Volume 2

Australian Beetles

Archostemata, Myxophaga, Adephaga, Polyphaga (part)

Editors: Adam Ślipiński and John F. Lawrence

CSIRO PUBLISHING

A catalogue record for this book is available from the National Library of Australia.

Published by:
CSIRO Publishing
Locked Bag 10
Clayton South VIC 3169
Australia

Telephone: +61 3 9545 8400
Email: publishing.sales@csiro.au
Website: www.publish.csiro.au

Cover images by Cate Lemann and Adam Ślipiński

Cover design by Andrew Weatherill
Typeset by Thomson Digital
Printed in China by Leo Paper Products Ltd.

Jul19_01

CONTENTS

PREFACE

As indicated in the preface to Volume 1 of this series, the chapters in the present volume and those following it, have been or will be written by specialists from various parts of the world, as well as by the two editors and other members of the CSIRO staff and students. Although the original plan was to include in this volume Australian families of Archostemata, Myxophaga, Adephaga, Hydrophiloidea, Staphylinoidea and Scarabaeoidea, the current availability of family specialists necessitated some reorganisation. As a result, Staphylinoidea will now be included in Volume 3, the families Scirtidae, Buprestidae and Tenebrionidae have been moved to the present volume, and chapter numbers will not be consistent with those used in Volume 1.

It is also well known that the family level classification of Coleoptera has changed considerably from that discussed and outlined in Volume 1, Chapter 2. This is due in part to several worldwide, in-depth studies, involving both adult and larval morphology, and in part to the increase in molecular sequence data and improvements in phylogenetic analysis. In the present volume, most Australian family limits have remained the same, with one major exception. Hydrophilidae is now broken up into several families, of which the following four have Australian representatives: Georissidae, Hydrochidae, Spercheidae and Hydrophilidae (including Sphaeridiinae); this is not a new concept and the change probably should have been made in the previous volume.

Below the level of family, the changes are often more considerable. The classification of Carabidae in Volume 1, for instance, included 13 Australian subfamilies, with most tribes included in Harpalinae. In Chapter 12 of the present volume, no subfamilies are recognised, and genera are placed in 37 tribes and numerous subtribes. Changes in subfamilial, tribal and generic concepts have occurred in various other families based on newly published data, and these are reflected in several chapters.

In addition to the Staphylinoidea and the remaining Polyphaga the Volume 3 will include new sets of family level keys to adult and larvae reflecting ongoing changes in the higher classification of beetles over the years.

DEDICATION

This volume is dedicated to the memory of

MARTIN BAEHR

March 10, 1943 – April 17, 2019

Dr Martin Baehr served his career as a curator at the Bavarian State Collection of Zoology (Zoologische Staatssammlung München) in Munich, Germany as a leading authority on the Carabidae of the Indo-Australian region. He first visited Australia and Tasmania for a year in 1972–73, and fell in love with the country, though then being interested in Onychophora, reptiles and arachnids. His PhD studies at Tübingen were on the functional structure of the thorax and legs of Carabidae and he began to work on their taxonomy, first on the European fauna, then a short 1976 paper on the Carabidae from his Tasmanian visit. A return visit to Queensland in 1981 allowed him to concentrate on carabid collecting and turned his interest to the Australian fauna. He returned for extended field and museum trips every two years for the rest of his life, exploring every corner of the continent and often spending weeks alone in remote areas like the Kimberley, Kakadu and Cape York. His prodigious collections were matched by his research output and he published 187 papers totalling 5789 pages on the Australian biota and described 1167 new Carabidae from Australia as well as 52 Araneae, 13 Hemiptera, 1 Onychophora and 33 Orthoptera. In parallel, he published another 183 papers on non-Australian Carabidae, mostly from SE Asia and the islands through to New Guinea. His final, and most comprehensive, work is the generic overview and synthesis of the whole Australian carabid fauna, written in collaboration with Kipling Will, which appears in this volume, and which he sadly did not live to see in print.

ACKNOWLEDGMENTS

We gratefully acknowledge The Australian National Insect Collection, CSIRO for support of research and for allowing the use of copyrighted images. We are also in debt to the many institutions and individuals who have supported our various research projects over the years by supplying specimens, literature and information. Anne Hastings is acknowledged for her artwork and graphics skills, and Cate Lemann for photography, testing the keys and continuous technical assistance.

We thank the following photographers for providing photos of living beetles: Kristi Ellingsen, Simon Grove, Owen Kelly, Zhenhua Liu, Jiří and Marie Lochman (Lochman Transparencies) and Nick Monaghan (Life Unseen).

CONTRIBUTORS TO VOLUME 2

Editors: Adam Ślipiński and John F. Lawrence
Australian National Insect Collection
CSIRO
GPO Box 1700
Canberra, ACT 2601, Australia

Martin Baehr
Zoologische Staatssammlung
Münchhausenstraße 21
81247 München, Germany

Alberto Ballerio
Viale Venezia 45
I-25123 Brescia, Italy

Hermes E. Escalona
Zoologisches Forschungsmuseum Alexander Koenig
Centre for Molecular Biodiversity Research
Adenauerallee 160
53113 Bonn, Germany

Martin Fikáček
Department of Entomology
National Museum Natural History
Cirkusová 1740 CZ-193 00 Praha 9 -
Horní Počernice, Czech Republic

Nicole L. Gunter
Department of Invertebrate Zoology
Cleveland Museum of Natural History
Cleveland, Ohio 44106, USA

W. Eugene Hall
University of Arizona Insect Collection
Department of Entomology
1140 E. South Campus Dr
Tucson, Arizona 85721, USA

Lars Hendrich
Zoologische Staatssammlung
Münchhausenstraße 21
81247 München, Germany

Paul M. Hutchinson
Quarantine WA
Department of Primary Industries and Regional Development,
Level 1, 24 Fricker Road
Perth Airport, WA 6105, Australia

Mengjie Jin
School of Life Sciences
Sun Yat-Sen University
Guangzhou 510275, China

Tomáš Lackner
Zoologische Staatssammlung
Münchhausenstraße 21
81247 München, Germany

Cate Lemann
Australian National Insect Collection
CSIRO
GPO Box 1700
Canberra, ACT 2601, Australia

Eric G. Matthews
South Australian Museum
North Terrace
Adelaide, SA 5000, Australia

Sławomir Mazur
Department of Forest Protection and Ecology
Warsaw University of Life Sciences
Nowoursynowska 159
02–776 Warszawa, Poland

Chris H. Moeseneder
Oceans and Atmosphere Flagship, CSIRO
Queensland Biosciences Precinct
306 Carmody Road
St. Lucia, Qld 4067, Australia

Chris A.M. Reid
Australian Museum
1 William Street
Sydney, NSW 2010, Australia

Owen D. Seeman
Queensland Museum
PO Box 3300
South Brisbane, Qld 4101, Australia

Chris H.S. Watts
South Australian Museum
North Terrace
Adelaide, SA 5000, Australia

Tom A. Weir
Australian National Insect Collection
CSIRO
GPO Box 1700
Canberra, ACT 2601, Australia

Kipling W. Will
Essig Museum of Entomology
University of California, Berkeley
Berkeley, California 94720, USA

Yu-Lingzi Zhou
Key Laboratory of Zoological Systematics and Evolution
Institute of Zoology, Chinese Academy of Sciences
1 Beichen West Road
Beijing 100101, China

Peter Zwick
Schwarzer Stock 9
D-36110 Schlitz, Germany

1. ARCHOSTEMATA KOLBE, 1908

John F. Lawrence and Hermes E. Escalona

The genera *Cupes* Latreille and *Omma* Newman were included in the suborder Adephaga by Ganglbauer (1903), Kolbe (1901, 1903) and Lameere (1903), based primarily on wing venation, while Kolbe (1908) excluded cupedids from Adephaga and placed them in a separate group, Archostemata, within what is now known as Polyphaga. Forbes (1926) considered Archostemata to be one of the three suborders of beetles, containing Cupedidae Laporte (including Ommatidae Sharp & Muir) and Micromalthidae Barber, based on wing folding, while by Böving & Craighead (1931) and Peyerimhoff (1933) supported this based on larval evidence. This classification was used in Crowson's (1955) classic work on the order. In a later paper, Crowson (1962) made the first attempt to integrate the work of paleontologists, such as A. Handlirsch, R. J. Tillyard, B. B. Rohdendorf and A. B. Martynov, on cupedid-like fossils with knowledge of the extant archostematan fauna known at that time. Ponomarenko (1964, 1966, 1968, 1969) published the first major work dealing with Archostemata, which was defined in a broad sense to include virtually all known Permian and Mesozoic fossils sufficiently complete to be described (11 families, 74 genera and 186 species) but lacking the distinctive features of either Polyphaga or Adephaga. The Lower Permian Tshekardocoleidae and several Upper Permian groups were placed in Protocoleoptera and Archecoleoptera, respectively, by Crowson (1975), but were considered to form a coleopterid stem group by Kukalová-Peck (1991). This was followed by Beutel (1997) and Beutel & Friedrich (2008), who added the Triassic Triadocupedidae to this paraphyletic stem group. Within the remaining Archostemata, relationships are still not clearly understood, particularly among the taxa with non-clathrate elytra (without window-punctures) such as Jurodidae, Ademosynidae, Schizophoridae and Catiniidae, and their possible affinities with the other three suborders, Adephaga, Myxophaga and Polyphaga. Numerous publications have appeared in the past 25 years on fossil Archostemata primarily from localities in Kazakhstan, Russia, Mongolia and northern China but also on those in the UK, USA, Spain, Australia and Argentina; included are Ashman *et al.* (2015); Jarzembowski & Wang (2016); Jarzembowski *et al.* (2012, 2013a, 2013b, 2015); Kirejtshuk (2005); Kirejtshuk *et al.* (2010a, 2010b, 2016); Lubkin (2003); Martin (2010); Martins-Neto *et al.* (2006); Ponomarenko (1994, 1997, 2000, 2004, 2006); Ponomarenko & Martinez-Delclòs (2000); Ponomarenko & Ren (2010); Ponomarenko *et al.* (2014); Ren & Tan (2006); Ren *et al.* (2006); Soriano & Delclòs (2006); Tan & Ren (2006a, 2006b, 2006c, 2007, 2009); Tan *et al.* (2007a); Tan *et al.* (2005); Tan *et al.* (2006a, 2006b, 2007b, 2007c); Tan *et al.* (2006c, 2013, 2012) and Yan *et al.* (2014).

In the study of Archostemata by Kirejtshuk *et al.* (2016), virtually all archostematans with clathrate or reticulate elytra are considered to belong to the single family Cupedidae, extending back to the Triassic, but without an adequate phylogenetic analysis and additional support from other members of the paleontological community, we continue to recognise the families Ommatidae and Cupedidae. Recent discoveries of Cupedidae in the Cretaceous amber of Myanmar may be found in the papers by Jarzembowski *et al.* (2017a, 2017b, 2017c).

Some important characters defining this suborder may be found in extant members of the Ommatidae and Cupedidae, but not in highly derived or little known groups like Micromalthidae (Beutel & Hörnschemeyer 2002a), Crowsoniellidae (Ge *et al.* 2011) or the doubtfully included Jurodidae (Yan *et al.* 2014). These include: (1) lack of cervical sclerites (also in Adephaga and Myxophaga); (2) external propleuron extending well in front of procoxal cavities (also in Adephaga, reduced in Myxophaga); (3) free, external protrochantin; (4) membranous joint between mesothorax and metathorax (also in some Polyphaga); (5) mesocoxal cavities broadly closed laterally by metanepisternum (narrowly so in a few Polyphaga); (6) well developed and exposed metatrochantins; (7) hind wing without a radial cell formed by meeting of RA_{1+2} and RA_{3+4} (also in Adephaga and Myxophaga); (8) hind wing with cross-veins between RP and MP_{1+2} forming an oblongum cell (also in Adephaga and Myxophaga), (9) hind wing with major fold crossing MP_{1+2} forming a sharp hinge (also in Adephaga and Myxophaga); (10) larva with 6-segmented legs and paired pretarsal claws (also in Adephaga); (11) larva with ligular sclerome.

There are relatively few extant species of Archostemata: the family Ommatidae contains four Australian species of *Omma* Newman and two species of *Tetraphalerus* Waterhouse, occurring primarily in Argentina and Bolivia, and the Cupedidae includes nine genera and 31 species occurring in North America, South America, the eastern Palaearctric and Oriental regions, the East Indies, New Guinea, Australia, New Caledonia, East Africa, South Africa and Madagascar (Neboiss 1984; Lawrence 1999; Hörnschemeyer 2009, 2016; Hörnschemeyer & Beutel 2016; Hörnschemeyer & Yavorskaya 2016). The group has become extinct in Europe, but several European species are known from the early Tertiary (Kirejtshuk 2005; Kirejtshuk *et al.* 2010a). Several morphological papers have been produced on both larvae (Beutel & Hörnschemeyer 2002a, 2002b; Hörnschemeyer *et al.* 2002; Grebennikov 2004; Yavorskaya *et al.* 2015) and adults (Baehr 1975; Hörnschemeyer *et al.* 2006; Beutel *et al.* 2008; Friedrich *et al.* 2009; Hünefeld *et al.* 2011). General summaries of extant Archostemata may be found in Hörnschemeyer (2016), Hörnschemeyer & Beutel (2016) and Hörnschemeyer & Yavorskaya (2016).

References

Ashman LG, Oberprieler RG, Ślipiński A (2015) *Rhopalomma stefaniae* gen. et sp. n., the first ommatid beetle from the Upper Jurassic in Australia (Coleoptera: Archostemata: Ommatidae). *Zootaxa* **3980**(1), 136–142. doi:10.11646/zootaxa.3980.1.8

Baehr M (1975) Skelett und Muskulatur des Thorax von *Priacma serrata* LeConte (Coleoptera, Cupedidae). *Zeitschrift für Morphologie der Tiere* **81**, 55–101. doi:10.1007/BF00290073

Beutel RG (1997) Über Phylogenese und Evolution der Coleoptera (Insecta), insbesondere der Adephaga. *Abhandlungen des Naturwissenschaftlichen Vereins in Hamburg* (NF) 31, 164 pp.

Beutel RG, Friedrich F (2008) The phylogeny of Archostemata (Coleoptera) and new approaches in insect morphology. *Entomologia Generalis* **31**(2), 141–154. doi:10.1127/entom.gen/31/2008/141

Beutel RG, Hörnschemeyer T (2002a) Description of the larva of *Rhipsideigma raffrayi* (Coleoptera: Archostemata), with phylogenetic and functional implications. *European Journal of Entomology* **99**, 53–66. doi:10.14411/eje.2002.011

Beutel RG, Hörnschemeyer T (2002b) Larval morphology and phylogenetic position of *Micromalthus debilis* LeConte (Coleoptera: Micromalthidae). *Systematic Entomology* **27**, 169–190. doi:10.1046/j.1365-3113.2002.00172.x

Beutel RG, Ge SQ, Hörnschemeyer T (2008) On the head morphology of *Tetraphalerus*, the phylogeny of Archostemata and the basal branching events in Coleoptera. *Cladistics* **24**, 270–298. doi:10.1111/j.1096-0031.2007.00186.x

Böving AG, Craighead FC (1931) An illustrated synopsis of the principal larval forms of the order Coleoptera. *Entomologica Americana* (N.S.) **11**(1930), 1–351.

Crowson RA (1955) *The Natural Classification of the Families of Coleoptera*. Nathaniel Lloyd, London.

Crowson RA (1962) Observations on the beetle family Cupedidae, with descriptions of two new forms and a key to the recent general. *Annals & Magazine of Natural History* **5**(51), 147–157. doi:10.1080/00222936208651227

Crowson RA (1975) The evolutionary history of Coleoptera, as documented by fossil and comparative evidence. *Atti del Congresso Nazionale Italiano di Entomologia* **10**, 47–90.

Forbes WTM (1926) The wing-folding patterns of the Coleoptera. *Journal of the New York Entomological Society* **34**, 42–115, pls 7–18.

Friedrich F, Farrell BD, Beutel RG (2009) The thoracic morphology of Archostemata and the relationships of the extent suborders of Coleoptera (Hexapoda). *Cladistics* **25**, 1–37. doi:10.1111/j.1096-0031.2008.00233.x

Ganglbauer L (1903) Systematisch-koleopterologische Studien. *Münchener Koleopterologische Zeitschrift* **1**, 271–319.

Ge S-Q, Hörnschemeyer T, Friedrich F, Beutel RG (2011) Is *Crowsoniella relicta* really a cucujiform beetle? *Systematic Entomology* **36**(1), 175–179. doi:10.1111/j.1365-3113.2010.00552.x

Grebennikov VV (2004) Review of larval morphology of beetles of the suborder Archostemata (Insecta: Coleoptera), including first-instar chaetotaxy. *European Journal of Entomology* **101**, 273–292. doi:10.14411/eje.2004.038

Hörnschemeyer T (2009) The species-level phylogeny of archostematan beetles—where do *Micromalthus debilis* and *Crowsoniella relicta* belong? *Systematic Entomology* **34**, 533–558. doi:10.1111/j.1365-3113.2009.00476.x

Hörnschemeyer T (2016) 5. Archostemata Kolbe, 1908, 5.1. Crowsoniellidae Iablokoff-Khnzorian, 1983. 5.3. Micromalthidae Barber, 1913. 5.5 Jurodidae Kirejtshuk, 1999 (= Sikhotealinidae). In *Handbuch der Zoologie/Handbook of Zoology. Band/Volume IV Arthropoda: Insecta. Teilband/Part 38. Coleoptera, Beetles. Volume 1: Morphology and Systematics (Archostemata, Adephaga, Myxophaga, Polyphaga partim).* 2nd edn. (Eds RG Beutel and RAB Leschen) pp. 41–43, 48–51, 57–62. Walter de Gruyter, Berlin.

Hörnschemeyer T, Beutel RG (2016) 5.4. Ommatidae Sharp & Muir, 1912. In *Handbuch der Zoologie/Handbook of Zoology. Band/Volume IV Arthropoda: Insecta. Teilband/Part 38. Coleoptera, Beetles. Volume 1: Morphology and Systematics (Archostemata, Adephaga, Myxophaga, Polyphaga partim).* 2nd edn. (Eds RG Beutel and RAB Leschen) pp. 52–56, 59–62. Walter de Gruyter, Berlin.

Hörnschemeyer T, Yavorskaya M (2016) 5.2. Cupedidae Laporte, 1832. Introduction and phylogeny. In *Handbuch der Zoologie/Handbook of Zoology. Band/Volume IV Arthropoda: Insecta. Teilband/Part 38. Coleoptera, Beetles. Volume 1: Morphology and Systematics (Archostemata, Adephaga, Myxophaga, Polyphaga partim).* 2nd edn. (Eds RG Beutel and RAB Leschen) pp. 44–47, 59–62. Walter de Gruyter, Berlin.

Hörnschemeyer T, Beutel RG, Pasop F (2002) Head structure of *Priacma serrata* LeConte (Coleoptera, Archostemata) inferred from X-ray tomography. *Journal of Morphology* **252**, 298–314. doi:10.1002/jmor.1107

Hörnschemeyer T, Goebbels J, Weidemann G, Faber C, Haase A (2006) The head morphology of *Ascioplaga mimeta* (Coleoptera: Archostemata) and the phylogeny of Archostemata. *European Journal of Entomology* **103**, 409–423. doi:10.14411/eje.2006.055

Hünefeld F, Marvaldi SE, Nüller B, Lawrence JF, Beutel RG (2011) The male postabdomen of the "ancestral" archostematan beetle *Tetraphalerus bruchi* Heller, 1913 (Ommatidae) and its phylogenetic significance. *Arthropod Structure & Development* **40**, 146–158. doi:10.1016/j.asd.2010.12.003

Jarzembowski EA, Wang B (2016) An unusual basal beetle from Myanmar (Coleoptera: Archostemata). *Alcheringia: An Australasian Journal of Palaeontology* **40**(2), 297–302. doi:10.1080/03115518.2016.1132493

Jarzembowski EA, Yan EV, Wang B, Zhang HC (2012) A new flying water beetle (Coleoptera: Schizophoridae) from the Jurassic Daohugou lagerstätte. *Palaeoworld* **21**(3), 160–166. doi:10.1016/j.palwor.2012.09.002

Jarzembowski EA, Yan EV, Wang B, Zhang H (2013a) Brochocolein beetles (Insecta: Coleoptera) from the Lower Cretaceous of northeast China and southern England. *Cretaceous Research* **44**, 1–11. doi:10.1016/j.cretres.2013.03.003

Jarzembowski EA, Yan EV, Wang B, Zhang H (2013b) Ommatin beetles (Insecta: Coleoptera) from the Lower Cretaceous of northeast China and southern England. *Terrestrial Arthropod Reviews* **6**, 135–161. doi:10.1163/18749836-06021062

Jarzembowski EA, Wang B, Zhang H, Fang Y (2015) Boring beetles are not necessarily dull: new notocupedins (Insecta: Coleoptera) from the Mesozoic of Eurasia and East Gondwana. *Cretaceous Research* **52**(B), 431–439.

Jarzembowski EA, Wang B, Zheng D (2017a) The first cupedine beetle from Burmese amber (Coleoptera: Cupedidae). *Comptes Rendus. Palévol* **16**, 241–247. doi:10.1016/j.crpv.2016.08.004

Jarzembowski EA, Wang B, Zheng D (2017b) A new reticulated beetle (Coleoptera: Cupedidae) with aedeagus preserved from mid-Cretaceous amber of Myanmar. *Cretaceous Research* **80**, 86–90. doi:10.1016/j.cretres.2017.08.015

Jarzembowski EA, Wang B, Zheng D (2017c) A new spiny rerticulated beetle (Coleoptera: Cupedidae) from Cretaceous Burmese amber. *Proceedings of the Geologist's Association.* doi:10.1016/j.pgeola.2017.07.003.

Kirejtshuk AG (2005) A revision of the genus *Cupes* Fabricius, 1801 from Baltic amber and some notes on taxonomy and composition of the family Cupedidae (Coleoptera, Archostemata). *Mitteilungen aus dem Geologisch-Paläontologischen Institut der Universität Hamburg* **89**, 55–84.

Kirejtshuk AG, Nel A, Collomb FM (2010a) New Archostemata (Insecta: Coleoptera) from the French Paleocene and early Eocene, with a note on the composition of the suborder. *Annales de la Société Entomologique de France* (N.S.) **46**(1–2), 216–227.

Kirejtshuk AG, Ponomarenko AG, Prokin AA, Chang H, Nikolajev GV, Ren D (2010b) Current knowledge of Mesozoic Coleoptera from Daohugou and Liaoning (northeast China). *Acta Geologica Sinica* **84**(4), 783–792[English Edition]. doi:10.1111/j.1755-6724.2010.00253.x

Kirejtshuk AG, Nel A, Kirejtshuk PA (2016) Taxonomy of the reticulate beetles of the subfamily Cupedinae (Coleoptera: Archostemata), with a review of the historical development. *Invertebrate Zoology* **13**(2), 61–190.

Kolbe HJ (1901) Vergleichend-morphologische Untersuchungen an Coleopteren nebst Grundlagen zu einem System und zur Systematik derselben. *Archiv für Naturgeschichte* **67** (Beiheft), 89–150, pls 2–3.

Kolbe HJ (1903) Zur Systematik der Coleopteren. *Allgemeine Zeitschrift für Entomologie* **1903**, 137–145.

Kolbe HJ (1908) Mein System der Coleopteren. *Zeitschrift für Wissenschaftliche Insektenbiologie* **4**, 116–123, 153–162, 219–226, 246–251, 286–294, 389–400.

Kukalová-Peck J (1991) 6. Fossil history and the evolution of hexapod structures. In *Insects of Australia: A Textbook for Students and Research Workers.* 2nd edn. (Ed. CSIRO) pp. 141–179. Melbourne University Press, Carlton, Victoria.

Lameere A (1903) Nouvelle notes pour la classification des Coléoptères. *Annales de la Société Entomologique de Belgique* **47**, 155–165.

Lawrence JF (1999) The Australian Ommatidae (Coleoptera: Archostemata): new species, larva and discussion of relationships. *Invertebrate Taxonomy* **13**, 369–390. doi:10.1071/IT99008

Lubkin SH (2003) *Paracupes svitkoi* (Coleoptera: Cupedidae) a new species from the Cretaceous of New Jersey. *Acta Zoological Cracoviensia* **46**, 189–194.

Martin SK (2010) Early Jurassic coleopterans from the Mintaja insect locality, Western Australia. *Acta Geologica Sinica* **84**(4), 925–953[English Edition]. doi:10.1111/j.1755-6724.2010.00276.x

Martins-Neto RG, Gallego OF, Mancuso AC (2006) The Triassic insect fauna from Argentina. Coleoptera from the Los Rastros Formation (Bermejo Basin), La Rioja Province. *Ameghiniana* **43**(3), 1–17.

Neboiss A (1984) Reclassification of *Cupes* Fabricius (s. lat.), with descriptions of new genera and species (Cupedidae: Coleoptera). *Systematic Entomology* **9**, 443–477. doi:10.1111/j.1365-3113.1984.tb00520.x

Peyerimhoff P de (1933) Les larves des Coléoptères d'après A. G. Böving et F. C. Craighead et les grands critériums de l'ordre. *Annales de la Société Entomologique de France* **102**, 77–106.

Ponomarenko AG (1964) New beetles of the family Cupedidae from the Jurassic of Karatau. *Палеонтологический журнал* **2**, 49–61[in Russian].

Ponomarenko AG (1966) New beetles of the family Cupedidae (Coleoptera) from Mesozoic deposits of Transbaikalia. *Entomologicheskoe Obozrenie* **1**, 138–143[in Russian].

Ponomarenko AG (1968) Preliminary review of the Jurassic beetles of the Karatau. In *Jurassic Insects of the Karatau.* (Ed. BB Rohdendorf) pp. 114–117. Nauka, Moscow.

Ponomarenko AG (1969) Historical development of the Coleoptera-Archostemata. *Trudy Paleontologicheskogo Instituta* **125**, 1–240 [in Russian].

Ponomarenko AG (1994) New Mesozoic cupedid beetles from Mongolia. Brochocoleini and Notocupedini. *Палеонтологический журнал* **28**(3), 102–115.

Ponomarenko AG (1997) New beetles of the family Cupedidae from the Mesozoic of Mongolia. Ommatini, Mesocupedini, Priacmini. *Paleontologicheskii Zhurnal* **31**(4), 45–55 [in Russian; translation in *Paleontological Journal* **31**(4), 389–399].

Ponomarenko AG (2000) Beetles of the family Cupedidae from the Lower Cretaceous locality of Semen, Transbaykalia. *Paleontological Journal* **34**(Suppl. 3), 317–322.

Ponomarenko AG (2004) Beetles (Insecta, Coleoptera) of the late Permian and early Triassic. *Paleontological Journal* **38**(Suppl. 2), S185–S196.

Ponomarenko AG (2006) On the types of Mesozoic archostematan beetles (Insecta, Coleoptera, Archostemata) in the Natural History Museum. London. *Paleontological Journal* **40**(1), 90–99. doi:10.1134/S0031030106010102

Ponomarenko AG, Martinez-Delclòs X (2000) New beetles (Insecta: Coleoptera) from Lower Cretaceous of Spain. *Acta Geologica Hispanica* **35**, 47–52.

Ponomarenko AG, Ren D (2010) First record of Notocupes (Coleoptera: Cupedidae) in locality Daohugou, Middle Jurassic of Inner Mongolia, China. *Annales Zoologici* **60**(2), 169–171. doi:10.3161/000345410X516812

Ponomarenko AG, Aristov DS, Bashkuev AS, Gubin YM, Khramov AV, Lukashevich ED, Popov YA, Pritykina LN, Sinitsa SM, Sinitshenkova ND, Sukatsheva ID, Vassilenko DV, Yan EV (2014) Upper Jurassic Lagerstätte Shar Teg, Southwestern Mongolia. *Paleontological Journal* **48**(14), 1573–1682. doi:10.1134/S0031030114140160

Ren D, Tan J (2006) A new cupedid genus (Coleoptera: Archostemata: Cupedidae) from Jehol Biota of western Liaoning, China. *Annales Zoologici* **56**(3), 457–464.

Ren D, Tan J, Ge S (2006) New fossil ommatid (Coleoptera: Archostemata: Ommatidae) from Jehol Biota of western Liaoning, China. *Progress in Natural Science* **16**(6), 639–643. doi:10.1080/10020070612330046

Soriano C, Delclòs X (2006) New cupedids from the lower Cretaceous of Spain and the palaeogeography of the family. *Acta Palaeontologica Polonica* **51**, 185–200.

Tan J, Ren D (2006a) Jurassic and Cretaceous Cupedomorpha (Insecta: Coleoptera) faunas of China. *Progress in Natural Science* **16**(Special Issue), 313–322.

Tan J, Ren D (2006b) New fossil Priacmini (Insecta: Coleoptera: Archostemata: Cupedidae) from the Jehol Biota of China. *Journal of Natural History* **40**(47–48), 2653–2661. doi:10.1080/00222930601121445

Tan J, Ren D (2006c) *Ovatocupes*: a new cupedid genus (Coleoptera: Archostemata: Cupedidae) from the Jehol Biota (Late Jurassic) of western Liaoning, China. *Entomological News* **117**, 223–232. doi:10.3157/0013-872X(2006)117[223:OANCGC]2.0.CO;2

Tan J, Ren D (2007) Two exceptionally well-preserved catiniids (Coleoptera: Archostemata: Catiniidae) from the Late Mesozoic of northeastern China. *Annals of the Entomological Society of America* **100**(5), 666–672. doi:10.1603/0013-8746(2007)100[666:TEWCCA]2.0.CO;2

Tan J, Ren D (2009) *Mesozoic Archostematan Fauna from China* (in Chinese with English summary). Science Press, Beijing.

Tan J, Ren D, Liu M (2005) New ommatids from the Late Jurassic of western Liaoning, China (Coleoptera: Archostemata). *Insect Science* **12**, 207–216. doi:10.1111/j.1005-295X.2005.00026.x

Tan J, Ren D, Shih C (2006a) First record of fossil *Priacma* (Coleoptera: Archostemata: Cupedidae) from the Jehol Biota of western Liaoning, China. *Zootaxa* **1326**, 55–68.

Tan J, Ren D, Shih C (2006b) New cupedids from the middle Jurassic of Inner Mongolia, China (Coleoptera: Archostemata). *Annales Zoologici* **56**(1), 1–6.

Tan J, Ren D, Shih C, Ge S (2006c) New fossil beetles of the family Ommatidae (Coleoptera: Archostemata) from the Jehol Biota of China. *Acta Geologica Sinica* **80**(4), 474–485.

Tan J, Huang D, Ren D (2007a) First record of fossil *Mesocupes* from China (Coleoptera: Archostemata: Cupedidae). *Acta Geologica Sinica* **81**(5), 688–696. doi:10.1111/j.1755-6724.2007.tb00993.x

Tan J, Ren D, Shih C (2007b) New beetles (Insecta: Coleoptera: Archostemata) from the Late Mesozoic of North China. *Annales Zoologici* **57**(2), 231–247.

Tan J, Ren D, Shih C (2007c) New ommatids of Ommatinae (Coleoptera: Archostemata: Ommatidae) from the Yixian Formation of western Liaoning, China. *Progress in Natural Science* **17**(7), 803–811. doi:10.1080/10002007088537476

Tan J, Wang Y, Ren D, Yang X (2012) New fossil species of ommatids (Coleoptera: Archostemata) from the middle Mesozoic of China illuminating the phylogeny of Ommatidae. *BMC Evolutionary Biology* **12**(113), 1–19.

Tan J, Ren D, Shih DC, Yang X (2013) New schizophorid fossils from China and possible evolutionary scenarios for Jurassic archostematan beetles. *Journal of Systematic Palaeontology* **11**(1), 47–62. doi:10.1080/14772019.2011.637515

Yan EV, Wang B, Ponomarenko AG, Zhang H (2014) The most mysterious beetles: Jurassic Jurodidae (Insecta: Coleoptera) from China. *Gondwana Research* **25**, 214–225. doi:10.1016/j.gr.2013.04.002

Yavorskaya MI, Kojima K, Machida R, Beutel RG (2015) Morphology of the first instar larva of *Tenomerga mucida* (Chevrolat, 1829) (Coleoptera: Archostemata: Cupedidae). *Arthropod Systematics & Phylogeny* **73**(2), 239–258.

2. OMMATIDAE SHARP AND MUIR, 1912

John F. Lawrence and Hermes E. Escalona

Fig. 2.1. *Omma mastersi* Macleay.

Introduction. Crowson (1962) recognised the cupedid subfamily Ommatinae (as Ommadinae of Sharp & Muir), including the genera *Omma* Newman and *Tetraphalerus* Waterhouse, each in its own tribe, and described the first fossil *Omma* from the Lower Jurassic of England. These were distinguished from the Cupedinae by short antennae, tridentate, vertically oriented mandibular apices, contiguous procoxae, simple fourth tarsomere, and non-overlapping abdominal ventrites. Ponomarenko (1964, 1966, 1968, 1969) considered these two genera, along with five other extinct taxa, to belong to the cupedid subfamily Ommatinae, and described eight more *Tetraphalerus* and five more *Omma* from the Lower Jurassic to the Lower Cretaceous of Russia, Kazakhstan and Kyrgyzstan. Since that time numerous other fossil taxa have been added to the Ommatinae *sensu* Ponomarenko from various deposits in Russia, Mongolia and China, as well as the UK, Spain and Australia. Lawrence (1999) suggested that several archostematan taxa be elevated to a higher rank in line with the current treatment of extant taxa (Triadocupedidae; Cupedidae with Mesocupedinae, Priacminae and Cupedinae; and Ommatidae with Lithocupedinae, Notocupedinae, Ommatinae and Tetraphalerinae). At least some palaeontologists (Kirejtshuk *et al.* 2010a, 2010b, 2016), however, consider the Ommatidae to be a subfamily of a broadly defined Cupedidae, which extends back to the Triassic. According to Tan *et al.* (2012) the group includes 13 genera and more than 100 species, but the relationships among extant and extinct species remain obscure. These workers presented a cladogram based on 27 characters in which the genera *Notocupoides* and *Eurydictyon* form a clade (unfortunately given an invalid name Pronotocupedini) sister to the remaining taxa, Notocupedini sister to the remaining four clades and Lithocupedini sister to (Brochocoleini (Ommatini and Tetraphalerini)). In the past 20 years, ommatids have been described from various sites in the UK (Jarzembowski *et al.* 2013a, 2013b; Ponomarenko 2006), Spain (Soriano & Delclòs 2006), China (Cai & Huang 2016; Jarzembowski *et al.* 2013a, 2013b; Ren *et al.* 2006; Tan *et al.* 2005, 2006, 2007, 2012), Mongolia (Ponomarenko 1994, 1997), Transbaikalia, Russia (Ponomarenko 2000), Myanmar (Jarzembowski & Wang 2016; Jarzembowski *et al.* 2017), and Australia (Ashman *et al.* 2015; Martin 2010).

Extant members of the family Ommatidae occur only in Australia (*Omma*) and in parts of southern South America (*Tetraphalerus*). The two groups differ in many respects and are placed in separate subfamilies. The South American *Tetraphalerus* Waterhouse is known from the relatively arid parts of northern Argentina and Bolivia but also occurs in southern Brazil. Species of *Tetraphalerus* differ from those of *Omma* in having a narrowly elongate, strongly flattened body almost entirely covered with apically flattened tubercles, each bearing a seta, head with sharply angulate temples and ventral antennal grooves, antennomeres 2–11 glabrous, prothorax with distinct lateral carinae, propleuron completely separated from sternum and extending almost to anterior edge of prothorax, and aedeagus with distinct phallobase and free parameres (Monrós & Monrós 1952; Crowson 1962; Vulcano & Pereira 1975; Beutel *et al.* 2007; Hünefeld *et al.* 2011). There are four species of *Omma* Newman in Australia. *Omma stanleyi* Newman occurs along the eastern coast but does not extend into the wet tropics, *O. mastersi* Macleay occurs mainly in the central and southern part of the continent, and *O. rutherfordi* Lawrence and *O. sagitta* Neboiss are southern and south-western in distribution (Hörnschemeyer & Beutel 2016).

Biology. An adult of *Omma stanleyi* was collected under bark of a *Eucalyptus* log in southern QLD, and it is likely that larvae of this species feeds in dead wood. An adult of *O. mastersi* was collected in southern QLD actively running on the trunk of a dead tree in full sun and its colour pattern was very similar to a clerid beetle (*Stigmatium gilberti* White) and a mutilid wasp (*Ephutomorpha* sp.). This is suggestive of a possible mimicry ring. An interesting parallel has been found in some fossil *Zygadenia* Handlirsch (Notocupedinae) from the Barremian (Early Cretaceous) of England; these beetles have elytral patterns similar to those in *Omma mastersi* and relatives and are likely to have been day-active (Jarzembowski *et al.* 2015). Males of *O. rutherfordi* were collected in flight-intercept traps at two sites in SA with deep sand and mixed mallee and the grass *Triodia*, while a female was collected in a swimming pool filter north of the Stirling Range in southern WA. The single known larva, which could belong either to *O. sagitta* or *O. rutherfordi*, was found in soil at Red Hill in southern WA and was likely feeding on underground roots or wood. It has been described by Lawrence (1999) and Grebennikov (2004).

Characteristics. Adults. 6–26 mm in length, and 2.4–3 times as long as wide, slightly flattened dorsally and distinctly so ventrally, with sides of pronotum and elytra slightly, independently rounded. Colour uniformly dark brown to black; upper surfaces tuberculate, (tubercles spine-like in *Omma rutherfordi*), and clothed with short, slightly thickened, decumbent setae in *O. stanleyi*, with flattened, ribbed scales which differ in colour and form a distinct pattern in other species of *Omma* (interspersed with stout, erect spines in *O. rutherfordi*). Head prognathous, always constricted posteriorly to form a neck, but without distinct temples in *O. rutherfordi*. Eyes entire and very finely facetted, without interfacetal setae. Antennal insertions lateral, barely concealed by frontal ridges; subantennal groove absent. Frontoclypeal suture absent. Labrum fused to clypeus. Antennae usually short, rarely exceeding posterior edge of pronotum, 11-segmented and filiform. Mandible tridentate with vertically aligned teeth, without mola or prostheca, its dorsal surface near base with setose cavity. Maxilla highly reduced and partly concealed by mentum, galea and lacinia slender, hyaline and setose, the latter without uncus. Apical maxillary palpomere expanded and securiform with small cavity bearing sensilla near outer edge of upper surface. Mentum about as long as wide, concealing small, rounded ligula and palpal insertions; apical labial palpomere expanded and securiform, with a cavity similar to that on the apical maxillary palpomere. Gular sutures well separated; gula longer than wide. Corpotentorium weakly developed or absent. Cervical sclerites absent. Prothorax usually distinctly narrower than distance across elytral humeri (less so in male *O. rutherfordi*), widest anteriorly with sides slightly sinuate; lateral pronotal carinae usually absent (irregular in *O. rutherfordi*). Prosternum in front of coxae moderately long, partly or completely fused to pleuron (pleurosternal suture incomplete or absent), without paired crural impressions. Prosternal process incomplete, narrowed apically and acute at apex. Notopleural suture complete to anterior edge, never joined by pleurosternal suture. Procoxae subglobular, not strongly projecting, without concealed lateral extensions, without or with reduced coxal plates; trochantin large, well sclerotised, broadly exposed and abutting sternum and pleuron. Procoxal cavities slightly transverse, contiguous, broadly open externally and internally. Scutellar shield flat, not elevated above adjacent sections of elytra. Elytra complete with independently rounded to acute apices and ten rows of circular or transversely oval window punctures; epipleura moderately broad and complete. Mesoventrite at middle with acute anterior projection separating paired procoxal rests extending onto mesanepisterna, mesoventral cavity absent; discrimen moderately long and transverse (mesokatepisternal) suture incomplete (not always visible externally). Mesoventral process divided into two short, acute processes. Mesocoxae more or less globular, not or slightly projecting, without or with short plates and broadly exposed trochantins. Mesocoxal cavities slightly transverse, contiguous, partly closed laterally by mesepimera and metanepisterna; meso- and metathoracic joint within coxal cavities membranous. Metaventrite strongly narrowed anteriorly; discrimen present; transverse (metakatepisternal) suture well developed. Metanepisternum completely exposed, widest near anterior end; exposed portion of metepimeron narrowly elongate and widest at posterior end; metatrochantins exposed. Metacoxae strongly transverse with subconical mesal projections, contiguous, extending laterally to meet elytra, with weak coxal plates. Metendosternite with a moderately long, slender stalk, short, broad ventrolateral processes, moderately long and distally slender lateral arms, moderately long, apically emarginate anterior process with tendons moderately widely separated on either side of emargination. Hind wing almost always well developed and moderately broad (reduced in males of *O. rutherfordi*), with very short bending zone near apex of radial bar (anterior wing strut); pterostigma absent; two cross-veins join RA_{1+2} and RA_{3+4} to form brachial cell; RP complete almost to wing base, joined to RA_{3+4} by four cross-veins; RP and MP joined by two cross-veins forming the oblongum cell. MP_{1+2} with very short bending zone or hinge and a very short, straight medial spur; medial field with four free veins and no medial fleck; wedge cell well developed and apically truncate; anal lobe well developed, without embayment; AP_{3+4} forked. Legs short and slender; tibia subequal in length to femur; protibia with small setose excavation at basal third in males (except *O. mastersi*); tibial spurs well developed, usually unequal and paired on all legs. Tarsi 5–5–5; tarsomeres without ventral lobes; pretarsal claws simple; empodium bisetose. Abdomen with five ventrites, which are flattened, more or less connate, separated by very narrow grooves and all on the same plane, with edges abutting. Functional spiracles on abdominal segments I–VIII. Anterior edge of segment IX in male subtruncate; apex broadly rounded; tergite IX apically rounded; segment X free, forming a cylindrical protrusion bearing the anus. Aedeagus of modified trilobate type, with parameres completely fused to phallobase and partially fused together, each with a mesal notch near apex. Testes of the tubular type in *Tetraphalerus* (Sarmiento 1969; Hünefeld *et al.* 2011) but not yet examined in *Omma*. Sternite VIII in female with long, articulated spiculum ventrale. Ovipositor narrowly elongate, lightly pigmented except for baculi; paraprocts much longer than gonocoxites, each of which with well developed apical gonostylus.

Larvae. Elongate, slightly flattened, widest at level of mesothorax and gradually narrowed posteriorly to apex of tergum IX. Head prognathous and slightly retracted; posterior edge of head capsule deeply emarginate; epicranial stem absent and frontal arms not apparent; endocarina thick and V-shaped with each arm forked at apex. Stemmata, four on each side, well separated. Frontoclypeal suture indistinct or absent. Labrum free, strongly transverse. Antennae moderately long, 4-segmented, with distinct antennifer; sensorium on antennomere 3 about half as long as antennomere 4. Mandibles tridentate, with large, slightly asymmetrical and irregular molae, which lack asperities or transverse ridges; prostheca absent. Ventral mouthparts retracted forming maxillolabial complex, with well developed maxillary articulating area. Galea articulated and palpiform, lacinia fixed and apically rounded with several stout spines and setae; maxillary palp 3-segmented with distinct palpifer. Labium consisting of prementum and postmentum, the former with a well developed ligular sclerome; labial palps 2-segmented and widely separated. Hypostomal rods long, converging for anterior third, then diverging and extending almost to posterior edge of head capsule; ventral epicranial ridges absent. Gular region reduced. Meso-and metatergum with strongly transverse ambulatory ampulla; thoracic venter without distinct armature. Legs reduced, 6-segmented, with two pretarsal claws; coxae widely separated. Abdomen with transverse tergal ampullae on first three segments; ampullae paired or reduced on posterior segments. Tergum IX subequal in length to VIII, apically rounded, without armature; sternite IX exposed, simple, not distinctly separated from segment X, which is posteriorly oriented. Anal opening Y-shaped with dorsal and paired ventrolateral lobes. Spiracles annular, with closing apparatus.

Classification of the Australian genera

Subfamily Ommatinae Sharp and Muir, 1912

Omma Newman, 1839 (Pl. 14F–J, M)

Type species. *Omma stanleyi* Newman, 1839.
Australian species. Four

Distribution. *Omma stanleyi* occurs in eastern Australia from VIC and SA to central QLD. *O. mastersi* Malceay is known from NSW and QLD but usually occurs inland from the coast. *O. rutherfordi* Lawrence is known from the mallee of SA and in the Stirling Range, WA, and *O. sagitta* Neboiss has been recorded from south-western WA and southern SA.
References. Neboiss (1968, 1989); Lawrence (1999).

Key to the species

1. Vestiture consisting of slender, recumbent setae (Fig. 2.4), which are yellowish-brown in colour; elytral window-punctures irregularly transversely oval (Figs 2.4, 2.6); length greater than 12 mm (Pl. 14F–G); eastern Australia .... ***O. stanleyi* Newman**

– Vestiture including scale-like setae (Figs 2.2–2.3), which may be yellowish-brown or white in colour, sometimes mixed with spine-like processes; elytral window-punctures circular (Figs 2.3, 2.5); length less than 12 mm **2**

2(1). Body slender; antennae longer, usually extending beyond basal third of elytra (Pl. 14M); head behind eyes gradually narrowed; elytral apices independently subacute; vestiture consisting of both scale-like setae and stout, erect, spine-like projections (Fig. 2.2); hind wings in male highly reduced; SA and WA........ ***O. rutherfordi* Lawrence**

– Body robust; antennae shorter, extending at most to basal third of pronotum (Pl. 14H–J); head behind eyes abruptly narrowed; elytral apices independently rounded; vestiture consisting of scale-like setae, without spines (Fig. 2.3); hind wings well developed in both sexes **3**

3(2). Body shorter and broader, elytra less than twice as long as wide; protibia without setose excavation at basal third in males; apical third of elytra with white scales bordering suture (Fig. 2.1; Pl. 14I–J); eastern and central Australia .... ***O. mastersi* Macleay**

– Body more elongate, elytra more than twice as long as wide; protibia with clear setose excavation at basal third in males; apical third of elytra without white scales bordering suture (Pl. 14H); WA .... ***O. sagitta* Neboiss**

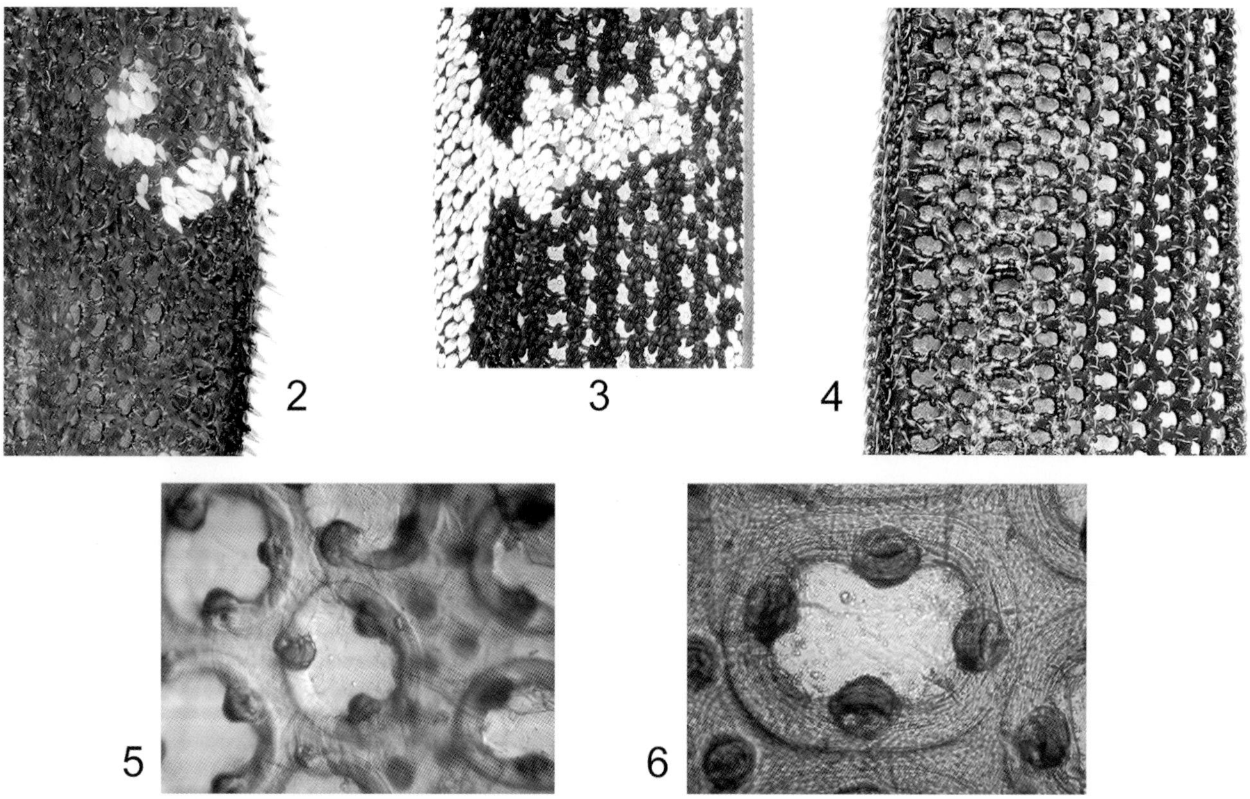

Figs 2.2–2.6. Australian Ommatidae. 2, *Omma rutherfordi* Lawrence, portion of elytron; 3, *O. mastersi* Macleay, portion of elytron; 4, *O. stanleyi* Newman, portion of elytron; 5, *Omma mastersi* Macleay, elytral window puncture; 6, *O. stanleyi* Newman, elytral window puncture.

References

Ashman LG, Oberprieler RG, Ślipiński A (2015) *Rhopalomma stefaniae* gen. et sp. n., the first ommatid beetle from the Upper Jurassic in Australia (Coleoptera: Archostemata: Ommatidae). *Zootaxa* **3980**(1), 136–142. doi:10.11646/zootaxa.3980.1.8

Beutel RG, Ge SQ, Hörnschemeyer T (2007) On the head morphology of *Tetraphalerus*, the phylogeny of Archostemata and the basal branching events in Coleoptera. *Cladistics* **23**, 1–29.

Cai C, Huang D (2016) *Omma daxishanense* sp. nov., a fossil representative of an extant Australian endemic genus recorded from the Late Jurassic of China (Coleoptera: Ommatidae). *Alcheringa: An Australasian Journal of Palaeontology* **3**, 1–7.

Crowson RA (1962) Observations on the beetle family Cupedidae, with descriptions of two new forms and a key to the recent generał. *Annals & Magazine of Natural History* **13**(51), 147–157. doi:10.1080/00222936208651227

Grebennikov VV (2004) Review of larval morphology of beetles of the suborder Archostemata (Insecta: Coleoptera), including first-instar chaetotaxy. *European Journal of Entomology* **101**, 273–292. doi:10.14411/eje.2004.038

Hörnschemeyer T, Beutel RG (2016) 5.4. Ommatidae Sharp & Muir, 1912. In *Handbuch der Zoologie/Handbook of Zoology. BandVolume IV Arthropoda: Insecta. Teilband/Part 38. Coleoptera, Beetles. Volume 1: Morphology and Systematics (Archostemata, Adephaga, Myxophaga, Polyphaga partim).* 2nd edn. (Eds RG Beutel and RAB Leschen) pp. 52–56. Walter de Gruyter, Berlin.

Hünefeld F, Marvaldi SE, Nüller B, Lawrence JF, Beutel RG (2011) The male postabdomen of the "ancestral" archostematan beetle *Tetraphalerus bruchi* Heller, 1913 (Ommatidae) and its phylogenetic significance. *Arthropod Structure & Development* **40**, 146–158. doi:10.1016/j.asd.2010.12.003

Jarzembowski EA, Wang B (2016) An unusual basal beetle from Myanmar (Coleoptera: Archostemata). *Alcheringia* **40**, 297–302. doi:10.1080/03115518.2016.1132493

Jarzembowski EA, Yan EV, Wang B, Zhang H (2013a) Brochocolein beetles (Insecta: Coleoptera) from the Lower Cretaceous of northeast China and southern England. *Cretaceous Research* **44**, 1–11. doi:10.1016/j.cretres.2013.03.003

Jarzembowski EA, Yan EV, Wang B, Zhang H (2013b) Ommatin beetles (Insecta: Coleoptera) from the Lower Cretaceous of northeast China and southern England. *Terrestrial Arthropod Reviews* **6**, 135–161. doi:10.1163/18749836-06021062

Jarzembowski EA, Wang B, Zhang H, Fang Y (2015) Boring beetles are not necessarily dull: New notocupedins (Insecta: Coleoptera) from the Mesozoic of Eurasia and East Gondwana. *Cretaceous Research* **52**(B), 431–439.

Jarzembowski EA, Wang B, Zheng D (2017) A new ommatin beetle (Insecta: Coleoptera) with unusual genitalia from mid-Cretaceous Burmese amber. *Cretaceous Research* **71**, 113–117. doi:10.1016/j.cretres.2016.10.010

Kirejtshuk AG, Nel A, Collomb FM (2010a) New Archostemata (Insecta: Coleoptera) from the French Paleocene and early Eocene, with a note on the composition of the suborder. *Annales de la Société Entomologique de France* (N.S.) **46**(1–2), 216–227.

Kirejtshuk AG, Ponomarenko AG, Prokin AA, Chang H, Nikolajev GV, Ren D (2010b) Current knowledge of Mesozoic Coleoptera from Daohugou and Liaoning (northeast China). *Acta Geologica Sinica* **84**(4), 783–792[English Edition]. doi:10.1111/j.1755-6724.2010.00253.x

Kirejtshuk AG, Nel A, Kirejtshuk PA (2016) Taxonomy of the reticulate beetles of the subfamily Cupedinae (Coleoptera: Archostemata), with a review of the historical development. *Invertebrate Zoology* **13**(2), 61–190.

Lawrence JF (1999) The Australian Ommatidae (Coleoptera: Archostemata): new species, larva and discussion of relationships. *Invertebrate Taxonomy* **13**, 369–390. doi:10.1071/IT99008

Martin SK (2010) Early Jurassic coleopterans from the Mintaja insect locality, Western Australia. *Acta Geologica Sinica* **84**(4), 925–953 [English Edition]. doi:10.1111/j.1755-6724.2010.00276.x

Monrós F, Monrós MM (1952) Las especies Argentinas de Cupedidae (Coleoptera). *Anales de la Sociedad Científica Argentina* **153**, 19–41.

Neboiss A (1968) The type of *Omma mastersi* Macleay (Coleoptera: Cupedidae). *Memoirs of the National Museum of Victoria* **28**, 21. doi:10.24199/j.mmv.1968.28.05

Neboiss A (1989) New species of Archostemata (Coleoptera, Ommatidae, Cupedidae). *Revue Française d'Entomologie* (N S.) **11**, 109–115.

Ponomarenko AG (1964) New beetles of the family Cupedidae from the Jurassic of Karatau. *Палеонтологический журнал* **2**, 49–61 [in Russian].

Ponomarenko AG (1966) New beetles of the family Cupedidae (Coleoptera) from Mesozoic deposits of Transbaikalia. *Entomologicheskoe Obozrenie* **1**, 138–143[in Russian].

Ponomarenko AG (1968) Preliminary review of the Jurassic beetles of the Karatau. In *Jurassic Insects of the Karatau.* (Ed. BB Rohdendorf) pp. 114–117. Nauka, Moscow.

Ponomarenko AG (1969) Historical development of the Coleoptera-Archostemata. *Trudy Paleontologicheskogo Instituta* **125**, 1–240[in Russian].

Ponomarenko AG (1994) New Mesozoic cupedid beetles from Mongolia. Brochocoleini and Notocupedini. *Палеонтологический журнал* **28**(3), 102–115.

Ponomarenko AG (1997) New beetles of the family Cupedidae from the Mesozoic of Mongolia. Ommatini, Mesocupedini, Priacmini. *Paleontologicheskii Zhurnal* **31**(4), 45–55 [in Russian; translation in *Paleontological Journal* **31**(4), 389–399].

Ponomarenko AG (2000) Beetles of the family Cupedidae from the Lower Cretaceous locality of Semen, Transbaykalia. *Paleontological Journal* **34**(Suppl 3), 317–322.

Ponomarenko AG (2006) On the types of Mesozoic archostematan beetles (Insecta, Coleoptera, Archostemata) in the Natural History Museum, London. *Paleontological Journal* **40**(1), 90–99. doi:10.1134/S0031030106010102

Ren D, Tan J, Ge S (2006) New fossil ommatid (Coleoptera: Archostemata: Ommatidae) from Jehol Biota of western Liaoning, China. *Progress in Natural Science* **16**(6), 639–643. doi:10.1080/10020070612330046

Sarmiento JAV (1969) El systema reproductor masculinoi y organos de copulation en Cupesidae (Coleoptera – Archostemata). *Revista de la Sociedad Entomológica Argentina* **31**(1–4), 43–48.

Soriano C, Delclòs X (2006) New cupedids from the lower Cretaceous of Spain and the palaeogeography of the family. *Acta Palaeontologica Polonica* **51**, 185–200.

Tan J, Ren D, Liu M (2005) New ommatids from the Late Jurassic of western Liaoning, China (Coleoptera: Archostemata). *Insect Science* **12**(3), 207–216. doi:10.1111/j.1005-295X.2005.00026.x

Tan J, Ren D, Shih C, Ge S (2006) New fossil beetles of the family Ommatidae (Coleoptera: Archostemata) from the Jehol Biota of China. *Acta Geologica Sinica* **80**(4), 474–485.

Tan J, Ren D, Shih C (2007) New ommatids of Ommatinae (Coleoptera: Archostemata: Ommatidae) from the Yixian formation of western Liaoning, China. *Progress in Natural Science* **17**(7), 803–811. doi:10.1080/10002007088537476

Tan J, Wang Y, Ren D, Yang X (2012) New fossil species of ommatids (Coleoptera: Archostemata) from the Middle Mesozoic of China illuminating the phylogeny of Ommatidae. *BMC Evolutionary Biology* **12**, 113, 19 pp.

Vulcano MA, Pereira FS (1975) Cupesidae (Coleoptera). *Arquivos do Instituto Biologico* **42**, 31–68.

3. CUPEDIDAE LAPORTE, 1836

John F. Lawrence and Hermes E. Escalona

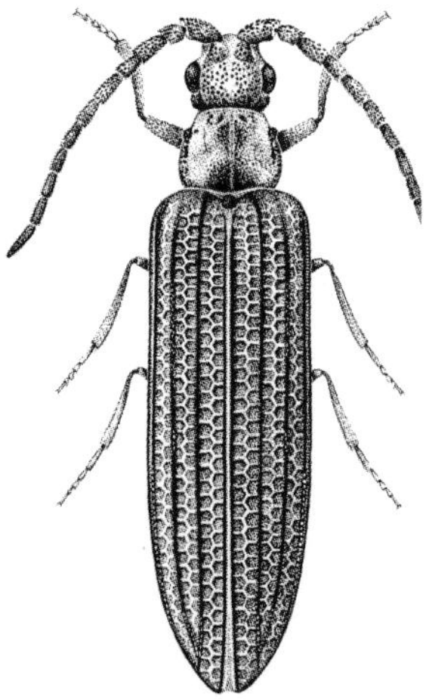

Fig. 3.1. *Adinolepis matthesonae* (Neboiss).

Common name. Reticulated beetles

Introduction. The family Cupedidae is a relictual group, containing 31 extant species in nine genera. According to Lawrence (1999), the modern cupedid fauna includes two subfamilies: Priacminae (Crowson 1962) with *Priacma serrata* (LeConte) from north-western North America and *Paracupes brasiliensis* (Kolbe) from Brazil and eastern Ecuador, and Cupedinae with *Cupes* Fabricius (eastern North America), *Prolixocupes* Neboiss (Chile-Argentina and south-western North America), *Tenomerga* Neboiss (Holarctic, Afrotropical and Oriental), *Rhipsideigma* Neboiss (Afrotropical), *Ascioplaga* Neboiss (Australia and New Caledonia) and the endemic Australian genera *Distocupes* Neboiss and *Adinolepis* Neboiss. Relationships of some of these genera are under scrutiny and some authors (Kirejtshuk 2005; Kirejtshuk *et al.* 2010a, 2010b, 2016) would dispense with the Priacminae, as well as the genera *Distocupes* and *Tenomerga* (= *Cupes*) and consider the families Cupedidae and Ommatidae to be subfamilies of Cupedidae extending back to the Lower Triassic. There have been two cladistic analyses of modern cupedids: one based primarily on head morphology (Hörnschemeyer *et al.* 2006) and another using a variety of morphological features (Hörnschemeyer 2009). In the first, a monophylum was recognised containing *Distocupes*, *Adinolepis* and *Ascioplaga* based mainly on the type of cuticular scales and the presence of a deep ventral mandibular cavity. This clade was sister to one containing *Rhipsidiegma*, *Prolixocupes*, *Cupes* and *Tenomerga*, which in turn, formed a clade with two basal lineages for *Priacma* and *Paracupes*. In the later study, there was a succession of clades for Priacmini, *Prolixocupes* and *Ascioplaga* + *Adinolepis*, with a large derived clade with *Distocupes* at the base and two main clades: the South African *Tenomerga leucophaea* + *Cupes capitatus* + *Rhipsideigma* sister to the remaining eleven *Tenomerga* species (Ge & Yang 2004; Hörnschemeyer & Yavorskaya 2016). Most references on Mesozoic fossils currently placed in this family may be found in the section on Archostemata; more recently, two new genera have been described from Cretaceous amber from Myanmar (Jarzembowski *et al.* 2017a, 2017b, 2017c).

Biology. Cupedid larvae are known to bore into dead wood that has been attacked by wood-rotting fungi. Snyder (1913) found larvae of the North American *Tenomerga cinerea* (Say) in wood infested with *Daedalea quercina* (Polyporaceae), while those of the Asian *T. mucida* (Chevrolat) were associated with a species of *Stromatoscypha* (Meripilaceae) (Fukuda 1941). Larvae of the Australian *Distocupes varians* (Lea) have been collected in structural timber and also in old rotten pine logs on the ground (Neboiss 1960, 1968, 1984, 1987; Lawrence 1999).

Characteristics. Adults. Elongate, strongly flattened, from 6 to 17 mm in length. Body surfaces entirely or mostly covered with scale-like setae. Head prognathous, constricted posteriorly to form a narrow neck and distinct temples. Eyes relatively large and more or less protuberant, entire, very finely facetted, without interfacetal setae. Antennal insertions dorsal and relatively narrowly separated, not concealed by frontal ridges. Head dorsally with one (*Adinolepis* and *Ascioplaga*) or two (*Distocupes*) pairs of subconical tubercles. Frontoclypeal suture and subantennal grooves absent. Labrum free, very small, apically truncate. Antennae moderately long, 11-segmented and filiform, with basal two to several antennomeres covered with scales. Mandible short and broad, unidentate with one or two subapical teeth; without mola or prostheca. Maxilla highly reduced and partly concealed by mentum, galea and lacinia hyaline and setose, the former very slender, the latter broad, without uncus. Terminal maxillary palpomere expanded and securiform with patch of sensilla lying on surface and not in deep cavity. Mentum distinctly transverse, concealing palpal insertions and prementum, the latter consisting of two trilobed, membranous processes; terminal labial palpomere expanded and secufirom. Gular sutures well separated; gula longer than wide. Corpotentorium weakly developed or absent. Pronotum widest at about middle, distinctly narrower at base than combined elytral bases. Lateral pronotal carinae complete, anterior angles usually slightly to moderately produced laterally (strongly produced forward in *Ascioplaga*). Prosternum in front of coxae moderately long and convex, with paired oblique crural impressions extending mesally almost to midline, where separated by a short carina, and laterally across the pleuron and onto the pronotal hypomeron. Notopleural and pleurosternal sutures separated posteriorly on either side of elongate, slender, anteriorly acute pleuron, the sutures uniting just before the anterior edge and continuing as a short notosternal suture. Prosternal process complete, narrowed posteriorly with narrowly rounded apex. Procoxae subglobular, not strongly projecting, without concealed lateral extensions, with distinct coxal plates; trochantin large, well sclerotised, broadly exposed and abutting sternum and pleuron. Procoxal cavities slightly transverse, narrowly separated, broadly open externally and internally. Scutellar shield abruptly elevated, anteriorly simple, posteriorly expanded and rounded to truncate. Elytra complete, usually with 5 longitudinal carinae (including lateral and sutural edges); epipleura narrow and complete. Mesoventrite with small cavity and short discrimen. Mesoventral process divided into two short, acute processes. Mesocoxae conical and projecting, with broadly exposed trochantins. Posterior portion of mesanepisterna, and entire mesepimeron concave, forming part of oblique crural impression for mid leg. Mesocoxal cavities contiguous, broadly closed laterally by mesepimera and metanepisterna. Metaventrite strongly narrowed anteriorly; discrimen more than half as long as ventrite and transverse (metakatepisternal) suture well developed. Metanepisternum completely exposed with anterior concavity and longitudinal groove forming part of crural impression for mid leg. Exposed portion of metepimeron narrowly elongate. Metacoxae strongly transverse with subconical mesal projections, contiguous, extending laterally to meet elytra, with well developed coxal plates. Metendosternite with a long, slender stalk, short, narrow ventrolateral processes, short, broad lateral arms, and a very long and slender anterior process bearing approximate anterior tendons. Hind wing moderately broad, with very short bending zone near apex of radial bar (anterior wing strut); apical field very short, with three radial extensions; pterostigma absent; RP complete to wing base, usually joined to RA_{3+4} by two cross-veins; RP and MP joined by two cross-veins forming the oblongum cell; MP_{1+2} with hinge crossed by fold and with short, straight medial spur; medial field with four or five free veins and no medial fleck, wedge cell well developed and apically truncate; anal lobe well developed, without embayment; AP_{3+4} forked. Legs short and slender; tibia distinctly longer than femur; tibial spurs small and paired on all legs. Tarsi 5–5–5; tarsomere 4 lobed beneath and tarsomeres 1–4 densely setose beneath; pretarsal claws simple; empodium not apparent. Abdomen with five ventrites, which are flattened but slightly overlapping; ventrite 1 distinctly shorter than 2, with well developed, narrowly acute intercoxal process. Functional spiracles on abdominal segments I–VIII. Anterior edge of segment IX in male broadly rounded; posterior edge narrowly rounded or acute; tergite IX at least partly fused to X. Aedeagus of modified trilobate type, with parameres completely fused to phallobase and partially fused together, each with a small hook near apex; penis with short basal struts. In *Distocupes*, testes are of the follicular type. Male diploid chromosome number is $2n = 19$, with a male sex chromosome system of the XO type and achiasmatic meiosis (Galián & Lawrence 1993). Sternite VIII in female with long, fixed spiculum ventrale. Ovipositor narrowly elongate, lightly pigmented except for bacula; paraprocts much longer than gonocoxites, which bear well developed gonostyli. (Hörnschemeyer & Yavorskaya 2016).

Larvae. Elongate, more or less parallel-sided, lightly pigmented except for mandibles and apex of tergum IX; surfaces smooth, clothed with scattered long setae. Head prognathous and protracted. Posterior edge of head capsule deeply emarginate; epicranial suture not apparent; median endocarina simple, not forked, extending as far as clypeus. Stemmata usually absent (one on each side in first instar of *Distocupes* Neboiss). Frontoclypeal suture indistinct. Labrum free, strongly transverse. Antennae in late instars moderately long and usually 4-segmented, with sensorium on antennomere 3 about half as long as antennomere 4; antennae of first instar 2-segmented with sensorium almost as long as apical antennomere. Mandibles tridentate; molae large, slightly asymmetrical and irregular with no asperities or transverse ridges; prostheca absent. Ventral mouthparts retracted forming maxillolabial complex, with well developed maxillary articulating area. Galea articulated and palpiform, lacinia fixed and apically rounded with several stout spines and setae. Maxillary palp 3-segmented with distinct palpifer. Labium consisting of prementum and postmentum, the former with a well developed ligular sclerome. Labial palps 2 segmented and widely separated. Hypostomal rods long and diverging; ventral epicranial ridges absent. Gular sutures separated; gula transverse. Meso- and metatergum with strongly transverse ambulatory ampulla. Thoracic venter with large field of asperities. Legs reduced, 6-segmented, with two unequal pretarsal claws or with claws fused together. Abdomen with transverse tergal ampulae present on most abdominal segments. Tergum IX slightly longer than VIII, apically narrowed to form narrow truncate process. Sternite IX exposed, simple; segment X posteroventrally oriented, with pair of large, globular projecting lobes. Anal opening Y-shaped, between these lobes and a weak transverse dorsal lobe. Spiracles annular, with closing apparatus. (Neboiss 1968; Grebennikov 2004; Yavorskaya *et al*. 2015).

Key to the Australian genera and species

1. Six or more basal antennomeres clothed with scales (Figs 3.4, 3.7); antennae shorter, extending at most to basal fourth of elytra (Pl. 14A–D. N–O); head with cylindrical, apically rounded scales intermixed with flattened scales (Figs 3.2, 3.4, 3.7); length usually less than 10 mm...***Adinolepis*** **2**

– Fewer than six basal antennomeres clothed with scales (Figs 3.5, 3.8–3.9); antennae longer, usually extending beyond basal third of elytra; head without cylindrical, apically rounded scales (Figs 3.3, 3.5); length almost always greater than 10 mm 4

2(1). Eyes much larger than temples and extending laterally beyond the edges of temples (Fig. 3.4; Pl. 14A–B); anterior edge of lateral pronotal carina with distinct tooth, so that pronotal width at this point is as great as maximum pronotal width; 6–9 mm (Pl. 14A–B); VIC and NSW to southern QLD .... ***A. yuanga*** **(Neboiss)**

– Eyes not or only slightly longer than temples and not extending laterally beyond their edges (Figs 3.2, 3.7; Pl. 14C–D, N–O); anterior edge of pronotal carina without or with very small tooth, so that width at this point is always less than maximum pronotal width **3**

3(2). Elytral scales more generally distributed, usually extending onto window punctures (Fig. 3.6); colour darker with white scales forming three irregular transverse bands; 6.3–7 mm (Pl. 14N–O); VIC, SA, and NSW .... ***A. eumana*** **(Neboiss)**

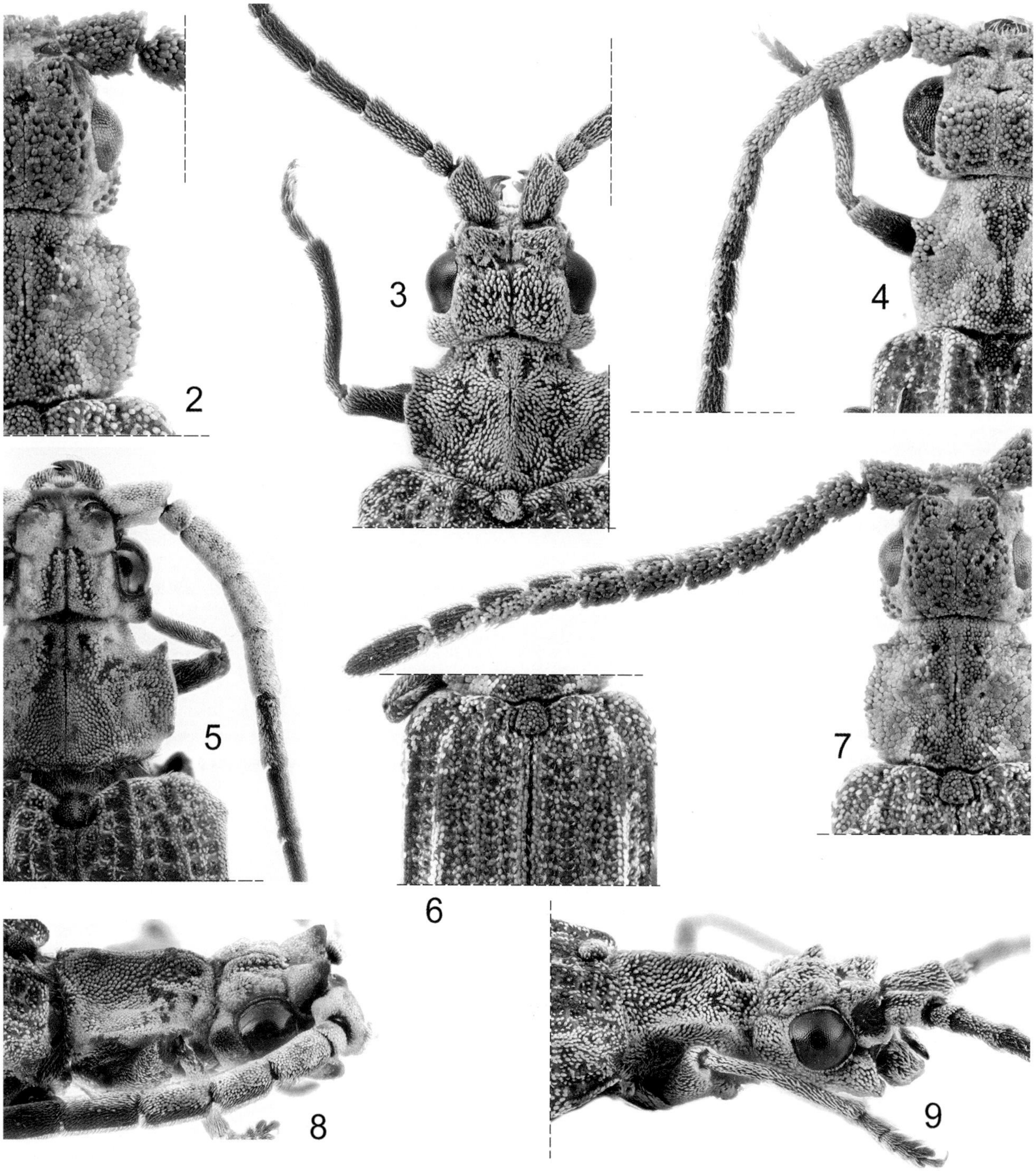

Figs 3.2–3.9. Australian Cupedidae. 2, *Adinolepis eumana* (Neboiss), head and prothorax, dorsal, showing less prominent eye and lack of distinct lateral pronotal tooth; 3, *Distocupes varians* (Lea), head and prothorax, dorsal, showing lack of apically rounded scales and presence of conical tubercles over antennal bases and eyes; 4, *Adinolepis youanga* (Neboiss), head and prothorax, dorsal, showing promient eyes and well developed lateral pronotal tooth; 5, *Ascioplaga scalena* (Neboiss), head and prothorax, dorsal, showing the first 5 antennomeres clothed with scales and a single conical tubercle over the antennal base; 6, *Adinolepis eumana* (Neboiss), anterior portion of elytra, dorsal, showing the more extensive covering of scales; 7, *Adinolepis eumana* (Neboiss), head and prothorax, dorsal, showing more than 5 antennomeres clothed with scales; 8, *Ascioplaga scalena* (Neboiss), head and prothorax, lateral, showing the first 5 antennomeres clothed with scales and a single conical tubercle over the antennal base; 9, *Distocupes varians* (Lea), head and prothorax, lateral, showing conical tubercles over both antennal bases and eyes.

– Elytral scales distributed mainly on the longitudinal costae and not or rarely extending onto the window punctures; colour lighter, except for dark areas alternating with white ones on costae (Fig. 3.1; Pl. 14); 7–10 mm (Pl. 14C–D); VIC, NSW and ACT 4
4(3). Elevated elytral interval 3 with three sections of pale scales; VIC, NSW and ACT .... ***A. mathesonae* (Neboiss)**
– Elevated elytral interval 3 with two sections of pale scales; WA *A. apodema* Neboiss
5(1). Two basal antennomeres clothed with scales (Fig. 3.9); head with two pairs of conical tubercles, one above antennal bases and one above eyes (Fig. 3.9); 10–14.5 mm (Pl. 14K–L); VIC, ACT, NSW, southern QLD .... ***Distocupes varians* (Lea)**
– Five basal antennomeres clothed with scales (Figs 3.5, 3.8); head with one pair of conical tubercles above antennal bases (Fig. 3.8); 16.2 mm (Pl. 14D); northern QLD .... ***Ascioplaga scalena* (Neboiss)**

Classification of the Australian genera

Distocupes Neboiss, 1984 (Pl. 14K–L; Figs 3.3, 3.9)

Type species. *Cupes varians* Lea, 1902.
Australian species. One.
Distribution. *Distocupes varians* (Lea) is known from TAS, VIC, NSW and QLD.
Biology. Larvae of this species have been taken in structural timber and in partly buried pine logs in a garden. As in *Tenomerga mucida* (Chevrolat) (Yavorskaya *et al.* 2015), the first instar of *D. varians* has 2-segmented antennae.
References. Neboiss (1960, 1968, 1984).

Adinolepis Neboiss, 1984 (Pl. 14A–D, N–O; Figs 3.2, 3.4, 3.6, 3.7)

Type species. *Cupes eumana* Neboiss, 1960.
Australian species. Four.
Distribution. Three of the four species of *Adinolepis* occur in the south-eastern part of the continent (VIC, NSW and QLD), while *A. apodema* Neboiss is known from southern WA. Hörnschemeyer (2009) suggested that *A. apodema* might be a colour variant of *A. mathesonae* Neboiss, which is known from the south-eastern part of the continent.
References. Neboiss (1960, 1984, 1987)

Ascioplaga Neboiss, 1960 (Pl. 14E; Figs 3.5, 3.8)

Type species. *A. mimeta* Neboiss, 1960.
Australian species. One.
Distribution. The genus *Ascioplaga* occurs in New Caledonia and Australia. *Ascioplaga scalena* (Neboiss) was described from northern QLD.
References. Neboiss (1984); Hörnschemeyer (2009).

References

Crowson RA (1962) Observations on the beetle family Cupedidae, with descriptions of two new forms and a key to the recent genera. *Annals & Magazine of Natural History* **5**(51), 147–157. doi:10.1080/00222936208651227

Fukuda A (1941) Some ecological studies on *Cupes clathratus*. *Transactions of the Natural History Society of Formosa* **31**, 394–399.

Galián J, Lawrence JF (1993) First karyotypic data on a cupedid beetle (Coleoptera: Archostemata) showing achiasmatic meiosis. *Entomological News* **104**(2), 83–87.

Ge SQ, Yang XK (2004) Two new Chinese species of *Tenomerga* Neboiss (Coleoptera: Cupedidae), with a world catalog of the genus. *Proceedings of the Entomological Society of Washington* **106**(3), 631–638.

Grebennikov VV (2004) Review of larval morphology of beetles of the suborder Archostemata (Insecta: Coleoptera), including first-instar chaetotaxy. *European Journal of Entomology* **101**, 273–292. doi:10.14411/eje.2004.038

Hörnschemeyer T (2009) The species-level phylogeny of archostematan beetles – where do *Micromalthus debilis* and *Crowsoniella relicta* belong? *Systematic Entomology* **34**, 533–558. doi:10.1111/j.1365-3113.2009.00476.x

Hörnschemeyer T, Yavorskaya M (2016) 5.2. Cupedidae Laporte, 1832. Introduction and phylogeny. In *Handbuch der Zoologie/Handbook of Zoology. Band/Volume IV Arthropoda: Insecta. Teilband/Part 38. Coleoptera, Beetles. Volume 1: Morphology and Systematics (Archostemata, Adephaga, Myxophaga, Polyphaga partim)*. 2nd edn. (Eds RG Beutel and RAB Leschen) pp. 44–47. Walter de Gruyter, Berlin.

Hörnschemeyer T, Goebbels J, Weidemann G, Faber C, Haase A (2006) The head morphology of *Ascioplaga mimeta* (Coleoptera: Archostemata) and the phylogeny of Archostemata. *European Journal of Entomology* **103**, 409–423. doi:10.14411/eje.2006.055

Jarzembowski EA, Wang B, Zheng D (2017a) The first cupedine beetle from Burmese amber (Coleoptera: Cupedidae). *Comptes Rendus. Palévol* **16**, 241–247. doi:10.1016/j.crpv.2016.08.004

Jarzembowski EA, Wang B, Zheng D (2017b) A new reticulated beetle (Coleoptera: Cupedidae) with aedeagus preserved from mid-Cretaceous amber of Myanmar. *Cretaceous Research* **80**, 86–90. doi:10.1016/j.cretres.2017.08.015

Jarzembowski EA, Wang B, Zheng D (2017c) A new spiny rerticulated beetle (Coleoptera: Cupedidae) from Cretaceous Burmese amber. *Proceedings of the Geologist's Association*. doi:10.1016/j.pgeola.2017.07.003.

Kirejtshuk AG (2005) A revision of the genus *Cupes* Fabricius, 1801 from Baltic amber and some notes on taxonomy and composition of the family Cupedidae (Coleoptera, Archostemata). *Mitteilungen aus dem Geologisch-Paläontologischen Institut der Universität Hamburg* **89**, 55–84.

Kirejtshuk AG, Nel A, Collomb FM (2010a) New Archostemata (Insecta: Coleoptera) from the French Paleocene and early Eocene, with a note on the composition of the suborder. *Annales de la Société Entomologique de France* (N.S.) **46**(1–2), 216–227.

Kirejtshuk AG, Ponomarenko AG, Prokin AA, Chang H, Nikolajev GV, Ren D (2010b) Current knowledge of Mesozoic Coleoptera from Daohugou and Liaoning (northeast China). *Acta Geologica Sinica* **84**(4), 783–792[English Edition]. doi:10.1111/j.1755-6724.2010.00253.x

Kirejtshuk AG, Nel A, Kirejtshuk PA (2016) Taxonomy of the reticulate beetles of the subfamily Cupedinae (Coleoptera: Archostemata), with a review of the historical development. *Invertebrate Zoology* **13**(2), 61–190.

Lawrence JF (1999) The Australian Ommatidae (Coleoptera: Archostemata): new species, larva and discussion of relationships. *Invertebrate Taxonomy* **13**, 369–390. doi:10.1071/IT99008

Neboiss A (1960) On the family Cupedidae, Coleoptera. *Proceedings of the Royal Society of Victoria* **72**, 12–20.

Neboiss A (1968) Larva and pupa of *Cupes varians* Lea, and some observations on its biology (Coleoptera; Cupedidae). *Memoirs of the National Museum of Victoria* **28**, 17–19. doi:10.24199/j.mmv.1968.28.04

Neboiss A (1984) Reclassification of *Cupes* Fabricius (s. lat.), with descriptions of new genera and species (Cupedidae: Coleoptera). *Systematic Entomology* **9**, 443–477. doi:10.1111/j.1365-3113.1984.tb00520.x

Neboiss A (1987) A new species of *Adinolepis* Neboiss from Western Australia (Cupedidae: Coleoptera). *Records of the Western Australian Museum* **13**, 323–325.

Snyder TE (1913) Record of the rearing of *Cupes concolor* Westw. (Coleoptera). *Proceedings of the Entomological Society of Washington* **15**, 30–31, pl. 1.

Yavorskaya MI, Kojima K, Machida R, Beutel RG (2015) Morphology of the first instar larva of *Tenomerga mucida* (Chevrolat, 1829) (Coleoptera: Archostemata: Cupedidae). *Arthropod Systematics & Phylogeny* **73**(2), 239–258.

4. MYXOPHAGA CROWSON, 1955

John F. Lawrence

The suborder Myxophaga contains four small families of minute beetles inhabiting aquatic or riparian habitats: Lepiceridae (Mexico to northern South America, Cretaceous of South-east Asia), Torridincolidae (South America, southern and central Africa, Madagascar, Japan and China), Sphaeriusidae (North to South America, Asia, Madagascar, Africa, Australia) and Hydroscaphidae (western North America, Mexico to South America, Eurasia, North Africa, South-east Asia and Madagascar) (Beutel 1999; Navarrete-Heredia *et al.* 2005; Anton & Beutel 2006; Kirejtshuk & Poinar 2006, 2013; Flowers *et al.* 2010; Ge *et al.* 2010; Sampaio & Ferreira 2013; Shepard *et al.* 2013; Short *et al.* 2015).

Adult subordinal autapomorphies include the absence of a maxillary galea, a movable tooth on the left mandible, a broad connection between the mesoventrite and metaventrite and long fringe-hairs lining the margins of the hindwings. The exposure of the propleuron, which extends anteriorly to the edge of the prothorax and posteriorly behind the procoxae, is a feature shared with both Archostemata and Adephaga, but in myxophagan adults, unlike those of the other two suborders, prothoracic trochantin is solidly attached to the coxa, separated by a suture in *Lepicerus* but solidly fused in the other myxophagan families to form a trochantinopleuron, as in Polyphaga. Wing venation and folding in Myxophaga, although highly reduced due to small size, is similar to that in Archostemata and Adephaga and very different from that in Polyphaga: (1) RA_{1+2} and RA_{3+4} diverge but never converge again to form a radial cell; and the latter extends into the apical field; (2) the posterior wing strut (MP_{1+2}) has a distinct medial hinge through which a fold passes; (3) an oblongum cell is almost always formed between RP and MP_{1+2}; (4) there is usually more than one cross-vein between RA and RP. The spiral rolling of the apical field is a character shared with Archostemata but not with Adephaga. Other features of adult Myxophaga include the small size (less than 2.8 mm but usually much smaller), antennae usually less than 10-segmented (11-segmented in *Delevea*, *Satonius* and *Sphaerius*), with the terminal antennomere more or less enlarged (with a compact 3-segmented club in *Sphaerius*), mandibular mola present; mesoventrite very short, mesocoxal cavities partly closed laterally by the mesepimeron and a narrow portion of the metanepisternum in Torridincolidae, but by the mesepimeron only in the other families. Internal features include the six free Malpighian tubules, tube-like, coiled testes and a telotrophic ovary (*Hydroscapha*).

Myxophagan larvae resemble those of basal Polyphaga in having the tibia and tarsus fused to form a tibiotarsus and microphagous mouthparts including a well developed, tuberculate or asperate mandibular mola. Although both of these features have been used to support a sister-group relationship between the two suborders, there are several structural differences in the mouthparts that suggest an independent development of microphagy in the two groups. Basal polyphagan larvae, for instance, have structures such as the hypopharyngeal bracon, hypopharyngeal sclerome and accessory ventral process on the mandible, which are absent in Myxophaga, while myxophagan larvae have sensory papillae on the labium. A distinctive feature of myxophagan larvae is the presence of three to six relatively large stemmata borne on a distinct lateral protuberance. The most prominent feature of almost all myxophagan larvae is the presence of spiracular gills, but the segmented structures of Torridincolidae differ significantly from the balloon-like gills of Hydroscaphiae and Sphaeriusidae. Tracheal gills are absent in the presumed larva of Lepiceridae (Lawrence *et al.* 2013), but there is some indirect evidence that this larva may have been misidentified. Antennae in most myxophagan larvae are 2-segmented, with the long, slender and more or less acute sensorium arising from antennomere 1 in Hydroscaphidae but from the middle or near the apex of antennomere 2 in Sphaeriusidae and Torridincolidae. Additional information on both larval and adult morphology may be found in Arce-Peréz *et al.* 2016, Beutel 1998, Beutel 2016, Beutel & Arce-Peréz 2016, Beutel & Haas 1998, Beutel & Vanin 2016, Beutel *et al.* 1999, Büning 2005, Vanin *et al.* 2016.

Crowson (1960) considered this taxon to be the sister group of Polyphaga based mainly on the fusion of the larval tibia and tarsus, and this conclusion was supported by Beutel & Haas (2000) and several other studies, but not by McKenna *et al.* (2015), who concluded that Archostemata + Myxophaga + Adephaga were sister to Polyphaga. Ponomarenko (1969) suggested that Recent Myxophaga may be descendants of the schizophoroid Archostemata (Ponomarenko 1969), especially the primarily Triassic and Jurassic Catiniidae, but at least two catiniid genera from the Jurassic and early Cretaceous, *Cervicatinus* Tan & Ren (2007) and *Forticatinius* Tan & Ren (2007), have been shown to be members of the family Trogossitidae (Polyphaga: Cucujiformia) (Kirejtshuk *et al.* 2010; Yu *et al.* 2012, 2015; Kirejtshuk 2017). Within the Myxophaga, morphological cladograms based on adults of all families and larvae of all except Lepiceridae (Beutel 1999) indicated that Lepiceridae was sister to Sphaeriusidae, Hydroscaphidae and Torridincolidae, but in cladograms based on molecular data (McKenna *et al.* 2015), Hydroscaphidae was sister to the remaining families and Sphaeriusidae + Lepiceridae sister to Torridincolidae. Fossil Lepiceridae from Burmese amber are in a confused state due in part to the poor condition of the types leading to morphological misinterpretations; at present, the fossils are placed in two genera, *Haplochelus* and *Lepiceroides*, by Kirejtshuk and Poinar (2006, 2013) but considered to be species of *Lepicerus* by Jałoszyński *et al.* (2017).

Myxophaga occur in various riparian or aquatic situations, such as in mud and sand in and at the edges of streams and ponds, in hot springs (some *Hydroscapha*) and on the surfaces of rocks which are covered with a moving film of water or constantly sprayed by waterfalls (hygropetric habitats). Although adults of Torridincolidae may have a ventral plastron (Hinton 1969) and larvae of Torridincolidae, Sphaeriusidae and Hydroscaphidae possess spiracular gills (tracheal gills) (Hinton 1967). The suborder is poorly represented in Australia by a single species of *Sphaerius* (see Chapter 5).

References

Anton E, Beutel RG (2006) On the head morphology of Lepiceridae (Coleoptera: Myxophaga) and the systematic position of the family and suborder. *European Journal of Entomology* **103**, 85–95. doi:10.14411/eje.2006.014

Arce-Pérez J, Navarrete-Heredia JL, Beutel RG (2016) 6.1. Lepiceridae Hinton, 1936 (= Cyathoceridae). In *Handbuch der Zoologie/Handbook of Zoology. BandVolume IV Arthropoda: Insecta. Teilband/Part 38. Coleoptera, Beetles. Volume 1: Morphology and Systematics (Archostemata, Adephaga, Myxophaga, Polyphaga partim).* 2nd edn. (Eds RG Beutel and RAB Leschen) pp. 65–67. Walter de Gruyter, Berlin.

Beutel RG (1998) Torridincolidae: II. Description of the larva of *Satonius kurosawai* (Satô, 1982) (Coleoptera). In *Water Beetles of China*. Vol. II. (Eds MA Jäch and L Ji) pp. 53–59. Zoologisch-Botanische Gesellschaft in Österreich & Wiener Coleopterologenverein, Vienna.

Beutel RG (1999) Phylogenetic analysis of Myxophaga (Coleoptera) with a redescription of *Lepicerus horni* (Lepiceridae). *Zoologischer Anzeiger* **237**(1998/1999), 291–308.

Beutel RG (2016) 6. Myxophaga Crowson, 1955. Introduction, phylogeny. In *Handbuch der Zoologie/Handbook of Zoology. BandVolume IV Arthropoda: Insecta. Teilband/Part 38. Coleoptera, Beetles. Volume 1: Morphology and Systematics (Archostemata, Adephaga, Myxophaga, Polyphaga partim).* 2nd edn. (Eds RG Beutel and RAB Leschen) pp. 63–64, 73–75. Walter de Gruyter, Berlin.

Beutel RG, Arce-Pérez R (2016) 6.3. Sphaeriusidae Erichson, 1845 (Jäch, 1999) (= Microsporidae). In *Handbuch der Zoologie/Handbook of Zoology. BandVolume IV Arthropoda: Insecta. Teilband/Part 38. Coleoptera, Beetles. Volume 1: Morphology and Systematics (Archostemata, Adephaga, Myxophaga, Polyphaga partim).* 2nd edn. (Eds RG Beutel and RAB Leschen) pp. 70–71. Walter de Gruyter, Berlin.

Beutel EG, Haas A (1998) Larval head morphology of *Hydroscapha natans* (Coleopptera, Myxophaga) with reference to miniaturization and the systematic position of Hydroscaphidae. *Zoomorphology* **118**, 103–116. doi:10.1007/s004350050061

Beutel RG, Haas F (2000) Phylogenetic relationships of the suborders of Coleoptera (Insecta). *Cladistics* **16**, 103–141.

Beutel RG, Vanin SA (2016) 6.2. Torridincolidae Steffan, 1964. In *Handbuch der Zoologie/Handbook of Zoology. BandVolume IV Arthropoda: Insecta. Teilband/Part 38. Coleoptera, Beetles. Volume 1: Morphology and Systematics (Archostemata, Adephaga, Myxophaga, Polyphaga partim).* 2nd edn. (Eds RG Beutel and RAB Leschen) pp. 67–70. Walter de Gruyter, Berlin.

Beutel RG, Maddison DR, Haas A (1999) Phylogenetic analysis of Myxophaga (Coleoptera) using larval characters. *Systematic Entomology* **24**, 171–192. doi:10.1046/j.1365-3113.1999.00075.x

Büning J (2005) The telotrophic ovary known from Neuroptera exists also in the myxophagan beetle *Hydroscapha natans*. *Development Genes and Evolution* **215**, 597–607. doi:10.1007/s00427-005-0017-8

Crowson RA (1960) The phylogeny of Coleoptera. *Annual Review of Entomology* **5**(1), 111–134. doi:10.1146/annurev.en.05.010160.000551

Flowers RW, Shepard WD, Mera RT (2010) A new species of *Lepicerus* (Coleoptrera: Lepiceridae) from Ecuador. *Zootaxa* **2639**, 35–39.

Ge SQ, Friedrich F, Beutel RG (2010) On the systematic position and taxonomic rank of the extinct myxophagan †*Haplochelus* (Coleoptera). *Insect Systematics & Evolution* **41**, 329–338. doi:10.1163/187631210X537385

Hinton HE (1967) On the spiracles of the larvae of the suborder Myxophaga (Coleoptera). *Australian Journal of Zoology* **15**, 955–959.

Hinton HE (1969) Plastron respiration in adult beetles of the suborder Myxophaga. *Journal of Zoology* **159**, 131–137, 3 pls.

Jałoszyński P, Yamamoto S, Takahashi Y (2017) Discovery of a new Mesozoic species of the ancient genus *Lepicerus* (Coleoptera: Myxophaga: Lepiceridae), with implications for the systematic placement of all previously described extinct "lepiceroids". *Cretaceous Research* **78**, 95–102. doi:10.1016/j.cretres.2017.06.001

Kirejtshuk AG (2017) Taxonomic notes on fossil beetles (Insecta: Coleoptera). *Russian Entomological Journal* **26**(1), 35–36.

Kirejtshuk AG, Poinar G (2006) Haplochelidae, a new family of Cretaceous beetles (Coleoptera: Myxophaga) from Burmese amber. *Proceedings of the Entomological Society of Washington* **108**(1), 155–164.

Kirejtshuk AG, Poinar G (2013) On the systematic position of the genera *Lepiceroides* gen. n. and *Haplochelus*, with notes on the taxonomy and phylogeny of the Myxophaga (Coleoptera). In *Insect Evolution in an Ambiferous and Stone Alphabet. Proceedings of the 6th International Congress on Fossil Insects, Arthropods and Amber.* (Eds D Azar, MS Engel, E Jarzembowski, L Krogmann, A Nel and J Santiago-Blay) pp. 55–69. Brill, Leiden.

Kirejtshuk AG, Ponomarenko AG, Prokin AA, Chang H, Nokolajev GV, Ren D (2010) Current knowledge of Mesozoic Coleoptera from Daohugou and Liaoning (Northeast China). *Acta Geologica Sinica* (English Edition) **84**(4), 783–792.

Lawrence JF, Ślipiński A, Beutel RG, Newton AF (2013) A possible larva of *Lepicerus inaequalis* Motschulsky (Coleoptera: Myxophaga: Lepiceridae) from Panama. *Zootaxa* **3701**(3), 393–400. doi:10.11646/zootaxa.3701.3.8

McKenna DD, Wild AL, Kanda K, Bellamy CL, Beutel RG, Caterino MS, Farnum CW, Hawks DC, Ivie MS, Jameson ML, Leschen RAB, Marvaldi AE, McHugh JV, Newton AF, Robertson JA, Thayer MK, Whiting MF, Lawrence JF, Ślipiński A, Maddison DR, Farrell BD (2015) The beetle tree of life reveals that Coleoptera survived end-Permian mass extinction to diversify during the Cretaceous terrestrial revolution. *Systematic Entomology* **40**(4), 835–880.

Navarrete-Heredia J, Cortes-Aguilar J, Beutel RG (2005) New findings on the enigmatic beetle family Lepiceridae (Coleoptera: Myxophaga). *Entomologische Abhandlungen* **62**(2), 193–201.

Ponomarenko AG (1969) Historical development of the Coleoptera-Archostemata. *Trudy Paleontologicheskogo Instituta* **125**, 1–240 [in Russian].

Sampaio BHL, Ferreira N, Jr (2013) A new species of *Iapir* Py-Daniel, Fonseca & Barbosa (Coleoptera: Myxophaga: Torridincolidae) from Brazil with key to species of the genus. *Zootaxa* **3753**(2), 196–200. doi:10.11646/zootaxa.3753.2.9

Shepard ED, Barr CB, Aguilar Julio CA (2013) The occurrence of Torridincolidae (Coleoptera: Myxophaga) in Paraguay and a world checklist of species. *Bolletin del Museo Nacional de Historia Natural del Paraguay* **17**(1), 76–82.

Short AEZ, Joly LJ, Garcia M, Wild AL, Bloom DD, Maddison DR (2015) Molecular phylogeny of the Hydroscaphidae (Coleoptera: Myxophaga) with description of a remarkable new lineage from the Guiana Shield. *Systematic Entomology* **40**(1), 214–229. doi:10.1111/syen.12097

Tan JJ, Ren D (2007) Two exceptionally well-preserved catiniids (Coleoptera: Archostemata: Catiniidae) from the Late Mesozoic of northeastern China. *Annals of the Entomological Society of America* **100**(5), 666–672. doi:10.1603/0013-8746(2007)100[666:TEWCCA]2.0.CO;2

Vanin SA, Beutel RG, Arce-Pérez R (2016) 6.4. Hydroscaphidae LeConte, 1874. In *Handbuch der Zoologie/Handbook of Zoology. BandVolume IV Arthropoda: Insecta. Teilband/Part 38. Coleoptera, Beetles. Volume 1: Morphology and Systematics (Archostemata, Adephaga, Myxophaga, Polyphaga partim).* 2nd edn. (Eds RG Beutel and RAB Leschen) pp. 71–73. Walter de Gruyter, Berlin.

Yu Y, Leschen RAB, Ślipiński A, Ren D, Pang H (2012) The first fossil bark-gnawing beetles from the middle Jurassic of Inner Mongolia, China (Coleoptera: Trogossitidae). *Annales Zoologici (Warszawa)* **62**(2), 245–252. doi:10.3161/000345412X652765

Yu Y, Leschen RAB, Ślipiński A, Ren D, Pang H (2015) New genera and species of bark-gnawing beetles (Coleoptera: Trogossitidae) from the Yixian Formation (Lower Cretaceous) of Western Liaoning, China. *Cretaceous Research* **53**(1), 89–97. doi:10.1016/j.cretres.2014.11.003

5. SPHAERIUSIDAE ERICHSON, 1845

W. Eugene Hall

Fig. 5.1. *Sphaerius ovensensis* (Oke)

Common name. Minute bog beetles

Family synonyms. Microsporidae Crotch, 1873; Sphaeriidae Erichson, 1845

Introduction. Sphaeriusidae has been recorded from North and Central America, Europe, Africa (incl. Madagascar), Asia and Australia (Horn 1868; Matthews 1888; Champion 1923; Lesne 1935, 1936, 1940; Oke 1954; Löbl 1995, 2003; Endrödy-Younga 1997; Hall 2000, 2003; Beutel & Raffaini 2003; Beutel & Arce-Pérez 2016; Mesaroš 2013). The family includes a single extant genus, *Sphaerius*, which contains ~25 species, with only one currently described from Australia. A fossil genus, *Burmasporum*, was recently described from Lower Cretaceous Burmese amber (Kirejtshuk 2009). Sphaeriusidae is currently under revision worldwide (W.E. Hall).

Sphaeriusids belong to the coleopteran suborder Myxophaga, a small monophyletic group including three other families: Hydroscaphidae, Lepiceridae and Torridincolidae (Lawrence & Newton 1982, 1995; Bouchard *et al.* 2011). Following a series of proposals and rulings (Clarke 1970; Spangler *et al.* 1981; ICZN 1985; Jäch 1999) the family-group name and type genus were conserved (ICZN 2000). Historical classification of Sphaeriusidae has placed the family within a variety of suborders and superfamilies before Myxophaga. The family, then known as Sphaeriidae, was placed by LeConte & Horn (1883) between Hydroscaphidae and Scaphidiidae, stating, 'The relations between this family and Trichopterygidae (= Ptiliidae) are so obvious as to require no further elucidation'. Matthews (1899) also placed Sphaeriusidae near the hydroscaphids, while at the turn of the 20th Century Kolbe (1901) and Ganglbauer (1903) transferred the family to Staphylinoidea, and this classification was retained by Leng (1920), Stickney (1923), and Williams (1938), the latter two authors commenting on shared characters of sphaeriusids with Ptiliidae. Based on hind-wing venation, Forbes (1926) transferred Sphaeriusidae to the adephagan superfamily Hydradephaga, while at the same time commenting that the sphaeriusid body lacked adephagan morphological characteristics. Sphaeriusidae was again transferred back to Staphylinoidea by Böving & Craighead (1931) and Jeannel & Paulian (1944). Based on morphological evidence, including the presence of notopleural sutures, Crowson (1955) erected the coleopteran suborder Myxophaga, at the time including the families Sphaeriusidae, Lepiceridae, Hydroscaphidae, and Calyptomeridae (calyptomerids are currently a subfamily within the polyphagan family Clambidae).

Recent molecular phylogenetic analysis suggests a possible Myxophaga + Archostemata sister-group relationship, and within Myxophaga the families Sphaeriusidae and Lepiceridae sister-groups to each other (McKenna *et al.* 2015). Previous subordinal and family analyses supported Myxophaga + Polyphaga and Sphaeriusidae + Hydroscaphidae sister-group relationships, respectively (Beutel 2016; Beutel & Haas 2000).

Characteristics. Adult. 0.5–1.20 mm in length, body subhemispherical, widest at middle, dorsally convex, flattened ventrally, black, subglabrous with short, sparse, dorsal setae. Head prognathous, broad, flattened, eyes well developed, entire, not protruding, antennal insertions exposed, groves below eyes antennae moderately long, 11-segmented with 3-segmented club, clypeal suture distinct, labrum well developed, mandible, short, broad, flattened, with distinct well developed mola, apically bidentate, prostheca well developed, membranous; left mandible with articulated lobe; maxilla reduced, partly concealed, with apical lobe; palps moderately developed, apical palpomere shorter and narrower than preapical one; gular sutures well separated; corpotentorium well developed; cervical sclerites absent. Pronotum widest at hind angles, angles slightly acute; posterior margin straight; disc convex. Prosternum short, prosternal process rounded at apex. Notopleural and pleurosternal sutures not meeting anteriorly. Procoxae subcylindrical, coxal plates absent, procoxal cavities strongly transverse, narrowly separated, closed internally by narrow bar. Scutellar shield small, posteriorly acute. Elytra complete, covering all abdominal segments, apices rounded, disc irregularly punctated. Mesoventral process reduced, truncate, fused to larger metaventrite. Mesocoxae widely separated, coxal plates absent. Metaventrite moderately transverse, flattened, parallel-sided. Metacoxae nearly contiguous, extending laterally to elytra, coxal plates present, large, covering hind femur. Metendosternite short, broad, lateral arms slender. Hind wing moderately reduced, membrane wide, rolled apically; radial bar short, extending mid-wing, bending-zone near apex; pterostigma absent; margin lined with setae. Legs short; protibia moderately expanded apically; pro- and mesofemora expanded at middle; tibia shorter than femur; tibial spurs single. Tarsi 3–3–3; tarsomeres slender, ventral lobes absent; pretarsal claws simple; empodium absent. Abdomen possessing three ventrites; ventrites 1 and 3 wider and longer than ventrite 2; coxal cavities large. Abdominal tergites weakly sclerotised, membranous;

spiracles on abdominal segments I–VIII functional; tergite IX divided into laterotergites; segment X free. Aedeagus trilobate, parameres fused to phallobase or absent; penis lacking anterior struts. Ovipositor short, broad (Britton 1966; Hinton 1967; Löbl 1995; Beutel *et al.* 1999; Hall 2000; Beutel & Arce-Pérez 2016; Lawrence & Ślipiński 2013).

Larva. Up to 1.5 mm in length, body oblong, flattened, widest anteriorly, narrowed posteriorly, dorsal surface more lightly pigmented than ventral, surface smooth with transverse serrate ridges. Head large, prognathous, protracted at base, broader than long, widest behind middle, moderately rounded laterally; epicranial stem absent. Stemmata four, lateral to antennae, clustered on projections. Frontoclypeal suture absent. Labrum free, large, broadly rounded. Antennae 2-segmented, short, directed posteriorly, well developed antennifer, conical sensorium at middle of apical antennomere. Mandible moderately endognathous, short, broad, unidentate, incisor serrate, large molae present; prostheca absent. Ventral mouthparts moderately retracted forming maxillolabial complex; stipes weakly elongate; maxillary palp 2-segmented. Labium composed of pre- and postmentum; ligula large, bilobed; labial palps 1-segemnted, widely separated. Gular region reduced. Prothorax broader than head, longer than meso- and metathorax combined, rounded laterally; posterior edge of all thoracic terga and abdominal terga I–VIII with fringe of posteriorly directed flattened setae. Legs short, 5-segmented, one simple pretarsal claw possessing a single seta, coxae widely separated. Abdomen nearly as broad as thorax, narrowed posteriorly. Tergum IX subequal to VIII; sternite IX exposed, simple; urogomphi absent; segment X possessing three membranous lobes, each with pair of hooks. Thoracic spiracles non-functional; abdominal segments I–VIII possessing lateral projections and balloon-shaped spiracular gills. (Britton 1966).

Pupa. Rests within the final larval exuviae.

Classification of the Australian genera

Sphaerius Waltl, 1838 (Fig. 5.1)

Type species. *Sphaerius acaroides* Waltl, 1838.
Microsporus Kolenati, 1846
Neosphaerius Oke, 1954

Australian species. *Sphaerius ovensensis* (Oke, 1954).

Distribution. The genus *Sphaerius* is widely distributed. In Australia, it is known from NSW and VIC.

Biology. Adult and larval sphaeriusids are found on sand and mud along edges of streams, creeks & rivers, under stones, mosses associated with bogs, on algae, among roots of plants, leaf litter further away from aquatic habitats, and most likely feed on algae (Löbl 1995; Hall 2000; Lawrence & Ślipiński 2013). A plastron that is present in other myxophagans (Hinton 1967; Messner & Joost 1984) is absent in Sphaeriusidae, but air is stored beneath the elytra.

Note. Oke (1954) described two new species of sphaeriusids from Australia under a proposed new genus *Neosphaerius* based on characters of the antennal club, but the genus presently is treated as a junior synonym of *Sphaerius*. While *S. ovensensis* (Oke 1954) is a valid species within the genus, *S. coenensis* (Oke 1954) is removed from the group as examination of the type material shows *S. coenensis* belongs in the polyphagan family Corylophidae (JF Lawrence and HE Escalona, personal communication).

References. Britton (1966); Lawrence & Ślipiński (2013); Oke (1954).

References

Beutel RG (2016) 6. Myxophaga Crowson, 1955. In *Handbuch der Zoologie/Handbook of Zoology. Band/Volume IV, Arthropoda: Insecta Teilband/Part 38. Coleoptera, Beetles. Volume 1. Morphology and Systematics (Archostemata, Adephaga, Myxophaga, Polyphaga partim).* 2nd edn. (Eds RG Beutel and RAB Leschen) pp. 63–65. W De Gruyter, Berlin.

Beutel RG, Arce-Pérez R (2016) 6.3. Sphaeriusidae Erichson, 1845 (Jäch 1999) (= Microsporidae). In *Handbuch der Zoologie/Handbook of Zoology. Band/Volume IV, Arthropoda: Insecta Teilband/Part 38. Coleoptera, Beetles. Volume 1. Morphology and Systematics (Archostemata, Adephaga, Myxophaga, Polyphaga partim).* 2nd edn. (Eds RG Beutel and RAB Leschen) pp. 63–65. W De Gruyter, Berlin.

Beutel RG, Haas F (2000) Phylogenetic relationships of the suborders of Coleoptera (Insects). *Cladistics* **16**, 103–141. doi:10.1111/j.1096-0031.2000.tb00350.x

Beutel RG, Raffaini GB (2003) First record of Sphaeriusidae for Argentina (Coleoptera: Myxophaga). *Koleopterologische Rundschau* **73**, 1–6.

Beutel RG, Maddison DR, Haas A (1999) Phylogenetic analysis of Myxophaga (Coleoptera) using larval characters. *Systematic Entomology* **24**, 171–192. doi:10.1046/j.1365-3113.1999.00075.x

Bouchard P, Bousquet Y, Davies AE, Alonso-Zarazaga MA, Lawrence JF, Lyal CHC, Newton AF, Reid CAM, Schmitt M, Ślipiński SA, Smith ABT (2011) Family-group names in Coleoptera (Insecta). *ZooKeys* **88**, 1–972. doi:10.3897/zookeys.88.807

Böving AG, Craighead FC (1931) An illustrated synopsis of the principal larval forms of the order Coleoptera. *Entomologica Americana* (N. S.) **11**, 1–351.

Britton EB (1966) On the larva of *Sphaerius* and systematic position of Sphaeriidae (Coleoptera). *Australian Journal of Zoology* **14**, 1193–1198. doi:10.1071/ZO9661193

Champion GC (1923) Some Indian Coleoptera. *Entomologist's Monthly Magazine* **59**, 45–53.

Clarke AH (1970) Sphaeriidae in Mollusca and Insecta: proposal for rectification of homonymy. Z.N (S.) 1892. *Bulletin of Zoological Nomenclature* **26**, 235–237. doi:10.5962/bhl.part.9909

Crowson RA (1955) *The Natural Classification of the Families of Coleoptera*. 187 pp. Nathaniel Lloyd, London.

Endrödy-Younga S (1997) Microsporidae (Coleoptera: Myxophaga), a new family for the African continent. *Annals of the Transvaal Museum* **36**(23), 309–311.

Forbes WTM (1926) The wing folding patterns of the Coleoptera. *Journal of the New York Entomological Society* **34**, 42–115, pls. 7–18.

Ganglbauer L (1903) Systematisch-koleopterologische Strudien. *Münchener Koleopterologische Zeitschrift* **1**, 271–319.

Hall WE (2000) Microsporidae Crotch, 1873. In *American Beetles. Vol. 1. Archostemata, Myxophaga, Adephaga, Polyphaga: Staphyliniformia.* (Eds RH Arnett, Jr. and MC Thomas) pp. 233–246. CRC Press, Boca Raton, Florida.

Hall WE (2003) Sphaeriusidae (Coleoptera). In *Water Beetles of China, Volume 3.* (Eds MA Jäch and L Ji) pp. 37–41. Wien: Zoologisch-Botanische Gesellschaft in Österreich and Wiener Coleopterologenverein, Vienna.

Hinton HE (1967) On the spiracles of the larvae of the suborder Myxophaga (Coleoptera). *Australian Journal of Zoology* **15**, 955–959. doi:10.1071/ZO9670955

Horn GH (1868) New species of Coleoptera from the Pacific District of the United States. *Transactions of the American Entomological Society* **2**, 129–140.

International Commission on Zoological Nomenclature (1985) Opinion 1331. Sphaeriidae Jeffreys, 1862 (1820) (Mollusca, Bivalvia) and Microsporidae Reichardt, 1976 (Insecta, Coleoptera): placed on the Official List. *Bulletin of Zoological Nomenclature* **42**, 230–232.

International Commission on Zoological Nomenclature (2000) Opinion 1957. *Sphaerius* Waltl, 1838 (Insecta, Coleoptera): conserved; and Sphaeriidae Erichson, 1845 (Coleoptera): spelling emended to Sphaeriusidae, so removing the homonymy with Sphaeriidae Deshayes, 1854 (1820) (Mollusca, Bivalvia). *Bulletin of Zoological Nomenclature* **57**, 182–184.

Jäch MA (1999) Case 3052. Sphaerius Waltl, 1838 and Sphaeriusidae Erichson, 1845 (Insecta, Coleoptera): proposed conservation by the partial revocation of Opinion 1331. *Bulletin of Zoological Nomenclature* **56**(2), 117–120. doi:10.5962/bhl.part.23046

Jeannel R, Paulian R (1944) Morphologie abdominale des coléoptères et systématique de l' ordre. *Revue Francaise d'Entomologie* **11**, 65–110.

Kirejtshuk AG (2009) A new genus and species of Sphaeriusidae (Coleoptera, Myxophaga) from Lower Cretaceous Burmese amber. *Denisia 26, zugleich Kataloge der oberösterreichischen Landesmuseen. Neue Serie* **86**, 99–102.

Kolbe HJ (1901) Vergleichend-morphologische Untersuchungen an Coleopteren nebst Grundlagen zu einem System und zur Systematik derselben. *Archiv für Naturgeschichte* **67**(Beiheft), 89–150.

Lawrence JF, Newton AF (1982) The evolution and classification of beetles. *Annual Review of Ecology and Systematics* **13**, 261–290. doi:10.1146/annurev.es.13.110182.001401

Lawrence JF, Newton AF, Jr (1995) Families and subfamilies of Coleoptera (with selected genera, notes, references and data on family-group names). In *Biology, Phylogeny, and Classification of Coleoptera. Papers Celebrating the 80th Birthday of Roy A. Crowson*. (Eds J Pakaluk and SA Ślipiński) pp. 779–1006. Muzeum i Instytut Zoologii Polska Akademia Nauk, Warszawa.

Lawrence JF, Ślipiński A (2013) *Australian Beetles: Morphology, Classification and Keys. Volume 1.* CSIRO Publishing, Melbourne, viii + 561pp.

LeConte JL, Horn GH (1883) Classification of the Coleoptera of North America. *Smithsonian Miscellaneous Collections* 26 (507), xxxviii + 567pp.

Leng CW (1920) *Catalogue of the Coleoptera of America, North of Mexico*. John D. Sherman, Jr., Mount Vernon, New York, X + 470pp.

Lesne P (1935) Les *Sphaerius* de France (Col. Sphaeriidae) (Note préliminaire). *Bulletin de la Société Entomologique de France* **40**, 215–217.

Lesne P (1936) Nouvelles données sur les coléoptères de la famille des Sphaeriidae. *Livre Jubilaire Bouvier*, Paris, 241–248.

Lesne P (1940) Entomological results from the Swedish expedition 1934 to Burma and British India. Coleoptera: Sphaeriidae et Bostrychidae recueillis par René Malaise. *Arkiv för Zoologi* **32B**(6), 1–4.

Löbl I (1995) New species of terrestrial *Microsporus* from Himalaya (Coleoptera: Microsporidae). *Entomologische Blätter* **91**(3), 129–138.

Löbl I (2003) Sphaeriusidae Erichson 1945. In *Catalogue of Palaearctic Coleoptera. Volume 1. Archostemata-Myxophaga-Adephhaga.* (Eds I Löbl and A Smetana). Apollo Books, Stenstrup.

Matthews A (1888) Sphaeriidae. In *Biologia Centrali-Americana, Insecta, Coleoptera 2, Part 1*, 156–158.

Matthews A (1899) *A Monograph on the Coleopterous Families Corylophidae and Sphaeriidae*. OE Janson & Son, London.

McKenna DD, Wild AL, Kanda K, Bellamy CL, Beutel RG, Caterino MS, Farnum CW, Hawks DC, Ivie MA, Jameson ML, Leschen RAB, Marvaldi A, McHugh JV, Newton AF, Robertson JA, Thayer MK, Whiting MF, Lawrence JF, Ślipiński A, Maddison DR, Farrell BD (2015) The beetle tree of life reveals that Coleoptera survived end-Permian mass extinction to diversify during the Cretaceous terrestrial revolution. *Systematic Entomology* **40**, 835–880. doi:10.1111/syen.12132

Mesaroš G (2013) Sphaeriusidae (Coleoptera, Myxophaga), a new beetle family to the fauna of Serbia. *Bulletin of the Natural History Museum* **6**, 71–74. doi:10.5937/bnhmb1306071M

Messner B, Joost W (1984) Die Plastronatmung von *Hydroscapha granulum* Imagines (Coleoptera, Hydroscaphidae). *Zoologische Jahrbuch der Anatomie* **112**(3), 269–278.

Oke CG (1954) Australian species of Sphaeriidae. *Proceedings of the Royal Society of Victoria (n.s.)* **65**, 57–59.

Spangler PJ, Starobogatov YI, Kuiper JGJ (1981) Sphaeriidae in Mollusca and Insecta: comments on proposals to remove the homonymy. Z.N.(S.) 1892. *Bulletin of Zoological Nomenclature* **38**, 157–158. doi:10.5962/bhl.part.8186

Stickney FS (1923) The head capsule of Coleoptera. *Illinois Biological Monographs* 8(1), 1–51, 26 pls.

Williams IW (1938) The comparative morphology of the mouthparts of the order Coleoptera treated from the standpoint of phylogeny. *Journal of the New York Entomological Society* **46**, 245–289.

6. ADEPHAGA SCHELLENBERG, 1806

John F. Lawrence

Adephaga is large, monophyletic group adults of which are distinguished from those of Myxophaga and Archostemata by the propleuron forming a large part of the ventrolateral wall of the prothorax but either not extending to the anterior edge of the prothorax or meeting it at a single point only, presence of a secondary mesal procoxal articulation, articulated but concealed protrochantin, relatively immobile metacoxae, six abdominal ventrites with the first three connate and the first divided by the metacoxae, and pygidial defence glands. Larvae have liquid-feeding mouthparts with a fused labrum and no mandibular molae, as well as 6 pairs of stemmata combined with 6-segmented legs (tibia and tarsus separate). The group is also characterised by having four Malpighian tubules, tubular testes and polytrophic ovarioles. Most Adephaga are predaceous as both larvae and adults, exceptions being the rhysodine Carabidae, which feed on Myxomycetes, algophagous Haliplidae and Noteridae and some phytophagous adults among the harpaline Carabidae. The earliest Adephaga in the fossil record are the Upper Permian Triaplidae (*Triaplus sibiricus* Volkov) (Volkov 2013), Gyrinidae (*Tunguskagyrus planus* Yan *et al.*) (Yan *et al.* 2018) and Trachypachidae (*Ademosynoides asiaticus* Martynov) (Ponomarenko & Volkov 2013).

In addition to the families covered in the following pages, the suborder includes Meruidae (northern South America), Aspidytidae (China and southern Africa), Amphizoidae (western North America, China and North Korea) and Trachypachidae (Holarctic and Chile). Further information on adult and larval moprphology, phylogeny and fossil history may be found in Ponomarenko (1977), Balke *et al.* (2005, 2008), Beutel *et al.* (2006, 2008, 2013), Maddison *et al.* (2009), Miller (2009), Dressler & Beutel (2010), Alarie *et al.* (2011), Short *et al.* (2012), Lawrence & Ślipiński (2013), Michat *et al.* (2014), Ponomarenko & Prokin (2015), Beutel & Ribera (2016), Toussaint *et al.* (2016), Gustafson *et al.* (2017), López-López & Vogler (2017).

References

Alarie Y, Short AEZ, Garcia M, Joly L (2011) Larval morphology of Meruidae (Coleoptera: Adephaga) and its phylogenetic implications. *Annals of the Entomological Society of America* **104**(1), 25–36. doi:10.1603/AN10054

Balke M, Ribera I, Beutel RG (2005) The systematic position of Asipdytidae, the diversification of Dytiscoidea (Coleoptera, Adephaga) and the phylogenetic signal of third codon positions. *Journal of Zoological Systematics and Evolutionary Research* **43**(3), 223–242. doi:10.1111/j.1439-0469.2005.00318.x

Balke M, Ribera I, Beutel R, Viloria A, Garcia M, Vogler AP (2008) Systematic placement of the recently discovered beetle family Meruidae (Coleoptera: Dytiscoidea) based on molecular data. *Zoologica Scripta* **37**(6), 647–650. doi:10.1111/j.1463-6409.2008.00345.x

Beutel RG, Ribera I (2016) 7. Adephaga Schellenberg, 1806. In *Handbuch der Zoologie/Handbook of Zoology. BandVolume IV Arthropoda: Insecta. Teilband/Part 38. Coleoptera, Beetles. Volume 1: Morphology and Systematics (Archostemata, Adephaga, Myxophaga, Polyphaga partim).* 2nd Edition. (Eds RG Beutel and RAB Leschen) pp. 77–79. Walter de Gruyter, Berlin.

Beutel RG, Balke M, Steiner WE, Jr (2006) The systematic position of Meruidae (Coleoptera, Adephaga) and the phylogeny of the smaller aquatic adephagan beetle families. *Cladistics* **22**, 102–131. doi:10.1111/j.1096-0031.2006.00092.x

Beutel RG, Ribera I, Bininda-Emonds ORP (2008) A genus-level supertree of Adephaga (Coleoptera). *Organisms, Diversity & Evolution* **7**, 255–269. doi:10.1016/j.ode.2006.05.003

Beutel RG, Wang B, Tan JJ, Ge SQ, Ren D, Yang XK (2013) On the phylogeny and evolution of Mesozoic and extant lineages of Adephaga (Coleoptera, Insecta). *Cladistics* **29**, 147–165. doi:10.1111/j.1096-0031.2012.00420.x

Dressler C, Beutel RG (2010) The morphology and evolution of the adult head of Adephaga (Insects: Coleoptera). *Arthropod Systematics & Phylogeny* **68**(2), 239–287.

Gustafson GT, Prokin AA, Bukontaite R, Bergsten J, Miller KB (2017) Tip-dated phylogeny of whirligig beetles reveals ancient lineage surviving on Madagascar. *Scientific Reports* **7**, 8619, 9 pp.

Lawrence JF, Ślipiński A (2013) *Australian Beetles: Morphology, Classification and Keys. Volume 1.* CSIRO Publishing, Melbourne, Victoria, viii + 561 pp.

López-López A, Vogler AP (2017) The mitogenome phylogeny of Adephaga (Coleoptera). *Molecular Phylogenetics and Evolution* **114**, 166–174. doi:10.1016/j.ympev.2017.06.009

Maddison DR, Moore W, Baker MD, Ellis TM, Ober KA, Cannone JJ, Gutell RR (2009) Monophyly of terrestrial adephagan beetles as indicated by three nuclear genes (Coleoptera: Carabidae and Trachypachidae). *Zoologica Scripta* **38**, 43–62. doi:10.1111/j.1463-6409.2008.00359.x

Michat MC, Alarie Y, Jia F, Xu S, Hajek J, Balke M (2014) Description of the second and third instars of *Aspidytes wrasei* Balke, Ribera & Beutel, 2003, with comments on the identification of larvae of *Aspidytes* Ribera, Beutel, Balke & Vogler, 2002 (Coleoptera: Aspidytidae), and phylogenetic considerations. *Zootaxa* **3881**, 362–372. doi:10.11646/zootaxa.3881.4.4

Miller KB (2009) On the systematics of Noteridae (Coleoptera: Adephaga: Hydradephaga): phylogeny, description of a new tribe, genus and species, and survey of female genital morphology. *Systematics and Biodiversity* **7**(2), 191–214. doi:10.1017/S1477200008002946

Ponomarenko AG (1977) Mesozoic Coleoptera. Composition and ecological characterstics of Mesozoic Coleoptera. Descriptions of new taxa. Suborder Adephaga. The Mesozoic stage in the evolution of Adephaga. Suborder Polyphaga. Polyphaga incertae *sedis*. Infraorder Staphyliniformia. *Trudy Paleontologicheskogo Institut Akademia Nauk SSSR* 161, 8–119 (in Russian; translation 1989. Smithsonian Institution Libraries and National Science Foundation, Washington, D. C.).

Ponomarenko AG, Prokin AA (2015) Review of paleontological data on the evolution of aquatic beetles (Coleoptera). *Paleontological Journal* **49**(13), 1383–1412.

Ponomarenko AG, Volkov AN (2013) *Ademosynoides asiaticus* Martynov, 1936, the earliest known member of an extant beetle family (Insects, Coleoptera, Trachypachidae). *Paleontological Journal* **47**(6), 601–606. doi:10.1134/S0031030113060063

Short AEZ, Alarie Y, Garcia M, Joly LJ (2012) Are noterids specialised meruids (Coleoptera, Adephaga)? A reply to Dressler *et al. Systematic Entomology* **37**(3), 417–419. doi:10.1111/j.1365-3113.2012.00626.x

Toussaint EFA, Beutel RG, Morinière J, Jia F, Shengquan X, Michat MC, Zhou X, Bilton DT, Ribera I, Hájek J, Balke M (2016) Molecular phylogeny of the highly disjunct cliff water beetles from South Africa and China (Coleoptera: Aspidytidae). *Zoological Journal of the Linnean Society* **176**, 537–546. doi:10.1111/zoj.12332

Volkov AN (2013) New species of Triaplidae from the Babii Kamen' locality (Kuznetsk Basin). *Paleontological Journal* **47**(1), 94–97. doi:10.1134/S0031030113010140

Yan E, Beutel R, Lawrence JF, Ponomarenko A (2018) Whirling in the Late Permian: ancestral Gyrinidae show early radiation of beetles before Ptermian-Triassic mass extinction. *BMC Evolutionary Biology* **18**(33), 1–10. doi:10.1186/s12862-018-1139-8

7. GYRINIDAE LATREILLE, 1810

Chris H.S. Watts

Fig. 7.1. *Macrogyrus australis* (Brulle).

Common name. Whirligig beetles

Introduction. Gyrinids are a successful cosmopolitan family of aquatic beetles most abundant in the tropics but also present in temperate regions. In Australia, they are commonest in the wetter northern and north-eastern coastal regions but two species reach Tasmania and three the south-west of Western Australia. Two species are also found in semi-permanent water in the arid centre, often in quite large numbers. By world standards the Gyrinid fauna of Australia is small with only 19 species in four geographically widespread genera: *Gyrinus* Geoffroy, *Aulonogyrus* Motschulsky, *Dineutus* MacLeay and *Macrogyrus* Regimbart. The species *Gyrinus convexiusculus* MacLeay, *Aulonogyrus strigosus* (Fabricius) and *Dineutus australis* (Fabricius) are found in much of South-east Asia as well as Australia but, apart from *Macrogyrus paradoxus* Regimbart which probably occurs in Timor, all other Australian species appear to be restricted to Australia.

The unusual morphology of both adult and larval Gyrinidae supports the widely held view that the family is monophyletic (Beutel & Roughley 1988; Miller & Bergsten 2012). Studies based predominately on morphology place Gyrinidae as the basal group within the Adephaga (Beutel *et al.* 2008). Molecular studies (based on 18S rRNA) place them as sister group to a clade comprising Haliplidae + Dytiscoidea (Noteridae, Hygrobidae, Amphizoidea) (Ribera *et al.* 2002).

The family Gyrinidae has recently been the subject of a comprehensive morphological and genetic study by Miller & Bergsten (2012) and the tribe Dineutini by Gustafson & Miller (2017). These studies included details of the female genitalia that provided several taxonomic informative characters. Unlike the female genitalia the male genitalia vary little and those of Australian species have been ignored by taxonomists but may repay study.

In Australia, the family as a whole has not been subject to taxonomic review since the studies by Ochs (1949, 1956). At the species level these works are essentially sound (Watts & Hamon 2010) and no additional taxa have since been discovered. However, the study by Gustafson & Miller (2017) provided no support for the subgeneric divisions within Australian *Macrogyrus* proposed by Ochs. The genus *Dineutes* was shown by Gustafson & Miller (2017) to be strongly monophyletic with an origin in the Eocene. The genus is well represented in South-east Asia, Africa and North America but has only two species in Australia. These Australian species appear to have dispersed from South-east Asia and New Guinea relatively recently. *Macrogyrus* with 13 Australian species appears to have originated in South America in the Palaeocene, where it is represented by the subgenus *Andogyrus* Ochs, and subsequently dispersed to Australia via Antarctica in the early Eocene with a later dispersal out of Australia to the Melanesian area (Gustafson & Miller 2017).

Biology. Adult beetles (other than the North American *Spanglerogyrus albiventris* Folkerts) are easily recognised by their unusual eyes which are completely divided into upper and lower portions (Miller & Bergsten 2012). When on the surface of the water the upper pair of eyes have an undisturbed view above the surface and the lower pair likewise below the water. Also characteristic are their streamlined form and stubby, paddle shaped mid and hind legs, which enable rapid movement both on the surface and below.

Wherever there is relatively permanent fresh water, gyrinids may be found. Mostly they favour still or slowly moving water such as ponds or pools in streams but some species e.g. *Macrogyrus angustatus* Regimbart in Western Australia, and *M. oblongus* Boisduval and *M. striolatus* (Guerin-Meneville) in Eastern Australia may be found in rivers and streams although mainly in areas of slower water. Many gyrinids are strong fliers and, particularly the inland species, are often found near lights at night. Strong dispersal ability lets inland species such as *Macrogyrus gibbosus* Ochs, exploit scattered desert rock pools and pools in drying creek beds. In northern Australia gyrinids are often the first beetles to colonise newly filled swamps after the start of the wet season. In southern Australia, most collections are made in the warmer months. This may reflect the preference of collectors and it is not known whether the beetles remain active all year or become inactive in the cooler months.

Adults are insectivorous and feed on insects and other small arthropods which they catch or scavenge on the water surface often after they have fallen or been blown onto the surface. Upwards in the food chain, the beetles themselves are used as food by fish and birds, although the unique eyes in the adults and their ability to move erratically and rapidly must limit the extent they are predated upon and allow them the luxury of a very visible life on the water surface.

Gyrinid larvae are long and thin with lateral tracheal gills. They have three larval instars. Most of their time is spent hunting among plant roots and debris at the bottom of the water for insect larvae such as those of chironomid flies. In open water, they swim strongly using a distinctive, dolphin-like swimming motion. The larvae pupate on land. *Macrogyrus australis* builds a small pupal cell out of mud on a solid surface a little above the water line (pers. observation). The larvae of *Aulonogyrus strigosus*, *Dineutus australis* and seven species of *Macrogyrus* are superficially known (Watts & Hamon 2010).

Gyrinids are seldom, if ever, found in polluted water bodies and appear sensitive to pollution, particularly surface pollution. Studies quantifying this are yet to be done but, on first principles, may make the beetles useful and easily seen predictors of water quality.

Characteristics. Adults. 4–19 mm long, elongate-oval, relatively flat, streamlined, shiny black with moderate to long prothoracic legs and short paddle-like meso- and metathoracic legs. Antennae short with a broad cup-shaped scape, offset subtriangular pedicel and compact flagellum with eight to 11 segments. Eyes divided into upper and lower portions. Mandibles short and compact, well hidden. Scutellum visible or not. Elytra often with longitudinal ridges and/or puncture rows, slightly truncated posteriorly, exposing one or two terminal tergites. Elytral margins with or without one or two subterminal spines. Ventral surface with short, narrow prosternal process. Mesoventrite extensive, flat, with or without discrimen. Metaventrite with discrimen. Fore legs long and slender, modified for grasping prey, protarsi in males with numerous small suckers. Distal part of mid and hind legs strongly shortened forming a paddle-like structure with strongly flattened and broadened tarsomeres. The male genitalia symmetrical, simple, flat with a near obsolete basal piece, a penis (median lobe) and two parameres (lateral lobes) which are fused at their bases. They vary only moderately between species. Gonocoxosterna VIII exposed, medially fused. Ovipositor short, broad, coxites well sclerotised, setose, without styli.

Larvae. Elongate, slender, head and pronotum relatively small, flattened, parallel-sided with six stemmata. Mandibles sickle-shaped, sucking channel present, short in Gyrinini long with narrow apical opening in Dineutini. Lateral tracheal gills on abdominal segments I–V. Segment IX with an additional pair of gills. Segment X with four apical hooks. Spiracles absent in first and second instars, present on abdominal segments I–III in third (final) instar.

Pupae. Membranous, compound eye divided into upper and lower portions, antennal sheaths short. Wings, legs and eight fully developed abdominal segments visible. Abdomen with eight segments visible ventrally; terminal ventrite represented by fused gonocoxosterna. Urogomphi absent. Gonopods present.

Key to the genera of Australian Gyrinidae

Adults

1.	Inner edges of elytra with slightly raised border (Gyrinini)	**2**
–	Inner edges of elytra without raised border (Dineutini)	**3**
2(1).	Length less than 5.0 mm. Elytra with rows of serial punctures, without linear grooves	***Gyrinus* Linnaeus**
–	Length greater than 6.0 mm. Elytra without rows of serial punctures, with well marked linear grooves	***Aulonogyrus* Régimbart**
3(2).	Scutellum visible	***Macrogyrus* Régimbart**
–	Scutellum not visible	***Dineutus* MacLeay**

Larvae*

1.	Lateral gills about two-thirds as long as width of abdominal segment	***Aulonogyrus* Régimbart**
–	Lateral gills longer than width of abdominal segment	**2**
2(1).	Prothorax short, square; front margin of head with one tooth	***Dineutus* MacLeay**
–	Prothorax longer, rectangular; front margin of head with four teeth	***Macrogyrus* Régimbart**

* The larva of the Australian species of *Gyrinus* is not known.

Classification of the Australian genera

Subfamily Gyrininae Latreille, 1810

Characteristics. Maxillary galea absent or one-segmented; without externally visible transverse metaventrite suture; complete fusion of the gonocoxosterna (forming an apically entire abdominal sternite V11) (Miller & Bergsten 2012).

Tribe Gyrinini Latreille, 1810

Characteristics. Eleven longitudinal grooves or series of punctures on each elytron (reduced or obscured in some taxa, especially medially); a one-segmented maxillary galea; scutellum exposed when elytra closed; nine antennomeres in the flagellum (Miller & Bergsten 2012).

Notes. The tribe Gyrinini is composed of the speciose and almost cosmopolitan genera *Gyrinus* and *Aulonogyrus* plus *Metagyrus* Brinck, a rare taxon from South-east Asia. Despite the tribes' success elsewhere it is represented in Australia by only one species of *Gyrinus* and two of *Aulonogyrus,* however these species have very successfully colonised Australia.

Gyrinus Geoffroy, 1762 (Fig. 7.3)

Type species. *Dytiscus natator* Linnaeus, 1758.

Characteristics. Small (Australian species 3.5–4.8 mm long), totally black, with visible scutellum, each elytron with 11 rows of small serial punctures, inner edges of the elytra with slightly raised borders, apex of elytron without spine(s).

Australian species. One, *Gyrinus convexiusculus* MacLeay.

Distribution. Coastal NT, QLD, NSW and occasionally VIC. There is a record of one specimen from Kangaroo Island in SA that needs to be confirmed. Outside of Australia, it has been recorded from Ceylon, India, China, Sumatra, Java, New Zealand, and New Caledonia. It almost certainly also occurs in New Guinea.

Biology. Most frequently seen in small to moderate sized groups in relatively shallow but extensive water bodies. Often these are flooded areas with a lot of emergent vegetation such as rushes and grasses.

Notes. The larvae are not known although there is a poorly preserved specimen in the South Australian Museum from north Queensland that probably belongs to this species. The aedeagus of a New Caledonian specimen is illustrated in Mazzoldi (2010).

Aulonogyrus Motschulsky, 1853 (Fig. 7.5)

Type species. *Gyrinus strigipennis* Suffrian, 1842.

Characteristics. Relatively small beetles (Australian species 5.5–6.5 mm long) which are most easily recognised by the raised inner margins of the elytra, deeply striate elytra and yellow margin to the prothorax.

Australian species. One, *Aulonogyrus strigosus* (Fabricius).

Distribution. All states and territories including Norfolk Island.

Biology. The Australian species, *A. strigosus*, is common and widespread and is found in a wide range of still water habitats – swamps, billabongs, dams and pools in slow flowing rivers and streams.

Notes. Larvae described by Watts & Hamon (2010). The aedeagus of a New Caledonian specimen is illustrated in Mazzoldi (2010).

Tribe Dineutini Desmarest, 1851

Characteristics. Defined more by the absence of obvious synapomorphies than the presence of them when compared with other tribes. Females have; spermatheca elongate, asymmetrical; a distinct elongate apodeme at the anterior end of the gonocoxa (Miller & Bergsten 2012).

Notes. *Dineutes* and *Macrogyrus* are the most speciose genera in the Dineutini and the only ones in Australia. Ochs (1926–27) placed them into different tribes based primarily on the presence or absence of a scutellum. This arrangement is not supported by recent morphological and genetic studies (Miller & Bergsten 2012). Of the two genera, *Dineutes* is widespread in Africa, South-east Asia and North America whereas *Macrogyrus* is restricted primarily to Australia and New Guinea where it has speciated considerably, and, as the subgenus *Andogyrus* Ochs, 1924, South America (Gustafson & Miller 2017).

Dineutus MacLeay, 1825 (Fig. 7.4)

Type species. *Dineutus politus* MacLeay, 1825.

Characteristics. Broad, squat, lacking a scutellum, lacking raised inner margins to the elytra, a strong spine near the apex of the elytron.

Australian species. Two.

Distribution. All States and Territories except TAS. A widespread and dominant genus in tropical areas of the world.

Biology. Found in dams, swamps, and pools in rivers and streams.

Notes. *Dineutus neohollandicus* Ochs is characterised by the yellow margins on the prothorax and elytra and has not been recorded outside of Australia. *D. australis* (Fabricius), which lacks the yellow margins, is perhaps the most abundant of all Australian Gyrinidae, being widespread in the northern two-thirds of the continent. It also occurs in China, The Philippines, Indonesia, New Guinea, Vanuatu, New Caledonia, Fiji and other Pacific Islands. The larva of *D. australis* is known (Watts & Hamon 2010). The aedeagus of a New Caledonian specimen of *Dineutus australis* is illustrated in Mazzoldi (2010).

Macrogyrus Régimbart, 1882 (Figs 7.1, 7.2, 7.6)

Type species. *Enhydrus howitti* Clark, 1866.

Characteristics. Moderate to large size (6.5–15.0 mm), scutellum visible, inner borders of elytra not raised. May have grooved elytra and one or two spines at apex of elytron.

Australian species. 15.

Distribution. All states and territories. New Guinea, Lesser Sunda Islands.

Biology. Widespread over most of the continent wherever there is reasonably permanent water, including rock pools in arid areas. Usually only one species is present but the large *Macrogyrus oblongus* Boisduval and *M. striolatus* Guerin-Meneville are often found together in deeply shaded pools in creeks in coastal NSW.

Key to species. Ochs (1949); Watts & Hamon (2010).

Notes. Ochs (1949) placed the Australian species into seven subgenera based on the shape of the side portion of

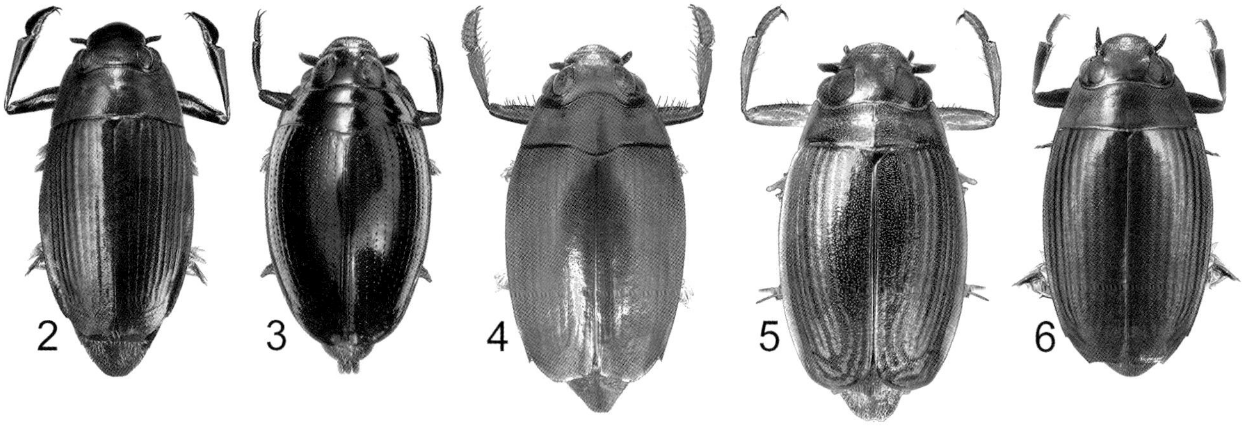

Figs 7.2–7.6. Australian Gyrinidae. 2, *Macrogyrus oblongus* (Boisduval); 3, *Gyrinus convexiusculus* Macleay; 4, *Dineutes australis* (Fabricius); 5, *Aulonogyrus strigosus* (Fabricius); 6, *Macrogyrus australis* (Brulle).

the metaventrite, the number of points on the tips of the elytra and the strength of the elytral grooves. This arrangement into subgenera was not followed by Watts & Hamon (2010) and was not supported by the major revision by Gustafson & Miller (2017) who placed all the Australian species in the sub genus *Macrogyrus* s.str. However, the characters Ochs used as well as body shape and colour can be useful in separating the species. Several the species recognised by Ochs (1949) may also prove not to be specifically distinct. In particular, the species *Macrogyrus australis, M. angustatus* and *M. finschi* Ochs are often hard to separate with confidence. In some species, notably *Macrogyrus striolatus* and *M. rivularis*, the front legs of the males are greatly elongate with strong sucker pads on their feet. The legs of the females are more normal in size. This character is not specifically diagnostic. Larger individuals also tend to have disproportionally longer front legs. The aedeagus of *Macrogyrus caledonicus* Fauvel from New Caledonia is illustrated in Mazzoldi (2010).

Acknowledgments

I would like to acknowledge my colleague, Howard Hamon, for his help in the production and selection of the images used here. Peter Hudson, collection manager in Entomology at the South Australian Museum, is thanked for facilitating access to the significant collection of Australian Gyrinids in his care.

References

Beutel RG, Roughley RE (1988) On the systematic position of the families Gyrinidae (Coleoptera: Adephaga). *Journal of Zoological Systematics and Evolutionary Research* **26**, 380–400. doi:10.1111/j.1439-0469.1988.tb00324.x

Beutel RG, Ribera I, Bininda-Emonds ORP (2008) A genus-level supertree of Adephaga (Coleoptera). *Organisms, Diversity & Evolution* **7**, 255–269. doi:10.1016/j.ode.2006.05.003

Gustafson GT, Miller KB (2017) Systematics and evolution of the whirligig beetle tribe Dineutini (Coleoptera: Gyrinidae: Gyrininae). *Zoological Journal of the Linnean Society* **20**, 1–33.

Mazzoldi P (2010) Gyrinidae (Coleoptera). In *Water Beetles of New Caledonia (part 1).* (Eds MA Jäch and M Balke) pp. 31–43. Zoologisch-Botanische Gesellschaft in Österreich and Wiener Coleopterologenverein (Monographs on Coleoptera. Vol. 3), Vienna.

Miller KB, Bergsten J (2012) Phylogeny and classification of whirligig beetles (Coleoptera: Gyrinidae) relaxed-clock model outperforms parsimony and time-free Bayesian analysis. *Systematic Entomology* **37**, 706–746. doi:10.1111/j.1365-3113.2012.00640.x

Ochs G (1949) A revision of the Gyrinidae. *Records of the South Australian Museum* **22**, 171–199. doi:10.3853/j.0067-1975.22.1949.599

Ochs G (1956) Additional remarks on Australian Gyrinidae. *Records of the Australian Museum* **24**, 31–36. doi:10.3853/j.0067-1975.24.1956.642

Ochs G (1926–27) Die Dineutini. 2. Tribus der unterfamilien Enhydrinae Fam. Gyrinidae (Col). *Entomologische Zeitschrift* **40**, 61–74, 112–126, 129–140, 190–197.

Ribera I, Hogan JE, Vogler AP (2002) Phylogeny of hydradephagan water beetles inferred from 18S rDNA sequences. *Molecular Phylogenetics and Evolution* **23**, 43–62. doi:10.1006/mpev.2001.1080

Watts CHS, Hamon H (2010) *Pictorial Guide to the Australian Whirligig Beetles*. http://www.samuseum.sa.gov.au/research/biological-sciences/terrestrial-invertebrates.

8. HALIPLIDAE AUBE, 1836

Chris H.S. Watts

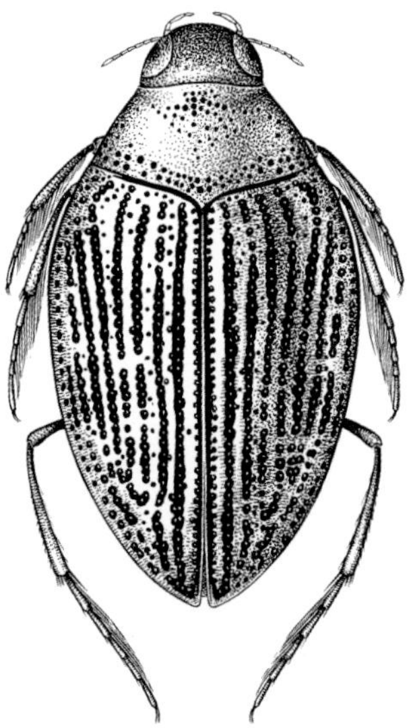

Fig. 8.1. *Haliplus (Liaphlus) australis* Clark

Common name. Crawling water beetles

Introduction. The family Haliplidae has a world-wide distribution and includes five genera only one of which, *Haliplus* Latreille, occurs in Australia, predominately in tropical areas. Nineteen species of *Haliplus* have been recorded from Australia; most are endemic, but three are also found in New Guinea, and one in New Caledonia. The species level taxonomy is relatively well known from the studies of Watts (1988) and van Vondel (1995). Additional species from the Pilbara region of Australia have been described by Watts & McRae (2010) and their larvae by van Vondel (2012).

The family is well characterised, monophyletic and is probably basal to both the Dytiscoidea and Geadephaga (Carabidae and close relatives) (Beutel *et al.* 2008). Within the family, *Haliplus, Algophilus* Zimmermann (from South Africa) and *Apteraliplus* Chandler (from North America) form a monophyletic group (van Vondel 2005). *Haliplus* is morphologically variable and does not appear to be monophyletic as presently defined. The relationships among its six subgenera and *Algophilus* and *Apteraliplus* are unclear. Australian species of *Haliplus* belong to three subgenera (van Vondel 1995): *Liaphlus* Guignot, *Neohaliplus* Netolitsky and *Phalilus* Guignot.

Biology. Little is known about the biology of Australian species. In both northern and southern areas of Australia they are strongly seasonal; in the north during the summer 'wet' season and in the south in spring and summer, in both areas occurring predominately in seasonally wet situations. Judging by collections made at light at the same locality in different years there is also a considerable yearly variation in population size (pers. obs.). The larvae are less well known than the adults and are seldom collected, but in the Pilbara area, which was well studied by the Western Australian Environment Department, larvae of all four species know to occur there were described by van Vondel (2012). Adults are omnivores feeding mainly on vegetation, notably algae, but small invertebrates, such as insect larvae, worms and crustaceans, are also included in their diet. They are air breathers and need to come to the surface periodically to replenish their air supplies. The long, thin, quite hard, larvae are herbivorous, living on filamentous algae or stoneworts (*Chara*). They feed by piercing individual plant cells with their small, hollow mandibles and sucking out the contents. Unlike the adults the larvae do not need to come to the surface to breath but obtain oxygen by diffusion through ventral and dorsal rows of gills and tubercles. After three moults (instars) the larvae leave the water and build cells in the bank in which to pupate. In Australia, because of the strong seasonality in adult emergence, either pupae or adults must over winter in the north, or over summer in the south, in the pupal cell before the adult emerges soon after the start of the rainy season. Adults, at least of the common northern species, are often attracted to lights at night and are strong flyers. In the north-east most adult specimens have been collected at light during January to April (the wet season). In southern and eastern Australia, assuming suitable swamps and ponds are available, adults and larvae have a longer active season during the warmer months.

Characteristics. Length 2.0–6.0 mm. Widest about middle. Head prognathous, eyes large, protuberant. Antennae 11-segmented, filiform, pedicel short and broad, not or little longer than scape. Mandible short and broad, unidentate. Maxillary palps 4-segmented with distinct palpifer and apical palpomere; labial palps 3-segmented. Pronotum ~0.5 times as long as wide, widest posteriorly. Lateral walls steep, almost vertical; propleuron ending well before anterior edge of prothorax. Prosternal process complete, parallel sided, its sharp lateral edges extending anteriorly to form a median shelf on a different plane than the remainder of the prosternum, its posterior edge truncate and abutting the metaventrite. Elytra ~1.3 times as long as wide and ~3.5 times as long as pronotum; disc with ten distinct puncture rows; epipleura moderately broad basally. Metacoxae greatly enlarged, slightly transverse, firmly attached to metaventrite but not fused together at midline, extending laterally to meet elytra with very large coxal plates concealing bases of femora and basal three ventrites. Legs relatively long and slender, fore legs shortest, with protibia slightly expanded at apex; femora distinctly widened at base and tibia slightly widened at apex; tarsi 5–5–5, tarsomeres long and slender, simple, with some swimming hairs; pretarsal claws simple; trochanters short and broad with transverse femoral joint. Abdomen with six ventrites, the first three solidly fused. Aedeagus trilobed, median lobe (penis) simple, lateral lobes (parameres) asymmetrical, articulating at base of penis. Ovipositor short and broad, coxites lightly sclerotised, setose, without styli (Lawrence & Slipinski 2013).

Key to the subgenera of Australian *Haliplus* Latreille

1. Pronotum without two basal grooves (plicae) (Fig. 8.2) .... ***Liaphlus* Guignot**
– Pronotum with two basal grooves (Fig. 8.3) **2**
2(1). Last abdominal ventrite without a central ridge .... ***Neohaliplus* Netolitzky**
– Last abdominal ventrite with a central longitudinal raised ridge .... ***Phalilus* Guignot**

Classification of the Australian genus and subgenera

Haliplus Latreille, 1802

Type species. *Dytiscus impressus* Fabricius, 1787.
Australian species. 18.

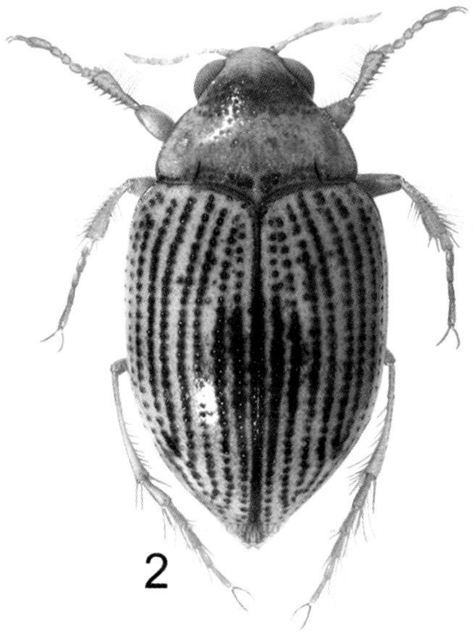

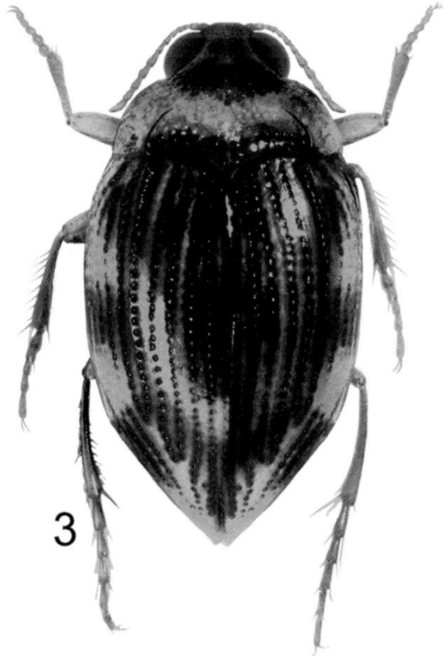

Figs 8.2–8.3. Australian Haliplidae. 2, *Haliplus (Neohaliplus) fuscatus* **Clark;** 3, *Haliplus (Phalilus) oberthuri* **Guignot.**

Distribution. All states and mainland territories.

Ecology. Species are found in still water or embayments at the edges of faster water. Strongly seasonal, particularly in the tropical north during the wet season and in the south in spring and summer. A favourite food of both adults and larva is the stonewort, *Chara,* which is typically found in 'hard' water. The little collection data available suggests that populations can vary considerably between years.

Notes. *Haliplus* is divided into six subgenera three of which occur in Australia (see below). However van Vondel (1995, 2012) cast doubt on the placement of some Australian species and further study is needed, particularly those now in *H.* (*Neohaliplus)* as they show strong differences from Palearctic species of *H.* (*Neohaliplus*) (van Vondel 1995). This was reinforced by the discovery of the larvae of *H.* (*N.*) *fortescueensis* Watts & McRae and *H.* (*N.*) *pinderi* Watts & McRae which showed that the assignment of these species to the subgenus was questionable (van Vondel 2012). Adults of most species have been collected at light.

References. Watts (1988); Watts & McRae (2010); van Vondel (1995, 2004, 2012).

Keys to species. Van Vondel (1995); Watts & Hamon (2015).

Subgenus *Liaphlus* Guignot, 1928

Type species. *Dytiscus fulvus* Fabricius, 1801

Characteristics. Pronotum without basal plicae (Fig. 8.1); with a setiferous striole on the dorsal side of the metatibia; right paramere with a small finger-like apical section (digitus).

Australian species. 10.

Distribution. Present in all states and mainland territories except Tasmania.

Notes. Present in all faunal regions of the world. Beutel & Ruhnau (1990) considered the subgenus to probably be polyphyletic.

References. Watts (1988); Watts & McRae (2010); van Vondel (1995, 2004).

Subgenus *Neohaliplus* Netolitzky, 1911 (Pl. 17K)

Type species. *Dytiscus lineatocollis* (Marsham, 1802)

Characteristics. Pronotum with basal plicae (Fig. 8.2), hind tibia without a setiferous striole on dorsal face. Last abdominal ventrite without a central ridge; right paramere without a solid digitus.

Australian species. Six.

Distribution. All states and mainland territories.

Notes. The two southern species, *H.* (*N.*) *fuscatus* Clark and *H.* (*N.*) *gibbus* Clark, can only be distinguished by examination of the male genitalia. Although mainly coastal in distribution these two species also occur in more inland areas than most Australian Haliplidae.

References. Watts (1988); Watts & McRae (2010); van Vondel (1995, 2004).

Subgenus *Phalilus* Guignot, 1935

Characteristics. Pronotum with two basal plicae (Fig. 8.3). Last ventrite with a central raised ridge; right paramere without a solid digitus.

Type species. *Haliplus oberthuri* Guignot, 1935

Australian species. Two.

Distribution. Coastal NT, QLD, NSW, New Caledonia.

Ecology. Both species are rare in collections with nothing known about their ecology apart from being collected from shallow swampy areas.

Notes. Very similar in general appearance to species of *Neohaliplus*, particularly the more common *H.* (*N.*) *bistriatus* Wehncke. The defining characteristic of a raised ridge at the apex of the abdomen is often difficult to see. *Haliplus* (*P.*) *oberthuri* also occurs in New Caledonia (Hendrich & van Vondel 2010).

References. Hendrich & van Vondel (2010); Watts & Hamon (2015); Watts (1988); Watts & McRae (2010); van Vondel (1995, 2004).

Acknowledgments

I would like to acknowledge the help of Alexis Tindall of the South Australian Museum for allowing me to utilise photographic equipment in her charge and for advice on its use. My colleague, Howard Hamon, contributed critically to the production of the images included here. Peter Hudson, collection manager in Entomology at the South Australian Museum, is thanked for facilitating access to the significant collection of Australian Haliplids in his care.

References

Beutel RG, Ruhnau S (1990) Phylogenetic analysis of the genera of Haliplidae (Coleoptera) based on characters of the adults. *Aquatic Insects* **12**, 1–17. doi:10.1080/01650429009361381

Beutel RG, Ribera I, Bininda-Emonds ORP (2008) A genus-level supertree of Adephaga (Coleoptera). *Organisms, Diversity & Evolution* **7**, 255–269. doi:10.1016/j.ode.2006.05.003

Hendrich L, van Vondel BJ (2010) Haliplidae (Coleoptera). In *Water Beetles of New Caledonia (Part 1)* (Eds MA Jäch and M Balke) pp. 237–242. Zoologisch-Botanische Gesellschaft in Österreich & Vienna Coleopterists Society (*Monographs on Coleoptera* 3), Vienna.

Lawrence JF, Slipinski A (2013) *Australian Beetles: Morphology, Classification and Keys*. Volume 1. CSIRO Publishing, Melbourne, Australia.

van Vondel BJ (1995) Revision of the Haliplidae (Coleoptera) of the Australian region and the Moluccas. *Records of the South Australian Museum* **28**, 61–102.

van Vondel BJ (2004) First description of larvae of *Haliplus* species from Australia (Coleoptera: Haliplidae). *Tijdschrift voor Entomologie* **147**, 57–61. doi:10.1163/22119434-900000140

van Vondel JB (2005) Haliplidae Aube, 1836. In *Handbuch der Zoologie/Handbook of Zoology. Band/Volume IV Arthropoda: Insecta. Teilband/Part 38. Coleoptera. Beetles. Volume 1. Morphology and Systematics (Archostemata, Adephaga, Myxophaga, Polyphaga partim).* (Eds RG Beutel and RAB Leschen) pp. 64–72. W. de Gruyter, Berlin.

van Vondel BJ (2012) Descriptions of larvae of four *Haliplus* species from Australia (Coleoptera: Haliplidae). *Tijdschrift voor Entomologie* **155**, 193–208. doi:10.1163/22119434-00002012

Watts CHS (1988) Revision of Australian Haliplidae (Coleoptera). *Records of the South Australian Museum* **22**, 21–28.

Watts CHS, Hamon H (2015) Pictorial guide to the Haliplidae (Crawling water beetles) of Australia. http://www.samuseum.sa.gov/research/biological-sciences/terrestrial-inverterbrates.

Watts CHS, McRae J (2010) The identity of *Haliplus* (Coleoptera: Haliplidae) from the Pilbara region of Australia, including the description of four new species. *Records of the Western Australian Museum* **25**, 387–398. doi:10.18195/issn.0312-3162.25(4).2010.387-398

9. NOTERIDAE THOMSON, 1860

Tom A. Weir and Cate Lemann

Fig. 9.1. *Sternocanthus australasiae* (Wehnke)

Family synonym. Phreatodytidae Uéno, 1957.

Common name. Burrowing water beetles.

Introduction. Noteridae is a rather small family of aquatic Adephaga, somewhat similar in appearance to Dytiscidae, of which they have been treated as a subfamily. They can be separated from dytiscids by the presence of a modified metacoxal process which often forms a ventral platform, together with the metaventrite (intercoxal process) and the prosternal process – the 'noterid platform' (Fig. 9.2) (Toledo 2010). According to the latest work on the family by Bacca *et al.* (2017) there are some 270 species in 17 genera worldwide. The most recent works on the family as a whole are those of Miller (2009), on the systematics and phylogeny using adult characters and Bacca *et al.* (2017), on the molecular phylogeny. Nilsson (2011) provided a world catalogue of Noteridae, while Dettner (2016) gave an overview of the family. Toledo (2010) provided information on the Noteridae of New Caledonia, Lawrence & Ślipiński (2013) gave an account of the Australian fauna, while Watts (2002) keyed and figured both adults and larvae of the Australian genera. Watts included the Noteridae in his 1985 faunal assessment of the Australian Hydradephaga (Watts 1985) and added a new species of *Canthydrus* (Watts 2001). The most recent work on the Australian fauna is that of Toledo & Hendrich (2006) in which they revised the genus *Hydrocanthus* (*Sternocanthus*) in the Australasian region, and added a new species, bringing the total Australian fauna to seven species in four genera.

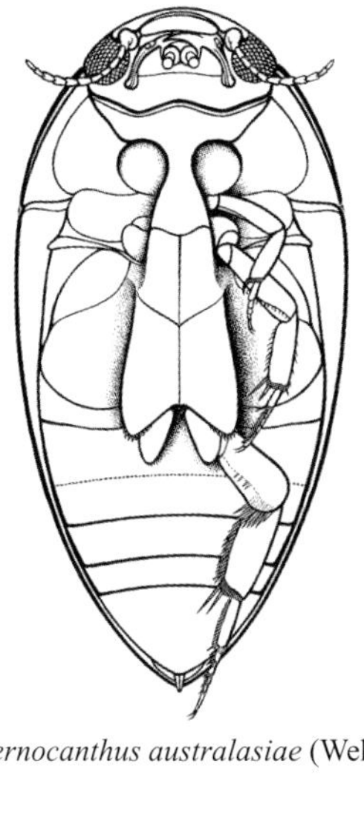

Fig. 9.2. *Sternocanthus australasiae* (Wehnke), ventral.

Diversity and distribution. Noterids are found worldwide with a predominately pantropical range and a few species in temperate areas (Toledo 2010). In Australia, the species occur mainly in the northern parts of the continent in WA, NT, QLD, with one extending to northern NSW and another species extending as far as the VIC/SA border.

Biology. Noterids usually occur along the margins of various sorts of shallow water bodies, both still and slow flowing where adults and larvae are commonly found among the roots of floating or emergent aquatic plants (Lawrence & Ślipiński 2013). Larson (1993, 1997) gives details of habitats for species in north Queensland. Young (1985) found *Hydrocanthus* to be associated with filamentous algae and *Typha*. According to Dettner (2016) widespread and abundant genera such as *Hydrocanthus* and *Noterus* show a higher degree of adaptation to the aquatic environment compared with representatives of more basal genera such as *Notomicrus*. Ribera *et al.* (2002a) suggest that in these basal groups, the middle and hind legs are moved alternately, whereas the 'higher noterids' such as *Sternocanthus* and *Canthydrus* move these legs simultaneously (advanced swimming abilities), while also having the front legs adapted for burrowing. Dettner (2016) states that both adults and larvae are carnivorous, whereas Lawrence & Ślipiński (2013) suggest that they are detritus feeders. Life cycles of most Noteridae are unknown, and only a few noterid larvae have been described (see Watts 2002 for figures of three of the Australian genera). Where known, larvae avoid the water surface and extract oxygen from aquatic plants with their last abdominal segment (Dettner 2016). Pupation, in the few known cases, is aquatic and occurs in air-filled cocoons attached to the roots of aquatic plants (Lawrence & Ślipiński 2013). Adults of some species may be attracted to light in great numbers.

Phylogeny and taxonomy. Noterid representatives have been included in more broadly focused morphological phylogenetic studies of Adephaga (Ribera *et al.* 2002b; Balke *et al.* 2008), but these have generally not had any bearing on relationships within Noteridae. These studies revealed that within the Dytiscoidea, families such as Amphizoidae, Hygrobiidae and Aspidytidae are more closely related to Dytiscidae than Noteridae. Miller

(2009) provided the most comprehensive morphological study of Noteridae and gave a revised classification with three subfamilies (Phreatodytinae, Notomicrinae and Noterinae), four tribes of Noterinae (Noterini, Neohydrocoptinae, Tonerini and Pronoterini) and a total of 16 genera. This classification was followed by Nilsson (2011) in his world catalogue of the family. Most recently Bacca *et al.* (2017) produced the first comprehensive phylogenetic reconstruction of Noteridae based on a multilocus molecular dataset, including all tribes and all but one genus. They provided the updated classification used herein which comprises only two subfamilies (Notomicrinae and Noterinae). Notomicrinae consists of two tribes, Phreatodytini (one genus) and Notomicrini (two genera), while Noterinae now has only Noterini after the synonymising of Neohydrocanthini, Tonerini and Pronoterini and contains 14 genera. Generic changes affecting the Australian fauna sees *Sternocanthus* raised from a subgenus of *Hydrocanthus* to full generic level.

Characteristics. Adults. Somewhat streamlined beetles, oval to elongate elliptical (Fig. 9.1), often teardrop-shaped, dorsally relatively convex and ventrally flattened, and having typical characteristics of the suborder Adephaga. Externally visible propleuron with both notopleural and pleurosternal sutures; eyes not completely divided into dorsal and ventral parts and usually in contact with anterior edge of pronotum; antennae 11-segmented, filiform or slightly serrate, scape constricted so that it appears 2-segmented, glabrous; metaventrite without transverse (katepisternal) suture; metacoxae immobile and fused medially to each other and to the metaventrite, extending laterally to meet elytra, without large plates concealing basal abdominal ventrites; metacoxal plates often markedly flattened and together with the metaventrite and prosternal process, more or less forming the characteristic so called 'noterid platform'; abdomen with 6 ventrites, first 3 solidly fused together and first ventrite distinctly divided by metacoxae; metatrochanter large and distinctly offset from the line of the metafemur; tarsi 5–5–5; with paired pygidial glands. Aquatic.

Larvae. Typical of Adephaga. Body either elongate and parallel sided, spindle shaped or teardrop shaped. Labrum and clypeus fused; mandibular mola absent; ligula without wedge-like sclerome; mandibles relatively short and broad at base, without distinct groove or internal perforation; head without endocarina, strongly transverse, without epicranial stem; antennae well developed and 4-segmented; legs relatively short, without swimming hairs, with 2 pretarsal claws; abdomen 8-segmented, urogomphi very short or apparently absent; segment 8 with or without a posterior prolongation. Aquatic.

Key to the genera of Australian Noteridae

1. Metacoxa and metaventrite fused laterally, suture obscured (Fig. 9.3); scutellum partly visible; elytra hyaline along outer edge in apical two thirds; very small size (< 1.5 mm in length); widely distributed across northern Australia and extending down east coast; WA, NT, QLD, NSW, VIC .... **NOTOMICRINAE: *Notomicrus* Sharp**

– Metacoxa and metaventrite not fused laterally, suture entire (Fig. 9.4); scutellum not visible; elytral colour entirely continuous to outer edge; size larger (> 2.0 mm in length)...**NOTERINAE** **2**

2(1). Protibia apically truncate, dorsoapical angle prominent; apical margin with several elongate spines; without one single apically curved spine (Fig. 9.5); without fringe of short setae along anterodorsal margin; dorsal microreticulation consisting of polygonal rounded cells; body parallel sided; 2.5 mm in length; widely distributed across northern Australia down to northern NSW; WA, NT, QLD, NSW .... ***Neohydrocoptus* Sato**

– Protibia with dorsoapical angle rounded, not prominent; apically with a single curved, elongate spine (Fig. 9.6); with fringe of short setae along anterodorsal margin; dorsal microreticulation very fine, arranged in wrinkles; body tear-drop shaped........ **3**

3(2). Metafemur and metatibia broad, longer metatibial spur serrulate; prosternum medially with uniform coverage of short setae; prosternal process about as long as wide (Fig. 9.7); uniformly black or reddish; 5.7 to 7.7 mm in length; widely distributed across northern Australia down to northern NSW; WA, NT, QLD, NSW .... ***Sternocanthus* Guignot**

– Metafemur and metatibia narrower, neither metatibial spur serrulate; prosternum medially with conspicuous longer setae; prosternal process longer than wide (Fig. 9.8); black with pale elytral spots 2.5 to 3.5 mm in length; northern Australia; WA, NT, QLD .... ***Canthydrus* Sharp**

Classification of the Australian genera

Here we follow the latest classification of Bacca *et al.* (2017). This differs from that of Miller (2009) and Nilsson (2011) in that Noterinae now contains only one tribe (Noterini) and Notomicrinae now contains Phreatodytini as well as Notomicrini. *Sternocanthus* is resurrected from synonymy with *Hydrocanthus* for the Old World species of *Hydrocanthus*. Generic names and synonyms applying to the Australian fauna are confirmed from Nilsson (2011) and Bacca *et al.* (2017).

Subfamily Noterinae Thomson, 1860

Characteristics. Metafurca and metacoxa internally are fused laterally in a narrow band forming a complete ring. Metacoxal plate ('noterid platform') extends anteriorly on to the metaventrite. Metacoxa and metaventrite not fused laterally, separated by a complete suture. Scutellum not visible. A subfamily with only one tribe.

Tribe Noterini Thomson, 1860

Distribution. 14 genera distributed worldwide.
Australian taxa. Three genera.
Key to genera. Watts (2002), Miller (2009).

Neohydrocoptus Sato, 1972 (Pl. 17H)

Type species. *Hydrocoptus bivittis* Motschulsky, 1859.
Hydrocoptus auctorum, nec Motschulsky, 1853

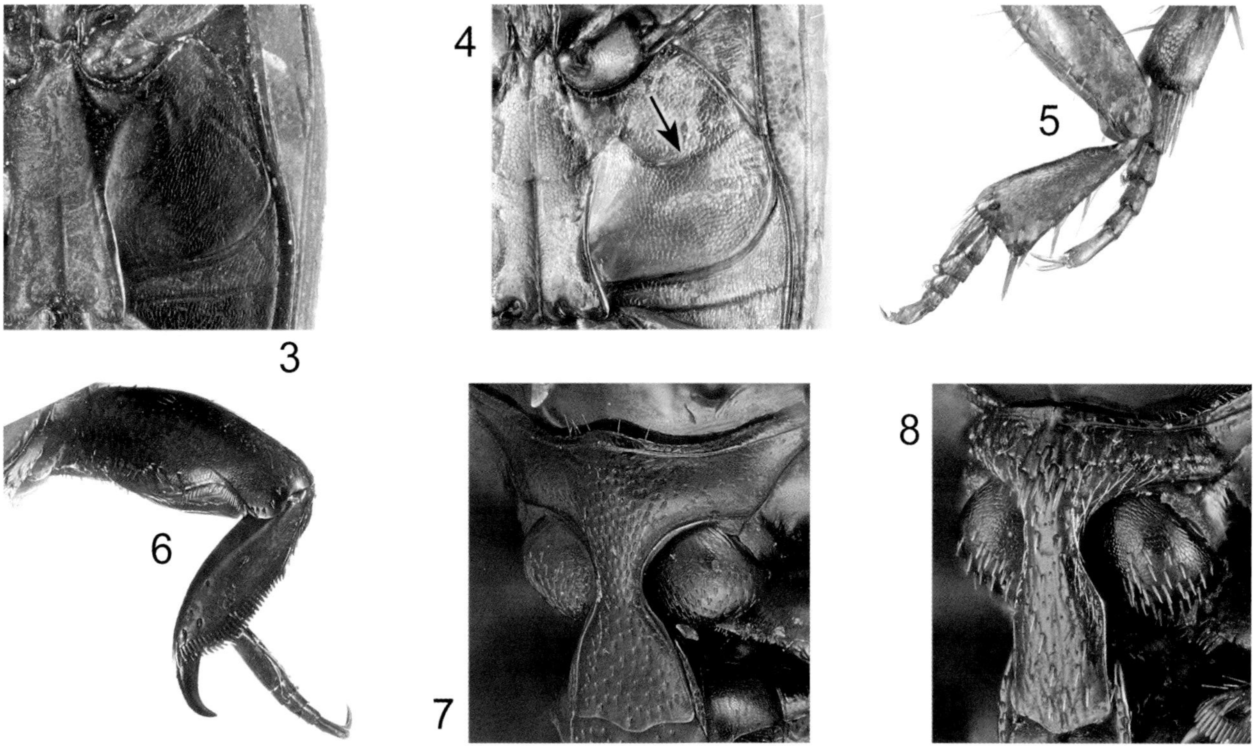

Figs 9.3–9.8. **Noteridae.** 3, *Notomicrus tenellus* Clark, left metacoxa and metaventrite; 4, *Neohydrocoptus subfasciatus* (Sharp), left metacoxa and metaventrite; 5, *Neohydrocoptus subfasciatus* (Sharp), protibial apex and protarsus; 6, *Sternocanthus australasiae* (Wehnke), protibial apex and protarsus; 7, *Sternocanthus australasiae* (Wehnke), prosternal process; 8, *Canthydrus bovillae* Blackburn, prosternal process.

Characteristics. Body length 2.5 mm in Australia, somewhat elongate. Dorsal surface covered with microreticulation of polygonal, rounded cells. Maxillary and labial palps conical, truncate distally, not bifid. Elytra with impressed and more or less well defined longitudinal series of dots; coloration testaceous, reddish or chestnut brown, often with paler markings more or less developed on elytra. Underside glabrous in most species, otherwise with setal punctation on metacoxal process; prosternal process lanceolate, bordered and strongly narrowed between coxae. Metacoxal process ('noterid platform') with lateral expansions more or less developed, depending on species; interlaminar bridge somewhat exposed on hind side, thus articulation with trochanter in part visible. Protibae apically expanded, truncate, dorsoapical angle prominent, with acuminate and straight spurs, at most curved but not hooked, without a fringe of short setae along anterodorsal margin. Males with slightly modified inner protarsal claws; penis in species from South-east Asia and the Australian region with elongate and stiff proximal portion, starting from a heavy robust base and widened apically. Female gonocoxae obliquely bifid and with a series of small teeth laterally (Miller 2009; Toledo 2010). Larvae keyed and figured by Watts (2002).

Australian species. One.

Distribution. Widely distributed in the Afrotropical, Oriental, Palaearctic regions, with some 29 described species worldwide, Nilsson (2011). *Neohydrocoptus subfasciatus subfasciatus* Sharp is widely distributed across northern Australia WA, NT, QLD and down to northern NSW; extralimital to New Caledonia and Indonesia.

Canthydrus Sharp, 1882 (Pl. 17I)

Type species. *Hydrocanthus guttula* Aube, 1838.

Characteristics. Body length: 2.5–3.5 mm in Australia, somewhat flattened ventrally. Dorsal surface shining, black, yellow or variegate in colouration, covered by a fine wrinkly reticulation. Protibia with an enlarged, curved, prominent spine. Dorsoapical angle of protibial rounded. Distinct setal fringe present along anterodorsal margin of pro- and mesotibia. Prosternal process flat, triangular, visibly longer than wide; together with metaventrite and metacoxal process densely setose. Metafemora with ventroapical tuft of setae. Metatibiae elongate, with apical spurs smooth, not serrulate. Males with few circular sucking hairs on basal pro- and mesotarsomeres. Penis broad and flat laterally, right side concave and right process of base strongly elongate and twisted distally. Left paramere somewhat blunt, more or less triangular with distinct, apical tuft of setae (Toledo 2010). Larvae keyed and figured by Watts 2002.

Australian species. Two.

Distribution. Widely distributed in the Afrotropical, Oriental and Palaearctic regions, with some 65 species worldwide. *Canthydrus bovillae* Blackburn and *Canthydrus ephemeralis* Watts both occur across northern Australia: NT, QLD, WA.

Key to species. Watts (2001), Toledo (2010)

Sternocanthus Guinot, 1948 (Pl. 17L)

Type species. *Hydrocanthus micans* Wehncke, 1883.

Characteristics. Body length: 5.7–7.7 mm in Australia, oblong, rarely strongly convex. Colouration evenly black, brown- or yellow-reddish; head and pronotum often paler than elytra. Elytra never variegate, at most with darker irrorations in some American species. Protibia with an enlarged, curved, prominent spine. Dorsoapical angle of protibia rounded. Distinct setal fringe present along anterodorsal margin of pro- and mesotibia. Sternal surface densely setose, as in *Canthydrus*, but setation often more scattered or even absent on prosternal process. Latter very broad, as wide as long, sometimes with depressions or even dimples on its surface. Metafemora with a ventroapical tuft of setae. Metatibiae broad and flat, external spur serrulate. Maxillary palps with terminal palpomere conical, bifid apically; terminal labial palpomere very broad, bifid apically. Males, in Oriental and Australian species, with penis more slender than in *Canthydrus*, basal processes unequal in length but not twisted. Left paramere elongate, with two, more or less distinct tufts of setae (Toledo 2010). Larvae (as *Hydrocanthus*) keyed and figured by Watts (2002) and figured by Lawrence & Ślipiński (2013).

Australian species. Three.

Distribution. Widely distributed across most regions of the world, with some 52 described species. *Sternocanthus australasiae* (Wehnke), *S. pederzanii* (Toledo & Hendrich) and *S. waterhousei* (Blackburn) are widely distributed across northern Australia: NT, QLD, WA with *S. australiasiae* extending down to northern NSW.

Key to species. Toledo & Hendrich 2006

Subfamily Notomicrinae Zimmermann, 1919

Characteristics. Metafurca and metacoxa internally are not fused laterally and do not form a complete ring. Metacoxal plate ('noterid platform') does not extend anteriorly on to the metaventrite. Metacoxa and metaventrite fused laterally, not separated by a suture. The female laterotergite is short and extends posteriorly well beyond the base of the gonocoxa (Miller 2009). Scutellum partly visible. A subfamily with two tribes, one of which occurs in Australia.

Distribution. Primarily tropical – south-eastern North America, much of South America, Australia and South-east Asia.

Tribe Notomicrini Zimmermann, 1919

Characteristics. As for subfamily. A tribe with two genera.

Australian taxa. One genus.

Key to genera. Miller 2009.

Notomicrus Sharp, 1882 (Pl. 17G)

Type species. *Notomicrus brevicornis* Sharp, 1882.

Characteristics. Body length very small: 1.0–1.3 mm. With respect to other noterids, this genus is characterised by fused metaventrite and metacoxal plates, absence of midgular apodeme and a partly exposed scutellum. The elytra have a hyaline area along the outer edge in the apical two thirds. The protibiae are expended apically, with several straight, not hooked spurs (Miller 2009; Toledo 2010). Larvae unknown.

Australian species. One.

Distribution. Widely distributed in the Australian, Nearctic, Neotropical, Oriental and Pacific regions with some 10 described species worldwide. *Notomicrus tenellus* (Clark) is widely distributed across northern Australia: WA, NT, QLD and extending down the east coast almost to the VIC/SA border; extralimital to Papua New Guinea, Solomon Islands, Indonesia, Malaysia, Philippines, Singapore and Samoa. However, Toledo (2010) suggests that there may be at least 4 species are actually merged under the name *N. tenellus*.

References

Bacca SM, Toussaint EFA, Miller KB, Short AEZ (2017) Molecular phylogeny of the aquatic beetle family Noteridae (Coleoptera: Adephaga) with an emphasis on data partitioning strategies. *Molecular Systematics and Evolution* **107**, 282–292.

Balke M, Ribera I, Beutel R, Viloria A, Garcia M, Vogler AP (2008) Systematic placement of the recently discovered beetle family Meruidae (Coleoptera: Dytiscoidea) based on molecular data. *Zoologica Scripta* **37**, 647–650. doi:10.1111/j.1463-6409.2008.00345.x

Dettner K (2016) Noteridae Thomson, 1857. In *Handbook of Zoology, Arthropoda: Insecta. Coleoptera, Beetles Volume 1 Morphology and Systematics (Archostemata, Adephaga, Myxophaga, Polyphaga partim)* 2nd edition. (Eds RG Beutel and RAB Leschen) pp. 98–107. W. De-Gruyter, Berlin.

Larson DJ (1993) Ecology of Tropical Australian Hydradephaga (Insecta: Coleoptera). Part 1. Natural History and distribution of northern Queensland species. *Proceedings of the Royal Society of Queensland* **103**, 47–63.

Larson DJ (1997) Habitat and Community Patterns of Tropical Australian Hydradephagan Water Beetles (Coleoptera: Dytiscidae, Gyrinidae, Noteridae). *Australian Journal of Entomology* **36**, 269–285. doi:10.1111/j.1440-6055.1997.tb01469.x

Lawrence JF, Ślipiński A (2013) *Australian Beetles. Volume 1. Morphology, Classification and Keys*. CSIRO Publishing, Melbourne, Victoria.

Miller KB (2009) On the systematics of Noteridae (Coleoptera: Adephaga: Hydradephaga): phylogeny, description of a new tribe, genus and species, and survey of female genital morphology. *Systematics and Biodiversity* **7**, 191–214. doi:10.1017/S1477200008002946

Nilsson AN (2011) A World Catalogue of the Family Noteridae, or Burrowing Water Beetles (Coleoptera, Adephaga). Version 16.VIII.2011. http://www2.emg.se/projects/biginst/andersen/WCN/wcn_index.htm

Ribera I, Beutel RG, Balke M, Vogler AP (2002a) Discovery of Aspidytidae, a new family of aquatic Coleoptera. *Proceedings. Biological Sciences* **269**, 2351–2356. doi:10.1098/rspb.2002.2157

Ribera I, Hogan JE, Vogler AP (2002b) Phylogeny of hydradephagan water beetles inferred from 18S rRNA sequences. *Molecular Phylogenetics and Evolution* **23**, 43–62. doi:10.1006/mpev.2001.1080

Toledo M (2010) Noteridae: Review of the species occurring east of the Wallace line (Coleoptera). In *Water Beetles of New Caledonia (part 1) Monographs on Coleoptera Vol. 3.* (Eds MA Jäch and M Balke) pp. 195–236. Zoologische-Botanische Gesellschaft, Vienna.

Toledo M, Hendrich L (2006) Taxonomic revision of the Austrtalasian *Hydrocanthus* Say 1833, with description of two new species. *Linzer Biologische Beitrage* **38**, 935–952.

Watts CHS (1985) A faunal assessment of Australian Hydradephaga. *Proceedings. Academy of Natural Sciences of Philadelphia* **137**, 22–28.

Watts CHS (2001) A new species of Australian *Canthydrus* Sharp with a key to the Australian species of Noteridae (Coleoptera). *Records of the South Australian Museum* **34**, 61–64.

Watts CHS (2002) *Checklist and guides to the identification, to genus, of adults and larval Australian water beetles of the families Dytiscidae, Noteridae, Hygrobiidae, Haliplidae, Gyrinidae, Hydraenidae and the superfamily Hydrophiloidea (Insecta-Coleoptera).* Cooperative Research Centre for Freshwater Ecology (Australia). Identification Ecology Guide, **43**, 110 pp.

Young FN (1985) A key to the American species of *Hydrocanthus* Say, with descriptions of new taxa (Coleoptera: Noteridae). *Proceedings. Academy of Natural Sciences of Philadelphia* **137**, 90–98.

10. HYGROBIIDAE RÉGIMBART, 1879

Lars Hendrich

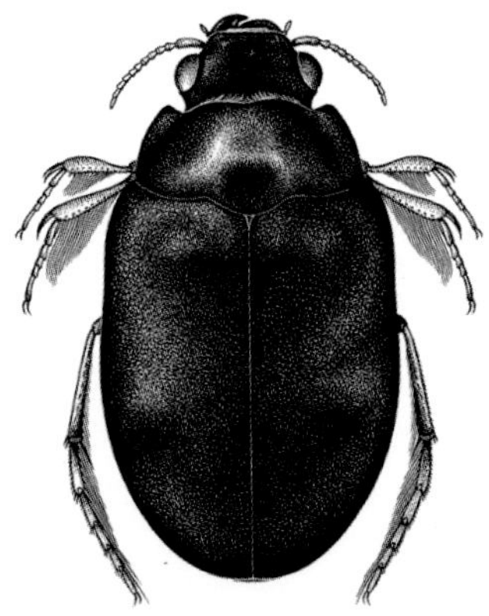

Fig. 10.1. *Hygrobia nigra* (Clark).

Common names. Screech beetles, Squeak beetles

Family synonym. Paelobiidae Erichson, 1837.

Introduction. The monotypic family Hygrobiidae is a small and ancient group of aquatic coleopterans commonly known as squeak beetles (Alarie *et al.* 2004; Hawlitschek *et al.* 2012). In term of phylogenetic relationships, there is a general agreement that Hygrobiidae is part of the adephagan clade Dytiscoidea along with Dytiscidae, Aspidytidae and Amphizoidae, and within this grouping it has been hypothesised to share a sister group relationship with Dytiscidae (Ruhnau 1986; Beutel & Haas 1996; Shull *et al.* 2001; Ribera *et al.* 2002; Alarie & Bilton 2005; Alarie *et al.* 2011) or with a clade formed by the families Dytiscidae, Aspidytidae and Amphizoidae (Balke *et al.* 2005; Hawlitschek *et al.* 2012). In using Hygrobiidae as family name, the present chapter follows Bouchard *et al.* (2011) (but see Nilsson 2005). *Hygrobia* Latreille, the only genus in the family, comprises six species worldwide (Hawlitschek *et al.* 2012): *H. hermanni* (Fabricius), relatively common and widespread in the Western Palearctic, *H. davidi* Bedel, found only in south-eastern China, and *H. wattsi* Hendrich, 2001, *H. australasiae* (Clark), *H. maculata* Britton, and *H. nigra* (Clark) endemic to Australia (Britton 1981; Hendrich 2001; Hawlitschek *et al.* 2012). With respect to the relationships among the Australian *Hygrobia*, a recent phylogenetic analysis based on adult morphological and molecular characters suggested a sister group relationship between *H. hermanni* and a clade formed by the Australian species, with *H. nigra* sister to *H. australasiae* (Hawlitschek *et al.* 2012).

Characteristics. Adult. Medium-sized (10–12 mm) beetles. Roundish-oval, dorsal and ventral sides very convex. Head very large, prognathous. Compound eyes not divided into ventral and dorsal portions, strongly protruding. Labrum short, transverse. Clypeus transverse, separated from frons by a suture. Antennae filiform, almost glabrous, 11-segmented; basal segment rather long. Mouthparts as in Haliplidae; distal segment of galea with a bifid apex. Parts of hind coxae not produced into plates concealing basal abdominal ventrites. Metaventrite with a transverse suture in front of hind coxae, separating an antecoxal plate from remaining part of metaventrite. Colour variable, black or black with reddish markings and light brown with black or dark-brown markings on elytra and pronotum. Hind legs moved alternately when swimming (Holmen 1987; Dettner 2016).

Larvae. Larvae fusiform, with large head and large prothorax in relation to the rest of the body. Tarsi with two claws and body with long ventral gill filaments. Abdomen with 8 well developed segments (segment 9 rudimentary). Descriptions or treatments of larvae of Australian Hygrobiidae have been presented by Alarie *et al.* (2004) and Michat *et al.* (2014). At the present time the larvae of three Australian species of *Hygrobia* are described (*H. australasiae*, *H. nigra* and *H. wattsi*).

Classification of the Australian genus

Hygrobia Latreille, 1804 (Pl. 17J; Figs 10.1, 10.2)

Type species. *Dytiscus hermanni* Fabricius, 1775.

Australian species. Four.

Distribution. *Hygrobia wattsi* is restricted to peatlands of south-western Australia. *Hygrobia nigra* extends from southern Victoria to the Atherton Tableland of northern Queensland, whereas *H. australasiae* extends into Tasmania and South Australia; these species have overlapping distributions and can be found in the same pond. The fourth species, H. *maculata*, is distributed in the river floodplains of the Northern Territory and the lowlands of northern Queensland (Cape York Peninsula) (Britton 1981; Hendrich 2001).

Biology. All Australian species occur in lowland areas and live deep in the mud, silt, and detritus of ponds of paperbark swamps, billabongs, small lakes and peatland areas often fringed by dense stand of larger sedges (e.g. *Baumea*) (Hendrich 2001). Both adults and larvae are preda-

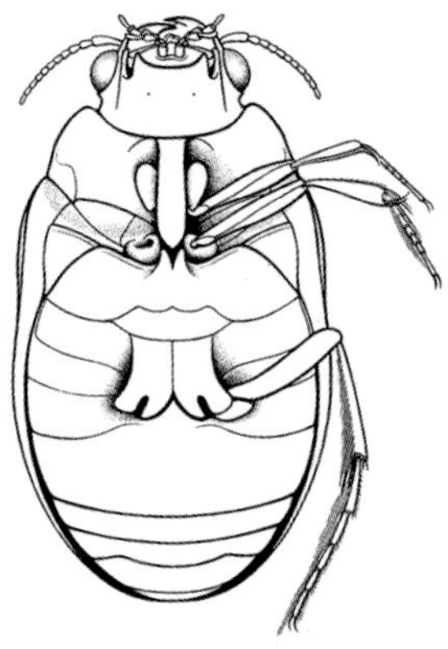

Fig. 10.2. *Hygrobia nigra* (Clark), ventral view.

tors, specialised on oligochaete worms, though adults of *H. hermanni* were also observed feeding on chironomid larvae (Balfour-Browne 1922). All Australian *Hygrobia* species are quite rare in collections and almost nothing is known of their biology, in contrast to the relatively well known West Palaearctic species (e.g. Dettner 1997; Cuppen 2000). It appears that they are rarely attracted to lights and that their population density is much lower than in *H. hermanni* (Hendrich 2001).

Keys to species. Britton (1981), Hendrich (2001).

References

Alarie Y, Bilton DT (2005) Larval morphology of Aspidytidae (Coleoptera: Adephaga) and its phylogenetic implications. *Annals of the Entomological Society of America* **98**, 417–430. doi:10.1603/0013-8746(2005)098[0417:LMOACA]2.0.CO;2

Alarie Y, Beutel RG, Watts CHS (2004) Larval morphology of three species of Hygrobiidae (Coleoptera: Adephaga: Dytiscoidea) with phylogenetic considerations. *European Journal of Entomology* **101**, 293–311. doi:10.14411/eje.2004.039

Alarie Y, Short AEZ, García M, Joly L (2011) Larval morphology of Meruidae (Coleoptera: Adephaga) and its phylogenetic implications. *Annals of the Entomological Society of America* **104**, 25–36. doi:10.1603/AN10054

Balfour-Browne F (1922) The life history of the water beetle *Pelobius tardus* Herbst. *Proceedings of the Zoological Society of London* **1922**, 79–97.

Balke M, Ribera I, Beutel RG (2005) The systematic position of Aspidytidae, the diversification of Dytiscoidea (Coleoptera, Adephaga) and the phylogenetic signal of third codon positions. *Journal of Zoological Systematics and Evolutionary Research* **43**, 223–242. doi:10.1111/j.1439-0469.2005.00318.x

Beutel RG, Haas A (1996) Phylogenetic analysis of larval and adult characters of Adephaga (Coleoptera) using cladistic computer programs. *Entomologica Scandinavica* **27**, 197–205. doi:10.1163/187631296X00043

Bouchard P, Bousquet Y, Davies AE, Alonso-Zarazaga MA, Lawrence JF, Lyal CHC, Newton AF, Reid CAM, Schmitt M, Ślipiński SA, Smith ABT (2011) Family-group names in Coleoptera (Insecta). *ZooKeys* **88**, 1–972. doi:10.3897/zookeys.88.807

Britton EB (1981) The Australian Hygrobiidae (Coleoptera). *Journal of the Australian Entomological Society* **20**, 83–86. doi:10.1111/j.1440-6055.1981.tb01004.x

Cuppen JGM (2000) Distribution, phenology, food and habitat of *Hygrobia hermanni* in The Netherlands (Coleoptera: Hygrobiidae). *Entomologische Berichten* **60**, 53–60.

Dettner K (1997) Insecta: Coleoptera: Hygrobiidae. In *Süßwasserfauna von Mitteleuropa* 20/2, 3, 4. (Eds J Schwoerbel and P Zwick) pp 127–147. Fischer, Stuttgart, Germany.

Dettner K (2016) 7.5 Hygrobiidae, Régimbart, 1879. In *Handbook of Zoology, vol. 4, Arthropoda: Insecta, Part 38, Coleoptera, vol. 1 (2nd edition): Morphology and Systematics (Archostemata, Adephaga, Myxophaga, Polyphaga partim)*. (Eds RG Beutel and RAB Leschen) pp. 112–118. W. de Gruyter, Berlin.

Hawlitschek O, Hendrich L, Balke M (2012) Molecular phylogeny of the squeak beetles, a family with disjunct Palearctic-Australian range. *Molecular Phylogenetics and Evolution* **62**, 550–554. doi:10.1016/j.ympev.2011.09.015

Hendrich L (2001) A new species of *Hygrobia* Latreille, from peatlands of south-western Australia (Coleoptera: Hygrobiidae). *Koleopterologische Rundschau* **71**, 17–25.

Holmen M (1987) *The aquatic Adephaga (Coleoptera) of Fennoscandia and Denmark. I: Gyrinidae, Haliplidae, Hygrobiidae and Noteridae. Fauna Entomologica Scandinavica 20.* E. J. Brill/Scandinavian Science Press, Leiden, Copenhagen.

Michat M, Alarie Y, Hendrich L (2014) Description of the third instar of *Hygrobia nigra* (Clark, 1862) (Coleoptera: Paelobiidae), with a key for the identification of mature larvae of *Hygrobia* Latreille and phylogenetic analysis. *Zootaxa* **3827**(3), 318–330. doi:10.11646/zootaxa.3827.3.2

Nilsson AN (2005) Family Paelobiidae. In *Amphizoidae, Aspidytidae, Haliplidae, Noteridae and Paelobiidae (Coleoptera, Adephaga). World Catalogue of Insects, 7.* (Eds AN Nilsson and BJ van Vondel) pp. 154–163. Apollo Books, Stenstrup, Denmark.

Ribera I, Beutel RG, Balke M, Vogler AP (2002) Discovery of Aspidytidae, a new family of aquatic Coleoptera. *Proceedings. Biological Sciences* **269**, 2351–2356. doi:10.1098/rspb.2002.2157

Ruhnau S (1986) Phylogenetic relationships within the Hydradephaga (Coleoptera) using larval and pupal characters. *Entomologica Basiliensia* **11**, 231–271.

Shull VL, Vogler AP, Baker MD, Maddison DR, Hammond PM (2001) Sequence alignment of 18S ribosomal RNA and the basal relationships of adephagan beetles: evidence for monophyly of aquatic families and the placement of Trachypachidae. *Systematic Biology* **50**, 945–969. doi:10.1080/106351501753462894

11. DYTISCIDAE LEACH, 1815

Lars Hendrich, Cate Lemann and Tom A. Weir

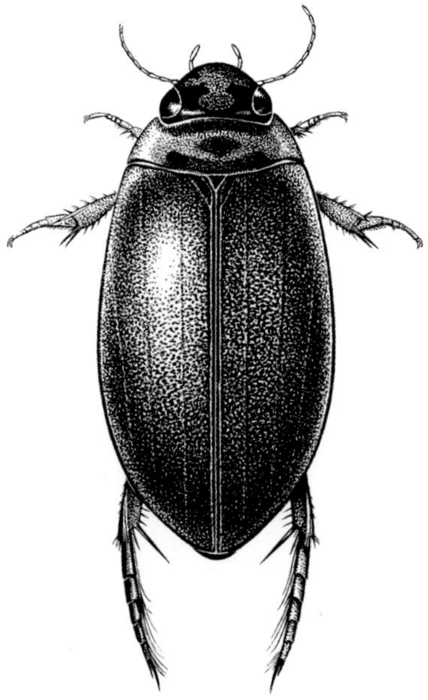

Fig. 11.1. *Rhantus suturalis* (W.S. Macleay, 1825).

Common name. Diving beetles.

Introduction. Diving beetles have had a long history of study by numerous taxonomists, ecologists and nature conservationists and might be one of the best known beetle families in the world today. There are more than 4440 species in some 182 genera of Dytiscidae currently known (Nilsson & Hájek 2018). The species diversity is probably highest in the tropical and subtropical regions, whereas the generic diversity is higher in the more temperate regions (Miller & Bergsten 2016). An overview of the systematics and biology of all Dytiscidae, including a key and habitus colour photos to all genera was recently published by Miller & Bergsten (2016). Another recent monograph on Dytiscidae (Yee 2014) contains several reviews on different aspects of diving beetle ecology, ethology and conservation biology. Nomenclature and classification of the Dytiscidae were reviewed by Nilsson *et al.* (1989), Pederzani (1995), Nilsson & Roughley (1997a, b), Miller (2001) and Miller & Bergsten (2014). An updated World Catalogue was presented by Nilsson & Hájek (2018). Changes to the tribal classification of Agabinae were recently made by Toussaint *et al.* (2016b). Currently, eleven subfamilies and 20 tribes of Dytiscidae are recognised (Miller & Bergsten 2016; Toussaint *et al.* 2016b).

Diversity and distribution. The Australian dytiscid fauna currently comprises 344 species in 39 genera. The altitudinal range in Australia is from at least 20 m underground up to ~1800 m in the Great Dividing Range. The Bassian (southern) region of Australia has the largest number of endemic genera, whereas the northern fauna is more representative of the Oriental fauna with only the genera *Neobidessodes* Hendrich & Balke, *Tiporus* Watts, *Austrodytes* Watts, *Sekaliporus* Watts and *Kakadudessus* Hendrich & Balke (Hendrich & Balke 2009; Hendrich *et al.* 2009) being endemic. There are several endemic hotspots with high numbers of short range endemics such as: the Pilbara region in the west, with at least seven endemic species in surface waters; the south-western corner of South-west Australia with 17 endemics (Hendrich *et al.* 2014); several regions in the south-east; the area around Brisbane, and the Atherton Tableland. The fauna of Tasmania has five endemics and 33 species which are widespread in the south-eastern mainland. The epigean Eyrean or inland fauna is dominated by mostly widespread species which are all capable to flight. The general distribution and species numbers of epigean Dytiscidae in Australia reflects the restriction of most species to the more humid areas of the continent. In contrast, the diverse stygofauna is especially rich in Australia's arid regions where groundwater provides an important refuge from arid surface conditions. The Yilgarn Region of Western Australia has a diverse subterranean dytiscid fauna (stygofauna) associated with calcrete (carbonate) aquifers, and here the dytiscid fauna shows an evolutionary pattern, with multiple calcretes containing 2–3 species in different size classes and cases of sympatric sister species, raising the possibility of their evolution by sympatric speciation. More than 100 species in three genera are stygobiontic and were taken from caves, wells or calcrete aquifers in western and central Australia (e.g. Watts & Humphreys 1999, 2000; Leys & Watts 2008; Watts & McRae 2013). It is likely that additional species are yet to be discovered.

Habitats. Australian Dytiscidae inhabit virtually every aquatic habitat. There are genera mainly occurring in springs (e.g. some *Platynectes* Régimbart and *Exocelina* Broun); rivers (even deep and fast flowing, all *Batrachomatus* Clark, some *Carabhydrus* Watts, *Sternopriscus* Sharp, *Tiporus* and *Laccophilus* Leach), small streams (few *Platynectes*, *Carabhydrus* and *Laccophilus*, *Australphilus* Watts, *Sekaliporus* Watts, *Tiporus*, some *Antiporus* Sharp), restpools of intermittent creeks (e.g. *Chostonectes* Sharp, *Necterosoma* W.J.Macleay, *Tiporus*, *Sternopriscus*, *Limbodessus* Guignot, *Platynectes* and *Exocelina*), and wet rock surfaces, i.e. hygropetric situations (e.g. the madicoulus *Petrodessus* Miller (2012) and a few *Platynectes*). In lotic situations they inhabit flooded meadows, peatland and sedge swamps (e.g. *Hyderodes* Hope, *Spencerhydrus* Sharp, some *Paroster* Sharp, many *Sternopriscus* and *Exocelina*, all *Gibbidessus* Watts and both *Brancuporus* Hendrich, Toussaint and Balke, some *Antiporus* and *Rhantus* Dejean), lakes, billabongs and farm dams (e.g. *Cybister* Curtis, *Onychohydrus* Schaum, *Sternhydrus* Brinck, some *Laccophilus*, *Lancetes* Sharp, *Hyphydrus* Illiger), and water filled rock pools on isolate granite outcrops, so called 'gnammas' (some *Paroster* and *Exocelina*). Some halotolerant species (e.g. *Necterosoma penicillatum* (Clark), *Allodessus bistrigatus* (Clark), *Megaporus howittii* (Clark)) are known from inner Australian salt lakes, whereas marine species are not known. Two species of *Paroster* (formerly *Terradessus* Watts) (Toussaint *et al.* 2016a) were described as terrestrial. Adults of those species have been sieved from damp soil in rainforests (Brancucci & Monteith 1997). As the immature stages of these species are still unknown, it is not possible yet to determine if their whole life cycle is really terrestrial.

Biology. Adult dytiscids need atmospheric air for respiration which is obtained by breaking the water surface with the tip of the abdomen. Continuous air transfer is then facilitated by a hydrofuge pubescence occurring dorsal on surface of the apical tergites (Larson *et al.* 2000) and the air is stored in the subelytral space. Underwater, this reservoir is kept in touch with the water via an air bubble held at the tip of the abdomen. This bubble functions as a physical gill, which must be renewed at the water surface from time to (Dettner & Peters 1999). The air supply and the rectal ampulla function as a hydrostatic control (Hicks & Larson 1991). Larval respiration differs from that of adults. Larvae of larger species (e.g. Cybistrinae and Dytiscinae), come to the water surface, rising the tip of the last abdominal ventrite out of the water. The large spiracles VIII, situated in a ventro-terminal position on that ventrite, are then used for respiration (Korschelt 1923). Air thus obtained is stored in the two main tracheal ducts. Cuticular respiration occurs at least in smaller larvae of other subfamilies as well as among stygobitic species (Balke & Hendrich 2015).

Dytiscid larvae are predators and scavengers and often cannibalistic, actively moving in the water, especially during the night and dawn. The kind of prey is more or less related to the body size of the larva and to its ability to catch a particular animal. Thus, small larvae feed on small crustaceans, dipteran larvae, etc., while large ones (*Sternhydrus*, *Onychohydrus*, *Cybister*) may well attack tadpoles and small fishes. Larvae of mostly small (Hydroporinae) and medium-sized species (*Platynectes, Copelatus* Erichson, *Exocelina*) are creepers, moving along the bottom of their habitat, usually close to the edge of the water, and occasionally with considerable speed. Larvae of the hygropetric *Platynectes tasmaniae* (Clark) creep along wet rock surfaces, where they are barely covered by a water film. The specific diet of most Australian species is still unexplored.

Most probably mating occurs under water. The eggs, the three instars and most of the adults are aquatic with but a few known exceptions. Time of development of eggs is strongly dependent on the temperature of the environment. The hatching time varies from ten days up to six weeks after oviposition in European *Dytiscus* Linnaeus species (Korschelt 1924). Interspecific variations were also observed between species of this genus. The life cycle of Australian dytiscids is likely more flexible and shorter, especially in the tropical north, than in colder regions and so larvae can be found at any time, as long as there is water. Some species of the genus *Paroster* adapted to arid conditions develop rather quickly which allows them to exploit ephemeral habitats such as rock pools on granite outcrops. It is likely that some *Paroster* belong to an ecological group of aquatic organisms which shows definite life history adaptations to summer drought and produce dormant eggs, as do many of the crustacean groups already recorded from gnammas (e.g. Pinder *et al.* 2000) and thus remain *in situ* in the dried out pools on the rocks, revitalising after rain (Hendrich & Fery 2008).

Pupation takes place on land, in oval cells formed by the third instar in mud or damp soil, often under stones, wood or mats of grass close to the water. The pupae of almost all Australian genera are unknown and not illustrated elsewhere. Pupation is believed to take place in air-filled crevices, although the possibility of underwater pupation cannot be fully excluded (Spangler & Decu 1998).

The species found living in groundwater generally show characteristic traits that are likely to be adaptive to a subterranean way of life, such as reduced pigmentation and reduction or loss of eyes (Watts *et al.* 2016). Sexual dimorphism occurs in many species, with shiny males and strongly reticulated, dull females (e.g. *Hyderodes* Hope) or males with highly modified and enlarged antennae (e.g. *Sternopriscus*), or having a little tooth (*Tiporus*) or a small notch (*Necterosoma*) on male protibia (Watts 1978, 2002) as well as modified pro- and mesotarsi.

Adults can roughly be separated into poor swimmers (usually running waters), crawlers (moving mostly within dense vegetation, e.g. *Hydrovatus* Motschulsky), manoeuvrability specialists (spherical species, i.e. most Hyphydrini) and high speed swimmers (e.g. *Rhantaticus* Klug, *Hydaticus* Leach) (Ribera & Nilsson 1995). Adults swim by simultaneously moving either middle (steering: Nachtigall 1960) and hind legs, or only the hind legs (Gewecke 1985). Some species are characterised by reduced swimming abilities. Australian subterranean bidessines run on the first and second pair of legs. Finally, some species evolved a remarkable jumping ability (*Australphilus*) which might be an escape behaviour (Beutel 1995).

Evolutionary biogeography. The Pleistocene Ice Ages were the most recent geohistorical event of major global impact, but their consequences for most parts of the southern hemisphere remain poorly known. Hawlitschek *et al.* (2012) investigated a radiation of ten species of *Sternopriscus* (*tarsalis*-group sensu Hendrich & Watts 2004), the most species-rich genus of epigean Australian diving beetles (Hendrich & Watts 2007). These species can be only identified by studying the male genitalia. They cannot be distinguished reliably by mtDNA (Hendrich *et al.* 2010b) and nDNA because of genotype sharing caused by incomplete lineage sorting. Their genetic similarity suggests a Pleistocene origin and that the repeated isolation of populations in glacial refugia might have led to divergent ecological adaptations and the fixation of morphological traits supporting reproductive isolation and therefore may have promoted speciation. The recent *Sternopriscus* radiation was the first described example of an aquatic insect species flock. The species of this group may represent a stage in speciation past the species flock condition because of their mostly broad and often non-overlapping ranges and preferences for different habitat types (Hawlitschek *et al.* 2012). The origin and evolution of all Australasian Hydroporini diving beetles was investigated in detail by Toussaint *et al.* (2016a) using an integrative approach combining DNA sequences, distributional data, fossil records, and latest methods in phylogenetics and likelihood-based approaches to diversification. The authors suggested that the aridification of Australia initiated in the Miocene might have played a cardinal role in shaping the extraordinary radiation of Australasian Hydroporini resulting in a striking diversity of species richness and types of colonised habitats, but also that later climatic adjustment, especially in seasonality of rainfall, contributed to a major wave of recent Pleistocene extinctions. Despite an astonishing adaptation to climate changes that occurred in the past million years, the group presents a boom-then-bust pattern of diversity dynamics, with a declining trajectory of diversity likely shaped by the on-going and increasing desertification that is occurring in Australia, which restricts the availability of suitable habitats and the likelihood of short-range dispersal events. As a result, this Australasian radiation has been influenced by climatic shifts and in particular the Quaternary climatic changes that likely opened new ecological opportunities, and it appears that whilst the climate continued to warm in the region, these beetles may have been the victims of the changes that once led to them thriving (Toussaint *et al.* 2016a).

Phylogeny and taxonomy. A comprehensive, combined morphological and molecular phylogeny by Miller & Bergsten (2014) led to a new classification of the family (see also Miller & Bergsten 2016). Australian Hydroporini, classified as subtribe Sternopriscina, were again retrieved monophyletic and sister to the *Graptodytes* Seidlitz and *Deronectes* Sharp groups (Siettitiina, Deronectina). *Hydrotrupes* Sharp was the sister of a lineage with the *Platynectes* group of genera and this clade, (Hydrotrupini), sister to the rest of Agabinae classified as Agabini. Toussaint *et al.* (2016b) resolved Platynectini as a third tribe of Agabinae to contain just *Platynectes* and *Andonectes* Guéorguiev. Dytiscinae in the traditional sense was indeed not monophyletic due to the sister group relationship of Laccophilinae + Cybistrinae, these placed as sister to Dytiscinae (without the *Cybister* group of genera). Matinae were retrieved as the sister of all other Dytiscidae.

Modern identification keys, monographs or taxonomic revisions, including comprehensive habitat information, are available for almost all Australian genera, except *Platynectes*, *Exocelina* and *Gibbidessus*, which will be revised in the near future. Furthermore, DNA Barcodes for more than 300 Australian species (Hendrich *et al.* 2010b) are available on Genbank.

Nature conservation. Despite the fact that many Australian dytiscid species are highly endemic to a particular river system or watershed (e.g. *Carabhydrus*, *Batrachomatus* Clark), peatland or wetland complex (e.g. *Gibbidessus*, many *Sternopriscus* and *Antiporus*), have been rarely collected and or seem to be restricted to small geographic regions, none has been listed on a regional or national list of endangered species. Their special ecological requirements (*Antiporus*, *Sternopriscus*, *Gibbidessus*, *Paroster*, *Exocelina*), and the reduced flight muscles in many if not all species of the genus *Carabhydrus*, greatly limit their dispersal capacity and the ability to colonise new habitats if environmental conditions have changed. The southern half of Australia is currently facing critical water supply and water quality problems in many freshwater ecosystems. These problems are expected to become worse over the coming decades due to drought, pollution, extraction and climate change.

In wetland management, Dytiscidae prove to be an important biomonitoring group as they inhabit virtually every kind of fresh- and brackish water habitat, from the smallest puddles up to large lakes and swamps and from streams to irrigation ditches and reservoirs. Due to their diversity in terms of variation in ecological niches they represent an ideal group for environmental impact assessments (EIAs), conservation assessments and biodiversity studies in a wider sense (e.g. Jäch & Balke 2008). The authors hope this chapter will prove useful for limnologists and technicians monitoring the biological functionality of rivers and wetlands in Australia.

Characteristics. Adults. Usually streamlined, dorsoventrally flattened beetles, with both dorsal and ventral surfaces convex, and having typical characteristics of the suborder Adephaga (Fig. 11.1). Large external propleuron and visible notopleural and pleurosternal sutures; eyes not completely divided into dorsal and ventral parts and usually in contact with anterior edge of pronotum; antennae with 11 antennomeres and filiform in most species, entirely or almost entirely glabrous; metaventrite without transverse (katepisternal) suture; metacoxae immobile and fused medially to each other and to the metaventrite, extending laterally to meet elytra, without

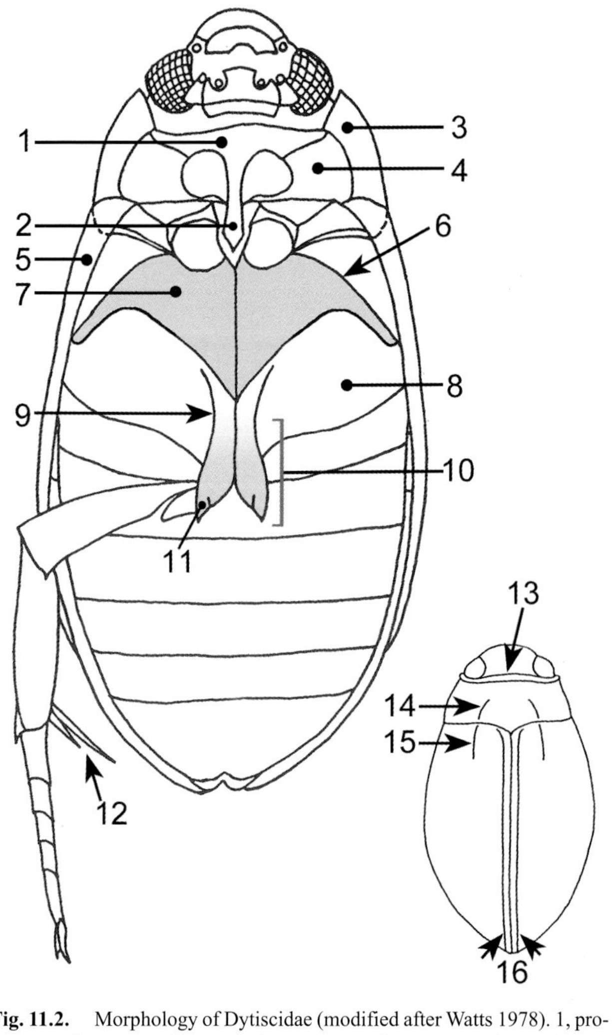

Fig. 11.2. Morphology of Dytiscidae (modified after Watts 1978). 1, prosternum; 2, prosternal process; 3, prothoracic hypomeron; 4, propleuron; 5, elytral epipleuron; 6, metaventral wing; 7, metaventrite; 8, hind coxa; 9, metacoxal line; 10, metacoxal process; 11, metacoxal lobe; 12, metatibial spurs; 13, cervical stria; 14, pronotal stria; 16, sutural stria.

large plates concealing basal abdominal ventrites or plates covering base of trochanters; abdomen with 6 ventrites, first 3 solidly fused together and first ventrite distinctly divided by the metacoxae; metatrochanter large and distinctly offset from the line of the metafemur; tarsi usually 5–5–5; pygidial glands paired (Fig. 11.2). Almost universally aquatic.

Larvae. Typical of Adephaga. Body elongate and elliptical in cross-section, widest at middle. Labrum and clypeus fused; mandibular mola absent; ligula without wedge-like sclerome; mandibles narrow and falcate, with open groove or internal perforation; head without endocarina, as long as or longer than wide, with well-developed epicranial stem, antennae well developed and 4-segmented; legs relatively long, 6-segmented, natatory or ambulatory with 2 protarsal claws; abdomen 8-segmented with articulated urogomphi composed of 1 or 2 segments or multiannulated; spiracles on segment 8 located ventrally or apicoventrally. Aquatic.

Keys to tribes and genera of Australian Dytiscidae

The majority of Australian dytiscids are found in lotic and lentic above ground environments and this key is for species found in these environments. In contrast to those found in subterranean and terrestrial environments, these species have well developed eyes, a more or less pigmented cuticle and in most cases fully developed metathoracic wings. Males are usually distinguished from females by the presence of more broadly expanded protarsomeres 1 to 3 which usually also have ventral adhesive setae. There may also be modifications of the antennae and legs and presence of stridulating devices.

Several genera come out more than once in the key and this is to allow for characters that are hard to see and variation of some characters within genera. The scutellum is hard to see in some species of *Carabhydrus,* the cervical stria (occipital line) is either present or absent in *Limbodessus* Guignot and *Gibbidessus,* and the transverse epipleural carina is either present or absent in *Limbodessus.*

Couplet 10 should be read with care as there are genera in both Bidessini and Hydroporini which exhibit characters of the other tribe: *Neobidessodes* in Bidessini and *Paroster* and *Antiporus gilbertii* (Clark) in Hydroporini.

Some morphological features of ventral and dorsal aspects are explained in Fig. 11.2. Characters of the male genitalia are used sparingly, but are sometimes necessary. All measurements refer to the Australian species in each genus.

Key to the epigean genera of Australian Dytiscidae

1. Scutellum not visible with elytra closed, or nearly completely obscured **2**
   Scutellum clearly visible with elytra closed **28**

2(1). Dorsally, lateral margins strongly discontinuous between pronotum and elytra; pronotum constricted basally, widest anterior to middle; body elongate; elytra each with two longitudinal grooves; 1.7 to 3.4 mm in length; eastern Australia; QLD, NSW, VIC, TAS; endemic .... **HYDROPORINI: *Carabhydrus* Watts (part)**

– Dorsally, lateral margins not strongly discontinuous between pronotum and elytra; pronotum not constricted basally, usually widest near posterior margin; elytra each with four or five longitudinal grooves or without grooves **3**

3(2). Each elytron with four or five longitudinal grooves; elytral colour black with yellow/orange patches; 4.2 to 4.5 mm in length; south-eastern Australia; NSW, VIC; endemic .... **HYDROPORINI: *Barretthydrus* Lea**

– Elytra without grooves but may have short elytral plicae in some species; elytral colour various **4**

4(3). Prosternum and prosternal process in same plane in lateral aspect; pro and meso-tarsi clearly pentamerous (5-segmented); first 4 segments of metatarsi with posteroapical margin lobed; metatarsi with single claw.... **LACCOPHILINAE** **5**

– Prosternum and prosternal process not in same plane in lateral aspect (variously arched) (Fig. 11.3); pro and mesotarsi usually pseudotetramerous (appearing 4 segmented, with tarsomere 3 enlarged and partially enclosing tarsomere 4), more evidently pentamerous in *Sternopriscus* and *Necterosoma*; metatarsomeres not lobed posteroapically; metatarsi with 2 claws... **HYDROPORINAE** **6**

5(4). Base of pronotum sinuate; metatibial spurs bifid at tip; metacoxal lines convering anteriorly, intralinear space not punctate or setose; 3.5 to 5.0 mm in length; northern Australia; WA, NT, QLD; found in all biogeographical regions .... ***Laccophilus* Leach**

– Base of pronotum relatively straight; metatibial spurs not bifid; metacoxal lines sub-parallel, intralinear space punctate and setose; 2.4 to 2.7 mm in length; south-eastern Australia; NSW, VIC, TAS; endemic .... ***Australphilus* Watts**

6(4). Metatarsal claws distinctly unequal in length, anterior shorter than posterior; metacoxal lobes absent; 3.0 to 4.5 mm in length; all states except VIC and TAS, with concentration in northern Australia; found in all biogeographical regions .... **HYPHYDRINI**: ***Hyphydrus* Illiger**

– Metatarsal claws equal or subequal in length; metacoxal lobes present **7**

7(6). Prosternal process short and apically broadly truncate, separating mesocoxae (Fig. 11.4); metacoxal processes apically broad with prominent lateral lobes, each with a deep medial emargination (Fig. 11.5); body broad and short, subglobose (except *Hydrovatus parallelus*); 2.0 to 4.0 mm in length; all states except TAS, with concentration in northern Australia; found in all biogeographical regions .... **HYDROVATINI: *Hydrovatus* Motschulsky**

– Prosternal process elongate and apically narrowly pointed or rounded; metacoxal processes various, but not broad with prominent lobes and deep medial emargination; body more elongate **8**

8(7). Pro and mesotarsi distinctly pentamerous (5-segmented), 4th tarsomere clearly visible; male protibiae ventrally with small notch or emargination **9**

– Pro- and mesotarsi pseudotetramerous, with very small 4th tarsomere lying within lobes of 3rd tarsomere; or with only 3 or 4 tarsomeres; male protibiae not ventrally emarginate **10**

9(8). Posterior margin of metacoxal processes indented in middle; metacoxal lobes rounded (Fig. 11.6); outer face of metatibiae virtually impunctate except for a row of setiferous punctures; elytra usually with distinct subapical spines; male antennomeres not elaborately expanded; 4.0 to 5.4 mm in length; widespread – all states; also found in New Caledonia and Timor .... ***Necterosoma* W.J.Macleay**

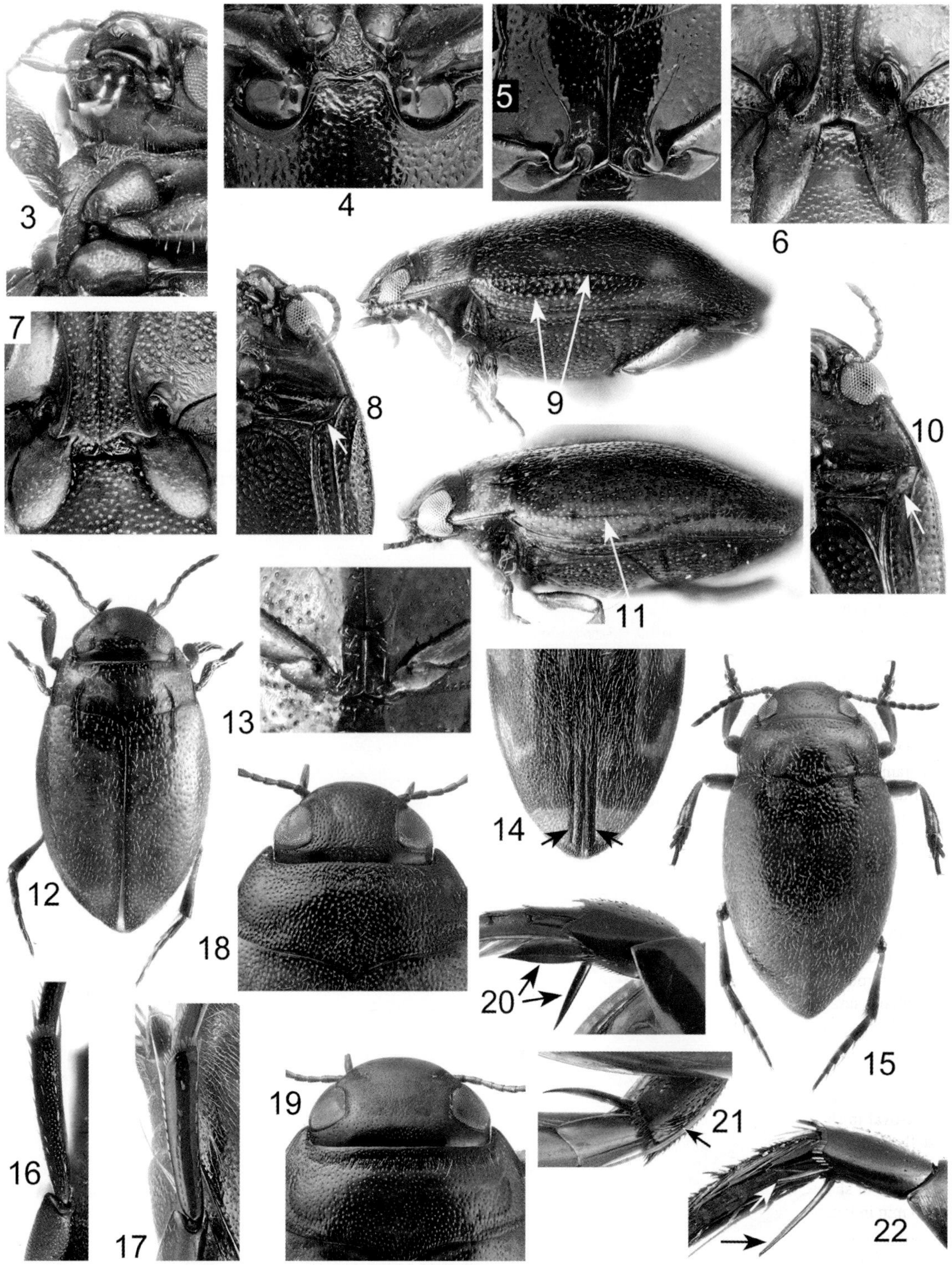

Figs 11.3–11.22. Dytiscidae. 3, *Necterosoma darwinii* (Babington), head, lateral; 4, *Hydrovatus nigrita* Sharp, prosternal process; 5, *Hydrovatus nigrita* Sharp, metacoxal lobes; 6, *Necterosoma darwinii* (Babington), metacoxal lobes; 7, *Sternopriscus aquilonaris* Hendrich & Watts, metacoxal lobes; 8, *Clypeodytes weiri* Hendrich & Wang, transverse epipleural carina; 9, *Clypeodytes weiri* Hendrich & Wang, body, lateral, two lateral carinae; 10, *Leiodytes migrator* (Sharp), epipleural carina absent; 11, *Leiodytes migrator* (Sharp), body, lateral, arrow - weak lateral carina; 12, *Gibbidessus chipi* Watts, habitus; 13, *Gibbidessus chipi* Watts, metacoxal lines; 14, *Hydroglyphus mastersii* (W.J.Macleay), elytral apices with sutural striae; 15, *Gibbidessus pictipes* (Lea), habitus; 16, *Tiporus undecimmaculatus* (Clark), outer face of metatibia; 17, *Chostonectes gigas* (Boheman), outer face of metatibia; 18, *Chostonectes gigas* (Boheman), head; 19, *Megaporus howittii* (Clark), head; 20, *Cybister tripunctatus* ssp. *temnenkii* Aubé, metatibial spurs; 21, *Onychohydrus scutellaris* (Germar), bifid spines on metatibia; 22, *Sandracottus bakewellii* (Clark), metatibial spurs.

– Posterior margin of metacoxal processes produced in the middle; metacoxal lobes somewhat truncate (Fig. 11.7); outer face of metatibiae densely and evenly punctate; elytra without distinct subapical spines; males often with elaborately expanded antennae; 2.00 to 4.5 mm in length; widespread – all states with a concentration in east, south-west and south-east Australia; endemic.................................. ***Sternopriscus* Sharp**

10(8). Pronotum with basal striae; elytra with basal striae (except *Neobidessodes* with pronotal striae weak or absent and elytral basal striae absent, but having prosternal process flat, parallel sided, margined and broadly rounded apically); metatibiae slender, gradually expanded apically; metacoxal lobes very small; metacoxal processes at same level as abdomen; male parameres two segmented....**BIDESSINI**..............................**11**

– Pronotum without basal striae (except *Antiporus gilbertii* with metatibiae densely and evenly punctured and pronotum strongly bordered); elytra without basal striae; metatibiae various, but not gradually expanded apically; metacoxal lobes larger, conspicuous; posterior margin of metacoxal process distinctly separated from level of abdomen (*Paroster* with metatibiae gradually expanded apically and metacoxal processes at same level as abdomen, but elytra with microreticulation of fine meshes); male parameres one segmented....**HYDROPORINI [major part]** **22**

11(10). Cervical stria (occipital line) present..**12**

– Cervical stria (occipital line) absent..**17**

12(11). Anterior clypeal margin modified, with distinct flattened, narrowly bordered margin; elytra with 1 or 2 lateral carinae; body short and oval **13**

– Anterior clypeal margin not modified as above; elytra without lateral carinae; body shape various **14**

13(12). Elytra with transverse epipleural carina at humeral angle, bordering basal cavity (Fig. 11.8); elytra with two lateral carinae forming a furrow (Fig. 11.9); 1.6 to 2.0 mm in length; northern and north-eastern Australia; WA, NT, QLD, NSW; also found in Africa, South-east Asia and New Guinea ....***Clypeodytes* Régimbart**

– Elytra without transverse epipleural carina at humeral angle (Fig. 11.10); elytra with a single weak lateral carina (Fig. 11.11); 1.8 to 2.0 mm in length; northern and north-eastern Australia; WA, NT, QLD, NSW; also found in Africa, South-east Asia and New Guinea............................ ....***Leiodytes* Guignot**

14(12). Clypeus with flattened area between antennal bases and anterior margin; body dorsoventrally compressed and flattened; pronotum parallel-sided, widest at middle; metaventrite with rows of punctures at midline; 2.2 to 2.5 mm in length; northern Australia; NT, QLD; endemic................. ....***Kakadudessus* Hendrich & Balke**

– Clypeus anteriorly without flattened area; body not dorsoventrally flattened as above; pronotum usually not parallel-sided, usually widest at base; metaventrite without rows of punctures at midline (except *Limbodessus capeensis* Watts & Leys)...**15**

15(12). Oblong-oval in shape; metacoxal lines short, not prolonged by row of punctures, separated by greater than half their own length (Fig. 11.13); less than 2.0 mm in length; south-west and south-east Australia; WA; NSW, TAS; endemic ....***Gibbidessus* Watts (part)**

– More elongate-oval in shape; metacoxal lines longer, sometimes prolonged by row of punctures, reaching nearly to metaventrite, separated by less than half their own length..**16**

16(15). Male parameres with apical segment simple, finger-like; antennomere 10 of female antennae with basal groove; metacoxal plates moderately punctured; female underside with black metathorax contrasting with yellow abdomen; male underside all black; 3.0 to 3.5 mm in length; widespread – (all states); also found sporadically in South-east Asia, New Caledonia and some other Pacific Islands.......***Allodessus* Guignot**

– Male parameres with apical segment strongly lobed, often hook-like; antennomere 10 of female antennae without basal groove; metacoxal plates virtually impunctate (strongly punctate in *Limbodessus capeensis*); underside of male and female more uniformly brown coloured; 1.8 to 3.0 mm in length; widespread – (all states); most species limited to Australia, one in New Caledonia.................***Limbodessus* Guignot (part)**

17(11). Anterior clypeal margin modified, flattened, beaded and protruding; 1.5 to 1.7 mm in length; north-east QLD; endemic........... ***Petrodessus* Miller**

– Anterior clypeal margin unmodified **18**

18(17). Elytra without basal striae; pronotal striae weak or absent; 2.3 to 4.2 mm in length; widespread across northern and eastern Australia; WA, NT, QLD, NSW, VIC; also found in New Guinea .... ***Neobidessodes* Hendrich & Balke**

– Elytra with basal striae; pronotal striae well marked **19**

19(18). Elytra with sutural striae (Fig. 11.14); 2.0 to 4.0 mm in length; widespread across northern and eastern Australia; WA, NT, QLD, NSW, VIC; also found throughout much of Africa, much of Europe, southern Asia and New Caledonia.. ***Hydroglyphus* Motschulsky**

– Elytra without sutural striae..**20**

20(19). Elytra with transverse epipleural carina at humeral angle, bordering basal cavity (*Limbodessus compactus* (Clark)); 1.6 to 2.5 mm in length); widespread – all Australian states; also found in New Guinea, Japan, South-east Asia, New Zealand, New Caledonia and several other Pacific Islands..***Limbodessus* Guinot (part)**

– Elytra without transverse epipleural carina at humeral angle, basal cavity not bordered..**21**

21(20). Elongate-oval in shape, depressed dorsoventrally; elytra unicolourous; metacoxal plates finely and densely punctured; sometimes found interstitially in gravel at edges of streams (*Limbodessus rivulus* (Larson); 1.8 to 2.2 mm in length; north-east QLD; *Limbodessus occidentalis* (Watts & Humphreys); 1.9 to 2.3 mm in length; Pilbara & Yilgarn regions of WA; endemic species) ....***Limbodessus* Guignot (part)**

– Oblong-oval in shape, not depressed dorsoventrally; elytra with distinct markings; metacoxal plates coarsely and sparsely punctured; less than 2.0 mm in length; south-west and south-east Australia; WA, NSW, TAS; endemic ....***Gibbidessus* Watts (part)**

22(10). Outer face of metatibiae densely and evenly punctured (Fig. 11.16); males with a single claw on protarsus..**23**

– Outer face of metatibiae virtually impunctate except for a row of setiferous punctures (Fig. 11.17); males with two claws on protarsus **26**

23(22). Lateral beaded edge of each elytron sharply bent at humeral angle before meeting pronotum; protarsi asymmetric, male protarsi with 3 tarsomeres; 3.3 to 5.0 mm in length; northern Australia; WA, NT, QLD; endemic ....***Tiporus* Watts**

– Lateral beaded edge of each elytron not sharply bent at humeral angle, either in a straight line or smooth curve before meeting pronotum; protarsi symmetric or asymmetric; male protarsi with 4 or 5 tarsomeres **24**

24(23). Lateral beaded edge of each elytron and lateral beaded edge of pronotum forming a nearly continuous straight line in combination; protarsi asymmetric, male protarsi with 4 tarsomeres; 3.2 to 3.5 mm in length; northern Australia; WA, NT, QLD; endemic....................***Sekaliporus* Watts**

– Lateral beaded edge of each elytron smoothly curved to meet pronotum at humeral angle; protarsi symmetric, male protarsi with 5 tarsomeres (pseudotetramerous); difficult to identify without dissection of male aedeagus (or DNA sequencing) **25**

25(24). Median lobe of aedeagus bilaterally symmetric; apical margins of elytra somewhat obliquely truncate and minutely serrate in most species; 3.5 to 6.8 mm in length; widespread – all states; also found in New Zealand ....***Antiporus* Sharp**

– Median lobe of aedeagus bilaterally asymmetric; apical margins of elytra not obliquely truncate or minutely serrate; female of one species (*Brancuporus pennifoldae* (Watts & Pinder) with apicolateral margins of elytra widely flanged; 3.0 to 3.5 mm in length; restricted to peatlands and seasonal swamps of south-west WA; endemic .... ***Brancuporus* Hendrich, Toussaint & Balke**

26(22). Outer edge of hind tarsal tarsomeres without fringe of stout spines; elytral epipleura narrowing abruptly at level of first abdominal ventrite and becoming very narrow in apical half; 1.8 to 4.2 mm in length; southern Australia; WA, SA, VIC, NSW; endemic .... ***Paroster* Sharp**

– Outer edge of hind tarsal tarsomeres with fringe of stout spines; elytral epipleura gradually narrowing to elytral apex **27**

27(26). Punctures on head well marked, not much smaller than those on pronotum (Fig. 11.18), or obscured by strong regular striations; metafemora slender with posteroventral angles rounded and posterodorsal angles slightly angulate; 4.0 to 6.0 mm in length; east and south-east Australia; QLD, NSW, VIC, SA, TAS; also found in New Guinea .... ***Chostonectes* Sharp**

– Punctures on head minute, much smaller than those on pronotum (Fig. 11.19); reticulation on head fine, regular, never with striations; metafemora stouter with more or less well defined angulate posteroventral angles and posterodorsal angles rounded; 5.8 to 7.2 mm in length; widespread – all states; also found in New Guinea, Fiji and New Caledonia .... ***Megaporus* Brinck**

28(2). Less than 4 mm in length; body outline with strong discontinuity at bases of pronotum and elytra; pronotum constricted basally, widest anterior to middle; body elongate; elytra each with two longitudinal grooves; 1.7 to 3.4 mm in length; eastern Australia; QLD, NSW, VIC, TAS; endemic .... **HYDROPORINI**: ***Carabhydrus* Watts (part)**

– Greater than 5 mm in length; dorsally, body outline without strong discontinuity at bases of pronotum and elytra; pronotum not constricted basally, usually widest near posterior margin; elytra each without grooves or incised lines **29**

29(28). Eye with anterior margin not emarginate above bases of antennae, outline fairly round; male with protarsomeres 1 to 3 greatly widened to form a large round or oval palette with ventral adhesive setae; aedeagus with symmetric median lobe of aedeagus **30**

– Eye with anterior margin emarginate above base of antenna, emargination formed by edge of clypeus; male with protarsomeres 1 to 3 moderately widened to form an elongate to elongate-oval palette with ventral adhesive setae; aedeagus with asymmetric median lobe of aedeagus **40**

30(29). Metatibial spurs different in size and shape, anterior spur acuminate, distinctly broader basally than posterior spur (Fig. 11.20); posteroapically with large dense cluster of bifid spines on inner side of metatibia (Fig. 11.21); adhesive setae of male protarsus with apices elongate oval, not circular.... **CYBISTRINAE** **31**

– Metatibial spurs similar in shape, but often different in length (Fig. 11.22); bifid spines on inner side of metatibia linear, not clustered (Fig. 11.23); adhesive setae of male protarsus with apices circular.... **DYTISCINAE** **36**

31(30). Prosternal process and prosternum with prominent longitudinal sulcus throughout **32**

– Prosternal process and prosternum without prominent longitudinal sulcus throughout, although margins variously bordered and anterior portion shallowly sulcate **34**

32(31). Metacoxal lines absent; claws on male protarsus equal; 14 to 17 mm in length; northern and eastern Australia; WA, NT, QLD, NSW; also found in New Guinea, the Moluccas and New Caledonia .... ***Sternhydrus* Brinck**

– Metacoxal lines present; claws on male protarsus unequal **33**

33(32). Metatarsus with 2 claws; larger claw of male protarsus strongly curved; 18 mm in length; south-west and south-east Australia; WA, VIC; endemic .... ***Spencerhydrus* Sharp**

– Metatarsus with 1 claw; larger claw of male protarsus almost straight (*Cybister weckwerthi* Hendrich; endemic); 16–17 mm in length; northern Australia; NT .... ***Cybister* Curtis (part)**

34(31). Metacoxal lines absent; metatarsus with 2 claws; 24 to 28 mm in length; south-west and eastern Australia – all states; also found in New Zealand .... ***Onychohydrus* Schaum**

– Metacoxal lines present; metatarsus with either 1 or 2 claws **35**

35(34). Metatarsus with 1 claw; lateral margins of prosternum and prosternal process not distinctly ridged throughout; mesotarsomeres with dorsal and ventral tufts of setae near apical margins; 22 to 36 mm in length; widespread – all states; found in all biogeographic regions .... ***Cybister* Curtis (part)**

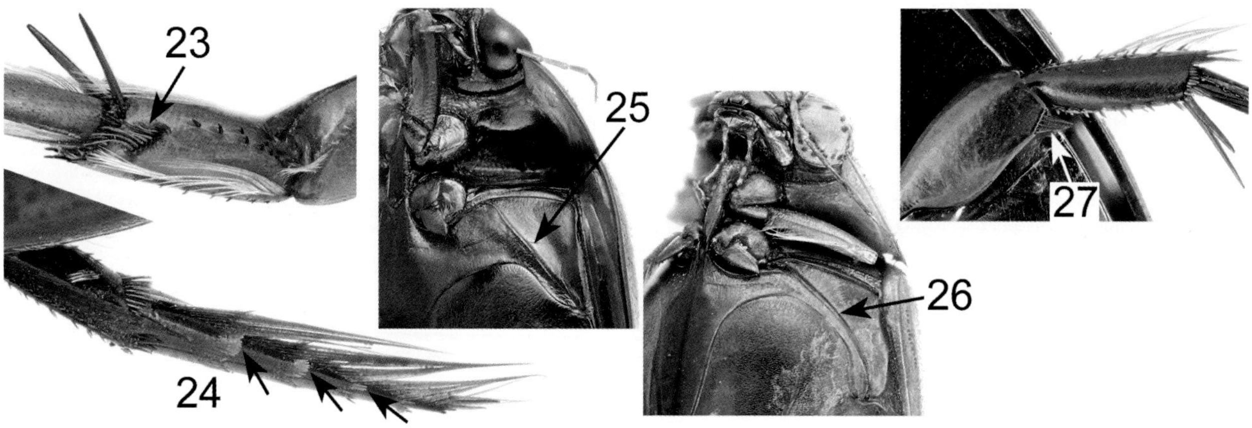

Figs 11.23–11.27. Dytiscidae. 23, *Eretes australis* (Erichson), bifid spines on inner side of metatibia; 24, *Rhantaticus congestus* (Klug) male hind tarsomere with adpressed setae; 25, *Hydaticus bihamatus* Aubé, metaventral wing; 26, *Eretes australis* (Erichson), metaventr wing; 27, *Platynectes* sp., metafemur with spine-like setae.

– Metatarsus with 2 claws; lateral margins of prosternum and prosternal process distinctly ridged throughout; mesotarsomeres with only ventral tufts of setae near apical margins; 16 to 20 mm in length; northern Australia, WA, NT, QLD; endemic.. ***Austrodytes* Watts**

36(30). Posterior margins of metatarsomeres 1–4 with coarse fringe of flat, adpressed golden setae (Fig. 11.24) **37**

– Posterior margins of metatarsomeres 1–4 without fringe of setae **39**

37(36). Apices of metatibial spurs simple, sharp; anterolateral margin of metaventral wing straight (Fig. 11.25); males with stridulatory apparatus on dorsal surfaces of protibiae and protarsi; 9 to 15 mm in length; widespread – all states except TAS; found in all biogeographical regions.. **HYDATICINI**: ***Hydaticus* Leach**

– Apices of metatibial spurs bifid; anterolateral margin of metaventral wing convexly curved (Fig. 11.26); males without stridulatory apparatus.... **ACILIINI** **38**

38(37). Posterolateral angle of tarsomeres 3 and 4 of mesotarsus with one long spine being much longer than width of tarsomeres; ventral setae on mesofemora with at least some as long as half of width of femora; elytra black with yellowish markings; 13 to 15 mm in length; northern, eastern and central Australia; WA, NT, QLD, NSW; also found in India, South-east Asia and New Guinea.. ***Sandracottus* Sharp**

– Posterolateral angle of tarsomeres 3 and 4 of mesotarsus with one long spine with at most same length as width of tarsomeres; ventral setae on mesofemora at most as long as one quarter of width of femora; elytra mottled brown with black markings; 8 to 10 mm in length; widely distributed in northern and eastern Australia; WA, NT, QLD; also found throughout Africa, Middle East, India, South-east Asia, New Guinea and New Caledonia.. ***Rhantaticus* Sharp**

39(36). Lateral margins of pronotum narrowly bordered; posterolateral margins of elytra with a series of short, acute spines; 9 to 15 mm in length; widespread – all states; also found in central America, Europe, Africa, Middle East, India, South-east Asia and New Guinea.. **ERETINI: *Eretes* Laporte**

– Lateral margins of pronotum broadly bordered; posterolateral margins of elytra without short spines; females sometimes with irregular rugose sculpturing on pronotum and elytra; 15 to 20 mm in length; south-west and south-east Australia; WA, SA, NSW, VIC; endemic.. **DYTISCINI: *Hyderodes* Hope**

40(29). Metafemora with distinct short linear series of stout spine-like setae on posteroventral angle (Fig. 11.27); 5.6 to 8.6 mm in length; widespread – all states; also found in South America, China, South-east Asia and New Guinea.. **PLATYNECTINI: *Platynectes* Régimbart**

– Metafemora without distinct short series of stout spine-like setae on posteroventral angle.. **41**

41(40). Prosternum and prosternal process together flat with a medial longitudinal groove, laterally bordered between procoxae and mesocoxae; 7.0 to 9.5 mm in length; south-west, northern and eastern Australia; WA, NT, QLD, NSW, VIC, TAS; endemic.. **MATINAE: *Batrachomatus* Clark**

– Prosternum and prosternal process together convex or slightly flattened, without medial longitudinal groove.. **42**

42(41). Metatarsal claws subequal in length; metacoxal lines closely approximated in front of metacoxal lobes then markedly diverging anteriorly; elytra sometimes with narrow longitudinal striae....**COPELATINAE** **43**

– Metatarsal claws markedly unequal in length; metacoxal lines not closely approximated in front of metacoxal lobes; elytra without longitudinal striae **44**

43(42). Elytra each with 4 or more sharply incised longitudinal striae; male without stout hook shaped seta on anteroventral angle of tarsomere 4 of protarsus; 4.5 to 8.0 mm in length; northern and eastern Australia; WA, NT, QLD; also found in Central and south America, Africa, India, China, South-east Asia, New Guinea, New Caledonia and other some Pacific Islands.. ***Copelatus* Erichson**

– Elytra without distinct longitudinal striae or with numerous very short striae; male with stout, hook shaped seta on anteroventral angle of tarsomere 4 of protarsus; 4.0 to 9.5 mm in length; widespread – all states; also found in China, Hawaii, New Guinea, New Zealand and New Caledonia.. ***Exocelina* Broun**

44(43). Elytra apically truncate, with distinct light and dark stripes; 7.5 to 12.0 mm in length; southern Australia; WA, SA, VIC, NSW; also found in South America, New Zealand and a few isolated islands.. **LANCETINAE**: ***Lancetes* Sharp**

– Elytra apically conjointly rounded, yellow to dark brown with dense dark black speckles; 10 to 13 mm in length; widespread – all states; found in all biogeographical regions.. **COLYMBETINAE**: ***Rhantus* Dejean**

Key to the subterranean and terrestrial genera of Australian Dytiscidae

This key is for those small species found in either subterranean (stygobitic) or terrestrial habitats. These species are highly convergent morphologically, general with eyes greatly reduced or absent, depigmented cuticle (subterranean species only), flightless (with metathoracic wings reduced or absent) and usually of very small size. Their classification is somewhat problematical and recent molecular analyses have helped clarify the clades in which they belong, even though they do not have many of the diagnostic characteristics of the epigean species in these clades. (Miller & Bergsten 2016). Some genera cannot be satisfactorily separated on morphological characters alone. Most genera belong to Hydroporinae: Bidessini and Hydroporini, one belongs to Copelatinae. All species are endemic to Australia.

Several species of Limbodessus and Exocelina with well-developed eyes are sometimes found interstitially in gravel at the edge of streams, but these are treated here as primarily epigean and will key out in that key.

1. Scutellum visible; subterranean.. **2**

– Scutellum not visible; subterranean or terrestrial.. **3**

2(1). Less than 3.0 mm in length (2–2.5 mm); elytra each with distinct longitudinal groove towards the side; metacoxal lines well separated, converging anteriorly; NSW (*Carabhydrus stephanieae* Watts, Hancock & Leys).. **HYDROPORINAE: HYDROPORINI: *Carabhydrus* Watts**

– Greater than 3.00 mm in length (3.5–4.5 mm); elytra without grooves; metacoxal lines close together, diverging anteriorly; NT, WA (*Exocelina abdita* (Balke, Watts, Cooper, Humphreys & Vogler) and *E. rasjadi* (Watts & Humphreys)).. **COPELATINAE**: ***Exocelina* Broun**

3(1). Terrestrial; pronotum broadly bordered, without basal striae; elytra with humeral angle broadly produced laterally, deflexed vertically and laterally in apical half, with distinct faint costae; legs without swimming hairs; 1.5 mm in length; collected in leaf litter above 750 m in Wet Tropics area of NQLD (*Paroster caecus* (Watts) and *P. anophthalmus* (Brancucci & Monteith) – formerly in *Terradessus*).. ....**HYDROPORINAE: HYDROPORINI: *Paroster* Sharp (part)**

– Subterranean; pronotum not bordered, with or without basal striae; elytra with humeral angle not produced laterally, smoothly rounded laterally, without costae; legs usually with at least a few long swimming hairs..**4**

4(3). Male parameres one-segmented; pronotum without basal striae; metatibiae approximately of same width throughout; 1.1 to 4.0 mm in length; WA, NT, NSW..**HYDROPORINAE: HYDROPORINI: *Paroster* Sharp (part)**

– Male parameres two-segmented (except *Limbodessus insolitus* Watts & Humphreys); pronotum usually with basal pronotal striae, although these sometimes weakly developed; metatibiae narrow basally then somewhat strongly expanded apically; 1.3 to 4.4 mm in length; WA................. **HYDROPORINAE: BIDESSINI: *Limbodessus* Guignot AND *Neobidessodes* Hendrich & Balke** [genera cannot be distinguished on morphological characters alone, but key to species in Watts & Humphreys (2009) can be utilised for individual species]

Classification of the Australian genera

Here we follow the classification of Miller and Bergsten (2016), except for the addition of the tribe Platynectini to the subfamily Agabinae (Toussaint *et al.* 2016b). Generic names and synonyms applying to the Australian fauna are confirmed from Nilsson & Hájek (2018).

Subfamily Agabinae Miller, 2001

Subfamily with three tribes worldwide (Toussaint *et al.* 2016b).

Tribe Platynectini Toussaint, Hendrich, Hájek & Balke, 2016

Characteristics. Medium-sized beetles (5.6–8.6 mm in Australia). Colour and sculpture variable. Body elongate to broadly oval. Elytron not striate except for two species in Australia (see below). Intralinear space of metacoxae relatively broad; metacoxal lines present in most genera. Metafemur ventrally with setiferous row at outer posterior angle in most species. Posterior (dorsal) face of metatibia without bifid setae. Metatarsal claws of the same length or not. The larva has been roughly described (Watts 1963, 2002).

Distribution. Neotropical, East Palearctic, Oriental and Australasian regions. Most diverse in the Indomalayan and Australasian realms.

Australian taxa. Only one genus: *Platynectes* Régimbart, 1879 with three subgenera.

Platynectes Régimbart, 1879 (Pl. 16F)

Type species. *Agabus decemnotatus* Aubé, 1838.

Leuronectes Sharp, 1882

Characteristics. As above for tribe. Medium-sized (5.6–8.6 mm), roundish to oval, rather flat, shiny and mostly black beetles, usually with at least two yellowish spots, markings or several yellowish stripes on elytra. Two species with rows of deep but small pronotal striae and seven to ten deep striae on elytron. Posterior angles of pronotum acute. Lateral wings of metaventrite narrow and almost parallel. Prosternal process sharp at apex. Parameres not strongly bent at middle. Metacoxal lines anteriorly obsolete, not reaching posterior margin of metaventrite. Median lobe of aedeagus somewhat asymmetric in ventral view, but with regular outline in lateral view.

Australian species. 17 described species; 10 new species will be described in the near future (Hendrich *et al.* unpublished).

Distribution. Distributed in Palearctic, Oriental and Australian realms. In Australia, one species in tropical northern Australia (WA, NT, QLD), one species in all states, three species endemic in south-western Australia and 12 species along the east coast, QLD, NSW, VIC, TAS, SA.

Biology. Most of Australian *Platynectes* are mainly lotic, found in small springs, streams, intermittent creeks and slow flowing rivers and its side and rest pools, often rich in rotten leaves, with sand or decaying plant material, from lowland areas near the coast up to 1000 m. At least two species are madicolous. Four species, two in the south-west and two in the south-east, including Tasmania, are restricted to lentic sites, such as seasonally flooded meadows, peatland habitats, coastal swamps and alpine grasslands.

References. Alarie *et al.* (2000); Guéorguiev (1972); Hendrich & Šťastný (2014); Miller & Bergsten (2016); Nilsson & Hájek (2018); Nilsson & Holmen (1995); Pederzani (1995); Toussaint *et al.* (2016b); Watts (1978, 2002).

Keys to species. Guéorguiev (1972); Watts (1978); Hendrich & Šťastný (2014).

Subfamily Colymbetinae Erichson, 1837

Characteristics. Medium-sized (10–13 mm in Australia), streamlined and more flattened diving beetles, narrowing towards apex. Scutellum fully exposed. Eyes anteriorly emarginat above bases of antennae, indentation caused by edges of clypeus. First three tarsomeres of fore tarsi of male broadened but not forming adhesive discs, although usually provided with plenty of very small adhesive tube or setae. Most taxa in this subfamily also with metatarsal claws unequal. Median lobe of aedeagus bilaterally asymmetric. Prosternum and prosternal process together in same plane and apices of elytra evenly rounded.

Distribution. A subfamily with eight genera worldwide (Miller & Bergsten 2016; Balke *et al.* 2017b).

Australian taxa. Only genus *Rhantus*.

Rhantus Dejean, 1833 (Pl. 1D, 16J)

Type species. *Colymbetes pulverosus* Stephens, 1828 (= *Colymbetes suturalis* W.S.Macleay, 1825).

Characteristics. Body form oval, with continuous outline. Pronoto-elytral angle negligible. Base of pronotum as wide as base of elytra. First four metatarsomeres apically lobate. Base of pronotum not strongly sinuate, with basal angles not acutely projecting posteriorly. Outer face of hind tibia with rows of setigerous punctures, otherwise almost unpunctured. Usually yellow to dark brown, with dense black speckles on elytra. Larvae of both Australian species of *Rhantus* have been described (Alarie *et al.* 2009; Balke *et al.* 2017b).

Australian species. Two.

Distribution. Worldwide (Balke *et al.* 2017b). In Australia, one species widespread (Balke *et al.* 2009), except tropical lowlands in northern Australia, and recorded from all states; one species endemic to south-western corner of WA, south of a line from Gingin to Albany.

Biology. Both species of Australian *Rhantus* are lotic. The common and widespread *R. suturalis* W.S.Macleay) is found in a large variety of standing waters such as flooded meadows, farm dams, ditches, peatlands, forest pools and rest pools of streams and creeks. The south-western *R. simulans* Régimbart is restricted to shaded, peaty and seasonal paperbark swamps.

References. Alarie *et al.* (2009); Balke *et al.* (2000, 2009, 2017b); Miller & Bergsten (2016); Nilsson & Hájek (2018); Pederzani (1995); Watts (1978, 2002).

Keys to species. Balke *et al.* (2000).

Subfamily Copelatinae Branden, 1885

Characteristics. Small to medium-sized, elongate or oblong-oval and more flattened diving beetles, narrowing towards apex. Scutellum visible. Elytron with or without longitudinal striae. Prosternum and its process on same plane or process slightly deflected upwards behind procoxae. Metepisternum extended to mesocoxal cavity. Prosternal process short. Metaventral wings relatively broad at level of mesocoxa and very narrow laterally. Metacoxal processes with apical cleft; intralinear space narrow or lines absent. Metafemur ventrally without a setiferous row at outer posterior angle. Tarsi with 5 distinct tarsomeres; male with protarsomeres 1–3 dilated but not together forming a round palette. Metatarsal claws of the same length (Nilsson & Holmen 1995). Larvae of *Exocelina* were roughly described by Watts (1963); all larvae of Australian *Copelatus* are unknown.

Distribution. A subfamily with eight genera worldwide but most diverse in tropical regions (Miller & Bergsten 2016; Nilsson & Hájek 2018).

Australian taxa. Two genera: *Copelatus* Erichson, 1832 and *Exocelina* Broun, 1886.

Copelatus Erichson, 1832 (Pl. 15D)

Type species. *Dytiscus posticatus* Fabricicus, 1801.

Characteristics. Small to medium-sized (4.5–8.0 mm in Australia), elongate or oblong-oval beetles. Elytron with longitudinal striae. Metacoxal lines present. Male pro- and mesotarsomeres 1–3 broadly dilated with 4 rows of larger adhesive discs bordered by long and thick setae. Median lobe aedeagus asymmetric.

Australian species. Eight.

Distribution. Worldwide (442 species) with greatest diversity in the tropics. In Australia, most species in the tropical north with two species extending along the east coast south to Brisbane (WA, NT, QLD).

Biology. Species of Australian *Copelatus* are lentic, found in small rock, rest and side pools of streams and intermittent creeks, in forest pools, and at flooded lake and river margins, often rich in submerged and emerged vegetation, and with sand or decaying plant material, mainly in lowland areas up to 200 m.

References. Nilsson & Holmen (1995); Nilsson & Hájek (2018); Watts (1978, 2002).

Keys to species. Watts (1978).

Exocelina Broun, 1886 (Pl. 15L)

Type species. *Exocelina advena* Broun, 1886.

Characteristics. As above for subfamily. Small to medium-sized (4.0–9.5 mm), elongate or oblong-oval beetles. Elytra without distinct longitudinal striae or with numerous very short striae. Protarsomere 4 bearing a hook on anterior angle (modified seta). Shiny, black, yellowish or brown in some species with pale basal marking on elytra.

Australian species. 16, 14 of which were formerly placed in *Copelatus*.

Distribution. 161 species recorded from Australia, New Guinea, New Caledonia, China, Hawaii, highest diversity in New Guinea and New Caledonia (Shaverdo *et al.* 2012, 2013; Wewalka *et al.* 2010. In Australia, 16 species, two from south-western Australia and 14 from QLD, NSW, VIC, SA and TAS. The genus is absent in north-western Australia (Kimberley region, Pilbara, west coast south to Geraldton) and most diverse in the south-east from around Brisbane to Melbourne.

Biology. Nine species of *Exocelina* are strongly lotic and restricted to small and often intermittent creeks, streams and its side pools, bottom with a layer of rotten leaves and debris, often situated in woodland or closed canopy forest sites. Four species are restricted to lentic sites such as forest pools and puddles, seasonally flooded meadows, marshes and peatland areas. Two species are stygobitic. From almost sea level to 1000 m.

References. Balke & Bergsten (2003); Miller & Bergsten (2016); Nilsson & Hájek (2018); Shaverdo *et al.* (2012, 2013); Watts (1963, 1978, 2002); Watts *et al.* (2016); Wewalka *et al.* (2010).

Keys to species. Watts (1978) (as *Copelatus*, couplets 9 to 20).

Subfamily Cybistrinae Miller & Bergsten, 2014

Characteristics. Large to extreme large diving beetles (15–50 mm). Scutellum visible. Eyes with anterior margin not emarginate above base of antenna, outline fairly rounded. Metatibial spurs different in size and shape, anterior spur wider, apically acuminate. Metatibia with large dense cluster of bifid spines posteroapically on inner side. Male with ventral surface of pro- and often mesotarsomeres 1 to 3 greatly widened to form a large round or oval palette equipped with ventral adhesive setae, apices elongate oval. Median lobe of aedeagus bilaterally symmetric, with distinct, elongate ventral sclerites. Females with single genital opening and gonocoxae fused dorsally. Larvae: Without frontal projection, anterior margin of clypeus prominently dentate. Anterior margin of prementum with distinct lobe without spinous setae. Last two abdominal ventrites with dense fringe of long setae. Larvae: Urogomphi very short or absent; described for four of the five Australian genera (except *Austrodytes*) (Watts 1963, 1964, 2002; Michat *et al.* 2015).

Distribution. A subfamily with seven genera worldwide occurring mainly in the tropic and subtropics (Miller *et al.* 2007; Miller & Bergsten 2016; Nilsson & Hájek 2018).

Australian taxa. Five genera (Nilsson & Hájek 2018).

Biology. Most cybistrines occur in lentic habitats with extensive emergent vegetation though some are common in turbid pools or those with mineral substrates. Larvae feed on other larger insects, small fish and anuran larvae while the adults feed on similar prey items or scavenge on freshly dead animals (Miller *et al.* 2007).

Keys to genera. Watts (1978, 2002); Miller *et al.* (2007); Miller & Bergsten (2016). Partial key to genera of larvae by Watts (2002).

Cybister Curtis, 1827 (Pl. 1C)

Type species. *Dytiscus lateralis* Fabricius, 1798 (= *Dytiscus tripunctatus* Olivier, 1795).

Characteristics. Large diving beetles (16–36 mm in Australia), olive green-black dorsally with yellow border to pronotum and elytra. Prosternum and prosternal process with or without a prominent longitudinal sulcus, lateral margins not distinctly ridged throughout. Metacoxal lines present. Unguicular cleft on metatarsus narrow and angular at base. Metatarsus with one claw. Larvae are known for two species (Watts 1964).

Australian species. Five.

Distribution. *Cybister* is a large world-wide genus with four subgenera and over 100 species (Miller *et al.* 2007). In Australia occur four species across northern Australia (WA, NT, QLD) and one is generally distributed over most of Australia (Watts 1978; Hendrich 1997).

Biology. All kinds of lentic and more permanent water bodies such as well vegetated billabongs, swamps, lakes, farm dams and pools. Most species, if not all, have been collected at light.

References. Watts (1963, 1964, 1978, 2002); Miller (2000); Miller *et al.* (2007); Hendrich (1997, 2003b); Miller & Bergsten (2016); Nilsson & Hájek (2018).

Keys to species. Watts (1978); Hendrich (1997).

Austrodytes Watts, 1978 (Pl. 15E)

Type species. *Cybister insularis* Hope, 1842.

Characteristics. Large size diving beetles (16–20 mm), olive green-brown dorsally with yellow border to pronotum and elytra. Prosternum and prosternal process without prominent longitudinal sulcus throughout, lateral margins distinctly ridged throughout. Metacoxal lines present. Metatarsus with two claws. Elytra with both large and small punctures. Larvae unknown. Endemic.

Australian species. Two.

Distribution. One species extends across northern Australia (WA, NT, QLD) and one is known only from the Pilbara (WA).

Biology. In contrast to all other members of Cybistrinae, *Austrodytes* inhabit permanent, clear streams and spring-fed rock pools (Larson 1993, 1997; Hendrich 2003b).

References. Watts (1978); Larson (1993, 1997); Hendrich (2003b); Nilsson & Hájek (2018).

Keys to species. Characters to separate the species are given by Hendrich (2003b).

Spencerhydrus Sharp, 1882 (Pl. 17F)

Type species. *Spencerhydrus latecinctus* Sharp, 1882.

Characteristics. Large size diving beetles (16–18 mm), olive green-brown dorsally with yellow border to pronotum and elytra. Prosternum and prosternal process with prominent longitudinal sulcus throughout, lateral margins not ridged throughout. Metacoxal lines present. Metatarsus with 2 claws. Claws on male protarsus unequal, larger claw strongly curved. Larvae known (Watts 2002). Endemic.

Australian species. Two.

Distribution. One species in south-western Australia (WA) and one in south-eastern Australia (VIC, TAS).

Biology. In permanent and temporary lentic habitats. Preferably in peatland and heathland swamps and flooded meadows. Both species are winter and early spring breeder. Rarely collected with nets, mostly with bottle traps and never found at light.

References. Watts (1978, 2002); Miller & Bergsten (2016); Nilsson & Hájek (2018).

Keys to species. Watts (1978).

Onychohydrus Schaum, 1847 (Pl. 16I)

Type species. *Cybister hookeri* White, 1846.

Homoeodytes Régimbart, 1878

Characteristics. Large size diving beetles (24–28 mm in Australia), olive green-black dorsally with yellow border to pronotum and elytra. Prosternum and prosternal process without prominent longitudinal sulcus throughout, lateral margins lightly ridged in anterior two thirds. Metacoxal lines absent. Unguicular cleft on metatarsus broad at base. Metatarsus with 2 claws. Inner apical angle of hind femora acute. Larvae known and roughly described (Watts 1963, 1964, 2002).

Australian species. One. A second species occurs in New Zealand.

Distribution. Small genus confined to Australia and New Zealand. *Onycohydrus scutellaris* is widely distributed in southern Australia (WA, SA, VIC, TAS, NSW, QLD).

Biology. All kinds of lentic and more permanent water bodies such as well vegetated swamps, lakes, farm dams and pools.

References. Miller & Bergsten (2016); Nilsson & Hájek (2018); Watts (1963, 1964, 1978, 1985, 2002).

Sternhydrus Brink, 1945 (Pl. 17E)

Type species. *Dytiscus atratus* Fabricius, 1801.

Characteristics. Medium size diving beetles (14–17 mm in Australia), olive green-black dorsally with yellow border to pronotum and elytra. Prosternum and prosternal process with prominent longitudinal sulcus throughout, lateral margins not ridged. Metacoxal lines absent. Inner apical angle of hind femora almost right. Claws on male protarsus equal. Larvae known (Michat *et al.* 2015).

Australian species. One.

Distribution. Small genus with four species known from Australia, New Guinea and Indonesia (Miller & Bergsten 2016; Nilsson & Hájek 2018). Further known from New Caledonia (Hendrich *et al.* 2010c) and Fiji Islands (Hendrich unpublished). *Sternhydrus atratus* (Fabricius) occurs across northern Australia (WA, NT, QLD), from Kimberley region in the north-west along the east coast south to Brisbane region, with an isolated population in the Pilbara (Hendrich 2003b)

Biology. All kinds of lentic and more permanent water bodies such as well vegetated billabongs, swamps, lakes, and rest pools of intermittent creeks and rivers.

References. Hendrich (2003b); Larson (1997); Miller & Bergsten (2016); Nilsson & Hájek (2018); Michat *et al.* (2015); Watts (1978, 2002).

Subfamily Dytiscinae Linnaeus, 1758

Characteristics. Medium to large size diving beetles. Scutellum visible. Eye with anterior margin not emarginate above base of antenna, outline fairly round. Prosternum and prosternal process in same plane. Metafemur without group of spine-like seta on posteroventral angle. Metatarsus with one or two claws which are equal or unequal in length. Male with protarsomeres 1 to 3 greatly widened to form a large round or oval palette equipped with varied number of ventral suckers; often with mesotarsomeres 1 to 3 widened and forming an elongate pad with numerous small suckers. Male with aedeagus symmetric. Females with single genital opening and gonocoxae fused dorsally.

Distribution. A subfamily of five tribes, four of which occur in Australia (Miller & Bergsten 2016; Nilsson & Hájek 2018).

Keys to tribes. Miller (2000); Miller & Bergsten (2016).

Tribe Hydaticini Sharp, 1882

Characteristics. Anteroapical and posteroapical metatarsal spurs similar in size and shape. Bifid spines on inner side of metatibia linear, not clustered. Adhesive setae of male protarsomeres 1 to 3 with apices circular. Anterolateral margin of metaventral wing straight. Posterior margins of metatarsomeres 1–4 with a coarse fringe of flat, adpressed golden setae. Apices of metatibial spurs simple, sharp. Larvae: Without frontal projection. Last two abdominal ventrites with dense fringe of long setae. Cerci present on last abdominal ventrite, lacking swimming hairs. Appendages on head relatively long, length of antenna about equal to width of head. Front pair of ocelli similar in size to rest. Apex of maxillary stipes with one small spine (Watts 2002).

Distribution. A worldwide tribe with one genus and 146 species (Nilsson & Hájek 2018).

Biology. In Australia hydaticines are mainly found in lentic habitats from small temporary pools with accumulated debris to large permanent ponds and reservoirs with emergent vegetation (Watts 1963, 2002; Larson 1993, 1997).

Hydaticus Leach, 1817 (Pl. 102J, K)

Type species. *Dytiscus transversalis* Pontoppidan, 1763.

Characteristics. Medium size diving beetles (9–15 mm in Australia), with variable dorsal colour patterns: yellow with small dark spots, densest and often contiguous medially; dark with pale lateral or sublateral longitudinal stripes; dark with pale sub-basal vita and apical vita or mottling; or dark with small lateral pale spots. Males with a stridulatory apparatus formed by a reticulate file on dorsal surface of male protarsomere 2 and short spines on dorsoproximal margin of protibia (Miller 2000; Miller & Bergsten 2016). Larvae known (Watts 1963, 2002).

Australian species. 12.

Distribution. Worldwide. Widely distributed in northern and eastern Australia (WA, NT, QLD, NSW, VIC), with several species occurring in central Australia (NT, SA).

References. Watts (1963, 1978, 2002); Daussin (1980); Larson (1993, 1997); Miller (2000); Miller & Bergsten (2016); Nilsson & Hájek (2018); Wewalka (1975, 1979, 2015); Hendrich (2003b).

Keys to species. Watts (1978), with subsequent descriptions and name changes and additions by Wewalka (1975, 1979, 2015), Daussin (1980) and Nilsson & Hájek (2018).

Tribe Aciliini Thomson, 1867

Characteristics. Anteroapical and posteroapical metatarsal spurs similar in size and shape. Bifid spines on inner side of metatibia linear, not clustered. Adhesive setae of male protarsomeres 1 to 3 with apices circular. Anterolateral margin of metaventral wing convexly curved. Posterior margins of metatarsomeres 1–4 with coarse fringe of flat, adpressed golden setae. Apices of metatibial spurs bifid. Larvae: Without frontal projection. Last two abdominal ventrites with dense fringe of long setae. Cerci present on last abdominal ventrite, lacking swimming hairs. Appendages on head short, length of antenna about half width of head (Watts 2002).

Distribution. A worldwide tribe with seven genera (Miller & Bergsten 2016; Nilsson & Hájek 2018).

Australian taxa. Two genera.

Biology. In Australia, aciliines are found in both lentic and lotic situations, from pools and reservoirs with emergent vegetation to large deep shaded pools with tangled roots (Larson 1993). Also in rest pools of temporary creeks and coastal heathland swamps.

Keys to genera. Watts (1978, 2002); Miller & Bergsten (2016).

Rhantaticus Sharp, 1882 (Pl. 16K)

Type species. *Hydaticus signatipennis* Laporte, 1835 (= *Hydaticus congestus* Klug, 1833).

Characteristics. Medium-sized diving beetles (8–10 mm), dorsally, basic colour yellow; head and posterior of pronotum with black markings; elytra with numerous small black maculae, coalescing to form broad central and supapical black fasciae, more or less distinct. Posterolateral angle of metatarsomeres 3 and 4 with one long spine having at most same length as width of tarsomeres. Larvae known (Watts 1963, 2002): ligula with two apical spines, spines on shaft of ligula small. A genus with only one species.

Australian species. One. The name of the single Australian species will be changed and another new species will be described (Bergsten pers. comm.).

Distribution. Widely distributed in the Afrotropical, Australian, Oriental and Palaearctic regions (Miller & Bergsten 2016). In Australia, widely distributed in northern Australia (WA, NT, QLD).

Biology. The species comes regularly to light.

References. Watts (1963, 1978, 2002); Larson (1993, 1997); Hendrich (2003b); Miller & Bergsten (2016); Nilsson & Hájek (2018).

Sandracottus Sharp, 1882 (Pl. 103C)

Type species. *Dytiscus fasciatus* Fabricius, 1775.

Characteristics. Medium-sized diving beetles (13–15 mm), dorsally black with variously sized yellow maculae on pronotum and elytra. Posterolateral angle of mesotarsomeres 3 and 4 with one long spine being much longer than width of tarsomeres. The larvae of the Australian species is partly figured in the key by Watts (2002): ligula with one apical spine, spines on shaft long. A small genus of 16 species.

Australian species. One.

Distribution. Oriental and Australasian regions (Miller & Bergsten 2016). In Australia, widely distributed across northern Australia (WA, NT, QLD) and central Australia (NT).

Notes. The central Australian populations are much darker than the northern forms, the yellow maculae being much reduced.

Biology. Prefers more permanent and at least partly shaded rest pools of creeks and backwaters of streams. In central Australia only known from shaded pools of gorges.

References. Watts (1978, 2002); Larson (1993, 1997); Miller & Bergsten (2016); Nilsson & Hájek (2018).

Tribe Eretini Crotch, 1873

Characteristics. Anteroapical and posteroapical metatarsal spurs similar in size and shape. Bifid spines on inner side of metatibia linear, not clustered. Adhesive setae of male protarsomeres 1 to 3 with apices circular. Posterior margins of metatarsomeres 1–4 without fringe of flat adpressed golden setae. Posterolateral margin of elytra with a linear series of short spines. Lateral margins of pronotum narrowly beaded. Long natatorial setae (swimming hairs) present along ventral margins of pro- and mesofemora. Lateral surfaces of apices of lateral lobes of aedeagus with short, proximally directed spines. Members of this tribe being rather lightly sclerotised and having elytra very thin and flattened. Larvae: Without frontal projection. Last two abdominal ventrites with dense fringe of long setae. Cerci present on last abdominal ventrite, lacking swimming hairs. Appendages on head relatively long, length of antenna about equal to width of head. Front pair of ocelli about twice of size of others. Apex of maxillary stipes with two strong spines (Watts 2002).

Distribution. A worldwide tribe with only one genus, basically circumtropical (Miller 2002; Miller & Bergsten 2016).

Biology. Eretines are associated with a wide variety of lentic environments, including those with ephemeral water sources where they can exploit the short-lived resources with their fast development. They are exceptional dispersers and can be found in small, very isolated pools far from other aquatic habitats (Miller 2002; Watts 2002; Larson 1993). They are often found in swimming pools even in suburban areas and are attracted to light, sometimes in huge numbers.

Eretes Laporte, 1833 (Pl. 2H, 15K)

Type species. *Dytiscus griseus* Fabricius, 1781.

Characteristics. Medium-sized diving beetles (9–15 mm), dorsally light yellow; with variously sized black maculae medially along posterior margin of head; elytra with very small black maculae, each set in a puncture; black maculae variably confluent especially at lateral black maculae and subapical black band. Larvae (Miller 2002; Watts 2002).

Australian species. Two.

Distribution. A worldwide genus with 4 species, two of which occur in Australia (Miller 2002; Miller & Bergsten 2016; Nilsson & Hájek 2018). In Australia, one species widely distributed (WA, NT, QLD, SA, VIC, NSW, ACT, TAS) the other confined to the extreme north (NT, WA).

Biology. One species (*E. australis* (Erichson)) in shallow temporary waters without any vegetation, where they often occur in great numbers. The second species (*E. griseus* (Fabricius)) only known from a few well vegetated billabongs in the north.

References. Watts (1963, 1978, 2002); Miller (2002); Larson (1993); Miller & Bergsten (2016); Nilsson & Hájek (2018).

Keys to species. Miller (2002).

Tribe Dytiscini Leach, 1815

Characteristics. Anteroapical and posteroapical metatarsal spurs similar in size and shape. Bifid spines on inner side of metatibia linear, not clustered. Adhesive setae of male protarsomeres 1 to 3 with apices circular. Posterior margins of metatarsomeres 1–4 without fringe of flat adpressed golden setae. Lateral margins of pronotum broadly beaded. Posterolateral margins of elytra without series of short spines. Median lobe of aedeagus covered with short stout setae. Parameres broad basally, thin and rounded apically. Larvae: Without frontal projection. Last two abdominal ventrites with dense fringe of long setae. Urogomphi present on last abdominal ventrite, swimming hairs present (Watts 2002).

Distribution. Disjunct distribution in Holarctic and Australian regions. A tribe with two genera (Miller & Bergsten 2016; Nilsson & Hájek 2018).

Australian taxa. In Australia this tribe contains a single genus.

Biology. Dytiscines tend to be found in lentic, often temporary habitats such as forest and seasonal heathland swamps, flooded meadows and slow flowing irrigation ditches, always rich in submerged vegetation. Occasionally in rest pools of intermittent creeks. Species tend to be winter and early spring breeders.

Hyderodes Hope, 1838 (Pl. 15G, H)

Type species. *Hyderodes shuckardi* Hope, 1838.

Characteristics. Medium-sized diving beetles (15–20 mm), dorsally almost entirely black-brown. Larvae known (Bertrand 1972; Watts 1964, 2002). Endemic.

Australian species. Two.

Distribution. One species in south-western Australia (WA) and one in south-eastern Australia (SA, NSW, VIC, TAS) (Watts 1978, 1985).

References. Bertrand (1972); Miller (2000); Miller & Bergsten (2016); Nilsson & Hájek (2018); Watts (1964, 1978, 2002).

Keys to species. Watts (1978).

Subfamily Hydroporinae Aube, 1836

Characteristics. Small to medium-sized beetles (1.1–6.3 mm in Australia). Very variable in body shape, size and colour. Here we follow Miller & Bergsten (2016) in the characterisation of the subfamily. Anteromedial portion of the prosternum is in a distinctly different plane than prosternal process (i.e. prosternal process is declivous with respect to prosternum); pro- and mesotarsi are pseudotetramerous with tarsomere 4 small and hidden within lobes of tarsomere 3, though some taxa, such as *Necterosoma* and *Sternopriscus*, have pro- and mesotarsi more distinctly pentamerous, and males of *Tiporus*, *Antiporus* and *Sekaliporus* have tarsi truly tetra- or trimerous; and scutellum is concealed with elytra closed, though *Carabhydrus* have a clearly visible scutellum and a few other taxa have scutellum partially visible. This group includes the great majority of subterranean diving beetles, which are difficult to diagnose using typical characters for epigean taxa (Miller & Bergsten 2016). Larvae of Australian species have been described from all four tribes (e.g. Biström 1996; Alarie & Watts 2005; Watts 2002; Michat 2006; Michat *et al*. 2011).

Distribution. A subfamily with 10 tribes, four subtribes and 120 genera worldwide. There are more species in this group than in any other diving beetles subfamily with over half of the total species diversity of diving beetles. In Australia four tribes are present (Miller & Bergsten 2016; Nilsson & Hájek 2018).

Keys to tribes. Miller & Bergsten (2016).

Tribe Bidessini Sharp, 1880

Characteristics. Body elongate to globular, unicoloured or with colour pattern; 1.0–3.5 mm in length (in Australia). Head sometimes with cervical line, joining posterior margin of eyes. Many species of sygofauna eyeless. Head frontally projecting downwards. Antennae generally slender, palpi fairly slender, apically generally finely bifid. Pronotum with basal striae, in some genera connected by transverse furrow. Elytra often with striae and in a few genera with elytral keels. Epipleura sometimes with basal cavity, posteriorly limited by a transverse carina. Metepisterna reaching mesocoxal cavities. Apex of prosternal process reaching metaventrite between fairly separated middle coxae. Base of metatrochanters totally visible and metacoxal processes almost at same level as and fused with first and second abdominal ventrites. Metacoxal lines distinct, generally longer than distance between them. Hind claws of almost equal length. Parameres formed by two or three segments and without long apical hairs, usually with few short setae apically. Median lobe aedeagus in Australian genera rather simple, curved and apically pointed (Biström 1988; Pederzani 1995).

Bidessini genera have to date been justified mainly on a diagnostic combination of structural features (Biström 1988; Miller & Short 2015) rather than apomorphies and this had to lead to recognition of genera that render others paraphyletic (Balke & Ribera 2004). Some of these features – such as presence or absence of an elytral plica or occipital line – have been shown to vary within clades of closely related species (Balke *et al.* 2015). In this context, the use of phylogenetic reconstructions based on DNA sequence data offer a source of information that helps to delineate monophyletic entities (Hendrich & Balke 2009; Balke *et al.* 2013).

Distribution. A worldwide tribe with 48 genera (Miller & Short 2015; Miller & Wheeler 2015; Miller & Bergsten 2016; Balke *et al.* 2017a; Nilsson & Hájek 2018).

Australian taxa. In Australia this tribe contains nine genera.

Biology. Most Bidessini inhabit shallow or very shallow lentic habitats, burrowing mud and among floating and submerged aquatic and emergent plants (Larson 1997), but some are lotic (Hendrich *et al.* 2009; Hendrich & Wang 2006) or even hygropetric (Miller 2012).

Keys to genera. Miller & Bergsten (2016).

Allodessus Guignot, 1953 (Pl. 102A)

Type species. *Hydroporus bistrigatus* Clark, 1862.

Characteristics. Small diving beetles (2.9–3.5 mm in Australia), elongate-oval and grey-yellowish. Head with cervical line and frontally bordered. Pronotum with basal striae but lacking transverse furrow between them. Elytral striae fairly distinct. Sutural lines totally absent. Elytra lack keels. Elytral punctuation not forming distinct rows. Epipleura lacking basal pit posteriorly limited by transverse carina. Prosternal process reaching metaventrite, its apex slightly broadened and laterally marginated. Metathorax lacking distinct lateral keels, not distinctly depressed posterior to mesocoxae. Metacoxal lines longer than distance between them posteriorly. Anteriorly metacoxal lines continued by rows of punctures. Metatrochanters and metafemora not strongly modified (Biström 1988). A genus with 5 species (Balke & Ribera 2004; Miller & Bergsten 2016). Larvae described by Michat *et al.* (2011).

Australian species. One.

Distribution. Australia, sporadically in areas of south-east Asia, Japan, Indonesia and certain oceanic islands. In Australia *Allodessus bistrigatus* is very common and widespread, more common in inland than in coastal area, but records from all states. The species is also known from New Zealand (Balke & Ribera 2004) and New Caledonia (Hendrich *et al.* 2010c).

Biology. In every type of lentic habitat, sometimes in slow flowing channels or ditches, too.

References. Watts (1978, 2002); Larson (1993, 1997); Biström (1988); Miller & Bergsten (2016); Nilsson & Hájek (2018).

Clypeodytes Régimbart, 1894 (Pl. 15B)

Type species. *Hydroporus cribrosus* Schaum, 1864.

Characteristics. Very small diving beetles (1.6–1.95 mm in Australia). Shape of body globular, head with cervical line, fore margin narrowly bordered. Pronotum with striae but lacking transverse furrow between them. Elytra with striae and lateral elytral furrow but lacking sutural lines. No distinct rows of punctures visible on elytra. Epipleuron anteriorly with a cavity, being posteriorly limited by transverse carina (Biström 1988). All Australian species with characteristic colour patterns on upper side (Hendrich & Wang 2006). A genus with 39 species (Miller & Bergsten 2016; Nilsson & Hájek 2018). Larvae unknown.

Australian species. Four.

Distribution. Africa, Europe, India, South-east Asia, Indonesia, New Guinea, Australia. In Australia, widely distributed in tropical northern and north-eastern Australia (WA, NT, QLD) with one species extending further south (NSW) (Hendrich & Wang 2006).

Biology. Three species occur in creeks, small springs and slowly flowing rivers and their rest pools, preferably with clear water. Two species occur in swamps, billabongs and rest pools of intermittent rivers and creeks. Larson (1997) and Hendrich & Wang (2006) provided details of habitats for all species.

References. Hendrich & Wang (2006); Miller & Bergsten (2016); Nilsson & Hájek (2018); Watts (1978).

Keys to species. Watts (1978); Hendrich & Wang (2006).

Gibbidessus Watts, 1978 (Pl. 102E; Figs 11.12, 11.15)

Type species. *Gibbidessus chipi* Watts, 1978

Characteristics. Very small diving beetles (1.1–1.9 mm). The smallest epigean dytiscids in Australia. Body oblong-oval and fairly compactly built. Head with or without cervical line, frontally not bordered. Palpi rather slender, apically very finely bifid. Pronotum with distinct striae but lacking transverse sutural lines and keels. Punctation of elytra not forming rows. Epipleura lacking basal cavity posteriorly limited by transverse carina. Prosternal process rather elongate, narrow, laterally distinctly marginated and with surface not medially excavated. Prosternal process reaching metaventrite; metaventrite not distinctly depressed posterior to mesocoxae. Metacoxal lines comparatively short, only slightly longer than distance between them posteriorly. Very fine punctures on each side of midline of metaventrite, but not forming distinct rows. Metatrochanters and metafemora not distinctly modified (Biström 1988). Larvae unknown. Endemic.

Australian species. One or two species (Watts 1978, 2002; Miller & Bergsten 2016; Nilsson & Hájek 2018) and five more undescribed (Hendrich *et al.* unpublished).

Notes. Based on the results of a phylogenetic reconstruction using DNA sequence data, *Uvarus pictipes* Lea will be transferred to the genus *Gibbidessus* (Hendrich *et al.* unpublished). The genus *Uvarus* is strictly African. This information must not be treated as nomenclatural act in the sense of the ICZN (1999: 103; disclaimer).

Distribution. Australia. Restricted to south-west and south-east Australia, including Tasmania.

Biology. In very shallow (1–5 cm depth) water of open or at least semi-exposed and ephemeral sedge and peatland swamps. Beetles can only be collected when using a net or strainer with very fine meshes.

References. Miller & Bergsten (2016); Nilsson & Hájek (2018); Watts (1978, 2002).

Hydroglyphus Motschulsky, 1853 (Pl. 102G, H)

Type species. *Dytiscus geminus* Fabricius, 1792

Guignotus Houlbert, 1934

Characteristics. Small (2–4 mm), elongate oval, fairly oblong, often with colour pattern on elytra. Head lacking cervical line, anteriorly unbordered. Palpi fairly slender, apically finely bifid. Pronotum with striae, but these not connected by transverse furrow. Elytra with striae and distinct sutural lines, but elytral keels absent. Elytral punctures not forming rows. Epipleura basally without cavity posteriorly limited by a transverse carina. Prosternal process reaching metaventrite, fairly elongate, narrow and with surface shallowly excavated. Metaventrite lacking lateral keels and not distinctly depressed posterior to mesocoxae. Metacoxal lines longer than distance between them posteriorly. On each side of metaventral midline fine rows of punctures visible. Metatrochanters and metafemora not distinctly modified (Biström 1988). A genus with 90 species (Miller & Bergsten 2016; Nilsson & Hájek 2018). Larvae known (Watts 2002; Michat *et al.* 2010).

Australian species. 10.

Distribution. Africa, most of Europe and Asia, Indonesia, New Guinea, Australia. In Australia, widely distributed across northern Australia (WA, NT, QLD), along east coast south to border of NSW and VIC. In WA south to Pilbara region (Hendrich 1999, 2003b).

Biology. Predominantly found in lentic sites. Larson (1997), Hendrich (1999) and Hendrich (2003b) provide details of habitats for most species in QLD and WA.

References. Watts (1978, 2002); Hendrich (1999, 2003b); Larson (1997); Biström (1988); Miller & Bergsten (2016); Nilsson & Hájek (2018).

Keys. Watts (1978); Hendrich (1999).

Kakadudessus Hendrich & Balke, 2009 (Pl. 102F)

Type species. *Kakadudessus tomweiri* Hendrich & Balke, 2009.

Characteristics. Very small diving beetles (2.2–2.25 mm in Australia). Habitus elongated, almost flat, with pale yellowish markings on elytra, head with cervical line, clypeus anteriorly bordered. Pronotum and elytra with well-developed striae. Elytra without keels, without sutural line and without accessory striae, posterior part of basal cavity of epipleura without transverse carina. Prosternal process reaching metaventrite, laterally margined and fairly elongate and narrow. Metaventrite with rows of punctures at midline, metacoxal lines longer than distance between them, strongly diverging anteriorly. Parameres two segmented, very thin, slender and elongate, only very slightly bifid anteriorly (Hendrich & Balke 2009). Larvae unknown. Endemic.

Australian species. One.

Distribution. Australia; Northern Territory (Kakadu area) and north eastern Queensland.

Biology. The single species of the genus inhabits small side and rest pools of shaded and intermittent streams and rivers with sandy or gravely bottom and without any vegetation (Hendrich & Balke 2009).

References. Hendrich & Balke (2009); Nilsson & Hájek (2018).

Leiodytes Guignot, 1936 (Pl. 16A)

Type species. *Hydroporus evanescens* Boheman, 1848.

Characteristics. Very small, yellowish diving beetles (1.75–2.0 mm in Australia) with black markings on elytra. Shape of body globular, more oval than in *Clypeodytes*, widest in middle, not flattened. Head with cervical line, fore margin narrowly and finely bordered. Pronotum with striae but lacking transverse furrow between them. Elytra with striae but lacking keels and sutural lines. No distinct rows of punctures visible on elytra. Epipleuron anteriorly lacking a basal cavity posteriorly limited by a transverse carina. Prosternal process reaching metaventrite, narrow and elongate and laterally marginated. Metacoxal lines subparallel and longer than distance between them posteriorly. Metaventrite on each side of midline with a row of punctures. Metatrochanters and metafemora not distinctly modified (Biström 1987, 1988). A genus with 27 species (Miller & Bergsten 2016; Nilsson & Hájek 2018). Larvae unknown.

Australian species. One (as *Clypeodytes migrator* (Sharp) in Watts 1978, 2002).

Notes. Based on the results of a phylogenetic reconstruction using DNA sequence data, *Clypeodytes migrator* will be transferred to the genus *Leiodytes* (Hendrich *et al.* unpublished). This information must not be treated as nomenclatural act in the sense of the ICZN (1999: 103; disclaimer).

Distribution. Africa, India, South-east Asia, Japan Indonesia New Guinea and Australia. Widely distributed in tropical northern and north-eastern Australia (WA, NT, QLD) and extending further south (NSW) (Watts 1978, 2002).

Biology. In swamps, billabongs and rest pools of intermittent rivers and creeks. Larson (1997) provided details of habitat for one species.

References. Biström (1987, 1988); Watts (1978, 2002); Larson (1997); Miller & Bergsten (2016); Nilsson & Hájek (2018).

Keys to species. Watts (1978) (in key to species of *Clypeodytes*).

Limbodessus Guignot, 1939 (Pl. 16G)

Type species. *Bidessus neoguineensis* Régimbart, 1892.
Boonguurus Larson, 1994
Kintingka Watts & Humphreys, 1999
Nirridessus Watts & Humphreys, 1999
Tjirtudessus Watts & Humphreys, 1999

Characteristics. Small (1.8–3.1 mm in Australia) elongate oval and fairly oblong species. Head with or without cervical line. Foremargin of head not bordered. Palpi fairly slender, apically finely bifid. Pronotum with striae, but without a transverse furrow connecting them. Elytra with striae but lacking keels and sutural lines. No rows of punctures on elytra. Prosternal process reaching metaventrite, fairly broad, laterally marginated and surface medially somewhat convex. Metaventrite lacking lateral keels and not distinctly depressed posterior to mesocoxae. Metacoxal lines longer than distance between them posteriorly. No signs of rows of punctures beside midline of metasternum. Metatrochanters and metafemora not distinctly modified (Biström 1988). A genus with 76 species (Miller & Bergsten 2016; Nilsson & Hájek 2018). Larvae known (Watts 2002).

Australian species. Nine epigean and 60 stygobitic species.

Distribution. Japan, South-east Asia, Indonesia, New Guinea, Australia and several Pacific Islands. In Australia, the epigean species are widely distributed, with more species along the east coast and in the south-east. Only one endemic species in the south-west. The stygofauna is concentrated in the Pilbara and Yilgarn areas in WA, and inland of NT and SA.

Biology. Mainly found in lentic sites (Larson 1997; Watts & Leys 2005) or subterranean in calcrete areas (Watts & Humphreys 1999, 2000, 2003, 2006, 2009).

References. Watts (2002); Watts & Humphreys (1999, 2000, 2003, 2006, 2009); Watts & Leys (2005); Larson (1997); Biström (1988); Leys *et al.* (2010); Miller & Bergsten (2016); Nilsson & Hájek (2017).

Keys to species. For epigean species Watts & Leys (2005), for the stygofauna Leys & Watts (2008); Watts & Humphreys (1999, 2000, 2003, 2006, 2009).

Neobidessodes Hendrich & Balke, 2009 (Pl. 103B)

Type species. *Bidessus denticulatus* Sharp, 1882.

Characteristics. A genus of small Bidessini (2.3–4.2 mm). Body elongate oval, basal pronotal striae in some species absent or fine, in others sharply incised, oblique, and not connected by a transverse groove. Elytra lacking basal striae and sutural striae, epipleura lacking transverse carina. Head lacking cervical line and its foremargin not bordered, prosternal process broad, surface distinctly excavated and marginated; inner margin of metacoxal wings strongly ridged. Hind margin of abdominal ventrites 3–5 without row of minor irregular dentate processes; well-developed finger-like apical lobe on distal segment of paramere lacking. All but two stygobitic species studied with more or less contrasting black/yellow surface. Ground pattern of these species including various yellow or reddish spots. Median lobe aedeagus simple and very elongate, in ventral view strongly tapered or rounded at tip (Hendrich *et al.* 2009). A genus with 10 species (Hendrich & Balke 2011). First instar larvae of *Neobidessodes limestonensis* (Watts & Humphreys) known (Michat *et al.* 2010).

Australian species. Nine.

Distribution. Australia and New Guinea. Widely distributed across northern Australia (WA, NT, QLD) and along the east coast south to VIC. In WA south to the Pilbara region.

Biology. All species occur in small streams, creeks, and pools of intermittent rivers or billabongs with sandy or gravelly bottom. Detailed descriptions of the habitats are provided by Hendrich *et al.* (2009) and Hendrich & Balke (2011).

Keys to species. Hendrich *et al.* (2009); Hendrich & Balke (2011).

References. Hendrich *et al.* (2009); Hendrich & Balke (2011); Nilsson & Hájek (2017); Michat *et al.* (2010).

Petrodessus Miller 2012 (Pl. 16D)

Type species. *Petrodessus conatus* Miller, 2012.

Characteristics. Very small diving beetles (1.5–1.7 mm). Head without cervical line, anterior clypeal margin strongly flattened, anteriorly produced, with broad anterior margin. Basal pronotal striae present, basally deeply impressed with a shallow, transverse groove connecting both striae. Basal elytral striae present, basally deeply impressed. Elytral sutural striae absent, elytron without longitudinal carinae. Epipleuron without transverse carina at humeral angle. Protibia broadly triangular, heavily spinous. Legs with relatively few natatorial setae, being robust and spinous (Miller 2012; Miller & Bergsten 2016). Larvae unknown. Endemic.

Australian species. One.

Distribution. Endemic to a few waterfalls from Tully Gorge and Paluma Range National Parks in north-eastern Queensland.

Biology. Hygropetric habitats near waterfalls, where thin layers of water seep over the surface of rocks.

References. Miller (2012); Miller & Bergsten (2016); Nilsson & Hájek (2018).

Tribe Hydroporini Aubé, 1836

Characteristics. Body elongate-oval, ratio length to width more than 1.6. Apex of prosternal process reaching metaventrite between fairly separated middle coxae. Hind claws almost of equal length. Parameres without long apical hairs, usually with a few short setae apically and formed by one segment. Metacoxal processes not in same plane as abdomen, without discontinuity or step in lateral view (Pederzani 1995). Larvae: Head with frontal projection; frontal projection with lateral notches and non spatulate, ~1.5 times as long as wide; body spindle shaped, with distinct colour pattern, often covered with small dark spots; cerci in older instars reaching beyond tip of siphon (Watts 2002; Michat 2006).

Distribution. A worldwide tribe with 36 genera (Miller & Bergsten 2016; Nilsson & Hájek 2018; Hendrich *et al.* 2014). The evolutionary history and recent diversification of the tribe in Australia have been studied by Hawlitschek *et al.* (2012) and Toussaint *et al.* (2015, 2016a).

Australian taxa. In Australia this tribe contains 11 genera all of which are included in the subtribe Sternopriscina.

Biology. Hydroporini inhabit all kind of shallow lentic and lotic habitats. Some genera are strictly lotic such as *Tiporus*, *Barretthydrus* and many *Sternopriscus* (Larson 1993, 1997; Hendrich 2001; Hendrich & Watts 2007, 2009; Hendrich & Balke 2015), others are strictly lentic such as *Paroster* (Hendrich & Fery 2008) and *Antiporus* (Hawlitschek *et al.* 2011; Hendrich & Watts 2010). A large number of *Paroster* are also part of the stygofauna (Watts & Humphreys 2003, 2006, 2009; Watts & Leys 2008).

Keys to genera. Miller & Bergsten (2016).

Antiporus Sharp, 1882 (Pl. 102B, C)

Type species. *Hydroporus femoralis* Boheman, 1858.

Characteristics. Small diving beetles (3.5–6.8 mm). Oval, reddish-yellow to black. One species, *A. gilberti*, has distinct dark lines on elytra. Posterior portion of epipleuron comparatively broad, hind tibiae evenly punctate. Fourth protarsomere scarcely visible. Lobes of protarsi more or less similar in size. Lateral edge of elytron curves smoothly upward before meeting pronotum. Males of all but one species with expansions to hind tibiae. Larvae known (Watts 2002; Michat 2006). A genus of 15 species.

Australian species. 14.

Distribution. Australia and New Zealand. Widely distributed in all wetter regions but with the highest number of local endemics in the south-west and the south-east. Only one species (*A. gilberti*) is distributed in the east and in the west part of the country, crossing the red centre.

Biology. Lentic and mainly in seasonal flooded swamps and peatlands. Only three species in NT, QLD and in the Pilbara, in rest pools along rivers and creeks.

References. Watts (1978, 1997a, 2002); Larson (1993, 1997); Hendrich & Watts (2010); Hendrich *et al.* (2014); Hawlitschek *et al.* (2011); Michat (2006); Miller & Bergsten (2016); Nilsson & Hájek (2018).

Keys to species. Watts (1978, 1997a); Watts & Pinder (2000); Hendrich & Watts (2010).

Barretthydrus Lea, 1927 (Pl. 15J)

Type species. *Barretthydrus geminatus* Lea, 1927.

Characteristics. Small diving beetles (4.2–4.5 mm). Elongate, black with distinct reddish or yellowish markings on elytra. Dorsal surface punctate; each elytron with 4–5 longitudinal grooves. Epipleuron relatively broad. Metatibia rugose punctate. Prosternal process broad, flat, weakly and broadly carinate; hind tarsi without lateral lobes on outside and with two claws, inner smaller than half length of outer. Larvae known (Watts 2002). Endemic.

Australian species. Three.

Distribution. Australia. Widely distributed along the south-east coast, in streams and rivers along the Great Dividing Range (VIC, NSW), absent from Tasmania.

Biology. Strictly lotic and mainly in permanent and oxygen rich rivers and larger streams with clear water, preferably in mountain or hilly regions.

Keys to species. Watts (1978).

References. Watts (1978, 2002); Toussaint *et al.* (2015); Miller & Bergsten (2016); Nilsson & Hájek (2018).

Brancuporus Hendrich, Toussaint & Balke, 2014 (Pl. 102D)

Type species. *Antiporus gottwaldi* Hendrich, 2001.

Characteristics. Small sized elongate-oval, reddish-brown beetles (3.0–3.35 mm). *Brancuporus* can be separated from all other Hydroporini by the combination of the following characters: Body elongate oval, reddish-brown, fourth tarsomere of protarsus scarcely visible; pronotum and elytron with narrow but well-marked lateral beading; elytra at least in females flanged; posterior part of epipleuron comparatively broad; humeral angle of elytron smoothly rounded; prosternal process blunt, sides weakly bowed, moderately ridged, and distinctly asymmetric median lobe of aedeagus. Separated from *Antiporus* by having a distinctly asymmetric median lobe of aedeagus, and flanged elytra, at least in females of both species (Hendrich *et al.* 2014). Larvae unknown. Endemic.

Australian species. Two.

Distribution. Australia. Restricted to the peatlands and seasonal swamps of the south-western corner of WA, from the Northcliffe area in the west to the Muir Lakes in the east.

Biology. Strictly lentic. Both species were collected in seasonal peatland swamps.

References. Watts (1978, 2002); Hendrich *et al.* (2014); Miller & Bergsten (2016); Nilsson & Hájek (2018).

Keys to species. Hendrich *et al.* (2014).

Carabhydrus Watts, 1978 (Pl. 15F)

Type species. *Carabhydrus monteithi* Watts, 1978.

Characteristics. Small diving beetles (1.7–3.4 mm), elongate and flattened beetles with comparatively long legs, surface colour black, dark brown or testaceous, in some species with reddish markings on elytra. Within tribe Hydroporini, *Carabhydrus* species are recognisable by their short and broad scutellum, strongly narrowed base of pronotum, and discontinuous body outline (as in many Carabidae). Pronoto-elytral angle fairly deep. Middle of prosternum in almost same plane as prosternal process. Metacoxal processes closely adpressed to abdominal first ventrite, not lobed. Each elytron with two wide more or less shallow longitudinal grooves. Metaventrite, metacoxae and abdominal ventrites 1 and 2 fused. Metatibia straight or arcuate, only weakly expanding towards apex, claws equal. Pro- and mesotarsi 5-segmented, fourth tarsomere extremely small. Hind wings of at least three epigean (*C. andreas*, *C. monteithi*, *C. mubboonus*) and the single stygobitic species (*C. stephanieae*) are reduced to very small stubs and are not functional (Hendrich & Watts 2009). A larval habitus illustration is given in Watts (2002). Endemic.

Australian species. 10.

Distribution. Australia. Eight species distributed from the Brisbane region in Queensland to the western part of the Great Dividing Range in Victoria. Only one species (*C. niger* Watts) occurs on Tasmania. The genus is absent in South Australia probably a consequence of the small amount of suitable habitat in that state. Two species (*C. mubboonus* Larson & Storey and *C. storeyi* Hendrich & Watts) are endemic to the tropical north-east of Queensland (Hendrich & Watts 2009).

Biology. At least nine if not all species are restricted to clean streams and rivers with sand, pebble and cobble beds, often situated in old growth or closed-canopy forest sites; one species (*C. stephanieae* Watts, Hancock & Leys) is recorded from groundwater only. It is probable that they are sensitive to low levels of dissolved oxygen and require cooler temperatures (Hendrich & Watts 2009).

References. Watts (1978, 2002); Zwick (1981); Larson & Storey (1994); Watts *et al.* (2007); Hendrich & Watts (2009); Miller & Bergsten (2016); Nilsson & Hájek (2018).

Keys to species. Hendrich & Watts (2009).

Chostonectes Sharp, 1880 (Pl. 15I)

Type species. *Hydroporus gigas* Boheman, 1858.

Characteristics. Small (4.0–6.1 mm), broadly oval, black or brownish beetles, with reddish spots on elytra. Fore and middle tarsi pseudotetramerous, with 4th tarsomere rudimentary, concealed or locked by apical lobes of 3rd tarsomere. Pronotum without latero-basal striae. Elytra without deeply incised geminate striae, at most with shallow longitudinal depression or longitudinal carinae. Epipleura with or without a basal cavity, posteriorly limited by a diagonal carina crossing near base; gradually tapering from base to apex, not abruptly narrowed at middle (Pederzani 1995). Larvae of no species described but genus keyed and figured in Watts (2002). A genus of 6 species.

Australian species. Five.

Distribution. Australia and New Guinea. In Australia, widely distributed along the east and south-east coast (QLD, NSW, AUS, VIC, SA, TAS). One species (*C. gigas* (Boheman)) also further inland in the McDonnell Range.

Biology. The habitat preference is lotic, the beetles are restricted to slow flowing rivers, streams and rest pools of creeks, very often in forested areas (Larson 1993, 1997; Wewalka 1994).

References. Watts (1978); Pederzani (1995); Larson (1993, 1997); Wewalka (1994); Miller & Bergsten (2016); Nilsson & Hájek (2018).

Keys to species. Watts (1978); Wewalka (1994).

Megaporus Brinck, 1943 (Pl. 16B)

Type species. *Hydroporus hamatus* Clark, 1862.

Macroporus Sharp, 1882

Characteristics. Medium-sized (5.8–7.2 mm) oval globular beetles. Colour reddish or black, sometimes with yellow and black colour pattern on elytra. *Megaporus* species distinguishable from all other Australasian Hydroporini by distinctly '4-segmented' protarsi and metatibia with only serial punctures on infero-external surface. Tarsi in fact pseudotetramerous, meaning 4th tarsomere very small and hardly visible. Terminal portion of epipleuron rather broad. Metafemur stout, postero-external angle sharply defined. Hind claws equal in length (Watts 1978; Miller & Bergsten 2016; Hendrich *et al.* 2010a). Larvae of no species described but genus keyed and figured in Watts (2002). A genus with 11 species.

Australian species. Eight.

Distribution. Australia, New Guinea, New Caledonia, Fiji. In Australia widely distributed in all wetter regions, but with the highest number of local endemics in the south-west and the south-east. Only one species (*M. howittii*) is distributed in the east and in the west part of the country, crossing the red centre.

Biology. Lentic and mainly in seasonal flooded swamps, peatlands, billabongs and stagnant rest pools of intermittent streams and rivers. Only three species in NT, QLD and in the Pilbara, in rest pools along rivers and creeks.

References. Watts (1978, 2002); Hendrich *et al.* (2010a); Larson (1993, 1997); Miller & Bergsten (2016); Nilsson & Hájek (2018).

Keys to species. Watts (1978).

Necterosoma W.J. MacLeay, 1871 (Pl. 16H)

Type species. *Necterosoma vittipenne* W.J.Macleay, 1871 (= *Hydroporus penicillatus* Clark, 1862).

Characteristics. Small to medium-sized beetles (4.0–5.4 mm), distinguished by distinctly 5-segmented protarsi, medially confluent metacoxal cavities and mesoventrite beeing in a more or less vertical position, at a pronounced angle with metaventrite. Males with front tibiae notched, often strongly so. Most species with distinct colour pattern on elytra, and raised elytral striae in three species. Larvae of no species havebeen described but genus was keyed and figured in Watts (2002). A genus of 12 species (Balke *et al.* 2013).

Australian species. Nine.

Distribution. Australia, New Caledonia, Timor. In Australia, widely distributed in all regions, but with the highest number of species along the east coast and in the south-east. One local endemic (*N. aphrodite* Watts) in the Grampians, one in the Kimberleys (*N. theonathani* Hendrich) the other species more widespread. One species, which tolerates high salinity (*N. penicillatum*) is distributed in the east and in the west part of the country, crossing the red centre.

Biology. Most *Necterosoma* appear to be restricted to slowly flowing rivers, streams and creeks and their backwaters, or to pools and puddles which form in their beds in the dry season. In these habitats up to two species of the genus are often abundant, and during the dry season aggregations of several hundred specimens in one spot are not unusual. *N. penicillatum* and *N. darwinii* (Babington) can be found near the shore of shallow lakes and in seasonal swamps, too.

References. Watts (1978, 2002); Zwick (1979); Larson (1993, 1997); Hendrich (2003a); Hendrich *et al.* (2010a); Balke *et al.* (2013); Miller & Bergsten (2016); Nilsson & Hájek (2018).

Keys to species. Watts (1978); Hendrich (2003a); Zwick (1979).

Paroster Sharp, 1882 (Pl. 16C, 16E, 17A)

Type species. *Hydroporus insculptilis* Clark, 1862.
Nirripirti Watts & Humphreys, 2001
Terradessus Watts, 1982

Characteristics. *Paroster* can be recognised by their exposed metatrochanter bases, evenly punctate metatibia and relatively strong microreticulation particularly in the females. Metacoxa are closely adpressed to first abdominal ventrite. Metacoxal cavities are exposed and well separated, metacoxal processes produced backwards in midline. Middle and posterior portions of epipleuron very narrow, and near shoulders crossed by an oblique carina (Watts 1985, 2002; Pederzani 1995; Hendrich & Fery 2008). Larva of an epigean species keyed and figured by Watts (2002). Endemic.

Notes. Within the last two decades, scientific research and biological surveys, often associated with environmental impact assessment of mining and other projects, have greatly increased knowledge about Australian stygofauna. Especially Western Australia is now known to be a world hotspot for such organisms. Within the genus the stygobitic species are morphologically very similar, with, apart from male sexual characters, only colour and size separating many of them (Watts & Humphreys 2003, 2006, 2009; Leys *et al.* 2010).

Australian species. 52 species, 16 epigean (Hendrich & Balke 2016) and 36 species belonging to the unique stygofauna in WA and NT (Miller & Bergsten 2016; Nilsson & Hájek 2018). Based on the results of a molecular study 2 species, formerly belonging to *Terradessus* and living in the moist leaf litter in Queensland´s rainforests, were transferred to *Paroster* (Toussaint *et al.* 2016a).

Distribution. Australia. The stygofauna is concentrated on the Yilgarn area in WA, inland of NT and SA. The epigean species are mainly distributed in south-western Australia and the southern parts of SA and VIC (Watts & Humphreys 2003, 2006, 2009; Leys *et al.* 2010).

Biology. The habitat preference of the epigean species is lentic living in temporary creeks, swamps, flooded areas and pan-gnammas (which occur on isolated granite outcrops in the south-west of Australia) that form in late winter and early spring but are dry by mid to late summer. In many places, particularly shallow gutters and pools, they are the dominant species and occur in considerable numbers for a relatively short time in spring and are usually gone well before the water has dried up. Breeding also occurs in these places but frequently adults are abundant yet no larvae are present. It would seem that breeding is highly synchronized and growth rapid, not unexpected in these short-term habitats where larval food can be plentiful and competition from other species is slight (Watts & Leys 2008).

References. Watts (1997b); Watts & Humphreys (2003, 2004, 2006, 2009); Watts & Leys (2008); Hendrich &

Fery (2008); Hendrich & Balke (2016); Leys *et al.* (2010); Miller & Bergsten (2016); Nilsson & Hájek (2017).

Keys to species. For epigean species see Watts & Leys (2008); for stygofauna see Watts & Humphreys (2003, 2004, 2006, 2009).

Sekaliporus Watts, 1997 (Pl. 17B)

Type species. *Sekaliporus kriegi* Watts, 1997.

Characteristics. Small (3.2–3.5 mm), oval and black beetles, with distinct colour pattern (yellowish spots on elytra and pronotum). From related genus *Tiporus*, having shoulders of elytra obliquely bent upwards, it can be distinguished by lateral edge of elytron and lateral edge of pronotum forming nearly continuous straight or slightly sinuate line in combination. Males with four distinctive asymmetric protarsomeres (Watts (1997b). Larvae of no species has been described but genus keyed and figured in Watts (2002). Endemic.

Australian species. Two.

Distribution. One species in northern Australia, from the inner Kimberley region in the north-west to north-eastern Queensland. The other species is endemic to the Kakadu area. The species are sympatric but never syntopic (Hendrich & Balke 2015).

Biology. The habitat preference is lotic, the beetles are restricted to slow flowing tropical rivers, streams and creeks very often in forested areas (Hendrich & Balke 2015).

References. Watts (1997b); Hendrich & Balke (2015).

Sternopriscus Sharp, 1880 (Pl. 17C)

Type species. *Hydroporus multimaculatus* Clark, 1862

Characteristics. Small (2.5–5.0 mm), elongate rugose-punctate species with a distinctive convex shape. Most species black to brown with variable yellowish markings dorsally. Pro- and mesotarsi distinctly 5-segmented. Fourth tarsomere of pro- and mesotarsi distinctly visible. Mesoventrite prominent between prosternum and metaventrite. Metacoxal cavities distinctly separated and metacoxal processes medially with posterior margin extending posteriorly. Metatibia rugose-punctate. Males with slight notch in protibia and with antennomeres often greatly expanded (Hendrich & Watts 2004; Miller & Bergsten 2016). Larvae keyed and figured by Watts (2002). Endemic.

Australian species. 29.

Distribution. The genus is restricted to Australia except for a dubious record from the Highlands of Irian Jaya, Indonesia. The genus is rather speciose in south-eastern Australia with 18 species of which four are endemic to Tasmania. Only two of these 18 species have a wider distribution over mainland Australia (*S. multimaculatus* (Clark) and *S. clavatus* Sharp). Five species are endemic to the south-east and four are distributed over the tropical north. Only one species is shared by two or more of these areas of endemism. This distribution reflects the restriction of all but *S. multimaculatus* to the more humid coastal areas of Australia (Hendrich & Watts 2004).

Biology. The species occur commonly in all but the most arid regions of Australia in both standing and running water. Ten species of the genus are rheophilic inhabiting rivers, streams and springs; eleven species are acidophilic living in seasonal or permanent swamps, ponds and pools of different types of peatlands, and seven species seem to be more or less eurytopic occurring in various water bodies, in open or forested country. The highest diversity is in lowland or coastal areas and hilly or low mountain ranges from 0–500 m. Only six species were collected up to 1000 m or above (Hendrich & Watts 2004).

References. Watts (1978); Hendrich & Watts (2004, 2007); Hawlitschek *et al.* (2012); Miller & Bergsten (2016); Nilsson & Hájek (2018).

Keys to species. Watts (1978); Hendrich & Watts (2004).

Tiporus Watts, 1985 (Pl. 17D)

Type species. *Hydroporus undecimmaculatus* Clark, 1862.

Hypodes Watts, 1978

Characteristics. Small (3.3–5.0 mm), oval, black, some with reddish spots on elytra and pronotum. From related genus *Sekaliporus* (with lateral edge of elytron and lateral edge of pronotum forming a nearly continuous straight or slightly sinuate line in combination), distinguishable by shoulders of elytra obliquely bent upwards; male protarsi 3-segmented. Larvae undescribed but keyed and figured by Watts (2002). Endemic.

Australian species. 13.

Distribution. Some *Tiporus* have a very wide distribution ranging from the Kimberley region in the west to the Arnhem Land in the north (e.g. *T. undecimmaculatus* (Clark), *T. centralis* (Watts), *T. emmae* Hendrich), others are endemic to a few creeks or river systems in the north-west (e.g. *T. moriartyensis* Watts and *T. georginae* Watts) or the Pilbara region (e.g. *T. tambreyi* (Watts) and *T. lachlani* Watts).

Biology. All *Tiporus* appear to be restricted to slowly flowing tropical rivers, streams and creeks and their backwaters, or to pools and puddles which form in their beds in the dry season. In these habitats up to five species of the genus are often abundant and during the dry season aggregations of several hundred specimens in one spot are not unusual.

References. Watts (1978, 2000a); Hendrich (2008); Hendrich *et al.* (2016); Miller & Bergsten (2016); Nilsson & Hájek (2017).

Keys to species. Watts (1978, 2000a); Hendrich *et al.* (2016).

Tribe Hydrovatini Sharp, 1880

Characteristics. Body broad and short, subglobose, ratio length to width less than 1.55. Apical margin of metacoxal processes on each side with deep sublateral emargination, making hind margin of metacoxal process conjointly tripartite – 2 widely separated narrow lateral lobes and a broad depressed middle region; lateral lobes partly covering bases of trochanters. Elytra with apices shortly produced. Prosternal process broadly triangular, separating mesocoxae. Larvae: Head with frontal projection; frontal projection without lateral notches and non spatulate, ~2 times as long as wide; body spindle shaped, without distinct colour pattern, often covered with small dark spots; cerci in older instars not reaching beyond tip of siphon which is sharply pointed (Watts 2002; Michat 2006; Biström 1996).

Distribution. A worldwide tribe with two genera (Biström 1996; Miller & Bergsten 2016; Nilsson & Hájek 2018).

Australian taxa. One genus.

Biology. Most hydrovatines inhabit shallow lentic habitats, burrowing in the bottom mud and silt among rooted aquatic and emergent plants (Larson 1993, 1997; Biström 1996).

Hydrovatus Motschulsky, 1853 (Pl. 102K)

Type species. *Hyphydrus cuspidatus* Kunze, 1818.

Characteristics. Small diving beetles (2–4 mm in Australia). Light brown to black dorsally, sometimes with vague colour pattern on elytra; parallel sided to subglobular. Most species with elytra having numerous punctures and microsculpture. Larvae known (Watts 2002; Biström 1996; Michat 2006). A genus of 213 species (Nilsson & Hájek 2018).

Australian species. Seven plus one undescribed species in the Pilbara (WA).

Distribution. Worldwide. In Australia widely distributed across northern and eastern Australia (WA, NT, QLD) with one species extending further south (NSW, VIC, SA).

References. Biström (1996); Larson (1993, 1997); Michat (2006); Miller & Bergsten (2016); Nilsson & Hájek (2018); Watts (1978, 2002).

Keys to species. Watts (1978); Biström (1996).

Tribe Hyphydrini Gistel, 1848

Characteristics. Body broad and short, subglobose. Clypeus distinctly margined and extended over base of labrum. Metacoxal processes not lobed or produced over base of trochanters; apex of metacoxal process adpressed to abdominal ventrite 2 and in approximately same plane. Elytral apices not produced. Metatarsus with inner claw ~0.25 or less than length of outer claw. Larvae (based on *Hyphydrus*): Head with frontal projection; frontal projection spatulate, without lateral notches; body spindle shaped; cerci in older instars reaching beyond tip of siphon (Watts 2002; Alarie & Watts 2005).

Distribution. A worldwide tribe with 14 genera (Miller & Bergsten 2016; Nilsson & Hájek 2018).

Australian taxa. One genus.

Hyphydrus Illiger, 1802 (Pl. 103E)

Type species. *Dytiscus gibbus* Fabricius, 1776.

Characteristics. Small diving beetles (3–4.5 mm in Australia). Light brown to yellowish with black maculae forming various patterns dorsally. Broadly oval and deep bodied. Elytra shiny or microreticulate and dull. Larvae of all four species known (Alarie & Watts 2005), with distinct yellow/dark colour pattern. A genus with more than 139 species in some 20 species groups.

Australian species. Four. Watts & Leys (2006) also discuss polymorphism in one species.

Distribution. Worldwide. In Australia, widely distributed in all states except VIC & TAS, with a concentration in northern Australia.

Biology. Species of *Hyphydrus* can be found in various types of lentic and slow moving lotic habitats with emergent and rooted aquatic vegetation.

References. Biström (1982); Watts (1978, 2002); Watts & Leys (2006); Larson (1993, 1997); Alarie & Watts (2005); Miller & Bergsten (2016); Nilsson & Hájek (2018).

Keys to species. Biström (1982); Biström *et al.* (1997); (Watts 1978); (Watts & Leys 2006).

Subfamily Laccophilinae Branden, 1885

A subfamily with two tribes and 14 genera worldwide (Nilsson & Hájek 2018; Miller & Bergsten 2016).

Tribe Laccophilini Nilsson, 1989

Characteristics. Medium-sized (3.0–4.5 mm in Australia), streamlined and more flattened diving beetles, narrowing towards apex. Fore and middle tarsi distinctly 5 segmented, 4th tarsomere approximately as long as 3rd. Posterior margins of first four tarsomeres of hind tarsi sinuate, their outer half produced into apical lobe and without swimming hairs at their posterior margins. Scutellum concealed. Prosternal process in same plane as prosternum, not projecting ventrally; its apex acuminate, either simple or three-pointed (Pederzani 1995).

Distribution. A worldwide tribe with 11 genera (Pederzani 1995; Miller & Bergsten 2016; Nilsson & Hájek 2018).

Australian taxa. Two genera.

Keys to genera. Watts (1978); Miller & Bergsten (2016).

Laccophilus Leach, 1815 (Pl. 103A)

Type species. *Dytiscus minutus* Linnaeus, 1758.

Characteristics. As above for tribe. Shiny, with clear colour pattern on elytra. Black, unicolorous or with testaceous or ferruginous markings on elytra. Prosternal process with one point. Basal angles of pronotum never spinose and not projecting. Longer spur of hind tibia apically notched or bifid (usually also the shorter one). Worldwide distribution (Brancucci 1983; Miller & Bergsten 2016; Nilsson & Hájek 2018). Larvae described by Alarie *et al.* (2000). A genus of 283 species.

Australian species. Nine.

Distribution. Worldwide. In Australia, most species in tropical northern Australia (WA, NT, QLD), with one species south to Sydney and the red centre of SA and NSW.

Biology. Species of Australian *Laccophilus* are mainly lotic, found in streams, creeks and slow flowing rivers and its side and rest pools (billabongs in the north), often rich in submerged vegetation, with sand or decaying plant material, mainly in lowland areas up to 200 m.

References. Alarie *et al.* (2000); Brancucci (1983); Nilsson & Hájek (2017); Miller & Bergsten (2016); Pederzani (1995); Watts (1978, 2002).

Keys to species. Watts (1978); Brancucci (1983).

Australphilus Watts, 1978 (Pl. 15A)

Type species. *Australphilus saltus* Watts, 1978.

Characteristics. As above for tribe but smaller (2.3–3.0 mm). Base of pronotum straight not sinuate, metatibial spines simple. Metacoxal process reticulate with round meshes; posterior margin bilobed. Shiny, black, with strong yellow markings on elytra. Larvae described by Alarie *et al.* (2000) and keyed and figured by Watts (2002). Endemic.

Australian species. Two.

Distribution. Both species in eastern mainland Australia (NSW, VIC) and one in Tasmania.

Biology. All species of *Australphilus* are strongly lotic and restricted to streams, creeks and rivers with sand, pebbles and cobble beds, often situated in woodland or closed canopy forest sites, most in lowland regions from almost sea level to 500 m.

References. Miller & Bergsten (2016); Nilsson & Hájek (2017); Watts (1978, 2002).

Keys to species. Watts (1978).

Subfamily Lancetinae Nilsson, 2016

Tribe Lancetini Branden, 1885

Characteristics. Medium-sized beetles (7.5–12 mm in Australia), body elongate and streamlined; apex of elytra sinuate or subtruncate; median lobe aedeagus asymmetric with distinct, elongate ventral sclerite; metartarsal claws unequal in length in both sexes. Dorsal surface yellowish, with elytra often irrorate or longitudinally fasciate (Miller & Bergsten 2016). A subfamily with only one tribe and genus containing 22 species. Larvae keyed and figured by Watts (2002).

Distribution. Southern part of South America, New Zealand and southern Australia (Nilsson & Hájek 2018; Miller & Bergsten 2016)

Australian taxa. One genus.

Lancetes Sharp, 1882 (Pl. 16L)

Type species. *Dytiscus varius* Fabricius, 1775

Characteristics. Yellowish with black basal marking on pronotum and on elytra longitudinally fasciate.

Australian species. One.

Distribution. Gondwanan distribution with 21 species in temperate southern South America and at high elevation in the Andes north through Peru. Some species occur in some of the most remote localities for any diving beetle including Tierra del Fuego, South Georgia Island or the King George Islands (Miller & Bergsten 2016). The single Australian species is distributed across southern Australia (WA, SA, NSW, ACT, TAS) and in New Zealand.

Biology. The Australian *Lancetes lanceolatus* (Clark) is mainly found in open, temporary or permanent, shallow lentic habitats, such as dune lakes, seasonal swamps, flooded meadows, farm dams and road side ditches. A good flyer, single specimens can be found in almost every type of habitat. In late summer the species often occurs in large numbers.

References. Miller & Bergsten (2016); Nilsson & Hájek (2017); Watts (1978, 2002).

Subfamily Matinae Branden, 1885

Tribe Matini Branden, 1885

Characteristics. Medium-sized (6.9–9.6 mm in Australia), elongate, flattened diving beetles. Scutellum visible. Eyes with anterior margin emarginate above base of antenna. Clypeus with anterior margin broadly and distinct emarginate medially. Prosternum and prosternal process in same plane, flat and with medial longitudinal groove. Metacoxal lines strong, parallel or diverging anteriorly, reaching metaventrite. Metafemur without group of setae on posteroapical angle. Metatarsal claws short, unequal in both sexes, outer claw curved. Male with protarsomeres 1–3 moderately widened, forming an elongate to elongate oval palette. Parameres abruptly narrowed near middle. Larvae of both genera have been described (Alarie *et al.* 2001; Alarie & Butera 2003; Hendrich & Balke 2013; Miller & Bergsten 2016).

Distribution. The single tribe contains only two genera: *Matus* Aubé in North America and *Batrachomatus* in Australia. Despite this highly disjunct distribution, members of the Matinae are postulated to share a monophyletic origin and to be relics of a once more extensively distributed taxon (Alarie *et al.* 2001; Alarie & Butera 2003; Hendrich & Balke 2013; Miller & Bergsten 2016).

Australian taxa. One genus.

Batrachomatus Clark, 1863 (Pl. 15C)

Type species. *Batrachomatus wingii* Clark, 1863.
Allomatus Mouchamps, 1964

Characteristics. Shiny, black, unicolorous or with testaceous or ferruginous markings on elytra. Dorsal surface with strong double reticulation and/or fine punctures.

Australian species. Five.

Distribution. Two species in northern Australia (WA, NT, QLD), two in eastern Australia (QLD, NSW, VIC, TAS) and one in south western Australia (WA).

Biology. All species of *Batrachomatus* are strongly lotic and restricted to permanent streams, creeks and rivers with sand, pebbles and cobble beds, often situated in woodland or closed canopy forest sites, most in lowland regions from almost sea level to 500 m (Hendrich & Balke 2013).

References. Alarie & Butera (2003); Alarie *et al.* (2001); Hendrich & Balke (2013); Miller & Bergsten (2016); Nilsson & Hájek (2018); Watts (1978, 2002).

Keys to species. Hendrich & Balke (2013).

References

Alarie Y, Butera FJ (2003) The larva of *Matus leechi* Young (Coleoptera: Adephaga: Dytiscidae): Implications for the larval ground plan of the genus *Matus* Aubé. *Aquatic Insects* **25**(1), 63–70. doi:10.1076/aqin.25.1.63.14017

Alarie Y, Watts CHS (2005) Larvae of four species of the *Hyphydrus lyratus* species-group (Coleoptera: Dytiscidae: Hydroporinae). *Australian Journal of Entomology* **44**(3), 244–251. doi:10.1111/j.1440-6055.2005.00481.x

Alarie Y, Nilsson AN, Hendrich L, Watts CHS, Balke M (2000) Larval morphology of four genera of Laccophilinae (Coleoptera: Adephaga: Dytiscidae) with an analysis of their phylogenetic relationships. *Insect Systematics & Evolution* **31**(2), 121–164. doi:10.1163/187631200X00354

Alarie Y, Watts CHS, Nilsson AN (2001) Larval morphology of the tribe Matini (Coleoptera: Dytiscidae, Colymbetinae): descriptions of *Batrachomatus daemeli*, *Matus bicarinatus*, and *Allomatus nannup* and phylogenetic relationships. *Canadian Entomologist* **133**, 165–196. doi:10.4039/Ent133165-2

Alarie Y, Michat MC, Nilsson AN, Archangelsky M, Hendrich L (2009) Larval morphology of *Rhantus* Dejean, 1833 (Coleoptera: Dytiscidae: Colymbetinae): descriptions of 22 species and phylogenetic considerations. *Zootaxa* **2317**, 1–102.

Balke M, Bergsten J (2003) *Copelatus* (*Papuadytes*) *shizong* sp.n. from Yünnan (China), the first member of *Papuadytes* Balke found west of the Wallace Line. In *Water Beetles of China*. Vol. 3 (Eds MA Jäch, L Ji) pp. 89–94. Zoologisch-Botanische Gesellschaft in Österreich and Wiener Coleopterologenverein, Vienna.

Balke M, Hendrich L (2015) *7.6 Dytiscidae Leach*. In *Handbuch der Zoologie/ Handbook of Zoology. Band/Volume IV Arthropoda: Insecta Teilband/ Part 38. Coleoptera, Beetles. Volume 1. Morphology and Systematics (Archostemata, Adephaga, Myxophaga, Polyphaga partim)*. 2nd Edition. (Eds Beutel RG and Leschen RAB) pp. 118–140. W. DeGruyter, Berlin.

Balke M, Ribera I (2004) Jumping across Wallace's line: *Allodessus* Guignot and *Limbodessus* Guignot revisited (Coleoptera: Dytiscidae, Bidessini) based on molecular-phylogenetic and morphological data. *Australian Journal of Entomology* **43**(2), 114–128. doi:10.1111/j.1440-6055.2004.00415.x

Balke M, Kovac D, Hendrich L, Flechtner G (2000) Rediscovery of the New Zealand diving beetle *Rhantus plantaris* Sharp, and notes on the south west Australian *R. simulans* Régimbart, with an identification key (Coleoptera: Dytiscidae). *New Zealand Journal of Zoology* **27**, 223–227. doi:10.1080/03014223.2000.9518229

Balke M, Ribera I, Hendrich L, Miller M, Sagata K, Posman A, Vogler AP, Meier R (2009) New Guinea highland origin of a widespread arthropod supertramp. *Proceedings of the Royal Society London* (Ser. B) **276**, 2359–2367.

Balke M, Toussaint EFA, Hendrich L, Hajek J (2013) A new species of the Australian genus *Necterosoma* from Timor (Coleoptera: Dytiscidae, Hydroporini). *Acta Entomologica Musei Nationalis Pragae* **53**(1), 65–74.

Balke M, Ruthensteiner B, Warikar E, Neven K, Hendrich L (2015) Two new species of *Limbodessus* diving beetles from New Guinea - short verbal descriptions flanked by online content (digital photography, µCT scans, drawings and DNA sequence data). *Biodiversity Data Journal* **3**, e7096 doi:10.3897/BDJ.3.e7096.

Balke M, Bergsten J, Wang LJ, Hendrich L (2017a) A new genus and two new species of Southeast Asian Bidessini as well as new synonyms for Oceanian species (Coleoptera: Dytiscidae). *ZooKeys* **647**, 137–151. doi:10.3897/zookeys.647.11231

Balke M, Hájek J, Hendrich L (2017b) Generic reclassification of species formerly included in *Rhantus* Dejean (Coleoptera, Dytiscidae, Colymbetinae). *Zootaxa* **4258**(1), 91–100. doi:10.11646/zootaxa.4258.1.7

Bertrand H (1972) *Larves et nymphes des coléoptères aquatiques du globe*. Paillait, Abeville, Paris.

Beutel RG (1995) The Adephaga (Coleoptera): phylogenetic analysis and thoughts on the evolutionary history. In *Biology, Phylogeny, and Classification of Coleoptera: Papers celebrating the 80th Birthday of Roy A. Crowson*. (Eds Pakaluk J and Ślipiński SA) pp. 173–217. Muzeum i Instytut Zoologii PAN, Warszawa.

Biström O (1982) A revision of the genus *Hyphydrus* Illiger (Coleoptera, Dytiscidae). *Acta Zoologica Fennica* **165**, 1–121.

Biström O (1987) Review of the genus *Leiodytes* in Africa (Coleoptera, Dytiscidae). *Annales Entomologici Fennici* **53**, 91–101.

Biström O (1988) Generic review of the Bidessini (Coleoptera, Dytiscidae). *Acta Zoologica Fennica* **184**, 1–41.

Biström O (1996) Taxonomic revision of the genus *Hydrovatus* Motschulsky (Coleoptera, Dytiscidae). *Entomologica Basiliensia* **19**, 57–584.

Biström O, Hendrich L, Kirk-Spriggs AH (1997) A new *Hyphydrus* species from Indonesia, with a list of and a key to the Oriental and Australasian species (Coleoptera, Dytiscidae). *The Raffles Bulletin of Zoology* **45**(2), 305–313.

Brancucci M (1983) Révision des espèces est-paléartiques, orientales et australiennes du genre *Laccophilus* (Col., Dytiscidae). *Entomologische Arbeiten aus dem Museum G. Frey* **31/32**, 241–426.

Brancucci M, Monteith GB (1997) A second *Terradessus* species from Australia (Coleoptera, Dytiscidae). *Entomologica Basiliensia* **19** (1996), 585–591.

Daussin GL (1980) A new name for *Hydaticus modestus* Watts, an Australian water beetle (Coleoptera: Dytiscidae). *Coleopterists Bulletin* **34**, 351.

Dettner K, Peters W (1999) *Lehrbuch der Entomologie*. Fischer, Stuttgart.

Gewecke M (1985) Swimming behaviour of the water beetle *Dytiscus marginalis* L. (Coleoptera, Dytiscidae) In *Insect Locomotion*. (Eds Gewecke M and Wendler G) pp. 111–120. Parey, Berlin.

Guéorguiev VB (1972) Notes sur les Agabini (Coleoptera, Dytiscidae). II. Révision des genres Platynectes Régimbart et Colymbinectes Falk. *Izvestija na Zoologitjeskija Institut s Musei Sofia* **34**, 33–62.

Hawlitschek O, Porch N, Hendrich L, Balke M (2011) Ecological niche modelling and nDNA sequencing support a new, morphologically cryptic beetle species unveiled by DNA barcoding. *PLoS One* **6**(2), e16662. doi:10.1371/journal.pone.0016662

Hawlitschek O, Hendrich L, Espeland M, Toussaint EFA, Genner MJ, Balke M (2012) Pleistocene climate change promoted rapid diversification of aquatic invertebrates in South-East Australia. *BMC Evolutionary Biology* **12**, 142. doi:10.1186/1471-2148-12-142

Hendrich L (1997) A new species of *Cybister* Curtis from the Kakadu National Park in northern Australia (Coleoptera: Dytiscidae). *Entomological Problems* **28**(2), 105–108.

Hendrich L (1999) A new species of *Hydroglyphus* Motschulsky, 1853 from Northern Australia (Coleoptera: Dytiscidae). *Linzer Biologische Beiträge* **31**(1), 63–69.

Hendrich L (2001) A new species of *Antiporus* Sharp, 1882 from peatland swamps of south-western Australia (Coleoptera: Dytiscidae). *Linzer Biologische Beiträge* **33**(1), 299–308.

Hendrich L (2003a) A new species of *Necterosoma* Sharp from the Kimberley region in north-western Australia (Coleoptera: Dytiscidae). *Entomologische Zeitschrift* **113**(5), 152–154 + figs on pp. 144–145.

Hendrich L (2003b) *Austrodytes plateni* sp.n., and a faunal analysis of the Hydradephaga of the Pilbara region, Western Australia (Coleoptera: Dytiscidae, Gyrinidae, Haliplidae). *Koleopterologische Rundschau* **73**, 43–58.

Hendrich L (2008) *Tiporus emmae* sp.n. from Northern Australia (Coleoptera: Dytiscidae, Hydroporinae). *Linzer Biologische Beiträge* **40**(1), 311–320.

Hendrich L, Balke M (2009) *Kakadudessus tomweiri*, a new genus and species of diving beetle from tropical northern Australia, based on molecular, phylogenetic and morphological data (Coleoptera, Dytiscidae, Bidessini). *Zootaxa* **2134**, 49–59.

Hendrich L, Balke M (2011) A simultaneous journal / wiki publication and dissemination of a new species description: *Neobidessodes darwiniensis* sp.n. from northern Australia (Coleoptera, Dytiscidae, Bidessini). *ZooKeys* **79**, 11–20. doi:10.3897/zookeys.79.803

Hendrich L, Balke M (2013) Taxonomic revision of Australian Matini diving beetles based on morphological and molecular data (Col.: Dytiscidae, Matinae), with description of a new species. *ZooKeys* **293**, 41–64. doi:10.3897/zookeys.293.4472

Hendrich L, Balke M (2015) Review of the genus *Sekaliporus* Watts, 1997 with description of a new species from northern Australia (Coleoptera, Dytiscidae: Hydroporini). *Zootaxa* **3981**(1), 107–116. doi:10.11646/zootaxa.3981.1.5

Hendrich L, Balke M (2016) A new epigean *Paroster* Sharp, 1882 from coastal New South Wales, Australia (Coleoptera: Dytiscidae, Hydroporinae). *Spixiana* **39**(2), 213–218.

Hendrich L, Fery H (2008) *Paroster baylyi* sp. n., *P. ursulae* sp. n. (Col. Dytiscidae, Hydroporinae) and the water beetle diversity of pan-gnammas on isolated granite outcrops in the Mallee of south-western Australia. *Zootaxa* **1704**, 27–41.

Hendrich L, Šťastný J (2014) Taxonomic revision of Australian *Platynectes* Régimbart, 1879 (part I) - Four new species from Queensland (Coleoptera: Dytiscidae, Agabinae). *Zootaxa* **3795**(1), 25–37. doi:10.11646/zootaxa.3795.1.3

Hendrich L, Wang LJ (2006) Taxonomic revision of Australian *Clypeodytes* Régimbart, 1894 (Coleoptera: Dytiscidae, Bidessini). *Entomological Problems* **37**(2), 1–12.

Hendrich L, Watts CHS (2004) Taxonomic revision of the Australian genus *Sternopriscus* Sharp, 1882 (Coleoptera: Dytiscidae, Hydroporinae). *Koleopterologische Rundschau* **74**, 75–142.

Hendrich L, Watts CHS (2007) Update of Australian *Sternopriscus* Sharp, 1882 with description of three new species (Coleoptera: Dytiscidae, Hydroporinae). *Koleopterologische Rundschau* **77**, 49–59.

Hendrich L, Watts CHS (2009) Taxonomic revision of the Australian predaceous water beetle genus *Carabhydrus* Watts, 1978 (Coleoptera: Dytiscidae, Hydroporinae). *Zootaxa* **2048**, 1–30.

Hendrich L, Watts CHS (2010) An endemic predaceous water beetle from the Murchison River in Western Australia – *Antiporus kalbarriensis* sp.n. (Coleoptera: Dytiscidae, Hydroporinae, Hydroporini). *Zootaxa* **2338**, 35–42.

Hendrich L, Hawlitschek O, Balke M (2009) The epigean Australasian species of *Neobidessodes* gen.n. diving beetles–a revision integrating morphology, cybertaxonomy, DNA taxonomy and phylogeny (Coleoptera: Dytiscidae, Bidessini). *Zootaxa* **2288**, 1–41.

Hendrich L, Balke M, Wewalka G (2010a) *Dytiscidae: Hydroporini (Coleoptera)*. In *Water Beetles of New Caledonia* (part 1). Monographs on Coleoptera 3. (Eds Jäch MA and Balke M) pp. 149–162. Zooligische-Botanische Gesellschaft in Österreich and Vienna Coleopterists Society (WCV), Vienna.

Hendrich L, Pons J, Ribera I, Balke M (2010b) Mitochondrial Cox1 sequence data reliably uncover patterns of insect diversity but suffer from high lineage-idiosyncratic error rates. *PLoS One* **5**(12), e14448 doi:10.1371/journal.pone.0014448.

Hendrich L, Balke M, Wewalka G (2010c) Dytiscidae: Annotated checklist of the Bidessini, Hydrovatini, Hyphydrini, Laccophilinae and Dytiscinae (Coleoptera). In *Water Beetles of New Caledonia* (part 1). Monographs on Coleoptera 3. (Eds Jäch MA and Balke M) pp. 171–194. Zooligische-Botanische Gesellschaft in Ősterreich and Vienna Coleopterists Society (WCV), Vienna.

Hendrich L, Toussaint EFA, Balke M (2014) A new genus of Hydroporini from south-western Australia (Coleoptera, Dytiscidae). *Spixiana* **37**(1), 103–109.

Hendrich L, Balke M, Watts CHS (2016) Description of a new *Tiporus* Watts, 1985 from northern Queensland, Australia (Coleoptera: Dytiscidae, Hydroporinae). *Zootaxa* **4189**, 174–182. doi:10.11646/zootaxa.4189.1.10

Hicks BJ, Larson DL (1991) The rectum as a hydrostatic organ in the predaceous diving beetle genus *Ilybius* Erichson (Coleoptera: Dytiscidae). *Coleopterists Bulletin* **45**(3), 274–278.

ICZN (1999) *International Code of Zoological Nomenclature*. 4th edn. The International Trust for Zoological Nomenclature, London, 306 pp.

Jäch M, Balke M (2008) Global diversity of water beetles (Coleoptera) in freshwater. *Hydrobiologia* **595**, 419–442.

Korschelt E (ed.) (1923) *Bearbeitung einheimischer Tiere. Erste Monographie. Der Gelbrand Dytiscus marginalis* L. Vol.1. Engelmann, Leipzig.

Korschelt E (ed.) (1924) *Bearbeitung einheimischer Tiere. Erste Monographie. Der Gelbrand Dytiscus marginalis* L. Vol. 2. Engelmann, Leipzig.

Larson DJ (1993) Ecology of tropical Australian Hydradephaga (Insecta: Coleoptera). Part 1. Natural history and distribution of northern Queensland species. *Proceedings of the Royal Society of Queensland* **103**, 47–63.

Larson DJ (1997) Habitat and community patterns of Tropical Australian hydradephagan water eetles (Coleoptera: Dytiscidae, Gyrinidae, Noteridae). *Australian Journal of Entomology* **36**, 269–285. doi:10.1111/j.1440-6055.1997.tb01469.x

Larson DJ, Storey RI (1994) *Carabhydrus mubboonus*, a new species of rheophilic water beetle (Coleoptera: Dytiscidae) from Queensland, Australia. *Canadian Entomologist* **126**, 895–906. doi:10.4039/Ent126895-3

Larson DJ, Alarie Y, Roughley RE (2000) *Predaceous diving beetles (Coleoptera: Dytiscidae) of the Nearctic Region, with emphasis on the fauna of Canada and Alaska*. NRC Research Press, Ottawa.

Leys R, Watts CHS (2008) Systematics and evolution of the Australian subterranean hydroporine diving beetles (Dytiscidae), with notes on *Carabhydrus. Invertebrate Systematics* **22**, 217–225. doi:10.1071/IS07034

Leys R, Roudnew B, Watts CHS (2010) *Paroster extraordinarius* sp. nov., a new groundwater diving beetle from the Flinders Ranges, with notes on other diving beetles from gravels in South Australia (Coleoptera: Dytiscidae). *Australian Journal of Entomology* **49**, 66–72. doi:10.1111/j.1440-6055.2009.00738.x

Michat MC (2006) The phylogenetic position of *Hydrovatus* Motschulsky: evidence from larval morphology of *H. caraibus* Sharp (Coleoptera: Dytiscidae: Hydroporinae). *Insect Systematics & Evolution* **37**(4), 419–432. doi:10.1163/187631206788831416

Michat MC, Alarie Y, Watts CHS (2010) Descriptions of the first instar larva of the hypogaeic species *Neobidessodes limestoneensis* (Watts and Humphreys) and the third instar larva of *Hydroglyphus balkei* Hendrich (Coleoptera: Dytiscidae: Bidessini) with phylogenetic considerations. *Zootaxa* **2658**, 38–50.

Michat MC, Alarie Y, Watts CHS (2011) Larval morphology of *Allodessus* Guignot (Coleoptera: Dytiscidae). *Aquatic Insects* **33**(1), 27–40. doi:10.1080/01650424.2011.574228

Michat MC, Alarie Y, Watts CHS (2015) Phylogenetic placement of the genus *Sternhydrus* (Coleoptera: Dytiscidae: Cybistrini) base don larval morphology, with description of *S. atratus*. *Annals of the Entomological Society of America* **108**(5), 881–892. doi:10.1093/aesa/sav067

Miller KB (2000) Cladistic analysis of the tribes of Dytiscinae and the phylogenetic position of the genus *Notaticus* Zimmermann (Coleoptera: Dytiscidae). *Insect Systematics & Evolution* **31**, 165–177. doi:10.1163/187631200X00363

Miller KB (2001) On the phylogeny of the Dytiscidae (Insecta: Coleoptera) with emphasis on the morphology of the female reproductive system. *Insect Systematics & Evolution* **32**, 45–89. doi:10.1163/187631201X00029

Miller KB (2002) Revision of the genus *Eretes* Laporte, 1833 (Coleoptera: Dytiscidae). *Aquatic Insects* **24**(4), 247–272. doi:10.1076/aqin.24.4.247.8238

Miller KB (2012) *Petrodessus conatus* sp. n., a new genus and species of Bidessini from hygropetric habitats in tropical Australia (Coleoptera: Dytiscidae: Hydroporinae). *Zootaxa* **3242**, 62–67.

Miller KB, Bergsten J (2014) The phylogeny and classification of predaceous diving beetles. In *Ecology, Systematics, and the Natural History of Predaceous Diving Beetles (Coleoptera: Dytiscidae)*. (Ed. Yee DA) pp. 49–171. Springer, Netherlands.

Miller KB, Bergsten J (2016) *Diving Beetles of the World. Systematics and Biology of the Dytiscidae*. Johns Hopkins University Press, Baltimore.

Miller KB, Short AEZ (2015) *Belladessus* Miller and Short (Coleoptera: Dytiscidae: Hydroporinae: Bidessini), New Genus for Two New Species from Northern South America: Parthenogenetic Diving Beetles? *Coleopterists Bulletin* **69**(3), 498–503. doi:10.1649/0010-065X-69.3.498

Miller KB, Wheeler Q (2015) *Zimpherus nancae* Miller and Wheeler (Coleoptera: Dytiscidae: Hydroporinae: Bidessini), New Genus and New Species from Venezuela. *Coleopterists Bulletin* **69**(3), 507–511. doi:10.1649/0010-065X-69.3.507

Miller KB, Bergsten J, Whiting MF (2007) Phylogeny and classification of diving beetles in the tribe Cybistrini (Coleoptera, Dytiscidae, Dytiscinae). *Zoologica Scripta* **36**, 41–59. doi:10.1111/j.1463-6409.2006.00254.x

Nachtigall W (1960) Über Kinematik, Dynamik und Energetik des Schwimmens einheimischer Dytisciden (Coleoptera, Dytiscidae). *Zeitschrift für Vergleichende Physiologie* **43**, 48–118.

Nilsson AN, Hájek J (2018) *A World Catalogue of the family Dytiscidae, or the Diving Beetles (Coleoptera, Adephaga)*. Version 1.1.2018, 306 pp. Available from: http://www.norrent.se & http://www.waterbeetles.eu

Nilsson AN, Holmen M (1995) *Fauna Entomologica Scandinavica 32. The aquatic Adephaga (Coleoptera) of Fennoscandia and Denmark. II. Dytiscidae*. E. J. Brill/Scandinavian Science Press, Leiden, Copenhagen.

Nilsson AN, Roughley RE (1997a) The genus- and family-group names of the Dytiscidae – additions and corrections. *Beiträge zur Entomologie* **47**(2), 359–364.

Nilsson AN, Roughley RE (1997b) A classification of the family Dytiscidae (Coleoptera). *Latissimus* **8**, 1–4.

Nilsson AN, Roughley RE, Brancucci M (1989) A review of the genus- and family-group names of the family Dytiscidae Leach (Coleoptera). *Entomologica Scandinavica* **20**, 287–316. doi:10.1163/187631289X00348

Pederzani F (1995) Keys to the identification of the genera and subgenera of adult Dytiscidae (sensu lato) of the world (Coleoptera Dytiscidae). *Atti dell'Accademia Roveretana degli Agiati* **224**, 5–83.

Pinder AM, Halse SA, Shiel RJ, McRae JM (2000) Granite outcrop pools in south-western Australia: foci of diversification and refugia for aquatic invertebrates. *Journal of the Royal Society of Western Australia* **83**, 149–161.

Ribera I, Nilsson AN (1995) Morphometric patterns among diving beetles (Coleoptera: Noteridae, Hygrobiidae, and Dytiscidae). *Canadian Journal of Zoology* **73**, 2343–2360. doi:10.1139/z95-275

Shaverdo HV, Surbakti S, Hendrich L, Balke M (2012) Introduction of the *Exocelina ekari*-group with descriptions of 22 new species from New Guinea (Coleoptera, Dytiscidae, Copelatinae). *ZooKeys* **250**, 1–76. doi:10.3897/zookeys.250.3715

Shaverdo HV, Hendrich L, Balke M (2013) *Exocelina baliem* sp.n., the only known pond species of New Guinea *Exocelina* Broun, 1886 (Coleoptera, Dytiscidae, Copelatinae). *ZooKeys* **304**, 83–99. doi:10.3897/zookeys.304.4852

Spangler PJ, Decu V (1998) Coleoptera aquatica. In *Encyclopaedia Biospeologica*. Tome II. (Eds Juberthie C. and Decu V) pp. 1031–1046. Société de Biospéologie, Moulis, Bucuresti.

Toussaint EFA, Condamine FL, Hawlitschek O, Watts CHS, Hendrich L, Balke M (2015) Unveiling the diversification dynamics of Australasian predaceous diving beetles in the Cenozoic. *Systematic Biology* **64**(1), 3–24. doi:10.1093/sysbio/syu067

Toussaint EFA, Hendrich L, Escalona H, Porch N, Balke M (2016a) Evolutionary history of a secondary terrestrial Australian diving beetle (Coleoptera, Dytiscidae) reveals a lineage of high morphological and ecological plasticity. *Systematic Entomology* doi:10.1111/syen.12182.

Toussaint EFA, Hendrich L, Hajek J, Michat MC, Panjaitan R, Short AEZ, Balke M (2016b) Evolution of Pacific rim diving beetles sheds light on Amphi-Pacific biogeography. *Ecography* **39**, 1–11.

Watts CHS (1963) The larvae of Australian Dytiscidae (Coleoptera). *Transactions of the Royal Society of South Australia* **87**, 23–40.

Watts CHS (1964) The larvae of Australian *Cybister* spp. Curt., *Homoeodytes* spp. Reg. and *Hyderodes shuckardi* Hope (Coleoptera: Dytiscidae). *Transactions of the Royal Society of South Australia* **88**, 145–156.

Watts CHS (1978) A revision of the Australian Dytiscidae (Coleoptera). *Australian Journal of Zoology Supplementary Series* **26**, 1–166. doi:10.1071/AJZS057

Watts CHS (1985) A faunal assessment of Australian Hydradephaga. *Proceedings. Academy of Natural Sciences of Philadelphia* **137**, 22–28.

Watts CHS (1997a) Four new species of *Antiporus* Sharp (Coleoptera, Dytiscidae) from Australia, with notes on *A. femoralis* (Boh.) and *A. interrogationis* (Clark). *Records of the South Australian Museum* **30**, 35–42.

Watts CHS (1997b) A new genus and species of Australian Dytiscidae (Coleoptera). *Records of the South Australian Museum* **29**, 121–123.

Watts CHS (2000a) Three new species of *Tiporus* Watts (Coleoptera: Dytiscidae) with re-descriptions of the other species of the genus. *Records of the South Australian Museum* **33**(2), 89–99.

Watts CHS (2000b) Six new species of *Nirridessus* Watts and Humphreys and *Tjirtudessus* Watts and Humphreys (Coleoptera: Dytiscidae) from underground waters in Australia. *Records of the South Australian Museum* **33**(2), 127–144.

Watts CHS (2002) *Checklist and guides to the identification, to genus, of adults and larval Australian water beetles of the families Dytiscidae, Noteridae, Hygrobiidae, Haliplidae, Gyrinidae, Hydraenidae and the superfamily Hydrophiloidea (Insecta-Coleoptera(Identification Ecology Guide 43)*. Cooperative Research Centre for Freshwater Ecology, Australia.

Watts CHS, Humphreys WF (1999) Three new genera and five new species of Dytiscidae (Coleoptera) from underground waters in Australia. *Records of the South Australian Museum* **32**, 121–142.

Watts CHS, Humphreys WF (2000) Six new species of *Nirridessus* Watts and Humphreys and *Tjirtudessus* Watts and Humphreys (Coleoptera: Dytiscidae) from underground waters in Australia. *Records of the South Australian Museum* **33**, 127–144.

Watts CHS, Humphreys WF (2003) Twenty-five new Dytiscidae (Coleoptera) of the genera *Tjirtudessus* Watts & Humphreys, *Nirripirti* Watts & Humphreys and *Bidessodes* Régimbart from underground waters in Australia. *Records of the South Australian Museum* **36**(2), 135–187.

Watts CHS, Humphreys WF (2004) Thirteen new Dytiscidae (Coleoptera) of the genera *Boonguurus* Larson, *Tjirtudessus* Watts & Humphreys and *Nirripirti* Watts & Humphreys, from underground waters in Australia. *Transactions of the Royal Society of South Australia* **128**, 99–129.

Watts CHS, Humphreys WF (2006) Twenty-six new Dytiscidae (Coleoptera) of the genera *Limbodessus* Guignot and *Nirripirti* Watts and Humphreys, from underground waters in Australia. *Transactions of the Royal Society of South Australia* **130**(1), 123–185. doi:10.1080/3721426.2006.10887055

Watts CHS, Humphreys WF (2009) Fourteen new Dytiscidae (Coleoptera) of the genera *Limbodessus* Guignot, *Paroster* Sharp, and *Exocelina* Broun from underground waters in Australia. *Transactions of the Royal Society of South Australia* **133**(1), 62–107. doi:10.1080/03721426.2009.10887112

Watts CHS, Leys R (2005) Review of the epigean species of Australian *Limbodessus* Guignot (Insecta: Coleoptera: Dytiscidae). *Transactions of the Royal Society of South Australia* **129**(1), 1–13.

Watts CHS, Leys R (2006) Notes on Australian *Hyphydrus* Illiger, including taxonomy, key to the species and sexual dimorphism. *Koleopterologische Rundschau* **76**, 15–21.

Watts CHS, Leys R (2008) Review of the epigean species of *Paroster* Sharp, 1882, with descriptions of three new species, and phylogeny based on DNA sequence data of two mitochondrial genes (Coleoptera: Dytiscidae: Hydroporinae). *Koleopterologische Rundschau* **78**, 9–36.

Watts CHS, McRae J (2013) *Limbodessus bennetti* sp. nov., the first stygobitic Dytiscidae (Coleoptera) from the Pilbara region of Western Australia. *Records of the South Australian Museum* **28**, 141–143. doi:10.18195/issn.0312-3162.28(2).2013.141-143

Watts CHS, Pinder A (2000) Two new species of *Antiporus* Sharp from Western Australia (Coleoptera: Dytiscidae). *Records of the South Australian Museum* **33**, 17–19.

Watts CHS, Hancock PJ, Leys R (2007) A stygobitic *Carabhydrus* Watts (Dytiscidae, Coleoptera) from the Hunter Valley in New South Wales, Australia. *Australian Journal of Entomology* **46**, 56–59. doi:10.1111/j.1440-6055.2007.00585.x

Watts CHS, Hendrich L, Balke M (2016) A new interstitial species of diving beetle from tropical northern Australia provides a scenario for the transition of epigean to stygobitic life (Coleoptera, Dytiscidae, Copelatinae). *Subterranean Biology* **13**, 23–29.

Wewalka G (1975) Revision der Artengruppe des *Hydaticus vittatus* (Fabricius), (Dytiscidae, Col.). *Koleopterologische Rundschau* **52**, 87–100.

Wewalka G (1979) Revision der Artengruppe des *Hydaticus* (*Guignotites*) *fabricii* (Mac Leay), (Col., Dytiscidae). *Koleopterologische Rundschau* **54**, 119–139.

Wewalka G (1994) A new species of *Chostonectes* from Australia. *Zeitschrift der Arbeitsgemeinschaft Österreichischer Entomologen* **46**, 140–142.

Wewalka G (2015) Revision of the *Hydaticus* (*Prodaticus*) *sexguttatus* species group, and resembling species from the Palaearctic, Oriental, Australian and Pacific Regions. *Koleopterologische Rundschau* **85**, 7–35.

Wewalka G, Balke M, Hendrich L (2010) Dytiscidae: Copelatinae (Coleoptera). In *Water Beetles of New Caledonia* (part 1). Monographs on Coleoptera 3. (Eds Jäch MA and Balke M) pp. 45–128. Zooligische-Botanische Gesellschaft in Ősterreich and Vienna Coleopterists Society (WCV), Vienna.

Yee DA (2014) *Ecology, Systematics and the Natural History of Predaceous Diving Beetles (Coleoptera: Dytiscidae).* Springer Netherlands, Dordecht.

Zwick P (1979) Notes on the genus *Necterosoma* (Col., Dytiscidae) with description *N. susanna* n.sp. from Australia. *Aquatic Insects* **1**(3), 179–184. doi:10.1080/01650427909360992

Zwick P (1981) *Carabhydrus andreas*, a new Australian dytiscid (Coleoptera, Dytiscidae). *Aquatic Insects* **3**, 167–170.

12. CARABIDAE LATREILLE, 1802

Martin Baehr and Kipling Will

Fig. 12.1. *Mecynognathus damelii* Macleay.

Common name. Ground beetles.

Family synonyms. Rhysodidae Laporte, 1840; Cicindellidae Latreille, 1802; Paussidae Latreille, 1806.

Introduction. Carabidae comprises one of the largest families of Coleoptera and by far the largest in the suborder Adephaga. In the sense of the present chapter (including 'Rhysodidae', 'Paussidae' and 'Cicindelidae'), the family consists of almost 1700 genera and ~40 000 described species and subspecies worldwide (Lorenz 2005). Currently, the Australian fauna includes 322 genera and more than 3100 described and valid species and subspecies (Baehr unpubl.) that span most of the major tribes in the family. However, in comparison to other faunal regions the composition of the Australian Carabid fauna is unique: whereas Broscini, Psydrini, and Pseudomorphini are diverse and abundant, Carabini, Platynini, Chlaeniini, and Brachinini are poorly represented. Also some tribes that are globally highly diverse and in some regions recognised as the groups with the greatest generic and species-level diversity, e.g. Cicindelini, Bembidiini, Trechini, Harpalini, these are comparatively less numerous in Australia, whereas groups often modestly represented in other regions, e.g. Scaritini, Licinini, and Helluonini, are numerous and highly diverse.

The Australian Carabid fauna is essentially a Southern Hemisphere one, therefore several tribes common in the northern continents are absent from Australia, e.g. Nebriini, Elaphrini, Loricerini, Notiophilini, Omophronini, Patrobini, Zabrini. On the other hand, several tribes have their, or one of their, centres of diversity in Australia, namely Migadopini, Broscini, Zolini, Psydrini, and Helluonini, or are endemic in Australia and do not occur elsewhere, e.g. Agonicini and Pamborini (with the single exception of the New Zealand *Maoripamborus* Brookes). The affinities of Australian taxa within these large, old groups is typically with southern South America and Africa, however, a significant connection to the Asian-Pacific fauna is also evident. Below the level of tribe, the Australian carabids are largely endemic, with several, major radiations of species in large genera that are wholly Australian, e.g. *Sphallomorpha* Westwood, *Carenum* Bonelli, and *Notonomus* Chaudoir.

The suprageneric systematics of Carabidae still is being resolved and several, quite different systems have been proposed. These classifications have different numbers of recognised subfamilies and strikingly different affiliations of tribes to subfamilies (e.g. Sloane 1923b; Jeannel 1941, 1942, 1949; Kryzhanovskij 1976; Moore *et al.* 1987; Erwin 1991; Lawrence & Britton 1991; Lawrence & Newton 1995; Liebherr & Will 1998; Ball 1979; Ball & Bousquet 2001; Löbl & Smetana 2003; Lorenz 2005; Ober & Maddison 2008; Bouchard *et al.* 2011; Arndt *et al.* 2016). Number and status of accepted tribes, as well as of subtribes, likewise is a focus of much discussion, and even the limits of Carabidae and the status of groups such as Trachypachidae, Cicindelinae, Paussinae, Brachininae, and Pseudomorphinae still is being debated. For practical reasons and because this chapter is not devoted to solving systematic or phylogenetic problems, we tried to use a simple, compromise system that is useful and yet reflects advances in our knowledge and the current state of affairs. Therefore, we do not use subfamilies, as these are problematic for many reasons, but employ tribes, which as suprageneric taxa go, have the greatest support from phylogenetic studies and are typically recognisable groups for many people. In the tribal and subtribal classification we largely, but not always follow what was proposed by Bouchard *et al.* (2011) and we try to adapt this system to the Australian fauna. As a consequence, we include in Carabidae Cicindelini, Paussini, Rhysodini, Brachinini, and Pseudomorphini as tribes.

Divergent opinions on the best higher classification of Carabidae draws attention to the need to erect the most general and objective system possible, one based on sound evolutionary hypotheses. To do so we need to develop a fuller understanding the phylogeny of Carabidae, with particular attention to its early evolution in Adephaga. Given we have not reached this stable, general classification; for the most part we have been restricted to the examination of extant forms and comparative morphology of external and genitalic characters. Present and future use of DNA sequence data promises to further elucidate the phylogeny and, in combination with morphology and fossil data, one day provide the evidence for a widely accepted classification system for the Carabidae (Assmann *et al.* 2008).

Life history, ecology and behaviour. Although their vernacular name 'ground beetles' denotes Carabidae as being a family of terrestrial beetles, at least in the tropics, large numbers of species are arbouricolous or planticolous, mainly as adults, and less commonly as larvae. In addition to species found on the ground, active adults can be found on trunks, twigs, and leaves of a variety of plants, and on or under the bark of standing trees and logs. In Australia, the number of corticolous (bark-inhabiting) carabid species is exceptionally high, and they live either under the bark of bark-shedding trees, such as various eucalypts, or in deep fissures in the bark of several eucalypt and non-eucalypt trees with ruggedly sculptured bark. This may be due to the preponderance of bark-shedding eucalypts over large parts of Australia, which is a microhabitat that provides excellent shelter against the very high ambient temperatures on the ground and frequent bushfires.

True ground living species are most common in forests, particularly in the more or less wet forests of eastern and southern Australia, whereas in the semiarid and arid areas of the interior, ground living species are rather scarce and mainly occur on the banks of rivers, shores of lakes and lagoons, and in seasonal wet places where they may be extremely abundant during the wet. There are other adaptations to the arid areas. Some carabids, mainly the Scaritini and Broscini, possess robust, dentate protibiae and the adult beetles are fossorial. They are particularly well represented in semiarid and arid regions, such as central Australia and interior Western Australia. Their larvae usually are entirely subterranean. Interstitial (edaphobitic) species, which occur deep in leaf litter and in the upper stratum of the soil, are represented by tiny species of the bembidiine subtribe Anillina, which has only recently received attention. In particular the Trechini include several true cavernicolous species that are depigmented and have reduced eyes or are completely blind. These cave species have been found mainly in southern mainland Australia and in Tasmania, but certainly the exploration of the Australian cave fauna is still insufficient. Amazing new species of Anillina and Zuphiini are being discovered in drill bores that allow for sampling to depths of 70 m (Baehr 2014e; Baehr & Main 2016). The subsurface fauna of Australian carabids represents another biodiversity horizon that remains poorly explored and is likely to be exceptionally rich in species.

Most of our knowledge of carabid beetle feeding habits is based on observations in Europe and North America (Lövei & Sunderland 1996; Larochelle 1990; Larochelle & Larivière 2003) but it is assumed, based on scattered observations, that the Australian species have a similar range of habits. Many adult Carabidae are general predators, but specialised feeding habits are common. The Northern Hemisphere tribe Cychrini includes many species that feed on land snails, some other Holarctic genera (Notiophini, Loricerini) hunt springtails (Collembola), some cavernicolous Trechini feed on the larvae of cave crickets, certain Lebiini on eggs and/or larvae of chrysomelids, species of chlaeniines and pterostichines feed on amphibians, some Brachinini on eggs of mole crickets, etc. However, several species, mainly in Harpalini, but probably also in other groups, secondarily became facultatively or entirely phytophagous, eating seeds, pollen, or even fruits. Probably, as has been shown in European species, the proportion of facultatively phytophagous species may be greater than is presently supposed.

Most carabid larvae are campodeiform, active general predators that probably move through the soil and leaf litter in search of prey. But those of the Cicindelini and of certain Pseudomorphini live in vertical burrows in the ground, anchored to the walls by hooks on their dorsal surface. There they lay waiting and ambush passing ants and other small invertebrates. The larvae of Paussini and of most Pseudomorphini live in the nests of ants or termites and have become physogastric, though in the later instars the hind body becomes more and more reduced. Also like Brachinini, Lebiini larvae that are ectoparasitic on eggs or pupae of other insects are also known to be physogastric.

Most genera of Pseudomorphini are ovoviviparous which means that the larvae hatch while in the female reproductive tract and emerge from the female as first instar larvae (Erwin 1981; Liebherr & Kavanaugh 1985; Baehr 1997b). These ovoviviparous species are myrmecophilous and living with ants or termites and so ovoviviparity is presumed to be a strategy to avoid egg predation by the ants. Maternal care for eggs and/or young larvae, or even feeding of larvae is known from several species, including certain typically predacious Pterostichini and phytophagous Harpalini (Brandmayr & Zetto Brandmayr 1979; Kavanaugh 1998; Horne 1990). While some *Notonomus* species brood larvae in shallow depressions in the soil they form under rocks and logs (Horne 1990), females of *Nurus* Motschulsky and some other large pterostichines make deep burrows in which the females nurture the young.

In general, the study of carabid larvae is not yet as well developed as that of the adults. This is even more pronounced in Australia where very little information has been published on habits and life cycles of Carabidae. Given the size of the fauna and commonness of the beetles, it is amazing that so little is known about the larvae and reproduction of Australian Carabidae. Moore (1964b, 1966b, 1966c, 1998), Moore & Lawrence (1994), Giachino (2005d) and Di Giulio & Moore (2004, 2009) described the larvae of single species of a several tribes, and Moore (1964b, 1974), Erwin (1981), Liebherr & Kavanaugh (1985), and Baehr (1997b) have described reproduction and larvae of certain pseudomorphine genera. The larvae of many presumably phylogenetically important, early branching, mostly southern, lineages have not been studied. However, based on larval morphology studies from other parts of the world, there is evidence that they can yield very useful character states for the reconstruction of carabid phylogeny (Arndt 1993, 1998). The limited knowledge of larval morphology was summarised by van Emden (1942) for tribes worldwide and by Thompson (1979) for the Nearctic tribes.

Carabidae, as adults and larvae have a large number of enemies from among vertebrates and various invertebrate groups. Typical vertebrate predators are insectivorous birds, several small, insectivorous mammals, small goannas, skinks, agamas, geckos, and most frogs and toads. For example, the introduced cane toad (*Rhinella marina* (L.)) is known to be a voracious predator of large carabid beetles (Turvey 2013). Among the invertebrates, spiders probably are the most important predators, some of which, e.g. the redback spider (*Latrodectus hasseltii* Thorell), can even catch very large Pterostichini, Broscini and Scaritini. Other likely important invertebrate predators include scorpions, centipedes, robber flies, and other carabids. Carabidae and their larvae may be victims of parasitic Hymenoptera or Diptera, and they also can be infested by parasitic fungi and horsehair worms (Nematomorpha) (Larochelle & Larivière 2007).

For protection from predation adult carabids rely on chemical secretions, nocturnal habits, protective colours, or aposematic patterns (Zetto Brandmayr *et al.* 2006). Carabids are preeminent among chemically defended arthropods. And the family includes beetles that produce a strikingly diverse array of defensive chemicals using a pair of homologous pygidial glands that discharge repellent fluids through orifices at the end of the abdomen. Carabid glandular products comprise 19 distinct classes of chemicals (representing > 250 individual compounds) including hydrocarbons, formic acid, aromatic aldehydes, sulphur compounds, cyanides and quinones (Moore & Wallbank 1968; Moore 1979; Will *et al.* 2000; Will & Attygalle unpubl.). While species in many carabid lineages produce a single class of chemicals, distantly related lineages of carabids also often produce the same defensive compounds presumably due to either parallel or convergent evolution. The chemical composition of these secretions is very diverse and can be used for phylogenetical studies (Moore & Wallbank 1968). Paussini and Brachinini include species commonly known as the bombardier beetles, which chemically blast defensive quinones at extremely hot temperatures (Eisner *et al.* 1977; Aneshansley *et al.* 1969).

Unfortunately, knowledge about the biology and ethology of Australian Carabidae is still very fragmentary and what is known, mostly relies on unpublished, personal experience of collectors, or simply is extrapolated from general trends and from the knowledge of carabid biology gained in the Northern Hemisphere. There the first general work was that of Lindroth (1945–1949) on the Fennoscandian Carabidae, which was extended by Thiele (1977), and many other European and American carabidologists, e.g. by Larochelle & Larivière (2003). Also the proceedings of the various International, European, and North American Symposia on Carabidology include a wealth of knowledge about many aspects of carabid biology, ecology, and ethology. Therefore, in the northern continents Carabidae are one of the most commonly used groups for eco-faunistic and environmental studies. Due to the limited knowledge of taxonomy, distribution, and ecology of the Australian Carabidae, presently this is possible only to a very restricted extent (New 1998).

Character systems. Flight is thought to be a key innovation in insect evolution and a well studied morphological feature in Carabidae is the flight wing size. The number of apterous species, those having greatly reduced metathoracic wings, is very high in species of wet forests, in particular in leaf litter inhabiting and montane species. Also cavernicolous species and most fossorial species are apterous. Many hygrophilous species, on the other hand, have full flight wings and are capable flyers. In particular, those that occur in periodically dry, seasonally wet, or disjunct habitat types tend to be capable fliers. Some species are dimorphic, with only a proportion of any given population apterous, while the rest have full wings and apparently are able to fly. Loss of flying ability and the short range endemism apparent in many apterous species makes Carabidae a preferred group for zoogeographical studies, e.g. Lindroth (1945–1949) for the European carabid fauna, Jeannel (1938) and Darlington (1961a,b, 1965, 1971) for the faunas of the southern continents and for the Australian fauna, and various island biogeography studies (Liebherr & Maddison 2013; Liebherr & Zimmerman 1998).

A very useful and widely used character set for carabid taxonomy and systematics is drawn from the structures of the male genitalia, in particular the form of the parameres, the shape of the phallus and the structure of the endophallus of the aedeagus all show variation that ranges in its utility from specific identification to higher-level grouping. Carabidae are thought to ancestrally have aedeagal asymmetry, in most species the true ventral surface of the aedeagal median lobe at rest (i.e. when retracted within the abdominal apex) faces to the left of the beetle and the true left side of the median lobe faces dorsally. When everted for mating, the median lobe and parameres rotate 90° counter clockwise back to the true anatomical orientation. Several carabid groups have evolved the mirror image situation where, when the median lobe is in repose it is right side up. Tribal level taxa like the Drimostomatini all share the derived condition, but in rare cases there is within-species dimorphism with mixed populations of left side up and right side up males (Liebherr & Will 2015). Three tribes found in Australia, Harpalini, Platynini and Abacetini, are known to have evolved monorchy, i.e. the absence of one of the testes in the male (Will *et al.* 2005). While the absent testis is consistently the left in Harpalini, Platynini, the Abacetini, including some of those in Australia, show variation with regard to which testis is absent. The female genitalia and reproductive tract, in general, are less divergent among closely related species, making it less useful for species-level identification, but characters are well suited for the characterisation of tribes and subtribes, and in some cases for genera and species groups (Liebherr & Will 1998; Deuve 1993). A taxonomy of the Australian and extra-Australian Cicindelini was almost completely based on the female genitalia (Rivalier 1963; Freitag 1979; McCairns *et al.* 1997). However, in many groups of Australian carabids female genitalia have not been studied in detail.

Chromosomes of Carabidae have been found to be surprisingly variable (Serrano 1981; Serrano & Galián 1998), with diploid numbers ranging from 8 to 69, the latter being the highest known in Coleoptera. The most frequently encountered number is 37 ($2n = 36 + X$) which is assumed to represent the ancestral condition (Galián & Moore 1994). Evidently numerous changes and reductions have occurred during evolution of the group and thus karyotype numbers are rather confusing and their use in phylogenetic studies seems to be rather problematic.

Molecular methods and DNA data are now widely used tools in taxonomy and phylogeny of carabids (Assmann *et al.* 2008) and these data and analyses have been applied at higher-levels for the world fauna (Maddison *et al.* 1999; Ober and Maddison 2008; Maddison *et al.* 2009), and to a few specific cases in the Australian fauna (e.g. Sota *et al.* 2005; Will 2015b), but have not yet been applied on a large scale to Australian Carabidae.

Geographic area. The area covered in this chapter includes mainland Australia, Tasmania, near offshore islands, including Torres Strait Islands, but also remote Norfolk, Lord Howe, and Christmas Islands. To better put genera in context, we have included summary information on their extra-Australian ('extraterritorial') distribution where appropriate.

For broader biogeographic patterns the standard names for the large biogeographic regions are used: the temperate

Palearctic and Nearctic Regions and the primarily tropical and subtropical Afrotropical, Oriental, Australian, and Neotropical Regions. The Afrotropical, Oriental, and Australian Regions are commonly combined into a single Paleotropical realm. While the Australian Region seems well circumscribed, the status of New Guinea and the Pacific Islands relative to it is debatable. Some authors have included them in the Australian Region, but more frequently they are kept as independent regions. This means any summary statement regarding distribution must be carefully interpreted and is best validated by reference to actual specimen data.

Keys. Prior to this work, no general keys to the Australian Carabidae were available. However, two publications are helpful for recognition of certain southern and south-eastern genera and species, namely the guide to the beetles of south-eastern Australia by B. P. Moore (1980–1989) which does not contain keys, but important information on tribes and some species; and the guide to the genera of beetles of South Australia by E. G. Matthews (1980) which contains illustrated keys to the subfamilies (= tribes) and genera of Carabidae that occur in South Australia. Both guides contain good pictures of several characteristic species of various genera.

The present chapter provides users, for the first time, a tool for the identification of all tribes, genera, and subgenera that are reported from Australia. The overarching goal is to offer a practical way to access the carabid biodiversity of the region. Generally we have confined the characters states used in the keys to the external morphology; those that are easily seen and do not require any dissections. We are aware that we have omitted several important distinctive characters, mainly of the male and female genitalia. However, including those characters would significantly complicate identification of supraspecific taxa. Moreover, for several genera information about their genitalia is lacking or so incomplete that the use of characters of the genitalia is not possible for identification. In certain large genera the morphological diversity of the genitalia is so great that a single example would be an insufficient representation of the group and obscure the characteristics of the respective genus.

We give keys to the tribes, subtribes, genera, and subgenera occurring in Australia. In some groups the keys to genera are found under the subtribes, in other groups, in which the identification of subtribes is more difficult or the subtribes are of questionable status, a complete key to all genera of the tribe is given. All keys and diagnoses only refer to the Australian fauna. We want to stress that for certain tribes and genera of which only single or few species occur in Australia, our keys and diagnoses may not be applicable to extra-Australian taxa.

Taxonomy. We provide characterizations of tribes and subtribes that include some general information about external and genitalic morphology (if this is recorded). For genera and subgenera we provide information about the number of included taxa, including undescribed ones (according to our present state of knowledge), distribution, and biology, if available. Unfortunately, generally very little has been recorded about habits and life histories of Australian Carabidae, so we strongly encourage filling of these knowledge-gaps though studies on ecology and life histories. We also suggest future systematic sampling in remote areas throughout Australia, not only to determine the actual species diversity, but also to more precisely fix the ranges of many species that are only known by single or few specimens or from a single locality.

Species identification. Although this publication is focused on identification and characterisation of supraspecific taxa, we have added references to papers devoted to species identification, monographs, revisions, or regional treatments when available. Because there is no comprehensive bibliography of the Australian Carabidae, we also enumerate older publications if they include keys to species that are still relevant and useful for the study of the respective supraspecific taxa.

The basis of any taxonomic work on the Australian Carabidae is still the catalogue of Moore *et al.* (1987) which includes, *inter alia*, some information about type depository, distribution, and habits. The general catalogues of Lorenz (1998, 2005, 2018) are very useful, but the present status of species has been taken from a provisional checklist that one of us (MB) has developed.

The references include those papers that contain the descriptions of tribes, subtribes, genera and subgenera, the most important citations of genera and subgenera, with special account to papers concerning neighbouring faunas, and those papers which contain keys, natural history information, or are otherwise helpful for the identification of species. We have included synonymies of genera and subgenera, but not of tribes and subtribes. These can be found in the treatment by Bouchard *et al.* (2011).

Publications about carabid faunas of adjacent areas. The carabid fauna of the Oriental Region has not yet received a general treatment. The work of Jedlicka (1963) on a part of the East Asian Carabidae is of some value, but the fauna of the Indonesian and Philippine archipelagos, which are much more important for the knowledge of the fauna of Australia, have not yet received a comprehensive treatment. Still of much value is the monumental work of Darlington (1958–1971) on the carabid fauna of New Guinea, although for several groups it is outdated because many additional species have been described and genera have been revised. The carabid fauna of Micronesia was studied by Darlington (1970). The catalogues of the New Zealand Carabidae of Larochelle & Larivière (2001, 2007) and the unpublished provisional checklist of the Carabidae of New Caledonia by one of us (KW) are useful for comparison and understanding the faunal relationships.

Illustrations. For all included genera, one species is illustrated. In some large and/or diverse genera we provide an image of more than one species to better demonstrate the diversity. However, we have omitted other illustrations, in particular those of male and female genitalia, because this would only be possible for a fraction of the genera at his point.

History of the study of Carabidae in Australia. Due to the late discovery of Australia by the European colonialists and the slow scientific exploration of the interior of the country, the history of carabidology in Australia has a relatively recent beginning. Single species were described in various works by Latreille, Dejean, W. S. Macleay, Hope, Erichson, Germar, and others in the first half of 19th Century, but the first major work was been done by F. L. Laporte (1867, 1868) who described a large number of species, and by W. Macleay (1863–1888) who described many species in a few papers focussing on particular carabid groups from selected geographic areas.

Almost all papers in the first half of 19th Century refer to species from the south-east of the continent including Tasmania, where the carabids received a special treatment by Erichson (1842). At that time the north, west and the interior of Australia were virtually unexplored by European naturalists or the newly arrived Australian colonists. It was W. Macleay's papers targeting particular areas in south-eastern and north-eastern Queensland that broadened the horizon for carabidologists. From ~1880 onwards, the Dodd family collected intensively in north Queensland, sending significant amounts of material for study to various researchers and collectors. So the north-eastern carabid fauna at the end of 19th Century was fairly broadly, though provisionally, explored. In contrast, Macleay's paper on the Carabidae of Kings Sound (Macleay 1888), which was the first to deal with the fauna of north-western Australia, was much more limited in scope. At the end of that century the first zoological explorations covered the interior of Australia and more material became available from the far north, the interior, and south-western and interior Western Australia.

A notable number of species were then described by the famous carabidologist Baron de Chaudoir in a series of papers (1843–1883), including, in some papers, keys to the taxa described. Perhaps most important was his monography on the Australian harpalines (Chaudoir 1878). The next large step forward was made by the Reverend T. Blackburn who between 1890 and 1912 described many species and also added keys to species to some of his descriptions.

Without a doubt, the greatest impact on our knowledge of Australian carabids are the many excellent papers of Thomas G. Sloane, mainly published between 1888 and 1923. He not only described hundreds of species but also prepared a large number of keys to species and genera, many of which are of major importance even now. Sloane also was the first to write regular revisions of genera and tribes, was an early adopter of Darwin's ideas on evolution, and published astute and seminal papers on the biogeography of Australia (Sloane 1915b; Ebach 2012). As a consequence, Sloane's work was and still is the basis for all further work done and to be done on Australian carabids.

Few publications on Australian carabids were produced after Sloane's last paper in the 1920s until P. J. Darlington from 1956 to 1958 spent almost two years in eastern Australia, systematically collecting the rainforests along the east coast and then publishing his results in ~15 revisionary papers. Darlington's papers in particular, greatly increased the knowledge of the rainforest inhabiting Pterostichini. From the 1960s onwards B. P. Moore devoted his attention to Australian Carabidae and, besides descriptions of species, he wrote important revisionary papers on the Psydrini, Pterostichini, and Trechini. He also described the carabid fauna of Norfolk and Lord Howe islands, various cave carabids (Moore 1962), and contributed early and important information on carabid beetle chemical defence (Moore & Wallbank 1968; Moore 1979). Arguably the most important of his contributions is his (1987) catalogue of the Australian Carabidae, which marks the basis for all further work.

Also from 1960s on researchers and affiliates of the prominent Australian museums, including the CSIRO supported Australian National Insect Collection (e.g. E. Britton, A. Calder, J. Lawrence, E. Matthews, R. Storey, M. Thayer, T. Weir and others), collected intensively in many regions, including the remote parts of the Northern Territory, Cape York Peninsula, and parts of the interior that had not been previously surveyed. At the same time G. Monteith of the Queensland Museum systematically collected in eastern Queensland and explored most of the mountains and tablelands of the Great Dividing Range. These 20th Century collectors extensively used sampling techniques that were unknown or little used in the 19th Century, namely light collecting, Berlese extraction, pitfall traps, Malaise traps, bark and canopy fogging, sifting ground litter, etc. Accordingly, many species that are rare or have special habits (e.g. arbouricolous or litter inhabiting species) were sampled for the first time.

While the entomological research greatly ramped up within Australia, some foreign visitors began collecting in selected geographic areas or habitats, e.g. H. Howden, A. Newton, S. Peck, and others, where all prolific collectors. From ~1985 M. Baehr collected in many parts of Australia and published a large number of papers with descriptions of many species and revisions of many genera and tribes. His most comprehensive papers are on the Australian Zuphiini, Pseudomorphini, Scaritini, Bembidiini, Psydrini, Pentagonicini, Odacanthini, and on certain genera of Lebiini. More recently P. M. Giachino revised some genera including the bembidiine Anillina and Tasmanian Trechini, and K. Will devoted his work mainly to the Pterostichine beetles. Due to the efforts of all these workers, a portion of the Australian carabid fauna now is fairly well, if provisionally, explored, but many groups and large parts of the fauna still are waiting for modern revisions that certainly will greatly increase the number of species.

Similarly, the systematic application of a variety of sampling techniques (e.g. light collecting, fogging canopy and bark, pitfall trapping, sifting ground litter, beating foliage, use of Malaise traps, window traps, etc.), even in areas generally collected, but not yet systematically so, can considerably augment both the number of species known from an area and increase our knowledge of their ranges, ecology, phenology and ethology. The increase of named species in the last 30 years from around 1900 species catalogued by Moore *et al.* (1987) to the current more than 3100 species and subspecies (Baehr unpubl.), is due to better exploration of some remote areas and the systematic application of the mentioned sampling techniques by several Australian and foreign collectors, and the follow-up efforts of scientists who have revised genera and tribes using the unidentified specimens housed in collections within and outside of Australia.

Geographical and ecological distribution of the Australian Carabidae. Due to the unequal exploration of the carabid fauna across Australia, our knowledge of the geographical distribution of many groups remains unsatisfactory. Many parts of Australia have not been systematically sampled; in particular the far north and north-west, and much of interior Western Australia, Northern Territory, and South Australia have not yet been properly sampled. A recent species inventory developed during a survey of the Pilbara region in northern Western Australia by the Western Australian Department of Environment and Conservation (Guthrie *et al.* 2010) found numerous undescribed carabid species from across the family and major range extension of some described species, demonstrating how much there is yet to be discovered regarding carabid beetles species distributions, taxonomy, and biology.

Carabidae are found in all areas and habitats in Australia, from tropical rainforest to arid areas, from the beaches to high mountains, and from the canopy of rainforest trees to caves. In comparison to the carabid faunas of other regions, the Australian fauna is highly distinctive due to the special geographical, geological, and botanical conditions within the continent, and because of the distinct biogeographical history of this southern island continent.

As a part of the former Gondwanan landmass, Australia through the Mesozoic was united with other southern continents and thus shares several uniquely austral carabid tribes. There are clear connections with the southern parts of South America, southern oceanic islands, New Zealand, New Caledonia, and New Guinea. A few groups seem to have affinities with southern Africa, South America, and India, some no doubt, as the result of old, vicariant events involving the separation of ancient continental plates and others as the result of more recent dispersal. But the isolation of the Australian landmass through much of the Tertiary and then its more recent contact to the South Asian Insular Belt fundamentally changed the composition of the fauna: some groups declined or perhaps died out, while others flourished and diversified greatly after this contact. Since the late Tertiary many groups have immigrated and presumably are still immigrating from the north. The hypothesised dispersal pattern is thought to generally originate in the Oriental fauna, moving southward via New Guinea and along the Cape York Peninsula. Hence, today Australia is home to a mixture of old indigenous tribes and genera, usually of southern origin, and of recently dispersing groups, most often of more northern origin.

Whereas the paleoendemic carabid fauna is generally composed of either cool and wet adapted, southern species that mainly inhabit temperate forests, or are more dry adapted species found in the sclerophyll forests, savannah woodlands, mallee and mulga scrubs, and more or less arid grasslands, the younger fauna of Oriental origin is mainly composed of warm adapted rainforest species or hygrophilous species that live on the banks of rivers, lagoons and in swamps.

Spencer's 19th Century biogeographic regionalisation of Australia (Spencer 1896) has served as the basis for three (or four) main faunal regions that been the descriptive core in biogeographical literature for the continent (Ebach 2012). These regions – Bassian, Torresian, and Eyrean – certainly can be, and have been, further subdivided, and are typically adjusted to suit the different age and history of various groups of animals. For example Sloane (1915b) adapted and subdivided Spencer's (1896) and Tate's (1889) biogeographic regionalisation in order to focus on the entomo-fauna of Australia. Even though no distinct borders exist between the regions or their faunas and so this is a rather rough division, it is suitable for our descriptive purposes here. In some areas, e.g. across the tropical latitudes, Torresian and Eyrean elements can be sympatric or intermixed where suitable microhabitats co-occur. In the wet forests of the Great Dividing Range of the east coast wide swathes of the Bassian and Torresian faunas are superposed with the lowland rainforest Torresian elements predominate, but on mountain tops and tablelands Bassian elements are numerous and diverse.

The Bassian faunal region, which covers the south-east of Australia, is home to mainly cool-adapted forest and sedgeland carabids. The Torresian faunal region covers the north-east and far north and includes groups inhabiting tropical rainforests, tropical savannahs, and wet or mesic tropical scrub. The Eyrean faunal region covers the arid centre and west of Australia. The carabid fauna of south-western Australia phylogenetically appears to belong to the Bassian fauna, but is commonly treated separately given the large number of endemic taxa.

In addition, the Australian carabid fauna has some special characteristics: the tropical rainforests possess a rich species-level fauna, which is mainly composed of Bassian faunal elements restricted to the mountain tops and tablelands, whereas the Torresian elements are represented by many distinct genera with relatively few species (Baehr 2003e). This probably is the result of the movement northward of the Australian plate during Tertiary that forced the cool-adapted Bassian fauna to retreat to these mountain refugia and of the subsequent contact of Australia with the Oriental Region and the movement of its fauna south into Australia.

Because eucalypt forests and woodlands represent a very important part of the Australian habitats a large number of the Australian carabid species inhabit or frequent eucalypts. These carabids are mainly found under the loose bark on trunks of bark-shedding eucalypts that are common in Australia. In this habitat they occur together with a very rich fauna of other beetles, insects and invertebrates (Baehr 1990d, 2018; Majer *et al.* 2003; Menzel *et al.* 2004) no doubt acting as both predators and prey in this ecosystem.

About two-thirds of Australia is arid and covered with arid bushland and various semiarid and arid grasslands. Carabids that live in these areas face extremely high surface temperatures that are a challenge for beetles to endure. They have several behaviours that allow them to persist under these extreme conditions. Primarily they avoid the brunt of the heat and dry by seeking microclimates in a tolerable range in spaces under bark or in deep bark cracks of trees or scrubs, or by going underground in fissures, interstitial spaces, burrows of other animals or digging. Adults have seasonally and daily periods of surface activity, emerging during wetter periods and at night. These habits help them avoid frequent bushfires in the sclerophyll forests and woodlands as well. As a consequence, about a quarter of the Australian carabid species live under or in the bark of trees, or wholly within the soil, a composition of habitat preferences very different from what is known from other continents. For example in most temperate or subtropical regions outside of Australia the majority of species are epigeal, frequently active in the leaf litter zone, while tropical regions have remarkable tree canopy diversity not seen in Australia. In Australia's open woodlands, bushlands, semiarid grasslands, and semideserts the proportion of trunk-inhabiting and digging species is even higher, and only very few species are true surface active, ground-living beetles, and most of these concentrate around water sources.

It is well known that the montane rainforests of eastern Australia are hotspots of species diversity (Baehr 1995a, 2003d; Harvey 2002; Yeates *et al.* 2002; Yeates & Monteith 2008; Austin *et al.* 2004; Staunton *et al.* 2014) with numerous micro-endemic carabid species occurring on single mountain tops or tablelands. Until recently, however, the extraordinary species diversity in the semiarid and arid areas of the west and the central regions has been overlooked. The high species diversity in the arid outback was first detected in some reptile groups, but similarly, exceptional diversity is also found for many carabid groups. Examples are the carenine scaritines (Moore *et al.* 1987) or the Lebiine subgenus *Nototarus* Chaudoir of the genus *Anomotarus* Chaudoir (Baehr 2009f). In these groups many species with apparently very restricted ranges occur in vast areas that today seem rather uniform with respect to their geology, climate, and vegetation. The best explanation of this high level of endemism is likely found in the climatic history of the late Tertiary and Quaternary. This time period is thought to have had repeated dry and wet cycles that induced repeated fluctuations of vegetation expansion and contraction. A similar example of micro-endemism is found in the many flightless species of the cicindeline subgenus *Rivacindela* van Nidek that occur on salt lakes in the interior of Western Australia and South Australia. Most are usually restricted to a single lake or at most to a group of neighbouring lakes. Overlaid on these patterns of narrow endemics are widespread, eurytopic species such as the strong-flying *Calosoma schayeri* Erichson and *Platycoelus melliei* (Montrouzier), leaf-litter dwelling, disturbance tolerant natives like *Mecyclothorax punctipennis* (Macleay) and a modest number of synanthropic adventive species, e.g. *Laemostenus complanatus* (Dejean).

In order to understand the detailed biogeographical patterns of carabid beetles, however, a more complete knowledge and description of species and a baseline understanding of their distributions is needed. This level of knowledge is achievable but can only be reached through efforts to better explore the unsampled areas of Australia (Baehr 1992c) and then to follow up with further revisionary work on taxa across the family.

Key to the tribes of Australian Carabidae

A few terms in the following keys and text are widely used by carabid specialists (Bell 1967) but may not be known to some users. *Open* or *closed procoxal cavities* (Fig. 12.2–12.3) refer to those which are externally closed by the meeting of the prosternal process with the two postcoxal bars arising from the proepimera and do not refer to an internal closure by the postcoxal bridges. A *disjunct mesocoxal cavity* (Fig. 12.2) is one which is partly closed laterally by the mesepimeron and may be referred to as a laterally open cavity in more general works on Coleoptera, while a *conjunct mesocoxal cavity* (Fig. 12.3) is one which is closed laterally by the meeting of the mesoventrite and metaventrite.

1\. Head with two deep oblique sulci, pronotum with three deep longitudinal sulci; antennomeres pearl-shaped, antenna short (Pl. 24H); elytra either with very coarsely grooved striae, or with three deep sulci ... **Rhysodini**

– Head and pronotum without such sulci; antennomeres when short not pearl-shaped; striae of elytra various, less coarsely grooved finely impressed or absent. At most moderately deeply impressed and sulcate, sulci shallower ... **2**

2(1). Scape of antenna not visible from above; head with deep antennal groove between eye and mandible (Fig. 12.4); legs more or less concealed beneath the body; body either dytiscid- or colydiid-like (Pl. 18A) ... **Pseudomorphini**

– Scape of antenna visible from above; head without deep antennal groove between eye and mandible; legs not concealed beneath the body; body variously shaped, but not distinctly dytiscid- or colydiid-like ... **3**

3(2). Abdomen with seven or eight ventrites (visible sternites); elytra truncate, one tergite visible from above; mandible with setiferous puncture in scrobe (Fig. 12.5); head with one pair of supraorbital setae above each eye; elytra multicarinate and black with yellow spots ... **Brachinini**

– Abdomen with six ventrites (visible sternites); other characters not in combination above ... **4**

4(3). Clypeus wider than distance between the insertions of the antennae (Fig. 12.6); eyes large and laterally protruding; mandibles with several elongate teeth (Fig. 12.6); procoxa protruding ventrally; legs very long and slender; protibia with two terminal spurs ... **Cicindelini**

– Clypeus narrower than distance between the insertions of the antennae; mandibles without or with fewer and less elongate teeth; procoxa not protruding ventrally; without other characters in combination ... **5**

5(4). Protibia with two terminal spurs, positioned at the same level and not in contact with the antennal cleaner groove (isochaete) (Fig. 12.7); mesocoxal cavities disjunct; elytra with posterolateral flange (Fig. 12.9, flange of Coanda) ... **6**

– Protibia with one apical spur and one spur located proximally distant from the apex and usually at or near the base of the antennal cleaner groove (anisochaete) (Fig. 12.8); mesocoxal cavities conjunct or disjunct; elytra without flange of Coanda ... **7**

6(5). Antenna filiform, 11 visible antennomeres; tibiae not flattened; elytra apically rounded covering tergites; antennal cleaner present; labial palps fusiform ... **Ozaenini**

– Antenna flattened (Fig. 12.10), 10 visible antennomeres, pedicel enclosed within the scape; tibiae flattened; elytra truncate, exposing one or two tergites; antennal cleaner absent; labial palps markedly securiform (Fig. 12.16) ... **Paussini**

7(5). Metepimeron not visible between metanepisternum and abdominal ventrite 1 (Fig. 12.2); mesocoxal cavities disjunct (Fig. 12.2) ... **8**

– Metepimeron visible between metanepisternum and abdominal ventrite 1 (Fig. 12.3); mesocoxal cavities conjunct or disjunct (Fig. 12.2–12.3) ... **10**

8(7). Procoxal cavities closed behind (Fig. 12.3); terminal maxillary palpomere fusiform; body length < 15 mm ... **Migadopini**

– Procoxal cavities open behind (Fig. 12.2); terminal maxillary palpomere truncate at apex; body length > 15 mm ... **9**

9(8). Mandibles with several teeth on the inner margin, their upper surface smooth; clypeus asetose ... **Pamborini**

– Mandibles not toothed on the inner margin, their upper surface rugose; clypeus with a seta at either side ... **Carabini**

10(7). Body pubescent and prothorax globular (Pl. 19B), without well defined lateral carina; palps long and slender ... **Apotomini**

– Body pubescent or glabrous except for typical fixed setae and prothorax form and lateral carina various; palps relatively short and robust ... **11**

11(10). Body pedunculate between prothorax and mesothorax (Pl. 20K), scutellum located entirely on the peduncle ... **12**

– Body not pedunculate, or if slightly so, then at least the tip of the scutellum extended beyond the peduncle and inserted between the elytral bases ... **15**

12(11). Mesocoxal cavities disjunct, partly closed by the mesepimeron (Fig. 12.2) .... **Scaritini**
– Mesocoxal cavities conjunct, laterally closed by meeting of meso- and metaventrite (Fig. 12.3) **13**
13(12). Mandible without seta in the scrobe; marginal series of elytral setae closely spaced in basal third and basal-most seta very near humeri **14**
– Mandible usually with a single seta in the scrobe (Fig. 12.5), rarely lacking; marginal series of elytral setae broadly spaced in basal third and basal-most seta very distant from humeri .... **Broscini**
14(14). Antennae filiform .... **Pterostichini (part)**
– Antennae moniliform .... **Cratocerini**
15(11). Mandible with a single large seta in the scrobe (Fig. 12.5) **16**
– Mandible without a single large seta in the scrobe, or scrobe with scattered sparse setae **21**
16(15). Base of elytra with complete margin (Fig. 12.17); marginal series of elytral setae composed of 12 equally spaced setae; terminal palpomere of maxillary palp fusiform (Fig. 12.13) .... **Pogonini**
– Basal elytral margin absent or variously developed; marginal series of elytral setae composed of eight or nine setae that are divided into two groups of four or five setae each; terminal maxillary palpomere conical or subulate (Fig. 12.14) **17**
17(16). Terminal maxillary palpomere very small and subulate (Fig. 12.14); frontal furrows short, not extended behind the eyes **18**
– Terminal maxillary palpomere as long as the penultimate palpomere and conical, not subulate, if small, frontal furrows elongate, extended behind the eyes (Fig. 12.21) **19**
18(17). Either lateroapical margin of protibia excised (Fig. 12.19) **and** elytra with recurrent stria at apex (Fig. 12.20) or lateroapical margin of protibia not excised (Fig. 12.18) **and** elytra without recurrent stria; elytral stria 5 not sulcate at base .... **Bembidiini**
– Lateroapical margin of protibia not excised (Fig. 12.18), elytra with recurrent stria at apex (Fig. 12.20); elytral stria 5 sulcate at base .... **Trechini**
19(17). Frontal furrows on head extending behind the eyes (Fig. 12.21); inner carina of the elytra not visible near apex, epipleura of elytra near apex without plica .... **Trechini**
– Frontal furrows of head not extending behind the eyes; inner carina of the elytra visible near apex, epipleura of elytra near apex with plica (Fig. 12.22) **20**
20(19). Elytra with lateral marginal series arranged as 4 (rarely 5) anterior and 5 (rarely 6) posterior setae, with a break between the series; male protarsus with two basal tarsomeres dilated and dentate on median side (Fig. 12.23) .... **Zolini**
– Elytra with lateral marginal series of 6–8 anterior and 6–7 (rarely 5) posterior setae, either a break between the series, or less commonly, a single seta near midpoint of hiatus; male protarsomere not so modified .... **Psydrini**
21(15). Labrum exceptionally large (Fig. 12.28), elongate, apex rounded or acute, never emarginate; tarsomeres dorsally setose; profemur at base usually with a ventral tubercle (Fig. 12.24) .... **Helluonini**
– Labrum usually not elongate, apex variously formed but typically transverse notably emarginate; tarsomeres dorsally glabrous or setose; profemur at base without a ventral tubercle **22**
22(21). Apex of elytra rounded, not truncate (Pl. 20G) (obliquely sinuate in some Platynini), usually abdomen completely covered by the elytra **23**
– Apex of elytra transversely or obliquely truncate (Pl. 23C,J), exposing the apex of the abdomen **40**
23(22). Head very wide, wider than pronotum and depressed, very narrowly constricted near occiput; tarsomeres very wide, deeply lobed (Fig. 12.30), and densely pilose beneath; body remarkably depressed; colour of elytra yellow with black apex .... **Hexagoniini**
– Head relatively narrower, not or only moderately constricted behind eyes; other characters variable or not in combination as above **24**
24(23). Head with one pair of supraorbital setiferous punctures **25**
– Head with two or more pairs of supraorbital punctures **29**
25(24). Antenna densely pubescent from middle of antennomere 3 .... **Harpalini**
– Antenna densely pubescent from antennomere 4 .... **26**
26(26). Inner carina of the elytra not visible near apex, epipleura without plica; terminal palpomere of labial palp markedly securiform .... **Agonicini**
– Inner carina of the elytra visible near apex, plica present (Fig. 12.22); terminal palpomere of labial palp not markedly securiform (Fig. 12.13) ... **27**
27(26). Elytral stria 8 forming a deep sulcus parallel to the margin, especially notable in the apical third of the elytra (Fig. 12.31) .... **Oodini (part)**
– Elytral stria 8 not forming a deep sulcus **28**
28(27). Elytra not pilose; colour uniformly black, prosternum cuneiform at base (Fig. 12.52) .... **Pterostichini (part)**
– Elytra usually pilose at least at margins; colour commonly green, frequently metallic, some with yellow patterning; prosternum not cuneiform at base .... **Chlaeniini**
29(24). Apical labial palpomere markedly securiform (Fig. 12.15); body coarsely punctate and densely pubescent .... **Panagaeini**
– Apical labial palpomere not markedly securiform, or if securiform then body not coarsely punctate and densely pubescent **30**
30(29). Elytra with sparse pubescence on the depressed marginal area lateral of stria 8; lateral margin of elytra widened towards apex (Fig. 12.32); small beetles (< 6 mm) with glossy surface and very shallowly impressed elytral striae .... **Perigonini**
– Elytra without pubescence lateral of stria 8; lateral margin of elytra not widened towards apex; elytral striae usually distinctly impressed **31**
31(30). Clypeus usually emarginate (Fig. 12.25), often deeply so; a pale membrane visible between clypeus and labrum; mandibles bifurcate at apex or right mandible with distinct tooth on inner margin in apical third .... **Licinini**
– Clypeus not emarginate; without a pale membrane between clypeus and labrum (except in the very large sized NQLD species of *Mecynognathus*, *Paranurus:* Pterostichini and *Hyperion*: Morionini (Fig. 12.26, 12.29); mandibles simple at apex **32**
32(31). Posterior tibiae with inner spur very long, nearly length of tarsomere 1 (Fig. 12.34) .... **Cyclosomini (part)**
– Posterior tibiae with spurs of normal length, about half or less than half the length of tarsomere 1 .... **33**
33(32). Elytral plica absent; parascutellar stria and angular base of stria 1 present (Fig. 12.33A,D,F); mental tooth entire-triangular or absent **34**
– Never with combination of characters as above; elytral plica present in the majority of species (Fig. 12.22), but frequently very small, narrow, or absent; parascutellar stria and/or angular base of stria one present or absent (Fig. 12.33); mental tooth form variable, often bifid **35**
34(33). Tarsal claws not pectinate, smooth or with a basal denticle .... **Platynini**
– Tarsal claws pectinate .... **Sphodrini**
35(33). Protibia fossorial, outer apical angle strongly produced; body form parallel-sided; antennae moniliform .... **Morionini**
– Protibia with outer apical angle not produced; other characters variable **36**

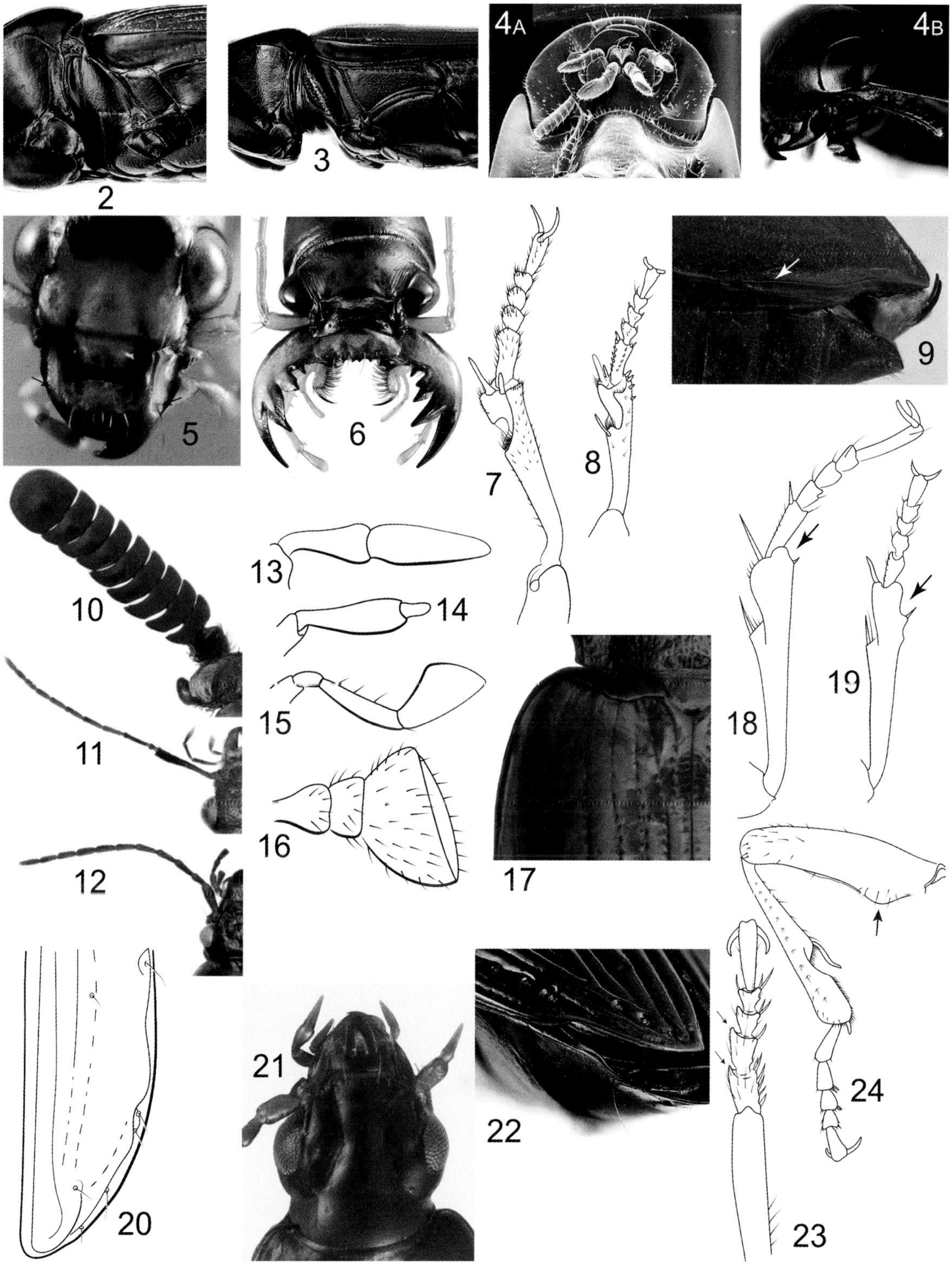

Figs 12.2–12.24. 2, *Pamborus alternans* Latreille, thoracic region; 3, *Platycoelus melliei* (Montrouzier); thoracic region; 4, *Adelotopus dytiscides* Newman, head: A, ventral, B anterodorsal; 5, *Pheropsophus verticalis* (Dejean), head, anterodorsal (arrow = scrobal seta); 6, *Megacephala* sp., head, anterodorsal; 7, *Mystropomus* sp., protibia and tarsus; 8, *Notonomus* sp., protibia and tarsus; 9, *Mystropomus regularis* Banninger, elytral apex, lateral (arrow = flange); 10–12, left antenna: 10, *Arthropterus* sp., 11, *Zuphium australe* Chaudoir; 12, *Pogonoglossus porosus* (Sloane); 13–16, labial palpomeres: 13, *Pogonus* sp., 14, *Polyderis* sp., 15, *Anomotarus* sp., 16, *Arthropterus* sp.; 17, *Pogonus* sp., elytral base; 18–19 protibia and tarsus: 18, *Bembidion* sp. (normal tibial apex), 19, *Paratachys* sp. (excised tibial apex); 20, *Paratachys* sp. elytral apex; 21, *Tasmanorites* sp., head, dorsal; 22. *Notonomus* sp., elytral apex showing plica; 23. *Sloaneana* sp., male protarsus (two basal tarsomeres dilated, dentate); 24, *Helluo* sp., male foreleg (arrow = femoral tubercle).

36(35). Antennae moniliform; elytron basal pore at base of stria 3 or absent (Fig. 12.33E) **37**
– Antennae filiform; elytron basal pore at base of stria 2 or absent (Fig. 12.33A–D,F) **38**
37(36). Elytron without basal pore; anteriolateral prothoracic setae positioned almost at anterior angles .... **Cratocerini**
– Elytron with basal pore at base of stria 3 (Fig. 12.33E); anteriolateral prothoracic setae ~2/5 prothoracic length behind anterior angles .... **Drimostomatini**
38(36). Elytral stria 8 forming a deep sulcus parallel to the margin (Fig. 12.31), especially notable in the apical third of the elytra .... **Oodini (part)**
– Elytral stria 8 not forming a deep sulcus **39**
39(38). Combined presence of all of the following characteristics: angular base of stria one absent (Fig. 12.33C); abdominal sterna smooth, rugulose or lightly punctate, but without deep punctures in transverse row or obvious transverse sulci; and colour black, piceous, rufous or pale brown, never with clear metallic colours .... **Abacetini**
– Without the combination of states above. Presence of **any one or more** of the counter states to the characteristics above: e.g. angular base of stria one present (Fig. 12.33A,B,D,E,F); abdominal sterna with deep punctures in transverse row or obvious transverse sulci (Fig. 12.48); and/or with clear metallic colours on all or part of the body .... **Pterostichini (part)**
40(22). Apical labial palpomere enormously enlarged; eye small; antenna short and moniliform (Fig. 12.83) .... **Zuphiini (Leleupidiina)**
– Apical labial palpomere not greatly enlarged; usually eye large, less commonly absent; antenna long, filiform or flattened **41**
41(40). Antenna densely pubescent from antennomere 1 or 2, scape at least 3x as long as pedicel (Fig. 12.11) **42**
– Antenna densely pubescent from antennomere 3 or 4, scape short, at most 2.5x as long as pedicel **44**
42(41). Prothorax subcylindrical (Pl. 23C), without or with an indistinct lateral carina; tarsomere 4 deeply bilobed .... **Dryptini**
– Prothorax dorsally depressed, with sharp lateral carina; tarsomere 4 not deeply bilobed **43**
43(42). Antennal scape shorter than the total length of following two antennomeres (Fig. 12.12); head sharply and narrowly constricted behind the eyes, with deep transverse sulcus; orbits usually swollen or ridged, with a row of setae; mandible elongate; elytral intervals convex but not costate .... **Physocrotaphini**
– Either antennal scape longer than the total length of following three antennomeres (Fig. 12.11), or elytra with interval 7 distinctly costate, or elytra with ~20 fine costae; neck without deep transverse sulcus; orbits not swollen or ridged .... **Zuphiini (part)**
44(41). Either prothorax subcylindrical, without or with indistinct lateral carinae, or prothorax elongate and head markedly narrowed to neck, or prothorax short and wide and elytra compact, brick red with dark pattern; tarsal claws never denticulate .... **Odacanthini (part)**
– Pronotum with distinct lateral carinae; when prothorax elongate then head not markedly narrowed to neck; tarsal claws typically denticulate 45
45(44). Prothorax subquadrate, slightly narrowed posteriorly; tarsomere 4 deeply bilobed; tarsal claws denticulate; surface uniformly dark yellow to pale brown with distinct golden or greenish lustre .... **Ctenodactylini**
– Without the characters in combination as above **46**
46(45). Posterior tibiae with inner spur very long, nearly length of tarsomere 1 (Fig. 12.34) .... **Cyclosomini (part)**
– Posterior tibiae with spurs of normal length, about half or less than half the length of tarsomere 1 .... **47**
47(46). Mentum and submentum fused, mental suture not evident; head markedly constricted behind very large eyes (Pl. 33C); mandible laminiform, scrobe not evident; tarsal claws not denticulate .... **Pentagonicini**
– Mentum and submentum not fused, mental suture evident; head not markedly constricted behind variously developed eyes; mandible with clearly evident scrobe **48**
48(47). Maxillary galea a single elongate segment; tarsal claws not denticulate; eytral punctures foveate, often encircled by a pale ring .... **Odacanthini**
– Maxillary galea comprising two subequal segments; tarsal claws usually denticulate; elytral punctures never foveate or encircled by a pale ring .... **Lebiini**

Classification of the Australian genera

Tribe Ozaenini Hope, 1838

Characteristics. Antenna typical filiform shape, not depressed or moniliform; procoxal cavities closed; mesocoxal cavities disjunct; a more or less conspicuous outward directed flange (Fig. 12.9) at the latero-apical margin of the elytra; oblongum cell of the metathoracic wing quadrate; both protibial spurs inserted at apex (isochaete) (Fig. 12.7); median lobe of the aedeagus, tubular though variously shaped, ostium typically dorsal or dorso-apical, median lobe base open; parameres slightly to very different in size, shape digitiform to plate-like, more or less densely setose at apex; gonocoxites one segmented, no medial articulation, quite varied in shape and sometime highly reduced, variously setose or asetose; abdomen nebridion type (Deuve 1993).

Notes. Ozaenini commonly are regarded the adelphotaxon of Paussini (and the rare Protopaussini) (e.g. Ball & Bousquet 2001), whereas Vigna Taglianti *et al.* (1998) considered Ozaenini plus Metriini to be the adelphotaxon of Paussini. Compared with Paussini, Ozaenini are plesiomorphic in several respects, e.g. the larvae of Paussini have a highly modified, disc-like abdominal apex (Di Giulio & Moore 2009) and ozaenines lack this modification. Only the single, endemic genus *Mystropomus* Chaudoir is found in Australia and is in its own subtribe, Mystropomina Horn.

Subtribe Mystropomina Horn, 1881

Characteristics. Antenna filiform, densely pilose from antennomere 5; mental tooth large, excised; scrobe of the mandible impilose; mesocoxae separated; latero-apical flange on the elytra inconspicuous; elytra with three raised or carinate intervals; flight wing reduced; protibia at the latero-apical margin oblique; male protarsus simple; aedeagus well sclerotised; parameres of notably different size and shape, the right very narrow, elongate,

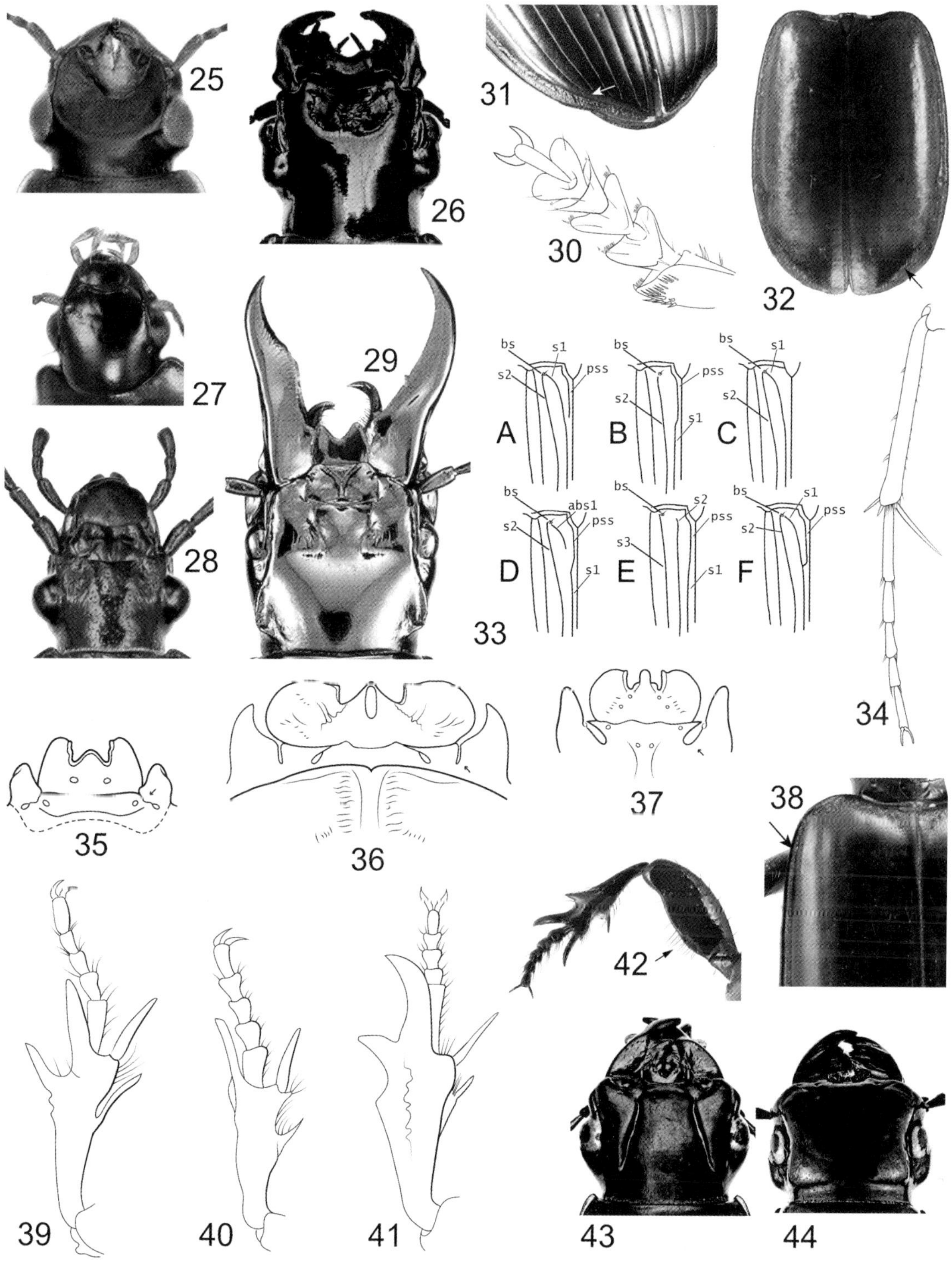

Figs 12.25–12.44. 25–29, dorsal view of heads: 25, *Physolaesthus caviceps* (Andrewes); 26, *Hyperion schroetteri* (Schreibers); 27, *Amblystomus* sp.; 28, *Helluarchus robustus* Sloane; 29, *Mecynognathus damelii* Macleay; 30, *Hexagonia* sp., protarsus, dorsal; 31. *Oodes* sp., elytron, apicolateral (arrow = deep sulcus parallel the margin); 32, *Perigona nigriceps* (Dejean), elytra, dorsal (arrow = preapical sulcus and marginal widening); 33A–F, Elytral striae: nomenclature (abs1 = angular base of stria 1; bs = basal seta; s1 = stria 1; s2 = stria 2; s3 = stria 3; pss = parascutellar stria); 34, *Tetragonoderus undatus* Dejean, metatibia and tarsus; 35–37, mentum and submentum: 35, *Carenum anthracinum* Macleay, 36, *Scaraphites rotundipennis* (Dejean), 37, *Geoscaptus laevissimus* Chaudoir; 38, *Scaraphites* sp. elytron, humeral region (arrow = marginal channel); 39–41, protibia and tarsus, dorsal: 39, *Carenum anthracinum* Macleay, 40, *Monocentrum convexum* (Sloane), 41, *Scaraphites rotundipennis* (Dejean); 42, *Trichocarenum cylindricum* Sloane, male foreleg (arrow = femoral setae); 43–44, head, dorsal: 43, *Conopterum leai* (Sloane), 44, *Monocentrum convexum* (Sloane).

the left broad, plate-like; gonocoxites asetose, gonocoxite very slender and elongate, markedly curved, glabrous.

Notes. This subtribe includes one genus, *Mystropomus* Chaudoir, which is restricted to Australia. Di Giulio & Moore (2009) described the larva of *Mystropomus*.

Mystropomus Chaudoir, 1848 (Pl. 30G; Figs 12.7, 12.9)

Type species. *Mystropomus subcostatus* Chaudoir, 1848, by monotypy.

Australian species. Two species and two subspecies.

Distribution. Eastern Australia from central eastern New South Wales to north-eastern Queensland.

Biology. Ground living in leaf litter and under logs in subtropical to tropical rainforests. Predacious and nocturnal. They are a type of bombardier beetle and can audibly eject quinones and hydrocarbons as a hot fluid (nearly 60°C) from their pygidial glands to defend against predation. One species has been reported from caves (W. Moore & G. Monteith, *in litt.*).

References. Csiki (1927); Bänninger (1927, 1940); Darlington (1961b); Moore *et al.* (1987); Lorenz (2005).

Key to species. Bänninger (1940).

Tribe Paussini Latreille, 1806

Characteristics. Antenna with less than 11 well developed antennomeres, antennomere 2 almost or completely reduced; procoxal cavities closed; mesepimeron disjunct; elytra with distinct latero-apical flange; median cell of the flight wing triangular or absent; procoxa strongly projected, contiguous; both protibial spurs inserted at apex (isochaete); aedeagus with ostium variously oriented, base broadly open; parameres slightly different in size, setose or not at apex; gonocoxites typically long, but quite varied in shape, sometimes reduced, setose or not.

Notes. The larvae are usually physogastric. Both Australian genera belong to the subtribe Cerapterina. They are inquilines and have been found with ants and termites. Some species stridulate in order to communicate with their ant hosts (Di Giulio *et al.* 2015). Paussine beetles are a type of bombardier beetle known to discharge hot defensive chemicals (Moore & Wallbank 1968) similar to true bombardier beetles in the tribe Brachinini. It has been proposed by some authors (e.g. Eisner *et al.* 1977) that this ability suggest a close relationship of these tribes. However, no studies of other character systems support this view.

Subtribe Cerapterina Billberg, 1820

Characteristics. Antenna with 10 well developed antennomeres, antennomere 2 small or very small; prothorax without trichome-bearing grooves.

Key to the genera of Australian Paussini

1. Antennomeres scarcely widened, about twice as long as wide; antennomere 2 small, but still visible at apex of 1; basal third of the pronotum constricted and not margined laterally .... ***Megalopaussus*** **Lea**

– Antennomeres markedly widened, much wider than long; antennomere 2 barely visible and nested within 1; pronotum lateral margins wide or narrow and reaching the base, or if ended anterad of base, then pronotum not notably constricted in the basal third .... ***Arthropterus*** **Macleay**

Arthropterus W. S. Macleay, 1838 (Pl. 19D; Figs 12.10, 12.16)

Type species. *Cerapterus macleaii* Donovan, 1805, by monotypy.

Phymatopterus Westwood, 1838
Telarthropterus Kolbe, 1924
Archarthropterus Kolbe, 1924
Euarthropterus Kolbe, 1924
Pelarthropterus Kolbe, 1924
Sticharthropterus Kolbe, 1924
Panarthropterus Kolbe, 1924

Characteristics. Antennomeres short and very wide; male and female genitalia not reported.

Australian species. 79 species and three subspecies.

Distribution. The whole of mainland Australia; extraterritorial in New Guinea.

Biology. Little has been recorded. Specimens have been sampled from under bark, under logs, in ant nests, but apparently mostly adults are found at lights. Most records are from more or less open forest or woodland. Probably all species are nocturnal and myrmecophilous. Two species were encountered in numbers moving about on the soil in the mallee of western Victoria in the company of ants of the genus *Camponotus* Mayr. Little else is known about habits, diet, and reproduction.

Notes. This genus urgently needs revision as the key in Luna de Carvalho (1987) is almost unusable. Probably many synonyms are concealed among the described species, but several undescribed species may exist as well. The first larva of the genus was described by Di Giulio & Moore (2004).

References. Sloane (1933); Darlington (1950, 1962a); Janssens (1953); Luna de Carvalho (1987); Moore *et al.* (1987); Lorenz (2005); Westwood (1838b).

Keys to species. Sloane (1933); Luna de Carvalho (1987).

Megalopaussus Lea, 1906 (Pl. 29I, K)

Type species. *Megalopaussus amplipennis* Lea, 1906, by monotypy.

Characteristics. Antennomeres longer than wide, small antennomere 2 visible; male and female genitalia not recorded.

Australian species. Only *M. amplipennis* Lea.

Distribution. North-eastern Queensland.

Biology. Little known. This is an extremely rarely sampled species. One of the few known specimens was collected at light. Nothing is known about habits, diet, and reproduction, beyond a specimen that was collected in a *Notostigma* Emery ant nest chamber (Lawrence & Ślipiński 2013).

References. Sloane (1933); Darlington (1950); Luna de Carvalho (1987); Moore *et al.* (1987); Lorenz (2005).

Tribe Cicindelini Latreille, 1802.

Characteristics. Abdomen with six visible sternites, cicindelidian type (Deuve 1993); clypeus wider than distance between the insertions of the antennae; eyes large and laterally protruding; mandibles with several elongate teeth; procoxa protruding ventrally; legs very long and slender; protibia with two terminal spurs (isochaete); aedeagus usually narrow and elongate, with the ostium dorsal; the endophallus usually with several sclerotised spiniform or flagelliform pieces, median lobe open at base, parameres long, narrow and nearly equal in form and size; female gonocoxites variously shaped, typically gonocoxite unisetose at median margin, gonocoxite-2 deeply divided, but of varied shape and length.

Cicindeline larvae are very characteristic, with large, rectangular, bent head and the mandibles directed upwards; they bear dorsal hooks with which they anchor themselves in deep, vertical burrows in the ground. From these burrows they ambush small invertebrates running along surface. Adults are often conspicuous, colourful and day-active. They are typically exceptionally fast runners and most fly readily. All appear to be predacious.

Notes. Tiger beetles are a highly automorphic and distinctive group. As such, they have frequently been treated as a family distinct from Carabidae. Studies of morphological and molecular data support the monophyly of the group and a placement within Carabidae. However, their sister group within Carabidae is unclear (Erwin and Sims 1984; Deuve 1993; Kavanaugh 1986; Liebherr and Will 1998; Maddison *et al.* 1999). For this reason we treat them as a tribe within carabids, consistent with the current evidence.

Key to the subtribes of Australian Cicindelini

1. Elytra fused; tarsomere 4 asymmetric; body ant-like; colour uniformly black ..... **Collyridina Brullé**
– Elytra usually free, rarely fused; tarsomere 4 symmetric; body not ant like; colour not uniformly black 2
2(1). Apical angles of pronotum projecting forward beyond apical margin of prosternum; head wide but eyes comparatively small and little produced; scutellum hidden, not dividing the elytra at base ..... **Megacephalina Latreille**
– Apical angles of pronotum not projecting beyond apical margin of prosternum; head relatively narrow, eyes large and prominently produced; scutellum visible, dividing the elytra at base ..... **Cicindelina Latreille**

Key to the genera of Australian Cicindelini

1. Elytra fused together; 4th tarsomeres asymmetric; body ant-like; colour uniformly black ..... ***Tricondyla* Latreille**
– Elytra usually free, rarely fused; 4th tarsomeres symmetric; body not ant-like; colour not uniformly black 2
2(1). Apical angles of pronotum projecting forward beyond apical margin of prosternum; head wide but eyes comparatively small and little produced; scutellum hidden, not dividing the elytra at base ..... ***Megacephala* Latreille**
– Apical angles of pronotum not projecting beyond apical margin of prosternum; head relatively narrow, eyes large and prominently produced; scutellum visible, dividing the elytra at base 3
3(2). Metanepisternum with a deep horseshoe-shaped groove; elytra with distinct punctate-undulate sculpture ..... ***Distipsidera* Westwood**
– Metanepisternum without a deep horseshoe-shaped groove; elytra usually with less distinct punctate-undulate sculpture 4
4(3). Pronotum with a large dorsal boss; labial palp very elongate ..... ***Rhysopleura* Sloane**
– Pronotum without a large dorsal boss, at most slightly convex in the middle; labial palp much shorter 5
5(4). Clypeus with two erect setae; 3rd palpomere of labial palp short and markedly thickened ..... ***Nickerlea* Horn**
– Clypeus without erect setae; 3rd palpomere of labial palp narrow and elongate ..... ***Cicindela* Linnaeus**

Subtribe Cicindelina Latreille, 1802

Characteristics. Rather narrow and elongate beetles; head narrow, eyes large and prominent; apical angles of pronotum not projecting further forward than apical margin of prosternum; scutellum visible, dividing the elytra at base; elytra usually free; 4th tarsomeres symmetric; gonocoxite-2 of varied size and length, more or less deeply divided.

Cicindela Linnaeus, 1758 (Pl. 21H, 36G)

Type species. *Cicindela campestris* Linnaeus, 1758, by subsequent designation by Latreille 1810.

Characteristics. Elytra typically free, but sometimes flight wings reduced; head narrow but eyes large and prominent; scutellum visible; metanepisternum without a horseshoe-shaped groove; pronotum without a large dorsal boss; labial palp not markedly elongate; clypeus without erect setae, 3rd palpomere of labial palp of normal length; gonocoxite-2 usually elongate, acute at apex, usually the lateral part much stouter and commonly also longer than the median part.

Australian species. 63 species and seven subspecies.

Distribution. The whole of mainland Australia; extraterritorial worldwide.

Biology. Ground dwelling, found on banks of rivers and creeks, shores of freshwater and salt lakes, temporary pools, at the seashore, and in open places in open forests, woodland and semidesert habitats. Predacious adults that have been observed catch small insects such as ants, beetles, and flies, and spiders. Mostly diurnal, agile runners with good eyesight, some species are extremely fast runners (Kamoun & Hogenhout 1996). Many species fly over more or less wide distances when disturbed and commonly land nearby orientated to face the source of the disturbance. Some species are commonly attracted to light. Several halophilous species are flightless. The larvae live in deep vertical burrows in the ground, where they ambush passing small invertebrates.

Note. In contrast to other authors the genus here is understood in a wide sense, and those taxa that are frequently regarded genera are treated as subgenera.

References. Horn (1926); Darlington (1962a); Freitag (1979); Matthews (1980); Sumlin (1981, 1992b, 1992c); Moore *et al.* (1987); Wiesner (1992); Lorenz (2005).

Keys to species. Freitag (1979); Sumlin (1981, 1992b, c).

Key to the subgenera of Australian *Cicindela*

1. Labrum about as long as wide, surface rather convex **2**
– Labrum much wider than long, surface rather depressed **4**
2(1). Antennomeres 5–8 distinctly dilated **or** surface of the elytra markedly dull, velvety .... ***Antennaria* Dokhtouroff**
– Antenna filiform **and** surface of the elytra not markedly dull **3**
3(2). Apical margin of labrum with three teeth .... ***Macfarlandia* Sumlin**
– Apical margin of labrum with seven teeth .... ***Micromentignatha* Sumlin**
4(1). Elytra considerably widened apicad; pale maculae usually confluent from humerus to apex; humeral area of elytra usually with white setae; pronotum setose; apex of elytra not microserrate .... ***Rivacindela* van Nidek**
– Combination of characters not as above **5**
5(4). Elytra glossy, deeply punctate; elytral pattern consisting of maculae along the lateral margin; body asetose except few setae on proepisternum and lateral margin of the metacoxa .... ***Archidela* Rivalier**
– Combination of characters not as above **6**
6(5). Propleura and sterna impilose **7**
– Propleura and sterna pilose **8**
7(6). Apex of elytra microserrate .... ***Euzona* Rivalier**
– Apex of elytra not microserrate .... ***Grandopronotalia* Horn**
8(7). Elytra usually with a sinuate median pale fascia; middle of abdominal sterna impilose (except for large, erect ambulatory setae); head dorso-ventrally compressed .... ***Hypaetha* LeConte**
– Combination of characters not as above **9**
9(8). Apex of labrum with a single tooth .... ***Cylindera* Westwood**
– Apex of labrum with three teeth .... ***Myriochila* Motschulsky**

Subgenus *Myriochila* Motschulsky, 1861

Type species. *Cicindela aegyptiaca* Klug, 1832 (= *Megacephala melancholica* Fabricius, 1798), by original designation.

Characteristics. Labrum wider than long, apex with three teeth; head not dorso-ventrally depressed; elytra rather parallel-sided; pronotum glabrous; pleura and abdominal sterna laterally pilose; elytra parallel-sided; full flight wing.

Australian species. Three species and three subspecies.

Distribution. Almost the whole of mainland Australia; extraterritorial in the Palearctic, Afrotropical, and Oriental regions.

Biology. Hygrophilous, near rivers, lakes, lagoons, temporary water bodies, and salt lakes. The adults are diurnally active, but also are commonly attracted to light. The diet may consist of various small invertebrates as flies, other small beetles, ants, etc. Two of the three species are the most commonly encountered cicindelines in Australia.

References. Horn (1926); Freitag (1979); Sumlin (1981, 1984); Moore *et al.* (1987); Wiesner (1992); Lorenz (2005).

Subgenus *Micromentignatha* Sumlin, 1981

Type species. *Cicindela leai* Sloane, 1905, by original designation.

Characteristics. Labrum as long a wide, apex with seven teeth; antenna filiform; surface of elytra not conspicuously dull; flight wing full.

Australian species. Two species and one subspecies.

Distribution. The northern tropical belt of Australia from north Queensland to northern Western Australia.

Biology. Hygrophilous, found near rivers, lakes, lagoons, and temporary water bodies. The adults are diurnal but sometimes are attracted to light. The diet may consist of various small invertebrates as flies, other small beetles, ants, etc.

References. Sumlin (1984); Moore *et al.* (1987); Wiesner (1992); Lorenz (2005).

Subgenus *Grandopronotalia* Horn, 1936 (Pl. 36G)

Type species. *Prothyma browni* Horn, 1836 (non *Cicindela browni* Sloane, 1913), = *Cicindela carnavona* Freitag, 1979.

Characteristics. A single rather large, stout beetle; pronotum and head glabrous; labrum short and wide, unidentate; mandibles rather elongate, with three subapical teeth; legs of moderate length; fore body green, elytra dark green with three yellow lateral spots; winged.

Australian species. Only *C.* (*G.*) *carnavona* Freitag.

Distribution. Only known from the vicinity of Carnarvon, central western Western Australia.

Biology. The species probably lives on the beach and on near shore river banks. Apparently diurnal.

References. Freitag (1979); Sumlin (1981); Moore *et al.* (1987); Wiesner (1992); Lorenz (2005).

Subgenus *Archidela* Rivalier, 1963

Type species. *Cicindela nigrina* Macleay, 1863, by original designation.

Characteristics. Labrum wider than long; elytra not markedly widened apicad; body largely asetose; elytra glossy, deeply punctate; full flight wing.

Australian species. Two species.

Distribution. The northern tropical belt of Australia from north Queensland to northern Western Australia.

Biology. Hygrophilous, found on the banks of rivers, lakes, and lagoons, also on the sea beach. Diurnal and predacious, but sometimes attracted to light. Little else is known about habits, diet, and reproduction.

References. Freitag (1979); Sumlin (1981, 1984); Moore *et al.* (1987); Wiesner (1992); Lorenz (2005).

Subgenus *Hypaetha* LeConte, 1860

Type species. *Cicindela quadrilineata* Fabricius, 1781, by original designation.

Characteristics. Labrum wider than long; antenna filiform; pleura and sterna pilose, but abdominal sterna impilose medially; elytra more or less widened apicad, apex microserrate; full flight wing.

Australian species. Four species and two subspecies.

Distribution. South-eastern Australia, eastern Australia, tropical northern Australia to central western Western Australia.

Biology. Diurnal and predacious found on sea beaches and near rivers and lakes close to the coast. The adults fly very well and often covering large distances, but are rarely attracted to light. Little else is known about habits, diet, and reproduction.

References. Horn (1926); Freitag (1979); Sumlin (1981, 1984); Moore *et al.* (1987); Wiesner (1992); Lorenz (2005).

Subgenus *Macfarlandia* Sumlin, 1981

Type species. *Cicindela arachnoides* Sumlin, 1981, by original designation.

Characteristics. A very small, extremely long-legged beetle; elytra oval-shaped; head and pronotum glabrous; labrum elongate, incised at the median setae; mandibles very elongate, with a single subapical tooth; fore body green, elytra yellow with a leaf-like dark central stripe; full flight wing.

Australian species. Only *C.* (*M.*) *arachnoides* Sumlin.

Distribution. Central Western Australia north of Northampton. Only recorded from the type locality.

Biology. Apparently a diurnal, riparian species.

References. Sumlin (1984); Moore *et al.* (1987); Wiesner (1992); Lorenz (2005).

Subgenus *Rivacindela* Brouerius van Nidek, 1973

Type species. *Cicindela blackburni* Sloane, 1906, by original designation.

Characteristics. Labrum wider than long; pronotum also setose on disc; lateral parts of thorax and abdomen densely setose; elytra distinctly widened apicad, apex not microserrate; commonly humerus setose; most species with reduced flight wings.

Australian species. 30 species and one subspecies.

Distribution. South-eastern Australia, northern Australia, and large parts of the interior. Most common in central Western Australia and central and northern South Australia.

Biology. Mainly on salt lakes, some species also on the sea shore or riparian habitats and near lakes. Most species in the interior are flightless, but are extremely agile runners on the dry surface of salt lakes (Kamoun & Hogenhout 1996). Little else is known about habits, diet, and reproduction.

References. Freitag (1979); Sumlin (1981, 1984, 1987); Moore *et al.* (1987); Wiesner (1992); Kamoun & Hogenhout (1996); Lorenz (2005).

Subgenus *Euzona* Rivalier, 1963

Type species. *Cicindela tetragramma* Boisduval, 1835, by original designation.

Characteristics. Labrum wider than long; pronotum impilose; pleura and abdominal sterna impilose; elytra not widened apicad, apex microserrate, full flight wing.

Australian species. Eight species.

Distribution. The northern tropical belt of Australia, at and close to the coast, from about Rockhampton in the east to Carnarvon in the west; one species in interior south-western Australia.

Biology. Typically found on the sea shore and on near rivers and lakes; one species on salt lakes. Strong flying, diurnal predators, actively catching flies and other small invertebrates. Little else is known about diet, habits, and reproduction.

References. Freitag (1979); Sumlin (1981, 1984); Moore *et al.* (1987); Wiesner (1992); Lorenz (2005).

Subgenus *Antennaria* Dokhtouroff, 1883

Type species. *Cicindela platycera* Gestro, 1879, by original designation.

Characteristics. Labrum about as long as wide, surface rather convex; commonly 5th–8th antennomeres distinctly dilated; prothorax and abdomen impilose; typically surface of the elytra markedly dull, velvety; narrow, elongate beetles; full flight wing.

Australian species. Four species.

Distribution. The northern tropical belt of Australia from north-east Queensland to the Kimberley Region.

Biology. Riparian, found near rivers, but occasionally near lakes, and lagoons. Diurnal and predacious and sometimes attracted to light. Little is known about habits, diet, and reproduction.

References. Horn (1926); Freitag (1979); Sumlin (1981, 1984); Moore *et al.* (1987); Wiesner (1992); Lorenz (2005).

Subgenus *Cylindera* Westwood, 1831

Type species. *Cicindela germanica* Linnaeus, 1758, by original designation.

Characteristics. Labrum wider than long, apex with a single tooth; head not dorso-ventrally depressed; pronotum glabrous; pleura and abdominal sterna laterally pilose; elytra rather parallel-sided; full flight wing.

Australian species. Only *Cicindela* (*C.*) *discreta* Schaum, 1863.

Distribution. The northern tropical belt of Australia from north-east Queensland to the Kimberley Region; extraterritorial in the whole Papuan, Oriental, Afrotropical, and Palearctic Regions.

Biology. Found near rivers, lakes, lagoons, and temporary pools. Diurnal predator, but commonly attracted to light. Little is known about habits, diet, and reproduction.

Note. *Cicindela discreta* Schaum is one of the few Australian cicindeline species which is widely distributed in the Oriental Region.

References. Horn (1926); Freitag (1979); Sumlin (1981, 1984); Moore *et al.* (1987); Wiesner (1992); Lorenz (2005).

Nickerlea Horn, 1899 (Pl. 31B)

Type species. *Nickerlea distypseroides* Horn, 1899, by monotypy.

Characteristics. Elytra free; head narrow but eyes large and prominent; labial palp not markedly elongate; clypeus with two erect setae, 3rd palpomere of labial palp short and thickened; pronotum without a large dorsal boss; scutellum visible; metanepisternum without a horseshoe-shaped groove; flight wing full; gonocoxite-2 elongate, acute at apex, the lateral part much stouter and also longer than the median part.

Australian species. Five species.

Distribution. Northern and central Western Australia.

Biology. Ground living, predacious species of open forest and semiarid woodland. Species where information about their habits is available seem to be hygrophilous. Little more is known about habits, diet, and reproduction.

References. Horn (1926); Sumlin (1985); Moore *et al.* (1987); Wiesner (1992); McCairns *et al.* (1997); Lorenz (2005).

Keys to species. McCairns *et al.* (1997); Sumlin (1997).

Distipsidera Westwood, 1837 (Pl. 24B)

Type species. *Distipsidera undulata* Westwood, 1837, by monotypy.

Characteristics. Elytra free; head narrow but eyes large and prominent; pronotum without a large dorsal boss; scutellum visible; elytra with distinct punctate-undulate sculpture; metanepisternum with a horseshoe-shaped groove; legs very elongate, flight wing full; gonocoxite-2 rather elongate, usually rather obtuse at apex, the lateral part much stouter and also longer than the median part.

Australian species. 12 species.

Distribution. Eastern Australia from northern New South Wales to the Cape York Peninsula in north Queensland; one species extraterritorial in New Guinea.

Biology. Arbouricolous, found on the trunks of trees in open and closed forests. Diurnal, predacious, have excellent vision and are very agile runners on tree trunks. Adults have been observed to catch flies and ants. Little more is known about habits, diet, and reproduction.

References. Horn (1926); Darlington (1962a); Moore *et al.* (1987); Wiesner (1992); Sumlin (1992a); McCairns *et al.* (1997); Lorenz (2005).

Key to species. McCairns *et al.* (1997).

Rhysopleura Sloane, 1906 (Pl. 37G)

Type species. *Distipsidera orbicollis* Sloane, 1904, by monotypy.

Characteristics. Very small beetles, easily distinguished from all other Australian genera by the humped pronotum; gonocoxite-2 very short, little excised, slightly obtuse at apex, the lateral part bifid at apex, much stouter and also longer than the median part.

Australian species. Only *R. orbicollis* (Sloane).

Distribution. North-eastern Queensland, mainly in the vicinity of Kuranda.

Biology. On trunks of rainforest trees. Very agile runner. Due to its small size and mottled elytral pattern this small beetle is extremely difficult to detect in its environment. Little more is known about habits, diet, and reproduction.

References. Horn (1926); Moore *et al.* (1987); Wiesner (1992); McCairns *et al.* (1997); Lorenz (2005).

Subtribe Collyridina Brullé, 1834

Characteristics. Elytra fused; 4th tarsomeres asymmetric; body ant-like; in the single Australian species the colour is uniformly black; gonocoxite-2 divided, both parts short and obtuse, the median part narrower and longer than the lateral part.

Note. In contrast to Bouchard *et al.* (2011) no separate subtribe is recognised for the genus *Tricondyla*.

Tricondyla Latreille, 1828 (Pl. 103J)

Type species. *Cicindela aptera* Olivier, 1790, by monotypy.

Characteristics. As for subtribe.

Australian species. Only *T. aptera* (Olivier).

Distribution. Northern half of Cape York Peninsula, north Queensland; extraterritorial in the Oriental Region and on New Guinea, New Hebrides, and Solomon Islands.

Biology. On logs and tree trunks in tropical rainforests. Probably a diurnal beetle and visual predator. Little more is recorded about habits, diet, and reproduction. Flight wings reduced.

References. Horn (1926); Darlington (1962a); Moore *et al.* (1987); Wiesner (1992); McCairns *et al.* (1997); Lorenz (2005).

Subtribe Megacephalina Laporte, 1834

Characteristics. 4th tarsomeres symmetrical; elytra usually free; apical angles of pronotum projecting further forward than apical margin of prosternum; head wide but eyes comparatively small and only slightly prominent; scutellum hidden, not dividing the elytra at base; usually rather bulky species, commonly green with or without some yellow spots; gonocoxite-2 elongate, deeply divided, usually the lateral part much more compact and typically longer than the inner part.

Megacephala, Latreille, 1802 (Pl. 2D, 29D; Fig. 12.6)

Type species. *Cicindela megacephala* Olivier, 1790, by subsequent designation by Latreille, 1810.

Tetracha Hope, 1838

Characteristics. As for subtribe.

Australian species. 29 species and two subspecies. Additional undescribed species are believed to occur mainly in Western Australia.

Distribution. Almost the whole of mainland Australia, but more common in the west; extraterritorial in subtropical and tropical parts of the Ethiopian and Neotropical Regions with single species in the southern Nearctic and Palearctic Regions. Two Australian species occur in south-eastern New Guinea.

Biology. Strictly nocturnal. In eastern and tropical northern Australia mainly on the banks of rivers, in the centre and west mainly at the shore of salt lakes, at some localities also in coastal or near coastal saline habitats. Predacious, during daytime usually hidden under cover or in small burrows in sand or mud. Very agile runners. A Victorian species has been observed at night devouring wood-lice (Isopoda) and dipteran larvae. Very large numbers of individuals of some species are commonly observed at night in some locations. Little else is known about habits, diet, and reproduction of the Australian species.

References. Horn (1926); Darlington (1962a); López-López *et al.* (2016); Matthews (1980); Moore *et al.* (1987); Wiesner (1992); McCairns *et al.* (1997); Sumlin (1997); Lorenz (2005).

Keys to species. McCairns *et al.* (1997), Sumlin (1997).

Key to the Australian subgenera of *Megacephala*

1. Elytra with a testaceous margin extending from humerus to apex .... ***Australicapitona* Sumlin**
– Elytra unicolourous or with testaceous apex only .... ***Pseudotetracha* Fleutiaux**

Subgenus *Pseudotetracha* Fleutiaux, 1894

Type species. *Megacephala cylindrica* Macleay, 1863, by monotypy.
Characteristics. As in key.
Australian species. 20 species and one subspecies.
Distribution. Almost the whole of mainland Australia, not yet recorded from Victoria. Most species are known from Western Australia.
Biology. As for the genus.
References. Horn (1926); Moore *et al.* (1987); Wiesner (1992); Lorenz (2005).

Subgenus *Australicapitona* Sumlin, 1992

Type species. *Megacephala australasiae* Hope, 1842, by original designation.
Characteristics. As in key.
Australian species. Nine species and one subspecies.
Distribution. Almost the whole of mainland Australia including northern and western Victoria. Most species are known from Western Australia.
Biology. As for the genus.
References. Lorenz (2005).

Tribe Carabini Latreille, 1802

Characteristics. Large beetles, typically exceeding 12 mm in size; procoxae open; mesocoxa disjunct; no seta in mandibular scrobe; protibial with external (lateral) terminal spur only slightly subapical (anisochaete, but only slightly so), subapical antennal cleaner, no distinct medial antenna cleaning organ; aedeagus tubular, base closed dorsally, parameres nearly the same shape and size, pointed, not setose; flight wing developed or reduced.
Notes. Most Carabini diversity is in the Palaearctic region where the genus *Carabus* (s.l.) has upward of 500 species. *Calosoma*, has close to 190 species and subspecies, of which only two are found in Australia. They are all highly predacious and *Calosoma* has the well earned common name of caterpillar hunters.

Calosoma Weber, 1801 (Pl. 1F, 20A)

Type species. *Carabus sycophanta* Linnaeus, 1758.
Callisoma Agassiz, 1846
Calodrepa Motschulsky, 1866
Callipara Motschulsky, 1866
Syncalosoma Breuning, 1927
Australodrepa Lapouge, 1929
Acalosoma Lafer, 1989
Characteristics. Large beetles, 15–30 mm long; elytra with 16 even intervals and well impressed striae; aedeagus with the ligula of the endophallus not sclerified; metaepisternum smooth.
Australian species. Two species and two subspecies.
Distribution. *Calosoma schayeri* Erichson is widely distributed in Australia, while *C. oceanicum* Perroud & Montrouzier occurs in northern Western Australia, Northern Territory and northern Queensland; extraterritorially in Timor, New Caledonia, and south-eastern Maluku Islands.
Biology. Australian species have a full flight wing and are strong fliers. They may be active during the day or night, but are most commonly night active and are frequently attracted to lights. Occasional incidents have recorded many hundreds to thousands of adults at a light in a single night. Adults are known from a range of more or less disturbed habitats and are fairly eurytopic, but are not typical in more extreme environments, e.g. closed rainforest or highly arid regions. Adults and larvae are predacious and feed on a variety of insects.
References. Moore *et al.* (1987); Lapouge (1929).

Tribe Pamborini Hope, 1838

Characteristics. Large beetles, typically exceeding 25 mm in size; procoxae open; mesocoxa disjunct; no seta in mandibular scrobe; protibial with external (lateral) terminal spur only slightly subapical (anisochaete, but only slightly so), subapical antennal cleaner, no distinct medial antenna cleaning organ; aedeagus base broadly open basally, parameres subequal in length; apex attenuate, glabrous; flight wing absent.
Notes. The only extraterritorial species in the tribe is *Maoripamborus fairburni* Brookes, 1944.

Pamborus Latreille, 1812 (Pl. 1H, 33D–E; Fig. 12.2)

Type species. *Pamborus alternans* Latreille, 1812, by monotypy.
Callimosoma Laporte, 1834
Characteristics. Large size beetles, 23–35 mm; pronotal hind angles notably produced; reduced flight wing; elytra oval or elongate oval; black and frequently with more or less brilliant metallic green or blue; mandibles are broad, and tightly curved at the tip; eyes prominent; head constricted behind the eyes; terminal palps securiform.
Australian species. 16 species and two subspecies.
Distribution. Endemic to coastal, eastern Australia from the mountains and tablelands near the southern end of the Cape York Peninsula to at least south of Sydney.
Biology. Mesic forests, mesic scrub and rainforests. Known to feed on snails and earthworms.
References. Takami & Sota (2006); Sota *et al.* (2005); Moore *et al.* (1987).

Tribe Migadopini Chaudoir, 1861

Characteristics. Procoxal cavities closed; mesocoxal cavities disjunct; protibial antennal cleaner absent; a single supraorbital seta above each eye; no lateral pronotal setae; 10 elytral stria; parascutellar stria impressed for entire length of the elytra; variously shaped male genitalia with large, setose parameres; female gonocoxites also variously shaped.

Notes. Migadopini are subdivided into Migadopina and Amarotypina, according to differences in the male and female genitalia (Deuve 1993; Liebherr & Will 1998; Roig-Juñent 2004; Baehr 2009a; Johns 2010). However, it is controversial how close their relationships are. Liebherr & Will (1998) stressed the differences between both groups and postulated that they should represent distinct tribes. Also Johns (2010) treated them as different tribes, but united them in a subfamily Migadopinae. The relationships and classification remains to be settled and herein we simply treat them as subtribes of Migadopini. For identification of genera and species of Australian Migadopini the male and female genitalia are of great value. Figures of these for all described species were published by Baehr (2009a, 2013a).

Key to the subtribes of Migadopini

1. Ostium of the aedeagus situated on the right side; left paramere setose at apex; gonocoxite-1 (in Australian genera) odd shaped and densely setose .... **Migadopina Chaudoir**
– Ostium of the aedeagus situated on the left side; left paramere asetose at apex; gonocoxite-1 straight and asetose .... **Amarotypina Erwin**

Key to the genera of Australian Migadopini

1. Apical angles of pronotum acute, far protruded, head deeply retracted in the prothorax **2**
– Apical angles of pronotum obtuse, little or not at all protruded, head not retracted in the prothorax **3**
2(1). Large, elongate species, body length > 11 mm; eye very large; pronotum with wide, thick lateral margin; elytra elongate, parallel-sided, striae distinctly crenulate .... ***Dendromigadops*** **Baehr**
– Small, wide beetles, body length < 8.5 mm; eyes small; pronotum with narrow lateral margin; overall form ovate; elytra very broad at base, tapering to apex, striae not or little crenulate .... ***Stichonotus*** **Sloane**
3(1). Small beetles, body length < 5 mm; elytra markedly oval-shaped, striae punctate .... ***Migadopiella*** **Baehr**
– Large beetles, body length > 8 mm; elytra not or far less oval-shaped, striae impunctate **4**
4(3). Surface greenish-violaceous, slightly metallic; mandibles elongate, straight; elytra elongate, parallel-sided .... ***Decogmus*** **Sloane**
– Surface brown or black, not metallic; mandibles shorter, rounded; elytra either relatively short or slightly oviform **5**
5(4). Large beetles, body length > 11 mm; eye depressed; pronotum not cordiform, lateral margin thick; elytra oblong, slightly oval-shaped, striae rather superficially impressed .... ***Calyptogonia*** **Sloane**
– Small beetles, body length c. 8 mm; eye laterally protruded; pronotum cordiform, lateral margin narrow and marginal sulcus wide; elytra short and wide, not oviform, striae deeply impressed .... ***Nebriosoma*** **Laporte**

Subtribe Migadopina Chaudoir, 1861

Characteristics. Highly variable in overall form; aedeagus with ostium invariably on the right side and, in Australian genera, both parameres setose, at least at their apex; gonocoxite-1 usually densely setose, gonocoxite-2 with a single apical seta.

Note. The subtribe also occurs with several genera in South America, on subantarctic islands, and in New Zealand.

Calyptogonia Sloane, 1920 (Pl. 20C)

Type species. *Calyptogonia atra* Sloane, 1920, by original designation.

Characteristics. Large, black; eye depressed; pronotum not cordiform with thick lateral margin; elytra oblong with shallow striae; aedeagus laterally depressed, curved, with large, excavated apex; gonocoxite-1 elongate, densely setose.

Australian species. Two species and one subspecies.

Distribution. Western and central Tasmania.

Biology. Ground living in dense forests, apparently restricted to montane areas. Probably predacious and nocturnal.

References. Csiki (1927); Moore *et al.* (1987); Lorenz (2005); Baehr (2013a).

Key to species. Baehr (2013a).

Decogmus Sloane, 1915 (Pl. 23F)

Type species. *Decogmus chalybaeus* Sloane, 1915, by monotypy.

Characteristics. Rather large, elongate, greenish-violaceous beetles; mandibles straight and elongate; elytra elongate, almost parallel-sided, deeply striate; aedeagus elongate, curved; gonocoxite-1 apically remarkably widened, densely setose.

Australian species. Only *D. chalybaeus* Sloane.

Distribution. Comboyne and Carrai Plateaus, eastern central New South Wales.

Biology. Specimens of this very rare species were collected under bark of a fallen tree and under a log in open forests. Both localities are in montane areas.

References. Csiki (1927); Moore *et al.* (1987); Lorenz (2005); Baehr (2013a).

Nebriosoma Laporte, 1867 (Fig. 12.110)

Type species. *Nebriosoma fallax* Laporte, 1867, by monotypy.
Characteristics. Medium-sized, black beetles; eye laterally protruded, prothorax cordiform; elytra short and wide, parallel-sided, with rather deep striae; aedeagus very curved, endophallus with dentate sclerites; gonocoxite-1 elongate, moderately setose, setae thick; gonocoxite-2 unusually large and wide.
Australian species. Only *N. fallax* Laporte.
Distribution. South-eastern New South Wales. Known only from the type locality, Cooma.
Biology. Very rare species of which only the two syntype specimens are known. Their collecting circumstances are unknown. Probably ground living in dense forests.
References. Lorenz (2005); Baehr (2013a).

Stichonotus Sloane, 1910 (Pl. 38J)

Type species. *Stichonotus leai* Sloane, 1910, by monotypy.
Characteristics. Rather small, oval-shaped species with trapezoidal pronotum; head deeply retracted in the prothorax; eyes moderately large, barely produced; aedeagus very elongate, markedly curved, without sclerotised internal structures; left paramere with one or few setae at apex; gonocoxite-1 narrow and elongate, densely setose.
Australian species. Four species.
Distribution. Tasmania.
Biology. Rather common in disturbed areas in several forest types (S. Grove and L. Forster (Hobart) *in litt.*). Ground living and probably nocturnal. Most specimens have been collected either from litter in wet forests, or in pitfall traps, commonly in rather high altitude.
References. Sloane (1910a, 1915a, 1920); Csiki (1927); Moore *et al.* (1987); Lorenz (2005); Baehr (2013a).
Keys to species. Sloane (1920); Baehr (2013a).

Dendromigadops Baehr, 2013 (Fig. 12.88)

Type species. *Dendromigadops alticola* Baehr, 2013, by original designation.
Characteristics. Large, wide, dark species with acute apical pronotal angles; head deeply retracted in the prothorax; eye very large; elytra elongate, parallel-sided with crenulate striae; aedeagus unknown; gonocoxite-1 widened at apex, densely setose, gonocoxite-2 asymmetrically inserted.
Australian species. Two species.
Distribution. South-eastern and north-eastern Queensland.
Biology. Arbouricolous on trunks of trees in upland tropical and subtropical rainforests. Apparently nocturnal.
References. Baehr (2013a).
Key to species. Baehr (2013a).

Subtribe Amarotypina Erwin, 1985

Characteristics. Usually small beetles; aedeagus with ostium on the left side and left paramere asetose; gonocoxites simple, straight, gonocoxite-1 asetose, gonocoxite-2 with a single apical seta.
Notes. So far this subtribe includes only two Australian species and *Amarotypus edwardsii* Bates, 1872 from New Zealand. However, Johns (2010) stated that in New Zealand several additional undescribed species are known to occur.

Migadopiella Baehr, 2009 (Pl. 29B)

Type species. *Migadopiella convexipennis* Baehr, 2009, by original designation.
Migadopidiella is an incorrect subsequent spelling introduced by Baehr (2013a).
Characteristics. Small, oval-shaped species, with punctate-striate elytra and without discal elytral setae.
Australian species. Two species.
Distribution. So far recorded from the northern parts of the central highland of Tasmania.
Biology. Largely unknown. According to label information the sampling circumstances of *M. octoguttata* Baehr, was pyrethrum fogged from low vegetation at high altitude.
References. Baehr (2009a, 2013a).
Key to species. Baehr (2009a).

Tribe Scaritini Bonelli, 1810

Characteristics. Procoxal cavities closed; mesocoxae disjunct; body pedunculate, with significant constriction of mesothorax and the bases of the elytra; scutellum visible, positioned well forward of elytral base; elytra margin complete, covering the abdomen and usually fixed tight laterally by a flange interlocking with the lateral margin of the abdomen; mandible without a seta in the scrobe; antennal insertions not visible from above; protibia usually fossorial, widened apicad and dentate at its external margin; with well developed antennal cleaner; aedeagus variously shaped, parameres rather similar in size and shape; gonocoxites also quite varied, usually rather setose.
Notes. The infratribal systematic has been much discussed. Liebherr & Will (1998) found no evidence for a close relationship of the non-clivinine Scaritini with Clivinina in their analysis of female genitalia; similarly the Scaritina and Clivinina differ in defensive secretions of the pygidial glands (Moore & Wallbank 1968; Moore 1979). However, a Scaritina plus Clivinina clade was found in analyses of DNA sequence data (Maddison *et al.* 1999). Due to conflicting evidence and an absence of a comprehensive study of the taxa involved, several significantly different classification exist (Lorenz 2005; Bouchard *et al.* 2011; Hogan 2012). Given this uncertainty, we use a conservative arrangement in which typically recognised groups of scaritine taxa are regarded subtribes.

Key to the subtribes of Australian Scaritini

1. Submentum contiguous with gena, not separated by posterior extension of buccal fissure (Fig. 12.35); second outer protibial spine proximal to the level of the apical spur (Fig. 12.39) or only one spine present (Fig. 12.40) .... **Carenina Macleay**
– Submentum separated on each side from gena by posterior extension of buccal fissure, fissure wide or very narrow (Fig. 12.36–37); second outer protibial spine at the level of the anterior spur (Fig. 12.41), two or more spines present **2**
2(1). Marginal channel of elytra granulate (Fig. 12.38) **3**
– Marginal channel of elytra not granulate **4**
3(2). Eyes minute, not visible from above, almost enclosed laterally by the dentiform orbits; body cylindrical; elytra deeply striate. **Scapterina Putzeys**
– Eyes large, visible from above and not enclosed laterally; body rather depressed; elytra at most finely striate .... **Scaritina Bonelli**
4(2). Marginal channel of the elytra with at most 3 subhumeral and two subapical, setose, umbilicate punctures; prothorax globose; in Australian species odd intervals with rows of erect setae .... **Dyschiriina Kolbe**
– Marginal channel of the elytra with an uninterrupted series of many setose, umbilicate punctures; prothorax more or less quadrate; odd intervals without rows of erect setae .... **Clivinina Rafinesque**

Subtribe Carenina Macleay, 1887

Characteristics. Head relatively large; frontal impressions deep and divergent; submentum contiguous with gena, not separated by posterior extension of buccal fissure; second outer protibial tooth proximal to the level of the anterior spur; body form either very elongate often cylindrical or subcylindrical or very wide with broad oval or nearly circular elytra; frequently with brilliant metallic colours; some species with prominent spines at the elytral apex.

Notes. Moore & Lawrence (1994) described the larvae of Carenina. These unique larvae differ from other scaritines in the short, modified antennae and lack of urogomphi, among other characters. The subtribe is in dire need of revision. There are many apparently undescribed species and the generic limits are very questionable and the key is only provisionally presented here.

Key to the genera of Australian Carenina

1. Lateral bead of the elytral humerus not dentate, uniformly rounded beyond the humerus **2**
– Lateral bead of the elytra dentate, tooth either low and rounded or prominent and humerus laterally and anteriorly expanded **4**
2(1). Three or four (rarely only two) setigerous punctures above each eye; prosternum with coarse, scattered, setigerous punctures; profemur with posterior-apical, ventral setae (Fig. 12.42) (rarely lacking) .... ***Trichocarenum* Blackburn**
– Two setigerous punctures above each eye; prosternum glabrous; profemur with or without posterior-apical, ventral setae **3**
3(2). Protibia with apical and subapical lateral spines (Fig. 12.39); head with frontal impressions deeply impressed (Fig. 12.43) .... ***Conopterum* Chaudoir**
– Protibia with apical spine only (Fig. 12.40) or at most a small denticle or low round subapical spine, much less than half the length of the apical spine; head with frontal impressions absent (Fig. 12.44), shallowly impressed or only deeply impressed posteriorly, between the eyes .... ***Monocentrum* Chaudoir**
4(1). Prominent subocular dentiform projection present (Fig. 12.45) .... ***Euryscaphus* Macleay**
– No subocular projection **5**
5(4). Elytra irregularly sculptured or with very large fovea **6**
– Elytra smooth or with very shallowly impressed and punctate striae and flat smooth intervals **7**
6(5). Elytra smooth, or more typically, irregularly sculptured and lacking rows of very large fovea; elytron with a sharp carina at the level of intervals 7–8 .... ***Philoscaphus* Macleay**
– Elytra with rows of very large fovea and no elytral carinae .... ***Laccopterum* Macleay**
7(6) Elytral humeral angles prominently produced .... ***Neoscaphus* Sloane**
– Elytral humeral angles flat, rounded denticle or very slightly produced **8**
8(7). Profemur with posterior-apical, ventral setae (Fig. 12.42); antennae filiform to moniliform .... ***Carenum* Bonelli**
– Profemur without posterior-apical, ventral setae; antennae moniliform **9**
9(8). Labial terminal palpomere very small, subsecuriform, only a little widening to apex .... ***Neocarenum* Laporte**
– Labial terminal palpomere very large and decidedly securiform, much widening to apex .... ***Epilectus* Blackburn**

Carenum Bonelli, 1813 (Pl. 2E, 20K; Figs 12.35, 12.39)

Type species. *Carenum bonellii* Westwood, 1842.
Arnidius Boisduval, 1835
Eutoma Newman, 1838
Carenoscaphus Macleay, 1887
Calliscapterus Macleay, 1887
Platythorax Macleay, 1887
Chariscapterus Sloane, 1888
Paliscaptus Sloane, 1888

Characteristics. Medium or large sized beetles. Elongate, cylindrical, or oval body form, elytra about a quarter, or more, longer than wide; protibial with two or three lateroapical spines; profemur with anterio-apical, ventral setae. Many *Carenum* are vividly metallic, typically green, blue, or violet, along the margins of the pronotum and elytra.

Australian species. 121 species described and many more undescribed.
Distribution. Throughout the whole of Australia, including Tasmania.
Biology. Flightless, fossorial, predacious. Found in a wide array of habitats, typically on sandy soils often open or with scrub or light forests.
References. Phylogeny, Hogan (2012); larvae, Moore & Lawrence (1994); classification, Moore *et al.* (1987).
Key to species. Species or groups variously covered by Sloane (1888, 1894, 1897, 1900b, 1917).

Philoscaphus Macleay, 1871 (Pl. 34H)

Type species. *Carenum tuberculatum* Macleay, 1863, subsequent designation by Bänninger (1940).
Characteristics. Large beetles, 14–30 mm, with the elytra irregularly sculptured with a sharp carina at the level of intervals 7–8; terminal palpomeres triangular or securiform, no subocular projection.
Australian species. Six species.
Distribution. From northern Queensland south to south-eastern New South Wales and west to Western Australia.
Biology. Nocturnally active, fossorial, predatory, flightless; low, open scrub or woodlands with sandy soils.
References. Phylogeny Hogan (2012); classification and ecology Moore *et al.* (1987); defensive chemistry Moore & Wallbank (1968).
Key to species. Bänninger (1940).

Conopterum Chaudoir, 1868 (Pl. 22E; Fig. 12.43)

Type species. *Conopterum insigne* Chaudoir, 1868, by original designation.
Carenidium Chaudoir, 1868
Characteristics. Narrow or elongate oval beetles; some with prominent, apical elytral spines; head with frontal impressions very deeply impressed, protibia with apical and subapical lateral spines; lateral bead of the elytra not dentate, uniformly rounded beyond the humerus.
Australian species. 22 species.
Distribution. The whole of mainland Australia, except for the deep interior.
Biology. Flightless, fossorial, predacious.
References. Classification, Moore *et al.* (1987).

Epilectus Blackburn, 1888 (Pl. 25D)

Type species. *Eurygnathus fortis* Blackburn, 1888, by monotypy.
Eurygnathus Blackburn, 1888
Carenarchus Sloane, 1905
Characteristics. Cylindrical, elongate, black beetles with greatly enlarged labial palpomeres; head with frontal impressions very deeply impressed.
Australian species. Two species.
Distribution. *Epilectus fortis* is known from South Australia and Victoria and *E. mastersi* (Macleay) is known from Victoria and Western Australia.
Biology. Fossorial, flightless, predacious, nocturnally active adults, typically in sandy soils.

Euryscaphus Macleay, 1865 (Pl. 25K; Fig. 12.45)

Type species. *Euryscaphus angulatus* Macleay, 1865 by subsequent designation (Bänninger, 1940).
Characteristics. Large beetles with broadly oval, almost circular elytra; head with frontal impressions very deeply impressed; elytral humeri with widely flattened margins.
Australian species. Seven species.
Distribution. The whole of mainland Australia.
Biology. Flightless, fossorial, predacious, and found in a variety of scrub, shrubland, and desert habitat. Typically on sandy soils.
References. Phylogeny, Hogan (2012); classification, Moore *et al.* (1987).
Key to species. Bänninger (1940). There is no key to all species and several very widespread species are highly variable in form.

Laccopterum Macleay, 1878 (Pl. 27J)

Type species. *Carenum spenceri* Macleay, 1878 (non Westwood, 1841) = *Laccopterum macleayi* Sloane, 1897; by original designation.
Laccoscaphus Sloane, 1905
Characteristics. Medium or large beetles elongate; head with frontal impressions very deeply impressed; elytral with large, deep foveate punctures; typically brightly metallic.
Australian species. 16 species.
Distribution. East and north coastal regions of Australia. North coastal and central Western Australia.
Biology. Flightless, fossorial, predacious, and found in a variety of scrub and open woodland habitats.
References. Classification, Moore *et al.* (1987).
Key to species. Most species are covered in Sloane's (1905b) key.

Monocentrum Chaudoir, 1868 (Pl. 30E; Figs 12.40, 12.44)

Type species. *Monocentrum grandiceps* Chaudoir, 1868, by original designation.
Teratidium Bates, 1874
Characteristics. Relatively narrow, elongate beetles; head with frontal impressions absent, shallowly impressed or

only deeply impressed posteriorly, between the eyes; protibial subapical spine absent or greatly reduced.
Australian species. 11 species.
Distribution. Western and northern mainland Australia.
Biology. Flightless, fossorial, predacious, and found in a variety of scrub and open woodland habitats with sandy soils.
References. Classification, Moore *et al.* (1987).
Key to species. Most species are covered in Sloane's (1905b) key.

Neocarenum Laporte, 1867 (Pl. 30I)

Type species. *Neocarenum singularis* Laporte, 1867:139. Here designated.
Characteristics. Elongate, narrow, parallel-sided body form; antennae moniliform; protibia with two apicolateral spines.
Australian species. 10 species.
Distribution. Primarily found in South Australia and Western Australia. One species, *N. elongatum*, is widespread.
Biology. Flightless, fossorial, predacious in open habitats on sandy soils.
References. Moore *et al.* (1987).
Key to species. Sloane (1897).

Neoscaphus Sloane, 1888 (Pl. 31F)

Type species. *Neoscaphus simplex* Sloane, 1888.
Characteristics. Elongate, narrow, parallel-sided body form; elytral humeral angles broadly and prominently angulate; palps securiform; head with frontal impressions very deeply impressed.
Australian species. Only *Neoscaphus simplexci* Sloane.
Distribution. *Neoscaphus simplex* is known only from Mulwala, NSW.
Biology. Flightless, fossorial, predacious, found on sandy soils.
References. Moore *et al.* (1987).

Trichocarenum Blackburn, 1892 (Pl. 39E; Fig. 12.42)

Type species. *Trichocarenum elderi* Blackburn, 1892, by monotypy.
Characteristics. Elongate, narrow, parallel-sided body form; elytral humeral rounded; head with frontal impressions very deeply impressed; typically three or four setigerous puncture above each eye.
Australian species. Three species.
Distribution. Found in Western Australia.
Biology. Flightless, fossorial, predacious, found on sandy soils.
References. Moore *et al.* (1987).
Key to species. Sloane (1892, 1897, 1905b).

Subtribe Scaritina Bonelli, 1810

Characteristics. Typically very large size beetles with very prominent mandibles; eyes large, visible from above; marginal channel of elytra granulate; submentum separated on each side from gena by posterior extension of buccal fissure; second outer protibial tooth at the level of the anterior spur; flight wing often full.
Notes. A large worldwide group that is questionably monophyletic (Hogan 2012).

Key to the genera of Australian Scaritina

1. Posterior extensions of buccal fissure wide, exposing the base of the maxilla (Fig. 12.37); sides of elytra more or less parallel .... ***Geoscaptus*** **Chaudoir**
– Posterior extension of buccal fissure present but very narrow (Fig. 12.36), not exposing the base of the maxilla; sides of elytra broadly rounded .... ***Scaraphites*** **Westwood**

Geoscaptus Chaudoir, 1855 (Pl. 26D; Fig. 12.37)

Type species. *Geoscaptus laevissimus* Chaudoir, 1855, by monotypy.
Characteristics. Medium or large beetles; parallel-sided body form; typically shiny black with elytral striae shallowly impressed or absent.
Australian species. Five species.
Distribution. Most species found from middle South Australia and Northern Territory eastward. Two species with scattered, mostly coastal records in Western Australia. One species *G. cacus* (Macleay) is also found in New Guinea.
Biology. Full flight wing, fossorial, predacious, found in wet habitats.
References. Classification, Moore *et al.* (1987); phylogeny, Hogan (2012).
Key to species. Sloane (1905b).

Scaraphites Westwood, 1842 (Pl. 37K; Figs 12.36, 12.38, 12.41)

Type species. *Scarites* (*Scaraphites*) *macleaii* Westwood, 1842, by monotypy.
Characteristics. Large beetles with broadly oval elytra; elytral striae absent or very shallowly impressed.

Australian species. Seven species.

Distribution. Throughout the continent, but mainly from western Victoria, Northern Territory and westward. One species, *S. insulanus* Sloane, is also known from Tasmania and Lord Howe Island.

Biology. Flightless, fossorial, predacious, found on sandy soils in a wide variety of open or woodland habitats.

References. Ecology, Guthrie (2001); classification, Moore *et al.* (1987); phylogeny, Hogan (2012).

Note. Hogan's (2012) study including on two species of *Scaraphites* for morphological characters placed the genus in a polytomy with Scaritina, Carenina, and Pasimachina. His molecular study using only the 18S gene and a single species each of *Scaraphites* and *Carenum* always placed these two as sister taxa. Hogan states that the position of terminal protibial spur relative to proximal spur is inconsistent within *Scaraphites*, which suggests the genus may not be monophyletic, but does not provide further description of the states. A more robust study is needed to confirm the position of *Scaraphites* and its placement in Scaritina, following Moore & Lawrence (1994) is provisional and Hogan's (2012) placement of the genus in Carenina may prove correct.

Key to species. Bänninger (1940).

Subtribe Scapterina Putzeys, 1867

Characteristics. Moderately large, rather cylindrical beetles; eyes very small, laterally more or less enclosed by the dentate orbits; elytra fused; flight wings reduced. Male and female genitalia only recorded from the Australian genus *Steganomma* Macleay; aedeagus very short and compact, with several sclerotised or denticulate pieces on the endophallus; gonocoxite-2 elongate, biseriately setose with rather stout setae.

Note. Of the two genera recorded from Australia the occurrence of one (*Parathlibops)* is doubtful.

Key to the genera of Australian Scapterina

1. Head wide, almost as wide as the prothorax; mandibles notably thick at base, scrobe markedly striolate, visible from above; pronotum short and wide, almost quadrate; elytra widened posteriorly, lacking dorsal punctures and setae .... ***Steganomma* Macleay**

– Head narrow, much narrower than the prothorax; mandibles moderately thick at base, scrobe little striolate, not visible from above; pronotum elongate, markedly widened posteriorly; elytra parallel-sided, with four setose dorsal punctures .... ***Parathlibops* Basilewsky**

Steganomma Macleay, 1887 (Pl. 38A)

Type species. *Steganomma porcatum* Macleay, 1887, by monotypy.

Characteristics. Eyes very small; mandibles thick at base and laterally markedly striolate; prothorax short and wide; elytra impunctate, deeply striate.

Australian species. Three species.

Distribution. North-eastern Queensland and central western part of Western Australia; the Western Australian record, however, is doubtful and may actually refer to a locality in Queensland.

Biology. Probably ground living or digging in the soil. Specimens are very rarely collected.

References. Macleay (1887); Sloane (1905b); Csiki (1927); Moore *et al.* (1987); Lorenz (2005); Baehr (2006b).

Key to species. Baehr (2006b).

Parathlibops Basilewsky, 1958 (Fig. 12.116)

Type species. *Scapterus dohrni* Chaudoir, 1868, by original designation.

Characteristics. Eyes very small; mandibles not very thick at base and laterally barely striolate; prothorax elongate, posteriorly widened; elytra quadripunctate, deeply striate. Male and female genitalia not recorded.

Australian species. The single known specimen is doubtfully attributed to *Parathlibops crenatus* (Chaudoir, 1863), described from the Oriental Region.

Distribution. Unknown, the single specimen recorded from Australia is only labelled 'Australia'. The occurrence of the single species of the genus in Australia is doubtful, as Baehr (2008j) discussed.

Biology. Unknown.

References. Basilewsky (1958); Lorenz (2005); Baehr (2008j).

Subtribe Clivinina Rafinesque, 1815

Characteristics. Small to rather large beetles; eyes large, laterally more or less protruded; prothorax quadrangular; marginal channel of elytra not granulate, with numerous, uninterrupted marginal elytral setae. Aedeagus form highly variable, usually with two sclerotised rods and with or without denticulate pieces on the endophallus; gonocoxites quite varied, but usually elongate and with elongate setae.

Notes. The subdivision of this subtribe and its extent has been much discussed and the number of genera that authors include is quite different. For some groups that do not occur in Australia separate subtribes are sometimes used.

Those genera and subgenera of which single species were doubtfully recorded from Australia (Baehr 2008j) are not included in the key below, because they seem to have been mislabelled and actually do not occur in Australia.

Key to the genera of Clivinina recorded from Australia

1. Maxillary palp narrowly securiform **and** antennomere 2 inserted eccentrically on scape; clypeofrontal region remarkably elongated anteriorly; elytra with distinct tooth between lateral margin and the carinate lateral part of basal margin; size small, body length < 4 mm .... ***Syleter* Andrewes**

– Maxillary palp not securiform; antennomere 2 usually not inserted excentrically on scape; clypeofrontal region usually not markedly elongated anteriorly; elytra usually without distinct tooth between lateral margin and basal margin, although the lateral part of the basal margin sometimes is carinate; size usually larger, rarely body length < 4 mm **2**

2(1). Peduncle between pronotum and elytra not sulcate; labrum 6-setose; body large, very elongate, convex; pronotum elongate and with many deep transverse furrows that produce a crenulate lateral margin .... ***Clivinarchus* Sloane**

Peduncle between pronotum and elytra sulcate; labrum 5-setose or 7-setose; body more or less elongate, size and shape various; pronotum without deep transverse furrows and without crenulate lateral margin **3**

3(2). Elytral stria 8 impressed medially of the setiferous pores of stria 9; meso and meta tibiae short; clypeus anteriorly deeply excised; head deeply sulcate; elytra with a single basal setiferous puncture on interval 3; body short and compact, with very coarsely punctate striae, and with wide, oval-shaped pronotum .... ***Rhysocara* Sloane**

– Elytral stria 8 not impressed; at least metatibia elongate; clypeus anteriorly not or less deeply excised; elytra usually with several setiferous punctures on interval 3; without combination of body and pronotum form, punctation and head structures above **4**

4(3). All tibiae wide and depressed; mesotibia with two subapical spines; metatibia externally remarkably crenulate; elytra bearing numerous marginal setiferous punctures (> 60); very large beetles, > 23 mm, with all striae free at base .... ***Platysphyrus* Sloane**

– All tibiae narrow and little depressed; mesotibia with only one spine or without spine; metatibia not crenulate externally; elytra usually (except *Pseudoclivina*) with fewer marginal setiferous punctures (about 30); body size much less than 12 mm, or if larger, then with only three or four inner striae free at base **5**

5(4). Mesotibia and metatibia on inner and outer surfaces densely hirsute; antennomeres 4–10 with a glabrous area on either side; elytra bearing numerous marginal setiferous punctures (> 60) .... ***Pseudoclivina* Kult**

– Mesotibia and metatibia on inner and outer surfaces not hirsute, only with a few spines; antennomeres 4–10 pubescent; elytra with only c. 30 marginal setiferous punctures **6**

6(5). Head and pronotum very densely punctate; head markedly convex in the middle, without clypeal suture; latero-basal border line of pronotum evenly curved, not interrupted; elytral interval 7 at humerus not raised but marginal border incurved and confluent with stria 4; elytral striae remarkably sulcate; no discal elytral setae present .... ***Rubidiclivina* Baehr**

– Head and pronotum less densely or only in parts punctate; head less convex, at least with traces of the clypeal suture; latero-basal border line of pronotum not evenly curved, interrupted or angulate at basal angle, 7th elytral interval at humerus raised, marginal border not incurved; elytral not or less sulcate; four discal elytral setae present .... ***Clivina* Latreille**

Clivina Latreille, 1802 (Pl. 22L)

Type species. *Scarites arenarius* Fabricius, 1792 (= *Tenebrio fossor* Linnaeus, 1758), by subsequent designation by Latreille, 1802.

Ceratoglossa W. J. Macleay, 1863

Scolyptus Putzeys, 1863

Characteristics. Small to very large beetles (in group), up to > 25 mm; usually rather parallel-sided; labrum 5- or 7-setose; meso- and metatibiae not hirsute; antennomeres 4–10 regularly pubescent; elytra with c. 30 marginal setiferous punctures and usually four setiferous punctures near stria 3; aedeagus of highly variable shape, but usually with two sclerotised rods and with denticulate or spinose folds in the endophallus; gonocoxite-2 usually elongate, variously setose.

Australian species. 343 species and 18 subspecies.

Distribution. The whole of Australia, including Tasmania, but more common in the tropical north; extraterritorial range is almost worldwide.

Biology. Most species live at the banks of rivers and creeks, the shore of pools, lakes, and billabongs, near and on salt lakes, in coastal and near coastal habitats, and in swamps. Some northern species, however, apparently live on the ground in more or less closed forests at some distance from water. The species seem to be nocturnal and are predacious, but very little is known about specific habits, diet, and reproduction of the Australian species. According to the very large numbers of specimens that are attracted to lights in northern Australia, the density of populations at certain localities must be very high.

In the Australian *Clivina* fauna the differences in size are unusually large (between 2.5 mm and almost 30 mm). Due to the poor representation of large, as well as of very small hygrophilous scaritine species of other genera in Australia, *Clivina* species apparently occupy the niches of both groups, which on other continents are represented by a multitude of large beetles of the genus *Scarites* Fabricius and its allies, and of the small beetles of the genus *Dyschirius* Bonelli (*sensu lato*).

Notes. The taxonomy of the genus *Clivina sensu lato* is much in discussion. Several authors who mostly have worked, or are working, the faunas of single continents, do not agree about a subdivision into subgenera or even genera. For this chapter and for the Australian fauna, a conservative and practical taxonomic approach is used. Most Australian species are included in the subgenus *Clivina* s. str., which has been subdivided in several species-groups that are believed to constitute monophyletic units (Baehr 2008j, 2015b, 2017a). A general and consistent system of the whole genus must await the revisions of the *Clivina* of other continents. Of this genus in Australia at least additional 50 species are known that await description, but most likely even more may exist and are still uncollected or have not been sorted from collection backlog of material. A general revision of the Australian species is in progress (MB). Those species recognised as undescribed belong to the subgenus *Clivina* s. str.

References. Lorenz (2005); Baehr (2008j, 2015b, 2017a).

Key to the Australian subgenera of *Clivina*

1. Proepisternum with a distinct submarginal ridge .... ***Semiclivina*** **Kult**
– Proepisternum without such submarginal ridge .... ***Clivina*** **Latreille**

Subgenus *Clivina* Latreille, 1802

Type species. *Scarites arenarius* Fabricius, 1792 (= *Tenebrio fossor* Linnaeus, 1758), by subsequent designation by Latreille 1802.

Characteristics. Proepisternum without a submarginal ridge; lateral margins of pronotum extended to the basal sulcus, no transverse prebasal sulcus present; elytra usually with four discal punctures and setae on interval 3.

Australian species. 342 species and 18 subspecies.

Note. The Australian species of this subgenus have been arranged preliminarily in ~25 species-groups (Baehr 2008j, 2015b) which are believed to constitute monophyletic units.

References. Putzeys (1846, 1863, 1866a, 1866b); Sloane (1896a, 1905); Csiki (1927); Andrewes (1929); Kult (1947, 1951, 1959); Darlington (1953, 1962a); Moore *et al.* (1987); Baehr (1989a, (2008j, 2015b, 2017a); Balkenohl (2001, 2003); Lorenz (2005); Ball (2001).

Keys to species. Sloane (1896a, 1905a, 1907a), Baehr (2008j, 2015b, 2017a).

Subgenus *Semiclivina* Kult, 1947

Type species. *Clivina dentipes* Dejean, 1825, by original designation.

Characteristics. Proepisternum with a submarginal ridge; lateral margins of pronotum extended to the basal sulcus, no transverse prebasal sulcus present; elytra with four discal punctures and setae on the interval 3.

Australian species. Only *C.* (*S.*) *doolani* Baehr, 2008.

Distribution. Known from a suburb of Sydney, where it has been collected repeatedly; extraterritorial in America.

Biology. Not recorded, probably hygrophilous.

Note. The single species recorded from Australia may have been accidentally introduced from the western hemisphere and, therefore, could be conspecific with an western hemisphere species. However, because identification of the numerous species of this western hemisphere subgenus presently is impossible without a thorough revision, at present this question cannot be solved and the Australian species has been provisionally described.

References. Lorenz (2005); Baehr (2008j).

Syleter Andrewes, 1941 (Pl. 38H)

Type species. *Ardistomis paradoxa* Putzeys, 1868, by original designation.

Psilus Putzeys, 1877. XLVI (*non Psilus* Panzer, 1806, *non Psilus* Fischer von Waldheim, 1813)

Characteristics. Small beetles, maxillary palp securiform; antennomere 2 inserted excentrically on scape; clypeofrontal region remarkably elongated anteriorly; elytra with distinct tooth between lateral margin and the carinate lateral part of basal margin; aedeagus markedly curved, without sclerotised rods; gonocoxite-2 elongate, with a single ensiform seta.

Australian species. Only *S. papua* Darlington, 1962.

Distribution. Extreme northern part of Cape York Peninsula; extraterritorial in New Guinea and the Oriental Region.

Biology. Largely unrecorded, most probably hygrophilous.

References. Basilewsky (1959); Darlington (1962a); Moore *et al.* (1987); Balkenohl (2001); Lorenz (2005); Baehr (2008j).

Clivinarchus Sloane, 1896 (Fig. 12.82)

Type species. *Clivinarchus perlongus* Sloane, 1896, by monotypy.

Characteristics. Large, very narrow and elongate beetles; labrum 6-setose; mandibles wide, depressed, and markedly striolate; pronotum elongate, parallel-sided, with many coarse, transverse furrows which produce a crenulate lateral margin; elytra with four setiferous punctures on stria 3; striae deep, coarsely punctate; aedeagus with two sclerotised rods; gonocoxite-2 elongate, sparsely setose.

Australian species. Only *C. perlongus* Sloane.

Distribution. Northern tropical Australia from north-eastern Queensland to extreme northern Western Australia.

Biology. Little recorded. A very rare species, of which few specimens have been ever recorded. One specimen apparently was collected at or from a termite mound.

References. Csiki (1927); Moore *et al.* (1987); Lorenz (2005); Baehr (2008j).

Platysphyrus Sloane, 1905 (Pl. 34K)

Type species. *Platysphyrus tibialis* Sloane, 1905, by monotypy.

Characteristics. Very large, convex beetles; labrum 7-setose; all tibiae wide and depressed; metatibia externally remarkably crenulate; elytra with > 60 marginal setiferous punctures; all striae free at base; stria 3 with 5 setiferous punctures; all tibiae wide and depressed; metatibia externally remarkably crenulate; male genitalia unknown; gonocoxite-2 short and compact, slightly crenulate, setose.

Australian species. Only *P. tibialis* Sloane.

Distribution. Roebuck Bay near Broome, north-western Western Australia.

Biology. One of the largest species of Clivinina. A very rare species of which apparently only two specimens have been ever collected. The collecting circumstances are not recorded. Nothing is known about habits, diet, and reproduction.

References. Csiki (1927); Moore *et al.* (1987); Lorenz (2005); Baehr (2008j).

Rhysocara Sloane, 1916 (Pl. 36D)

Type species. *Rhysocara crassa* Sloane, 1916, by monotypy.

Characteristics. Short and compact beetles; clypeus anteriorly deeply excised; labrum 7-setose; head deeply sulcate; pronotum wide, oval-shaped; elytra with 8th stria impressed; with a single setiferous puncture near base of stria 3, striae coarsely punctate; meso- and meta-tibiae short; aedeagus with two sclerotised rods in endophallus, gonocoxite-2 elongate, slightly crenulate, sparsely setose.

Australian species. Only *R. crassa* Sloane.

Distribution. Northern tropical Australia from north Queensland to extreme northern Western Australia. Most commonly collected in the northern parts of Northern Territory.

Biology. Little recorded, because most recorded specimens were collected at light.

References. Csiki (1927); Moore *et al.* (1987); Lorenz (2005); Baehr (2008j).

Pseudoclivina Kult, 1947 (Fig. 12.127)

Type species. *Clivina grandis* Dejean, 1826, by original designation.

Characteristics. Moderately large, rather convex beetles; labrum 7-setose; antennomeres 4–10 with a glabrous area on either side; elytra with five striae free at base, with > 60 marginal setiferous punctures; stria 3 4-punctate; meso- and metatibia on inner and outer surfaces densely hirsute; aedeagus with two setose rods in endophallus; gonocoxite-2 elongate, rather sparsely setose.

Australian species. Only *P. australiana* Baehr, 2008.

Distribution. The single species refers to two old specimens labelled 'Australia' which are different from all known Oriental species of this genus.

Note. The occurrence of the single species in Australia still is uncertain, because the two recorded specimens are old records without a definite locality except 'Australia'.

References. Kult (1959); Balkenohl (2001); Lorenz (2005); Baehr (2008j).

Rubidiclivina Baehr, 2015 (Fig. 12.129)

Type species. *Rubidiclivina punctatissima* Baehr, 2015, by monotypy.

Characteristics. Medium-sized, rather convex beetles; colour uniformly rufous; labrum 7-setose; Head without any trace of the clypeal suture; latero-basal border line of pronotum evenly curved, not interrupted; 7th elytral interval at humerus not raised but incurved, attaining the base of interval 3; elytral margin serrate in basal half; elytral striae sulcate; discal setae absent; marginal elytral setae interrupted in the middle; apical setae of terminal sternum widely distant; pronotum and elytra very densely punctate; male genitalia unknown; gonocoxite-2 elongate, rather sparsely setose.

Australian species. Only *R. punctatissima* Baehr.

Distribution. Northern part of Cape York Peninsula, north Queensland.

Biology. The female holotype was sampled at light.

References. Baehr (2015b).

Subtribe Dyschiriina Kolbe, 1880

Characteristics. Small sized beetles; prothorax globose; elytra with at most three subhumeral and two subapical marginal punctures and setae; disc with variable number of setiferous punctures; aedeagus variously shaped, with a narrow, very elongate, sclerotised flagellum in the endophallus and usually additional sclerotised pieces; parameres elongate, differently shaped, usually with a single elongate apical seta; gonocoxites narrow and elongate, with a very elongate apical seta.

Note. Only the endemic genus *Setodyschirius* occurs in Australia.

Setodyschirius Fedorenko, 1996 (Pl. 37A)

Type species. *Dyschirius zonatus* Putzeys, 1868, by original designation.

Characteristics. Disc of elytra with rows of setiferous punctures on odd intervals; aedeagus variously shaped with more or less distinctly sclerotised folds in the endophallus; female genitalia not recorded.

Australian species. 13 species.

Distribution. Almost the whole of mainland Australia, including Kangaroo Island, but most common in the tropical North. One specimen of an Australian species is labelled 'Sumbawa', but probably was mislabelled.

Biology. Most specimens were collected at light at or near wet environments. swamps, rivers, lakes, and lagoons. Matthews (1980) suggests that one southern species may feed on small Staphylinidae (rove beetles) as certain European *Dyschirius* species are known to do.

References. Lorenz (2005); Bulirsch (2011).

Key to species. Bulirsch (2011).

Tribe Rhysodini Laporte, 1840

Characteristics. Transverse metaventral (metakatepisternal) suture absent; metacoxae widely separated so that large portion of ventrite 1 is visible between them; antennae relatively thick and moniliform; neck narrow and condyliform; mentum elongate and more or less fused to genae concealing remaining mouthparts in ventral view; dorsal surface of head with deep longitudinal groove, often connected to system of grooves; apices of metatibiae in male with outwardly directed tooth (calcar); pronotum often and elytra sometimes with longitudinal grooves, which may be lined with micropapillae (appearing as dull greyish surface).

Key to the Australian subtribes of Rhysodini

1. Mentum separated laterally from gena by a cleft 2
– Mentum completely fused laterally to gena 3
2(1). Shorter setae on subapical antennomere in broad band; pronotum with paramedian grooves narrow, complete; median groove with anterior and posterior pits; pronotum much narrower than elytra; elytral humerus not dentate .... **Leoglymmiina Bell & Bell**
– Shorter setae on subapical antennomere forming subapical ring; pronotum with rounded basal impressions, without paramedian grooves; median groove linear, not expanded at either end, without anterior or posterior median pits; pronotum broad, almost as wide as elytra; elytral humerus dentate .... **Sloanoglymmiina Bell & Bell**
3(2). Shorter setae on subapical antennomeres forming a tuft on ventral side OR shorter setae entirely absent, except on terminal antennomere .... **Clinidiina Bell & Bell**
– Shorter setae on subapical antennomere forming subapical ring (usually only one seta in width) 4
4(3). Median lobe of head elongate, extending to neck constriction, widely separating temporal lobes .... **Rhysodina Laporte**
– Median lobe of head not as elongate, not separating temporal lobes .... **Omoglymmiina Bell & Bell**

Subtribe Leoglymmiina Bell & Bell, 1978

Leoglymmius Bell & Bell, 1978 (Pl. 1E; Fig. 12.103)

Type species. *Rhysodes lignarius* Olliff, 1885, by original designation.

Characteristics. Large, stout rhysodine beetles; labrum with one pair of setae; one or two minute temporal setae, in isolated punctures posteromedial to eye; temporal lobe otherwise impunctate.

Australian species. Only *Leoglymmius lignarius* (Olliff).

Distribution. Central and southern New South Wales, Australian Capital Territory, Victoria and Tasmania.

Reference. Bell & Bell (1991).

Subtribe Sloanoglymmiina Bell & Bell, 1991

Sloanoglymmius Bell & Bell, 1991 (Fig. 12.132)

Type species. *Rhysodes planatus* Lea, 1904, by original designation.

Characteristics. Medium size beetles; antennal segments 5–10 with a single row of minor setae; eyes large; large round temporal lobes with one temporal seta; pronotum is much broader than the head and has rectangular hind angles; the elytron has seven punctate striae; meso- and metatibiae each have two small equal apical spurs. The hind wing is fully developed

Australian species. Only *Sloanoglymmius planatus* (Lea).

Distribution. New South Wales, Australian Capital Territory and Victoria.

Reference. Bell & Bell (1991).

Subtribe Rhysodina Laporte, 1840

Kaveinga Bell & Bell, 1978 (Pl. 31I; Fig. 12.101)

Type species. *Rhysodes abbreviatus* Lea, 1904.

Characteristics. The antennal stylet is absent; antennomeres VI-X with minor setae; the medial margins of the temporal lobes are curved or oblique; they do not closely parallel the margins central lobe of the head; the orbital groove, if present, is terminated at the posterior margin of the eye. In most species, the pronotum has complete paramedian grooves. In two species, these are represented anteriorly by very coarse puncture rows which fail to reach the anterior margin. The prosternum has precoxal carinae. The humeral tubercle is well developed. The middle and hind tibiae each have only one spur.

Distribution. New South Wales, Australian Capital Territory and Victoria.

Reference. Bell & Bell (1991).

Key to the Australian subgenera of *Kaveinga*

1. Paramedian grooves linear, slightly sinuate; pollinosity of paramedian grooves largely or entirely limited to punctures......... ***Angekiva* Bell & Bell**
– Paramedian grooves broad, deep, entirely pollinose, not visibly punctate .... ***Kaveinga* Bell & Bell**

Subgenus *Angekiva* Bell & Bell, 1979 (Fig. 12.101)

Type species. *Rhysodes frontalis* Grouvelle, 1903.
Australian species. Three species.
Distribution. Eastern Australia from northern Queensland to Tasmania.
Reference. Bell & Bell (1991).

Subgenus *Kaveinga* Bell & Bell, 1979 (Pl. 31I)

Type species. *Rhysodes abbreviatus* Lea, 1904.
Australian species. Only *Kaveinga* (*K.*) *abbreviata* (Lea).
Distribution. Northern Queensland; extraterritorial in New Guinea, the Solomon Is. and Bismark Archipelago west to Sulawesi and Mindanao.
Reference. Bell & Bell (1991).

Subtribe Clinidiina Bell & Bell, 1978

Rhyzodiastes Fairmaire, 1895 (Fig. 12.128, Pl. 24H)

Type species. *Rhyzodiastes parumcostatus* Fairmaire, 1895 (Brazil).
Characteristics. The elytra have fewer than six striae; the paramedian grooves are complete; eyes reduced; flight wings are reduced.
Australian species. Three species.
Distribution. Widespread: South America, South-east Asia, New Guinea, Australia, New Zealand, Fiji and intervening islands.
Reference. Bell & Bell (1991).

Key to the Australian subgenera of *Rhyzodiastes*

1. Paramedian grooves closer together at midline than at base or apex; outer carina much wider at middle than at either end; tufts of short setae on antennomeres 6–10 or 7–10.. ***Rhyzoarca* Bell & Bell**
– Paramedian grooves closer together at ends; outer carina not widened at middle; tufts of short setae on segments 5–10 (in Australian species)........ ....***Temanoa* Bell & Bell**

Subgenus *Rhyzoarca* Bell & Bell, 1985

Type species. *Rhysodes montrouzieri* Chevrolat, 1875 (New Caledonia).
Australian species. *Rhyzodiastes* (*R.*) *burnsi* (Oke) and *R.* (*R.*) *ovicollis* Bell & Bell.
Distribution. Southern Queensland, New South Wales and Victoria; extraterritorial in New Zealand and New Caledonia.
Reference. Bell & Bell (1991).

Subgenus *Temanoa* Bell & Bell, 1985 (Fig. 12.128; Pl. 24H)

Type species. *Clinidium spissicorne* Fairmaire, 1895 (Malaysia).
Australian species. Only *Rhyzodiastes* (*Temanoa*) *mirabilis* (Lea, 1904).
Distribution. Northern Queensland; extraterritorial in Caroline Is. and Solomon Is. west to Taiwan, Andaman Is. and eastern India.
Reference. Bell & Bell (1991).

Subtribe Omoglymmiina Bell & Bell, 1978

Omoglymmius Ganglbauer, 1892 (Fig. 12.112)

Type species. *Rhysodes germari* Ganglbauer, 1892.
Characteristics. The central lobe of the head is short; the temporal lobes usually have distinct medial angles and come closest to one another posterior to the apex of the median lobe; frontal grooves are deep and complete; minor setae form rings on the outer segments of the antenna.
Australian species. Five species.
Distribution. Almost cosmopolitan, but absent from Madagascar and New Zealand and poorly represented in Australia, Africa and South America.
Reference. Bell & Bell (1978, 1982, 1991).

Key to the Australian subgenera of *Omoglymmius*

1. Antennomere 11 with apical stylet; eyes small, diameter less than 0.3 times depth of head; medial edge of subapical elytral tubercle aligned with stria 2 .... ***Caeconavitia* Bell & Bell**
– Antennomere 11 without apical stylet; eyes large, diameter more than 0.5 times depth of head; medial edge of subapical tubercle aligned with stria 4 .... ***Omoglymmius* Ganglbauer**

Subgenus *Caeconavitia* Bell & Bell, 1982

Type species. *Omoglymmius* (*Nitiglymmius*) *zimmermani* Bell & Bell, 1978.
Australian species. Only *O.* (*C.*) *okei* Bell & Bell, 1991.
Distribution. Northern Queensland; extraterritorial in Fiji.
Reference. Bell & Bell (1978, 1991).

Subgenus *Omoglymmius* Ganglbauer, 1892

Type species. *Rhysodes germari* Ganglbauer, 1891 by monotypy.
Australian species. Three species.
Distribution. East coastal region of Queensland, extraterritorially diverse in New Guinea, east to the Caroline Islands, and north-west into Eurasia.
Reference. Bell & Bell (1978, 1982, 1991).

Tribe Apotomini LeConte, 1853

Characteristics. Metepimeron visible between metanepisternum and ventrite 1; procoxal cavities closed; mesocoxal cavities conjunct; body pedunculate between prothorax and elytra, the scutellum located entirely on the peduncle; body pilose; mentum edentate; mandible with a seta in the lateral scrobe; pronotum without distinct lateral margin; aedeagus curved, with the ostium on the dorsal surface; both parameres rather similarly sized and shaped, polysetose; gonocoxite-1 asetose, gonocoxite-2 rather elongate, curved, with two short subapical setae.

Apotomus Illiger, 1807 (Pl. 19B)

Type species. *Scarites rufus* Rossi, 1790, by monotypy.
Australian species. Two species.
Distribution. South-eastern, eastern, and northern Australia, including northern Western Australia; recently the genus was recorded from Tasmania; extraterritorial in the Oriental, Afrotropical, and Palearctic Regions.
Biology. Ground living and hygrophilous at the banks of rivers, lakes, and lagoons, also in swamps and wet meadows. Probably nocturnal, because specimens are commonly attracted to light.
References. Laporte (1867, 1868); Macleay (1871); Csiki (1928); Darlington (1962a); Matthews (1980); Moore *et al.* (1987); Baehr (1990a); Lorenz (2005).
Key to species. Baehr (1990a).

Tribe Broscini Hope, 1838

Characteristics. Metepimeron visible between metanepisternum and ventrite 1; procoxal cavities closed; mesocoxal cavities conjunct; body pedunculate between prothorax and elytra, the scutellum located entirely on the peduncle; body glabrous; pronotum with well defined lateral margin; mandible commonly, but not always, with a seta in the lateral scrobe. Aedeagus variously shaped, ostium on the dorsal surface; both parameres of rather similar size and shape, usually polysetose; gonocoxites short and stout, gonocoxite-2 obtusely triangular with or without a few subapical setae.
Notes. In the classification of genera we largely follow the catalogue of broscine genera of Häckel *et al.* (2010) which for certain genera differs from the arrangement in the catalogue of Moore *et al.* (1987). However, some synonymies created by the reconfiguring of genera need revaluation and careful revision. In general, the Australian Broscini urgently need a thorough revision as identification of species in most genera is extremely difficult or impossible.
A cladistic analysis by Roig-Juñent (1998, 2000) examining the relationships of the tribe and among the included genera found it to be monophyletic and most closely related to Apotomini and Melaenini. Broscini are subdivided in several subtribes (Roig-Juñent 2000), of which two occur in Australia. The larvae of two Australian genera (*Eurylychnus* Bates and *Promecoderus* Dejean) were described by Moore (1964b).

Key to the subtribes of Broscini recorded from Australia

1. Mentum with bifid tooth; mentum with two circular foveae .... **Nothobroscina Roig-Juñent**
– Mentum with unidentate tooth, or edentate; mentum without circular foveae .... **Creobiina Jeannel**

Key to the Australian genera of Broscini

1. Suborbital ridge distinct and complete **2**
– Suborbital ridge indistinct or incomplete **6**
2(1). Prosternum strongly projected backwards behind the procoxae .... ***Brithysternum* Macleay**
– Prosternum not projected backwards behind the procoxae **3**
3(2). Protibia markedly dentate at the latero-apical margin; elytra usually with very large foveae .... ***Gnathoxys* Westwood**
– Protibia not strongly dentate at the latero-apical margin; elytra without large foveae **4**
4(3). Mentum edentate **5**
– Mentum with a distinct median tooth **6**

5(4).	Male with securiform terminal palpomeres; mesotarsus in male lacking squamose setae beneath; usually dorsally convex	***Adotela*** **Laporte**
–	Male with cylindrical terminal palpomeres; two or three basal tarsomeres of mesotarsus in male with squamose setae beneath; usually dorsally depressed beetles	***Cerotalis*** **Laporte**
6(1,4).	Elytra with four to six marginal punctures	**7**
–	Elytra with a large number of marginal punctures	**9**
7(6).	Male protarsus with squamose setae beneath	**8**
–	Male protarsus without squamose setae beneath	***Anheterus*** **Putzeys**
8(7).	Male mesotarsus with squamose setae beneath	***Promecoderus*** **Dejean**
–	Male mesotarsus without squamose setae beneath	***Acallistus*** **Sharp**
9(6).	Head with a distinct transverse impression behind eyes	**10**
–	Head without a distinct transverse impression behind eyes	**11**
10(9).	Mandible with a seta in the scrobe	***Eurylychnus*** **Bates**
–	Mandible without a seta in the scrobe	***Chylnus*** **Sloane**
11(9).	Head with a single supraorbital seta above each eye; antenna moniliform	***Percolestus*** **Sloane**
–	Head with several supraorbital setae above each eye; antenna longer, filiform	***Percosoma*** **Schaum**

Subtribe Creobiina Jeannel, 1941

Characteristics. Medium-sized to very large beetles; head with one supraorbital seta above each eye; mentum unidentate or edentate, without circular foveae; glossa bisetose or quadrisetose; male protarsus with squamose setae beneath; aedeagus with elongate ostium; left paramere usually not setose; gonocoxites usually asetose.

Promecoderus Dejean, 1829 (Pl. 35I)

Type species. *Promecoderus brunnicornis* Dejean, 1829, by monotypy.

Characteristics. Medium-sized beetles; mentum with a distinct tooth; glossa quadrisetose; a distinct suborbital ridge present; head with a single supraorbital seta above each eye; with a distinct suborbital ridge; mandible with a seta in the scrobe; elytra usually with only four marginal punctures; male protarsus and mesotarsus with squamose setae beneath; aedeagus without a large, sclerotised tooth on the endophallus; left paramere asetose.

Australian species. 40 species.

Distribution. The whole of Australia including Tasmania.

Biology. Ground living in a variety of habitats, from forest to grassland, also at high altitude above tree-line, not in arid environments. Nocturnal and predacious.

Notes. Most species of this genus presently are almost undeterminable. The old key (Sloane 1890) includes only species groups, the more recent key (Sloane 1920) refers only to Tasmanian species and, moreover, includes the species that presently are placed in the genus *Acallistus* Sharp.

Not all species included in *Promecoderus* have the marginal series of the elytra consisting of only four punctures. The Tasmanian *P. viridiaeneus* Sloane has up to 8 punctures. The genus, therefore, not only needs a taxonomic revision, but also a critical examination of the infra-generic grouping. Probably the genus is not monophyletic and should be further dismembered. Roig-Juñent (2000) figured some external characters and the male and female genitalia.

References. Laporte (1868); Putzeys (1868); Sloane (1890, 1898, 1920); Csiki (1928); Matthews (1980); Moore *et al.* (1987); Roig-Juñent (2000); Lorenz (2005); Häckel *et al.* (2010).

Keys to species. Sloane (1890, 1920).

Acallistus Sharp, 1886 (Pl. 23H; Fig. 12.70)

Type species. *Acallistus simplex* Sharp, 1886, by monotypy.

Characteristics. Medium-sized beetles; mentum with a distinct tooth; glossa quadrisetose; head with a single supraorbital seta above each eye; in most species without a distinct suborbital ridge; mandible with a seta in the scrobe; elytra with only four marginal punctures; only the male protarsus with squamose setae beneath; aedeagus without a large, sclerotised tooth on the endophallus; left paramere asetose.

Australian species. Four species.

Distribution. Tasmania.

Biology. Ground living species that occur in a variety of habitats. Nocturnal and predacious.

Notes. The evidence for this genus is doubtful. In the paper by Moore *et al.* (1987) it is arranged as synonymous with *Promecoderus*, and indeed, the differences between the genera do not appear significant. Roig-Juñent (2000) figured some external characters and the male and female genitalia.

References. Csiki (1928); Britton (1949); Moore *et al.* (1987); Roig-Juñent (2000); Lorenz (2005); Häckel *et al.* (2010).

Key to species. Sloane (1920), as a part of *Promecoderus*.

Anheterus Putzeys, 1868 (Pl. 19A)

Type species. *Promecoderus gracilis* Germar, 1848, by monotypy.

Characteristics. Rather small beetles; mentum with a distinct tooth; glossa quadrisetose; head with a single supraorbital seta above each eye; in most species without a distinct suborbital ridge; mandible with a seta in the scrobe; elytra with only four marginal punctures; in the male neither the protarsus nor the mesotarsus is with squamose setae beneath; aedeagus without a large, sclerotised tooth on the endophallus; left paramere asetose.

Australian species. Three species.

Distribution. Widely distributed in southern and inland mainland Australia.

Biology. Ground living in more or less open country. Nocturnal and predacious.

Notes. The justification of this genus is doubtful; previously it was included in the genus *Promecoderus*. Roig-Juñent (2000) figured some external characters and the male and female genitalia.

References. Csiki (1928); Moore *et al.* (1987); Roig-Juñent (2000); Lorenz (2005); Häckel *et al.* (2010).

Key to species. None.

Adotela Laporte, 1867 (Pl. 18E)

Type species. *Adotela concolour* Laporte, 1867, subsequent designation (Roig-Juñent 2000).

Parroa Laporte, 1867

Characteristics. Medium-sized to large beetles; mentum edentate; glossa bisetose; head with a single supraorbital seta above each eye; with a distinct suborbital ridge; mandible with a seta in the scrobe; male palps securiform; elytra with more than four marginal punctures; only the male protarsus with squamose setae beneath; aedeagus with a large, sclerotised tooth on the endophallus; left paramere asetose.

Australian species. 17 species.

Distribution. The whole southern half of mainland Australia, including the interior.

Biology. Ground living in a variety of habitats, but most species in rather dry environments. Nocturnal and predacious.

Notes. The justification of the synonymy of *Parroa* is doubtful and a thorough revision of the genus is needed to clarify the status. In general this genus (or both genera) urgently need(s) a revision, because identification of species is nearly impossible. Moreover some species formerly placed in *Parroa* possess a very small mental tooth, whereas in included species originally described as *Cerotalis* virtually no sign of the tooth is visible. Roig-Juñent (2000) figured some external characters and the male and female genitalia.

References. Laporte (1868); Putzeys (1868, 1873); Sloane (1890); Csiki (1928); Moore *et al.* (1987); Roig-Juñent (2000); Häckel *et al.* (2010).

Keys to species. Sloane (1890, 1893), but the first key is incomplete and the latter key only works for some species that were included in *Parroa*.

Cerotalis Laporte, 1867 (Pl. 21E)

Type species. *Cerotalis substriata* Laporte, 1867, by subsequent designation by Roig-Juñent (2000).

Characteristics. Medium-sized to moderately large beetles; mentum edentate; glossa bisetose; head with a single supraorbital seta above each eye; with a distinct suborbital ridge; mandible with a seta in the scrobe; male palps cylindrical; elytra with more than four marginal punctures; male protarsus and mesotarsus with squamose setae beneath; aedeagus with a large, sclerotised tooth on the endophallus; left paramere at least in some species sparsely setose.

Australian species. Seven species.

Distribution. Southern half of mainland Australia, excluding the wetter, eastern parts.

Biology. Ground living in rather dry to notably xeric environments. Nocturnal and predacious.

Notes. The genus urgently needs a revision, as identification of species is almost impossible. Roig-Juñent (2000) figured some external characters and the male and female genitalia.

References. Laporte (1868); Putzeys (1868); Sloane (1890); Csiki (1928); Matthews (1980); Moore *et al.* (1987); Roig-Juñent (2000); Häckel *et al.* (2010).

Brithysternum Macleay, 1873 (Pl. 20E)

Type species. *Brithysternum calcaratum* Macleay, 1873, by monotypy.

Characteristics. Very large beetles; mentum edentate; head with a single supraorbital seta above each eye; glossa bisetose; with a distinct suborbital ridge; mandible with a seta in the scrobe; prosternal process remarkably elongate, surpassing the procoxae; elytra with more than four marginal punctures; neither the male protarsus nor the mesotarsus with squamose setae beneath; aedeagus with a large, sclerotised tooth on the endophallus; left paramere asetose.

Australian species. Three species.

Distribution. Across interior Queensland, central and northern parts of Northern Territory, and north-western Western Australia.

Biology. Ground living in rather dry to notably xeric environments. Nocturnal and predacious.

Note. Roig-Juñent (2000) figured some external characters and the male and female genitalia.

References. Sloane (1910a); Csiki (1928); Moore *et al.* (1987); Roig-Juñent (2000); Lorenz (2005); Häckel *et al.* (2010).

Key to species. Sloane (1910a).

Gnathoxys Westwood, 1842 (Pl. 26L)

Type species. *Gnathoxys granularis* Westwood, 1842, by subsequent designation (Roig-Juñent 2000).

Characteristics. Variously sized beetles; mentum edentate; glossa bisetose; head with a single supraorbital seta above each eye; with a distinct suborbital ridge; mandible with a seta in the scrobe; elytra short and dorsally convex, either ocellate or with very rugose surface; with more than four marginal punctures; neither male protarsus nor mesotarsus with squamose setae beneath; aedeagus without a large, sclerotised tooth on the endophallus; left paramere asetose.

Australian species. 17 species.

Distribution. Across the majority of Australia, except the east and north-east.

Biology. Ground living in rather dry to notably xeric environments, from open sclerophyll forests to woodlands, mallee and mulga, to semiarid grassland. Some species have been observed digging rather deep holes in the ground where they stay during daytime. Nocturnal and predacious.

Note. Roig-Juñent (2000) figured some external characters and the male and female genitalia.

References. Putzeys (1868); Sloane (1898); Csiki (1928); Matthews (1980); Moore *et al.* (1987); Roig-Juñent (2000); Lorenz (2005); Häckel *et al.* (2010).

Subtribe Nothobroscina Roig-Juñent, 2000

Characteristics. Medium-sized to very large beetles; mentum with bifid tooth, with two circular foveae; glossa bisetose; head with one or several supraorbital setae above each eye; male protarsus without squamose setae beneath; aedeagus with short ostium; left paramere setose or not; gonocoxites usually sparsely setose.

Eurylychnus Bates, 1891 (Pl. 2G, 25A)

Type species. *Eurylychnus olliffi* Bates, 1891, by monotypy.

Characteristics. Medium-sized beetles; mentum dentate; head with a single supraorbital seta above each eye; without a suborbital ridge; with a deep transverse impression behind the eyes; mandible with a seta in the scrobe; labial palps slightly widened; elytra with more than four marginal punctures; neither male protarsus nor mesotarsus with squamose setae beneath, but in both sexes the tarsomeres of the protarsus laterally produced; left paramere asetose.

Australian species. Eight species.

Distribution. Eastern Victoria, New South Wales, and Tasmania.

Biology. Ground living in a variety of habitats from rainforest to open woodland, also at high altitude. Nocturnal and predacious.

Notes. Identification of species is difficult, because the key by Sloane (1892) covers only a few species and for the additional species only descriptions are available. Roig-Juñent (2000) figured some external characters and the male and female genitalia.

References. Sloane (1892, 1915a, 1916a); Csiki (1928); Moore (1960b); Moore *et al.* (1987); Roig-Juñent (2000); Lorenz (2005); Häckel *et al.* (2010).

Keys to species. Sloane (1892, 1915a).

Chylnus Sloane, 1920 (Pl. 21L)

Type species. *Lychnus ater* Putzeys, 1868, by monotypy.

Lychnus Putzeys, 1868 (*non Lychnus* Matheson, 1832)

Characteristics. Medium-sized beetles; mentum dentate; head with a single supraorbital seta above each eye; without a suborbital ridge; with a deep transverse impression behind the eyes; mandible without a seta in the scrobe; labial palps slightly widened; elytra with more than four marginal punctures; neither male protarsus nor mesotarsus with squamose setae beneath; in both sexes the tarsomeres of the protarsus laterally but little produced; left paramere asetose.

Australian species. Three species.

Distribution. Eastern Victoria and Tasmania.

Biology. Ground living in a variety of forest types. Nocturnal and predacious.

Notes. Häckel *et al.* (2010) transferred some species to *Chylnus* that in Moore's *et al.* (1987) catalogue were included in *Percosoma*. However, at least one of these transferred species, namely *Percosoma substriatum* Moore, 1960, possesses several supraorbital setae and thus actually belongs in *Percosoma* and not *Chylnus* and it is herewith re-transferred to the latter genus. Roig-Juñent (2000) figured some external characters and the male and female genitalia.

References. Csiki (1928); Moore *et al.* (1987); Roig-Juñent (2000); Lorenz (2005); Häckel *et al.* (2010).

Key to species. Sloane (1920).

Percolestus Sloane, 1892 (Pl. 34I)

Type species. *Percolestus blackburni* Sloane, 1892, by monotypy.

Characteristics. Moderately large beetles; mentum dentate; head with a single supraorbital seta above each eye; without a suborbital ridge; without a deep transverse impression behind the eyes; mandible without a seta in the scrobe; labial palps slightly widened; elytra with more than four marginal punctures; neither male protarsus nor mesotarsus with squamose setae beneath; left paramere asetose.

Australian species. Only *Percolestus blackburni* Sloane.

Distribution. Eastern Victoria.

Biology. Ground living at high altitude in subalpine habitats. Nocturnal and predacious.

Note. Roig-Juñent (2000) figured some external characters and the male and female genitalia.

References. Csiki (1928); Moore *et al.* (1987); Roig-Juñent (2000); Lorenz (2005); Häckel *et al.* (2010).

Percosoma Schaum, 1858 (Pl. 35D)

Type species. *Broscus carenoides* White, 1846, by monotypy.

Characteristics. Rather large to very large beetles; mentum dentate; head with several supraorbital setae above each eye; without a suborbital ridge; without a deep transverse impression behind the eyes; mandible without a seta in the scrobe; labial palps slightly widened; elytra with more than four marginal punctures; neither male protarsus nor mesotarsus with squamose setae beneath; in both sexes the tarsomeres of the protarsus laterally little produced; left paramere setose.

Australian species. Three species.

Distribution. Tasmania, eastern Victoria; extraterritorial one species described from New Caledonia, but the generic affiliation of the recorded species is doubtful and it is most likely an undescribed genus near *Percolestus* or possible a species in that genus (D. Seldon in litt.).

Biology. Ground living in closed forests, the single mainland species at high altitude. Nocturnal and predacious.

Notes. Some of the species that Moore *et al.* (1987) included in *Percosoma* were transferred to the genus *Chylnus* by Häckel *et al.* (2010). For discussion see under that genus. Roig-Juñent (2000) figured some external characters and the male and female genitalia.

References. Putzeys (1868); Sloane (1892, 1920); Csiki (1928); Moore *et al.* (1987); Roig-Juñent (2000); Lorenz (2005); Häckel *et al.* (2010).

Keys to species. Sloane (1892, 1920).

Tribe Zolini Sharp, 1886

Characteristics. Procoxal cavities closed; mesocoxae conjunct; head with one or two supraorbital setae above each eye; mandibles with a seta in the scrobe; palps not subulate; penultimate palpomere of the maxillary palp setose; frontal furrows short and not extended behind the eyes; basal margin of the elytra incomplete; series of marginal setiferous punctures of the elytra interrupted; elytral epipleura with a plica; male protarsus with two basal tarsomeres dilated and dentate on median side; aedeagus with large ostium, usually with a more or less sclerotised or denticulate piece on the endophallus; parameres of different size, triangular, with several elongate apical setae; gonocoxites elongate, gonocoxite-2 curved, with a various number of setae.

Key to the genera of Zolini recorded from Australia

1. Eyes very small, depressed; pronotum narrow, without anterior submarginal sulcus; legs unusually slender and elongate .... ***Idacarabus* Lea**
– Eyes large, laterally protruding; prothorax wide, with or without anterior submarginal sulcus; legs relatively robust, not notably slender and elongate **2**

2(1). Head with two supraorbital setae over each eye **3**
– Head with a single supraorbital seta over each eye .... ***Pterocyrtus* Sloane**

3(2). Pronotum without an anterior submarginal sulcus **and** without the posterior latero-marginal seta .... ***Thayerella* Baehr**
– Pronotum with an anterior submarginal sulcus; if body oval shaped then the posterior marginal seta present; if body more parallel-sided the posterior marginal seta absent **4**

4(3). Pronotum with posterior marginal seta; body oval-shaped .... ***Sloaneana* Csiki**
– Pronotum without posterior marginal seta; body rather parallel-sided .... ***Percodermus* Sloane**

Idacarabus Lea, 1910 (Pl. 27C)

Type species. *Idacarabus troglodytes* Lea, 1910, by monotypy.

Characteristics. Rather elongate, yellow to light, reddish-brown beetles; head with two supraorbital setae above each eye; eyes minute; antenna very elongate; pronotum narrow, without an anterior submarginal sulcus; legs very elongate; aedeagus compact, endophallus with a sclerotised piece; parameres moderately dissimilar, rather elongate, triangular, with 4–5 setae.

Australian species. Three species.

Distribution. Tasmania.

Biology. Cave inhabiting. Found in various limestone caves in Tasmania. Predacious.

References. Sloane (1920); Csiki (1928); Moore (1967, 1978, 1994); Moore *et al.* (1987); Lorenz (2005); Eberhard & Giachino (2011).

Keys to species. Moore (1967, 1978).

Percodermus Sloane, 1920 (Pl. 34C)

Type species. *Percodermus niger* Sloane, 1920, by monotypy.

Characteristics. Head with two supraorbital setae above each eye; eyes of normal size; antenna not unusually elongate; prothorax with distinct anterior submarginal sulcus; without the basal latero-marginal seta; elytra rather parallel-sided, with 4–5 setiferous punctures on stria 3; aedeagus compact, endophallus with a sclerotised piece; parameres moderately dissimilar, rather stout, triangular, with 4–5 setae; gonocoxites not recorded.

Australian species. Only *P. niger* Sloane.

Distribution. Central highlands of Tasmania.

Biology. Ground living in closed forests. Eberhard & Giachino (2011) mention specimens as being sieved from litter in tall *Nothofagus* forests. Probably nocturnal and predacious.

References. Csiki (1928); Moore *et al.* (1987); Lorenz (2005); Eberhard & Giachino (2011).

Pterocyrtus Sloane, 1920 (Pl. 36L)

Type species. *Pterocyrtus globosus* Sloane, 1920, by original designation.

Characteristics. Head with a single supraorbital seta above each eye; eyes of normal size; antenna not unusually elongate; pronotum with distinct anterior submarginal sulcus, basal marginal seta present; elytra rather oval-shaped, with no more than three setiferous punctures on stria 3; aedeagus variously shaped, endophallus with a sclerotised piece; parameres rather dissimilar, more or less elongate, triangular, with 4–5 setae.

Australian species. Six species.

Distribution. Tasmania and eastern Victoria.

Biology. Ground living. Most species in leaf litter in various habitats at various altitudes, one species in caves. Nocturnal and predacious.

References. Csiki (1928); Moore (1994); Moore *et al.* (1987); Lorenz (2005); Eberhard & Giachino (2011).

Keys to species. Sloane (1920), Eberhard & Giachino (2011).

Sloaneana Csiki, 1933 (Pl. 38G; Fig. 12.23)

Type species. *Brachydema tasmaniae* Sloane, 1915, by monotypy.

Brachydema Sloane, 1915. 452 (non *Brachydema* Fairmaire, 1881)

Characteristics. Head with two supraorbital setae above each eye; eyes of normal size; antenna not unusually elongate; pronotum wide, widest near base, with distinct anterior submarginal sulcus; with the basal latero-marginal seta; elytra rather oval-shaped, depressed, with 2–4 setiferous punctures on stria 3; aedeagus moderately short, endophallus with a sclerotised piece; parameres rather dissimilar, more or less elongate, triangular, with 2–3 setae.

Australian species. Four species.

Distribution. Tasmania and south-eastern Australia from eastern Victoria to south-eastern Queensland.

Biology. Ground living in leaf litter of closed forests, commonly at high altitude, at least in south-eastern Queensland. Nocturnal and predacious.

References. Moore *et al.* (1987); Baehr (2002c); Lorenz (2005); Eberhard & Giachino (2011).

Key to species. Baehr (2002c, 2015a).

Thayerella Baehr, 2016 (Fig. 12.145)

Type species. *Thayerella newtoni* Baehr, 2016, by original designation.

Characteristics. Head with two supraorbital setae above each eye; eyes of normal size; antenna not unusually elongate; pronotum without anterior submarginal sulcus, without the basal latero-marginal seta; elytra rather oval-shaped, with two setiferous punctures on the stria 3; pronotum and elytra with conspicuous, yellow margin; aedeagus unknown.

Australian species. Only *T. newtoni* Baehr.

Distribution. Southern Victoria.

Biology. Ground living in leaf litter in wet sclerophyll forests.

References. Baehr (2016b).

Tribe Trechini Bonelli, 1810

Characteristics. A very species rich and diverse tribe; procoxal cavities closed; mesocoxae conjunct; head with two supraorbital setae above each eye; mandibles with a seta in the scrobe; palps usually not subulate but elongate-acute, except in the genus *Tasmanitachoides* Erwin, 1972 as discussed below; frontal furrows deep and elongate, usually curved; basal margin of the elytra incomplete; series of marginal setiferous punctures of the elytra interrupted; elytral epipleura without plica; very commonly apex of the elytral stria 3 recurrent and hooked; aedeagus quite varied, parameres of different size, triagonal, with several elongate apical setae; gonocoxites more or less elongate, gonocoxite-2 usually curved, with or without various numbers of setae.

Notes. This tribe includes many species and they have a large diversity of body shape and morphological structures, including many small-eyed or eyeless, usually more or less depigmented subterranean and cave beetles.

The subtribal classification is presently in flux and a phylogenetical survey of the whole Trechine-Bembidiine complex is in progress (Maddison *in litt.*; Faille *in litt.*). In particular the phylogenetically early branching Trechodina may not be monophyletic, and the status of the genus *Tasmanitachoides* is still unclear. According to Maddison (*in litt.*) the genus *Tasmanitachoides* and some South American relatives will be combined to a new tribe that probably will be sister to the whole Trechine-Bembidiine complex (Trechitae *sensu* Jeannel). For the moment, however, we include *Tasmanitachoides* in Trechini without a definite subtribal status.

The key below is based largely on the revision of the Australian Trechini of Moore (1972).

Key to the Australian subtribes of Trechini

1. Palps subulate, apical palpomere of maxillary palp much thinner than the preapical palpomere .... **Trechini *incerta sedis*: *Tasmanitachoides* Erwin**
– Palps not subulate, apical palpomere elongate, conical **2**
2(1). Surface pilose; eyes pilose; very small size, parallel-sided, depressed form .... **Perileptina Sloane**
– Surface not pilose; eyes glabrous; usually larger, less parallel-sided, less depressed form **3**
3(2). Ostium of aedeagus (i.e. the dorsal surface) open; aedeagus without a basal bulbus .... **Trechodina Jeannel**
– Ostium of aedeagus (i.e. the dorsal surface) closed; aedeagus with basal bulbus .... **Trechina Bonelli**

Key to the Australian genera of Trechini

1. Palps subulate, apical palpomere of maxillary palp much thinner than the preapical palpomere .... ***Tasmanitachoides* Erwin**
– Palps not subulate, apical palpomere elongate, conical **2**
2(1). Surface pilose; eyes pilose; very small, parallel-sided, depressed form .... ***Perileptus* Schaum**
– Surface not pilose; eyes glabrous; usually larger, less parallel-sided, less depressed form **3**
3(2). Ostium of aedeagus (i.e. the dorsal surface) open, aedeagus without basal bulbus; elytra usually sericeous and very glossy **4**
– Ostium of aedeagus (i.e. the dorsal surface) closed, aedeagus with basal bulbus; elytra usually less sericeous and glossy **7**
4(3). Elytra with six striae; basal border interrupted at position of interval .... ***Trechobembix* Jeannel**
– Elytra with at most three distinct striae; basal border not interrupted **5**
5(4). Elytra with only the sutural stria **6**
– Elytra with three distinct striae .... ***Paratrechodes* Jeannel**
6(5). Body shape wide and depressed; eyes large, prominently protruding .... ***Trechodes* Blackburn**
– Body shape narrow and dorsally convex; eyes smaller, only slightly protruding .... ***Cyphotrechodes* Jeannel**
7(3). Recurrent stria at apex hooked or ending abruptly in a deep puncture or sulcus **8**
– Recurrent stria at apex not hooked, either uniting with a discal stria, or gradually diminishing **14**
8(7). Elytra only with a sinuate postmedian fascia .... ***Trechiella* Jeannel**
– Elytra unicolourous or with different pattern of lighter colours **9**
9(8). Recurrent stria ending at the posterior discal puncture; eyes reduced .... ***Mimanillus* Moore**
– Recurrent stria not in contact with a discal puncture; eyes well developed **10**
10(9). Frontal furrows duplicated near each eye (Fig. 12.46) .... ***Eutrechopsis* Moore**
– Frontal furrows simple **11**
11(10). Antenna short, antennomeres moniliform .... ***Mimotrechus* Moore**
– Antenna longer, antennomeres filiform **12**
12(11). Elytra with distinct pale pattern **13**
– Elytra unicolourous, without distinct pattern .... ***Trechimorphus* Jeannel**
13(12). Discal punctures of elytra large, foveiform; elytral pattern variegated .... ***Bothynotrechus* Moore**
– Discal punctures of elytra small; elytra with a distinct humeral spot and a postmedian fascia .... ***Eutrechus* Moore**
14(7). The posterior discal puncture of the elytra located far back on the apical declivity, inside the recurrent stria; stria 2 ending near this puncture **16**
– The posterior discal puncture of the elytra located in front of the apical declivity, in front of the recurrent stria, or the puncture is absent; stria 2 ending far from this puncture when present **15**
15(14). Discal punctures of elytra interval 3 small, not in contact with the adjacent striae; colour black, sometimes with indistinct maculae .... ***Tasmanorites* Jeannel**
– Discal punctures of elytra interval 3 large, in contact with the adjacent striae; elytral pattern variegated .... ***Sloanella* Jeannel**
16(14). Mentum concave and fused to the gula, the gular suture indistinct; discal punctures of elytra very large, in contact with the adjacent striae; yellow species with variegated elytra or, rarely, concolorous .... ***Tropidotrechus* Jeannel**
– Mentum depressed, the gular suture distinct; discal punctures of elytra smaller, not or barely in touching the adjacent stria; colour and elytral pattern different **17**
17(16). Humeral border of the elytra incurved, rectilinear and forming a distinct basal margin **18**
– Humeral border of the elytra sharply curved apically at the level of stria 3 or 4 .... **19**
18(17). Apex of labium prominent and triangularly produced .... ***Pogonoschema* Jeannel**
– Apex of labium almost truncate, not prominent .... ***Austrotrechus* Moore**
19(17). Elytra with two discal punctures, the anterior one lacking .... ***Tasmanotrechus* Moore**
– Elytra with three discal punctures **20**
20(19). Eyes reduced or absent .... ***Goedetrechus* Moore**
– Eyes well developed **21**
21(20). Mandibles slender, elongate and straight .... ***Nototrechus* Moore**
– Mandibles short and curved .... ***Trechistus* Moore**

Subtribe Trechina Bonelli, 1810

Characteristics. Eyes variously developed, reduced or absent; if present, eyes glabrous; palps not subulate; dorsal surface typically glabrous, but in some cavernicolous species pilose; shape of head, prothorax and elytra quite varied; elytra with various number of discal setae; aedeagus dorsally closed, ostium small, aedeagus with basal bulbus, highly variably shaped and structured; gonocoxite-2 curved, usually with one dorsal seta, two ventral setae, and one or two subapical setae.

Notes. This is by far the largest subtribe of Trechini, with a high diversity of shapes, structures and colours. The subtribe includes many subterranean or cave inhabiting species throughout the world. Without a doubt the Australian cavernicolous trechine fauna still is inadequately recorded and additional species, perhaps even distinct genera, may be detected in future by systematic exploration of Tasmanian and southern mainland caves.

Tasmanorites Jeannel, 1927 (Pl. 38L; Figs 12.21, 12.142)

Type species. *Trechus nitens* Putzeys, 1874, by original designation.

Characteristics. Medium-sized to large beetles; frontal furrows complete, simple; apex of labrum emarginate; recurrent stria at apex not hooked; posterior discal puncture of elytra located in front of the apical declivity; stria 2 continued beyond level of posterior discal puncture; discal punctures small; elytra uniformly black or indistinctly maculate; flight wings reduced; aedeagus well sclerotised, tubular.

Australian species. 23 species and one subspecies.

Distribution. Tasmania.

Biology. Ground living, typically in leaf litter of montane forests and woodlands. According to Moore (1972) some species were found in the near running water or in other wet places. Nocturnal and predacious. Little else is known about habits, diet, and reproduction.

References. Csiki (1928); Moore (1972, 1983); Moore *et al.* (1987); Lorenz (2005); Eberhard & Giachino (2011).

Keys to species. Moore (1972, 1983); Eberhard & Giachino (2011).

Sloanella Jeannel, 1927 (Fig. 12.131)

Type species. *Trechus simsoni* Blackburn, 1894, by original designation.

Characteristics. Medium-sized beetles; frontal furrows complete, simple; apex of labrum emarginate; recurrent stria at apex not hooked; elytra short and rounded; posterior discal puncture of elytra located in front of the apical declivity; stria 2 continued beyond level of posterior discal puncture; discal punctures large, in contact with stria 3, which wraps around the puncture; elytra distinctly maculate; flight wings reduced; aedeagus well sclerotised, tubular.

Australian species. Five species.

Distribution. Tasmania.

Biology. Ground living, typically in leaf litter of montane forests. Nocturnal and predacious. Little else is known about habits, diet, and reproduction.

References. Csiki (1928); Moore (1972, 1983); Moore *et al.* (1987); Lorenz (2005); Eberhard & Giachino (2011).

Keys to species. Moore (1972, 1983), Eberhard & Giachino (2011).

Tropidotrechus Jeannel, 1927 (Pl. 39H)

Type species. *Trechus victoriae* Blackburn, 1894, by original designation.

Characteristics. Medium-sized to large beetles; frontal furrows complete, simple; apex of labrum emarginate; recurrent stria at apex not hooked; elytra short and rounded; posterior discal puncture of elytra located far back on the apical declivity, inside the recurrent stria; stria 2 ending near this puncture; discal punctures large, impinging on stria 3; elytra often distinctly maculate; flight wings reduced; aedeagus stout, strongly sclerotised, membranous over much of upper surface, endophallus with a large copulatory piece.

Australian species. Four species.

Distribution. Eastern Victoria in montane areas.

Biology. Ground living in leaf litter of montane forests and alpine woodlands, mostly at very high altitude. Nocturnal and predacious. Little else is known about habits, diet, and reproduction.

References. Csiki (1928); Moore (1972); Moore *et al.* (1987); Lorenz (2005).

Key to species. Moore (1972).

Pogonoschema Jeannel, 1927 (Fig. 12.123)

Type species. *Trechus robustus* Sloane, 1920, by original designation.

Characteristics. Large beetles; frontal furrows complete, simple or interrupted; apex of labrum truncate; recurrent stria at apex not hooked; elytra rather short and rounded; humeral border incurved rectilinearly, forming a distinct basal margin; posterior discal puncture of elytra located far back on the apical declivity, inside the recurrent stria; stria 2 ending near this puncture; discal punctures fairly large, but not in contact with stria 3 or not altering the stria's path; elytra unicolourous black; flight wings reduced; aedeagus well sclerotised, endophallus with a distinct copulatory piece.

Australian species. Four species.

Distribution. Northern Tasmania.

Biology. Ground living in leaf litter of closed forest at various altitudes. Nocturnal and predacious. Little else is known about habits, diet, and reproduction.

References. Csiki (1928); Moore (1972); Moore *et al.* (1987); Lorenz (2005).

Key to species. Moore (1972).

Austrotrechus Moore, 1972 (Fig. 12.79)

Type species. *Trechus kosciuszkoanus* Sloane, 1923, by original designation.

Characteristics. Medium-sized beetles; frontal furrows complete, simple; apex of labrum triangularly produced; recurrent stria at apex not hooked; elytra rather short and rounded; humeral border incurved rectilineary, forming a distinct basal margin; posterior discal puncture of elytra located far back on the apical declivity, inside the recurrent stria; stria 2 ending near this puncture; discal punctures fairly large, but not impinging on stria 3; elytra not maculate; flight wings reduced; aedeagus stout, strongly sclerotised, ostium elongate, endophallus with or without a copulatory piece.

Australian species. Two species.

Distribution. Eastern Victoria, south-eastern New South Wales.

Biology. Ground living in leaf litter at the alpine zone, usually in areas with snow gum. Nocturnal and predacious. Little else is known about habits, diet, and reproduction.

References. Moore *et al.* (1987); Lorenz (2005).

Key to species. Moore (1972).

Tasmanotrechus Moore, 1972 (Fig. 12.143)

Type species. *Trechus leai* Sloane, 1923, by original designation.

Characteristics. Rather large beetles; frontal furrows complete, simple; apex of labrum truncate; eyes of normal size or most commonly more or less reduced; recurrent stria at apex not hooked; elytra rather elongate and oval-shaped; humeral border of elytra suddenly curved apically at the level of stria 3 or 4; posterior discal puncture of elytra located far back on the apical declivity, inside the recurrent stria; stria 2 ending near this puncture; only two discal punctures present; elytra not maculate; flight wings reduced; aedeagus moderately sclerotised, ostium short, endophallus with a small copulatory piece.

Australian species. 11 species.

Distribution. Tasmania.

Biology. Ground living, typically in leaf litter in montane forests or wet open areas; some species cavernicolous. Nocturnal and predacious. Little else is known about habits, diet, and reproduction.

References. Moore (1983, 1994); Moore *et al.* (1987); Lorenz (2005); Eberhard & Giachino (2011).

Keys to species. Moore (1983, 1994); Eberhard & Giachino (2011).

Goedetrechus Moore, 1972 (Fig. 12.93)

Type species. *Goedetrechus talpinus* Moore, 1972, by original designation.

Characteristics. Medium-sized to rather large beetles; frontal furrows complete, simple; apex of labrum truncate; eyes very small or absent; recurrent stria at apex not hooked; elytra rather elongate and oval-shaped; humeral border of elytra suddenly curved apically at the level of stria 3 or 4; posterior discal puncture of elytra located far back on the apical declivity, inside the recurrent stria; stria 2 ending near this puncture; three discal punctures present; elytra not maculate; flight wings reduced; aedeagus small, weakly sclerotised, ostium very elongate, endophallus with a small copulatory piece.

Australian species. Seven species.

Distribution. Tasmania.

Biology. Endogeous or in leaf litter of montane forests, but most species cavernicolous. Nocturnal and predacious.

References. Moore *et al.* (1987); Lorenz (2005); Eberhard & Giachino (2011).

Keys to species. Moore (1972); Eberhard & Giachino (2011).

Nototrechus Moore, 1972 (Fig. 12.111)

Type species. *Nototrechus unicolor* Moore, 1972, by original designation.

Characteristics. Small beetles; frontal furrows complete, simple; apex of labrum truncate; eyes of normal size; mandibles slender, straight, and elongate; recurrent stria at apex not hooked; elytra rather short and oval-shaped; humeral border of elytra suddenly curved apically at level of stria 3 or 4; posterior discal puncture of elytra located far back on the apical declivity, inside the recurrent stria; stria 2 ending near this puncture; three discal punctures present; elytra unicolourous piceous; flight wings reduced; only the tarsomere 1 of male protarsus widened and with squamose setae beneath; aedeagus weakly sclerotised, dorsally open, endophallus without a copulatory piece.

Australian species. Only *N. unicolor* Moore.

Distribution. Eastern Victoria.

Biology. Ground living in leaf litter of subalpine woodland. Nocturnal and predacious. Little else is known about habits, diet, and reproduction.

References. Moore *et al.* (1987); Lorenz (2005).

Trechistus Moore, 1972 (Fig. 12.149)

Type species. *Trechistus humicola* Moore, 1972, by original designation.

Characteristics. Medium-sized beetles; frontal furrows complete, simple; apex of labrum truncate; eyes of normal size; mandibles short and curved; recurrent stria at apex not hooked; elytra rather elongate and oval-shaped; humeral border of elytra suddenly curved apically at level of stria 3 or 4; posterior discal puncture of elytra located far back on the apical declivity, inside the recurrent stria; stria 2 ending near this puncture; three discal punctures present; elytra maculate or rarely unicolourous; flight wings reduced; aedeagus weakly sclerotised, tubular, endophallus without a copulatory piece.

Australian species. Six species.

Distribution. Tasmania.

Biology. Ground living in leaf litter and in moss of montane and subalpine habitats. Little else is known about habits, diet, and reproduction.

References. Moore *et al.* (1987); Lorenz (2005); Eberhard & Giachino (2011).

Keys to species. Moore (1972); Eberhard & Giachino (2011).

Trechiella Jeannel, 1927 (Fig. 12.147)

Type species. *Trechus subornatellus* Blackburn, 1901, by original designation.

Characteristics. Medium-sized beetles; frontal furrows complete, simple; apex of labrum truncate; eyes of normal size; recurrent stria at apex deep and hooked; elytra short and oval-shaped, deeply striate; base of elytra not margined; posterior discal puncture of elytra located far back on the apical declivity, inside the recurrent stria; three discal punctures present; elytra unifasciate in apical half; flight wings reduced; aedeagus weakly sclerotised, dorsally open, endophallus without a copulatory piece.

Australian species. Two species.

Distribution. Eastern Victoria, south-eastern Queensland. Little else is known about habits, diet, and reproduction.

Biology. Ground living in leaf litter of subapine woodlands and temperate rainforests. Little else is known about habits, diet, and reproduction.

References. Moore (1972); Moore *et al.* (1987); Lorenz (2005).

Key to species. Moore (1972).

Mimanillus Moore, 1972 (Fig. 12.107)

Type species. *Mimanillus gracilis* Moore, 1972, by original designation.

Characteristics. Small beetles; frontal furrows complete, simple; apex of labrum deeply emarginate; eyes very small; recurrent stria at apex touching the 3rd discal puncture; elytra fused, fairly elongate and oval-shaped, rather faintly striate; base of elytra not margined; three discal punctures present; colour pale reddish; flight wings reduced; aedeagus weakly sclerotised, tubular, endophallus with a distinct copulatory piece.

Australian species. Only *M. gracilis* Moore, 1972.

Distribution. Western Tasmania.

Biology. Largely unknown. Probably ground living in leaf litter of woodlands.

References. Moore *et al.* (1987); Lorenz (2005).

Eutrechopsis Moore, 1972 (Figs 12.46, 12.89)

Type species. *Eutrechopsis ovalis* Moore, 1972, by original designation.

Characteristics. Small beetles; frontal furrows duplicated near the eye (Fig. 12.46; the anterior supraorbital puncture situated in a deep, elongate groove; apex of labrum truncate; eyes of normal size; recurrent stria deep, abruptly terminated in a deep pit; elytra short and oval-shaped, moderately striate; base of elytra not margined; three discal punctures present; elytra maculate; flight wings reduced; aedeagus small, weakly sclerotised, tubular, ostium short, endophallus without a copulatory piece.

Australian species. Only *E. ovalis* Moore.

Distribution. Extreme south-eastern New South Wales.

Biology. Ground living in leaf litter of tree ferns at median altitude.

Note. According to Moore (1972) the single species apparently is only known from the type locality.

References. Moore *et al.* (1987); Lorenz (2005).

Mimotrechus Moore, 1972 (Fig. 12.108)

Type species. *Trechus australiensis* Sloane, 1923, by original designation.

Characteristics. Small beetles; frontal furrows simple; apex of labrum emarginate; eyes of moderate size; antenna short, median antennomeres moniliform; recurrent stria deep, hooked at apex; elytra rather elongate and more or less parallel-sided, rather deeply striate; base of elytra not margined; three large discal punctures present; elytra unicolourous or maculate; flight wings usually reduced, but variable; aedeagus moderately sclerotised, tubular, ostium large, endophallus without a copulatory piece.

Australian species. Four species.

Distribution. Eastern Victoria, south-eastern New South Wales, central Tasmania.

Biology. Ground living in leaf litter of tree ferns and snow gums, also 'in damp moss and litter' (Moore 1972), usually at high altitude. Little else is known about habits, diet, and reproduction.

References. Moore *et al.* (1987); Lorenz (2005).

Key to species. Moore (1972).

Bothynotrechus Moore, 1972 (Pl. 35H; Fig. 12.80)

Type species. *Trechus castelnaui* Sloane, 1920, by original designation.

Characteristics. Medium-sized beetles; frontal furrows simple; apex of labrum truncate; eyes of normal size; antenna elongate, middle antennomeres filiform; recurrent stria deep, hooked at apex, terminating in a deep pit; elytra short and wide, deeply striate; base of elytra not margined; three discal punctures present, the anterior two large and deep, the posterior one small and situated far back on the declivity; elytra maculate; flight wings reduced; aedeagus moderately sclerotised, tubular but upper surface largely membranous, endophallus without a copulatory piece; setosity of parameres reduced in number.

Australian species. Two species.

Distribution. Eastern Victoria and adjacent extreme south-eastern New South Wales.

Biology. Ground living in leaf litter of forests at median altitude, also in 'litter around forest pools' (Moore 1972). Little else is known about habits, diet, and reproduction.

References. Moore *et al.* (1987); Lorenz (2005).

Key to species. Moore (1972).

Eutrechus Moore, 1972 (Fig. 12.90)

Type species. *Sloanella otwayensis* Moore, 1960, by original designation.

Characteristics. Medium-sized to fairly large beetles; frontal furrows simple; apex of labrum emarginate; eyes large; antenna elongate, median antennomeres filiform; recurrent stria deep, not hooked at apex, but terminating in a deep, curved groove; elytra short and wide, superficially striate; base of elytra not margined; three small discal punctures present, the posterior one situated far back on the declivity; elytra maculate; flight wings reduced; aedeagus well sclerotised, tubular but upper surface largely membranous, endophallus with a large copulatory piece.

Australian species. Four species.

Distribution. Eastern Australia from southern Victoria to New England Tableland in north-eastern New South Wales.

Biology. Little recorded, probably ground-living in leaf litter of forests.

References. Moore *et al.* (1987); Lorenz (2005).

Key to species. Moore (1972).

Trechimorphus Jeannel, 1927 (Fig. 12.148)

Type species. *Trechus diemenensis* Bates, 1878, by original designation.

Characteristics. Medium-sized to fairly large beetles; frontal furrows simple; apex of labrum slightly emarginate; eyes large; antenna elongate, median antennomeres filiform; recurrent stria deep, more or less distinctly hooked at apex, terminating in a deep channel; elytra rather elongate, more or less deeply striate; base of elytra not margined inside of stria 4; two or three small discal punctures present, the posterior one situated far back on the declivity; elytra unicolourous; flight wings varied; aedeagus well sclerotised, tubular, ostium apical, endophallus with a small copulatory piece.

Australian species. Six species.

Distribution. Eastern Australia from southern Victoria to south-eastern Queensland, Tasmania, and south-western Australia.

Biology. Ground living in leaf litter of various forest and woodland types. Some species are common and very widely distributed. In some populations fully winged and brachypterous forms occur. Little else is known about habits, diet, and reproduction.

References. Csiki (1928); Moore (1972); Moore *et al.* (1987); Lorenz (2005); Eberhard & Giachino (2011).

Key to species. Moore (1972).

Subtribe Trechodina Jeannel, 1926

Characteristics. Eyes of varied size, glabrous; palps not subulate; surface not pilose; ostium of aedeagus open, aedeagus without basal bulbus; gonocoxites as in Trechina. Usually the elytra are sericeous and very glossy.

Note. Trechodina may not represent a monophyletic group, and in particularly, the genus *Trechobembix* Jeannel seems rather remotely related to the other included genera.

Trechodes Blackburn, 1901 (Fig. 12.151)

Type species. *Bembidion secaloides* Blackburn, 1890, by subsequent designation by Jeannel 1926.

Characteristics. Rather small beetles; frontal furrows simple but deep; apex of labrum deeply emarginate; eyes very large, prominently produced laterally, orbits short; antenna elongate, middle antennomeres filiform; base of prothorax laterally excised; recurrent stria barely developed; elytra wide and short, rather parallel-sided, unistriate; base of elytra completely margined; three discal punctures present, the posterior one situated at base of declivity; elytra unicolourous or bicolorous; flight wings fully developed; aedeagus dorsally open, endophallus with a small, elongate copulatory piece.

Australian species. Four species.

Distribution. Eastern and northern Australia from Victoria to north-eastern Queensland, and in the northern part of Northern Territory; extraterritorial in the Oriental and Afrotropical Regions.

Biology. Hygrophilous, found on or near river banks, shores of lakes and lagoons, and swamps. All species are fully winged and commonly are attracted to light. Predacious and most probably nocturnal. Little else is known about habits, diet, and reproduction.

References. Csiki (1928); Jeannel (1926); Moore (1972); Moore *et al.* (1987); Lorenz (2005).

Key to species. Moore (1972).

Paratrechodes Jeannel, 1926 (Fig. 12.117)

Type species. *Trechus macleayi* Sloane, 1920, by original designation.

Characteristics. Medium-sized beetles; frontal furrows simple; apex of labrum almost truncate; eyes large, prominently produced laterally, orbits short; antenna very elongate, middle antennomeres filiform; base of prothorax not laterally excised; recurrent stria developed but rather short; elytra wide and short, rather parallel-sided, with three distinct striae; base of elytra completely margined; three discal punctures present, the posterior one situated far back on the declivity; elytra unicolourous but margins slightly paler; flight wings fully developed; aedeagus dorsally open, endophallus with a small, elongate copulatory piece.

Australian species. Only *P. macleayi* (Sloane).

Distribution. Victoria and Tasmania.

Biology. Little recorded, according to Moore (1972) hygrophilous and found on the banks of rivers. The single species is fully winged. Little else is known about habits, diet, and reproduction.

References. Csiki (1928); Moore (1972); Moore *et al.* (1987); Lorenz (2005).

Cyphotrechodes Jeannel, 1926 (Fig. 12.87)

Type species. *Trechodes gibbipennis* Blackburn, 1901, by original designation.

Characteristics. Small beetles; frontal furrows simple; apex of labrum slightly emarginate; eyes rather large, but only moderately produced laterally, orbits short; antenna moderately elongate, median antennomeres filiform; pronotum convex, pedunculate, base laterally oblique; recurrent stria barely developed; elytra moderately elongate, unistriate; base of elytra completely margined; three discal punctures present, the anterior situated in a transverse groove, the posterior one situated far back on the declivity; elytra unicolourous black; flight wings varied; aedeagus rudimentary, endophallus without a copulatory piece.

Australian species. Only *C. gibbipennis* (Blackburn).

Distribution. Victoria, south-eastern New South Wales, and Tasmania.

Biology. Very hygrophilous, according to Moore (1972) on wet ground, 'in *Juncus* beds, besides slow-moving or standing water'. The single species is dimorphic in the flight wings. Little else is known about habits, diet, and reproduction.

References. Csiki (1928); Moore (1972); Moore *et al.* (1987); Lorenz (2005); Eberhard & Giachino (2011).

Trechobembix Jeannel, 1926 (Fig. 12.150)

Type species. *Trechus baldiensis* Blackburn, 1894, by original designation.

Characteristics. Small to medium-sized, quite variable species; frontal furrows simple; apex of labrum almost truncate; eyes moderately large, only slightly produced laterally; antenna moderately elongate, median antennomeres filiform; pronotum narrowed to base, base laterally excised and basal angles produced backwards; recurrent stria barely developed; shape of elytra varied, sex-striate; basal margin interrupted at position of interval 3; three discal punctures present, the posterior one situated far back on the declivity; elytra black and very glossy, apex more or less distinctly rufous; flight wings varied; aedeagus elongate, dorsal surface open, endophallus with an elongate, folded copulatory piece.

Australian species. Only *T. baldiensis* (Blackburn) with three subspecies.

Distribution. Eastern Australia from South Australia to north-eastern Queensland, south-western part of Western Australia, and Tasmania.

Biology. Ground living in a variety of habitats. Predacious and probably nocturnal. All populations are wing-dimorphic. Little else is known about habits, diet, and reproduction.

Notes. The single species has been divided into three subspecies that are endemic in south-eastern Australia, north-eastern Queensland, and south-western Australia, respectively.

References. Csiki (1928); Moore (1972); Moore *et al.* (1987); Matthews (1980); Lorenz (2005).

Key to subspecies. Moore (1972).

Subtribe Perileptina Sloane, 1903

Characteristics. Very small, elongate, parallel-sided, and depressed beetles; eyes pilose; palps not subulate, conical; surface pilose; aedeagus weakly sclerotised, tubular, with small, apical ostium; basal with bulbus, endophallus without a copulatory piece; both parameres triangular, setose at apex; gonocoxites rather short, gonocoxite-2 curved, acute, with one dorsal seta, two ventral setae, and a rather elongate subapical seta.

Perileptus Schaum, 1860 (Fig. 12.119)

Type species. *Carabus areolatus* Creutzer, 1799, by monotypy.

Characteristics. Small, narrow, and rather depressed beetles; surface pilose; margins of pronotum distinct; discal punctures of the elytra very small; protibia without a sulcus at the lateral surface.

Australian species. Six species.

Distribution. South-eastern and eastern Australia, northern tropical Australia, in the north-west south to the Pilbara, and also in central Australia; extraterritorial in the Old World and on Caribbean Islands.

Biology. Hygrophilous in sand and small gravel on the banks of rivers and the shore of lakes and lagoons. Commonly attracted to light. Predacious and nocturnal. Little else is known about habits, diet, and reproduction.

Note. The Australian species are divided into two subgenera, according to their body shape and surface structure.

References. Csiki (1928); Jeannel (1926); Andrewes (1935a); Moore (1966a, 1972); Darlington (1962a); Moore *et al.* (1987); Baehr (1987b, 1997a); Lorenz (2005).

Keys to species. Moore (1966a), Baehr (1987b).

Key to the subgenera of *Perileptus* recorded from Australia

1. Body convex; microreticulation inconspicuous but punctures of pronotum coarse; median tooth of the apical border of the labrum indistinct; elytra with 10–12 setae on either side .... ***Perileptus* Schaum**
– Body depressed; microreticulation usually more conspicuous, punctures of pronotum fine; median tooth of the apical border of the labrum distinct; elytra with at least 16 setae on either side .... ***Pyrrhotachys* Sloane**

Subgenus *Perileptus* Schaum, 1860

Type species. *Carabus areolatus* Creutzer, 1799, by monotypy.

Characteristics. As in the key; pronotum and elytra dorsally rather convex, with sparser pilosity.

Australian species. Only *P. cylindricollis* Baehr, 1997.

Distribution. Northern tropical Australia from northern Queensland to northern Western Australia to the southern margin of the Kimberley; extraterritorial in the Old World and on Caribbean Islands.

Biology. As for genus; most specimens of the single species were sampled at light near the sandy banks, or shores, of rivers and lagoons. Predacious and nocturnal.

References. Csiki (1928); Jeannel (1926); Andrewes (1935a); Moore (1966a, 1972); Darlington (1962a); Moore *et al.* (1987); Baehr (1987b, 1997a); Lorenz (2005).

Subgenus *Pyrrotachys* Sloane, 1896

Type species. *Pyrrotachys constricticeps* Sloane, 1896, by monotypy.

Characteristics. As in the key; depressed beetles with denser pilosity on the elytra.

Australian species. Five species.

Distribution. Eastern, northern, and central Australia; extraterritorial dispersed in the Oriental and Afrotropical Regions.

Biology. Hygrophilous in sand, gravel, or pebbles on the banks of rivers and the shore of lakes and lagoons. Most recorded specimens, however, were collected at light. Predacious and probably nocturnal. At some localities three *Perileptus* species occur together. Little else is known about habits, diet, and reproduction.

References. Sloane (1903); Jeannel (1926); Csiki (1928); Moore (1966a, 1972); Moore *et al.* (1987); Baehr (1987b); Lorenz (2005).

Keys to species. Moore (1966a), Baehr (1987b).

Subtribe incertae sedis

Tasmanitachoides Erwin, 1972 (Fig. 12.141)

Type species. *Bembidion hobarti* Blackburn, 1901, by original designation.

Characteristics. Small to very small, *Perileptus*-like beetles; frontal furrows rather deep; eyes pilose; palpomeres subulate; base of elytra not margined; basal margin curved into stria 5, which is deeply engraved at base; recurrent stria short, not hooked; elytra glabrous, variously striate; with three discal punctures; aedeagus sclerotised, without basal bulbus, ostium rather elongate, endophallus with a few sclerotised pieces; gonocoxites narrow and elongate, gonocoxite-2 usually with a very elongate apical seta.

Note. Due to the subulate palps this genus originally was described as belonging to the subtribe Tachyina of Bembidiini. The larval morphology (Grebennikov 2008) and molecular analyses (Maddison *in litt.*), however support its inclusion into, or a position as sister to, the whole Trechine-Bembidiine complex (Maddison *in litt.).*

Australian species. 25 species.

Distribution. Eastern Australia including Tasmania, to northern Queensland south of Cape York Peninsula, northern tropical Australia, and in the west to the Pilbara region.

Biology. Hygrophilous in sand and gravel of the banks of rivers, creeks and lagoons. In the south-east and east rather montane and in more or less dense forests, in the north and north-west even in semiarid areas. Most known specimens were collected at light, in particular those from northern Australia. Predacious and probably nocturnal. Grebennikov (2008) described the larva as being rather similar to those of Perileptina. Little else is known about habits, diet, and reproduction.

References. Moore *et al.* (1987); Baehr (1990b, 2001c, 2008b, 2008c, 2009c, 2010f, 2013c); Lorenz (2005).

Keys to species. Baehr (2001c, 2010f, 2013c).

Tribe Bembidiini Stephens, 1827

Characteristics. An extremely species-rich and diverse tribe worldwide; procoxal cavities closed; mesocoxae conjunct; head with two supraorbital setae above each eye; mandibles with a seta in the scrobe; palps subulate; frontal furrows short; basal margin of the elytra incomplete; series of marginal setiferous punctures of the elytra interrupted; elytral epipleura without plica; apex of elytral stria 3 recurrent or not; aedeagus varied, endophallus usually with one or several sclerotised pieces; parameres more or less dissimilar, triagonal, with several elongate apical setae; gonocoxites more or less elongate, gonocoxite-2 usually curved, with or without a varied number of ventral, dorsal, and subapical setae.

Note. Some Australian tachyine species at present cannot be attributed to any of the accepted tachyine genera. Therefore, they are inserted in the key to the genera as *Tachys* (*incert. sed.*) with their specific name or under a group name.

Key to the Australian subtribes of Bembidiini

1. External apical margin of protibia straight, not excised or oblique (Fig. 12.18) .... **Bembidiina Stephens**

– External apical margin of protibia excised or oblique (Fig. 12.19) **2**

2(1). Eyes absent; elytra not or barely striate but with more or less seriate pilosity; body size tiny, usually < 2 mm, rarely larger (in some Western Australian species) .... **Anillina Jeannel**

– Eyes almost always well developed (except a single species of *Lymnastis*); elytra at least with the sutural stria; body usually not pilose (except genus *Lymnastis* and a new, not yet described genus); body size commonly larger **3**

3(2). Protibia barely excised (less so than Fig. 12.19) on outer apical margin; body rather convex; terminal palpomeres remarkably elongate .... **Xystosomina Erwin**

– Protibia with distinct excision (Fig. 12.19) on outer apical margin, body commonly more depressed; terminal palpomeres not unusually elongate .... **Tachyina Motschulsky**

Key to the genera of Bembidiina, Xystosomina, and Tachyina*

*For the genera of Anillina a separate key is provided, see below.

1. Eyes absent **2**

– Eyes well developed **6**

2(1). Elytra striate or not, but with more or less seriate pilosity; if striate, elytra narrow and markedly parallel sided (Fig. 12.76); body size varied **3**

– Elytra striate, pilosity not seriate; body size > 2.2 mm .... ***Lymnastis* Motschulsky (part)**

3(2). Elytra with four distinct striae, prothorax and elytra very elongate, narrow, and parallel sided; body size 2.4 mm .... ***Angustanillus* Baehr [Anillina]**

– Elytra only with the sutural stria; body shape usually different, when very narrow and elongate, body size much smaller **4**

4(3). Elytra with a distinct longitudinal sulcus extending between the scutellary puncture and the 9th puncture of the marginal series **5**

– Elytra without such sulcus .... ***see key to Anillina***

5(4). Pronotum cordiform, elytra short and markedly oval shaped .... ***Illaphanus* Macleay [Anillina]**

– Pronotum not cordiform, elytra elongate and rather parallel sided; Western Australian genera .... ***see key to Anillina***

6(5). External apical margin of protibia straight, not excised or oblique .... ***Bembidion* Latreille**

– External apical margin of protibia excised or oblique **7**

7(6). Dorsal surface finely pilose **8**

– Dorsal surface glabrous, except for fixed discal setae **9**

8(7). Pilosity sparse, decumbent; eyes small, depressed .... ***Lymnastis* Motschulsky (part)**

– Pilosity dense, erect; eyes very large, prominently protruded .... ***Setitachys* Baehr**

9(8). Recurrent stria of elytra straight and elongate, situated very close to margin; tarsal claws denticulate .... ***Tachyta* Kirby**

– Recurrent stria of elytra differently shaped, rarely situated very close to margin but then very short; tarsal claws not denticulate **10**

10(9). Mentum with two deep foveae **11**

– Mentum without deep foveae **20**

11(10). Pronotum with punctate transverse basal sulcus, but without a deep fovea in middle; elytra with coarsely punctate or punctate-striate striae .... ***Pericompsus* LeConte**

– Not all characters together present **12**

12(11). Protibia barely excised on outer apical margin; body rather convex; terminal palpomeres remarkably elongate .... ***Philipis* Erwin**

– Protibia with distinct excision on outer apical margin, body commonly more depressed; terminal palpomeres not unusually elongate .... ***Tachys* Stephens (*sensu lato*) 13**

13(12). Recurrent stria very short **14**

– Recurrent stria elongate, more or less incurved at its end, or confluent with stria 3 .... **15**

14(13). Very small, < 2 mm long, unicolourous reddish to piceous species; basal pronotal angles wide .... ***Polyderis* Motschulsky**

– Larger, 3–4 mm long, quadrimaculate species; basal pronotal angles rectangular and acute .... ***Tachys macleayi* Sloane**

15(13). Large, wide, depressed species with very wide pronotum .... ***Tachys ectromoides*-group**

– Usually smaller, less wide species with narrower pronotum **16**

16(15). Frontal sulci absolutely straight, oblique, prolonged onto clypeus, enclosing a triangular median area .... ***Tachys mitchelli* Sloane**

– Frontal sulci not absolutely straight, more or less parallel, not enclosing a triangular median area **17**

17(16). The posterior discal puncture located on disk in front of the recurrent stria **18**

– The posterior discal puncture located within the recurrent stria, or attached to its end **19**

18(17). Recurrent stria at apex incurved, not confluent with stria 3 .... ***Tachys blackburni* Sloane**

– Recurrent stria at apex confluent with stria 3 .... ***Tachys mulwalensis* Sloane**

19(17). The posterior discal puncture located within the recurrent stria; the anterior group of marginal punctures of the elytra not in a furrow; the anterior discal puncture located at 4th or 5th stria and behind midel of the elytra .... ***Paratachys* Casey**

– The posterior discal puncture attached to the end of the recurrent stria; the anterior group of marginal punctures of the elytra situated in a furrow; the anterior discal puncture located at 3rd stria and in middle of the elytra .... ***Tachys* (s. str.) Dejean**

20(10). Frontal sulci absolutely straight, oblique, prolonged onto clypeus, enclosing a triangular median area .... ***Sphaerotachys* Müller**

– Frontal sulci not absolutely straight, more or less parallel, not enclosing a triangular median area **21**

21(20). Elytra short, oval, and dorsally convex, with a single discal puncture, stria 8 only in apical half present .... ***Elaphropus* Motschulsky**

– Elytra longer, less oval, and dorsally depressed, with two discal punctures, stria 8 complete **22**

22(21). Humerus dentate or at least produced; prebasal transverse sulcus of pronotum absent .... ***Tachylopha* Motschulsky**

– Humerus not dentate; prebasal transverse sulcus of pronotum present .... ***Tachyura* Motschulsky**

Subtribe Bembidiina Stephens, 1827

Characteristics. No recurrent stria on the elytra; straight, not excised latero-apical margin of the protibia; aedeagus quite variably shaped and structured, with or without varied sclerotised structures on the endophallus; parameres more or less dissimilar, always setose; gonocoxites varied, variously setose.

Note. Following Toledano (2005) only the genus *Bembidion* is used for the Australian Bembidiina and all included taxa are treated as subgenera.

Bembidion Latreille, 1802 (Pl. 20J; Fig. 12.18)

Type species. *Cicindela quadrimaculata* Linnaeus, 1761, by original designation.

Characteristics. Body shape, surface structure, and colouration very diverse; aedeagus variously shaped, with a wide diversity of sclerotised internal structures; parameres usually elongate, triagonal, bi- or trisetose; gonocoxites more or less elongate, gonocoxite-2 usually little curved, with 1–4 elongate ventral and 1–2 rather elongate subapical setae, with or without a dorsal seta.

Australian species. 12 species.

Distribution. The whole of Australia including Tasmania; extraterritorial worldwide.

Biology. Ground living and usually hygrophilous at the banks of rivers and creeks and margins of pools, lakes, and billabongs. Some species also in saline habitats and/or at the sea shore. Predacious, devouring small insects and their larvae, spiders, and worms. Diurnal or nocturnal. Some species are commonly attracted at light. Little else is known about habits, diet, and reproduction in Australia.

Note. Two species were introduced into Australia: *B. (Notaphus) brullei* Gemminger & Harold, 1868 from South America and *B. (Peryphus) tetracolum* Say, 1823, probably from Europa.

References. Csiki (1928); Netolitzky (1942, 1943); Darlington (1962a, 1962b); Matthews (1980); Moore *et al.* (1987); Lorenz (2005); Toledano (2005).

Key to species. Toledano (2005).

Key to the Australian subgenera of the genus *Bembidion*

1. Elytra with large, yellow humeral and apical spots that are separated along suture and laterally, forming a cruciate dark pattern .... ***Peryphus*** **Stephens**

– Colour pattern of elytra different **2**

2(1). Head very large; mandibles elongate and straight; elytra elongate and depressed .... ***Desarmatocillenus*** **Netolitzky**

– Head smaller, if rather large, eyes markedly protruded and mandibles short; elytra usually less elongate **3**

3(2). Surface dull from distinct microreticulation; elytra with a pale apical lunula; head large with normal shaped mandibles and large, markedly protruded eyes **4**

– Surface various; if dull, head of normal size and eyes less produced; elytral pattern various **5**

4(3). Pronotum cordiform, constricted posteriorly, base laterally not excised .... ***Notaphocampa*** **Netolitzky**

– Base of pronotum laterally deeply excised .... ***Sloanephila*** **Netolitzky**

5(4). Body size large, > 5 mm; pronotum transverse, depressed, and rectangular; elytra with several pale spots .... ***Notaphus*** **Stephens**

– Body size smaller, < 4.5 mm; pronotum narrower, convex, and more cordiform; elytra unicolourous or with a single apical lunule **6**

6(5). Body size larger, > 3.6 mm, typically larger; elytra elongate and rather parallel-sided; basal angles of pronotum very obtuse; two basal tarsomeres of the protarsus markedly widened .... ***Ananotaphus*** **Netolitzky**

– Body size smaller, < 3.6 mm, usually smaller; elytra shorter and laterally more rounded; basal angles of pronotum rather angulate; two basal tarsomeres of the protarsus normally widened **7**

7(6). Elytra with additional setiferous punctures on intervals 3 and 5; basal angles of pronotum less angulate .... ***Gondwanabembidion*** **Toledano**

– Elytra without additional setiferous punctures on intervals 3 and 5; basal angles of pronotum angulate .... ***Australoemphanes*** **Toledano**

Subgenus *Desarmatocillenus* Netolitzky, 1942

Type species. *Cillenus yokohamae* Bates, 1883, by original designation.

Characteristics. Medium-sized to rather large beetles; head very large with straight, elongate mandibles; elytra elongate, more or less parallel-sided, dorsally depressed, without cruciate pattern.

Australian species. Four species.

Distribution. Coastal in northern and eastern Australia, in the south apparently to northern Tasmania; extraterritorial from Japan and eastern China through the Philippine and Indonesian Insular belts, to Fiji Islands.

Biology. Halophilous, predacious, nocturnal species of the sea shore and near-shore saline habitats, where they may hunt small insects and crustaceans. All Australian species are very rarely collected, and mostly at light.

References. Lindroth (1980); Moore *et al.* (1987); Baehr (1995b); Lorenz (2005); Toledano (2005).

Key to species. Toledano (2005).

Subgenus *Ananotaphus* Netolitzky, 1931

Type species. *Bembidion errans* Blackburn, 1888, by original designation.

Characteristics. Medium-sized beetles; head normal shaped with moderate mandibles; pronotum rather cordiform, with very obtuse basal angles; elytra fairly elongate, without cruciate pattern; two basal tarsomeres of protarsus markedly widened.
Australian species. Two species.
Distribution. Southern Australia from southern Western Australia to south-eastern New South Wales, Tasmania; extraterritorial in New Zealand.
Biology. The species are ground living and hygrophilous, and occur close to the coast in coastal heath or in estuarine habitats; one species is only known from saline mound springs in central South Australia. Predacious and apparently at least partly diurnal.
References. Csiki (1933); Netolitzky (1942, 1943); Darlington (1962b; Moore *et al.* (1987); Lorenz (2005); Toledano (2005).
Key to species. Toledano (2005).

Subgenus *Sloanephila* Netolitzky, 1931 (Pl. 20J)

Type species. *Bembidion jacksoniense* Guérin-Méneville, 1830, by original designation.
Characteristics. Rather small to medium-sized beetles; head normal shaped with moderate mandibles; eyes large and laterally well produced; pronotum rather cordiform, with laterally deeply excised base; elytra fairly elongate, greenish with pale apical lunule, microreticulation distinct; two basal tarsomeres of protarsus normally widened.
Australian species. Only *B.* (*S.*) *jacksoniense* Guérin-Méneville.
Distribution. Almost the whole of mainland Australia; extraterritorial in the Holarctic Region.
Biology. Ground living, hygrophilous, very common at the margins of all sorts of water bodies, including saline habitats. Apparently at least partly diurnal, because commonly running in bright sunshine on wet sandy banks or mudflats. Predacious, specimens have been observed to chase small flies, but may hunt various other small insects, spiders, and small crustaceans. Little else is known about habits, diet, and reproduction.
References. Csiki (1933); Netolitzky (1942, 1943); Moore *et al.* (1987); Toledano (2005).

Subgenus *Notaphocampa* Netolitzky, 1914

Type species. *Bembidion niloticum* Dejean, 1831, by original designation.
Notaphomimus Netolitzky, 1931
Characteristics. Medium-sized beetles; head normal shaped head with moderate mandibles; eyes large and laterally well produced; pronotum cordiform, with angulate basal angles, base laterally not excised; elytra fairly elongate, greenish with pale apical lunule, microreticulation distinct; two basal tarsomeres of protarsus normally widened.
Australian species. Only *B.* (*N.*) *opulentum riverinae* Sloane, 1895.
Distribution. Most of mainland Australia; extraterritorial in the Ethiopian and Oriental regions, and on New Caledonia, Solomon Islands, and Fiji Islands.
Biology. Ground living, hygrophilous, common at the margins of all sorts of water bodies, including saline habitats. At least partly diurnal, because specimens commonly are observed running in bright sunshine on wet sand or mudflats. Also rather commonly attracted to light. Predacious. Little else is known about habits, diet, and reproduction.
Note. The single Australian species previously was regarded a subspecies of the Oriental species *B. opulentum* Nietner.
References. Csiki (1928); Darlington (1962b); Moore *et al.* (1987); Lorenz (2005); Toledano (2005).

Subgenus *Notaphus* Stephens, 1828

Type species. *Carabus varius* Olivier, 1795, by subsequent designation by Westwood (1838a).
Characteristics. Rather large sized, depressed beetles; head normal shaped, with moderate mandibles prominent; eyes large and laterally well produced; pronotum wide, cordiform, with angulate basal angles, base not laterally excised; elytra fairly elongate, parallel-sided, and depressed, bronzed with several pale spots, microreticulation moderate; two basal tarsomeres of protarsus normally widened.
Australian species. Only *B.* (*N.*) *brullei* Gemminger & Harold, 1868.
Distribution. South eastern Australia from eastern South Australia to New South Wales, also on Lord Howe Island; extraterritorial in the Holarctic Region and in South America, and New Zealand.
Biology. Ground living, hygrophilous, at the margins of pools, lakes, and slow flowing rivers. Predacious and probably at least partly diurnal.
Note. The single Australian species was introduced from South America and apparently first recorded in 1957 (Moore *et al.* 1987).
References. Csiki (1928); Netolitzky (1942, 1943); Darlington (1962b); Moore *et al.* (1987); Lorenz (2005); Toledano (2005).

Subgenus *Peryphus* Stephens, 1828

Type species. *Carabus litoralis* Olivier, 1795 (= *Carabus ustulatus* auctorum = *Bembidion tetracolum* Say, 1823), by subsequent designation by Jeannel (1941).
Characteristics. Medium-sized rather convex beetles; head normal shaped head with moderately prominent mandibles; eyes moderately large and laterally well produced;

pronotum cordiform, with rectangular basal angles; elytra rather short and wide, dorsally convex, yellow with dark cruciate pattern.

Australian species. Only *B.* (*P.*) *tetracolum* Say, 1823.

Distribution. Victoria; extraterritorial in the Holarctic, Ethiopian, and Neotropical Regions.

Biology. Ground living and hygrophilous, usually at the sandy or gravely shores of rivers and lakes. In Europe a common species that is partly diurnal, seen chasing small flies, other small insects, their larvae, and spiders. The larvae live under gravel and stones and can be found hunting at night.

Note. The single Australian species was introduced probably from Europe.

References. Csiki (1928); Netolitzky (1942, 1943); Moore *et al.* (1987); Lorenz (2005); Toledano (2005).

Subgenus *Australoemphanes* Toledano, 2005

Type species. *Bembidion blackburni* Csiki, 1928, by original designation.

Characteristics. Rather small, convex beetles; head normal shaped head with moderately prominent mandibles; eyes rather large and moderately produced laterally; pronotum narrow, cordiform, with angulate basal angles, base not laterally excised; elytra fairly elongate, parallel-sided, dorsally convex; without additional setae on intervals 3 and 5; black with or without small apical pale spot, glossy; two basal tarsomeres of protarsus normally widened.

Australian species. Only *B.* (*A.*) *ateradustum* Liebherr, 2008.

Distribution. Southern Australia from South Australia to New South Wales, Tasmania.

Biology. Hygrophilous, ground living, apparently as well in estuarine habitats as at high altitude, where it occurs also on wet, open heaths near and above tree line (Darlington 1962b). Predacious and apparently at least partly diurnal, because Darlington reported specimens 'running in sunlight on the ground in open places'.

References. Stephens (1828); Csiki (1928); Andrewes (1925); Darlington (1962a, 1962c, 1963a); Matthews (1980); Moore *et al.* (1987); Sciaky & Vigna Taglianti (2003); Lorenz (2005).

Key to species. Sloane (1921).

Subgenus *Gondwanabembidion* Toledano, 2005

Type species. *Bembidion proprium* Blackburn, 1888, by original designation.

Characteristics. Rather small, convex beetles; head normal shaped head with moderately prominent mandibles; eyes rather large and moderately produced laterally; pronotum narrow, moderately cordiform, with rather obtuse basal angles, base laterally not excised; elytra fairly elongate, parallel-sided, dorsally convex; with additional setae on intervals 3 and 5; black or brown with a small apical spot, glossy; two basal tarsomeres of protarsus normally widened.

Australian species. *B.* (*G.*) *proprium* Blackburn.

Distribution. Southern Australia from South Australia to New South Wales, and Tasmania.

Biology. Hygrophilous, ground living, apparently mainly in estuarine habitats. Predacious, probably at least partly diurnal.

References. Toledano (2005).

Subtribe Tachyina Motschulsky, 1862

Characteristics. Terminal palpomeres not unusually elongate; eyes typically well developed (except in a single cavernicolous species of *Lymnastis* Motschulsky from Christmas Island); usually a recurrent stria on the elytra present; protibia with oblique moderately to deeply excised latero-apical margin (Fig. 12.19); aedeagus variously sized and shaped, endophallus commonly with one or more, variously sclerotised pieces; parameres more or less dissimilar, variously shaped, 2–5-setose; gonocoxites elongate, gonocoxite-2 most commonly with a single ventral seta and a subapical seta, sometimes also with a dorsal seta (but female genitalia recorded only from few species).

Notes. The Australian Tachyina urgently needs a thorough revision, because many additional species and some that are divergent enough that they may require an additional genus are known to exist. Hence, apart from the genera *Pericompsus, Sphaerotachys, Tachyura, Tachylopha*, and *Tachyta*, identifications presently are almost impossible. Moreover, several Australian species have not yet been placed in genera and so are species *incertae sedis*.

Although there are several papers by recent authors (Erwin 1974a; Baehr 1988a, 1989b, 2017e; Bruneau de Miré 1991; Ball & Bousquet 2001; Kopecky 2003; Sciaky & Vigna Taglianti 2003; Lorenz 2005) the generic and subgeneric taxonomy of the former genus '*Tachys sensu latissimo*' still is remarkably controversial. Here we try to find an intermediate and practical system regarding genera and several subgenera as used by previous authors. We do that in order to avoid confusion, to establish putative monophyletic units, and to facilitate the use of the present chapter for identification. We cannot provide the needed analysis-based phylogeny as a basis for the classification. At present this is impossible, given that some Australian tachyine species still are of uncertain systematic status and, moreover, because a large part of the recognised Australian species urgently needs a thorough revision. Moreover, as long as the taxonomy of the Oriental, New Guinean, and Afrotropical Tachyina has not been evaluated and adjusted, the taxonomy of the Australian Tachyina cannot be placed into a generally accepted or acceptable system.

Those species of the subtribe Tachyina that presently are not placed to genus (as the genera are given in this chapter), trace in the keys to '*Tachys incert. sed.*' or as '*ectromoides*-group', respectively.

The genera *Tachys s. str., Paratachys,* and *Polyderis* currently are being revised (MB). Therefore affiliations of some species to one of these genera in the present paper deviates from those in some recent publications, e.g. Lorenz (2005).

Tachys Dejean, 1821 (Fig. 12.138)

Type species. *Tachys scutellaris* Stephens, 1828, by subsequent designation by Hope (1834).

Characteristics. Medium-sized beetles; mentum with two deep pits; elytra wide and depressed; bisetose on disc; recurrent stria elongate, situated in the middle of the elytra, at apex incurved; the posterior discal puncture located within the recurrent stria; tarsal claws not denticulate; aedeagus short and stout, endophallus with a sclerotised piece; parameres elongate, rather dissimilar, triagonal, bisetose, gonocoxites in one Australian species elongate, gonocoxite-2 with a ventral and a subapical seta.

Australian species. Seven species.

Distribution. Almost the whole of Australia, extraterritorial almost worldwide.

Biology. All Australian species are hygrophilous and live at the banks of rivers and brooks, at the shores of lakes and lagoons, also on the sea shore. Predacious and probably mainly nocturnal.

Notes. Darlington (1962a) mentioned the species of this genus as the *Tachys quadrillum* group. The species presently are being revised and a couple of additional species is known to exist.

References. Stephens (1828); Csiki (1928); Andrewes (1925); Baehr (1989b); Darlington (1962a, 1962c, 1963a); Matthews (1980); Moore *et al.* (1987); Sciaky & Vigna Taglianti (2003); Lorenz (2005).

Key to species. Sloane (1921).

Paratachys Casey, 1918 (Figs 12.19, 12.20, 12.115)

Type species. *Paratachys austinicus* Casey, 1918, by original designation.

Characteristics. Moderately small to medium-sized, rather depressed beetles; mentum with two deep pits; elytra bisetose on disc; recurrent stria elongate, situated in the middle of the elytra, at apex little or not incurved; the posterior discal puncture situated within the recurrent stria, but usually near the apex; elytra moderately depressed; tarsal claws not denticulate; aedeagus variously sized and shaped, endophallus with a sclerotised piece; parameres moderately elongate, rather dissimilar, triagonal, usually trisetose; gonocoxites elongate, gonocoxite-2 most commonly with one ventral, one dorsal, and one subapical seta, but varied.

Australian species. Eight species.

Distribution. Almost the whole of Australia; extraterritorial almost worldwide.

Biology. The described Australian species are hygrophilous and live at the banks of rivers and brooks, at the shores of lakes and lagoons, including saline habitats. Predacious and probably mainly nocturnal. Many species are commonly attracted to light. Little else is known about habits, diet, and reproduction.

Notes. Darlington (1962a) mentioned the species of this genus as the *Tachys fasciatus* group. Identification of species presently is difficult. The genus presently is being revised and several additional species are known to occur. The key in Baehr (2012i) only covers species related to *T. mastersi* Sloane.

References. Csiki (1928); Sciaky & Vigna Taglianti (2003); Lorenz (2005); Baehr (2012i, 2017e).

Keys to species. Sloane (1921); Baehr (2012i).

Polyderis Motschulsky, 1862 (Figs 12.14, 12.47, 12.124)

Type species. *Tachys brevicornis* Chaudoir, 1846, by subsequent designation by Jeannel 1941.

Characteristics. Very small beetles; body shape various; mentum with or without two deep pits; elytra bisetose on disc; basal angles of pronotum rather wide; recurrent stria very short or almost invisible; tarsal claws not denticulate; aedeagus very short and stout, endophallus with a sclerotised piece; parameres rather short dissimilar, triagonal, bisetose. Female genitalia not yet recorded.

Several Australian species lacks the mental foveae, a character which so far was known only from the Oriental-New Guinean species *P. truncatus* (Nietner).

Australian species. Six species.

Distribution. Almost the whole of Australia; extraterritorial almost worldwide.

Biology. The Australian species are hygrophilous and live at the banks of rivers and shores of lakes and lagoons. Predacious and probably mainly nocturnal.

Notes. Darlington (1962a) mentioned the species of this genus as the *Tachys truncatus* group. Identification of species presently is extremely difficult, because for this genus the key of Sloane (1921) is almost unserviceable. The genus is currently being revised and several additional species are known to exist.

References. Csiki (1928); Andrewes (1925); Moore *et al.* (1987); Giachino (2003b); Sciaky & Vigna Taglianti (2003); Lorenz (2005).

Key to species. Sloane (1921).

ectromoides-group

Characteristics. Comparatively large beetles; mentum with two deep pits; elytra wide and very depressed, bisetose on disc; recurrent stria elongate, situated in the middle of the elytra, at apex incurved; tarsal claws not denticulate; aedeagus short and compact, endophallus with a sclerotised piece; paremeres moderately dissimilar, with 4–5 setae; gonocoxite-2 narrow and elongate, with two fairly elongate ventral setae, one short dorsal seta, and one very elongate subapical seta.

Australian species. Nine species.

Distribution. Southern and eastern Australia from southern Western Australia to Windsor Tableland in north Queensland; extraterritorial in New Guinea.

Biology. All species seem to live either under loose bark of trees, or on moss covered bark, as well in rainforests, as in open forests and woodlands. Predacious and probably nocturnal.

Note. This group of species at present cannot be attributed to any of the presently described genera.

References. Baehr (1989a, 1991a, 2003b, 2016g, 2017e).

Keys to species. Baehr (1991a, 2003b, 2016g, 2017e).

Setitachys Baehr, 2016 (Fig. 12.130)

Type species. *Setitachys macrops* Baehr, 2016, by original designation.

Characteristics. Small, narrow and elongate, pale beetles; the whole surface with dense, erect pilosity; mentum with two deep pits; eyes very large, promiently protruded; elytra bisetose on disc; pilosity on the elytra seriate; recurrent stria elongate, situated in the middle of the elytra, at apex incurved; the posterior discal puncture situated within the recurrent stria; tarsal claws not denticulate; aedeagus very short and compact, endophallus with a coiled, sclerotised piece near base; parameres elongate, rather dissimilar, triagonal, bisetose; gonocoxites moderately elongate, gonocoxite-2 triangular, with one ventral and one subapical seta.

Australian species. Only *S. macrops* Baehr.

Distribution. Northern Queensland: Cape York Peninsula and South Wellesley Islands in the Gulf of Carpenteria.

Biology. Hygrophilous and at the banks of rivers. Predacious and probably mainly nocturnal, one specimen attracted to light.

References. Baehr (2016a).

Pericompsus LeConte, 1852 (Pl. 27L; Fig. 12.118)

Type species. *Bembidion ephippiatum* Say, 1834, by subsequent designation by LeConte 1859.

Characteristics. Rather small, dorsally fairly convex beetles; mentum with two deep pits; pronotum with punctate transverse basal sulcus, but without a deep fovea in the middle; elytra with coarsely punctate or punctate-striate striae, bisetose on disc; recurrent stria short, situated more close to the lateral margin than to the middle of the elytra, at apex barely incurved; the posterior discal puncture situated within the recurrent stria; tarsal claws not denticulate; aedeagus short and stout, endophallus usually with a weakly sclerotised piece; parameres fairly elongate, moderately dissimilar, triagonal, with 3–5 setae; gonocoxites rather elongate, gonocoxite-2 with a ventral seta, without or with a dorsal seta, and with a subapical seta.

Australian species. Nine species.

Distribution. Almost the whole of Australia, including Tasmania and Lord Howe Island; extraterritorial in the Neotropical and Nearctic Regions. One Australian species is likely a human mediated adventive in New Zealand (Darlington 1963a; Larochelle & Larivière 2001).

Biology. Hygrophilous at the margins of various water bodies, including saline habitats. The species are nocturnal, because they are commonly attracted to light. Predacious.

Notes. Darlington (1963a) mentioned the species of this genus as the *Tachys australis* group. All Australian species belong in the subgenus *Upocompsus* Erwin, 1974.

References. LeConte (1859); Andrewes (1925); Csiki (1928); Moore *et al.* (1987); Erwin (1974a); Lorenz (2005); Baehr (2017e).

Key to species. Erwin (1974a); Baehr (2017e).

Subgenus *Upocompsus* Erwin, 1974

Type species. *Tachys australis* Schaum, 1863, by original designation.

Characteristics. As for the genus; distinguished from the two other American subgenera by the 8th stria that is sulcate only in the apical third, and consists of a row of punctures in the basal two-thirds.

Australian species. Nine species.

Distribution. Only in Australia. One introduced species in New Zealand.

References. Andrewes (1925); Csiki (1928); Jeannel (1932); Andrewes (1925, 1935a); Darlington (1962a); Moore *et al.* (1987); Lorenz (2005); Baehr (2008d, 2017e).

Keys to species. Jeannel (1932); Darlington (1962a); Baehr (2008c, 2017e).

Lymnastis Motschulsky, 1862 (Fig. 12.105)

Type species. *Lymnaeum indicum* Motschulsky, 1851, by subsequent designation by Jeannel 1932.

Characteristics. Small, very elongate, parallel-sided, depressed, pale beetles; mentum with two deep pits; eye variously sized, but eye in one species from Christmas Island absent; antenna moniliform; surface pilose; elytra completely striate, bisetose on disc; recurrent stria short, united with stria 3; tarsal claws not denticulate; aedeagus short and stout, endophallus without a sclerotised piece; parameres very dissimilar, the left one much larger than right, both trisetose; gonocoxites very elongate, gonocoxite-2 curved, acute, with 1–2 ventral setae, apparently without dorsal seta, with or without a short subapical seta (but female genitalia recorded only from very few species).

Australian species. Two species.

Distribution. Eastern and northern tropical Australia from eastern Victoria to northern parts of Northern Territory and Western Australia, also on Christmas Island; extraterritorial in the Palearctic, Ethiopian, Oriental, Papuan Regions, and Hawaii.

Biology. Hygrophilous, but ecology unrecorded. Probably strictly nocturnal, because almost all specimens were collected at light, usually in or close to wet environments. The species from Christmas Island is blind and lives in volcanic lava tube caves.

Note. Additional species are known to exist and await their description.

References. Andrewes (1925); Csiki (1928); Jeannel (1932); Andrewes (1925, 1935a); Darlington (1962a); Moore *et al.* (1987); Lorenz (2005); Baehr (2008d).

Keys to species. Jeannel (1932); Darlington (1962a); Baehr (2008c).

Elaphropus Motschulsky, 1839 (Pl. 103G)

Type species. *Elaphropus caraboides* Motschulsky, 1839, by original designation.

Characteristics. Small, dorsally convex beetles; mentum without pits; frontal sulci rather parallel-sided; elytra oval shaped, unistriate; with a single seta on disc; recurrent stria rather short, situated in the middle of the elytra; 8th stria not impressed in the basal half; tarsal claws not denticulate; aedeagus rather short and stout, endophallus with a sclerotised piece; parameres moderately elongate, triagonal, bisetose. Female genitalia not recorded.

Australian species. Only *E. latissimus* (Motschulsky, 1851).

Distribution. Eastern and northern tropical Australia, Tasmania; extraterritorial in the Palearctic, Afrotropical, Oriental, and Papuan Regions.

Biology. Hygrophilous at the banks of rivers and shores of lakes and lagoons. Predacious and probably mainly nocturnal.

Notes. Darlington (1962a) mentioned the species of this genus as the *Tachys haliploides* group. The single Australian species was formerly known under the name *Tachys bifoveatus* (Macleay, 1871).

References. Andrewes (1925); Csiki (1928); Sciaky & Vigna Taglianti (2003); Lorenz (2005).

Sphaerotachys Müller, 1926 (Fig. 12.133)

Type species. *Carabus hoemorrhoidalis* Ponza, 1805, by subsequent designation by Müller (1926).

Characteristics. Rather small beetles; mentum without pits; frontal sulci straight and oblique, enclosing a convex, triangular field; elytra rather short and convex, bipunctate and bistriate; recurrent stria situated in the middle of the elytra; 8th stria complete; tarsal claws not denticulate; aedeagus short and stout, endophallus with a sclerotised piece; parameres moderately elongate, triagonal, trisetose; gonocoxites elongate, gonocoxite-2 with one ventral, one dorsal, and one subapical seta.

Australian species. Only *S. fumicatus* (Motschulsky, 1851).

Distribution. Eastern and northern tropical Australia; extraterritorial in the Palearctic, Afrotropical, Oriental, and Papuan Regions.

Biology. Hygrophilous at the banks of rivers and shores of lakes and lagoons. A fairly common species. Predacious and at least partly diurnal, as specimens sometimes can be seen running in sunshine on wet ground.

Notes. Darlington (1962a) mentioned the species of this genus as the *Tachys fumicatus* group. In Moore *et al.* (1987) the single Australian species is still kept under the name *T. curticollis* Sloane, 1896. The genus *Sphaerotachys* as used here does not fully correspond to the name as used by Sciaky & Vigna Taglianti (2003).

References. Csiki (1933); Baehr (1988a); Sciaky & Vigna Taglianti (2003); Lorenz (2005).

Tachyura Motschulsky, 1862 (Pl. 25G)

Type species. *Elaphrus quadrisignatus* Duftschmid, 1812, by subsequent designation by Motschulsky (1862).

Characteristics. Moderately small to fairly large, usually rather convex beetles; mentum without pits; frontal sulci rather parallel sided; base of pronotum with a transverse sulcus; elytra variously striate, in one species even the striae are duplicated; bisetose; recurrent stria situated in the middle of the elytra; 8th stria complete and sulcate; tarsal claws not denticulate; aedeagus rather short and stout, endophallus with a sclerotised piece; parameres moderately elongate, triagonal, bi- or trisetose; gonocoxites elongate, number of setae varied, most commonly gonocoxite-2 with one ventral and one subapical seta.

Australian species. 26 species.

Distribution. Almost the whole of mainland Australia but more common in the north; extraterritorial in the Palearctic, Afrotropical, Oriental, and Papuan Regions.

Biology. Hygrophilous at the banks of rivers and shores of lakes and lagoons. Some species are very common on wet sand or mud and may roam about even in bright sunshine, but they are also very commonly attracted to light. They are predacious and have been observed chasing small flies.

Notes. The genus *Tachyura* as used in the present chapter is equivalent to the *politus*-group of Andrewes (1925) and Darlington (1962a). It does not fully correspond to the name as used by Sciaky & Vigna Taglianti (2003).

References. Andrewes (1925); Csiki (1928); Baehr (1988a, 2017e); Sciaky & Vigna Taglianti (2003); Lorenz (2005).
Key to species. Baehr (1988a, 2014d, 2017e).

Tachylopha Motschulsky, 1862 (Fig. 12.137)

Type species. *Lopha ovata* Motschulsky, 1862, by monotypy.
Characteristics. Moderately large, usually rather convex beetles; mentum without pits; frontal sulci rather parallel-sided; base of pronotum without a transverse sulcus; humerus dentate or at least prominent; elytra unistriate or bistriate, bisetose; recurrent stria situated in the middle of the elytra; 8th stria complete and sulcate; tarsal claws not denticulate; aedeagus moderately short and stout, with slightly prolonged apex, endophallus with a sclerotised piece; parameres moderately elongate, triagonal, trisetose; gonocoxites elongate, gonocoxite-2 with one ventral and one subapical seta, apparently without a dorsal seta.
Australian species. Three species.
Distribution. The northern half of Australia and south-eastern Australia from mid-eastern New South Wales, also in the Interior; extraterritorial in the Afrotropical and Oriental Regions.
Biology. Hygrophilous at the banks of rivers and shores of lakes and lagoons. Predacious and probably mainly nocturnal, because specimens are commonly attracted to light.
References. Andrewes (1925); Csiki (1928); Baehr (1988a); Sciaky & Vigna Taglianti (2003); Lorenz (2005).
Key to species. Baehr (1988a).

Tachyta Kirby, 1837 (Fig. 12.139)

Type species. *Tachyta picipes* Kirby, 1837 (= *Tachys nanus inornatus* Say, 1823), by monotypy.
Characteristics. Rather small beetles; mentum without pits; body shape various, either wide, rather parallel-sided, depressed, and distinctly microreticulate, or oval-shaped, dorsally convex, and glossy; elytra bisetose on disc; recurrent stria elongate, straight, situated close to the margin of the elytra, at apex barely incurved; tarsal claws denticulate; aedeagus short and stout, endophallus with a sclerotised piece; parameres moderately dissimilar, elongate, trisetose; gonocoxites elongate, gonocoxite-2 acute, slightly curved, with two large ventral setae and two elongate subapical setae.
Australian species. Six species.
Distribution. Tropical northern Australia from northern Queensland to northernmost Western Australia; extraterritorial almost worldwide.
Biology. Most Australian species live under the loose bark of dead and living trees, as well in rainforests or more open forests. Sometimes they have been observed in large aggregations under loose bark of dead trees. Predacious and probably nocturnal.
Notes. Darlington (1962a) mentioned the species of this genus as the *Tachys nana* group.
The genus includes three subgenera, two of which are represented in Australia. An additional species presently is being described (MB).
References. Schaum (1863); Csiki (1928); Andrewes (1925); Erwin (1975); Baehr (1986a, 2006f, 2012h); Moore *et al.* (1987); Moore (2000); Sciaky & Vigna Taglianti (2003); Lorenz (2005).
Key to species. Baehr (2012h).

Key to the Australian subgenera of *Tachyta* Kirby

1. Dorsal surface with visible, though in some species rather superficial microreticulation; elytra depressed, with relatively finely punctate striae and punctate intervals .... ***Tachyta* Kirby**
– Dorsal surface without microreticulation; elytra moderately convex, with very coarsely punctate striae and impunctate intervals .... ***Australotachyta* Baehr**

Subgenus *Tachyta* Kirby, 1837

Type species. *Tachyta picipes* Kirby, 1837 (= *Tachys nanus inornatus* Say, 1823), by monotypy.
Characteristics. Body more or less elongate, depressed; elytra with distinct, though in some species superficial microreticulation and punctate intervals.
Australian species. Five species.
Distribution. Tropical northern Australia from northern Queensland to northernmost Western Australia; extraterritorial almost worldwide.
Biology. All species live under the loose bark of trees, as well in rainforests and in more open forests.
References. Schaum (1863); Csiki (1928); Andrewes (1925); Erwin (1975); Baehr (1986a, 2006f, 2012h); Moore *et al.* (1987); Moore (2000); Sciaky & Vigna Taglianti (2003); Lorenz (2005).
Key to species. Baehr (2012h).

Subgenus *Australotachyta* Baehr, 2012

Type species. *Tachyta punctipennis* Baehr, 2012, by original designation.
Characteristics. Body dorsally convex; elytra oval shaped, lacking microreticulation; intervals impunctate, but striae very coarsely punctate.

Australian species. Only *T. (A.) punctipennis* Baehr.

Distribution. McIllwraith Range, central Cape York Peninsula, north Queensland, known only from the type locality.

Biology. The holotype of this species was collected by fogging from a *Pandanus* tree in an open forest.

References. Baehr (2012h).

Subtribe Xystosomina Erwin, 1994

Characteristics. Rather small, usually fairly convex beetles; mentum with two deep pits; apical palpomeres markedly elongate; pronotum with transverse basal sulcus, with a deep fovea in the middle; basal grooves elongate, oblique, laterally bordered by a distinct carina; elytra variously striate, bisetose on disc; recurrent stria elongate, situated in the middle of the elytra, at apex barely incurved; laterobasal part of the protibia only slightly excised (less than as in Fig. 12.19); tarsal claws not denticulate; aedeagus with a coiled sclerite on the endophallus; parameres rather similar, triagonal, with four or five apical setae; gonocoxites elongate, gonocoxite-2 ensiform with ventral, dorsal, and subapical setae.

Notes. The evidence of this subtribe is doubtful due to the paucity of significant character differences among this subtribe and putative near relatives. The subtribe is represented in South America by several genera and a multitude of species; in Australia only by the genus *Philipis*.

Philipis Erwin, 1974 (Fig. 12.120)

Type species. *Tachys trunci* Darlington, 1963, by original designation.

Characteristics. Flight wings present or more or less reduced, most species bear an elytral pattern of spots or fasciae.

Australian species. 44 species.

Distribution. Eastern Australia from extreme south-eastern Victoria to Windsor and Carbine Tablelands in north-eastern Queensland, but most common in north-eastern Queensland.

Biology. Ground living or on the lower parts of moss covered rainforest trees, usually in montane rainforests, rarely at low altitude. Some species possess very small ranges at the uppermost tops of high mountains in north Queensland. Predacious and apparently nocturnal. In most species almost all specimens have been sampled by fogging trunks of rainforest trees.

Note. An additional undescribed species is known and is presently being described (MB).

References. Baehr (1995a, 2002b, 2008g); Lorenz (2005).

Keys to species. Baehr (1995a, 2002b, 2008g).

Subtribe Anillina Jeannel, 1937

Characteristics. Tiny, depressed, depigmented, pilose beetles; eyes absent; elytra most commonly unistriate, commonly with an elongate sulcus running down at the middle of the elytra; flight wings absent; legs short, tarsi not denticulate; aedeagus usually with a more or less darkly sclerotised sclerite on the endophallus, parameres triagonal, usually with two apical setae; female gonocoxites of Australian species only recorded from a few Western Australian genera (Baehr & Main 2016). Gonocoxite-2 elongate, curved, acute with or without one dorso-median and ventro-lateral ensiform seta, but always with an elongate subapical nematiform seta.

Note. An additional new genus is known to exist and currently is being described (MB).

Key to the genera of Australian Anillina

1. Elytra with four distinct striae; both, pronotum and elytra, very elongate, ratio width/length of pronotum 0.86, ratio length/width of elytra > 2.2; surface of pronotum and elytra with distinct, deep and large punctures. North-west WA: Pilbara .... ***Angustanillus* Baehr**

– Elytra with at most a faint parasutural stria; pronotum and elytra less elongate, ratio width/length of pronotum > 0.97, ratio length/width of elytra < 2.15; surface of pronotum without or with less deep and large punctures **2**

2(1). Elytra with a distinct longitudinal sulcus extending between the scutellary puncture and the 9th puncture of the marginal series (in one species from WA only visible in basal third) **3**

– Elytra without longitudinal sulcus **5**

3(2). Body size large, always > 2 mm; base of elytra wide, humerus almost rectangular; elytra with three discal punctures. North-west WA: Pilbara .... ***Magnanillus* Baehr**

– Body size smaller, < 1.75 mm (usually much less); base of elytra and humerus variously shaped; elytra with at most two discal punctures **4**

4(3). Elytra more or less oval-shaped, decidedly narrowed to base, humerus oblique or rounded; prothorax markedly cordiform. Eastern Australia .... ***Illaphanus* Macleay**

– Elytra parallel-sided or almost so, little or not narrowed to base, humerus almost rectangular; prothorax usually rather quadrate. North-west WA: Pilbara .... ***Gracilanillus* Baehr**

5(2). Body size large, > 2.4 mm **and** pronotum almost as long as wide **and** base much wider than apex **and** humerus oblique-convex; apex of aedeagus slightly bent down. North-west WA: Pilbara .... ***Pilbaranillus* Baehr**

– Body size smaller, < 1.8 mm; **either** pronotum rather quadrate with base about as wide as apex, and with elongate elytra with almost rectangular humerus; **or** pronotum cordiform with much narrower base than apex, and with more or less oval-shaped elytra with oblique or rounded humerus; apex of aedeagus not bent down **6**

6(5). Pronotum not cordiform, base as wide as apex; elytra elongate with almost rectangular humerus. South-east NSW ....***Austranillus* Giachino**
– Pronotum decidedly cordiform, base considerably narrower than apex; elytra more or less oval-shaped or with oblique-convex humerus 7
7(6). Elytra with an additional marginal puncture, situated slightly inside of 2nd marginal puncture (Fig. 12.141) **and** elytra distinctly oval-shaped and with oblique-convex humerus. TAS .... ***Tasmanillus* Giachino**
– Elytra without an additional marginal puncture (Fig. 12.126); elytra oval shaped or not; not in Tasmania 8
8(7). Body size smaller, length < 1.26 mm; pronotum very wide, ratio width/length > 1.35; elytra shorter, ratio length/width < 1.55. Eastern NSW, south-east QLD.. ***Pseudillaphanus* Giachino**
– Body size larger, length > 1.35 mm; pronotum usually narrower; **when** ratio width/length > 1.30, body size larger, length c. 1.50 mm, and elytra longer, ratio length/width 1.75; elytra longer, ratio length/width > 1.65. North-west and interior southern WA 9
9(8). Body size larger, length > 1.50 mm; eye area markedly prominent, head almost as wide as pronotum; pronotum wider, ratio width/length > 1.18. North-west WA: Pilbara .... ***Hesperanillus* Baehr**
– Body size smaller, length 1.35 mm; eye area not prominent, head decidedly narrower than pronotum; pronotum narrower, ratio width/length 1.05. interior southern WA..***Externanillus* Baehr**

Illaphanus Macleay, 1865 (Fig. 12.100)

Type species. *Illaphanus stephensii* Macleay, 1865, by monotypy.

Characteristics. Mentum with or without tooth; elytra with a longitudinal sulcus between the parascutellar puncture and the 9th puncture of the marginal series; without an additional marginal puncture at level of 2nd marginal puncture; aedeagus very variably shaped, endophallus with a more or less feebly sclerotised piece; parameres varied, triagonal, both bisetose. Female genitalia not yet recorded.

Australian species. 30 species.

Distribution. South-eastern and eastern Australia, from eastern Victoria to north Queensland, south of Cape York Peninsula, and on Norfolk and Lord Howe Islands.

Biology. Endogeous in ground leaf litter in rainforests and closed sclerophyll forests.

Note. Two additional species are currently being described (MB).

References. Csiki (1928); Jeannel (1937); Moore *et al.* (1987); Lorenz (2005); Giachino (2005a).

Key to species. Giachino (2005a).

Pseudillaphanus Giachino, 2005 (Fig. 12.126)

Type species. *Pseudillaphanus barringtoni* Giachino, 2005, by original designation.

Characteristics. Mentum without tooth; pronotum cordiform, with narrow base; elytra without a longitudinal sulcus between the parascutellar puncture and the 9th puncture of the marginal series; without an additional marginal puncture at level of 2nd marginal puncture (Fig. 12.126); aedeagus rather elongate, endophallus with a feebly sclerotised piece; parameres elongate, triagonal, both bisetose. Female genitalia not recorded.

Australian species. Two species.

Distribution. Central eastern New South Wales and south-eastern Queensland

Biology. Endogeous in ground leaf litter in rainforests and closed sclerophyll forests.

References. Giachino (2005a).

Key to species. Giachino (2005a).

Tasmanillus Giachino, 2005 (Fig. 12.140)

Type species. *Tasmanillus daccordii* Giachino, 2005, by original designation.

Characteristics. Elytra without a longitudinal sulcus between the parascutellar puncture and the 9th puncture of the marginal series; mentum with large tooth; pronotum cordiform, with narrow base; elytra without a longitudinal sulcus between the parascutellar puncture and the 9th puncture of the marginal series; with an additional marginal puncture at level of 2nd marginal puncture (Fig. 12.141); aedeagus short and stout, endophallus with a feebly sclerotised piece; parameres short, triagonal, both bisetose. Female genitalia not recorded.

Australian species. Two species.

Distribution. Central and south-eastern Tasmania.

Biology. Endogeous in ground litter in temperate rainforests.

References. Giachino (2005a).

Key to species. Giachino (2005a).

Austranillus Giachino, 2005 (Fig. 12.78)

Type species. *Illaphanus macleayi* Lea, 1906, by original designation.

Characteristics. Mentum without tooth; pronotum not cordiform, with very wide base; elytra without a longitudinal sulcus between the parascutellar puncture and the 9th puncture of the marginal series; without an additional marginal puncture at level of 2nd marginal puncture; male genitalia unknown, female genitalia not recorded.

Australian species. Only *A. macleayi* (Lea).

Distribution. South-eastern New South Wales.

Biology. Endogeous, but specific habitat unknown.

References. Giachino (2005a).

Magnanillus Baehr, 2017 (Fig. 12.106)

Type species. *Macranillus bennetti* Baehr in Baehr & Main, 2016, by original designation.

Macranillus Baehr, 2016

Characteristics. Very large, elongate, parallel-sided species. Elytra not striate but with a longitudinal sulcus between the scutellary puncture and the 9th puncture of the marginal series; mentum without tooth; pronotum not cordiform, with very wide base; elytra with almost rectangular humerus and with three discal punctures. Aedeagus fairly elongate, markedly curved, with short, rather obtuse apex, internal sac with two or three moderately sclerotised pieces; parameres elongate, both with two elongate setae; female gonocoxite-2 elongate with one dorso-median and one ventro-lateral ensiform seta, and with an elongate subapical nematiform seta.

Australian species. Five species.

Distribution. Pilbara, north-western Western Australia.

Biology. Endogeous, all species have been sampled by 'trog scrape' (i.e. sampling in bore holes) in stratified sandstone at depths up to 65 m. Nothing else in known about habits, diet, and reproduction.

References. Baehr (2017d); Baehr & Main (2016).

Key to species. Baehr & Main (2016).

Pilbaranillus Baehr in Baehr & Main, 2016 (Fig. 12.122)

Type species. *Pilbaranillus latibasis* Baehr in Baehr & Main, 2016 by original designation.

Characteristics. Large, elongate species with slightly ovoid elytra. Elytra not striate and without a longitudinal sulcus; mentum without tooth; pronotum not cordiform, with very wide base, rather quadrate; elytra with slightly oblique humerus and with three discal punctures. Aedeagus fairly elongate, markedly curved, with short, rather obtuse apex, internal sac with three narrow, sclerotised pieces; parameres elongate, both with two elongate setae; female genitalia unknown.

Australian species. Only *P. latibasis* Baehr & Main.

Distribution. Pilbara, north-western Western Australia.

Biology. Endogeous, the single species has been sampled by 'trog scrape' (i.e. sampling in bore holes) in stratified sandstone at 71 m. Nothing else in known about habits, diet, and reproduction.

References. Baehr & Main (2016).

Gracilanillus Baehr in Baehr & Main, 2016 (Fig. 12.96)

Type species. *Gracilanillus longulus* Baehr in Baehr & Main, 2016 by original designation.

Characteristics. Small, elongate, parallel-sided species. Elytra not striate but with a weakly developed longitudinal sulcus between the scutellary puncture and the 9th puncture of the marginal series; mentum with a faint tooth; pronotum not cordiform, elongate, with wide base, rather quadrate; elytra with almost rectangular humerus and with three discal punctures. Aedeagus compact, in basal part markedly curved, with short, rather obtuse apex; internal sac with one moderately sclerotised piece; parameres fairly differently shaped, right paramere slenderer than left, both with two elongate apical setae; female gonocoxite-2 narrow and elongate, slightly curved, with acute apex, without ensiform setae, but with an elongate subapical nematiform seta.

Australian species. Six species.

Distribution. Pilbara, north-western Western Australia.

Biology. Endogeous, all species have been sampled by 'trog scrape' (i.e. sampling in bore holes) in stratified sandstone at depths up to 39 m. Nothing else in known about habits, diet, and reproduction.

References. Baehr & Main (2016).

Key to species. Baehr & Main (2016).

Hesperanillus Baehr in Baehr & Main, 2016 (Fig. 12.97)

Type species. *Hesperanillus scanloni* Baehr in Baehr & Main, 2016 by original designation.

Characteristics. Medium-sized, moderately elongate, not parallel sided species. Elytra not striate and without a longitudinal sulcus; mentum with faint tooth; eye area markedly prominent; pronotum rather cordiform, with fairly narrow base; elytra slightly to fairly ovoid, with more or less oblique humerus and with one or two discal punctures. Aedeagus rather compact, lower surface in basal part markedly curved, in apical three fourths straight, apex moderately short, slightly produced, narrow, with obtusely triangular tip; internal sac with one slightly sclerotised piece; parameres fairly differently shaped, right paramere slenderer than left, both with two elongate apical setae; female gonocoxite-2 rather elongate, slightly curved, with acute apex, without ensiform setae, with an elongate subapical nematiform seta.

Australian species. Three species.

Distribution. Pilbara, north-western Western Australia.

Biology. Endogeous, all species have been sampled by 'trog scrape' (i.e. sampling in bore holes) in stratified sandstone at depths up to 59 m. Nothing else in known about habits, diet, and reproduction.

References. Baehr & Main (2016).

Key to species. Baehr & Main (2016).

Externanillus Baehr in Baehr & Main, 2016 (Fig. 12.91)

Type species. *Externanillus mcraeae* Baehr in Baehr & Main, 2016 by original designation.

Characteristics. Small, elongate, slightly ovoid species. Elytra not striate and without a longitudinal sulcus; mentum without tooth; eye area not prominent, pronotum gently cordiform, with moderately wide base; elytra gently oval-shaped, with rounded humerus and with one hree discal puncture. Male genitalia unknown; female gonocoxite-2 elongate, acute, without ensiform setae, with an extremely elongate subapical nematiform seta.

Australian species. Only *E. mcraeae* Baehr & Main.

Distribution. Inland southern Western Australia.

Biology. Endogeous, the species has been sampled by 'trog scrape' (i.e. sampling in bore holes) in stratified sandstone at 50 m. Nothing else in known about habits, diet, and reproduction.

References. Baehr & Main (2016).

Angustanillus Baehr in Baehr & Main, 2016 (Fig. 12.76)

Type species. *Angustanillus striatipennis* Baehr in Baehr & Main, 2016 by original designation.

Characteristics. Large, remarkably narrow and elongate, parallel-sided species. Elytra distinctly striate (four median striae), but without a longitudinal sulcus; mentum without tooth; pronotum narrow and elongate, not cordiform, with wide base; elytra extremely elongate, parallel-sided, with almost rectangular humerus and with two discal punctures. Male and female genitalia unknown.

Australian species. Only *A. striatipennis* Baehr & Main.

Distribution. Pilbara, north-western Western Australia.

Biology. Endogeous, the species has been sampled by 'trog scrape' (i.e. sampling in bore holes) in stratified sandstone at unknown depth. Nothing else in known about habits, diet, and reproduction.

References. Baehr & Main (2016).

Tribe Pogonini Laporte, 1834

Characteristics. Procoxal cavities closed; mesocoxae conjunct; head with two supraorbital setae above each eye; mandibles with a seta in the scrobe; palps not subulate; frontal furrows short and not extended behind the eyes; basal margin of the elytra complete; marginal series of setiferous punctures uninterrupted; elytral epipleura without plica; aedeagus with a variously shaped sclerite on the endophallus; parameres dissimilar in size, triagonal, with 2–3 apical setae; gonocoxites variously shaped, gonocoxite-2 sparsely setose or asetose.

Key to the Australian genera of Pogonini

1. Elytra without microreticulation; humerus sharply angulate; protibia with several rows of elongate spines on dorsal and lateral surfaces that occupy almost the whole length; pronotum notably trapezoidal, widest at the anterior angles .... ***Syrdenoidius* Baehr & Hudson**

– Elytra with distinct microreticulation; humerus not so angulate; protibia with a short row of moderate length spines on lateral surface only; pronotum shape not trapezoidal, widest far behind anterior angles .... ***Pogonus* Nicolai**

Pogonus Nicolai, 1822 (Pl. 103D; Figs 12.13, 12.17)

Type species. *Carabus chalceus* Marsham, 1802, by subsequent designation by Curtis 1824.

Characteristics. Body shape average, not markedly slender and depressed; mandibles moderately elongate; widest diameter of pronotum well behind the anterior angles; humerus moderately dentate; elytra microreticulate; protibia with only a single row of elongate spines; flight wings usually present; aedeagus fairly elongate; gonocoxite-2 elongate.

Australian species. 17 species.

Distribution. Throughout mainland Australia; extraterritorial almost worldwide, except the Neotropical region.

Biology. Ground living, hygrophilous, and halophilous, found on large parts of the coast and in saline habitats of the interior. These predacious, nocturnal species hide under debris, algae and stones, on salt lakes they also dig burrows into the soil surface. A single species (*P. perovalis* Baehr & Hudson, 2001) is flightless, a very uncommon state in the tribe. The diet may consist of small insect adults and larvae, and small crustaceans. Nothing else is known about diet, habits, and reproduction.

Note. Several additional undescribed species are known from on or near salt lakes in central South Australia and interior south-western Australia (P. Hudson *in litt.*).

References. Sloane (1895b); Csiki (1928); Moore (1977a); Moore *et al.* (1987); Matthews (1980); Baehr & Hudson (2001); Lorenz (2005); Baehr (2010b).

Keys to species. Baehr & Hudson (2001), Baehr (2010b).

Syrdenoidius Baehr & Hudson, 2001 (Fig. 12.136)

Type species. *Syrdenoidius spinipes* Baehr & Hudson, 2001, by original designation.

Characteristics. Body shape narrow, elongate, dorsally depressed; mandibles very elongate; prothorax trapezoidal; humerus sharply dentate; elytra not microreticulate; protibia with several rows row of elongate spines; flight wings absent; aedeagus short; gonocoxite-2 short, obtuse, with a single elongate apical seta.

Australian species. Only *S. spinipes* Baehr & Hudson.

Distribution. Only recorded from Lake Gairdner, central South Australia.

Biology. A nocturnal, predacious species that hides in burrows on the surface of dry salt lakes. The absence of flight wings is exceptional within the tribe and is shared only with the Australian *Pogonus perovalis* Baehr & Hudson, 2001 and the dimorphic European *P. chalceus* (Marsham, 1802) (Dhuyvetter *et al.* 2007).

References. Lorenz (2005).

Tribe Psydrini LeConte, 1853

Characteristics. A remarkably diverse tribe in terms of overall shape and morphological structure; procoxal cavities closed; mesocoxae conjunct; head with two supraorbital setae above each eye; mandibles with a seta in the scrobe; palps not subulate; penultimate palpomere of the maxillary palp glabrous; frontal furrows short and not extended behind the eyes; basal margin of the elytra incomplete; series of marginal setiferous punctures of the elytra interrupted; elytral epipleura with evident plica; male protarsus not much modified; male and female genitalia quite diverse in shape and structure; parameres setose at apex or not; gonocoxites usually rather elongate, with a variable number of setae.

Notes. The subtribal classification has been controversial and much discussed in recent years. We here follow the classification proposed by Baehr (1999a, 2005c), but downgrading tribes to subtribes. The relationships of the subtribe Psydrina to the other subtribes are unsettled (Baehr 1999a; Maddison *et al.* 1999; Liebherr 2011). Perhaps Psydrina form a separate tribe that is more closely related to the Holarctic tribe Patrobini than to other Psydrini. The larva of one Australian genus (*Melisodera*) was described (Moore 1964b) and compared with the putative larva of *Nomius pygmaeus* (Dejean). The differences likewise suggest a separate taxonomic position of Psydrina within or even outside of the rest of the Psydrini.

Key to the Australian subtribes of Psydrini

1.	8th elytral interval laterally compressed to form a carina at its apex	**2**
–	8th elytral interval simple	**4**
2(1).	Mesoventrite very narrow between coxae	**Amblytelina Blackburn**
–	Mesoventrite much broader	**3**
3(2).	Metanepisternum elongate; metaventrite elongate between meso- and metacoxae	**Moriomorphina Sloane**
–	Metanepisternum transverse or quadrate; metaventrite very short	**Tropopterina Sloane**
4(3).	Tarsi setose dorsally	**Psydrina LeConte**
–	Tarsi glabrous dorsally	**5**
5(4).	Labrum and glossa at apex truncate; elytral interval 3 with three setiferous punctures	**Mecyclothoracina Jeannel**
–	Labrum at apex deeply emarginated; glossa conical; elytral interval 3 asetose	**Meonidina Sloane**

Subtribe Psydrina LeConte, 1853

Characteristics. Antenna short, moniliform; pronotal margins plurisetose, 8th elytral interval not carinate, dorsal surface of the tarsi setose; aedeagus elongate, open at the base and small ostium, parameres elongate, rather similar, both densely setose; gonocoxites elongate, gonocoxite-2 little curved, acute, with three ventral setae, one dorsal seta, and one rather elongate subapical seta.

Note. Apart from the Australian genus *Laccocenus* Sloane, 1890 this subtribe includes the Nearctic genus *Psydrus* LeConte and the Holarctic and Ethiopian genus *Nomius* Laporte.

Laccocenus Sloane, 1890 (Pl. 27A)

Type species. *Laccocenus ambiguus* Sloane, 1890, by monotypy.

Characteristics. Elongate, parallel-sided, rather depressed beetles; head with distinct supraocular ridge; labial palp plurisetose.

Australian species. Two species.

Distribution. North-eastern New South Wales, south-eastern Queensland.

Biology. One species lives under logs and under bark of logs in montane subtropical to temperate rainforests, the other is cavernicolous. Predacious and probably nocturnal.

References. Csiki (1929); Moore (1963b, 2004); Moore *et al.* (1987); Baehr (2003d); Lorenz (2005).

Key to species. Moore (2004).

Subtribe Moriomorphina Sloane, 1890

Characteristics. Mesoventrite wide between the coxae; 8th elytral interval compressed and carinate; metanepisternum and metaventrite elongate; aedeagus variously shaped, with various sclerotised pieces on the endophallus, parameres moderately dissimilar, elongate, most commonly asetose, rarely minutely setose; gonocoxites more or less elongate, with variable numbers of ventral, dorsal, and subapical setae.

Note. This subtribe is restricted to Australia.

Key to the Australian genera of Moriomorphina

1.	Antenna elongate, filiform; legs long and slender	***Rhaebolestes* Sloane**
–	Antenna short, moniliform; legs short and stout	**2**
2(1).	Basal impressions of pronotum bordered by a distinct submarginal ridge; eyes markedly protruded laterally	**3**
–	Basal impressions of pronotum not bordered by a submarginal ridge; eyes slightly protruded laterally, orbits oblique	**4**

3(2). Margins of pronotum plurisetose; flight wings much reduced; male protarsus without squamose setae beneath .... ***Celanida* Laporte**
– Margins of pronotum unisetose; flight wings fully developed; male protarsus with squamose setae beneath .... ***Melisodera* Westwood**
4(2). Mesotibia markedly arcuate; tarsi laterally compressed .... ***Moriomorpha* Laporte**
– Mesotibia straight; tarsi not compressed .... ***Moriodema* Laporte**

Celanida Laporte, 1867 (Pl. 21A)

Type species. *Celanida montana* Laporte, 1867, by monotypy.

Characteristics. Large beetles; eyes globular, strongly protruded laterally; submarginal ridge of pronotum distinct, margin plurisetose; flight wings reduced; tibiae not curved; tarsi not laterally compressed; male protarsus simple; aedeagus with parameres elongate, according to Moore (1963b) both minutely setose; gonocoxites unknown.

Australian species. Only *C. montana* Laporte.

Distribution. Eastern Victoria.

Biology. According to Moore *et al.* (1987) ground living in tall forest and hygrophilous. Predacious and apparently montane.

References. Laporte (1868); Csiki (1929); Moore (1963b); Moore *et al.* (1987); Lorenz (2005).

Rhaebolestes Sloane, 1903 (Pl. 36A)

Type species. *Rhaebolestes walkeri* Sloane, 1903, by monotypy.

Characteristics. Large, elongate beetles; antenna elongate, filiform; prothorax cordiform, laterally bisetose; flight wings fully developed; legs slender, elongate; tibiae not curved; tarsi not laterally compressed; male protarsus widened and with squamose setae beneath; aedeagus moderately elongate, endophallus with feebly sclerotised pieces; parameres moderately dissimilar, asetose; female genitalia unknown.

Australian species. Two species.

Distribution. South-eastern New South Wales, south-eastern Queensland.

Biology. Probably ground living in forests. The species from southern Queensland was apparently collected in montane rainforests. Predacious and probably nocturnal.

Note. Both Australian species are extremely rare and only the holotypes are recorded.

References. Csiki (1929); Moore (1963b); Moore *et al.* (1987); Lorenz (2005); Baehr (2011c).

Key to species. Baehr (2011c).

Melisodera Westwood, 1835 (Pl. 29A)

Type species. *Melisodera picipennis* Westwood, 1835, by monotypy.

Characteristics. Large, rather broad beetles; antenna moniliform; eyes globular, strongly protruded laterally; prothorax laterally bisetose, with distinct submarginal ridge; flight wings fully developed; legs rather short; tibiae not curved; tarsi not laterally compressed; male protarsus widened and with squamose setae beneath; aedeagus little known, both parameres apparently asetose; gonocoxite-2 elongate, almost straight; with three elongate ventral setae, two dorsal setae, and two fairly elongate subapical setae.

Australian species. Two species.

Distribution. South-eastern Australia from eastern Victoria to New England Tableland in north-eastern New South Wales

Biology. Probably ground living or inside of logs and stumps in open forests. One sample was collected 'from inside eucalypt log, feeding on lucanid larvae' at very high, subalpine altitude.

References. Csiki (1929); Moore (1963b); Moore *et al.* (1987); Lorenz (2005); Baehr (2011c).

Key to species. Baehr (2011c).

Moriomorpha Laporte, 1867 (Pl. 30A)

Type species. *Moriomorpha victoriae* Laporte, 1867, by subsequent designation by Moore (1963b).

Characteristics. Large, rather wide beetles; antenna moniliform; eyes moderately protruding laterally; prothorax laterally bisetose, without distinct submarginal ridge; flight wings fully developed; legs rather short; tibiae curved; tarsi laterally compressed; male protarsus widened and with squamose setae beneath; aedeagus elongate, endophallus with or without sclerotised pieces; parameres moderately dissimilar, elongate, both asetose; gonocoxite-1 asetose at apex; gonocoxite-2 rather elongate, with two very elongate ventral setae, one similarly sized dorsal seta, and a rather short subapical seta.

Australian species. Six species.

Distribution. South-eastern and eastern Australia from eastern South Australia to Paluma Range in north-eastern Queensland.

Biology. Ground living under or in hollow logs and under bark in open forests and rainforests. Specimens from north Queensland occur in montane rainforests. Predacious and probably nocturnal.

Note. For most species only the holotype specimen is known. Those that are known from more material seem to be widely distributed.

References. Laporte (1868); Csiki (1929); Moore (1963b); Moore *et al.* (1987); Lorenz (2005); Baehr (2011b).

Key to species. Baehr (2011b).

Moriodema Laporte, 1867 (Pl. 30C)

Type species. *Moriodema mcoyei* Laporte, 1867 (= *Moriodema paramattensis* Laporte, 1867), by subsequent designation by Moore 1963.

Characteristics. Moderately large, rather slender beetles; antenna moniliform; eyes slightly protruding laterally; prothorax laterally bisetose, without distinct submarginal ridge; flight wings fully developed; legs rather short; tibiae straight; tarsi not laterally compressed; male protarsus widened and with squamose setae beneath; aedeagus rather elongate, endophallus with several elongate sclerotised pieces; parameres moderately dissimilar, elongate, both asetose; gonocoxite-2 rather elongate, with two elongate ventral setae, one elongate dorsal seta, and a rather short subapical seta.

Australian species. Two species.

Distribution. South-eastern Australia from eastern Victoria to south-eastern Queensland.

Biology. Ground living, usually in open forests, in the north of its range at high altitude. One specimen was collected from under bark. Predacious and probably nocturnal.

References. Laporte (1868); Sloane (1903); Csiki (1929); Moore (1963b); Moore *et al.* (1987); Lorenz (2005); Baehr (2011c).

Key to species. Baehr (2011c).

Subtribe Tropopterina Sloane, 1898

Characteristics. Mesoventrite wide between the coxae; metaventrite very short and metanepisternum transverse or quadrate; 8th elytral interval compressed and carinate; flight wings reduced; aedeagus variously shaped, with various sclerotised pieces on the endophallus, parameres more or less dissimilar, elongate, the right one and sometimes also the left one minutely setose; gonocoxites more or less elongate, with variable numbers of ventral, dorsal, and subapical setae.

Note. Most genera of this subtribe are Australian, one genus each occurs in New Zealand and southern South America.

Key to the Australian genera of Tropopterina

1.	Eyes reduced	***Trephisa*** **Moore**
–	Eyes fully developed, but sometimes rather small	**3**
2(1).	Posterior pronotal seta absent; antennomere 3 sparsely setose	***Pterogmus*** **Sloane**
–	Posterior pronotal seta present; antennomere 3 glabrous	**3**
3(2).	Antenna short, moniliform; legs short and stout; lateral mental lobes obtusely rounded at apex	***Teraphis*** **Laporte**
–	Antenna longer, filiform; legs slender; lateral mental lobes angulate at apex	**4**
4(3).	Pronotum trapezoidal, the median basal impression punctiform, the lateral impression absent; body short and ovoid	***Sitaphe*** **Moore**
–	Pronotum cordiform, the median basal impression elongate, the lateral impression visible; body elongate	***Theprisa*** **Moore**

Pterogmus Sloane, 1920 (Pl. 36E)

Type species. *Pterogmus rufipes* Sloane, 1920, by monotypy.

Characteristics. Small beetles; eyes fully developed; antennomere 3 sparsely setose; pronotum without the basal marginal seta; disc of elytra tripunctate; legs moderately short; tarsi setose above; aedeagus elongate, endophallus with sclerotised pieces; parameres moderately dissimilar, both sparsely setose, but the right one with slightly denser setosity; female genitalia not recorded.

Australian species. Only *P. rufipes* Sloane.

Distribution. Tasmania.

Biology. Ground living in a variety of habitats. Predacious and probably nocturnal. Little else is known about habits, diet, and reproduction.

References. Csiki (1929); Moore (1963b); Moore *et al.* (1987); Lorenz (2005); Eberhard & Giachino (2011).

Teraphis Laporte, 1867 (Pl. 39L)

Type species. *Teraphis melbournensis* Laporte, 1867, by subsequent designation by Moore 1963.

Phersita Sloane, 1903 (unnecessary replacement name).

Characteristics. Rather small beetles; eyes fully developed; antennomere 3 asetose; pronotum bisetose; disc of elytra uni- or bipunctate; legs short and stout; tarsi glabrous above aedeagus rather elongate, endophallus with sclerotised pieces; parameres rather similar, the left one asetose, the right one with extremely short, sparse setosity; female genitalia not recorded.

Australian species. Six species.

Distribution. South-eastern Australia from eastern Victoria to south-eastern New South Wales, Tasmania.

Biology. Ground living in closed forests, one species cavernicolous. Little else is known about habits, diet, and reproduction.

References. Laporte (1868); Csiki (1929); Moore (1963b); Moore *et al.* (1987); Baehr (2003d; Lorenz (2005).

Key to species. Sloane (1920).

Theprisa Moore, 1963 (Pl. 39A)

Type species. *Phersita convexa* Sloane, 1920, by original designation.

Characteristics. Rather small, elongate-oval beetles; eyes fully developed; antenna filiform, antennomere 3 asetose; pronotum subcordiform, bisetose; disc of elytra bipunctate; legs short but slender; tarsi glabrous above; aedeagus rather compact, endophallus with sclerotised pieces; parameres fairly similar form, the left one asetose, the right one with extremely short, sparse setosity; female genitalia not recorded.

Australian species. Three species.

Distribution. Eastern Victoria, Tasmania.

Biology. Ground living in closed forests. Predacious and probably nocturnal. Little else is known about habits, diet, and reproduction.

Note. The genus needs a revision as examined specimens show variation that suggest there are undescribed species.

References. Moore *et al.* (1987); Lorenz (2005); Eberhard & Giachino (2011).

Key to species. Sloane (1920).

Trephisa Moore, 1963 (Fig. 12.152)

Type species. *Trephisa parallela* Moore, 1963, by original designation.

Characteristics. Rather small, elongate, almost parallel-sided beetles; eyes much reduced; antenna short, almost moniliform, antennomere 3 asetose; pronotum square, bisetose; disc of elytra bipunctate; legs fairly elongate; tarsi setose above; aedeagus fairly compact, endophallus with sclerotised pieces; parameres moderately dissimilar, the left one asetose, the right one with extremely short, sparse setosity; female genitalia not recorded.

Australian species. Only *T. parallela* Moore.

Distribution. South-eastern Queensland, known only from Lamington Plateau.

Biology. Endogeous in closed montane rainforests. Predacious and probably nocturnal.

References. Moore *et al.* (1987); Baehr (2003d); Lorenz (2005).

Sitaphe Moore, 1963 (Pl. 38D)

Type species. *Sitaphe rotundata* Moore, 1963, by original designation.

Characteristics. Rather small, short, markedly ovoid beetles; eyes fully developed; antenna rather short, but filiform, antennomere 3 asetose; pronotum trapezoidal with punctiform median basal impression, bisetose; disc of elytra impunctate; legs short but slender; tarsi glabrous above; aedeagus length notably variable, endophallus always with some sclerotised pieces; parameres dissimilar, the left one asetose, the right one with extremely short, sparse setosity; gonocoxite-1 asetose at apex; gonocoxite-2 rather short, with two elongate ventral setae, one elongate dorsal seta, and a short subapical seta.

Australian species. Eight species.

Distribution. Eastern Queensland, in the north to Windsor and Carbine Tablelands.

Biology. Probably ground living in rainforests, usually at high altitude. However, because most recorded specimens have been collected by pyrethrum knockdown at the bases of trunks of moss covered trees, the species also may live on the lower parts of rainforest trees and on logs. Predacious. Apparently strictly nocturnal, because almost never seen during daytime. Little else is known about habits, diet, and reproduction.

Note. Some species occur in an extremely restricted range at the uppermost tops of certain rainforest capped mountains.

References. Moore *et al.* (1987); Baehr (2003d); Lorenz (2005).

Key to species. Baehr (2003d).

Subtribe Amblytelina Blackburn, 1892

Characteristics. Mesoventrite narrow between the coxae; metanepisternum and metaventrite elongate; 8th elytral interval compressed and carinate; elytra usually rather elongate and depressed, and commonly with distinct colour pattern; aedeagus variously shaped, usually with complexly folded, sclerotised pieces on the endophallus; parameres moderately dissimilar, with elongate apex, usually rather sparsely setose, the right one more than the left one; gonocoxites fairly elongate, gonocoxite-1 setose at apical rim, gonocoxite-1 with 2–5 setae at apex, gonocoxite-2 very similar in all genera, invariably with two rather large ventral setae, one elongate dorsal seta, and usually two short subapical setae.

Note. This subtribe is restricted to Australia.

Key to the Australian genera of Amblytelina

1. Fourth tarsomeres neither markedly widened nor deeply cleft, without squamose setae beneath .... ***Epelyx* Blackburn**
– Fourth tarsomeres **either** markedly widened **or** deeply cleft, at least the lower surface of protarsus and mesotarsus with dense squamose setae beneath **2**
2(1). Fourth tarsomeres of protarsus and mesotarsus markedly widened though not deeply cleft, those of metatarsus barely widened and not or only with sparse squamose setae beneath **3**
– Fourth tarsomeres of all tarsi markedly widened **and** deeply cleft **and** with dense squamose setae beneath **4**
3(2). Brachypterous flight wings, elytra considerably narrowed at base; prothorax cordiform, both lateral setae present; 4th tarsomere large, very asymmetrically excised, half as long as 5th tarsomere .... ***Trichamblytelus* Baehr**

– Full flight wing, elytra little narrowed at base; prothorax wide, not cordiform, anterior lateral seta absent; 4th tarsomere smaller, less asymmetrically explanate, a third as long as 5th tarsomere .... ***Paratrichothorax*** **Baehr**

4(3). Both marginal setae of pronotum present .... ***Amblytelus*** **Erichson**

– Either anterior or posterior marginal seta of pronotum absent **5**

5(4). Posterior marginal seta of pronotum absent, pronotum disciform (discus shaped), without distinct basal angles, with depressed, widely explanate margins .... ***Pseudamblytelus*** **Baehr**

– Anterior marginal seta of pronotum absent, pronotum not disciform, with distinct basal angles, with upturned, less explanate margins .... ***Dystrichothorax*** **Blackburn**

Amblytelus Erichson, 1842 (Pl. 19C)

Type species. *Carabus curtus* Fabricius, 1801, by monotypy.

Characteristics. Pronotum not disciform, with distinct basal angles, bisetose; 4th tarsomeres wide, deeply cleft, and with dense squamose setae beneath; flight wings well developed; many species bear a distinct colour pattern on the elytral; aedeagus variously shaped, endophallus usually with sclerotised pieces; parameres more or less dissimilar, mostly with very elongate apex, usually setose, the right one usually more densely than the left one; gonocoxites as for subtribe.

Australian species. 44 species and four subspecies.

Distribution. Southern and eastern Australia from southern Western Australia to Windsor and Carbine Tablelands in north-eastern Queensland, Tasmania.

Biology. Under bark of bark-shedding eucalypts, in deep fissures of a variety of trees, in open forest and woodland, on the bark of moss-covered trees in rainforests. Apparently strictly nocturnal. Predacious but diet and habits not recorded. Some species can be very common under bark of shedding eucalypts.

References. Blackburn (1891); Sloane (1920); Csiki (1929); Moore (1963b); Moore *et al.* (1987); Matthews (1980); Baehr (2005c, 2006g, 2008f, 2016d); Lorenz (2005); Eberhard & Giachino (2011).

Keys to species. Baehr (2005c, 2008f, 2016d).

Dystrichothorax Blackburn, 1892 (Pl. 24A)

Type species. *Amblytelus amplipennis* Macleay, 1871, by original designation.

Characteristics. Pronotum not disciform, with distinct basal angles, unisetose, the anterior marginal seta absent; 4th tarsomeres wide, deeply cleft, and with dense squamose setae beneath; flight wings well developed; some species bear a distinct elytral colour pattern; aedeagus very differently shaped, endophallus usually with sclerotised pieces; parameres more or less dissimilar, mostly with very elongate apex, usually very sparsely setose, sometimes even asetose; gonocoxites as for subtribe.

Australian species. 50 species and three subspecies.

Distribution. Southern and eastern Australia from eastern South Australia to Windsor and Carbine Tablelands in north Queensland; Tasmania.

Biology. Under bark of bark-shedding eucalypts, in deep fissures of a variety of trees, in open forest and woodland, on the bark of moss-covered trees in rainforests. Apparently strictly nocturnal. Predacious but diet and habits not recorded.

References. Sloane (1920); Csiki (1929); Moore (1963b); Moore *et al.* (1987); Baehr (2005c, 2006i, 2008f, 2016d); Lorenz (2005); Eberhard & Giachino (2011).

Keys to species. Baehr (2005c, 2006i, 2008f, 2016d).

Epelyx Blackburn, 1892 (Pl. 25J)

Type species. *Epelyx lindensis* Blackburn, 1892, by subsequent designation by Moore 1987.

Characteristics. Pronotum rather cordiform, with distinct basal angles, bisetose; 4th tarsomeres not widened, not deeply cleft, and without squamose setae beneath; flight wings well developed; elytra without distinct pattern; aedeagus moderately elongate, endophallus with sclerotised pieces; parameres rather similar, with short apex, usually sparsely setose, the right one commonly with slightly more setae; gonocoxites as for subtribe.

Australian species. Five species.

Distribution. Southern mainland Australia from south-western Western Australia through southern South Australia to south-eastern New South Wales.

Biology. Under bark of bark-shedding eucalypts and in deep fissures of various non bark-shedding trees, in open forests, woodlands, and mallee. Predacious and probably nocturnal.

References. Csiki (1929); Moore (1963b); Moore *et al.* (1987); Matthews (1980); Baehr (2005c); Lorenz (2005).

Key to species. Baehr (2005c).

Pseudamblytelus Baehr, 2005 (Fig. 12.125)

Type species. *Pseudamblytelus orbicollis* Baehr, 2005, by original designation.

Characteristics. Pronotum disciform, without distinct basal angles, unisetose, the posterior marginal seta absent; 4th tarsomeres wide, deeply cleft, and with dense squamose setae beneath; flight wings well developed; elytra without colour pattern. Male and female genitalia unknown.

Australian species. Only *P. orbicollis* Baehr.

Distribution. South-eastern Victoria, only recorded from Fern Tree Gully east of Melbourne.

Biology. Probably arbouricolous.

References. Lorenz (2005).

Paratrichothorax Baehr, 2005 (Pl. 33B)

Type species. *Paratrichothorax brevistylus* Baehr, 2005, by original designation.

Characteristics. Pronotum wide, not disciform, with distinct basal angles, unisetose, the anterior marginal seta absent; 4th tarsomeres of protarsus and mesotarsus wide but not deeply cleft, with squamose setae beneath, 4th tarsomere of metatarsus narrow and without squamose setae beneath; flight wings well developed; elytra without distinct colour pattern; aedeagus short and wide, endophallus with some sclerotised pieces; both parameres short and wide, with very short apex, right paramere more densely setose the left one; gonocoxites as for subtribe.

Australian species. Only *P. brevistylus* Baehr.

Distribution. South-western part of Western Australia.

Biology. Under bark of bark-shedding eucalypts, particularly on karri (*Eucalyptus diversicolor*). Predacious and apparently strictly nocturnal. Sometimes very common under loose bark.

References. Lorenz (2005).

Trichamblytelus Baehr, 2005 (Fig. 12.153)

Type species. *Trichamblytelus ovalipennis* Baehr, 2005, by original designation.

Characteristics. Pronotum narrow, cordiform, with distinct basal angles, bisetose; elytra oval-shaped with narrow base; 4th tarsomeres of protarsus and mesotarsus wide but not deeply cleft, with squamose setae beneath, 4th tarsomere of metatarsus narrow and without squamose setae beneath; flight wings absent; elytra without distinct colour pattern; aedeagus short and wide, endophallus with some sclerotised pieces; both parameres short and wide, with short apex, right paramere very densely setose; female genitalia unknown.

Australian species. Only *T. ovalipennis* Baehr.

Distribution. South-western part of Western Australia, known only from Walpole National Park.

Biology. The single known specimen was collected in a malaise trap, although the specimen is flightless.

References. Lorenz (2005).

Subtribe Meonidina Sloane, 1898

Characteristics. Eyes moderately large, but laterally little produced; labrum deeply emarginate; mandibles elongate and straight; antenna filiform, pilose from antennomere 4; pronotum more or less cordate; metanepisternum and metaventrite elongate; elytra ovoid, disc impunctate; elytral interval 8 not carinate; metathorac wings absent; tarsi glabrous above; three basal tarsomeres of male protarsus with squamose setae beneath; aedeagus variously shaped, usually with sclerotised or denticulate pieces on the endophallus; parameres rather dissimilar, usually setose, the right one more densely so than the left one; gonocoxite-2 with two very elongate ventral setae, one elongate dorsal seta, and one short subapical setae.

Note. This subtribe includes the genus *Selenochilus* Chaudoir from New Zealand and probably also the South American genus *Bembidiomorphum* Champion.

Key to the Australian genera of Meonidina

1. Base of elytra not bordered; scutellum very small; paraglossae setose; lacinia densely setose and without a terminal hook; aedeagus narrow and long .... ***Meonis* Laporte**

– Base of elytra weakly bordered; scutellum normal; paraglossae glabrous; lacinia spinose and with a strong terminal hook; aedeagus short and compact .... ***Raphetis* Moore**

Meonis Laporte, 1867 (Pl. 29F)

Type species. *Meonis niger* Laporte, 1867, by subsequent designation of Moore 1963.

Characteristics. Rather large beetles; lacinia densely setose; paraglossae setose; pronotum bisetose; scutellum very small; elytra with a most five striae; base of elytra not bordered; aedeagus usually narrow and elongate, endophallus with sclerotised or markedly dentate pieces; gonocoxite-2 acute, with two elongate ventral setae, a very elongate dorsal seta, and a small subapical seta.

Australian species. 16 species and six subspecies.

Distribution. Eastern Australia from south-eastern New South Wales to south-eastern Queensland.

Biology. Ground living in more or less closed forests, under rocks and woody debris. Because the species are flightless, they usually occupy restricted, allopatric ranges. Predacious and apparently nocturnal. It has been speculated that the glossy surface and the long, pincer-like mandibles might be well suited for eating snails.

References. Laporte (1868); Sloane (1900a, 1911, 1915a, 1916a); Csiki (1929); Moore (1963b); Moore *et al.* (1987); Lorenz (2005); Baehr (2003d, 2007b).

Key to species. Baehr (2007b).

Key to the Australian subgenera of *Meonis* Laporte

1. Prothorax cordiform; basal margin of elytra short; prosternal process asetose; aedeagus narrow and elongate; without markedly dentate pieces; left paramere narrower, setose .... ***Meonis* Laporte**
– Prothorax subcordiform, with wide base; basal margin of elytra elongate; prosternal process setose; aedeagus wider; with two markedly dentate pieces; left paramere wide, asetose .... ***Meonidius* Baehr**

Subgenus *Meonis* Laporte, 1867

Type species. *Meonis niger* Laporte, 1867, by subsequent designation of Moore 1963.

Characteristics. Prothorax cordiform, considerably narrowed basad; humerus of elytra barely angulate, basal margin of elytra short; prosternal process asetose; microreticulation of elytra fine though visible, consisting of transverse meshes; aedeagus narrow and elongate; with some sclerotised rods, but without markedly dentate pieces; left paramere less wide, rather triangular, more or less sparsely setose.

Australian species. 15 species and five subspecies.

Distribution. Eastern Australia from south-eastern New South Wales to south-eastern Queensland.

References. Laporte (1868); Sloane (1900a, 1911, 1916a); Csiki (1929); Moore (1963b); Moore *et al.* (1987); Lorenz (2005); Baehr (2007b).

Subgenus *Meonidius* Baehr, 2007

Type species. *Meonis uncinatus* Baehr, 2007, by original designation.

Characteristics. Prothorax subcordiform, with wide base; humerus of elytra moderately angulate, basal margin elongate; prosternal process setose; microreticulation of elytra barely recognisable, consisting of very fine, transverse lines; aedeagus wider; without sclerotised rods on the endophallus, but with two markedly dentate pieces; left paramere very wide, not triangular, asetose.

Australian species. Only *M. uncinatus* Baehr, with one subspecies.

Distribution. South-eastern Queensland (Lamington Plateau) and immediately adjacent north-eastern New South Wales (Mt. Warning).

References. Baehr (2007b).

Raphetis Moore, 1963 (Pl. 36C)

Type species. *Raphetis darlingtoni* Moore, 1963, by original designation.

Characteristics. Medium-sized to rather small beetles; lacinia spinose; paraglossae glabrous; pronotum unisetose, the posterior marginal seta absent; scutellum of normal size; elytra shallowly, but completely striate; base of elytra with scarcely evident border; gonocoxite-2 of various length, with two very elongate ventral setae, elongate dorsal seta, and one short subapical seta.

Australian species. Three species and four subspecies.

Distribution. Eastern Queensland on Lamington Plateau, Eungella Plateau, and Atherton and Windsor Tablelands.

Biology. Ground living in and under logs and on the bases of moss covered tree trunks in montane rainforests. Predacious and apparently strictly nocturnal. It has been speculated that the long, pincer-like mandibles might be well suited for eating snails.

References. Moore *et al.* (1987); Baehr (2003d; Lorenz (2005).

Key to species. Baehr (2003d).

Subtribe Mecyclothoracina Jeannel, 1944

Characteristics. Eyes moderately large, laterally moderately produced; mandibles or normal size; antenna filiform, pilose from antennomere 4; pronotum variously shaped; metanepisternum and metaventrite elongate; elytra ovoid, disc variously punctate; elytral interval 8 not carinate; metathorac wings usually present; tarsi glabrous above; three basal tarsomeres of male protarsus widened and with squamose setae beneath; aedeagus variously shaped, usually with sclerotised and/or dentate pieces on the endophallus; parameres moderately dissimilar, setose, the right one more densely so; gonocoxites rather elongate, Gonocoxite-1 with without or with few setae at apex; gonocoxite-2 with variable numbers of ventral, dorsal, and subapical setae.

Notes. This subtribe is distributed from Java through Sulawesi, Timor, New Guinea, New Caledonia, Australia, New Zealand, Polynesia to the Hawaiian Islands, it also occurs on certain subantarctic islands.

Key to the Australian genera of Mecyclothoracina

1. Terminal palpomeres glabrous; pronotum rather cordate .... ***Mecyclothorax* Sharp**
– Terminal palpomeres finely pilose; pronotum subrectangular .... ***Neonomius* Moore**

Mecyclothorax Sharp, 1903 (Pl. 29G)

Type species. *Cyclothorax punctipennis* Macleay, 1871, by monotypy.

Cyclothorax W. J. Macleay, 1871 (non *Cyclothorax* Frauenfeld, 1867).

Characteristics. Rather small beetles; terminal palpomeres glabrous; prothorax more or less cordate or oval-shaped; male and female genitalia as for tribe, but male genitalia rotated 180° in some specimens of *Mecyclothorax punctipennis* (Liebherr & Will 2015); genital structures of both sexes rather varied in shape and setosity.

Australian species. 21 species and three subspecies.

Distribution. Southern and eastern Australia, Tasmania, Lord Howe and Norfolk Islands; extraterritorial on Java, Sulawesi, Timor, New Guinea, New Caledonia, New Zealand, and, especially numerous, on Society and Hawaiian Islands.

Biology. Ground living and in and under bark of trees in a variety of habitats, in upland rainforest as well as in open forests, woodlands, and disturbed areas. Predaceous and nocturnal. Commonly found in large numbers under loose eucalypt bark and in leaf litter in mesic spots. Ground dwelling species are nocturnally active and can be found in mixed species aggregations.

Notes. The Australian species of this genus are in need of a revision as identifications are difficult or even impossible without comparison to types. Without a doubt there are undescribed species.

References. Sloane (1894, 1920); Csiki (1929); Moore (1963b, 1984, 1985, 1992); Moore *et al.* (1987); Matthews (1980); Baehr (2000b, 2003d, 2009g, 2016f); Lorenz (2005); Eberhard & Giachino (2011).

Keys to species. Sloane (1894); Moore (1984); Baehr (2003d). Keys cover only a part of the recorded Australian species.

Neonomius Moore, 1963 (Pl. 31G)

Type species. *Mecyclothorax laevicollis* Sloane, 1915, by original designation.

Characteristics. Small beetles; terminal palpomeres finely pilose; prothorax almost parallel-sided with rectangular basal angles; aedeagus rather elongate, parameres moderately dissimilar, the left one asetose, the right one moderately setose; female genitalia not recorded.

Australian species. Three species.

Distribution. South-western part of Western Australia, south-eastern Australia from eastern Victoria to south-eastern New South Wales.

Biology. Ground living, in leaf litter of various habitats, from closed forests to alpine habitats. Predacious and apparently nocturnal.

References. Moore *et al.* (1987); Lorenz (2005).

Key to species. Moore (1963b).

Tribe Brachinini Bonelli, 1810

Characteristics. Males with eight visible abdominal sternites, females with seven; elytral apex truncate; head with a single supraorbital seta above each eye; mandible with a seta in the scrobe; antenna densely pilose from antennomere 3; suture between mentum and submentum distinct; procoxal cavities open or closed; mesocoxal cavities conjunct; protibia with one apical and one subapical spur (anisochaete) with distinct antenna cleaner; aedeagus without basal bulb, broadly open; the parameres are usually balteate, the left is larger than the right, and both are wrapped around the basal portion of the median lobe, ostium usually directed dorsad; both gonocoxites asetose.

Notes. The systematic position of the tribe has been widely discussed and its sister-group is still unclear. Relationships to Paussini as well as to Pseudomorphini have been postulated, but published phylogenetic analyses place the tribe near the subfamily Harpalinae (Deuve 1988; Liebherr & Will 1998; Beutel 1993; Maddison *et al.* 1999; Ober 2002).

Brachinini are well known for their explosive chemical defence mechanism. They discharge a hot mixture of quinones and hydrocarbons that is primarily heated by an enzymatic reaction involving hydroquinone and hydrogen peroxide in a sclerotised reaction chamber portion of the pygidial glands. The fluid is ejected at temperatures of up to 100°C and can be sprayed over a distance of more than 50 cm. Because the beetles are commonly gregarious, the group defensive is highly successful against predators and even against man.

Pheropsophus Solier, 1833 (Pl. 2F, 34E; Fig. 12.5)

Type species. *Carabus complanatus* Fabricius, 1775, by subsequent designation by Hope, 1838.

Characteristics. Elytra distinctly costate; mentum edentate; apical palpomeres moderately widened; ligula at apex bisetose, paraglossae somewhat surpassing the glossa; colour either completely black or black and yellow.

Australian species. Seven species.

Distribution. Almost the whole of mainland Australia, except the in the extreme southern area. The highest diversity is the northern tropical belt; extraterritorial worldwide in tropical and subtropical regions.

Biology. Hygrophilous, found near rivers, lakes, pools, lagoons. Sometimes occurring in very large aggregations. Strictly nocturnal, during the day hidden under stones, grass or in small burrows in wet sand or mud. The adults are rather omnivorous, feeding on worms, insect larvae, etc. The larvae of Australian species are unknown, but larvae of extra-Australian species apparently feed exclusively on eggs of mole crickets (Habu 1967; Frank 1994).

Note. All Australian species belong to the subgenus *Stenaptinus* Maindron, 1906.

References. Csiki (1933); Jedlicka (1963); Habu (1967); Darlington (1968); Matthews (1980); Moore *et al.* (1987); Giachino (2003a, 2005b); Lorenz (2005); Baehr (2012d).

Keys to species. Giachino (2003a), Baehr (2012d).

Tribe Pterostichini Bonelli, 1810

Characteristics. Medium to very large size beetles, typically 10–30 mm, but the full range is from 3–70 mm; procoxal cavities are closed, mesocoxa conjunct; protibia anisochaete with a well developed antennal cleaning groove; two supraorbital setae above each eye (rarely only one or more than two); mandibular scrobe lacking seta, rarely plurisetose; elytral plica typically present, but occasionally elytra with only an internal ridge or no evident structure; flight wing full or very frequently, variously reduced in size or absent; abdomen of the harpalidian type (Deuve 1993); aedeagus with well developed basal bulb, foramen small or base open; aedeagus with parameres glabrous, highly variable in shape, but typically dissimilar in form with the left larger; female reproductive tract usually with gland on spermatheca or bursa; gonocoxites with a few setae or glabrous, usually gonocoxite-2 elongate, rarely completely reduced.

Notes. Pterostichines are usually treated as a tribe, but have been ranked anywhere from subtribe to family by various authors. The group has, in part, been treated in restricted faunal reviews (e.g. Moore (1965) for Australia) and organised by a mental analysis or using no explicit data at all to compose various classifications. Studies that include cladistic analyses are few and very limited in taxonomic scope (Sasakawa & Kubota 2007; Casale & Ribera 2008; Will & Gill 2008). All attempts to cover the world fauna long predate cladistics analysis (e.g. papers by T. Tschitschérine, T. Sloane, M. Chaudoir, R. Jeannel, S. Straneo, etc.), and are based on limited samples of taxa. Subsequent catalogues and checklists have followed these, often arbitrary, established placements (Csiki 1932; Moore *et al.* 1987; Bousquet 2012; Lorenz 2005), with catalogue authors doing what they can to merge disparate ideas into a single classification. Given the scope of the present work, we also provide a compromise arrangement as reflected in the keys. To the degree possible, we take in to account ongoing, but unpublished work on the tribe as a whole that includes DNA sequence data and morphological data (KWW unpubl.). We treat separately several taxa as tribes (Morionini, Cratocerini, Drimostomatini, Abacetini) that have been lumped in a larger concept of pterostichines by some previous authors (e.g. Moore *et al.* 1987; Darlington 1962a). Though these tribes are each monophyletic groups, there is no evidence that they share a very close relationship to Pterostichini. In the classification used here only one subtribe is recognised in the Australian fauna, Euchroina, which constitutes a monophyletic group worldwide. The remaining taxa do not constitute a monophyletic group. The use of a subtribe Pterostichina would obscure the lineage-level structure in the Australian pterostichine fauna, but until a more thorough study is completed there no benefit to inserting further structure or using formal taxa. Moore's (1965) series for pterostichites captured some of that lineage information and here we present a set of informal series that are likely to reflect the natural affinities of the taxa, or at least draw attention to important similarities. Like Moore's series, these are presented as provisional groups and as an aide to identification. As many characters of these series are found on structures that can only be examined by dissection and often are only available in one sex, all genera and subgenera of the Australian Pterostichini are included in a single key below. Those genera that are in Euchroina, an informal series, or incertae sedis are marked as such.

The genus *Geobaenus* Dejean, 1829, represented by the single species *G. australasiae* Guérin-Méneville, 1830, was included in Pterostichitae by Moore *et al.* (1987) but it is not included in the present work. The species is only known from the original description, the type locality is unknown, no additional specimens have been found, and several attempts to locate the holotype have failed. If this is a species of *Geobaenus*, then it may have been mislabelled or its provenance is incorrectly reported. Possibly it was a human introduction but no population was established. There is also a high probability that *G. australasiae* is not a *Geobaenus*, but rather it belongs in an existing Australian genus. Otherwise, this genus is found in southern Africa and the affinities of *Geobaenus* are unclear. They may be related to Platynini, Oodini or Melanchitonini, but there is no evidence they are closely related to any pterostichine group.

Similarly, *Cyrtoderus australasiae* Hope, 1841 is the basis for a monotypic genus for which the type is apparently lost and no additional specimens can be attributed to the species based on the brief description. For these reasons, and also because it is nearly certain, given the little information in the description and the type locality reported, a described species of *Delinius* Westwood, we follow Moore's (1965) recommended synonymy. We have not included this genus in the keys.

Key to the subtribes and informal series of Australian Pterostichini

1. All of the following character states: abdominal ventrites transversely sulcate (Fig. 12.48); size < 18 mm, gula narrow (Fig. 12.51), elytral intervals flat or convex, never costate or carinate; propleuron smooth; and elytral interval 3 with one or more punctures .... **Euchroina (part)**

– Not the combination above; lacking **any one or more** of those characteristics **2**

2(1). Elytra without punctures in interval 3 and punctures of last ventrite large, deep fovea .... **Euchroina: *Setalis* Laporte**

– Not the combination of character states above .... various pterostichine taxa. See characteristics of series and genera below.

Key to the genera and subgenera of Australian Pterostichini

1. Abdominal ventrites transversely sulcate (Fig. 12.48) 2
– Abdominal ventrites without sulci 16

2(1). Prosternal process variously shaped at apex, rounded or truncate, never with a rearward projected cuneiform apex; body size under 25 mm; widely distributed 3
– Prosternal process with a rearward projected cuneiform apex (Fig. 12.52); large beetles, 25 mm and larger; restricted to Western Australia ..... ***Cuneipectus* Sloane**

3(2). Gula narrow, its width about half the width between the buccal fissures or less (Fig. 12.51); black, iridescent or rarely subtly metallic coloured beetles; flight wing various, but usually reduced; widely distributed 4
– Gula wide, its width as or nearly as wide as the width between the buccal fissures (Fig. 12.50); metallic green beetles with full-sized flight wings; restricted to Queensland ..... ***Lesticus* Dejean**

4(3). Elytral intervals convex or flat, never sharply carinate 5
– Elytral intervals 3, 5, & 7 sharply carinate for all or most of their length ..... ***Loxogenius* Sloane**

5(4). Frontal impressions elongate and well marked, either linear or angulate 10
– Frontal impressions absent or punctiform 6

6(5). Proepisternum smooth; widely distributed 7
– Proepisternum distinctly, transversely rugose; restricted to Lord Howe Island ..... ***Eurystomis* Chaudoir**

7(6). Mandibles short, heavily built; labrum not or only slightly emarginate; elytra with one or more punctures in interval 3 on the disc; marginal series of elytral setae ended near humerus 8
– Mandibles slender, porrect; labrum deeply emarginate; elytra without punctures on the disc; marginal series of elytral setae very distant from humerus ..... ***Darodilia* Laporte**

8(7). Metatrochanter short, less than half the length of the femur 9
– Metatrochanter long, 2/3 the length of the femur or more ..... ***Gastrogmus* Sloane [Euchroina]**

9(8). Pronotum with posterior latero-marginal seta positioned very near the margin; elytral interval 3 with three or less punctures, rarely four ..... ***Simodontus* s.str. [Euchroina]**
– Pronotum with posterior latero-marginal seta positioned very distant from the margin, the distance more than 2x the puncture width; elytral interval 3 with four punctures ..... ***Simodontus (Trochoglymmus)* Straneo [Euchroina]**

10(5). Propleuron smooth or shallowly punctate 11
– Propleuron distinctly rugose ..... ***Rhytisternus* Chaudoir (part)**

11(10). Elytral disc with one or more punctures in interval 3 14
– Elytral disc without punctures 12

12(11). Elytral interval 6 costate at its base; pronotum with a single inner basal impression ..... ***Liopasa* Tschitschérine**
– Elytral interval 6 not costate at base; pronotum with inner and outer basal impressions evident 13

13(12). Small size beetles, < 10 mm; frontal impressions very deeply impressed and linear; eyes are strongly protruded ..... ***Setalis* Laporte [Euchroina]**
– Large size beetles, > 15 mm; frontal impressions very shallowly impressed; eyes only moderately protruded or nearly flat ..... ***Cratogaster* Blanchard**

14(11). Last ventrite with typical sized subapical pores 15
– Last ventrite with large, foveate subapical pores (Fig. 12.49) ..... ***Setalimorphus* Sloane [Euchroina]**

15(14). Elytral interval 3 with two or more punctures on the disc ..... ***Prosopogmus* Chaudoir [Euchroina]**
– Elytral interval 3 with one puncture on the disc ..... ***Phaenaulax* Tschitschérine [Euchroina]**

16(1). Stipes with two or more setae (Fig. 12.53–54) 17
– Stipes with one seta near the base (Fig. 12.55) 26

17(16). Marginal series of elytral setae with three pores or more between the primary and the humerus (Fig. 12.56B), check both sides, rarely unilaterally reduced 18
– Marginal series of elytral setae with two pores between the primary and the humerus (Fig. 12.56A) 25

18(17). Head without postocular ridges 19
– Head with paramedial, postocular longitudinal ridges (Fig. 12.29) ..... ***Mecynognathus* Macleay**

19(18). Elytra without punctures on interval 7; with or without punctures on intervals 3 and/or 5 20
– Elytra with punctures on intervals 3, 5 & 7 24

20(19). Elytral interval 7 flat or slightly convex near the apex of the elytra, may be raised above adjacent intervals but is not carinate 21
– Elytral interval 7 carinate near the apex from near the apical 1/3 to the suture 23

21(20). Mandibles very long with scrobe occupying only half the total length; parameres of male aedeagus variable; eastern Australia 22
– Mandibles shorter, scrobe nearly always half the total length or more; parameres of male aedeagus markedly dissimilar in form, the left shorter, conchoid and the right styloid and/or falcate; primarily eastern Australia with one species in south-western Western Australia ..... ***Trichosternus* Chaudoir**

22(21). Elytral form relatively short, broad and convex; parameres of male aedeagus markedly dissimilar in form, the left shorter, conchoid and the right styloid and/or falcate ..... ***Nurus* Motschulsky**
– Elytral form elongate, ovoid or parallel-sided, depressed; parameres of male aedeagus similarly conchoid, the right smaller than the left ..... ***Castelnaudia* Tschitschérine**

23(20). Elytral striae broadly and indistinctly impressed; intervals 3, 5, & 7 noticeably more prominent than other intervals; head proportionally very large, as wide as the pronotum ..... ***Paranurus* Tschitschérine**
– Elytral striae narrowly and distinctly impressed; elytral intervals all equally convex head large, but of more typical proportions, narrower than pronotum ..... ***Notabax* Moore**

24(19). Antennomere 1 with a dorsal depression (Fig. 12.57); prosternal process setose at the tip ..... ***Cratoferonia* Tschitschérine**
– Antennomere 1 without a dorsal depression; prosternal process glabrous at the tip ..... ***Megadromus* Motschulsky**

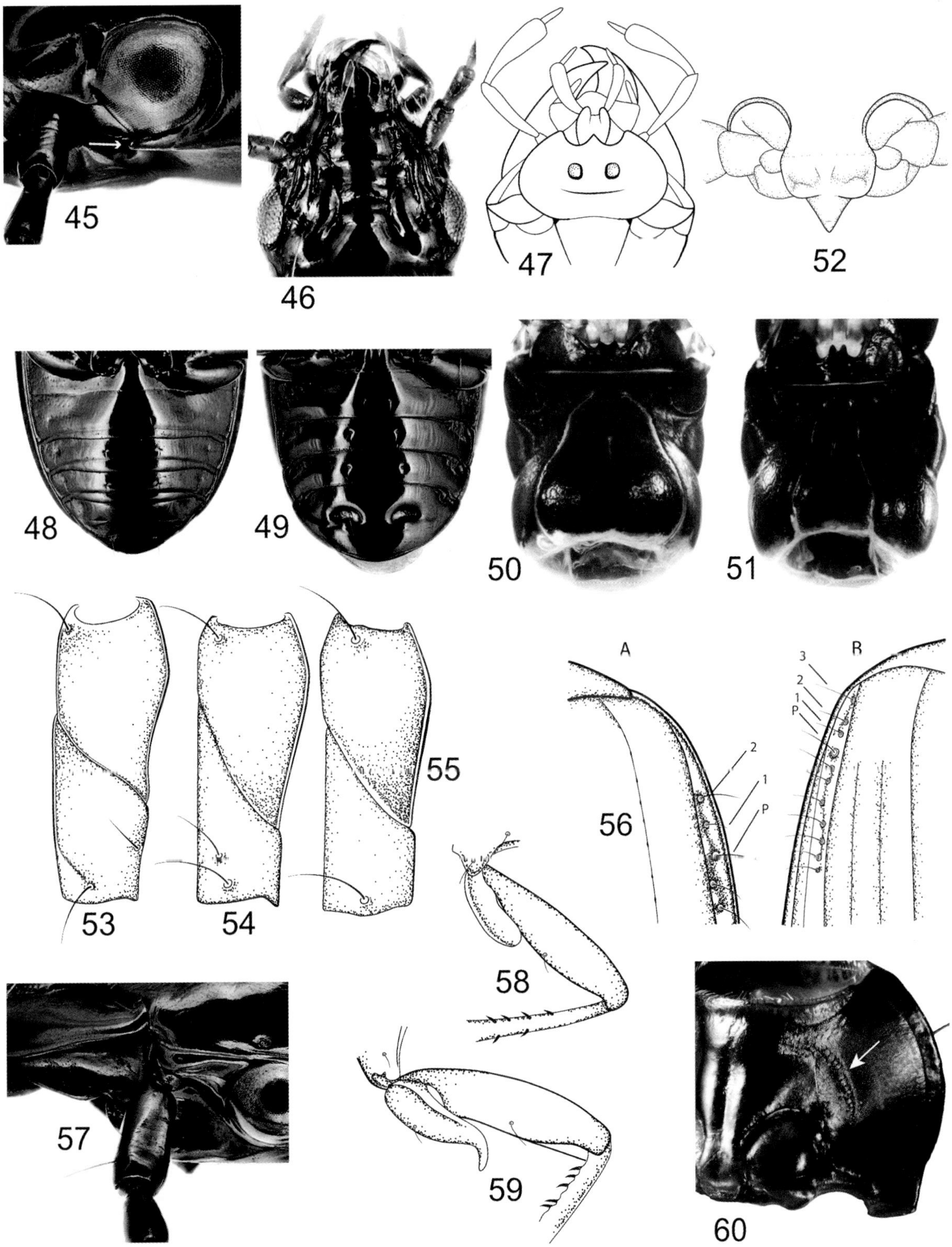

Figs 12.45–12.60. 45, *Euryscaphus dilatatus Macleay*, head, lateral (arrow = subocular projection); 46, *Eutrechopsis ovalis* Moore, head, dorsal; 47, *Polyderis* sp., head, ventral; 48–49, abdominal ventrites: 48, *Prosopogmus chalybaeipennis* (Chaudoir), 49, *Setalimorphus punctiventris* Sloane; 50–51, ventral view of head: 50, *Notonomus (Leiradira) aurifer* (Darlington), 51, *Notonomus hephaestus* Will; 52, *Cuneipectus foveatus* Sloane, prosternal process; 53–55, maxillary stipes: 53, *Cuneipectus foveatus*, 54, *Castelnaudia* sp., 55, *Cratogaster* sp.; 56A–B, humeral region: scheme for elytral umbilicate setigerous punctures (P, primary seta, typically larger and positioned lateral to series, often touching or in stria 8); 57, *Cratoferonia phylarchus* (Sloane), head, lateral; 58–59, metafemur and trochanter: 58, *Cratogaster* sp., 59, *Teropha besti* (Sloane); 60, *Abacetus* sp., prothorax, ventral (arrow = stridulatory file).

25(17). Metatrochanter distinctly hooked at the apex (Fig. 12.59) ... ***Teropha* Laporte**
– Metatrochanter blunt at apex (Fig. 12.58) ... ***Secatophus* Laporte**
26(16). Elytra with eight to 10 apparent intervals, widths variable, but never more than a 5-fold difference between the wide and narrow interval widths; colour variable ... **27**
– Elytra with only five apparent intervals; intervals 1, 3, & 5 massively wide, more than five time the width of 2 & 4; brilliant coppery colour ... ***Notolestus* Sloane**
27(26). Length between the hind margin of the mesocoxa and the anterior margin of the metacoxa noticeably less than diameter of the mesocoxa; flight wing reduced ... **31**
– Length between the hind margin of the mesocoxa and the anterior margin of the metacoxa equal to or greater than the diameter of the mesocoxa; flight wing usually full ... **28**
28(27). Antennae pubescent from antennomere 5; size very large, 25–65 mm ... ***Catadromus* Macleay**
– Antennae pubescent from antennomere 4; small to medium-large size, < 20 mm ... **29**
29(28). Parascutellar stria present (Fig. 12.33A–B,D–F); propleuron smooth or shallowly punctate but never with an indication of rugosities ... **30**
– Parascutellar stria absent (Fig. 12.33C); propleuron distinctly rugous or, in a few species, rugosities are apparent but extremely shallowly impressed ... ***Rhytisternus* Chaudoir (part)**
30(29). Mental tooth entire triangular ... ***Pseudoceneus* Sloane**
– Mental tooth bifid or emarginate ... ***Platycoelus* Blanchard**
31(27). Propleuron without a deep transverse sulcus; last labial palp form various, but typically fusiform or narrowly securiform ... **32**
– Propleuron with a deep transverse sulcus; last labial palp broadly securiform ... ***Delinius* Westwood**
32(31). Lateral margins of the pronotum broadly explanate posteriorly ... **33**
– Lateral margin of the pronotum not explanate, typically narrowed posteriorly ... **35**
33(32). Hind angles of the pronotum broadly obtuse or rounded and not produced over the humeri; tarsi dorsally smooth ... **34**
– Hind angles of the pronotum right angled and produced over the humeri; tarsi dorsally striolate ... ***Rhabdotus* Chaudoir**
34(33). All or nearly all elytral striae moderate to shallowly impressed ... ***Sarticus* s.str.**
– Elytral with only striae 1,3, and 4 impressed and those deeply sulcate ... ***Sarticus (Coronocanthus)* Macleay**
35(32). Gula very wide (Fig. 12.50), reaching or surpassing the buccal fissure laterally, sutures often very lightly marked ... ***Notonomus* (*Leiradira*) Laporte (part)**
– Gula narrow (Fig. 12.51), not as wide or up to just a little wider than the apices of the mental epilobes, sutures usually clearly evident ... **36**
36(35). Eyes and postocular region at most slightly prominent; metatrochanters short and moderately pointed or blunt ... **37**
– Eyes and postocular region greatly enlarged and very prominent; metatrochanters very long and pointed ... ***Notonomus* (*Acanthoferonia*) *ferox* Moore**
37(36). Elytra intervals convex or flat, not narrowly costate; lustre of elytra somewhat dull to shiny ... **38**
– Elytral intervals narrowly costate; elytral lustre very dull ... ***Notonomus* (*Loxodactylus*) Chaudoir**
38(37). Body form exceptionally elongate cylindrical; elytra with internal ridge and no external plica ... ***Notonomus* (*Conchitella*) *clivinoides* Moore**
– Body form various, but not elongate cylindrical; elytral plica and internal ridge variously developed, but plica usually present ... ***Notonomus* Chaudoir (part)**

Subtribe Euchroina Chaudoir, 1874

Characteristics. Small to medium size beetles, 5–17 mm; majority are black or brown, often iridescent, but metallic colours or metallic reflex are common, especially among species in the tropics; eyes well developed, often very prominent; abdominal ventrites 3–6 transversely sulcate in most; gula narrow; flight wing full or reduced; most with 1–3 punctures in elytral interval 3; two umbilicate punctures apicad the primary in the margin at the elytral humerus; elytral plica evident; lateral margins of pronotum typically with two setae; aedeagus with ostium dorsal, left paramere conchoid, right small; gonocoxite-1 with sparse setae or glabrous; and spermathecal gland with duct diverticulum present in most species.

Notes. A subtribe of ~30 genera worldwide, primarily found in South America, Africa and Australia, although a few species are known from southern parts of North America and Europe (Will 2002).

Gastrogmus Sloane, 1915 (Fig. 12.92)

Type species. *Gastrogmus ischialis* Sloane, 1915, by monotypy.

Characteristics. Robust, elongate oval, piceous beetles; clypeo-ocular sulcus complete; mentum tooth bifid; elytral plica present; angular base of stria 1 present; flight wing full; abdominal ventrites 4–6 transversely sulcate; meta trochanters very long, about two-thirds the length of the femur; male protarsomeres expanded, with squamous setae ventrally; aedeagus with ostium dorsal; parameres short, left conchoid, right small.

Australian species. Only *Gastrogmus ischialis* Sloane.

Distribution. Known only from near Albany, Western Australia.

Biology. Holotype was collected in December, 'under a log, in a very damp place, beside a swamp' (Sloane 1915a).

Prosopogmus Chaudoir, 1865 (Pl. 35A; Fig. 12.48)

Type species. *Feronia* (*Prosopogmus*) *impressifrons* Chaudoir, 1865, by monotypy.

Characteristics. As for subtribe and mentum tooth emarginate; frontal sulci usually well impressed and elongate, sometimes short; pronotum cordiform to trapezoidal; punctures of last ventrite not deeply foveate.

Australian species. 23 species.

Distribution. Primarily eastern mainland Australia, rarely in interior and Western Australia, Tasmania, and Lord Howe Island; extraterritorially in Maluku Islands, New Caledonia, New Guinea, New Zealand and possibly Fiji.

Biology. Eurytopic in mesic to wet habitats. Open and sometimes disturbed habitat, riparian, and more or less closed forests. Nocturnally active, ground dwelling and predacious. Flight-capable species are sometimes taken at lights.

References. Fauvel (1882); Lorenz (2005); Moore *et al.* (1987); Moore & Wallbank (1968); Moore (1965); Sloane (1895a, 1898, 1900a, 1911, 1915a, 1920); Straneo (1936, 1937); Tschitschérine (1902); Will (2011); Yeates & Monteith (2008).

Keys to species. For some Tasmanian species there is a key by Sloane (1920).

Simodontus Chaudoir, 1843 (Pl. 38B)

Type species. *Simodontus aeneipennis* Chaudoir, 1843, by monotypy.

Leptopodus Blackburn, 1890 (nec Leptopodus Chaudoir, 1874)

Pseudoceneus Tschitschérine, 1890 (non *Pseudoceneus* Sloane, 1903)

Characteristics. As for subtribe and frequently reminiscent of Harpalini in general form, i.e. head relative very broad; pronotum short and wide; elytra convex, short and oval, but many have the same form as typical *Prosopogmus*; mandibles short; mentum tooth emarginate or entire; frontal sulci not impressed; pronotum trapezoidal; flight wing reduced; punctures of last ventrite not deeply foveate.

Australian species. 19 species.

Distribution. Southern Queensland south to Tasmania and west to south-western coastal South Australia; in southern and west coastal Western Australia. Scattered locations further inland in the south and west.

Biology. Eurytopic, ground dwelling and more common in wet and mesic habitats. In more arid habitats usually near temporary water sources. Found in tall closed forests, woodlands and open habitats, including some disturbed habitats like roadside ditches. Most are probably predacious, but the broad heads and thick mandibles in some species suggest seed feeding habits similar to harpalines.

References. Lorenz (2005); Moore (1965); Moore *et al.* (1987); Sloane (1895a, 1898, 1900a); Straneo (1936, 1937, 1957); Tschitschérine (1898).

Keys to species. For some Western Australian species Sloane (1898) and Tasmanian species Sloane (1920).

Subgenus *Trochoglymmus* Straneo, 1937

Type species. *Simodontus aeneipennis* Chaudoir, 1843, by monotypy.

Characteristics. As for subtribe and with form as very similar to typical *Prosopogmus*; mandibles short; head relatively large; mentum tooth emarginate or entire; frontal sulci not impressed; pronotum trapezoidal; flight wing reduced; punctures of last ventrite not deeply foveate.

Australian species. Three species.

Distribution. Tasmania and southern Victoria north at least to Mt. Baw Baw.

Biology. More or less closed, wet forests, nocturnal.

Note. Given variation in characteristics of *Simodontus* s.str. and *Prosopogmus* this is a weakly supported subgenus based only on chaetotaxy of the pronotum and elytra.

References. Straneo (1936, 1937); Moore *et al.* (1987).

Phaenaulax Tschitschérine, 1898 (Pl. 34J)

Type species. *Phaenaulax stenomorphus* Tschitschérine, 1898 (= *Setalimorphus nanus* Sloane, 1895), by monotypy.

Characteristics. As for subtribe and mentum tooth entire; frontal sulci impressed; pronotum elongate rectangular, lateral margin with one or two setae (posterior seta sometimes absent); flight wing reduced; punctures of last ventrite not deeply foveate; elytral interval 3 with one puncture; legs stout.

Australian species. Only *Phaenaulax nanus* (Sloane).

Distribution. South-eastern coastal Victoria.

Biology. Under woody debris in forests. Nocturnal.

References. Lorenz (2005); Moore (1965); Moore *et al.* (1987); Tschitschérine (1898); Sloane (1907a).

Setalimorphus Sloane, 1895 (Pl. 37I; Fig. 12.49)

Type species. *Setalimorphus punctiventris* Sloane, 1895, by original designation.

Characteristics. As for tribe and small beetles, parallel-sided form, similar to typical *Prosopogmus*; head with frontal impressions deep; antennae stout, submoniliform; mentum tooth broad, truncate; pronotum quadrate; elytral interval 3 with one puncture; angular base of stria 1 absent; flight wing reduced; punctures of last ventrite foveate.

Australian species. Two species.

Distribution. South-east coastal Australia, inland at least to Mt. Buffalo; New South Wales, Victoria.

Biology. More or less closed forests, alpine; found under woody debris. Nocturnal, probably predacious.

References. Lorenz (2005); Moore *et al.* (1987); Moore (1965).

Setalis Laporte, 1867 (Pl. 37L)

Type species. *Setalis niger* Laporte, 1867, by monotypy.

Loxogmus Sloane, 1890.

Characteristics. As for tribe and medium size beetles; head with frontal impressions very deep; eyes prominently produced; antennae stout; mentum tooth broad, truncate; pronotum quadrate; elytral interval 3 without punctures; angular base of stria 1 absent; flight wing reduced; punctures of last ventrite foveate.

Australian species. Three species.

Distribution. Northern coastal Queensland to northern New South Wales.

Biology. Restricted to wet closed forests. Presumed predatory, nocturnal.

References. Lorenz (2005); Moore *et al.* (1987).

Keys to species. Moore (1965).

Trichosternus series

Stipes with two or more setae; three umbilicate punctures apicad of the primary at the elytral humeral margin (Fig. 12.56B); ostium of the median lobe of the aedeagus deflected to the right; female reproductive tract with spermathecal gland duct diverticulum, diverticulum often ramifying; pygidial gland reservoirs lacking a dorsal lobe.

Castelnaudia Tschitschérine, 1890 (Pl. 20F; Fig. 12.54)

Type species. *Feronia* (*Omalosoma*) *nitidicollis* Laporte, 1867, by original designation.

Omocycla Tschitschérine, 1902.

Characteristics. Very large beetles; black or frequently with striking metallic colours (usually green, blue or purple); pronotum generally cordiform, lateral margins with two or more setae; pro- and mesosterna glabrous or setose; elytra fused, flight wing reduced; plica not developed, internal ridge present; elytral intervals costate or carinate; parascutellar striae absent or obscured; anterior tarsi of male usually with two or three basal segments dilated with squamose setae ventrally, setae sometimes lacking; aedeagus median lobe with ostium deflected to the right; parameres conchoid, the right smaller than the left.

Australian species. 17 species.

Distribution. Restricted to eastern Queensland from the southern edge of the tropics northward to near Cooktown.

Biology. Most species are found in temperate and tropical rainforest, a few in wet sclerophyll or mesic scrub. Predacious, nocturnal and often found under or in woody debris.

References. Moore & Wallbank (1968); Staunton *et al.* (2014); Galián & Moore (1994).

Keys to species. Darlington (1961c) covered most species.

Cratoferonia Tschitschérine, 1902 (Pl. 23D; Fig. 12.57)

Type species. *Feronia regalis* Laporte, 1867, by original designation.

Characteristics. Extremely large beetles, typically around 35–45 mm long; black with a more or less clearly evident, blue, copper, or purple metallic cast; pronotum broadly cordiform, with two setae on the lateral margin; elytra with parascutellar stria present; elytral intervals convex, with setigerous punctures on the third, fifth and seventh interval; flight wing reduced to a small strap; male protarsomeres with three basal segments dilated, with squamose setae ventrally; aedeagus with ostium deflected to the right; parameres dissimilar, the left conchoid, the right elongate, subfalcate.

Australian species. Three species.

Distribution. East coastal mountains from the Illawarra Range in New South Wales to the border region between New South Wales and Queensland.

Biology. Wet montane forests, nocturnal, and predacious.

References. Moore & Wallbank (1968).

Keys to species. Differential diagnoses published by Garetto & Giachino (2003).

Mecynognathus Macleay, 1873 (Pl. 1G, 29H; Figs 12.1, 12.29)

Type species. *Mecynognathus daemeli* Macleay, 1873, by monotypy.

Characteristics. Extremely large beetles, typically around 50 mm long or longer; black; pronotum broad; elytra dull, striae very shallowly impressed, intervals impunctate; elytra without plica; head with prominent postocular ridges; labrum and clypeus strongly emarginate; flight wing reduced; male protarsomeres unmodified; male mandibles long and asymmetrical; aedeagus with ostium deflected to the right; parameres dissimilar, the left conchoid, the right elongate, subfalcate.

Australian species. Only *Mecynognathus daemelii* Macleay (Fig. 12.1).

Distribution. Northern part of the Cape York Peninsula, Queensland.

Biology. Nocturnal, predacious or scavenging, in open forests and heathlands. Attracted to trap with rotten fish.

Note. Sexually dimorphic with males showing allometric development of the adult head and mandibles. Because of the asymmetrical mouthparts and deeply emarginate labrum and clypeus, various authors have placed the genus in Licinini. However, most characters including those listed for the *Trichosternus* series, male genitalia (Moore 1965), female reproductive tract, and DNA data (K. Will unpubl.) all clearly place this species in Pterostichini.

References. Moore (1965).

Megadromus Motschulsky, 1865 (Pl. 29E)

Type species. *Megadromus viridilimbatis* Motschulsky, 1865 (= *Feronia* (*Trichosternus*) *antarctica* Chaudoir, 1865), by monotypy.

Nesopterostichus Tschitscherine, 1902

Protodromus Moore, 1965

Characteristics. Large beetles, typically around 23 mm long; black with a more or less clearly evident, copper, metallic cast at least near the elytral margins; pronotum cordiform, with two setae on the lateral margin; elytra with parascutellar stria present; elytral intervals slightly convex, with setigerous punctures on the first, third, fifth and seventh interval; flight wing reduced; male protarsomeres with three basal segments dilated, with squamose setae ventrally; aedeagus with ostium deflected to the right; parameres dissimilar, the left conchoid, the right elongate, falcate.

Australian species. Two Australian species.

Distribution. Known only from Mount Royal Range and New England National Park, New South Wales and Australia; extraterritorially New Zealand (27 species).

Biology. Nocturnal, predatory, flightless, in temperate, closed rainforest.

Note. The subgenus *Protodromus* erected for the two Australian species is quite distinct from all New Zealand species and may represent a separate genus. However, there are no recent collections of *M. australicus* (Sloane) and the type specimen is significantly damaged, hindering conclusive study.

Keys to species. Differential description by Moore (1965).

Notabax Moore, 1976 (Pl. 31C)

Type species. *Notabax monteithi* Moore, 1976, by original designation.

Characteristics. Large beetles, typically around 20 mm long; black, relatively shiny; pronotum subcordiform, with two setae on the lateral margin; elytra with parascutellar stria present; elytral intervals slightly convex, without setigerous punctures; flight wing reduced; male protarsomeres with three basal segments slightly dilated, with squamose setae ventrally; aedeagus with ostium deflected to the right; parameres dissimilar, the left conchoid, the right short, peg-like.

Australian species. Only *N. monteithi* Moore.

Distribution. Restricted to the Iron Range area of the Cape York Peninsula, Queensland.

Biology. Nocturnal, predacious, in closed forests.

Nurus Motschulsky, 1865 (Pl. 32E)

Type species. *Nurus brevis* Motschulsky, 1865, by monotypy.

Nuridius Sloane, 1890

Pachymelas Tschitschérine, 1902

Characteristics. Large, robustly built beetles, typically around 20 mm long; black, relatively shiny; pronotum cordiform, with two setae on the lateral margin; elytral with small setigerous punctures on third interval; flight wing reduced; male protarsomeres with two basal segments slightly dilated and with squamose setae ventrally or, typically, unmodified; aedeagus with ostium deflected to the right; parameres dissimilar, the left conchoid, the right falcate.

Australian species. 14 species.

Distribution. East coastal New South Wales and Queensland.

Biology. Flightless, burrowing, predacious, in closed forests.

Note. Species generally have highly restricted ranges and *Nurus atlas* (Laporte) and *N. brevis* are formally listed as an endangered species in New South Wales. They are currently only known from selected locations near Lismore, Mallanganee, and Alstonville, NSW. All females build extensive burrows and brood eggs and larvae (Charley & Andren 2018).

References. Moore (1965); Darlington (1961a, b); Staunton *et al.* (2014); Greenslade (1994).

Keys to species. Darlington (1961b); Will & Monteith (2018).

Paranurus Tschitschérine, 1901 (Pl. 33H)

Type species. *Feronia* (*Trichosternus*) *dilaticeps* Chaudoir, 1865, by original designation.

Characteristics. Large beetles, typically around 35 mm long or longer; black; head massive relative to body; pronotum broad; elytra duller than pronotum, striae very broadly impressed, intervals impunctate; elytra seventh interval carinate, without plica; flight wing reduced; male protarsomeres unmodified; male mandibles long and asymmetrical; aedeagus with ostium deflected to the right; parameres dissimilar, the left conchoid, the right elongate, subfalcate.

Australian species. Three species.

Distribution. Cape York Peninsula, Queensland from south near Cooktown north to include the islands in the Torres Strait.

Biology. Nocturnal, predacious or scavenging, in open forests and heathlands.

Note. The three named species have been insufficiently studied and appear to be rather polymorphic. It is very likely there are additional species-level taxa to be recognised in this complex.

Key to species. Tschitschérine (1902), but see Sloane (1903).

Secatophus Laporte, 1867 (Pl. 37D)

Type species. *Acinopus australis* Hope, 1845, by original designation.

Hoploprion Tschitschérine, 1903.

Characteristics. Heavily built beetles, legs stout; tarsomeres thick; black, typically with a coppery or red metallic border on the pronotum and elytra; head massive relative to body; overall body form is convex, subcylindrical; striae very shallowly impressed, intervals flat; elytra with prominent plica; flight wing reduced; male protarsomeres with three basal segments dilated, with squamose setae ventrally; aedeagus with ostium deflected to the right; parameres dissimilar, the left conchoid, the right small.

Australian species. Only *Secatophus australis* Hope.

Distribution. South Australia in the mountain and hill regions along the southern Mount Lofty Ranges.

Biology. Flightless, predacious, in open woodland.

Teropha Laporte, 1867 (Pl. 39D; Fig. 12.59)

Type species. *Platysema flindersii* White, 1859, by original designation.

Morphnos Schaufuss, 1867

Characteristics. Massive, heavily built beetles; legs stout, tarsomeres thick; black; head very massive relative to body; overall body form is convex; striae well impressed; elytra with low plica; flight wing reduced; hind trochanters elongate, hooked at apex, especially elongate in males; male protarsomeres unmodified; aedeagus with ostium deflected to the right; parameres dissimilar, the left conchoid, the right small.

Australian species. Two species.

Distribution. South-east coastal Victoria to the Eyre Basin, South Australia.

Biology. Flightless, predacious, in open scrub and open woodlands or light forests.

References. Moore & Wallbank (1968); Moore *et al.* (1987).

Key to species. Differential description by Sloane (1901).

Trichosternus Chaudoir, 1865 (Pl. 39K)

Type species. *Omalosoma vigorsi* Gory, 1833, by subsequent designation by Moore *et al.* 1987.

Omalosoma Gory, 1833

Homalosoma Agassiz, 1847

Homalosome Chaudoir, 1865

Characteristics. Large to very large beetles; black or frequently with striking metallic colours (usually green, blue or purple); pronotum generally cordiform, lateral margins with two or more setae; prosternum setose and mesosterna glabrous or setose; elytra fused, flight wing reduced; plica very small; elytral intervals typically costate; parascutellar striae small; anterior tarsi of male usually with two or three basal segments dilated with squamose setae ventrally, setae sometimes lacking; aedeagus median lobe with ostium deflected to the right; parameres conchoid, the right longer than the left falcate or styloid.

Australian species. 15 species described and several recognised in collections but undescribed.

Distribution. All but one species are found along the east coastal region from near the level of Port Douglas, Queensland, south to just north of Sydney. One species, *T. relictus* (Darlington), is found in the Margaret River region of Western Australia.

Biology. Most species are found in tropical rainforest, a few in wet sclerophyll or mildly mesic scrub. Predacious, nocturnal and often found under or in woody debris or rocks.

References. Moore & Wallbank (1968); Staunton *et al.* (2014); Constantine & Seeman (2014).

Keys to species. Darlington (1961b).

Notonomus series

Ostium of the median lobe of the aedeagus dorsal; left paramere of the aedeagus conchoid; female reproductive tract without spermathecal duct digital diverticulum; pygidial gland reservoirs with large dorsal lobe.

Notonomus Chaudoir, 1865 (Pl. 31A, L, 32A–B, F–G, I, L; Figs 12.8, 12.22, 12.50, 12.51)

Type species. *Notonomus triplogenioides* Chaudoir, 1865, by subsequent designation (Moore, 1965).

Orbitus Motschulsky, 1865.

Neuropates Motschulsky, 1865.

Ternox Motschulsky, 1865.

Adetipa Laporte, 1867.

Characteristics. Medium to large (10–35 mm) generally elongate ovoid beetles, typically black, but often with metallic colours; flight wing reduced; anterior tarsi of male usually with two or three basal segments dilated with squamose setae ventrally, setae sometimes lacking and tarsomeres unmodified; aedeagus median lobe with ostium dorsal; parameres left conchoid, the right small.

Australian species. The genus includes 132 named species and many undescribed species.

Distribution. In the east the beetles in the genus are found from Cooktown, Queensland south to coastal Victoria, the islands of the Bass Strait, Tasmania, and west to Mt. Gambier, South Australia. Only *Notonomus mediosulcatus* (Chaudoir) is known from south-west Western Australia.

Biology. The majority of species are found in more or less closed canopy forests and dense scrub. A few species are found in the alpine and open grass woodlands. All are flightless, nocturnal, and presumed to be predatory. Some species are known to make shallow burrows under woody debris where females brood eggs and larvae.

References. Will (2015a); Staunton *et al.* (2014); Nash *et al.* (2008a); Nash *et al.* (2008b); Horne (1990, 1992); Moore & Wallbank (1968).

Keys to species. Species-group, species keys and descriptions were published by Sloane (1913, 1916a, 1923a) and Moore (1960a, 1961, 1963c). See also references under subgenera below.

Subgenus *Notonomus* Chaudoir, 1865 (Pl. 32F–G, I)

Type species. *Notonomus triplogenioides* Chaudoir, 1865, by subsequent designation (Moore, 1965).

Characteristics. As for the genus.

Australian species. The genus includes 108 named species and many undescribed species.

Distribution. In the east the beetles in the genus are found from Cooktown, Queensland south to coastal Victoria, the islands of the Bass Strait, Tasmania, and west to Mt. Gambier, South Australia. Only *Notonomus mediosulcatus* (Chaudoir) is known from south-west Western Australia.

Biology. The majority of species are found in more or less closed canopy forests and dense scrub. A few species are found in the alpine and open grass woodlands. All are flightless, nocturnal, and presumed to be predatory. Some species are known to make shallow burrows under woody debris where females brood eggs and larvae.

References. Will (2015a); Staunton *et al.* (2014); Nash *et al.* (2008a); Nash *et al.* (2008b); Horne (1990, 1992); Moore & Wallbank (1968).

Keys to species. Species-group, species keys and descriptions were published by Sloane (1913, 1916a, 1923a) and Moore (1960a, 1961, 1963c). See also references under subgenera below.

Subgenus *Acanthoferonia* Moore, 1965 (Pl. 31L)

Type species. *Acanthoferonia ferox* Moore, 1965, by original designation.

Characteristics. Characteristics as for the genus but with relatively very large head, post-ocular orbits expanded; legs relatively stout and male metatrochanters elongate with pointed apex; male with three basal segments dilated with squamose setae ventrally.

Australian species. Only *Notonomus* (*Acanthoferonia*) *ferox* Moore.

Distribution. Known only from a few locations in the Mount Lofty Ranges, Fleurieu Peninsula, South Australia.

Biology. Open woodland and scrub.

References. Will (2015a); MLR Southern Emu-wren & Fleurieu Peninsula Swamps Recovery Team (2007).

Subgenus *Conchitella* Moore, 1962 (Pl. 31A)

Type species. *Conchitella clivinoides* Moore, 1962, by original designation.

Characteristics. Characteristics as for the genus but overall body form distinctly cylindrical, legs short and stout.

Australian species. Only *Notonomus* (*Conchitella*) *clivinoides* Moore.

Distribution. Found only in the Alpine National Park region of Victoria.

Biology. Possibly fossorial, from open or lightly forested alpine habitat.

References. Will (2015a).

Subgenus *Leiradira* Laporte, 1867 (Pl. 32A, L; Fig. 12.50)

Type species. *Leiradira auricollis* Laporte, 1867, by subsequent designation (Darlington 1961c).

Metadira Darlington, 1961

Stomimorphus Straneo, 1953

Characteristics. Characteristics as for the genus and typically metallic coloured; most species with more or less geniculate antennae, mentum tooth absent, and wide gula, others as in typical *Notonomus* for these characters.

Australian species. 20 species.

Distribution. Subtropical northern New South Wales, north to Cooktown, Queensland.

Biology. Rainforests or closed canopy mesic forest, frequently found under woody debris.

References. Yeates *et al.* (2002); Staunton *et al.* (2014).

Keys to species. Darlington (1961d); Will (2015a).

Subgenus *Loxodactylus* Chaudoir, 1865 (Pl. 32B)

Type species. *Feronia* (*Loxodactylus*) *carinulata* Chaudoir, 1865, by subsequent designation by Moore, 1965.

Characteristics. Characteristics as for the genus and all elytral intervals costate, third interval with three punctures; legs slender; parameres with denticulate apices.

Australian species. Only *Notonomus* (*Loxodactylus*) *carinulatus* (Chaudoir). However, several subspecies have been recognised as species in the past and their status remains unclear. Will (2015a) listed '*australiensis* (Sloane, 1895)' as a species in *Loxodactylus*; however, this was a lapsus (name refers to a *Loxandrus* species) and no such combination exists.

Distribution. South-east Victoria.

Biology. Found in open forests and scrub at the edge of forests.

References. Will (2015a).

Sarticus Motschulsky, 1865 (Pl. 37F)

Type species. *Sarticus orbicollis* Motschulsky, 1865 (= *Feronia* (*Steropus*) *obesula* Chaudoir, 1865), by subsequent designation by Moore *et al.* (1987).

Characteristics. Small to medium (5–20 mm) generally elongate ovoid beetles; pronotum is orbiculate or nearly so, hind angles rounded and reflexed, typically broadly expanded; black or brown, sometimes iridescent and rarely with metallic colours along the margin of the elytra; flight wing reduced; anterior tarsi of male with three basal segments dilated with squamose setae ventrally, setae sometimes lacking and tarsomeres unmodified; aedeagus median lobe with ostium dorsal; parameres left conchoid, the right small.

Australian species. 23 species described and many undescribed from throughout the range.

Distribution. Throughout the continent except for the deep interior.

Biology. Flightless, nocturnal, predacious, found in a variety of open grassland, scrub, forest, and agricultural habitat.

References. Will (2015a); Moore & Wallbank (1968).

Keys to species. Many species covered in the key by Sloane (1889).

Subgenus *Sarticus* Motschulsky, 1865

Type species. *Sarticus orbicollis* Motschulsky, 1865 (= *Feronia* (*Steropus*) *obesula* Chaudoir, 1865), by subsequent designation by Moore *et al.* 1987.

Characteristics. As in the genus.

Australian species. 21 species described and many undescribed from throughout the range.

Distribution. Throughout the continent except for the deep interior.

Biology. Flightless, nocturnal, predacious, found in a variety of open grassland, scrub, forest, and agricultural habitat.

References. Will (2015a); Moore & Wallbank (1968).

Keys to species. Many species covered in the key by Sloane (1889).

Subgenus *Coronacanthus* Macleay, 1878.

Type species. *Coronacanthus sulcatus* Macleay, 1878 by monotypy.

Characteristics. Characteristics as for the genus and distinguished by the deeply sulcate elytral intervals.

Australian species. Two species.

Distribution. Northern coast of Northern Territory.

Biology. Open forest.

Keys to species. Moore (1963c).

Pseudoceneus Sloane, 1903 (Pl. 36J)

Type species. *Feronia iridipennis* Laporte, 1867 by subsequent destination by Giachino 2005e.

Leptopodus Chaudoir, 1874. *Nomem nudum.*

Poeciliridius Straneo, 1937.

Pseudoceneus Tschitschérine, 1890.

Characteristics. Small, black beetles, often with an iridescent lustre; eyes prominent; elytra striate with three puncture in third interval; parascutellar stria long and well impressed; elytral plica present; flight wing full; anterior tarsi of male with three basal segments dilated with squamose setae ventrally; aedeagus median lobe with ostium dorsal; parameres left conchoid, the right small, narrow, pointed.

Australian species. Five species.

Distribution. Widespread from Adelaide, South Australia to southern Queensland, primarily in the coastal region, Tasmania, and Norfolk Island. Scattered records in Western Australia may represent introductions from the east; extraterritorially a single species is described from New Caledonia (Fauvel 1903) but that may be an Australian species that was introduced (Will 2011).

Biology. Good flier often attracted to lights, hygrophilus, often found along streams and billabongs.

References. Giachino (2005e); Straneo (1937).

Keys to species. Giachino (2005e).

Platycoelus series

Propleuron smooth or shallowly punctate, not rugose; parameres of the aedeagus attenuate with long narrow apex or broad and apically pointed, both nearly of equal length; female reproductive tract without spermathecal duct digital diverticulum, without spermathecal gland duct diverticulum.

Delinius Westwood, 1864 (Pl. 23G)

Type species. *Delinius essingtoni* Westwood, 1864, by monotypy.

Characteristics. Medium to large beetles (16–26 mm), stout, black; head and mandibles somewhat elongate; antennae often geniculate; mentum transverse, very shallowly emarginate; flight wing reduced; elytra with raised odd intervals, convex or carinate at least in the apical half, no discal punctures; male protarsomeres not modified; aedeagus with dorsoapical ostium; both parameres narrow with long narrow apex.

Australian species. Nine species.

Distribution. North coastal Northern Territory including coastal islands, west to eastern Western Australia.

Biology. Flightless, predacious, found in open woodlands, e.g. reported from *Callitris intratropica* and *Eucalyptus tetrodonta* dominated habitats.

References. Baehr (2008m); Giachino (2003c); Giachino & Picciau (2005).

Keys to species. Baehr (2008m).

Platycoelus Blanchard, 1843 (Pl. 35K; Fig. 12.3)

Type species: *Platycoelus depressus* Blanchard, 1853, by monotypy.

Platycaelus Blanchard, 1843

Psegmatopterus Chaudoir, 1878

Chlaenioidius Chaudoir, 1865

Hypherpinus Straneo, 1938

Dalbertisia Straneo, 1939

Characteristics. Medium size black, metallic or iridescent beetles; mental tooth emarginate (sometimes shallowly); stipes with a single setae near base; elytra usually with eight more or less impressed striae; when impressed, the parascutellar stria is present and intervals are of more or less equal width; abdominal ventrites without transverse sulci; flight wing full length or very short; anterior tarsi of male with three basal segments expanded, ventrally squamous; aedeagus ostium dorsal; parameres attenuate with long narrow apex or broad and apically pointed, both nearly of equal length, left wider than right.

Australian species. 11 species.

Distribution. Throughout continental Australia and near shore islands. One species in Tasmania; extraterritorially widespread or endemic species are found in New Guinea, New Caledonia, New Zealand, the Moluccas, and New Brittan Island.

Biology. Predacious, strong fliers or flightless. Flight capable species are frequently attracted to lights. Eurytopic species found in open, often disturbed habitats but usually associated with water or rainforest leaf litter.

Keys to species. Will (2015c).

Rhytisternus series

All members of the *Rhytisternus* series share a very similar form of male aedeagus with the ostium dorsally oriented, a thin elongate blade, left paramere subconchoid with a denticulate apex, and right paramere with a long attenuated apex. Punctures of the elytral disc are lacking and plica is well developed. Protarsomeres of the male with three basal segments expanded (narrowly in *Cratogaster*), and ventrally squamous. All but *Rhytisternus* have transversely sulcate abdominal ventrites (shallowly impressed laterally only in *Eurystomis*) and many species have the proepisterna transversely wrinkled. Preliminary analyses of DNA data (Will unpubl.) consistently place these five genera together.

Cratogaster Blanchard, 1853 (Pl. 23B; Figs 12.55, 12.58)

Type species. *Cyphosoma unicolour* Hope, 1842, by monotypy.

Cyphosoma Hope, 1842 (not Mannerheim, 1837).

Pachidius Chaudoir, 1865.

Tibarisus Laporte, 1867.

Characteristics. Robust, medium-size, black beetles; mentum shallowly emarginate; pronotum cordiform or suborbicular, margins bisetose; prosterna glabrous; proepisterna smooth; flight wing reduced; elytra fully striate; elytral apex declivous; abdominal sternites transversely sulcate; aedeagus slender, ostium dorsal; parameres subconchoid, the right with a slender point.

Australian species. Five species.

Distribution. All states and Northern Territory near coasts and inland, except the very dry centre.

Biology. Mesic to somewhat arid habitats in open forests or scrub. Often in coarsely rocky or boulder-strewn habitat. Flightless, predatory, and nocturnally active.

References. Moore & Wallbank (1968); Driscoll & Weir (2005).

Darodilia Laporte, 1867 (Pl. 23K)

Type species. *Darodilia mandibularis* Laporte, 1867, by monotypy.

Aulacocoelius Chaudoir, 1870.

Characteristics. Small or medium-size (5–10 mm), black beetles; frequently very shiny; mentum shallowly emarginate; eyes often prominent and mandibles elongate; pronotum orbicular or suborbicular, margins bisetose; prosterna glabrous; proepisterna distinctly rugose, shallowly rugose, or smooth; flight wing typically full or rarely reduced; elytra striate on disc only, often reduced to only striae 1–3; elytral apex declivous; abdominal sternites transversely sulcate.

Australian species. 10 species.

Distribution. East coastal Victoria, New South Wales, coastal Queensland, Northern Territory, coastal Western Australia, south to the southern Pilbara region; extraterritorially New Caledonia.

Biology. Adults of most species are able to fly, one species flightless. Nocturnal, and predacious, attracted to lights in open shrubland typically near streams or found in riparian flotsam.

Keys to species. Sloane (1900a) covers five of the species.

Eurystomis Chaudoir, 1878 (Pl. 25C)

Type species. *Eurystomis castelnaui* Chaudoir, 1878, by monotypy.

Characteristics. Compact, black, moderately shiny beetles; mandibles prominently elongate; pronotum hind angles rounded and base narrowing; anterior tarsi of male with three basal segments expanded and squamose beneath; propleuron distinctly, transversely rugose; abdominal ventrites shallowly transversely sulcate only near the margins.

Australian species. Only *Eurystomis castelnaui* Chaudoir.

Distribution. Restricted to Lord Howe Island.

Biology. Flightless, predacious, found in open forests under rocks.

References. Moore (1992).

Liopasa Tschitschérine, 1901 (Pl. 28B)

Type species. *Liopasa crepera* Tschitschérine, 1901, by monotypy.

Characteristics. Elongate, black, shiny beetles; head relatively very large and eyes small; sixth elytral interval raised and costate at base, pronotum elongate-cordiform; abdominal sternites transversely sulcate; aedeagus, with dorsal ostium; parameres broad right attenuate, longer than the left.

Australian species. Only *Liopasa crepera* Tschitschérine.

Distribution. McPherson range in Queensland and New South Wales.

Biology. Flightless, predacious, found under rocks in closed mesic forests.

Loxogenius Sloane, 1907 (Pl. 28E)

Type species. *Homalosoma opacipennis* Macleay, 1887, by original designation.

Characteristics. Black, dull except for head, which is somewhat shiny; Pronotum cordiform, margins plurisetose; elytra with third, fifth and seventh intervals sharply carinate; abdominal sternites transversely sulcate; anterior tarsi of male with three basal segments expanded and squamose beneath; aedeagus slender, ostium very slightly displaced to the right; parameres broad but attenuate, the right longer than the left.

Australian species. Only *Loxogenius opacipennis* Macleay.

Distribution. Northern coastal Queensland between Cooktown and Kuranda.

Biology. Flightless, predacious, nocturnally active. Found under rocks and woody debris in closed rain forests.

References. Staunton *et al.* (2014).

Rhytisternus Chaudoir, 1865 (Pl. 37E)

Type species. *Feronia* (*Rhytisternus*) *liopleura* Chaudoir, 1865, by subsequent designation (Britton 1940).

Amastus Motschulsky, 1865.

Characteristics. Black, somewhat or very shiny, frequently more or less prominent spectral iridescence; pronotum cordiform, sometimes suborbiculate; margins bisetose; proepisterna transversely rugous in nearly all species, in a few the wrinkles are scarcely visible; elytra striate only on disc with outer striae more or less effaced; plica well developed; angular base of stria 1 absent; no discal pores; flight wings fully developed; anterior tarsi of male with three basal segments expanded and squamose beneath; aedeagus slender, ostium dorsal; parameres subconchoid, apices attenuate.

Australian species. 23 species.

Distribution. Throughout continental Australia and Tasmania; extraterritorially one species, *R. laevis* (Macleay) is known from New Guinea and two species *R. miser* (Chaudoir) and *R. liopleurus* (Chaudoir) established in New Zealand. One species *R. externus* Baehr is known from Timor.

Biology. Adults of most species are capable fliers, nocturnal, scavengers and predators. Many are found in somewhat or significantly disturbed habitats. Most or somewhat hygrophilous and are found near flowing and still water in leaf litter and woody debris. They are common at lights at night.

Note. There arc numerous undescribed species and the genus is in need of revision.

References. Horne & Edward (1998); Moore & Wallbank (1968); Baehr & Reid (2017).

Keys to species. Many species covered by Sloane (1895a).

Pterostichini subtribe incertae sedis

Catadromus Macleay, 1825 (Pl. 20H)

Type species. *Carabus tenebrioides* Olivier, 1790, by monotypy.

Characteristics. Large to very large. Individuals may be up to 90 mm long, but are rarely above 60 mm. Black, somewhat shiny, with green, metallic borders on the pronotum and elytra. First four antennomeres glabrous; mentum tooth bifid; mandibles broad, strongly curved; pronotal margins bisetose; prosternum glabrous; elytra fully striate; parascutellar striae long; third intervals unipunctate; flight wings full; anterior tarsi of male with three basal segments expanded and squamose beneath; aedeagus slender, ostium on dorsum or slightly displaced to the right; parameres dissimilar, the left conchoid (pointed in some species), the right peg-like.

Australian species. Seven species.

Distribution. Throughout mainland Australia, Tasmania; extraterritorially *C. tenebrioides* is recorded from New Guinea, Sulawesi, Java, and the Moluccas.

Biology. Predatory, known to feed on frogs and various invertebrates; nocturnal, strong flier, found along riparian forests and wooded savannas. Attracted to lights at night.

References. Spencer & Richards (2010).

Keys to species. Giachino (2005c).

Cuneipectus Sloane, 1907 (Pl. 23I; Figs 12.52, 12.53)

Type species. *Cuneipectus frenchi* Sloane, 1907, by monotypy.

Characteristics. Large, dull black beetles; flightless; first three antennomeres glabrous; paraglossae with apex setose; labrum very large; number of supraorbital setae variable, one or more over each eye, rarely absent; pronotal lateral margin with three or four setae; flight wing reduced; elytral striae very deeply punctate, nearly fenestrate; umbilicate punctures of the elytral margin small, variable in number; abdominal ventrites transversely sulcate; pygidial gland reservoirs without a dorsal lobe; female reproductive tract with spermathecal gland duct diverticula.

Australian species. Two species.

Distribution. Central and southern Western Australia.

Biology. Arid, shrubland and heath; nocturnal; predacious, adults found feeding on termites at night.

Note. *Cuneipectus* has been treated as a separate tribe (Sloane 1907a) or variously placed in association with non-pterostichine taxa such as Chlaeniini and Licinini. However, characteristics are typical of Australian Pterostichini, including the presence of the spermathecal gland duct diverticulum in the female (Liebherr & Will 1998). The spermathecal gland duct diverticulum is found in various pterostichines including members of the

Trichosternus series, with which *Cuneipectus* also shares a plurisetose stipes. Setose paraglossae are also known in some *Megadromus* species. However, the sulcate ventrites and left facing ostium of the aedeagus medium lobe are distinct from the *Trichosternus* series taxa. Preliminary analyses of DNA data (Will unpubl.) consistently places *Cuneipectus* with Australian Pterostichini but not within the *Trichosternus* series.

References. Will (2015b); Sloane (1907a); Moore *et al.* (1987); Lorenz (2005).

Lesticus Dejean, 1828 (Pl. 28I)

Type species. *Lesticus janthinus* Dejean, 1828, by monotypy.

Characteristics. Large, metallic green to slightly blue beetle; head broad; mandibles notably curved and sharply pointed; gula very wide (Fig. 12.50); mentum short, transverse; third intervals with three punctures; flight wings full; anterior tarsi of male with three basal segments expanded and squamose beneath; abdominal sternites transversely sulcate; left paramere with transverse apophysis.

Australian species. The only Australian species, *L. chloronotus chloronotus*, is included in the subgenus *Triplogenius* Chaudoir. *Lesticus* (*Triplogenius*) *chloronotus dahli* Kuntzen is a subspecies found on New Britain Island.

Distribution. Northern Queensland, northward along the Cape York Peninsula; extraterritorially the genus is distributed across the Oriental region and oceanic islands.

Biology. Nocturnally active on the ground in tropical, more or less closed forests. Predatory. Capable flier occasionally taken at lights.

Keys to species. Dubault *et al.* (2008).

Notolestus Sloane, 1895 (Pl. 31H)

Type species. *Abax sulcipennis* Macleay, 1873, by monotypy.

Characteristics. Large, robust and brilliantly metallic beetles; mentum tooth bifid; elytral plica present; elytral with only four discal striae; flight wing absent; aedeagus with ostium dorsal; right paramere styloid, left conchoid; three umbilicate punctures anterad primary at elytral humerus;

Australian species. Only *Notolestus sulcipennis* Macleay.

Distribution. Forests of the eastern ranges of southern Queensland and northern New South Wales.

Biology. Tall, subtropical rainforests; nocturnal, predaceous.

Note. This species is more closely related to the New Zealand genus *Holcaspis* Chaudoir than it is any Australian taxa. Among the Australian taxa it is most closely related to the *Trichosternus* series species with which it shares the plurisetose stipes and three umbilicate puncture anterad primary puncture in striae 8.

Rhabdotus Chaudoir, 1865 (Pl. 36H)

Type species. *Feronia* (*Rhabdotus*) *reflexa* Chaudoir, 1865, by monotypy.

Characteristics. Medium-sized beetles; black, often with a more or less apparent metallic sheen; mentum median tooth bifid; antennae filiform, with three basal segments glabrous; postocular orbits small; pronotum broadly cordiform; lateral margins bisetose; pro- and mesosterna glabrous; elytra fused, fully striate, with evident plica; para scutellar striae present; third interval with discal pores; hind wings vestigial; anterior tarsi of male with three basal segments dilated with squamose setae ventrally; aedeagus with orifice on dorsum; parameres dissimilar, the left conchoid, with a pointed, membranous tip, the right a small strap.

Australian species. Two species.

Distribution. Tasmania.

Biology. Wet forest and scrubland. Possibly larval brooders.

Note. Giachino (2005d) described the larva based on field-collected larvae that were associated with a single male *Rhabdotus reflexus* (Chaudoir). Given this unusual situation of 16 larvae with a single male and no female, the apparent brooding behaviour remains to be confirmed in *Rhabdotus*. Larval brooding is common in *Notonomus* (Horne 1990, 1992, KWW numerous species with unpub lished observations of brooding).

Keys to species. Differential diagnoses by Giachino (2005d).

Tribe Abacetini Chaudoir, 1873

Characteristics. Small to medim size beetles (3–17 mm long, typically ~8 mm); most are shiny, have spectral iridescence, piceus to deep brown colour, frequently have contrastingly paler legs; triangular mental tooth; lack the angular base of stria 1 (Fig. 12.33B,E); elytral plica absent in many species, though present in most; species from the Australian region rarely have distinct pale markings on the elytra, which is common among Neotropical species; in general form, most are rather typical looking Harpalitae, frequently resembling an average Platynini, Harpalini or Pterostichini species; aedeagus with ostium dorsal; parameres conchoid, the right smaller than the left.

Australian species. There are 31 named species in the tribe from Australia, however, all of the genera in the group are in need of significant revision and a great many species await description.

Biology. Abacetine beetles are typically hygrophilous and are most common near water or in areas with very moist soil and significant leaf litter cover. They can be seasonally very abundant at temporary water sources and are frequently attracted to lights near these wet areas. In Australia they can be found from sea-level to moderately high elevations throughout their range. They are common along lakes and streams well into the interior and are fairly common in wet areas in lowland scrub, present but less common at higher elevations in rainforest mountain ranges in the tropics.

Distribution. The tribe has representatives throughout Australia, New Zealand, Africa, South America, Central America, North America, southern Europe and subtropical and tropical Asia. In Australia they are found everywhere except for the driest parts of the central desert. The greatest species diversity is found in the border region between New South Wales and Queensland, while very few species are found in southern Western Australia.

References. Bousquet (2012); Darlington (1968, 1971).

Notes. The possible association of loxandrines (e.g. *Loxandrus*, *Zeodera*, *Pediomorphus* and *Cerabilia*) as recognised by Allen & Ball (1980) as a '*fundamentally distinct group*' from Pterostichini, with abacetines (e.g. Abacetini of Jeannel (1942, 1948), and Abacetides of Chaudoir (1873)) has been suggested in several papers (Arndt 1988; Bousquet & Larochelle 1993; Bousquet 1985, 1999; van Emden 1949; Straneo 1991) and found in a preliminary cladistic analysis of morphological characters (Will 2000). Bousquet (2012) adopted the use of Abacetini as the inclusive taxa though he states that Abacetini in his sense is '*inadequately characterized*' and that the '[m]*onophyly of this tribe has not yet been demonstrated*'. Nevertheless, he extends the Abacétides of Chaudoir (1873) to include Celioschesini of Jeannel (1948) and all Loxandrini *sensu* Erwin and Sims in this concept of Abacetini. A multi-gene and morphological data analysis strongly supports a clade of Abacetina + Loxandrina separate from Pterostichini as presented here (KWW unpubl.).

Key to the Australian genera of Abacetini

1. Propleuron without stridulatory ridges or pegs, smooth, punctate or longitudinally strigose; pedicel of antenna symmetrically, or nearly symmetrically, inserted onto scape 2
– Propleuron with stridulatory ridges or pegs (Fig. 12.60); pedicel of antenna notably asymmetrically inserted onto scape (Fig. 12.61) .... ***Abacetus*** **Dejean**

2(1). Elytral interval 3 without dorsal puncture 3
– Elytral interval 3 with one dorsal puncture 6

3(2). Penultimate labial palpomere similar diameter to adjacent palpomeres; other features variable, but never in the combination below 4
– Penultimate labial palpomere enlarged relative to adjacent palpomeres; palpomeres plurisetose, bodyform parallel-sided; small sized beetles (4–8 mm) that are pale fulvous or testaceous colour .... ***Pediomorphus*** **Chaudoir**

4(3). Flight wing usually full; metanepisternum long, anterior margin half the length or less, of the lateral margin .... ***Cosmodiscus rubripictus*** **Sloane**
– Flight wing completely reduced; metanepisternum very short, anterior margin equal the length of the lateral margin 5

5(4). Male with protarsomeres moderate or notably widely, asymmetrically expanded; head frontal impressions punctiform; eastern Queensland .... ***Cerabilia*** **Laporte**
– Male with protarsomeres narrowly, symmetrically expanded; head frontal impressions absent; central Western Australia (Cue) .... ***Litarthrum browni*** **Sloane**

6(2). Elongate, parallel-sided or ovoid body form; pronotum variously formed, typically subquadrate or elongate .... ***Loxandrus*** **LeConte**
– Broad, abaci-form body form; pronotum broadly trapezoidal .... ***Zeodera atra*** **Laporte**

Abacetus Dejean, 1828 (Pl. 18B; Figs 12.60, 12.61)

Type species. *Abacetus gagates* Dejean, 1828, by monotypy.

Characteristics. Small beetles (5–7 mm) with well developed eyes and notably excentrically inserted antennomere 2 (pedicel) (Fig. 12.61). Australian species are black or rarely deep brown and paler near the apex of the elytra. Legs are contrasting paler; proepisternum with minute stridulatory pegs or ridges (Fig. 12.60).

Australian species. Eight species are known from Australia.

Distribution. The genus is distributed across northern Australia through tropical and subtropical latitudes except in the dry interior; extraterritorially it is found across Asia, Africa and southern Europe.

Biology. Very hygrophilous, frequently found in muddy flotsam along streams and lakes. Frequently attracted to lights.

References. Straneo (1958).

Cosmodiscus Sloane, 1907 (Pl. 22A, C)

Type species. *Cosmodiscus rubripictus*, Sloane, 1907, by monotypy.

Characteristics. Broadly ovoid, convex and shiny beetles, with short mandibles, and prominent eyes, 6–7 mm long. In the only Australian species, the colour is piceous with an irregular set of light red patches on each elytron. Patches vary from small spots near the humeri, laterally in the apical third and on interval 1 in the apical third, to large, continuous facia connecting the entire red pattern.

Australian species. *Cosmodiscus rubripictus* is the only species known from Australia

Distribution. Northern Queensland; extraterritorially species of *Cosmodiscus* are found in South-east Asia, west to India and north-east to Japan.

Biology. Based on New Guinea specimens, these beetles are found in rainforest leaf litter (Darlington 1962a).

References. Darlington (1962a).

Cerabilia Laporte, 1867, (Pl. 25F)

Type species. *Cerabilia maori* Laporte, 1867, by monotypy.
Zabronothus Broun, 1893
Nelidus Chaudoir, 1878
Australomasoreus Baehr, 2007

Subgenus *Feronista* Moore, 1965

Type species. *Feronista amaroides* Moore, 1965, by original designation.

Characteristics. Very small to moderately small beetles (3–12 mm), body form elongate-ovoid to notably ovoid. Colour is uniformly piceus to castaneous, rarely with paler regions on stria 1 and near the apex. Flight wing entirely reduced, thorax very compact.

Australian species. Currently there are five named species in Australia (see notes below). However, in a revision of the group that is in preparation there are 26 undescribed species (KWW unpubl.).

Distribution. Eastern Queensland; extraterritorially species of other subgenera are found in New Zealand and New Caledonia.

Biology. *Cerabilia* in Australia are all found in rainforest leaf litter. They are rarely encountered during the day, but are active at night.

Notes. All Australian species are in the subgenus *Feronista* with the exception of *Cerabilia australis* (= *Nelidus australis* Chaudoir, 1878), which is known only from the damaged holotype. The type of this species is published as being from the Paroo River area, but based on its characteristics and the absence of any specimens of this species or any *Cerabilia* species from the Paroo River area, it seem more likely to be a mislabelled. The type is very similar to some New Zealand species belonging to *Cerabilia* s.str. and so was transferred to that genus (Will 2015b).

Key to species. Moore (1965).

Litarthrum Sloane, 1915 (Fig. 12.104)

Type species. *Litarthrum browni* Sloane, 1915, by monotypy.

Characteristics. Small, castaneous brown beetle, ovoid shaped, no dorsal elytral punctures, no angular base of stria 1; abdominal ventrites without transverse sulci, elytra without plica.

Australian species. Only *Litarthrum browni* Sloane.

Distribution. Known only from the type locality, Cue, Western Australia.

Note. Known only from the single, highly damaged holotype specimen. The general form is like a species of *Simodontus*, but the absence of the elytral plica, lack of the angular base of stria 1, and absence of sulci on abdominal ventrites are states not known in that genus but are all consistent with a placement in Abacetini. Sloane's (1915a) original description, based on the intact specimen, suggested an affinity to *Pediomorphus*. Though very different in form, *Cerabilia*, which is very similar to *Litarthrum*, was not known to Sloane. The status of this genus should be examined when more material is found, but the provisional placement in Abacetini is consistent with the evidence currently at hand.

Pediomorphus Chaudoir, 1878 (Pl. 33G, K)

Type species. *Pediomorphus planiusculus* Chaudoir, 1878, by monotypy.

Characteristics. Small (4–8 mm) pale brown or rarely castaneous coloured beetles with a parallel sided and dorso-ventrally depressed body form and relatively large pronotum; penultimate segment of labial palpi enlarged, subglobose.

Australian species. There are currently four species of *Pediomorphus* described, but nine more undescribed species are known from collections (KWW unpubl.).

Distribution. Because specimens are rarely collected the distribution is poorly known, but records are scattered throughout Australia in mesic and dry sclerophyl woodlands and grasslands, especially in the north.

Biology. Beetles have been found along temporary pools of water in open grassland and at lights in the vicinity of riparian areas and billabongs.

Key to species. Sloane (1900b).

Zeodera Laporte, 1867 (Pl. 39I)

Type species. *Zeodera atra* Laporte, 1867, by monotypy.

Characteristics. Large beetles (18–22 mm), deep black and shiny. Their broad body form and trapezoidal pronotum is distinctive.

Australian species. Only *Zeodera atra* Laporte.

Distribution. This species is only found in suitable habitat in the New South Wales and Queensland border region.

Biology. They are flightless, leaf litter dwelling beetles that are nocturnally active and are found in rainforest and wet sclerophyll forests.

Loxandrus LeConte, 1852 (Pl. 28A)

Type species. *Feronia recta* Say, 1823, by subsequent designation by Casey 1918.

Characteristics. Medium size beetles (in the Australian fauna 6–15 mm, most ~8 mm) that are usually black, occasionally brown, typically shiny with a distinct spectral iridescence (rarely dull) and contrastingly pale legs. Antennae and legs are relatively long and lightly built. Most species have a full flight wing, though some reduction of the wing has occurred in a few species.

Australian species. Currently 13 named species are recognised as valid, however at least 25 undescribed Australian species are known from museum collections (KWW unpubl. data).

Distribution. The genus has representatives throughout Australia, New Guinea, Sulawesi, South America, Central America, and North America. In Australia they are found everywhere except for the driest parts of the central desert and the far south-west of Western Australia. The greatest species diversity is found in the border region of New South Wales and Queensland. *Loxandrus gagatinus* (Laporte) is described from Tasmania; however there are no recent records of this species form Tasmania and it may represent an introduction that did not establish from the mainland or a mislabelled specimen.

Biology. Most species are decidedly hygrophilous. They are often common near water that is permanent or seasonal, flowing or still. Most are associated with leaf litter, flotsam and muddy areas, while a few are restricted to gravel and cobble along rivers and streams. A few species have been found in caves in the Northern Territory, but as they have full-sized and apparently functional eyes, this is probably facultative.

References. Allen & Ball (1980); Allen (1982); Darlington (1962a); Straneo (1991).

Tribe Sphodrini Laporte, 1834

Characteristics. Widely distributed tribe, mostly in the Northern Hemisphere that is quite variable in form, but in general form is similar to platynines and pterostichines.

Note. The only species from this tribe in Australia was an early adventive introduction.

Laemostenus Bonelli, 1810 (Pl. 28D)

Type species. *Carabus venustus* Dejean, 1828, by subsequent designation (Madge 1975).

Pristonychus Dejean, 1828.

Characteristics. Large, black beetles; bluish reflection on the elytra; pubescent upper surface of the tarsomeres; pronotum cordiform; elytra with feebly punctate striae; interval 3 without dorsal punctures; claws denticulate at base; flight wing full; left paramere broad, with membranous apical tip, right narrow, elongate, not hooked at apex.

Australian species. Only *Laemostenus complanatus* (Dejean, 1828).

Distribution. Coastal New South Wales, west along near-coast areas to South Australia and in Western Australia; extraterritorially globally subcosmopolitan.

Biology. Adventive, synanthropic and eurytopic in more or less disturbed habitats.

References. Casale (1988); Sloane (1910b).

Tribe Morionini Brullé, 1835

Characteristics. A tribe with a worldwide distribution in tropical and subtropical regions. Species in Australia all share the following: moniliform antennomeres with the first four antennomeres glabrous, prominent eyes, elytral plica not externally visible, expanded apex of the protibia with a apicolaterally produced spine, and projecting dentiform process above the insertion point of the antenna, deep black to piceus brown, rather shiny, elongate, parallel-sided and somewhat depressed and subpedunculate; aedeagus with ostium dorsal; parameres small, the left conchoid, the right small, peg-like.

Distribution. Found in tropical and subtropical habitats in Australia and worldwide. Significantly diverse in Africa and Asia.

References. Will (2003); Ober (2002).

Key to the genera of Australian Morionini

1. Temples very large, nearly as prominent as the eyes (Fig. 12.26); size very large (length up to 75 mm) .... ***Hyperion* Laporte**
– Temples not large, half or less as prominent as the eyes; size moderate (length less than 30 mm) .... ***Morion* Latreille**

Hyperion Laporte, 1834 (Pl. 27G; Fig. 12.26)

Type species. *Scarites schroetteri* Schreibers, 1802, by monotypy.

Heteroscelis Boisduval, 1835

Campylocnemis Westwood, 1842

Characteristics. Massive, black, shiny, beetles, parallel-sided form with large head, as broad as pronotum; antennae short; flight wings full; hind tibia arcuate.

Australian species. Only *Hyperion schroetteri* (Schreibers).

Distribution. Occurs in Victoria, South Australia and New South Wales.

Biology. Reported as a predator of dynastine grubs and associated with the rotten centre of large eucalypts; nocturnal; attracted to lights. The largest adults are nearly 75 mm long and only challenged in size by the largest individuals of the pterostichine *Catadromus*.

References. Moore (1965).

Morion Latreille, 1810 (Pl. 30H)

Type species. *Harpalus monilicornis* Latreille, 1810, by original designation.

Characteristics. Medium to large size, black, shiny, beetles, parallel-sided form with relatively large head, nearly as broad as pronotum; antennae short, moniliform; flight wings full.

Australian species. 10 species.

Distribution. In Australia it is known from coastal regions in north-eastern Victoria, New South Wales and Queensland. Most species are found in north Queensland, several are known from there and extraterritorially in New Guinea. The genus is found throughout the range of the tribe.

Biology. Predatory, nocturnal, strong fliers, associated with dead wood, often found under bark. Frequently attracted to lights at night.

References. Will *et al.* (2000); Moore *et al.* (1987).

Tribe Drimostomatini Chaudoir, 1872

Characteristics. Small size, < 8 mm, oval or subovale, generally black or piceous beetles, very shiny; pronotal basal impressions linear, often more or less convergent anteriorly; elytra with deep, more or less crenulate striae; angular base stria 1 absent; elytral basal setae at the level of the third stria; aedeagus right side up in repose; parameres variable, but typically right larger and usually conchoid.

Caelostomus W. S. Macleay, 1825 (Pl. 20G)

Type species. *Anaulaucus (Caelostomus) picipes* W. S. Macleay, 1825, by monotypy.

Characteristics. Small, convex, shiny black or brown beetles; antennae relatively short; mandibles porrect, sharp; pronotum with deep, linear basal impressions; anterior pronotal setae a little before the midpoint; basal puncture of the elytra at the level of the third stria; aedeagus inverted, right-side up in repose.

Australian species. Three species.

Distribution. In Australia restricted to the far north, north-east coastal, Northern Territory and Queensland; extraterritorially distributed across the Ethiopian and Oriental Regions, north-east to Japan.

Biology. Typically under bark of decaying logs or in rotting vegetation and litter. Possibly brought to Australia by humans in wood and vegetable materials.

References. Straneo (1938); Darlington (1962a).

Keys to species. Straneo (1938).

Tribe Cratocerini Lacordaire, 1854

Characteristics. Head broad, eyes very prominent; antennae stout or moniliform; tergite VIII divided medially into hemitergites, each hemitergite subdivided into a median tergite and an epitergite.

Note. Genera now considered to be in this tribe have been variously placed in Harpalinae the past, though typically they have been positioned near or within the Pterostichini. Lorenz (2005) combined cratocerines with catapieseines and drimostomatines within his subfamily Pterostichinae. The tribe Cratocerini is considered to be composed of the two subtribes, the New World Catapiesina (*Catapiesis* Solier and *Homalomorpha* Brulle) and Cratocerina s. str. (*Cratocerus*, *Brachidius* Chaudoir and *Oxyglychus* Straneo; New World tropics, tropical Asia and Australia, and Japan, respectively). Bouchard *et al.* (2011) placed these as tribes Cratocerini and Catapiesini within the subfamily Harpalinae. Synapomorphic modifications of tergite VIII, shared defensive chemicals and DNA sequence data all support the monophyly of Cratocerini (Grzymala & Will 2014; Will *et al.* 2000, Will unpubl.). However, the sister group of the tribe is unknown, though preliminary DNA data suggests that they are related to Morionini. Like morionines, all species are strongly associated with decaying wood.

Brachidius Chaudoir, 1852 (Pl. 20D)

Type species. *Brachidius crassicornis* Chaudoir, 1852, by monotypy.

Characteristics. Small, stout beetles; eyes very prominent; antennae short, moniliform; mandibles broad; flight wing full; elytral without puncture at base; scutellum anterad elytral base; anterior seta of the pronotum far forward near apical angle; aedeagus with dorsal ostium; median lobe left side up in repose; left paramere conchoid, right small, blunt.

Australian species. Only *Brachidius crassicornis* Chaudoir.

Distribution. Restricted to north-east costal Queensland; extraterritorially from Solomon Islands, New Guinea, north-west to Myanmar and east at least to the Philippines.

Biology. Nocturnal, found at lights. Associated with rotten wood, under bark of logs in lowland tropical forests.

References. Darlington (1962a); Will & Kavanaugh (2012); Grzymala & Will (2014); Will *et al.* (2000).

Tribe Platynini Bonelli, 1810

Characteristics. Worldwide diversity of the group is difficult to summarise but Australia's small number of species are all of the typical form: lightly built; legs relatively thin and long; head narrow; pronotum moderately wide, elytra relatively wide; antennae long; parascutellar and angular base of stria 1 present; small to medium size beetles, typically 5–10 mm; procoxal cavities are closed, mesocoxa conjunct; protibia anisochaete with a well developed antennal cleaning groove; two supraorbital setae above each eye; mandibular scrobe lacking seta; elytral plica absent; flight wing full or very frequently, variously reduced in size or absent; abdomen of the harpalidian type (Deuve 1993); aedeagus with well developed basal bulb, foramen small; aedeagus with ostium usually dorsal, parameres glabrous, highly variable in shape, but typically dissimilar in form with the left larger; female reproductive tract usually with gland on spermatheca or bursa; usually gonocoxite-2 elongate, gonocoxite-1 with setae; pygidial gland reservoirs with dorsal lobe.

Key to the Australian genera of Platynini

1. Anterior pronotal seta and the anterior discal elytral seta absent .... ***Altagonum* Darlington**
– All regular pronotal and elytral setae present **2**
2(1). Tarsal claws with an acute tooth at base .... ***Dicranoncus* Chaudoir**
– Tarsal claws smooth, without basal tooth **3**
3(2). Mesepimeron broadly triangular, lateral edge ~1/2 length of anterior edge .... ***Euplynes* Schmidt-Göbel**
– Mesepimeron narrow, more or less parallel-sided for most of its length, expanded laterally so the lateral edge is much less than 1/2 length of anterior edge **4**
4(3). Tarsomeres 1–4 with only two ventrolateral rows of setae and two-four apical setae; three elytral punctures on disc in interval 3 notably foveate .... ***Lorostema* Motschulsky**
– Tarsomeres 1–4 with numerous setae ventrally between and/or laterally above the two ventrolateral rows of setae; three elytral punctures variable **5**
5(4). Brilliant metallic purple; long spines at elytra apices; tarsi asymmetrically bilobed .... ***Violagonum* Darlington**
– Without the combination of characters above; typically more sombre colours or metallic green; elytral apices with small or no spines; tarsi variable, but usually only emarginate **6**
6(5). Elytra distinctly metallic green .... ***Metacolpodes* Jeannel**
– Elytra black, brown, bronze or deep blue **7**
7(6). Metathoracic wings short; elytra rather oval-shaped; metanepisternum little longer than wide **8**
– Metathoracic wings fully developed; elytra rather parallel-sided; metanepisternum about twice as long as wide .... ***Notagonum* Darlington**
8(7). Microreticulation isodiametric, on head and pronotum distinct, on elytral very coarse, elytra dull; upper surface of tarsomeres strongly quadrisulcate; female terminal sternum 8(7)-setose .... ***Paraplatynus* Baehr**
– Microreticulation on head and pronotum very superficial, on elytral distinct but fine, elytra rather glossy; upper surface of tarsomeres moderately bisulcate; female terminal sternum 4- or 6-setose **9**
9(8). Antenna dark; lateral margin of pronotum near base concave; microreticulation of elytra very transverse; female terminal sternum 6-setose .... ***Mooreagonum* Baehr**
– Antenna pale red; lateral margin of pronotum near base not concave; microreticulation of elytra almost isodiametric; female terminal sternum 4-setose; restricted to Lord Howe and Norfolk Islands .... ***Notoplatynus* Moore**

Dicranoncus Chaudoir, 1850 (Pl. 24L)

Type species. *Dicranoncus femoralis* Chaudoir, 1850, by monotypy.

Characteristics. Slender, brown beetles, ~7 mm long; elytral sutural angles denticulate; flight wings full; with an acute tooth at base; males with only one testis.

Australian species. Only *Dicranoncus queenslandicus* (Sloane, 1903).

Distribution. North-east coastal Queensland; extraterritorial east to Solomon Islands, north to the Philippines, west to India.

Biology. Found in relatively open areas with tall grass, low scrub or more forested areas.

References. Sloane (1910b); Darlington (1952, 1971); Will *et al.* (2005).

Keys to species. Darlington (1956).

Euplynes Schmidt-Göbel, 1846 (Pl. 25E)

Type species. *Euplynes cyanipennis* Schmidt-Göbel, 1846, by monotypy.

Xatis Fairmaire, 1901.

Anarmosta Peringuey, 1896.

Characteristics. Small, 5–7 mm, lightly-built, broad beetles; elytra with a light, metallic blue reflex over the rufous ground colour; elytra wide, well rounded humeri; striae complete and finely impressed; aedeagus with broad, dorsal ostium; parameres conchoid, left larger than right.

Australian species. Only *Euplynes decoloratus* Baehr, 2000.

Distribution. In Australia known from the central Cape York Peninsula; extraterritorially other species of the genus are widespread across the Afrotropical and Oriental Regions, including New Guinea.

Biology. Lowland rainforest, collected by pyrethrum spraying of logs and tree trunks.

References. Darlington (1952, 1971).

Lorostema Motschulsky, 1864 (Pl. 28C)

Type species. *Lorostema alutacea* Motschulsky, 1864, by monotypy.

Feanus Bates, 1889

Lorostemmoides Habu, 1978

Characteristics. Fairly small, 6–7 mm, brown to piceous beetles; paler at the margins of the pronotum and elytra, dorsally with a slightly aeneous reflex; eyes large and prominent; three dorsal elytral setae in wide depressed foveae that are as wide or wider than their intervals; elytra without apical denticle;

Australian species. Only *Lorostema bothriophora* (Redtenbacher, 1867)

Distribution. East coastal Queensland; extraterritorially *L. bothriophora* is known from in Vanuatu, New Caledonia; Tahiti, Samoa, and Rapa. The genus is more widely distributed, reaching the Oriental Region at least north-east to Japan

Biology. Found in flood debris (Darlington 1956).

References. Darlington (1952, 1971); Liebherr (2005).

Keys to species. Darlington (1956) (as *Lorostemma cooki* (Sloane) [sic!]); Liebherr (2005).

Metacolpodes Jeannel, 1948 (Pl. 22I)

Type species. *Colpodes buchanani* Hope, 1831, by original designation.

Characteristics. Relatively large beetles, around 14 mm; flight wing full; shiny; piceous; often metallic or at least with a slight aeneous reflex laterally on elytra; elytral striae faintly to obsoletely punctulate, intervals flat to slightly convex; elytral apex with small, sutural denticles; male and female genitalia figured by Liebherr (2005).

Australian species. Only *Metacolpodes truncatellus* (Fairmaire 1881).

Distribution. In Australia found in north-east, coastal Queensland; extraterritorial *M. truncatellus* is found from the Moluccas, across New Guinea, the Solomon Islands, the Santa Cruz Islands, Vanuatu, and Fiji. The genus is distributed widely in the Pacific-Asian region and *M. buchanani* is adventive in western North America.

Biology. Predacious, nocturnal, collected at lights.

Keys to species. Liebherr (2005).

Notagonum Darlington, 1952 (Pl. 31K)

Type species. *Notagonum externum* Darlington, 1952, by original designation.

Characteristics. Small or medium-sized, 4.5–9.5 mm; black or piceous coloured, rarely notably iridescent; flight wing full; elytral basal margin entire; humeri rounded or very obtusely angulate; elytral apices variable, rounded, denticulate, angulate, or rarely spined; male and female genitalia figured by Liebherr (2005).

Australian species. 10 species and one subspecies.

Distribution. Most species are found in the region from north-east New South Wales to north-east Queensland. *Notagonum marginellum* (Erichson) is found in South Australia along the gulf coast, Tasmania and Norfolk Island. Extraterritorial the genus is widespread in the Pacific including New Guinea, Fiji, Indonesia; Vanuatu; New Caledonia, and New Zealand.

Biology. Ground dwelling in moist habitats especially among leaf litter. From low, open forests to montane rainforests.

References. Sloane (1910b); Baehr (2016i); Liebherr (2005); Darlington (1952, 1956).

Keys to species. Baehr (2016i), Darlington (1963b) cover species from tropical north Queensland.

Notoplatynus Moore, 1985 (Pl. 32J)

Type species. *Notoplatynus darlingtoni* Moore, 1985, by monotypy.

Characteristics. Small, piceous beetles with ovoid elytra, pronotum broadly explanate along lateral margins; emarginate tarsomere 4; male genitalia figured by Moore (1985).

Australian species. Two species.

Distribution. Restricted to Norfolk and Lord Howe Islands.

Biology. Flightless, predators in closed forests.

References. Moore (1985, 1992).

Keys to species. Baehr (2016i).

Violagonum Darlington, 1956 (Pl. 39G)

Type species. *Harpalus violaceus* Chaudoir, 1844, by original designation.

Characteristics. Rather wide form with purple or purplish-blue metallic colour. Elytra with spine at apex; third elytral interval with three punctures; bi-lobed 4th tarsomere with outer lobe larger than inner.

Australian species. Only *Violagonum violaceum* (Chaudoir).

Distribution. North Queensland; extraterritorially New Guinea, Solomon Islands.

Biology. Arboreal, strong flier, in closed tropical forests.

Mooreagonum Baehr, 2016 (Fig. 12.109)

Type species. *Mooreagonum brevipenne* Baehr, 2016, by monotypy.

Characteristics. Moderately sized, dark piceous beetles with short, ovoid elytra, pronotum dorsally convex, lateral margin near base distinctly concave; lateral margin upturned; wings shortened, metepisternum little longer than wide; tarsi faintly bisulcate; male and female genitalia figured by Baehr (2016i), female terminal sternum six-setose.

Australian species. Only *M. brevipenne* Baehr.

Distribution. South-eastern New South Wales.

Biology. Flightless, predators in closed forests, probably a montane species.

References. Baehr (2016i).

Paraplatynus Baehr, 2016 (Fig. 12.114)

Type species. *Paraplatynus sulcatipes* Baehr, 2016, by monotypy.

Characteristics. Moderately sized, pale rufous beetles with rather short, ovoid, markedly depressed elytra, pronotum dorsally depressed, lateral margin near base distinctly concave; wings almost completely reduced, metepisternum little longer than wide; tarsi very coarsely qudrisulcate; pronotum and elytra with very coarse, isodiametric microreticulation; male genitalia unknown, female genitalia figured by Baehr (2016i), female terminal sternum eight-setose.

Australian species. Only *P. sulcatipes* Baehr.

Distribution. Southern central Northern Territory.

Biology. Flightless, predator, but ecology is unknown.

References. Baehr (2016i).

Altagonum Darlington, 1952 (Fig. 12.72)

Type species. *Altagonum caducum* Darlington, 1952, by original designation.

Characteristics. This is a morphologically very diverse genus of mainly New Guinean species, thus far not yet recorded from Australia. Variously sized and shaped bettles, which have in common the absence of certain of the ambulatory setae on either head, either prothorax, either elytra. The great amount of variation in external and genitalic morphology makes it a genus of convenience, which in the course of a revision likely will be divide into two or even more genera. The single Australian species is closely related to two small species from New Guinea, therefore it is provisionally included in *Altagonum*, as long as the genus has not yet been revised. The small species has fully developed wings; the male genitalia are figured in Baehr (2016i), the aedeagus bears a densely denticulate fold in the apical part of the endophallus.

Australian species. Only *Altagonum kurandae* Baehr 2016.

Distribution. North-eastern Queensland near Kuranda.

Biology. Predator, holotype collected in upland rain forest from leaf litter.

References. Darlington (1952, 1971); Baehr (2016i).

Tribe Perigonini Horn, 1881

Characteristics. Procoxal cavities closed; mesocoxae conjunct; head with two supraorbital setae above each eye; mandibles without a seta in the scrobe; terminal palpomeres markedly elongate, fusiform; elytra epipleura without plica; marginal sulcus of the elytra apicad widened (Fig. 12.32) and the latero-apical margin pilose; aedeagus commonly with extremely complexly shaped median lobe, sclerotised rods on the endophallus; parameres short, wide, asetose; gonocoxites rather short, gonocoxite-2 curved, with a few ventral, dorsal, and subapical setae.

Perigona Laporte, 1835 (Pl. 34L; Fig. 12.32)

Type species. *Perigona pallida* Laporte, 1834, by monotypy.

Siltopia Laporte, 1867

Australian species. 17 species.

Distribution. Eastern and northern tropical Australia, but most common and most diverse in north-eastern Queensland; extraterritorial almost worldwide.

Biology. Ground living in leaf litter and on trunks of moss covered rainforest trees, perhaps also in nests of birds and small mammals. Predacious and probably nocturnal. Rather commonly attracted to light, but some species can be also sieved from ground litter.

References. Sloane (1904); Csiki (1931); Darlington (1964, 1968); Moore *et al.* (1987); Lorenz (2005); Baehr (2013d).

Keys to species. Darlington (1964); Baehr (2013d).

Key to the Australian subgenera of *Perigona* Laporte

1. Setiferous punctures in the subapical marginal channel form a straight line .... ***Perigona*** **Laporte**
– Setiferous punctures in the subapical marginal channel form a triangle .... ***Trechicus*** **LeConte**

Subgenus *Perigona* Laporte, 1835

Type species. *Perigona pallida* Laporte, 1834, by monotypy.

Characteristics. Elongate and rather parallel-sided, depressed beetles; subapical setiferous punctures form a straight line.

Australian species. Eight species.

Distribution. Eastern Australia, eastern Queensland to north-eastern New South Wales.

References. Sloane (1904); Csiki (1931); Darlington (1964, 1968); Moore *et al.* (1987); Lorenz 2005); Baehr 2013d.

Subgenus *Trechicus* LeConte, 1853

Type species. *Trechicus umbripennis* LeConte, 1853

Characteristics. Rather short and oval-shaped, somewhat convex beetles; subapical setiferous punctures arranged in a triangle.

Australian species. Nine species.

Distribution. Eastern Queensland, extreme northern parts of Northern Territory and of Western Australia.

Notes. One species, *Perigona nigriceps* (Dejean, 1831), is a globally subcosmopolitan adventive and is thought to have been transported in shipments of crops and soil.

References. Csiki (1931); Moore *et al.* (1987); Lorenz 2005); Baehr 2013d.

Tribe Harpalini Bonelli, 1810

Characteristics. Highly variable in body form and size, but most are recognisable from the head typically being relatively broad and mandibles heavily built, legs often heavily built, spinose and somewhat fossorial, except in the subtribe Pelmatellina, which often has relatively narrow head, a platynine-like body form, and lightly built legs. Procoxal cavities are closed, mesocoxa conjunct; protibia anisochaete with a well-developed antennal cleaning groove; male protarsi may be expanded with dense pads, biseriately setose, or lack ventral setae; antennae elongate; one supraorbital setae above each eye; mandibular scrobe lacking seta; elytral plica absent; parascutellar and angular base of stria 1 usually present; flight wing full or frequently, variously reduced in size or absent; abdomen of the harpalidian type (Deuve 1993); aedeagus with well-developed basal bulb, foramen small; aedeagus with ostium usually dorsal, parameres glabrous, highly variable in shape, but typically dissimilar in form with the left larger; female reproductive tract usually with gland on spermatheca or bursa; usually gonocoxite-2 elongate, gonocoxite-1 with setae; pygidial gland reservoirs with dorsal lobe; male with only one testis.

Key to the Australian subtribes of Harpalini

1. Penultimate labial palpomere with usually two or more rarely, three setae along the anterior margin **2**
– Penultimate labial palpomere plurisetose (4 or more) along the anterior margin or rarely with only three setae **4**
2(1). Male protarsomeres ventrally with dense pads of setae having a spongy appearance (Fig. 12.62) **3**
– Male protarsomeres ventrally biseriate squamulous (Fig. 12.63) .... **Stenolophina Kirby**
3(2). Clypeo-ocular sulci not impressed or punctiform, shallow and obscure, not reaching the margin of the eye .... **Anisodactylina Lacordaire (part)**
– Clypeo-ocular sulci well impressed and reaching or nearly reaching the margin of the eye .... **Pelmatellina Bates**
4(1). Male protarsomeres ventrally with dense pads of setae having a spongy appearance (Fig. 12.62) .... **Anisodactylina Lacordaire (part)**
– Male protarsomeres ventrally biseriate squamulous (Fig. 12.63) **5**
5(4). Parascutellar stria separate from stria 1; stria 1 continuously impressed to the base (Fig. 12.33A); clypeus at apex emarginate, a pale membrane visible between clypeus and the usually asymmetric labrum (Fig. 12.27) .... **Amblystomina Fauvel**
– Parascutellar stria joined to stria 1; angular base of stria 1 separate from apical portion of stria or absent (Fig. 12.33D); apex of clypeus transverse or subemarginate, no membrane visible between clypeus and the symmetric labrum .... **Harpalina Bonelli**

Key to the Australian genera of Harpalini

1. Males with ventrally setose protarsomeres; protarsomeres typically expanded with dense pads or biseriately arranged squamulous setae (Fig. 12.62–63) **2**
– Without ventrally setose protarsomeres, either male or female **20**
2(1). Protarsomeres ventrally with dense pads of setae having a spongy appearance (Fig. 12.62) **3**
– Protarsomeres ventrally biseriate squamulous (Fig. 12.63) **10**
3(2). Penultimate labial palpomere bisetose along anterior margin; occasionally one or a few small setae may be present apically or in other areas **4**
– Penultimate labial palpomere plurisetose (> 3 setae) along anterior margin; often with other, smaller setae as well **5**
4(3). Body form various but not as described below; typically overall shape elongate ovoid; eyes relatively large; subocular gena much narrower than the width of the first antennomere; widespread and eurytopic, but most commonly found in leaf litter .... ***Lecanomerus* Chaudoir (part)**
– Body form slender, elongate, reminiscent of a typical platynine; pronotum distinctly narrower than elytra; eyes relatively small, not prominent, with subocular gena about the width of the first antennomere; known from caves in the southern part of Victoria and South Australia .... ***Notospeophonus* Moore (part)**
5(3). Elytra with one or no punctures in interval 3 and intervals 5 and 7 impunctate **6**
– Elytra with more than one puncture in interval 3, intervals 5 and 7 with or without punctures .... ***Gnathaphanus* Macleay (part)**
6(5). Mentum tooth present; pronotum form various, but not as below **7**
– Mentum tooth absent; pronotum suborbiculate (Pl. 21G), lateral depression prominently flattened and sharply delimited from disc .... ***Cenogmus* Sloane (part)**
7(6). Head large, but of more typical relative size for Harpalini; metepimeron form and sternal setation variable; widespread genera of usually larger sized beetles **8**
– Head relatively enormous, as long and nearly as wide as pronotum; metepimeron very short; abdominal sterna glabrous except for paramedial setae; relatively small sized beetles, 6–8 mm; restricted to south-western Western Australia .... ***Nornalupia* Kataev (part)**
8(7). Metacoxa with < 8 setae along anterior margin, typically 1–2 setae, more in some *Hypharpax* **9**
– Metacoxa with fringe of > 10 setae along anterior margin .... ***Phorticosomus* Schaum (part)**
9(8). First metatarsomere relatively short; shorter than or only slightly longer than the combined length of 2 and 3 .... ***Hypharpax* Macleay (part)**
– First metatarsomere relatively very long; nearly as long as the combined length of 2,3 and 4 .... ***Notiobia* Perty (part)**
10(2). Penultimate labial palpomere plurisetose (> 3 setae) **11**
– Penultimate labial palpomere bisetose **15**

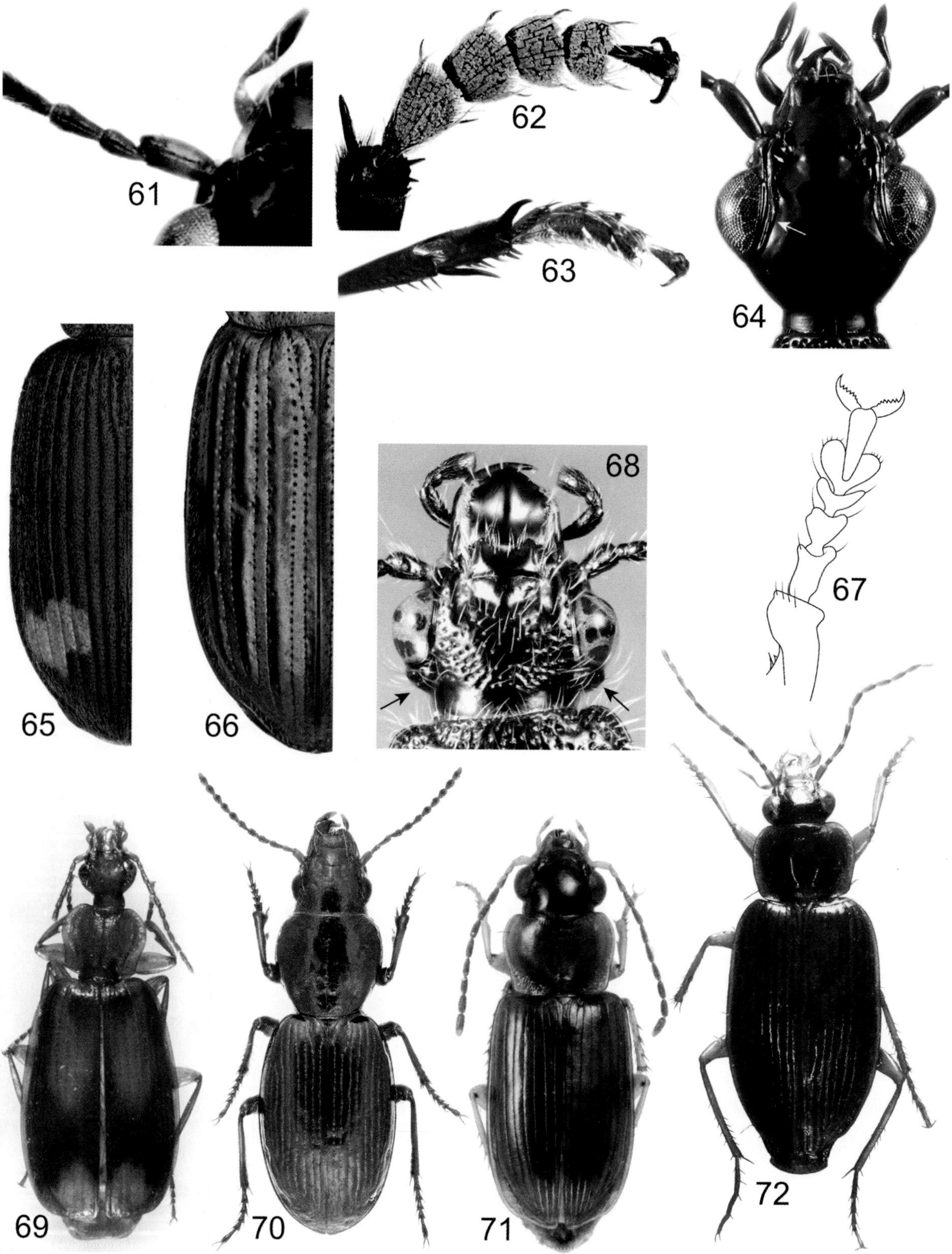

Figs 12.61–12.72. 61, *Abacetus* sp., basal antennomeres, dorsal; 62–63, male protarsomeres, ventral: 62, *Gnathaphanus philippensis* (Chevrolat), 63, *Harpalus fulvicornis* (Thunberg); 64, *Dicraspeda obscura* (Laporte), head, dorsal; 65–66, elytron: 65, *Chlaenius flaviguttatus* W. S. Macleay, 66, *Chlaenius greyanus* White; 67, *Demetrida* sp.; protarsomeres; 68, *Dicranoglossus resplendens* (Laporte), head, dorsal; 69–72, habitus: 69, *Abaditicus meyeri* Ball & Hilchie, 70, *Acallistus cuprescens* (Sloane), 71, *Acupalpus brunnicolor* (Sloane), 72, *Altagonum kurandae* Baehr.

11(10). Elytra with one or no punctures in interval 3 and intervals without pubescence **12**

– Elytra with more than one puncture in interval 3 or intervals densely pubescent on the disc **14**

12(11). Mentum tooth present, prominent or only slightly produced; parascutellar stria joined to stria 1 (Fig. 12.33A); angular base of stria 1 separate from apical portion of stria or absent; apex of clypeus, no membrane visible between clypeus and the symmetric labrum; typically larger sized beetles, > 5 mm **13**

– Mentum tooth absent; parascutellar stria separate from stria 1 (Fig. 12.33D); stria 1 continuously impressed to the base; clypeus at apex emarginate, a pale membrane visible between clypeus and the usually asymmetric labrum (Fig. 12.27); small size beetles, 3–5 mm .... ***Amblystomus* Erichson (part)**

13(12). Prosternum at base and abdominal sterna with short scattered pubescence; north-eastern Queensland .... ***Trichotichnus* Morawitz (part)**

– Prosternum at base and abdominal sterna glabrous; south-western Western Australia .... ***Harpalus* Latreille (part)**

14(11). Elytra with intervals densely pubescent on the disc; elytral apices not denticulate .... ***Parophonus javanus* (Gory) (part)**

– Elytra with interval 3 with a series of > 3 punctures but otherwise intervals are glabrous on the disc; elytral apices denticulate .... ***Coleolissus papua* Darlington (part)**

15(10). Mentum tooth absent or at most indicated by a very low, rounded prominence **16**

– Mentum tooth well developed and triangular .... ***Euthenarus* Bates (part)**

16(15). Suture between mentum and submentum entire and evident **17**

– Mentum and submentum fused and not evident at least laterally .... ***Loxoncus marginatus* (Macleay)**

17(16). Abdominal sterna with short pubescence **18**

– Abdominal sterna glabrous except for paramedial fixed setae .... ***Egadroma* Motschulsky (part)**

18(17). Prosternal process with long setae at the apex **19**

– Prosternal process without long setae at the apex .... ***Stenolophus* Dejean (part)**

19(18). Ligula with two setae; pronotum usually with posteriolateral angles and sides rounded .... ***Acupalpus* Latreille (part)**

– Ligula with four setae; pronotum usually with posteriolateral angles angulate, sides straight or sinuate .... ***Anthracus Motschulsky* (part)**

20(1). Penultimate labial palpomere plurisetose (> 3 setae) **21**

– Penultimate labial palpomere bisetose **31**

21(20). Elytra with one or no punctures in interval 3 and intervals 5 and 7 impunctate; elytra with intervals not pubescent on the disc **22**

– Elytra with more than one puncture in interval 3, intervals 5 and 7 with or without punctures or elytra with intervals densely pubescent on the disc **29**

22(21). Metacoxa with < 8 setae along anterior margin; typically 1–2, occasionally more in some *Hypharpax*; male protarsomeres with or without ventral vestiture of setae **23**

– Metacoxa with fringe of > 10 setae along anterior margin; male protarsomeres without ventral vestiture of setae .... ***Phorticosomus* Schaum (part)**

23(22). Mentum tooth present, rarely only slightly produced (in *Harpalus*) **24**

– Mentum tooth absent **28**

24(23). Abdominal sterna glabrous; metatarsomere 1 relatively very long, clearly longer than the combined length of 2 and 3 .... **25**

– Abdominal sterna with short pubescence or rarely glabrous; when glabrous metatarsomere 1 relatively short, shorter than or only slightly longer than the combined length of 2 and 3 .... **26**

25(24). Head large, but of more typical relative size for Harpalini; metapimeron elongate; larger sized beetles, > 8 mm; mentum tooth low and rounded .... ***Harpalus* Latreille (part)**

– Head relatively enormous, as long and nearly as wide as pronotum; metapimeron very short; relatively small size beetles, 6–8 mm; mentum tooth triangular and produced .... ***Nornalupia* Kataev (part)**

26(24). Metatarsomere 1 relatively very long; nearly as long as the combined length of 2, 3 and 4 .... **27**

– Metatarsomere 1 relatively short, shorter than or only slightly longer than the combined length of 2 and 3 .... ***Hypharpax* Macleay (part)**

27(26). Eyes very large and prominent; subocular gena much narrower than the width of the antennomere 1; if eye somewhat smaller, then suture between clypeus and frons very deeply impressed; clypeus notably tumescent; clypeo-ocular sulcus deep and complete .... ***Trichotichnus* Morawitz (part)**

– Eyes typical size and not notably prominent; subocular gena about as wide or notably wider than the width of antennomere 1; suture between clypeus and frons very shallowly impressed or not evident; clypeus not tumescent; clypeo-ocular sulcus incomplete, absent or shallow and complete .... ***Notiobia* Perty (part)**

28(23). Clypeal apex emarginate exposing base of labrum (Fig. 12.27); small size beetles, 3–5 mm; pronotum quadrate or somewhat cordate, without distinctly flattened lateral depression .... ***Amblystomus* Erichson (part)**

– Clypeal apex not emarginate; larger size beetles, > 5 mm; pronotum suborbiculate (Pl. 21B), lateral depression prominently flattened and sharply delimited from disc .... ***Cenogmus* Sloane (part)**

29(21). Elytra with intervals glabrous on the disc except for 3 or more setigerous punctures in interval 3 .... **30**

– Elytra with intervals densely pubescent on the disc .... ***Parophonus javanus* (Gory) (part)**

30(29). Mentum tooth present; elytral apices denticulate .... ***Coleolissus papua* Darlington (part)**

– Mentum tooth absent; elytral apices not denticulate .... ***Gnathaphanus* Macleay (part)**

31(20). Mentum tooth absent **32**

– Mentum tooth present **37**

32(31). Suture between mentum and submentum entire and evident **33**

– Mentum and submentum fused and not evident, at least laterally .... ***Loxoncus marginatus* (Macleay)**

33(32). Elytral outer striae well impressed throughout their length or only slightly shallower in the basal third **34**

– Elytral outer striae, typically 6–9, very shallow in the basal half, usually 7–8 completely effaced in the basal third .... ***Haplaner* Chaudoir**

34(33). Abdominal sterna with short pubescence **35**
– Abdominal sterna glabrous except for paramedial fixed setae or at most a few, short setae are present medioposterad metacoxae .... ***Egadroma* Motschulsky (part)**
35(34). Prosternal process with long setae at the apex **36**
– Prosternal process without long setae at the apex .... ***Stenolophus* Dejean (part)**
36(35). Ligula with two setae; pronotum usually with posteriolateral angles and sides rounded .... ***Acupalpus* Latreille (part)**
– Ligula with four setae; pronotum usually with posteriolateral angles angulate, sides straight or sinuate .... ***Anthracus* Motschulsky (part)**
37(31). Abdominal sterna glabrous except for paramedial fixed setae **38**
– Abdominal sterna with short pubescence .... ***Euthenarus* Bates (part)**
38(37). Head and pronotum without punctures or punctation fine, shallow and more or less dense **39**
– Head and pronotum with scattered, deep, coarse punctures (Pl. 20B) .... ***Batoscelis oblongus* (Dejean)**
39(38). Body form various but not as described below; typically overall shape ovoid or subcylindrical with prominent eyes; widespread and eurytopic, but most commonly found in leaf litter .... ***Lecanomerus* Chaudoir (part)**
– Body form slender, elongate, reminiscent of a typical platynine; pronotum distinctly narrower than elytra; known from caves in the southern areas of Victoria and South Australia .... ***Notospeophonus* Moore (part)**

Subtribe Pelmatellina Bates, 1882

Characteristics. Mentum and submentum separated by complete transverse suture. Penultimate labial palpomere with two or three setae on anterior margin. Apex of prosternal lobe without setae. Protarsomeres of males with one or more of first four articles laterally expanded and with ventral spongy pubescence.

Lecanomerus Chaudoir, 1850 (Pl. 28H)

Type species. *Lecanomerus insidiosus* Chaudoir, 1850, by monotypy.
Thenarotes Bates, 1878
Nemaglossa Sloane, 1920
Odontagonum Darlington, 1956
Veradia Laporte, 1867
Characteristics. Small to medium size beetles (3–11 mm, most around 4–5 mm); not dorsally pubescent, typically more or less shiny, often with contrastingly light coloured elytral patterns. Head somewhat narrower than typical harpalines, but general form of Australian species highly varied from oval, oodine-like to more elongate, platynine-like.
Australian species. 25 species.
Distribution. Coastal regions throughout Australia and further inland in more mesic habitats in the east; extraterritorially New Guinea, New Caledonia, and New Zealand.
Biology. Flying or flightless species, presumed primarily predatory, nocturnal, found in leaf litter. Often hygrophilous, in a range of forest types, also known from caves.
References. Jaeger (2016); Will (2015b).

Notospeophonus Moore, 1962 (Pl. 32H)

Type species. *Notospeophonus castaneus* Moore, 1962, by original designation.
Characteristics. Castaneous to reddish-brown beetles. Form of an elongate *Lecanomerus* with very long legs, long antennae, and cordiform pronotum.
Australian species. Three species, two with subspecies.
Distribution. Southern Victoria, eastern New South Wales, and south-eastern Western Australia.
Biology. Flightless cave-dwelling and deep-soil beetles.

Subtribe Harpalina Bonelli, 1810

Characteristics. Mentum and submentum separated by complete transverse suture; penultimate labial palpomere with three or more setae on anterior margin; Protarsomeres of males of most species with some articles laterally expanded and with ventral biseriate vestiture or unmodified.

Coleolissus Bates, 1892 (Pl. 22H)

Type species. *Hypolithus perlucens* Bates, 1878, by subsequent designation by Andrewes (1939).
Characteristics. Broad medium-sized beetles; shiny, iridescent; head with eyes relatively large; elytra sutural intervals denticulate at apices; sutural striae long; interval 3 seriate-punctate; flight wing full.
Australian species. Single, *Coleolissus (Coleolissus) papua* Darlington.
Distribution. North coastal Queensland; extraterritorially New Guinea.
Keys to species. Darlington (1968).

Harpalus Latreille, 1802 (Pl. 26I; Fig. 12.63)

See Lorenz (2005) for generic synonyms.
Type species. *Carabus proteus* Paykull, 1790 (= *Carabus affinis* Schrank, 1781), by subsequent designation by Andrewes (1935b).
Characteristics. Species introduced into Australia are typical of Northern Hemisphere *Harpalus* with stout form, broad head and heavily built mandibles. Adults are brown, fairly dull, and legs paler, castaneous. Frontal fovea not or shallowly impressed; eyes not prominent; elytra without plica.
Australian species. Two species.
Distribution. Both species are adventive and are species native to South Africa. One, *Harpalus (Harpalus) fulvicornis* (Thunberg) is known from Western Australia,

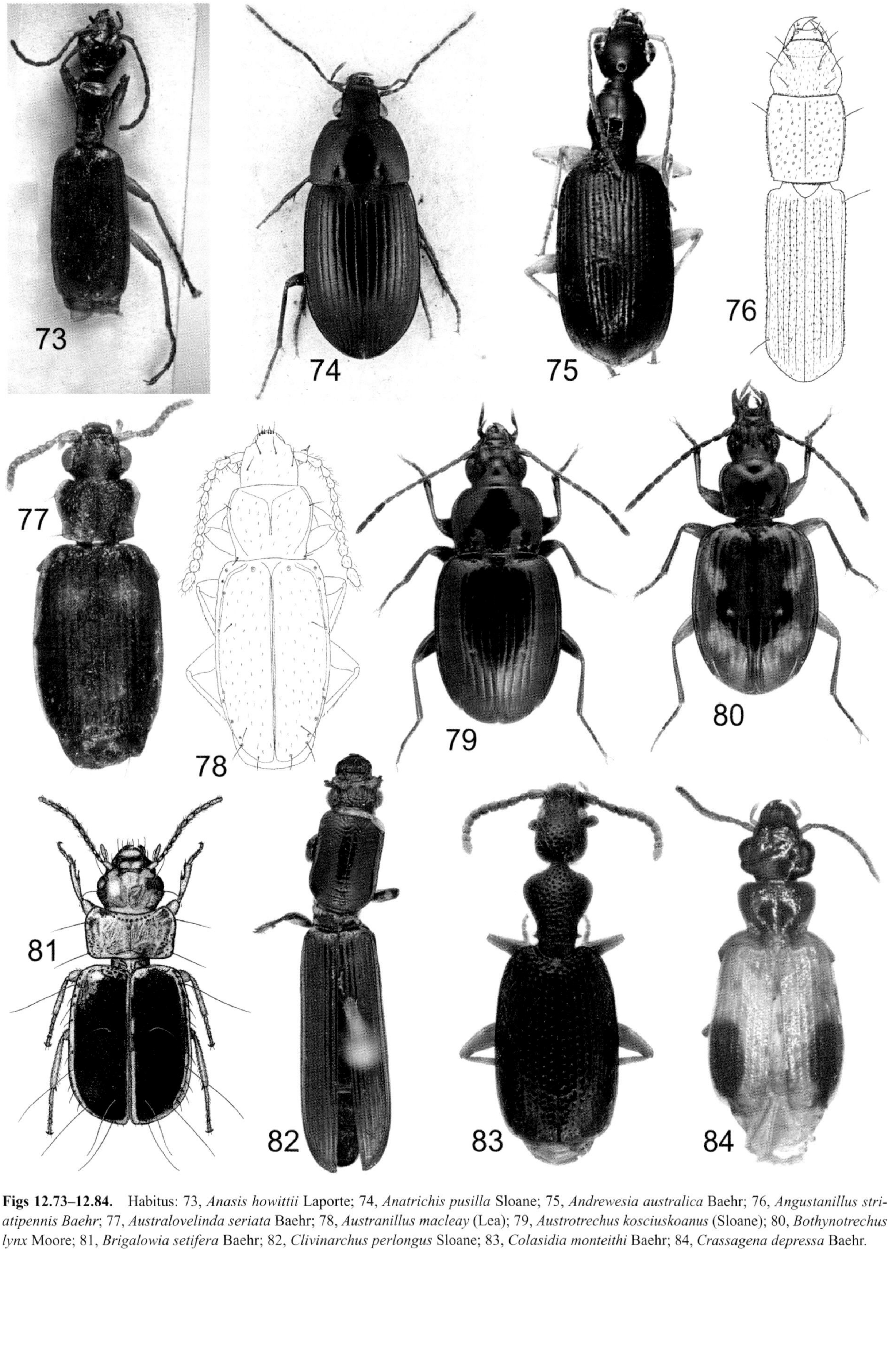

Figs 12.73–12.84. Habitus: 73, *Anasis howittii* Laporte; 74, *Anatrichis pusilla* Sloane; 75, *Andrewesia australica* Baehr; 76, *Angustanillus striatipennis Baehr*; 77, *Australovelinda seriata* Baehr; 78, *Austranillus macleay* (Lea); 79, *Austrotrechus kosciuskoanus* (Sloane); 80, *Bothynotrechus lynx* Moore; 81, *Brigalowia setifera* Baehr; 82, *Clivinarchus perlongus* Sloane; 83, *Colasidia monteithi* Baehr; 84, *Crassagena depressa* Baehr.

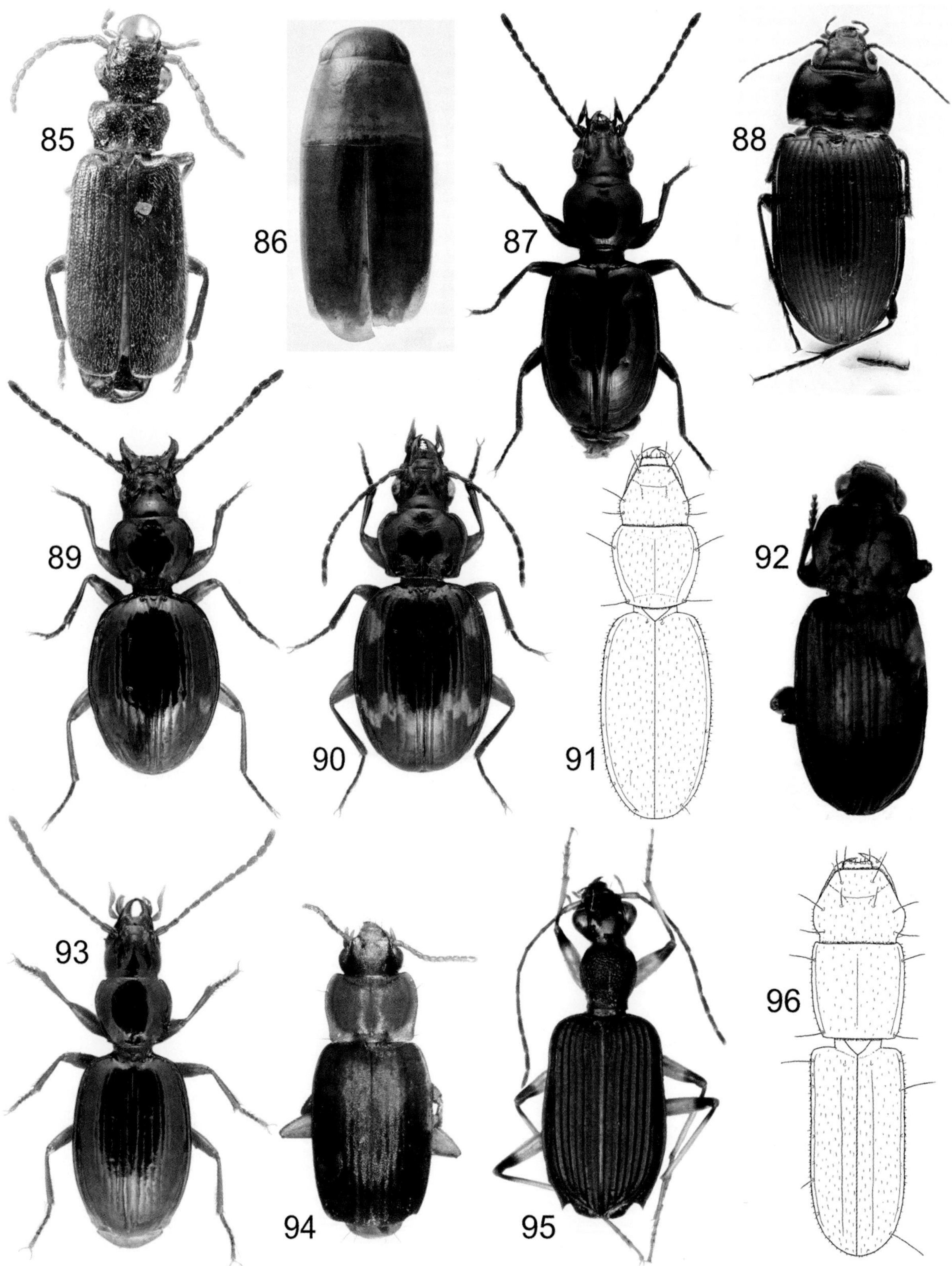

Figs 12.85–12.96. Habitus: 85, *Creagris labrosus* Nietner; 86, *Cryptocephalomorpha australica* Baehr; 87, *Cyphotrechodes gibbipennis* (Blackburn); 88, *Dendromigadops gloriosus* Baehr; 89, *Eutrechopsis ovalis* Moore; 90, *Eutrechus otwayensis* (Moore); 91, *Externanillus mcraei* Baehr; 92, *Gastrogmus ischialis* Sloane; 93, *Goedetrechus talpinus* Moore; 94, *Geoffreyella lamingtonensis* Baehr; 95, *Giachinoana carinipennis* Baehr; 96, *Gracilanillus longulus* Baehr.

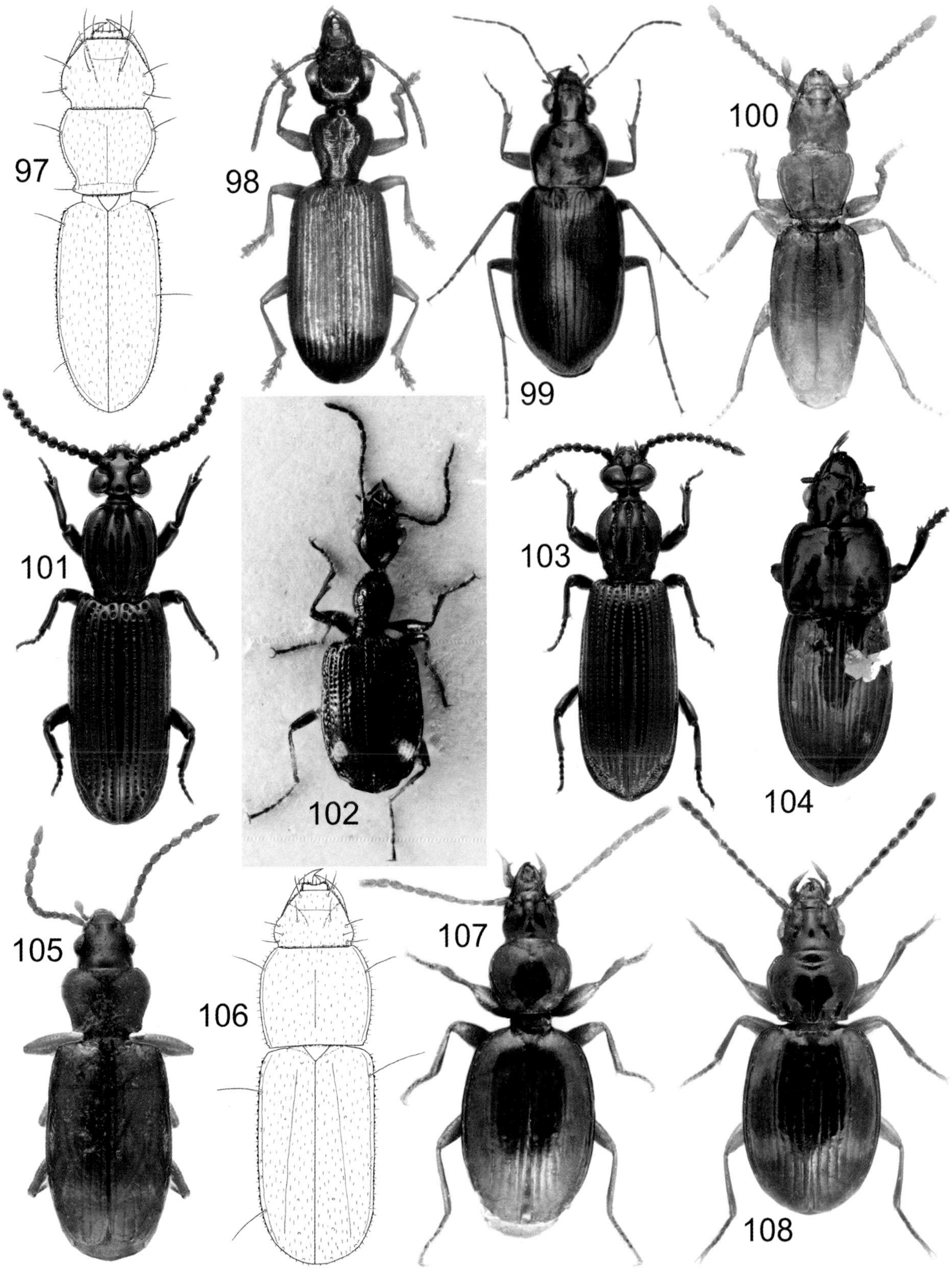

Figs 12.97–12.108. Habitus: 97, *Hesperanillus scanloni* Baehr; 98, *Hexagonia bilyi* Baehr; 99, *Hololeius ceylanicus* (Nietner); 100, *Illaphanus* sp.; 101, *Kaveinga* (*Angekiva*) *frontalis* (Grouvelle); 102, *Lachnothorax tokkia* Gestro; 103, *Leoglymmius lignarius* (Olliff); 104, *Litarthrum browni* Sloane; 105, *Lymnastis* sp.; 106, *Magnanillus magnus* (Baehr) 107, *Mimanillus gracilis* Moore; 108, *Mimotrechus australiensis* (Sloane).

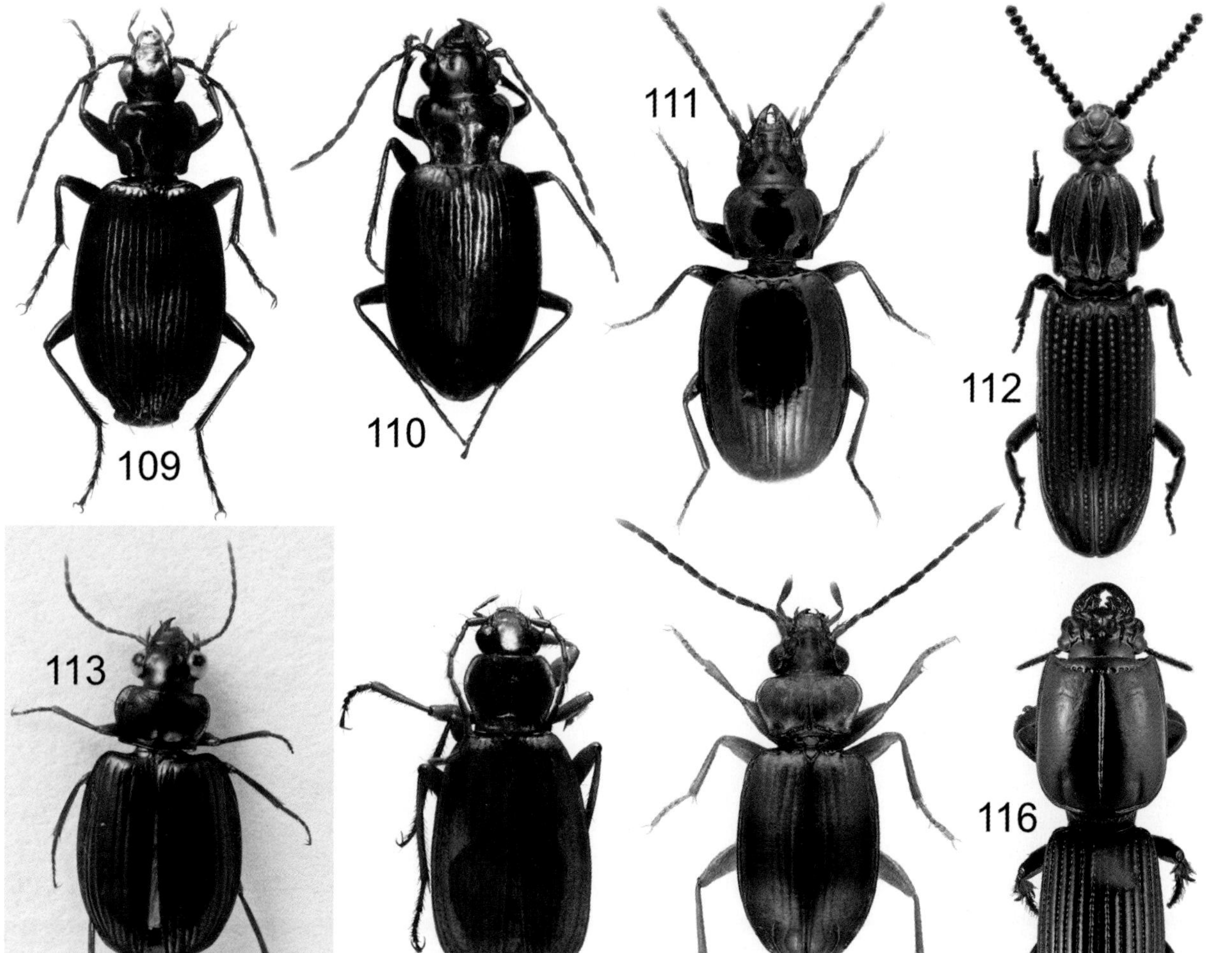

Figs 12.109–12.120. Habitus: 109, *Mooreagonum brevipenne* Baehr; 110, *Nebriosoma fallax* Laporte; 111, *Nototrechus unicolor* Moore; 112, *Omoglymmius* (*Omoglymmius*) *ichthyocephalus* (Lea); 113, *Oxyodontus tripunctatus* Chaudoir; 114, *Paraplatynus sulcatipes* Baehr; 115, *Paratachys fasciatus* Motschulsky; 116, *Parathlibops crenatus* (Chaudoir); 117, *Paratrechodes macleayi* (Sloane); 118, *Pericompsus australis* (Schaum); 119, *Perileptus sloanei* Moore; 120, *Philipis alpina* Baehr.

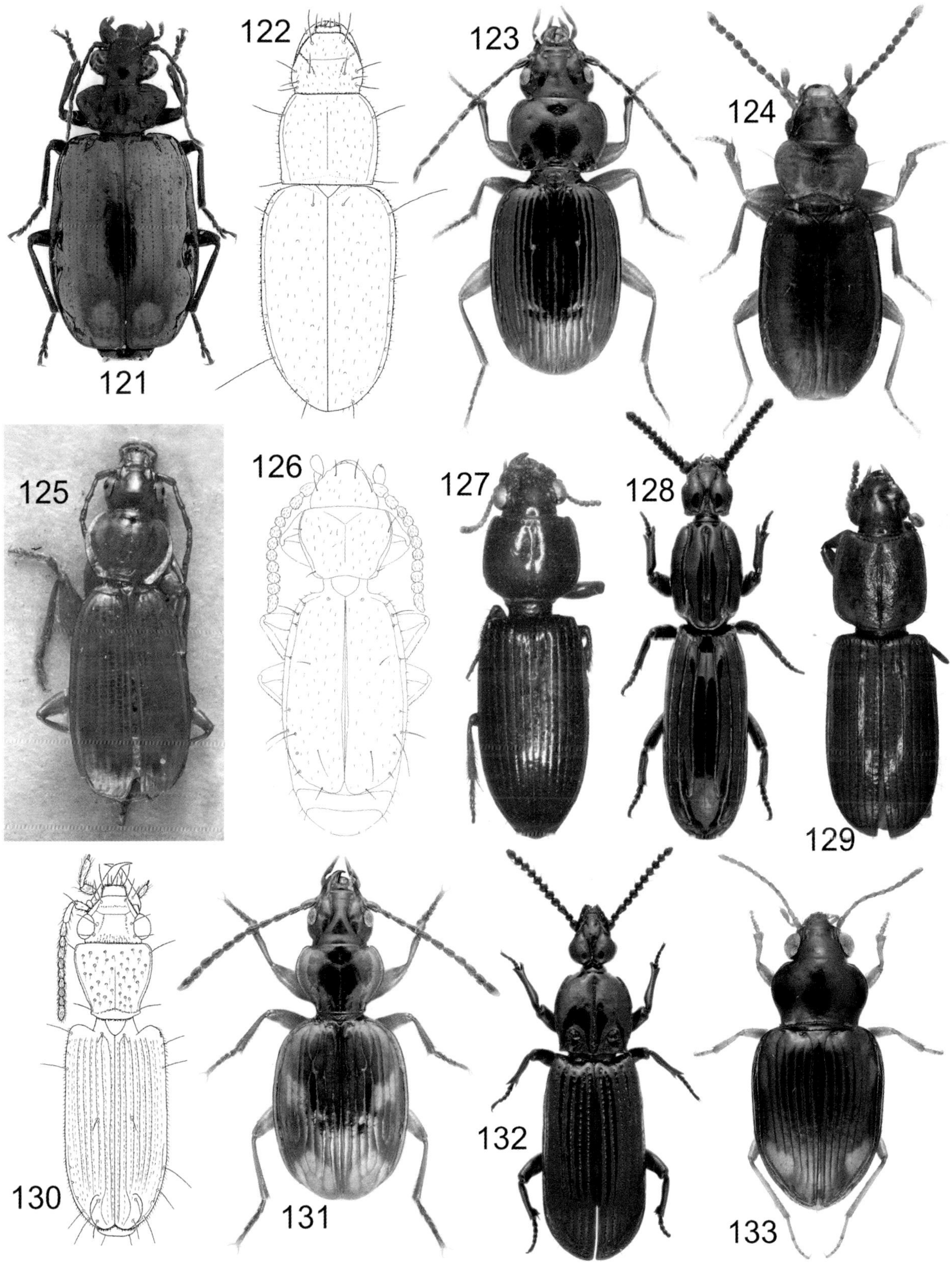

Figs 12.121–12.133. Habitus: 121, *Physodera bacchusi* Darlington; 122, *Pilbaranillus latibasis* Baehr; 123, *Pogonoschema pallipes* Moore; 124, *Polyderis captus* (Blackburn); 125, *Pseudamblytelus orbicollis* Baehr; 126, *Pseudillaphanus barringtoni* Giachino; 127, *Pseudoclivina australiana* Baehr; 128, *Rhyzodiastes* (*Temoana*) *mirabilis* (Lea); 129, *Rubidiclivina punctatissima* Baehr; 130, *Setitachys macrops* Baehr; 131, *Sloanella simsoni* (Blackburn); 132, *Sloanoglymmius planatus* (Lea); 133, *Sphaerotachys curticollis* Sloane.

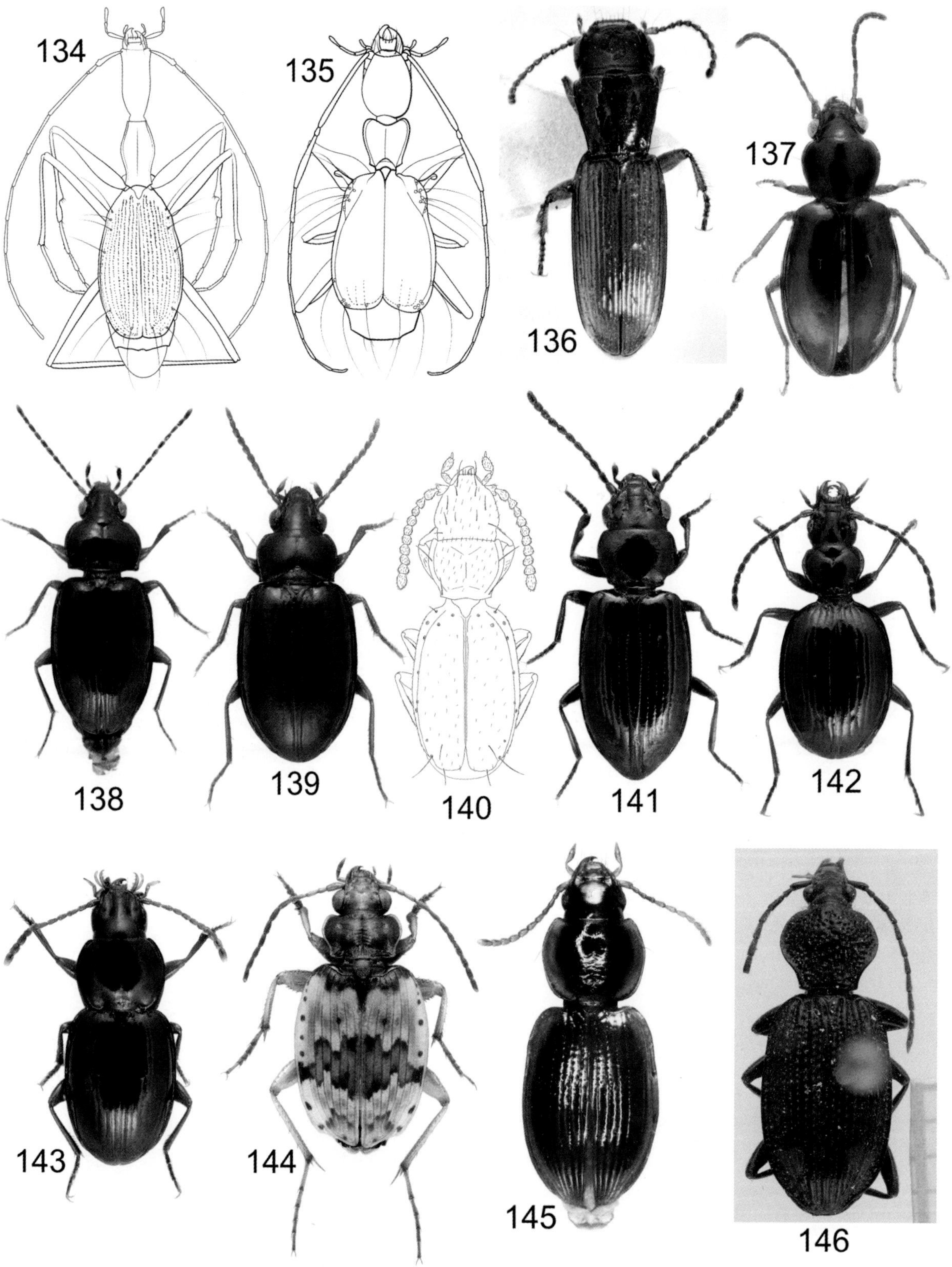

Figs 12.134–12.146. Habitus: 134, *Speothalpius grayi* Moore; 135, *Speozuphium poulteri* Moore; 136, *Syrdenoidius spinipes* Baehr & Hudson; 137, *Tachylopha iaspidea* (Sloane); 138, *Tachys mulwalensis* Sloane; 139, *Tachyta rexensis* Moore; 140, *Tasmanillus daccordii* Giachino; 141, *Tasmanitachoides wattsense* (Blackburn); 142, *Tasmanorites flavipes* (Lea); 143, *Tasmanotrechus compactus* Moore; 144, *Tetragonoderus undatus* (Dejean); 145, *Thayerella newtoni* Baehr; 146, *Tinognathus parviceps* Chaudoir.

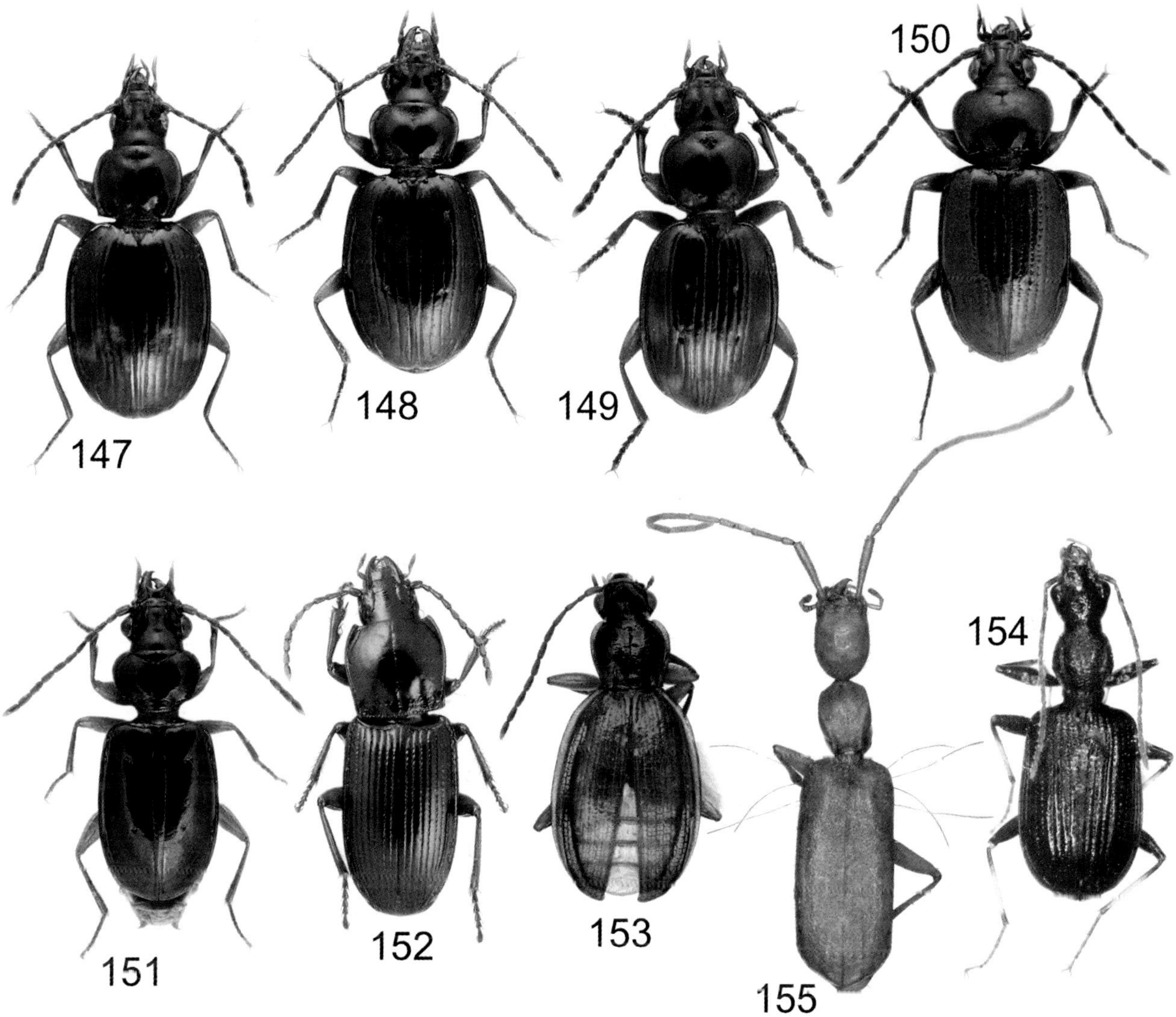

Figs 12.147–12.155. Habitus: 147, *Trechiella subornatella* (Blackburn); 148, *Trechimorphus diemenensis* (Bates); 149, *Trechistus humicola* Moore; 150, *Trechobembix baldiensis* (Blackburn); 151, *Trechodes secalioides* (Blackburn); 152, *Trephisa parallela* Moore; 153, *Trichamblytelus ovalipennis* Moore; 154, *Tricharnhemia browni* Baehr; 155, *Typhlozuphium longipenne* Baehr.

and the other, *Harpalus (Harpalus) parvulus* Dejean, is recorded from New South Wales and South Australia.

Biology. These species are found in human modified habitats.

References. Kataev (2015); Moore (1977b).

Parophonus Ganglbauer, 1892 (Pl. 33F)

Type species. *Carabus maculicornis* Duftschmid, 1812, by subsequent designation (Jeannel, 1942).

Hypolithus Dejean, 1829 (not Eschscholtz, 1829)

Characteristics. Dorsally punctate beetles, head micropunctate, glabrous; pronotum punctate more coarsely along margins, particularly latero-basally; elytra completely, densely and regularly punctate with at least outer intervals pubescent; intervals 3, 5 and 7 with discal pores, punctation of intervals more fine and dense, with six to seven punctures in interval 3; last ventrite in female rounded, in male slightly emarginate at apex; flight wing full; male with protarsi with squamous setae ventrally.

Australian species. Only *Parophonus (Hyparpalus) javanus* (Gory, 1833)

Distribution. Coastal Western Australia, Northern Territory, and Queensland; extraterritorially distributed widely across Indonesia.

References. Kataev (2010).

Keys to species. Kataev (2010).

Trichotichnus Morawitz, 1863 (Pl. 39C)

Type species. *Trichotichnus longitarsis* Morawitz, 1863, by monotypy.
Asmerinx Tschitschérine, 1898
Bellogenus Clarke, 1971
Carbanus Andrewes, 1937
Lyter Darlington, 1968
Pseudotrichotichnus Habu, 1973
Pteropalus Casey, 1914
Velimius Jedlička, 1952
Characteristics. Normal looking Harpalini without dorsal pubescence, mentum with tooth, labial penultimate palpomere plurisetose; males usually with pro- and mesotarsi with squamous setae ventrally
Australian species. Five species.
Distribution. Found in northern Queensland; extraterritorially in eastern North America, Eurasia, and south to the Malay Archipelago and New Guinea
Biology. Leaf litter in temperate and tropical forests. In Australia, flight capable beetles found in tropical rainforest leaf litter. Taken at lights at night.
References. Baehr (1990a); Darlington (1968).
Keys to species. Baehr (1997f).

Key to the Australian subgenera of *Trichotichnus*

1 Clypeo-ocular sulci only well impressed near the clypeus and ended before reaching the margin of the eye .... ***Trichotichnus* Morawitz**
– Clypeo-ocular sulci well impressed and reaching the margin of the eye .... ***Bottchrus* Jedlička**

Subgenus *Trichotichnus* Morawitz, 1863

Type species. *Trichotichnus longitarsis* Morawitz, 1863, by monotypy.
Australian species. Four species.
Distribution. Widespread outside of Australia. Species in Australia are restricted to north-east Queensland.

Subgenus *Bottchrus* Jedlička, 1935

Type species. *Bottchrus philippinus* Jedlička, 1935, by monotypy
Bellogenus Clarke, 1971
Characteristics. The clypeo-ocular furrows are deep throughout, reaching the margin of eyes; mentum with acute median tooth; male pro- and mesotarsi with squamous setae ventrally; median lobe of aedeagus with ostium slightly shifted left.
Australian species. One species, *T. maculipennis* Baehr.
Distribution. Northern Queensland; extraterritorially in south-eastern Asia from Sri Lanka southward.
References. Kataev (2016).

Subtribe Amblystomina Fauvel, 1889

Amblystomus Erichson, 1837 (Pl. 18F; Fig. 12.27)

Type species. *Acupalpus mauritanicus* Dejean, 1829.
Hispalis Rambur, 1838
Artizoum Gistel, 1857
Megaristerus Nietner, 1858
Notophilus Blackburn, 1888
Thenarotidius Sloane, 1898
Psilonothus Sloane, 1900
Entomorrhinus Jeannel, 1948
Characteristics. Small sized beetles (3–5 mm), head relatively very large; clypeus deeply emarginate, exposing membranous region, parascutellar stria not joined to stria 1; pronotum typically rounded, suborbiculate, occasionally cordate;
Australian species. 16 species.
Distribution. In Australia found throughout the continent, except for central deserts, and Tasmania; extraterritorially most of the warmer regions of the Old World, especially in the tropical and subtropical areas.
Biology. Terrestrial leaf litter in a variety of habitats.
References. Moore *et al.* (1987); Lorenz (2005).

Subtribe Anisodactylina Lacordaire, 1854

Characteristics. Mentum and submentum separated by complete transverse suture, fused laterally but separated medially, or completely fused. Penultimate labial palpomere with three or more setae on anterior margin. Protarsomeres 1(2)-4 laterally expanded and with ventral spongy pubescent vestiture.

Cenogmus Sloane, 1898 (Pl. 21B)

Type species. *Cenogmus castelnaui* Csiki (= *Harpalus rotundicollis* Laporte, 1867), by subsequent designation by Noonan, 1973.
Characteristics. More or less parallel-sided or cylindrical beetles with the mentum tooth absent; pronotum suborbiculate, usually a bit wider than long, lateral depression prominently flattened and sharply delimited from disc. Complete transverse suture between mentum and submentum.
Australian species. Three species.
Distribution. Widespread across the Australian continent.
Biology. Omnivorous, strong fliers, found in open habitats often near water. Frequently come to lights at night.
References. Moore (1966c).

Gnathaphanus Macleay, 1825 (Pl. 26G; Fig. 12.62)

Type species. *Trechus* (*Gnathaphanus*) *vulneripennis* Macleay, 1825, by monotypy.
Pachauchenius W. J. Macleay, 1864
Mirosarus Bates, 1878
Iwosiopelus Nakane & Ishida, 1959
Characteristics. Medium size harpaline beetles, typically black, some with metallic green or cupreous reflex, somewhat shiny; male pro- and mesotarsi broadly expanded with squamous setae ventrally; with complete transverse suture between mentum and submentum; some species with massively broad head.
Australian species. 17 species.
Distribution. In Australia they are found across the continent and in Tasmania; extraterritorially found from the Oriental region and throughout the Pacific region.
Biology. Omnivorous, regularly seed or pollen feeding, strong fliers,
References. Guthrie *et al.* (2010).
Keys to species. Some species covered by Darlington (1968).

Haplaner Chaudoir, 1878 (Pl. 26C)

Type species. *Harpalus velox* Laporte, 1867, by monotypy.
Characteristics. Small sized harpaline beetles (3–4 mm) with the elytral outer striae very shallow or absent.
Australian species. Two species.
Distribution. North-east coastal Queensland and across the south from south western Western Australia, South Australia, and Victoria.
Biology. Flight capable, omnivore.
Note. The genus and putative near relatives need to be reviewed as there appears to be considerable confusion in the published literature and among specimens in museums regarding characteristics of this genus. Users of the key presented here may find *Haplaner* or *Haplaner*-like specimens that fail to key properly.
References. Possible confusion with *Haplanister* Moore, 1996; Larochelle & Larivière (2007).

Hypharpax Macleay, 1825 (Pl. 27I)

Type species. *Harpalus* (*Hypharpax*) *lateralis* Macleay, 1825, by monotypy.
Sagraemerus Redtenbacher, 1867.
Characteristics. This genus is quite variable in size, form and colour. Typically they are medium size, black (some more or less metallic), often shiny beetles with relatively broad heads and the pronotal hind angles are rounded, pronotum nearly orbiculata in some; flight wing full; complete transverse suture between mentum and submentum; male pro- and mesotarsi broadly expanded with squamous setae ventrally.
Australian species. 25 species.
Distribution. This genus is distributed from Tasmania, throughout continental Australia; extraterritorially New Guinea, Sulawesi, Java and Sumatra, and New Zealand.
Biology. Omnivores, sometimes feeding on seeds and fruit, many strong fliers, found in a wide variety of open habitats. Frequently attracted to lights at night.
Note. This genus is in need of revision as species identifications can only be done with recourse to types and original descriptions. The boundaries of the genus relative to other anisodactyline genera is questionable.

Nornalupia Kataev, 2002 (Pl. 103K)

Type species. *Nornalupia megacephala* Kataev, 2002, by original designation.
Characteristics. Relatively massive head; eyes small; mentum with prominent medial tooth; mentum and submentum separated by complete transverse suture; elytron with one discal pore on interval 3; no parascutellar stria.
Australian species. Two species.
Distribution. Southern Western Australia.
Biology. Found in leaf litter in various eucalypt forest; nocturnal.
References. Kataev (2002a, 2007).

Notiobia Perty, 1830 (Pl. 31D)

Type species. *Notiobia nebrioides* Perty, 1830, by monotypy.
Anisotarsus Chaudoir, 1837
Diaphoromerus Chaudoir, 1843
Eurytrichus LeConte, 1847
Harpalodes Motschulsky, 1864
Stilbolidus Casey, 1914
Characteristics. Complete transverse suture between mentum and submentum
Australian species. 26 species. All apparently in the subgenus *Anisotarsus*.
Distribution. In Australia these beetles are known from south-west coastal Western Australia, south coastal South Australia, throughout most of Victoria, New South Wales and Queensland except in deep interior; extraterritorially the genus is found worldwide and the subgenus *Anisotarsus* is found throughout the North and South America.
Biology. Omnivorous, strong fliers found in a variety of open, scrub and forest habitats.

Phorticosomus Schaum, 1863 (Pl. 34D)

Type species. *Phorticosomus felix* Schaum, 1863, by monotypy.
Characteristics. Medium to large size, concolorous brown beetles; metacoxa with a fringe of setae on anterior margin; metatrochanter with a cluster of medially setae and

fringe of setae on the ventral surface; abdominal ventrites with fringe of setae on apical margins and dense cluster of setal on the intercoxal process; male pro- and mesotarsi not expanded and lacking squamous setae ventrally, or rarely, 3rd and 4th protarsomeres with biseriate vestiture.

Australian species. 18 species.

Distribution. Found across mainland Australia and several island groups in the Gulf of Carpentaria, and the Timor and Arafura Seas.

Biology. Nocturnal, omnivorous, seed feeding. Primarily semi-arid to arid areas, open grasslands, savannahs and open woodlands. Capable fliers frequently attracted to lights at night.

References. Martínez-Navarro *et al.* (2005). This genus currently is being revised by N. Guthrie, WA.

Subtribe Stenolophina Kirby, 1837

Mentum and submentum separated by complete transverse suture; mental tooth present or absent; Penultimate labial palpomere with two or three setae on anterior margin; Protarsomeres of male with articles 1(2)- 4 laterally expanded and with ventral biseriate vestiture or unmodified; median lobe of male genitalia with dorsal membranous area long, extended to basal bulb in most; ostium dorsal in position in most; generally small beetles, less than 10 mm in length.

Acupalpus Latreille, 1829 (Pl. 18K; Fig. 12.71)

Type species. *Carabus meridianus* Linnaeus, 1760

Manicellus Motschulsky, 1864

Palcuapus Habu, 1973

Pseudanthracus Habu, 1973

Characteristics. Small size (3.5–4 mm) pale brown; mentum tooth absent; ligula with two setae; clypeo-ocular sulcus well impressed; pronotum with rounded, obtuse hind angles.

Australian species. One species, *Acupalpus brunnicolor* (Sloane).

Distribution. In Australia found only in northern parts of Queensland, Northern Territory and Western Australia; extraterritorially the genus is distributed nearly worldwide.

Biology. Omnivorous, seed-feeding, strong flier, found in open habitat. Frequently attracted to lights at night.

References. Jaeger (2016).

Anthracus Motschulsky, 1850 (Pl. 18G)

Type species. *Carabus consputus* Duftschmid, 1812

Balius Schiodte 1861

Characteristics. Small size (3–4 mm) brown; mentum tooth absent; ligula with four setae; clypeo-ocular sulcus well impressed; pronotum cordiform, margins sinuate, with well marked about right or obtuse hind angles.

Australian species. Two species.

Distribution. Found in northern Queensland; extraterritorially the genus is known from Nearctic, Palearctic, and across the Old World tropics, Indonesia, and New Caledonia.

References. Jaeger (2016).

Batoscelis Dejean, 1836 (Pl. 20B)

Type species. *Agonoderus oblongus* Dejean, 1831, by subsequent designation by Noonan, 1976.

Systenognathus Putzeys, 1863

Leptocellus Muller, 1942

Characteristics. Small harpaline beetles that are elongate, parallel-sided; antennae are relatively short and stout; eyes prominent; pronotal disk with numerous coarse punctures, especially along the margins, but also on the disc, though more sparsely.

Australian species. Only *Batoscelis (Agonoderus) oblongus* Dejean.

Distribution. Northern Australia; extraterritorially the genus occurs from Africa and across Asia.

Euthenarus Bates, 1874 (Pl. 25L)

Type species. *Euthenarus brevicollis* Bates, 1874, by subsequent designation by Noonan, 1976.

Characteristics. Small beetles (3–6 mm); eyes relatively large; mentum and submentum separated by laterally incomplete transverse suture, mentum with a medial tooth; ventrites 5–6 with numerous short setae, in addition to paired ambulatory setae; apex of prosternal process pubescent.

Australian species. Six species.

Distribution. Tasmania, Norfolk Island, and mainland Australia; extraterritorially the genus is found in New Zealand.

Biology. Nocturnal, capable fliers, found in a wide variety of habitats including disturbed areas and open habitat. Often near water or attracted to lights at night.

Key to species. Some species covered by Larochelle & Larivière (2005).

Loxoncus Schmidt-Göbel, 1846 (Pl. 28K)

Type species. *Loxoncus elevatus* Schmidt-Göbel, 1846, by monotypy.

Anoplogenius Chaudoir, 1852

Megrammus Motschulsky, 1857

Lepithrix Nietner, 1857

Neolissus Landin, 1955

Characteristics. Mentum and submentum fused at least laterally, gonocoxite-2 flat and with numerous spines along single outer margin, elytra weakly sinuate before apex and lacking parascutellar stria, and abdominal sternites with only large ambulatory setae.

Australian species. Only *Loxoncus marginatus* (Macleay, 1888).

Distribution. In Australia, the one species is transcontinental in the north near the coast, from King Sound, Western Australia to Townsville, Queensland; extraterritorially the genus is found from Africa, the Palearctic to Sumatra and southward.

Biology. Capable flier, omnivore, typically found near water in more or less open habitat.

Keys to species. Kataev (2002b).

Stenolophus Dejean, 1821

Type species. *Carabus vaporariorum* Linnaeus, *sensu* Fabricius, 1787 (= *Carabus teutonus* Schrank, 1781, by subsequent designation by Westwood, 1838a.

Notiocharis Gistel, 1856

Characteristics. Small beetles (5–6 mm); Eyes moderately prominent; clypeo-ocular sulcus reaching near eye; mentum lacking medial tooth; pronotal margins straight or sinuate before hind angles. Habitus as in *Egadroma* (Pl. 24I).

Australian species. Two species.

Distribution. Northern Queensland; extraterritorially the genus is widespread in the Northern Hemisphere and is distributed from the Oriental region south through the Malay Archipelago and New Guinea.

Biology. Omnivorous, flight capable.

Note. The identity of the two New Guinea and Australian species is discussed by Darlington (1968).

Keys to species. Darlington (1968).

Egadroma Motschulsky, 1855 (Pl. 24I)

Type species. *Carabus smaragdulus* Fabricius, 1798, by monotypy.

Characteristics. Small beetles (4–6 mm); Eyes moderately prominent; clypeo-ocular sulcus reaching near eye; mentum lacking medial tooth; pronotal margins rounded, straight or subsinuate before hind angles.

Australian species. Eight species.

Distribution. In Australia found in coastal regions of South Australia, Victoria, New South Wales, and Queensland. Distributed further inland in north Queensland and across the Northern Territory; extraterritorially found in the Palaearctic region, across Asia and south into New Guinea and New.

Biology. Omnivorous, flight capable.

Note. *Egadroma* is frequently treated as a subgenus of *Stenolophus*. The rank, relationships, and monophyly of these groups remains to be adequately tested.

Keys to species. Some species covered by Darlington (1968) and Larochelle & Larivière (2005).

Tribe Agonicini Sloane, 1920

Characteristics. Procoxal cavities closed; mesocoxae conjunct; head with a single supraorbital seta above each eye; labrum at apex emarginate; labial palps securiform; mandible straight and elongate, without a seta in the scrobe; body somewhat pedunculate; elytra oval-shaped, without visible plica; apex barely excised; flight wings reduced; aedeagus with ostium situated largely on the dorsal surface, parameres large, rounded, of almost similar size and shape, asetose; gonocoxite-2 short to moderately elongate, obtuse at apex, without or with few ensiform and/or nematiform setae.

Note. This tribe was regarded a subtribe of Peleciini by Straneo & Ball (1989). Peleciini and Agonicini are closely related and probably Agonicini is the adelphotaxon of Peleciini. Compared to Peleciini, Agonicini appears to have retained more putative plesiomorphic characters states. We prefer the status of a tribe. Agonicini is a tribe endemic to Australia.

Key to the genera of Agonicini

1. Vertex of head convex; frontal sulci well impressed; elytra with distinctly impressed striae; endophallus of aedeagus without denticulate or spinose folds .... ***Agonica*** **Sloane**
– Vertex of head depressed; frontal sulci barely indicated; elytra indistinctly or not striate, if present striae very shallow or not impressed; endophallus of aedeagus with denticulate or spinose folds .... ***Pseudagonica*** **Moore**

Agonica Sloane, 1920 (Pl. 18J)

Type species. *Agonica simsoni* Sloane, 1920, by subsequent designation by Moore 1963a.

Characteristics. Frontal sulci distinct; elytra clearly striate; endophallus of the aedeagus without any denticulate or spinose folds.

Australian species. Four species.

Distribution. South-eastern Victoria and Tasmania.

Biology. Ground living in closed forests. Specimens of *Agonica* are rarely collected and the best sampling method may be sifting of leaf litter in rainforests or Berlese extraction of litter. Certain peleciine species have been reported to eat millipedes (Straneo & Ball 1989). Hence, this may be also the diet of agonicine species.

References. Csiki (1931); Moore (1960a, 1963a); Moore *et al.* (1987); Straneo & Ball (1989); Lorenz (2005); Baehr (2012n).

Keys to species. Moore (1963a); Baehr (2012n).

Pseudagonica Moore, 1960 (Pl. 36B)

Type species. *Pseudagonica nitida* Moore, 1960, by original designation.

Characteristics. Frontal sulci indistinct; elytra almost non-striate; endophallus of the aedeagus with a denticulate or spinose fold.

Australian species. 14 species and five subspecies.

Distribution. Eastern Victoria to south-eastern New South Wales and the Australian Capital Territory.

Biology. Ground living, and perhaps in leaf litter, in closed forests, commonly in montane areas. Specimens are very rarely collected, and of most species only single or very few specimens are known. Nothing else is recorded about habits and life histories. Due to the very small size of several species, we do not have any idea about their diet.

References. Moore (1963a); Moore *et al.* (1987); Straneo & Ball (1989); Lorenz (2005); Baehr (2012n).

Key to species. Baehr (2012n).

Tribe Hexagoniini Horn, 1881

Characteristics. Procoxal cavities closed; mesocoxae conjunct; form very depressed; head wide, quadrate, constriction behind the eyes forming a narrow neck; mandibles with distinct scrobe, without seta in the scrobe; mentum with triangular tooth; glossa at apex excised, paraglossae far surpassing glossa; apical part of the inner lobe of the maxilla movable; First three antennomeres glabrous; elytra entire and depressed, completely covering the abdomen; tarsi very wide and underneath densely pilose, 4th tarsomeres deeply bilobed; tarsal claws denticulate; aedeagus with simple folding, endophallus usually with a sclerotised piece; parameres asetose at apex; gonocoxite-1 with several elongate setae on the apical rim, gonocoxite-2 with many short ensiform setae on the ventral surface and with some nematiform setae near apex.

Notes. The relationships of the tribe are uncertain. Habu (1967) and Darlington (1968) placed it between Pentagonicini and Odacanthini, but without any explanation. Lorenz (2005), following Jeannel (1949), includes it in the tribe Ctenodactylini. We believe that a position close to Ctenodactylini is more appropriate, at the present state of knowledge. The moveable inner lobe of the maxilla is unique within Carabidae.

Hexagonia Kirby, 1825 (Figs 12.30, 12.98)

Type species. *Hexagonia terminata* Kirby, 1825, by monotypy.

Characteristics. Extremely flat beetles; head wide, neck very narrow; elytra elongate and very depressed, completely striate; Australian species possess red elytra with more or less extended black apex.

Note. Darlington (1968) had already noted the presence of an undescribed species in North Queensland.

Australian species. Two species.

Distribution. North-eastern Queensland; northern parts of Northern Territory; extraterritorial in the Afrotropical and Oriental Regions, and in New Guinea.

Biology. Largely unrecorded. Some species were observed on grass and reed vegetation near water, where specimens have been found between or under the leaf sheets. Predacious. Darlington (1968) suggested that the species are diurnal, but a specimen recently recorded from Australia was sampled at light. Nothing else is known about habits, diet, and reproduction,

References. Csiki (1932); Habu (1967); Darlington (1968, 1971); Lorenz (2005); Baehr (2012m).

Key to species. Baehr (2012m).

Tribe Ctenodactylini Laporte, 1834

Characteristics. Procoxal cavities closed; mesocoxae conjunct; mandibles with distinct scrobe, without seta in the scrobe; mentum with distinct tooth; antenna elongate, pilose from antennomere 4; pronotum elongate; elytra slightly sinuate at apex, not completely covering the abdomen; legs slender and elongate; 4th tarsomeres wide and very deeply cleft; four basal tarsomeres with biseriate squamose setae beneath; tarsal claws denticulate; male and female genitalia not recorded in Australian species.

Plagiotelum Solier, 1849 (Pl. 35G)

Type species. *Plagiotelum irinum* Solier, 1849, by monotypy.

Characteristics. Elongate, yellow-greenish, glossy metallic beetles; 4th tarsomeres deeply excised and bilobed, pilose beneath; tarsal claws denticulate.

Australian species. Only *P. opalescens* Olliff, 1885.

Distribution. Tasmania; extraterritorial in the southern Neotropical Region.

Biology. An arbouricolous species that has been found on *Leptospermum* flowers. Predacious and probably diurnal.

References. Lacordaire (1854); Olliff (1885); Csiki (1932); Liebke (1938); Moore *et al.* (1987); Lorenz (2005).

Tribe Odacanthini Laporte, 1834

Characteristics. Procoxal cavities closed; mesocoxae conjunct; mandibles with distinct scrobe; mentum usually with rather elongate tooth; antenna elongate, usually pilose from antennomere 4, but sometimes the whole antenna pilose; pronotum usually elongate; elytra either complete or at apex transverse or slightly sinuate, then not completely covering the abdomen; legs slender and elongate; tarsal claws simple; male aedeagus slender, with the ostium dorsal, usually without distinctly sclerotised parts on the endophallus; parameres asetose at apex; female gonocoxite-1 usually with several elongate setae at the apical rim, gonocoxite-2 more or less elongate and curved, with two to four ensiform setae at the latero-ventral surface and with one short nematiform seta near the apex. The male and female genitalia of all Australian species, even across genera, are very similar in shape and structure.

Notes. The Australian Odacanthini species are exceptionally diverse in their external morphology and their habits as well. They include the probably most plesiotypic genera of Odacanthini, as well as some highly apotypic ones. Therefore, several genera do not exhibit the 'typical' odacanthine habitus. i.e, with elongate pronotum, elongate elytra, and vivid elytral pattern. All species seem to be strictly nocturnal, but their ecology is quite variable. Although all seem to be hygrophilous, they may live among pebbles on river banks, on muddy ground in swamps, as well as on grassy and reedy vegetation. The larva of one Australian species (*Eudalia macleayi* Bates) was described by Moore (1966c) and it was stated that it is rather similar to the larva of the genus *Ophionea*.

Key to the Australian genera of Odacanthini

1. 4th tarsomeres deeply excised in the middle and markedly bilobed 2
– 4th tarsomeres not deeply excised in the middle and not markedly bilobed 4

2(1). Head elongate, more or less distinctly triangular; prothorax tubular, without distinct lateral sulcus; elytra narrow, upper surface depressed, red with black or blue pattern and white spots .... ***Ophionea*** **Klug**
– Head not elongate, convex; prothorax not tubular, with distinct lateral sulcus; elytra rather short and wide, upper surface not markedly depressed, uniformly black or dark piceous, elytra sometimes with narrow yellow apical margin 3

3(2). Head with a distinct ridge mesal to each eye (Fig. 12.64); pronotum with distinct lateral sulcus; elytra without yellow apical margin .... ***Dicraspeda*** **Chaudoir (part)**
– Head without a ridge mesal to each eye; pronotum without lateral sulcus; elytra with narrow yellow apical margin .... ***Andrewesia*** **Liebke**

4(1). Head not distinctly narrowed behind eyes; colour of surface brick-red, elytra with serrate, cruciate dark pattern .... ***Porocara*** **Sloane**
– Head distinctly narrowed behind eyes; colour of surface different, elytra with different pattern, or unicolourous 5

5(4). Antennomere 3 very elongate, as long as 4 and 5 together .... ***Clarencia*** **Sloane**
– Antennomeres 3 and 4 of about equal length 6

6(5). Elytra elongate, parallel-sided; upper surface remarkably depressed; apex almost transverse, without any excision .... ***Anasis*** **Laporte**
– Elytra usually shorter and less parallel, when elongate, then upper surface not markedly depressed and apex oblique, usually more or less distinctly excised 7

7(6). Head with distinct longitudinal sulcus and ridge inside of each eye 8
– Head without or with indistinct sulcus and ridge inside of each eye 13

8(7). Odd elytral intervals cariniform; external angle of elytra spiniform .... ***Giachinoana*** **Baehr**
– Odd elytral intervals not cariniform; external angle of elytra not spiniform, at most slightly angulate 9

9(8). Pronotum with indistinct lateral margin, or margin not medially bordered by a deep sulcus; elytra glossy black with two or four small white spots .... ***Archicolliuris*** **Liebke**
– Pronotum with very conspicuous ridge-like margin, margin medially bordered by a deep sulcus; elytra piceous or black, without white spots 10

10(9). Elytra deeply punctate-striate in basal third, barely striate in apical two-thirds; with deep transverse sulcus in basal third; elytra rather narrow and elongate; lateral margins of pronotum conspicuously sinuate .... ***Basistichus*** **Sloane**
– Elytra fully striate, or striation becoming gradually weaker towards apex; without or with only shallow transverse sulcus in basal third, in latter case elytra more depressed and rather wide; lateral margins of pronotum barely sinuate 11

11(10). Surface with dense, elongate, erect pilosity; tibiae dark with conspicuous yellow ring 12
– Surface without pilosity; tibiae uniformly yellow or dark, without yellow ring .... ***Dicraspeda*** **Chaudoir (part)**

12(11). Elytra with shallow transverse depression near apex; interval 7 tumid in apical third; antennomere 3 sparsely setose .... ***Neoeudalia*** **Baehr**
– Elytra without transverse depression near apex; interval 7 not tumid; antennomere 3 glabrous except for apical setae .... ***Tricharnhemia*** **Baehr**

13(7). Antennomere 3 impilose 14
– Antennomere 3 and usually also 1 and 2, pilose 17

14(13). Elytral striae impunctate, rather sulcate; head markedly triangular behind eyes .... ***Aulacolius*** **Sloane**
– Elytral striae punctate, not sulcate; head convex behind eyes 15

15(14). Lateral margin of pronotum medially bordered by a deep sulcus; **all** odd intervals with a row of many (> 10) erect setae .... ***Renneria*** **Baehr**
– Lateral margin of pronotum medially not bordered by a deep sulcus; at most elytral intervals 3,5, and 7 with a row of fewer (< 5) erect setae 16

16(15). Elytra ampliate, apicad considerably widened and externally angulate or spinose; antenna very elongate, antennomere 4 impilose .... ***Gestroania*** **Liebke**
– Elytra not ampliate, apicad not or little widened and externally not angulate; antenna shorter, antennomere 4 pilose .... ***Eudalia*** **Laporte**

17(13). Elytral striae almost complete; prothorax densely punctate and pilose on disc **18**
– Elytral striae incomplete, either only the basal third striate, or only two median striae complete; prothorax impunctate in the middle, either glabrous or coarsely, transversely rugose, impilose on disc .... ***Myrmecodemus* Sloane**
18(17). Lateral margin of pronotum medially bordered by a deep sulcus; elytral striae punctate and impressed .... ***Deipyrodes* Bousquet**
– Lateral margin of pronotum medially not bordered by a deep sulcus; elytral striae only coarsely punctate, but not impressed .... ***Lachnothorax* Motschulsky**

Anasis Laporte, 1867 (Fig. 12.73)

Type species. *Anasis howittii* Laporte, 1867, by monotypy.

Characteristics. Elongate, parallel-sided, markedly depressed beetles; head without paraorbital ridge; first three antennomeres impilose; lateral margin of pronotum distinct, without a marginal sulcus; elytra elongate, parallel, depressed, completely striate, microreticulate, impilose; apex transverse, not excised; 4th tarsomeres narrow, not bilobed; colour uniformly dark; male genitalia unknown; gonocoxite-1 with comparatively few and short setae at apex; gonocoxite-2 unusually elongate, with 3 ventral setae.

Australian species. Only *A. howittii* Laporte.

Distribution. Victoria, only known from the type locality, Geelong.

References. Laporte (1868); Csiki (1932); Liebke (1938); Moore *et al.* (1987); Baehr (2005b); Lorenz (2005).

Andrewesia Liebke, 1938 (Fig. 12.75)

Type species. *Odacantha apicalis* Chaudoir, 1872, by original designation.

Characteristics. Moderately elongate, fairly convex beetles; head convex, deeply constricted, without paraorbital ridge; first thee antennomeres impilose; lateral margin of pronotum indistinct, without a marginal sulcus; elytra rather short, fairly convex, completely striate, microreticulate, impilose; apex oblique, not excised; 4th tarsomeres deeply excised; elytra black with small apical spot; aedeagus elongate, straight, endophallus without sclerotised pieces; gonocoxite-1 with rather many elongate setae at apex; gonocoxite-2 fairly elongate, with three ventral setae.

Australian species. Only *A. australica* Baehr, 2009.

Distribution. Extreme north of Northern Territory; extraterritorial from southern mainland Asia through the Indonesian insular belt to the Moluccas.

Biology. The Australian species was collected 'on mud in saltmarsh' close to the sea.

References. Jedlicka (1963); Darlington (1968); Baehr (2005b, 2009e); Lorenz (2005).

Archicolliuris Liebke, 1931 (Pl. 19G)

Type species. *Casnoia bimaculata* Kollar & Redtenbacher, 1842, by original designation.

Colliuris DeGeer, 1774

Characteristics. Elongate, narrow beetles; head elongate, triangular, with a distinct paraorbital ridge; first three antennomeres impilose, antennomere 3 not markedly elongate; pronotum elongate, without distinct margin, without a marginal sulcus; elytra rather elongate, slightly widened apicad, incompletely striate, not microreticulate, impilose; apex oblique, slightly excised; 4th tarsomeres narrow, not excised; elytra black, variously spotted; aedeagus moderately elongate, straight, endophallus without distinctly sclerotised pieces; gonocoxite-1 with rather many moderately elongate setae at apex; gonocoxite-2 fairly short, with two ventral setae.

Australian species. Three species.

Distribution. Cape York Peninsula, north-eastern Queensland; extraterritorial in the Oriental Region and in New Guinea.

Biology. Probably hygrophilous at the banks of rivers, shores of lagoons, and in swamps. Usually only collected at light.

References. Csiki (1932); Liebke (1938); Jedlicka (1963); Habu (1967); Baehr (2005b, 2009e); Lorenz (2005).

Key to species. Baehr (2005b).

Aulacolius Sloane, 1923 (Pl. 19E)

Type species. *Aulacolius triordinatus* Sloane, 1923, by monotypy.

Characteristics. Rather stout beetles; head elongate, triangular, without a paraorbital ridge; first three antennomeres impilose; pronotum with distinct margin, without a marginal sulcus; elytra short and wide, incompletely striate but striae deeply sulcate, not microreticulate, impilose; odd intervals with several erect setae; apex oblique, not excised; 4th tarsomeres narrow, not excised; elytra black, with small apical spot; aedeagus narrow, elongate, with elongate, wide apex, endophallus without distinctly sclerotised pieces; gonocoxite-1 with rather many short setae at apex; gonocoxite-2 fairly elongate, with two small ventral setae.

Australian species. Only *A. triordinatus* Sloane.

Distribution. Northern parts of Northern Territory to Cape York Peninsula, north Queensland.

Biology. Most specimens were collected at light. Very likely hygrophilous.

References. Csiki (1932); Liebke (1938); Moore *et al.* (1987); Baehr (2005b, 2009e); Lorenz (2005).

Basistichus Sloane, 1917 (Pl. 19H)

Type species. *Odacantha micans* Macleay, 1864, by original designation.

Characteristics. Rather narrow beetles; head short, triangular, with produced eyes, with a paraorbital ridge; first three antennomeres impilose; pronotum with distinct margin, with a marginal sulcus; elytra rather elongate, parallel-sided, deeply impressed in basal third, incompletely striate, not microreticulate, impilose; apex oblique, slightly excised; 4th tarsomeres narrow, not excised; elytra black, paler in basal third; aedeagus narrow, elongate, with fairly wide apex, endophallus without distinctly sclerotised pieces; gonocoxite-1 with rather few short setae at apex; gonocoxite-2 fairly short, with two small ventral setae.

Australian species. Only *B. micans* (Macleay).

Distribution. Eastern Queensland, northern parts of Northern Territory, and extreme north of Western Australia; extraterritorial in New Guinea.

Biology. Hygrophilous and ground living at the banks of rivers, shores of lakes and lagoons, and in swamps. Apparently nocturnal. Sometimes collected from among pebbles.

References. Sloane (1923a); Csiki (1932); Liebke (1938); Darlington (1968); Moore *et al.* (1987); Baehr (2005b); Lorenz (2005).

Clarencia Sloane, 1917 (Pl. 21J)

Type species. *Casnonia aliena* Pascoe, 1860, by original designation.

Characteristics. Narrow and elongate beetles; head elongate, triangular, with a paraorbital ridge; first three antennomeres impilose, antennomere 3 very elongate; pronotum with distinct margin, without a marginal sulcus; elytra elongate, parallel-sided, incompletely striate, not microreticulate, very shortly pilose, odd intervals with erect setae; apex oblique, more or less deeply excised; 4th tarsomeres narrow, not excised; elytra black, variously spotted; aedeagus narrow, rather elongate, with short, slightly curved apex, endophallus without distinctly sclerotised pieces; gonocoxite-1 with rather few elongate setae at apex; gonocoxite-2 fairly short, with three ventral setae.

Australian species. Four species.

Distribution. Tropical northern Australia from northern Queensland to extreme north-eastern Western Australia; extraterritorial in New Guinea.

Biology. Hygrophilous, probably mainly on low vegetation at the banks of rivers, shores of lakes and lagoons, and in swamps. Probably nocturnal. Most commonly sampled at light.

References. Sloane (1923a); Csiki (1932); Liebke (1938); Darlington (1968); Moore *et al.* (1987); Baehr (2005b, 2009e); Lorenz (2005).

Key to species. Baehr (2005b).

Deipyrodes Bousquet, 2002 (Pl. 23E)

Type species. *Lachnothorax palustris* Sloane, 1910, by original designation.

Deipyrus Liebke, 1938 (*non Deipyrus* Champion, 1908)

Characteristics. Compact beetles; head rather elongate, triangular, with very narrow neck, without a paraorbital ridge; antenna pilose from antennomere 1; pronotum with distinct margin, with a deep marginal sulcus; elytra short and wide, completely striate, not microreticulate, densely pilose; apex convex; 4th tarsomeres narrow, not excised; elytra black, inconspicuously paler at apex; aedeagus narrow, rather elongate, with short, fairly wide apex, endophallus without distinctly sclerotised pieces; gonocoxite-1 with rather few elongate setae at apex; gonocoxite-2 fairly short, with two small ventral setae.

Australian species. Two species.

Distribution. North-eastern Queensland, northern parts of Northern Territory. The range of the second species (*D. inops* Baehr) is unknown.

Biology. Hygrophilous, at wet places, but habits little known. Most recorded specimens were collected at light.

References. Lorenz (2005); Baehr (2006a, 2009e); Liebke (1938); Csiki (1932); Moore *et al.* (1987); Baehr (2005b); Lorenz (2005).

Key to species. Baehr (2005b).

Dicraspeda Chaudoir, 1862 (Pl. 24G, J; Fig. 12.64)

Type species. *Dicraspeda brunnea* Chaudoir, 1862, by monotypy.

Macrocentra Chaudoir, 1869a

Loxocara Sloane, 1907b

Philemonia Liebke, 1938

Characteristics. Rather compact beetles; head short, triangular, with produced eyes, with a paraorbital ridge; first three antennomeres impilose; pronotum with distinct margin, with a variously shaped marginal sulcus; elytra varied, rather parallel-sided, completely striate, microreticulate or not, impilose; apex oblique, more or less distinctly excised, in extra-Australian species even bi- or quadridentate; 4th tarsomeres varied; elytra black or dark piceous; aedeagus narrow, rather elongate, with rather elongate, slightly curved apex, endophallus without distinctly sclerotised pieces; gonocoxite-1 with rather few short setae at apex; gonocoxite-2 fairly short, with 2–3 ventral setae.

Australian species. Nine species.

Distribution. Eastern and northern Australia; extraterritorial from the Moluccas through New Guinea to Solomon Islands.

Biology. Hygrophilous and ground living at the banks of rivers, shores of lakes and lagoons, and in swamps; in New Guinea, according to Darlington (1968), probably also away from water in lowland rainforest on the ground. Probably strictly nocturnal, because beetles are commonly attracted to light.

Note. This genus is quite diverse in its external character states, hence different authors have dismembered it into several genera. However, because intermediate character states are known in several aspects, we retain a single genus.

References. Sloane (1923a); Csiki (1932); Liebke (1938); Jedlicka (1963); Darlington (1968); Moore *et al.* (1987); Baehr (2000a, 2003c, 2005b, 2006a, 2009e, 2012g); Lorenz (2005).

Keys to species. Baehr (2003c, 2005b, 2006a, 2009e, 2012g).

Eudalia Laporte, 1867 (Pl. 25B)

Type species. *Odacantha latipennis* Macleay, 1864, by original designation.

Characteristics. Rather compact beetles; head short, triangular, with produced eyes, without a paraorbital ridge; first three antennomeres impilose; pronotum with distinct margin, without a marginal sulcus; elytra various, completely striate, microreticulate or not, pilose or impilose; apex oblique-convex, not excised; 4th tarsomeres narrow, not excised; elytra black or piceous, rarely with slightly paler apex; aedeagus narrow, rather elongate, with curved and/or upturned apex, endophallus without distinctly sclerotised pieces; gonocoxite-1 with numerous, very elongate setae at apex; gonocoxite-2 rather short, with 3–4 elongate ventral setae.

Australian species. Fourteen species and three subspecies.

Distribution. South-eastern, eastern and northern tropical Australia, one species was recently recorded from Tasmania.

Biology. Hygrophilous and ground living at the banks of rivers, shores of lakes and lagoons, and in swamps. Some found among pebbles and gravel on montane rivers and streams. Nocturnal. Beetles were observed devouring larvae and stranded adults of mayflies on a river bank.

References. Laporte (1868); Sloane (1917, 1923a); Csiki (1932); Liebke (1938); Darlington (1968); Moore *et al.* (1987); Baehr (2005b, 2006a, 2009e, 2016c); Lorenz (2005).

Keys to species. Baehr (2005b, 2009e, 2016c).

Gestroania Liebke, 1938 (Pl. 26B)

Type species. *Casnonia amplipennis* Gestro, 1875, by monotypy.

Characteristics. Rather compact beetles; head short, triangular, with produced eyes, without a paraorbital ridge; first three antennomeres impilose; pronotum with distinct margin, without a marginal sulcus; elytra short and rather wide, widened apicad, depressed, completely striate, microreticulate, impilose, but odd intervals usually with several erect setae; apex rather deeply excised, external angles dentate; 4th tarsomeres narrow, not excised; elytra black; aedeagus narrow, elongate, with short, barely curved apex, endophallus without distinctly sclerotised pieces; gonocoxite-1 with rather few short setae at apex; gonocoxite-2 short, with 2–3 ventral setae.

Australian species. Four species.

Distribution. Northern Australia from north Queensland to the Pilbara in northern Western Australia.

Biology. Hygrophilous and ground living at the banks of rivers, shores of lakes and lagoons, and other wet places. Probably nocturnal. One species has been found in large numbers on muddy ground around a shallow pool under cover of leaves and timber.

References. Csiki (1932); Moore *et al.* (1987); Baehr (2005b, 2009e); Lorenz (2005).

Key to species. Baehr (2005b).

Giachinoana Baehr, 2003 (Fig. 12.95)

Type species. *Giachinoana carinipennis* Baehr, 2003, by original designation.

Characteristics. Rather compact beetles; head short, triangular, with produced eyes, with a paraorbital ridge; first three antennomeres impilose; pronotum with distinct margin, without a marginal sulcus; elytra short and rather wide, parallel-sided, depressed, completely striate, microreticulate, impilose; odd intervals carinate; apex deeply excised, internal and external angles sharply dentate; 4th tarsomeres narrow, not excised; elytra black; aedeagus narrow, elongate, with short, barely curved apex, endophallus without distinctly sclerotised pieces; gonocoxite-1 with rather few short setae at apex; gonocoxite-2 short, with three ventral setae.

Australian species. Three species.

Distribution. Northern parts of Northern Territory, adjacent northernmost Western Australia; one species from the islands of the Pellew Group off the north coast of Northern Territory.

Biology. Hygrophilous and probably ground living at the banks of rivers, shores of lakes and lagoons, and other wet places. Apparently nocturnal.

References. Baehr (2003a, 2005b, 2009e, 2012g, 2016c); Lorenz (2005).

Key to species. Baehr (2012g, 2016c).

Lachnothorax Motschulsky, 1862 (Fig. 12.102)

Type species. *Lachnothorax biguttatus* Motschulsky, 1862, by monotypy.

Characteristics. Short and compact beetles; whole surface pilose; head rather short, triangular, with produced eyes, without a paraorbital ridge; antenna pilose from antennomere 1; pronotum with distinct margin, without a marginal sulcus; elytra short and wide, convex, completely striate, microreticulate, with dense pilosity; apex oblique, faintly excised; 4th tarsomeres narrow, not excised; elytra black, plagiate near apex; aedeagus narrow, elongate, curved, with rather short, curved apex, endophallus without distinctly sclerotised pieces; gonocoxite-1 with rather few moderate setae at apex; gonocoxite-2 rather short, with 2–3 ventral setae.

Australian species. Only *L. tokkia* Gestro, 1875.

Distribution. Central Cape York Peninsula, north Queensland; extraterritorial from mainland South Asia through the Indonesian and Philippine insular belts to New Guinea.

Biology. Hygrophilous and ground living at the banks of rivers, shores of lakes and lagoons, and other wet places. Probably nocturnal.

References. Sloane (1917, 1923a); Csiki (1932); Liebke (1938); Jedlicka (1963); Darlington (1968); Baehr (2000a, 2005b); Lorenz (2005).

Myrmecodemus Sloane, 1923 (Pl. 30B)

Type species. *Casnonia riverinae* Sloane, 1890, by original designation.

Characteristics. Small beetles; head short, triangular, posteriorly strongly narrowed, with produced eyes, without a paraorbital ridge; antenna pilose from antennomere 1; pronotum with distinct margin, without a marginal sulcus; elytra short, oval-shaped, incompletely striate, with variable microreticulation, pilose or not; apex oblique, barely excised; 4th tarsomeres narrow, not excised; elytra with various colour patterns; aedeagus narrow, elongate, with elongate, widened apex, endophallus without distinctly sclerotised pieces; gonocoxite-1 with few very short and stout setae at apex; gonocoxite-2 short, without or with two ventral setae.

Australian species. Five species.

Distribution. Northern Australia from lower Cape York Peninsula through northern parts of Northern Territory and Western Australia; also in the interior of New South Wales and in northern Victoria.

Biology. Hygrophilous and ground living at the banks of rivers, shores of lakes and lagoons, and other wet places; one species found in reedy swamps beside large inland rivers. Predacious and probably nocturnal.

Note. The genus includes two subgenera which differ in the pilosity of head, structure of the surface of the pronotum, and the striation of the elytra.

References. Csiki (1932); Liebke (1938); Moore *et al.* (1987); Baehr (2005b, 2009e); Lorenz (2005).

Key to species. Baehr (2005b).

Key to the subgenera of *Myrmecodemus*

1. Head, pronotum, and elytra (except for fixed setae) glabrous; pronotum without transverse wrinkles; elytra without striae in apical two-thirds .... ***Myrmecodemus* Sloane**

– Head densely punctate and pilose; pronotum with coarse transverse wrinkles in the middle; elytra with very deep, punctate and pilose impressions in basal third, with two punctate and pilose striae along suture .... ***Trichodemus* Baehr**

Subgenus *Myrmecodemus* Sloane, 1923

Type species. *Casnonia riverinae* Sloane, 1890, by original designation.

Characteristics. Surface glabrous, impilose; pronotum not rugose; apical two-thirds of the elytra not striate; gonocoxite-2 with two ventral setae.

Australian species. Four species.

Distribution. Northern Australia from lower Cape York Peninsula through northern parts of Northern Territory and Western Australia; also in the interior of New South Wales and in northern Victoria.

References. Csiki (1932); Liebke (1938); Moore *et al.* (1987); Baehr (2005b, 2009e); Lorenz (2005).

Subgenus *Trichodemus* Baehr, 2005

Type species. *Myrmecodemus pilosellus* Baehr, 2005, by original designation.

Characteristics. Head punctate, pilose; pronotum rugose, two median elytral striae present; gonocoxite-2 without ventral setae.

Australian species. Only *M.* (*Trichodemus*) *pilosellus* Baehr.

Distribution. Lower Cape York Peninsula, north Queensland.

References. Lorenz (2005).

Neoeudalia Baehr, 2005 (Pl. 31E)

Type species. *Eudalia nigra* Sloane, 1910, by original designation.

Characteristics. Rather compact beetles; head short, triangular, with produced eyes, with a paraorbital ridge; first two antennomeres impilose, antennomere 3 sparsely pilose; pronotum with distinct margin, with a deep marginal sulcus; elytra rather short and wide, convex, completely striate, not microreticulate, pilose; with a shallow transverse depression near apex and with tumid 7th interval in the apical third; apex oblique-convex, not excised; 4th tarsomeres narrow, not excised; elytra black; aedeagus narrow, elongate, with fairly elongate, slightly curved apex, endophallus without distinctly sclerotised pieces;

gonocoxite-1 with rather few very elongate setae at apex; gonocoxite-2 short, with 3 ventral setae.

Australian species. Only *N. nigra* (Sloane).

Distribution. Large parts of eastern and northern Australia.

Biology. Hygrophilous and ground living at the banks of rivers, shores of lakes and lagoons, and other wet places. Probably nocturnal.

References. Lorenz (2005); Baehr (2009e).

Ophionea Klug, 1821 (Pl. 32D)

Type species. *Attelabus indicus* Thunberg, 1784, by subsequent designation of Hope 1834.

Ophionaea Eschscholtz, 1829

Casnoidea Laporte, 1834

Characteristics. Narrow, elongate beetles; head various, with rather produced eyes, with a paraorbital ridge; first three antennomeres impilose; pronotum tubular, without distinct margin, without a marginal sulcus; elytra elongate, parallel-sided, completely striate, microreticulate; apex oblique, barely excised; 4th tarsomeres wide, deeply excised; elytra black or blue and red, with white spots; aedeagus rather short to moderately elongate, with short, usually knobbed and asymmetric apex, endophallus without distinctly sclerotised pieces; gonocoxite-1 with a varied number of moderately elongate setae at apex; gonocoxite-2 elongate, with 2–4 ventral setae of varied size.

Australian species. Five species.

Distribution. Eastern and northern Australia; extraterritorial in the Oriental Region and from New Guinea to Solomon Islands.

Biology. Hygrophilous in and on low vegetation near water. Nocturnal, commonly attracted to light.

Note. The Oriental species *Ophionea indica* (Thunberg) is an adventive species which was recorded at the harbour of Geraldton on a Japanese ship.

References. Sloane (1917, 1923a); Csiki (1932); Liebke (1938); Habu (1967); Baehr (2005b, 2009e); Lorenz (2005); Laporte (1834); Csiki (1932); Jedlicka (1963); Darlington (1968); Moore *et al.* (1987); Baehr (1996b).

Keys to species. Baehr (1996b, 2005b).

Porocara Sloane, 1917 (Pl. 35J)

Type species. *Porocara punctata* Sloane, 1917, by monotypy.

Characteristics. Compact beetles; head large, wide, with wide base and produced eyes, without a paraorbital ridge; first three antennomeres impilose; pronotum quadrangular, coarsely punctate, with distinct margin, without a marginal sulcus; elytra short and wide, very compact, completely striate, microreticulate, impilose; apex convex; 4th tarsomeres narrow, not excised; elytra brick-red, with dark, serrate-cruciate pattern; aedeagus narrow, elongate, with elongate, slightly knobbed or upturned apex, endophallus without distinctly sclerotised pieces; gonocoxite-1 with rather few moderately elongate setae at apex; gonocoxite-2 rather elongate, with 2 small ventral setae.

Australian species. Five species and two subspecies.

Distribution. Northern Australia from north-eastern Queensland to Gascoyne River in mid-Western Australia.

Biology. Hygrophilous and ground living on sandy or pebbly banks of rivers. Strictly nocturnal and extremely agile runners. Beetles were observed hunting larvae and adults of mayflies on a riverbank.

Note. Species of this genus may represent the earliest branching lineage of Odacanthines. Their unusual habits that resemble those of Chlaeniines are consistent with this hypothesis.

References. Sloane (1923a); Csiki (1932); Liebke (1938); Baehr (1986b, 1996a, 2005b, 2009e); Moore *et al.* (1987); Lorenz (2005).

Keys to species. Baehr (1986b, 1996a, 2005b, 2009e).

Renneria Baehr, 1999 (Pl. 36I)

Type species. *Renneria kamouni* Baehr, 1999, by original designation.

Characteristics. Short and wide beetles; head triangular, with produced eyes, without a distinct paraorbital ridge; first three antennomeres impilose; pronotum quadrangular, coarsely punctate, with distinct margin, with a deep marginal sulcus; elytra short and wide, widened apicad, depressed, completely striate, microreticulate, impilose; odd intervals plurisetose; apex oblique, slightly excised; 4th tarsomeres narrow, not excised; elytra black or dark piceous; aedeagus very narrow and elongate, with fairly elongate, curved apex, endophallus without distinctly sclerotised pieces; gonocoxite-1 with rather few moderately elongate setae at apex; gonocoxite-2 fairly elongate, with 2 small ventral setae.

Australian species. Only *R. kamouni* Baehr.

Distribution. Northern parts and interior of Northern Territory, northern Western Australia to the northern margin of the Pilbara Region.

Biology. Hygrophilous and ground living at the banks of rivers and shores of lakes and lagoons. Apparently nocturnal, as almost all known specimens were sampled at light.

References. Baehr (2005b, 2009e); Lorenz (2005).

Tricharnhemia Baehr, 2009 (Fig. 12.154)

Type species. *Tricharnhemia browni* Baehr, 2009, by original designation.

Characteristics. Compact beetles; head large, moderately triangular, with a paraorbital ridge; first three antennomeres impilose; pronotum with distinct margin, with a deep marginal sulcus; elytra moderately elongate, compact, completely striate, not microreticulate, pilose; apex oblique, barely excised; 4th tarsomeres narrow, not excised; elytra black; aedeagus narrow, elongate, with short, barely curved apex, endophallus without distinctly sclerotised pieces; gonocoxite-1 with rather few elongate setae at apex; gonocoxite-2 elongate, with two ventral setae.

Australian species. Only *T. browni* Baehr.

Distribution. Arnhem Land, extreme north of Northern Territory.

Biology. Unknown, but most probably hygrophilous and ground living.

Note. This species is only known from the holotype.

References. Baehr (2009e).

Tribe Pentagonicini Bates, 1873

Characteristics. Procoxal cavities closed; mesocoxae conjunct; mandibles flat and relatively wide, without distinct scrobe; eyes large or very large, very convex; pronotum transverse, commonly rather pentagonal, usually laterally markedly angulate; elytra at apex transverse or slightly sinuate, usually not completely covering the abdomen; antenna elongate, pilose from antennomere 4 or 5; legs slender and elongate; claws simple; male genitalia with the ostium on the upper surface, usually without distinctly sclerotised parts on the endophallus; parameres asetose at apex; female gonocoxite-1 usually with some setae at the apical rim, gonocoxite-2 more or less elongate and curved, with 2–5 ensiform setae at the latero-ventral surface and with one or two short nematiform setae near the apex.

Note. Liebherr (1988) in the course of a cladistic analysis united Pentagonicini with Odacanthini and included these in a supertribe Odacanthitae based on a shared, derived bipartite spermatheca. A placement that is also supported by characters of the larva of one species (*Scopodes simplex* Blackburn), described by Moore (1966c) as similar to known odacanthine larvae. Also supporting this relationship is an analysis of DNA sequence data that placed them as sister taxa (Ober & Maddison 2008), though with modest support. Ball & Bousquet (2001) placed Pentagonicini close to Oodini, stressing the differences of the defensive secretions of the tribes. The chemical data is derived from one species, *Scopodes boops* Erichson that produces a saturated acid (isovaleric acid; Moore 1979), as compared to formic acid that was found in two species of odacanthine (*Colliuris pensylvanicus* L. and *Odacantha melanura* L.; Will *et al.* 2000; Schildknecht *et al.* 1968). Isovaleric acid has been found in 14 different tribes ranging from Opisthiini to Anthiini, but this compound is not known from Odacanthini or Oodini. *Pentagonica picticornis* Bates is now known to produce formic acid (Will & Attygalle unpubl.) and also secondary compounds shared with odacanthines and *Scopodes*. In sum, the evidence is significant that Odacanthini and Pentagonicini are closely related. However, there is currently no consensus on the relationship between them or their reciprocal monophyly and so we keep them as separate tribes.

Key to the Australian genera of Pentagonicini

1. Pronotum decidedly pentagonal; surface of elytra neither sericeous, nor metallic, nor with foveate setiferous punctures .... ***Pentagonica* Schmidt-Göbel**

– Pronotum not decidedly pentagonal; surface of elytra either sericeous, or metallic, or with foveate setiferous punctures **2**

2(1). Eyes not very large; elytra with setiferous punctures encircled by a light ring .... ***Homethes* Newman**

– Eyes very large; elytra with setiferous punctures not encircled by a light ring but sometimes brightly coloured **3**

3(2). Elytra metallic though not sericeous, regularly punctate-striate, without conspicuous, foveate punctures on interval; mentum with short and wide, bidentate tooth; glossa not swollen; male protarsus with biseriate squamous setae .... ***Parascopodes* Darlington**

– Elytra either sericeous, or striate, but then striae rarely markedly punctate, and usually with conspicuous, foveate punctures on interval 3; mentum without distinct tooth; glossa swollen; male protarsus with uniseriate squamous setae .... ***Scopodes* Erichson**

Homethes Newman, 1842 (Pl. 27E)

Type species. *Homethes elegans* Newman, 1842, by monotypy.

Characteristics. Elongate, depressed beetles; base of head; not markedly narrowed; eyes moderately large and protruded; lateral margins of the pronotum not deeply sinuate; surface of elytra very sericeous; setiferous punctures on the elytra encircled by distinct pale rings; aedeagus without distinctly sclerotised pieces on the endophallus; apical rim of gonocoxite-1 setose; gonocoxite-2 fairly elongate and rather acute, with two large ensiform setae on the latero-ventral rim.

Australian species. Nine species.

Distribution. Southern and eastern Australia from southern Western Australia to about Mackay in central eastern Queensland, also sporadically found in the interior, Tasmania; extraterritorial on the Moluccas.

Biology. Ground living, but sometimes also in bark fissures at the base of tree trunks in a variety of habitats, from closed forest through open woodland to mallee. Predacious and probably nocturnal. Almost nothing else is known about diet, habits, and reproduction.

Notes. In the catalogue of Moore *et al.* (1987) the genus is still included in Platynini, but Liebherr (1991) demonstrated that *Homethes* and the related Oriental genus *Aeolodermus* Andrewes, 1929 belong to Pentagonicini. This genus is much in need of a revision.

References. Blackburn (1892); Sloane (1920); Csiki (1931); Darlington (1956); Moore *et al.* (1987); Liebherr (1991); Lorenz (2005); Baehr (2010g, 2012a).

Key to species. Sloane (1920), but this key applies only to Tasmanian species.

Pentagonica Schmidt-Göbel, 1846 (Pl. 33C)

Type species. *Pentagonica ruficollis* Schmidt-Göbel, 1846, by subsequent designation by Andrewes (1939).

Rhombodera Reiche, 1842
Didetus LeConte, 1853
Elliota Nietner, 1856
Trichothorax Montrouzier, 1860
Xenothorax Wollaston, 1867
Wakefieldia Broun, 1880

Characteristics. Variously sized beetles; mentum edentate; glossa and paraglossae not separated; eyes very large and, strongly produced laterally; neck narrow; prothorax markedly pentagonal; elytra neither sericeous, nor metallic; setiferous punctures on interval 3 small, inconspicuous; aedeagus mostly without sclerotised parts on the endophallus; apical rim of gonocoxite-1 markedly setose, gonocoxite-2 short, obtuse, with several short ensiform setae on the ventral surface.

Australian species. 13 species, and the occurrence of one additional species is doubtful.

Distribution. Eastern and northern Australia, Tasmania, Lord Howe and Norfolk Islands; extraterritorial almost worldwide in tropical and subtropical regions.

Biology. Ground living and on tree trunks in a variety of habitats, from rainforest to open woodland. Predacious and apparently nocturnal. Darlington (1968) mentioned New Guinean species as being collected 'by sweeping low vegetation'. Most Australian specimens, however, apparently were collected at light, by sieving leaf litter, or in pitfall traps.

References. Chaudoir (1877); Sloane (1898); Dupuis (1913); Csiki (1932); Britton (1941); Jedlicka (1963); Habu (1967); Darlington (1968); Moore (1985, 1992); Moore *et al.* (1987); Larochelle & Larivière (2001, 2007); Bousquet (2003); Lorenz (2005); Baehr (2012a).

Key to species. Baehr (2012a).

Parascopodes Darlington, 1968 (Pl. 33L)

Type species. *Scopodes cyaneus* Sloane, 1907, by original designation.

Characteristics. Mentum bidentate; glossa short and not swollen; paraglossae elongate; eyes extremely large, strongly produced laterally; prothorax not markedly pentagonal; elytra bright metallic; setiferous punctures on interval 3 small, inconspicuous; male protarsus with biseriate squamose setae beneath; ventral and lateral surfaces of the aedeagus conspicuously striolate; apical rim of gonocoxite-1 not extensively setose; gonocoxite-2 fairly elongate and rather acute, with two large ensiform setae on the latero-ventral rim.

Australian species. Two species and two subspecies.

Distribution. North-eastern Queensland including Cape York Peninsula and Torres Strait Islands; extraterritorial in Papua New Guinea.

Biology. Ground living in rainforest and tropical Savannah Woodland. Predacious, but unknown whether diurnal or nocturnal. Darlington (1968) noted a specimen as being collected 'in flooded grassland'.

References. Moore *et al.* (1987); Lorenz (2005); Baehr (2010g).

Key to species. Baehr (2010g).

Scopodes Erichson, 1842 (Pl. 37H)

Type species. *Scopodes boops*, Erichson, 1842, by monotypy.

Molpus Newman, 1842
Helaeotrechus White, 1846
Periblepusa Redtenbacher, 1867

Characteristics. Mentum edentate; glossa short and swollen; paraglossae elongate; eyes extremely large, strongly produced laterally; prothorax not markedly pentagonal; elytra variously coloured; setiferous punctures on interval 3 usually foveate and conspicuous; male protarsus with uniseriate squamose setae beneath; lower surface of the aedeagus not striolate; apical rim of gonocoxite-1 not extensively setose; gonocoxite-2 fairly elongate and rather acute, with two large ensiform setae on the latero-ventral rim.

Australian species. 55 species and two subspecies.

Distribution. Almost the whole of Australia, including the interior, Tasmania, and Lord Howe Island; extraterritorial in New Guinea, New Britain, New Ireland, New Caledonia, New Zealand, and Java.

Biology. Ground living and on the trunks of trees and on logs in various habitats. Tropical rainforests, various types of open forests and woodlands, bogs, banks of rivers and creeks, shores of lakes and lagoons, and in semiarid areas on wet ground around temporary pools. Probably mostly diurnal, as certain species have been seen running in bright sunshine on wet ground, in bare dirt between grass, around temporary pools or on logs in forests. Very rarely seen at light. Beetles have been observed hunting small flies at the margin of a temporary pool. Little else is known about habits, diet, and reproduction.The larva of *Scopodes simplex* Blackburn was described by (Moore 1966c).

References. Lacordaire (1854); Blackburn (1894a, 1894b, 1895); Fauvel (1903); Sloane (1903, 1907a, 1920); Csiki (1932); Andrewes (1933); Britton (1941); Darlington (1968); Moore (1962, 1992); Moore *et al.* (1987); Matthews (1980); Baehr (1994d, 2010g); Larochelle & Larivière (2001); Lorenz (2005).
Key to species. Baehr (2010g).

Tribe Panagaeini Bonelli, 1810

Characteristics. Small to large beetles with very prominent eyes; convex form; frequently black with bold patterns of red or yellow, or solid metallic; pronotum and elytral striae deeply and coarsely punctate; male protarsi with two basal tarsomeres slightly expanded and squamose beneath or unmodified.

Key to the Australian genera of Panagaeini

1. Labrum with medial pair of setae positioned notably forward of lateral pair; paraglossae not prolonged beyond apex of ligula; colour solid bluish or greenish black with contrastingly pale rufous legs .... ***Trichisia*** **Motschulsky**
– Labrum with medial pair of setae not notably forward of lateral pair; paraglossae prolonged well beyond apex of ligula; usually with yellow or red spots, at least the pronotum laterally paler near the base **2**
2(1). Lateral margins of the pronotum and legs, at least femur, pale; metatarsomere 4 deeply bilobed or emarginate .... ***Adischissus*** **Fedorenko**
– Lateral margins of the pronotum and legs dark, black or piceus; metatarsomere 4 emarginate, but not bilobed **3**
3(2). Mentum tooth narrow, elongate; mandibles elongate .... ***Tinognathus*** **Chaudoir**
– Mentum tooth broad and truncate; mandibles short and abruptly curved .... ***Craspedophorus*** **Hope**

Craspedophorus Hope, 1838 (Pl. 2C, 22G)

Type species. *Carabus reflexus* Fabricius, 1781 (not *C. reflexus* Fabricius, 1801), by subsequent designation by Basilewsky, 1953.
Camptoderus Hope 1838
Eudema Laporte 1840
Epicosmus Chaudoir 1846
Isotarsus Laferté-Sénectere 1851
Brachyonychus Chaudoir 1879
Acanthocosmus Jeannel 1949
Brachycosmus Jeannel 1949
Characteristics. Medium to large beetles with elytra boldly marked with four yellow or reddish-yellow maculae; male protarsi unmodified; flight wing reduced.
Australian species. 12 species.
Distribution. Members of this genus can be found in tropical regions of the Old World from Africa to the eastern Palearctic, through Asia south to New Guinea and the Australian continent.
Biology. Nocturnal, dry forests.
References. Sloane (1903); Kirschenhofer (2000); Häckel & Farkač (2012).
Keys to species. Sloane (1923a).

Adischissus Fedorenko, 2015 (Pl. 24C, 29L)

Type species. *Carabus notulatus* Fabricius, 1801, by original designation.
Characteristics. Small to medium size panagaeine beetles with four pale maculae on dark elytra; pale explanate pronotal margins; tarsomere 4 bilobed, flight wing full; antennal scape clearly longer than width of eye in dorsal view.
Australian species. Two species.
Distribution. The genus is distributed from Africa, across Asia and south through New Guinea to Australia. In Australia, found in the north-east coastal region of Queensland including the Cape York Peninsula.
Biology. Predatory, capable flier, open forest and grassland.
Note. Fedorenko (2015) provisionally moved *Microcosmodes quadrimaculatus* (Csiki) (Pl. 29L) to *Adischissus*.
References. Fedorenko (2015); Darlington (1968).

Tinognathus Chaudoir, 1879 (Fig. 12.146)

Type species. *Tinognathus parviceps* Chaudoir, 1878, by monotypy.
Characteristics. Around 11 mm long, head constricted behind eyes, eyes very prominent; mentum tooth narrow, elongate; mandibles relatively long and straight; elytral faint red maculae.
Australian species. Only *Tinognathus parviceps* Chaudoir.
Distribution. South coastal Queensland and north-east coastal New South Wales.
Biology. Flightless, predacious.

Trichisia Motschulsky, 1864 (Pl. 39B)

Type species. *Trichisia cyanescens* Motschulsky, 1864, by monotypy.
Characteristics. Medium size beetles (9–12 mm); striking blue to purple metallic colour, legs contrastingly pale.
Australian species. Only *Trichisia azurea* (Chaudoir, 1861).
Distribution. The genus is distributed from Africa, across Asia and south through New Guinea to Australia. In Australia found in north-east coastal Queensland around Innisfail, south to northern New South Wales.

Biology. Flight capable, nocturnal, predacious.

References. Moore (1979).

Tribe Chlaeniini Brullé, 1834

Characteristics. Medium to large sized beetles that are typically metallic coloured (frequently brilliantly so) and in most notably pubescent; procoxal cavities closed; mesocoxae conjunct; head with a single supraorbital seta above each eye; labial palps fusiform; mandible without a seta in the scrobe; mental tooth present, emarginate or entire; elytra oval-shaped, with or without visible plica; apex barely excised; flight wings usually full; aedeagus with ostium situated largely on the dorsal surface, parameres large, rounded, of almost similar size and shape, asetose; gonocoxite-2 short to moderately elongate, obtuse at apex, without or with few ensiform and/or nematiform setae.

Key to the Australian genera of Chlaeniini

1. Head shallowly rugulose, not punctate; pronotum with few, shallow scattered punctures; prosternum without punctures anterior to coxae .... *Hololeius* **LaFerté-Sénectère**
– Head punctate; pronotum coarsely, densely punctate; prosternum with punctures anterior to coxae .... ***Chlaenius*** **Bonelli**

Chlaenius Bonelli, 1810 (Pl. 21C, I, K; Figs 12.65, 12.66)

Type species. *Carabus festivus* Fabricius, 1810, by subsequent designation (Madge 1975).

Characteristics. Medium to large sized beetles variously metallic coloured, many with pale elytral maculations; elytra usually extensively pubescent, only marginally in *C. greyanus* White (Fig. 12.65–66); flight wings full.

Australian species. Nine species.

Distribution. The genus has a cosmopolitan distribution. In Australia they are found across the continent. Nearly half of the species in Australia are extralimital from Asia and the Pacific.

Biology. Hygrophilous, strong fliers, predacious but probably also scavenge.

References. Darlington (1968); Moore (1979).

Notes. The Australian species are arrayed in five different subgenera in current classifications (Lorenz 2005).

Keys to species. Sloane (1910b).

Hololeius LaFerté-Sénectère, 1851 (Pl. 12.99)

Type species. *Chlaenius nitidulus* Dejean, 1826, by monotypy. *Hololius* Chaudoir 1857

Characteristics. An ovate and convex beetles that lack apparent microsculpture on the pronotum and elytra and so are very shiny and metallic green. Elytral pubescence sparse.

Australian species. Only *Hololeius ceylanicus* (Nietner, 1856).

Distribution. The genus is found in the Oriental region, south through New Guinea and in Australia coastally in Victoria, New South Wales, Queensland, and the Northern Territory.

Biology. Under leaf litter and other cover along riparian flood zones and backwaters. Strong flier, attracted to lights at night.

References. Darlington (1968); Sloane (1910b).

Tribe Oodini LaFerté-Sénectère, 1851

Characteristics. Small to medium-sized beetles that are typically brown or black though some are metallic, in a few cases brilliantly so. The typical body form is elongate-ovate, ovate, or even nearly round. Head with one or two pairs of supraorbital setae. Elytral stria 8 is distinctly deepened and in its apical part (~1/3) delimited internally by a strong carina reaching suture (or almost so). The epipleura are crossed near apex; metepisternum is laterally coadunate with the elytral epipleuron. The Oodines are general glabrous, setation and punctuation is often reduced or absent. They are strongly hygrophilous, some even semiaquatic, but many are found in wet soil or wet leaf litter in forests far from bodies of water. Flight wing is fully developed in most, but frequently reduced in terrestrial, leaf litter species.

Notes. Key to genera given here from Borislav Guéorguiev (in litt.). Generic limits for the tribe as a whole, including many of the Australian genera, are not clearly supported by character evidence. A revision at the world level is much needed.

Key to the Australian genera of Oodini

1. Abdominal ventrites 4–5 with paramedial ambulatory setae **2**
– Abdominal ventrites 4–5 without paramedial ambulatory setae .... ***Coptocarpus*** **Chaudoir**
2(1). Clypeus with one pair of setae; three male protarsomeres with ventral pads of setae .... ***Oodes*** **Bonelli**
– Clypeus without setae; two male protarsomeres with ventral pads of setae **3**

3(2). Submentum with one pair of setae; mesocoxa without posterior seta (but lateral spike presenting); male ventrite VI lacking setae; small size beetles, 5.7 mm or less in length .... ***Anatrichis* LeConte**
– Submentum with two pairs of setae; mesocoxa with posterior seta (besides the lateral spike); male ventrite VI with two setae; relatively larger size beetles, 5.7 mm or more in length .... ***Nanodiodes* Bousquet**

Coptocarpus Chaudoir, 1857 (Pl. 22B)

Type species. *Oodes australis* Dejean, 1831, by original designation.

Characteristics. Medium to large oodine beetles; clypeus with one or no setae at lateral corners; submentum with one seta at the lateral edge; without ambulatory setae on the abdominal sterna; flight wing full or reduced; three male protarsomeres in variously expanded, sometimes remarkably wide and asymmetrical, ventrally with dense pads of setae having a spongy appearance, ventral setae only on a third of tarsomere 3.

Australian species. 14 species.

Distribution. Coastal regions in eastern Victoria and New South Wales, tropical Queensland, Northern Territory, and Western Australia, and far south-western Western Australia. One additional species is found in New Guinea.

Biology. Predaceous, hygrophilous or in more terrestrial rainforests or mesic forest habitats.

References. Sloane (1910b); Bousquet (2012); Moore (1979); Staunton *et al.* (2014); Baehr (2017c).

Keys to species. Erwin (1974b); Baehr (2017c).

Oodes Bonelli, 1810 (Pl. 32C; Fig. 12.31)

Type species. *Carabus helopioiodes* Fabricius, 1792, by subsequent designation by Samouelle, 1819.

Characteristics. Small to large oodine beetles, typically black or with aeneus lustre; male protarsomeres ventrally with dense pads of setae having a spongy appearance.

Australian species. 13 species.

Distribution. The genus is found in all biogeographical regions worldwide.

Biology. Predaceous. Very hygrophilous, sometimes semi-aquatic beetles typically found in swampy areas and riparian flood-zones near pools with well vegetated shorelines.

References. Moore (1979).

Keys to species. Sloane (1910b).

Anatrichis LeConte, 1853 (Fig. 12.74)

Type species. *Oodes minutus* Dejean, 1831, by original designation.

Characteristics. Small, black oodine beetles without setae on the edge of the clypeus, submentum with a single seta at the lateral edge; mesocoxa without a posteriomedial seta; metatrochanter without a seta; male protarsi with two or three basal tarsomeres slightly expanded and squamose beneath.

Australian species. Only *A. pusilla* Sloane, 1910.

Distribution. The genus is found extraterritorially in the Nearctic, Neotropical, and Oriental Regions in temperate and tropical areas.

Biology. Predaceous, hygrophilous.

References. Darlington (1968); Erwin (1974b); Spence (1983); Guéorguiev (2014); Guéorguiev & Morita (2017).

Nanodiodes Bousquet, 1996 (Pl. 30K)

Type species. *Oodes piceus* Nietner, 1856, by original designation.

Nanodes Habu, 1956 (not Schönherr, 1825).

Characteristics. Small, black or slightly aeneous oodine beetles without setae on the edge of the clypeus, submentum with a pair of setae at the lateral edge; mesocoxa with a posterior seta; metatrochanter without a seta; male protarsi with two or three basal tarsomeres slightly expanded and squamose beneath.

Australian species. Three species.

Distribution. The genus is found from southern Asia to Australia, where it is known from coastal regions throughout the continent.

Biology. Predaceous, hygrophilous.

References. Bousquet (1996); Will (2015b).

Keys to species. Sloane (1910b) as *Anatrichis* LeConte.

Tribe Licinini Bonelli, 1810

Characteristics. Small to large size beetles; procoxal cavities are closed, mesocoxa conjunct; protibia anisochaete with a well developed antennal cleaning groove; mandibular scrobe lacking seta; elytral plica absent, apical margin not truncate; flight wing full or very frequently, variously reduced in size or absent; abdomen of the harpalidian type (Deuve 1993); aedeagus with moderately well developed basal bulb, foramen small or base open; aedeagus with parameres glabrous, highly variable in shape, but typically dissimilar in form with the left larger; female reproductive tract usually with gland on spermatheca or bursa; gonocoxites with a few setae or glabrous, usually gonocoxite-2 elongate.

Key to the Australian genera of Licinini

1.	One mandible deeply notched with a prominent tubercle behind the notch	2
–	Neither mandible with notch; tubercle lacking	4
2(1).	Left mandible notched; only pedicle of antenna glabrous	***Badister*** **Clairville**
–	Right mandible notched; first three antennomeres glabrous	3
3(2).	Elytra without lateroapical spine; small size, 4–7 mm	***Physolaesthus*** **Chaudoir**
–	Elytra with lateroapical spines; large size, 11–15 mm	***Omestes*** **Andrewes**
4(1).	Apex of mandible bidentate	5
–	Apex of left mandible simple, right mandible with subapical tooth on inner margin	***Dicrochile*** **Guérin-Méneville**
5(4).	Antenna pubescent from antennomere 3	***Lestignathus*** **Erichson**
–	Antenna pubescent from antennomere 4 (some scattered pubescence may be present at the extreme apical end of 3 in *Platylytron*)	6
6(5).	Metepisternum short, lateral and anterior edges equal in length or anterior edge longer	7
–	Metepisternum elongate, lateral edge slightly or much longer than anterior edge	8
7(6).	Labrum very deeply emarginate; prosternal process not margined	***Lacordairia*** **Laporte**
–	Labrum very shallowly sinuate emarginate with a low medial tooth or nearly straight; prosternal process margined at least laterally	***Siagonyx*** **Macleay**
8(6).	Apical margin of elytra notably sinuate	***Hormacrus*** **Sloane**
–	Apical margin of elytra rounded or very slightly sinuate	9
9(8).	Antennomere 4 shorter than 3 and 5; large size, 18–20 mm	***Platylytron*** **Macleay**
–	Antennomere 4 longer than 3 and nearly equal to length of 5; smaller size	10
10(9).	Reflexed margin of pronotum, lateral margins of elytra and the region around each dorsal setae contrasting pale on the black ground colour; labrum slightly emarginate	***Dilonchus*** **Andrewes**
–	Without the colour pattern as above; margins sometimes paler, but not so distinctly contrasting and narrowly pale; labrum typically deeply emarginate	***Microferonia*** **Blackburn**

Badister Clairville, 1806 (Pl. 103I)

Type species. *Carabus bipustulatus* Fabricius, 1792 (= *Carabus bullatus* Schrank, 1798), by monotypy.

Characteristics. Small beetles (5–6 mm), black, slightly iridescent, prothorax and elytra paler at the lateral margins; elongate ovate form.

Australian species. Only *Badister (Baudia) sundaicus* Andrewes, 1926.

Distribution. Distributed worldwide. *Badister sundaicus*, which is widespread in the Pacific region, is only found in Australia in north-east costal Queensland.

Biology. Strong fliers, predacious. Found among leaf litter and vegetation in very wet areas and swamps.

References. Moore (1979); Darlington (1968).

Dicrochile Guérin-Méneville, 1846 (Pl. 24E)

Type species. *Rhembus goryi* Boisduval, 1832, by subsequent designation by Moore (1985).

Dicronochilus Guérin-Méneville, 1846

Stomatocoelus W. J. Macleay, 1864

Pedalopia Laporte, 1867

Characteristics. Medium size beetles, black, either shiny iridescent or dull from microsculpturing; broad elytra and depressed form; pronotum typically sub-orbiculate; palpomeres fusiform.

Australian species. 10 species.

Distribution. The genus is found in New Zealand, New Guinea, Moluccas, Solomons, New Caledonia, Tasmania, Norfolk Island, and in coastal regions in continental Australia, further inland in the south-east.

Biology. Flight capable or flightless, usually hygrophilous, wet forests, or more open habitat near wetlands.

References. Moore & Wallbank (1968); Ball (1959).

Dilonchus Andrewes, 1936 (Pl. 24F)

Type species. *Dilonchus bidens* Andrewes, 1936, by monotypy.

Characteristics. Small, lightly-built beetles; brownish-black with contrastingly pale legs and lateral margins of the pronotum and elytra.

Australian species. Only *Dilonchus pictus* Darlington, 1963.

Distribution. The genus is known from one species in India and *D. pictus* described from north-east coastal Queensland.

Biology. Flight capable, predacious, rainforest, leaf litter.

References. Darlington (1963b).

Hormacrus Sloane, 1898 (Pl. 27F)

Type species. *Hormacrus latus* Sloane 1898, by monotypy

Characteristics. Medium size black, shiny beetles, form wide, especially across the elytra; body little convex; head and pronotum relatively small; elytra striate, third interval with two punctures.

Australian species. Two species.

Distribution. Endemic, found only in south-west coastal Western Australia and gulf region of South Australia.

Biology. Flightless, predacious, found in temperate coastal forest.

Keys to species. Sloane (1898).

Lacordairia Laporte, 1867 (Pl. 28G)

Type species. *Lacordairia cychroides* Laporte, 1867, by subsequent designation by Moore (1985).

Characteristics. Medium size, black beetles; often iridescent; pronotum typically elongate; elytra elongate ovate; terminal palpomeres expanded, pubescent.

Australian species. 12 species.

Distribution. Endemic, Tasmania, Norfolk Island, Lord Howe Island, and continental Australia from south coastal Victoria, New South Wales, and north-east costal Queensland.

Biology. Flightless, predacious, hygrophilous, in leaf litter of closed forests.

References. Staunton *et al.* (2014).

Lestignathus Erichson, 1842 (Pl. 28L)

Type species. *Lestignathus cursor* Erichson, 1842, by monotypy.

Characteristics. Small to medium size beetles, elongate ovoid form; black or brown with paler, sometimes reddish-brown margins of the pronotum and elytra; head relatively small; labrum emarginate with four setae at the margin.

Australian species. Four species.

Distribution. Endemic to Tasmania.

Biology. Flightless, predacious, hygrophilous, in leaf litter of closed forests.

Keys to species. Baehr (2000b).

Microferonia Blackburn, 1890 (Pl. 29C)

Type species. *Microferonia adelaide* Blackburn, 1890, by monotypy.

Genycerus Andrewes, 1933

Characteristics. Small size beetles, brown, piceous, or black, typically with paler margins on the pronotum and elytra; frequently iridescent.

Australian species. Five species.

Distribution. New Guinea, Java, Lord Howe Island, and continental Australia in south-western Western Australia, coastal regions of South Australia, Victoria, New South Wales, and Queensland.

Biology. Flight capable or flightless, predacious, found in open woodlands.

References. Moore *et al.* (1987).

Omestes Andrewes, 1933 (Pl. 32K)

Type species. *Omestes torta* Andrewes, 1933, by monotypy.

Characteristics. Clypeus deeply emarginate, exposing the basal membrane of the labrum, which is rather narrow, emarginate nearly to its base, margin with six setae; mandibles short, stout, compressed and blunt at apex, right one deeply notched at middle; mentum with only a vestige of a tooth; maxillary palpomeres long and slender, labial palpomeres shorter; pronotum subcordate, with two lateral setae on each side. Elytra depressed apex subtruncate, laterad of slightly emargination with a short, sharp spine.

Australian species. Only *Omestes torta* Andrewes.

Distribution. Found in the Philippines, New Guinea, Sulawesi, the Moluccas, and Australia.

References. Darlington (1968).

Physolaesthus Chaudoir, 1850 (Pl. 34B, G; Fig. 12.25)

Type species. *Physolaesthus australis* Chaudoir, 1850, by monotypy.

Physolaesthus Lacordaire, 1854

Characteristics. Small, smooth black or brown beetles, some with pale vittae; body form depressed, head wide; pronotum transverse; mentum lacking medial tooth.

Australian species. Seven species.

Distribution. New Guinea, Java, the Philippines, New Zealand, and continental Australia in south-western Western Australia, coastal regions of South Australia, New South Wales, Queensland, and Northern Territory.

Biology. Predacious, flight capable, hygrophilous.

References. Darlington (1968).

Platylytron Macleay, 1873 (Pl. 35F)

Type species. *Platylytron amplipenne* Macleay, 1873, by monotypy.

Platelytron Rye, 1875

Characteristics. Large size (18–20 mm) black, somewhat shiny beetles; eyes prominent; pronotum broadly rounded and explanate laterally; elytra relatively very broad; terminal palpomeres enlarged, securiform.

Australian species. Only *Platylytron amplipenne* Macleay.

Distribution. South-western Western Australia.

Biology. Flightless, predacious.

Siagonyx Macleay, 1871 (Pl. 38C)

Type species. *Siagonyx amplipennis* Macleay, 1871, by subsequent designation by Moore *et al.* (1987).

Characteristics. Medium size black, typically shiny, iridescent beetles; slightly convex and elongate ovoid form; legs and antennae relatively long; neither mandible notched, apices bidentate; labrum very shallowly sinuate emarginate with a low medial tooth or nearly straight; prosternal process margined at least laterally; metanepisternum short, lateral and anterior edges equal in length.

Australian species. Three species.

Distribution. Coastal regions of southern Queensland and New South Wales. Coastal and further inland in the southeast and across Victoria.

Biology. Flightless, predacious, hygrophilous.

References. Moore & Wallbank (1968).

Tribe Cyclosomini Laporte, 1834

Characteristics. Procoxal cavities closed; mesocoxae conjunct; mandibles with distinct scrobe; glossa united with paraglossae; mentum dentate or edentate; first three antennomeres impilose; elytra either entire, covering the abdomen, or slightly shortened; tibial spurs of all legs, but particularly of the metatibia, extremely elongate, commonly microserrulate; 4th tarsomeres variously shaped; tarsal claws denticulate or not; aedeagus with simple folding; parameres asetose at apex; female gonocoxites variously shaped, variously setose.

Note. The limits of the tribe as well as the infratribal grouping are much in discussion (e.g. Jedlicka 1963; Darlington 1968; Ball & Bousquet 2001). Also the use of the names given to the various named subgroups differ much between authors. Following Bouchard *et al.* (2011) we recognise two subtribes. Cylosomina and Masoreina. In the appointment of genera to subtribes we follow Ball & Bousquet (2001).

Key to the Australian subtribes of Cyclosomini

1. Maxillae at the external apical margin pilose .... **Cyclosomina Laporte**
– Maxillae at the external apical margin impilose .... **Masoreina Chaudoir**

Key to the Australian genera of the tribe Cyclosomini

1. Mentum not toothed; rather wide, black species with sericeous elytra .... ***Aephnidius* Macleay**
– Mentum toothed; either very small, short and wide, uniformly brownish species, or larger, more elongate, usually vividly patterned species **2**
2(1). Body size very small, < 3 mm, body short and wide, uniformly more or less dark brown .... ***Caphora* Schmidt-Göbel**
– Body size larger; body longer, elytra usually with rather vivid pattern **3**
3(2). Tarsal claws denticulate; 4th tarsomeres of pro and meso legs usually wide and bilobed, and at least with remnants of ventral squamose setae, very rarely simple; elytra with various patterns, but not with many dark spots on yellow ground .... ***Sarothrocrepis* Chaudoir**
– Tarsal claws not denticulate; 4th tarsomeres of all legs narrow and simple; elytra with many dark spots on yellow ground (in Australia) .... ***Tetragonoderus* Dejean**

Subtribe Cyclosomina Laporte, 1834

Characteristics. Maxillae at the external apical margin pilose; usually depressed beetles with transverse apex of the elytra which do not completely cover the abdomen; elytra usually with distinct pattern.

Sarothrocrepis Chaudoir, 1876 (Pl. 37B–C)

Type species. *Carabus corticalis* Fabricius, 1801, by original designation.

Lebiomorpha Chaudoir, 1876.

Ectroma Blackburn, 1890.

Characteristics. Medium-sized to large, depressed beetles with large, strongly projecting eyes; 4th tarsomeres usually wide and deeply cleft, very rarely narrow and simple; tarsal claws denticulate; elytra usually with distinct colour pattern that does not consist of many small dark spots on yellow ground; aedeagus with the ostium mainly on the left side; without distinctly sclerotised, but sometimes finely denticulate pieces in the endophallus; parameres very dissimilar, asetose; both gonocoxites very short, asetose, gonocoxite-2 apicad widened, with wide, membranous apex.

Australian species. 115 species.

Distribution. Australia including Tasmania and Lord Howe Island; extraterritorial in New Guinea, New Caledonia and on the Philippine and Indonesian insular belts.

Biology. The Australian species live under bark or in deep fissures in the bark of a multitude of trees from rainforests to semiarid areas and can be found there sometimes in great numbers. Many species occur together with certain lebiine species that possess similarly depressed body shape and comparable colour patterns. They are best distinguished from these lebiine species by the very elongate and microserrulate tibial spurs. Probably strictly nocturnal, because specimens commonly are attracted to light.

Note. Some of the described species either belong to other tribes or are of uncertain status.

References. Baehr (2018); Blackburn (1901); Sloane (1917); Csiki (1932); Jedlicka (1963); Darlington (1968); Matthews (1980); Moore *et al.* (1987); Moore (1992); Baehr (2004c, 2010a); Lorenz (2005).

Keys to species. Baehr (2018) published a key to all species. Blackburn (1901), Sloane (1917, 1920) keys are either outdated and do not work properly for the described species, or they cover only Tasmanian species (Sloane 1920).

Tetragonoderus Dejean, 1829 (Pl. 22D; Figs 12.34, 12.144)

Type species. *Carabus quadrum* Fabricius, 1792, by subsequent designation by Jeannel 1949.

Characteristics. Medium-sized, rather wide beetles with moderately large, rather projecting eyes; 4th tarsomeres narrow and simple; tarsal claws not denticulate; aedeagus with the ostium mainly on the left side; without distinctly sclerotised pieces on the endophallus; parameres very dissimilar, asetose at apex; both gonocoxites very short, asetose, gonocoxite-2 more or less rectangular, with very wide, membraneous apex; in the single Australian species the elytra bear many small dark spots on yellow ground.

Australian species. Only *T. undatus* (Dejean, 1829), introduced from South America.

Distribution. Eastern New South Wales; extraterritorial in the Oriental, Ethiopian, Nearctic, and Neotropical Regions.

Biology. The single Australian species occurs on the ground at the banks of rivers and creeks.

References. Csiki (1932); Jedlicka (1963); Moore (1976); Moore *et al.* (1987); Lorenz (2005).

Subtribe Masoreina Chaudoir, 1871

Characteristics. Maxillae at the external apical margin impilose, usually rather short and wide beetles with entire elytra which almost completely cover the abdomen; elytra without distinct colour pattern.

Aephnidius W. S. Macleay, 1825 (Pl. 18H)

Type species. *Anaulacus adeloides* W. S. Macleay, 1825, by monotypy.

Characteristics. Wide, medium-sized, oval shaped, black species (in Australia); mentum not toothed; elytral surface markedly sericeous; aedeagus with or without sclerotised spines on the endophallus; parameres dissimilar, asetose at apex; both gonocoxites asetose, gonocoxite-2 straight, with very obtuse apex.

Australian species. Only *A. adeloides* Macleay.

Distribution. Northern tropical Australia from north Queensland to north-eastern Western Australia; extraterritorial in the Oriental Region and in New Guinea.

Biology. Most specimens were collected at light. Probably ground living in open forest and woodland, partly near water.

Notes. Ball & Shpeley (2002) consider *Aephnidius* a subgenus of *Anaulacus* Macleay, 1825. We think that both are sufficiently diferent to consider them separate genera. Habu (1967) reported the larva as being described.

References. Csiki (1932); Jedlicka (1963); Habu (1967); Darlington (1968); Moore *et al.* (1987); Ball & Shpeley (2002); Lorenz (2005).

Caphora Schmidt-Göbel, 1846 (Pl. 20L)

Type species. *Caphora humilis* Schmidt-Göbel, 1846, by monotypy.

Characteristics. Very small, wide, and in Australia, dark brown beetles; mentum sharply dentate; aedeagus with the ostium on the dorsal surface; without distinctly sclerotised pieces on the endophallus; parameres very dissimilar, asetose at apex; both gonocoxites short, asetose, gonocoxite-2 more or less rectangular, with membranous apex.

Australian species. Only *C. humilis* Schmidt-Göbel.

Distribution. Cape York Peninsula, north Queensland; extraterritorial in the Oriental Region and in New Guinea.

Biology. Apparently ground living in leaf litter in open and closed forests. Darlington (1968) found the species in northern Cape York Peninsula 'in leaf litter mixed with bird droppings under a large tree in which colonial birds have nested'.

References. Csiki (1932); Jedlicka (1963); Darlington (1968); Moore *et al.* (1987); Lorenz (2005).

Tribe Pseudomorphini Hope, 1838

Characteristics. Head with a deep antennal groove, antennomere 1 not visible from above; labium without suture between mentum and submentum; procoxal cavities closed; mesocoxal cavities conjunct; protibia with one apical and one subapical spur (anisochaete) with distinct antenna cleaner; abdomen with six visible sterna; elytra apex truncate, last abdominal tergite visible from above.

Notes. Deuve (1988) refuted the relationships with Brachinini or Scaritini that some authors had previously postulated. Adaptations to myrmecophilous life can constrain morphological evolution such that it obscures evidence of relationships. Erwin & Geraci (2008) suggested a close relationship to the likewise myrmecophilous or termitophilous Orthogoniini, a relationship that is supported by DNA sequence data analyses (Ober 2002; Ober & Maddison 2008).

From what is known, most genera of this tribe are unique in that they are ovoviviparous (Liebherr & Kavanaugh 1985; Baehr 1997b), meaning that the larvae hatch within the female oviducts. Apparently this is an adaptation to the myrmecophilous life probably shared by all pseudomorphine species (Baehr 1992a, 1994a; Erwin 1981). Other adaptations to this mode of life are conspicuous morphological modifications in adults towards development of a compact form, shortening and flattening of antennae and legs, reduction of chaetotaxy, somewhat thicker cuticle, and the trend towards physogastric larvae (Erwin 1981).

Key to the Australian genera of Pseudomorphini

1. Eyes dorsal **and** delimited ventrally by a continuous raised bead (Fig. 12.4B)........ ***Adelotopus*** **Hope**
– Eyes dorsal or lateral, when more or less dorsal, **not** delimited ventrally by a continuous raised bead........ **2**
2(1). Maxilla laterally developed into a large, conspicuous lobe; with a conspicuous triangular process or a large concavity beneath the eye; tibiae distinctly compressed; dorsal surface very coarsely punctate........ **3**
– Lateral part of maxilla not so developed; no conspicuous process or concavity beneath eye; tibiae not distinctly compressed; surface finely punctate or at most, with scattered coarse punctures, then each puncture bearing an elongate seta........ **4**
3(2). Triangular process below eye; tibiae moderately compressed; tarsi slightly thickened........ ***Cainogenion*** **Notman**
– Large circular concavity below eye; tibia very compressed; tarsi thick........ ***Paussotropus*** **Waterhouse**
4(2). Head hypognathus; labrum not visible; mandibles relatively minute; body size very small, < 5 mm, body very convex, oviform........ ***Cryptocephalomorpha*** **Ritsema**
– Head more or less prognathous; labrum visible; mandibles typical size; body size usually larger, either elongate and cylindrical, or wide and rather depressed........ **5**
5(4). Margins of elytra with a fringe of elongate setae; dorsal surface normally with scattered, elongate setae; mental tooth elongate, very acute........ ***Austropseudomorpha*** **Baehr**
– Margins of elytra without a fringe of elongate setae; dorsal surface without elongate setae, at most with short pilosity; mental tooth either widely rounded, or slightly bidentate, or absent........ ***Sphallomorpha*** **Westwood**

Sphallomorpha Westwood, 1837 (Pl. 38I)

Type species. *Sphallomorpha decipiens* Westwood, 1837, by monotypy.

Silphomorpha Westwood, 1837.

Characteristics. Body wide, more or less depressed; head prognathous; antenna elongate; marginal setiferous punctures on the elytra numerous, no elongate setae along the lateral margins or on the disc of the elytra; legs elongate with tibiae not compressed and tarsi wide or short; male terminal abdominal sternum excised; aedeagus with basal bulbus and small enclosed foramen, parameres dissimilar shapes, left typically conchoid and right more elongate, median lobe variously shaped; gonocoxites pilose, gonocoxite-2 dentiform.

Australian species. 151 species.

Distribution. The whole of mainland Australia, one species recorded from Tasmania; extraterritorial in New Guinea.

Biology. Under bark and in deep fissures in the bark of various eucalypt and non-eucalypt trees; from closed forest to open woodland, bushland (mallee, mulga) to semiarid and arid steppe areas where some tree growth is present. Very few species have been found in rainforests. All species probably are nocturnal and predacious, but nothing is known about their diet and feeding mode. The species of this genus are oviparous and the larvae (at least of the few species of which larval habits have been observed) appear to live in or around ant nests and to feed on ants (Moore 1964b, 1974, 1983). In spite of their wide body shape the beetles are very agile runners and can fly well.

References. Notman (1925); Csiki (1933); Darlington (1968); Matthews (1980); Moore *et al.* (1987); Baehr (1992a, 1993a, 1994c, 2002a, 2006h; 2008e, 2009d, 2014a); Lorenz (2005).

Keys to species. Baehr (1992a, 1993a, 1994c, 2002a, 2005d, 2006h, 2008e, 2009d, 2016e).

Austropseudomorpha Baehr, 1997 (Pl. 36K)

Type species. *Adelotopus insignis* Sloane, 1910, by original designation.

Characteristics. Head with mouthparts orthognathous; terminal palpomere of labial palp markedly securiform; antenna moderately elongate, not widening in the middle; body shape cylindrical, parallel-sided; number of setiferous marginal punctures of the elytra reduced; elytra fused, with elongate setae along the lateral margins and usually also on the disc; flight wings absent; legs elongate with tibiae not compressed and tarsi not much widened or shortened; aedeagus very elongate; both gonocoxites very narrow and elongate, asetose except one short apical seta at gonocoxite-2.

Australian species. Three species and one subspecies.

Distribution. South-western and south-eastern Australia, in moderately dry areas.

Biology. Little known. The few recorded specimens have been collected in open forests, woodland, and dry bushland.

Note. The genus originally was described as a subgenus of *Pseudomorpha* Kirby, 1823, but herewith is raised to generic rank, because the genus *Pseudomorpha* has been divided into several genera by Erwin & Geraci (2008) and *Austropseudomorpha* is so different from those species that what now remains in *Pseudomorpha* s. str. appears to be generically distinct.

References. Lorenz (2005); Erwin & Geraci (2008); Baehr (2009d).

Key to species. Baehr (1997b).

Adelotopus Hope, 1834 (Pl. 18A, D; Fig. 12.4)

Type species. *Adelotopus gyrinoides* Hope, 1834, by monotypy.

Characteristics. Body shape moderately wide and convex to cylindrical, head with mouth parts rather orthognathous; eyes situated on the dorsal surface of the head with a continuous raised bead along the ventral border; terminal palpomere of labial palp markedly securiform; antenna short, widened in the middle; elytra free; number of setiferous marginal punctures of the elytra much reduced; flight wing full; legs moderately elongate with tibiae slightly compressed and tarsi shortened; most tactile setae on head, prothorax, and abdomen absent; aedeagus moderately elongate, with very large ostium; gonocoxites-1 and -2 fused and characteristically foliaceous, with a few short apical setae.

Australian species. 133 species and 10 subspecies.

Distribution. The whole of mainland Australia, few species in Tasmania; extraterritorial in New Guinea, Solomon Islands, Moluccas, Java, and Malaysia; one species introduced into New Zealand.

Biology. Under bark and in deep fissures in the bark of various eucalypt and non-eucalypt trees; from closed forest to open woodland, bushland (mallee, mulga) to semiarid and arid steppe areas where river gums occur. Few species were recorded from rainforests. All species probably are nocturnal and predacious, but nothing is known about their diet and feeding mode. The species of this genus are ovoviviparous and the larvae live in ant nests. For a few species, adults were also found in ant nests under the bark of trees. In spite of their rather clumsy looking body shape they are agile runners, and can fly well.

Note. Additional species are known to exist and presently are being described.

References. Notman (1925); Csiki (1933); Darlington (1968); Matthews (1980); Moore *et al.* (1987); Baehr (1997b, 2002a, 2008e, 2009d); Lorenz (2005).

Keys to species. Baehr (1997b, 2002a, 2005d, 2007c, 2008e, 2009d, 2016e, 2016h).

Cainogenion Notman, 1925 (Pl. 20I)

Type species. *Adelotopus ipsoides* Westwood, 1837, by original designation.

Characteristics. Body shape parallel-sided and dorsally depressed; head with mouthparts rather orthognathous; lateral part of the maxilla is developed into a large, conspicuous lobe; eyes situated on the dorsal surface of the head, with a triangular process beneath the eye; terminal palpomere of labial palp markedly securiform; antenna short, widened in the middle; elytra free; number of setiferous marginal punctures of the elytra much reduced; flight wing full; legs moderately elongate with tibiae distinctly compressed and tarsi shortened; most tactile setae on head, prothorax, and abdomen absent; aedeagus moderately elongate, with very large ostium; gonocoxites-1 and -2 fused and characteristically foliaceous, with a few short apical setae.

Australian species. 12 species and two subspecies.

Distribution. Inland, eastern and northern mainland Australia.

Biology. All species seem to live under bark or in deep fissures in the bark of various trees, mainly in sclerophyll forests and open woodland. Nothing else is known about habits and diet. The species of this genus are ovoviviparous and the larvae may live in ant nests.

Note. Two subgenera are recognised.

References. Csiki (1933); Matthews (1980); Moore *et al.* (1987); Baehr (1997b, 2002a, 2008e); Lorenz (2005).

Key to species. Baehr (1997b).

Key to the subgenera of *Cainogenion*

1. Basal angles of pronotum rounded, base without any sinuosity; microreticulation of pronotum and elytra superficial, surface rather glossy; apex of glossa narrow, bisetose .... ***Procainogenion* Baehr**
– Basal angles of pronotum obtuse to angulate, base with a distinct sinuosity near basal angles; microreticulation of pronotum and elytra very distinct, surface dull; apex of glossa wider, multisetose .... ***Cainogenion* Notman**

Subgenus *Cainogenion* Notman, 1925

Type species. *Adelotopus ipsoides* Westwood, 1837, by original designation.

Australian species. Eleven species and two subspecies.

Distribution. Inland, eastern and northern mainland Australia.

References. Csiki (1933); Moore *et al.* (1987); Baehr (1997b); Lorenz (2005).

Key to species. Baehr (1997b).

Subgenus *Procainogenion* Baehr, 1997

Type species. *Adelotopus ephippiatus* Newman, 1856, by original designation.

Australian species. Only *C.* (*P.*) *ephippiatum* (Newman).

Distribution. The single species occurs in south-eastern Australia from eastern South Australia to southern and central eastern Queensland.

References. Lorenz (2005).

Paussotropus Waterhouse, 1877 (Pl. 33I)

Type species. *Paussotropus parallelus* Waterhouse, 1877, by monotypy.

Characteristics. Body elongate and parallel-sided and dorsally depressed; head with mouthparts orthognathous; lateral part of the maxilla is developed into a large, conspicuous lobe; eyes situated on the dorsal surface of the head, with a large circular concavity beneath the eye; terminal palpomere of labial palp markedly securiform; antenna very short, widened in the middle; elytra free; number of setiferous marginal punctures of the elytra much reduced; flight wing full; legs short with tibiae very compressed and tarsi extremely short and thick; most tactile setae of head, prothorax, and abdomen absent; aedeagus moderately elongate, wide, with wide apex; gonocoxites-1 -2 fused and characteristically foliaceous, with a few short apical setae.

Australian species. Only *P. parallelus* Waterhouse.

Distribution. Widely distributed in interior and northern Australia.

Biology. Almost all recorded specimens were sampled at light, therefore it is not even settled whether the species is corticolous. Nothing is known about habits and diet. The single species is ovoviviparous and the remarkably physogastric larva may live in ant nests.

References. Notman (1925); Csiki (1933); Darlington (1968); Moore *et al.* (1987); Baehr (1997b, 2002a, 2008e); Lorenz (2005).

Cryptocephalomorpha Ritsema, 1875 (Fig. 12.86)

Type species. *Cryptocephalomorpha gaverei* Ritsema, 1875.

Characteristics. Size very small, < 5 mm, body shape convex, surface glossy; head ventrally deflected; labrum hidden; mandibles minute, in the single Australian species conspicuously dentate on their lateral margin; eyes situated on the dorsal surface of the head; terminal palpomere of labial palp very large and markedly securiform; antenna very short, widened in the middle; elytra free; number of setiferous marginal punctures of the elytra much reduced; flight wing full; legs short with tibiae very compressed and tarsi short; most tactile setae on head, prothorax, and abdomen absent; aedeagus variously shaped; gonocoxites very elongate and narrow, asetose except for a single apical seta on gonocoxite-2.

Australian species. Only *C. australica* Baehr, 1997.

Distribution. Iron Range, central Cape York Peninsula, north Queensland; the single Australian species is known only from the type locality and the holotype; extraterritorial in New Guinea, Solomon Islands, southern Asia, and South Africa.

Biology. Probably ground living, because the holotype of the single species was sampled in a pitfall trap. A related species in New Guinea was found with ants (Darlington 1968). Nothing else is known about habits, diet, and mode of reproduction, and the larvae are unknown.

References. Notman (1925); Csiki (1933); Darlington (1968); Baehr (1997b, 2002a); Lorenz (2005).

Tribe Lebiini Bonelli, 1810

Characteristics. Procoxal cavities closed; mesocoxae conjunct; mandibles with distinct scrobe; pronotum usually wider than long; apex of elytra transverse or more or less deeply sinuate, or even dentate or spined, not completely covering the abdomen; tibiae with normal sized tibial spurs, these not microserrulate; male aedeagus quite variously shaped, with the ostium on the dorsal or left surface, with or without distinctly sclerotised parts on the endophallus; parameres asetose at apex; gonocoxites highly variably shaped, often very small.

Notes. This is one of the largest and most diverse carabid tribes and it includes hundreds of genera worldwide that cover a huge range of character states. Therefore the subdivision of the tribe into subtribes is particularly controversial, see e.g. Jeannel (1949), Basilewsky (1984), Casale (1998), Ball & Bousquet (2001), Lorenz (2005). In the present paper we largely follow the system proposed by Bouchard *et al.* (2011), except that we regard Lichnasthenina a separate subtribe. Because the subtribal concepts are so controversial, all recorded Australian lebiine genera are included in a general key. In Lebiini, the female gonocoxites are variably shaped and their form and features are commonly characteristic for subtribes. Therefore the subdivision of Lebiini to a large part makes use of the shape and structure of the gonocoxites.

Key to the Australian subtribes of Lebiini

1.	Tarsomere 4 wide, deeply excised and with squamose setae beneath (Fig. 12.67)	2
–	Tarsomere 4 narrow, not excised, without squamose setae beneath	5
2(1).	Base of pronotum distinctly produced in the middle, laterally excised	**Lebiina Bonelli**
–	Base of pronotum not produced in the middle, not excised laterally	3
3(2).	Apical angle or the entire lateral margin of pronotum with several additional setae	**Physoderina Chaudoir**
–	Apical angle of pronotum without additional setae	4

4(3). Antenna short; tarsi short and wide; body compact, dorsally rather convex; gonocoxite-2 quadrate, with a few stout setae at apex .... **Metallicina Basilewsky**
– Antenna elongate; tarsi narrow and elongate; body elongate, dorsally rather depressed; gonocoxite-2 elongate, with a tuft of fine setae at apex .... **Calleidina Chaudoir (part)**
5(1). Head with coarse and regular longitudinal sulci and ridges; aedeagus with an elongate filum that stands out from the ostium .... **Apenina Ball**
– Head without regular longitudinal sulci and ridges; aedeagus without such filum **6**
6(5). Labrum longer than wide **and** labial palp narrowed apicad **7**
– Labrum either distinctly wider than long **or** labial palp distinctly widened apicad, or even securiform **8**
7(6). Base of pronotum not produced in the middle, not excised laterally .... **Pericalina Hope**
– Base of pronotum distinctly produced in the middle, laterally excised .... **Somotrichina Mateu**
8(6). Labial palp distinctly widened apicad, or even securiform .... **Calleidina Chaudoir (part)**
– Labial palp not widened apicad, labrum much wider than long **9**
9(8). Dorsal surface very rugose and elytral intervals with a single, dense row of setae .... **Lichnasthenina Thomson**
– Dorsal surface not so rugose; elytra either glabrous or with more or less dense, decumbent setae **10**
10(9). Tarsal claws finely denticulate; apex of elytra shortened, transverse or slightly excised and not covering the apex of the abdomen .... **Dromiusina Bonelli**
– Tarsal claws not denticulate; apex of elytra evenly rounded and almost covering the apex of the abdomen .... **Celaenephina Habu**

Key to the Australian genera of Lebiini

1. Tarsomere 4 deeply excised (Fig. 12.67) and with squamose setae beneath **2**
– Tarsomere 4 not deeply excised and without squamose setae beneath **9**
2(1). Upper surface of tarsi pilose **3**
– Upper surface of tarsi glabrous **6**
3(2). Base of pronotum markedly produced in the middle, laterally excised **4**
– Base of pronotum not produced in the middle, laterally not excised **5**
4(3). Dorsal surface very rugose; elytra deeply striate, with elongate, erect pilosity .... ***Lachnoderma* Macleay**
– Dorsal surface smooth, barely striate, glabrous .... ***Physodera* Eschscholtz**
5(3). Lateral margins of pronotum with many elongate setae; elytra with decumbent setae .... ***Endynomena* Chaudoir**
– Lateral margins of pronotum bisetose; elytra usually glabrous, rarely with sparse pilosity .... ***Demetrida* White**
6(2). Base of pronotum markedly produced in the middle, laterally excised .... ***Lebia* Latreille**
– Base of pronotum not markedly produced in the middle, laterally not excised **7**
7(6). Lateral apical angles of elytra sharply dentate; pronotum semicircular .... ***Aristolebia* Bates**
– Lateral apical angles of pronotum not sharply dentate, pronotum more or less cordiform **8**
8(7). Body compact, dorsally rather convex; antenna short; legs rather short; elytra either piceous or broadly fasciate with blue .... ***Parena* Motschulsky**
– Body longer and more slender, dorsally rather depressed; antenna elongate; legs longer; elytra with pale discal spots, commonly also with pale apical spots .... ***Trigonothops* Macleay**
9(1). Tarsal claws virtually not denticulate, commonly claws large, if claws small, species small and regularly pilose **10**
– Tarsal claws commonly smaller, usually distinctly denticulate, at least with traces of denticles **16**
10(9). Small beetles, body length < 4.5 mm; surface with dense, decumbent pilosity; elytra with indistinct colour pattern **11**
– Large beetles, body length > 6 mm, usually more; surface glabrous; colour reddish-piceous, or uniformly black, or metallic **12**
11(10). Base of pronotum distinctly produced in the middle, laterally excised; mentum edentate; surface reddish-brown, elytra with distinct, wide, dark, and transverse band in the middle .... ***Somotrichus* Seidlitz**
– Base of pronotum absolutely straight; mentum with distinct triangular tooth; pronotum reddish, elytra dark piceus or black with faint, paler, reddish base .... ***Geoffreyella* Baehr**
12(10). Body slender, parallel-sided; colour black **13**
– Body usually wider, when rather parallel-sided, then colour metallic **14**
13(12). Head with a single supraorbital seta above each eye, eye small, laterally not produced, orbit elongate .... ***Miscelus* Klug**
– Head with two supraorbital setae above each eye, eye large, laterally produced, orbit short .... ***Celaenephes* Schmidt-Göbel**
14(12). Eye large, produced laterally; elytra rather parallel-sided and dorsally convex, if rather wide and depressed, pronotum with several anterior marginal setae; colour usually bright metallic .... ***Catascopus* Kirby**
– Eye usually smaller, less produced laterally; elytra wider and dorsally depressed; pronotum with a single anterior marginal seta; colour not bright metallic, at most with slight violaceous lustre **15**
15(14). Eye small, not longer than orbit; colour rufo-testaceous to brown .... ***Speotarus* Moore**
– Eye larger, longer than orbit; colour piceous-black or black, with or without slight violaceous lustre .... ***Coptoglossus* Chaudoir**
16(9). Dorsal surface extremely rugose; each elytral interval with a single row of setae; dentation of tarsal claws very inconspicuous; body length c. 2.5 mm .... ***Australovelinda* Baehr**
– Dorsal surface less rugose; elytra either glabrous or with dense pilosity; dentation of tarsal claws more distinct; body length usually greater....... **17**
17(16). Apex of elytra deeply excised; body rather parallel-sided; colour metallic .... ***Holcoderus* Chaudoir**
– Apex of elytra differently shaped; when deeply excised then body not parallel-sided, and colour black **18**

18(17). Apex of elytra spined at sutural and external angles; large, broad beetles with scattered erect setae on the elytra .... ***Stricklandiana* Bousquet**

– Apex of elytra unarmed, or only spined at sutural angle; surface without erect setae, but sometimes with dense, short, usually decumbent pilosity **19**

19(18). Surface with dense, short pilosity, or with sparse, erect pilosity **20**

– Surface impilose **28**

20(19). Head with large postocular prominences (Pl. 19K) **21**

– Head without large postocular prominences (doubtful genera under both couplets) **24**

21(20). Elytra with sparse, erect setae; flight wings reduced .... ***Anomotarus* Chaudoir (part)**

– Elytra with dense, decumbent pilosity; flight wings fully developed **22**

22(21). Head without longitudinal strioles; elytra with short and decumbent pilosity; colour of elytra either yellow with a large, dark spot laterally in apical half, or dark with base and centre reddish **23**

– Head with distinct longitudinal strioles near eyes; elytra with longer and irregularly erect pilosity; colour of elytra reddish-black to piceous with indistinctly lighter base .... ***Rugitarus* Baehr**

23(22). Head and pronotum black, elytra yellow with a large, black, contrasting spot laterally in apical half; apical palpomere of labial palp not markedly widened .... ***Crassagena* Baehr**

– Head black, pronotum reddish-brown, elytra dark piceous apico-laterally, reddish-brown along base and disc; apical palpomere of labial palp considerably widened .... ***Anomotariella* Baehr**

24(20). Head very large with large, semicircular eyes **and** pronotum short and very wide **and** anterior lateral pronotal seta situated at or in front of apical third **and** elytra wide, markedly depressed, quadrate **and** pilosity of pronotum and elytra very short, regular, and markedly depressed .... ***Minuthodes* Andrewes**

– Not all these characters in combination **25**

25(24). Pronotum rather narrow and elongate, markedly cordiform; pilosity of surface extremely short but erect; colour either uniformly black or regularly quadrimaculata .... ***Mochtherus* Schmidt-Göbel**

– Pronotum usually wider, not as cordiform; pilosity of surface usually slightly longer and inclined; colour usually not regularly quadrimaculata **26**

26(25). Penultimate palpomere of labial palp longer than apical palpomere; usually larger species > 7.5 mm .... ***Philophloeus* Chaudoir**

– Penultimate palpomere of labial palp not longer than apical palpomere; usually smaller species < 7.5 mm **27**

27(26). Elytra dorsally and laterally convex, reversely oviform; pronotum dorsally convex, barely cordiform and with more or less obtuse basal angles .... ***Cheilagona* Baehr**

– Elytra dorsally depressed, laterally less convex, not reversely oviform; pronotum dorsally more or less depressed, cordiform and with angulate or acute basal angles .... ***Agonocheila* Chaudoir**

28(19). Whole frons with regular and elongate longitudinal sulci and ridges .... ***Habutarus* Ball & Hilchie**

– Frons without regular and elongate longitudinal sulci and ridges, at most with some elongate punctures **29**

29(28). Orbit of eyes short or barely indicated (Pl. 35L), eye large, greatly protruding laterally **30**

– Orbit of eyes distinct, at least half as long as eye; or if less, swollen as postocular prominence (Pl. 19K) **38**

30(29). All tarsi short and thick; colour unicolourous reddish-brown to pale brown .... ***Plochionus* Dejean**

– All tarsi slender and elongate; colour various **31**

31(30). Mentum with distinct tooth **32**

– Mentum edentate **36**

32(31). Eye very large, conspicuously protruding laterally, neck narrow; elytra usually with cruciate pattern .... ***Phloeocarabus* Macleay**

– Eye smaller, less protruding laterally, neck not unusually narrow; elytra either uniformly dark or quadrimaculata **33**

33(32). Mentum with bifid or slightly notched tooth; base of pronotum distinctly produced in the middle, laterally excised .... ***Syntomus* Hope**

– Mentum with unidentate tooth; **either** base of pronotum slightly produced and elytra quadrimaculate, **or** base of pronotum not produced and elytra uniformly black **34**

34(33). Elytra conspicuously quadrimaculate .... ***Barrymooreana* Baehr**

– Elytra uniformly black **35**

35(34). Elytra markedly microreticulate, dull, interval 3 with a single puncture close to apex .... ***Mochtheroides* Andrewes**

– Elytra with barely noticeable microreticulation, glossy, interval 3 tripunctate .... ***Oxyodontus* Chaudoir**

36(31). Labrum wider than long; base of pronotum produced in the middle, laterally distinctly excised .... ***Microlestodes* Baehr**

– Labrum longer than wide; base of pronotum not produced in the middle, laterally not excised **37**

37(36). Setiferous punctures on interval 3 distinct .... ***Coptodera* Dejean**

– Setiferous punctures on interval 3 small, fine, barely visible .... ***Dolichoctis* Schmidt-Göbel**

38(29). Postocular prominences remarkably produced; small, very wide and depressed beetles with extremely elongate discal elytral setae .... ***Brigalowia* Baehr**

– Postocular prominences little produced; moderately wide and depressed beetles with short discal elytral setae **39**

39(38). Orbit rather elongate, but not developed as postocular prominence; labial palp never securiform **40**

– Orbit more or less elongate, always developed as postocular prominence; labial palp usually securiform .... ***Anomotarus* Chaudoir (part)**

40(39). Head sharply constricted behind the eyes to form a distinct neck; microreticulation of elytra isodiametric .... ***Abaditicus* Ball & Hilchie**

– Head not constricted behind the eyes; microreticulation of elytra transverse .... ***Diabaticus* Bates**

Subtribe Lebiina Bonelli, 1810

Characteristics. 4th tarsomeres deeply excised and with squamose setae beneath; upper surface of tarsi impilose; eyes very large and semicircular; base of pronotum produced in the middle, laterally excised; external apex of the elytra not dentate; endophallus of aedeagus commonly with sclerotised parts; gonocoxites usually asetose, gonocoxite-2 short and wide, variously shaped.

Aristolebia Bates, 1892 (Pl. 19J)

Type species. *Aristolebia quadridentata* Bates, 1892, by monotypy.

Characteristics. Eyes very large and laterally remarkably protruding; pronotum semicircular; lateral apical angles of the elytra sharply angulate or even dentate; 4th tarsomeres wide, deeply excised, and with squamose setae beneath; endophallus of the aedeagus with sclerotised parts; gonocoxites large and wide, both gonocoxites usually asetose, gonocoxite-2 short and irregularly trigonal.

Australian species. Two species.

Distribution. North-eastern Queensland; extraterritorial in the whole Oriental Region to New Guinea,

Biology. Little recorded. Probably on and under bark of tree trunks and logs in tropical rainforests. Predacious. Darlington (1968) says that the New Guinean species are diurnal, but specimens have been sampled at light. Therefore they probably are at least partly nocturnal. Nothing else is recorded about habits, diet and reproduction.

References. Csiki (1932); Jedlicka (1963); Darlington (1968); Moore *et al.* (1987); Baehr (2004c, 2010a); Lorenz (2005).

Key to species. Baehr (2010a).

Lebia Latreille, 1802 (Pl. 28F)

Type species. *Carabus haemorrhoidalis* Fabricius, 1792 (= *Buprestis marginatus* Geoffroy in Fourcroy, 1785), by subsequent designation by Andrewes (1935b).

Eulebia W. J. Macleay, 1871

Characteristics. Eyes usually large and markedly protruding laterally; 4th tarsomeres wide, deeply excised, and with squamose setae beneath; pronotum more or less cordiform; lateral apical angles of the elytra not or barely angulate; endophallus of the aedeagus with sclerotised or dentate parts; gonocoxites large and wide, both gonocoxites usually asetose, gonocoxite-2 with widely rounded apex.

Australian species. 12 species.

Distribution. Humid eastern and tropical northern Australia from mid-New South Wales to Cape York Peninsula in north Queensland, and to north-western Northern Territory; extraterritorial almost worldwide.

Biology. Most known specimens were sampled at light. They may live on the ground, on or below bark of standing trees and logs, on twigs and foliage in open, and in closed forests. Pesticide fogging and light samples from the canopy suggest, however, that at least some species live in the canopy of rainforest trees and may be quite common there. In other regions species are known as predators of chrysomelid beetle eggs.

Note. This is an extremely diverse genus that has been divided into several subgenera, but the subgeneric concepts are different in various faunal regions. Therefore no subgeneric concept has been applied to the Australian species.

References. Csiki (1932); Jeannel (1942); Jedlicka (1963); Habu (1967); Darlington (1968); Matthews (1980); Moore *et al.* (1987); Baehr (2007a, 2012j); Lorenz (2005).

Keys to species. Baehr (2004b, 2007b, 2010a, 2012j).

Subtribe Physoderina Chaudoir, 1877

Characteristics. Eyes large and semicircular; base of pronotum slightly convex in the middle, or produced and laterally excised; lateral margin of pronotum with several setae; upper surface of tarsi pilose; 4th tarsomeres deeply excised and with squamose setae beneath; endophallus of aedeagus commonly with sclerotised parts; gonocoxites elongate, with tufts of setae at apex.

Physodera Eschscholtz, 1829 (Fig. 12.121)

Type species. *Physodera dejeani* Eschscholtz, 1829, by monotypy.

Characteristics. Base of pronotum produced in the middle, laterally excised; apical angles of pronotum with several setae; elytra dorsally convex, almost glabrous, in Australia with a small subapical pale spot. Genitalia of the Australian species not recorded.

Australian species. Only *P. bacchusi* Darlington, 1971.

Distribution. North-eastern Queensland; extraterritorial in the Oriental Region and in New Guinea.

Biology. Rainforest living species. Probably nocturnal.

References. Csiki (1932); Jedlicka (1963); Darlington (1971); Moore *et al.* (1987); Lorenz (2005).

Lachnoderma Macleay, 1873 (Pl. 28J)

Type species. *Lachnoderma cinctum* Macleay, 1873, by monotypy.

Characteristics. Eyes large and semicircular; base of pronotum produced in the middle, laterally excised; surface very rugose and with conspicuous, erect pilosity; tarsal claws clearly dentate; aedeagus short, with sclerotised pieces on the endophallus; gonocoxites elongate, straight; gonocoxite-2 at apex usually oblique, sparsely setose.

Australian species. Two species.

Distribution. Eastern Australia from north-eastern New South Wales to the tip of Cape York Peninsula in north Queensland; extraterritorial in the Oriental Region and on New Guinea.

Biology. Ground living and on logs in subtropical and tropical rainforests. Predacious and probably nocturnal.

References. Sloane (1915a); Csiki (1932); Jedlicka (1963); Habu (1967); Darlington (1968); Moore *et al.* (1987); Lorenz (2005).

Key to species. Sloane (1915a).

Endynomena Chaudoir, 1872 (Pl. 25H)

Type species. *Plochionus pradieri* Fairmaire, 1849, by monotypy.

Characteristics. Eyes large and almost semicircular; base of pronotum not produced and nor excised laterally; surface not rugose but with dense and short pilosity; tarsal claws strongly dentate; aedeagus with sclerotised pieces on the endophallus; gonocoxites elongate, straight; gonocoxite-2 at apex rather densely setose.

Australian species. Only *E. pradieri* (Fairmaire).

Distribution. Christmas Island; extraterritorial widely distributed in the Oriental and Papuan Regions, also occurring on New Caledonia, Fiji, Tonga, Samoa, Society, and Hawaii Islands.

Biology. Little recorded, probably ground living in a variety of habitats. In other areas, this species is commonly sampled at light. The good flying ability of these beetles may explain the large range, but they may also have been transported by human activities.

References. Csiki (1932); Jedlicka (1963); Habu (1967); Darlington (1968); Lorenz (2005).

Subtribe Metallicina Basilewsky, 1984

Characteristics. Eyes large and almost semicircular; base of pronotum not produced in the middle; apical angles without additional setae; external apex of the elytra not dentate; 4th tarsomeres deeply excised and with squamose setae beneath; tarsal claws strongly dentate; endophallus of aedeagus commonly with sclerotised or dentate parts; gonocoxite-1 asetose, gonocoxite-2 variously shaped, usually short, with few of setae at apex.

Parena Motschulsky, 1860 (Pl. 33A)

Type species. *Parena bicolor* Motschulsky, 1860, by monotypy.

Phloeodromius W. J. Macleay, 1871

Crossoglossa Chaudoir, 1872a

Characteristics. Legs short and stout; upper surface of tarsi sparsely pilose; aedeagus short and stout, with a varied pattern of sclerotised or denticulate pieces on the endophallus; gonocoxites short and wide; gonocoxite-2 usually short, about quadrate, with a few stout setae at apex,

Australian species. Two species.

Distribution. Eastern and north-eastern Queensland; extraterritorial in the Oriental and Papuan Regions.

Biology. Both Australian species apparently live on logs and trunks in more or less close forests. Predacious and probably nocturnal.

References. Andrewes (1927); Csiki (1932); Jedlicka (1963); Habu (1967); Darlington (1968); Moore *et al.* (1987); Lorenz (2005).

Key to species. No key available.

Subtribe Pericalina Hope, 1838

Characteristics. Head without coarse and regular longitudinal sulci and ridges; labrum longer than wide; labial palp narrowed apicad, not securiform; base of pronotum not produced in the middle, laterally not excised; apex of elytra variously shaped; 4th tarsomeres narrow, not excised, and without squamose setae beneath; tarsal claws dentate or not; aedeagus variously shaped, with or without distinctly sclerotised pieces on the endophallus, but without an elongate filum that stands out from the ostium; gonocoxites variously shaped, gonocoxite-2 short or elongate, with a variable number of ventral and dorsal setae.

Coptoglossus Chaudoir, 1869 (Pl. 22F)

Type species. *Coptoglossus sulcatulus* Chaudoir, 1869, by monotypy.

Characteristics. Rather large, uniformly dark beetles; labrum elongate; apex of the elytra rounded, not in any way excised; tarsi narrow and elongate; tarsal claws not denticulate; aedeagus without distinctly sclerotised pieces; gonocoxite-2 curved, with two large ventral setae and one elongate dorsal seta.

Notes. Due to body shape and structure of the apex of the elytra this genus was, previously included in the tribe Platynini. *Colpodes porphyriacus* (Sloane, 1910) was still placed in Platynini in the catalogue by Moore *et al.* (1987) and was transferred from Platynini to *Coptoglossus* in Lebiini by Baehr (2012b).

Australian species. Five species.

Distribution. Eastern Victoria to south-eastern Queensland.

Biology. Arbouricolous at and under bark of various trees in sclerophyll habitats and rainforests. Predacious and probably nocturnal.

References. Csiki (1932); Darlington (1963b; Moore *et al.* (1987); Lorenz (2005); Baehr (2012b, 2013b).

Key to species. Baehr (2012b, 2013b).

Miscelus Klug, 1834 (Pl. 30J)

Type species. *Miscelus javanus* Klug, 1834, by monotypy.

Characteristics. Large, narrow, parallel-sided, black beetles; a single supraorbital seta present in the Australian species; tarsi slender; tarsal claws large, not denticulate; aedeagus without distinctly sclerotised pieces; gonocoxite-2 curved, with two large ventral setae and one elongate dorsal seta.

Australian species. Only *M. unicolor* Putzeys, 1845.

Distribution. North-eastern Queensland; extraterritorial in the Oriental Region and on New Guinea.

Biology. On and under bark of logs and trunks in tropical rainforests. Predacious and nocturnal.

References. Lacordaire (1854); Chaudoir (1861a); Csiki (1932); Jedlicka (1963); Darlington (1968); Moore *et al.* (1987); Lorenz (2005).

Catascopus Kirby, 1825 (Pl. 21G)

Type species. *Catascopus hardwickii* Kirby, 1825, by monotypy.

Characteristics. Variously sized and shaped, but commonly elongate and dorsally rather convex, mostly metallic beetles; labrum elongate; eyes large, distinctly protruding laterally; tarsi narrow and elongate; tarsal claws not denticulate; aedeagus usually without distinctly sclerotised pieces; gonocoxite-2 comparatively short, curved, usually with two large ventral setae and one elongate dorsal seta. Most species are bright metallic. The apex of the elytra can be simple or dentate at the lateral and/or sutural angles.

Australian species. Seven species and one subspecies.

Distribution. Eastern and northern Australia from north-eastern New South Wales to Cape York Peninsula and the northern part of Northern Territory; extraterritorial in the Oriental region, New Guinea, the Bismarck Archipelago, Solomon Islands, and less diverse, in the Afrotropical and Neotropical Regions.

Biology. Arbouricolous on logs, trunks and branches of trees in rainforests and wet sclerophyll forests. Probably largely diurnal predators. In other regions beetles have been observed stalking flies and other beetles on logs during daytime.

References. Chaudoir (1861a); Sloane (1910a); Csiki (1932); Jedlicka (1963); Habu (1967); Darlington (1968); Moore *et al.* (1987); Straneo (1994); Baehr (1997e, 2000a, 2012l); Shpeley & Ball (2000); Lorenz (2005).

Keys to species. Straneo (1994); Baehr (2012l).

Coptodera Dejean, 1825 (Pl. 22J–K)

Type species. *Coptodera emarginata* Dejean, 1825, by subsequent designation by Blanchard 1842.

Belonognatha Chaudoir, 1843

Eucalyptocolia W. J. Macleay, 1871

Ectinochila Chaudoir, 1883

Characteristics. Labrum elongate; eyes large and laterally protruding; mentum edentate; tarsi narrow and elongate; elytra rather wide and depressed, glabrous, with 2–4 punctures; discal punctures distinct; tarsal claws denticulate; aedeagus with conspicuous, multidentate pieces on the endophallus; gonocoxite-2 narrow and elongate, with about four strong ventral setae.

Australian species. 11 species and two subspecies.

Distribution. Eastern Australia from central eastern New South Wales to north-eastern Queensland; extraterritorial in the whole Oriental Region, New Guinea, Solomon Islands, and in the Afrotropical and Neotropical Regions.

Biology. On and under bark of various tree species in sclerophyll forests and rainforests, also on and under bark of fallen logs. Predacious and probably nocturnal.

Note. All Australian species belong to the subgenus *Coptoderina* Jeannel, 1949.

References. Blanchard (1842); Sloane (1907a); Csiki (1932); Jedlicka (1963); Habu (1967); Darlington (1968); Moore *et al.* (1987); Shpeley & Ball (1993); Lorenz (2005); Baehr (2008j).

Key to species. Baehr (2008k).

Dolichoctis Schmidt-Göbel, 1846 (Pl. 24D)

Type species. *Dolichoctis striata* Schmidt-Göbel, 1846, by monotypy.

Characteristics. Labrum elongate; eyes large and laterally protruding; mentum edentate; elytra rather wide and depressed, glabrous, bipunctate; discal punctures very fine; apex either oblique and unarmed or spinose at sutural angle; tarsi narrow and elongate; tarsal claws denticulate; aedeagus with variable denticulate or spinose pieces on the endophallus; gonocoxite-2 rather short, curved, usually with two strong ventral setae.

Australian species. Three species.

Distribution. North-eastern Queensland; extraterritorial in the whole Oriental Region and from New Guinea to Solomon Islands.

Biology. Arbouricolous on logs and on tree trunks in rainforests, according to Darlington (1968) also on twigs and leaves in the understorey. Predacious and probably nocturnal, although Darlington suggested that the species are predominantly diurnal. However, specimens are commonly attracted to light. Little else is known about habits, diet, and reproduction.

Notes. The genus was divided into three subgenera two of which occur in Australia (Baehr 1999b).

The specimens of all Australian species enumerated in Moore *et al.* (1987) belong to other species.

References. Lacordaire (1854); Csiki (1932); Jedlicka (1963); Habu (1967); Darlington (1968); Moore *et al.* (1987); Baehr (1999b, 2006j, 2007d, 2013e); Lorenz (2005).

Keys to species. Baehr (2007d, 2013e).

Key to the Australian subgenera of *Dolichoctis*

1. Sutural angle of elytra not dentate or spined; elytra quadrimaculate .... ***Dolichoctis*** **Schmidt-Göbel**
– Sutural angle of elytra dentate or spined; elytra unicolourous dark piceous to black .... ***Spinidolichoctis*** **Baehr**

Subgenus *Dolichoctis* Schmidt-Göbel, 1846

Type species. *Dolichoctis striata* Schmidt-Göbel, 1846, by monotypy.

Characteristics. Sutural angle of the elytra not dentate or spinose; elytra quadrimaculate.

Australian species. Only *D. tetrastigma macleayi* (Sloane, 1901).

Distribution. North-eastern Queensland; extraterritorial in the whole Oriental Region and from New Guinea to Solomon Islands. Other subspecies of *D. tetrastigma* are also widely distributed in New Guinea and the southern Oriental Region.

Note. Moore *et al.* (1987) noted *D. striata* for Australia. However, that species does not occur in Australia.

References. Baehr (1999b, 2013e).

Subgenus *Spinidolichoctis* Baehr, 1999

Type species. *Dolichoctis aculeata* Chaudoir, 1869, by original designation.

Characteristics. Sutural angle of the elytra dentate or spinose; elytra unicolourous black or dark piceous.

Australian species. Two species.

Distribution. North-eastern Queensland; extraterritorial in New Guinea, Kei Islands, the Bismarck Archipelago, and Solomon Islands.

References. Lorenz (2005); Baehr (2007d).

Key to species. Baehr (2007d).

Mochtherus Schmidt-Göbel, 1846 1846 (Pl. 30D)

Type species. *Mochtherus angulatus* Schmidt-Göbel, 1846 (= *Dromius tetraspilotus* W. S. Macleay, 1825), by monotypy.

Characteristics. Labrum elongate; eyes large and laterally protruding; mentum obtusely toothed; prothorax rather narrow and elongate, markedly cordiform, with upturned lateral margin; elytra rather wide and depressed, with fine and dense, erect pilosity, bipunctate; apex oblique, barely excised; tarsi narrow and elongate; tarsal claws denticulate; aedeagus with a finely denticulate piece on the endophallus; gonocoxite-2 rather short, curved, with two heavy ventral setae.

Australian species. Two species.

Distribution. North-eastern Queensland; extraterritorial in the Oriental and Papuan Regions. Recently *Mochtherus tetraspilotus* (W. S. Macleay) was found in to have been introduced in North America (Choate 2001).

Biology. At and under bark of logs and tree trunks in tropical rainforests. Predacious and probably nocturnal. In the southern Oriental Region and in New Guinea the species is commonly attracted to light. Little else is known about habits, diet, and reproduction.

Note. This genus commonly is included as a subgenus in *Dolichoctis*. The pilosity of the dorsal surface, however, suggests that it should be regarded a separate genus.

References. Csiki (1932); Jedlicka (1963); Habu (1967); Darlington (1968); Moore *et al.* (1987); Baehr (2000a); Lorenz (2005).

Key to species. Two species occurring in Australia are easily distinguished, as *M. tetraspilotus* has quadrimaculate elytra and those of *M. obscurus* (Sloane, 1907) are unicolourous.

Mochtheroides Andrewes, 1923 (Pl. 30F)

Type species. *Masoreus sericans* Schmidt-Göbel, 1846, by original designation.

Characteristics. Labrum elongate; eyes large and laterally protruding; mentum obtusely toothed; prothorax rather wide, cordiform; elytra rather wide and depressed, impilose, densely microreticulate, with a single discal puncture close to apex; apex oblique, little excised; tarsi narrow and elongate; tarsal claws denticulate; male and female genitalia not recorded.

Australian species. Only *M. niger* Jedlicka, 1934.

Distribution. North-eastern Queensland; extraterritorial in South Asia, New Guinea, and New Britain.

Biology. Under bark of trees and logs in rainforests. Predacious and probably nocturnal.

References. Csiki (1932); Jedlicka (1963); Darlington (1968); Lorenz (2005).

Minuthodes Andrewes, 1941 (Pl. 30L)

Type species. *Platia lineella* Chaudoir, 1869, by subsequent designation by Andrewes (1939).

Platia Chaudoir, 1869b (*non Platia* Hübner, 1820).

Characteristics. Labrum elongate; head very large; eyes very large and laterally protruding; prothorax short and very wide; anterior lateral pronotal seta situated at or in front of apical third; elytra wide, quadrate, depressed, with fine and dense, depressed pilosity, tripunctate; tarsi narrow and elongate; tarsal claws denticulate; aedeagus without strongly sclerotised pieces; gonocoxite-2 rather short, curved, with two remarkably stout and elongate ventral setae and a similarly elongate dorsal seta.

Australian species. Seven species.

Distribution. Eastern and northern Australia from eastern South Australia to north Queensland, and in the northern tropical belt to north-western Australia north of Great Sandy Desert; extraterritorial from Sulawesi and the Moluccas through New Guinea to Solomon Islands.

Biology. Under bark of bark-shedding eucalypts and on logs in rainforests, open eucalypt woodlands, and even in semiarid savannah woodlands. In the Northern Territory and in north-western Australia most commonly sampled under bark of River Eucalypt. Predacious and, because specimens are commonly attracted to light, presumably nocturnal.

References. Darlington (1968); Baehr (1990c, 1994b, 2001a, 2006c); Lorenz (2005).

Keys to species. Baehr (1990c, 1994b, 2001a).

Oxyodontus Chaudoir, 1869 (Fig. 12.113)

Type species. *Oxyodontus tripunctatus* Chaudoir, 1869, by monotypy.

Characteristics. Labrum elongate; head large; eyes large and laterally protruding; mentum distinctly toothed; prothorax rather short and wide, moderately cordiform; anterior lateral pronotal seta situated behind apical third; elytra wide, widened apicad, almost impilose, tripunctate; apex oblique, barely excised; tarsi narrow and elongate; tarsal claws denticulate; male and female genitalia not recorded.

Australian species. Only *O. tripunctatus* Chaudoir.

Distribution. Mid-Cape York Peninsula, north-eastern Queensland; extraterritorial in the southern parts of the Oriental Region and in New Guinea.

Biology. At and under bark of logs and trunks in rainforests, probably also on low vegetation. Predacious and probably nocturnal.

References. Csiki (1932); Darlington (1968); Baehr (2000a); Lorenz (2005).

Stricklandiana Bousquet, 2003 (Pl. 38E)

Type species. *Stricklandia pericalloides* Macleay, 1886, by monotypy.

Stricklandia W. J. Macleay, 1886 (non *Stricklandia* Billings, 1859).

Characteristics. Large, wide, black beetles; labrum elongate; head large; eyes large and laterally protruding; mentum barely toothed; prothorax wide, very cordiform, with several lateral pronotal setae; elytra wide, intervals with many erect setae, tripunctate; apex spined at sutural and external angles; tarsi narrow and elongate; tarsal claws denticulate; aedeagus elongate, with a sclerotised, spinose piece; gonocoxites short, gonocoxite-2 curved, with two very elongate ventral setae and a similarly sized dorsal seta.

Australian species. Only *S. nigra* (Sloane, 1907).

Distribution. North-eastern Queensland; extraterritorial in New Guinea, New Britain, and the Moluccas.

Biology. On logs and trunks of rainforest trees. Predacious and probably nocturnal.

References. Baehr (2011d); Csiki (1932); Darlington (1968); Moore *et al.* (1987); Baehr (1997c); Lorenz (2005).

Holcoderus Chaudoir, 1869 (Pl. 27D)

Type species. *Holcoderus praemorsus* Chaudoir, 1869, by monotypy.

Characteristics. Elongate, parallel-sided, metallic beetles; labrum elongate; eyes large and laterally protruding; mentum toothed; prothorax rather narrow, with sulcate median line and with one or two setae at apical angles; elytra narrow, glabrous, with 4–5 discal setae; apex deeply excised, lateral apical angles angulate; tarsi narrow and elongate; tarsal claws denticulate; aedeagus without distinctly sclerotised pieces; gonocoxite-2 short, curved, with two very elongate ventral setae and a similarly sized dorsal seta.

Australian species. Only *H. caeruleipennis* Sloane, 1910.

Distribution. North-eastern Queensland; extraterritorial in the Oriental Region and in New Guinea and the Bismarck Archipelago.

Biology. Apparently arboreal on logs and trunks of trees in rainforests, perhaps also on understorey foliage and in the canopy. Predacious and probably nocturnal. In New Guinea specimens were repeatedly sampled at light.

References. Sloane (1910a); Csiki (1932); Jedlicka (1963); Darlington (1968); Moore *et al.* (1987); Lorenz (2005).

Philophloeus Chaudoir, 1844 (Pl. 34A)

Type species. *Cymindis australis* Dejean, 1826, by original designation.

Gomelina Blanchard, 1853

Idius Chaudoir, 1869c

Characteristics. Rather large, depressed beetles, commonly with distinct elytral pattern of longitudinal stripes; labrum elongate; eyes large and laterally protruding; mentum toothed; penultimate palpomere of labial palp longer than apical palpomere; prothorax rather wide, moderately cordiform; elytra wide, depressed, densely pilose, pilosity decumbent; apex little excised, unarmed; tarsi narrow and elongate; tarsal claws denticulate; aedeagus in the few species with recorded male genitalia, with some distinctly sclerotised pieces; gonocoxite-2 very short, triangular, with two rather large ventral setae and one dorsal seta.

Australian species. 31 species.

Distribution. Southern and eastern Australia, including Tasmania.

Biology. On and under bark of various eucalypt and non-eucalypt trees, in various forests and woodlands.

In some areas of eastern and southern Australia species of *Philophloeus* are common under bark of bark-shedding eucalypts. Predacious and probably nocturnal. However, little is known about habits, diet, and reproduction.

References. Chaudoir (1869c); Sloane (1898, 1920); Csiki (1932); Matthews (1980); Moore *et al.* (1987); Lorenz (2005).

Keys to species. Chaudoir (1869c); Sloane (1888, 1920). All keys are outdated and do not cover several the described species.

Agonocheila Chaudoir, 1848 (Pl. 18L)

Type species. *Agonocheila guttata* Chaudoir, 1848, by monotypy.

Agonochila Chaudoir, 1869c (misspelling)

Characteristics. Rather small, depressed beetles, commonly with various sorts of elytral patterns; labrum elongate; eyes large and laterally protruding; mentum toothed; penultimate palpomere of labial palp shorter than apical palpomere; prothorax variously shaped, more or less cordiform; elytra fairly wide, depressed, densely pilose, pilosity inclined; apex little excised, unarmed; tarsi narrow and elongate; tarsal claws denticulate; aedeagus in the few species with recorded male genitalia with distinctly sclerotised pieces; gonocoxite-2 very short, triangular, with two rather large ventral setae and one elongate dorsal seta.

Australian species. 28 species.

Distribution. The whole of Australia including Tasmania, but apparently more numerous and diverse in the East and South. Contrary to Darlington (1968), *Agonocheila* does not occur in New Guinea, as all species referred to this genus by Darlington have been transferred to other genera (Baehr 2006c). One Australian species occurs in New Zealand (Larochelle & Larivière 2001).

Biology. All species are arbouricolous in sclerophyll forests, semiarid woodlands, and in the moss of tree trunks in rainforests. They live under bark or in deep fissures in the bark of many eucalypt and non-eucalypt trees. *Agonocheila* is part of the rich cortical and subcortical carabid community of Australia and many species belong to the commonest and most numerous inhabitants in the mentioned habitats. Predacious and probably nocturnal. However, little else is known about habits, diet, and reproduction.

References. Blackburn (1892, 1894b); Sloane (1898); Matthews (1980); Moore *et al.* (1987); Baehr (1990c, 2006c); Lorenz (2005); Blackburn (1895); Sloane (1920); Csiki (1932); Darlington (1968).

Key to species. The key in Sloane (1920) only covers Tasmanian species. For the mainland species no usable key is available.

Cheilagona Baehr, 2006 (Pl. 21D)

Type species. *Agonocheila gressitti* Darlington, 1968, by original designation.

Characteristics. Small, dorsally and laterally convex species, with various elytral patterns; labrum elongate; eyes rather large and laterally protruded; mentum toothed; penultimate palpomere of labial palp shorter than apical palpomere; prothorax wide, not cordiform, with obtuse basal angles; elytra ovoid, convex, densely pilose, pilosity decumbent; apex barely excised, unarmed; tarsi narrow and elongate; tarsal claws denticulate; aedeagus with sclerotised, commonly dentate pieces; gonocoxite-2 rather short, curved, with two remarkably stout and elongate ventral setae and a similarly elongate dorsal seta.

Note. Two Australian species were transferred from *Agonocheila* to *Cheilagona* by Baehr (2006c), but additional ones may exist, either still included in *Agonocheila*, or ones yet to be described.

Australian species. Two species.

Distribution. North-eastern Queensland; extraterritorial in New Guinea.

Biology. Arbouricolous on or under bark of rainforest trees. Predacious and probably nocturnal.

References. Baehr (2006c).

Crassagena Baehr, 2006 (Fig. 12.84)

Type species. *Crassagena depressa* Baehr, 2006, by original designation.

Characteristics. Small, wide, very depressed beetles; eyes large, with very wide, quadrangular orbits; labrum elongate; mentum with shallow tooth; prothorax wide, trapezoidal; elytra wide and very depressed, sparsely pilose, with rounded apex; tarsi short, impilose, with narrow 4th tarsomeres; tarsal claws denticulate; male genitalia unknown; gonocoxite-2 short, curved, with obtuse apex, with two stout ventral setae and a similar dorsal seta. The colouration of the elytra with two conspicuous dark, latero-apical spots is very characteristic.

Australian species. Only *C. depressa* Baehr.

Distribution. North-eastern Queensland, known only from Mt. Webb near Cooktown.

Biology. Probably a bark inhabiting species.

References. Baehr (2006e).

Subtribe Apenina Ball, 1983

Characteristics. Head usually with coarse and regular longitudinal sulci and ridges; labrum longer than wide; labial palp usually more or less widened, or securiform, in particular in males; base of pronotum not produced in the middle, laterally not excised; apex of elytra oblique, little excised; 4th tarsomeres narrow, not excised and without squamose setae beneath; tarsal claws dentate; aedeagus variously shaped, but always with an elongate filum that stands out from the ostium; gonocoxites variously shaped, gonocoxite-2 usually short, with a varied number of ventral and dorsal setae.

Habutarus Ball & Hilchie, 1983 (Pl. 26E, J)

Type species. *Nototarus papua* Darlington, 1968, by monotypy.

Characteristics. Moderately large, usually unicolourous beetles; head with very distinct and regular longitudinal furrows; eyes fairly large with convex orbits; mentum with large tooth; labial palp in females widened, in males securiform; prothorax cordiform; elytra rather short, slightly ovoid, deeply striate, glabrous, with oblique apex; flight wings reduced; tarsi dorsally pilose; 4th tarsomeres narrow; tarsal claws denticulate; aedeagus elongate, without distinct sclerotised pieces on the endophallus, but with a long filum; gonocoxite-1 with a spinose projection, gonocoxite-2 short, with a single large ventral and dorsal seta each.

Australian species. 19 species.

Distribution. Eastern Australia from northern New South Wales to Cape York Peninsula, northern parts of Northern Territory; extraterritorial in New Guinea.

Biology. Ground living in wet sclerophyll forests and rainforests, commonly in montane areas. Predacious and probably nocturnal. Flightless. Most species seem to be very rare, or they are rarely sampled for whatever reason.

Notes. Ball and Hilchie (1983) regarded this genus a subgenus of the Oriental genus *Cymindoidea* Laporte, 1833. However, we think that the morphological differences are large enough to regard *Habutarus* a valid genus. Two of the species enumerated by Moore *et al.* (1987) in the genus *Nototarus* Chaudoir, 1875 belong to *Habutarus*. All Australian species belong in the nominate subgenus.

References. Ball & Hilchie (1983) (as subgenus of *Cymindoidea* Laporte, 1833); Lorenz (2005); Baehr (2008i).

Key to species. Baehr (2008i).

Subtribe Calleidina Chaudoir, 1872

Characteristics. Labrum wider than long; labial palp distinctly widened, or securiform, in particular in males; base of pronotum not produced in the middle, laterally not excised; apex of elytra oblique, variously shaped; 4th tarsomeres wide, deeply excised and with squamose setae beneath, or narrow, without squamose setae beneath; tarsal claws commonly dentate; aedeagus variously shaped, but always without an elongate filum that stands out from the ostium; gonocoxites usually elongate, gonocoxite-2 usually straight, with a tuft of setae at apex, rarely asetose.

Notes. Casale (1998) conducted a cladistic analysis of morphological characters to reconstruct the phylogenetic relationships for species in the subtribe. The extent of the subtribe is much in discussion in the literature. Habu (1967) separated Anomotarina as a subtribe and included Metallicina and Physoderina in Calleidina, Lorenz (2005) united Calleidina and Anomotarina in the subtribe Agrina, Ball & Bousquet (2001) separated Calleidina and Agrina, and Bouchard *et al.* (2011) likewise separated Calleidina and Agrina but included Anomotarina in Calleidina. All authors except Habu separated Physoderina and Metallicina from Calleidina. We take an intermediate position in uniting Anomotarina with Calleidina, but regarding Metallicina and Physoderina separate subtribes.

Demetrida White, 1846 (Pl. 23J; Fig. 12.67)

Type species. *Demetrida lineella* White, 1846, by subsequent designation by Moore 1987.

Xanthophoea Chaudoir, 1848

Characteristics. Usually elongate body shape, but in colour and elytral pattern quite variable; head with more or less large eyes; labrum short; mentum toothed; labial palp in females widened, in males securiform; prothorax not or little cordiform; base not produced nor laterally excised; elytra elongate, striate, glabrous or with erect pilosity; apex either oblique with more or less distinct excision, dentate or spinose in various ways; tarsi dorsally pilose; 4th tarsomeres wide and excised, with squamose setae beneath; tarsal claws denticulate; aedeagus not known for most species, with or without distinctly sclerotised pieces; gonocoxite-2 in the few recorded species short, curved, asetose or with few setae.

Australian species. 39 species.

Distribution. The whole of Australia, including Tasmania; extraterritorial in New Caledonia, New Guinea and the Bismarck Archipelago.

Biology. Most species live on and under bark of various tree species from rainforests to semiarid areas where species can be found on river gums; some, however, seem to be ground living. In New Guinea, according to Darlington (1968), many species may live in the understorey and in the foliage of trees. Predacious and probably nocturnal, because specimens sometimes are attracted to light. All Australian species seem to possess unarmed elytra.

Notes. We are unable to distinguish between *Demetrida* and *Xanthophoea*, which are sometimes regarded subgenera (Lorenz 2005). Therefore we regard *Xanthophoae* simply a synonym of *Demetrida*.

References. Lacordaire (1854); Csiki (1932); Britton (1941); Darlington (1968); Matthews (1980); Moore *et al.* (1987); Larochelle & Larivière (2001); Lorenz (2005).

Key to species. Sloane (1917), but this key covers only a relatively small part of the then known species.

Trigonothops Macleay, 1864 (Pl. 39J)

Type species. *Calleida pacifica* Erichson, 1842, by original designation.

Characteristics. Usually rather elongate, depressed, variously patterned beetles; eyes very large, laterally much produced; mentum toothed; labial palp in females widened, in males securiform; prothorax more or less cordiform; base not produced or laterally excised; elytra elongate, striate, glabrous; apex oblique, barely excised; flight wings present; tarsi dorsally impilose; 4th tarsomeres wide and excised, with squamose setae beneath; tarsal claws denticulate; aedeagus elongate, endophallus with an elongate, sclerotised rod; gonocoxites elongate, gonocoxite-2 straight, with a tuft of setae at apex.

Australian species. 10 species.

Distribution. Almost the whole of Australia including Tasmania; extraterritorial in New Guinea.

Biology. On and under the bark of various trees in sclerophyll forests and woodlands, even in semiarid regions. Predacious and probably nocturnal. In some areas in northern Australia common under bark-shedding, gum-type eucalypts. Little else is known about habits, diet, and reproduction.

Notes. Ball & Hilchie (1983) and subsequently Moore *et al.* (1987) included in *Trigonothops* several other taxa, e.g. *Phloeocarabus* Macleay, *Abaditicus* Ball & Hilchie, *Diabaticus* Bates, and *Speotarus* Moore, that we regard separate genera as they differ in significant character states.

References. Chaudoir (1877); Macleay (1888); Blackburn (1892); Csiki (1932); Darlington (1968); Matthews (1980); Ball & Hilchie (1983); Moore *et al.* (1987); Lorenz (2005).

Phloeocarabus Macleay, 1871 (Pl. 34F)

Type species. *Phloeocarabus mastersii* Macleay, 1871, by monotypy.

Notoxena Chaudoir, 1877

Characteristics. Usually rather elongate, depressed, variously patterned beetles; eyes very large, laterally much produced; mentum toothed; labial palp in females widened, in males securiform; prothorax more or less cordiform; base produced and laterally excised; elytra elongate, striate, glabrous; apex oblique, not excised; flight wings present; tarsi dorsally impilose; 4th tarsomeres narrow and not excised, without squamose setae beneath; tarsal claws denticulate; aedeagus elongate, endophallus with an elongate, sclerotised rod; gonocoxites elongate, gonocoxite-2 straight, with a tuft of setae at apex.

Australian species. Seven species.

Distribution. Eastern and northern Australia; extraterritorial in New Guinea.

Biology. On and under bark of various eucalypt and non-eucalypt trees in various sorts of eucalypt woodlands. In New Guinea perhaps also in rainforests. Predacious and probably nocturnal, because specimens sometimes are attracted to light. In eucalypt woodlands of the Far North quite common under bark. Little else is known about habits, diet, and reproduction.

Notes. Ball & Hilchie (1983) and Moore *et al.* (1987) include the genus in *Trigonothops* Macleay, 1864. We think, however, that the morphological differences are large enough for the status of a separate genus.

References. Sloane (1898); Csiki (1932); Darlington (1968); Matthews (1980); Moore *et al.* (1987); Lorenz (2005).

Abaditicus Ball & Hilchie, 1983 (Fig. 12.69)

Type species. *Diabaticus collaris* Blackburn, 1901, by original designation.

Characteristics. Rather elongate, depressed, unicolourous beetles; eyes very large, laterally much produced; neck very constricted; mentum toothed; labial palp in females widened, in males securiform; prothorax narrow and more or less cordiform; base not produced and not laterally excised; elytra elongate, striate, glabrous; apex oblique, not excised; flight wings present; tarsi dorsally impilose; 4th tarsomeres narrow and not excised, without squamose setae beneath; tarsal claws denticulate; aedeagus elongate, endophallus with an elongate, sclerotised rod; gonocoxites elongate, gonocoxite-2 straight, with a tuft of setae at apex.

Australian species. Two species.

Distribution. South-eastern Australia from eastern Victoria to south-eastern New South Wales, in the north to at least Barrington Tops.

Biology. Both species live under bark of several bark-shedding eucalypts. Predacious and probably nocturnal. Little else is known about habits, diet, and reproduction.

Note. This genus was described as a subgenus of *Trigonothops* Macleay, 1864, by Ball & Hilchie (1983). However, we believe the differences between the two species of *Abaditicus* and the species of the nominate subgenus are significant and their separation as a genus seems justified.

References. Ball & Hilchie (1983); Moore et al. (1987); Lorenz (2005).

Keys to species. Ball & Hilchie (1983).

Diabaticus Bates, 1878 (Pl. 23L)

Type species. *Plochionus australis* Erichson, 1842, by monotypy.

Characteristics. Rather elongate, more or less unicolourous species; eyes variously sized; neck constricted or not; mentum toothed; labial palp widened; prothorax more or less cordiform; base not produced and not laterally excised; elytra elongate, striate, glabrous or pilose; apex oblique, not excised; metathoracic wings present or absent; tarsi dorsally pilose or impilose; 4th tarsomeres narrow and not excised, not squamose beneath; tarsal claws denticulate; aedeagus elongate, internal sac with an elongate, sclerotised rod; gonocoxites elongate, gonocoxite 2 straight, with a tuft of setae near apex and sometimes with an apical hyaline seta.

Australian species. Five species.

Distribution. South-eastern Australia from Victoria to north-eastern New South Wales, Tasmania.

Biology. Little recorded. One species definitely is arbouricolous and occurs under bark of bark shedding eucalypts, in other species the habits are not recorded and doubtful. Predacious and probably nocturnal. Nothing else is known about habits, diet, and reproduction.

Notes. This genus was regarded a subgenus of *Trigonothops* by Ball & Hilchie (1983), but the differences in many characters are so significant that Baehr (2010c, 2012f) regarded it a separate genus. However, the genus is very diverse in shape and structure and likely could be subdivided into subgenera or even genera. Because one peculiar Tasmanian species is not well known, we presently refrain from undertaking any nomenclatorial decisions.

References. Csiki (1932); Moore *et al.* (1987); Lorenz (2005).

Keys to species. Baehr (2010c, 2012f).

Speotarus Moore, 1964 (Pl. 38F)

Type species. *Speotarus lucifugus* Moore, 1964, by original designation.

Characteristics. Elongate, rufo-brunneous beetles; eyes moderately large, but laterally little produced, orbits elongate; neck not much constricted; mentum dentate; labial palp moderately widened; prothorax narrow; base not produced and laterally not excised; elytra striate, glabrous; apex convex, not excised; flight wings present; 4th tarsomeres narrow and not excised, without squamose setae beneath; tarsal claws not denticulate; aedeagus elongate, endophallus with an elongate, sclerotised rod; gonocoxites elongate, gonocoxite-2 straight, with tapering apex, with a tuft of short subapical setae and a larger dorsal and ventral seta each near apex.

Australian species. Only *S. lucifugus* Moore with two subspecies.

Distribution. Southern Western Australia to southern New South Wales.

Biology. Cavernicolous in a variety of caves. Predacious. Little else is known about habits, diet, and reproduction.

Note. Although both subspecies are apparently true cavernicoles, they still possess fully developed flight wings. An additional species is currently being described from inland Australia (MB). It is not known if this species is also cavernicolus.

References. Matthews (1980); Moore *et al.* (1987); Lorenz (2005).

Key to species. Moore (1964a).

Anomotarus Chaudoir, 1875 (Pl. 19K; Fig. 12.15)

Type species. *Anomotarus olivaceus* Chaudoir, 1875, by monotypy.

Uvea Fauvel, 1881

Characteristics. Quite variously shaped, structured, and patterned species; eyes variously sized; neck constricted or not; labrum short; mentum dentate or not; labial palp widened to securiform; prothorax more or less cordiform; base more or less produced in the middle but laterally not distinctly excised; elytra variously shaped, striate, glabrous or pilose; apex oblique, variously or not excised; flight wings present or absent; tarsi dorsally impilose; 4th tarsomeres narrow and not excised, without squamose setae beneath; tarsal claws denticulate or not; aedeagus elongate, endophallus with an elongate, sclerotised rod; gonocoxites elongate, gonocoxite-2 straight, asetose or with a tuft of setae at apex.

Australian species. 109 species and seven subspecies.

Distribution. The whole of Australia, but the subgenera are quite differently distributed. The nominate subgenus occurs from southern and south-eastern mainland Asia through the Indonesian and Philippine insular belts, New Guinea, the Bismarck Archipelago to Australia, Solomon Islands, New Hebrides, New Caledonia, and Loyalty Islands. All other subgenera are restricted to Australia.

Biology. All species are ground-living and they occur from rainforests to semiarid and arid areas, and from coastal dunes to high mountains. Most species are strictly nocturnal and during daytime hide under stones, logs, debris, and in cracks in the soil, sometimes they are found under or in the bark at the base of trees. Some southern species, however, are occasionally seen during the daytime. Very little is known about habits, diet, and reproduction.

Notes. The genus presently includes five subgenera, some of which formerly were regarded separate genera. However, Ball & Hilchie (1983) and Baehr (2009f) demonstrated that the characters formerly used for separation of *Nototarus* Chaudoir, 1875 from *Anomotarus* are very weak and not usable to separate *Nototarus* as a distinct genus. The whole genus was revised by Baehr (2004a, 2006d, 2009f). A few additional species were described more recently.

References. Sloane (1898, 1917); Csiki (1932); Jedlicka (1963); Habu (1967); Darlington (1968); Matthews (1980); Ball & Hilchie (1983); Moore (1992); Moore *et al.* (1987); Baehr (2004a, 2006d, 2009f, 2011a, 2012e); Lorenz (2005).

Keys to species. Baehr (2006d, 2009f, 2011a, 2012e).

Key to the subgenera of *Anomotarus* Chaudoir recorded from Australia

1. Body surface covered with elongate, erect pilosity 2
– Body surface impilose, or pilosity very inconspicuous and depressed 3

2(1). Whole surface without microreticulation, very glossy; pronotum coarsely punctate, but not transversely sulcate; colour blue or green .... ***Lithostrotus*** **Blackburn**
– At least elytra very coarsely microreticulate, remarkably dull; pronotum coarsely punctate and transversely sulcate; colour ferruginous to reddish-brown .... ***Rubritarus*** **Baehr**

3(2). Mentum distinctly toothed; rarely flight wings reduced (mainly species from mountain tops in south-eastern Australia) .... ***Anomotarus*** **Chaudoir**
– Mentum without distinct tooth; flight wings always reduced 4

4(3). Elytra very wide, remarkably oviform, > 2 × as wide as pronotum, apical margin deeply excised, lateral apical angles protruded; intervals with coarse, conspicuous punctures; interval 3 with several setae .... ***Amplitarus*** **Baehr**
– Elytra narrower, not or far less oviform, < 2 × as wide as pronotum, apical margin barely or not excised, lateral apical angles not protruded; punctures of intervals commonly barely visible, rarely with conspicuous punctation; interval 3 bisetose .... ***Nototarus*** **Chaudoir**

Subgenus *Anomotarus* Chaudoir, 1875

Type species. *Anomotarus olivaceus* Chaudoir, 1875, by monotypy.

Characteristics. Variously shaped and patterned species; eyes moderately large, orbits fairly large; neck not much constricted; mentum dentate; labial palp securiform; prothorax rather cordiform; base more or less produced in the middle but laterally not distinctly excised; elytra variously shaped, striate, glabrous; apex oblique, little excised; flight wings usually present, rarely absent; tarsal claws denticulate; aedeagus elongate, endophallus with an elongate, sclerotised rod; gonocoxites elongate, gonocoxite-2 straight, usually with a tuft of setae at apex.

Australian species. 43 species and seven subspecies.

Distribution. The whole of Australia, including Tasmania, but most species occur in the east and south-east and in the tropical north of Australia. Only few species are found in the arid interior and most of Western Australia; extraterritorial across the whole Oriental Region and in New Guinea, the Bismarck Archipelago, Solomon Islands, New Hebrides, New Caledonia, and the Loyalty Islands.

Biology. Ground living from rainforests through any sort of sclerophyll forests, to semiarid and arid areas, also above tree line on high mountains and tablelands of south-eastern Australia. One Tasmanian species was even sampled on coastal dunes. Populations of one species were found in caves. Most species living on high mountains possess brachypterous wings, all other species are able to fly and fly well. As strictly nocturnal beetles, they are commonly attracted to light. During the daytime they hide under stones, logs, moss, ground litter, debris, and in cracks in the soil. Predacious. Many eastern and northern species seem to prefer moist habitats and they are typically encountered near swamps and water bodies.

References. Sloane (1898, 1917); Csiki (1932); Jedlicka (1963); Habu (1967); Darlington (1968); Matthews (1980); Ball & Hilchie (1983); Moore (1992); Moore *et al.* (1987); Baehr (2004a, 2006d, 2009f, 2011a, 2012e); Lorenz (2005).

Keys to species. Baehr (2006d, 2011a, 2012e).

Subgenus *Lithostrotus* Blackburn, 1894

Type species. *Lithostrotus caerulescens* Blackburn, 1894, by monotypy.

Lestianthus Sloane, 1895a

Characteristics. Rather small, unicolourous green or blue beetles, without microreticulation; with erect pilosity; eyes moderately large, orbits fairly large; neck not much constricted; mentum dentate; labial palp securiform; prothorax rather cordiform; base more or less produced in the middle but laterally not distinctly excised; elytra deeply punctate-striate, pilose; apex oblique, little excised; flight wings reduced; tarsal claws denticulate; aedeagus elongate, endophallus with an elongate, sclerotised rod; gonocoxites elongate, gonocoxite-2 straight, with a tuft of short setae at apex.

Australian species. Only *A. (L.) caerulescens* (Blackburn).

Distribution. The single species occurs in montane south-eastern Australia from eastern Victoria to south-eastern New South Wales, and in the Australian Capital Territory.

Biology. Ground living in temperate, wet and dry sclerophyll forests, usually at rather high altitude. Predacious and apparently nocturnal.

Note. The single species was included in *Nototarus* Chaudoir, 1875 by Sloane (1898), but was transferred to *Lithostrotus* as a separate subgenus by Baehr (2009f).

References. Ball & Hilchie (1983); Baehr (2009f).

Subgenus *Nototarus* Chaudoir, 1875

Type species. *Nototarus australis* Chaudoir, 1875, by monotypy.

Characteristics. Variably shaped, structured, and patterned species; eyes moderately large, orbits usually large; neck not much constricted; mentum edentate; labial palp securiform; prothorax rather cordiform; base more or less produced in the middle but laterally not distinctly excised; elytra variously striate, impilose or very inconspicuously pilose; apex oblique, little excised; flight wings reduced; tarsal claws usually not or barely denticulate; aedeagus elongate, endophallus with an elongate, sclerotised rod; gonocoxites elongate, gonocoxite-2 straight, asetose.

Australian species. 61 species.

Distribution. Western and northern Australia with single species as far east as north-western Queensland, Cape York Peninsula, and eastern South Australia, but the bulk of the species occurs in western and the western part of central Western Australia and western South Australia.

Biology. Predacious, ground living from savannah woodlands to semiarid and arid habitats, commonly in dry savannah, mulga, and mallee country, but also in Spinifex and Mitchell grass plains. As strictly nocturnal and generally short-winged beetles, specimens are best collected by use of pitfall traps. Single specimens were recorded in bark cracks at lower parts of trunks of Acacias and mallee trees. One of the smaller species was found in a cave.

Note. For a long time this subgenus was regarded a separate genus. Ball & Hilchie (1983) first demonstrated that the differencs between *Nototarus* and *Anomotarus* (s. str.) are not sufficient for separation of these genera, and Baehr (2009f), when revising *Nototarus*, came to the same conclusion.

References. Sloane (1898, 1917); Csiki (1932), Darlington (1968); Ball & Hilchie (1983); Moore *et al.* (1987); Lorenz (2005); Baehr (2009f, 2012e).

Key to species. Baehr (2009f, 2012e).

Subgenus *Amplitarus* Baehr, 2009

Type species. *Nototarus angusticollis* Sloane, 1915, by original designation.

Characteristics. Large, black beetles with wide, ovoid elytra; eyes moderately large, orbits large; neck not much constricted; mentum edentate; labial palp securiform; prothorax rather cordiform; base more or less produced in the middle but laterally not distinctly excised; elytra deeply punctate-striate, interval 3 with seriate setiferous punctures; apex oblique, deeply excised; flight wings reduced; tarsal claws not denticulate; aedeagus elongate, endophallus with an elongate, sclerotised rod; gonocoxites elongate, gonocoxite-2 straight, asetose.

Australian species. Two species.

Distribution. Central Western and north-western Western Australia south of Great Sandy Desert.

Biology. Ground living in semiarid open savannah to mulga woodland and Spinifex grassland. Predacious. As strictly nocturnal and generally short-winged beetles, specimens are usually collected by use of pitfall traps. Almost nothing is known about habits, diet, and life history.

References. Baehr (2009f, 2012e).

Key to species. Baehr (2009f).

Subgenus *Rubritarus* Baehr, 2009

Type species. *Anomotarus setiferus* Baehr, 2009, by original designation.

Characteristics. Ferruginous coloured beetles with erect pilosity on the dorsum; eyes moderately large, orbits large; neck not much constricted; mentum edentate; labial palp securiform; prothorax rather cordiform; base more or less produced in the middle but laterally not distinctly excised; elytra deeply striate, with erect setosity; apex oblique, slightly excised; flight wings reduced; tarsal claws not denticulate; aedeagus elongate, endophallus with an elongate, sclerotised rod; gonocoxites elongate, gonocoxite-2 straight, with a tuft of short setae at apex.

Australian species. Two species.

Distribution. North-western Australia, apparently restricted to the Pilbara Region.

Biology. Ground living in semiarid open savannahs, Mulga woodlands and Spinifex grasslands. Predacious. As strictly nocturnal and short-winged beetles, specimens were only collected by use of pitfall traps. Nothing is known about habits, diet, and life history.

References. Baehr (2009f).

Key to species. Baehr (2009f).

Rugitarus Baehr, 2009 (Pl. 37J)

Type species. *Rugitarus puellarum* Baehr, 2009, by original designation.

Characteristics. Depressed, rufous to black, pilose beetles; head depressed with several irregular longitudinal sulci mediad of the eyes; eyes moderately large, but laterally little produced, orbits large, convex; neck much constricted; mentum dentate; labial palp securiform; prothorax wide, slightly cordiform; base not produced and laterally not excised; elytra striate, pilose; apex convex, not excised; flight wings full; 4th tarsomeres narrow, not excised, without squamose setae beneath; tarsal claws denticulate; aedeagus elongate, endophallus with a very elongate, sclerotised rod; gonocoxites rather short, gonocoxite-2 trapezoidal, with oblique apex, with two apical setae.

Australian species. Only *R. puellarum* Baehr, with additional six subspecies.

Distribution. Southern, eastern and northern Australia, but apparently most common in semiarid areas of the southern half of Australia.

Biology. In and under bark of various trees, including Mallee and Mulga, in open forests and woodlands. Predacious and probably nocturnal, but specimens are rarely seen at light. Little else is known about habits, diet, and reproduction. Almost all known specimens were sampled by pyrethrum spraying bark of trees with deeply fissured bark.

References. Baehr (2009f).

Key to subspecies. Baehr (2009f).

Anomotariella Baehr, 2012 (Pl. 19F)

Type species. *Anomotariella hippocrepis* Baehr, 2012, by original designation.

Characteristics. Small, depressed beetles with horseshoe-shaped elytral pattern; eyes moderately large, but laterally little produced, orbits large, tumescent; neck moderately constricted; mentum dentate; labial palp markedly widened; prothorax cordiform; base produced in the middle, laterally slightly excised; elytra striate, pilose, pilosity very short, erect; apex oblique, barely excised; flight wings full; 4th tarsomeres narrow, not excised, without squamose setae beneath; tarsal claws denticulate; aedeagus elongate, endophallus with an elongate, sclerotised rod; gonocoxites elongate, gonocoxite-2 straight, with a dense tuft of setae at apex.

Australian species. Only *A. hippocrepis* Baehr.

Distribution. North-eastern Queensland: Atherton and Windsor Tablelands.

Biology. The single species was collected by pyrethrum spraying the bark of tree trunks, mainly Bunya Pines (*Araucaria bidwilli*), at high altitude.

References. Baehr (2012c).

Plochionus Dejean, 1821 (Pl. 35L)

Type species. *Carabus pallens* Fabricius, 1775, by monotypy.

Characteristics. Compact, uniformly brown beetles; eyes large, laterally rather produced, orbits small; neck not constricted; palps moderately widened; antenna short; prothorax not cordiform; base not produced and laterally not excised; elytra striate, impilose; apex convex, not excised; flight wings full; tarsi short and thick, glabrous above; 4th tarsomeres not widened or excised, without squamose setae beneath; tarsal claws denticulate; aedeagus without distinctly sclerotised folds; female genitalia not recorded.

Australian species. Only *P. pallens* (Fabricius).

Distribution. North-eastern Queensland; extraterritorial almost worldwide, mainly in and near harbours; also widely distributed on Pacific islands.

Biology. Little recorded. Probably ground living in rainforests.

Note. The single species recorded from Australia probably is a South American species which was carried by humans over most of the world, including many Pacific islands.

References. Csiki (1932); Jedlicka (1963); Darlington (1968); Lorenz (2005).

Subtribe Somotrichina Mateu, 1963

Characteristics. Head without coarse and regular longitudinal sulci and ridges; labrum longer than wide; labial palp narrowed apicad, not securiform; base of pronotum produced in the middle, laterally excised; apex of elytra oblique; 4th tarsomeres narrow, not excised, without squamose setae beneath; tarsal claws denticulate; aedeagus variously shaped, without distinctly sclerotised pieces on the endophallus; gonocoxites rather elongate, gonocoxite-2 varied, with a variable number of ventral and dorsal setae.

Somotrichus Seidlitz, 1887 (Pl. 103H)

Type species. *Carabus elevatus*, Fabricius, 1792 (= *Lebia unifasciata* Dejean, 1831), by original designation.

Characteristics. Small, wide, depressed, yellow beetles with a wide, transverse, dark band on the elytra; whole surface with short, erect pilosity; pronotum wide, with additional setae at apical and basal angles; aedeagus without distinctly sclerotised pieces; gonocoxite-2 fairly elongate, slightly curved, with one dorsal and two ventral subapical setae.

Australian species. Only *S. unifasciatus* (Dejean).

Distribution. Only recorded from Atherton, north-eastern Queensland; extraterritorial almost worldwide distributed.

Biology. Little recorded. The single species is commonly carried by humans with vegetables or seeds and is most commonly encountered in harbours. It is probably a rainforest living species.

Notes. The single Australian species apparently comes from tropical Africa, but is now distributed almost worldwide. Thus far the record from Atherton seems to be unique in Australia and it is unclear if the species has established.

References. Csiki (1932); Jedlicka (1963); Habu (1967); Darlington (1968); Kabak (2003); Lorenz (2005); Baehr (2007a).

Subtribe Celaenephina Habu, 1982

Characteristics. Head without coarse and regular longitudinal sulci and ridges; labrum wider than long; mentum edentate; labial palp narrowed apicad, not securiform; base of pronotum not produced in the middle but slightly convex; apex of elytra convex and almost covering the abdomen; 4th tarsomeres narrow, not excised, without squamose setae beneath; tarsal claws smooth; aedeagus without distinctly sclerotised pieces on the endophallus; parameres almost similarly sized and shaped; gonocoxites rather elongate, gonocoxite-1 at apex densely setose; gonocoxite-2 with rounded apex, with one dorsal and two ventral subapical setae and with a very elongate apical seta.

Celaenephes Schmidt-Göbel, 1846 (Pl. 21F)

Type species. *Celaenephes parallelus* Schmidt-Göbel, 1846, by subsequent designation by Lacordaire 1858.

Taromorpha Blackburn, 1894b.

Characteristics. Moderately large, elongate, depressed, black beetles; apex of elytra almost completely covering the abdomen.

Australian species. Two species.

Distribution. Eastern and northern Australia from eastern Victoria to Cape York Peninsula and far Northern Territory; extraterritorial in the whole Oriental Region and in New Guinea, the Bismarck Archipelago, Solomon Islands, Fiji, and Samoa.

Biology. On and under bark of trees in rainforests and more open forests. Predacious and apparently nocturnal, because specimens are commonly attracted to light. Nothing else is known about their habits, diet, and reproduction.

References. Lacordaire (1854); Csiki (1932); Jedlicka (1963); Habu (1967); Darlington (1968); Moore *et al.* (1987); Shpeley *et al.* (1985); Kabak (2003); Lorenz (2005).

Key to species. Shpeley *et al.* (1985).

Subtribe Dromiusina Bonelli, 1810

Characteristics. Head without coarse and regular longitudinal sulci and ridges; labrum wider than long; mentum dentate or not; labial palp narrowed apicad, not securiform; base of pronotum produced in the middle or not; apex of elytra oblique, variously excised; 4th tarsomeres narrow, not excised, without squamose setae beneath; tarsal claws dentate or edentate; aedeagus usually with distinctly sclerotised pieces on the endophallus; parameres differently sized and shaped; gonocoxites rather short; gonocoxite-2 variously shaped, with or without few apical setae.

Syntomus Hope, 1838 (Pl. 38K)

Type species. *Carabus truncatellus* Linnaeus, 1761, by subsequent designation by Andrewes 1934.

Metabletus Schmidt-Göbel, 1846.

Characteristics. Small beetles; eyes large, laterally rather produced, orbits small; neck not constricted; mentum bidentate; paraglossae barely surpassing the glossa; labial palp not widened; prothorax slightly cordiform; base produced and laterally excised; elytra striate, impilose; apex oblique, little excised; flight wings full; tarsi narrow and elongate; 4th tarsomeres not widened or excised, without squamose setae beneath; tarsal claws denticulate; aedeagus rather stout, endophallus usually with a variously shaped sclerotised piece; gonocoxites rather short, gonocoxite-2 trapezoidal, with oblique apex, asetose.

Australian species. Only *S. quadripunctatus* (Schmidt-Göbel, 1846).

Distribution. North-eastern Queensland; extraterritorial almost worldwide. *Syntomus quadripunctatus* is widely distributed in the Oriental Region.

Biology. Ground living in leaf litter in wet environments. Predacious and probably nocturnal. Little is known about habits, diet, and reproduction.

Note. Lorenz (2005) placed the genus *Syntomus* in a separate tribe Lionychina Jeannel, 1949. We do not consider the differences between Dromiusina and Lionychina sufficiently important to justify their separation.

References. Csiki (1932); Jeannel (1942); Jedlicka (1963); Habu (1967); Darlington (1968); Baehr, (1987a); Moore *et al.* (1987); Lorenz (2005).

Microlestodes Baehr, 1987 (Pl. 29J)

Type species. *Dromius humeralis* Macleay, 1871 (= *Microlestodes macleayi* Csiki, 1932), by original designation.

Characteristics. Small beetles; eyes large, laterally rather produced, orbits small; neck not constricted; mentum edentate; paraglossae well surpassing the glossa; labial palp not widened; prothorax more or less cordiform; base produced and laterally excised; elytra striate, impilose; apex oblique, little excised; flight wings full or reduced; tarsi narrow and elongate; 4th tarsomeres not widened or excised, without squamose setae beneath; tarsal claws denticulate; aedeagus rather stout, endophallus usually with a variously shaped sclerotised piece; gonocoxites rather short, gonocoxite-2 trapezoidal, with oblique apex, asetose.

Australian species. 14 species.

Distribution. Almost the whole continent, including Tasmania; extraterritorial in New Guinea.

Biology. Ground living in a variety of habitats, but most commonly in humid environments or in arid areas near water. Predacious. The species are strictly nocturnal and thus are almost always sampled at light.

References. Baehr (1990e, 2009b, 2010e); Lorenz (2005).

Keys to species. Baehr (1987a, 2009b, 2010e).

Key to the subgenera of *Microlestodes*

1. Flight wings fully developed; metanepisternum elongate; basal and apical groups of marginal elytral punctures barely separated .... ***Microlestodes* Baehr**

– Flight wings reduced; metanepisternum quadrate; basal and apical groups of marginal elytral punctures distinctly separated .... ***Cyclolestodes* Baehr**

Subgenus *Microlestodes* Baehr, 1987

Type species. *Dromius humeralis* Macleay, 1871 (= *Microlestodes macleayi* Csiki, 1932), by original designation.

Characteristics. Flight wings well developed; metanepisternum elongate.

Australian species. 13 species.

Distribution. Almost the whole continent, including Tasmania; extraterritorial in New Guinea.

References. Lorenz (2005).

Keys to species. Baehr (1987a, 1990e, 2009b, 2010e).

Subgenus *Cyclolestodes* Baehr, 1987

Type species. *Microlestodes ovatus* Baehr, 1987, by original designation.

Characteristics. Flight wings reduced; metepisterna short; hind body rather ovate.

Australian species. Only *M.* (*Cyclolestodes*) *ovatus* Baehr.

Distribution. South-eastern Australia from eastern Victoria to south-eastern New South Wales, a record from the New England Tableland in north-eastern New South Wales needs to be verified.

Biology. Apparently ground living in montane areas. Probably mainly in more or less open sclerophyll forests.

References. Lorenz (2005).

Barrymooreana Baehr, 1997 (Pl. 19L)

Type species. *Mooreana quadrimaculata* Baehr, 1987, by original designation.

Mooreana Baehr, 1987a (non *Mooreana* Evans, 1926).

Characteristics. Small, quadrimaculate beetles; eyes large, laterally rather produced, orbits small; neck not constricted; mentum unidentate; paraglossae well surpassing the glossa; labial palp not widened; prothorax wide, cordiform; base slightly produced in the middle; elytra striate, impilose; apex oblique, little excised; flight wings full; tarsi narrow and elongate; 4th tarsomeres not widened or excised, without squamose setae beneath; tarsal claws denticulate; aedeagus moderately stout, endophallus with a sclerotised piece; gonocoxites rather short, gonocoxite-2 trapezoidal, with oblique apex, asetose.

Australian species. Only *B. quadrimaculata* (Baehr).

Distribution. Central and southern Cape York Peninsula, north-eastern Queensland.

Biology. Probably ground living in rainforests. Specimens were collected either at light or by Berlese extraction from ground litter. Predacious and probably nocturnal.

References. Lorenz (2005).

Brigalowia Baehr, 2006 (Fig. 12.81)

Type species. *Brigalowia setifera* Baehr, 2006, by original designation.

Characteristics. Small, very wide and depressed beetles; head wide; eyes of moderate size, laterally little produced; orbits large, oblique, pilose; labrum short and wide; mentum strongly dentate; pronotum very wide, not cordiform; base not produced; elytra short and wide, quadrate, very depressed; all pronotal, discal elytral and most marginal setae extremely elongate; tarsi narrow and elongate, dorsally pilose; 4th tarsomeres narrow, not excised; tarsal claws denticulate; aedeagus rather short, with several sclerotised pieces; gonocoxites elongate; gonocoxite-2 straight, with several extremely short setae at apex.

Australian species. Only *B. setifera* Baehr.

Distribution. Central western and north-western Queensland, adjacent eastern Northern Territory.

Biology. The single known species was collected by pyrethrum spraying of the deep fissured bark of Brigalow and Gidgee Acacias.

References. Baehr (2006k).

Geoffreyella Baehr, 2012 (Fig. 12.94)

Type species. *Geoffreyella holoserica* Baehr, 2012, by original designation.

Characteristics. Small, wide, depressed, ferruginous coloured, inconspicuously patterned beetles; eyes large, but laterally little produced, orbits small; neck not constricted; mentum with elongate tooth; paraglossae barely surpassing the glossa; labial palp not widened; prothorax wide, slightly cordiform; base not produced in the middle; elytra striate, densely pilose; apex oblique, little excised; flight wings full; tarsi narrow and elongate; 4th tarsomeres not widened or excised, without squamose setae beneath; tarsal claws very large, not denticulate; aedeagus elongate, endophallus without sclerotised parts but with several parallel folds; gonocoxite-2 very short, convex at apex, with two large ventral setae and a large dorsal seta.

Australian species. Two species.

Distribution. South-eastern and north-eastern Queensland.

Biology. Arbouricolous on moss covered trees in upland rainforests. All recorded specimens were collected by pyrethrum spraying the lower trunks of trees. Probably predacious and nocturnal.

References. Baehr (2012c).

Key to species. Baehr (2012c).

Subtribe Lichnasthenina Thomson, 1858

Characteristics. Head without coarse and regular longitudinal sulci and ridges; whole dorsal surface extremely rugose, with erect pilosity; eyes large and laterally well produced; labrum wider than long; mentum dentate or not; labial palp widened apicad, but not securiform; base of pronotum not distinctly produced in the middle; apex of elytra oblique, shallowly excised; 4th tarsomeres narrow, not excised, without squamose setae beneath; tarsal claws denticulate; aedeagus with a small, sclerotised piece on the endophallus; parameres very differently sized and shaped; gonocoxites rather short, gonocoxite-2 oblique at apex, with a dense brush of setae on the apical rim and with a single elongate seta.

Notes. The justification of this subtribe is controversial. Whereas Lorenz (2005) regards it as a separate subtribe, Bouchard *et al.* (2011) included it in Dromiusina, Jeannel (1949) included it in Singilina, together with the genus *Somotrichus*, which in the present paper is placed in a separate subtribe, Somotrichina. However, the female genitalia of Lichnasthenina are very similar to what is known in Dromiusina. Additional study of the genera of Lichnasthenina probably will demonstrate that they should be included in Dromiusina. Schüle & Lorenz (2008) published a key to the genera, only lacking the recently described Australian genus.

Australovelinda Baehr, 2012 (Fig. 12.77)

Type species. *Australovelinda seriata* Baehr, 2012, by original designation.

Characteristics. Very small, extremely rugose, vaguely quadrimaculate beetles with hirsute pilosity; eyes very large and laterally greatly produced; mentum dentate; labial palp moderately widened but not securiform; pronotum cordiform; elytra with rather sparse, erect pilosity; apex oblique, shallowly excised; tarsal claws inconspicuously denticulate; aedeagus elongate, with a denticulate, sclerotised piece in the middle of the endophallus; female genitalia unknown.

Australian species. Only *A. seriata* Baehr.

Distribution. Eastern central Queensland. Known only from a single locality.

Biology. The single recorded specimen was collected by pyrethrum spraying of tree trunks in 'Bendee scrub' (*Acacia catenulata*). Nothing else is recorded about habits, diet and reproduction.

References. Baehr (2012c).

Tribe Dryptini Bonelli, 1810

Characteristics. Body form is recognisably elongate, parallel-sided, pronotum, head and mandibles narrow-elongate; pubescent dorsal surface; antennomere 1 very long and 2 very short; elytra without raised lateral margins.

Key to the genera of Dryptini

1. Tarsal claws smooth .... ***Drypta*** **Latreille**
– Tarsal claws pectinate .... ***Dendrocellus*** **Schmidt-Göbel**

Drypta Latreille, 1797 (Pl. 24K)

Type species. *Carabus emarginatus* Gmelin, 1790 (= *Carabus dentatus* P. Rossi, 1790)

Australian species. Three species.

Distribution. The genus is known from many tropical regions worldwide. In Australia it is found in the east coastal regions of South Australia, Victoria, New South Wales, Queensland, and also in the Northern Territory.

Biology. Strong fliers, possibly diurnal, predacious, found in open, often grassy habitats and swampy areas.

References. Darlington (1968); Moore *et al.* (1987); Liang *et al.* (2004).

Dendrocellus Schmidt-Göbel 1846 (Pl. 23C)

Type species. *Dendrocellus discolor* Schmidt-Göbel, 1846 (= *Desera nepalensis* Hope, 1831), subsequent designation by Andrewes (1939).

Desera Hope, 1831

Characteristics. Medium size beetles with a distinct blue or green lustre; legs bicolored; claws pectinate.

Australian species. Two species.

Distribution. The genus is found from Africa, through Asia and south to Australia. North Queensland and Northern Territory.

Biology. Strong fliers, possibly diurnal, predacious, found in open, often grassy habitats.

References. Liang & Kavanaugh (2007).

Key to species. Liang & Kavanaugh (2007).

Tribe Zuphiini Bonelli, 1810

Characteristics. Procoxal cavities closed; mesocoxae conjunct; head large, with well developed neck; eyes variously shaped, sometimes absent; two supraorbital setae above each eye; antenna with elongate scape, pubescent from base; pronotum usually cordiform; elytra variously shaped, at apex transverse or slightly sinuate, not completely covering the abdomen; whole surface variously pilose; male protarsus not or little widened, biseriate squamose setae beneath; tarsal claws smooth; aedeagus variously shaped, with the ostium on the dorsal surface, with two sclerotised ligulae; parameres very dissimilar, asetose at apex; gonocoxite-2 variously shaped, with or without subapical setae.

Notes. The subdivision of the tribe is somewhat controversial, in particular the position of Planetina. Planetines either are included in the tribe Galeritini (Bouchard *et al.* 2011), or regarded a separate subtribe (Habu 1967), or included in Zuphiini (Lorenz 2005). We use the latter classification.

Key to the Australian subtribes of Zuphiini

1. Antenna short, median antennomeres barely longer than wide; elytra oval shaped and dorsally rather convex .... **Leleupidiina Basilewsky**
– Antenna elongate, median antennomeres at least 2 × as long as wide; elytra elongate, either dorsally depressed, or parallel-sided **2**
2(1). Elytra fused together at suture, with raised or carinate 7th interval and with erect pilosity; large, black, apterous species .... **Dicrodontina**
– Elytra free, intervals not carinate, convex, or depressed; pilosity decumbent or depressed; if black form very depressed, or elytra with 18 striae and intervals **3**
3(2). Elytra with no more than nine striae and intervals .... **Zuphiina Bonelli**
– Elytra with 18 striae and intervals .... **Planetina Jedlicka**

Key to the Australian genera of the tribe Zuphiini

1. Antenna short, median antennomeres barely longer than wide; elytra oval shaped and dorsally rather convex .... ***Colasidia* Basilewsky**
– Antenna elongate, median antennomeres at least 2 × as long as wide; elytra elongate, either dorsally depressed, or parallel-sided **2**
2(1). Elytra with no more than nine striae and intervals **3**
– Elytra with 18–19 striae and intervals .... ***Planetes* Macleay**
3(2). Eye absent **4**
– Eye present **6**
4(3). Head large, oval shaped; prothorax cordiform; elytra short, apicad considerably widened .... ***Speozuphium* Moore**
– Head elongate, parallel-sided; prothorax narrow and elongate; elytra elongate **5**
5(4). Head without supraorbital setae above each eye; elytra with distinct, punctate striae .... ***Speothalpius* Moore**
– Head with one supraorbital seta above each eye; elytra very shallowly striate .... ***Typhlozuphium* Baehr**
6(3). Pronotum with several anterior marginal setae; elytra with erect setae on all odd intervals, 7th interval carinate or at least more convex than other intervals; flight wings absent, elytra fused together at suture .... ***Acrogenys* Macleay**
– Pronotum with a single anterior marginal seta; elytra without erect setae on all odd intervals, 7th interval not carinate or more convex than other intervals; flight wings present, elytra free **7**
7(6). Antennomere 1 less than 1.5 × as long as 2 and 3 together; base of head convex, neck more than half as wide as head between eyes; elytra dorsally convex, intervals convex .... ***Pseudaptinus* Laporte**
– Antennomere 1 more than 1.5 × as long as 2 and 3 together; base of head quadrangular, neck less than half as wide as head between eyes; elytra dorsally depressed, intervals barely convex **8**
8(7). Antennomere 1 with a single seta near apex; elytra with distinct striae and intervals .... ***Zuphium* Latreille**
– Antennomere 1 with several setae; elytra with indistinct striae and intervals .... ***Parazuphium* Jeannel**

Subtribe Zuphiina Bonelli, 1810

Characteristics. Antenna elongate, median antennomeres more than 2 times as long as wide; mentum usually dentate; palps usually at least slightly widened; elytra depressed, with no more than nine striae and intervals; aedeagus rather short and compact; gonocoxite-2 variously shaped, setose or not.

Pseudaptinus Laporte, 1834 (Pl. 36F)

Type species. *Pseudaptinus albicornis* Laporte, 1834, by original designation.

Diaphorus Dejean, 1831.

Characteristics. Antennomere 1 less than 1.5 × as long as 2 and 3 together; base of head convex, neck more than half as wide as head between eyes; elytra dorsally not markedly depressed, with no more than nine striae and intervals; striae more or less coarsely punctate, intervals convex; aedeagus comparatively elongate, endophallus with sclerotised folds; gonocoxite-2 trapezoidal, with several subapical setae.

Distribution. Australia; extraterritorial in South America and southern parts of North America.

Biology. Ground living in various habitats. Little is known about the ecology of the American species.

Note. All Australian species belong to the subgenus *Thalpius* LeConte, 1851. The nominate subgenus occurs only in the Neotropical Region.

References. Csiki (1932); Liebke (1934); Moore *et al.* (1987); Baehr (1985a, 1986d, 1988c, 1995c, 2001b, 2008l); Lorenz (2005).

Subgenus *Thalpius* LeConte, 1851

Type species. *Helluo pygmaeus* Dejean, 1826, by monotypy.

Zuphiosoma Laporte, 1867.

Characteristics. Prothorax not markedly cordiform; elytra rather depressed.

Australian species. 11 species.

Distribution. Eastern and northern Australia from eastern South Australia through eastern and central New South Wales and Queensland, northern parts of Northern Territory to the Pilbara Region in northern Western Australia; extraterritorial in the Neotropical Region and the southern Nearctic Region.

Biology. Probably ground living in woodland and semiarid areas. Most specimens were sampled at light in non-forested country, which means that the species may be preferably nocturnal. A few specimens were found under stones or in cracks in the soil, commonly in wet environments or near water. Probably they also live in leaf litter. Predacious. Little else is known about habits, diet, and reproduction.

Note. Additional undescribed species are known and await description.

References. Csiki (1932); Moore *et al.* (1987); Baehr (1985a, 2008l); Lorenz (2005).

Keys to species. Baehr (1985a, 2008l).

Zuphium Latreille, 1806 (Pl. 39F; Fig. 12.11)

Type species. *Carabus olens* Rossi, 1790, by subsequent designation by Jeannel (1942).

Characteristics. Moderately large to large, depressed, flavous to black beetles; eyes present but usually moderately sized to rather small; antenna elongate with very elongate scape; antennomere more than 1.5 times as long as 2 and 3 together, with a single seta near apex; elytral striae distinct but shallow; aedeagus short and compact; gonocoxite-2 falciform, asetose.

Australian species. Seven species and three subspecies.

Distribution. Almost the whole of mainland Australia, but most common in the east and the tropical North; extraterritorial almost worldwide, but mainly in tropical and subtropical regions.

Biology. Ground living in a variety of woodlands and in semi arid areas. However, most specimens were sampled at light in non-forested country, which means that the species seem to be strictly nocturnal. A few specimens were found under stones or in cracks in the soil, commonly near water. The depressed body shape indicates that the species are adapted to live in fissures in the ground or under various tight cover. Little else is known about habits, diet, and reproduction.

References. Csiki (1932); Jeannel (1942); Jedlicka (1963); Darlington (1968); Matthews (1980); Moore *et al.* (1987); Baehr (1986c, 1986d, 1988c, 2001b, 2010d); Lorenz (2005).

Keys to species. Baehr (1986c, 2001b, 2010d).

Parazuphium Jeannel, 1942 (Pl. 33J)

Type species. *Zuphium chevrolati* Laporte, 1833.

Characteristics. Rather small, depressed, usually flavous to pale brown beetles; eyes present, moderately sized to rather small; antenna elongate with elongate scape; antennomere 1 more than 1.5 times as long as 2 and 3 together, with several setae; elytral striae indistinct; aedeagus short and compact; gonocoxite-2 falciform, asetose.

Australian species. Nine species.

Distribution. Most of mainland Australia, except the southwest; extraterritorial in the Oriental, Afrotropical, and Palearctic Regions.

Biology. Little recorded, ground living and probably in leaf litter or in cracks in the soil and in various sorts of more or less open woodlands and in semiarid areas. Because most specimens were collected at light, the species may be preferably nocturnal. A few specimens were found under stones or in cracks in the soil, commonly near water. One subterranean species of the subgenus *Austrozuphium* Baehr, 1985 occurs in the Pilbara Region and has been found by drilling in the ground. Little else is known about habits, diet, and reproduction.

References. Baehr (1985b, 1986d, 1988c, 2001b, 2014c); Lorenz (2005).

Key to species. Baehr (1985b).

Key to the Australian subgenera of *Parazuphium*

1. Head with two setae behind the eye; elytral striae usually more distinct; apex of elytra more incurved towards suture .... ***Parazuphium* Jeannel**
– Head with a single seta behind the eye; elytral striae usually indistinct; apex of elytra far less incurved towards suture .... ***Austrozuphium* Baehr**

Subgenus *Parazuphium* Jeannel, 1942

Type species. *Zuphium chevrolati* Laporte, 1833.

Characteristics. Head with two setae behind the eye; elytral striae fairly distinct; apex of elytra deeply incurved towards suture.

Australian species. Three species.

Distribution. Northern tropical Australia from north Queensland to northern Western Australia north of Great Sandy Desert; extraterritorial in the Oriental, Afrotropical, and Palearctic Regions.

References. Baehr (1985b, 1986d, 1988c, 2001b); Lorenz (2005).

Subgenus *Austrozuphium* Baehr, 1985

Type species. *Zuphium mastersii* Laporte, 1867, by original designation.

Characteristics. Head with a single seta behind the eye; elytral striae very feeble; apex of elytra moderately incurved towards suture.

Australian species. Six species.

Distribution. Eastern and northern Australia from eastern South Australia to the Pilbara in northern Western Australia, not yet recorded from the interior and from southwestern Australia.

Note. In the Pilbara Region a small-eyed species has been found at the depth of 17 m in a bore in the ground.
References. Baehr (2014c); Lorenz (2005).

Typhlozuphium Baehr, 2014 (Fig. 12.155)

Type species. *Typhlozuphium humicolum* Baehr, 2014, by original designation.
Characteristics. Eyeless, depigmented, narrow beetles; antenna and legs very elongate; head elongate and parallel, with a single supraorbital seta above each eye; prothorax elongate, slightly cordiform, without marginal setae; elytra elongate and parallel-sided, striae very feeble. The male and female genitalia are figured in Baehr (2014c).
Australian species. Two species.
Distribution. Pilbara Region in northern Western Australia.
Biology. Both recorded species are endogeous and have been sampled in bores at depths up to 60 m.
References. Baehr (2014c).

Speozuphium Moore, 1995 (Fig. 12.135)

Type species. *Speozuphium poulteri* Moore, 1995, by original designation.
Characteristics. Eyeless, depigmented beetles; head large, oval shaped; antenna very elongate; pronotum cordiform; elytra short, apicad widened, almost lacking elytral striae; marginal elytral setae very elongate; setae on head and pronotum absent.
Australian species. Only *S. poulteri* Moore.
Distribution. The single species was recorded from Weebubbie Cave in the Nullarbor Plain, Western Australia.
Biology. Not recorded, except that it is cavernicolous.
References. Lorenz (2005).

Speothalpius Moore, 1995 (Fig. 12.134)

Type species. *Speothalpius grayi* Moore, 1995, by original designation.
Characteristics. Eyeless, depigmented beetles; head narrow, elongate; antenna very elongate; pronotum narrow, elongate; elytra elongate, oval shaped, densely pilose; striae complete, distinct, punctate; marginal elytral setae elongate; setae on head and pronotum absent.
Australian species. Only *S. grayi* Moore.
Distribution. The single species was recorded from Nurina Cave in the Nullarbor Plain, Western Australia.
Biology. Little recorded, except that it is cavernicolous and that it was found near cracks in the soil in the far end of the cave.
References. Lorenz (2005).

Subtribe Dicrodontina Machado, 1992

Characteristics. Antennomere 1 not much shorter than 2 and 3 together; whole antenna densely pilose; mentum dentate; palps slightly widened; elytra distinctly striate, with erect or slightly decumbent pilosity; flight wings reduced; gonocoxite-2 elongate, with some elongate setae on the medio-lateral margin.
Note. The Australian genus *Acrogenys* Macleay is doubtfully included in this subtribe, mainly due to the moderately short antennomere 1 and the reduction of the flight wings. Machado (1992), when erecting the subtribe, discussed the status of *Acrogenys*.

Acrogenys Macleay, 1864 (Pl. 18I)

Type species. *Acrogenys hirsuta* Macleay, 1864, by monotypy.
Characteristics. Rather large, dorsally fairly convex, black, flightless beetles; whole dorsal surface densely pilose; eye rather large, with distinct orbit; antenna pilose from base; mentum dentate; palps slightly widened; pronotum markedly cordiform, with several anterior marginal setae; elytra tightly connected; with rows of erect setae on the odd intervals; 7th interval carinate or more raised then the other intervals; aedeagus rather compact, usually with hooked apex; endophallus without sclerotised pieces; gonocoxite-2 falciform, elongate, variously setose.
Australian species. 12 species.
Distribution. Eastern and northern Australia from eastern Queensland through the northern parts of Northern Territory and Western Australia to the Pilbara; also in Central Australia.
Biology. Ground living, in open sclerophyll forests and woodlands, also in semiarid areas. Predacious and nocturnal. Specimens are rarely collected, usually under stones or logs, and in debris, most commonly in wet environments. Little else is known about habits, diet, and reproduction.
References. Gestro (1875); Csiki (1932); Baehr (1984, 1992b, 2001b, 2008h, 2012k); Moore *et al.* (1987); Lorenz (2005).
Keys to species. Baehr (1984, 2008h, 2012k).

Key to the subgenera of *Acrogenys*

1. 7th interval not carinate, sutural interval not raised, therefore disc of elytra not depressed .... ***Paracrogenys* Baehr**
– 7th interval carinate, sutural interval raised, therefore disc of elytra depressed .... ***Acrogenys* Macleay**

Subgenus *Acrogenys* W. Macleay, 1864

Type species. *Acrogenys hirsuta* Macleay, 1864, by monotypy.

Characteristics. 7th interval carinate; sutural interval raised; surface of elytra depressed or concave; aedeagus rather compact with shorter apical part, and with upturned terminal knob.

Australian species. 11 species.

Distribution. Eastern and northern Australia from eastern Queensland through the northern parts of Northern Territory and Western Australia to the Pilbara Region; also in Central Australia.

References. Gestro (1875); Csiki (1932); Baehr (1984, 1986d, 1988c, 2001b, 2008h, 2012k); Moore *et al.* (1987); Lorenz (2005).

Keys to species. Baehr (2008h, 2012k).

Subgenus *Paracrogenys* Baehr, 1984

Type species. *Acrogenys longicollis* Gestro, 1875, by original designation.

Characteristics. 7th interval not carinate; sutural interval not raised; surface of elytra convex; aedeagus slender, with longer apical part, and with strongly hooked but not knobbed apex.

Australian species. Only *A.* (*Paracrogenys*) *longicollis* (Gestro).

Distribution. Eastern Queensland.

References. Moore *et al.* (1987); Lorenz (2005); Baehr (2008h).

Subtribe Planetina Jedlicka, 1941

Characteristics. Medium-sized, depressed, black or bimaculate beetles; head with distinct neck; mentum dentate; palps slightly widened; antenna moderately elongate; antennomere 1 less than 1.5 × as long as 2 and 3 together; elytra depressed, with 18–19 striae and carinate intervals; aedeagus rather elongate without distinctly sclerotised pieces; gonocoxite-2 moderately elongate, curved, with obtusely rounded apex, with two large ventral setae, one similarly sized dorsal seta, and a sort subapical seta.

Note. Some authors, e.g. Bouchard *et al.* (2011), include this subtribe in the tribe Galeritini. However, the suprageneric taxonomy of the zuphiine-galeritine complex still is unsettled.

Planetes W. S. Macleay, 1825 (Pl. 35B)

Type species. *Planetes bimaculatus* W. S. Macleay, 1825, by monotypy.

Characteristics. Moderately large, depressed, parallel-sided, full flight wing, in Australia dark piceous to black species; surface pilose; elytra with 18–19 striae and costate intervals, doubling due to presence of secondary costae; male and female genitalia as in Characteristics of subtribe.

Australian species. Four species.

Distribution. Eastern Queensland from about Rockhampton through northern parts of Northern Territory and Western Australia to the Pilbara Region; extraterritorial in New Guinea and the Oriental and Afrotropical Regions.

Biology. Most probably ground living, in rather wet envi ronments. Most specimens were collected at light, but almost always near water. Probably the species live in debris or under dense vegetation in swamp areas and near the banks of rivers and shores of lakes and lagoons. Predacious and nocturnal.

References. Csiki (1932); Jedlicka (1963); Habu (1967); Darlington (1968); Moore *et al.* (1987); Baehr (1986d, 1988c); Lorenz (2005).

Key to species. Baehr (1986d).

Subtribe Leleupidiina Basilewsky, 1951

Characteristics. Small, dorsally convex species with seriate pilosity; head oval shaped, neck very constricted; eyes various, laterally not or little produced; mentum dentate; labial palp very large; antenna short, moniliform. Antennomere 1 short; pronotum cordiform; elytra ovoid, usually rather convex; striae coarsely punctate and with erect pilosity; apex transverse or convex; aedeagus variously shaped, usually with distinct but various sclerotised parts on the endophallus; gonocoxite-1 very narrow and elongate; gonocoxite-2 curved, with elongate ventral and dorsal setae.

Colasidia Basilewsky, 1954 (Fig. 12.83)

Type species. *Colasidia malayica* Basilewsky, 1954, by monotypy.

Characteristics. Elytra with regular rows of large punctures; apex of aedeagus not hooked. The single Australian species is exceptional in its very elongate head and extremely small eyes.

Australian species. Only *C. monteithi* Baehr, 1987.

Distribution. Iron Range, mid Cape York Peninsula, northern Queensland; extraterritorial in the southern Oriental Region and in New Guinea.

Biology. Ground living in leaf litter in rainforests. Only the holotype of the single species is known.

References. Baehr (1988c, 1991b, 1997d); Lorenz (2005).

Tribe Physocrotaphini Chaudoir, 1862

Characteristics. Procoxal cavities closed; mesocoxae conjunct; head large, with well developed neck; eyes large, orbits large, convex or even dentate; two supraorbital setae present; labrum very large, covering most of the mandibles; mentum dentate; antenna pubescent from base; pronotum cordiform; elytra parallel-sided and depressed, at apex transverse or slightly sinuate, not completely covering the abdomen; male protarsus little widened; biseriate squamose setae beneath; tarsal claws simple; aedeagus compact, with the ostium on the dorsal surface and with two sclerotised ligulae; without sclerotised pieces on the endophallus; parameres very dissimilar; gonocoxites short, gonocoxite-2 curved and acute at apex, with or without short subapical setae.

Note. Moore (1998) discussed the probable relationships of this tribe. These are still doubtful, because adult and larval characters suggest different relationships.

Pogonoglossus Chaudoir, 1862 (Pl. 35E; Fig. 12.12)

Type species. *Pogonoglossus validicornis* Chaudoir, 1862, by monotypy.

Characteristics. Medium-sized to rather large, depressed, parallel-sided, more or less dark beetles; surface pilose; head large, orbits convex or angulate; mandibles elongate, straight, only at apex incurved; mentum dentate; pronotum cordiform; elytra depressed, striate, pilose; gonocoxite-2 elongate, curved, somewhat falciform, with or without a few short ventral and dorsal setae.

Australian species. Eight species.

Distribution. Central New South Wales, eastern Queensland, northern parts of Northern Territory and Western Australia to the Pilbara Region; extraterritorial in the Oriental Region and in New Guinea.

Biology. Australian specimens have been sampled from under bark of logs in rainforests, but mostly collected at light. The beetles may live under bark or in narrow fissures in the ground in various types of forests and woodlands. One species was described from a cave. Two additional species were captured from bores in the ground down to 43 m. According to Moore (1998) the larvae live under bark. Predacious and probably nocturnal.

References. Csiki (1932); Andrewes (1937); Jedlicka (1963); Darlington (1968); Moore *et al.* (1987); Baehr (1988b, 1993b, 2008a, 2014b); Moore (1998); Lorenz (2005).

Keys to species. Baehr (1988b, 1993b, 2008a, 2014b).

Tribe Helluonini Hope, 1834

Characteristics. Procoxal cavities closed; mesocoxae conjunct; head large, with well developed neck; eyes large; one or two supraorbital setae present; antenna pubescent from base, but more densely from antennomere 5; pronotum transverse, cordiform; elytra wide and elongate, parallel-sided, at apex transverse or slightly sinuate, not completely covering the abdomen; male protarsus not or little widened, biseriate squamose setae beneath; tarsal claws simple; aedeagus short and wide, with the ostium on the dorsal surface and with two sclerotised ligulae; with or without sclerotised pieces on the endophallus; parameres very dissimilar, asetose at apex; female gonocoxites elongate, gonocoxite-2 curved and acute at apex, with or without short subapical setae.

Note. For a general description of the tribe, but with special emphasis on the Neotropical fauna, see Reichardt (1974).

Key to the Australian subtribes of Helluonini

1. Profemur without a small triangular prominence near base; labium laterally asetose; head usually with two supraorbital setae .... **Omphrina Jedlicka**
– Profemur with a small triangular prominence near base (Fig. 12.24); labium laterally setose; head almost always with one supraorbital seta .... **Helluonina Hope**

Key to the Australian genera of Helluonini

1. Head with two supraocular setae **2**
– Head with a single supraocular seta **3**
2(1). Profemur with a small prominence on lower surface near base; body size c. 14 mm .... ***Epimicodema* Sloane**
– Profemur without a prominence on lower surface near base; body size c. 10 mm .... ***Creagris* Nietner**
3(1). Head with large postocular prominences (Fig. 12.68) **4**
– Head without distinct postocular prominences **8**
4(3). Labrum at apex pointed, with more than four setae; apex of clypeus emarginated **5**
– Labrum at apex widely rounded, with at most four setae; apex of clypeus truncate **6**
5(4). Mentum with a triangular tooth; labium convex at apex .... ***Helluonidius* Chaudoir**
– Mentum edentate; labium deeply excised at apex .... ***Dicranoglossus* Chaudoir**
6(4). Prothorax constricted and laterally very sinuate near base; labium narrow, elongate, and convex at apex .... ***Helluosoma* Laporte**
– Prothorax narrowed to basis, but not markedly constricted; labium wide, either widely rounded or emarginate at apex **7**
7(6). Body very depressed; labium convex at apex .... ***Platyhelluo* Baehr**
– Body rather convex; labium emarginate at apex .... ***Helluodema* Laporte**

8(3). Flight wing reduced, metaventrite shorter than metacoxa; metanepisternum short; 9th interval of elytra not considerably narrower than 8th **9**
– Full flight wing, metaventrite much longer than metacoxa; metanepisternum elongate; 9th interval of elytra considerably narrower than 8th....... **11**
9(8). Elytra with odd intervals carinate or costate; mental tooth elongate, triangular; lateral lobes of mentum elongate and acute at apex .... ***Helluo* Bonelli**
– Intervals not carinate nor costate; mental tooth short and wide; lateral lobes of mentum not elongate, obtuse at apex **10**
10(9). Apex of elytra rounded; elytra depressed on disc .... ***Helluarchus* Sloane**
– Apex of elytra truncate, disc of elytra rather convex .... ***Helluapterus* Sloane**
11(8). Labium wide, apex rounded; 8th interval pluripunctate **12**
– Labium deeply excised at apex; 8th interval biseriate punctate .... ***Ametroglossus* Sloane**
12(11). Mentum with a short, triangular tooth; prothorax not markedly hexagonal; colour black .... ***Gigadema* Thomson**
– Mentum edentate; **either** prothorax markedly hexagonal **or** colour blue-violaceous **13**
13(12). Colour blue-violaceous; prothorax transverse, slightly narrowed posteriorly .... ***Aenigma* Newman**
– Colour black; prothorax markedly hexagonal, strongly narrowed posteriorly .... ***Neohelluo* Sloane**

Subtribe Helluonina Hope, 1834

Characteristics. Head usually with a single supraorbital setae (except genus *Epimicodema* Sloane, 1914); profemur with a small prominence near base. Male and female genitalia have not been studied for most Australian genera.

Note. The subtribe is restricted to Australia and New Guinea.

Aenigma Newman, 1836 (Pl. 18C)

Type species. *Aenigma iris* Newman, 1836, by monotypy.

Characteristics. Rather large, blue beetles; orbit small, not produced; mentum edentate; glossa wide, at apex rounded; prothorax transverse; 8th elytral interval pluripunctate; flight wings fully developed.

Australian species. Only *A. iris* Newman.

Distribution. Eastern Australia from northern New South Wales to northern Queensland.

Biology. Subcortical under the bark of trees in open and closed forests, nocturnal, predacious.

References. Sloane (1914); Csiki (1932); Moore *et al.* (1987); Lorenz (2005).

Ametroglossus Sloane, 1914 (Pl. 19I)

Type species. *Gigadema atrum* Macleay, 1887, by original designation.

Characteristics. Rather large, black beetles; orbit small, not produced; mentum edentate; glossa at apex deeply excised; prothorax cordiform; 8th elytral interval biseriate punctate; flight wings fully developed.

Australian species. Only *A. ater* (Macleay, 1887).

Distribution. North-eastern Queensland.

Biology. Subcortical under the bark of trees in open and closed forests, nocturnal, predacious.

References. Csiki (1932); Moore *et al.* (1987); Lorenz (2005).

Dicranoglossus Chaudoir, 1872 (Pl. 23A; Fig. 12.68)

Type species. *Helluosoma resplendens* Laporte, 1867, by monotypy.

Characteristics. Comparatively small, blue or greenish beetles; orbits large, produced; glossa deeply excised; mentum edentate; labrum at apex pointed, polysetose; clypeus truncate at apex; flight wings fully developed.

Australian species. Only *D. resplendens* (Laporte).

Distribution. Eastern Queensland, northern parts of Northern Territory and Western Australia.

Biology. Subcortical under bark of eucalypt and non-eucalypt trees, predacious, probably mainly nocturnal. In some areas common in open eucalypt woodland or at River Gums. Occasionally specimens are seen running in sunshine on the bark of eucalypts.

References. Sloane (1914); Csiki (1932); Moore *et al.* (1987); Lorenz (2005).

Epimicodema Sloane, 1914 (Pl. 25I)

Type species. *Helluosoma mastersii* Macleay, 1871, by original designation.

Characteristics. Medium-sized, black beetles; distinguished from all genera of the subtribe Helluonina by presence of two supraorbital setae on the head; flight wings fully developed.

Australian species. Only *E. mastersii* (Macleay).

Distribution. Eastern Australia from New South Wales to central eastern Queensland.

Biology. Subcortical under bark of eucalypt and non-eucalypt trees. Predacious and probably nocturnal.

References. Csiki (1932); Moore *et al.* (1987); Lorenz (2005).

Gigadema Thomson, 1859 (Pl. 26K)

Type species. *Gigadema titanum* Thomson, 1859, by monotypy.

Penichrodema Gestro, 1875

Characteristics. Mostly large or very large, depressed, black beetles; orbits short, not produced; mentum dentate; glossa at apex rounded; 8th elytral interval pluripunctate; flight wings fully developed.

Australian species. 13 species.

Distribution. Almost the whole of mainland Australia, but more common in central and western parts.

Biology. Little recorded, because mostly sampled at light, more rarely sampled in pitfall traps in open forests and woodlands. Specimens also were collected from under bark of eucalypt trees. Nocturnal, predacious.

References. Sloane (1914); Csiki (1932); Darlington (1968); Matthews (1980); Moore *et al.* (1987); Lorenz (2005).

Key to species. Sloane (1914).

Helluo Bonelli, 1813 (Pl. 26H)

Type species. *Helluo costatus* Bonelli, 1813, by monotypy.

Characteristics. Large, rather stout, black beetles; orbits short, not produced; mentum with elongate tooth; lateral lobes of mentum elongate and very acute; odd intervals of elytra costate; 9th interval almost as wide as 8th; flight wings reduced; metathorax short.

Australian species. Two species.

Distribution. South-eastern mainland Australia from western Victoria to central Queensland.

Biology. Ground living under logs and slabs of rock in more or less open forests, nocturnal, predacious.

References. Sloane (1914); Csiki (1932); Moore *et al.* (1987); Lorenz (2005).

Key to species. Sloane (1914).

Helluapterus Sloane, 1914 (Pl. 26A)

Type species. *Helluapterus niger* Sloane, 1914, by original designation.

Characteristics. Large, compact, rather convex, black beetles; orbits short, not produced; mentum dentate, but tooth short and wide; lateral lobes of mentum short, obtuse; elytra dorsally convex; apex truncate; intervals not costate; 9th interval almost as wide as 8th; flight wings reduced; metathorax short.

Australian species. Only *H. niger* Sloane.

Distribution. Central Western Australia.

Biology. Ground living in open woodland to semiarid scrub, nocturnal, predacious.

References. Csiki (1932); Moore *et al.* (1987); Lorenz (2005).

Helluarchus Sloane, 1914 (Pl. 26F; Fig. 12.28)

Type species. *Helluarchus robustus* Sloane, 1914, by original designation.

Characteristics. Large, compact, rather depressed, black beetles; orbits short, not produced; mentum dentate, but tooth short and wide; lateral lobes of mentum short, obtuse; elytra dorsally depressed; apex rounded; intervals not costate; 9th interval almost as wide as 8th; flight wings reduced; metathorax short.

Australian species. Two species.

Distribution. Central Western Australia, adjacent north-western South Australia and central Northern Territory.

Biology. Ground living in semiarid open woodlands and shrublands, nocturnal, predacious.

References. Csiki (1932); Matthews (1980); Moore *et al.* (1987); Lorenz (2005).

Key to species. Sloane (1914).

Helluodema Laporte, 1867 (Pl. 27H)

Type species. *Helluomorpha batesii* Thomson, 1857, by original designation.

Simoglossus Chaudoir, 1872b

Characteristics. Medium-sized, narrow, black beetles; orbits elongate, laterally produced; glossa at apex emarginate; labrum at apex rounded, at most quadrisetose; clypeus at apex truncate; prothorax not cordiform; elytra elongate, parallel-sided, dorsally rather convex; flight wings fully developed.

Australian species. Two species.

Distribution. Eastern and northern Australia from eastern New South Wales through eastern Queensland to northern parts of Northern Territory.

Biology. Apparently ground living in more or less open forests, nocturnal, predacious. Commonly attracted to light.

References. Laporte (1868); Sloane (1914); Csiki (1932); Darlington (1968); Moore *et al.* (1987); Lorenz (2005).

Key to species. A short differential diagnosis in Sloane (1914).

Helluonidius Chaudoir, 1872 (Pl. 27K)

Type species. *Aenigma cyanipennis* Hope, 1842, by monotypy.

Characteristics. Medium-sized to fairly large, narrow, black or blue beetles; orbits elongate, laterally produced; mentum dentate; glossa at apex rounded; labrum at apex pointed, polysetose; clypeus at apex emarginate; prothorax rather cordiform; elytra elongate, parallel-sided, dorsally rather depressed; flight wings fully developed;

Australian species. Four species.

Distribution. Eastern and northern Australia from north-eastern New South Wales through northern parts of Queensland, the Northern Territory, and Western Australia; extraterritorial in New Guinea.

Biology. Subcortical under bark of trees in open forest and woodland, nocturnal, predacious. Commonly attracted to light.

References. Sloane (1914); Csiki (1932); Darlington (1968); Moore *et al.* (1987); Lorenz (2005).

Key to species. Sloane (1914).

Helluosoma Laporte, 1867 (Pl. 27B)

Type species. *Helluosoma ater* Laporte, 1867, by subsequent designation by Moore *et al.* (1987).

Helluodema Chaudoir, 1872b (non *Helluodema* Laporte, 1867, as *Helluodoma*).

Characteristics. Medium-sized to fairly large, narrow, black beetles; orbits elongate, laterally produced; mentum dentate; glossa narrow, at apex rounded; labrum at apex rounded, at most quadrisetose; clypeus at apex truncate; prothorax distinctly cordiform; elytra elongate, parallel-sided, dorsally rather depressed; flight wings fully developed.

Australian species. Seven species.

Distribution. Northern Australia from central eastern Queensland through the northern parts of Queensland, the Northern Territory, and Western Australia; extraterritorial in Papua New Guinea.

Biology. Apparently ground living on open forest to semiarid woodland, nocturnal, predacious. Commonly attracted to light.

References. Laporte (1868); Gestro (1875); Sloane (1914); Csiki (1932); Darlington (1968); Moore *et al.* (1987); Baehr (2005a, 2017b); Lorenz (2005).

Key to species. Baehr (2005a, 2017b).

Neohelluo Sloane, 1914 (Pl. 31J)

Type species. *Neohelluo angulicollis* Sloane, 1914, by monotypy.

Characteristics. Medium-sized to fairly large, narrow, black beetles; orbits short, post-ocular not produced; mentum edentate; glossa wide, at apex rounded; labrum at apex rounded; prothorax hexagonal, markedly narrowed apicad; elytra elongate, parallel-sided, dorsally rather depressed; 8th elytral interval pluripunctate; flight wings fully developed;

Australian species. Only *N. angulicollis* Sloane.

Distribution. North-eastern Queensland.

Biology. Corticolous on or under the bark of trees in more or less closed forests, nocturnal, predacious.

References. Csiki (1932); Moore *et al.* (1987); Lorenz (2005).

Platyhelluo Baehr, 2005 (Pl. 35C)

Type species. *Platyhelluo weiri* Baehr, 2005, by original designation.

Characteristics. Medium-sized, very depressed, black beetles; orbits elongate, laterally produced; glossa at apex convex; labrum at apex rounded, with at most four setae; clypeus at apex truncate; prothorax not cordiform; elytra rather elongate, dorsally very depressed; flight wings fully developed.

Australian species. Only *P. weiri* Baehr.

Distribution. Northern part of Northern Territory.

Biology. Probably subcortical and nocturnal, sampled at light, predacious.

References. Lorenz (2005).

Subtribe Omphrina Jedlicka, 1941

Characteristics. Head always with two supraorbital setae above each eye; profemur without prominence.

Creagris Nietner, 1858 (Fig. 12.85)

Type species. *Creagris labrosa* Nietner, 1858, by monotypy. *Pseudohelluo* Laporte, 1867.

Characteristics. Rather small, black or bimaculate beetles; two supraorbital setae present; labium laterally asetose; mentum with an elongate, acute tooth; profemur without prominence; 4th tarsomeres widened and deeply excised. The elongate mental tooth and the bilobed tarsomeres distinguish the genus from extra-Australian genera. Both Australian species are unicolourous.

Australian species. Two species.

Distribution. Northern Queensland; extraterritorial in southern and south-eastern Asia.

Biology. Under bark of trees, preferably eucalypts, in open forest and woodland. Nocturnal, predacious.

References. Gestro (1875); Sloane (1914); Csiki (1932); Jedlicka (1963); Darlington (1968); Moore *et al.* (1987); Lorenz (2005).

Key to species. Sloane (1914).

Acknowlegments

Our sincere thanks are particularly due to the following persons, who kindly provided support and reference materials: Cate Lemann and Tom Weir (ANIC), Geoffrey Monteith (QM), and Max Barclay (NHM, London). Special thanks are also due to Jürgen Wiesner (Wolfsburg) for his assistance in taxonomic questions in Cicindelinae and to Wolfgang Lorenz (Tutzing) for providing various information about recent taxonomic and nomenclatorial changes. We thank Nadine Guthrie (Perth), and James Liebherr (CUIC) for useful comments, corrections, and suggestions on various versions of the keys. Adam Slipinski (ANIC) deserves special thanks for his patience and efforts to help us complete this work.

Image sources: We thank Cate Lemann (ANIC) for superbly photographing the bulk of the figures. The majority of the specimens used for images were carefully prepared by the late Dr Barry P. Moore. The genus-level coverage of habitus images was made possible by Moore's lifetime of dedicated work building a reference collection that was donated to ANIC. Some line drawings were done by Erica Pang (Riverside, CA) and Connie Kim (UC, Berkeley, CA). Some important images were graciously provided by the following people and institutions: Australian National Insect Collection, CSIRO, Canberra, ANIC (12.74, 80, 82, 89, 92, 95, 104, 111, 120, 143, 149, 153); Dmitri Fedorenko, A.N. Severtsov Institute of Ecology and Evolution, Moscow (12.116); P. Giachino, Torino, Italy (121,78, 126, 140); James Liebherr, Cornell University Insect Collection, CUIC, Ithaca (12.152); David Maddison, Oregon State University, Corvallis (12.13–14); Geoff Monteith, Queensland Museum, Brisbane, QM (12.81, 86, 113); Wendy Moore, University of Arizona, Tuscon (12.9); Cezary Rojewski, Sydney (12.1, 29); D. Shpleley, University of Alberta (12.69); Museum of Victoria, Melbourne (12.73, 125); Muséum national d'Histoire naturelle, Paris, MNHN (12.146); South Australian Museum, Adelaide (12.136); Danny Shpeley, University of Alberta, Edmonton (12.69);

Udo Schmidt (12.10); Figures 121.35–37 and 12.39–41 after Moore & Lawrence (1994); *Tricondyla aptera* and *Somotrichus unifasciatus* images by Udo Schmidt licensed under the Creative Commons Attribution-Share Alike 2.0 Generic license.

Funding support: We heartily thank our supporting organizations. MB was supported by the German Scientific Organisation (Deutsche Forschungsgemeinschaft, DFG) for repeated collecting trips to Australia and visits to various Australian Museums by several grants. KW received support from the US, National Science Foundation Grant DEB 0444726 for research trips to Australia and museum visits.

References

Allen RT (1982) The Species of *Nebrioferonia* and *Homalonesiota* in New Guinea (Coleoptera: Carabidae: Pterostichini). *Coleopterists Bulletin* **36**, 402–411.

Allen RT, Ball GE (1980) Synopsis of Mexican taxa of the *Loxandrus* series (Coleoptera: Carabidae: Pterostichini). *Transactions of the American Entomological Society* **105**[1979], 481–575.

Andrewes HE (1923) On the types of Carabidae described by Schmidt-Göbel in his Faunula Coleopterorum Birmaniae. *Transactions of the Royal Entomological Society of London* **1923**, 1–63.

Andrewes HE (1925) A revision of the Oriental species of the genus *Tachys. Annali del Museo Civico di Storia Naturale Giacomo Doria* **51**, 327–502.

Andrewes HE (1927) Coleoptera: Carabidae. *Insects of Samoa* **4**, 1–14.

Andrewes HE (1929) Coleoptera. Carabidae. *Vol. I. – Carabinae. The Fauna of British India, including Ceylon and Burma.* Taylor & Francis, London.

Andrewes HE (1933) On some new species of Carabidae, chiefly from Java. *Treubia* **14**, 274–286.

Andrewes HE (1934) On the genotypes of British Carabidae. *Entomologist's Monthly Magazine* **70**, 200–201.

Andrewes HE (1935a) *The Fauna of British India including Ceylon and Burma. Coleoptera. Carabidae. Vol. II. – Harpalinae – I.* Taylor & Francis, London.

Andrewes HE (1935b) On the genotypes of British Carabidae. – II. *Annals and Magazine of Natural History* (series 10) **16**, 12–25.

Andrewes HE (1937) On the species of *Pogonoglossus* found in Java and Sumatra. *Bulletin de la Société Entomologique de France* **42**, 152–156.

Andrewes HE (1939) Papers on Oriental Carabidae. XXXV. On the types of some Indian genera. *Annals & Magazine of Natural History* (series 11) **3**(13), 128–139. doi:10.1080/03745481.1939.9723582

Andrewes HE (1941) Papers on the Oriental Carabidae-XXXVII. *Annals & Magazine of Natural History* (series 11) **7**(39), 307–317. doi:10.1080/03745481.1941.9727934

Aneshansley DJ, Eisner T, Widom JM, Widom B (1969) Biochemistry at 100°C: explosive secretory discharge of Bombardier Beetles (Brachinus). *Science* **165**, 61–63. doi:10.1126/science.165.3888.61

Arndt E (1988) Beschreibung der larve von *Abacetus villiersianus* Straneo (Coleoptera, Carabidae, Pterostichini). *Entomologische Nachrichten Berichte* **32**, 169–173.

Arndt E (1993) Phylogenetische Untersuchungen larvalmorphologischer Merkmale der Carabidae (Insecta, Coleoptera). *Stuttgarter Beiträge zur Naturkunde.* Ser. A, Nr. **488**, 1–56.

Arndt E (1998) Phylogenetic investigation of Carabidae (Coleoptera) using larval characters. In *Phylogeny and classification of Caraboidea* (Eds Ball GE, Casale A & Vigna Taglianti A). *Atti de Museo regionale di Scienze naturale de Torino* **5**, 171–190.

Arndt E, Beutel RG, Will KW (2016) 7.10. Carabidae Latreille, 1802. In *Handbuch der Zoologie/Handbook of Zoology. BandVolume IV Arthropoda: Insecta. Teilband/Part 38. Coleoptera, Beetles. Volume 1: Morphology and Systematics (Archostemata, Adephaga, Myxophaga, Polyphaga partim).* 2nd Edition. (Eds RG Beutel and RAB Leschen) pp. 162–190. Walter de Gruyter, Berlin.

Assmann T, Buse J, Drees C, Habel J, Härdtle W, Matern A, von Oheimb G, Schuldt A, Wrase DW (2008) From Latreille to DNA systematics – towards a modern synthesis for carabidology. In *Back to the Roots and back to the future: towards a new synthesis between taxonomic, ecological, and biogeographical approaches in carabidology. Proceedings of the XIII European Carabidologists Meeting, Blagoevgrad, August 20–24, 2007.* (Eds L Penev, TL Erwin and T Assmann) pp. 41–76. Pensoft Publishers, Sofia.

Austin A, Yeates DK, Cassis G, Fletcher MJ, La Salle J, Lawrence JF, McQuillan PB, Mound LA, Bickel DJ, Gullan PJ, Hales DF, Taylor GS (2004) Insects "Down Under" – Diversity, endemism and evolution of the Australian insect fauna: examples from selected orders. *Australian Journal of Entomology* **43**, 216–234. doi:10.1111/j.1326-6756.2004.00448.x

Baehr M (1984) Revision der australischen Zuphiinae. 1. Gattung *Acrogenys* Macleay (Insecta, Coleoptera, Carabidae). *Spixiana* **7**, 115–134.

Baehr M (1985a) Revision of the Australian Zuphiinae. 3. The genus *Pseudaptinus* Laporte de Castelnau (Insecta, Coleoptera, Carabidae). *Spixiana* **8**, 33–57.

Baehr M (1985b) Revision of the Australian Zuphiinae. 4. The genus *Parazuphium* Jeannel (Insecta, Coleoptera, Carabidae). *Spixiana* **8**, 295–321.

Baehr M (1986a) Review of the Australian species of the genus *Tachyta* Kirby (Coleoptera, Carabidae, Bembidiinae). *Entomofauna* **7**, 305–314.

Baehr M (1986b) Revision of the Australian ground-beetle genus *Porocara* Sloane (Coleoptera: Carabidae: Odacanthinae). *Australian Journal of Zoology* **34**, 717–731. doi:10.1071/ZO9860717

Baehr M (1986c) Revision of the Australian Zuphiinae. 5. The genus *Zuphium* Latreille (Insecta, Coleoptera, Carabidae). *Spixiana* **9**, 1–23.

Baehr M (1986d) Revision of the Australian Zuphiinae. 6. The genus *Planetes* Macleay. Supplement to the other genera (Insecta, Coleoptera, Carabidae). *Spixiana* **9**, 151–168.

Baehr M (1987a) Revision of the Australian Dromiine ground beetles, formerly placed in the genus *Microlestes* SCHMIDT-GÖBEL (Coleoptera, Carabidae, Lebiinae). *Entomologische Arbeiten aus dem Museum G. Frey Tutzing bei München* **35/36**, 21–65.

Baehr M (1987b) The Australian species of the Carabid genus *Perileptus* (Coleoptera: Carabidae: Trechinae). *Invertebrate Taxonomy* **1**, 1–16. doi:10.1071/IT9870001

Baehr M (1988a) A review of the Australian Tachyine beetles of the subgenera *Tachyura* Motschoulsky and *Sphaerotachys* Müller, with special regard to the tropical fauna (Insecta, Coleoptera, Carabidae, Bembidiinae). *Spixiana* **10**(1987), 225–269.

Baehr M (1988b) Revision of the Australian species of the genus *Pogonoglossus* Chaudoir (Insecta: Coleoptera: Carabidae: Helluodinae). *Invertebrate Taxonomy* **2**, 961–972. doi:10.1071/IT9880961

Baehr M (1988c) Revision of the Australian Zuphiinae 2. *Colasidia monteithi* sp. nov. from north Queensland, first record of the tribe Leleupidiini in Australia (Insecta: Coleoptera: Carabidae). *Memoirs of the Queensland Museum* **25**(1987), 135–140.

Baehr M (1989a) On new and rare *Clivina* LATR. from northern and western Australia 1. (Coleoptera, Carabidae, Scaritinae). *Mitteilungen der Münchner Entomologischen Gesellschaft* **79**, 5–30.

Baehr M (1989b) A new species of the *Tachys ectromoides*-group from Western Australia (Coleoptera, Carabidae, Bembidiinae). *Spixiana* **12**, 279–283.

Baehr M (1990a) Revision of the Australian species of the genus *Apotomus* Illiger (Insecta, Coleoptera, Carabidae, Apotominae). *Invertebrate Taxonomy* **3**, 619–627. doi:10.1071/IT9890619

Baehr M (1990b) Revision of the Australian Ground-beetle genus *Tasmanitachoides* Erwin (Insecta: Coleoptera: Carabidae: Bembidiinae), with special regard to the tropical species. *Invertebrate Taxonomy* **4**, 867–894. doi:10.1071/IT9900867

Baehr M (1990c) A review of the Australian species of *Minuthodes* Andrewes, with the description of two new species (Coleoptera, Carabidae, Lebiinae). *Spixiana* **13**, 33–41.

Baehr M (1990d) The carabid community living under the bark of Australian eucalypts. In *The Role of Ground Beetles in Ecological and Environmental Studies.* (Ed. NE Stork) pp. 3–11. Intercept, Andover.

Baehr M (1990e) A new species of *Microlestodes* Baehr from Western Australia (Insecta, Coleoptera, Carabidae, Lebiinae). *Spixiana* **13**, 183–185.

Baehr M (1990f) Two new species of *Trichotichnus* Morawitz from North Queensland (Coleoptera: Carabidae. Harpalinae). *Memoirs of the Queensland Museum* **28**, 383–388.

Baehr M (1991a) *Tachys windsorensis*, spec. nov. from north Queensland, a further new species of the *Tachys ectromoides*-group (Insecta, Coleoptera, Carabidae, Bembidiinae). *Spixiana* **14**, 189–192.

Baehr M (1991b) On new and rare Leleupidiini from the Oriental and Australian regions (Coleoptera, Carabidae, Zuphiinae). *Mitteilungen der Münchner entomologischen Gesellschaft* **81**, 193–202.

Baehr M (1992a) Revision of the Pseudomorphinae of the Australian Region 1. The previous genera *Sphallomorpha* Westwood and *Silphomorpha* Westwood. Taxonomy, phylogeny, zoogeography (Insecta, Coleoptera, Carabidae). *Spixiana Supplemente* **18**, 1–440.

Baehr M (1992b) A new *Acrogenys* Macleay from Central Australia. Supplement to the revision of the Australian Zuphiinae (Insecta, Coleoptera, Carabidae). *Spixiana* **15**, 75–80.

Baehr M (1992c) An Introduction to the Biogeography of the Carabidae of Montane Refugia in Northern Australia (Coleoptera: Carabidae). In *The Biogeography of Ground Beetles of Mountains and Islands* (Eds GR Noonan, GE Ball and NE Stork) pp. 67–78. Intercept, Andover.

Baehr M (1993a) New species and new records of the genus *Sphallomorpha* Westwood from Australia. Supplement to the "Revision of the Pseudomorphinae of the Australian Region 1." (Insecta, Coleoptera, Carabidae). *Spixiana* **16**, 25–42.

Baehr M (1993b) A new species of *Pogonoglossus* Chaudoir from Australia (Insecta, Coleoptera, Carabidae, Helluodinae). *Spixiana* **16**, 141–144.

Baehr M (1994a) Phylogenetic relations and biogeography of the genera of Pseudomorphinae (Coleoptera, Carabidae). In *Carabid Beetles: Ecology and Evolution* (Eds K Desender, M Dufrene, M Loreau, ML Luff and J-P Malfait) pp. 11–17. Series Entomologica 51, Kluwer, Dordrecht, Boston, London.

Baehr M (1994b) A new species of *Minuthodes* Andrewes from Australia (Insecta, Coleoptera, Carabidae, Lebiinae). *Spixiana* **17**, 37–41.

Baehr M (1994c) New species and new records of the genus *Sphallomorpha* Westwood from Australia and New Guinea. 3rd Supplement to the "Revision of the Pseudomorphinae of the Australian Region 1." (Insecta, Coleoptera, Carabidae). *Spixiana* **17**, 215–235.

Baehr M (1994d) Revision of the genus *Scopodes* Erichson from New Guinea (Insecta, Coleoptera, Carabidae, Pentagonicinae). *Spixiana* **17**, 97–155.

Baehr M (1995a) Revision of *Philipis* (Coleoptera: Carabidae: Bembidiinae), a genus of arboreal Tachyine beetles from the rainforests of eastern Australia. Taxonomy, phylogeny, and biogeography. *Memoirs of the Queensland Museum* **38**, 315–381.

Baehr M (1995b) Three new species of *Bembidion* (s.l.), subgenus *Desarmatocillenus* Netolitzky, from the Philippines and Australia (Insecta, Coleoptera, Carabidae, Bembidiinae. *Stuttgarter Beiträge zur Naturkunde. Serie A: Biologie* **525**, 1–8.

Baehr M (1995c) A new species of *Pseudaptinus* Laporte de Castelnau from Australia. 2nd supplement to the "Revision of the Australian Zuphiinae" (Insecta, Coleoptera, Carabidae, Zuphiinae). *Koleopterologische Rundschau* **65**, 15–18.

Baehr M (1996a) The Australian ground beetle genus *Porocara* Sloane. Second revision (Insecta, Coleoptera, Carabidae, Odacanthinae). *Spixiana* **19**, 253–265.

Baehr M (1996b) The ground beetle genus *Casnoidea* Laporte de Castelnau. Taxonomy, phylogeny, zoogeography (Insecta, Coleoptera, Carabidae, Odacanthinae). *Invertebrate Taxonomy* **10**, 1041–1084. doi:10.1071/IT9961041

Baehr M (1997a) A new species of *Trichotichnus* Morawitz from northern Australia (Insecta, Coleoptera, Carabidae, Harpalinae). *Spixiana* **20**, 131–135.

Baehr M (1997b) Revision of the Pseudomorphinae of the Australian Region 2. The genera *Adelotopus* Hope, *Cainogenion* Notman, *Paussotropus* Waterhouse, *Pseudomorpha* Kirby, and *Cryptocephalomorpha* Ritsema. Taxonomy, phylogeny, zoogeography (Insecta, Coleoptera, Carabidae). *Spixiana Supplemente* **23**, 1–510.

Baehr M (1997c) The genus *Stricklandia* Macleay in New Guinea (Carabidae, Lebiinae). *Coleoptera* **1**, 117–136.

Baehr M (1997d) Leleupidiini from the Oriental Region. 1. New species of the genus *Colasidia* Basilewsky (Insecta, Coleoptera, Carabidae, Zuphiinae). *Revue Suisse de Zoologie* **104**, 611–659. doi:10.5962/bhl.part.80014

Baehr M (1997e) Review of the Australian-Papuan *laevigatus*-group of the genus *Catascopus* Kirby, with description of a new species (Insecta, Coleoptera, Carabidae, Lebiinae). *Spixiana* **20**, 227–234.

Baehr M (1997f) A new species of *Trichotichnus* Morawitz from northern Australia (Insecta, Coleoptera, Carabidae, Harpalinae). *Spixiana* **20**, 131–135.

Baehr M (1999a) A preliminary survey of the classification of the Psydrinae (Coleoptera: Carabidae). Phylogeny and classification of Caraboidea. *Atti di Museo regionale di Scienze naturale de Torino* **5**, 359–368.

Baehr M (1999b) The genus *Dolichoctis* Schmidt-Göbel in New Guinea and surrounding areas (Insecta, Coleoptera, Carabidae, Lebiinae). *Coleoptera* **2**, 121–169 (1998).

Baehr M (2000a) Some genera and species of ground beetles new to Australia (Coleoptera, Carabidae). *Memoirs of the Queensland Museum* **46**, 9–14.

Baehr M (2000b) A new species of the genus *Lestignathus* Erichson from Tasmania. With a note on the Tasmanian species of *Mecyclothorax* Sharp (Insecta: Coleoptera: Carabidae: Licininae, Psydrinae). *Records of the South Australian Museum* **33**, 123–126.

Baehr M (2001a) A new species of the genus *Minuthodes* Andrewes from north Queensland, Australia (Insecta, Coleoptera, Carabidae, Lebiinae). *Spixiana* **24**, 171–175.

Baehr M (2001b) New species and new records of Zuphiinae from Australia (Coleoptera: Carabidae). *Special Bulletin of the Japan Coleopterology Society* **1**, 83–101.

Baehr M (2001c) *Tasmanitachoides* Erwin *glabellus* n. sp. from north Queensland, Australia with a note on *Tasmanitachoides lutus* (Darlington) (Insecta, Coleoptera, Carabidae, Bembidiinae). *Animal Biodiversity and Conservation* **24**, 1–7.

Baehr M (2002a) New species and new records of Australian Pseudomorphinae. 4th Supplement to the "Revision of the Pseudomorphinae of the Australian Region." (Insecta, Coleoptera, Carabidae). *Spixiana* **25**, 101–129.

Baehr M (2002b) New species and additional records of the tachyine genus *Philipis* Erwin (Coleoptera: Carabidae: Bembidiinae). *Memoirs of the Queensland Museum* **48**, 1–7.

Baehr M (2002c) Two new species of *Sloaneana* Csiki from southern Queensland (Coleoptera, Carabidae, Merizodinae). *Memoirs of the Queensland Museum* **48**, 9–15.

Baehr M (2003a) A peculiar new genus of Odacanthinae from northern Australia (Insecta, Coleoptera, Carabidae). *Mongraphie de Museo Regionale di Scienze naturali di Torino* **35**, 99–110.

Baehr M (2003b) A further new species of the *Tachys (s.l.) ectromoides*-group from Queensland, Australia (Insecta, Coleoptera, Carabidae, Bembidiinae). *Spixiana* **26**, 143–147.

Baehr M (2003c) A revision of the *brunnea*-group of the genus *Dicraspeda* Chaudoir (Coleoptera, Carabidae, Odacanthinae). *Spixiana* **26**, 249–267.

Baehr M (2003d) Psydrine ground beetles (Coleoptera: Carabidae: Psydrinae) excluding Amblytelini, of eastern Queensland rainforests. *Memoirs of the Queensland Museum* **49**, 65–109.

Baehr M (2003e) Australia's subantarctic Tropics – a contradiction? Abstract of a talk held at Forum Herbulot, 2003. *Spixiana* **26**, 206.

Baehr M (2004a) A revision of the Oriental, New Guinean, and Pacific species of the ground beetle genus *Anomotarus* Chaudoir (Coleoptera: Carabidae: Lebiinae). *Coleoptera* **7**, 27–78 (2003).

Baehr M (2004b) The genus *Lebia* Latreille in the Australian Region (Insecta, Coleoptera, Carabidae, Lebiinae). *Spixiana* **27**, 205–246.

Baehr M (2004c) A new species of the genus *Aristolebia* Bates from Thailand, with notes on some Papuan and Australian species (Insecta, Coleoptera, Carabidae, Lebiinae). *Spixiana* **27**, 247–251.

Baehr M (2005a) A new genus and three new species of helluonine beetles from Australia (Insecta, Coleoptera, Carabidae, Hellunoniae). *Spixiana* **28**, 21–31.

Baehr M (2005b) A revision of the Australian odacanthine ground beetles, including Checklists for Australia and the Papuan Subregion (Insecta: Coleoptera: Carabidae). *Memoirs of the Queensland Museum* **50**, 133–194.

Baehr M (2005c) The Amblytelini. A tribe of corticolous ground beetles from Australia. Taxonomy, phylogeny, biogeography (Insecta, Coleoptera, Carabidae, Psydrinae). *Coleoptera* **8**, 1–286.

Baehr M (2005d) New species and new records of Australian Pseudomorphinae. 6th Supplement to the "Revision of the Pseudomorphinae of the Australian Region." (Insecta, Coleoptera, Carabidae). *Spixiana* **28**, 259–269.

Baehr M (2006a) New species and new records of the genera *Dicraspeda* Chaudoir and *Eudalia* Castelnau from the Papuan and Australian regions, with a nomenclatorial note on *Deipyrus* Liebke (Insecta, Coleoptera, Carabidae, Odacanthinae). *Spixiana* **29**, 51–72.

Baehr M (2006b) The Australian scaritine genus *Steganomma* Macleay (Coleoptera: Carabidae: Scaritinae). *Mitteilungen der Münchner Entomologischen Gesellschaft* **95**, 59–66 (2005).

Baehr M (2006c) Revision of the genera *Agonocheila* Chaudoir and *Minuthodes* Andrewes in New Guinea (Insecta, Coleoptera, Carabidae, Lebiinae). *Spixiana* **29**, 103–145.

Baehr M (2006d) Revision of the Australian species of the subgenus *Anomotarus* Chaudoir s. str. (Insecta, Coleoptera, Carabidae, Lebiinae). *Coleoptera* **9**, 11–107 (2005).

Baehr M (2006e) A peculiar new genus of lebiine ground beetles from Australia (Coleoptera: Carabidae: Lebiini). *Koleopterologische Rundschau* **76**, 1–5.

Baehr M (2006f) A new species of the genus *Tachyta* Kirby from northern Queensland, Australia, with the first record of *T. brunnipennis* (Macleay) from New Guinea (Coleoptera, Carabidae, Bembidiinae). *Mitteilungen der Münchner entomologischen Gesellschaft* **96**, 81–85.

Baehr M (2006g) A peculiar new species of the genus *Amblytelus* Erichson from southern Queensland, Australia (Insecta, Coleoptera, Carabidae, Psydrinae). *Spixiana* **29**, 243–246.

Baehr M (2006h) A new and another remarkable species of the genus *Sphallomorpha* Westwood from central Queensland, Australia (Insecta, Coleoptera, Carabidae, Pseudomorphinae). *Mitteilungen der Münchner entomologischen Gesellschaft* **96**, 119–124.

Baehr M (2006i) Two new species of the genus *Dystrichothorax* Blackburn from Queensland, Australia (Carabidae, Psydrinae). *Coleoptera* **10**, 1–9.

Baehr M (2006j) New taxa of the lebiine genus *Dolichoctis* and of a related new genus, from the Papuan subregion (Coleoptera: Carabidae: Lebiinae). *Folia Heyrovskyana* **14**, 37–66.

Baehr M (2006k) A peculiar new genus of arboricolous lebiine ground beetles from interior Australia (Insecta: Coleoptera: Carabidae: Lebiinae). *Memoirs of the Queensland Museum* **52**, 1–6.

Baehr M (2007a) A further new *Lebia* Latreille of the *karenia*-group from northern Queensland, Australia, with the first Australian record of *Somotrichus elevatus* (Fabricius) (Insecta, Coleoptera, Carabidae, Lebiinae). 2nd supplement to "The genus *Lebia* Latreille in the Australian-Papuan Region". *Spixiana* **30**, 169–172.

Baehr M (2007b) Revision of the genus *Meonis* Laporte de Castelnau (Insecta, Coleoptera, Carabidae, Psydrinae). *Annali del Museo Civico di Storia Naturale de Genova* **99**, 563–662.

Baehr M (2007c) A new species of the genus *Adelotopus* Hope from northern Queensland, Australia (Insecta, Coleoptera, Carabidae, Pseudomorphinae). *Spixiana* **30**, 25–28.

Baehr M (2007d) New taxa of the genus *Dolichoctis* Schmidt-Göbel and of related genera from the Australian–Papuan Region (Insecta, Coleoptera, Carabidae, Lebiinae). *Tijdschrift voor Entomologie* **150**, 77–86. doi:10.1163/22119434-900000214

Baehr M (2008a) A new species of the genus *Pogonoglossus* Chaudoir from Australia (Carabidae, Physocrotaphinae). *Coleoptera* **11**(2007), 105–111.

Baehr M (2008b) A new species of the tachyine genus *Tasmanitachoides* Erwin from New South Wales, Australia (Coleoptera, Carabidae, Bembidiinae, Tachyini). *Mitteilungen der Münchner entomologischen Gesellschaft* **98**, 121–126.

Baehr M (2008c) Two new species of the genus *Tasmanitachoides* Erwin from north Queensland, Australia (Insecta, Coleoptera, Carabidae, Bembidiinae). *Annals of the Carnegie Museum* **77**, 13–19. doi:10.2992/0097-4463-77.1.13

Baehr M (2008d) A new, cavernicolous, eyeless species of the genus *Lymnastis* Motschulsky from Christmas Island, Australia (Insecta, Coleoptera, Carabidae, Bembidiinae). *Records of the Western Australian Museum* **24**, 199–203. doi:10.18195/issn.0312-3162.24(2).2008.199-203

Baehr M (2008e) New and rare pseudomorphine species from Western Australia (Insecta, Coleoptera, Carabidae, Pseudomorphinae). 8th Supplement to the "Revision of the Pseudomorphinae of the Australian Region.". *Records of the Western Australian Museum* **24**, 151–180. doi:10.18195/issn.0312-3162.24(2).2008.151-180

Baehr M (2008f) New species and new records of the amblyteline genera *Amblytelus* Erichson and *Dystrichothorax* Blackburn from eastern Australia (Insecta, Coleoptera, Carabidae, Psydrinae). *Spixiana* **31**, 263–276.

Baehr M (2008g) New species of the tachyine genus *Philipis* Erwin from southeastern Australia (Insecta, Coleoptera, Carabidae, Bembidiinae). *Folia Heyrovskyana*, Ser. A, **15**(2007), 17–26.

Baehr M (2008h) New species of the zuphiine genus *Acrogenys* Macleay from Australia (Insecta, Coleoptera, Carabidae, Zuphiinae). *Coleoptera* **11**(2007), 113–124.

Baehr M (2008i) Revision of the genus *Habutarus* Ball & Hilchie from the Australian region (Insecta, Coleoptera, Carabidae, Lebiinae). *Memoirs of the Queensland Museum* **52**, 13–48.

Baehr M (2008j) The Australian Clivinini 1. The genera *Ancus* Putzeys, *Aspidoglossa* Putzeys, *Clivinarchus* Sloane, *Platysphyrus* Sloane, *Pseudoclivina* Kult, *Rhysocara* Sloane, *Syleter* Andrewes, the subgenera *Paraclivina* Kult, *Semiclivina* Kult, and the *atrata-, biplagiata-, brevicornis-, coronata-, coryzoides-, cribrosa-, denticollis-, grandiceps-, incerta-, lobata-, obliquata-, obsoleta, orbitalis-, planiceps-, sulcaticeps-, tranquebaria-*, and *wurargae*-groups of the genus *Clivina* Latreille. With a note on a record of the genus *Parathlibops* Basilewsky (Scapterini) (Carabidae, Scaritinae). *Coleoptera* **12**, 1–220.

Baehr M (2008k) The Australian species of the genus *Coptodera* Dejean (Insecta, Coleoptera, Carabidae, Lebiinae). *Coleoptera* **11**, 175–199 (2007).

Baehr M (2008l) Two new species of the genus *Pseudaptinus* Laporte de Castelnau from northern Australia (Insecta: Coleoptera: Carabidae: Zuphiini). *The Beagle. Records of the Museum and Art Gallery of the Northern Territory* **24**, 55–62.

Baehr M (2008m) A review of the genus *Delinius* Westwood from northern Australia, with description of five new species. (Insecta, Coleoptera, Carabidae, Pterostichini). *Coleoptera* **12**, 227–240.

Baehr M (2009a) A new genus and two new species of the subfamily Migadopinae from Tasmania (Coleoptera, Carabidae). *Folia Heyrovskyana* **17**, 95–103.

Baehr M (2009b) A new species of the genus *Microlestodes* Baehr from Southeast Queensland, Australia (Coleoptera, Carabidae, Lebiinae). *Mitteilungen der Münchner Entomologischen Gesellschaft* **99**, 5–9.

Baehr M (2009c) A new species of the tachyine genus *Tasmanitachoides* from the Kimberley Division, Western Australia (Coleoptera, Carabidae). *Records of the Western Australian Museum* **25**, 159–164. doi:10.18195/issn.0312-3162.25(2).2009.159-164

Baehr M (2009d) New and rare species of the pseudomorphine genera *Sphallomorpha* Westwood, *Adelotopus* Hope, and *Pseudomorpha* Kirby mainly from far Northern Territory, Australia (Insecta, Coleoptera, Carabidae, Pseudomorphinae). 12th supplement to the "Revision of the Pseudomorphinae of the Australian Region". *Spixiana* **32**, 231–253.

Baehr M (2009e) New genera and species and new records of odacanthine carabid beetles from the Australian, Papuan, and Oriental Regions (Insecta, Coleoptera, Carabidae, Odacanthini). *Spixiana* **32**, 201–229.

Baehr M (2009f) Revision of the species of the genus *Anomotarus* (s. l.) Chaudoir from the Australian region, formerly alluded to the genus *Nototarus* Chaudoir (Insecta, Coleoptera, Carabidae, Lebiinae). *Coleoptera* **13**, 33–200.

Baehr M (2009g) A new species of the genus *Mecyclothorax* Sharp from New South Wales (Insecta: Coleoptera, Carabidae: Psydrinae). *Records of the Australian Museum* **61**(1), 89–92. doi:10.3853/j.0067-1975.61.2009.1519

Baehr M (2010a) New species of the genera *Lebia* Latreille and *Aristolebia* Bates from Indonesia (Coleoptera: Carabidae: Lebiini). 3nd supplement to "The genus *Lebia* Latreille in the Australian–Papuan Region". *Stuttgarter Beiträge zur Naturkunde* Ser. *ANS (Nijmegen)* **3**, 111–121.

Baehr M (2010b) Two new species of the genus *Pogonus* Nicolai from northern and western Australia (Insecta, Coleoptera, Carabidae, Pogonini). *Records of the Western Australian Museum* **25**, 441–448. doi:10.18195/issn.0312-3162.25(4).2010.441-448

Baehr M (2010c) A new species of the genus *Diabaticus* Bates from eastern Australia (Coleoptera, Carabidae, Lebiini). *Spixiana* **33**, 59–63.

Baehr M (2010d) A new species of the genus *Zuphium* Latreille from northern central Queensland, Australia (Coleoptera, Carabidae, Zuphiinae). *Spixiana* **33**, 65–68.

Baehr M (2010e) A new species of the genus *Microlestodes* Baehr from Arnhem Land, Northern Territory, Australia (Insecta: Coleoptera: Carabidae: Lebiinae). *The Beagle. Records of the Museum and Art Gallery of the Northern Territory* **26**, 103–107.

Baehr M (2010f) Two new species of the bembidiine genus *Tasmanitachoides* Erwin from northern Australia (Coleoptera, Carabidae, Trechitae). *Entomologische Blätter und Coleoptera* **106**, 25–39.

Baehr M (2010g) Revision of the genus *Scopodes* Erichson in Australia (Insecta, Coleoptera, Carabidae, Pentagonicinae). *Coleoptera* **14**, 1–188.

Baehr M (2011a) New species and new records of the genus *Anomotarus* Chaudoir, subgenus *Anomotarus* s. str., from the Oriental and Australian Regions. Supplement to the "Revisions of the species of the subgenus *Anomotarus* Chaudoir s. str." (Insecta, Coleoptera, Carabidae, Lebiinae). *Spixiana* **34**, 21–32.

Baehr M (2011b) New species and new records of the genus *Moriomorpha* Laporte de Castelnau from Australia (Coleoptera, Carabidae, Psydrinae, Moriomorphini, Melisoderina). *Spixiana* **34**, 199–213.

Baehr M (2011c) New species of the Genera *Melisodera* Westwood, *Rhaebolestes* Sloane, and *Moriodema* Laporte de Castelnau from Australia (Coleoptera: Carabidae: Psydrini). *Australian Entomologist* **38**, 129–144.

Baehr M (2011d) A new species of the genus *Stricklandiana* Macleay from New Guinea (Insecta, Coleoptera, Carabidae, Lebiinae). *Tijdschrift voor Entomologie* **154**, 167–171. doi:10.1163/004074912X13397496981382

Baehr M (2012a) The genus *Pentagonica* Schmidt-Göbel in the Oriental, Papuan, and Australian Regions (Insecta, Coleoptera, Carabidae, Pentagonicini). *Coleoptera* **15**, 1–200.

Baehr M (2012b) The genus *Coptoglossus* Chaudoir in eastern Australia (Insecta, Coleoptera, Carabidae, Lebiinae). *Memoirs of the Queensland Museum – Nature* **56**, 85–96.

Baehr M (2012c) Three peculiar new genera of lebiine carabid beetles from Queensland, Australia (Insecta, Coleoptera, Carabidae, Lebiini). *Memoirs of the Queensland Museum – Nature* **56**, 99–115.

Baehr M (2012d) A new species of the genus *Pheropsophus* Solier (Coleoptera: Carabidae: Brachininae) from northern Australia. *Records of the Western Australian Museum* **27**, 62–67. doi:10.18195/issn.0312-3162.27(1).2012.062-067

Baehr M (2012e) New species and new records of the genus *Anomotarus* Chaudoir from Australia. 2nd supplement to the "Revisions of the species of the genus *Anomotarus* Chaudoir (Coleoptera: Carabidae: Lebiini). *Transactions of the Royal Society of South Australia* **136**, 16–34. doi:10.1080/03721426.2012.10887159

Baehr M (2012f) Two new species of the genus *Diabaticus* Bates from Australia (Insecta, Coleoptera, Carabidae, Lebiini). *Transactions of the Royal Society of South Australia* **136**, 35–44. doi:10.1080/03721426.2012.10887160

Baehr M (2012g) New species and new records of odacanthine carabid beetles from the Australian, Papuan, and southern Oriental Regions (Coleoptera: Carabidae: Odacanthini). *Folia Heyrovskyana* **19**, 129–138.

Baehr M (2012h) New species and a new subgenus of the genus *Tachyta* Kirby from Australia (Coleoptera: Carabidae: Bembidiini). *Entomofauna* **34**, 65–77.

Baehr M (2012i) The species of the Australian *Tachys* (*s. l.*) *mastersi* complex (Coleoptera, Carabidae, Bembidiini, Tachyina). *Mitteilungen der Münchner Entomologischen Gesellschaft* **102**, 51–63.

Baehr M (2012j) Three new species of the genus *Lebia* Latreille from the Australian-Papuan Region. 4th supplement to "The genus *Lebia* Latreille in the Australian-Papuan Region". With nomenclatorial notes on *Lebia trivittata* Baehr, 2004 and on the genus *Pseudoplatia* Baehr, 2006 (Insecta, Coleoptera, Carabidae, Lebiini). *Mitteilungen der Münchner Entomologischen Gesellschaft* **102**, 41–49.

Baehr M (2012k) Three new species of the genus *Acrogenys* Macleay, 1864 from Australia (Insecta, Coleoptera, Carabidae, Zuphiini). *Entomologische Blätter und Coleoptera* **108**, 1–10.

Baehr M (2012l) Revision of some species-groups of the genus *Catascopus* Kirby from the Oriental and Australian Regions (Coleoptera, Carabidae, Lebiini). *Entomologische Blätter und Coleoptera* **108**, 11–108.

Baehr M (2012m) Two new Australian species of the genus *Hexagonia* Kirby, first record of the tribe Hexagoniini from Australia (Coleoptera, Carabidae). *Spixiana* **35**, 29–34.

Baehr M (2012n) A revision of the Australian subtribe Agonicina (Coleoptera, Carabidae, Peleciini). *Spixiana* **35**, 209–236.

Baehr M (2013a) A revision of the carabid tribe Migadopini in Australia (Insecta, Coleoptera, Carabidae, Migadopini). *Memoirs of the Queensland Museum – Nature* **56**(2), 279–304.

Baehr M (2013b) A new species of the genus *Coptoglossus* Chaudoir from Australia (Insecta, Coleoptera, Carabidae, Lebiini). *Memoirs of the Queensland Museum – Nature* **58**, 1–6.

Baehr M (2013c) New species and new records of the genus *Tasmanitachoides* Erwin from Australia (Insecta, Coleoptera, Carabidae, Trechini). *Mitteilungen der Münchner Entomologischen Gesellschaft* **103**, 85–94.

Baehr M (2013d) The species of the genus *Perigona* Castelnau from New Guinea, Sulawesi, Halmahera, and Australia, and of the *parvicollis-pygmaea*-lineage (Carabidae, Perigonini). *Entomologische Blätter und Coleoptera* **109**, 1–132.

Baehr M (2013e) The *Dolichoctis striata* complex (Insecta, Coleoptera, Carabidae, Lebiini). *Entomologische Blätter und Coleoptera* **109**, 133–222.

Baehr M (2014a) A peculiar new species of the genus *Sphallomorpha* Westwood from northern Australia (Coleoptera, Carabidae, Pseudomorphini). *Entomofauna* **36**, 441–452.

Baehr M (2014b) Two new species of the genus *Pogonoglossus* Chaudoir from the Pilbara, northwestern Australia (Insecta: Coleoptera: Carabidae: Physocrotaphini). *Records of the Western Australian Museum* **29**, 89–94. doi:10.18195/issn.0312-3162.29(2).2014.089-094

Baehr M (2014c) Three new species and a new genus of subterranean Zuphiini from the Pilbara region of northwestern Australia (Coleoptera: Carabidae: Harpalinae). *Records of the Western Australian Museum* **29**, 95–104. doi:10.18195/issn.0312-3162.29(2).2014.095-104

Baehr M (2014d) New Tachyine species from the Oriental, Papuan, and Australian Regions (Coleoptera, Carabidae, Bembidiini, Tachyina). *Entomologische Blätter und Coleoptera* **110**, 1–31.

Baehr M (2015a) A new species of the genus *Sloaneana* Csiki from Tasmania (Insecta, Coleoptera, Carabidae, Zolini). *Spixiana* **38**, 231–235.

Baehr M (2015b) Revision of the Australian Clivinini 2. The *ambigua-, bataviae-, bullata-, cava-, emarginata-, heterogena-, impressiceps-,* and *sloanei*-groups of the genus *Clivina* Latreille, the new genus *Rubidiclivina*, and additions to the 1st part (Coleoptera, Carabidae, Scaritinae). *Entomologische Blätter und Coleoptera* **111**, 59–446.

Baehr M (2016a) A peculiar new genus of the subtribe Tachyina from north Queensland, Australia (Insecta, Coleoptera, Carabidae, Bembidiini, Tachyina). *Transactions of the Royal Society of South Australia* **140**, 96–106. doi:10.1080/03721426.2016.1147116

Baehr M (2016b) A new genus and species of the tribe Zolini from southeastern Australia (Coleoptera: Carabidae: Zolini). *Australian Entomologist* **43**, 39–45.

Baehr M (2016c) New species and new records of odacanthine carabid beetles from the Oriental, Papuan, and Australian Regions (Coleoptera: Carabidae: Odacanthini). *Entomologische Blätter und Coleoptera* **112**, 25–51.

Baehr M (2016d) New species of the genera *Amblytelus* Erichson, 1842 and *Dystrichothorax* Blackburn, 1892 from Queensland and New South Wales, Australia (Insecta: Coleoptera: Carabidae: Psydrini, Amblytelina). *Memoirs of the Queensland Museum – Nature* **60**, 13–27.

Baehr M (2016e) New species of the genera *Sphallomorpha* Westwood, 1837 and *Adelotopus* Hope, 1834 from Queensland, Australia (Insecta: Coleoptera: Carabidae: Pseudomorphini). *Memoirs of the Queensland Museum – Nature* **60**, 29–41.

Baehr M (2016f) A new subspecies of *Mecyclothorax punctatus* (Sloane) from south-western Australia (Insecta, Coleoptera, Carabidae, Psydrini, Mecyclothoracina). *Spixiana* **39**, 93–97.

Baehr M (2016g) A new species of the *Tachys* (*s. l.*) *ectromoides*-group from south-eastern Queensland, Australia (Insecta, Coleoptera, Carabidae, BembidiinI). *Australian Entomologist* **43**, 69–74.

Baehr M (2016h) Two new species of the genus *Adelotopus* Hope, 1834 from North Queensland, Australia (Insecta, Coleoptera, Carabidae, Pseudomorphini) 15th supplement to the "Revision of the Pseudomorphinae of the Australian Region". *Spixiana* **39**, 257–263.

Baehr M (2016i) The *Notagonum marginellum-submetallicum*-complex. With description of new genera and species of Platynini from Australia (Insecta, Coleoptera, Carabidae, Platynini). *Entomologische Blätter und Coleoptera* **112**, 1–24.

Baehr M (2017a) Revision of the Australian Clivinini 3. The *procera*- and *elegans* -groups of the genus *Clivina* Latreille (Coleoptera, Carabidae, Scaritinae). *Spixiana* **40**, 197–304.

Baehr M (2017b) New species of the helluonine genus *Helluosoma* Castelnau, 1867 from Australia (Insecta, Coleoptera, Carabidae, Helluonini). *Mitteilungen der Münchner Entomologischen Gesellschaft* **107**, 23–31.

Baehr M (2017c) A new species of the genus *Coptocarpus* Chaudoir, 1857 from northern Australia (Insecta, Coleoptera, Carabidae, Oodini). *Spixiana* **40**, 17–40.

Baehr M (2017d) A replacement name for *Macranillus* Baehr, 2016 (Insecta: Coleoptera: Carabidae: Bembidiini: Anillina). *Records of the Western Australian Museum* **32**, 237. doi:10.18195/issn.0312-3162.32(2).2017.237

Baehr M (2017e) New Tachyine species from the Australian and Oriental Regions (Coleoptera, Carabidae, Bembidiini, Tachyina). *Entomologische Blätter und Coleoptera* **113**, 1–53.

Baehr M (2018) The cyclosomine genus *Sarothrocrepis* Chaudoir (Coleoptera, Carabidae, Cyclosomini), *Spixiana* **41**, 211–336.

Baehr M, Hudson P (2001) New species and a new genus of the subfamily Pogoninae from South Australia (Insecta: Coleoptera: Carabidae). *Records of the South Australian Museum* **34**, 1–18.

Baehr M, Main D (2016) New genera and species of Anilline Bembidiini from the Pilbara, north–western Australia (Insecta: Coleoptera: Carabidae: Bembidiini: Anillina). *Records of the Western Australian Museum* **31**, 59–89. doi:10.18195/issn.0312-3162.31(2).2016.059-089

Baehr M, Reid CAM (2017) On a collection of Carabidae from Timor Leste, with descriptions of nine new species (Insecta: Coleoptera, Carabidae). *Records of the Australian Museum* **69**, 421–450. doi:10.3853/j.2201-4349.69.2017.1660

Balkenohl M (2001) Key and Catalogue of the Tribe Clivinini from the Oriental Realm, with Revisions of the genera *Thliboclivina* Kult, and *Trilophidius* Jeannel (Insecta, Coleoptera, Carabidae, Scaritinae, Clivinini). *Pensoft Publishers*, Sofia–Moscou.

Balkenohl M (2003) Subfamily Scaritinae Bonelli, 1810. In *Catalogue of Palaearctic Coleoptera*, Vol. 1. Archostemata – Myxophaga – Adephaga (Eds I. Löbl and A. Smetana) pp. 219–234. Apollo Books, Stenstrup.

Ball GE (1959) A taxonomic study of the North American Licinini with notes on the Old World species of the genus *Diplocheila* Brulle (Coleoptera). *Memoirs of the American Entomological Society* **16**, 1–258.

Ball GE (1979) Conspectus of carabid classification: history, holomorphology, and higher taxa. In *Carabid Beetles: Their Evolution, Natural History, and Classification.* (Eds TL Erwin, GE Ball, DR Whitehead and AL Halpern) pp. 63–111. W. Junk, The Hague.

Ball GE (1983) Evaluation of the Baron Maximilien de Chaudoir's contribution to classification of cymindine Lebiini and Masoreimorphini (Coleoptera: Carabidae). *Coleopterists Bulletin* **36**, 513–530.

Ball GE (2001) The Subgenera of *Clivina* Latreille in the Western Hemisphere, and a Revision of Subgenus *Antroforceps* Barr (new status), with Notes about Evolutionary Aspects (Coleoptera: Carabidae: Clivinini). *Special Publication of the Japan Coleopterological Society Osaka* **1**, 129–156.

Ball GE, Bousquet Y (2001) Family 6. Carabidae Latreille, 1810. In *American Beetles **1** Archostemata, Myxophaga, Adephaga, Polyphaga, Staphyliformia.* (Eds RH Arnett Jr & MC Thomas) pp. 32–132. *CRC Press*, Boca Raton, London, New York, Washington, D.C.

Ball GE, Hilchie GJ (1983) Cymindine Lebiini of Authors: Redefinition and Reclassification of Genera (Coleoptera: Carabidae). *Quaestiones Entomologicae* **19**, 93–216.

Ball GE, Shpeley D (2002) The Neotropical Subgenera and Species of The Pantropical Genus *Anaulacus* MacLeay (*sensu novo*) (Coleoptera: Carabidae: Masoreini): a Taxonomic Revision, with Notes about Way of Life, Evolution, and Geographical History. *Transactions of the American Entomological Society* **128**, 265–343.

Bänninger M (1927) Die Ozaenini. *Deutsche Entomologische Zeitschrift* **1927**, 177–216.

Bänninger M (1940) On Australian Pamborini, Ozaenini and Scaritini (Coleoptera, Carabidae). *Novitates Zoologicae* **42**, 203–214.

Basilewsky P (1951) *Leleupidia luvubuana*, nov. gen. et nov. sp. (Col. Carabidae). *Revue de Zoologie et de Botanique Africaines* **44**, 175–179.

Basilewsky P (1954) Un genre nouveau de Leleupidiini de la presqueîle de Malacca. *Revue Francaise d'Entomologie* **21**, 213–216.

Basilewsky P (1958) Coléoptères Carabiques africaines nouveaux IX. *Revue de Zoologie et de Botanique Africaines* **57**, 321–334.

Basilewsky P (1959) Sur les genres *Syleter* Andrewes, *Psilidius* Jeannel et *Leleuporella* Basilewsky (Col. Carabidae Scaritinae). *Revue Francaise d'Entomologie* **26**, 5–16.

Basilewsky P (1984) Essai d'une classification supragénerique naturelle des carabides lébiens d'Afrique et de Madagascar (Coleoptera Carabidae Lebiinae). *Revue de Zoologie Africaine* **98**, 525–559.

Bates HW (1873) On the geoadephagous Coleoptera of Japan. *The Transactions of the Entomological Society of London* **1873**, 219–322.

Bates HW (1878) New genera and species of Carabidae from Tasmania. *Cistula Entomologica* **2**, 317–326.

Bates HW (1891) Notes on three Australian Carabidae. *Entomologist's Monthly Magazine* **2**, 285–287.

Bates HW (1892) Viaggio di Leonardo Fea in Birmania e regione vicine. XLIV. List of the Carabidae. *Annali del Museo Civico di Storia Naturale di Genova* **12**(32), 267–428.

Bell RT (1967) Coxal cavities and the classification of the Adephaga (Coleoptera). *Annals of the Entomological Society of America* **60**(1), 101–107. doi:10.1093/aesa/60.1.101

Bell RT, Bell JR (1978) Rhysodini of the world. Part I. A new classification of the tribe, and a synopsis of *Omoglymmius*, subgenus *Nitiglymmius*, new subgenus (Coleoptera: Carabidae or Rhysodidae). *Quaestiones Entomologicae* **14**, 43–88.

Bell RT, Bell JR (1982) Rhysodini of the world. Part III. Revision of *Omoglymmius* Ganglbauer (Coleoptera: Carabidae or Rhysodidae) and substitutions for preoccupied generic names. *Quaestiones Entomologicae* **18**, 127–259.

Bell RT, Bell JR (1991) The Rhysodini of Australia (Insecta: Coleoptera: Carabidae or Rhysodidae). *Annals of the Carnegie Museum* **60**(3), 179–210.

Beutel RG (1993) Phylogenetic analysis of Adephaga (Coleoptera) based on characters of the larval head. *Systematic Entomology* **18**, 127–147. doi:10.1111/j.1365-3113.1993.tb00658.x

Billberg GJ (1820) Enumeratio Insectorum in Museo Gust. Joh. Billberg. *Gadel*, Stockholm.

Blackburn T (1890) Notes on Australian Coleoptera, with descriptions of new species. Part IV. *Proceedings of the Linnean Society of New South Wales* **4**, 707–746. doi:10.5962/bhl.part.15065

Blackburn T (1891) Notes on Australian Coleoptera, with descriptions of new species. Part IX. *Proceedings of the Linnean Society of New South Wales* **5**, 775–790. doi:10.5962/bhl.part.18665

Blackburn T (1892) Notes on Australian Coleoptera, with descriptions of new species. Part XI. *Proceedings of the Linnean Society of New South Wales* **7**, 65–151. doi:10.5962/bhl.part.26043

Blackburn T (1894a) Further notes on Australian Coleoptera, with descriptions of new genera and species. XVI. *Transactions of the Royal Society of South Australia* **18**, 200–240.

Blackburn T (1894b) Notes on Australian Coleoptera, with descriptions of new species. XV. *Proceedings of the Linnean Society of New South Wales* **9**, 85–108. doi:10.5962/bhl.part.18095

Blackburn T (1895) Further notes on Australian Coleoptera, with descriptions of new genera and species XVIII. *Transactions of the Royal Society of South Australia* **19**, 1–258.

Blackburn T (1901) Further notes on Australian Coleoptera, with descriptions of new genera and species. XXIX. *Transactions of the Royal Society of South Australia* **25**, 99–131.

Blanchard CE (1842) Les Insectes. Atlas pls 20–21. In: Cuvier, G (1836–1849): Le Regne Animal accompagnée de planches gravées representant les types de tous les genres par une reunion de disciples de Cuvier. *Masson*, Paris.

Blanchard CE (1853) Description des Insectes. In: Dumont d'Urville, JS *Voyage au Pôle Sud et dans l'Océanie sur les corvettes l'Astrolabe et la Zélée; pendant les années, 1837–40*. Zoologie. Vol. IV. Gide & Baudry, Paris.

Bonelli FA (1810) Tabula synoptica exhibens genera carabicorum in sectiones et stirpes disposita. Galetti, Torino.

Bonelli FA (1813) Observations entomologiques. Deuxuième partie. *Memorie della Reale Accademia delle Scienze di Torino* **20**, 433–484.

Bouchard P, Bousquet Y, Davies AE, Alonso-Zarazaga MA, Lawrence JF, Lyal CHC, Newton AF, Reid CAM, Schmitt M, Slipinski SA, Smith ABT (2011) Family-group names in Coleoptera (Insecta). *ZooKeys* **88**(Special Issue), 1–972. doi:10.3897/zookeys.88.807

Bousquet Y (1985) Morphologie comparée des larves de Pterostichini (Coleoptera: Carabidae): descriptions et tables de détermination des espèces du nord-est de l'Amerique du nord. *Naturaliste Canadien* **112**, 191–251.

Bousquet Y (1996) Taxonomic revision of Nearctic, Mexican, and West Indian Oodini (Coleoptera: Carabidae). *Canadian Entomologist* **128**(3), 443–537. doi:10.4039/Ent128443-3

Bousquet Y (1999) Supraspecific classification of the Nearctic Pterostichini (Coleoptera: Carabidae). *Fabreries* **9**(Supplement), 1–292.

Bousquet Y (2002) Additions and corrections to the world catalogue of genus-group names of Geadephaga (Coleoptera) published by Wolfgang Lorenz (1998). *Folia Heyrovskyana* **9**(Supplementum), 1–78.

Bousquet Y (2003) Tribe Pentagonicini Bates, 1873. In *Catalogue of Palaearctic Coleoptera, Vol. 1. Archostemata – Myxophaga – Adephaga* (Eds Löbl I & Smetana A) p. 448. Apollo Books, Stenstrup.

Bousquet Y (2012) Catalogue of Geadephaga (Coleoptera, Adephaga) of America, north of Mexico. *ZooKeys* **245**, 1–1722. doi:10.3897/zookeys.245.3416

Bousquet Y, Larochelle A (1993) Catalogue of the geodephaga (Coleoptera: Trachypachidae, Rhysodidae, Carabidae including Cicindelini) of America north of Mexico. *Memoirs of the Entomological Society of Canada* **167**, 1–397.

Brandmayr P, Zetto Brandmayr T (1979) The evolution of parental care phenomena in Pterostichinae ground beetles with special reference to the genera *Abax* and *Molops* (Coleoptera, Carabidae). *Miscellaneous Papers of the Landbouwhogeschool Wageningen* **18**, 135–149.

Britton EB (1940) The Carabidae (Coleoptera) of New Zealand Part 1. PTEROSTICHINI. *Transactions and Proceedings of the Royal Society of New Zealand* **69**, 473–508.

Britton EB (1941) The Carabidae (Coleoptera) of New Zealand. Part II. Tribe s Lebiini and Pentagonicini. *Proceedings of the Royal Entomological Society of London* **10**, 185–196.

Britton EB (1949) The Carabidae (Coleoptera) of New Zealand. Part III. *Transactions of the Royal Society of New Zealand* **77**, 533–581.

Brouerius van Nidek CMC (1973) A new genus and a new species from the Australian region (Cicindelidae: Coleoptera). *Entomologische Blätter für Biologie und Systematik der Käfer* **69**, 61–63.

Broun T (1880) Manual of the New Zealand Coleoptera. J. Hughes, Wellington.

Brullé GA (1834) Coléoptères. I. Histoire naturelle des insects, traitant de leur organisation et de leurs moeurs en general, par M. V. Audouin, et comprenant leur classification et la description des espèces, par M. A. Brullé. Le tout accompagne de planches gravées sur acier, d'après des peintures executées pour cette edition sur la collection du Muséum de Paris. Tome IV. Pillot, Paris.

Bruneau de Miré P (1991) Note sur le genre *Elaphropus* Motschulsky, 1839 (Coleoptera Carabidae). *Bulletin de la Société Entomologique de France* **96**, 361–364.

Bulirsch P (2011) A taxonomic revision of the Australian genus *Setodyschirius* with description of six new species (Coleoptera: Carabidae: Scaritinae). *Acta Entomologica Musei Nationalis Pragae* **51**, 55–81.

Casale A (1988) *Revisione degli Sphodrina (Coleoptera, Carabidae, Sphodrini)*. Museo Regionale di Scienze Naturali, Torino.

Casale A (1998) Phylogeny and biogeography of Calleidina (Coleoptera: Carabidae: Lebiini): a preliminary survey. Phylogeny and classification of Caraboidea. *Atti de Museo regionale di Scienze naturale de Torino* **5**, 381–428.

Casale A, Ribera I (2008) Are Molopina of Euro-Mediterranean region related to the Madagascan, South African and Australian Pterostichini? (Coleoptera, Carabidae). *Biogeographia* **29**, 33–44.

Charley D, Andren M (2018) The distribution and conservation of the endangered atlas rainforest ground-beetle, 'Nurus atlas' (Laporte, 1867) (Coleoptera: Carabidae), in north-east New South Wales. *Australian Entomologist* **45**(4), 389–402.

Chaudoir M (1843) Genres nouveaux de la famille des Carabiques. (Continuation). *Bulletin de la Société Impériale des Naturalistes de Moscou* **16**, 383–427.

Chaudoir M (1844) Trois mémoires sur la famille des Carabiques. *Bulletin de la Société Impériale des Naturalistes de Moscou* **17**, 415–479.

Chaudoir M (1848) Mémoire sur la famille des Carabiques. 1e partie. *Bulletin de la Société Impériale des Naturalistes de Moscou* **21**, 3–134.

Chaudoir M (1861a) Beitrag zur Kenntnis einiger Carabicinen-Gattungen. *Berliner Entomologische Zeitschrift* **5**, 116–131.

Chaudoir M (1861b) Matériaux pour servir à l'étude des Cicindélètes et des Carabiques (continuation). *Bulletin de la Société Impériale des Naturalistes de Moscou* **34**, 491–576.

Chaudoir M (1862) Matériaux pour servir à l'étude des Carabiques. 3e partie. *Bulletin de la Société Impériale des Naturalistes de Moscou* **35**, 274–320.

Chaudoir M (1869a) Descriptions des Cicindélètes et de Carabiques nouveaux par le Baron de Chaudoir (Suite). *Revue et magasin de zoologie pure et appliquée* **21**, 203–208.

Chaudoir M (1869b) Mémoire sur les Thyréoptérides. *Annales de la Société Entomologique de Belgique* **12**, 113–162.

Chaudoir M (1869c) Mémoire sur les Coptodérides. *Annales de la Société Entomologique de Belgique* **12**, 163–256.

Chaudoir M (1871) Monographie des Lébiides. *Bulletin de la Société Impériale des Naturalistes de Moscou* **43**, 111–255.

Chaudoir M (1872a) Monographie des Callidides. *Annales de la Société Entomologique de Belgique* **15**, 97–204.

Chaudoir M (1872b) Descriptions d'espèces nouvelle de Carabiques de la tribu des Troncatipennes. *Revue et Magazin de Zoologie* **23**, 101–107, 138–143, 168–172, 212–221, 241–250.

Chaudoir M (1873) Essai monographique sur les drimostomides et les cratocérides et description d'un genre nouveau de morionides. *Annales de la Société Entomologique de Belgique* **15**.

Chaudoir M (1875) Genres aberrants du groupe des Cymindides. *Bulletin de la Société Impériale des Naturalistes de Moscou* **44**, 1–61.

Chaudoir M (1876) Étude monographique des Masoréides, des Tetragonodérides et du genre *Nematotarsus*. *Bulletin de la Société Impériale des Naturalistes de Moscou* **51**, 1–84.

Chaudoir M (1877) Genres nouveaux et espèces inédites de la famille des Carabiques. *Bulletin de la Société Impériale des Naturalistes de Moscou* **52**, 186–268.

Chaudoir M (1878) Les Harpaliens d'Australie d'après la collection de M. le Compte de Castelnau at la mienne. *Annali di Museo Civico di Storia Naturale di Genova* **12**, 475–517.

Chaudoir M (1883) Descriptions des Carabiques nouveaux. In: *Coleopterorum novitiates. Recueil spécialement consacré à l'étude des Coléoptères* **1**(1), 48 pp. Oberthür, Rennes.

Choate PM (2001) An Asian Ground Beetle, *Mochtherus tetraspilotus* (MacLeay), in Florida (Coleoptera: Carabidae: Lebiini). Fla. Dept. Agric. & Consumer Services, Division of Plant Industry. Entomology Circular No. 404, 2 pp.

Constantine RA, Seeman O (2014) Three new species of *Eutarsopolipus* (Acari: Podapolipidae) from Australian carabid beetles (Coleoptera: Carabidae). *Systematic and Applied Acarology* **19**(1), 87–109. doi:10.11158/saa.19.1.8

Csiki E (1927) Coleopterorum Catalogus. Carabidae: Carabinae II. Pars 92. W. Junk, Berlin.

Csiki E (1928) Coleopterorum Catalogus. Carabidae: Mormolycinae, Harpalinae I. Pars 97. W. Junk, Berlin.

Csiki E (1929) Coleopterorum Catalogus. Carabidae: Harpalinae III. Pars 104. W. Junk, Berlin.

Csiki E (1931) Coleopterorum Catalogus. Carabidae: Harpalinae V. Pars 115. W. Junk, Berlin.

Csiki E (1932) Coleopterorum Catalogus. Carabidae: Harpalinae VII. Pars 124. W. Junk, Berlin.

Csiki E (1933) Coleopterorum Catalogus. Carabidae: Harpalinae VIII. Pars 126. W. Junk, Berlin.

Darlington PJ, Jr (1950) Paussid beetles. *Transactions of the American Entomological Society* **76**, 47–142.

Darlington PJ, Jr (1952) The carabid beetle of New Guinea. Part 2. The Agonini. *Bulletin of the Museum of Comparative Zoology* **107**(3), 89–252.

Darlington PJ, Jr (1953) Australian carabid beetles I. Some *Clivina* from Western Australia. *Psyche* **60**, 52–61. doi:10.1155/1953/97605

Darlington PJ, Jr (1956) Australian carabid beetles III. Notes on the Agonini. *Psyche* **63**, 1–10. doi:10.1155/1956/57940

Darlington PJ, Jr (1961a) Australian carabid beetles V. Transition of wet forest faunas from New Guinea to Tasmania. *Psyche* **68**, 1–24. doi:10.1155/1961/64175

Darlington PJ, Jr (1961b) Australian Carabid beetles VI. The tropical and some subtropical species of *Pamborus, Mystropomus*, and *Nurus*. *Breviora* **142**, 1–13.

Darlington PJ (1961c) Australian carabid beetles VII. *Trichosternus*, especially the tropical species. *Psyche* **68**, 113–130.

Darlington PJ (1961d) Australian Carabid beetles VIII. *Leiradira*, especially the tropical species. *Breviora* **147**, 1–11.

Darlington PJ, Jr (1962a) The Carabid Beetles of New Guinea. Part I. Cicindelinae, Carabinae, Harpalinae through Pterostichini. *Bulletin of the Museum of Comparative Zoology* **126**, 319–564.

Darlington PJ, Jr (1962b) Australian carabid beetles X. *Bembidion*. *Breviora* **162**, 1–12.

Darlington PJ, Jr (1962c) Australian carabid beetles XI. Some *Tachys*. *Psyche* **69**, 117–128. doi:10.1155/1962/71454

Darlington PJ, Jr (1963a) Australian carabid beetles XII. More *Tachys*. *Psyche* **70**, 22–33. doi:10.1155/1963/96060

Darlington PJ, Jr (1963b) Australian carabid beetles XIII. Further notes on Agonini, and a genus of Licinini new to Australia. *Breviora* **183**, 1–10.

Darlington PJ, Jr (1964) Australian carabid beetles XIV. Perigona. *Psyche* **71**, 125–129. doi:10.1155/1964/68570

Darlington PJ, Jr (1965) Biogeography of the Southern End of the World. Harvard University Press, Cambridge, Mass.

Darlington PJ, Jr (1968) The Carabid Beetles of New Guinea. Part III. Harpalinae (continued): Perigonini to Pseudomorphini. *Bulletin of the Museum of Comparative Zoology* **137**, 1–253.

Darlington PJ, Jr (1970) Coleoptera, Carabidae including Cicindelinae. *Insects of Micronesia* **15**(1), 1–49.

Darlington PJ, Jr (1971) The carabid beetles of New Guinea. Part IV. General considerations; analysis and history of fauna; taxonomic supplement. *Bulletin of the Museum of Comparative Zoology* **142**, 129–337.

DeGeer C (1774) Memoires pour Servir a l'Histoire des Insectes. Tome quadtrième. P. Hesselberg, Stockholm.

Dejean PFMA (1821) Catalogue de la collection de coléoptères de M. le Baron Dejean. Crevot, Paris.

Dejean PFMA (1825) Species général des coléoptères, de la collection de M. le Compte Dejean. Tome première. Crevot, Paris.

Dejean PFMA (1829) Species général des coléoptères, de la collection de M. le Compte Dejean. Tome quadrième. Mequinon–Marvis, Paris.

Dejean PFMA (1831) Species général des coléoptères, de la collection de M. le Compte Dejean. Tome cinquième. Mequinon–Marvis.

Deuve T (1988) Étude phylogénétique des Coléoptères Adephaga: redefinition de la famille des Harpalidae, sensu novo, et position systématique des Pseudomorphinae et Brachininae. *Bulletin de la Société Entomologique de France* **92**, 161–182.

Deuve T (1993) L'abdomen et les genitalia des femelles de Coléoptères Adephaga. *Mémoires du Muséum National d'Histoire Naturelle, Paris* **155**, 1–184.

Dhuyvetter H, Hendrickx F, Gaublomme E, Desender K (2007) Differentiation between two salt marsh beetle ecotypes: evidence for ongoing speciation. *Evolution* **61**, 184–193. doi:10.1111/j.1558-5646.2007.00015.x

Di Giulio A, Moore W (2004) The first-instar larva of the genus *Arthropterus* (Coleoptera: Carabidae: Paussinae): implications for evolution of myrmecophily and phylogenetic relationships within the subfamily. *Invertebrate Systematics* **18**, 101–115. doi:10.1071/IS03028

Di Giulio A, Moore W (2009) The first known larva of the Australian genus *Mystropomus* Chaudoir (Coleoptera: Carabidae: Paussinae). *Australian Journal of Entomology* **48**, 140–148. doi:10.1111/j.1440-6055.2009.00698.x

Di Giulio A, Maurizi E, Barbero F, Sala M, Fattorini S, *et al.* (2015) The Pied Piper: a parasitic beetle's melodies modulate ant behaviours. *PLoS ONE* **10**(7), e0130541. doi.1371/journal.pone.0130541

Dokhtouroff W (1883) Essai sur la subdivision du genre *Cicindela. Revue Mensuelle d'Entomologie Pure et Appliquée* **1**, 66–70.

Driscoll DA, Weir T (2005) Beetle Responses to Habitat Fragmentation Depend on Ecological Traits, Habitat Condition, and Remnant Size. *Conservation Biology* **19**(1), 182–194. doi:10.1111/j.1523-1739.2005.00586.x

Dubault G, Lassalle B, Roux P (2008) Le groupe de *Lesticus chloronotus* Chaudoir, 1868 (Coleoptera, Pterostichidae, Trigonotomi). *Bulletin Mensuel de la Societe Linneenne de Lyon* **77**, 165–181. doi:10.3406/linly.2008.13692

Dupuis P (1913) Coleoptera Adephaga (Fam. Carabidae, subfam. Pentagonicinae). *Genera Insectorum* **145**, 1–4, pl. 1.

Ebach MC (2012) A history of biogeographical regionalisation in Australia. *Zootaxa* **34**, 1–34.

Eberhard S, Giachino PM (2011) Tasmanian Trechinae and Psydrinae (Coleoptera, Carabidae): a taxonomic and biogeographic synthesis, with description of new species and evaluation of the impact of Quaternary climate changes on evolution of the subterranean fauna. *Subterranean Biology* **9**, 1–72. doi:10.3897/subtbiol.9.2516

Eisner T, Jones TH, Aneshansley DJ, Tschinkel WR, Silberglied RE, Meinwald J (1977) Chemistry of defensive secretions of bombardier beetles (Brachinini, Metriini, Ozaenini, Paussini). *Journal of Insect Physiology* **23**, 1383–1386. doi:10.1016/0022-1910(77)90162-7

Erichson WF (1842) Beitrag zur Insecten-Fauna von Vandiemensland, mit besonderer Berücksichtigung der geographischen Verbreitung der Insecten. *Archiv für Naturgeschichte* **8**, 83–247.

Erwin TL (1972) Two new genera of Bembidiine Carabid beetles from Australia and South America with notes on their phylogenetic and zoogeographic significance (Coleoptera). *Breviora* **383**, 1–19.

Erwin TL (1974a) Studies of the Subtribe Tachyina (Coleoptera: Carabidae: Bembidiini). Part II: A revision of ther New World-Australian genus *Pericompsus* LeConte. *Smithsonian Contributions to Zoology* **162**, 1–96.

Erwin TL (1974b) The genus *Coptocarpus* Chaudoir of the Australian Region with notes on related African species (Coleoptera: Carabidae: Oodini). *Smithsonian Contributions to Zoology* **184**, 1–25.

Erwin TL (1975) Studies of the Subtribe Tachyina (Coleoptera: Carabidae: Bembidiini). Part III: Systematics, phylogeny, and zoogeography of the genus *Tachyta* Kirby. *Smithsonian Contributions to Zoology* **208**, 1–68. doi:10.5479/si.00810282.208

Erwin TL (1981) A synopsis of the immature stages of Pseudomorphini (Coleoptera: Carabidae) with notes on tribal affinities and behaviours in relation to life with ants. *Coleopterists Bulletin* **35**, 53–68.

Erwin TL (1985) The taxon pulse: a general pattern of lineage radiation and extinction among carabid beetles. In *Taxonomy, phylogeny, and zoogeography of beetles and ants* (Ed Ball GE) pp. 437–472. W. Junk, Dordrecht.

Erwin TL (1991) The Ground-Beetles of Central America (Carabidae), Part II: Notiophilini, Loricerini, and Carabini. *Smithsonian Contributions to Zoology* **501**, 1–30. doi:10.5479/si.00810282.501

Erwin TL (1994) Arboreal beetles of tropical forests: the Xystosomi group, subtribe Xystosomina (Coleoptera: Carabidae: Bembidiini). Part I. Character analysis, taxonomy, and distribution. *Canadian Entomologist* **126**, 549–666. doi:10.4039/Ent126549-3

Erwin TL, Geraci CJ (2008) New genera of Pseudomorphini (Insecta: Coleoptera, Carabidae) in the Western Hemisphere. In *Back to the Roots and back to the future: towards a new synthesis between taxonomic, ecological, and biogeographical approaches in carabidology. Proceedings of the XIII European Carabidologists Meeting, Blagoevgrad, August 20–24, 2007.* (Eds L Penev, TL Erwin and T Assmann) pp. 77–100. Pensoft Publishers, Sofia.

Erwin TL, Sims LL (1984) Carabid beetles of the West Indies (Insects: Coleoptera): a synopsis of the genera and checklists of tribes of Caraboidea, and of the West Indian species. *Quaestiones Entomologicae* **20**, 351–466.

Eschscholtz JF (1829) Zoologischer Atlas, enthaltend Abbildungen und Beschreibungen neuer Thierarten, während des Flottcapitains von Kotzebue zweiter Reise um die Welt, auf der Russisch-Kaiserlichen Kriegsschalupp Predpriaetië in den Jahren 1823–1826. Zweites Heft. Reimer, Berlin.

Fauvel A (1882) Les Coléoptères de la Nouvelle-Calédonie er dépendances avex descriptions, notes et synonymies nouvelles. *Revue d'Entomologie* **1**, 217–273.

Fauvel A (1903) Faune analytique des Coléoptères de la Nouvelle-Calédonie. *Revue d'Entomologie* **22**, 203–243.

Fedorenko DN (1996) Reclassification of world Dyschiriini, with a revision of the Palearctic fauna (Coleoptera, Carabidae). *Series Faunistica* **4**, 4–224. Pensoft Publishers, Sofia, Moscow, St. Petersburg.

Fedorenko DN (2015) Notes on the genera *Dischissus* and *Microcosmodes* (Coleoptera, Carabidae, Panagaeini) from the Oriental Region, with description of a new genus and a new species. *Russian Entomological Journal* **24**, 271–279.

Fleutiaux E (1894) Les Megacephalidae d'Australie. *Revue d'Entomologie* **13**, 123–128.

Frank JH (1994) Inoculative biological control of moe crickets. In *Integrated Pest Management for Turf and Ornamentals* (Ed A Leslie) pp. 467–475. CRS Press.

Freitag R (1979) Reclassification, phylogeny and zoogeography of the Australian species of *Cicindela* (Coleoptera: Cicindelidae). *Australian Journal of Zoology. Supplementary Series* **66**, 1–99.

Galián J, Moore BP (1994) Chromosome numbers and sex-determining mechanism in Australian Carabidae (Coleoptera). *Coleopterists Bulletin* **48**(3), 226–235.

Garetto R, Giachino PM (2003) A new species of *Cratoferonia* Tschitcherine 1902 (Coleoptera, Carabidae) from New South Wales. Results of the Zoological Missions to Australia of the Regional Museum of Natural Sciences of Turin, Italy. I. *Monographie di Museo Regionale di Scienze Naturali Torino* **35**, 85–98.

Gestro R (1875) Note sopra alcuni Carabici appartenenti al Museo Civico di Genova con descrizione di species nuove. *Annali del Museo Civico di Storia Naturale di Genova* **7**, 850–894.

Giachino PM (2003a) The genus *Pheropsophus* Solier, 1833 in Australia (Coleoptera, Carabidae). *Monografie de Museo Regionale di Scienze naturali di Torino* **35**, 29–64.

Giachino PM (2003b) A new *Polyderis* species from the Northern Territory, Australia (Coleoptera, Carabidae, Bembidiinae). *Monografie de Museo Regionale di Scienze naturali di Torino* **35**, 65–72.

Giachino PM (2003c) Some data on the genus *Delinius* Westwood, 1864 (Coleoptera, Carabidae) with the description of a new species from the Northern Territory. Results of the Zoological Missions to Australia of the Regional Museum of Natural Sciences of Turin, Italy. I. *Monografie di Museo Regionale di Scienze Naturali Torino* **35**, 73–84.

Giachino PM (2005a) Revision of the Australian Anillina (Coleoptera, Carabidae, Bembidiini). *Monografie di Museo Regionale di Scienze naturali di Torino* **42**, 137–238.

Giachino PM (2005b) The genus *Pheropsophus* Solier, 1833 in New Guinea (Insecta, Coleoptera, Carabidae, Brachininae). *Spixiana* **28**, 223–257.

Giachino PM (2005c) Revision of the genus *Catadromus* W.S. Macleay, 1825 (Coleoptera, Carabidae). Results of the Zoological Missions to Australia of the Regional Museum of Natural Sciences of Turin, Italy. II. *Monografie di Museo Regionale di Scienze Naturali Torino* **42**, 239–268.

Giachino PM (2005d) Revision of the genus *Rhabdotus* Chaudoir, 1865 and description of the larva of R. reflexus (Chaudoir, 1865) (Coleoptera, Carabidae). Results of the Zoological Missions to Australia of the Regional Museum of Natural Sciences of Turin, Italy. II. *Monografie di Museo Regionale di Scienze Naturali Torino* **42**, 279–294.

Giachino PM (2005e) Notes on the genus *Pseudoceneus* Sloane, 1903, with the description of a new species from Tasmania (Coleoptera, Carabidae). Results of the Zoological Missions to Australia of the Regional Museum of Natural Sciences of Turin, Italy. II. *Monografie di Museo Regionale di Scienze Naturali Torino* **42**, 295–316.

Giachino PM, Picciau L (2005) A new *Delinius* species from the Northern Territoy, Australia (Coleoptera, Carabidae). Results of the Zoological Missions to Australia of the Regional Museum of Natural Sciences of Turin, Italy. II. *Monografie di Museo Regionale di Scienze Naturali Torino* **42**, 269–278.

Grebennikov VV (2008) *Tasmanitachoides* belongs to Trechini (Coleoptera: Carabidae): discovery of the larva, its phylogenetic implications and revised key to Trechitae genera. *Invertebrate Systematics* **22**, 479–488. doi:10.1071/IS07051

Greenslade P (1994) Heritage listing of invertebrate sites in southeastern Australia. *Memoirs of the Queensland Museum* **36**, 67–76.

Grzymala TL, Will KW (2014) Taxonomic review of *Cratocerus* Dejean, 1829 (Coleoptera, Carabidae) with description of six new species. *ZooKeys* **416**, 77–112. doi:10.3897/zookeys.416.6455

Guéorguiev B (2014) Two new related oodine genera in the Oriental Region, with remarks on the systematic position of the genera *Hololeius* and *Holosoma* (Coleoptera, Carabidae). *Deutsche Entomologische Zeitschrift* **61**, 87–104. doi:10.3897/dez.61.7754

Guéorguiev B, Morita S (2017) Two new species of *Anatrichis* LeConte (Coleoptera: Carabidae: Oodini) from Japan, with notes on the taxonomic position of *Oodes infimus* Andrewes. *Zootaxa* **4226**, 249–263. doi:10.11646/zootaxa.4226.2.5

Guthrie NA (2001) Aspects of the taxonomy and ecology of the ground beetle (Coleoptera: Carabidae) assemblage of the Swan Coastal Plain: (with particular reference to habitat fragmentation and the Quindalup dune system). Masters Thesis. Edith Cowan University.

Guthrie NA, Weir T, Will KW (2010) Localised and regional patterns in ground-dwelling beetle assemblages in a semi-tropical arid zone environment. *Records of the Western Australian Museum* **78**(Supplement), 169–184. doi:10.18195/issn.0313-122x.78(1).2010.169-184

Habu A (1967) Fauna Japonica: Carabidae. Truncatipennes group (Insecta: Coleoptera). Biogeographical Society of Japan, 338 pp.

Habu A (1982) Revised and supplementary notes on and descriptions of the Truncatipennes group of Japan (I). (Coleoptera, Carabidae). *Entomological Review of Japan* **36**, 85–142.

Häckel M, Farkač J (2012) A checklist of the subfamily Panagaeinae Hope, 1838 of the World (Coleoptera: Carabidae). *Studies and reports of District Museum Prague-East (taxonomical series)* **8**(1–2), 67–116.

Häckel M, Farkač J, Wrase DW (2010) A check-list of the tribe Broscini Hope, 1838 of the World. *Studies and Reports. Taxonomical Series* **6**, 43–84.

Harvey MS (2002) Short-range endemism among the Australian fauna: some examples from non-marine environments. *Invertebrate Systematics* **16**, 555–570. doi:10.1071/IS02009

Hogan JE (2012) Taxonomy, Systematics and Biogeography of the Scaritinae (Insecta, Coleoptera, Carabidae). PhD Thesis. Oxford Brookes University.

Hope FW (1834) Characters and descriptions of several new genera and species of coleopterous insects. *Transactions of the Zoological Society of London* **1**, 91–112. doi:10.1111/j.1096-3642.1835.tb00607.x

Hope FW (1838) The coleopterist's manual, part the second, containing the predaceous land and water beetles of Linneus and Fabricius. Henry G. Bohn, London, pp. 169.

Horn GH (1881) On the genera of Carabidae with special reference to the fauna of boreal America. *Transactions of the American Entomological Society and Proceedings of the Entomological Section of the Academy of Sciences* **9**, 91–196.

Horn W (1899) Über das System der Cicindeliden. *Deutsche Entomologische Zeitschrift* **1899**, 33–51.

Horn W (1926) Carabidae, Cicindelinae. Coleopterorum Catalogus. Pars **86**, W. Junk, Berlin.

Horn W (1936) Über die erste australische *Prothyma*-Art sowie über die Fauna von Haiti. *Entomologische Blätter Für Biologie Und Systematik Der Käfer* **32**, 22–23.

Horne PA (1990) Parental care in *Notonomus* Chaudoir (Coleoptera: Carabidae: Pterostichinae). *Australian Entomologist* **17**(2), 65–69.

Horne PA (1992) Comparative life histories of 2 Species of *Notonomus* (Coleoptera, Carabidae) in Victoria. *Australian Journal of Zoology* **40**(2), 163–171.

Horne PA, Edward CL (1998) The effect of tillage on pest and beneficial beetles. *Australian Journal of Entomology* **37**, 60–63. doi:10.1111/j.1440-6055.1998.tb01545.x

Illiger JWC (1807) Vorschlag zur Aufnahme im fabrischen Systeme fehlender Käfergattungen. *Magazin für Insektenkunde* **6**, 318–349.

Jaeger B (2016) Two new species of the *Anthracus annamensis* group from Australia and New Caledonia, and notes on identity and generic placement of *Acupalpus angulatus* MacLeay, 1871 and *Acupalpus trapezus* Fauvel, 1882 (Coleoptera, Carabidae, Harpalini, Stenolophina, Pelm). *Linzer Biologische Beiträ ge* **48**, 1295–1312.

Janssens E (1953) Cupesidae – Paussidae. (Editio Secunda). Coleopterorum Catalogus Supplementum Pars **5**. W. Junk, The Hague.

Jeannel R (1926) Monographie des *Trechinae*. Morphologie comparé et distribution géographique d'un groupe des coléoptères. (Première livraison). *L'Abeille: Journal d'Entomologie* **32**, 221–550.

Jeannel R (1927) Monographie des Trechinae. Morphologie comparé et distribution géographique d'un groupe des coléoptères. (Deuxième livraison). *L'Abeille: Journal d'Entomologie* **33**, 1–592.

Jeannel R (1932) Révision du genre *Limnastis* (Coleoptera, Carabidae). *Livre de Centenaire*: 167–187. *Société Entomologique de France*, Paris.

Jeannel R (1937) Les Bembidiides endogées (Col. Carabidae). Monographie d'une lignée gondwanienne. *Revue Francaise d'Entomologie* **3**, 241–396.

Jeannel R (1938) Les Migadopides (Coleoptera Adephaga), une lignée subantarctique. *Revue Francaise d'Entomologie* **5**, 1–55.

Jeannel R (1941) Coléoptères Carabiques. Première Partie. *Faune de France* **39**, 1–571.

Jeannel R (1942) Coléoptères Carabiques. Deuxième Partie. *Faune de France* **40**, 572–1173.

Jeannel R (1948) Coléoptères carabiques de la région Malgache (deuxième partie). Faune de l'Empire francais, vol. X, 765 pp.

Jeannel R (1949) Coléoptères Carabiques de la region Malgache, Part 3. *Faune de l'Empire Français* **11**, 767–1148.

Jedlicka A (1941) Versuch einer Monographie der pal. Carabiden-Gattungen mit abgesetzten Flügeldecken (Truncatipennen) mit Berücksichtigung der indischen Fauna. A. Jedlicka, Prague.

Jedlicka A (1963) Monographie der Truncatipennen aus Ostasien. Lebiinae – Odacanthinae – Brachyninae (Coleoptera, Carabidae). *Entomologische Abhandlungen und Berichte aus dem Staatlichen Museum für Tierkunde in Dresden* **28**, 269–579.

Johns PM (2010) Migadopini (Coleoptera: Carabidae; Migadopinae) of New Zealand. *Records of the Canterbury Museum* **24**, 39–63.

Kabak I (2003) Tribe Lebiini Bonelli, 1810. In *Catalogue of Palaearctic Coleoptera*, Vol. 1. Archostemata – Myxophaga – Adephaga (Eds I Löbl and A. Smetana) pp. 488–439. Apollo Books, Stenstrup.

Kamoun S, Hogenhout SA (1996) Flightlessness and rapid terrestrial locomotion in tiger beetles of the *Cicindela* L. subgenus *Rivacindela* van Nidek from saline habitats of Australia (Coleoptera, Cicindelidae). *Coleopterists Bulletin* **50**, 221–230.

Kataev BM (2002a) A new genus and species of the subtribe Anisodactylina from south-western Australia (Coleoptera: Carabidae: Harpalini). *Acta Zoologica Academiae Scientiarum Hungaricae* **48**, 173–179.

Kataev BM (2002b) Revision of the genus *Loxoncus* Schmidt-Goble, 1846 from the Palearctic, the Oriental Region and Australia (Coleoptera: Carabidae: Harpalini). *Russian Entomological Journal* **11**, 351–382.

Kataev BM (2007) A new species of the genus *Nornalupia* from the Stirling Range Nature Park, Western Australia (Coleoptera: Carabidae: Harpaline). *Zoosystematica Rossica* **16**, 33–36.

Kataev BM (2010) A taxonomic review of the subgenus *Hyparpalus* (genus *Parophonus*) of the Oriental and Australian regions (Coleoptera: Carabidae: Harpalini). *Zoosystematica Rossica* **19**, 277–300.

Kataev BM (2015) New data on distribution of ground-beetles of the tribe Harpalini in the Palaearctic, Oriental Region and in Australia (Coleoptera: Carabidae: Harpalini). *Entomological Review* **95**, 536–543. doi:10.1134/S0013873815040181

Kataev BM (2016) On taxonomic status of *Bottchrus*, with a description of a new brachypterous species of the genus *Trichotichnus* (Coleoptera: Carabidae) from the Philippines. *Zootaxa* **4061**, 68–76. doi:10.11646/zootaxa.4061.1.7

Kavanaugh DH (1986) A systematic review of am- phizoid beetles (Amphizoidae: Coleoptera) and their phylogenetic relationships to other Adephaga. *Proceedings of the California Academy of Sciences* **44**, 67–109.

Kavanaugh DH (1998) Field observations confirming brood care in Percus passerinii in the Tuscan Apennines, Italy (Coleoptera Carabidae). *Bollettino della Societa Entomologica Italiana* **130**, 55–63.

Kirby W (1825) A description of such genera and species of insects alluded to in the "Introduction to Entomology" of Messrs. Kirby and Spence, as appear not to have been before sufficiently noted or described. *Transactions of the Linnean Society of London* **14**, 563–572. doi:10.1111/j.1095-8339.1823.tb00103.x

Kirby W (1837) Insects. In *Fauna Boreali-Americana; or the zoology of the northern parts of British America*. Part **4**. (Ed. J. Richardson) pp. 1–325. London, Longmans.

Kirschenhofer E (2000) Neue und wenig bekannte Panagaeini der östlichen Paläarktis sowie der Orientalis. (Coleoptera, Carabidae). *Entomofauna* **21**(29), 321–372.

Klug JCF (1821) Entomologiae Brasilianae specimen. *Nova Acta Academiae Leopoldo Carlinae* **10**, 277–324.

Klug JCF (1834) Jahrbücher der Insektenkunde mit besonderer Rücksicht auf die Sammlung im Königl. Museum in Berlin herausgegeben. Band **1**. Enslin, Berlin.

Kolbe HJ (1880) Natürliches Sytem der carnivoren Coleoptera. *Deutsche Entomologische Zeitschrift* **24**, 258–280.

Kopecky T (2003) Tachyini. In *Catalogue of Palaearctic Coleoptera*, Vol. 1. Archostemata – Myxophaga – Adephaga. (Eds I Löbl and A Smetana) pp. 273–280. Apollo Books, Stenstrup.

Kryzhanovskij OL (1976) An attempt at a revised classification of the family Carabidae (Coleoptera) [in Russian]. Entomologicheskoe Obozrenie 55, 80–91 [An English translation was issued in Entomological Review 55(1), 56–64.]

Kult K (1947) The 3rd study to the knowledge of Tribus Clivinini (Col. Carab.). *Acta Societatis Entomologicae Cechoslovakiae* **44**, 26–37.

Kult K (1951) Revision of the genus *Clivina* Latr. from Oriental region. *Acta Societatis Entomologicae Cechoslovakiae* **48**, 16–32.

Kult K (1959) Revision of the African species of the old genus *Clivina* Latr. (Col. Carabidae). *Revue de zoologie et de botanique africaines, Wavre Belgium* **60**, 172–225.

Lacordaire JT (1854) Histoire Nouvelle des Insectes. genera des Coléoptères Tom. 1 Contenant les familles des Cicindélètes, Carabiques, Dytiscides, Gyrinides et Palpicornes. Libraire Encyclopédique de Roret, Paris.

Laporte FL (1834) Études Entomologiques, ou Description d'Insectes Nouveaux et Observations sur leur Synonymie. *Méquinon–Marvis*, Paris.

Laporte FL (1835) Études Entomologiques, ou Description d'Insectes Nouveaux et Observations sur leur Synonymie. Première partie. Méquinon–Marvis, Paris.

Laporte FL (1867) Notes on Australian Coleoptera. *Transactions and Proceedings of the Royal Society of Victoria* **139**, 1–139.

Laporte FL (1868) Notes on Australian Coleoptera. *Transactions and Proceedings of the Royal Society of Victoria* **8**, 30–38, 95–225.

Lapouge GV (1929) Sous-tribu des Calosomina. *Miscellanea Entomologica* **32**(3), 1–10.

Larochelle A (1990) The food of carabid beetles (Coleoptera: Carabidae, including Cicindelinae). *Fabreries* **5**(Supplement), 1–132.

Larochelle A, Larivière M-C (2001) Carabidae (Insecta: Coleoptera): catalogue. *Fauna of New Zealand* **43**, 285 pp.

Larochelle A, Larivière M-C (2003) A natural history of the ground beetles (Coleoptera: Carabidae) of America north of Mexico. Pensoft, Sofia-Moscow.

Larochelle A, Larivière M-C (2005) Harpalini (Insecta: Coleoptera: Carabidae: Harpalinae). Fauna of New Zealand **53**, 160 pp.

Larochelle A, Larivière M-C (2007) Carabidae (Insecta: Coleoptera): synopsis of supraspecific taxa. *Fauna of New Zealand* **60**:188 pp.

Latreille PA (1802) Histoire naturelle, générale et particulière des crustacés et des insectes. Ouvrage faisant suite à l'histoire naturelle générale et particulière, composée par Leclerc de Buffon, et rédigée par C. S. Sonnini, membre de plusieurs sociétés savantes. Familles naturelles des genres. Tome troisième. F. Dufart, Paris.

Latreille PA (1806) Genera crustaceorum et insectorum secundum ordinem naturalem in familias dispositas, iconibus exempliusque plurimis explicata. Tomus secundus. Armand Koenig, Paris.

Latreille PA (1810) Considérations générales sur l'ordre natural des animaux composant les classes de crustacés, des arachnides, et des insectes avec une tableau méthodique de leurs genres, disposés en familles. F. Schoell, Paris.

Lawrence JF, Britton EB (1991) Coleoptera (Beetles). In *The Insects of Australia. A textbook for students and research workers*. Second Ed. Vol. II. pp. 543–683. Melbourne University Press.

Lawrence JF, Newton AF (1995) Families and subfamilies of Coleoptera (with selected genera, notes, references and data on family-group names). In *Biology, phylogeny, and classification of Coleoptera: Papers Celebrating the 89th Birthday of Roy A. Crowson* (Eds J Pakaluk J and SA Slipinski) pp. 779–1005. Muzeum I Instytut Zoologii PAN, Warszawa.

Lawrence JF, Ślipiński A (2013) Australian Beetles. Morphology, Classification and Keys. Volume 1. CSIRO Publishing, Collingwood, Vic.

LeConte JL (1851) Descriptions of new species of Coleoptera from California. *Annals of the Lyceum of Natural History of New York* **5**, 125–184. doi:10.1111/j.1749-6632.1852.tb00123.x

LeConte JL (1852) Descriptions of new species of Coleoptera from California. *Annals of the Lyceum of Natural History of New York* **5**, 185–216.

LeConte JL (1853) Notes on the classification of the Carabidae of the United States. *Transactions of the American Philosophical Society* **10**, 363–403. doi:10.2307/1005287

LeConte JL (1859) The Complete Writings of Thomas Say on the Entomology of North America, Vol. II. Balière, New York.

LeConte JL (1860) Revision on the Cicindelidae of the United States. *Transactions of the American Philosophical Society* **11**, 27–64. doi:10.2307/3231925

Liang HB, Kavanaugh DH (2007) Review of the genus *Dendrocellus* Schmidt-Gobel (Coleoptera: Carabidae: Dryptini), with descriptions of seven new species. *Coleopterists Bulletin* **61**, 1–39. doi:10.1649/934.1

Liang HB, Kavanaugh DH, Tian MY (2004) Notes on *Drypta longicollis* MacLeay and the status of the genus-group name *Desera* Dejean, 1825 (Coleoptera: Carabidae: Dryptini). *Proceedings of the California Academy of Sciences* (Series 4) **55**, 377–383.

Liebherr JK (1988) Redefinition of the supertribe Odacanthitae and revision of the West Indian Lachnophorini (Coleoptera: Carabidae). *Quaestiones Entomologicae* **24**, 1–42.

Liebherr JK (1991) A new tribal placement for the Australasian genera *Homethes* and *Aeolodermus* (Coleoptera: Carabidae: Odacanthini). *The Pan-Pacific Entomologist* **66**, 312–321.

Liebherr JK (2005) Platynini (Coleoptera: Carabidae) of Vanuatu: Miocene diversification on the Melanesian Arc. *Invertebrate Systematics* **19**, 263–295. doi:10.1071/IS04032

Liebherr JK (2011) Cladistic assessment of subtribal affinities within the tribe Moriomorphini with description of *Rossjoycea glacialis*, gen. n. and sp. n. from the South Island, and revision of Meonochilus Liebherr and Marris from the North Island, New Zealand (Coleoptera, Carabidae). *ZooKeys* **147**, 277–335. doi:10.3897/zookeys.147.1898

Liebherr JK, Kavanaugh DH (1985) Ovoviviparity in carabid beetles of the genus *Pseudomorpha* (Insecta: Coleoptera). *Journal of Natural History* **19**, 1079–1086. doi:10.1080/00222938500770681

Liebherr JK, Maddison DR (2013) Colonisation of the Pacific by Bembidion beetles (Coleoptera: Carabidae), with description of Bembidion tahitiense, sp. nov., from Tahiti, French Polynesia. *Invertebrate Systematics* **27**, 439. doi:10.1071/IS13003

Liebherr JK, Will KW (1998) Inferring phylogenetic relationships within Carabidae (Insecta, Coleoptera) from characters of the female reproductive tract. In *Phylogeny and classification of Caraboidea* (Eds GE Ball, A Casale and A Vigna Taglianti). *Atti de Museo regionale di Scienze naturale de Torino* **5**, 107–170.

Liebherr JK, Will KW (2015) Antisymmetric male genitalia in Western Australian populations of *Mecyclothorax punctipennis* (Coleoptera: Carabidae: Moriomorphini). *Insect Systematics & Evolution* **46**, 393–409. doi:10.1163/1876312X-45042124

Liebherr JK, Zimmerman EC (1998) Cladistic analysis, phylogeny and biogeography of the Hawaiian Platynini (Coleoptera: Carabidae). *Systematic Entomology* **23**, 137–172. doi:10.1046/j.1365-3113.1998.00044.x

Liebke M (1931) Die afrikanischen Arten der Gattung *Colliuris* Degeer (Col. Car.). *Revue de Zoologie et de Botanique Africaines* **20**, 280–301.

Liebke M (1934) Die Arten der Gattung *Pseudaptinus* Cast. (Col. Carabidae). *Revista de Etologia* **4**, 372–388.

Liebke M (1938) Denkschrift über die Carabiden-Tribus Colliurini. *Festschrift zum 60. Geburtstage von Professor Dr. Embrik Strand* **4**, 37–141.

Lindroth CH (1980) A revisionary study of the taxon *Cillenus* Samouelle, 1819 and related forms. (Coleoptera. Carabidae: Bembidiini). *Entomologica Scandinavica* **11**, 179–205. doi:10.1163/187631280X00509

Lindroth CH (1945–1949) Die fennoskandischen Carabidae. Eine tiergeographische Studie. I–III. *Göteborgs Kungliga Vetenskaps och Vitterhets Samhälles Handlingar* B **4**(3), 1–911.

Löbl I, Smetana A (Eds) (2003) Catalogue of Palaearctic Coleoptera, Vol. 1. Archostemata – Myxophaga – Adephaga. Apollo Books, Stenstrup.

López-López A, Hudson P, Galián J (2016) Islands in the desert: species delimitation and evolutionary history of *Pseudotetracha* tiger beetles (Coleoptera: Cicindelidae: Megacephalini) from Australian salt lakes. *Molecular Phylogenetics and Evolution* **101**, 279–285. doi:10.1016/j.ympev.2016.05.017

Lorenz W (1998) Systematic List of extant Ground Beetles of the World (Insecta Coleoptera "Geadephaga": Trachpachidae and Carabidae incl. Paussinae, Cicindelinae. Rhysodidae). Printed by the author, Tutzing.

Lorenz W (2005) Systematic List of extant Ground Beetles of the World (Insecta Coleoptera "Geadephaga": Trachpachidae and Carabidae incl. Paussinae, Cicindelinae. Rhysodidae). 2nd edn. Printed by the author, Tutzing.

Lorenz W (2018) CarabCat-Checklist. Available at: http://www.carabidfauna.com/checklist.php [accessed 22 October 2018].

Lövei GL, Sunderland KD (1996) Ecology and Behavior of Ground Beetles. *Annual Review of Entomology* **41**, 231–256. doi:10.1146/annurev.en.41.010196.001311

Luna de Carvalho E (1987) Essai monographique des Coléoptères Protopaussines et Paussines. Contenant des déscritions et iconographie des taxa actuelles et fossiles avec des clefs dichotomiques de toutes les espèces. *Memórias do Instituto de Investigação. Científica Tropical, Lisboa* **70**, 1027 pp.

Machado A (1992) Monografia de los Carábidos de las Islas Canarias (Insecta, Coleoptera). *Instituto de Estudios Canarios, La Laguna*.

Macleay WS (1825) Annulosa Javanica, or an attempt to illustrate the natural affinities and analogies of the insects collected in Java by Thomas Horsfield, M.D.F.L. & G.S. and deposited by him in the museum of the hounourable East-India Company. Number 1. Kingsbury, Parbury & Allen, London.

Macleay WS (1838) Illustrations of the Annulosa of South Africa; being a portion of the objects of the natural history chiefly collected during an expedition into the interior of South Africa, under the direction of Dr. Andrew Smith, in the years 1834, 1835, and 1836. Smith, Elder & Co., London, Macleay WJ (1863) On the Scaritidae of New Holland. *Transactions of the Entomological Society of New South Wales* **1**, 55–74.

Macleay WJ (1864) Descriptions of new genera and species of Coleoptera from Port Denison. *Transactions of the Entomological Society of New South Wales* **1**, 106–130.

Macleay WJ (1865) Description of a new genus of carabideous insects. *Transactions of the Entomological Society of New South Wales* **1**, 155–157.

Macleay WJ (1871) Notes on a collection of insects from Gayndah. *Transactions of the Entomological Society of New South Wales* **2**, 79–205.

Macleay WJ (1873) Miscellania entomologica. *Transactions of the Entomological Society of New South Wales* **2**, 319–370.

Macleay WJ (1886) The insects of the Fly river, New Guinea, "Coleoptera. *Proceedings of the Linnean Society of New South Wales* **1**, 136–157.

Macleay WJ (1887) The insects of the Cairns district, northern Queensland. *Proceedings of the Linnean Society of New South Wales* **2**, 213–238. doi:10.5962/bhl.part.29175

Macleay WJ (1888) The insects of King's Sound and its vicinity. *Proceedings of the Linnean Society of New South Wales* **3**, 443–480.

Maddison DR, Baker MD, Ober KA (1999) Phylogeny of carabid beetles as inferred from 18S ribosomal DNA. *Systematic Entomology* **24**, 103–138. doi:10.1046/j.1365-3113.1999.00088.x

Maddison DR, Moore W, Baker MD, Ellis TM, Ober KA, Cannone JJ, Gutell RR (2009) Monophyly of terrestrial adephagan beetles as indicated by three nuclear genes (Coleoptera: Carabidae and Trachypachidae). *Zoologica Scripta* **38**, 43–62. doi:10.1111/j.1463-6409.2008.00359.x

Madge RB (1975) The type-species of Bonelli's genera of Carabidae (Coleoptera). *Quaestiones Entomologicae* **11**, 579–586.

Maindron M (1906) Description d'une nouvelle espèce de *Pheropsophus* et remarques sur ce genre (Col.). *Bulletin de la Société Entomologique de France* **1906**, 15–16.

Majer JD, Recher HF, Graham R, Gupta R (2003) Trunk invertebrate faunas of Western Australian forests and woodlands: influence of tree species and season. *Austral Ecology* **28**, 629–641. doi:10.1046/j.1442-9993.2003.01320.x

Martínez-Navarro EM, Galián J, Serrano J (2005) Phylogeny and molecular evolution of the tribe Harpalini (Coleoptera, Carabidae) inferred from mitochondrial cytochrome-oxidase I. *Molecular Phylogenetics and Evolution* **35**, 127–146. doi:10.1016/j.ympev.2004.11.009

Mateu J (1963) Notas sobre tres séries filéticas de Lebiidae (Lichnasthenini Thomson, Singilini Jeannel, Somotrichini nov.) (Coleoptera-Carabidae) y rectificaciones sinonìmicas. *Annali del Museo Civico di Storia Naturale Giacomo Doria* **74**, 122–139.

Matthews EG (1980) A guide to the genera of Beetles of South Australia. Part 1: Archostemata and Adephaga: 68 pp. *Special Educational Bulletin Series, South Australian Museum,* Adelaide.

McCairns RF, Freitag R, Rose HA, Donald FJD (1997) Taxonomic revision of the Australian Cicindelidae (Coleoptera), excluding species of *Cicindela*. *Invertebrate Taxonomy* **11**, 599–687. doi:10.1071/IT94011

Menzel F, Kitching RL, Boulter SL (2004) Host specificity or habitat structure? – The epicortical beetle assemblages in an Australian subtropical rainforest. *European Journal of Entomology* **101**, 251–259. doi:10.14411/eje.2004.035

MLR Southern Emu-wren & Fleurieu Peninsula Swamps Recovery Team (2007) Recovery Plan for the Mount Lofty Ranges Southern Emu-wren *Stipiturus malachurus intermedius*: 2006–2011. Conservation Council of South Australia, Adelaide.

Moore BP (1960a) Studies on Australian Carabidae (Coleoptera) – 1. New species of the tribes *Agonicini, Trechini* and *Pterostichini. Proceedings of the Royal Entomological Society of London* **29**, 165–169[B].

Moore BP (1960b) Notes on Australian Carabidae (Col.) I. A new species of *Percosoma* Schaum from Victoria. *Entomologist's Monthly Magazine* **95**, 173–174.

Moore BP (1961) Notes on Australian Carabidae (Col.). II *Notonomus atridermis* Sloane. *Entomologist's Monthly Magazine* **96**, 134–136.

Moore BP (1962) Notes on Australian Carabidae (Col.) V. New species of *Scopodes* Erichson (Pentagonicini). *Entomologist's Monthly Magazine* **98**, 238–240.

Moore BP (1963a) Studies on Australian Carabidae (Coleoptera) 2. – The Agonicinae. *Proceedings of the Royal Entomological Society of London* **32**, 20–24[B].

Moore BP (1963b) Studies on Australian Carabidae (Coleoptera) 3. – The Psydrinae. *Transactions of the Royal Entomological Society of London* **115**, 277–290. doi:10.1111/j.1365-2311.1963.tb00810.x

Moore BP (1963c) New or little known Australian Carabidae in the Frey Museum (Col.). *Entomologischen Arbeiten aus dem Museum Georg Frey* **14**, 435–444.

Moore BP (1964a) New cavernicolous Carabidae (Coleoptera) from mainland Australia. *Journal of the Entomological Society of Queensland* **3**, 69–74. doi:10.1111/j.1440-6055.1964.tb00624.x

Moore BP (1964b) Australian larval Carabidae of the subfamilies Broscinae, Psydrinae, and Pseudomorphinae (Coleoptera). *Pacific Insects* **6**, 242–246.

Moore BP (1965) Studies on Australian Carabidae (Coleoptera) 4. – The Pterostichinae. *Transactions of the Royal Entomological Society of London* **117**, 1–32. doi:10.1111/j.1365-2311.1965.tb00042.x

Moore BP (1966a) The genus *Perileptus* in Australia (Coleoptera, Trechinae). *Bulletin of the National Science Museum, Tokyo* **9**, 317–320.

Moore BP (1966b) The larva of *Pamborus* (Coleoptera: Carabidae) and its systematic position. *Proceedings of the Royal Entomological Society of London* **35**, 1–4[B].

Moore BP (1966c) Australian larval Carabidae of the subfamilies Harpalinae, Licininae, Odacanthinae, and Pentagonicinae (Coleoptera). *Proceedings of the Linnean Society of New South Wales* **90**, 157–163.

Moore BP (1967) New Australian cave Carabidae (Coleoptera). *Proceedings of the Linnean Society of New South Wales* **91**, 179–184.

Moore BP (1972) A revision of the Australian Trechinae. *Australian Journal of Zoology. Supplementary Series* **18**, 1–61.

Moore BP (1974) The larval habits of two species of *Sphallomorpha* Westwood (Coleoptera: Carabidae: Pseudomorphinae). *Journal of the Australian Entolomogical Society* **13**, 179–183. doi:10.1111/j.1440-6055.1974.tb02171.x

Moore BP (1976) Notes on two South American Carabid species (Coleoptera) established in Australia. *Australian Entomological Magazine* **2**, 126–129.

Moore BP (1977a) New or little known Pogoninae (Coleoptera, Carabidae) from Lake Eyre, South Australia. *Australian Entomological Magazine* **4**, 63–67.

Moore BP (1977b) *Harpalus fulvicornis* Thunberg: a South African Carabid Beetle Established in Western Australia. *Australian Entomological Magazine* **4**(1), 7–10.

Moore BP (1978) A new species of the Tasmanian cave carabid genus *Idacarabus* (Coleoptera). *Australian Entomological Magazine* **5**, 23–25.

Moore BP (1979) Chemical defense in carabids and its bearing on phylogeny. In *Carabid Beetles: Their Evolution, Natural History, and Classification.* (Eds Erwin TL, Ball GE, Whitehead DR and Halpern AL) pp. 193–203. *W. Junk*, The Hague.

Moore BP (1980–1989) *A Guide to the Beetles of Southeastern Australia.* Fasc. 1–9. Australian Entomological Press.

Moore BP (1983) New Tasmanian Trechini (Coleoptera: Carabidae). *Australian Entomological Magazine* **10**, 2–6.

Moore BP (1984) Taxonomic notes on some Australasian *Mecyclothorax* (Coleoptera: Carabidae: Psydrini) and descriptions of new species. *Journal of the Australian Entomological Society* **23**, 161–166. doi:10.1111/j.1440-6055.1984.tb01935.x

Moore BP (1985) The Carabidae of Norfolk Island. In *Taxonomy, Phylogeny and Zoogeography of beetles and ants* (Ed GE Ball) pp. 237–256. W. Junk, Dordrecht.

Moore BP (1992) The Carabidae of Lord Howe Island (Coleoptera: Carabidae). In *The biogeography of Ground Beetles of Mountains and Islands* (Eds GR Noonan, GE Ball and NE Stork) pp. 159–173. Intercept, Andover.

Moore BP (1994) New species and new records of Tasmanian cave Carabidae (Coleoptera). *Australian Entomologist* **21**, 75–80.

Moore BP (1998) The enigmatic tribe Physocrotaphini Chaudoir (Helluodini auct.) (Coleoptera: Carabidae). In *Phylogeny and classification of Caraboidea* (Eds GE Ball, A Casale and A Vigna Taglianti). *Atti de Museo regionale di Scienze naturale de Torino* **5**, 369–380.

Moore BP (2000) A new species of *Tachyta* Kirby (Coleoptera: Carabidae: Bembidiini) from tropical Queensland. *Australian Entomologist* **27**, 97–100.

Moore BP (2004) A new cave–dwelling species of *Laccocenus* Sloane (Coleoptera: Carabidae: Psydrini) from southern New South Wales. *Australian Entomologist* **31**, 45–48.

Moore BP, Lawrence JF (1994) The extraordinary larval characters of *Carenum* Bonelli and their bearing on the phylogeny of the Scarititae (Coleoptera: Carabidae). *Canadian Entomologist* **126**, 503–514. doi:10.4039/Ent126503-3

Moore BP, Wallbank BE (1968) Chemical composition of the defensive secretions in carabid beetles and its importance as a taxonomic character (Coleoptera: Carabidae). *Proceedings of the Royal Entomomological Society of London* **37**, 62–72[B].

Moore BP, Weir TA, Pyke JE (1987) Rhysodidae and Carabidae. *Zoological Catalogue of Australia.* Volume 4, Australian Government Publications Service, Canberra.

Motschulsky V (1839) Coléoptères du Caucase et des provinces transcaucasiennes (Continuation). *Bulletin de la Société Impériale des Naturalistes de Moscou* **12**, 68–93.

Motschulsky V (1860) Insectes des Indies orientales, et des contrées analogues. Études Entomologiques **8**, 25–118.

Motschulsky V (1861) Entomologie spéciale. Insectes du Japon. Études Entomologiques **10**, 1–24.

Motschulsky V (1862) Entomologie speciale. Remarques sur la collection d'insectes de V. de Motschulsky. Coléotères. Études Entomologiques **11**, 15–55.

Müller J (1926) I coleotteri della Venezia Guilia. Catalogo ragionato con descrizioni e tabelle per la classificazione delle specie e dei generi meno noti compilato con la cooperatione degli entomologi Triestini. *Studia Entomologica* **1**(2), 1–306.

Nash MA, Thomson LJ, Hoffmann AA (2008a) Effect of remnant vegetation, pesticides, and farm management on abundance of the beneficial predator *Notonomus gravis* (Chaudoir) (Coleoptera: Carabidae). *BioControl* **46**(2), 83–93.

Nash MA, Thomson LJ, Horne PA, Hoffmann AA (2008b) *Notonomus gravis* (Chaudoir) (Coleoptera: Carabidae) predation of *Deroceras reticulatum* Müller (Gastropoda: Agriolimacidae), an example of fortuitous biological control. *Biological Control* **47**, 328–334. doi:10.1016/j.biocontrol.2008.08.013

Netolitzky F (1914) Die Bembidiini in Winklers Catalogus. *Entomologische Blätter zur Biologie und Systematik der Käfer* **10**, 164–176.

Netolitzky F (1931) Überprüfungen afrikanischer und australischer Bembidien. *Coleopterologisches Centralblatt* **6**, 1–3.

Netolitzky F (1942) Bestimmungstabelle der Bembidion-Arten des paläarctischen Gebietes. *Koleopterologische Rundschau* **28**, 29–124.

Netolitzky F (1943) Bestimmungstabelle der Bembidion-Arten des paläarctischen Gebietes. *Koleopterologische Rundschau* **29**, 1–70.

New TR (1998) The role of ground beetles (Coleoptera: Carabidae) in monitoring programmes in Australia. *Annales Zoologici Fennici* **35**, 163–171.

Nicolai EA (1822) Dissertatio inauguralis medica sistens Coleopterorum species agri Halensis quam consensu illustrissimi medicorum ordinis in celebrrima Academia Fridericiana Halensi et Vitebergensi consociata pro summis in medicina et chirurgia honoribus rite obtinendis die X. mensis septembris MDCCCXXII publice defendet. Grunert, Halae.

Nietner J (1856) Entomological papers, being descriptions of new Ceylon Coleoptera with such observations on their habits as appear in any way interesting. *Journal of the Asiatic Society of Bengal* **25**, 381–394, 523–554.

Nietner J (1858) Entomological papers, being chiefly descriptions of new Ceylon Coleoptera with such observation on their habits, etc., as appear in any way interesting. *Journal of the Asiatic Society of Bengal* **26**, 132–156.

Notman H (1925) A review of the beetle family Pseudomorphindae, and a suggestion for a rearrangement of the Adephaga, with descriptions of a new genus and new species. *Proceedings of the United States National Museum* **67**(2586), 1–34. doi:10.5479/si.00963801.2586

Ober KA (2002) Phylogenetic relationships of the carabid subfamily Harpalinae (Coleoptera) based on molecular sequence data. *Molecular Phylogenetics and Evolution* **24**, 228–247. doi:10.1016/S1055-7903(02)00251-8

Ober KA, Maddison DR (2008) Phylogenetic relationships of tribes within Harpalinae (Coleoptera: Carabidae) as inferred from 28S ribosomal DNA and the wingless gene. *Journal of Insect Science* **8**(63), 1–32. doi:10.1673/031.008.6301

Olliff AS (1885) Contributions towards a knowledge of the Coleoptera of Australia. *Proceedings of the Linnean Society of New South Wales* **10**, 467–472. doi:10.5962/bhl.part.17941

Putzeys J (1846) Monographie des *Clivina* et genres voisins précédée d'un tableau synoptique des genres de la tribu des Scaritides. *Mémoirs de la Société royal des Sciences de Liège* **2**, 521–663.

Putzeys J (1863) Postscriptum ad Clivindarum monographiam atque de quibusdam aliis. *Mémoirs de la Société royal des Sciences de Liège* **18**, 1–78.

Putzeys J (1866a) Révision des Clivinides d'Australie. *Stettiner Entomologische Zeitschrift* **27**, 33–42.

Putzeys J (1866b) Révision générale des clivinides. *Annales de la Société Entomologique de Belgique* **10**, 1–242.

Putzeys J (1867) Supplément à la révision générale des Clivinides. *Annales de la Société Entomologique de Belgique* **11**, 7–22.

Putzeys J (1868) Les Broscides. *Stettiner Entomologische Zeitung* **29**, 305–379.

Putzeys J (1873) Révision des Broscides de l'Australie d'après la collection de M. le Compte de Laporte de Castelnau. *Annali del Museo Civico di Storia Naturale di Genova* **4**, 307–341.

Putzeys J (1877) Description des quelques Clivinides de l'Inde. *Annales de la Société Entomologique de Belgique* **20**, Comptes Rendues XL–XLVII.

Rafinesque CS (1815) Analyse de la nature ou tableau de l'univers et des corps organisés. *C. S. Rafinesque*, Palermo.

Redtenbacher L (1867) Zoologischer Theil. Zweiter Band. 1. Abtheilung A 1. Coleoptera. In *Reise der österreichischen Fregatte Novara um die Erde in den Jahren 1857, 1858, 1859 unter dem Befehl des Commodore B. von Wüllerstorf-Urbair* 249 pp. Karl Gerold's Sohn, Wien.

Reichardt H (1974) Monograph of the Neotropical Helluonini, with notes and discussions of Old World forms (Coleoptera Carabidae). *Studia Entomologica* **17**, 211–302.

Reiche L (1842) Coléoptères de Colombie. *Revue Zoologique* **5**, 307–314.

Ritsema C (1875) *Tijdschrift voor Entomologie*, Verslag XCII–XCII.

Rivalier E (1963) Démembrement du genre *Cicindela* Linné. V. Faune Australienne. *Revue Francaise d'Entomologie* **30**, 30–48.

Roig-Juñent S (1998) Cladistic relationships of the tribe Broscini (Coleoptera: Carabidae). In *Phylogeny and classification of Caraboidea* (Eds Ball GE, Casale A & Vigna Taglianti A). *Atti de Museo regionale di Scienze naturale de Torino* **5**, 343–358.

Roig-Juñent S (2000) The subtribes and genera of the tribe Broscini (Coleoptera: Varabidae): Cladistic analysis, taxonomic treatment, and biogeographical considerations. *Bulletin of the American Museum of Natural History* **255**, 1–90. doi:10.1206/0003-0090(2000)255<0001:TSAGOT>2.0.CO;2

Roig-Juñent S (2004) South American Migadopini (Coleoptera: Carabidae): description of male and female genitalia and phylogenetic and biogeographic considerations. *Acta entomologica Chilena* **28**, 7–29.

Sasakawa K, Kubota K (2007) Phylogeny and genital evolution of carabid beetles in the genus *Pterostichus* and its allied genera (Coleoptera, Carabidae) inferred from two nuclear gene sequences. *Annals of the Entomological Society of America* **100**, 100–109. doi:10.1603/0013-8746(2007)100[100:PAGEOC]2.0.CO;2

Schaum HR (1858) Cicindelidae und Carabidae. *Naturgeschichte der Insecten Deutschlands* **1**, 353–552 [Nicolai, Berlin.].

Schaum HR (1860) Cicindelidae und Carabidae. *Naturgeschichte der Insecten Deutschlands* **2**, 583–791 [Nicolai, Berlin.].

Schaum HR (1863) Beiträge zur Kenntniss einiger Carabicinen-Gattungen. *Berliner Entomologische Zeitschrift* **7**, 67–92. doi:10.1002/mmnd.18630070107

Schildknecht H, Maschwitz U, Winkler H (1968) Über arthropodenabwehrstoffe, XXXII. Zur evolution der Carabiden-wehrdrüsensekrete. *Naturwissenschaften* **55**, 112–117. doi:10.1007/BF00624238

Schmidt-Göbel HM (1846) Faunula Coleopterorum Birmaniae, adjectis nonullis Bengaliae indignis. Haase, Prag.

Schüle P, Lorenz W (2008) A new genus of lebiine ground beetles, *Velindastus* gen. nov. (Carabidae: Lebiinae: 'Lichnasthenina'), from South Africa. *Annals of the Transvaal Museum* **45**, 91–99.

Sciaky R, Vigna Taglianti A (2003) Observations on the systematics of the tribe Tachyini (Coleoptera Carabidae). *Bollettino della Società Entomologica Italiana* **135**, 79–96.

Seidlitz G (1887) Fauna Baltica. Die Kaefer (Coleoptera) der deutschen Ostseeprovinzen Russlands. Zweite neu bearbeitete Auflage. Hartungsche Verlagsdruckerei, Königsberg.

Serrano J (1981) Chromosome numbers and karyotypic evolution of Caraboidea. *Genetica* **55**, 51–60. doi:10.1007/BF00134005

Serrano J, Galián J (1998) A review of karyotypic evolution and phylogeny of carabid beetles (Coleoptera). In *Phylogeny and classification of Caraboidea* (Eds GE Ball, A Casale and A Vigna Taglianti). *Atti de Museo regionale di Scienze naturale de Torino* **5**, 191–228.

Sharp D (1886) On New Zealand Coleoptera, with descriptions of new genera and species. *Scientific Transactions of the Royal Dublin Society* **3**, 351–456.

Sharp D (1903) Coleoptera Caraboidea. Coleoptera pt. II. In *Fauna Hawaiiensis*. Volume 3, part 3 pp. 175–292. Cambridge University Press, Cambridge.

Shpeley D, Ball GE (1993) Classification, reconstructed phylogeny and geographical history of the New World species of *Coptodera* Dejean (Coleoptera. Carabidae: Lebiini). *Proceedings of the Entomological Society of Ontario* **124**, 3–182.

Shpeley D, Ball GE (2000) A taxonomic review of the subtribe Pericalina (Carabidae: Lebiini) in the Western hemisphere, with descriptions of new species and notes about classification and zoogeography. *Insecta Mundi* **14**, 1–185.

Shpeley D, Madge RB, Ball GE (1985) The genus *Celaenephes* Schmidt–Göbel: Characteristics, species, and relationships (Coleoptera: Carabidae: Lebiini). *Proceedings of the Entomological Society of Washington* **87**, 283–307.

Sloane TG (1888) A note on the Carenides, with descriptions of new species. *Proceedings of the Linnean Society of New South Wales* **3**, 1101–1122.

Sloane TG (1889) Studies in Australian entomology. No. I. Review of the genus *Sarticus* (Carabidae). *Proceedings of the Linnean Society of New South Wales* **4**, 501–512. doi:10.5962/bhl.part.15053

Sloane TG (1890) Studies in Australian Entomology. No. III. On *Promecoderus* and closely allied genera (Carabidae). *Proceedings of the Linnean Society of New South Wales* **5**, 189–242. doi:10.5962/bhl.part.18632

Sloane TG (1892) Studies in Australian Entomology. No. V. Notes on the Subfamily Broscini (Carabidae). with descriptions of new Species. *Proceedings of the Linnean Society of New South Wales* **7**, 45–64. doi:10.5962/bhl.part.26042

Sloane TG (1893) Broscides (sub-fam.) of Coleoptera. Scientific results of the Elder Expedition. *Transactions of the Royal Society of South Australia* **16**, 203–213.

Sloane TG (1894) A second note on the Carenides, with descriptions of new species. *Proceedings of the Linnean Society of New South Wales* **8**, 448–483.

Sloane TG (1895a) Studies in Australian Entomology. No. VII. New Genera and Species of Carabidae (Including Some Notes on Previously Described Species, and Synoptic Lists of Genera and Species. *Proceedings of the Linnean Society of New South Wales* **9**, 393–455. doi:10.5962/bhl.part.18118

Sloane TG (1895b) Notes on Carabidae from Lake Callabonna, Central Australia. *Transactions of the Royal Society of South Australia* **19**, 124–137.

Sloane TG (1896a) On the Australian Clivinides (Fam. Carabidae). Revision of the Australian species of the genus *Clivina* with the description of a new genus, *Clivinarchus. Proceedings of the Linnean Society of New South Wales* **21**, 143–257. doi:10.5962/bhl.part.8469

Sloane TG (1896b) On the Australian Bembidiines referable to the genus *Tachys*, with the description of a new allied genus *Pyrrhotachys. Proceedings of the Linnean Society of New South Wales* **21**, 355–377. doi:10.5962/bhl.part.8477

Sloane TG (1897) On the Carenides (fam. Carabidae). No. III. *Proceedings of the Linnean Society of New South Wales* **22**, 190–229.

Sloane TG (1898) On Carabidae from west Australia, sent by Mr. A. M. Lea (with descriptions of new genera and species, synoptic tables, &c.). *Proceedings of the Linnean Society of New South Wales* **23**, 444–520.

Sloane TG (1900a) Studies in Australian Entomology. No. IX. New Species of Carabidae (with Notes on some previously described species, and synoptic lists of species). *Proceedings of the Linnean Society of New South Wales* **24**, 553–584. doi:10.5962/bhl.part.7682

Sloane TG (1900b) On the Carenides (fam. Carabidae). No. IV. *Proceedings of the Linnean Society of New South Wales* **25**, 361–388. doi:10.5962/bhl.part.12158

Sloane TG (1901) Studies in Australian entomology. No. X. *Proceedings of the Linnean Society of New South Wales* **25**, 632–636. doi:10.5962/bhl.part.12180

Sloane TG (1903) Studies in Australian Entomology. No. XII. New Carabidae (Panagaeini, Platysmatini, Platynini, Lebiini, with Revisional lists of Genera and Species, some Notes on Synonymy, etc.). *Proceedings of the Linnean Society of New South Wales* **28**, 566–642.

Sloane TG (1904) Studies in Australian Entomology. No. XIV. New Species of Geodephagous Coleoptera from Tropical Australia. Cicindelidae (3), and Carabidae (5). *Proceedings of the Linnean Society of New South Wales* **29**, 527–538 [Platysmatini, Morioni, Perigonini, Masoreini, and Physocrotaphini]. doi:10.5962/bhl.part.20170

Sloane TG (1905a) Revisional notes on Australian Carabidac. Part I. *Proceedings of the Linnean Society of New South Wales* **29**, 699–733. doi:10.5962/bhl.part.20177

Sloane TG (1905b) Revisional notes on Australian Carabidae. Part II. *Proceedings of the Linnean Society of New South Wales* **30**, 103–135. doi:10.5962/bhl.part.12891

Sloane TG (1907a) Studies in Australian Entomology. No. XV. New genera and species of Carabidae, with some notes on synonymy (Clivinini, Scaritini, Cuneipectini, Trigonotomini and Lebiini). *Proceedings of the Linnean Society of New South Wales* **32**, 346–381. doi:10.5962/bhl.part.19572

Sloane TG (1907b) New Carabidae from German New Guinea and its dependencies (Col.). *Deutsche Entomologische Zeitschrift* **1907**, 177–185.

Sloane TG (1907c) Further Carabidae from German New Guinea and its dependencies (Col.). *Deutsche Entomologische Zeitschrift* **1907**, 467–474.

Sloane TG (1910a) Studies in Australian Entomology. No. xvi. New Species of Carabidae. *Proceedings of the Linnean Society of New South Wales* **35**, 378–406.

Sloane TG (1910b) Revisional notes on Australian Carabidae, Part iii. *Proceedings of the Linnean Society of New South Wales* **35**, 435–458.

Sloane TG (1911) Carabidae from Dorrigo, N.S.W.; with an appendix on Tenebrionidae from Dorrigo by H. J. Carter. *Proceedings of the Linnean Society of New South Wales* **35**, 823–848.

Sloane TG (1913) Revisional notes on Australian Carabidae. Part iv. The genus *Notonomus*. *Proceedings of the Linnean Society of New South Wales* **38**, 404–449.

Sloane TG (1914) Revisional notes on Australian Carabidae. Part V. Tribe Helluonini. *Proceedings of the Linnean Society of New South Wales* **39**, 568–614. doi:10.5962/bhl.part.2293

Sloane TG (1915a) Studies in Australian Entomology. No. XVII. New genera and species of Carabidae. (Pamborini, Migadopini, Broscini, Cuneipectini, Nomiini, Pterostichini, Platynini, Oodini, Harpalini, Lebiini). *Proceedings of the Linnean Society of New South Wales* **40**, 438–473. doi:10.5962/bhl.part.18880

Sloane TG (1915b) On the faunal subregions of Australia. *Proceedings of the Royal Society of Victoria* **28**, 139–148.

Sloane TG (1916a) Carabidae from the Upper Williams River, N.S.W. (Coleoptera). *Proceedings of the Linnean Society of New South Wales* **41**, 196–208. doi:10.5962/bhl.part.15312

Sloane TG (1916b) New Australian species of Carabidae belonging to the tribe Scaritini (Coleoptera). *Proceedings of the Linnean Society of New South Wales* **41**, 597–630.

Sloane TG (1917) Carabidae from Tropical Australia. (New genera and species, notes on synonymy, and synoptic tables. Tribe s Scaritini, Harpalini, Odacanthini, Lebiini, and Helluonini). *Proceedings of the Linnean Society of New South Wales* **42**, 406–438. doi:10.5962/bhl.part.4857

Sloane TG (1920) The Carabidae of Tasmania. *Proceedings of the Linnean Society of New South Wales* **45**, 113–178. doi:10.5962/bhl.part.19535

Sloane TG (1921) Revisional notes on Australian Carabidae. Part VI. Tribe Bembidiini. *Proceedings of the Linnean Society of New South Wales* **46**, 192–208. doi:10.5962/bhl.part.14008

Sloane TG (1923a) Studies in Australian Entomology. No. XVIII. New genera and species of Carabidae. (Scaritini, Pterostichini, Merizodini, Bembidiini, Trechini, Odacanthini, Panagaeini, Licinini, and Lebiini). *Proceedings of the Linnean Society of New South Wales* **48**, 17–39.

Sloane TG (1923b) The Classification of the family Carabidae. *Transactions of the Royal Entomological Society of London* **1923**, 234–250.

Sloane TG (1933) Notes on the Australian species of the family Paussidae (Coleoptera). *Proceedings of the Linnean Society of New South Wales* **58**, 396–404.

Solier AJJ (1833) Observations sur les deux genres *Brachinus* et *Aptinus* du spécies de M. le Compte Dejean; et description d'une nouvelle espèce de *Gyrinus*. *Annales de la Société Entomologique de France* **2**, 459–463.

Solier AJJ (1849) Order III. Coleopteros [Pentamera, Heteromera I]. In *Historia Fisica y Politica de Chile* Vol. **4**. (Ed C Gay) pp. 105–380. Paris.

Sota T, Takami Y, Monteith GB, Moore BP (2005) Phylogeny and character evolution of endemic Australian carabid beetles of the genus *Pamborus* based on mitochondrial and nuclear gene sequences. *Molecular Phylogenetics and Evolution* **36**, 391–404. doi:10.1016/j.ympev.2005.02.015

Spence JR (1983) Taxonomic status, relationships, and biogeography of *Anatrichis* LeConte and *Oodinus* Motschulsky (Carabidae: Oodini). *The Coleopterists Bulletin* **36**[1982], 567–580.

Spencer CP, Richards K (2010) The Green-lined Ground Beetle, *Catadromus lacordairei*, in Tasmania. *Tasmanian Naturalist* **132**, 15–19.

Spencer WB (1896) Report on the work of the Horn Scientific Expedition to Central Australia: Part I – Introduction, narrative, summary of results, supplement to zoological Report, map. Melville, Mullen & Slade, Melbourne, 220 pp.

Staunton KM, Robson SKA, Burwell CJ, Reside AE, Williams SE (2014) Projected distributions and diversity of flightless ground beetles within the australian wet tropics and their environmental correlates. *PLoS One* **9**, 1–16. doi:10.1371/journal.pone.0088635

Stephens JF (1827) Illustrations of British entomology; or, a synopsis of indigenous insects: containing their generic and specific distinctions; with an account of their metamorphoses, times of appearance, food, end economy, as far as practicable. Mandibulata. Vol. 1. Baldwin & Craddock, London, 186 pp.

Stephens JF (1828) Illustrations of British entomology; or, a synopsis of indigenous insects: containing their generic and specific distinctions; with an account of their metamorphoses, times of appearance, food, end economy, as far as practicable. Mandibulata. Vol. 2. Baldwin & Craddock, London, 200 pp.

Straneo SL (1936) Sui tipi dei Platysmatini (Coleopt. Carabid.) australiani della collezione Laporte de Castelnau nel Museo Civico di Genova. *Museo civico di storia naturale Giacomo Doria* (Genoa, Italy) **59**, 246–259.

Straneo SL (1937) Sui tipi dei Platysmatini (Coleop. Carabid.) australiani della collezione Laporte de Castelnau II. *Museo civico di storia naturale Giacomo Doria* (Genoa, Italy) **59**, 461–475.

Straneo SL (1938) Studi sulle specie orientali del genere Caelostomus MacL. (Coleopt. Carabid.). *Museo civico di storia naturale Giacomo Doria (Genoa, Italy)* **60**, 5–100.

Straneo SL (1957) Elenco di Carabidi entrati recentemente nelle Collezioni del Museo Frey. *Entomologische Arbeiten aus dem Museum G. Frey Tutzing bei München* **11**, 416–428.

Straneo SL (1958) Coleoptera: Carabidae Pterostichinae. In *South African Animal Life, Vol. V.* (Eds B. Hanstrom, P. Brinck, and G. Rudebeck) pp. 318–455. Almquist and Wiksell, Stockholm.

Straneo SL (1991) South American species of Loxandrus LeConte, 1852 (Coleoptera: Carabidae: Pterostichini). *Annals of the Carnegie Museum* **60**, 1–62.

Straneo SL (1994) Sulle species orientali del genere *Catascopus* Kirby, 1825 (Coleoptera: Carabidae). *Elytron* **8**, 141–172.

Straneo SL, Ball GB (1989) Synopsis of the Genera and Subgenera of the Tribe Peleciini, and Revision of the Neotropical and Oriental Species (Coleoptera: Carabidae). *Insecta Mundi* **3**(2), 73–178.

Sumlin WD (1981) Studies in the Australian Cicindelidae II: new taxa from Australia (Coleoptera). *Coleopterologists Bulletin* **35**, 273–280.

Sumlin WD (1984) Studies on the Australian Cicindelidae III: observations on the Australian Members of the genus *Cicindela* L. (Coleoptera). *Entomological News* **95**(5), 189–199.

Sumlin WD (1985) Studies on the Australian Cicindelidae IV: a review of the genus *Nickerlea* (Coleoptera). *Entomological News* **96**(2), 63–68.

Sumlin WD (1987) Studies on the Australian Cicindelidae VI: a review of the subgenus *Rivacindela* of the genus *Cicindela* (Coleoptera). *Entomological News* **98**, 117–134.

Sumlin WD (1992a) Studies on the Australian Cicindelidae VII: observations on seven *Distipsidera* species from Queensland. *Cicindelidae: Bulletin of worldwide research* **4**(1), 2–6.

Sumlin WD (1992b) Studies on the Australien Cicindelidae VIII: notes on *Megacephala* and *Cicindela* with descriptions of new taxa (Coleoptera). *Cicindelidae: Bulletin of worldwide research* **4**(1), 7–26.

Sumlin WD (1992c) Studies on the Australien Cicindelidae X: notes on *Megacephala* and *Cicindela* with the description of a new species (Coleoptera). *Cicindelidae: Bulletin of worldwide research* **4**(1), 27–29.

Sumlin WD (1997) Studies on the Australian Cicindelidae XII: additions to *Megacephala*, *Nickerlea* and *Cicindela* with notes (Coleoptera). *Cicindelidae: Bulletin of worldwide research* **4**(4), 1–56.

Takami Y, Sota T (2006) Four new species of the Australian *Pamborus* Latreille (Coleoptera, Carabidae) carabid beetles. *Australian Journal of Entomology* **45**, 44–54. doi:10.1111/j.1440-6055.2006.00495.x

Tate R (1889) On the influence of physiological changes in the distribution of life in Australia. *Report of the First Meeting of the Australian Association for the Advancement of Science*, pp. 312–326.

Thiele HU (1977) Carabid beetles in their environments: a study on habitat selection by adaptations in physiology and behavior. Springer, Berlin, 369 pp.

Thompson RG (1979) Larvae of North American Carabidae with a key to the tribes. In *Carabid Beetles: Their Evolution, Natural History, and Classification.* (Eds Erwin TL, Ball GE, Whitehead DR and Halpern AL) pp. 209–291. W. Junk, The Hague.

Thomson J (1858) Voyage au Gabon. Deuxième partie. Histoire naturelle proprement dite. Insectes. I. Ordre coléptères, livr. **13–14**: 29–80. In: Thomson, J. (ed.): Archives entomologiques ou recueil continent des illustrations d'insectes nouveaux ou rares. Tome deuxième. *Société Entomologique de France*: 469 pp.

Thomson J (1859) Description des deux espèces nouvelles de Carabiques. *Arcana Naturae* **1**, 93–94.

Toledano L (2005) The Australian species of *Bembidion* Latreille, 1802: a taxonomic treatment, with notes about Gondwana as an evolutionary source area (Coleoptera, Carabidae, Bembidiini). *Monografie de Museo Regionale di Scienze naturali di Torino* **42**, 73–136.

Tschitschérine T (1898) Matériaux pour servir à l'étude des Féroniens IV. *Horae Societatis Entomologicae Rossicae* **32**, 1–224.

Tschitschérine T (1902) Notes sur les Platysmatini de l'Australie. *Horae Societatis Entomologicae Rossicae* **35**, 502–534.

Turvey N (2013) *Cane Toads, A Tale of Sugar, Politics and Flawed Science.* Sydney University Press, Sydney.

van Emden FI (1942) A key to the genera of larval Carabidae. *Transactions of the Royal Entomological Society of London* **92**, 1–99. doi:10.1111/j.1365-2311.1942.tb03318.x

van Emden FI (1949) LXIX. New and little-known Neotropical Carabidae. *Journal of Natural History* **12**(2), 861–893.

Vigna Taglianti A, Santarelli F, di Guiulio A, Oliverio M (1998) Phylogenetic implications of larval morphology in the tribe Ozaenini. In *Phylogeny and classification of Caraboidea.* (Eds GE Ball, A Casale and A Vigna Taglianti). *Atti de Museo regionale di Scienze naturale de Torino* **5**, 273–295.

Waterhouse CO (1877) Descriptions of twenty new species of Coleoptera from various localities. *The Transactions of the Entomological Society of London* **1877**, 1–13.

Westwood JO (1838a) Synopsis of the genera of British insects. In *An introduction to the modern classification of insects* (Ed. JO Westwood) pp. 1–48. Longman, London.

Westwood JO (1838b) On the genus *Cerapterus* of Swederus. *The Entomolocical Magazine* **5**, 500–505.

Westwood JO (1842) On the Scaritideous beetles of New Holland. In *Arcana Entomologica or illustrations of new rare and interesting insects* I. (Ed. JO Westwood) pp. 81–90. Smith, London.

White A (1846) Insects. In *The Zoology of the voyage of H.M.S. Erebus and Terror, under the command of Capt. Sir J. C. Ross*. Vol. 4. (Eds J Richardson and JE Gray) pp. 1–24. E. W. Janson, London.

Wiesner J (1992) Checklist of the Tiger Beetles of the World. Verlag Erna Bauer, Kaltern-Weiler.

Will KW (2000) Systematics and Zoogeography of Abaryform genera (Coleoptera: Carabidae: Pterostichini), and a phylogenetic hypothesis for pterostichine genera. PhD Thesis. Cornell University.

Will KW (2002) Revision of the New World Abariform Genera *Neotalus* N. Gen. and *Abaris* Dejean (Coleoptera: Carabidae: Pterostichini (Auctorum). *Annals of the Carnegie Museum* **71**, 143–213.

Will KW (2003) Review and cladistic analysis of the generic-level taxa of Morionini Brulle (Coleoptera: Carabidae). *The Pan-Pacific Entomologist* **79**, 212–229.

Will KW (2011) Taxonomic review of the Pterostichini and Loxandrini fauna of New Caledonia (Coleoptera, Carabidae). *ZooKeys* **147**, 337–397. doi:10.3897/zookeys.147.1943

Will KW (2015a) A multigene phylogenetic analysis results in a redefinition of the genus *Notonomus* Chaudoir (Coleoptera, Carabidae) and descriptions of new species of the subgenus *Leiradira* Laporte de Castelnau. *Invertebrate Systematics* **29**, 313–344. doi:10.1071/IS14021

Will KW (2015b) Resolution of taxonomic problems in Australian Harpalini, Abacetini, Pterostichini, and Oodini (Coleoptera, Carabidae). *ZooKeys* **545**, 131–137. doi:10.3897/zookeys.545.6752

Will KW (2015c) A taxonomic review, new species and a key to species of *Platycoelus* Blanchard, 1843 (Coleoptera: Carabidae: Pterostichini). *Zootaxa* **4034**, 291–308. doi:10.11646/zootaxa.4034.2.3

Will KW, Gill AS (2008) Phylogeny and classification of *Hypherpes* auctorum (Coleoptera: Carabidae: Pterostichini: Pterostichus). *Annals of the Carnegie Museum* **77**, 93–127. doi:10.2992/0097-4463-77.1.93

Will KW, Kavanaugh DH (2012) A new species of *Lesticus* Dejean, 1828 (Coleoptera, Carabidae) from the Finisterre Range, Papua New Guinea and a key to the genera of pterostichine-like Harpalinae of New Guinea. *ZooKeys* **246**, 27–37. doi:10.3897/zookeys.246.4112

Will KW, Monteith GB (2018) A multi-locus phylogeny, taxonomic review and description of new species of Australian Nurus (sensu stricto) Motschulsky, 1865 (Coleoptera: Carabidae: Pterostichini). *Australian Entomologist* **45**, 353–388.

Will KW, Attygalle AB, Herath K (2000) New defensive chemical data for ground beetles (Coleoptera: Carabidae): interpretations in a phylogenetic framework. *Biological Journal of the Linnean Society. Linnean Society of London* **71**, 459–481. doi:10.1111/j.1095-8312.2000.tb01269.x

Will KW, Liebherr JK, Maddison DR, Galián J (2005) Absence asymmetry: the evolution of monorchid beetles (Insecta: Coleoptera: Carabidae). *Journal of Morphology* **264**, 75–93. doi:10.1002/jmor.10319

Wollaston TV (1867) Coleoptera Hesperidum, being an enumeration of the coleopterous insects of the Cape Verde archipelago. J. Van Voorst, London.

Yeates DK, Monteith GB (2008) The invertebrate fauna of the wet tropics: diversity, endemism and relationships. In *Biodiversity and Environmental Change: Monitoring, Challenges and Direction*. (Eds D Lindenmayer, E Burns, N Thurgate and A Lowe) pp. 178–191. CSIRO Publishing, Melbourne.

Yeates DK, Bouchard P, Monteith GB (2002) Patterns and levels of endemism in the Australian Wet Tropics rainforest: evidence from flightless insects. *Invertebrate Systematics* **16**, 605–619. doi:10.1071/IT01032

Zetto Brandmayr T, Bonacci T, Massolo A, Brandmayr P (2006) What is going on between aposematic carabid beetles? The case of *Anchomenus dorsalis* (Pontoppidan 1763) and *Brachinus sclopeta* (Fabricius 1792) (Coleoptera Carabidae). *Ethology Ecology & Evolution* **18**, 335–348. doi:10.1080/08927014.2006.9522700.

13. POLYPHAGA EMERY, 1886

John F. Lawrence

The suborder Polyphaga represents the crown group of beetles, containing at least 166 families and more than 300 000 species, most of modern beetles. The group is usually distinguished from almost all members of the other three suborders by (1) fusion of the protrochantin and propleuron and the reduction and at least partial internalisation of the resulting trochantinopleuron, so that it barely extends in front of the coxal cavity in only a few Derodontidae and Scirtoidea and never extends behind the coxal cavity; (2) presence of lateral cervical sclerites (secondarily reduced in several lineages), (3) formation of a radial cell in the hind wing due to the separation and subsequent fusion of RA_{1+2} and RA_{3+4} (secondary reduction or loss of the cell in various lineages); (4) absence of an abrupt hinge in the posterior wing strut or medial bar (MP_{1+2}), (5) mesocoxal cavities rarely partly closed laterally by a small segment of the metanepisternum and never broadly closed by this sclerite. Larval polyphagans are distinguished from adephagans and archostematans by the fusion of tibia and tarsus to form a tibiotarsus, fusion of the two pretarsal claws to form a single claw and the presence of a microphagous feeding system in all basal lineages. The last of these, discussed in detail by Betz *et al.* (2003) for Staphylinoidea, apparently occurs also in the suborder Myxophaga, a fact which has been used to support a Myxophaga–Polyphaga sister group relationship (Anton & Beutel 2006). As noted in Chapter 4, however, the microphagous feeding system differs in the two groups and may have evolved independently.

On the whole Polyphaga appears to a well defined, monophyletic assemblage (Beutel & Leschen 2016), but in recent years the subordinal distinctions have become somewhat blurred. One example is the family Jurodidae, including four species of *Jurodes* Ponomarenko from the Middle and Upper Jurassic of northern China (Inner Mongolia) and the Transbaikalian region of Russia and the Recent species *Sikhotealinia zhiltovae* Lafer from the Sikhoe-Alin Range in Far Eastern Russia (Lafer 1996; Kirejtshuk 1999; Yan *et al.* 2014). Although this group is included in the Archostemata by almost all authors, the wing venation resembles that in Polyphaga in several respects, such as the presence of a distinct radial cell and lack of a hinge in the posterior wing strut. Fedorenko (2009, p. 212) admits that *Sikhotealinia* could be polyphagan and, if so, possibly closest to Scirtidae, and Lawrence (2016) tentatively included the family in Polyphaga. In the available illustrations and photographic images of the unique type of *Sikhotealinia zhiltovae*, the external propleuron and vislble metatrochantin are not clearly indicated, although both are shown in illustrations of *Jurodes* species. The other very unusual feature of this group is the presence of three ocelli on the head, which are found in no other beetle (Leschen & Beutel 2004).

Recent palaeontological studies have also cast doubt on the distinction and composition of the four currently recognised suborders. Yan *et al.* (2017) redefined the family Ademosynidae (Upper Triassic to Lower Jurassic) and removed it from the Archostemata, along with the Jurodidae and various genera formerly included in Schizophoridae, Catiniidae and Ademosynidae; the suborder Archostemata was restricted to the families Ommatidae and Cupedidae. Yan *et al.* (2018a) proposed a new family, Peltosynidae containing *Peltosyne* Ponomarenko and two new genera from the Triassic. Unlike the previous work, Jurodidae was placed in the Archostemata. In the most recent paper by Yan *et al.* (2018b), a new genus *Ponomarenkia* was described from Late Permian of Australia, which the authors concluded could be in the stem group of one of the modern suborders or the stem group of a clade representing more than one of them. As more new fossil data emerges it is likely that the entire higher classification of Coleoptera will have to be reexamined.

References

Anton E, Beutel RG (2006) On the head morphology of Lepiceridae (Coleoptera: Myxophaga) and the systematic position of the family and suborder. *European Journal of Entomology* **103**, 85–95. doi:10.14411/eje.2006.014

Betz O, Thayer MK, Newton AF (2003) Comparative morphology and evolutionary pathways of the mouthparts in spore-feeding Staphylinoidea (Coleoptera). *Acta Zoologica* **84**, 179–238 [Stockholm]. doi:10.1046/j.1463-6395.2003.00147.x

Beutel RG, Leschen RG (2016) 8. Polyphaga Emery, 1886. In *Handbuch der Zoologie/Handbook of Zoology. BandVolume IV Arthropoda: Insecta. Teilband/Part 38. Coleoptera, Beetles. Volume 1: Morphology and Systematics (Archostemata, Adephaga, Myxophaga, Polyphaga partim).* 2nd Edition. (Eds RG Beutel and RAB Leschen) pp. 199–200. Walter de Gruyter, Berlin.

Fedorenko DN (2009) *Evolution of the Beetle Hind Wing, with Special Reference to Folding (Insecta, Coleoptera).* 336 pp. Pensoft, Sofia, Moscow.

Kirejtshuk AG (1999) *Sikhotealinia zhiltzovae* (Lafer, 1996) – recent representative of the Jurassic coleopterous fauna (Coleoptera, Archostemata, Jurodidae). *Proceedings of the Zoological Institute Russian Academy of Sciences* **281**, 21–26.

Lafer GS (1996) Sikhotealiniidae Lafer, fam. n. In *Key to the Insects of the Russian FarEast. Vol. III. Coleoptera. Part 3.* (Ed. P. A. Ler) pp. 390–396 (in Russian).

Lawrence JF (2016) 2. Classification (families and subfamilies). In *Handbuch der Zoologie/Handbook of Zoology. BandVolume IV Arthropoda: Insecta. Teilband/Part 38. Coleoptera, Beetles. Volume 1: Morphology and Systematics (Archostemata, Adephaga, Myxophaga, Polyphaga partim).* 2nd Edition. (Eds RG Beutel and RAB Leschen) pp. 13–22. Walter de Gruyter, Berlin.

Leschen RAB, Beutel RG (2004) Ocellar atavism in Coleoptera: plesiomorphy or apopmorphy? *Journal of Zoological Systematics and Evolutionary Research* **42**(1), 63–69. doi:10.1046/j.0947-5745.2003.00241.x

Yan EV, Wang B, Ponomarenko AG, Zhang H (2014) The most mysterious beetles: Jurassic Jurodidae (Insecta: Coleoptera) from China. *Gondwana Research* **25**, 214–225. doi:10.1016/j.gr.2013.04.002

Yan EV, Beutel RG, Ponomarenko AG (2017) Ademosynidae (Insecta: Coleopter): a new concept for a coleopteran key taxon and its phylogenetic affinities to the extant suborders. *Palaeontologia Electronica* 20.2.31A, 1–22.

Yan EV, Beutel RG, Ponomarenko AG (2018a) †Peltosynidae, a new beetle family from the Middle-Late Triassic of Kyrgyzstan: its affinities with Polyphaga (Insecta, Coleoptera) and the groundplan of this megadiverse suborder. *Journal of Systematic Palaeontology* **16**, 1–16 [online].

Yan EV, Lawrence JF, Beattie R, Beutel RG (2018b) At the dawn of the great rise: †*Ponomarenkia belmontensis* (Insecta: Coleoptera), a remarkable new Late Permian beetle from the Southern Hemisphere. *Journal of Systematic Palaeontology* **16**, 1–9 [online].

14. SUPERFAMILY SCIRTOIDEA FLEMING, 1821

John F. Lawrence

The superfamily Scirtoidea is more or less equivalent to the Eucinetiformia of Crowson (1981, p. 695), with the families Scirtidae, Eucinetidae and Clambidae, but with the addition of the eastern Palaearctic family Decliniidae (Nikitsky *et al.* 1994; Lawrence *et al.* 1995; Sakai & Satô 1996; Beutel & Leschen 2016; Bocak 2016; Lawrence 2016a, b). Adults are usually distinguished by having a relatively short prothorax, the prosternum very short in front of the coxae, and a strongly declined head, which is ventrally concave between a pair of sharp subgenal ridges, which often lie against the procoxae. These are usually combined with plesiomorphic features, such as a mesothoracic discrimen, metakatepisternal suture, and the mesocoxal cavities partly closed by the metanepisternum and unusual features, such as the presence of six cuticular rings in the rectum. The group was included in an expanded Elateriformia by Lawrence *et al.* (1995) and Lawrence & Newton (1995), but subsequent molecular (Caterino *et al.* 2002; Hunt *et al.* 2007; Maddison *et al.* 2009); and morphological (Lawrence 2001; Lawrence & Yoshitomi 2007) evidence suggested that these families lie at the base of the polyphagan clade, sister to the remaining Polyphaga. In more recent molecular analyses, however, the family Derodontidae also forms part of this basal clade (Bocak *et al.* 2014; McKenna *et al.* 2015; Zhang *et al.* 2018). In the morphological analysis by Lawrence *et al.* (2011) Clambidae forms a basal polyphagan clade, while Eucinetidae, Scirtidae and Decliniidae, together with Derodontidae and Nosodendridae form a sister group to the combined superfamilies Cleroidea, Cucujoidea, Coccinelloidea, Tenebrionoidea, Chrysomeloidea and Curculionoidea. Although this last analysis was based on over 500 larval and adult characters, its results were in some respects at odds with those of several molecular analyses, including that based on most of the same taxa (McKenna *et al.* 2015).

Although scirtoids form one or more clades at the base of the polyphagan lineage according to recent molecular studies, they appear relatively late in the fossil record. As noted in Chapter 15, fossil Scirtidae have been described from Baltic amber (upper Eocene to Oligocene) and Fusan amber (lower Eocene). In addition, Kirejtshuk & Nel (2008, 2013) described two species of *Cyphon* and a species of *Scirtes* from lower Eocene French amber. The single scirtid known from Burmese amber (mid Cretaceous) is *Mesernobius anawrahtai* Engel (2010), described in the family Ptinidae but moved to Scirtidae by Peris *et al.* (2015). Otherwise, Scirtidae appear to be uncommon among Burmite fossils, and no others have been described. The family Clambidae is represented by the genus *Eoclambus* Kirejtshuk & Azar from the lower Cretaceous Lebanese amber (Kirejtshuk & Azar 2008) and both Clambidae and Eucinetidae are common and diverse in Burmite, but not yet described (C. Cai *in litt.*) The oldest known Scirtoidea, placed in the family Mesocinetidae by Kirejtshuk & Ponomarenko (2010), are known as impression fossils from Mongolia, China, Kazakhstan and Transbaikalia, Russia; five genera and ten species have been described, but undescribed forms are also known. Members of this group share various features with both Scirtidae and Eucinetidae.

Several other fossil groups must be considered which may or may not be related to Scirtoidea. The first of these is the family Elodophthalmidae, containing the genus *Elodophthalmus* Kirejtshuk & Azar, 2008, with two species from Lebanese amber (Lower Cretaceous). The authors considered this group to be close to Scirtoidea, but several features are unusual for the superfamily, such as the lack of metacoxal plates in a group which does not appear to be saltatorial and the distinct difference between the mesotarsi and metatarsi (the latter with a very long tarsomere 1 and apparently only 4 tarsomeres). Although a complete examination of the type would be needed to confirm this, I think, based on published figures, that *Elodophthalmus harmonicus* Kirejtshuk & Azar, 2008 (length 1.4 mm) may well be a member of the tenebrionoid family Aderidae.

A second possible scirtoid is the lower Eocene French amber (Oise) specimen *Boleopsis polinae* Kirejsthuk & Nel (2013), which was included in the Cleroidea as a new family Boleopsidae. The placement in Cleroidea may have been based in part on the relatively soft-bodied adult with 5–5–5 tarsi and a lobed tarsomere 4. In any case, the presence of subgenal ridges and well developed metacoxal plates (at least mesally) must have argued against this placement. Although the head in the amber specimen was porrect, the presence of subgenal ridges make it likely that it was capable of considerabe declination. The very short prothorax (shorter than the head) also supports an alternative placement in the Scirtoidea. As in the above case, a definitive statement concerning its placement would require an examination of the type.

A final problematic fossil taxon is *Ptisma zasukhae* Kirejtshuk & Azar (Kirejtshuk *et al.* 2016a, b) from Lower Cretaceous Lebanese amber, which was made the type of a family Ptismidae. Although this taxon was placed in the Staphylinoidea, it is far more likely to be related to the family Clambidae and a part of the superfamily Scirtoidea.

The most perplexing problem concerning this superfamily is the small number of Cretaceous Scirtidae and a similar lack of impression fossils from Jurassic localities, given the very basal position of this family within Polyphaga. This could be explained in part by the relatively thin cuticle and loose-joined body, but it is also possible that some mesocinetids are actually basal scirtids or that fossil scirtids have been placed in other families, as suggested above.

References

Beutel RG, Leschen RG (2016) 8. Polyphaga Emery, 1886. In *Handbuch der Zoologie/Handbook of Zoology. BandVolume IV Arthropoda: Insecta. Teilband/Part 38. Coleoptera, Beetles. Volume 1: Morphology and Systematics (Archostemata, Adephaga, Myxophaga, Polyphaga partim).* 2nd edn. (Eds RG Beutel and RAB Leschen) pp. 199–200. Walter de Gruyter, Berlin.

Bocak L (2016) 9. Scirtiformia Fleming, 1821. In *Handbuch der Zoologie/ Handbook of Zoology. BandVolume IV Arthropoda: Insecta. Teilband/*

Part 38. Coleoptera, Beetles. Volume 1: Morphology and Systematics (Archostemata, Adephaga, Myxophaga, Polyphaga partim). 2nd edn. (Eds RG Beutel and RAB Leschen) pp. 201–202. Walter de Gruyter, Berlin.

Bocak L, Barton C, Crampton Platt A, Chester D, Ahrens D, Vogler AP (2014) Building the Coleoptera tree-of-like for >8000 species: composition of public DNA data and fit with Linnaean classification. *Systematic Entomology* **39**, 97–110. doi:10.1111/syen.12037

Caterino MS, Shull VS, Hammond PM, Vogler AP (2002) Basal relationships of Coleoptera inferred from 18S rDNA sequences. *Zoologica Scripta* **31**, 41–49. doi:10.1046/j.0300-3256.2001.00092.x

Crowson RA (1981) *The Biology of the Coleoptera.* Academic Press, London.

Engel MS (2010) A primitive anobiid beetle in mid-Cretaceous amber from Myanmar (Coleoptera: Anobiidae). *Alavesia* **3**, 31–34.

Hunt T, Bergsten J, Levkanicova Z, Papadopoulou A, St. John O, Wild R, Hammond PM, Ahrens D, Balke M, Caterino MS, Gómez-Zurita J, Ribera I, Barraclough TG, Bocakova M, Bocak L, Vogler AP (2007) A comprehensive phylogeny of beetles reveals the evolutionary origins of a superradiation. *Science* **318**, 1913–1916. doi:10.1126/science.1146954

Kirejtshuk AG, Azar D (2008) New taxa of beetles (Insecta, Coleoptera) from Lebanese amber with evolutionary and systematic comments. *Alavesia* **2**, 15–46.

Kirejtshuk AG, Nel A (2008) New beetles of the suborder Polyphaga from the Lowermost Eocene French amber (Insecta: Coleoptera). *Annales de la Société Entomologique de France* (N. S.) 44 (4), 419–442.

Kirejtshuk AG, Nel A (2013) Current knowledge of Coleoptera (Insecta) from the Lowermost Eocene Oise amber. *Insect Systematics & Evolution* **44**, 175–201. doi:10.1163/1876312X-04402007

Kirejtshuk AG, Ponomarenko AG (2010) A new coleopterous family Mesocinetidae fam. nov. (Coleoptera: Scirtoidea) from the Late Mesozoic and notes on fossil remains from Shar-Teg (Upper Jurassic, South-Western Mongolia). *Zoosystematica Rossica* **19**(2), 301–325.

Kirejtshuk AG, Chetverikov PE, Azar D, Kirejtshuk PA (2016a) Ptismidae fam. nov. (Coleoptera, Staphyliniformia) from the Lower Cretaceous Lebanese amber. *Cretaceous Research* **59**, 201–213. doi:10.1016/j.cretres.2015.10.027

Kirejtshuk AG, Chetverikov PE, Azar D, Kirejtshuk PA (2016b) Corrigendum to "Ptismidae fam. nov. (Coleoptera, Staphyliniformia) from the Lower Cretaceous Lebanese amber". *Cretaceous Research* **61**, 275[Cretac Res 59(2016) 201–213]. doi:10.1016/j.cretres.2016.02.004

Lawrence JF (2001) A new genus of Valdivian Scirtidae (Coleoptera) with commentws on Scirtoidea and the beetle suborders. In *Sukunohikona. Special Publication No. 1.* (Eds K Morimoto, K Mizuno, Y. Hayashi, T. Ito, K. ando, M. tanikado, S. Shiyake) pp. 351–361. The Japan Coleopterological Society, Osaka.

Lawrence JF (2016a) 2. Classification (families and subfamilies). In *Handbuch der Zoologie/Handbook of Zoology. BandVolume IV Arthropoda: Insecta. Teilband/Part 38. Coleoptera, Beetles. Volume 1: Morphology and Systematics (Archostemata, Adephaga, Myxophaga, Polyphaga partim).* 2nd Edition. (Eds RG Beutel and RAB Leschen) pp. 13–22. Walter de Gruyter, Berlin.

Lawrence JF (2016b) 10.1 Decliniidae Nikitsky, Lawrence, Kirejtshuk & Gratshev, 1994. In *Handbuch der Zoologie/Handbook of Zoology. BandVolume IV Arthropoda: Insecta. Teilband/Part 38. Coleoptera, Beetles. Volume 1: Morphology and Systematics (Archostemata, Adephaga, Myxophaga, Polyphaga partim).* 2nd Edition. (Eds RG Beutel and RAB Leschen) pp. 201–204. Walter de Gruyter, Berlin.

Lawrence JF, Newton AF Jr (1995) Families and subfamilies of Coleoptera (with selected genera, notes, references and data on family-group names). In *Biology, Phylogeny, and Classification of Coleoptera. Papers Celebrating the 80th Birthday of Roy A. Crowson.* (Eds J Pakaluk and SA Ślipiński) pp. 779–1006. Muzeum I Instytut Zoologii PAN, Warsaw.

Lawrence JF, Yoshitomi H (2007) *Nipponocyphon*, a new genus of Japanese Scirtidae and its phylogenetic significance. *Elytra* **35**(2), 507–527[Tokyo].

Lawrence JF, Nikitsky NB, Kirejtshuk AG (1995) Phylogenetic position of Decliniidae (Coleoptera: Scirtoidea) and comments on the classification of Elateriformia (*sensu lato*). In *Biology, Phylogeny, and Classification of Coleoptera. Papers Celebrating the 80th Birthday of Roy A. Crowson.* (Eds J Pakaluk and SA Ślipiński) pp. 375–410. Muzeum i Instytut Zoologii PAN, Warszawa.

Lawrence JF, Ślipiński A, Seago AE, Thayer MK, Newton AF, Marvaldi AE (2011) Phylogeny of the Coleoptera based on morphological characters of adults and larvae. *Annales Zoologici* **61**(1), 1–217. doi:10.3161/000345411X576725

Maddison DR, Moore W, Baker MD, Ellis TM, Ober KA, Cannone JJ, Gutell RR (2009) Monophyly of terrestrial adephagan beetles as indicated by three nuclear genes (Coleoptera: Carabidae and Trachypachidae). *Zoologica Scripta* **38**, 43–62. doi:10.1111/j.1463-6409.2008.00359.x

McKenna DD, Wild AL, Kanda K, Bellamy CL, Beutel RG, Caterino MS, Farnum CW, Hawks DC, Ivie MA, Jameson ML, Leschen RAB, Marvaldi AE, McHugh JV, Newton AF, Robertson JA, Thayer MK, Whiting MF, Lawrence JF, Ślipiński A, Maddison DR, Farrell BD (2015) The beetle tree of life reveals that Coleoptera survived end-Permian mass extinction to diversity during the Cretaceous terrestrial revolution. *Systematic Entomology* **40**, 835–880. doi:10.1111/syen.12132

Nikitsky NB, Lawrence JF, Kirejtshuk AG, Gratshev VG (1994) A new beetle family, Decliniidae fam. n., from the Russian Far Easat and its taxonomic relationships (Coleoptera, Polyphaga). *Russian Entomological Journal* **2**(5–6), 3–10.

Peris D, Philips TK, Delclòs X (2015) Ptinid beetles from the Cretaceous gymnosperm-dominated forests. *Cretaceous Research* **52**, 440–452. doi:10.1016/j.cretres.2014.02.009

Sakai M, Satô M (1996) The coleopteran family Decliniidae (Elateriformia, Scirtoidea) new to Japan, with description of its second representative. *Elytra* **24**(1), 103–111[Tokyo].

Zhang SQ, Che LH, Li Y, Liang D, Pang H, Ślipiński A, Zhang P (2018) Evolutionary history of Coleoptera revealed by extensive sampling of genes and species. *Nature Communications* **9**, 205.

15. SCIRTIDAE FLEMING, 1821

Chris H.S. Watts and Peter Zwick

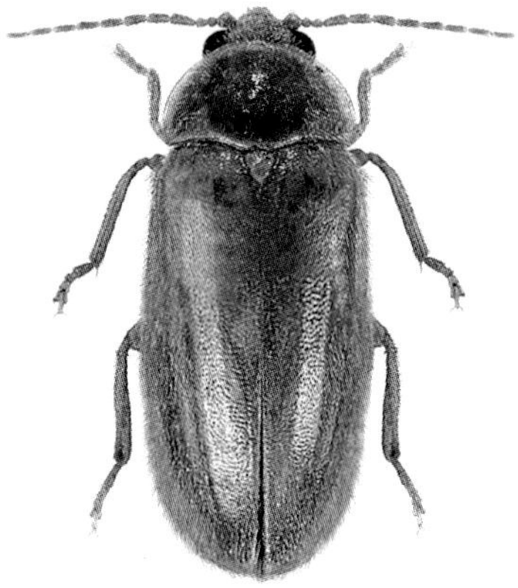

Fig. 15.1. *Pseudomicrocara orientalis* Armstrong.

Common name. Marsh beetles

Family synonyms. Cyphonidae Stephens, 1830; Elodidae Shuckard, 1840; Helodidae Agassiz, 1847.

Introduction. Scirtids occur world-wide except in arid areas. Australia has 308 named species and at least 20 undescribed. Adults are rather nondescript, narrowly oval to round, flattish to deep-bodied and range in size from 1.5 mm to 12 mm long. Most are a reddish brown to black in colour with a very few with strong yellow/black or red/black dorsal colour. The adults are terrestrial but the larvae are aquatic which restricts Scirtidae to mesic regions of Australia, mainly in the east. Four species live in permanent waters in the dry centre e.g. Stanley Chasm. Biodiversity is high and comparable between tropical areas of Queensland and cool temperate Tasmania. All but one species (*Contacyphon putonii* (Brisout de Barneville)) and most genera are endemic to the Australasian Region, with minimal overlap between Australia and New Guinea. A few taxa with flightless females may be related to New Zealand genera (Zwick unpubl.). A few genera (see below) are widespread in South-east Asia and beyond. The Australian genus *Pseudomicrocara* Armstrong (Fig.15.1) has also been recorded from temperate South America, New Caledonia and New Guinea (Libonatti & Ruta 2013).

Within Australia a relatively large group of mainly species-poor genera are restricted to higher rainfall regions of eastern Australia. This group includes all the saproxylic (living in water-saturated logs) genera and ones breeding in phytotelmata (water filled tree-holes) as well as several short range endemics restricted to areas of wet forest. Most are endemic to Australia (see generic accounts for details). Other genera such as the speciose *Scirtes* Illiger, *Pseudomicrocara* Armstrong, *Austrocyphon* Zwick and *Nothocyphon* Zwick inhabit generally dryer areas often with seasonal rainfall. Other than the cosmopolitan genus *Scirtes* these genera and a few *Pseudomicrocara* species are also endemic but, unlike the first grouping, have many species that are widespread across Australia. The tropical north is home to genera such as *Ypsiloncyphon* Klausnitzer, *Contacyphon* Gozis, *Calvarium* Pic and *Ora* Clark, all genera with members in countries to the north. These would seem to be more recent arrivals. The far south-west is an exception since, although its total species count is comparatively low (seven genera and 16 species nearly all endemic to the region) it contains species related to members of all the above loose groupings (i.e. *Heterocyphon* Armstrong and *Accolabass* Watts – wet forest; *Scirtes* – widespread; and *Papuacyphon* Zwick – New Guinea) as well as an endemic, *Cygnocyphon* Zwick, and even one species, *Contacyphon putonii* otherwise known only from central and SW Europe, primarily Spain and France (Zwick 2015b).

According to Lawrence & Newton (1995), the superfamily Scirtoidea contains the four families Decliniidae, Scirtidae, Eucinetidae and Clambidae, although in some more recent molecular phylogenies (Bocák *et al.* 2014; McKenna *et al.* 2015), the latter two families are more closely related to Derodontidae. Relations with other Coleoptera were analysed by Lawrence *et al.* (2011), and monophyly of Scirtidae and its subfamilies was discussed by Lawrence (2001) and Lawrence & Yoshitomi (2007). Most Australian species belong to the subfamily Scirtinae, all members of which have irregularly punctate elytra and male terminalia conforming to a common groundplan (see below). No Scirtinae have the three-lobed aedeagus present in both Nipponocyphoninae and Stenocyphoninae, and all scirtines have paired tibial keels, which are absent in these two groups.

The slender long-legged Stenocyphoninae (only genus *Stenocyphon* Lawrence) are represented by one species each in Chile (Lawrence 2001) and New Zealand (Ruta *et al.* 2011), and one undescribed species from Tasmania. In the Nipponocyphoninae (only the Japanese *Nipponocyphon nakanei* Lawrence & Yoshitomi 2007) elytral punctures are arranged in regular lines; *Nipponocyphon* is not addressed here.

Several fossil Scirtidae are known mostly from Baltic amber, variously dated from the lower Oligocene to the upper Eocene (Alekseev 2013). Most were (by habitus) assigned to extant genera but several to genera not known from the modern fauna (Yablokov-Khnzoryan 1961; Klausnitzer 2012; Alekseev 2013). They are also known from early Eocene amber from Fushun China (Hong 2002; Ross 2014) and from a single specimen of a new genus from Burmese amber dated to the mid-Cretaceous (Engel 2010; Peris *et al.* 2015). The only Australian fossils are some well preserved unidentified larvae from a lower Createaceous fossil bed near Koonwarra thought to be from a shallow lake-side environment (Jell & Duncan 1986).

Knowledge of Australian Scirtidae began with Blackburn (1892) but their taxonomy has only recently been intensely studied, based mainly on adults (Watts 2014 and previous; Zwick 2016 and previous). Studies of the larvae (Watts 2014) and molecular data (Cooper *et al.* 2014) are still incomplete and the relationships between the various genera are not yet well understood.

Biology. Oviposition, eggs and early larval development have been observed in a few European species of *Contacyphon, Elodes* Latreille, *Prionocyphon* Redtenbacher and *Scirtes*, but only in the laboratory (Zwick & Zwick 2008a, 2010). Here between 20–50 eggs were laid near the water's edge or into clefts of wet material. The soft whitish eggs developed immediately, larvae hatched after 9–10 days at 21°C. Newborn larvae of *Elodes* had air in the tracheae and fed, while those of *Scirtes haemisphaericus* (Linnaeus) and *Prionocyphon serricornis* (Müller) had some yolk in their guts. After 2–3 days, at the first moult, mouth and anus of the latter species opened and air first appeared in the tracheae (Zwick & Zwick 2008a and unpubl. observations). From the little gleaned from rearing studies most Australian species are similar (Watts pers. obs.). The large range of morphologies of the apical portion of the ovipositor in Australian taxa imply ovipositing onto or into a range of different substrates (see below).

Scirtinae develop in very different water bodies, from stream and lake shores to puddles, freshwater crayfish burrows (Horwitz 1989), water-filled tree holes (phytotelmata), or water-filled cells in rotting logs (saproxylic) (Watts 2014). The 'smooth cells formed in the damp earth underneath a deeply imbedded log' in which the New Zealand *Veronatus tricostellus* White was found (Hudson 1934) may be similar.

Larvae breathe atmospheric air at the water surface (Beier 1949; Treherne 1952). In older literature the (mostly five) rectal papillae were called blood gills but their primary function is osmoregulation (Treherne 1954; Wichard & Komnick 1974). This may be particularly important in small water bodies subject to evaporation and hence concentration of salts i.e. phytotelmata. It is notable that it is species from these habitats that have multibranched anal papillae. However, although their principle function is osmoregulation some gas exchange across the very delicate cuticle is inevitable and may be important for inhabitants of poorly aerated waters. *Exochomoscirtes* Pic spp. from phytotelmata and Australian larvae from rotting logs have highly branched rectal papillae (Zwick 2011a; Watts 2014). *Prionocyphon serricornis* in phytotelmata rhythmically retracts and extends its rectal papillae (Zwick, pers. obs.). Larvae of *Contacyphon* found in large numbers in deep groundwater (Klausnitzer & Pospisil 1991) must practice cutaneous respiration as is probably the case in early instars of most species.

Larvae are unable to bite and except for a few unusual species (e.g. saproxylic ones) their mandibles lack teeth during most instars (Fig. 15.7e). Late in development, mostly at the last moult, the mandible developes a sharp tip, called the incisivus (Beier 1949, 1952), which is used to build a pupal cell (Fig. 15.7f). Scirtinae larvae belong to the functional feeding type of gatherers and collectors. With dense brushes of fine hair and comb-shaped setae on maxillae and mandibles (Hannappel & Paulus 1987) they indiscriminately sweep deposited fine particulate matter from submerged surfaces, at rates of 70–80 sweeps per minute (Beier 1949, s.e.m. illustrations in Wichard *et al.* 2002). Rasp-like surfaces on the molae grind the material. Excess water is squeezed out in a press of the hypopharynx (therefore, larvae are sometimes called filterers, e.g. Klausnitzer 2006) and a food pellet is formed. Guts contain very fine particulate matter, organic as well as mineral. The deposited fine particulate organic matter ingested consists (at least in streams) mainly of refractory macromolecules which cannot be processed by insects but are accessible to microbes. As in other limnic fine particle consumers, readily digested microbes probably constitute an important part of the diet (e.g. Edwards & Meyer 1990; Findlay *et al.* 1984; Hall & Meyer 1998). Short gut passage times of only about one to two hours (Zwick 2011b: fig. 15.7.77, *Elodes*) are in line with this.

Most papers assume that larvae pass through five instars. However, laboratory rearing of some European species documented 9–11 instars (Zwick & Hecht 2008; Zwick & Zwick 2008a, 2010). Most Scirtinae pupate on land in a cell built from locally available particles or within masses of decaying damp leaves or sheets of drying algae. Mandible structure and mode of pupation are correlated. The few Scirtinae known to pupate under water, e.g. *Hydrocyphon* Redtenbacher (not Australian) have a bluntly pointed tip to their mandible in the last larval instar. In contrast, *Scirtes haemisphaericus* lacerates live submerged plants with its multidentate mandibles (e.g. Fig. 15.7g). The pupa attaches itself to the aerenchyma, using the plant as a snorkel (Zwick & Zwick 2008b). It is possible that the Australian *Scirtes auratus* Watts which is phylogenetically close, and has a similar mandible, may behave similarly (Watts 2014). However, *S. haemisphaericus* can also pupate between dense masses of filamentous algae and probably also on land as *S. auratus* is known to do. Late instars of all known saproxylic larvae developing in wet cavities of rotting logs have large, often pluridentate incisivi (Watts 2014) and pupate in damp areas in the log above the wet cavities inhabited by the larvae. The role of these unusually large and robust mandibles in late instars of all known saproxylic species, both in Australia and elsewhere (Klausnitzer 2006; Ruta *et al.* 2018) is unknown. Watts (2014) suggested competition between individual larvae for scarce pockets of water. We add, defence against predators or to help with movement through their very restricted microhabitat. The content of the gut – mineral and vegetable matter – precludes predation (Hudson 1934).

Most of the whitish, soft pupae of Scirtinae have 2 or 4 horn-like spines on the corners of the pronotum and 2–4 anchor organs at the abdominal tip (e.g. *Austrocyphon charon* Zwick, *Nothocyphon amita* Zwick, *Pseudomicrocara spilota* (Blackburn)), also those few known from saproxylic species. Numbers, size and detail of structure vary between species (our data on European and personal observations of various Australian species). However, the only described Australian pupa, *Prionocyphon niger* Kitching and Allsopp, has neither horns nor anchor organs (Kitching & Allsopp 1987).

Observations in the laboratory show that pupation generally takes 2–6 days (Watts 2014), the shortest periods being in *Scirtes* species. If a cell has been built a further 2–7 days is spent in it before emergence.

Adults of most species are most often collected in late spring or summer months from riparian vegetation, particularly flowering shrubs, or at light and intercept traps. Other collecting sites/methods include pyrethrum fogging of trees or fungusy logs, under bark, also sifting debris on the forest floor or on banks. Worldwide there is no information on what the adults feed. Mandibular structure varies considerably from weak and simple to robust and multitoothed (Fig. 15.4a, b). Adult diet is likely to be just as varied. Most flower-visiting species have simple mandibles and have been found with pollen grains in their guts. Fungal hyphae and spores have been found in the guts of species with more robust mandibles. Guts of adult *Elodes* and *Odeles* Klausnitzer species (both Palaearctic) never contain particulate matter (Zwick pers. obs.).

Copulating pairs are rarely seen, in the few that have the male mounts the female to quickly establish a tail-to-tail position (Nyholm 1972; Zwick pers obs.).

The only field study of Australian Scirtid life cycles was of *Prionocyphon niger* inhabiting tree-holes in south-east Queensland (Kitching & Allsopp 1987). Tree-holes were sampled monthly over 17 months. Larvae, often in large numbers and many instars, were present most of the time indicating continuous breeding if conditions were suitable. In Europe, based on the age distributions of larvae, life cycles appear to be univoltine, plurivoltinism has not been recorded. A univoltine life cycle was documented by monthly benthos samples combined with emergence trap records of adult *Elodes minuta* (Linnaeus) and *Odeles marginata* (Fabricius) for more than a year (Rasmussen 1978; Zwick 2011b). Adult emergence of these European species from headwater streams was poorly synchronised. In contrast, in Australia, species that inhabit seasonal creeks and swamps can have highly synchronised emergence times, e.g. *Leptocyphon furcalonga* Zwick and some species of *Scirtes* and *Pseudomicrocara* (Watts pers. obs.), usually in late spring or early summer. Species inhabiting habitats with more permanent water are less seasonal with both adults and larvae present most of the time although fewer are present in winter in colder regions. The life stage which survives dry periods is unknown. Since neither eggs nor larvae can survive desiccation and there is no evidence from laboratory observations of delayed emergence of pupae (Watts pers. obs.), it seems most likely that it is the adults which survive dry periods by aestivating in sheltered places.

Parasites of Scirtidae. Other than a mermithid (Nematoda) (Kaiser & Klausnitzer 2000) and a single record of water mites (Müller 2015), parasites of Scirtidae are not known.

Characteristics. Adults (see also Lawrence 2016; Lawrence & Ślipiński 2013, as well as *Explanatory Notes* below). Length ~1.5–12 mm. Body broadly ovate to narrowly elongate, strongly convex to strongly flattened; sides of prothorax and elytra usually slightly to strongly, independently rounded, rarely continuously rounded or subparallel, usually finely pubescent (subglabrous in *Macrohelodes*) (Pl. 41A–D). Head declined. Eyes entire. Mandibles well developed, varying from delicate and toothless to robust with several strong teeth on inner edge. Antennal insertions exposed and widely separated; subgenal ridge well developed (Fig. 15.2). Pronotum ~0.25–0.70 times as long as wide, usually widest posteriorly or at middle; sides slightly to strongly curved, often explanate. Pronotal epipleura broad, flat or weakly concave, bounded interiorly by a carina of variable length and strength. Procoxal cavities strongly transverse, narrowly separated, externally broadly open, without narrow lateral extensions, internally open. Trochantin broadly exposed and very narrow. Elytra ~0.85–1.95 times as long as wide and 2.9–8.5 times as long as pronotum, entire and conjointly rounded at apex, regularly or irregularly punctate, occasionally apunctate, without scutellary striole; epipleura relatively narrow but more or less complete. Mesoventral discrimen usually present but anteriorly obscured by mesoventral cavity. Mesoventral process apically divided. Metaventrite flat to strongly convex, with short to very long discrimen. Metafemora greatly enlarged in some genera. Outer edges of tibiae with paired longitudinal carinae which may be more or less spinose. Tibial spurs double, glabrous or pubescent, hind pair of equal or unequal length. Tarsi 5–5–5 with ventral lobe on penultimate tarsomere; claws simple. Abdomen with five ventrites, with basal two more or less connate. Anterior edge of sternite VIII in male with or without median strut; anterior edge of tergite VIII with paired lateral struts. Anterior edge of sternite IX in male often with median strut; tergite IX with paired lateral struts, which never unite with sternite to form subgenital plate or spiculum gastrale. Aedeagus highly variable but never a tubular penis with phallobase. Instead, the ejaculatory duct opens between loosely connected dorsal and ventral sclerites, often with lateral appendages (see below, Figs 15.9–15.12). Ovipositor and internal female tract complex (Fig. 15.13).

Larvae. (Pl. 2) Body elongate to broadly elliptical and slightly flattened. Surfaces and ventral abdominal surfaces lightly to more or less heavily sclerotised and evenly pigmented except for saproxylic species which are uniformly cream coloured, generally smooth; vestiture consisting of fine hairs, setae or short blunt spines. Head prognathous, epicranial stem very short or absent; frontal arms V-shaped or U-shaped, contiguous or distinctly separated at base. Stemmata as many as 5 on each side, but often tightly clustered, sometimes absent. Antennae usually long with 14–150 annuli, with or without sensorium on antennomere 2. Clypeus intimately fused with labrum, combined structure free (Fig. 15.7a); epipharynx with complex armature. Mandibles symmetrical, somewhat flattened, apex somewhat rounded in earlier instars but tip usually unidentate and acute but occasionally bidente or multidentate in last instar. Ventral mouthparts retracted; maxillary articulating area present but more or less concealed. Maxillary palp 3-segmented or 4-segmented with reduced apical palpomere, the apical or subapical palpomere often with transverse rows of microspines; palpifer usually well developed. Labium consisting of prementum and postmentum, the latter large, subquadrate or trapezoidal, with blunt posterior edge, not divided longitudinally, laterally expanded to conceal maxillary articulating area. Ligula very short and broad or absent. Labial palps 2-segmented or 3-segmented with reduced apical palpomere. Hypopharyngeal region complex, consisting of a series of teeth, combs, plates, and patches of simple and complex hairs (Fig. 15.7). Thoracic segments about equal in length or with prothorax slightly larger. Prothoracic venter usually with anterolateral cervical sclerites but without other armature. Thoracic legs 5-segmented, including pretarsus, which is acute, claw-like and bi- to multisetose; mesocoxae large, more or less conical and projecting, narrowly to moderately widely separated; femur and tibiotarsus with stout setae or spines. Abdomen less than twice as long as thorax. Abdominal apex with respiratory chamber formed from terga VIII and IX and enclosing pair of enlarged, posteriorly oriented, annular spiracles (Fig. 15.8), reduced in size in saproxylic genera and *Leptocyphon*; thoracic

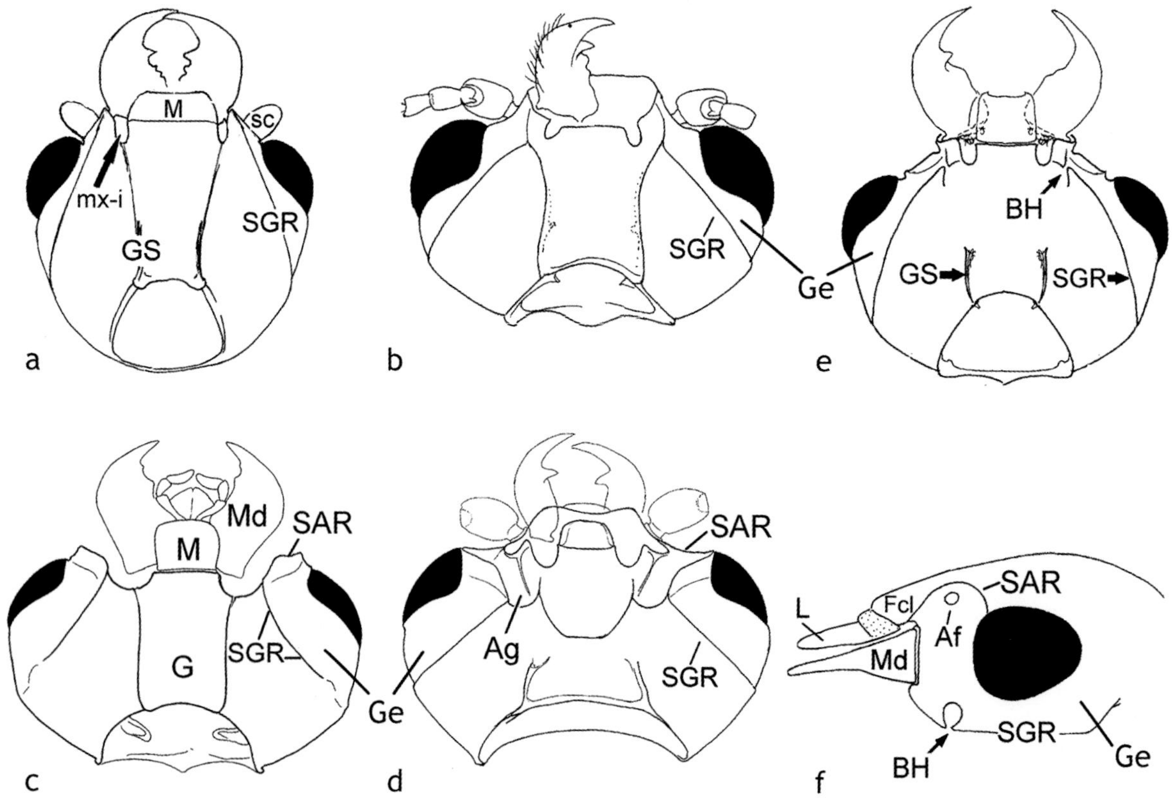

Fig. 15.2. Diagrammatic views of head capsules; a–e, ventral views; f, lateral view. a, *Heterocyphon cooloolaensis* Watts; b, *Nothocyphon lindensis* (Blackburn); c, *Pachycyphon serratus* Zwick; d, *Calvarium* (*Calvariellum*) sp.; e,f, *Contacyphon coarctatus* (Paykull). Af, antennal foramen; Ag, antennal groove; BH, button hole; Fcl, frontoclypeus; Ge, gena; GS, gular suture; M, mentum; Md, mandible; mx-I, insertion of maxilla; SAR, supra-antennal ridge; sc, scapus; SGR, subgenal ridge. Modified from Zwick (2013a, b, 2016).

and anterior abdominal spiracles reduced and non-functional or absent. Abdominal segment IX shorter than VIII, sometimes concealed from above by tergum VIII; tergum IX completely dorsal, without paired urogomphi. Sternum IX simple without hinged operculum. Segment X membranous and more or less fused to IX. Anal region posteriorly or terminally oriented, often with five or more, simple or branched, anal papillae or gills (Fig. 15.8); anal hooks absent.

Explanatory notes to facilitate use of keys and descriptions of adults

Head capsule (Fig. 15.2). Characteristic patterns result from different configurations of the subgenal ridge (SGR) and supraantennal ridge (SAR) (Fig. 15.2). The SGR is a sharp fold starting dorsolaterally on the occiput and running forward beneath the eye. The SAR extends outward from the anterolateral edge of the frontoclypeus. It surrounds the base of antenna and in most Scirtinae ends where it meets the inner edge of the eye (Fig. 15.2a, b). In a few genera the SAR turns ventrally before it reaches the eye, forms a crest in front of it and connects to the SGR beneath the eye (Fig. 15.2c, d). In this way, a continuous ridge delimiting an antennal groove is formed.[1] These genera may be close relatives of New Zealand genera, e.g. *Atopida* White (Zwick, unpubl.) which have a similar structure. In most Australian Scirtinae the front end of the SGR joins the sclerotised edge of the oral cavity near the mandibular insertion, sometimes together with an extension of the gular suture (Fig. 15.2a,b). However, in some genera the SGR ends freely on a raised cone at some distance from the oral cavity. A distinct gap in the lower edge of the head is then apparent in side view ('buttonhole', Fig. 15.2e,f). In taxa with a wide and very short head (most Australian *Prionocyphon*), the buttonhole is visible only in antero-lateral view. It is exposed in *Contacyphon, Cygnocyphon, Nanocyphon* Zwick, and *Petrocyphon* Watts.

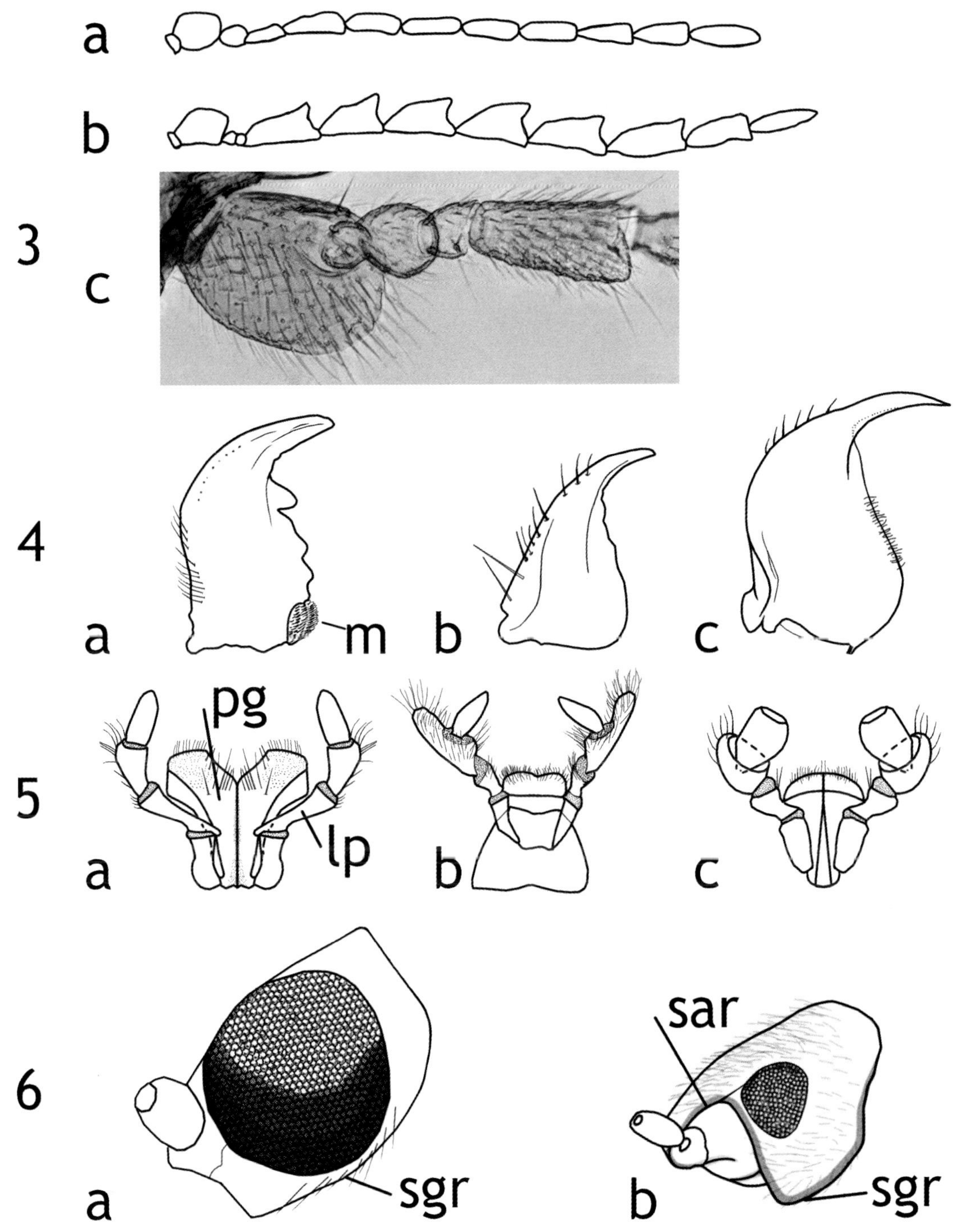

Figs 15.3–15.6. 3a–e. Male antennae: a, filiform antenna of *Eurycyphon fulvus* Watts; b, serrated antenna of *Macrodascillus scalaris* (Lea); c, base of antenna of *Prionocyphon serricornis* (Müller). (a, b, From Watts 2011; c, modified from Zwick 2016.) 4a–c. Mandibles: a, *Heterocyphon cooloolaensis* Watts; b, *Eurycyphon fulvus* Watts; c, *Ypsilocyphon longus* Zwick, m, mola. (From Watts 2009, 2011 and Zwick 2014b.) 5a–c. Labial palpi: a, simple form, *Heterocyphon cooloolaensis* Watts; b, forked form, *Scirtes kaytae* Watts; c, cup-like terminal segment, *Spaniosdascillus elongatus* Watts. (From Watts 2009, 2011.) lp, lapial palpi; pg, paraglossa. 6a, b. Lateral view of head illustrating relative size of eyes: a, *Peneveronatus australis* Armstrong; b, *Dasyscyphon hadrostiktos* Watts. (From Watts 2009, 2011.) SAR, supraantennal ridge; SGR, subgenal ridge.

Head appendages (Figs 15.3–15.5c).

Antenna. Eleven-segmented, filiform (e.g. Fig. 15.3a), in a few taxa (*Macrodascillus* (Fig. 15.3b), *Spaniosdascillus* Watts, some *Prionocyphon* spp.) the flagellum may be more or less serrate, especially in males. The proportions between the 11 segments vary (see basal segments in Fig. 15.3). Only *Prionocyphon* exhibits structural modifications at the antennal base (Fig. 15.3c): the scape is flattened and almost disc-like, usually with a narrow front edge; the smaller pedicle inserts near the edge of the disc, mostly on its underside, causing the antenna, at rest, to extend straight back under the body; flagellomere 1 (= segment 3) is very small, almost ring-like. In several Australian *Prionocyphon* the frontoclypeus forms lobes projecting over the bases of the mandibles, medially from the disc-like scape.

Mandible. Broad based, curving inwards and narrowing to a sharp point, outer edge thicker with a fringing row of long setae, inner edge thinner sharp with 0–3 teeth and in some species a patch of dense small spines on the mola area at the base (e.g. Fig. 15.4a). The general robustness of the mandible varies from delicate (e.g. many small *Pseudomicrocara* species, Fig. 15.4b) to robust (e.g. *Heterocyphon* Armstrong, Fig. 15.4a) presumably reflecting diet. In most cases the variation in structure is genus-specific. An exception being the genus *Pseudomicrocara*, probably reflecting its polyphyletic nature (see later).

Maxilla. Varies little between genera in adult Scirtinae and conforms to the basic Coleopteran structure of a basal cardo, stipes, lacinia, galea and a maxillary palpus usually with four segments.

Labial palpus. Three segmented, usually slightly curved (Fig. 15.5a). Last segment of variable size and shape, often standing at an angle to the rest of the palpus and inserted baso-laterally on the penultimate segment. Differences between most taxa are gradual and not easily quantified. An extreme variant in which the palpus appears to be forked (Fig. 15.5b) reveals that actually the 2nd segment is prolonged/extended on the outside where it bears large setae. This extreme condition is widespread among, e.g. *Prionocyphon* spp. and in genera near *Scirtes*, also in northern hemisphere genera, e.g. *Elodes, Odeles, Microcara* Thomson. In some Australian taxa segment 2 is approximately cup-shaped with an almost oblique end but large setae on the outside (Fig. 15.5c) suggesting it may be a reduction of the forked morph.

Prosternal process and mesoventral groove and process. The pronotal process is very narrow between the procoxae which it narrowly separates. Behind the procoxae the pronotal process varies from thin and sharply pointed (e.g. *Scirtes*) to broad with rounded apex (e.g. *Daploeuros*). Its posterior end in most Australian species reaches the mesoventrite which in most cases has a central groove which accommodates the apex of the pronotal process (absent or nearly so in some, e.g. *Tectocyphon* Zwick). This groove varies from short and transverse (e.g. *Petrocyphon*) to long and narrow (e.g. *Scirtes*). The backward extension of the mesoventrite, which varies from short and wide (e.g. *Dasyscyphon* Watts) to long and narrow (e.g. *Scirtes*), separates the mesocoxae and reaches a small forward-directed triangular midline extension of the metacoxal plate in all Australian species except *Eximiocyphon excisus* Zwick.

Larval morphology (Figs 15.7, 15.8). All known scirtid larvae feed by scraping wet surfaces to obtain deposited vegetable matter and associated microorganisms. However their actual microhabitat varies from flowing water to films of water in the interior of water-saturated forest logs. These different microhabitats are strongly reflected in their anatomy, less so in their phylogenetic relationships (Cooper *et al.* 2014). Those inhabiting logs (saproxylic) are pale, sparsely covered with setae, some have sensory structures on their tergites, have a greatly reduced number of flagellum segments (14–30), multilobed anal papillae, robust mandibles multidentate in later instars and a highly modified hypopharynx and clypeus (Watts 2014) (Fig. 15.7a–c). In comparison larvae living in surface waters are more uniform in body structure although differences, usually reflecting different genera, are present in setation, morphology of the apical abdominal segments and mandible structure (most unidentate but bidentate in some *Scirtes* and *Pseudomicrocara*). The terminal abdominal segments are complex (Fig. 15.8). Dorsally the two large tracheae which run down the body end in a pair of spiracles which enter an atrium formed between segments 8 and 9. In most species the spiracles have a protective grille of short setae (Fig. 15.8e), although these are absent in *Austrocyphon* (Fig. 15.8f). Below the atrium and above the anus are anal papillae (Fig. 15.8a, b) which can be extruded or retracted, presumably by hydrostatic pressure changes in the haemolymph. A few, notably those living in phytotelmata, have multilobed anal papillae, although mostly the basic five-lobed morphology is retained (8–10 in *Scirtes auratus*) (Watts 2014).

Male terminalia. Scirtidae: Stenocyphoninae. Males of the unnamed Tasmanian species are unknown. Males of *S. sasajii* Lawrence, 2001 (Chile) and of *S. neozealandicus* Ruta *et al.* 2011 (New Zealand) have similar genitalia including a trilobed aedeagus with a phallobase and are very different from Scirtinae.

The groundplan of Scirtinae male terminalia (Figs 15.9–15.10)

Scirtidae: Scirtinae. The male genitalia differ considerably from the summary descriptions of beetle genitalia given in Lawrence & Ślipiński (2013). At the same time, there is global agreement between genera in the involved structural components as well as in their relative positions, there is a common groundplan. In Scirtinae, the terminalia comprise the invaginated segments 8 and 9 and the genitalia proper. In the Scirtinae groundplan (Zwick 2015c and Figs 15.9–15.10), segments 8 and 9 are complete, both comprising sclerotised tergites and sternites. The tergites possess paired anterior apodemes. The plate of tergite 8 is covered with setae and/or microtrichia, its caudal edge often with a regular fringe. The soft bare plate of tergite 9 lies over the anus. Sternite 8 is a wide sclerotised bracket supporting a largely membranous plate with a few setae at its caudolateral corners. Its front is attached to the inside of sternite 7 and, unlike sternite 9, it cannot be freely everted. Sternite 9 is basally soft, caudally it is bilobed and bears socketed setae. The genitalia proper consist of a dorsal tegmen (te), which may be a plate, or a band-like sclerite lying across the penis base. The tegmen is

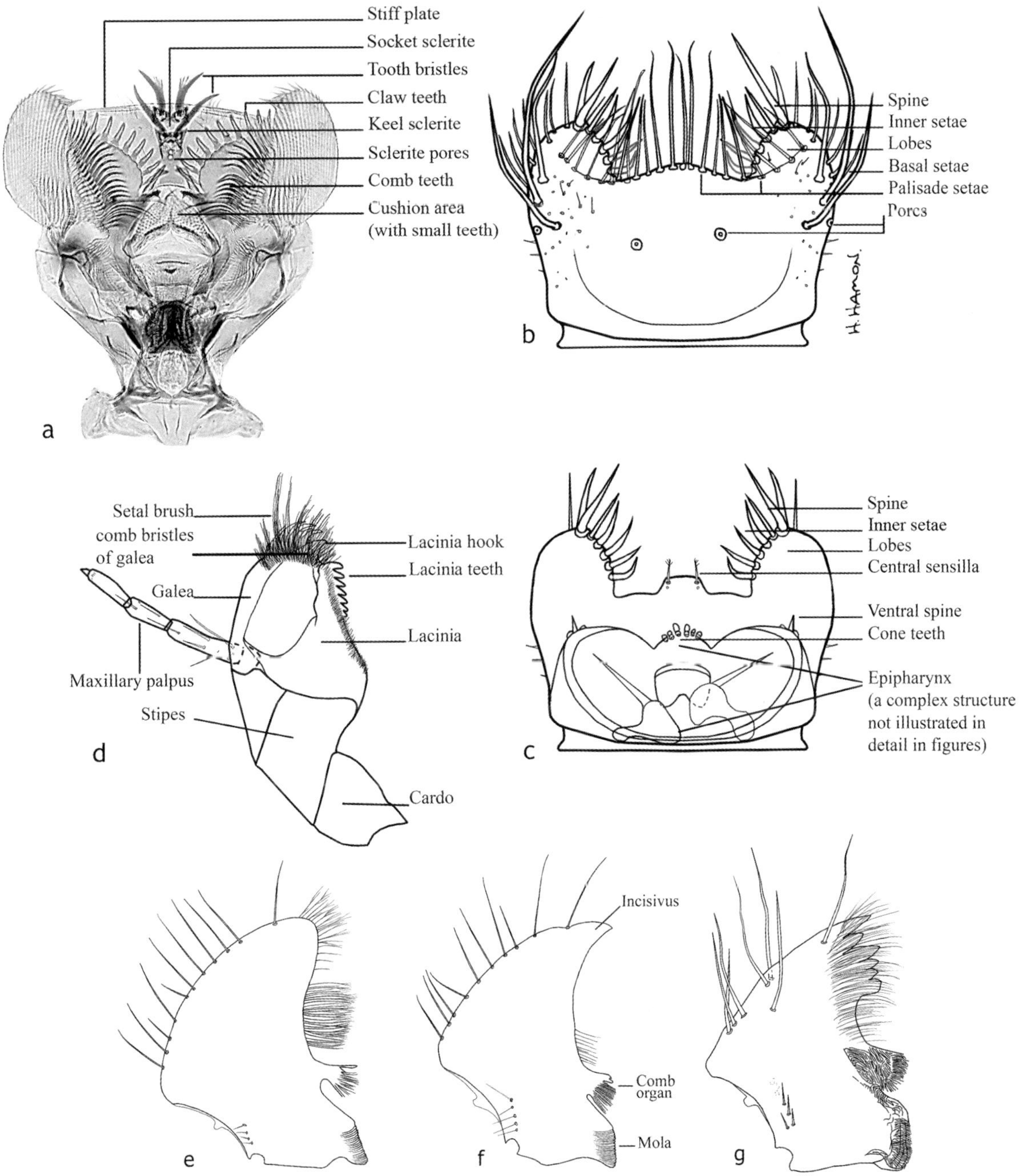

Fig. 15.7. Larval mouthparts: a, dorsal view of hypopharynx of *Austrocyphon pictus* (Blackburn); b, ditto, dorsal view of clypeus; c, ditto, ventral view of clypeus; d, maxilla of *Heterocyphon tasmanicus* Watts. (From Watts 2014). Dorsal view of left mandibles, to similar size: *Scirtes exoletus* Waterhouse, penultimate instar (e) without, last instar (f) with incisivus; *Scirtes auratus* Watts, last instar with pluridentate incisivus (g), the European *S. haemisphaericus* (Linnaeus) is similar.

caudolaterally divided into parameres (pe) wrapping distally around the penis beneath. There is no phallobase. The base of the penis, the pala (pa), is surrounded by a sclerite which caudolaterally ends in a pair of immobile parameroids (pd) with sensilla at their tips. Between the parameroids is the median trigonium (tr). It can be bent down by muscles in the pala and anchors the male during copula. There may be paired styli (st) located baso-ventrally from the penis.

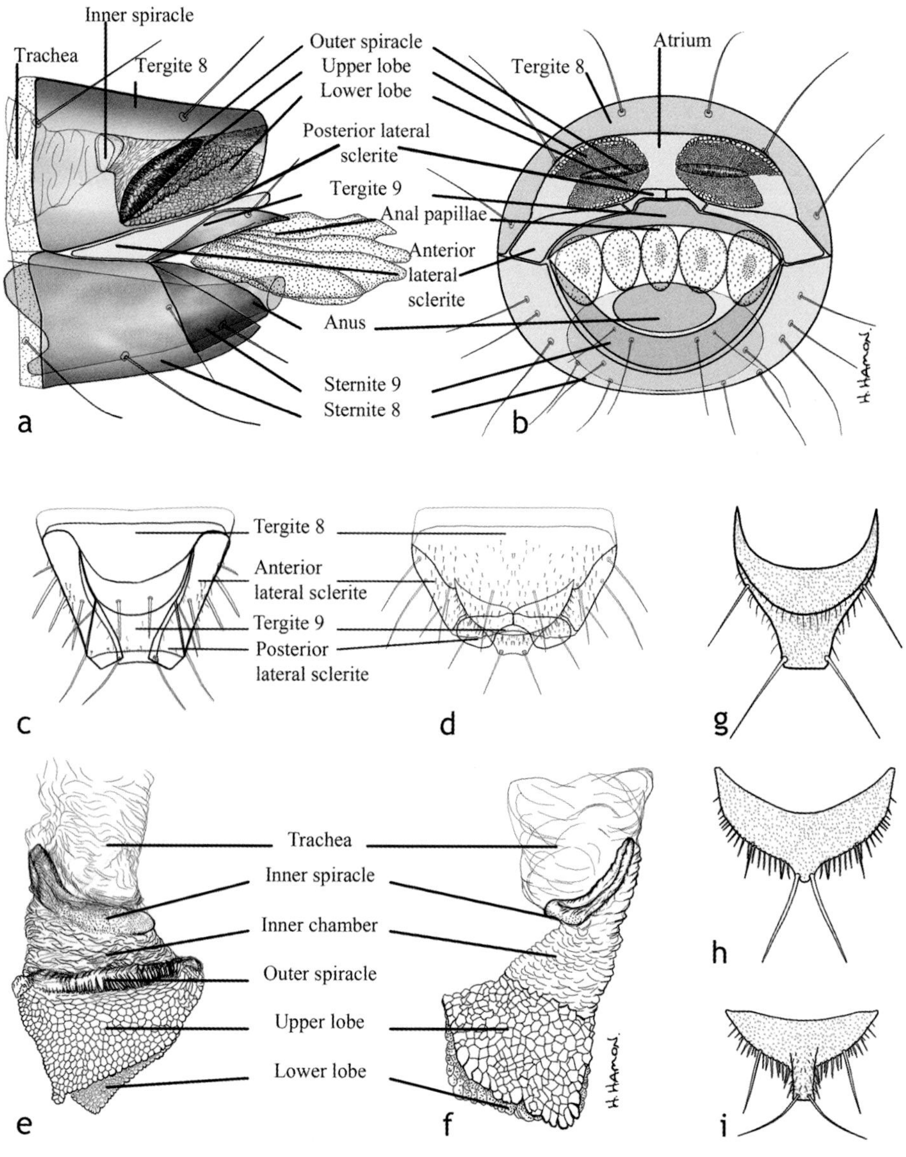

Fig. 15.8. Semidiagrammatic views of terminal abdominal segments of larvae: a, lateral view of abdominal segments 8 and 9; b, ditto, caudal view; c, lateral sclerites of tergite 8 of *Scirtes helmsi* Blackburn; d, ditto of *Pseudomicrocara orientalis* Armstrong; e, spiracle apparatus of *Pseudomicrocara atkinsoni* (Waterhouse); f, ditto of *Austrocyphon adelaidae* (Blackburn); g, tergite 9 of *Scirtes emmaae* Watts; h, ditto *Austrocyphon adelaidae*; i, ditto *Nothocyphon lindensis* (Blackburn). (From Watts 2014.)

Modifications of the Scirtinae groundplan

Genitalia conforming to the groundplan include all structures listed above and occur in most Australian genera. In several they are morphologically very simple, e.g. *Accolabass, Macrohelodes* (Fig. 15.10), and some *Pseudomicrocara*, but also in, e.g. the Northern hemisphere genus *Microcara* (Zwick 2015c). The hypothetical common ancestor of Scirtinae may also have had structurally simple genitalia. However, even in genitalia conforming to the groundplan individual genital components can take an amazing variety of complex shapes and forms including asymmetries. Such seem to be apomorphic character expressions within the groundplan.

Tergite 8 is rarely modified, if so, it is usually simply notched. Sternite 8 can have its sclerites reduced to various degrees, e.g. thin Y- or V-shaped rods. Partial sclerotisation and multiple modifications of tergite 9 can occur and sternite 9 can be transformed in many ways, e.g. it can become an unpaired spine (e.g. group of *Austrocyphon stylatus* Zwick), be deeply divided into long blades of different shape and length (e.g. *Ypsiloncyphon brevis* Zwick) (Fig. 15.11b). In some,

9

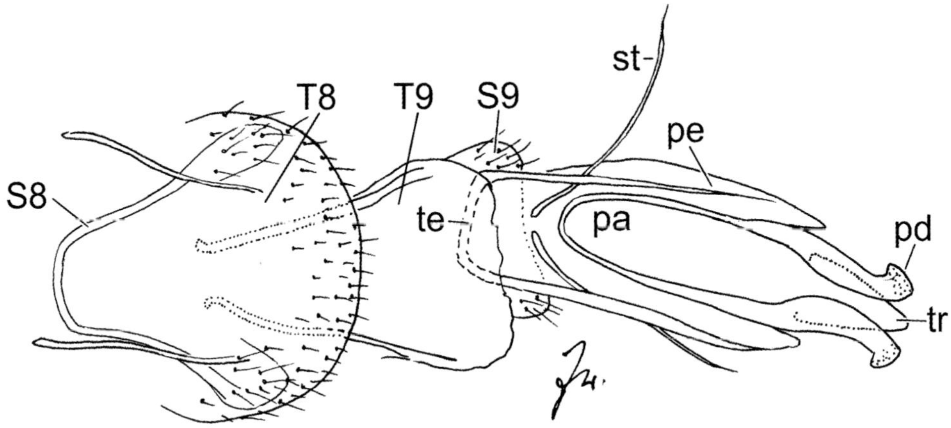

10

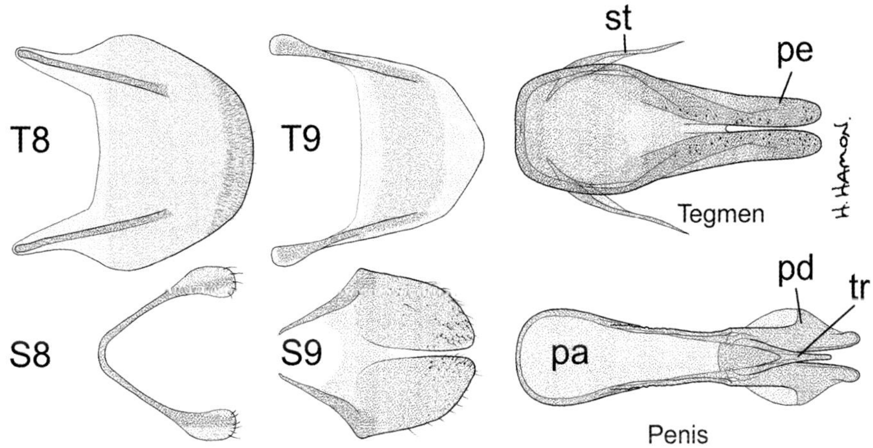

Figs 15.9–15.10. Examples of male terminalia exhibiting all structural elements present in the groundplan of Scirtinae. 9, *Pseudomicrocara* nr. *spilota* (Blackburn); 10, *Macrohelodes crassus* Blackburn (expanded). pa, pala; pd, parameroids; pe, parameres; S (with ordinal number), sternite; st, stylus; T (with ordinal number), tergite; tr, trigonium.

e.g. *Peneveronatus* the trigonium is complex (Fig. 15.11d). In others the tegmen can have the parameres heavily armed with spines and hooks, and/or be subdivided into several lobes (e.g. some *Pseudomicrocara*) (Fig. 15.11f).

Taxa not conforming with the groundplan

When some components are missing, or others transformed into completely new structures taxa are considered not to conform to the groundplan. For example, in *Contacyphon* sclerites of sternite 8 are completely lost and there is also no trigonium, its clasping function is performed by paired caudal processes of the penis, the prosthemes (Fig. 15.11c) (Nyholm 1972). Sternite 9 may form a large trough with paired caudal processes and accommodate other parts of the genitalia (e.g. *Contacyphon mobula* (Zwick) (Fig. 15.11c)).

Parts can also be connected to form some new structure. For example, the parameroids of *Austrocyphon* are connected to form a dorsal plate above the trigonium, with a transparent central foramen (Fig. 15.11a). In *Nanocyphon* (Fig. 15.11g) the parameroids merge and form a plate looking like a trigonium while the true trigonium is spine-like and situated beneath that plate. In some species of *Ora* (Fig. 15.12a) and *Scirtes* (Fig. 15.12b) the tegmen and penis are fused in a way that makes it difficult to identify the component structures and in some Australian *Prionocyphon* spp. (e.g. *P. warra* Watts, Fig. 15.11h), homology of the transformed genitalia can only be inferred by comparison to various less strongly modified congeners.

In several genera homology with elements from the groundplan is unclear, e.g. the origin of the large hook apparatus (uncifer) which *Papuacyphon* ssp. possess (Fig. 15.12c), in addition to the normal genital structures. The homology of the genitalia of several other Australian genera is completely obscure, e.g. *Calvarium* spp. with large claspers articulating on a common base (Fig. 15.12d), *Eximiocyphon excisus* Zwick with long cigar-shaped genitalia (Fig. 15.12e), *Paracyphon avicularis* Zwick with two-layered claspers attached to a single tendon (Fig. 15.12f), or *Leptocyphon abnormis* Zwick. *Petrocyphon antarcticus* Watts and its closest relatives as well as several *Prionocyphon* spp. (Fig. 15.12h–i) possess large spines in the terminal section of the endophallus which may

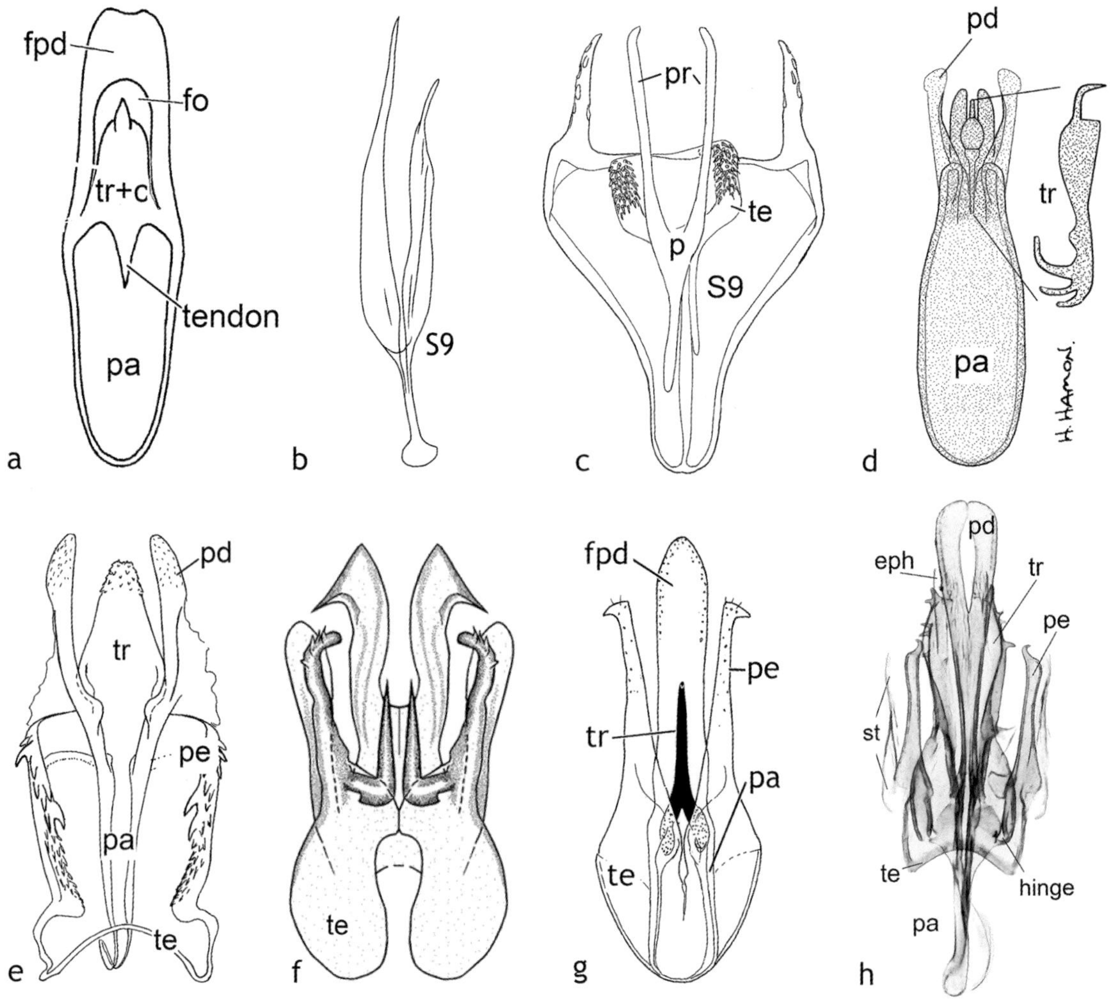

Fig. 15.11. Male genitalia and terminalia: a, penis of *Austrocyphon* sp. (modified from Zwick 2013c); b, sternite 9 of *Ypsiloncyphon brevis* Zwick (from Zwick 2014b); c, sternite 9 and genitalia of *Contacyphon mobula* (Zwick) (from Zwick 2013b); d, penis of *Peneveronatus macedonensis* (Armstrong) (from Watts 2009); e, genitalia of *Nothocyphon lanceolatus* Zwick (from Zwick 2015a); f, multilobed tegmen of *Pseudomicrocara serrata* Watts (from Watts 2007); g, genitalia of *Nanocyphon australicus* Zwick (from Zwick 2013b); h, genitalia of *Prionocyphon warra* Watts (from Zwick 2016). c, centema; eph, endophallus; fo, foramen; fpd, fused parameroids; p, penis; pa, pala; pd, parameroid; pe, paramere; pr, prostheme; S9, sternite 9; st, stylus; te, tegmen; tr, trigonium.

be visible even when the genitalia are contracted (Fig. 15.12h). None of these are considered to conform to the groundplan.

Female terminalia and genitalia (Figs 15.13–15.15)

The invaginated terminal segments 8 and 9 are weakly sclerotised. Tergite 8 has long freely projecting apodemes. Sternite 8 is membranous, with embedded lateral supporting rods (Fig. 15.13a, b). Tergite 9 and sternite 9 are assumed to be indistinguishably fused into an ovipositor (Fig. 15.13b) whose apodemes extend through the entire abdomen, or sometimes beyond. Subterminally, a small recurrent branchlet may support associated membranes (Fig. 15.13a). Caudally, a condylus of the apodeme articulates with an elongate gonocoxite which narrows caudally and supports a terminal tubular gonostyle with sensilla. Between the gonopods a delicate end plate of tergite 9 may be visible. Taxonomically useful variations concern the front end of the sternite 8 rods, the gonocoxites and the gonostyles.

Several Australian Scirtidae have modified ovipositors (Fig. 15.14a–d). In *Dasyscyphon* the gonocoxites are slender with very small terminal gonostyles furnished with long stiff bristles which suggests that particular sensory cues are required to oviposit, possibly into fine crevices (Fig. 15.14a). In *Heterocyphon* (Fig. 15.14b) and *Pseudomicrocara variegata* (Carter) the gonocoxae are robust and triangular with sub-terminal gonostyles. Oviposition is into decaying forest logs and into dense leaf-packs in shallow water respectively. The most extreme modification is shown in *Chameloscyphon* (Fig. 15.14c) and *Pachycyphon* (Fig. 15.14d) which have unusually thick and long apodemes supporting triangular, hard, shovel-like gonocoxites with rudimentary or absent gonostyles

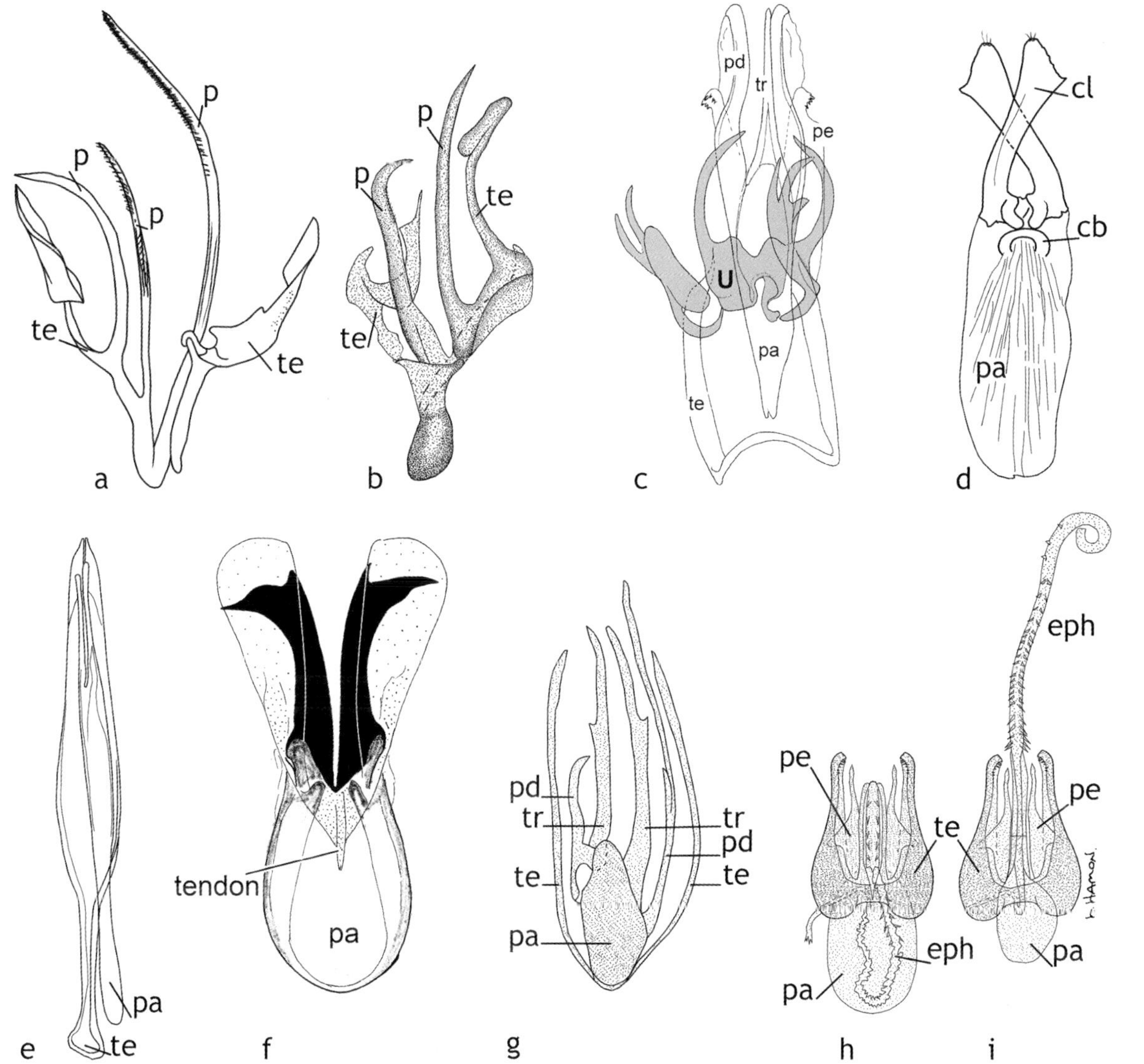

Fig. 15.12. Male genitalia: a, *Ora floccosa* Watts (from Watts 2004); b, *Scirtes albamaculatus* Watts; c, *Papuacyphon darwini* Zwick (from Zwick 2014b); d, *Calvarium australiense* Zwick (from Zwick 2014b); e, *Eximiocyphon excisus* Zwick (from Zwick 2015b); f, *Paracyphon avicularis* Zwick (from Zwick 2015b); g, *Petrocyphon monga* Watts (from Watts 2011); h, *Prionocyphon quasichameleon* Watts, with endophallus retracted; i, ditto with endophallus extended (from Watts 2010b). cb, common base; cl, clasper; eph, endophallus; mf, membranous flap; p, penis; pa, pala; pd, parameroid; pe, paramere; sh, sclerotised hook; te, tegmen; tr, trigonium; u, uncifer.

respectively. The larval habitat is unknown but the robustness of the ovipositor suggests ovipositing into relatively solid substrates.

The terminology of the female genitalia evolved successively (Nyholm 1948, 1969, 1972, 2002). In cleared preparations the mesodermal paired ovaries are lost. The terminal section of the genitalia, the vagina (= vulva), in some taxa contains a vulvar sclerite, at the level of the invaginated segments 8 and 9. The entry of the delicate cuticular common oviduct separates the vagina from the bursa in front. The bursa often houses a prehensor (Figs 15.13c, 15.15a–d) grasping the spermatophore during mating (Nyholm 1969). The anteriormost part of the bursa, the bursella, functions as a seminal receptacle. There is often also a very large accessory gland. Additional structures occur in some taxa, e.g. a comb organ in the oviduct of *Contacyphon* (Zwick 2013b), a network of sclerotised meshes ('dictyon', Zwick 2014b) in *Ypsiloncyphon*, and a pair of cuticular rings in some *Petrocyphon* (Zwick 2012), all of unknown function. At the entrance of the accessory gland there may be a complex bursal sclerite. Additional small glands and appendages may

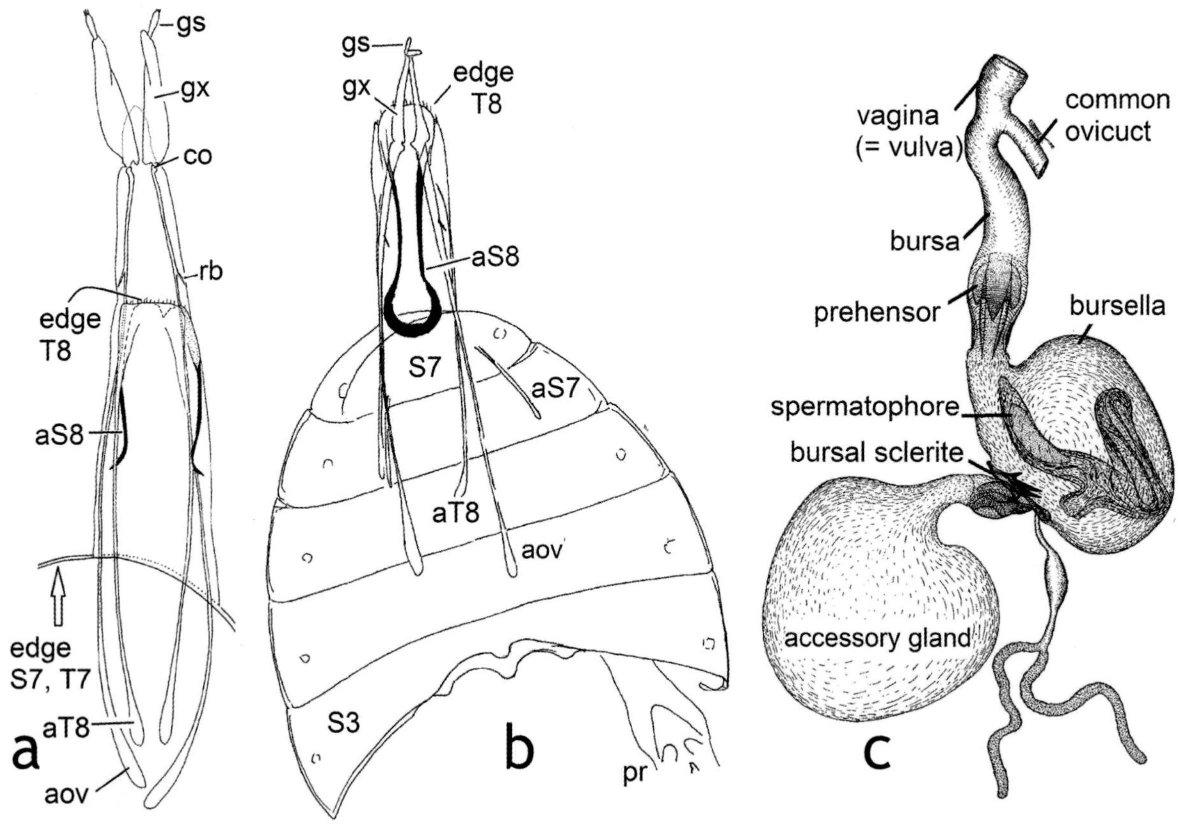

Fig. 15.13. Female terminalia and genitalia: a, *Austrocyphon* sp.; b, *Nothocyphon ypsilon* Zwick, cleared abdomina with overview of genitalia, apodemes of S8 shown in black; c, genital tract of *Scirtes haemisphaericus* (Linnaeus). (a and b, From Zwick (2013c, 2015a); c, modified (with permission) from Nyholm 2002.) aov, apodeme of ovipositor; aS7, apodeme of S7 (or T7, S8, T8, respectively); co, condylus; gs, gonostyle; gx, gonocoxite; pr, prehensor; rb, recurrent branchlet; S,T, sternite or tergite, respectively, with ordinal number; sp, sclerotised pocket; sr, sclerite ring, tp, tooth-bearing plate.

occur. All mentioned sclerites and the prehensors vary and are taxonomically important. In cleared preparations, prehensors and bursal sclerites can lie far in front in the abdomen or, being attached to long ducts, even in front of the abdomen.

No species has all of the sclerotised structures, some have none (e.g. *Austrocyphon* spp.), many possess only a prehensor. Prehensor and bursal sclerites may occur in combination in e.g. *Scirtes* (Fig. 15.13c).

Key to the genera of Australian Scirtinae based on adult characters

(using elements from the keys in Watts 2011 and Zwick 2016)

Abbreviations: BH = buttonhole; BL = body length (= combined length of pronotum and elytra, excluding head); SAR = supraantennal ridge; SGR = subgenal ridge (Fig. 15.2).

Key characters consider only Australian species and may not apply to exotic species of widespread genera, e.g. Contacyphon. For genera with a single Australian species its name is given; for others, the number of Australian species is given in the generic descriptions below.

1. Saltatorial; hind femora thick, tibial spurs unequal, one about as long as basal hind tarsal segment 2

– Unable to jump, legs unmodified, slender 4

2(1). Male tergite 7 with small projecting nipple. Male genitalia membranous, penis bag-shaped with an unpaired median process, no distinct parameroids; tegmen with lateral parameres plus a median appendage (e.g. Zwick 2011a: Fig. 15.29; appendage missing in Watts 2004: Fig. 15.15). Females with narrow tube-shaped prehensor, no large bursal sclerite .... ***Exochomoscirtes ruforotundus* (Watts)**

– Male tergite 7 without caudal nipple, rounded. Male genitalia sclerotised; tegmen with one or two large sclerotised parameres, no median appendage. Females with ill-defined prehensor and a large, toothed bursal sclerite 3

3(2). Hind coxae separated, or touching each other only in basal portion, posteriorly diverging (Ruta & Yoshitomi 2010). Articulation of hind femur exposed .... ***Ora* Clark**

– Hind coxae meeting along full length of median line (Ruta & Yoshitomi 2010). Metacoxa variable, an extension often concealing the articulation of the femur .... ***Scirtes* Illiger**

4(1). SAR short, ending where it meets the medial edge of the eye; no crest in front of eye; SAR and SGR not connected beneath the eye (Fig. 15.2a–b, e–f) 9

14
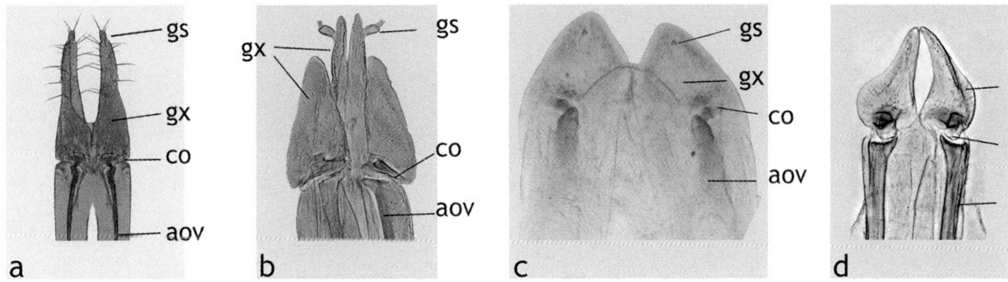

15
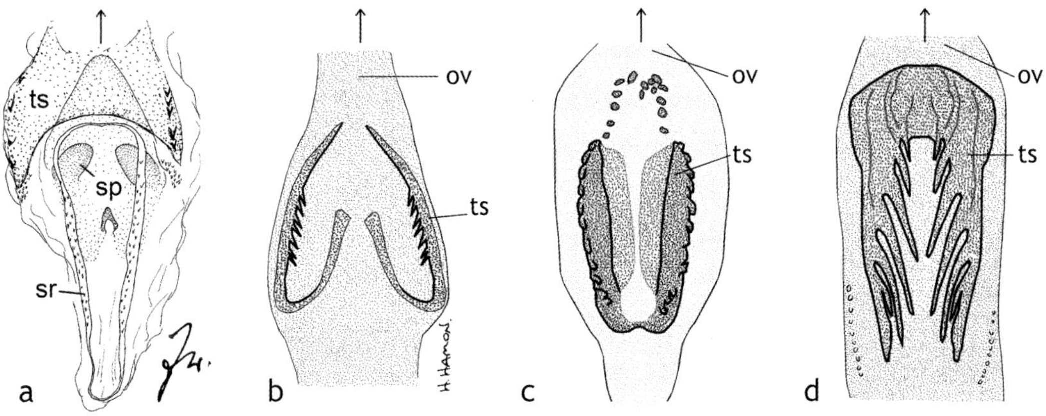

Figs 15.14–15.15. 14. Modified ovipositors: a, *Dasyscyphon hadrostiktos* Watts; b, *Heterocyphon cooloolaensis* Watts; c, *Chameloscyphon huonensis* Watts; d, *Pachycyphon serratus* Zwick. (From Watts 2009, 2011 and orig., Zwick.) aov, apodeme of ovipositor; co, condylus; gs, gonostyle; gx, gonocoxite.15. Prehensors in gonoducts: a, *Nothocyphon ypsilon* Zwick; b, *Chameloscyphon huonensis* Watts; c, *Eurycyphon aquilus* Watts; d, *Dasyscyphon victoriaensis* Watts. (From Zwick 2015a and Watts 2011.) ov, oviduct; sp, sclerotised pocket; sr, sclerite ring; ts, tooth-bearing sclerite; arrow, direction of vagina.

– SAR long, turning ventrally in front of eye and forming a crest separating the eye from the concave antennal sulcus; SAR and SGR connected beneath the eye (Fig. 15.2c–d) **5**

5(4). Stout convex beetles. Head transverse, fitted tightly into front of prothorax. Eyes of normal size. Mesoventral process short, not longer than wide. Mesoventral groove triangular, large, occupying half or more of length of mesoventrite **6**

– Elongate beetles (Fig. 15.1). Head hanging freely in front of prothorax. Eyes very small. Mesoventral process longer than wide. Mesoventral groove triangular, moderate size, occupying less than half length of mesoventrite **7**

6(5). Scape enlarged, a flat disc; pedicel cylindrical, a little shorter than scape. Gena visible only in ventral view, its medially projecting edge delimits a deep and narrow antennal groove (Fig. 15.2d). Ventrites 3–6 not connate, abdomen not rigid. Ovipositor unmodified (e.g. Fig. 15.13a). Both sexes fully winged .... ***Calvarium* Pic**

– Scape not enlarged, not disc-like, ovoid; pedicel similar but shorter. Gena freely visible from side, antennal groove wide, open. Ventrites 3–6 connate, abdomen rigid. Ovipositor fossorial, bacula very strong, gonocoxae triangular, shovel-like, without gonostyle (Fig. 15.14d). Females micropterous .... ***Pachycyphon* Zwick**

7(5). Mandibles strongly projecting. Pronotum with projecting front angles (Pl. 40J). Prosternal process behind coxae broad, apex broadly rounded. Mesepisternum rugose. Gonocoxites normal. Females fully winged **8**

– Mandibles not strongly projecting. Front angles of pronotum rounded. Prosternal process behind coxae broad with pointed apex. Mesepisternum smooth. Gonocoxites triangular, shovel-like, the minute gonostyle in a slot on underside of gonocoxite (Fig. 15.14c). Female micropterous .... ***Chameloscyphon huonensis* Watts**

8(7). Elytral punctures much larger than those on pronotum. Female ventrite 4 without transverse row of setae .... ***Dasyscyphon* Watts**

– Elytral punctures similar to those on pronotum. Female ventrite 4 with transverse row of setae .... ***Daploeuros* Watts**

9(4). Scape enlarged and disc-like, flat; pedicle much smaller than scape, cylindrical, inserting on lower face of flat scape or directly on its sharp edge (Fig. 15.3c); segment 3 smaller than pedicle, wider than long; antennae often at least partially serrate in males (e.g. Fig. 15.3b) **10**

– Scape normal size; pedicle and segment 3 not as above (Fig. 15.3a), male antennae not serrate **12**

10(9). Elytra with smooth surface without ribs. Front corners of clypeus extended forward in many species. Small, broadly oval .... ***Prionocyphon* Redtenbacher**

– Elytra with ribs or very uneven surface. Front corners of clypeus not expanded **11**

11(10). Elongate. Labial palpus with apical segment arising from inner margin of penultimate, broad, scoop-like (Fig. 15.5c). (Female unknown) .... ***Spaniosdascillus elongatus* Watts**

– Broad-bodied. Labial palpus unmodified .... ***Macrodascillus* Carter**

12(9). Body almost circular, no gap between elytra and pronotum. Elytral epipleura as wide as mesanepisternum, caudally abruptly narrowed **13**

– Never almost circular, a distinct gap between base of elytra and pronotum in dorsal view. Epipleura not widened **15**

13(12). Frontoclypeus anterolaterally extended around the antennal insertions. Prosternal process a thin blade not projecting beyond the front coxae, not reaching mesosternum, no receiving groove on mesoventrite. Mesoventral process a short triangle; mesocoxae partly contiguous .... ***Eximiocyphon excisus* Zwick**

– Antennal insertions not surrounded by frontoclypeus. Prosternal process, mesoventral groove, and mesoventral process well developed; mesocoxae separate **14**

14(13). Pronotum with blunt front angles (Pl. 1: 4), head partly visible from above. Elytral epipleura with longitudinal furrow .... ***Eurycyphon* Watts**

– Pronotum semicircular, no front angles, head concealed from above. Elytral epipleura flat .... ***Tectocyphon* Zwick**

15(12). SGR ending beneath eye, in side view with buttonhole (Fig. 15.2e–f). Apical segment of maxillary palpus twice as long as penultimate .... ***Cygnocyphon ibex* Zwick**

– SGR not ending beneath eye, without buttonhole. Apical and penultimate segments of maxillary palpus subequal in length **16**

16(15). Pronotal front angles distinct, square or projecting to variable degrees, front margin between angles straight or recessed. Body length usually ~3 mm or less **17**

– Pronotal front margin rounded, front angles obtuse, not projecting, or no angles at all; if with front angles extended then dorsal surface devoid of setae, shiny, impunctate or nearly so (some *Macrohelodes*). Size variable **26**

17(16). Maxillary and labial palpi with thickened penultimate and short conical terminal segments. Mandible with apical third abruptly narrowing and slender, tip sharply pointed, toothless (Fig. 15.4c) **18**

– Penultimate palpus segments slender (Fig. 15.5a), last segment of labial palpus bean-shaped, often standing at an angle to penultimate one. Mandible not abruptly narrowing, bluntly pointed, often toothed **19**

18(17). Mesoventral groove wide, almost triangular; mesoventral process widening caudally, truncate. Subgenal ridge without a buttonhole beneath eye. Male: tegmen and penis separate, both and also S9 with narrow handle-like front end (Fig. 15.11b); female with network of sclerotised meshes (dictyon) on bursella .... ***Ypsiloncyphon* Klausnitzer**

– Mesoventral groove short and transverse between two tiny cones; mesoventral process parallel, end finely incised. Subgenal ridge forms a buttonhole beneath eye (resembling Fig. 15.2e–f). Male: tegmen and penis connected in a functional unit with spine-like trigonium (Fig. 15.11g); female without dictyon .... ***Nanocyphon* Zwick**

19(17). Prosternal process reduced, narrow, parallel sided; mesosternal groove indistinct (WA, NT) .... ***Leptocyphon* Zwick**

– Prosternal process well developed, broadening caudally; mesosternal groove well developed **20**

20(19). SGR with buttonhole, ridge ending beneath eye (e.g. Fig. 15.2e–f). Mesosternal groove rhomboid **21**

– SGR without buttonhole, front end of ridge connects to edge of oral cavity. Mesosternal groove elongate, U-shaped. Body shape oval to elongate oval. (*Study of male terminalia required in couplets 22–25*) **22**

21(20). Slender, elytra parallel-sided, wider than the transverse pronotum. Head often projecting and with bulging eyes (Pl. 40F). Male: several species with prolonged and thickened antennae. Sternite 8 present; penis with trigonium and parameroids (Fig. 15.12g), often asymmetrical .... ***Petrocyphon* Watts**

– Body usually oval, elytra not strikingly wider than pronotum. Head bent against fore coxae, eyes not strongly projecting, antennae not sexually dimorphic. Male: without sternite 8, penis without trigonium, with paired prosthemes (Fig. 15.11c) .... ***Contacyphon* de Gozis**

22(20). Male terminalia conform with groundplan of Scirtinae (p. 226), all elements fully developed, no additional structures present. However, character expressions, e.g. presence of spines, varies a lot between species .... ***Nothocyphon* Zwick**

– Male terminalia differ from groundplan of Scirtinae in various ways such as transformation or absence of standard components, or presence of additional elements **23**

23(22). Penis sac-shaped, no trigonium, with pair of divergent double-layered appendages (Fig. 15.12f) .... ***Paracyphon avicularis* Zwick**

– Penis not sac-like, with median trigonium, without additional appendages **24**

24(23). Stout convex beetles. Male genitalia with a complex structure (uncifer) comprising very large hooks located above sternite 9, tergites 8 and 9 unmodified, with fully developed plates, parameroids separate, unmodified (Fig. 15.12c) .... ***Papuacyphon darwini* Zwick**

– Oval to elongate-oval relatively flat beetles. Male genitalia with no hook-bearing uncifer; plate of male tergite 9 variously modified and partly reduced. Penis slender, trigonium distinct, parameroids fused, forming an elongate median shield over the trigonium **25**

25(24). Mandible toothed. Prosternal process caudally widened. Tergite 9 complex, at least partly sclerotised. Tegmen narrowly V-shaped, parameres are simple thin struts. Penis simple; trigonium a finger with nail-like centema in front of a central pale window in the fused parameroids (Fig. 15.11a) .... ***Austrocyphon* Zwick**

– Mandible toothless. Prosternal process blade-like, tergite 9 membranous. Tegmen and parameres well developed. Trigonium and parameroids of complex shape .... ***Tasmanocyphon heideae* Zwick**

26(16). Dorsal surface of elytra and pronotum without setae .... ***Macrohelodes* Blackburn**

– Pronotum and elytra with setae (except ♀ *Pseudomicrocara hangayi* Watts) **27**

27(26). Without dorsal pupillate punctures, without pronotal grooves **28**

– Pupillate punctures present on dorsal surface of anterior body, may have pronotal grooves **29**

28(27). Pronotal base about as wide as base of elytra, surface smooth .... ***Pseudomicrocara* Armstrong**

– Pronotum at base ~15% narrower than elytra, surface flat, wrinkled .... ***Peneveronatus* Armstrong**

29(27). Base of pronotum with paired grooves. Gonocoxae strongly sclerotised, styli small, subapical. Trigonium completely or partially bifid .... ***Heterocyphon* Armstrong**

– Base of pronotum without paired grooves. Gonocoxae normal. Trigonium simple, not bifid .... ***Accolabass* Watts**

Key to the genera of Australian Scirtidae based on last instar[2] larval characters

(Modified from Watts 2014).
* Potentially new genera (2018)

1(0). Live in wet logs on forest floor (saproxylic); white to pale testaceous; antenna short with 14–30 annuli, not reaching beyond prothorax; eyes vestigial (Pl. 41B) **18**
– Do not live within wet logs on forest floor; generally dark coloured; generally with longer antennae, with 16–150 annuli; with normal sized eyes **2**
2(1). The pair of long apical setae on tergite 9 close together (Fig. 15.8h), often with tergite 9 narrowly extended; maxillary palpus with three segments; body less than 7 mm long **3**
– The pair of long apical setae on tergite 9 more separated (Fig. 15.8g), tergite 9 either not extended or broadly so; maxillary palpus four segmented (segment 4 can be small; three segments in *Scirtes auratus*); body length variable **5**
3(2). Dorsal setae on abdominal segments short, less than 1/4 width of segment; antenna relatively short, flagellum with 15–30 annuli .... ***Nothocyphon frater*** **(Blackburn) and related species**
– Dorsal setae on abdominal segments sparse and long, most more than 1/2 width of segment; antenna usually longer, flagellum with 16–80 annuli **4**
4(3). Tergite 9 abruptly extended into a long narrow portion .... ***Nothocyphon lindensis*** **(Blackburn) and related species**
– Tergite 9 not so strongly extended .... ***Austrocyphon*** **Zwick**
5(2) Pretarsus with two parallel rows of ~10 stout setae .... ***Prionocyphon lamingtonensis*** **Watts***
– Pretarsus with 2–4 stout setae **6**
6(5). Socket and keel bristles on hypopharynx separate (Fig. 15.7a) **7**
– Socket and keel bristles on hypopharynx with bases touching **8**
7(6). Body long, cylindrical, strongly punctate; hypopharynx with 20–30 pairs of claw teeth, stiff plate setae strong and numerous .... ***Peneveronatus*** **Armstrong**
– Body flatter, sparsely punctate; hypopharynx with 11–12 pairs of claw teeth, stiff plate setae absent or very fine .... ***Pseudomicrocara variegata*** **(Carter) ***
8(6). Clypeolabrum with bases of lobes nearly touching, spine and inner setae concentrated in a group at apex of lobes **9**
– Clypeolabrum lobes with bases well separated, inner setae spread out along inner edge of lobes **10**
9(8). Mandible with distinct tip .... ***Exochomoscirtes ruforotundus*** **(Watts)**
– Mandible with small apical tubercle .... ***Prionocyphon niger*** **Kitching and Allsopp**
10(8). Clypeolabrum without palisade setae; dorsal setae long but sparse .... ***Contacyphon mobula*** **(Zwick)**
– Clypeolabrum with well developed palisade setae; dorsal setae variable but never long and sparse **11**
11(10). Tergite 9 broadly extended backwards, most dorsal setae strong, dorsal shape tending spindle-shaped **14**
– Tergite 9 not so extended; most dorsal setae short, fine, relatively dense, body shape tending parallel-sided **12**
12(11). Tip of mandible not bifid in last instar **13**
– Tip of mandible in last instar bifid .... ***Pseudomicrocara*** **Armstrong, in part**
13(12). Length more than 10 mm; maxillary palpus with segment 4 less than 1/4 length of segment 3; clypeolabrum with five inner setae; tuft of long setae at base of stiff plate on hypopharynx; live in ponds in or beside creek beds .... ***Pseudomicrocara davidsoni*** **(Carter)***
– Length 5–8 mm long; maxillary palpus with segment 4 more than 1/4 length of segment 3; clypeolabrum with seven inner setae; stiff plate on hypopharynx with dense band of small setae inwards from claw teeth; live in phytotelmata .... ***Prionocyphon kurandaensis*** **Watts and related species**
14(11). Maxillary palpus with segment 4 as long as segment 3; tip of mandible blunt (Yoshitomi 2005) .... ***Ora*** **Clark**
– Maxillary palpus with segment 4 less than 1/4 as long as segment 3; tip of mandible sharply pointed **15**
15(14). Dorsal abdominal setae peg-like, dorsal cuticle scale-like .... ***Scirtes microrotundus*** **Watts**
– Dorsal abdominal setae thin, pointed, dorsal cuticle not scale-like or with very slight scales **16**
16(15). Tip of mandible bifid or multipointed **17**
– Tip of mandible a single point .... ***Scirtes*** **Illiger, in part**
17(16). Tip of mandible bifid .... ***Scirtes emmae*** **Watts,** ***S. kaytae*** **Watts,** ***S. tindaleensis*** **Watts**
– Tip of mandible multipointed (resembles Fig. 15.7g) .... ***Scirtes auratus*** **Watts**
18(1). Mandible antler-like .... ***Macrodascillus scalaris*** **(Lea)**
– Mandible simple or with a tooth on inner margin **19**
19(18). Mandible elongate, projecting well beyond front of head; clypeolabrum with lobes virtually lacking, spine and inner setae undifferentiated .... ***Prionocyphon insolitus*** **Watts ***
– Mandible broader, apex not projecting well beyond front of head; clypeolabrum with moderately developed lobes, spine and inner setae distinct **20**
20(19). Apex of mandible bifid; lacinia teeth very strong; hypopharynx with bifid socket bristles .... ***Heterocyphon australis*** **(Erichson)**
– Apex of mandible simple; lacinia teeth relatively weak; hypopharynx with bristle teeth not bifid **21**
21(20). Top of head with a pair of distinct horns; keel bristles long, curved .... **? genus**
– Head without horns; keel bristles short, peg-like .... ***Pseudomicrocara thunguttiensis*** **Watts***

Classification of the Australian genera

Subfamily Stenocyphoninae Lawrence and Yoshitomi, 2007

Type genus and species. *Stenocyphon sasajii* Lawrence, 2001.

Characteristics. Frontal clypeal suture distinctly impressed; subgenal ridge distinct; apex of mandible bidentate; lacinia with uncus; protrochantin large, quadrate and broadly exposed; hind wing with four free veins in medial field; outer edge of tibia without longitudinal carinae; aedeagus laterally compressed with distinct phallobase and articulated parameroids.

Australian species. There is one undescribed species in Australia (TAS).

Biology. Two female specimens, possibly in the genus *Stenocyphon,* are known from temperate closed forest dominated by *Nothofagus* in south-western Tasmania. Apart from this nothing is known of their biology.

Notes. The subfamily has a Gondwanian/transantarctic distribution, occurring in Tasmania, Chile (*S. sasajii*) and New Zealand (*S. neozealandicus* Ruta *et al.* 2011).

References. Lawrence (2001); Ruta *et al.* (2011); Watts (2011).

Subfamily Scirtinae Fleming, 1821

Type genus. *Scirtes* Illiger, 1807.

Characteristics. Frontoclypeal suture vaguely impressed or absent; subgenal ridge distinct; apex of mandible unidentate; lacinia without uncus; protrochantin slender and usually more or less concealed by strongly declined head; elytra irregularly punctate; hind wing with three or fewer veins in medial field; outer edges of tibiae with paired longitudinal carinae; aedeagus dorsoventrally flattened, without phallobase, parameroids fixed or absent.

Accolabass Watts, 2009 (Pl. 40G)

Type species. *Helodes maculatus* Waterhouse, 1877.

Characteristics. Small to medium size. Body elongate, in some species semicylindrical. Eyes some distance from subgenal ridge. Antenna simple. Subgenal ridge unmodified. Labial palpus unmodified. Mandibles stout, with one large and several smaller teeth, mola with small area of small spines. Pronotum slightly narrower than elytral base, 1.5 × as wide as long, anterolateral angles not projecting forwards, posterior margin without pits, epipleura strongly bounded inside by raised carina. Punctures on dorsal and ventral surfaces rather small, dense, often rugose. Apical portion of prosternal process relatively broad, spatulate. Male terminalia conform to Scirtinae groundplan: without or with poorly developed stylus. Ovipositor with well sclerotised, elongate gonocoxites, gonostyli small, terminal; gonoduct without sclerites. Sternite 6 in females often with transverse row of long setae associated with a glandular area. Length 4.0–8.0mm.

Australian species. Six.

Distribution. TAS, southern VIC, south-west WA.

Biology. Most species are common. The single Victorian species is confined to wet forest areas of the Dandenong Ranges. The single WA species (and specimen) came from an area of sand-plain south of Perth. Tasmanian species have been most commonly collected from flowering shrubs in subalpine areas in the south-west. In several Tasmanian species only males are known to visit flowers; with both sexes collected by other means such as in flight intercept traps. Despite the abundance of adults the larvae are unknown, suggesting that they may be semi-terrestrial in wet soil/logs etc. rather than in shallow water bodies.

Notes. Endemic to Australia. Watts (2009) placed the species into three groups on morphological characters. This arrangement was consistent with sequence data in Cooper *et al.* (2014). Sequence data (Cooper *et al.* 2014) clearly places the species *Pseudomicrocara rufusensis* Watts in this genus.

References. Watts (2009).

Key to species. Watts (2009).

Austrocyphon Zwick, 2013 (Pl. 40P, 41B, larva)

Type species. *Austrocyphon charon* Zwick, 2013.

Characteristics. Small size; habitus and overall structure not distinctive. Dorsally light to dark brown, a few with patterned elytra. Most species are oval and relatively flat, not highly domed. Eyes large, in side view extending down to subgenal ridge or almost so; subgenal ridge unmodified. Head fitted into slightly recessed front of pronotum. Antennae slender, unmodified. Mandibles short, weakly toothed. Palpi unmodified; apical segment of maxillary palpus tapering, slightly concave; labial palpus bent, with bean-shaped terminal segment. Pronotum with blunt anterolateral angles projecting a little, sides curved, the rounded base a little narrower than base of elytra. Prosternal process drop-shaped, slender. Ventral side of pterothorax and abdomen unmodified. Aedeagus (Fig. 15.11a) and T9 greatly modified, T8 rarely modified. Ovipositor unmodified, gonoduct without sclerites. Length 1.8–3.2 mm.

Australian species. 39.

Distribution. All Australian states and mainland territories. One species shared with New Guinea.

Biology. Adults are riparian, living in riparian vegetation often together with species of *Nothocyphon*. They can be common on flowering shrubs close to water suggesting that they might feed on parts of flowers. Larvae live among detritus and stones at the sides of creeks and streams particularly in places with running water. Pupae have been found beneath mats of drying algae on rocks in streams.

Notes. Males are easily recognised and identified to species by the modified penis and T9 (see key). In contrast females have simple unmodified genitalia and can usually not even be distinguished from other genera with similar generalised habitus, e.g. *Nothocyphon*. Cooper *et al.* (2014) showed that the 17 taxa that they studied formed a monophyletic group sister to some *Nothocyphon* species.

Reference. Zwick (2013c, 2013d).

Key to species. Zwick (2013c).

Calvarium Pic, 1918

Type species. *Calvarium maxi* Pic, 1918 (designated by Ruta 2010).

Characteristics. Small size. The nominate subgenus is oval and fairly convex, subgenus *Calvariellum* Zwick, 2014b is smaller (BL 1.7–2.2 mm), more slender, some specimens almost parallel-sided, and less convex. Dorsally brownish, sometimes with lighter edges, no distinctive patterns. Subantennal groove on the underside of the flat, deflexed head, deep and narrow. Antennal scape large, flat, pedicle and 3rd segment large, slender. Labial palpus forked in the nominate subgenus, less strongly modified in *Calvariellum*. Mandible symmetrical, slender with a large subterminal tooth, and bare molar area. Mesoventral process short, barely longer than wide. Male terminalia very different between the subgenera: *C.* (*Calvarium*) with paired caudal processes articulating on a common cupule-like sclerite (Fig. 15.12d); homology of components doubtful. Male terminalia of *Calvariellum* agree with the scirtid groundplan (Zwick 2014b). Ovipositor unmodified, gonoduct with variably developed prehensor. Length 2.3–4.5 mm.

Australian species. *Calvarium* (*Calvarium*), 2; *Calvarium* (*Calvariellum*), 4.

Distribution. *Calvarium* (*Calvarium*) is widespread in Africa and SE Asia, the two Australian species occur on near-coastal islands of NT. Species of *C.* (*Calvariellum*) live in tropical QLD, one extending south to ~27°S.

Biology. One specimen of *Calvarium* (s.str.) was collected from *Podocarpus*, there is no information on the other three individuals known. Most *Calvariellum* seem to have been taken near streams, with Malaise or light traps, some by sweeping. Adults of *C. (Calvariellum) hamifer* Zwick have been collected from riparian vegetation in riverine tropical rainforest (CHSW pers obs.). The larvae have not yet been found but are presumed to live in the streams close to where adults are found.

Notes. The genus is best identified by the deep, narrow subantennal groove, and by the short mesoventral process.

References. Ruta (2010); Zwick (2014b).

Key to species. Zwick (2014b).

Chameloscyphon Watts, 2011 (Pl. 40L)

Type species. *Chameloscyphon huonensis* Watts, 2011.

Characteristics. Small size. Elongate; strongly punctured. Eyes small; large distance from bottom edge of eye to subgenal ridge. Antenna simple. Subantennal groove moderately well developed, bounded by raised carina. Mandibles each with a strong tooth. Apical segment of labial palpus arising from inner portion of apical edge of penultimate segment. Pronotum moderately flanged, anterolateral angles rounded, posterior edge in females with two large excavations. Male terminalia conform to Scirtinae groundplan: aedeagus with very broad trigonium, its posterior angles each with a downwardly curved, spine-like hook; without stylus. Females micropterous; disc of pronotum deeply excavated either side of midline posteriorly, excavations open posteriorly; gonocoxites broad; gonostyli vestigial, subapical (Fig. 15.14c); gonoduct with moderately developed, spinose, bursal sclerite. Length 3.0–4.4 mm.

Australian species. One.

Distribution. South-west TAS.

Biology. The only known species occurs in a limited area of wet Eucalypt forest in south-west Tasmania. The females are almost apterous and have only been collected in pitfalls on the forest floor. The larvae are unknown. Watts (2011) suggested that they might live in wet soil – an atypical habitat for Scirtid larvae.

Notes. Endemic to Australia. Cooper *et al.* (2014), using molecular sequencing, grouped this genus distantly with *Daploeuros,* another genus from wet forest with flightless females.

References. Watts (2011); Cooper *et al.* (2014).

Contacyphon Gozis, 1886

Cyphon Paykull, 1799, a junior synonym of *Elodes* Latreille, 1796.

Type species. *Cantharis variabilis* Thunberg, 1787.

Characteristics. Small size, mostly relatively flat, oval, brownish beetles without colour pattern. Subgenal ridge with distinct buttonhole. Mandibles asymmetrical, only right one toothed, both toothless in the tiny *C. forcipatus* Zwick. Labial palpus curved, middle segment slightly thicker than the others. Mesoventral groove transverse, rhomboid, surrounded by little ridges. Males lack a sclerotised S8, penis strongly modified, without trigonium, trigonium replaced by paired caudal appendages (prosthemes) (Fig. 15.11c). Ovipositor unmodified; gonoduct with a large prehensor and a dense patch of hollow, curved, hair-like outgrowths of unknown function – the comb organ. Length 1.7–3.5 mm.

Australian species. Four indigenous and one apparently introduced.

Distribution. Cosmopolitan, with hundreds of species (Zwick *et al.* 2013). Native Australian species in NSW, QLD (*C. mobuloides* (Zwick) shared with New Guinea), one introduced species (*C. putonii* (Brisout)) in WA.

Biology. The tropical species were taken in light traps near shallow rivers or swamps. The specimen from temperate NSW was found on *Alphitonia* blossoms. The pale coloured larvae of *Contacyphon mobula* (Zwick) have been collected within mats of dead vegetation in shallow areas at the edge of swamps (CHSW, pers. obs.).

Notes. The genitalia of both sexes are very diverse and provide excellent specific characters. By its location in the genital tract, the comb organ, which is typical of the genus (Zwick 2013b) may be homologous to the bursal sclerites. The indigenous Australian *Contacyphon mobula, C. mobuloides,* and *C. reductus* (Zwick) are distinguished by large horn-like caudal outgrowths of sternite 9 and a delicate Y-shaped penis (Fig. 15.11c). They have close relatives in Irian Jaya. However *C. forcipatus* Zwick from Queensland has no known close relatives. How the apparently vital population of the European *C. putonii* got to WA is unknown.

References. Zwick (2013b, 2015b).

Cygnocyphon Zwick, 2015

Type species. *Cygnocyphon ibex* Zwick, 2015.

Characteristics. Slender, ~2 mm long, head with buttonhole configuration, mandibles with three teeth. The maxillary palpus has a long outwardly curved end segment; labial palpus straight, unmodified. Of the antenna only the unmodified scape and pedicel known. Immediately recognised by the short and stout cupule-like trigonium between long outwardly curved parameroids resembling goat's horns.

Australian species. One.

Distribution. WA.

Biology. Unknown.

Notes. Known from a single historical specimen taken near the Swan River by A. M. Lea. Endemic to Australia, affinities doubtful, no similar species known world-wide.

Reference. Zwick (2015b).

Daploeuros Watts, 2011 (Pl. 40J)

Type species. *Daploeuros lamingtonensis* Watts, 2011.

Characteristics. Large size, elongate, robust, strongly punctured. Eyes small: large distance from bottom edge of eye to subgenal ridge. Antenna simple. Subantennal groove well developed, bounded by raised carina. Mandibles each with a strong tooth. Apical segment of labial palpus arising from centre of apical edge of penultimate segment. Pronotum with anterolateral angles strongly produced forward. Male terminalia conform to Scirtinae groundplan; penis short, broad, trigonium consisting of two curved parts; without stylus. Females micropterous, sternite 6 with row of long setae near posterior margin; gonocoxite elongate; gonostylus apical; gonoduct with well developed, multipronged prehensor. Length 7.0–11.0 mm.

Australian species. Four.

Biology. A little known genus from wet forests of south-east Australia and Tasmania. Females of *Daploeuros tasmanicus* are micropterous and it is expected that those of the other three species are also. The larvae are unknown but both male and female specimens have been collected from log emergence traps in south-west Tasmania suggesting that the larvae are saproxylic (Yee *et al.* 2006).

Distribution. South-east QLD, Coastal NSW, TAS.

Notes. Endemic to Australia. Cooper *et al.* (2014), using sequence data, placed *Daploeuros lamingtonensis* distantly with *Chameloscyphon huonensis*.

References. Watts (2011); Cooper *et al.* (2014).

Dasyscyphon Watts, 2011

Type species. *Dasyscyphon hadrostiktos* Watts, 2011.

Characteristics. Medium size, robust, setose, punctures strong, legs short, stout. Head recumbent; eyes small; distance from bottom edge of eye to subgenal ridge large (Fig. 15.6b). Antenna simple. Subantennal groove well developed, margined above by a well developed supra-antennal ridge and posteriorly by a strong subgenal ridge. Mandibles stout, each with one strong and two to three weaker teeth. Labial palpus with broad penultimate segment; apical segment arising from centre of apical edge of penultimate segment in *D. victoriaensis* Watts, and from the inner corner of a triangularly shaped, penultimate segment in *D. hadrostiktos*. Anterolateral angles of pronotum projecting weakly forward. Apex of pronotal process broad. Male terminalia conform to Scirtinae groundplan; without styli. Females micropterous, elongate; gonocoxite thin, well sclerotised: gonostylus very short, apical (Fig. 15.14a), gonoduct with well developed multipronged prehensor. Length 3.0–4.5 mm.

Australian species. Two.

Distribution. TAS, southern VIC.

Biology. The collection of flightless females from wet forest litter suggests that the habitat of the species is the ground of wet forests. The larvae are unknown.

Notes. Endemic to Australia. The small eyes, strong subantennal groove, flightless females with well developed prehensor, suggest a phylogenetic relationship with *Chameloscyphon* and *Daploeuros*. No sequence data is available.

References. Watts (2011).

Key to species. Watts (2011).

Eurycyphon Watts, 2011 (Pl. 40B)

Type species. *Eurycyphon fulvus* Watts, 2011.

Characteristics. Small to medium size, broadly oval, flattish, weakly punctate. Eyes of moderate size. Antenna simple. Mandible simple. Labial palpus with apical segment arising from inside apical corner of penultimate segment. Pronotum considerably narrower than elytra; anterolateral angles rounded. Elytral epipleura very wide in anterior half, as wide as pronotal epipleura, grooved with a strongly raised basal carina. Genital characters variable (see Notes). Length 3.3–5.0 mm.

Australian species. 10.

Distribution. From southern VIC north through NSW into tropical N.QLD.

Biology. The two originally described species were collected together in an area of *Nothofagus* forest near Warburton, Victoria. Specimens of several other species were also taken beside creeks or rivers. The larvae are unknown.

Notes. Endemic to Australia. The genus is not homogenous. In males of *E. fulvus* and five close relatives (from S.VIC to N.QLD) the pala is much longer than the distal portion of the penis, and the claw-shaped parameroids are shorter than the band-shaped or almost spherical trigonium between them. In addition, the parameres have a weakly developed appendage, and their tips are complex and mostly convoluted. Females in the group (those of only two species known) have a large conical projection on the apex of sternite 7, an unmodified ovipositor, and the gonoduct has a well developed prehensor.

A group of four other species are only 1.9–3.5 mm long, have less complex male genitalia, except for the appendages to the parameres which are more strongly developed. The two known females of these species have unmodified sternites but well developed prehensors.

Cooper *et al.* (2014) suggest a distant phylogenetic relationship (of *E. aquilus* Watts and *E. fulvus*) with some Australian species of 'typical Australian *Prionocyphon*' – as defined by Zwick (2015b). (See later under *Prionocyphon*)

References. Watts (2011); Zwick (2015b).

Key to species. Watts (2011); Zwick (2015b).

Eximiocyphon Zwick, 2015

Type species. *Eximiocyphon excisus* Zwick, 2015.

Characteristics. Stoutly oval and domed, with slender walking legs. Frontoclypeus anterolaterally extended around the base of the slender, unmodified antenna. The sharply pointed mandible is toothless, the maxillary palpus unmodified, the labial palpus forked although the lateral prolongation of segment 2 is short (resembling Fig. 15.5b). Pronotum with rounded front, no angles. Mesocoxae partially contiguous. S7 with a deep semicircular notch. Tegmen and penis are superimposed and form a cigar-shaped unit about half as long as the beetle (Fig. 15.12e). Female unknown. Length ~2 mm.

Australian species. One.

Distribution. NE QLD, the single specimen was taken near a stream at 1150m on the Carbine Tableland.

Biology. Unknown.

Notes. Endemic to Australia. Several characters are reminiscent of *Scirtes* species but the walking legs speak against a close relationship. The genitalia are unique. No similar species is known in Australia or elsewhere. The species name refers to the deep notch on S7.

Reference. Zwick (2015b).

Exochomoscirtes Pic, 1916 (Pl. 40A)

Type species. *Exochomoscirtes binotatus* Pic, 1916 (designated by Ruta & Yoshitomi 2010).

Characteristics (Australian species). Medium size, broad, almost circular, dorsal surface red or red and black. Maxillary palpus swollen. Saltorial hind legs, femur incrassate, tibia with long unequal spurs. Hind coxae meeting along full length of median line, articulation of hind trochanters covered. Male tergite 7 with projecting nipple; penis small with oval pala and narrow filiform parameroids; tegmen trilobed with long thin parameres. Ovipositor unmodified; gonoduct with both a sclerotised tube-shaped prehensor and a weak bursal sclerite. Length 3.2–5.0 mm.

Australian species. One: *Exochomoscirtes ruforotundus* (Watts 2007).

Distribution. QLD, NT, PNG. SE Asia.

Biology. Sparsely distributed in tropical open woodland including orchards. Larvae found in phytotelmata often in dense concentrations.

Notes. The genus is widespread in South-east Asia and includes many vividly coloured and variously patterned species. The sole Australian representative belongs to an Oriental group but differs from its closest relatives most distinctly in the shape of sternite 9. The saltorial hind legs ally it taxonomically with the genera *Scirtes* and *Ora*, a position confirmed by sequence data (Cooper *et al.* 2014).

References. Watts (2004); Ruta & Yoshitomi (2010); Zwick (2011a).

Heterocyphon Armstrong, 1953 (Pl. 40K)

Type species. *Cyphon australis* Erichson, 1842.

Characteristics. Medium to large size, elongate; uniformly testaceous; setae short and dense. Antenna simple. Eyes large; bottom edge of eye close to subgenal ridge (resembling Fig. 15.6a). Mandibles stout with three to four teeth. Labial palpus straight. Pronotum flanged, anterolateral angles rounded, posterior margin with two well marked basal pits. Male terminalia conform to Scirtinae groundplan; aedeagus relatively complex, trigonium varying from deeply forked to totally bilobed; tegmen usually deeply bilobed, lobes often with serrated edges; with or without stylus. Female with relatively large rectangular or oval area of long setae (? glandular) towards posterior margin of sternite 6. Gonocoxites broadly triangular; gonostyli small, arising from side near to but not at apex of gonocoxites (Fig. 15.14b); apical portions of sternites 8 and 9 often strongly sclerotised; gonoduct with weakly sclerotised prehensor. Length 5.0–9.0 mm.

Australian species. 14.

Distribution. South-west WA, Qld, NSW, VIC, TAS.

Biology. Specimens can be abundant in patches of *Nothofagus* rainforest in eastern Australia with many individual species seemingly restricted to a separate isolated area of wet forest. Most specimens, predominantly males, have been collected at light traps. More even sex ratios have been collected from malaise traps and beating vegetation such as tree ferns. The larvae of the Tasmanian *Heterocyphon tasmanicus* Watts are known from the interior of water-saturated forest logs. Pupation occurs in areas of damp friable wood (Watts 2014).

Notes. Endemic to Australia. Sequence data (Cooper *et al.* 2014) shows *Heterocyphon* to be monophyletic with the exception of *H. spencei* (Armstrong) and to be a member of a large group of taxa loosely related to *Pseudomicrocara.* Individual species are best identified by characters of the male genitalia.

References. Watts (2009, 2014); Cooper *et al.* (2014).

Key to species. Watts (2009).

Leptocyphon Zwick, 2015 (Pl. 40H)

Type species. *Leptocyphon furcalonga* Zwick, 2015.

Characteristics. Parallel sided, relatively small, slender beetles, habitus resembling some small *Pseudomicrocara* (in the wide sense). No buttonhole configuration, head unmodified, antenna and palpi straight. Right mandible with small tooth. Male terminalia of the type species conform to the Scirtinae groundplan. The penis and tegmen vary considerably between species. Female (only that of *L. furcalonga* known): ovipositor rigidly attached to bacula; gonocoxites slender, triangular; gonostyles small, terminal, wart-like; gonoduct with prehensor. Length 2.5–3.5 mm.

Australian species. Three.

Distribution. WA, NT.

Biology. *Leptocyphon furcalonga* can be common in spring on flowering bushes close to small semi-permanent streams with marginal swampy areas in forested areas of south-west Western Australia. The larvae live in cracks and crevises around the water-line in rotting part-submerged logs in creeks and swampy areas near where the adults are collected. Pupation takes place further up the log. No cell is built. The little known Northern Territory *L. abnormis* Zwick is known from similar habitat near Darwin.

Notes. Endemic to Australia. The genus was established because of numerous differences between the two species from WA and the type species of *Pseudomicrocara,* e.g. the very narrow prosternal process, weak or absent mesosternal groove and exceptional male and female genitalia. However, the included species also differed between themselves. Doubts about the monophyly of *Leptocyphon* increased with the discovery of the NT species. Morphological and preliminary sequence evidence (Watts pers. obs.) places *L. furcalonga* and *L. abnormis* as distant from each other and from other Australian Scirtids but probably within the *Pseudomicrocara* group (Cooper *et al.* (2014). The larva of *L. furcalonga* is unique among species living in free water in having greatly reduced apical spiracles and associated structures. Both this genus and *Pseudomicrocara* appear to be polyphyletic and require revision.

References. Zwick (2015b, 2016).

Key to species. Zwick (2015b) (WA species only).

Macrodascillus Carter, 1935 (Pl. 40N, 41A, larva)

Type species. *Helodes scalaris* Lea, 1895 (= *Macrodascillus denticornis* Carter, 1935)

Characteristics. Large size, elongate oval. Eyes large. Antenna with enlarged scape and segments 2 & 3 very small, other segments serrate. Mandibles stout, each with one strong tooth. Pronotum with anterolateral angles projecting strongly forwards. Labial palpus with apical segment arising from inner apical angle of penultimate segment; apical segment broader and sucker-like in *M. pannuceus* Watts. Male genitalia conform to Scirtinae groundplan; trigonium waisted, caudal angles with small sharp extensions; with styles. Gonocoxite elongate; gonostylus terminal, gonoduct with prehensor. Length 5.0–12.0 mm.

Australian species. Two.

Distribution. QLD, NSW, VIC.

Biology. Uncommon. Restricted to wet sclerophyll forests of eastern Australia. Larvae of *M. scalaris* are saproxylic living in the small amount of water that collects in old insect burrows etc. in water-saturated Eucalypt logs on the forest floor.

Notes. Endemic to Australia. The enlarged scape, small pedicle and small segment 3 of the antenna somewhat resemble *Prionocyphon* spp. some of which also have a serrate flagellum. However, the scape lacks the narrow front edge of *Prionocyphon* and *Macrodascillus* also lacks the heterogeneous setation on the abdomen, a synapomorphy found in most Australian *Prionocyphon* species (Zwick 2015b). Sequence data (Cooper *et al.* 2014) suggests a relationship with the morphologically atypical *Prionocyphon*; *P. insolitus* and *P. lamingtonensis.* The strength of this relationship requires further study.

References. Watts (2011, 2014).

Key to species. Watts (2011).

Macrohelodes Blackburn, 1892 (Pl. 5A, 40I)

Type species. *Helodes princeps* Blackburn, 1892.

Characteristics. Medium to large size, oval to round, deep bodied, moderate size; dorsal surface shiny, without setae. Antennae simple. Mandibles each with one strong tooth. Labial palpus unmodified. Male terminalia conform to the Scirtinae groundplan; stylus present in some species (Fig. 15.10). Female with area of atypical setae in centre of sternite 6. Gonocoxites moderately to quite strongly sclerotised, elongate; gonostyles small, subapical; gonoduct lacking sclerotised structures in *M. montanus* and *M. lucidus,* while *M. crassus* has a moderate-sized bilobed prehensor. Length 3.2–11.5 mm.

Australian species. 16.

Distribution. Coastal eastern Australia. QLD, NSW, VIC, TAS.

Biology. One group of species within the genus are colourful, active during the day and are commonly found on flowers such as *Leptospermum*. A second group are black or black and red and are only active at night. This latter group are restricted to wet forest. The larvae are unknown although adults have been caught in emergence traps placed over large Eucalypt logs, suggesting that the larvae are saproxylic.

Notes. Endemic to Australia. Watts (2010a) divided the genus into two morphological groups – see above. This was supported by sequence data (Cooper *et al.* 2014) which also suggested that Tasmanian and Mainland populations of the common *Macrohelodes crassus* Blackburn may be separate species.

Key to species. Watts (2010a).

References. Watts (2010a).

Nanocyphon Zwick, 2013

Type species. *Nanocyphon australicus* Zwick, 2013.

Characteristics. Fairly slender, Australian species almost cylindrical. Head wide, eyes relatively small. Genal ridge with buttonhole. Base of antenna unmodified; flagellar segments less than twice as long as wide. Mandibles sharply pointed, toothless. Labial palpus with apical segment short, conical, arising centrally from the apical margin of the swollen penultimate segment. Mesoventral groove a small transverse rhomboid between two raised granules; mesoventral process short but not widened, finely incised. Aedeagus with tegmen and penis firmly connected, the parameroids forming a slender plate dorsally from the spine-like trigonium (Fig. 15.11g). Ovipositor unmodified; gonoduct without sclerotised structures. Length 1.2–2.0 mm.

Australian species. Two, but possibly more undescribed.

Distribution. QLD, NSW, ACT, TAS. New Guinea.

Biology. Adults were taken in light traps, found on flowers, or in dry leaf litter. Larvae unknown.

Notes. Endemic to the Australasian region. Several slightly bulkier species occur in New Guinea. The Australian species resemble each other very much but exhibit puzzling resemblances in individual details with New Guinean species, e.g. the presence of ultra-fine spinules on the penis in one New Guinean species and in some Australian specimens. *Nanocyphon tasmanicus* Zwick and a species in Irian Jaya differ in habitus but their penes share a rhomboid apex and are hard to distinguish.

References. Zwick (2013b, 2015b).

Nothocyphon Zwick, 2015 (Pl. 40M, 41C)

Type species. *Helodes (Cyphon?) lindensis* Blackburn, 1892.

Characteristics. Small size, oval, with a generalised body structure, most species brownish but a few have characteristic dorsal pigment marks. Head inserted between blunt, slightly projecting anterolateral angles of pronotum. No antennal groove. Subgenal ridge without buttonhole, meeting the edge of the oral cavity (Fig. 15.2b–c). Mandible with one tooth, or none. Antenna and maxillary palpus unmodified. Labial palpus with bean-shaped apical segment arising eccentrically from apical margin of penultimate segment. Male terminalia complex (Fig. 15.11e), very diverse. The few identified females known have unmodified ovipositors: the front ends of the sclerites of sternite 8 are mostly connected by a ring or a transverse bracket; prehensors vary from small and indistinct to large, complex (Figs 15.13b, 15.15a) or heavily toothed. Length 1.5–3.5 mm.

Australian species. 39 plus several undescribed ones.

Distribution. SA, TAS, VIC, ACT, NSW, QLD.

Biology. Adults are riparian, most typically found beside small, usually permanent, streams in eucalypt-dominated forested areas of southern and eastern Australia. They commonly occur on flowering shrubs close to water which suggests that they feed on flowers, probably pollen. In cooler months some species may be found under the bark of eucalypt trees in or close to water. Larvae live among stones and dead leaves in the shallow side-pools of streams or in the beds of drying streams, mainly in areas of still or slowly flowing water. All instars are usually present indicating a continuous breeding season if water is present. Adults are often found together with species of the genus *Austrocyphon* whose larvae favour packs of dead leaves in areas of streams with more rapid water movement. Exceptions are the larvae of *Nothocyphon frater* and related species which typically live at the sides of larger permanent rivers and are morphologically distinct (Watts 2014).

Notes. Endemic to Australia. *Nothocyphon* will probably prove to be polyphyletic, the included species lack any common derived character expressions to unite them as are found in similar genera. However the male terminalia are species-specific and very diverse, permitting the distinction of several informal species groups which appear monophyletic. Unfortunately too few representatives were available to Cooper *et al.* (2014) for reliable conclusions about phylogeny to be made. Female *N. lanceolatus* Zwick and *N. armstrongi* Zwick possess a pair of porefields on abdominal sternite 7.

References. Zwick (2015a); Watts (2014).

Keys to species. Zwick (2015a).

Ora Clark, 1865 (Pl. 40D)

Type species. *Scirtes trobertii* Guérin-Méneville 1861 (by subsequent designation of Champion 1918).

Characteristics (Australian species). Moderate size, oval to round, flat. Saltorial hind legs. Hind coxae diverge caudally, touching only in front, articulation of hind trochanter exposed. Aedeagi of *O. floccosa* Watts (Fig. 15.12a) and *O. juxtafloccosa* Watts very complex with penis and tegmen fused, that of *O. improtecta* Watts is much simpler with penis and tegmen separate. Female genitalia of *O. improtecta* with large bursal sclerite resembling some *Scirtes* spp. Length 2.5–4.5 mm.

Australian species. Three.

Distribution. Northern QLD, Northern NT, Pantropical.

Biology. Nothing known about the Australian species other than their capture sites in or near wet forest. Larvae of Japanese species are known from phytotelmata (Yoshitomi 2005). The North American *O. texana* Champion inhabits marshes (Epler 2010).

Notes. Globally the species assigned to *Ora* have very different genitalia and the genus appears heterogeneous. Unfortunately, there is no detailed description of the Neotropical type species. Of the three Australian species, *Ora floccosa* and *O. juxtafloccosa* are morphologically very similar, *O. improtecta* is smaller and differently coloured. *Ora improtecta* is present on Halmahera, Indonesia (SA Museum specimen). Sequence data (Cooper *et al.* 2014) places *Ora floccosa* as sister to both *Exochomoscirtes ruforotundus* (Watts) and several species of Australian *Scirtes.*

References. Watts (2004); Libonatti (2014).

Pachycyphon Zwick, 2013

Type species. *Pachycyphon turgidus* Zwick, 2013a.

Characteristics. Small size, strongly convex, in dorsal view broadly to elongate oval, the micropterous females more slender than males. Head wide; eyes small; with a subantennal groove bordered externally by a ridge formed by the extended supraantennal ridge which connects with the subgenal ridge, forming a single sinuous ridge (Fig. 15.2c). Antennae unmodified. Labial palpus gently curved, all segments slender. Pronotal anterolateral angles project distinctly forward. Abdominal sternites connate, rigid except the wide sternite 7. Male abdominal segments 8 and 9 complete but soft; tegmen, parameres and penis large, complex, all strongly sclerotised. Female micropterous; sternite 6 with a porous pilose area; ovipositor with exceptionally strong rigid bacula supporting shovel-shaped gonocoxae that lack gonostyles (Fig. 15.14d); two types of well developed prehensors occur. Length 2.2–3.4 mm.

Australian species. 10.

Distribution. Wet forests of tropical QLD.

Biology. Not known. Females apparently oviposit into substrata, perhaps soil or rotting wood, the presumably terrestrial larvae are unknown.

Notes. Endemic to Australia. Head structure, female micropterism and ovipositor resemble the Tasmanian *Chameloscyphon* from which it differs in its stout convex shape, finely punctate shining dorsal surface, small mandibles, connate abdominal sternites and male genitalia.

Reference. Zwick (2013a).

Key to species. Zwick (2013a).

Papuacyphon Zwick, 2014

Type species. *Papuacyphon globulus* Zwick, 2014.

Characteristics. Stout, strongly convex beetles. Antenna unmodified; no subantennal groove. Subgenal ridge without buttonhole, continues to oral cavity. Mandibles symmetrical, with subterminal tooth. Labial palpus bean-shaped, terminal segment inserted terminally on thickened penultimate segment. Pronotum with projecting anterolateral angles. Prosternal process narrow. Mesoventral groove narrow, long. Male abdominal segments 8 & 9 complete. Tegmen with elongate parameres wrapping around the slender penis; between the tegmen and sternite 9 lies a complex sclerite, the uncifer, supporting two pairs of huge hooks or rods, the shape of which differs between species (Fig. 15.12c). Ovipositor unmodified; gonoduct very slender, the prehensor an oval sclerite ring and two divergent, heavily toothed ribs opposite it. Length 2.1–2.9 mm (exotic species up to 3.5 mm).

Australian species. One.

Distribution. Endemic to the Australasian region, disjunct: south-west corner of WA, several species on high mountains in Irian Jaya.

Biology. Adults have been collected from flowering shrubs in or close to shallow swamps. The larvae are assumed to live in the swamps.

Notes. Phylogenetic relationships are unclear.

Reference. Zwick (2014a, b).

Paracyphon Zwick, 2015

Type species. *Paracyphon avicularis* Zwick, 2015.

Characteristics. Small size, elongate-oval, anterolateral angles of pronotum weakly projecting, in general facies resembling *Austrocyphon* and *Nothocyphon*. Genal ridge without buttonhole. Antennae and palpi unmodified, terminal segment of labial palpus bean-shaped, obliquely inserted on the end of the penultimate segment. The male genitalia are unique: the oval penis base houses large muscles attaching to the common base of a pair of caudal, two-layered appendages, there is no trigonium (Fig. 15.12f); tegmen and parameres form a loop with enlarged flaps at the end. Female unknown. Length 2.3–2.5 mm.

Australian species. One.

Distribution. VIC and NSW.

Biology. The two known specimens were caught in wet sclerophyll forest and subalpine forest, respectively.

Notes. Endemic to Australia. Close relatives are not known. The pair of large mediobasally connected caudal processes of the male genitalia superficially resemble those of *Calvarium* (Fig. 15.12d). However, in *Paracyphon* there is no subantennal groove and no connection between the supraantennal ridge and the subgenal ridge as there is in *Calvarium*. *Paracyphon* shares with *Contacyphon* the trigonium being replaced by paired caudal appendages. However the absence of a buttonhole in the subgenal ridge in *Paracyphon* and its presence in *Contacyphon* speaks against close affinities. Also in *Contacyphon* the male sternite 8 is completely missing, while it is distinct and V-shaped in *Paracyphon*.

Reference. Zwick (2015b).

Peneveronatus Armstrong, 1953 (Pl. 40O, 41D; Fig. 15.6a)

Type species. *Peneveronatus australis* Armstrong, 1953.

Characteristics. Large size, elongate, relatively flat, uniformly testaceous. Pronotum flat, surface uneven, noticeably narrower than base of elytra, with rounded anterolateral angles, without basal pits. Mandible with one weak tooth; mola region with well developed area of small spines. Male terminalia conform to Scirtinae groundplan; trigonium elongate, often split in the dorso-ventral/sagittal plane or strongly curved downwards (Fig. 15.11d); without styli. Gonocoxites elongate/triangular; gonostyli apical; S8 variably shaped between species, quite strongly sclerotised; gonoduct not known. Length 5.0–10.0 mm.

Australian species. Four.

Distribution. ACT, Coastal NSW, VIC and southern TAS.

Biology. Found in wet forest. Presumptive larvae are found among decaying leaves in shallow, swampy side areas of small creeks in deep shade (Watts 2014). Unusual for Australian scirtids most adult specimens of all species have been collected in winter. Supporting a winter emergence the last instars of the presumptive larvae of the Victorian *P. macedonensis* (Armstrong) are not present until early autumn (Watts pers obs.).

Notes. Endemic to Australia. Rarely collected. Based on the rounded front angles to the pronotum and sequence data from presumptive larvae the genus is a member of the *Pseudomicrocara* group of taxa.

References. Watts (2009, 2014); Cooper *et al.* (2014).

Key to species. Watts (2009).

Petrocyphon Watts, 2011 (Pl. 40F)

Type species. *Petrocyphon antarcticus* Watts, 2011.

Characteristics. Small size, elongate oval, some species sexually dimorphic with antennae more prominent in the male; eyes of moderate size. Subgenal ridge with buttonhole configuration. Mandible with one strong tooth. Labial palpus with apical segment arising from inner apical angle of penultimate segment. Pronotum with anterolateral angles slightly projecting forward, with shallow groove along the posterior margin. Apex of pronotal process broad. Mesoventral groove short and broad. Male terminalia conform to Scirtinae groundplan; trigonium, parameroids and parameres complex, often divided (Fig. 15.12g); without styli. Gonocoxites elongate; gonostyli apical; prehensors large and complexly structured but without major sclerites. Several species with a pair of elaborate cuticular rings on the bursella. Length 1.8–2.6 mm.

Australian species. 19.

Distribution. QLD, NSW, VIC, TAS.

Biology. Adults are riparian, living in vegetation beside and overhanging streams and rivers in typically *Nothofagus* dominated closed forest. The larvae are unknown. Most patches of closed forest in coastal eastern Australia and Tasmania have their own species. In a few localities in Tasmania two species are present but this appears to be an exception.

Notes. Endemic to Australia. Morphology and DNA sequence data (Cooper *et al.* 2014), suggest that the genus is phylogenetically isolated.

References. Watts (2011); Zwick (2012, 2016).

Key to species. Zwick (2012, 2016).

Prionocyphon Redtenbacher, 1858 (Pl. 40E)

Type species. *Cyphon serricornis* Müller, 1821.

Characteristics. Small to medium size, oval to round. Antennae with scape enlarged, flat, with sharp edge; pedicle and especially segment 3 small (Fig. 15.3c); male antennae of several species serrate. Anterolateral corners of clypeus often greatly extended. Subgenal ridge with buttonhole configuration. Labial palpi forked. Male terminalia varying from simple and conforming to the Scirtinae groundplan to highly modified with fused penis and tegmen (Fig. 15.11h) or an extendable endophallus (Fig. 15.12i); with styli. Ovipositor unmodified; gonoduct of the few

Australian females known without or with poorly developed prehensor and/or bursal sclerite. Length 2.0–5.0 mm, some of the problematic species up to 7.2 mm.

Australian species. 20, plus three probably misplaced in the genus.

Distribution. WA, NT, QLD, NSW (including Lord Howe Island), VIC, TAS, South-east Asia, Holarctic, Neotropics.

Biology. Found in wet forested areas. The known larvae live in phytotelmata. Exceptions are the larvae of some Tasmanian populations of *P. niger* which live in forest ponds as well as phytotelmata; *P. lamingtonensis* which inhabit rock pools beside rivers, and *P. insolitus* which is saproxylic, living in the interiors of wet logs. Both of these latter species are probably taxonomically misplaced.

Notes. The original description of the European type species focused on the flat disc-like scape and the minuteness of antennal segment 3. Australian species conforming to this diagnosis seem to be closely interrelated, and also related to the type species, although perhaps distantly (Cooper *et al.* 2014). There are Asian genera similar to *Prionocyphon* in several respects (Zwick 2015d). A global solution for the *Prionocyphon* complex is needed.

References. Watts (2010b, 2014); Cooper *et al.* (2014); Zwick (2016).

Key to Australian species. Watts (2010b), added species in Zwick (2016).

Pseudomicrocara Armstrong, 1953 (Fig. 15.1)

Type species. *Pseudomicrocara orientalis* Armstrong, 1953.

Characteristics. Small to large size. A very heterogeneous genus best characterised by: pronotum as wide as base of elytra, or nearly so, with anterolateral angles not projecting forwards, without pits on posterior margin. All species, other than female *P. hangayi*, dorsally well covered with setae. Antennae simple. Mandibles slight to robust, with or without teeth. Elytral epipleura anteriorly narrower than pronotal epipleura. Prosternal process well developed, reaching mesoventrite; mesoventral groove weakly to well developed. Male terminalia conform to the Scirtinae groundplan; penis variable between species from simple trilobed ones (Fig. 15.9), to highly modified structures, tegmen (e.g. Fig. 15.11f) and female genitalia are likewise very variable. Length 2.4–9.0 mm.

Australian species. 41 described, 6–8 undescribed.

Distribution. WA, NT, QLD, NSW, VIC, TAS, SA. Also in New Guinea, New Caledonia, South America.

Biology. Most species inhabit areas of open sclerophyll forest with the larvae typically found at the side of ponds, swampy areas and in slower moving water at the side of creeks and rivers. Several species have larvae which are saproxylic, living inside water-saturated Eucalypt logs on the floor of wet forest. Adults of some species can be abundant on flowering shrubs beside water such as *Leptospermum*; others, despite being relatively abundant, have only been collected at light.

Notes. Watts (2007) on morphological evidence grouped the 41 species into five groups and three unplaced species. Work since then involving DNA sequencing (Cooper *et al.* 2014) and larval morphology (Watts 2014) has shown that most of these groupings cannot be maintained and that the genus is polyphyletic. However the sequencing also showed that all the taxa studied were phylogenetically quite closely related and linked with species in the genera *Accolabass*, *Heterocyphon, Macrohelodes,* and *Peneveronatus* to form an indigenous radiation.

Several species of *Pseudomicrocara,* which have all the characteristics given above, have recently been recorded and described from southern South America (Libonatti & Ruta 2013), however Cooper *et al.* (2014) suggest that the presence of the genus in South America needs to be confirmed, preferably by sequence data. Contributing to the taxonomic difficulty of this complex of taxa is the preponderance of what appear to be plesiomorphic characters. (See also Zwick (2015c) in regard to the North American genus *Sarabandus* Leach, thought by its describer to be close phylogenetically to *Pseudomicrocara*). Further study, both morphological and molecular will be needed to confirm the nature of the relationship between the Australian and South American representatives of *Pseudomicrocara*.

References. Watts (2007, 2014); Libonatti & Ruta (2013).

Key to Australian species. Watts (2007).

Scirtes Illiger, 1807 (Pl. 40C)

Type species. *Chrysomela haemisphaerica* Linnaeus, 1767.

Characteristics. Small to medium size, oval to round. Saltatorial hind legs; hind coxae meeting along full length of midline; articulation of hind trochanters not exposed. Mouthparts variably developed, labial palpus in some species strongly forked (Fig. 15.5b). Male terminalia very variable. Gonoduct with soft or weakly sclerotised prehensors and well sclerotised, complex, bursal sclerites. Length 2.0–5.0 mm.

Australian species. 30, plus three undescribed.

Distribution. All Australian states and mainland territories. Cosmopolitan, more abundant in tropical areas.

Biology. Common and widespread in Australia, with one species, *Scirtes exoletus* Waterhouse, occurring in all States and Territories. Larvae typically live in ponds and swamps although a few species inhabit gravel at the sides of creeks and small rivers. Adults are found in emergent vegetation and can occur in large numbers on flowering shrubs adjacent to water. Often strongly attracted to light traps placed near water.

Notes. Based on the morphology of the penis most Australian species belong in the *Scirtes helmsi* group (Watts 2007).

This group is known to occur only in Australia but, judging from the illustrations of male genitalia in Ruta & Yoshitomi (2010), also occurs in New Caledonia. One Australian species, *Scirtes auratus* Watts, appears, on both larval and adult morphology (Watts 2014; Zwick 2016) and DNA sequence data (Cooper *et al.* 2014), to be closely related to the European *S. haemisphaericus*. The little sequence data available (Cooper *et al.* 2014) shows that the Australian species of *Scirtes*, *Ora* and *Exochomoscirtes* studied are phylogenetically close. Many of the individual species of Australian *Scirtes* are morphologically superficially similar. Specific identification is based mainly on characters of the penis. The form of the female bursal sclerite is also useful in the identification of species (see update in Watts *et al.* 2017).

References. Watts (2007, 2014); Cooper *et al.* (2014); Ruta & Yoshitomi (2010).

Key to Australian species. Watts (2007).

Spaniosdascillus Watts, 2011

Type species. *Spaniosdascillus elongatus* Watts, 2011.

Characteristics. Large size, elongate; eyes large, protuberant. Antennae with scape moderately large; pedicle and segment 3 small; segments 4–7 strongly serrate (only male known). Labial palpus with penultimate segment broad, scoop-like; apical segment broad, arising from anterior portion of the inner edge of penultimate segment (Fig. 15.4c). Mandible with one strong tooth. Pronotum with anterolateral angles not projecting forward. Penis elongate, parameroids much longer than trigonium. Other male terminalia, the female and the larva not known. Length 7.3 mm.

Distribution. NSW.

Australian species. One.

Biology. Nothing known. The only known specimen was collected from eucalypt woodland.

Notes. Endemic to Australia. Watts (2011) when describing the only species (and specimen) suggested, based on the form of the antennal base, that it might have a relatively close phylogenetic relationship with *Macrodascillus*. However, it differs noticeably in the unusual form of the labial palpus and the rounded anterolateral angles to the pronotum.

References. Watts (2011).

Tasmanocyphon Zwick, 2013

Type species. *Tasmanocyphon heideae* Zwick, 2013c.

Characteristics. Small size, elongate-oval; elytra brown with two pale spots. Habitus and general structure resemble *Austrocyphon,* as do the male genitalia with fused parameroids lying over the penis, but all the components are more complexly shaped than in that genus. Length 2.7–3.3 mm.

Australian species. One.

Distribution. TAS.

Biology. Adults are riparian living in vegetation beside permanent rivers, streams and waterfalls. Less common than species in the genera *Austrocyphon* and *Nothocyphon* which are found in the same habitat. Larvae are unknown.

Notes. Endemic to Tasmania, sister genus to *Austrocyphon*.

References. Zwick (2013c).

Tectocyphon Zwick, 2015

Type species. *Tectocyphon microphallus* Zwick, 2015b.

Characteristics. Small size, body rather flat, oval, no gap between elytra and pronotum. Head concealed under pronotum; no subantennal grove. Subgenal ridge continues to edge of oral cavity. Antennae unmodified, setiform. Mandible tip sharply pointed, toothless. Maxillary palpus unmodified. Labial palpus strongly bent, the bean-shaped apical segment stands on the inside of the penultimate one. Pronotum short, strongly narrowing from a wide base to the rounded front margin which lacks anterolateral angles. Anterior portion of elytra with sides slightly flanged above the wide and flat epipleura. Prosternal process slender, drop-shaped. Mesoventral groove rhomboid. Male sternites 8 and 9 unmodified; tegmen spade-shaped, parameres long with a strongly toothed basolateral appendage; trigonium either a simple small triangle, or a transverse scale with a serrate caudal edge, always shorter than the parameroids. Female with pairs of finely pilose pore plates on sternites 4–6. Ovipositor unmodified; gonoduct with a large prehensor with a partly cleft sclerite, bearing big hooks along sides. Length 3.0–3.6 mm.

Australian species. Three.

Distribution. ACT, NSW, VIC.

Biology. Little known. The few available adults were taken at elevations of 1400–1900 m in alpine forest or closed forest. In Victoria *Tectocyphon victoriae* Zwick has been collected from tree-fern fronds beside a strongly flowing creek.

Notes. Endemic to Australia. Habitus in dorsal view reminiscent of *Eurycyphon* spp. which occur in similar habitat but *Tectocyphon* can be separated from that genus by e.g. the pronotum whose base is hardly narrower than the elytra and whose convexly projecting front edge completely conceals the head, epipleura not concave, and genital characters of both sexes. Larvae are unknown.

References. Zwick (2015b).

Key to Australian species. Zwick (2015b).

Ypsiloncyphon Klausnitzer, 2009

Type species. *Cyphon chlorizans* Klausnitzer, 1973 (Philippines).

Characteristics. Small size, oval, females of some species elongate; blackish, usually with yellow appendages. Antenna unmodified. Without buttonhole on subgenal ridge. Mandible sharply pointed, toothless (Fig. 15.4c). Maxillary palpus with a short conical apical segment. Mesoventral process very short, widening caudally. Except for *Y. virgulifer* Zwick, in the Australian and New Guinean males the symmetrical penis, the asymmetrical sternite 9 (Fig. 15.11b) and the hairpin-like tegmen plus parameres each have a slender handle-like base (capulus). Sternite 9 is deeply divided into flat blades which differ in shape, one often being extremely long. Ovipositor unmodified, gonoduct with a large vulvar sclerite but without prehensor. The spherical bursella is armed with a pair of large cuticular meshworks, the dictyon. Length 1.4–2.3 mm.

Australian species. Seven, plus probably several undescribed.

Distribution. Northern WA, NT, tropical and subtropical QLD, New Guinea.

Biology. Adults of *Y. longus* Zwick have been collected from emergent vegetation around small swamps (CHSW pers obs), otherwise nothing is known of their ecology.

Notes. Palpi and mandibles of Australasian *Ypsiloncyphon* resemble *Nanocyphon* but there is no buttonhole, the mesoventrum and the genitalia of both sexes are also very different. DNA sequence data (Cooper *et al.* 2014) placed *Ypsiloncyphon* (represented by *Y. longus*) as an isolated lineage within Australian Scirtidae. The monophyly of the Australasian species assemblage, as compared with the other species in the genus, is probable but unconfirmed. They do not share several derived characters observed in most Asian species of the genus (Zwick 2016).

References. Zwick (2014b, 2016).

Key to Australasian species. Zwick (2014b).

Acknowledgments

We would like to acknowledge the work of Howard Hamon who greatly contributed to the production of many of the images in both this and earlier publications by C. H. S. Watts on Australian Scirtidae and Rafal Ruta who drew our attention to the literature on fossil Scirtidae.

Endnotes

1 Lawrence and Yoshitomi (2007) called the same structure 'subocular carina'.

2 The last instar is identified by having a sharply pointed tip rather than a blunt one, to the mandible (Fig. 15.7e, f). Earlier instars of saproxylic larvae may also have sharply pointed mandibles, but this will not affect the working of the key.

References

Alekseev VI (2013) The beetles (Insecta: Coleoptera) of Baltic amber: the checklist of described species and preliminary analysis of biodiversity. *Zoology and Ecology* **23**, 5–12. doi:10.1080/21658005.2013.769717

Beier M (1949) Körperbau und Lebensweise der Larve von *Helodes hausmanni* Gredler (Col. Helodidae). *Eos: Revista Española de Entomología* **25**, 49–100.

Beier M (1952) Bau und Funktion der Mundwerkzeuge bei den Helodiden-Larven (Col.). *Transactions of the Ninth International Congress of Entomology* **1**, 135–138.

Blackburn T (1892) Notes on Australian Coleoptera with descriptions of new species. Part X. *Proceedings of the Linnean Society of New South Wales* **6** (1891), 479–550.

Bocák L, Barton C, Crampton-Platt A, Chesters D, Ahrens D, Vogler AP (2014) Building the Coleoptera tree-of-life for >8000 species: composition of public DNA data and fit with Linnaean classification. *Systematic Entomology* **39**, 97–110.

Champion CG (1918) New and little-known saltatorial Dascillidae. *Entomologist's Monthly Magazine* **54**, 93–103, 139–149, 188–198, 219–225, 256–273.

Cooper SJB, Watts CHS, Saint KM, Leijs R (2014) Phylogenetic relationships of Australian Scirtidae (Coleoptera) based on mitochondrial and nuclear sequences. *Invertebrate Systematics* **28**, 628–642 doi:10.1071/IS13046.

Edwards RT, Meyer JL (1990) Bacterivory by deposit-feeding mayfly larvae (*Stenonema* spp.). *Freshwater Biology* **24**, 453–462. doi:10.1111/j.1365-2427.1990.tb00724.x

Engel MS (2010) A primitive anobiid beetle in mid-Cretaceous amber from Myanmar (Coleoptera: Anobiidae). *Alavesia* **3**, 31–34.

Epler J (2010) *The Water Beetles of Florida – an identification manual for the families Chrysomelidae, Curculionidae, Dryopidae, Dytiscidae, Elmidae, Gyrinidae, Haliplidae, Helophoridae, Hydraenidae, Hydrochidae, Hydrophilidae, Noteridae, Psephenidae, Ptilodactylidae and Scirtidae* (PDF, 152 MB). http://publicfiles.dep.state.fl.us/dear/labs/biology/biokeys/beetles10.pdf

Findlay S, Meyer JL, Smith PJ (1984) Significance of bacterial biomass in the nutrition of a freshwater isopod (*Lirceus* sp.). *Oecologia* **63**, 38–42. doi:10.1007/BF00379782

Hall RO, Meyer JL (1998) The trophic significance of bacteria in a detritus-based stream food web. *Ecology* **79**, 1995–2012. doi:10.1890/0012-9658(1998)079[1995:TTSOBI]2.0.CO;2

Hannappel U, Paulus HF (1987) Arbeiten zu einem phylogenetischen System der Helodidae (Coleoptera) - Feinstrukturuntersuchungen an europäischen Larven. *Zoologische Beiträge. Neue Folge* **31**, 77–150.

Hong Y-C (2002) *Amber insects of China*. Beijing Scientific and Technological publishing House.

Horwitz P (1989) The faunal assemblage (or pholeteros) of some freshwater crayfish burrows in southwest Tasmania. *Bulletin of the Australian Society of Limnology* **12**, 57–60.

Hudson GV (1934) *New Zealand beetles and their larvae. An elementary introduction to the study of our native Coleoptera*. Ferguson and Osborn, Wellington, 236 pp.

Jell AP, Duncan PM (1986) Invertebrates, mainly insects, from the freshwater lower Cretaceous, Koonwarra fossil bed (Korumburra Group), South Gippsland, Victoria. *Memoirs of the Association of Australian Palaeontologists* **3**, 111–205.

Kaiser H, Klausnitzer B (2000) *Scirtimermis praematura* n. gen. n. sp. (Nematoda, Mermithidae) ein neuer Parasit aus *Cyphon variabilis* (Thunberg, 1787) (Coleoptera, Scirtidae). *Entomologische Blätter* **96**, 219–227.

Kitching RL, Allsopp PG (1987) *Prionocyphon niger* sp. n. (Coleoptera: Scirtidae) from water-filled tree holes in Australia. *Journal of the Australian Entomological Society* **26**, 73–79. doi:10.1111/j.1440-6055.1987.tb00264.x

Klausnitzer B (2006) Zur Kenntnis der Larven einiger Scirtidae aus Neuseeland und Borneo mit wahrscheinlich terrestrischer Lebensweise (Coleoptera). *Entomologische Nachrichten und Berichte* **50**, 141–152.

Klausnitzer B (2012) Neue Arten der Familie Scirtidae (Coleoptera) aus Baltischem Bernstein (Teil 3) (172. Beitrag zur Kenntnis der Scirtidae). *Linzer biologische Beiträge* **44**, 313–318.

Klausnitzer B, Pospisil P (1991) Larvae of *Cyphon* sp. (Col., Helodidae) in Ground Water. *Aquatic Insects* **13**, 161–165. doi:10.1080/01650429109361437

Lawrence JF (2001) A new genus of valdivian Scirtidae with comments on Scirtoidea and beetle suborders. In *Sukanahikoma, Special Publication of the Japan Coleopterological Society Osaka* 1 (Eds K Morimoto, K Mizuno, Y Hayashi, T Ito, N Ito, K Ando, M Tanikado, S Shiyake) pp. 351–361. Japan Coleopterological Society, Osaka.

Lawrence JF (2016) 10.4. Scirtidae Fleming, 1821.In *Handbuch der Zoologie/Handbook of Zoology. BandVolume IV Arthropoda: Insecta. Teilband/Part 38. Coleoptera, Beetles. Volume 1: Morphology and Systematics (Archostemata, Adephaga, Myxophaga, Polyphaga partim).* 2nd Edition. (Eds RG Beutel and RAB Leschen) pp. 215–225. Walter de Gruyter, Berlin.

Lawrence JF, Newton AF, Jr (1995) Families and subfamilies of Coleoptera (with selected genera, notes, references and data on family-group names). In *Biology, Phylogeny, and Classification of Coleoptera: papers celebrating the 80th Birthday of Roy A. Crowson* (Eds J Pakaluk and SA Ślipiński) pp. 779–1006. Muzeum i Instytut Zoologii PAN, Warszawa.

Lawrence JF, Ślipiński A (2013) *Australian Beetles Morphology, Classification and Keys*. Vol. 1, VIII+561 pp., CSIRO Publishing, Collingwood.

Lawrence JF, Yoshitomi H (2007) *Nipponocyphon*, a new genus of Japanese Scirtidae (Coleoptera) and its phylogenetic significance. *Elytra* **35**(2), 507–527.

Lawrence JF, Ślipiński A, Seago AE, Thayer MK, Newton AF, Marvaldi AE (2011) Phylogeny of the Coleoptera based on morphological characters of adults and larvae. *Annales Zoologici (Warsaw)* **61**, 1–217. doi:10.3161/000345411X576725

Libonatti ML (2014) A revision of the genus *Ora* Clark, 1865 (Coleoptera: Scirtidae) in Argentina (part I) – descriptions of new species. *Zootaxa* **3884**(1), 27–44. doi:10.11646/zootaxa.3884.1.2

Libonatti ML, Ruta R (2013) Review of the Argentinean species of *Pseudomicrocara* Armstrong (Coleoptera: Scirtidae). *Zootaxa* **3718**, 137–157. doi:10.11646/zootaxa.3718.2.3

McKenna DD, Wild AL, Kanda K, Bellamy CL, Beutel RG, Caterino MS, Farnum CW, Hawks DC, Ivie MA, Jameson ML, Leschen RAB, Marvaldi AE, McHugh JV, Newton AF, Robertson JA, Thayer MK, Whiting MF, Lawrence JF, Ślipiński A, Maddison DR, Farrell BD (2015) The beetle tree of life reveals that Coleoptera survived end-Permian mass extinction to diversity during the Cretaceous terrestrial revolution. *Systematic Entomology* **40**, 835–880.

Müller H (2015) Erster Nachweis parasitischer Wassermilbenlarven an Sumpfkäfern: *Piersigia intermedia* Williamson, 1912 (Acari: Hydrachnidia: Piersigiidae) parasitiert *Contacyphon laevipennis* (Insecta: Coleoptera: Scirtidae) / First record of watermites parasitic on marsh beetles: *Piersigia intermedia* Williamson, 1912 (Acari: Hydrachnidia: Piersigiidae) parasitizes *Contacyphon laevipennis* (Insecta: Coleoptera: Scirtidae). *Lauterbornia* **80**, 112–114.

Nyholm T (1948) Studien über die Familie Helodidae III. Neue paläarktische Arten der Gattung *Cyphon* Payk. *Arkiv för Zoologi* **41A**(4), 1–7 [pls. I–V.].

Nyholm T (1969) Über Bau und Funktion der Kopulationsorgane bei den Cyphones (Col. Helodidae) Studien über die Familie Helodidae. X. *Entomologisk Tidskrift* **90**(3–4), 233–271 [plates I–Vii].

Nyholm T (1972) Zur Morphologie und Function des Helodiden-Adoeagus (Col). *Entomologica Scandinavica* **3**, 81–119. doi:10.1163/187631272X00193

Nyholm T (2002) *Scirtes japonicus* Kiesenwetter and its allies, with description of *Scirtes ussuriensis* n. sp. (Coleoptera, Scirtidae). *Entomologische Blätter* **98**, 49–60.

Peris D, Phillips KT, Delcios X (2015) Ptinid beetles from the Cretaceous gymnosperm-dominated forests. *Cretaceous Research* **52**, 440–452. doi:10.1016/j.cretres.2014.02.009

Rasmussen K (1978) Livscyklus og habitat praeference hos *Helodes marginata* Fabr. og *Helodes minuta* L. (Coleoptera), og de to artes udbredelse i Jylland. *Flora og Fauna* **82**, 80–84.

Ross A (2014) Palaeontology: Chinese amber insects bridge the gap. *Current Biology* **24**(14), R642–R643.

Ruta R (2010) Revision of the genus *Calvarium* Pic, 1918 (Coleoptera: Scirtidae). Part 1. Redescription of the genus and catalogue of described taxa. *Annales Zoologici* **60**(3), 341–350[Warszawa]. doi:10.3161/000345410X535343

Ruta R, Yoshitomi H (2010) Scirtidae: The Genus *Scirtes* Illiger (Coleoptera). In *Water Beetles of New Caledonia. Part 1.* (Eds MA Jäch and M Balke) pp. 403–438 Natural History Museum Vienna, Austria.

Ruta R, Thorpe S, Yoshitomi H (2011) *Stenocyphon neozealandicus*, a new species from New Zealand of a previously monotypic subfamily from Chile (Coleoptera: Scirtidae: Stenocyphoninae). *Zootaxa* **3113**, 65–68.

Ruta R, Klausnitzer B, Prokin A (2018) South American terrestrial larva of Scirtidae (Coleoptera: Scirtoidea): the adaptation of Scirtidae larvae to saproxylic habitat is more common than expected. *Australian Entomologist* **57**, 50–61. doi:10.1111/aen.12270

Treherne JE (1952) The respiration of the larva of *Helodes minuta* (Col.). *Transactions of the ninth International Congress of Entomology* **9**, 311–314.

Treherne JE (1954) Osmotic regulation in the larvae of *Helodes* (Coleoptera: Helodidae). *Transactions of the Royal Entomological Society of London* **105**, 117–130. doi:10.1111/j.1365-2311.1954.tb00779.x

Watts CHS (2004) Revision of Australian *Scirtes* Illiger and *Ora* Clark (Scirtidae: Coleoptera). *Transactions of the Royal Society of South Australia* **128**, 131–168.

Watts CHS (2007) Revision of Australian *Pseudomicrocara* Armstrong (Scirtidae: Coleoptera). *Transactions of the Royal Society of South Australia* **131**, 1–82. doi:10.1080/03721426.2007.10887068

Watts CHS (2009) Revision of *Heterocyphon* Armstrong, *Peneveronatus* Armstrong and *Accolabass* gen. nov. (Scirtidae: Coleoptera). *Transactions of the Royal Society of South Australia* **133**, 108–149. doi:10.1080/03721426.2009.10887113

Watts CHS (2010a) Revision of the genus *Macrohelodes* (Scirtidae: Coleoptera; Insecta). *Transactions of the Royal Society of South Australia* **134**, 19–52. doi:10.1080/3721426.2010.10887130

Watts CHS (2010b) Revision of Australian *Prionocyphon* Redtenbacher (Scirtidae: Coleoptera). *Transactions of the Royal Society of South Australia* **134**, 53–88. doi:10.1080/3721426.2010.10887131

Watts CHS (2011) Revision of Australian Scirtidae of the genera *Chameloscyphon* gen. nov., *Daploeuros* gen. nov., *Dasyscyphon* gen. nov., *Eurycyphon* gen. nov., *Macrodascillus* Carter, *Petrocyphon* gen. nov., and *Spaniosdascillus* gen. nov. (Coleoptera). *Transactions of the Royal Society of South Australia* **135**, 66–110. doi:10.1080/03721426.2011.11877485

Watts CHS (2014) The larvae of some Australian Scirtidae (Coleoptera) with a key to known genera. *Transactions of the Royal Society of South Australia* **138**, 1–91. doi:10.1080/03721426.2014.10887191

Watts CHS, Cooper SJB, Saint KM (2017) Review of Australian *Scirtes* Illiger, *Ora* Clark and *Exochomoscirtes* Pic (Coleoptera: Scirtidae) including descriptions of new species, new groups and a multi-gene molecular phylogeny of Australian and non-Australian species. *Zootaxa* **4347**, 511–532 https://doi.org/10.11646/zootaxa.4347.3.5. doi:10.11646/zootaxa.4347.3.5

Wichard W, Komnick H (1974) Feinstruktur und Funktion der Analpapillen von aquatischen Käferlarven (Elodidae). *International Journal of Insect Morphology & Embryology* **3**(3–4), 335–341. doi:10.1016/0020-7322(74)90027-0

Wichard W, Arens W, Eisenbeis G (2002) *Biological Atlas of Aquatic Insects*. 338 pp., Apollo Books, Stenstrup.

Yablokov-Khnzoryan SM (1961) Notes on the family Helodidae (Coleoptera) from Baltic amber. *Палеонтологический журнал* **1**, 108–116 [in Russian].

Yee M, Grove SJ, Richardson AMM, Mohammed CL (2006) Brown rot in inner heartwood: Why large logs support characteristic saproxylic beetle assemblages of conservation concern. In Insect biodiversity and dead wood: proceedings of a symposium for the 22nd International Congress of Entomology. (Eds SJ Grove, J Hanula and J James) pp. 42–56. *General Technical Report SRS-93, Ashville, NC: US Department of Agriculture Forest Service, Southern Research Station*, 109 pp.

Yoshitomi H (2005) Systematic revision of the family Scirtidae of Japan, with phylogeny, morphology and bionomics (Insecta: Coleoptera, Scirtoidea). *Japanese Journal of Systematic Entomology monograph series* **3**, 1–212.

Zwick P (2011a) Some Scirtidae (Coleoptera) from Palawan (the Philippines), mainly from phytotelmata. *Aquatic Insects* **33**(3), 233–252. doi: 10.1080/01650424.2011.614257

Zwick P (2011b) 7.6 Arthropoda. In *Central European Stream Ecosystems. The Long Term Study of the Breitenbach* (Eds R Wagner *et al.*) pp.216–327. Wiley-Blackwell, Weinheim.

Zwick P (2012) Australian Marsh Beetles (Coleoptera: Scirtidae). 1. Additions to genus *Petrocyphon*. *Entomologische Blätter und Coleoptera* **108**, 159–180.

Zwick P (2013a) Australian Marsh Beetles (Coleoptera: Scirtidae). 2. *Pachycyphon*, a new genus of presumably terrestrial Australian Scirtidae. *Zootaxa* **3626**, 326–344. doi:10.11646/zootaxa.3626.3.2

Zwick P (2013b) Australian Marsh Beetles (Coleoptera: Scirtidae). 3. A restricted concept of genus *Cyphon*, Australian species of *Cyphon* s. str. and the new Australasian genus *Nanocyphon*. *Genus* **24**, 163–189.

Zwick P (2013c) Australian Marsh Beetles (Coleoptera: Scirtidae). 4. Two new genera, *Austrocyphon* and *Tasmanocyphon*. *Zootaxa* **3706**, 1–74. doi:10.11646/zootaxa.3706.1.1

Zwick P (2013d) ERRATUM Peter Zwick (2013) Australian Marsh Beetles (Coleoptera: Scirtidae) 4. Two new genera, *Austrocyphon* and *Tasmanocyphon*. Zootaxa, 3706(1), 001–074. *Zootaxa* **3710**(6), 600. doi:10.11646/zootaxa.3710.6.6

Zwick P (2014a) Australian Marsh Beetles (Coleoptera: Scirtidae). 5. New species of genus *Cyphon* (s.str.), and *Papuacyphon* gen. nov. from New Guinea. *Stuttgarter Beiträge zur Naturkunde. Neue Serie* **7**, 131–152.

Zwick P (2014b) Australian Marsh Beetles (Coleoptera: Scirtidae). 6. Genera *Calvarium* Pic, *Papuacyphon* Zwick, and *Ypsiloncyphon* Klausnitzer. *Zootaxa* **3846**, 1–41. doi:10.11646/zootaxa.3846.1.1

Zwick P (2015a) Australian Marsh Beetles (Coleoptera: Scirtidae). 7. Genus *Nothocyphon*, new genus. *Zootaxa* **3981**, 301–359. doi:10.11646/zootaxa.3981.3.1

Zwick P (2015b) Australian Marsh Beetles (Coleoptera: Scirtidae). 8. The new genera *Cygnocyphon, Eximiocyphon, Paracyphon, Leptocyphon, Tectocyphon*, and additions to *Contacyphon* de Gozis, *Nanocyphon* Zwick and *Eurycyphon* Watts. *Zootaxa* **3981**(4), 451–490. doi:10.11646/zootaxa.3981.4.1

Zwick P (2015c) To the knowledge of *Sarabandus robustus* (LeConte) (Col.: Scirtidae: Scirtinae), and on the groundplan of male marsh beetle genitalia. *Linzer biologische Beiträge* **47**(2), 1439–1449.

Zwick P (2015d) Three new Marsh Beetles (Col.: Scirtidae) from New Guinea and Java. *Linzer biologische Beiträge* **47**(2), 1885–1895.

Zwick P (2016) Australian Marsh Beetles (Coleoptera: Scirtidae). 9. The relations of Australasian *Ypsiloncyphon* species to their Asian congeners, additions, mainly to *Petrocyphon* and *Prionocyphon*, and a key to Australian genera of Scirtinae. *Zootaxa* **4085**(2), 151–198. doi:10.11646/zootaxa.4085.2.1

Zwick P, Hecht M (2008) Life history and instar number of *Elodes minuta*, with notes on additional species (Coleoptera: Scirtidae). *Lauterbornia* **62**, 79–88.

Zwick P, Zwick H (2008a) Number of larval instars, early instar structure, and life history of *Scirtes haemisphaericus* (Coleoptera: Scirtidae) in central Germany. *Lauterbornia* **63**, 87–99.

Zwick P, Zwick H (2008b) *Scirtes hemisphaericus* uses macrophyte snorkels to pupate under water. With notes on pupae of additional European genera of Scirtidae (Coleoptera). *Aquatic Insects* **30**, 83–95. doi:10.1080/01650420701882244

Zwick P, Zwick H (2010) The number of larval instars in some Central European Marsh Beetles (Col.: Scirtidae). *Entomologische Blätter* **106**, 431–441.

Zwick P, Klausnitzer B, Ruta R (2013) *Contacyphon* Gozis, 1886 removed from synonymy (Coleoptera: Scirtidae) to accommodate species so far combined with the invalid name, *Cyphon* Paykull, 1799. *Entomologische Blätter und Coleoptera* **109**, 337–353.

16. SUPERFAMILY HYDROPHILOIDEA LEACH, 1815

Martin Fikáček

Introduction. The superfamily Hydrophiloidea contains ~6600 described species distributed worldwide (Archangelsky *et al.* 2005), mostly associated with decaying organic material (leaf litter, decaying wood, excrement) or aquatic habitats. Ten families are recognised, grouped into two undoubtedly monophyletic clades: the hydrophiloid lineage (often also called Hydrophiloidea s.str.) with six families (Helophoridae, Georissidae, Epimetopidae, Hydrochidae, Spercheidae and Hydrophilidae) and histeroid lineage (often treated as a separate superfamily Histeroidea) with four families (Synteliidae, Sphaeritidae, Histeridae and the extinct Cretohisteridae). Only five of these families occur in Australia, remaining ones have more restricted distribution: Helophoridae are confined to the Holarctic region, northern India and Africa; Epimetopidae are distributed worldwide in tropical and subtropical areas but absent from Australia; Synteliidae are confined to East Asia and Central America; Sphaeritidae are Holarctic; and Cretohisteridae are only known from the Lower Cretaceous Yixian Formation in China (Zhou *et al.* 2018).

The monophyly of Hydrophiloidea (s.l.) was recognised by most recent studies based both on adult and/or larval morphology (e.g. Hansen 1997; Beutel & Komarek 2004; Beutel & Leschen 2005; Lawrence *et al.* 2011) and molecular markers (McKenna *et al.* 2015a,b; Zhang *et al.* 2018). Some molecular studies however revealed Hydrophiloidea s.l. paraphyletic, indicating closer relationship of the histeroid lineage to some staphylinoid clades than to the hydrophiloid lineages (Bocák *et al.* 2014). All analyses support the position that Hydrophilioidea s.l. stand close to staphylinoid and scarabaeoid lineages in the Coleoptera tree of life; the precise position of the clade, however, varies considerably in published studies according to data and analytical methods used and hopefully will be resolved by upcoming transcriptomic or genomic studies.

Adults of the superfamily are characterised by the following synapomorphies: (1) spiracles on eigth abdominal segment atrophied; (2) the antennomere preceding the antennal club cupuliform (so-called cupule, e.g. Fig. 20.2D–E, except Georissidae); (3) antenna with 3-segmented club (e.g. Figs 20.2D–F, 21.59; in some groups the antennomeres are fused and the club is one- or two-segmented only: e.g. Figs 17.2D, 21.53); (4) mandibles with tubular glands (reduced in some groups). [Adapted from Lawrence & Ślipiński 2013].

Larvae are characterised by: (1) prognathous or hyperprognathous (back-tilted) head (Fig. 20.3A); (2) frons, clypeus and labrum completely fused, i.e. anterior margin of head simple (Fig. 19.3D–E) or with median teeth and lateral epistomal lobes (e.g. Figs 17.3D, 18.3C–D, 20.3D); (3) ecdysial lines reduced in the final instar; (4) antennal sensorium/sensoria situated laterally (e.g. Fig. 20.3E); (5) mandibular mola reduced or absent (Fig. 20.3F); (6) mandible with penicillus (lost in Hydrochidae, Spercheidae and Hydrophilidae); (7) lacinia absent (inner projection of stipes in Spercheidae and Hydrochidae are not homologous with lacinia, as proved by chaetotaxy); (8) galea located on palpifer (e.g. Figs 19.3H, 20.3H); (9) abdomen largely membranous; and (10) predatory habits. [Adapted from Lawrence & Ślipiński 2013].

Key to the Australian families of Hydrophiloidea

Both larval and adult keys are constructed to allow the assignment of Australian taxa to families, but not all characters used are diagnostic at world-wide scale and the keys hence cannot be used for taxa outside of Australia.

Adults

1. Metacoxal cavities widely transverse, only very narrowly separated from each other by a narrow abdominal process (e.g. Figs 18.2C, 19.2C, 20.2B–C). Antennae with 7–9 antennomeres, antennal club 3-segmented (Figs 18.2E, 20.2D–F) or seemingly 4-segmented (Fig. 19.2D), loosely segmented or compact. 3

– Metacoxal cavities rounded, widely separated by a wide abdominal process (e.g. Figs 17.2C, 21,2B). Antennae with 7–11 antennomeres, antennal club always compact, with 3, 2 or 1 antennomere only (e.g. Fig. 28.2D). 2

2(1). Prosternum totally concealed under large procoxal plates (Fig. 17.2C). Antenna with 7 antennomeres, antennal club one-segmented (Fig. 17.2D). Parts of pronotum and head with granulate microsculpture arranged to characteristic patterns (Fig. 17.2H–I). Tiny beetles (1.0–1.5 mm). .... **Georissidae**

– Prosternum always clearly visible between procoxae, forming a highly elevated and very apparent plate (Fig. 21.63). Antenna with 9–11 antennomeres, antennal club one- to three-segmented (Figs 21.53, 21.59). Dorsal surface of head and pronotum without granulate patterns. Body size variable .... **Histeridae**

3(2). Antenna with 7 antennomeres; antennal pedicel enlarged, much thicker than antennomeres 1 and 3, pubescent (Fig. 19.2D). Mentum very large (Fig. 19.2C). Tarsal formula 4–4–4 (Fig. 19.2H). Sides of pronotum denticulate (Fig. 19.2F–G). Clypeus bilobate, sexually dimorphic (Figs 19.2I–J). .... **Spercheidae**

– Antenna with 7–9 antennomeres, pedicel only slightly thicker than subsequent antennomeres, never pubescent (Figs 18.2E, 20.2D–F). Tarsal formula 5–5–5. Sides of pronotum smooth. Clypeus simple (e.g. Figs 18.2D, 20.4C, 20.11F–I, 20.19D–F) or emarginate medially with exposed membrane between clypeus and labrum (e.g. Figs 20.6M, 20.8F,P,a, 20.12F–G), with same morphology in males and females (except *Eurygmus* when the dimorphism concerns the extent of upturning, not the shape in dorsal view: Figs 20.14Z,a–b). 4

4(3) Procoxal cavities closed posteriorly (Fig. 18.2F). Ventral surface with numerous pubescent pits and depressions (Fig. 18.2C). Eyes large, globular, projecting laterally; head strongly constricted in front of eyes (Fig. 18.2D). Legs without swimming hairs. .... **Hydrochidae**

– Procoxal cavities open posteriorly (e.g. Fig. 20.2B–C). Ventral surface densely pubescent to bare, never with apparent pubescent pits and depressions. Eyes not largely protruding from head outline (e.g. Figs 20.4C, 20.8F, 20.11F–I, 20.19D–F), if largely protruding and head strongly constricted in front of eyes (Fig. 20.5O), then tibiae and tarsi of meso- and metathoracic legs with dense series of swimming hairs (Fig. 20.5E). .... **Hydrophilidae**

Larvae

1. Apex of abdomen (segments 8–10) modified, forming a spiracular atrium into which enlarged last pair of spiracles opens (Fig. 20.3I). Urogomphi reduced (1-segmented but hard to see: Fig. 20.3I). Labium with or without ligula. Second antennomere with a single large sensorium on apex (e.g. as Fig. 20.2E). **3**

– Abdominal segment 8–10 not modified, last pair of spiracles on segment 8 of the same morphology as spiracles on abdominal segments 1–7 (e.g. Fig. 17.3A). Urogomphi present and distinct (two- or one-segmented) or reduced to absent. Labium always without ligula. Second antennomere with two large sensoria on apex **2**

2. Ocular area with 6 distinct stemmata (Fig. 17.3A–B). Legs reduced, 3-segmented, without separate claw (Fig. 17.3E). Abdominal segment 8 differing from segment 7 by bearing distinct dorsal plate (Fig. 17.3A). Urogomphi seemingly absent (1-segmented but hard to see). Maxillary palps 3-segmented, labial palps 2-segmented. .... **Georissidae**

– Ocular area with a single stemma, or eyes totally absent. Legs normally developed or partly reduced, but always 5-segmented with well developed claw. Abdominal segment 8 with same structure as segment 7, without additional dorsal plate. Urogomphi distinct (2-segmented or 1-segmented), or present as small remanants and hence seemingly absent. Maxillary palps 3–4-segmented, labial palpi 2–3-segmented. .... **Histeridae**

3. Gula present, wide (Fig. 19.3C). Ocular area with 5 stemmata (Fig. 19.3F). Labium with fused mentum and prementum (Fig. 19.3I). Maxillary stipes with large median projection bearing series of setae and cuticular projections (Fig. 19.3H). General habitus of the larva myrmeleonid-like (Fig. 19.3A). .... **Spercheidae**

– Gula absent, only single gular suture present (Figs 18.3E–F, 20.3C). Ocular area with 6 distinct stemmata (e.g. Fig. 20.12d) which may be fused to 2–3 ocular spots only (Figs 20.17i–j, 20.18g) or to a single one (Figs 18.3D, 20.19e); never with 5 distinct stemmata. Mentum and prementum always separated (e.g. Figs 18.3L, 20.3G). Maxillary stipes conical (Fig. 20.3H) or with small median projection only (Fig. 18.3K). General habitus of larva elongate, campodeiform. **4**

4. First antennomere long and wide, antennomeres 2–3 minute (Fig. 18.3D, J). Submentum free, not fused to head capsule (Fig. 18.3E–F). Stipes with small median projection (Fig. 18.3K). Mandibles sickle-shaped (Fig. 18.3G–I). .... **Hydrochidae**

– First antennomere not exceptionally long and wide, antennomeres 2–3 well developed, only slightly thinner than antennomere 1 (Fig. 20.3E). Submentum fused to anterior margin of head capsule (Fig. 20.3C). Stipes conical, without inner projection (Fig. 20.3H). Mandibles variable in shape, but never sickle-shaped. .... **Hydrophilidae**

References

Archangelsky M, Beutel RG, Komarek A (2005) 10. Hydrophiloidea. Introduction, phylogeny. pp. 157–158. In: Beutel RG, Leschen RAB (eds) Handbook of Zoology, Band/Volume IV Arthropoda: Insecta, Teilband/Part 28 Coleoptera, Beetles. Volume 1: Morphology and Systematics (Archostemata, Adephaga, Myxophaga, Polyphaga partim). Walter de Gruyter, Berlin, 567 pp.

Beutel RG, Komarek A (2004) Comparative study of thoracic structures of adults of Hydrophiloidea and Histeroidea with phylogenetic implications (Coleoptera, Polyphaga). *Organisms, Diversity & Evolution* **4**, 1–34. doi:10.1016/j.ode.2003.10.001

Beutel RG, Leschen RAB (2005) Phylogenetic analysis of Staphyliniformia (Coleoptera) based on characters of larvae and adults. *Systematic Entomology* **30**, 510–548. doi:10.1111/j.1365-3113.2005.00293.x

Bocák L, Barton C, Crampton-Platt A, Chesters D, Ahrens D, Vogler AP (2014) Building the Coleoptera tree-of-life for >8000 species: composition of public DNA data and fit with Linnean classification. *Systematic Entomology* **39**(1), 97–110. doi:10.1111/syen.12037

Hansen M (1997) Phylogeny and classification of the staphyliniform beetle families (Coleoptera). *Biologiske Skrifter* **48**, 1–339.

Lawrence JF, Ślipiński A (2013) *Australian Beetles. Volume 1. Morphology, Classification and Keys.* CSIRO Publishing, Melbourne.

Lawrence JF, Ślipiński A, Seago AE, Thayer MK, Newton AF, Marvaldi AE (2011) Phylogeny of the Coleoptera based on morphological characters of adults and larvae. *Annales Zoologici* **61**(1), 1–217 [Warszawa]. doi:10.3161/000345411X576725

McKenna DD, Farrell BD, Caterino MS, Farnum CW, Hawks DC, Maddison DR, Seago AE, Short AEZ, Newton AF, Thayer MK (2015a) Phylogeny and evolution of Staphyliniformia and Scarabaeiformia: forest litter as stepping stone for diversification of nonphytophagous beetles. *Systematic Entomology* **40**(1), 35–60. doi:10.1111/syen.12093

McKenna DD, Wild AL, Kanda K, Bellamy CL, Beutel RG, Caterino MS, Farnum CW, Hawks DC, Ivie MS, Jameson ML, Leschen RAB, Marvaldi AE, McHugh JV, Newton AF, Robertson JA, Thayer MK, Whiting MF, Lawrence JF, Ślipiński A, Maddison DR, Farrell BD (2015b) The beetle tree of life reveals that Coleoptera survived end-Permian mass extinction to diversify during the Cretaceous terrestrial revolution. *Systematic Entomology* **40**(4), 835–880. doi:10.1111/syen.12132

Zhang SQ, Che LH, Li Y, Liang D, Pang H, Ślipiński A, Zhang P (2018) Evolutionary history of Coleoptera revealed by extensive sampling of genes and species. *Nature Communications* **9**(205), 1–11. doi:10.1038/s41467-017-02644-4

Zhou YL, Caterino MS, Ślipiński A, Cai CY (2018) Cretohisteridae, a new beetle family from the Early Cretaceous of China, and its implications for the early evolution of the basal group of Histeroidea (Coleoptera). *Systematic Entomology* **43**, 716–728.

17. GEORISSIDAE CASTELNAU, 1840

Martin Fikáček

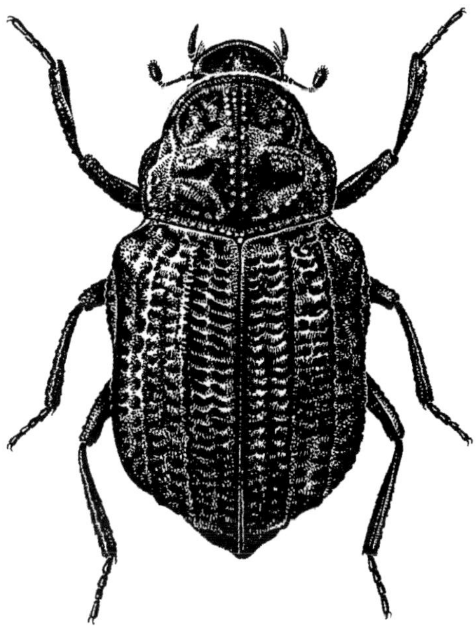

Fig. 17.1. *Georissus kingii* MacLeay

Common name. Minute mudloving beetles.

Introduction. The family Georissidae contains a single genus *Georissus* Latreille with 83 described species. The genus is distributed world-wide, but the highest diversity is known from the Old World tropics (Delève 1967a–c, 1972; Hansen 1999). In contrast, the diversity in New World is very low (2 species in North America, 2–3 species in the Neotropics).

Due to its highly modified adult morphology not much resembling other groups of the Hydrophiloidea, *Georissus* was considered as closely related to different beetle families (e.g. Dryopidae, Elmidae, Byrrhidae, Latridiidae, Thorictidae) by historical authors. Moreover, the Neotropical genus *Lepicerus* Motschulsky (nowadays in Myxophaga: Lepiceridae) was originally described and for a long time listed within the Georissidae. The close proximity of *Georissus* to Hydrophiloidea was recognised by Crowson (1950) based on adult morphology, and confirmed after the discovery of *Georissus* larvae by van Emden (1956). Since then, *Georissus* is consistently placed in the Hydrophiloidea and this placement was also confirmed by all recent phylogenetic analyses performed, both morphological and molecular (e.g. Beutel & Leschen 2005; Lawrence *et al.* 2011; Bocák *et al.* 2014; McKenna *et al.* 2015).

The relationship of Georissidae to remaining families of Hydrophiloidea s.str. is not properly understood so far. Studies based on different character sets revealed incongruent results. Hansen (1991) placed the Georissidae into the 'helophorid-lineage' of the Hydrophiloidea (together with Helophoridae, Epimetopidae and Hydrochidae). Although the monophyly of the lineage as a whole was not confirmed by subsequent studies, most phylogenetic studies performed recently place Georissidae close to Helophoridae and/or Epimetopidae (e.g. Archangelsky 1998; Beutel and Komarek 2004; Bernhard *et al.* 2009; Fikáček *et al.* 2012a), which seems to correspond well especially with larval morphology of these groups.

Taxonomy of the family is rather poorly known in most parts of the world except for the Afrotropical region, where it was revised taxonomically by Delève (1967a–c) (two additional species were described later from Madagascar by Calamandrei & Mascagni 1993). In other parts of the world many species remain undescribed and a detailed taxonomic revision is needed. Published papers only contain descriptions of individual new species from SW Asia (van Emden 1934; Delève 1969, 1972, 1974a,b; Mascagni 1996; Fikáček 2012), New Guinea (Delève 1969; Mascagni 1996), Himalaya (Champion 1923), Near East (Fikáček & Trávníček 2009; Fikáček & Falamarzi 2010; Fikáček *et al.* 2012b), Japan (Satô 1970, 1972, 1981; Nakane 1995) and Australia (King 1865; MacLeay 1871; Carter 1932).

Immature stages (Fig. 17.3) are described for three non-Australian species only: European *G. crenulatus* (Rossi) (van Emden 1956; Hansen 2000: egg case, larva, pupa), Nearctic *G. pusillus* LeConte (Archangelsky 1997: larva) and Nearctic *G. californicus* LeConte (Shepard 2003: larva, pupa).

The fossil record of Georissidae is very poor – two fossils from the Cretaceous and Eocene of Near East and Europe were recently assigned to the fossil genus *Georissites* Ponomarenko by Ponomarenko (2008). They are known only as isolated elytra and their assignment to Georissidae is therefore very doubtful (see also Haupt 1956; Hörnschemeyer *et al.* 1995). Additional fossils and subfossils assigned to *Georissus* are known from Miocene to Pleistocene of Europe and Alaska (Piton 1939; Hopkins *et al.* 1971; Matthews *et al.* 2003; Green *et al.* 2006; Ashworth and Nelson 2014).

Biology. Biology was studied only in few European and one North American species of *Georissus* (Messner 1965, 1972; Shepard 2003). Adults may be found on the surface of moist clay-sand deposits along banks of streams and rivers (more rarely of standing waters or at puddles and on open ground in salines). Adults process the substrate they live in as the food (Messner 1965, 1972), likely feeding on the algae and organic remains in the substrate. Larvae live in small cavities in the clay-sand deposits at the same places as adults. They are predaceous, preying on small invertebrates (nematodes and larvae of Diptera are supposed as a probable prey: Hansen 2000). Larval development includes two or three instars (two instars are recorded for European *G. crenulatus*, three for North American *G. californicus*: Hansen 2000; Shepard 2003) and takes about four weeks in Europe (Hansen 2000). Eggs are laid in an egg case (Fig. 17.3F) burrowed into the soil from which it only projects by a short mast (Hansen 2000). Pupation (Fig. 17.3G–H) takes place in a pupal chamber constructed in moist soil.

In *Georissus* (s.str.), alive adult beetles are often camouflaged by soil particles which they actively attach to the pronotum and elytra (Bameul 1989; Shepard 2003). However, alive specimens of species of other subgenera lack this camouflage (e.g. Fikáček

& Falamarzi 2010) and the muddy cover is absent also in most species in the collections. This suggests that the camouflage behaviour is likely present only in some species (probably only in the subgenus *Georissus* s.str.) and is not generally present in all Georissidae, as it is sometimes stated.

Few species in Africa and SE Asia are not associated with water at all, and they inhabit tropical forest leaf litter (Fikáček 2012; H. Yoshitomi, pers. comm.).

Characteristics. Adult. (Fig. 17.2). Length 1.0–2.2 mm (Australian species 1.0–1.5 mm). Body ~1.5–2.7 times as long as wide, sides not evenly rounded (lateral margin of pronotum and elytra does not form a continuous line), body well sclerotised and moderately to strongly convex; colour brown or black; dorsal surface tuberculate (with granules each bearing apical sensillum) and often costate; ventral surfaces without hydrofuge pubescence.

Head moderately to strongly declined, largely covered by anterior lobe of pronotum in dorsal view. Eyes well developed, not protuberant, circular and entire anteriorly, finely facetted, without interfacetal setae. Antennal insertions exposed from above; subantennal groove absent. Frontoclypeal and mid-cranial sutures impressed or indistinct; clypeus large, rounded anteriorly. Labrum transverse, well sclerotised, exposed in dorsal view, rounded apically. Antennae short, 7-segmented in Australian species (9-segmented in non-Australian *Georissus*), with basal antennomeres glabrous and 1-segmented (in Australian *Georissus*) or 3-segmented (in non-Australian species) pubescent club, antennomere preceding antennal club is not cup-like. Mandibles large and partly concealed beneath clypeus and labrum; apex unidentate or bidentate; mola well developed and often asymmetrical; prostheca well developed, membranous, pubescent and not articulated. Maxilla usually with setose galea and lacinia, the former rounded, sclerotised basally, apically membranous with series of long curved setae, the latter series of long setae, apical ones stout; maxillary palps slender, ca. as long as antennae; apical palpomere fusiform. Mentum triangular; ligula bilobed; apical labial palpomere expanded. Posterior tentorial pits and gular sutures confluent. Corpotentorium very narrow; anterior tentorial arms slender, not expanded into laminatentoria. Cervical sclerites present.

Pronotum ~0.7–0.8 times as long as wide, widest ca. at midlength, narrower than combined elytral bases; sides sinuate and sometimes slightly explanate; anterior edge strongly produced forward and rounded. Lateral pronotal carinae complete and simple; anterior angles absent; posterior angles rounded, not embracing elytral bases; posterior edge slightly arcuate or slightly angulate, with underlying ventral ridge, which interlocks with elytra and scutellar shield. Prosternum completely concealed by procoxae, shorter than shortest diameter of procoxal cavity, without median carina, anterior margin strongly concave. Prosternal process absent. Notosternal sutures absent. Hypomeron not divided into shiny lateral and pubescent mesal portions; with anterior groove for reception of antennal club. Procoxal cavities circular, widely separated, open externally, without narrow anterolateral extension.

Scutellar shield abruptly elevated, minute, sometimes not visible, posteriorly narrowly rounded. Elytra ~1.0–1.2 times as long as combined width and 1.7–2.2 times as long as pronotum; sides moderately curved or subparallel, apices conjointly rounded; humeri well developed; disc with ten puncture rows, without scutellary striole, alternate or all elytral intervals weakly to highly costate; epipleura absent. Mesoventrite separated by complete sutures from mesanepisterna, anterior portion of mesoventrite on different plane than metaventrite; without or with very weakly defined procoxal rests; posteromesal portion with elevated subpentagonal plate; mesoventral cavity absent; discrimen absent. Mesanepisterna widely separated from each other anteriorly. Mesocoxal cavities subcircular, widely separated, closed laterally by mesepimeron. Metaventrite distinctly transverse, flat to slightly convex; discrimen and transverse (metakatepisternal) suture reduced, weakly visible; metanepisternum not exposed ventrally, concealed beneath elytra. Metacoxae slightly transverse, widely separated, extending laterally to meet elytra. Metendosternite with short wide stalk and long lateral arms. Hind wing well developed, or in some species/populations reduced in size (brachypterous) or absent (apterous); when fully developed, wing narrowly elongate with apical field forming ca. two thirds of wing area; veins weakly sclerotised, ScA and RA reaching nearly to anterior hidge, radial cell unpigmented, RP not developed basally, proximally joint with MP and forming the R-M loop, basal cell present, with AA3 projection from it nearly to posterior margin of the wing, other veins of median are (braches of M, Cu and A) absent; anal lobe absent; wing margin with long sparsely arranged setae; wings of brachypterous species/specimens with ScA, RA and R-M loop present, but apical, median and anal areas reduced.

Legs. Protrochantins concealed; procoxae plate-like; endopleuron not fused to notum. Procoxae large, plate-like, completely concealing prosternum; protrochanter partly fused to protibia. Meso- and metacoxae slightly transverse; meso- and metatrochanter well developed, attached to posterobasal part of femur, femoral base hence oblique and anteriorly in contact with coxa. Protibiae slightly flattened and bisinuate on outer face; meso- and metatibiae cylindrical to slightly flattened, laterally more or less denticulate, without swimming hairs; tibial spurs short, indistinct. Tarsal 5–5–5 in majority of species examined (all non-Australian), 5–4–4 in Australian species; tarsomeres simple, tarsomere 1 small, globular and partly concealed by apex of tibia, and tarsi hence apparently tetramerous; in meso- and metatarsi of all Australian species the globular small first tarsomere is completely absent; pretarsal claws simple, arcuate; empodium small, asetose.

Abdomen with five free ventrites; basal 2 ventrites connate; ventrite 1 without median carina; intercoxal process wide and truncate. Functional spiracles on abdominal segments I–VI; abdominal tergites relatively lightly sclerotised, each with patches of wing-folding asperities. Anterior edge of sternite VIII in male without median strut. Segment IX in male U-shaped, with weakly delimited long lateral struts. Aedeagus of the trilobate type, symmetrical or with asymmetrical parameres, with narrowly elongate phallobase and simple parameres; penis with basal struts. Ovipositor moderately elongate, lightly sclerotised; paraprocts much shorter than coxites, which are subdivided with broad proximal lobe, slender distal lobe and small terminal styli.

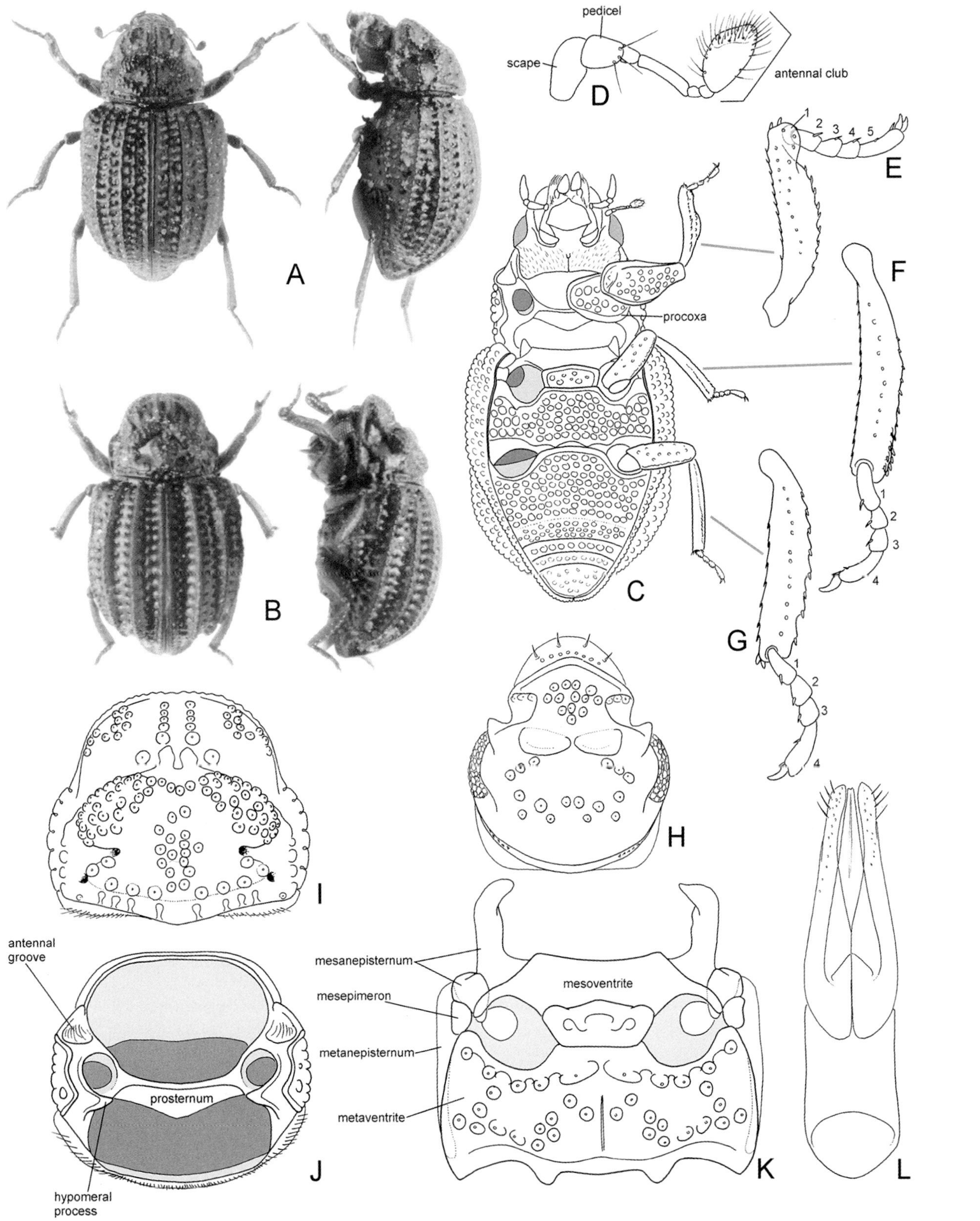

Fig. 17.2. Australian Georissidae, genus *Georissus*. A–B, general habitus in dorsal and lateral views: A, *Georissus* cf. *kingii* MacLeay; B, *Georissus* sp. (undescribed species). C–L, general morphology: C, ventral morphology; D, antenna; E–G, pro-, meso- and metatibiae and tarsi, ventral views; H, head in dorsal view; I, pronotum with dorsal granulation; J, pronotum, ventral morphology; K, morphology of mesothorax, ventral view; L, aedeagus. C, J, *Georissus* cf. *kingi* MacLeay; D–G, H–I, K–L, *Georissus* sp. (undescribed species).

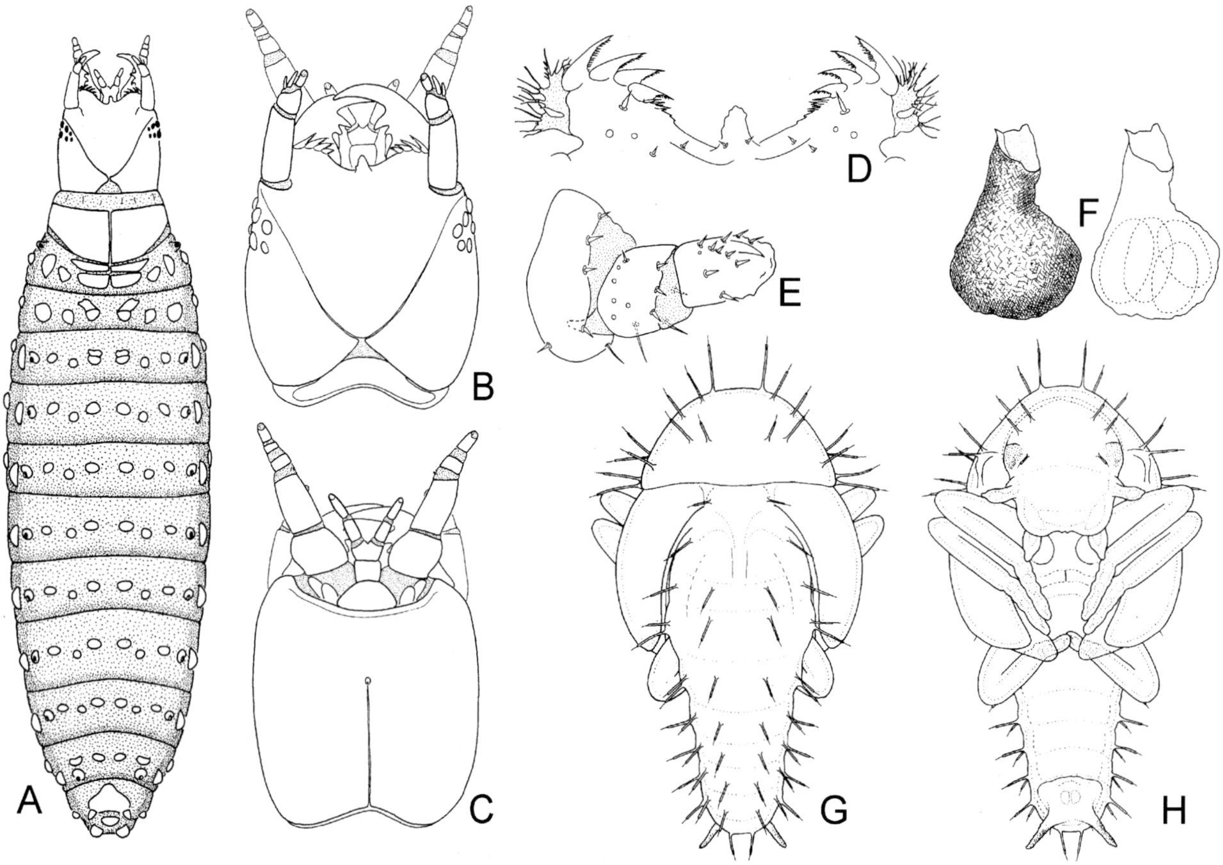

Fig. 17.3. Immature stages of *Georissus* (based on non-Australian species). A, general habitus of the larva of *G. pusillus* LeConte; B–C, larval head of *G. californicus* LeConte: B, dorsal view; C, ventral view. D, anterior margin of the larval head capsule of *G. pusillus* LeConte. E, prothoracic leg of *G. pusillus* LeConte (from Archangelsky 1997). F, egg case of *G. crenulatus* (Rossi). G–H, pupa of *G. crenulatus* (Rossi). A, D and E from Archangelsky (1997), F–H from Hansen (2000).

Larvae (based on non-Australian species, Fig. 17.3). Body elongate, nearly parallel-sided, slightly wider at middle; head and protergum well sclerotised, meso- and metathorax with terga divided into several small sclerites; dorsum of abdominal segments I-VIII with a transverse row sclerotised asperities; dorsum of abdominal segment IX with a sclerotised plate; membranous surfaces with asperites; lateral projections absent in thorax and abdomen.

Head prognathous to slightly hyperprognathus, ca. as long as wide, with sides only slightly curved; posterior edge distinctly emarginate; epicranial stem absent; frontal arms V-shaped with bases contiguous; median endocarina absent. Each side of head with 6 stemmata, well separated from each other, forming two vertical rows. Frontoclypeal suture absent; labrum fused to head capsule and forming clypeolabrum; clypeolabrum with median narrow projection (nasale) and large symmetrical paired adnasalia (= epistomal lobes); adnasalia with few wide flat setae with ciliate inner margin, and with membranous ciliate lateral portions. Antennae well developed, 3-segmented; antennomere 1 longest, antennomere 2 with laterally placed two sensoria. Mandibles symmetrical, moderately broad at base with narrow, strongly curved and unidentate apex, each with 2 retinacula; mola or prostheca absent; basal penicillus present. Ventral mouthparts protracted, maxillary articulating area absent; maxilla hexamerous, with segment-like palpifer: cardines distinct, each divided into 2 sclerites; stipes very short and cylindrical, palpifer long and cylindrical; galea (= inner appendage) membranous, not projecting; lacinia absent; palp 3-segmented. Labium consisting of prementum and postmentum; palps 2-segmented; ligula absent. Ventral tentorial pits and gular sutures fused.

Prothorax longer than meso- or metathorax each, with single tergal plate divided by median ecdysial line; meso- and metaterga with several separated sclerites. Prothoracic venter with two pairs of sclerites submesally, not sclerotised mesally. Legs 3-segmented, with pretarsus claw-like but fused to tibiotarsus; procoxae rather close to each other, meso- and metacoxae widely separated. Abdominal segments I–VIII usually lightly sclerotised, each with transverse row of small, round sclerites; without lateral projections or gills; abdominal sterna without prolegs. Abdominal apex without spiracular atrium, segment VIII is not terminal, tergum IX is completely visible with 1-segmented urogomphi, and segment X is terminal. Functional spiracles present on mesothorax and abdominal segments I–VIII.

Classification of the Australian taxa

Characters marked by an asterisk (*) apply only for Australian species, non-Australian species may bear different character states.

Georissus Latreille, 1809 (Pl. 43A, B; Figs 17.1–17.3)

Type species. *Pimelia pygmaea* Fabricius, 1798 (= *Georissus crenulatus* (Rossi, 1794)).

Characteristics. Adults (Fig. 17.2) strongly sclerotised, widely oval; head largely concealed beneath anterior lobe of the pronotum; dorsal surface of head with species-specific pattern of granules (tubercles with apical pore); labrum small but well sclerotised and exposed anterior to clypeus, semicircular; antennae 7-segmented with 1-segmented club*; mentum triangular, apical labial palpomeres enlarged and widely triangular; prothorax narrower than elytra combined; pronotum bearing species-specific pattern of granules and impressions; prosternum completely hidden beneath procoxae; scutellum very small or completely hidden; elytra moderately to highly costate at least on alternate intervals, with granules in some species; mesocoxal cavities widely separated by the broadly subpentagonal mesoventral plate; metaventrite transverse, with species-specific pattern of granules; hind wings well developed or reduced in size; abdomen with 5 ventrites, first two connate, all ventrites with species-specific pattern of granules; legs long, protibia sinuate on outer face, tarsal formula 5–4–4*; aedeagus with narrowly elongate phallobase; no sexual dimorphism; brown to black; 1.0–1.5 mm. Based on the combination of small body size, characteristic dorsal habitus and unique morphology of antenna and prothorax, *Georissus* cannot be confused with any other Australian beetle.

Larvae (based on non-Australian species, Fig. 17.3): V-shaped frontal arms, epicranial stem absent, posterior margin of head capsule excised dorsally; anterior margin of head symmetrical, with narrowly projecting median nasale and large epistomal lobes (= adnasalia) with flat wide ciliate setae; submentum not inserted into anteroventral margin of head capsule, hence submental line straight; each segment of abdomen with transverse row of sclerites, membranous parts with numerous asperites; spiracular atrium absent, urogomphi short, 1-segmented; legs short, reduced, with 3 segments only.

Australian species. Six to seven species are present in the Australian material examined so far. Three species are described at the moment: *G. australis* King, *G. kingii* MacLeay and *G. occidentalis* Carter. A taxonomic revision of the Australian *Georissus* is in progress (M. Fikáček, in prep.).

Distribution. The genus is nearly world-wide in distribution (see the Introduction to the family for details). The distribution in Australia is not well known, I saw specimens from Northern Territory, Queensland, New South Wales, Victoria and Western Australia.

Note. There are three subgenera recognised in *Georissus* at the moment, defined largely by the sculpture of the pronotum and elytra: *Nipponogeorissus* Satô is characterised by uniformly granulate pronotum and includes two species from Japan and New Guinea; *Georissus* s.str. is characterised by pronotum containing both smooth and granulate areas and smooth elytra without granules and costae, it is distributed in the Palaearctics and its border zones with the Oriental Region, and in Americas. *Neogeorissus* Satô is characterised by pronotum containing both smooth and granulate areas, but elytra with costate and/or partially granulate intervals (for details see Satô 1979; Hebauer 2004). Based on these characters, the Australian species agree with *Neogeorissus* and are classified in this subgenus until now. However, all Australian *Georissus* differ from the non-Australian representatives of the genus by the 7-segmented antennae with the antennal club consisting of a single segment (in contrast to 9-segmented antenna with 3-segmented club in non-Australian species), and in meso- and metatarsi with 4 tarsomeres (in contrast to pentamerous meso- and metatarsi in non-Australian *Georissus* – the basal tarsomere is small and not easily seen in non-Australian species, but it is always present). Based on these characters, and especially on the morphology of the antenna, it is clear that Australian *Georissus* form a monophyletic clade and differ strongly from the non-Australian species. For that reason, the Australian species cannot be classified within *Neogeorissus*, and will be assigned in a separate subgenus (M. Fikáček, in prep.). It is interesting, that the possible separate generic status of Australian members of Georissidae, indicated by the unusual antennal morphology, was mentioned already by King (1865) in the description of *G. australis*. The unusual antennal morphology was not commented by later authors, and the separate position of Australian *Georissus* hence remained overlooked until now.

Acknowledgments

I am indebted to L. Hendrich (Zoologische Staatssammlung München, Germany) and C. H. S. Watts (South Australian Museum) for the possibility to examine their material of Australian *Georissus*, to V. Grebennikov (Canadian Food Inspection Agency, Ottawa, Canada) for the long-term loan of his slide of the larvae of *G. californicus*, and to M. Archangelsky (CONICET, Esquel, Argentina) for the permission to use his drawings of larval *Georissus* in this chapter and the comments on the early draft of this chapter. J. Lawrence kindly helped with adapting the description of adult morphology.

References

Archangelsky M (1997) Studies on the biology, ecology, & systematics of the immature stages of New World Hydrophiloidea (Coleoptera: Staphyliniformia). *Bulletin of the Ohio Biological Survey. New Series* **12**(1), 1–207.

Archangelsky M (1998) Phylogeny of Hydrophiloidea (Coleoptera: Staphyliniformia) using characters from adult and preimaginal stages. *Systematic Entomology* **23**, 9–24. doi:10.1046/j.1365-3113.1998.00039.x

Ashworth AC, Nelson RE (2014) The paleoenvironment of the Olympia beds based on fossil beetles from Discovery Park, Seattle, Washington, U.S.A. *Quaternary International* **341**, 243–254. doi:10.1016/j.quaint.2013.09.022

Bameul F (1989) Description du comportement de camouflage d'un Coléoptère: le déguisement actif de *Georissus crenulatus* (Coleoptera Georissidae), et proposition d'une nouvelle classification des déguisements chez les invertebrates. *Comptes Rendus de l'Academie des Sciences (Paris), Serie III* **309**, 351–356.

Bernhard D, Ribera I, Komarek A, Beutel RG (2009) Phylogenetic analysis of Hydrophiloidea (Coleoptera: Polyphaga) based on molecular data and morphological characters of adults and immature stages. *Insect Systematics & Evolution* **40**, 3–41. doi:10.1163/187631209X416741

Beutel RG, Komarek A (2004) Comparative study of thoracic structures of adults of Hydrophiloidea and Histeroidea with phylogenetic implications (Coleoptera, Polyphaga). *Organisms, Diversity & Evolution* **4**, 1–34. doi:10.1016/j.ode.2003.10.001

Beutel RG, Leschen RAB (2005) Phylogenetic analysis of Staphyliniformia (Coleoptera) based on characters of larvae and adults. *Systematic Entomology* **30**, 510–548. doi:10.1111/j.1365-3113.2005.00293.x

Bocák L, Barton C, Crampton-Platt A, Chesters D, Ahrens D, Vogler AP (2014) Building the Coleoptera tree-of-life for >8000 species: composition of public DNA data fit with Linnean classification. *Systematic Entomology* **39**, 97–110. doi:10.1111/syen.12037

Calamandrei S, Mascagni A (1993) Contributo alla conoscenza dei Georissidae dei Madagascar con descrizione de due nuice species. *Bollettino della Società Entomologica Italiana* **125**, 109–112.

Carter HJ (1932) New Guinea and Australian Coleoptera. *Proceedings of the Linnean Society of New South Wales* **57**(3–4), 101–115.

Champion GC (1923) Some Indian Coleoptera. 11. *Entomologist's Monthly Magazine* **59**, 165–179.

Crowson RA (1950) The classification of the families of British Coleoptera. *Entomologist's Monthly Magazine* **86**, 149–171, 274–288, 324–344.

Delève J (1967a) Notes sur les *Georissus* d'Afrique et descriptions d'espèces nouvelles (Coleoptera Georissidae). *Bulletin de l'Institute Royal des Sciences Naturelles de Belgique* **43**(26), 1–23.

Delève J (1967b) Les *Georissus* de Madagascar (Coleoptera Georissidae). *Bulletin et Annales de la Société Entomologique de Belgique* **103**, 233–254.

Delève J (1967c) The scientific results of the Hungarian Soil Zoological Expedition to the Brazzaville-Congo. 18. Espèces de la familie Georyssidae (Coleoptera). *Opuscula Zoologica* **7**, 117–123[Budapest].

Delève J (1969) Quelques *Georissus* Latreille noveaux de l'Indonésie, Vietnam et de Nouvelle Guinée (Coleoptera: Georissidae). *Acta Zoologica Academiae Scientiarum Hungaricae* **15**, 23–30.

Delève J (1972) Les Georissidae (Coleoptera) de Ceylan. *Bulletin et Annales de la Société Entomologique de Belgique* **108**, 149–165.

Delève J (1974a) *Georissus capitatus* (Coleoptera, Georissidae), une nouvelle espèce du Vietnam. *Annales Historico-Naturales Musei Nationalis Hungarici* **66**, 173–174.

Delève J (1974b) Un *Georissus* nouveau de Malaisie (Coleoptera). *Revue Suisse de Zoologie* **81**, 487–488. doi:10.5962/bhl.part.76016

Fikáček M (2012) A new species of leaf-litter inhabiting *Georissus* from Bornao (Coleoptera: Hydrophiloidea: Georissidae). *Zootaxa* **3316**, 63–68.

Fikáček M, Falamarzi Sh (2010) *Georissus persicus* sp. nov. from Iran, with notes on the West-Palaearctic species of the *G. laesicollis* group (Coleoptera: Georissidae). *Acta Entomologica Musei Nationalis Pragae* **50**, 107–116.

Fikáček M, Trávníček D (2009) Georissidae of the United Arab Emirates. In *Arthropod fauna of the UAE, volume 2* (Ed A van Harten) pp. 145–148. Multiply Marketing Consultancy Services, Abu Dhabi.

Fikáček M, Prokin A, Angus RB, Ponomarenko A, Yue Y, Ren D, Prokop J (2012a) Phylogeny and fossil record of the Helophoridae reveal Jurassic origin of extant hydrophiloid lineages (Coleoptera: Polyphaga). *Systematic Entomology* **37**, 420–447. doi:10.1111/j.1365-3113.2012.00630.x

Fikáček M, Delgado JA, Gentili E (2012b) The Hydrophiloid beetles of the Socotra Island (Coleoptera: Georissidae, Hydrophilidae). *Acta Entomologica Musei Nationalis Pragae* **52**(supplementum 2), 107–130.

Green CP, Branch NP, Coope GR, Field MH, Keen DH, Wells JM, Schwenninger JL, Preece RC, Schreve DC, Canti MG, Gleed-Owen CP (2006) Marine Isotope Stage 9 environments of fluvial deposits at Hackney, north London, UK. *Quaternary Science Reviews* **25**, 89–113. doi:10.1016/j.quascirev.2004.10.011

Hansen M (1991) The hydrophiloid beetles. Phylogeny, classification and a revision of the genera (Coleoptera, Hydrophiloidea). *Biologiske Skrifter* **40**, 1–367.

Hansen M (1999) *Wold Catalogue of Insects. Volume 2. Hydrophiloidea (s.str.) (Coleoptera).* Apollo Books, Stenstrup, 416 pp.

Hansen M (2000) Observations on the immature stages of Georissidae (Coleoptera: Hydrophiloidea), with remarks on the evolution of the hydrophiloid egg cocoon. *Invertebrate Taxonomy* **14**, 907–916. doi:10.1071/IT00016

Haupt H (1956) Beitrag zur Kenntnis der eozänen Arthropodenfauna des Geiseltales. *Nova Acta Leopoldina* **128**, 1–90.

Hebauer F (2004) Systematic and zoogeographic notes on the genus *Georissus* Latreille, 1809 (Coleoptera: Hydrophiloidea). *Acta Coleopterologica* **22**(1), 3–6.

Hopkins DM, Matthews JV, Wolfe JA, Silberman ML (1971) A Pliocene flora and insect fauna from the Bering Strait region. *Palaeogeography, Palaeoclimatology, Palaeoecology* **9**, 211–231. doi:10.1016/0031-0182(71)90032-0

Hörnschemeyer T, Tröster G, Wedmann S (1995) Die oezänen Käferfauna des Geiseltals und der Grube Messel - ein Vergleich und systematischen und paläoökologischen Gesichtspunkten. *Hallesches Jahrbuch der Geowissenschaften* **17**, 107–119.

King RL (1865) Description of Australian species of georyssides and parnides. *Transactions of the Entomological Society of New South Wales* **1**, 158–161[+ pl. xiv.].

Lawrence JF, Ślipiński A, Seago AE, Thayer MK, Newton AF, Marvaldi AE (2011) Phylogeny of the Coleoptera based on morphological characters of adults and larvae. *Annales Zoologici* **61**(1), 1–217 [Warszawa]. doi:10.3161/000345411X576725

MacLeay W (1871) Notes on a collection of insects from Gyandah. *Transactions of the Entomological Society of New South Wales* **2**, 79–205.

Mascagni A (1996) Descrizione dei maschi di *Georissus biroi* Deleve e *G. emdeni* Deleve (Coleoptera Georissidae). *Bollettino della Società Entomologica Italiana* **127**, 227–230.

Matthews JV, Westgate JA, Ovenden L, Carter LD, Fouch T (2003) Stratigraphy, fossils, and age of sediments at the upper pit of the Lost Chicken gold mine: new information on the late Pliocene environment of east central Alaska. *Quaternary Research* **60**, 9–18. doi:10.1016/S0033-5894(03)00087-5

McKenna DD, Farrell BD, Caterino MS, Farnum CW, Hawks DC, Maddison DR, Seago AE, Short AEZ, Newton AF, Thayer MK (2015) Phylogeny and evolution of Staphyliniformia and Scarabaeiformia: forest litter as a stepping stone for diversification of nonphytophagous beetles. *Systematic Entomology* **40**, 35–60. doi:10.1111/syen.12093

Messner B (1965) Zur Biologie der Georyssiden (Coleopt.). *Entomologische Berichte* **1964**(3), 97–100.

Messner B (1972) Zur Biologie einheimischer Käferfamilien. 9. Georyssidae. *Entomologische Berichte* (Berlin) **(1972)**, 1–4.

Nakane T (1995) Notes on some little-known beetles (Coleoptera) in Japan. 13. *Kita-kyushû no Konchû* **42**(1), 41–46.

Piton L (1939) Note complémentaire sur les insectes fossiles des cinérites pliocenes du lac Chambon (Puy-de-Dome). *Revue des Sciences Naturelles de l'Auvergne (N.S.)* **5**, 102–108.

Ponomarenko AG (2008) The beetles. Order Coleoptera (=Scarabaeida). In *Plant-Arthropod interactions in the early angiosperm history: evidence from the Cretaceous of Israel* (Eds V Krassilov and A. Rasnitsyn) pp. 193–196. Pensoft, Sofia-Moscow.

Satô M (1970) Une nouvelle espèce de genre *Georyssus* Latreille au Japon (Coleoptera: Georyssidae). *Journal of Nagoya Women's College* **16**, 199–200.

Satô M (1972) The Georissid Beetles of Japan. *Journal of Nagoya Women's College* **18**, 207–213.

Satô M (1979) Ergebnisse der Bhutan-Expedition 1972 und Indien-Nepal-Expeditionen 1975–1977 des Natur-historischen Museums in Basel. Coleoptera: Fam. Georissidae, Hydraenidae, Hydrophilidae und Ptilodactylidae. *Entomologica Basiliensia* **4**, 43–67.

Satô M (1981) An outline of Japanese Georissidae. *Coleopterist's News* **53**, 1–4.

Shepard WD (2003) The biology of *Georissus californicus* LeConte (Coleoptera, Hydrophilidae, Georissinae). *Special Bulletin of the Japanese Society of Coleopterology* **6**, 121–128 [Tokyo].

van Emden FI (1934) Von H. Overbeck gefangene neue und bekannte Arten der Carabidae, Hydrophilidae, Georyssidae und Clavicornia (Col.). *Entomologische Blätter* **30**(5), 193–200.

van Emden FI (1956) The Georyssus larva – a hydrophilid. *Proceedings of the Royal Entomological Society, London* **31**(1–3), 20–24.

18. HYDROCHIDAE THOMSON, 1859

Martin Fikáček

Fig. 18.1. *Hydrochus horni* Blackburn

Common name. Elongated water scavenger beetles.

Introduction. The family Hydrochidae contains a single genus *Hydrochus* Leach (Fig. 18.1) with 180 described species. It is distributed world-wide (except some islands, e.g. New Zealand) and fairly diverse both in tropical and temperate regions (Hansen 1999; Short & Fikáček 2011).

Hydrochidae is rather homogenous morphologically and without any doubt monophyletic; both adults and larvae are characterised by many unique characters. Nevertheless, the position of the family within Hydrophiloidea s.str. remains unclear. The earliest morphological analyses (Hansen 1991; Beutel 1994) were influenced by the absence of relevant information on larval morphology. Larval characters were re-examined and more correctly interpreted by Archangelsky (1997, 1998) and Beutel (1999), but the knowledge on larval characters still remains very incomplete. Two principal hypotheses about phylogenetic position of hydrochids were proposed based on morphology. The first one considers Hydrochidae as a part of the 'hydrophilid lineage' together with Hydrophilidae and Spercheidae (Archangelsky 1998; Beutel and Leschen 2005). The second hypothesis places Hydrochidae into the 'helophorid lineage' together with Helophoridae, Georissidae and Epimetopidae; it seems to be supported mainly by adult characters (Hansen 1991; Anton and Beutel 2004). Both hypotheses were supported by molecular analyses based on different datasets and gene selections: those by Korte *et al.* (2004) and Hunt *et al.* (2007) support the first hypothesis and place hydrochids close to Spercheidae; those of Caterino *et al.* (2005), Bernhard *et al.* (2006, 2009) and McKenna *et al.* (2015) support the second hypothesis and place Hydrochidae close to Georissidae and Helophoridae.

The taxonomy of the Hydrochidae is still rather poorly known. Larger modern revisions were published only for faunas of South America (Oliva 1992) and Australia (Watts 1999). Fauna of the Palaearctic region is also rather well known, despite the fact that it was never treated as a whole (Angus 1977; Berge Henegouwen 1988; Berge Henegouwen & Sainz-Cantero 1992; Castro & Delgado 1998, 1999; Valladares 1988; Valladares *et al.* 1999; Hidalgo-Galiana *et al.* 2010; Shatrovskiy 1989, 1993). Knowledge about the Oriental, Afrotropical and Nearctic faunas remains very inadequate. The interest in the taxonomy of the Hydrochidae was unfortunately largely decimated by the activities of D. Makhan who published several dozens of species descriptions of very poor formal quality (see e.g. Hansen 1998, 1999; Oliva 2000; Short & Hebauer 2006 for more detailed critique).

Phylogenetic relationships within Hydrochidae were not studied so far. The only exception is a study by Hidalgo-Galiana & Ribera (2011) focused on western-European species; a single Australian representative was included in the analysis and revealed as sister to few South and North American species, but all these species were included as outgroups only and better taxon sampling is necessary to understand the position of Australian *Hydrochus* properly. The division of *Hydrochus* into nine genera was proposed by Makhan (1994, 1998, 2001a,b, 2004a), but all newly erected genera were later synonymized with *Hydrochus* for more or less formal reasons (Hansen 1998, 1999; Short & Hebauer 2006) and are not accepted by other authors than D. Makhan. Some of the Makhan's genera were partly based on characters which are very likely phylogenetically informative (e.g. number of antennomeres of the antennal club, morphology of male genitalia) but a wider comparative study across the family in phylogenetic context is necessary to understand their relevance.

Immature stages (Fig. 18.3) are very little known, and were described in detail only for the Nearctic *H. squamifer* LeConte and *H. rufipes* Melsheimer (Richmond 1920; Archangelsky 1997) and the European *H. elongatus* (Schaller) (Beutel 1999, head morphology only). Brief notes on mandibular morphology of five European species were provided by Makhan (2004b), some additional photographs of larval morphology of the European species and the Neotropical *H. piroei* Makhan were published by Makhan *et al.* (2012). Few details on morphology of egg case and larvae of European species were provided by Angus (2015). Larva of Australian species is unknown. Pupae of Hydrochidae remain unknown.

Fossil record of Hydrochidae is very poor. Two fossil species of *Hydrochus* have been described by Scudder (1890): *Hydrochus relictus* Scudder from the Eocene Green River Formation, and *H. amictus* Scudder from the Pleistocene deposits in Ohio; their identity is unclear and requires a revision. Subfossil remains of *Hydrochus* are also known from the Pleistocene peat bog deposits in the Great Britain (Angus 1977); all were attributed to species living today. Re-examination of the Early Cretaceous genus *Cretohelophorus* Ponomarenko revealed that it may be related to *Hydrochus*; a detailed study is in progress (Ponomarenko 1987, Fikáček *et al.*, unpublished data).

Biology. Both adults and larvae are aquatic, inhabiting densely vegetated standing waters, leaf debris and detritus at edges of standing and slowly running waters, and/or gravel or sandy banks of streams and rivers. Adults are slowly moving and it is supposed that they feed on algae (Archangelsky 1997). As in the Hydrophilidae, adult hydrochids breathe atmospheric oxygen which is stored on pubescent ventral surface; the contact of the ventral bubble with the atmosphere needed for restoring the air supply is gained by antennae (Hrbáček 1950). Eggs are laid in silky egg cases with or without mast attached to the substrate (usually plants), each egg case contains one or two eggs (Richmond 1920; Angus 2015). Larvae likely have three larval instars, similarly to most other members of Hydrophiloidea s.str. (J. Balfour-Browne, unpublished data), but little is known about their biology. First instars are easy to obtain from egg cases laid by adults, but higher instars were only rarely obtained by rearing. Larval diet is unknown, but the unusual morphology of mandibles suggests it may be highly specialised. Makhan (2004b) reported a mixed diet for the European *H. elongatus* consisting of algae ingested when the larval head is submerged, and aquatic oligochaetes which are digested extraorally with the head held up above water surface as in Hydrophilidae (see there for details). Makhan (2004b) also reports that larvae ingest air bubbles which are then visible in the intestines; same behaviour is also known in many hydrophilid larvae where they increase buyonancy and facilitate surfacing. Nothing is known about the length of larval development. Larvae breathe atmospheric oxygen which they pump into tracheal system through large spiracles of abdominal segment VIII situated in the spiracular atrium. Pupa and pupation habits are unknown.

Characteristics. Adults (Fig. 18.2). Length 0.9–5.4 mm. Body ~2.5–3.0 times as long as wide, slightly flattened, with sides more or less parallel-sided; colour reddish to black, often with metallic sheen, in some species with spotted elytra; dorsal surfaces usually smooth and glabrous, in some species with elevated granules; ventral surfaces with hydrofuge pubescence (plastron). Head prognathous, deeply inserted into prothorax. Eyes well developed, semiglobular and strongly protuberant, circular and entire anteriorly, finely facetted, with interfacetal setae. Antennal insertions weakly concealed from above; subantennal groove present. Frontoclypeal and midcranial sutures distinct as impressions only, frontoclypeal suture angulate; clypeus large, truncate anteriorly. Labrum exposed, strongly transverse, well sclerotised, rounded anteriorly, widest at base. Antennae short, with 5–7 antennomeres, basal antennomeres glabrous pubescent club usually with 3 antennomeres, in some species with two or one antennomere only; the club preceded by moderately to strongly asymmetrical cup-like antennomere (cupule). Mandibles large, partly exposed laterally and anteriorly of clypeus; apex bidentate; mola well developed, asymmetrical; prostheca reduced, inner face of mandible only with long pubescence, without membranous or sclerotised appendages. Maxilla with setose galea and lacinia, galea sclerotised basally, slightly flattened, with apical curved setae arranged in regular rows, lacinia with moderately small uncus; maxillary palps slender, ca. as long as antennae; with 4 palpomeres, apical palpomere long and asymmetrical. Mentum slightly wider than long, angulate on anterior margin; ligula bilobed, each half with large transversely oval sclerite; labial palpus with 3 palpomeres, apical one expanded, globular. Corpotentorium moderately narrow; laminatentoria well developed, sometimes meeting at midline. Cervical sclerites small. Pronotum ca. as long as wide, widest anteriorly, narrower than elytral bases at posterior margin; sides straight, slightly curved or sinuate, not explanate laterally; anterior edge arcuate, not produced forwards; lateral pronotal carinae complete, simple or denticulate; anterior angles not produced, subangulate; posterior angles right to rounded; posterior edge straight to slightly arcuate, with underlying ventral ridge, which interlocks with elytra and scutellar shield. Prosternum long, in front of coxa ca. as long as diameter of procoxal cavity, flat and with pubescent impressions. Prosternal process complete, apically expanded and truncate. Notosternal sutures absent. Hypomeron with narrow glabrous lateral portion and pubescent mesal portion separated by ridge; antennal grooves minute on anterolateral portion of hypomeron. Procoxae subglobular and projecting, with very short concealed lateral extensions; trochantins concealed; endopleuron fused to notum. Procoxal cavities subcircular, narrowly separated by prosternal process, externally closed, with distinct closed anterolateral extension, internally closed. Scutellar shield abruptly elevated, minute, much longer than wide, rounded posteriorly. Elytra ~1.6–2.0 times as long as combined width and 2.4–3.0 times as long as pronotum; sides subparallel with apices subtruncate; humeri well developed; disc with ten distinct puncture rows or striae, without scutellary stria, even intervals often costate; epipleura usually incomplete (rarely developed throughout), always narrow, not divided into bare and pubescent portions. Mesoventrite wide at anterior margin, separated from mesanepisterna by complete sutures, without procoxal rests; mesal portion flat, without carinae or elevations, with numerous setiferous impressions; mesoventral cavity absent; discrimen absent. Mesocoxae subglobular with concealed trochantins. Mesocoxal cavities narrowly separated, laterally closed by meso- and metaventrite. Metaventrite usually ca. as long as wide, flat to slightly convex, with numerous pubescent impressions; discrimen weakly developed in posterior portion, metakatepisternal suture present mesally. Metacoxae moderately transverse and subcontiguous, extending laterally to meet elytra; plates absent. Metendosternite relatively small, with short stalk and rather short lateral arms; anterior process either very short and broadly emarginate, so that anterior tendons are on widely separated lobes. Hind wing with long radial bending zone at the end of which is a distinct hinge; apical field with three radial extensions; cross-vein r3 absent; cross-vein r4 complete; RP obliterated basally of r3, well sclerotised distally of it; medial spur very short, straight; medial field usually with four free veins; MP_{3+4} with basal cross-vein; basal cell and wedge cell well developed, basal cell posteriorly with long AA3 nearly reaching elytral margin; anal lobe well developed, with deep notch-like embayment; AP present and undivided. Femoral attachment of mesotrochanter oblique, anterobasal corner of not facing trochanter. Tibiae cylindrical, lined with spines and several series of long (swimming?) hairs; tibial spurs very short. Tarsi 5–5–5, basal tarsomere always very short, globular; each tarsomere with

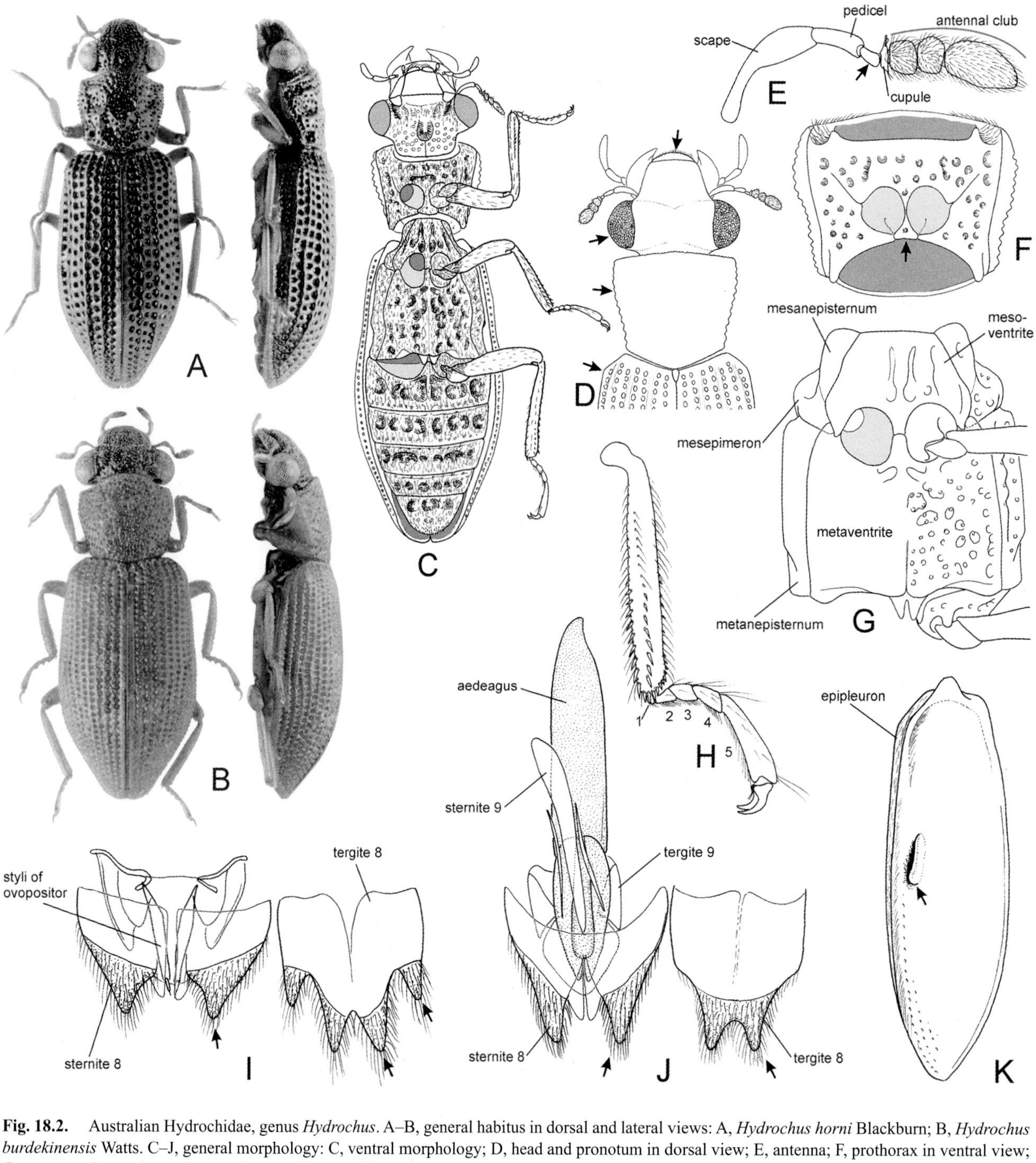

Fig. 18.2. Australian Hydrochidae, genus *Hydrochus*. A–B, general habitus in dorsal and lateral views: A, *Hydrochus horni* Blackburn; B, *Hydrochus burdekinensis* Watts. C–J, general morphology: C, ventral morphology; D, head and pronotum in dorsal view; E, antenna; F, prothorax in ventral view; G, meso- and metathorax in ventral view; H, mesotibia and mesotarsus; I, female genitalia and associated sclerites; J, male genitalia and associated sclerites; K, elytron in ventral view.

or without few long swimming hairs dorsally; pretarsal claws simple; empodium well developed, bisetose. Abdomen with five free ventrites; ventrite 1 without median carina; intercoxal process rather acute. Abdominal tergites relatively lightly sclerotised, with patches of wing-folding asperities. Sternite VIII and tergite VIII of both sexes with pubescent triangular lobes on posterior margin, tergite VIII sexually dimorphic, with two lobes in males and four lobes in females; in some species (non-Australian ones), the lobes of tergite 8 can be prolonged and articulated (see Hansen 1998 for details); anterior edge of sternite VIII without median strut. Segment IX in male without spiculum gastrale; sternite IX tongue-like with lateral struts, tergite IX emarginate; tergite X free. Aedeagus of the trilobate type, symmetrical or with asymmetrical phallobase, genital morphology highly variable among species-groups, often phallobase very long and parameres very short, in some non-Australian species whole aedeagus extremely long and slender; parameres simple; penis with short basal struts or without struts. Ovipositor moderately elongate, lightly sclerotised; gonocoxites subdivided with broad proximal lobe, slender distal lobe and small terminal styli [adapted from Lawrence & Ślipiński 2013; see also Hansen 1998; Watts 1999].

Larvae (Fig. 18.3). Body elongate, parallel-sided, slightly wider at middle; head and protergum well sclerotised, meso- and metatergum each with pair of large tergites; abdomen membranous in first instars, higher instars with two pairs of weakly sclerotised tergites on each abdomen [observed on very transparent slide-mounted 3rd instar of *H. angustatus* from J. Balfour-Browne collection, also weakly seen on the photo of higher instar of *H. elongatus* in Makhan *et al.* (2012)]; abdominal segment VIII with sclerotised tergum which is entire or subdivided medially; abdomen without lateral abdominal projections.

Head prognathous, strongly transverse, more or less parallel-sided; posterior edge straight to weakly emarginate; coronal line (= epicranial stem) absent; frontal arms converging posteriorly, bases widely separated, not meeting each other; median endocarina absent. Stemmata aggregated [six stemmata mentioned by Richmond (1920) and Archangelsky (1997), single stemma by Beutel (1999), none of that can be confirmed based on material examined]. Frontoclypeal suture absent and labrum fused to head capsule forming clypeolabrum; medial nasale projecting or not, with small denticles; epistomal lobes (= adnasalia) present, rather large, symmetrical. Antennae

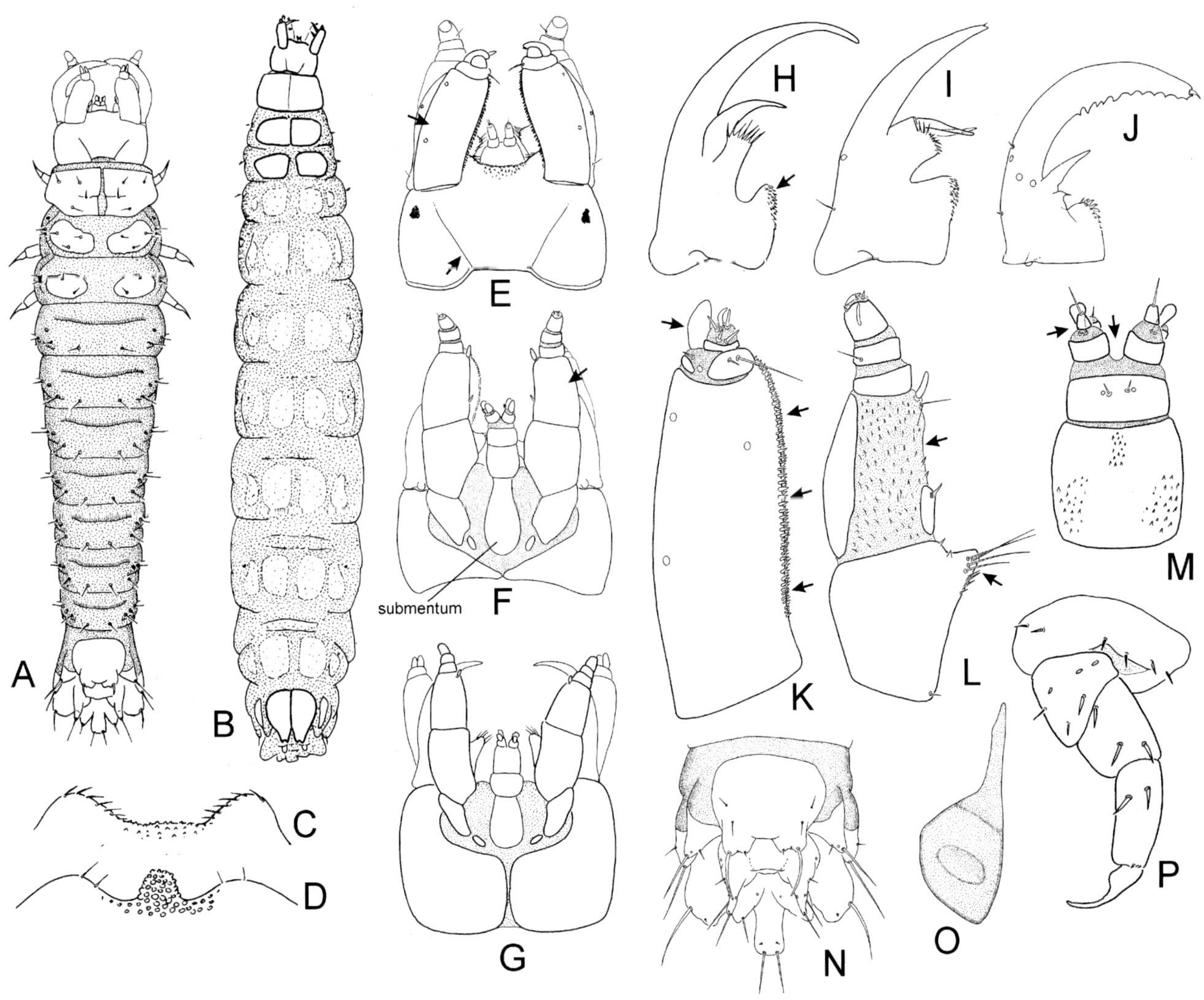

Fig. 18.3. Immature stages of Hydrochidae, genus *Hydrochus* (on non-Australian species). A–B, general habitus (A, first instar of *H. rufipes* Melsheimer; B, third instar of *H. angustatus* Germar). C–D, anterior margin of head (C, *H. crenatus* (Fabricius); D, *H. megaphallus* Berge Henegouwen). E, head of *H. crenulatus* in dorsal view; F–G, head in ventral view (F, *H. megaphallus*; G, *H. angustatus*). H–J, mandible (H, *H. angustatus*; I, *H. crenatus*; J, *H. rufipes*). K–M, head appendages of *H. megaphallus* (K, antenna; L, maxilla in dorsal view; M, mentum and prementum in dorsal view); N, abdominal apex of *H. rufipes*. O, egg case of *H. squamifer* LeConte. P, metathoracic leg of *H. angustatus*. A, J, N–O from Archangelsky (1997).

well developed, 3-segmented; antennomere 1 very long and stout, whole inner face with spines or bare with spinose mesal projection (*in H. rufipes*), antennomeres 2–3 minute, antennomere 2 with laterally placed large sensorium. Mandibles symmetrical, with basal denticulate mola-like structure, moderately broad at base with narrow, strongly curved unidentate apex, with 1–2 inner teeth (= retinacula); prostheca absent; basal penicillus absent or present (penicillus-like structure found in *H. crenatus*, homology with Helophoridae and Georissidae is unclear). Ventral mouthparts distinctly retracted and with articulating area; maxilla hexamerous: cardines distinct, large, subdivided basally; stipes rather short and wide; lacinia (= inner projection of stipes) short, rounded and bearing setae; palp 3-segmented, with very long segment-like palpifer; galea (= inner appendage) small, digitiform, strongly sclerotised. Labium with submentum elongate, not fused with head capsule; mentum subquadrate, prementum short, palps 2-segmented, palpomere 1 apically with two large sensilla ventrally and dorsally of palpomere 2, hence palpomere 2 virtually trilobate; ligula absent. Hypopharyngeal sclerome and hypostomal rods absent; ventral epicranial ridges absent. Gular region usually very short, longer in some species (e.g. *H. angustatus*), gular sutures nearly fused, gular region membranous [gular sutures considered as well separated and gula present by most authors, I cannot confirm it based on material examined, as only a very narrow membranous area is present between the margins of left and right epicranium in all examined larvae]. Prothorax as long as meso- or metathorax, with single tergal plate divided by median ecdysial line; meso- and metaterga with paired plates. Prothoracic venter without distinct presternal plate. Legs well developed, 5-segmented with pretarsus forming distinct claw, without swimming hairs; coxae moderately widely separated on all thoracic segments. Abdominal segments I–VII lightly sclerotised, in examined third instar of *H. angustatus* with two pairs of moderately large probably weakly sclerotised tergites, examined first instar larvae without such tergites; abdominal sterna III–VII without prolegs or asperities. Abdominal apex with spiracular atrium formed by modifications of segments VIII–X (tergum VIII forming sclerotised dorsal valve at apex of abdomen, segment IX trilobed and forming ventral valve, which bears a pair of small 1-segmented urogomphi, posterolateral portions of segments 8–9 in some species with several pairs of sclerotised finger-like projections; segment X reduced and ventral. Functional spiracles absent except for an enlarged annular pair in spiracular atrium [based on non-Australian larvae described or illustrated by Richmond 1920, Archangelsky 1997, Beutel 1999 and Makhan *et al.* 2012; I also examined first and third instar larvae of *H. angustatus* Germar on slides of J. Balfour-Browne deposited in the Natural History Museum, London, and first instars of *H. megaphallus*, *H. crenatus* and *H. ignicollis* reared from egg cases laid by adults and deposited in National Museum, Prague].

Classification of the Australian taxa

Characters marked by an asterisk (*) apply only for Australian species of the respective taxon, non-Australian species may bear different character states.

Hydrochus Leach, 1817 (Pl. 43C–D; Figs 18.2, 18.3)

Type species. *Silpha elongata* Schaller

Characteristics. Body elongate, depressed to moderately convex; dorsal surface in some species with elevated granules; whole ventral surface with deep pubescent impressions; head strongly constricted in front and behind eyes; eyes semiglobular, strongly protruding laterally; clypeus straight anteriorly; labrum strongly sclerotised, exposed; antenna with 7 antennomeres*, with 3-segmented* pubescent antennal club; pronotum subquadrate, with or without shallow impressions; prosternum very long in front of procoxae, procoxal cavities closed; elytral base wider than base of pronotum; elytron with 10 punctural striae, with or without ridges/keels on every second interval; epipleuron very narrow; anapleural sutures distinct, converging but not meeting anteriorly; mesoventrite very wide anteriorly, mesanepisterna not each other; metaventrite completely pubescent, metanepisternum narrow, laterally excised subbasally; abdomen with five visible ventrites, all covered with dense hydrofuge pubescence; ventrite 5 entire on posterior margin; tarsi 5-segmented, of the same shape in male and female; hydrofuge pubescence confined to basal portion of meso- and metafemora; coloration pale reddish to dark brown or black, often with greenish metallic sheen, in some species with spotted elytra; 0.9–4.7 mm. Easily recognisable from all other Australian beetles by the combination of elongate body, large globular eyes, clubbed antenna, subquadrate pronotum and ventrum with deep setiferous impressions. Among water beetles, it may resemble *Berosus* (Hydrophilidae) by head shape, large eyes, long legs and antennae with 7 antennomeres, but differs from it by long prosternum, anteriorly wide mesoventrite, elytra without scutellary stria and ventral surface with pubescent impressions.

Larvae (based on non-Australian species, Fig. 18.3): Head transverse; anterior margin with distinct epistomal lobes and with or without median projection (nasale), median area denticulate; gula absent; antenna with long and stout first antennomere armed by numerous spines on inner face, antennomeres 2–3 min, sensorium large; mandibles symmetrical, with one or two inner teeth and basal spinose mola; maxilla rather short stipes with mesal setose projection, palpifer longest; inner process small, strongly sclerotised, finger-like; labium with submentum not fused to head capsule, elongate; ligula absent; labial palpus with two large appendages on palpomere 1, hence palpomere 2 virtually trilobed; prothorax with large dorsal tergite, meso- and metathorax each with pair of smaller sclerites, abdominal segments without projections, largely membranous or with two pairs of weakly sclerotised plates; legs short, with well developed claw; abdominal apex with posteriorly bilobed dorsal plate (may be subdivided medially), stigmatic atrium present, often with several pairs of lateral projections; ventral prolegs absent. Easily recognisable by characteristicly transverse head shape, masive internally spinose antennae and unique shape of the mandible [based on non-Australian larvae, see under larval morphology of the whole family for sources of information].

Australian species. 25 species were recognised in the revision by Watts (1999). Two additional new species were mentioned subsequently by Makhan (2004a,b), one of which is *nomen nudum* as it does not fulfill the requirements of the International Code of Zoological Nomenclature, and the second (*H. hellenae* Makhan) evidently fits Watts' (1999) concept of *H. interioris* Blackburn based on comments provided by Makhan (2008).

Distribution. The genus is world-wide in distribution. In Australia it is distributes through the continent with the exception of central Western Australia, and is also present in Tasmania. 21 Australian species are endemic to Australia, three (*H. gitaraiae* Makhan, *H. imamkhani* Makhan and *H. obscuroaeneus* Fairmaire) also occur in New Guinea, and one (*H. aschnakiranae* Makhan) in Solomon Islands.

Biology. All species are aquatic, found in various types of standing and running waters: among submerged vegetation, among leaves and detritus and water edge, in gravel or sand or among stones at side (or in side pools) of streams or rivers (Watts 1999).

Note. Makhan (2004a, b) considers the Australian fauna to consist of two genera: *Hydrochus* and *Satishius* Makhan; part of species group 2 *sensu* Watts (1999) is placed into the latter genus. Except of the fact that the type species *Satishius* is *nomen nudum* and the genus does not fulfill the requirements of the International Code of the Zoological Nomenclature, no diagnosis from *Hydrochus* was provided for the genus. Hence, *Satishius* is here not considered as valid genus, in agreement with Short & Hebauer (2006). See also Introduction for more details.

References. Hansen (1991) (adult morphology); Archangelsky (1997) (larval morphology); Watts (1999) (revision of Australian fauna).

Key to species. Watts (1999).

Acknowledgments

I am indebted to R. Angus and C. Taylor (Natural History Museum, London) for the loan of the larval slides of *Hydrochus angustatus* from the collection of J. Balfour-Browne and to M. Archangelsky (CONICET, Esquel, Argentina) for the permission to use his drawings of larval *Hydrochus* in this chapter. J. Lawrence (ANIC, Canberra) kindly helped to adapt the description of adult morphology.

References

Angus RB (1977) A re-evaluation of the taxonomy and distribution of some European species of *Hydrochus* Leach (Col.: Hydrophilidae). *Entomologist's Monthly Magazine* **112**, 177–201.

Angus RB (2015) Egg cocoons and larvae of *Hydrochus* Leach (Coleoptera: Hydrophiloidea. *Acta Entomologica Musei Nationalis Pragae* **55**, 862.

Anton E, Beutel RG (2004) On the head morphology and systematic position of *Helophorus* (Coleoptera: Hydrophiloidea: Helophoridae). *Zoologischer Anzeiger* **242**, 313–346.

Archangelsky M (1997) Studies on the biology, ecology, & systematics of the immature stages of New World Hydrophiloidea (Coleoptera: Staphyliniformia). *Bulletin of the Ohio Biological Survey. New Series* **12**(1), 1–207.

Archangelsky M (1998) Phylogeny of Hydrophiloidea (Coleoptera: Staphyliniformia) using characters from adult and preimaginal stages. *Systematic Entomology* **23**, 9–24. doi:10.1046/j.1365-3113.1998.00039.x

Berge Henegouwen A (1988) *Hydrochus megaphallus*, a new and widespread European water beetle described from the Netherlands (Coleoptera, Hydrophilidae). *Balfour-Browne Club Newsletter* **42**(Supplement), 18–21.

Berge Henegouwen A, Sainz-Cantero CE (1992) *Hydrochus nooreinus*, a new species from Spain (Coleoptera, Hydrochidae). *Storkia* **1**, 26–28.

Bernhard D, Schmidt C, Korte A, Fritzsch G, Beutel RB (2006) From terrestrial to aquatic habitats and back again - molecular insights into the evolution and phylogeny of Hydrophiloidea (Coleoptera) using multigene analyses. *Zoologica Scripta* **35**, 597–606. doi:10.1111/j.1463-6409.2006.00251.x

Bernhard D, Ribera I, Komarek A, Beutel RG (2009) Phylogenetic analysis of Hydrophiloidea (Coleoptera: Polyphaga) based on molecular data and morphological characters of adults and immature stages. *Insect Systematics & Evolution* **40**, 3–41. doi:10.1163/187631209X416741

Beutel RG (1994) Phylogenetic analysis of the Hydrophiloidea based on characters of the head of adults and larvae (Coleoptera: Staphyliniformia). *Koleopterologische Rundschau* **64**, 103–131.

Beutel RG (1999) Morphology and evolution of the larval head structures of Hydrophiloidea and Histeroidea (Coleoptera: Staphylinoidea). *Tijdschrift voor Entomologie* **142**, 9–30. doi:10.1163/22119434-99900016

Beutel RG, Leschen RAB (2005) Phylogenetic analysis of Staphyliniformia (Coleoptera) based on characters of larvae and adults. *Systematic Entomology* **30**, 510–548. doi:10.1111/j.1365-3113.2005.00293.x

Castro A, Delgado JA (1998) Notes on the presence of *Hydrochus grandicollis* Kiesenwetter, 1870 in the Iberian Peninsula (Coleoptera, Hydrochidae). Notas sobre la presencia de *Hydrochus grandicollis* Kiesenwetter, 1870 en la Peninsula Iberica (Coleoptera, Hydrochidae). *Boletin de la Asociacion Espanola de Entomologia* **22**(1–2), 145–149.

Castro A, Delgado JA (1999) *Hydrochus aljibensis* sp. n., una nueva especies del sur de la Peninsula Iberica (Coleoptera, Hydrochidae). *Boletin de la Asociacion Espanola de Entomologia* **23**, 25–28.

Caterino MS, Hunt T, Vogler AP (2005) On the constitution of Staphyliniformia (Insecta: Coleoptera). *Molecular Phylogenetics and Evolution* **34**, 655–672. doi:10.1016/j.ympev.2004.11.012

Hansen M (1991) The hydrophiloid beetles. Phylogeny, classification and a revision of the genera (Coleoptera, Hydrophiloidea). *Biologiske Skrifter* **40**, 1–367.

Hansen M (1998) The systematic status of *Kiransus* Makhan, 1994, with the description of three new species of *Hydrochus* Leach, 1817 (Coleoptera: Hydrochidae). *Entomologica Scandinavica* **29**, 223–232. doi:10.1163/187631298X00302

Hansen M (1999) *Wold Catalogue of Insects. Volume 2. Hydrophiloidea (s.str.) (Coleoptera).* Apollo Books, Stenstrup.

Hidalgo-Galiana A, Ribera I (2011) Late Miocene diversification of the genus *Hydrochus* (Coleoptera, Hydrochidae) in the west Mediterranean area. *Molecular Phylogenetics and Evolution* **59**, 377–385. doi:10.1016/j.ympev.2011.01.018

Hidalgo-Galiana A, Jäch MA, Ribera I (2010) *Hydrochus farsicus* sp.n. from Iran and notes on other Palaearctic species of the genus (Coleoptera: Hydrophiloidea: Hydrochidae). *Zootaxa* **2344**, 61–64.

Hrbáček J (1950) On the morphology and function of the. antenna of the Central European Hydrophilidae (Coleoptera). *Transactions of the Royal Entomological Society of London* **101**(7), 239–256. doi:10.1111/j.1365-2311.1950.tb00452.x

Hunt T, Bergsten J, Levkanovicova Z, Papadopoulou A, St. John O, Wild R, Hammond PM, Ahrens D, Balke M, Caterino MS, Goméz-Zurita J, Ribera I, Barraclough TG, Bocáková M, Bocák L, Vogler AP (2007) A comprehensive phylogeny of beetles reveals the evolutionary origins of a superradiation. *Science* **318**, 1913–1916. doi:10.1126/science.1146954

Korte A, Ribera I, Beutel RG, Bernhard D (2004) Interrelationships of staphyliniform groups inferred from 18S and 28S rDNA sequences, with special emphasis on Hydrophiloidea (Coleoptera, Staphyliniformia). *Journal of Zoological Systematics and Evolutionary Research* **42**, 281–288. doi:10.1111/j.1439-0469.2004.00282.x

Lawrence JF, Ślipiński A (2013) *Australian Beetles. Volume 1. Morphology, classification and keys.* CSIRO Publishing, Collingwood.

Makhan D (1994) Descriptions of a new species in a new genus, *Kiransus*, and five new species of *Hydrochus* (Coleoptera, Hydrophilidae). *Zoological Studies (Taipei, Taiwan)* **33**(2), 160–164.

Makhan D (1998) Three new genera and two new species of Hydrochidae from different parts of the World (Coleoptera). *Annales Historico-Naturales Musei Nationalis Hungarici* **90**, 139–150.

Makhan D (2001a) A new genus *Rishihydroius* and three new *Hydrochus* species (Coleoptera: Hydrochidae) from Africa. *Bangladesh Journal of Zoology* **29**(1), 83–95.

Makhan D (2001b) A new genus, *Soesilius*, and a new species of Hydrochidae (Coleoptera) from America. *Russian Entomological Journal* **10**(4), 389–393.

Makhan D (2004a) Hydrochidae of the World, Dryopidae and Hydrophilidae (Coleoptera). *Calodema* **2**, 11–26.

Makhan D (2004b) *Hydrochidae of the World. Dryopidae and Hydrophilidae.* Rembrand, Suriname.

Makhan D (2008) *Hydrochus interioris* Blackburn, 1896 from Australia (Coleoptera: Hydrochidae). *Calodema. Supplementary Paper* **93**, 1–3.

Makhan D, Eftekharzadeh G, Soheili A, Khani M, Banihashemi M (2012) Additional observations on the morphology of the larvae from the genera *Aschius* Makhan *Hydrochus* Leach and *Rishihydroius* Makhan (Coleoptera, Hydrochidae). *Calodema* **213**, 1–7.

McKenna DD, Farrell BD, Caterino MS, Farnum CW, Hawks DC, Maddison DR, Seago AE, Short AEZ, Newton AF, Thayer MK (2015) Phylogeny and evolution of Staphyliniformia and Scarabaeiformia: forest litter as a stepping stone for diversification of nonphytophagous beetles. *Systematic Entomology* **40**, 35–60. doi:10.1111/syen.12093

Oliva A (1992) The species of *Hydrochus* (Coleoptera; Hydrochidae; Hydrophiloidea) described from South America. *Bulletin & Annales de la Société Royale Belge d'Entomologie* **128**, 87–104.

Oliva A (2000) *Hydrochus* synonymies. *Latissimus* **12**, 14.

Ponomarenko AG (1987) Novye mezozoickiye vodnye zhestkokrylye (Insecta, Coleoptera) iz Azii. *Paleontologicheskiy Zhurnal* **1987**, 83–97.

Richmond AE (1920) Studies on the biology of aquatic Hydrophilidae. *Bulletin of the American Museum of Natural History* **42**, 1–92.

Scudder SH (1890) The fossil insects of North America with notes on some European species. Volume 2. The Tertiary insects of North America. *Report of the United States Geological Survey of the Territories* **13**, 1–734 + 28 pls + 1 map.

Shatrovskiy AG (1989) Hydraenidae, Hydrophilidae. Pp. 260–293. In: Ler, PA (Ed.) *Opredelitel' nasekomykh Dal'nego Vostoka SSSR v shesti tomakh. Tom 3. Zhestkokrylye ili zhuki.* Nauka, Leningrad, 572 pp.

Shatrovskiy AG (1993) On new and little known hydrophilids of the genus *Hydrochus* Leach (Coleoptera, Hydrophilidae). *Entomologicheskoe Obozrenie* **72**, 827–829.

Short AEZ, Fikáček M (2011) World catalogue of Hydrophiloidea (Coleoptera): additions and corrections II (2006–2010). *Acta Entomologica Musei Nationalis Pragae* **51**, 83–122.

Short AEY, Hebauer F (2006) World Catalogue of Hydrophiloidea - additions and corrections, 1 (1999–2005) (Coleoptera). *Koleopterologische Rundschau* **76**, 315–359.

Valladares LF (1988) Descripcion de *Hydrochus angusi* n. sp. del norte de España (Coleoptera, Hydrophilidae). *Nouvelle Revue d'Entomologie* **5**, 83–87.

Valladares LF, Díaz JA, Delgado JA (1999) *Hydrochus ibericus* sp.n. (Coleoptera: Hydrochidae) from the Iberian Peninsula. *Aquatic Insects* **21**(2), 81–87. doi:10.1076/aqin.21.2.81.4529

Watts CHS (1999) Revision of Australian *Hydrochus* (Coleoptera: Hydrochidae). *Records of the South Australian Museum* **32**(1), 1–43.

19. SPERCHEIDAE ERICHSON, 1837

Martin Fikáček

Fig. 19.1. *Spercheus platycephalus* MacLeay

Common name. Filterfeeding water scavenger beetles.

Introduction. The family Spercheidae contains a single genus *Spercheus* Kugelann with 18 described species and two subspecies. It is distributed world-wide except the Nearctic, but most species occur in the Afrotropical and Oriental Regions. Only two species are distributed in South America and Australia, and the Palearctic Region only hosts one widespread species (Hebauer 1997, 1999; Hansen 1999; Archangelsky 2001).

Spercheidae are morphologically very distinctive and without any doubt form a monophyletic group, characterised by many unique adult and larval characters. Most of them seem to be related to the filter-feeding habits and are found on the head and alimentary canal (Beutel 1994, 1999; Beutel *et al.* 2001). The larval head, in particular, resembles in many aspects that of basal staphylinoid families. This is also the reason why two alternative hypotheses were proposed for the phylogenetic position of the Spercheidae. The first hypothesis considers spercheids as a derived clade within the Hydrophiloidea s.str., considering the morphology of larval head as secondarily derived from the predaceous hydrophiloid ancestor. Spercheidae is then placed as a sister-group of Hydrophilidae (Hansen 1991; Archangelsky 1998; Beutel & Komarek 2004; Lawrence *et al.* 2011) or Hydrochidae (Beutel & Leschen 2005). The second hypothesis considers Spercheidae as a sister-group to all other Hydrophiloidea s.str. (Beutel 1994) or the whole Hydrophiloidea + Histeroidea (Beutel 1999), and the spercheid-like head morphology is then considered as ancestral for the whole group. Molecular analyses confirm both above hypotheses when using different markers. The sister relationship to Hydrophilidae was revealed by Korte *et al.* (2004) (two nuclear rDNA), Bernhard *et al.* (2009) (mitochondrial protein-coding genes and rDNA combined with morphology) and McKenna *et al.* (2015a) (six nuclear protein coding genes and two rDNA). Spercheids as sister to all Hydrophiloidea s.str. were revealed by Bernhard *et al.* (2006) (mitochondrial protein-coding and rDNA) and McKenna *et al.* (2015b) (one nuclear protein-coding gene and one rDNA).

Hebauer (1997) revised the Spercheidae and provided the key to species, and two additional species were described subsequently by Hebauer (1999) and Archangelsky (2001). The revision is however rather superficial and based on small subset of characters, and additional studies are clearly necessary.

Immature stages (Fig. 19.2) are described for three non-Australian species: European *Spercheus emarginatus* (Schaller) (e.g. Archangelsky 1997), the Neotropical *S. halophilus* Archangelsky (Archangelsky 2001) and partly for the African-Near East *S. cerisyi* Guérin-Méneville (not illustrated, diagnosed in Archangelsky 2001). The pupa was never properly described and was only incompletely illustrated for *S. emarginatus* by Cussac (1852).

The fossil record of Spercheidae goes back to the Late Jurassic – the assignment of the well preserved fossil of *Prospercheus cristatus* Prokin from the Shar-Teg deposit in Mongolia to Spercheidae may be confirmed by the morphology of mouthparts of the fossil species (Prokin 2009; Fikáček *et al.*, unpublished data). The other Mesozoic fossils assigned to or close to Spercheidae either belong to non-hydrophiloid beetles (*Wollastonia* Heer to Archostemata; *Paraspercheus* Ponomarenko to Coleoptera *incertae sedis*, possibly to Scarabaeoidea) or seem to not belong to Coleoptera at all (*Orphnospercheus* Hong likely belonging to Heteroptera) (Heer 1852; Ponomarenko 1977; Hong 1992; Fikáček, unpublished data). Undescribed fossils (isolated elytra assigned to *Spercheus*) are know from the Eocene-Oligocene deposits of Bembridge Marl Beds, Isle of Wight, UK (A. Ponomarenko, personal communication).

Biology. Both adults and larvae are aquatic, inhabiting densely vegetated standing waters. Biology was discussed for the European species *Spercheus emarginatus* (Cussac 1852; Fowler 1882; Schlick 1887; Bukh 1910; Böving and Henriksen 1938), notes on biology of the South American *S. halophilus* were provided by Archangelsky (2001). Adults live in a cryptic way among submerged vegetation, and are able to walk on the underside of water surface and filter plant remnants and decaying organic material from it (Rothmeier and Jäch 1986). As in Hydrophilidae, they breathe atmospheric oxygen which is held in an air bubble on the ventral side of the head, thorax and first abdominal ventrite; antennae are used to establish a connection between atmosphere and the air bubble on the venter once the beetle needs to renew the air supply (Hrbáček 1950). Eggs are laid in a bag-shaped egg case which is attached to metatibiae and carried by female on the ventral side of the abdomen (Fig. 19.3L). Embryonic development takes ca. 4–5 days in *S. halophilus*. Larvae are slowly moving, stay close to water surface and can also walk on the underside of it. They feed on drifting detritus, algae and detritus grazed from the substrate

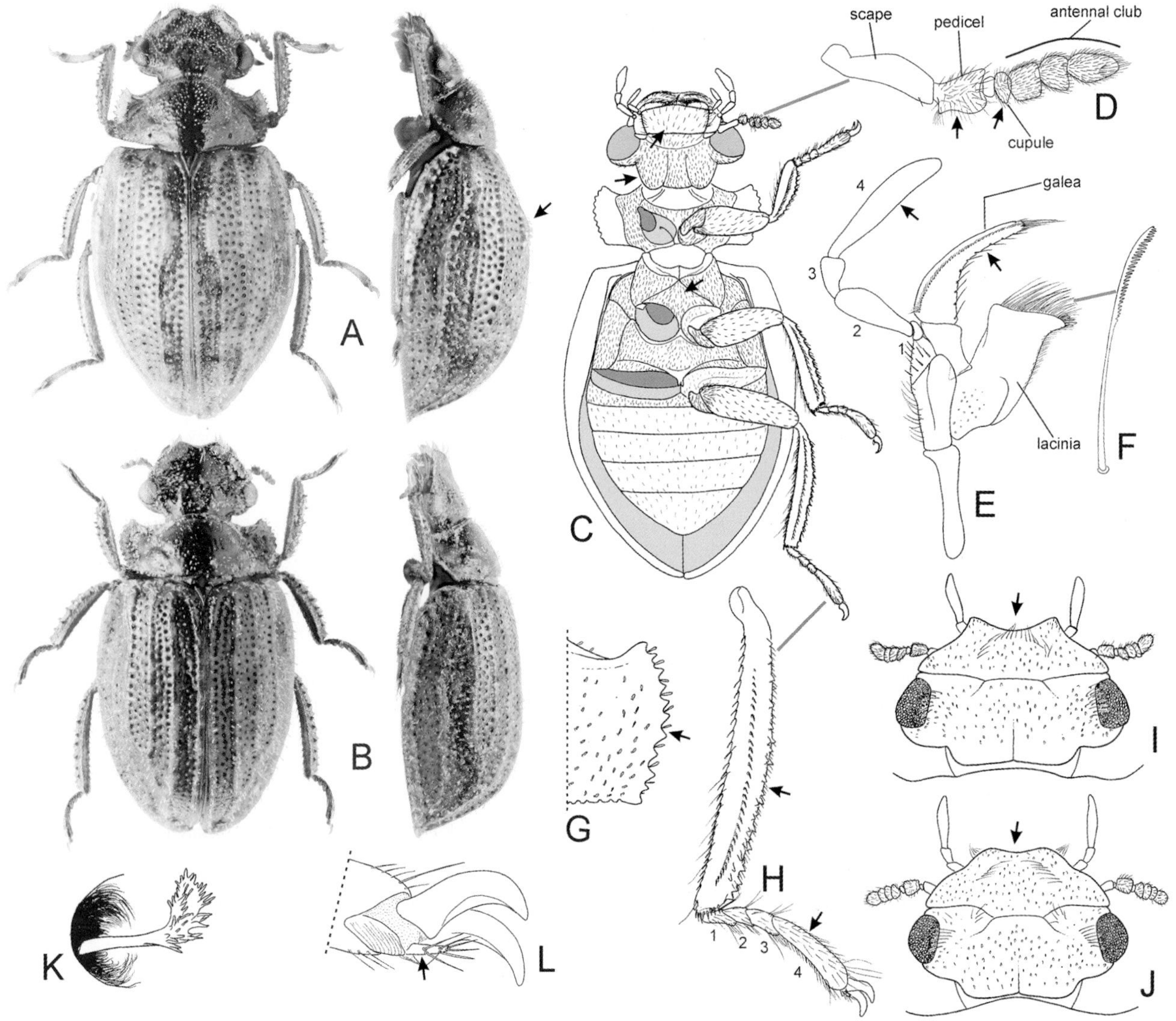

Fig. 19.2. Australian Spercheidae, genus *Spercheus*. A–B, general habitus in dorsal and lateral views: A, *Spercheus platycephalus* MacLeay; B, *S. wattsi* Hebauer, 1999. C–J, general morphology: C, ventral morphology; D, antenna; E, maxilla in ventral view; F, detail of lacinial comb-like seta; G, lateral margin of pronotum of *S. platycephalus*; H, metatibia and metatarsus; I-H, head of *S. platycephalus* in dorsal view (I, male; J, female); K, detail of the club-like seta of dorsal surface; L, pretarsus with claws and empodium. Drawings D–F, H, K–L based on non-Australian *S. senegalensis* Castelnau.

and on dead invertebrate prey. Some studies (Fowler 1882; Böving & Henriksen 1938) indicate that *S. emarginatus* larvae are also predaceous, feeding on small invertebrates and being cannibalistic, which was not observed in *S. halophilus*. All examined *Spercheus* larvae have a large muscular denticulate proventriculus (Beutel 1999) which probably helps in processing the detritus-based food. Additional studies on food preferences are necessary, but from available data it seems that larvae may be opportunistic and combine small invertebrates and detritus particles as food sources. Larval development includes three larval instars and takes ca. 30–35 days in *S. emarginatus* (Bukh 1910). Archangelsky (2001) mentions that the first instar takes 4–6 days and the second one 6–7 days in *S. halophilus*. Pupation takes place outside of water, in a pupal chamber built up of the soil and plant particles attached to surfaces (plants?) few centimetres above the water surface or situated among plant debris at the water edge; pupal stage is short, taking ca. 5–7 days (Bukh 1910; Cussac 1852; M. Soviš, personal observation).

Morphology. Adults (Figs 19.1, 19.2). Length ~1.9–7.0 mm (Australian species 3.2–4.2 mm). Body ~1.8–2.1 times as long as wide, with sides more or less evenly rounded, well sclerotised and moderately to strongly convex; yellowish to dark brown, elytra often with a pattern of dark and pale spots; dorsal surfaces smooth and glabrous, with stout erect club-like setae (Fig. 19.2K); elytra costate, costae in some species with elevated tubercles; ventral surfaces completely covered with

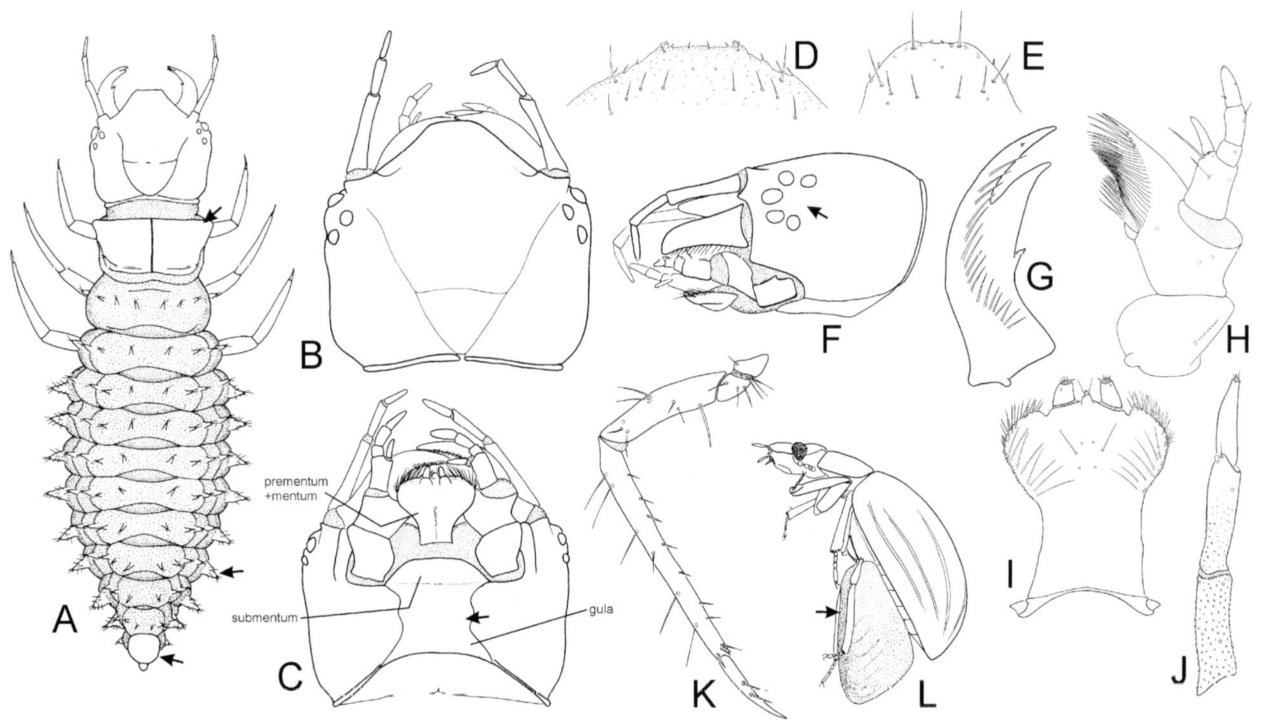

Fig. 19.3. Immature stages of Spercheidae, genus *Spercheus*. A–J, larva: A, general habitus; B, head, dorsally; C, head, ventrally; D-F, anterior margin of head (D, *S. emarginatus* (Schaller); E, *S. halophilus* Archangelsky); F, head, laterally; G, mandible; H, maxilla; I, labium; H, antenna; K, leg; L, female of *S. emarginatus* carrying egg case. A–D, F–J based on non-Australian *S. emarginatus*; E on *S. halophilus*; K on *S. belli* Champion. A, D, H from Archangelsky (1997); E from Archangelsky (2001); K redrawn and adapted from Darilmaz & Kiyak (2011).

hydrofuge pubescence (plastron) in *S. emarginatus*, without dense pubescence on abdominal ventrites 2–5 and on epipleura in other *Spercheus* species.

Head more or less prognathous, deeply inserted into prothorax, strongly constricted behind eyes. Eyes well developed, slightly protuberant, circular and entire, finely facetted, without interfacetal setae. Antennal insertions concealed from above; subantennal groove present. Frontoclypeal suture obsolete or partly distinct, nearly straight, mid-cranial suture distinct only posteriorly in some species; clypeus large, sexually dimorphic, deeply emarginate and with angulate lobes in males (Fig. 19.2I), weakly emarginate and with rounded lobes in females (Fig. 19.2J); sexual dimorphism very weak in *S. fimbriicollis* Bruch and *S. halophilus*. Labrum strongly transverse, well sclerotised, truncate or sinuate on anterior margin, not narrowed at base, retracted under clypeus and not visible from above. Antennae (Fig. 19.2D) short, 7-segmented, with pubescent widely conical pedicel, cupule and 3-segmented antennal club, pedicel and cupule separated by a small glabrous antennomere. Mandibles moderately large, concealed beneath clypeus; apex bidentate; mola well developed, slightly asymmetrical; prostheca well developed, membranous, without articulated or sclerotised processes. Maxilla (Fig. 19.2E) with setose galea and lacinia, galea long, slender and finger-like, with two series of setae; lacinia with a small uncus hidden among long setae, setae simple or comb-like (Fig. 19.2F); maxillary palps with 4 palpomeres, slender, ca. as long as antennae; apical maxillary palpomere long and asymmetrical. Mentum very large, transverse, subrectangular; prementum bilobed, pubescent, each lobe with a transversely oval sclerite; labial palps with three palpomeres, moderately long, situated on anterolateral corners of mentum. Corpotentorium narrow; laminatentoria well developed, sometimes meeting at midline. Cervical sclerites well developed.

Pronotum ~0.4–0.5 times as long as wide, widest at middle (incl. Australian species) or in anterior third, about as wide as combined elytral bases (narrower than elytral bases in some non-Australian species); sides strongly curved, explanate. Surface flat with club-like setae, or with median and submedian impressions (in *S. spangleri* Hebauer, *S. fimbriicollis* and *S. halophilus*). Anterior edge strongly bisinuate with median portion slightly produced forward. Lateral pronotal carinae complete, denticulate (Fig. 19.2G) or with finger-like lobes (in non-Australian species); anterior angles strongly produced and acute; posterior angles right to rounded; posterior edge slightly sinuate to straight, with underlying ventral ridge which interlocks with elytra and scutellar shield. Prosternum flat or weakly elevated medially, in front of coxae shorter than shortest diameter of procoxal cavity. Prosternal process complete and slightly expanded apically in most *Spercheus*, incomplete in *S. emarginatus*. Notosternal sutures complete.

Hypomeron with wide glabrous and shiny lateral portion separated by ridge from pubescent mesal portion; without antennal grooves. Procoxae more or less conical and projecting, with short to moderately long concealed lateral extensions; trochantins more or less exposed; endopleuron not fused to notum. Procoxal cavities slightly transverse, narrowly separated (most *Spercheus*) or contiguous posteriorly (*S. emarginatus*), in most species closed externally, open in *S. emarginatus*, with short and wide anterolateral extension, internally closed.

Scutellar shield abruptly elevated, triangular, acute posteriorly. Elytra ~1.2–1.5 times as long as combined width and 3.0–3.5 times as long as pronotum; sides usually moderately curved and apices conjointly rounded; humeri well developed, somewhat impressed where overlapped by pronotum; disc without distinct puncture rows, punctation irregular, surface with four more or less distinct costae (absent in few non-Australian species). Epipleura wide throughout, explanate, in most species not divided into pubescent and bare portions (with wide pubescent portion basally in *S. emarginatus*). Mesoventrite separated from mesanepisterna by complete sutures, flat, without any carinae or projections, without procoxal rests, without median discrimen, mesoventral cavity absent; mesanepisterna contiguous at midline, rather widely contacting. Mesocoxae slightly transverse with exposed trochantins. Mesocoxal cavities contiguous, closed laterally by mesepimeron and mesanepisternum. Metaventrite distinctly transverse, flat, discrimen absent, transverse (metakatepisternal) suture present and very distinct; exposed portion of metanepisternum moderately broad, not concealed under elytra. Metacoxae strongly transverse and subcontiguous, extending laterally to meet elytra; plates absent. Metendosternite with short, stalk and long, lateral arms; anterior process absent. Hind wing with long radial bending zone at the end of which is a distinct hinge; apical field usually with three radial extensions; cross-vein r3 absent; cross-vein r4 complete; RP moderately long, obsolete basally, well sclerotised distally and together with MP_{1+2} forming R-M loop, medial spur very short, straight and ending well before wing margin; MP_{3+4} absent; basal and wedge cells present, with partly obsolete veins on their posterior margins, AA_3 mostly absent, in *S. emarginatus* long and reaching wing margin; AP absent.

Femoral attachment of trochanter oblique with base of femur free and not contacting trochanter anteriorly in middle leg, contacting trochanter in hind leg. Tibiae broadly cylindrical, with rows of spines, in many species also with cuticular teeth especially along lateral margin (Fig. 19.2H), swimming hairs present on meso- and metatibiae in many species; tibial spurs very short. Tarsi usually 4–4–4 (5–5–5 in *S. emarginatus*), tarsomeres simple, with few swimming hairs dorsally in some species; pretarsal claws simple; empodium well developed, multisetose (Fig. 19.2L).

Abdomen with five free ventrites; ventrite 1 without median carina; intercoxal process acute. Functional spiracles on abdominal segments I–VI; abdominal tergites relatively lightly sclerotised, usually with patches of wing-folding asperities on one or more segments. Anterior edge of sternite VIII in male without median strut. Anterior edge of segment IX in male broadly rounded or truncate, without spiculum gastrale; tergite IX emarginate; tergite X free. Aedeagus of the trilobate type, symmetrical, with very short phallobase and simple parameres; penis with basal struts. Ovipositor moderately elongate, lightly sclerotised; paraprocts much shorter than coxites, which are subdivided with broad proximal lobe, slender distal lobe and small terminal styli. [Adapted from Lawrence and Ślipiński 2013; see also Hebauer 1997; Archangelsky 2001; Archangelsky *et al.* 2005].

Larvae (Fig. 19.3). Body elongate, widest at middle of abdomen, slightly flattened; head and protergum well sclerotised, remaining surfaces more or less membranous, tergum VIII with a sclerotised tergite; surfaces pubescent, abdomen with lateral projections.

Head prognathous, wider than long, more or less parallel-sided; posterior edge straight dorsally, slightly emarginate ventrally; epicranial stem (= coronal line) absent; frontal arms V-shaped with bases contiguous; median endocarina absent. Each side of head with 5 stemmata well separated from each other (Fig. 19.3F). Frontoclypeal suture absent and labrum fused to head capsule forming clypeolabrum, its anterior edge arcuate or angulate, without producing median portion (= nasale) and epistomal lobes (= adnasalia) (these parts seem to be secondarily reduced based on chaetotaxy). Antennae well developed, slender and elongate, 3-segmented, antennomeres subequal in length or antennomeres 2–3 prolonged (in non-Australian *S. halophilus*), antennomere 2 with laterally placed sensorium which is minute in size and hence very indistinct. Mandibles symmetrical, moderately broad at base with narrow, curved unidentate apex, with 2 inner teeth (= retinacula) of which the distal one is situated close to mandibular apex; the mandible is hence seemingly bidentate, homology of the distal inner tooth with that of other hydrophiloids proved by chaetotaxy; ventral face of mandible with row of setae (most species) or irregular row of small teeth and asperities (in *S. halophilus*); mola and prostheca absent; basal penicillus absent. Ventral mouthparts distinctly retracted, with articulating area; maxilla hexamerous, cardo distinct, consisting of single large sclerite; stipes about as long as wide and open mesally; galea (= inner appendage) long, well sclerotised, finger-like, articulated and located at apex of palpifer; lacinia (= inner projection of stipes homologous with inner face of stipes in other hydrophiloids based on the position of stipital sensilla) broad at base, apically acute and lined with setae/spines; palp 3-segmented, with segment-like palpifer (considered as first palpomere by some authors). Labium consisting of submentum fused to head capsule, and free plate-like structure (prementomentum) formed by fused mentum and prementum; palps 1-segmented, articulated (in most species) or fused with prementomentum (in *S. halophilus*); ligula usually present, absent in *S. halophilus*. Hypopharyngeal sclerome and hypostomal rods absent; ventral epicranial ridges absent. Gular region wide, gular sutures well separated.

Prothorax as long as meso- or metathorax, with single tergal plate divided by median ecdysial line; mesothorax with a pair of narrow tergites anteriorly, metathorax without sclerotised plates. Prothoracic venter without presternal plate, only with sclerotised precoxal regions. Legs well developed, long and slender, 5-segmented with pretarsus forming distinct claw, with few long setae on tibiotarsus; coxae widely separated.

Abdominal segments lightly sclerotised, segments I-VII each with transverse row of small finger-like setiferous projections dorsally, segments I-VIII with short conical lateral projections; abdominal sterna without prolegs, sternum 7 with finger-like projections in *S. halophilus*. Abdominal apex with spiracular atrium formed by modifications of segments VIII–X (tergum VIII forming dorsal valve with sclerotised plate at apex of abdomen, segment IX trilobed and forming ventral valve, which bears a pair of small 1-segmented urogomphi; large posteriorly oriented spiracles located between the two, and segment X reduced and ventral. Functional spiracles absent except for an enlarged annular pair in spiracular atrium. [Adapted from Lawrence and Ślipiński 2013 according to Archangelsky 1997, 2001; first instar larvae of *S. belli* Champion and *S. crenulatus* deposited in National Museum, Prague were also checked].

Classification of the Australian taxa

Characters marked by an asterisk (*) apply only for Australian species, non-Australian species may bear different character states.

Spercheus Kugelann, 1798 (Pl. 43E; Fig. 19.2)

Type species. *Dytiscus emarginatus* Schaller, 1783.

Characteristics. Adults. Body oval, moderately convex; head strongly constricted behind eyes; eyes weakly protruding; clypeus sexually dimorphic*, deeply excised and with sharp lobes in male, shallowly excised with rounded lobes in female; labrum concealed under clypeus; antenna with 7 antennomeres, pedicel, cupule and three-segmented club pubescent, pedicel large; pronotum transverse, explanate laterally; lobate on lateral margins*; prosternum short in front of procoxae, procoxal cavities closed*; elytra without punctural series, with four keels, mesal keel with (*S. platycephalus*) or without (*S. wattsi*) elevated bulge on anterior portion; epipleuron wide throughout, without mesal pubescent portion*; lateral margin of elytra smooth*; anapleural sutures distinct, converging anteriorly, mesanepisterna contacting each other anteriorly, mesoventrite flat; metaventrite flat and nearly completely pubescent, metanepisternum parallel-sided and wide; abdomen with five visible ventrites, only ventrite 1 with dense hydrofuge pubescence; ventrite 5 entire on posterior margin; tarsi 4-segmented*, of the same shape in male and female; hydrofuge pubescence confined to basalmost portion of meso- and metafemora; meso- and metatibiae with denticles on outer face; coloration pale to dark brown, mesal portions of pronotum and head darker, elytra with variable pattern of dark and pale spots; 3.2–4.2 mm. Easily recognisable from all other Australian beetles by the combination of neck-like constricted head behind eyes, emarginate clypeus, explanate and lobate lateral margin of pronotum, costate elytra and meso- and metatibae with denticles on outer face; also characteristic by the unique antennal morphology. Females carry egg cases on abdomen, which otherwise only occurs in *Helochares* (Hydrophilidae) in Australia.

Larvae (based on non-Australian species, Fig. 19.3). General habitus myrmeleonid-like, head large, transverse, with 5 stemmata on each side; anterior margin arcuate to slightly angulate; gula present and wide; antenna long and slender, with very small and indistinct sensorium; mandibles symmetrical, with two inner teeth and ventral series of setae*; maxilla with large setose projection of stipes and long finger-like galea (= inner appendage on palpifer); labrum with mentum and prementum fused into a transversely oval, laterally setose plate bearing one-segmented palps and small ligula*; prothorax with large dorsal tergite, meso- and metathorax largely membranous; abdominal segments with lateral projections and transverse series of small sclerotised cuticular sclerites; legs very long, slender; abdominal apex with dorsal oval plate, stigmatic atrium present, ventral prolegs absent. Easily recognisable by characteristic body shape and morphology of head venter with plate-like prementomentum and wide gula.

Australian species. Two species, both endemic to Australia.

Distribution. The genus is nearly world-wide in distribution (absent in North America and distant oceanic islands). In Australia, *Spercheus platycephalus* Macleay is known from all territories with most records from northern and north-eastern parts of the continent, and *S. wattsi* Hebauer is so far known from northern Western Australia and Northern Territory.

Biology. Aquatic species inhabiting various types of well vegetated pools and ponds; also comes to light.

Notes. The genus was considered as very uniform by most authors, but detailed studies revealed many differences especially in thoracic and leg morphology between most *Spercheus* and the European *S. emarginatus*. In addition, both South American species and the Oriental *S. spangleri* also differ from remaining species by many characters, most probably including characters of the larva (described for the South American *S. halophilus*). The phylogeny of the genus is under study at the moment in order to reveal the relevance of these differences and their correspondence to the DNA-based phylogeny (Fikáček and Short, unpublished data).

References. Hansen (1991) (adult morphology); Archangelsky (1997) (larval morphology); Hebauer (1997, 1999) (species revision including description of Australian species).

Key to species. Diagnosis of both Australian species provided by Hebauer (1999); both species can be easily distinguished by presence/absence of the elevated bulge in anterior portion of the first elytral keel (compare Figs 19.2A–B).

Acknowledgments

I am indebted to A. Ślipiński (ANIC, Canberra) for the loan of the types of *Spercheus wattsi*, to M. Archangelsky (CONICET, Esquel, Argentina) for the permission to use his drawings of larval *Spercheus* in this chapter, and to M. Soviš (Prague, Czech Republic) for providing his observations of the pupa of *Spercheus emarginatus*. J. Lawrence (ANIC, Canberra) kindly helped to adapt the description of adult morphology.

References

Archangelsky M (1997) Studies on the biology, ecology, & systematics of the immature stages of New World Hydrophiloidea (Coleoptera: Staphyliniformia). *Bulletin of the Ohio Biological Survey. New Series* **12**(1), 1–207.

Archangelsky M (1998) Phylogeny of Hydrophiloidea (Coleoptera: Staphyliniformia) using characters from adult and preimaginal stages. *Systematic Entomology* **23**, 9–24. doi:10.1046/j.1365-3113.1998.00039.x

Archangelsky M (2001) A new Neotropical species of *Spercheus* Kugelann, and its larval stages (Coleoptera, Hydrophiloidea: Spercheidae). *Studies on Neotropical Fauna and Environment* **36**(3), 199–204. doi:10.1076/snfe.36.3.199.2128

Archangelsky M, Beutel RG, Komarek A (2005) 10.1. Hydrophilidae. In *Handbook of Zoology. Volume IV Arthropoda: Insecta. Part 38 Coleoptera, beetles Vol. 1: Morphology and systematice (Archostemata, Adephaga, Myxophaga, Polyphaga partim)* (Eds RG Beutel and RAB Leschen) pp. 158–183. Walter de Gruyter, Berlin-New York.

Bernhard D, Schmidt C, Korte A, Fritzsch G, Beutel RB (2006) From terrestrial to aquatic habitats and back again - molecular insights into the evolution and phylogeny of Hydrophiloidea (Coleoptera) using multigene analyses. *Zoologica Scripta* **35**, 597–606. doi:10.1111/j.1463-6409.2006.00251.x

Bernhard D, Ribera I, Komarek A, Beutel RG (2009) Phylogenetic analysis of Hydrophiloidea (Coleoptera: Polyphaga) based on molecular data and morphological characters of adults and immature stages. *Insect Systematics & Evolution* **40**, 3–41. doi:10.1163/187631209X416741

Beutel RG (1994) Phylogenetic analysis of the Hydrophiloidea based on characters of the head of adults and larvae (Coleoptera: Staphyliniformia). *Koleopterologische Rundschau* **64**, 103–131.

Beutel RG (1999) Morphology and evolution of the larval head structures of Hydrophiloidea and Histeroidea (Coleoptera: Staphylinoidea). *Tijdschrift voor Entomologie* **142**, 9–30. doi:10.1163/22119434-99900016

Beutel RG, Komarek A (2004) Comparative study of thoracic structures of adults of Hydrophiloidea and Histeroidea with phylogenetic implications (Coleoptera, Polyphaga). *Organisms, Diversity & Evolution* **4**, 1–34. doi:10.1016/j.ode.2003.10.001

Beutel RG, Leschen RAB (2005) Phylogenetic analysis of Staphyliniformia (Coleoptera) based on characters of larvae and adults. *Systematic Entomology* **30**, 510–548. doi:10.1111/j.1365-3113.2005.00293.x

Beutel RG, Anton E, Bernhard D (2001) Head structures of adults of *Spercheus* (Coleoptera: Spercheidae): their function and possible significance to staphyliniform phylogeny. *Annales Zoologici* **51**(4), 473–484.

Böving AG, Henriksen KL (1938) The developmental stages of the Danish Hydrophilidae (Ins., Coleoptera). *Videnskabelige Meddelelser fra Danks Naturhistorik Forening i Khobenhavn* **102**, 27–162.

Bukh F (1910) Lebensweise und Entwicklung von *Spercheus emarginatus*. *Entomologische Rundschau* **27**, 127–128.

Cussac ME (1852) Moeurs et metamorphoses du *Spercheus emarginatus* et de l'*Helochares lividus*. *Annales de la Société Entomologique de France* **10**, 617–627[+ pl. XII.].

Darilmaz M, Kiyak S (2011) A study of the family Spercheidae (Coleoptera) from Turkey. *Turkish Journal of Zoology* **35**, 441–444.

Fowler WW (1882) A contribution to the life-history of *Spercheus emarginatus*. *Entomologist's Monthly Magazine* **19**, 79.

Hansen M (1991) The hydrophiloid beetles. Phylogeny, classification and a revision of the genera (Coleoptera, Hydrophiloidea). *Biologiske Skrifter* **40**, 1–367.

Hansen M (1999) *Wold Catalogue of Insects. Volume 2. Hydrophiloidea (s.str.) (Coleoptera)*. Apollo Books, Stenstrup.

Hebauer F (1997) Revision der Arten der Familie Spercheidae Erichson, 1837 (Coleoptera, Hydrophiloidea). *Entomologische Blätter* **93**, 9–42.

Hebauer F (1999) *Spercheus wattsi* sp. n. - a second Australian species of the genus (Coleoptera, Hydrophiloidea). *Acta Coleopterologica* **15**(2), 5–6.

Heer O (1852) Die Lias-Insel des Aargau's. In (*Zwei geologische Vorträge gehalten im März 1852 von Oswald Heer und A. Escher von der Linth*. (Eds O Heer, A Escher von der Linth) 15 pp, 1 pl. E. Kiesling, Zürich.

Hong YC (1992) [*Paleontological Atlas of Jilin Province*]. Jilin Science and Technology Press, Jilin, China

Hrbáček J (1950) On the morphology and function of the. antenna of the Central European Hydrophilidae (Coleoptera). *Transactions of the Royal Entomological Society of London* **101**(7), 239–256. doi:10.1111/j.1365-2311.1950.tb00452.x

Korte A, Ribera I, Beutel RG, Bernhard D (2004) Interrelationships of Staphyliniform groups inferred from 18S and 28S rDNA sequences, with special emphasis on Hydrophiloidea (Coleoptera, Staphyliniformia). *Journal of Zoological Systematics and Evolutionary Research* **42**(4), 281–288. doi:10.1111/j.1439-0469.2004.00282.x

Lawrence JF, Ślipiński A (2013) *Australian Beetles. Volume 1. Morphology, classification and keys*. CSIRO Publishing, Collingwood.

Lawrence JF, Ślipiński A, Seago AE, Thayer MK, Newton AF, Marvaldi AE (2011) Phylogeny of the Coleoptera based on morphological characters of adults and larvae. *Annales Zoologici* **61**(1), 1–217. doi:10.3161/000345411X576725

McKenna DD, Wild AL, Kanda K, Bellamy CL, Beutel RG, Caterino MS, Farnum CW, Hawks DC, Ivie MA, Jameson ML, Leschen RAB, Marvaldi AE, McHugh JV, Newton AF, Robertson JA, Thayer MK, Whiting MF, Lawrence JF, Ślipiński A, Maddison DR, Farrell BD (2015a) The beetle tree of life reveals that Coleoptera survived end-Permian mass extinction to diversify during the Cretaceous terrestrial revolution. *Systematic Entomology* **40**(4), 835–880. doi:10.1111/syen.12132

McKenna DD, Farrell BD, Caterino MS, Farnum CW, Hawks DC, Maddison DR, Seago AE, Short AEZ, Newton AF, Thayer MK (2015b) Phylogeny and evolution of Staphyliniformia and Scarabaeiformia: forest litter as a stepping stone for diversification of nonphytophagous beetles. *Systematic Entomology* **40**, 35–60. doi:10.1111/syen.12093

Ponomarenko AG (1977) Infraotryad Staphyliniformia. [Infraorder Staphyliniformia]. In *Mezozoiskie zhestkokrylye*. [*Mesozoic Coleoptera*]. (Eds LV Arnoldi, VV Zherikhin, LM Nikritin, AG Ponomarenko) pp. 106–119. Nauka, Moskva.

Prokin AA (2009) New water scavenger beetles (Coleoptera: Hydrophilidae) from the Mesozoic of Mongolia. *Paleontological Journal* **43**, 660–663. doi:10.1134/S0031030109060094

Rothmeier G, Jäch MA (1986) Spercheidae, the only filter-feeders among Coleoptera. *Proceedings of the 3rd European Congress of Entomology (Amsterdam)*, pp. 133–137.

Schlick W (1887) Yngleforhold hos *Spercheus emarginatus*. *Entomologiske Meddelelser* **1**, 26–27.

20. HYDROPHILIDAE LEACH, 1815

Martin Fikáček

Fig. 20.1. *Helochares* (*Hydrobaticus*) *percyi* Watts

Common name. Water Scavenger Beetles

Introduction. The family Hydrophilidae contains ~2900 described species distributed worldwide (Short & Fikáček 2011, 2013). Five of the six subfamilies are most diverse in the tropics and subtropics; Cylominae mostly inhabit temperate regions of the southern hemisphere. Representatives of subfamilies Hydrophilinae, Chaetarthriinae, Enochrinae and Acidocerinae are mostly aquatic, inhabiting both standing and running waters as well as some specialised habitats (wet rocks, waterfalls, phytotelmata). Representatives of Cylominae and Sphaeridiinae are mostly terrestrial, inhabiting various types of decaying plant material (usually leaf litter or dung).

The monophyly of the family is strongly supported both by morphological synapomorphies of adults and larvae (Hansen 1991; Archangelsky 1998; Beutel & Komarek 2004; Archangelsky *et al.* 2005) and molecular data (e.g. Short & Fikáček 2013; Bocák *et al.* 2014). The relationships of Hydrophilidae to remaining families of Hydrophiloidea *s.str.* remain unclear, as different character sets reveal conflicting results. Based on morphological characters, Hansen (1991) considered Hydrophilidae as sister to Spercheidae, together forming the 'hydrophilid-lineage'. Close relationship of Hydrophilidae and Spercheidae, sometimes also including Hydrochidae, was revealed by some analyses based on morphology (Archangelsky 1998; Beutel & Komarek 2004; Lawrence *et al.* 2011) and molecules (McKenna *et al.* 2015a). In contrast, other analyses, morphological (e.g. Beutel 1994), molecular (Bernhard *et al.* 2006; Bocák *et al.* 2014; McKenna *et al.* 2015b) and combined (Bernhard *et al.* 2009) contradicted the sister relationship of Hydrophilidae with spercheids and/or hydrochids.

The species-level taxonomy of the family was quite intensively studied in last decades, and especially aquatic species are rather well known in many parts of the world (Hansen 1999a; Short & Hebauer 2006; Short & Fikáček 2011). Despite of that, many undescribed species are expected especially in tropical regions. Bloom *et al.* (2014) estimated that ~25% of aquatic species and ~40% of terrestrial ones remain undescribed. Australian fauna is rather well known at species level, and most genera (both terrestrial and aquatic) were more or less recently revised by C. Watts, E. Gentili, A. Komarek and M. Hansen (see under respective genera for references to their papers). Only few terrestrial Australian genera were never revised: *Pseudohydrobius* (Cylominae), all omicrine genera, and *Notocercyon*, *Cryptopleurum* and *Merosoma* (all Megasternini). Discoveries of new species may be, however, expected even in previously revised genera, especially in tropical regions of northern Australia (e.g. Watts 1996; Gentili 2006).

Higher-level systematics and phylogeny was analysed recently by Short & Fikáček (2013) based on a multigene dataset. The results stand in contrast to previously used morphology-based classification by Hansen (1991, 1999a). In most cases, morphological support was however found for newly revealed or redefined clades, which allowed the use of the results for establishing a new classification at tribe and subfamily levels, which is also adopted in this chapter. Subsequent phylogenetic analyses focused on particular clades within the family, adding more representative taxon sampling and/or endemic genera of unclear position which were not available in DNA quality quality before: Fikáček & Vondráček (2014) for Anacaenini, Toussaint *et al.* (2016) and Toussaint & Short (2018) for Laccobiini; Short *et al.* (2017) and Toussaint & Short (2017) for Hydrobiusini; Toussaint *et al.* (2017) for Hydrophilini; and Deler-Hernández *et al.* (2018) for Coelostomatini. Additional studies are currently in progress for the phylogeny of the Berosini (Clarkson-Mattos, unpublished), Chaetarthriinae (Clarkson-Mattos *et al.*, unpublished), Acidocerinae (Short, unpublished), Cylominae (Seidel *et al.*, unpublished), Omicrini (Fikáček *et al.*, unpublished) and Megasternini (Arriaga-Varela *et al.*, unpublished). Among Australian genera, *Phelea* and *Petasopsis* are those which would deserve the test of their phylogenetic position by means of DNA, but are not available at the moment. Little is known about phylogenetic relationships at lower level, and the monophyly of some larger genera and their subgenera, as well as relationships among species need to be tested in future studies. For example, some subgenera of large genera *Helochares*, *Enochrus*, *Laccobius* and *Cercyon* are very likely polyphyletic. Despite of that, I am using subgenera in the current concept within this chapter, as they are mostly defined by easy-to-observe characters and may be helpful for species identification.

203 species of Hydrophilidae are recorded from Australia at the moment, of which 165–172 species are endemic to Australia (i.e. 82–86%; the uncertainty is caused by limited information about Omicrini which were not revised in Australia

so far). Of the non-endemic species, ~18 are distributed in areas adjacent to Australia only (New Guinea, New Caledonia, New Zealand, Pacific Islands), and only 20 species are more widespread (of these, 6 species of sphaeridiine genera *Cercyon* and *Dactylosternum* are adventive, at the moment of world-wide distribution). All main hydrophilid clades but one (Protosternini) are represented in Australia, although some of them (Laccobiini, Hydrobiusini, Omicrini and Coelostomatini) by a very limited number of genera and species relative to their world-wide diversity. Forty-eight genera are recorded from Australia, of which 15 are endemic for the continent (i.e. 31%). Most of them belong to the southern temperate subfamily Cylominae (6 genera) and the sphaeridiine tribe Megasternini (6 genera); remaining endemic genera belong to Anacaenini (2 genera) and Hydrobiusini (1 genus). Australian fauna of Cylominae and partly also Megasternini is related to those of other continents of southern hemisphere: cylomines otherwise occur in New Zealand, austral South America and South Africa; the Australian *Austrotypus* has a sister vicariant species in high altitude forests of Peru (Fikáček *et al.* 2014a); and the *Oosternum* group of genera of the Megasternini (to which most Australian genera belong) is nearly exclusively confined to South and Central America and Australia+New Guinea. A southern disjunct distribution occurs in the anacaenine genus *Crenitulus* (one species in Australia, otherwise distributed and diverse in Neotropical region; Fikáček & Vondráček 2014). Part of the Australian fauna is shared with (and likely related to) Asia and/or Africa (e.g. genera *Regimbartia*, *Hydrophilus*, *Hydrobiomorpha*, *Chaetarthria*, *Anacaena*, *Helochares* (*Hydrobaticus*), *Agraphydrus* and *Sphaeridium*, tribes Omicrini, Coelostomatini, and few genera of Megasternini). Additional phylogenetic studies are however necessary to reveal the relationships of Australian species to those occurring outside Australia.

Immature stages are known for most aquatic genera, especially thanks to detailed studies by M. Archangelsky (e.g. Archangelsky 1997) and Y. Minoshima (e.g. Minoshima & Hayashi 2011a). The knowledge of the larval morphology of the terrestrial genera still remains rather limited, and identification is usually possible to tribes or groups of genera only. In most cases, the knowledge of larval morphology is based on non-Australian species of the respective genera. In Australia, only three papers dealing with larval morphology were published so far: Anderson (1976) describing immature stages of few Acidocerinae and Enochrinae species, Watts (2002) describing larvae of aquatic genera including the Australian-endemic *Hybogralius*, and Fikáček *et al.* (2014a) describing the larva of the terrestrial cylomine genus *Austrotypus*. For basic introduction into larval morphology and the characters diagnosing the genera or larger clades, we refer the reader to the papers of Archangelsky (1997) and Minoshima & Hayashi (2011a).

Fossil record of the Hydrophilidae is rather rich, with the oldest fossils known from the Late Jurassic and Early Cretaceous (Fikáček *et al.* 2014b). Many fossils are known from Cenozoic deposits, all representing recent genera (Fikáček, unpubl. data). In the Late Jurassic and Early Cretaceous, the Hydrophilidae were already widespread and fossils are known from both northern and southern hemispheres. Some of the Cretaceous fossils clearly represent modern tribes (e.g. Hydrobiusini, Anacaenini: Fikáček *et al.* (2014b, 2017a). Two of the Mesozoic fossils are Australian – the Late Jurassic *Protochares brevipalpis* from Talbragar Fish Beds which is at the moment the oldest definite hydrophilid beetle, and the Early Cretaceous *Cretoxenus australis* from Koonwarra (Fikáček *et al.* 2014b). Triassic fossils assigned to Hydrophilidae (Tillyard 1916; Dunstan 1923) either represent other beetle groups (mostly Archostemata: Ponomarenko 1969; Jell 2004) or cannot be assigned to any beetle family as they are only preserved as isolated elytra. So far, no Cenozoic fossils of Hydrophilidae were found in Australia. The fossil record is currently under revision, see e.g. Fikáček & Schmied (2013) for references to recently published papers.

Biology. Adult and larvae of Hydrophilinae, Chaetarthriinae, Acidocerinae and Enochrinae are mostly aquatic, inhabiting various types of water bodies from extremely temporary ones to permanent well vegetated pools or forest streams. Some representatives of these families are found in hygropetric habitats as well sun-exposed rocks at sides of waterfalls or on seepage habitats (e.g. non-Australian genera *Oocyclus* Sharp, *Tritonus* Mulsant, some *Laccobius* Erichson, and the New Zealand endemic *Horelophus* Orchymont; e.g. Short 2008; Clarkson & Short 2012; Gentili & Fikáček 2009; Fikáček *et al.* 2012). Few representatives of these subfamilies even colonised terrestrial habitats, though still highly humid: e.g. the New Zealand *Tormus* Sharp (tribe Laccobiini) inhabits mosses in forest floor (Fikáček *et al.* 2013), one species of *Limnoxenus* Motschulsky (tribe Hydrobiusini) in Hawaii including its larva was found in tree mosses (Short & Liebherr 2007), and the Neotropical genus *Quadriops* Hansen (Acidocerinae) was found in rotten fruits, dead wood and tree sap (Hansen 1999b; Girón & Short 2017). In contrast, subfamilies Cylominae and Sphaeridiinae are predominately terrestrial, and we suppose that their common ancestor left water and colonised terrestrial habitats (even in this case high humidity is still required, due to sensitivity of larvae to drying up; Archangelsky 1999a). Many groups of Cylominae and Sphaeridiinae inhabit various kinds of decaying organic material, and can be usually found in forest leaf litter or dung, some of them are also found on fungi or myxomycetes. Few groups returned to aquatic habitats, either to usual ones (e.g. *Coelostoma* in standing waters and at sides of streams and rivers) or to special ones (phytotelmata in inflorescences of gingers and heliconias, water tanks in bromeliads or water accummulated in broken bamboo culms, e.g. genera *Lachnodacnum* Orchymont (Coelostomatini) and *Pelosoma* Mulsant, *Cycreon* Orchymont and *Pseudocercyon* Orchymont (Megasternini)) (e.g. Jia *et al.* 2014; Clarkson *et al.* 2014; Arriaga-Varela *et al.* 2018). Some genera of Megasternini are specialised inhabitants of beach wrack and are capable to survive occasional submersion in sea water (Australian genera *Cercyodes* Broun and *Ercycodes* Hansen, and several species of *Cercyon* in Europe, western coast of North America, eastern Palaearctic and South Africa; Smetana 1978; Ôhara and Jia 2006; Hansen 1987, 1990b). The adults of cylomine genera *Pseudohydrobius* Blackburn and *Rygmodus* White can be found on flowers (Hansen 1990b; Minoshima *et al.* 2018); larvae of *Rygmodus* inhabit sides of the streams (Minoshima *et al.* 2018) unlike all other hydrophilids in which larvae inhabit the same habitat as adults.

Adults are scavengers, feeding on various kinds of decaying organic matter and in some taxa probably also algae (seepage taxa) and higher plants (e.g. *Hydrophilus*). Detailed knowledge on food preferences are not known, the experience from laboratory rearings show that at least ocassional protein-rich diet (dog food or fish dried food) is necessary for successful reproduction. This may explain the occasional records of adult feeding on animal corpses or even alive animals. Experiments with dung-inhabiting *Sphaeridium* showed that the adults feed on small detritus particles of specific size effectively selected from the substrate (Holter 2004). Flower-visiting adults of New Zealand genus *Rygmodus* are feeding on pollen; the mola of their mandibles is grinder-like and not sieve-like as in all other studied Hydrophilidae (Minoshima *et al.* 2018). Larvae are ambush-type predators (e.g. Formanowicz & Brodie 1988), feeding on various invertebrates as insect larvae (incl. mosquitoes), annelids or crustaceans; larvae of *Hydrophilus* are specialised predators of planorbiid and lymnaeid snails able to open the shell of these snails using their mandibles (Archangelsky 1997; Inoda *et al.* 2003, 2015). Larvae of larger aquatic species may occasionally feed on small vertebrates (fish, tadpoles). Larvae of dung-inhabiting *Sphaeridium* feed on fly larvae in the specific succession phase of the dung (Sowig *et al.* 1997). Larvae are often cannibalistic; cannibalism can be prevented by short dispersal phase after hatching (absent in some dung-inhabiting Sphaeridiinae probably due to time constraints; Archangelsky 1999a). Larval food preferences are not known for most genera.

Larvae digest the prey extraorally, i.e. digestive fluids from alimentary canal are spit on the prey, the food is then processed by the mouthparts and antennae (and sometimes legs; Archangelsky 2008), and fluids from pre-digested prey are subsequently sucked into the intestines; mandibles play role only for catching the prey and its processing and in most groups do not help with ingestion). To prevent the dissolution of digestive fluids by water, many aquatic taxa hold the prey above water level during processing it while raising the head into vertical position (hence the hyperprograthous, i.e. back-leaned, head). Benthic larvae (genera *Berosus*, *Laccobius*, *Hybogralius*) are able to process food when submerged, due to the ability to ingest the predigested prey using the grooved left mandible covered by enlarged left epistomal lobe (Archangelsky 1997, 1999a; Minoshima & Hayashi 2015). This morphology is always accompanied by reductions of labium and labial palps (probably because the prey is not manipulated above water surface) and evolved several times independently in the family; Archangelsky (2008); Fikáček *et al.* (2018a). *Hydrophilus* larvae also process food underwater, as digestive fluids are prevented from dissolution by the shell of the eaten snail (Fikáček, personal observation). Terrestrial larvae often do not raise the head during prey processing (Archangelsky 1999a), as predigested prey cannot be dissolved by water in terrestrial environment, or as larger prey (fly eggs or larvae) are consumed. The ingestion of pre-digested food is moreover facilitated by densely pubescent mouthparts in many groups (hypopharyngeal lobe: *Austrotypus*, some *Dactylosternum*, *Sphaeridium* and all Megasternini; maxillae and labium: *Tormus*).

Adults of all species and larvae of most groups breathe atmospheric air. In adults, antennae are used to establish a connection between atmosphere and the air bubble on the venter (Hrbáček 1950). The bubble (plastron) is held by hydrophobic pubescence which in most aquatic taxa covers the majority of ventral surface, often including femora. The only exception are many *Hydrophilus* in which pubescence on the abdomen is largely reduced (Fig. 20.8b), probably as the ventral bubble would hamper swimming abilities. Many terrestrial taxa retain the ventral pubescence (although it may be reduced e.g. on median portion of the metaventrite), and only in few genera the venter is completely or nearly completely bare (e.g. *Noteropagus*, *Cryptopleurum*, *Pachysternum*). Most larvae are metapneustic, i.e. have all spiracles but the last pair closed (but exceptions seem to exist, and it would require further study to confirm that thoracic and abdominal spiracles 1–7 are really non-functional; see Fikáček *et al.* 2017b). The last pair, placed on abdominal segment 8 is large, annular and connected to the main tracheal trunks, and is enclosed in a spiracular atrium (Fig. 20.3I) formed by abdominal segments 8–10; the atrium opens when the abdomen is in contact with atmosphere and closes when the larva is submerged. The spiracular atrium is present also in all examined terrestrial larvae, as it may be probably useful at ocassional flooding of the very humid substrate in which the larvae live. Benthic larvae of *Berosus*, *Hybogralius* and *Laccobius* (*Yateberosus*) are apneustic, i.e. even the last pair of spiracles is reduced and the spiracular atrium is absent. Tracheal gills are present in *Berosus* and *Laccobius* (*Yateberosus*) to facilitate the gas exchange from water; in *Hybogralius* no tracheal gills are present. Partly reduced posterior spiracles and spiracular atrium is present in larvae of *Hemiosus*, a sister group of *Berosus* (Archangelsky 2000, 2008). Many aquatic larvae ingest air bubbles which are then visible in their intestines; they increase the buyonancy and facilitate to reach the water level for breathing and/or prey processing. Bubbles are not present in benthic larvae, good swimmers and in terrestrial ones (Minoshima & Hayashi 2015).

Stridulation signals produced by moving elytral plectrum (Fig. 20.2M) across stridulatory file on laterosternite 3 (Fig. 20.2Q) are involved before mating in many species of Hydrophilinae and were best studied in genera *Berosus* and *Tropisternus* (different type of stridulation is used as defensive sounds in these species; e.g. Van Tassell 1965, Ryker 1972, 1976). Eggs are laid in silky egg cases woven by the ovipositor styli in most taxa; multiple egg cases may be produced without repeated mating (Archangelsky 1997; Fikáček, personal observation). In most cases, each egg case contains one to few eggs enclosed in a bag-shaped part of the egg case which continues into a collar-, ribbon- or mast-shaped projection (Figs 20.3J–L) usually reaching above water and likely facilitating gas exchange (Archangelsky 1999a). Egg cases are usually stuck on plants or submerged surfaces (i.e. in *Berosus* and *Enochrus*, e.g. Figures 20.3J–K). Egg cases of many terrestrial species are cup-shaped, without mast (Fig. 20.3N). In few genera of Acidocerinae (*Helochares*, *Helobata*, *Radicitus*), egg case is bag-shaped and is carried by female until hatching of larvae (Fig. 20.3Q; Archangelsky 1997; Short and García 2014). In hydrophiline genera *Hydrophilus*, *Hydrobiomorpha* and *Hydrochara*, egg cases are floating on water surface, with the mast projecting upwards (Fig. 20.3P; Archangelsky *et al.* 2004; van Berge Henegouwen 1982; Minoshima *et al.* 2012). Eggs are deposited in a loosely woven silky web (i.e. without proper egg case) in *Anacaena* and some *Enochrus* (attached to plants, Fig. 20.3M; Archangelsky & Fikáček 2004; Minoshima *et al.* 2016), *Chasmogenus* (in a mossy hollow in wet soil; Fig. 20.3O; Anderson 1976) and some *Cymbiodyta* (Archangelsky 1997). Larvae usually hatch after 4–8 days after ovoposition (2–3 days after ovoposition in dung-inhabiting Sphaeridiinae) (Archangelsky 1997).

Larval development comprises three larval instars and is relatively quick, taking 40–60 days in most aquatic forms (rarely more, e.g. ~90 days in *Sperchopsis*; Archangelsky 1997). In studied Sphaeridiinae (*Dactylosternum*, *Sphaeridium*, *Cercyon*), larval development is quicker, taking ~7–20 days (Hafez 1939; Archangelsky 1997; Sowig *et al.* 1997). Larval instars in field collected larvae may be identified by width of the head, presence of additional setae on the head and mouthparts in 2nd and 3rd instars in most groups (first instar with primary chaetotaxy only, see Fikáček *et al.* 2008 and Minoshima & Hayashi 2011a; in some Cylominae additional setae are present already in first instar larvae: Minoshima *et al.* 2015), and absence of frontal lines on the head in 3rd instar larvae; first instars of some Sphaeridiinae genera may be also recognised by the presence of egg bursters on the head. Before pupation, a short prepupal stage is present, in which the larva actively seeks for pupation place and does not feed.

Pupation takes place usually in a pupal chamber constructed by the prepupa in a moist substrate, aquatic taxa pupate outside of water in wet soil (Archangelsky 1997). Terrestrial species living in decaying plant substrate build pupal chamber often with chewed plant tissue (Archangelsky 1999a). Numerous styli present on dorsal and pleural portions of the pupa (Fig. 20.3R–S) prevent the pupa from contact with the moist walls of pupal chamber. Pupal stage usually takes 5–7 days (Archangelsky 1997). Free pupation is only known in *Derallus* and *Amphiops* whose pupae are freely attached to aquatic plants (Archangelsky 1997; Watts 2002). Adaptations to free pupation include coloration (green in *Derallus*, dark in *Amphiops*, whitish in all other hydrophilids), modification of styli (extremely long possibly cryptic in *Derallus*, very short in *Amphiops*: Fig. 20.4h) and shortened pupal stage (~2 days in *Derallus*, unknown for *Amphiops*).

Newly emerged, often still teneral adults of many aquatic and dung-inhabiting species are in a dispersal phase, actively flying during the day and shortly after sunset. Flying adults localise new water bodies as well as fresh dung by means of polarised light reflected from water surface or wet substrate surface (Schwind 1991; Kriska *et al.* 2006); for that reason adults are occasionally attracted to red or black car polish.

Characteristics. Adults (Fig. 20.2). Length ~0.9–46 mm (most species ~1.5–10 mm, only Hydrophilini much larger). Body ~1.2–2.5 times as long as wide, with sides more or less evenly rounded, well sclerotised, moderately to strongly convex; colour yellow to brown or black, occasionally bicolored (Pl. 43F–N, 44–47), in some genera (*Paracymus*) with metallic sheen; dorsal surfaces usually smooth and glabrous, sometimes with fine pubescence in Berosini, Cylominae, and Megasternini; clypeus, frons, pronotum and alternate elytral intervals in most species with groups of trichobothria (having ring-like socket and bearing long seta; 20.2G); ventral surfaces with hydrofuge pubescence (plastron) which is partly reduced in many Sphaeridiinae, some Hydrophilinae and Chaetarthriini, and totally absent from some Omicrini (e.g. *Noteropagus*, Fig. 20.16F) and Megasternini (e.g. *Cryptopleurum* and *Pachysternum*, Figs 20.21C–D).

Head more or less prognathous, deeply inserted into prothorax. Eyes usually well developed, protuberant or not, circular and entire to slightly emarginate anteriorly (deeply emarginate in some Coelostomatini and *Sphaeridium*, Figs 20.17H, 20.18G), occasionally divided into dorsal and ventral parts (*Amphiops*, Fig. 20.4B–C), finely facetted, without interfacetal setae. Antennal insertions concealed from above (but more or less exposed from above due to constriction of clypeus before eyes in Megasternini and Omicrini); subantennal groove usually present. Frontoclypeal suture often distinct, angulate, with attached mid-cranial suture; clypeus large, truncate, rounded or emarginated anteriorly. Labrum strongly transverse, slightly narrowing at the base; exposed and strongly sclerotised in most Hydrophilinae, Chaetarthriini, Enochrinae, most Acidocerinae, some Cylominae, most Omicrini and many Coelostomatini, partly to largely membranous and partly to completely concealed from above in *Amphiops*, some Anacaenini, few Acidocerinae, most Cylominae, *Psalitrus*, many Coelostomatini and all Megasternini. Antennae short, usually 8- or 9-segmented (7-segmented in *Berosus* and few *Laccobius*), with glabrous basal antennomeres and 3-segmented pubescent club preceded by cup-like, often asymmetrical, antennomere (cupule). Mandibles usually large (partly reduced in some Omicrini) and more or less concealed beneath clypeus and labrum; apex unidentate or bidentate; mola well developed and often asymmetrical; prostheca well developed, usually membranous, sometimes with sclerotised process (articulated e.g. in *Berosus*, non-articulated e.g. in *Laccobius*). Maxilla with setose galea and lacinia, galea mostly flattened, partly sclerotised, with apical curved setae arranged in regular rows (most groups) or irregularly arranged (some Cylominae, *Sphaeridium* and Megasternini); lacinia short, largely membranous, without uncus, sometimes with apical setiferous membranous to slightly sclerotised projection (e.g. in Cylominae and some Sphaeridiinae); maxillary palps slender, in some aquatic groups (Hydrophilini, Enochrinae, Acidocerinae) longer than antennae, or about as long as antennae (Berosini, Laccobiini, Hydrobiusini, Chaetarthriinae), in most terrestrial groups (Cylominae, Sphaeridiinae) distinctly shorter than antennae; consisting of 4 palpomeres, basal palpomere minute, apical palpomere asymmetrical or fusiform. Mentum usually transverse and apically truncate or bisinuate (except some Cylominae); ligula bilobed, setiferous; labial palp with 3 palpomeres, basal one minute, second one with few long setae (Hydrophilinae, Chaetarthriinae, Enochrinae, Acidocerinae) or multiple setae (*Tormus*, *Chaetarthria*, Cylominae, Sphaeridiinae). Corpotentorium narrow; laminatentoria well developed, sometimes meeting at midline. Cervical sclerites well developed.

Pronotum ~0.25–0.65 times as long as wide, widest posteriorly or at middle, about as wide as combined elytral bases (slightly narrower e.g. in *Horelophus* and some *Berosus*); sides slightly to strongly curved, sometimes slightly explanate, deflexed to ventral side in some Megasternini (e.g. *Cryptopleurum*, Fig. 20.2C). Lateral pronotal carinae complete

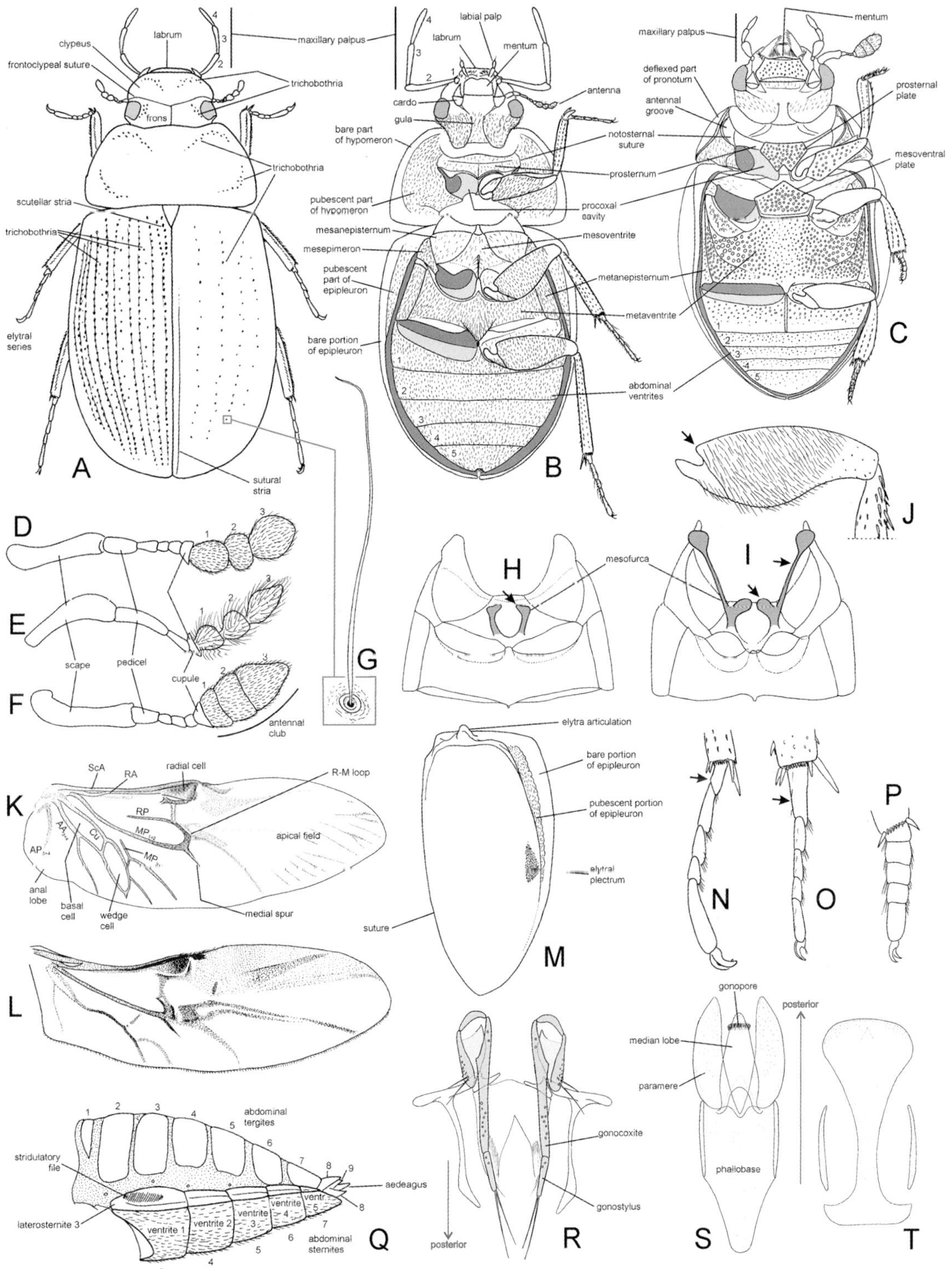

Fig. 20.2. General morphology of adult Hydrophilidae. A–C, general habitus (A, generalised hydrophilid, dorsal; B, *Helochares* sp., ventral; C, *Cryptopleurum* sp., ventral). D–F, antenna (D, *Enochrus* sp.; E, *Berosus* sp.; F, Megasternini gen. sp.). G, detail of trichobothria. H–I, internal view of ventral portion of meso- and metathorax with mesofurca (H, *Chaetarthria* sp.; I, *Amphiops* sp.). K–L, metathoracic wing (K, *Austrotypus nothofagi* Fikáček, Minoshima and Newton; L, Megasternini: *Kanala* sp.). M, ventral view of elytron of *Laccobius* sp. N–P, posterior tarsus (N, *Helochares* sp.; O, *Coelostoma* sp.; P, *Noteropagus* sp.). Q, abdomen in lateral view; R, female genitalia of Megasternini: *Kanala* sp. S–T, male genitalia of *Austrotypus nothofagi* (S, aedeagus; T, sternite 9). H–I, N–O adapted from Short & Fikáček (2013); K, S–T from Fikáček *et al.* (2014a); L and R adapted from Fikáček (2010c); M and Q adapted from Hansen (1991).

and simple; anterior angles not or slightly produced forward and rounded; posterior angles right to rounded and sometimes embracing elytral bases; posterior edge straight to slightly sinuate, with underlying ventral ridge, which interlocks with elytra and scutellar shield. Prosternum in front of coxae about as long as to much shorter than shortest diameter of procoxal cavity, usually tumid, with or without median carina, in some groups (some Laccobiini, Anacaenini and Enochrinae) with a transverse groove, anterior edge sometimes produced and angulate or acute; in many Megasternini the median portion of prosternum is delimited from lateral portions by ridges (e.g. Figs 20.20A, C–D, F, K) or is elevated into a plate (e.g. Figs 20.21C–D); in Berosini the prosternum is largely reduced and narrowly exposed only between coxae. Prosternal process weakly developed, apically pointed, or narrow and slightly widened behind procoxae (this portion is usually not visible until coxae are detached). Notosternal sutures complete, rarely somehow obsolete. Hypomeron with glabrous and shiny lateral portion and pubescent mesal portion, often separated by a fine ridge; usually without anterior groove for reception of antennal club (with antennal groove in some Omicrini and most Megasternini). Procoxae more or less conical and projecting, with short to moderately long concealed lateral extensions; trochantins concealed or exposed; endopleuron not fused to notum. Procoxal cavities subcircular to slightly transverse, contiguous in most groups, narrowly separated by prosternal plate or process in some Megasternini, *Sphaeridium* and *Hydrophilus*, broadly open externally, often with distinct narrow anterolateral extension, internally closed.

Scutellar shield abruptly elevated, posteriorly acute. Elytra ~1.0–1.65 times as long as combined width and 1.0–5.0 times as long as pronotum; sides weakly to moderately curved, apices conjointly rounded (not closely fitting to each other at apex in *Sphaeridium*); humeri well developed, often somewhat impressed where overlapped by pronotum; disc usually with ten distinct puncture rows or striae (less frequently with nine rows, e.g. in some Megasternini, or 11 rows in some *Dactylosternum*), with (in most Hydrophilinae, Enochrinae, Acidocerinae) or without (Chaetarthriinae, Cylominae, Sphaeridiinae) scutellary stria, punctation sometimes irregular or with puncture rows indistinct; epipleura largely reduced posteriorly in most groups, but complete and very wide e.g. in many Omicrini, Coelostomatini and few Megasternini, sometimes obliquely or vertically explanate basally, divided into glabrous external and pubescent internal portion; in some Megasternini (e.g. *Cryptopleurum*, 20.2C) largely reduced throughout. Mesoventrite usually separated by complete anapleural sutures from mesanepisterna, which are usually narrowly separated or contiguous at midline (mesanepisterna widely separated at midline in Amphiopini, some Berosini and Chaetarthriini, partly or completely fused with mesoventrite in some Hydrophilinae, Laccobiini: *Afrotormus*, Cylominae: *Petasopsis* and Sphaeridiinae), anterior edge often on different plane than metaventrite, anapleural sutures divergent posteriorly (most groups) or subparallel anteriorly (some Chaetarthriinae, some Enochrinae and some Acidocerinae); procoxal rests (= grooves for reception of procoxae) usually absent (well defined in Megasternini and some Omicrini, weakly defined in some Coelostomatini); mesal portion of mesoventrite flat to slightly bulged (e.g. *Hybogralius*, some Anacaenini, some Acidocerinae, most Cylominae, *Sphaeridium*), but more frequently with median or posteromedian carina (e.g. Berosini, some Hydrobiusini, Enochrinae, Acidocerinae, Cylominae: *Coelostomopsis*, Megasternini, in Hydrophilini highly elevated and fused to metaventral carina) or elevation of variable shape (transverse or arrow-like: most Laccobiini, some Chaetarthriinae and Cylominae, many Coelostomatini and Protosternini; plate-like: Amphiopini, Omicrini, some Coelostomatini, Megasternini); mesoventral cavity absent; discrimen absent. Mesocoxae slightly to strongly transverse with exposed trochantins. Mesocoxal cavities usually narrowly separated, widely separated by meso- and metaventral projections or mesoventral plate in some Amphiopini and many Sphaeridiinae, contiguous in *Regimbartia*; closed laterally by mesepimeron or mesepimeron and mesanepisternum. Mesendosternite (mesofurca) consisting of a pair of arms originated separately, each arm short and slightly widened apically in Chaetarthriinae, Enochrinae, Acidocerinae, Cylominae and Sphaeridiinae (Fig. 20.2H), or with a long thin apical extension widened apically and nearly fused to dorsal body wall in Hydrophilinae (Fig. 20.2I). Mesometaventral junction consisting of abutting or overlapping ventrites. Metaventrite distinctly transverse, sometimes much shorter than mesoventrite, flat to slightly convex, in Sphaeridiinae often with mesal portion somewhat raised; discrimen and transverse (metakatepisternal) suture present or absent; exposed portion of metanepisternum usually moderately broad, very narrow and concealed beneath elytra in some Megasternini (e.g. *Cryptopleurum*, Fig. 20.2C). Metacoxae strongly transverse and subcontiguous, extending laterally to meet elytra; plates absent. Metendosternite (metafurca) usually with long, slender stalk and long, lateral arms; anterior process either absent or very short and broadly emarginate, so that anterior tendons are on widely separated lobes. Hind wing (Figs 20.2K–L) usually with long radial bending zone at the end of which is a distinct hinge; apical field usually with three radial extensions, and with a small sclerite just beyond radial cell, which is short and broad, partly enclosing pterostigma and with basal edge incomplete or absent; cross-vein r3 usually short or absent; cross-vein r4 complete; RP moderately long, obsolete basally, well sclerotised distally and together with MP_{1+2} forming R-M loop, medial spur straight and ending well before wing margin, sometimes very short; MP_{3+4} with basal cross-vein; both basal and wedge cells often well developed, AA_3 mostly long and reaching wing margin; veins of the medial field (especially MP_{3+4} and AA_{3+4}) reduced in Omicrini and Megasternini, basal cell reduced in size and wedge cell missing in these taxa; anal lobe well developed to reduced, sometimes with notch-like embayment; AP undivided or absent; in many terrestrial taxa wings may be reduced in size (with reduced apical field e.g. in some specimens of *Micramphiops*, narrow in shape with preserved remnants of longitudinal veins in many Omicrini, reduced to short scales in some Megasternini) or completely absent (e.g. *Tormus*). Femoral attachment of trochanter oblique with base of femur free, not contacting trochanter (Fig. 20.2J). Tibiae cylindrical to broadly flattened, often lined with teeth or spines, in some aquatic groups swimming hairs present on mesotibiae (*Amphiops*) or meso- and metatibiae (Berosini);

tibial spurs well developed. Tarsi almost always 5–5–5 (5–5–4 in Enochrinae: *Cymbiodyta*, *Helocombus*), tarsomeres simple, tarsomere 1 usually shorter than tarsomere 2 (Fig. 20.2N), in Sphaeridiinae and some Cylominae tarsomere 1 thick and long (Fig. 20.2O), in some terrestrial taxa (e.g. Cylominae: *Petasopsis*, Omicrini: *Noteropagus*, *Psalitrus*) tarsus is reduced in length, with all tarsomeres subequal in length (Fig. 20.2P); pretarsal claws simple; empodium usually well developed, bisetose.

Abdomen almost always with five free ventrites (6 ventrites in *Laccobius*, 4 ventrites in *Regimbartia* and *Allocotocerus*); ventrite 1 sometimes with median carina; intercoxal process acute. Functional spiracles on abdominal segments I–VI; abdominal tergites relatively lightly sclerotised, usually with patches of wing-folding asperities on one or more segments; laterosternite 3 with stridulatory file consisting in most groups of separate bird-head-shaped asperites, which are arranged in rows or totally fused into longitudinal ridges in several clades of Hydrophilinae. Anterior edge of sternite VIII in male usually without median strut, with a median strut in some Chaetarthriini and Megasternini. Anterior edge of segment IX in male broadly rounded or truncate, without spiculum gastrale; tergite IX emarginate; tergite X free. Aedeagus of the trilobate type, usually symmetrical (or at most with slightly to strongly asymmetrical phallobase), with short and broad to elongate phallobase and simple parameres; penis (= median lobe) with basal struts. Ovipositor moderately elongate, lightly sclerotised; paraprocts shorter than coxites, which are subdivided with broad proximal lobe, slender distal lobe and small terminal styli. (Adapted from Lawrence & Ślipiński 2013.)

Most hydrophilids do not exhibit external sexually dimorphic characters, and sex is recognisable only by genitalia dissections (sometimes apices of genitalia are protruding from abdominal apex in alcohol-preserved specimens, females may then my recognised by a pair of styli; Fig. 20.2R, gonocoxites + gonostyli). External sexually dimorphic characters are present in males of some genera only; they are likely structures facilitating the attachement of the male on the female during mating (tarsal claws of modified shape or enlarged size in Hydrophilini, some *Paracymus*, *Eurygmus* and *Sphaeridium*; expanded male protarsi with special setae or lamellae in *Berosus*, *Laccobius*, some *Paracymus*, *Hydrophilus*, *Eurygmus* and *Sphaeridium*; protibial suckers or lamellae in some *Chaetarthria* (see Spangler 1977); suckers on ventral surface of labrum (sometimes called 'googles' by some authors) in many *Laccobius* species; maxillary suckers in males of *Sphaeridium* and Megasternini) or with pre-mating communication (enlarged maxillary and labial palps of some *Hydrophilus* and *Hydrobiomorpha*). In addition, males of some Cylominae (*Eurygmus*, some *Rygmodus*) have expanded and upturned anterolateral margin of the clypeus, males of some Megasternini (e.g. the Australian species of *Merosoma*) bear additional setae and impression on the metaventrite, and males and females of *Berosus* often differ in the shape of the emargination on posterior margin of abdominal ventrite 5, in the shape of the spines on elytral apices, or in the character of dorsal microsculpture of pronotum and elytra; in *Bourdonnaisia mahensis* Scott (Coelostomatini), males have antennae with 8 antennomeres, and females with 9 antennomeres.

Larvae (Fig. 20.3A–H). Body elongate, parallel-sided or slightly wider at middle, conical to slightly flattened; head and protergum well sclerotised, meso- and metatergum with a pair of sclerites which can be large (in most groups) to rather small and subdivided (e.g. in Chaetarhriini and Omicrini); abdomen membranous, with sclerotised plate or pair of plates on tergum VIII (absent in *Hybogralius*) and/or small tergal sclerites on most abdominal segments; surfaces usually pubscent and sometimes granulate or asperate, or with lateral abdominal projections.

Head hyperprognathous; epicranial stem short or absent; frontal arms V-shaped with bases contiguous (in Amphiopini, Hydrobiusini, Hydrophilini, Enochrinae, Acidocerinae, Cylominae, Coelostomatini and Protosternini) or not contiguous (in Berosini, Laccobiini, Chaetarthriinae, Omicrini, Sphaeridiini and Megasternini); median endocarina absent. Each side of head with 6 stemmata, usually well separated in two vertical rows, sometimes fused to form two (in Omicrini, Sphaeridiini and some Protosternini and Coelostomatini) or one group (in Chaetarthriini and Megasternini); stemmata largely reduced in some Coelostomatini, totally absent in *Sphaerocetum* (Protosternini) and some Omicrini. Frontoclypeal suture absent and labrum fused to head capsule forming clypeolabrum consisting of a medial nasale and paired epistomal lobes (adnasalia), nasale genus-specific in shape, usually with one to 6 teeth, rarely multidentate (e.g. *Regimbartia*, *Enochrus*) or without teeth (*Hydrobiomorpha*, *Hydrophilus*, some *Berosus*, Sphaeridiini and Megasternini), epistomal lobes not projecting in Sphaeridiini and Megasternini. Antennae well developed, almost always 3-segmented (4-segmented in second and third instars of *Hydrophilus*); antennomere 1 usually longest, antennomere 2 with single laterally placed sensorium. Mandibles symmetrical or asymmetrical, moderately broad at base with narrow, strongly curved and unidentate apex, with 1 to 3 retinacula (inner teeth; absent in some Sphaeridiinae); mola, prostheca and basal penicillus absent. Ventral mouthparts usually protracted with maxillary articulating area absent; maxilla hexamerous: cardo in form of a single sclerite (sometimes associated by few more or less distinct sclerites of unclear homology, see Minoshima 2013); stipes elongate and cylindrical, its apical portion (palpifer) segment-like (considered as first segment of maxillary palp by many authors, e.g. Minoshima & Hayashi 2011a); galea (inner appendage by some authors) small, palpiform, articulated and located at apex of palpifer (membranous and reduced to various extent in Sphaeridiini and Megasternini); lacinia (= inner projection of stipes) absent; palp 3-segmented (not counting the palpifer). Labium with submentum largely (with submental line still present) or completely (e.g. in *Laccobius*) fused with head-capsule, submental line V-shaped when present; mentum and prementum well divided, mentum with asymmetrical pubescent membranose hypopharyngeal lobe in Cylominae: *Austrotypus*, some Coelostomatini, *Sphaeridium* and Megasternini; palps 2-segmented; ligula present or absent. Hypopharyngeal sclerome and hypostomal rods absent; ventral epicranial ridges absent. Gular region long, with gular sutures fused and sometimes largely obsolete (Protosternum).

Prothorax usually slightly longer than meso- or metathorax, with single tergal plate divided by median ecdysial line; meso- and metaterga often with paired plates. Prothoracic venter

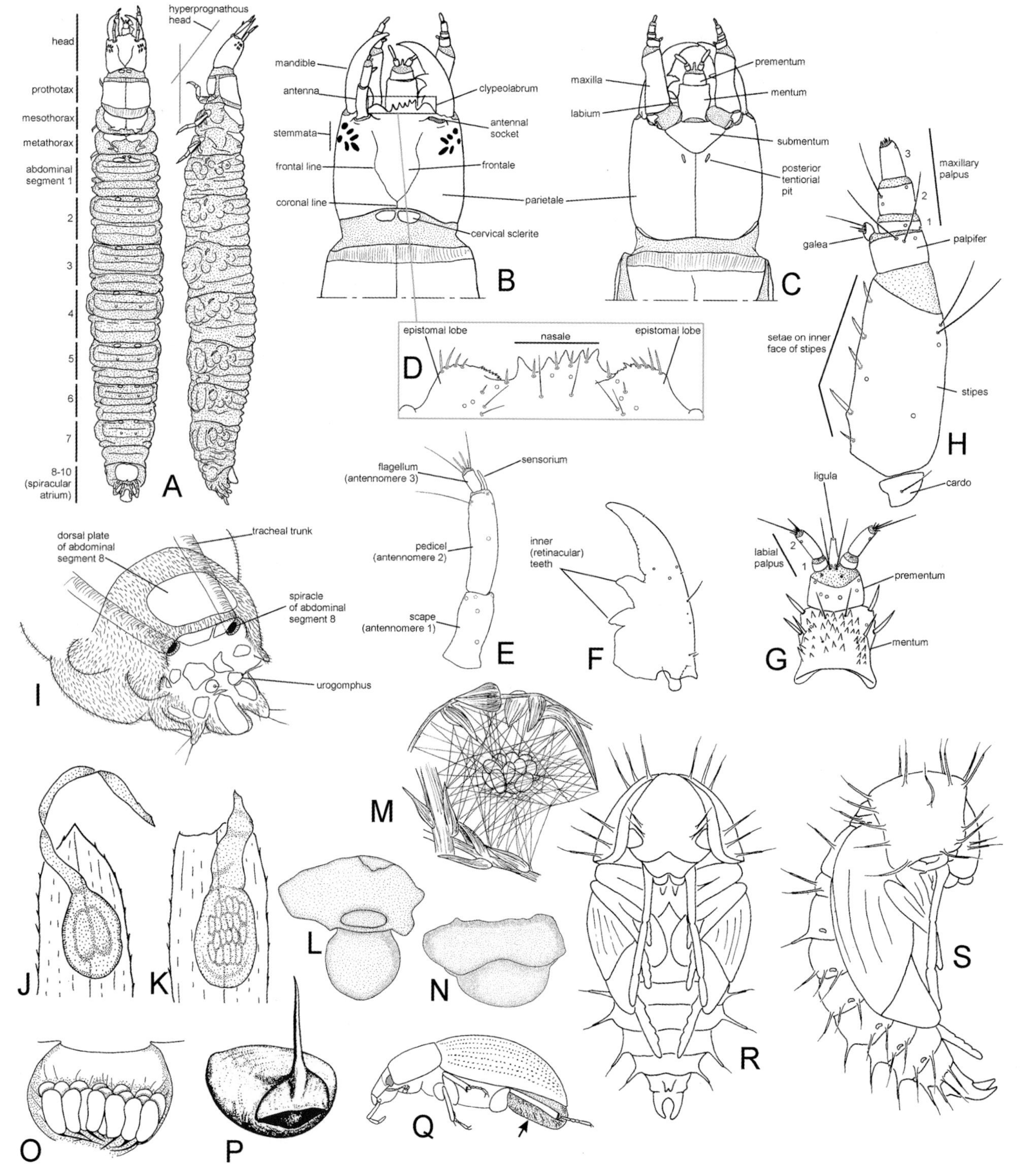

Fig. 20.3. Immature stages of the Hydrophilidae. A–I, larval morphology: A, general larval habitus dorsally and laterally; B, larval head dorsally; C, larval head ventrally; D, details of clypeolabrum; E, antenna; F, mandible; G, labium in dorsal view; H, maxilla in ventral view; I, abdominal apex with spiracular atrium. J–Q, egg cases: J–K, *Berosus* sp.; L, *Limnohydrobius melaenus* (Germar); M, *Anacaena lutescens* (Stephens); N, *Cercyon* sp.; O, *Chasmogenus nitescens* (Fauvel); P, *Hydrophilus piceus* (Linnaeus); Q, *Helochares* sp. R–S, *Chasmogenus nitescens*, pupa (R, ventral view; S, lateral view). Adopted and adapted from: A–D, F–H, Minoshima & Hayashi (2011a); E, Minoshima & Hayashi (2012); J–N, O, R–S, Archangelsky (1997); M, Archangelsky and Fikáček (2004); P, van Berge Henegouwen (1982). I, redrawn and adapted from Clarkson *et al.* (2014).

usually with divided or undivided prosternal plate. Legs usually well developed and 5-segmented with pretarsus forming distinct claw (reduced in size in some taxa: 5-segmented with reduced claw in *Sphaeridium*, 3-segmented and without pretarsus in *Chaetarthria*, Fig. 20.10f; gradually reduced in Megasternini: 3-segmented in *Pelosoma* and some *Cercyon*, 2-segmented in *Oosternum*, as a small spinose lobe in *Cryptopleurum* and some *Cercyon*, totally absent in *Merosoma*; Archangelsky 1999a, 2016; Arriaga-Varela *et al.* 2017, this chapter), with long swimming hairs on femur in some Hydrophilini (Fig. 20.9T–U); procoxae widely, meso- and metacoxae narrowly separated. Abdominal segments I–VIII weakly sclerotised, with or without lateral membranous projection, each with a pair or transverse row of minute sclerites; terga I–VII sometimes with transverse plicae; segments I–VII with long, lateral projections in *Allocotocerus* and *Regimbartia*, and long tracheal gills in *Berosus*; abdominal sterna III–VII in *Enochrus*, *Agraphydrus*, *Lachnodacnum* and some *Dactylosternum* with paired, asperity-bearing prolegs. Abdominal apex usually with spiracular atrium formed by modifications of segments VIII–X (tergum VIII forming sclerotised dorsal plate at apex of abdomen, segment IX trilobed and forming ventral valve of spiracular atrium, which bears a pair of small 1-segmented urogomphi; large posteriorly oriented spiracles located within spiraculr atrium; segment X reduced and ventral); spiracular atrium as well as urogomphi absent in *Berosus*, *Hybogralius*, and *Laccobius* (*Yateberosus*), largely reduced in *Hemiosus*. Functional spiracles absent except for an enlarged annular pair in spiracular atrium in most taxa, all spiracles non-functional in species without spiracular atrium). (Adapted from Lawrence & Ślipiński 2013; Archangelsky 1997, 2016; Minoshima & Hayashi 2011a; Minoshima 2013; Fikáček *et al.* 2015.)

Pupae (Fig. 20.3R–S). Exarate, adectic with free appendages, usually whitish in coloration (greenish in free-pupating *Derallus*, dark in free-pupating *Amphiops*), eyes turning red in pharate imago. Dorsal surface of head, pronotum, meso- and metanotum and dorsal and pleural parts of abdominal segments I–VII or I–VIII with cuticular projections (styli) each bearing a seta of variable length; styli very short and lacking apical seta in *Amphiops*. Abdominal segment IX with a pair of cerci. Functional spiracles present on abdominal segment II–VI (Archangelsky 1997; Watts 2002).

Key to the genera of Australian Hydrophilidae (adults)

The key contains all genera recorded until now from Australia, plus also the megasternine genus *Platycyon* Hansen which is endemic for New Guinea and its presence in northernmost Queensland and Northern Territory cannot be excluded. The key does not follow the classification and phylogenetic position of the respective genus, for key to subfamilies and tribes see Short & Fikáček (2013). In few cases, single genus is keyed twice, either in case of morphologically diverse genera or in case when the used key character may be easily misinterpreted.

1. Medium-sized to large species (body length 10–46 mm). Meso- and metaventrite raised into a common sternal keel extending posteriorly into a long or short spine (Fig. 20.8A, E, J, O, U, Y–Z). Meso- and metatarsi with dense series of swimming hairs on inner face (Fig. 20.8A, J, U). **Hydrophilini** **2**

– Small to medium-sized species (1–11 mm). Metaventrite flat or with two median bulges, never with an elevated median keel projecting into a spine posteriorly; mesoventrite of variable morphology. Meso- and metatarsi with or without swimming hairs **4**

2(1). Middle portion of prosternum deeply emarginate, forming two lobes receiving the anterior portion of ventral keel (Fig. 20.8W). Anterior part of ventral keel not notched in lateral view (Fig. 20.8X). Medium-sized to large species (body length 18–46 mm) .... ***Hydrophilus* Geoffroy**

– Middle portion of prosternum simply carinate or tectiform (Fig. 20.8C, M–L). Anterior portion of median keel with a small notch when observed from lateral view (Fig. 20.8D, N). Medium-sized species (10–18 mm) **3**

3(2). First segment of antennal club globular (Fig. 20.8B). All femora with hydrofuge pubescence basally (Fig. 20.8A). Anterior tip of prosternal carina with tuft of setae (20.8D). Anterior margin of clypeus entire or narrowly excised (Fig. 20.8F–G) .... ***Sternolophus* Solier**

– First segment of antennal club deeply split into two lobes (Fig. 20.8K). Meso- and metafemora glabrous (Fig. 20.8J). Anterior tip of prosternal carina without tuft of setae (Fig. 20.8N). Anterior margin of clypeus widely concave or excised (Fig. 20.8P–R) .... ***Hydrobiomorpha* Blackburn**

4(1). Eyes completely divided into dorsal and ventral portion (Fig. 20.4A–C). Body highly convex (Fig. 20.4B). Mesotibiae with dense fringe of long swimming hairs, metatibiae and all tarsi without swimming hairs (Fig. 20.4D). Metanepisterna very wide. Antennae with 8 antennomeres (Fig. 20.4E) .... ***Amphiops* Erichson**

– Eyes not divided into dorsal and ventral portion. Body of variable shape, highly convex to depressed. Legs either without swimming hairs, or swimming hairs present on tibiae and/or tarsi of meso- and metathoracic legs. Metanepisterna narrow. Antennae with 7–9 antennomeres **5**

5(4). Meso- and metatibiae and tarsi dorsally with fringe of long swimming hairs (Fig. 20.5A, C, E). Body rather highly convex (Fig. 20.5P–Q), head capable of being markedly deflexed (Pl. 43K). Antennae with 7–8 antennomeres (Fig. 20.5K–L) ... **Berosini** **6**

– Tibiae always without swimming hairs, meso- and metatarsi with or without swimming hairs. Body of different shape, head not capable to being markedly deflexed. Antennae with 8–9 antennomeres **8**

6(5). Eyes not protuberant, head wide anteriorly (Fig. 20.5M–N). Antennae with 8 antennomeres (Fig. 20.5L). Body completely black and shiny dorsally (Pl. 43G–H), without dorsal pubescence. Abdomen with four visible ventrites, the fifth is retracted in the abdomen (Fig. 20.5A, C) **7**

– Eyes protuberant, head narrow anteriorly (Fig. 20.5O). Antennae with 7 antennomeres (Fig. 20.5K). Body usually yellowish with black patterns (Pl. 43I–K), with or without distinct dorsal pubescence; if the dorsal coloration is largely black, there are at least small yellowish spots at bases of elytra and dorsal punctation is very coarse. Abdomen with five visible ventrites (Fig. 20.5E) .... ***Berosus* Leach**

7(6). Body narrowly drop-shaped in dorsal view, elytra attenuate posteriorly (Pl. 43H). Elytra with very distinct striae (Fig. 20.5Q). Metaventrite with a low median ridge throughout its length (Fig. 20.5C) .... ***Regimbartia* Zaitzev**

– Body widely and shortly globular, elytra not attenuate posteriorly (Pl. 43G). Elytral punctation uniform, without clearly defined series of larger punctures (sometimes except near lateral margin; Fig. 20.5P). Metaventrite with two median tooth-like projections (anterior and posterior, best seen in lateral view; Fig. 20.5B) .... ***Allocotocerus* Kraatz**

8(5). First abdominal ventrite with fringe of long setae arising from its basal margin (Fig. 20.10A). Antennal pedicel large, globular (Fig. 20.10C). Labrum strongly sclerotised, widely exposed anterior to clypeus (Fig. 20.10B). Mesoventrite very wide at anterior margin (Fig. 20.10A). Body small (1.5–2.5 mm), globular, completely black; elytra without puncture rows, only with sharp sutural stria (Pl. 44H) .... ***Chaetarthria* Stephens**

– First abdominal ventrite without series of long setae. Antennal pedicel conical, never apparently large and globular (e.g. Figs 20.6O–Q, 20.11J–M, 20.17I–K, 20.19O–P). Labrum strongly sclerotised and exposed, or membranous and largely retracted under clypeus. Body small to medium-sized (1–10 mm), depressed to globular, pale to black; elytra with or without puncture rows and with or without sutural stria **9**

9(8). Metatrochanters large, detached from metafemora, larger and of different shape than mesotrochanters (Fig. 20.6A). Abdomen with six exposed ventrites (Fig. 20.6A, R). Metatibia straight or distinctly arcuate (Fig. 20.6D–E). Body always pale, with variable pattern of dark/metallic spots (Pl. 43L–M). Antenna with 8 antennomeres (Fig. 20.6O). Male protarsus with tarsomeres 2–3 enlarged (Fig. 20.6C) .... ***Laccobius* Erichson**

– Metatrochanters of the same size and shape as mesotrochanters. Abdomen with five exposed ventrites. Metatibia straight. Body pale or dark. Antenna with 8–9 antennomeres. Male protarsus enlarged or not **10**

10(9). Metatarsomere 1 distinctly shorter than metatarsomere 2 (e.g. Figs 20.2N, 20.7A, E, 20.14F, O, T, V). Antennal club loosely segmented (e.g. Figs 20.7D, 20.11J–M, 20.14L). Mesoventrite always separated from mesanepisterna by distinct anapleural sutures **11**

– Metatarsomere 1 either distinctly longer than metatarsomere 2 (Figs 20.2O, 20.14C, K, 20.16D, 20.18A, 20.20–21), or metatarsi shortened and all metatarsomeres more or less of the same length (e.g. Figs 20.2F, 20.14R, 20.16H, O). Antennal club compact or loosely segmented. Mesoventrite with distinct anapleural sutures, or mesoventrite fused with mesanepisterna and anapleural sutures absent **30**

11(10). Meso- and metatarsi with very apparent dense series of swimming hairs dorsally (Fig. 20.7A). Mesoventrite with high subquadrate lamina mesally (Fig. 20.7C). Body medium-sized (body length 8–11 mm), elongate oval, dorsal coloration completely black, sometimes with greenish stripes on elytra (Pl. 44B) .... ***Limnoxenus* Motschulsky**

– Meso- and metatarsi without swimming hairs. Mesoventrite flat or at most with median protuberance or tooth-like projection, never with high subquadrate lamina. Body small to medium-sized (1–9 mm), of variable shape and coloration **12**

12(11). Elytra with 10 very distinct and regular rows consisting of punctures larger than surrounding punctation (do not confuse real serial punctures with serially arranged dark spots!) (Pl. 44I–J, 45E, H, J, M). Mesoventrite always flat, without any median protuberance, elevation or tooth-like projection **13**

– Elytral punctuation either totally irregular, or with serial punctures different than above (Pl. 43N, 44K, M, 45L, N–O). If elytral punctures (or part of them) are more or less serially arranged, then there are much less or many more than 10 rows, or the serially arranged punctures are irregular and/or difficult to distinguish from surrounding punctuation (Pl. 44L, 45A–D, F). Mesoventrite flat or with median protuberance, tooth-like projection or high elevation **19**

13(12). Elytron with short but distinct scutellary stria (e.g. Figs 20.7J, 20.13J). Elytral intervals with or without scattered punctures which are much larger than normal interval punctation (trichobothria; Figs 20.2A,G, 20.7J) **14**

– Elytron never with distinct scutellary stria or enlarged scattered punctures on elytral intervals **15**

14(13). Body rather depressed (Pl. 45B), maxillary palps very long, often as long as or longer than the width of the head (Fig. 20.13F). Last abdominal ventrite with apical emargination containing stiff setae (Fig. 20.13M). Nearly whole ventral surface of all femora (except the extreme apex) covered by very dense pubescence (Fig. 20.13F). Mesoventrite widely parallel-sided anteriorly (Fig. 20.13F) .... ***Helochares*: *Hydrobaticus* MacLeay**

– Body rather highly convex (Pl. 44A), maxillary palps much shorter than the width of the head (Fig. 20.7E). Posterior margin of the last abdominal ventrite entire, without apical emargination or group of stiff setae (Fig. 20.7E). Meso- and metafemora at most with very sparsely arranged long setae basally (Fig. 20.7E). Mesoventrite narrowly triangular anteriorly (Fig. 20.7E) .... ***Hybogralius* Orchymont**

15(13). Pronotum and elytra with very apparent long pubescence (Fig. 20.14h). Body widely oval, elytra widely explanate laterally (Pl. 45H). Posterior margin of the last abdominal ventrite bears a group of stiff setae mesally (Fig. 20.14G) .... ***Borborophorus* Hansen**

– Dorsal surface without apparent long pubescence. Elytra not explanate laterally. Body elongate to widely oval (Pl. 44I–J, 45 J, M). Posterior margin of the last abdominal ventrite never with the median group of stiff setae **16**

16(15). Antennae with 8 antennomeres (Fig. 20.11L–M). Labrum largely exposed (Fig. 20.11H–I) **17**

– Antennae with 9 antennomeres (Fig. 20.14L). Labrum membranous, completely or nearly completely concealed under clypeus (Fig. 20.14Z, e) **18**

17(16). Head short and very wide, clypeus subrectangular in shape, not narrowing anteriorly (Fig. 20.11I). Labrum large, inclined, straight on anterior margin. Mesofemur pubescent only at extreme base, otherwise bare; metafemur bare, abdomen pubescent only on ventrites 1–3 (Fig. 20.11E) .... ***Phelea* Hansen**

– Head narrowing anteriorly. Labrum small, partly concealed by clypeus, anterior margin of labrum more or less emarginate (Fig. 20.11H). Mesofemur with dense pubescence covering whole ventral surface except the extremely distal portion; metafemur largely pubescent or bare, abdomen completely pubescent (Fig. 20.11D) .... ***Notohydrus* Balfour-Browne**

18(16). Body small (body length 2.5–2.7 mm). Anterior margin of clypeus upturned (Fig. 20.14Z,b). Protibia with a pair of large subapical spines on outer margin (Fig. 20.14N). Male pro- and mesotarsi with dilated tarsomere 5 and enlarged more curved claws than in female (Fig. 20.14M, i–j) .... ***Eurygmus* Hansen**

– Body large (body length 5.5–6.5 mm). Anterior margin of clypeus flat. Protibia without large subapical spines on outer margin. Male tarsi not modified .... ***Pseudohydrobius* Blackburn (part)**

19(12). Posterior margin of the last abdominal ventrite with small but distinct notch containing stiff setae (Figs 20.6T, 20.13K–M) **20**

– Posterior margin of the last abdominal segment entire, without distinct notch **24**

20(19). Metafemora nearly completely bare, with dense pubescence present at most at very base (Fig. 20.6G). Body small (body length 1.5–3.0 mm), black or partly metallic (Pl. 43N) .... ***Paracymus* Thomson (part)**

– Metafemora always nearly completely covered with dense short pubescence, only its extreme apex bare (Figs 20.12A, D, 20.13A–B, D). Body small to medium-sized (body length 1.4–9.0 mm), coloration variable, but never metallic **21**

21(20). Elytra without sutural stria (Fig. 20.13G, I). Body either medium-sized (6.1–8.2 mm) and brown/black (Pl. 45F), or small (1.4–2.6 mm), yellowish, with dark head and central portion of pronotum (Pl. 45D) **22**

– Elytra with very distinct sutural stria reaching basal half of elytra (e.g. Fig. 20.13H). Body small to medium-sized (2–9 mm), of variable coloration; if small and pale, then pronotum never with dark central spot **23**

22(21). Body medium-sized (6.1–8.2 mm), brown/black (Pl. 45F). Maxillary palps very long, much longer than the width of the head. Posterior portion of mesoventrite with medina carina (Fig. 20.13D) .... ***Helochares*: *Helochares* Mulsant**

– Body small (1.4–2.6 mm), yellowish, with dark head and dark central spot on pronotum (Pl. 45D). Maxillary palps rather short, equal to or shorter than the width of the head. Mesoventrite without median carina (Fig. 20.13A). .... ***Agraphydrus* Régimbart**

23(21). Maxillary palpomere 2 (i.e. first visible) straight or distinctly curved outwards. Maxillary palpomere 3 (i.e. second visible) with small basal notch on inner face (Fig. 20.12J–K). Anapleural sutures strongly curved (Fig. 20.12A,D), mesoventrite with high tooth-like process or keel (Fig. 20.12B–C, E). Body middle-sized to small, coloration pale to black (Pl. 45A–B) .... ***Enochrus* Thomson**

– Maxillary palpomere 2 (i.e. the first visible) straight or curved inwards. Maxillary palpomere 3 (i.e. the second visible) without basal notch on inner face (Fig. 20.13P). Anapleural sutures nearly straight (Fig. 20.13B), mesoventrite with low median keel (Fig. 20.13B–C). Body medium-sized, completely black dorsally (Pl. 45C) .... ***Chasmogenus* Sharp**

24(19). Labrum membranous, largely concealed under clypeus (Figs 20.14f–g). Prosternum completely flat, without median carina or transverse ridge (Fig. 20.14S, U). Meso- and metafemora with dense pubescence covering the whole surface except the extreme tip (Fig. 20.14S, U). All abdominal ventrites completely covered with dense pubescence. Body medium-sized (3.5–5.0 mm), dorsal coloration entirely yellowish or reddish, or with yellowish anterior and black posterior half (Pl. 45L–O) **25**

– Labrum well sclerotised and exposed in front of clypeus, well visible at least in anterior view (Figs 20.6M, 20.11F–I). Prosternum flat or with fine longitudinal median carina or distinct transverse groove (Figs 20.6G, 20.11A, C, D, E). At least metafemora with reduced extent of dense pubescence, which is either totally missing (Fig. 20.11E), or limited to the basal half or anterobasal portion (Figs 20.6G, 20.11A, C, D). Abdominal ventrites 4–5 with or without dense pubescence. Small to very small beetles (1.5–3.8 mm), dorsal coloration uniformly pale, pale with dark patterns or largely dark **26**

25(24). Anterior margin of clypeus slightly uplifted. Labrum with dense series of long setae projecting anteriorly (Fig. 20.14g). Elytra uniformly reddish or bicolored, yellowish anteriorly and black posteriorly (Pl. 45N–O) .... ***Rygmostralia* Orchymont,**

– Anterior margin of clypeus not uplifted. Labrum without densely arranged long setae projecting anteriorly (Fig. 20.14f). Elytra uniformly yellowish (Pl. 45L–M) .... ***Pseudohydrobius* Blackburn (part)**

26(24). Head short and very wide, clypeus subrectangular in shape, not narrowing anteriorly (Fig. 20.11I). Prosternum flat (Fig. 20.11E). Abdominal ventrites 4–5 shiny, lacking dense pubescence (Fig. 20.11E). Claws strong, with basal tooth. Mesoventrite only indistinctly elevated mesally. Mesofemur with pubescence only at extreme base, metafemur bare (Fig. 20.11E) .... ***Phelea* Hansen**

– Head distinctly narrowing anteriorly (Figs 20.6M, 20.11F–H). Prosternum with fine longitudinal carina and/or distinct transverse groove (Figs 20.6G, 20.11A, C). All abdominal ventrites with dense pubescence. Claws small, arcuate, without distinct basal tooth. Mesoventrite flat, slightly elevated or bearing large tooth-like projection. Meso- and metafemur with or without dense pubescence **27**

27(26). Clypeus anteriorly with wide angular excision (Fig. 20.6M). Mesoventrite with high tooth-like projection (Fig. 20.6G–H) .... ***Paracymus* Thomson (part)**

– Anterior margin of clypeus straight or weakly concave medially (Fig. 20.11F–H). Mesoventrite flat, weakly elevated or with high tooth-like projection (Fig. 20.11A–D) **28**

28(27). Antenna with 8 antennomeres (Fig. 20.11J, L). Eyes excised anteriorly. Pubescence of metafemur confined to extreme base or basal third (Fig. 20.11A, D). Mesoventrite without median projection (Fig. 20.11D) or with tooth-like transverse projection (Fig. 20.11A–B) **29**

– Antenna with 9 antennomeres (Fig. 20.11K). Eyes not excised anteriorly. Metafemur pubescent on the whole basal half (Fig. 20.11C). Mesoventrite without median projection (Fig. 20.11C) .... ***Crenitulus* Winters**

29(28). Median portion of mesoventrite with high tooth-like projection (Fig. 20.11A–B). Labrum largely concealed under clypeus, anterior margin of labrum not distinctly emarginate (Fig. 20.11F) .... ***Anacaena* Thomson**

– Median portion of mesoventrite flat or very weakly bulged (Fig. 20.11D). Labrum partly concealed under clypeus, anterior margin of labrum more or less emarginate (Fig. 20.11H) .... ***Notohydrus* Balfour-Browne (part)**

30(10). Each elytron with 9–11 distinct puncture rows or striae (Pl. 45G, K, 46C–D, G–I, 46L–N, 47A–F, H–O) **37**

– Elytron punctation more or less irregular, never with easy-to-see 9–11 puncture rows or striae, at most with sharply impressed sutural stria. If serially arranged punctures are seen laterally, they are always absent in anteromesal portion of elytra, or the number or series is much larger than 11 (Pl. 45I, 46A, C–D, J–K, 47G) **31**

31(30). Antennal club compact (e.g. Figs 20.18E, 20.19P) **32**

– Antennal club loosely articulated (e.g. Figs 20.14L, 20.16N, 20.17I) **35**

32(31). Medium-sized beetles (length 5–8 mm). Scutellar shield much longer than wide. Eyes deeply emarginated anteriorly (Fig. 20.18F–G). Median portion of prosternum and procoxae with large spines (Fig. 20.18A). Meso- and metatibiae with large spines on both inner and outer faces (Fig. 20.18D). Protarsi and protarsal claws of males enlarged (Fig. 20.18B) .... ***Sphaeridium* Fabricius**

– Small to medium-sized beetles (length 1.1–5.9 mm). Scutellar shield as long as wide, widely triangular. Eyes not deeply emarginated anteriorly. Prosternum and procoxae without large spines. Meso- and metatibiae without very large spines mesally or laterally. Protarsi and protarsal claws not sexually dimorphic, of the same size in males and females **33**

33(32). Body widely rounded in dorsal view, highly convex, lateral margins of elytra widely explanate (Pl. 46A, C, D). Prosternum very short in front of procoxae (Fig. 20.16A, I, M). Labrum sclerotised, exposed anterior to clypeus (Fig. 20.16P, R). Epipleura wide throughout (Fig. 20.16A, I). First abdominal ventrite with or without median longitudinal carina **34**

– Body narrowly elongate in dorsal view, compressed, elytra not explanate laterally (Fig. 20.19C). Prosternum long anterior of procoxae (Fig. 20.20G). Labrum membranous, concealed by clypeus (as Fig. 20.19D). Epipleura narrow in posterior half (Fig. 20.20G). First abdominal ventrite with fine median longitudinal carina .... ***Cercyodes* Broun**

34(33). First abdominal ventrite with median carina (Fig. 20.16I) .... ***Paromicrus* Scott**
– First abdominal ventrite without median carina (Fig. 20.16A) .... ***Mircogioton* Orchymont**
35(31). Small beetles (length 1.1–2.0 mm). Median portion of mesoventrite with a subpentagonal plate (Fig. 20.16M) .... ***Psalitrus* Orchymont**
– Medium-sized beetles (length 4.5–5.9 mm). Mesoventrite never with subpentagonal plate, bearing either narrow posteromesal keel (Fig. 20.14H–J) or rather wide arrowhead-shaped elevation (Fig. 20.17A–B, D) **36**
36(35). Mesoventrite posteromedially with very narrow keel (either high and with a tooth and hence more apparent, or low and without the tooth, and hence rather inconspicuous; Fig. 20.14H–J). Anterior portion of mesoventrite flat, without anteromedian pit-like groove. Metaventrite densely pubescent on its whole surface or with a very small bare median area .... ***Coelostomopsis* Hansen**
– Mesoventrite posteromedially with high and wide arrowhead-shaped elevation which is facing the wide anterior projection of the metaventrite (Fig. 20.17A). Anteromedian portion of mesoventrite with pit-like groove. Median portion of metaventrite largely bare, without dense pubescence .... ***Coelostoma* Brullé**
37(30). Antennal club loosely articulated (Figs 20.14L, 20.17K) **38**
– Antennal club compact (Fig. 20.17K, 20.19P) **40**
38(37). Small (length 1.8–2.2 mm) highly sculptured beetles (Pl. 45K). Body widely rounded in dorsal view. Lateral parts of elytra explanate, elytral margin finely denticulate (Fig. 20.14d). Clypeus very wide, only very weakly narrowing anterior of eyes (Fig. 20.14c). Tarsi shortened, all tarsomeres of similar length (Fig. 20.14R). First abdominal ventrite without median carina (Fig. 20.14P) .... ***Petasopsis* Hansen**
– Medium-sized (length 3.5–5.2 mm), dorsal surface smooth (Pl. 45G, 46G–I). Body widely rounded or elongate in dorsal view. Lateral parts of elytra explanate or not, but lateral margin never denticulate. Clypeus narrowing anterior of eyes (Fig. 20.17F–H). Metatarsomere 1 distinctly longer than metatarsomere 2. First abdominal ventrite with or without median carina **39**
39(38). First abdominal ventrite with median longitudinal carina. Posteromedian portion of mesoventrite with high arrowhead-shaped elevation facing anteromesal projection of metaventrite (Fig. 20.17D) .... ***Dactylosternum* Wollaston (part)**
– First abdominal ventrite without median carina. Posteromedian portion of mesoventrite only very weakly elevated, not forming a high arrowhead-shaped elevation (Fig. 20.14A) .... ***Austrotypus* Fikáček, Minoshima and Newton**
40(37). Lateral portions of clypeus not excised anterior of eyes, overlapping antennal insertions. Eyes emarginate anteriorly (Fig. 20.17F). Elytron with 10–11 rows of punctures .... ***Dactylosternum* Wollaston (part)**
– Lateral portions of clypeus anterior of eyes excised above antennal insertions (Figs 20.16P–S, 20.19D–E). Eyes not emarginated anteriorly. Elytron with 9–10 rows of punctures **41**
41(40). Mesoventral plate widely pentagonal or triangular, as wide as long or wider than long (Figs 20.16F, 20.20A, 20.21C–D). Grooves for reception of procoxae large, reaching anterior margin of mesocoxal cavities or nearly so **42**
– Mesoventral plate elongate oval (Figs 20.16A–B, 20.20C–K, 20.21A–B, E–H), triangular or pentagonal (Fig. 20.16I–J), always much narrower than wide. Grooves for reception of procoxae absent or small, never reaching close to mesocoxal cavities **45**
42(41). Mesoventral pentagonal plate with narrow to wide median carina (Fig. 20.20A–B). Antennal grooves never reaching lateral margin of prothorax (Fig. 20.20A) .... ***Australocyon* Hansen**
– Mesoventral pentagonal plate flat, without median carina or ridge. Antennal grooves large, always reaching lateral margin of prothorax. (Figs 20.16F, 20.21C–D) **43**
43(42). Median portion of prosternum elevated to a plate without median carina (Fig. 20.21C–D). Labrum membranous and concealed below clypeus (Fig. 20.19E). Metanepisternum very narrow anteriorly, distinctly widening posteriorly, exposed or partly concealed by elytra (Fig. 20.21C–D). Lateral portion of metaventrite with a ridge arising from anterolateral corner and facing posteromesally (Fig. 20.21C–D) **44**
– Median portion of prosternum highly roof-like (Fig. 20.16F). Labrum well sclerotised and not concealed below clypeus (Fig. 20.16Q). Metanepisterna narrow, but of the same width throughout and never concealed by elytra. Lateral portion of metaventrite without ridges, only with very coarse punctures. (Fig. 20.16F) .... ***Noteropagus* Orchymont**
44(43). Mesoventral plate about as long as wide, triangular in shape (Fig. 20.21D). Outer margin of anterior tibia with emargination at midlength, and hence sinuate in shape (Fig. 20.19N). Mentum with anterolateral corners subrectangular .... ***Pachysternum* Motschulsky**
– Mesoventral plate wider than long, pentagonal in shape (Fig. 20.21C). Outer margin of anterior tibia arcuate, without any trace of emargination at midlength (Fig. 20.19M). Mentum with anterolateral corners projecting anteriorly .... ***Cryptopleurum* Mulsant**
45(41). Anterolateral corners of metaventrite demarcated from rest of metaventrite by an arcuate ridge more or less corresponding to arc described by the hindfemoral tip (Figs 20.20F, K, 20.21B) **46**
– Metaventrite without the ridge demarcating anterolateral corners (Figs 20.20C–D, G–J, 20.21A, E–H) **50**
46(45). Median portion of prosternum elevated, demarcated from lateral portions by a pair of sharp edges (Fig. 20.20C, F, K) **47**
– Median portion simply carinate medially, without elevated portion demarcated from lateral portions (Fig. 20.21B) **49**
47(46). Anteromedian portion of metaventrite with deep cavity bridged mesally by meso- and metaventral plates, i.e. forming a pair of deep pits posterolaterally of mesoventral plate; pits are usually filled by whitish secretion. Femoral lines of metaventrite absent. (Fig. 20.20F) .... ***Cenebriophilus* Hansen**
– Metaventrite without a pair of deep pits filled by whitish secretion situated laterally of meso-metaventral bridge. Femoral lines on metaventrite present or absent (Fig. 20.20C, K) **48**
48(47). Mesoventral plate elongate oval, pointed posteriorly. Elevated median portion of metaventrite projecting anteriorly by single flat tubercle. Femoral lines of metaventrite absent. Ventral tentorial pits very large. (Fig. 20.20K). Elytra oval basally (Pl. 47K) .... ***Notocercyon* Blackburn**
– Mesoventral plate narrowly subpentagonal, truncate or slightly concave posteriorly. Elevated median portion of metaventrite with two flat tubercles anteriorly. Femoral lines of metaventrite present. Ventral tentorial pits never enlarged. (Fig. 20.20C). Elytra rectangular basally (Pl. 47H) .... ***Ceronocyton* Hansen**
49(46). Dorsal surface of elytra densely pubescent. Body small, 1.4–1.8 mm long. Elytral intervals costate, elytral series deeply impressed (Pl. 47M). Lateral elytral margin finely serrate. Male maxilla without sucking disc .... ***Pseudoosternum* Hansen**
– Dorsal surface of elytral bare. Body small to medium-sized (2.1–4.1 mm). Elytral intervals flat, elytral series fine, slightly impressed or not impressed at all. Lateral elytral margin smooth. Male maxilla with sucking disc .... ***Platycyon* Hansen (not yet recorded from Australia)**

50(45). Antennal grooves of prothorax present (Figs 20.20D–E, J, 20.21A–B, E–H) **51**
– Antennal grooves of prothorax absent (Figs 20.16I, 20.20H–I) **54**
51(50). Ventral face of meso- and metatibiae with long dense pubescence (Fig. 20.21A). Antennal cupule small or large (Fig. 20.19O–P) .... ***Pilocnema* Hansen**
– Tibiae without long pubescence on ventral surface. Antennal cupule small (as in Fig. 20.19P) **52**
52(51). Median portion of prosternum demarcated from lateral portions (Fig. 20.20D–E) .... ***Chledocyon* Hansen**
– Median portion of prosternum not demarcated from lateral portions (Figs 20.20J, 20.21E–H) **53**
53(52). Male sternite IX tongue-like medially (Fig. 20.19K). Sternite VIII without anteromedian projection (Fig. 20.19L). Median lobe reaching deeply into phallobase, freely moving anteriorly and posteriorly (Fig. 20.19J) .... ***Cercyon* Rey**
– Male sternite IX crescent-like medially (Fig. 20.19H). Sternite VIII with anteromedian projection (Fig. 20.19I). Median lobe not reaching deeply into phallobase, firmly attached to the base of parameres and not freely movable (Fig. 20.19G) .... ***Merosoma* Balfour-Browne**
54(50). Elytra widely explanate posterolaterally (Pl. 46C–D), epipleuron very wide throughout (Fig. 20.16I). Prosternum very narrow in front of procoxae Fig. 20.16I). Very small beetles (length 1.0–1.7 mm) .... ***Paromicrus* Scott**
– Elytra not explanate posterolaterally (Pl. 47F, I), epipleuron very narrow posteriorly (Fig. 20.20H–I). Prosternum rather long in front of procoxae (Fig. 20.20H–I). Moderately large to large beetles (length 2.0–4.9 mm) **55**
55(54). Prosternum highly carinate medially (Fig. 20.20I). Mesoventral plate slightly widened posteriorly (Fig. 20.20I). Meso- and metatibiae widened and with stout spines distally (Fig. 20.19S). Body small (length 2.0–2.6 mm) .... ***Ercycodes* Hansen**
– Prosternum without distinct median carina (Fig. 20.20H). Mesoventral plate narrowed posteriorly (Fig. 20.20H). Meso- and metatibiae not markedly widened distally (Fig. 20.20H). Body medium-sized (length 4.4–5.0 mm) .... ***Cetiocyon* Hansen**

Key to the larvae of Australian hydrophilid genera

The knowledge of the larval morphology of Australian hydrophilid fauna is still rather limited, as only three studies describing larvae of Australian species were published so far (Anderson 1976; Watts 2002; Fikáček *et al.* 2014a). In *Paracymus*, *Chaetarthria*, *Anacaena*, *Crenitulus*, *Agraphydrus*, *Coelostoma* and *Sphaeridium*, larvae of Australian species are not known, but larvae are described for non-Australian species. I used the data on non-Australian species for constructing the key and for larval illustrations, but the reader should be aware that some characters may vary within genera and that the key is hence necessarily tentative until Australian species of the respective genera will be described. Larvae of *Phelea*, *Notohydrus*, most Cylominae, Omicrini and native Megasternini are not known so far. When constructing the key, I also examined some field-collected larvae from Australia associated with adults of *Anacaena*, *Borborophorus*, *Noteropagus*+*Paromicrus* and *Merosoma*, and the larvae collected with adults of *Dactylosternum dytiscoides* in New Guinea (see respective genera for details).

The use of the key should be possible without any special dissections or preparation of permanent slide mounts. However, it is still needed to examine details of the head and mouthparts, which may be best done by placing the larva into a drop of the medium (e.g. alcohol, glycerol, water), cover it or only its anterior part by a cover glass, and observe it under higher magnification of a binocular microscope or small magnification of compound microscope. The key should work for all three larval instars, only head sulci (frontal and coronal sulcus) are usually largely obsolete in the third instar; use other characters in that case.

1. Mandibles strongly asymmetrical, each mandible with different number of inner teeth (Figs 20.5l, 20.6e, 20.7j, 20.12 g, 20.13k, 20.17k, 20.18c, 20.19k–l) **2**
– Both mandibles similar in shape, with the same number of inner teeth (e.g. Figs 20.4G, 20.5d, 20.6k, 20.7k) **11**
2(1). Left mandible with a complex set of spines on the inner teeth, right mandible with 2–3 simple inner teeth. Left epistomal lobe large, with a dense row of stout setae on inner margin (Figs 20.5l, 20.6e, 20.7j) **3**
– Left mandible differs from the right one in having fewer inner teeth (one or none), or in having the basal inner tooth subdivided into two (dorsal and ventral). All inner teeth simple, without a complex set of spines **5**
3(2). Abdominal segments with long tracheal gills laterally (Fig. 20.5k) .... ***Berosus* Leach**
– Abdominal segments without long tracheal gills laterally **4**
4(3). Right epistomal lobe bearing setae (Fig. 20.7h). Antennal sensorium minute, much shorter than antennomere 3 (as in Fig. 20.7d). Frontal sulci converging posteriorly, coronal sulcus present (Fig. 20.7i). Prementum with short but distinct ligula (Fig. 20.7k) .... ***Hybogralius* Orychmont**
– Right epistomal lobe bare, without setae (Fig. 20.6a). Antennal sensorium large, nearly as long as antennomere 3. Frontal sulci not converging posteriorly, reaching posterior margin of the head, coronal sulcus absent (Fig. 20.6c). Ligula absent (Fig. 20.6f) .... ***Laccobius* Erichson**
5(2). Inner face of stipes with a series of many (more than 5) stout setae (Figs 20.17r, 20.18e, 20.19m). Left mandible lacking inner teeth or with a very minute basal spine only, right mandible with one or two large inner teeth (Figs 20.17k, 20.18c, 20.19k–l). Legs short or partly reduced, or absent. Terrestrial species **6**
– Inner face of stipes with 5 stout setae (e.g. Fig. 20.12i). Left mandible with at least one large tooth (e.g. Fig. 20.12g). Legs always well developed. Aquatic species **8**
6(5). Abdomen with two finger-like projections on each side of segment 8 (Fig. 20.18h–i). Ligula present (Fig. 20.18b). Legs 5-segmented with reduced claw (Fig. 20.18a) .... ***Sphaeridium* Fabricius**
– Abdomen completely without lateral projections or with projections on segments 1–7 (Figs 20.17e, 20.19a); segment 8 always without projections. Ligula absent (Figs 20.17q, 20.19n). Legs either 5-segmented with well developed claw, or largely reduced (at most 3-segmented and lacking claw; Fig. 20.19c–d) **7**
7(6). Anterior margin of head with well developed epistomal lobe and projecting nasale, hence trilobate (Fig. 20.17b–c, i). Legs with well developed claw .... ***Dactylosternum* Wollaston (part, incl. *D. abdominale*)**
– Anterior margin of head without distinct epistomal lobes or projecting nasale, straight or sinuate, with a pubescent emargination on the left side (Fig. 20.19f–h). Legs highly reduced, at most with 3 recognisable minute segments and without claw, sometimes totally absent (Fig. 20.19b–d) .... **tribe Megasternini**

8(5). Mentum conspicuously wider than prementum (e.g. Fig. 20.5f). Frontal sulci not converging posteriorly (e.g. Figs 20.5e, 20.9N). Maxilla slender and very long, antenna-like in general form (e.g. Figs 20.5e, 20.9N). Anterior margin of head straight, either completely without teeth (Figs 20.5a, 20.9O), or with numerous spines (Fig. 20.5b). Ventral portion of abdominal segments without spinose prolegs **16**

– Mentum only indistinctly wider than prementum (e.g. Figs 20.12f, 20.13l). Frontal sulci V–shaped, converging posteriorly (Figs 20.12d, 20.13b–c). Maxilla stout, not resembling antenna in morphology, with stipes much thicker than antenna (Figs 20.12d, 20.13b–c). Anterior margin of head always with well defined epistomal lobes and slightly to strongly asymmetrical nasale bearing several distinct teeth (Figs 20.12b–c, 20.13e–g). Abdominal segments 3–7 each with a pair of spinose prolegs........................**9**

9(8). Nasale with 6 distinct teeth (Fig. 20.13e–g)........................**10**

– Nasale with marginal large teeth but with the serrate median portion (Fig. 20.12b)........................***Enochrus* (sg. *Methydrus* Rey)**

10(9). Nasale strongly asymmetrical, its right portion projecting further than left one (Fig. 20.13e)........................***Agraphydrus* Régimbart**

– Nasale symmetrical, not projecting further on one side (Fig. 20.13i)........................***Chasmogenus* Sharp**

11(1). Inner face of stipes with a series consisting of many stout setae (many more than 5; Figs 20.15F, 20.16m, 20.17r). Stipes always much longer than the following segment (palpifer)........................**12**

– Inner face of stipes with less than 7 stout setae. Stipes shorter, as long as or longer than the following segment (palpifer)........................**15**

12(11). Setal pattern differs between right and left epistomal lobe (right lobe with a series two to many densely arranged stout setae, left one without stout setae; Figs 20.15B–C, 20.17b). Abdominal segment 8 with two pairs of finger-like projections (Fig. 20.15A) or without such projections...**13**

– Left and right epistomal lobe with the same pattern of few stout setae (Figs 20.16K, 20.17a). Abdominal segment 8 without lateral projections..**14**

13(12). Abdominal segment 8 with two pairs of lateral finger-like projections (Fig. 20.15A). Left epistomal lobe with many stout setae (Fig. 20.15B–C). Hypopharyngeal lobe present (Fig. 20.15G)........................***Austrotypus* Fikáček, Minoshima & Newton**

– Abdomen without lateral projections (as in Fig. 20.17e). Left epistomal lobe with two stout setae only (Fig. 20.17b). Hypopharyngeal lobe absent ....***Dactylosternum* Wollaston (part, incl. *D. dytiscoides*)**

14(12). Ligula distinctly developed (as in Fig. 20.17p). Nasale more or less simply triangular (20.17a). Aquatic........................***Coelostoma* Brulle**

– Ligula absent (Fig. 20.16n). Nasale with 3–4 teeth (Fig. 20.16c–d). Terrestrial........................**tribe Omicrini (part)**

15(11). Prementum conspicuously narrower than mentum, mentum with or without projecting anterolateral corners (e.g. Figs 20.4H, 20.5f, 20.9Q). Antenna and maxilla long and slender, maxillary stipes always much longer than the following segment (palpifer). Anterior margin of head smooth (e.g. Figs 20.5a, 20.9O) or with distinct teeth........................**16**

– Prementum only indistinctly narrower than mentum, mentum never with projecting anterolateral corners (e.g. Figs 20.6l, 20.7f, 20.10e, 20.11g). Antenna and maxilla not extremely long and slender, maxillary stipes may be nearly as long as the following segment (palpifer) or much longer. Anterior margin of head always with distinct set of teeth (e.g. Figs 20.6g, 20.7b, 20.10E, 20.11b–c, 20.12b–c)........................**22**

16(15). Antennomere 1 with a finger-like projection on a distal portion of inner face (Fig. 20.5h–i). Meso- and metathorax and abdominal segments with numerous setiferous projections (as in Fig. 20.5c, f). Anterior margin without distinct nasale and epistomal lobes, totally smooth or (in higher instars) with numerous spines (Fig. 20.5a–b). Prementum much longer than mentum (Fig. 20.5g). Mandibles slender and falcate; small basal tooth bifid in left mandible and simple in right one (Fig. 20.5d)........................**17**

– Antennomere 1 never with inner finger-like projection (e.g. Figs 20.4e, 20.9H, M, R–S). Thorax and abdomen without setiferous projections. Anterior margin of head with well developed epistomal lobes, nasale well developed with 5 to 8 distinct teeth (Figs 20.4F, 20.9E, 20.13 g) or more or less smooth (Fig. 20.9J,O). Prementum about as long as mentum or shorter. Mandibles falcate or robust, never with basal tooth bifid on the left mandible and simple on the right one........................**18**

17(16). Inner projection of antennomere 1 pointed apically; apex of antennomere 2 extending mesally (Fig. 20.5i)........................***Allocotocerus* Kraatz**

– Inner projection of antennomere 1 rounded apically; apex of antennomere 2 not extending mesally (Fig. 20.5h)........................***Regimbartia* Zaitzev**

18(16). Anterior margin of head without distinct large teeth, either smooth or serrate (Fig. 20.9J, O). Legs long, with dense series of swimming hairs both dorsally and ventrally (Fig. 20.9U). Apex of abdomen with a pair of long membranous projections (Fig. 20.9V). Large larvae........................**19**

– Anterior margin of head always with 5–8 distinct teeth. Legs never with dense series of setae dorsally. Apex of abdomen without a pair of long membranous projections. Small to medium-sized larvae........................**20**

19(18). Head elongate, frontal sulci V-shaped (Fig. 20.9D). Mandibles narrowly falcate (Fig. 20.9K). Antenna always with 3 antennomeres (Fig. 20.9M) ....***Hydrobiomorpha* Blackburn**

– Head transversely rounded, frontal sulci widely U-shaped (Fig. 20.9N). Mandibles robust (Fig. 20.9P). Antenna with 3 antennomeres in first instar, but with 4 antennomeres in second and third instars (Fig. 20.9R–S)........................***Hydrophilus* Geoffroy**

20(19). Nasale with 5 teeth (Figs 20.4F, 20.9E)........................**21**

– Nasale with more than 5 teeth (Fig. 20.13g)........................***Helochares* Mulsant (part, incl. *H. foveicollis*)**

21(20). Antennal sensorium very small, rather indistinct (Fig. 20.9H). Mentum with one large triangular projection on each anterolateral portion, prementum well developed, about as long as mentum, ligula present (Fig. 20.9G). Antennomere 1 bare (lower instars) or with a 'ring' or long setae in distal third (higher instars; Fig. 20.9H)........................***Sternolophus* Solier**

– Antennal sensorium well developed, at least 1/3 times as long as antennomere 3 (Fig. 20.4e). Mentum with two large triangular projections on each anterolateral portion; prementum small, shorter than mentum; ligula absent (Fig. 20.4f). Antennomere 1 without distal 'ring' of setae in all instars........................***Amphiops* Erichson**

22(15). Ligula absent. Median lobe of spiracular atrium deeply bilobate or trilobate........................**tribe Omicrini (part)**

– Ligula present. Median lobe of spiracular atrium entire........................**23**

23(22). Nasale with fewer than 5 teeth (Figs 20.6g, 20.10E)........................**24**

– Nasale with 5 or more teeth (e.g. Figs 20.7b, 20.11b–c, 20.15K)........................**25**

24(23). Ligula large, rounded, overlapping labial palps (Fig. 20.10e). Nasale with three teeth, lateral ones may be minute (Fig. 20.10E). Mandibles each with two inner teeth (Fig. 20.10a)........................***Chaetarthria* Stephens**

– Ligula narrow, not overlapping labial palps (Fig. 20.6l). Nasale with 4 teeth (Fig. 20.6 g). Mandibles each with three inner teeth (Fig. 20.6k) ....***Paracymus* Thomson**

25(23). Ligula long and narrow, as long as labial palps or longer (Fig. 20.11g). Labial palps directed slightly laterally (Fig. 20.11 g). Stipes short, as long as or shorter than the following segment (Fig. 20.11e). Pronotum with more or less developed lateral bulges (Fig. 20.11h). Nasale with 5 teeth (Fig. 20.11b–c) **26**

– Ligula not reaching the tip of labial palps (e.g. Figs 20.7f, 20.15P). Labial palps directed anteriorly, subparallel to each other (e.g. Figs 20.7f, 20.15P). Stipes much longer than following segment (e.g. Fig. 20.15N). Pronotum without conspicuous bulges. Nasale with 5 or more teeth **27**

26(25). Nasale strongly asymmetrical, left-most tooth the lowest, teeth gradually more projecting to the right (Fig. 20.11c) .... ***Crenitulus* Winters**

– Nasale slightly asymmetrical, with median tooth the lowest, and lateral ones projecting further (Fig. 20.11b) .... ***Anacaena* Thomson**

27(25). Nasale with 5 teeth (Figs 20.7b, 20.15K) **28**

– Nasale with more than 5 teeth (Figs 20.12c, 20.13f) **29**

28(27). Each mandible with 3 inner teeth but basal most tooth small (Fig. 20.7e). Nasale with the left-most tooth larger and widely isolated from other teeth (Fig. 20.7b) .... ***Limnoxenus* Motschulsky**

– Each mandible with 2 inner teeth (Fig. 20.15O). Nasale with both lateral-most teeth large and widely isolated from mesal three teeth (Fig. 20.15K) .... **Cylominae (part, incl. ?*Borborophorus*)**

29(27). Ligula short, not overlapping the first labial palpomere (as in Fig. 20.12f). Base of stipes bare on inner face, without group of cuticular spines (as in Fig. 20.12i) .... ***Enochrus* Thomson (part, sg. *Hydatotrephis*)**

– Ligula overlapping the first labial palpomere (Fig. 20.13l). Stipes with a group of cuticular spines basaly on inner face (Fig. 20.13d)... ***Helochares* Mulsant (part, incl. sg, *Hydrobaticus*)**

Classification of the Australian taxa

Characters marked by an asterisk (*) apply only for Australian species of the respective taxon, non-Australian species may bear different character states.

Subfamily Hydrophilinae Latreille, 1802

Characteristics. The unique synapomorphy of the subfamily is the long mesofurcal arm reaching the dorsal body wall (Fig. 20.2I), this character is however not seen from outside in complete specimens. At least some of the Australian members of this subfamily can be recognised by tibial swimming hairs combined with body length exceeding 15 mm.

Australian taxa. 11 genera, 85 species. All species are aquatic.

Tribe Amphiopini Kuwert, 1890 (Fig. 20.4)

Characteristics. Body hemispherical, head with the capacity to curl up to ventral face of prothorax; eyes completely divided into dorsal and ventral part; labrum not exposed*; antennae with 8 antennomeres; prosternum extremely reduced anteriorly of procoxae; anapleural sutures distinct*; mesoventrite wide anteriorly, with posterior elevation reaching the level of metaventral process; mesocoxal cavities widely separated*; metanepisternum very wide*; scutellar shield longer than wide; mesotibiae with swimming hairs*; apex of abdominal ventrite 5 without emargination.

Larvae are only known for *Amphiops*. They are characterised by nasale with 5 teeth of the same size, low epistomal lobes, long coronal sulcus, narrowly falcate mandibles symmetrical, with two large and a minute inner teeth, mentum much wider than prementum and with large triangular projections anterolaterally, and maxillary palpomere 1 extremely short; characteristic is also the yellowish head with few small black spots dorsally (Watts 2002; Minoshima & Hayashi 2012).

A small ancient group only occurring in Australia, Asia and Africa, containing two genera: the widespread aquatic *Amphiops* Erichson and the tiny terrestrial *Micramphiops* Short endemic to Madagascar (Short 2009; Short & Fikáček 2013).

Australian taxa. Five species of the genus *Amphiops*.

Amphiops Erichson, 1843 (Pl. 43F; Fig. 20.4)

Type species. *Hydrophilus gibbus* Illiger, 1801

Characteristics. See the tribal characteristics. Body uniformly brown to black; length 2.7–5.3 mm. It may be only confused with the genus *Allocotocerus*, from which it differs by eyes completely divided by a lateral canthus into dorsal and ventral portion, labrum completely hidden by clypeus, very wide metanepisterna, five abdominal ventrites and swimming hairs confined to mesotibiae only.

Larvae: See the tribal characteristics. It may be only confused with *Sternolophus*, from which it differs by very short prementum and very wide mentum with two large triangular teeth on each side.

Australian species. Five species, all endemic to Australia.

Distribution. Most species occur in wet tropical (mostly coastal) areas of Australia (Western Australia, Northern Territory, Queensland), two species reach south-east Queensland.

Biology. All species inhabit various kinds of standing water, especially well vegetated swamps and pools.

References. Hansen (1991) (adult morphology); Watts (1998a) (revision of Australian fauna); Watts (2002) (larval morphology); Minoshima & Hayashi (2012) (larval morphology).

Key to species. Watts (1998a).

Tribe Berosini Mulsant, 1844 (Fig. 20.5)

Characteristics. Body conspicuously convex to almost hemisphaerical; head with the capacity to curl-up; antennae with 7–8 antennomeres; prosternum very short; mesothorax with distinct anapleural sutures, mesoventrite wide to narrow anteriorly, with variably shaped median projection/elevation; meso- and metatibiae with

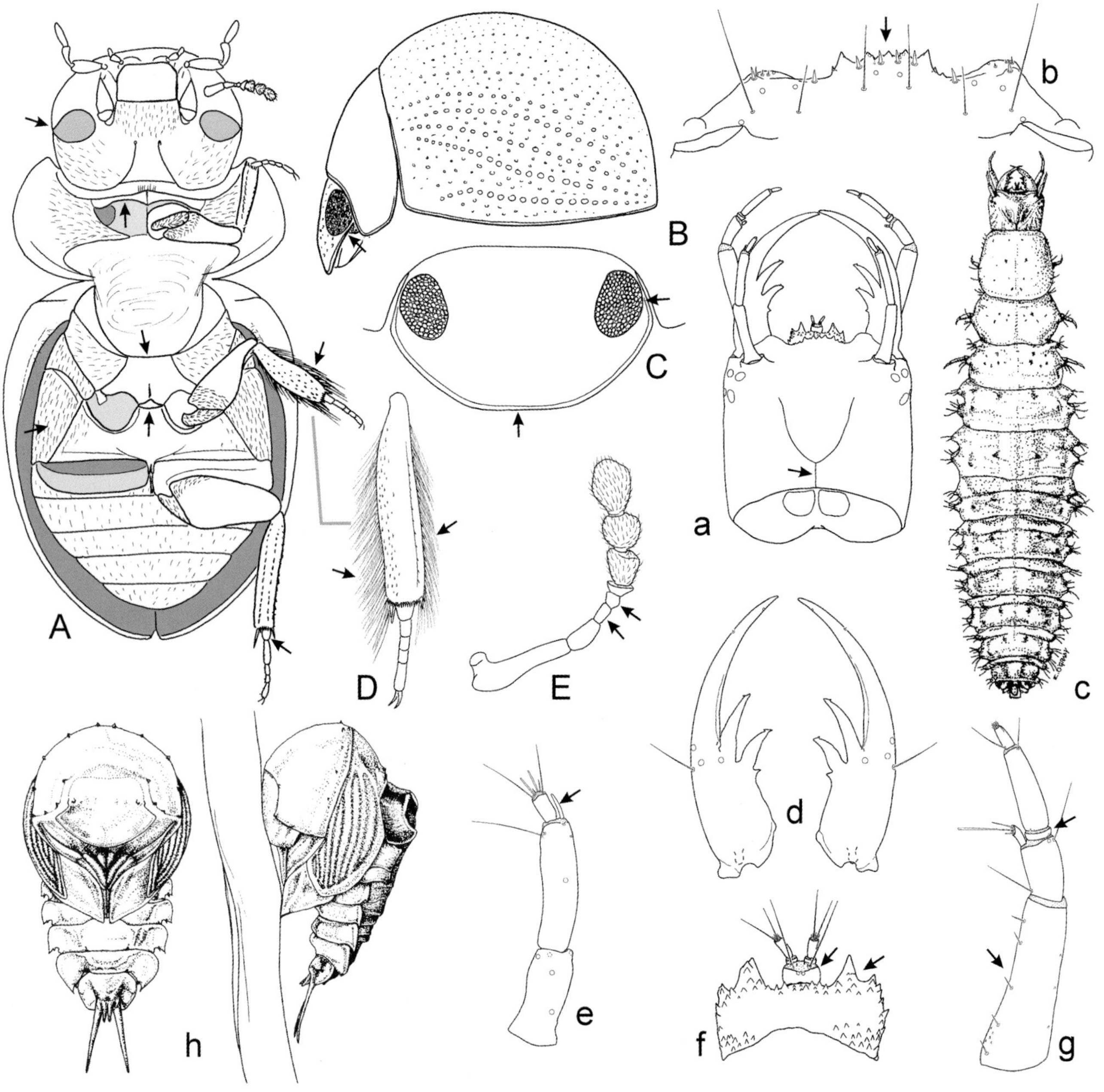

Fig. 20.4. Tribe Amphiopini, genus *Amphiops*. A–E, adult: A, ventral view; B, lateral view; C, head in dorsal view; D, mesotibia and tarsus; E, antenna. a–g, larva: a, larval head; b, anterior margin of larval head; c, general habitus; d, mandibles; e, antenna; f, labium; g, maxilla. h, pupa in ventral and lateral views. a–b, d–g from Minoshima & Hayashi (2012) (larva of non-Australian *A. mater* Sharp); c, h from Watts (2002) (larva and pupa of *A. queenslandicus* Balfour-Browne).

fringe of long swimming hairs on dorsal face; abdomen with four or five visible ventrites; ventrite 5 (exposed or concealed by ventrite 4) in most species with median rectangular emargination; laterosternite 3 with organised stridulatory file (only seen in detached abdomen at high magnification).

Larvae are known for all five genera of the tribe (Archangelsky & Durand 1992; Archangelsky 1994, 1997, 2002a,b, 2008; Watts 2002; Minoshima & Hayashi 2015). Two different larval forms exist within the tribe (benthic ones with strongly asymmetrical clypeolabrum and mandibles in *Berosus* and *Hemiosus*, metapneustic ones with symmetrical clypeolabrum and falcate nearly symmetrical mandibles in *Allocotocerus*, *Regimbartia* and *Derallus*), all of them are however characterised by a projection on distal inner part of antennomere 1 (developed as membra-

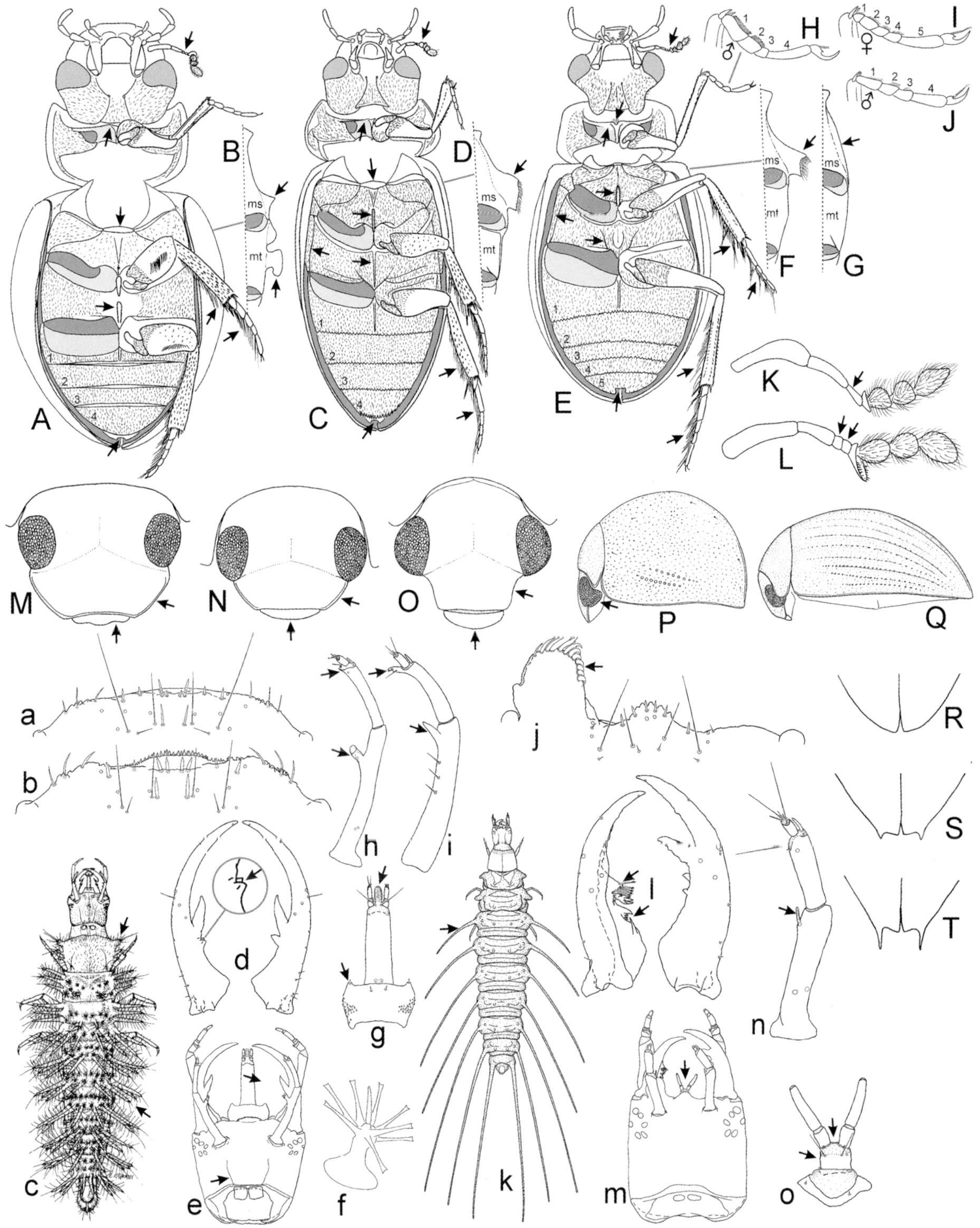

Fig. 20.5. Tribe Berosini, adults (A–T) and larvae (a–n). A–G, ventral view and detail of meso- and metathorax in lateral view (A–B, *Allocotocerus tibialis* (Balfour-Browne); C–D, *Regimbartia attenuata* (Fabricius); E–F, *Berosus queenslandicus* Blackburn; G, *B. gibbae* Watts). H–J, protarsi (H–I, *B. australiae* Mulsant and Rey; J, *B. involutus* (MacLeay)). K–L, antenna (K, *Berosus*; L, *Allocotocerus*). M–N, head, dorsal view (M, *Allocotocerus*; N, *Regimbartia*, O, *Berosus*). P–Q, body, laterally (P, *Allocotocerus*; Q, *Regimbartia*). R–T, elytral apices of different *Berosus* species. a–e, g–h, larva of *Regimbartia attenuata* (a, anterior margin of head, first instar; b, same, third instar; c, general habitus; d, mandibles; e, head dorsally; h, labrium; h, antenna). f, i, larva of *Allocotocerus punctatus* (Blackburn): f, abdominal projection, i, antenna. j–o, larva of *Berosus* (based on non-Australian species) (j, anterior margin of head; k, general habitus; l, mandibles; m, head dorsally; n, antenna; o, labium. S–T adapted from Watts (1987); a–b, d–e, g–h, j–o from Minoshima & Hayashi (2015) (*B. japonicus* Sharp); c, i from Watts (2002); f from Archangelsky (2008).

nous lobe, sclerotised projection or even articulated projection), absence of coronal line, and frontal sulci reaching posterior head margin separately.

A large group (5 genera, ~380 described species, all aquatic) of world-wide distribution. Many widespread and common aquatic beetles belong to this tribe.

Australian taxa. Three genera, 40 species.

Berosus Leach, 1817 (Pl. 43I–K; Fig. 20.5E–K, O, R–T)

Type species. *Dytiscus luridus* Linnaeus, 1761.

Characteristics. Body narrowly to widely elongate, elytron in some species with subapical spine; head constricted in front of eyes, eyes conspicuously protruding; anterior margin of clypeus straight; labrum exposed, large; antenna with 7 antennomeres; elytra with distinct striae incl. scutellar striole, bare or in few species densely pubescent; mesoventrite with low to high median lamina; metaventrite without a median ridge throughout its length; abdomen with five visible ventrites, posterior margin of ventrites with or without spines; posterior margin of ventrite 5 usually with median emargination (missing in few species); male protarsi with 4 tarsomeres only, some of which are slightly enlarged and with stiff setae ventrally; meso- and metafemora with hydrofuge pubescence covering at least basal portion; body length 1.5–9.0 mm; dorsal coloration usually reddish to yellowish with variable pattern of dark spots, rarely uniformly dark/black. The genus is very characteristic by its head with protruding eyes and in most species also by pale-dark coloration, and is hence difficult to be confused with other Australian hydrophilids. The shape of the head resembles that of the genus *Hydrochus* (Hydrochidae) from which it differs by pronotum posteriorly as wide as elytra, very short prosternum, presence of scutellar stria on elytron, abdominal ventrite 5 emarginate, mesoventrite longitudinally elevated medially, and ventrum without numerous pubescent impressions.

Larvae (Fig. 20.5k–o): Very characteristic, with highly modified head morphology, reduced stigmatic atrium but distinct tergite on abdominal segment 8 (i.e. larva apneustic) and with a pair of long tracheal gills on each side of all or most abdominal segments. Head capsule with left epistomal lobe very large and bearing series of stout spines on anterior margin, right epistomal lobe absent*; nasale bears several small teeth or lacks teeth totally; mandibles asymmetrical, left one with two inner teeth each bearing comb-like projections, right one falcate with 2–3 inner teeth; antennomere 1 with inner membranous projection; labium reduced in size, with very short mentum and prementum and without ligula; coronal sulcus absent, frontal sulci not converging posteriorly. The larva of *Berosus* is similar to that of *Laccobius* and *Hybogralius* in the highly modified head morphology, but differs from both in absence of right epistomal lobe (present and large in the other genera) and in presence of abdominal tracheal gills (absent in the other genera except Laccobius (Yateberosus)).

Australian species. 36 species, additional ones are expected in tropical parts of northern Australia (Watts 1987, 1996; Schödl 1993). All species are endemic to Australia, with the exception of *B. pulchellus* MacLeay which is widely distributed in Oriental region.

Distribution. *Berosus* is one of the largest hydrophilid genus, with ~300 described species distributed world-wide. In Australia members of *Berosus* occur throughout the continent.

Biology. Aquatic species inhabiting various types of pools and ponds, many of them frequently come to light. Some species are characteristic for side pools of stony rivers or forest streams. Larvae are adapted to benthic lifestyle, i.e. are able to process the food when submerged and breathe oxygen dissolved in water and hence do not need to draw in atmospheric air.

Notes. Three subgenera are recognised within *Berosus* at the moment: *Enoplurus* Hope for species with elytra each bearing subapical spine, *Berosus* s.str. for species without such spine, and *Phelerosus* Sharp containing the New Zealand species. This subgeneric classification was not applied for Australian fauna, which was instead divided into five species groups (Watts 1987). Unpublished studies by B. Clarkson (São Paulo) based on morphology and molecular data indicate that the subgeneric division of *Berosus* into three main clades corresponding to current subgenera may be adequate. The presence/absence of elytral spines is however not a character distinguishing these clades, and some Australian species are close to the New Zealand one and should belong to the subgenus *Phelerosus*.

References. Hansen (1991) (adult morphology); Watts (1987) (revision of Australian fauna, definition of species groups); Schödl (1993) (comments on *B. pulchellus* and *B. devisi* as separate species); Watts (1996) (description of three additional species); Archangelsky (1997) (larval morphology); Watts (2002) (larval morphology); Minoshima & Hayashi (2015) (larval morphology).

Key to species. Watts (1987); Watts (1996); Schödl (1993).

Regimbartia Zaitzev, 1908 (Pl. 43H; Fig. 20.5C–D, N, Q)

Type species. *Volvulus inflatus* Brullé, 1835.

Characteristics. Body narrowly drop-shaped in dorsal view; elytra attenuate posteriorly; head not constricted in front of eyes, eyes not protruding; anterior margin of clypeus weakly concave; labrum exposed, small; antenna with 8 antennomeres; elytra with distinct striae, bare; mesoventrite with high median lamina; metaventrite with a fine median ridge throughout its length; abdomen with four visible ventrites, ventrite 5 retracted under ventrite 4; posterior margin of ventrites with spines; posterior margin of ventrite 5 with median emargination; male protarsi with 5 tarsomeres, not dilated; hydrofuge pubescence confined to basalmost portion of meso- and metafemora; body length 3.6–5.0 mm; dorsal coloration uniformly black. A genus very characteristic in its body shape and black coloration, and hence hardly to be confused with any other Australian hydrophilid.

Larvae (Fig. 20.5a–e, g–h): Head capsule wider than long, anterior margin of head smooth, at most with numerous minute teeth; mandibles falcate, symmetrical or nearly so, each with two inner teeth, basal teeth of left mandible bifid; antennomere 1 with apically rounded inner projection distally; mentum much wider than prementum; prementum very long; ligula as long as labial palps; abdominal segments with numerous finger-like projections bearing setae. The larva of *Regimbartia* is very similar to that of *Allocotocerus*, and may be distinguished from it by details of antennal morphology (see the key, couplet 17). It may also resemble larvae of *Amphiops* by general habitus and falcate mandibles, but differs by very long prementum, nasale without 5 large teeth and presence of inner projection on antennomere 1 and absence of coronal sulcus. Smooth anterior margin of clypeolabrum of *Regimbartia* resembles larvae of *Hydrobiomorpha* and *Hydrophilus*, see under these genera for diagnostic characters.

Australian species. Single species, *R. attenuata* (Fabricius).

Distribution. The genus occurs in Old World and Australia, with 10 described species. *Regimbartia attenuata* has a very wide range reaching from Iran and Arabian Peninsula through tropical and subtropical Asia to Japan and New Guinea and Australia. In Australia, it is recorded from Western Australia through Northern Territory and Queensland to north-east New South Wales.

Biology. Aquatic species inhabiting various types of standing waters, especially well vegetated pools and swamps.

References. Hansen (1991) (adult morphology); Watts (1998a) (revision of Australian fauna); Watts (2002) (larval morphology); Minoshima & Hayashi (2015) (larval morphology).

Key to species. Watts (1998a).

Allocotocerus Kraatz, 1883 (Pl. 43G; Fig. 20.5A–B, M, P)

Type species. *Globaria leachii* Hope, 1838.

Characteristics. Body widely globular/hemispheric; head not constricted in front of eyes, eyes weakly protruding; anterior margin of clypeus concave; labrum exposed, small (bifid anteriorly in male of one species); antenna with 8 antennomeres; elytra without distinct striae sometimes except laterally, bare; mesoventrite with high spine-like median projection; metaventrite with a high median ridge divided into anterior and posterior one; abdomen with four visible ventrites, ventrite 5 retracted under ventrite 4; posterior margin of ventrites bare; posterior margin of ventrite 5 with median emargination; male protarsi with 5 tarsomeres, not dilated; hydrofuge pubescence confined to basalmost portion of meso- and metafemora; meso- and/or metafemora with groups of long goldish setae in some species (sometimes in males only); body length 3.5–5.0 mm; dorsal coloration uniformly black. It may be only confused with *Amphiops*, from which it differs by large eyes without any trace of division in dorsal and ventral portion, exposed labrum, extremely narrow metanepisterna hidden under elytra, four abdominal ventrites, and swimming hairs on tibiae and tarsi of middle and hind legs.

Larvae (Fig. 20.5h): Very similar to the larva of *Regimbartia* (see there for details) from which it differs by sharply pointed mesal process of the antennomere 1 (blunt apically in *Regimbartia*) and additional process on apex of antennomere 2 (without such process in *Regimbartia*).

Australian species. Three species, all endemic to Australia.

Distribution. The genus is distributed in Australian, Oriental and Afrotropical Regions. In Australia it occurs in tropical and subtropical wet regions in Western Australia, Northern Territory and Queensland.

Biology. Aquatic species inhabiting various types of pools and ponds.

References. Hansen (1991) (adult morphology); Watts (1998a) (revision of Australian species); Watts (2002) (larval morphology).

Key to species. Watts (1998a).

Tribe Laccobiini Houlbert, 1922 (Fig. 20.6)

Characteristics. Body weakly to moderately convex*; antennae with 7–9 antennomeres; labrum largely exposed, well sclerotised; prosternum moderately long*, longitudinally carinate or at least elevate (roof-like) mesally; mesothorax with distinct anapleural sutures, mesoventrite narrow anteriorly, with variably shaped median projection/elevation; elytron without scutellary stria, with or without serial punctures; ridge dividing bare and pubescent portions of epipleuron consisting of small arcs; anterior tarsi of males widened or not; tibiae without swimming hairs; meso- and metatarsi with sparse long swimming hairs dorsally*; abdomen with five to six visible ventrites; ventrite 5 with or without median emargination; laterosternite 3 with organised stridulatory file (only seen in detached abdomen at high magnification); body uniformly black or yellowish with variable pattern of dark spots, often with metallic median/discal dark spot on pronotum.

Larvae are known for five genera of the tribe (*Laccobius* and *Oocyclus*: Archangelsky 1997; *Paracymus*: Archangelsky 1997, 1999b; Archangelsky & Fikáček 2004; *Tormus*: Fikáček *et al.* 2013; *Tritonus*: Fikáček *et al.* 2017b) and largely differ between the *Paracymus*- and *Laccobius*-clade. Hence, see under

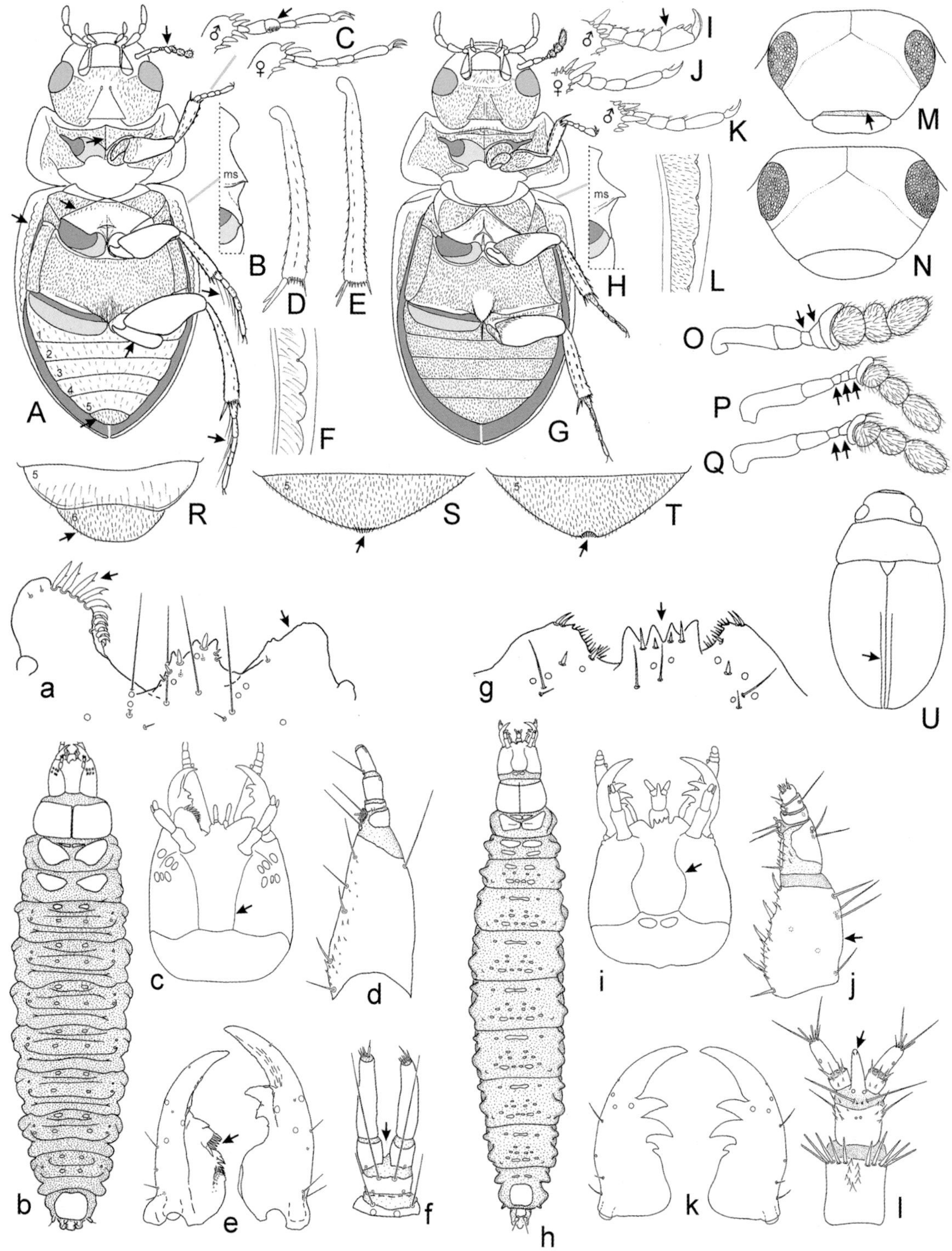

Fig. 20.6. Tribe Laccobiini, adults (A–U) and larvae (a–l). A–F, *Laccobius*: A, body in ventral view; B, mesoventrite in lateral view; C, protarsus of male and female; D, hind tibia of sg. *Microlaccobius*; E, hind tibia of *L. zietzi*; F, detail of arcs on epipleuron. G–L, *Paracymus*: G, body in ventral view; H, mesoventrite in lateral view; I–J, protarsus of *P. pygmaeus* (MacLeay) (male and female); K, male protarsus of *P. spenceri* Blackburn; L, detail of epipleuron with arcs. M–N, head dorsally: M, *Paracymus*; N, *Laccobius*. O–Q, antenna: O, *Laccobius*; P, *Paracymus pygmaeus*; Q, *P. spenceri*. R–T, abdominal apex: R, *Laccobius*; S, *Paracymus pygmaeus*; T, *P. spenceri*. U, *Paracymus* in dorsal view. a–f, larva of *Laccobius* (based on non-Australian species): a, anterior margin of head; b, habitus; c, head dorsally; d, maxilla; e, mandibles; f, labium. g–l, larva of *Paracymus* (based on non-Australian species): g, anterior margin of head; h, habitus; I, head dorsally; j, maxilla; k, mandibles; l, labium. a, d–f from Minoshima (2013) (*L. kunashiricus* Shatrovskiy); b–c from Archangelsky (1997) (*L. minutoides* Orchymont); g–l from Archangelsky (1997) (*P. subcupreus* (Say)).

Laccobius and *Paracymus* for list of diagnostic characters of each genus.

The tribe was originally revealed as paraphyletic (Short & Fikáček 2013), but confirmed as a monophylum after including a more representative taxon sampling by Toussaint *et al.* (2016) and Toussaint & Short (2018). It consists of two large clades, the *Laccobius*-clade and the *Paracymus*-clade. *Laccobius*-clade is a large group (5 genera and nearly 400 species) of world-wide distribution, most of the confined to running waters and seepages. *Paracymus*-group is a rather small group formed by four small sourthern hemispere genera (*Tormus*, *Afrotormus*, *Scoliopsis*, *Tritonus*) and a world-wide genus *Paracymus* (with ~80 species), all except *Afrotormus* and *Tormus* are aquatic.

Australian taxa. Two genera, 21 species.

Laccobius Erichson, 1837 (Pl. 43L–M; Fig. 20.6A–F, N–O, R)

Type species. *Chrysomela minuta* Linnaeus, 1758.

Characteristics. Body oval, weakly to moderately convex; head not constricted in front of eyes, eyes large; anterior margin of clypeus widely concave; labrum largely exposed, inserted into clypeal emargination, no membrane exposed between clypeus and labrum; antenna with 8 antennomeres; elytra with more or less regular series of punctures or without distinct series; mesoventrite with or without median projection of variable shape; metaventrite flat; abdomen with six visible ventrites; posterior margin of ventrite 5 without median emargination; male protarsomere 2–3 dilated; metatrochanters enlarged, detached from metafemora; hydrofuge pubescence absent in meso- and metafemora; metatibiae straight or curved; body length 1.6–4.0 mm; dorsal coloration yellowish with dark metallic median portion of pronotum and head and in some species with variable dark pattern on elytra. It may be confused with *Anacaena*, *Crenitulus* and *Paracymus*, but easily differs from all these genera by the combination of large exposed labrum, metatrochanter detached from metafemur and abdomen with 6 ventrites.

Larvae (Fig. 20.6a–f): Frontal sulci not converging posteriorly, coronal sulcus absent; nasale irregularly rounded without distinctly defined teeth; epistomal lobes asymmetrical, overlapping nasale, left one with series of stout setae, right one bare; mandibles asymmetrical, left one with numerous spines on inner teeth, right one with 2–3 inner teeth (basalmost may be minute); palpifer much shorter than maxillary palpomere; labium very small, ligula absent; abdomen without spiracular gills, stigmatic atrium on segment 8 present*. The larva of *Laccobius* is very characteristic by the morphology of nasale and mandibles, in which it closely resembles larvae of *Berosus* and *Hybogralius*. It may be distinguished from the larva of *Berosus* by the presence of right epistomal lobe (absent in most *Berosus*), and abdomen without tracheal gills* and with stigmatic atrium* (without stigmatic atrium and with tracheal gills in *Berosus*). It differs from *Hybogralius* by the frontal sulci reaching posterior margin of head separately (converging posteriorly and with coronal sulcus present in *Hybogralius*) and by totally absent ligula (small but present in *Hybogralius*).

Subgenera. The genus is divided into nine subgenera, five or which occur in Australia. Monophyly of the subgenera was never tested, and assignment of some Australian species moreover varies in different papers (Gentili 1981, 2005; Hansen 1999a). The following key and the assignment of the species to subgenera follows Gentili (2005):

1. Metatibia straight (Fig. 20.6E). Mesoventrite with weak median keel or bulge. Larger species (3.0–3.9 mm) .... **sg. *Notoberosus* Blackburn** (1 sp., *L. zietzi* Blackburn)
– Metatibia curved (Fig. 20.6D). Mesoventrite with distinct median keel or projection. Small to larger species (1.6–3.9 mm) **2**

2(1). Elytra with at most four more or less irregular series .... **sg. *Hydroxenus* Wollaston** (1 sp., *L. clarus* Gentili)
– Elytra with ~20 series of punctures **3**

3(2). Elytra with two alternating types of series, those with more serially arranged or larger punctures and those with less organised or smaller punctures **4**
– All elytral series consisting of punctures of about the same size and arrangement .... **sg. *Laccobius* s.str.** (2 spp.)

4(3). Series of smaller punctures alternating with more or less scattered larger punctures. Mesoventral keel with two lateral spurs .... **sg. *Dimorpholaccobius* Zaitzev** (2 spp.)
– Series of larger and more impressed punctures alternating with irregular series of smaller punctures. Mesoventral keel simple, without lateral spurs .... **sg. *Microlaccobius* Gentili** (6 spp.)

Comments. Two species currently classified in subgenus *Dimorpholaccobius* (*L. brittoni* Gentili and *L. collium* Gentili) are rather similar to species endemic to New Zealand by extrenal morphology as well as by the three-dimensional aedeagus. Additional studies are needed to test whether they form an endemic Australian-New Zealand clade without close relationship to true *Dimorpholaccobius* which is endemic to the Palaearctic.

Australian species. 12 species in Australia, all but one (*L. roseiceps* Régimbart) endemic to Australia.

Distribution. The genus is distributed world-wide except of South America and belongs to the largest hydrophilid genera (with ~260 described species). In Australia it occurs in all regions except Tasmania.

Biology. Aquatic both as adults and larvae, inhabiting banks of various types of standing and running waters.

References. Gentili (1981) (revision of Australian fauna); Gentili (1989) (description of *L. billi* and *L. quantulus* and new records of other Australian species); Gentili (1996a) (description of *L. tantillus*); Archangelsky (1997) (larval morphology of *Laccobius*); Watts (2002) (description of larva of *L. decipiens*); Gentili (2005) (review of all species of the Australian Region and neighbouring areas).

Key to species. Gentili (2005).

Paracymus Thomson, 1867 (Pl. 43N; Fig. 20.6G–M, P–Q, S–U)

Type species. *Hydrophilus aeneus* Germar, 1824.

Characteristics. Body oval, moderately to highly convex; head not constricted in front of eyes, eyes large; anterior margin of clypeus widely emarginate, narrowly exposing the membrane between clypeus and labrum; labrum largely exposed; antenna with 8–9 antennomeres; elytra without distinct striae except sutural stria; mesoventrite narrow anteriorly, with median projection of variable shape; metaventrite flat; abdomen with five visible ventrites; posterior margin of ventrite 5 with or without median emargination; male protarsi dilated or not; metacoxae not enlarged; hydrofuge pubescence confined to basal portion of mesofemora and very base of metafemora; metatibiae straight; body length 1.6–3.0 mm; dorsal coloration uniformly dark (often with metallic sheen) or yellowish with dark metallic median portion of pronotum. *Paracymus* may be confused with *Anacaena* and *Crenitulus* from which it differs by triangularly converging anapleural sutures and mesoventrite with median projection (sutures parallel-sided in *Anacaena* and mesoventrite without projection in *Crenitulus*) and by widely excised clypeus exposing membrane (not excised and without exposed membrane in *Anacaena*, very narrowly excised in *Crenitulus*); some *Paracymus* also resembles *Laccobius* by coloration (see under *Laccobius* for diagnostic characters).

Larvae (Fig. 20.6g–l): Frontal sulci not converging posteriorly, coronal sulcus absent; nasale with 4 teeth; epistomal lobes slightly asymmetrical, not reaching further than nasale, each with few setae; mandibles symmetrical, each with three inner teeth (basalmost small); palpifer slightly longer than maxillary palp; mentum only slightly wider than prementum; prementum as long as wide; ligula shorter than labial palps. The larva of *Paracymus* is very characteristic by nasale bearing 4 teeth and epistomal lobes not overlapping nasale, and cannot be hence confused with any other Australian genus with known larva (based on larva of Nearctic *P. subcupreus*; Archangelsky 1997).

Australian species. Nine species, all but one endemic to Australia, one (*P. pygmaeus*) also occurring in New Zealand (Gentili 2000).

Distribution. The genus is world-wide in distribution, with 82 described species. Australian species occur in whole Australia except central Western Australia.

Biology. Aquatic both as adults and larvae, inhabiting banks of various types of standing and running waters. Often attracted at light.

References. Winterbourn (1973) (description of larva of *Paracymus pygmaeus* from New Zealand, under name *Anacaena tepida*); Archangesky (1997) (larval morphology of *Paracymus*); Gentili (2000) (revision of Australian fauna).

Key to species. Gentili (2000).

Tribe Hydrobiusini Mulsant, 1844 (Fig. 20.7)

Characteristics. Body moderately convex; antennae with 9 antennomeres; labrum exposed, well sclerotised; prosternum moderately long, with or without median longitudinal carina; mesothorax with distinct anapleural sutures, mesoventrite narrow anteriorly, flat or with high median keel; elytron with 10 series, scutellary stria present (serial punctures may be very small and indistinct!), alternate elytral intervals with large trichobothria; anterior tarsi without sexual dimorphism; tibiae without swimming hairs; meso- and metatarsi each with fringe of long swimming hairs dorsally or without such setae; abdomen with five visible ventrites; ventrite 5 with or without median emargination; laterosternite 3 with or without organised stridulatory file (only seen in detached abdomen at high magnification); medium-sized species (5.5–12.0 mm).

Larvae: Known for all genera except the New Caledonian endemic *Limnocyclus* (Archangelsky 1997; Minoshima & Hayashi 2011a, b); larvae are known for both Australian genera but largely differ from each other due to adaptation to benthic lifestyle in *Hybogralius*. Hence, see below for list of diagnostic characters of each genus.

A small group (8 genera, 45 described species) distributed especially in temperate zones of both northern and southern hemispheres. The phylogeny of the tribe was analysed by Short *et al.* (2017), the origin of its fragmented distribution by Toussaint & Short (2017).

Australian taxa. Two genera, two species.

Limnoxenus Motschulsky, 1853 (Pl. 44B; Fig. 20.7A–D, G–H)

Type species. *Hydrophilus oblongus* Herbst, 1797.

Characteristics. Body elongate oval, moderately convex; head not constricted in front of eyes, eyes moderately large; anterior margin of clypeus straight; labrum exposed;

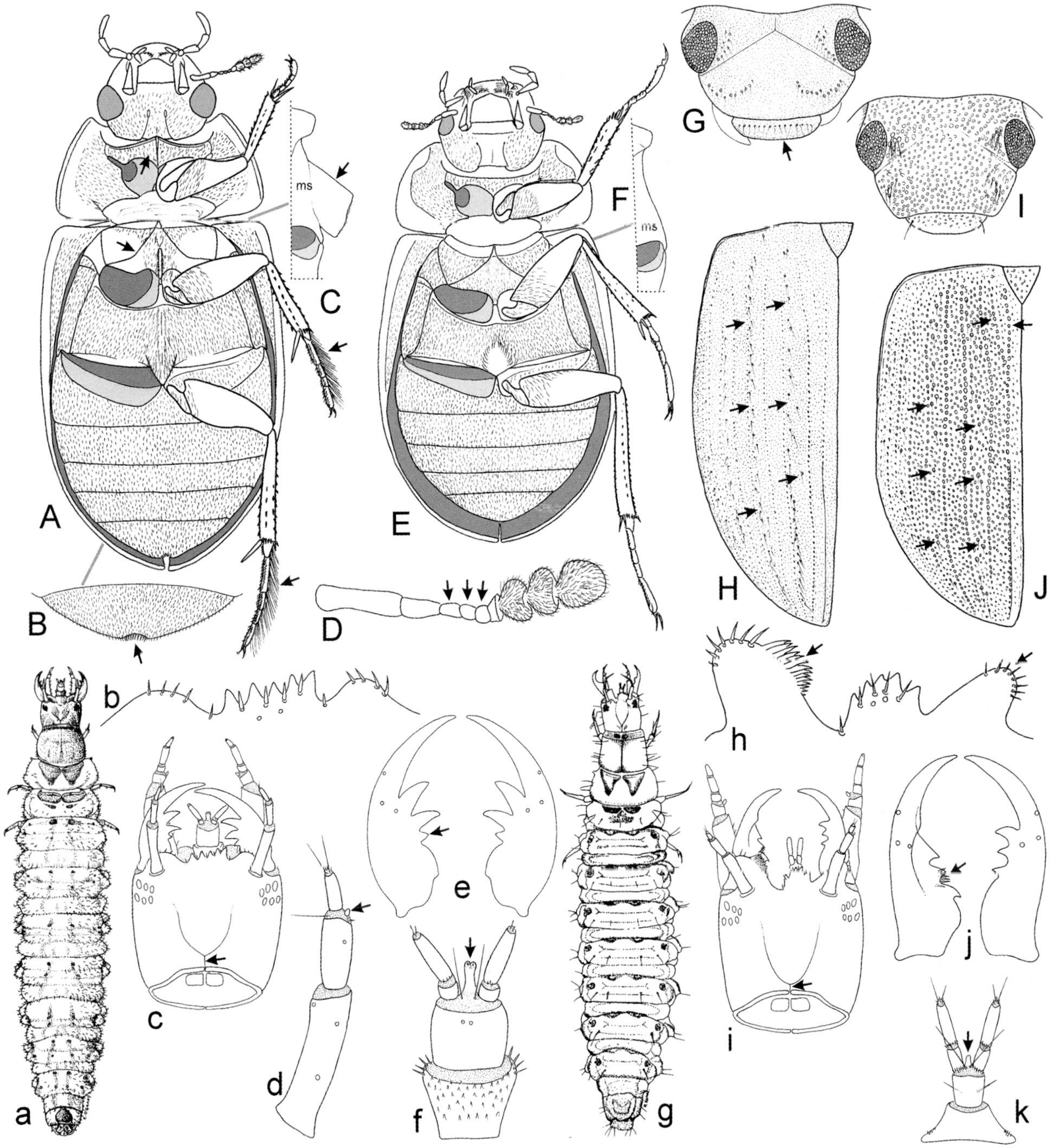

Fig. 20.7. Tribe Hydrobiusini, adults (A–J) and larvae (a–k). A–D, *Limnoxenus zealandicus* (Broun): A, body in ventral view; B, abdominal apex; C, mesoventrite in lateral view; D, antenna. E–F, *Hybogralius hartmeyeri* (Régimbart): E, body in ventral view; F, mesoventrite in lateral view. G–H, *L. zealandicus* (G, head; H, elytron); I–J, *H. hartmeyeri* (I, head; J, elytron). a–f, larva of *Limnoxenus zealandicus*: a, habitus; b, anterior margin of head; c, head dorsally; d, antenna; e, mandibles; f, labium. g–k, larva of *Hybogralius hartmeyeri*: g, habitus; h, anterior margin of head; i, head dorsally; j, mandibles; k, labium. a from Watts (2002); h, j, k redrawn from Watts (2002); i based on multiple drawings of Watts (2002).

antenna with 9 antennomeres; elytra 10 series of punctures which are very fine and indistinct mesally and more distinct laterally, scutellary stria present; prosternum carinate medially; mesoventrite with high median keel (subquadrate in lateral view); metaventrite flat, posteriorly projecting between metacoxae; abdomen with five completely pubescent ventrites; posterior margin of ventrite 5 with median emargination; hydrofuge pubescence confined to basal portion of meso- and metafemora; tarsi with fringe of swimming hairs dorsally; body length 8.7–11.2 mm;

dorsal coloration uniformly black or black with alternate greenish stripes on elytra and irregular greenish spots on pronotum. By the coloration and body form, *Limnoxenus* may be confused with *Sternolophus* (see there for diagnostic characters) and with *Enochrus mastersi* and some *Helochares* (differs from them by the high rectangular mesoventral keel and meso- and metafemoral pubescence confined to the femoral base).

Larvae (Fig. 20.7a–f): Frontal sulci V-shaped; coronal sulcus present; nasale with 5 teeth, left lateral-most separated from others by wider gap; epistomal lobes small, rounded, symmetrical; antennal scape of all instars without numerous setae; mandibles slender, symmetrical, with three inner teeth (basalmost very small); mentum about as wide as prementum, without projecting anterolateral corners. The larva is very similar to that of *Sternolophus* in most characters, but easily differs from it by the form of the labium (mentum about as wide as prementum) and in higher instars by antennal chaetotaxy without subapical fringe of setae.

Australian species. Single species, *Limnoxenus zealandicus* (Broun), also occurring in New Zealand and New Caledonia.

Distribution. The genus has two species in Europe, one in South Africa, one in Australian Region and underwent a small radiation in Hawaii (7 spp., Short & Liebherr 2007). *Limnoxenus zealandicus* is known from all parts of Australia, but the majority of records are from southern half of the continent.

Biology. Aquatic both as adults and larvae, inhabiting various types of standing waters.

References. Hansen (1991) (adult morphology of *Limnoxenus*); Watts (2002) (larval morphology); Short (2010a) (photo, diagnosis); Short *et al.* (2017) (phylogeny); Toussaint & Short (2017) (biogeography).

Hybogralius Orchymont, 1942 (Pl. 44A; Fig. 20.7E–F, I–J)

Type species. *Hydrobius hartmeyeri* Régimbart, 1908

Characteristics. Body oval, highly convex; head not constricted in front of eyes, eyes moderately large; anterior margin of clypeus straight; labrum exposed; antenna with 9 antennomeres; elytra with 10 series of large punctures, scutellary stria present, alternate intervals with very large trichobothria; prosternum flat, without carina or grooves; mesoventrite flat, without median keel or projection; metaventrite flat, slightly projecting between metacoxae posteriorly; abdomen with five completely pubescent ventrites; tarsi without swimming hairs; posterior margin of ventrite 5 entire; meso- and metafemora with sparse long pubescence basally; body length 5.7–6.0 mm; dorsal coloration uniformly brown to black, head metallic. By body form, size (and striate elytra), *Hybogralius* may resemble some *Helochares* and *Enochrus mastersi*, but differs from both by the largely reduced meso- and metafemoral pubescence and entire abdominal ventrite 5. By the rather highly convex body and coarsely punctate dorsal surface *Hybogralius* may also resemble some *Berosus*, from which it may be easily distinguished by the form of the head and lack of swimming hairs on tibiae and tarsi.

Larvae (Fig. 20.7g–k): Frontal sulci V-shaped, coronal sulcus present; nasale irregularly rounded with five teeth; epistomal lobes slightly asymmetrical, largely overlapping nasale, left one with series of stout setae, right one bare; mandibles asymmetrical, left one with numerous spines on inner teeth, right one with 3 inner teeth; palpifer much shorter than maxillary palp; labium small, ligula present; abdomen without spiracular gills, stigmatic atrium as well as tergite on segment 8 absent. The larva of *Hybogralius* is very characteristic by the morphology of nasale and mandibles, in which it closely resembles larvae of *Berosus* and *Laccobius*. It may be distinguished from larva of *Berosus* by the presence of right epistomal lobe (absent in *Berosus*), and abdomen without tracheal gills (with tracheal gills in *Berosus*). For difference from *Laccobius* see under that genus.

Australian species. Single species, *Hybogralius hartmeyeri* (Régimbart).

Distribution. The genus is endemic to Western Australia. It is only known from a small area on the slopes of Darling Ranges east of Perth.

Biology. The species inhabits winter and spring streams. The strong resemblance of the larvae to those of *Berosus* and *Laccobius* is due to a parallel adaptation to the benthic lifestyle.

Note. The adult and larval morphology of *Hybogralius* is very unusual for the Hydrobiusini, and its assignment to the tribe was hence sometimes questioned; the molecular analysis by Short & Fikáček (2013) and Short *et al.* (2017) confirmed that it belongs to the Hydrobiusini.

References. Hansen (1991) (adult morphology); Watts (2002) (larval morphology, biology); Archangelsky (2008) (notes on morphology of the larva); Short & Fikáček (2013) (confirmation of its assignment to the Hydrobiusini); Short *et al.* (2017) (phylogenetic position).

Tribe Hydrophilini Latreille, 1802 (Figs 20.8–20.9)

Characteristics. Members of this tribe are easily recognisable by meso- and metaventrite fused into a raised ventral keel extending posteriorly in more or less distinct spine. The tribe includes medium-sized to large beetles (body length 10–48 mm). Additional diagnostic characters: body elongate oval, moderately convex; antenna 9 antennomeres; labrum exposed, well sclerotised; prosternum moderately long, longitudinally carinate or with cavity to receive anterior tip of ventral keel; mesothorax with distinct to obsolete anapleural sutures; elytron with few irregular series of fine setiferous punctures (trichobothria), but without 10 distinct elytral series and without su-

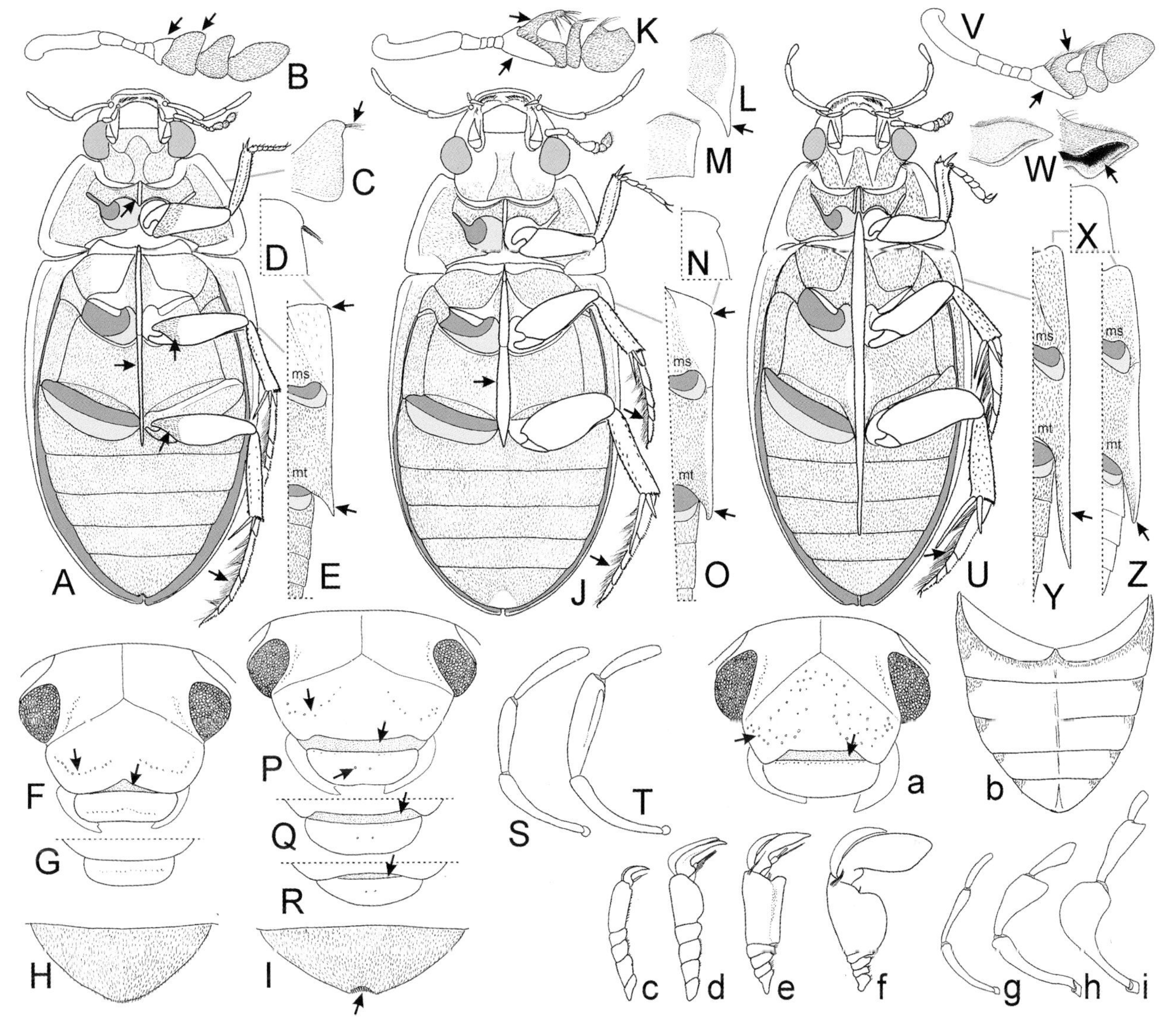

Fig. 20.8. Tribe Hydrophilini, adults. A–I, *Sternolophus* (A–F, I, *S. marginicollis* (Hope); G–H, *S. rufipes* (Fabricius)): A, body in ventral view; B, antenna; C, prosternum in lateral view; D, detail of anterior portion of mesoventral keel in lateral view; E, meso-metaventral keep in lateral view; F–G, head and clypeus; H–I, abdominal apex. J–T, *Hydrobiomorpha* (J–L, N–P, *H. helenae* Blackburn): J, body in ventral view; K, antenna; L–M, prosternum in lateral view (L, *H. helenae*; M, *H. microspina* Watts); N, detail of anterior portion of mesoventral keel; O, meso-metaventral keel in lateral view; P–R, head and clypeus (P, *H. helenae*; Q, *H. debbae* Watts; R, *H. troxi* Watts); S–T, maxillary palpomere (S, *H. microspina*; T, *H. troxi*). U–I, *Hydrophilus*: U, body of *H. bilineatus* (MacLeay) in ventral view; V, antenna; W, prosternum in lateral and posterolateral views; X, detail of anterior portion of mesoventral keel; Y–Z, meso-metaventral keel in lateral view (Y, *H. bilineatus*; Z, *H. macronyx* (Régimbart)); a, head in dorsal view; b, abdominal ventrites of *H. macronyx*; c–f, male protarsus (c, *H. wattsi* Hansen; d, *H. pedipalpus* (Bedel); e, *H. novaeguineae* Watts; f, *H. macronyx*); g–i, male maxillary palpomere (g, *H. brevispina* Fairmaire; h, *H. latipalpus* Castelnau; i, *H. pedipalpus*). Q–R and S–T adapted from Watts (1990), c–i adapted from Watts (1988).

tural stria*; anterior tarsi of males widened or not; tibiae without swimming hairs; meso- and metatarsi flattened, with dense fringe of long swimming hairs mesally; abdomen with five visible ventrites; ventrite 5 with or without median emargination; laterosternite 3 with organised stridulatory file (only seen in detached abdomen at high magnification); body uniformly dark*.

Larvae are known for all genera except *Brownephilus* and *Protistolophus*, including all Australian genera of the tribe (Archangelsky 1997; Watts 2002; Minoshima & Hayashi 2011a). They share these characters: antennal scape with numerous setae at least second and third instars (also in the first in *Hydrophilus*); antennal sensorium extremely small; maxilla long and slender, of similar shape as antenna; mentum about twice as wide as prementum, with anterolateral corners projecting anteriorly; ligula well developed; legs with dense swimming hairs on trochanters and femora.

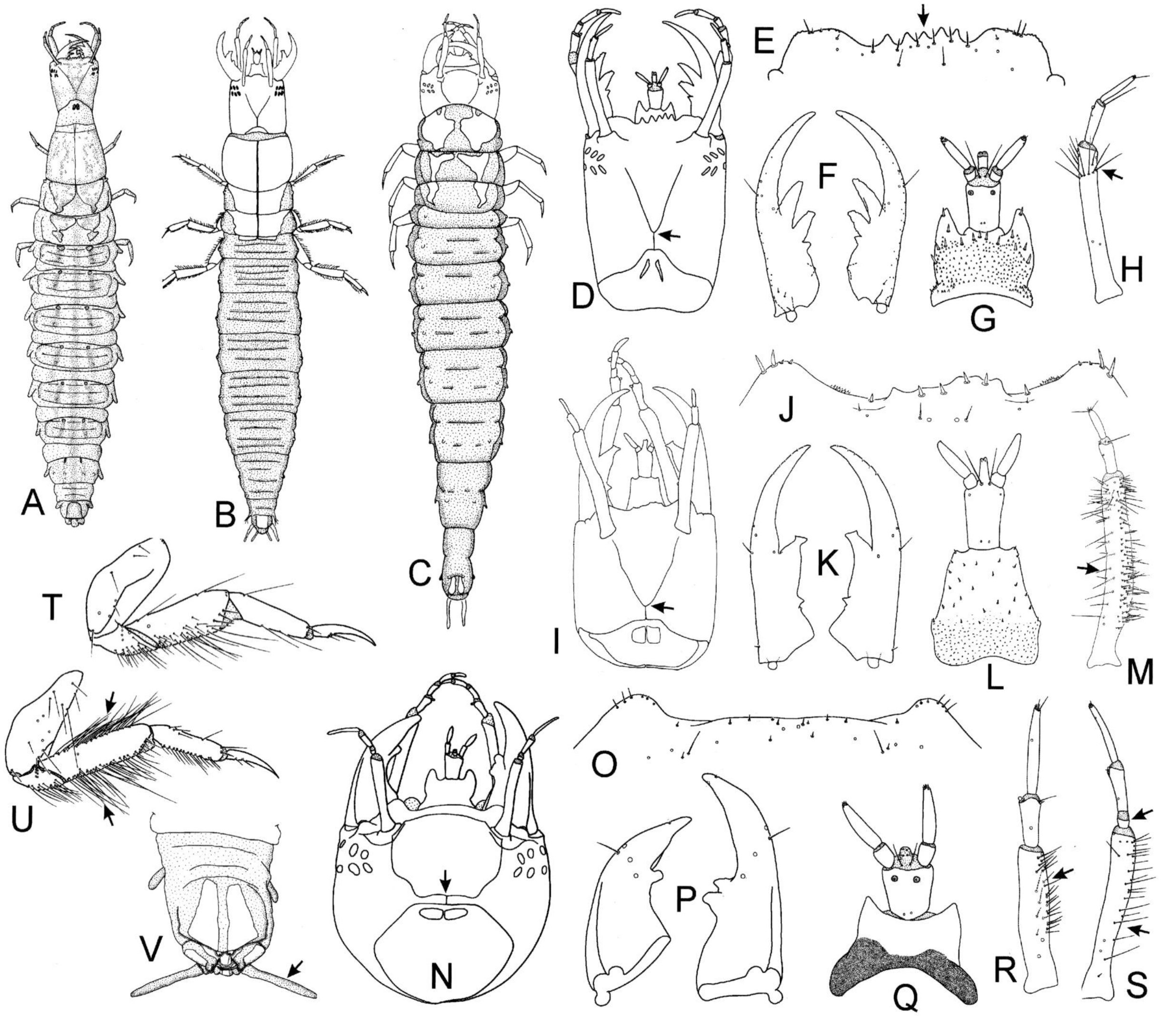

Fig. 20.9. Tribe Hydrophilini, larvae (figures based on non-Australian species, for Australian ones see Watts 2002). A–C, larval habitus (A, *Sternolophus*; B, *Hydrobiomorpha*; C, *Hydrophilus*). D–H, larva of *Sternolophus rufipes* (Fabricius): D, head dorsally; E, anterior margin of head; F, mandibles; G, labium; H, antenna of the third instar. I–M, *Hydrobiomorpha casta* (Say): I, head dorsally; J, anterior margin of head; K, mandibles; L, labium; M, antenna of the third instar. N–S, *Hydrophilus acuminatus* Motschulsky: N, head dorsally; O, anterior margin of head; P, mandibles; Q, labium; R, antenna of the first instar; S, antenna of the third instar. T–U, mesothoracic legs (T, *Sternolophus*; U, *Hydrophilus*). V, abdominal apex of *Hydrophilus acuminatus*. A, D–H and N–V from Minoshima & Hayashi (2011a), J–M from Archangelsky (1997).

A moderately large group (7 genera, ~200 described species) of world-wide distribution. All species are aquatic and many are widespread and commonly collected. Phylogeny and genus-level systematics was revised by Short (2010b) and Toussaint *et al.* (2017).

Australian taxa. Three genera, 16 species.

Hydrobiomorpha Blackburn, 1888 (Pl. 44F–G; Fig. 20.8J–T)

Type species. *Hydrobiomorpha bovilli* Blackburn, 1888.

Characteristics. Anterior margin of clypeus emarginate, exposing the membrane between clypeus and labrum; labrum exposed, with two pits mesally; first antennomere of antennal club divided in two lobes and bearing long setae; maxillary palps long, sexually dimorphic in some species; prosternum simply carinate mesally, with or without posterior spine; ventral keel with a small notch anteriorly (in lateral view); abdomen with five visible ventrites; posterior margin of ventrite 5 entire; all ventrites completely pubescent except of a glabrous patch on ventrite 5; male protarsi slightly widened, with claws of different shape than in female; meso- and metafemur without pubescence; aedeagus very characteristic, with very narrow parameres and median lobe with a hook in most species; body length 11–18 mm; dorsal coloration uniformly dark. It may be only confused with other genera of the Hydrophilini (see the key, couplets 2–3), from other medium-sized hydrophilids it differs by the distinct ventral keel.

Larvae (Fig. 20.9B, I–M): Frontal sulci V-shaped, coronal sulcus present; nasale without distinct teeth; epistomal lobes present, symmetrical, projecting slightly further than nasale; antennal scape with numerous setae on inner face; antenna of all instars with 3 antennomeres; mandibles symmetrical, each with three inner teeth (basalmost small), the distalmost teeth with quadrate apex; prementum longer than wide; abdominal apex with an additional pair of long membranous projections (prostyli). The combination of nasale without teeth, symmetrical mandibles, legs with swimming hairs and presence of prostyli easily distinguishes the larva from all other Australian hydrophilids.

Australian species. Five species, all endemic to Australia.

Distribution. The genus occurs in all continents except of Europe (55 known species) and is confined to tropics and subtropics. Australian species occur in whole Australia except Tasmania.

Biology. Aquatic both as adults and larvae, inhabiting especially well vegetated pools.

References. Watts (1990) (revision of Australian fauna); Archangelsky (1997) (description of larva); Watts (2002) (description of larva); Short (2010b) (diagnosis of the genus); Toussaint *et al.* (2017) (phylogeny).

Key to species. Watts (1990).

Hydrophilus Geoffroy, 1762 (Pl. 44D–E; Fig. 20.8U–Z, a–i)

Type species. *Dytiscus piceus* Linnaeus, 1758.

Characteristics. Anterior margin of clypeus emarginate, exposing the membrane between clypeus and labrum; labrum exposed, without mesal pits*; first antennomere of antennal club divided in two lobes and bearing long setae; maxillary palps long, sexually dimorphic in some species; prosternum mesally elevated, hood-like with a cavity to receive anterior tip of ventral keel*; ventral keel without anterior notch (in lateral view); abdomen with five visible ventrites; posterior margin of ventrite 5 entire; either all ventrites completely pubescent, or median portion of ventrites lacking pubescence; male protarsi weakly to largely widened; anterior claws with or without sexual dimorphism (if present, enlarged and of different shape in males); meso- and metafemur without pubescence; aedeagus of simple form without narrow parameres and hooks on median lobe; body length 18–46 mm; dorsal coloration uniformly dark, black or greenish. Large species cannot be confused with any other hydrophilid genus simply due to the body size, smaller species may resemble other genera of the Hydrophilini, but are easily diagnosed from the by the hood-like form of the prosternum.

Larvae (Fig. 20.9C, O–R, U–V): Frontal sulci U-shaped, coronal sulcus present but short; nasale absent, anteromesal portion of head straight, without teeth; epistomal lobes small, symmetrical; antennal scape with numerous setae on inner face; antenna of second and third instars with four antennomeres; mandibles robust, asymmetrical, left with one, right with two teeth; prementum about as long as wide; abdominal apex with an pair of long membranous projections (prostyli). The combination of nasale without teeth, asymmetrical robust mandibles, legs with swimming hairs, presence of prostyli and body size easily distinguishes the larva from all other Australian hydrophilids.

Australian species. Eight species, all but one (*H. bilineatus* (MacLeay)) endemic to Australia.

Distribution. The genus is distributed world-wide (49 species). Australian species occur throughout all regions of the continent.

Biology. Aquatic both as adults and larvae, inhabiting especially well vegetated pools. Larvae of some species are specialised predators of aquatic snails.

References. Watts (1988) (revision of Australian fauna); Archangelsky (1997) (description of larva); Watts (2002) (description of larva); Short (2010b) (diagnosis of the genus); Toussaint *et al.* (2017) (phylogeny).

Key to species. Watts (1988).

Sternolophus Solier, 1834 (Pl. 44C; Fig. 20.8A–I)

Type species. *Sternolophus rufipes* Solier, 1834 (= *S. solieri* Laporte, 1840).

Characteristics. Anterior margin of clypeus straight or with small median notch; labrum exposed, without mesal pits; first antennomere simple, not divided in two lobes, without long setae; maxillary palps never sexually dimorphic; prosternum simply carinate mesally, with tuft of setae anteriorly; ventral keel with anterior notch with setae (in lateral view); metafemora pubescent at extreme base; abdomen with five visible ventrites; posterior margin of ventrite 5 entire or narrowly emarginate; all abdominal ventrites completely pubescent; protarsi not sexually dimorphic, male anterior claws at most of very slightly different shape than in female; aedeagus of simple form without narrow parameres and hooks on median lobe; body length 10–15 mm; dorsal coloration uniformly dark, black or greenish. It may be confused by the size, coloration and body form with *Limnoxenus zealandicus* and *Enochrus mastersi* (differs from both by the presence of the ventral keel and the absence of sutural stria), and may also resemble some small *Hydrobiomorpha* (see couplet 3 in the key for diagnostic characters).

Larvae (Fig. 20.9A, D–H): Frontal sulci V-shaped; coronal sulcus present, quite long; nasale with 5 teeth, left lateralmost one separated from others by wider gap; epistomal lobes small, rounded, symmetrical; antennal scape of second and third instars with fringe of setae subapically; antenna on all instars with 3 antennomeres; mandibles slender, symmetrical, with three inner teeth (basalmost very small); distal tooth never rectangularly quadrate at apex; additional pair of membranous projections (prostyli) abdominal apex reduced (hence not apparent). By the combination of the shape of nasale, slender mandibles and body size closely resembling *Limnoxenus zealandicus* (see under *Limnoxenus* for diagnostic characters).

Subgenera. The genus is divided into two subgenera which differ in the form of the anterior margin of clypeus (straight in *Sternolophus* s.str., emarginate in *Neosternolophus* Zaitzev). The subgenera were revealed as reciprocally monophyletic in the analysis of adult morphology (Nasserzadeh *et al.* 2017), the monophyly of *Sternolophus* s.str. was not revealed by molecular phylogenetic analysis by Toussaint *et al.* (2017). The subgenera were omitted in the revision of the genus by Nasserzadeh & Komarek (2017), but all Australian species would belong to *Neosternolophus* based on clypeal morphology.

Australian species. Four species, three endemic to Australia and one (*S. marginicollis*) also occurring in New Guinea, Ceram, Sulawesi, Philippines, Solomon Islands, New Caledonia and Fiji (Nasserzadeh & Komarek 2017). *Sternolophus rufipes* Fabricius, which was indicated by Watts (1989) as possibly reaching Australia, is not occurring in Australia (Nasserzadeh & Komarek 2017).

Distribution. The genus is distributed in tropical and subtropical regions of Asia, Africa and Australian regions and comprises 16 described species.

Biology. Aquatic both as adults and larvae, inhabiting especially well vegetated pools.

References. Watts (1989) (revision of Australian fauna); Watts (2002) (description of larva); Short (2010b) (diagnosis of the genus); Minoshima & Hayashi (2011a) (description of larva); Nasserzadeh & Komarek (2017) (taxonomic revision); Toussaint *et al.* (2017) (molecular phylogeny); Nasserzadeh *et al.* (2017) (morphological phylogeny).

Key to species. Watts (1989); Nasserzadeh & Komarek (2017).

Subfamily Chaetarthriinae Bedel, 1881

Characteristics. No unique adult synapomorphy known. Characterised by the combination of following characters: maxillary palpomere IV with digitiform sensilla on dorsolateral surface (seen under electron microscope only); elytron without scutellar stria; anapleural sutures of mesothorax always present; mesofurca without long dorsal arms (Fig. 20.2H; not seen from outside); posterior margin of abdominal ventrite never with apical emargination.

Larvae known for six genera of the subfamily, all share the following characters: coronal sulcus absent; frontal lines not converging; antennal sensorium as long as antennomere 3 or slightly longer; mandibles symmetrical, with 2–3 inner teeth; labium with long ligula, which exceeds labial palps in length.

Moderately large group, containing 14 genera and ~250 described species distributed world-wide. Most species are aquatic or semiaquatic, only few minute clades lost the association with aquatic habitats (in Australia this may be the case of *Phelea*).

Australian taxa. Six genera, 16 species, most of them aquatic. Both tribes of Chaetarthriinae are represented in Australia.

Tribe Chaetarthriini Bedel, 1881 (Fig. 20.10)

Characteristics. All genera of the tribe have a fringe of long setae on the base of abdominal ventrite 1 which covers a depression of the ventrite usually filled with a hyaline mass. For additional characters helpful for identification, see under *Chaetarthria*.

Larvae are known for two genera (*Chaetarthria* and *Guyanobius*; Archangelsky 1997, 2002b), they both have all characters of larval Chaetarthriinae but otherwise do not share any obvious character which would be a candidate for the larval synapomorphy of the tribe.

A small group containing seven genera and ~75 species distributed world-wide, but with very few species in Australian and Afrotropical Regions. Most species are aquatic or semiaquatic, few Asian species are possibly inhabiting forest leaf litter. Majority of species are rather tiny in body size.

Australian taxa. One genus with a single species.

Chaetarthria Stephens, 1835 (Pl. 44H; Fig. 20.10A–C)

Type species. *Hydrophilus seminulum* Herbst, 1797.

Characteristics. Body widely rounded, semiglobular; head narrowing anteriorly; labrum very large, well sclerotised and exposed; antennae with 8 antennomeres, pedicel very large and globular; gular sutures fused; prosternum extremely short; elytra without series of punctures, only with sharply impressed sutural stria; mesothorax with well developed anapleural sutures; mesoventrite wide anteriorly, elevated to the level of metaventrite posteromesally; abdomen with 5 ventrites, first ventrite covered with fringe of long yellowish setae; posterior margin of ventrite 5 without emargination; dorsal coloration uniformly black; body size 1.5–2.5 mm. The genus may be confused with *Anacaena*, *Crenitulus* and *Paracymus* or with terrestrial members of the tribe Omicrini based on tiny size, globular body and uniformly dark coloration. It may be easily distinguished from all these taxa by a very characteristic and unique morphology of head and antenna, and by the presence of fringe of yellowish setae over abdominal ventrite 1.

Larvae (Fig. 20.10a–g): Nasale symmetrical with 3 teeth, median one large, lateral ones tiny; epistomal lobes low

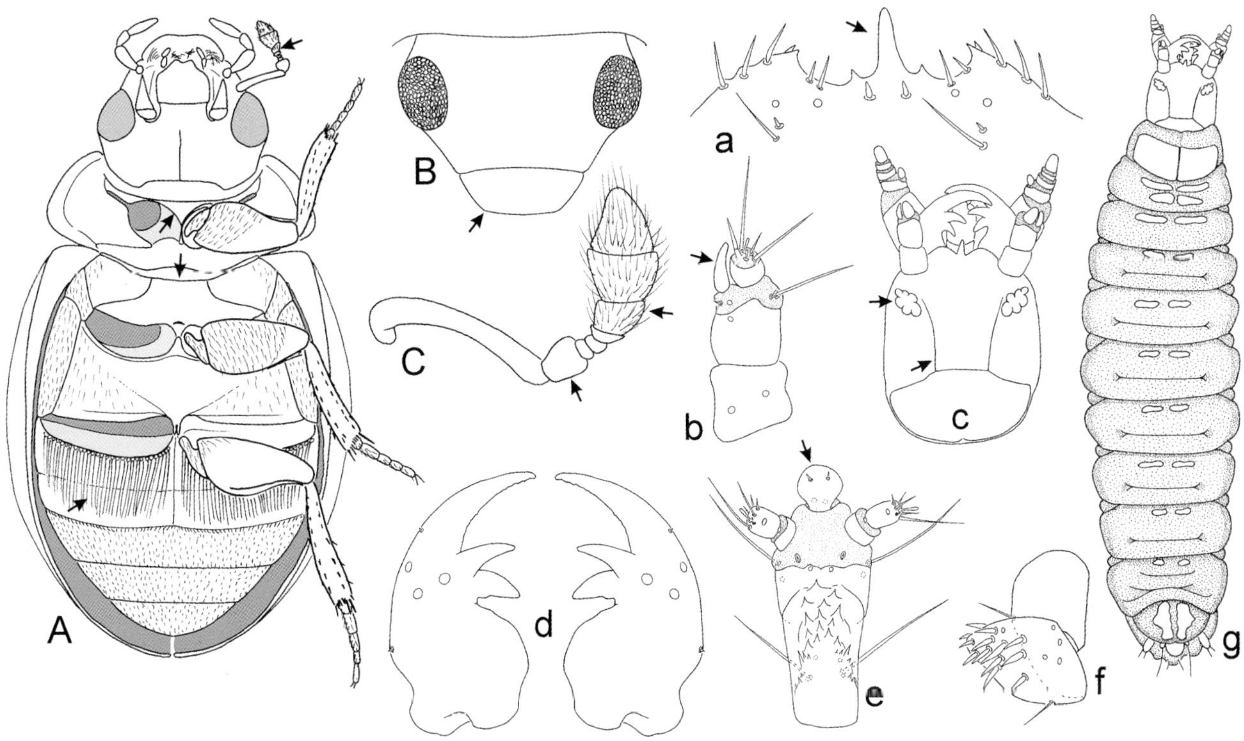

Fig. 20.10. Tribe Chaetarthriini, *Chaetarthria*. A–C, adult of *C. nigerrima* Blackburn: A, body in ventral view; B, head in dorsal view; C, antenna. a–g, larva of *Chaetarthria*: a, anterior margin of head; b, antenna; c, larval head; d, mandibles; e, labium; f, prothoracic leg; g, general habitus. a–e, g from Archangelsky 1997 (larva of *Chaetathria* sp. from USA), f from Archangelsky 2002b (larva of *C. bruchi* Balfour-Browne).

and symmetrical; frontal sulci not converging posteriorly; stemmata closely aggregated; mandibles symmetrical, each with two inner teeth; ligula overlapping labial palps, widely rounded; labial palps directed laterally; legs reduced in size, without claws; thorax and abdomen without conspicuous lateral lobes (based on known Palearctic/Neotropical larvae, larva of the Australian species is unknown; Archangelsky 1997, 2002b; Fikáček unpubl. data).

Australian species. One species, *C. nigerrima* (Blackburn), endemic to Australia.

Distribution. The genus is worldwide, comprising ~50 species, most of them occurring in the Neotropical and Oriental Regions. *Chaetarthria nigerrima* is widespread in Australia, occurring from Pilbara region (Western Australia) to eastern Victoria, but usually it is only locally common.

Biology. Collected from sand and gravel edges of rivers or pools, sometimes coming to light.

References. Hansen (1991) (adult morphology); Watts (2000) (taxonomy and distribution of Australian species); Fikáček (2010a) (short comparison with Asian species).

Key to species. Watts (2000).

Tribe Anacaenini Hansen, 1991 (Fig. 20.11)

Characteristics. Prosternum flat or with transverse groove; anapleural sutures well developed; mesoventrite flat, bearing a tooth or transverse ridge; abdomen with 5 visible ventrites; posterior margin of ventrite 5 without median notch or group of stout setae; elytron always without scutellary stria and with sutural stria; organised stridulatory file on abdominal laterosternite 5 absent. Most species are rather small (under 3.7 mm) and moderately convex, *Phelea* is more depressed.

Larvae are known for four genera only (*Anacaena*, *Crenitulus*, *Crenitis*: Archangelsky 1997; Archangelsky & Fikáček 2004; *Pseudorygmodus*: Spangler 1979 (as *Cylorygmus*)). All known larvae have very short stipes and extremely long palpifer, which is unique for the tribe within the Hydrophilidae.

A moderately large world-wide group containing seven genera and ~170 species. The clade is so far most diverse in the Oriental Region due to many species of the genus *Anacaena* occurring there. Many anacaenine genera including the early-branching ones are largely restricted to the Southern Hemisphere, only three of them (*Crenitis*,

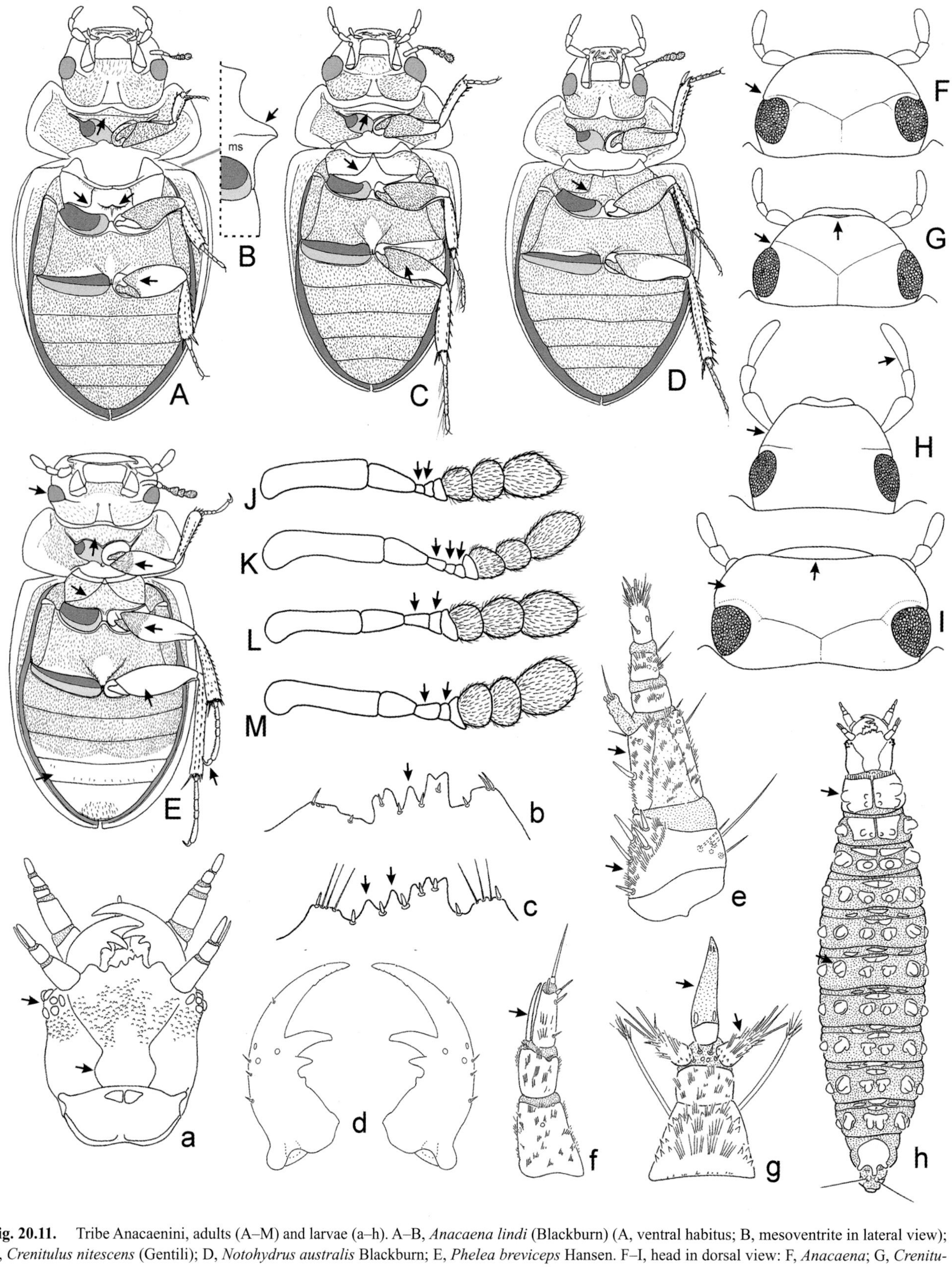

Fig. 20.11. Tribe Anacaenini, adults (A–M) and larvae (a–h). A–B, *Anacaena lindi* (Blackburn) (A, ventral habitus; B, mesoventrite in lateral view); C, *Crenitulus nitescens* (Gentili); D, *Notohydrus australis* Blackburn; E, *Phelea breviceps* Hansen. F–I, head in dorsal view: F, *Anacaena*; G, *Crenitulus*; H, *Notohydrus*; I, *Phelea*. J–M, antenna: J, *Anacaena lindi*; K, *Crenitulus nitens*; L, *Notohydrus*; M, *Phelea*. a–g, larvae (based in non-Australian species: a, c–h, *Crenitulus suturalis* (LeConte), b, *Anacaena lutescens* (Stephens). a, head in dorsal view; b–c, anterior margin of head; d, mandibles; e. maxilla; f, antenna; g, labium; g, general habitus. a–b from Fikáček and Vondráček (2014); c–h from Archangelsky and Fikáček (2004).

part of *Anacaena* and few *Crenitulus*) reach northern hemisphere (of these, only *Anacaena* and *Crenitis* are species-diverse there).

Australian taxa. Four genera, 15 species. All species are aquatic or semiaquatic. *Crenitis* was recorded from Australia by Gentili (1996b), but the species in fact belongs to *Pseudohydrobius* and *Crenitis* is hence not a member of Australian fauna (Fikáček & Watts 2015).

Anacaena Thomson, 1859 (Pl. 44L–M; Fig. 20.11A–B, F, J)

Type species. *Hydrophilus globulus* Paykull, 1798.

Characteristics. Elongate to widely oval, moderately convex; labrum largely retracted under clypeus, straight on anterior margin; eyes moderately large, not protruding, emarginated anteriorly; antenna with 8 antennomeres*, antennal club loosely segmented; prosternum moderately long, without median carina or with short median carina in anterior half, with weak transverse ridge submedially; anapleural sutures distinct, nearly parallel-sided anteriorly; mesoventrite always with high posteromesal projection; elytra not explanate laterally; elytral punctures irregularly or subserially arranged, in latter case with alternating larger (primary) and smaller (secondary) series; epipleuron extremely narrow posteriorly; mesofemur nearly completely pubescent except for extreme apex; metafemur pubescent at base and along anterior margin; tibiae and tarsi without swimming hairs; metatarsomere 1 much shorter than metatarsomere 2; abdomen with 5 ventrites; ventrite 1 not carinate medially; posterior margin of ventrite 5 entire; body length 1.7–3.0 mm; dorsal coloration yellowish (sometimes with various pattern of dark spots) to dark brown or black. It may be easily confused with *Crenitulus* and *Paracymus* but differs from both easily by characteristic shape of the mesoventrite (wide and with anapleural sutures nearly parallel-sided anteriorly). For additional differential characters see under *Crenitulus* and *Paracymus*.

Larvae (Fig. 20.11b): Head subquadrate with slightly projecting ocular region; frontal sulci not converging posteriorly; nasale with five teeth (two at the right and left sides reaching further than the median one); epistomal lobes moderately large and nearly symmetrical, coronal sulcus absent, stemmata not closely aggregated; mandibles symmetrical with two large and one minute inner teeth each; maxillary stipes very short, with five stout setae on inner face, palpifer much longer than stipes; mentum not much wider than prementum, with large ligula overlapping labial palps; legs well developed, short, not visible in dorsal view; proscutum with lateral lobes (based on known Palearctic/Nearctic larvae, larvae of Australian species are unknown; Richmond 1920 (as *Paracymus*); d'Orchymont 1940; Fikáček & Vondráček 2014).

Australian species. Seven species, all of them endemic to Australia (Komarek 2007).

Distribution. The genus is world-wide in distribution (~130 known species). Australian species occur in whole Australia, highest species diversity is known from Queensland and New South Wales.

Biology. Aquatic to semiaquatic both as adults and larvae, inhabiting banks of various types of streams, lakes and lagoons, usually found in submerged vegetation, moss or accumulated wet debris.

Notes. The generic assignment of Australian species varied through time, they were originally described as *Paracymus*, later transferred to a separate genus *Paranacaena* (Gentili 1993), which was later synonymized with *Anacaena* (Komarek & Beutel 2007). One species originally treated in *Anacaena* was recently transferred to *Crenitulus* (Fikáček & Vondráček 2014).

References. Gentili (1993) (resurrection of *Paranacaena*, description of new species); Komarek & Beutel (2007) (synonymisation of *Paranacaena*); Komarek (2007) (revision of Australian fauna); Richmond (1920), d'Orchymont (1940) (description of larvae of Northern Hemisphere species); Fikáček & Vondráček (2014) (adapted generic concept, larval morphology).

Key to species. Komarek (2007).

Crenitulus Winters, 1926 (Pl. 44K; Fig. 20.11C, G, K)

Type species. *Limnebius suturalis* LeConte, 1866.

Characteristics. Body elongate oval with elytra attenuate posteriorly, moderately convex; labrum exposed, well sclerotised, straight on anterior margin; clypeus weakly concave, narrowly exposing membrane; eyes large, globular, not emarginated anteriorly; antenna with 9 antennomeres, antennal club loosely segmented; prosternum very short, ecarinate, with transverse ridge submedially; anapleural sutures distinct, converging anteriorly; mesoventrite flat, without protuberance*; elytra not explanate laterally; elytral punctures arranged subserially, with alternating larger (primary) and smaller (secondary) series; epipleuron extremely narrow posteriorly; meso- and metafemora pubescent ca. in basal two thirds; tibiae without swimming hairs; meso- and metatarsi with few fine long swimming hairs dorsally; metatarsomere 1 much shorter than metatarsomere 2; abdomen with 5 ventrites; ventrite 1 not carinate medially, posterior margin of ventrite 5 entire; body length 1.9–2.1 mm; dorsal coloration dark brown to black, with paler margins of pronotum. It may be confused with *Anacaena* or *Paracymus* by general body shape and small body size. It differs from *Anacaena* by antennae with 9 antennomeres (8 antennomeres in *Anacaena*), eye not emarginated anteriorly (anteriorly emarginated by lateral portion of clypeus in *Anacaena*), flat mesoventrite (mesoventrite with median tooth in *Anacaena*) and phallobase widely rounded basally (projecting into narrow pointed manubrium basally in Australian *Anacaena*). It differs from *Paracymus* by medially ecarinate prosternum (medially elevated to slightly carinate in *Paracymus*), flat mesoventrite (always with median projection in *Paracymus*), hind femur with dense hydrofuge pubescence in basal two thirds (with dense pubescence only at very base in *Paracymus*) and elytra with subserially arranged punctures (with irregularly arranged punctures in *Paracymus*).

Larvae (Fig. 20.11a, c–h): Head subquadrate with slightly projecting ocular region; nasale with five teeth (right three projecting further than left two); epistomal lobes moderately large and nearly symmetrical; frontal sulci not converging posteriorly; coronal sulcus absent; stemmata not closely aggregated; mandibles symmetrical, with two large and one minute inner teeth each; maxillary stipes very short, with only five stout setae on inner face, palpifer about twice as long as stipes; mentum not much wider than prementum, without distinct teeth, with large ligula overlapping labial palps; prosternum incompletely divided medially; legs short, not visible in dorsal view; proscutum with large lateral lobes; plate on abdominal segment 8 trilobate posteriorly (based on Neotropical/Nearctic *C. suturalis* (LeConte), larva of Australian species unknown; Archangelsky & Fikáček 2004).

Australian species. Single species, *C. nitens* (Gentili, 1993).

Distribution. Australian *C. nitens* occurs in northern Queensland (Komarek 2007). Outside Australia, there are 19 recent Neotropical species, one of which reaches even the United States, and one Miocene fossil from Dominican amber (Fikáček & Engel 2011; Fikáček & Vondráček 2014; Santana *et al.* 2017; García 2018).

Biology. Aquatic both as adult and larva, but more detailed data from Australia are absent.

Note. The generic assignment of the Australian species was very unstable since its description. It was first assigned to *Paranacaena* Blackburn (Gentili 1993), later transferred to its own genus *Gentilina* (Hebauer 2003), then to *Anacaena* (Komarek 2007) and only recently moved to the resurrected genus *Crenitulus* (Fikáček & Vondráček 2014).

References. Gentili (1993) (description of *C. nitens*); Hebauer (2003) (comments on generic placement); Komarek (2007) (redescription and illustration of diagnostic characters); Fikáček & Vondráček (2014) (resurrection of *Crenitulus*).

Key to species. Komarek (2007).

Notohydrus Balfour-Browne, 1939 (Pl. 44I; Fig. 20.11D, H, L)

Type species. *Laccobius australis* Blackburn, 1891.

Characteristics. Widely oval, moderately convex; head narrow and elongate; labrum largely exposed, sinuate on anterior margin; eyes moderately large, weakly protruding, emarginated anteriorly; antenna with 8 antennomeres, antennal club loosely segmented; maxillary palps apparently long; prosternum moderately long, without carina or transverse ridge; anapleural sutures distinct, subparallel anteriorly; mesoventrite only with a low not very distinct posteromesal projection; elytra not explanate laterally; elytra with 10 elytral series and impressed sutural stria; epipleuron extremely narrow posteriorly; mesofemur nearly completely pubescent except for extreme apex; metafemur pubescent at base and along anterior margin; tibiae and tarsi without swimming hairs; metatarsomere 1 much shorter than metatarsomere 2; abdomen with 5 ventrites; ventrite 1 not carinate medially; posterior margin of ventrite 5 entire; body length 2.3–3.7 mm; dorsal coloration yellowish to brownish often with slightly darker discs of elytra and pronotum and central portion of head. It may be confused with *Anacaena* but differs from it easily by the elongate head with very long maxillary palps and largely exposed labrum, and by elytra with 10 distinct series. It may also resemble some Cylominae, but differs from them by anteriorly subparallel anapleural sutures and antenna with 8 antennomeres.

Larva unknown.

Australian species. The genus is endemic to Australia, with 6 described species.

Distribution. Endemic to the south-eastern part of Australia, confined to mountain regions of Victoria, New South Wales and Australian Capital Territory.

Biology. Found in wet debris and among wet leaves at the margins of forests streams; occassionally collected using flight intercept traps (Gentili 1992, 2006).

References. Balfour-Browne (1939a) (description of the genus); Gentili (1981) (redescription); Gentili (1992) (species revision); Gentili (2006) (description of *N. shorti*).

Key to species. Gentili (1992, 2006).

Phelea Hansen, 1999b (Pl. 44J; Fig. 20.11E, I, M)

Type species. *Phelea breviceps* Hansen, 1999.

Characteristics. Widely elongate oval, depressed; labrum large, exposed and well sclerotised, inclined, straight on anterior margin; clypeus subquadrate, not narrowed anteriorly, its anterior margin weakly concave; eyes small, not protruding laterally, weakly emarginated anteriorly; antenna with 8 antennomeres, antennal club loosely segmented; prosternum very long, ecarinate and without transverse ridge submedially; anapleural sutures present, converging anteriorly; mesoventrite with very low and indistinct posteromedian bulge; elytra narrowly explanate laterally; serially arranged punctures of the same size as interval punctation, hence elytral series rather indistinct; sutural stria present; epipleuron wide even posteriorly. mesofemora pubescent on extreme base; metafemora not pubescent; tibiae and tarsi without swimming hairs; claws large, with basal tooth; metatarsomere 1 much shorter than metatarsomere 2; abdomen with 5 ventrites; ventrite 1 not carinate medially; posterior margin of ventrite 5 entire; dense pubescence present only on ventrites 1–3; body length 3.5 mm; dorsal coloration dark brown to black. *Phelea* can be hardly confused with any other hydrophilid genus, there is no other genus known from Tasmania which would combine the coarsely punctate depressed body, quadrate head with large inclined labrum, long prosternum without median carina, meso- and metafemora without dense pubescence and abdomen with ventrites 4–5 lacking dense pubescence.

Larva unknown.

Australian species. Endemic to Australia (Tasmania), with single species, *P. breviceps* Hansen.

Distribution. The genus and its only species is endemic to Tasmania, all specimens were collected in a small area around Mt. Cradle in Tasmania between 900–1300 m of altitude (last record is known from 1959).

Biology. A short series collected in 1959 was found in small tarns on the plateau at the base of Cradle Mt., indicating that the species may be aquatic. The reduced hydrofuge pubescence on abdominal ventrites 4–5 may indicate that the species is hygropetric.

References. Hansen (1999b) (description of the genus); Fikáček & Watts (2015) (redescription, biology).

Subfamily Enochrinae Short and Fikáček, 2013 (Fig. 20.12)

Characteristics. Australian species of the subfamily may be easily recognised by the characteristic form of the maxillary palps, with the third (second visible from above) palpomere excised mesally at the base, and the palps with a 'zig-zag' appearance (Fig. 20.12J–K). See under *Enochrus* for additional characters. No unique adult synapomorphy is known for the subfamily, but both principal clades are easily recognisable: *Enochrus*-clade by the unique morphology of maxillary palps mentioned above, *Cymbiodyta*-clade (not in Australia) by meso- and metatarsi with four tarsomeres only.

Larvae are known for all genera except *Enochrella* Hansen (Archangelsky 1997; Watts 2002; Minoshima & Hayashi 2011a). They are all characterised by nasale with several small weakly defined teeth, sometimes present along with 6 or 7 well defined ones (in all cases it is hence very difficult to count the real number of teeth on nasale), mandibles symmetrical (with 2 inner teeth each) or asymmetrical (left with one, right with two inner teeth), frontal lines converging posteriorly, maxillary stipes with 5 stout spines on inner face, labium with distinct ligula (but always much shorter than labial palps). See under *Enochrus* for additional characters of the genus.

A moderately large group containing five genera and ~280 species distributed world-wide, of which most (~235) belong to the genus *Enochrus*. All species are aquatic.

Australian taxa. One genus and 14 species.

Enochrus Thomson, 1859 (Pl. 45A–B; Fig. 20.12A–L)

Type species. *Hydrophilus bicolor sensu* Gyllenhal (= *Hydrophilus melanocephalus* Olivier, 1792).

Characteristics. Body elongate oval, moderately convex; eyes moderately large; anterior margin of clypeus concave with membrane exposed between clypeus and labrum; labrum exposed; antenna with 9 antennomeres; maxillary palpomere 2 (first visible) straight or curved outwards (in the latter case palps 'zig-zag' in appearance), third palpomere with a small notch at basal inner face; elytra with sharply impressed sutural stria, otherwise without series of large punctures or at most with few irregular series of setiferous punctures (trichobothria); prosternum flat or with transverse groove; mesoventrite triangular (sg. *Hydatotrephis*) or wide and parallel-sided anteriorly (sg. *Methydrus*) with median keel or cone-like projection; abdomen with five ventrites; tarsi with few long dorsal setae or totally without swimming hairs; posterior margin of ventrite 5 emarginated mesally*; all femora with dense pubescence ventrally except at extreme apex; body length 1.6–9.1 mm; pale with dark head and central part of pronotum to completely black. Large dark species of *Enochrus* may be confused with *Limnoxenus* (which differs by elytra with fine but distinct 10 series of punctures and meso- and metatarsi with dense fringe of swimming hairs on dorsal surface and lacks *Enochrus* form of maxillary palp), medium-sized species resemble *Chasmogenus* (lacks typical *Enochrus*-form of maxillary palps, mesoventrite is wide but not parallel-sided anteriorly). Genera *Helochares* and *Agraphydrus* which are also slightly similar may be distinguished from *Enochrus* by absence of sharply impressed sutural stria. In all cases, *Enochrus* is unique in Australia in the straight or outwardly curved second palpomere 2 and basal notch on palpomere 3, by which it can be easily distinguished from all other Australian genera.

Larvae (Fig. 20.12a–i): Frontal sulci V- or U-shaped, coronal sulcus present but short; nasale usually slightly asymmetrical, with 6 distinct to many small irregular and rather indistinct teeth; epistomal lobes small, asymmetrical, not overlapping nasale; mandibles either each with two inner teeth (sg. *Hydatotrephis*) or with one inner tooth on left one and two in right one (sg. *Methydrus*); palpifer much shorter than maxillary palp; labium with ligula present but small, much shorter than maxillary palps; abdomen without spiracular gills, stigmatic atrium on segment 8 present; abdominal segments 2–7 each with pair of ventral prolegs (sg. *Methydrus*) or without such prolegs (sg. *Hydatotrephis*).

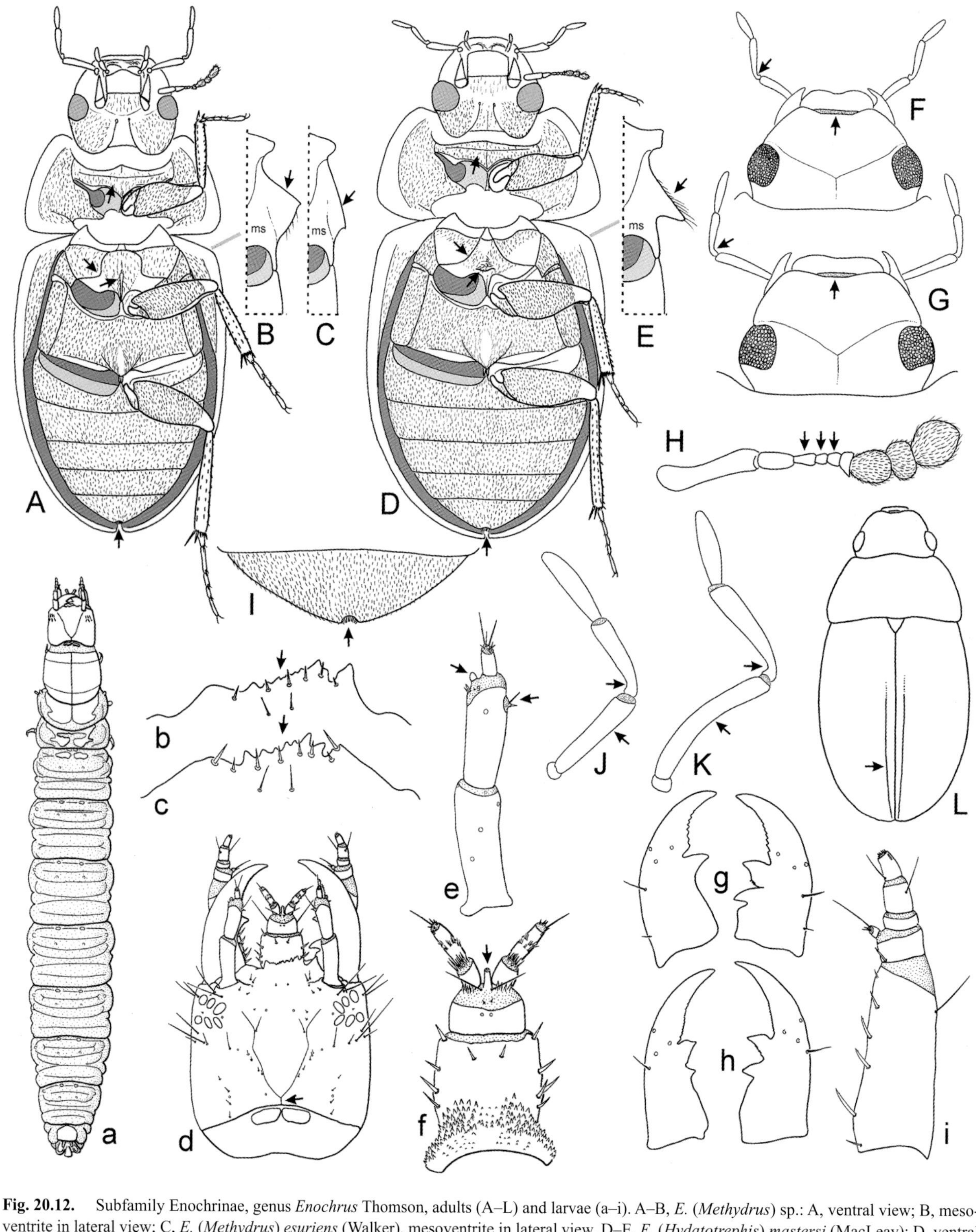

Fig. 20.12. Subfamily Enochrinae, genus *Enochrus* Thomson, adults (A–L) and larvae (a–i). A–B, *E.* (*Methydrus*) sp.: A, ventral view; B, mesoventrite in lateral view; C, *E.* (*Methydrus*) *esuriens* (Walker), mesoventrite in lateral view. D–E, *E.* (*Hydatotrephis*) *mastersi* (MacLeay): D, ventral view; E, mesoventrite in lateral view. F, *E.* (*Methydrus*), head dorsally; G–J, *E.* (*Hydatotrephis*) *mastersi*: G, head dorsally; H, antenna; I, abdominal apex; J, maxillary palpomere. K–L, *E.* (*Methydrus*): K, maxillary palpomere; L, dorsal habitus. a, general habitus of the larva of *E.* (*Methydrus*). b–c, anterior margin of head: b, *E.* (*Methydrus*) *eyrensis* (Blackburn); c, *E.* (*Hydatotrephis*) *mastersi*. d–f, *E.* (*Methydrus*): d, head dorsally; e, antenna; f, labium; g–h, mandibles: g, *E.* (*Methydrus*) *eyrensis*; h, *E.* (*Hydatotrephis*) *mastersi*. I, maxilla of *E.* (*Methydrus*). b–c, g–h based on Watts (2002); a, d–f, i based non-Australian *E. japonicus* (Sharp) by Minoshima & Hayashi (2011a), i slightly adapted.

Subgenera. The genus is divided into six subgenera, three of which occur in Australia. The monophyly of all subgenera has to be, however, tested in the future, as many of them are defined arbitrarily by the combination of one/two characters only and there are species (not in Australia) which cannot be assigned to any subgenus based on their current definition (e.g. Jia & Short 2013). The following key follows Hansen (1990a) and Watts (1998b):

1. Ultimate and penultimate maxillary palpomeres subequal in length (Fig. 20.12J). Body length 3.0–9.2 mm **2**
– Ultimate maxillary palpomere distinctly shorter than penultimate (Fig. 20.12K). Body length 1.6–8.2 mm ..... **sg. *Methydrus* Rey** (12 spp.)
2(1). Body medium-sized (around 3.2 mm long). Mesoventrite with median keel ..... **sg. *Enochrus* s.str.** (1 sp.)
– Body large (8.0–9.1 mm). Mesoventrite with median cone-shaped projection (Fig. 20.12E) ..... **sg. *Hydatotrephis* MacLeay** (1 sp.)

Australian species. 14 species belonging to three subgenera: *Hydatotrephis* MacLeay (1 sp., *H. mastersi* (MacLeay)), *Enochrus* s.str. (1 sp., *E. peregrinus* Knisch) and *Methydrus* Rey (12 spp.).

Distribution. The genus is world-wide in distribution. Of the species occurring in Australia, 10 are endemic for the continent, two more widespread in western Pacific (*E. elongatus* (MacLeay), *E. maculiceps* (MacLeay)) and two reaching Australia and Pacific from Asia (*E. esuriens* (Walker), *E. malabarensis* (Régimbart)).

Biology. Common in various types of waters including small pools, river banks, lagoon etc. Frequently coming to light.

References. Anderson (1976) (descriptions of Australian larvae); Hansen (1990a) (position of *Hydatotrephis*, rede scription of *E. mastersi*); Hansen (1991) (adult morphology); Watts (1998b) (revision of Australian fauna); Watts (1999) (rediscovery of *E. peregrinus*); Watts (2002) (descriptions of Australian larvae).

Key to species. Watts (1998b).

Subfamily Acidocerinae Zaitzev, 1908 (Fig. 20.13)

Characteristics. Antennae with 9 antennomeres*; Prosternum moderately to very long, with transverse groove; anapleural sutures well developed; mesoventrite wide or narrow anteriorly, with low median keel or posteromedian elevation; abdomen with 5 visible ventrites; posterior margin of ventrite 5 with or without median notch; elytron either lacking elytral series (sometimes except of few irregularly arranged series of setiferous punctures) or with 10 distinct series and a scutellary stria; sutural stria absent or present; organised stridulatory file on abdominal laterosternite 5 absent. Adults are generally rather similar to those of Enochrinae (*Enochrus*), from which they easily differ by the morphology of the maxillary palp. Females of some members of this group carry egg case attached to abdomen (Fig. 20.3Q) similarly as morphologically dissimilar Spercheidae.

Larvae are known for five genera only, of which some (*Chasmogenus*) are only incompletely described and others (*Helobata*) are known from first instar only (Anderson 1976; Archangelsky 1997; Watts 2002; Minoshima & Hayashi 2011a; Minoshima *et al.* 2013). Australian ones are characterised by asymmetrical nasale with 6 well defined teeth or symmetrical nasale with 5 to multiple teeth, mandibles symmetrical (with 2 inner teeth each) or asymmetrical (left with one, right with two inner teeth), frontal lines converging posteriorly, maxillary stipes with 5 stout spines on inner face, labium with distict ligula (but always much shorter than labial palps). They are generally very similar to those of Enochrinae (*Enochrus* in Australia) which differ by serrate (multidentate) asymmetrical nasale.

A large world-wide group containing 17 genera and ~300 described species. Most species of the subfamily are aquatic (Girón & Short 2018).

Australian taxa. Three genera, 13 species. All species are aquatic.

Agraphydrus Régimbart, 1903 (Pl. 45D; Fig. 20.13A, G, N)

Type species. *Agraphydrus punctatellus* Régimbart, 1903.

Characteristics. Body oval, weakly convex; eyes moderately large; anterior margin of clypeus straight; labrum exposed; antenna with 9 antennomeres*; maxillary palps slightly longer than width of head; prosternum moderately long with transverse groove; elytra without sutural stria and without series of punctures (at most with few irregular series of setiferous trichobothria); prosternum long; mesoventrite without median projection*; abdomen with five ventrites; abdominal ventrite 5 with median notch surrounded by stout setae; all femora ventrally with dense pubescence in basal two thirds; body length 1.4–2.6 mm; pale with dark head and central part of pronotum. It may be confused with small pale *Enochrus esuriens* as well as with some other small aquatic hydrophilids, but differs from all of them by the lack of sharply impressed sutural stria on elytron. *Helochares pallens* also has elytra without elytral series and without sutural stria and is of similar body form and coloration, but can be easily separated by anteriorly parallel-sided mesoventrite and conical (not trilobate) aedeagus.

Larvae (Fig. 20.13b, e, k; based on larva of *A. narusei*): Frontal sulci V-shaped, coronal sulcus present but short; nasale strongly asymmetrical (projecting more anteriorly at right side), with 6 teeth; epistomal lobes small, slightly

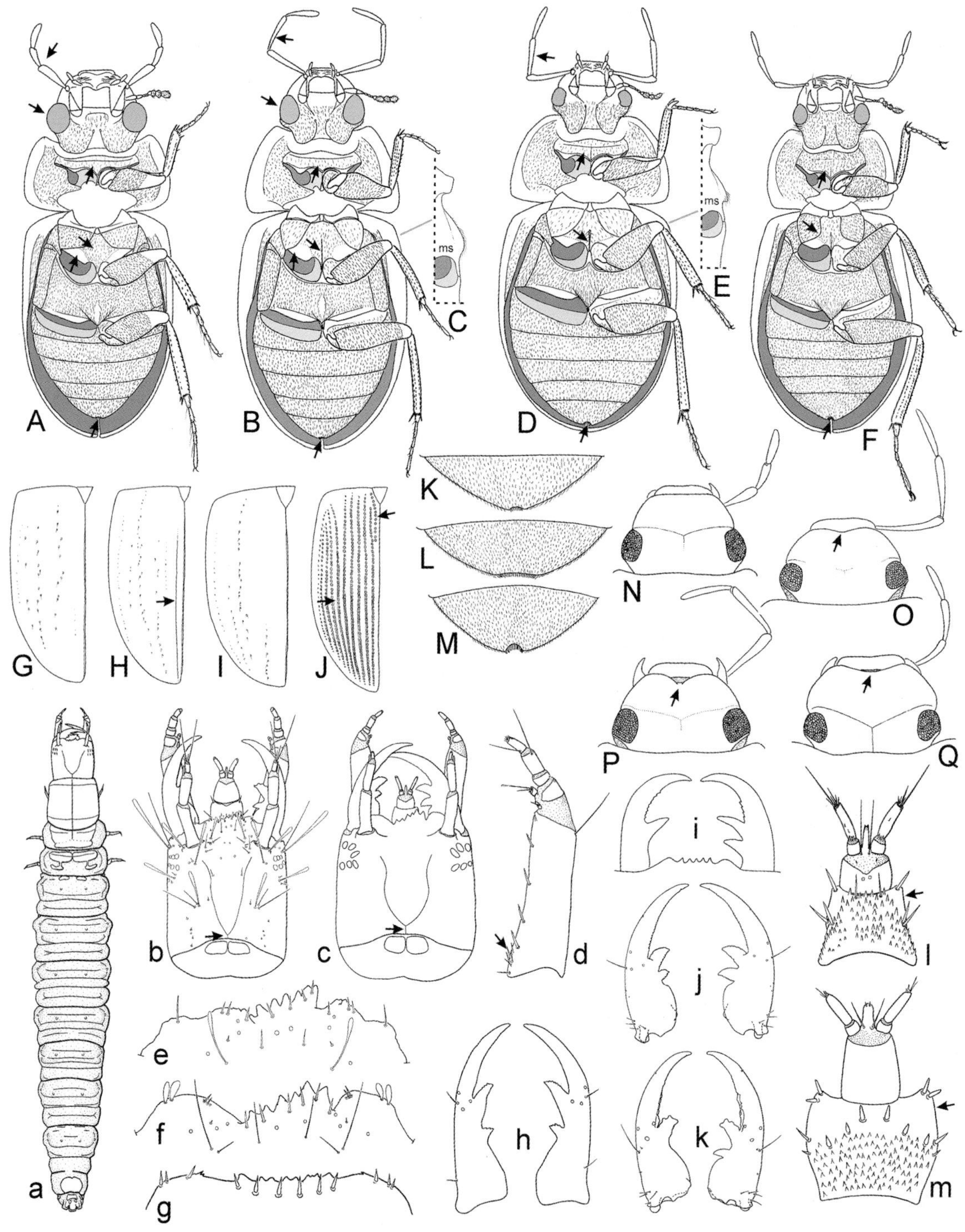

Fig. 20.13. Subfamily Acidocerinae, adults (A–Q) and larvae (a–m). A–F, ventral view and mesoventrite in lateral view: A, *Agraphydrus coomani* (Orchymont); B–C, *Chasmogenus nitescens* (Fauvel); D–E, *Helochares* (s.str.) *foveicollis* (Montrouzier); F, *Helochares* (*Hydrobaticus*) sp. G–J, elytron: G, *Agraphydrus* coomani; H, *Chasmogenus nitescens*; I, *Helochares* (s.str.) *foveicollis*; J, *Helochares* (*Hydrobaticus*) sp. K–M, abdominal apex: K, *Chasmogenus nitescens*; L, *Helochares foveicollis*; M, *Helochares* (*Hydrobaticus*) sp. N–Q, head in dorsal view: N, *Agraphydrus coomani*; O, *Helochares foveicollis*; P, *Chasmogenus nitescens*; Q, *Helochares* (*Hydrobaticus*) *percyi* Watts. a, habitus of *Helochares* sg. *Hydrobaticus*; b–c, larval head: b, *Agraphydrus*; c, *Helochares* sg. *Hydrobaticus*. d, maxilla of *Helochares* sg. *Hydrobaticus*; e–g, anterior margin of head: e, *Agraphydrus*; f, *Helochares* sg. *Hydrobaticus*; g, *Helochares* (s.str.) *foveicollis*. h–k, mandibles: h, *Helochares* (s.str.) *foveicollis*; i, mandibles and anterior margin of head of *Chasmogenus nitescens*; j, *Helochares* sg. *Hydrobaticus*; k, *Agraphydrus*. l–m, labium: l, *Helochares* sg. *Hydrobaticus*; m, *Helochares* (s.str.) *foveicollis*. a–f and j–l from Minoshima & Hayashi (2011a) based on non-Australian species *Agraphydrus narusei* (Satô) and *Helochares* (*Hydrobaticus*) *nipponicus* Hebauer; g–h and m adapted from Watts (2002), i from Anderson (1976).

asymmetrical, not overlapping nasale; mandibles asymmetrical, left with one inner tooth, right with two inner teeth; head with some setae scale-like; palpifer much shorter than maxillary palp; inner face of stipes with 5 setae; labium with ligula present but small, much shorter than maxillary palps; abdomen without spiracular gills, stigmatic atrium on segment 8 present; abdominal segments 3–7 with spinose ventral prolegs; body narrowly elongate, whitish. By asymmetrical mandibles and distinct teeth of nasale most similar to larva of *Chasmogenus*, from which it differs by asymmetrical nasale with 6 teeth; by asymmetrical mandibles similar to some *Enochrus* which differs by serrate nasale (rather than bearing 6 distinct teeth).

Australian species. Single species, *Agraphydrus coomani* (Orchymont) (confirmed by A. Komarek, pers. comm. 2015).

Distribution. The genus occurs from Africa through Arabian Peninsula to Asia and includes very many undescribed species. *Agraphydrus coomani* occurs from continental SE Asia through New Guinea to Australia (common in northern part of the continent).

Biology. Most species of the genus live at sides of stony streams or rivers, under partly submerged stones. Biology of Australian species is unknown.

References. Hansen (1991) (adult morphology); Watts (1995) (Australian records); Minoshima & Hayashi (2011a) (larval description of *Agraphydrus*).

Key to species. Watts (1995).

Chasmogenus Sharp, 1882 (Pl. 45C; Fig. 20.13B–C, H, K, P)

Type species. *Chasmogenus fragilis* Sharp, 1882.

Characteristics. Body elongate oval, weakly convex; eyes moderately large; anterior margin of clypeus excised, with exposed membrane between clypeus and labrum; labrum exposed; antenna with 9 antennomeres; maxillary palps much longer than width of head; elytra with sharply impressed sutural stria, without series of punctures (only with irregularly arranged setiferous trichobothria); prosternum long with transverse carina; mesoventrite widely rhomboid, weakly carinate posteriorly; abdomen with five ventrites; abdominal ventrite 5 with median notch; all femora ventrally with dense pubescence except on extreme apex; body length 2.5–5.2 mm; dorsal surface uniformly black. It may be easily confused with medium-sized *Enochrus* but differs from it by very long maxillary palps with second (first visible) palpomere curved invards and third (second visible) without distinct basal notch, and by different shape of the excision of the clypeus (compare with Figs 20.12F–G). The wide rhomboid mesoventrite is characteristic for *Chasmogenus* and not found in any other Australian hydrophilid.

Larvae (Fig. 20.13i; based on incomplete description by Anderson 1976): frontal sulci V-shaped, coronal sulcus present but short; nasale nearly symmetrical with 5 teeth; epistomal lobes small, nearly symmetrical, about as high as nasale; mandibles asymmetrical, left with one inner tooth, right with two inner teeth; labium with ligula present but small, much shorter than maxillary palps; abdomen without spiracular gills, stigmatic atrium on segment 8 present; abdominal segments with spinose ventral prolegs; body narrowly elongate. The larva seems to be generally quite similar to *Agraphydrus* (except of the shape of the nasale), see under that genus for more details.

Australian species. Only a single species, *C. nitescens* (Orchymont), was recorded from Australia by Watts (1995). Specimens from Northern Territory studied during the preparation of this chapter are, however, surely not conspecific with *C. nitescens* from New Caledonia (where type locality of *C. nitescens* is situated) and differ from it easily by larger body size and different morphology of male genitalia. These specimens also differ from *C. irianus* Hebauer from New Guinea by genital morphology. Additional studies are hence necessary to properly understand Australian fauna of *Chasmogenus*.

Distribution. The genus occurs world-wide in tropical areas. *Chasmogenus nitescens* occurs in Australia, New Guinea and Pacific Islands (New Caledonia, Solomon Islands, Fiji), in Australia it was recorded from tropical parts (Queensland, Northern Territory, New South Wales).

Biology. Aquatic, usually collected from submerged vegetation or leaf packs in vegetated side pools of streams or in pools of swampy meadows, also flying to light (Short 2010c).

References. Anderson (1976) (description of the larva, as *Helochares nitescens*); Hansen (1991) (adult morphology of the genus); Watts (1995) (summary of Australian records); Hebauer (2001) (description of *C. irianus* and records of *C. nitescens* from New Guinea); Short (2010c) (diagnosis of *C. nitescens*, genitalia drawing).

Helochares Mulsant, 1844 (Pl. 45E–F; Fig. 20.13D–F, I–J, L–M, O, Q)

Type species. *Dytiscus lividus* Forster, 1771.

Characteristics. Body elongate oval, weakly convex; eyes moderately large; anterior margin of clypeus concave, not or only very narrowly exposing membrane between clypeus and labrum; labrum exposed; antenna with 9 antennomeres; maxillary palps much longer than width of head; elytra without traces of elytral series and without sutural stria (sg. *Helochares* s.str.) or with 10 distinct series of punctures and scutellar stria (sg. *Hydrobaticus*); prosternum long; mesoventrite triangular (*H. foveicollis*) or moderately wide and parallel-sided anteriorly (sg. *Hydrobaticus*), with low median keel or low posteromedian bulge; abdomen with five ventrites; abdominal ventrite 5 with median notch surrounded by stout setae; all femora ventrally with dense pubescence except in extreme apex; body length 3.4–8.2 mm; dorsal body uniformly brown-black, or with mesally black head and pronotum. *Helochares* s.str. may be confused with some *Enochrus*, *Chasmogenus* or with smaller specimens of *Limnoxenus*, but differ from all the them by absence of sutural stria and maxillary palps nearly twice as long as width of the head. *Helochares* (*Hydrobaticus*) are very characteristic by coarsely striate elytra and very long maxillary palps, and by the habitus may be possibly confused only with *Hybogralius* which is however much more convex, has shorter maxillary palps, nearly completely bare posterior femora and lacks apical notch on abdominal ventrite 5. Adult females of *Helochares* carry egg cases attached to the abdomen, and are hence easily recognisable in the field (in Australia only *Spercheus* (Spercheidae) is also carrying egg cases, see there for comparison). *Helochares pallens* may be confused with *Agraphydrus*, see there for diagnostic characters.

Larvae (Fig. 20.13a, c–d, f–h, j, l–m): of two different types, both with frontal sulci V-shaped, coronal sulcus present but short; mandibles symmetrical or nearly so, each with two inner teeth; head without scale-like setae; palpifer much shorter than maxillary palp; inner face of stipes with 5 setae, base with group of cuticular spines; labium with ligula present but small, much shorter than maxillary palps; abdomen without spiracular gills, stigmatic atrium on segment 8 present; abdominal segments without ventral prolegs. Larvae of sg. *Hydrobaticus* with asymmetrical nasale with 6 teeth, maxillary palp not extremely elongate, stout mandibles and mentum not markedly wider than prementum (hence except of the shape of nasale very similar to larvae of *Enochrus* and other Acidocerinae); larvae *H. foveicollis* with multidentate straight anterior margin of head, extremely elongate maxilla and slender mandible, and mentum much wider than prementum (by this resembling larvae of *Regimbartia*, *Allocotocerus*, *Sternolophus* and *Hydrobiomorpha*, see under each genus for diagnostic characters).

Subgenera. *Helochares* is currently divided in five subgenera, two of which are reported from Australia (both are the only subgenera with world-wide distribution, remaining three include few species only and are restricted to Africa or South America). The subgenera are defined on a combination of few arbitrarily selected adult characters, but other character sets (genitalia, larval morphology) indicate they may be polyphyletic. Of the Australian taxa this concerns *H. foveicollis* which is now placed in sg. *Helochares* s.str. originally defined for European species, but both its genital morphology and larval characters indicate it is not related to those. The following key allows to distinguish the subgenera/species as defined at the moment:

1. Mesoventrite wide and parallel-sided anteriorly (as in Fig. 20.13F). Aedeagus of cylindrical form. Elytra with or without puncture rows. Small to medium-sized species **2**

– Metaventrite triangularly narrowing anteriorly (Fig. 20.13D). Aedeagus of trilobate type. Elytra finely punctate, without elytral series and sutural stria (Fig. 20.13I). Large species (6.1–8.2 mm) .... ***H.* (s.str.) *foveicollis***

2(1). Elytra either very coarsely punctate (hence elytral series indistinct) or with 10 series of punctures (Fig. 20.13J). Body medium-sized (3.4–7.2 mm) .... **sg. *Hydrobaticus* MacLeay** (12 spp.)

– Elytra finely punctate, without elytral series (but in some species with dark stripes). Small (2.3–3.3 mm) .... ***H.* (s.str.) *pallens*** (not yet recorded)

Australian species. 13 species. One additional widely distributed tiny species, *H.* (s.str.) *pallens* (MacLeay) reaches New Guinea and its occurrence in northernmost Australia cannot be excluded.

Distribution. The genus is world-wide in distribution (~200 described species). Of the Australian species, four are distributed in Australia plus New Guinea and/or New Caledonia, nine are endemic to Australia. Most species occur in the tropical northern part, few reach more to the south (*H. tristis* to Victoria) and one (*H. tenuistriatus*) is endemic to a smaller region around Perth in Western Australia.

Biology. Aquatic species, occurring in various types of waters, especially common in well vegetated pools. Frequently coming at light.

References. Anderson (1976) (larva of *H. tristis*); Hansen (1991) (adult morphology); Watts (1995) (revision of Australian fauna); Watts (2002) (larvae of *H. tristis* and *H. foveicollis*); Hebauer & Hendrich (1999) (description of two additional species, updated key to Australian *Hydrobaticus*); Minoshima & Hayashi (2011a) (detailed larval descriptions of *H. pallens* and *Hydrobaticus*).

Key to species. Watts (1995); Hebauer & Hendrich (1999).

Subfamily Cylominae Zaitzev, 1908 (Figs 20.14–20.15) = Rygmodinae Orchymont, 1916

Characteristics. Antenna with 9 antennomeres*, antennal club loosely segmented*; gula wide; prosternum moderately long, without median carina*;scutellary stria absent; anapleural sutures well developed (except in *Petasopsis*); mesoventrite triangular, narrow anteriorly, flat or with low/small/narrow posteromedian elevation; anteromedian pit on mesoventrite absent or weakly developed (*Petasopsis*); abdomen with 5 ventrites; first abdominal ventrite without median carina*; abdomen without apical notch (but widened setae at abdominal apex are present in

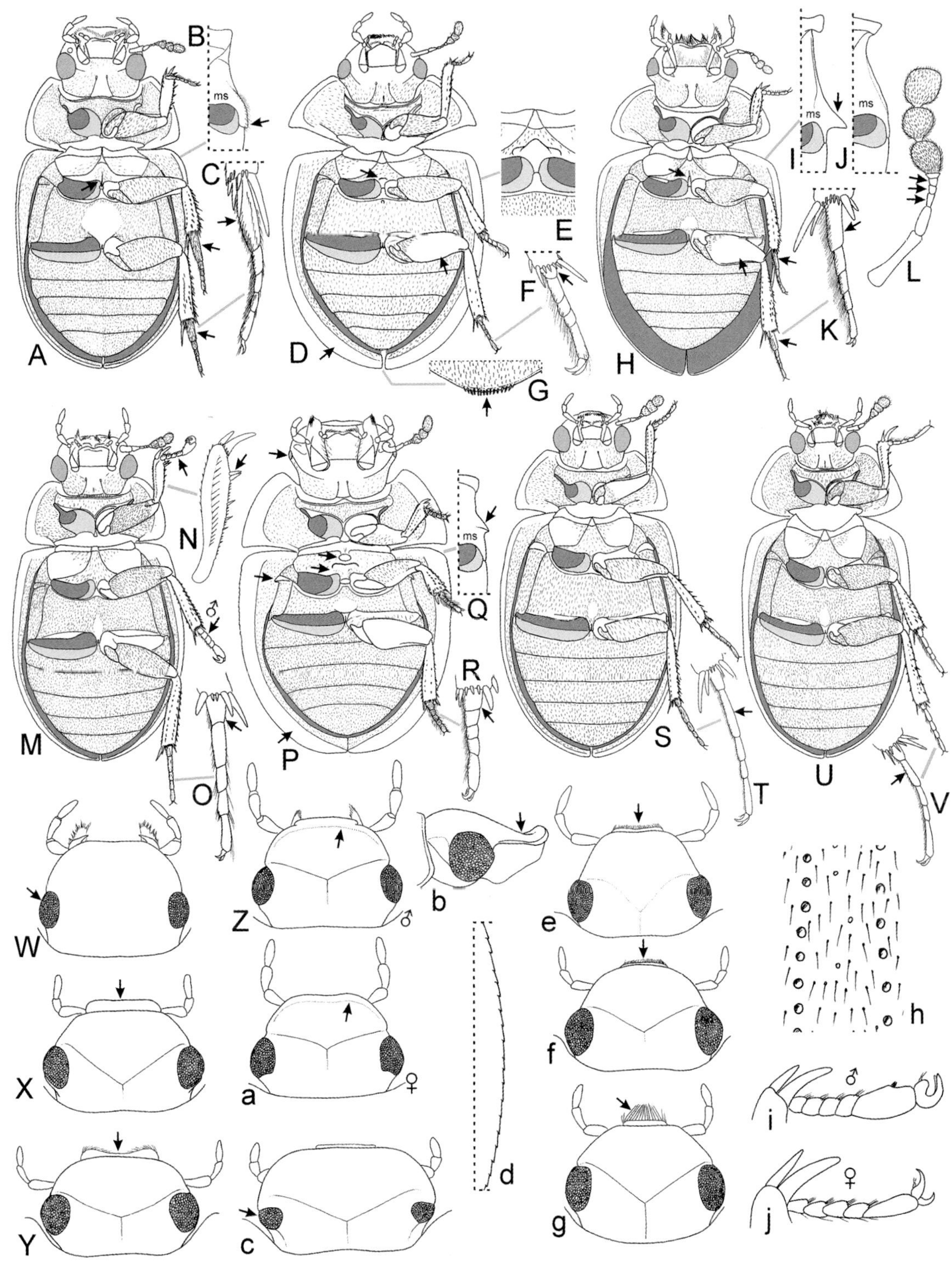

Fig. 20.14. Subfamily Cylominae, adults (A–Z, a–j). A–C, *Austrotypus nothofagi* Fikáček, Minoshima and Newton (A, ventral view; B, mesoventrite laterally; C, metatarsus). D–G, *Borborophorus* Hansen (D, ventral view of *B. tuberculus* Hansen; E, mesoventrite of *B. punctatus* Hansen; F, metatarsus; G, abdominal apex). H–L, *Coelostomopsis* Hansen (H–I, ventral view and mesoventrite laterally of *C. picea* Hansen; J, mesoventrite laterally of *C. major* Hansen; K, metatarsus; L, antenna). M–O, *Eurygmus helocharoides* Hansen (M, ventral view; N, protibia ventrally; O, metatarsus). P–R, *Petasopsis brevitarsis* Hansen (P, ventral view; Q, mesoventrite laterally; R, metatarsus), S–T, *Pseudohydrobius* cf. *flavus* Lea (S, ventral view; T, metatarsus). U–V, *Rygmostralia brunnea* Orchymont (U, ventral view; V, metatarsus). W–a, head dorsally (W, *Austrotypus*; X, *Borborophorus*; Y, *Coelostomopsis*; Z, *Eurygmus*, male; a, *Eurygmus*, female. b, male head of *Eurygmus*, lateral view. d, *Petasopsis*, lateral margin of elytron. e–g, head dorsally (e, *Pseudohydrobius floricola* Blackburn; f, *P.* cf. *flavus*; g, *Rygmostralia brunnea*). h, *Borborophorus*, elytral punctation and pubescence. i–j, protarsus of *Eurygmus* (i, male; j, female).

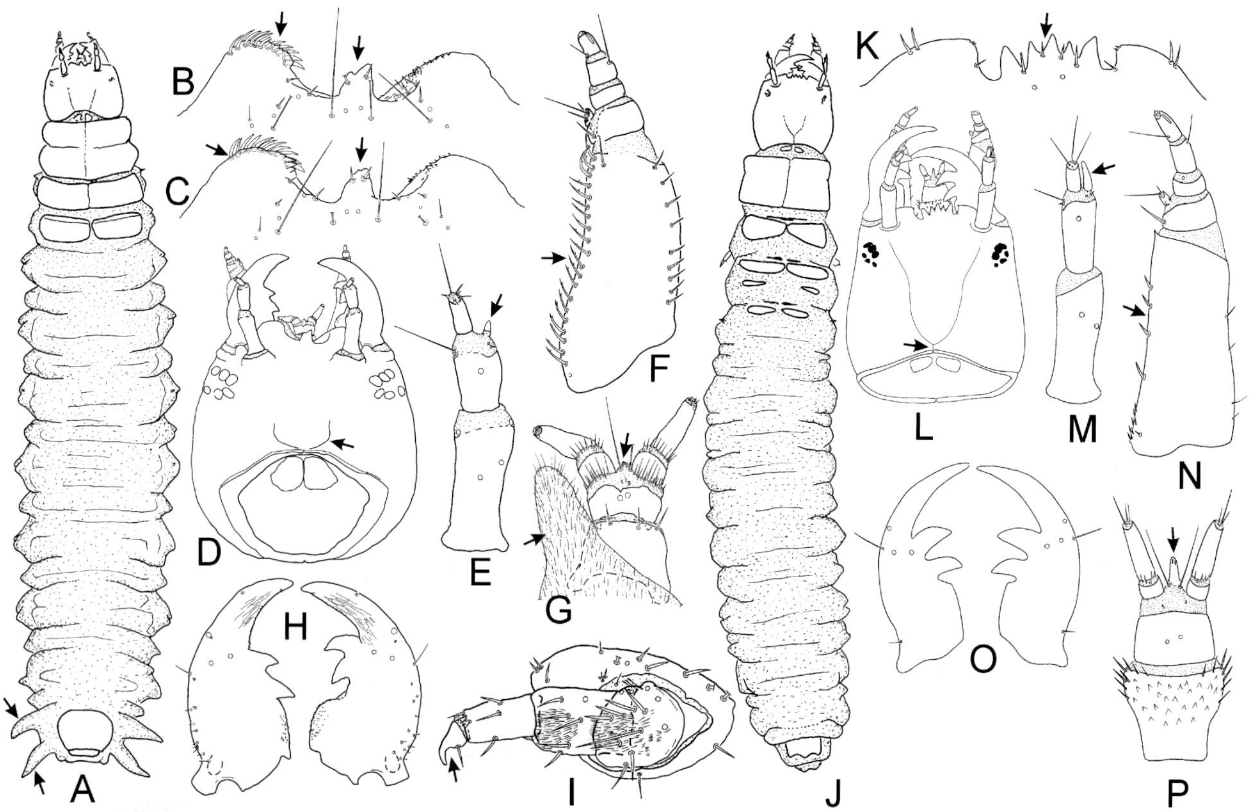

Fig. 20.15. Subfamily Cylominae, larvae. A–I, *Austrotypus nothofagi* Fikáček, Minoshima and Newton, 2014 (A, habitus; B, anterior margin of head, third instar; C, anterior margin of head, second instar; D, head dorsally E, antenna; F, maxilla; G, labium with hypopharyngeal lobe; H, mandibles; I, metathoracic leg). J–P, *Borborophorus* sp. (J, habitus; K, anterior margin of head; L, head dorsally; M, antenna; N, maxilla; O, mandibles; P, labium. A–I from Fikáček *et al.* (2014a).

Borborophorus); basal meso- and metatarsomere shorter than tarsomere 2 (much longer than following tarsomeres in *Austrotypus* and *Coelostomopsis*).

Larvae: described for few genera only (*Andotypus*, *Austrotypus*: Fikáček *et al.* 2014a; *Tormissus* and *Cyloma*: Ordish 1974; *Anticura* and *Cylomissus*: Minoshima *et al.* 2015; Cylorygmus: Seidel *et al.* 2018) and known and to be described also for all New Zealand genera except *Exydrus* (Minoshima and Fikáček, in prep.). All known larvae share the following characters: Frontal sulci converging posteriorly, coronal sulcus present and short or absent; nasale with one median projection or with 2–5 teeth, symmetrical to slightly asymmetrical; epistomal lobes small to large, symmetrical to slightly asymmetrical (in *Austrotypus* with asymmetrical chaetotaxy); mandibles symmetrical, with two inner teeth; palpifer shorter than maxillary palp; inner face of stipes with 5 to many setae; labium with ligula (may be very small: *Austrotypus*), short to longer than maxillary palps; abdomen with or without lateral projections, stigmatic atrium on segment 8 present, without spinose ventral prolegs; body narrowly elongate, whitish. Australian cylomines have two distinct larval types, resembling either Coelostomatini and Sphaeridiini (*Austrotypus*, see there for diagnosis) or Hydrobiusini (*Borborophorus* and likely most other Australian genera, see under *Borborophorus* for diagnosis); unknown larva of *Eurygmus* may differ from both these types based on the phylogenetic position of the genus (Fikáček, unpubl. data).

A moderately large group (19 genera, ~60 species) confined to southern hemisphere (Australia, New Zealand, Chile, South Africa; Seidel *et al.* 2016, 2018), non-Australian species are confined to temperate zone. Most species are terrestrial, few non-Australian ones are aquatic (*Anticura*, *Cylomissus*; Minoshima *et al.* 2015).

Note. Previous authors used the name Rygmodinae for the subfamily, but the transfer of the New Zealand genus *Cyloma* to the subfamily made by Short & Fikáček (2013) made this name a junior synonym of Cylominae. See Seidel *et al.* (2016) for details.

Australian taxa. There are seven genera in Australia, all but one (*Austrotypus*) endemic to Australia. All of them are terrestrial, some of them (*Pseudohydrobius*) visit flowers.

Austrotypus Fikáček, Minoshima and Newton, 2014 (Pl. 45G; Fig. 20.14A–C, W)

Type species. *A. nothofagi* Fikáček, Minoshima and Newton, 2014.

Characteristics. Elongate oval, moderately convex; anterior margin of clypeus not upturned in both sexes; eyes large ventrally, rather small dorsally; labrum not exposed; elytra with fine but distinct elytral series, epipleura wide only in anterior half; mesoventrite elevated slightly above metaventral process posteromesally; pro- and mesotarsi not widened in both sexes; metatarsomere 1 very long (about as long as metatarsomeres 2–4 combined); dorsum not distinctly pubescent; body usually dark to black with basal reddish spots/stripe on elytra; moderately convex; 3.6–5.2 mm. May be most easily confused with *Dactylosternum* from which it differs by well developed anapleural sutures, weakly elevated mesoventrite lacking anteromedian pit and first abdominal ventrite without median carina.

Larvae (Fig. 20.15A–I): Nasale simple to slightly bifid; epistomal lobes moderately large, nearly symmetrical, right one with numerous spines, left one bare; antenna with minute sensorium; maxillary stipes with numerous stout setae on inner face; mandibles slightly asymmetrical, each with 2 inner teeth; labium with indistinct ligula and large hypopharyngeal lobe; abdominal segment 8 with two pairs or lateral finger-like projections. in general habitus most similar to larva of *Sphaeridium*, may be distinguished by morphology of nasale and epistomal lobes, number of inner mandibular teeth and morphology of legs (well developed in *Austrotypus*, with reduced claw in *Sphaeridium*). By head morphology very similar to some *Dactylosternum*, which however always lack the two pairs of projections on abdominal segment 8.

Australian species. One species endemic for Australia, *A. nothofagi* Fikáček, Minoshima and Newton.

Distribution. Genus of widely vicariant distribution, known from Australia and high altitude forest in the Andes in Peru. Australian species locally common in moist forests in north-eastern New South Wales and extreme south-eastern Queensland.

Biology. Terrestrial species. Most commonly collected in pitfall traps baited with squid carrion, fermenting malt or human dung, or associated with rotting fruits on the ground; also captured in flight intercept traps. Found in moist forest habitats including relict cool temperate forests of *Nothofagus moorei* and nearby warm-temperate and subtropical rainforests. Larvae found with adults in rotting fruits on the ground and in rotting mushrooms in early December and late February. Adults probably saprophagous, associated with and probably feeding on decaying carrion, dung, fruits and fungi.

References. Newton (1989) (key to genera, as 'undescribed genus and species'); Hansen (1990b) (key to genera, as 'new genus, to be described by A. F. Newton'); Hansen (1991) (key to genera, morphology, as 'Genus A'); Fikáček *et al.* (2014a) (description of genus and species, morphology of larva, key to species).

Borborophorus Hansen, 1990 (Pl. 45H; Fig. 20D–G, X, h)

Type species. *Borborophorus pubescens* Hansen, 1990.

Characteristics. Body widely oval, moderately convex; clypeus not expanded anterior of eyes, not upturned anteriorly; labrum well sclerotised, exposed; eyes large and slightly protruding; anapleural sutures present; mesoventrite without anteromedian pit, with low to high and wide transverse posteromedal elevation; elytra widely explanate laterally, with distinct elytral series, elytral intervals flat; epipleura very wide throughout, lateral elytral margin straight; pro- and mesotarsi not enlarged in both sexes; metatarsomere 1 slightly shorter than metatarsomere 2; dorsum markedly pubescent; body reddish to dark brown; 3.1–3.6 mm. Very characteristic by widely oval body with explanate elytra and pubescent dorsum, hardly to be confused with another hydrophilid genus. Potentially it may be confused with genera of few coccinelloid or cucujoid families, but always easy to assigned to Hydrophilidae by head and antennal morphology.

Larvae (Fig. 20.15J–P): Frontal sulci U-shaped, coronal sulcus present but very short; nasale slightly asymmetrical, with 5 teeth; epistomal lobes symmetrical about as long as nasale; mandibles symmetrical with two inner teeth; palpifer much shorter than maxillary palp; inner face of stipes with 5 setae; ligula present but short, shorter than maxillary palps; abdomen without spiracular gills, stigmatic atrium on segment 8 present; body narrowly elongate, whitish. By nasale with 5 teeth and symmetrical mandibles resembling *Amphiops*, *Limnoxenus* and *Sternolophus*, but can be distinguished by mandibles with two inner teeth, mentum not apparently wider than prementum and antennal sensorium about as long as antennomere 3. (Diagnosis and figures based on larvae associated with adults from Brindle Creek, Wiangarie State Forest, collected by S. and J. Peck, deposited in ANIC Canberra; association considered as correct as the larvae agree in main characters with some New Zealand cylomine larvae realiably identified by DNA).

Australian species. Three species endemic to Australia.

Distribution. The genus is endemic to Australia, it is known from rainforests in north-eastern Queensland and *Nothofagus* forests southern Queensland and in New South Wales.

Biology. Terrestrial species. Available specimens were found in rainforest leaf litter, dung and at carrion, quite many were collected by pyrethrum fogging from branches and dead wood with fungi; the aforementioned larva and associated adults were sifted from log and bark litter.

References. Hansen (1990b) (genus description, key to species); Hansen (1991) (key to genera, morphology); Short & Fikáček (2013) (transferred to Cylominae).
Key to species. Hansen (1990b).

Coelostomopsis Hansen, 1990 (Pl. 45I; Fig. 20.14H–L, Y)

Type species. *Coelostomopsis picea* Hansen, 1990.
Characteristics. Anterior margin of clypeus not or very slightly upturned in both sexes; labrum membranous but partly exposed, sinuate on anterior margin; eyes large and slightly protruding; anapleural sutures present; mesoventrite with low posteromedian keel or sharp teeth-like carina; pro- and mesotarsi not widened in both sexes; elytra without distinct elytral series, only with sutural stria; epipleura wide only in anterior half; pro- and mesotarsi not enlarged in both sexes; metatarsomere 1 very long and wide (about as long as metatarsomeres 2–4 combined); dorsum not distinctly pubescent; body dark brown to black, without pale spots; moderately convex; 3.9–5.9 mm. By dorsal habitus it may be confused with *Coelostoma* from which it differs by well developed anapleural sutures, carinate mesoventrite lacking anteromedian pit, exposed anterior concave labrum and largely pubescent mesofemora. By tarsal morphology it resembles *Austrotypus*, but differs from it by elytra without puncture rows and by partly exposed labrum.
Larva unknown.
Australian species. Two species, both endemic for Australia.
Distribution. The genus is endemic to Australia, both species are known from rainforests in north-eastern Queensland
Biology. Terrestrial species. Little is known about ecology, most specimens were collected in various types of rainforest using flight intercept traps (incl. baited ones), few were found while brushing sticks or sieving leaf litter.
References. Hansen (1990b) (original description, key to genera and species); Hansen (1991) (generic morphology); Short & Fikáček (2013) (transferred to Cylominae).
Key to species. Hansen (1990b).

Eurygmus Hansen, 1990 (Pl. 45J; Fig. 20.14M–O, Z, a–b, i–j)

Type species. *Eurygmus helocharoides* Hansen, 1990.
Characteristics. Elongate oval, weakly convex; anterior margin of clypeus very distinctly upturned in both sexes (more markedly in male in which clypeus is angularly projecting anteriorly); eyes moderately large; anapleural sutures present; mesoventrite flat, without elevations or pits; elytra with distinct elytral series; dorsum not distinctly pubescent; epipleura moderately wide throughout; pro- and mesotarsi of males with enlarged tarsomere 5 and claws; protibia with two closely standing stout spine on outer margin; metatarsomere 1 shorter than metatarsomere 2; body yellowish to brown; 2.5–2.8 mm. By body shape it may be most easily confused with small Acidocerinae or Enochrinae (*Agraphydrus,* small *Helochares* or small *Enochrus*) from which it differs by very short maxillary palps (moreover not zig-zag and with basal notch in contrast to *Enochrus*), absence of scutellary stria, and combination of upturned anterior clypeal margin and enlarged male pro- an mesotarsus (both latter characters absent from aforementioned genera).
Larva unknown.
Australian species. Single species, *E. helocharoides* Hansen, endemic to Australia.
Distribution. The genus is endemic for Australia. Known from rainforests in north-west Queensland.
Ecology. Terrestrial species. Very little is known about biology, available specimens were collected in flight intercept traps.
References. Hansen (1990b) (original description, key to genera); Hansen (1991) (key to genera, morphology).

Petasopsis Hansen, 1990 (Pl. 45K; Fig. 20.14P–R, d)

Type species. *Petasopsis brevitarsis* Hansen, 1990.
Characteristics. Body widely oval, moderately convex; clypeus widely expanded anterior of eyes, not upturned anteriorly; eyes very small; prosternal process moderately wide; anapleural sutures absent; mesoventrite with anteromedian pit and transverse posteromedian ridge; metanepisternum with a teeth fitting a small cleft of epipleuron; elytra widely explanate laterally, with distinct elytral series, elytral intervals convex; epipleura very wide throughout, lateral elytral margin denticulate; hind wings absent; pro- and mesotarsi not enlarged in both sexes; metatarsomere 1 as long as metatarsomere 2 (tarsus very short); dorsum not markedly pubescent; body yellowish to brown; 1.9–2.0 mm. By small body size and some ventral characters it may be confused with some sphaeridiine genera, but differs from them by the combination of loosely segmented antennal club, mesoventrite without well defined elevated plate and the head shape which is unique among Australian hydrophilids.
Larva unknown.
Australian species. Single species, *P. brevitarsis* Hansen, endemic to Australia.
Distribution. The genus is endemic to Australia, occurring in rainforests in north-west Queensland, at elevations 1400–1600 m.
Biology. Terrestrial species. Available specimens were sifted from rainforest leaf litter.
Note. Hansen (1990b) described *Petasopsis* as close to *Borborophorus*, but discussed also the possibility that it may stand closer to some Sphaeridiinae. Later, Hansen (1991)

included it to the Borborophorini (now Cylominae) based on body form, wide epipleura and few other characters shared with *Borborophorus*, even though he mentioned that similarly looking taxa also occur in sphaeridiine tribe Coelostomatini. The morphology of *Petasopsis* is very derived and partly similar to some Coelostomatini, and the presence of the anteromedian pit on the mesoventrite is also more typical for Coelostomatini (present in most genera) than Cylominae (present in New Zealand genera *Adolopus* and *Cyloma* only). The assignment of *Petasopsis* to Cylominae needs to be tested by molecular data, but DNA-grade specimens were not available for the studies until now.

References. Hansen (1990b) (original description, key to genera); Hansen (1991) (key to genera, morphology); Short & Fikáček (2013) (transferred to Cylominae).

Pseudohydrobius Blackburn, 1898 (Pl. 45L–M; Fig. 20.14S–T, e–f)

Type species. *Pseudohydrobius floricola* Blackburn, 1898.

Characteristics. Body elongate oval, moderately convex; clypeus not expanded anterior of eyes, not upturned anteriorly; labrum membranous and largely covered by clypeus, without long setae on anterior margin; mandibles with well developed or reduced apex; eyes moderately large to large, slightly protruding; anapleural sutures present; mesoventrite flat without anteromedian pit or posteroventral elevation; elytra not explanate laterally, with elytral series which may be very fine and indistinct (*P. flavus* group), elytral intervals slightly convex to flat; epipleura moderately wide posteriorly, lateral elytral margin straight; pro- and mesotarsi not enlarged in both sexes; metatarsomere 1 much shorter than metatarsomere 2; dorsum not markedly pubescent; body reddish to brown; 3.5–6.5 mm. By body form and pale coloration it may be confused with some *Enochrus*, but differs from it by completely flat mesoventrite, short maxillary palps not being zig-zag and without basal notch on palpomere 3, and largely concealed labrum.

Larva unknown.

Australian species. The genus was not revised until now. Four species were described: larger *P. floricola* Blackburn bearing often reddish-dark pattern (Pl. 45M), and smaller uniformly yellowish/reddish species which belong to the *P. flavus* species complex (Pl. 45L): *P. flavus* Lea, *P. barrettii* Deane, 1936 (now considered as synonym of the latter) and *P. neogallicus* (Gentili). The *P. flavus* species complex contains ~5 valid species, some of which need to be described (Fikáček, unpubl. data).

Distribution. The genus is endemic for Australia, occurring in eastern parts of the continent (Queensland, New South Wales, Victoria).

Biology. Terrestrial species. Little is known about biology, except for the fact that both *P. floricola* and representatives of the *P. flavus* complex visit flowers (e.g. *Leptospermum*). Preliminary studies show that *Pseudohydrobius* is related to the New Zealand *Rygmodus* White which also visits flowers as adult but which larvae are aquatic (Minoshima *et al.* 2018). The larvae of *Pseudohydrobius* may be hence also aquatic.

References. Blackburn (1898) (original description); Lea (1919) (description of *P. flavus*); Deane (1936) (description of *P. barrletti*); Hansen (1990b) (generic diagnosis); Hansen (1991) (key to genera, morphology); Gentili (1996b) (description of *Crenitis neogallica*); Fikáček & Watts (2015) (transfer of *Crenitis neogallica* to *Pseudohydrobius*).

Rygmostralia Orchymont, 1933 (Pl. 45N–O; Fig. 20.14U–V, g)

Type species. *Rygmostralia brunnea* Orchymont, 1933.

Characteristics. Body elongate oval, moderately convex; clypeus not expanded anterior of eyes, slightly upturned anteriorly; labrum membranous and largely covered by clypeus, with long setae on anterior margin; mandibles with well developed apex; eyes large, slightly protruding; anapleural sutures present; mesoventrite flat without anteromedian pit or posteroventral elevation; elytra not explanate laterally, with fine elytral series or without elytral series, elytral intervals flat; epipleura narrow posteriorly, lateral elytral margin straight; pro- and mesotarsi not enlarged in both sexes; metatarsomere 1 much shorter than metatarsomere 2; dorsum not markedly pubescent; body uniformly reddish to brown or bicolored (pale anteriorly, dark posteriorly); 4.1–4.9 mm. Externally very similar to *Pseudohydrobius*, from which it differs by long setae on labrum and possibly also by well developed mandibles. From all other genera which may be habitually similar, *Rygmostralia* differs by completely flat mesoventrite, short maxillary palps and largely concealed labrum.

Larva unknown.

Australian species. Single described species, *R. brunnea* Orchymont (Pl. 45N). At least one additional species (Pl. 45O) has to be described.

Distribution. Genus endemic to Australia, *R. brunnea* described from New South Wales, the undescribed species from northern Queensland.

Biology. Terrestrial species, ecology unknown. Specimens of the undescribed species from Queensland were collected at light.

Note. *Rygmostralia* is very similar and closely related to *Pseudohydrobius*, and the relationships between these two genera need to be examined in detail.

References. d'Orchymont (1933) (original description); Hansen (1990b) (key to genera, basic information about the genus); Hansen (1991) (generic diagnosis).

Subfamily Sphaeridiinae Latreille, 1802 (Figs 20.16–20.21)

Characteristics. Morphologically rather diverse group, but sharing following characters: metatarsomere 1 distinctly longer than metatarsomere 2 (as long as metatarsomere 2 in taxa with shortened tarsi, e.g. some Omicrini); mesoventrite partly to completely fused to mesanepisterna (i.e. anapleural sutures obsolete or absent); mesoventrite with posteromedian elevation or plate (except *Sphaeridium*); scutellary stria absent; antennal club loosely segmented or compact.

Larvae are poorly known and no synapomorphies are recognised, but known larvae share following characters: nasale always with less than five teeth (often only with single median projection or totally without teeth); legs short, in many groups even with reduced claws or other segments, or totally absent; stemmata aggregated to one or two groups, or largely reduced to totally absent (Hayashi 1986; Archangelsky 1997; Hansen & Richardson 1998; Short & Fikáček 2013; Fikáček *et al.* 2015, 2018b).

Large and ecologically and morphologically diverse group (~90 genera and nearly 1000 described species) of terrestrial species mostly living in various kinds of decaying organic substrates including forest leaf litter and dung; few genera are associated with ants or termites (none known from Australia), few secondarily returned to aquatic environment (in Australia only genus *Coelostoma*).

Australian taxa. 21 genera and at least 56 species; several groups were still not revised in Australia and several species remains undescribed.

Tribe Omicrini Smetana, 1975 (Fig. 20.16)

Characteristics. Small to minute species (most Australian species 0.9–1.7 mm long, only *Microgioton* slightly larger), widely rounded and moderately to highly convex species characterised especially by modified head morphology (Fig. 20.16P–S): eyes small, situated on angulate sides of the head (large in *Psalitrus*) clypeus excised before eyes (antennal bases exposed) and oblique or vertical anteriorly, with laterally projecting lobes partly or completely divided from median portion by ridges and covering head from sides; frontal sutures not visible. Additional characters: labrum usually exposed (except *Psalitrus*), antenna with 8–9 antennomeres; antennal club usually compact (loosely segmented in *Psalitrus*); prosternum very short in front of procoxae; epipleuron moderately to very wide throughout.

Larvae: only known for three genera: *Peratogonus* (incomplete description by Hayashi 1986), *Omicrus* (Hansen and Richardson 1998) and unidentified larva associated with *Noteropagus* and *Paromicrus* (illustrated in Figs 20.16b–b, e, m–o, based on specimens from Lake Barrine, Northern Queensland, collected on 29.vii.1982 by S. and J. Peck in rainforest bark and log litter, deposited in ANIC). They may be characterised as follows: nasale with 3–4 teeth, symmetrical to partly asymmetrical; epistomal lobes well developed, reaching about as far as nasale, symmetrical; mandibles symmetrical, with 2 inner teeth (inner face sometimes angulate basally and forming additional small basal tooth); maxilla with long stipes, inner face with 5 stout setae or series of many stout setae; inner appendage of palpifer well developed; antenna with sensorium as long as antennomere 3; labium only weakly protruding, ligula absent, hypopharyngeal lobe absent; stemmata reduced and aggregated in two small groups or totally absent; frontal sulci not converging posteriorly, coronal sulcus absent; sclerites of meso- and metathorax narrow and transverse; abdomen without larger tergites except on segment 8; stigmatic atrium present, its median lobe bilobed or trilobed; legs short but with all segments distinct and a well developed claw.

Poorly known group with 15 genera and ~110 described species distributed world-wide in the tropics, mostly living in leaf litter, sometimes associated with decaying banana trunks (Jia *et al.* 2015) or bromeliads (Hansen & Richardson 1998; Albertoni & Fikáček 2014), *Tylomicrus* and *Oreomicrus* living in termite nests (Fikáček, unpubl. data). Omicrines are sometimes characterised by highly derived mouthparts with adapted labium and reduced mandibles, although this was so far examined in few genera only (e.g. *Aculomicrus*, *Nannomicrus*, *Psalitrus*: Smetana 1975; Bameul 1992), additional studies are necessary to confirm that the modification is present and same in all omicrines.

Australian taxa. Four genera are present in Australia. The fauna was not revised, hence number of species living in Australia is unknown.

Mircogioton Orchymont, 1937 (Pl. 46A; Fig. 20.16A–E, P)

Type species. *Mircogioton coomani* Orchymont, 1937.

Characteristics. Body widely oval, moderately convex; eyes small; lateral lobes of clypeus large and not divided by complete ridge; labrum exposed, angulate or bisinuate on anterior margin; antenna with 9 antennomeres, antennal club compact; prosternum carinate medially; antennal grooves absent; elytra with very fine (and quite indistinct) series of punctures and without sutural stria, epipleuron moderately wide throughout; mesoventral plate narrowly oval (in some species projecting posteriorly over metaventrite); mesocoxal cavities narrowly separated; metaventrite short, pubescent; abdominal ventrite 1 without median carina; metatarsomere 1 distinctly longer than metatarsomere 2; body uniformly brownish; 2.4–2.7 mm. May be confused with some megasternine genera with narrowly oval mesoventral plate, differs from them by

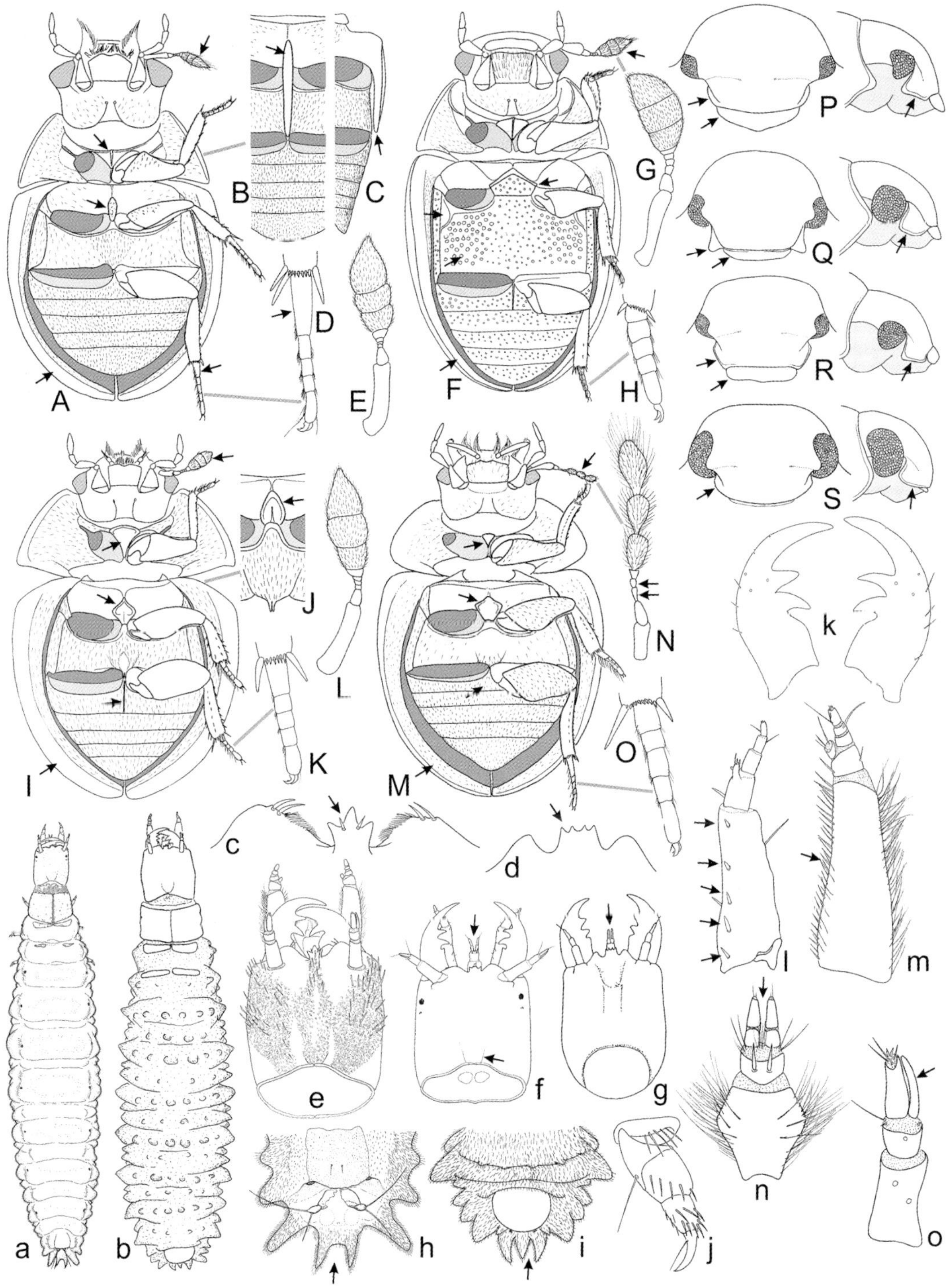

Fig. 20.16. Tribe Omicrini, adults of Australian genera (A–S) and larvae of Australian and non-Australian genera (a–o). A–E, *Microgioton* sp. (A, ventral habitus; B–C, ventral and lateral meso- and metathorax of *Mircogioton* cf. *huonus* Hebauer; D, metatarsus; E, antenna). F–H, *Noteropagus* sp. (F, ventral habitus; G, antenna; H, metatarsus). I–L, *Paromicrus* sp. (I, ventral habitus; J, ventral view of meso- and metathorax of another species; K, metatarsus; L, antenna). M–O, *Psalitrus* sp. (M, ventral view; N, antenna; O, metatarsus). P–S, head in dorsal and lateral view (P, *Mircogioton* sp.; Q, *Noteropagus* sp.; R, *Paromicrus* sp.; S, *Psalitrus* sp.). a–b, larval habitus (a, *Omicrus ingens* Hansen and Richardson; b, unidentified larva asociated with *Noteropagus* and *Paromicrus*). c–d, anterior margin of head (c, *Noteropagus*/*Paromicrus*; d, *Peratogonus*). e–g, head in dorsal view (e, *Noteropagus*/*Paromicrus*; f, *Omicrus*; g, *Peratogonus*). h–i, abdominal apex (h, *Omicrus*; j, *Noteropagus*/*Paromicrus*). j–k, larva of *Noteropagus*/*Paromicrus* (j, mesothoracic leg; k, mandibles). l, maxilla of *Omicrus ingens*. m–o, larva of *Noteropagus*/*Paromicrus* (m, maxilla; n, labium; o, antenna). a, f, h, l from Hansen and Richardson (1998); d, g adapted from Hayashi (1986).

wide epiepleura, very short prosternum in front of procoxae, absence of antennal grooves and abdominal ventrite 1 without median carina. The posteriorly projecting mesoventral plate of some species is unique, and these species hence cannot be confused with any other hydrophilid genera.

Larvae: Unknown.

Australian species. Australian fauna is not revised, but at least two species are present in material examined. They are likely close to the New Guinean species which were revised by Hebauer (2006), one of the species has posteriorly projecting mesoventral plate as in Fig. 20.16B–C, and is likely close to *M. huonus* Hebauer.

Distribution. The genus is known from the Oriental Region, few species were described from New Guinea. Examined Australian specimens were collected in northern Queensland.

Biology. Examined specimens were collected from under bark on log with fungi and by spraying the bark. Oriental *M. coomani* Orchymont was collected from decaying banana trunks (Jia *et al.* 2015).

References. Hansen (1990b) (key to genera of Australian Sphaeridiinae); Hansen (1991) (generic diagnosis); Hebauer (2006) (description of New Guinean species and key to species).

Key to species. Not available for Australian species. Key to described species provided by Hebauer (2006).

Noteropagus Orchymont, 1919 (Pl. 46B; Fig. 20.16F–H, Q)

Type species. *Noteropagus politus* Orchymont, 1919.

Characteristics. Body oval, moderately convex; eyes moderately large; lateral lobes of clypeus large, divided by complete or incomplete ridge; labrum exposed, arcuate on anterior margin; antenna with 9 antennomeres, antennal club compact; prosternum roof-like and carinate medially; antennal grooves present and large; elytra with 10 distinct puncture rows or striae; epipleuron moderately wide throughout, pubescent portion reduced posteriorly; mesoventral plate widely pentagonal, wider than long; mesocoxal cavities widely separated; metaventrite long, base with coarse punctures laterally, with anterolateral ridges; abdominal ventrite 1 with median carina; all tarsomeres of metatarsus subequal in length; body uniformly brown to black; 0.9–1.2 mm By the ventral morphology, it may be easily confused with megasternine genera *Cryptopleurum* and *Pachysternum*, but differs from them by roof-like carinate prosternum, lateral portions of pronotum not deflexed and not visible in ventral view, moderately wide epipleuron and metanepisternum of the same width throughout.

Larvae: Unknown. Larva illustrated on Fig. 20.16 was associated with adults of *Noteropagus* and *Paromicrus*.

Australian species. Australian fauna is not revised. The taxonomy of the genus is little studied and even non-Australian species cannot be identified easily.

Distribution. The genus is known from SE Asia, Malay Archipelago and New Guinea, and from islands in the Pacific (Hawaii, Samoa, Fiji) and Indian oceans (Madagascar, Mascarene Is.). Examined Australian specimens were collected in northern Queensland.

Biology. Examined specimens were collected from under bark on log with fungi and in rainforest bark and fungi litter. Oriental species were repeatedly collected in longer series in cut off standing decaying banana trunks (Jia and Fikáček, unpubl. data). Hawaiian and Mascarene specimens were reported from wet decaying banana trunks, papaya stems and agave, under bark of mango trees and palms, and in garden compost (Bameul 1986; Hansen 1995).

References. d'Orchymont (1932) (key to known species); Hansen (1990b) (key to genera of Australian Sphaeridiinae); Hansen (1991) (generic diagnosis).

Key to species. Not available for Australian species. Identification key to known species was provided by d'Orchymont (1932).

Paromicrus Scott, 1913 (Pl. 46C–D; Fig. 20.16I–L, R)

Type species. *Paromicrus carinatus* Scott, 1913.

Characteristics. Body oval, moderately to highly convex; eyes small; lateral lobes of clypeus large, divided by incomplete ridge; labrum exposed, straight to sinuate on anterior margin; antenna with 9 antennomeres, antennal club compact; prosternum without median carina; antennal grooves absent; elytra with 10 puncture rows or without series, narrowly to widely explanate laterally; epipleuron moderately to very wide throughout; mesoventral plate subtriangular or subpentagonal, about as long as wide or longer than wide; mesocoxal cavities narrowly separated; metaventrite short, largely pubescent; abdominal ventrite 1 with median carina; metatarsomere 1 slightly longer than metatarsomere 2 (tarsi shortened); body brown to dark brown, sometimes paler laterally; 1.0–1.7 mm. By widely oval explanate elytra it may resemble *Borborophorus* or some *Dactylosternum*, but easily differs from them by small body size, clypeus excised above antennal bases and with large lateral lobes, and compact antennal club. From *Psalitrus*, which has similar mesoventral plate, it differs by exposed labrum and compact antennal club.

Larvae: Unknown. Larva illustrated on Fig. 20.16 was associated with adults of *Noteropagus* and *Paromicrus*.

Australian species. Australian fauna is not revised, but at least three likely undescribed species were recognised in material examined by Hansen (1990b). One of them differs from other known species by subtriangular mesoventral plate (Fig. 20.16J) and widely explanate elytra (Pl. 46C). The taxonomy of the genus is little studied and even non-Australian species cannot be identified easily.

Distribution. The genus is distributed in Mascarene Islands, South-east Asia, Malay Archipelago and New Guinea. Australian specimens are from northern New South Wales and Queensland.

Biology. Australian specimens as well as those from Sulawesi were collected from various types of forest leaf litter and decaying logs (Hansen 1990b; Bameul 1993). Specimens from Seychelles were collected among leaf bases of various palms and under bark of partly decayed tree trunk (Scott 1913).

References. Scott (1913) (description of the genus); Hansen (1990b) (notes on Australian species); Hansen (1991) (generic diagnosis).

Key to species. Not available for Australian species. Bameul (1993) provided key to the species from Sulawesi; no other areas close to Australia were revised.

Psalitrus Orchymont, 1919 (Pl. 46E–F; Fig. 20.16M–O, S)

Type species. *Psalitrus vandenbosscheae* Orchymont, 1919.

Characteristics. Body oval, moderately to highly convex, eyes rather large; lateral lobes of clypeus moderately large, divided by incomplete ridge; labrum concealed by clypeus; antenna with 8 antennomeres, antennal club loosely segmented; prosternum without median carina, largely reduced except median triangular portion; antennal grooves absent; elytra with more than 10 serially arranged puncture rows (and no interval punctures!), narrowly explanate laterally, epipleuron very wide throughout; mesoventral plate subpentagonal, about as long as wide; mesocoxal cavities moderately separated; metaventrite short, with reduced pubescence; abdominal ventrite 1 without median carina; metatarsomeres subsequal in length (tarsi shortened); body dark brown or pale with a dark spot on each elytron; 1.0–1.4 mm. The combination of loosely segmented antennal club, pentagonal mesoventral plate and wide epipleura is unique in Hydrophilidae and *Psalitrus* is hence difficult to be confused with other genera. For diagnosis from somewhat similar *Paromicrus* see there. From *Borborophorus*, which also has loosely segmented antenna and wide epipleuron, it differs by largely reduced prosternum, bare dorsum and clypeus excised above antennal bases.

Larvae: Unknown.

Australian species. Australian fauna is not revised, but at least three likely undescribed species were recognised in material examined by Hansen (1990b).

Distribution. The genus is widely distributed in Afrotropical and Oriental Regions and reaches to New Guinea, Australia and New Caledonia, with 36 described species. In Australia *Psalitrus* is known from Northern Territory, Queensland and New South Wales (Hansen 1990b).

Biology. Australian specimens were mainly collected from rainforest log and leaf litter, but also under rotten bark and in moss (Hansen 1990b). Indian, Sri Lankan and New Caledonian species were mostly sifted from rainforest litter (Bameul 1992; Fikáček 2010b).

References. Hansen (1990b) (notes on Australian species); Hansen (1991) (generic diagnosis).

Key to species. Not available for Australian species. Key to New Caledonian species was provided by Fikáček (2010b); New Guinean and Malayan species were not revised.

Tribe Coelostomatini Heyden, 1891 (Fig. 20.17)

Characteristics. Australian species easy to distinguish by medium-sized body and characteristic arrow-head shape of mesoventral plate with distinct anteromedian pit. Additional characters: body moderately widely rounded* and weakly to highly convex species; head with more or less distinct frontal lines; clypeus not excised laterally, covering antennal bases, not excised anteriorly; eyes small to large, sometimes slightly protruding, excised anteriorly or not; antenna with 9 antennomeres*, antennal club loosely segmented to compact; maxillary palps short; labrum usually largely retracted under clypeus, sometimes exposed (some *Dactylosternum*); prosternum short to moderately long, flat, carinate or with median elevation; anapleural suture obsolete or completely absent; mesoventrite with anteromedian groove*, posteromedian elevation arrow-head-shaped; elytron with 10–11 puncture rows, or without series and with impressed sutural stria*; epipleuron present throughout, narrow to wide posteriorly; abdomen with 5 ventrites, first one carinate or not, last one entire on apex*; femora without dense pubescence; metatarsomere 1 much longer than metatarsomere 2.

Larvae: known for few genera only: *Hydroglobus* (Archangelsky & Fernández 1994, as *Phaenonotum*), *Phaenonotum* (Archangelsky 1997; Archangelsky *et al.* 2016), *Coelostoma* (Hrbáček 1943; Fikáček unpubl. data), *Dactylosternum* (Costa *et al.* 1988; Archangelsky 1997; De Marzo 2000; Archangelsky *et al.* 2016) and *Lachnodacnum* (Clarkson *et al.* 2014). Larvae are rather uniform in habitus, but have rather variable head morphology, especially in the genus *Dactylosternum* (see under that genus for details). Larvae may be characterised as follows: frontal sulci convergning posteriorly, coronal sulcus very short or absent; nasale in form of single subtriangular or subrectangular symmetrical to slightly asymmetrical projection, without teeth; epistomal lobes moderately large, symmetrical to slightly asymmetrical; head appendages variable, but maxilla always with long stipes with row of many stout setae on inner face, inner projection of palpifer well developed; legs short but with well developed

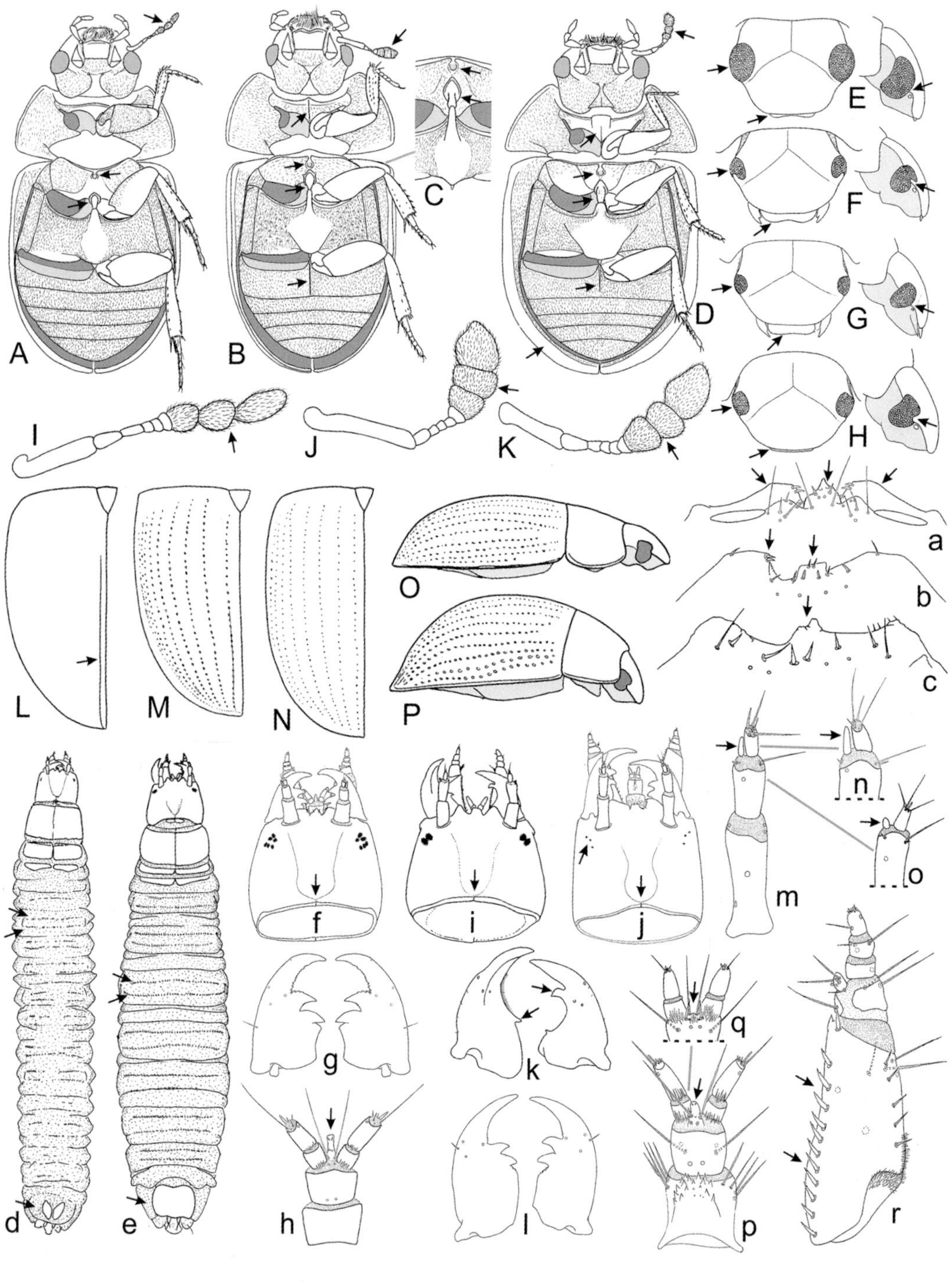

Fig. 20.17. Tribe Coelostomatini, adults of Australian species (A–P) and larvae of Australian and non-Australian species (a–r). A–D, ventral morphology (A, *Coelostoma fabricii* (Montrouzier); B, *Dactylosternum abdominale* (Fabricius); C, *D. hydrophiloides* (MacLeay); D, *D. marginale* (Sharp)). E–F, head in dorsal and lateral views (E, *Coelostoma fabricii*; F, *Dactylosternum abdominale*; G, *D. dytiscoides* (Fabricius); H, *D. marginale*). I–J, antenna (I, *Coelostoma fabricii*; J, *Dactylosternum abdominale*; K, *D. marginale*). L–N, elytron (L, *Coelostoma fabricii*; M, *Dactylosternum marginale*; N, *D. hydrophiloides*). O–P, body in lateral view (O, *Dactylosternum abdominale*; P, *D. marginale*). a–c, anterior margin of larval head (a, *Coelostoma orbiculare* (Fabricius); b, *Dactylosternum dytiscoides*; c, *D. cacti* (LeConte)). d–e, habitus (d, *Coelostoma orbiculare*; e, *Dactylosternum abdominale*). f–h, *Coelostoma orbiculare* (f, head; g, mandibles; h, labium). i–l, larvae of *Dactylosternum* (i, head of *D. abdominale*; j, head of *D. dytiscoides*; k, mandibles of *D. abdominale*; l, mandibles of *D. dytiscoides*). m–o, antenna (m, *Dactylosternum* sp. B; n, *Coelostoma orbiculare*; o, *Dactylosternum cacti*). p–q, labium of *Dactylosternum* (p, *D.* sp. B; q, *D. cacti*). r, maxilla of *Dactylosternum* sp. B. c, m, o, p–r from Archangelsky (1997); e, i, k adapted from De Marzo (2000).

segments and claw; abdomen membranous, often with two transverse series of asperities on each segment ventrally and/or dorsally, or with spinose prolegs ventrally on segments 2–7; tergite of abdominal segment 8 entire or subdivided mesally (may vary between instars: entire in first instar, divided in higher instars; Clarkson *et al.* 2014); spiracular atrium present.

A moderately large group of 17 genera and ~230 described species distributed world-wide, most diverse in tropical areas. Most species are found in various types of decaying plant material (rotten banana and tree trunks, garden compost), some inhabit forest leaf litter, some genera are secondarily aquatic (e.g. *Coelostoma*), in some cases specialised in phytotelmata (bromeliads: *Lachnodacnum*, heliconia inflorescences and watered mesenchyme of living banana trunks: some *Dactylosternum*; Clarkson *et al.* 2014; Fikáček unpubl. data). Some species associated with rotten plant material are often accidentally introduced e.g. with plant and fruit transports, few *Dactylosternum* species easily establish and are widely distributed at the moment.

Australian taxa. In Australia, the fauna of Coelostomatini is very poor and mostly includes widely distributed species. Two genera and five species are recorded.

Coelostoma Brullé, 1835 (Pl. 46J; Fig. 20.17A, E, I, L)

Type species. *Hydrophilus orbicularis* Fabricius, 1775.

Characteristics. Body oval, moderately convex; eyes large, protruding, not excised anteriorly; labrum membranous and largely retracted under clypeus; antennal club loosely segmented; elytron without puncture rows, with sharply impressed sutural stria; abdominal ventrite 1 without median carina; uniformly black; 5.7–5.9 mm. By medium size, black coloration and elytra without puncture rows but with sutural stria it resembles *Coelostomopsis* and some *Enochrus* (especially *E. mastersi*). For diagnosis from *Coelostomopsis* see under that genus, from *Enochrus* it differs by non-exposed labrum, entire clypeus, short maxillary palps without basal notch on palpomere 3, anteriorly obsolete anapleural sutures and arrow-head shaped mesoventral projection.

Larvae (Fig. 20.17a, d, f–h, n): coronal sulcus short; stemmata well developed and well separated; nasale triangular, nearly symmetrical, epistomal lobes symmetrical; mandibles symmetrical, with two inner teeth; antennal sensorium as long as antennomere 3; labium without hypopharyngeal lobe, ligula well developed, longer than labial palpomere 1; abdomen without prolegs and with rather indistinct transverse rows of asperites [based on European *C. orbiculare*, larva of Australian species unknown]. Based on triangular nasale and stipes with many stout setae on inner face easily recognisable from other Australian hydrophilid larvae.

Australian species. Single species, *C. fabricii* Montrouzier.

Subgenera. Four subgenera are recognised at the moment, defined by combination of characters of femora, abdominal apex and claws (for key see Mouchamps 1958 and Jia *et al.* 2014). The only Australian species is classified in *Coelostoma* s.str. Preliminary molecular studies show that this subgenus is likely paraphyletic in respect to some other subgenera (Sýkora and Fikáček, unpubl. data).

Distribution. The genus is widespread in Old World, with many species in Afrotropical and Oriental regions, and few species reaching Palaearctic. *Coelostoma fabricii* is widespread in New Guinea, New Caledonia, Vanuatu and Australia (recorded from all territories), also recorded from Malay Archipelago (Sumatra, Borneo) and continental Asia (but continental records need revision; Hansen 1999a; Jia *et al.* 2014).

Biology. Aquatic, inhabiting various types of well vegetated standing waters, frequently comming to light.

References. Hansen (1990b) (comments on Australian species); Hansen (1991) (generic diagnosis); Fikáček (2010b) (redescription of *C. fabricii*, illustration of genitalia).

Dactylosternum Wollaston, 1854 (Pl. 46G–I; Fig. 20.17B–D, F H, J–K, M–P)

Type species. *Dactylosternum roussetii* Wollaston, 1854 (= *Sphaeridium abdominale* Fabricius).

Characteristics. Morphologically a very diverse genus, Australian species largely differing from each other: body oval to widely oval, rather depressed (*D. abdominale*) to moderately convex; eyes small, not protruding, emarginate anteriorly or not (*D. dytiscoides*); labrum well sclerotised and exposed (*D. dytiscoides*) or membranous and largely retracted under clypeus (*D. marginale*); antenna with 9 antennomeres, antennal club more or less compact; prosternum moderately long, simply carinate or with median elevation (*D. marginale*); anapleural sutures usually absent; mesoventrite with anteromedian pit and arrow-head shaped posteromedian elevation; anterior metaventral process usually elevated above surface of metaventrite (not elevated in *D. marginale*); elytron with 10–11 elytral series (fine and rather indistinct in *D. dytiscoides*), narrowly to widely (in *D. marginale*) explanate; epipleura very narrow (*D. abdominale*) to very wide posteriorly (*D. marginale*); abdominal ventrite 1 with median carina; abdominal apex without emargination; metatarsomere 1 usually much longer than metatarsomere 2, only slightly longer in *D. marginale*; body uniformly brown to black, or bicolored (in *D. dytiscoides*); 4.0–7.7 mm. By combination of medium size, arrow-head shape mesoventral plate, anteromedian pit on mesoventrite and elytra with puncture rows easily recognisable from other Australian hydrophilids; by body form resembling *Austrotypus*, see under that genus for diagnostic characters.

Larvae (Fig. 20.17b–c, e, i–m, o, p–r): morphologically rather diverse especially in head morphology: coronal sulcus short or absent; stemmata aggregated in two groups or well separated but largely reduced; nasale subtriangular or subrectangular, slightly asymmetrical, epistomal lobes symmetrical or slightly asymmetrical; mandibles symmetrical with two inner teeth or asymmetrical, with left mandible without teeth except small basal denticle, and right mandible with one inner tooth; antennal sensorium as long as or much shorter than antennomere 3; stipes long with row of many stout setae on inner face; labium with or without hypopharyngeal lobe, ligula well developed and longer than labial palpomere 1, or absent; with transverse rows of asperites dorsally (based on descriptions of three American species by Archangelsky 1997, Costa *et al.* 1988 and Archangelsky *et al.* 2016, incomplete description of *D. abdominale* by De Marzo 2000, and the larvae collected from rotten banana trunks in association with adults of *D. dytiscoides* in Indonesia, Papua, Jayapura distr., Sentani env. by J. Hájek, deposited in Prague Museum; larvae of *D. marginale* and *D. hydrophiloides* unknown). Larvae of the morphotype with symmetrical mandibles (in Australia represented by *D. dytiscoides*) very similar to those of *Coelostoma*, but seem to differ in reduced or largely reduced stemmata, subrectanglar nasale and slightly asymmetrical epistomal lobes. Larvae of the morphotype with asymmetrical mandibles, reduced ligula and hypopharyngeal lobe (in Australia represented by *D. abdominale*) resemble larvae of *Sphaeridium* and Megasternini, but differ from them by well developed legs (incl. claw), frontal sulci converging posteriorly, and well defined nasale and epistomal lobes without setose area on the left side.

Australian species. Five species are known from Australia: *D. marginale* (Sharp), *D. hydrophiloides* (MacLeay), *D. dytiscoides* (Fabricius), *D. abdominale* (Fabricius), and *D. subquadratum* (Fairmaire).

Distribution. *Dactylosternum* is distributed in world-wide tropics and subtropics, with 77 described species. Of the five Australian species, two are probably native: *D. marginale* is widespread in eastern Australia (New South Wales, Queensland, later introduced to New Zealand and Norfolk Island; Newton 1989), and *D. dytiscoides* (widespread in Malay Archipelago, New Guinea and adjacent islands, in Australia in northern Queensland; Newton 1989). *Dactylosternum abdominale* is currently worldwide synantropic species probably accidentally introduced to Australia at the begining of 20th Century at the latest and now recorded from all regions except Tasmania and Northern Territory (Newton 1989; Hansen 1999a). *Dactylosternum hydrophiloides* was probably purposely introduced to Australia (Hansen 1999a). *Dactylosternum subquadratum* is a species distributed in New Guinea, adjacent islands and Pacific islands, which is currently known from Australia by a single historical specimen from Gyandah (coll. Institute Royal des Sciences Naturelles in Brussels, Belgium).

Biology. All species collected from various types of decaying plant material, e.g. decaying banana and papaya trunks (adults and larvae of *D. dytiscoides* described illustrated here were collected from banana trunks), decaying logs, and in garden compost; *D. marginale* occurs in these habitats in native tropical, subtropical and temperate forests; *D. abdominale* is found synantropically, sometimes also in dung and poultry droppings.

Notes. Despite the morphologically diverse adults and larvae, preliminary molecular analyses did not confirm the expected polyphyly of the genus so far. Additional studies are needed to understand the morphological evolution of *Dactylosternum*. *D. marginale* seems to stand close to endemic Pacific/New Caledonian species based on these studies (Sýkora & Fikáček unpubl. data).

References. Newton (1989) (revision of Australian fauna); Hansen (1991) (generic diagnosis); Hansen (1999a) (*D. hydrophiloides* recorded from Australia).

Key to species. Australian species keyed by Newton (1989), but *D. hydrophiloides* and *D. subquadratum* are not included; the key to Hawaiian *Dactylosternum* (Hansen 1995) contains all Australian species except *D. marginale*. Genitalia of *D. marginale*, *D. abdominale* and *D. hydrophiloides* are illustrated by Newton (1989), Smetana (1978) and Satô (1979), respectively.

Tribe Sphaeridiini Latreille, 1802 (Fig. 20.18)

Characteristics. Body widely oval, rather depressed; clypeus covering antennal bases; eyes small, anteriorly excised; labrum exposed, well sclerotised; antenna with 8 antennomeres, scape very long, antennal club compact; male maxilla with ventral sucking disc; prosternum moderately long, without median carina, prosternal process widely triangular and bearing spines; antennal grooves absent; anapleural sutures absent; mesoventrite bulged and with numerous spines posteromesally; metaventrite densely pubescent laterally, with distinct median discrimen; elytra separately rounded posteriorly, with 10 rather indistinct puncture rows; epipleura narrow; abdomen with 5 ventrites, ventrite 1 without median carina, abdominal apex narrowly exposed behind elytral tip, ventrite 5 without notch on posterior margin; tibiae with numerous long spines; protarsus sexually dimorphic (except in Brazilian *S. braziliense* van Berge Henegouwen), shortened and with expanded tarsomere 5 and one claw in males; metatarsomere 1 much longer than metatarsomere 2; body black with yellow apical thirds of elytra and pronotal margins* (in non-Australian species often also with reddish humeral spot); Australian species 5.5–7.6 mm long.

Larvae (Fig. 20.18a–i): Frontal sulci not converging posteriorly; nasale in form of small projection, epistomal lobes

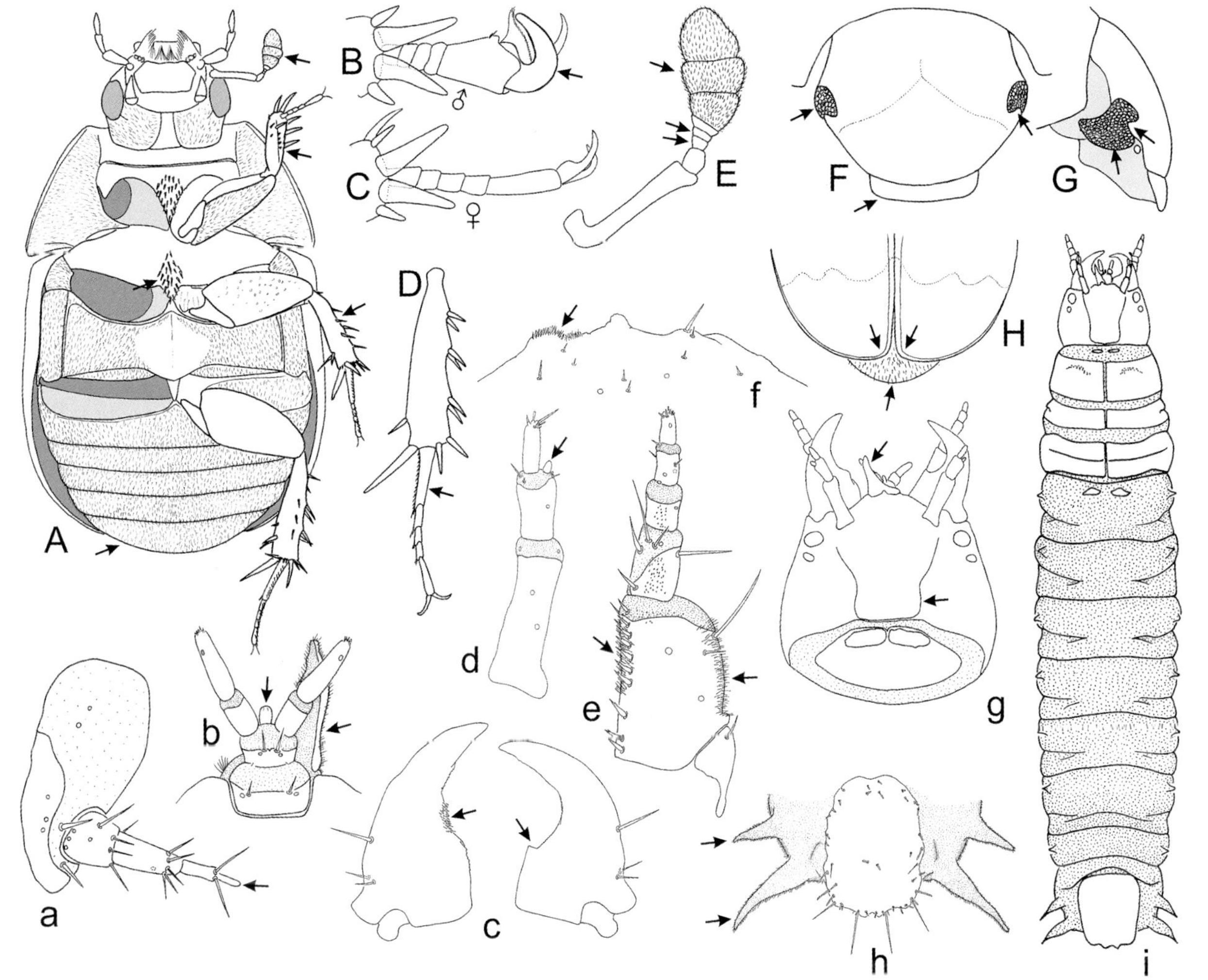

Fig. 20.18. Tribe Sphaeridiini, genus *Sphaeridium* Fabricius, adults (A–H, based on *S. discolor* Orchymont) and larva (a–i, based on non-Australian *S. scarabaeoides* (Linnaeus)). A, ventral view; B–C, male and female protarsus; D, metatibia and tarsus; E, antenna; F, head in dorsal view; G, head in lateral view; H, apical portion of elytra. a, prothoracic leg of larva; b, larval labium; c, larval mandibles; d, larval antenna; e, larval maxilla; f, anterior margin of larval head; g, halrval head; g, abdominal apex; I, habitus of the larva. a–i from Archangelsky (1997).

very low, i.e. anterior margin of head weakly lobate, without teeth, left part with setose area; stemmata aggregated into two groups; mandibles asymmetrical, left without any tooth, right with small basal denticle only; antennal sensorium small, much shorter than antennomere 3; maxilla with stipes rather short, with series of many stout setae on inner face; palpifer rather long, inner process reduced and membranous; labium with large memranous asymmetrical hypopharyngeal lobe, ligula present but small; abdominal segment 8 with two pairs of lateral projections; legs short, with reduced claw (based on larvae of *Sphaeridium scarabaeoides*, *S. bipustulatum* and *S. caffrum* (Prins 1984; Archangelsky 1997; De Marzo 2000), larva of Australian species unknown).

Australian taxa. Single species of the genus *Sphaeridium* Fabricius.

Sphaeridium Fabricius, 1775 (Pl. 46K; Fig. 20.18A–H)

Type species. *Dermestes scarabaeoides* Linnaeus, 1758.

Characteristics. As for the tribe. Adults may be easily recognised from all other Australian hydrophilids by medium body size, characteristic coloration, legs with spinose tibiae and males with enlarged anterior claws.

Larvae (Fig. 20.18a–i): See the tribe for diagnostic characters. By general habitus, *Sphaeridium* larvae resemble those of *Austrotypus*, from which they differ by mandibles without teeth, stemmata aggregated in two groups, maxillary stipes only about twice as long as palpifer, and leg with reduced claw.

Australian species. Single species, *S. discolor* Orchymont. Two additional species are known from New Guinea, but were so far not recorded from Australia (see van Berge Henegouwen 1986a).

Distribution. The genus is nearly exclusively confined to Old World, in New World it only has one native Brazilian species and few European species introduced to North America (Smetana 1978; van Berge Henegouwen 1986b). The highest number of species is known from tropical Africa. *Sphaeridium discolor* is widespread in the Oriental Region, in Australia it is recorded from northern Western Australia, Northern Australia, Queensland and New South Wales.

Biology. *Sphaeridium* nearly exclusively inhabit excrements of large herbivorous mammals and feed in dung particles (Holter 2004). Larvae inhabit the same habitat and are predators of dipteran larvae (Sowig *et al.* 1997).

References. Hansen (1990b) (notes on Australian species); Hansen (1991) (adult morphology); Archangelsky (1997) (larval morphology).

Key to species. Not available. *Sphaeridium discolor* differs from other species of the genus by coloration, obliquely cut off posterolateral corners of pronotum, and rounded apex of the median lobe of the aedeagus (illustrated by d'Orchymont 1933).

Tribe Megasternini Mulsant, 1844 (Figs 20.19–20.21)

Characteristics. Antenna with 9 antennomeres*, antennal club compact; clypeus excised anterior of eyes, hence antennal bases exposed in dorsal view (weakly excised only in *Ercycodes*); male maxilla each with small rounded sucker ventrally (seen only in good microscope and in specimens with open maxillae, absent in *Pseudooster-num* and *Ceronocyton*); prosternum carinate medially, or with elevated plate; hypomeron with antennal grooves (rarely absent); elytron with 9–10 series (rarely without series), scutellary stria absent; anapleural sutures absent; mesoventrite with median elongate or pentagonal plate; anteromedian pit on mesoventrite absent; abdomen with 5 ventrites; first abdominal ventrite carinate*; abdomen without apical notch*; femora never with dense pubescence ventrally; basal meso- and metatarsomere longer than tarsomere 2. By the combination of compact antennal club, mesoventrite with median plate and clypeus excised in front of eyes easily recognisable from most other hydrophilid groups except of some Omicrini, which differ by exposed well sclerotised labrum, clypeus with large lateral lobes and epipleura wide even posteriorly.

Larvae: only known for few genera so far (Archangelsky 1997, 2016, 2018; the only Australian larva examined was collected in association with *Merosoma* sp., see under that genus for details). All larvae are generally very similar to each other: frontal sulci not converging posteriorly; stemmata fused into one eye spot on each side; anterior margin of head smooth, without tooth, with small pubescent notch on the left side; mandibles asymmetrical, left without inner tooth, right usually with one inner tooth (with two teeth in *Merosoma*); maxilla with stipes bearing many stout setae on inner face, inner appendage of palpifer reduced and membranous; labium small, with large asymmetrical hypopharyngeal lobe, ligula absent; legs short with reduced number of segments (3 in *Pelosoma*, 2 in some *Cercyon*; Archangelsky 1997, 1999a, 2016; Arriaga-Varela *et al.* 2017) or totally absent (spinose lobe present on the place of reduced legs in *Cryptopleurum* and some *Cercyon*, Oosternum and Motonerus; Archangelsky 1997, 2016, 2018; Arriaga-Varela *et al.* 2017); abdomen without finger-like lateral projections, stigmatic atrium present. The general morphology slightly resembles that of *Sphaeridium* and *Austrotypus*. *Austrotypus* differs in symmetrical mandibles, shape of anterior margin of the head, presence of two pairs of long projections on abdominal segment 8 (see under that genus for details); *Sphaeridium* differs in abdominal segment 8 with two pairs of projections, stemmata aggregated in two groups, rather short stipes and legs with 4 preserved segments and labium with distinct ligula.

A very diverse group containing 51 genera and ~550 described species. In Australia represented by rich fauna of endemic genera and species. Majority of species are terrestrial, inhabiting various types of decaying organic material, most frequently rainforest leaf litter. Outside Australia the tribe also includes some secondarily aquatic species, inhabiting edges of standing waters or phytotelmata, and some species associated with ants.

Generic groups and genital morphology. Based on the preliminary phylogenetic analysis (Short & Fikáček 2013), there are two principal clades within the tribe that can be recognised from each other by the morphology of the male genitalia and associated structures. The majority of Australian genera belong to the *Oosternum* group of genera, a clade largely confined to Australia and South America, characterised by crescent-like median portion of abdominal sternite 9, medially projecting sternite 8, and the median lobe of the aedeagus firmly joint with bases of parameres (hence, it does not reach deeply into phallobase and cannot freely move anteriorly or posteriorly; Fig. 20.19G–I). The *Cercyon* group of genera, in Australia only containing the introduced genera *Cercyon* and *Pachysternum* and the native *Cryptopleurum* and *Pilocnema*, is characterised by tongue-like median part of sternite 9, sternite 8 without median projections, and median lobe freely movable within tegmen, not firmly joint to parameres and hence reaching deeply into phallobase (Fig. 20.19J–L). In some cases, these characters are the only way to recognise similarly looking genera each of which belong to the different clade, which is in Australia the case of genera *Cercyon* and *Merosoma*.

Australian taxa. There are 14 genera in Australia, of which two (*Cercyon* and *Pachysternum*) are introduced, two are native but widespread outside Australia (*Cryptopleurum*, *Australocyon*), three also occur in New Guinea (*Cetiocyon*, *Merosoma*, *Pilocnema*) and one is also present in New Zealand (*Cercyodes*). All Australian species are terrestrial.

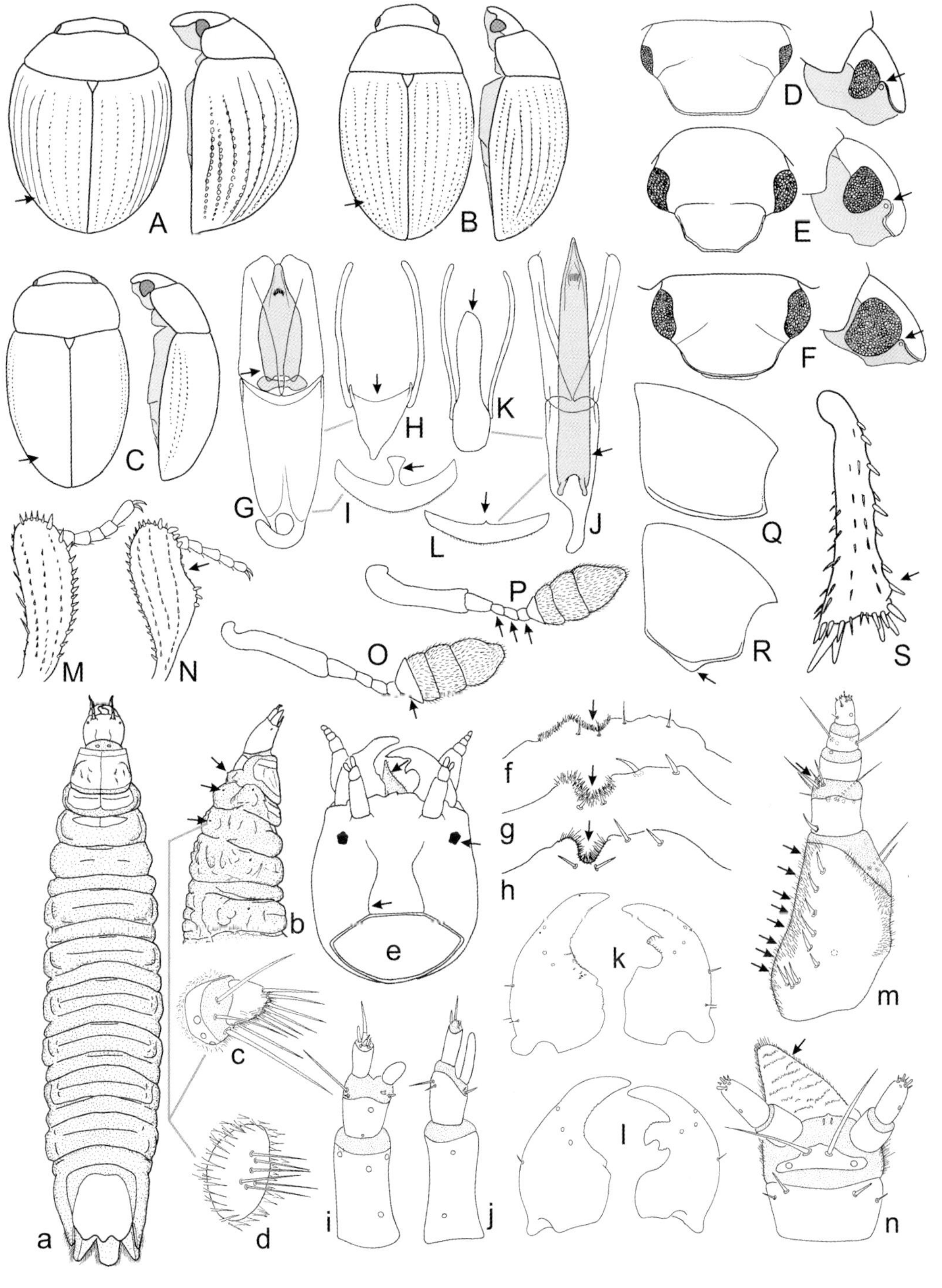

Fig. 20.19. Tribe Megasternini. A–C, adults in dorsal and ventral views (A, *Pilocnema maculiapex* Hansen; B, *Cercyon haemorrhoidalis* (Fabricius); C, *Cercyodes kingensis* (Blackburn)). D– F, head in dorsal and lateral view (D, *Ercycodes fossus* (Blackburn); E, *Cryptopleurum* sp.; F, *Cercyon* (*Paracycreon*) *laminatus* Sharp). G–L, male genitalia and associated structures: G–H, *Merosoma* sp. (G, aedeagus; H, sternite 9; I, sternite 8); J–L, *Cercyon haemorrhoidalis* (J, aedeagus; K, sternite 9; L, sternite 8). M–N, anterior tibiae (M, *Cryptopleurum* sp.; N, *Pachysternum capense* (Mulsant)). O–P, antenna of *Pilocnema* species (O, *P. maculiapex*; P, *P. nigra* Hansen). Q–R, pronotum in lateral view (Q, *Cercyon haemorrhoidalis*; R, *Cryptopleurum* sp.). S, middle tibia of *Ercycodes fossus*. a–m, larvae of non-Australian species: a, *Cercyon* sp., dorsal habitus; b, same species, head and thorax in lateral view; c, *Cercyon praetextatus* (Say), prothoracic leg; d, *Motonerus obscurus* Hansen, prothoracic lobe with completely reduced leg; e, *Megasternum* sp., head in dorsal view; f–h, nasale (f, *Megasternum* sp.; g, *Cercyon* sp.; h, *Cryptopleurum* sp.); i–j, antenna (I, *Cercyon* sp.; j, *Merosoma* sp.); k–l, mandibles (k, *Cercyon* sp.; l, *Merosoma* sp.); m, *Cercyon* sp., maxilla; n, *Cercyon* sp., labium. O–P, S adapted from Hansen (1990b), a–b from Minoshima (2013), c, e–i, k, m–n from Archangelsky (1997).

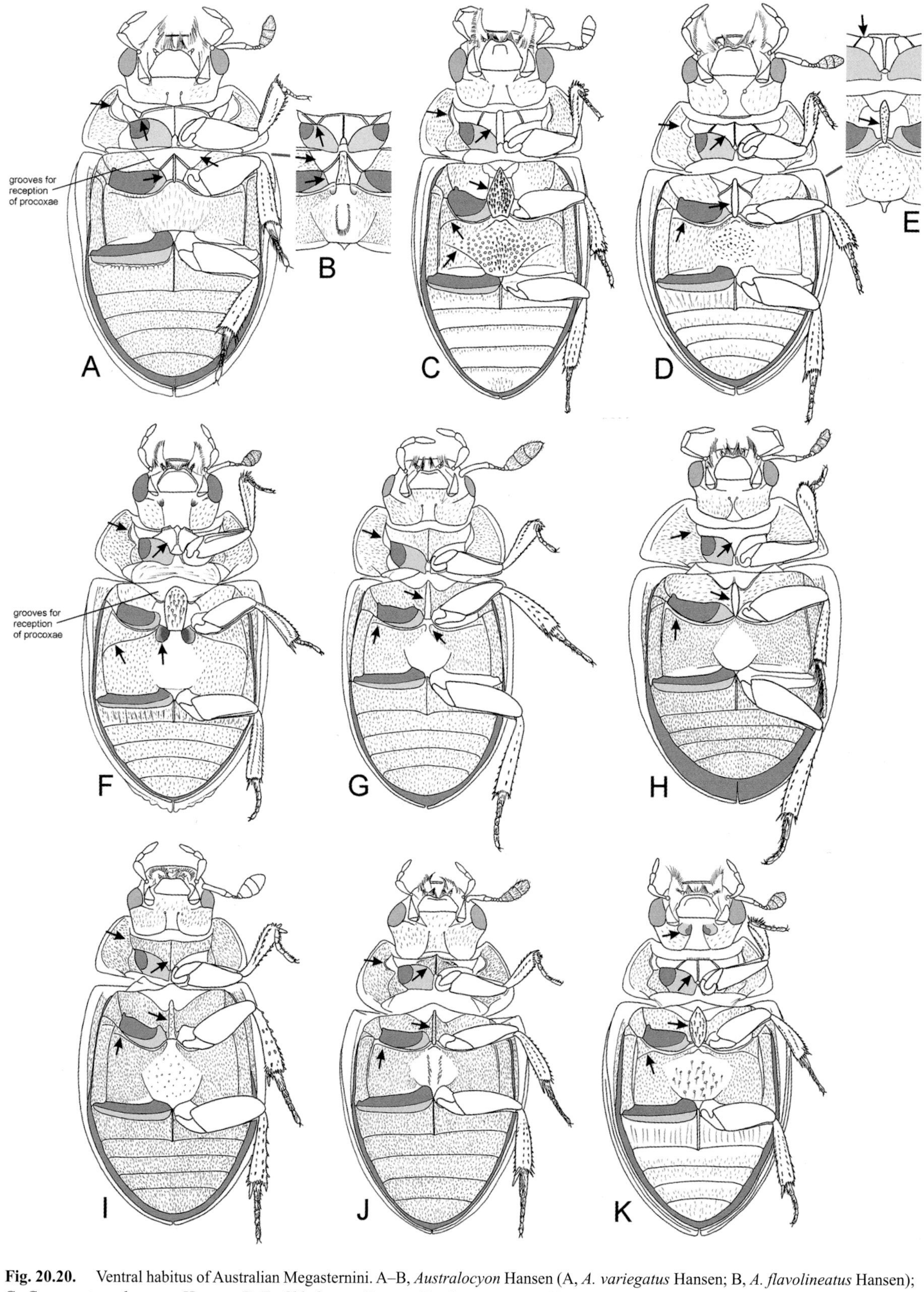

Fig. 20.20. Ventral habitus of Australian Megasternini. A–B, *Australocyon* Hansen (A, *A. variegatus* Hansen; B, *A. flavolineatus* Hansen); C, *Ceronocyton obscurum* Hansen; D–E, *Chledocyon* Hansen (D, *C. marmoratus* Hansen; E, *C. semiopacus* Hansen); F, *Cenebriophilus costatus* Hansen; G, *Cercyodes kingensis* (Blackburn); H, *Cetiocyon hanseni* Fikáček and Short (New Guinean species); I, *Ercycodes fossus* (Blackburn); J, *Merosoma* sp.; K, *Notocercyon ornatus* Blackburn.

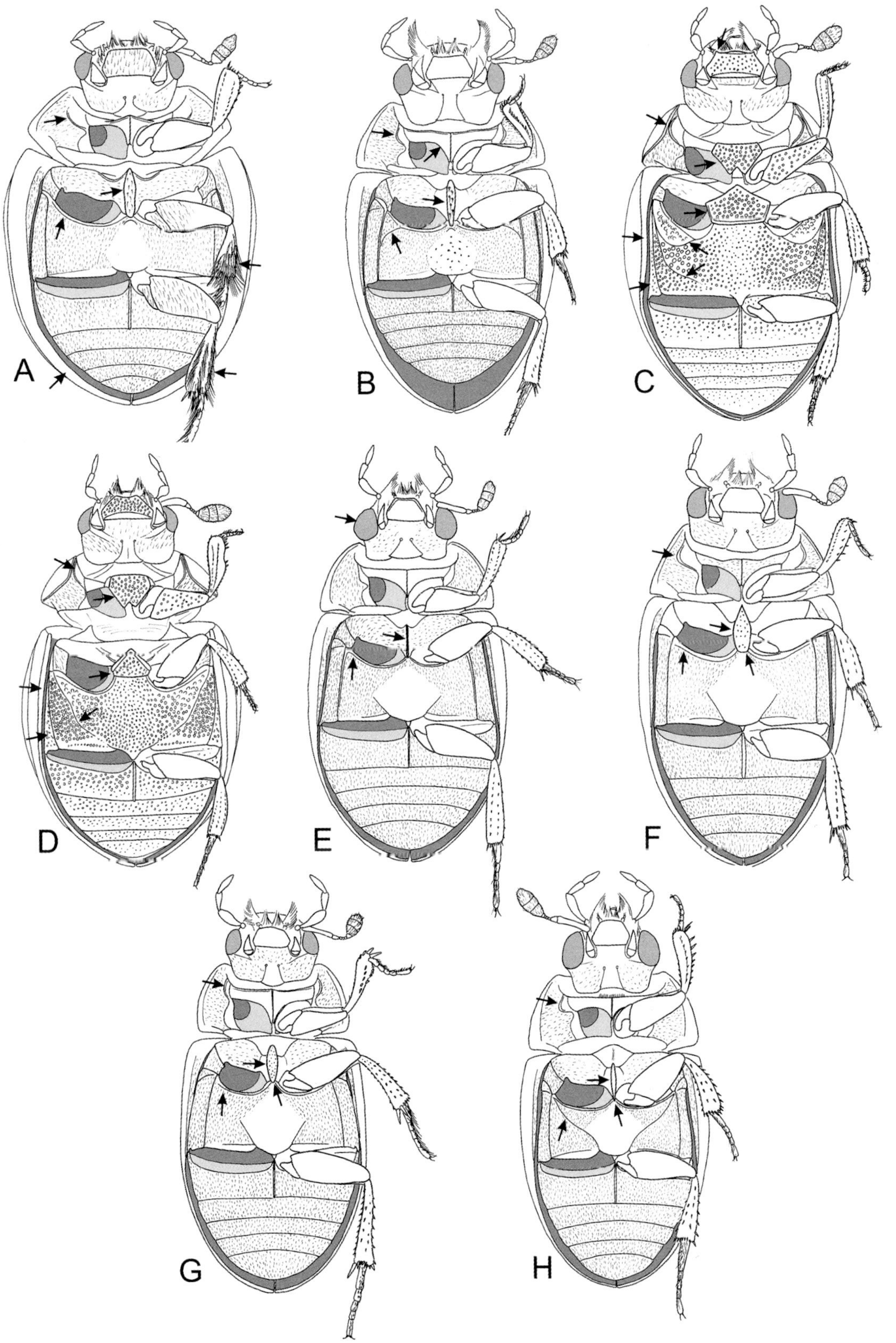

Fig. 20.21. Ventral habitus of Australian Megasternini. A, *Pilocnema maculiapex* Hansen; B, *Pseudoosternum maculatum* Hansen; C, *Cryptopleurum evansi* Balfour-Browne; D, *Pachysternum capense* (Mulsant); E, *Cercyon* (*Paracycreon*) *laminatus* Sharp; F, *Cercyon* (*Paracercyon*) *analis* (Paykull); G, *Cercyon* (s.str.) *maritimus* Knisch; H, *Cercyon* (s.str.) *nigriceps* (Marsham).

Australocyon Hansen, 1990 (Pl. 47A–B; Fig. 20.30A–B)

Type species. *Australocyon variegatus* Hansen, 1990.

Characteristics. Body rather widely oval, highly convex; head with small tentorial pits; prosternum with median plate carinate mesally; antennal grooves present, not reaching lateral margin of hypomeron; mesoventral plate pentagonal, widely contacting metaventrite, bearing narrow or wide median keel (except in *A. weiri*!); grooves for reception of procoxae deep, reaching nearly to mesocoxae; metaventrite without or with weakly developed anterolateral ridges; metaventrite medially flat without depressions or pits (*A. variegatus* group) or with anteromedian pits in both sexes and a deep setiferous impression in males (*A. nanus* group), without femoral lines; metanepisternum not extremely narrow; anterior tibiae not excised; metatibiae without dense long setae ventrally; dorsal punctuation setiferous; elytron with 10 elytral series, series not impressed, intervals flat; male sucking disc on maxilla present; male sternite 9 crescent-like; median lobe fixed to base of parameres; body brown to black, often with various pattern of paler spots on elytra; body length 1.4–2.5 mm. Easy-to-recognise by combination of prosternum with carinate median plate and mesoventrite with carinate pentagonal plate, by which it differs from similar genera of Megasternini as well as from similarly looking omicrines. Species with less apparent (*A. calderi*) or absent (*A. weiri*) median ridge of mesoventral plate may be distinguished from similarly looking omicrine genera by compact antennal club (from *Psalitrus*), concealed labrum (from *Mircogioton*, *Noteropagus* and *Paromicrus*), narrow epipleura, presence of antennal grooves (from *Mircogioton*, *Paromicrus* and *Psalitrus*) and carinate abdominal ventrite 1 (from *Mircogioton* and *Psalitrus*). Unique among Australian Megasternini in deep grooves for reception of procoxae nearly reaching mesocoxal cavities.

Larvae: Unknown.

Australian species. Six species classified in two species groups.

Distribution. The genus in current understanding is distributed in Australia, South America and SE Asia (Hansen 2003), but at least the Asian species actually represent a separate, as yet undescribed genus not related to *Australocyon*. In Australia the genus occurs in various types of rainforests in Queensland and New South Wales.

Biology. Most frequently collected by sifting leaf litter (most species), in dung or carcass (incl. baited pitfall traps, *A. nanus* species group) or flight intercept traps.

References. Hansen (1990b) (description of the genus, revision on Australian fauna); Hansen (2003) (discovery of *Australocyon* outside Australia).

Key to species. Hansen (1990b).

Cenebriophilus Hansen, 1990 (Pl. 47C; Fig. 20.20F)

Type species. *Cenebriophilus costatus* Hansen, 1990.

Characteristics. Body widely oval with subrectangular elytral bases, highly convex; head with small tentorial pits; prosternum with median plate widely carinate mesally and excised anteriorly; antennal grooves present, small, not reaching lateral margin of hypomeron; mesoventral plate elongate oval, widely contacting metaventrite, flat (without median keel); grooves for reception of procoxae small, not reaching to mesocoxae; metaventrite with very distinct arcuate anterolateral ridges; metaventrite anteromedially with a pair of deep pits, without femoral lines; metanepisternum very narrow; anterior tibiae not excised; metatibiae without dense long setae ventrally; dorsal punctation setiferous; elytron with 9 elytral series, series deeply impressed, intervals costate; male sucking disc on maxilla present; male sternite 9 crescent-like; median lobe fixed to base of parameres, not reaching into phallobase; body uniformly brown to black, sometimes with paler elytral base; 1.5–2.0 mm. Easy-to-recognise by combination of costate elytra, prosternum with carinate median plate with anterior excision, mesoventrite with elongate oval plate and the presence of anteromedian pits of metaventrite (in this similar to some *Australocyon* from which it differs by mesoventral morphology). By general habitus resembling *Ceronocyton* and *Pseudoosternum*, but differs from both by presence of prosternal plate.

Larvae: Unknown.

Distribution and Australian species. Endemic to Australia (Queensland, New South Wales), with two species: *C. costatus* Hansen and *C. subcostatus* Hansen.

Biology. Collected from carrion (most specimens of *C. costatus*), dung traps and forest leaf litter (most specimens of *C. subcostatus*) in rainforests.

References. Hansen (1990b) (description of the genus, key to species).

Key to species. Hansen (1990b).

Cercyodes Broun, 1886 (Pl. 47G; Fig. 20.19C, 20.20G)

Type species. *Cercyodes laevigatus* Broun, 1886.

Characteristics. Unique among Australian Megasternini by the absence of elytral series (except short vestiges laterally). Except of that characterised by the following characters: body elongate oval, rather depressed; head with small tentorial pits; prosternum simply carinate, without median plate; antennal grooves minute, virtually absent; mesoventral plate narrowly elongate, narrowly contacting metaventrite; grooves for reception of procoxae small, not reaching to mesocoxae; metaventrite without anterolateral ridges, without pits or impressions, femoral lines absent; metanepisternum moderately wide; anterior tibiae not excised; metatibiae without dense long setae ventrally;

dorsal punctuation without setae; male sucking disc of maxilla present; male sternite 9 crescent-like; median lobe fixed to base of parameres; body uniformly dark brown; 2.3–2.6 mm. Unique among Australian megastenines by the absence of elytral series, virtually absent antennal grooves, very narrow mesoventral plate and characteristic body shape, and hence cannot be confused with any other Australian hydrophilid. In the beachwrack, it may occur syntopically with *Ercycodes*, but differs from it easily by absence of elytral series.

Larvae: Unknown.

Australian species. Single species, *C. kingensis* (Blackburn).

Distribution. The genus is endemic for Australian Region. It includes two species, *C. kingensis* is endemic to Tasmania incl. King Island, and *C. laevigatus* Broun is endemic to New Zealand (occurs in both main islands and Chatham Islands).

Biology. Both species of the genus are most frequently collected from/under accumulated decaying seaweed on beaches, rarely from forest leaf litter.

References. Hansen (1990b) (description of the genus and redescription of *C. kingensis*).

Key to species. Hansen (1990b).

Cercyon Leach, 1817 (Pl. 46L–N; Figs 20.20B, F, J–L, Q, 20.21E–H)

Type species. *Dermestes melanocephalus* Linnaeus, 1758.

Characteristics. Body elongate oval, weakly to moderately convex; head with small tentorial pits; prosternum carinate mesally, without median plate; antennal grooves present, minute to moderately large, not reaching lateral margin of hypomeron; mesoventral plate elongate oval narrowly to widely (sg. *Paracercyon*) contacting metaventrite, or simply carinate (sg. *Paracycreon*); grooves for reception of procoxae small, not reaching to mesocoxae; metaventrite without arcuate anterolateral ridges*; metaventrite without pits, with or without femoral lines; metanepisternum moderately wide; anterior tibiae not excised; metatibiae without dense long setae ventrally; dorsum bare; elytron with 10 elytral series, series impressed or not, intervals never costate; male sucking disc of maxilla present; male sternite 9 tongue-like; median lobe freely movable, reaching into phallobase; body pale yellowish to black, sometimes with darker elytral base and sutural interval and/or paler elytral apex; 1.3–4.2 mm. Easily recognisable from most Australian genera by the combination of simply carinate prosternum, narrowly oval or simply carinate mesoventral plate and metaventrite without anterolateral ridges, in these characters only resembling *Pilocnema* and *Merosoma*; from *Pilocnema* it differs by meso- and metatibiae without dense long pubescence and epipleura narrow posteriorly, from *Merosoma* recognisable by *Cercyon*-like male genitalia (Fig. 20.19J–L) (see also under *Merosoma* for additional characters).

Larvae (Fig. 20.19a–c, g, k, m–n): Corresponding to tribal diagnosis, legs present as small 3- or 2-segmented vestiges or totally reduced, right mandible with single tooth (based larvae described by Hafez 1939; Prins 1984; Schulte 1985; Hayashi 1986; Archangelsky 1997, 2016; De Marzo 2000; Hayashi & Hayama 2009; and Arriaga-Varela *et al.* 2017, including larvae of *C. haemorrhoidalis*, *C. quisquilius, C. maritimus* and *C. laminatus* which occur in Australia).

Australian species. Six species known to occur in Australia, all of them introduced, belonging to two subgenera: *Cercyon* (s.str.) *haemorrhoidalis*, *C.* (s.str.) *quisquilius*, *C.* (s.str.) *nigriceps* and *C.* (s.str.) *maritimus*, *C.* (*Paracycreon*) *laminatus* and *C.* (*P.*) *subsolanus*. Two additional species are expected, as they are already introduced and established e.g. in New Zealand: *C.* (*Paracercyon*) *analis* and C. (s.str.) *depressus*. Hansen (1990b) mentions the first four above species and 'at least eight as yet unidentified species'. At least one of them actually belongs to *Merosoma* (see below), but no attempt was made to identify all these species, hence they may represent additional species of *Cercyon* as well as of *Merosoma*.

Generic concept and subgenera. *Cercyon* is the largest genus of Megasternini (as well as of Sphaeridiinae), with ~260 described species. It was hence divided into 8 subgenera, two of which occur in Australia and the third is expected. First results of molecular phylogenetic studies (Short & Fikáček 2013, Arriaga-Varela unpubl. data) however suggest that *Cercyon* is polyphyletic in respect to other megasternine genera. Additional studies are necessary to clarify the systematics of the genus. In the current concept, the subgenera occurring or expected in Australia may be identified as follows:

1.	Metaventrite with complete femoral lines reaching anterolateral corners (Fig. 20.21H). Body small, with dark head (length 1.3–1.8 mm)	***C.* (s.str.) *nigriceps***
–	Metaventrite without or with incomplete femoral lines. Body larger	**2**
2(1).	Mesoventrite simply laminate, not forming a plate (Fig. 20.21E). Body pale with dark head (Pl. 46L), eyes large (Fig. 20.19F)	**sg. *Paracycreon* (2 spp.)**
–	Mesoventrite in a form of narrowly elongate plate. Body larger to smaller, head and pronotum always dark, eyes not apparently large	**3**

3(2). Mesoventral plate wide posteriorly, abutting moderately wide metaventral process (Fig. 20.21F). Elytra attenuate posteriorly (Pl. 46N) .... ***C.* (*Paracercyon*) *analis***

– Mesoventral plate narrow posteriorly, narrowly contacting metaventrite (Fig. 20.21G). Elytra not apparently attenuate posteriorly (Pl. 46M) ....**sg. *Cercyon* s.str. (part)** (3 species and one expected, of which *C. quisquilius* has yellowish elytra, and *C. haemorrhoidalis* dark reddish elytra with black T at elytral base and along suture)

Distribution. *Cercyon* is distributed world-wide, but it is not native in Australian Region (no native species are known from Australia, and even the species described as *Cercyon* from New Guinea by Hebauer (2001) do not belong there but close to *Platycyon* and *Merosoma* in *Oosternum* genus group). Most species recorded from Australia are today of world-wide distribution (*C. nigriceps*, *C. haemorrhoidalis*, *C. quisquilius*, *C. laminatus*), *C. maritimus* is native in Republic of South Africa and *C. subsolanus* widespread in Oriental region. In Australia the species are mostly recorded from southern (temperate) regions, *C. maritimus* was recorded from southern Western Australia and New South Wales and recently collected in King Island, and examined specimens of *C. laminatus* and *C. subsolanus* were collected in Northern Queensland (Granite Gorge and Malanda, respectively).

Biology. Most species recorded from Australia inhabit dung of herbivorous mammals (cow, horse, sheep). *Cercyon maritimus* (and also *C. depressus* which discovery in Australia is expected) are found under decaying wet to partly dry beach wrack. *Cercyon analis* which is also expected in Australia is usually found in garden compost both in Europe and New Zealand. Most species come frequently to light.

Key to species. Hansen (1987); Forster *et al.* (2014) (keys to Scandinavian/UK fauna which includes all species so far recorded or expected in Australia except *C. maritimus* and *C. subsolanus*); Ryndevich (2008) (key to some species of *Paracycreon* incl. *C. laminatus* and *C. subsolanus*).

Ceronocyton Hansen, 1990 (Pl. 47H; Fig. 20.20C)

Type species. *Ceronocyton obscurum* Hansen, 1990.

Characteristics. Body elongate oval, moderately convex; head with tentorial pits small to moderately large; prosternum with median plate widely carinate mesally, not excised anteriorly; antennal grooves minute, virtually absent; mesoventral plate narrowly subpentagonal, rather widely contacting metaventrite, flat (without median keel); grooves for reception of procoxae small, not reaching to mesocoxae; metaventrite with very distinct anterolateral arcuate ridges; metaventrite without pits or impressions, with distinct femoral lines; metanepisternum moderately wide; anterior tibiae not excised; metatibiae without dense long setae ventrally; dorsal punctuation setiferous; elytron with 10 elytral series, series impressed, intervals convex; male sucking disc of maxilla absent; male sternite 9 crescent-like; median lobe fixed to base of parameres; body uniformly brown to black; 1.7–2.3 mm. By prosternum with median plate, elongate mesoventral plate and presence of arcuate ridge on metaventrite similar to *Cenebriophilus*, but differing from it by presence of 10 elytral series, presence of femoral lines, and absence of deep pits anteromesally on the metaventrite.

Larvae: Unknown.

Distribution and Australian species. The genus is endemic to Australia (Queensland, New South Wales, Victoria), with three described species

Biology. Collected from leaf litter, dung-baited pitfall traps and flight intercept traps in rainforests.

References. Hansen (1990b) (description of the genus, key to species).

Key to species. Hansen (1990b).

Cetiocyon Hansen, 1990 (Pl. 47F; Fig. 20.20H)

Type species. *Cercyon papuensis* Orchymont, 1924.

Characteristics. Body widely oval, highly convex; head with small tentorial pits; prosternum simple, without median carina*; antennal grooves absent; mesoventral plate narrowly oval, contacting metaventrite in single point; grooves for reception of procoxae small, not reaching to mesocoxae; metaventrite without anterolateral arcuate ridges; metaventrite without pits, impressions or femoral lines; metanepisternum moderately wide; anterior tibiae not excised; metatibiae without dense long setae ventrally; dorsal punctuation without apparent setae; elytron with 10 elytral series, series sharply impressed*, intervals slightly convex; male sucking disc of maxilla present; male sternite 9 crescent-like; median lobe fixed to base of parameres; body uniformly brown to black*; 4.4–4.9 mm*. Differs from other Megasternini by large body size and prosternum without median plate, without median carina and without antennal grooves. By the morphology of meso- and metaventrite it resembles *Cercyon*, *Pilocnema* and *Merosoma*, from which it differs by the above prosternal characters, plus from *Cercyon* and *Pilocnema* also by the crescent-like median part of male sternite 9, and from *Pilocnema* by bare posterior tibiae.

Larvae: Unknown.

Australian species. Single undescribed species; it was identified as *C. papuensis* (Orchymont) by Hansen (1990b), but likely belongs to an undescribed species close to *C. papuensis* and *C. hanseni* Hebauer (Fikáček & Short 2010). Unfortunately, it is known only from females which is why it remains undescribed.

Distribution. The genus is nearly exclusively confined to New Guinea where it has 18 described species

(Szczepański *et al.* 2018). Outside of New Guinea it has one species is northernmost Queensland, and one aberrant species in northern South America (Suriname) which may be not congeneric.

Biology. Terrestrial, but little is known about the lifestyle. The Australian specimens were extracted from rainforest leaf litter using Berlese funnels, New Guinean species were mostly collected in flight intercept traps.

References. Hansen (1990b) (description of the genus, review of Australian specimens); Fikáček & Short (2010) (revision of the genus, corrected identity of Australian specimens); Szczepański *et al.* (2018) (description of additional New Guinean species, updated key to species).

Key to species. Fikáček & Short (2010); Szczepański *et al.* (2018).

Chledocyon Hansen, 1990 (Pl. 47D–E; Fig. 20.20D–E)

Type species. *Chledocyon semiopacus* Hansen, 1990.

Characteristics. Body widely oval, highly convex; head with tentorial pits small to enlarged (*C. venustus*); prosternum with median portion divided by ridges (and in some species elevate), carinate medially, not excised anteriorly; antennal grooves small, mesoventral plate narrowly elongate to subpentagonal with median narrow plate, narrowly contacting metaventrite; grooves for reception of procoxae small, not reaching mesocoxae; metaventrite with distinct anterolateral arcuate ridges; metaventrite without pits or impressions, without femoral lines; metanepisternum moderately wide; anterior tibiae not excised; metatibiae without dense long setae ventrally; dorsal punctuation setiferous; elytron with 10 elytral series, series impressed, intervals convex; male sucking disc of maxilla present; male sternite 9 crescent-like; median lobe fixed to base of parameres; body brown to black, in some species with pattern of pale spots on elytra; 1.7–2.3 mm. By prosternum with median plate, elongate mesoventral plate and presence of arcuate ridge on metaventrite similar to *Cenebriophilus*, but differing from it by presence of 10 elytral series, presence of femoral lines, and absence of deep pits anteromesally on the metaventrite.

Larvae: Unknown.

Distribution and Australian species. *Chledocyon* is endemic to Australia, comprising five species classified in two species groups; four species are endemic to south-west Australia, one (*C. queenslandicus*) occurs in northern Queensland.

Biology. Terrestrial, collected from forest leaf and log litter and using carrion traps.

References. Hansen (1990b) (description of the genus, key to species).

Key to species. Hansen (1990b).

Cryptopleurum Mulsant, 1844 (Pl. 47N; Figs 20.19E, M, R, 20.21C)

Type species. *Sphaeridium minutum* Fabricius, 1775.

Characteristics. Body widely oval, moderately convex; head with tentorial pits small; clypeus angulate, with nearly complete transverse groove; prosternum with wide ecarinate coarsely punctate plate; antennal grooves large, reaching lateral portions of hypomeron (hence lateral margin of pronotum angulate in lateral view); mesoventral plate subpentagonal, wider than long; mesocoxal cavities widely separated; grooves for reception of procoxae small, not reaching mesocoxae; metaventrite without anterolateral arcuate ridges, but with distinct anterior portion of femoral lines and additional postcoxal ridges; metaventrite without pits or impressions, not densely pubescent but coarsely punctate laterally; metanepisternum extremely narrow, narrowing anteriorly; anterior tibiae not excised; metatibiae without dense long setae ventrally; dorsal punctation setiferous; elytron with 10 elytral series, series impressed, intervals convex to costate; male sucking disc of maxilla present; male sternite 9 tongue-like; median lobe freely movable, reaching into phallobase; body brown, sometimes with pale elytral apex; 1.9–2.1 mm. By the combination of large prosternal plate, widely pentagonal mesoventral plate and coarsely punctate metaventrite resembling *Pachysternum* and *Noteropagus*; from *Pachysternum* it differs by arcuate (not sinuate) outer margin of anterior tibiae; from *Noteropagus* it differs by ecarinate prosternum and labrum retracted under clypeus.

Larvae (Fig. 20.19h): Corresponding to tribal diagnosis, legs totally absent, right mandible with a single inner tooth (based on larva of *C. minutum* described by Archangelsky (1997), larva of the Australian species unknown).

Australian species. Single species; examined Australian specimens seem conspecific with a specimen of *C. evansi* Balfour-Browne from Hawaii identified by M. Hansen in coll. National Museum Prague.

Distribution. The genus is widely distributed in Palaearctic, Nearctic, Afrotropical and Orietal Regions, but does not reach to South America, and is poorly represented in New Guinea and Pacific (Hebauer 2001; Hansen 1995). Examined Australian specimens are from northern Queensland.

Biology. Terrestrial, usually found in dung or compost, some species collected from forest leaf litter. Australian specimens were collected by baited flight intercept traps and in yellow pan traps.

References. Hansen (1990b) (comments on Australian species); Hansen (1991) (genetic diagnosis).

Key to species. Not available for Australia. Oriental species were keyed by d'Orchymont (1926) and Jia & Zhang (2017), New Guinean ones were diagnosed by Hebauer (2001), and the Pacific *C. evansi* is diagnosed by Hansen (1995).

Ercycodes Hansen, 1990 (Pl. 47I; Figs 20.19D, S, 20.20I)

Type species. *Cercyon fossum* Blackburn, 1888.

Characteristics. Body widely oval, moderately convex; clypeus widely angulate, very weakly excised above antennal bases; prosternum rather long, simple, carinate medially; antennal grooves absent; mesoventral plate very narrow, narrowly contacting metaventrite; grooves for reception of procoxae small, not reaching mesocoxae; metaventrite without anterolateral arcuate ridges, pits, impressions or femoral lines; metanepisternum moderately wide; anterior tibiae arcuate on outer margin; meso- and metatibiae without dense long setae ventrally, but widened distally; dorsal punctation without apparent setae; elytron with 10 elytral series, series weakly impressed, intervals slightly convex; male sucking disc on maxilla present; male sternite 9 crescent-like; median lobe fixed to base of parameres; uniformly black dorsally; 2.0–2.6 mm. Differs from most other Australian Megasternini by absence of antennal grooves (by this resembling *Cercyodes* which lacks elytral series, and *Cetiocyon* which has ecarinate prosternum). It is unique among Australian megasternine species by distally widened tibiae with stout spines.

Larvae: Unknown.

Distributuion and Australian species. The genus is endemic to Australia. It includes two species: *E. fossus* (Blackburn) in mainland Australia (New South Wales, Victoria) and *E. tasmanicus* Hansen (Tasmania).

Biology. Both species were found in litter and under seaweed on the beaches.

References. Hansen (1990b) (description of the genus, key to species).

Key to species. Hansen (1990b).

Merosoma Balfour-Browne, 1939 (Pl. 47J; Figs 20.19G–I, 20.20J)

Type species. *Pelosoma eremita* Knisch, 1925.

Characteristics. Body moderately wide, moderately convex; head with tentorial pits small; prosternum with median plate slithgly elevated and with different sculpture, but not divided by ridges, carinate mesally; antennal grooves moderately large, not reaching lateral margin of hypomeron; mesoventral plate narrowly elongate, narrowly contacting metaventrite; grooves for reception of procoxae small, not reaching to mesocoxae; metaventrite without anterolateral arcuate ridges, pits, impressions or femoral lines, median portion with two submedian longitudinal series of stout setae; metanepisternum moderately wide; anterior tibiae not excised; metatibiae without dense long setae ventrally; dorsal punctuation setiferous (but setae sparse and rather indistinct); elytron with 10 elytral series, series impressed, intervals convex; male sucking disc of maxilla present; male sternite 9 crescent-like; median lobe fixed to base of parameres; body pale reddish with dark central spot on pronotum and along elytral series; 2.2–2.4 mm. By external characters crresponding with *Cercyon*, but differing from Australian *Cercyon* by characteristic coloration, slightly elevate median portion of prosternum and series of stout setae on median portion of metaventrite, and male genitalia of *Oosternum* form (Fig. 20.19G–I).

Larvae (Fig. 20.1): Corresponding to tribal diagnosis, legs completely absent, right mandible with two inner teeth, abdominal segmentes lobate laterally (based on larvae collected in association with udescribed *Merosoma* in New England NP, 26.viii.1982, collected by S. and J. Peck (event. SBP114), deposited in ANIC).

Distribution and Australian species. Single, yet undescribed species known from New England National Park in New South Wales. Additional species may be present and overlooked in the collections among unidentified *Cercyon*.

Biology. Examined specimens were collected in Malaise traps, the sample including adults and larvae was sifted from *Nothofagus* litter.

Notes. Examined specimens were found among material of ANIC, Canberra identified by M. Hansen as *Cercyon* (see also under *Cercyon* for details); the species was hence overlooked and left undescribed. Examination of the male genitalia confirms it does not belong to *Cercyon* but to the *Oosternum* group of genera, and does not fit any genus of this group recorded until now from Australia. It may however very likely stand close to New Guinean *Cercyon*-like species which were described in *Cercyon* by Hebauer (2001) but actually belong to *Oosternum* group of genera based on genital morphology. Similar situation occurs in the New Guinean *Pelosoma* which were separated to a separate genus *Merosoma* by Balfour-Browne (1939b) (but later synonymized again with *Pelosoma* by Hansen (1991)): despite the similarity to *Pelosoma*, *Merosoma* has the *Oosternum* type of male genitalia, and cannot hence be a synonym of *Pelosoma* which belongs to *Cercyon* group. Despite slight differences of mesoventral morphology, I suppose that New Guinean '*Cercyon*' and '*Pelosoma*' (= *Merosoma*) are likely rather closely related. For this reason, I temporarily place the Australian *Cercyon*-like species with *the Oosternum* type of genitalia to the genus *Merosoma*.

References. Balfour-Browne (1939b) (description of the genus).

Notocercyon Blackburn, 1898 (Pl. 47K; Fig. 20.20K)

Type species. *Notocercyon ornatum* Blackburn, 1898.

Characteristics. Body elongate oval moderately convex; head with ventral tentorial pits forming huge holes; prosternum with a plate, carinate medially; antennal grooves small, not reaching lateral margin of hypomeron; mesoventral plate elongate oval, contacting metaventrite in single point; grooves for reception of procoxae small, not reaching to mesocoxae; metaventrite with anterolateral arcuate ridges, metaventral pits, impressions or femoral lines absent; metanepisternum moderately wide; anterior tibiae arcuate on outer margin; meso- and metatibiae without dense long setae ventrally; dorsal punctation without apparent setae; elytron with 10 elytral series, series weakly impressed, intervals slightly convex; male sucking disc on maxilla present; male sternite 9 crescent-like; median lobe fixed to base of parameres; body length 1.8–2.4 mm; body dark with pale spots on elytra or pale with dark elytral spots. *Notocercyon* is most similar to *Pseudoosternum* and *Chledocyon* by general habitus and ventral morphology; it differs from *Pseudoosternum* by median portion of prosternum delimited by sharp ridges and by enlarged tentorial pits on the head, and from *Chledocyon* by presence of the arcuate anterolateral ridge on the metaventrite (it also differs from most *Chledocyon* except *C. venustus* in enlarged tentorial pits on the head).

Larvae: Unknown.

Distribution and Australian species. *Notocercyon* is endemic to south-eastern Australia. There are two described species and ~8 undescribed species (based on material presorted by M. Hansen in ANIC, Canberra). All known specimens are from Tasmania, Victoria and New South Wales.

Biology. Terrestrial, found in various kinds of decaying organic material, main in leaf and log litter, but also in dung and fungi (Hansen 1990b).

References. Hansen (1990b) (redescription, notes on Australian fauna); Hansen (1991) (generic diagnosis).

Pachysternum Motschulsky, 1863 (Pl. 47O; Figs 20.19N, 20.21D)

Type species. *Pachysternum nigrovittatum* Motschulsky, 1863.

Characteristics. Body widely oval, moderately convex; head with tentorial pits minute; prosternum with large plate without carina and with coarse punctures; antennal grooves large, reaching lateral margin of hypomeron (hence lateral margin of pronotum angulate in lateral view); mesoventral plate widely triangular, widely contacting metaventrite, without median carina; grooves for reception of procoxae large, nearly reaching mesocoxae; metaventrite without anterolateral arcuate ridges or pits and impressions, but with very distinct femoral lines reaching anterolateral corner of metaventrite; metanepisternum extremely narrow, virtually absent anteriorly; protibiae sinuate on outer margin metatibiae without dense long setae ventrally; dorsal punctuation without apparent setae; elytron with 10 elytral series, series impressed*, intervals flat; male sucking discs present; male sternite 9 tongue-like; median lobe freely movable and reaching deeply into phallobase; body length 2.3–2.5 mm*; body black with reddish pattern on elytral disc and yellowish spot at elytral apex. Based on general habitus and ventral morphology similar to *Cryptopleurum*, from which it differs by proportions of mesoventral plate and shape of the protibia (see the Identification key). Ventral morphology also resembles *Noteropagus*, see under that genus for diagnostic characters. It is unique among Australian Megasternini by its coloration.

Larvae: Unknown.

Australian species. Single introduced species, *Pachysternum capense* Mulsant.

Distribution. The genus is native in Oriental and Afrotropical Regions. *Pachysternum capense* is native and widely distributed in sub-Saharan Africa, but frequently introduced worldwide and recently established in Europe (Lőkkös *et al.* 2014). It was introduced to Western Australia, where it was collected only twice at two close localities (in 1969 at Kojonup and in 1981 in Rottnest Island) and did likely not establish.

Biology. In Africa, *P. capense* is a common inhabitant of excrements of herbivorous mammals. No data are available about the habitat from which the Australian specimens were collected.

References. Fikáček & Boukal (2004) (redescription of *P. capense*); Lőkkös *et al.* (2014) (summary of current distribution of *P. capense* outside Africa).

Pilocnema Hansen, 1990 (Pl. 47L; Figs 20.19A, O–P, 20.21A)

Type species. *Pilocnema maculiapex* Hansen, 1990.

Characteristics. Body widely oval and highly convex; head with minute tentorial pits; antennal cupule small or markedly enlarged; prosternum simply carinate medially; antennal grooves moderately large, not reaching lateral margin of hypomeron; mesoventral plate narrowly oval, rather narrowly contacting metaventrite; grooves for reception of procoxae small, not reaching mesocoxae; metaventrite without anterolateral arcuate ridges, pits, impressions or femoral lines; metanepisternum moderately wide; protibiae arcuate on outer margin; meso- and metatibiae with dense long setae ventrally; dorsal punctuation without apparent setae; elytron with 9 elytral series, series sharply impressed laterally, intervals flat to slightly convex; epipleura very wide throughout; male sucking disc present on maxilla; male sternite 9 tongue-like; median lobe freely movable, deeply inserted into phallobase; body uniformly dark (*P. nigra*) or with bicolored elytra (*P. maculiapex*); 2.0–2.5 mm;. Unique among Australian Megasternini in densely pubescent ventral face of meso- and metatibiae, very wide epipleura and in *P. maculiapex* also by enlarged antennal cupule; besides that easily recognisable by highly convex body with sharply impressed lateral elytral series. By ventral morphology it resembles *Cercyon*, *Cetiocyon* and *Merosoma*; except of aforementioned characters differs from them by tongue-like median portion of male sternite 9 (from *Merosoma* and *Cetiocyon*) and by carinate prosternum (from *Cetiocyon*).

Larvae: Unknown.

Australian species. Two species, *P. nigra* Hansen and *P. maculiapex* Hansen.

Distribution. The genus is confined to Australia, New Guinea (15 spp.) and Philippines (1 sp.). The two Australian species are both endemic to Australia, occurring in rainforests of northern Queensland (*P. maculiapex*) and southern Queensland and northern New South Wales (*P. nigra*).

Biology. Most specimens sifted from rainforest leaf litter, less frequently collected using flight intercept traps.

References. Hansen (1990b) (description of the genus, revision of Australian fauna); Hansen (2003) (discrovery of *Pilocnema* in New Guinea and Philippines, key to species); Hebauer (2004) (few more species from New Guinea).

Key to species. Hansen (1990b).

Pseudoosternum Hansen, 1990 (Pl. 47M; Fig. 20.21B)

Type species. *Pseudoosternum maculatum* Hansen, 1990.

Characteristics. Body elongate oval, moderately convex; head with minute tentorial pits; prosternum simply carinate medially; antennal grooves small, not reaching lateral margin of hypomeron; mesoventral plate elongate oval to narrowly lamellar, rather narrowly contacting metaventrite; grooves for reception of procoxae small, not reaching mesocoxae; metaventrite with anterolateral arcuate ridges, but without pits, impressions or femoral lines; metanepisternum moderately wide; protibiae arcuate on outer margin; meso- and metatibiae without dense long setae ventrally; dorsal punctuation setiferous; elytron with 10 elytral series, series impressed, intervals flat to convex; male sucking disc of maxilla absent; male sternite 9 crescent-like; median lobe joint to parameres; body uniformly reddish brown (*P. brunneum*) or dark brown with pale basal spots on elytra (*P. maculatum*); 1.3–1.8 mm. By the simple prosternum and narrow mesoventral plate resembling *Pilocnema, Cercyon*, *Cetiocyon*, *Cercyodes* and *Ercycodes*, but differing from all of them (besides their unique characters) by metaventrite with anterolateral arcuare ridge. By body size and habitus similar to *Cenebriophilus* and *Ceronocyton*, which differ by prosternum with median portion divided by ridges.

Larvae: Unknown.

Distribution and Australian species. Endemic to Australia, with two described species distributed in tropical areas of Queensland, Northern Territory and Western Australia (*P. brunneum*) or only in Queensland (*P. maculatum*).

Biology. Terrestrial, usually collected from leaf and log litter in rainforests, also using flight intercept traps.

References. Hansen (1990b) (description of the genus, review of species).

Key to species. Hansen (1990b).

Acknowledgments

I am indebted to A. Ślipiński (ANIC Canberra), Maxwell Barclay (Museum of Natural History, London), Alexey Solodovnikov (Zoologisk Museum, Copenhagen, Denmark) and L. Hendrich (Zoologische Staatssammlung München, Germany) for loans of part of the specimens which were used for preparation of this chapter. Thanks are due to M. Archangelsky, B. Clarkson, E. Gentili, M. Jäch, F.-L. Jia, A. Komarek, Y. Minoshima, A. Short, C. Watts and T. Weir for consultations during preparation of this chapter and/or comments and critique of the early version of the manuscript. M. Archangelsky and Y. Minoshima allowed me to use many of their drawings, including some unpublished ones. I also apologise to all my collaborators and colleagues for big delays in our projects caused by the work on this chapter, and to my ex-partner M. Pokoš for a year without holidays and proper Christmas.

References

Albertoni FF, Fikáček M (2014) A new bromeliad-inhabiting species of *Omicrus* Sharp from South Brazil (Coleoptera, Hydrophilidae, Sphaeridiinae). *Spixiana* **37**, 111–122.

Anderson JME (1976) Aquatic Hydrophilidae (Coleoptera). The biology of some Australian species with descrptions of immature stages reared in the laboratory. *Journal of the Australian Entomological Society* **15**, 219–228. doi:10.1111/j.1440-6055.1976.tb01696.x

Archangelsky M (1994) Description of immature stages of three Nearctic species of the genus *Berosus* Leach (Coleoptera: Hydrophilidae). *Internationale Revue der Gesamten Hydrobiologie* **79**(3), 357–372. doi:10.1002/iroh.19940790305

Archangelsky M (1997) Studies on the biology, ecology, and systematics of the immature stages of New world Hydrophiloidea (Coleoptera: Staphyliniformia). *Bulletin of the Ohio Biological Survey. New Series* **12**(1), 1–207.

Archangelsky M (1998) Phylogeny of Hydrophiloidea (Coleoptera: Staphyliniformia) using characters from adult and preimaginal stages. *Systematic Entomology* **23**, 9–24. doi:10.1046/j.1365-3113.1998.00039.x

Archangelsky M (1999a) Adaptations of immature stages of Sphaeridiinae (Staphyliniformia, Hydrophiloidea: Hydrophilidae) and state of knowledge of preimaginal Hydrophilidae. *Coleopterists Bulletin* **53**(1), 64–79.

Archangelsky M (1999b) Immature stages of *Paracymus rufocinctus* Bruch, 1915 and *Enochrus* (*Methydrus*) *vulgaris* (Steinheil, 1869) (Coleoptera: Hydrophilidae: Hydrophilinae). *Elytron* **13**, 87–99.

Archangelsky M (2000) Immature stages of Neotropical Hydrophilidae (Coleoptera): *Hydramara argentina* (Knisch, 1925) and *Hemiosus bruchi* Knisch, 1924. *Proceedings of the Entomological Society of Washington* **102**(2), 280–291.

Archangelsky M (2002a) Immature stages of Neotropical *Berosus* (Coleoptera, Hydrophilidae): *B. toxacanthus* Oliva, 1989, *B. coptogonus* Jensen-Haarup, 1910, *B. cornicinus* Knisch, 1922 and *B. auriceps* Boheman, 1859. *Tijdschrift voor Entomologie* **145**, 19–28. doi:10.1163/22119434-900000096

Archangelsky M (2002b) Nuevas larvas de Hydrophilidae (Coleoptera: Hydrophiloidea): *Hemiosus multimaculatus* y *Chaetarthria bruchi*. *Revista de la Sociedad Entomológica Argentina* **61**(3–4), 89–97.

Archangelsky M (2008) Phylogeny of Berosini (Coleoptera: Hydrophilidae. Hydrophilinae) based on larval and adult characters, and evolutionary scenarios related to habitat shift in larvae. *Systematic Entomology* **33**, 635–650. doi:10.1111/j.1365-3113.2008.00425.x

Archangelsky M (2016) Chaetotaxy and larval morphometry of *Cercyon praetextatus* (Say) and *Cercyon quisquilius* (Linnaeus) (Coleoptera: Hydrophilidae: Sphaeridiinae) and its phylogenetic implications. *Arthropod Systematics & Phylogeny* **74**(2), 177–193.

Archangelsky M (2018) Larval chaetotaxy and morphometry of *Oosternum costatum* (Coleoptera: Hydrophilidae) including a discussion on larval characters with phylogenetic relevance. *Acta Entomologica Musei Nationalis Pragae* **58**(2), 499–511. doi:10.2478/aemnp-2018-0038

Archangelsky M, Durand ME (1992) Description of the preimaginal stages of *Derallus angustus* Sharp, 1882 (Coleoptera: Hydrophilidae, Berosinae). *Aquatic Insects* **14**(3), 169–178. doi:10.1080/01650429209361479

Archangelsky M, Fernández LA (1994) Description of the preimaginal stages and biology of *Phaenonotum* (*Hydroglobus*) *puncticolle* Bruch (Coleoptera: Hydrophilidae). *Aquatic Insects* **16**, 55–63.

Archangelsky M, Fikáček M (2004) Descriptions of the egg case and larva of *Anacaena* and a review of the knowledge and relationships between larvae of Anacaenini (Coleoptera: Hydrophilidae: Hydrophilinae). *European Journal of Entomology* **101**, 629–636. doi:10.14411/eje.2004.086

Archangelsky M, Torres PLM, Michat MC (2004) Immature stages of *Hydrobiomorpha spinosa* (Orchymont, 1928) (Coleoptera: Hydrophilidae). *Aquatic Insects* **26**(3–4), 253–264. doi:10.1080/0165-0420400000593

Archangelsky M, Beutel RG, Komarek A (2005) 10.1. Hydrophilidae. Pp. 158–183. In: Beutel, RG, Leschen, RAB (eds) *Handbook of Zoology. Volume IV Arthropoda: Insecta. Part 38 Coleoptera, beetles Vol. 1: Morphology and systematice (Archostemata, Adephaga, Myxophaga, Polyphaga partim)*. Walter de Gruyter, Berlin-New York, 567 pp.

Archangelsky M, Rodriguez G, Torres PLM (2016) Primary chaetotaxy and larval morphometry of *Phaenonotum exstriatum* and *Dactylosternum cacti* (Coleoptera: Hydrophilidae). *Acta Entomologica Musei Nationalis Pragae* **56**(1), 167–193.

Arriaga-Varela E, Seidel M, Deler-Hernández A, Senderov V, Fikáček M (2017) A review of the *Cercyon* Leach (Coleoptera, Hydrophilidae, Sphaeridiinae) of the Greater Antilles. *ZooKeys* **681**, 39–93. doi:10.3897/zookeys.681.12522

Arriaga-Varela E, Wong SY, Kirejtshuk A, Fikáček M (2018) Review of flower-inhabiting water scavenger beetle genus *Cycreon* (Coleoptera, Hydrophilidae), with description of new species and comments on its biology. *Deutsche Entomologische Zeitschrift* **65**(1), 99–115. doi:10.3897/dez.65.26261

Balfour-Browne J (1939a) Contribution to the study of the Palpicornia II. *Entomologist's Monthly Magazine* **75**, 1–8.

Balfour-Browne J (1939b) On the aquatic Coleoptera of the New Hebrides and Banks Islands. Dytiscidae, Gyrinidae and Palpicornia. *Annals and Magazine of Natural History, Series 11* **3**, 370–376. doi:10.1080/03745481.1939.9723612

Bameul F (1986) Les Hydrophiloidea des îles Mascareignes (Coleoptera). *Revue Suisse de Zoologie* **93**(3), 875–910. doi:10.5962/bhl.part.79518

Bameul F (1992) Revision of the genus *Psalitrus* d'Orchymont from Southern India and Sri Lanka (Coleoptera: Hydrophilidae: Omicrini). *Systematic Entomology* **17**, 1–20. doi:10.1111/j.1365-3113.1992.tb00317.x

Bameul F (1993) Omicrini from Sulawesi Utara, Indonesia (Coleoptera: Hydrophilidae: Sphaeridiinae). *Invertebrate Taxonomy* **7**, 751–778. doi:10.1071/IT9930751

Bernhard D, Schmidt C, Korte A, Fritzsch G, Beutel RG (2006) From terrestrial to aquatic habitats and back again – molecular insights into the evolution and phylogeny of Hydrophiloidea (Coleoptera) using multigene analyses. *Zoologica Scripta* **35**, 597–606.

Bernhard D, Ribera I, Komarek A, Beutel RG (2009) Phylogenetic analysis of Hydrophiloidea (Coleoptera: Polyphaga) based on molecular data and morphological characters of adults and immature stages. *Insect Systematics & Evolution* **40**, 3–41. doi:10.1163/187631209X416741

Beutel RG (1994) Phylogenetic analysis of Hydrophiloidea based on characters of the head of adults and larvae. *Koleopterologische Rundschau* **64**, 103–131.

Beutel RG, Komarek A (2004) Comparative study of thoracic structures of adults of Hydrophiloidea and Histeroidea with phylogenetic implications (Coleoptera, Polyphaga). *Organisms, Diversity & Evolution* **4**, 1–34. doi:10.1016/j.ode.2003.10.001

Blackburn T (1898) Further notes on Australian Coleoptera, with descriptions of new genera and species XXIV. *Transactions of the Royal Society of South Australia* **22**(1897–1898), 221–233.

Bloom DD, Fikáček M, Short AEZ (2014) Clade age and diversification rate variation explain disparity in species richness among water scavenger beetle (Hydrophilidae) lineages. *PLoS One* **9**(6), e98430 doi:10.1371/journal.pone.0098430.

Bocák L, Barton C, Crampton-Platt A, Chesters D, Ahrens D, Vogler AP (2014) Building the Coleoptera tree-of-life for >8000 species: composition of public DNA data and fit with Linnaean classification. *Systematic Entomology* **39**, 97–110. doi:10.1111/syen.12037

Clarkson B, Short AEZ (2012) Revision of the *Oocyclus* Sharp of Brazil with description of ten new species (Coleoptera: Hydrophilidae: Laccobiini). *Zootaxa* **3183**, 1–35.

Clarkson B, Albertoni FF, Fikáček M (2014) Taxonomy and biology of the bromeliad–inhabiting genus *Lachnodacnum* (Coleoptera: Hydrophilidae: Sphaeridiinae). *Acta Entomologica Musei Nationalis Pragae* **54**, 157–194.

Costa C, Vanin SA, Casari-Chen SA (1988) *Larvas de Coleoptera do Brasil*. Museo di Zoologia-Universidade de Sao Paulo, Sao Paulo, 282 pp.

d'Orchymont A (1926) Contribution à l'étude des Hydrophilides VI. *Bulletin et Annales de la Société Entomologique de Belgique* **66**, 201–248.

d'Orchymont A (1932) Zur Kenntniss der Kolbenwasserkäfer (Palpicornia) von Sumatra, Java und Bali. In: Tropische Binnengewässer, Band II. *Archiv für Hydrobiologie* **Supplement 9**, 623–714.

d'Orchymont A (1933) Contribution à l'étude des Palpicornia VIII. *Bulletin et Annales de la Société Entomologique de Belgique* **73m**, 271–314, pl. 5.

d'Orchymont A (1940) Les Palpicornia des iles Atlantiques. *Memoires du Musee Royal d'Histoire Naturelle de Belgique, 2e série* **20**, 1–86.

De Marzo L (2000) Larvae di coleotteri indetriti vegetali di origine agricola: lineamenti morfologici e presenza stagionale (Polyphaga: 20 famiglie). *Entomologica* **34**, 65–131[Bari].

Deane C (1936) Notes on micro-Hydrophilidae. *Victorian Naturalist* **52**, 206–207.

Deler-Hernández A, Sýkora V, Seidel M, Cala-Riquelme F, Fikáček M (2018) Multiple origins of the *Phaenonotum* beetles in the Greater Antilles (Coleoptera: Hydrophilidae): phylogeny, biogeography and systematics. *Zoological Journal of the Linnean Society* **183**, 97–120. doi:10.1093/zoolinnean/zlx071

Dunstan B (1923) *Mesozoic Insects of Queensland. Part I. Introduction and Coleoptera*. A.J. Cumming, Government Printer, Brisbane.

Fikáček M (2010a) Hydrophilidae: The genus *Chaetarthria* Stephens – description of a new species from New Caledonia and notes in *C. nigerrima* (Blackburn) from Australia (Coleoptera). In: Jäch, MA, Balke M (eds) Water Beetles of New Caledonia. Part 1. *Monographs in Coleoptera* **3**, 263–269.

Fikáček M (2010b) Hydrophilidae: Sphaeridiinae (Coleoptera). In: Jäch MA, Balke M (eds) Water Beetles of New Caledonia. Part 1. *Monographs in Coleoptera* **3**, 323–364.

Fikáček M, Boukal M (2004) *Pachysternum capense*, a new genus and species for Europe, and updated key to genera and subgenera of European Sphaeridiinae (Coleoptera: Hydrophilidae). *Klapalekiana* **40**, 1–12.

Fikáček M, Engel MS (2011) An aquatic water scavenger beetle in Early Miocene amber from the Dominican Republic (Coleoptera: Hydrophilidae). *Annales Zoologici* **61**(4), 621–628. doi:10.3161/000345411X622462

Fikáček M, Schmied H (2013) Insect fauna of the Late Miocene locality of Öhningen (Germany) less diverse than reported: an example of the hydrophilid beetles (Coleoptera). *Journal of Paleontology* **87**(3), 427–443. doi:10.1666/12-101.1

Fikáček M, Short AEZ (2010) Taxonomic revision and phylogeny of the genus *Cetiocyon* and its discovery in the Neotropical region (Insecta: Coleoptera: Hydrophilidae). *Arthropod Systematics & Phylogeny* **68**(3), 309–329.

Fikáček M, Vondráček D (2014) A review of *Pseudorygmodus* (Coleoptera: Hydrophilidae), with notes on the classification of the Anacaenini and on distribution of genera endemic to southern South America. *Acta Entomologica Musei Nationalis Pragae* **54**, 479–514.

Fikáček M, Watts CHS (2015) Notes on Australian Anacaenini (Coleoptera: Hydrophilidae): description of male of *Phelea breviceps* Hansen and unravelling the identity of *Crenitis neogallica* Gentili. *Zootaxa* **3980**(3), 427–434. doi:10.11646/zootaxa.3980.3.6

Fikáček M, Archangelsky M, Torres PLM (2008) Primary chaetotaxy of the larval head capsule and head appendages of the Hydrophilidae (Coleoptera) based on larva of *Hydrobius fuscipes* (Linnaeus, 1758). *Zootaxa* **1874**, 16–34.

Fikáček M, Leschen RAB, Newton AF, Gunter N (2012) *Horelophus walkeri* rediscovered: adult morphology and notes on biology (Coleoptera: Hydrophilidae). *Acta Entomologica Musei Nationalis Pragae* **52**(1), 129–146.

Fikáček M, Minoshima Y, Vondráček D, Gunter N, Leschen RAB (2013) Morphology of adults and larvae and integrative taxonomy of southern hemisphere genera *Tormus* and *Afrotormus* (Coleoptera: Hydrophilidae). *Acta Entomologica Musei Nationalis Pragae* **53**, 75–126.

Fikáček M, Minoshima YN, Newton AF (2014a) A review of *Andotypus* and *Austrotypus* gen. nov., rygmodine genera with an austral disjunction (Hydrophilidae: Rygmodinae). *Annales Zoologici* **64**, 557–596. doi:10.3161/000345414X685893

Fikáček M, Prokin A, Yan E, Yue Y, Wang B, Ren D, Beattie R (2014b) Modern hydrophilid clades present and widespread in the Late Jurassic and Early Cretaceous (Coleoptera: Hydrophiloidea: Hydrophilidae). *Zoological Journal of the Linnean Society* **170**, 710–734. doi:10.1111/zoj.12114

Fikáček M, Maruyama M, Komatsu T, von Beeren C, Vondráček D, Short AEZ (2015) Protosternini (Coleoptera: Hydrophilidae) corroborated as monophyletic and its larva described for the first time: a review of the myrmecophilous genus *Sphaerocetum*. *Invertebrate Systematics* **29**(1), 23–36. doi:10.1071/IS14026

Fikáček M, Minoshima YN, Komarek A, Short AEZ, Huang D, Cai C (2017a) *Cretocrenis burmanicus*, the first Mesozoic amber inclusion of a water scavenger beetle (Coleoptera: Hydrophilidae). *Cretaceous Research* **77**, 49–55. doi:10.1016/j.cretres.2017.04.017

Fikáček M, Gustafson GT, Short AEZ (2017b) On wet rock with snorkels: immature stages of *Tritonus* cascade beetles with unusual modification of spiracles (Hydrophilidae: Laccobiini). *Annales Zoologici* **67**(1), 91–107. doi:10.3161/00034541ANZ2017.67.1.011

Fikáček M, Minoshina YN, Jäch MA (2018a) Larval morphology of *Yateberosus*, a New Caledonian endemic subgenus of *Laccobius* (Coleoptera: Hydrophilidae), with notes on parallel evolution of 'Berosus-like' larval morphology in Hydrophiloidea. *Acta Entomologica Musei Nationalis Pragae* **58**(1), 195–206. doi:10.2478/aemnp-2018-0017

Fikáček M, Liang WR, Hsiao Y, Jia FL, Vondráček D (2018b) Biology and morphology of immature stages of banana-associated *Protosternum* beetles, with comments on the status of Taiwanese endemic *P. abnormale* (Coleoptera: Hydrophilidae). *Zoologischer Anzeiger* **277**, 85–100. doi:10.1016/j.jcz.2018.10.001

Formanowicz DR, Brodie ED (1988) Prey density and ambush site changes in *Tropisternus lateralis* larvae (Coleoptera: Hydrophilidae). *Journal of the Kansas Entomological Society* **61**(4), 420–427.

Forster GN, Bilton DT, Friday LE (2014) Keys to adults of the water beetles of Britain and Ireland (part 2)(Coleoptera: Polyphaga: Hydrophiloidea – both aquatic and terrestrial species). *Handbooks for the Identification of British Insects* **4**(5b): i–vi+1–126.

García M (2018) El género Crenitulus en Venezuela (Coleoptera: Hydrophilidae) con descripción de doce nuevas especies. *UNED Research Journal* **10**(2), 304–329.

Gentili E (1981) The genera *Laccobius* and *Notohydrus* (Coleoptera, Hydrophiloidea) in Australian and New Zealand. *Records of the South Australian Museum* **18**(7), 143–154.

Gentili E (1989) Alcune novità sul genere *Laccobius* (Coleoptera, Hydrophilidae). *Annuario Osservatorio di Fisica terresre e Museo Antonio Stoppani del Seminario Arcivescovilledi Milano (N.S.)* **10**(1987), 31–39.

Gentili E (1992) The *Notohydrus* of Australia (Coleoptera Hydrophilidae). *Bollettino della Società Entomologica Italiana* **124**(1), 21–26.

Gentili E (1993) *Paranacaena* Blackburn, 1889: a valid genus (Coleoptera: Hydrophilidae). *Giornale Italiano di Entomologia* **6**, 285–296.

Gentili E (1996a) *Laccobius* nuovi o poco noti di Sierra Leone, Pakistan, Vietnam, Australia, Sumatra, Filippine e Nuova Guinea (Coleoptera Hydrophilidae). *Memorie Società Entomologica Italiana* **74**(1995), 21–32.

Gentili E (1996b) Notes on the Anacaenini from Australia and Melanesia with description of new species (Coleoptera, Hydrophilidae). *Giornale Italiano di Entomologia* **8**, 177–189.

Gentili E (2000) The *Paracymus* of Australia (Coleoptera, Hydrophilidae). *Records of the South Australian Museum* **33**(2), 101–122.

Gentili E (2005) The genus *Laccobius* Erichson, 1837 in the Australian region (Coleoptera, Hydrophilidae). Pp. 317–370. In: Daccordi M, Giachino PM (eds) Results of the zoological missions to Australia of the Regional Museum of Natural Sciences of Turin, Italy. II. *Monografie del Museo Regionale di Scienze Naturali, Torino* **42**, 1–643.

Gentili E (2006) *Notohydrus* Balfour-Browne, 1939: description of a new species and new distribution data (Coleoptera, Hydrophilidae). *Giornale Italiano di Entomologia* **53**, 179–182.

Gentili E, Fikáček M (2009) Taxonomic notes on *Laccobius*, subgenus *Glyptolaccobius*, with new records and descriptions of four new species (Coleoptera: Hydrophilidae). *Acta Entomologica Musei Nationalis Pragae* **49**, 607–623.

Girón JC, Short AEZ (2017) Revision of the Neotropical water scavenger beetle genus *Quadriops* Hansen, 1999 (Coleoptera, Hydrophilidae, Acidocerinae). *ZooKeys* **705**, 115–141. doi:10.3897/zookeys.705.19815

Girón JC, Short AEZ (2018) Three new genera of acidocerine water scavenger beetles from tropical South America (Coleoptera, Hydrophilidae, Acidocerinae). *ZooKeys* **768**, 113–158. doi:10.3897/zookeys.768.24423

Hafez M (1939) The external morphology of the full grown larva of *Cercyon quisquilius* L. (Coleoptera: Hydrophilidae). *Bulletin de la Société Fouad Ier d'Entomologie* **23**, 339–343.

Hansen M (1987) The Hydrophiloidea (Coleoptera) of Fennoscandia and Denmark. *Fauna Entomologica Scandinavica* **18**, 1–254.

Hansen M (1990a) *Hydatotrephis* Macleay, a subgenus of *Enochrus* Thomson (Coleoptera: Hydrophilidae). *Entomologica Scandinavica* **21**, 71–76. doi:10.1163/187631290X00058

Hansen M (1990b) Australian Sphaeridiinae (Coleoptera: Hydrophilidae): a taxonomic outline with descriptions of new genera and species. *Invertebrate Taxonomy* **4**, 317–395. doi:10.1071/IT9900317

Hansen M (1991) The hydrophiloid beetles. Phylogeny, classification and a revision of the genera (Coleoptera, Hydrophiloidea). *Biologiske Skrifter* **40**, 1–367.

Hansen M (1995) A review of the Hawaiian Hydrophilidae (Coleoptera). *Pacific Science* **49**(3), 266–288.

Hansen M (1999a) *Wold Catalogue of Insects. Volume 2. Hydrophiloidea (s.str.) (Coleoptera).* Apollo Books, Stenstrup, 416 pp.

Hansen M (1999b) Fifteen new genera of Hydrophilidae (Coleoptera), with remarks on generic classification of the family. *Entomologica Scandinavica* **30**, 121–172. doi:10.1163/187631200X00228

Hansen M (2003) Discovery of *Australocyon* Hansen and *Pilocnema* Hansen (Coleoptera, Hydrophilidae) outside the Australian region. Pp. 53–84. In: Cuccodoro, G, Leschen, RAB (eds) Systematics of Coleoptera:

Papers Celebrating the Retirement of Ivan Löbl. Memoirs of Entomology International, Associated Publisher.

Hansen M, Richardson B (1998) A new species of *Omicrus* Sharp (Coleoptera: Hydrophilidae) from Puerto Rico and its larva, the first known larva of Omicrini. *Systematic Entomology* **23**, 1–8. doi:10.1046/j.1365-3113.1998.00036.x

Hayashi N (1986) Hydrophilidae. In: Morimoto, K, Hayashi, N (eds) The Coleoptera of Japan in Color, Vol. I. Hoikusha Publishing Co. Ltd., Tokyo, vi+323 pp. + 113 pls.

Hayashi M, Hayama T (2009) Notes on the immature stages of *Cercyon aptus* Sharp (Coleoptera: Hydrophilidae). *Coleopterists News* **165**, 1–4.

Hebauer F (2001) Beitrag zur Kenntniss der Hydrophilidae von Neuguinea. Ergebnisse der zoologischen Forschungsreisen von. M. Balke und L. Hendrich nas West Neuguinea (Irian Jaya) in den Jahren 1990–1998 (Results of the German Hydroentomological Mission No. 4 [in part].) sowie Nachweise aus früheren Expeditionen. *Acta Coleopterologica* **17**(1), 3–72.

Hebauer F (2003) A new genus of water beetle *Gentilina* gen. nov. from Australia (Coleoptera: Hydrophilidae). *Records of the South Australian Museum* **36**, 109–114.

Hebauer F (2004) New species of the genus *Pilocnema* Hansen from New Guinea (Coleoptera: Hydrophilidae). *Acta Coleopterologica* **20**(2), 43–50.

Hebauer F (2006) Three new species of the genus *Mircogioton* d'Orchymont, 1937 from Papua New Guinea (Coleoptera: Hydrophilidae: Sphaeridiinae: Omicrini). *Acta Coleopterologica* **22**(2), 25–30.

Hebauer F, Hendrich L (1999) Two new species of *Helochares* from Northern Australia (Coleoptera: Hydrophilidae). *Entomological Problems* **30**(1), 47–51.

Holter P (2004) Dung feeding in hydrophilid, geotrupid and scarabaeid beetles: examples of parallel evolution. *European Journal of Entomology* **101**(3), 365–372. doi:10.14411/eje.2004.051

Hrbáček, J (1943) Dvě nové larvy Hydrophilidů. Beschreibung zweier neuen Hydrophilidenlarven. Časopis Československé Společnosti Entomologické **40**, 98–105.

Hrbáček J (1950) On the morphology and function of the. antenna of the Central European Hydrophilidae (Coleoptera). *Transactions of the Royal Entomological Society of London* **101**(7), 239–256. doi:10.1111/j.1365-2311.1950.tb00452.x

Inoda T, Hirata Y, Kamimura S (2003) Asymmetric mandibles of water–scavenger larvae improve feeding effectiveness on right–handed snails. *American Naturalist* **162**(6), 811–814. doi:10.1086/378903

Inoda T, Inoda Y, Rullan JK (2015) Larvae of the water scavenger beetle, *Hydrophilus acuminatus* (Coleoptera: Hydrophilidae) are specialist predators of snails. *European Journal of Entomology* **112**(1), 145–150.

Jell PA (2004) The fossil insects of Australia. *Memoirs of the Queensland Museum* **50**, 1–124.

Jia FL, Short AEZ (2013) *Enochrus algarum* sp. nov., a new hygropetric water scavenger beetle from China (Coleoptera: Hydrophilidae: Enochrinae). *Acta Entomologica Musei Nationalis Pragae* **53**, 609–614.

Jia FL, Zhang RJ (2017) A review of the genus *Cryptopleurum* from China (Coleoptera: Hydrophilidae). *Acta Entomologica Musei Nationalis Pragae* **57**(2), 577–592. doi:10.1515/aemnp-2017-0090

Jia FL, Aston P, Fikáček M (2014) Review of the Chinese species of the genus *Coelostoma* Brullé, 1835 (Coleoptera: Hydrophilidae: Sphaeridiinae). *Zootaxa* **3887**(3), 354–376.

Jia FL, Lin RC, Li BJ, Fikáček M (2015) A review of the omicrine genera *Omicrogiton, Mircogioton* and *Peratogonus* of China (Coleoptera, Hydrophilidae, Sphaeridiinae). *ZooKeys* **511**, 99–116. doi:10.3897/zookeys.511.8980

Komarek A (2007) Taxonomic revision of *Anacaena* Thomson, 1859. IV. Australia (Coleoptera: Hydrophilidae). *Koleopterologische Rundschau* **77**, 147–170.

Komarek A, Beutel RG (2007) Phylogenetic analysis of Anacaenini (Coleoptera: Hydrophilidae: Hydrophilinae) based on morphological characters of adults. *Systematic Entomology* **32**, 205–226. doi:10.1111/j.1365-3113.2006.00359.x

Kriska G, Csabai Z, Boda P, Malik P, Horváth G (2006) Why do red and dark-coloured cars lure aquatic insects? The attraction of aquatic insects to car paintwork explained by reflection-polarization signals. *Proceedings. Biological Sciences* **273**, 1667–1671. doi:10.1098/rspb.2006.3500

Lawrence JF, Ślipiński A (2013) *Australian Beetles. Volume 1. Morphology, classification and keys.* CSIRO Publishing, Collingwood, 561 pp.

Lawrence JF, Ślipiński A, Seago AE, Thayer MK, Newton AF, Marvaldi AE (2011) Phylogeny of the Coleoptera based on morphological characters of adults and larvae. *Annales Zoologici* **61**(1), 1–217. doi:10.3161/000345411X576725

Lea AM (1919) Notes on some miscellaneous Coleoptera with descriptions of new species - part V. *Transactions of the Royal Society of South Australia* **43**, 166–261[+ pls. xxv–xxvii.].

Lőkkös A, Boukal M, Fikáček M (2014) The hydrophilid beetle *Pachysternum capense* (Mulsant)(Coleoptera: Hydrophilidae) became a world-wide adventive species: a summary of distribution with new records from Europe, Australia, and South America. *North-Western Journal of Zoology* **10**(2), 333–336.

McKenna DD, Farrell BD, Caterino MS, Farnum CW, Hawks DC, Maddison DR, Seago AE, Short AEZ, Newton AF, Thayer MK (2015a) Phylogeny and evolution of Staphyliniformia and Scarabaeiformia: forest litter as a stepping stone for diversification of nonphytophagous beetles. *Systematic Entomology* **40**, 35–60. doi:10.1111/syen.12093

McKenna DD, Wild AL, Kanda K, Bellamy CL, Beutel RG, Caterino MS, Farnum CW, Hawks DC, Ivie MA, Jameson ML, Leschen RAB, Marvaldi AE, McHugh JV, Newton AF, Robertson JA, Thayer MK, Whiting MF, Lawrence JF, Ślipiński A, Maddison DR, Farrell BD (2015b) The beetle tree of life reveals that Coleoptera survived end-Permian mass extinction to diversify during the Cretaceous terrestrial revolution. *Systematic Entomology* **40**, 835–880. doi:10.1111/syen.12132

Minoshima Y (2013) Systematic study on the Japanese Hydrophilidae based on the larval morphology (Coleoptera). Unpublished Ph.D. thesis at Systematic Entomology, Division of Environmental Resources, Graduate School of Agriculture, Hokkaido University, Sapporo.

Minoshima Y, Hayashi M (2011a) Larval morphology of the Japanese species of the tribes Acidocerini, Hydrobiusini and Hydrophilini (Coleoptera: Hydrophilidae). *Acta Entomologica Musei Nationalis Pragae* **51**(supplementum), 1–118.

Minoshima Y, Hayashi M (2011b) Larval morphology of the genus *Hydrocassis* Fairmaire (Coleoptera: Hydrophilidae). *Journal of Natural History* **45**(45–46), 2757–2784. doi:10.1080/00222933.2011.602805

Minoshima Y, Hayashi M (2012) Larval morphology of *Amphiops mater mater* Sharp (Coleoptera: Hydrophilidae: Chaetarthriini). *Zootaxa* **3351**, 47–59.

Minoshima YN, Hayashi M (2015) Description of the larval stages of the berosine genera *Berosus* and *Regimbartia* based on the Japanese species *B. japonicus* and *R. attenuata* (Coleoptera: Hydrophilidae). *Acta Entomologica Musei Nationalis Pragae* **55**(1), 47–83.

Minoshima Y, Iwata Y, Hayashi M (2012) Morphology of the immature stages of *Hydrochara libera* (Sharp) (Coleoptera, Hydrophilidae). *Elytra. New Series* **2**(2), 285–302.

Minoshima Y, Hayashi M, Kobayashi N, Yoshitomi H (2013) Larval morphology and phylogenetic position of *Horelophopsis hanseni* Satô et Yoshitomi (Coleoptera, Hydrophilidae, Horelophopsinae). *Systematic Entomology* **38**, 708–722. doi:10.1111/syen.12025

Minoshima YN, Fikáček M, Gunter N, Leschen RAB (2015) Larval morphology and biology of New Zealand-Chilean vicariant genera *Cylomissus* and *Anticura* (Coleoptera: Hydrophilidae: Rygmodinae). *Coleopterists Bulletin* **69**(4), 687–712. doi:10.1649/0010-065X-69.4.687

Minoshima YN, Iwata Y, Hayashi M (2016) Egg-case and pupa of *Enochrus* (*Methydrus*) *japonicus* (Sharp) (Coleoptera: Hydrophilidae). *Coleopterists Bulletin* **70**(3), 562–565. doi:10.1649/0010-065X-70.3.562

Minoshima YN, Seidel M, Wood JR, Leschen RAB, Gunter NL, Fikáček M (2018) Morphology and biology of flower-visiting water scavenger beetle genus *Rygmodus* (Coleoptera, Hydrophilidae). *Entomological Science* **21**, 363–384. doi:10.1111/ens.12316

Mouchamps R (1958) Notes sur quelques *Coelostoma* (Brullé) principalement africains (12me note). *Bulletin de l'Institut Royal des Sciences Naturelles de Belgique* **34**(41), 1–36.

Nasserzadeh H, Komarek A (2017) Taxonomic revision of the water scavenger beetle genus *Sternolophus* Solier, 1834 (Coleoptera: Hydrophilidae). *Zootaxa* **4282**(2), 201–254. doi:10.11646/zootaxa.4282.2.1

Nasserzadeh H, Alipanah H, Gilasian E (2017) Phylogenetic study of the genus *Sternolophus* Solier (Coleoptera, Hydrophilidae) based on adult morphology. *ZooKeys* **712**, 69–85. doi:10.3897/zookeys.712.14085

Newton AF (1989) Review of *Dactylosternum* Wollaston species of Australia and New Zealand (Coleoptera: Hydrophilidae). *Australian Entomological Magazine* **16**(3), 49–58.

Ôhara M, Jia FL (2006) Terrestrial hydrophilid beetles of the Kuril Archipelago (Coleoptera, Hydrophilidae). *Biodiversity and Biogeography of the Kuril Islands and Sakhalin* **2**, 129–150.

Ordish RG (1974) Arthropoda of the subantarctic islands of New Zealand (3). Coleoptera: Hydrophilidae. *Journal of the Royal Society of New Zealand* **4**(3), 307–314. doi:10.1080/03036758.1974.10419398

Ponomarenko AG (1969) Istoricheskoe razvitie zhestkokrylykh – Arkhostemat. *Trudy Paleontologicheskogo Instituta Akademii Nauk SSSR* **125**, 1–238 + 14 pls.

Prins AJ (1984) Morphological and biological notes on some South African arthropods associated with decaying organic matter. Part 2. The predatory families Carabidae, Hydrophilidae, Histeridae, Staphylinidae and Silphidae (Coleoptera). *Annals of the South African Museum* **92**(4), 295–356.

Richmond EA (1920) Studies on the biology of the aquatic Hydrophilidae. *Bulletin of the American Museum of Natural History* **42**(1), 1–93 + 16 Pls.

Ryker LC (1972) Acoustic behaviour of four sympatric species of water scavenger beetles (Coleoptera, Hydrophilidae, *Tropisternus*). *Occasional Papers of the Museum of Zoology, University of Michigan* **666**, 1–19.

Ryker LC (1976) Acoustic behaviour of *Tropisternus ellipticus, T. columbianus,* and *T. lateralis limbalis* in western Oregon (Coleoptera: Hydrophilidae). *Coleopterists Bulletin* **30**(2), 147–156.

Ryndevich SK (2008) Review of species of the genus *Cercyon* Leach, 1817 of Russia and adjacent regions. IV. The subgenera *Paracycreon* Orchymont, 1942 and *Dicyrtocercyon* Ganglbauer, 1904 (Coleoptera: Hydrophilidae). *Zoosystematica Rossica* **17**(2), 89–97.

Santana L, Benetti CJ, Pes AM (2017) A new species of *Crenitulus* Winters, 1926 (Coleoptera: Hydrophilidae: Chaetarthriinae: Anacaenini) from northern South America. *Zootaxa* **4319**(1), 169–176. doi:10.11646/zootaxa.4319.1.9

Satô M (1979) Ergebnisse der Bhutan-Expedition 1972 und Indien-Nepal Expeditionen 1975–1977 des Naturhistorisches Museums in Basel. Coleoptera: Fam. Georissidae, Hydraenidae, Hydrophilidae und Ptilodactylidae. *Entomologica Basiliensia* **4**, 43–67.

Schödl S (1993) Revision der Gattung *Berosus* Leach. 3. Teil: Die paläarktischen und orientalischen Arten der Untergattung *Berosus* s.str. (Coleoptera: Hydrophilidae). *Koleopterologische Rundschau* **63**, 189–233.

Schulte F (1985) Eidonomie, Ethökologie und Larvalsystematik dungbewohnender *Cercyon*-Species (Coleoptera: Hydrophilidae). *Entomologia Generalis* **11**(1–2), 47–55. doi:10.1127/entom.gen/11/1985/47

Schwind R (1991) Polarization vision in water insects and insects living on moist substrate. *Journal of Comparative Physiology. A, Neuroethology, Sensory, Neural, and Behavioral Physiology* **169**, 531–540. doi:10.1007/BF00193544

Scott MA (1913) No. X.–Coleoptera; Hydrophilidae, Histeridae. In: The Percy Sladen Trust Expedition to the Indian ocean in 1905, under leadership of Mr. J. Stanley Gardiner, M.A. Vol. V. *Transactions of the Linnean Society of London, Second Series Zoology* **16**, 193–235 + pl. 14.

Seidel M, Arriaga-Varela E, Fikáček M (2016) Establishment of Cylominae Zaitzev, 1908 as a valid name for the subfamily Rygmodinae Orchymont, 1916 with an updated list of genera (Coleoptera: Hydrophilidae). *Acta Entomologica Musei Nationalis Pragae* **56**(1), 159–165.

Seidel M, Minoshima YN, Arriaga-Varela E, Fikáček M (2018) Breaking a disjunct distribution: a review of the southern hemisphere genera Cylorygmus and Relictorygmus gen. nov. (Hydrophilidae: Cylominae). *Annales Zoologici* **68**(2), 375–402 [Warszawa]. doi:10.3161/00034541ANZ2018.68.2.011

Short AEZ (2008) Revision of the Afrotropical genus *Tritonus* Mulsant (Coleoptera: Hydrophilidae). *Zootaxa* **1855**, 41–55.

Short AEZ (2009) Description of *Micramphiops* gen.n. from Madagascar (Coleoptera: Hydrophilidae). *Koleopterologische Rundschau* **79**, 189–195.

Short AEZ (2010a) Hydrophilidae: Hydrobiusina (Coleoptera). In: Jäch, MA, Balke M (eds) Water Beetles of New Caledonia. Part 1. *Monographs in Coleoptera* **3**, 319–322.

Short AEZ (2010b) Phylogeny, evolution and classification of the giant water scavenger beetles (Coleoptera: Hydrophilidae: Hydrophilini: Hydrophilina). *Systematics and Biodiversity* **8**(1), 17–37. doi:10.1080/14772000903529375

Short AEZ (2010c) Hydrophilidae: Review of the subtribe Acidocerina of the Southwest Pacific Islands (Coleoptera). In: Jäch MA, Balke M (eds) Water Beetles of New Caledonia. Part 1. *Monographs in Coleoptera* **3**, 297–318.

Short AEZ, Fikáček M (2011) World catalogue of the Hydrophiloidea (Coleoptera): additions and corrections II (2006–2010). *Acta Entomologica Musei Nationalis Pragae* **51**, 83–122.

Short AEZ, Fikáček M (2013) Molecular phylogeny, evolution and classification of the Hydrophilidae (Coleoptera). *Systematic Entomology* **38**(4), 723–752. doi:10.1111/syen.12024

Short AEZ, García M (2014) A new genus of egg case-carrying water scavenger beetle from the Guiana shield (Coleoptera: Hydrophilidae: Acidocerinae). *Zootaxa* **3835**, 251–262. doi:10.11646/zootaxa.3835.2.5

Short AEZ, Hebauer F (2006) World catalogue of Hydrophiloidea – additions and corrections, 1 (1999–2005) (Coleoptera). *Koleopterologische Rundschau* **76**, 315–359.

Short AEZ, Liebherr JK (2007) Systematics and biology of the endemic water scavenger beetles of Hawaii (Coleoptera: Hydrophilidae, Hydrophilini). *Systematic Entomology* **32**, 601–624. doi:10.1111/j.1365-3113.2007.00403.x

Short AEZ, Cole J, Toussaint EFA (2017) Phylogeny, classification and evolution of the water scavenger beetle tribe Hydrobiusini inferred from morphology and molecules (Coleoptera: Hydrophilidae: Hydrophilinae). *Systematic Entomology* **42**(4), 677–691. doi:10.1111/syen.12239

Smetana A (1975) Revision of the New World genera of the tribe Omicrini trib. nov. of the hydrophilid subfamily Sphaeridiinae (Coleoptera). *Studies on the Neotropical Fauna* **10**, 153–182. doi:10.1080/01650527509360490

Smetana A (1978) Revision of the subfamily Sphaeridiinae of America north of Mexico (Coleoptera: Hydrophilidae). *Memoirs of the Entomological Society of Canada* **110**(S105), 1–292. doi:10.4039/entm110105fv

Sowig P, Himmelsbach R, Himmelsbach W (1997) Predator-prey relationship between insect larvae: growth of *Sphaeridium* larvae (Coleoptera: Hydrophilidae) under time constraints through predation on *Musca autumnalis* maggots (Diptera: Muscidae). *Canadian Journal of Zoology* **75**, 2069–2076. doi:10.1139/z97-841

Spangler PJ (1977) Three new Ecuadorian species of the aquatic beetle genus *Chaetarthria* (Coleoptera: Hydrophilidae). *Proceedings of the Biological Society of Washington* **90**, 566–578.

Spangler PJ (1979) Description of the larva and pupa of *Cylorygmus lineatopunctatus* (Coleoptera: Hydrophilidae: Rygmodini). *Proceedings of the Biological Society of Washington* **92**, 743–752.

Szczepański WT, Vondráček D, Seidel M, Wardhaugh C, Fikáček M (2018) High diversity of *Cetiocyon* beetles (Coleoptera: Hydrophilidae) along an elevational gradient on Mt. Wilhelm, New Guinea, with new records from the Bird's Head Peninsula. *Arthropod Systematics & Phylogeny* **76**(2), 323–347.

Tillyard RJ (1916) *Mesozoic and Tertiary Insects of Queensland and New South Wales*. A.J. Cumming, Government Printer, Brisbane.

Toussaint EFA, Short AEZ (2017) Biogeographic mirages? Molecular evidence for dispersal-driven evolution in Hydrobiusini water scavenger beetles. *Systematic Entomology* **42**(4), 692–702. doi:10.1111/syen.12237

Toussaint EFA, Short AEZ (2018) Transoceanic stepping-stones between Cretaceous waterfalls? The enigmatic biogeography of pantropical *Oocyclus* cascade beetles. *Molecular Phylogenetics and Evolution*, doi: doi:10.1016/j.ympev.2018.04.023.

Toussaint EFA, Fikáček M, Short AEZ (2016) India-Madagascar vicariance explains cascade beetle biogeography. *Biological Journal of the Linnean Society. Linnean Society of London* **118**(4), 982–991. doi:10.1111/bij.12791

Toussaint EFA, Bloom D, Short AEZ (2017) Cretaceous West Gondwana vicariance shaped giant water scavenger beetle biogeography. *Journal of Biogeography* **44**(9), 1952–1965. doi:10.1111/jbi.12977

van Berge Henegouwen AL (1982) Notes on the larval stages of some East African Hydrophilinae (Coleoptera: Hydrophilidae). *Entomologische Berichten* **42**(1), 11–16[Amsterdam].

van Berge Henegouwen AL (1986a) *Sphaeridium huijbregtsi*, a new species from Halmahera, Irian Jaya, Papua New Guinea and the Solomon Islands (Coleoptera, Hydrophilidae). *Bulletin et Annales de la Société Royale Belge d'Entomologie* **122**, 253–257.

van Berge Henegouwen AL (1986b) *Sphaeridium braziliense*, a new species from Brazil (Coleoptera: Hydrophilidae). *Entomologische Berichten* **46**, 36–38[Amsterdam].

Van Tassell ER (1965) An audiospectrographic study of stridulation as an isolating mechanism in the genus *Berosus* (Coleoptera: Hydrophilidae). *Annals of the Entomological Society of America* **58**(4), 407–413. doi:10.1093/aesa/58.4.407

Watts CHS (1987) Revision of Australian *Berosus* Leach (Coleoptera: Hydrophilidae). *Records of the South Australian Museum* **21**(1), 1–28.

Watts CHS (1988) Revision of Australasian *Hydrophilus* Muller, 1764 (Coleoptera: Hydrophilidae). *Records of the South Australian Museum* **22**(2), 117–130.

Watts CHS (1989) Revision of Australasian *Sternolophus* Solier (Coleoptera: Hydrophilidae). *Records of the South Australian Museum* **23**(2), 89–95.

Watts CHS (1990) Revision of Australian *Hydrobiomorpha* Blackburn (Coleoptera: Hydrophilidae). *Records of the South Australian Museum* **24**(1), 35–42.

Watts CHS (1995) Revision of the Australasian genera *Agraphydrus* Regimbart, *Chasmogenus* Sharp and *Helochares* Mulsant (Coleoptera: Hydrophilidae). *Records of the South Australian Museum* **28**, 113–130.

Watts CHS (1996) Three new *Berosus* Leach (Coleoptera: Hydrophilidae) from Australia. *Records of the South Australian Museum* **29**(2), 147–152.

Watts CHS (1998a) Revision of Australian *Amphiops* Erichson, *Allocotocerus* Kraatz and *Regimbartia* Zaitzev (Coleoptera: Hydrophilidae). *Records of the South Australian Museum* **30**(2), 93–106.

Watts CHS (1998b) Revision of Australian *Enochrus* Thomson (Coleoptera: Hydrophilidae). *Records of the South Australian Museum* **30**(2), 137–156.

Watts CHS (1999) Rediscovery of *Enochrus peregrinus* in Australia (Coleoptera, Hydrophilidae). *Records of the South Australian Museum* **32**(1), 119.

Watts CHS (2000) Revision of Australian *Chaetarthria* Stephens (Coleoptera: Hydrophilidae). *Records of the South Australian Museum* **33**(1), 29–31.

Watts CHS (2002) The larvae of some Australian aquatic Hydrophilidae (Coleoptera: Insecta). *Records of the South Australian Museum* **35**(2), 105–138.

Winterbourn MJ (1973) The larva of *Anacaena tepida* (Coleoptera: Hydrophilidae) from a Rotorua hot spring. *New Zealand Entomologist* **5**, 171–174. doi:10.1080/00779962.1973.9722992

21. HISTERIDAE GYLLENHAL, 1808

Yu-Lingzi Zhou, Sławomir Mazur, Tomáš Lackner and Adam Ślipiński

Fig. 21.1. *Hister ambulator* Thayer

Common name. Clown beetles

Introduction. Histeridae is a moderately large, cosmopolitan family with ~400 genera and 3900 species (Mazur 2011) currently placed in eleven subfamilies. Described Australian taxa are currently placed in 9 subfamilies, 48 genera and 260 species. The family is well defined morphologically and there is little doubt that Histeridae with two small northern Hemisphere families, Synteliidae and Sphaeritidae, form a monophyletic group, often considered to be an independent superfamily Histeroidea (Beutel 2016). The phylogenetic relationships of histeroid families are not well established. Morphology-based phylogenetic research (Hansen 1997; Ślipiński & Mazur 1999; Caterino & Vogler 2002) recovered a basal Sphaeritidae, with Synteliidae more closely related to Histeridae, but molecular studies (Caterino *et al.* 2005; Bocák *et al.* 2014; McKenna *et al.* 2015) recovered Sphaeritidae as more closely related to Histeridae. Three Burmese amber (~99 Ma) inclusions constitute the oldest records of the family Histeridae (Poinar & Brown 2009; Caterino *et al.* 2015; Caterino & Maddison 2018).

The subfamily and tribal classification of Histeridae is still controversial, since relationships of major clades have not been properly established. Ôhara (1994), Ślipiński & Mazur (1999), Caterino & Vogler (2002) and McKenna *et al.* (2015) studied histerid phylogeny based on morphological and/or molecular evidence and proposed several, largely incongruent patterns of relationship among recognised clades/subfamilies. The major controversial points are (1) placement of cylindrical-bodied bark beetle predators (Niponiinae, Trypanaeinae, and Trypeticinae), (2) delimitation of Onthophilinae and Tribalinae, and (3) subdivisions of Histerinae (Ślipiński & Mazur 1999; Kovarik & Caterino 2016).

Except for revisions of the subfamilies Chlamydopsinae (Caterino 2003, 2007; Caterino & Dégallier 2007; Dégallier & Caterino 2005a, 2005b) and Saprininae (Lackner & Leschen 2017), a revision of the genus *Trypeticus* (Kanaar, 2003) and several papers dealing with Australian 'micro-Histeridae' (mostly Acritini and Bacaniini) (Gomy (1984, 1991, 1994, 1995, 2009a), taxonomy of Australian Histeridae has not been studied in detail since the papers by Lea (1910, 1912, 1914a, b, 1918, 1919, 1925a–c). The Australian species of Histeridae were also catalogued by Mazur (1984, 1997b, 2011).

Several histerids have been introduced into Australia as part of biological control programmes aimed mostly at control of *Haematobia exigua* de Meijere (buffalo fly) and *Musca vetustissima* Walker (bush fly) that breed in the dung of introduced herbivores. Apparently *Hister calidus* Erichson, *Hister cruentus* Erichson and *Pachylister caffer* Erichson from Africa, and *Hololepta quadridentata* (Olivier) and *Plaesius javanus* Erichson from Java have not established (Waterhouse & Sands 2001).

Biology. Histerid adults and their larvae are mostly predators, feeding on small arthropods, particularly the larvae of cyclorrhaphan flies. Histerids attracted to dung or carrion belong to Histerini and Saprinini, mostly *Hister*, *Notosaprinus*, *Saprinus*, *Chalcionellus*, *Gnathoncus* and *Hypocaccus*. *Gnathoncus* and *Carcinops* are likewise routinely attracted to poultry manure. Several species of the genus *Tomogenius* are found in caves where they prey on fly larvae occurring in bat guano (Lackner & Leschen 2017). Members of the genus *Hypocaccus*, as well as those of *Halacritus*, are found on beaches where they most likely prey on arthropod larvae and adults occurring on or under wrack (Lackner & Leschen 2017). Micro-histerids (*Acritus*, *Acritomorphus*, *Australanius*, *Bacanius*) most likely prey on mites, while several sub-cortical taxa (*Epiechinus*, *Epierus*, *Parepierus*, *Tribalus*) have their mouthparts adapted to consume the fungal spores, as well as the larvae of tiny arthropods (Kovarik & Caterino 2016). The Australian histerid fauna consists of a comparatively large number of myrmecophilous taxa, mostly classified in the largest subfamily Chlamydopsinae (107 Australian species in 6 genera; Mazur 2011). However, apparent symphiles have also been found in Saprininae; *Iridoprinus* lives inside the nests of *Iridiomyrmex* ants (Lackner & Leschen 2017) and two specialised termitophilous species, *Arbolister myrmecophilus* Mazur and *Saprinus rarus* Lackner & Leschen, were found in arboreal nests of *Nasutitermes* Dudley (Lackner & Leschen 2017). The biology of Chlamydopsinae, which are obligate inquilines of social insects (mainly ants) was summarised by Caterino & Dégallier (2007). These beetles exhibit a wide range of bizarre morphological characteristics associated with their lifestyle (Caterino & Dégallier 2007). Dorso-ventrally flattened members of the Australian Hololeptini and Platysomatini (Histerinae) and several Paromalini (Dendrophilinae) live mostly under bark preying on small insects and other arthropods. The cylindrical species of Niponiinae, Trypeticinae, and Teretriini occur in the galleries of bark or Ambrosia beetles, where they prey upon larvae and sometimes also upon adults (Kanaar 2003; Kovarik & Caterino 2016).

Characteristics. Adults of Histeridae are usually easily recognised by their compact, heavily sclerotised bodies, retracted head with large projecting mandibles, short geniculate and clubbed antennae and truncate elytra usually exposing two abdominal tergites (propygidium and pygidium). The beetles are usually uniformly black or brown (bicolored in some Saprininae and Histerinae) and glabrous with their dorsal surfaces feebly shining to strongly iridescent. Chlamydopsinae are mostly dull, with dorsal setae and distinct microsculpture, often brick-red or ochre coloured. Body length varies from 0.8 mm (*Aeletes, Acritus, Bacaniomorphus*) to more than 10 mm (*Nasaltus*). A more complete adult description with emphasis on the Australian fauna may be found in Volume 1 (Lawrence & Ślipiński 2013).

The histerid body plan is relatively constant and several critical morphological characters are used extensively in adult classification and identification (Wenzel 1962; DeMarzo & Vienna 1982a, 1982b, 1982c; Kukalová-Peck & Lawrence 1993; Ôhara 1994; Kovarik & Caterino 2001; Yélamos 2002; Kovarik & Tishechkin 2004; Kovarik & Caterino 2005, 2016; Kanaar 1997, 2003; Lackner 2010; Caterino & Tishechkin 2014). The main diagnostic features used in the keys to identification are illustrated on Fig. 21.2 and the terms are briefly explained below.

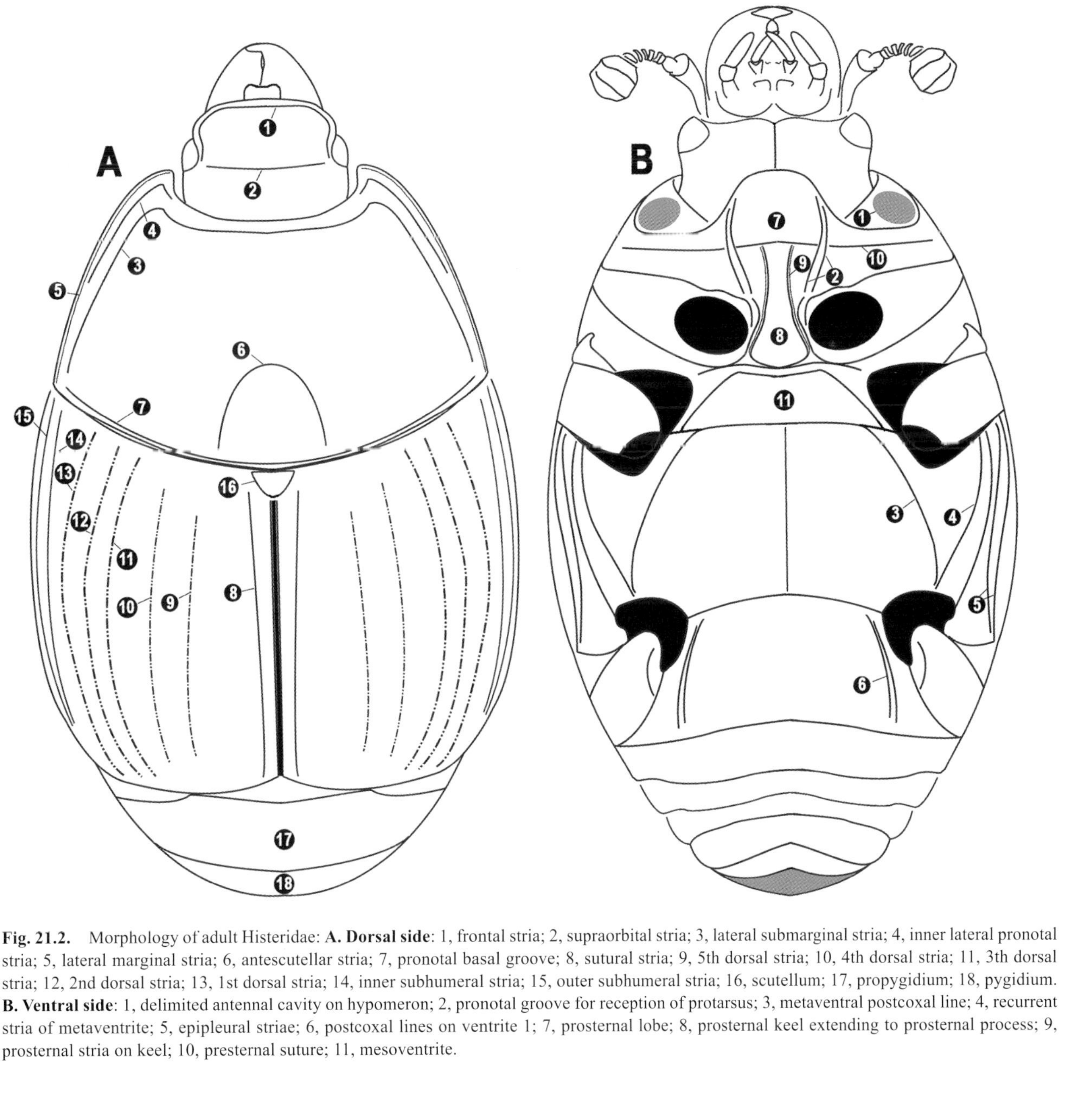

Fig. 21.2. Morphology of adult Histeridae: **A. Dorsal side**: 1, frontal stria; 2, supraorbital stria; 3, lateral submarginal stria; 4, inner lateral pronotal stria; 5, lateral marginal stria; 6, antescutellar stria; 7, pronotal basal groove; 8, sutural stria; 9, 5th dorsal stria; 10, 4th dorsal stria; 11, 3th dorsal stria; 12, 2nd dorsal stria; 13, 1st dorsal stria; 14, inner subhumeral stria; 15, outer subhumeral stria; 16, scutellum; 17, propygidium; 18, pygidium. **B. Ventral side**: 1, delimited antennal cavity on hypomeron; 2, pronotal groove for reception of protarsus; 3, metaventral postcoxal line; 4, recurrent stria of metaventrite; 5, epipleural striae; 6, postcoxal lines on ventrite 1; 7, prosternal lobe; 8, prosternal keel extending to prosternal process; 9, prosternal stria on keel; 10, presternal suture; 11, mesoventrite.

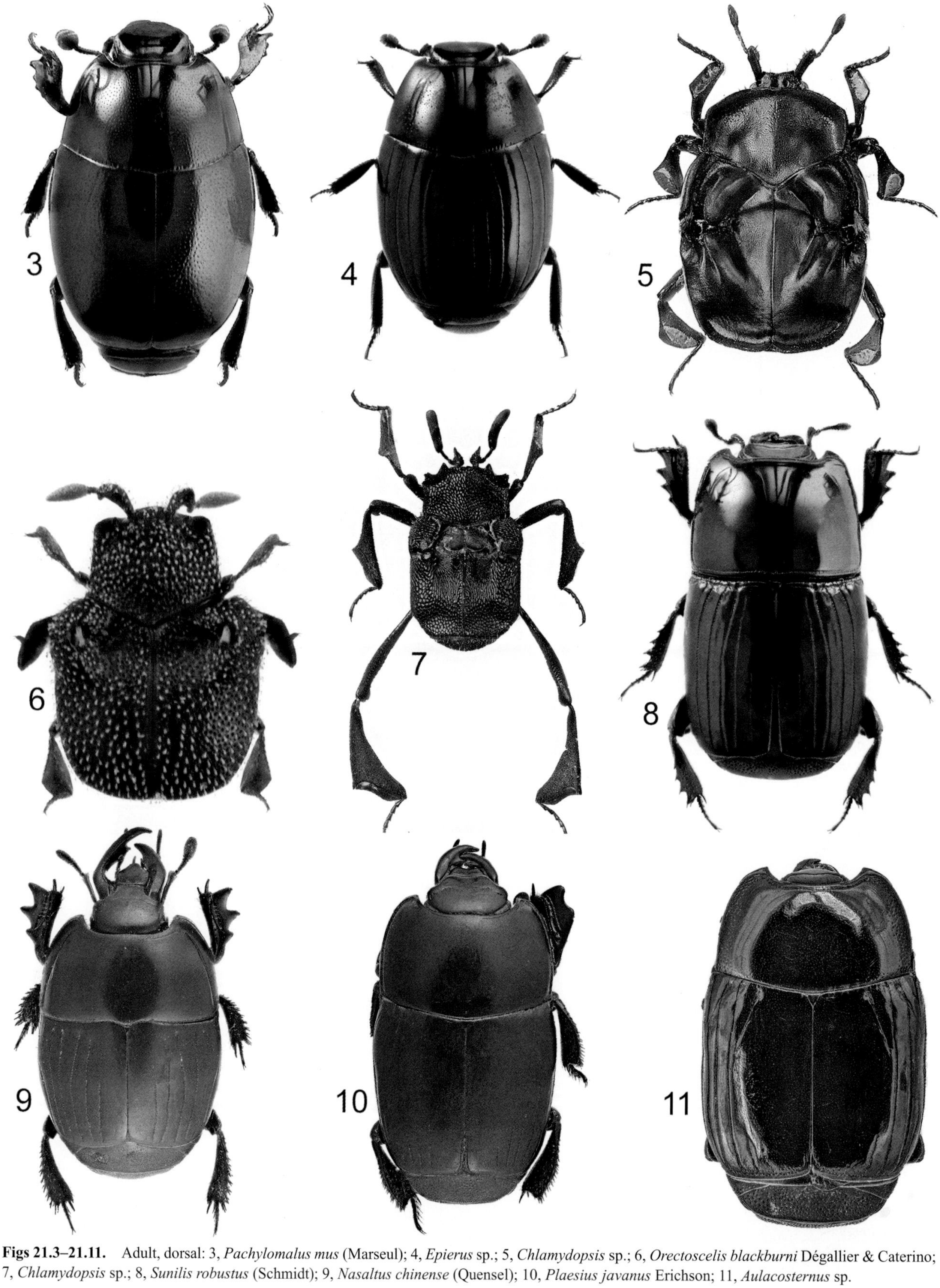

Figs 21.3–21.11. Adult, dorsal: 3, *Pachylomalus mus* (Marseul); 4, *Epierus* sp.; 5, *Chlamydopsis* sp.; 6, *Orectoscelis blackburni* Dégallier & Caterino; 7, *Chlamydopsis* sp.; 8, *Sunilis robustus* (Schmidt); 9, *Nasaltus chinense* (Quensel); 10, *Plaesius javanus* Erichson; 11, *Aulacosternus* sp.

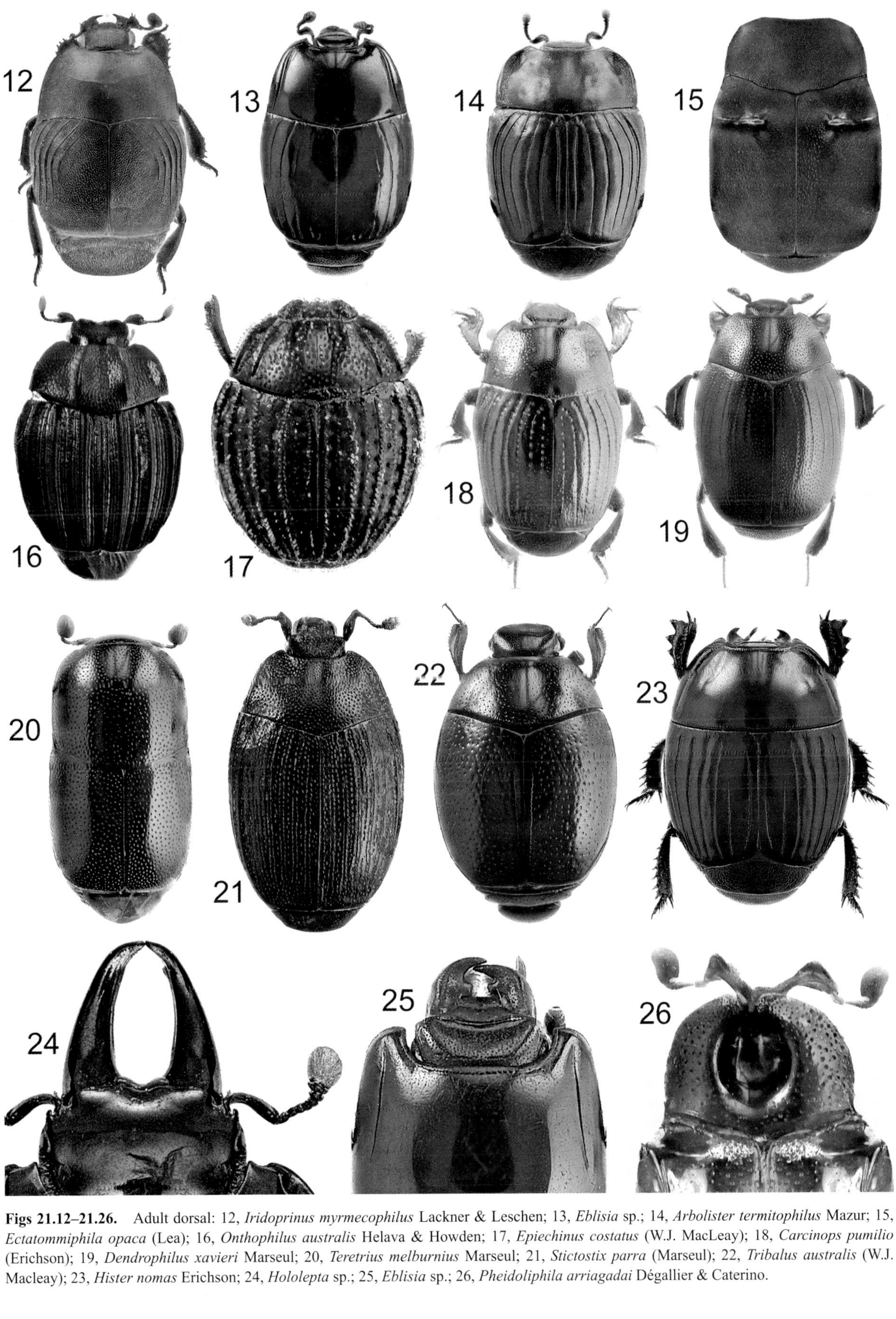

Figs 21.12–21.26. Adult dorsal: 12, *Iridoprinus myrmecophilus* Lackner & Leschen; 13, *Eblisia* sp.; 14, *Arbolister termitophilus* Mazur; 15, *Ectatommiphila opaca* (Lea); 16, *Onthophilus australis* Helava & Howden; 17, *Epiechinus costatus* (W.J. MacLeay); 18, *Carcinops pumilio* (Erichson); 19, *Dendrophilus xavieri* Marseul; 20, *Teretrius melburnius* Marseul; 21, *Stictostix parra* (Marseul); 22, *Tribalus australis* (W.J. Macleay); 23, *Hister nomas* Erichson; 24, *Hololepta* sp.; 25, *Eblisia* sp.; 26, *Pheidoliphila arriagadai* Dégallier & Caterino.

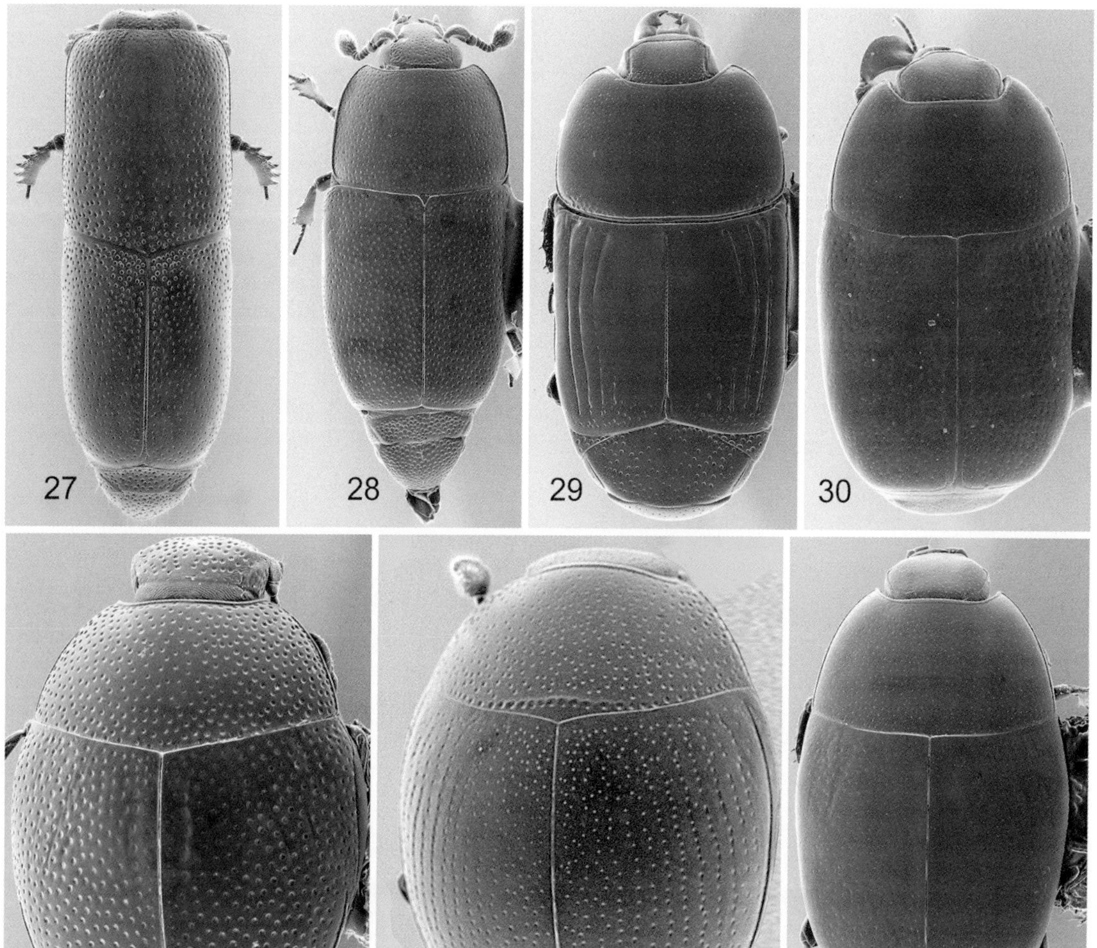

Figs 21.27–21.36. Adult dorsal: 27, *Trypeticus monteithi* Kanaar; 28, *Acritomorphus* sp.; 29, *Asterix parallelus* (Lewis); 30, *Platylomalus terraereginae* (Blackburn); 31, *Bacanius suturalis* (Lea); 32, *Australanius* sp.; 33, *Eulomalus miliaris* (Marseul); 34, *Acritus* sp.; 35, *Halacritus lividus* (Lea); 36, *Aeletes franzi* Gomy.

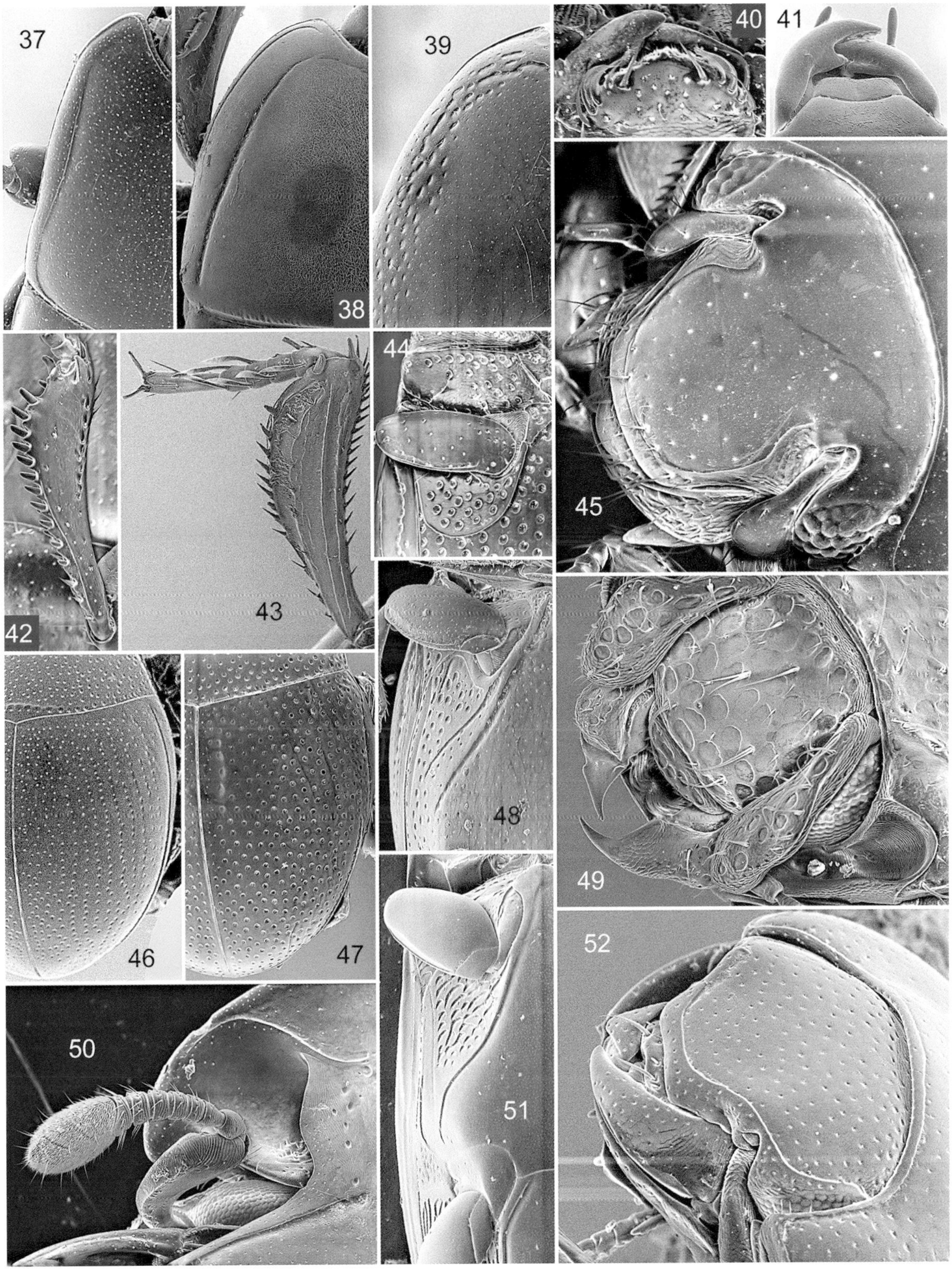

Figs 21.37–21.52. Adult characters: 37, *Platysoma bipunctatum* Lewis, pronotum, lateral margin; 38, *Nasaltus chinese* (Quensel), pronotum, lateral margin; 39, *Aulacosternus* sp., pronotum, lateral margin; 40, *Halacritus lividus* (Lea); 41, *Plaesius javanus* Erichson, labrum and mandibles; 42, *Halacritus lividus* (Lea), protibia; 43, *Acritus* sp., protibia; 44, *Bacanius suturalis* (Lea), metaventral postcoxal line; 45, *Aeletes franzi* Gomy, head; 46, *Australanius* sp.; 47, *Bacanius suturalis* (Lea), subhumeral stria; 48, *Carcinops troglodytes* (Paykull), paired metaventral postcoxal lines; 49, *Chlamydopsis mallee* Caterino, head; 50, *Pacifister urvillei* (Le Guillou), antenna and hypomeral antennal cavity; 51, *Eulomalus miliaris* (Marseul), metaventral postcoxal line; 52, *Platylomalus terraereginae* (Blackburn), head.

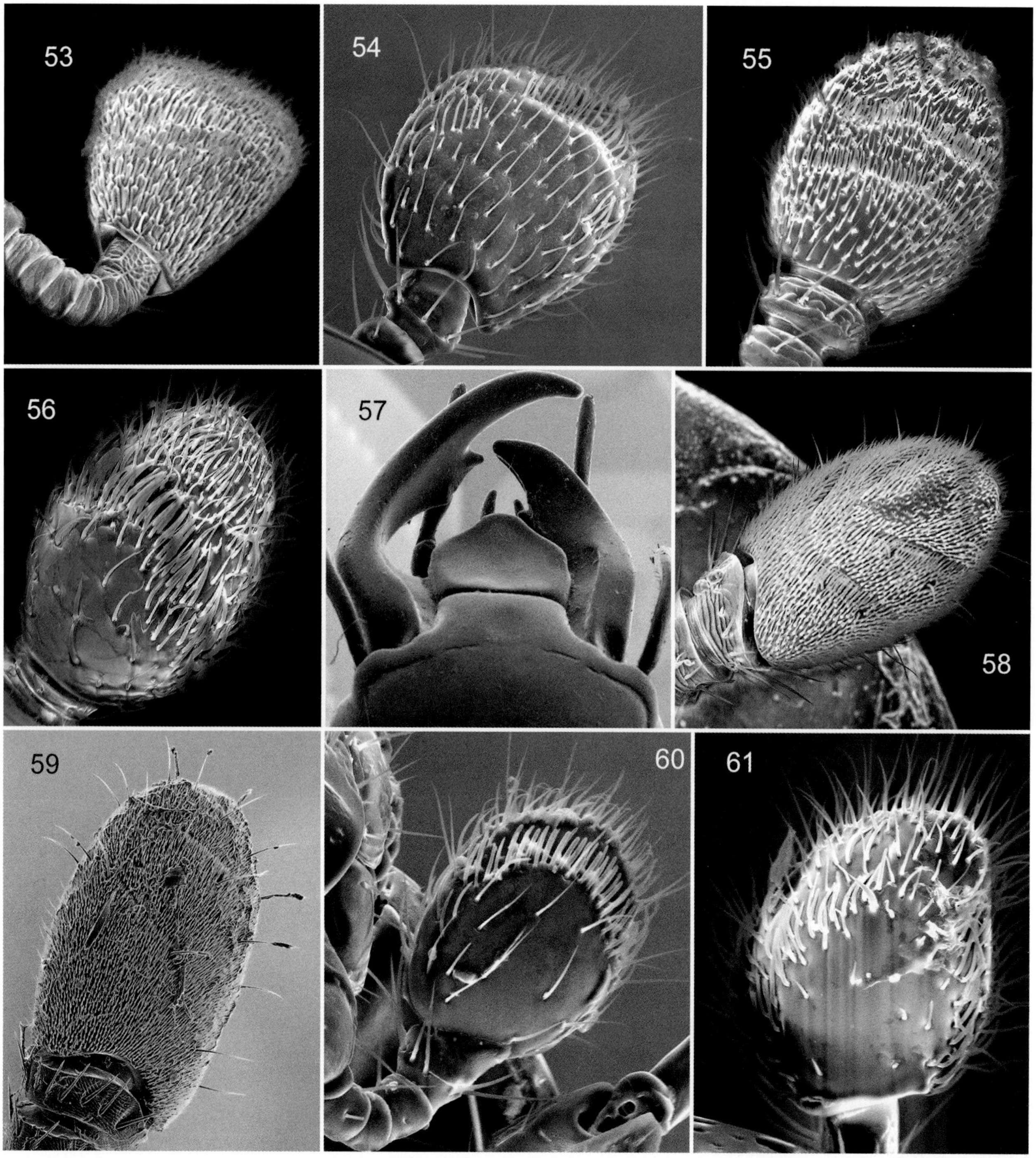

Figs 21.53–21.61. 53–56, 58–61, antennal club: 53, *Tribalus australis* (W.J. Macleay); 54, *Eulomalus miliaris* (Marseul); 55, *Carcinops troglodytes* (Paykull); 56, *Acritomorphus* sp.; 57, *Nasaltus chinense* (Quensel), labrum and mandibles; 58, *Pacifister urvillei* (Le Guillou); 59, *Nasaltus chinense* (Quensel); 60, *Acritus* sp.; 61, *Australanius* sp.

Head: shape, setation and visibility of labrum in dorsal aspect (Figs 21.40, 21.41, 21.49, 21.57); frontal stria (usually beginning along inner margin of eyes and continuing across frontoclypeus, Fig. 21.52); supraorbital stria (often concealed by the anterior part of pronotum); frontoclypeal; suture (usually not visible, but obvious in *Bacanius*); insertion point of the antennae in relation to the eye and its dorsal visibility (Figs 21.45, 21.49, 21. 52); length and shape of the antennal scape and the subdivision (sutures) of the antennal club, which is always composed of three antennomeres (Figs 21.53–21.57, 21.58–21.61); in Saprininae club segments are solidly fused and bear specialised sensory plaques called Reichardt's organs (Fig. 21.82).

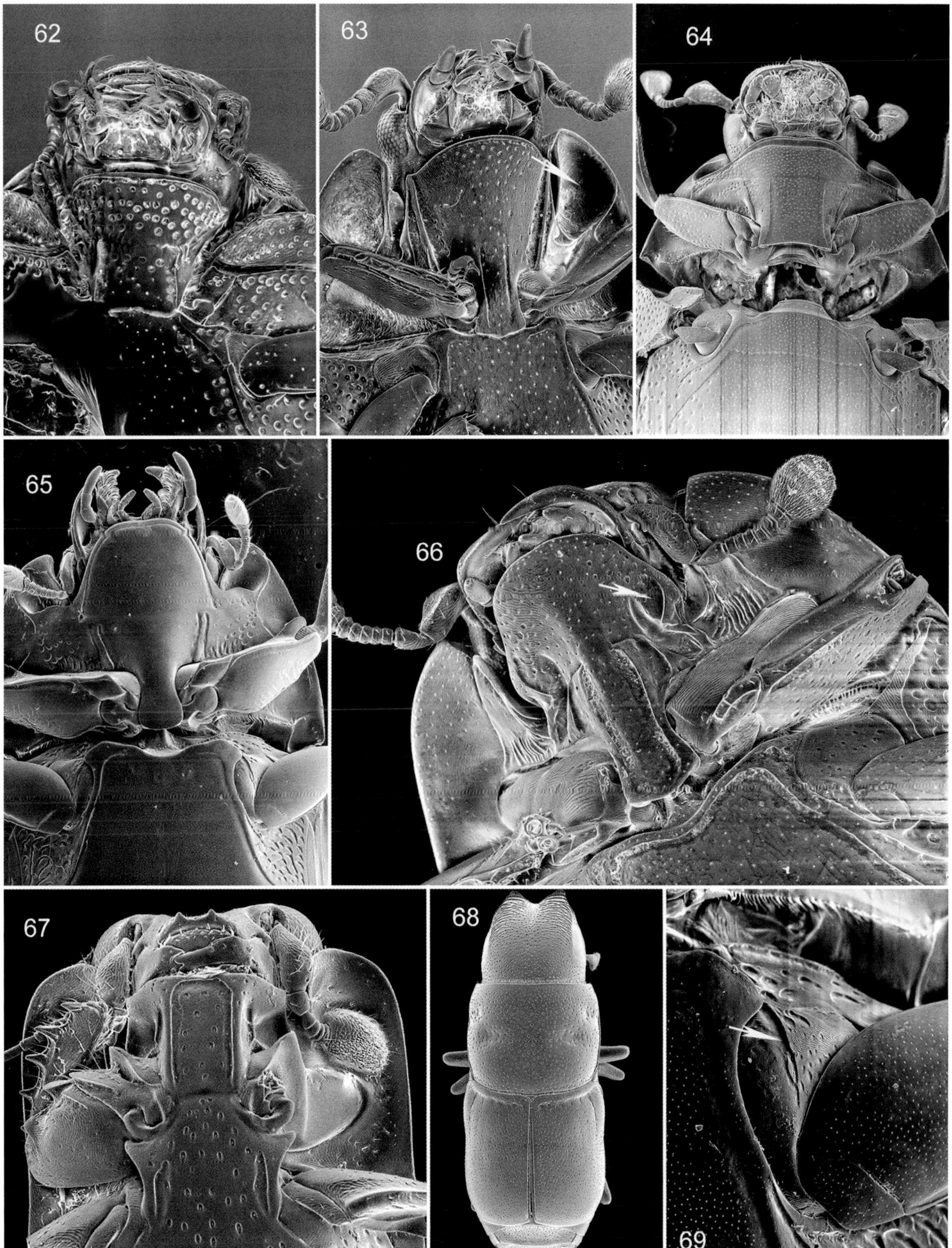

Figs 21.62–21.69. 62, *Bacanius suturalis* (Lea), prosternum; 63, *Acritomorphus* sp., prosternum; 64, *Tribalus australis* (W.J. Macleay), prosternum; 65, *Pacifister urvillei* (Le Guillou), prosternum; 66, *Carcinops troglodytes* (Paykull), prosternum; 67, *Trypeticus monteithi* Kanaar, prosternum and mesoventrite; 68, *Niponius* sp., India; 69, *Platysoma bipunctatum* Lewis, mesocoxa.

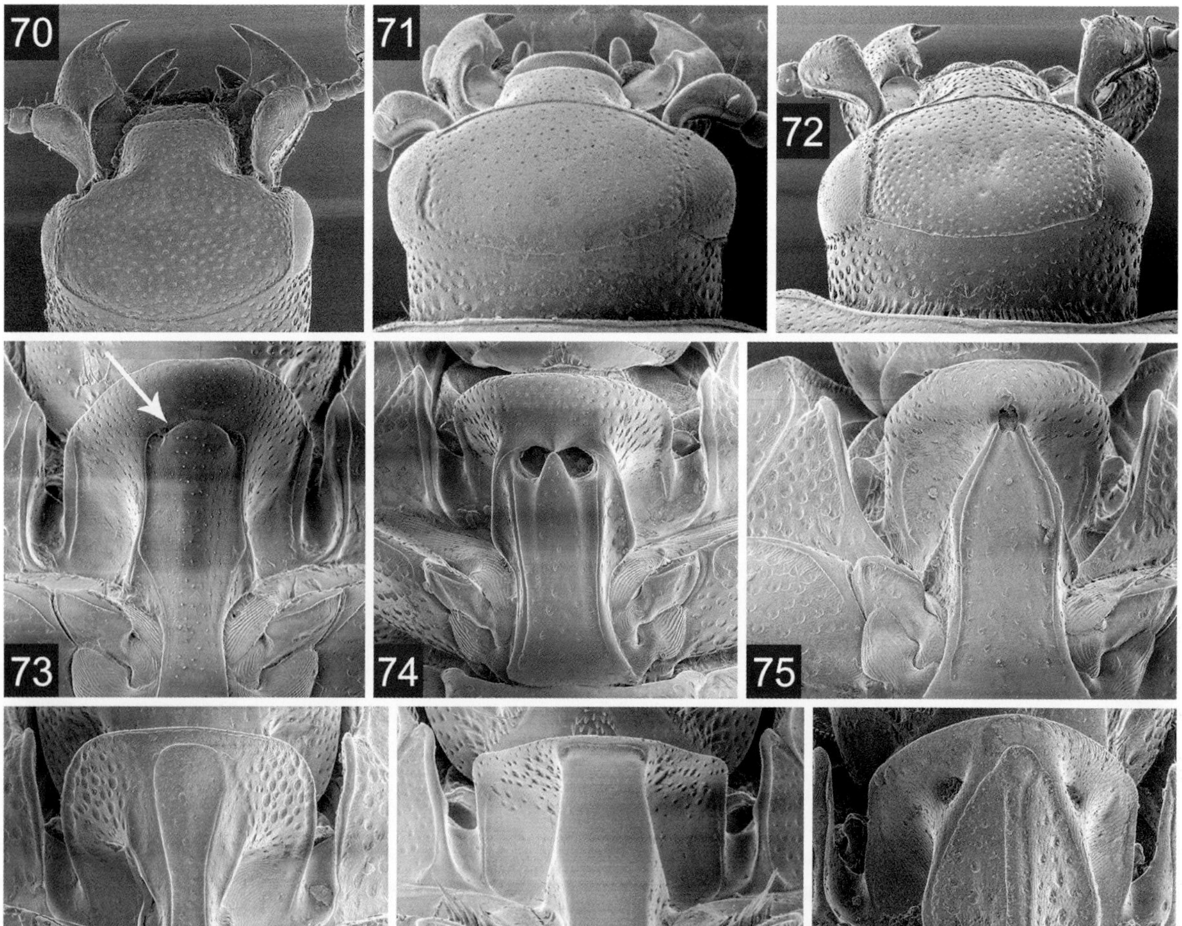

Figs 21.70–21.82. Saprininae adult characters: 70, *Tomogenius motocola* Mazur, head; 71, *Chalcionellus aeneovirens* (Schmidt), head; 72, *Saprinus cyaneus* (Fabricius), head; 73, *Euspilotus (Neosaprinus) rubriculus* (Marseul), prosternum; 74, *Tomogenius ripicola* (Marseul), prosternum; 75, *Gnathoncus brevisternus* Lewis, prosternum; 76, *Saprinodes falcifer* Lewis, prosternum; 77, *Notosaprinus irinus* (Marseul), prosternum; 78, *Hypocacus interpunctatus* (Schmidt), prosternum; 79, *Chalcionellus aeneovirens* (Schmidt), prosternum; 80, *Saprinus amethystinus* Lewis, protibia; 81, *Saprinodes distinctus* Dégallier, protibia; 82, *Tomogenius incisus* (Erichson), antennal club.

Prothorax dorsal: marginal pronotal stria (forming a bead along the edge of pronotum; Fig. 21.39); lateral submarginal pronotal stria (situated along anterior pronotal margin; Fig. 21.37); inner lateral pronotal stria (rarely present, incomplete and located near anterior angles between marginal and submarginal pronotal striae, Fig. 21.38); antescutellar stria, complete or incomplete arcuate line in front of scutellum, and marginal stria along the base of pronotum.

Prothorax ventral: development of the anterior part of prosternum and associated structures (lateral alae and lateral notch) to accommodate scape and protect antennal club in repose (Figs 21.62–21.67); prosternal lobe (anteriorly projecting median part of prosternum often delimited posteriorly by fine stria (= presternal suture)); prosternal keel (elevated median part of prosternum); prosternal process (posterior part of prosternum between and behind procoxae); prosternum usually has lateral striae and often median pair of carinal striae located on prosternal keel (Fig. 21.63).

Elytra: dorsal elytral striae 1–5, with numbering starting from the most external stria located at the inner side of humeral tubercle; sutural stria along the elytral suture (usually incomplete anteriorly); subhumeral and epipleural striae (on external side of each elytron beyond first dorsal elytral stria).

Pterothorax and abdomen ventral: anterior margin of mesoventrite and marginal mesoventral stria (Figs 21.62–21.67); shape and end length of metaventral postcoxal line (= lateral metaventral stria; sometimes double, Figs 21.44, 21.48, 21.51); recurrent metaventral stria. Abdominal ventrite 1 with single or double postcoxal lines (= abdominal ventral striae).

Histerid larvae are elongate, lightly sclerotised and flattened with prognathous head and usually 2-segmented urogomphi. They are carnivorous and resemble some Hydrophilidae but are distinguished by having only a single stemma on each side of the head, mandible lacking a mola but with a basal penicillus and a maxilla apparently lacking a cardo (Newton 1991). Kovarik (1995) described in detail the eclosion of larva from egg, as well as feeding habits of larvae, and according to him all histerid larvae are liquid feeders and digestion is extraoral. Clown beetles have only two larval instars, and according to Kovarik & Caterino (2005) the shortest development is found with the taxa associated with dung or carcasses, while the subcortical taxa undergo slightly longer larval life. For further information on larval cycle, morphology and development the reader is referred to the works Kovarik & Passoa (1993); Kovarik (1995); Beutel (1999); Kovarik & Caterino (2005); and Caterino & Tishechkin (2006).

Notes. We have not been able to establish an identity of *Apobletes errans* Marseul, 1860, subsequently made the type species of *Diabletes* Reichardt, 1933 (= *Diaphorus* Marseul, 1860). This is probably not an Australian species and Marseul (1860) only provisionally indicated Tasmania as a type locality. This species and the genus is treated as *incertae sedis* and excluded from the Australian fauna, pending further research.

Key to the genera of Australian Histeridae

1.	Body distinctly cylindrical (Figs 21.20, 21.27); elytra never with complete striae	**2**
–	Body oval or flattened; IF somewhat parallel-sided THEN elytra with some complete striae (Fig. 21.8)	**5**
2(1).	Head prognathous (horizontal); clypeus produced to form two anterior horns (Fig. 21.68)	**Niponiinae: *Niponius* Lewis**
–	Head oblique or vertical; clypeus not produced into horns	**3**
3(2).	Antennal scape strongly expanded, triangular (Fig. 21.49); scutellum concealed; tarsi short; head not retractable	**Chlamydopsinae: *Teretriopsis* Caterino & Dégallier**
–	Antennal scape elongate; scutellum visible, triangular; tarsi very long and slender	**4**
4(3).	Prosternal process incised apically receiving projecting mesoventrite	**Abraeinae: *Teretrius* Erichson**
–	Prosternal process not incised apically (Fig. 21.67)	**Trypeticinae: *Trypeticus* Marseul**
5(1).	Elytra with humeral trichomes and gland openings (Figs 21.5–21.7); eyes large, extending on frontal surface; antennal scape very large, triangular, completely covering eye in repose (Fig. 21.49)	**Chlamydopsinae…6**
–	Elytra without humeral trichomes and gland openings; eyes small and located laterally; antennal scape expanded apically, not triangular, not completely covering eye in repose (Fig. 21.52)	**10**
6(5).	Scutellum externally visible	**7**
–	Scutellum concealed, either by posterior margin of pronotum, or by junction of two elytra	**9**
7(6).	Dorsum flat; humeral trichomes transverse (Fig. 21.15)	***Ectatommiphila* Lea**
–	Dorsum convex, often irregular (Fig. 21.5); trichomes, never transverse	**8**
8(7).	Anterior margin of pronotum flat; humeral trichomes bearing elongate hairs, as long as the elytra	***Eucurtia* Mjöberg**
–	Anterior margin of pronotum raised in most species (Fig. 21.5); trichomes with short hairs	***Chlamydopsis* Westwood**
9(6).	Pronotum usually with prominent, bifid anterior marginal processes (Fig. 21.26)	***Pheidoliphila* Lea**
–	Pronotal processes usually paired and separate (Fig. 21.6)	***Orectoscelis* Lewis**
10(5).	Elytra with raised costae (Fig. 21.16)	**11**
–	Elytra with striae or punctures only (Fig. 21.18)	**13**
11(10).	Pronotum with pair of admedian costae (Fig. 21.16); each elytron with four strong costae	***Onthophilus* Leach (Onthophilinae)**
–	Pronotum flat; elytral costae very weak (Fig. 21.17)	**12**
12(11).	Dorsum with rows of short and stout setae (Fig. 21.17)	***Epiechinus* Lewis (Onthophilinae)**
–	Dorsum without apparent setae (Fig. 21.21)	***Stictostix* Marseul, part (Tribalinae)**
13(10).	Length less than 1.4 mm. Body strongly convex, globular (Fig. 21.31)	**14**
–	Length more than 1.5 mm. Body variable	**20**

14(13). Scutellum reduced, not visible externally **15**
– Scutellum very small but visible **19**
15(14). Antescutellar stria on pronotum complete .... ***Bacaniomorphus* Mazur (Dendrophilinae)**
– Antescutellar stria of pronotum absent **16**
16(15). Body elongate-oval and weakly convex (Fig. 21.33); metaventral postcoxal line long, descending and reaching almost end of ventrite (Fig. 21.51); prosternal process with parallel carinae. Length ~1.3–1.5 mm .... ***Eulomalus* Cooman, part (Dendrophilinae)**
– Body oval and strongly convex; metaventral postcoxal line short and usually recurved (Fig. 21.44); prosternal process without carinae. Length usually less than 1.2 mm **17**
17(16). Frontal stria absent. Frontoclypeus with trace of darker fronto-clypeal suture; tarsi 5–5–5. Protibia broad, widest in the middle and with fine spines externally. Antennal club tomentose, much denser on apical 1/3 (Fig. 21.61) **18**
– Frontal stria thin, but complete along anterior margin of clypeus to inner side of eye margins (Fig. 21.45). Frontoclypeus without traces of dark suture; tarsi 5–5–4. Protibia widened apically and without spines externally. Antennal club with basal 3/4 shiny and with few visible setae (Fig. 21.60) .... ***Aeletes* Horn (Abraeinae)**
18(17). Elytra with irregular punctures (Fig. 21.31); elytral subhumeral stria absent or visible only in apical half (Fig. 21.47) .... ***Bacanius* LeConte (Dendrophilinae)**
– Elytra with punctures forming 6–7 regular rows, sometimes weakly impressed forming rudimentary striae (Fig. 21.32); subhumeral stria sharp and complete (Fig. 21.46) .... ***Australanius* Gomy (Dendrophilinae)**
19(14). Protibia distinctly expanded apically and spinose along its outer margin (Fig. 21.42). Body outline subquadrate (Fig. 21.35) .... ***Halacritus* Schmidt, part (Abraeinae)**
– Protibia slender and setose along outer margin (Fig. 21.43). Body outline more elongate-oval (Fig. 21.34) .... ***Acritus* LeConte (Abraeinae)**
20(13). Anterior margin of prosternum with pair of deep lateral notches receiving antennal funicle in repose (Figs 21.66, 21.78) **21**
– Anterior margin of prosternum without lateral notches, continuous on each side with hypomeron (Figs 21.64, 21.65) **36**
21(20). Smaller beetles, length at most 3.4 mm; elytra without or with strongly reduced striae (Figs 21.3, 21.19); prosternal keel broad and flat (Fig. 21.66); antennal cavity, if present, located on prothoracic hypomeron (Fig. 21.63); antenna variable but always without Reichardt's organs **22**
– Larger beetles, length at least 3.8 mm; elytra with characteristic pattern of striae, dense punctures and smooth lustrous areas on elytra (Fig. 21.12); dorsum shiny, often iridescent; prosternal keel narrow and strongly elevated medially with deep antennal cavity at level of antennal notch (Fig. 21.76); antenna short with solid club bearing Reichardt's organs (Fig. 21.82)… **Saprininae** **28**
22(21) Elytral disc with impressed punctate striae (Fig. 21.18) **23**
– Elytral disc punctate without impressed striae except for some vague rudiments (Fig. 21.3) **24**
23(22). Frontal and supraorbital striae distinctly reduced; metaventral postcoxal lines single .... ***Dendrophilus* Leach (Dendrophilinae)**
– Frontal and supraorbital striae complete and obvious; metaventral postcoxal lines double (Fig. 21.48) .... ***Carcinops* Marseul (Dendrophilinae)**
24(22). Scutellum visible (Fig. 21.28) **25**
– Scutellum externally not visible (Fig. 21.30) **26**
25(24). Body elongate oval and flat (Fig. 21.28), dark brown; tarsal formula 5–5–5 .... **Wenzel (Abraeinae)**
– Body short, oval (Fig. 21.35) and more convex, light brown; tarsal formula 5–5–4 .... ***Halacritus* Schmidt (Abraeinae)**
26(24). Pronotal base with antescutellar stria, often interrupted in the middle (Fig. 21.3); metaventral postcoxal line short and recurved meeting metanepisternum; propygidium with anterior transverse stria .... ***Pachylomalus* Schmidt (Dendrophilinae)**
– Pronotal base without antescutellar stria; metaventral postcoxal line long and descending posteriorly along metanepisternum (Fig. 21.51); propygidium without anterior stria **27**
27(26) Frontal stria widely interrupted anteriorly .... ***Eulomalus* Coomann, part (Dendrophilinae)**
– Frontal stria complete anteriorly (Fig. 21.52) .... ***Platylomalus* Cooman (Dendrophilinae)**
28(21). Protibia arcuate devoid of teeth or denticles externally, apically narrowed and prolonged into single apical tooth (Fig. 21.81) .... ***Saprinodes* Lewis**
– Protibia normal with teeth and apical spurs, not terminating in single tooth (Fig. 21.80) **29**
29(28). Anterior part of frontal stria totally absent (Fig. 21.70) **30**
– Anterior part of frontal stria usually present (Fig. 21.72), sometimes partly reduced in *Saprinus* **32**
30(29). Carinal striae on prosternum joined anteriorly by deep sulcus (Fig. 21.73); elytral disc between fourth dorsal elytral and sutural elytral striae without hooked appendix; marginal elytral stria single .... ***Euspilotus rubriculus* (Marseul)**
— Carinal striae on prosternum not joined anteriorly by deep sulcus; elytral disc between fourth dorsal and sutural elytral striae with short hooked appendix; marginal elytral stria double **31**
31(30). Prosternum anteriorly with two large median foveae separated by apex of triangular process (Fig. 21.74) .... ***Tomogenius* Marseul**
– Prosternum anteriorly with single small median fovea (Fig. 21.75) or without foveae [in *G. communis* (Marseul)] .... ***Gnathoncus* Jacquelin-Duval**
32(29). Prosternum without carinal striae (Fig. 21.77) (occasionally vestiges of carinal prosternal striae present between procoxae) .... ***Notosaprinus* Kryzhanovskij**
– Prosternum always with carinal striae **33**
33(32). Elytra with strongly carinate elytral striae; sutural stria absent (Fig. 21.12) .... ***Iridoprinus* Lackner & Leschen**
– Elytral striae never carinate; sutural elytral stria always present, sometimes shortened basally **34**
34(33). Prosternum anteriorly without foveae .... ***Saprinus* Erichson**
– Prosternum anteriorly with foveae (Fig. 21.78) **35**
35(34). Prosternal foveae (Fig. 21.79) large and deep, anteriorly connected by marginal prosternal stria; frontal stria not carinate, somewhat weakened medially (Fig. 21.71) .... ***Chalcionellus* Reichardt**
– Prosternal foveae (Fig. 21.78) anteriorly not connected by marginal prosternal stria; frontal stria often carinate .... ***Hypocaccus* Thomson**
36(20). Labrum with setigerous punctures (Fig. 21.40); antennal scape normal length, club with oblique furrows (Fig. 21.59), never with V-shaped furrows; middle leg elongate with small spines on outer surface.... **Tribalinae** **37**

– Labrum without setigerous punctures (Fig. 21.41); antennal scape relatively long, club usually with V-shaped furrows (Fig. 21.58); middle leg distinctly broad, usually with coarse spines on outer surface.... **Histerinae** ... **40**

37(36). Elytral striae strongly reduced, except for sutural stria (Fig. 21.22) ... ***Tribalus* Erichson**

– Elytral striae well developed (Fig. 21.4) ... **38**

38(37). Elytral intervals with rows of punctures (Fig. 21.21); antennal scape very long and slender, almost twice as long as antennal club ... ***Stictostix* Marseul**

– Elytral intervals smooth or with punctures not forming rows; antennal scape stouter and about as long as antennal club ... **39**

39(38). Antennal insertion visible at least frontally; antennal scape triangular, simple; elytral striae not reaching apical margin of elytra ... ***Parepierus* Bickhardt**

– Antennal insertions concealed and not visible dorsally or frontally; antennal scape emarginate externally to accommodate part of antennal club in repose; elytral striae reaching the elytral apex (Fig. 21.4), at most 5th and sutural stria abbreviated basally ... ***Epierus* Erichson**

40(36). Antennal cavities on hypomeron absent or open, not delimited posteriorly (Fig. 21.67) ... **41**

– Antennal cavities on hypomeron completely closed posteriorly (Fig. 21.50) ... **43**

41(40). Labrum very small; mandible arched without teeth along inner margin (Fig. 21.24); mentum strongly expanded filling entire space between mandibular bases ventrally and covering maxillae; body very flat ... ***Hololepta* Paykull**

– Labrum large; mandible with one or two teeth along inner edge (Fig. 21.41); mentum narrow, maxillary cardines well visible; body more convex ... **42**

42(41). I nner lateral pronotal stria (between lateral marginal stria and submarginal stria) absent; prosternal lobe broadly rounded, strongly projecting forward; antennal cavity on hypomeron partially delimited; prosternal process broad with pair of striae; labrum transverse (Fig. 21.41) ... ***Plaesius* Erichson**

– Inner lateral pronotal stria (between lateral marginal stria and submarginal stria) present (Fig. 21.38); prosternal lobe very short, and pointed apically; antennal cavity on hypomeron absent; prosternal process keel like medially and without striae; labrum quadrate or longer than broad and projecting anteriorly (Fig. 21.57) ... ***Nasaltus* Mazur & Węgrzynowicz**

43(40). Inner lateral pronotal stria (between lateral marginal stria and submarginal stria) present; metaventral postcoxal lines strongly recurved ... ***Hister* Linnaeus**

– Inner lateral pronotal stria absent; metaventral postcoxal lines usually strongly descending and oblique ... **44**

44(43). Lateral submarginal pronotal stria absent (Fig. 21.39) ... **45**

– Lateral submarginal pronotal stria present (Fig. 21.37) ... **47**

45(44). Length less than 2.5 mm; body brown and very flat (Fig. 21.29 ... ***Asterix* Mazur**, part

– Length more than 5 mm; body black and convex ... **46**

46(45). Pronotum with distinct longitudinal impressions along lateral margins; each elytron with 6 complete striae (Fig. 21.14) ... ***Arbolister* Mazur**

– Pronotum without lateral impressions replaced by multiple rows of punctures situated on submarginal pronotal region (Fig. 21.39); each elytron with maximum 3 complete or incomplete striae (Fig. 21.11) ... ***Aulacosternus* Marseul**

47(44). Dorsal elytral striae connected by a transverse line (Fig. 21.8), pronotal base with deep basal groove ... ***Sanilis* Mazur & Ôhara**

– Dorsal elytral striae not connected by a transverse line; pronotal base without or at most with very fine stria ... **48**

48(47). Lateral submarginal stria broadly separated from lateral marginal stria (Fig. 21.25) ... ***Eblisia* Lewis**

– Lateral submarginal stria very close to lateral marginal stria (Fig. 21.37) ... **49**

49(48). Abdominal ventrite 2 with short stria on each side; length over 6 mm ... ***Pacifister* Mazur & Ôhara**

– Abdominal ventrite 2 without striae; length less than 5 mm ... **50**

50(49). Length less than 2.5 mm; prosternal process very broad, paired prosternal striae long, broadly separately from each other ... ***Asterix* Mazur**, part

– Length more than 3 mm; usually without prosternal striae; if present, much shorter and closer to each other ... **51**

51(50). Mesocoxa without cariniform stria ... ***Eurylister* Bickhardt**

– Mesocoxa with distinct cariniform stria (Fig. 21.69) ... ***Platysoma* Leach**

Classification of the Australian Genera

Subfamily Niponiinae Fowler, 1912

Characteristics. Body elongate, cylindrical (Fig. 21.68). Head slightly narrower than pronotum, 1.5 times as long as wide, not retracted into prothorax, prognathous. Frontoclypeus anteriorly projecting into horns or knobs. Labrum not visible dorsally. Mandibles deflexed, moving in a plane at right angles to the long axis of the head. Antenna 10-segmented with 3-segmented club. Prothorax longer than wide. Prosternum straight anteriorly, without prosternal plate.

Niponius Lewis, 1885 (Fig. 8.68)

Type species. *Niponius impressicollis* Lewis, 1885.
Arunus Sengupta & Pal, 1995

Characteristics. Cylindrical, slender to moderately stout, shiny. Frontoclypeal horn or knob carinate apically and usually with one or two postapical carinae. Antennal groove below each eye, receiving antennal scape in repose. Eyes well removed from anterior margin of pronotum. Propygidium and pygidium usually with foveae.

Australian species. One: *N. interstitialis* Lewis.

Distribution. Northern QLD, East and South-east Asia, from Russian Far East and Japan to India, Malaysia, Indonesia.

Biology. *Niponius* species occur in association with bark beetles (Coleoptera: Curculionidae: Scolytinae) and are most likely predacious upon them.

References. Gardner (1935).

Subfamily Chlamydopsinae Bickhardt, 1914

Characteristics. Antennal cavities closed beneath by the prosternal alae, situated in the anterior prothoracic angles.

Antennal scape strongly angulate and usually greatly expanded covering eyes in repose, antennal club strongly elongate, at least three times as long as broad. Elytral humeri often with trichomes.

Australian taxa. Six genera are known to occur in Australia.

Chlamydopsis Westwood, 1869 (Pl. 5E, 41I, 41K; Figs 21.5, 21.7)

Type species. *Chlamydopsis striatella* Westwood, 1869.
Byzenia King, 1869

Characteristics. Body compact, rectangular in outline. Head completely retracted into prothorax. The species belonging here are very diverse in morphology, all share a visible scutellum and upturned anterior pronotal margin, at least above the antennal cavities, or along the entire margin. None are entirely flattened dorsally or have the strikingly elongate trichome setae.

Australian species. 68.

Distribution. *Chlamydopsis* is distributed throughout Australia with few species known from New Guinea, New Caledonia and Vanuatu.

Biology. Myrmecophilous.The most common hosts appear to be ants in the subfamily Ponerinae, although hosts from Dolichoderinae, Formicinae and Myrmicinae ants have also been reported.

References. Caterino (2003); Caterino & Dégallier (2007).

Ectatommiphila Lea, 1914 (Pl. 41F–G; Fig. 21.15)

Type species. *Chlamydopsis glabra* Lea, 1910.

Characteristics. Easily recognised by its large size (length ~4 mm) and flat dorsum. Elytra is flat with small trichomes opening into transverse slits which are not raised above the level of the elytra.

Australian species. Two: *E. glabra* (Lea) and *E. opaca* (Lea).

Distribution. Endemic to Australia: QLD, NSW, VIC.

Biology. Myrmecophilous. The only known host is a ponerinae ant (*Rhytidoponera metallica* (Smith)).

References. Caterino & Dégallier (2007).

Eucurtia Mjöberg, 1912

Type species: *Eucurtia paradoxa* Mjöberg, 1912 (= *Chlamydopsis comata* Blackburn, 1901).

Characteristics. Body oval, dorsum smooth and shiny. Antennal scape about half as broad as long, arcuate. Elytra with dense, elongate tufts of setae arising from the anterior and posterior elevations of the elytral trichomes.

Australian species. One.

Distribution. *Eucurtia comata* (Blackburn), the only representative of the genus, occurs in South Australia.

Biology. Termitophilous, but the host is unknown. A fluid was observed to be secreted from the apices of the humeral trichome setae. This substance attracted the termites, which apparently consumed the secretion.

References. Caterino & Dégallier (2007).

Orectoscelis Lewis, 1903 (Pl. 42C; Fig. 21.6)

Type species: *Orectoscelis humeralis* Lewis, 1903.

Characteristics. Body oblong, slightly convex; scutellum completely hidden. Elytra without striae; humeral trichomes usually obliquely elevated, fringed by a single continuous setal margin.

Australian species: Nine.

Distribution. *Orectoscelis* has an Australasian distribution: from northern Australia (QLD, NT) through New Guinea, Indonesia (Sulawesi, Halmahera) and Japan (Ryukyu Archipelago).

Biology. Unknown, but presumably myrmecophilous.

References. Dégallier & Caterino (2005a); Caterino & Dégallier (2007).

Pheidoliphila Lea, 1914 (Pl. 42K)

Type species. *Pheidoliphila minuta* Lea, 1914.

Characteristics. The genus contains small species (less than 2 mm long), characterised by the presence of bifid projection of anterior pronotal margin. Trichomes (when present) small, oblique and found in the extreme anetrolateral corners of each elytron.

Australian species. 25.

Distribution. Eastern Australia, Tasmania, New Guinea.

Biology. Myrmecophilous species, associated with myrmicine ants of the genus *Pheidole* Westwood.

References. Dégallier & Caterino (2005b); Caterino & Dégallier (2007).

Teretriopsis Caterino & Dégallier, 2007

Type species. *Teretriopsis theryi* Caterino & Dégallier, 2007.

Characteristics. Body subcylindrical, without pronotal processes and humeral trichomes.

Australian species. One.

Distribution. *Teretriopsis theryi* Caterino & Dégallier, the only included species, is known only from Queensland.

Biology. Collected in flight intercept traps.

References. Caterino & Dégallier (2007).

Subfamily Onthophilinae W.S. MacLeay, 1819

Characteristics. Labrum setose; antennal cavities at least partially covered from below. Elytra, and often pronotum with distinct longitudinal costae. Dorsum densely and coarsely punctured. Mesoventrite short, often with metaventrite jointly depressed at sides.

Australian taxa. Two genera are represented in the Australian fauna.

Epiechinus Lewis, 1891 (Pl. 41H; Fig. 21.17)

Type species. *Onthophilus costipennis* Fåhraeus, 1851.

Characteristics. Dorsum more or less setose, setae short and squamiform, most conspicuous on the costae and in the thoracic punctures. Prosternal keel flat and broad, weakly emarginate apically. Meso- and metaventrites deeply foveolate, foveae varying much in shape, there is also sometimes a median sulcus at metaventrite.

Australian species. Two.

Distribution. South Eastern Europe, Japan, SouthEast Asia including India, tropical Africa, and Australia. *E. costatus* (W.J. MacLeay) and *E. notogaeus* Bickhardt are both known from NSW.

Biology. Found under bark of older dead trees, in rotting wood and in piles of sawdust. The adults feed mainly on fungal spores.

Onthophilus Leach, 1817 (Pl. 42F; Fig. 21.16)

Type species. *Scolytus punctatus* O.F. Müller, 1776.

Characteristics. Body oval, moderately convex. Antennal scape expanded or strongly angulate. Prosternal plate short, not completely covering head from below; antennal cavities partially exposed. Prothorax with 2 costae, each elytron with 8 striae and 4 costae; entire body with strongly sculptured appearance.

Australian species. One.

Distribution. Genus broadly distributed in Northern Hemisphere, represented in both, Eurasia and North America. *O. australis* Helava & Howden is reported from Queensland.

Biology. Probably the species of *Onthophilus* are predators of fly larvae or of other small arthropods in leaf litter, dung, carrion and in the nests and faecal chamber of some rodents and other small mammals. *O. australis* was collected by the traps with decaying fungi or with human faeces at a low, sandy hillside with a grassy-herbaceous cover and scattered eucalypts.

References. Helava & Howden (1977).

Subfamily Tribalinae Bickhardt, 1914

Characteristics. Labrum with at least two setae. Pronotum lacking costae, elytra usually normally striated, sometimes striae reduced or outer ones replaced by low carinae. Prosternal keel short, its lateral extension (prosternal ala) on each side without notch for reception of antennal funicle.

Australian taxa. Four genera present in Australia.

Epierus Erichson, 1834 (Pl. 41L; Fig. 21.4)

Type species. *Hister fulvicornis* Fabricius, 1801.

Characteristics. Body elongate-oval. Prosternal plate distinctly projecting forward, covering gular part of head from below. Prosternal keel elevated, narrowed medially with paired narrowly separated striae. Elytron with dorsal striae, reaching the elytral apex.

Australian species. One.

Distribution. Most species are known from North and South America; few species are distributed in the Palaearctic, Oriental and Australasian Regions. *Epierus beccarii* Marseul is known from Sarawak, Sumatra, New Guinea and northern QLD.

Biology. Adults are both predators and microphagous spore specialists; inhabiting humus accumulating beneath the bark of decaying tree trunks.

References. Bousquet & Laplante (2006).

Parepierus Bickhardt, 1913 (Pl. 42J)

Type species. *Epierus amandus* Schmidt, 1892.

Characteristics. Body broadly oval. Prosternnal plate short, not covering head from below; prosternal keel weakly elevated with pair broadly separated striae. Dorsal elytral striae abbreviated apically.

Australian species. One, *P. faederatus* (Lewis).

Distribution. South-eastern Asia from southern China and Taiwan through to Indo-Malaya and Indonesia to Australia and New Zealand.

Biology. The species are mostly found in dead and rotten wood where they feed mainly on fungal spores.

References. Zhang & Zhou (2007b).

Stictostix Marseul, 1870 (Pl. 71H; Fig. 21.21)

Type species. *Epierus parra* Marseul, 1870.

Characteristics. Antenna longer than the head width. Dorsal elytral striae 1–3 and subhumeral stria of the elytron often represented by low carinae.

Australian species. Five.

Distribution. Western North America and Australia (NSW).

References. Bousquet & Laplante (2006).

Tribalus Erichson, 1834 (Pl. 71I; Fig. 21.22)

Type species. *Hister capensis* Paykull, 1811.

Characteristics. Body shortly oval. Elytra without dorsal striae (except the sutural stria) occasionally with rudiments of subhumeral striae.

Australian species. Two: *T. australis* (W.J. MacLeay) and *T. marseuli* Gomy.

Distribution. Afrotropical and Oriental Regions, New Guinea, Australia (NSW, QLD).

Biology. Found under bark of dead trees and in rotting wood, feeding on fungal spores.

Subfamily Histerinae Gyllenhal, 1808

Characteristics. Labrum sclerotised without setae. Antennal scape cylindrical or slightly expanded apically, antennal club round or oval, setose. Dorsal elytral striae distinct in most species. Prosternum with prosternal lobe, usually covering head from below; lateral extension (alae) without incision for the reception of antennal funicle; antennal cavities located near anterior angles of prothorax and usually covered by alae from below. Protibia with apical and marginal spines or denticles.

References. Gomy (1985).

Tribe Exosternini Bickhardt, 1914

Characteristics. Antennal club with straight sutures; frontal stria, if present, always reaching the antennal base; aedeagus simple and tubular.

Australian taxa. Two Australian genera are placed in this tribe.

Arbolister Mazur, 1990 (Pl. 41E; Fig. 21.14)

Type species. *Arbolister termitophilus* Mazur, 1990.

Characteristics. Antennal club small with a transverse subapical annulus; antennal funicle thickening apically. Pronotum without submarginal striae, strongly longitudinally impressed laterally. Elytron with 6 complete dorsal elytral striae. Protarsal grooves sharply defined, S-shaped. Prosternum with short prosternal plate, weakly emarginate anteriorly; antennal cavities not delimited posteriorly; paired prosternal striae present, but short and shallow.

Australian species. One: *Arbolister termitophilus* Mazur.

Distribution. Queensland.

Biology. *Arbolister termitophilus* was found in an arboreal nest of *Nasutitermes graveolus* Hill.

References. Mazur (1990); Dégallier (1993).

Asterix Mazur, 1993 (Pl. 41M; Fig. 21.29)

Type species. *Apobletes appendiculatus* Schmidt, 1893.

Characteristics. Body flat and elongate. Dorsal elytral striae present, sometimes reduced or abbreviated. Prosternal process very broad, with long broadly separated carinae. Mesoventrite shallowly emarginate, not margined.

Australian species. Three.

Distribution. Malaysia, Indonesia, New Guinea, Australia (QLD).

Biology. Found under the bark of trees.

References. Mazur (1993).

Note. *Pseudister nigropunctatus* Lewis, 1899 probably belongs to this genus but we were not able to confirm its identity.

Tribe Histerini Gyllenhal, 1808

Characteristics. Body oval, convex, heavily sclerotised. Head with large mandibles; labrum sclerotised well visible; frontal stria complete. Pronotum with complete lateral submarginal stria and short inner sublateral stria near anterior corners. Prosternal lobe triangular, moderately strongly developed, lateral alae reduced delimiting shallow antennal grooves. Anterior margin of mesoventrite truncate or emarginate, not bisinuate. Tarsal groove of protibia straight, distinctly limited at inner margin only. Mid- and hind tibiae with two rows of strong spinules.

Australian taxa. Two genera have been recorded from Australia.

Hister Linnaeus, 1758 (Pl. 42A–B; Fig. 21.23)

Type species. *Hister unicolor* Linnaeus, 1758.

Characteristics. Body oblong, convex heavily sclerotised and uniformly black. Head with complete frontal carina. Inner lateral pronotal stria always present near anterior angles; lateral submarginal stria complete. Prosternum with triangular, medially pointed pronotal lobe; antennal cavity shallow; prosternal keel without striae. Elytron with 4–5 dorsal striae and almost complete sutural stria. Mesoventrite straight. Metaventral postcoxal line recurved meeting metanepisternum.

Australian species. Two: a native species, *H. ambulator* Thayer (= *H. walkeri* Lewis), and *H. nomas* Erichson, introduced from Africa.

Distribution. *Hister* has a worldwide distribution. In Australia recorded from ACT, NSW, QLD, VIT and WA.

Biology. All species probably feed on fly larvae found primarily in the dung of large mammals, vertebrate carrion and in rotting fruit and various types of fungi.

References. Caterino (1999); Bousquet & Laplante (2006).

Nasaltus Mazur & Węgrzynowicz, 2008 (Fig. 21.9)

Type species. *Hister orientalis* Paykull, 1811.

Characteristics. Body oblong, convex heavily sclerotised and uniformly black. Head with complete frontal carina. Mandibles elongate and asymmetrical. Inner lateral pronotal stria always present near anterior angles; lateral submarginal stria complete, broadly separated from lateral margin. Prosternum with triangular, medially pointed pronotal lobe; antennal cavity very shallow; prosternal process narrowed medially, without carinal striae. Elytron with 4–5 dorsal striae, sutural stria absent. Mesoventrite deeply emarginate medially. Metaventral postcoxal line recurved meeting metanepisternum.

Australian species. One: *N. chinensis* (Quensel), introduced from the Oriental region.

Distribution. Native to Oriental Region; introduced to Fiji, Samoa, Vanuatu, Solomon Is., Hawaii, Trinidad, French Guyana and Australia.

Biology. Predacious, usually occurring in fresh cow dung where it feeds on dung inhabiting fly larvae.

References. Mazur & Węgrzynowicz (2008).

Tribe Hololeptini Hope, 1840

Characteristics. Body distinctly flattened with head distinctly narrower than pronotum, prognathous. Frontal stria absent. Pronotal hypomera not ciliate, with a groove for reception of the antennae. Elytra with dorsal striae strongly reduced. Propygidium long, situated at same level as elytra.
Australian taxa. Single genus, *Hololepta*, present in Australia.

Hololepta Paykull, 1811 (Pl. 42G; Fig. 21.24)

Type species. *Hololepta humilis* Paykull, 1811.
Characteristics. Body black and shiny. Labrum small and glabrous, emarginate arteriorly. Mandibles well developed without teeth on inner edges, with a deep groove for reception the maxillary palp ventrally. Prosternal lobe very short not covering head from below and not extending laterally; antennal cavities weakly impressed on hypomeron, not delimited. Anterior margin of mesoventrite shallowly emarginate, metaventrite broad and flat. Tarsal groove of protibia S-shaped, well impressed.
Australian species. Five.
Distribution. Worldwide.
Biology. Predators, found under bark of decaying trees.
References. Bousquet & Laplante (2006).

Tribe Platysomatini Bickhardt, 1914

Characteristics. Body more or less depressed, rarely narrow and more convex. Prosternal lobe well developed extending laterally to hypomeron and usually covering antennal cavities from below. Lateral metaventral stria never curved outwardly, extending or not, to the apical margin. Tarsal groove on protibia usually S-shaped and well delimited. Meso- and metatibia with larger teeth but usually without dense spines.
Australian taxa. Seven genera present in Australia.
Notes. This is very large but morphologically uniform group of small and medium-sized histerids, distributed worldwide and mostly living under bark of various trees. The generic classification of Platysomatini is largely artificial, highly volatile, and requires thorough molecular approach to resolve this taxonomically difficult group. In the key to genera and generic review below we have adopted very conservative approach, recognising broadly defined *Platysoma* that may be divided into numerous subgenera, often regarded as separate genera (see Mazur 1999; Ôhara & Mazur 2000, 2002; and Mazur & Ôhara 2009).

Aulacosternus Marseul, 1853 (Pl. 71D; Fig. 21.11)

Type species. *Aulacosternus zelandicus* Marseul, 1853.
Sternaulax Marseul, 1863
Characteristics. Body oblong oval, moderately convex; length 6–7 mm. Pronotal keel with paired striae. Pronotum without lateral submarginal stria; outer subhumeral elytral stria well marked.
Australian species. One, undescribed.
Distribution. New Caledonia, New Zealand, Australia (QLD, NSW).
Biology. A New Zealand species has been found in damaged fronds of the Nikau palm (*Rhopalostylis sapida*).
References. Wenzel (1955); Ôhara & Mazur (2000).

Eblisia Lewis, 1889 (Pl. 41Q; Fig. 21.13)

Type species. *Phelister lunaticus* Marseul, 1864.
Chronus Lewis, 1914 (subgenus)
Characteristics. Body convex, oval in outline; length less than 5 mm. Both lateral marginal striae and submarginal stria complete, with lateral submarginal stria broadly separated from lateral marginal stria. Prosternal lobe margined, marginal stria mostly complete, situated close to anterior margin of prosternal lobe. Mesoventrite emarginate, with distinct striae on each corner.
Australian species. Several, all undescribed.
Distribution. Asia, New Guinea, Australia: QLD.
Biology. Predators, occurring beneath the bark of decaying trees.
References. Mazur (1990, 2007); Mazur and Ôhara (2009).

Plaesius Erichson, 1834 (Fig. 21.10)

Type species. *Plaesius javanus* Erichson, 1834.
Characteristics. Body large (over 10 mm long), oval, weakly depressed. Prosternal lobe long, covering the gular part, laterally reduced; antennal cavities not completely delimited posteriorly. Prosternal keel distinctly projecting forward, with paired striae. Anterior margin of mesoventrite with a notch medially. Protibia strongly bidentate, its ventral surface covered with short spinules on outer margin. Mid- and hind tibia broadly expanded, each with two rows of spinules at outer margin.
Australian species. *Plaesius javanus* was introduced into Australia but probably has not become established.
Distribution. Asia, New Guinea, Australia: QLD.
Biology. The species of *Plaesius* are known as predators of the banana weevil (*Cosmopolites sordidus* Quensel) and used for biological control. *P. javanus* has been introduced to Taiwan, Fiji, Haiti, Mauritius, Trinidad and Jamaica.
References. Ôhara & Mazur (2000); Mazur & Ôhara (2009).

Eurylister Bickhardt, 1920 (Pl. 41O)

Type species: *Platysoma sincerum* Schmidt, 1892.
Characteristics. Pronotum with lateral submarginal stria close to lateral margin. Mesocoxa without carina. Elytral striae reduced, sutural stria absent.
Australian species. Four described and several undescribed.
Distribution. North America (two species), Asia, New Guinea, New Zealand, Australia: QLD, NSW, TAS, Norfolk Is.

Biology. Found under bark of dead trees, also in polypores.
References. Mazur (1990, 2007).

Pacifister Mazur & Ôhara, 2009 (Pl. 42I; Fig. 21.9)

Type species. *Hololepta urvillei* Le Guillou, 1844.
Characteristics. Body oblong, somewhat convex; length 6–7 mm. Pronotum with distinct lateral marginal stria, and with basal groove, at least visible laterally. Elytral striation reduced to two complete dorsal elytral striae. Elytra without subhumeral or sutural striae. Prosternal process without carinal striae. Mesoventrite shallowly emarginate anteriorly. Postcoxal line on metaventrite long, comparatively close to metacoxa. Abdominal ventrite 1 with at least 8 grooves laterally; ventrite 2 with stria on each side.
Australian species. One: *P. urvillei* (Le Guillou).
Distribution. Mariana Is., Moluccas, New Guinea, Vanuatu, New Caledonia, Fiji, Tahiti, Madagascar, Australia: QLD.
Biology. Poorly known, found in rotten fruits.

Platysoma Leach, 1817 (Pl. 42D, L–M, O–P)

Type species. *Hister compressum* Herbst, 1783.
Kanaarister Mazur, 1999 (subgenus)
Popinus Mazur, 1999 (subgenus)
Platylister Lewis, 1892 (subgenus)
Characteristics. Body variable but usually elongate oval and flattened; length less than 5 mm. Frontal stria present. Pronotum with lateral marginal stria complete laterally, reaching anterior pronotal angles. Elytral striation present, variable; subhumeral elytral striae absent, occasionally inner subhumeral stria indistinctly marked. Post mesocoxal striae long, parallel to lateral metaventral stria.
Australian species. 35.
Distribution. World-wide.
Biology. Members of this genus are found under the bark of dead trees, often in galleries of Scolytinae weevils (bark beetles).
References. Mazur (1999); Ôhara & Mazur (2002); Mazur & Ôhara (2009).
Notes. As mentioned above, the generic concept of *Platysoma* adopted here is conservative pending further research to establish validity of multiple genera/subgenera recognised in this complex, as we were unable to provide a key to distinguish Australian species classified in these genera.

Sunilis Mazur & Ôhara, 2009 (Pl. 71K; Fig. 21.8)

Type species. *Platysoma robustum* Schmidt, 1892.
Characteristics. Body oblong, somewhat subcylindrical; length 6–7 mm; frontal stria present. Mandibles with single dent on inner mandibular margin. Pronotum impunctate, lateral submarginal stria complete and deeply impressed; pronotal base with deep basal groove. Elytral striae 1–4 complete; sutural stria absent, dorsal elytral striae basally connected by a transverse line. Prosternal lobe broad and anteriorly margined; prosternum narrow, with carinal striae around base. Mesoventrite weakly emarginate anteriorly.
Australian species. One: *Sunilis robustus* (Schmidt).
Distribution. Endemic to Australia: QLD.
References. Mazur & Ôhara (2009).
Note. The placement of *Platysoma incongruum* Lea, 1925 provisionally assigned to this genus by Mazur & Ôhara (2009) requires further research.

Subfamily Dendrophilinae Reitter, 1909

Characteristics. Labrum with setae; prosternal lobe broad, with deep lateral notch for reception of antennal funicle and basal groove receiving long apical spine of protibia at rest. Antennal cavities open, poorly defined, usually located in front of coxa at middle of hypomeron. Elytra with or without dorsal elytral striae, these occasionally represented only as vague rudiments near basal edge. Protibia enlarged; lateral protibial edge curved.
Australian taxa. Out of four recognised Dendrophilinae tribes three are present in Australia.

Tribe Bacaniini Kryzhanovskij, 1976

Characteristics. Very small and globular beetles, length usually below 1.2 mm. Frontoclypeus with angulate dark line on level of antennal insertions. Scutellum not visible externally. Protibia without apical spurs. Propygidium at least partially covered by elytra.
Australian taxa. Three genera in Australia.

Australanius Gomy, 2009 (Fig. 21.32)

Type species. *Australanius verrucosus* Gomy, 2009a.
Characteristics. Elytron with complete lateral stria along edge. Dorsal elytral striae weakly impressed and mostly composed of strial punctures.
Australian species. Three.
Distribution. Endemic Australian genus known from Queensland.
Biology. Known specimens were collected in pitfall traps and fight intercept traps.
References. Gomy (2009a, 2009b).

Bacaniomorphus Mazur, 1989

Type species. *Abraeomorphus semiellipticus* Thérond, 1965.
Characteristics. Body less than 1 mm long, somewhat elongate, very finely punctulate dorsally. Prosternal lobe very short and shallowly emarginate. Mesoventrite medially with transverse punctiform stria. Sexually dimorphic taxon: forehead of male distinctly excavated, covered with dense tomentosum.

Australian species. One: *Bacaniomorphus semiellipticus* (Thérond).
Distribution. Malaysia, New Guinea, Solomon Is., Tonga, Fiji, Australia: QLD.
Biology. Specimens were collected in pitfall traps, flight intercept traps and sifting litter.

Bacanius LeConte, 1853 (Fig. 21.31)

Type species. *Bacanius tantillus* LeConte, 1853.
Characteristics. Body globular, less than 1.5 mm. Frontoclypeus usually with dark transverse line close to eye level. Elytron usually coarsely punctured, without striae, except for inner subhumeral stria. Prosternum short, striate laterally. Protibia expanded, flat with several small denticles laterally.
Australian species. Four.
Distribution. Pantropical. In Australia recorded from QLD, NSW, TAS.
Biology. Found in rotten wood, under bark of trees, in litter and among decaying plant material, occasionally in ant nests.
References. Bousquet & Laplante (2006).

Tribe Dendrophilini Reitter, 1909

Characteristics. Elytra with well developed striae.
Australian taxa. One genus.

Dendrophilus Leach, 1817 (Pl. 41J; Fig. 21.19)

Type species. *Hister punctatus* Herbst, 1791.
Characteristics. Body elongate-oval, convex and densely punctate dorsally. Frontal stria absent. Prosternal lobe projecting forward; prosternal process with pair of striae. Dorsal elytral striae 1–4 fine and almost complete, sutural stria incomplete to absent; scutellum visible. Mid- and hind tibiae flattened, wide, with curved lateral edge.
Australian species. One, introduced.
Distribution. Holarctic. *D. xavieri* Marseul from Japan and Siberia has been introduced to ACT and NSW.
Biology. Found in rotting vegetable matter, pig and hen manure, carrion and nests of birds and rodents.
References. Ôhara (1994); Bousquet & Laplante (2006).

Tribe Paromalini Reitter, 1909

Characteristics. Clypeus broad, not distinctly narrowing anteriorly. Frontal stria present, sometimes interrupted medially.
Australian taxa. Four genera.

Carcinops Marseul, 1855 (Pl. 41N; Fig. 21.18)

Type species. *Paromalus pumilio* Erichson, 1834.
Characteristics. Body oval, slightly depressed dorsoventrally. Frontal stria nearly complete, shortly interrupted medially. Dorsal elytral striae 1–5 more or less complete. Scutellum small, distinct. Pronotal lobe projecting forwards. Metaventrite with double postcoxal lines. Meso- and metatibiae rather narrow.
Australian species. Three.
Distribution. Most of the species of *Carcinops* inhabit the New World but many species are known from Asia, Africa and Australia.
Biology. Beetles are found in decaying matter of animal and plant origin, in bird and rodent nests; often collected also in anthropogenic settings as pigsties, chicken coops, etc.
References. Ôhara (1994); Bousquet & Laplante (2006).

Eulomalus Cooman, 1937 (Fig. 21.33)

Type species. *Eulomalus vermicipygus* Cooman, 1937.
Characteristics. Body depressed, small, 0.9–2.5 mm. Frontal stria broadly interrupted medially. Elytra punctate without striae. Prosternum with pair of carinal prosternal striae. Meso- and metatibiae short, dilated towards apex. Metaventral postcoxal line single and long, nearly extending to metacoxa. A sexually dimorphic taxon: pygidium of female variably sculptured.
Australian species. Four.
Distribution. Asia, Australia: QLD, NSW, Lord Howe Island.
Biology. Predators, living under the bark of trees.
References. Cooman (1937); Zhang & Zhou (2007a).

Pachylomalus Schmidt, 1897 (Pl. 42E; Fig. 21.3)

Type species. *Paromalus leo* Marseul, 1879.
Characteristics. Body oval, convex. Frontal stria broadly interrupted medially. Pronotal base with shallow antescutelar stria interrupted medially. Scutellum absent. Elytra punctate, dorsal elytral striae at most visible as light traces laterally. Propygidium with transverse stria. All tibiae dilated. Prosternum with carinal striae. Metaventral postcoxal line recurved to lateral side. Ventrite 1 with double postcoxal lines.
Australian species. One: *Pachylomalus mus* (Marseul).
Distribution. Asia, New Guinea, Australia: QLD.
Biology. Poorly known, beetles are usually found under stones or in the forest litter.
References. Cooman (1941).

Platylomalus Cooman, 1948 (Pl. 42H; Fig. 21.30)

Type species. *Hister complanatus* Panzer, 1797.
Characteristics. Frontal stria complete, not interrupted medially. Prosternum with a pair of carinal striae. Elytra punctate, dorsal elytral striae absent or visible as light traces laterally. Scutellum invisible. Metaventral postcoxal line single and long, extending to metacoxa. Propygid-

ium without transverse stria. Ventrite 1 with single postcoxal lines. Pygidium even or with weak groove in male, with numerous deep grooves in female of most species.
Australian species. Eight.
Distribution. Most species are known from Asia but the genus extends to Europe North America and Australia.
Biology. Found under bark and in decaying organic matter.
References. Zhang & Zhou (2007a).

Subfamily Saprininae Blanchard, 1845

Characteristics. Moderately-sized, rectangular oval to ovoid-shaped, not depressed beetles; cuticle in many cases metallic green, blue or purple. Antennal insertions visible from above, labrum usually setose. Antennal club with specialised sensory apparatus (Reichardt's organ) in form of sensory patches, slit-like orifices and internal vesicles, or combinations of above-mentioned (Fig. 21.1). Prosternum usually with prosternal keel bearing carinal striae, apex of prosternum in some species with median fovea or with paired prosternal foveae. Antennal cavities located on sides of prosternum. Scutellum distinct, elytra usually striae, dense punctures and smooth lustrous areas. Propygidium and pygidium distinct.
Australian taxa. Nine genera.

Chalcionellus Reichardt, 1932

Type species. *Saprinus amoenus* Erichson, 1834.
Izpaniolus Mazur, 1972
Eosaprinus Kozminykh, 2000
Characteristics. Rather small, iridescent, ovoid beetle; elytra lighter than pronotum. Frontal stria complete. Most of pronotal surface punctate, punctures coarser in lateral and anterior area. Dorsal elytral striae well developed. Prosternum with prosternal foveae, linked by a transverse stria. Protibia with 7–8 moderately large teeth topped by denticles.
Australian species. One. Introduced in Western Australia.
Distribution. Old World. *Chalcionellus aeneovirens* (Schmidt) has been introduced into Western Australia.
Biology. Most of *Chalcionellus* species are generalist predators attracted to carrion or dung. Australian species is found mostly on cow pads, occasionally also on carrion.
References. Reichardt (1941); Mazur (2011); Lackner (2014); Lackner & Leschen (2017).

Euspilotus Lewis, 1907

Type species. *Euspilotus zonalis* Lewis, 1907.
Neosaprinus Bickhardt, 1909 (subgenus)
Myrmeosaprinus Mazur, 1974
Characteristics. Anterior part of frontal stria and supraorbital striae absent; eyes visible from above. Cuticle dark brown to black. Elytral striae 1–4 well developed. Prosternal striae on keel divergent on apical half, united anteriorly by deep sulcus.
Australian species. One.
Distribution. Almost exclusively in the New World; one species occurs in the Palaearctic and one in the Oriental Region. *Euspilotus* (*Neosaprinus*) *rubriculus* (Marseul) has been introduced into Australia.
Biology. Most of the species are typical generalist predators and saprobionts.
References. Mazur (2011); Lackner (2014); Lackner & Leschen (2017).

Gnathoncus Jacquelin-Duval, 1858

Type species. *Hister rotundatus* Kugelann, 1792.
Characteristics. Cuticle brown to black, never iridescent. Anterior part of frontal stria and supraorbital striae absent. Elytral fourth dorsal stria never connected with sutural stria; with characteristic hooked appendix between fourth dorsal and sutural striae at elytral base. Prosternum without prosternal foveae or with median fovea; prosternal striae on keel strongly convergent anteriorly, united under sharp angle; prosternal process flattened, broad.
Australian species. Two, introduced.
Distribution. Distributed mostly in the Holarctic, one species occurs in the Afrotropical and one in the Oriental Region. *G. rotundatus* (Kugelann) and *G. communis* (Marseul) have been introduced into Australia.
Biology. Most of the species live inside the nests of birds or ground mammals; several species are widely distributed across the globe by human activity.
References. Mazur (2011); Lackner (2010, 2014); Vienna & Ratto (2013); Lackner & Leschen (2017).

Hypocaccus C. Thomson, 1867

Type species. *Hister quadristriatus* Hoffmann, 1803.
Rhytidoprinus Houlbert & Monnot, 1923
Nessus Reichardt, 1932 (subgenus)
Baeckmanniolus Reichardt, 1932 (subgenus)
Characteristics. Small to median size; dorsum often iridescent. Frontal stria usually well developed, straight; frons coarsely punctate or with several to multiple short to long transverse rugae. Pronotal disc either smooth or punctate; hypomeron glabrous; prosternal foveae present on each side of keel, often well developed. Protibia on outer margin with 5–11 teeth topped by denticle; metatibia on outer margin with two or three rows of denticles.
Australian species. Two: native *Hister ambulator* Thayer and *H. nomas* Erichson introduced from tropical Africa.
Distribution. Large genus with over 200 species distributed almost exclusively in the Old World.
Biology. Most of the species are generalist predators found in open xerothermic landscapes, on sandy soils, and are often found on beach or sandy riverbanks hunting arthropod larvae and adults occurring under wrack.

References. Mazur (2011); Lackner (2010, 2014); Lackner & Leschen (2017).

Iridoprinus Lackner & Leschen, 2017 (Fig. 21.12)

Type species. *Iridoprinus myrmecophilus* Lackner & Leschen, 2017.
Characteristics. Cuticle light brown, elytra with strong blue iridescent metallic lustre. Frontal stria prolonged far onto clypeus. Pronotal disc on apical two-thirds with leathery appearance with punctures forming elongate wrinkles. Five dorsal elytral striae curved and carinate; inner subhumeral and sutural elytral striae absent. Prosternal foveae absent; metanepisternum with deep elongate groove for reposing mesotarsus. Pro- and mesotibiae slightly dilated.
Australian species. One: *Iridoprinus myrmecophilus* Lackner & Leschen, 2017.
Distribution. Endemic to Australia: NT and NSW.
Biology. Found in the nest of a meat ant, *Iridomyrmex purpureus* (Smith).
References. Lackner & Leschen (2017).

Notosaprinus Kryzhanovskij, 1972

Type species. *Saprinus irinus* Marseul, 1862.
Characteristics. Body comparatively large, length 4.2–6.0 mm. Pronotum with bronze, elytra with blue metallic lustre. Frontal stria prolonged onto clypeus. Labrum in males with large median projection. Dorsal elytral striae reduced, usually only fourth dorsal elytral stria discernible.
Australian species. One: *Notosaprinus irinus* (Marseul).
Distribution. Endemic to Australia: QLD, NSW, WA.
Biology. Predator found mostly on carrion.
References. Mazur (2011); Lackner & Leschen (2017).

Saprinodes Lewis, 1891

Type species. *Saprinodes falcifer* Lewis, 1891.
Characteristics. Frontal stria broadly interrupted medially. Prosternal keel with complete striae, joined by broad loop anteriorly; prosternal foveae absent. Protibia very slender and elongate without teeth or denticles with large apical spur (Fig. 21.81).
Australian species. Two: *Saprinodes distinctus* Dégallier and *S. falcifer* Lewis.
Distribution. Australia: QLD and NSW.
Biology. Specimens have been collected using pitfall traps.
References. Dégallier (1993); Mazur (2011); Lackner & Leschen (2017).

Saprinus Erichson, 1834 (Pl. 71L)

Type species. *Hister nitidulus* Fabricius, 1801.
Plesioprinus Houlbert & Monnot, 1923
Syrinus Erichson, 1841
Characteristics. Dorsum usually with strong metallic lustre that can become dull and dark with older specimens. Frontal stria complete. Prosternal keel with complete striae, joined anteriorly by broad loop (striae absent in one species); prosternal foveae absent.
Australian species. Nine native, two introduced.
Distribution. Large genus distributed worldwide.
Biology. Members of *Saprinus* normally prefer open xeric landscapes; they are typical generalist predators, but some species are also attracted to flowers, bird nests, or rotting fungi.
References. Mazur (2011); Lackner (2014); Lackner & Leschen (2017).

Tomogenius Marseul, 1862

Type species. *Saprinus incisisternus* Marseul, 1862.
Characteristics. Cuticle brown to black. Head with frontal and supraorbital striae absent. Elytra with short hooked appendix between fourth dorsal and sutural striae. Prosternum with carinal atria parallel to lateral margins incomplete anteriorly, ending in deep foveae, separated by apex triangular process.
Australian species. Three.
Distribution. Australia, New Guinea, New Zealand.
Biology. Australian species occur mostly in caves where they prey upon arthropod larvae living in guano but are occasionally found also on carrion or in pit traps.
References. Mazur (2011); Lackner (2014); Lackner & Leschen (2017).

Subfamily Abraeinae MacLeay, 1819

Characteristics. Body small, generally round, oval rarely depressed. Antennal insertions visible from above, located near inner edge of eye. Labrum with setae. Prosternum with lateral notch for reception of the antennal funicle. Antennal cavities open, located either in front of procoxa or at level of anterior angle of prothorax. Elytron without dorsal striae (with vague impressions in some species).
Australian taxa. Three tribes.

Tribe Acritomorphini Wenzel, 1944

Characteristics. Body elongate and relatively flattened. Frontal stria absent. Prothoracic hypomeron with well developed antennal cavity. Scutellum triangular, relatively large. Metaventral postcoxal line oblique and short. Tarsi 5-segmented.
Australian taxa. This tribe contains a single genus.

Acritomorphus Wenzel, 1944 (Fig. 21.28)

Type species. *Acritomorphus praecursor* Wenzel, 1944.
Macroabraeus Ôhara, 1999
Characteristics. Length 1.5–1.6 mm. Prosternum long, with a comparatively flat narrow keel; prosternal striae

broadly separated and incomplete. Protibia expanded apically, slender basally, denticulate. Meso- and metatibiae with several marginal spinules in a single row.

Australian species. One: *Acritomorphus silvestris* Mazur.

Distribution. Japan, Philippines, Australia: QLD, Lord Howe Is.

Biology. Australian specimens were collected in Malaise traps.

References. Mazur (1997a); Ôhara (1999).

Tribe Acritini Wenzel, 1944

Characteristics. Beetles very small, usually less than 1.5 mm. Body oval, moderately convex. Head rather wide, clypeus not separated from frons. Elytra truncate at apex, the dorsal and subhumeral striae absent. Pygidium usually vertical. Prosternum moderately wide, marginal and carinal striae well developed.

Australian taxa. Three genera are present in Australia.

Acritus LeConte, 1853 (Fig. 21.34)

Type species. *Hister nigricornis* Hoffmann, 1803.

Pycnacritus Casey, 1916 (subgenus)

Characteristics. Scutellum visible, sometimes very minute; clypeus and pygidium not margined. Carinal prosternal striae divergent both apically and basally, rarely parallel. Mesoventrite straight or weakly emarginate anteriorly, without a row of large punctures. Protibia slender and setose along outer margin.

Australian species. 20.

Distribution. Widely distributed through the world.

Biology. *Acritus* species live under bark of trees, in leaf litter, in rotten vegetation and trees, and tree holes.

References. Mazur (1987); Gomy (1984); Bousquet & Laplante (2006).

Aeletes Horn, 1873 (Fig. 21.36)

Type species. *Acritus politus* LeConte, 1853.

Characteristics. Frontal stria complete. Scutellum not visible externally. Pygidium usually with a continuous stria along lateral and apical. Mesoventral disc usually with a row of parallel, longitudinal punctures or sulci.

Australian species. One: *A. franzi* Gomy.

Distribution. Europe, North and South America, Hawaii, Fiji and Australia.

Biology. Found under bark and in leaf litter.

References. Gomy (1984); Yélamos (1998).

Halacritus Schmidt, 1893 (Fig. 21.35)

Type species. *Abraeus punctum* Aubé, 1842.

Characteristics. Body elongate and quadrangular in shape, from light to dark brown in colour. Scutellum visible. Clypeus and pygidium not margined. Protibia distinctly expanded and spinose along its outer margin.

Australian species. Two.

Distribution. Pantropical genus.

Biology. Species are found on beaches in and under decaying seaweed. Specific information regarding their food is lacking, but it is possible that they may be scavengers rather than predators.

References. Wenzel (1944); Gomy (1984).

Tribe Teretriini Bickhardt, 1914

Characteristics. Body subcylindrical. Antennal club with medial longitudinal sclerotization. Mesepimeron produced dorsally between pronotum and elytron, visible from above. Mesoventrite pointed medially, fitting into incision of the prosternal process. Elytral striae obsolete.

Australian taxa. One genus.

Teretrius Erichson, 1834 (Pl. 71J; Fig. 21.20)

Type species. *Teretrius fabricii* Mazur, 1972.

Neotepetrius G. Müller, 1937 (subgenus)

Cyclosternum G. Müller, 1937

Characteristics. Frons flat or evenly concave. Pronotum at least three-fourths the length of elytra in most species. Elytron punctate with a smooth humeral region. Pygidium regularly convex. Outer margin of protibia with 7–9 large spinules.

Australian species. Nine.

Distribution. Australia: QLD, VIC, TAS.

Biology. Found under the bark on bare trunks of dead trees, often in galleries of xylophagous beetles: Bostrichidae, Ptiniidae and Eucnemidae.

References. Yélamos (2002); Bousquet & Laplante (2006).

Subfamily Trypeticinae Bickhardt, 1913

Characteristics. Body cylindrical; head vertical in repose; antennae consisting of seven segments and a club. Anterior margin of prosternal alae with rather deep, longitudinal incisions for the reception of the antennal funicle in repose.

Australian taxa. One genus is present Australia.

Trypeticus Marseul, 1864 (Fig. 21.27)

Type species. *Trypeticus gilolous* Marseul, 1864.

Characteristics. Cylindrical, shiny. Frontoclypeus without visible suture, usually more or less prolonged into a rostrum. Labrum transverse, setose; in frontal view often hidden by the rostral tip. Mandibles short and wide. Eyes large, strongly convex. Pronotum with marginal stria laterally. Scutellum small, triangular. Elytra without striae, elytral apices rounded, forming a rather shallow sutural angle.

Australian species. Two: *T. gilolous* Marseul and *T. monteithi* Kanaar.

Distribution. Members of the genus *Trypeticus* occur predominantly in the Indo-Australian subregion; in Australia they are known only from QLD.

Biology. Species of *Trypeticus*, live in the burrows of bark and ambrosia beetles and other wood-boring insects, where they prey upon the larvae and sometimes also upon the adults.

References. Kanaar (2003).

Acknowledgments

This research was supported by the National Natural Science Foundation of China Grants No. 31402008, and International Postdoctoral Exchange Fellowship No. 20150064 to the senior author. Cate Lemann is sincerely thanked for technical support and digital imaging.

References

Beutel RG (1999) Morphology and evolution of the larval head of Hydrophiloidea and Histeroidea (Coleoptera: Staphyliniformia). *Tijdschrift voor Entomologie* **142**, 9–30. doi:10.1163/22119434-99900016

Beutel RG (2016) Introduction, phylogeny of Histeroidea Gyllenhal, 1808. In *Handbuch der Zoologie/Handbook of Zoology. BandVolume IV Arthropoda: Insecta. Teilband/Part 38. Coleoptera, Beetles. Volume 1: Morphology and Systematics (Archostemata, Adephaga, Myxophaga, Polyphaga partim).* 2nd edn. (Eds RG Beutel and RAB Leschen) p. 273. Walter de Gruyter, Berlin.

Bocák L, Barton C, Crampton-Platt A, Chesters D, Ahrens D, Vogler AP (2014) Building the Coleoptera tree-of-life for >8000 species: composition of public DNA data and fit with Linnean classification. *Systematic Entomology* **39**(1), 97–110. doi:10.1111/syen.12037

Bousquet Y, Laplante S (2006) *The insects and Arachnids of Canada, Part 24. Coleoptera: Histeridae.* NCR Research Press, Ottawa.

Caterino MS (1999) The taxonomy and phylogenetics of the *coenosus* group of *Hister* Linnaeus (Coleoptera: Histeridae). *University of California Publications in Entomology* **119**, viii+75 pp.

Caterino MS (2003) New species of Chlamydopsinae (Histeridae: Chlamydopsinae), with a review and phylogenetic analysis of all known species. *Memoirs of the Queensland Museum* **49**, 159–235.

Caterino MS (2007) Two new species of *Chlamydopsis* Westwood (Coleoptera: Histeridae). *Coleopterists Bulletin* **61**(3), 327–333. doi:10.1649/0010-065X(2007)61[327:TNSOCW]2.0.CO;2

Caterino MS, Dégallier N (2007) A review of the biology and systematics of Chlamydopsinae (Coleoptera: Histeridae). *Invertebrate Systematics* **21**, 1–28. doi:10.1071/IS06017

Caterino MS, Hunt T, Vogler AP (2005) On the constitution and phylogeny of Staphyliniformia (Insecta: Coleoptera). *Molecular Phylogenetics and Evolution* **34**, 655–672.

Caterino MS, Maddison DR (2018) An early and mysterious histerid inquiline from Cretaceous Burmese amber (Coleoptera, Histeridae). *Zookeys* **733**, 119–129.

Caterino MS, Tishechkin AK (2006) DNA identification and morphological description of the first confirmed larvae of Hetaeriinae (Coleoptera: Histeridae). *Systematic Entomology* **31**, 405–418. doi:10.1111/j.1365-3113.2006.00338.x

Caterino MS, Tishechkin AK (2014) New genera and species of Neotropical Exosternini (Coleoptera: Histeridae). *ZooKeys* **381**, 11–78. doi:10.3897/zookeys.381.6772

Caterino MS, Vogler AP (2002) The phylogeny of the Histeroidea (Coleoptera: Staphyliniformia). *Cladistics* **18**, 394–415. doi:10.1111/j.1096-0031.2002.tb00158.x

Caterino MS, Wolf-Schwenninger K, Bechly G (2015) *Cretonthophilus tuberculatus*, a remarkable new genus and species of hister beetle (Coleoptera: Histeridae) from Cretaceous Burmese amber. *Zootaxa* **4052**, 241–245.

Cooman A (1937) Etude sur les genres *Paromalus* Er. et *Eulomalus* n. g. (Col. Histeridae). Avec description d'espèces nouvelles. *Notes d'Entomologie Chinoise* **4**, 89–167.

Cooman A (1941) Coléoptères Histeridae d'Extrême Orient, principalement du Tonkin. *Notes d'Entomologie Chinoise* **8**, 291–333.

Dégallier N (1993) Description des *Saprinodes distinctus* sp. nov., désignation d'un lectotype pour *S. falcifer* Lewis, 1891 et notes sur un autre Histeridae (Coleoptera) d'Australie. *Nouvelle Revue d'Entomologie* (n.s.) **10**, 47–53.

Dégallier N, Caterino MS (2005a) Notes taxonomiques sur les Chlamydopsinae et descriptions d'espèces nouvelles. I. – Genres *Ceratohister* Reichensperger, *Eucurtiopsis* Silvestri et *Orectoscelis* Lewis (Coleoptera: Histeridae). *Bulletin de la Société Entomologique de France* **110**(3), 299–326.

Dégallier N, Caterino MS (2005b) Notes taxonomiques sur les Chlamydopsinae et descriptions d'espèces nouvelles. - II. Genre *Pheidoliphila* Lea (Coleoptera, Histeridae). *Bulletin de la Société Entomologique de France* **110**(4/5), 463–494.

DeMarzo L, Vienna P (1982a) Osservazioni morfologiche e ultrastrutturali su particolari organi di senso della clave antennali in Insteridi della subf. Saprininae e considerazioni sistematiche. *Entomologica* **57**, 53–77.

DeMarzo L, Vienna P (1982b) Osservazioni morfologiche e ultrastrutturali su un particolari apparato di senso della clave antennali di Platysomatini e Hololeptini (Coleoptera: Histeridae) e considerazioni sistematiche. *Entomologica* **57**, 79–87.

DeMarzo L, Vienna P (1982c) Studio morfologico della spermatheca in Coleotteri Isteridi con particolare attenzione alla tribu Saprinini. *Entomologica* **57**, 163–179.

Gardner JCM (1935) Fam. Histeridae, subfam. Niponiinae. In *Genera Insectorum, fasc. 202.* (Ed. P Wytsman) pp. 1–6. L. Desmet-Verteneuil, Bruxelles.

Gomy Y (1984) Révision des Acritini d'Australie et de Tasmanie (Coleoptera, Histeridae). *Annales de la Société Entomologique de France* (n.s.) **20**, 171–201.

Gomy Y (1985) Contribution à la connaissance des *Eutribalus* Bickhardt de la faune australo-papouae (Coleoptera, Histeridae). *Nouvelle Revue d'Entomologie* (n.s.) **2**, 377–383.

Gomy Y (1991) Description de deux nouveaux *Bacanius* Leconte de la faune Australienne (I.) (Coleoptera, Histeridae). Nouvelle Revue d'Entomologie (N.S.) **9**(1), 75–81.

Gomy Y (1994) Désignation du Lectotype de *Platysoma planiceps* MacLeay, 1871 d'Australie. Nouvelle Revue d'Entomologie (N.S.) **11**(3), 306–307.

Gomy Y (1995) Contribution à la connaisance des Micro-Histeridae australiens (Coleoptera). *Bulletin Mensuel de la Société Linnéenne de Lyon* **64**, 373–380. doi:10.3406/linly.1995.11088

Gomy Y (2009a) Un nouveau genre et trois nouvelles espèces d'Histeridae Bacaniini d'Australie (Coleoptera). *Bulletin de la Société Entomologique de France* **114**, 189–193.

Gomy Y (2009b) Désignation de l'espèce-type du genre *Australanius* Gomy, 2009 (Col., Histeridae) et lieu de conservation des holotypes. *Bulletin de la Société Entomologique de France* **114**, 361–364.

Hansen M (1997) Phylogeny and classification of the staphyliniform beetle families (Coleoptera). *Biologiske Skrifter* **48**, 1–339.

Helava J, Howden HF (1977) The genus *Onthophilus* in Australia, with description of a species (Coleoptera: Histeridae). *Journal of the Australian Entomological Society* **16**, 82–86.

Kanaar P (1997) Revision of the genus *Paratropus* Gerstaecker (Coleoptera: Histeridae). *Zoölogische Verhandelingen* **315**, 1–185.

Kanaar P (2003) Revision of the genus *Trypeticus* Marseul (Coleoptera: Histeridae). *Zoölogische Verhandelingen* **342**, 3–318.

Kovarik PW (1995) Development of *Epierus divisus* Marseul (Coleoptera: Histeridae). *Coleopterists Bulletin* **49**, 253–260.

Kovarik PW, Caterino MS (2001) Histeridae Gyllenhal, 1808. In *American beetles: Archostemata, Myxophaga, Adephaga, Polyphaga: Staphyliniformia. Volume 1.* (Eds RHJr Arnett and MC Thomas) pp. 212–227. CRC Press, Boca Raton.

Kovarik PW, Caterino MS (2005) Histeridae Gyllenhal, 1808. In *Coleoptera, Beetles. Volume 1: Morphology and Systematics (Archostemata, Adephaga, Myxophaga, Polyphaga partim.).* (Volume eds RG Beutel and RAB Leschen). In *Handbook of Zoology: A Natural History of the Phyla of the Animal Kingdom. Volume IV. Arthropoda: Insecta. Part 38.* (Eds NP Kristensen and RG Beutel) pp. 190–222. Walter de Gruyter, Berlin/New York.

Kovarik PW, Caterino MS (2016) Histeridae Gyllenhal, 1808. In *Handbook of Zoology, Coleoptera, Beetles. Morphology and Systematics (Archostemata, Adephaga, Myxophaga, Polyphaga partim) Vol. 1.* (Eds RG Beutel and RAB Leschen) pp. 281–314. De Gruyter, Berlin/Boston.

Kovarik PW, Passoa S (1993) Chaetotaxy of larval Histeridae based on a description of *Onthophilus nodatus* LeConte. *Annals of the Entomological Society of America* **86**, 560–576. doi:10.1093/aesa/86.5.560

Kovarik PW, Tishechkin AK (2004) A new genus and species of Hetaeriinae (Coleoptera: Histeridae) from southwestern North America. *Coleopterists Bulletin* **58**, 317–327. doi:10.1649/617

Kukalová-Peck J, Lawrence JF (1993) Evolution of the hind wing in Coleoptera. *Canadian Entomologist* **125**, 181–258. doi:10.4039/Ent125181-2

Lackner T (2010) Review of the Palaearctic genera of Saprininae (Coleoptera: Histeridae). *Acta Entomologica Musei Nationalis Pragae* **50**(Supplementum), 1–254.

Lackner T (2014) Phylogeny of the Saprininae subfamily reveals interesting ecological shifts in the history of the subfamily (Coleoptera: Histeridae). *Zoological Journal of the Linnean Society* **172**, 521–555.

Lackner T, Leschen RAB (2017) A monograph of the Australopacific Saprininae (Coleoptera, Histeridae). *ZooKeys* **689**, 1–263. doi:10.3897/zookeys.689.12021

Lawrence JF, Ślipiński SA (2013) *Australian Beetles. Volume 1. Morphology, Classification and Keys.* CSIRO Publishing, Melbourne.

Lea AM (1910) Australian and Tasmanian Coleoptera inhabiting or resorting to the nests of ants, bees and termites. *Proceedings of the Royal Society of Victoria.* New Series **23**, 116–230.

Lea AM (1912) Australian and Tasmanian Coleoptera inhabiting or resorting to the nests of ants, bees and termites. *Proceedings of the Royal Society of Victoria.* New Series **25**, 31–78.

Lea AM (1914a) On Australian and Tasmanian Coleoptera, with descrip[tions of new species. Part II. *Proceedings of the Royal Society of Victoria.* New Series **26**, 211–227

Lea AM (1914b) Notes on some miscellaneous Coleoptera, with descriptions of new species. *Transactions and Proceedings of the Royal Society of South Australia* **38**, 249–344.

Lea AM (1918) On Australian Coleoptera. Part 1. *Records of the South Australian Museum* **1**, 83–104.

Lea AM (1919) Notes on some miscellaneous Coleoptera, with descriptions of new species. Part V. *Transactions and Proceedings of the Royal Society of South Australia* **43**, 166–261.

Lea AM (1925a) Descriptions of new species of Australian Coleoptera. Part XVIII. *Proceedings of the Linnean Society of New South Wales* **50**, 414–431.

Lea AM (1925b) On Australian Histeridae (Coleoptera). *The Transactions of the Entomological Society of London* **1925**, 239–262.

Lea AM (1925c) Results of Dr. E. Mjoberg's Swedish Scientific Expeditions to Australia 1910–1913. 42. Histeridae. *Arkiv för Zoologi* **17A**(17), 1–3.

Marseul SM (1860) Supplément a la monographie des histérides. *Annales de la Société entomologique de France* (3) **8**, 835–866.

Mazur S (1984) A world catalogue of Histeridae. *Polskie Pismo Entomologiczne* **54**(3–4), 1–376.

Mazur S (1987) Description of a new *Pholioxenus* from Morocco with additional histerid notes (Col., Histeridae). *Entomologische Blätter* **83**, 28–32.

Mazur S (1990) Notes on Oriental and Australian Histeridae (Coleoptera). *Polskie Pismo Entomologiczne* **59**, 743–759.

Mazur S (1993) Two new histerids from Namibia and two new genera from the *Platysoma* complex (Coleoptera, Histeridae). *Stuttgarter Beiträge zur Naturkunde* (A) **492**, 8 pp.

Mazur S (1997a) New species of histerid beetles (Coleoptera, Histeridae). *Annales Zoologici* **38**, 27–32 [Warszawa].

Mazur S (1997b) A world catalogue of the Histeridae (Coleoptera: Histeridae). *Genus* (Supplement), 1–373.

Mazur S (1999) Preliminary studies upon the *Platysoma* complex (Col. Histeridae). *Annals of Warsaw Agricultural University-SGGW* **39**, 3–29.

Mazur S (2007) On new and little known histerids (Coleoptera: Histeridae) from Taiwan with additional notes on the species composition and zoogeography. *Formosan Entomologist* **27**, 67–81.

Mazur S (2011) *A Concise Catalogue of the Histeridae (Insecta: Coleoptera).* Warsaw University of Life Sciences, SGGW Press, Warsaw.

Mazur S, Ôhara M (2009) Notes on the genus *Eblisia* Lewis, 1899 in relation to Platysomatini, with description of four new genera (Coleoptera: Histeridae). *Studies and Reports of District Museum Prague-East* **5**, 233–248.

Mazur S, Węgrzynowicz P (2008) Notes on the genus *Santalus* Lewis, 1906 and *Pachylister* Lewis, 1904, with the description of *Nasaltus* gen. n. (Coleoptera, Histeridae). *Deutsche Entomologische Zeitschrift* **55**, 185–197. doi:10.1002/mmnd.200800016

McKenna DD, Farrell BD, Caterino MS, Farnum CW, Hawks DC, Maddison DR, Seago AE, Short AEZ, Newton AF, Thayer MK (2015) Phylogeny and evolution of Staphyliniformia and Scarabaeiformia: forest litter as stepping stone for diversification of nonphytophagous beetles. *Systematic Entomology* **40**(1), 35–60. doi:10.1111/syen.12093

Newton AF (1991) Histeridae (Hydrophiloidea) (including Niponiidae) In *Immature Insects, Vol. 2.* (Ed. FW Stehr) pp. 361–363. Kendall/Hunt, Dubuque, Iowa.

Ôhara M (1994) A revision of the superfamily Histeroidea of Japan (Coleoptera). *Insecta Matsumurana (N. S.)* **51**, 1–238.

Ôhara M (1999) A revision of the superfamily Histeroidea of Japan (Coleoptera). Supplementum I. *Insecta Matsumurana* (n.s.) **55**, 75–132.

Ôhara M, Mazur S (2000) A revision of the genera of the tribe Platysomatini (Coleoptera, Histeridae, Histerinae) Part 3: Redescriptions of the type species of *Althanus, Caenolister, Idister, Diister, Placodes, Plaesius, Hyposolenus* and *Aulacosternus. Insecta Matsumurana* (n.s) **57**, 1–37.

Ôhara M, Mazur S (2002) A revision of the genera of the tribe Platysomatini (Coleoptera, Histeridae, Histerinae) Part 4: Redescriptions of the type species of *Heudister, Platysoma, Cylister, Cylistus, Nicotikis, Mesostrix* and *Desbordesia. Insecta Matsumurana* (n.s.) **59**, 1–28.

Poinar G, Brown AE (2009) *Pantostictus burmanicus*, a new genus and species of Cretaceous Beetles (Coleoptera: Hydrophiloidea: Histeridae) in Burmese amber. *Proceedings of the Entomological Society of Washington* **111**, 38–46.

Reichardt A (1941) Semeystva Sphaeritidae i Histeridae (Vol. 1). [Families Sphaeritidae and Histeridae]. In *Fauna SSSR, Nasekomye Zhestokrylye, V, 3.* Nauka, Moskva-Leningrad, xiii + 419 pp. [in Russian].

Ślipiński SA, Mazur S (1999) *Epuraeosoma*, a new genus of Histerinae and phylogeny of the family Histeridae (Coleoptera: Histeroidea). *Annales Zoologici* **49**, 209–230.

Vienna P, Ratto G (2013) Un nuovo *Gnathoncus* Jacquelin du Val, 1858 dell'Iran (Insecta, Coleoptera, Histeridae). *Bolletino Dell Museo Civico di Storia Naturale Di Venezia* **64**, 29–37.

Waterhouse DF, Sands DPA (2001) *Classical biological control of arthropods in Australia.* ACIAR Monograph No. 77, Canberra.

Wenzel RL (1944) On the classification of the histerid beetles. *Fieldiana. Zoology* **28**, 51–151.

Wenzel RL (1955) The histerid beetles of New Caledonia (Coleoptera: Histeridae). *Fieldiana. Zoology* **37**, 601–634.

Wenzel RL (1962) Histeridae, the Hister beetles. In *Beetles of the United States*, fasc. 26. (Ed RHJr Arnett) pp. 372–379. Catholic University of America Press, Washington D.C., USA.

Yélamos T (1998) The *Aeletes* of the Hawaiian Islands (Coleoptera: Histeridae). *Bishop Museum Occasional Papers* **54**, 1–63.

Yélamos T (2002) *Fauna Iberica. Vol.17, Coleoptera Histeridae.* Museo Nacional de Ciencias Naturales Consejo Superior de Investigaciones Cientificas, Madrid.

Zhang YJ, Zhou HZ (2007a) Taxonomy of the Paromalini Reitter (Coleoptera: Histeridae, Dendrophilinae) from China. *Zootaxa* **1544**, 1–40.

Zhang YJ, Zhou HZ (2007b) Description of three new species of *Parepierus* Bickhardt from South China (Coleoptera, Histeridae, Tribalinae). *Deutsche Entomologische Zeitschrift* **54**, 253–260. doi:10.1002/mmnd.200700022

22. SUPERFAMILY SCARABAEOIDEA LATREILLE, 1802

John F. Lawrence, Nicole L. Gunter and Tom A. Weir

Introduction. The superfamily Scarabaeoidea is one of the three most stable and obviously monophyletic groups of Polyphaga, the others being Chrysomeloidea and Curculionoidea. Scarabaeoids were recognised as Lamellicornes by Latreille (1825) and most early workers, although separated into Pectinicornes (passalids and lucanids) and Lamellicornes in Lacordaire's (1856) classification based on antennal differences and briefly combined with the hydrophiloid family Synteliidae as the Actinorhabda of Kolbe (1903). Scarabaeoidea is recognised as a monophyletic group in virtually all modern works (Crowson 1955; Lawrence & Newton 1995; Smith 2006; Smith *et al.* 2006; Scholtz & Grebennikov 2016), including those based on cladistic analyses of morphological and/or molecular characters (Hunt *et al.* 2007; Lawrence *et al.* 2011; McKenna *et al.* 2015a, 2015b). The relationship of this series to other polyphagan groups is, however, still somewhat uncertain. Böving & Craighead (1931) considered the family Dascillidae to be sister group to Scarabaeoidea based on larval features, and this was followed by Crowson (1960, 1971) who noted the division of the penis into dorsal and ventral lobes in both Dascilloidea and *Pleocoma* LeConte (Pleocomidae) and placed the superfamilies Dascilloidea and Scarabaeoidea into a series Scarabaeiformia. Lawrence & Newton (1982) claimed that similarities between scarabaeoid and dascillid larvae were either plesiomorphic or associated with a soil-dwelling habit, and this was supported more recently by Grebennikov & Scholtz (2004). An earlier concept, originating with Kolbe (1908) combined the lamellicorns with Staphylinoidea (including Histeridae) and Actinorhabda (scarabaeoids plus Synteliidae) into Haplogastra, based primarily on characters of the abdomen, although hydrophiloids were still treated as an unrelated group, Palpicornia. Forbes (1926), in his study of the hind wing, also recognised the Haplogastra, but with the addition of the hydrophiloids. The concept of Haplogastra was revived in works on the hind wing by Kukalová-Peck & Lawrence (1993) and Scholtz *et al.* (1994) and in an analysis of both adult and larval characters by Hansen (1997). A combined molecular and morphological analysis by Caterino *et al.* (2005) produced a monophyletic Haplogastra, with Scarabaeoidea derived from within Staphyliniformia. A morphological analysis based on more than 300 beetle families and subfamilies and more than 500 characters of both adults and larvae (Lawrence *et al.* 2011) unexpectedly recovered a Dascillidae + Scarabaeoidea clade near the base of a clade consisting of all remaining Elateroidea, with the exception of Rhipiceridae and Rhinorhipidae; however a molecular analysis based on almost the same set of taxa (McKenna *et al.* 2015b) resulted in Hydrophiloidea (*s. lat.*) + Scarabaeoidea sister to Staphylinoidea (Bayesian) or Scarabaeoidea sister to Hydrophiloidea (*s. lat.*). + Staphylinoidea (Maximum Likelihood). Similar results were found with a larger selection of staphyliniform and scarabaeiform taxa but with fewer outgroups (McKenna *et al.* 2015a).

Although numerous classification schemes have existed, the classifications of Lawrence and Newton (1995), and Browne and Scholtz (1995, 1999) that includes thirteen families is almost universally accepted and has remained relatively stable. The one exception is that Ceratocanthinae is now considered a subfamily within Hybosoridae and not a distinct family, which has been supported by both larval and adult characters (Grebennikov *et al.* 2004; Ocampo 2006) and molecular analyses. Bouchard *et al.* (2011) also list two additional family names, proposed by Genise (2004) that represent extinct ichnotaxa. The superfamily is diverse with over 2500 genera and 30 000 species globally, and ~320 genera and 2650 species from six families recorded within Australia equivalent to ~10% of the beetle diversity of the continent. Additional information on the biology and distribution of the Australian scarabaeoid fauna is provided in the following chapters.

Characteristics. Adults. Scarabaeoidea are usually distinguished by the following features: (1) a lamellate antennal club; (2) a highly modified, burrowing prothorax with large procoxae (almost always with concealed trochantins and closed cavities); 3) usually dentate tibiae with a single spur; (4) no metacoxal plates; (5) hind wings with reduced venation and a strong radial bending zone coupled with a hinge; (6) abdominal sternite II represented by a pair more or less concealed lateral sclerites; (7) tergite VIII forming a pygidium and not concealed by tergite VII; and (8) four Malpighian tubules. More information on adult morphology may be found in Scholtz & Grebennikov (2016).

Larvae. Almost always lightly pigmented, grub-like and C-shaped with well developed antennae and legs, rarely a single pair of stemmata, usually with a distinct epicranial suture, often with underlying median endocarina, large mandibles with well developed, asymmetrical molae, abdomen with ten segments, most of which are usually divided by 2 to 4 folds, a simple ninth segment without urogomphi, a more or less terminal anus and usually cribriform spiracles (biforous in some Geotrupidae and Trogidae). Trends in scarabaeoids include the reduction of adult spiracles on segment VIII and movement of terminal spiracles dorsally or ventrally, loss of the larval spiracular closing apparatus, reduction of the number of adult antennomeres, and increase in the number of larval antennomeres from three to four. The group appears to be primitively adapted for burrowing in soil, with secondary trends in other directions. Loss of the spiracular closing apparatus may have accompanied the development of cribriform spiracles, although in the family Trogidae, the two structures may occur together. All scarabaeoid larvae, but in particular members of the more derived scarabaeid subfamilies (Melolonthinae, Rutelinae, Dynastinae, Cetoniinae) have complex structures on the underside of the labrum (epipharynx) and the ventral portion of the tenth abdominal segment (raster). Both character complexes have been very useful in the identification of scarabaeoid larvae, and a set of terms for each was devised by Hayes (1929) and modified and improved by Böving (1936) and Ritcher (1966). Modified versions of Boving's figures are shown in Figs 22.51–22.56. Further information on larval scarabaeoids may be found in Medvedev (1952), Ritcher (1966), McQuillan (1985), Grebennikov & Scholtz (2004), Scholtz & Grebennikov (2016).

Key to the families of Scarabaeoidea occurring in Australia

Adults

1. Antennae 11-segmented; apical 3 antennomeres broad, circular or oval, flattened and capable of close apposition, forming a cupuliform club (Fig. 23.7); body highly convex and hemispherical, often with extensive armature on head and prothorax in male (Fig. 23.19) .... **GEOTRUPIDAE**

– Antennae with fewer than 11 segments **2**

2(1). Mesocoxal cavities closed laterally by meeting of mesoventrite and metaventrite **3**

– Mesocoxal cavities open laterally (partly closed by mesepimeron) **4**

3(2). Head hypognathous, not or only partly visible from above, with concealed mouthparts; abdomen with 5 ventrites; mesocoxal cavities contiguous; elytra dull and more or less tuberculate, without striae (Fig. 25.1); apical 3 antennomeres asymmetrically widened .... **TROGIDAE**

– Head prognathous, completely visible from above, with exposed mouthparts (Fig. 24.17); abdomen with 5 or 6 ventrites (Figs 24.14–15); mesocoxal cavities distinctly separated; elytra shining and distinctly striate; apical 3 to 6 antennomeres asymmetrically widened and antennae curled in repose (Fig. 24.17) .... **PASSALIDAE**

4(2). Lateral expansions of apical antennomeres relatively narrow, thick and not usually capable of close apposition (Fig. 26.2); abdomen with 5 ventrites; antennae with long scape, usually elbowed (Fig. 26.15); mandibles of males often greatly enlarged (Fig. 26.13); head and prothorax rarely modified in males .... **LUCANIDAE**

– Lateral expansions of apical antennomeres usually thin, often broad and always capable of close apposition (Fig. 27.6); abdomen with 6 ventrites; mandibles in male not larger than those in female; male armature (if present) involving head and/or prothorax **5**

5(4). Body strongly contractile, capable of being rolled into shape of a ball (Fig. 27.14); eye completely divided into two parts by ocular canthus; metatibia and to a lesser extent protibia and mesotibia expanded and flattened, with a carinate outer edge; body length less than 5 mm .... **HYBOSORIDAE: Ceratocanthinae**

– Body not contractile OR IF moderately so THEN eye not completely divided, metatibia not flattened, without carinate outer edge; body length variable **6**

6(5). Labrum and mandibles clearly visible from above, the latter more or less opposable; antennae 10-segmented; antennal club 3-segmented and cupuliform (proximal segment circular and concave and two distal ones fitting into it); metatibiae without obliquely transverse ridges or groups of spines .... **HYBOSORIDAE: Hybosorinae and Liparochrinae**

– IF labrum visible from above THEN antennal club not cupuliform and metatibiae with obliquely transverse ridges or groups of spines; IF mandibles visible from above and with opposable apices THEN antennae 9-segmented .... **SCARABAEIDAE**

LARVAE

1. Antennae 2-segmented (Fig. 22.1); abdomen cylindrical, neither tapered posteriorly nor curled in repose (C-shaped) (Fig. 22.2); concavity and scar of abdominal spiracles directed posteriorly; metathoracic leg highly reduced, unsegmented with several teeth along anterior edge and serving as stridulatory plectrum acting against patch of fine, parallel carinae (stridulitrum) on mesocoxa (Fig. 22.3); mandibles without accessory ventral process (Fig. 22.4); living in association with adults in rotting timber .... **PASSALIDAE**

– Antennae 3- or 4-segmented (Fig. 22.5); larvae almost always curled in repose (C-shaped) (Fig. 22.6); concavity and scar of abdominal spiracles directed anteriorly, anteroventrally or ventrally (Fig. 22.7); usually with all three pairs of legs equally developed (Fig. 22.8); metathoracic leg, if reduced, with 2 or more segments (Fig. 22.9); mandibles with accessory ventral process (Fig. 22.10) **2**

2(1). Anterior edge of labrum serrate (Fig. 22.11) or with 3 truncate lobes (Fig. 22.12); stridulatory organ involving fore and mid legs OR labial palps 1-segmented (Fig. 22.13); head with dorsal endocarina extending anteriorly onto frontal plate (Fig. 22.14); apical antennomere with large sensory spot (Fig. 22.15) **3**

– Anterior edge of labrum rounded to weakly trilobed but never with serrations or truncate lobes; stridulatory organ, if present, involving mid and hind legs (Fig. 22.16); labial palps 2-segmented (Fig. 22.17); IF dorsal endocarina extending onto frontal plate THEN apical antennomere without large sensory spot **4**

3(2). Antennae 3-segmented due to fusion of antennomeres 3 and 4 (Fig. 22.18); stridulatory organ formed from procoxa and mesotibiotarsus (Fig. 22.19); anterior edge of labrum with 3 truncate lobes (Fig. 22.20); labial palps 2-segmented .... **HYBOSORIDAE: Hybosorinae (*Phaeochrous*)**

– Antennae 4-segmented, but appearing 3-segmented in ventral view due to partial fusion of antennomeres 1 and 2 (Fig. 22.21); stridulatory organ absent; anterior edge of labrum serrate (Fig. 22.22); labial palps 1-segmented (Fig. 22.23) .... **HYBOSORIDAE: Ceratocanthinae (*Cyphopisthes*)**

4(2). All legs more or less equally developed and 5-segmented (Fig. 22.24), with stridulatory organ formed by mid and hind legs (mesocoxa and metatrochanter) (Fig. 22.25); antennae 3- or 4-segmented, with apical antennomere much shorter and narrower than penultimate one (Fig. 22.26); abdominal terga not plicate; anal opening Y-shaped or longitudinal, lying between two fleshy lobes (Fig. 22.27) .... **LUCANIDAE**

– IF stridulatory organ formed by mid and hind legs THEN hnd legs reduced and 3-segmented (Fig. 22.28); abdominal terga with 2 or 3 transverse plicae or annulets (Fig. 22.29); anal opening transverse (Fig. 22.30), angulate or Y-shaped (Fig. 22.31), not lying between two fleshy lobes **5**

5(4). Antennae 4-segmented (Fig. 22.32) or apparently 5-segmented (first antennomere vaguely subdivided, Fig. 22.33); anterior abdominal terga with 3 plicae or annulets (Fig. 22.34); galea and lacinia usually completely fused (Fig. 22.35) or at least fitting very tightly together (Fig. 22.36) and tormae not united mesally (Fig. 22.37); IF galea and lacinia distinctly separated (Fig. 22.38) and tormae united mesally (Fig. 22.39) THEN body hump-backed (with mid abdominal terga greatly inflated) (Fig. 22.40) and legs 3-segmented with reduced pretarsal claw (Fig. 22.41) OR body less than 15 mm in length) .... **SCARABAEIDAE**

– Antennae 3-segmented (Figs 22.42–43); galea and lacinia distinctly separated (Fig. 22.44); tormae always united mesally (Fig. 22.45); body usually greater than 15 mm in length **6**

6(5). Anterior abdominal terga with 3 plicae or annulets clothed with conspicuous sharp setae and longer hairs (Fig. 22.46); frontoclypeal suture present (Fig. 22.47); legs all 5-segmented with well developed pretarsal claw; head capsule usually heavily pigmented, with one pair of stemmata .... **TROGIDAE**

– Anterior abdominal terga without or with 2 plicae or annulets, without sharp setae (Figs 22.48–49); frontoclypeal suture absent (Fig. 22.50); IF legs 5-segmented THEN pretarsal claw reduced; head capsule lightly pigmented, without stemmata .... **GEOTRUPIDAE**

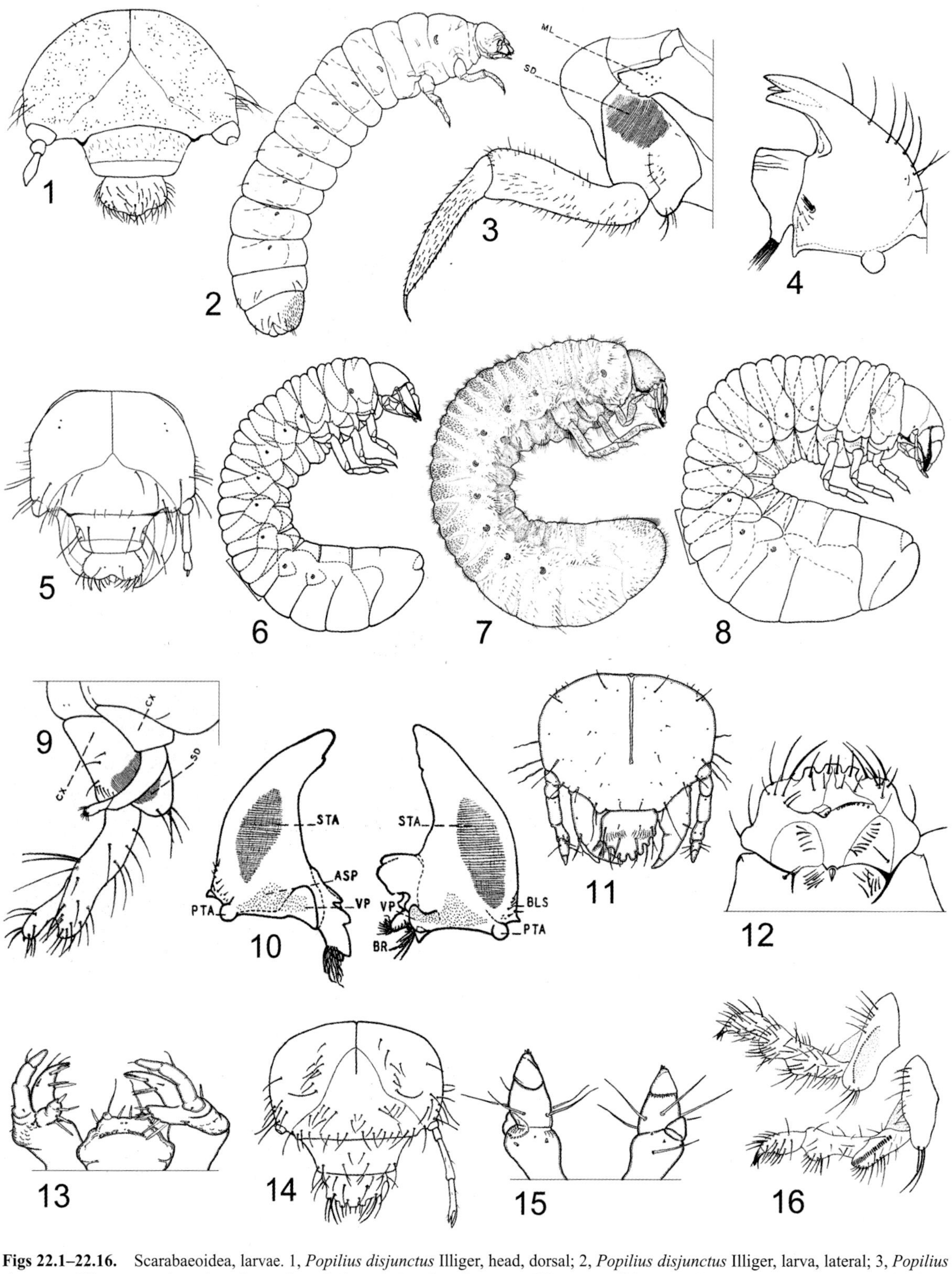

Figs 22.1–22.16. Scarabaeoidea, larvae. 1, *Popilius disjunctus* Illiger, head, dorsal; 2, *Popilius disjunctus* Illiger, larva, lateral; 3, *Popilius disjunctus* Illiger, mesothoracic and metathoracic legs; 4, *Popilius disjunctus* Illiger, mandible, ventral; 5, *Omorgus suberosus* (Fabricius), head, dorsal; 6, *Anomala innuba* (Fabricius), larva, lateral; 7, *Anoplognathus pindarus* Carne, larva, lateral; 8, *Cyclocephala immaculata* Olivier, larva, lateral; 9, *Geotrupes blackburni* Beauvois, metathoracic and mesothoracic leg; 10, *Anomala innuba* (Fabrficius), mandibles, ventral; 11, *Cyphopisthes descarpentriesi* Paulian, head, dorsal; 12. *Phaeochrous emarginatus* Laporte, abrum-epipharynx, ventral; 13, *Cyphopisthes descarpentriesi* Paulian, maxillolabial complex, ventral; 14, *Phaeochrous emarginatus* Laporte, head, dorsal; 15, *Cyphopisthes descarpentriesi* Paulian, antennal apex, dorsal & ventral; 16, *Dorcus parallelus* (Say), meso- and metathoracic legs.

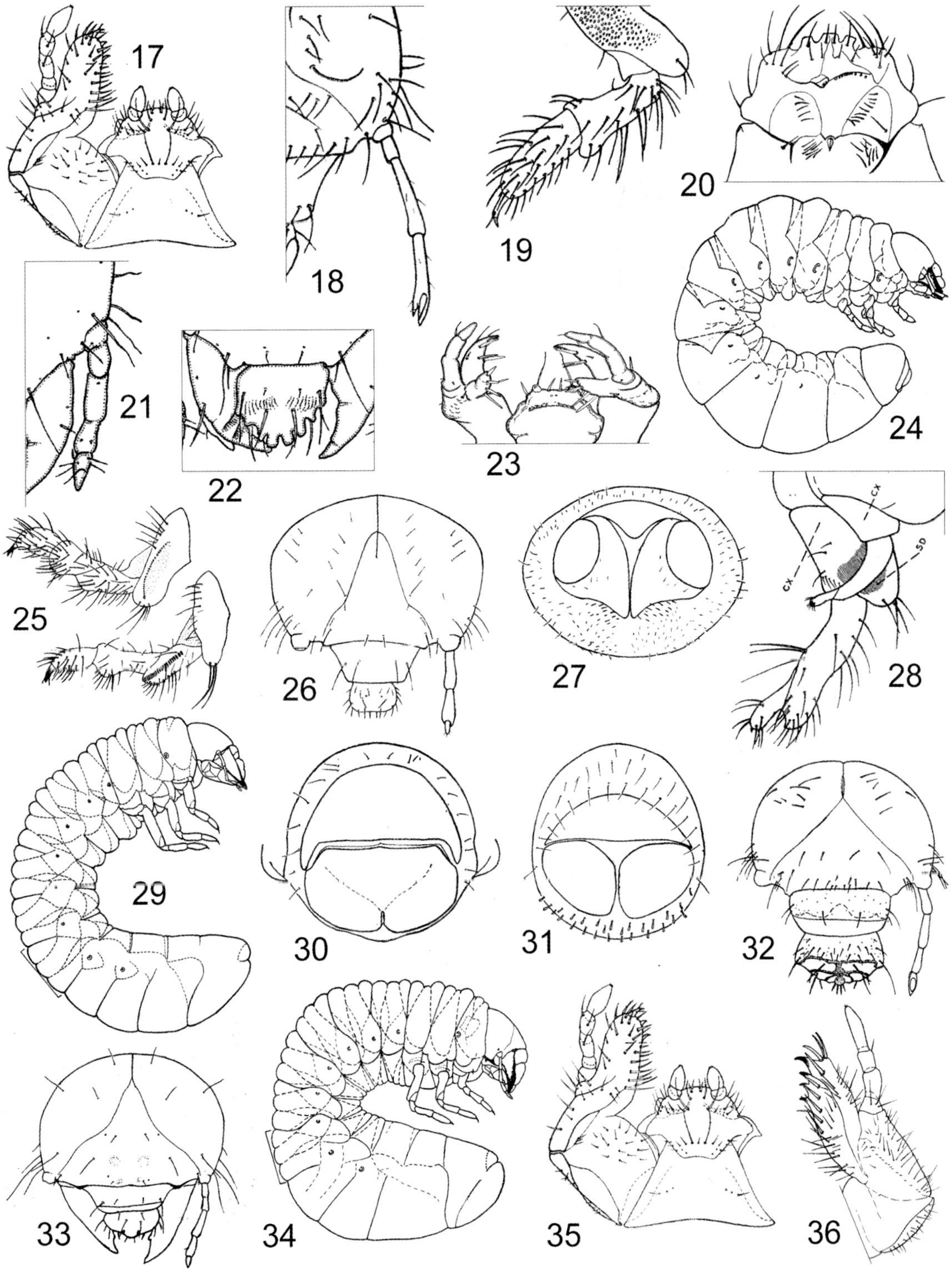

Figs 22.17–22.36. Scarabaeoidea, larvae.17, *Anomala innuba* (Fabrficius), maxilla and labium, ventral; 18, *Phaeochrous emarginatus* Laporte, head, dorsal; 19. *Hybororus orientalis* Westwood, prothoracic leg; 20, *Phaeochrous emarginatus* Laporte, epipharynx; 21, *Cyphopisthes descarpentriesi* Paulian, antenna; 22, *Cyphopisthes descarpentriesi* Paulian, labrum; 23, *Cyphopisthes descarpentriesi* Paulian, labium & maxillae; 24, *Dorcus parallelus* (Say), larva, lateral; 25, *Dorcus parallelus* (Say), meso- and metathoracic legs; 26, *Platycerus oregonensis* Westwood, head, dorsal; 27, *Platycerus oregonensis* Westwood, abdominal apex; 28, *Geotrupes blackburni* Beauvois, meta- and mesothoracic legs; 29, *Anomala innuba* (Fabricius), larva, lateral; 30, *Aphodius haemorrhoidalis* (Linnaeus), abdominal apex; 31, *Psammodius oregon*ensis Cartwright, abdominal apex; 32, *Plectris aliena* Chapin, head, dorsal; 33, *Aphodius hamatus* Say, head, dorsal; 34, *Cyclocephala immaculata* Olivier, larva, lateral; 35, *Anomala innuba* (Fabricius), labium & maxilla, ventral; 36, *Polyphylla decimlineata* (Say), maxilla, dorsal.

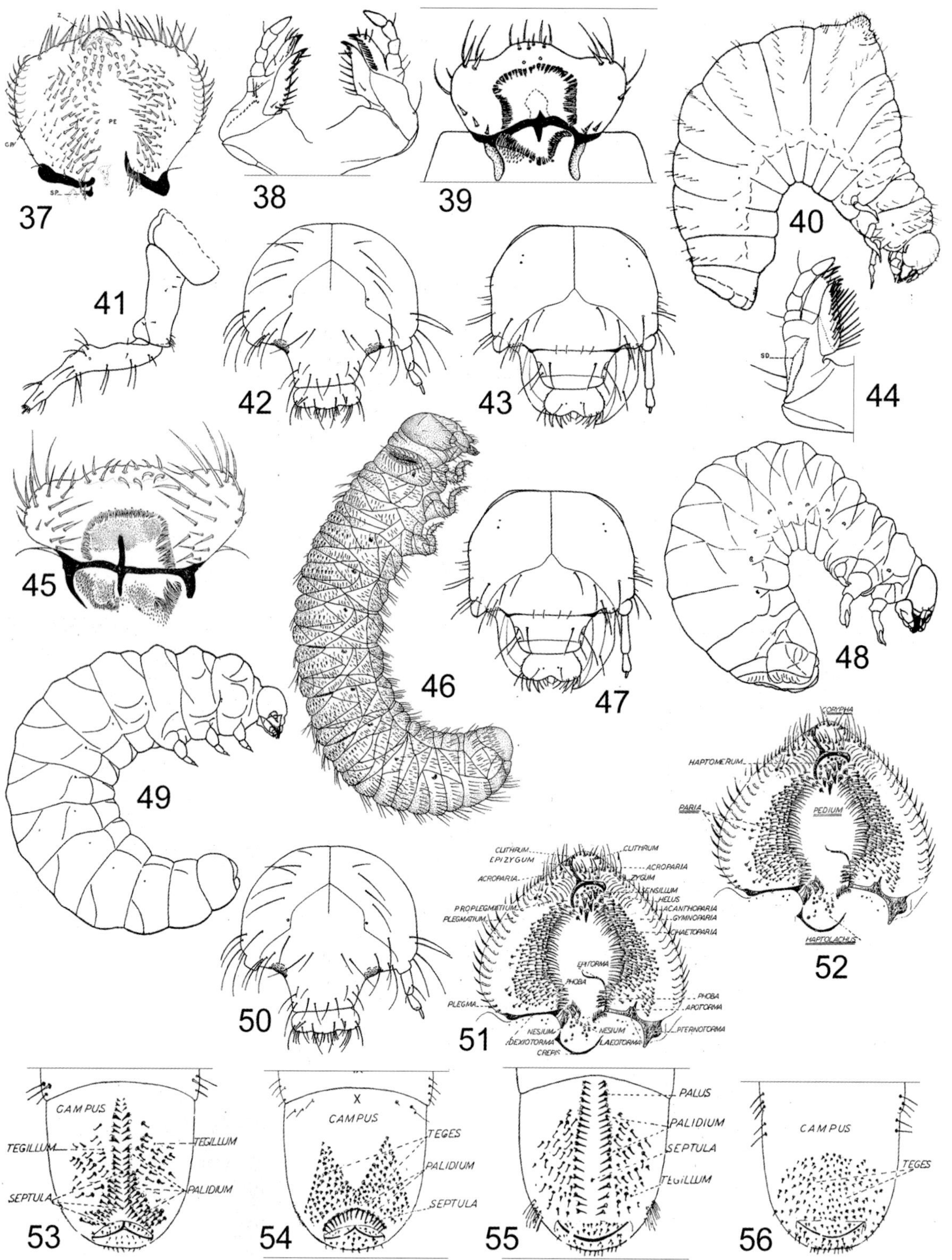

Figs 22.37–22.56. Scarabaeoidea, larvae. 37, *Parastasia brevipes* (LeConte), epipharynx; 38, *Aphodius hamatus* Say, maxillae, dorsal & ventral; 39, *Onthophagus hecate* (Panzer), epipharynx; 40, *Onthophagus hecate* (Panzer), larva, lateral; 41, *Dichotomius carolina* (Linnaeus), mesothoracic leg; 42, *Geotrupes blackburni* Beauvois, head, dorsal; 43, *Omorgus suberosus* (Fabricius), head, dorsal; 44, *Omorgus suberosus* (Fabricius), maxilla, dorsal; 45, *Geotrupes blackburni* Beauvois, epipharynx; 46, *Omorgus suberosus* (Fabricius), larva, lateral; 47, *Omorgus suberosus* (Fabricius), head, dorsal; 48, *Geotrupes blackburni* Beauvois, larva, lateral; 49, *Eucanthus lazarus* (Fabricius), larva lateral; 50, *Geotrupes blackburni* Beauvois, head, dorsal; 51–52, Epipharyngeal nomenclature from Böving (1936); 53–56, Raster types from Böving (1936).

Acknowledgments

The authors are indebted to Oregon State University, Corvallie, Oregon, for allowing us to use the images in Figs 22.1–22.10, 22.12–22.20, 22.24–22.45, 22.47–22.50 from Ritcher (1966), to Museu de Zoologia, Universidade de Sao Paulo, Brazil for allowing us to use Fig. 46 from Costa *et al.* (1988), and to Vasily Grebennikov, Canadian Food Inspect Agency, Ottawa, for sending images used in Figs 22.11, 22.13, 22.15, 22.21 22.23, which were also published in Grebennikov *et al.* (2002).

References

Bouchard P, Bousquet Y, Davies AE, Alonso-Zarazaga MA, Lawrence JF, Lyal CHC, Newton AF, Reid CAM, Schmitt M, Slipinski SA, Smith ABT (2011) Family names in Coleoptera (Insecta). *ZooKeys* **88**, 1–972. doi:10.3897/zookeys.88.807

Böving AG (1936) Description of the larva of *Plectris aliena* Chapin and explanation of new terms applied to the epipharynx and raster. *Proceedings of the Entomological Society of Washington* **38**(8), 169–185, pls 9–10.

Böving AG, Craighead FC (1931) An illustrated synopsis of the principal larval forms of the order Coleoptera. *Entomologica Americana* (N.S.) **11**, 1–351.

Browne DJ, Scholtz CH (1995) Phylogeny of the families of Scarabaeoidea (Coleoptera) based on characters of the hindwing articulation, hindwing base and wing venation. *Systematic Entomology* **20**, 145–173. doi:10.1111/j.1365-3113.1995.tb00089.x

Browne DJ, Scholtz CH (1999) A phylogeny of the families of Scarabaeoidea (Coleoptera). *Systematic Entomology* **24**, 51–84. doi:10.1046/j.1365-3113.1999.00067.x

Caterino MS, Hunt T, Vogler AP (2005) On the constitution and phylogeny of Staphyliniformia (Insecta: Coleoptera). *Molecular Phylogenetics and Evolution* **34**, 655–672.

Costa C, Vanin SA, Casari-Chen S (1988) *Larvas de Coleoptera do Brasil.* Museu de Zoologia, Universidade de São Paulo, São Paulo, Brazil.

Crowson RA (1955) The Natural Classification of the Families of Coleoptera. Nathaniel Lloyd, London.

Crowson RA (1960) The phylogeny of Coleoptera. *Annual Review of Entomology* **5**, 111–134.

Crowson RA (1971) Observations on the superfamily Dascilloidea (Coleoptera: Polyphaga), with the inclusion of Karumiidae and Rhipiceridae. *Zoological Journal of the Linnean Society* **50**, 11–19.

Forbes WTM (1926) The wing folding patterns of the Coleoptera. *Journal of the New York Entomological Society* **34**, 42–115, pls. 7–18.

Genise JF (2004) Ichnotaxonomy and ichnostraitigraphy of chambered trace fossils in paleosols attricbuted to coleopterans, ants and termites. In *The application of ichnology to palaeoenvironmental and stratigraphic analysis.* (Ed. D McIlroy) pp. 419–453. Geology Society, special publication 228, London.

Grebennikov VV, Scholtz CH (2004) The basal phylogeny of Scarabaeoidea (Insecta: Coleoptera) inferred from larval morphology. *Invertebrate Systematics* **18**, 321–348. doi:10.1071/IS03013

Grebennikov VV, Ballerio A, Scholtz CH (2002) Larva and pupa of *Cyphopisthes descarpentriesi* Paulian (Coleoptera: Scarabaeoidea: Ceratocanthidae) and their phylogenetic implications. *Australian Journal of Entomology* **41**, 367–374. doi:10.1046/j.1440-6055.2002.00307.x

Grebennikov VV, Ballerio A, Ocampo FC, Scholtz CH (2004) Larvae of Ceratocanthidae and Hybosoridae (Coleoptera: Scarabaeoidea): study of morphology, phylogenetic analysis and evidence of paraphyly of Hybosoridae. *Systematic Entomology* **29**, 524–543. doi:10.1111/j.0307-6970.2004.00257.x

Hansen M (1997) Phylogeny and classification of the staphyliniform beetle families (Coleoptera). *Biologiske Skrifter* **48**, 1–340.

Hayes WP (1929) Morphology, taxonomy and biology of larval Scarabaeoidea. *Illinois Biological Monographs* **12**(2), 1–119.

Hunt T, Bergsten J, Levkanicova Z, Papadopoulou A, St. John O, Wild R, Hammond PM, Ahrens D, Balke M, Caterino MS, Gómez-Zurita J, Ribera I, Barraclough TG, Bocakova M, Bocak L, Vogler AP (2007) A comprehensive phylogeny of beetles reveals the evolutionary origins of a superradiation. *Science* **318**, 1913–1916. doi:10.1126/science.1146954

Kolbe HJ (1903) Zur Systematik der Coleopteren. *Allgemeine Zeitschrift für Entomologie* **1903**, 137–145.

Kolbe H (1908) Mein System der Coleopteren. *Zeitschrift für Wissenschaftliche Insektenbiologie* **4**, 116–123, 153–162, 219–226, 246–251, 286–294, 389–400.

Kukalová-Peck J, Lawrence JF (1993) Evolution of the hind wing in Coleoptera. *Canadian Entomologist* **125**, 181–258.

Lacordaire T (1856) *Histoire Naturelle des Insectes. Genera des Coléoptères. Tome 3.* Libraire Encyclopédique de Roret, Paris.

Latreille PA (1825) *Familles Naturelles du Règne Animal, Exposées Succinctement et dans un Order, Analytique, avec Indication de leurs Genres.* J.-B. Baillière, Paris.

Lawrence JF, Newton AF (1982) Evolution and classification of beetles. *Annual Review of Ecology and Systematics* **13**, 261–290. doi:10.1146/annurev.es.13.110182.001401

Lawrence JF, Newton AF, Jr (1995) Families and subfamilies of Coleoptera (with selected genera, notes, references and data on family-group names). In *Biology, Phylogeny, and Classification of Coleoptera. Papers Celebrating the 80th Birthday of Roy A. Crowson.* (Eds J Pakaluk and SA Ślipiński) pp. 779–1006. Muzeum I Instytut Zoologii Polska Akademia Nauk, Warsaw.

Lawrence JF, Ślipiński A, Seago AE, Thayer MK, Newton AF, Marvaldi AE (2011) Phylogeny of the Coleoptera based on morphological characters of adults and larvae. *Annales Zoologici* **61**(1), 1–217. doi:10.3161/000345411X576725

McKenna DD, Farrell BD, Caterino MS, Farnum CW, Hawks DC, Maddison DR, Seago AE, Short AEZ, Newton AF, Thayer MK (2015a) Phylogeny and evolution of Staphyliniformia and Scarabaeiformia: forest litter as a stepping stone for diversification of nonphytophagous beetles. *Systematic Entomology* **40**, 35–60. doi:10.1111/syen.12093

McKenna DD, Wild AL, Kanda K, Bellamy CL, Beutel RG, Caterino MS, Farnum CW, Hawks DC, Ivie MA, Jameson ML, Leschen RAB, Marvaldi AE, McHugh JV, Newton AF, Robertson JA, Thayer MK, Whiting MF, Lawrence JF, Ślipiński A, Maddison DR, Farrell BD (2015b) Beetles survived the end Permian mass extinction to diversify during the Cretaceous terrestrial revolution. *Systematic Entomology* **40**, 835–880. doi:10.1111/syen.12132

McQuillan PB (1985) The identification of root-feeding cockchafer larvae (Coleoptera: Scarabaeidae) found in pastures in Tasmania. *Australian Journal of Zoology* **33**, 509–546. doi:10.1071/ZO9850509

Medvedev SI (1952) *Larvae of Scarabaeid Beetles of the USSR Fauna. Classification Keys to the Fauna of the USSR, Publication of the Zoological Institute, Academy of Sciences, USSR, 47.* (in Russian). Academy of Sciences, Moscow, Leningrad.

Ocampo FC (2006) Phylogenetic analysis of the scarab family Hybosoridae and monographic revision of the New World subfamily Anaidinae (Coleoptera: Scarabaoidea). 3. Phylogenetic analysis of the subfamily Anaidinae (Coleoptera: Scarabaoidea). *Bulletin of the University of Nebraska State Museum* **19**, 13–177.

Ritcher PO (1966) *White Grubs and their Allies. A Study of North American Scarabaeoid Larvae.* Oregon State University Press (Studies in Entomology No. 4), Corvallis, Oregon.

Scholtz CH, Grebennikov VV (2016) 15. Scarabaeoidea Latreille, 1802. Introduction, phylogeny. In *Handbuch der Zoologie/Handbook of Zoology. BandVolume IV Arthropoda: Insecta. Teilband/Part 38. Coleoptera, Beetles. Volume 1: Morphology and Systematics (Archostemata, Adephaga, Myxophaga, Polyphaga partim).* 2nd Edition. (Eds RG Beutel and RAB Leschen) pp. 443–465, 505–525. Walter de Gruyter, Berlin.

Scholtz CH, Browne DJ, Kukalová-Peck J (1994) Glaresidae, archaeopteryx of the Scarabaeoidea (Coleoptera). *Systematic Entomology* **19**, 259–277. doi:10.1111/j.1365-3113.1994.tb00590.x

Smith ABT (2006) A review of the family-group names for the superfamily Scarabaeoidea (Coleoptera) with corrections to nomenclature and a current classification. –. *Coleopterists Society Monographs* **5**, 144–204.

Smith ABT, Hawks DC, Heraty JM (2006) An overview of the classification and evolution of the major scarab beetle clades (Coleoptera: Scarabaeoidea) based on preliminary molecular analyses. *Coleopterists Society Monographs* **5**, 35–46.

23. GEOTRUPIDAE LATREILLE, 1802

John F. Lawrence and Mengjie Jin

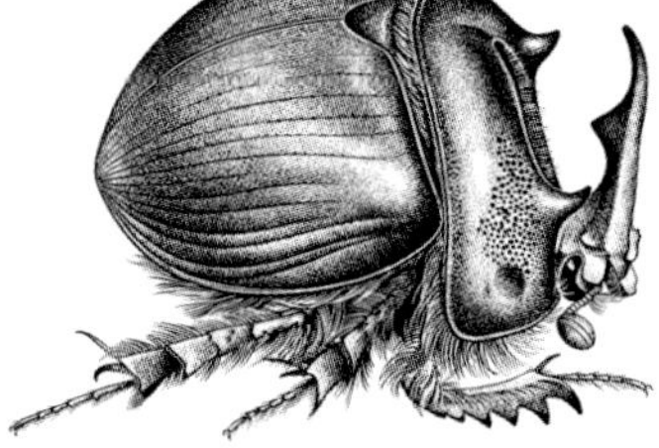

Fig. 23.1. *Blackburnium cavicolle* (Macleay), male, lateral.

Common names. Earth-boring scarab beetles; dor beetles; musician beetles.

Introduction. Geotrupidae, defined here in the broad sense of Smith (2006), is a widely distributed family containing ~65 genera and 500 species placed in the three subfamilies and 13 extant tribes (based in part on the works of Nikolajev 1970, 1990, 2002, 2003 and Zunino 1984a, 1984b): Taurocerastinae, Bolboceratinae (including Bolboceratini, Athyreini, Bolbochromini, Gilletinini, Eubolbitini, Bolbelasmini, Australobolbini, Eucanthini and Stenaspidiini) and Geotrupinae (including Geotrupini, Lethrini, Chromogeotrupini and Ceratotrupini). The family is considered by Browne & Scholtz (1996, 1999 and references therein) and Scholtz & Grebennikov (2016) to be polyphyletic based on a series of larval and adult characters, with an emphasis on the axillary region of the hind wing. According to these authors Bolboceratinae belongs to a major scarabaeoid lineage also containing Glaphyridae, Pleocomidae, Trogidae, Diphyllostomatidae, Lucanidae and Passalidae, while Geotrupinae, Lethrinae and Taurocerastinae belong to a separate lineage also containing Hybosoridae and Ochodaeidae. In a morphological analysis of 359 beetle taxa based on over 500 adult and larval characters (Lawrence *et al.* 2011) *Odonteus* Samouelle (= *Bolboceras* Kirby), *Geotrupes* and *Taurocerastes* Philippi formed a monophyletic group, but in a molecular analysis of most of the same beetle taxa (McKenna *et al.* 2015) *Geotrupes* fromed a clade with Trogidae, *Taurocerastes* formed a clade with Lucanidae and *Odonteus* (= *Bolboceras*) formed a clade with Passalidae. A monophyletic Geotrupidae, including Bolboceratinae and Taurocerastinae, was supported in a preliminary molecular study by Smith *et al.* (2006) and a morphological analysis based on larvae (Verdú *et al.* 2004).

Characteristics. Adults. Short, broad and strongly convex, ~5–25 mm in length, with upper surfaces more or less glabrous but undersurfaces and legs setose; colour yellowish-brown to black, with metallic blue or violet undersurface in *Geotrupes*. Head deeply inserted into prothorax, usually abruptly constricted immediately behind eyes forming very short temples and a short, broad neck (not in *Geotrupes*). Eyes well developed, usually strongly protuberant, deeply emarginate or completely divided by canthus, finely facetted, without interfacetal setae. Antennal insertions concealed from above; subantennal grooves present. Frontoclypeal suture absent in Bolboceratinae (well developed and angulate in *Geotrupes*). Males sometimes with median elevation on frons and with dorsally or anteriorly projecting horn on anterior portion of clypeus. Labrum strongly transverse and truncate to slightly emarginate. Antennae short, 11-segmented with strong, 3-segmented, compact, lamellate club; scape slightly curved or asymmetrically expanded. Mandibles with carinate outer edge, unidentate or weakly bidentate apex, well developed molae and setose prostheca. Maxilla usually with 1-segmented, subfalcate galea and lacinia with apical and mesal unci (in *Geotrupes* with 2-segmented, apically expanded, densely setose galea and a densely setose lacinia bearing a weak, bidentate uncus); maxillary palp 4-segmented with apical palpomere fusiform. Mentum with bilobed ligula; labial palps 3-segmented with apical palpomere fusiform. Gular usually elongate. Pronotum usually wider at base than combined elytral bases with strongly curved sides, complete lateral carinae and usually a simple disc with various depressions and elevations (horns) in some males. Prosternum in front of coxae very short, biconcave or slightly tumid, the anterior edge sometimes produced to form chin piece. Prosternal process complete but sometimes interrupted at middle, distinctly expanded and truncate to angulate apically. Hypomeron usually concave anteriorly forming pockets for the reception of the eyes when head is retracted. Procoxae large with mesal portion projecting below prosternum. Procoxal cavities strongly transverse, very narrowly separated or contiguous, closed externally and open internally. Elytra with sides moderately curved and apices conjointly rounded concealing all abdominal tergites or exposing part of pygidium only; disc with 10–15 puncture rows or striae and no scutellary striole; epipleura relatively narrow and complete or almost complete. Mesoventrite without paired procoxal rests, sometimes with median prosternal rest or broad collar; mesoventral cavity absent; discrimen absent. Mesocoxae strongly transverse and slightly oblique, with exposed trochantins. Mesocoxal cavities usually widely separated (narrowly separated in *Eucanthus* and *Elephastomus* and contiguous in *Geotrupes*), partly closed laterally by mesepimeron. Mesometaventral junction usually absent due to fusion meso- and metaventrites (ventrites abutting or overlapping in *Elephastomus* and *Eucanthus*, not quite meeting in *Geotrupes*). Metaventrite strongly transverse and strongly convex; transverse (metakatepisternal) suture short and curved towards posterior edge of ventrite; discrimen complete, extending to anterior edge of ventrite; exposed portion of metanepisternum relatively short and broad. Metacoxae strongly transverse, subcontiguous, extending laterally to meet elytra; plates absent. Metendosternite with long, slender stalk, long, transverse lateral arms and short to moderately long anterior process with narrowly separated tendons. Hind wing with long radial bending zone combined with distinct hinge at end of radial bar; apical field long, with as many as 4 radial extensions; radial cell incomplete, without base; basal portion of RP relatively short and apical extension extending to wing margin; medial field with 4 free veins; MP_3 and MP_4 not connected basally; wedge cell absent; anal lobe highly reduced, separated by wide, deep embayment, AP_{3+4} not forked, sometimes absent. Legs rather stout, with expanded and flattened femora; trochanterofemoral joint strongly oblique with femur in contract with coxa. Protibia apically expanded with one spur, its outer edge with a variable number of lobes or teeth. Meso- and metatibiae usually apically expanded and outer edge with at least one transversely oblique subapical ridge; spurs well developed and usually asymmetrical. Tarsi 5–5–5; tarsomeres relatively long and slender, simple; pretarsal claws simple; empodium well developed and bisetose. Abdomen with 7 free ventrites (sternites II–VIII), but sternite II divided by coxae into 2 small lateral sclerites; abdominal intercoxal process absent. Functional spiracles usually on segments I–VII, located in sternal notches (in *Geotrupes* on abdominal segments I–VIII, located mainly in pleural membrane). Anterior edge of sternite VIII in male without median strut. Anterior edge of segment IX with broadly rounded subgenital plate; tergite IX almost completely divided into two laterotergites (united by a slender strip of membrane); tergum X free. Aedeagus of modified trilobate type, with large, well sclerotised phallobase; parameres free, fused to phallobase and/or fused to one another, symmetrical or asymmetrical; penis mostly membranous, sometimes with paired struts; endophallus with or without armature. Sternite VIII in female without spiculum ventrale. Sternite IX in female divided into two parts; proctiger usually sclerotised. Ovipositor short and broad; paraprocts slightly shorter than coxites, without bacula; coxites undivided, setose, without styli.

Larvae. Subcylindrical and strongly curved ventrally (C-shaped); relatively weakly sclerotised and lightly pigmented. Head protracted, hypognathous, about as long as wide or slightly transverse, strongly rounded laterally and slightly flattened, with straight posterior edge and no stemmata. Epicranial stem moderately long; frontal arms V-shaped or U-shaped, with bases contiguous; median endocarina coincident with or extending anterior to epicranial stem. Frontoclypeal suture absent; clypeus sometimes asymmetrical; labium free, transverse, broadly rounded or weakly trilobed; tormae united mesally. Antennae less than half as long as head width, 3-segmented; antennomere 2 usually with 2 to 4 or 5 short, conical sensoria (in *Geotrupes* with a single, dome-like sensorium). Mandibles large and more or less asymmetrical, unidentate or weakly bidentate, with accessory ventral process, well developed mola and no prostheca; somewhat reduced in Bolboceratinae. Ventral mouthparts slightly retracted, with well developed maxillary articulating area; cardines more or less oblique, divided; stipes as long as wide or slightly transverse, its dorsal surface with stridulatory teeth. Galea and lacinia well separated, relatively sparsely setose, each with terminal spine in *Geotrupes*, variously reduced in Bolboceratinae; palp 3-segmented with segment-like palpifer. Labium with ligula more or less truncate; widely separated. Hypopharyngeal bracon absent; hypopharynx with paired symmetrical or asymmetrical oncyli. Hypostomal rods moderately long and diverging or absent; ventral epicranial ridges absent. Gular region absent. Thoracic segments more or less equal in length, without distinct tergal plates. Prothoracic venter without distinct sclerites. Pro- and mesothoracic legs in *Geotrupes* moderately long, 4-segmented with trochanter absent and pretarsus claw-like but minute; metathoracic leg reduced and stridulatory organ formed by pars stridens on posterior surface of the mesocoxa and a longitudinal row of teeth (plectrum) on the anterior edge of the metafemur. Legs in Bolboceratinae reduced, subequal and 3-segmented with pretarsal claw absent in some; coxae moderately widely separated. Abdominal terga without sclerotised plates, terga I–VIII usually transversely divided forming 2 plicae. Tergum IX about as long as VIII, not extending onto ventral surface, without urogomphi; segment X visible from above, more or less cylindrical in Bolboceratinae, laterally expanded and obliquely flattened in *Geotrupes*; anal opening Y-shaped, surrounded by single dorsal lobe and paired ventral lobes. Functional spiracles usually cribriform and slightly reniform with spiracular scar located ventrally (biforous in *Eucanthus*), those on segment VIII slightly reduced; closing apparatus absent (Ritcher 1966; Houston 2016).

Biology. Adult geotrupids are usually nocturnal or fly only at dusk and dawn; they are often attracted to lights, but may also be caught in flight intercept traps. Adults dig deep burrows in the soil, and usually remain there during the day. Although many exotic species, especially from mesic or humid forested areas, provision larval tunnels with vegetation, leaf litter, humus or dung (Howden 1955; Woodruff 1973), this is not the case with the majority of Australian species. The one exception is the introduced *Geotrupes spiniger* Marsham, which provisions its larval burrows with dung and was introduced for the control of cattle dung. Carne (1965) reported that specimens of *Elephatomus meraldus* Carne in the ACT utilised cattle dung, but according to Howden *et al.* (2007), Carne observed burrows beside or through dung piles but did not excavate burrows to determine if dung was actually the food source. Subsequent examination by the Howdens of many *Elephastomus* burrows in a similar patch of dung piles found no dung. Howden *et al.* (2007) reported finding bits of unidentified fungi in the burrows of some species of *Blackbolbus*, *Blackburnium* and *Bolborhachium* species. Houston & Bougher (2010) found that two species of *Blackbolbus* had fed on different species of hypogean fungi, based on examination of faeces and gut contents, while adults of *Blackbolbus frontalis* (Guérin-Méneville) were found in burrows with hypogean sporocarps (truffles) of the basidiomycote genera *Amarrendia* (Amanitaceae, Justo *et al.*, 2010), *Hysterangium* (Hysterangiaceae) and *Scleroderma* (Sclerodermataceae). Some species of *Blackburnium*, *Bolboleaus* and *Bolborhachium*, on the other hand, were found to have ingested mycelium and spores of a mycorrhizal glomeralean (Glomeromycota, Oehl *et al.* 2011), along with quantities of soil. But none of these records involved bolboceratine larvae. A concerted effort was made by T. F. Houston (2007, 2011, 2016) to determine just what larval bolboceratines feed on in Australia. Observations in the field and laboratory (Houston 2011) revealed that specimens of four Western Australian *Bolborhachium*, *Blackburnium* and *Blackbolbus* laid one gigantic egg at a time, those of *Blackburnium reichei* (Guérin-Méneville) weighing up to 56% as much as the female that laid them. Howden (1985) and Howden & Cooper (1977) reported that the deep larval cells

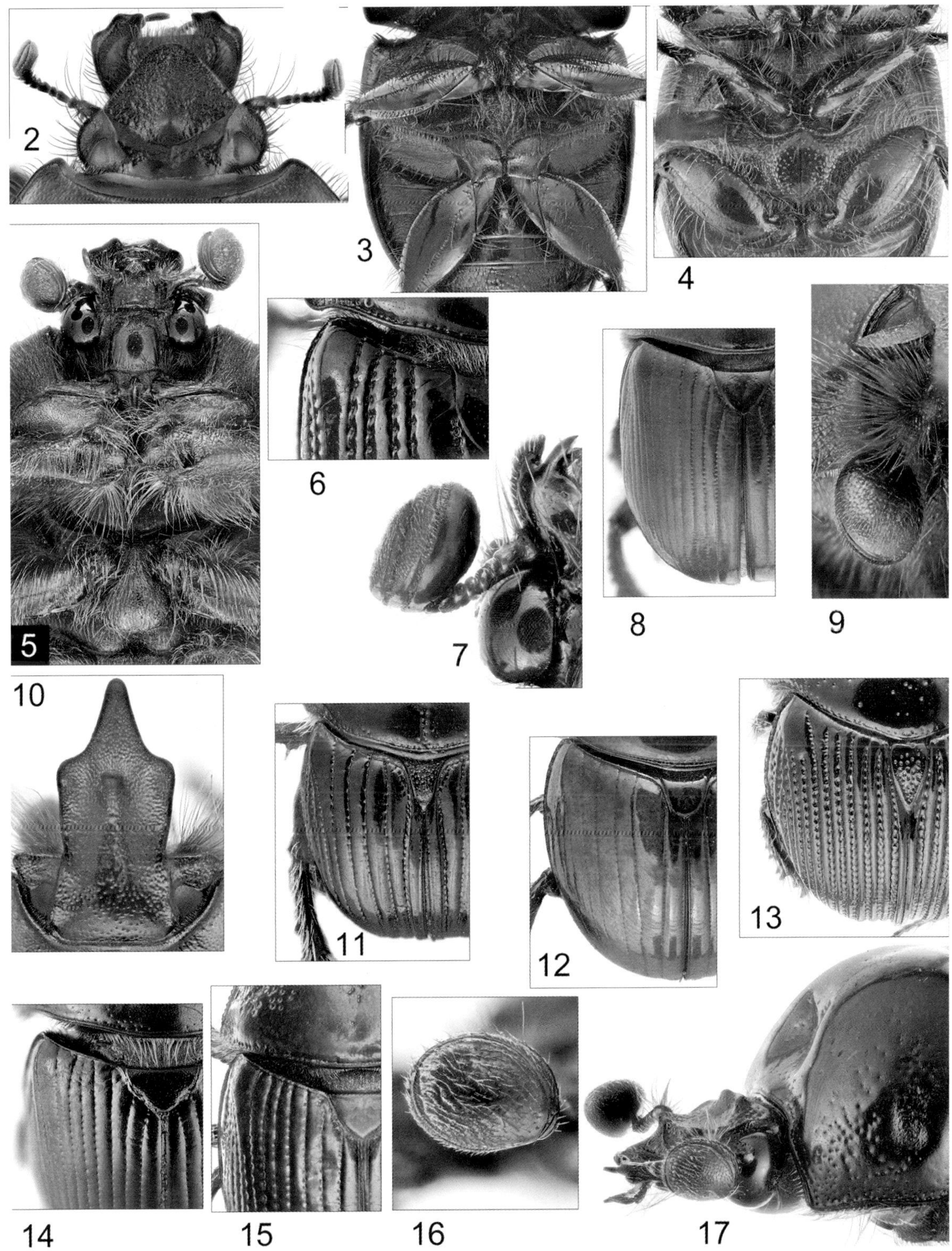

Figs 23.2–23.17. Geotrupidae adults. 2, *Geotrupes spiniger* Marsham, head, dorsal; 3, *Geotrupes spiniger* Marsham, mesothorax, ventral; 4, *Eucanthus felskei* Boucomont, meso- and metathorax, ventral; 5, *Blackburnium cavicolle* (Macleay), head and thorax, ventral; 6, *Eucanthus felskei* Boucomont, anterior portion of left elytron; 7, *Eucanthus felskei* Boucomont, antenna; 8, *Elephastomus meraldus* Carne, left elytron; 9, *Elephastomus meraldus* Carne, antennal club; 10, *Elephastomus meraldus* Carne, head, dorsal; 11, *Stenaspidius albosetosus* Howden, left elytron, dorsal; 12, *Bolbaineus planiceps* (Malceay), left elytron; 13, *Gilletinus corrugatus* (Lea), left elytron; 14, *Bolborhachium recticorne* (Guérin-Méneville), left elytron; 15, *Blackbolbus taurus* (Westwood), left elytron;16, *Australobolbus fenestratus* (Blackburn), antennal apex. 17, *Australobolbus fenestratus* (Blackburn), female, head and pronotum, lateral.

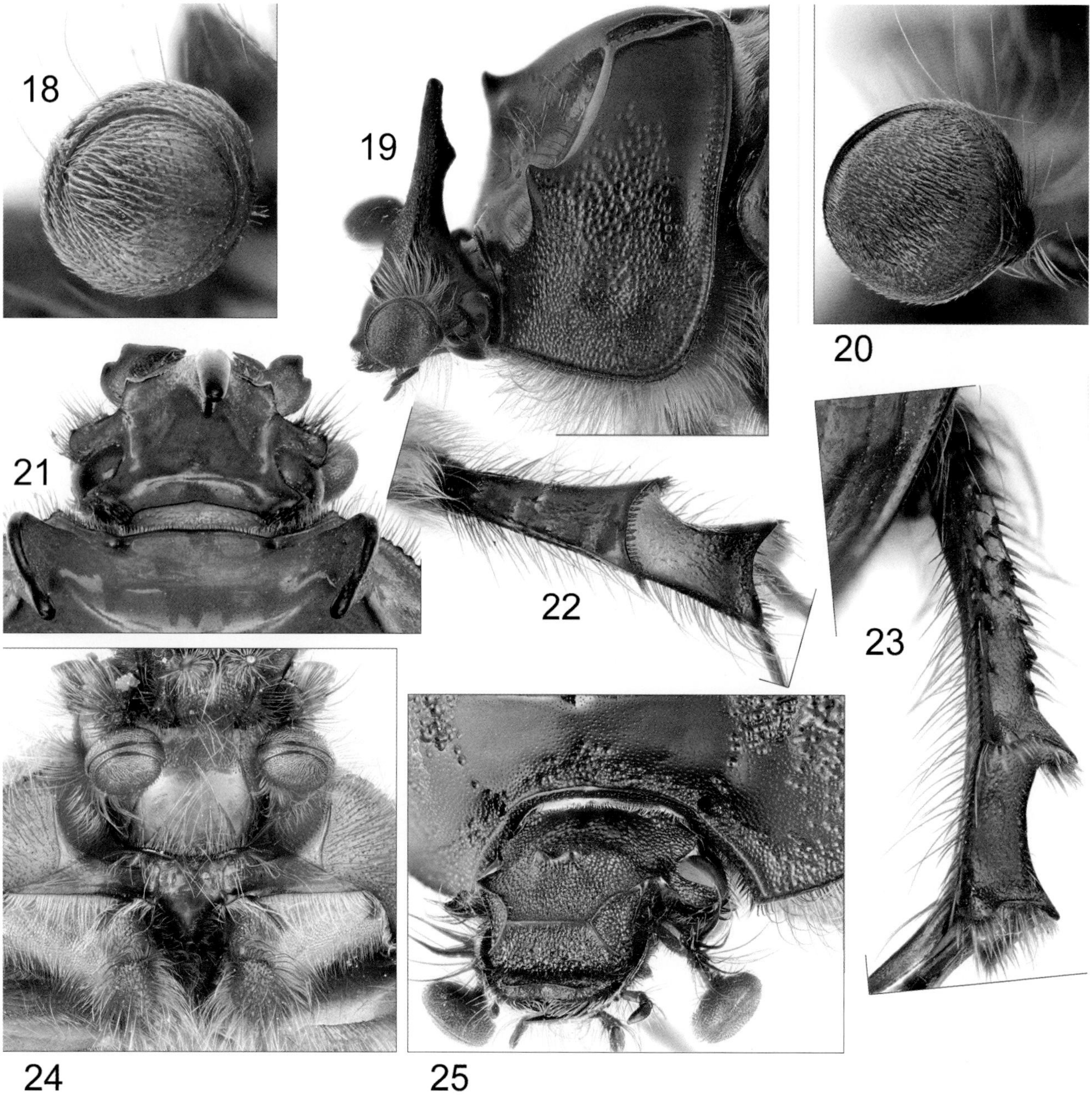

Figs 23.18–23.25. Geotrupidae adults. 18, *Blackbolbus bispinicollis* (Lea), antennal apex; 19, *Blackburnium cavicolle* (Macleay), male, head and prothorax, lateral; 20, *Blackburnium cavicolle* (Macleay), antennal club; 21, *Blackburnium angulicorne* (Macleay), head and pronotum; 22, *Blackburnium cavicolle* (Macleay), metatibia; 23, *Bolboleaus truncatus* (Blackburn), metatibia; 24, *Blackbolbus bispinicollis* (Lea), head and prosternum, ventral; 25, *Bolboleaus froggatti* (Blabkburn), head and prothorx, anterodorsal.

of two species of *Bolborhachium* appeared to be provisioned with humus-rich soil or finely divided humus obtained at or near the soil surface. However, Houston (2011) found no evidence of this provisioning; four laid eggs, a larva, a pupa and two teneral adults of *B. reichei* were found in relatively small bare cells with no trace of faecal material. Also, no sign of brood cell construction were noted, even in borrows where heavily gravid females of *Bolborhachium*, *Blackburnium* and *Blackbolbus* were present. This study provided strong circumstantial evidence that mass provisioning of brood cells does not occur in all bolboceratines and at least some of these develop in the absence of food. In a later paper (Houston 2016) it was suggested that ingestion of salts and/or humid and fulvic acidsmight enable the larvae to absorb water, and soil bacteria were also suggested as a possible food source. Adult geotrupids are capable of producing disturbance sounds (Alexander *et al.* 1963; Palestrini *et al.* 1990), but among the Australian species only *Geotrupes* is capable of stridulation in the larval stage.

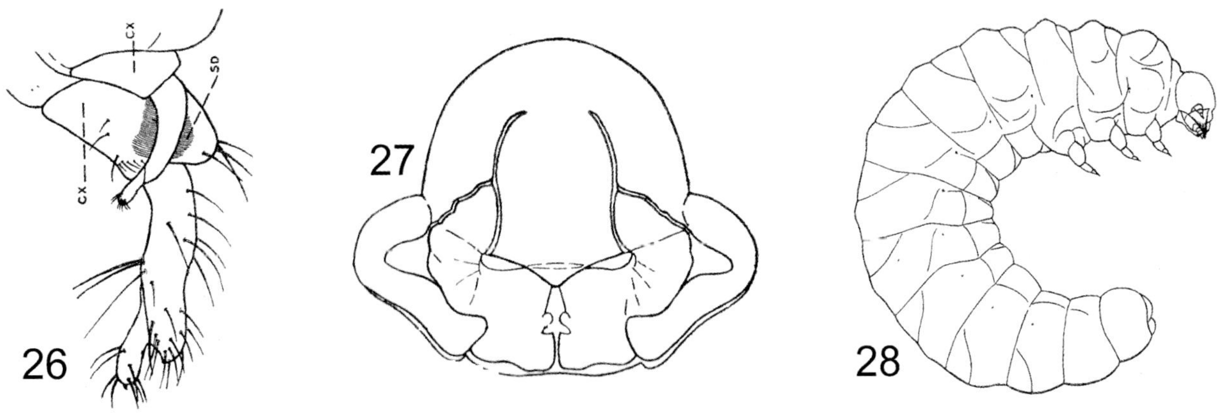

Figs 23.26–23.28. Geotrupidae larvae. 26, *Geotrupes blackburni* (Fabricius), larva, meso- and metathoracic legs; 27, *Geotrupes blackburni* (Fabricius), larva, abdominal apex; 28, *Eucanthus lazarus* (Fabricius), larva, lateral.

Key to the Australian genera

1. Frontoclypeal suture angulate (Fig. 23.2); antennal club about half as large as remaining antennomeres combined; mesocoxae contiguous (Fig. 23.3); galea expanded apically and densely setose; ventral surfaces metallic blue (Pl. 48F, 76C) .... ***Geotrupes* Latreille**

– Frontoclypeal suture absent; antennal club about as long as remaining antennomeres combined; mesocoxae not contiguous (Figs 23.4–23.5); galea falcate and not densely setose; ventral surfaces not metallic blue **2**

2(1). Mesocoxae subcontiguous, separated by less than 0.1 times the shortest diameter of a mesocoxal cavity (Fig. 23.4) **3**

– Mesocoxae separated by more than 0.33 times the shortest diameter of a mesocoxal cavity (Fig. 23.5) **4**

3(2). Elytron with 5 striae between suture and humeral umbone (Fig. 23.6); inner (proximal) segment of antennal club almost entirely glabrous (Fig. 23.7); clypeus not produced anteriorly; length less than 15 mm (Pl. 48E); WA and SA .... ***Eucanthus* Westwood**

– Elytron with 7 striae between suture and humeral umbone (Fig. 23.8); inner (proximal) segment of anntennal club only partly glabrous (Fig. 23.9); clypeus of males produced anteriorly (Fig. 23.10); length usually greater than 16.5 mm (Pl. 49E–G); eastern and southern Australia, including TAS .... ***Elephastomus* W.S. Macleay**

4(2). Scutellum distinctly longer than its basal width; elytral intervals strongly convex (Fig. 23.11) **5**

– Scutellum shorter than or as long as its basal width; elytral intervals flat to slightly convex (Fig. 23.12) **6**

5(4). Elytron with 7 striae between suture and humeral umbone; strial punctures large, separated by less than 2 puncture diameters (Fig. 23.13); base of elytron without carina (Pl. 49C); New Guinea and north-eastern Australia .... ***Gilletinus* Boucomont**

– Elytron with 5 striae between suture and humeral umbone; strial punctures small, usually separated by more than 2 puncture diameters (Fig. 23.11); base of elytron with carina; length 6–11 mm (Pl. 49A–B); south-eastern, south-western and northern coastal Australia .... ***Stenaspidius* Westwood**

6(4). Sutural stria anteriorly curving inwardly and terminating near apex of scutellar shield (Fig. 23.14), which is sharply deflected at base, with lateral edge internally arcuate or curving slightly inwardly at base; 9–22 mm (Pl. 48G–H, N–P); western, southern and eastern Australia .... ***Bolborhachium* Boucomont**

– Sutural stria anteriorly curving outwardly around scutellar shield and terminating at elytral base (Fig. 23.12); base of scutellar shield rounded or gradually sloping downwards, its lateral edge curving slightly outwardly at base **7**

7(6). Elytron with 5 striae between suture and humeral umbone (Fig. 23.12); 8–10 mm (Pl. 48A); northern north-eastern and north-western coastal regions and some dry inland localities south to SA. .... ***Bolbobaineus* Howden & Cooper**

– Elytron with 7 striae between suture and humeral umbone (Fig. 23.15) **8**

8(7). Antennal club segments ovoid in outline (Fig. 23.16); metatarsomeres 2–4 without subapical setae or with no more than 2 subapical setae on 3 and 4; pronotum with punctures just above lateral fovea not noticeably larger than those below it (Fig. 23.17); little sexual dimorphism; length 5–15 mm (Pl. 48M); New Guinea and most of Australia, but rare in the west .... ***Australobolbus* Howden & Cooper**

– Antennal club segments circular in outline (Fig. 23.18) (rarely ovoid with outer/distal segment greatly enlarged); metatarsomeres 2–4 each usually with 2 or more subapical setae; pronotum with punctures just above lateral fovea 2 to 4 times as large as those below it (Fig. 23.19); sexual dimorphism considerable; length 12–25 mm **9**

9(8). Diameter of outer/distal segment of antennal club ~0.80 times that of inner/proximal segment (Fig. 23.18); prosternum anteriorly flat or concave (rarely with transverse clefts) between paramedian ridges (Fig. 23.24); metatarsomeres 2–4 with subapical setae located laterally (rarely with a few ventral setae also present); 13–23 mm (Pl. 48K–L); common in WA but rare elsewhere .... ***Blackbolbus* Howden & Cooper**

– Diameter of outer/distal segment of antennal club only slightly less than that of inner/proximal segment (Fig. 23.20); prosternum anteriorly convex or with median longitudinal ridge between paramedian ridges (Fig. 23.5); metatarsomeres 2–4 with subapical setae located ventrally or ventrally and laterally **10**

10(9). Metatibia with series of short, transverse carinae above subapical carina (Fig. 23.22); pronotum anteriorly on each side behind eye with relatively small, sharply defined, deep fovea or pit, and edge surrounding fovea not heavily punctate (Fig. 23.21); metatibia with series of short, transverse carinae above subapical carina (Fig. 23.22); 15–25 mm; New Guinea and most of continental Australia (Pl. 48B–D) .... ***Blackburnium* Boucomont**

– Metatibia with series of tooth-like projections above subapical carina (Fig. 23.23); pronotum anteriorly on each side behind eye usually without sharply delimited pit, OR IF shallow fovea present, THEN the edge surrounding it heavily punctate (Fig. 23.25); 10–26 mm (Pl. 48I–J); northern, eastern and south-eastern Australia .... ***Bolboleaus* Howden & Cooper**

Key to the subfamilies of Geotrupidae based on larvae

1. Prothoracic and mesothoracic legs moderately long and 4-segmented, metathoracic legs reduced and 3-segmented; stridulatory organ formed by mesocoxa and metafemur (Fig. 23.26); abdominal segment X short and obliquely flattened; anal opening surrounded by complex, flap-like lobes (Fig. 23.27); spiracles cribriform .... **Geotrupinae**
– Legs more or less equally developed but relatively short; stridulatory organ absent (Fig. 23.28); abdominal segment X neither reduced nor flattened; anal lobes simple; spiracles biforous .... **Bolboceratinae**

Classification of Australian taxa

Subfamily Bolboceratinae Mulsant, 1842

Tribe Australobolbini Nikolajev 1990

Australobolbus Howden & Cooper, 1977 (Pl. 48M)

Type species. *Bolboceras obscurium* Blackburn, 1904.
Australian species. 45.
Distribution. *Australobolbus* species are widely distributed in Australia.
Biology. Specimens of *A. gayndahensis* (Macleay) were taken in numbers at light in NSW (Steinbauer & Weir 2007), and some adults have been taken in borrows, but without food items. Larvae are unknown.
References. Howden (1992); Howden & Cooper (1977); Cassis & Weir (1992); Howden *et al.* (2007).

Blackbolbus Howden & Cooper, 1977 (Pl. 48K–L)

Type species. *Bolboceras taurus* Westwood, 1848.
Australian species. 31.
Distribution. *Blackbolbus* species occur primarily in WA, SA, NT, western NSW and western QLD.
Biology. An immobile final instar larva of *Blackbolbus hoplocephalus* (Lea) found in a burrow by Houston (2011) had an unusual globose body with vestigial legs and mouthparts; the pupa of this species was also photographed.
References. Howden (1985, 1993); Howden & Cooper (1977); Cassis & Weir (1992); Howden *et al.* (2007).

Blackburnium Boucomont, 1911 (Pl. 48B–D)

Type species. *Bolboceras reichei* Guérin-Méneville, 1829.
Australian species. 29.
Distribution. *Blackburnium* species are widely distributed in Australia.
Biology. The larva of *Blackburnium reichei* (Guérin-Méneville) was photographed by Houston (2011), and a male of this species was reared from an egg by Houston (2016), who photographed the pupa.
References. Howden (1979); Howden & Cooper (1977); Cassis & Weir (1992); Howden *et al.* (2007).

Bolbobaineus Howden & Cooper, 1977 (Pl. 48A)

Type species. *Bolboceras planiceps* Macleay, 1873.
Australian species. One.
Distribution. *Bolbobaineus planiceps* is known from northern QLD and NT.
Biology. Specimens are usually collected at lights and larvae are unknown.
References. Howden (1992); Howden & Cooper (1977); Cassis & Weir (1992); Howden *et al.* (2007).

Bolboleaus Howden & Cooper, 1977 (Pl. 48I–J)

Type species. *Bolboceras truncatum* Blackburn, 1904.
Australian species. 14.
Distribution. *Bolboleaus* species are widely distributed in Australia.
Biology. Adults of *B. quadriarmigerus* (Howden) were found in burrows without food items. An egg of *Bolboleaus hiaticollis* Howden was reared to the third instar, and a late instar was reared through to the pupal stage by Houston (2016). The larva, pupa and details of the head capsule were photographed.
References. Howden (1985); Howden & Cooper (1977); Cassis & Weir (1992); Howden *et al.* (2007).

Bolborhachium Boucomont, 1911 (Pl. 48G–H, N–P)

Type species. *Bolboceras reticorne* Guérin-Méneville, 1829.
Australian species. 25.
Distribution. *Bolborhachium* species are widely distributed in Australia.
Biology. Eggs and larvae of *Bolborhachium annae* Howden and *B. recticorne* (Guérin-Méneville) were described by Howden (1985). The larva of *B. trituberculatum* (Bainbridge) was illustrated by Houston (2011). Houston (2016) obtained eggs, larvae and pupae of *B. recticorne* producing photographs and developmental details. Also,

a late instar of *B. inclinatum* Howden was reared in the laboratory to produce a minor male.

References. Howden (1985); Howden & Cooper (1977); Cassis & Weir (1992); Howden *et al.* (2007).

Elephastomus W.S. Macleay, 1819 (Pl. 49E–G)

Type species. *Scarabaeus proboscideus* Schreibers, 1802.

Australian species. Seven.

Distribution. *Elephastomus* species occur in TAS, VIC, ACT, NSW and QLD.

Biology. Adults of several species have been found in burrows feeding on bits of fungi, but immature stages are as yet unknown.

References. Carne (1965); Howden & Cooper (1977); Cassis & Weir (1992); Howden *et al.* (2007).

Tribe Eucanthini Nikolajev 2002

Eucanthus Westwood, 1848 (Pl. 48E)

Type species. *Scarabaeus meliboeus* Fabricius, 1792.

Australian species. One.

Distribution. *E. felschei* Boucomont occurs in SA and southern WA.

Biology. A few specimens have been collected in burrows in an area of coastal heath with rich layer of black humus; however there are no records of larvae. Three larvae of the North American *E. lazarus* (Fabricius) were collected from soil in a vineyard in Arkansas, USA, and two were reared to the adult stage (Ritcher 1966). There is no record of food items.

References. Browne (1991); Howden (1992); Cassis & Weir (1992); Howden & Cooper (1977); Howden *et al.* (2007).

Tribe Gilletinini Nikolajev 1990

Gilletinus Boucomont, 1932 (Pl. 49C)

Type species. *Bolboceras multicostatum* Lansberge, 1885.

Australian species. Five.

Distribution. The genus *Gilletinus* includes 12 described species distributed from eastern Indonesia and Papua New Guinea south into QLD and northern NSW.

Biology. Adults are usually collected at light or in flight intercept traps. Larvae are unknown.

References. Howden (1990, 1992); Howden & Cooper (1977); Cassis & Weir (1992); Howden *et al.* (2007); Telnov (2011).

Tribe Stenaspidiini Nikolajev 2003

Stenaspidius Westood, 1848 (Pl. 49A–B)

Type species. *Bolboceras* (*Stenaspidius*) *nigricornis* Westwood, 1848.

Australian species. 10.

Distribution. *Stenaspidius* species occur in VIC, NSW, QLD, SA and WA.

Biology. Many specimens have been collected in burrows or flight intercept traps. Adults of *S. matthewsi* Howden were collected in the same area with *Eucanthus felscheii.* Larvae are unknown.

References. Howden (1974, 1993); Howden & Cooper (1977); Cassis & Weir (1992); Howden *et al.* (2007).

Subfamily Geotrupinae Latreille, 1802

Geotrupes Latreille, 1796 (Pl. 48F, 76C)

Type species. *Scarabaeus stercorarius* Linnaeus, 1758.

Australian species. One (introduced).

Distribution. The genus *Geotrupes* is endemic to the Nearctic and Palaearctic regions. *Geotrupes spiniger* Marsham was introduced from France and occurs in NSW, ACT, VIC and TAS.

Biology. *Geotrupes spiniger* is the only Australian geotrupid to provision its larval burrows with dung, usually taken from cattle dung pads in pastures. The nesting behaviour of the European species *G. spiniger* Marsham has been studied in detail by Klemperer (1979), and the same author (1980) reported kleptoparasitism by *Aphodius rufipes* (Linnaeus) in nests of the same species.

References. Howden (1955); Ritcher (1966); Klemperer (1979, 1980); Cassis & Weir (1992).

Acknowledgments

The authors are grateful to Oregon State University for permission to use Figs 23.6–23.8.

References

Alexander RD, Moore TE, Woodruff RE (1963) The evolutionary differentiation of stridulatory signals in beetles (Insecta: Coleoptera). *Animal Behaviour* **11**(1), 111–115. doi:10.1016/0003-3472(63)90018-6

Browne J (1991) Wing structure of the genus *Eucanthus* Westwood: confirmation of the primitive nature of the genus (Scarabaeidae: Geotrupidae: Bolboceratidae). *Journal of the Entomological Society of Southern Africa* **54**, 221–230.

Browne DJ, Scholtz CH (1996) The morphology of the hind wing articulation and wing base of the Scarabaeoidea (Coleoptera) with notes on phylogenetic trends. *Bonner Zoologische Monographien* **40**, 1–200.

Browne J, Scholtz CH (1999) A phylogeny of the families of Scarabaeoidea (Coleoptera). *Systematic Entomology* **24**, 51–84. doi:10.1046/j.1365-3113.1999.00067.x

Carne PB (1965) A revision of the genus *Elephastomus* Macleay (Coleoptera: Geotrupidae). *Journal of the Entomological Society of Queensland* **4**, 3–13.

Cassis G, Weir TA (1992) Geotrupidae. In *Zoological Catalogue of Australia. Volume 9. Coleoptera: Scarabaeoidea.* (Ed. WWK Houston) pp. 41–64. Australian Government Printing Office, Canberra.

Houston TF (2007) Geotrupid beetles – what role do they play in the dispersal of mycorrhizal fungi? *Western Australian Insect Study Society Newsletter*, February 2007, 6–8.

Houston TF (2011) Egg gigantism in some Australian earth-borer beetles (Coleoptera: Geotrupidae: Bolboceratinae) and its apparent association with reduction or elimination of larval feeding. *Australian Journal of Entomology* **50**, 164–173. doi:10.1111/j.1440-6055.2010.00794.x

Houston TF (2016) Brood cells, life-cycle stages and development of some earh-borer beetles in the genera *Bolborhachium, Blackburnium* and *Bolboleaus* (Coleoptera: Geotrupidae), with notes on captive rearing and a discussion of larval diet. *Australian Entomologist* **55**, 49–62. doi:10.1111/aen.12151

Houston TF, Bougher NL (2010) Records of hypogeous mycorrhizal fungi in the diet of some Western Australian bolboceratine beetles (Coleoptera: Geotrupidae, Bolboceratinae). *Australian Journal of Entomology* **49**(1), 49–55. doi:10.1111/j.1440-6055.2009.00720.x

Howden HF (1955) Biology and taxonomy of North American beetles of the subfamily Geotrupinae with revision of the genera *Bolbocerosoma, Eucanthus, Geotrupes* and *Peltotrupes* (Scarabaeidae). *Proceedings of the United States National Museum* **104**, 151–319. doi:10.5479/si.00963801.104-3342.151

Howden HF (1974) A revision of the Australian genus *Stenaspidius* Westwood (Coleoptera, Scarabaeidae, Geotrupinae). *Records of the South Australian Museum* **17**(2), 11–21.

Howden HF (1979) A revision of the Australian genus *Blackburnium* Boucomont (Coleoptera: Scarabaeidae: Geotrupinae). *Australian Journal of Zoology* **72**(Supplement), 1–88.

Howden HF (1985) A revision of the Australian beetle genera *Bolboleaus* Howden & Cooper, *Backbolbus* Howden & Cooper, and *Bolborhachium* Boucomont (Scarabaeidae: Geotrupinae). *Australian Journal of Zoology* **111**(Supplement), 1–179.

Howden HF (1990) The Geotrupinae of New Guinea (Coleoptera: Scarabaeidae). *Invertebrate Taxonomy* **3**(1989), 261–289.

Howden HF (1992) A revision of the Australian beetle genera *Eucanthus* Westwood, *Bolbaineus* Howden & Cooper, *Australobolbus* Howden & Cooper and *Gilletinus* Boucomont (Scarabaeidae: Geotrupinae). *Invertebrate Taxonomy* **6**, 605–717. doi:10.1071/IT9920605

Howden HF (1993) New Bolboceratini from Western Australia (Coleoptera: Scarabaeidae: Geotrupinae). *Journal of the Australian Entomological Society* **32**, 379–386. doi:10.1111/j.1440-6055.1993.tb00608.x

Howden HF, Cooper JB (1977) The generic classification of the Bolboceratini of the Australian region, with descriptions of four new genera (Scarabaeidae: Geotrupinae). *Australian Journal of Zoology* **50**(Supplement), 1–50.

Howden H, Howden A, Holloway G (2007) Digging down under: Australian Bolboceratini, their habits and a list of species (Coleoptera: Scarabaeoidea: Geotrupidae). *Zootaxa* **1499**, 47–59.

Justo A, Morgenstern I, Hallen-Adams HE, Hibbett DS (2010) Convergent evolution of sequestrate forms in *Amanita* under Mediterranean climate conditions. *Mycologia* **102**(3), 675–688. doi:10.3852/09-191

Klemperer HG (1979) An analysis of the nesting behavior of *Geotrupes spiniger* MArsham (Coleoptera, Scarabaeidae). *Ecological Entomology* **4**(2), 133–150. doi:10.1111/j.1365-2311.1979.tb00569.x

Klemperer HG (1980) Kleptoparasitic behavior of *Aphodius rufipes* (L.) larvae in nests of *Geotrupes spiniger* Marsh. (Coleoptera, Scarabaeidae). *Ecological Entomology* **5**(2), 143–151. doi:10.1111/j.1365-2311.1980.tb01135.x

Lawrence JF, Ślipiński A, Seago AE, Thayer MK, Newton AF, Marvaldi AE (2011) Phylogeny of the Coleoptera based on morphological characters of adults and larvae. *Annales Zoologici* **61**(1), 1–217.

McKenna DD, Wild AL, Kanda K, Bellamy CL, Beutel RG, Caterino MS, Farnum CW, Hawks DC, Ivie MS, Jameson ML, Leschen RAB, Marvaldi AE, McHugh JV, Newton AF, Robertson JA, Thayer MK, Whiting MF, Lawrence JF, Ślipiński A, Maddison DR, Farrell BD (2015) The beetle tree of life reveals that Coleoptera survived end-Permian mass extinction to diversify during the Cretaceous terrestrial revolution. *Systematic Entomology* **40**(4), 835–880.

Nikolajev GV (1970) The taxonomic rank of the groups within the subfamily Geotrupinae (Coleoptera, Scarabaeidae). In *Proceedings of the II Scientific Conference of Young Researchers and Postgraduates, dedicated to the 100th Anniversary of the Birth of V. I. Lenin and to the 50th Anniversary of Kazakhstan.* pp. 31–34. Almaty, Kazakhstan [in Russian].

Nikolajev GV (1990) The usage of particularities of wing venation for clarification of the system of Lamellicornia beetles. In *Advances in Entomology in USSR. Coleoptera. Proceedings of the X Meeting of the All-Soviet Entomology Society, Lenigrad, September 11–15, 1989.* (Eds GS Medvedev and BA Korotaev) pp. 98–99. USSR Academy of Sciences Zoology Institute, Leningrad [in Russian].

Nikolajev GV (2002) A description of the larva of the genus *Eubolbitus* Reitter (Coleoptera, Scarabaeidae, Bolboceratinae). *Evraziatskii Entomologicheskii Zhurnal* **1**, 207–209 [in Russian].

Nikolajev GV (2003) The taxonomic composition of the subfamily Bolboceratinae (Coleoptera, Scarabaeidae) from Palaearctic faunistic region. *Tethys Entomological Research* **8**, 187–206 [in Russian].

Oehl F, Sieverding E, Palenzuela J, Ineichen K, Alves da Silva G (2011) Advances in Glomeromycota taxonomy and classification. *IMA Fungus* **2**(2), 191–199. doi:10.5598/imafungus.2011.02.02.10

Palestrini C, Zunino M, Zucchelli M (1990) Sound production in the larvae of *Geotrupes spiniger* (Marsham) (Coleoptera: Geotrupidae). *Bioacoustics* **2**(3), 209–216. doi:10.1080/09524622.1990.9753133

Ritcher PO (1966) *White Grubs and their Allies. A Study of North American Scarabaeoid Larvae.* Oregon State University Press (Studies in Entomology No. 4), Corvallis, Oregon.

Scholtz CH, Grebennikov VV (2016) 15. Scarabaeoidea Latreille, 1802. In *Handbuch der Zoologie/Handbook of Zoology. BandVolume IV Arthropoda: Insecta. Teilband/Part 38. Coleoptera, Beetles. Volume 1: Morphology and Systematics (Archostemata, Adephaga, Myxophaga, Polyphaga partim).* 2nd Edition. (Eds RG Beutel and RAB Leschen) pp. 443–525. Walter de Gruyter, Berlin.

Smith ABT (2006) A review of the family-group names for the superfamily Scarabaeoidea (Coleoptera) with corrections to nomenclature and a current classification. *Coleopterists Society Monographs* **5**, 144–204.

Smith ABT, Hawks DC, Heraty JM (2006) An overview of the classification and evolution of the major scarab beetle clades (Coleoptera: Scarabaeoidea) based on preliminary molecular analyses. *Coleopterists Society Monographs* **5**, 35–40.

Steinbauer MJ, Weir TA (2007) Summer activity patterns of nocturnal Scarabaeoidea (Coleoptera) of the southern tablelands of New South Wales. *Australian Journal of Entomology* **46**, 7–16. doi:10.1111/j.1440-6055.2007.00579.x

Telnov D (2011) New *Gilletinus* Boucomont, 1932 (Coleoptera: Geotrupidae: Bolboceratinae) from Misool, with the key to known species. In *Biodiversity, Biogeography and Nature Conservation in Wallacea and New Guinea. Vol. 1.* (Ed. D Telnov) pp. 287–290, pls. 38–39. The Entomological Society of Latvia, Riga.

Verdú JR, Galante E, Lumaret JP, Cabrero-Sañudo FJ (2004) Phylogenetic analysis of Geotrupidae (Coleoptera, Scarabaeoidea) based on larvae. *Systematic Entomology* **29**, 509–523. doi:10.1111/j.0307-6970.2004.00256.x

Woodruff RE (1973) *Arthropods of Florida and Neighboring Land Areas. Volume 8. The Scarab Beetles of Florida (Coleoptera: Scarabaeidae). Part I. The Laparosticti (Subfamilies: Scarabaeinae, Aphodiinae, Hybosorinae, Ochodaeinae, Geotrupinae, Acanthocerinae).* Florida Dept. of Agriculture and Consumer Services, Division of Plant Industry, Gainesville.

Zunino M (1984a) Analisi sistematica e zoogeografica della sottofamiglia Taurocerastinae Germain (Coleoptera, Scarabaeoidea: Geotrupidae). *Bollettino del Museo Regionale di Scienze Naturali* **2**(2), 445–464.

Zunino M (1984b) Sistematica genetica dei Geotrupinae (Coleoptera, Scarabaeoidea: Geotrupidae), filogenesi della sottogamiglia e considerazioni biogeografiche. *Bollettino del Museo Regionale di Scienze Naturali* **2**, 9–162.

24. PASSALIDAE LEACH, 1815

Eric G. Matthews and Owen D. Seeman

Fig. 24.1. *Mastachilus quaestionis* (Kuwert).

Common name. The bess beetles.

Introduction. Passalids are pantropical, with some dispersal into adjacent temperate regions both north and south. In Australia their distribution suggests invasion from the tropical north mainly along the eastern forested region with just one species of *Pharochilus* Kaup penetrating as far as Tasmania.

There is no question about the monophyly of the Passalidae (e.g. Grebennikov & Scholtz 2004; Boucher 2006). Most studies regard Passalidae as a basal branch in a group of seven families including them plus Diphyllostomatidae, Lucanidae, Glaphyridae, Trogidae, Pleocomidae and Bolboceratidae, which radiated during the Mesozoic (Browne & Scholtz 1996; Scholtz & Grebennikov 2005). On the other hand, Passalidae is one of the only two families of Scarabaeoidea which have not been found in the rich Mesozoic fossil record of this superfamily (Nikolajev 2007; Bai *et al.* 2012a) (there are a few Cenozoic fossils). This seems to correlate with the molecular phylogeny of McKenna *et al.* (2015) which places Passalidae among the most derived scarabs. Bai *et al.* (2012b) report a fossil from the Lower Cretaceous of China which they characterise as a 'missing link' between Passalidae, Lucanidae and Diphyllostomatidae. Boucher (2006) states that a variety of biogeographic and ecological data strongly supports Trogidae and Chironidae as sister groups, while Boucher *et al.* (2016) describe some very well preserved specimens of an apparent sister family of Passalidae, the Passalopalpidae, from mid-Cretaceous Burmese amber. Based on molecular data, Smith *et al.* (2006) and McKenna *et al.* (2015) group Passalidae with Bolboceratidae and Pleocomidae, and Ahrens *et al.* (2014) with Bolboceratidae and Trogidae.

The taxonomy of the Australian members of the family has received little recent attention and requires the examination of all types in order to establish synonymies and define new species. Percheron (1835) and Kaup (1868, 1871) were the first to monograph the world fauna. Gravely (1918) provided the first modern treatment. Kuwert (1891, 1898) described 18 species for Australia, and Dibb (1938) reviewed the Australian fauna. The key to genera adopted here is modified from that of Dibb (1938), while the suprageneric classification follows that of Reyes-Castillo (1970) as adopted by Cassis & Weir (1992), with no tribal divisions in Australia and just two subfamilies: Aulacocyclinae and Passalinae. Reyes-Castillo (1970) details the history of the classification of passalid groups, which at one time numbered up to 38 subfamilies. According to Boucher (2006) there are two tribes in Aulacocyclinae and five in Passalinae, of which only one in each subfamily, the Aulacocyclini and the Passalini respectively, occurs in Australia. Monophyly of the subfamily Aulacocyclinae was supported by the phylogenetic analysis of Hosoya *et al.* (2008). Fonseca *et al.* (2011) proposed a classification of the family based on the structure of the hind gut which again elevates some groups to subfamily level. They arrived at the following cladogram: (Aulacocyclinae (Solenocyclinae (Proculinae (Passalinae, Macrolininae)))).

Biology. Passalids feed on decaying wood and its associated microorganisms in both the larval and adult stages. They are known for their highly developed social behaviour which has been investigated extensively in the Western Hemisphere and Japan, but there is only one brief observation published on an Australian species (Fearn 1988). What is known about the behaviour of the New World taxa was surveyed by Reyes-Castillo & Halffter (1984) based on references cited therein, which will not be repeated here. Their information can be summarised as follows, with some more recent observations added.

The beetles form colonies which live inside chambers chewed out of rotten wood. A founding pair initially bores a short vertical tunnel which is then expanded horizontally in the wood to form a large chamber or branched gallery. The more cylindrical passalids, such as *Mastachilus* Kaup, bore their chamber in the duramen or heartwood, while the more flattened, like *Analaches* Kuwert, tend to make a subcortical gallery. Copulation by the founding pair is usually inside the gallery in an end-to-end position because of the restricted head space (providing an explanation for the rotated aedeagus mentioned below), but there is one observation of two pairs copulating during nuptial flights (MacGown & MacGown 1996). In flight adults emit a continuous buzzing sound produced by the wings alone. Copulation takes place after an elaborate ceremonial courtship involving frequent stridulation. Eggs may be laid anywhere in the gallery over a period of two or more months, but are then gathered by the pair in their mandibles and deposited in a single 'nest' made of frass. Hatching takes place over a period of time with the result that the colony contains first, second and third instar larvae, pupae, and teneral adults, as well as the founding pair. Only the latter have the mandibular strength necessary to scrape away at the walls of the chamber, consequently the larvae, and perhaps the teneral adults, are unable to eat unless they are offered a triturated pabulum by the fully sclerotised parents. Larvae also need to ingest the faeces

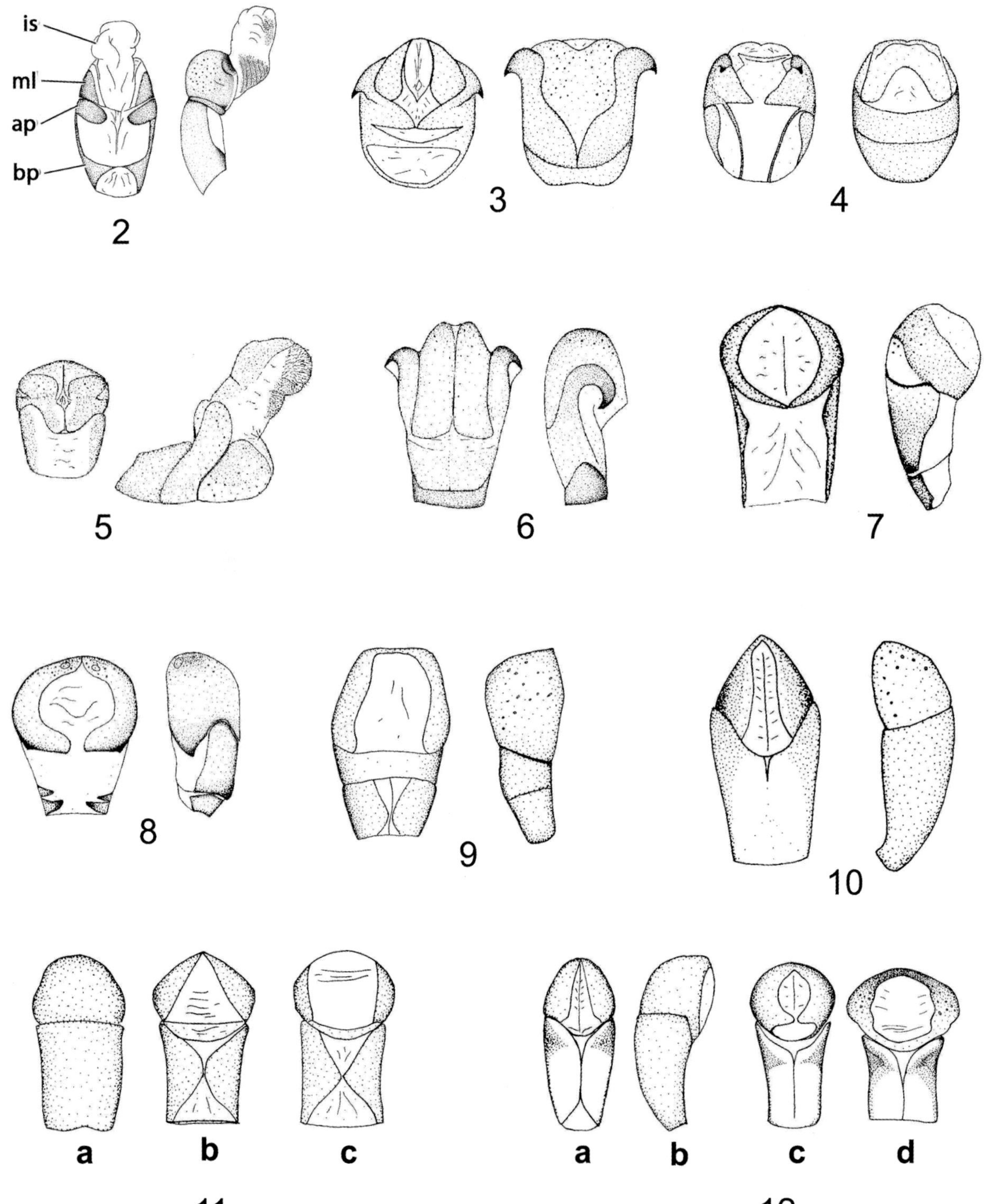

Figs 24.2–24.12. Aedeagi. 2, *Mastachilus polyphyllus* (Macleay), dorsal (L) and lateral (R): ap, apical piece; bp, basal piece; is, internal sac (everted); ml, median lobe; 3, *Mastachilus australasicus* (Percheron), dorsal (L) and lateral (R); 4, *Mastachilus quaestionis* (Kuwert), dorsal (L) and ventral (R); 5, *Pharochilus dilatatus* (Dalman), dorsal (L) and lateral with internal sac everted (R); 6, *Pharochilus rugiceps* (Hope & Westwood), ventral (L) and lateral (R); 7, *Austropassalus hultgreni* Mjöberg, dorsal (L) and lateral (R); 8, *Labienus inaequalis* Gravely, dorsal (L) and lateral (R); 9, *Analaches australiensis* (Stoliczka), dorsal (L) and lateral (R); 10, *Aulacocyclus aliicornis* Kuwert, dorsal (L) and lateral (R); 11, *Aulacocyclus fracticornis* Kuwert: a, ventral view of a specimen from Mt Baldy (QLD); b, dorsal view of same; c, dorsal view of a specimen from Cape Tribulation (QLD); 12, *Aulacocyclus kaupii* Macleay: a, dorsal view of typical form (Jimna State Forest, QLD); b, lateral view of same; c, dorsal view of specimen from Bunya Mountains; d, dorsal view of specimen from Blackdown Tableland.

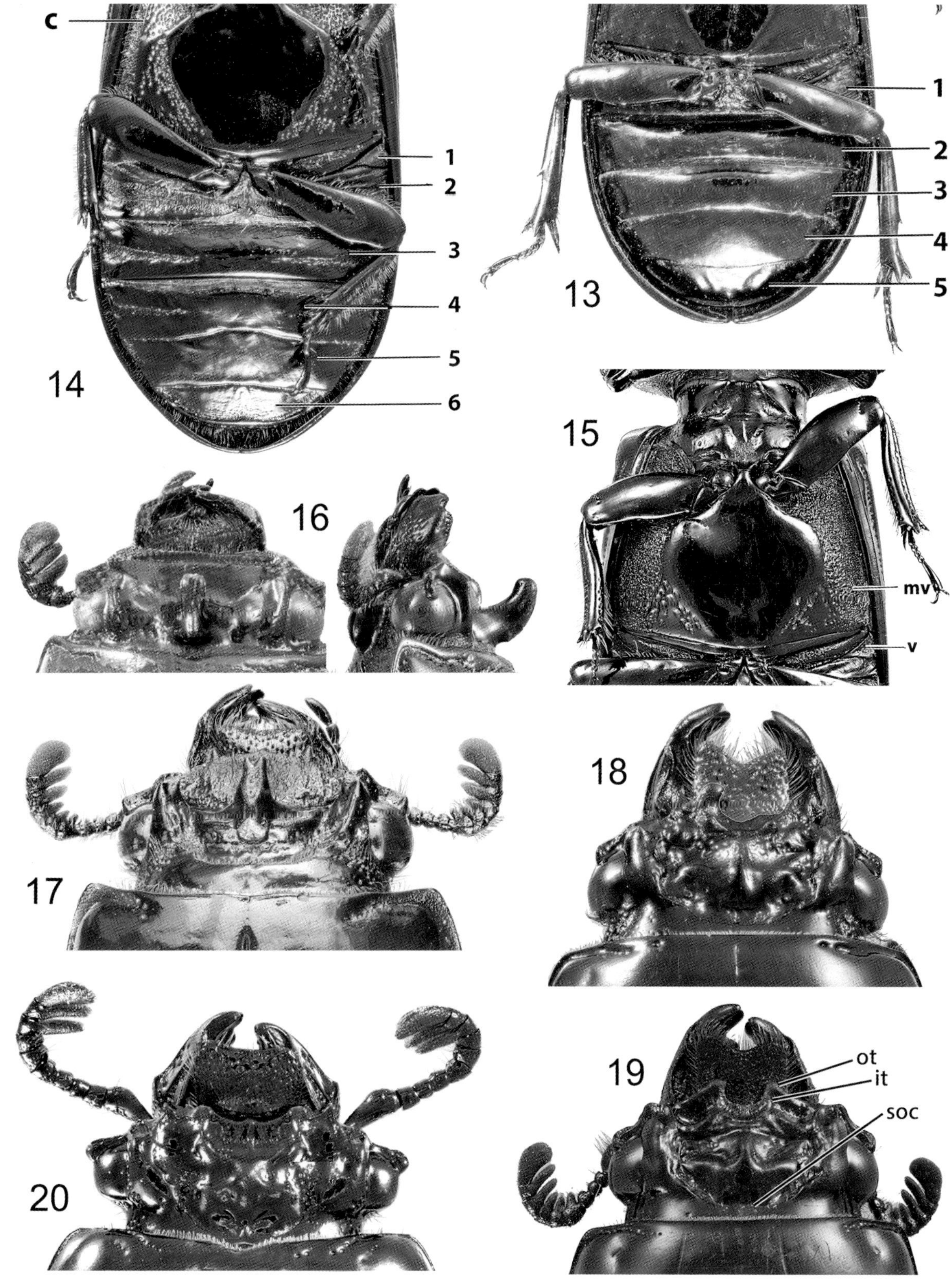

Figs 24.13–24.20. 13, *Aulacocyclus* sp., venter, ventrites numbered; 14, *Gonatas* sp., venter, ventrites numbered, c – carina on metaventrite; 15, *Protomocoelus australis* (Boisduval), part of venter: mv – metaventrite without carina; v – visible part of first abdominal ventrite; 16, *Aulacocyclus* sp., head, dorsal and lateral; 17, *Austropassalus hultgreni* Mjöberg, head, dorsal; 18, *Labienus inaequalis* Gravely, head, dorsal; 19, *Protomocoelus australis* (Boisduval), head, dorsal: it – inner tubercle; ot – outer tubercle; soc – supra-occipital carina; 20, *Leptaulax dentatus* (Fabricius), head, dorsal.

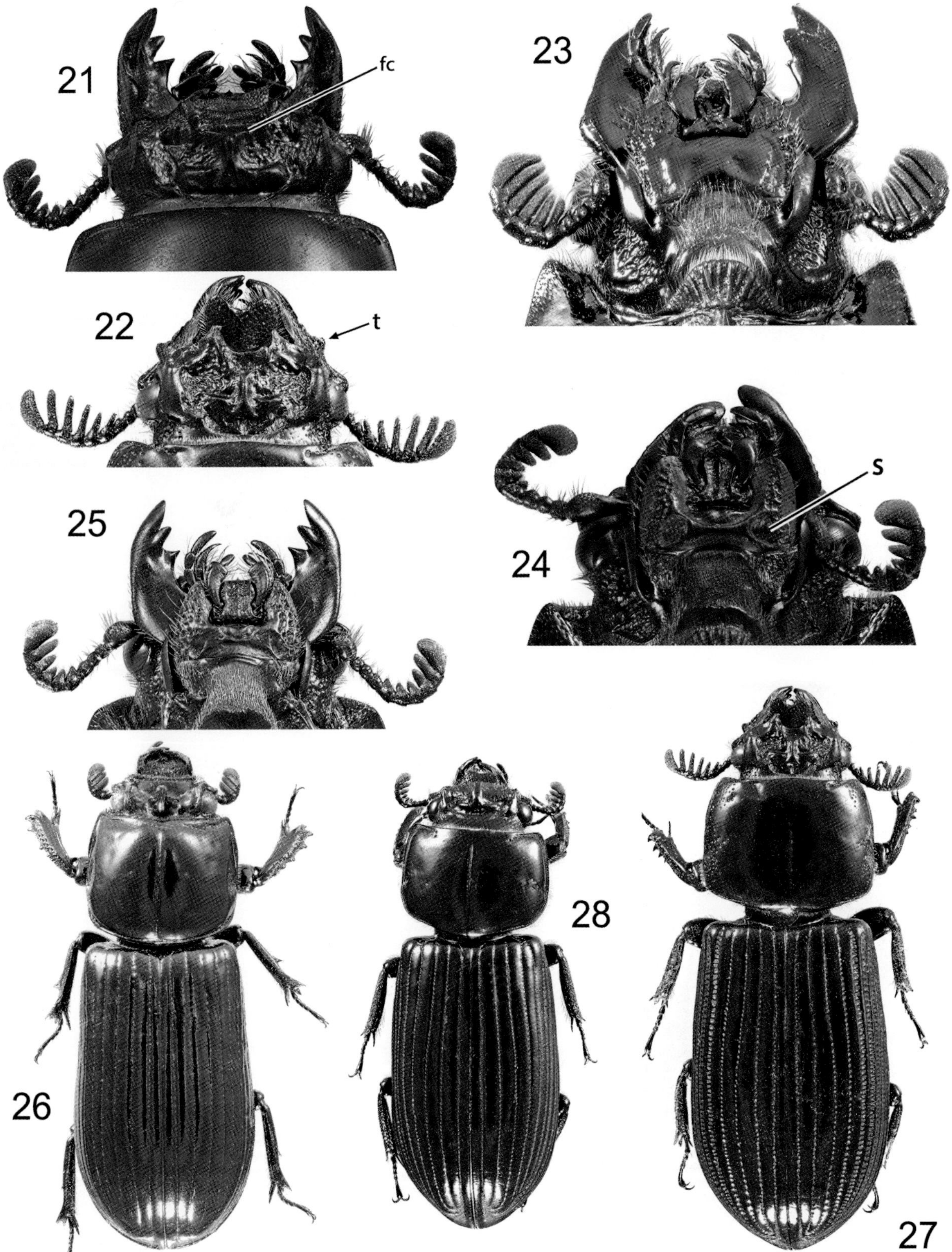

Figs 24.21–24.28. 21, *Mastachilus* sp., head, dorsal: fc – frontal carina; 22, *Analaches australiensis* (Stoliczka), head, dorsal: t – outer tooth of mandible; 23, *Gonatas* sp., head, ventral; 24, *Pharochilus* sp., head, ventral: s – scar on mentum; 25, *Mastachilus* sp., head, ventral; 26, *Aulacocyclus* sp., habitus; 27, *Analaches australiensis* (Stoliczka), habitus; 28, *Austropassalus hultgreni* Mjöberg, habitus.

Figs 24.29–24.34. Passalidae habitus. 29, *Gonatas* sp.; 30, *Labienus inaequalis* Gravely; 31, *Leptaulax dentatus* (Fabricius); 32, *Mastachilus australasicus* (Percheron); 33, *Pharochilus dilatatus* (Dalman); 34, *Protomocoelus australis* (Boisduval).

of adults in order to acquire the microorganisms necessary to digest wood (Valenzuela-González 1992). The pulped wood and fungal diet is supplemented in both adults and larvae by proteins apparently derived from cannibalised sibling and filial individuals. Pupation takes place in a cocoon made of frass, constructed with the help of both the founding pair and the filial adults. There is some evidence that the pupal cocoon is different in Aulacocyclinae and Passalinae (G.B. Monteith, pers. comm. 2016): the former form a discreet capsular cocoon while the latter merely pile frass around the pupa.

Reyes-Castillo & Halffter (1984) cite cases where numerous related adults may coexist for a time, forming hierarchies of dominant and submissive individuals (the latter sometimes flightless), equivalent to castes and suggesting a division of labour. Together with the high degree of parental investment, these observations point to a level of eusociality among passalids which is perhaps unique in the Coleoptera (Schuster & Schuster 1985). Passalids have highly developed mushroom bodies in their brains, a characteristic otherwise seen only in social insects (Panov 2010).

Activities in the colony are constantly accompanied by stridulation in both adults and larvae. Adults stridulate using an abdomino-alary apparatus consisting of specialised areas of abdominal tergite 6 (the pars stridens) rubbing against a patch of spinules on both metathoracic wings (the plectrum). In larvae, claws of the much reduced metathoracic leg scrape along a striated area of the mesocoxa to produce a sound. Oscillograms of adult signals showed that up to 17 different types of sound could be produced within a single species, depending on the behavioural context. Resident adults of either sex will confront conspecific intruders by engaging in head-to-head combat accompanied by continuous stridulation (further investigated by Wicknick & Miskelly 2009). Stridulation also has a defensive function: experiments with crows as predators showed that the birds hesitated to attack stridulating beetles as opposed to silenced ones. Furthermore, stridulation signals are species-specific; this specificity appears to serve the purpose of avoiding interspecies interactions when several passalid species occupy the same log (see also Palestrini *et al.* 2003). There is evidence of a supplementary chemical signalling.

More recent observations on an aulacocycline species occurring at high altitudes in Japan, *Cylindrocaulus patalis* (Lewis), show some departures from the general pattern outlined above. In the first place, only a single offspring is raised to maturity in a given colony during the breeding season, although several eggs are laid. Invariably, fatal combat occurs among the hatched larvae until only one, usually the senior one, survives (Ento *et al.* 2003, 2010). In the second place, infertile trophic eggs are offered directly to third-instar larvae by the parent female as a supplement to the normal protein-poor diet (Ento *et al.* 2008).

Numerous symbiotic organisms associate with passalid beetles. Mites are the most diverse of these with 24 families, of which six are specific to passalids. A few of these mite families are parasites or feed from the beetle, but typically the relationship is phoretic (Hunter 1993; Seeman 2001). The immature mites live within the galleries and most probably prey on nematodes and other mites, then move to beetles upon adulthood (Seeman 2000). The ectoparasitic fungi of the Laboulbeniomycetes infest beetles and their entourage of mites (Seeman & Nahrung 2000). The passalid gut is host to a diverse microfauna (e.g. trichomycete fungi, nematodes), the most important of which are the cellulolytic yeasts which inhabit numerous caecae in the adult gut (absent in larvae) where fermentation takes place (Houseknecht *et al.* 2011). The walls of the brood chamber are lined with a pasty mixture of triturated wood and faeces of both larvae and adults. The microbial activity which takes place in this lining is equivalent to that which occurs in the gut pouches of other saproxylophagous scarab larvae. Hence the passalid chamber has been termed an 'external rumen'. Ceratocanthine hybosorids are regularly found in passalid galleries, particularly *Germarostes* Paulian in the Americas (A. Ballerio, pers. comm. 2013) and *Pterorthochaetes* Gestro in Malaysia (Kon *et al.* 2010). It is not clear whether the presence of these ceratocanthines represents a special association with the passalids or merely a fortuitous co-occurrence in rotten wood.

The only published observations on an Australian species are those of Fearn (1988) on *Pharochilus rugiceps* (Hope and Westwood) (as *P. politus* Burmeister) in Tasmania. Most colonies had 2–5 adults and ~10 larvae, but one contained 21 adults and 67 larvae in a chamber gnawed out along the interface between a eucalypt stump and the surrounding soil.

Characteristics. Adults. Elongate, generally large (~20–55 mm in length), subcylindrical or more or less flattened and shiny black (young adults brown), with a narrow waist or pedicel between prothorax and hind body. Head prominent and dorsally bearing various tubercles and ridges, often also a small median horn. Mandibles large and projecting forward. Antennae relatively short, 10-segmented, curved forwards and not geniculate, with a terminal club of three to six segments not capable of close apposition. Mentum very deeply excised anteriorly. Scutellum hidden. Mesocoxal cavities closed by thoracic ventrites. Wings with one or two disconnected posterior branches of vein MP between MP_{1+2} and $CuA+AA_3$, and a stridulatory area in region of r4 crossvein where the wing folds. First abdominal ventrite either hidden in Aulacocyclinae (Fig. 24.13 where the number 1 corresponds to the second ventrite of Passalinae) or represented by two very small sections visible at sides of metacoxae in Passalinae (Figs 24.14 (1), 24.15 (v). Abdominal tergite 6 with median stridulatory area in form of two tubercles. Aedeagus in form of a characteristic inverted semi-globose capsule (Figs 24.2–24.12). No sexual dimorphism.

The aedeagus of Passalidae is unusually abbreviated compared to that of other scarabs and consists of a short tegmen which may or may not be divided into a basal piece and an apical piece (the latter also called lateral lobes or parameres). Distal to these is a globose sclerotised median lobe which is usually wider and sometimes longer than the tegmen. Inside the aedeagus is a membranous internal sac or penis which is eversible. The aedeagus of Passalidae is permanently rotated 180° so that the morphological ventral surface is dorsal in position (Bührnheim 1978). A two-part aedeagus, consisting of the median lobe and a one-piece tegmen, is characteristic of many Aulacocyclinae including *Aulacocyclus*, but not all of them (Hosoya *et al.* 2008), whereas a three-part aedeagus with a divided tegmen is seen in all other Passalidae examined (Cano 1994).

Larvae. Only slightly curved, unlike those of most other scarabs, appearing at first glance to have only two pairs of legs, the third pair being greatly reduced and one-segmented, ending in apical teeth which scrape along a ridged patch on the mesocoxae to produce a sound. Larval antennae are short and two-segmented.

Key to the subfamilies and genera of Australian Passalidae

1. Prosternal process sunk between protruding coxae; 5 visible abdominal ventrites (Fig. 24.13); frontal area of head without inner or outer tubercles or supra-occipital carinae, usually with central horn with slightly bilobed apex (Fig. 24.16); antennal club with only 3 lamellae; pronotum with complete marginal furrows .... **AULACOCYCLINAE: *Aulacocyclus* Kaup**

– Prosternal process on similar plane with coxae; 6 visible abdominal ventrites, first one consisting of two small widely separated sections outside metacoxae (Figs 24.14 **1**, 24.15 **v**); frontal area with inner and outer tubercles and supra-occipital carina (Fig. 24.19 **soc**); antennal club with more than 3 lamellae except in *Leptaulax*; pronotum with lateral marginal furrows only ... **PASSALINAE** 2

2(1). Antennal club with 3 lamellae; 4 tubercles on anterior margin of head (Fig. 24.20) .... ***Leptaulax* Kaup**

– Antennal club with more than 3 lamellae; 2–3 tubercles on anterior margin of head 3

3(2). Anterior margin of head with 3 tubercles; central tubercle of head horn-like, proclinate with bifid apex (Fig. 24.17). Lateral 5 elytral intervals matt .... ***Austropassalus* Mjöberg**

– Anterior margin of head with 2 tubercles or teeth; central tubercle variable. All elytral intervals glossy except in *Analaches* 4

4(3). Lateral and intermediate areas of metaventrite without a carina dividing the two regions (Fig. 24.15 **mv**) 5

– Lateral and intermediate areas of metaventrite separated by a distinct carina (Fig. 24.14 **c**) 6

5(4). Median pronotal sulcus entire; left outer tubercle of head much larger than right, directed inwards (Fig. 24.18) .... ***Labienus* Kaup**

– Median pronotal sulcus obsolete; right and left outer tubercles of head of equal length, but right one slightly more attenuated (Fig. 24.19 **ot**) .... ***Protomocoelus* Zang**

6(4). Mentum with 2 broad shallow depressions without distinct lateral margins (Fig. 24.23). Restricted to far northern QLD .... ***Gonatas* Kaup**

– Mentum with depressions or scars usually defined in part by distinct margins (Figs 24.24 **s**, 24.25). Of more widespread occurrence 7

7(6). Outer edge of mandibles with prominent angulate tooth (Fig. 24.22 **t**); lateral 5 elytral intervals matt .... ***Analaches* Kuwert**

– Outer edge of mandibles rounded or feebly angulate; all elytral intervals glossy unless worn 8

8(7). Mentum laterally with a moderate to wide depressed matt border (Fig. 24.24), or entirely matt .... ***Pharochilus* Kaup**

– Mentum laterally with a very narrow depressed matt border, or matt border absent (Fig. 24.25) .... ***Mastachilus* Kaup**

Classification of the Australian Genera

Subfamily Aulacocyclinae Kaup, 1868

Characteristics. Frontal area of head without tubercles, usually with prominent apically divided central horn. Antennal club with three segments. Mandibles without mobile tooth. Pronotum usually with complete marginal furrow. Procoxae covering central area of prosternum. Wings with two free branches of MP. Abdomen with five visible ventrites. Body subcylindrical. Male genital capsule of two sclerotised parts: undivided tegmen and median lobe, latter varying in form between species or even populations (Figs 24.10–24.12). Length 20–40 mm. [Dibb 1933; Hosoya *et al.* 2008].

Distribution. Palaearctic, Oriental and Australian regions.

Australian taxa. One genus.

Aulacocyclus Kaup, 1868 (Pl. 72B; Figs 24.10–12, 24.13, 24.16, 24.26)

Type species. *Passalus edentulus* W.S. Macleay, 1826.

Caulifer Kaup, 1871

Australian species. Fourteen named of which one is probably a synonym (van Doesberg 1992; Cassis & Weir 1992; Reyes-Castillo & Jiménez-Ferbans 2016).

Distribution. Eastern Australia in forested areas as far south as Victoria, also north of NT and of WA, India, Malay Peninsula, Philippines, New Guinea and neighbouring islands, New Caledonia.

Subfamily Passalinae Leach, 1815

Characteristics. Frontal area of head with paired inner and outer tubercles and central tubercle or horn. Antennal club with more than three segments except in *Leptaulax*. Mandibles with a mobile tooth. Pronotum without marginal furrow on anterior and posterior edges. Procoxae not meeting medially, leaving prosternum visible. Wings with only one free branch of MP. Abdomen with six visible ventrites, the first represented by two small widely separated sections outside hind coxae (Figs 24.14 (1), 24.15 (v). Male genital capsule of three sclerotised parts: a tegmen divided into basal and apical pieces, and a median lobe (Figs 24.2–24.9).

Distribution. Pantropical.

Australian taxa. There are seven confirmed genera of Passalinae in Australia and one, *Leptaulax*, known from Christmas Island, treated here in alphabetical order (number of named Australian species in parentheses): *Analaches* (1), *Austropassalus* (1), *Gonatas* (5), *Labienus* (1), *Leptaulax* (2?), *Mastachilus* (5), *Pharochilus* (6), and *Protomocoelus* (1). *Analaches* and *Pharochilus* are treated as subgenera of *Mastachilus* in Gravely's (1918) world revision.

Analaches Kuwert, 1891 (Pl. 51A; Figs 24.9, 24.22, 24.27)

Type species. *Analaches puberilis* Kuwert, 1891.

Characteristics. Antennal lamellae long. Outer edge of mandibles with a prominent angle often forming an acute tooth. Outer tubercles of head asymmetrical, left larger than right. Central tubercle small. Frontal carinae present. Mentum entirely glossy, with sharply defined scars. Median sulcus of pronotum fine, obsolescent anteriorly. Lateral and intermediate areas of metaventrite separated by distinct carina. Lateral five elytral striae consisting of matt bands separated by nitid costae, each band with a series of nitid convexities. Lateral lobes of genital capsule dorsally forming ring about base of median lobe, ventrally fusing with median lobe (Fig. 24.9). Body flattened. Length 26–30 mm.

Australian species. One: *A. australiensis* (Stoliczka).

Distribution. Eastern NSW, QLD and New Guinea.

Austropassalus Mjöberg, 1917 (Pl. 72J; Figs 24.7, 24.17, 24.28)

Type species. *Austropassalus hultgreni* Mjöberg, 1917.

Characteristics. Antennal lamellae short. Outer tubercles of head symmetrical. Central tubercle pedunculate and proclinate, flattened dorsally with median depression and bifid apex. Frontal carinae absent. Mentum with broad lateral matt edges. Lateral and intermediate areas of metaventrite with carina separating them. Lateral five elytral striae consisting of matt bands separated by nitid convex intervals. Genital capsule with small basal piece, apical piece forming middle segment, median lobe rounded. Body flattened. Length 26–30 mm.

Australian species. There is only one known species in the genus: *A. hultgreni.*

Distribution. Northern QLD.

Gonatas Kaup, 1871 (Pl. 72D; Figs 24.14, 24.23, 24.29)

Type species. *Passalus naviculator* Percheron, 1844.

Characteristics. Antennal lamellae short or long. Outer tubercles of head asymmetrical, left larger than right. Frontal carinae present. Central tubercle small. Mentum with two shallow depressions without sharp edges. Median sulcus of pronotum entire. Lateral and intermediate areas of metaventrite separated by a distinct carina. Genital capsule not examined. Body flattened. Length 18–33 mm.

Australian species. Five.

Distribution. North-eastern QLD, New Guinea, Melanesia and the Oriental region.

Labienus Kaup, 1871 (Pl. 72C; Figs 24.8, 24.18, 24.30)

Type species. *Eriocnemis ptox* Kaup, 1868 (New Guinea).

Characteristics. Antennal lamellae fairly short. Outer tubercles of head asymmetrical, left directed inwards, much larger than right. Central tubercle small. Frontal carinae present. Median sulcus of pronotum entire. Mentum entirely glossy, with sharply defined scars. Lateral and intermediate areas of metaventrite without a carina separating them. Mid-tibiae with dense clothing of orange pilosity, also present beneath hind angles. Genital capsule with small basal piece; apical piece forming middle segment; median lobe large, enclosing circular depression ventrally, posteriorly extended on dorsal and lateral surfaces (Fig. 24.8). Body subcylindrical. Length 32–41 mm.

Australian species. One: *L. inaequalis* Gravely.

Distribution. North-eastern QLD, New Guinea and the Oriental Region.

Leptaulax Kaup, 1868 (Pl. 72I; Figs 24.20, 24.31)

Type species. *Passalus dentatus* Fabricius, 1792.

Characteristics. Antennae with only three short lamellae. Anterior margin of head with four tubercles. Central tubercle small. Frontal carinae present. Mentum with sharply defined scars. Median sulcus of pronotum entire. Lateral area of metaventrite very narrow, with carina separating it from intermediate area. Genital capsule not examined. Body flattened. Length 21–26 mm.

Australian species. Doubtfully two widespread Fabrician species which have numerous synonyms (see Dibb 1938).

Distribution. Gravely (1918) states that the distribution of this genus 'possibly' extends to Australia from New Guinea and the Oriental region, but there are no mainland Australian specimens in collections. However, the genus was recorded from Christmas Island by Gahan (1888) and is represented from there in the Australian Museum by specimens which are possibly *L. dentatus* (Fabricius) (C. Reid pers. comm. 2014), one of which is illustrated here.

Mastachilus Kaup, 1868 (Pl. 72K; Figs 24.1–4, 24.21, 24.25, 24.32)

Type species. *Passalus polyphyllus* W.S. Macleay, 1826.

Mastochilus Kaup, 1868

Episphenoides Kuwert, 1891

Characteristics. Antennal lamellae long. Outer tubercles of head symmetrical. Central tubercle variable. Frontal carinae present. Mentum uniformly glossy, with sharply defined scars. Median sulcus of pronotum either obsolete, obsolescent, or entire. Lateral and intermediate areas of metaventrite separated by a distinct carina. Genital capsule with lateral lobes of apical piece modified into spines

(Figs 24.3, 24.4) except in *M. polyphyllus* (Macleay, 1826) (Fig. 24.2). Body subcylindrical. Length 35–55 mm.

Distribution. Coastal NSW, QLD, and doubtfully New Guinea.

Australian species. Five named and at least one unnamed.

Pharochilus Kaup, 1868 (Pl. 72A; Figs 24.5, 24.6, 24.24, 24.33)

Type species. *Passalus dilatatus* Dalman, 1817.

Characteristics. Antennal lamellae short. Outer tubercles of head symmetrical. Central tubercle small. Frontal carinae present. Mentum laterally with depressed matt border, with sharply defined scars. Median sulcus of pronotum entire. Lateral and intermediate areas of metaventrite separated by distinct carina. Genital capsule with lateral lobes modified into whorls (Fig. 24.6) or arms (Fig. 24.5). Body flattened. Length 28–37 mm.

Australian species. Six.

Distribution. Eastern Australia including Tasmania. The genus is not known outside Australia.

Note. *Pharochilus* and *Mastachilus* are closely related and often not clearly separable.

Protomocoelus Zang, 1905 (Pl. 72L; Figs 24.15, 24.19, 24.34)

Type species. *Passalus australis* Boisduval, 1835.

Characteristics. Antennal lamellae short. Right outer tubercle equal in length to left but more attenuated. Central tubercle small. Frontal carinae present. Mentum with deep V-shaped scar. Dentition of both mandibles reduced, especially that of left one. Median sulcus of pronotum obsolete. Lateral and intermediate areas of metaventrite without carina separating them. Mesotibiae with dense clothing of orange pilosity. Genital capsule not examined. Body subcylindrical. Length 36–45 mm.

Australian species. One, *P. australis*.

Distribution. North-eastern QLD. The genus also inhabits New Guinea and adjacent islands, Ceram, the Solomon Islands, and New Caledonia. The type specimens of *P. australis* are from Indonesia. The Queensland Museum has 19 specimens of this species from Iron Range plus one from an unspecified locality in Lamington National Park which appears to be a different species, but more specimens are required to confirm this.

Acknowledgments

The authors wish to thank C.A.M. Reid of the Australian Museum for valuable advice on preparing this chapter and for sending a specimen of *Leptaulax* collected on Christmas Island, G. Monteith of the Queensland Museum for information on pupae, G. Thompson of the Queensland Museum for assistance with photography, and P. Hudson of the South Australian Museum for help with literature search.

References

Ahrens D, Schwarzer J, Vogler A (2014) The evolution of scarab beetles tracks the sequential rise of angiosperms and mammals. *Proceedings of the Royal Society B* **281**, 20141470. doi:10.1098/rspb.2014.1470

Bai M, Ahrens D, Yang X, Ren D (2012a) New fossil evidence of the early diversification of scarabs. *Alloioscarabaeus cheni* (Coleoptera: Scarabaeoidea) from the Middle Jurassic of Inner Mongolia, China. *Insect Science* **19**, 159–171. doi:10.1111/j.1744-7917.2011.01460.x

Bai M, Ren D, Yang X (2012b) *Prosinodendron krelli* from the Yixian Formation, China: a missing link among Lucanidae, Diphyllostomatidae and Passalidae (Coleoptera: Scarabaeoidea). *Cretaceous Research* **34**, 334–339. doi:10.1016/j.cretres.2011.11.017

Boucher S (2006) Évolution et phylogénie des coléoptères Passalidae (Scarabaeoidea). Les taxons du groupe famille. La tribu néotropicale des Proculini et son complexe *Veturius*. *Annales de la Société Entomologique de France* **41**(3–4) [2005], 238–599.

Boucher S, Bai M, Wang B, Zhang W, Yang X (2016) Passalopalpidae, a new family from the Cretaceous Burmese amber, as the possible sister group of Passalidae Leach (Coleoptera: Scarabaeoidea). *Cretaceous Research* **64**, 67–78. doi:10.1016/j.cretres.2016.03.017

Browne DJ, Scholtz CH (1996) The morphology of the hind wing articulation and wing base of the Scarabaeoidea (Coleoptera) with some phylogenetic implications. *Bonner Zoologische Monographien* **40**, 1–200.

Bührnheim PF (1978) O edeago na sistemática de passalídeos americanos (Coleoptera: Passalidae). *Acta Amazonica* **8**(1, suppl. 2), 1–57.

Cano EB (1994) Estado actual sobre el conocimiento del edeago en Passalidae (Coleoptera), con la descripción del edeago en seis especies del Viejo Mundo. *Acta Zoológica Mexicana* (n.s.) **61**, 21–34.

Cassis G, Weir TA (1992) Passalidae. In *Zoological Catalogue of Australia. Coleoptera: Scarabaeoidea* Vol. 9. (Ed. WWK Houston) pp. 20–29. AGPS, Canberra.

Dibb JR (1933) Notes on the Aulacocyclinae (Col. Passalidae). *Entomologist's Monthly Magazine* **69**, 195–200.

Dibb JR (1938) Synopsis of Australian Passalidae (Coleoptera). *Transactions of the Royal Entomological Society of London* **87**, 103–124. doi:10.1111/j.1365-2311.1938.tb00092.x

Ento K, Araya K, Kon M (2003) The life history and colony composition of *Cylindrocaulus patalis* (Coleoptera: Passalidae). *Sociobiology* **42**(3), 795–806.

Ento K, Araya K, Kudo S (2008) Trophic egg provisioning in a passalid beetle (Coleoptera). *European Journal of Entomology* **105**(1), 99–104. doi:10.14411/eje.2008.014

Ento K, Araya K, Kudo S (2010) Laboratory observations of siblicide with hatching asynchrony in an insect with parental provisioning. *Journal of Ethology* **28**(2), 405–407. doi:10.1007/s10164-010-0208-2

Fearn S (1988) Notes on an unusually large colony of the passalid *Pharochilus politus* (Scarabaeoidea, Passalidae) in Tasmania. *Tasmanian Naturalist* **92**, 6.

Fonseca CRV, Barbosa MML, Souza Fernández MF (2011) A hypothetical evolutionary history of passalid beetles narrated by the comparative anatomy of the hindgut (Coleoptera: Passalidae). *Zootaxa* **3012**, 1–20.

Gahan CJ (1888) On the Coleoptera of Christmas Island. *Proceedings of the Zoological Society of London* **56**, 538–541. doi:10.1111/j.1469-7998.1888.tb06731.x

Gravely FH (1918) A contribution towards the revision of the Passalidae of the world. *Memoirs of the Indian Museum* **7**, 1–144.

Grebennikov VV, Scholtz CH (2004) The basal phylogeny of Scarabaeoidea (Insecta: Coleoptera) inferred from larval morphology. *Invertebrate Systematics* **18**(3), 321–348. doi:10.1071/IS03013

Hosoya T, Araya K, Masahiro K (2008) Phylogenetic relationships of Aulacocyclinae (Coleoptera, Passalidae) inferred from morphological characters. *Elytra* **36**(2), 301–309.

Houseknecht JL, Hart EL, Suh S-O, Zhou J (2011) Yeasts in the *Sugiyamaella* clade associated with wood-ingesting beetles and the proposal of *Candida bullrunensis* sp. nov. *International Journal of Systematic and Evolutionary Microbiology* **61**(7), 1751–1756. doi:10.1099/ijs.0.026427-0

Hunter PE (1993) Mites associated with New World passalid beetles (Coleoptera: Passalidae). *Acta Zoológica Mexicana* (n.s.) **58**, 1–37.

Kaup JJ (1868) Prodromus zu einer Monographie der Passaliden. *Coleopterologische Hefte* **4**, 1–31.

Kaup JJ (1871) Monographie der Passaliden. *Berliner Entomologische Zeitschrift* **15**, 1–125. doi:10.1002/mmnd.18710150306

Kon M, Johki Y, Ballerio A (2010) The ceratocanthid beetle *Pterorthochaetes haroldi* (Coleoptera, Ceratocanthidae) collected from a gallery of the passalid beetle, *Leptaulax planus* (Coleoptera, Passalidae), in Sabah, Malaysia. *Kogane* **1**, 93–95.

Kuwert AF (1891) Systematische Übersicht der Passaliden-Arten und Gattungen. *Deutsche Entomologische Zeitschrift* **1891**, 161–192.

Kuwert AF (1898) Die Passaliden dichotomisch bearbeitet. *Novitates Zoologicae* **5**, 259–349. doi:10.5962/bhl.part.1649

MacGown J, MacGown M (1996) Observation of a nuptial flight of the horned passalus beetle, *Odontotaenius disjunctus* (Illiger) (Coleoptera: Passalidae). *Coleopterists Bulletin* **50**(3), 201–203.

McKenna DD, Wild AL, Kanda K, Bellamy CL, Beutel RG, Caterino MS, Farnum CW, Hawks DC, Ivie MA, Jameson ML, Leschen RAB, Marvaldi AE, McHugh JV, Newton AF, Robertson JA, Thayer MK, Whiting MF, Lawrence JF, Ślipiński A, Maddison DR, Farrell BD (2015) The beetle tree of life reveals that Coleoptera survived end-Permian mass extinction to diversify during the Cretaceous terrestrial revolution. *Systematic Entomology* **40**, 835–880 doi:10.1111/syen.12132.

Mjöberg E (1917) Results of Dr. E. Mjöberg's Swedish Scientific Expedition to Australia 1910–1913. Cetoniidae, Rutelidae, Passalidae, Chrysomelidae: Subfam. Sagrinae, Cassidinae, Hispinae. *Arkiv för Zoologi* **11**, 1–19.

Nikolajev GV (2007) The Mesozoic stage of evolution of the scarabaeoid beetles (Insecta: Coleoptera: Scarabaeoidea). Kazak Universiteti, Almaty. [In Russian]

Palestrini C, Pensati F, Barbero E, Zunino M (2003) Differences in distress signals of adult passalid beetles (Coleoptera, Passalidae). *Bollettino della Società Entomologica Italiana* **135**(1), 45–53.

Panov AA (2010) Structure of the mushroom bodies in Scarabaeoidea (Insecta: Coleoptera). 1. Basal families and coprophagous Scarabaeidae. *The Biological Bulletin* **37**(5), 502–510. doi:10.1134/S1062359010050109

Percheron AR (1835) *Monographie des Passales et des genres qui en ont été séparés; accompagnée de planches dessinées par l'auteur, ou toutes les espèces ont été figurées.* Mercklein, Paris.

Reyes-Castillo P (1970) Coleoptera: Passalidae: morfología y división en grandes grupos; géneros americanos. *Folia Entomológica Mexicana* **20–22**, 1–240.

Reyes-Castillo P, Halffter G (1984) La estructura social de los Passalidae (Coleoptera: Lamellicornia). *Folia Entomológica Mexicana* **61**, 49–72.

Reyes-Castillo P, Jiménez-Ferbans L (2016) *Aulacocylus yorkensis* a new species of Passalidae (Coleoptera: Scarabaeoidea) from Australia, with a key to the identification of Australian species of the genus. *Revista Brasileira de Entomológica* **60**(3), 211–213. doi:10.1016/j.rbe.2016.05.003

Scholtz CH, Grebennikov VV (2005) 12. Scarabaeiformia Crowson, 1960. In *Handbook of Zoology. A Natural History of the Phyla of the Animal Kingdom. Volume IV – Arthropoda: Insecta. Part 38. Coleoptera, Beetles.* Volume 2: Systematics. Part 2. (Eds RG Beutel and RAB Leschen) pp. 345–365. Walter de Gruyter, Berlin.

Schuster JC, Schuster LB (1985) Social behavior in passalid beetles (Coleoptera: Passalidae): cooperative brood care. *The Florida Entomologist* **68**(2), 266–272. doi:10.2307/3494359

Seeman OD (2000) The immature life stages of the Fedrizziidae (Mesostigmata: Fedrizzioidea). *Acarologia* **41**, 39–52.

Seeman OD (2001) Myriad Mesostigmata associated with log-inhabiting arthropods. In *Acarology: Proceedings of the 10th International Congress.* (Eds RB Halliday, DE Walter, H Proctor, RA Norton and M Colloff) pp. 272–276. CSIRO Publishing, Melbourne.

Seeman OD, Nahrung HF (2000) Mites as fungal vectors? The ectoparasitic fungi of mites and their arthropod associates in Queensland. *Australasian Mycologist* **19**, 3–9.

Smith ABT, Hawks DC, Heraty JM (2006) An overview of the classification and evolution of the major scarab beetle clades (Coleoptera: Scarabaeoidea) based on preliminary molecular analyses. *Coleopterists' Society Monographs. Patricia Vaurie Series* **5**, 35–46.

Valenzuela-González JE (1992) Adult-juvenile alimentary relationships in Passalidae (Coleoptera). *Folia Entomológica Mexicana* **85**, 25–37.

van Doesberg PH (1992) A new species of *Aulacocyclus* from Australia (Coleoptera: Passalidae). *Zoölogische Mededeelingen* **66**(29), 413–415.

Wicknick JA, Miskelly SA (2009) Behavioral interactions between non-cohabiting bess beetles, *Odontotaenius disjunctus* (Illiger) (Coleoptera:Passalidae). *Coleopterists Bulletin* **63**(1), 108–116. doi:10.1649/0010-065X-63.1.108

25. TROGIDAE MACLEAY, 1819

John F. Lawrence

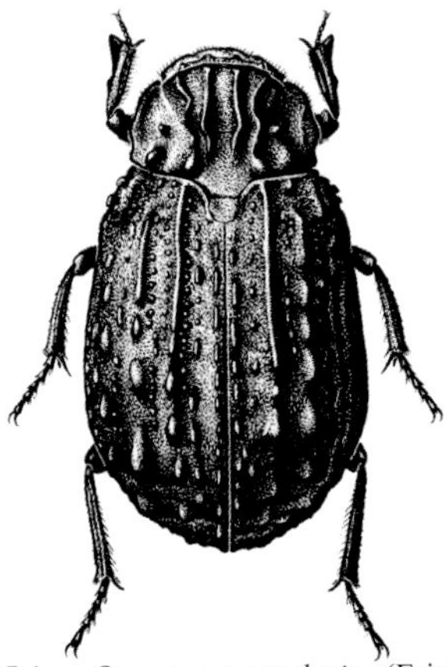

Fig. 25.1. *Omorgus australasiae* (Erichson).

Introduction. Trogidae is a relatively large, more or less cosmopolitan family, containing ~300 species in the genera *Trox* Fabricius (including subgenera *Afromorgus* Scholtz and *Haroldomorgus* Scholtz), *Omorgus* Erichson (including subgenus *Phoberus* Macleay) and *Polynoncus* Burmeister. The Australian fauna includes 53 native species of *Omorgus* (*Omorgus*) and two introduced New World species, *Omorgus suberosus* (Fabricius) and *Trox scaber* (Linnaeus) (Houston 1992; Zidek 2013).

Biology. Adults and larvae apparently feed on dry animal remains, being one of the last in the succession of insects that invade carcasses; however gut contents of several adults of the wingless species *Omorgus rotundulus* (Haaf) contained several insect parts, particularly of termites and ants (Houston *et al.* 2010). Trogid larvae live in vertical burrows in the soil beneath the carcass. Adults stridulate by rubbing an abdominal plectrum against a file along the internal edges of the elytra (Alexander & Moore 1963). The group appears to flourish in the more arid parts of Australia. (Scholtz 1986a, 1986b, 1991; Strümpher & Scholtz 2011).

Characteristics. Adults. Length ~5–30 (usually 7–20) mm and body oblong; moderately to strongly convex above but relatively flat below; colour brown to grey or black; dorsal surfaces dull, often granulate to tuberculate and sometimes encrusted with dirt; vestiture of short to very short, erect, bristles. Head hypognathous, deeply inserted into prothorax, with low, curved occipital carina, which extends anteriorly on each side to meet supraocular carinae. Eyes well developed, usually strongly protuberant, shallowly emarginate, finely facetted, without interfacetal setae; ommatidium usually of exocone or duocone type (eucone in *Trox*). Antennal insertions concealed from above by frontal carina, which is continuous with supraocular carina on each side and extends to midline; frontoclypeal suture very weak or incomplete mesally; frontoclypeal region abruptly declined anterior to frontal carina (reflexed area very short in *Trox*). Labrum well sclerotised, strongly transverse and slightly emarginate. Antennae very short, 10-segmented, with 3-segmented, lamellate club with pubescent segments capable of close apposition; scape usually more than twice as long as wide, more than twice as long as pedicel, more or less curved and clothed with stout bristles. Mandibles short and broad, unidentate; mola weakly developed and slightly subbasal; prostheca a membranous lobe. Maxilla with galea short, 2-segmented, more or less truncate and clothed with stout setae; lacinia only slightly narrower than galea, with densely setose inner edge, apicolateral 2- or 3-dentate uncus and apicomesal unidentate uncus; apical palpomere fusiform. Labial palps with bases concealed from below by mentum and apical palpomere fusiform. Pronotum ~0.55–0.8 times as long as wide, widest at about middle but almost always sharply constricted posteriorly, so that base is distinctly narrower than combined elytral bases. Lateral pronotal carinae complete, simple, undulate or crenulate; pronotal disc with various impressions and elevations, sometimes forming longitudinal costae. Prosternum in front of coxae relatively short, biconcave, oblique to almost vertical and on different plane than prosternal process, which is complete, narrow to laminate at base but abruptly expanded and subtruncate apically. Notosternal sutures absent; hypomera with large anterior pockets for reception of head. Procoxae large, transverse, projecting slightly below prosternum, with moderately long concealed lateral extension; trochantin concealed; endopleuron not fused to notum. Procoxal cavities moderately transverse, narrowly separated or subcontiguous, broadly closed externally by postcoxal hypomeral projections fitting into lateral cavities near apex of prosternal process, internally open. Elytra ~1.1–1.3 times as long as combined width and 2.0–3.2 times as long as pronotum, their apices broadly, conjointly rounded concealing all abdominal tergites; disc usually without distinct puncture rows or striae, but usually with tubercles and/or longitudinal costae, sometimes completely fused together; epipleura moderately broad and complete. Mesoventrite with median prosternal rest or broad collar formed with mesanepisterna; mesoventral cavity and discrimen absent. Mesocoxae more or less globular, with concealed trochantins; mesocoxal cavities contiguous, broadly closed laterally by meeting of meso- and metaventrites. Metaventrite without transverse (metakatepisternal) suture; discrimen usually short to moderately long, sometimes absent. Metacoxae strongly transverse, subcontiguous, extending laterally to meet elytra; plates absent. Metendosternite usually with moderately long and narrow stalk, long, transverse lateral arms and moderately long, anterior process with narrowly separated tendons. Hind wing with long radial bending zone combined with distinct hinge at end of radial bar; apical field long, with 2 or 3 radial extensions; radial cell complete, shorter than wide, more or less oblique, with lumen completely filled by pterostigma; cross-vein r3 absent; r4 more or less complete; R-M loop forming a narrowly acute angle; basal portion of RP short, apical extension interrupted basally but extending to wing margin; medial spur not curved posteriorly, extending to wing margin;

medial field with 4 free veins; MP_3 and MP_4 not connected basally; wedge cell absent; AA_3 not joining Cu; anal lobe well developed, with very small, notch-like embayment, AP_{3+4} not forked; hind wings sometimes absent. Legs relatively long and slender; trochanterofemoral joint strongly oblique with femur in contact with coxa. Profemur expanded and flattened with carinate anterior edge (concealing tibiae, tarsi and prosternum in repose); all tibiae more or less expanded apically; meso- and metatibiae in *Trox* with one or two teeth at outer apical angle and a longitudinal row of spines along outer edge, those in *Omorgus* without teeth but often with socketed spines; outer edge and outer apical angle of protibia with several large teeth; meso- and metatibial spurs well developed and more or less asymmetrical; protibia one large, apical spur. Tarsi 5–5–5; tarsomeres slender, simple; pretarsal claws simple; empodium absent or not apparent. Abdomen usually with 5 connate ventrites (sternites III–VII), with ventrite 1 strongly inclined and much shorter than 2, and ventrites 2–5 all on the same plane and separated by distinct grooves; lateral portions of sternite II completely concealed by coxae. Functional spiracles usually on abdominal segments I–VIII (on I–VII in *Trox*) and located in pleural membrane. Abdominal tergites (including tergite VIII) relatively lightly sclerotised, with patches of wing-folding asperities on tergites V and VI. Anterior edge of sternite VIII in male without median strut. Anterior edge of segment IX with broadly rounded subgenital plate; tergite IX almost completely divided into two laterotergites; tergum X membranous with paired struts. Aedeagus of trilobate type, usually symmetrical, with short broad phallobase, slender, simple parameres and broad, somewhat flattened penis without basal struts. Sternite VIII in female without spiculum ventrale. Sternite IX in female divided into two parts; proctiger membranous with lateral struts. Ovipositor very short and broad; paraprocts much shorter than coxites, without baculi; coxites undivided, with strongly oblique basal bacula, with or without apical styli (Tanner 1927; Scholtz 1986a, 1986b, 1990; D'Hotman & Scholtz 1990; Caveney & Scholtz 1993; Jameson 2002; Scholtz & Grebennikov 2016).

Larvae. Body subcylindrical and strongly curved ventrally (C-shaped, Fig. 25.6); relatively weakly sclerotised and lightly pigmented, except for heavily pigmented head and weakly pigmented prothoracic plate; vestiture relatively dense, consisting of long and short setae and short spines. Head hypognathous, distinctly transverse, strongly rounded laterally and slightly flattened; posterior edge straight. Epicranial stem moderately long; frontal arms V-shaped or U-shaped, with bases contiguous; median endocarina coincident with epicranial stem. Stemmata one on each side. Frontoclypeal suture usually present; clypeus symmetrical. Labrum free, transverse, emarginate; tormae united mesally. Antennae moderately long, 3-segmented; antennomere 3 min; antennomere 2 with dome-like sensorium. Mandibles large, unidentate, with accessory ventral process; mola well developed, somewhat asymmetrical, with small

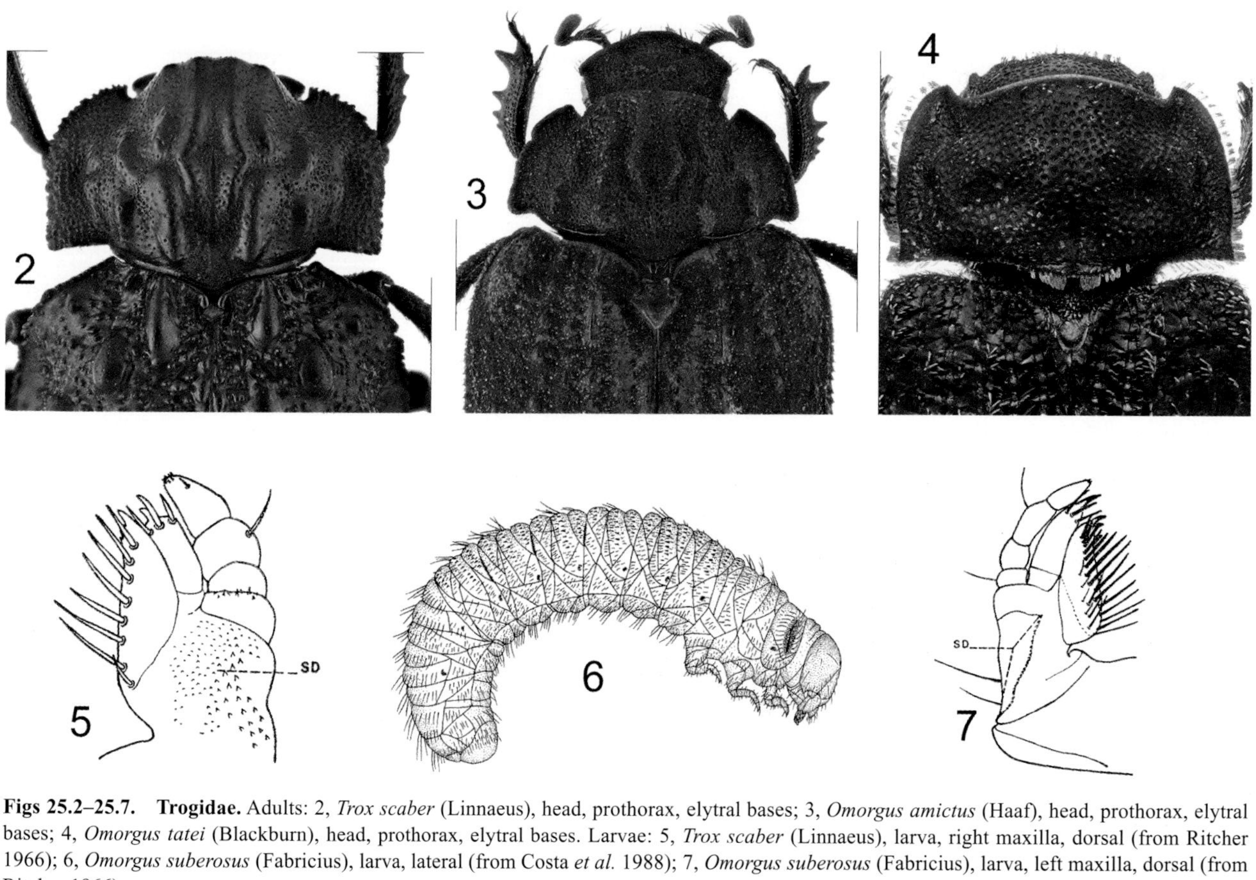

Figs 25.2–25.7. Trogidae. Adults: 2, *Trox scaber* (Linnaeus), head, prothorax, elytral bases; 3, *Omorgus amictus* (Haaf), head, prothorax, elytral bases; 4, *Omorgus tatei* (Blackburn), head, prothorax, elytral bases. Larvae: 5, *Trox scaber* (Linnaeus), larva, right maxilla, dorsal (from Ritcher 1966); 6, *Omorgus suberosus* (Fabricius), larva, lateral (from Costa *et al.* 1988); 7, *Omorgus suberosus* (Fabricius), larva, left maxilla, dorsal (from Ritcher 1966).

sub-basal tuft of hairs; prostheca absent. Ventral mouthparts slightly retracted, with well developed maxillary articulating area; cardines more or less oblique, internally divided; stipes elongate its dorsal surface with row of stridulatory teeth. Galea and lacinia well separated, the former articulated, 2-segmented, with few setae and a terminal spine, the latter with bidentate uncus and stout setae along inner edge; palp 3-segmented with segment-like palpifer. Labium with ligula weakly developed and truncate; palps widely separated. Hypopharyngeal bracon absent; hypopharynx with slender undulate transverse bar with 2 pairs of distal projections. Hypostomal rods short and very slightly diverging; ventral epicranial ridges absent. Gular region absent. Thoracic segments more or less equal in length, with distinct tergal plate on prothorax. Legs moderately long, slender, 5-segmented with pretarsus claw-like and bisetose; stridulatory mechanism absent; coxae moderately widely separated. Abdominal terga without sclerotised plates, terga I–VI or VII transversely divided forming 3 plicae, each with at least one transverse row of short, spine-like setae, in addition to longer setae. Tergum IX about as long as VIII, not extending onto ventral surface, without urogomphi; segment X visible from above, more or less cylindrical and terminal, more or less evenly clothed with short and long, stiff hairs; anal opening Y-shaped, surrounded by single dorsal lobe and paired ventral lobes. Functional spiracles usually cribriform and slightly reniform with spiracular scar located anteroventrally on both thorax and abdomen (spiracles biforous in *Trox*); closing apparatus present (Böving & Craighead 1931; Ritcher 1966; Medvedev & Nikolayev 1972; Scholtz & Peck 1990; Carlson 1991; Scholtz 1991; Jameson 2002; Scholtz & Grebennikov 2016).

Key to the Australian genera

1. *Adult*: scutellar shield widest at base, slightly longer than wide and broadly rounded at apex; length always less than 8 mm; surface of pronotal disc without longitudinal ridges (Fig. 25.2). *Larva*: spiracles biforous; maxilla with stridulatory area consisting of a patch of minute teeth (Fig. 25.5) .... ***Trox* Fabricius**

– *Adult:* scutellar shield hastate, widest beyond base and narrowly rounded or acute at apex (Fig. 25.3); length almost always greater than 8 mm OR IF less THEN pronotal disc with two pairs of undulating longitudinal ridges (Figs 25.3–4). *Larva*: spiracles cribriform; maxilla with stridulatory area consisting of a row of minute teeth (Fig. 25.7) .... ***Omorgus* Erichson**

Classification of Australian genera

Omorgus Erichson, 1847 (Pl. 49H–I)

Type species. *Trox suberosus* Fabricius, 1775.
Australian species. 57.
Distribution. The genus *Omorgus* occurs primarily in the Neotropical, southern Nearctic, Afrotropical and Australian regions. The genus is widely distributed in Australia.

Trox Fabricius, 1775 (Pl. 49D)

Type species. *Scarabaeus sabulosus* Linnaeus, 1758.
Australian species. One (introduced).
Distribution. The genus is endemic to the Nearctic and Palaearctic regions. The introduced species *Trox scaber* (Linnaeus) occurs in NSW, VIC and SA.

Acknowledgments

The authors are grateful to Oregon State University and Museu de Zoologia, Universidade de Sao Paulo for permission to use Figs 25.5, 25.7 and 25.6, respectively.

References

Alexander RD, Moore TE (1963) The evolutionary differentiation of stridulatory signals in beetles (Insecta: Coleoptera). *Animal Behaviour* **11**(1), 111–115. doi:10.1016/0003-3472(63)90018-6

Böving AG, Craighead FC (1931) *An illustrated synopsis of the principal larval forms of the order Coleoptera. Entomologica Americana* (New Series) **11**, 1–351.

Carlson CD (1991) Lucanidae (Scarabaeoidea), Passalidae (Scarabaeoidea), Scarabaeidae (Scarabaeoidea) (including Acanthoceridae, Ceratocanthidae, Cetoniidae, Diphyllostomatidae, Geotrupidae, Glaphyridae, Hybosoridae, Pleocomidae, Trogidae, etc.). In *Immature Insects. Volume 2.* (Ed. FW Stehr) pp. 372–384. Kendall/Hunt, Dubuque, Iowa.

Caveney S, Scholtz CH (1993) Evolution of ommatidium structure in the Trogidae (Coleoptera). *Systematic Entomology* **18**, 1–10. doi:10.1111/j.1365-3113.1993.tb00652.x

Costa C, Vanin SA, Casari-Chen SA (1988) *Larvas de Coleoptera do Brasil.* Museu de Zoologia, Universidade de São Paulo, São Paulo.

D'Hotman D, Scholtz CH (1990) Phylogenetic significance of the structure of the external male genitalia in the Scarabaeoidea (Coleoptera). *Republic of South Africa, Department of Agricultural Development. Entomology Memoirs* **77**, 1–51.

Houston WWK (Ed.) (1992) *Zoological Catalogue of Australia. Volume 9. Coleoptera: Scarabaeoidea.* xii + 544 pp. Australian Government Publication Service, Canberra.

Houston TF, Zhang J, Hanich BP (2010) Diet of the flightless trogid beetle *Omorgus rotundulus* (Haaf) (Coleoptera: Trogidae) in the Little Sandy Desert of Western Australia. *Australian Entomologist* **36**(4), 207–212.

Jameson ML (2002) 27. Trogidae MacLeay 1819. In *American Beetles. Volume 2. Polyphaga: Scarabaeoidea through Curculionoidea.* (Eds RH Arnett Jr, MC Thomas, PE Skelley and JH Frank) pp. 17–19. CRC Press, Boca Raton, Florida.

Medvedev SI, Nikolayev GV (1972) Descriptions of the immature stages of beetles of the genera *Trox* F. and *Lethrus* Scop. (Coleoptera, Lamellicornia) and notes on their biology. *Entomologicheskoye Obozreniye* 51, 617–624 (in Russian; translation in *Entomological Review* 51, 371–375).

Ritcher PO (1966) *White Grubs and Their Allies. A Study of North American Scarabaeoid Larvae.* Oregon State University Press (Studies in Entomology No. 4), Corvallis, Oregon.

Scholtz CH (1986a) Phylogeny and systematics of the Trogidae (Coleoptera: Scarabaeoidea). *Systematic Entomology* **11**, 355–363. doi:10.1111/j.1365-3113.1986.tb00186.x

Scholtz CH (1986b) Revision of the genus *Trox* Fabricius (Coleoptera: Trogidae) of the Australasian Region. *Australian Journal of Zoology* **125**(Supplement), 1–99.

Scholtz CH (1990) Revision of the Trogidae of South America (Coleoptera: Scarabaeoidea). *Journal of Natural History* **24**(6), 1391–1456. doi:10.1080/00222939000770841

Scholtz CH (1991) Descriptions of larvae of Australian *Omorgus* Erichson, with implications for the phylogeny of the Trogidae (Insecta: Coleoptera). *Invertebrate Taxonomy* **5**, 827–835. doi:10.1071/IT9910827

Scholtz CH, Grebennikov VV (2016) 15.3. Trogidae Macleay, 1819. In *Handbuch der Zoologie/Handbook of Zoology. BandVolume IV Arthropoda: Insecta. Teilband/Part 38. Coleoptera, Beetles. Volume 1: Morphology and Systematics (Archostemata, Adephaga, Myxophaga, Polyphaga partim).* 2nd Edition. (Eds RG Beutel and RAB Leschen) pp. 470–471. Walter de Gruyter, Berlin.

Scholtz CH, Peck S (1990) Description of a *Polynoncus* Burmeister larva, with implications for phylogeny of the Trogidae (Coleoptera: Scarabaeoidea). *Systematic Entomology* **15**, 383–389. doi:10.1111/j.1365-3113.1990.tb00072.x

Strümpher WP, Scholtz CH (2011) New species of Trogidae (Coleoptera: Scarabaeoidea) from Australia. *Australian Journal of Entomology* **50**, 139–143. doi:10.1111/j.1440-6055.2010.00804.x

Tanner VM (1927) A preliminary study of the genitalia of female Coleoptera. *Transactions of the American Entomological Society* **53**, 5–50.

Zidek J (2013) Checklist and bibliography of the Trogidae (Coleoptera: Scarabaeoidea). *Insecta Mundi* **0314**, 1–38.

26. LUCANIDAE LATREILLE, 1804

Chris Reid

Fig. 26.1. *Lamprima aurata* Latreille

Common name. Stag beetles.

Introduction. Stag beetles are mostly large beetles, popular in the public imagination (Sprecher-Uebersax 2008), and particularly popular amongst insect collectors (Brock 2006; Raffles 2010), therefore their taxonomy and distribution is relatively well known. There is a huge literature on this group, including two recent reviews of all Australian species (Hangay & de Keyzer 2017; Bartolozzi *et al.* 2017). Stag beetles are widespread globally but almost restricted to forested areas, predominantly in the tropics and the cool temperate southern hemisphere. There are ~110 genera and 1300 described species. These numbers are approximate because this family is notorious for weakly defined taxa in both ranks, including numerous infrageneric and infraspecific names, with the status of these taxa being controversial. The cataloguer Parry was already complaining about excess genera in 1870. As far as possible the approach used here has been to eliminate use of subspecies and subgenera. The greatest diversity of Lucanidae is in Asia, particularly tropical Asia. The Australian fauna is diverse at generic level but relatively poor in species, however several species in collections remain to be described. The 19 genera and 91 species are restricted to Tasmania and the margins of the east, north and south-west of the mainland, one species of *Figulus* occurring up to 450km inland in the east. Lucanidae are absent from the centre and west of Australia, except the south-west. Lucanidae are also found on remote islands such as Christmas Island, Lord Howe and Norfolk Island. Nine genera are endemic to Australia.

The Australian Lucanidae were catalogued by Moore & Cassis (1992) but species continue to be discovered, revised and described. Recent revisions include: *Figulus* (Monte *et al.* 2016), *Safrina* (Reid & Beatson 2016) and *Lamprima* (Reid *et al.* 2018b). Two endemic Australian genera have been synonymised (Reid 1999). All Australian described species are illustrated, with diagnoses and natural history observations, by Hangay & de Keyzer (2017), a work which this chapter attempts to complement. All species, including many primary types, are illustrated in large format by Bartolozzi *et al.* (2017).

One genus is added here to the Australian fauna: *Mitophyllus* Parry. This New Zealand genus, described in detail by Holloway (1998, 2007), is defined by several features, including elytral punctures and scales, tibial structure, male and female genitalia. These features are shown by two Australian species hitherto placed in *Ceratognathus*: *M. minutus* (Lea, 1929) **comb. nov.** and *M. ocularis* (Carter, 1925) **comb. nov.** A third species, only known from its description, also fits *Mitophyllus* and is therefore placed in this genus: *M. abdominalis* (Parry, 1870) **comb. nov.**

Serrognathus Motschulsky was formerly recorded from Australia, with two local species (Moore & Cassis 1992). This genus is often treated as a synonym of *Dorcus* (e.g. Mizunuma & Nagai 1994; Kim & Farrell 2015) but a detailed analysis of lucanine morphology has provided support for validity of the genus (Huang & Chen 2013). Molecular evidence provides additional support for validity of *Serrognathus*, as the only species sampled (as a *Dorcus*) is sister to the remaining *Dorcus* species (Kim & Farrell 2015). However, the two Australian species formerly placed in *Serrognathus* (Moore & Cassis 1992) lack its morphological characteristics (Holloway 2007) and are better placed in the revised concept of *Dorcus* used by Huang & Chen (2013). The two species have recently been transferred to *Dorcus* (Hangay & de Keyzer 2017).

The key to the remaining genera provided here is new and based on examination of almost all described species.

Conservation. Stag beetles are important in conservation in several different ways: (i) rare or localised species, requiring protection; (ii) species restricted to endangered habitats, therefore useful indicator species; (iii) barometers of environmental change.

Rarity. 20 species of world Lucanidae are on the IUCN Red List (International Union for Conservation of Nature and Natural Resources 2014), 6 from Europe and 14 from South Africa. Several Australian species should qualify for inclusion. A survey published in 1991 of Australian insects in the export trade listed only six lucanid species, with concern expressed for the trade in *Phalacrognathus muelleri*, the most spectacular Australian lucanid (Hawkeswood *et al.* 1991). Prices for it at that time were in the range $500–1150. Since 1991 interest in collecting Lucanidae has increased greatly, especially in Japan (Brock 2006). *Phalacrognathus muelleri* is now known to be a widespread and common species in north Queensland rainforests (Wood *et al.* 1996), and prices have dropped to $50–200, but it has also been so heavily exported to, and reared in, Japan that it may have formed feral colonies there, like several other imported species (Goka *et al.* 2004; Okushima 2004; Brock 2006). The two most obvious Australian candidates for protection are the distinctive and attractive *Homolamprima crenulata* and *Lamprima imberbis*, which are only known from early collections and have not been rediscovered despite extensive searching in the last 50 years (Hangay & de Keyzer 2017; Reid *et al.* 2018b). Several other species are severely restricted in range, leaving them vulnerable to climate change, collection pressure and other disturbances from human activity. Best known of these are the three *Hoplogonus* species, confined to small areas of Tasmania, where they have been extensively surveyed (Meggs *et al.* 2003; Meggs *et al.* 2004; Munks *et al.* 2004; Spencer & Richards 2013). Two *Hoplogonus* species occupy ranges of less than 100 km^2 (Munks *et al.* 2004), and at least one fits the Vulnerable category (International Union for Conservation of Nature and Natural Resources 2001). Several species of *Lissapterus*, *Lissotes* and *Safrina* show the same narrow ranges but only a few species of *Lissotes* have been carefully surveyed (Meggs & Taylor 1999; Meggs & Munks 2003). Studies of *Hoplogonus* and *Lissotes* species were undertaken from a commercial forest management point of view, in response to a study showing the importance of old growth forests for Tasmanian lucanid conservation (Michaels & Bornemissza 1999). While concern is expressed for species whose ranges are largely outside reserves (Munks & Taylor 2000), occurrence in protected reserves is no impediment to collectors. Collection pressure is a particular problem in Lucanidae, as collectors include dealers who are only interested in sales, not long-term viability of populations (New 2005; Brock 2006). Furthermore, as numbers diminish, the increased rarity leads to increased collecting pressure threatening extinction of populations (Huang 2014). For example, two Japanese dealers were apprehended in 2002 in possession of ~1000 adult and larval *Lamprima insularis*, endemic to the lowlands of Lord Howe (total area ~11 km^2), a world heritage listed site (UNESCO 2015). As the population of this species had not been adequately surveyed it was impossible to determine the impact of these thefts, but it was likely to be significant, for both beetles and habitat. Single males of *Lamprima insularis* had been priced at up to $250 at the time of the arrest (2002) but were being offered for up to $1850 four years later (Brock 2006), suggesting that the arrest blocked supply. It is therefore sadly ironic that publications describing stag beetles lead to further pressure by collectors (Brock 2006; Holloway 2007; Switala *et al.* 2014).

Habitat indicators and population monitoring. Lucanidae may be useful indicators for habitat conservation because they are popular with collectors (therefore with well known distributions), taxonomically well known and relatively poor dispersers. However, in Australia many species are not well defined taxonomically and have poorly known ranges. Rainforest species of Lucanidae were used, along with other invertebrate groups, for assessing the significance of the remnant rainforests of eastern Australia (Williams 2002) and they showed high local endemism. In central Europe, Lucanidae are recommended for habitat quality assessment in beech forests, as the family has the highest percentage of indicator species of all saproxylic beetles assessed (Lachat *et al.* 2012). The usefulness of Lucanidae is partly due to their low dispersal. A Swedish population of *Ceruchus* moved less than 10 m after 17 years (Karlsson *et al.* 2013), and even actively volant species, such as *Lucanus cervus* in Europe, move relatively small distances, less than 800 m in females (Rink & Sinsch 2007, 2011). Monitoring of lucanids may be problematic, as surveying may be habitat destructive. In Europe, the stridulation of larvae of *Lucanus cervus* can be detected outside logs (Harvey *et al.* 2011). The same authors also found that road kills and lures could be used to monitor adult populations. However, in Tasmania, road kills are seen as a potential threat to *Hoplogonus simsoni* (Spencer & Richards 2013).

Environmental change. The study of impact of environmental change on lucanids has focussed on the European species *Lucanus cervus*. Dispersive flight in this species is correlated with temperature, but high temperatures lead to rapid weight loss which may affect reproductive performance (Rink & Sinsch 2011). Male mandible size varies seasonally, decreasing over the active period of 3 months (Hardersen *et al.* 2011), suggesting that temperature is negatively correlated with nutritional value of larval food. In Ukraine, radiation near the Chernobyl reactor is causing much higher than normal levels of male mandible asymmetry in *Lucanus cervus*, leading to failed pairing (Moller 2002).

There have been no studies of the effects of fire regime on lucanids in Australia, despite fire being such a significant factor in the landscape. Many species, for example some *Figulus*, clearly exist in high fire-frequency habitats. In Switzerland, no significant effect of differing fire regimes was found on lucanids (Moretti *et al.* 2004).

Several lucanid species have limited ranges because they are confined to small islands (for example, *Lamprima aenea* on Norfolk Island). Here they are particularly vulnerable to climate change but also to environmental change through the introduction of exotic species. In the Seychelles Islands, the exotic lucanid *Aegus chelifer* appears to be displacing or eradicating the native *Figulus* species (Carpanato et al. 2010).

Biology. *Life history*. Notes on the life histories of most Australian genera and many species are provided by Hangay & de Keyzer (2017). Most female lucanids tunnel into rotting wood to lay their eggs. In *Lamprima*, each egg is laid in its own chamber (Fearn 1996), whereas in *Phalacrognathus* the eggs may be laid together in adult or larval pulp (Wood *et al.* 1996). Lucanidae larvae are primarily saproxylic, feeding in or on the surface of rotting, fungus-infected, wood. Larval guts have mycophagous enzymes and larvae can survive for at least 14 days on a pure diet of fungi (Tanahashi *et al.* 2009; Tanahashi & Kubota 2013). The fungal associations may not be specific; at least six species of fungi in two families were associated with the rotting wood hosts of *Phalacrognathus* (Wood *et al.* 1996), but species do seem to be either white rot or brown rot specialists. White rot fungi decompose lignin whereas brown rot fungi leave the lignin (Pandey & Pitman 2003). Adult female Lucanidae carry mycangia (Tanahashi & Hawes 2016). Many of the Australian lucanids seem to prefer development in or under partially buried logs (Hangay & de Keyzer 2017), where they may be protected from bushfires. Some Australian lucanids are known to have wide larval host ranges, including exotic species; these beetles tend to be widespread, commonly occurring in gardens (Fearn 1996;

Holloway 2007). Moisture content and presence of fungi may be as important for oviposition site as host species. Other lucanid species may have narrow host ranges (Munks *et al.* 2004). Larvae of an exotic *Figulus* species appear to be tended by adults (Mori & Chiba 2009), although the sociality is not as developed as in Passalidae (Chapter 24). Survivorship of *Figulus* larvae is lowest when related adults are in attendance and highest when larvae are alone, but intermediate for larvae with unrelated adults, which sometimes prey on the larvae (Mori & Chiba 2009). Cannibalism between larvae of a single species is common in laboratory conditions but appears to be rare in the field (Tanahashi & Togashi 2009). Pupation is usually in the host wood, but may be in adjacent soil, the site varying within a single species (Fearn 1996). The life span varies according to climate and quality of food. In cool temperate Tasmania, *Lamprima aurata* takes 3 years to develop from egg to mature adult (Fearn 1996), whereas in tropical Queensland the related *Phalacrognathus muelleri* can take one year, but as much as two years if the food quality is poor (Wood *et al.* 1996). Adults mature for several months, immobile in their pupal cells, before emerging. Mating of the actively volant species *Lamprima aurata* occurs on flowers (Fearn 1996), but many Australian species are flightless and copulation is likely to occur on or near the oviposition site. Adult longevity of Australian species is only known for captive adults of *Lamprima* (6 months: Hangay & de Keyzer 2017), *Lissapterus* (6 months: Hangay & de Keyzer 2017), *Phalacrognathus* (18 months: Wood *et al.* 1996) and *Safrina* (1 month: Hangay & de Keyzer 2017), but it may be up to three years in mark and recapture studies of wild European and Japanese *Dorcus* species (Fremlin & Hendriks 2013). Mark and recapture studies also show that small males of *Lucanus cervus* in Italy survive longer than large males, possibly due to lower predation (Chiari *et al.* 2014).

Parasites and parasitoids of overseas Lucanidae include tachinid Diptera in larvae (O'Hara 2013), canestriniid mites on adults (Okabe *et al.* 2012) and diplogastrid nematodes (Kanzaki *et al.* 2014). In Australia, larvae of *Phalacrognathus muelleri* are parasitised by a tachinid and a scoliid wasp (Wood *et al.* 1996). The same tachinid genus (*Amphibolia*) parasitises *Lamprima* larvae in New South Wales (label data in Australian Museum).

Adult Lucanidae seem to be mostly liquid feeding but there are few data for Australian species. *Phalacrognathus* adults feed at floral nectaries, sap flows on tree-trunks and fruit (Wood *et al.* 1996). Overseas studies show that males and females of some lucanid species have different diets; males feed on sap flows and females on sap and fungoid wood. In *Lamprima*, only males snip off shoot tips but both sexes drink the ensuing sap (Fearn 1996, 2015). Females of Japanese species have mycangia with yeast-like microbes that help to digest wood but these are lacking in males of the same beetles (Tanahashi *et al.* 2010). Females chew their way into wood to lay eggs. Female fecundity can be improved by feeding them pulped insects (Fremlin & Hendriks 2013). This may reflect cannibalism in the wild.

Adults of some genera in the Australasian region have well developed wings and are good dispersers, particularly *Aegus* (Christmas Island: Gahan 1888), and *Lamprima*, which appears to have dispersed to remote islands off the east Australian coast where it has speciated (Reid *et al.* 2018b). Individual male lucanids may disperse further and more often than females; in the case of *Lucanus cervus* males flew frequently and up to 2km, females flew once, up to 700m (Rink & Sinsch 2007).

Behaviour. Adult male Lucanidae are generally notable for their elongated mandibles, hence 'stag beetles', however this may be a derived feature within the family, with at least two separate origins (Hosoya & Araya 2005; Kim & Farrell 2015). Enlarged mandibles are used to defend spatially restricted critical resources (Emlen 2008), which for lucanid males may be feeding sites (sap flows), females, or oviposition sites (logs). Emlen argues that enlarged male lucanid mandibles are particularly associated with controlling access to sap flows. The evidence is more complicated. At feeding sites, larger males win intraspecific male:male combats (Fearn 1996, 2015, 2016) but early male occupiers are usually unchallenged, even if small (Inoue & Hasegawa 2013). Interspecific male:male combat can be induced in the laboratory, where the strength and type of the grip is more important than the size of the mandibles for victory (Hongo & Okamoto 2013). Male mandibles are also used in copulation, as females are often resistant to male attentions (Okada & Hasegawa 2005). Male to male combat is often prolonged and involves tipping over or disabling rivals (Goyens *et al.* 2015), and the grip may be strong, to the extent of drawing blood on human fingers (*pers. obs.* with *Lucanus cervus* in the UK). The enlarged muscles for greater grip require lateral expansion of the head (Goyens *et al.* 2014), typical of males in most sexually dimorphic species. Larger males have greater success in intraspecific male: male combat (Lagarde *et al.* 2005) but surprisingly have lower success in copulation as small males fight resistant females more actively (Okada & Hasegawa 2005). If the latter is true of lucanids generally, it partly accounts for the typical tendency towards male dimorphism in most species (Okada *et al.* 2008). Species with several male morphs may have several different mating strategies (Matsumoto & Knell 2017). There also appears to be a longevity cost to large males due to higher predation rates (Chiari *et al.* 2014). Single-species studies have shown mandible development is due to a combination of juvenile hormone production stimulated by larval nutrition (Gotoh *et al.* 2011), which correlates with size of the adult, and inherited allometry (Gotoh *et al.* 2012). Mandible size and shape is controlled by a complex of the genes associated with development of arthropod appendages (Gotoh *et al.* 2017). There are natural limits to mandible size (Knell *et al.* 2004), which may be due to the mechanical limitations imposed on other structures, such as the carrying capacity of wings (Kawano 1997). However, the effect of allometry is useful for dealers who rear the beetles on artifical diets to produce large males, as proportionally large mandibles sell for a premium. Some genera include species without obvious sexual dimorphism, for example *Figulus*, and there is evidence that these are subsocial and therefore cooperative rather than combative, not requiring large male mandibles (Mori & Chiba 2009). Small male mandibles and subsocial behaviour seem to be linked, as phylogenetic analysis suggests *Figulus* and related genera had ancestors with large male mandibles (Hosoya & Araya 2005).

Morphologically distinct sister species may be similar when allopatric but easily distinguishable when sympatric (Kawano 2003). It has been suggested that this avoids competition and copulation in interspecifics. The potential for interspecific combat is also reduced by sympatric species having different activity periods (Hongo 2014).

Stag beetles as pests. Lucanids are rare timber pests, causing significant damage only when fungal infection has already occurred in structural timber, as recorded for *Syndesus cornutus* in a wooden bridge, Canberra (Lawrence 1981a), or *Lamprima* in telegraph poles in Tasmania (Fearn 1996). Introduced species of Lucanidae may become environmental pests, although their effects have never been quantified. No exotic species have established in Australia (but see below under *Dorcus*), but at least two Australian species have established in New Zealand: *Ryssonotus nebulosus* and *Syndesus cornutus* where they are common in suburban habitat, mostly breeding in non-native trees (Holloway 2007). However, in Japan so many exotic species have been released in the wild that there are problems with (i) competitive exclusion of natives by exotics; (ii) hybridisation; and (iii) the introduction of associated parasites, particularly mites (Goka *et al.* 2004). The invasion of the Seychelles Islands by *Aegus chelifer* is claimed to have resulted in loss of the endemic *Figulus* species (Carpanato *et al.* 2010).

Stag beetles as food. Australian Lucanidae are relatively small and rare and not recorded as human food. Elsewhere, the only record of human consumption is of several large species of *Sphaenognathus* in Colombia, South America (Salazar-E *et al.* 2010).

Systematics, taxonomy & biogeography. The Lucanidae are scarabaeoids but their position in this superfamily is unclear. Lucanidae may be most closely related to Diphyllostomatidae and Hybosoridae (molecular evidence: Kim & Farrell 2015), or Diphyllostomatidae, Trogidae and Glaresidae (molecular evidence: Smith *et al.* 2006), or Diphyllostomatidae, Passalidae and Trogidae (morphological evidence: Browne & Scholtz 1999), or Passalidae, Diphyllostomatidae and Pleocomidae (morphological evidence: Lawrence *et al.* 2011). Lucanidae are generally considered monophyletic, although there is debate about the position of Diphyllostomatidae. Molecular analysis suggests that the small North American genus *Diphyllostoma*, split off from Lucanidae (Holloway 1972), should be reunited (Kim & Farrell 2015), but the latter authors did not include a large range of outgroups in their study. Mizunuma & Nagai (1994) included *Diphyllostoma* in Lucanidae without comment. The family Lucanidae is obviously old, with several genera or tribes distributed in Australasia and South America (Moore 1978; Reid 1999; Holloway 2007; Paulsen 2010; Reid & Beatson 2016) and several Mesozoic fossils (Bai *et al.* 2012; Kim & Farrell 2015). The classification of subfamilies remains controversial (Howden & Lawrence 1974; Hosoya & Araya 2005; Holloway 2007; Paulsen 2013; Kim & Farrell 2015), with the two last, molecular, studies providing completely conflicting relationships (Paulsen 2013; Kim & Farrell 2015). Monophyly of Aesalinae seems well established, with inclusion of former subfamilies Nicaginae and Ceratognathinae (the latter used for the Australian aesalines in Volume 1 of this series: Lawrence & Ślipiński 2013). There are four distinct generic groups in Australia, in subfamilies Aesalinae, Lampriminae, Lucaninae and Syndesinae. Most worldwide and Australian generic diversity is in Lucaninae, which is divided into numerous tribes, many without clear distinguishing features or evidence of monophyly and which add nothing useful to the classification of Lucanidae (Holloway 2007). Conservatively, there is evidence from morphological and molecular analyses for the following monophyletic groups which can be treated as tribes: Casignetini, Chiasognathini, Lucanini (including Cladognathini, Dorcini, Figulini & Penichrolucanini) and Platycerini (including Platyceroidini) (Bouchard *et al.* 2011; Huang & Chen 2013; Kim & Farrell 2015). It should be noted that several nomenclaturally unavailable names are used by Kim & Farrell, which merely adds to the confusion. Another tribe Sclerostomini is supported by some authors but has not been morphologically defined or generically circumscribed and some genera have also been treated as Platycerini, based largely on the larval stridulatory organ (Lawrence 1981b; Richards & Spencer 2014). If valid, Sclerostomini may include Australian genera such as *Lissotes* and *Lissapterus* which are treated here under Lucanini.

Three tribes of Lucaninae are present in Australia: Casignetini, Chiasognathini and Lucanini. Casignetini are represented by a single genus. There are 3 genera of Australian Chiasognathini. In Australian Lucanini, there are several distinct and apparently unrelated generic groups, which are placed in different tribes by some authors: (i) *Aegus*; (ii) *Dorcus*, *Dorculus*, *Prospocoilus* & *Pseudodorcus*; (iii) *Hoplogonus*, *Lissotes*; (iv) *Figulus*; (v) *Lissapterus*. The genera of the second group are poorly distinguished, and Arrow (1943) thought they should all be placed in *Dorcus*. However, there is evidence from morphological and molecular studies for two different lineages in this group (Huang & Chen 2013; Kim & Farrell 2015). The genus most obviously absent from the Australian fauna is *Cyclommatus*, which is widespread from Asia to the Solomons (Mizunuma & Nagai 1994; Bartolozzi *et al.* 2011; Huang & Chen 2013) and therefore might be expected in the northern tropics of Australia.

Species in Lucanidae have traditionally been delimited from male secondary sexual characters. Male genitalia are useful (Kawano 2004; Holloway 2007) and now widely used (Paulsen 2010), but there has been little use of female genitalia although these may be equally useful (Holloway 2007). Clades in Australia have not been dated, so it is difficult to determine the ages of species. In an old, long-isolated, clade, speciation of South African *Colophon* occurred 14–5.5 MY (Switala *et al.* 2014). On the other hand, island speciation in Taiwan has been dated to 1.5 MY for *Lucanus* (Huang & Lin 2010) or 1 MY for *Neolucanus* (Tsai *et al.* 2014). Speciation is ongoing in the *Neolucanus* of Taiwan, where populations of three 'species' have been geographically but not genetically isolated by Pleistocene glaciations, leading to a call for more conservative taxonomy in the genus, i.e. lumping (Tsai *et al.* 2014). Tasmania, with its numerous morphologically similar endemic species of *Lissotes*, has had a similar Pleistocene history to Taiwan and

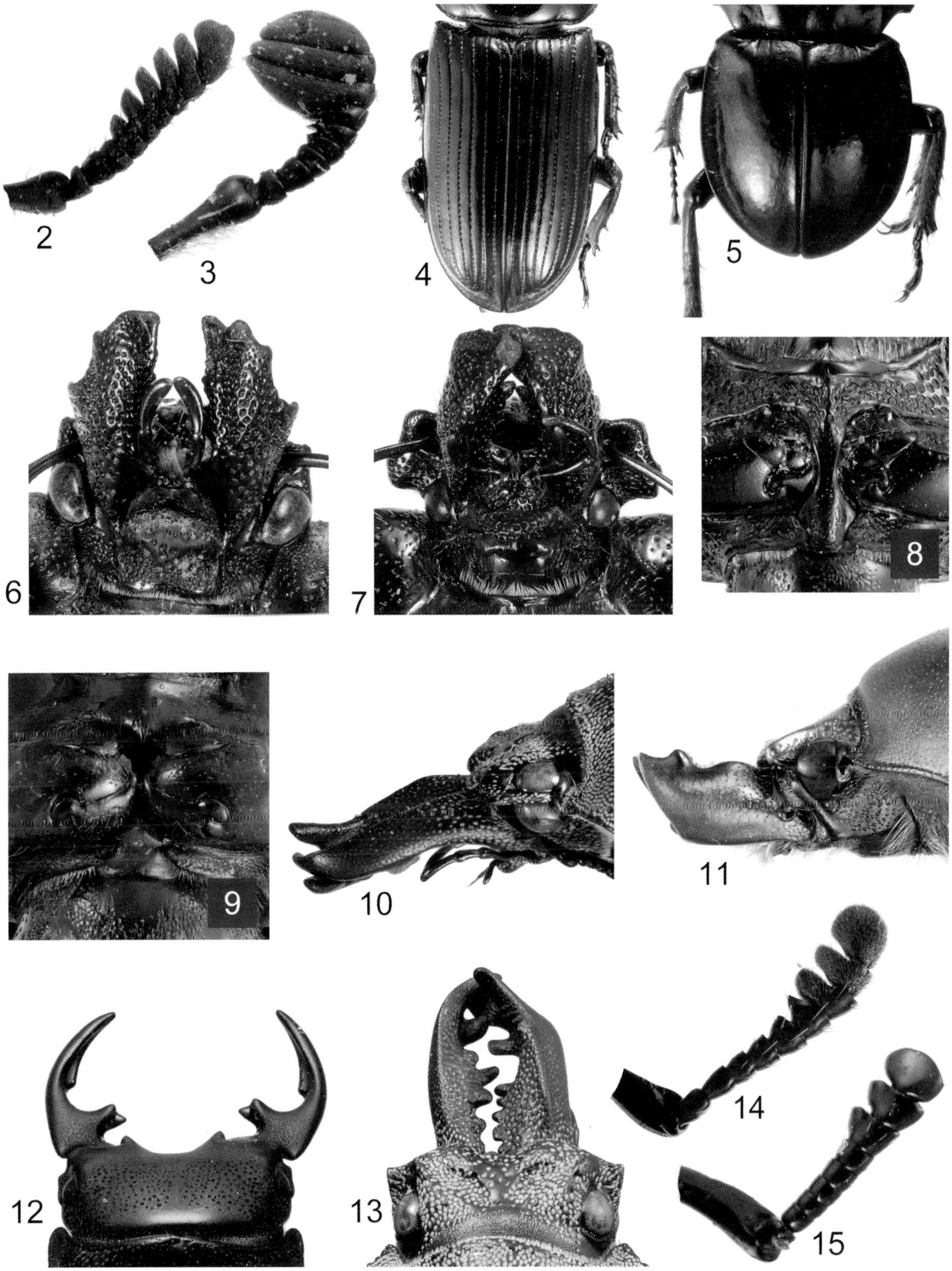

Figs 26.2– 26.15. 2, antenna, *Safrina moorei* Reid & Beatson; 3, antenna, *Lamprima aurata* Latreille; 4, *Figulus nitens* Waterhouse, elytra; 5, *Safrina moorei* Reid & Beatson, elytra; 6, ventral mouthparts, *Ryssonotus nebulosus* (Kirby); 7, ventral mouthparts, *Safrina moorei* Reid & Beatson; 8, prothoracic venter, *Figulus regularis* Waterhouse; 9, prothoracic venter, *Safrina moorei* Reid & Beatson; 10, lateral of head, *Cacostomus squamosus* Newman; 11, lateral of head, *Lamprima aenea* (Fabricius); 12, head, *Aegus jansoni* Boileau; 13, head, *Cacostomus squamosus* Newman; 14, antenna, *Aegus jansoni* Boileau; 15, antenna, *Figulus regularis* Waterhouse.

Figs 26.16– 26.28. 16, base of elytra, *Lissapterus pelorides* (Westwood); 17, base of elytra, *Figulus nitens* Waterhouse; 18, lateral, *Lamprima aurata* Latreille; 19, lateral, *Hoplogonus simpsoni*, Parry; protibia, *Lamprima aenea* (Fabricius); 21, protibia, *Homolamprima crenulata* Macleay, 22, head, lateral, *Phalacrognathus muelleri* Macleay; 23, head, lateral, *Lamprima aurata* Latreille; 24, ventral, *Ceratognathus niger* Westwood; 25, *Lissotes pelorides* (Westwood); 26, head, *Ceratognathus macrognathus* Boileau; 27, head, *Mitophyllus ocularis* (Carter); 28, lateral base of elytra, *Lissotes pelorides* (Westwood).

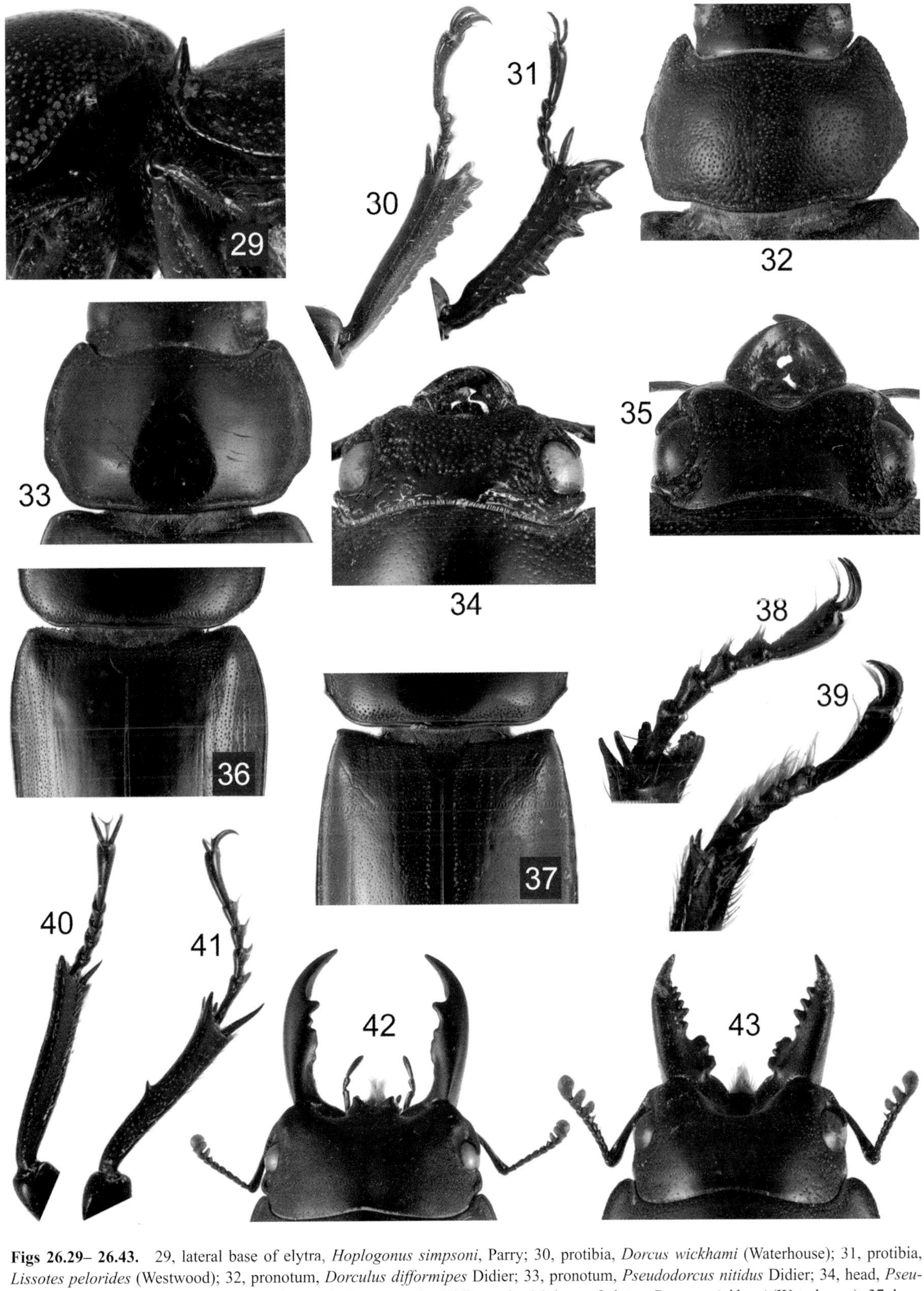

Figs 26.29– 26.43. 29, lateral base of elytra, *Hoplogonus simpsoni*, Parry; 30, protibia, *Dorcus wickhami* (Waterhouse); 31, protibia, *Lissotes pelorides* (Westwood); 32, pronotum, *Dorculus difformipes* Didier; 33, pronotum, *Pseudodorcus nitidus* Didier; 34, head, *Pseudodorcus nitidus* Didier female; 35, head, *Pseudodorcus nitidus* Didier male; 36, base of elytra, *Dorcus wickhami* (Waterhouse); 37, base of elytra, *Prosopocoilus torresensis* Deyrolle; 38, metatarsi, *Pseudodorcus nitidus* Didier; 39, metatarsi, *Dorcus wickhami* (Waterhouse); 40, metatibia, *Prosopocoilus torresensis* Deyrolle; 41, metatibia, *Pseudodorcus nitidus* Didier; 42, head, *Dorcus wickhami* (Waterhouse); 43, head, *Prosopocoilus torresensis* Deyrolle.

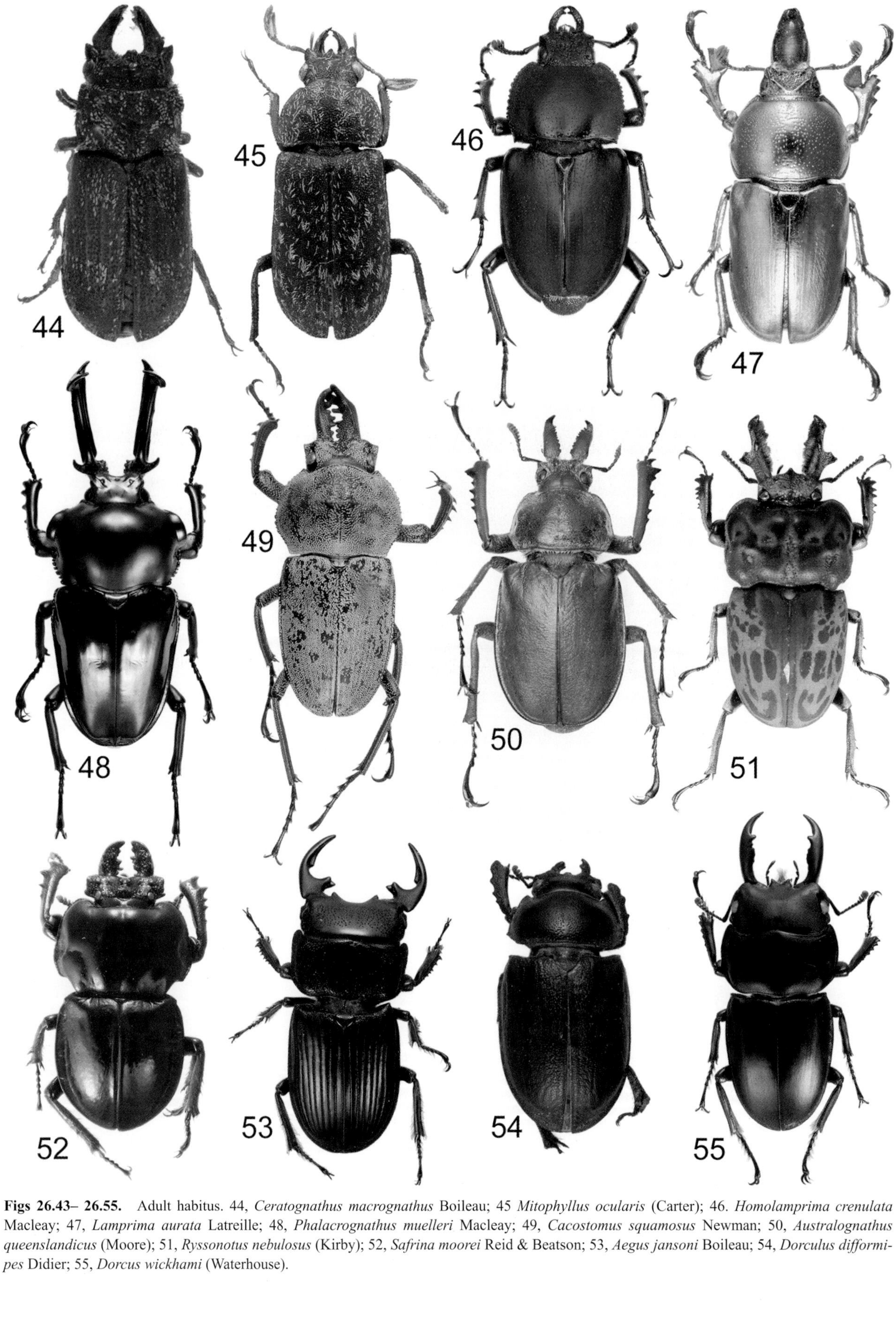

Figs 26.43– 26.55. Adult habitus. 44, *Ceratognathus macrognathus* Boileau; 45 *Mitophyllus ocularis* (Carter); 46. *Homolamprima crenulata* Macleay; 47, *Lamprima aurata* Latreille; 48, *Phalacrognathus muelleri* Macleay; 49, *Cacostomus squamosus* Newman; 50, *Australognathus queenslandicus* (Moore); 51, *Ryssonotus nebulosus* (Kirby); 52, *Safrina moorei* Reid & Beatson; 53, *Aegus jansoni* Boileau; 54, *Dorculus difformipes* Didier; 55, *Dorcus wickhami* (Waterhouse).

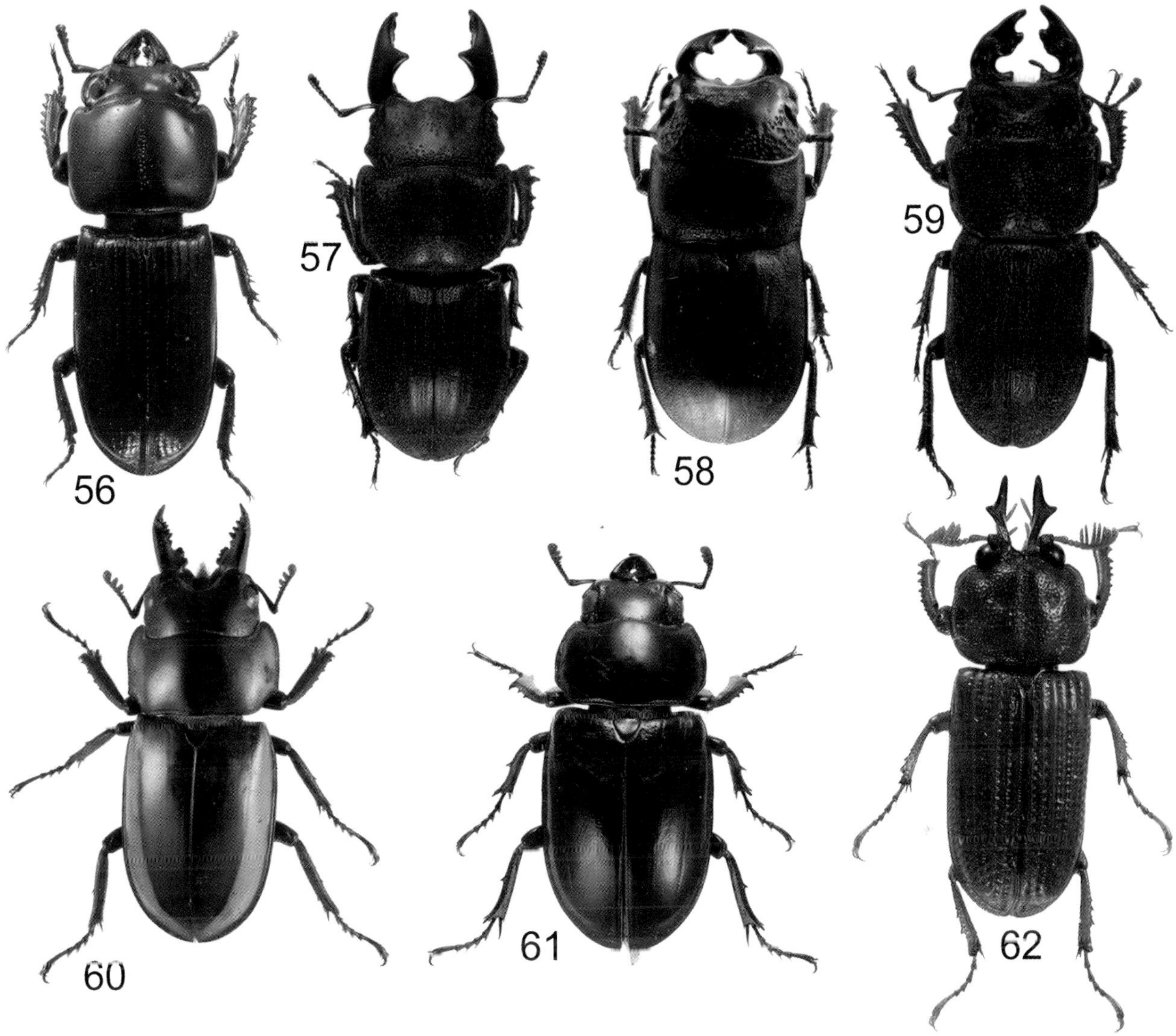

Figs 26.56– 26.62. Adult habitus: 56, *Figulus nitens* Waterhouse; 57, *Hoplogonus simpsoni* Parry; 58, *Lissapterus pelorides* (Westwood); 59, *Lissotes darlingtoni* Benesh; 60, *Prosopocoilus torresensis* (Deyrolle); 61, *Pseudodorcus nitidus* Didier; 62, *Syndesus cornutus* (Fabricius).

molecular studies may elucidate similar problems of species differentiation. In the northern tropics of Australia, a few supposed Australian endemics may be synonymous with New Guinean species. Considerable taxonomic work is still required on the Australian lucanid fauna.

Characteristics (Australian taxa only). Medium-sized to very large beetles, length: 4–75 mm. Elongate-ovate and flattened to cylindrical; usually dull brown to black, but shining metallic in subfamily Lampriminae and dark metallic green in some Lucaninae; dorsal vestiture absent, or with short stiff setae, or with scales; head prominent, usually with relatively small lateral eyes (not *Syndesus*), which are often divided by a genal lobe; labrum fused to clypeus, with suture usually obscure, and this part of head often hidden from above by projection of frontoclypeus; mandibles large and anteriorly directed, usually larger in male; antennal insertions hidden from above by expanded genal area of head; antennae 9- or 10-segmented, short, less than 1/3 body length, often geniculate, scape longer than antennomeres 2 and 3; antennal terminal club of 3–7 segments, which may be well separated or closely apposed, usually with densely setose lateral lobes or concave areas; mentum not bilobed (except *Syndesus*); prothorax without horns (median tubercle in *Syndesus*); procoxal cavities closed by hypomeral lobes reaching apex of prosternal process; scutellum visible (long and narrow to apparently absent in *Figulus*); anterior tibiae fossorial, relatively expanded, with lateral teeth; first abdominal ventrite complete, not divided by metacoxae; apical abdominal tergite without stridulatory area. Larva: C-shaped; antennae 3-segmented; all legs well developed; stridulatory mechanisms (pars stridens) on mesocoxa and metatrochanter; transverse folds on abdominal terga absent; apex of abdomen with two oval smooth areas either side of anus.

Key to the Australian genera of Lucanidae

Adults

1. Antennal club with 5–7 segments, each at least partly densely setose and/or with sensory surface (Fig. 26.2) **2**
– Antennal club with 3–4 segments, if 4, only last 3 with densely sensory surfaces (7th antennomere always smooth, without sensory surface) (Fig. 26.3) **5**
2(1). Elytra non-striate (Fig. 26.5); 5–6 antennal club segments (Fig. 26.2); labium broad **3**
– Elytra strongly striate (Fig. 26.4); 7 antennal club segments; labium narrow (length 7–15 mm) ..... ***Syndesus* Macleay**
3(2). Antennal club 5 segmented; mentum membranous & smooth (Fig. 26.6) (length 15–33 mm) ..... ***Ryssonotus* Macleay**
– Antennal club 6 segmented (Fig. 26.2); mentum well sclerotised, punctate (Fig. 26.7) **4**
4(3). Prosternal process visible between procoxae (Fig. 26.8); profemur without anterior ridge; pregula not abruptly raised and thickened; male head without dorsal tubercles (length 33–48 mm) ..... ***Australognathus* Chalumeau & Brochier**
– Prosternal process hidden at middle by procoxae (Fig. 26.9); profemur with anterior ridge; pregula abruptly raised & thickened; male head with dorsal tubercles (length 15–28 mm) ..... ***Safrina* Reid & Beatson**
5(1). Eyes split by extension of genal lobe, with separate dorsal and ventral parts (Fig. 26.10) **6**
– Eyes not entirely split, showing some continuity of ommatidia from dorsal to ventral surfaces (Fig. 26.11) **9**
6(5). Without dorsal scales (ie broadly flattened setae, usually recumbent, semirecumbent or recurved) (Fig. 26.12); mesoventral process wide at tip, mesocoxae broadly separated **7**
– Dorsal scales present, at least on sides of vertex [but may be rubbed off] (Fig. 26.13); mesoventral process tapered to a point or at least mesocoxae narrowly separated; (mentum semicircular; surface of last antennomere entirely sensory; length 7–25 mm) ..... ***Cacostomus* Newman**
7(6). Apical antennomere entirely sensory, without differentiated smooth and sensory areas (Fig. 26.14); antennomere 7 at least slightly cupuliform, shape unlike 6, with long apicolateral setae (tarsal empodium prominent, apex apparently bisetose, in two diverging clumps of 2–6 fused or apparently fused setae; elytra strongly striate; length 13–24 mm) ..... ***Aegus* Macleay**
– Apical antennomere smooth and shining except sensory area in hollowed or clearly differentiated apex (Fig. 26.15); antennomere 7 not cupuliform, shaped like 6, without long apicolateral setae **8**
8(7). Scutellum semicircular to almost equilaterally triangular (Fig. 26.16); tarsal empodium prominent, apex apparently bisetose, in two diverging clumps of 3–6 fused or apparently fused setae; bases of ventrites 2–5 not visibly grooved; elytra only partly striate (length 17–35 mm) ..... ***Lissapterus* Deyrolle**
– Scutellum strongly elongate-triangular to absent (Fig. 26.17); tarsal empodium hidden from below by apicoventral lobe of 5th tarsomere, apex with single seta; bases of ventrites 2–5 distinctly transversely grooved; elytra entirely & evenly striate (length 5–19 mm) ..... ***Figulus* Macleay**
9(5). Mesometaventrite junction abruptly anteriorly elevated (Fig. 26.18); (prosternal process narrowed to a thin ridge between procoxae, often hidden) **10**
– Mesometaventrite junction gradually raised or flat, not abruptly elevated (Fig. 26.19) **12**
10(9). Mesometaventrite junction anteriorly triangular; apex prosternal process flat; male fore tibia with broad spines (Fig. 26.20) **11**
– Mesometaventrite junction anteriorly bilobed; apex prosternal process elevated; male fore tibia with large narrow spines (Fig. 26.21) (spur simple; length 19 mm) ..... ***Homolamprima* Macleay**
11(10). Anterior angles of head without prominent genae, eyes anteriorly convex; male mandible with prominent basal dorsal tooth (Fig. 26.22), female mandible with three dorsal cusps; prosternal process narrow but visible between procoxae; male protibia with simple spur (as female); male meso– and metatibiae smooth and cylindrical, without spines; protibia with subsidiary teeth between large teeth on outer margin (length 23–75 mm) ..... ***Phalacrognathus* Macleay**
– Genae prominent anterior to eyes, eyes anteriorly concave; male mandible without basal dorsal tooth (Fig. 26.23), female mandible with only one dorsal cusp; prosternal process hidden between procoxae; protibia with spur expanded as a flat blade (except *L. imberbis*); male meso– and metatibiae with lateral spines; only large teeth present on protibial outer margin (length 15–37 mm) ..... ***Lamprima* Latreille**
12(9). Procoxae touching at midline, obscuring middle of prosternal process (Fig. 26.24); eyes entire, without genal lobe; elytral punctures ocellate, with raised centres (dorsal surface scaled; pronotal lateral margins slightly crenulated) **13**
– Procoxae separated, prosternal process entirely visible (Fig. 26.25); eyes with narrow genal lobe; elytral punctures simply concave **14**
13(12). Head with median tubercle or paired tubercles in both sexes, but often obscure in female (Fig. 26.26); anterior margin of head depressed at middle, truncate or with short lobe; basal half of outer edge of protibia with mixed smaller and larger teeth; inner margin of male metatibia excavate (except *C. macrognathus*) (length 8–14 mm) ..... ***Ceratognathus* Westwood.**
– Head without tubercles (Fig. 26.27); anterior margin of head flat with long median lobe between bases of mandibles; basal half of outer margin of protibia with uniformly small teeth; inner margin of male metatibia simple (length 4–8 mm) ..... ***Mitophyllus* Parry**
14(12). Base of upper margin of elytral epipleuron simple, at most with right-angled tubercle (Fig. 26.28) **15**
– Base of upper margin of elytral epipleuron with elongate erect spine (Fig. 26.29) (bases of ventrites grooved; short genal lobe present on eye; length 18–28 mm) ..... ***Hoplogonus* Parry**
15(14). Dorsally glabrous, or with minute stubble at sides of pronotum and elytra; usually larger, length 15–50mm; external apex protibiae extended as a broad multitoothed lobe (Fig. 26.30) (except *Pseudodorcus nitidus*) **16**
– Dorsally scaled & setose (but may be rubbed off); usually smaller, length 9–23mm; external apex protibiae obliquely produced with single tooth (Fig. 26.31) (anterior female head truncate; antennomere 7 not laterally expanded) ..... ***Lissotes* Westwood**

16(15). Diameters of punctures on pronotal disc much less than width of basal beading and punctures separated by 4–10 diameters, or punctures obliterated by rugose microsculpture (Fig. 26.33); female protibiae with > 5 teeth on outer edge; meso–and metatibiae thin, length of metatibia at least 4x apical width; apex of metatibia without smooth semicircular surface external to tarsal insertion, usually with three distinct teeth; (male mandibles usually longer than head) **17**

– Diameters of punctures on pronotal disc about equal to width of basal beading and punctures separated by 1–5 diameters (Fig. 26.32); protibiae with 5 or less teeth on outer edge in both sexes; meso– and metatibiae thickened, length of metatibia at most 3x apical width; apex of metatibia truncate with smooth semicircular surface external to tarsal insertion, bounded by two short teeth (female first metatarsomere elongate; male mandibles shorter than head; male head shining and strongly punctured; length 15–23 mm) .... ***Dorculus* Didier**

17(16). Inner edge of mandibles with single median dorsal tooth (Fig. 26.34) (left mandible may have pair of median teeth arising from one swelling, one dorsal, one ventral); mandibles as long as or shorter than head; temples shorter than half eye length; head never densely microreticulate [females] **18**

– Inner edge of mandibles with more than one tooth (Fig. 26.35) (one minor male with second tooth weakly expressed); mandibles usually longer than head; temples usually longer than eye; head usually densely microreticulate [males] **20**

18(17). Basal half of elytra without distinct stria near suture (Fig. 26.36); protibia shorter, with 0–2 subsidiary teeth between relatively crowded large teeth at apical half of outer edge **19**

– Basal half of elytra with distinct strongly punctured stria near suture, contrasting with fine background punctures (Fig. 26.37); protibia longer, with 2–4 small subsidiary teeth between sparsely placed larger teeth on apical half of outer edge (head broadest at canthus; canthus half eye length or less; elytral humeri angulate) .... ***Prosopocoilus* Hope**

19(18). Metatarsomere 1 elongate, much longer than 2 and almost as long as 5 (Fig. 26.38); apical metatibial spines longer, more than half apical width tibia; head usually broadest at temples, sometimes width at canthus equal to width at temples; canthus shorter, less than half eye length; elytral humeri evenly rounded .... ***Pseudodorcus* Parry**

– Metatarsomere 1 short, about as long as 2 and at most half length of 5 (Fig. 26.39); apical metatibial spines shorter, less than half apical width tibia; head usually broadest at canthus, sometimes width at temples equal to width at canthus; elytral humeri at least slightly angulate .... ***Dorcus* Macleay**

20(17). Dorsum of head densely microreticulate, obscuring or almost obscuring fine and sparse punctures; mandibles longer than head (one minor male only slightly so); temples at least as long as eyes; metatibial median spine small (much less than half mesotibial width) or absent (Fig. 26.40); **21**

– Dorsum of head shining, with strong punctures, close anteriorly, mandibles shorter than head; temples less than half eye length; metatibial median spine large and prominent, almost third or more mesotibial width (Fig. 26.41); (anterior of head shallowly arcuately concave; antennomeres 2–6 not elongate) .... ***Pseudodorcus* Parry**

21(20). Anterior of head almost flat, not deeply excavate (Fig. 26.42); temples usually laterally produced; mandibles without large internal basal tooth or lobe; prosternal process flat; antennomeres 2–6 elongate, 7 slightly expanded; .... ***Dorcus* Macleay**

– Anterior of head with a deep concave excavation (Fig. 26.43); temples not laterally produced; base of internal edge of mandibles with massive blunt lobe; prosternal process strongly elevated to truncate apex; antennomeres 2–6 quadrate to transverse, 7 strongly laterally expanded as thin lobe .... ***Prosopocoilus* Hope**

Larvae of Australian subfamilies

1. Mesocoxal pars stridens with one or more distinct rows of large dark granules, or a granular carina; apex of abdomen without a well defined median-dorsal smooth and glabrous swelling, with or without paired apical swellings (anal lobes) **2**

– Mesocoxal pars stridens without either rows of granules or a carina, the granules randomly scattered and often indistinct; apex of abdomen with a well defined median-dorsal smooth and glabrous swelling, also a pair of ovate smooth glabrous swellings below this (anal lobes) **3**

2(1). Apex of abdomen with contiguous anal lobes, which are finely setose; sternite X with raster confined to apex; left mandible without incisor teeth between 2–3 apical teeth and mola (pars stridens with a single row of larger granules; tarsungulus absent or minute) .... **Lampriminae**

– Apex of abdomen with pair of well separated smooth glabrous anal lobes; sternite X with raster occupying most of ventral surface; left mandible with at least 1 incisor tooth between apical teeth and mola (pars stridens often with multiple rows of larger granules; tarsungulus often well developed) .... **Lucaninae**

3(2). Each side of pronotum produced anteriorly into a sclerotised lobe; plectrum with 6–15 well separated carinae; tarsungulus strongly attenuated at apex, not curved; outer edges of anal lobes without adjacent setae .... **Syndesinae**

– Without lateral pronotal lobes; plectrum with > 30 dense carinae; tarsungulus simple, apex long and curved; outer edges of anal lobes with adjacent setae .... **Aesalinae**

Classification of the Australian genera

Subfamily Aesalinae Macleay, 1819

Characteristics. Dorsal surface with distinct scales; eyes entire, without genal lobe; antennae not geniculate, with 3-segmented club; middle of prosternal process hidden between procoxae; elytral punctures ocellate; profemora not anteriorly ridged; abdominal ventrites 2–5 laterally finely margined; bases of ventrites 2–5 transversely grooved (often hidden).

Australian genera. Two.

Ceratognathus Westwood, 1838 (Pl. 3E, 4C, 50A; Fig. 26.44)

Type species. *Ceratognathus niger* Westwood, 1838.

Characteristics. Length 8–16 mm; dorsal surface with irregular patches of narrow, acutely pointed, scales; male head anteriorly truncate or slightly produced; male genae approximately right-angled or rounded anteriorly; temples slightly angularly produced to straight in both sexes, pronotum distant from eyes; middle of male head with two small tubercles (raised shiny patches) or these fused into a single tubercle; female head usually distinctly narrowed, quadrate to transverse; eye entire, slightly transverse, with slight anterior excavation, without canthus; male mandibles shorter than to slightly longer than head length; pregula flat; labium broad ~1/3 head width; mentum solid, punctured, apex slightly convex; antennae 10-segmented, with 3-segmented club, flabellate in males, surface of apical segment entirely sensory; antennomere 7 not cupuliform but slightly asymmetric; prosternal process hidden between procoxae, apex of process flat; scutellum semicircular; mesoventral process tapered, junction with metaventrite not abruptly raised; elytral humeri not spined; elytra non-striate, punctures ocellate, with raised centres and surface between punctures smooth or finely granulose; profemur without anterior ridge; male protibia with simple apical spur; protibial outer edge with mixture of large and small teeth; male metatibia with setose excavation (except *C. macrognathus*); tarsal empodium prominent between claws, with paired bifurcating apical setae.

Australian species. 11.

Distribution. *Ceratognathus* species are found in near-coastal south-east Australia. The only species described from outside that region, *C. flabellatus* Boileau, is almost certainly based on a mislabelled specimen (Hangay & de Keyzer 2017).

Biology. *Ceratognathus* species occur in closed forest to open woodland. The following notes are from a study of *C. niger* Westwood in Victoria which includes a description of the larva and pupa (Alderson 1975b). *Ceratognathus niger* breeds in both *Acacia* and *Eucalyptus* logs, with preference for the first. Mature larvae may pupate in either spring or late summer. Adults of this species are nocturnal but other species may be diurnal as well as nocturnal (Moore 1984–1986; pers. obs.). Hangay & de Keyzer (2017) note that *Ceratoganthus* larvae are found in logs with white rot, and adults are short-lived and active fliers.

Note. Three species have been removed to *Mitophyllus*.

Mitophyllus Parry, 1843 (Pl. 76L; Fig. 26.45)

Type species. *Mitophyllus irroratus* Parry, 1843.

Characteristics. Length 4–8 mm; dorsal surface with irregular patches of narrow acutely pointed scales; male head with strong anteromedian lobe; male genae prominently anteriorly angled; no dorsal head tubercles; female head narrow, slightly elongate; male without temples, contracted behind eye, female head parallel-sided behind eye; eye entire, rounded or with slight posterior excavation, without canthus; male mandibles shorter than head length; pregula flat; labium broad ~1/3 head width; mentum solid, punctured, with slightly convex apical margin; antennae 10-segmented, with 3-segmented club, flabellate in males, surface of apical segment entirely sensory; antennomere 7 not cupuliform; prosternal process hidden between procoxae, apex of process flat; scutellum triangular; mesoventral process tapered, junction with metaventrite not abruptly raised; elytral humeri not spined; elytra non-striate, punctures ocellate, with raised centres and surface between punctures smooth or finely granulose; profemur without anterior ridge; male protibia with simple apical spur; protibial outer edge with two large triangular teeth, at apex and middle, and small even-sized teeth between these and to base; male metatibia without setose excavation; tarsal empodium prominent between claws, with paired bifurcating apical setae.

Australian species. Three.

Distribution. South-west and south-east near-coastal Australia, including Tasmania, and New Zealand.

Biology. Hangay & de Keyzer (2017) have published the only biological information for this genus in Australia. *Mitophyllus minutus* (Lea) and *M. ocularis* (Carter) are both considered nocturnal, although only the latter is attracted to light. Adults are most easily found in or on host logs, from December to February. The larvae breed in rotting sapwood of eucalypt or acacia (only *M. minutus*) logs. The life cycle is approximately one year in artificial conditions. The larva of a New Zealand species has been described (Hudson 1934).

Notes. Two of the three species transferred here to *Mitophyllus* agree exactly with the diagnostic characters of New Zealand species (Holloway 1998, 2007), except for the incomplete ventral suture on the basal piece of the aedeagus and the narrower scales. Holloway regards the basal piece as an important diagnostic character for separation of *Mitophyllus* from the Australian genus *Ceratognathus* and the New Zealand genus *Holloceratognathus*, but her descriptions and drawings of two New Zealand species of *Mitophyllus* show that this suture can be reduced to an imperceptible trace (Holloway 2007: figs 596, 603) therefore it may not be a reliable diagnostic feature. The South American genus *Hilophyllus* Paulsen & Mondaca, 2006, is similar to *Mitophyllus* but has short male mandibles, broad galeae, symmetrical parameres and basal piece without any groove or division. The Australian species have elongate-triangular galeae, as in the New Zealand species *M. foveolatus* (Broun, 1880). Both *Hilophyllus* and the Australian species have similar narrow scales which are centrally placed on the elytral punctures, which Paulsen & Mondaca treat as diagnostic for *Hilophyllus*, but in at least three New Zealand species, including the type *M. irroratus* Parry, 1843, the elytral scales are variable in width and many are centred on the punctures. The partly dissected female *M. ocularis* I have seen has laterally hooked styli, like some New Zealand species of *Mitophyllus* (Holloway 2007: figs

70, 71). The third species, *M. abdominalis* Parry, is only known from the original description of a female (holotype illustrated by Hangay & de Keyzer 2017), which also fits the characters of *Mitophyllus*. *Mitophyllus abdominalis* is similar to *M. minutus* and it is possible that these are synonyms.

Subfamily Lampriminae Macleay, 1819

Characteristics. Without dorsal scales; eyes undivided, without genal lobe; antennae not geniculate, with 3-segmented club; prosternal process narrowed to a thin ridge between procoxae, middle often hidden; metaventrite abruptly elevated anteriorly; profemora not anteriorly ridged; abdominal ventrites 2–5 finely margined or without margins; bases of ventrites 2–5 without transverse grooves.

Australian genera. Three.

Homolamprima Macleay, 1885 (Pl. 50B; Fig. 26.46)

Type species. *Hololamprima crenulata* Macleay, 1885.

Characteristics. Length 19 mm; dorsal surface without scales or visible setae; male without dorsal tubercles on head, anterior of head truncate, slightly medially produced; male genae obtuse-angled not produced; female head not narrowed compared with male; eyes almost entire, transverse, with shallow excavation of posterior margin; antennae not geniculate, 10-segmented, with 3-segmented club, the lobes of the club segments densely setose and capable of being closely appressed; antennomere 7 slightly asymmetric, 6 symmetrical; male mandibles about as long as head; labium broad, approximately one third width of head; mentum solid and punctate; pregula flat; prosternal process narrow but visible between procoxae, apex produced and elevated; elytra non-striate, with scattered simple shallow concave punctures; surface of interspaces smooth but dull from microsculpture; base of elytral epipleuron not spined; mesoventral process abruptly elevated, the anterior face deeply concave for retention of apex of prosternal process, slightly tapered to junction with metaventrite; profemur without anterior ridge; male protibia with simple apical spur; female protibia without subsidiary teeth between large teeth on outer edge; male metatibia without setose excavation; tarsal empodium prominent with paired divergent thin tufts of apical setae.

Australian species. *Hololamprima crenulata* Macleay.

Distribution. *Homolamprima crenulata* was originally described from Clarence River, which in the 1880s might have been anywhere within 50 km of Grafton, north-east New South Wales. Lea (1915) noted a specimen in the Carter collection from Dorrigo plateau, ~90 km south of Grafton.

Biology. Nothing is known of the biology of this species, but it is most likely to occur in rainforest. The probably closely related genus *Streptocerus* Fairmaire (South America) is associated with *Nothofagus* (Paulsen 2010).

Notes. This is an endemic monotypic genus, known from less than 5 specimens, all collected more than 100 years ago. *Homolamprima* is flat, dull surfaced and with relatively delicate appendages, compared with the other genera of Lampriminae. It may be most closely related to the South American genus *Streptocerus* (Paulsen 2010).

Lamprima Latreille, 1804 (Pl. 7C, 50G, 50K, 72H; Figs 26.1, 26.47)

Type species. *Lethrus aeneus* Fabricius, 1792.

Neolamprima Gestro, 1875

Characteristics. Length 13–38 mm; dorsal surface without scales or visible setae; male without dorsal tubercles on head, anterior of head concave or truncate; male and female genae not laterally and at most slightly anteriorly angled; female head not narrowed compared with male; temples short but slightly angulate, pronotum almost reaching eyes; eyes reniform, with shallow anterior and posterior excavations; antennae not geniculate, 10-segmented, with 3-segmented club, the club segments entirely densely setose and often closely appressed; antennomere 7 slightly cupuliform, with thin lateral lobe and 6 slightly asymmetric (male) or with thin lobe (female); male mandibles longer than head (only slightly so in *L. aurata* from semi-arid areas) and without basal dorsal tooth; female mandibles with one cusp; labium broad, approximately one third width of head; mentum solid and punctate, apex truncate; pregula flat; prosternal process hidden between procoxae, apex flat; scutellum almost equilateral triangular; elytra non-striate but may have irregular shallow grooves, with scattered simple shallow concave punctures, surface of interspaces smooth and shining to dull and wrinkled; elytral humeri not spined; mesoventral process almost parallel-sided to junction with metaventrite, abruptly elevated anterior to this; profemur without anterior ridge; male protibia with expanded blade– or fan-like spur (except *L. imberbis*, with simple spur); female protibia without subsidiary teeth between large teeth on outer edge; male meso– and metatibiae with lateral spines; male metatibia without setose excavation; tarsal empodium prominent with paired divergent thin tufts of apical setae.

Australian species. Four.

Distribution. *Lamprima aurata* Latreille is widespread in Australia, occupying almost the entire eastern edge of the continent from north Queensland to Tasmania and South Australia, at up to 400 km inland, and almost the entire southern edge, absent only from the

Nullarbor Plain (Reid *et al.* 2018b). The other mainland eastern species, *Lamprima imberbis* Carter, is only known from a single collecting event in New South Wales more than 85 years ago (Carter 1926). There are endemic species on the remote oceanic islands of Lord Howe (*L. insularis* Macleay) and Norfolk (*L. aenea* (Fabricius)). The genus also occurs in New Guinea (one species).

Biology. The following notes mostly concern the commonest species, *L. aurata* (Fearn 1996, 2015, 2016; Reid *et al.* 2018b). *Lamprima* adults are commonly diurnal and are often found on flowers, where they may mate. Adults feed on living plant material, in *L. aurata* usually *Eucalyptus* species, the males cutting off the tips of shoots and lapping up the exuded fluids. They may also eat flowers. However, female mandibles are non-functional and this sex relies on males to create sap flows. Feeding sites may be used for male–male aggression and for copulation. Fearn (1996) noticed a 3:1 male to female sex ratio in the field, which is supported by the strongly male-biased material in collections (pers. obs.). Oviposition is usually underground, the females tunnelling into soil around partly buried decaying wood with a great variety of host species (Fearn 1996). The larva of *L. aurata* has been described (Alderson 1975a: as *L. varians*). *Lamprima aurata* larvae in Tasmania are most commonly in decaying roots but prefer standing timber in north Queensland rainforest (Wood *et al.* 1996). *Lamprima insularis* larvae mostly inhabit fallen timber (Reid *et al.* 2018b). Evidently in wetter areas the larvae are better able to survive above ground. The larvae commonly bore upwards. Pupation is in a chamber, usually just beneath the wood surface but also in adjacent soil (Fearn 1996). In Tasmania the entire life cycle is at least 3 years. *Lamprima* larvae feed in a wide variety of hosts, including exotic species.

Notes. *Lamprima* species are relatively convex. Until recently the genus included numerous species in eastern Australia based on variations in colour, size and shape, but with identical genitalia. Male characters in these populations appear to show clinal variation according to latitude and longitude (Reid *et al.* 2018b). These mainland populations or varieties have all been placed under the name *L. aurata* (Reid et al. 2018b).

Phalacrognathus Macleay, 1885 (Pl. 50E; Fig. 26.48)

Type species. *Lamprima muelleri* Macleay, 1885.

Characteristics. Length 23–75 mm (Wood *et al.* 1996; Hangay & de Keyzer 2017); dorsal surface without scales or visible setae; male without dorsal tubercles on head, anterior of head shallowly concave; male and female genae not laterally or anteriorly angled; female head not narrowed compared with male; temples absent in both sexes, pronotum reaching eye; eyes almost entire, with convex anterior and shallow excavation of posterior margin; antennae not geniculate, 10-segmented, with 3-segmented club, the club segments entirely densely setose and often closely appressed; antennomere 7 not asymmetric (male) or slightly asymmetric (female); male mandibles longer than head, with prominent basal dorsal tooth; female mandibles with three dorsal cusps; labium broad, approximately one third width of head; mentum solid and punctate, slightly medially produced; pregula flat; prosternal process narrow but visible between procoxae, apex flat; scutellum almost equilateral triangular; elytra non-striate, with scattered punctures, which are sharp edged and enclose a small tubercle and seta; surface of interspaces smooth and shining; base of elytral epipleuron not spined; mesoventral process slightly tapered to junction with metaventrite, abruptly elevated anterior to junction; profemur without anterior ridge; male protibia with simple apical spur; female protibia with small subsidiary teeth between large teeth on outer edge; male meso– and metatibiae smooth and cylindrical, without lateral spines; male metatibia without setose excavation; tarsal empodium prominent with paired divergent thin tufts of apical setae.

Australian species. *Phalacrognathus muelleri*.

Distribution. *Phalacrognathus* is an endemic monotypic genus, confined to the wet tropical rainforests of north-east Queensland (Wood *et al.* 1996). Hyperparasites of the larvae include a tachinid fly and a scoliid wasp (illustrated in Wood *et al.* 1996).

Biology. The following notes are from the detailed study by Wood *et al.* (1996). Adults rarely come to light and they feed on flowers, sap flows and fruit, usually diurnally. Oviposition occurs throughout the year. Females create a tunnel in rotten wood of at least 24 species of hosts and lay up to 50 eggs, in the wood pulp. The white wood rot required is associated with at least 6 species of fungi. The larvae are described by Wood *et al.* (1996) and Molino-Olmedo (2014). The larvae occur in standing or fallen rotting wood and may travel several metres. There are 3 larval instars and pupation is in a cell. Development from egg to adult varies from 1–4 years and adults can live for up to 18 months. Males use their mandibles both for dislodging rivals and for moving females.

Subfamily Lucaninae Latreille, 1804

Characteristics. Eyes completely or partly divided by genal lobe; antennae geniculate, with 3– to 6-segmented club; ventrites 2–5 laterally variable in shape and basally with or without grooves.

Australian genera. 13.

Notes. The tribal classification of Lucaninae is poorly supported by phylogenetic analysis (Huang & Chen 2013;

Kim & Farrell 2015) and the tribes have never been carefully morphologically defined. Furthermore, many tribal names in current use are unavailable as they do not conform to the rules of nomenclature (International Commission on Zoological Nomenclature 1999; Bouchard *et al.* 2011). The Australian genera are arranged in their putative tribes below, for reference, but no attempt is made to provide diagnoses.

Tribe Casignetini Reid, 1999

Note. The tribe Casignetini is treated as valid here but it appears to be sister to Chiasognathini and could therefore be synonymised with the latter (Costa *et al.* 1988; Kim & Farrell 2015; Reid & Beatson 2016). The genera of Casignetini show classic late-Gondwanan distribution (Reid 1999). There is a single genus in Australia.

Cacostomus Newman, 1840 (Pl. 49J, 50C; Fig. 26.49)

Type species. *Cacostomus squamosus* Newman, 1840.

Eucarteria Lea, 1914

Lepidodes Westwood, 1841

Characteristics. Length 6–26 mm; at least dorsal surface of head with recumbent or adpressed scales (entire dorsum in the commonest species, *C. squamosus*); male head without dorsal tubercles, anterior margin concave or truncate; male genae strongly right-angled; female head not strongly narrowed compared with male; temples absent, head of both sexes excavate behind eye and pronotum almost reaching eye; eyes completely divided by posterior extension of genal lobe, dorsal part elongate-ovate to circular, slightly smaller than circular ventral part; antennae geniculate, 10-segmented, with 3-segmented club, the club segments almost entirely densely setose and often closely appressed; antennomere 7 slightly asymmetric, 6 symmetrical; male mandibles longer than head; labium broad, approximately one quarter width of head; mentum solid and punctate, semicircular in shape; pregula flat; prosternal process narrow between procoxae, but convex and visible, apex flat; scutellum semicircular; elytra non-striate, with scattered simple shallowly concave punctures, surface of interspaces shining and smooth or slightly wrinkled; elytral humeri not spined; mesoventral process flat, slightly to strongly tapered to junction with metaventrite; profemur without anterior ridge; male protibia with simple apical spur; female protibia without subsidiary teeth between large teeth on outer edge; male metatibia without setose excavation; tarsal empodium prominent, with paired divergent thin tufts of apical setae; bases of abdominal ventrites 2–5 not grooved; sides of abdominal ventrites finely margined (*C. squamosus*) or with the edges slightly laterally flanged.

Australian species. Three.

Distribution. *Cacostomus* is endemic to Australia. *Cacostomus floralis* (Lea) and *C. subvittatus* (Moore) are endemic to isolated uplands in central eastern New South Wales, but *C. squamosus* Newman is widespread from southern New South Wales to northern Queensland.

Biology. Adults of two species, *C. floralis* and *C. subvittatus* are diurnal and occasionally found in abundance on foliage (Williams 2002) or grass tussocks, where they copulate (Krake 1998), but the third species is most commonly collected at light. Larvae of *C. squamosus* are associated with rotting log 'detritus' and do not appear to be associated with a particular fungus (Wood *et al.* 1996). Larvae of the Australian species have not been described but are likely to be similar to *Casignetus* (Costa *et al.* 1988).

Note. *Cacostomus* includes two species formerly placed in *Eucarteria* (Reid 1999).

Tribe Chiasognathini Burmeister, 1847

Note. In the strictest sense, this Gondwanan tribe includes two genera from South America and the three listed below from Australia (Reid & Beatson 2016). In the broadest sense suggested by Kim and Farrell (2015) it might include Casignetini and the South African genus *Colophon* Gray.

Australognathus Chalumeau & Brochier, 1993 (Pl. 49L; Fig. 26.50)

Type species. *Sphaenognathus queenslandicus Moore*, 1978.

Characteristics. Length 33–48 mm; dorsal surface without scales, but head, pronotum and margins of elytra with long setae; male head with a pair of dorsal tubercles, anterior of head almost truncate but with median & lateral emarginations; male genae right-angled; female head not narrowed compared with male; eyes completely divided by posterior extension of genal lobe, dorsal part smaller, circular, ventral part ovate; temples absent, head excavate behind posterior margin of genal lobe but pronotum distant from eyes; antennae geniculate, 10-segmented, with 6-segmented club, the club segment lobes short, densely setose and not closely adpressed (club pectinate); male mandibles longer than head; labium broad, 1/3–1/4 of head width; mentum solid and punctured, with concave apex; pregula sinuate in profile, evenly convex at base, with transverse depression before elevated junction with mentum, pair of small pits at base; prosternal process thin but visible and slightly convex between procoxae, apex flat; scutellum semicircular; elytra non-striate, with dense minute shallow punctures, surface of interspaces microreticulate and dull, slightly wrinkled; elytral humeri not spined; mesoventral proces strongly convexly elevated at base, almost parallel-sided to junction with metaventrite; profemur without anterior ridge on apical half; male protibia with simple apical spur; female protibia without subsidiary teeth between large teeth on outer edge; male metatibia without setose excavation; tarsal empodium prominent with paired divergent thin tufts of 3–4 apical setae; bases of abdominal ventrites 2–5 not grooved; sides of ventrites 2–5 without margins.

Australian species. Two, *Australognathus queenslandicus* (Moore), *A, munchowae* (Moore & Monteith).

Distribution. The two species of this endemic Australian genus are confined to isolated upland plateaux in northern and central Queensland.

Biology. The following notes are from Moore & Monteith (2004), who described the biology of *A. munchowae*. Larvae occur under deeply embedded rotten logs of the genera *Casuarina* and *Eucalyptus*. The larvae do not penetrate deeply into the logs but create a cell in the soil. Several generations occur together, as larval development is slow, up to 4 years at least. Larvae stridulate but the sound is not audible to humans. Pupation is also in soil. Adults of *A. queenslandicus* are attracted to light.

Notes. The first discovered Australian species, *A. queenslandicus*, was placed in the hitherto South American genus *Sphaenognathus* (Moore 1978), which it strongly resembled. Chalumeau & Brochier (1993) erected *Australognathus* for this species based largely on differences in the male mandibles from *Sphaenognathus* and *Chauliognathus*, another South American genus. Their new genus was downgraded to a subgenus in the description of the second species, *A. munchowae* (Moore & Monteith 2004). Morphologically *Australognathus* and *Sphaenognathus* are almost identical. However phylogenetic analyses of Lucanidae, particularly Chiasognathini, support recognition of *Australognathus* as a valid genus and also suggest that it has been isolated from its sister genera for almost 50 million years, a remarkable example of morphological stasis (Kim & Farrell 2015). The differing biologies of *Australognathus* and *Sphaenognathus* also support their generic separation (Moore & Monteith 2004).

Ryssonotus Macleay, 1819 (Pl. 4E–G, 49K; Fig. 26.51)

Type species. *Lucanus nebulosus* Kirby, 1818.

Characteristics. Length 15–33 mm; dorsal surface without scales, but elytra with minute erect clavate setae (visible ′60); male head with dorsal tubercles, anterior of head concave or truncate; male genae right-angled; female head not narrowed compared with male; temples absent in both sexes, head deeply excavate behind eyes and pronotum almost touching eyes; eyes completely divided by posterior extension of genal lobe, dorsal part smaller, ovate, ventral part transverse; antennae geniculate, 10-segmented, with 5-segmented club, the club segment lobes short, densely setose and not closely adpressed (club pectinate); male mandibles longer than head; labium broad, approximately one third head width; mentum membranous and impunctate, with bilobed apex; pregula not elevated, with pair of broad depressions at base; prosternal process thin but visible and strongly convex between procoxae, apex flat; scutellum semicircular; elytra non-striate, with scattered minute shallow punctures, surface of interspaces microgranulate and dull, not wrinkled; elytral humeri not spined; mesoventral proces abruptly elevated from base, almost parallel-sided to junction with metaventrite; profemur with anterior ridge on apical half; male protibia with simple but robust apical spur; female protibia without subsidiary teeth between large teeth on outer edge; male metatibia without setose excavation; tarsal empodium prominent with paired divergent thin tufts of 1–2 apical setae; bases of abdominal ventrites 2–5 not grooved; sides of ventrites 2–5 without margins.

Australian species. Only *Ryssonotus nebulosus* Kirby.

Distribution. *Ryssonotus nebulosus* is widespread in eastern Australia, up to 400 km inland (Reid & Beatson 2016), and has been accidentally introduced to New Zealand (Holloway 2007).

Biology. Adults are nocturnal and commonly attracted to lights, in the suburbs of eastern Australian towns. They breed in a variety of hosts, including exotic species in gardens (Holloway 2007; pers. obs.). Larvae of *R. nebulosus* have been collected in the cores of rotting logs associated with white rot in north Queensland rainforest (Wood *et al.* 1996). The larva was partially described by Lawrence (1981b) and Reid & Beatson (2016).

Notes. The genus is endemic to Australia and monotypic. Holloway (2007) described the antennal club as 6-segmented, but only antennomeres 6–10 have densely setose lateral lobes, whereas antennomere 5 lacks a lateral lobe and has scattered setae. Holloway thought *Ryssonotus* was closely related to *Aegus*, but it is much closer to genera of Chiasognathini (Kim & Farrell 2015) and has been placed in that tribe (Reid & Beatson 2016).

Safrina Reid & Beatson 2016 (Pl. 76A; Fig. 26.52)

Type species. *Rhyssonotus laticeps* Macleay, 1885.

Characteristics. Length 15–28 mm; dorsal surface without scales, elytra with minute erect pointed setae and often longer simple setae at apex; male head with dorsal tubercles, anterior of head truncate; male genae rounded, right-angled or laterally produced, genae in both sexes considerable laterally expanded; female head often narrowed compared with male; temples absent in both sexes, head deeply excavate behind eyes and pronotum almost touching eyes; eyes completely divided by posterior extension of genal lobe, both ovate, dorsal part slightly smaller; antennae geniculate, 10-segmented, with 6-segmented club, the club segment lobes short, densely setose and not closely adpressed (club pectinate); male mandibles about as long as or slightly longer than head; labium narrow, 1/5–1/7 head width; mentum solid, punctured, with convex apex; pregula strongly elevated as a transverse ridge; prosternal process thin, not elevated, middle hidden by procoxae, apex flat; scutellum transverse with truncate apex to semiovate; elytra non-striate, with scattered shallow punctures, punctures sometimes in irregular lines between elongate ridges, surface of interspaces microreticulate and dull or smooth and shining, not wrinkled; elytral humeri not spined; mesoventral process abruptly elevated at base, strongly narrowed to ridge-like junction with metaventrite; profemur with anterior ridge on apical half; male protibia with simple apical spur; female protibia without subsidiary teeth between large teeth on outer edge; male metatibia without setose excavation; tarsal empodium prominent with paired divergent thin tufts of 1–2 apical setae; bases of abdominal ventrites 2–5 not grooved; sides of ventrites 2–5 without margins.

Australian species. Eight.

Distribution. *Safrina* is endemic to Australia and is confined to the eastern seaboard from south central Queensland to south-west Victoria, and there is little overlap in species' distributions. *Safrina* species are montane or at least upland in distribution.

Biology. Adults are nocturnal but have not been recorded in flight or at light. Most specimens have been collected under logs or in pitfall traps, or have been reared from larvae. Hosts of *Safrina* species include *Nothofagus* and *Eucalyptus* infected with brown rot. Preferred habitat varies according to species, from eucalypt woodland (*S. laticeps* (Macleay)) to rainforest (*S. polita* (Carter))

Notes. *Safrina* was recently named for a group of species formerly placed in *Ryssonotus*. There is also molecular support for the separation of these two genera (Kim & Farrell 2015).

Tribe Lucanini Latreille, 1804 *sensu lato*

Note. The majority of former Lucaninae tribes are placed under this name (Bouchard *et al.* 2011). In Australia this includes genera placed in Cladognathini, Dorcini, Figulini and Sclerostomini, which probably represent separate generic lineages (Kim & Farrell 2015).

Aegus Macleay, 1819 (Pl. 50J; Fig. 26.53)

Type species. *Aegus chelifer* Macleay, 1819.

Paraegus Gahan, 1888

Characteristics. Length 13–30 mm; dorsal surface without scales, with minute erect setae more visible at sides; male head without dorsal tubercles, flat, anterior concave to almost truncate; male genae rounded to almost right-angled; female head not greatly narrowed compared with male; temples straight to strongly laterally produced in male, in female not or slightly produced, pronotum not reaching eyes; eyes completely divided by posterior extension of gena, upper part smaller, elongate-ovate, lower more rounded; antennae geniculate, 10-segmented, with 3-segmented club, club segments entirely densely setose and often closely appressed; antennomere 7 strongly asymmetric with truncate apex and 6 slightly asymmetric (male) or antennomere 7 with thin tapered lobe and 6 strongly asymmetric (female); male mandibles longer than head; labium broad, approximately half to one third head width; mentum solid and punctate, apex slightly excavate; pregula flat except raised junction with mentum; prosternal process flat and relatively broad between procoxae, apex flat; scutellum semicircular to almost equilateral triangular; elytra striate, striae consisting of often coalescent shallowly concave punctures in grooves, intervals with dense concave punctures and smooth and shining or finely microsculptured dull interspaces; humeri not spined; mesoventral process slightly raised from base, tapering to junction with metaventrite; profemur with complete anterior ridge; male protibia with simple apical spur; female protibia with 0–2 subsidiary teeth between large external teeth; male metatibia without setose excavation; tarsal empodium prominent, with divergent thin paired tufts of apical setae; bases of abdominal ventrites 2–5 with transverse grooves which are usually hidden; sides of ventrites 2–5 strongly laterally lobed.

Australian species. Two: *Aegus jansoni* Boileau, *A. listeri* (Gahan).

Distribution. *Aegus* is a widespread and evidently dispersive genus in South-east Asia, Australasia and the west Pacific, including Vanuatu, Fiji and Samoa (Boucher 1991; Bomans 1993). In Australia, there is an endemic species on Christmas Island (Gahan 1888), and an endemic species in north Queensland (Moore & Cassis 1992).

Biology. Larvae of *A. jansoni* are associated with powdery brown rot fungal infection of rotting logs (Wood *et al.* 1996). Surprisingly, the larva of *Aegus* is undescribed.

Notes. *Aegus* species are distinguished from other lucanids by the dull, flat, body combined with striate elytra. Holloway (2007) suggested that *Aegus* was related to *Ryssonotus* (Chiasognathini) but molecular evidence supports the placement of *Aegus* deeply within Lucanini (Kim & Farrell 2015).

Dorculus Didier 1930 (Pl. 76I; Fig. 26.54)

Type species. *Dorculus bouvier* Didier, 1930.

Characteristics. Length 15–23 mm; dorsal surface without scales, without obvious erect setae; male head without dorsal tubercles, flat, anterior concave but with short truncate median lobe; male genae rounded; female head not greatly narrowed compared with male; temples reduced to narrow unswoolen strip between eye and pronotum in both sexes; eyes globular, with posterior extension of gena dividing anterior half and slight invagination on posterior margin of eye; antennae geniculate, 10-segmented, with 3-segmented club, club segments entirely densely setose and often closely appressed; antennomere 7 cupuliform with thin blade-like lateral lobe, 6 asymmetric and strongly transverse; male mandibles shorter than head; labium broad, slightly less than one third head width; mentum solid and punctate, with convex apex; pregula flat except narrowly raised at junction of mentum; prosternal process convex and relatively broad between procoxae, apex strongly elevated; scutellum almost equilateral-triangular; elytra non-striate but may have shallow elongate grooves, both sexes with dense simple concave punctures, interspaces smooth and shining to dull and microgranulate; humeri not spined; mesoventral process flat, tapering or parallel-sided to junction with metaventrite; profemur with complete anterior ridge; male protibia with simple apical spur; female protibia without subsidiary teeth between large external teeth; male metatibia without setose excavation; tarsal empodium prominent, with divergent thin paired tufts of apical setae; bases of abdominal ventrites 2–5 with well defined transverse grooves but these usually hidden; ventrites 2–5 strongly laterally lobed.

Australian species. Single, *Dorculus difformipes* Didier.

Distribution. *Dorculus* species are found in eastern Indonesia and the northern tropics of Australia.

Biology. Nothing is known about the biology of *Dorculus*, however this small lucanine is notable for its strongly thickened tibia and relatively cylindrical shape, which suggests a more fossorial behaviour than other lucanines. *Dorculus* has rarely been collected in Australia.

Notes. The genus *Dorculus* was reviewed by Okuda (2006), however the genitalia of the two Indonesian species are structurally different from the Australian species (the type species of *Dorculus*) and it is probable that two genera are confused under this name. The Australian *Dorculus* is like a small fossorial *Pseudodorcus*, and both genera are similar to female *Prosopocoilus*.

Dorcus Macleay, 1819 (Pl. 72G; Fig. 26.55)

Type species. *Scarabaeus parallelipipedus* Linnaeus, 1758.

Characteristics. Length 21–50 mm; dorsal surface without scales, females often with short erect setae at sides; male head without dorsal tubercles, flat, anterior concave but with short truncate median lobe; male genae rounded; female head not greatly narrowed compared with male; temples long and laterally swollen in male, short and slightly excavate behind eyes in female with pronotum almost reaching eyes; eyes globular, with posterior extension of gena dividing anterior half and slight invagination on posterior margin of eye; antennae geniculate, 10-segmented, with 3-segmented club, club segments entirely densely setose and often closely appressed; antennomere 7 asymmetric with truncate apex and 6 symmetrical; male mandibles longer than head; labium broad, approximately one third head width; mentum solid and punctate, apex truncate; pregula flat; prosternal process flat and relatively broad between procoxae, apex slightly elevated; scutellum almost equilateral-triangular; elytra non-striate, in female with scattered simple concave punctures, larger and denser at sides, interspaces smooth and shining, in male with minute punctures obscured by dull microgranulate surface; humeri not spined; mesoventral process flat, tapering to junction with metaventrite; profemur with complete anterior ridge; male protibia with simple apical spur; female protibia with 0–2 subsidiary teeth between large external teeth; male metatibia without setose excavation; tarsal empodium prominent, with divergent thin paired tufts of apical setae; bases of abdominal ventrites 2–5 with transverse grooves which are usually hidden; sides of ventrites 2–5 strongly laterally lobed.

Australian species. Two: *Dorcus intermedius* Gestro, *D. wickhami* (Waterhouse).

Distribution. The genus *Dorcus*, as currently understood, is found in the Holarctic, south Asia and east to Australia and the Solomon Islands. The Australian species of *Dorcus* are confined to the northern tropics of Queensland and the Northern Territory where they are rarely collected.

Biology. There is a large amount of information about the temperate members of this genus but little on the tropical species and nothing for the genus in Australia. *Dorcus* was not found in the survey of rainforest Lucanidae by Wood *et al.* (1996), which suggests that it inhabits drier forest. None of the material examined for this study included detailed locality data but most specimens seem to have been collected on or near the coast. *Dorcus intermedius* (as *D. pilosipes*) is a common pest of young coconut trees in the Solomon Islands (Froggatt 1911) and has a wide host range in lowland forest and woodland in New Guinea (Hawkeswood & Bomans 1992).

Notes. The supposedly endemic Australian species *D. wickhami* (formerly in genus *Eurytrachelus*) may be synonymous with the widespread New Guinean species *D. arfakianus* (Lansberge, 1880), formerly in *Serrognathus*. The two species are supposedly separated by minor differences in the male mandibles. The definition of *Dorcus* has varied enormously; the recent taxonomic treatment is summarised here in regard to *D. arfakianus* and *D. wickhami*. *Serrognathus arfakianus* and *Eurytrachelus wickhami* were first placed in *Dorcus* by Mizinuma & Nagai (1994). In contrast, Holloway (2007), in a discussion of the generic limits of *Serrognathus*, suggested that *D. arfakianus* was related to *Lucanus saiga* Olivier, 1789 (type species of *Telodorcus* Didier, 1931), and that these species belonged to neither *Dorcus* nor *Serrognathus*. However the antennal setal characters used by Holloway to distinguish genera are not even consistent amongst specimens of the type species of *Serrognathus*, *S. titanus* Boisduval, 1835 (*pers. obs.*). Bartolozzi *et al.* (2011) studied *S. arfakianus* and *L. saiga* and placed them in *Dorcus*. Huang & Chen (2013) concluded that *Serrognathus* and *Dorcus* were valid genera from their cladistic study of Lucaninae, and that *S. arfakianus* and *L. saiga* properly belonged in *Dorcus*. They synonymised *Dorcus* and *Eurytrachelus*. Hao & Chen redefined *Serrognathus* and reduced it to a few east and south-east Asian species. There are clearly several unresolved issues in the analysis presented by Hao & Chin, for example: at least one of the synapomophies defining *Serrognathus* is incompatible with a synapomorphy for the clade to which it belongs and several characters are duplications giving biased weighting to the clades they define. Nevertheless, Huang & Chen (2013) were able to define genera morphologically and provide a workable key to the lucanine genera they examined, therefore these authors are followed here.

The second Australian species listed, *Dorcus intermedius*, was originally described as *Serrognathus australicus* Bomans, 1985. The synonymy was proposed by Hangay & de Keyzer (2017) and is confirmed here. Males of both species show: (i) elongate mandibles with multiple median teeth but without a basal tooth; (ii) laterally prominent eyes with canthus about half length of eye; (iii) lateral margin pronotum abruptly excavate anterior to middle; (iv) tarsomeres 1–4 short with long ventral setal tufts. The Australian material was originally described from two specimens collected by university students in an unlikely location (the low rainfall temperate climate of Tamworth, New South Wales). Repeated searches in this area by Australian collectors have failed to find any further specimens and this record should be regarded as an accidental introduction or a result of mislabelling. However, there is a specimen collected from the Australian lowland tropics at Rockhampton, Queensland, in ANIC. This may represent an established population. On the other hand, *D. intermedius* is a common pest species in Melanesia (generally under the name *D. pilosipes* (Waterhouse)) therefore it is likely that Australian specimens represent accidental introductions.

Figulus Macleay, 1819 (Pl. 50I; Fig. 26.56)

Type species. *Lucanus striatus* Olivier, 1789.

Characteristics. Length 5–19 mm; dorsal surface without scales; male head without dorsal tubercles or with 2–4 pairs of low tubercles; anterior of head concave, truncate or medially produced; male genae rounded; female head not narrowed compared with male; head excavate posterior to extended genal lobe, but pronotum distant from eyes; eyes completely divided by posterior extension of broad genal lobe, dorsal part smaller, reniform, ventral part circular; antennae geniculate, 9- (*F. lilliputanus*) or 10-segmented, with 3-segmented club, the club segments with short lateral lobes (e.g. *F. trilobus*) or flattened-triangular (most species), shining except for apical well defined densely sensillate areas; antennomeres 6–7 symmetrical or almost so; male mandibles as long as or shorter than head; labium broad, approximately one third head width; mentum solid and punctate, apex broadly concave; pregula not elevated; prosternal process distinct and flat between procoxae, apex flat to slightly elevated; scutellum elongate-triangular to absent; elytra strongly punctate-striate, strial punctures deeply concave, surface of intervals smooth and shining; elytral humeri not spined; scutellum distinct but elongate (*F. trilobus*), or very thin, or apparently absent (*F. lilliputanus*); mesoventral process anteriorly slightly elevated, slightly tapered to junction with metaventrite, profemur with anterior ridge on apical half; male protibia with simple apical spur; female protibia without subsidiary teeth between large teeth on outer edge; male metatibia without setose excavation; tarsal empodium not exserted between claws or if so hair-like, with single apical seta; bases of abdominal ventrites 2–5 deeply transversely grooved, grooves not hidden; sides of ventrites 2–5 strongly laterally expanded.

Australian species. Seven (Monte *et al.* 2016).

Distribution. *Figulus* may be the most drought tolerant genus of lucanids in Australia. Species are widespread in northern, eastern and south-western Australia, including the semi-arid inland Murray–Darling basin (< 450 mm per annum). Rainforest associated species also occur, including *F. nitens* Waterhouse in near-coastal eastern Australia and on Lord Howe Island. *Figulus rossi* Gahan is endemic to Christmas Island and Cocos Keeling Islands. *Figulus regularis* Waterhouse and *F. sulcicollis* Hope & Westwood also occur in New Guinea (Monte *et al.* 2016).

Biology. *Figulus* species inhabit a wide range of timbers in a wide range of habitats. Several tropical rainforest species in New Guinea prefer trees with viscous sap in family Euphorbiaceae, including rubber (Hawkeswood & Bomans 1992). The biology of Australian *Figulus* species is little known (Hangay & de Keyzer 2017).

Notes. The Australian species of this genus have recently been revised, with several new synonyms, reducing the fauna from 11 (Moore & Cassis 1992) to seven species (Monte *et al.* 2016). The genus is easily recognised amongst Australian Lucanidae by the flat, subparallel body, broad and flat genal lobe around eye, short mandibles, almost quadrate pronotum, minute scutellum and strongly striate elytra.

Hoplogonus Parry, 1875 (Pl. 50F; Fig. 26.57)

Type species. *Hoplogonus simpsoni* Parry, 1875.

Characteristics. Length 15–30 mm; dorsal surface without scales, but at least sides of elytra with stubble of short clavate setae, or with long hair-like setae; male head without dorsal tubercles, flat, anterior concave but with short triangular anteroventral lobe; male genae rounded; female head narrowed compared with male; temples long and laterally produced in male, long and straight in female, eyes distant from pronotum; eyes small and flat, with posterior extension of gena dividing anterior half and slight invagination on posterior margin of eye; antennae geniculate, 10-segmented; with 3-segmented laterally lobed club, sensory areas on first two club segments confined to apices of lobes, last segment mostly densely setose, segments often closely appressed; antennomere 7 almost symmetrical and 6 symmetrical; male mandibles longer than head; labium broad, 1/4–1/3 head width; mentum solid and punctate, apex truncate to weakly convex; pregula flat; prosternal process flat and relatively broad between procoxae, apex slightly elevated; scutellum transverse semi-ovate; elytra semistriate, with some punctures in irregular lines and shallow elongate grooves; interspaces smooth and shining or irregularly strigose; humeri spined, with erect acute-angled tubercle at base of upper margin of epipleuron; mesoventral process slightly elevated and tapered to junction with metaventrite; profemur with anterior ridge on apical half; male protibia with simple apical spur; female protibia with 0–1 subsidiary teeth between large external teeth; male metatibia without setose excavation; tarsal empodium prominent, with divergent thin paired 1–2 apical setae; bases of abdominal ventrites 2–5 with well defined grooves, usually visible; sides of ventrites 2–5 laterally lobed.

Australian species. *Hoplogonus bornemisszai* Bartolozzi, *H. simsoni* Parry and *H. vanderschoori* Bartolozzi.

Distribution. All *Hoplogonus* species are endemic to moderate to high rainfall forests in the north-east of Tasmania (Meggs *et al.* 2003; Munks *et al.* 2004).

Biology. All three *Hoplogonus* species are patchy within their ranges, and usually occur in low densities (Meggs *et al.* 2003; Munks *et al.* 2004). *Hoplogonus simsoni* evidently survives clear felling and wildfire, presumably because the larvae are in root systems, like many other lucanids in Australia. The larvae of all species have been described (Richards & Spencer 2014).

Notes. *Hoplogonus* is distinguished from *Lissotes* only by the longer spine-like tubercle at the elytral humeri; the larvae are almost identical (Richards & Spencer 2014). *Hoplogonus* is probably a clade of species within *Lissotes* and could be merged with that genus.

Lissapterus Deyrolle, 1870 (Pl. 4D, 50D; Fig. 26.58)

Type species. *Dorcus howittanus* Westwood, 1863.

Characteristics. Length 15–42 mm; dorsal surface without scales, or short erect setae; male head without dorsal tubercles, flat, but anterior deeply concave, lower edge of cavity with short narrow median lobe; male genae rounded to almost right-angled; female head narrowed compared with male; temples strongly convexly produced in both sexes, pronotum distant from eyes; eyes completely divided by posterior extension of gena, each part of eye small and round, lower larger than upper (males) or similar sized (female); antennae geniculate, 10-segmented, with 3-segmented club, club segments flat, triangular-trapezoid, smooth and shining except for well defined sensory areas on apical surfaces; antennomeres 7 and 6 symmetrical; male mandibles about as long as head; labium narrow relative to head, ~1/5 head width; mentum solid and punctate with concave apex; pregula flat; prosternal process flat and relatively broad between procoxae, apex slightly elevated; scutellum semicircular to almost equilateral-triangular; elytra semi-striate, female with lines of separated punctures on basal half but these obscured by similar sized interstrial punctures, male with short lines of punctures on only basal 1/5 with disc of elytra impunctate but shallowly grooved; punctures simple, shallow, concave; interspaces smooth but finely microreticulate in both sexes; humeri not spined; mesoventral process slightly elevated and slightly tapering to junction with metaventrite; profemur with indistinct anterior ridge on apical half; male protibia with simple apical spur; female protibia without subsidiary teeth between large external teeth; male metatibia without setose excavation; tarsal empodium prominent, with divergent thin paired tufts of apical setae; bases of abdominal ventrites 2–5 with transverse groove, usually hidden; sides of ventrites 2–5 laterally lobed.

Australian species. Eight.

Distribution. *Lissapterus* is endemic to Australia. The 8 species show almost no overlap in distribution and occur in north Queensland (*L. darlingtonia* Bomans, *L. ogivus* Bomans) and south Queensland to Victoria but are curiously absent from Tasmania where they are replaced by *Lissotes* species.

Biology. Most species are montane. The larva of *L. howittanus* (Westwood) has been described (Alderson 1975a). Adults and larvae are found under partly buried logs (Hangay & de Keyzer 2017).

Notes. The species '*L. notestinei* Bomans' was named for the collector Mary Notestine, a woman, and the name should therefore be *L. notestineae*.

Lissotes Westwood, 1855 (Pl. 3F, 50H; Fig. 26.59)

Type species. *Lissotes menalcas* Westwood, 1855.

Characteristics. Length 8–24 mm; dorsal surface without scales, but at least sides of elytra with stubble of short clavate setae, or with long hair-like setae; male head without dorsal tubercles, flat, anterior concave but with short triangular anteroventral lobe; male genae rounded; female head narrowed compared with male; male temples long and straight to strongly convex, female temples short and straight with pronotum almost reaching eye; eyes flat, with posterior extension of gena dividing anterior half and slight invagination on posterior margin of eye; antennae geniculate, 10-segmented; with 3-segmented laterally lobed club, sensory areas on first two club segments confined to apices of lobes, last segment mostly densely setose, segments often closely appressed; antennomere 7 almost symmetrical and 6 symmetrical; male mandibles shorter than to slightly longer than head; labium broad, 1/4–1/3 head width; mentum solid and punctate, apex truncate to slightly concave; pregula flat; prosternal process flat and relatively broad between procoxae, apex slightly elevated; scutellum semicircular; elytra non-striate, with scattered deep simple concave punctures, to semistriate, with some punctures in irregular lines and shallow elongate grooves; interspaces smooth and shining; humeri not spined, but with erect right-angled tubercle at base of upper margin of epipleuron; mesoventral process slightly elevated and tapered to junction with metaventrite; profemur with anterior ridge on apical half; male protibia with simple apical spur; female protibia with 0–1 subsidiary teeth between large external teeth; male metatibia without setose excavation; tarsal empodium prominent, with divergent thin paired 1–2 apical setae; bases of abdominal ventrites 2–5 with well defined groove; sides of ventrites 2–5 laterally lobed.

Australian species. 29.

Distribution. *Lissotes* is endemic to south-east Australia, where it has undergone an extraordinary radiation of species, especially in Tasmania (25 species). Most of these species have small ranges. The remaining four species are confined to upland areas of Victoria (Bartolozzi *et al.* 2014).

Biology. *Lissotes* larvae are usually found beneath rather than within logs (Moore 1984–1986) and they do not appear to be host-specific. The larvae of *L. furcicornis* Westwood and *L. rudis* Lea have been described (Alderson 1975a; Lawrence 1981b; Richards & Spencer 2014). They inhabit wood with brown rot (Hangay & de Keyzer 2017). Most species are found in wetter forests but some species tolerate wildfire and clearfelling, like related *Hoplogonus* (Meggs & Taylor 1999; Meggs & Munks 2003).

Notes. The seven New Zealand species formerly placed in this genus have been split off as *Paralissotes* Holloway (Holloway 2007). The synonymy of *L. desmaresti* Deyrolle and *L. punctatus* Lea was by Bartolozzi (2003). Many of the Tasmanian species have doubtful validity, as noted by Hangay & de Keyzer (2017).

Prosopocoilus Hope, in Hope & Westwood, 1845 (Pl. 4H, 72F; Fig. 26.60)

Type species. *Lucanus cavifrons* Hope & Westwood, 1845.

Characteristics. Length 18–50 mm; dorsal surface without scales or visible short setae; male head without dorsal tubercles, flat, anterior deeply concave without median lobe; male genae rounded to almost right-angled; female head not greatly narrowed compared with male; male temples long, almost parallel-sided, female temples short and flat but pronotum not reaching eye; eyes globular, with posterior extension of gena dividing anterior half and slight invagination on posterior margin of eye; antennae geniculate, 10-segmented, with 3-segmented club, club segments entirely densely setose and often closely appressed; antennomere 7 cupuliform, with thin lateral lobe, 6 slightly asymmetric; male mandibles longer than head; labium broad, almost one third head width; mentum solid and punctate, semicircular; pregula flat, slightly elevated at junction with mentum; prosternal process broad and almost flat between procoxae, apex elevated; scutellum almost equilateral-triangular; elytra non-striate, in female with scattered simple concave punctures, larger and denser at sides, interspaces smooth and shining, in male with minute punctures obscured by dull microgranulate surface (minor males shinier and smoother); humeri not spined; mesoventral process convexly elevated at base, slightly tapered to junction with metaventrite; profemur with complete anterior ridge; male protibia with simple apical spur; female protibia with 2–4 subsidiary teeth between larger external teeth; male metatibia without setose excavation; tarsal empodium prominent, with divergent thin paired tufts of apical setae; bases of abdominal ventrites 2–5 with transverse basal groove, usually hidden; sides of ventrites 2–5 laterally lobed.

Australian species. *Prospocoilus torresensis* (Deyrolle).

Distribution. *Prospocoilus* is a widespread genus in the Old World tropics from western Africa to east Asia and the Solomon Islands. The single Australian species occurs in the lowland tropics from Darwin to north Queensland as far south as Townsville area. *Prosopocoilus torresensis* appears to be identical to the widespread Melanesian species *P. bison* (Olivier) but lacks the major males of that species.

Biology. Larvae of this species are associated with white rot in rotting logs in rainforest (Wood, Hasenpusch & Storey 1996).

Notes. Two species were listed by Moore & Cassis (1992): *P. bison* and *P. torresensis*. Bartolozzi *et al.* (2011) list *P. bison* as widespread from eastern Indonesia to the Solomon Islands but absent from Australia and all Australian records of *P. bison* are now referred to *P. torresensis*, which is supposedly endemic, although it is likely that this is a synonym of *P. bison* (Hangay & de Keyzer 2017). In Australia, the name *Prospocoilus bison* was used for specimens with clear yellow margins and *P. torresensis* for dark specimens, but they are structurally identical. The dark form occurs further south than the paler form but they overlap around Cooktown. *Prospocoilus torresensis* was probably wrongly named; in a footnote attached to the original description, Parry notes that the collector Mniszech indicated that it was found in 'extreme northern' Australia, not necessarily the Torres Straits (Parry 1870: 80). The subspecies *P. torresensis suzukii* Fujita, 2010, described from Darwin, is within the range of variation of North Queensland material.

Pseudodorcus Perry, 1870 (Pl. 76B; Fig. 26.61)

Type species. *Lucanus hydrophiloides* Hope & Westwood, 1845.

Characteristics. Length 17–33 mm; dorsal surface without scales, without visible setae; male head without dorsal tubercles, flat, anteriorly deeply concave but with short narrow median lobe; male genae rounded; female head not greatly narrowed compared with male; both sexes with temples reduced to short swelling between eye and pronotum; eyes globular, with posterior extension of gena dividing anterior half and slight invagination on posterior margin of eye; antennae geniculate, 10-segmented, with 3-segmented club, club segments entirely densely setose and often closely appressed; antennomere 7 cupuliform (more so in female), asymmetric with thin lateral lobe, 6 slightly asymmetric; male mandibles shorter than head; labium broad, 1/3–1/4 head width; mentum solid and punctate, semicircular; pregula raised as thick flat ridge, with deep narrow transverse groove posteriorly; prosternal process slightly convex and relatively broad between procoxae, apex strongly elevated; scutellum almost equilateral-triangular; elytra non-striate but may have shallow elongate grooves, both sexes with fine but dense simple concave punctures, interspaces shining, smooth or microreticulate; humeri not spined; mesoventral process gradually elevated from base, almost parallel-sided to junction with metaventrite; profemur with complete anterior ridge; male protibia with simple apical spur; female protibia with 0–2 subsidiary teeth between large external teeth; male metatibia without setose excavation; tarsal empodium prominent, with divergent thin paired tufts of apical setae; bases of abdominal ventrites 2–5 with transverse basal groove, usually hidden; sides of ventrites 2–5 laterally lobed.

Australian species. *Pseudodorcus hydrophiloides* (Hope & Westwood), *P. nitidus* Didier.

Biology. Nothing is known about the biology of *Pseudodorcus* but the few known localities suggest preference for relatively dry forests.

Notes. The two species are feebly distinguished and probably synonymous. Both sexes of *Pseudodorcus* are similar to females of *Dorcus* and *Prosopocoilus*.

Subfamily Syndesinae Macleay, 1819

Characteristics. Cylindrical; dorsal surface without scales; eyes entire, sphaerical, without genal lobe; antennae not geniculate, with 7-segmented lamellate club; middle of prosternal process narrow but convex and visible between procoxae; elytra striate, punctures deep and sharply rimmed; bases of ventrites 2–5 without transverse groove; sides of ventrites 2–5 unmargined.

Australian genus. Only *Syndesus*.

Syndesus Macleay, 1819 (Pl. 3G, 72E; Fig. 26.62)

Type species. *Sinodendron cornutum* Fabricius, 1801.

Characteristics. Length 7–15 mm; dorsal surface without scales; male head without dorsal tubercles, anterior of head deeply concave, therefore genae anteriorly projecting; eyes entire but posterior margins slightly excavate; temples absent, head abruptly excavate behind eyes, pronotum reaching eyes; female head not narrowed compared with male; antenna not geniculate, 10-segmented, with 7-segmented apical club, club segments lamellate, often closely adpressed, and surfaces entirely densely setose; male mandibles longer than head; labium narrow, < 1/4 head width; mentum small, solid, punctured and apically deeply bilobed; pregula raised as a low transverse ridge; prosternal process convex and visible between coxae but narrow and blade-like and apex flat; scutellum semicircular; elytra striate, punctures deeply concave, surface of interspaces smooth and shining; elytral humeri not spined; mesoventral process tapered to apex and junction with metaventrite not abruptly elevated; profemur without anterior ridge; male protibia with simple apical spur; female protibia with mixed large and small teeth along outer edge; mesotibial setae not in rows, without smooth lines between; male metatibia without setose excavation; tarsal empodium projecting but short, apex with two diverging setae; bases of abdominal ventrites not grooved and sides unmargined.

Australian species. *Syndesus cornutus* (Fabricius), *S. macleayi* Boileau.

Distribution. There are two species native to Australia and one in New Caledonia, and one Australian species has been introduced to New Zealand (Holloway 2007). *Syndesus* is widespread in South America (Holloway 2007; Paulsen 2014). The two Australian species are restricted to SW Western Australia (*S. macleayi*) and near-coastal eastern Australia, from north Queensland to Tasmania and South Australia (*S. cornutus*). The unremarked record for oceanic Norfolk Island (Moore & Cassis 1992) and recent record from Lord Howe Island (Reid *et al.* 2018a), presumably represent accidental introductions of this occasional timber pest.

Biology. Larvae of *S. cornutus* are associated with powdery brown rot fungal infection of logs (Wood *et al.* 1996). *Syndesus cornutus* is an occasional pest of timber which has been attacked by this brown rot fungus (Lawrence 1981a), which may explain its accidental introduction to New Zealand (Holloway 2007). The larva of *S. cornutus* has been described (Alderson 1975a; Lawrence 1981b).

Notes. The South American *Psilodon* Perty 1830, with 8 species, is included here as a synonym of *Syndesus*, following Holloway (2007), not Paulsen (2014). The two genera are supposedly distinguished by antennal segment number but this overlaps in the species of both genera (Paulsen 2014). Paulsen notes lack of morphological differences but argues that the two genera should be maintained because of their long isolation. If they are all treated as one genus, the discovery of a *Syndesus* species in Miocene amber in the West Indies (Woodruff 2009), not far from known extant species in northern South America, is not unexpected. *Syndesus* species are strongly cylindrical compared with other Australian Lucanidae. Their uniquely shaped heads also provide easy recognition.

Acknowledgments

I am particularly grateful to Roger de Keyzer & George Hangay for continued discussions of all things lucanid and for loaning material. I am also grateful to John Lawrence for information about lucanids. Almost all the photographs were taken by my colleague Max Beatson, for which many thanks; the others were taken by Cate Lemann in Canberra, for which I am also very grateful. The following curators kindly loaned or provided access to specimens: Max Barclay (Natural History Museum, London), Rob Blackburn (Macleay Museum, Sydney), Peter Hudson (South Australian Museum, Adelaide), Geoff Monteith (Queensland Museum, Brisbane), Adam Ślipiński & Cate Lemann (Australian National Insect Collection, Canberra).

References

Alderson J (1975a) Descriptions of the larvae of four species of Lucanidae (stag beetle). *Victorian Naturalist* **92**, 71–79.

Alderson J (1975b) Descriptions of the larvae of *Ceratognathus niger* (Westw.). Coleoptera: Lucanidae (stag beetle). *Victorian Naturalist* **92**, 217–221.

Arrow GJ (1943) On the genera and nomenclature of the lucanid Coleoptera, and descriptions of a few new species. *Proceedings of the Royal Entomological Society Series B* **12**(9–10), 133–143.

Bai M, Ren D, Yang X (2012) *Prosinodendron krelli* from the Yixian formation, China: a missing link among Lucanidae, Diphyllostomatidae and Passalidae (Coleoptera: Scarabaeoidea). *Cretaceous Research* **34**, 334–339. doi:10.1016/j.cretres.2011.11.017

Bartolozzi L (2003) Contribution to the knowledge of the Tasmanian lucanids, with a description of two new species of *Lissotes* Westwood, 1855 (Coleoptera: Lucanidae). *Museo Regionale di Scienze Naturali Monographie (Turin)* **35**, 329–342.

Bartolozzi L, Cianferoni F, Monte C (2011) Checklist of Lucanidae (Insecta: Coleoptera) from the Indo-Australian transition zone. In *Biodiversity, Biogeography and Nature Conservation in Wallacea and New Guinea. Volume 1*. (Ed. D Telnov) pp. 43–58. The entomological Society of Latvia, Riga.

Bartolozzi L, de Keyzer R, Zilioli M (2014) Contribution to the knowledge of the Australian stag beetles, with description of a new species of *Lissotes* Westwood, 1855 (Coleoptera: Lucanidae). *Coleopterists Bulletin* **68**(3), 345–350. doi:10.1649/072.068.0301

Bartolozzi L, Zilioli M, de Keyzer R (2017) *The Stag Beetles of Australia, New Zealand, New Caledonia and Fiji (Coleoptera, Lucanidae)*. Taita Publishers, Hradec Králové, Czech Republic.

Bomans HE (1993) Considerations sur le genre *Aegus* et descriptions d'espèces nouvelles (Coleoptera: Lucanidae). *Lambillionea* **93**(2), 193–210.

Bouchard P, Bousquet Y, Davies AE, Alonso-Zarazaga MA, Lawrence JF, Lyal CHC, Newton AF, Reid CAM, Schmitt M, Ślipiński SA, Smith ABT (2011) Family group names in Coleoptera (Insecta). *ZooKeys* **88**, 1–972. doi:10.3897/zookeys.88.807

Boucher S (1991) Coleoptera Lucanidae de Nouvelle-Caledonie. Un nouvel *Aegus* McLeay [sic] endemique. *Mémoires du Muséum National d'Histoire Naturelle Serie A Zoologie* **149**, 161–165.

Brock RL (2006) Insect fads in Japan and collecting pressure on New Zealand insects. *Weta* **32**, 7–15.

Browne J, Scholtz CH (1999) A phylogeny of the families of Scarabaeoidea (Coleoptera). *Systematic Entomology* **24**, 51–84. doi:10.1046/j.1365-3113.1999.00067.x

Carpanato GM, Bartolozzi L, Mazzei P, Pimpinelli I, Viglioglia V (2010) *Aegus chelifer* Macleay 1819, an Asian stag beetle (Coleoptera Lucanidae) invading the Seychelles Islands: a threat for endemic saproxylic species? *Tropical Zoology (Florence)* **23**(2), 173–180.

Carter HJ (1926) Revision of the Australasian species of *Anilara* (Fam. Buprestidae) and *Helmis* (Fam. Dryopidae) with notes and descriptions of other Australian Coleoptera. *Proceedings of the Linnean Society of New South Wales* **51**(2), 50–71.

Chalumeau F, Brochier B (1993) Un nouveau genre de Chiasognathinae Australien (Coleoptera: Lucanidae). *Bulletin de la Société Sciences Naturelles* **79**, 16–17.

Chiari S, Zauli A, Audisio P, Campanaro A, Donzelli PF, Romiti F, Svensson GP, Tini M, Carpaneto GM (2014) Monitoring presence, abundance and survival probability of the stag beetle, *Lucanus cervus*, using visual and odour-based capture methods: implications for conservation. *Journal of Insect Conservation* **18**(1), 99–109. doi:10.1007/s10841-014-9618-8

Costa C, Vanin SA, Casari-Chen SA (1988) *Larvas de Coleoptera do Brasil*. Museu de Zoologia, São Paulo.

Emlen D (2008) The evolution of animal weapons. *Annual Review of Ecology, Evolution and Systematics* **39**, 387–413.

Fearn S (1996) Observations of the life history and habits of the stag beetle *Lamprima aurata* (Latreille) (Coleoptera: Lucanidae) in Tasmania. *Australian Entomologist* **23**(4), 133–138.

Fearn S (2015) A new adult host tree record for the green and gold stag beetle *Lamprima aurata* (Scarabaeoidea: Lucanidae) in Tasmania. *Tasmanian Naturalist* **137**, 2–4.

Fearn S (2016) New ecological and behavioural observations on the green and gold stag beetle *Lamprima aurata* (Scarabaeoidea: Lucanidae) in coastal Tasmania. *Tasmanian Naturalist* **138**, 57–61.

Fremlin M, Hendriks P (2013) Lesser stag beetles *Dorcus parallelipipedus* (L.) (Coleoptera: Lucanidae) longevity – at least three years. *Bulletin of the Amateur Entomological Society* **72**, 48–56.

Froggatt WW (1911) Pests and diseases of the coconut palm. *Department of Agriculkture of New South Wales, Science Bulletin* **2**, 1–20.

Gahan CJ (1888) On the Coleoptera of Christmas Island. *Proceedings of the Zoological Society of London* **37**, 538–541.

Goka K, Kojima H, Okabe K (2004) Biological invasion caused by commercialization of stag beetles in Japan. *Global Environmental Research* **8**(1), 67–74.

Gotoh H, Cornette R, Koshigawa S, Okada Y, Levine LC, Emlen DJ, Miura T (2011) Juvenile hormone regulates extreme mandible growth in male stag beetles. *PLoS One* **6**(6), 1–5. doi:10.1371/journal.pone.0021139

Gotoh H, Fukaya K, Miura T (2012) Heritability of male mandible length in the stag beetle *Cyclommatus metallifer. Entomological Science* **15**(4), 430–433. doi:10.1111/j.1479-8298.2012.00527.x

Gotoh H, Zinna RA, Ishikawa Y, Miyakawa H, Ishikawa A, Sugime Y, Emlen DJ, Lavine LC, Miura T (2017) The function of appendage patterning genes in mandible development of the sexually dimorphic stag beetle. *Developmental Biology* **422**(1), 24–32. doi:10.1016/j.ydbio.2016.12.011

Goyens J, Dirckx J, Dierick M, Van Hoorebeke L, Aerts P (2014) Biomechanical determinants of bite force dimoprhism in *Cyclommatus metallifer* stag beetles. *The Journal of Experimental Biology* **217**(7), 1065–1071. doi:10.1242/jeb.091744

Goyens J, Dirckx J, Aerts P (2015) Stag beetle behavior and its associated anatomical adaptations. *Journal of Insect Behavior* **28**(3), 227–244. doi:10.1007/s10905-015-9495-3

Hangay G, de Keyzer R (2017) *A Guide to Stag Beetles of Australia*. CSIRO Publishing, Melbourne.

Hardersen S, Macagno ALM, Sacchi R, Toni I (2011) Seasonal constraints on the mandible allometry of *Lucanus cervus* (Coleoptera: Lucanidae). *European Journal of Entomology* **108**(3), 461–468. doi:10.14411/eje.2011.059

Harvey DJ, Hawes CJ, Gange AC, Finch P, Chesmore D, Farr I (2011) Development of non-invasive monitoring methods for larvae and adults of the stag beetle *Lucanus cervus. Insect Conservation and Diversity* **4**(1), 4–14. doi:10.1111/j.1752-4598.2009.00072.x

Hawkeswood TJ, Bomans HE (1992) Notes on a small collection of Lucanidae from northern Papua New Guinea (insecta, Coleoptera). *Spixiana* **15**(2), 137–142.

Hawkeswood T, Callister DJ, Antram F (1991) Collection and export of Australian insects. *Traffic Bulletin* **12**(3), 41–48.

Holloway BA (1972) The systematic position of the genus *Diphyllostoma* Fall (Coleoptera: Scarabaeoidea). *New Zealand Journal of Science* **15**, 31–38.

Holloway BA (1998) A re-evaluation of the genera of New Zealand aesaline stag beetles (Coleoptera: Lucanidae). *Journal of the Royal Society of New Zealand* **28**(4), 641–656. doi:10.1080/03014223.1998.9517578

Holloway BA (2007) Lucanidae (Insecta: Coleoptera). *Fauna of New Zealand* **61**, 1–254.

Hongo Y (2014) Interspecific relationship between the Japanese horned beetle and two Japanese stag beetle species. *Entomological Science* **17**(1), 134–137. doi:10.1111/ens.12040

Hongo Y, Okamoto K (2013) Interspecific contests between males of two Japanese stag beetle species, *Lucanus maculifemoratus* and *Prospocoilus inclinatus*: what overcomes a body size disadvantage? *Behaviour* **150**(1), 39–59. doi:10.1163/1568539X-00003036

Hosoya T, Araya K (2005) Phylogeny of Japanese stag beetles (Coleoptera: Lucanidae) inferred from 16S mtrRNA gene sequences, with reference to the evolution of sexual dimorphism of mandibles. *Zoological Science* **22**, 1305–1318. doi:10.2108/zsj.22.1305

Howden HF, Lawrence JF (1974) The New World Aesalinae, with notes on the North American lucanid subfamilies (Coleoptera, Lucanidae). *Canadian Journal of Zoology* **52**(12), 1505–1510.

Huang H, Chen C-C (2013) Phylogeny and systematics of Lucaninae (sensu Holloway, 1960 & 1968) from China. In *Stag beetles of China II*. (Eds H Huang and C–C Chen) pp. 41–113. Formosa Ecological Company, Taiwan.

Huang J-P (2014) Modeling the effects of anthropogenic exploitation and climate change on an endemic stag beetle, *Lucanus miwai* (Lucanidae), of Taiwan. *Journal of Asia-Pacific Entomology* **17**(3), 423–429. doi:10.1016/j.aspen.2014.03.009

Huang J-P, Lin C-P (2010) Diversification in subtropical mountains: phylogeography, Pleistocene demographic expansion, and evolution of polyphenic mandibles in Taiwanese stag beetle, *Lucanus formosanus. Molecular Phylogenetics and Evolution* **57**(3), 1149–1161. doi:10.1016/j.ympev.2010.10.012

Hudson GV (1934) *New Zealand Beetles and Their Larvae*. Ferguson & Osborn Ltd, Wellington.

Inoue A, Hasegawa E (2013) Effect of morph types, body size and prior residence on food-site holding by males of the male-dimorphic stag beetle *Prospocoilus inclinatus* (Coleoptera: Lucanidae). *Journal of Ethology* **31**(1), 55–60. doi:10.1007/s10164-012-0350-0

International Commission on Zoological Nomenclature (1999) *International Code of Zoological Nomenclature, 4th edition*. International Trust for Zoological Nomenclature, London.

International Union for Conservation of Nature and Natural Resources (2001) *Red List Categories and Criteria. Version 3.1* (second edition). http://www.iucnredlist.org/technical-documents/categories-and-criteria/2001-categories-criteria

International Union for Conservation of Nature and Natural Resources (2014) *The IUCN Red List of Threatened Species. Version 2014.3*. <http://www.iucnredlist.org>. Downloaded on 2 February 2015.

Kanzaki N, Ragsdale EJ, Hermann M, Sommer RJ (2014) Two new and two recharacterized species from a radiation of *Pristionchus* (Nematoda: Diplogastridae) in Europe. *Journal of Nematology* **46**(1), 60–74.

Karlsson M, Jonsell M, Eriksson P (2013) En inplanterad population av svartoxe (*Ceruchus chrysomelinus*) 17 ar senare: etablering, vedval och framtid. *Entomologisk Tidskrift* **134**(4), 207–220.

Kawano K (1997) Cost of evolving exaggerated mandibles in stag beetles (Coleoptera: Lucanidae). *Annals of the Entomological Society of America* **90**(4), 453–461. doi:10.1093/aesa/90.4.453

Kawano K (2003) Character displacement in stag beetles (Coleoptera: Lucanidae). *Annals of the Entomological Society of America* **96**(4), 503–511. doi:10.1603/0013-8746(2003)096[0503:CDISBC]2.0.CO;2

Kawano K (2004) Developmental stability and adaptive variability of male genitalia in sexually dimorphic beetles. *American Naturalist* **163**(1), 1–15. doi:10.1086/379796

Kim SI, Farrell BD (2015) Phylogeny of world stag beetles (Coleoptera: Lucanidae) reveals a Gondwanan origin of Darwin's stag beetle. *Molecular Phylogenetics and Evolution* **86**, 35–48. doi:10.1016/j.ympev.2015.02.015

Knell RJ, Pomfret JC, Tomkins JL (2004) The limits of elaboration: curved allometries reveal the constraints on mandible size in stag beetles. *Proceedings of the Royal Society Biological Sciences Series B* **271**, 523–528. doi:10.1098/rspb.2003.2641

Krake GJ (1998) Description of the female of *Eucarteria subvittata* Moore (Coleoptera: Lucanidae). *Australian Entomologist* **25**(3), 93–95.

Lachat T, Wermelinger B, Gossner MM, Bussier H, Isacsson G, Mueller J (2012) Saproxylic beetles as indicator species for dead-wood amount and temperature in European beech forests. *Ecological Indicators* **23**, 323–331. doi:10.1016/j.ecolind.2012.04.013

Lagarde F, Corbin J, Goujon C, Poisbleau M (2005) Polymorphisme et performances au combat chez les males de lucane cerf-volant (*Lucanus cervus*). *Revue d'Ecologie la Terre et la Vie* **60**(2), 127–137.

Lawrence JF (1981a) The occurrence of *Syndesus cornutus* (F.) in structural timber (Coleoptera: Lucanidae). *Journal of the Australian Entomological Society* **20**, 171–172. doi:10.1111/j.1440-6055.1981.tb01024.x

Lawrence JF (1981b) Notes on larval Lucanidae (Coleoptera). *Journal of the Australian Entomological Society* **20**, 213–219. doi:10.1111/j.1440-6055.1981.tb01035.x

Lawrence JF, Ślipiński A (2013) *Australian Beetles. Volume 1: Morphology, Classification and Keys.* CSIRO Publishing, Melbourne.

Lawrence JF, Ślipiński A, Seago AE, Thayer MK, Newton AF, Marvaldi AE (2011) Phylogeny of the Coleoptera based on morphological characters of adults and larvae. *Annales Zoologici* **61**(1), 1–217. doi:10.3161/000345411X576725

Lea AM (1915) Descriptions of new species of Australian Coleoptera. Part X. *Proceedings of the Linnaean Society of New South Wales* **39** (1914), 650–694, plate 76.

Matsumoto K, Knell RJ (2017) Diverse and complex male polymorphisms in Odontolabis stag beetles (Coleoptera: Lucanidae). *Scientific Reports* **7**(16733), 1–11.

Meggs JM, Munks SA (2003) Distribution, habitat characteristics and conservation requirements of a forest-dependent threatened invertebrate *Lissotes latidens* (Coleoptera: Lucanidae). *Journal of Insect Conservation* **7**(3), 137–152. doi:10.1023/A:1027333323362

Meggs JM, Taylor RJ (1999) Distribution and conservation status of the Mt Mangana stag beetle, *Lissotes menalcas* (Coleoptera: Lucanidae). *Papers and Proceedings of the Royal Society of Tasmania* **133**, 23–28.

Meggs JM, Munks SA, Corkrey R (2003) The distribution and habitat characteristics of a threatened lucanid beetle *Hoplogonus simsoni* in north-east Tasmania. *Pacific Conservation Biology* **9**(3), 172–186. doi:10.1071/PC030172

Meggs JM, Munks SA, Corkrey R, Richards K (2004) Development and evaluation of predictive habitat models to assist the conservation planning of a threatened lucanid beetle, *Hoplogonus simsoni*, in north-east Tasmania. *Biological Conservation* **118**(4), 501–511. doi:10.1016/j.biocon.2003.10.001

Michaels K, Bornemissza G (1999) Effects of clearfell harvesting on lucanid beetles (Coleoptera: Lucanidae) in wet and dry sclerophyll forests in Tasmania. *Journal of Insect Conservation* **3**(2), 85–95. doi:10.1023/A:1009696130694

Mizunuma T, Nagai S (1994) *The Lucanid Beetles of the World.* Mushi-Sha, Nakano.

Molino-Olmedo F (2014) Redescripcion de la larva de *Phalacrognathus muelleri* (Macleay, 1855 [sic]) (Coleoptera, Lucanidae). *Lambillionea* **114**(3), 184–188.

Moller AP (2002) Developmental instability and sexual selection in stag beetles from Chernobyl and a control area. *Ethology* **108**(3), 193–204. doi:10.1046/j.1439-0310.2002.00758.x

Monte C, Zilioli M, Bartolozzi L (2016) Revision of the Australian species of *Figulus* Macleay, 1819 (Coleoptera: Lucanidae). *Zootaxa* **4189**(3), 447–484. doi.org/10.11646/zootaxa.4189.3.2

Moore BP (1978) A new Australian stag beetle (Coleoptera: Lucanidae) with neotropical affinities. *Journal of the Australian Entomological Society* **17**, 99–103. doi:10.1111/j.1440-6055.1978.tb01751.x

Moore BP (1984–1986) Family Lucanidae (stag beetles). In *A Guide to the Beetles of South-eastern Australia. Fascicles 6* [1984] *& 7* [1986]. (Ed. BP Moore) pp. 98–103. Australian Entomological Press, Greenwich.

Moore BP, Cassis G (1992) Lucanidae. *Zoological Catalogue of Australia* **9**, 4–19.

Moore BP, Monteith GB (2004) A second Australian species of the Gondwanan stag beetle genus *Sphaenognathus* Buquet (Coleoptera: Lucanidae). *Memoirs of the Queensland Museum* **49**(2), 693–699.

Moretti M, Obrist MK, Duelli P (2004) Arthropod diversity after forest fires: winners and losers in the winter fire regime of the southern Alps. *Ecography* **27**(2), 173–186. doi:10.1111/j.0906-7590.2004.03660.x

Mori H, Chiba S (2009) Sociality improves larval growth in the stag beetle *Figulus binodulus* (Coleoptera: Lucanidae). *European Journal of Entomology* **106**(3), 379–383.

Munks SA, Taylor RJ (2000) Conserving threatened fauna in production forests: the Tasmanian process. In *Conservation in Production Environements: Managing the Matrix.* (Eds JL Craig, N Mitchell and DA Saunders) pp. 670–674. Surrey Beatty & Sons, Chipping Norton.

Munks S, Richards K, Meggs J, Wapstra M, Corkrey R (2004) Distribution, habitat and conservation of two threatened stag beetles, *Hoplogonus bornemisszai* and *H. vanderschoori* (Coleoptera: Lucanidae) in north-east Tasmania. *Australian Zoologist* **32**(4), 586–596. doi:10.7882/AZ.2004.004

New TR (2005) 'Inordinate fondness': a threat to beetles in south east Asia? *Journal of Insect Conservation* **9**(3), 147–150. doi:10.1007/s10841-005-1309-z

O'Hara JE (2013) Taxonomic and host catalogue of the Tachinidae of America north of Mexico. http://www.nadsdiptera.org/Tach/Nearctic/CatNAmer/Home/CatNAmerhome.html. Accessed 3 February 2015.

Okabe K, Masuya H, Kanzaki N, Taki H (2012) Regional collapse of symbiotic specificity between lucanid beetles and canestriniid mites. *Naturwissenschaften* **99**(11), 959–965. doi:10.1007/s00114-012-0979-0

Okada Y, Hasegawa E (2005) Size-dependent precopulatory behavior as mate-securing tactic in the Japanese stag beetle, *Prospocoilus inclinatus* (Coleoptera: Lucanidae). *Journal of Ethology* **23**(2), 99–102. doi:10.1007/s10164-004-0134-2

Okada Y, Fujisawa H, Kimura Y, Hasegawa E (2008) Morph-dependent form of asymmetry in mandibles of the stag beetle *Prospocoilus inclinatus* (Coleoptera: Lucanidae). *Ecological Entomology* **33**(5), 684–689. doi:10.1111/j.1365-2311.2008.01023.x

Okuda N (2006) A new species of the genus *Dorculus* and a new species of the genus *Prosopocoilus* (Coleoptera, Lucanidae) from the Lesser Sunda Islands, Indonesia. *Gekkan-Mushi* **426**, 16–19.

Okushima Y (2004) *Phalacrognathus muelleri* (Coleoptera, Lucanidae) collected from Okayama Prefecture, western Honshu, Japan. *Bulletin of the Kurashiki Museum of Natural History* **19**, 87–88.

Pandey KK, Pitman AJ (2003) FTIR studies of the changes in wood chemistry following decay by brown-rot and white-rot fungi. *International Biodeterioration & Biodegradation* **52**(3), 151–160. doi:10.1016/S0964-8305(03)00052-0

Parry FJ (1870) A revised catalogue of the lucanoid Coleoptera; with remarks on the nomenclature, and descriptions of new species. *Transactions of the Entomological Society of London* **1870**(1), 53–118.

Paulsen MJ (2010) The stag beetles of southern South America (Coleoptera: Lucanidae). *Bulletin of the University of Nebraska State Museum* **24**, 1–148.

Paulsen MJ (2013) A new genus for the Neotropical species of *Aesalus* Fabricius, with desciptions of eight new species (Coleoptera: Lucanidae: Aesalinae). *Insecta Mundi* **325**, 1–25.

Paulsen MJ (2014) Generic guide to New World scarab beetles. *Psilodon* Perty 1830: 53. http://museum.unl.edu/research/entomology/Guide/Scarabaeoidea/Lucanidae/SYN/PSI/Psilodon.html. Accessed 3 February 2015.

Raffles H (2010) *Insectopedia.* Pantheon Books, New York.

Reid CAM (1999) A new generic synonym in the Australian Lucanidae (Coleoptera). *Coleopterists Bulletin* **53**(2), 175–177.

Reid CAM, Beatson M (2016) Revision of the stag beetle genus *Ryssonotus* MacLeay (Coleoptera: Lucanidae) with descriptions of a new genus and three new species. *Zootaxa* **4150**(1), 1–39. doi:10.11646/zootaxa.4150.1.1

Reid CAM, Jenkins Shaw J, Jensen A (2018a) The Australian Museum 2017 expeditions to Lord Howe: Coleoptera. *Technical Reports of the Australian Museum* **26**, 53–67.

Reid CAM, Smith K, Beatson M (2018b) Revision of the genus *Lamprima* MacLeay in Australia (Coleoptera: Lucanidae). *Zootaxa* **4446**(2), 151–202.

Richards K, Spencer CP (2014) Descriptions and key to the larvae of the Tasmanian endemic genus *Hoplogonus* Parry (Coleoptera: Lucanidae), and comparison with the sympatric *Lissotes rudis* Lea. *Zootaxa* **3884**, 347–359. doi:10.11646/zootaxa.3884.4.4

Rink M, Sinsch U (2007) Radio-telemetric monitoring of dispersing stag beetles: implications for conservation. *Journal of Zoology* **272**(3), 235–243. doi:10.1111/j.1469-7998.2006.00282.x

Rink M, Sinsch U (2011) Warm summers negatively affect duration of activity period and condition of adult stag beetles (*Lucanus cervus*). *Insect Conservation and Diversity* **4**(1), 15–22. doi:10.1111/j.1752-4598.2009.00073.x

Salazar-E JA, Rios-Malaver C, Martinez Agudelo JF (2010) Coleoptera (VII). Noticias sobre los lucanidos colombianos *Sphaenognathus pubescens* (Warterhouse [sic]), *S. bordoni* (Brochier & Chalumeau); *S. bellicosus* (Boileau y *S. nobilis colombiensis* (Moxey) (Coleoptera: Lucanidae). *Boletin Cientifico Museo de Historia Natural Universidad de Caldas* **14**(2), 141–154.

Smith ABT, Hawks DC, Heraty JM (2006) An overview of the classification and evolution of the major scarab beetle clades (Coleoptera: Scarabaeoidea) based on preliminary molecular analyses. *The Coleopterists Society Monographs* **5**, 35–46.

Spencer CP, Richards K (2013) Are invertebrate pedestrians threatened? Observations of *Hoplogonus simsoni* from road line transects in northestern Tasmania. *Tasmanian Naturalist* **135**, 28–40.

Sprecher-Uebersax E (2008) The stag beetle *Lucanus cervus* (Coleoptera, Lucanidae) in art and mythology. *Revue d'Ecologie la Terre et la Vie* **10**(Supplement), 153–159.

Switala AK, Sole CL, Scholtz CH (2014) Phylogeny, historical biogeography and divergence time estimates of the genus *Colophon* Gray (Coleoptera: Lucanidae). *Invertebrate Systematics* **28**(3), 326–336. doi:10.1071/IS13054

Tanahashi M, Hawes CJ (2016) The presence of a mycangium in European *Sindendron cylindricum* (Coleoptera: Lucanidae) and the associated yeast symbionts. *Journal of Insect Science* **16**, 76. doi:10.1093/jisesa/iew054

Tanahashi M, Kubota K (2013) Utilization of the nutrients in the soluble and insoluble fractions of fungal mycelium by larvae of the stag beetle, *Dorcus rectus* (Coleoptera: Lucanidae). *European Journal of Entomology* **110**(4), 611–615. doi:10.14411/eje.2013.083

Tanahashi M, Togashi K (2009) Interference competition and cannibalism by *Dorcus rectus* (Motschulsky) (Coleoptera: Lucanidae) larvae in the laboratory and field. *Coleopterists Bulletin* **63**(3), 301–310. doi:10.1649/1143.1

Tanahashi M, Matsushita N, Togashi K (2009) Are stag beetles fungivorous? *Journal of Insect Physiology* **55**(11), 983–988. doi:10.1016/j.jinsphys.2009.07.002

Tanahashi M, Kubota K, Matsushita N, Togashi K (2010) Discovery of mycangia and the associated xylose-fermenting yeasts in stag beetles (Coleoptera: Lucanidae). *Naturwissenschaften* **97**(3), 311–317. doi:10.1007/s00114-009-0643-5

Tsai C-L, Wan X, Yeh W-B (2014) Differentiation in stag beetles, *Neolucanus swinhoei* complex (Coleoptera: Lucanidae): four major lineages caused by periodical Pleistocene glaciations and separation by a mountain range. *Molecular Phylogenetics and Evolution* **78**, 245–259. doi:10.1016/j.ympev.2014.05.004

UNESCO (2015) Lord Howe Island Group. http://whc.unesco.org/en/list/186 (accessed June 2015).

Williams G (2002) A taxonomic and biogeographic review of the invertebrates of the Central Eastern Rainforest Reserves of Australia (CERRA) World Heritage Area, and adjacent regions. *Technical Reports of the Australian Museum* **16**, 1–208. doi:10.3853/j.1031-8062.16.2002.1353

Wood GA, Hasenpusch J, Storey RI (1996) The life history of *Phalacrognathus muelleri* (Macleay) (Coleoptera: Lucanidae). *Australian Entomologist* **23**(2), 37–48.

Woodruff RE (2009) A new fossil species of stag beetle from Dominican Republic amber, with Australasian connections (Coleoptera: Lucanidae). *Insecta Mundi* **98**, 1–10.

27. HYBOSORIDAE ERICHSON, 1847

Eric G. Matthews and Alberto Ballerio

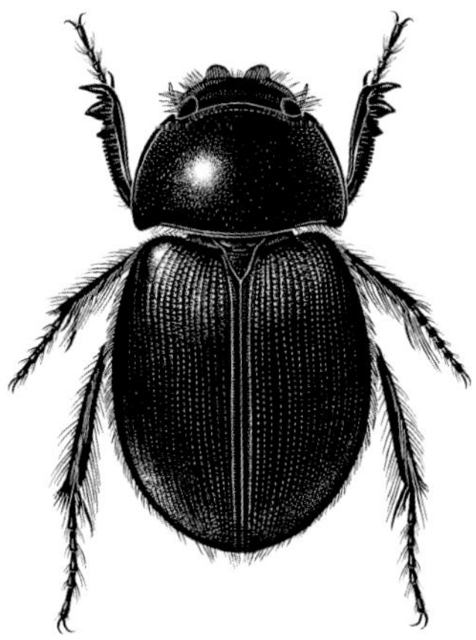

Fig. 27.1. *Phaeochrous emarginatus* Laporte.

Common names. Scavenger scarabs; pill scarabs.

Introduction. The Hybosoridae are basically a pantropical family which achieves its greatest diversity in the Neotropical region. The classification adopted here is that of Ocampo & Ballerio (2006), based on the phylogenetic analyses of Ocampo (2006) and Ocampo & Hawks (2006) which support the monophyly of the family providing it includes Ceratocanthinae, previously considered to be a separate family. Five extant subfamilies, 79 extant genera comprising ~600 species worldwide (see Zoological Record, Ballerio 2013, Ballerio & Grebennikov 2016 and Frolov *et al.* 2017 for post 2006 references) are recognised, of which three subfamilies (Hybosorinae, Liparochrinae and Ceratocanthinae), five named genera and 48 species occur in Australia. The most probable sister taxon of Hybosoridae is the Holarctic Ochodaeidae (Scholtz *et al.* 1988; Ocampo & Hawks 2006). Within Hybosoridae, Ceratocanthinae is a derived group which is the sister taxon of Liparochrinae (Ocampo & Hawks 2006).

Recent revisions of the Australian fauna include that of *Liparochrus* Erichson of Australia and New Guinea by Paulian (1980), that of the species of *Phaeochrous* Laporte by Kuijten (1978) with an additional species described by Keith (2002), and that of the species of Ceratocanthinae by Ballerio (2013). Paulian (1980) divided *Liparochrus* into four subgenera (one restricted to New Guinea) but made some errors of nomenclature which were corrected by Allsopp (1982, 1984) and further corrected by Cassis & Weir (1992). Collection records for nine species were published by Barbero & Palestrini (2003).

The classification of the Australian taxa adopted here is the one currently in use, but it is very likely that a future revision will elevate the subgenera of *Liparochrus* to generic status and divide *Antiochrus* Sharp into two separate genera, bringing the Australian total to eight genera.

The family Hybosoridae has an unusually long fossil history, much of it revealed recently in the Mesozoic fossil beds of Kazakhstan, southern Russia, and Inner Mongolia (Nikolajev 2007, 2010a; Yan *et al.* 2012). Impressions date back to the Middle Jurassic, pointing to divergence times much earlier than those inferred from morphology-based trees where hybosorids are on terminal branches (e.g. Scholtz & Grebennikov 2005), as pointed out by Bai *et al.* (2012). In the Early Cretaceous the taxonomic diversity of the family considerably increases to the point where hybosorids make up 10% of the known number of scarabaeoid species (Nikolajev 2010b; Bai *et al.* 2016). Today they represent less than 2% of scarab diversity. All subfamilies of Hybosoridae are said to be represented as fossils, but the affinities of most of these fossils are very doubtful due to poor preservation (Yan *et al.* 2012; Ballerio & Grebennikov 2016). The first known fossil which is a well preserved whole specimen, rather than a compression, was found in mid-Cretaceous Burmese amber (Bai *et al.* 2016) and can be placed in the extant genus *Hybosorus* Macleay. Liparochrinae, which is of special interest here since it is mainly an Australian group, is said to be represented in Lower Cretaceous Lebanese amber (Kirejtshuk *et al.* 2011), but there is nothing in these authors' description of this fossil which points specifically to Liparochrinae. Similarly, the Ceratocanthinae are recorded from the Lower Cretaceous of Middle Mongolia with the genus *Mesoceratocanthus* Nikolajev *et al.*, 2010, which clearly shares some characters with Ceratocanthinae, but at the same time displays a habitus completely different from the ground plan of Ceratocanthinae and therefore its placement in the subfamily must be regarded as doubtful.

Biology. The Hybosorinae, Liparochrinae and most Anaidinae are scavengers known to feed on carrion, faeces and other rotting organic matter. The Neotropical Anaidinae are tropical forest dwellers found in leaf litter and are supposed to be mycophagous (Ocampo 2006), while Hybosorinae, Liparochrinae and Pachyplectrinae have a much wider ecological range, occurring also in drier environments such as Mediterranean and steppe biomes. *Hybosorus illigeri* Reiche is known to be predatory on small insects (Scholtz & Grebennikov 2005). Activity is nocturnal at least in *Phaeochrous* (Kuijten 1978). Monteith (2009) reported that in dry far-inland north Queensland near Longreach a pig which had been shot during the day was found early that evening to be covered by a vast, seething mass of beetles which proved to be *Phaechrous*, probably *P. emarginatus* Laporte. In a matter of just 2–3 h since nightfall the beetles had already stripped the flesh and skin from the jaw of the pig, and eight days later only a cleaned skeleton remained. The carcass was surrounded by thousands of burrows dug by the beetles, into which they had dragged fragments of flesh to feed their larvae. Initial aggregations of this sort may give rise to huge masses of dead beetles which are self-sustaining over many years as more beetles are continuously attracted to the decaying previous ones. G. Monteith (*in litt.* 2013) has also observed that *Phaeochrous* are not attracted to dung or mushroom baits, suggesting that they are carrion specialists. Species of *Liparochrus* have frequently been trapped in pitfalls baited with human faeces in forests. Williams & Williams (1983) found an

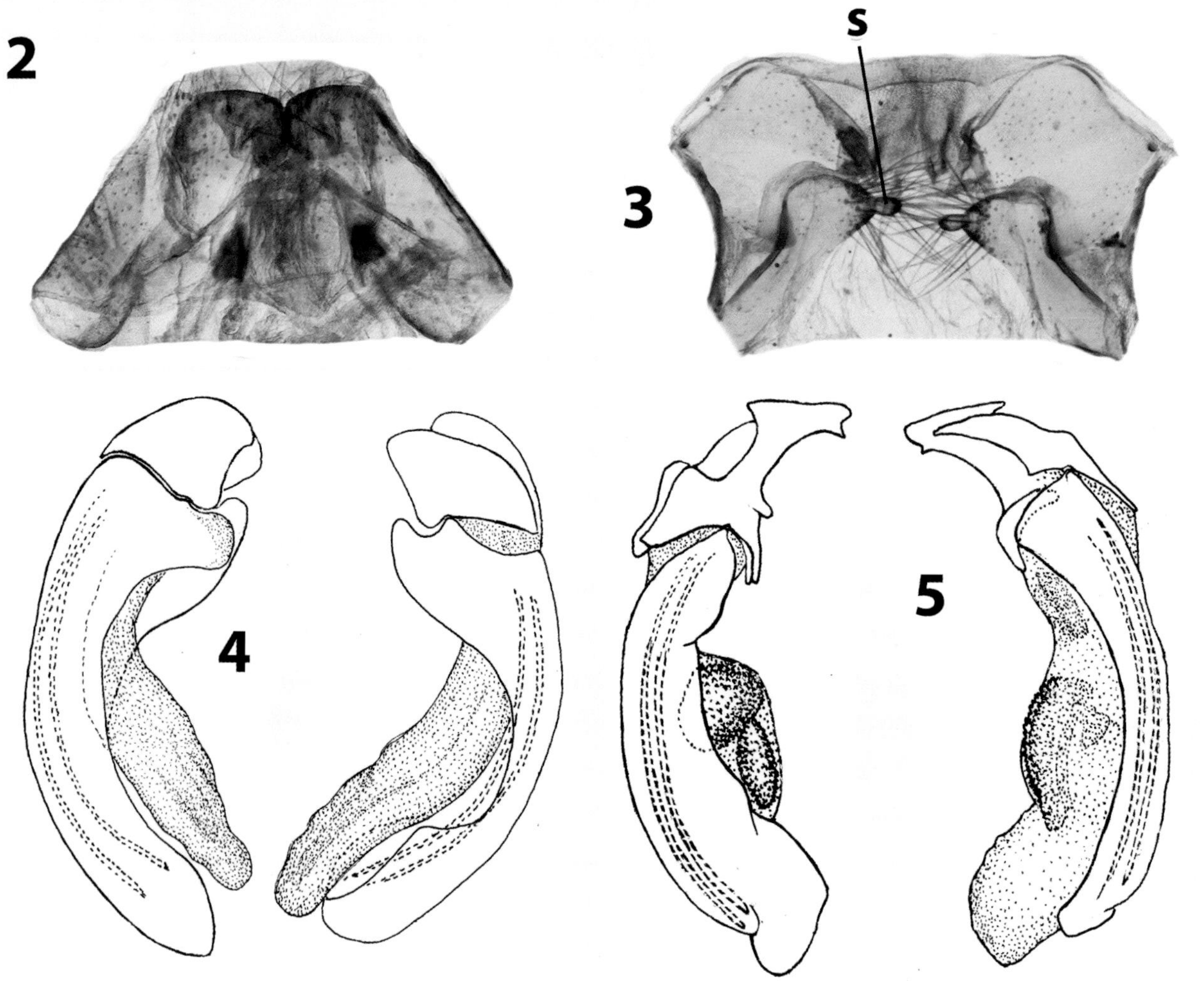

Figs 27.2–27.5. 2, *Phaeochrous emarginatus* Laporte, ovipositor; 3, *Liparochrus (Ranidichrus) crenatulus* Fairmaire, ovipositor, s – stylus; 4, *Liparochrus (Liparochrus) hackeri* Blackburn, aedeagus both sides; 5, *Liparochrus (Ranidichrus) crenatulus* aedeagus both sides.

ecological separation between *L. silphoides* Harold, which was trapped only in rainforest, and *L. fossulatus* Westwood, which occurred only in wet sclerophyll forest. Both species came to faecal traps, while the latter also came to carrion. Allsopp (1977) found that two species of *Liparochrus* came commonly to traps baited with fish bones. When G. Monteith (*in litt.* 2013) set traps baited with rotten mushrooms and dung, he found that *Liparochrus* came more frequently to the mushroom traps. Specimens of *Antiochrus* in Queensland have not been found in wet environments near the coast, only inland in semi-arid country, but not in the desert proper (G. Monteith *in litt.* 2013).

The Ceratocanthinae are possibly mycophagous and display a diverse biology (Ballerio & Grebennikov 2016), with flightless species occurring in leaf litter, volant species, often metallic in colour, living in the lower canopy of forests (there is also a record of an Australian 'Ceratocanthinae' collected in the canopy, Stork & Grimbacher 2006), species living under bark and in rotten wood and species living in termite nests. The Australian *Cyphopisthes descarpentriesi* Paulian was found as adults and larvae in galleries of *Mastotermes darwiniensis*; the other Australian *Cyphopisthes* Gestro and all Australian *Pterorthochaetes* Gestro were collected through sifting of leaf litter or at flight-intercept traps. Elsewhere *Pterorthochaetes* have been collected under bark of dead logs, sometimes associated with Passalidae (Kon *et al.* 2010; Ballerio & Maruyama 2010), or in termite nests. The majority of Ceratocanthinae occur in tropical rainforests, although a few species inhabit temperate forests, seasonal tropical rainforests, and dry forests (such as the interesting cases of the Australian *Cyphopisthes monteithi* Ballerio which occurs in brigalow scrub, and *Cyphopisthes descarpentriesi* which occurs in open eucalypt woodland).

Characteristics. The adults of Australian Hybosoridae are shiny, reddish brown to black and range in size from 4 mm or less (Ceratocanthinae) to 5–12 mm (other subfamilies). Labrum and mandibles are exposed. Eyes are divided by a canthus (Ceratocanthinae) or not (Hybosorinae and Liparochrinae). Antennae of Hybosorinae and Liparochrinae have the first segment of the three-segmented club cupuliform, enclosing

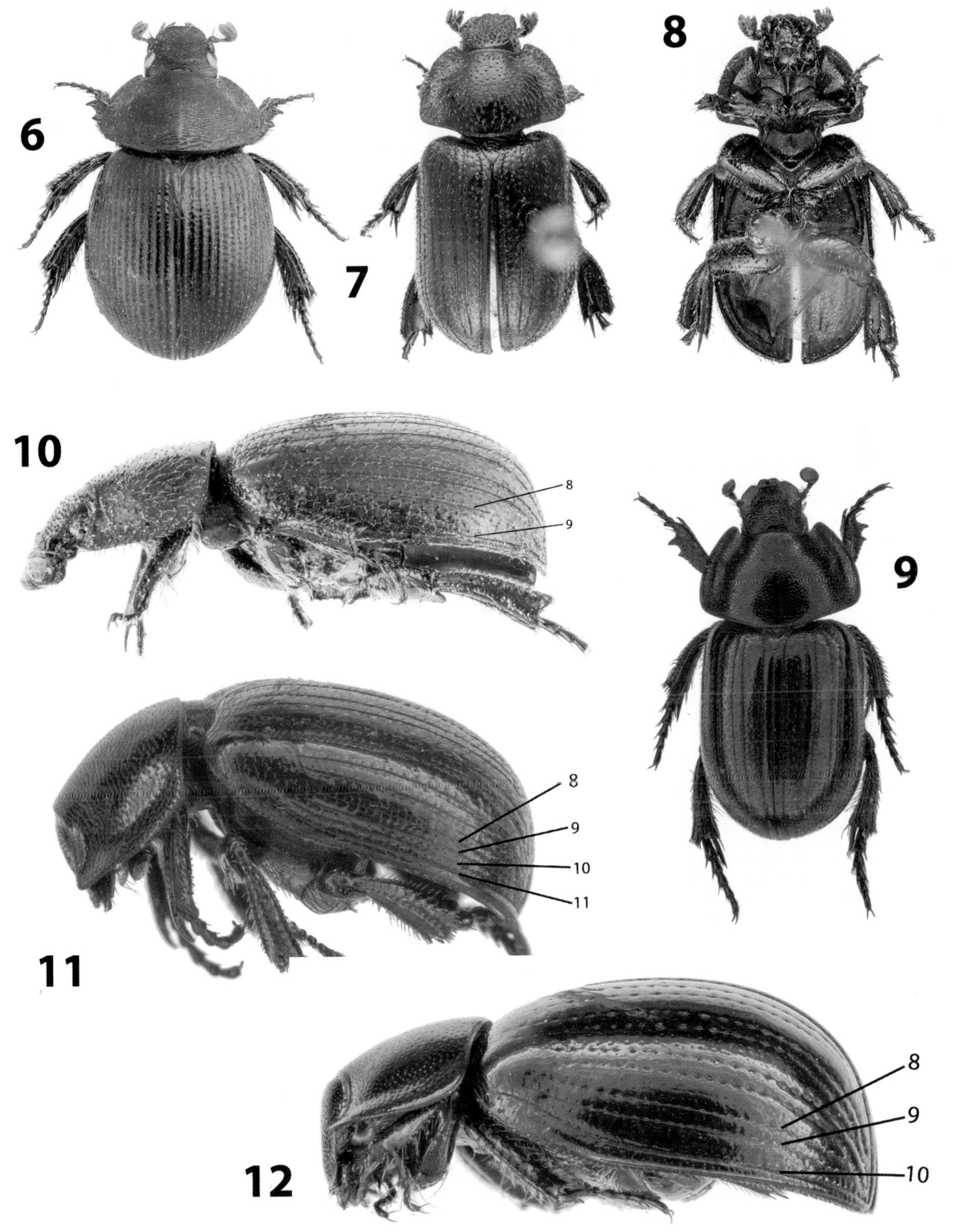

Figs 27.6–27.12. 6, *Liparochrus (Ranidichrus) septemdecimlineatus* Petrovitz dorsal; 7, *Antiochrus brunneus* Sharp holotype dorsal, © MNHN, A. Mantilleri; 8. *Antiochrus brunneus* holotype ventral, © MNHN, A. Mantilleri; 9, *Antiochrus* (*auctt.*) *politulus* (Macleay) dorsal; 10, *Antiochrus brunneus* holotype lateral, © MNHN, A. Mantilleri; 11, *Antiochrus* (*auctt.*) sp. Lateral; 12, *Liparochrus (Liparochrus) bimaculatus* (Macleay) lateral.

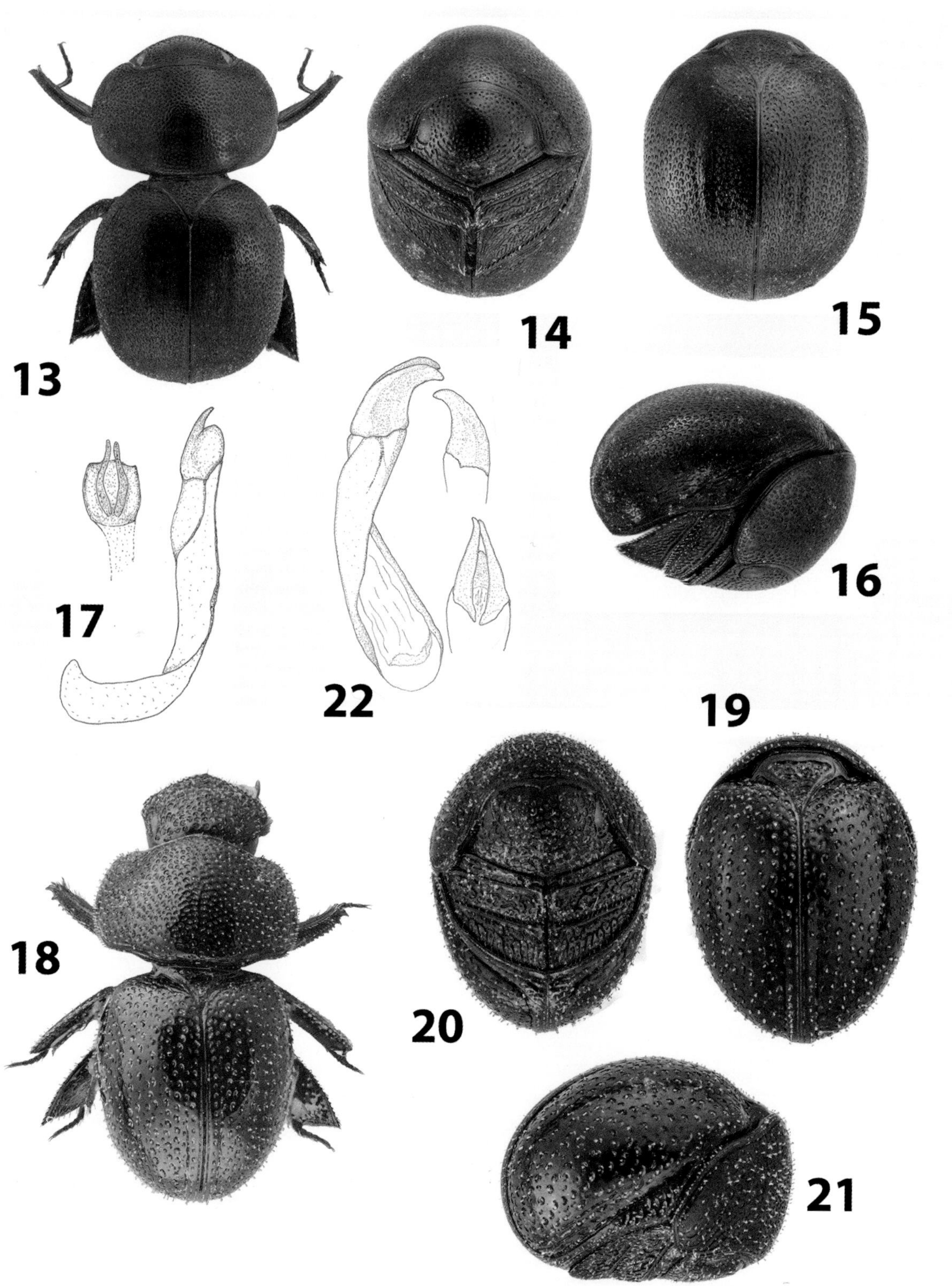

Figs 27.13–27.22. 13, *Cyphopisthes monteithi* Ballerio habitus dorsal extended, © A. Ballerio; 14, *Cyphopisthes monteithi* habitus ventral retracted, © A. Ballerio; 15, *Cyphopisthes monteithi* habitus dorsal retracted, © A. Ballerio; 16, *Cyphopisthes monteithi* habitus lateral retracted, © A. Ballerio; 17, *Cyphopisthes monteithi* aedeagus lateral and parameres dorsal, © A. Ballerio; 18, *Pterorthochaetes storeyi* Ballerio habitus dorsal extended, © A. Ballerio; 19, *Pterorthochaetes storeyi* habitus dorsal retracted, © A. Ballerio; 20, *Pterorthochaetes storeyi* habitus ventral retracted, © A. Ballerio; 21, *Pterorthochaetes storeyi* habitus lateral retracted, © A. Ballerio; 22, *Pterorthochaetes cribricollis* Gestro aedeagus lateral and parameres lateral and dorsal. © A. Ballerio.

the remaining two. The pygidium is concealed by the elytra. Tibiae are without transverse carinae. Ceratocanthinae can roll up into a compact ball, with all involved parts perfectly matching together, and some Liparochrinae are also able to curl up but to a lesser extent. Descriptions of the larvae of only some Hybosorinae, Anaidinae and Ceratocanthinae are available (including the Australian *Cyphopisthes descarpentriesi*, see Grebennikov *et al.* 2002). They are C-shaped as are most other scarab larvae (Ceratocanthinae larvae are remarkably elongate and narrow), have a hypognathous cranium, and the labrum with fore margin dentate. The larval synapomorphies defining the family, according to Grebennikov *et al.* (2004), are: dorsal medial endocarina on cranium extended anteriorly into frontal sclerite, presence of a large membranous spot on apical antennomere, labium dorsally with four pores in centre (secondarily reduced to two pores in some groups), and presence of a stridulatory organ on fore- and middle legs (secondarily reduced in some groups).

Key to the subfamilies, genera and subgenera of Australian Hybosoridae

1. Mesoventrite with a deep fossa just in front of mesocoxae. Tarsal claws split or with basal tooth. Ventral insertion of protarsus at or distad to second tibial tooth. Shape not globose or spherical, head not able to be deflexed. Ovipositor without styli (Fig. 27.2) .... **HYBOSORINAE: *Phaeochrous* Laporte**
– Mesoventrite without fossa. Tarsal claws simple. Ventral insertion of protarsus at or basad to second tibial tooth. Shape sometimes globose or nearly spherical (when head and pronotum deflexed), but may be oblong, head able to be partially or fully deflexed. Ovipositor with or without styli **2**

2(1). First segment of antennal club not cupuliform. Head able to be fully deflexed, form globose (body capable of rolling up to form a compact ball with all parts matching together) (Figs 27.16, 27.21). Ovipositor without styli ... **CERATOCANTHINAE** **3**
– First segment of antennal club cupuliform. Head partially deflexed, form occasionally globose but never forming a perfect ball with all parts matching together. Ovipositor with styli (Fig. 27.3 **s**) ... **LIPAROCHRINAE** **4**

3(2). Antennae 10-segmented, labrum subtruncate, elytra with a distinct pseudepipleuron .... ***Cyphopisthes* Gestro**
– Antennae 9-segmented, labrum not truncate, somewhat depressed distally, elytra without a distinct pseudepipleuron .... ***Pterorthochaetes* Gestro**

4(2). Elytral striae numbering 10 or 17–20. Body form generally oval ... ***Liparochrus* Erichson** **5**
– Elytral striae numbering 9 or 11 (rarely 12). Body form oblong, parallel-sided in middle ... ***Antiochrus sensu lato*** **7**

5 (4). Elytral striae uniformly spaced, 17 or 18 on each elytron (Fig. 27.6). Aedeagus with valves asymmetrical, right one more or less hypertrophied, apically with toothed spatula (Fig. 27.5) .... Subgenus ***Ranidichrus* Allsopp**
– Elytral striae unevenly spaced, 10 or 20 on each elytron. Aedeagus with valves symmetrical or, if asymmetrical, simple (Fig. 27.4) **6**

6(5). 10 striae on each elytron, partly geminate (Fig. 27.12), discal interstriae not carinate .... Subgenus ***Liparochrus* Erichson**
– 20 striae on each elytron, none geminate, 4 discal interstriae carinate .... Subgenus ***Ropalichrus* Allsopp**

7(4). Lateral elytral striae after number 8 reduced to just the one along epipleuron so that the total is 9, with stria 8 shortened and a wide gap between 8 and 9 (Fig. 27.10). Meso- and metatibiae extremely widened (Figs 27.7, 27.8) .... ***Antiochrus* Sharp**
– Striae after 8 crowded together (Fig. 27.11) and very finely carinate, with the total equal to 11 (rarely 12). Meso- and metatibiae moderately widened (Fig. 27.9) .... **'*Antiochrus*' *auctorum***

Classification of the Australian Genera

Subfamily Hybosorinae Erichson, 1847

Characteristics. Head not able to be deflexed. Beetle oblong, not capable of rolling up. Volant. Antennae 10-segmented, club with first segment cupuliform, enclosing second and third segments. Eyes undivided. Ventral insertion of protarsi at or apical to second tibial tooth. Mesosternum with a deep median fossa. Elytra striate, without longitudinal carinae. Tarsal claws with ventral split or tooth. Ovipositor without styli (Fig. 27.2). Length 10–12 mm.

Australian taxa. One genus.

Phaeochrous Laporte, 1840 (Pl. 49O; Figs 27.1, 27.2)

Type species. *Phaeochrous emarginatus* Laporte, 1840.

Silphodes Westwood, 1845

Australian species. Three named (Kuijten 1978; Keith 2002).

Distribution. Northern and western Australia, New Guinea and adjacent archipelagos, Asia, Africa.

Subfamily Liparochrinae Ocampo, 2006

Characteristics. Beetle oblong to globose, more or less capable of rolling up with head and pronotum able to be deflexed. Antennae 10-segmented, club with first segment cupuliform. Eyes undivided. Ventral insertion of protarsi at or basad to second tibial tooth. Mesosternum without fossa. Elytra striate, with or without longitudinal carinae. Tarsal claws simple. Ovipositor with styli (Fig. 27.3s). Length 5–15 mm.

Australian taxa. Two named genera and three subgenera.

Liparochrus Erichson, 1848 (Pl. 49M–N; Figs 27.3, 27.4–27.6, 27.12)

Type species. *Liparochrus geminatus* Westwood, 1852.

Ranidia Paulian, 1980

Parolichrus Allsopp, 1982 (subgenus) (New Guinea)

Ranidichrus Allsopp, 1982 (subgenus)

Ropalichrus Allsopp, 1982 (subgenus)

Paulichrus Allsopp, 1984 (*lapsus calami* for *Ranidichrus*)

Characteristics. Form suboval, sides of elytra more or less convex. Surface of lateral edges of elytral disc similar to that of rest of disc, striae numbering 10 or 16–20. Middle and hind tibiae moderately widening apically. Dorsal surface smooth or finely tuberculate, glabrous or with more or less dense short erect setae. Most species winged in both sexes, but at least one species flightless (*L. laevissimus* Paulian), and some flightless in females only (subgenus *Ropalichrus*). Aedeagus variable (see key). Most species without sexual dimorphism, but some with protibiae externally bidentate in males and tridentate in females, and others with protarsus of males dilated and/or inner claw enlarged. Length 5–15 mm.

Australian taxa. Three subgenera and 31 named species. Paulian (1980) presents keys to all the species of *Liparochrus*.

Distribution. Throughout mainland Australia mainly in the north (one species on Lord Howe Island), New Guinea, Indonesia, New Caledonia, Loyalty Islands, possibly Vanuatu (Paulian 1980).

Antiochrus Sharp, 1873 (Pl. 49P; Figs 27.7–27.11)

Type species. *Antiochrus brunneus* Sharp, 1873.

Characteristics. Form oblong, sides of elytra parallel, body length ~1.5–2 times as long as wide. Volant. Elytra either with 9 striae, of which the 8th is effaced posteriorly, with a wide gap between 8 and 9 (Fig. 27.10), or with 11 (rarely 12) striae of which the last three form approximated fine raised lines (Fig. 27.11). Dorsal surfaces and legs usually covered with short to moderately long curved setae, but some species glabrous. Meso- and metatibiae more or less widened and flattened. Aedeagus with valves asymmetrical, left one tapering to a rounded apex, right one shorter and truncate. Length 6–8 mm.

Australian species. There are seven named species, but some specimens in Australian collections from the Elder Expedition misidentified as *Liparochrus rufus* Blackburn belong to *Antiochrus* as noted by Paulian (1980), and represent an additional, unnamed species.

Distribution. Throughout mainland Australia, mostly in the west and north.

Notes. All species placed in *Antiochrus* by authors subsequent to Sharp (1873) only agree with the nominotypical species *A. brunneus* in being oblong with parallel sides and in having wider meso- and metatibiae than *Liparochrus*, but do not agree with what appears to be the most important character: just nine elytral striae, with stria eight shortened (Fig. 27.10), therefore it is likely that there are two different genera presently under the name *Antiochrus*, separated by the characters described in couplet 7 of the key. Sharp mentioned the reduced number of striae (he counted eight, probably omitting the one along the epipleuron) and stressed the strongly dilated last four tibiae (Fig. 27.8) otherwise seen only in Ceratocanthinae, but these features were ignored by subsequent authors, who concentrated on the longer body form (von Harold 1874; Petrovitz 1963, 1968, 1971) or a setose dorsal surface (Schmidt 1913). Blackburn (1905) introduced the important character of the lateral elytral striae appearing as approximated raised lines and this is the main diagnostic feature of all species other than *A. brunneus*.

At the time of writing no specimens of the distinctive *A. brunneus* have been found in Australian collections, hence the species is known only from the unique type in the Muséum National d'Histoire Naturelle in Paris (Figs 27.7, 27.8, and 27.10), from a locality given simply as Nova Hollandia occidentalis. The type of *Liparochrus oblongus* Harold, 1873 which von Harold (1874) subsequently synonymised with *A. brunneus*, could not be found in Paris (A. Mantilleri *in litt.* 2013), but from Harold's remarks it does not appear that he examined Sharp's type, and *L. oblongus* is most probably not a synonym.

Subfamily Ceratocanthinae Martínez, 1968

Characteristics. Body globose, capable of rolling up to form a compact ball with all parts perfectly matching together. Antennal club with first segment not cupuliform, shaped like the remaining two. Eyes divided by a canthus. Body dorsally covered by puncturation, mostly horseshoe-shaped. Elytra without any striation. Ovipositor without styli. Length 4–5 mm. [Ballerio 2013; Ocampo & Ballerio 2006].

Australian taxa. Two genera.

Cyphopisthes Gestro, 1898 (Figs 27.13–27.17)

Type species. *Synarmostes amphicyllis* Sharp, 1875.

Characteristics. Reddish-brown to black. Volant. Dorsum glabrous (30x). Elytra somewhat flattened dorsally and forming a distinct pseudepipleuron laterally. Antennae 10-segmented. Labrum subtruncate. Protibiae sexually dimorphic (female with two outer apical teeth, male with only one). Mesotibiae: inner apical spur of males curved inwards, in females both apical spurs straight. Male metatibiae with two straight apical spurs. Aedeagus with parameres short, dorsally flattened and symmetrical (Fig. 27.17). Length 4 mm.

Australian species. Three named, all of them endemic and occurring in eastern Queensland (rainforests of Iron Range, open eucalypt woodland near Townsville, and brigalow scrub around Brisbane).

Distribution. From India to Australia (Queensland) through Malaysia, Indonesia, Philippines and New Guinea, with a doubtful record for New Caledonia.

Pterorthochaetes Gestro, 1898 (Pl. 56E; Figs 27.18–27.22)

Type species. *Synarmostes gestroi* Harold, 1874.

Characteristics. Dark brown to black. Volant. Dorsum setose (setae short, thick and clavate). Elytra regularly convex, without any distinct pseudepipleuron. Antennae 9-segmented. Labrum not truncate, somewhat depressed distally. Male mesotibiae with 2 apical spurs, inner one curved inwards, females with 1 apical spur. Male metatibiae with one twisted apical spur and one straight. Aedeagus with asymmetrical parameres (Fig. 27.22). Female genitalia with bursa copulatrix with two paired symmetrical/asymmetrical sclerites. Length 4–5 mm.

Australian species. Four, all occurring in north-east Queensland (rainforests of Cape York Peninsula and Wet Tropics), one of them shared with New Guinea.

Distribution. From India and Sri Lanka through Nepal, southern China, the whole Indochinese peninsula, Indonesia, Philippines, New Guinea, Solomon Islands, Vanuatu islands and Australia (Queensland).

Acknowledgments

The authors wish to thank Antoine Mantilleri of the Muséum National d'Histoire Naturelle, Paris, for sending images of the holotype of *Antiochrus brunneus* Sharp; Geoff Monteith of the Queensland Museum for valuable field observations on Hybosorinae and Liparochrinae and for donating representative liparochrines to the South Australian Museum; Georgiy Nikolajev of Al-Farabi Kazak University, Almaty, Kazakhstan for information on fossil Scarabaeoidea; Cate Lemann of the Australian National Insect Collection for sending information on *Liparochrus rufus* Blackburn, and Peter Hudson of the South Australian Museum for help with literature search.

References

Allsopp PG (1977) Further records of dung beetles (Coleoptera: Scarabaeidae) from the Toowoomba District. *Australian Entomological Magazine* **3**(8), 101–103.

Allsopp PG (1982) Corrections to Paulian (1980): "Révision taxonomique des *Liparochrus* Erichson, genre Australo-Mélanésien de Coléoptères Hybosoridae, Scarabaeoidea. *Journal of the Australian Entomological Society* **21**(3), 199–200. doi:10.1111/j.1440-6055.1982.tb01794.x

Allsopp PG (1984) Checklist of the Hybosorinae (Coleoptera: Scarabaeidae). *Coleopterists Bulletin* **38**, 105–117.

Bai M, Ahrens D, Yang X, Ren D (2012) New fossil evidence of the early diversification of scarabs: *Alloioscrabaeus cheni* (Coleoptera: Scarabaeoidea) from the Middle Jurassic of Inner Mongolia, China. *Insect Science* **19**, 159–171. doi:10.1111/j.1744-7917.2011.01460.x

Bai M, Zhang W, Ren D, Shih C, Yang X (2016) *Hybosorus ocampoi*: the first hybosorid from the Cretaceous Myanmar amber. *Organisms, Diversity & Evolution* **16**, 233–240. doi:10.1007/s13127-015-0245-3

Ballerio A (2013) Revision of the Australian Ceratocanthinae (Coleoptera, Scarabaeoidea, Hybosoridae). *ZooKeys* **339**, 67–91. doi:10.3897/zookeys.339.6033

Ballerio A, Grebennikov VV (2016) Rolling into a ball: phylogeny of the Ceratocanthinae (Coleoptera: Hybosoridae) inferred from adult morphology and single origin of an unique body enrollment coaptation in terrestrial arthropods. *Arthropod Systematics & Phylogeny* **74**, 23–52.

Ballerio A, Maruyama M (2010) The Ceratocanthinae of Ulu Gombak: high species richness at a single site, with descriptions of three new species and an annotated checklist of the Ceratocanthinae of Western Malaysia and Singapore (Coleoptera, Scarabaeoidea, Hybosoridae). In Ratcliffe B, Krell F-T (eds) Current advances in Scarabaeoidea research. *Zookeys* **34**, 77–104.

Barbero E, Palestrini V (2003) Coleoptera Trogidae, Geotrupidae, Hybosoridae, Scarabaeidae (Scarabaeinae and Dynastinae) collected in Australia. In Daccordi M and Giachino PM (Eds) Results of the zoological mission to Austeralia of the Regional Museum of Natural History of Turin, Italy I. *Museo Regionale di Scienze Naturali di Torino, Monografie* **35**, 303–328.

Blackburn T (1905) Further notes on Australian Coleoptera, with descriptions of new genera and species. *Transactions of the Royal Society of South Australia* **29**, 270–332.

Cassis G, Weir TA (1992) Ceratocanthidae, Hybosoridae. *Zoological Catalogue of Australia* **9**, 65–75.

Frolov AV, Ocampo FC, Akhmetova LA, Vaz de Mello F (2017) A new genus and species of the termitophilous Neotropical Hybosorinae (Coleoptera: Scarabaeoidea: Hybosoridae) associated with *Cornitermes* (Isoptera: Termitidae) in the Cerrado ecoregion in Brazil. *Journal of Natural History* **51**, 1759–1765. doi:10.1080/00222933.2017.1353150

Grebennikov VV, Ballerio A, Scholtz CH (2002) Larva and pupa of *Cyphopisthes descarpentriesi* Paulian (Coleoptera: Scarabaeoidea: Ceratocanthidae) and their phylogenetic implications. *Australian Journal of Entomology* **41**, 367–374. doi:10.1046/j.1440-6055.2002.00307.x

Grebennikov VV, Ballerio A, Ocampo FC, Scholtz CH (2004) Larvae of Ceratocanthidae and Hybosoridae (Coleoptera: Scarabaeoidea): study of morphology, phylogenetic analysis and evidence of paraphyly of Hybosoridae. *Systematic Entomology* **29**, 524–543. doi:10.1111/j.0307-6970.2004.00257.x

Keith D (2002) *Phaeochrous nicolasi*, nouvelle espèce d'Australie (Col. Scarabaeoidea Hybosoridae). *Bulletin Mensuel de la Société Linnéenne de Lyon* **71**, 355–356. doi:10.3406/linly.2002.13421

Kirejtshuk AG, Azar D, Montreuil O (2011) First Mesozoic representative of the subfamily Liparochrinae (Coleoptera: Hybosoridae) from the Lower Cretaceous Lebanese amber. *Zoosystematica Rossica* **20**(1), 62–70.

Kon M, Johki Y, Ballerio A (2010) The ceratocanthid beetle *Pterorthochaetes haroldi* (Coleoptera, Ceratocanthidae), collected from a gallery of the passalid beetle, *Leptaulax planus* (Coleoptera, Passalidae) in Sabah, Malaysia. *Kogane* **11**, 93–95.

Kuijten PJ (1978) Revision of the Indo-Australian species of the genus *Phaeochrous* Castelnau, 1840 (Coleoptera: Scarabaeidae: Hybosorinae), with notes on the African species. *Zoölogische Verhandelingen* **185**, 1–40.

Monteith GB (2009) Beetles bringing home the bacon. *News Bulletin of the Entomological Society of Queensland* **37**(5), 87–88.

Nikolajev GV (2007) The Mesozoic stage of evolution of the scarabaeoid beetles (Insecta: Coleoptera: Scarabaeoidea). Kazak Universiteti, Almaty. [In Russian]

Nikolajev GV (2010a) A new genus of the subfamily Anaidinae from the Mesozoic of Transbaikalia. *Paleontological Journal* **44**(2), 192–194. doi:10.1134/S0031030110020115

Nikolajev GV (2010b) On the Mesozoic taxa of Scarabaeoid beetles of the family Hybosoridae (Coleoptera: Scarabaeoidea). *Paleontological Journal* **44**(6), 649–653. doi:10.1134/S0031030110060067

Nikolajev GV, Wang B, Liu Y, Zhang H (2010) First record of Mesozoic Ceratocanthinae (Coleoptera, Hybosoridae). *Acta Palaeontologica Sinica* **49**(4), 443–447.

Ocampo FC (2006) Phylogenetic analysis of the scarab family Hybosoridae and monographic revision of the New World subfamily Anaidinae (Coleoptera: Scarabaeoidea). 3. Phylogenetic analysis of the subfamily Anaidinae (Coleoptera: Scarabaeoidea). *Bulletin of the University of Nebraska State Museum* **19**, 13–177.

Ocampo FC, Ballerio A (2006) Phylogenetic analysis of the scarab family Hybosoridae and monographic revision of the New World subfamily Anaidinae (Coleoptera: Scarabaeoidea). 4. Catalog of the subfamilies Anaidinae, Ceratocanthinae, Hybosorinae, Liparochrinae, and Pachyplectrinae (Scarabaeoidea: Hybosoridae). *Bulletin of the University of Nebraska State Museum* **19**, 178–209.

Ocampo FC, Hawks DC (2006) Phylogenetic analysis of the scarab family Hybosoridae and monographic revision of the New World subfamily Anaidinae (Coleoptera: Scarabaeoidea). 2. Molecular phylogenetics and systematic placement of the family Hybosoridae (Coleoptera: Scarabaeioidea). *Bulletin of the University of Nebraska State Museum* **19**, 7–12.

Paulian R (1980) Révision taxonomique des *Liparochrus* Erichson, genre Australo-Mélanésien de Coléoptères Hybosoridae, Scarabaeoidea. *Annales de la Société Entomologique de France* (N.S.) **16**, 389–433.

Petrovitz R (1963) Neue Aegialiinae, Orphninae und Hybosorinae (Col.). *Entomologische Arbeiten aus dem Museum G. Frey* **14**, 118–125.

Petrovitz R (1968) Bekannte und unbekannte Scarabaeidae (Hybosorinae, Troginae, Orphninae, Dynamopinae, Geotrupinae, Aegialiinae). *Entomologische Arbeiten aus dem Museum G. Frey* **19**, 179–187.

Petrovitz R (1971) Neue australische Hybosoridae (Col.). *Entomologische Arbeiten aus dem Museum G. Frey* **22**, 134–135.

Schmidt A (1913) Hybosoridae. *Genera Insectorum* **150**, 1–87.

Scholtz CH, Grebennikov VV (2005) Scarabaeoidea. In *Handbuch der Zoologie, Band IV Arthropoda: Insecta, Part. 38: Coleoptera, Beetles, Volume 1: Morphology and Systematics*. (Eds RG Beutel and RAB Leschen) pp. 367–425. W De Gruyter, Berlin.

Scholtz CH, D'Hotman D, Evans AV, Nel A (1988) Phylogeny and systematics of the Ochodaeidae (Insecta: Coleoptera: Scarabaeoidea). *Journal of the Entomological Society of Southern Africa* **51**, 207–240.

Sharp D (1873) Descriptions de nouveaux genres et de nouvelles espèces de coléoptères lamellicornes. *Revue et Magasin de Zoologie* (ser. 3) **1**, 262–273.

Stork NE, Grimbacher PS (2006) Beetle assemblages from an Australian tropical rainforest show that the canopy and the ground strata contribute equally to biodiversity. *Proceedings of the Royal Society* **273**, 1969–1975[B]. doi:10.1098/rspb.2006.3521

von Harold E (1874) Zur Kenntnis der kugelförmigen Trogiden. *Coleopterologische Hefte* **12**, 26–51.

Williams GA, Williams T (1983) A survey of the Aphodiinae, Hybosorinae and Scarabaeinae (Coleoptera: Scarabaeidae) from small wet forests of coastal New South Wales, Part 4: Lansdowne State Forest. *Victorian Naturalist* **100**, 146–154.

Yan Z, Bai M, Ren D (2012) A new genus and species of fossil Hybosoridae (Coleoptera: Scarabaeoidea) from the Early Cretaceous Yixian Formation of Liaoning, China. *Alcheringa: An Australasian Journal of Palaeontology* doi:10.1080/03115518.2012.716231

28. SCARABAEIDAE LATREILLE, 1802

John F. Lawrence and Tom A. Weir

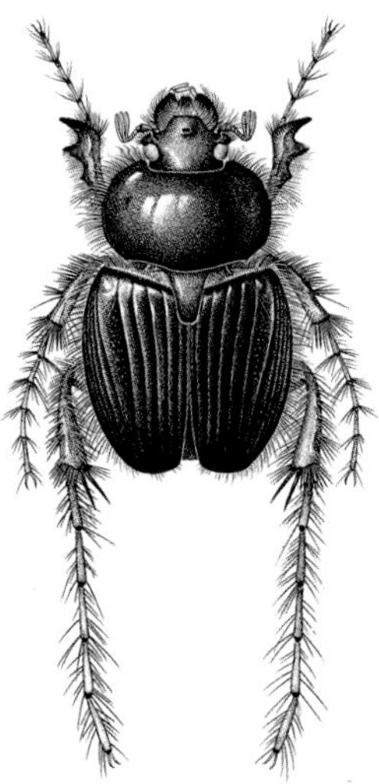

Fig. 28.1. *Phaenognatha* sp.

Introduction. Scarabaeidae is the largest and most diverse scarabaeoid family, with ~1600 genera and 27 000 species; however, it is not as clearly defined as the superfamily. According to Scholtz & Grebennikov (2016), the monophyly of Scarabaeidae is supported by numerous apomorphies of the wing base and wing venation (detailed in Browne & Scholtz 1995, 1998). In most cladograms based on either morphological or molecular data, the scarab subfamilies Melolonthinae, Rutelinae, Dynastinae and Cetoniinae form a monophyletic group, while Scarabaeinae and Aphodiinae form a separate clade, sometimes closer to other scarabaeoid families, such as Hybosoridae, Glaphyridae or Ochodaeidae (Smith *et al.* 2006; Hunt *et al.* 2007; Lawrence *et al.* 2011; Bocak *et al.* 2014; McKenna *et al.* 2015a, 2015b). The results of a larval analysis by Grebennikov & Scholtz (2004) led to the conclusion that larval morphology alone could not address basal relationships in this group (Smith 2006; Grebennikov & Scholtz 2004; Scholtz & Grebennikov 2016).

Keys to subfamilies and some tribes of Australian Scarabaeidae

ADULTS

1. Pretarsal claws unequal (at least on hind legs and usually on all legs) (Figs 28.2, 28.3) with one claw being more slender than the other and either both claws simple, one claw toothed or both claws toothed; claws movable in one plane and often held at different angles in dead specimens; tarsomere 5 variously ventrally cleft to allow movement of one or both claws; IF metapretarsal claws appear simple, almost equal in thickness and not held at different angles (e.g. *Mesystoechus, Chlopocha* – Rutelinae), THEN protibial spur shorter than first protarsomere OR absent 2

– Pretarsal claws equal, more or less fixed and held at same angle in dead specimens; tarsomere 5 not cleft ventrally; IF metapretarsal claws appear held at different angles (e.g. some *Teinogenys* – Dynastinae) THEN protibial spurs longer than first protarsomere 3

2(1). Both claws on all legs toothed (bifid) (Fig. 28.3); metatarsi at least 1.5 times as long as metatibiae; metatarsomere 1 as long as or longer than 2 and 3 combined; spurs on metatibial apex reduced to mere stubs in males; uniform yellowish brown species 10–14 mm in length with a covering of recumbent and erect setae on pronotum and elytra; one introduced species *Plectris aliena*; eastern NSW .... **MELOLONTHINAE: Macrodactylini**

– Not more than one claw toothed (bifid); metatarsi short, not longer than metatibiae; metatarsomere 1 short, about same length as 2; spurs on metatibial apex longer, clearly visible; cuticle often brightly coloured and metallic .... **RUTELINAE**

3(1). Abdominal spiracles entirely concealed beneath elytra and not visible in intact specimen (Fig. 28.4 – with elytron raised) OR very small and difficult to see (last spiracle EITHER on dorsal part of penultimate ventrite OR in pleural membrane) 4

– At least the last pair of abdominal spiracles exposed and CLEARLY visible in intact specimen (Figs 28.5, 28.6) (last spiracle associated with pleural suture or groove on penultimate ventrite) 7

4(3). Labrum exposed, acuminate, and about as long as wide; mandibles foliaceously exposed with rounded, non-opposable apices; head in male with small horn (Fig. 28.7); tarsi of male slender and very long, metatarsi at least 3 times as long as tibiae; body of male somewhat wedge shaped with individually rounded elytra (females little known) .... **MELOLONTHINAE: Phaenognathini**

– Labrum concealed or exposed and transverse, sometimes fused to clypeus; mandibles concealed or slightly exposed, never foliaceously expanded, apices opposable or not; head with armature or not; tarsi shorter, metatarsi never more than 2 times as long as metatibiae; body of different shape 5

5(4). Labrum exposed, transverse, situated either below clypeus (fused to it in Automoliini, most Maechidiini, some Liparetrini and some Heteronychini) or attached to apical margin of clypeus, fused or not (Sericini, Phyllotocidiini, Systellopini, Pachytrichini and *Diphucephala*); head and pronotum always unarmed; mesocoxae slightly to moderately oblique and narrowly separated; last abdominal spiracle located on dorsal part of penultimate sternite (lift elytra to see); tarsal claws simple, toothed or bifid; male antennal club sometimes with more than 3 segments; metatibae with 0, 1 or 2 spurs, if with 2 then these usually placed one above and one below tarsal articulation (except Diphucephalini, Diplotaxini, Pachytrichini Macrodactylini, Melolonthini which have 1 or 2 below) .... **MELOLONTHINAE (major part)**

– Labrum concealed (except *Saprus* – Aphodiinae – which has a typical aphodiine habitus); head and pronotum often armed; mesocoxae strongly oblique or longitudinal and variously separated; last abdominal spiracle located in pleural membrane and not visible even on lifting elytra; tarsal claws usually simple (toothed or subdentate in some Canthonini – Scarabaeinae); antennal club never with more than 3 segments; metatibiae with 1 or 2 spurs usually placed below tarsal articulation (except a few Aphodiinae which have 1 above and 1 below) 6

6(5). Metatibia with 2 apical spurs (very small in Rhyparini and Stereomerini), both located below tarsal articulation (Fig. 28.8) (except in Aphodiini which have 1 above and 1 below (Fig. 28.9)); mesocoxae strongly oblique and separated by less than their width (separated by one width in *Odontolochus*); elytra exposing pygidium or not; scutellum exposed (except Rhyparini and Stereomerini); abdominal sternites together about as long as metaventrite; anterior hypomeral carina from procoxae to lateral edge or anterior angles of pronotum absent (except *Rhyparus*); more or less elongate beetles lacking cephalic or pronotal armature; length usually less than 7 mm (as long as 14 mm in some *Acrossidius*).. **APHODIINAE**

– Metatibia with 1 apical spur located below tarsal articulation (Fig. 28.10); mesocoxae longitudinal and separated by more than their width (Fig. 28.11); elytra always leaving pygidium exposed; scutellum almost always absent (present, mostly small in native *Thyregis,* and *Coptodactyla,* and in introduced *Onitis, Euoniticellus* and *Liatongus*); abdominal ventrites narrowed in midline so that together they are shorter than metaventrite; anterior hypomeral carina from procoxae to lateral edge or anterior angles of pronotum present, often deliminating a cavity for reception of profemur (except in introduced Onitini and some *Canthonosoma*); more or less rounded beetles often with cephalic or pronotal armature; length from 2 mm to 30 mm ....**SCARABAEINAE**

7(3). Procoxae conical, produced ventrally (Fig. 28.12); mesepimera usually visible from above between pronotum and elytra (except in Valgini which are less than 5 mm long, covered with scales and have propygium exposed); antennal insertions exposed from above (clypeus with sides constricted just anterior to eyes); mandibles concealed; hind tibial apices without articulated spines (few in some Valgini); dorsal surfaces somewhat flattened; usually brightly coloured..**CETONIINAE**

– Procoxae transverse, not produced ventrally (Fig. 28.13); mesepimera not visible from above; antennal insertions not normally visible from above (IF slightly exposed THEN mandibles also exposed); mandibles concealed or slightly exposed; metatibial apices usually with articulated spines (except some Dynastinae); dorsal surfaces more convex..**8**

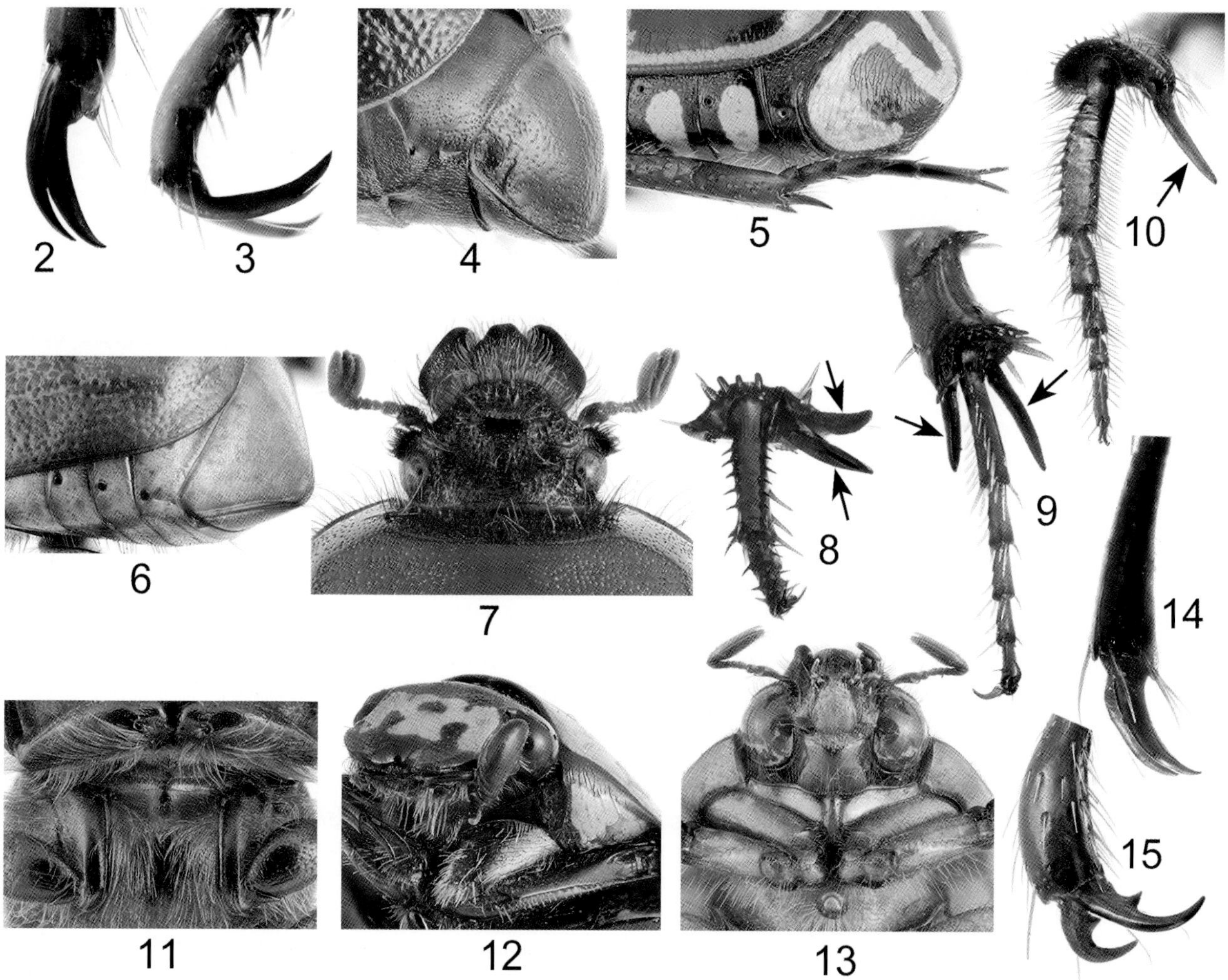

Figs 28.2–28.15. Adults. 2, Rutelinae, *Anoplognathus porosus* (Dalman), hind tarsal claw; 3, Melolonthinae, *Plectris aliena* Chapin, hind tarsal claw; 4, Melolonthinae, *Antitrogus plaiceps* Blackburn, abdominal spiracles; 5, Cetoniinae, *Eupoecila australasiae* (Donovan), abdominal spiracles; 6, Dynastinae, *Cyclocephala signaticollis* Burmeister, abdominal spiracles; 7, Melolonthinae, *Phaenognatha aequistriata* Arrow, abdominal spiracles; 8, Aphodiinae, *Ataenius picinus* Harold, hind tarsus; 9, Aphodiinae, *Aphodius cincticulus* Hope, hind tibia and tarsus; 10, Scarabaeinae, *Onthophagus* sp., hind tibia and tarsus; 11, Scarabaeinae, *Onthophagus* sp., mesocoxae; 12, Cetoniinae, *Eupoecila australasiae* (Donovan), prosternum; 13, Dynastinae, *Cyclocephala signaticollis* Burmeister, prosternum; 14, Dynastinae, *Teinogenys dives* (Blackburn), apical protarsomere; 15, Melolonthinae, *Antitrogus planiceps* (Blackburn), apical protarsomere.

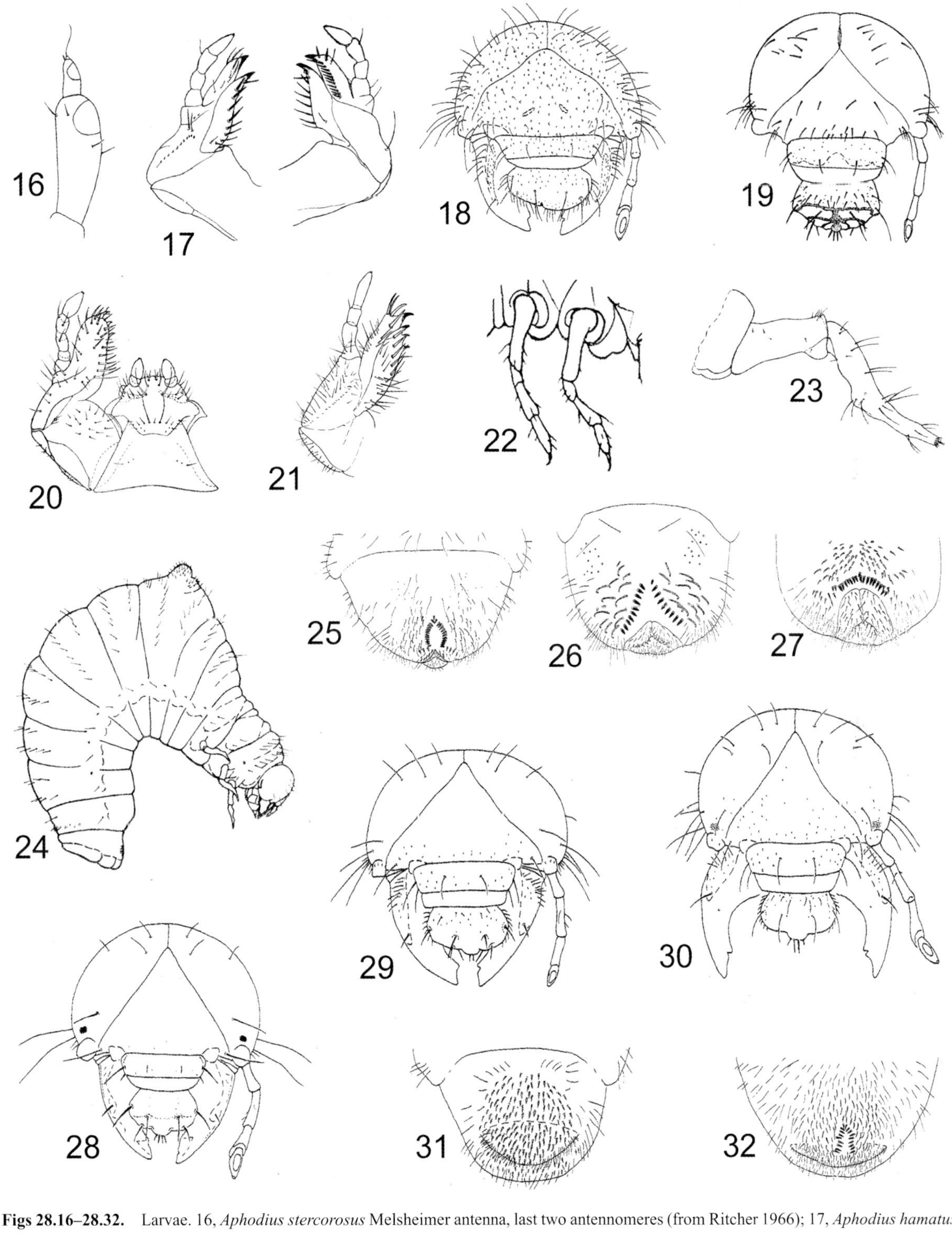

Figs 28.16–28.32. Larvae. 16, *Aphodius stercorosus* Melsheimer antenna, last two antennomeres (from Ritcher 1966); 17, *Aphodius hamatus* Say, maxillae, dorsal and ventral (from Ritcher 1966), 18, *Cheiroplatys latipes* (Guérin-Méneville), head, dorsal (from Mcquillan 1985); 19, *Plectris aliena* Chapin, head, dorsal (from Ritcher 1966); 20, *Anomala innuba* (Fabricius), maxilla and labium, ventral (from Ritcher 1966); 21, *Polyphylla decimlineata* (Say), maxilla, ventral (from Ritcher 1966); 22, *Aphodius sparsus* LeConte, larva, lateral (from Ritcher 1966); 23, *Dichotomius carolinus* (Linnaeus), larval, mesothoracic leg (from Ritcher 1966); 24, *Onthophagus hecate* (Panzer), larva, lateral (from Ritcher 1966); 25, *Colpochila obesa* Boisduval, raster (from Mcquillan 1985); 26, *Diphucephala smaragdula* (Boisduval), raster (from Mcquillan 1985); 27, *Heteronyx tasmanicus* Blackburn raster (from Mcquillan 1985); 28, *Telura* sp., head capsule (from Mcquillan 1985); 29, *Scitala sericans* Erichson, head capsule (from Mcquillan 1985); 30, *Heteronyx tasmanicus* Blackbuirn, head capsule (from Mcquillan 1985); 31, *Cheiroplatys latipes* (Guérin-Méneville), raster (from Mcquillan 1985); 32, *Saulostomus villosus* (Waterhouse), raster (from Mcquillan 1985).

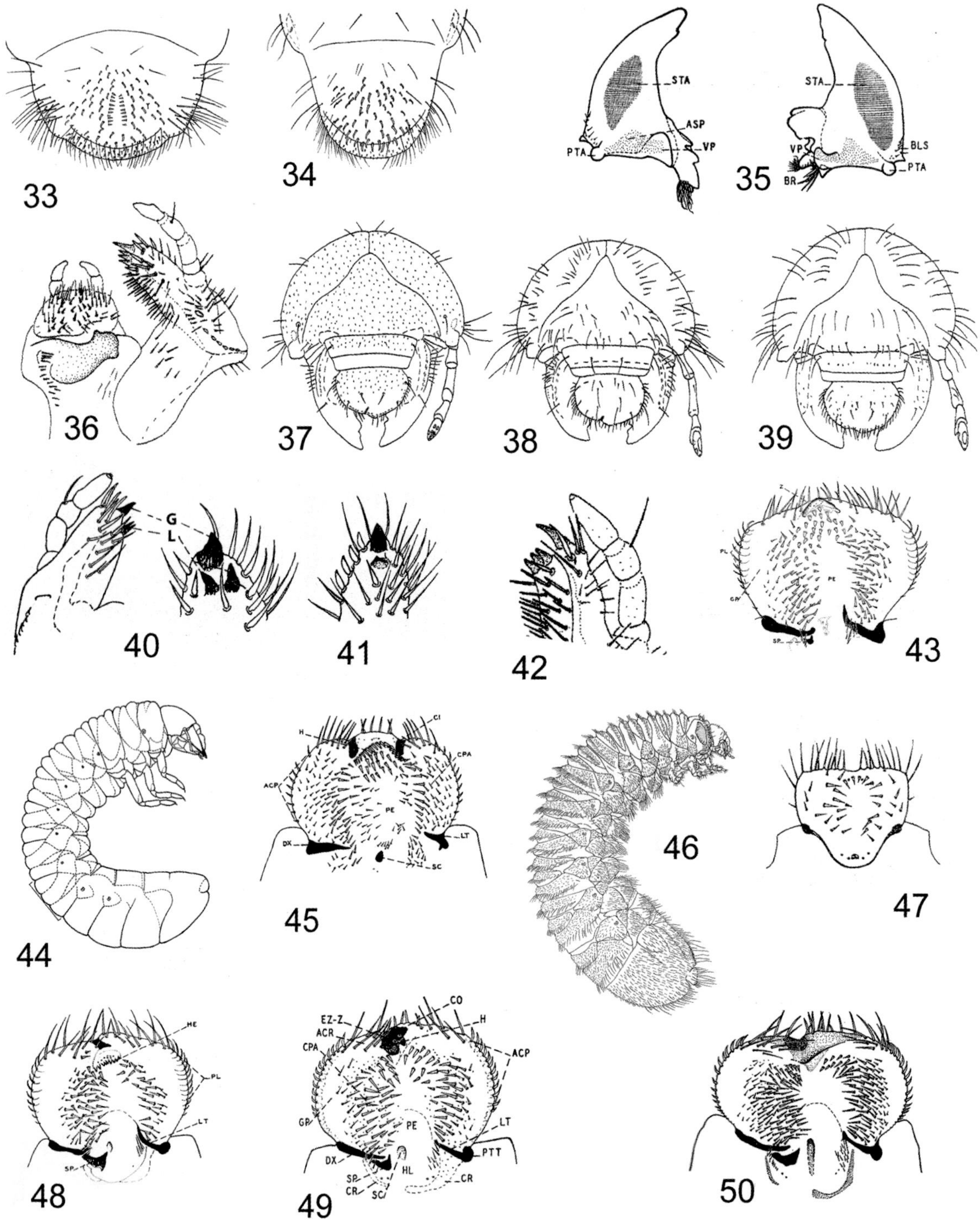

Figs 28.33–28.50. Larvae. 33, *Anomala innuba* (Fabricius), abdominal apex (from Ritcher 1966); 34, *Cyclocephala immaculata* Olivier, abdominal apex (from Ritcher 1966); 35, *Anomala innuba* (Fabricius), mandible, dorsal (from Ritcher 1966); 36, *Saulostomus villosus* (Waterhouse), maxilla and labium, dorsal (from Mcquillan 1985); 37, *Pimelopus nothus* Burmeister, head capsule (from Mcquillan 1985); 38, *Saulostomus villosus* (Waterhouse), head capsule (from Mcquillan 1985); 39 *Anoplognathus suturalis* Boisduval, head capsule (from Mcquillan 1985); 40 *Cotalpa lanigera* (Linnaeus), maxilla, dorsal, (from Ritcher 1966); 41, *Parastasia brevipes* (LeConte), maxillary apex, (from Ritcher 1966); 42, *Adoryphorus couloni* (Burmeister) maxilla, dorsal (from Mcquillan 1985); 43, *Parastasia brevipes* (LeConte), epipharynx(from Ritcher 1966); 44, *Anomala innuba* (Fabricius), larva, lateral(from Ritcher 1966); 45, *Cotinus nitidus* (Linnaeus), epipharynx (from Ritcher 1966); 46, *Hoplopyga brasiliensis* (Gory & Percheron), larva, lateral (from Costa *et al.* 1988); 47, *Valgus canaliculatus* (Fabricius), epipharynx (from Ritcher 1966); 48, *Adoretus sinicus* Burmeister, epipharynx (from Ritcher 1966); 49, *Cyclocephala immaculata* Olivier, epipharynx (from Ritcher 1966); 50, *Xyloryctes jamaicensis* (Drury), epipharynx (from Ritcher 1966).

8(7). Pretarsal claws of meso- and metatibiae simple (Fig. 28.14); mandibles usually at least partly visible from above; males and females always with 3 segmented antennal club (enlarged in males of some genera); labrum almost always concealed, IF exposed THEN not strongly transverse; head and pronotum in males (and sometimes females) often with tubercles, horns or complex elevations; often more than one pair of abdominal spiracles visible in intact specimen .... **DYNASTINAE**

– Pretarsal claws of meso- and metatibiae toothed or bifid (Fig. 28.15) (except *Zietzia* which has an 8 segmented antennal club in male); mandibles always concealed from above; males often with an enlarged antennal club of more than 3 segments; labrum exposed, strongly transverse; head and pronotum in males and female unarmed; only one pair of abdominal spiracles visible in intact specimen; mostly Melolonthini but some Pachtrichini and others may come out here as well .... **MELOLONTHINAE (part)**

LARVAE

1. Apical antennomere reduced, at most half as wide as preapical one, the apex of which bears a pale oval sensorium on inner side (Fig. 28.16); galea and lacinia distinctly separated for their entire lengths (Fig. 28.17) **2**

– Apical antennomere about as wide as preapical one, which lacks a sensorium (Figs 28.18–28.19); galea and lacinia entirely fused (Fig. 28.20), fused at base or fitting very tightly together (Fig. 28.21) **3**

2(1). Legs 5-segmented, including well developed pretarsal claw (Fig. 28.22); head capsule dark brown; body in lateral view more or less evenly C-shaped, without dorsal hump (as in Fig. 28.44) .... **APHODIINAE**

– Legs 2-segmented, without or with very small pretarsal claws (Fig. 28.23); head usually yellow; body in lateral view asymmetrical, with prominent dorsal hump (Fig. 28.24) .... **SCARABAEINAE**

3(1). Anal opening almost always angulate or Y-shaped and abdominal sternite IX usually with distinct raster composed of paired, curved, longitudinal rows or a curved, transverse row of stout setae (pali) (Figs 28.25–28.27); mandible with ventral stridulatory area absent or consisting of a patch of minute granules; galea and lacinia fused proximately or fitting very tightly together (Fig. 28.21); apical antennomere with single, large, oblong sensory spot (Figs 28.28–28.30) .... **MELOLONTHINAE**

– Anal cleft transverse (Figs 28.31–28.34); sternite IX rarely with raster composed of curved rows of stout setae (Fig. 28.32); mandible with ventral, oval stridulatory area consisting of a few to many transverse granular ridges (Fig. 28.35); galea and lacinia often entirely fused to form mala (Figs 28.36, 28.20); apical antenomere with one or more sensory spots (Figs 28.37–28.39) **4**

4(3). Lacinia with 1 or 2 well developed terminal unci (Fig. 28.41) or rarely with uncus almost absent **5**

– Lacinia with 3 well developed terminal unci (Figs 28.40, 28.42) **7**

5(4). Labrum (Figs 28.43) usually asymmetrical with anterior edge strongly rounded and epipharynx without clithra and never with haptomerum consisting of conspicuous, transverse, curved row of setae; legs long, more than half as long as thoracic width (Fig. 28.44) **RUTELINAE (major part)**

– Labrum (Fig. 28.45) symmetrical with anterior edge almost always trilobed and epipharynx with sclerite (clithrum) on either side of middle lobe and with haptomerum often with consisting of transverse, curved row of stout setae; legs usually short, less than half as long as thoracic width (Fig. 28.46) **6**

6(5). Anterior edge of labrum trilobed and apex of epipharynx with pair of clithra (Fig. 28.45); apical antennomere with 2 or more sensory spots; segments IX and X partly fused, the suture between them incomplete dorsally; dorsal surfaces clothed with short, stout spines mixed with longer setae; larva capable of back-crawling (Fig. 28.46); length usually greater than 15 mm .... **CETONIINAE (major part)**

– Anterior edge of labrum subtruncate or slightly concave and epipharynx without clithra (Fig. 28.47); apical antennomere with 1 sensory spot; segments IX and X not fused dorsally; larva not capable of back-crawling; length less than 15 mm .... **CETONIINAE: Valgini**

7(4). Epiphaynx with plegmatia and with haptomerum consisting of a dense, transverse row of 6 to 9 heli (Fig. 28.48); length less than 25 mm; known only from Melville I, NT .... **RUTELINAE: Adoretini**

– Epipharynx without plegmatia and with haptomerum consisting of a raised, bilobed, tooth-like or ridge-like process (Figs 28.49–28.50); length usually greater than 25 mm; widely distributed .... **DYNASTINAE**

Descriptions and illustrations of various scarabaeid larvae occurring in Australia may be found in Mcquillan (1985), and exotic species are treated in Medvedev (1952) and Ritcher (1966).

Acknowledgments

The authors are indebted to Oregon State University for allowing us to use illustrations from Ritcher (1966), to the Museu de Zoologia de Universidade de São Paulo, Brazil for the use of Fig. 28.46 from Costa *et al.* (1988) and to CSIRO and Peter Mcquillan for the use of images from Mcquillan (1985).

References

Bocak L, Barton C, Crampton-Platt A, Chesters D, Ahrens D, Vogler AP (2014) Building the Coleoptera tree-of-life for >8000 species: composition of public DNA data and fit with Linnaean classification. *Systematic Entomology* **39**, 97–110. doi:10.1111/syen.12037

Browne J, Scholtz CH (1995) Phylogeny of the families of Scarabaeoidea (Coleoptera) based on characters of the hindwing articulation, hindwing base and wing venation. *Systematic Entomology* **20**, 145–173. doi:10.1111/j.1365-3113.1995.tb00089.x

Browne J, Scholtz CH (1998) Evolution of the scarab hindwing articulation and wing base: a contribution toward the phylogeny of the Scarabaeidae (Scarabaeoidea: Coleoptera). *Systematic Entomology* **23**, 307–326. doi:10.1046/j.1365-3113.1998.00059.x

Costa C, Vanin SA, Casari-Chen SA (1988) *Larvas de Coleoptera do Brasil.* Museu de Zoologia, Universidade de São Paulo, São Paulo.

Grebennikov VV, Scholtz CH (2004) The basal phylogeny of Scarabaeoidea (Insecta: Coleoptera) inferred from larval morphology. *Invertebrate Systematics* **18**, 321–348. doi:10.1071/IS03013

Hunt T, Bergsten J, Levkanicova Z, Papadopoulou A, St. John O, Wild R, Hammond PM, Ahrens D, Balke M, Caterino MS, Gómez-Zurita J, Ribera I, Barraclough TG, Bocakova M, Bocak L, Vogler AP (2007) A comprehensive phylogeny of beetles reveals the evolutionary origins of a superradiation. *Science* **318**, 1913–1916. doi:10.1126/science.1146954

Lawrence JF, Ślipiński A, Seago AE, Thayer MK, Newton AF, Marvaldi AE (2011) Phylogeny of the Coleoptera based on morphological characters of adults and larvae. *Annales Zoologici* **61**(1), 1–217. doi:10.3161/000345411X576725

McKenna DD, Farrell BD, Caterino MS, Farnum CW, Hawks DC, Maddison DR, Seago AE, Short AEZ, Newton AF, Thayer MK (2015a) Phylogeny and evolution of Staphyliniformia and Scarabaeiformia: forest litter as a stepping stone for diversification of nonphytophagous beetles. *Systematic Entomology* **40**, 35–60. doi:10.1111/syen.12093

McKenna DD, Wild AL, Kanda K, Bellamy CL, Beutel RG, Caterino MS, Farnum CW, Hawks DC, Ivie MA, Jameson ML, Leschen RAB, Marvaldi AE, McHugh JV, Newton AF, Robertson JA, Thayer MK, Whiting MF, Lawrence JF, Ślipiński A, Maddison DR, Farrell BD (2015b) Beetles survided the end Permian mass extinction to diversify during the Cretaceous terrestrial revolution. *Systematic Entomology* **40**, 835–880. doi:10.1111/syen.12132

Mcquillan PB (1985) The identification of root-feeding cockchafer larvae (Coleoptera: Scarabaeidae) found in pastures in Tasmania. *Australian Journal of Zoology* **33**, 509–546. doi:10.1071/ZO9850509

Medvedev SI (1952) *Larvae of Scarabaeid Beetles of the USSR Fauna. Classification Keys to the Fauna of the USSR,* (in Russian). Academy of Sciences, Moscow, Leningrad.

Ritcher PO (1966) *White Grubs and their Allies. A Study of North American Scarabaeoid Larvae*. Oregon State University Press (Studies in Entomology No. 4), Corvallis, Oregon.

Scholtz CH, Grebennikov VV (2016) 15. Scarabaeoidea Latreille, 1802. Inroduction, phylogeny. In *Handbuch der Zoologie/Handbook of Zoology. BandVolume IV Arthropoda: Insecta. Teilband/Part 38. Coleoptera, Beetles. Volume 1: Morphology and Systematics (Archostemata, Adephaga, Myxophaga, Polyphaga partim).* 2nd Edition. (Eds RG Beutel and RAB Leschen) pp. 443–525. Walter de Gruyter, Berlin.

Smith ABT (2006) A review of the family-group names for the superfamily Scarabaeoidea (Coleoptera) with corrections to nomenclature and a current classification. *Coleopterists Society Monographs* **5**, 144–204.

Smith ABT, Hawks DC, Heraty JM (2006) An overview of the classification and evolution of the major scarab beetle clades (Coleoptera: Scarabaeoidea) based on preliminary molecular analyses. *Coleopterists Society Monographs* **5**, 35–46.

29. SCARABAEIDAE: APHODIINAE LEACH, 1815

Tom A. Weir, John F. Lawrence, Cate Lemann and Nicole L. Gunter

Fig. 29.1. *Acrossidius tasmaniae* (Hope)

Introduction. Aphodiinae is a large cosmopolitan group containing ~3300 species, but there is debate over the number of genera, these ranging from 178 (Dellacasa *et al.* 2001; Scholtz & Grebennikov 2016) to 280 (Skelley 2008) to over 300 (Cave & Ratcliffe 2008). The tribal classification is complicated but currently there are 15 recognised tribes: Aegialini (Holarctic, Australia, South America), Aphodiini (cosmopolitan), Aulonocnemini (Africa, Madagascar), Chironini (Africa, Asia). Corythoderini (Central & South Africa, India, Burma), Eupariini (cosmopolitan), Lomanoxiini (South America), Odochilini (Oriental), Odontolochini (nearly Pantropical), Proctophanini (South Africa, Australia, New Zealand, Moluccas), Psammodiini (cosmopolitan), Rhyparini (Pantropical), Stereomerini (Peru, Singapore, Borneo, Australia), Termitoderini (Africa) and Termitotrogini (India, Africa) (Forshage 2002). Scholtz & Grebennikov (2016) provide details of tribal revisional works of recent authors. The Australian fauna includes six genera in Aphodiini, three in Proctophanini, eight in Psammodiini, two in Odontolochini, and seven in Eupariini, with a single genus of Rhyparini and Aegialiini and two genera of Stereomerini, giving a total fauna of 183 species in 30 genera and eight tribes (Stebnicka 2009). Here we use *Aphodius* (*sensu lato*) and retain the subgenera as proposed by Stebnicka (2009).

There are eight genera endemic to Australia and one each to Lord Howe Island and Christmas Island. The genera *Ataenius* and *Saprosites* are pantropical, the former being highly conservative morphologically and constituting the central core of taxa from which the other genera have been recognised (Stebnicka 2009). Approximately 12% or 21 species are definite introductions from parts of Africa, Asia and America with climatic conditions similar to the target regions in Australia. The native Australian species show a high level of endemicity (Allsopp 1999) and tend to be concentrated around the periphery of the continent. Most species occur in the east and the north of the continent with fewer numbers in the arid central zone, south-west of Western Australia, South Australia and Tasmania (Stebnicka 2009). In very broad terms, the Aphodiinae of Australia have a close relationship with those in South America (Stebnicka & Howden 1997; Smith & Skelley 2007; Cabrero-Sañudo & Lobo 2009).

The age and origin of the subfamily remain debated. A study of aphodiine biogeography suggested the subfamily may date back as far as the early Jurrassic (Cabrero-Sañudo & Lobo 2009), however even the oldest molecular divergence analyses calibrated with numerous Mesozoic scarabaeoid fossils estimate a Lower Cretaceous origin for Aphodiinae (Ahrens *et al.* 2014; Gunter *et al.* 2016). The phylogenetic relationships of the group are also understudied and while numerous studies have confirmed a sister-relationship with the Scarabaeinae (Browne & Scholtz 1998; Philips *et al.* 2004; Smith *et al.* 2006; Monaghan *et al.* 2007; Ahrens *et al.* 2014; Gunter *et al.* 2016), taxon sampling is too limited within the Aphodiinae. Some evidence suggests the group represents a grade rather than a monophylum (see review of Smith *et al.* 2006). The composition of natural groups within the subfamily is contentious and compounded by species richness, intermediate character states between taxonomic groups and authors specialising in faunas from limited areas (Stebnicka 2001).

As in most other subfamilies in the Scarabaeidae, the early fossil record of Aphodiinae is poor with partial preservation, ambiguous morphological characters or unresolved relationships complicating the interpretation of the earliest records. For example, there is significant debate regarding the taxonomic placement of Aegialiini Laporte, which is either considered a tribe within Aphodiinae or a separate subfamily. While the oldest aegialiine fossil dates to the Lower Createaceous (130 MYA), the oldest known, unambiguous aphodiine fossil *Aphodius charauxi* Piton is recorded from the Eocene (56 MYA) (Krell 2007). The Oligocene and Miocene fossil record is substantially richer and although over 30 fossil species have been identified, none are known from Australia (Krell 2007).

The early taxonomy of the Australian species consisted mainly of isolated descriptions with Hope describing five species in 1846 and 1847, Macleay, eight species in 1871 and 1888, Blackburn, 36 species between 1892 and 1904 and Lea, a further 30 species between 1904 and 1926. Another 30 years elapsed before Petrovitz added 30 more species between 1958 and 1975, while Rakovic produced papers on the Australian species of Psamodiini between 1981 and 1984. Matthews & Stebnicka (1986) later removed *Demarziella* Balthasar to the subfamily Scarabaeinae. Dellacasa (1988) produced a comprehensive world catalogue and Cassis & Weir (1992) covered all the known Australian species in the *Zoological Catalogue of Australia*. Complete revisions of the Australian fauna by Stebnicka & Howden (1994, 1995, 1996, 1997), Howden & Storey (1992, 2000) and Storey & Howden (1996) added 74 new taxa, while Stebnicka (1998, 2001) detailed the neighbouring faunas of New Guinea and New Zealand respectively. Stebnicka (2009) published a comprehensive Iconography of the Australian Aphodiinae containing colour images of almost all of the 183 species and 30 genera found on the Australian mainland as well as Tasmania, Lord Howe Island, Norfolk Island, Christmas Island and Cocos-Keeling Islands.

Biology. Aphodiines are diverse in their habits, feeding on detritus, dung, carrion, fungi or living plants, with several groups living with termites or ants (Vårdal & Forshage 2010). The majority of species, however, feed on humus or dung. Depending on the structure of the mouthparts, they can be divided into hard consumers (Aegailiini) or soft consumers (most of the rest) of Cambefort (1991). Alternately, Stebnicka (2009) uses the terms 'hard saprophagy' or 'soft saprophagy or coprophagy' to describe these feeding habits. Examples of 'soft saprophagy' include most species of *Saprosites, Phycocus, Proctophanes* and *Ataenius,* while true coprophagy occurs in most species of Aphodiini, some Eupariini and Psammodiini and some members of other tribes (Stebnicka 2001). Several introduced species of *Aphodius* are attracted to dung, as is the native *Acrossidius tasmaniae* (Hope), but the latter and its close relative *A. pseudotasmaniae* Given are phytophagous in the larval stage. Their larvae construct vertical burrows from which they emerge at night and drag leaves into their burrows for consumption (Carne 1950, 1957). The cosmopolitan Psammodiini inhabit sandy areas and feed on plant roots. Many species of *Ataenius* and *Saprosites* are found under bark or in rotten wood or decaying vegetation. (Stebnicka 2009). A generalised pattern between larval biology and climate has been reported, tropical species generally breed in decomposing plant material while temperate species breed in dung (Stebnicka 2001). Interestingly, the adults of many of these dung-breeding temperate species are saprophagous, the adults of some species like the aforementioned *Acrossidius* spp. are coprophagous with saprophagous larvae and some adults have a mixed diet (Stebnicka 2001). It has been suggested that the variable diet of aphodiines has enabled them to adapt to so many specialised niches (Stebnicka 2001). The most specialised adaptations include kleptoparasitism, and living in the mounds of ants and termites, nests of birds and burrows of small mammals (Stebnicka 2001).

In terms of dung beetle communities measured by species richness and total abundance, aphodiines dominate only in the southern/temperate regions of Australia where they contribute greatly to nutrient cycling through dung removal (Doube *et al.* 1991; Steinbauer & Weir 2007; Cabrero-Sañudo & Lobo 2009). Conversely, scarabaeines dominate dung beetle communities in the northern regions of the continent. While scarabaeines display a higher degree of variability in dung feeding and breeding ecology, coprophagous aphodiines live and reproduce within the dung source. Females lay an average of 20–25 eggs, and when hatched the larvae are free-living within the dung, receiving no parental care (Stebnicka 2001). There is some evidence for species level spatial niches within dung pats (e.g. top-, bottom- or periphery- dwellers *v.* those species that occur evenly throughout) (Holter 1982). The generation time, development cycle, emergence pattern are variable, however most species have two generations per year (Stebnicka 2001).

Typical methods for collecting aphodiines, such as baited pitfall trapping, litter and log sifting, exploit their biologically association with dung and plant debris, while flighted species are regularly collected at light and in flight intercept traps. Although aphodiines are usually collected on the ground, fully-winged species regularly frequent the canopy (Stork & Grimbacher 2006; Davis *et al.* 2011).

Characteristics. Adults. Australian species vary from 2 to 14 mm in length, although some exotic species of Chironini are more than 30 mm in length (Huchet 2000). Most species are black or brown in colour, but *Aphodius fimetarius* (Linnaeus) has black head and pronotum with red elytra, while *Podotenus* species often have yellow elytra with black markings. The body is usually more or less parallel-sided, although oblong to ovoid in a few genera. The labrum is concealed by the clypeus in almost all Australian species (exposed only in *Saprus*) and the mandibles are usually flattened and plate-like. A combination of characters separating aphodiines from other scarabaeid subfamilies includes tarsi of equal length, elytra concealing the entire abdomen, pygidium partially or completely exposed, abdominal spiracles concealed in pleural membrane, mesocoxae strongly oblique and usually separated by less than their width, mesoscutellum almost always present (minute or absent in Rhyparini and Stereomerini), labrum almost always concealed (slightly exposed in *Saprus*), and metatibia with 2 apical spurs (very small in Rhyparini and Stereomerini) (Stebnicka 2009; Lawrence & Slipinski 2013).

Larvae. May be distinguished from those of other scarabaeoids by the 4- or apparently 5-segmented antennae with a reduced and narrow apical antennomere, distinctly separated galea and lacinia, anal opening usually surrounded by anal lobes, 5-segmented legs, each with a distinct, claw-like pretarsus, a well sclerotised head capsule, a more or less cylindrical, C-shaped body (without dorsal hump) (Carne 1950; Jerath 1960; Ritcher 1966).

Key to the tribes and genera of Australian Aphodiinae

Some species of *Australammoecius* and *Ataenius* are difficult to distinguish at the generic level. The key tries to allow for these differences by bringing out these species in couplets 23, 24, 31 and 32.

1. Labrum well sclerotised and visible from above; mandibular apices heavily sclerotised and visible just beyond apex of labrum; 2.7–4.5 mm; TAS, VIC, NSW .... **AEGIALIINI:** ***Saprus*** **Blackburn**

– Labrum and mandibular apices more or less membranous and concealed beneath clypeus **2**

2(1). Pronotum and elytra longitudinally costate or carinate; head with longitudinal grooves or costae; mesoscutellum absent **3**

– Pronotum never longitudinally costate or carinate; head without longitudinal grooves or costae; mesoscutellum rarely absent **5**

3(2). Elytra more than twice as long as pronotum, parallel-sided with sharply raised longitudinal costae separated by double rows of large punctures; subsutural costa with trichome near posterior end; pronotal costae sharp and more or less continuous; prosternal process hastate, expanded subapically and acute at apex; body surfaces subglabrous; tarsi 5-5-5; metatarsomere 1 longer than 2 and 3 combined; NSW, QLD .... **RHYPARINI:** ***Rhyparus*** **Westwood**

– Elytra less than twice as long as pronotum; elytral costae not separated by double rows of large punctures and without trichomes; pronotal costae rounded, separated by narrow grooves and interrupted by transverse impressions; prosternal process not hastate; body surfaces covered with flattened scales; tarsi 5–5–4 or apparently 5–3–3; metatarsomere 1 shorter than 2 and 3 combined.....**STEREOMERINI**..............................**4**

4(3). Lateral head grooves S-shaped; anterior edge of clypeus not vertical, in same plane as clypeal base; elytra with strong transverse striations, widest at base and narrowing posteriorly; mesotibia and metatibia oval in cross-section, not flattened; 2.3–2.6 mm; northern QLD .... ***Daintreeola*** **Howden & Storey**

– Head grooves straight and parallel; anterior edge of clypeus vertical, in different plane than clypeal base; elytra with transverse striations weak or absent, widest at about middle; mesotibia and metatibia strongly flattened; 2.2–3.2 mm; northern QLD, NT..................***Australoxenella*** **Howden & Storey**

5(2). Metatibial spurs separated by at least a few setae and located on each side of tarsal insertion so that basal tarsomere passes between them in moving (Fig. 29.2) (or nearly so in *Candezeollus*); pronotum evenly convex, rarely with median longitudinal line or groove; elytral base without marginal basal bead; pygidium more or less oblique and usually concealed in intact specimens, smooth, without roughened area, carina or notch; mesocoxae contiguous and oblique; abdominal ventrites not fused, freely moveable...**APHODIINI**..............................**6**

– Metatibial spurs more or less contiguous, located on medial side of tarsal insertion, so that basal tarsomere passes above them when moving (Figs 29.3, 29.4); pronotum evenly convex or with impressions and swellings; elytral base with or without marginal basal bead; pygidium usually vertical and at least partially exposed in intact specimens, usually carinate or with roughened area or dorsal (basal) notch; mesocoxae contiguous or separated, oblique or parallel; abdominal ventrites usually fused..............................**11**

6(5). Elytra usually with distinct preapical umbone and with odd intervals differing, at least slightly, from even ones either in height, width or in tubercles or setae; posterior portion of pronotal disc usually with median longitudinal groove or finely impressed line; sides of pronotum often emarginate or excavate posteriorly; outer edge of protibia with 2 teeth in male and 3 teeth in female; upper surfaces setose; head supplied with readily visible abundant long or short setae; sides of pronotum with fringe of long or short setae; 4.0–8.0 mm; widely distributed.. ***Podotenus*** **Schmidt**

– Elytra without preapical umbone; odd elytral intervals not differing from even ones, without tubercles or swellings; pronotum rarely with median longitudinal groove; sides of pronotum rarely emarginate or excavate posteriorly; outer edge of protibia with 3 teeth in both sexes; upper surfaces setose or glabrous; head either glabrous (*Aphodius, Ozodius*), or with extremely minute setae (*Candezeollus,* most *Acrossidius*) or with a few long upright setae (*Lorditomaeus*); sides of pronotum with or without fringe of setae [*Acrossidius yorkensis* has a few moderate setae on the head, but the clypeus has a marked fringe of close set setae, pronotum is very convex, and is 10–14 mm in length]..............................**7**

7(6). Elytra with striae 9 distinctly curving anteriorly to meet striae 10 (Fig. 29.5); sides of pronotum either with complete fringe of short setae (longer in *Aphodius paradivisus*) or without setae; head sometimes with carinae or tubercles; 3.0–7.5 mm; widely distributed.......... ***Aphodius*** **Illiger**

– Elytra with striae 9 separate from striae 10 anteriorly; sides of pronotum with complete or partial fringe of long setae or without setae; head never with carinae or tubercles..............................**8**

8(7). Elytra with striae 7 and 8 meeting anteriorly; head with large punctures bearing long, upright setae; disc of pronotum with moderate setose punctures and scattered larger punctures bearing long, upright setae; sides of pronotum with fringe of setae confined to posterior angles; elytra arcuate at sides; metatarsal segment 1 subequal to next 4 segments combined; 5–6 mm; SA.................................... ***Lorditomaeus*** **Peringuey**

– Elytra with striae 7 and 8 separate anteriorly; head either glabrous or with extremely minute setae or with a few moderate setae; disc of pronotum either glabrous or with a few setae in anteroloateral angles; sides of pronotum with fringe of setae complete or confined to anterior angles, or without setae; elytra more or less parallel-sided; metatarsal segment 1 at most just longer than next 3 segments combined..............................**9**

9(8). Size small, 2.5–4.0 mm; disc of pronotum with punctures of two distinct sizes, moderate and very small; sides of pronotum without setae; metanepisternum with a longitudinal carinae where it abuts elytra (Fig. 29.6), pygidium pointed apically, edge with 2 or 4 setae; elytra with ninth interval somewhat costate; NSW, ACT, VIC, WA.. ***Ozodius*** **Skelley**

– Size larger, 5.0 mm or more; disc of pronotum with punctures basically of one size; sides of pronotum with fringe of setae complete or confined to anterior angles; metanepisternum without longitudinal carina, rounded where it abuts elytra; pygidium less pointed apically, edge with at least 6 setae; elytra with ninth interval not costate..............................**10**

10(9). Length 5.5–7.5 mm; body EITHER with head and pronotum brown and elytra yellowish with darker spots OR deep reddish brown with explanate pronotum (*Candezeollus bicrenulatus*); sides of head converging anteriorly; clypeal margin distinctly emarginate medially; elytra with epipleural flange weak in apical third; VIC, SA, WA .... ***Candezeollus*** **Stebnicka & Howden**

– Length 8–14 mm; body unicoloured, reddish brown to black; sides of head more or less semicircular; clypeal margin weakly emarginate medially (females) or truncate (males); elytra with epipleural flange strong in apical third; NSW, VIC, SA, TAS .... ***Acrossidius*** **Schmidt**

11(5). Body globular; head granulate; metatibiae twice as wide as mesotibiae; metatarsi very short, tarsomeres triangular; claws setaceous; strongly associated with sand dunes on beaches...**PSAMMODIINI (part)**..............................**12**

– Body elongate (except *Leiopsammodius newcastlensis*); head granulate or not; metatibiae slender, not twice as broad as mesotibiae; metatarsi long or moderate in length; claws corneous; not normally associated with sand dunes on beaches..............................**13**

12(11). Pronotum very short, less than 0.5 times as long as wide, impunctate, smooth; elytra less than 1.15 times as long as wide, without striae; elytral base without marginal basal bead; outer side of metatibia with irregularly spaced denticles; eyes absent; 2.5–3.0 mm; TAS......... ***Phycocus*** **Broun**

– Pronotum longer, more than 0.5 times as long as wide, punctate, with vestiges of transverse furrows; elytra more than 1.15 times as long as wide, striate; elytral base with marginal basal bead; outer side of metatibiae with a strong transverse ridge; eyes vestigial; 2.8–3.2 mm; TAS .... ***Tesarius*** **Rakovic**

13(11). Elytral base lacking marginal basal bead, without distinct epipleural denticle below humeral area; pygidium not modified, without deep basal longitudinal groove to receive elytral sutural process (although shallow groove maybe present) (Figs 29.7, 29.8, 29.9); pygidium never with transverse carina and/or central notch; femora never grooved on anterior or posterior margins; transverse ridges of metatibia strongly developed (less so in *Harmogaster*)...**PROCTOPHANINI**..............................**14**

– Elytral base with marginal basal bead: if weak or hard to see, then epipleural denticle present below humeral area (Fig. 29.24, arrow); pygidium modified, with deep basal longitudinal groove that receives an elytral sutural process at rest (may need to raise elytra to see) (Figs 29.10, 29.11, 29.12); pygidium with transverse carina and/or central notch (absent in *Saprosites* and *Cnemoplatys*); femora often with grooved margins; transverse ridges of metatibiae less developed, or absent..............................**16**

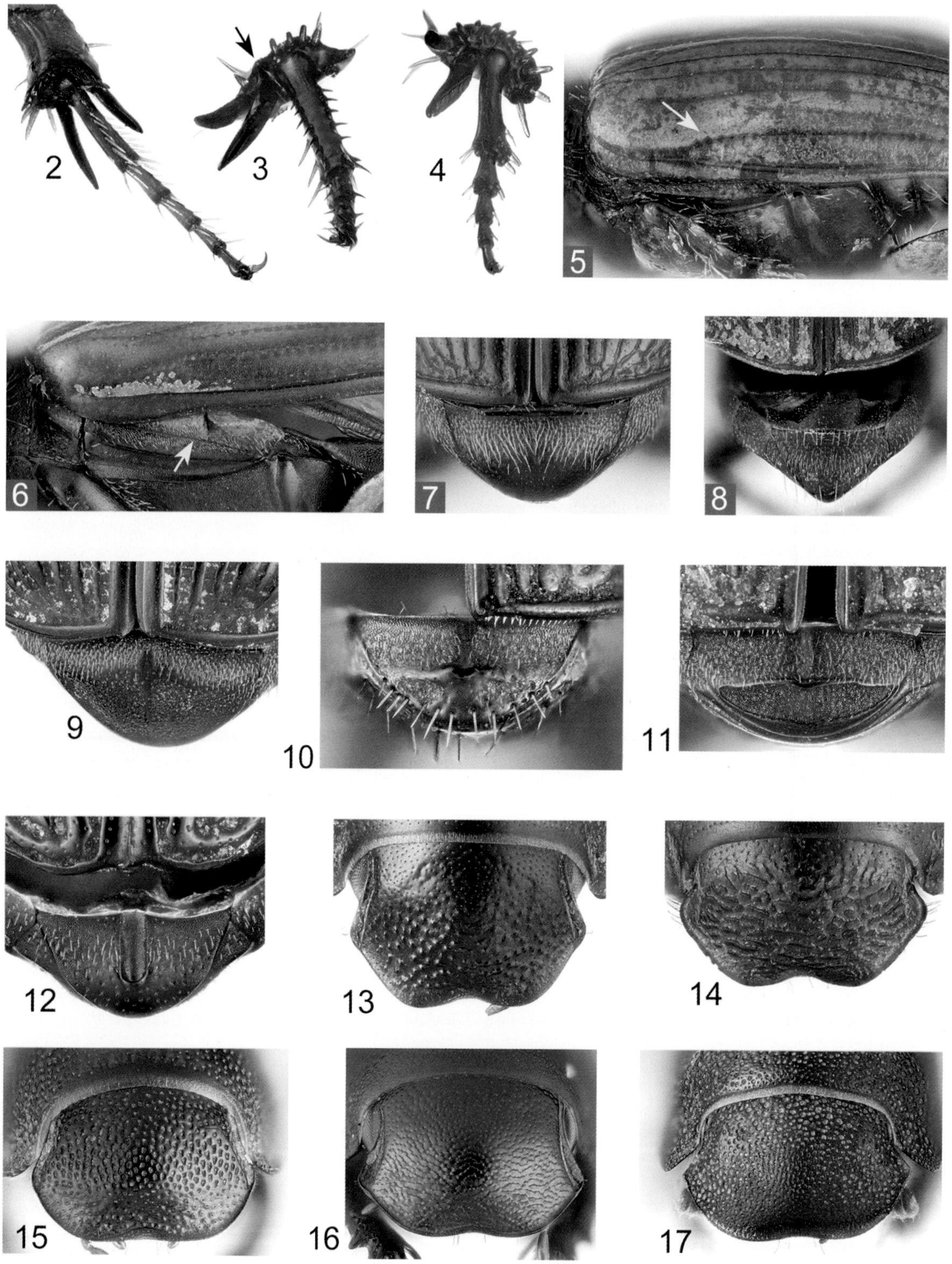

Figs 29.2–29.17. Aphodiinae. 2, *Aphodius cincticulus* Hope, metatibial apex and metatarsus; 3, *Ataenius picinus* Harold, metatibial apex and metatarsus; 4, *Australammoecius goyderensis* (Blackburn), metatibial apex and metatarsus; 5, *Aphodius cincticulus* Hope, elytron, lateral; 6, *Ozodius neglectus* (Schmidt), metanepisternum and elytron, lateral; 7, *Proctophanes distinguendus* (Schmidt), pygidium; 8, *Harmogaster geminata* Schmidt, pygidium; 9, *Proctophanes sculptus* (Hope), pygidium; 10, *Aphodopsammobius rugicollis* (Macleay), pygidium; 11, *Ataenius australasiae* (Boheman), pygidium; 12, *Saprodites mendax* (Blackburn), pygidium; 13, *Australammoecius goyderensis* (Blackburn), head, anterodorsal; 14, *Parataenius simulator* (Harold), head, anterodorsal; 15, *Ataenius edungalbae* Stebnicka & Howden, head, anterodorsal; 16, *Ataenius picinus* Harold, head, anterodorsal; 17, *Airapus obscurus* (Macleay), head, anterodorsal.

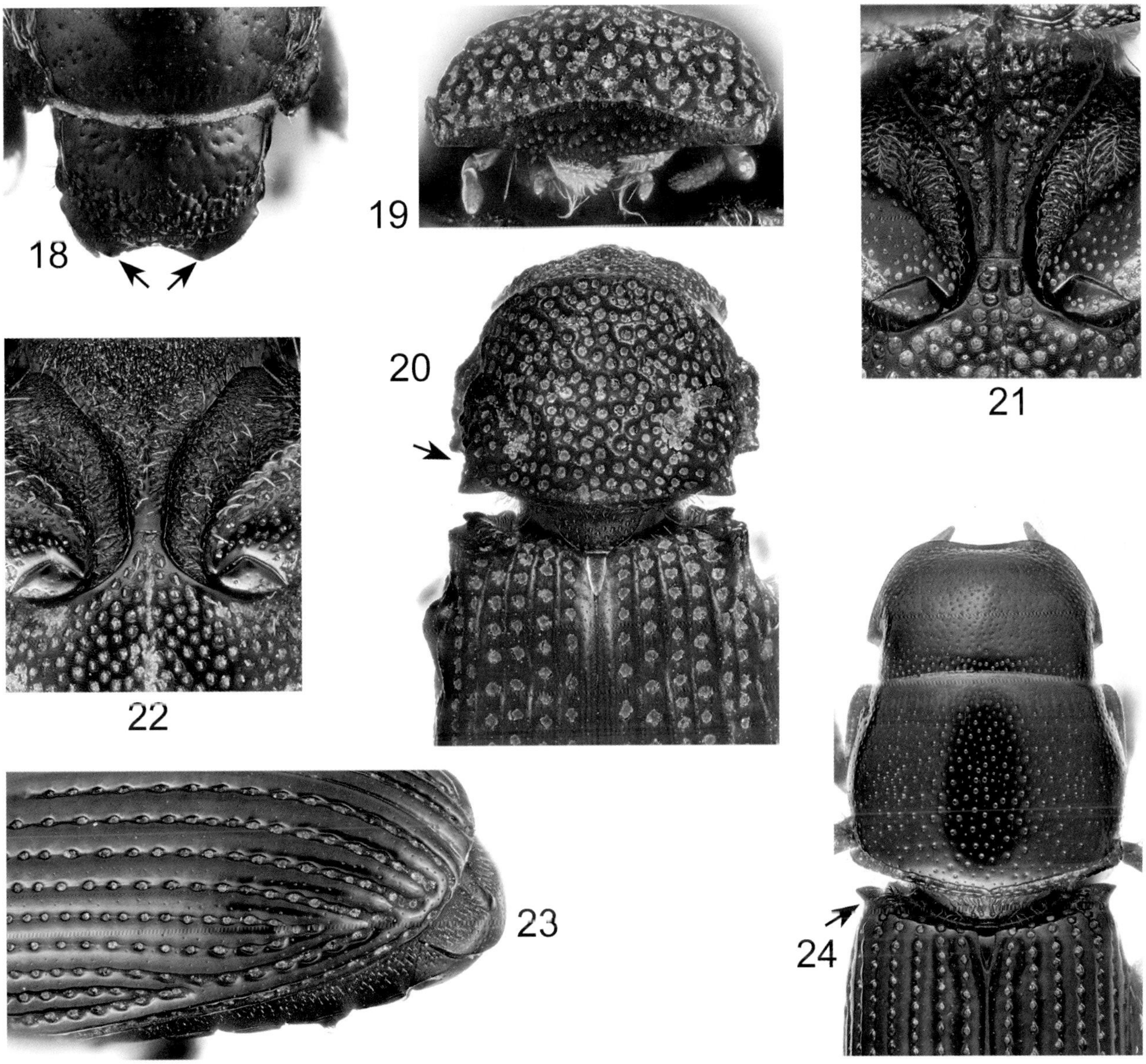

Figs 29.18–29.24. Aphodiinae. 18, *Psammorpha lawrencei* Stebnicka, head, dorsal; 19, *Odontolochus weiri* Stebnicka & Howden, clypeus, frontal; 20, *Odontolochus weiri* Stebnicka & Howden, pronotum and anterior portion of elytra; 21, *Airapus obscurus* (Macleay), meso-metaventral junction (with suture); 22, *Ataenius angulatus* Petrovitz, meso-metaventral junction (without suture); 23, *Saprosites mendax* (Blackburn), elytral apex, dorsolateral; 24, *Saprosites mendax* (Blackburn), head, pronotum and anterior portion of elytra, dorsal.

14(13). Body short, oval convex; base of pronotum without marginal border; mesocoxae separated by distance about as great as width of mesofemur; head sometimes with carinae or tubercles; 2.5–5.8 mm; widely distributed .... ***Proctophanes*** **Harold**

– Body elongate oval, moderately convex; base of pronotum with marginal border; mesocoxae separated by distance less than one-fifth as great as width of mesofemur; head never with carinae or tubercles **15**

15(14). Each elytral stria composed of 3 undulate lines and 2 rows of punctures; 4.2–4.8 mm; NSW, SA, QLD .... ***Harmogaster*** **Harold**

– Each elytral stria composed of 1 row of punctures; 3.2–4.5 mm; TAS, VIC, SA, WA .... ***Australophodius*** **Balthasar**

16(13) Clypeus, from anterior margin to front of eyes distinctly granulate (Fig. 29.13), or coarsely or finely transversely wrinkled, at least laterally (Fig. 29.14) **17**

– If clypeus distinctly granulate (Fig. 29.15), or transversely wrinkled (Fig. 29.16), then this confined to a narrow anterior band; otherwise whole of clypeus impunctate, simply punctate (Fig. 29.17), subgranulate, rugulose, scabrous or with longitudinal punctures **25**

17(16) Pronotum EITHER with up to 5 distinct transverse ridges and furrows OR their vestiges frequently marked by rows of large punctures, at least visible near eyes OR pronotum evenly convex with only lateral foveae (*Aphodopsammobius*); longitudinal furrow or impression often present, although sometimes only marked by a row of punctures; metatarsi shorter than metatibiae and basal tarsal segment shortened and widened (except *Aphodopsammobius zeitzi*); clypeus always granulate (Fig. 29.18)... **PSAMMODIINI (part)** **18**

– Pronotum without transverse ridges and furrows or their vestiges or lateral foveae; clypeus either granulate, or transversely wrinkled ...**EUPARIINI (part)** **23**

18(17). Pronotum with 5 distinct transverse ridges and furrows or with furrow 3 more distinct and deeper than others; with median longitudinal groove dividing at least ridges 4 and 5 .... **19**

– Pronotum without 5 distinct transverse ridges and furrows; median longitudinal groove represented at most by a row of punctures **20**

19(18). Elytra glabrous; 3.0–3.8 mm; NT, QLD, SA, WA, Christmas Island .... ***Rhyssemus* Mulsant**

– Elytra with setae; 3.5–3.8 mm; Christmas Island, Cocos-Keeling Is. .... ***Neotrichiorhyssemus* Rakovič & Král**

20(18). Clypeus angulate either side of anterior emargination (Fig. 29.18 – arrows); head characteristically elongate and parallel sided; eyes vestigial (Fig. 29.18); elytral humeri each with 2 denticles; pronotum with vestiges of transverse and longitudinal furrows represented by lines of punctures; pygidium granulate; 3.8 mm; known only from Christmas Island .... ***Psammorpha* Stebnicka**

– Clypeus not angulate, evenly rounded either side of anterior emargination; head not elongate; eyes present or vestigial; elytral humeri without denticles; pronotum with or without vestiges of transverse and longitudinal furrows; pygidium not granulate **21**

21(20). Meso- and metatibiae externally with at least vestiges of 2 transverse ridges; pronotum without vestiges of transverse furrows, with random large punctures and lateral foveae; 2.3–4.5 mm; widely distributed .... ***Aphodopsammobius* Endrodi**

– Meso- and metatibiae externally without transverse ridges, with at least a few denticles; pronotum with vestiges of transverse furrows represented by rows of punctures **22**

22(21). Lateral edges of pronotum with a fringe of setae crenulating margin; meso and metatibiae with rows of denticles; clypeofrontal suture impressed or represented by a dark line; elytra with striae 8 shorter than 7, not reaching basal bordered margin anteriorly; 2.6–3.6 mm; NSW, SA, WA, NT .... ***Leiopsammodius* Rakovic**

– Lateral edges of pronotum smooth, glabrous; meso and metatibiae with a few denticles, not in rows; frontoclypeal suture indistinct; elytra with striae 8 reaching basal bordered margin anteriorly; 2.7–4.0 mm; SA, WA, QLD, Lord Howe Island; .... ***Platytomus* Mulsant**

23(17). Clypeus granulate; edges of pronotum smooth, without a fringe of setae; metatarsi shorter than metatibiae; apex of meso- and metatibiae with only a fringe of spines, lacking any projections (Fig. 29.4); outer side of metatibiae with incomplete transverse ridges; 3.2–5.0 mm; WA, SA, QLD, wNT .... ***Australammoecius* Petrovitz (part)**

– Clypeus transversely wrinkled; edges of pronotum smooth or crenulated, with a fringe of setae; metatarsi about as long as metatibiae; as well as a fringe of spines, apex of meso- and metatibiae with a small to very small projection near shorter spur and separate from it by one spine (Fig. 29.3 – arrow), and another projection on the outer corner; outer side of metatibiae without transverse ridges; introduced species **24**

24(23). Clypeus coarsely transversely wrinkled (Fig. 29.14); metafemora enlarged, about as wide as profemora, without marginal grooves; 3.8–5.9 mm; NSW, VIC, QLD; (*Parataenius simulator*) .... ***Parataenius* Balthasar**

– Clypeus finely transversely wrinkled; metafemora not enlarged, somewhat parallel sided, narrower than profemora; with marginal grooves, at least near apex; 3.5–6.0 mm; widespread including Christmas Island .... ***Ataenius* Harold (part)**

25(16). Clypeus with anterior edge broadly inflexed and extending from gena to gena, forming a strongly transverse horizontal plate (Fig. 29.19); pronotum sharply declivous laterally, dentate, somewhat emarginate posteriorly, with distinct median longitudinal furrow and/or lateral sulci; head able to be deflexed at nearly 90 degrees to horizontal, forming, with anterior portion of pronotum, a cavity for forelegs... **ODONTOLOCHINI** **26**

– Clypeus with anterior edge not broadly inflexed, sometimes with a small triangular inflexed process at middle; pronotum laterally without denticles, IF slightly angulate and posteriorly emarginate THEN anterior pronotal angles explanate, without distinct median longitudinal furrow or lateral sulci; head deflexion less than 90 degrees...**EUPARIINI (part)** **27**

26(25). Upper surfaces more or less glabrous; sides of pronotum angulate and posteriorly emarginate (Fig. 29.20); elytra less than 1.3 times as long as wide; mesocoxae separated by more than shortest diameter of one coxa; mesoventrite without median carina; elytral striae wider than intervals; pygidium with longitudinal as well as transverse carinae; 2.0–4.5 mm; NSW, QLD .... ***Odontolochus* Stebnicka & Howden**

– Upper surfaces clothed with short, stout, yelolow bristles; sides of pronotum dentate; elytra more than 1.3 times as long as combined width; mesocoxae separated by less than shortest diameter of one coxa; mesoventrite with median, longitudinal carina; elytral striae narrower than intervals; pygidium with only transverse carina; 5.9–6.5 mm; northern QLD .... ***Gongrolophus* Stebnicka & Howden**

27(25). Sides of pronotum anteriorly explanate; distinct meso-metaventral suture present (Fig. 29.21); mesocoxae widely separated, parallel or subparallel; mesoventrite subdepressed or flat with various calosities **28**

– Sides of pronotum at most slightly reflexed; meso-metaventral suture absent or indistinct (Fig. 29.22); mesocoxae slightly separated or contiguous, oblique; mesoventrite evenly convex or with median longitudinal elevation, without calosities **29**

28(27). Body glabrous or setose; elytral intervals smooth or irregularly swollen; meso- and metatibiae straight, with 2 to 4 longitudinal serrate or setigerous lines; 3.5–8.0 mm; NSW, QLD, NT .... ***Airapus* Stebnicka & Howden**

– Body covered with scales and curled setae; elytral intervals strongly tuberculate; meso- and metatibiae slightly sinuate, without longitudinal lines; 4.8–5.8 mm; endemic to Lord Howe Island .... ***Euparotrix* Stebnicka & Howden**

29(27). Body elongate, dorsum somewhat flattened; elytra more or less parallel-sided; elytral interval 10 curving around and meeting interval 2 at apex (Fig. 29.23); head almost as wide as anterior edge of pronotum (Fig. 29.24), usually finely punctuate; pygidium with transverse carina weak or absent, with simple punctures **30**

– Body less elongate, dorsum convex; elytra often more or less rounded laterally, widest at middle; elytral interval 10 ending well before apex and not meeting interval 2; head distinctly narrower than anterior edge of pronotum, surface variously sculptured; pygidium with obvious transverse carina and apical area usually scabrous or roughened **31**

30(29). Mesocoxae narrowly separated by metaventrite forming a median carina; meso- and metatibiae almost parallel-sided to apex, excised at basal joint; tibial spurs strongly curved; tarsomeres thick; elytra more than 1.75 times as long as combined width; 4.5–5.0 mm; northern QLD .... ***Cnematoplatys* Schmidt**

– Mesocoxae not separated by distinct carina, metaventrite usually flat; meso- and metatibiae expanded apically, not excised at joint; tibial spurs not or slightly curved; tarsomeres slender; elytra less than 1.7 times as long as combined width; 2.1–5.8 mm; widely distributed .... ***Saprosites* Redtenbacher**

31(29) Clypeus EITHER with heavily or lightly impressed longitudinal punctures and/or lines of united longitudinally confluent punctures that are always well defined at least laterally OR strongly scabrous; 2.5–6.0 mm; widespread .... ***Ataenius* Harold (part)**

– Clypeus EITHER impunctate, simply punctate OR with ill-defined granulations, subgranulations or rugulose areas anteriorly **32**

32(31) With at least one of the following characters: elytral intervals on disc carinate or convex with their margins crenulate; profemora with posterior faces rough or wholly or partly distinctly punctate; abdominal ventrites densely punctate; 3.0–5.5 mm .... ***Ataenius* Harold (part) and *Australammoecius* Petrovitz (part) (*Australammoecius persimilis*)**

– With all of the following characters: elytral intervals on disc flat; profemora posterior face smooth and shiny; abdominal ventrites with a few punctures, mostly lateral; 2.2–4.8 mm... ***Ataenius* Harold (part) (*Ataenius gibbus, Ataenius nudus, Ataenius parvus*) and *Australammoecius* Petrovitz (part) (*Australammoecius occidentalis, Australammoecius peckorum*)**

Classification of Australian genera

Here we follow the classification in Stebnicka (2009). We take a conservative view of the genus *Aphodius* and retain the subgenera.

Tribe Aegialiini Laporte, 1840

Characteristics. Small body size, length of most less than 5 mm. Head flat or slightly convex, surface granulate or punctate. Mandibles and labrum not covered by the clypeus. Aegialiini is most closely related to the Eupariini-Psammodiini complex of species.

Distribution. Mainly Holarctic with a few members in Australia and South America, with the world fauna comprising approximately eight genera and 65 species.

Australian taxa. One genus.

Saprus Blackburn, 1904 (Pl. 51G)

Type species. *Saprus griffithi* Blackburn, 1904.

Australian species. Four.

Distribution. Endemic to Australia, occurring in TAS, VIC and NSW.

Biology. Occurring in leaf litter in temperate and subtropical rainforest.

References. Stebnicka & Howden (1995); Stebnicka (2009).

Tribe Aphodiini Leach, 1815

Characteristics. Elytra with basal margin not bordered. Femora not grooved at front or hind margins. Middle and hind tibiae with two or more transverse carinae. Pygidium smooth, unmodified, usually covered by the elytra. Metatibial apical spurs set apart so that tarsus can articulate between them. Abdominal ventrites not fused, with intersegmental membranes. Aphodiini forms the core of the family.

Distribution. Worldwide (but a paucity in South America), with ~170 genera and 2000 species (Skelley 2008).

Australian taxa. The current fauna in Australia consists of two elements: endemic species and those accidentally introduced from Africa, South America and the Oriental region. Six genera.

Acrossidius Schmidt, 1914 (Pl. 52H, 53B)

Type species. *Aphodius tasmaniae* Hope, 1847.

Australian species. Four.

Distribution. Endemic to Australia and introduced to New Zealand. *Acrossidius pseudotasmaniae* (Given) is restricted to TAS and *A. tasmaniae* (Hope) is known from SA, VIC, NSW, TAS and New Zealand. *A. brittoni* Stebnicka & Howden and *A. yorkensis* (Blackburn) both occur in SA.

Biology. Both *A. tasmaniae* and *A. pseudotasmaniae* are pasture pests and the larva is usually referred to as the black-headed pasture cockchafer. Larvae construct vertical burrows from which they emerge at night to feed on leaves. Larvae were described by Carne (1950).

References. Stebnicka & Howden (1995); Stebnicka (2009).

Aphodius Illiger, 1798 (Pl. 52L)

Type species. *Scarabaeus fimetarius* Linnaeus, 1758.

Australian species. Eight species in six subgenera.

Distribution. Worldwide. All Australian species, except *A. albertisi* Harold from NT and QLD, have been introduced from various parts of the world. *A. granarius* (Linnaeus) is cosmopolitan, *A. fimetarius* (Linnaeus) is primarily Palearctic, two species are primarily Oriental and two apparently introduced from South Africa. *Aphodius cinctulus* Hope occurs in the Nearctic, Neotropical, Afrotropical and Oceanic regions and is one of the very few *Aphodius* found in the dry interior.

Biology. Most species occur in dung of various types although debris, decaying vegetation, compost, carrion are also utilised. Several species are attracted to lights. Larvae have been described by Jerath (1960).

References. Stebnicka & Howden (1995); Stebnicka (2009).

Candezeolus Stebnicka & Howden, 1995 (Pl. 52E)

Type species. *Aphodius candezei* Harold, 1868.

Australian species. Three.

Distribution. Endemic to Australia. Species known from VIC, SA and WA.

References. Stebnicka & Howden (1995); Stebnicka (2009).

Lorditomaeus Peringuey, 1901

Type species. *Aphodius deplanatus* Roth, 1851.

Australian species. One.

Distribution. Native to South Africa; introduced into Australia. *L. bifidus* Schmidt recorded from SA.

Biology. Outside Australia, this species is commonly collected in various kinds of excrement.
References. Stebnicka (2009).

Ozodius Skelley, 2007 (Pl. 51K)

Type species. *Aphodius neglectus* Schmidt, 1912.
Drepanocanthoides (in part): Stebnicka & Howden 1995.
Australian species. Four.
Distribution. Endemic to Australia with species in NSW, ACT, VIC and WA.
Biology/Ecology. Collected in dung traps and litter berlesates.
References. Stebnicka & Howden (1995); Skelley (2007); Stebnicka (2009).

Podotenus Schmidt, 1913 (Pl. 53C)

Type species. *Aphodius insignior* Blackburn, 1904.
Trogaphodius Balthasar, 1965.
Australian species. 33.
Distribution. Australia and southern South America. Australian species occur in all states in the western, southern, eastern and north-eastern areas of the continent, but absent from the dry interior. There is a concentration in the east in wooded areas including rainforest.
Biology. Most species are attracted to human and bovine dung, but they depend mostly on the pellet-like dung of marsupials. Some are attracted to carrion. Specimens collected in various baited pitfall traps, flight intercept traps and at light.
References. Stebnicka & Howden (1994); Stebnicka (2009).

Tribe Eupariini Schmidt, 1910

Characteristics. Elytra with basal margin usually margined. Femora usually grooved at front and hind margins. Middle and hind tibiae without transverse carinae. Abdominal ventrites usually separated by transverse band of short, longitudinal carinae (fluted). Pygidium modified, with longitudinal basal groove and often transverse carina.
Distribution. Worldwide, with ~45 genera and 565 species.
Australian taxa. Seven genera.

Airapus Stebnicka & Howden, 1996 (Pl. 59D)

Type species. *Aulonocnemis sumatrae* Fairmaire, 1896.
Australian species. Five.
Distribution. Indonesia (Sumatra, Papua), Papua-New Guinea, Solomon Islands, NT, QLD and NSW.
Biology. Most specimens collected at light.
References. Stebnicka & Howden (1996); Stebnicka (2009).

Ataenius Harold, 1867 (Pl. 52J)

Type species. *Ataenius scutellaris* Harold, 1867.
Australian species. 52 endemic and four introduced.
Distribution. Pantropical. Widely distributed in Australia and also found on Christmas Island.
Biology. Most species are saprophagous and may be found in excrement, carrion, compost heaps or flowers. Many species are readily attracted to light. Larvae have been described by Jerath (1960) and Verdú & Galante (1999).
References. Stebnicka & Howden (1997); Stebnicka (2009).

Australammoecius Petrovitz, 1958 (Pl. 51D)

Type species. *Australammoecius demarzi* Petrovitz, 1958 (= *Ataenius coloratus* Blacklburn, 1904).
Australian species. Six.
Distribution. Endemic to Australia but mainly limited to western and northern areas (WA, NT, QLD, SA).
Biology. Most specimens collected at light.
References. Stebnicka & Howden (1996); Stebnicka (2009).

Cnematoplatys Schmidt, 1913 (Pl. 52F)

Type species. *Cnemoplatys benardi* Schmidt, 1913.
Neosaprosites Endrödi, 1951.
Australian species. Two.
Distribution. The genus occurs in Nepal, Thailand, Sumatra, Moluccas, New Guinea, Solomon Islands, Micronesia and northern Australia. *Cnematoplatys yorkensis* Stebnicka and *C. tozerensis* Stebnicka & Howden both occur on Cape York Peninsula, QLD.
Biology. Species of *Cnematoplatys* live under bark and in rotten fermenting logs or trunks.
References. Stebnicka & Howden (1996); Stebnicka (2006, 2009).

Euparotrix Stebnicka & Howden, 1996 (Pl. 53A)

Type species. *Euparia squamosa* Lea, 1923.
Australian species. One.
Distribution. *Euparotrix squamosa* (Lea) is endemic to Lord Howe Island.
Biology/Ecology. Specimens have been collected in moss, fungi, rotting wood and leaf litter.
References. Stebnicka & Howden (1996); Stebnicka (2009).

Parataenius Balthasar, 1961 (Pl. 52I)

Type species. *Parataenius mirabilis* Balthasar, 1961 (= *Aphodius derbesis* Solier, 1851).
Australian species. One.
Distribution. South-eastern USA, Neotropical region; introduced into Australia, New Zealand, Africa, Europe. The New World species, *P. simulator* (Harold) has been introduced into NSW, QLD and VIC.

Biology. Specimens have been collected at light and in log and leaf litter berlesates. Larvae have been described by Verdú & Galante (1999).
References. Stebnicka & Howden (1996); Stebnicka (2009).

Saprosites Redtenbacher, 1858 (Pl. 52A)

Type species. *Saprosites peregrinus* Redtenbacher, 1858.
Australian species. 11.
Distribution. Pantropical. Widely distributed in Australia.
Biology. Saprosites species are found in a variety of habitats, including lawn compost, leaf litter, rotten wood, and under bark. Larvae have been described by Jerath (1960). They are good examples of the 'soft saprophagy' life style, with adults probably feeding on the liquid fractions of rotten wood.
References. Stebnicka & Howden (1996); Stebnicka (2009).

Tribe Odontolochini Stebnicka & Howden, 1996

Characteristics. Head broad, able to be deflected at nearly 90 degree to pronotum. Clypeal margin usually broad and double edged. Pronotum swollen in front and sharply declivous to sides, anterior portion with clypeus forming cavity for fore legs; lateral margin rarely visible from above and usually emarginated near base, often barining denticles. Mesocoxae widely separated by flattened meso-metasternal suture. Odontolochini is most closely related to some genera of the Eupariini complex associated with social insects.
Distribution. Nearly pantropical, Australia, Africa and New World, with seven genera and 33 species.
Australian taxa. Two genera.

Gongrolophus Stebnicka & Howden, 1996 (Pl. 53D)

Type species. *Gongrolophus storeyi* Stebnicka & Howden 1996.
Australian species. One.
Distribution. Endemic to Australia with *G. storeyi* being described from northern QLD.
Biology. Adults have been collected from flight intercept traps and berlesates from leaf and log litter.
References. Stebnicka & Howden (1996); Stebnicka (2009).

Odontolochus Schmidt, 1916 (Pl. 51I)

Type species. *Odontoderus gestroi* Clouët, 1900.
Odontoderus Clouët, 1900 (not Schwarz, 1894).
Anodontolochus Paulian, 1942.
Loeblietus Endrödi, 1979.
Australian species. Two.
Distribution. The genus *Odontolochus* is known from Africa and Australia. The Australian species *O. monteithi* Stebnicka & Howden and *O. weiri* Stebnicka & Howden both occur in NSW and QLD.
Biology. Adults of both Australian species have been collected in berlesates of leaf and log litter.
References. Stebnicka & Howden (1996); Stebnicka (2009).

Tribe Psammodini Mulsant, 1842

Characteristics. Head usually distinctly granulate. Pronotum with up to 5 transverse furrows separated by swellings, sometimes reduced with only traces of impressions remaining; a longitudinal furrow usually present. Elytra with basal margin usually bordered. Pygidium with longitudinal basal groove. Mesotarsomeres often triangularly expanded, especially the basal segment. Psammodiini is most closely related to Eupariini.
Distribution. Worldwide, with ~26 genera and 400 species.
Australian taxa. Eight genera.

Aphodopsammobius Endrödy, 1964 (Pl. 52B)

Type species. *Aphodius mirabilis* Endrödy, 1964.
Australian species. Four.
Distribution. The genus is known from Africa and Australia. Species of *Asphodopsammobius* are widely distributed in continental Australia.
Biology/Ecology. Most specimens collected at light.
References. Stebnicka & Howden (1996); Stebnicka (2009).

Leiopsammodius Rakovič, 1981 (Pl. 51C)

Type species. *Psammodius laevicollis* Klug, 1845.
Australian species. Five.
Distribution. Worldwide. Australian species are known from NSW, SA, WA and NT.
Biology. Specimens have been collected at light.
References. Stebnicka & Howden (1996); Stebnicka (2009).

Neotrichiorhyssemus Rakovič & Kral, 1997 (Pl. 51H)

Type species. *Trichiorhyssemus hirsutus* Clouët, 1901.
Australian species. One.
Distribution. *Neotrichiorhyssemus hirsutus* occurs in Indonesia, Papua New Guinea, Christmas Island and Cocos-Keeling Islands.
References. Stebnicka (2009).

Phycocus Broun, 1886

Type species. *Phycocus graniceps* Broun, 1886.
Australian species. One.
Distribution. *Phycocus graniceps* Broun is endemic to New Zealand and has been introduced into TAS.
Biology. These beetles are associated with coastal sand dunes and are commonly found under debris or logs.
References. Stebnicka & Howden (1996); Stebnicka (2009).

Platyomus Mulsant, 1842 (Pl. 52G)

Type species. *Scarabaeus tibialis* Fabricius, 1798.
Australian species. Three.
Distribution. *Platyomus* is worldwide in distribution. *P. tibialis* (Fabricius) occurs in the New World, Eurasia, Africa and the Pacific Islands; it has been introduced into QLD and WA. *P. obscurior* (Blackburn) occurs in SA and WA and *P. pachypus* (Lea) is known from Lord Howe Island and New Caledonia.
References. Stebnicka & Howden (1996); Stebnicka (2009).

Psammorpha Stebnicka, 1994 (Pl. 51L)

Type species. *Psammorpha lawrencei* Stebnicka, 1994.
Australian species. One.
Distribution. *Psammorpha lawrencei* is known only from Christmas Island.
Biology. The single specimen was berlesed from leaf and log litter.
References. Stebnicka (1994, 2009); Stebnicka & Howden (1996).

Rhyssemus Mulsant, 1842 (Pl. 51J)

Type species. *Ptinus germanus* Linnaeus, 1767.
Australian species. Two.
Distribution. Worldwide. *R. inscitus* (Walker) occurs in the Afrotropical, Oriental and Australian regions and has been recorded from QLD, NT and Christmas Island. *R. blackburni* Clouët is known from SA, and WA.
Biology. Most specimens collected at light.
References. Stebnicka & Howden (1996); Stebnicka (2009).

Tesarius Rakovič, 1981 (Pl. 55P)

Type species. *Phycocus sulcipennis* Lea, 1904.
Australian species. One.
Distribution. Nearctic region, New Zealand, Australia. The Australian species *T. sulcipennis* (Lea) is endemic to TAS and introduced into New Zealand.
Biology. The species is found on coastal sand dunes.
References. Stebnicka & Howden (1996); Stebnicka (2009).

Tribe Proctophanini Stebnicka & Howden, 1995

Characteristics. Body usually dark brown to black, robust. Elytra with basal margin not bordered. Femora not grooved at front and hind margins. Meso- and metatibiae with weak transverse carinae. Pygidium fully or partially exposed, sometimes with median longitudinal carina. Metatibial apical spurs set close together, so metatarsus can only articulate below them. Proctophanini is primarily an Australian-African tribe representing a traditional link between Aphodiini and Eupariini.
Distribution. Africa, Australia, New World (introduced), with ~5 genera and 30 species.
Australian taxa. Three genera.

Australaphodius Balthasar, 1942 (Pl. 51E)

Type species. *Aphodius melbournicus* Balthasar 1942 (= *Aphodius frenchi* Blackburn 1892).
Australian species. One.
Distribution. *Australaphodius frenchi* (Blackburn) is native to South Africa, but introduced into Australia, New Zealand and Chile. In Australia, it has been recorded from TAS, VIC, SA and WA.
Biology. Adults have been collected in pitfall traps in pastures.
References. Stebnicka & Howden (1995); Stebnicka (2009).

Harmogaster Harold, 1861 (Pl. 52D)

Type species. *Harmogaster exarata* Harold, 1861.
Australian species. One.
Distribution. *Harmogaster geminata* Schmidt, a native of southern Africa, has been introduced into NSW, QLD and SA.
References. Stebnicka & Howden (1995); Stebnicka (2009).

Proctophanes Harold, 1861 (Pl. 53L)

Type species. *Aphodius sculptus* Hope, 1847.
Australian species. Eight.
Distribution. Australia, New Zealand (introduced), Moluccas. The genus is widely distributed in Australia.
Biology. Adults have been collected in dung.
References. Stebnicka & Howden (1995); Stebnicka (2009).

Tribe Rhyparini Schmidt, 1910

Characteristics. Head with 4 basal tubercles. Clypeal disc convex. Ringed with a groove; clypeal margin broad, with double edge. Pronotum with longitudinal costae and grooves transected by a single median transverse groove or set of deep fassae. Elytra usually distinctly costate on alternaste intervals, with apical bulbous protuberances with trichomes. Prosternal process projecting posteriorly, hastate. Metaventrite usually with a posterior lateral projection. All protibial teeth located in apical quarter. Meso- and metatibiae with apical spurs reduced. Rhyparini share affinities with Seroemerini.
Distribution. Nearly pantropical, with only one species in Africa. Worldwide there are 11 genera and 77 species.
Australian taxa. One genus.

Rhyparus Westwood, 1845 (Pl. 52C)

Type species. *Rhyparus desjardinsii* Westwood, 1845.
Ryparus Westwood 1845 (not Spinola, 1844).
Ryhparus Agissiz, 1846 (unjustified emendation, but in prevailing usage and so considered a justified emendation). See Bouchard *et al.* 2011 [241].
Australian species. Two.
Distribution. *Rhyparus* is a Pantropical genus. *R. helophoroides* is widely distributed throughout the Philippines,

Indonesia, Vanuatu and New Guinea and extends into QLD and NSW. *R. ironensis* Stebnicka & Howden occurs in northern QLD and has a close relative in south-eastern Papua New Guinea.

Biology. *Rhyparus* species are oftern considered to be termitophilous based on the unusual morphology, but there is no direct evidence of termitophily.

References. Stebnicka & Howden (1996); Stebnicka (2009); Bouchard *et al.* (2011).

Tribe Stereomerini Howden & Storey, 1992

Characteristics. Body flattened, broadly oval, small, less than 3 mm, surface usually covered with scale-like microsetae. Pronotum and elytra modified with unusual patterns of ridges and grooves. Prosternal process projecting posteriorly, hastate. Apical tibial spurs greatly reduced. Stereomerini share affinities with Rhyparini.

Distribution. Primarily in South-east Asia, Indonesia and Australia with one species from Peru. Worldwide there are 7 genera and 17 species.

Australian taxa. Two genera.

Australoxenella Howden & Storey, 1992 (Pl. 51F)

Type species. *Australoxenella humptydooensis* Howden & Storey, 1992.

Australian species. 11.

Distribution. *Australoxenella* is endemic with species occurring in northern QLD and NT.

Biology. As with the Rhyparini, species of Stereomerini are highly modified and have the appearance of inquilines, but there is no direct data supporting termitophily or myrmecophily. Most specimens were collected in flight intercept traps or berlesates.

References. Howden & Storey (1992, 2000); Storey & Howden (1996).

Daintreeola Howden & Storey, 2000

Type species. *Daintreeola grovei* Howden & Storey, 2000.

Australian species. One.

Biology. Specimens taken in flight intercept traps on tree trunks in rainforest.

Distribution. *Daintreeola* is endemic with *D. grovei* occurring in northern QLD.

References. Howden & Storey (2000).

References

Ahrens D, Schwarzer J, Vogler AP (2014) The evolution of scarab beetles tracks the sequential rise of angiosperms and mammals. *Proceedings of the Royal Society B Biological Series* **281**, 20141470.

Allsopp PG (1999) How localized are the distributions of Australian scarabs (Coleoptera, Scarabaeoidea)? *Diversity & Distributions* **5**, 143–149. doi:10.1046/j.1472-4642.1999.00050.x

Bouchard P *et al.* (2011) Family-group names in Coleoptera (Insecta). *ZooKeys* **88**, 1–972. doi:10.3897/zookeys.88.807

Browne DJ, Scholtz CH (1998) Evolution of the scarab hind wing articulation and wing base: a contribution towards the phylogeny of the Scarabaeoidea (Coleoptera). *Systematic Entomology* **23**, 307–326. doi:10.1046/j.1365-3113.1998.00059.x

Cabrero-Sañudo FJ, Lobo JM (2009) Biogeography of Aphodiinae dung beetles based on the regional composition and distribution patterns of genera. *Journal of Biogeography* **36**, 1474–1492. doi:10.1111/j.1365-2699.2009.02093.x

Cambefort Y (1991) From saprophagy to coprophagy In *Dung Beetle Ecology.* (Eds I Hanski and Y Cambefort) pp. 22–35. Princeton University Press, Princeton, NJ.

Carne PB (1950) The morphology of the immature stages of *Aphodius howitti* Hope (Col. Scarab. Aphodiinae). *Proceedings of the Linnean Society of New South Wales* **75**, 158–166.

Carne PB (1956) An ecological study of the pasture scarab *Aphodius howitti* Hope. *Australian Journal of Zoology* **4**, 259–314. doi:10.1071/ZO9560259

Cassis G, Weir TA (1992) Aphodiinae. In *Zoological Catalogue of Australia. Volume 9. Coleoptera: Scarabaeidae.* (Ed. WWK Houston) pp. 81–105. AGPS, Canberra.

Cave RD, Ratcliffe BC (2008) Scarab Beetles (Coleoptera: Scarabaeoidea). In *Encyclopedia of Entomology.* (Ed. JL Capinera) pp. 3273–3286. Springer, Heidelberg. doi:10.1007/978-1-4020-6359-6_4045

Davis AJ, Sutton SL, Brendell MJD (2011) Vertical distribution of beetles in a tropical rainforest in Sulawesi: the role of canopy in contributing to biodiversity. *Sepilok Bulletin* 13 & 14, 59–83.

Dellacasa GP, Bordat P, Dellacasa M (2001) A revisional essay of world genus-group taxa of Aphodiinae (Coleoptera: Scarabaeoidea: Aphodiinae). *Memorie della Societa Entomologica Italiana* **79**, 1–482.

Dellacasa M (1988) Contribution to a world wide catalogue of Aegialiidae, Aphodiidae, Aulonocnemidae, Termitotrogidae (Coleoptera: Scarabaeoidea). *Memorie della Societa Entomologica Italiana* **67**, 3–455.

Doube BM, Macqueen A, Ridsdill-Smith TJ, Weir TA (1991) Native and introduced dung beetles in Australia. In *Dung Beetle Ecology*. (Ed. I Hanski and Y Cambefort) pp. 255–178. Princeton University Press, Princeton, NJ.

Forshage M (2002) *State of knowledge of dung beetle phylogeny – a review of phylogenetic hypotheses regarding Aphodiinae (Coleoptera; Scarabaeidae).* Uppsala University (Master's thesis, Biology), Uppsala, Sweden: https://www.researchgate.net/publication/237625313_State_of_knowledge_of_dung_beetle_phylogeny-a_review_of_phylogenetic_hypotheses_regarding_Aphodiinae_Coleoptera_Scarabaeidae.pdf.

Gunter NL, Weir TA, Slipinski A, Bocak L, Cameron SL (2016) If dung beetles (Scarabaeindae: Scarabaeinae) arose in association with dinosaurs, did they also suffer a mass co-extinction at the K-PG boundary? *PLOSONE* **11**, e0153570. doi:10.1371/journal.pone.0153570

Holter P (1982) Resource utilization and local co-existence in a guild of scarabaeid dung beetles (*Aphodius* spp.). *Oikos* **39**, 213–227. doi:10.2307/3544488

Howden HF, Storey RI (1992) Phylogeny of the Rhyparini and the new tribe Stereomerini, with descriptions of new genera and species (Coleoptera; Scarabaeidae; Aphodiinae). *Canadian Journal of Zoology* **70**, 1810–1823. doi:10.1139/z92-248

Howden HF, Storey RI (2000) New Stereomerini and Rhyparini from Australia, Borneo and Fiji (Coleoptera: Scarabaeidae: Aphodiinae). *Memoirs of the Queensland Museum* **46**(1), 175–182.

Huchet JB (2000) Scission du genre *Chiron* Macleay, 1819 et description du deux nouveaux genres de Chironidae (Coleoptera: Scarabaeidae). *Annales de Societe d'Entomologie* **36**, 3–28.

Jerath ML (1960) Notes on larvae of nine genera of Aphodiinae in the United States (Coleoptera: Scarabaeidae). *Proceedings of the United States National Museum* **111**, 43–94. doi:10.5479/si.00963801.111-3425.43

Krell FT (2007) Catalogue of fossil Scarabaeoidea (Coleoptera: Polyphaga) of the Mesozoic and Tertiary. Available at http://museum.unl.edu/research/entomology/fossilrecord-web.pdf

Lawrence JF, Slipinski A (2013) *Australian Beetles Volume 1: Morphology, Classification and Keys.* CSIRO Publishing, Melbourne.

Matthews EG, Stebnicka Z (1986) A review of *Demarziella* Balthasar, with a transfer from Aphodiinae to Scarabaeinae (Coleoptera: Scarabaeidae). *Australian Journal of Zoology* **34**, 449–461. doi:10.1071/ZO9860449

Monaghan MT, Inward DJG, Hunt T, Vogler AP (2007) A molecular phylogenetic analysis of the Scarabaeinae (dung beetles). *Molecular Phylogenetics and Evolution* **45**, 674–692. doi:10.1016/j.ympev.2007.06.009

Philips TK, Pretorius E, Scholtz CH (2004) A phylogenetic analysis of dung beetles (Scarabaeinae: Scarabaeidae): unrolling an evolutionary history. *Invertebrate Systematics* **18**, 53–88. doi:10.1071/IS03030

Ritcher PO (1966) *White Grubs and Their Allies. A Study of North American Scarabaeoid Larvae. (Studies in Entomology Number Four).* Oregon State University Press, Corvallis, OR.

Scholtz CH, Grebennikov VV (2016) 15.13.1. Aphodiinae Leach, 1815. In *Handbuch der Zoologie/Handbook of Zoology. BandVolume IV Arthropoda: Insecta. Teilband/Part 38. Coleoptera, Beetles. Volume 1: Morphology and Systematics (Archostemata, Adephaga, Myxophaga, Polyphaga partim).* 2nd Edition. (Eds RG Beutel and RAB Leschen) pp. 487–490. Walter de Gruyter, Berlin.

Skelley PE (2007) *Ozodius* n. gen. for the Australian members of the genus *Drepanocanthoides* Schmidt (Coleoptera: Scarabaeidae: Aphodiinae). *Insecta Mundi* **15**, 1–2.

Skelley PE (2008) *Guide to New World Scarab Beetles. Scarabaeidae: Aphodiinae.* University of Nebraska State Museum – Division of Entomology. http://museum.unl.edu/research/entomology/Guide/Scarabaeoidea/Scarabaeidae/Aphodiinae/Aphodiinae-Overview/AphodiinaeO.html

Smith ABT, Skelley P (2007) A review of the Aphodiinae (Coleoptera: Scarabaeidae) of southern South America. *Zootaxa* **1458**, 1–80.

Smith ABT, Hawks DC, Heraty JM (2006) An overview of the classification and evolution of the major scarab beetle clades (Coleoptera: Scarabaeiodea) based on preliminary molecular analyses. *Coleopterists Society Monograph* **5**, 35–46.

Stebnicka ZT (1993) A new genus of modified *Psammodiini* from Christmas Island *(Coleoptera: Scarabaeoidea: Aphodiinae). Acta Zoologica Cracoviensia* **37**, 65–69.

Stebnicka ZT (1994) A new genus of unusually modified Psammodiini from Christmas Island (Coleoptera: Scarabaeoidea: Aphodiinae). *Acta Zoologica Crakoviensia* **37**, 71–80.

Stebnicka ZT (1998) The Aphodiinae of New Guinea, the Bismarck Archipelago and the Solomon Islands (Coleoptera: Scarabaeoidea). *Invertebrate Taxonomy* **12**, 833–895. doi:10.1071/IT97006

Stebnicka ZT (2001) *Fauna of New Zealand. Number 42. Aphodiinae (Insecta: Coleoptera: Scarabaeidae).* Manaaki Whenua Press, Lincoln, Canterbury, New Zealand.

Stebnicka ZT (2006) Revision of the Indo-Australian genus *Cnemoplatys* Schmidt with description of a new species from Queensland (Coleoptera: Scaebaeidae: Aphodiinae: Eupariini). *Australian Journal of Zoology* **45**, 168–175.

Stebnicka ZT (2009) *Aphodiinae of Australia (Coleoptera: Scarabaeoidea): Iconography.* Institute of Systematics & Evolution of Animals, Polish Academy of Sciences, Kraków, Poland.

Stebnicka ZT, Howden HF (1994) A Revision of the Australian genus *Podotenus* A. Schmidt (Coleoptera: Scarabaeidae: Aphodiini). *Invertebrate Taxonomy* **8**, 17–62. doi:10.1071/IT9940017

Stebnicka ZT, Howden HF (1995) Revision of Australian genera in the tribes Aphodiini, Aegialiini and Proctophanini (Coleoptera: Scarabaeidae: Aphodiinae). *Invertebrate Taxonomy* **9**, 709–766. doi:10.1071/IT9950709

Stebnicka ZT, Howden HF (1996) Australian genera and species in the tribes Odontolochini, Psammodiini, Rhyparini, Stereomerini and part of the Eupariini (Coleoptera: Scarabaeoidea: Aphodiinae). *Invertebrate Taxonomy* **10**, 97–170. doi:10.1071/IT9960097

Stebnicka ZT, Howden HF (1997) Revision of the Australian species of *Ataenius* Harold (Coleoptera: Scarabaeoidea: Aphodiinae: Eupariini. *Invertebrate Taxonomy* **11**, 735–821. doi:10.1071/IT95046

Steinbauer MJ, Weir TA (2007) Summer activity patterns of nocturnal Scarabaeoidea (Coleoptera) of the southern tablelands of New South Wales. *Australian Journal of Entomology* **46**, 7–16. doi:10.1111/j.1440-6055.2007.00579.x

Storey RI, Howden HF (1996) Revision of *Australoxenella* Howden & Storey in Australia (Coleoptera: Scarabaeidae: Aphodiinae). *Memoirs of the Queensland Museum* **39**, 365–380.

Stork NE, Grimbacher PS (2006) Beetle assemblages from an Australian tropical rainforest show that the canopy and the ground strata contribute equally to biodiversity. *Proceedings. Biological Sciences* **273**, 1969–1975. doi:10.1098/rspb.2006.3521

Vårdal H, Forshage M (2010) A new genus and species and a revised phylogeny of Stereomerini (Coleoptera, Scarabaeidae, Aphodiinae), with notes on assumedly termitophilic aphodiines. *ZooKeys* **34**, 55–76. doi:10.3897/zookeys.34.264

Verdú JR, Galante E (1999) Larvae of *Ataenius* (Coleoptera: Scarabaeidae: Aphodiinae): generic characteristics and species descriptions. *European Journal of Entomology* **96**, 57–68.

30. SCARABAEIDAE: SCARABAEINAE LATREILLE, 1802

Nicole L. Gunter, Cate Lemann and Tom A. Weir

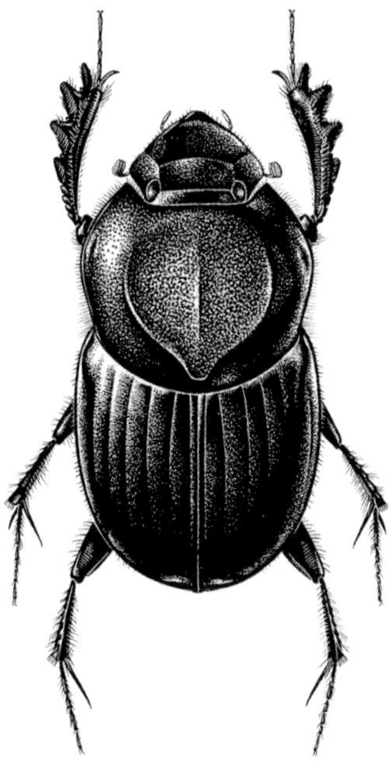

Fig. 30.1. *Onthophagus declivis* Harold.

Common name. Dung Beetles

Introduction. The subfamily Scarabaeinae has a worldwide distribution and contains over 6200 described species in ~270 genera (Tarasov & Génier 2015). The greatest diversity is recorded from the Afrotropical region and together with the Neotropical and Oriental, these three regions represent 80% of the total diversity of Scarabaeinae (Davis *et al.* 2017). The Australian fauna is considered relatively depauperate with only 351 described species representing ~5% of the described diversity.

The monophyly of Scarabaeinae is supported by morphological synapomorphies of the female reproductive system (Richter & Baker 1974), wing articulation (Browne & Scholtz 1998) and larvae characters (Edmonds & Halffter 1978; Grebennikov & Scholtz 2004). Phylogenetic analyses of morphological (e.g. Montreuil 1998; Philips *et al.* 2004; Vaz-de-Mello 2007; Tarasov & Génier 2015) and molecular (e.g. Monaghan *et al.* 2007; Sole & Scholtz 2010; Ahrens *et al.* 2014; McKenna *et al.* 2015; Gunter *et al.* 2016; Tarasov & Dimitrov 2016) data has supported the monophyly of the subfamily and examined the relationships between tribes. The close relationship between Scarabaeinae and Aphodiinae has been recovered in all phylogenetic analyses and they are accepted as sister groups (e.g. Browne & Scholtz 1995, 1998; Grebennikov & Scholtz 2004; Ahrens 2006; Smith *et al.* 2006; Hunt *et al.* 2007; Ahrens & Vogler 2008; Bocak *et al.* 2013; Ahrens *et al.* 2014; Lawrence *et al.* 2011; McKenna *et al.* 2015; Gunter *et al.* 2016). Scarabaeinae and Aphodiinae are the major saprophagous subfamilies in the family Scarabaeidae, and together they are considered sister to the phytophagous scarab subfamilies or 'pleurostict scarabs' (Ratcliffe *et al.* 2002; Ahrens 2006; Ahrens & Vogler 2008; Gunter *et al.* 2016). Other studies reject the monophyly of the family, instead suggesting Scarabaeinae and Aphodiinae are more closely related to the scarabaeoid families Glaphyridae, Hybosoridae and Ochodaeidae (Smith *et al.* 2006; Hunt *et al.* 2007; Ahrens *et al.* 2014).

Historically, dung beetles have been divided into tribes characterised by morphology and behavioural ecology. The first conception of a tribal classification divided the subfamily into six tribes including four subtribes within Coprini and five subtribes within Scarabaeini (Janssens 1949). Currently, twelve tribes are generally recognised that are broadly divided into two behavioural groups, tunnellers: Dichotomiini, Onitini, Phanaeini, Coprini, Oniticellini and Onthophagini, and rollers: Canthonini, Scarabaeini, Gymnopleurini, Eucraniini, Eurysternini and Sisyphini (Scholtz *et al.* 2009). Phylogenetic studies have demonstrated the paraphyly, polyphyly, and artificial nature of many of the tribes. Montreuil (1998) was the first to propose changes to the classification based on phylogenetic results but the most significant changes redefined the diagnoses of Deltochilini, Dichotomiini and Coprini (Tarasov & Dimitrov 2016). As a result of this revision 125 genera previously classified within these three tribes were treated as *incertae sedis* (Tarasov & Dimitrov 2016). With almost half of the known genera currently unplaced, systematists must work towards an updated classification that reflects natural groupings.

Despite the instability in tribal classification, both morphological and molecular studies are relatively congruent and support lineages with some geographic structure (Davis *et al.* 2002). Most endemic Australian genera, including those also found in New Guinea, form a relationship with endemic genera from New Zealand and New Caledonia. This lineage is referred to as the Australasian endemic clade and is morphologically diverse. The included genera were previously placed in the tribes Coprini and Deltochilini, but now all genera are unplaced to tribe. Australian *Onthophagus* are nested in the cosmopolitan clade of Onthophagini. Of the genera included in phylogenetic analyses, only the phylogenetic position of *Boletoscapter* remains unresolved but may represent a third distinct element of the Australian fauna (Tarasov & Dimitrov 2016; Gunter *et al.* 2018).

Twenty-one native genera are recorded from Australia and six genera have been introduced (discussed below). Fifteen genera are endemic to the continent while *Onthophagus* Latreille has a cosmopolitan distribution, *Amphistomus* Lansberge is recorded from New Guinea and the Moluccas (Huijbregts & Krikken 2007), and *Coptodactyla* Burmeister, *Lepanus* Balthasar, *Temnoplectron* Westwood and *Tesserodon* Hope are known only from Australia and New Guinea (Matthews 1974; Paulian 1985; Reid 2000; Reid & Storey 2000). Dung beetles have been recorded from all Australian bioregions, although only *Onthophagus* are recorded in central Australia and Tasmania. Richness is highest on the eastern coast within the mesic zone forests.

Species-level taxonomy was studied in detail by Matthews (1972, 1974, 1976) who documented 284 native species. In the last forty years, most taxonomic work has been conducted by a handful of authors N. Gunter, E. Matthews, G. Monteith, C. Reid, Z. Stebnicka, R. Storey and T. Weir. Recently *Coptodactyla* and *Temnoplectron* have been revised (Reid 2000; Reid & Storey 2000) and the genera *Matthewsius* Gunter & Weir and *Monteithocanthon* Gunter & Weir described (Gunter & Weir 2017). These works, together with a few smaller works on *Aptenocanthon* Matthews (Storey 1984; Storey & Monteith 2000), *Aulacopris* White (Storey 1986), *Demarziella* Balthasar (Matthews & Stebnicka 1986), *Lepanus* Balthasar (Matthews & Weir 2002; Weir & Monteith 2010), *Onthophagus* (Storey 1977; Storey & Weir 1990; Monteith & Storey 2013) and *Tesserodon* (Storey 1991) include all Australian species descriptions since Matthews' major revisions (1972, 1974, 1976). As of 2017, 351 described species are recorded from Australia, however over 100 undescribed species are known from museum collections (Monteith 2015). To aid curation an informal nomenclature for undescribed species was developed by G. Monteith and T. Weir, these represent 25% of scarabaeine species in collections. Tarasov & Génier (2015) estimated that between 30% and 50% of dung beetle species remain undescribed, while ~70% of Australian insects may be undescribed (Chapman 2009). It is likely that further discoveries of new species, particularly of flightless genera with apparently broadly distributed species in the fragmented mesic zone, may bring the total Australian fauna to 500–600 species in line with the estimates of Tarasov & Génier (2015).

Species distribution was studied in detail by Matthews (1972, 1974) and Allsopp (1995, 1999), while Ridsdill-Smith *et al.* (1989) provided a field guide to the scarabs of south-western WA. Matthews (1972, 1974) grouped species into 11 broad patterns based on habitat and geographic range, while Allsopp (1995, 1999) analysed distribution based geographical range as reflected by proportions of total continental surface area. In general, species in endemic genera (referred to as 'Scarabaeini' in the study) have localised distributions in rainforests or sclerophyll forests while *Onthophagus* spp. show little preference to either habitat and have both restricted and wide range distributions (Allsopp 1999). The differing dispersal ability and environmental tolerance of the endemic genera and Onthophagini may be related to the differing evolutionary histories confirmed in phylogenetic analyses (Gunter *et al.* 2016; Tarasov & Dimitrov 2016; Breeschoten et al. 2016).

The native Onthophagini have been recorded from all Australian IBRA regions, although fewer species occupy the driest areas of the continent. Most species occupy open forest (i.e. sclerophyll forest) or the more inland savannah woodlands with ~40% of species found in each of these habitats. Only 15% of species are found in closed rainforests and truly arid species are rare (Matthews 1972). In comparison, the endemic genera are mostly confined to the warmer and wetter areas of Australia and generally absent from areas that receive less than 760 mm of rainfall (Matthews 1974). These genera are notably absent from Tasmania which has been attributed to the low temperatures and not the lack of rainfall (Matthews 1974). Approximately half the species are found in closed forests (i.e. rainforest), a third are found in open forest (i.e. sclerophyll forest) and the remainder are distributed in savannah woodlands, paperbark swamps and desert biomes (Matthews 1974).

Distributional data highlights large gaps in geographic ranges of all endemic genera that corresponds to regions of drier savannah grasslands and eucalypt forests that divide the Australian mesic zone forests (Gunter *et al.* 2018). The absence of endemic genera in these drier regions is not an artefact of sampling effort as native *Onthophagus* occur within these regions, but instead suggests such habitats create significant geographic barriers for forested restricted taxa with limited dispersal (Gunter *et al.* 2018). Furthermore, phylogenetic evidence suggests these endemic genera represent part of the *Southern* faunal element of Australia (*sensu* Yeates & Cassis 2017) and when distributional data and species relationships are considered, evidence supports a broad ancestral distribution of endemic genera when climate and habitat were more suitable with diversification linked to fragmentation of the mesic zone forests (Gunter *et al.* 2018).

While such studies support the Gondwanan origins of the Australasian endemic clade and thus a Cretaceous age, the origins of Scarabaeinae remain contentious due to various methodologies (see reviews by Davis *et al.* 2017 and Gunter *et al.* 2018). The fossil record of Scarabaeinae was reviewed by Tarasov *et al.* (2016). Only 35 fossils reported as Scarabaeinae are known, of which only 21 can be confidently assigned to the subfamily (Tarasov *et al.* 2016). The oldest fossil known to date is *Lobateuchus parisii* Montreuil, Génier & Nel from Oise amber of Eocene age (~53 Ma), while most of fossils are Miocene in age (Tarasov *et al.* 2016). Five extinct dung beetle genera have been described from fossils, however due to poor preservation and lack of preserved characters four genera cannot be unequivocally placed in the subfamily. Nine extant genera representing at least four tribes are represented in the fossil record (Tarasov *et al.* 2016). Fossilised brood balls have been recorded from South America, North America, Africa and Asia; most are of Eocene age (Laza 2006), although Carvalho *et al.* (2009) reported a single specimen from a Cretaceous formation in Brazil (see review Genise *et al.* 2016). Other trace fossils attributed to dung beetles are burrows preserved in Upper Cretaceous hadrosaur coprolites (Chin and Gill 1996), although the identity of such ichnofossils remains purely speculative. No Australian Scarabaeinae fossils are known.

Feeding biology. Adults are terrestrial and may be found in leaf litter, perching on vegetation and foraging at food sources (Matthews 1972, 1974, 1976; Howden *et al.* 1991). Some species have adapted prehensile tarsal claws that allow them to cling directly to mammal fur such as a few species of Australian *Onthophagus* that cling to macropod fur (Halffter & Matthews 1966; Matthews 1972) and three Neotropical genera that are phoretically associated with sloths (Ratcliffe 1980). Dung beetles are well known for their association and exploitation of animal faeces, particularly mammal dung, with both adults and larvae feeding on faeces (Scholtz *et al.* 2009). Most species feed on dung of herbivorous mammals but alternative food sources are also well documented. Their mouthparts are adapted for feeding on wet dung but also other liquid food with similar nutritional content such as rotting fruit, detritus and carrion (Bai *et al.* 2015). Only one genus, *Pachysoma* Macleay is known to feed on solid detritus (Holter *et al.* 2009). Alternative food sources include vertebrate and invertebrate (millipede, ants, snails, squid) carrion, rotting fruit, fungi, detritus from vegetable matter, fallen flowers and other plant parts. Many dung beetles that feed on alternative food sources are generalist feeders and the largest numbers are known from the tropics (Gill 1991; Hanski & Krikken 1991; Scholtz *et al.* 2009).

The feeding biology of Australian dung beetles is intriguing. Although all species are coprophagous in part, a high proportion of them utilise food resources other than mammalian dung. In addition to feeding on dung, many species feed on mushrooms and some species feed on carrion and vegetable matter (Matthews 1974; Hill 1996). *Cephalodesmius* Westwood also collects vegetable matter for brood ball construction (Monteith & Storey 1981). There are accumulating observations of the smallest of Australian species that live in leaf litter feeding on bird droppings; these species are rarely collected at mammal dung or mushroom baited pitfall traps. Within Australia, only one study has documented the food preferences of dung beetles from multiple genera within a localised area. In a comparative study, Hill (1996) documented the dung beetle abundance attracted to agile wallaby (*Macropus agilis* Gould) dung-, mushroom-, banana- and liver-baited pitfall traps as well as collected in flight intercept traps and at light in Kirrama State Forest, northern Queensland. Of the 10 most abundant species collected in the study, eight were reported from multiple baits including four reported at all four baits, indicating that these species may have a generalist diet, while only two were recorded from a single bait thus appearing to specialise on either dung (*Demarziella interupta* (Carter)) or liver (*Lepanus latheticus* Matthews) (Hill 1996). Abundance data from pitfall traps cleared just after dusk and again after dawn indicated that the majority of dung beetles collected in their study were night active (Hill 1996).

New Zealand dung beetles also have a generalist diet, feeding on mammal, bird, reptile and invertebrate dung, and are commonly collected in mushroom and squid-baited pitfall traps (Hodge *et al.* 2010; Seldon & Beggs 2010; Jones *et al.* 2012;). Mushroom feeding is common among species in the Australasian clade of dung beetles, the *latro* and *dunningi* groups of *Onthophagus* (Matthews 1972) and has been reported for species from most Australian genera (Matthews 1974; Hill 1996). Few studies have addressed food preference to date but no species has been reported as an obligate mushroom feeder (Hill 1996). However, *Boletoscapter* Matthews is rarely attracted to dung baits and appears to be the only Australian genus that may be an obligate mushroom feeder. Genera from some of the most primitive dung beetle lineages also feed on mushrooms and it has been proposed that these genera never deviated from their ancestral fungus feeding habits (Mlambo *et al.* 2015). The common association with fungus feeding in Australian endemic genera may reflect persistence of the ancestral state or reversal/adaptation from dung feeding ancestors.

The introduction of agricultural livestock to Australia presented an unusual ecological problem due to limited invertebrate nutrient cycling. In the absence of most placental mammals (with the exception of bats and rodents), native dung beetles had adapted to feeding on marsupial dung. Although many species are generalists, few feed on dung of cattle, sheep and horses and those that do have little impact on it (Edwards *et al.* 2015). Since 1968, 25 of the introduced dung beetle species from Africa, Asia and Europe have established in Australia specifically to assist in breakdown of agricultural dung through feeding and burial in nest provisioning (Edwards *et al.* 2015).

Breeding biology. Scarabaeine species may be divided into three behavioural groups: tunnellers, rollers, and dwellers (Halffter & Edmonds 1982). Tunnellers are the most diverse and excavate burrows, often immediately below the food source, before provisioning the burrow; rollers remove dung or alternative sources for food and breeding and then bury it, and dwellers live and breed directly inside the dung source (Scholtz *et al.* 2009).

For rollers, food is removed from the original source by carving, dragging and rotating movements that results in a rough spherical shape (Matthews 1974). Ball rolling occurs either head up using its front legs to pull the ball while walking backwards, or head down using its front legs to push against the ground while its middle and hind legs guide the ball backwards (Matthews 1974). Other groups cannot shape their own balls but instead roll an already formed pellet without any alteration to the dung. Collected dung is used as a food source or as a breeding resource in which its larva will feed (Matthews 1974).

Few beetle groups build nests and dung beetles are the only group that provision their nesting burrow with dung (Scholtz *et al.* 2009). Nesting behaviour, brood ball construction and parental cooperation varies among species. Cooperation between a paired couple is common in rollers and tunnellers, although some mated females can nest unaided by a male (Kryger *et al.* 2006; Sato 1998; Scholtz *et al.* 2009). In general, female tunnellers dig the burrow and males forage and provision the nest, while in rollers pairs cooperate in ball-formation, rolling and nesting (see Scholtz *et al.* 2009 for detailed review).

One or multiple brood balls can be constructed from the collected dung that is reshaped into smaller ball-, pear- or sausage-like masses in which an egg is laid (Scholtz *et al.* 2009). These brood balls may or may not be coated in a clay shell. Most males leave the nest after eggs are laid and females may leave once brood balls are complete and only sometimes do females or pairs remain in the nest (Scholtz *et al.* 2009). Once hatched the developing larvae feeds from within and in rare cases parents may continue to provision the brood ball. Most Scarabaeinae develop from egg to adult in ~30–50 days however some species take over a year to develop (Halffter & Matthews, 1966). This complex nesting behaviour results in low juvenile mortality, however fecundity is also low given the small number of offspring produced in each breeding event (Scholtz *et al.* 2009).

Breeding biology of only a few Australian species is well documented: *Onthophagus compositus* Lea in kangaroo dung (Bornemissza 1971a), *Onthophagus dunningi* Harold in mushrooms (Bornemissza 1971b), and *Cephalodesmius* (Monteith & Storey 1981). In *Cephalodesmius*, bonded pairs build a burrow together and the male forages for decaying vegetable matter to construct a large brood mass (Monteith & Storey 1981). The female constantly works on the brood mass until it is 3–4 times her size, adding foraged material and the pair's frass to the outside, and working on its consistency with her protibial teeth and mouthparts. This large ball is allowed to compost for up to two weeks before the female shapes it into the egg balls (Monteith & Storey 1981). Copulation occurs usually before egg ball construction. Each egg ball will be hollowed out to a cup-like shape with a flared rim, then it is positioned on her abdomen, egg laying takes place and then it is sealed and rounded. Usually one egg ball is made per day and up to seven balls are formed (Monteith & Storey 1981). Eggs hatch within a week and as the larvae grows the female will add more material from the brood mass to the egg ball ensuring there is enough food for the developing larvae (Monteith & Storey 1981). The male will continue to forage or wait within the nest. *Cephalodesmius* is the only known genus of dung beetle that form bonded pairs for their lifetime but breeding only occurs once (Monteith & Storey 1981). Although not well documented it appears *Canthonosoma* Macleay is also a progressive provisioner as initial egg balls are too small for larvae to mature in (Monteith 2015).

This strategy of larvae developing within a spherical chamber within a brood ball has attributed to superficially similar C-shaped larvae amongst all Scarabaeinae. Larvae are eyeless and have well sclerotised head capsule, well developed mouthparts, a district 'coprine' hump (i.e. enlarged dorsal portions of at least the first 6 abdominal segments), a specially flattened anal segment known as Fabre's trowel and reduced legs. The legs play no role in locomotion, instead the 'coprine' hump and Fabre's trowel allow the larvae to move within the brood ball. The larvae use these to features to wedge themselves in position, allowing either their fore body to move for feeding or hind body to swing in to a new position in the chamber. The well developed cutting and crushing mouthparts of the larvae are different from the adults suggesting larvae are able to consume more than liquids suspended in the dung.

Development comprises three larval instars. Pupation occurs in a pupal chamber constructed by the prepupa. In general, the pupal chambers of rollers are either formed by an uneaten proportion of the dung or the clay shell of the brood ball, while the prepupa of tunnellers construct the pupal chamber around itself from remnant dung using its mouthparts and Fabre's trowel. Pupae generally have projections issuing from the middle of the pronotum; meso- and meta- thorax and from the abdomen. Placement of projections varies between species. These projections are thought to reduce contact between pupae and the chamber wall that may increase survival due to lower exposure to pathogens.

Newly emerged adults are physically immature and must feed before reaching sexual maturity. 'Maturation feeding' varies among species but may relate to the dung available in the larval stages, i.e. larger species in small brood balls may require longer periods of maturation feeding (Scholtz *et al.* 2009) but also the availability of dung. Dung may be patchy within an environment and competition for dung is also dependent on the dispersal ability of species. Over 40% of species of the endemic genera are flightless, but there is no pattern in flight ability that correlate with habitat or general range of species. Nine Australian genera are entirely flightless, nine other genera have flighted and flightless species and all species of only four genera are capable of flight. The majority of flightless species are short-range endemics, with only few species recorded with apparently broad distributions. Some fully winged species are also short range endemics thus dispersal ability cannot necessarily be correlated with range. Dispersal ability not only affects species ability to find food but also to find mates. These factors, coupled with the relative abundance and stability of food, ultimately shape the effective population size of species.

There are many ecological benefits associated with the breeding biology of scarabaeines including but not limited to nutrient cycling, soil fertilization and aeration, secondary seed dispersal and reduction in pest flies and enteric parasites associated with the dung (Andresen & Feer 2005).

Conservation. Like many invertebrates, the conservation status of many dung beetles is unknown primarily due to the limited distributional data available. Of the 624 Scarabaeinae assessed world-wide by IUCN, 246 are listed as data deficient, 331 are of least concern, 20 are near threatened, 14 are vulnerable and 12 are endangered species (IUCN 2017). Less than 20% of Australian species have been assessed and to date, no Australian dung beetle species is considered endangered and only *Aulacopris matthewsi* Storey is listed as vulnerable given its extremely restricted distribution (Weir 2013). However, it is likely that once more data becomes available that additional species may be listed in categories of conservation concern, particularly species that are restricted to narrow ranges or those restricted in threatened habitats. Coggan (2012) highlighted a concern for arid and semi-arid zone Australian native dung beetles due to the sharp decline in marsupials occupying this habitat but noted that dung specificity is a major assumption underpinning a correlation in decline of marsupials and dung beetle assemblages. However, a study of dung beetle assemblages associated with dung of the five most abundant small mammals in Lamb Range, northern Queensland, demonstrated that of the species that were collected more than twice, no preference for dung was apparent (Vernes *et al.* 2005). Given the apparent generalist diet Australian dung beetles, the decline in native marsupials may not result in extinction of generalist species.

Habitat loss due to land-use and environmental change may be a more critical factor in conservation of scarabaeines. Few long-term studies have monitored a single locality, the most notable monitored dung beetles at Welder Wildlife Refuge, Texas in 1974, 1985 and 1999 (Howden & Howden 2001). Over the three surveys a decline in species richness and shift in species composition was recorded which may be attributed to historical changes in vegetation, climate or the establishment of the imported red fire ant *Solenopsis invicta* Buren (Howden & Howden 2001). A study of elevational distribution of Bolivian dung beetles from the tribe Phanaeini reported that short-range endemic species with narrow elevational ranges, especially on low-elevation mountains are at greatest risk of climate-change related extinction (Herzog *et al.* 2013). The Great Dividing Range (GDR) is a complex of mountain ranges, plateaus, escarpments and upland areas that stretches over 3500 km along the eastern coast of Australia. There are almost 100 peaks in the GDR that range from ~300–2200 m in height. Much of the diversity of Australian scarabaeines is known from the GDR and also from other low elevation coastal and inland ranges within Australia, and like Bolivian phanaeines, the endemic species with narrow elevational ranges may be of conservation concern. Flightless dung beetles of the rainforests of the Wet Tropics, northern Queensland have been identified as being vulnerable to climate change, as the majority of species are endemic to a single or few-mountain tops, often restricted to narrow elevational ranges, and no flightless species were recorded from lowland subregions (Aristophanous 2014).

Dung beetles have also been proposed as ideal candidates as invertebrate models for conservation assessment. Spector (2006) reviewed the benefits of dung beetles as surrogates for monitoring broader patterns in biodiversity, highlighting their broad geographic distribution, multi-species communities, varying sensitivity to environmental disturbances, ecological and economic importance and correlation with other taxa through their association with vertebrate dung. Monitoring dung beetles through pitfall trapping requires minimal effort with replicable surveying at both spatial and temporal scales and as such are ideal surrogates compared to other invertebrate species. Within Borneo, dung beetles have been used as indicators of impact on invertebrate communities following forest disturbance (Davis & Sutton 1998; Davis 2000; Davis *et al.* 2001) and in ecosystem evaluation in Europe and Africa (Koch *et al.* 2000; Verdú *et al.* 2000; Davis 2002; Davis *et al.* 2004).

Recently, it has been proposed that due to the lifecycle and association with agricultural dung, that dung feeding beetles could play a role in reducing greenhouse gases (Nichols *et al.* 2008; Penttilä *et al.* 2013; Iwasa *et al.* 2015; Slade *et al.* 2016). A Finnish study reported a 0.05–0.13% reduction in overall greenhouse emissions from cattle production associated with reduced methane and nitrous oxide fluxes directly from pats utilised by dung beetles and at the pasture level through nutrient cycling (Slade *et al.* 2016). Slade *et al.* (2016) suggested that the reduction in greenhouse gas emissions may be greater in tropical countries due to increased dung beetle activity in warmer climates.

Characteristics. Adults. Suboval to oblong scarabs, 2 to 30 mm in length, often with pronotal or cephalic armature. Clypeus expanded, covering mouthparts, with or without apical teeth. Labrum membranous, concealed. Mandibles lamelliform, mostly membranous. Antennae 8- or 9-segmented, with 3-segmented lamellate club. Anterior hypomeral carina usually present from procoxa to lateral edge or anterior angles of pronotum. Pretarsal claws always equal. Protibiae flattened and usually with three teeth. Scutellum almost always invisible. Mesocoxae longitudinal and separated by more than their width. Metatibia with a single spur. Elytra exposing pygidium. Abdominal spiracles entirely concealed beneath elytra. Abdomen with six visible fused ventrites, compressed medially and together shorter than metaventrite.

Larvae. C-shaped with characteristic dorsal hump formed from enlarged dorsal parts of at least the first six abdominal segments. Head well sclerotised and usually yellow. Mouthparts well developed, galea and lacinia distinctly separated for their entire lengths. Apical antennomere reduced, at most half as wide as preapical one, the apex of which bears a pale oval sensorium on the inner side. Legs reduced, 2-segmented, with or without very small pretarsal claws.

Key to the tribes and genera of Australian Scarabaeinae

(modified from Matthews 1972, 1974, 1976 and various other sources)

The key contains all genera recorded until now from Australia, including introduced genera. Several genera are keyed out twice, allowing for the morphological diversity found in these genera. Sexual dimorphism in scarabaeine genera varies greatly. As a general rule the last abdominal ventrite in males is shorter medially than in females, but this is often hard to determine in isolation. Almost every part of the body can show some difference between male and female, including the armature of head and/or pronotum (Onthophagini, some Onitini, Oniticellini, Coprini), foretibiae (Onitini, many *insertae sedis* genera), metathoracic legs (Sisyphini, Onitini, some *insertae sedis* genera), clypeal teeth (some *insertae sedis* genera), metaventrite (some *insertae sedis* genera), pygidium (some *insertae sedis* genera) and forespur (some *insertae* sedis genera).

1. Body compressed laterally with sides of pronotum flattened and vertical; pronotal hypomeron not excavated anteriorly for reception of profemur; sides of metaventrite vertical; middle coxae very widely separated, forming lateral margin of body; antennae 8-segmented; metafemora and metatibiae greatly elongated, the latter curved; elytra markedly tapered to apex, triangular in shape; 6 to 10 mm in length; eastern Australia – QLD, NSW .... **Sisyphini: *Neosisyphus* Muller**

– Body not so compressed laterally; sides of pronotum not flattened and vertical; IF sides of metaventrite vertical and middle coxae widely separated and forming lateral margin of body, THEN antennae 9-segmented and pronotal hypomeron excavated anteriorly for the reception of profemur (*Amphistomus*); antennae 8- or 9-segmented; metafemora and metatibiae may be elongate with the latter curved; elytra otherwise shaped (tapered in some *Pseudignambia* which are less than 3 mm in length) **2**

2(1). Pronotum with a small depression on each side of midline near base (Fig. 30.2); protarsi lacking (Fig. 30.3) in both sexes (2 introduced genera).... **Onitini** **3**

– Pronotum without basal pits near midline; protarsi usually present (except *Coptodactyla*) **4**

3(2). Prosternum with a rounded projection behind procoxae (Fig. 30.4); scutellum not visible; male with paired head horns and carinate projection on pronotum (Fig. 30.5); female with a carina on head (Fig. 30.6) and transverse carina on pronotum; 12 to 16 mm in length; southern Australia – WA, SA, VIC, NSW .... ***Bubas* Mulsant**

– Prosternum without a rounded projection behind procoxae; scutellum visible; male and female with only carinae and tubercles on head; pronotum without projections or carina in both sexes; 11 to 22 mm in length; widespread – all states except TAS; .... ***Onitis* Fabricius**

4(2). Scutellum visible between bases of elytra (small to very small and often hard to see in *Coptodactyla,* which has protarsi absent) **5**

– Scutellum not visible between bases of elytra **8**

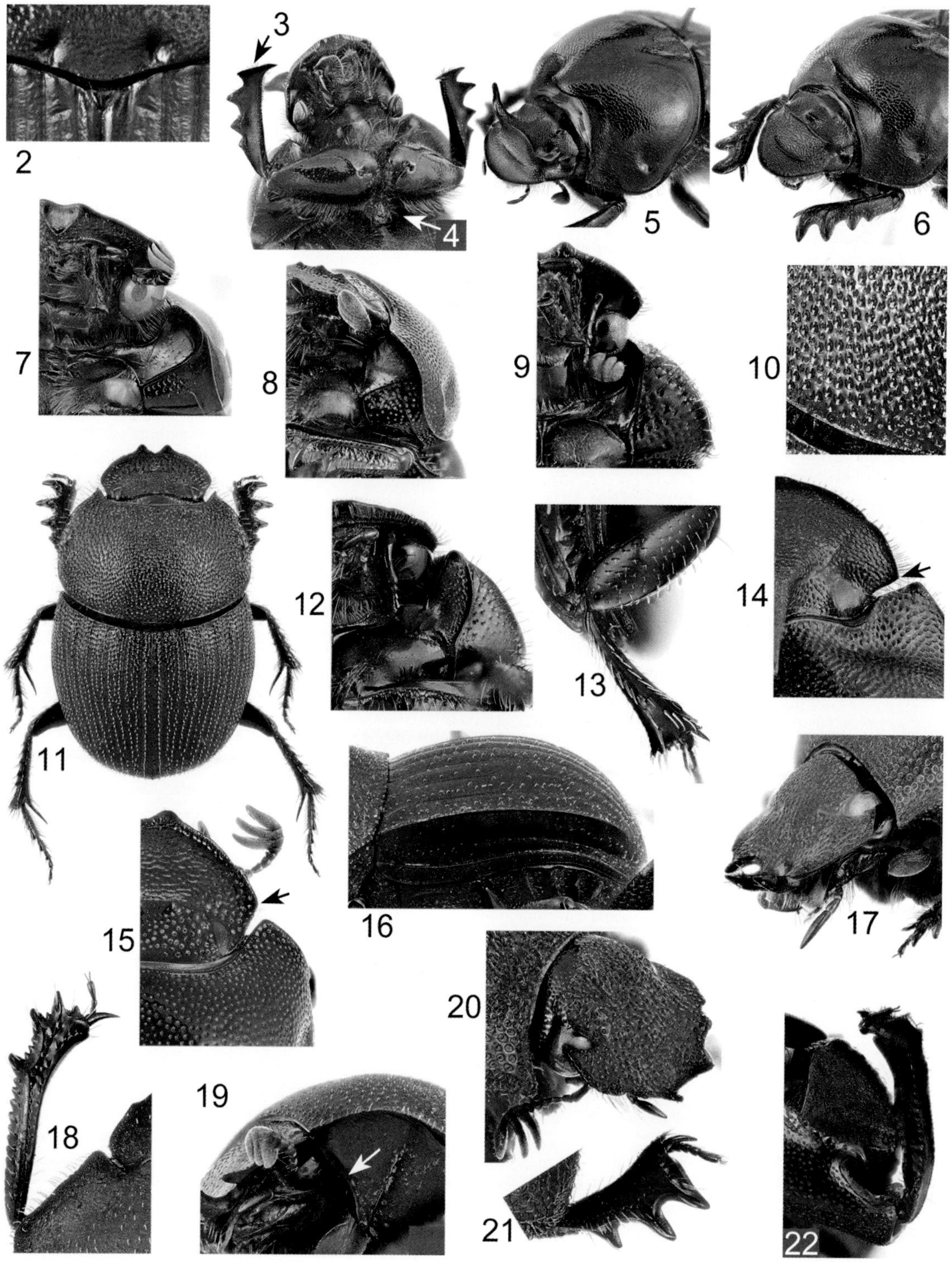

Figs 30.2–30.22 2, *Onitis alexis* Klug, female, pronotal depressions; 3–4, *Bubas bison* (Linnaeus), male head and prothorax, ventral: 3, protibia, 4, prosternal process; 5, *Bubas bison* (Linnaeus), male head and pronotum, anterolateral; 6, *Bubas bison* (Linnaeus), female head and prothorax, anterolateral; 7, *Coptodactyla glabricollis* (Hope), prothorax, ventral; 8, *Thyregis kersawi* Blackburn, prothorax, ventral; 9, *Hamonthophagus depressus* (Harold), prothorax, ventral; 10, *Hamonthophagus depressus* (Harold), pronotal punctures and setae; 11, *Hamonthophagus depressus* (Harold); 12, *Digitonthophagus gazella* (Fabricius), prothorax, ventral; 13, *Demarziella yorkensis* Matthews & Stebnicka, hind tibia; 14, *Copris hispanus* Linnaeus, eye canthus; 15, *Thyregis kershawi* Blackburn, eye canthus; 16, *Cephalodesmius armiger* Westwood, elytra lateral; 17, *Cephalodesmius armiger* Westwood, eye canthus; 18, *Cephalodesmius armiger* Westwood, protibia; 19, *Cephalodesmius armiger* Westwood, hypomeral ridge; 20, *Labroma umbratilis* Matthews, head eye canthus; 21, *Labroma umbratilis* Matthews, protibia; 22, *Aulacopris reichei* White, profemur.

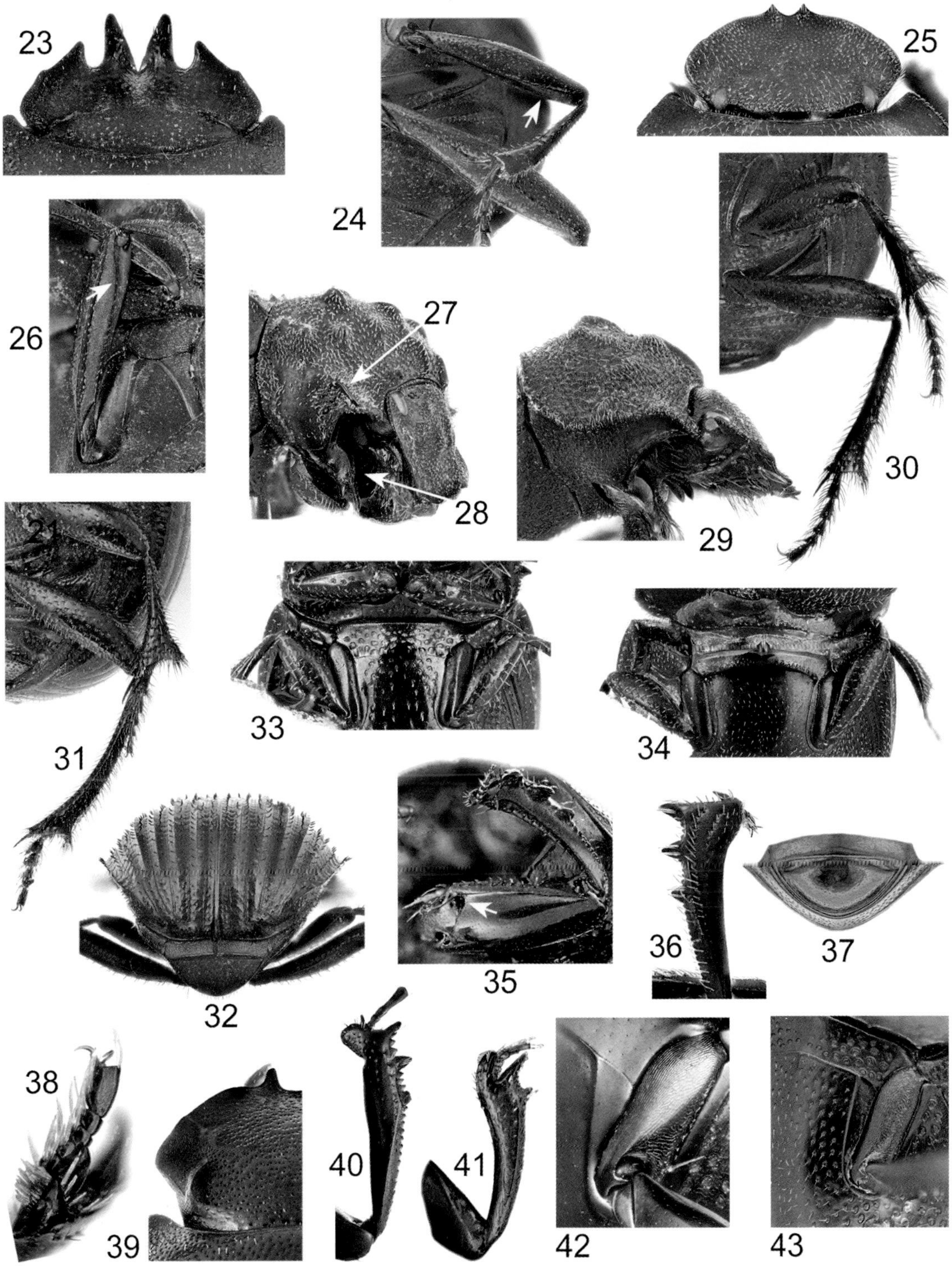

Figs 30.24–30.43. 23, *Cephalodesmius armiger* Westwood, male head; 24, *Cephalodesmius armiger* Westwood, mesofemur; 25, *Canthonosoma macleayi* (Harold), head; 26, *Canthonosoma macleayi* (Harold), mesofemur; 27–28, *Amphistomus tuberosus* Matthews: 27, incomplete lateral carina, 28, hypomeral cavity; 29, *Labroma umbratilis* Matthews, prothorax, lateral; 30, *Mentophilus hollandiae* (Castelnau), meso- and metatibia; 31, *Coproecus hemisphaericus* (Guérin-Méneville), meso- and metatibia; 32, *Mentophilus hollandiae* (Castelnau), elytra; 33, *Tesserodon novaehollandiae* (Fabricius), mesoventrite; 34, *Boletoscapter cornutus* (Macleay), mesoventrite; 35, *Matthewsius illawarrensis* (Matthews), profemoral pit; 36, *Matthewsius illawarrensis* (Matthews), male protibia; 37, *Matthewsius penelopae* (Matthews & Weir), pygidium; 38, *Matthewsius illawarrensis* (Matthews), protibial comb; 39, *Sauvagesinella becki* (Paulian), head; 40, *Sauvagesinella becki* (Paulian), protibia; 41, *Lepanus ovatus* Balthasar, protibia; 42, *Monteithocanthon arator* (Matthews), metaventrite and hind coxa; 43, *Lepanus villosus* Matthews, metaventrite and hind coxa.

5(4). Dorsal surfaces matt with at least a few setae laterally on elytra; dark brown to yellowish in colour; meso- and metatarsi with segment 1 longer than next 3 segments combined; protibiae with 4 teeth; apical edges of meso- and metatibiae straight with a fringe of stout setae/spines; antennae with 8 segments ....**Oniticellini**..**6**

– Dorsal surfaces usually shiny and always glabrous; black in colour; meso- and metatarsi with segment 1 shorter than next 3 segments combined; protibiae with 3 teeth; apical edges of meso- and metatibiae scalloped, with tufts of setae; antennae with 9 segments........................**7**

6(5). Sides of head angulate behind frontoclypeal suture; apical margin of elytra with fringe of long setae, distinct from other elytral pilosity; body colour dorsally light brown with some mottling; 7 to 13 mm in length; widespread – all states except TAS........................***Euoniticellus* Janssens**

– Sides of head rounded, not angulate behind frontoclypeal suture; apical margin of elytra lacking distinct fringe of setae; body colour otherwise – dark brown pronotum with yellow at the sides, elytra light brown with broken black stripes; 8 to 10 mm in length; northern and eastern Australia – NT, QLD, NSW........................***Liatongus* Reitter**

7(5). Protarsi absent; pronotal hypomeron anteriorly excavated for the reception of profemur (Fig. 30.7), excavation shallow but well defined by carina from procoxae to lateral margin; mesoscutellum small to very small, often hard to see; 9 to 18 mm in length; northern and eastern Australia – NT, QLD........................***Coptodactyla* Burmeister**

– Protarsi present; pronotal hypomeron without well defined anterior excavation for reception of profemur, carina from procoxae to lateral margin incomplete (Fig. 30.8); mesoscutellum larger, clearly visible; 8 to 12 mm in length; south-eastern and south-western Australia – NSW, VIC, WA........................***Thyregis* Blackburn (part)**

8(4). Meta- and usually mesotarsi with segment 1 longer than or subequal to the following 3 segments combined....**Onthophagini**........................**9**

– Meta- and usually mesotarsi with segment 1 at most as long as following 2 segments combined (mesotarsi may have segment 1 slightly longer than following 2 segments in some *Amphistomus*)........................**11**

9(8). Pronotal hypomeron with anterior carina from procoxa to anterior angle flange-like, enclosing a pocket for reception of antenna (Fig. 30.9); anterior angles of pronotum sharply acute and turned outwards (Fig. 30.11); pronotum characteristically densely granulate-punctate with each granule bearing a very short, thick, truncated, backwardly directed seta (Fig. 30.10); elytral intervals with 2 to 4 rows of setae, similar to those on pronotum; 6 to 9 mm in length; NSW, southern QLD........................***Hamonthophagus depressus* (Harold)**

– Pronotal hypomeron with anterior carina from procoxa to anterior angle not flange-like and often incomplete, not forming a pocket for the reception of antennae; anterior angles of pronotum variously angled; pronotal granulation and/or punctuation not as above; if elytral intervals with setae, then setae never as above........................**10**

10(9). Pronotal hypomeron with anterior carina reaching lateral margin at anterior angle and forming a small rounded lobe, this lobe being the anterior part of the hypomeron and visible from above (Fig. 30.12); male with protibiae elongate with strong interoapical tooth, curved down at apex; pronotum dark brown with yellow lateral and posterior edges; elytra light brown with lighter basal and apical areas; meso and metafemora pale with dark patches in middle; 8 to 12 mm in length; widespread – all states except TAS................***Digitonthophagus gazella* (Fabricius)**

– Not with the exact combination of characters described above; 2.5 to 23 mm in length; widespread – all Australian states..***Onthophagus* Latreille**

11(8). Metatibiae relatively short, markedly widened apically; ratio of length:width less than 3:1........................**12**

– Metatibiae elongate, slender, often curved, usually gradually widened to apex (except *Mentophilus* and *Coproecus* in which they are elongate but abruptly widened apically AND which have elytra with pseudepipleura); ratio of length:width greater than 3:1........................**14**

12(11). Metatibiae without distinct transverse carinae (Fig. 30.13); pronotal hypomeron excavated anteriorly for reception of profemur; elytra with sparse stout setae; 3 to 5 mm in length; northern and eastern Australia – NT, QLD, NSW, WA;........................***Demarziella* Balthasar**

– Metatibiae with well developed transverse carinae; pronotal hypomeron without anterior excavation for reception of profemur; elytra glabrous....**13**

13(12). Elytra with 9 striae visible basally, striae 9 along epipleural edge; protarsi short, not extending past end of apical protibial tooth; genal angles sub-quadrate or acute (Fig. 30.14); 16 to 26 mm in length; eastern and south-western Australia........................**Coprini: *Copris* Geoffroy**

– Elytra with 10 striae visible basally, striae 10 along epipleural edge; protarsi longer, with segments 4 and 5 extending past apical protibial tooth; genal angles rounded, obtuse (Fig. 30.15); 7 to 8 mm in length; VIC........................***Thyregis* Blackburn (part)**

14(11). Elytra with pseudepipleura (Fig. 30.16): an abrupt deflection of elytral surface in same plane as epipleura, marked by a sharp ridge, rounded fold or row of tubercles just outside seventh striae (short and confined to anterior quarter in *Aulacopris matthewsi* which has a ventral tooth on the profemur); at least parts of striae 8, 9 or 10 usually visible on pseudepipleura (although may be faint); tarsal claws simple........................**15**

– Elytra without pseudepipleura, surface evenly rounded to epipleura, OR, if short pseudepipleura present (*Diorygopyx*), elytra are deflexed along striae 8 and pseudepipleura occupy only basal third of elytra or less; tarsal claws simple, or with basal angulation or toothed........................**25**

15(14). Eye canthus (Fig. 30.17) complete (need to extrude head to see), broadly joining occipital margin, completely dividing eye into dorsal and ventral parts; protibiae with the two anterior teeth very close together at outer apex (Fig. 30.18); pronotal hypomeron with a ridge along notosternal suture, the ridge continuing forward along free edge of hypomeron (Fig. 30.19); elytra with 9 striae, 7 on disc and 2 on pdeudepipleura; length greater than 6 mm........................**15**

– Eye canthus (Fig. 30.20) incomplete (need to extrude head to see), remote from occipital edge or just touching it, usually leaving a more or less narrow strip of eye surface joining dorsal and ventral parts, or eye canthus completely absent; eye occasionally largely absent dorsally; protibiae with the two anterior teeth not very close together at outer apex, teeth more evenly spaced (Fig. 30.21); pronotal hypomeron without a ridge along notosternal suture (except in *Labroma*); elytra with 8, 9 or 10 striae, 7 on disc and 1, 2 or 3 on pseudepleura; length variable............**18**

16(15). Profemur with a strong tooth on ventral margin (Fig. 30.22); pronotum with longitudinal nitid carinae and tubercles (one poorly developed median carina in *Aulacopris matthewsi*); elytra tuberculate; male with a median tubercle and depression on metaventrite; pronotal hypomeron not excavated anteriorly to receive profemur; 8 to 30 mm in length; eastern Australia – QLD, NSW, VIC........................***Aulacopris* White**

– Profemur without a tooth on ventral margin; pronotum with at most a pair of low nitid tubercles; elytra smooth, non tuberculate; pronotal hypomeron excavated anteriorly for reception of profemur or not........................**17**

17(16). Clypeal margin with 4 to 8 teeth or rounded projections (Fig. 30.23), the middle pair often very prominent; protibiae with a supplementary tooth between spur and outer apical tooth (except some males); flattened posterior face of mesofemur without a longitudinal carina in the middle (Fig. 30.24); pronotal hypomeron anteriorly excavated anteriorly for reception of profemur or not; 6 to 19 mm in length; eastern Australia – QLD, NSW........................***Cephalodesmius* Westwood**

– Clypeal margin with 2 small obtuse teeth (Fig. 30.25), protibiae without a supplementary tooth; flattened posterior face of mesofemur with a sharply defined longitudinal carina slightly behind the middle (Fig. 30.26); pronotal hypomeron not excavated anteriorly for reception of profemur; 13 to 16 mm in length; eastern Australia – QLD.......... ***Canthonosoma* Macleay**

18(15). Dorsal surfaces tuberculate and with numerous curved setae capable of holding layer of soil; meso- and metatibiae with distinct small transverse carinae or spines on outer face; pygidium without a basal groove.......... **19**

– Dorsal surfaces of elytra nontuberculate; with or without setae of various sorts; meso- and metatibiae without transverse carinae or spines on outer face (suspicion of carinae in some *Mentophilus* which have tectate elytra); pygidium with or without a basal groove.......... **20**

19(18). Lateral edges of pronotum carinate or beaded only anteriorly (Fig. 30.27), posteriorly rounded; pronotal hypomeron deeply excavated anteriorly for reception of profemur (Fig. 30.28); mesometaventral suture straight; 4 to 10 mm in length; northern and eastern Australia – NT, QLD, NSW, WA.......... ***Amphistomus* Lansberge**

– Lateral edges of pronotum fully carinate (Fig. 30.29); pronotal hypomeron at most only shallowly excavated anteriorly for reception of profemur; mesometaventral suture slightly arched with vague swellings either side of midpoint; 8 to 12 mm in length; south-western WA.......... ***Labroma* Sharp**

20(18). Protibiae with 1 or 2 teeth; small size, less than 3 mm in length; setae on elytra and pronotum long, curved or squamate, tending to form tufts or clumps; elytra somewhat constricted laterally and tapered from midpoint to apex; wet tropics of north QLD.......... ***Pseudignambia* Paulian & Pluot-Sigwalt**

– Protibiae with 3 teeth; size larger, 4 mm or greater in length; setae on elytra of various sorts but never in tufts or clumps; elytra not constricted, laterally a smooth curve.......... **21**

21(20). Meso- and metatibiae abruptly expanded apically (Figs 30.30, 30.31); metatarsi with segment 1 longer than segment 2; elytra markedly convex; flightless – hind wings atrophied.......... **22**

– Meso- and metatibiae not abruptly expanded apically, gradually widened or somewhat parallel-sided apically; metatarsi with segment 1 shorter than or equal to segment 2; elytra less convex; flightless or not.......... **23**

22(21). Elytral intervals strongly tectate (Fig. 30.32); ridge defining pseudepipleuron ending before elytral apex, not meeting epipleuron; 6 to 10 mm in length; western WA.......... ***Mentophilus* Castelnau**

– Elytral intervals not tectate; ridge defining pseudepipleuron continuous to elytral apex and meeting epipleuron just before apex; 10 to 16 mm in length; north-western WA.......... ***Coproecus* Reiche**

23(21). Dorsal surfaces of elytra either glabrous or with minute setae visible with high power; pronotum glabrous with fine round somewhat superficial punctures; labial palpi with basal segment shorter than or equal to second segment; pygidium with or without a basal groove; 4 to 6 mm in length; eastern Australia – QLD, NSW.......... ***Aptenocanthon* Matthews**

– Dorsal surfaces of elytra with obvious curved or erect setae; pronotum densely and often deeply punctuate with oval or elongate punctures bearing setae; labial palpi with basal segment longer than second segment; pygidium with or without a basal groove.......... **24**

24(23). Mesoventrite 3–4 times as wide as long with large annular or cicatricose (scar-like) punctures each bearing a seta (Fig. 30.33); head without deep medial depression posterior to clypeal teeth; male clypeal teeth separate; pygidium with a basal groove; setae on elytra usually short and erect; 4 to 9 mm in length; western, northern, eastern and central Australia – WA, NT, QLD.......... ***Tesserodon* Hope**

– Mesoventrite very short, ~10 times as wide as long, with very small punctures without setae (Fig. 30.34); head with a deep medial depression posterior to clypeal teeth; male with clypeal teeth coalesced into a conspicuous process; pygidium without a basal groove; setae on elytra short and curved; 5 to 8 mm in length; north-eastern Australia – QLD.......... ***Boletoscapter* Matthews**

25(14). Meso- and metatibiae apically bearing a single large curved spur on inner angle which is about equal in length to first 3 tarsal segments combined; inner edge of profemur bearing a spine or tubercle; 3 to 7 mm in length; northern and eastern Australia – WA, NT, QLD, NSW.......... ***Monoplistes* Lansberge**

– Meso- and metatibiae apically with much shorter spurs; mesotibiae with 2 unequal spurs (some male *Temnoplectron* have a prolongation of the apex of the metatibiae, but this is not analogous to the spur described above, but mesotibiae still have 2 unequal spurs); inner edge of profemur without a spine or tubercle.......... **26**

26(25). Elytra with 8 striae; striae 8 located along the somewhat flange-like dorsal edges of the epipleura; protibiae with 2 or 3 teeth; tarsal claws either simple, or with a basal angulation or distinctly toothed; pygidium either simple or with a basal groove or a median depression or a modified central area; generally smaller species 2 to 6.5 mm in length (NB if some or all of the elytral striae are hard to see or effaced, then the somewhat flange-like dorsal edge of the epipleuron is evident).......... **27**

– Elytra with 9 or 10 striae, some of which may be short; striae 8 not located along edge of epipleura which are never somewhat flange-like; protibiae with 3 teeth (third tooth may be small); tarsal claws distinctly toothed; pygidium either simple or with a basal groove or transverse impressed line which may be interrupted in the middle; generally larger species 3.5 to 13 mm in length (NB if some or all of the elytral striae are hard to see or effaced, then stria 8 is not located along a somewhat flange-like dorsal edge of the epipleuron).......... **31**

27(26). Proleg with trochanterofemoral anterior pit (Fig. 30.35), situated in the apical border of trochanter; protibiae with 3 teeth, each bearing dorsally 2 small brushes of setae (Fig. 30.36); mesepimeron punctate; pygidium (Fig. 30.37) with a basal groove, extending from side to side; male protibiae inner apex bearing a comb of stiff bristles and a ventral brush of setae as well as a short spur (Fig. 30.38); tarsal claws simple; hindwings absent; 2.4 to 5.6 mm in length; NSW, SA WA.......... ***Matthewsius* Gunter and Weir**

– Proleg without trochanterofemoral anterior pit; protibiae with 2 or 3 teeth, each bearing dorsally a single brush of setae; mesepimeron impunctate; pygidium variously modified or not, if with a basal groove, then groove not extending from side to side; male protibiae either as above (*Matthewsius* type) or not; tarsal claws either simple, or with a basal angulation or toothed; hindwings present or absent.......... **28**

28(27). Tarsal claws large, distinctly toothed; metatarsi with segment 5 the longest, longer than segment 2 and as long as or longer than segments 3 and 4 combined; eyes (Fig. 30.39) with raised inner edge; protibiae with 3 teeth; mesocoxae strongly oblique; pygidium simple, punctuate and/or shagreened, setose; dorsal carina of protibiae (Fig. 30.40) from third tooth to base exposing large lateral basal area (view from above); profemora with a flattened posterior face; metafemora with a reticulate, flattened posterior face bearing 1, 2 or 3 carinae; male protibiae of *Matthewsius* type; 2.5 to 5 mm in length; south-west WA,SA.......... ***Sauvagesinella* Paulian**

– Tarsal claws smaller, simple or with a basal angulation or toothed; metatarsi with segment 5 subequal or shorter than segment 2 and shorter than segments 3 and 4 combined; eyes without raised inner edge; protibiae with 2 or 3 teeth; mesocoxae less oblique; pygidium variously modified or not, if with a basal groove, then groove not extending from side to side; dorsal carina of protibiae usually located more laterally or poorly defined, exposing less of lateral basal area (view from above); profemora without a flattened posterior face; metafemora usually without a reticulate, flattened posterior face; male protibiae of *Matthewsius* type or not **29**

29(28). Protibiae with 2 teeth, apical edge usually V-shaped between apical tooth and inner apical end (Fig. 30.41), occasionally approximately at right angles to the inner edge; pygidium either simple, or with a depression or a central modified area, but without a distinct basal groove; male protibiae never of the *Matthewsius* type; 1.8 to 5.2 mm in length; mainly eastern Australia: QLD, NSW, VIC, WA, NT .... ***Lepanus* Balthasar (part)**

– Protibiae with 3 teeth, apical edge usually at approximately right angles to the inner edge between apical tooth and inner apical end, occasionally V-shaped; pygidium variously modified or with a basal groove; male protibiae of the *Matthewsius* type or not **30**

30(29). Pygidium with a distinct curved or straight basal groove which does not extend from side to side; tarsal claws toothed; protibiae always with apical edge at approximately right angles to the inner edge between apical tooth and inner apical end; male protibiae of the *Matthewsius* type; median lobe of metaventrite with lateral margined area half or less than length of mesocoxae (Fig. 30.42); 2.1 to 3.2 mm in length; QLD, south-west WA .... ***Monteithocanthon* Gunter and Weir**

– Pygidium either simple, or with a depression or a central modified area; tarsal claws simple or with basal angulation; protibiae with apical edge either at approximately right angles to the inner edge or V-shaped (Fig. 30.41); male protibiae never of the *Matthewsius* type, either with inner apical end extended to a long or short apical digit or slightly expanded with a comb of stiff bristles; median lobe of metaventrite with lateral margined area longer than half length of mesocoxae (Fig. 30.43); 1.6 to 6.5 mm in length; QLD, NSW,WA, NT .... ***Lepanus* Balthasar (part)**

31(26). Pygidium with a basal groove or impressed line which may be interrupted in the middle; elytral epipleura basally strongly curved inwards and elytra deflexed along striae 8, here forming short basal pseudepipleura (except *cuspidatus*); hypomeral striae absent (except a short one in *cuspidatus*); metatibiae nearly straight, relatively short; elytra with 9 striae (all effaced in *cuspidatus*); male with protibiae greatly broadened apically; flightless, hindwings vestigial or reduced (except *tibialis* which is often fully winged); 5 to 9 mm in length; eastern Australia – QLD, NSW .... ***Diorygopyx* Matthews**

– Pygidium simple; elytral epipleura evenly curved; no basal pseudepipleura; hypomeral striae present (minute in some *diversicolle*); Metatibiae relatively long, parallel for at least distal half, compressed and blade-like, strongly curved in male; elytra with 10 striae, if with only 9 striae, then striae 8 absent (distal striae may be hard to see or effaced); male protibiae not greatly broadened apically; capable of flight or flightless, hind wings fully developed or vestigial; 3.5 to 13 mm in length; northern and eastern Australia – NT, QLD .... ***Temnoplectron* Westwood**

Classification of the Australian genera

Characters marked by an asterisk (*) apply only for Australian species of the respective taxon, non-Australian species may bear different character states.

Subfamily Scarabaeinae Latreille, 1802

The Australian fauna consists of 30 genera and some 382 described species. Of these, *Onthophagus* contains both native and introduced species, *Digitonthophagus* has one introduced species, *Copris, Onitis, Bubas, Euoniticellus, Liatongus* and *Neosisyphus* contain introduced species and *Hamonthophagus* has one accidently introduced species. The genera are distributed amongst 5 defined tribes and a large *Incerte sedis* group which contains almost all of the native genera.

Tribe Coprini Leach, 1815

Characteristics. Tribe Coprini *sensu novo* of Tarasov and Dimitrov (2016). Body robust and convex. Head and/or pronotum often with horns. Elytra with 10 striae, striae 9 and 10 close together along epipleural carina. Scutellum invisible. Anterior hypomeral carina stretches towards margin of hypomeron. Metatibiae relatively short, marked widened apically.

Distribution. Afrotropical, Oriental, Palaearctic, Nearctic, Neotropical.

Australian taxa. One genus.

References. Tarasov & Dimitrov (2016).

Copris Geoffroy, 1762 (Pl. 6D, 69C)

Type species. *Scarabaeus lunaris* Linnaeus, 1758.

Australian species. Two introduced.

Distribution. *Copris elphenor* Klug is a native to eastern Africa and established in eastern Queensland, while *C. hispanus* Linnaeus is native to Europe and northern Africa and established in southern Western Australia.

Biology. Burrows underneath dung pad, beetles fly at night, activity highest in autumn to early winter.

References. Edwards *et al.* (2015).

Tribe Oniticellini Kolbe, 1905

Characteristics. Body elongate, usually flattened. Antennae 8 segmented. Mesoscutellum visible. Pronotum without basal pits either side of midline. Meso- and metatarsi with segment 1 longer than segments 2–4 combined.

Distribution. Africa, Asia.

Australian taxa. Two genera.

References. Janssens (1953).

Euoniticellus Janssens, 1953 (Pl. 70G, H)

Type species. *Scarabaeus fulvus* Goeze, 1777.

Characteristics. As for tribe. Sides of head angulate behind frontoclypeal suture. Apical margin of elytra with a fringe of long setae, distinct from other elytral pilosity.

Australian species. Four introduced.

Distribution. *Euoniticellus africanus* (Harold) is native to South Africa and established from southern Queensland to central New South Wales, *E. intermedius* (Reiche) is native to Africa and widespread in Australia, while *E. fulvus* (Goeze) and *E. pallipes* (Fabricius) are native from Europe to India and established in southern Australia.
Biology. Nest below dung pad, beetles fly during the day, activity highest spring to autumn.
References. Edwards *et al.* (2015).

Liatongus Reitter, 1893 (Pl. 70I)

Type species. *Onthophagus phanaeroides* Westwood, 1840.
Characteristics. As for tribe. Sides of head rounded, not angulate behind frontoclypeal suture. Apical margin of elytra without a distinct fringe of long setae.
Australian species. One introduced.
Distribution. *Liatongus militaris* (Castelnau) is native to southern and eastern Africa and established in eastern Queensland, northern New South Wales and around Darwin, Northern Territory.
Biology. Burrows underneath dung pad, mostly day flying, activity highest spring to autumn.
References. Edwards *et al.* (2015).

Tribe Onitini Laporte, 1840

Characteristics. Body oblong, robust. Antennae 9 segmented. Pronotum with basal pits either side of midline. Mesoscutellum small and visible or invisible. Protarsi absent in both sexes. Elytra each with a somewhat prominent lateral carina.
Distribution. Africa, Palaearctic, Oriental.
Australian taxa. One genus.
References. Janssens (1937).

Bubas Mulsant, 1842 (Pl. 69A, F, H, L)

Type species. *Scarabaeus bison* Linnaeus, 1767.
Characteristics. Prosternum with a rounded projection behind procoxae. Scutellum not visible externally. Male with paired head horns and carinate projection on pronotum, female with a carina on head and transverse carina on pronotum.
Australian species. Two introduced.
Distribution. *Bubas bison* (Linnaeus) and *B. bubalus* (Olivier) are native to Europe and established in southern Australia.
Biology. Burrows underneath dung pad, mostly diurnal, activity highest in autumn to early winter.
References. Edwards *et al.* (2015).

Onitis Fabricius, 1798 (Pls 8G, 70B–D, J, K)

Type species. *Scarabaeus inuus* Fabricius, 1775.
Characteristics. Prosternum without a rounded projection behind procoxae. Mesoscutellum small but visible. Male and female with only carinae and tubercles on head. Pronotum without projections or carinae in both sexes.
Australian species. Six introduced.
Distribution. All introduced species native to Africa. *Onitis alexis* Klug is widespread in Australia, *O. aygulus* (Fabricius) is established in southern Australia, *O. caffer* Boheman is established in south-western Western Australia, and south-eastern Queensland to northern New South Wales, *O. pecuaris* Lansberge is established southern Queensland to Victorian border, *O. vanderkelleni* Lansberge is established near Brisbane and Cairns, Queensland, and *O. viridulus* Boheman is widespread in Queensland and in northern Northern Territory.
Biology. Burrows underneath dung pad, mostly diurnal, species activity varies.
References. Edwards *et al.* (2015).

Tribe Onthophagini Burmeister, 1846

Characteristics. Body not compressed laterally. Pronotum without basal pits either side of midline. Mesoscutellum not visible. Meta- and usually mesotarsi with segment 1 longer than or sub equal segments 2–4 combined.
Distribution. Cosmopolitan.
Australian taxa. Three genera.

Onthophagus Latreille, 1802 (Pl. 7B, 70A, E, L, O)

Type species. *Scarabaeus taurus* Schreber, 1759.
Monapus Erichson, 1848
Psilax Erichson, 1848
Macropocopris Arrow, 1920
Characteristics. Small to large sized, oblong to oval beetles; labial palpi with basal segment shorter than second segment; mesoscutellum not visible; elytra with 8 striae without intercalary striae; protibiae with 4 outer teeth; meso- and metatibia without transvers carinae, strongly dilated apically; metatarsi elongate and straight.
Australian species. 203 (6 introduced)
Distribution. Cosmopolitan distribution with over 2300 species globally. Australian endemic species distributed in all IBRA regions of Australia. The majority of species are known from broad distributions with only ~35% with a restricted distribution. Almost 60% of species are recorded from multiple states or territories. Over 150 species known from Queensland, over 60 species known from NSW, over 50 species known from the Northern Territory, over 45 species are known from Western Australia, over 20 species are known from South Australia, Victoria and the Australian Capital Territory, and ~10 species are known from Tasmania. 15 species are shared between New Guinea and Australia (Krikken & Huijbregts 2012).
Biology. Terrestrial, usually collected from baited pitfall traps in all major habitats, also using flight intercept and light traps.

References. Matthews (1972); Storey (1977); Storey & Weir (1990); Cassis & Weir (1992); Monteith & Storey (2013).

Key to species. Matthews (1972); Storey (1977); Storey & Weir (1990); Monteith & Storey (2013); Edwards *et al.* (2015); Monteith & Rossini (2017).

Digitonthophagus Balthasar, 1959 (Pl. 9G, 69B, E, M, R)

Type species. *Scarbaeus bonasus* Fabricius, 1775.

Characteristics. Similar to *Onthophagus* but with anterior portion of hypomeron produced anteriorly and visible from above, with the setae that would be located along the anterior edge in a ventral position now in a dorsal position. Male protibia elongate with a distinctly shaped interoapical tooth which is ventrally flat and obliquely oriented downwards.

Distribution. Africa, India, South-east Asia.

Australian species. One introduced species, *D. gazella* (Fabricius), widespread, mainly northern, all states except VIC, TAS.

References. Génier & Moretto (2017).

Hamonthophagus Roggero, 2016 (Fig. 30.11)

Type species. *Onthophagus bituberculatus* Olivier, 1789.

Characteristics. Similar to *Onthophagus,* but body somewhat flattened. Anterior margin of clypeus bidentate. Anterior hypomeral carina from procoxae to anterior pronotal angles flange-like, enclosing pocket for reception of antenna. Pronotum with sharp anterior angles, surface densely granulo-punctate with each granule bearing a short, thick, truncated, backwardly directed seta.

Distribution. Afrotropical.

Australian species. One accidentally introduced species, *H. depressus* (Harold) found in NSW and southern QLD.

References. Roggero *et al.* (2016).

Tribe Sisyphini Mulsant, 1842

Characteristics. Body laterally compressed with sides of pronotum and metaventrite vertical. Antennae 8-segmented. Mesoscutellum not visible. Mesocoxae very widely separated, forming lateral margin of body. Meso- and metatibiae elongate, curved, barely widened at apex.

Distribution. Afrotropical, Oriental, Palaearctic, Neotropical.

Australian taxa. One genus.

References. Tarasov & Dimitrov (2016).

Neosisyphus Muller, 1942 (Pl. 71F)

Type species. *Sisyphus atratus* Klug, 1855.

Australian species. Two introduced species.

Distribution. *Neosisyphus rubrus* (Paschalidis) and *N. spinipes* (Thunberg) are both native to Europe and widespread in QLD to northern NSW.

Biology. Rollers that build burrows away from dung pad, activity highest spring to autumn.

References. Edwards *et al.* (2015) [as *Sisyphus*].

Genera *Insertae sedis*

Characteristics. The following genera (except *Boletoscapter*) form what is called 'the Australasian endemic clade'. Although molecular and morphological studies have loosely defined this group, no concrete set of morphological characters has yet been assembled.

Distribution. Australia, New Guinea, New Caledonia, New Zealand.

Australian taxa. 21 genera, 154 species.

References. Tarasov & Génier (2015); Gunter *et al.* (2018).

Amphistomus Lansberge, 1874 (Pl. 68I)

Type species. *Merodontus calcaratus* Macleay, 1871.

Merodontus Macleay, 1871 (nec Jekel, 1854)

Platyphymatia Waterhouse, 1874

Characteristics. Small sized, oblong beetles trapezoidal to convex in cross section, length 4–10 mm; uniform in colour usually reddish brown to black; dorsal surfaces tuberculate and carinate with dense curved setae that often collect soil; head with 6 teeth, the 2 medial teeth longest; eye canthus incomplete; mentum broader than long; labial palpi with basal segment longer or subequal to second segment; pronotal disc strongly convex with tubercles; pronotal hypomeron deeply excavated anteriorly for reception of the profemur; mesoscutellum absent; elytra with 10 striae, 7 on disc and 3 on pseudepipleura; tubercles present on odd intervals; lateral edge of elytra with teeth, tubercles or udulations; mesoventrite 2–5 times as long often with median longitudinal ridge; trochanterofemoral pit absent; protibia with distal edge straight at right angle to inner edge and 3 large, evenly spaced, apical tooth with 2 tufts of setae dorsally, others each with 1 tuft of setae; protibial spur small and simple; mesotibiae with 2 transverse carinae on outer face; metatibia elongate and not abruptly expanded; tarsi short ~1/3 length of tibia; metatarsi with segment 1 longer than segment 2; claws small and simple; ventrites foreshortened, medially crowded together; pygidium flat to moderately convex, finely bordered with up to 7 tubercles on disc; hind wings fully developed, reduced or absent; male parameres symmetrical.

Australian species. 18.

Distribution. Australasian distribution with 18 species from Australia, seven from New Guinea and one from the Indonesian Moluccas. Australian endemic species primarily in the east coast mesic zone forests of Queensland and New South Wales almost to Victorian border, *A. tuberosus* Matthews is known from north coast Western Australia and near Darwin, Northern Territory; *A. tuberculatus* (Lansberge) is shared between Australia and New Guinea). Six species, *A. fasciculatus* (Gillet), *A. respiciens* (Kolbe), *A. monticola* Paulian, *A. kukali* Paulian, *A. speculate* (Gillet) and *A. uncinatus* Paulian are endemic to New Guinea, and *A. alfurorum* Huijbregts and Krikken is endemic to the Moluccas

Biology. Terrestrial, usually collected from baited pitfall traps in open and closed forests, also using flight intercept and light traps. Have been observed rolling balls.

References. Matthews (1974); Paulian (1985) (New Guinea only); Cassis & Weir (1992); Rortais (2003); Huijbregts & Krikken (2007).

Key to species. Matthews (1974).

Aptenocanthon Matthews, 1974 (Pl. 68E)

Type species. *Panelus hopsoni* Carter, 1936.

Characteristics. Small sized, oval beetles, convex, length 3.8–6.2 mm; uniform in colour usually brownish black; surfaces glabrous or with minute setae; head with 2 small teeth; eye canthus incomplete; mentum subquadrate; labial palpi with basal segment subequal or shorter than second segment; pronotum densely punctate with punctures bearing setae; pronotal disc shallowly convex; pronotal hypomeron excavated anteriorly for reception of the profemur; mesoscutellum absent; elytra with 9 or 10 striae and pseudepipleura outside 7th striae; mesoventrite 4 times as wide as long; trochanterofemoral pit present; protibia with distal edge at right angle to inner edge, 3 large evenly spaced teeth usually with 2 tufts of setae dorsally on each; protibial spur short, inserted near inner edge; meso- and metatibia not abruptly expanded, feebly curved, tarsal segments short; metatarsi with segment 1 shorter than segment 2 which is equal in length to 5th segment; claws small and simple; 5th sternite fused with 6th; pygidium transverse, punctured, simple or with a basal groove in *A. woolumbin* Storey and Monteith, and *A. kaburra* Storey and Monteith; hind wings absent; male parameres asymmetrical, left paramere longest.

Australian species. Six.

Distribution. Endemic to Australia. *Aptenocanthon kaburra* Storey and Monteith, *A. monteithi* Storey, *A. winyar* Storey and Monteith, *A. wollumbin* Storey and Monteith are from North Queensland, *A. hopsoni* (Carter) is from central coastal New South Wales, and *A. jimara* Storey and Monteith is from Northern Territory.

Biology. Terrestrial, usually collected from baited pitfall traps in closed forest, also using flight intercept and light traps.

References. Matthews (1974); Storey (1984); Storey & Monteith (2000); Cassis & Weir (1992); Gunter & Weir (2017).

Key to species. Storey & Monteith (2000). Note two species, *A. rossi* Matthews and *A. speewah* Storey and Monteith, have been removed to *Matthewsius*.

Aulacopris White, 1859 (Pl. 68J)

Type species. *Aulacopris reichei* White, 1859.

Characteristics. Large sized, oblong beetles, length 16–30 mm, convexly rounded; uniform in colour usually brownish-black; dorsal surfaces covered with short, stout setae, nitid carinae and tubercles; head wider than long with 2 small median teeth; eye canthus complete; mentum subquadrate; labial palpi with basal segment rectangular longer than second segment ovoid and minute third segment; pronotum heart-shaped, deflexed portion not excavated anteriorly but strongly beaded along outer edge; pronotal hypomeron not excavated anteriorly for reception of the profemur; mesoscutellum absent; elytra with 9 striae and sharply deflexed pseudepipleura outside 7th striae; intervals feebly undulate and odd numbered intervals tuberculate on disc only; mesoventrite 4–5 times as wide as long; trochanterofemoral pit absent; protibia with 2–3 very small teeth crowded near apex, each with 1 or 2 tufts of setae dorsally; protibial spur very small inserted at inner apical angle; profemora with strong tooth on ventral margin; meso- and metatibia elongate, slender and moderately strongly curved; meso- and metafemora with longitudinal ventral ridge with beaded hind edge; tarsi short; meso- and metatibial spurs longer than 1st tarsomere; metatarsi with segment 1 shorter than segment 2 and subequal to 3rd and 4th; claws moderate and simple; ventrites of normal proportion, basally with numerous large punctures bearing setae; pygidium feebly convex, strongly bordered with numerous punctures bearing setae; hind wings fully developed; male parameres asymmetrical, left one large.

Australian species. Three.

Distribution. Endemic to Australia, *A. matthewsi* Storey from North Queensland, *A. maximus* Matthews from southern Queensland to central coastal New South Wales and *A. reichei* White from southern New South Wales to Victoria.

Biology. Terrestrial, usually collected from baited pitfall traps in various forest types. Found in association with bat guano including in caves and in hollow tree trunks of roosts.

References. Matthews (1974); Storey (1986); Cassis & Weir (1992).

Key to species. Storey (1986).

Boletoscapter Matthews, 1974 (Pl. 68K, L)

Type species. *Cephalodesmius cornutus* Macleay, 1887.

Characteristics. Small sized, suboval to oblong beetles, length 5–8 mm; uniform in colour usually reddish brown; head with 2 prominent median teeth coalesced into a conspicuous process; eye canthus incomplete; mentum subquadrate; labial palpi with basal segment longer than second segment; pronotum punctate; pronotal hypomeron excavated anteriorly for reception of the profemur; mesoscutellum absent; elytra with 10 striae and incomplete pseudepipleura outside 7th striae, terminating 2/3 of way down elytral length; short and curved elytral setae; mesoventrite extremely short; trochanterofemoral pit absent; protibia with 3 evenly spaced teeth, apical tooth with 1 seta dorsally, others each with 2 setae; protibial spur long, medially located on distal edge, distal angle prolonged to falsely give appearance of a broad spur; metatibia elongate and gradually expanded; tarsi short; metatarsi with segment 1 equal to segment 2; claws small and simple; ventrites of normal proportion; pygidium transverse, feebly convex and evenly punctate; hind wings fully developed; male parameres symmetrical.

Australian species. Two.

Distribution. Endemic to Australia, both species *B. cornutus* (Macleay) and *B. furcutus* Matthews from Queensland.

Biology. Terrestrial, usually collected from mushroom baited pitfall traps in various forest types.

References. Matthews (1974); Cassis & Weir (1992).

Key to species. Matthews (1974).

Canthonosoma Macleay, 1871 (Pl. 69K)

Type species. *Canthonosoma mastersi* Macleay, 1871.
Homodesmius Sharp, 1873

Characteristics. Large sized, suboval to oblong beetles, length 13–16 mm; uniform in colour usually brownish black; head with 2 small triangular teeth; dorsal surfaces matt; eye canthus complete; mentum subquadrate; labial palpi with basal segment oblong slightly longer than second segment; pronotum with pair of impunctate spots present on either side of the midline in basal 1/3; pronotal disc strongly convex, lateral edge beaded, posterior edge with small tubercles; pronotal hypomeron not excavated anteriorly for reception of the profemur; mesoscutellum absent; elytra smooth with 9 striae and very sharp pseudepipleura outside 7th striae, pseudepipleura broad and deflected with 2 striae; mesoventrite 3–4 times as wide as long; trochanterofemoral pit absent; all femora with longitudinal carina on ventral side; protibia with 3 teeth, the distal 2 close together at outer apex, each tooth with 2 tufts of setae dorsally; protibial spur short; mesotibia strongly curved and subequal to length of femora; metatibia elongate, slender and strongly bowed in distal half; tarsi short and broad; metatarsi with segment 1 shorter to segments 2, 3 and 4; claws small and simple; ventrites of normal proportion and impunctate; pygidium semicircular, moderately convex and evenly punctate with short cirri; hind wings vestidual ~1/3 abdominal length; male parameres asymmetrical, left prolonged and bent under right.

Australian species. Three.

Distribution. Endemic to Australia, all species *C. castelnaui* (Harold), *C. macleayi* (Harold), *C. mastersi* Macleay are from southern Queensland.

Biology. Terrestrial, usually collected from mushroom baited pitfall traps in various forest types.

References. Matthews (1974); Cassis & Weir (1992).

Key to species. Matthews (1974).

Cephalodesmius Westwood, 1841 (Pl. 69G)

Type species. *Cephalodesmius armiger* Westwood, 1841.

Characteristics. Small to medium-sized, oblong beetles with broadened prothorax, length 6–19 mm; uniform in colour usually blackish-brown; surface matt with very fine setae that may hold small amounts of soil; head with 4–8 teeth medial 2 long and acute, 2nd pair shorter but prominent, others minute; eye canthus complete; mentum broader than long; labial palpi with basal segment subequal to second segment; pronotum moderately to strongly convex; pronotal hypomeron with a ridge along notosternal suture and anteriorly excavated for reception of the profemur or not; mesoscutellum absent; elytra with 9 striae, elytral disc relatively flat with 7 striae, a sharp pseudepipleural edge with 2 striae on pseudepipleura; mesoventrite 4–5 times as wide as long; trochanterofemoral pit absent; profemora uniformly punctate on ventral surface; protibia with distal edge straight approximately at right angle to inner edge, 3 teeth, the distal 2 closer together and usually supplementary tooth between spur and outer apical tooth, apical tooth with 1 or 2 tufts of setae dorsally, others each with 2 tufts of setae; protibial spur large and moveable inserted near inner apical angle; metatibia elongate and not abruptly expanded; tarsi normal proportion; metatarsi with segment 1 subequal or longer than to segment 2; claws normal and simple; ventrites not unusually foreshortened; pygidium subtriangular, flat and margined; hind wings reduced or absent; male parameres symmetrical.

Australian species. Three.

Distribution. Endemic to Australia, *C. quadridens* Macleay from southern Queensland, *C. laticollis* Pascoe from southern Queensland to Northern New South Wales, *C. armiger* Westwood from southern Queensland to central New South Wales, one undescribed species from central Queensland.

Biology. Terrestrial, usually collected from baited pitfall traps in various forest types. Pairs bond for life building a burrow that is provisioned with decaying vegetable matter.

References. Matthews (1974); Monteith & Storey (1981); Cassis & Weir (1992); Dalgleish & Elgar (2005).
Key to species. Matthews (1974).

Coproecus Reiche, 1841 (Pl. 69N)

Type species. *Circellium hemisphaericum* Guérin-Méneville, 1830.
Characteristics. Large sized, rotund beetles, very convex length 13–17 mm; uniform in colour usually brownish black; dorsal surfaces wrinkled or of leathery appearance without tubercles or carinae; head with 2 triangular teeth that form a V-shaped excision; eye canthus incomplete; mentum subquadrate; labial palpi with 1st and 3rd segment longer but narrower than second ovoid segment; pronotum slightly narrower than hind body; pronotal disc strongly convex and densely punctate; sides of the pronotum subparallel in basal half but convergent in apical half, anterior edge beaded; pronotal hypomeron deeply excavated anteriorly for reception of the profemur; mesoscutellum absent; elytra very broad with 10 striae and broad sharp pseudepipleura outside 7th striae, 8th striae only present as posterior traces, intervals moderately convex with 2 rows of simple punctures; mesoventrite 3 times as wide as long; trochanterofemoral pit absent; protibia widened apically with distal edge at right angle to inner edge and 3 large, evenly spaced teeth, each with multiple setae dorsally; protibial spur small; mesocoxae subparallel; meso- and metafemora beaded on posterior edge; mesotibia greatly expanded at apex; metatibia elongate and slighly expanded at apex, spurs as long as tarsomere 1; tarsi subtriangular; metatarsi with segment 1 subequal to segment 2; claws small and simple; ventrites of normal proportion; pygidium strongly transvers feebly convex, fully bordered with a transverse basal groove; hind wings greatly reduced; male parameres asymmetrical left slightly longer.
Australian species. One.
Distribution. Endemic to Australia, only known species *C. hemisphaericus* Guérin-Méneville from central coast Western Australia.
Biology. Terrestrial, usually collected from baited pitfall traps. Does not appear to be able to make its own dung ball but instead observed rolling and burying small faecal pellets.
References. Matthews (1974); Cassis & Weir (1992).
Key to species. Matthews (1974).

Coptodactyla Burmeister, 1846 (Pl. 69P, Q)

Type species. *Copris glabricollis* Hope, 1841.
Boucomontia Paulian, 1933
Characteristics. Small to medium-sized, oblong beetles, feebly to strongly convex, length 9–18 mm; uniform in colour usually blackish-brown; surface largely glabrous; head with 2 medial teeth and edges usually with marginal beading; eye canthus incomplete; mentum broader than long, more or less excavated apically; underside of clypeus with a transverse ridge before mouthparts; labial palpi with basal segment smaller than the second segment, the third smallest but distinct; pronotum moderately to strongly convex, often armed in male; pronotal hypomeron deflexed in areas and shallowly excavated anteriorly for reception of the profemur, the excavation posteriorly bordered by a well defined carina extending from procoxae to the lateral margin; a longitudinal carina about half the prothoracic length or more running from posterior edge; small to minute mesoscutellum visible between elytral bases; elytra with 10 striae including 1 along the epipleural margin, 8th stria occasionally effaced anteriorly, 9th always well separated from the 10th for the anterior third (which may be partly or wholly effaced here) then merging abruptly for the posterior thirds; mesoventrite 3.5–4 times as wide as long; trochanterofemoral pit present; protarsi absent; protibia with distal edge transversely truncate, 3 distinct, evenly-spaced teeth, apical two teeth with 1 or 2 tufts of setae dorsally; protibial spur long but quickly worn and moveable inserted near inner apical angle; apical edges of meso- and meta tibia scalloped with tufts of setae; meso- and metatibia abruptly expanded distally, with a transverse carina on the outer edge; meso- and metatarsi short with subtriangular segments decreasing in size distally, basal segment slightly shorter than preceding 2 combined; claws small, moderately curved and simple; ventrites strongly contracted medially and fused; pygidium transverse, strongly and completely margined, simple with a fine deep groove all around the disc; hind wings fully developed; male parameres symmetrical.
Australian species. 14.
Distribution. Endemic to Australia and New Guinea, *C. glabricollis* (Hope) shared between coastal Northern Territory and Queensland, Australia and New Guinea, *C. merdeka* Reid from Indonesia, West Papua, *C. papua* Lansberge and *C. tuberculata* Gillet from Papua New Guinea, *C. depressa* Paulian, *C. ducalis* Blackburn, *C. nitida* Paulian, *C. monstrosa* Felsche, *C. matthewsi* Reid, *C. onitoides* Gillet, *C. storey* Reid, *C. subaenea* Harold, *C. torresica* Matthes from North Queensland, *C. lesnei* Paulian and *C. stereocera* Gillet from northern Northern Territory, *C. brooksi* Matthews from inland central Queensland, and *C. meridionalis* Matthews from southern Queensland.
Biology. Terrestrial, usually collected from baited pitfall traps in various forest types.
References. Matthews (1976); Reid (2000); Cassis & Weir (1992).
Key to species. Reid (2000).

Diorygopyx Matthews, 1974 (Pl. 69J)

Type species. *Temnoplectron tibiale* Macleay, 1871.

Characteristics. Medium-sized, broadly oval beetles, more or less convex, length 5–9.5 mm; uniform in colour, black; surfaces entirely glabrous and nitid; head with 2 small teeth that are close together or merged into 1 in male *cuspidatus*, edge of teeth and between them greatly thickened; eye canthus incomplete; mentum subquadrate or slightly longer than wide; labial palpi with basal segment longer than second segment; pronotum evenly convex, sides of the pronotum straight in basal 2/3; pronotal hypomeron excavated anteriorly for reception of the profemur; mesoscutellum absent; elytra with 9 superficial striae, surface strongly deflexed along 8th striae to form a short basal pseudepipleuron (except *cuspidatus*); striae 8 and 9 greatly shortened and at edge of deflexed humeral portion; mesoventrite 5 times as wide as long, greatly shortened medially; trochanterofemoral pit present; protibia greatly broadened apically with straight transverse distal edge, 3 evenly spaced small teeth apical tooth with 2 tufts of setae dorsally, others each with 1 tuft of setae; protibial spur short, inserted near inner apical angle; protarsi very small; pro- and metafemora beaded along anterior and posterior edges; mesofemora beaded only along anterior edge; meso- and metatibia relatively short and nearly straight and not abruptly expanded, distal edge transverse; metatarsi with segment 1 shorter and broader than segment 2; claws large and toothed; ventrites of normal proportion 5th and 6th clearly demarcated; pygidium transverse, finely bordered with thickened border at base, disc strongly convex, glabrous and finely punctate with a basal groove or impressed line that may be interrupted in the middle; hind wings vestigial or slightly reduced in *tibialis*; male parameres asymmetrical.

Australian species. Eight.

Distribution. Endemic to Australia, *D. cuspidatus* Matthews from central Queensland, *D. incomptus* Matthews, *D. simpliciclunis* Matthews and *D. tibialis* (Macleay) from southern Queensland to Northern New South Wales, *D. asciculifer* Matthews, *D. duplodentatus* Matthews, *D. incrassatus* Matthews, *D. niger* Matthews from northern New South Wales.

Biology. Terrestrial, usually collected from baited pitfall traps in closed forest. Individuals of some species observed rolling balls.

References. Matthews (1974); Cassis & Weir (1992).

Key to species. Matthews (1974).

Demarziella Balthasar, 1961 (Pl. 68C, D)

Type species. *Demarziella mirifica* Balthasar, 1961.
Notopedaria Matthews, 1976

Characteristics. Small sized, oblong beetles, feebly convex, length 2.9–5 mm; uniform in colour often bronzed and occasionally metallic; dorsally stout, surface with decumbent bristles combined with some very small setae; head subcircular with 2 medial teeth, widely separated by an arcuate emargination; eye canthus incomplete; mentum subquadrate; underside of clypeus with a strong angular ridge; labial palpi with basal segment subequal to second segment and both subtriangular and very wide, the third segment minute; pronotum subquadrate, not strongly convex, entirely unarmed but with various shallow depressions and elevations, and sparse, stout bristles often arranged in a transverse row; pronotal hypomeron deflexed and deeply excavated anteriorly for reception of the profemur, the excavation with a well defined posterior carina and a second carina paralleling posterior portion of lateral edge; mesoscutellum absent; elytra with 10 striae including 1 along the epipleural margin, 8th striae represented only by a short anterior section, 9th only present in posterior half and very close to 10th striae which is greatly enlarged and deepened anteriorly; striae usually unevenly spaced with odd intervals wider and often higher than even ones; mesoventrite 4 times as wide as long; trochanterofemoral pit present; protarsi present, short, held in transverse grooves in tibial truncation; protibia with distal edge transversely truncate, 3 distinct, evenly-spaced teeth, apical tooth with 2 tufts of setae dorsally, others each with 1 tuft of setae; protibial spur small to minute inserted near inner apical angle; meso- and metatibia abruptly expanded distally, distal edges more or less straight and transverse, with a fringe of very short, stout bristles and some longer of setae; mesotibiae with 2 short, unequal terminal spurs and metatibiae with 1 spur; meso- and metatarsi short with first 2 segments subequal then decreasing in size distally; claws small and simple; all leg segments with short bristles; ventrites fused, basal sternite longer medially than preceding 4; pygidium transverse, strongly margined, with a groove along the dorsal edge, setose; hind wings fully developed; male parameres assymmetrical.

Australian species. 14.

Distribution. Endemic to Australia, *D. geminata* (Macleay), *D. imitatrix* Balthasar, *D. pratensis* (Matthews), *D. storeyi* (Matthews), *D. tropicalis* (Matthews) and *D. yorkensis* Matthews and Stebnicka from northern Queensland, *D. interrupta* (Carter) from northern Queensland to northern New South Wales, *D. metallica* (Carter) from southern Queensland to northern New South Wales, *D. alternata* (Lea) and *D. planitarsis* Petrovitz from coastal Northern Territory including Groote Eylandt, *D. scarpensis* (Matthews) from central New South Wales, *D. eungella* (Matthews) from central Queensland, and *D. mirifica* Balthasar from coastal northern Australia.

Biology. Terrestrial, usually collected from baited pitfall traps in various forest types.

References. Matthews (1976); Matthews & Stebnicka (1986); Cassis & Weir (1992).
Key to species. Matthews & Stebnicka (1986).

Labroma Sharp, 1873 (Pl. 70M)

Type species. *Labroma horrens* Sharp, 1873.
Characteristics. Medium-sized, oblong beetles moderately convex, length 8–12 mm; uniform in colour usually brownish-black; dorsal surfaces with short curved setae that often collect soil, carinate on pronotum and tuberculate on elytra; head with 6 short teeth, clypeal edge medially broadened with U-shaped excision; eye canthus incomplete; mentum nearly semicircular and feebly excised distally; labial palpi with basal segment ovoid and subequal to second segment, 3rd segment about half the size; pronotum moderately convex with various longitudinal carinae, lateral edges fully carinate; pronotal hypomeron at most shallowly excavated anteriorly for reception of the profemur; mesoscutellum absent; elytra moderately convex with 10 striae and very sharp pseudepipleura outside 7th striae; edge of pseudepipleura tuberculate or crenulate, terminating at hind angle, no tubercles on pseudepipleura; mesoventrite 2 1/2 times as wide as long; trochanterofemoral pit absent; protibia normal and flattened, distal edge at approximate right angle to inner edge and 3 large, evenly spaced teeth, apical tooth with 2 unequal tufts of setae dorsally, others each with 1 tuft of setae; protibial spur slender and inserted near inner apical angle; protarsi relatively long; mesotibiae and metatibia elongate, slender, straight and with 1–2 transverse carinae or spines on outer face; tarsi very long, longer than spurs; metatarsi with segment 1 subequal to segment 2 and 3 combined; claws moderate and simple; ventrites normal in proportion; pygidium finely bordered laterally with punctures; hind wings reduced; male parameres asymmetrical, left one longer.
Australian species. Three.
Distribution. Endemic to Western Australia, *L. horrens* Sharp and *L. umbratilis* Matthews from south-western Western Australia, *L. tuberculata* (Waterhouse) from central coastal Western Australia.
Biology. Terrestrial, mostly collected in baited pitfall traps in jarrah and marri forests.
References. Matthews (1974); Cassis & Weir (1992).
Key to species. Matthews (1974).

Lepanus Balthasar, 1966 (Pl. 9H, 70P)

Type species. *Lepanus ovatus* Balthasar, 1961.
Characteristics. Small sized, broadly oval beetles, convex, length 1.6–6.5 mm; colour variable usually uniformly coloured occasionally biocoloured, colours range from brown to black; surfaces usually glabrous and nitid; head with 2 small teeth that are close together and separated by a broad or normal U-shape; eye canthus incomplete; mentum subquadrate; labial palpi with second segment twice as long as 1st and 3rd segments; pronotum strongly convex; pronotal edges usually beaded; pronotal hypomeron deeply excavated anteriorly for reception of the profemur; mesoscutellum absent; elytra with 8 striae without pseudepipleura or any carina, striae 8 located on a somewhat flanged-dorsal edge of the epipleura; epipleura broad basally; intervals flat; mesoventrite 4 times as wide as long; trochanterofemoral pit absent; protibia with concave distal edge, 2–3 evenly spaced small teeth each bearing 1 small tuft of setae dorsally, apical edge usually V-shaped between apical tooth and inner apical end when 2 tibial teeth present; protibial spur short to long, inserted near inner apical angle but projecting forward and downwards particularly in males; femora finely beaded on anterior and posterior edges; mesotibiae with 2 unequal spurs; meso- and metatibia not abruptly expanded apically but generally broadened, feebly to moderately curved with very short spurs; metatarsi with segment 1 shorter and slightly broader than segment 2, segment 5 subequal or shorter than segment 2 and shorter than segment 3–4 combined; claws usually subdentate; ventrites of normal proportion 5th and 6th connate; pygidium rarely strongly transverse, fully bordered, disc variable often with a pit or groove, sometimes simple; hind wings usually fully developed except absent in *occidentalis*; male parameres asymmetrical, left paramere only slightly longer.
Australian species. 20.
Distribution. Endemic to Australia and New Guinea, *L. dichrous* (Gillet), *L. furcifer* Matthews, *L. globulus* (Macleay), *L. latheticus* Matthews, *L. monteithi* Matthews, *L. nitidus* Matthews, *L. palumensis* Matthews, *L. parapisoniae* Matthews, *L. pisoniae* (Lea), *L. villosus* Matthews from northern Queensland, *L. vestitus* Matthews and *L. gelasinus* Matthews central coastal Queensland, *L. storeyi* Weir and Monteith, and *L. niger* (Lansberge) (status unknown) from south-east Queensland, *L. ustulatus* (Lansberge) and *L. politus* (Carter) from coastal central Queensland to central New South Wales, *L. australis* Matthews from coastal central Queensland to coastal Victoria, *L. bidentatus* (Wilson) from coastal southern Queensland to central New South Wales, *L. pygmaeus* (Macleay) from northern coastal Australia, *L. occidentalis* Matthews from south-western Western Australia. *L. ovatus* Balthasar, *L. howdeni* Paulian, *L. gressiti* (Balthasar), *L. fossulatus* Paulian and *L. papuanus* Paulian from New Guinea.
Biology. Terrestrial, usually collected from baited pitfall traps in various forest types, also using flight intercept and light traps. Some species observed rolling balls. Several species recorded on bird droppings.
References. Matthews (1972); Cassis & Weir (1992); Gunter & Weir (2017).
Key to species. Matthews (1974) (in part); Paulian (1985) (New Guinea only).

Matthewsius Gunter & Weir, 2017 (Pl. 70F)

Type species. *Matthewsius stricklandensis* Gunter & Weir, 2017.

Characteristics. Small to medium-sized, broadly oval beetles, convex, length 3.4–5.6 mm; uniformly coloured brownish black; surfaces usually glabrous and nitid, yellow antennae; head with 2 teeth separated by a broad U-shape; eye canthus incomplete; mentum subquadrate; labial palpi with basal segment subequal to 2nd; pronotal disc convex; pronotal edges usually beaded; pronotal hypomeron excavated anteriorly for reception of the profemur and excavation posteriorly; mesoscutellum absent; elytra with 8 striae without pseudepipleura or any carina, striae 8 located on a somewhat flanged-dorsal edge of the epipleura and in a deep groove at least apically; mesoventrite 5–6 times as wide as long; trochanterofemoral pit present situated in the apical border of the trochanter; protibia with distal edge at right angle to inner edge, 3 small teeth each bearing 2 small tufts of setae dorsally, apical teeth closer together; male protibiae inner apex bearing a comb of stiff bristles and a ventral brush of setae, apical digit and broad, short triangular spur; metafemora with beaded anterior and posterior edges and a posterior longitudinal carina; long dense setae on all tarsomeres; mesotibiae with 2 unequal spurs and males with dense setae on inner edge; meso- and metatibia gradually widened apically with short spurs; male metatibiae with small blunt tooth on inner apical angle; metatarsi with segment 1 widest, segment 2 equal in length to segment 5; claws small and simple (sometimes subdentate in *illawarrensis*); ventrites of normal proportion; pygidium transverse, with a deep groove right across the base and a fine groove around the rest of the edge; hind wings absent; male parameres asymmetrical, right paramere apically bent downwards and expanded.

Australian species. Six.

Distribution. Endemic to Australia. *M. illawarrensis* (Matthews), *M. rossi* (Matthews) and *M. stricklandensis* Gunter and Weir from coastal New South Wales, *M. penelopae* (Matthews and Weir) from South Australia, *M. vanleeuweni* Gunter and Weir from north-western Western Australia, and *M. speewah* (Storey and Monteith) from northern Queensland.

Biology. Terrestrial, usually collected from baited pitfall traps in various forest types, also using flight intercept and light traps.

References. Gunter & Weir (2017).

Key to species. Gunter & Weir (2017).

Mentophilus Castelnau, 1840 (Pl. 70Q)

Type species. *Mentophilus hollandiae* Castelnau, 1840.
Aulacium Reiche, 1841
Minthophilus (Mintophilus) Lacordaire, 1856
Menthophilus Gemminger & Harold, 1968

Characteristics. Medium-sized, rotund beetles, moderately convex length 6–10 mm; uniform in colour usually brownish black; antennal clubs black; dorsal surfaces strongly textured and sculptured; head with 2 prominent teeth flanked by deep notches, with smaller notches at clypeogenal sutures; eye canthus incomplete; mentum somewhat transverse; labial palpi with 1st and 3rd segment longer but narrower than second ovoid segment; pronotum as broad as hind body, disc convex; sides of the pronotum very strongly expanded, lateral edges sinuate, posterior edge finely beaded, other edges clearly beaded; pronotal hypomeron poorly excavated anteriorly for reception of only antenna, excavation poorly defined posteriorly; mesoscutellum absent; elytra short and broad with 9 striae and very broad pseudepipleura marked by sharp outer edge just outside 7th striae and ending before elytral apex, not meeting epipleuron; intervals strongly tectate; mesoventrite 4 times as wide as long; trochanterofemoral pit present/absent; protibia with distal edge at right angle to inner edge and 3 large, evenly spaced teeth, each with multiple setae dorsally; protibial spur short and inserted near apical angle; large protarsi; meso- and metafemora finely beaded on posterior edge; meso- and metatibia abruptly expanded apically, spurs as long as tarsomere 1; tarsi subtriangular; metatarsi with segment 1 slightly longer than segment 2; claws moderate and simple; ventrites somewhat foreshortened; pygidium transverse with a deep basal groove and fully bordered; hind wings reduced to minute vestiges; male parameres asymmetrical left slightly longer, terminally downturned and spatulate.

Australian species. Two.

Distribution. Endemic to Western Australia, both species *M. hollandiae* Castelnau and *M. subsulculatus* Sharp from central coastal Western Australia.

Biology. Terrestrial usually collected from baited pitfall traps. Observed rolling faecal pellets and making balls.

References. Matthews (1974); Cassis & Weir (1992).

Key to species. Matthews (1974).

Monoplistes Lansberge, 1874 (Pl. 70N)

Type species. *Monoplistes haroldi* Lansberge, 1874.

Characteristics. Small to medium-sized, broadly oval beetles, moderately convex, length 3–7 mm; uniform in colour, black or occasionally brownish black; surfaces entirely glabrous; head with 2 or 4 small teeth an additional angulation at edge of clypeogenal suture and a small median tooth on lower part of anterior edge; eye canthus incomplete; mentum subquadrate; labial palpi with basal segment longer than second segment, all edges of pronotum very finely beaded; pronotal disc strongly downturned at sides, lateral edge remains hidden in dorsal view; pronotal hypomeron excavated anteriorly for reception of the profemur; mesoscutellum absent; elytra without pseudepipleura, 9 distinct striae but incomplete posteriorly, striae 8 effaced anteriorly and close to striae 9, intervals flat; mesoventrite 4.5 times as wide as length along midline; trochanterofemoral pit absent; inner edges of protibia and profemur bearing a spine or tubercle; protibia greatly broadened apically with oblique distal edge, 3 evenly spaced small teeth, each with 2 tufts of setae dorsally; protibial spur very broad, quadrate or with inner angle produced; protarsi very small; mesocoxae strongly oblique; meso- and metafemora stout, beaded along anterior edge of metafemora only; mesofemora beaded only along anterior edge; meso- and metatibia strongly curved with spinulose outer edge bearing a short spur-like prolongation of outer angle and on inner angle a single large curved spur about equal in length to first 3 tarsomeres combined; tarsi relatively long, segments subcylindrical; metatarsi with segment 1 much shorter than segment 2; claws large and subdentate; ventrites of normal proportion, glabrous, 5th and 6th clearly demarcated; pygidium transverse, finely bordered, glabrous, with a deep basal groove; hind wings fully developed; male parameres symmetrical except in *haroldi*.

Australian species. Six.

Distribution. Endemic to Australia, *M.* curvipes (Lea) and *M. tropicus* Lea from northern Queensland, *M. haroldi* Lansberge and *M. phanophilus* Lea from northern coastal Australia, *M. occidentalis* (Macleay) from eastern Queensland, *M. leai* Paulian from southern Queensland to central New South Wales.

Biology. Terrestrial, collected in flight intercept traps, baited pitfall traps; one species observed perching on leaves in rainforest at night; one sighting on bird dropping.

References. Matthews (1974); Cassis & Weir (1992).

Key to species. Matthews (1974).

Monteithocanthon Gunter & Weir, 2017 (Pl. 103F)

Type species. *Lepanus arator* Matthews, 1974.

Characteristics. Small sized, broadly oval beetles, convex, length 2.1–3.2 mm; uniformly coloured reddish brown to brownish black; surfaces usually glabrous and nitid; head with 2 small teeth separated by a broad U-shape; eye canthus incomplete; mentum subquadrate, labial palpi with basal segment smaller than 2nd; pronotal disc convex; pronotal edges beaded or not; pronotal hypomeron excavated anteriorly for reception of the profemur and excavation posteriorly; mesoscutellum absent; elytra with 8 striae without pseudepipleura or any carina, striae 8 located on a somewhat flanged-dorsal edge of the epipleura; mesoventrite 5–6 times as wide as long; trochanterofemoral pit absent; protibia with distal edge at right angle to inner edge, 3 small teeth each bearing 1 small tuft of setae dorsally, apical tooth teeth closer together, male protibiae inner apex bearing a comb of stiff bristles and a ventral brush of setae, apical digit virtually nonexistent with a very short spur, metafemora with beaded anterior and posterior edges and a posterior longitudinal carina; mesotibiae with 2 unequal spurs and males with dense setae on inner edge; meso- and metatibia gradually widened apically, strongly curved with short spurs, male metatibiae generally bent with an extension (except straight in *peckorum*); metatarsi with segment 1 shorter than segment 2, segment 5 longest and shorter than 3–4 combined; claws small and toothed, ventrites of normal proportion; pygidium transverse, with a transverse groove not reaching basal angle; hind wings fully developed, reduced or absent; male parameres asymmetrical, right paramere apically bent downwards and more or less apically expanded.

Australian species. Six.

Distribution. Endemic to Australia. *M. arator* (Matthews), *M. elliotensis* Gunter and Weir, *M. koombooloomba* Gunter and Weir, and *M. paraarator* Gunter and Weir from northern Queensland, *M. glaber* (Matthews) from south-eastern Queensland, *M. peckorum* Gunter and Weir from south-western Western Australia.

Biology. Terrestrial, usually collected from baited pitfall traps in various forest types.

References. Gunter & Weir (2017).

Key to species. Gunter & Weir (2017).

Pseudignambia Paulian & Pluot-Sigwalt, 1985 (Pl. 71A)

Type species. *Ignambia mimerops* Matthews, 1974.

Characteristics. Small sized, oval beetles, convex, length 2.5–2.8 mm; uniform in colour brownish black; dorsal surfaces covered in thick setae tending to form tufts or clumps; head with 2 small, close set teeth; eyes without any dorsal separation; eye canthus completely absent, mentum twice as wide as long; labial palpi with basal segment quadrate, shorter than second trapezoidal segment; labial palpi and galea of maxilla with long tufts if curved setae; pronotum as slightly narrower than hind body, disc strongly convex, lateral edges curved, edges not beaded; pronotal hypomeron deeply excavated anteriorly for reception of the profemur; mesoscutellum absent; elytra constricted laterally from midpoint to apex with 9 striae and distinct pseudepipleura outside 7th striae, epipleura narrow basally disappearing at apex; mesoventrite 4 times as wide as long; trochanterofemoral pit present/absent (hard to see); protibia with distal edge oblique and inner apical angle produced forward and with 1 or 2 small teeth, each with a few setae dorsally; protibial spur short and inserted near apical angle; meso- and metafemora finely beaded on anterior and posterior edges; meso- and metatibia gradually expanded apically, spurs extremely small; tarsi short and cylindrical; metatarsi with segment 1 subequal to segments 2 to 4; claws minute and simple, all ventrites free; pygidium simple, nearly in horizontal plane; hind wings reduced to minute vestiges; male parameres asymmetrical left slightly longer, terminally downturned and spatulate.

Australian species. Two.

Distribution. Endemic to Australia, both species *P. mimerops* (Matthews) and *P. squamata* (Matthews) from northern Queensland.

Biology. Terrestrial, usually collected from baited pitfall traps in closed forest. May be myrmecophilous.

References. Matthews (1974) (as *Ignambia*); Paulian & Pluot-Sigwalt (1985); Cassis & Weir (1992).

Key to species. Matthews (1974) (as *Ignambia*).

Sauvagesinella Paulian, 1934 (Pl. 71G)

Type species. *Sauvagesinella monstrosa* Paulian, 1934.

Characteristics. Small sized, broadly oval beetles, convex, length 2.5–5.5 mm; uniformly coloured brownish black with black antennal clubs, surfaces variable but without distinct carinae or tubercles; head rounded with 2 small slender teeth separated by a broad U-shape; eyes with raised inner edge; eye canthus incomplete; mentum subquadrate; labial palpi with basal segment smaller than 2nd; pronotal disc convex; pronotal edges usually beaded; pronotal hypomeron excavated anteriorly for reception of the profemur; mesoscutellum absent; elytra with 8 striae without pseudepipleura or any carina, striae 8 located on a somewhat flanged-dorsal edge of the epipleura; mesoventrite 5–6 times as wide as long; trochanterofemoral pit absent; protibia with distal edge at right angle to inner edge, 3 evenly spaced small teeth each bearing 1 small tuft of setae dorsally; protibia with flattened posterior surface, male protibiae inner apex bearing a comb of stiff bristles and a ventral brush of setae as well as short spur; protarsi large inserted below spur; metafemora with ventral longitudinal carina and finely beaded on anterior and posterior edges; mesotibiae with 2 unequal spurs; meso- and metatibia gradually widened apically, strongly curved with short spurs; metatibiae with a reticulate, flattened posterior surface bearing 1, 2 or 3 carinae; metatarsi with segment 1 subequal segment 2, segment 5 longest subequal or longer than segments 3–4 combined; claws large and strongly toothed; ventrites of normal proportion; pygidium strongly transverse, simple, punctate and fully bordered; hind wings fully developed, reduced or absent; male parameres asymmetrical, left paramere only longer.

Australian species. Four.

Distribution. Endemic to Australia. *S. becki* (Paulian), *S. monstrosa* Paulian, *S. palustris* Matthews from south-western Western Australia, and *S. loftyensis* (Matthews and Weir) from South Australia.

Biology. Terrestrial, usually collected from baited pitfall traps in various forest types. Observed ball-making and rolling.

References. Matthews (1974); Cassis & Weir (1992); Gunter & Weir (2017).

Key to species. Gunter & Weir (2017).

Temnoplectron Westwood, 1841 (Pl. 71C)

Type species. *Temnoplectron rotundum* Westwood, 1841.

Characteristics. Small to medium-sized, oval beetles, convex, length 3.5–13 mm; uniform in colour, brownish black to black; surfaces entirely glabrous and nitid or covered with small wrinkles; head with 2 small teeth that are close together; ventral part of eyes usually very large; eye canthus incomplete; mentum subquadrate; labial palpi with basal segment longer than second segment; pronotal disc evenly convex; pronotal hypomeron excavated anteriorly for reception of the profemur; mesoscutellum absent; elytra without pseudepipleura and with 9 or 10 striae- if only 9 striae then striae 8 is absent, if 10 striae then striae 8 effaced anteriorly and posteriorly; epipleura narrow and evenly curved; mesoventrite 2.5–4 times as wide as long and subparallel; trochanterofemoral pit present; protibia relatively short with transversely truncated distal edge, 3 evenly spaced large teeth, each with 2 tufts of setae dorsally; protibial spur large; protarsi relatively long or very short, femora beaded along anterior and posterior edges; mesotibia short and gradually widened to an apex with 2 articulates spurs; metatibia very long and curved, very strongly compressed and blade-like, bearing dense setae on most of the inner surface, 1 or 2 fringes of shorted setae along upper and lower edges of tibiae; metatarsi with segment 1 subequal or shorter than segment 2; claws large and toothed; ventrites of normal proportion, compressed, 5th and 6th clearly demarcated; pygidium triangular, finely bordered all around, simple; hind wings fully developed to vestigial; male parameres symmetrical or asymmetrical.

Australian species. 16.

Distribution. Endemic to Australia and New Guinea.

Thirteen species, *T. aenopiceum* Matthews, *T. bournemisszai* Matthews, *T. cooki* Reid and Storey, *T. diversicolle* Blackburn, *T. disruptum* Matthews, *T. finnigani* Reid and Storey, *T. involucre* Matthews, *T. leave* (Castelnau), *T. lewisense* Reid and Storey, *T. monteithi* Reid and Storey, *T. politulum* Macleay, *T. reyi* Paulian, and *T. subvolitans* Matthews from northern Queensland, *T. rotundum* Westwood from coastal Queensland and Northern Territory *T. boucomonti* Paulian from coastal *Queensland*, Northern Territory, Western Australia and New Guinea, *T. major* Paulian from northern Queensland and New Guinea. Three species, *T. aeneolum* Lansberge, *T. atropolitum* Gillet and *T. wareo* Reid and Storey are endemic to New Guinea.

Biology. Terrestrial, usually collected from baited pitfall traps in various forest types, also using flight intercept and light traps; several species observed perching on leaves in rainforest at night.

References. Matthews (1974); Cassis & Weir (1992); Reid (2000); Bell *et al.* (2004).
Key to species. Reid (2000).

Tesserodon Hope, 1837 (Pl. 71E)

Type species. *Scarabaeus novaehollandiae* Fabricius, 1775
Tessarodon Reiche, 1842
Characteristics. Small sized, oval beetles, shallowly convex, length 4–9 mm; uniform in colour usually brownish black; surfaces with numerous short setae; head with 2 triangular median teeth flanked by V-shaped excision; eye canthus incomplete; mentum subquadrate; labial palpi with basal segment longer than second segment; pronotum densely punctate with punctures bearing setae; pronotal disc shallowly convex, anterior and lateral edges beaded; pronotal hypomeron excavated anteriorly for reception of the profemur, sides if pronotum subparallel on basal 2/3; mesoscutellum absent; elytra with 10 striae and incomplete pseudepipleura outside 7th striae, terminating 3/4 of way down elytral length, elytral setae generally short and erect; mesoventrite 3 times as wide as long; trochanterofemoral pit present; protibia with distal edge at right angle to inner edge, 3 large, evenly spaced teeth, each with 2 tufts of setae and/or multiple setae dorsally, protibial spur inserted near inner edge; meso- and metatibia short and not abruptly expanded but sometimes widening before or at apex, feebly curved, tarsal segments triangular; metatarsi with segment 1 shorter than segment 2; claws small and simple. 5th sternite fused with 6th with suture still visible but fine; pygidium transverse, convex and with a deep basal groove; hind wings fully developed except absent in *tenebroides* and *pilicrepus*; male parameres asymmetrical, left paramere longest.
Australian species. 13.
Distribution. Endemic to Australia, New Guinea and Indonesian Moluccas. *T. erratum* Storey, *T. intricatum* Lea, and *T. variolosum* Macleay from coastal Northern Territory, *T. feehani* Storey, *T. gestroi* Lansberge, and *T. henryi* Storey from northern Queensland, *T. hilleri* Storey from central Queensland, *T. novaehollandiae* (Fabricius), *T. simplicipunctatum* Storey from northern Australia, *T. granulatum* Matthews from central Australia,*T. pilcrepus* Matthews and *T. tenebroides* Matthews from coastal central Western Australia,*T. angulatum* Westwood from south-western Western Australia. *T. elongatus* Lansberge, *T. heurni* Paulian, and *T. setolusum* Balthasar are endemic to New Guinea. *T. seramicum* Krikken and Huijbrets is endemic to Indonesian Moluccas.
Biology. Terrestrial, usually collected from baited pitfall traps in a variety of open forests. Burrows generally observed beside or directly under food source.
References. Matthews (1974); Paulian (1985) (New Guinea only); Storey (1991); Cassis & Weir (1992); Krikken & Huijbregts (2009).
Key to species. Storey (1991).

Thyregis Blackburn, 1904 (Pl. 71B)

Type species. *Thyregis kershawi* Blackburn, 1904.
Arrowianella Paulian, 1933
Characteristics. Small sized, oblong beetles, strongly convex, length 7–12 mm; uniform in colour usually reddish-brown to blackish-brown; surface largely glabrous; head with 2 medial teeth and edges without distinct or with very narrow marginal beading; eye canthus incomplete; mentum subquadrate; underside of clypeus with a transverse ridge before mouthparts; labial palpi with basal segment subequal to or smaller than second segment, the third smallest distinct; antennae with large club; pronotum moderately to strongly convex, entirely unarmed and without carinae, but may have an anterior depression; pronotal hypomeron deflexed in areas and not excavated anteriorly for reception of the profemur; usually an incomplete longitudinal carina about half the prothoraic length running from posterior edge; small mesoscutellum visible between elytral bases or absent; elytra with 10 striae including 1 along the epipleural margin, 9th and 10th striae gradually merging posteriorly; mesoventrite 4 times as wide as long; trochanterofemoral pit absent; protarsi present, longer than length of apical tooth; protibia with distal edge oblique, 3 distinct, evenly-spaced teeth, apical tooth with 2 tufts of setae dorsally, others with 1 tuft of setae dorsally; protibial spur long and moveable, inserted near inner apical angle; apical edges of meso- and meta tibia scalloped with tufts of setae; meso- and metatibia abruptly expanded distally, with a transverse carina on the outer edge; mesotibiae with 2 strong, unequal terminal spurs and metatibiae with 1 spur; meso- and metatarsi long with segments decreasing in size distally, basal segment slightly shorter than preceding 2 combined; claws small and slender, moderately curved and simple; ventrites strongly contracted medially and fused, basal sternite longer medially than preceding 4 combined; pygidium transverse, strongly margined, simple; hind wings fully developed or reduced; male parameres symmetrical.
Australian species. Four.
Distribution. Endemic to Australia. *T. monteithi* Matthews and *T. relictus* Matthews from northern New South Wales, *T. kershawi* Blackburn from southern New South Wales to Victoria, and *T. tarsatus* (Paulian) from central Western Australia.
Biology. Terrestrial, usually collected from baited pitfall traps in various forest types.
References. Matthews (1976); Cassis & Weir (1992).
Key to species. Matthews (1976).

References

Ahrens D (2006) The phylogeny of Sericini and their position within the Scarabaeidae based on morphological characters (Coleoptera: Scarabaeidae). *Systematic Entomology* **31**, 113–144. doi:10.1111/j.1365-3113.2005.00307.x

Ahrens D, Vogler AP (2008) Towards the phylogeny of chafers (Sericini): Analysis of alignment-variable sequence and the evolution of segment numbers in the antennal club. *Molecular Phylogenetics and Evolution* **47**, 783–798. doi:10.1016/j.ympev.2008.02.010

Ahrens D, Schwarzer J, Vogler AP (2014) The evolution of scarab beetles tracks the sequential rise of angiosperms and mammals. *Proceedings of the Royal Society B Biological Series* **281**, 20141470.

Allsopp PG (1995) Biogeography of the Australian Dynastinae, Rutelinae, Scarabaeinae, Melolonthinini, Scitalini and Geotrupidae (Coleoptera: Scarabaeoidea). *Journal of Biogeography* **22**, 31. doi:10.2307/2846071

Allsopp PG (1999) How localized are the distributions of Australian scarabs (Coleoptera: Scarabaeoidea)? *Diversity & Distributions* **5**, 143–149. doi:10.1046/j.1472-4642.1999.00050.x

Andresen E, Feer F (2005) The role of dung beetles as secondary seed dispersers and their effect on plant regeneration in tropical rainforests In *Seed fate: predation, dispersal and seedling establishment.* (Eds PM Forget, JE Lambert, PE Hulme, and SB Vander Wall) pp. 331–349. CABI International, Oxon, United Kingdom.

Aristophanous M (2014) *Understanding patterns of endemic dung beetle (Coleoptera: Scarabaeidae: Scarabaeinae) biodiversity in the Australian Wet Tropics*. James Cook University (PhD thesis), Townsville QLD.

Bai M, Li S, Lu Y, Yang H, Tong Y, Yang X (2015) Mandible evolution in the Scarabaeinae (Coleoptera: Scarabaeidae) and adaptations to coprophagous habits. *Frontiers in Zoology* **12**(30), 1–10. doi:10.1186/s12983-015-0123-z

Bell KL, Yeates DK, Moritz C, Monteith GB (2004) Molecular Phylogeny and biogeography of the dung beetle genus *Temnoplectron* Westwood (Scarabaeidae: Scarabaeinae) from Australia's Wet Tropics. *Molecular Phylogenetics and Evolution* **31**, 741–753. doi:10.1016/j.ympev.2003.09.010

Bocák L, Barton C, Crampton-Platt A, Chesters D, Ahrens D, Vogler AP (2013) Building the Coleoptera tree-of-life for >8000 species: composition of public DNA data and fit with Linnaean classification. *Systematic Entomology* **39**, 97–110. doi:10.1111/syen.12037

Bornemissza GF (1971a) A new variant of the paracopric nesting type in the Australian dung beetle, *Onthophagus compositus. Pedobiologica* **11**, 1–10.

Bornemissza GF (1971b) Mycetophagous breeding in the Australian dung beetle, *Onthophagus dunningi. Pedobiologica* **11**, 133–142.

Breeschoten T, Doorenweerd C, Tarasov S, Vogler AP (2016) Phylogenetics and biogeography of the dung beetle genus *Onthophagus* inferred from mitochondrial genomes. *Molecular Phylogenetics and Evolution* **105**, 86–95. doi:10.1016/j.ympev.2016.08.016

Browne DJ, Scholtz CH (1995) Phylogeny of the families Scarabaeoidea (Coleoptera) based on characters of the hind wing articulation, hindwing base and wing venation. *Systematic Entomology* **20**, 145–173. doi:10.1111/j.1365-3113.1995.tb00089.x

Browne DJ, Scholtz CH (1998) Evolution of the scarab hind wings articulation and wing base: a contribution towards the phylogeny of the Scarabaeidae (Coleoptera: Scarabaeoidea). *Systematic Entomology* **23**, 307–326. doi:10.1046/j.1365-3113.1998.00059.x

Carvalho IS, Gracioso DE, Fernandes ACS (2009) Uma câmara de coleóptero *(Coprinisphaera)* do Cretáceo Superior, Bacia Bauru. *Revista Brasileira de Geociencias* **39**, 679–684. doi:10.25249/0375-7536.2009394679684

Cassis G, Weir TA (1992) Scarabaeinae. In. *Zoological Catalogue of Australia. Volume 9. Coleoptera: Scarabaeoidea.* (Ed. WWK Houston) pp 106–173. AGPS, Canberra, ACT.

Chapman AD (2009) Numbers of living species in Australia and the world. Toowomba, Australia: 2nd edition. Australian Information Services.

Chin K, Gill B (1996) Dinosaurs, dung beetles and conifers: participants in a Cretaceous food web. *Palaios* **11**, 280–285. doi:10.2307/3515235

Coggan N (2012) Are native dung beetle species following mammals in the Critical Weight Range towards extinction? *Proceedings of the Linnean Society of New South Wales* **134**, A5–A9.

Dalgleish EA, Elgar MA (2005) Breeding ecology of the rainforest dung beetle *Cephalodesmius armiger* (Scarabaeidae) in Tooloom National Park. *Australian Journal of Zoology* **53**, 95–102. doi:10.1071/ZO04067

Davis ALV (2000) Does reduced-impact logging help preserve biodiversity in tropical rainforests? A case study from Borneo using dung beetles (Coleoptera: Scarabaeoidea) as indicators. *Environmental Entomology* **29**, 467–475. doi:10.1603/0046-225X-29.3.467

Davis ALV (2002) Dung beetle diversity in South Africa: influential factors, conservation status, data inadequacies and survey design. *African Entomology* **10**, 53–65.

Davis ALV, Sutton S (1998) The effects of rainforest canopy loss on arboreal dung beetles in Borneo: implications for measurement of biodiversity in derived tropical ecosystems. *Diversity & Distributions* **4**, 167–173. doi:10.1046/j.1472-4642.1998.00017.x

Davis ALV, Holloway JD, Huijbregts H, Krikken J, Kirk-Spriggs A, Sutton S (2001) Dung beetles as indicators of change in forests of northern Borneo. *Journal of Applied Ecology* **38**, 593–616. doi:10.1046/j.1365-2664.2001.00619.x

Davis ALV, Scholtz CH, Phillips TK (2002) Historical biogeography of scarabaeine dung beetles. *Journal of Biogeography* **29**, 1217–1256. doi:10.1046/j.1365-2699.2002.00776.x

Davis ALV, Scholtz CH, Dooley P, Bham N, Kryger U (2004) Scarabaeine dung beetles as indicators of biodiversity, habitat transformation and pest control chemicals in agro-ecosystems. *South African Journal of Science* **100**, 415–424.

Davis ALV, Scholtz CH, Sole CL (2017) Biogeographical and co-evolutionary origins of scarabaeine dung beetles: Mesozoic vicariance versus Cenozoic dispersal and dinosaur versus mammal dung. *Biological Journal of the Linnean Society. Linnean Society of London* doi:10.1111/bij.12893.

Edmonds WD, Halffter G (1978) Taxonomic review of immature dung beetles of the subfamily Scarabaeinae (Coleoptera: Scarabaeinae). *Systematic Entomology* **3**, 307–331. doi:10.1111/j.1365-3113.1978.tb00002.x

Edwards P, Wilson P, Wright J (2015) *Introduced Dung Beetles in Australia: A Pocket Field Guide*. CSIRO Publishing, Clayton, Victoria.

Génier F, Moretto P (2017) *Digitonthophagus* Balthasar, 1959: taxonomy, systematics, and morphological phylogeny of the genus revealing an African species complex (Coleoptera: Scarabaeidae: Scartabaeinae). *Zootaxa* **4248**, 1–110. doi:10.11646/zootaxa.4248.1.1

Genise JF, Bedatou E, Bellosi ES, Sarzetti LC, Sanchez MV, Krause JM (2016) The Phanerozoic four revolutions and evolution of paleosol ichnofacies. In *The Trace-fossil Record of Major Evolutionary Events. Volume 2. Mesozoic and Cenozoic.* (Eds M Gabriela Mangano and LA Buatois) pp. 301–370. Springer.

Gill BD (1991) Dung beetles in tropical American forests. In *Dung Beetle Ecology.* (Eds I Hanski and Y Camberfort) pp. 211–229. Princeton University Press, Princeton, New Jersey.

Grebennikov VV, Scholtz CH (2004) The basal phylogeny of Scarabaeoidea (Insecta: Coleoptera). *Invertebrate Systematics* **18**, 321–348. doi:10.1071/IS03013

Gunter NL, Weir TA (2017) Two new genera of Australian dung beetle (Scarabaeidae: Scarabaeinae) with the description of six new species and transfer of six described species. *Zootaxa* **4290**, 201–243. doi:10.11646/zootaxa.4290.2.1

Gunter NL, Weir TA, Slipinski A, Bocak L, Cameron SL (2016) If dung beetles (Scarabaeidae: Scarabaeinae) arose in association with dinosaurs, did they also suffer a mass co-extinction at the K-PG boundary? *PLOSOne* **11**, e0153570. doi:10.1371/journal.pone.0153570

Gunter NL, Monteith GB, Cameron SL, Weir TA (2018) Evidence from Australian mesic zone dung beetles support Gondwanan vicariance and Mesozoic origin of the Scarabaeinae (Coleoptera: Scarabaeidae). *Invertebrate Systematics and Evolution* doi:10.1163/1876312x-00002171.

Halffter G, Edmonds WD (1982) The nesting behavior of dung beetles (Scarabaeinae): An ecological and evolutionary approach. Instituto de Ecología: DF, Mexico City, Mexico.

Halffter G, Matthews EG (1966) The natural history of dung beetles of the subfamily Scarabaeinae (Coleoptera: Scarabaeidae). *Folia Entomologica Mexicana* **12–14**, 1–312.

Hanski I, Cambefort Y (Eds) (1991) *Dung Beetle Ecology*. Princeton University Press, Princeton, New Jersey.

Hanski I, Krikken J (1991) Dung beetles in tropical forests in south-east Asia. In *Dung Beetle Ecology*. (Eds I Hanski and Y Cambefort) pp 179–197. Princeton University Press. Princeton, New Jersey.

Herzog SK, Hamel-Leigue AC, Larsen TH, Mann DJ, Soria-Auza RW, Gill BD, Edmonds WD, Spector S (2013) Elevational distribution and conservation biogeography of phanaeine dung beetles (Coleoptera: Scarabaeinae) of Bolivia. *PLoS One* **8**, e64963. doi:10.1371/journal.pone.0064963

Hill CJ (1996) Habitat specificity and food preference of an assemblage of tropical Australian dung beetles. *Journal of Tropical Ecology* **12**, 449–460. doi:10.1017/S026646740000969X

Hodge S, Marshall SA, Oliver H, Berry J, Marris J, Andrew I (2010) Preliminary survey of insects collected using mushroom baits in native and exotic New Zealand woodlands. *New Zealand Entomologist* **33**, 43–54. doi:10.1080/00779962.2010.9722191

Holter P, Scholtz CH, Stenseng L (2009) Desert detritivory: feeding biology of a dung beetle (*Pachysoma glentoni*) subsisting on plant litter in arid South African sand dunes. *Journal of Arid Environments* **73**, 1090–1094. doi:10.1016/j.jaridenv.2009.04.009

Howden HF, Howden AT (2001) Changes through time: a third survey of the Scarabaeinae (Coleoptera: Scarabaeinae) at Welder Wildlife Reserve. *Coleopterists Bulletin* **55**, 356–362. doi:10.1649/0010-065X(2001)055[0356:CTTATS]2.0.CO;2

Howden HF, Howden AT, Storey RI (1991) Nocturnal perching of scarabaeine dung beetles (Coleoptera: Scarabaeinae) in an Australian tropical rain forest. *Biotropica* **23**, 51–57. doi:10.2307/2388687

Huijbregts J, Krikken J (2007) First record of the genus *Amphistomus* from the Moluccas (Coleoptera: Scarabaeidae: Scarabaeinae). *Tijdschrift voor Entomologie* **150**, 31–37. doi:10.1163/22119434-900000210

Hunt T, Bergensten J, Levkanicova Z, Papadopoulou A, John OS, Wild R, Hammond PM, Ahrens D, Balke M, Caterino MS, Gómez-Zurita J, Ribera I, Barraclough TG, Bocakoca M, Bocak L, Vogler AP (2007) A comprehensive phylogeny of beetles reveals the evolutionary origins of superradiation. *Science* **318**, 1913–1916. doi:10.1126/science.1146954

IUCN (2017) The IUCN red list of threatened species. Version 2017–3. http://www.iucnredlist.org. Downloaded on 30 March 2017.

Iwasa M, Moki Y, Takahashi J (2015) Effects of the activity of coprophagous insects on greenhouse gas emissions from Cattle dung pats and changes in amounts of nitrogen, carbon and energy. *Environmental Entomology* **44**, 106–113. doi:10.1093/ee/nvu023

Janssens A (1937) Revision des Onitides. *Memoiresdu Musee Royal d'Histoire Naturelle Belgique* **11**, 3–200.

Janssens A (1949) Contibution à l'étude des Coléptères Lamellicornes. XIII. Table synoptique et essai de classification pratique des Coléptères Scarabaeidae. *Bulletin de la Institut Royal de Science Natural Belgique* **25**, 1–30.

Janssens A (1953) Oniticellini (Coleoptera Lamellicornia) *in* Exploration du Parc National de l'Upemba, Mission G.F. De Witte. Institu des Parcs Nationaux du Congo Belge. Fascicle **11**, 1–118.

Jones AG, Forgie SA, Scott DJ, Beggs JR (2012) Generalist dung attraction response in New Zealand dung beetle that evolved with an absence of mammalian herbivores. *Ecological Entomologist* **37**, 124–133. doi:10.1111/j.1365-2311.2012.01344.x

Koch SO, Chown SL, Davis ALV, Endrody-Youunga S, van Jaarsveld AS (2000) Conservation strategies for poorly surveyed taxa: a dung beetle (Coleoptera: Scarabaeidae) case study from southern Africa. *Journal of Insect Conservation* **4**, 45–56. doi:10.1023/A:1009634318926

Krikken J, Huijbregts J (2012) Taxonomy of Papuasian *Onthophagus*: twenty new species and their relatives (Coleoptera: Scarabaeidae: Scarabaeinae). *Tijdschrift voor Entomologie* **155**, 73–131. doi:10.1163/22119434-00002013

Krikken J, Huijbregts J (2009) A new *Haroldius* Boucomont and a new *Tesserodon* Hope from the Moluccas (Coleoptera: Scarabaeidae: Scarabaeinae). *Koleopterologische Rundschau* **79**, 257–266.

Kryger U, Coles KS, Tukker R, Scholtz CH (2006) Biology and ekology of *Circellium bakchus* (Fabricius 1781)(Coleoptera Scarabaeidae), a South African dung beetle of conservation concern. *Tropical Zoology* **19**, 185–207.

Lawrence JF, Ślipiński A, Seago AE, Thayer MK, Newton AF, Marvaldi AE (2011) Phylogeny of the Coleoptera based on morphological characters of adults and larvae. *Annales Zoologici* **61**, 1–217. doi:10.3161/000345411X576725

Laza JH (2006) Dung-beetle fossil brood balls: the ichnogenera *Coprinisphaera* Sauer and *Quirogaichnus* (Coprinisphaeridae). *Ichnos* **13**, 217–235. doi:10.1080/10420940600843641

Matthews EG (1972) A revision of the scarabaeine dung beetles of Australia. I. Tribe Onthophagini. *Australian Journal of Zoology. Supplementary Series* **19**, 1–330.

Matthews EG (1974) A revision of the scarabaeine dung beetles of Australia. II. Tribe Scarabaeini. *Australian Journal of Zoology. Supplementary Series* **24**, 1–211.

Matthews EG (1976) A revision of the scarabaeine dung beetles of Australia. III. Tribe Coprini. *Australian Journal of Zoology. Supplementary Series* **38**, 1–52.

Matthews EG, Stebnicka Z (1986) A review of *Demarziella* Balthasar, with a transfer from Aphodiinae to Scarabaeinae (Coleoptera: Scarabaeidae). *Australian Journal of Zoology* **34**, 449–461. doi:10.1071/ZO9860449

Matthews EG, Weir TA (2002) Two new species of the genus *Lepanus* Balthasar from South Australia (Coleoptera: Scarabaeidae). *Records of the South Australian Museum* **35**, 79–84.

McKenna DD, Farrell BD, Caterino MS, Farnum CW, Hawks DC, Maddison DR, Seago AE, Short AEZ, Newton AF, Thayer MK (2015) Phylogeny and evolution of Staphyliniformia and Scarabaeiformia: forest litter as a stepping stone for diversification of nonphytophagous beetles. *Systematic Entomology* **40**, 35–60. doi:10.1111/syen.12093

Mlambo S, Sole CL, Scholtz CH (2015) A molecular phylogeny of the African Scarabaeinae (Coleoptera: Scarabaeidae). *Arthropod Systematics & Phylogeny* **73**, 303–321.

Monaghan MT, Inward DJG, Hunt T, Vogler AP (2007) A molecular phylogenetic analysis of the Scarabaeinae (dung beetles). *Molecular Phylogenetics and Evolution* **45**, 674–692. doi:10.1016/j.ympev.2007.06.009

Monteith GB (2015) Australian native dung beetles. *Entomology Society of Queensland* **43**, 24–32.

Monteith GB, Rossini M (2017) A new name for the Australian dung beetle *Onthophagus bicornis* Macleay, 1888 (Coleoptera: Scarabaeidae), with notes on tytpe locality, distribution and biology. *Australian Entomologist* **44**, 161–171.

Monteith GB, Storey RI (1981) The biology of *Cephalodesmius*, a genus of dung beetles that synthesizes its dung from plant material (Coleoptera: Scarabaeidae: Scarabaeinae). *Memoirs of the Queensland Museum* **20**, 253–277.

Monteith GB, Storey RI (2013) A review of the *Onthophagus posticus*-group (Coleoptera: Scarabaeinae) of Australian dung beetles with five new species. *Memoirs of the Queensland Museum* **58**, 55–72.

Montreuil O (1998) Analyse phylogénétique et paraphylie des Coprini et Dichotomiini (Coleoptera: Scarabaeidae). Scénario Biogéographique. *Annales de la Société Entomologique de France* (N.S.) **34**, 135–148.

Nichols E, Spector S, Louzada J, Larsen T, Amezquita S, Favilla ME, The Scarabarinae Research Network (2008) Ecological functions and ekosystém services of Scarabaeine dung beetles: a review. *Biological Conservation* **141**, 1461–1474. doi:10.1016/j.biocon.2008.04.011

Paulian R (1985) Les Coléoptères Scarabaeidae canthonines de Nouvell-Guinée. *Annales de la Société Entomologique de France* **21**, 219–238.

Paulian R, Pluot-Sigwalt D (1985) Les Canthonines de Nouvelle-Caledonie (Coleoptera, Scarabaeidae) Etude systematique et biogeographue. *Bulletin du Museum National d'Histoire Naturelle* (section A zoologie biologie et ecologie animals) **6**(4) 1984, 1091–1133.

Penttilä A, Slade EM, Simojoki A, Riutta T, Minkkinen K, Roslin T (2013) Quantifying beetle-mediated effects on gas fluxes from dung pats. *PLoS One* **8**, e71454. doi:10.1371/journal.pone.0071454

Philips TK, Pretorius E, Scholtz CH (2004) A phylogenetics analysis of dung beetles (Scarabaeinae: Scarabaeidae): unrolling an evolutionary history. *Invertebrate Systematics* **18**, 53–88. doi:10.1071/IS03030

Ratcliffe BC (1980) New species of Coprini (Coleoptera: Scarabaeidae: Scarabaeinae) taken from the pelage of three toed sloths (*Bradypus tridactylus* L.) (Edentata: Brachypodidae) in central Amazonia, with a brief commentary on scarab-sloth relationships. *Coleopterists Bulletin* **34**, 337–350.

Ratcliffe BC, Jameson ML, Smith ABT (2002) 34 Scarabaeidae Latreille 1802. In *American Beetles, Volume 2.* (Eds RH Arnett, M Thomas, P Skelley and JH Frank) pp. 39–81. CRC Press, Boca Raton.

Reid CAM (2000) A complex of cryptic species in the genus *Coptodactyla* Burmeister (Coleoptera: Scarabaeidae: Coprini). *Memoirs of the Queensland Museum* **46**, 231–251.

Reid CAM, Storey RI (2000) Revision of the dung beetle genus *Temnoplectron* Westwood (Coleoptera: Scarabaeidae: Scarbaeini). *Memoirs of the Queensland Museum* **46**, 253–297.

Richter PO, Baker CW (1974) Ovariole numbers in Scarabaeoidea (Coleoptera: Lucanidae, Passalidae, Scarabaeidae). *Proceedings of the Entomological Society of Washington* **76**, 480–494.

Ridsdill-Smith TJ, Hall GP, Weir TA (1989) A field guide to the dung beetles (Scarabaeidae: Scarabeinae and Aphodiinae) common in pastures in South-western Australia. *Journal of the Royal Society of Western Australia* **71**, 49–58.

Roggero A, Dierkens M, Barbero E, Palestrini P (2016) Combined phylogenetic analysis of two new Afrotropical genera of Onthophagini (Coleoptera, Scarabaeidae). *Zoological Journal of the Linnean Society* **2016**, 1–23.

Rortais A (2003) Distribution of two endemic and flightless tropical dung beetles *Amphistomus pectoralis* Matthews and *Temnoplectron involucre* Matthews (Coleoptera: Scarabaeinae), at a rainforest edge. *Australian Journal of Entomology* **42**, 212–213. doi:10.1046/j.1440-6055.2003.00349.x

Sato H (1998) Payoffs of the two alternative nesting tactics in the African dung beetle, *Scarabeus catenatus*. *Ecological Entomology* **23**, 62–67. doi:10.1046/j.1365-2311.1998.00099.x

Scholtz CH, Davis ALV, Kryger U (2009) *Evolutionary Biology and Conservation of Dung Beetles.* Pensoft, Sofia.

Seldon DS, Beggs JR (2010) The efficacy of baited and live capture pitfall traps in collecting large-bodied forest carabids. *New Zealand Entomologist* **33**, 30–37. doi:10.1080/00779962.2010.9722189

Slade EM, Riutta T, Roslin T, Tuomisto HL (2016) The role of dung beetles in reducing greenhouse gas emissions from cattle farming. *Scientific Reports* **6**, 18140. doi:10.1038/srep18140

Smith ABT, Hawks DC, Heraty JM (2006) An overview of the classification and evolution of the major scarab beetle clades (Coleoptera: Scarabaeoidea) based on preliminary molecular analyses. *Coleopterists Society Monograph* **5**, 35–46.

Sole CL, Scholtz CH (2010) Did dung beetles arise in Africa? A phylogenetics hypothesis based on five gene regions. *Molecular Phylogenetics and Evolution* **56**, 631–641. doi:10.1016/j.ympev.2010.04.023

Spector S (2006) Scarabaeine dung beetles (Coleoptera: Scarabaeidae: Scarabaeinae): an invertebrate focal taxon for biodiversity research and conservation. *Coleopterists Bulletin* **60**, 71–83. doi:10.1649/0010-065X(2006)60[71:SDBCSS]2.0.CO;2

Storey RI (1977) Six new species of *Onthophagus* Latreille (Coleoptera: Scarabaeidae) from Australia. *Journal of the Australian Entomological Society* **16**, 313–320. doi:10.1111/j.1440-6055.1977.tb00111.x

Storey RI (1984) A new species of *Aptenocanthon* Matthews from north Queensland (Coleoptera: Scarabaeidae: Scarabaeinae). *Memoirs of the Queensland Museum* **21**, 387–390.

Storey RI (1986) A new flightless species of *Aulacopris* White from north Queensland (Coleoptera: Scarabaeidae: Scarabaeinae). *Memoirs of the Queensland Museum* **22**, 197–303.

Storey RI (1991) New species and new records of *Tesserodon* Hope (Coleoptera: Scarabaeidae) from northern Australia. *Memoirs of the Queensland Museum* **30**, 577–588.

Storey RI, Monteith GB (2000) Five new species of *Aptenocanthon* Matthews (Copeotera: Scarabaeidae: Scarabaeinae) from tropical Australia, with notes on distribution. *Memoirs of the Queensland Museum* **46**, 349–358.

Storey RI, Weir TA (1990) New species of *Onthophagus* Latreille (Copeotera: Scarabaeidae) from Australia. *Invertebrate Taxonomy* **3**, 783–815. doi:10.1071/IT9890783

Tarasov S, Dimitrov D (2016) Multigene phylogenetics analysis redefines dung beetles relationships and classification (Coleoptera: Scarabaeidae: Scarabaeinae). *BMC Evolutionary Biology* **16**, 257. doi:10.1186/s12862-016-0822-x

Tarasov S, Génier F (2015) Innovative Bayesian and parsimony phylogeny of dung beetles (Coleoptera: Scarabaeindae: Scarabaeinae) enhanced by ontology-based partitioning of morphological characters. *PLoSONE* **10**, e0116671.

Tarasov S, Vaz-de-Mello FZ, Krell F, Dimitar D (2016) A review and phylogeny of scarabaeine dung beetle fossils (Coleoptera: Scarabaeidae: Scarabaeinae) with the description of two *Canthochilium* species from Dominican amber. *PeerJ* **4**, e1998.

Vaz-de-Mello FZ (2007) Revisión taxonómica y análisis filogenético de la tribu Ateuchini (Coleoptera: Scarabaeidae: Scarabaeinae). *Intituto de Ecologia. A.C.* 1–238.

Verdú JR, Crespo MB, Galante E (2000) Conservation strategy of a nature reserve in Mediterranean ecosystems: the effects of protection from grazing on biodiversity. *Biodiversity and Conservation* **9**, 1707–1721. doi:10.1023/A:1026506725251

Vernes K, Pope LC, Hill CJ, Bärlocher F (2005) Seasonality, dung specificity and competition in dung beetle assemblages in the Australian Wet Tropics, north-eastern Australia. *Journal of Tropical Ecology* **21**, 1–8. doi:10.1017/S026646740400224X

Weir TA (2013) *Aulacopris matthewsi.* The IUCN Red List of Threatened Species 2013: e.T137526A526573. doi:10.2305/IUCN.UK.2013-2.RLTS.T137526A526573.en. Downloaded on 01 May 2017.

Weir TA, Monteith GB (2010) *Lepanus storeyi,* a new species of dung beetle (Coleoptera: Scarabaeidae: Scarabaeinae) from southeast Queensland. *Australian Entomologist* **36**, 235–241.

Yeates DK, Cassis G (2017) Australian Insect Biogeography: beyond faunal provinces and elements towards processes. In *Handbook of Australasian Biogeography.* (Ed. MC Ebach) pp. 213–239. CRC Press.

31. SCARABAEIDAE: MELOLONTHINAE LEACH, 1819

Tom A. Weir, John F. Lawrence, Cate Lemann and Nicole L. Gunter

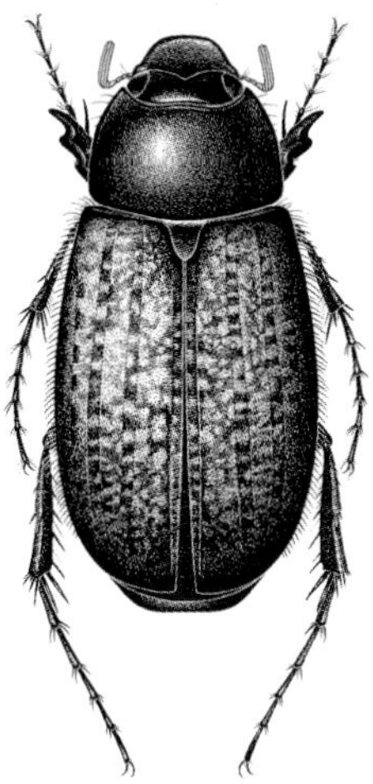

Fig. 31.1. *Sericesthis geminata* Boisduval.

Common name. Chafers.

Introduction. The Melolonthinae is the largest of the scarabaeid subfamilies, containing ~750 genera and 11 000 species, and widely distributed throughout the world. According to Smith (2006), there are 28 extant tribes included in this subfamily, 16 of which occur in Australia, but we have included the Phaenognathini, formerly placed in the subfamily Aclopinae (Ocampo & Mondaca 2012). The most recent global review of the subfamily is outdated and unreliable as it was published over a century ago (Dalla Torre 1913), however the faunas of Australia and the New World have been revised (Britton 1957–2000; Evans 2003). As a whole, melolonthine classification remains largely neglected, most tribes are inadequately defined within the original descriptions, many of which have not been reviewed since (Smith 2006). Furthermore, diagnostic characters have been inconsistently applied, resulting in erroneous placement of genera especially during the earliest taxonomic efforts when tribes were defined by a single diagnostic character which we now known to occur across multiple tribes. The richness and diversity of the subfamily makes study of the natural groupings logistically challenging and while cladistic analyses have attempted to explore relationships, sampling has been insufficient to comment on the relationships between or within most tribes (Smith *et al*. 2006; Coco-Abia 2007). A handful of authors have made significant progress towards understanding the internal relationships of the tribes Sericini and Hoplini (Ahrens 2006; Ahrens & Vogler 2008; Ahrens *et al*. 2011; Liu *et al*. 2015), but the majority remained unstudied.

Despite the issues regarding classification, major clades are consistently recovered in phylogenetic analysis of the Scarabaeidae (Smith *et al*. 2006; Ahrens *et al*. 2014; Gunter *et al*. 2016; Šípek *et al*. 2016). The subfamily is not monophyletic, instead the major melolonthine clades form the earliest branching lineages within the 'Pleurosticti', represented by the phytophagous scarab subfamilies Melolonthinae, Cetoniinae, Rutelinae and Dynastinae. There is strong evidence to support the monophyly of some tribes, however others are polyphyletic indicating numerous classification changes are warranted especially within a large clade that contains Australasian melolonthines. Preliminary evidence suggests the major melolonthine lineages had already diverged by the middle of the Cretaceous (Ahrens *et al*. 2014) and as early as the Lower Cretaceous (Gunter *et al*. 2016; Toussaint *et al*. 2017). This Lower Cretaceous age is supported by the fossil record, however there is some uncertainty regarding classification of the fossil taxa *Cretomelolontha* and *Lithanomala,* which are considered melolonthines and the earliest record for the family (Krell 2007).

The Australian fauna currently includes 118 genera and ~1250 species, but many species are as yet undescribed, especially in genera such as *Heteronyx* Guérin-Méneville. Gunter *et al*. (2016) included 32 Australian genera within a larger phylogeny of the Scarabaeidae, confirming the relationship of *Phaenognatha* (Phaenognathini) within the Melolonthinae as suggested by Ocampo & Mondaca (2012) and Lawrence & Ślipiński (2013) and recovering *Dermolepida* Arrow and *Lepidiota* Kirby within the primarily Palaearctic and Oriental tribe Melolonthini congruent with Britton's (1978) hypothesis on their northern origins. All other genera were recovered in a well supported large clade. In this analysis the tribes Liparetini, Scitalini, Phyllotocini and Xylonychini were each represented by at least three genera and were all recovered as polyphyletic, with genera from tribes Automollini, Diphucephalini, Heteroncycini and Maechidiini also scattered throughout the clade. Genera currently placed in Scitalini and Colymbomorphini also occur in South America, however the relationship between these neotropical genera and the predominately Australian clade has yet to be tested using molecular methods. It should be noted that the most widely followed classification of Scarabaeidae (i.e. Smith 2006; Bouchard *et al*. 2011) includes the Australian 'Sericini' within the subtribe Phyllotocina, which are distantly related to a monophyletic clade that contains the remaining sericine subtribes Sericina and Trochalina (Ahrens 2006). This phylogenetic relationship is more congruent with the classification as the tribe Phyllotocini (Ahrens 2006; Ahrens & Vogler 2008; Ahrens *et al*. 2014).

The taxonomic effort and description of Australian melolonthines has been sporadic, with most of species described by a handful of authors. The most prolific were Blanchard (1846–1853), Macleay (1864–1888), Blackburn (1888–1912), Lea (1895–1930) and Britton (1955–2000), who together described of over 85% of species (1053) that remain valid today. Britton's taxonomic contribution extended beyond species description, and he published the first and most recent keys to the Australian tribes (Britton 1957, 1978, 1990) and revisions of Automoliini, Maechidiini, Phyllotocidiini,

Sericini and Colymbomorphini (as Xylonychini) (Britton 1957), Melolonthini (Britton 1978), *Liparetrus* (Liparetrini) (Britton 1980), *Colpochila* (Liparetrini) (Britton 1986), Scitalini and Comophorinini (Britton 1987a), *Heteronyx* and other Heteronycini (Britton 1988, 2000), Liparetrini (Britton 1990) and Diphucephalini (Britton 1995b). Allsopp also contributed to the revisionary work of the tribes Systellopini and Pachytrichini (Allsopp 1989a, 1990a). The richness of Australian species is unevenly distributed between tribes and genera. Liparetini is the most diverse and speciose Australian tribe with 32 genera and 428 species, and the richness of Heteronycini (11 genera, 236 species), Scitalini (17 genera, 128 species), and Melolonthini (16 genera, 119 species) are also notable. On the other hand, the endemic tribes Phyllotocidiini, Comophorini and Pachytrichini are each represented by a single genus with 1–7 species. Given that the majority of endemic and austral tribes are paraphyletic within the large Australiasian clade, indicating that the current tribal classification does not reflect natural groupings but instead a geographic radiation, several genera may be synonymised once the fauna is revised. Macrodactylini is also represented by a single genus and species within Australia, with *Plectris aliena* Chapin being introduced from the Neotropics.

Three hyper-diverse genera represent almost 50% of the diversity (e.g. *Liparetrus* Guérin-Méneville with 240 species, *Heteronyx* with 207 species and *Colpochila* Erichson with 133 species), *Maechidius* Macleay (71 species) and *Lepidiota* Kirby (62 species) each account for another 5% of diversity and the remaining 40% placed within 113 genera. All but six of the genera are endemic to Australia; five of the non-endemic genera occur on other Australasian land masses (New Guinea, New Caledonia, New Zealand or Moluccas), while *Lepidiota* Kirby extends from Australia into New Guinea and Indonesia and through South-east Asia to India. Melolonthines occur in all regions of Australia. While few genera are regionally endemic, many display regional centers of diversity that may reflect radiation following temporal isolation (Britton 1986; Allsopp 1995; Lawrence & Ślipiński 2013).

Biology. Adult and larval Melolonthinae are almost always phytophagous, but some adult Melolonthini (e.g. *Antiotrogus* and *Rhopaea*) apparently do not feed at all. The life cycles of most species are poorly known, but the adult life span usually varies from a few days to a few weeks, whereas larvae may live as long as two years. Most adults are nocturnal or crepuscular and may fly to lights at night, but there are conspicuous metallic or bioloured diurnal adults in genera such as *Diphucephala* or *Phyllotocus*, respectively. Many adults, including species of *Automolius*, *Diphucephala*, *Heteronyx*, *Liparetrus*, *Phyllotocus* and *Sericesthes*, feed on the leaves of *Eucalyptus* and other trees, and the concentration of adults facilitates mating, but males of non-feeding species locate females by means of a sex pheromones produced by the latter (Soo Hoo & Roberts 1965). Eggs are laid in the soil, and pupation occurs in a subterranean cell, but adults usually remain inactive in the pupal cell until the soil is softened by rain. According to Britton (1986, 2000), the fact that adults remain in their pupal cells until released by rain, combined with the short adult and long larval life spans, and the short distances the adult males fly, can lead to temporal isolation and speciation, especially in a relatively arid habitat where rainfall is low but isolated thunderstorms fairly common. During an isolated storm, a portion of a widespread population of adults in their pupal cells will emerge, mate and return to the soil to lay eggs, while adults in other areas remain in their cells. Another portion of this population in a different area may emerge at another time in response to a different storm, and some adults in the general population may not survive at all in a drought. In either case, this would lead to the genetic isolation of the limited population, and given the limited mobility of adults, isolation might last long enough for random genetic drift to occur, leading to mutual sterility. Prolonged drought common in Australia, could reduce a widespread population of a single species to a few isolated, minute enclaves where local conditions remain favourable.

Although many leaf-feeding adults in genera mentioned above may be defoliators of eucalypts and other trees, the larvae of some melolonthines may be serious pests of some crops and grasslands. Larvae of *Sericesthis* feed on young roots of grasses and clovers, as well as organic matter in the soil (Ridsdill-Smith & Roberts 1976). *Heteronyx piceus* Blanchard is apparently a serious pests of peanuts in southern Queensland (Rogers *et al.* 1992), and larvae of *Dermolepida albohirtum* (Waterhouse) are the most serious pests of sugarcane in northern Queensland (Britton 1978; Roberts *et al.* 1982a, 1982b).

Characteristics. Adults. Members of the subfamily Melolonthinae may be distinguished by the position and alignment of the abdominal spiracles, which are located in the upper parts of the sternites and do not diverge outwardly towards the abdominal apex, and the equal, fixed and divergent pretarsal claws, which may be simple, bifid or toothed. Antennae may be 7- to 10-segmented, the elytra sufficiently short to expose the pygidium, the protibiae flattened with an apical tooth and usually one or two teeth on the outer edge, pronotum, without horns or pronounced depressions, the labrum sclerotised, the mandibles concealed from above, the metatarsus articulated either between between the two tibial spurs or below them and the colour usually yellowish- or reddish-brown, rarely black, with bicoloured elytra in Phyllotyocini, and with metallic colours in *Diphucephala*, *Phyllotocidium* and some Xylonychini.

Larvae. Melolonthine larvae possess the following distinguishing characters in combination: Apical antennomere about as wide as preapical antennomere and bearing a single large oblong sensory spot, preapical antennomere lacking a sensorium, galea and lacinia fused proximally or fitting very tightly together, anal opening almost always angulate or Y-shaped and abdominal sternite IX usually with a distinct raster composed of paired, curved, longitudinal rows or a curved transverse rowof stout setae (pali) and mandible with ventral stridulating are absent or consisting of a patch of minute granules (Ritcher 1966; McQuillan 1985; Cribb *et al.* 1998; Miller & Allsopp 2000).

Key to the tribes and some genera of Australian Melolonthinae

(modified from Britton 1990)

Male characters are used sparingly throughout these keys as necessary. Males often have enlarged antennal clubs compared to females, this is not always easy to determine in isolation. Differences may also occur in the shape of the abdominal ventrites, but there is no one rule to cover all species. Differences in the shape of the pretarsal claws are also used throughout the keys: (a) bifid claws are split so that there are two rami basically parallel to each other as in Figs 31.8, 31.10; (b) toothed claws have a tooth on the inner concave side, usually at right angles to the shaft as in Figs 31.9, 31.10; (c) appendiculate claws have any other type of immoveable structure attached to the concave side as in Figs 31.4, 31.85, 31.86; (d) there may also be membranous pulvilli beneath the claws as in Fig. 31.13.

1. Mandibles and labrum projecting well in front of and on same plane as clypeus; labrum free, about as long as wide, apically rounded to acute and extending to mandibular apices (Fig. 31.2); protrochantin exposed; metatarsi in males at least 3 times as long as metatibiae; abdomen at midline shorter than metaventrite, spiracles concealed in intact specimens; antennae 9-segmented; body of male somewhat wedge shaped, with individually rounded elytra; females little known; length 7.0–21.2 mm; central to northern QLD, NT .... **Phaenognathini**: ***Phaenognatha* Hope**

– Mandibles never exposed; labrum, IF exposed, either fused to clypeus or distinctly transverse and on different plane than clypeus; protrochantin concealed; metatarsi almost never more than 2 times as long as metatibiae (slightly more in some Systellopini); abdomen at midline usually longer than metaventrite, with last pair of spiracles exposed or concealed in intact specimens; antennae 7- to 10-segmented **2**

2(1). Metatibia with 2 spurs placed one above and one below tarsal articulation (Fig. 31.3) so that the basal tarsomere in moving tends to pass between the spurs (some Maechidiini have 1 spur and *Phyllotocidium* has 1 or 0 spurs, but simple claws); tarsal claws simple or with small tooth, sclerotised lamina (Fig. 31.4) or membranous pulvillus (Fig. 31.5) on inner side of each claw **3**

– Metatibia with 2 spurs (very short in *Plectris*, Fig. 31.7) placed close together below tarsal articulation (Fig. 31.6) so that the basal tarsomere in moving can only pass above the spurs (some Diphucephalini have 1 or 0 spurs, but bifid claws); tarsal claws bifid (Figs 31.8, 31.10) or with large tooth on inner curve (Figs 31.9, 31.10), or rarely without a tooth (*Zietzia*) **18**

3(2). Anterior pretarsal claws of male strongly asymmetrical (Fig. 31.11); each procoxa, as seen from in front, projecting to a distance greater than width of its coxal cavity (Fig. 31.12); metacoxae large, usually longer than metanepisternum (except *Adossa, Ancyclonyx, Cheiragra*); abdomen with not more than 5 ventrites visible from below; prosternum usually with long, slender, setose process arising from mid posterior margin (except *Cheiragra* and some *Ancylonyx*); maxilla with lacinia usually setose and without teeth (except in *Sphaeroscelis*); labrum attached to apical margin of clypeus, either fused to it or separated by a suture; length less than 10 mm .... **Phyllotocini**

– Anterior pretarsal claws of male equal (Fig. 31.13); each procoxa, as seen from in front projecting to a distance less than width of its coxal cavity (Fig. 31.14); metacoxae usually shorter than meteanpisternum, if longer then male pretarsal claws and procoxae not as above (eg *Colymbomorpha*, which has median keel extending forwards between procoxae); abdomen often with part of 6th ventrite visible from below; prosternum with or without a prosternal process; maxilla with lacinia toothed on inner side and rarely setose; labrum variously attached to clypeus; length often greater than 10 mm **4**

4(3). Mesoventrite and metaventrite together forming prominent, median, anteriorly directed process or lamina which often passes between procoxae (short in some *Xylonichus*) (*Homolotropus taylori* [Scitalini] has this character but is slightly metallic brown in colour and the elytra are densely clothed in decumbent white setae and with numerous long erect setae as well); legs slender, digging adaptations reduced; colour usually greenish or partly metallic .... **Colymbomorphini**

– Mesoventrite and metaventrite without trace of median process; legs usually well adapted for digging **5**

5(4). Surfaces of body and legs metallic green or cupreus, at least in part; protibiae with 1 or 2 teeth on outer edge, including apical; metatibiae with 0, 1 or 2 spurs; male pro- and mesotarsal claws with long membranous pulvilli beneath (Fig. 31.13); labrum transverse, attached to apical margin of clypeus and separated from it by a suture (Figs 31.14, 31.15); antennal club with 3 segments; 4.5–7 mm in length; VIC, NSW, WA .... **Phyllotocidiini**: ***Phyllotocidium* Blackburn**

– Surfaces of body and legs not metallic; protibiae with 1, 2 or 3 teeth on outer edge, including apical; tarsal claws without pulvilli beneath (except *Odontonyx* [Heteronychini] and *Nosphisthis* [Liparetrini]); labrum variously attached to clypeus; antennal club often with more than 3 segments **6**

6(5). Labrum and clypeus fused, at most separated by faint indication of suture at each side **7**

– Labrum and clypeus distinct and clearly separated **11**

7(6). Protibiae with 3 teeth on outer edge, including apical; proximal tooth small to very small, acute or rounded, situated in proximal half of tibiae, closely applied to edge of tibiae, apically directed and separated from edge of tibiae by an acute angled notch (Fig. 31.16); labrum in same plane as anterior face of clypeus (Fig. 31.17); penultimate abdominal tergite and sternite fused .... **Automoliini**

– Protibiae with 1, 2 or 3 teeth on outer edge, including apical; if protibiae with 3 teeth, then proximal tooth obtuse and usually situated either in distal half or approximately half way (slightly more proximal in *Microcoenus* [Maechidiini]), without acute angled notch as above **8**

8(7). Pretarsal claws toothed .... **Heteronycini (part – *Acheilo,* some *Eurychelus,* some *Webbella*)**

– Pretarsal claws simple **9**

9(8). Form depressed, metacoxae strongly transverse, narrow (Fig. 31.18); elytra usually with rows of characteristic annular setiferous punctures (extra long setae as well in *Epholcis bilobiceps*); abdominal ventrites with longitudinal ridge, fitting to the elytra (Fig. 31.19) .... **Maechidiini (part – *Maechidius, Epholcis*)**

– Form not unusually depressed; metacoxae less transverse, wider; elytra without rows of characteristic annular setiferous punctures; abdominal ventrites with or without longitudinal ridge **10**

10(9). Suture between penultimate abdominal tergite and sternite absent (Fig. 31.20) so that the segments are fused (not to be confused with longitudinal abdominal ridge if present) .... **Maechidiini (part – *Microthopus, Harpechys, Microcoenus*)**

– Suture between penultimate abdominal tergite and sternite present (not to be confused with longitudinal abdominal ridge if present) **11**

11(6,10). Form depressed; labrum situated under clypeus and separated by semicircular suture (*Proborhinus* (Heteronycini) also has a semicircular suture here but has an upturned apical edge to the clypeus); sutural edge of labrum fringed with short, stout spines (Fig. 31.21); lateral edges of clypeus deeply sinuate, anterior edge concave with sharp lateral angles (Fig. 31.22); pronotum without well defined lateral edges and margins; metacoxae strongly transverse, narrow .... **Maechidiini (part – *Termitophilus*)**

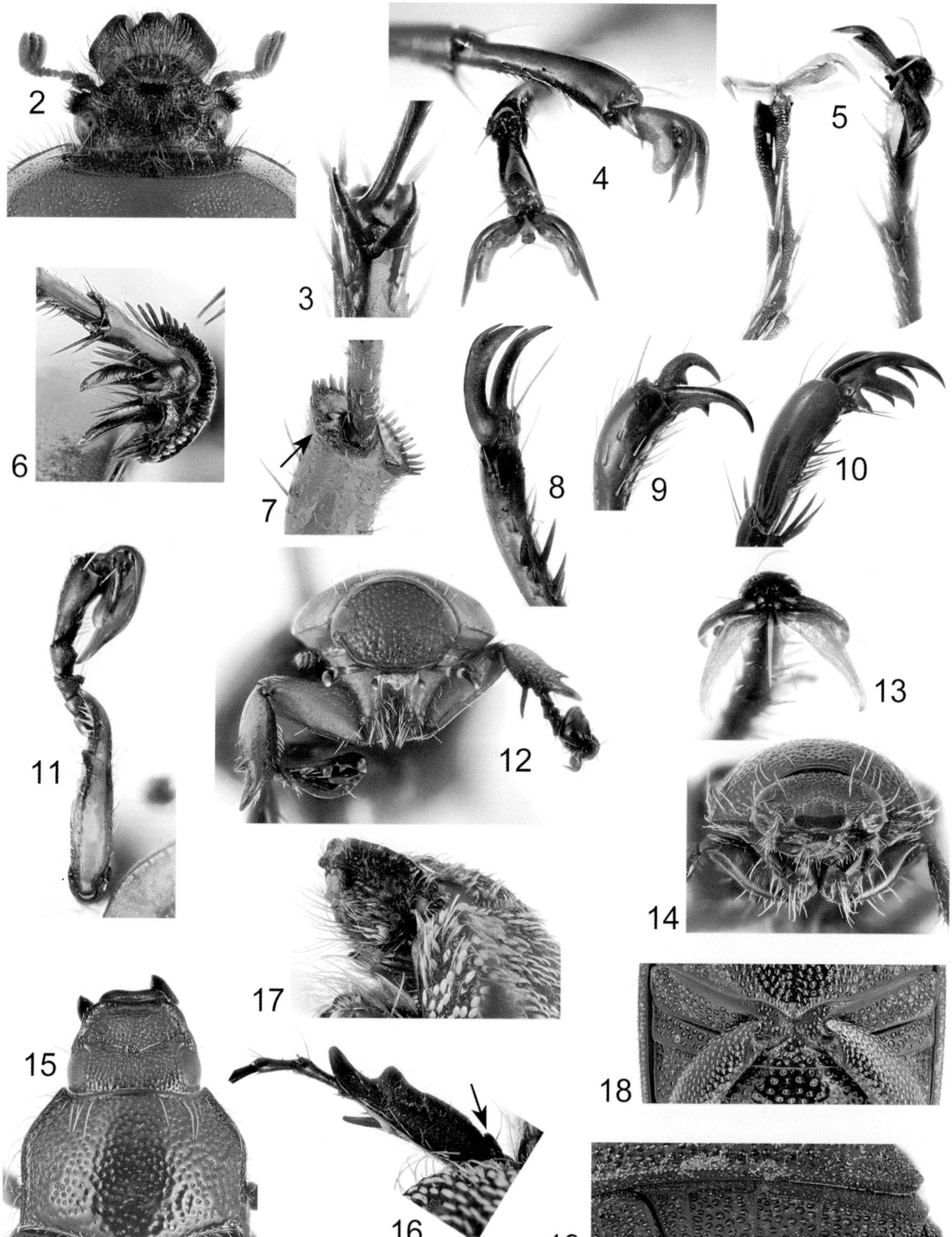

Figs 31.2–31.19. Melolonthinae. 2, *Phaenognatha aequistriata* Arrow, head, dorsal; 3, *Cheiragra ruficollis* Macleay, apex of metatibia; 4, *Ancyclonyx* sp., metatarsal claws; 5, *Cheiragra ruficollis* Macleay, metatarsal claws; 6, *Pachytricha castanea* Hope, apex of metatibia; 7, *Plectris aliena* Chapin, apex of metatibia; 8, *Plectris aliena* Chapin, protarsal claws; 9, *Antitrogus plaiceps* (Blackburn), protarsal claws; 10, *Pachytricha castanea* Hope, metatarsal claws; 11, *Cheiragra ruficollis* Macleay, pretarsal claws; 12, *Cheiragra ruficollis* Macleay, procoxae; 13, *Phylloctocidium pictum* Lea, pretarsal claws; 14, *Phylloctocidium pictum* Lea, head, frontal; 15, *Phylloctocidium pictum* Lea, head and pronotum; 16, *Maechidinus sculptilis* Britton, protibia; 17, *Maechidinus sculptilis* Britton, labrum, laterally; 18, *Maechidius hopeanus* Westwood, metacoxae; 19, *Maechidius hopeanus* Westwood, abdomen, lateral.

– Form not unusually depressed; labrum otherwise differentiated from clypeus or fused or partly fused to it; sutural edge of labrum never with fringe of spines; pronotum with well defined lateral edges and margins; metacoxae less transverse, wider **12**

12(11). Pretarsal claws simple **13**

– Pretarsal claws each with small, often minute, sclerotised tooth or membranous appendage on concave side (difficult to see in *Callabonica* [Heteronychini] which is flightless, 11 mm in length, black and has the elytra immovably locked at the suture in intact specimen) **16**

13(12). At least the first 3 abdominal ventrites with glabrous or setose longitudinal ridge (Fig. 31.23) on each side (raise edge of elytron if necessary) [a few species of *Colpochila* and *Sciton* (Liparetrini) will also come out here, but this is accounted for in the key to genera of Scitalini] .... **Scitalini**

– Abdominal ventrites without longitudinal ridge on each side (raise edge of elytron if necessary), although there may be a line of setae in this position **14**

14(13). Metatibia produced into an obliquely tapered process at distal edge on lower side and without a trace of a transverse spinose ridge near the middle (Fig. 31.24); clypeus with fringe of stout, erect, pointed spines on anterior edge (Fig. 31.25); outer edge of protibia with 1 tooth (apical process only) .... **Comophorinini: *Comophorina* Strand**

– Metatibia not produced into tapered process at apex and with at least a partial transverse spinose ridge near middle; clypeus without fringe of spines on anterior edge; protibia with 2 or 3 teeth on outer edge including apical process **15**

15(14). Labrum large, extended in front of clypeus and in same plane (Figs 31.26, 31.27); metacoxae inflated, usually without defined lateral margin; metafemora inflated; metatibiae (Fig. 31.28) very short and stout (length/width = 0.9–2.0); antennal club consisting of 5 or 6 long lamellae (Fig. 31.26); lateral edge of clypeus deeply emarginate before ocular canthus .... **Systellopini**

– Labrum usually small, below level of clypeus; metacoxae flat, not inflated, usually with well defined lateral margin or angular edge separating ventral and lateral faces (Fig. 31.29); antennal club with 3 to 7 lamellae; lateral edges of clypeus not deeply emarginate before ocular canthus [a few species of *Scitala* and *Sericesthis* (Scitalini) will also come out here but these will key out in the key to genera of Liparetrini] .... **Liparetrini (major part)**

16(12). Pretarsal claws without membranous pulvilli .... **Heteronycini (major part)**

– Each pretarsal claw with membranous pulvillus beneath (Fig. 31.30) **17**

17(16). Each pretarsal claw with large tooth on concave side (Fig. 31.30) .... **Heteronycini (part – *Odontonyx*)**

– Pretarsal claws simple, without tooth on concave side .... **Liparetrini (part – *Nosphisthis*)**

18(2). Pretarsal claws with a tooth beneath or simple (tooth very small in *Microrhopaea, Holorhopaea, Nanorhopaea*) .... **Melolonthini**

– Pretarsal claws bifid, with or without a basal tooth **19**

19(18). Labrum with a deep, V-shaped emarginate, extending in front of and on same plane as clypeus; mouthparts covered with woolly pubescence (Fig. 31.31); length more than 30 mm; metatarsus about twice as long as tibia; pretarsal claws with a distinct basal tooth as well as the bifurcation (Fig. 31.10); WA .... **Pachytrichini: *Pachytricha* Hope**

– IF labrum emarginate (some *Diphucephala*), THEN emargination in the shape of a broad U; mouthparts not covered with woolly pubescence; length less than 15 mm; metatarsus at most 1.5 times as long as metatibiae; pretarsal claws without a basal tooth in addition to the bifurcation **20**

20(19). Body surfaces metallic green, blue or cupreous; metatibiae with 2, 1 or no apical spurs; pronotum with a distinct fovea on each side near mid lateral angle (Fig. 31.32); length 10mm or less .... **Diphucephalini**

– Body surfaces yellowish brown, dark brown or black; metatibiae each with 2 spurs (very short in *Plectris*); pronotum without fovea; length 8 to 14 mm **21**

21(20). Abdominal sternite VII (ventrite 5) and tergite VII (propygidium) partly separated by a suture extending from spiracle to posterior edge (Fig. 31.33), abdomen with 6 distinct ventrites (Fig. 31.34); dorsal surfaces dull yellowish brown and densely setose; pretarsal claws unequal, long, capable of independent movement in one plane (Fig. 31.8), those of metatarsus approximately equal in length to tarsal segment 5; one introduced species *Plectris aliena*; 10 to 14 mm in length; eastern NSW .... **Macrodactylini: *Plectris* Lepeletier & Serville**

– Abdominal sternite VII (ventrite 5) and tergite VII (propygidium) completely fused together, without suture between them; abdomen appearing to have 5 ventrites, ventrite 6 only narrowly visible beyond ventrite 5 and retracted into it (Fig. 31.35); dorsal surfaces shiny dark brown or black, glabrous; pretarsal claws shorter, not capable of independent movement (Fig. 31.36), those on metatarsus approximately half the length of tarsal segment 5; 8 to 10 mm in length; northern Australia QLD, NT .... **Diplotaxini: *Apogonia* Kirby**

Classification of the Australian Genera of Melolonthinae

The tribal classification used here follows that of Smith (2006) and Bouchard *et al.* (2011) except for the transfer of Phaenognathini from Aclopinae as per Ocampo & Mondaca (2012) and Lawrence & Ślipiński (2013) and the raising to tribal level of Phyllotocini for the Australian genera placed in Sericini as per Ahrens (2006), Ahrens & Vogler (2008) and Ahrens *et al.* (2014). Generic placements within these tribes follow the various works of Britton from 1957 to 2000.

Tribe Automoliini Britton, 1978

Characteristics. Metatibiae with 2 spurs placed one above and one below tarsal articulation. Tarsal claws simple or with membranous pulvilli beneath, never toothed or bifid. Labrum completely fused to clypeus. Penultimate abdominal tergite and ventrite fused. Protibiae with 3 teeth, proximal tooth small to very small, situated in proximal half of tibiae, closely applied to edge of tibiae, apically directed and separated from edge of tibiae by an acute angled notch. Abdominal ventrites with or without a longitudinal ridge on each side. Whole body usually setose, the setae arising from characteristic large annular punctures.

Distribution. Australia.

Australian taxa. Five genera.

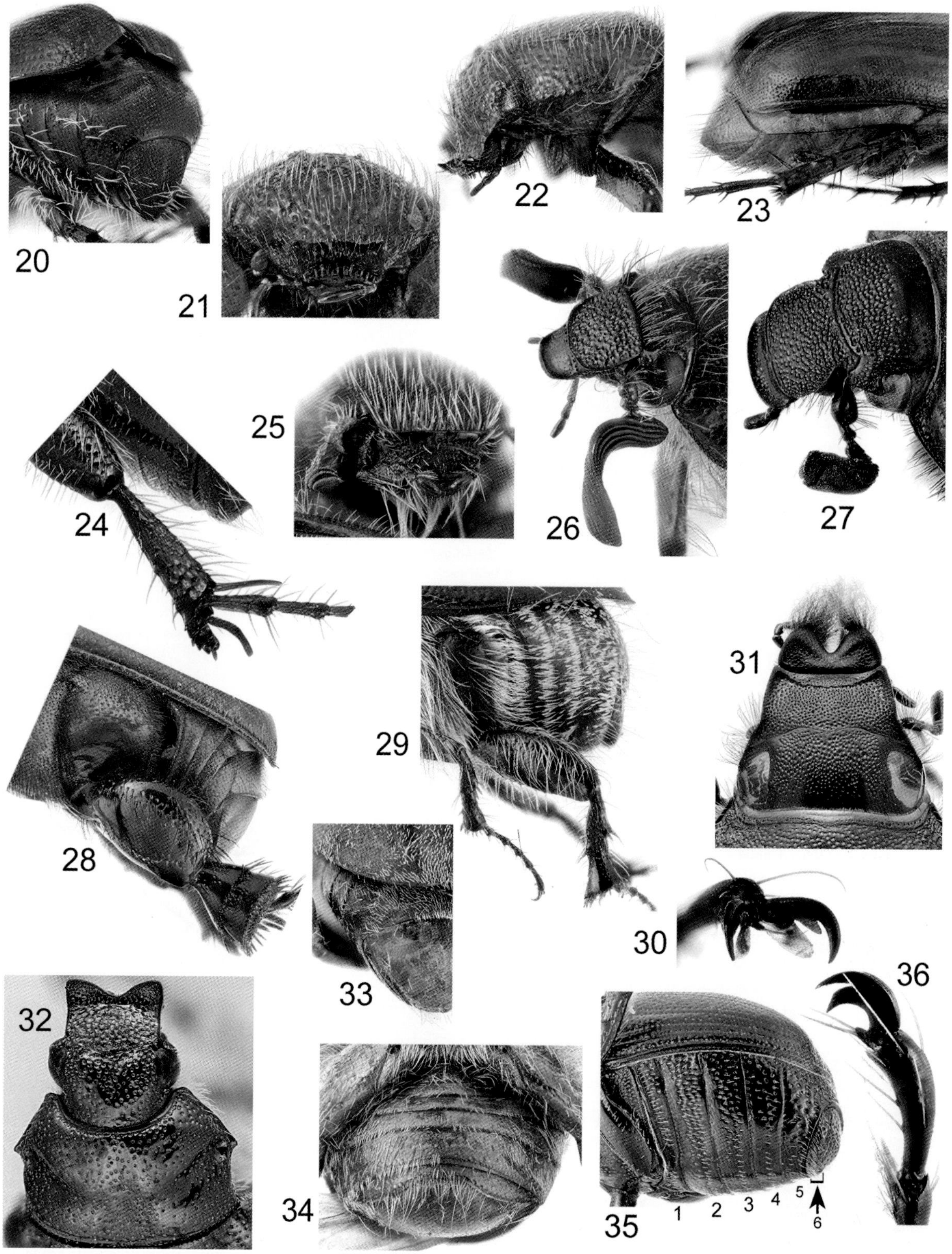

Figs 31.20–31.36. Melolonthinae. 20, *Microthopus castanopterus* Burmeister, pygidium; 21, *Termitophilus spadix* Britton, head, frontal; 22, *Termitophilus spadix* Britton, head, lateral; 23, *Hadrops flavus* (Lea), abdomen, lateral; 24, *Comphorina testaceipennis* (Blanchard), metatibia; 25, *Comphorina testaceipennis* (Blanchard), head, frontal; 26, *Sphyrocallus brunneus* Sharp, head; 27, *Systellopus obtusus* Sharp, head; 28, *Systellopus obtusus* Sharp, hind leg; 29, *Liparetrus lepidopygus* Lea, hind leg; 30, *Odontonyx ruficeps* Lea, pretarsal claws; 31, *Pachytricha castanea* Hope, head, dorsal; 32, *Diphucephala elegans* Blackburn, male head and pronotum; 33, *Plectris aliena* Chapin, pygidium; 34, *Plectris aliena* Chapin, abdomen; 35, *Apogonia destructor* Bos, abdomen; 36, *Apogonia destructor* Bos; pretarsal claws.

Key to the genera of the Australian Automoliini

(modified from Britton 1957)

1. Elytra shortened, exposing part or all of the propygidium in intact specimen (Pl. 58E, 59F); abdominal ventrites without a longitudinal ridge on either side (partial ridge in *Maechidinus sculptilis*); antennae with 7, 8 or 9 segments **2**

– Elytra not shortened, long enough to cover propygidium in intact specimen; abdominal ventrites usually with a longitudinal ridge on each side fitting to the elytra in intact specimen; antennae with 8 or 9 segments with 3 segmented club **3**

2(1). Pronotal hypomera produced forward as a flange to form pockets for reception of antennae (Fig. 31.37); body clothed with expanded, flattened, scale-like setae (Pl. 59F); antennae 9 segmented; 5 to 9.5 mm in length; south-west WA .... ***Maechidinus* Lea**

– Pronotal hypomera not produced forward, at most a ridge where it meets prosternum (Fig. 31.38); body clothed with semi erect setae (Pl. 58E); antennae 7, 8 or 9 segmented; 3.5 to 8 mm in length; widespread – all Australian States .... ***Automolius* Britton**

3(1). Mentum inflated, almost hemispherical, densely clothed with long setae (Fig. 31.39); metafemora strongly inflated; metatibiae short, strongly expanded apically (Fig. 31.40); antennae 8 segmented; 5 to 8 mm in length; WA .... ***Brittonius* Özdikmen & Demir**

– Mentum flat or slightly convex, anteriorly bare or with few setae; metafemora not inflated, relatively slender; metatibiae longer, slender and much less expanded apically; antennae 8 or 9 segmented **4**

4(3). Antennae 9 segmented (Fig. 31.41); body vestiture of either long or short setae or small scales; setae or scales on disc of elytra often arranged in longitudinal bands with relatively narrow spaces between them (Pl. 58F); striae not obvious; 5 to 12 mm in length; widespread but mainly WA .... ***Aplopsis* Blanchard**

– Antennae 8 segmented (Fig. 31.42); body vestiture only of long or short setae; setae on disc of elytra not obviously arranged in longitudinal bands (Pl. 61B); striae obvious, at least in parts; 7.5 to 10 mm in length; TAS .... ***Deuterocaulobius* Dalla Torre**

Aplopsis Blanchard, 1850 (Pl. 58F)

Type species. *Aplopsis lineoligera* Blanchard, 1850.
Ocnodus Burmeister, 1855
Haplopsis Gemminger & Harold, 1869
Atermonichila Blackburn, 1907

Australian species. 22.
Distribution. Endemic. Widely distributed in Australia but most species from WA.
References. Britton (1957, 1987a).

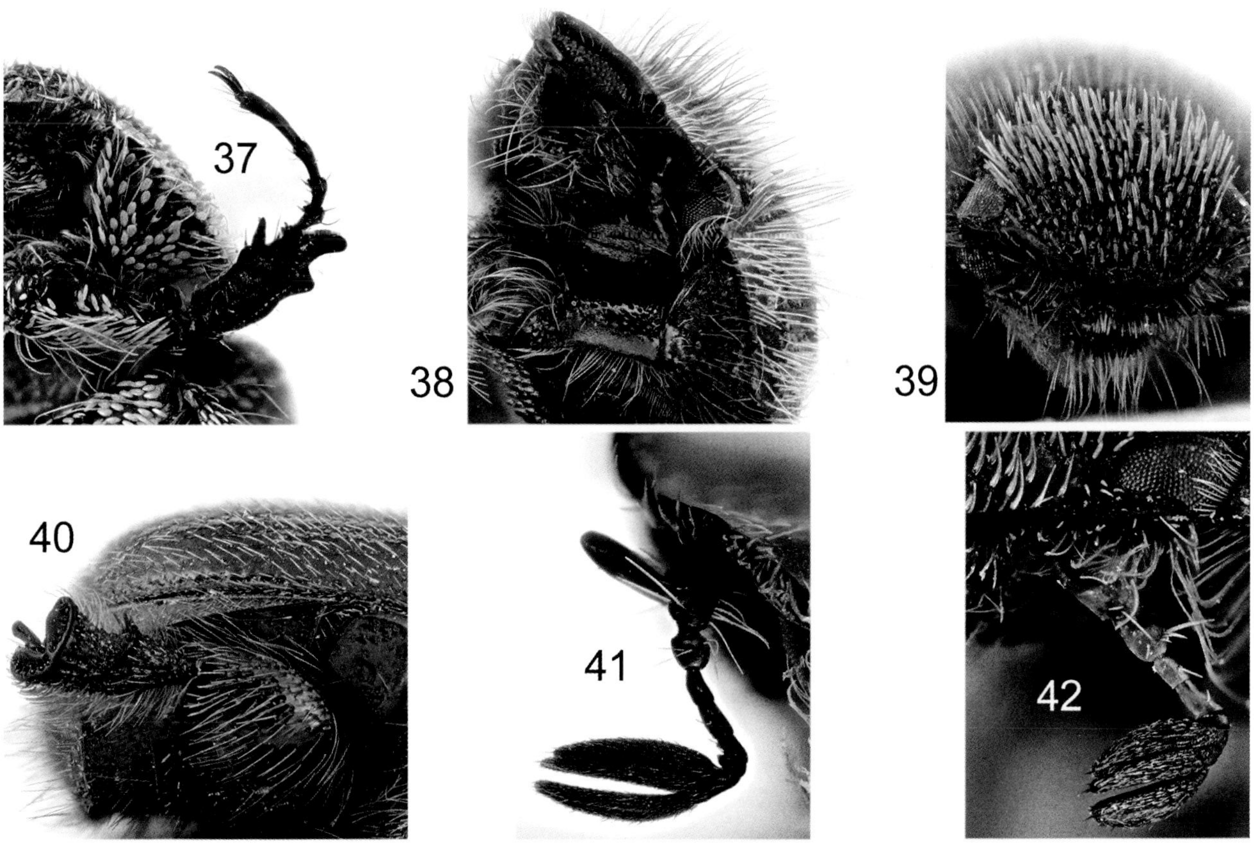

Figs 31.37–31.42. Melolonthinae: Automoliini. 37, *Maechidinus sculptilis* Britton, pronotal hypomeron; 38, *Automolius depressus* (Blanchard), pronotal hypomeron; 39, *Brittonius castanea* (Britton), head, frontal; 40, *Brittonius castanea* (Britton), hind leg; 41, *Aplopsis lineoligera* Blanchard, antenna; 42, *Deuterocaulobius villosus* (LeGuillou), antenna.

Automolius Britton, 1978 (Pl. 58E)

Type species. *Automolus angustulus* Burmeister 1855.
Automolus Burmeister 1855, not Reichenbach, 1853
Australian species. 30
Distribution. Endemic. Widely distributed in Australia.
Biology. The larva of *A. depressus* (Blanchard) from TAS was described by McQuillan (1985).
References. Britton (1957, 1978).

Brittonius Özdikmen & Demir, 2008 (Pl. 58K)

Type species. *Bryantella castanea* Britton 1957.
Bryantella Britton 1957 (not Chickering 1946)
Australian species. One.
Distribution. Endemic. *B. castanea* Britton occurs in WA.
References. Britton (1957); Özdikmen & Demir (2008)

Maechidinus Lea, 1919 (Pl. 59F)

Type species. *Maechidinus latericollis* Lea, 1919.
Australian species. Three.
Distribution. Endemic. *Maechidinus* species are restricted to southern WA.
References. Britton (1957).

Deuterocaulobius Dalla Torre, 1912 (Pl. 61B)

Type species. *Caulobius villosus* Guillou, 1844
Phyllochlaenia Blanchard, 1846, not *Phyllochloenia* Dejean, 1833 (misspelling – see Smith & Evans 2005 for explanation)
Caulobius Guillou, 1844, not Duponchel, 1838.
Australian species. Two.
Distribution. Endemic. *Deuterocaulobius rufescens* (Blanchard) and *D. villosus* (Guillou) are known only from TAS.
References. Britton (1957); Smith & Evans (2005)

Tribe Colymbomorphini Blanchard, 1850

Characteristics. Metatibiae with 2 spurs placed one above and one below tarsal articulation. Tarsal claws simple. Labrum distinct from clypeus. Mesoventrite and metaventrite together forming a prominent, median, anteriorly directed process or lamina which often passes between the procoxae. Abdominal ventrites with or without a longitudinal ridge on each side. Pronotum with posterior edge strongly sinuate and posterior angles usually acute. Body colour usually greenish and slightly metallic.
Distribution. Australia, New Zealand, South America.
Australian taxa. Six genera.

Key to the Australian genera of Colymbomorphini

(modified from Britton 1957, 1979)

1. Each elytron with 14 punctate striae. (Pl. 60A, 62c) **2**
– Each elytron with at most 9 punctate striae (Pl. 60L, 63A) **4**
2(1). Antennae 8-segmented; antennal club 4-lamellate in both sexes; protibiae with 1 tooth (apical) or with a very obtuse trace of a second tooth; elytra (Pl. 62C) without setae or scales; 16–18 mm in length; coastal heath of southern QLD ..... ***Exostethus* Britton**
– Antennae 9-segmented; antennal club with 3 1/2, 5 or 6 lamellae; protibia with 2 or 3 obvious teeth (including apical); elytra with setae or scales or both (e.g. Pl. 60A) **3**
3(2). Antennal club with 3 1/2 lamellae (Fig. 31.43); tarsal segment 5 without stout spines beneath; meso- metaventral process elongate and produced forwards, so that its anterior end is level with anterior faces of procoxae (Fig. 31.44); 14–16 mm in length; northern QLD ..... ***Chariochilus* Moser**
– Antennal club with 5 or 6 lamellae (Fig. 31.45); tarsal segment 5 bearing 4 to 6 stout, pointed spines beneath; meso- metaventral process not so elongate, not or just reaching posterior faces of procoxae (Fig. 31.46); 16–22 mm in length; TAS, VIC, southern NSW and montane northern NSW ..... ***Xylonichus* Boisduval**
4(1). Meso-metaventral process in form of a thin, vertical lamina which extends forwards between procoxae (Fig. 31.47); abdominal ventrites with a longitudinal ridge on each side (Fig. 31.48); 6–9 mm in length; south-western coastal WA ..... ***Colymbomorpha* Blanchard**
– Meso-metaventral process elongate, subcylindrical or in the form of a thick vertical ridge, which does not reach procoxae; abdominal ventrites without a longitudinal ridge on each side (Fig. 31.49) **5**
5(4). Meso-metaventral process elongate, subcylindrical, extending forwards so that its rounded anterior end is at least level with the posterior faces of procoxae (Fig. 31.50); labrum visible in front (Fig. 31.51), but below level of clypeus; labrum and clypeus with deep emargination; antennal club 3-lamellate; 11–13 mm in length; south-western WA ..... ***Phyllococerus* Waterhouse**
– Meso-metaventral process in the form of a short, thick, vertical ridge which does not reach procoxae (Fig. 31.52); labrum visible, extended in front of and on same plane as surface of clypeus; labrum and clypeus not emarginate (Fig. 31.53); antennal club 6-lamellate in male and 5-lamellate in female length 9–11 mm; northern NSW ..... ***Nitorellus* Britton**

Chariochilus Brenske, 1898 (Pl. 60A)

Type species. *Chariochilus metallescens* Brenske, 1898.
Lepidostethaspis Arrow, 1924
Australian species. One.
Distribution. Endemic. *Chariochilus costipennis* Moser occurs in northern QLD.
References. Britton (1957, 1979).

Colymbomorpha Blanchard, 1850 (Pl. 60L)

Type species. *Colymbomorpha lineata* Blanchard, 1850.
Australian species. Three.
Distribution. Endemic. *Colymbomorpha* species all occur in southern WA.
References. Britton (1957, 1979).

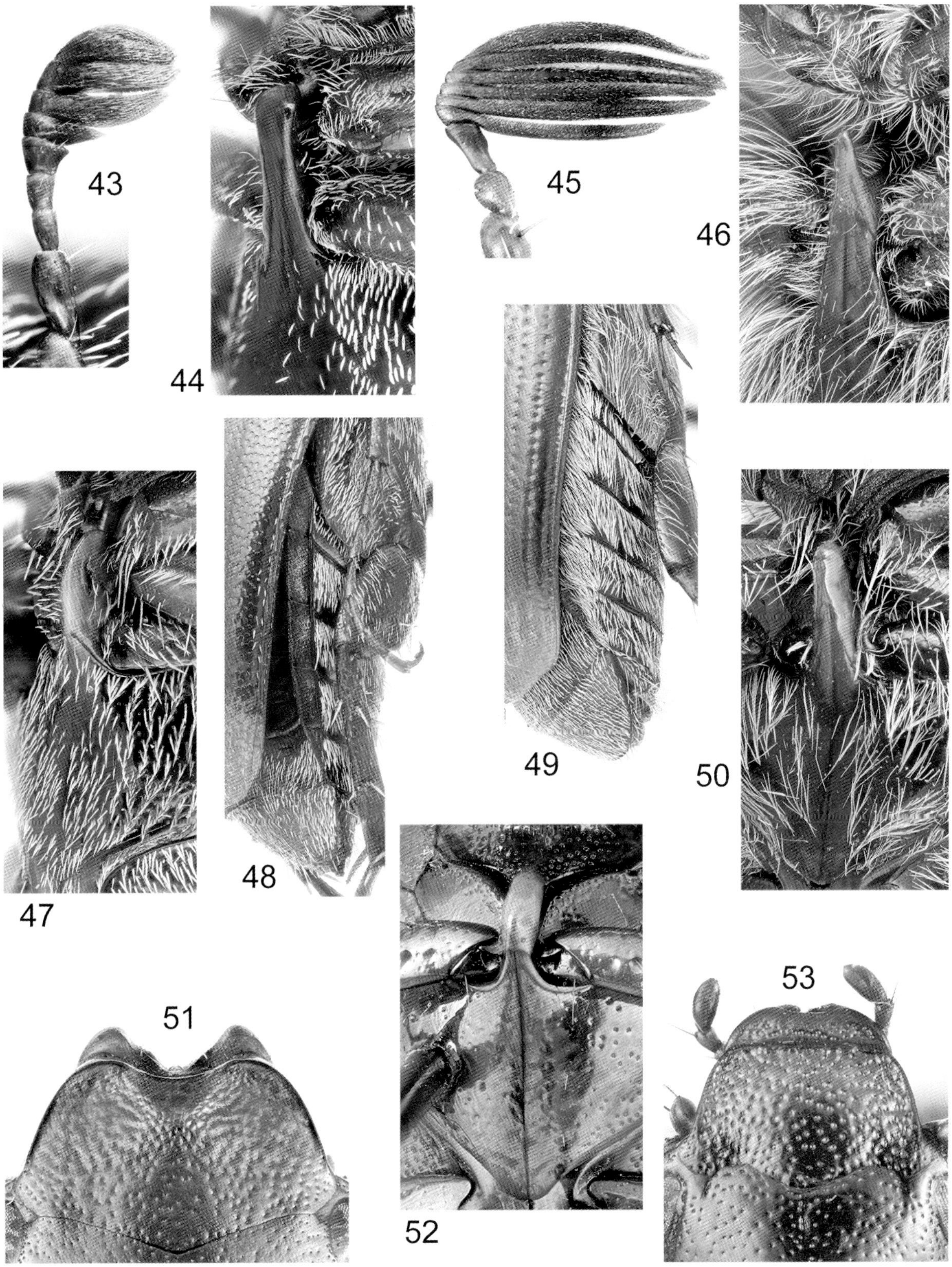

Figs 31.43–31.53. Melolonthinae: Colymbomorphini. 43, *Chariochilus costipenis* Moser, antennal club; 44, *Chariochilus costipenis* Moser, meso-metaventral process; 45, *Xylonychus eucalypti* (Boisduval), antennal club; 46, *Xylonychus eucalypti* (Boisduval), meso-metaventral process; 47, *Colymbomorpha vittata* Britton, meso-metaventral process; 48, *Colymbomorpha vittata* Britton, abdomen, lateral; 49, *Phyllococerus purpurescens* Waterhouse, abdomen, lateral; 50, *Phyllococerus purpurescens* Waterhouse, meso-metaventral process; 51, *Phyllococerus purpurescens* Waterhouse, labrum and clypeus; 52, *Nitorellus splendidus* (Lea), meso-metaventral process; 53, *Nitorellus splendidus* (Lea), labrum and clypeus.

Exostethus Britton, 1979 (Pl. 62C)

Type species. *Exostethus brunneus* Britton, 1979.
Australian species. One.
Distribution. Endemic. *Exostethus brunneus* occurs in northern QLD.
References. Britton (1979).

Nitorellus Britton, 1957 (Pl. 63A)

Type species. *Colymbomorpha splendida* Lea, 1919.
Australian species. One.
Distribution. Endemic. *Nitorellus splendidus* (Lea) is known from northern NSW.
References. Britton (1957, 1979).

Phyllococerus Waterhouse, 1876 (Pl. 64I)

Type species. *Phyllococerus purpurascens* Waterhouse, 1876.
Australian species. One.
Distribution. Endemic. *Phyllococerus purpurascens* is known from southern WA.
References. Britton (1957, 1979).

Xylonichus Boisduval, 1835 (Pl. 66L)

Type species. *Xylonichus eucalypti* Boisduval, 1835.
Stethaspis Hope, 1837
Australian species. Three.
Distribution. Endemic. *Xylonichus* species are known from TAS, VIC and NSW.
References. Britton (1957, 1979).

Tribe Comophorinini Britton, 1957

Characteristics. Metatibiae with 2 spurs placed one above and one below tarsal articulation. Metatibiae produced into an obliquely tapered process at distal edge on lower side and without a trace of a transverse spinose ridge near the middle. Tarsal claws simple. Labrum distinct from clypeus. Clypeus with a fringe of stout, erect, pointed spines on anterior edge. Outer edge of protibiae with only one apical tooth. Propygidium very large.
Distribution. Australia – one genus.

Comophorina Strand, 1928 (Pl. 61H)

Type species. *Comophorus testaceipennis* Blanchard, 1850.
Comophorus Blanchard, 1850 (not Agassiz 1846)
Comophus Britton, 1978
Australian species. One.
Distribution. Endemic. *Comophorina testaceipennis* (Blanchard) is known from NSW, ACT and SA.
References. Britton (1957, 1978, 1987a).

Tribe Diphucephalini Laporte, 1840

Characteristics. Metatibiae with 1 or 2 spurs placed below tarsal articulation or without spurs. Tarsal claws strongly bifid. Labrum either extended in front of clypeus and fused with it, visible from above or located below the anterior face of the clypeus. Pronotum with a distinct fovea on each side near mid lateral angle. Body colour metallic green red or blue.
Distribution. Australia
Australian taxa. Three genera.

Key to the Australian genera of Diphucephalini

(modified from Britton 1995b)

1. Labrum extended in front of clypeus, visible from above, fused with clypeus but separated by transverse ridge or declivity or rarely by lateral traces of a suture; clypeolabrum obviously sexually dimorphic (Fig. 31.54): expanded and angulate in male (Fig. 31.55); ventral surfaces usually clothed with adpressed, usually flat, white setae (Fig. 31.56); metatibia with 2, 1 or 0 spurs; 4 to 10 mm in length; widespread eastern and western Australia .... ***Diphucephala* Dejean**

– Labrum below anterior face of clypeus, not visible from above, separated from clypeus by curved, dorsally convex groove (Fig. 31.57); ventral surfaces with or without long, flat, adpressed setae; metatibia with 1 or 0 spurs **2**

2(1). Antennae 8-segmented; pretarsal claws testaceous, with basal ramus of bifurcation wider at base than apical one and usually truncate (Fig. 31.58); clypeus broadly rounded at sides and constricted at base before ocular canthi; frons between large eyes narrower (ratio of head across eyes to minimum distance between eyes >1.8:1; Fig. 31.59); parameres expanded apically; 4.5 to 8.5 mm in length; high altitude rainforest NSW and QLD .... ***Watkinsia* Britton**

– Antennae 8- or 9-segmented; pretarsal claws black, with basal ramus of bifurcation acute and as wide at base as apical one (Fig. 31.60); clypeus paralled sided and not constricted before ocular canthi; frons between normal size eyes broader (ratio of head across eyes to minimum distance between eyes 1.5:1; Fig. 31.61); parameres not expanded apically; 5.5 to 7.5 mm in length; WA and NSW .... ***Cunderdinia* Lea**

Cunderdinia Lea, 1916 (Pl. 61F)

Type species. *Cunderdinia variabilus* Lea, 1916.
Australian species. Three.
Distribution. Endemic. *Cunderinia* species have been described from NSW and southern WA.
References. Britton (1995b).

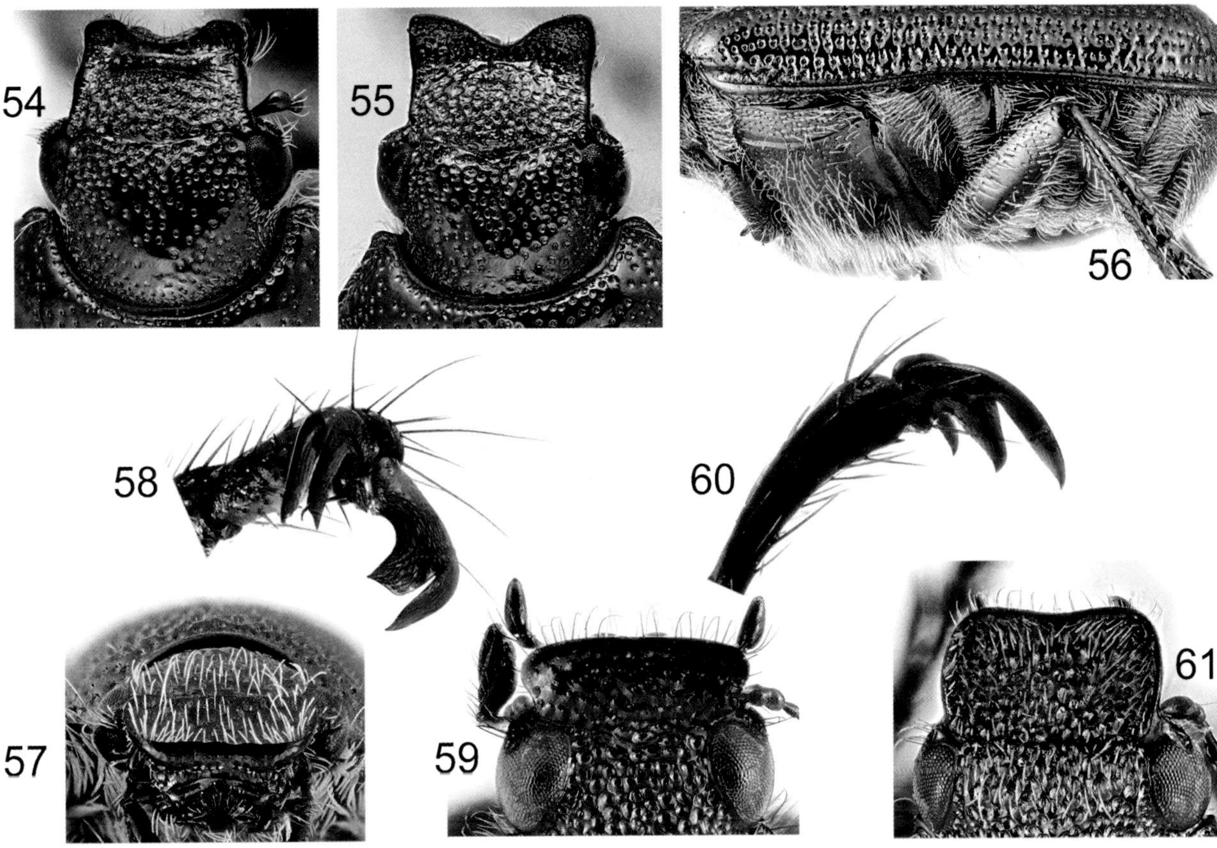

Figs 31.54–31.61. Melolonthinae: Diphucephalini. 54, *Diphucephala elegans* Blackburn, female head; 55, *Diphucephala elegans* Blackburn, male head; 56, *Diphucephala elegans* Blackburn, elytra and abdomen, lateral; 57, *Cunderdinia variabilis* Lea, head, frontal; 58, *Cunderdinia variabilis* Lea, pretarsal claws; 59, *Watkinsia bella* Britton, head, dorsal; 60, *Watkinsia bella* Britton, pretarsal claws; 61, *Cunderdinia variabilis* Lea, head, dorsal.

Diphucephala Dejean, 1821 (Pl. 5H, 6F, 61G, L)

Type species. *Melolontha sericea* Kirby, 1818.
Australian species. 68
Distribution. Endemic. Widely distributed in Australia.
Biology. The larva of *D. colasidoides* (Gyllenhal) from TAS was described and illustrated by McQuillan (1985).
References. Britton (1995b).

Watkinsia Britton, 1995 (Pl. 66B, G)

Type species. *Watkinsia bella* Britton, 1995.
Australian species. Eight.
Distribution. Endemic. *Watkinsia* species occur in NSW and QLD.
References. Britton (1995b).

Tribe Diplotaxini Kirby, 1837

Characteristics: Metatibae with 2 spurs placed below the tarsal articulation. Tarsal claws bifid. Labrum distinct, transverse, located below clypeus. Antennae 9 or 10 segmented with 3 segmented club. Penultimate abdominal ventrite (5) and tergite (propygidium) completely fused together, without a suture between them. Abdominal ventrite 6 nearly completely retracted beneath ventrite 5.
Distribution: Worldwide.
Australian taxa: One genus.

Apogonia Kirby, 1818 (Pl. 59A)

Type species. *A. gemellata* Kirby, 1818 (= *Melolontha rauca* Fabricius, 1781).
Australian species. Three known in collections at generic level.
Distribution. Mainly Ethiopian and Oriental regions, Indonesia, NewGuinea. In Australia *Apogonia* species are known from northern NT and northern QLD.
References. Brown *et al.* (2000); Bezdåk (2004).

Tribe Heteronycini Lacordaire, 1856

Characteristics. Metatibiae with 2 spurs placed one above and one below tarsal articulation. Tarsal claws each with a lamina on the concave side, lamina ending distally in an acute or rounded tooth or sharp declivity. Anterior tarsal claws equal. Labrum distinct (absent in *Acheilo*), often reflexed upwards, never with a fringe of short spines. Elytral stiae almost always indisdinct or absent. Abdominal ventrites with a longitudinal ridge on each side. Antennae 7–9 segmented, club with 3 very short lamellae. Body testaceous, castaneous, piceous or black, never metallic.

Distribution. Basically Australian, with *Heteronyx* also found in Indonesia, New Guinea and New Caledonia and *Neoheteronyx* also recorded from Indonesia and New Guinea.

Australian taxa. 11 genera.

Key to the Australian genera of Heteronycini

(modified from Britton 1988; *Cubidens* for *Paraheteronyx* from Britton 1992)

1. Hind wings absent or reduced, at most half as long as elytra, not folded; elytra immovably locked at suture in intact specimen; body black or dark brown **2**

– Hind wings fully developed and folded; elytra free; body usually testaceous or castaneous, rarely black or piceous **3**

2(1). Labrum reflexed upwards, anterior face flat, visible in front of clypeus (Fig. 31.62) when viewed from above middle of pronotum; antennae 8-segmented; elytra with pale, membranous apical edges (Fig. 31.63); pretarsal claws on concave side with longitudinal lamina (Fig. 31.64) ending in distal declivity (hard to see); prosternum without setose projection on mid-posterior margin; 11 mm in length; SA .... ***Callabonica* Blackburn**

– Labrum not reflexed, broadly and deeply concave (Fig. 31.65), not visible in front of clypeus when viewed from above middle of pronotum; antenna 8- or 9-segmented; elytra without pale, membranous apical edges; pretarsal claws (Fig. 31.66) on concave side with longitudinal lamina ending distally in acute tooth; prosternum usually with median, setose projection on mid-posterior margin (Fig. 31.67); 10 to 15 mm in length; QLD, NSW, VIC, TAS, ACT .... ***Pseudoheteronyx* Blackburn**

3(1). Clypeus extended forwards and upwards as a curved, narrowed lamina (Fig. 31.68); terminal maxillary palpomere securiform (ratio of length to apical width 1.5–1.7:1; Fig. 31.68, white arrow); pronotum and elytra uniformly clothed with short pubescence; antennae 9-segmented; body length less than 7 mm **4**

– Clypeus not extended forwards and upwards as a curved, narrow lamina; terminal maxillary palpomere not securiform; pronotum and elytra with or without iniform, short pubescence; antennae 7- to 9-segmented; body length 3.5 to 17 mm **5**

4(3). Labrum clearly distinguishable below clypeus (Fig. 31.69); clypeus not expanded apically, its posterolateral angles sharply angled and extended laterally beyond level of eyes (Fig. 31.70); pronotum and elytra coarsely punctured (75–100 per square mm); 4 to 5 mm in length; northern WA .... ***Proborhinus* Britton**

– Labrum fused to and curved upwards with clypeus; labral apex visible just before clypeal apex (as viewed from above; Fig. 31.71); clypeus sometimes expanded apically; posterolateral angles of clypeus rounded and not extending laterally beyond level of eyes (Fig. 31.71); pronotum and elytra finely and densely punctured (~200 per square mm); 5 to 7 mm in length; northern WA and northern QLD .... ***Acheilo* Britton**

5(3). Protibia with 1 or 2 teeth, including apical, within distal third of its length and a very small, acute tooth very close to base (Fig. 31.72), apically directed and separated from edge of tibiae by an acute angled notch; metacoxae narrow, its length about half that of metanepisternum; antennae 8-segmented **6**

– Protibia almost always with 3 teeth, including apical, the proximal one not as above (Figs 31.73, 31.74); metacoxae wider, its length much greater than half that of metanepisternum and often equal to it in length; antennae 7- to 9-segmented **7**

6(5). Dorsal surfaces of pronotum and elytra without setae; pretarsal claws appendiculate and with multisetose empodium present between claws (Fig. 31.75); pronotum without narrow anterior and posterior margins (Fig. 31.76); male pro- and mesotarsi expanded and with dense spongiose setae beneath; 5 to 8 mm in length; NT, QLD, NSW, TAS, also found in Indonesia and New Guinea .... ***Neoheteronyx* Blackburn**

– Dorsal surfaces of pronotum and elytra densely setose; pretarsal claws bifid and strongly curved, without multisetose empodium; pronotum with narrow defined anterior margin; length 7–7.5 mm .... ***Cubidens* Britton**

7(5). Antennae 7-segmented; abdominal ventrites without lateral longitudinal ridges on each side (Fig. 31.77); dorsal surface covered with fine reticulation; 7 mm; TAS .... ***Nepytis* Erichson**

– Antennae 8- or 9-segmented; abdominal ventrites with or without lateral longitudinal ridges on each side; dorsal surface otherwise **8**

8(7). Pretarsal claws each with large, pale, membranous pulvillus beneath, in addition to an acute tooth (Fig. 31.78); elytral striae equally spaced, impressed, strongly and regularly punctured; elytra without setae or pubescence on disc; abdominal ventrites without longitudinal ridges on each side (Fig. 31.79); antennae 9-segmented; 6 to 14 mm in length; QLD, NSW .... ***Odontotonyx* Macleay**

– Pretarsal claws without pulvilli, bifid or appendiculate; elytral striae indistinct or absent; elytra usually with setae or pubescence on disc; abdominal ventrites usually with longitudinal ridges on each side; antennae 8- or 9-segmented **9**

9(8). Elytral pubescence mottled (with small distinct patches of whitish and brown setae); antennae 8-segmented; pretarsal claws bifid (Fig. 31.80); labrum not visible from above (Fig. 31.81), suture between it and clypeus faint or absent (Fig. 31.82); 8 to 12 mm in length; NSW, ACT, VIC .... ***Eurychelus* Blanchard**

– Elytral pubescence, if present, uniformly coloured; antennae 8- or 9-segmented; pretarsal claws bifid or appendiculate; labrum often visible from above, suture between it and clypeus distinct, faint or absent **10**

10(9). Labrum with median carina in anterior emargination, suture between it and clypeus sometimes faint or absent in middle (Fig. 31.83); pretarsal claws bifid (Fig. 31.84); 5 to 10 mm in length; NSW, ACT, VIC .... ***Webbella* Britton**

– Labrum without median carina in anterior emargination, suture between it and clypeus always distinct; pretarsal claws toothed, bifid or appendiculate (Figs 31.85–31.88); 3.5 to 17 mm in length; widespread, also found in Indonesia, New Guinea and New Caledonia .... ***Heteronyx* Guérin-Méneville**

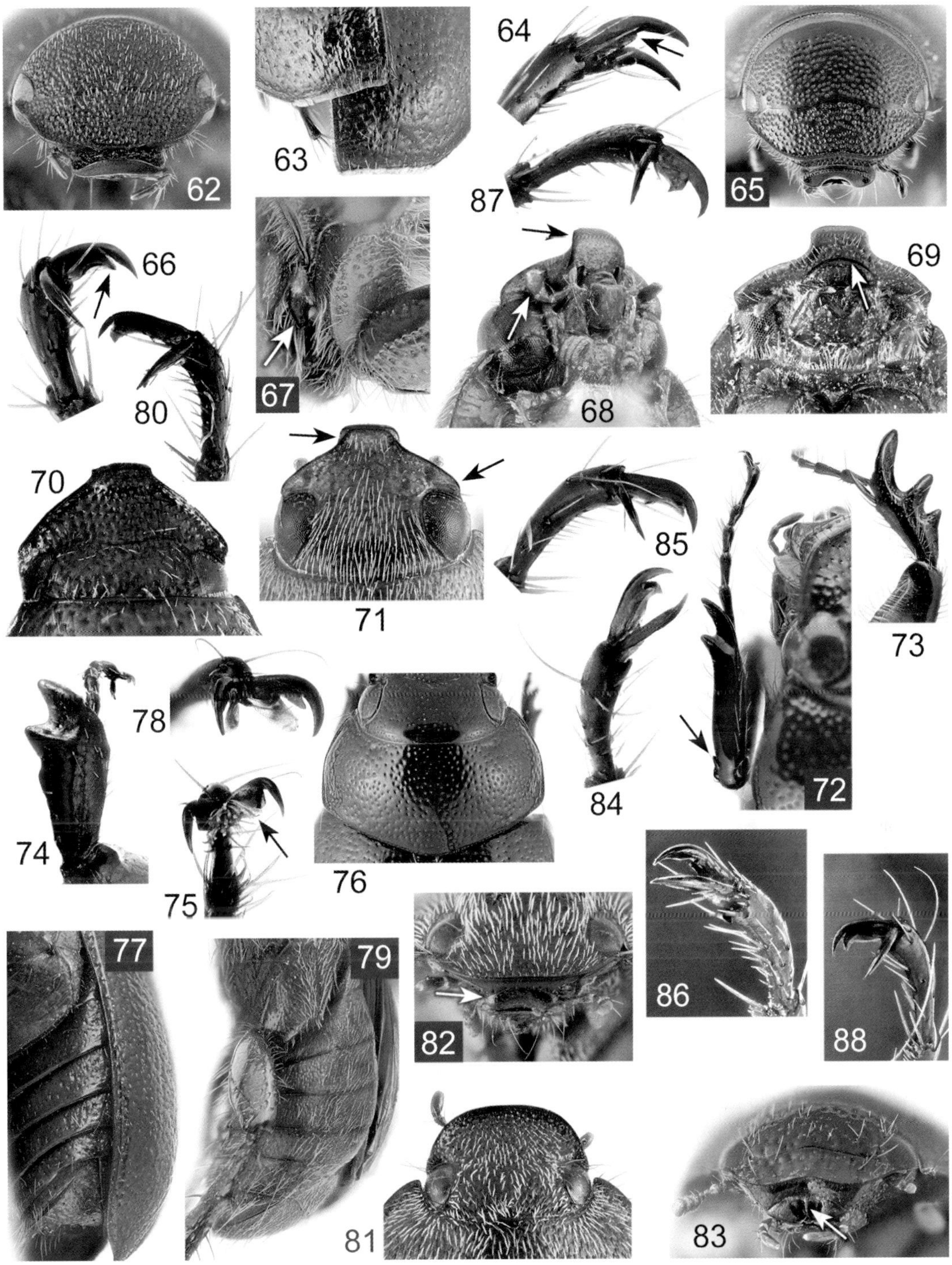

Figs 31.62–31.88. Melolonthinae: Heteronychini. 62, *Callabonnica propria* Blackburn, head, anterodorsal; 63, *Callabonnica propria* Blackburn, elytral apex with membrane; 64, *Callabonnica propria* Blackburn, pretarsal claws; 65, *Pseudoheteronyx creber* Blackburn, head, anterodorsal; 66, *Pseudoheteronyx creber* Blackburn, pretarsal claws; 67, *Pseudoheteronyx creber* Blackburn, prosternal process; 68, *Acheilo clypeatus* Britton, head, ventral; 69, *Proborhinus cornutus* (Blackburn), head, ventral; 70, *Proborhinus cornutus* (Blackburn), head, dorsal; 71, *Acheilo clypeatus* Britton, head, dorsal; 72, *Neoheteronyx brevicollis* (Blackburn), protibia; 73, *Heteronyx cribriceps* Blackburn, protibia; 74, *Nepytis* sp, protibia; 75, *Neoheteronyx brevicollis* (Blackburn), pretarsal claws; 76, *Neoheteronyx brevicollis* (Blackburn), head and pronotum; 77, *Nepytis* sp., abdomen, lateral; 78, *Odontonyx ruficeps* Lea, pretarsal claws; 79, *Odontonyx ruficeps* Lea, abdomen, lateral; 80, *Eurychelus marmoratus* Blanchard, pretarsal claw; 81, *Eurychelus marmoratus* Blanchard, head and pronotum; 82, *Eurychelus marmoratus* Blanchard, head, frontal; 83, *Webbella firma* (Blackburn), head, frontal; 84, *Webbella firma* (Blackburn), pretarsal claws; 85, *Heteronyx australis* Guérin-Méneville, pretarsal claws; 86, Heteronyx constans Blackburn, pretarsal claws; 87, *Heteronyx cribriceps* Blackburn, pretarsal claws; 88, *Heteronyx darwini* Blackburn, pretarsal claw.

Acheilo Britton, 1988 (Pl. 56I)

Type species. *Heteronyx capitalis* Blackburn 1910.
Australian species. Three.
Distribution. Endemic. *Acheilo* species are known from northern QLD and northern WA.
References. Britton (1988).

Callabonica Blackburn, 1895 (Pl. 58C)

Type species. *Callabonica propria* Blackburn, 1895.
Australian species. One.
Distribution. Endemic. *Callabonica propria* occurs in SA.
References. Britton (1995b).

Cubidens Britton, 1992

Type species. *Heteronyx unguiculata* Burmeister, 1855.
Paraheteronyx Britton, 1988, not Moser, 1924
Australian species. One.
Distribution. Unknown. Known only from the types labelled 'Neu Holland'
References. Britton (1988, 1992).

Eurychelus Blanchard, 1850 (Pl. 61E)

Type species. *Eurychelus marmoratus* Blanchard, 1850.
Australian species. One.
Distribution. Endemic. *Eurychelus marmoratus* is known from VIC and NSW.
References. Britton (1988).

Heteronyx Guérin-Méneville, 1831 (Pl. 62F)

Type species. *Heteronyx australis* Guérin-Méneville, 1831.
Cotidius Boisduval, 1835
Silopa Erichson, 1842
Hostilina Blanchard, 1850
Anacheirotus Blackburn, 1900
Australian species. More than 220.
Distribution. Widely distributed throughout Australia, with exotic species known from Indonesia, New Guinea and New Caledonia.
Biology. Larvae of *Heteronyx* species, like those of most other melolonthines live in the soil and feed on roots and humus. According to Britton (2000) adults probably do not respond to pheromones (as evidenced by the small antennal clubs) but rather congregate on tree foliage to mate. They emerge from the soil at sunset and fly at night, with emergence dependent more on moisture than temperature. Most specimens attracted to lights are males, suggesting that females have a brief emergence from the soil. Larvae of many species may coexist in the soil and form an important food resource for ground feeding birds. Swarming behaviour was described by Morgan (1977) for *H. piceus* Blanchard in SA; swarming began at dusk and continued until an hour before dawn and beetles congregated on the top of trees and extremities of branches, where copulation took place. Swarming beetles have been known to destroy irrigated crops of lucerne and sunflower, and according to Rogers *et al.* (1992), *H. piceus* is a major pest of peanuts in southern QLD. The larva of *H. tasmanicus* Blackburn was described and Illustrated by McQuillan (1985).
Note. Species of *Heteronyx* described from South America have been transferred to the genus *Blepharotoma* Blanchard (Smith 2008).
References. Britton (1988, 2000).

Neoheteronyx Blackburn, 1890 (Pl. 59K)

Type species. *Neoheteronyx lividus* Blackburn 1890.
Australian species. Nine.
Distribution. *Neoheteronyx* species are known from northern QLD, NT, New Guinea and Indonesia.
References. Britton (1988, 2000).

Nepytis Erichson, 1842

Type species. *Nepytis russula* Erichson, 1842.
Australian species. One.
Distribution. Endemic. *Nepytis russula* occurs in TAS.
References. Britton (1988).

Odontotonyx W. J. Macleay, 1871 (Pl. 63J)

Type species. *Odontotonyx brunneipennis* W. J. Macleay, 1871.
Australian species. Two.
Distribution. Endemic. *Odontotonyx ruficeps* Lea occurs in NSW and *O. brunneipennis* in southern QLD.
References. Britton (1988).

Proborhinus Britton, 1988 (Pl. 64H)

Type species. *Heteronyx cornutus* Blackburn, 1910.
Australian species. One.
Distribution. Endemic. *Proborhinus cornutus* (Blackburn) occurs in northern WA.
References. Britton (1988).

Pseudoheteronyx Blackburn, 1892 (Pl. 59H)

Type species. *Heteronyx baldiensis* Blackburn, 1892.
Australian species. Seven.
Distribution. Endemic. *Pseudoheteronyx* species occur in VIC, NSW, QLD and TAS.
References. Britton (1988); McQuillan & Michaels (1997).

Webbella Britton, 1988 (Pl. 66D)

Type species. *Heteronyx firmus* Blackburn, 1909.
Australian species. Two.

Distribution. Endemic. *Webbella firma* (Blackburn) and *W. labralis* (Blackburn) occur in NSW.

References. Britton (1988).

Tribe Liparetrini Burmeister, 1855

Characteristics. Metatibiae with 2 spurs placed one above and one below tarsal articulation. Tarsal claws simple, without a tooth on the concave side. Labrum distinct. Elytral striae paired or approximately equally spaced. Abdominal ventrites without a longitudinal ridge on each side. Prosternum usually without a long slender process arising from the mid-posterior margin. Antennae 8, 9 or 10 segmented with a 3–7 lamellate club. Body colour testaceous, castaneous, or black.

Distribution. Australia

Australian taxa. 32 genera.

Key to the Australian genera of Liparetrini

(modified from Britton 1990; *Evbrittonia* added from Szito 1993; *Luftius* (= *Colpomorpha*) added from Szito 1994a)

The genera *Cheilo, Sciton, Parasciton, Leonotus, Stenochelyne* and *Colobostoma* appear twice in the key to allow for possible alternate decisions at couplets 1 and 25.

1. Elytral striae grouped into pairs so that adjacent intervals are very different in width (ratio of width of interval 4 to that of interval 5 near middle greater than 1.5:1) or striae indistinct (except sutural stria), the strial punctures being confused with interstrial punctures **2**

– Elytral striae not grouped into obvious pairs, adjacent intervals about equal in width (ratio of width of interval 4 to that of interval 5 near middle less than 1.5:1); striae obvious, number of intervals always 10 .... **25**

2(1). Posterior and sometimes lateral edges of metacoxae pale and membranous (Fig. 31.89); elytra usually abbreviated (exposing part of propygidium, Fig. 31.90); less than 13 mm in length; widespread .... ***Liparetrus* Guérin-Méneville (part)**

– Posterior and lateral edges of metacoxae sclerotised and of same dark colour as rest of coxal surface **3**

3(2). Penultimate abdominal segment without lateral longitudinal sutures separating tergite (propygidium) and ventrite; elytra abbreviated (ratio of length to width of one elytron 2.2:1) and propygidium partly exposed; outline of eye approximately semicircular; less than 13 mm in length; widespread .... ***Liparetrus* Guérin-Méneville (part)**

Penultimate abdominal segment with obvious lateral longitudinal sutures separating tergite (propygidium) and ventrite; elytra not abbreviated (ratio of length to width of one elytron greater than 2.2:1); outline of eye more than semicircular (only slightly truncate behind); body length often greater than 13 mm **4**

4(3). Labrum and clypeus fused in front (separated by fine suture on each side but with suture obliterated at middle, Fig. 31.91) **5**

– Labrum and clypeus clearly separated both at middle and at sides **7**

5(4). Labrum reflexed upwards in front (Fig. 31.92), visible from above beyond anterior edge of clypeus and fused with upper surface of clypeus at middle (Fig. 31.91); all edges of pronotum except mid-anterior with fringes of long setae; 14 to 18 mm in length; WA .... ***Aphanesia* Britton**

– Labrum not reflexed, placed below clypeus, not visible from above **6**

6(5). Antennae 8-segmented; protibiae with 2 teeth, basal one acute (Fig. 31.93); anterior face of clypeolabrum densely setose (Fig. 31.94); outline of clypeus semielliptical; posterior pronotal angles absent, lateral and posterior edges united in uniform curve; elytral disc without setae (Pl. 56J); 5 mm in length; WA (Montebello Islands) .... ***Allara* Britton**

– Antennae 9-segmented; protibiae with 2 teeth, basal one obtuse (Fig. 31.95); anterior face of clypeolabrum shining, without setae; outline of clypeus almost semicircular; posterior angles of pronotum obvious, approximately rectangular; elytral disc with sparse, long, erect setae (Pl. 61A); 5 mm in length; WA .... ***Dysphanochila* Blackburn**

7(4). Labrum fully visible from above, projecting very strongly from beneath clypeus and strongly reflexed (Figs 31.96, 31.97); antennal club 3-lamellate; 13 to 16 mm in length; WA .... ***Glossocheilifer* Blackburn**

– Labrum not or scarcely visible from above, if slightly projecting then not strongly reflexed; antennal club 3- to 7-lamellate **8**

8(7). Antennal club 6-lamellate (basal 1 or 2 lamellae may be shortened) **9**

– Antennal club 3-, 4-, 5- or 7-lamellate **10**

9(8). Protibiae with 2 very broad teeth, the basal one occupying half of tibiae; clypeus semicircular in outline (Pl. 63B), with dorsal surface very deeply concave; 19 to 22 mm in length; NSW, SA .... ***Pachygastra* Germar**

– Protibiae with 3 teeth; clypeus various; 22 to 29 mm in length for the 4 species listed here (*C. antennalis, C. laminata, C.pilosa, C. vanga*) .... ***Colpochila* Erichson (part)**

10(8). Clypeus truncate; antennal club always 3-lamellate **11**

– Clypeus rounded; antennal club 3- to 7-lamellate **13**

11(10). Pronotal hypomeron with a densely setose ridge along notosternal suture, continuing on to anterior edge of hypomeron as a flange overlying the eye (Fig. 31.98); anterior face of clypeus bearing setae; parameres of aedeagus without setae; 12 mm in length; NT .... ***Cheilo* Britton (part)**

– Pronotal hypomeron without such a ridge or flange (Fig. 31.99); anterior face of clypeus without setae; parameres of aedeagus with at least a few setae **12**

12(11). Prosternum with a usually long, stout, conical, laterally flattened, usually apically setose process arising from mid-posterior margin (Figs 31.99, 31.100); parameres of aedeagus bearing many setae; 10 mm in length: QLD, NSW, VIC, SA, WA .... ***Sciton* Blackburn (part)**

– Prosternum without such a process; parameres of aedeagus bearing only 2 or 3 very short setae; 10 mm in length; NT .... ***Parasciton* Britton (part)**

13(10). Metacoxae densely setose; metatibiae markedly expanded towards apex; metafemora with a setose ridge or row of strong setae (Fig. 31.101) parallel to hind margin (can be obscured by other longer setae if present) **14**

– Metacoxae virtually glabrous, with at most a few setae (Fig. 31.102); metatibiae little expanded towards apex, mostly parallel sided; metatibiae without a setose ridge parallel to hind margin, but there may be a few setae in this position **24**

14(13). Anterior face of labrum shallower than face of clypeus **15**
– Anterior face of labrum (including any excavation) as deep as or deeper than face of clypeus **19**
15(14). Antennae 9 segmented with a 4-lamellate club **16**
– Antennae 9 segmented with a 3-lamellate club **17**
16(15). Pronotal disc without setae (Pl. 58B); clypeus semicircular, basal edge without a ridge; 12 mm in length; SA, WA .... ***Aneucomides* Blackburn**
– Pronotal disc densely setose (Pl. 56F); clypeus disciform, basal edge marked by an obvious ridge; 16 mm in length; WA .... ***Evbrittonia* Szito**
17(15). Protibiae with 2 teeth; labrum small, not projecting in front of clypeus; clypeus markedly tapered, semi-elliptical, margin strongly reflexed (Fig. 31.103); parameres of aedeagus with many setae; 9 to 11 mm in length; WA .... ***Stenochelyne* Britton (part)**
– Protibiae with 3 teeth; labrum larger, projecting to a more or less degree in front of clypeus; clypeus rounded or semi-elliptical, if the latter then margin not strongly reflexed; parameres of aedeagus without setae **18**
18(17). Posterior pronotal angles distinct, broadly rounded; metatibiae very short and strongly expanded to apex (ratio of length to width ~1.7:1); transverse spinose ridge near middle of outer side of metatibiae not or only minutely interrupted at middle (Fig. 31.101); head with dark setae about eyes; pronotum with disc glabrous and fringes of dark setae on all sides; 16 mm in length; SA .... ***Ictigaster* Britton**
– Posterior pronotal angles absent, the lateral and posterior edges united in a smooth curve; metatibiae elongate and less expanded (ratio of length to width more than 2:1); transverse ridge near middle of outer side of metatibia broadly interrupted at middle (Fig. 31.104); head (Fig. 31.105), pronotum and scutellum in male clothed with long, very dense, fine, pale setae; 12 to 15 mm in length; SA .... ***Leonotus* Britton (part)**
19(14). Pronotal hypomeron with distinct ridge along notosternal suture, the ridge continuing forward along the free edge of the hypomeron as a narrow flange (Fig. 31.106); antennae 8 segmented with a 3-lamellate club **20**
– Pronotal hypomeron without ridge along notosternal suture; antennae 8 or 9 segmented with a 3- to 7-lamellate club **21**
20(19). Labrum with vertical ridge in middle of deep, wide anterior cavity; 10 mm in length; SA .... ***Astibicola* Britton**
– Labrum without vertical ridge in middle of anterior cavity (Fig. 31.107); 9 mm in length; WA .... ***Lutfius* Őzdikmen**
21(19). Antennae 8 segmented with a 4-lamellate club (*Colpochila affinis* has this antennal arrangement, but the 4th segment is sub-lamellate and the pronotum is glabrous **22**
– Antennae 8 or 9 segmented with a 3-, 5- or 7-lamellate club **23**
22(21). Posterior angles of pronotum distinct, nearly rectangular, the lateral edges behind the middle about parallel; protibiae with 3 teeth; disc of pronotum densely setose (Pl. 66A); labrum with anterior edge concave, visible beyond clypeus (Fig. 31.108); 15 to 17 mm in length; WA .... ***Xyridea* Britton**
– Posterior angles of pronotum absent, the lateral and posterior edges united in a smooth curve; protibiae with 2 teeth; disc of pronotum sparsely setose (Pl. 61C); labrum with anterior edge strongly arched, not visible beyond clypeus (Figs 31.109, 31.110); 10 mm in length; WA .... ***Dikellites* Britton**
23(21). Meso- and metatarsomeres 1–3 in male and mesotarsomeres 1–3 in female each with dense tuft of pale yellowish setae beneath; metatarsomeres 4 and 5 with at most a few setae beneath (Fig. 31.111), beside apical setae; ocular canthus unusually narrow (Fig. 31.112); antennae 9-segmented with a 3-lamellate club; 8 to 10 mm in length; NT .... ***Petinopus* Blackburn**
– Tarsomeres without dense tufts of setae beneath; metatarsomeres 4 and 5 with many setae beneath; ocular canthus of normal width; antennae 8- or 9-segmented with a 3-, 5- or 7-lamellate club; 11 to 33 mm in length; widespread .... ***Colpochila* Erichson (major part)**
24(13). Antenna 8 segmented with a 3- or 4-lamellate club (Fig. 31.113) .... **Scitalini: *Sericesthis* Boisduval (part)**
– Antenna 9 segmented with a 3-lamellate club .... **Scitalini: *Scitala* Erichson (part)**
25(1). Pronotal hypomeron with low ridge or flange along notosternal suture and continuing onto anterior edge of hypomeron **26**
– Pronotal hypomeron without such a ridge or flange **31**
26(25). Protibia with 2 very broad teeth (Pl. 62H), the basal one occupying half of tibiae; metatibia very short and strongly expanded to apex (ratio of length to width at apex ~1.8:1; Fig. 31.114); 12 to 18 mm in length; VIC, WA .... ***Hadropechys* Britton**
– Protibia with 2 or 3 teeth, if 3 basal tooth sometimes very small, if 2 then arrangement not as above; metatibia less strongly expanded to apex (ratio of length to width at apex ~2.6–2.9:1) **27**
27(26). Clypeus viewed from above truncate, its anterior face flat or very slightly concave (Pl. 60I); labrum not projecting; pronotum without posterior angles **28**
– Clypeus viewed from above broadly rounded (Fig. 31.115); labrum projecting slightly beyond level of face of clypeus; pronotum with distinct posterior angles (Pl. 60E) **29**
28(27). Metatarsi long (ratio of length of tarsus to that of tibia ~2:1); lamellae of male antennal club long (~1.9 mm); pronotum without defined posterior margin; eyes very large and convex; 10 mm in length; northern QLD .... ***Scitaloides* Britton**
– Metatarsi short (ratio of length of tarsus to that of tibia ~1:1); lamellae of male antennal club short (~0.65 mm); pronotum with narrow, continuous, raised posterior margin; eyes not unusually convex; 12 mm in length; NT .... ***Cheilo* Britton (part)**
29(27). Labrum shallow (ratio of mid depth of anterior face of labrum to that of clypeus less than 1:1); terminal labial palpomere thick, short, fusiform (Fig. 31.116); hypomeral flange not strongly developed, straight as seen from side; 9 to 11 mm in length; NSW, QLD .... ***Colobostoma* Blanchard (part)**
– Labrum deep (ratio of mid depth of anterior face of labrum to that of clypeus greater than 1:1); terminal labial palpomere elongate (ratio of length to width ~2.8:1; Figs 31.117, 31.118) and parallel-sided; hypomeral flange strongly developed, often curved as seen from side **30**
30(29). Abdominal ventrites each with complete transverse row of well developed, regularly spaced setae (Fig. 31.119); pronotum with fine, continuous posterior margin; lateral edges of pronotum somewhat angulate in the middle, usually straight or slightly concave before posterior angles, which are approximately square; 11 to 16 mm in length; QLD, NSW, VIC, SA, WA .... ***Frenchella* Blackburn**
– Abdominal ventrites 2, 3 and 4 without or with very reduced, sparse setae in middle, with well developed setae only at sides (Fig. 31.120); pronotum without continuous posterior margin or with fine margin in middle only; lateral edges of pronotum more evenly curved and convergent to obtuse posterior angles; 10 to 14 mm in length; QLD, NSW, NT .... ***Neso* Blackburn**
31(25). Antennal club 5- or 6-lamellate **32**
– Antennal club 3-lamellate **34**

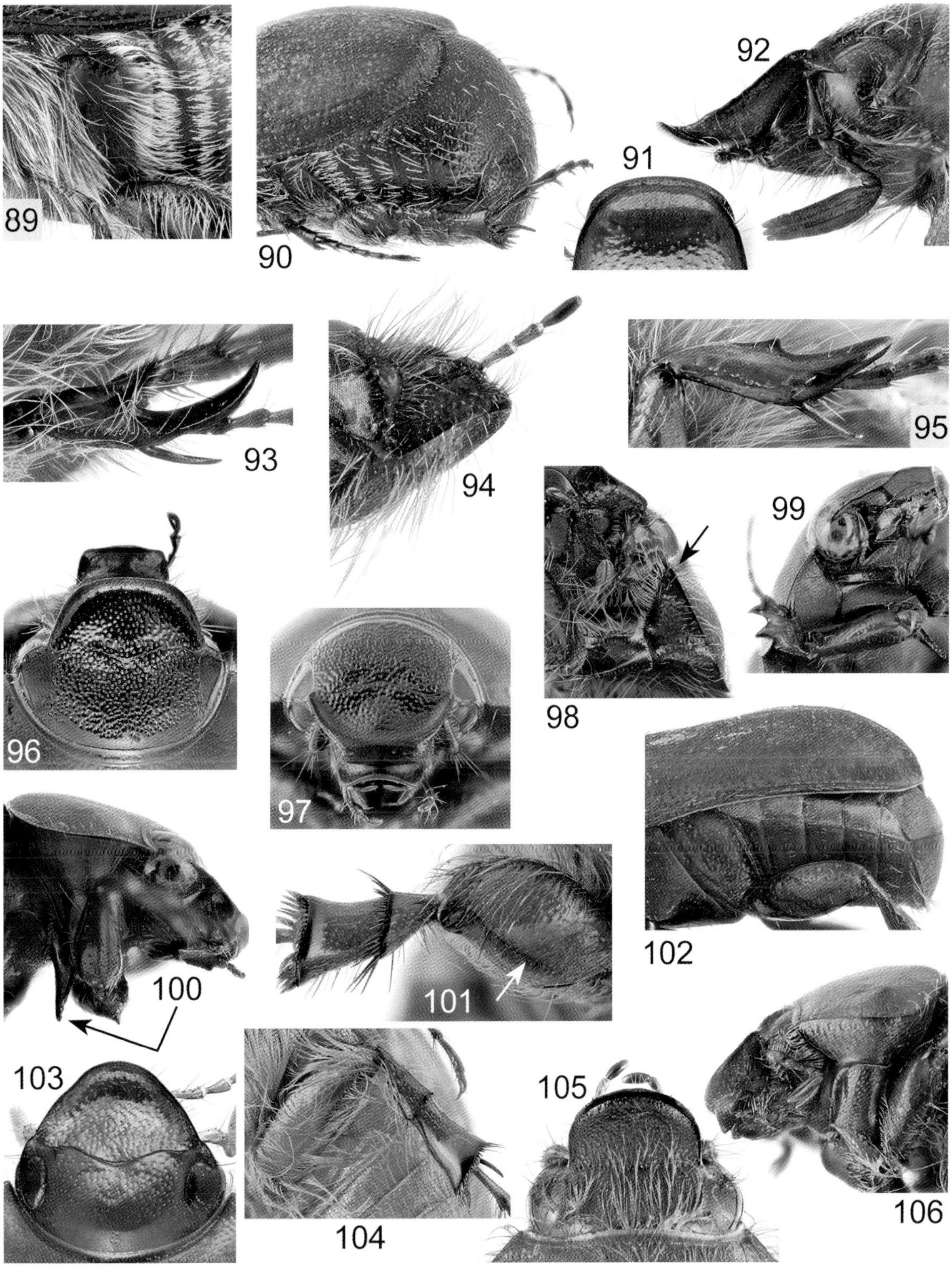

Figs 31.89–31.106. Melolonthinae: Liparetrini. 89, *Liparetrus lepidopygus* Lea, hind leg; 90, *Liparetrus hirpex* Britton, abdominal apex; 91, *Aphanesia greyi* Britton, labrum and clypeus; 92, *Aphanesia greyi* Britton, head, lateral; 93, *Allara insularis* Britton, protibia; 94, *Allara insularis* Britton, head, lateral; 95, *Dysphanochila pilosipennis* Blackburn, protibia; 96, *Glossocheilifer labialis* Blackburn, head, dorsal; 97, *Glossocheilifer labialis* Blackburn, head, anterolateral; 98, *Chelio liparetroides* Britton, prothoracic hypomeron; 99, *Sciton* sp., prothoracic hypomeron; 100, *Sciton* sp., prothorax, lateral; 101, *Ictigaster ruficollis* (Lea), hind leg; 102, *Scitala hirticeps* (Blackburn) [Scitalini], abdomen, lateral; 103, *Stenochelyne noctis* Britton, male head, dorsal; 104, *Leonotus pilosicollis* (Lea), hind leg; 105, *Leonotus pilosicollis* (Lea), head, dorsal; 106, *Luftius parva* (Szito), prothorax, lateral.

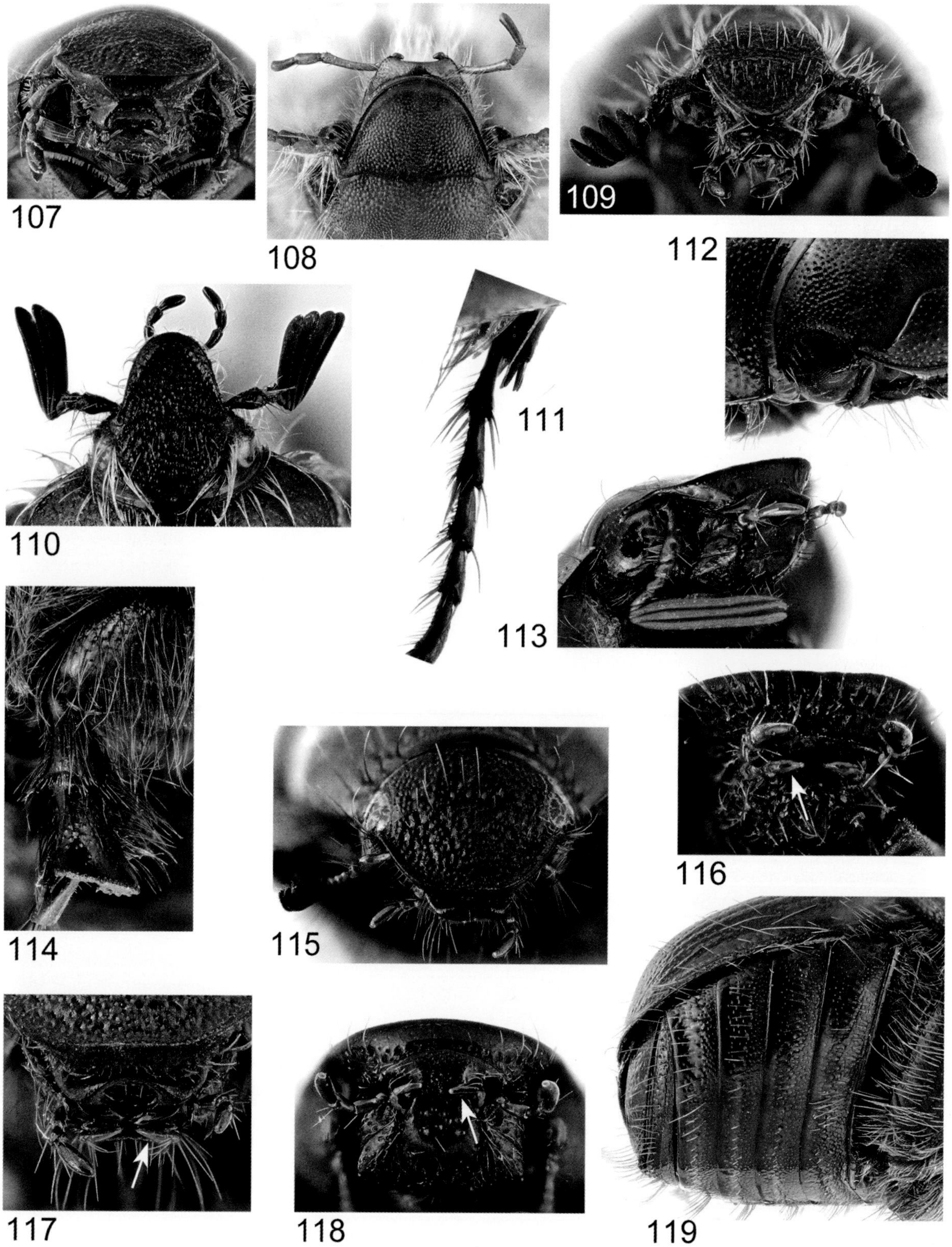

Figs 31.107–31.119. Melolonthinae: Liparetrini. 107, *Luftius parva* (Szito), head, frontal; 108, *Xyridea hirticollis* (Blackburn), head, dorsal; 109, *Dikellites abditus* Britton, male head, frontal; 110, *Dikellites abditus* Britton, male head, dorsal; 111, *Petinopus aegrotus* Blackburn, mesotarsus; 112, *Petinopus aegrotus* Blackburn, eye; 113, *Sericesthis geminata* Boisduval [Scitalini], antenna; 114, *Hadropechys victoriae* (Blackburn), metatibia; 115, *Colobostoma hirsuta* (Frey), head, anterodorsal; 116, *Colobostoma hirsuta* (Frey), labial palps; 117, *Frenchella hirticollis* Blackburn, labial palps; 118, *Neso planicollis* Blackburn, labial palps; 119, *Frenchella hirticollis* Blackburn, abdominal ventrites.

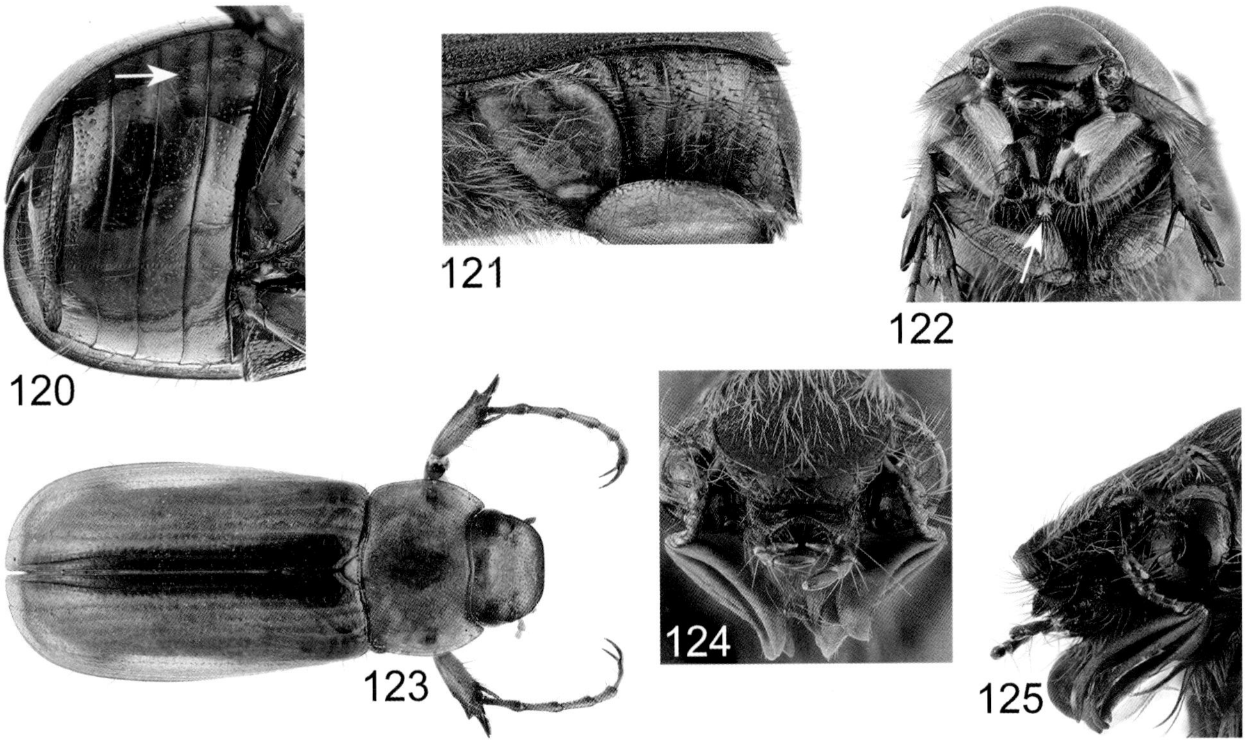

Figs 31.120 31.125. Melolonthinae: Liparetrini. 120, *Neso flavipennis* (Macleay), abdominal ventrites; 121, *Biphyllocera kirbyana* White, metacoxa and abdominal ventrites; 122, *Biphyllocera kirbyana* White, prosternum; 123, *Teluroides* sp., habitus; 124, *Leonotus pilosicollis* (Lea), head, frontal; 125, *Leonotus pilosicollis* (Lea), head, lateral.

32(31). Antennae 10-segmented; antennal club 5- or 6-lamellate; dorsal surfaces of head, pronotum and elytra sparsely clothed with very long (~1 mm) erect setae (Pl. 64L); 11 to 12 mm in length; NSW.......... ***Paronyx* Britton**

– Antennae 9-segmented; antennal club 5- or 6-lamellate; dorsal surfaces of head, pronotum and elytra without sparse, long erect setae.......... **33**

33(32). Disc of pronotum and elytra densely (~100 per square mm) and uniformly clothed with pale, short, recumbent setae; metacoxa (Fig. 31.121) longer than metanepisternum (ratio of length of coxa to that of metanepisternum 1.25:1); prosternum with long, slender, cylindrical, spinose process arising from mid-posterior edge (Fig. 31.122); antennal club in male 6-lamellate and fimbriate beneath, that of female 5-lamellate and non fimbriate; body length 14–16 mm; WA.......... ***Biphyllocera* White**

– Disc of pronotum and elytra without clothing of setae; metacoxa shorter than metanepisternum; prosternum with a short spinose process arising from mid-posterior edge; antennal club 5-lamellate, not fimbriate beneath; body length 6–8 mm; WA.......... ***Macleayella* Britton**

34(31). Pretarsal claws curved, each with pale, membranous pulvillus beneath; lamellae of antennal club in both male and female very short (less than 0.4 mm) and thick (ratio of length to width of lamella ~3:1); 10 to 15 mm in length; NSW & QLD.......... ***Nosphisthis* Blackburn**

– Pretarsal claws less curved, without membranous pulvilli; lamellae of antennal club longer (more than 0.6 mm).......... **35**

35(34). Protibiae with only 2 distinct teeth.......... **36**

– Protibiae with 3 teeth.......... **38**

36(35). Clypeus tapered towards apex (Fig. 31.103); posterior angles of pronotum sharply defined; parameres of aedeagus with many setae; 9 to 11 mm in length; WA.......... ***Stenochelyne* Britton (part)**

– Clypeus more or less rounded; posterior angles of pronotum rounded; parameres of aedeagus without setae.......... **37**

37(36). Elytra elongate and narrow (Fig. 31.123); anterior face of labrum equal in depth to face of clypeus; eyes very large; pale testaceous with dark brown sutural interval; 9 mm in length; WA.......... ***Teluroides* Britton**

– Elytra of normal proportions; anterior face of labrum much shallower than face of clypeus; eyes normal; head, pronotum and scutellum black, elytra dark brown; 6 to 8 mm in length; northern QLD.......... ***Anacanthodes* Britton**

38(35). Anterior face of clypeus truncate, smooth, shining, without setiferous punctures in middle (Fig. 31.99); aedeagus with many or just a few setae on parameres.......... **39**

– Anterior face of clypeus rounded, with transverse row of setiferous punctures not interrupted in middle (Figs 31.124, 31.125); aedeagus without setae on parameres.......... **40**

39(38). Prosternum with a usually long, stout, conical, laterally flattened, usually apically setose process arising from mid-posterior margin (Fig. 31.100); parameres of aedeagus bearing many setae; 10 mm in length; QLD, NSW, VIC, SA, WA.......... ***Sciton* Blackburn (part)**

– Prosternum without such a process; parameres of aedeagus bearing only 2 or 3 very short setae; 10 mm in length; NT.......... ***Parasciton* Britton (part)**

40(38). Lateral edges of pronotal disc straight and parallel behind middle; posterior pronotal angles approximately square (Pl. 65F); 9 to 11 mm in length; NSW, QLD .... ***Colobostoma* Blanchard (part)**
– Lateral edges of pronotal disc convergent behind middle; posterior pronotal angles broadly rounded **41**
41(40). Body length 7–8 mm; pronotum without setae on disc (Pl. 61I); protibiae with teeth very acute; southern QLD .... ***Engyopsina* Britton**
– Body length 14–17 mm; entire pronotal disc very densely clothed with long, fine, pale setae in male (Pl. 56H); base of pronotum only clothed with fine setae in female; protibiae with teeth blunt, rounded and black at their apices; SA .... ***Leonotus* Britton (part)**

Allara Britton, 1955 (Pl. 56J)

Type species. *Allara insularis* Britton, 1955.
Australian species. Two.
Distribution. *Allara insularis* Britton was described from the Monte Bello Islands, WA.
References. Britton (1955, 1990).

Anacanthodes Britton, 1990 (Pl. 56B)

Type species. *Anacanthopus inermis* Blackburn 1898.
Anacanthopus Blackburn 1898, not Montandon, 1894.
Australian species. One.
Distribution. Endemic. *Anacanthodes inermis* (Blackburn) is known from northern QLD.
References. Britton (1990).

Aneucomides Blackburn, 1898 (Pl. 58B)

Type species. *Aneucomides coloratus* Blackburn, 1898.
Australian species. One.
Distribution. Endemic. *Aneucomides coloratus* is known from SA and WA.
References. Britton (1990).

Aphanesia Britton, 1990 (Pl. 58A)

Type species. *Aphanesia greyi* Britton, 1990.
Australian species. One.
Distribution. Endemic. *Aphanesia greyi* occurs in southern WA.
References. Britton (1990).

Astibicola Britton, 1990 (Pl. 58L)

Type species. *Astibicola bicolor* Britton, 1990.
Australian species. One.
Distribution. Endemic. *Astibicola bicolor* is known from SA.
References. Britton (1990).

Biphyllocera White, 1841 (Pl. 58H)

Type species. *Biphyllocera kiryana* White, 1841.
Australian species. Two.
Distribution. Endemic. *Biphyllocera fabriciana* White and *B. kirbyana* White both occur in southern WA.
References. Britton (1990).

Cheilo Britton, 1990 (Pl. 60I)

Type species. *Cheilo liparetroides* Britton, 1990.
Australian species. One.
Distribution. Endemic. *Cheilo liparetroides* is known from NT.
References. Britton (1990).

Colobostoma Blanchard, 1851 (Pl. 60E, 65F)

Type species. *Sericesthis rufipennis* Boisduval, 1835.
Platydesmus W. J. Macleay, 1887, not Lucas 1843
Pteroplatydesmus Dalla Torre, 1912
Australian species. Eight.
Distribution. Endemic. *Colobostoma* species occur in NSW and QLD.
References. Britton (1990).

Colpochila Erichson, 1843 (Pl. 8D, 60H)

Type species. *Sericesthis gouldii* Hope, 1842.
Haplonycha Blanchard, 1850
Australian species. 133.
Distribution. Endemic. *Colpochila* is widely distributed in Australia.
Biology. The larva of *C. obesa* Boisduval from TAS was described and illustrated by McQuillan (1985).
References. Britton (1986, 1990); Szito (1994b, 1995).

Dikellites Britton, 1990 (Pl. 61C)

Type species. *Dikellites abditus* Britton, 1990.
Australian species. One.
Distribution. Endemic. *Dikellites abditus* Britton is known from WA.
References. Britton (1990).

Dysphanochila Blackburn, 1898 (Pl. 61A)

Type species. *Dysphanochila pilosipennis* Blackburn, 1898.
Australian species. One.
Distribution. Endemic. *Dysphanochila pilosipennis* Blackburn occurs in southern WA.
References. Britton (1990).

Engyopsina Britton, 1990 (Pl. 61I)

Type species. *Engyops spectans* Blackburn 1898.
Engyops Blackburn 1898 (not Bauer & Bergenstamm, 1890)
Australian species. One.

Distribution. Endemic. *Engyopsina spectans* (Blackburn) occurs in southern QLD.
References. Britton (1990).

Evbrittonia Szito, 1993 (Pl. 56F)

Type species. *Evbrittonia longitarsa* Szito, 1993.
Australian species. One.
Distribution. Endemic. *Evbrittonia longitarsa* occurs in WA.
References. Szito (1993).

Frenchella Blackburn, 1892 (Pl. 62K)

Type species. *Frenchella lubrica* Blackburn, 1892.
Australian species. Seven.
Distribution. Endemic. *Frenchella* species are known from VIC, NSW, QLD, SA and WA.
References. Britton (1990).

Glossocheilifer Blackburn, 1898 (Pl. 59G)

Type species. *Glossocheilifer labialis* Blackburn, 1898.
Australian species. Two.
Distribution. Endemic. *Glossocheilifer labialis* and G. *bidentatus* Lea occur in southern WA.
References. Britton (1990).

Hadropechys Britton, 1990 (Pl. 62H)

Type species. *Pachygastra victoriae* Blackburn, 1897.
Australian species. One.
Distribution. Endemic. *Hadropechys victoriae* (Blackburn) is known from VIC.
References. Britton (1990).

Ictigaster Britton, 1986 (Pl. 62D)

Type species. *Haplonycha ruficollis* Lea, 1917.
Australian species. One.
Distribution. Endemic. *Ictigaster ruficollis* (Lea) is known from SA.
References. Britton (1986, 1990).

Leonotus Britton, 1986 (Pl. 56H)

Type species. *Haplonycha pilosicollis* Lea, 1930.
Australian species. One.
Distribution. Endemic. *Leonotus pilosicollis* (Lea) occurs in SA.
References. Britton (1986, 1990).

Liparetrus Guérin-Méneville, 1831 (Pl. 4A, 6G, 57J)

Type species. *Liparetrus discipennis* Guérin-Méneville, 1831.
Liparetra Hope, 1837
Australian species. 239.
Distribution. Endemic. *Liparetrus* species are widely distributed in Australia.
Biology. Larvae of a TAS species was described by McQuillan (1985).
References. Britton (1980, 1990).

Luftius Özdikmen, 2009 (Pl. 62G)

Type species. *Colpomorpha parva* Szito, 1994.
Colpomorpha Szito, 1994, not Meyrick, 1929
Australian species. One.
Distribution. Endemic. *Luftius parva* (Szito) is known from WA.
References. Szito (1994a), Özdikmen (2009).

Macleayella Britton, 1990 (Pl. 57K)

Type species. *Macleayia singularis* Blackburn, 1888.
Macleayia Blackburn 1888, not Haswell, 1830.
Australian species. One.
Distribution. Endemic. *Macleayella singularis* (Blackburn) is known from WA.
References. Britton (1990).

Neso Blackburn, 1898 (Pl. 63I)

Type species. *Neso usta* Blackburn 1898.
Australian species. Six.
Distribution. Endemic. Species of *Neso* are known from NSW, QLD and NT.
References. Britton (1990).

Nosphisthis Blackburn, 1898 (Pl. 63G)

Type species. *Nosphisthis parvicornis* Blackburn, 1898.
Australian species. Two.
Distribution. Endemic. *Nosphisthis parvicornis* Blackburn occurs in NSW and *N. perkinsi* Blackurn in QLD.
References. Britton (1990).

Pachygastra Germar, 1848 (Pl. 63B)

Type species. *Pachygastra tasmanica* Germar, 1848.
Australian species. One.
Distribution. Endemic. *Pachygastra tasmanica* occurs in SA.
References. Britton (1990).

Parasciton Britton, 1990

Type species. *Parasciton inermis* Britton, 1990.
Australian species. One.
Distribution. Endemic. *Parasciton inermis* is known from NT.
References. Britton (1990).

Paronyx Britton, 1990 (Pl. 64L)

Type species. *Paronyx setifera* Britton, 1990.
Australian species. One.
Distribution. Endemic. *Paronyx setifera* is known from NSW.
References. Britton (1990).

Petinopus Blackburn, 1898 (Pl. 64A)

Type species. *Petinopus aegrotus* Blackburn, 1898.
Australian species. One.
Distribution. Endemic. *Peltinopus aegrotus* is known from NT.
References. Britton (1990).

Scitaloides Britton, 1990 (Pl. 65K)

Type species. *Scitaloides malanda* Britton, 1990.
Australian species. One.
Distribution. Endemic. *Scitaloides malanda* is known from northern QLD.
References. Britton (1990).

Sciton Blackburn, 1892 (Pl. 65J)

Type species. *Sciton ruber* Blackburn, 1892.
Australian species. Four.
Distribution. Endemic. Species of *Sciton* are known from QLD, NSW, VIC, SA and WA.
References. Britton (1990).

Stenochelyne Britton, 1990 (Pl. 65I)

Type species. *Stenochelyne noctis* Britton, 1990.
Australian species. One.
Distribution. Endemic. *Stenochelyne noctis* is known from southern WA.
References. Britton (1990).

Teluroides Britton, 1990 (Fig. 31.123)

Type species. *Telura suturalis* Lea, 1920.
Australian species. Two.
Distribution. Endemic. *Teluroides suturalis* (Lea) and *T. clypealis* (Lea) were both described from WA.
References. Britton (1990).

Xyridea Britton, 1990 (Pl. 66A)

Type species. *Aneucomides hirticollis* Blackburn, 1912.
Australian species. One.
Distribution. Endemic. *Xyridea hirticollis* (Blackburn) is known from WA.
References. Britton (1990).

Tribe Macrodactylini Kirby, 1837

Characteristics. Metatibiae with 2 very short spurs placed below the tarsal articulation. Tarsal claws unequal, bifid, without a tooth on the concave side, capable of independent movement in one plane. Abdomen with 6 visible ventrites, fifth longer than fourth. Penultimate abdominal ventrite and tergite partly separated by a suture extending from spiracle to posterior edge. Dorsal surfaces dull yellowish brown and densely setose.
Distribution. Neotropical and Nearctic.
Australian taxa. One introduced genus.

Plectris Saint-Fargeau & Serville, 1828 (Pl. 64F)

Type species. *Plectris tomentosa* Saint-Fargeau & Serville, 1828.
Australian species. One.
Distribution. *Plectris* species are endemic to the Neotropical region, with one species extending into the south-eastern USA and introduced into NSW.
Biology/Ecology. *Plectris aliena* Chapin is known to damage roots of grasses and crops in north-east coastal NSW.
References. Böving (1936); Roberts (1968); Katovich (2008).

Tribe Maechidiini Burmeister, 1855

Characteristics. Metatibiae with 2 spurs placed one above and one below tarsal articulation, sometimes with only one spur. Tarsal claw without a tooth on the concave side. Sometimes with a pale membranous pulvillus beneath each claw. Labrum and clypeus fused. Metacoxae strongly transverse and narrow. Penultimate abdominal tergite and ventrite wholly or partly fused. Abdominal ventites with or without a longitudinal ridge on each side. Antennae 8 or 9 segmented with a 3–5 lamellate club.
Distribution. Australia.
Australian taxa. Six genera.

Key to the genera of Maechidiini

(modified from Britton 1957)

1. Protibia with only 1 tooth (apical; Fig. 31.126); disc of frons, pronotum and elytra without setae (Pl. 63C); propygidium nearly fully exposed in intact specimen; abdominal ventrites without a longitudinal ridge on each side (Fig. 31.127); metafemora strongly inflated; metatibiae strongly expanded apically, with a transverse spinose ridge on both upper and lower sides; form not depressed; 7 to 10 mm in length; south-west WA .... ***Microthopus* Burmeister**

– Protibia with 2 or 3 teeth; disc of frons, pronotum and elytra usually bearing setae; propygidium usually completely covered by elytra in intact specimen; abdominal ventrites usually with a longitudinal ridge on each side; metafemora slender; metatibiae not strongly expanded apically, without a transverse spinose ridge or with the ridge only on the upper side; form depressed or not **2**

2(1). Labrum situated under clypeus and separated by a semicircular suture; sutural edge of labrum fringed with short, stout spines (Fig. 31.128); pronotum without well defined lateral edges and margins (Fig. 31.129); metatibial apex of male with elongate process on inner side bearing a long flattened spur (Fig. 31.130); female with just long flattened spur (Fig. 31.131); form depressed; 5.5 to 9 mm in length; south-west WA .... ***Termitophilus* Britton**

– Labrum and clypeus fused; labrum without a row of short, stout spines; pronotum with well defined lateral edges and margins; metatibial apex normal, with 1 or 2 spurs, spurs not flattened; form depressed or not **3**

3(2). Protibia with 2 elongated teeth (Fig. 31.132); pronotum highly convex with crenulate lateral margins (Fig. 31.132); metatibiae without a transverse spinose ridge (Fig. 31.133); claws without pulvilli (Fig. 31.132); antennae 8 segmented; form not unusually depressed; 6.5 mm in length; northern SA .... ***Harpechys* Britton**

– Protibia with 3 teeth, first two elongated or not; pronotum not highly convex, lateral edges crenulated or not; metatibiae with or without transverse spinose ridge; claws with or without pulvilli; antennae 7, 8 or 9 segmented; form depressed or not **4**

4(3). Pronotal hypomera more or less produced forward as a flange to form pockets for reception of antennae (Fig. 31.134), the free edge angulate or emarginate anteriorly; antennae 7, 8 or 9 segmented; claws with pulvilli or not; 5 to 13 mm in length; widespread – all Australian States except TAS .... ***Maechidius* Macleay**

– Pronotal hypomera not produced forward, at most a straight ridge where it meets prosternum, not forming pockets for reception of antennae (Fig. 31.135); antennae 8 or 9 segmented; claws with pulvilli or not **5**

5(4). Claws with pulvilli (Fig. 31.136); protibiae with first two teeth not elongated (Fig. 31.137); metatibiae with transverse ridge slight or absent; setae or scales on disc of elytra (Pl. 61J) often arranged in longitudinal bands with relatively narrow spaces between them, striae not obvious; antennae 9 segmented; 5 to 9 mm in length; QLD, NSW, NT .... ***Epholcis* Waterhouse**

– Claws without pulvilli; protibiae with first two teeth elongated (Fig. 31.138); metatibiae with an obvious transverse ridge on the upper side (Fig. 31.139); setae on disc of elytra not in bands, striae obvious; antennae 8 segmented; 3 to 3.5 mm in length; WA .... ***Microcoenus* Britton**

Epholcis Waterhouse, 1875 (Pl. 61J)

Type species. *Epholcis divergens* Waterhouse, 1875.
Australian species. Five.
Distribution. Endemic. Species of *Epholcis* occur in NSW, QLD and NT.
References. Britton (1957).

Harpechys Britton, 1957 (Pl. 62B)

Type species. *Harpechys chilo* Britton, 1957.
Australian species. One.
Distribution. Endemic. *Harpechys chilo* was described from the River Diamantina, northern SA.
References. Britton (1957).

Maechidius W. S. Macleay, 1819 (Pl. 59L)

Type species. *Trox spurius* Kirby, 1818.
Geobatus Boisduval, 1835
Australian species. 70.
Distribution. Endemic. *Maechidius* species are widely distributed throughout continental Australia.
Biology. *Maechidius chadwicki* Britton has been collected in nests of *Coptotermes lacteus* Froggatt.
References. Britton (1957, 1963).

Microcoenus Britton, 1957 (Pl. 63E)

Type species. *Microcoenus nanus* Britton, 1957.
Australian species. One.
Distribution. Endemic. *Microcoenus nanus* occurs in WA.
References. Britton (1957).

Microthopus Burmeister, 1855 (Pl. 63C)

Type species. *Microthopus castanopterus* Burmeister, 1855.
Australian species. One.
Distribution. Endemic. *Microthopus castanopterus* occurs in southern WA.
References. Britton (1957).

Termitophilus Britton, 1957 (Pl. 66I)

Type species. *Termitophilus spadix* Britton, 1957.
Australian species. One.
Distribution. Endemic. *Termitophilus spadix* is known from southern WA.
Biology. This species has been collected in mounds of *Nasutitermes exidiosus* (Hill).
References. Britton (1957).

Tribe Melolonthini Leach, 1819

Characteristics. Metatibiae with 2 spurs placed below the tarsal articulation. Tarsal claws almost always with a tooth on the concave side. Labrum separate from clypeus and located below it. Antennae 9 or 10 segmented with a 3–8 lamellate club. Body surface not metallic.
Distribution. World-wide.
Australian taxa. 16 genera.

Key to the Australian genera of Melolonthini

(modified from Britton 1978; *Megacoryne* added from Britton 1987b)

The genera *Lepidiota, Antitrogus* and *Metatrogus* appear more than once, mainly to allow for possible alternate decisions on scales and setae in couplet 12.

1. Antennae 9-segmented; antennal club 3-lamellate **2**

– Antennae 10-segmented; antennal club with 3- to 8-lamellate **4**

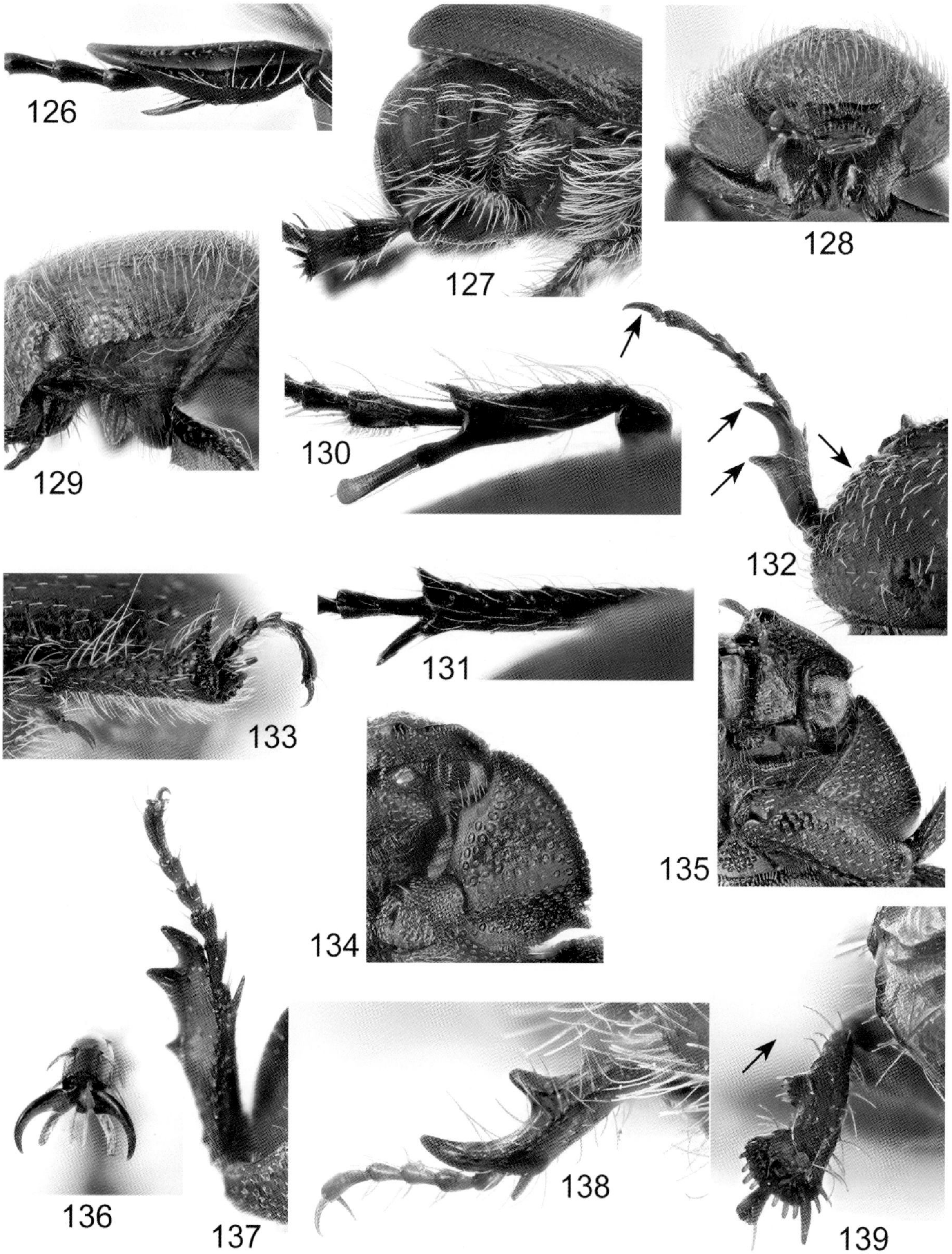

Figs 31.126–31.139. Melolonthinae: Maechidiini. 126, *Microthopus castanopterus* Burmeister, protibia; 127, *Microthopus castanopterus* Burmeister, abdomen, lateral; 128, *Termitophilus spadix* Britton, head, frontal; 129, *Termitophilus spadix* Britton, pronotum, lateral; 130, *Termitophilus spadix* Britton, male hind tibia; 131, *Termitophilus spadix* Britton, female hind tibia; 132, *Harpechys chilo* Britton, front leg and lateral margin of pronotum; 133, *Harpechys chilo* Britton, hind tibia; 134, *Maechidius hopeanus* Westwood, prothoracic hypomeron; 135, *Epholcis divergens* Waterhouse, prothoracic hypomeron; 136, *Epholcis divergens* Waterhouse, pretarsal claws and pulvili; 137, *Epholcis divergens* Waterhouse, protibia, ventral; 138, *Microcoenus nanus* Britton, front tibia; 139, *Microcoenus nanus* Britton, hind tibia.

2(1). Dorsal surfaces clothed with broadly ovate, sometimes sparse, white scales (Fig. 31.140) .... ***Lepidiota* Kirby (part)**

– Dorsal surfaces clothed with inclined, fine, acute, yellowish setae or without setae or scales **3**

3(2). Metafemur strongly distended; metatibia strongly expanded at apex (Fig. 31.141); metatarsomere 1 as long as 2 and 3 combined; clypeus semicircular in outline (Fig. 31.142); pretarsal claws with minute tooth at base (Fig. 31.143); posterior angles of pronotum rounded (Pl. 62A); 16 mm in length; northern NT .... ***Holorhopaea* Britton**

– Metafemur not unusually distended; metatibia not strongly expanded at apex (Fig. 31.144); metatarsomere 1 as long as 2 only; clypeus strongly transverse (Fig. 31.145); pretarsal claws with small tooth near base (Fig. 31.146); Posterior angles of pronotum prominent; 10 mm in length; northern WA .... ***Nanorhopaea* Britton**

4(1). Pretarsal claws without a tooth; antennal club composed of 8 long, equal lamellae; frontoclypeal suture raised and angulate laterally (Fig. 31.147); 14 to 17 mm in length; central Australia: SA, NT .... ***Zietzia* Blackburn**

– Pretarsal claws with a tooth, either prominent and near middle or small and close to base; antennal club 3- to 8-lamellate; frontoclypeal suture not raised or angulate **5**

5(4). Antennal club 3-lamellate **6**

– Antennal club 4- to 8-lamellate **8**

6(5). Anterior face of clypeus deep (ratio of greatest width to mid depth less than 5); and bearing numerous scattered setiferous punctures (Fig. 31.148); lamellae of antennal club in male at least as long as remainder of antenna .... ***Antitrogus* Blackburn (part)**

– Anterior face of clypeus shallow (ratio of greatest width to mid depth more than 7); and with single line of setiferous punctures across middle or with shining unpunctured area in middle (Fig. 31.149); lamellae of antennal club in male usually shorter than remainder of antenna **7**

7(6). Anterior edge of clypeus, as seen from above the head, concave (Fig. 31.150) or straight in middle; clypeus of various shapes; body usually bearing adpressed white scales (Pl. 6H, 62I) of various sizes and shapes (the scales sometimes minute and contained within the punctures); 11 to 38 mm in length; widespread northern WA, NT, QLD, eastern Australia to northern NSW .... ***Lepidiota* Kirby (major part)**

– Anterior edge of clypeus, as seen from above the head, uniformly convex or straight, transverse, sharply curved at each side and constricted at base towards ocular canthi (Fig. 31.151); body with very fine scales; 22 to 25 mm in length; northern QLD .... ***Metatrogus praeceps* Britton**

8(5). Junctions between abdominal ventrites 1–5 as obvious at middle as at sides; antennal club of 6, 7 or 8 long lamellae **9**

– Junctions between abdominal ventrites much less obvious at middle than at sides, some or all of the ventrites being connate or partly fused in middle (Fig. 31.152); antennal club often with fewer than 6 lamellae **12**

9(8). Tooth on pretarsal claws minute and very close to base; elytra with sparse white scales; clypeus, frons and pronotum clothed with long backwardly directed white setae; antennal club 7-lamellate; 7 mm in length; northern NT .... ***Microrhopaea* Lea**

– Tooth on pretarsal claws obvious (Fig. 31.153); body length 12–17 mm; elytra without white scales; if clypeus, frons and pronotum have long setae, then these are brown **10**

10(9). Antennal club of male 8-lamellate; inner edges of lamellae serrate; metafemur (Fig. 31.154) strongly inflated (ratio of length to greatest width ~1.6); elytral striae indistinct (Pl. 62E); 12 to 14 mm in length; northern WA .... ***Megacoryne* Britton**

– Antennal club of male 6- or 7-lamellate; inner edges of lamellae not serrate; metafemur not unusually inflated (ratio of length to greatest width ~2.5 or 3.5); at least a few elytral striae obvious **11**

11(10). Metatibial spurs short, broad and flattened (Fig. 31.155); surface of pronotum without long, erect setae (Pl. 63L); pygidium uniformly clothed with short setae (Fig. 31.155); antennal club 6-lamellate; 13 to 17 mm in length; NSW, southern QLD .... ***Othnonius* Olliff**

– Metatibial spurs long and uniformly tapered (Fig. 31.156); surface of pronotum bearing long, erect setae, at least on anterior and posterior margins (Pl. 60B); pygidium either shagreened with sparse very short setae or densely punctate with adpressed white scales (Fig. 31.156); antennal club 6- or 7-lamellate, 11 to 16 mm in length, northern QLD .... ***Allothonius* Britton**

12(8). Pronotal hypomera, anterior faces of profemora and protibiae and/or pronotum and scutellum setose (Fig. 31.158) or bare, but without white scales or whitish, flattened, adpressed setae **13**

– Pronotal hypomera, anterior faces of profemora and protibiae and/or pronotum and scutellum bearing AT LEAST A FEW broad or elongate, adpressed white scales or flattened adpressed setae **16**

13(12). Body densely clothed with recumbent setae, disc of pronotum may have longer setae as well (Fig. 31.157); posterior pronotal angles often rounded; 19 to 30 mm in length; QLD, NSW, ACT, VIC, SA .... ***Rhopaea* Erichson**

– Dorsal surfaces bare or sparsely clothed with setae, if appearing recumbent on elytra then disc of pronotum with only long setae; posterior pronotal angles obvious **14**

14(13). Antenna with 7-lamellate club; upper surface of clypeus less transverse (ratio of greatest width to mid length less than 2.3), the surface usually strongly concave (Fig. 31.159); labrum strongly projecting, deflexed; 15 to 26 mm in length; central and western Australia: SA, WA .... ***Pararhopaea* Blackburn**

– Antenna with 3-, 6- or 7-lamellate club, if 7-lamellate, then clypeus transverse (ratio of greatest width to mid length usually more than 2.5) and the surface not strongly concave; labrum not strongly projecting and not deflexed **15**

15(14). Anterior face of clypeus deep (ratio of greatest width to mid length less than 6) and bearing numerous scattered setiferous punctures (Fig. 31.148); 16 to 26 mm in length; eastern Australia: QLD, NSW, ACT, VIC, TAS .... ***Antitrogus* Blackburn (major part)**

– Anterior face of clypeus shallow and wide (ratio of greatest width to mid length ~10) and with single row of setiferous punctures across middle or without punctures (Fig. 31.160); 23 to 30 mm in length; southern QLD, northern NSW .... ***Metatrogus* Britton (part)**

16(12). Anterior edge of metafemur concave near base (Fig. 31.161); body colour black or dark brown, with few or many white, elongate, adpressed, scale-like setae on dorsal surface (Pl. 61K); length 23–33 mm **17**

– Anterior edge of metafemur not concave near base (Fig. 31.162) **18**

17(16). Frons and anterior margin of pronotum bearing long, yellowish, backwardly curved setae (Fig. 31.163); 24 to 33 mm in length; northern QLD .... ***Dermolepida* Arrow**

– Frons and anterior margin of pronotum without long setae; 23 to 30 mm in length; southern QLD, northern NSW .... ***Metatrogus* Britton (part)**

Figs 31.140–31.152. Melolonthinae: Melolonthini. 140, *Lepidiota rothei* Blackburn, habitus; 141, *Holorhopaea sagata* Britton, hind leg; 142, *Holorhopaea sagata* Britton, head, dorsal; 143, *Holorhopaea sagata* Britton, pretarsal claws; 144, *Nanohropaea testacea* Britton, hind leg; 145, *Nanohropaea testacea* Britton, head, dorsal; 146, *Nanohropaea testacea* Britton, pretarsal claws; 147, *Zeitzia geologa* Blackburn, head, dorsal; 148, *Antitrogus gubbi* Allsopp, head, ventral; 149, *Lepidiota sororia* Moser, head, frontal; 150, *Lepidiota sororia* Moser, head, dorsal; 151, *Metatrogus praeceps* Britton, head, dorsal; 152, *Rhopaea magnicornis* Blackburn, abdomen.

18(16). Antennal club 8-lamellate in male (maximum length of lamellae 4.0–5.6 mm), 7-lamellate in female (maximum length of lamellae ~2 mm); labrum projecting far beyond clypeus (Pl. 62L, Fig. 31.164); 33 to 37 mm in length; semi-desert: southern SA, southern WA .... ***Megarhopaea* Britton**

– Antennal club of 7 or fewer full length lamellae; labrum much less projecting; 30 mm or less in length **19**

19(18). Antennal club in both sexes with 6 or 7 lamellae (length of lamella of segment 5 at least 0.7 times length of lamella of segment 6) **20**

– Antennal club in both sexes with 5 or fewer lamellae (length of lamella of segment 5 at most 0.6 times length of lamella of segment 6) **22**

20(19). Labrum deeply excavated anteriorly (Fig. 31.165); male antennal club with 6 or 7 full length lamellae; pygidium truncate apically (Fig. 31.166); upper surface of clypeus deeply concave; 11 to 20 mm in length; northern Australia: WA. QLD .... ***Pseudoholophylla* Blackburn**

– Labrum not or only slightly excavated anteriorly; male antennal club with 6 full length lamellae; pygidium rounded apically; upper surface of clypeus concave or not **21**

21(20). Pronotum without a well defined anterior margin; terminal abdominal ventrite of male truncate and lobed, visible from above at sides (Fig. 31.168); frons and clypeus with long setae; head, pronotum, scutellum and elytra reddish brown; 16 to 20 mm in length; central QLD .... ***Antitrogus setifrons* Britton**

– Pronotum with a well defined anterior margin; terminal abdominal ventrite of male rounded, not visible from above (Fig. 31.170); frons and clypeus without long setae; head, pronotum and scutellum dark brown/black, elytra pale yellowish-brown (Pl. 59E); 12 mm in length; northern QLD .... ***Hypolepida* Britton**

22(19). Anterior face of clypeus deep (ratio of greatest width to mid depth less than 5) and with numerous scattered setiferous punctures (Figs 31.167, 31.171); disc of pronotum bare or with very short setae; male antennal club of 3 to 5 full length lamellae; 20 mm in length; QLD, NSW .... ***Antitrogus* Britton (part)**

– Anterior face of clypeus shallow (ratio of greatest width to mid depth more than 7) and with a single transverse line of setiferous punctures or a bare (Figs 31.172, 31.173), smooth, median area; disc of pronotum with rounded or elongate white scales; male antennal club of 5 full length lamellae; 17 to 20 mm in length .... ***Lepidota ciliata* Britton** (NSW) & ***Lepidiota quinaria* Britton** (Northern QLD)

Allothnonius Britton, 1978 (Pl. 60B)

Type species. *Allothnonius brooksi* Britton, 1978.
Australian species. Two.
Distribution. Endemic. *Allothnonius brooksi* and *A. barretti* Britton occur in northern QLD.
References. Britton (1978).

Antitrogus Burmeister, 1855 (Pl. 58I)

Type species. *Antitrogus tasmanicus* Burmeister, 1955.
Australian species. 23.
Distribution. Endemic. *Antitrogus* species occur from TAS to northern QLD and in SA.
Biology. This genus includes several major sugarcane pests, as well as pasture pests, with larvae feeding on roots of grasses and related monocots. Adult *Antitrogus* fly in spring and early summer following rain; adults do not feed and females are poor dispersers. Mating occurs near emergence sites, and males fly little, are attracted by pheromones but not to light.
References. Britton (1978); Allsopp (2003); Allsopp & Lambkin (2006); Miller & Allsopp (2000).

Dermolepida Arrow, 1941 (Pl. 61K)

Type species. *Lepidoderma albohirtum* Waterhouse 1875. *Lepidoderma* Waterhouse 1875, not Reuss, 1855
Australian species. Two.
Distribution. *Dermolepida* species occur in eastern Indonesia, Papua New Guinea and northern QLD.
References. Britton (1978); Miller & Allsopp (2000).

Holorhopaea Britton, 1978 (Pl. 62A)

Type species. *Holorhopaea sagata* Britton, 1978.
Australian species. One.
Distribution. Endemic. *Holorhopaea sagata* is known from NT.
References. Britton (1978).

Hypolepida Britton, 1978 (Pl. 59E)

Type species. *Hypolepida wilsoni* Britton, 1978.
Australian species. One.
Distribution. Endemic. *Hypolepida wilsoni* Britton occurs in northern QLD.
References. Britton (1978).

Lepidiota Kirby, 1828 (Pl. 6H, 62I)

Type species. *Melolontha stigma* Fabricius 1798.
Australian species. 61.
Distribution. The genus *Lepidiota* is distributed from India to China and south to the Philippines, Indonesia, New Guinea and Australia. Australian species are known from NSW, QLD, NT and WA.
Biology. Several *Lepidiota* species are pests, especially of sugarcane.
References. Britton (1978, 1985); Allsopp (1989b, 1990b, 1999); Allsopp & Watkins (1995); Miller & Allsopp (2000).

Megacoryne Britton, 1987 (Pl. 62E)

Type species. *Megacoryne carnabyorum* Britton, 1987.
Australian species. One.
Distribution. Endemic. *Megacoryne carnabyorum* occurs in northern WA.
References. Britton (1987b).

Megarhopaea Britton, 1978 (Pl. 62L)

Type species. *Pararhopaea gigas* Lea, 1916.
Australian species. One.

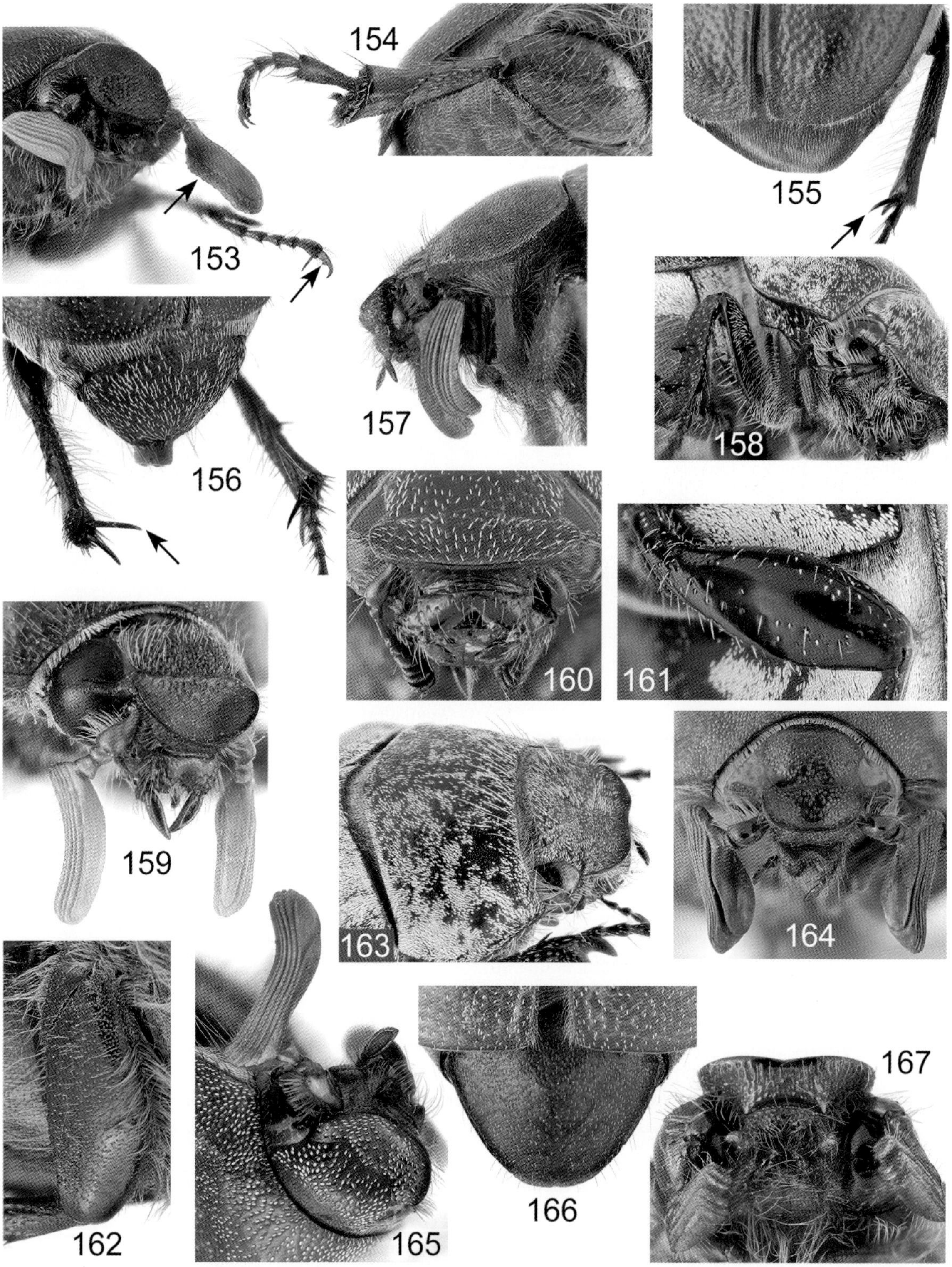

Figs 31.153–31.167. Melolonthinae: Melolonthini. 153, *Megacoryne carnabyorum* Britton, head and protarsus; 154, *Megacoryne carnabyorum* Britton, hind leg; 155, *Othnonius batesi* Oliff, pygidium and hind leg; 156, *Allothnonius brooksi* Britton, pygidium and hind leg; 157, *Rhopaea magnicornis* Blackburn, pronotum, lateral; 158, *Dermolepida albohirtum* (Waterhouse), head and pronotum lateral; 159, *Pararhopaea callbonensis* (Blackburn), head; 160, *Metatrogus lukei* Allsopp, head, frontal; 161, *Dermolepida albohirtum* (Waterhouse), metafemur; 162, *Megarhopaea gigas* (Lea), metafemur; 163, *Dermolepida albohirtum* (Waterhouse), head and pronotum; 164, *Megarhopaea gigas* (Lea), head, anterodorsal; 165, *Pseudoholophylla cavifrons* (Lea), head, anterolateral; 166, *Pseudoholophylla cavifrons* (Lea), pygidium; 167, *Antitrogus setifrons* Britton, head, ventral.

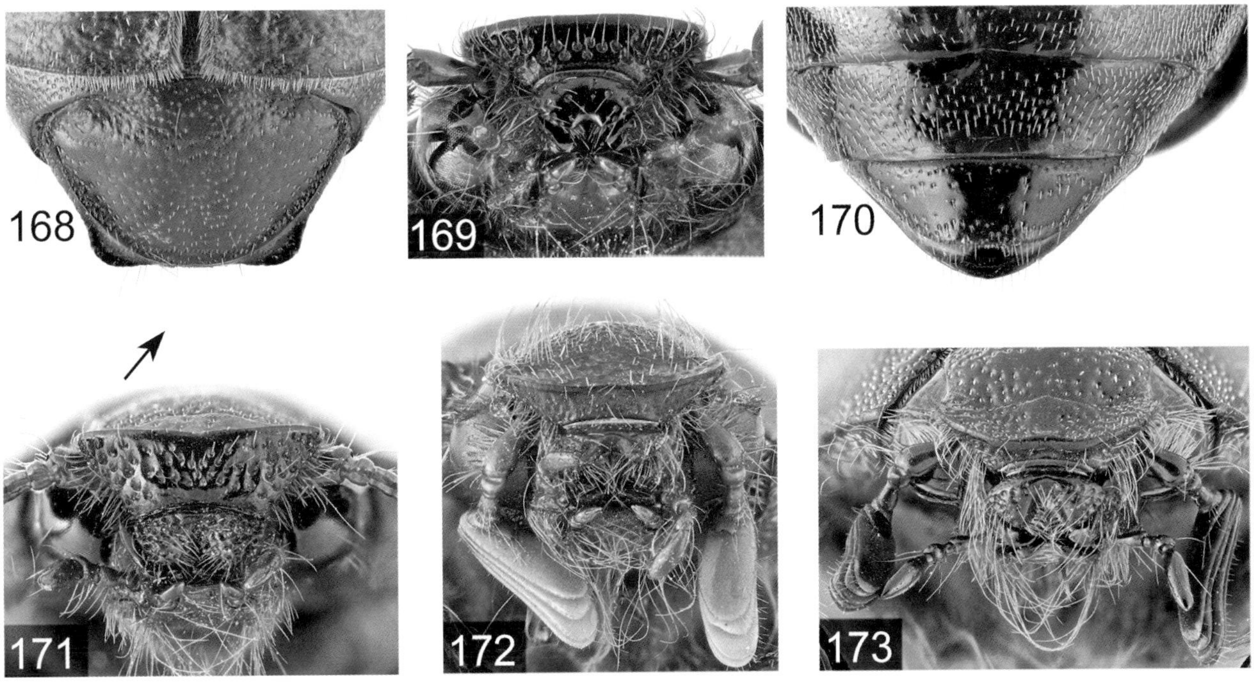

Figs 31.168–31.173. Melolonthinae: Melolonthini. 168, *Antitrogus setifrons* Britton, pygidium; 169, *Hypolepida wilsoni* Britton, head, ventral; 170, *Hypolepida wilsoni* Britton, abdominal apex; 171, *Antitrogus nox* Britton, head, frontal; 172, *Lepidiota ciliata* Britton, head, frontal; 173, *Lepidiota quinaria* Britton, head, frontal.

Distribution. Endemic. *Megarhopaea gigas* (Lea) is known from SA and WA.
References. Britton (1978).

Metatrogus Britton, 1978 (Pl. 63K)

Type species. *Metatrogus septuosus* Britton, 1978.
Australian species. Four.
Distribution. Endemic. Species of *Metatrogus* are known from NSW and QLD.
References. Britton (1978), Allsopp (1999).

Microrhopaea Lea, 1920

Type species. *Microrhopaea flavipennis* Lea, 1920.
Australian species. One.
Distribution. Endemic. *Microrhopaea flavipennis* is known from NSW.
References. Britton (1978).

Nanorhopaea Britton, 1978

Type species. *Nanorhopaea testacea* Britton, 1978.
Australian species. One.
Distribution. Endemic. *Nanorhopaea testacea* occurs in northern WA.
References. Britton (1978).

Othnonius Olliff, 1891 (Pl. 63L)

Type species. *Othnonius batesii* Olliff, 1891.
Australian species. One.
Distribution. Endemic. *Othnonius batesii* occurs in NSW and QLD.
References. Britton (1978).

Pararhopaea Blackburn, 1911 (Pl. 64J)

Type species. *Rhopaea callabonnensis* Blackburn, 1894.
Australian species. Four.
Distribution. Endemic. *Pararhopaea* species are known from NSW, QLD, NT, SA and WA.
References. Britton (1978); Allsopp (1990b).

Pseudholophylla Blackburn, 1911 (Pl. 64K)

Type species. *Holophylla furfuracea* Burmeister, 1855.
Holophylla Burmeister 1855, not Erichson, 1847
Paralepidiota Blackburn, 1911
Australian species. Five.
Distribution. Endemic. Species of *Pseudholophylla* are known from northern QLD and northern WA.
References. Britton (1978).

Rhopaea Erichson, 1847 (Pl. 65L)

Type species. *Rhopaea verrauxii* Blanchard, 1951.
Australian species. 10.

Distribution. Endemic. *Rhopaea* species are known from VIC, ACT, NSW, QLD and SA.
Biology/Ecology. Larvae of some *Rhopaea* are pasture pests, especially in northern NSW.
References. Britton (1978); Miller & Allsopp (2000).

Zietzia Blackburn, 1894 (Pl. 66F)

Type species. *Zietzia geologa* Blackburn, 1894.
Australian species. One.
Distribution. Endemic. *Zietzia geologa* is known from SA and NT.
References. Britton (1978).

Tribe Pachytrichini Burmeister, 1855

Characteristics. Metatibia with 2 spurs placed below the tarsal articulation. Tarsal claws with a large tooth on concave side. Labrum deeply emarginated in middle, extended in front of, and at same level as, clypeus. Mouthparts covered with dense woolly pubescence. Metatarsi about twice as long as metatibiae. Body surface not metallic.
Distribution. Australia
Australian taxa. One genus.

Pachytricha Hope, 1841 (Pl. 64E)

Type species. *Pachytricha castanea* Hope, 1841.
Australian species. Five.
Distribution. Endemic. All five Australian *Pachytricha* occur in WA.
References. Allsopp (1990a).

Tribe Phaenognathini Iablokoff-Khnzorian, 1977

Characteristics. Labrum prominent, in same plane as clypeus, tapering to a rounded or pointed apex. Mandibles visible from above, expanded beside and same length as labrum. Clypeus with a prominent horn (males) or carina (females) and covered with long dense setae. Pronotum of males with a shallow anterior excavation, marking an almost vertical surface. Metatarsi of males at least three times as long as metatibiae.
Distribution. Australian and Neotropical.
Australian taxa. One genus.

Phaenognatha Hope, 1842 (Pl. 51B)

Type species. *Phaenognatha erichsoni* Hope, 1841.
Australian species. Eight.
Distribution. Endemic. *Phaenognatha* species occur in central to northern QLD and NT.
Biology. Females are known for a single species, *P. jenae* Allsopp; reduction of the hind wings and metatarsi and inflation of the abdomen suggest that female *Phaenognatha* live below ground, coming to the surface for a short time to mate.
References. Allsopp (1981, 1983).
Note. In most works on scarabaeids, including Smith (2006), *Phaenognatha* and its New World counterpart *Neophaenognatha* Allsopp are placed in the subfamily Aclopinae, but they were excluded from that subfamily by Ocampo & Mondaca (2012). They are here returned to Melolonthinae, where they were originally placed by Lacordaire (1856) pending further study.

Tribe Phyllotocidiini Britton, 1957

Characteristics. Metatibiae with 1 or 2 spurs placed below tarsal articulation or without spurs. Pro- and mesotarsal claws of male with long membranous pulvilli beneath. Labrum transverse, attached to apical margin of clypeus and separated from it by a suture. Prosternum without a setose process. Protibiae with 1 or 2 teeth, including apical. Surfaces of body and legs metallic green or cupreous, at least in part.
Distribution. Australia.
Australian taxa. One genus.

Phyllotocidium Blackburn, 1898 (Pl. 65C)

Type species. *Cheiragra macleayi* Blackburn, 1892.
Australian species. Four.
Distribution. Endemic. Species of *Phyllotocidium* are known from VIC, NSW and WA.
References. Britton (1957).

Tribe Phyllotocini Burmeister, 1855

Characteristics. Metatibiae with 2 spurs placed one above and one below tarsal articulation. Tarsal claws simple, those of meso- and metatarsi sometimes elongate and thin, protarsal claws of male strongly asymmetrical and sometimes toothed. Labrum attached to apical margin of clypeus, either fused to it or separated by a suture. Procoxae more vertical than horizontal, projecting to a distance greater than the width of the coxal cavity. Metacoxae large, usually longer than metanepisternum. Prosternum usually with a long, slender, setose process arising from mid posterior margin.
Distribution. Australia.
Australian taxa. 10 genera.

Key to the Australian genera of Phyllotocini

(modified from Britton 1957 (as Sericini); *Adossa* added from Britton 1995)

1. Metapretarsal claws sharply curved, each with a tooth, membranous process or sclerotised lamina on underside 2
– Metapretarsal claws almost straight or evenly curved, without tooth, process or lamina beneath 3

2(1). Pretarsal claws with membranous pulvilli beneath (Fig. 31.174); labrum and clypeus fused; anterior edge of head as seen from above (Pl. 60F) forming a fairly uniform approximately semicircular curve; 3 to 5 mm in length; NSW & QLD .... ***Cheiragra*** **Macleay**

– Pretarsal claws without pulvilli (Fig. 31.175) but each with sclerotised tooth or lamina beneath; labrum separated from clypeus by transverse suture; anterior edge of head not uniformly curved; 5 to 7 mm in length; nothern QLD .... ***Ancyclonyx*** **Britton**

3(1). Labrum and clypeus fused, not separated by suture (Fig. 31.176); upper surfaces of body (Pl. 65D) clothed with pale, recumbent, flattened setae; maxilla with lacinia toothed on inner side; propretarsal claws of male greatly enlarged and strongly asymmetrical; 4 to 6 mm in length; WA .... ***Sphaeroscelis*** **Burmeister**

– Labrum separated from clypeus by suture (labrum very narrow in *Cheirora*); body bare or with normal erect setae; maxilla with lacinia pubescent, not toothed **4**

4(3). Outline of head approximately semicircular; labrum very narrow, just visible beyond anterior edge of clypeus (Fig. 31.177); pronotum and elytra with uniform, rather sparse clothing of erect setae; elytra tapered from middle to apex; protibia with 1 or 2 teeth (including apical), the basal tooth when present very obtuse; 5 mm in length; northern QLD .... ***Cheirora*** **Britton**

– Outline of head not semicircular; labrum obvious; pronotum and elytra bare or IF with setae THEN protibia with 2 or 3 teeth (including apical).... **5**

5(4). Pronotum almost circular in outline (except for anterior truncation), sides and base united in one almost uniform curve (Pl. 63D); labrum somewhat elongate, sometimes pointed or arched, usually with a median longitudinal ridge (Fig. 31.178); maxillary palps usually elongate; 4 to 7 mm in length; WA, western QLD. .... ***Neophyllotocus*** **Blackburn**

– Pronotum not circular in outline, the posterior angles apparent but sometimes broadly rounded; labrum more transverse, not arched, rarely pointed or ridged; maxillary palps normal **6**

6(5). Posterior pronotal angles distinct, obtuse or square (Figs 31.179, 31.180), not rounded, closely applied to elytral humeri; posterior edge of pronotum slightly convex in middle and straight or slightly concave at each side (Pl. 60D) **7**

– Posterior pronotal angles absent; lateral and posterior edges of pronotal disc meeting in a curve; posterior edge of pronotum uniformly curved and not closely applied to elytral humeri (Pl. 63F) **8**

7(6). Anterior edge of labrum almost straight (Fig. 31.181); metacoxae very large, longer than metanepisternum; elytral disc without setae and not mottled brown or testaceous; mid-posterior process of prosternum long, slender, cylindrical; 4 to 6 mm in length; NSW & southern QLD .... ***Cheirodontus*** **Britton**

– Anterior edge of labrum obviously concave in middle (Fig. 31.182); metacoxae shorter than metanepisternum; elytral disc, especially on lateral declivities, with patches of long, shining, white, adpressed setae and surface mottled brown or testaceous; mid-posterior process of prosternum short, compressed; 7.5 to 10 mm in length; TAS .... ***Adossa*** **Britton**

8(6). Clypeus with transverse ridge near anterior edge (Fig. 31.183); protibia with 2 distinct teeth (including apical); 6 to 7 mm in length; NSW............ .... ***Opsitocus*** **Britton**

– Clypeus without transverse ridge; protibia with 2 or 3 teeth (including apical) **9**

9(8). Surface of frons, clypeus and usually labrum with shallow longitudinal concavity (view from behind parallel to surface, Fig. 31.184); inner propretarsal claw of male divided longitudinally, the inner half usually flattened (Fig. 31.185); frons with setae near anterior edge; 4 to 6 mm in length; NSW & QLD .... ***Anthotocus*** **Britton**

– Surface of frons, clypeus and labrum usually slightly convex (view from behind parallel to surface, Fig. 31.186); inner propretarsal claw in male enlarged but not divided (Fig. 31.187); frons with or without setae; 4 to 10 mm in length; widespread .... ***Phyllotocus*** **Fischer**

Adossa Britton, 1995 (Pl. 56K)

Type species. *Adossa sylvicola* Britton, 1995.
Australian species. Two.
Distribution. Endemic. The two *Adossa* species occur in TAS.
References. Britton (1995a).

Ancylonyx Britton, 1957 (Pl. 58D)

Type species. *Phyllotocus vittatus* Macleay, 1887.
Australian species. Three.
Distribution. Endemic. The three *Ancylonyx* species occur in northern QLD.
References. Britton (1957).

Anthotocus Britton, 1957 (Pl. 58G)

Type species. *Cheiragra lurida* Macleay, 1864.
Australian species. Five.
Distribution. Endemic. *Anthotocus* species occur in NSW and QLD.
References. Britton (1957).

Cheiragra W. J. Macleay, 1864 (Pl. 60F)

Type species. *Phyllotocus pusillus* Blanchard, 1850.
Australian species. Five.
Distribution. Endemic. *Cheiragra* species are known from NSW to northern QLD.
References. Britton (1957).

Cheirodontus Britton, 1957 (Pl. 60D)

Type species. *Phyllotocus basicollis* Lea, 1919.
Australian species. Two.
Distribution. Endemic. *Cheirodontus bilobus* (Lea) is known from NSW and *C. basicollis* (Lea) from NSW and QLD.
References. Britton (1957).

Cheirora Britton, 1957 (Pl. 60C)

Type species. *Cheirrhamphica interstitialis* Blackburn, 1898.
Australian species. One.
Distribution. Endemic. *Cheirora interstitialis* (Blackburn) occurs in northern QLD.
References. Britton (1957).

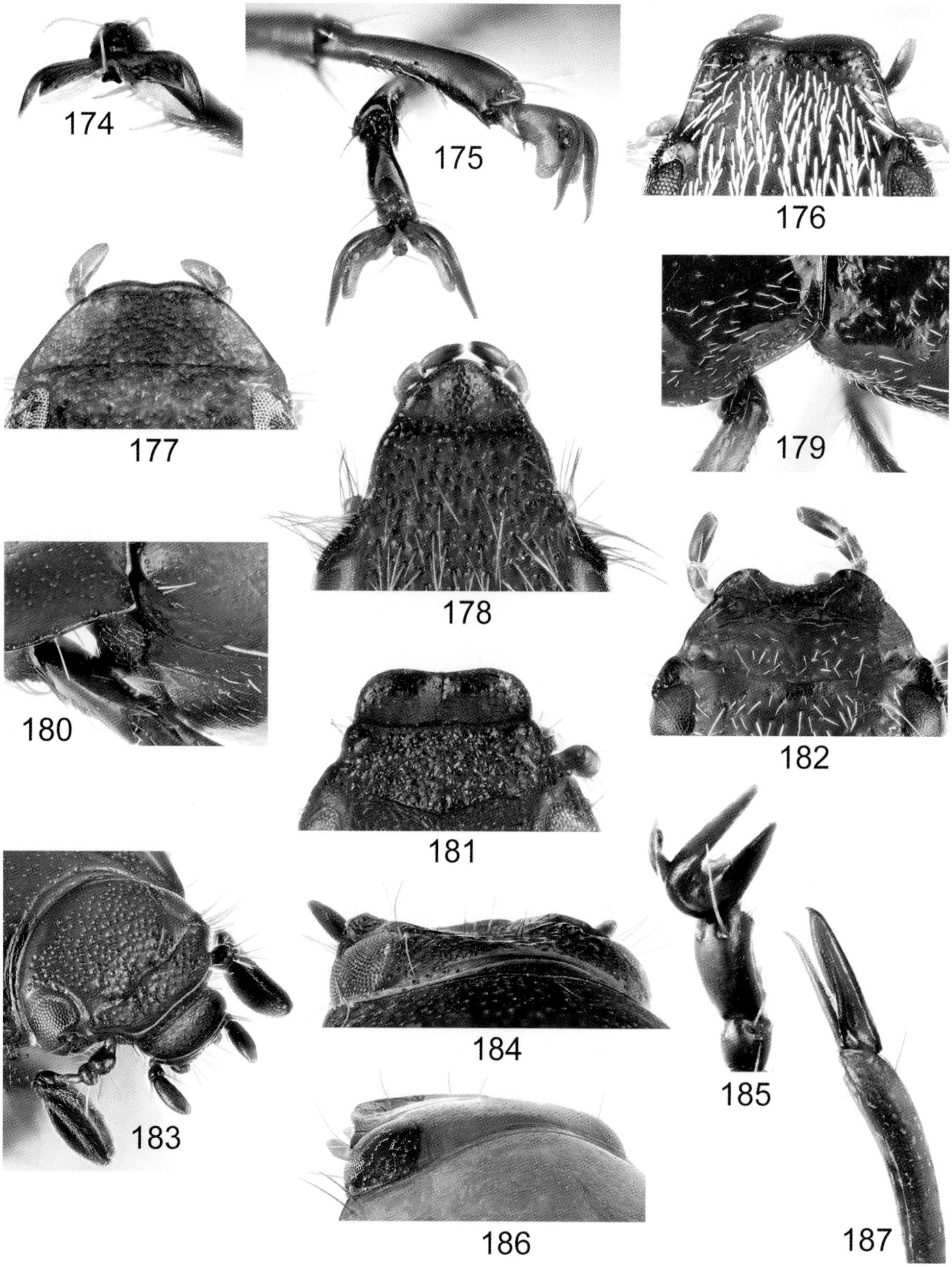

Figs 31.174–31.187. Melolonthinae: Phyllotocini. 174, *Cheiragra ruficollis* Macleay, metatarsal claws; 175, *Ancyclonyx* sp., metatarsal claws; 176, *Sphaeroscelis pectoralis* Burmeister, labrum and clypeus; 177, *Cheirora interstitialis* (Blackburn), labrum and clypeus; 178, *Neophyllotocus dispar* (Blackburn), labrum and clypeus; 179, *Adossa sylvicola* Britton, posterior angle of pronotum; 180, *Cheirodontus bilobus* (Lea), posterior angle of pronotum; 181, *Cheirodontus bilobus* (Lea), labrum and clypeus; 182, *Adossa sylvicola* Britton, labrum and clypeus; 183, *Opsitocus obscurus* (Lea), head; 184, *Anthotocus luridus* (Macleay), head from behind; 185, *Anthotocus luridus* (Macleay), protarsal claws; 186, *Phyllotocus bimaculatus* Erichson, head from behind; 187, *Phyllotocus bimaculatus* Erichson, protarsal claws.

Neophyllotocus Blackburn, 1898 (Pl. 63D)

Type species. *Macrothops pallidipennis* W. J. Macleay, 1864.
Macrothops W. J. Macleay, 1864, not Boisduval, 1835.
Australian species. Seven.
Distribution. Endemic. Most *Neophyllotocus* species occur in WA, but *N. undarus* Allsopp was described from western QLD.
References. Britton (1957); Allsopp (1976).

Opsitocus Britton, 1957 (Pl. 63F)

Type species. *Phyllotocus obscurus* Lea, 1920.
Australian species. One.
Distribution. Endemic. *Opsitocus obscurus* (Lea) was described from NSW.
References. Britton (1957).

Phyllotocus Fischer, 1823 (Pl. 8A, 64B)

Type species. *Phyllotocus macleayi* Fischer 1823.
Macrothops Boisduval, 1835.
Australian species. 28.
Distribution. Endemic. *Phyllotocus* species are widely distributed in Australia.
Biology/Ecology. Larvae of *P. macleayi* Fischer and three other species from TAS were described and illustrated by McQuillan (1985).
References. Britton (1957).

Sphaeroscelis Burmeister, 1855 (Pl. 65D)

Type species. *Sphaeroscelis pectoralis* Burmeister, 1855.
Australian species. One.
Distribution. Endemic. *Sphaeroscelis pectoralis* occurs in WA.
References. Britton (1957).

Tribe Scitalini Britton, 1957

Characteristics. Metatibiae with 2 spurs placed one above and one below tarsal articulation. Tarsal claws simple, without a tooth on the concave side. Labrum distinct from clypeus. Abdominal ventrites almost always with a longitudinal ridge on each side. Protibiae with 2 or 3 teeth including apical. Antennae 7 to 9 segmented with 3–6 lamellate club.
Distribution. Basically Australian, with *Gnaphalopoda* also found in New Zealand and New Caledonia.
Australian taxa. 17 genera.

Key to the Australian genera of Scitalini

(modified from Britton 1987a)

The genus *Gnaphalopoda* appears twice to allow for possible alternate decisions on the flange at couplet 1. The genera *Colpochilodes*, *Scitala* and *Sericesthis* appear twice to take account of both conditions in couplet 10.

1. Pronotal hypomeron with a ridge along notosternal suture, the ridge continuing forward along the free edge of the hypomeron as a narrow flange (Fig. 31.188) (sometimes slight) 2
– Pronotal hypomeron without ridge along notosternal suture or flange (Fig. 31.189) 7
2(1). Posterior edge of propygidium with fringe of setae (Fig. 31.190); pronotum with groove defining anterior margin continuous from side to side (Fig. 31.191); elytral striae not grouped in pairs, intervals of approximately equal width; antennae 8- or 9-segmented 3
– Posterior edge of propygidium without setae (Fig. 31.192); pronotum with groove defining anterior margin usually present only towards sides (Fig. 31.193); elytral striae associated in pairs so that intervals 3, 5 and 7 are much narrower than adjacent ones, OR striae and intervals not obvious; antennae 7- 8- or 9-segmented 5
3(2). Antennae 8-segmented, with 3-laminate club (Fig. 31.188); setae on posterior edge of propygidium stout, dense; apical edges of elytra with dense fringe of stout setae (Fig. 31.190); 11 to 15 mm in length; QLD, NSW, ACT .... ***Ophropyx* Britton**
– Antennae 9-segmented; setae on posterior edge of propygidium dense or not; apical edges of elytra with dense fringe of setae or not 4
4(3). Antennal club 3-lamellate; posterior edge of propygidium (Fig. 31.195) with dense fringe of setae; lateral and apical edges of elytra with dense fringe of setae (Fig. 31.195); protibiae with 2 teeth (Fig. 31.196); 12 mm in length; northern QLD .... ***Xyrodes* Britton**
– Antennal club 6-lamellate in male (Fig. 31.197), 4.5 lamellate in female; posterior edge of propygidium with fringe of fine, sparse setae (Fig. 31.198); lateral edges of elytra with sparser and finer setae and apical edges without fringe of setae; protibiae with 3 teeth (Fig. 31.199); 12 to 14 mm in length; NSW, QLD .... ***Xyrine* Britton**
5(2). Pronotum without groove defining posterior margin (Fig. 31.193); protibiae with 3 teeth; 11 to 16 mm in length; northern QLD ...... ***Xyroa* Britton**
– Pronotum with fine, raised posterior margin defined by continuous groove (Fig. 31.194); protibiae with 2 or 3 teeth 6
6(5). Antennae 7-segmented, third segment sometimes bearing 2 constrictions as if formed by fusion of 3 segments; protibiae with 2 teeth; tarsomeres with setae beneath, but not in dense tufts (Fig. 31.201); clypeus with 2 transverse swellings on upper surface (Fig. 31.200); body pale brown; 7 to 10 mm in length; northern QLD .... ***Colpochelyne* Britton**
– Antennae 9-segmented; protibiae with 3 teeth; tarsomeres 1–4 each with dense tuft of setae beneath (Fig. 31.202); clypeus without transverse swellings on upper surface (Fig. 31.194); body usually black; 7 to 13 mm in length; widespread, also found in New Zealand and New Caledonia .... ***Gnaphalopoda* Reiche (part)**

7(1). Labrum with small tooth, ridge or tubercle as well as two setae in middle of anteroventral concavity (Fig. 31.203) AND elytral striae not grouped in pairs, intervals of uniform width AND longitudinal abdominal ridge glabrous; eyes unsually large **8**

– Labrum usually without tooth, ridge or tubercle in middle of anteroventral concavity (some *Homolotropus* have a ventrally directed tooth, but lack the two setae and have elytral striae grouped in pairs and longitudinal abdominal ridges setose); elytral striae EITHER grouped in pairs, OR not grouped in pairs OR intervals not clearly defined; longitudinal abdominal ridge EITHER glabrous OR setose; eyes normal **10**

8(7). Clypeus with long, erect seta arising from large puncture on each side of midline near frontoclypeal suture (Fig. 31.203); anterior face of clypeus not visible from above; elytral apices truncate; elytral edges and disc of pygidium with many long erect setae; 8 to 11 mm in length; northern QLD .... ***Hadrops* Britton**

– Clypeus without setigerous punctures on upper surface; anterior face of clypeus posteriorly inclined and visible from above; elytral apices rounded; elytral edges with not more than 9 long erect setae; disc of pygidium glabrous or with fewer and shorter setae **9**

9(8). Antennal club in male 5-lamellate (Fig. 31.204); labrum barely extending in front of clypeus when viewed from above; elytral edges with up to 9 widely spaced long erect setae; 9 to 13 mm in length; NSW, VIC, ACT, TAS .... ***Telura* Erichson**

– Antennal club in male 3-lamellate (Fig. 31.205); labrum markedly extending in front of clypeus when viewed from above (Fig. 31.206); elytral edges with only short setae; 6 to 9 mm in length; northern QLD .... ***Protelura* Britton**

10(7). Abdominal ventrites with lateral ridges on each side bearing fringes of close-set setae (Fig. 31.207) **11**

– Abdominal ventrites with lateral ridges on each side without setae (Figs 31.208, 31.209) **16**

11(10). Antennae 9 segmented with a 5- or 6-lamellate club (basal lamellae may be small); pronotal disc with at least 1 very long, erect or semi-erect seta on each side (Fig. 31.210); pygidium with long erect setae as well as decumbent setae (Fig. 31.207); surface of body and legs with at least a trace of metallic colour; 15 to 20 mm in length; rainforest in QLD, northern NSW .... ***Homolotropus* Macleay**

– Antennae 8 or 9 segmented with a 3-lamellate club (Fig. 31.211); pronotal disc usually without erect or semi-erect setae on each side few short setae in *Scitala hirticeps*); pygidium without setae; body colour not metallic **12**

12(11). Antennae 8 segmented with a 3-lamellate club (Fig. 31.211) .... ***Sericesthis* Boisduval (part – few species)**

– Antennae 9 segmented with a 3-lamellate club **13**

13(12). Prosternum with a long, stout, conical, laterally compressed setose process arising from mid-posterior margin (Fig. 31.212); clypeus truncate, anterior face without setae .... **Liparetrini: *Sciton* Blackburn (part – few species)**

– Prosternum without such a process; clypeus rounded or truncate, if the latter then anterior face with at least a few fine setae **14**

14(13). Metacoxae without setae or with a few sort setae (Fig. 31.209); pronotal disc with a few erect setae on each side; 7 to 10 mm in length; NSW .... ***Scitala hirticeps* (Blackburn)**

– Metacoxae densely setose with long setae (Fig. 31.208); pronotal disc without erect setae **15**

15(14). Pronotum without narrow anterior and posterior margins defined by grooves (Fig. 31.213); 13 to 20 mm in length; WA, SA, VIC, TAS .... ***Colpochilodes* Blackburn (part)**

– Pronotum with narrow anterior (wholly or partly) and posterior (wholly) margins defined by grooves (Fig. 31.214) .... **Liparetrini: *Colpochila* Erichson (part – few species)**

16(10). Antennae 9-segmented **17**

– Antennae 8-segmented **21**

17(16). Apex of labrum turned upwards and visible from above in an emargination of anterior edge of clypeus (Fig. 31.215); body black, heavily sclerotised; frons with transverse ridge parallel to and close to frontoclypeal suture; body coarsely punctured; 9 to 13 mm in length; WA, SA, VIC .... ***Byrrhomorpha* Blackburn**

– Apex of labrum not reflexed and not visible from above; anterior edge of clypeus at most slightly concave in middle (Fig. 31.213); colour brown to black; frons without ridge close to frontoclypeal suture; body usually not coarsely punctured **18**

18(17). Tarsomeres 1–4 each with dense tuft or pad of yellow setae beneath (Fig. 31.202); body usually black; 7 to 13 mm in length; widespread, also found in New Zealand and New Caledonia .... ***Gnaphalopoda* Reiche (part)**

– Tarsomeres with setae beneath, but not in dense tufts (Fig. 31.216); colour usually brown **19**

19(18). Antennal club 5- or 6-lamellate; elytra striae not grouped in pairs, distinct; lateral angles of pronotal disc concave before posterior angles (Fig. 31.217); 8 mm in length; northern QLD .... ***Idanastes* Britton**

– Antennal club 3-lamellate; elytral striae grouped in pairs or indistinct; lateral angles of pronotal disc not concave before posterior angle **20**

20(19). Pronotum without anterior and posterior margins defined by grooves (Fig. 31.213); 13 to 20 mm in length; WA, SA, VIC, TAS .... ***Colpochilodes* Blackburn (part)**

– Pronotum with anterior or posterior margin or both, defined by a groove and usually continuous from side to side; 5 to 17 mm in length; widespread .... ***Scitala* Erichson (major part)**

21(16). Anterior edge of clypeus with deep median emargination (Fig. 31.218); apical lamella of antennal club with pale longitudinal groove on outer side (Fig. 31.219); pygidium with broad, shallow median longitudinal depression; body piceous or black; pronotum without groove defining anterior margin; 14 mm in length; NSW .... ***Phorine* Britton**

– Anterior edge of clypeus without emargination; apical lamella of antennal club without pale longitudinal groove on outer side; pygidium without median longitudinal depression; body rarely black; pronotum usually with anterior margin defined by groove **22**

22(21). Labrum and clypeus fused (Fig. 31.220); protibia with 2 teeth, basal one small and obtuse (Fig. 31.221); surfaces of pronotum and elytra with sparse, erect setae; elytral striae and intervals not obvious; colour pale brown; 9 mm in length; WA .... ***Synchilus* Britton**

– Labrum and clypeus separated by suture (Fig. 31.222); protibia with 2 or 3 teeth; surfaces of pronotum and elytra usually without setae; elytra usually obvious, grouped in pairs; colour various; 8 to 19 mm in length; widespread .... ***Sericesthis* Boisduval (major part)**

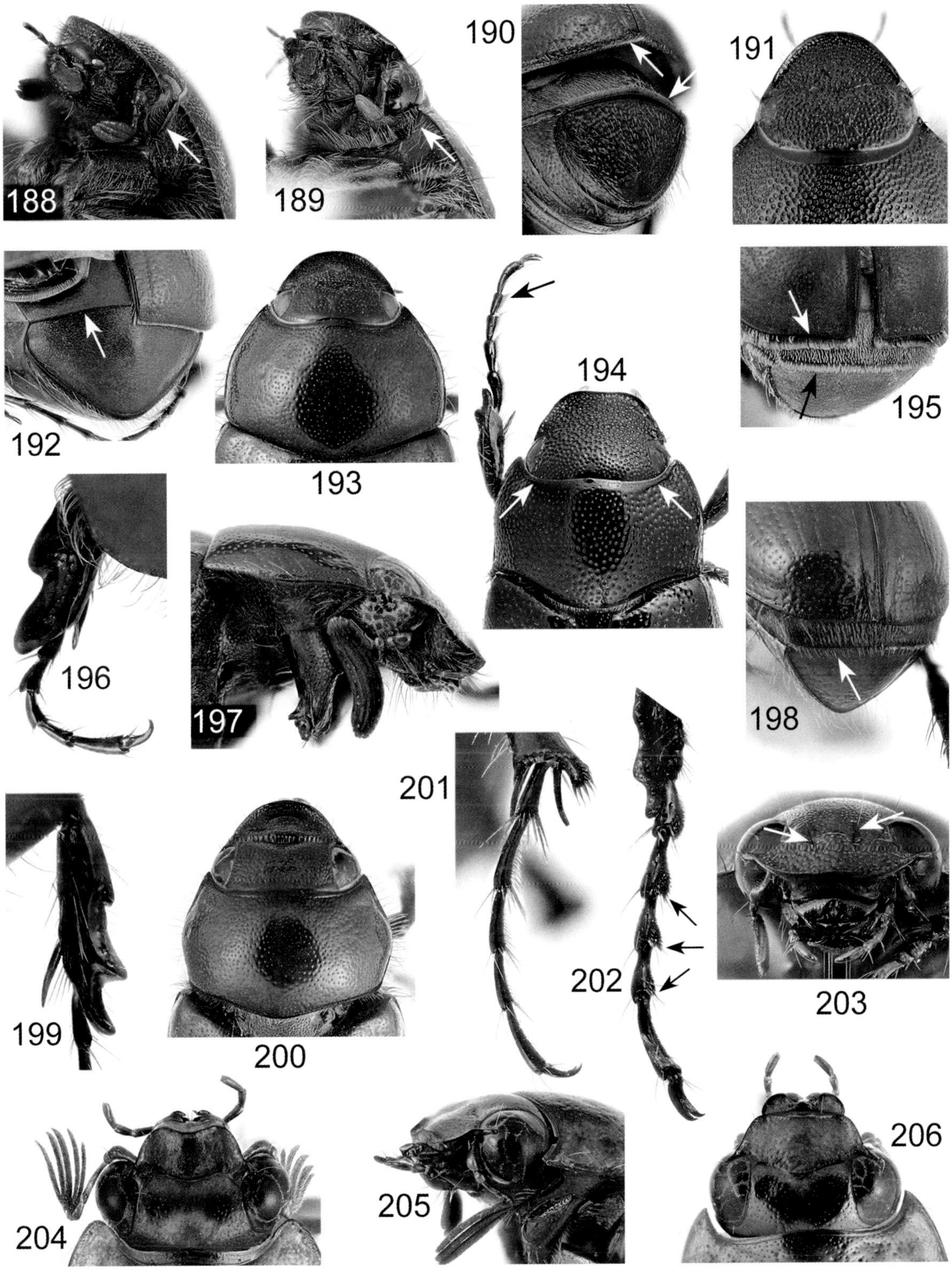

Figs 31.188–31.206. Melolonthinae: Scitalini. 188, *Ophropyx ciliata* (Boisduval), head and pronotum, lateral; 189, *Homolotopus sericeus* Britton, head and pronotum, lateral; 190, *Ophropyx ciliata* (Boisduval), pygidium; 191, *Ophropyx ciliata* (Boisduval), head and pronotum; 192, *Xyroa polita* Britton, female pygidium; 193, *Xyroa polita* Britton, female head and pronotum; 194, *Gnaphalopoda carnei* Britton, head and pronotum and leg; 195, *Xyrodes calorata* (Blackburn), pygidium; 196, *Xyrodes calorata* (Blackburn), protibia; 197, *Xyrine inusitatus* Britton, head and pronotum, lateral; 198, *Xyrine inusitatus* Britton, male pygidium; 199, *Xyrine inusitatus* Britton, male protibia; 200, *Colpochelyne dux* Britton, head, pronotum; 201, *Colpochelyne dux* Britton, hind tarsus; 202, *Gnaphalopoda biloba* (Lea), fore tarsus; 203, *Hadrops flavus* (Lea), head, frontal; 204, *Telura vitticollis* Erichson, head, dorsal; 205, *Protelura guttata* Britton, head, lateral; 206, *Protelura guttata* Britton, male head, dorsal.

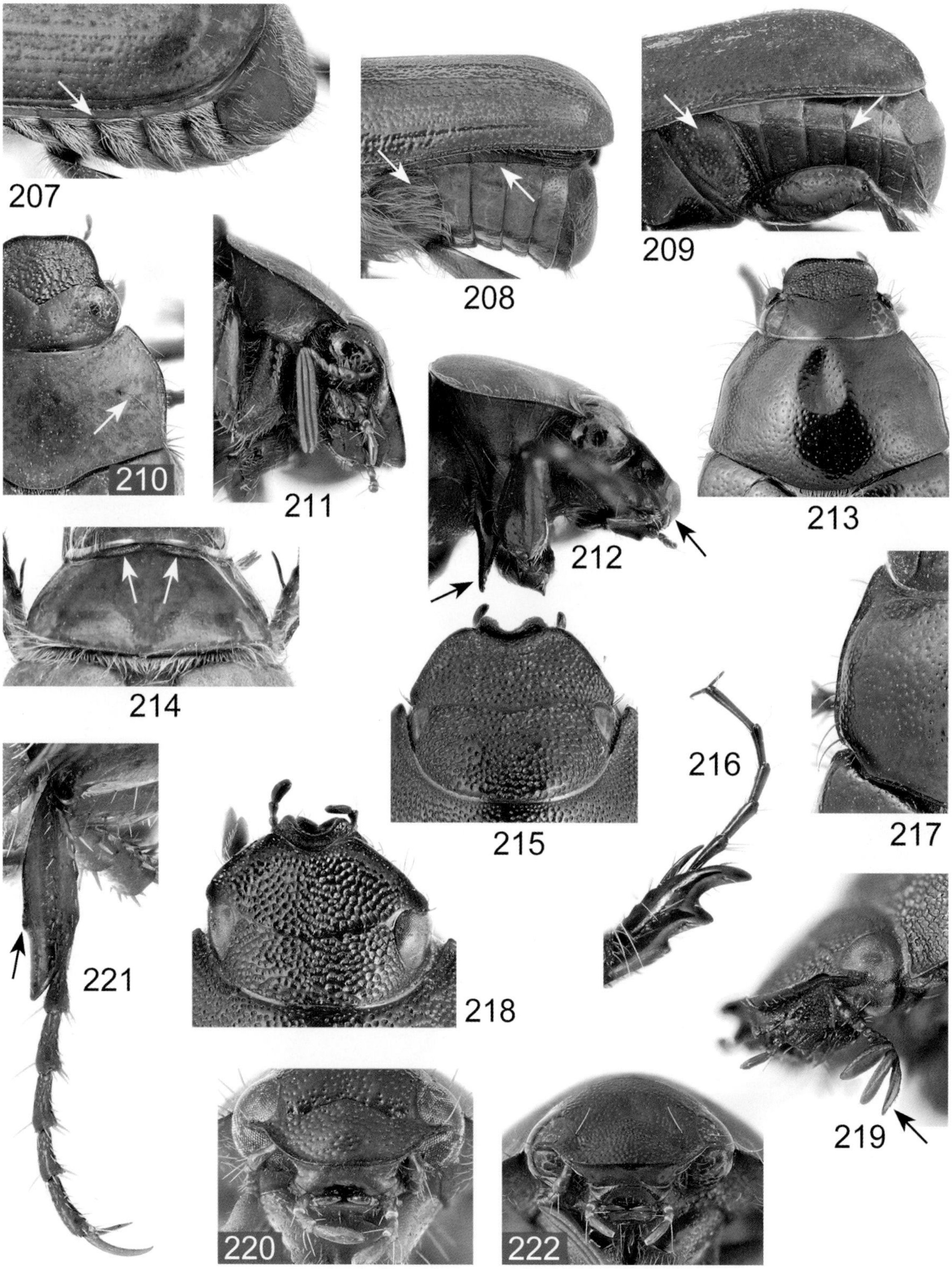

Figs 31.207–31.222. Melolonthinae: Scitalini. 207, *Homolotopus sericeus* Britton, abdomen and elytra, lateral; 208, *Colpochilodes peregrina* Britton, hind coxa and abdomen, lateral; 209, *Scitala hirticeps* (Blackburn), hind coxa and abdomen, lateral; 210, *Homolotopus sericeus* Britton, head and pronotum; 211, *Sericesthis geminata* Boisduval, head and pronotum, lateral; 212, *Sciton* sp. [Liparetrini], head and prothorax, lateral; 213, *Colpochilodes peregrina* Britton, head and pronotum; 214, *Colpochila andersoni* Britton [Liparetrini], pronotum; 215, *Byrrhomorpha verres* Blackburn, head, dorsal; 216, *Idanastes abditus* Britton, protibia and protarsus; 217, *Idanastes abditus* Britton, lateral margin of pronotum; 218, *Phorine anomala* (Blackburn), head, dorsal; 219, *Phorine anomala* (Blackburn), antenna; 220, *Synchilus gisleni* Britton, head, frontal view; 221, *Synchilus gisleni* Britton, protibia and protarsus; 222, *Sericesthis geminata* Boisduval, head, frontal view.

Byrrhomorpha Blackburn, 1892 (Pl. 58J)

Type species. *Byrrhomorpha verres* Blackburn, 1892.
Australian species. Three.
Distribution. Endemic. *Byrrhomorpha* are known from VIC, SA and WA.
References. Britton (1987a).

Colpochelyne Britton, 1987 (Pl. 60J)

Type species. *Colpochelyne dux* Britton, 1987.
Australian species. One.
Distribution. Endemic. *Colpochelyne dux* is known from northern QLD.
References. Britton (1987a).

Colpochilodes Blackburn, 1898 (Pl. 60K)

Type species. *Colpochilodes raucipennis* Blackburn, 1898.
Australian species. Two.
Distribution. Endemic. *Colpochilodes peregrina* Britton occurs in TAS, VIC, SA and southern WA, while *C. raucipennis* is known only from southern WA.
References. Britton (1987a).

Gnaphalopoda Reiche, 1860 (Pl. 62J)

Type species. *Rhisotrogus leptopoda* Montrouzier, 1860.
Xylostygnus Broun, 1886
Ocnodus Blackburn, 1892
Australian species. 21.
Distribution. The genus *Gnaphalopoda* is known from Australia, New Zealand and New Caledonia. Most Australian species occur in SA, WA and NT, but *G. carnei* Britton occurs in VIC and ACT, and both *G. lepida* (Blackburn) and *G. lesouefi* Britton have been recorded from QLD or NSW.
References. Britton (1987a).

Hadrops Britton, 1987 (Pl. 56D)

Type species. *Hadrops halei* Britton, 1987.
Australian species. Two.
Distribution. Endemic. *Hadrops flavus* (Lea) and *H. halei* are known from northern QLD.
References. Britton (1987a).

Homolotropus W. J. Macleay, 1871 (Pl. 57H)

Type species. *Homolotropus luridipennis* W. J. Macleay, 1871.
Australian species. Five.
Distribution. Endemic. *Homolotropus* species are known from NSW to northern QLD.
References. Britton (1987a).

Idanastes Britton, 1987 (Pl. 59B)

Type species. *Idanastes abditus* Britton, 1987.
Australian species. One.
Distribution. Endemic. *Idanastes abditus* occurs in northern QLD.
References. Britton (1987a).

Ophropyx Britton, 1987 (Pl. 63H)

Type species. *Frenchella hispida* Blackburn, 1898.
Australian species. Three.
Distribution. Endemic. *Ophropyx* species are known from VIC, NSW and QLD.
References. Britton (1987a).

Phorine Britton, 1987 (Pl. 64D)

Type species. *Byrrhomorpha anomala* Blackburn, 1907.
Australian species. One.
Distribution. Endemic. *Phorine anomala* is known from NSW.
References. Britton (1987a).

Protelura Britton, 1987 (Pl. 64C)

Type species. *Protelura nana* Britton, 1987.
Australian species. Two.
Distribution. Endemic. *Protelura guttata* Britton and *P. nana* occur in northern QLD.
References. Britton (1987a).

Scitala Erichson, 1842 (Pl. 65A)

Type species. *Scitala sericans* Erichson, 1842.
Australian species. 22.
Distribution. Endemic. *Scitala* species are widely distributed in Australia.
Biology. Larvae of *S. sericans* Erichson from TAS was described by McQuillan (1985).
References. Britton (1987a).

Sericesthis Boisduval, 1835 (Pl. 65B)

Type species. *Sericesthis geminata* Boisduval, 1835.
Anodontonyx Sharp, 1890
Australian species. 52.
Distribution. Endemic. *Sericesthis* species are widely distributed in Australia.
Biology. The larvae of *S. nigra* (Lea) and *S. nigrolineata* (Boisduval) from TAS were described and illustrated by McQuillan (1985).
References. Britton (1987a).

Synchilus Britton, 1955 (Pl. 65G)

Type species. *Synchilus gisleni* Britton, 1955.
Australian species. One.
Distribution. Endemic. *Synchilus gisleni* occurs in southern WA.
References. Britton (1987a).

Telura Erichson, 1842 (Pl. 66E)

Type species. *Telura vitticollis* Erichson, 1842.
Australian species. Five.
Distribution. Endemic. *Telura* species occur in TAS, VIC, ACT and NSW.
Biology. The larva of *T. vitticollis* Erichson was described by McQuillan (1985).
References. Britton (1987a).

Xyrine Britton, 1987 (Pl. 66C)

Type species. *Xyrine carnei* Britton, 1987.
Australian species. Two.
Distribution. Endemic. *Xyrine carnei* occurs in NSW while *X. inusitatus* Britton apparently extends into QLD.
References. Britton (1987a).

Xyroa Britton, 1987 (Pl. 66J)

Type species. *Xyroa polita* Britton, 1987.
Australian species. Four.
Distribution. Endemic. All *Xyroa* species occur in northern QLD.
References. Britton (1987a).

Xyrodes Britton, 1987 (Pl. 66K)

Type species. *Frenchella calorata* Blackburn, 1907.
Australian species. One.
Distribution. Endemic. *Xyrodes colorata* occurs in northern QLD.
References. Britton (1987a).

Tribe Systellopini Sharp, 1877

Characteristics. Metatibiae with 2 spurs placed one above and one below tarsal articulation. Tarsal claws simple, without a tooth on the concave side. Labrum large, extended in front of clypeus and in the same general plane, usually separated by a complete suture. Clypeus with lateral margins deeply emarginate before ocular canthus. Metacoxae inflated, usually without defined lateral margin. Metafemora inflated. Metatibiae very short and stout. Antennal club with 5–7 long lamellae. Body surface not metallic.
Distribution. Australia.
Australian taxa. Six genera.
References. Allsopp (1989a, 1993).

Key to the genera of Systellopini

(modified from Allsopp 1989a, 1993)
This key applies to males only as females are rare.

1. Antennal club 5- or 6-lamellate, lamellae with long, pilose filaments (Figs 31.223, 31.224); 10 to 13.5 mm in length; WA, NT .... ***Sarothromerus* Blackburn**
– Antennal club 5-, 6- or 7-lamellate, lamellae setose or glabrous 2
2(1). Labrum less than 0.33 times as long as wide (Fig. 31.225); metafemora strongly dilated (Fig. 31.227), less than 1.3 times as long as wide; pronotum basally swollen; dorsal and ventral surfaces entirely black; 14.5 to 27 mm in length; south-central, coastal WA .... ***Systellopus* Sharp**
– Labrum more than 0.33 times as long as wide; metafemora less strongly dilated, more than 1.4 times as long as wide; pronotum not basally swollen; at least ventral surfaces not entirely black 3
3(2). Antennal club 5- or 6-lamellate, inner faces of lamellae glabrous; disc of pygidium rough, rugose-punctate with numerous erect setae (Fig. 31.226); inner apical angle of elytra rounded, usually densely spinose 4
– Antennal club 5-, 6- or 7-lamellate, inner faces of lamellae setose; disc of pygidium smooth, punctate or not and with at most a few scattered setae; inner apical angles of elytra angulate, not densely spinose, although a few setae may be present 5
4(3). Labrum with transverse ridge within anterior one-third of length of labrum (Fig. 31.228); pronotal disc glabrous (Pl. 60G); 11.5 to 16 mm in length; northern QLD, northern NT .... ***Chilodiplus* Sharp**
– Labrum without a transverse ridge, with reflexed anterior and lateral margins forming a continuous ridge (Fig. 31.229); pronotal disc with at least a few long setae; 11.5 to 15.5 mm in length; northern WA, central Australia: NT, SA .... ***Sphyrocallus* Sharp**
5(3). Pronotal disc glabrous (Pl. 64G); antennal club 5-, 6- or 7-lamellate; 11.5 to 20 mm in length; south-western WA (*Enamillus septus* will come out here – antennal club with 7 lamellae) .... ***Prochelyna* Erichson**
– Pronotal disc densely setose with long setae (Pl. 61D) (except bicoloured *Enamillus septus*); antennal club 5- or 6-lamellate; 9 to 16 mm in length; SA and south-western WA .... ***Enamillus* Sharp**

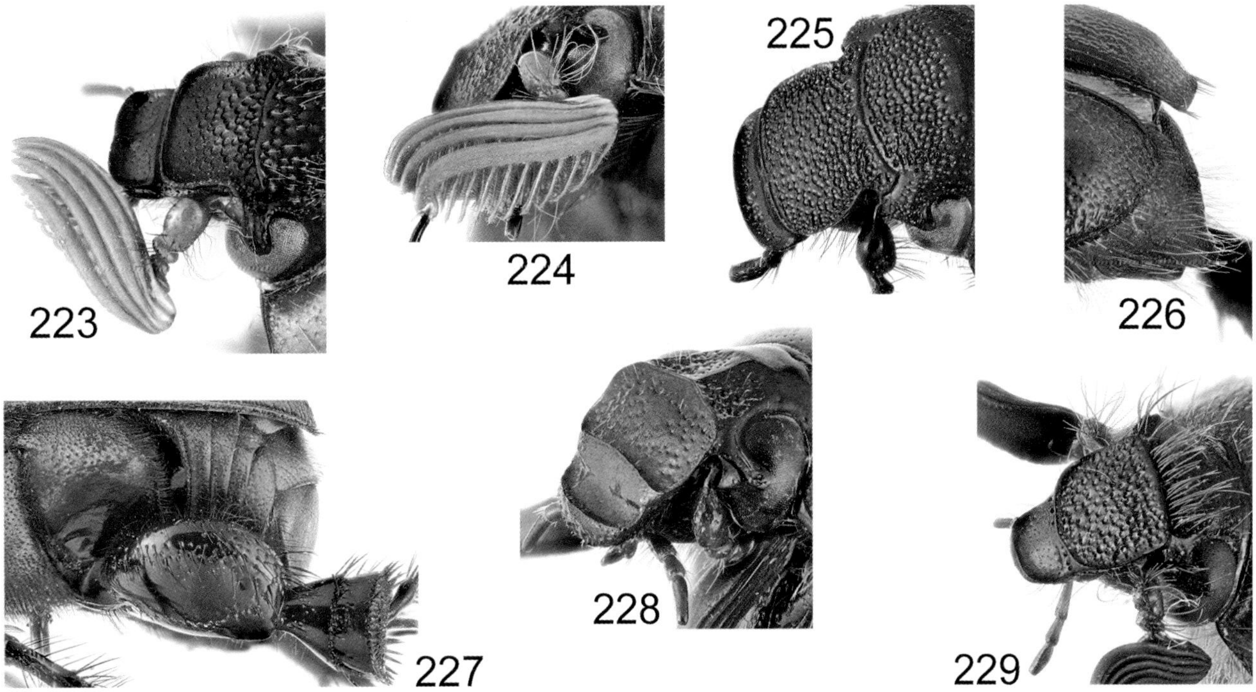

Figs 31.223–31.229. Melolonthinae: Systellopini. 223, *Sarothromerus aranda* Allsopp, head, dorsal; 224, *Sarothromerus aranda* Allsopp, antenna; 225, *Systellopus obtusus* Sharp, head; 226, *Chilodiplus weiri* Allsopp, pygidium; 227, *Systellopus obtusus* Sharp, hind leg; 228, *Chilodiplus weiri* Allsopp, head; 229, *Sphyrocallus brunneus* Sharp, head.

Chilodiplus Sharp, 1877 (Pl. 60G)

Type species. *Chilodiplus albertisi* Sharp, 1877.
Australian species. Four.
Distribution. Endemic. *Chilodiplus* species occur in northern QLD and NT.
References. Allsopp (1989a, 1993).

Enamillus Sharp, 1877 (Pl. 61D)

Type species. *Enamillus striatus* Sharp, 1877.
Australian species. Eight.
Distribution. Endemic. *Enamillus* species are known from SA and southern WA.
References. Allsopp (1989a).

Prochelyna Erichson, 1847 (Pl. 64G)

Type species. *Prochelyna heterodoxa* Burmeister, 1855.
Atholerus Sharp, 1877
Tosotarsus Sharp, 1877
Australian species. Three.
Distribution. Endemic. *Prochelyna* species all occur in southern WA.
References. Allsopp (1989a).

Sarothromerus Blackburn, 1907 (Pl. 65E)

Type species. *Enamillus sharpi* Blackburn, 1905.
Australian species. Three.
Distribution. Endemic. *Sarothromenus* species occur in northern WA and NT.
References. Allsopp (1989a).

Sphyrocallus Sharp, 1877 (Pl. 65H)

Type species. *Sphyrocallus brunneus* Sharp, 1877.
Australian species. Three.
Distribution. Endemic. *Sphyrocallus* species occur in NSW, NT, SA and northern WA.
References. Allsopp (1989a).

Systellopus Sharp, 1877 (Pl. 66H)

Type species. *Systellopus obtusus* Sharp, 1877.
Australian species. One.
Distribution. Endemic. *Systellopus obtusus* Sharp is known from northern WA.
References. Allsopp (1989a).

Genera of Melolonthinae excluded from Australian fauna

***Neolepidiota* Blackburn, 1890.** Type species. *N. obscura* Blackburn, 1890. Arrow (1917) has shown that the damaged type specimen, described without locality, in fact belongs to an Indian species *Holotrichia serrata* (Fabricius).

***Metascelis* Westwood, 1842**. Type species. *M. flexilis* Westwood, 1842. According to Allsopp (1989a) this is not Australian genus and does not belong in Systellopini.

References

Ahrens D (2006) The phylogeny of Sericini and their position within the Scarabaeidae based on morphological characters (Coleoptera: Scarabaeidae). *Systematic Entomology* **31**, 113–144. doi:10.1111/j.1365-3113.2005.00307.x

Ahrens D, Vogler AP (2008) Towards the phylogengy of Chafers (Sericini): Analysis of alignment-variable sequences and the evolution of segment numbers in the antennal club. *Molecular Phylogenetics and Evolution* **47**, 783–798. doi:10.1016/j.ympev.2008.02.010

Ahrens D, Scott M, Vogler AP (2011) The phylogeny of monkey beetles basedon mitochondrial and ribosomal RNA genes (Coleoptera: Scarabaeidae: Hopliini). *Molecular Phylogenetics and Evolution* **60**, 408–415. doi:10.1016/j.ympev.2011.04.011

Ahrens D, Schwarzer J, Vogler AP (2014) The evolution of scarab beetles tracks the sequential rise of angiosperms and mammals. *Proceedings. Biological Sciences* **281**(20141470), 1–10 10.1098/rspb.2014.1470.

Allsopp PG (1976) *Neophyllotocus undarus* (Coleoptera: Scarabaeidae: Melolonthinae), a new species from Queensland. *Journal of the Australian Entomological Society* **14**, 327–328. doi:10.1111/j.1440-6055.1975.tb02047.x

Allsopp PG (1981) Revision of the Australian species of *Phaenognatha* Hope (Coleoptera: Scarabaeidae: Aclopinae). *Journal of the Australian Entomological Society* **20**, 185–195. doi:10.1111/j.1440-6055.1981.tb01028.x

Allsopp PG (1983) *Neophaenognatha*, a new genus for the Neotropical species of *Phaenognatha* Hope (Scarabaeidae: Aclopinae) with the description of *N. capella* n. sp. and designation of lectotypes. *Coleopterists Bulletin* **37**, 208–211.

Allsopp PG (1989a) Revision of the Systellopini (Coleoptera: Scarabaeidae: Melolonthinae). *Invertebrate Taxonomy* **3**, 197–227. doi:10.1071/IT9890197

Allsopp PG (1989b) Two new species of *Lepidiota* Kirby (Coleoptera: Scarabaeidae: Melolonthinae) from Australia with noted on *L. noxia* Britton. *Journal of the Australian Entomological Society* **28**, 39–43. doi:10.1111/j.1440-6055.1989.tb01190.x

Allsopp PG (1990a) Revision of the Pachytrichini (Coleoptera: Scarabaeidae: Melolonthinae). *Invertebrate Taxonomy* **4**, 753–761. doi:10.1071/IT9900753

Allsopp PG (1990b) Two new species and distribution records of Australian melolonthinae (Coleoptera: Scarabaeidae). *Coleopterists Bulletin* **44**(2), 217–223.

Allsopp PG (1993) Three new species of *Chilodiplus* Sharp (Coleoptera: Scarabaeidae: Melolonthinae) from northern Australia. *Australian Entomologist* **20**, 49–55.

Allsopp PG (1995) Biogeography of the Australian Dynastinae, Rutelinae, Scarabaeinae, Melolonthini, Scitalini and Geotrupidae (Coleoptera: Scarabaeoidea). *Journal of Biogeography* **22**, 31–48. doi:10.2307/2846071

Allsopp PG (1999) Three new species of Melolonthinae (Coleoptera: Scarabaeidae) from Australia. *Memoirs of the Queensland Museum* **43**(2), 453–458.

Allsopp PG (2003) Synopsis of *Antitrogus* Burmeister (Coleoptera: Scarabaeidae: Melolonthini). *Australian Journal of Entomology* **42**, 159–178. doi:10.1046/j.1440-6055.2003.00340.x

Allsopp PG, Lambkin CL (2006) Canegrubs and cladistics: what story do adult, larval and ecological characters tell? *Australian Journal of Entomology* **45**, 55–66. doi:10.1111/j.1440-6055.2006.00511.x

Allsopp PG, Watkins SG (1995) *Lepidiota brittoni*, a new species from coastal New South Wales (Coleoptera: Scarabaeidae: Melolonthinae). *Australian Entomologist* **22**(3), 79–82.

Arrow GJ (1917) Some systematic notes on melolonthine Coleoptera. *Annals & Magazine of Natural History* Series 8, **19**(109), 59–65. doi:10.1080/00222931709486912

Bezdåk A (2004) Catalogue of Diplotaxini (Coleoptera: Scarabaeidae: Melolonthinae) of the Old World. *Zootaxa* **463**, 1–90. doi:10.11646/zootaxa.463.1.1

Bouchard P, Bousquet Y, Davies AE, Alonso-Zarazaga MA, Lawrence JF, Lyal CHC, Newton AF, Reid CAM, Schmitt M, Ślipiński SA, Smith ABT (2011) Family group names in Coleoptera (Insecta). *ZooKeys* **88**, 1–972. doi:10.3897/zookeys.88.807

Böving AG (1936) Description of the larva of *Plectris aliena* Chapin and explanation of new terms applied to the epipharynx and raster. *Proceedings of the Entomological Society of Washington* **38**, 169–185.

Britton EB (1955) Coleoptera: Scarabaeidae: melolonthinae and Dynastinae from the Monte Bello Islands. *Proceedings of the Linnaen Society of London* **165**(2), 124–126. doi:10.1111/j.1095-8312.1955.tb00725.x

Britton EB (1957) *A revision of the Australian Chafers (Coleoptera: Scarabaeidae: Melolonthinae). Volume I.* British Museum (Natural History), London, viii + 185 pp., 42 pls.

Britton EB (1963) A new species of *Maechidius* (Coleoptera: Melolonthinae) from Northern Queensland. *Entomologischen Arbeiten aus dem Museum Georg Frey.* **14**, 126–127.

Britton EB (1978) A revision of the Australian chafers (Coleoptera: Scarabaeidae: Melolonthinae). Vol. 2. Tribe Melolonthini. *Australian Journal of Zoology. Supplementary Series* **60**, 150 pp.

Britton EB (1979) New Australian Melolonthinae (Coleoptera: Scarabaeidae). *Journal of the Australian Entomological Society* **18**, 193–197. doi:10.1111/j.1440-6055.1979.tb00836.x

Britton EB (1980) A revision of the Australian chafers (Coleoptera: Scarabaeidae: Melolonthinae). Vol. 3. Tribe Liparetrini: genus *Liparetrus. Australian Journal of Zoology. Supplementary Series* **76**, 209 pp.

Britton EB (1985) *Lepidiota noxia* sp. n. (Coleoptera: Scarabaeidae: Melolonthinae), a pest of sugarcane in Queensland. *Journal of the Australian Entomological Society* **24**, 117–119. doi:10.1111/j.1440-6055.1985.tb00202.x

Britton EB (1986) A revision of the Australian chafers (Coleoptera: Scarabaeidae: Melolonthinae). Vol. 4. Tribe Liparetrini: genus *Colpichila Australian Journal of Zoology. Supplementary Series* **118**, 135 pp.

Britton EB (1987a) A revision of the Australian chafers (Coleoptera: Scarabaeidae: Melolonthinae). Vol. 3. Tribes Scitalini and Comophorini. *Invertebrate Taxonomy* **1**, 685–799. doi:10.1071/IT9870685

Britton EB (1987b) *Megacoryne carnabyorum*, a new genus and species of Melolonthinae (Coleoptera: Scarabaeidae) from Australia. *Journal of the Australian Entomological Society* **26**, 41–43. doi:10.1111/j.1440-6055.1987.tb00256.x

Britton EB (1988) Synopsis of the genera of Australian Heteronycini (Coleoptera: Scarabaeidae: Melolonthinae*). Journal of the Australian Entomological Society* **27**, 27–36. doi:10.1111/j.1440-6055.1988.tb01139.x

Britton EB (1990) A synopsis of the Australian genera of Liparetrini (Coleoptera: Scarabaeidae: Melolonthinae). *Invertebrate Taxonomy* **4**, 159–195. doi:10.1071/IT9900159

Britton EB (1992) Appendix III. Taxonomic decisions made in this work. New Replacement Name. In *Zoological Catalogue of Australia. Vol. 9. Coleoptera: Scarabaeoidea* (WWK Houston ed.) pp 458–463. AGPS, Canberra.

Britton EB (1995a) *Adossa* gen. n. and two new species of Sericini (*A. sylvicola* and *A. infera*) (Coleoptera: Scarabaeidae: Melolonthinae) from Tasmania. *Journal of the Australian Entomological Society* **34**, 85–89. doi:10.1111/j.1440-6055.1995.tb01289.x

Britton EB (1995b) A synopsis of the genera of Diphucephalini (Coleoptera: Scarabaeidae: Melolonthinae) including *Watkinsia*, gen. nov. et spp. nov. *Invertebrate Taxonomy* **9**, 115–128. doi:10.1071/IT9950115

Britton EB (2000) A review of *Heteronyx* Guérin-Méneville (Coleoptera: Scarabaeidae: Melolonthinae). *Invertebrate Taxonomy* **14**, 465–589. doi:10.1071/IT98009

Brown H, Chin D, Smith ESC, Brown GR (2000) Northern Territory Economic Insect Reference Collection. In *1999–2000 Technical Annual*

Report Department of Primary Industries and Fisheries, pp 234–235. NT Government, Darwin.

Coco-Abia MM (2007) Phylogenetic relationships of the subfamily Melolonthinae (Coleoptera: Scarabaeidae). *Insect Systematics & Evolution* **38**, 447–472. doi:10.1163/187631207794760921

Cribb BW, Hull CD, Moore CJ, Miller LJ, Yeates DK (1998) Structure of raster in melolonthine larvae. *Annals of the Entomological Society of America* **91**, 202–210. doi:10.1093/aesa/91.2.202

Dalla Torre KW (1913) Coleopterorum catalogus. Vol. 20, pars 50, Scarabaeidae. *Melolonthidae* **IV**, 291–450.

Evans AV (2003) A checklist of the New World Chafers (Coleoptera: Scarabaeidae: Melolonthinaer). *Zootaxa* **211**, 1–458. doi:10.11646/zootaxa.211.1.1

Gunter NL, Weir TA, Ślipiński A, Bocak L, Cameron L (2016) If dung beetles (Scarabaeidae: Scarabaeinae) arose in association with dinosaurs, did they also suffer a mass co-extinction at the K-Pg boundary? *PLoS One* **11**(5), e0153570 doi:10.1371/journal.pone.0153570.

Katovich K (2008) A generic-level phylogenetic review of the Macrodactylini)Coleoptera: Scarabaeidae: Melolonthinae). *Insecta Mundi* **0023**, 1–78.

Krell F-T (2007) Fossil Record and Evolution of Scarabaeoidea Coleoptera: Polyphaga). *Coleopterists Bulletin* **60**, 120–143.

Lacordaire JT (1856) *Histoire naturelle des insectes. Genera des Coléoptères, ou exposé méthodique et critique de tous genmres proposés jusquici dans cet ordre d'insectes. Tome Troisième. Contenant les familles des Pectinicornes et Lamellicornes*. Libraire Encyclopaedique de Roret, Paris, 594 pp.

Lawrence JF, Ślipiński A (2013) *Australian Beetles, Volume 1: Morphology, classification and Keys.* CSIRO Publishing, Collingwood. viii + 561 pp.

Liu W-G, Eberle J, Bai M, Yang X-K (2015) A phylogeny of Sericini with particular reference to Chinese species using mitochondrial ribosomal DNA (Coleoptera: Scarabaeidae). *Organisms, Diversity & Evolution* **15**, 343–350. doi:10.1007/s13127-015-0204-z

McQuillan PB (1985) The identification of root-feeding cockchafer larvae (Coleoptera: Scarabaeidae) found in pastures in Tasmania. *Australian Journal of Zoology* **33**, 509–546.

McQuillan PB, Michaels K (1997) First record of *Pseudoheteronyx* (Coleoptera: Scarabaeidae: Melolonthinae) from Tasmania, with description of *Pseudoheteronyx littoralis* sp.n. from sandy beaches. *Australian Journal of Entomology* **36**, 117–120. doi:10.1111/j.1440-6055.1997.tb01443.x

Miller LJ, Allsopp PG (2000) Identification of Australian canegrubs (Coleoptera: Scarabaeidae: Melolonthini). *Invertebrate Taxonomy* **14**, 377–409. doi:10.1071/IT98019

Morgan DF (1977) Swarming behaviour in the Australian beetle, *Heteronyx obesus*, with notes on related species. *Ecological Entomology* **2**, 91–96. doi:10.1111/j.1365-2311.1977.tb00868.x

Ocampo FC, Mondaca J (2012) Revision of the scarab subfamily Aclopinae Blanchard (Coleoptera: Scarabaeidae) in Argentina and Chile. *Zootaxa* **3409**, 1–29.

Özdikmen H (2009) Nomenclatural changes for five preoccupied scarab beetle genus group names (Coleoptera: Scarabaeidae). *Munis Entomology and Zoology Journal* **4**(1), 139–147.

Özdikmen H, Demir H (2008) Replacement names for two preoccupied beetle genera (Coleoptera: Scarabaeidae: Melolonthinae). *Entomological News* **119**(3), 315–317. doi:10.3157/0013-872X(2008)119[315:RNFTPB]2.0.CO;2

Ridsdill-Smith TJ, Roberts RL (1976) Insect density effects in root feeding by larvae of *Sericesthis nigrolineata* (Coleoptera: Scarabaeidae). *Journal of Applied Ecology* **13**, 423–428. doi:10.2307/2401791

Ritcher PO (1966) *White Grubs and their Allies. A Study of North American Scarabaeoid Larvae*. Oregon State University Press (Studies in Entomology No. 4), Corvallis, Oregon.

Roberts RJ (1968) An introduced pasture beetle, *Plectris aliena* Chapin (Scarabaeidae: Melolonthinae). *Journal of the Australian Entomological Society* **7**, 15–20. doi:10.1111/j.1440-6055.1968.tb00693.x

Roberts RJ, Ridsdill-Smith TJ, Porter MR, Sawtell NL (1982a). Fluctuations in the abundance of pasture scarabs over an 18-year period of light trapping. In *Proceedings of the Third Australasian Conference on Grassland Invertebrate Ecology, Adelaide 30 Nov.–4 Dec. 1981* (Lee KE, ed.) pp. 75–79. Government Printer, Adelaide.

Roberts RJ, Campbell AJ, Porter MR, Sawtell NL (1982b). The distribution and abundance of pasture scarabs in relation to *Eucalyptus* trees. In *Proceedings of the Third Australasian Conference on Grassland Invertebrate Ecology, Adelaide 30 Nov.–4 Dec. 1981* (Lee KE, ed.) pp. 207–214. Government Printer, Adelaide.

Rogers DJ, Brier HB, Houston KJ (1992) Scarabaeidae (Coleoptera) associated with peanuts in southern Queensland. *Journal of the Australian Entomological Society* **31**, 177–181. doi:10.1111/j.1440-6055.1992.tb00477.x

Šípek P, Fabrizi S, Eberle J, Ahrens D (2016) A molecular phylogeny of rose chafers (Coleoptera: Scarabaeidae: Cetoniinae) reveals a complex and contorted morphological evolution related to their flight mode. *Molecular Phylogenetics and Evolution* **101**, 163–175. doi:10.1016/j.ympev.2016.05.012

Smith ABT (2006) A review of the family-group names for the superfamily Scarabaeoidea (Coleoptera) with corrections to nomenclature and a current classification. *Coleopterists Society Monographs* **5**, 144–204.

Smith ABT (2008) South American Melolonthinae (Coleoptera: Scarabaeidae) classification and nomenclature: some problems and solutions. *Insecta Mundi* **0060**, 1–28.

Smith ABT, Evans AV (2005) A supplement to the checklist of the New World chafers (Coleoptera: Scarabaeidae: Melolonthinae) with notes on their tribal classification. *Zootaxa* **1032**, 29–60.

Smith ABT, Hawks DC, Heraty JM (2006) An overview of the classification and evolution of the major scarab beetle clades (Coleoptera: Scarabaeoidea) based on preliminary molecular analyses. *Coleopterists Society Monographs* **5**, 35–46.

Soo Hoo CF, Roberts RJ (1965) Sex attraction in *Rhopaea* (Coleoptera: Scarabaeidae). *Nature* **205**, 724–725. doi:10.1038/205724a0

Szito A (1993) *Evbritonia longitarsa*, a new genus and species of Melolonthinae (Coleoptera: Scarabaeidae) from Western Australia. *Journal of the Australian Entomological Society* **32**, 141–143. doi:10.1111/j.1440-6055.1993.tb00562.x

Szito A (1994a) *Colpomorpha parva*, a new genus and species of Liparetrini (Coleoptera: Scarabaeidae: Melolonthinae) from Western Australia. *Journal of the Australian Entomological Society* **33**, 363–365. doi:10.1111/j.1440-6055.1994.tb01247.x

Szito A (1994b) New species of *Colpochila* Erichson from Western Australia (Coleoptera: Scarabaeidae: Melolonthinae: Liparetrini). *Journal of the Australian Entomological Society* **33**, 1–4. doi:10.1111/j.1440-6055.1994.tb00904.x

Szito A (1995) New species of *Colpochila* Erichson from Western Australia II (Coleoptera: Scarabaeidae: Melolonthinae: Liparetrini). *Journal of the Australian Entomological Society* **34**, 153–157. doi:10.1111/j.1440-6055.1995.tb01309.x

Toussaint EFA, Seidel M, Arriaga-Varela E, Hájek J, Král D, Sekerka L, Short AEZ, Fikáček M (2017) The peril of dating beetles. *Systematic Entomology* **42**(1), 1–10 https://doi.org/10.1111/syen.12198. doi:10.1111/syen.12198

32. SCARABAEIDAE: RUTELINAE MACLEAY, 1919

Tom A. Weir, John F. Lawrence, Cate Lemann and Nicole L. Gunter

Fig. 32.1. *Calloodes grayianus* (White).

Introduction. The subfamily Rutelinae includes ~200 genera and over 4000 species worldwide. There are currently six recognised tribes: Anatistini (Central and South America), Anoplognathini (Australia, Central and South America), Geniatini (Central and South America), Rutelini (widely distributed but primarily Neotropical), Anomalini (widely distributed but primarily in the Old World) and Adoretini (exclusively in the Old World). Phylogenetic evidence indicates the subfamily is not monophyletic, instead a larger clade comprised of a grade of ruteline tribes and the subfamily Dynastinae is consistently recovered in molecular analyses. Despite varying or limited taxon sampling, phylogenetic results are relatively congruent indicating Anomalini and Adoretini are distinct lineages from the Rutelini and/or Anoplognathini, and that future classification changes are necessary (Smith *et al.* 2006; Ahrens *et al.* 2011; Gunter *et al.* 2016).

The Australian fauna consists of 23 genera and 114 species, all in the tribe Anoplognathini, with the exception of three species of *Anomala* (Anomalini) and a single *Adoretus* species (Adoretini) (Ohaus 1904, 1918, 1934; Machatschke 1957, 1965, 1972; Carne 1954, 1955, 1956, 1957, 1958, 1961; Cassis & Weir 1992; Smith 2006). Most of species were described by a handful of authors including Macleay, Waterhouse, Lea, Ohaus, Carne and Allsopp who together described almost three-quarters (84 spp.) of the Australian species. As was the case with dynastines, much of the native fauna was described within the earliest descriptions of the Australian beetles, including 50 species described before 1900 with *Repsimus aeneus* Fabricius 1775 representing one of the first described Australia's insect species. Carne (1954, 1955, 1956, 1957, 1958) revised the subtribes of Anoplognathini, retaining 74 valid species and describing another 20. Ohaus' recognised five subtribes within Anoplognathini, including two Australian subtribes: Anoplognathina and Schizognathina and although Carne commented on the validity of the Australian subtribes, no formal changes where implemented. Preliminary molecular evidence indicates that Anoplognathini is monphyletic, and Anoplognathina and Schizognathina are paraphyletic with respect to each other (Gunter *et al.* 2016), which is largely congruent with Carne's views on the validity of these subtribes.

Rutelines are known from all regions of Australia. Although regional endemicity is highest in the more arid inland region of the continent, ruteline richness and abundance is higher in coastal regions and most species are distributed within two to three adjacent regions. Three major clusters are apparent: a strong south-eastern cluster, a loose eastern tropical and semi-arid cluster and a loose western, north-western and arid cluster (Allsopp 1995). Arguably, *Anoplognathus* spp. are one of the most recognisable Australian insect groups due to their large size, attractive metallic sheen, wide distribution and abundance from November through February which gave rise to their common name, Christmas beetles. The intense lustrous and metallic colours of rutelines are generally produced by the cuticular reflection of circularly polarised light (Pye 2010). While most detailed studies examine structural colour in the New World genus *Chrysina* Kirby, strong polarization has also been confirmed in some *Anoplognathus* spp. (Sharma *et al.* 2009; Pye 2010).

Biology. Adult Rutelinae feed on leaves and flowers, with the exception of *Paraschizognathus* which do not feed as adults. Most Australian species feed on *Eucalyptus* or *Acacia* and feeding damage is often visible as jagged, ripped areas radiating inwards from the edges of leaves. Given the gregarious nature of adults, some rutelines are considered pest species and can cause severe defoliation as they feed on both mature and immature leaves in eucalypts. Some primarily exotic Anomalini, such as the Japanese Beetle, *Popilia japonica* Newman, the Garden Chafer, *Phyllopertha horticola* (Linnaeus) and various species of *Anomala* are agricultural pests.

Following mating, females lay their eggs in soil where the larvae develop, feeding on roots, compost and decaying vegetation. While most species feed on native vegetation and grasses, they are considered pests of pastures and lawns causing yellowing and die-back of grasses. In general pupation and emergence is linked to weather, especially springtime rains and despite the enlarged, weakly fossorial protibiae, most Australian rutelines generally pupate close to the soil surface as if the soil remains too dry newly emerged adults are unable to burrow great distances and may die underground.

Characteristics. Adults. Elongate-oval beetles, variable in colour but often metallic or patterened. Antenna 9- or 10-segmented with 3-segmented club. Scutellar shield exposed. Mesotibial apex with 2 adjacent spurs mesal to tarsal articulation. Pretarsal claws on all legs independently movable, unequal in length or size and frequently split at apex; onychium laterally compressed. Pygidium exposed beyond elytral apices. Male genitalia frequently very slender and sometimes with penis exserted ventrally between parameres, which may be partly or completely fused dorsally. Differing from Australian Dynastinae in having mentum not or only slightly contracted towards ligula, mandibles rarely produced beyond clypeus, labrum visible and sclerotised, head and pronotum unarmed, elytra with sides often lined by membrane, protibiae weakly fossorial with lateral teeth commonly crowded into distal half

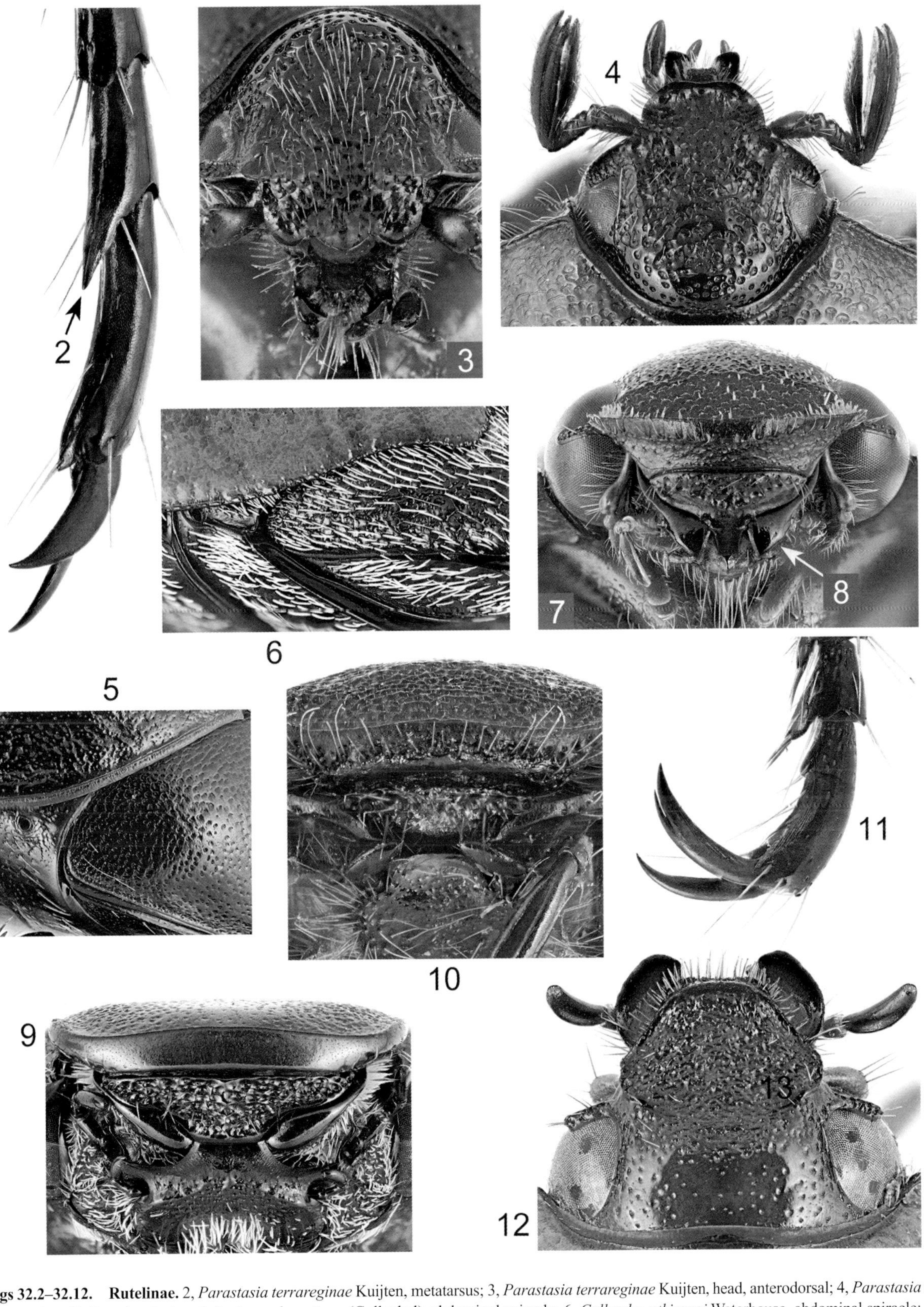

Figs 32.2–32.12. Rutelinae. 2, *Parastasia terrareginae* Kuijten, metatarsus; 3, *Parastasia terrareginae* Kuijten, head, anterodorsal; 4, *Parastasia terrareginae* Kuijten, head, dorsal; 5, *Anomala antiqua* (Gyllenhal), abdominal spiracle; 6, *Calloodes atkinsoni* Waterhouse, abdominal spiracle; 7, *Adoretus melvillensis* Lea, head, frontal; 8, *Adoretus melvillensis* Lea, mandible; 9, *Calloodes atkinsoni* Waterhouse, head, frontal; 10, *Amblyterus simplicitarsus* Carne, head, frontal; 11, *Amblyterus simplicitarsus* Carne, metatarsal claws; 12, *Dungoorus frater* Smith, head, dorsal.

of tibia, meso- and metatibiae not or weakly dilated posteriorly, with carinae lacking or poorly defined, mesocoxae frequently separated by an evident mesoventral process, pretarsal claws of mid and hind legs asymmetrical, movable, frequently toothed, usually very large, propygidium without stridulatory apparatus and pygidium not exhibiting strong sexual dimorphism.

Larvae. Mandibles with stridulatory area consisting of transverse ridges. Maxilla with row of anteriorly directed, sharp, pointed, stridulatory teeth; lacinia with 1–3 unci. Epipharynx with haptomerum, with or without plegmata, without proplegmata, two nesia usually present; haptomerum with two or more prominent heli (beak-like or mound-like) and with 15 or more prominent spine-like setae. Pretarsus bisetose. Dorsal surface of abdominal segments 9 and 10 never fused. Raster with or without palidia; anal slit transverse, slightly curved (Ritcher 1966; Mcquillan 1985).

Key to the genera of Australian Rutelinae

Male characters are used at various points in the key. Males can usually be distinguished from females by having various extensions, constrictions or reflexions of the clypeus whereas the female clypeus is more rounded. The protarsi of the females usually have segment 1 elongate, as long as segments 2–4 combined and often as long as segment 5 and often the large claw of the protarsi is apically toothed. Males of some genera also have much enlarged antennal clubs.

There may be difficulties at couplet 6 separating some species of *Paraschizognathus,* which runs to couplet 7 and species of *Trioplognathus* and *Schizognathus* which run to couplet 13. In *Paraschizognathus*, the scissorial (apical) area of the mandibles is simple, whereas in *Trioplognathus* and *Schizognathus*, the scissorial (apical) area of the mandibles is bidentate and thinner.

1. Meso- and metatarsomere 4 each with sharply pointed apicoventral projection bearing 4 long setae (Fig. 32.2); anterior edge of clypeus, as seen from above, at least slightly produced to form pair of tubercles or teeth; labrum horizontal, projecting well in front of clypeal apex; mandibles exposed beyond clypeus, parallel-sided with truncate or rounded, slightly upturned apices (Figs 32.3, 32.4); frontoclypeal suture absent; lateral edge of elytron without membranous border; 10–19 mm; northern QLD, Christmas Island. .... **RUTELINI**: ***Parastasia* Westwood**

– Meso- and metatarsomere 4 with other, variously shaped projections and spines; anterior edge of clypeus, as seen from above, evenly rounded, truncate or emarginate but without paired teeth or tubercles; labrum vertical or horizontal; mandibles, if exposed, of different shape; frontoclypeal suture almost always present, at least laterally; lateral edge of elytron with or without membranous border.. **2**

2(1). Terminal abdominal spiracle (VII) free, not contiguous with suture or groove (Fig. 32.5); antennae 9-segmented; lateral edge of elytron with membranous border extending to elytral suture; 13–18 mm; northern Australia (WA, NT, QLD), Christmas Island, Cocos-Keeling Islands .... **ANOMALINI:** ***Anomala* Samouelle**

– Terminal abdominal spiracle (VII) contiguous with suture or groove (Fig. 32.6); antennae 9- or 10-segmented; lateral edge of elytron with or without membranous border..**3**

3(2). Labrum distinctive, vertical and contiguous with clypeus, its apex strongly produced, truncate, with transverse row of denticles, its apex overlapping that of labium (Fig. 32.7); mandibles exposed and bent ventrally (Fig. 32.8); 10–12 mm; northern NT .... **ADORETINI:** ***Adoretus* Dejean**

– Labrum vertical or horizontal, rounded or truncate, not strongly produced, with at most a small anteroventral process contiguous with labium; IF mandibles exposed THEN their apices not bent ventrally...**ANOPLOGNATHINI**..**4**

4(3). Mentum laterally compressed, strongly convex and apically setose; labrum horizontal and exposed beyond clypeus; mandibles exposed beyond clypeus; antennal club longer than shaft; length less than 14 mm...**5**

– Mentum flat or at most slightly convex in lateral view; without other features combined..**6**

5(4). Edges of clypeus strongly recurved; elytral intervals 1, 3 and 5 neither raised nor costate; metafemur expanded with dorsal edge markedly curved; protibia with long, often curved setae; 8–13 mm; central Australia (SA, NT) .... ***Clilopocha* Lea**

– Edges of clypeus not strongly recurved; elytral intervals 1, 3 and 5 raised or costate; metafemur not expanded, somewhat parallel-sided; protibia with intermingled long and short setae; 9–12 mm; southern QLD, northern NSW.. ***Mesystoechus* Waterhouse**

6(4). Labium with anteromedian process vertical, curved into mouth cavity and contacting anterior process of labrum (Fig. 32.9); when viewed from the front, maxillary galeae mostly appearing large, filling most of the space between labrum and labium at sides and often grooved for reception of mandibles **7**

– Labium without such a curved process, usually flat or gently convex, its apex rounded, bilobed or with an obtuse median prominence (Fig. 32.10), rarely (*Exochogenys*) with a flat, truncate median process; when viewed from the front, maxillary galeae less obvious and space between labrum and labium narrower..**13**

7(6). Meso-metaventral process produced to form anterior projection, ranging from a short, rounded lobe scarcely surpassing anterior edges of mesocoxae to a large, curved, robust prong lying beneath procoxae; IF metaventral process not surpassing anterior edges of mesocoxae, THEN either body brilliant metallic green (*Anoplognathus parvulus* Waterhouse, northern QLD) OR pygidium clothed with white decumbent setae AND elytra without membranous border (*A. blackdownensis* Carne, central QLD and *A. debaari* Allsopp, southern QLD); elytra usually without externally visible membranous border in apical half..**8**

– Meso-metaventral process not produced as above; elytra with externally visible narrow membranous border in apical half..................................**10**

8(7). Outer edge of protibia usually with 3 teeth; IF basal tooth small or absent THEN colour not dark purplish black or dark green without elytral discal striae AND sutural stria not deeply impressed apically; disc of mentum either flat or with basal depression wholly contained within mentum; 12–32 mm; widely distributed (all states with concentration in north and east)... ***Anoplognathus* Leach**

– Outer edge of protibia with 1 or 2 teeth (basal tooth sometimes small); disc of mentum either with basal depression extending onto submentum or anteriorly slightly swollen..**9**

9(8). Outer edge of protibia with 1 tooth (apical); mentum with basal depression extending onto submentum; sutural stria not deeply impressed apically; metatarsomere 5 with ventral tooth; hind legs of males not greatly enlarged or modified; 18–31 mm; northern Australia (WA, NT, QLD........ .... ***Calloodes* White**

– Outer edge of protibia with 2 teeth (basal tooth sometimes small); mentum slightly swollen anteriorly, without distinct basal depression; sutural stria deeply impressed apically; metatarsomere 5 without ventral tooth, although ventral ridge may be present; hind legs of male grossly enlarged and modified; 15–27 mm; eastern Australia (QLD, NSW, VIC) .... ***Repsimus* Macleay**

10(7). Prosternal process apically flattened; epipleural setae virtually absent from apical half of elytra; head and pronotal disc glabrous (except for a few setae near eyes); males with EITHER patch of specialised setae on mentum OR adpressed white scales on lateral portions of mentum; often brightly but variably coloured; 19–28 mm; northern QLD, northern NT, central and south-western WA .... ***Anoplostethus* Brullé**

– Prosternal process, if present, rounded or acute, not apically flattened; usually at least a few epipleural setae present on apical half of elytra; head usually setose (except in some *Paraschizognathus*); pronotal disc setose or not; males without specialised setae on mentum **11**

11(10). Pronotal disc coarsely punctate and clothed with long, erect setae; larger metatarsal claws toothed apically18–21 mm; south-western WA .... ***Epichrysus* White**

– Pronotal disc more finely punctate and glabrous; larger metatarsal claws not toothed apically **12**

12(11). Clypeus half as long as wide with sides diverging anteriorly and apex broadly curved with small median indentation; 18–19 mm; southern QLD .... ***Wambo* Allsopp**

– Clypeus usually more than half as long as wide with sides slightly curved or straight and parallel or converging anteriorly; 13–22 mm; eastern Australia (southern QLD, NSW, VIC) .... ***Paraschizognathus* Ohaus**

13(6). Larger meso- and metatarsal claws toothed, sometimes finely so, more marked in females than in males (Fig. 32.11); dark reddish-brown or reddish black; 13–23 mm; southern QLD, NSW .... ***Amblyterus* Macleay (part)**

– Larger meso- and metatarsal claws not toothed **14**

14(13). Pronotum and elytra covered partly or completely with decumbent, elongate, white or yellow scale-like setae or scales, sometimes sparse **15**

– Pronotum and elytra without decumbent, elongate, white or yellow scale-like setae or scales **17**

15(14). Total length 20–25 mm; metaventral process anteriorly produced, rounded and extending to anterior edges of mesocoxae; white scale-like setae on elytra not arranged in rows; southern QLD, NSW .... ***Trioplognathus* Ohaus**

– Total length 7–16 mm; metaventral process not anteriorly produced; setae on elytra arranged in rows or not **16**

16(15). Elytra with elongate, white, scale-like setae not arranged in rows or patches; ventralsurfaces clothed in long, erect white setae; male antennal club enlarged, leaf-like, much longer than shaft; outer edge of protibia with 3 well defined teeth; 13–16 mm; eastern NSW .... ***Saulostomus brunneoviridis* Lea**

– Elytra with white or yellow scale-like setae or scales arranged either in rows or patches (sparse and often rubbed off in *Mimadoretus leucothyreus*); ventral surfaces clothed in either long, erect, yellow setae or scales; male antennal club not as enlarged, not leaf-like, at most as long as shaft; outer edge of protibia with 2 or 3 teeth, if the latter then basal tooth ill defined; 7–15 mm; northern QLD .... ***Mimadoretus* Arrow**

17(14). Labium with distal edge produced (in same plane) to form conspicuous, flat, apically truncate process; anterior edge of clypeus slightly concave; all black species or with elytra reddish-black; 12–16 mm; eastern Australia (southern QLD, NSW, VIC) .... ***Exochogenys* Carne**

– Labium with distal edge bilobed, gently rounded or with obtuse median lobe, never with distinct truncate process; anterior edge of clypeus straight or convex; colour various but not as above **18**

18(17). Labrum more or less produced ventrally in middle, in same vertical plane, often overlapping labium **19**

– Labrum not produced ventrally in same plane, appearing horizontal, with any small apical protuberance directed back into mouth cavity **22**

19(18). Ocular canthus short, triangular, subvertical, hardly extending onto surface of eye; pronotum clothed with long, erect setae; labial palps 2-segmented; male eyes protuberant; frons and clypeus compressed and elongate with clypeus expanded anteriorly; head, pronotum and scutellum dark metallic green, elytra light brown; 11–15 mm; southern QLD, northern NSW .... ***Amblochilus* Blanchard**

– Ocular canthus extending well onto surface of eye; pronotum glabrous with at most a few setae near anterior angles; labial palps 3-segmented; male eyes not protuberant; frons and clypeus neither compressed nor elongate; colour pattern different **20**

20(19). Elytra coarsely punctate, strongly wrinkled, with intervals strongly costate; apical maxillary palpomere small, not more than one-third (male) or one half (female) as long as antennal club; clypeus markedly concave, sides reflexed; pronotum and scutellum black, elytra light brown; 10–13.5 mm; eastern Australia (QLD, NSW, VIC) .... ***Bilobatus* Machatschke**

– Elytra more finely punctate, sometimes slightly wrinkled but never costate; apical maxillary palpomere enlarged or not, at least half as long as antennal club; male clypeus not as concave, parallel-sided or constricted anteriorly; colour other than above **21**

21(20). Apical maxillary palpomere as long as antennal club, with very large, deeply impressed sensorium; pygidium with long, erect setae only; male clypeus more or less parallel-sided; metaventral process not anteriorly produced; antennae 9- or 10-segmented; 12–16 mm; eastern Australia (QLD, NSW, VIC) .... ***Pseudoschizognathus* Ohaus**

– Apical maxillary palpomere one-half to two-thirds as long as antennal club, with sensorium not deeply impressed; pygidium clothed with small to minute, decumbent setae (sometimes with a few long erect setae near edges as well); EITHER male clypeus strongly constricted before reflexed anterior edge OR metaventral process produced anteriorly; 14–22 mm; eastern Australia (QLD, NSW, VIC) .... ***Schizognathus* Fischer**

22(18). Clypeus strongly narrowed anteriorly, truncate; mandibles plate-like, projecting well beyond clypeus, their lateral edges strongly reflexed (Fig. 32.12); apical maxillary palpomere in male as long as or longer than antennal club, with very deeply impressed sensorium; antennae 9-segmented; 11–13 mm; central Australia (QLD, NT) .... ***Dungoorus* Carne**

– Clypeus not strongly narrowed anteriorly, sides parallel or slightly rounded; IF mandibles projecting beyond clypeus THEN clypeus anteriorly rounded; apical maxillary palpomere in male at least slightly shorter than antennal club; antennae 9- or 10-segmented **23**

23(22). Meso- and metatarsomeres 1–4 combined at least twice as long as 5; dorsal surfaces glabrous or at most with sparse setae near anterior angles of lateral margins of pronotum; male antennal club enlarged and leaf-like; 11–16 mm; eastern and southern Australia (QLD, SA, WA) .... ***Eosaulostomus* Carne**

– Meso- and metatarsomeres 1–4 combined equal to or scarcely longer than 5; dorsal surfaces with conspicuous setae at least on head; male antennal club may be elongate but never leaf-like **24**

24(23). Elytral striae grouped in pairs, so that intervals 2 and 4 are different in width to intervals 3 and 5 OR striae indistinct; head, pronotum and elytra clothed in long, erect setae (may be rubbed off); antennae 10-segmented; 8–16 mm; eastern Australia (QLD, NSW, VIC, TAS) .... ***Saulostomus* Waterhouse (part)**

– Elytral striae not grouped in obvious pairs, adjacent intervals about equal in width; elytral striae deeply punctate; pronotum and sides of elytra with shorter, often decumbent setae (except in *A. paradoxus* Carne, which has 9-segmented antennae); 10–14 mm; eastern Australia (QLD, NSW).. ***Amblyterus* Macleay (part)**

Classification of the Australian genera

The tribal classification used here follows that of Smith (2006) and Bouchard *et al.* (2011). For convenience, in Anoplognatini, we retain the distribution of genera between the subtribes Anoplognathina and Schizognathina as in Carne (1958), even though this may prove to be somewhat artificial.

Tribe Adoretini Burmeister, 1844

Characteristics. Frontoclypeal suture present. Terminal abdominal spiracle (VII) contiguous with suture or groove. Labrum vertical and contiguous with clypeus, its apex strongly produced, truncate, with transverse row of denticles, its apex overlapping that of labium. Mandibles exposed and bent ventrally. Antennae with 10 segments. Inner protibial spur subapical. Elytra without membranous border.

Distribution. Palaearctic, Ethiopian and Oriental regions

Australian taxa. One genus.

Adoretus Dejean, 1833

Type species. *Melolontha nigricornis* Steven, 1809.

Australian species. One.

Distribution. The genus *Adoretus* is distributed throughout the Palaearctic, Afrotropical and Oriental regions and extends into Papua New Guinea. *Adoretus melvillensis* Lea was described from Melville Island, NT.

Biology. The exotic Chinese Rose Beetle, *Adoretus sinicus* Burmeister is a serious pest in Hawaii, attacking ~55 different species of cultivated crops and ornamental plants.

References. Carne (1958); Habeck (1964); Krell (2007)

Tribe Anomalini Streubel, 1839

Characteristics. Frontoclypeal suture present, at least laterally. Terminal abdominal spiracle (VII) free, not contiguous with suture or groove. Labrum horizontal with respect to the clypeus. Mandibles concealed. Antennae with 9 segments. Inner protibial spur subapical. Elytra with membranous border extending to elytral suture.

Distribution. Worldwide.

Australian taxa. One genus.

Anomala Samouelle, 1819 (Pl. 67L)

Type species. *Melolontha frischii* Fabricius, 1775 (= *Scarabaeus dubia* Scopili, 1763).

Australian species. Three.

Distribution. The genus *Anomala* is more or less cosmopolitan in distribution. Of the three Australian species, *A. aeneotincta* Fairmaire occurs in QLD and WA, as well as New Britain and Papua New Guinea, *A. antiqua* (Gyllenhal) occurs in northern QLD, NT and northern WA, as well as the Oriental and Palaearctic regions, and *A. katherine* Jin *et al.* occurs in NT.

Biology. Adults of *A. antiqua* are nectar feeders and have been reported from the flowers of the Leichardt Tree, *Nauclea orientalis* (Rubiaceae) and *Lantana* sp. (Verbenaceae). Larvae occur in piles of organic refuse.

References. Carne (1958); Jin *et al.* (2014).

Tribe Anoplognathini Macleay, 1819

Characteristics. Frontoclypeal suture present, at least laterally. Terminal abdominal spiracle (VII) contiguous with suture or groove. Labrum horizontal or vertical with respect to clypeus, sometimes projecting in front of clypeus. Mandibles concealed or exposed. Membranous border of elytra present or absent. Antennae with 9 or 10 segments. Inner protibial spur subapical or absent.

Distribution. Australasian and Neotropical.

Australian taxa. 20 genera.

Subtribe Anoplognathina Macleay, 1819

Anoplognathus Leach, 1815 (Pl. 3A–B, H, 7F, 67B, E, H, K, 68H, 76D, E, J–K, H)

Type species. *Melolontha viridiaenea* Donovan, 1825.

Paranonca Laporte, 1840

Australian species. 37.

Distribution. *Anoplognathus* species are widely distributed in Australia and occur in Papua New Guinea.

Biology. Adult *Anoplognathus* often occur in large numbers and defoliate a variety of trees, but especially *Eucalyptus* and related genera. They are not considered to be major pests, but might cause serious damage to ornamental trees in parks and streets, windbreaks or nurseries. Larvae occur in soil under natural pasture or sometimes under rotten logs, feeding on soil, organic matter, grass roots, leaf mould or finer roots of eucalypts. The larva of *A. suturalis* Boisduval was described and illustrated by Mcquillan (1985).

References. Carne (1957, 1958); Carne *et al.* (1974).

Anoplostethus Brullé, 1837 (Pl. 67F)

Type species. *Anoplostethus opalinus* Brullé, 1837.

Anoplosternus Guérin-Méneville, 1838

Panschizus Blackburn, 1888

Australian species. Three.

Distribution. Endemic to Australia. *Anoplostethus* species occur in northern QLD, NT, northern and southern WA.

References. Carne (1958).

Calloodes White, 1845 (Pl. 6A, 67A, C–D, I)

Type species. *Calloodes grayianus* White, 1845.

Australian species. Four.

Distribution. *Calloodes* species are known from northern QLD and northern WA, and one species, *C. atkinsoni* Waterhouse, extends into Papua New Guinea.
Biology. Adults of *C. atkinsoni* and *C. grayanus* White were recorded from the foliage of Red Bloodwood, *Corymbia gummifera* and larvae were found in pure sand in a river bed.
References. Carne (1958).

Epichrysus White, 1841 (Pl. 67G)

Type species. *Brachysternus* (*Epichrysus*) *lamprimoides* White, 1841.
Australian species. One.
Distribution. Endemic to Australia. *Epichrysus lamprimoides* occurs in southern WA.
References. Carne (1958).

Paraschizognathus Ohaus, 1904

Type species. *Schizognathus prasinus* Boisduval, 1835.
Australian species. 13.
Distribution. Endemic to Australia. *Paraschizognathus* species are known from VIC, ACT, NSW and QLD.
References. Ohaus (1904); Carne (1958).

Repsimus W. S. Macleay, 1819 (Pl. 5B, 67J, 68F–G)

Type species. *Anoplognathus brownii* W. S. Macleay, 1819 (= *Rutela manicata* Swartz, 1817).
Australian species. Two.
Distribution. Endemic to Australia. *Repsimus* species occur in VIC, ACT, NSW and QLD.
Biology. Adults of *R. aeneus* (Fabricius) were observed feeding on foliage of two myrtaceous trees, *Corymbia gummifera* and *Syzygium tierneyana*.
References. Carne (1958).

Wambo Allsopp, 1988 (Pl. 57C)

Type species. *Wambo puticasus* Allsopp, 1988.
Australian species. One.
Distribution. Endemic to Australia. *Wambo puticasus* occurs in southern QLD.
References. Allsopp (1988).

Subtribe Schizognathina Ohaus, 1918

Amblochilus Blanchard, 1851 (Pl. 57F)

Type species. *Amblochilus bicolor* Blanchard, 1851.
Australian species. One.
Distribution. Endemic to Australia. *Amblochilus bicolor* is known from QLD.
References. Carne (1958).

Amblyterus W. S. Macleay, 1819 (Pl. 57G)

Type species. *Amblyterus geminatus* W. S. Macleay, 1819.
Australian species. Eight.
Distribution. Endemic to Australia. Species of *Amblyterus* are known from NSW and QLD.
References. Carne (1958, 1961).

Bilobatus Machatschke, 1970 (Pl. 57B)

Type species. *Homotropus luridipennis* Waterhouse, 1878.
Homotropus Waterhouse, 1878 (not Foerster, 1868)
Australian species. Two.
Distribution. Endemic to Australia. *Bilobatus testaceipennis* (Ohaus) occurs in VIC and *B. luridipennis* (Waterhouse) is known from southern QLD and VIC.
Biology. Carne (1958); Machatschke (1972).
References. Carne (1958); Machatschke (1972).

Clilopocha Lea, 1914

Type species. *Clilopocha whiteae* Lea, 1914.
Dynastomorphus Carne, 1954.
Australian species. Five.
Distribution. Endemic to Australia. *Clilopocha* species are known from SA and NT.
References. Carne (1954, 1955, 1958).

Dungoorus Carne, 1958 (Pl. 57D)

Type species. *Dungoorus murrumbullus* Carne, 1958.
Australian species. One.
Distribution. Endemic to Australia. *Dungoorus murrumbullus* was described from QLD.
References. Carne (1958).

Eosaulostomus Carne, 1956 (Pl. 68B)

Type species. *Eosaulostomus excisus* Carne, 1956.
Australian species. Six.
Distribution. Endemic to Australia. Species of *Eosaulostomus* are known from SA, WA and QLD.
References. Carne (1956, 1958).

Exochogenys Carne, 1958 (Pl. 57I)

Type species. *Schizognathus nigripennis* Blanchard, 1851.
Australian species. One.
Distribution. Endemic to Australia. *Exochogenys nigripennis* occurs in VIC, NSW and southern QLD.
References. Carne (1958).

Mesystoechus Waterhouse, 1878 (Pl. 68A)

Type species. *Mesystoechus ciliatus* Waterhouse, 1878.
Australian species. Two.

Distribution. Endemic to Australia. *Mesystoechus* species are known from NSW and QLD.
References. Carne (1958).

Mimadoretus Arrow, 1901 (Pl. 8B, 57E, N)

Type species. *Popillia flavimaculata* Macleay, 1887.
Popillia Macleay, 1887, not Leach 1826.
Australian species. Three.
Distribution. Endemic to Australia. *Mimadoretus* species occur in northern QLD.
References. Carne (1958).

Pseudoschizognathus Ohaus, 1904 (Pl. 57L)

Type species. *Pseudoschizognathus schonfeldti* Ohaus, 1904.
Australian species. Three.
Distribution. Endemic to Australia. *Pseudoschizognathus* species are known from VIC, NSW and QLD.
References. Carne (1958).

Saulostomus Waterhouse, 1878 (Pl. 57M)

Type species. *Saulostomus villosus* Waterhouse, 1878.
Australian species. Five.
Distribution. Endemic to Australia. *Saulostomus* species occur in TAS, VIC, NSW and QLD.
Biology. The larva of *S. villosus* Waterhouse was described and illustrated by Mcquillan (1985).
References. Carne (1956, 1958).

Schizognathus Fischer, 1823 (Pl. 57P)

Type species. *Schizognathus macleayi* Fischer, 1823.
Australian species. Eight.
Distribution. Endemic to Australia. *Schizognathus* species are known from VIC, ACT, NSW and QLD.
References. Carne (1958).

Trioplognathus Ohaus, 1904 (Pl. 57O)

Type species. *Anoplognathus griseopilosus* Ohaus, 1901.
Australian species. One.
Distribution. Endemic to Australia. *Trioplognathus griseopilosus* occurs in NSW.
References. Carne (1958).

Tribe Rutelini Macleay, 1819

Characteristics. Frontoclypeal suture absent. Labrum horizontal, projecting well in front of clypeus. Terminal abdominal spiracle (VII) contiguous with suture or groove. Mandibles exposed beyond clypeus, upturned. Antennae 10 segmented. Inner protibial spur apical. Membranous border of elytra absent.
Distribution. Worldwide.
Australian taxa. One genus.

Parastasia Westwood, 1841 (Pl. 57A)

Type species. *Parastasia canaliculata* Westwood, 1841.
Australian species. Three.
Distribution. South-east Asia, Philippines, Indonesia, New Guinea, Australia, Pacific Islands and eastern North America (one species). *Parastasia montrouzieri* Fairmaire and *P. terrareginae* Kuijten occur in northern QLD and *P. christmasensis* Wada occurs on Christmas Island.
References. Kuijten (1992); Jameson (2002); Wada (2008).

References

Ahrens D, Scott M, Vogler AP (2011) The phylogeny of monkey beetles based on mitochondrial and ribosomal RNA genes (Coleoptera: Scarabaeidae: Hopliini). *Molecular Phylogenetics and Evolution* **60**, 408–415. doi:10.1016/j.ympev.2011.04.011

Allsopp PG (1988) *Wambo puticasus* gen. et sp. nov., a new ruteline from south Queensland (Coleoptera: Scarabaeidae). *Memoirs of the Queensland Museum* **25**, 255–258.

Allsopp PG (1995) Biogeography of the Australian Dynastinae, Rutelinae, Scarabaeinae, Melolonthini, Scitalini and Geotrupidae (Coleoptera: Scarabaeoidea). *Journal of Biogeography* **22**, 31–48. doi:10.2307/2846071

Bouchard P, Bousquet Y, Davies AE, Alonso-Zarazaga MA, Lawrence JF, Lyal CHC, Newton AF, Reid CAM, Schmitt M, Slipinski SA, Smith ABT (2011) Family names in Coleoptera (Insecta). *ZooKeys* **88**, 1–972. doi:10.3897/zookeys.88.807

Burmeister H (1844) *Handbuch der Entomologie*, vol. 4, part 1. Enslin, Berlin. 586 pp.

Carne PB (1954) Notes on the Australian Rutelinae (Coleoptera) and description of a new genus. *Proceedings of the Royal Entomological Society of London* **23**, 36–40[Series B].

Carne PB (1955) Notes on the Australian Rutelinae (Scarabaeidae, Coleoptera). Suppression of a generic name under *Clilopocha* Lea. *Proceedings of the Linnean Society of New South Wales* **80**(2), 137.

Carne PB (1956) A revision of *Saulostomus* Waterhouse and description of a new ruteline genus (Scarabaeidae, Coleoptera). *Proceedings of the Linnean Society of New South Wales* **81**(1), 62–70.

Carne PB (1957) A revision of the ruteline genus *Anoplognathus* Leach (Coleoptera: Scarabaeidae). *Australian Journal of Zoology* **5**(1), 88–143. doi:10.1071/ZO9570088

Carne PB (1958) A review of the Australian Rutelinae (Coleoptera: Scarabaeidae). *Australian Journal of Zoology* **6**, 162–240. doi:10.1071/ZO9580162

Carne PB (1961) Supplementary note to a revision of the Australian Rutelinae (Coleoptera: Scarabaeidae). *Proceedings of the Linnean Society of New South Wales* **86**(1), 126–127.

Carne PB, Greaves RTG, Mcinnes RS (1974) Insect damage to plantation grown eucalypts in north coastal New South Wales, with particular reference to Christmas beetles (Coleoptera: Scarabaeidae). *Journal of the Australian Entomological Society* **13**, 189–206. doi:10.1111/j.1440-6055.1974.tb02173.x

Cassis G, Weir TA (1992) Rutelinae. In *Zoological Catalogue of Australia. Vol. 9. Coleoptera: Scarabaeoidea.* (Eds G Cassis, WWK Houston, TA Weir and BP Moore) pp. 359–382. Australian Government Publishing Service, Canberra.

Gunter NL, Weir TA, Slipinski A, Bocak L, Cameron SL (2016) If dung beetles (Scarabaeidae: Scarabaeinae) arose in association with dinosaurs, did they also suffer a mass co-extinction at the K-Pg boundary? *PLoS One* **11**(5), e0153570. doi:10.1371/journal.pone.0153570

Habeck DH (1964) Notes on the biology of the Chinese rose beetle, Adoretus sinicus Burmeister in Hawaii. *Proceedings of the Hawaiian Entomological Society* **18**(3), 399–403. http://scholarspace.manoa.hawaii.edu/bitstream/handle/10125/10891/18_399-404.pdf

Jameson ML (2002) IV Rutelinae MacLeay 1819. In *American Beetles. Volume 2. Polyphaga: Scarabaeoidea through Curculionoidea*. (Eds RH Arnett, MC Thomas, PE Skelley and JH Frank) pp. 60–64. CRC Press, Boca Raton, Florida.

Jin M, Weir T, Slipinski A, Pang H (2014) Description of a new species of *Anomala* Samouelle (Coleoptera: Scarabaeidae, Rutelinae) from Northern Territory, Australia. *Zootaxa* **3872**(5), 591–600. doi:10.11646/zootaxa.3872.5.9

Krell F-T (2007) Author, date and type species of *Adoretus* (Scarabaeidea: Rutelinae). *Coleopterists Bulletin* **61**(1), 125–126. doi:10.1649/952.1

Kuijten PJ (1992) A revision of the genus *Parastasia* in the Indo-Australian region (Coleoptera: Scarabaeidae: Rutelinae). *Zoologische Verhandelingen*, Leiden **275**, 1–207, 1 pl.

Laporte FL (1840) *Histoire Naturelle des Insectes Coléoptères* (volume 2 of Histoire Naturelle des Animaux Articulés). P. Duménil, Paris. 564 pp.

Machatschke JW (1957) *Genera Insectorum. Fascicle 199B. Coleoptera Lamellicornia Fam. Scarabaeidae. Subfam. Rutelinae. Zweiter Teil.* Genera Insectorum, Crainhem, Belgium, 219 pp., 6 pls.

Machatschke JW (1965) *Genera Insectorum. Fascicle 199C. Coleoptera Lamellicornia Fam. Scarabaeidae. Subfam. Rutelinae. Dritter Teil.* Genera Insectorum, Crainhem, Belgium, 145 pp., 2 pls.

Machatschke JW (1972) Scarabaeoidea: Melolonthidae, Rutelinae. *Coleopterum Catalogus Supplementa* **66**(1), 1–361.

MacLeay WS (1819) *Horae Entomologicae: or Essays on the Annulose Animals*, volume 1, part 1. S. Bagster, London. 160 pp.

Mcquillan PB (1985) The identification of root-feeding cockchafer larvae (Coleoptera: Scarabaeidae) found in pastures in Tasmania. *Australian Journal of Zoology* **33**, 509–546. doi:10.1071/ZO9850509

Ohaus F (1904) Revision der Anoplognathiden (Coleoptera lamellicornia). *Stettiner Entomologische Zeitung* **65**, 57–175.

Ohaus F (1918) Scarabaeidae: Euchirinae, Phaenomerinae, Rutelinae. *Coleopterorum Catalogus* **20**, 1–241[dated 1915].

Ohaus F (1934) *Genera Insectorum de P. Wytsman. Fascicule 199A. Coleoptera Lamellicornia Fam. Scarabaeidae Subfam. Rutelinae. Erster Teil.* Genera Insectorum, Tervuren, Louis Desmet-Verteneuil, Brussels, 172 pp., 6 pls.

Pye JD (2010) The distribution of circularly polarized light reflection in the Scarabaeoidea (Coleoptera). *Biological Journal of the Linnean Society. Linnean Society of London* **100**, 585–596. doi:10.1111/j.1095-8312.2010.01449.x

Ritcher PO (1966) *White Grubs and Their Allies. A Study of North American Scarabaeoid Larvae*. (Studies in Entomology No. 4). Oregon State University Press, Corvallis, Oregon.

Sharma V, Crne M, Park JO, Srinivasarao M (2009) Structural origin of Circularly Polarised Iridescence in Jeweled Beetles. *Science* **325**(5939), 449–451. doi:10.1126/science.1172051

Smith ABT (2006) A review of the family-group names for the superfamily Scarabaeoidea (Coleoptera) with corrections to nomenclature and a current classification. *Coleopterists Society Monographs* **5**, 144–204.

Smith ABT, Hawks DC, Heraty JM (2006) An overview of the classification and the evolution of the major scarab beetle clades (Coleoptera: Scarabaeoidea) based on preliminary molecular analyses. *Coleopterists Society Monographs* **5** (2006), pp. 35–46.

Wada K (2008) Five new species of the genus *Parastasia* from the Oriental region. *Elytra, Tokyo* **36**(1), 87–99.

33. SCARABAEIDAE: DYNASTINAE MACLEAY, 1919

Tom A. Weir, John F. Lawrence, Cate Lemann and Nicole L. Gunter

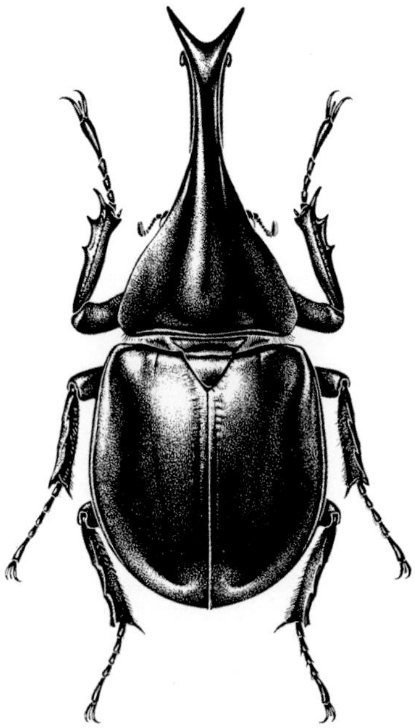

Fig. 33.1. *Xylotrupes australicus australicus* Thomson.

Common name. Rhinoceros beetles.

Introduction. The subfamily Dynastinae, including ~225 genera and 1500 species, is worldwide in distribution, although most diverse in tropical regions. There are eight included tribes, the largest of which, Pentodontini, is worldwide in distribution and includes most Australian species. Other dynastine tribes include Agaocephalini (New World tropics), Hexodontini (Madagascar), Cyclocephalini (primarily New World with one African genus and a single introduced Australian species), Oryctoderini (Oriental, with three genera extending into Australia), Oryctini (widely distributed but with only one genus extending into northern Australia), Phileurini (primarily pantropical, but including the widely distributed Australian genus *Cryptodus*) and Dynastini (with several Asian genera, one in Africa, four in tropical America and two extending into Australia) (Endrödi 1985; Dechambre 2005a). Native species are known from all States in Australia with species richness highest in south-east and north-east Queensland, south-west Western Australia and inland Australia (Allsopp 1995). Over a third of species in south-west Western Australia and inland Australia are endemic to these regions while in eastern Australia species few species are endemic to a single region but instead can be divided into two more broadly distributed groups, one ranging from south-eastern Queensland to south-eastern New South Wales and the other ranging from north-eastern Queensland and north-western Queensland to south-western New South Wales, west of the Great Dividing Range (Allsopp 1995).

Only a few studies have explored the phylogenetic relationships of the subfamily but despite the limited attention, relationships are relatively congruent. In all studies Dynastinae forms a clade with Rutelinae (Howden 1982; Browne & Scholtz 1998; Jameson 1998; Smith *et al.* 2006; Hunt *et al.* 2007; McKenna *et al.* 2015; Gunter *et al.* 2016) or Rutelinae and Pachydemini in part (Melolonthinae) (Ahrens *et al.* 2014). Although the exact relationship between these subfamilies differ depending on taxon sampled, molecular studies with comprehensive sampling within the family Scarabaeidae support the monophyly of Dynastinae (Hunt *et al.* 2007; Ahrens *et al.* 2014; Gunter *et al.* 2016). The tribal relationships within the subfamily have received limited attention. Two studies of the Dynastini both recovered two clades consisting of the Oriental/Australian fauna and Afrotropical + Neotropical/Nearctic fauna (Rowland and Miller 2012; Jin *et al.* 2016). The estimated divergence of these clades occurred after the break-up of Pangea and the distribution of the tribe is attributed to dispersal in the Upper Cretaceous (Jin *et al.* 2016). Divergence date estimates suggest Dynastinae originated ~100–115 Ma (Gunter *et al.* 2016; Jin *et al.* 2016) and significantly predate the earliest known fossils which are recorded from the Oligocene (Krell 2007).

Some 192 dynastine species in 39 genera have been recorded in Australia. Six species are classed as introduced, of which *Cyclocephala cygnaticollis* and *Heteronychus arator* are established and widespread. The earliest efforts to describe the fauna consisted of descriptions of one or two species in the publications Fabricius (1775), Kirby (1818), W.S. Macleay (1819), Guérin-Méneville (1830, 1831), Boisduval (1835), Hope (1837, 1845), Castelnau (1840), Westwood (1841, 1856), White (1841), Erichson (1842), Germar (1848), Reiche (1860), Montrouzier (1860), Olliff (1889) and Sternberg (1910). Larger efforts by a handful of taxonomists saw the description of 7 species described by Burmeister (1847) and Fairmaire (1877), 9 species described by Macleay (1871, 1887, 1888), 12 species described by Sharp (1873, 1875), 37 species described by Blackburn (1888, 1889, 1890a–e, 1892a–b, 1894, 1895, 1896, 1897). Almost 50% of species (95) were described before 1900 which can be attributed to their large size and the charismatic ornamentation of the males. In fact, the first formally described Australian insect species was *Haploscapanes barbarossa,* which also holds the record as Australia's heaviest beetle. Later taxonomic effort included Arrow (1908, 1911, 1937) and Lea (1917, 1919a–b, 1920, 1926) who described another 9 and 20 species respectively, before the first major revision of the fauna by Carne (1957b) which described a further 45 species. Since then a few isolated works have seen the description of the remaining 19 species (Allsopp 1993, 2000; Allsopp & Carne 1986; Carne 1976, 1978, 1980, 1981, 1985a; Carne & Allsopp 1987; Dechambre 2003a, 2003b; Endrödi 1971a, 1971b, 1974, 1977, 1978; Howden & Vladislav 2005; Silvestre 2012).

Biology. Little is known of the biology of most dynastines. Adults are generally nocturnal occurring on the ground and not feeding on foliage. Members of the genus *Cryptodus* live in association with termites and ants. Introduced species, such as *Heteronychus arator* and *Cyclocephala signaticollis*, are regarded as pasture pests and lawn pests, but several native species in the genera

Carneodon, *Metanastes*, *Adoryphorus* and *Dasygnathus* have been implicated in pasture damage (Cumpston 1940; Carne 1957b; Mcquillan 1985). *Oryctes rhinoceros* is regarded as a serious coconut pest overseas. Despite their size, adults typically consume less than larvae and as such they produce little waste. Furthermore, they are relatively easy to maintain in culture and are becoming increasingly popular in insect zoos and as pets. When disturbed some adults generate a warning through stridulating their abdomen against their elytra which produces a squeaking or hissing sound.

Although the larger species in the genera *Xylotrupes* and *Haplanoscapanes* have a two or three year life cycles, most are probably univoltine. Larval dynastines, in general, feed on roots, in decaying vegetable matter in soil or in rotten logs. Two distinctive types of life cycle are known. In the first type, exhibited by species of *Cyclocephala, Cryptodus*, and probably members of the Pseudoryctina, eggs are laid in mid to late summer, larvae overwinter and pupate in spring, and adults emerge in early summer; this is correlated, at least in the pseudoryctines, with marked sexual dimorphism, vestigial or non-functional mouthparts and sedentary females that are rarely collected. In the second type, exhibited by members of the Cheiroplatina and Pentodontina, eggs are laid in early summer, larvae develop rapidly, and adults either emerge in autumn or remain in their pupal cells until the following spring. These species have strongly developed, functional mouthparts, a minimum of male secondary sexual armature and active females collected as commonly as males.

Characteristics. Adults. Length 9–60 mm. Stout-bodied beetles usually moderately convex, (flattened in *Cryptodus*), ranging in colour from reddish-yellow to reddish-brown, reddish-black or black, occasionally with elytra differing in colour from head and pronotum and very rarely yellow with distinct colour pattern. Antennae almost always 10-segmented, rarely with 8 or 9 antennomeres; club 3-segmented. Mentum usually contracted towards ligula (broad concealing all mouthparts in *Cryptodus*). Mandibles simple or toothed, usually exposed beyond clypeus, or visible at sides. Labrum rarely exposed and always membranous. Pronotum usually armed and with impression or excavation. Elytra without membranous lateral edges. Mesocoxae united or almost so, never conspicuously separated by mesometaventral process. Protibia strongly fossorial, with lateral teeth evenly spaced; meso- and metatibiae normally dilated posteriorly, with 1 or more commonly 2 ciliate carinae; apical spurs set to one side of tarsal articulation. Pretarsal claws immovable, usually symmetrical and never toothed, those of fore leg sometimes asymmetrical in male. Last 3 abdominal spiracles distinctly diverging apically (unlike Melolonthinae). Propygidium often with stridulatory ridges, either dispersed or organised into bands, the sclerite frequently enlarged. Pygidium characteristically large, transverse, more convex in male (except in some *Heteronychus*, *Metanastes*, *Oryctes* and *Crytpodus*). Male genitalia stout with penis exserted dorsally between parameres, the latter never fused along dorsal edges.

Larvae. Apical antennomere as wide as preapical one. Galea and lacinia fused to form mala. Anal cleft transverse. Sternum IX without raster consisting of longitudinal rows of spines. Mandible with ventral stridulatory area. Labrum symmetrical with a weak, pigmented notch on the right side of midline. Legs long, with metacoxa larger than clypeal width; dorsal stridulatory teeth on dorsal surface of maxilla bluntly rounded. Head usually strongly punctate (Ritcher 1966; Cumpston 1940; Mcquillan 1985).

Key to the tribes, subtribes and genera of Australian Dynastinae

Male characters are used throughout the key. Males can be distinguished from females in most cases by having the last abdominal ventrite about as long as the preceding one and emarginate to various degrees apically. Females have the last abdominal ventrite longer than the preceding one, somewhat triangular in shape and apically rounded. In some genera the male protarsi are more or less thickened and the pretarsal claws asymmetrical OR male protarsi elongated. Characters involving the propygidium require lifting of one of the elytra. Characters involving the maxillary galea usually require mouthpart dissection, although the maxillae are somewhat exposed in a few Pentodontina.

1. Mentum dilated, concealing at least bases of labial palps and usually most of maxillae and labium ....**Phileurini**....2
– Mentum normal, more or less narrow, not concealing bases of maxillary palps 3

2(1). Mandibles produced beyond clypeus, which is apically narrowed; mentum not concealing maxillae (Fig. 33.2) and not produced at suture with submentum; antennal scape not expanded and wedge-like, not concealing funicular segments; 20 to 26 mm in length; northern QLD;.... **Phileurina: *Eophileurus* Arrow**
– Mandibles not visible beyond clypeus, which is broadly rounded or truncate at apex; mentum greatly enlarged, concealing all mouthparts (Fig. 33.3), and produced and often notched at suture with submentum; antennal scape apically dilated and wedge-like, concealing all funicular segments; 12 to 27 mm in length; widespread – all Australian States, Lord Howe Island;.... **Cryptodina: *Cryptodus* Macleay**

3(1). Apex of metatibia scalloped (Fig. 33.4) or with longer or shorter fixed teeth, never with articulated spines 4
– Apex edge of metatibia truncate or very slightly produced, almost always with at least a few (Oryctoderini) or many articulated spines; IF spines lacking THEN head with broad, more or less vertical, rugosely punctate, frontoclypeal plate (*Temnorrhynchus*)....6

4(2). Propygidium enlarged and bearing transverse stridulating ridges (Fig. 33.5); apical pygidial ridge setose; metatarsomere 1 produced apically (Fig. 33.6); sexual dimorphism of armature, slight, head with median horn in both sexes; 37 to 47 mm in length; northern QLD, northern NT, Cocos-Keeling Islands .... **Oryctini: *Oryctes* Illiger**
– Propygidium not enlarged, without stridulating ridges; apical pygidial ridge glabrous; metatarsomere 1 not produced apically; sexual dimorphism of armature great, head in female at most with tubercle....**Dynastini**....5

5(4). Meso- and metatibiae with apices obtusely scalloped and intermediate and basal carinae non-ciliate (Fig. 33.7); clypeal apex bituberculate; mandibles strongly bidentate (Fig. 33.8); male with bifid cephalic horn and median pronotal process; metatibial spurs extremely short; female with dorsal surfaces conspicuously punctate and head with traces of cephalic tubercles; 29 to 61 mm in length; QLD, NSW, NT ....***Xylotrupes* Hope**
– Meso- and metatibiae with apices strongly bispinose and intermediate and basal carinae ciliate (Fig. 33.9); clypeal apex truncate or rounded; mandibles untoothed (Fig. 33.10); metatibial spurs elongate in both sexes; 26 to 56 mm in length; QLD, NSW, NT, WA ....***Haploscapanes* Arrow**

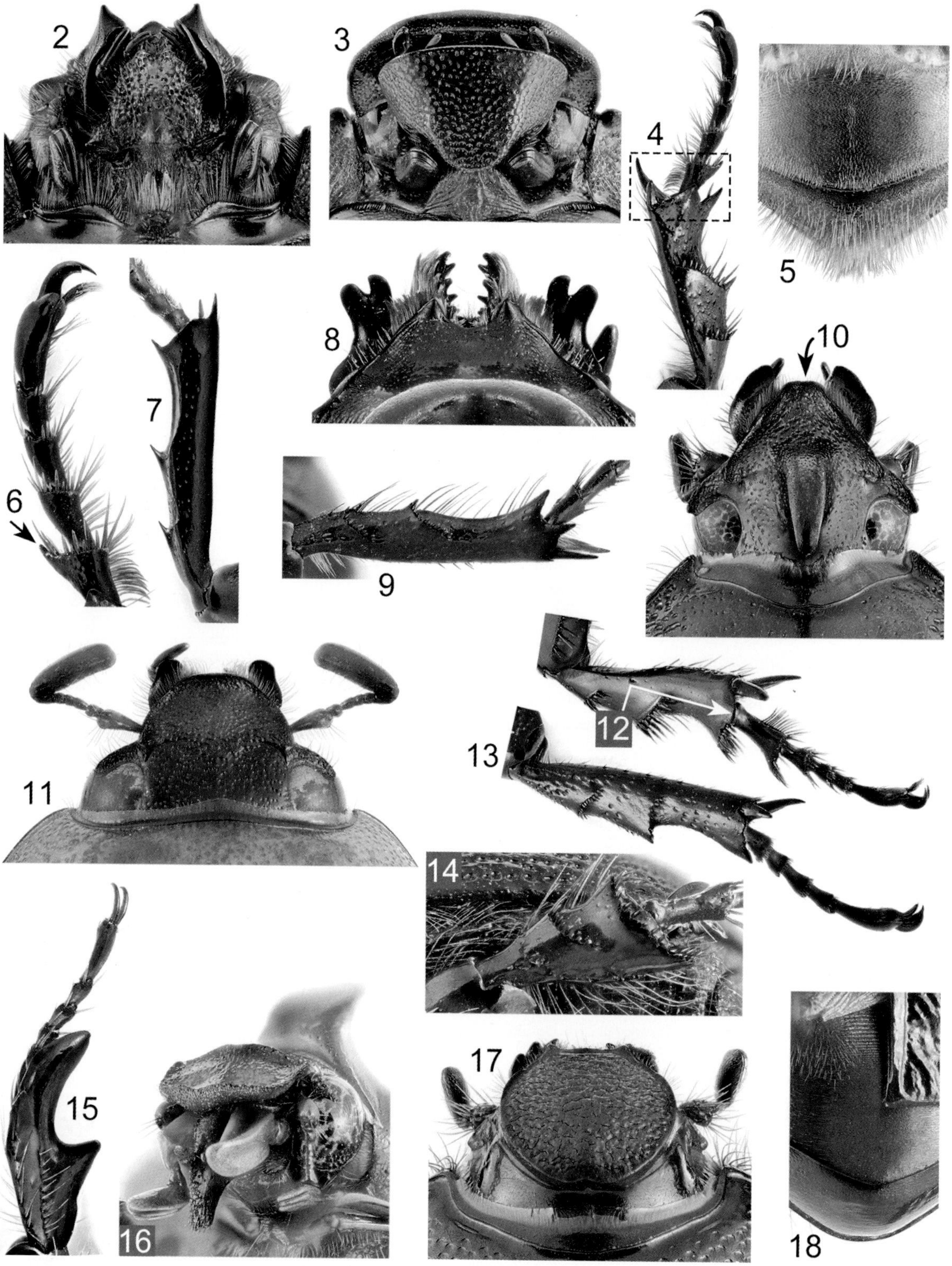

Figs 33.2–33.18. Dynastinae. 2, *Eophileurus dentatus* (Blackburn), head, ventral; 3, *Cryptodus tasmanianus* Westwood, head, ventral; 4, *Oryctes rhinoceros* (Linnaeus), metatibia; 5, *Oryctes rhinoceros* (Linnaeus), pygidium; 6, *Oryctes rhinoceros* (Linnaeus), metatarsus; 7, *Xylotrupes australicus australicus* Thomson, hind tibia; 8, *Xylotrupes australicus australicus* Thomson, clypeal apex and mandibles; 9, *Haploscapanes australicus* (Arrow), metatibia; 10, *Haploscapanes australicus* (Arrow), head; 11, *Cyclocephala signaticollis* Burmeister, head, dorsal; 12, *Oryctoderus latitarsis* Boisduval, metatibia; 13, *Onychionyx australicus* Dechambre, metatibia; 14, *Neocnecus punctata* (Sharp), metatibia; 15, *Neocnecus punctata* (Sharp), protibia; 16, *Nephrodopus* sp., head, ventral; 17, *Temnorrhynchus retusus* (Fabricius), head, dorsal; 18, *Dipelicus* sp., pygidium.

6(1). Head and pronotum unarmed in both sexes, the latter simply convex; ratio of head width across eyes to interocular space at eye canthus greater than 1.7:1 (Fig. 33.11); mandibles extending beyond clypeus, outer sides straight and apically upturned; male protarsi thickened, with large asymmetrical claws, protarsomere 5 as long as 1–4 combined; metatibia not greatly enlarged, with slender, sharp spurs in male and short blunt spurs in female; propygidium without stridulatory area; light brownish-yellow, shining, pronotum and elytra with symmetrical and sometimes vaguely indicated dark brown markings; length 12–15 mm; south-eastern NSW, south-eastern QLD, ACT, SA........ **Cyclocephalini: *Cyclocephala signaticollis* Burmeister**

– Head and pronotum with more or less distinct frontoclypeal suture, tubercles, horns, impressions, or hollows; IF unarmed or armature not obvious, THEN ratio of head width across eyes to interocular space at eye canthus less than 1.4: 1; AND EITHER propygidium with stridulating area (*Heteronychus*) OR elytral striae very obvious (*Neocorynophyllus, Adoryphorus,* some *Anomalomorpha*) OR sides of clypeus deeply emarginate before ocular canthi (*Oryctoderus*); mandibles various; male protarsi thickened with asymmetrical claws or not; propygidium with or without stridulatory area; colour more or less uniform, brownish to black, without pronotal and elytral colour pattern, sometimes bicoloured; length often greater than 15 mm........ **7**

7(6). Meso- and metatibial apices EITHER with slight median projection separating spines from spurs (Fig. 33.12) OR with lateral projection and few spines (Fig. 33.13); head with one or two tubercles OR IF without, THEN sides of clypeus deeply emarginate in front of ocular canthi; pronotum simply convex or with small depression and tubercle; elytra impunctate or with random punctation; propygidium without stridulatory area **...Oryctoderini**........ **8**

– Meso- and metatibial apices truncate or only slightly produced, spines continuous with spurs or rarely entirely absent (Fig. 33.14); head and pronotum with more or less distinct frontoclypeal suture, tubercles, horns, impressions or hollows; IF unarmed or armature not obvious THEN EITHER propygidium with stridulating area OR elytral striae very obvious; propygidium with or without stridulating area.... **Pentodontini**........ **10**

8(7). Clypeus with 2 distinct teeth; head with 2 low tubercles; meso- and metatibial apices with a lateral projection and only a few spines (Fig. 33.13); elytra with random ocellate punctures; 20mm; north QLD........ ***Onychionyx* Arrow**

– Clypeus rounded and truncate, with or without slight emargination; head with single tubercle or without tubercle; meso- and metatibial apices with slight median projection separating spines from spurs (Fig. 33.12); elytra impunctuate or with some horseshoe shaped punctures........ **9**

9(8). Clypeus deeply emarginate in front of ocular canthi, that of male deeply impressed; frons of male without tubercle; that of female with frontal tubercle feebly developed or not; pronotum without pit or tubercle in either sex; male protarsi thickened; elytra impunctate, smooth; 35–40 mm; northern QLD........ ***Oryctoderus* Boisduval**

– Clypeus not emarginate in front of ocular canthi; frons in both sexes with tubercle; pronotum in both sexes with tubercle and in males with small but distinct impression; male protarsi not thickened; elytra somewhat randomly punctuate, with some horseshoe shaped punctures; 30–32 mm; northern QLD........ ***Oryctoderinus* Endrödi**

10(7). Protibia bidentate, apical tooth elongate (Fig. 33.15); antenna 9-segmented with club scarcely longer in male than in female; mentum not compressed; pronotum and elytra conspicuously punctate; legs very short and stout; metafemora bisulcate (Fig. 33.14); black species; 8–12 mm in length; WA........ ***Neocnecus* Sharp**

– Protibia tridentate, OR IF occasionally bidentate (some *Adoryphorus and Cheiroplatys*), THEN apical tooth broadly rounded and poorly separated from second tooth; antennae 9 or 10 segmented, with club of various lengths; mentum compressed or not; other characters variable........ **11**

11(10). Mentum and ligula compressed to form thin vertical lamina divided by ligular suture; apical maxillary palpomere remarkably enlarged and specialised (Fig. 33.16); elytra with a conical process near scutellum; pronotum with anterior horn and pair of lateral horns or tubercles, with an excavation extending to base; scutellum very large; hind legs stout, the metafemora unisulcate; yellow, reddish-brown or black species, often bicoloured; 12–16 mm in length; female unknown; QLD, NT;........ ***Nephrodopus* Sharp**

– Mentum flat or strongly compressed but not produced to form vertical lamina; apical maxillary palpomere not enlarged; elytra unarmed; pronotum various........ **12**

12(11). Meso- and metatibial apices without spines; head plate-like, flattened, more or less vertical, and coarsely, somewhat rugosely punctate (Fig. 33.17); clypeus not separated from frontal plate; pronotum declivous in front; 15–18 mm; NSW, WA........ ***Temnorrhynchus* Hope**

– Meso- and metatibial apices with spines continuous with spurs; IF head plate-like and vertical THEN smooth and impunctate........ **13**

13(12). Propygidium enlarged, its posterior edge convex and its disc covered with numerous fine transverse stridulating ridges (Fig. 33.18); head plate-like, vertical and smooth, forming a horn in males (Fig. 33.19); apical labial palpomere globular; glossy reddish-brown to reddish-black species; 12–32 mm in length; NSW, QLD, NT, WA........ **Dipelicina: *Dipelicus* Hope**

– Propygidium not enlarged, with or without stridulating ridges; if ridges present, then these are either scattered or form a pair of paramedian bands; head without vertical plate; apical labial palpomere cylindrical........ **14**

14(13). Males and females sexually dimorphic with regard to the antennal club; male antennal club enlarged and segments leaf-like and flattened, longer than shaft; female antennal club (where known) shorter than shaft; male head and pronotum often with bizarre armature; female head and pronotum (where known) umarmed; females either represented in collections by few specimens or unknown; only males can be successfully keyed out in couplets 15 to 25....**Pseudoryctina (part)**........ **15**

– Males and females not sexually dimorphic with regard to the antennal club; antennal club subequal in both sexes, not enlarged as above, segments not as flattened, shorter than or equal in length to shaft, usually shorter than funicle; male and female pronotum similar or different; females and males more or less equally represented in collections; combined male and female characters used from couplet........ **26**

15(14). Frontoclypeal suture absent; clypeus only slightly narrowed anteriorly, with thickened bead all around; setae on mentum very long, dense and extending well past front edge of clypeus; antennal lamellae widest near apex; male protarsal claws asymmetrical; black species; 14 mm in length; female unknown; northern QLD;........ ***Neocorynophyllus* Carne**

– Frontoclypeal suture usually well developed, often giving rise to tubercles or horns of various sizes; IF suture weak or not evident THEN clypeus narrowed to reflexed apex; setae on mentum shorter; antennal lamellae usually widest at middle; male claws symmetrical........ **16**

16(15). Clypeus narrowed to reflexed apex; frontoclypeal suture subobsolete; apical pygidial ridge glabrous; pronotum with basal ridge continuous and with anteriomedian tubercle and impression; dorsal surface conspicuously punctuate; black species; 13–19 mm in length; WA........ ***Epironastes* Carne**

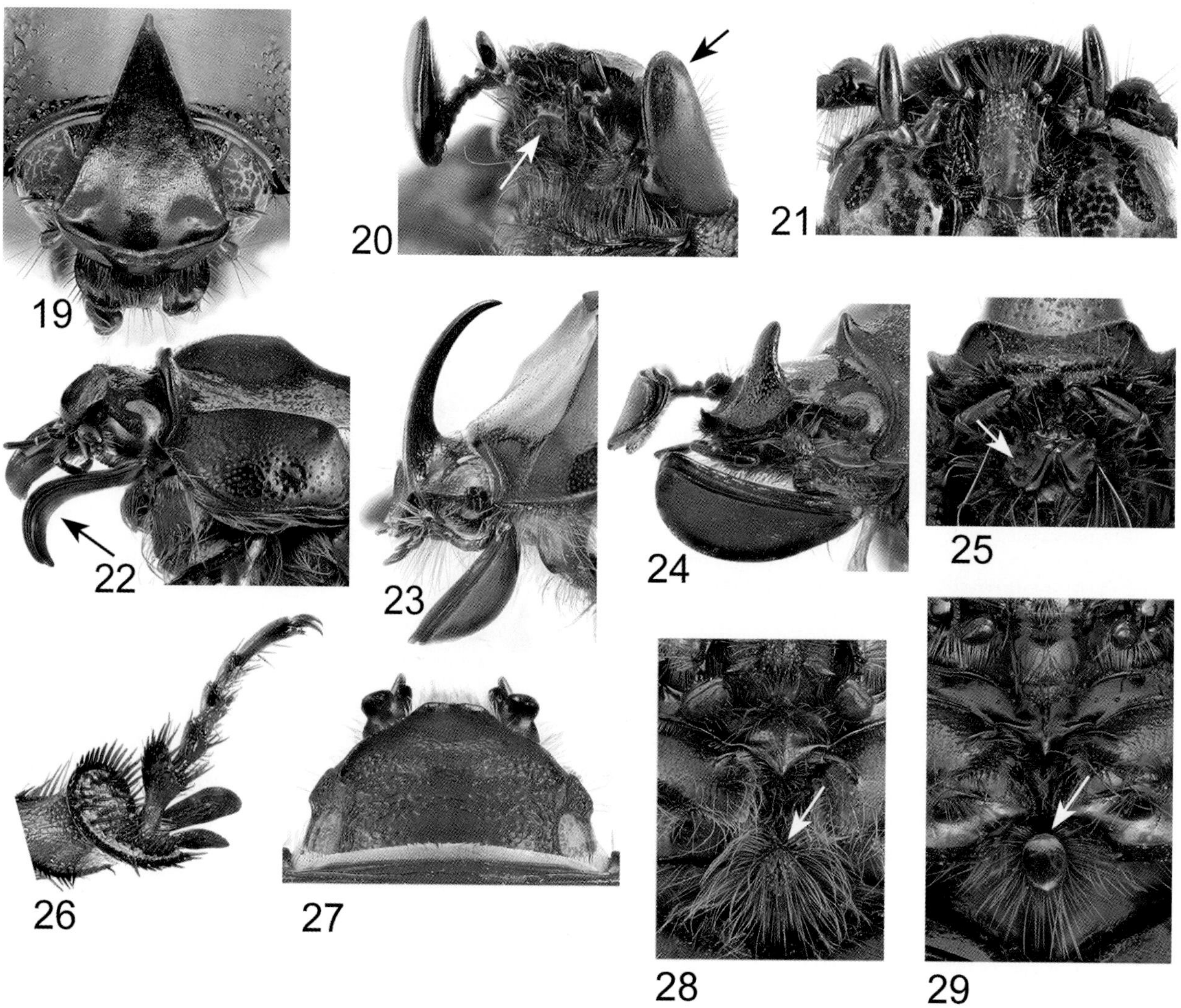

Figs 33.19–33.29. Dynastinae. 19, *Dipelicus* sp., head, anterodorsal; 20, *Cavonus armatus* Sharp, head, ventral, white arrow – mentum, black arrow – antennal club; 21, *Teinogenys dives* (Blackburn), head, ventral; 22, *Pseudocavonus antennalis* Blackburn, head and pronotum, lateral; 23, *Corynophyllus curvicornis* Lea, head, lateral; 24, *Corynophyllus major* Sharp, antennal club; 25, *Dasygnathus trituberculatus* Blackburn, mentum; 26, *Hyphoryctes maculatus* Blackburn, metatibia and metatarsus; 27, *Pimelopus nothus* Burmeister, head, dorsal; 28, *Carneodon glauerti* (Carne), prosternal process; 29, *Metanastes vulgivagus* (Olliff), prosternal process.

– Clypeus subrectangular, rounded or slightly narrowed, only rarely apically reflexed; with carinate or raised frontoclypeal suture, median tubercle or horn; other characters variable **17**

17(16). Head with raised or carinate frontoclypeal suture, median tubercle or small to medium horn; IF with horn THEN pronotum with anteriomedial tubercle and impression not extending to posterior edge and not limited by an elevated transverse ridge **18**

– Head armed with a conspicuous horn; pronotum with EITHER a major excavation extending to posterior edge OR an excavation limited by a sharp elevated transverse ridge or sharp elevations **24**

18(17). Head with carinate frontoclypeal suture only **19**

– Head with raised frontoclypeal suture, median tubercle or small to medium horn **25**

19(18). Anterior edge of pronotum narrowly membranous, at least on either side median third (where this is occupied by a horn-like process), as well as bearing a fringe of setae **20**

– Anterior edge of pronotum entirely sclerotised, only with fringe of setae **22**

20(19). Mandibles concealed beneath clypeus, the latter with sides convex; antennal club not much longer than funicle; legs moderately stout; metatibia contracted subapically; frontoclypeal ridge transverse, elevated; dorsal surfaces shining and impunctate; pronotum with small, anteromedian tubercle and shallow depression; sutural stria linear; black species ~14 mm in length; female unknown; SA .... ***Orthocavonus* Carne**

– Mandibles exposed beyond clypeus; antennal club much longer than funicle; other characters variable **21**

21(20). Clypeus with sides rounded or sinuately narrowed towards apex; IF rounded THEN epipleural setae sparse, not extending on to apical third of elytra; antennal club not much longer than shaft, kidney shaped (Fig. 33.20); mentum flat or compressed anteriorly, strongly reflexed (Fig. 33.20); metatibia strongly contracted subapically; pronotum with anteromedian horn or tubercle and impression, sometimes bounded by lateral horns; black species 12–21 mm in length; SA, WA, VIC, NSW .... ***Cavonus* Sharp**

– Clypeus rounded; antennal club much longer than shaft, its sides more or less subparallel; mentum highly compressed (Fig. 33.21); metatibia evenly dilated from femoral articulation to distal edge; pronotum with small, rarely strongly elevated, anteromedian tubercle plus an impression OR with surface evenly convex; epipleural setae dense, extending or nearly extending to elytral apices; reddish-brown to black species 11–17 mm in length; widely distributed in arid regions..***Teinogenys* Sharp (part)**

22(19). Maxillary palps very short and inconspicuous, not much longer than labial palps; mandibles and labrum concealed; lateral pronotal horns irregular and turret-like; metatibia unicarinate; spurs straight and subequal in length; metatarsomere 1 strongly dilated and twice the length of metatarsomere 2; reddish-brown, bicolored or black species; 14–24 mm in length; northern Australia;...***Pseudoryctes* Sharp**

– Maxillary palps elongate, much longer than labial palps; mandibles and labrum more or less exposed; pronotum with simple, lateral horns or lateral elevated angles; metatibia bicarinate; metatarsomere 1 not dilated and equal in length to metatarsomere...**23**

23(22). Antennal club very large, sinuate, bitexturous (Fig. 33.22); pronotum with lateral elevated angles (Fig. 33.22) and median third of basal edge strongly produced over scutellum; dark reddish-brown species 16–17 mm in length; female unknown; northern WA ....***Pseudocavonus* Blackburn**

– Antennal club not strongly sinuate, unitexturous; pronotum with simple lateral horns and basal edge normally rounded; reddish-brown, bicoloured or black species 12–27 mm in length; widely distributed ....***Carneoryctes* Ozdikmen**

24(17). Head with horn extending back over more than half of pronotum (Fig. 33.23); pronotum with excavation not extending to posterior edge and bounded posteriorly by a sharp elevated transverse ridge or sharp elevations; basal area of pygidium glabrous; QLD; .... ***Corynophyllus* Hope (part)**

– Head with horn not or only just extending back over pronotum; pronotum with excavation extending to posterior edge and bounded laterally by blunt lateral elevations; basal area of pygidium markedly setose; WA, NT, SA..***Phylliocephala* Blackburn**

25(18). Clypeus evenly rounded with moderately deep, truncate face; frontoclypeal ridge transverse or posteriorly angulate, usually giving rise to median tubercle which is in line with or behind anterior edge of eye; pronotum strongly transverse, non-tuberculate, evenly convex or with slight, submarginal impression in front; antennal club with sides more or less subparallel; apical pygidial ridge glabrous; black or reddish-brown species 11–17 mm in length; NSW, WA...***Teinogenys* Sharp (part)**

– Clypeus elongate, narrowed anteriorly with truncate apex OR IF rarely evenly rounded THEN without truncate apical face; frontoclypeal ridge transverse, giving rise to median elevation, tubercle or distinct, sometimes bifurcate, horn all of which are in front of anterior edge of eye; antennal club greatly enlarged and of characteristic shape (Fig. 33.24); pronotal margin with median tubercle or node; pronotum with anteromedian tubercle and impression; apical pygidial ridge setose; reddish-brown, black or bicolored species 12–20 mm in length; eastern Australia .... ***Corynophyllus* Hope (part)**

26(14). Mandibles either evenly rounded, slightly scalloped or straight sided, exposed beyond or beside clypeus or concealed; propygidium without a trace of stridulating ridges; last abdominal ventrite sometimes with well defined transverse sulcus, especially in female**....Cheiroplatina...........27**

– Mandibles almost always conspicuously sinuate, bidentate or tridentate and exposed beyond or beside clypeus; propygidium with either short scattered stridulating ridges or a pair of paramedian stridulatory bands; last abdominal ventrite without transverse sulcus....**Pentodontina.......36**

27(26). Clypeus (viewed dorsally) strongly contracted towards apex with sides concave; concave sides with characteristic vertical faces; mandibles somewhat straight sided and exposed beyond or beside clypeus **28**

– Clypeus (viewed dorsally) broadly triangular, subrectangular or trapezoidal, with sides straight; mandibles rounded or slightly scalloped, exposed beyond or beside clypeus or concealed **29**

28(27). Frontoclypeal suture posteriorly angulate, giving rise to triangular elevation or large horn in male; pronotum somewhat pentagonal, especially in female, its sides contracted towards anterolateral angles; anterior edge of pronotum membranous only across median third; pronotum in male with large impression or excavation and without anteromedian tubercle; antennae 10-segmented; base of pygidium setose; aedeagus with simple parameres; reddish-brown to reddish-black species; larger species 16–25 mm in length; widespread QLD, NSW, VIC, SA, WA, NT..***Novapus* Sharp**

– Frontoclypeal suture usually straight, rarely tuberculate; pronotum with sides more evenly curved towards anterolateral angles; anterior edge of pronotum wholly membranous; pronotum in male usually with anteromedian tubercle and small impression; antennae usually 9-segmented (very rarely with 8 or 10 segments); base of pygidium glabrous; aedeagus with complex parameres; dark reddish-brown to black species; smaller species 9–16 mm in length; mainly WA with one species in south-eastern Australia..***Trissodon* Burmeister**

29(27). Clypeus broadly triangular, sides straight; mandibles exposed beyond or beside clypeus or concealed; galea strongly developed and conspicuously toothed; palps short **30**

– Clypeus broadly subrectangular or trapezoidal, sides straight or rounded; mandibles usually concealed beneath clypeus (exposed at sides in some *Neodasygnathus*); galea reduced, either weakly toothed or having only a small, conical process terminating in a pencil of hairs; palps elongate or not...**31**

30(29). Head quite unarmed; clypeus with oblique, sometimes serrate, anterior edge; galea strongly toothed; mandibles concealed or slightly exposed at sides of clypeus; pronotum with anterior tubercle and (usually) deep impression in male; armature reduced but still discernible in most females; protibia with three or two blunt teeth; first metatarsomere widened, flattened and with distinct longitudinal carina externally; deep reddish-brown to black species 15–28 mm in length; widespread QLD, NSW, VIC, TAS, SA ....***Cheiroplatys* Hope**

– Head armed with low tubercle or small horn; clypeus with anterior edge slightly upturned, never oblque; galea weakly toothed; mandibles strongly exposed beyond clypeus; pronotum with anterior and posterior foveae developed in both sexes, without anterior tubercle; protibia with three sharp teeth; first metatarsomere more rounded, not carinate externally; elytra sometimes costate; reddish-black to black species; 12–26 mm in length; widespread QLD, NSW, VIC, SA, WA, NT...***Semanopterus* Hope**

31(29). Large species, 17–32 mm in length; head armed with tubercle or small horn on straight or recurved frontoclypeal suture; IF tubercle hard to see or absent (some *Dasygnathus*), THEN base of mentum deeply hollowed or cleft; pronotum sometimes with a submarginal groove; elytra never conspicuously guttered; galea weakly toothed; pronotum of male with excavation of various development, sometimes with anteromedian impression and tubercle in both sexes...**32**

– Smaller species, 9–20 mm in length; head EITHER unarmed, OR with small swelling on vertex (*Adoryphorus coulonii*), OR with small tubercle on recurved frontoclypeal suture; IF with small tubercle on recurved frontoclypeal suture THEN pronotum with a submarginal groove and elytra conspicuously guttered and epipleura horizontal in anterior half (some *Anomalomorpha*); base of mentum never deeply cleft; galea with only a small conical process terminating in a pencil of hairs; pronotum EITHER evenly convex with micropuntate surface (*Adoryphorus, Anomalomorpha*) OR with small anterior tubercle and discernable punctures (*Enracius*)..33

32(31). Head with preocular armature of tubercle or small horn, always anterior to a line joining the front of the eyes; frontoclypeal suture straight; mandibles exposed at side of clypeus or not; base of mentum not hollowed or cleft; pronotum with anteromedian tubercle in both sexes; elytral punctation heavily impressed; size and development of male armature not exhibiting marked variation; reddish brown to pitch black species 20–32 mm in length; eastern states NSW, TAS..***Neodasygnathus* Carne**

– Head with interocular armature; frontoclypeal suture recurved, giving rise to tubercle or small horn usually posterior to a line joining the front of the eyes; mandibles almost always concealed; base of mentum hollowed or deeply cleft (Fig. 33.25); pronotum never with anteromedian tubercle, variously excavated, often bounded posteriorly by sharp points; elytral punctation lightly impressed; extremely variable in size and development of male armature; reddish brown to reddish black species 17–32 mm in length; eastern states QLD, NSW, VIC, SA..***Dasygnasthus* Macleay**

33(31). Pronotum and elytra guttered with epipleura horizontal in anterior half; head with recurved frontoclypeal suture often terminating in a tubercle; males with all tarsi elongate, protarsomere 5 much longer than 1; claws enlarged, robust and not much curved; dark brown or black species with yellow legs; 13–20 mm in length; eastern Australia QLD, NSW, VIC and north-west WA........................***Anomalomorpha* Arrow (part)**

– Pronotum and elytra not guttered, epipleura vertical; head EITHER unarmed OR with small swelling on vertex; frontoclypeal suture straight; males with tarsi elongate or not, protarsomere 5 much longer than 1 or not; claws enlarged or not..34

34(33). Clypeus setose; pronotum with discernable punctures; male pronotum with small tubercle on anterior edge; head unarmed; frontoclypeal suture carinate; tarsi elongate in both male and female; male protarsomere 5 much longer than 1; claws enlarged and only slightly curved; brownish black species 12–15 mm in length; southern WA;..***Enracius* Dechambre**

– Clypeus without setae; pronotum evenly convex with micropunctate surface; frontoclypeal suture carinate or not; male tarsi elongate or not; protarsomere 5 much longer than 1 or not; claws enlarged or not..35

35(34). Anterior edge of pronotum entirely narrowly membranous; male with tarsi elongate, protarsomere 5 much longer than 1; claws enlarged, robust and only slightly curved; elytral striae vaguely impressed; female penultimate abdominal ventrite with row of setae removed from posterior edge; parameres with inner edges neither spinose nor toothed; brown species 12–14 mm in length; south-eastern Australia NSW, VIC..***Anomalomorpha* Arrow (part)**

– Anterior edge of pronotum with narrow membrane not extending to anterior angles; male tarsi not so elongate, protarsomere 5 about same length as 1; claws not enlarged and more strongly curved; elytral striae more deeply impressed and distinctly punctate; female penultimate abdominal ventrite with row of setae very close to posterior edge; parameres with inner edges spinose or toothed; brownish to black species 9–17 mm in length; eastern Australia QLD, NSW, VIC, SA..***Adoryphorus* Blackburn**

36(26). Protibiae with more than 3 teeth externally, EITHER with 3 main teeth and smaller basal tooth OR with 3 main teeth and denticles between second and third tooth and behind third tooth; male protarsomere 5 enlarged with asymmetrical claws; black species..37

– Protibiae clearly with only 3 teeth externally; male protarsomere 5 enlarged or not and claws symmetrical or not; light brown, reddish brown, reddish black or black species..39

37(36). Propygidium with 2 well defined fine stridulatory bands; head unarmed; pygidium impunctate and highly convex, more so in female; pronotum evenly convex; black species 9–15 mm in length; widespread QLD, NSW, VIC, SA, WA..***Heteronychus* Burmeister**

– Propygidium with dispersed stridulatory ridges, not forming well defined bands; head armed with either 1 or 2 tubercles or horns; pygidium variously punctuate and less convex; pronotum EITHER evenly convex OR with a depression anteriorly..**38**

38(37). Head armed with two tubercles or horns; pronotum evenly convex; clypeus narrowed and truncate anteriorly with 2 small upturned teeth; smaller black species 12–22 mm in length; eastern states QLD, NSW, VIC..***Metanastes* Arrow (part)**

– Head armed with a horn (male) or median tubercle (female); pronotum with anterior rugose depression, bounded posteriorly by a tubercle in both sexes; clypeus more broadly tuncate anteriorly with both anterior angles bearing a sharp tooth; larger black species 22–28 mm in length; north QLD..***Papuana* Arrow**

39(36). Apical truncate faces of meso- and metatibiae densely setose (Fig. 33.26); frontoclypeal suture strongly marked, posteriorly angulate, and giving rise to a median tubercle; clypeus narrowed to a slightly upturned apex, sides concave; mandibles conspicuously broadly bidentate; pronotum unarmed, evenly convex; male protarsal claws asymmetrical; reddish-brown to reddish-yellow species with little or no elytral punctation; 15–20 mm in length; northern Australia QLD, NT, WA..***Hyphoryctes* Blackburn**

– Apical truncate faces of meso- and metatibiae glabrous; frontoclypeal suture straight or obscurely recurved with 1 or 2 low tubercles or a horn; clypeus various; mandibles bidentate or tridentate; pronotum EITHER unarmed OR with a depression and/or tubercle anteriorly; male protarsal claws asymmetrical or not; reddish brown, reddish black or black species..40

40(39). Mandibles bidentate with inner apical tooth much narrower than outer one (Fig. 33.27); pronotum either completely unarmed or with a major excavation bounded posteriorly by a ridge (one species); frontoclypeal suture obscure, somewhat recurved, giving rise to an obtuse median swelling, tubercles or horn; clypeus more or less rounded, apex obliquely truncate with lower portion of truncate face slightly produced beyond upper sinuate edge; propygidium with dispersed stridulating ridges; reddish-brown, highly convex rather featureless species 12–26 mm in length; widely distributed QLD, NSW, VIC, TAS, SA, WA, Norfolk Island, Lord Howe Island..***Pimelopus* Erichson**

– Mandibles variously tridentate; pronotum EITHER completely unarmed OR with anteromedian tubercle and impression; frontoclypeal suture straight, giving rise to 2 tubercles or none; clypeus narrowed to apex which is either somewhat rounded or bears two small tubercles or teeth; propygidium with either dispersed stridulating ridges or ridges organised into short, coarse paramedian bands..41

41(40). Pygidium evenly and heavily punctuate and glabrous; clypeus somewhat rounded and slightly upturned at apex; protibiae with distal tooth somewhat elongate, basal tooth small; mandibles with inner apical tooth markedly produced; propygidium with dispersed stridulating ridges; black species 12–14 mm in length; northern Australia QLD, NT..***Neonastes* Carne**

– Pygidium EITHER impunctate OR punctate and densely setose; clypeus with apex bearing 2 small tubercles or teeth; protibiae with teeth more evenly sized; mandibles with inner apical tooth not so produced; propygidium with either dispersed stridulating ridges or ridges organised into short, coarse paramedian bands..42

42(41). Propygidium with more or less well defined short, coarse stridulating bands (except *laevicollis*); male pronotum armed with anteromedian tubercle and impression (except *laevis*), female sometimes with slight tubercle or swelling; frontoclypeal suture carinate or bituberculate; clypeus apically with 2 small upturned tubercles or teeth; metatibial spurs more or less rounded apically, spatulate; prosternal process more or less rounded apically (Fig. 33.28); IF male propretarsal claws asymmetrical THEN protarsomere 5 not enlarged; reddish brown to reddish black species 10–17 mm in length; widespread QLD, NSW, VIC, SA, WA, NT, Cocos-Keeling Islands .... ***Carneodon* Ozdikmen**

– Propygidium with only dispersed stridulating ridges; pronotum with vague tubercle in male, evenly convex in female; frontoclypeal suture bituberculate; clypeus apically with 2 small forwardly projecting teeth; metatibial spurs sharp; prosternal process expanded and flattened apically (Fig. 33.29); male propretarsal claws strongly asymmetrical and protarsomere 5 enlarged; black species 17–20 mm in length; eastern Australia QLD, NSW .... ***Metanastes* Arrow** (part)

Classification of the Australian genera

The tribal and subtribal classification used here follows that of Smith (2006) and Bouchard *et al.* (2011). Endrödi (1985) used a similar tribal classification, but did not use subtribes. It differs from those used by Carne (1957b) and Dechambre (2005a).

Tribe Cyclocephalini Laporte, 1840

Characteristics. Head and pronotum of both sexes similar, simply convex, at most with two indistinct tubercles behind frontal suture. Outer side of mandibles simple, without teeth. Protarsi of males usually thickened. Apex of metatibiae truncated. Basal segment of metatarsi not triangular. Propygidium without stridulatory area.

Distribution. Neotropical and Nearctic regions with one genus from Africa. Endrödi (1985) gives the world fauna as 301 species in 14 genera.

Australian taxa. One genus.

Cyclocephala Dejean, 1821 (Pl. 55K)

Type species. *Melolontha amazona* Linnaeus, 1767.

Australian species. One.

Distribution. *Cyclocephala* species are endemic to the Nearctic and Neotropical regions. *Cyclocephala signaticollis* Burmeister has been introduced into southern NSW.

Biology. This species is a serious pest in paspalum pastures in Sydney.

References. Carne (1957a, 1957b).

Tribe Dynastini Macleay, 1819

Characteristics. Large size, usually with very strong sexual dimorphism. Horns of male head and pronotum often very long, head and pronotum of females usually without ornamentation. Prolegs of males much longer than those of female. Mandibles usually excised at apex, outer side straight of lobed. Antennal club in both sexes short. Elytra often different between the sexes: smooth in male and sculptured in female. Propygidium with or without a stridulatory area. Basal segment of metatarsi cylindrical in both sexes.

Distribution. Worldwide. Endrödi (1985) gives the world fauna as 60 species in 10 genera, while Rowland and Miller (2012) add a further 3 genera.

Australian taxa. Two genera.

Haploscapanes Arrow, 1908 (Pl. 53E, G, 55L)

Type species. *Scarabaeus barbarossa* Fabricius, 1775.

Liteupatorus Prell, 1911

Australian species. Three.

Distribution. Endemic to Australia. *Haploscapanes* species occur in NSW, QLD, NT and WA.

References. Carne (1957b); Dechambre & Drumont (2004).

Xylotrupes Hope, 1837 (Pl. 8F, 9F, 53H, J, 56G)

Type species. *Scarabaeus gideon* Linnaeus, 1767.

Endebius Lansberge, 1880

Australian species. One.

Distribution. The genus *Xylotrupes* occurs throughout South-east Asia, from India and Sri Lanka to Malaysia and the Philippines, is widely distributed throughout Indonesia, Papua New Guinea and the Solomon Islands and extends into northern Australia. The Australian species is *X. australicus* Thomson with two subspecies: *X. a. australicus* from QLD and NSW and *X. a. darwinia* Rowland from the NT. *Xylotrupes macleayi* Montrouzier is recorded from Vanuatu.

Biology. *Xylotrupes* adults have been observed feeding on bark, tunnelling into ripe pineapple and boring into spathes of young palms.

References. Carne (1957b); Silvestre (2003); Rowland (2011); Rowland & Miller (2012).

Tribe Oryctini Mulsant, 1842

Characteristics. Large size, body elongate with sides convex or nearly parallel. Sexual dimorphism pronounced or not, with armature usually on both head and pronotum. Propygidium usually with well developed stridulatory area. Tibiae short and strong, apex of metatibiae usually with two or three sharp teeth. Basal segment of metatarsi triangularly dilated.

Distribution. Australian, Oriental, Ethiopian, Neotropical, Nearctic regions. Endrödi (1985) gives the world fauna as over 200 species in 26 genera.

Australian taxa. One genus.

Oryctes Illiger, 1798

Type species. *Scarabaeus nasicornis* Linnaeus, 1758.

Australian species. Two.

Distribution. Species of *Oryctes* occur primarily in the Palaearctic, Afrotropical and Oriental regions but the generic range has expanded throughout the Australo-Pacific region, primarily due to the introduction from Asia of *O. rhinoceros* (Linnaeus), the well known pest of coconut palm. *O. rhinoceros* has been reported from Thursday Island, QLD and Bathurst Island, NT, and was reported from the Cocos-Keeling Islands in 1940. *Oryctes centaurus* Sternberg, a New Guinea species, was recorded from Cape York, QLD.

References. Szent-Ivany (1969); Carne (1957b).

Tribe Oryctoderini Endrödi, 1966

Characteristics. Clypeus truncated and upturned or with one to four teeth. Frons with one or two tubercles. Mandibles covered by or projecting beside clypeus. Pronotum usually simply convex, rarely slightly hollowed or with tubercle. Elytra usually without distinct rows of punctures. Propygidium with or without stridulatory area. Metatibiae with one or two transverse carinae, apex truncate or with one to four obtuse triangular teeth.

Distribution. Australia, New Guinea, New Caledonia. Endrödi (1985) gives the world fauna as 24 species in 10 genera.

Australian taxa. Three genera.

Onychionyx Arrow, 1914 (Pl. 59C)

Type species. *Onychionyx opacipennis* Arrow, 1914.

Australian species. One.

Distribution. New Guinea. *Onychionyx australicus* Dechambre occurs in northern QLD.

References. Endrödi (1971a); Dechambre (2003b).

Oryctoderinus Endrödi, 1978 (Pl. 54A)

Type species. *Oryctoderinus walfordorum* Endrödi, 1978.

Australian species. One.

Distribution. Endemic to Australia. *Oryctoderinus walfordorum* Endrödi occurs in northern QLD.

References. Endrödi (1978).

Oryctoderus Boisduval, 1835 (Pl. 54L, 56A)

Type species. *Oryctoderus latitarsis* Boisduval, 1835.

Australian species. One.

Distribution. *Oryctoderus latitarsis* Boisduval is a New Guinea species occurring in northern QLD.

References. Carne (1957b); Endrödi (1971a); Allsopp (1990).

Tribe Pentodontini Mulsant, 1842

Characteristics. Mentum narrowed towards ligula, the insertions of labial palps visible. Head and pronotum commonly with armature in male. Postcoxal process of prosternum forming a more or less vertical columnar or tapering process. Protarsi subequal in both sexes, the fifth segment rarely twice as long as first. Meso- and metalegs with claws and onychia small, tibiae with ciliate (or less commonly digitate) apices and carinae and first tarsal segment sometimes dilated but not strongly dilated. Anterior claws of male often asymmetrical. Propygidium usually with stridulatory area.

Distribution. Worldwide. Largest tribe with over 500 species in more than 100 genera (Endrödi 1985).

Australian taxa. 30 genera distributed in four subtribes.

Subtribe Pseudoryctina Carne, 1957

Characteristics. Antennal club sexually dimorphic, enlarged in male. Maxillae with galeae always reduced to small more or less untoothed pieces, the palps often elongate and very conspicuous. Mentum characteristically compressed, transversely convex, reflexed to simple ligula. Mandibles and labrum both commonly exposed beyond clypeus, the latter often with a deep truncated face and often at a considerable angle to plane of frons. Head and pronotum of male with conspicuous armature. Legs usually slender but occasionally stout, tibiae ciliate, claws always simple. Females are in collections.

Australian taxa. 12 genera.

Carneoryctes Özdikmen, 2009 (Pl. 55J)

Type species. *Pseudoryctes tectus* Blackburn, 1892.

Cryptoryctes Carne, 1957 (not Reed, 1954)

Australian species. 14.

Distribution. *Carnoryctes* species are known from NSW, VIC, QLD, SA, WA and NT. Also found in New Guinea.

References. Carne (1957b, 1981); Howden & Vladislav (2005); Özdikmen (2009).

Cavonus Sharp, 1875 (Pl. 55C)

Type species. *Cavonus armatus* Sharp, 1875.

Australian species. Seven.

Distribution. Endemic to Australia. *Cavonus* species are known from VIC, SA and WA.

References. Carne (1957b).

Corynophyllus Hope, 1845 (Pl. 55G)

Type species. *Corynophyllus fortnumi* Hope, 1845.

Australian species. 16.

Distribution. Endemic to Australia. *Corynophyllus* species are known from VIC, NSW, QLD and SA.

References. Carne (1957b); Dechambre (2003a; 2004).

Epironastes Carne, 1957 (Pl. 54O)

Type species. *Epironastes nigrosetosus* Carne, 1957.

Australian species. Three.

Distribution. Endemic to Australia. The three species of *Epironastes* occur in southern WA.

References. Carne (1957b).

Neocnecus Sharp, 1878 (Pl. 54P)

Type species. *Neocnemis punctata* Sharp, 1875.
Neocnemis Sharp, 1875, not Crotch, 1867
Neatocnemis Prell 1936
Australian species. Two.
Distribution. Endemic to Australia. *Neocnecus punctata* (Sharp) and *N. minor* (Carne) occur in both southern and northern WA.
References. Carne (1957b); Daniels (1999).

Neocorynophyllus Carne, 1985 (Pl. 54G)

Type species. *Neocorynophyllus aputaringus* Carne, 1985.
Australian species. One.
Distribution. Endemic to Australia. *N. aputaringus* Carne is known from northern QLD.
References. Carne (1985b).

Nephrodopus Sharp, 1873

Type species. *Nephrodopus enigma* Sharp, 1873.
Australian species. One.
Distribution. Endemic to Australia. *N. enigma* occurs in northern QLD and NT.
References. Carne (1957b).

Orthocavonus Carne, 1957

Type species. *Neocavonus occidentalis* Blackburn, 1888.
Australian species. One.
Distribution. Endemic to Australia. *O. occidentalis* (Blackburn) occurs in SA.
References. Carne (1957b).

Phylliocephala Blackburn, 1890 (Pl. 54N, 55A)

Type species. *Phylliocephala nigrohirta* Blackburn, 1890.
Xynedria Blackburn, 1894
Australian species. Two.
Distribution. Endemic to Australia. *Phylliocephala nigrohirta* is known from southern WA while *P. interioris* (Blackburn) is known from central Australia NT, SA.
References. Carne (1957b), Dechambre (2003a).

Pseudocavonus Blackburn, 1890 (Pl. 55H)

Type species. *Pseudocavonus antennalis* Blackburn, 1890.
Australian species. One.
Distribution. Endemic to Australia. *P. antennalis* Blackburn occurs in northern WA.
References. Carne (1957b).

Pseudoryctes Sharp, 1873 (Pl. 53F)

Type species. *P. mullerianus* Sharp, 1873.
Australian species. 10.
Distribution. Endemic to Australia. *Pseudoryctes* species are known from NSW, QLD, NT and WA.
References. Carne (1957b, 1980).

Teinogenys Sharp, 1873 (Pl. 54F)

Type species. *Teinogenys nitidus* Sharp, 1873.
Australian species. 16.
Distribution. Endemic to Australia. Species of *Teinogenys* are known from VIC, NSW, QLD, SA, WA and NT.
References. Carne (1957b); Endrödi (1971b); Allsopp & Carne (1986).

Subtribe Cheiroplatina Carne, 1957

Characteristics. Mandibles evenly rounded distally, often concealed beneath clypeus. Antennae not sexually dimorphic. Ocular canthi usually setose, clypeofrontal ridge transverse or posteriorly arcuate, often giving rise to a median tubercle or horn, never to paired aramature. Pronotum sometimes unarmed but more often with impressions or excavations. Claws of male protarsi rarely asymmetrical. Propygidium without stridulatory area, apical pygidial ridge usually setose. Last abdominal ventrite with a transverse sulcus.
Australian taxa. Nine genera.

Adoryphorus Blackburn, 1889 (Pl. 54B)

Type species. *Dasygnathus coulonus* Burmeister, 1847.
Australian species. Three.
Distribution. Endemic to Australia. *Adoryphorus* species are known from TAS, VIC, NSW, QLD and SA. *A. couloni* (Burmeister) has been recorded from New Zealand.
Biology. The larva of *A. couloni* was described and illustrated by Mcquillan (1985).
References. Carne (1957b).

Anomalomorpha Arrow, 1908 (Pl. 54J)

Type species. *Anomalomorpha anthracina* Arrow, 1908.
Australian species. Three.
Distribution. Endemic to Australia. *Anomalomorpha* species occur in VIC, ACT and NSW to northern QLD.
Biology. Allsopp (2000).
References. Carne (1957b).

Cheiroplatys Hope, 1837 (Pl. 55O)

Type species. *Scarabaeus latipes* Guérin-Méneville, 1838
Australian species. 10.
Distribution. Endemic to Australia. *Cheiroplatys* species are known from TAS, VIC, NSW, QLD and SA.
Biology. The larva of *C. latipes* (Guérin-Méneville) was described and illustrated by Mcquillan (1985).
References. Carne (1957b, 1976).

Dasygnathus W. S. Macleay, 1819 (Pl. 55B)

Type species. *D. dejeani* W. S. Macleay, 1819.
Australian species. Four.
Distribution. Endemic to Australia. Species of *Dasygnathus* are known from VIC, ACT, NSW, QLD and SA. *D. dejeani* Macleay has been recorded from New Zealand.
References. Carne (1957b, 1978).

Enracius Dechambre, 1999 (Pl. 54I)

Type species. *Enracius longipes* Dechambre, 1999.
Australian species. One.
Distribution. Endemic to Australia. *Enracius longipes* Dechambre is known from southern WA.
References. Dechambre (1999a).

Neodasygnathus Carne, 1957 (Pl. 55M)

Type species. *Scarabaeus juba* Kirby, 1818.
Australian species. Three.
Distribution. Endemic to Australia. Species have been described from TAS and NSW. *Neodasygnathus juba* (Kirby) has been recorded from Vanuatu.
References. Carne (1957b).

Novapus Sharp, 1875 (Pl. 55E, N)

Type species. *Novapus crassus* Sharp, 1875.
Australian species. 12.
Distribution. Endemic to Australia. *Novapus* species have been described from both southern and northern WA, SA, VIC, NSW, QLD and NT.
References. Carne (1957b); Carne & Allsopp (1987).

Semanopterus Hope, 1847 (Pl. 55I)

Type species. *Phileurus subcostatus* Laporte, 1840.
Australian species. Seven.
Distribution. Endemic to Australia. *Semanopterus* species are widely distributed in Australia.
References. Carne (1957b).

Trissodon Burmeister, 1847 (Pl. 54D)

Type species. *Isodon sustralasiae* Hope, 1837.
Australian species. 12.
Distribution. Endemic to Australia. *Trissodon* species have been described from VIC, NSW, QLD, SA and WA.
References. Carne (1957b); Allsopp (1993).

Subtribe Dipelicina Carne, 1957

Characteristics. Last segment of labial palps securiform, labrum visible between mandibles, clypeus truncate, ocular canthi setose. Protibiae simply tridentate, lacking supernumerary denticles. Male with cephalic horn and retusely excavated pronotum, the latter usually smooth and glossy, anterior claws simple. Metatibiae with apical margin ciliate, first segment of metatarsi strongly dilated. Propygidium with posterior margin convex, disc with fine transverse stridulating ridges. Pygidium with apical ridge glabrous, disc glabrous on both sexes.
Australian taxa. One genus.

Dipelicus Hope, 1845 (Pl. 55D)

Type species. *Dipelicus cantori* Hope, 1845.
Australian species. Four.
Distribution. The genus *Dipelicus* occurs throughout the Oriental region and extends to New Guinea and Australia, where species are known from NSW, QLD, NT and northern WA. *Dipelicus montrouzieri* (Reiche) has been recorded from New Caledonia and Vanuatu.
References. Carne (1957b); Silvestre (2006).

Subtribe Pentodontina Mulsant, 1842

Characteristics. Mandibles toothed, visible beyond clypeus. Antennae not sexually dimorphic. Ocular canthi usually glabrous, clypeus narrowed, truncate, anterior margin bisinuate or bidentate, clypeofrontal ridge more or less transverse, usually elevated, bituberculate, rarely giving rise to a median cephalic horn. Pronotum usually evenly convex or with an anteromedian tuibercle and impression, rarely with a major excavation. Claws of male protarsi often asymmetrical. Propygidium with stridulatory ridges in most species. Sexes equally abundant in collections, adults with functional mouthparts.
Australian taxa. Eight genera.

Carneodon Özdikmen, 2009 (Pl. 54E)

Type species. *Cheiroplatys pecuarius* Reiche. 1860.
Neodon Carne, 1957 (not Horsfield, 1841)
Australian species. 10.
Distribution. Endemic to Australia. *Carneodon* species occur in VIC, NSW, QLD, SA, WA and NT.
Biology. *C. pecuarius* (Reiche) is known to attack a variety of crops in QLD and NSW, including tomatoes, potatoes, carrots and sugarcane.
References. Carne (1957b); Özdikmen (2009).

Heteronychus Burmeister, 1847 (Pl. 53K)

Type species. *Scarabaeus arator* Fabricius, 1775.
Australian species. One (introduced).
Distribution. Species of *Heteronychus* occur primarily in the Afrotropical region, but *H. arator* (Fabricius) has been introduced into southern Brazil, New Zealand and Australia; *H. arator* was introduced into NSW, ACT, SA

and WA. Carne (1957b) records a single old specimen of the Palearctic *H. parumpuntatus* Burmeister from 'Victoria', but no further specimens were found in collections.
Biology. The larva of *Heteronychus arator* (Fabricius) was described and illustrated by Cumpston (1940).
References. Carne (1957b).

Hyphoryctes Blackburn, 1895 (Pl. 55F)

Type species. *Hyphoryctes maculatus* Blackburn, 1895.
Australian species. One.
Distribution. Endemic to Australia. *H. maculatus* Blackburn is known from northern QLD, NT and northern WA.
References. Carne (1957b).

Metanastes Arrow, 1911 (Pl. 54K, 59I)

Type species. *Heteronychus australis* Fauvel, 1862.
Australian species. Four.
Distribution. Species of *Metanastes* are known from Australia, New Caledonia and Papua New Guinea, with the Australian species occurring in VIC, NSW, QLD and NT.
Biology. The larva of *M. vulgivagus* (Olliff) was described and illustrated by Cumpston (1940).
References. Carne (1957b).

Neonastes Carne, 1957 (Pl. 54C)

Type species. *Isodon glabricollis* W. J. Macleay, 1887.
Australian species. Two.
Distribution. Endemic to Australia. *Neonastes glabricollis* (Macleay) and *N. uptoni* Carne occur in northern QLD and NT.
References. Carne (1957b, 1985a).

Papuana Arrow, 1911 (Pl. 53I)

Type species. *Papuana semistriata* Arrow, 1911.
Australian species. One.
Distribution. The genus *Papuana* is known from the Oriental region, Indonesia, Melanesia and Papua New Guinea, with *P. woodlarkiana* (Montrouzier) extending into northern QLD. Two species are recorded from Vanuatu.
References. Carne (1957b); Dechambre (2005b).

Pimelopus Erichson, 1842 (Pl. 59J)

Type species. *Pimelopus porcatus* Erichson, 1842.
Australian species. Six.
Distribution. Widely distributed in Australia and on Norfolk Island and Lord Howe Island. *Pimelopus fischeri* (Montrouzier) has been recorded from New Caledonia.
References. Carne (1957b).

Temnorrynchus Hope, 1837 (Pl. 54H)

Type species. *Geotrupes retusus* Fabricius 1781.
Australian species. One.
Distribution. *Temnorhynchus retusus* (Fabricius) is native to eastern and southern Africa but has been introduced into southern WA and southern NSW. It has also been recorded from Hawaii.
References. Carne (1957b); Allsopp (1987); Krell & Hangay (1998); Jameson *et al.* (2009).

Tribe Phileurini Burmeister, 1847

Characteristics. Mentum strongly dilated, covering bases of labial palps. Sexual dimorphism weak to strong. Outer side of mandibles simply curved or with two or three teeth or lobes. Frons usually armed with one or two tubercles or horns. Femora never strongly thickened. Transverse carinae of meso and metatibiae usually well developed. Body mostly flattened. Stridulatory area of propygidium present or absent.
Distribution. Australian, Oriental, Ethiopian and Neotropical regions. Endrödi (1985) gives the world fauna as 183 species in 35 genera.
Australian taxa. Two genera, one in each of two subtribes.

Subtribe Cryptodina Burmeister & Schaum, 1840

Characteristics. Clypeus rounded or slightly narrowed and truncate, usually emarginated, apex never acuminate. Mandibles small, bidentate, completely concealed. Antennal scape apically dilated and wedge shaped, concealing all funicular segments.
Australian taxa. One genus.

Cryptodus W. S. Macleay, 1819 (Pl. 52K)

Type species. *Cryptodus paradoxus* W. S. Macleay, 1819.
Australian species. 21.
Distribution. Widely distributed in Australia including Lord Howe Island and also occurring in New Guinea and New Caledonia.
References. Carne (1957b).

Subtribe Phileurina Burmeister, 1847

Characteristics. Clypeus strongly narrowed, acuminate, apex reflexed. Mandibles large, usually pointed and untoothed, projecting beyond clypeus. Antennal scape not expanded and wedge-like, not concealing funicular segments.
Australian taxa. One genus.

Eophileurus Arrow, 1908 (Pl. 56C)

Type species. *Geotrupes planatus* Weidemann, 1823.
Australian species. Two.
Distribution. The genus *Eophileurus* is distributed throughout the eastern Palaearctic and Oriental regions and

extends into Melanesia, Papua New Guinea and Australia. *E. dentatus* (Blackburn) occurs in northern QLD and *E. convexus* (Arrow) is known only from Christmas Island.

References. Carne (1957b).

Genera excluded from the Australian fauna

Pentodon Hope, 1837

Type species. *Geotrupes punctatus* Villers, 1789.

Note. The Palaearctic species *Pentodon algerinum* (Herbst) was recorded by Carne (1957b) from a specimen only labelled 'Victoria' (as *Pentodon bispinosus* Kuster). No other specimens have been found from Australia in collections.

References. Carne (1957b); Dechambre (2005a).

Phyllognathus Eschscholtz, 1830

Type species. *Scarabaeus silenus* Fabricius, 1775 (= *Scarabaeus excavatus* Forster, 1771).

Note. *Phyllognathus carnei* Allsopp, described from northern QLD, is synonymous with the West African *P. degener* Fairmaire, 1891 (= *P. burmeisteri* Arrow, 1911), and it is likely that the Queensland specimen has been mislabelled (Dechambre 1999b). No other specimens have been found in collections.

References. Allsopp (1990); Dechambre (1999b, 2005a).

References

Ahrens D, Schwarzer J, Vogler AP (2014) The evolution of scarab beetles tracks the sequential rise of angiosperms and mammals. *Proceedings of the Royal Society B Biological Series* **281**, 20141470.

Allsopp PG (1987) An additional exotic dynastine *Temnorrhynchus retusus* (F.) (Coleoptera: Scarabaeidae) in Australia. *Journal of the Australian Entomological Society* **26**, 189–191. doi:10.1111/j.1440-6055.1987.tb00283.x

Allsopp PG (1990) *Phyllognathus carnei* sp. n. and *Oryctoderus latitarsis* Boisduval (Coleoptera: Scarabaeidae: Dynastinae), two additions to the Australian fauna. *Journal of the Australian Entomological Society* **29**, 229–232. doi:10.1111/j.1440-6055.1990.tb00353.x

Allsopp PG (1993) *Trissodon howdenorum* and *T. bidentatus* spp. n. from Western Australia with further records of other *Trissodon* spp. (Coleoptera: Scarabaeidae: Dynastinae). *Journal of the Australian Entomological Society* **32**, 201–205. doi:10.1111/j.1440-6055.1993.tb00574.x

Allsopp PG (1995) Biogeography of the Australian Dynastinae, Rutelinae, Scarabaeinae, Melolonthini, Scitalini and Geotrupidae (Coleoptera: Scarabaeoidea). *Journal of Biogeography* **22**, 31–48. doi:10.2307/2846071

Allsopp PG (2000) Revision of the Australian genus *Anomalomorpha* Arrow (Coleoptera: Scarabaeidae: Dynastinae) with a new species from the wet tropics of Queensland. *Memoirs of the Queensland Museum* **46**(1), 1–7.

Allsopp PG, Carne PB (1986) *Teinogenys mooniensis* sp. n. and new locality records for other species of *Teinogenys* (Coleoptera: Scarabaeidae: Dynastinae). *Journal of the Australian Entomological Society* **25**, 89–93. doi:10.1111/j.1440-6055.1986.tb01081.x

Arrow GJ (1908) A contribution to the classification of the Coleopterous family Dynastidae. *The Transactions of the Entomological Society of London* **1908**, 321–358.

Arrow GJ (1911) Notes on the Coleopterous subfamily Dynastinae, with descriptions of new genera and species. *Annals & Magazine of Natural History* **8**, 151–176. doi:10.1080/00222931108693008

Arrow GJ (1937) Systematic notes on beetles of the subfamily Dynastinae, with descriptions of a few new species in the British Museum Collection (Coleoptera). *Transactions of the Royal Entomological Society of London* **86**, 35–57. doi:10.1111/j.1365-2311.1937.tb00246.x

Blackburn T (1888) Notes on Australian Coleoptera, with descriptions of new genera and species. *Proceedings of the Linnean Society of New South Wales* **3**, 805–875.

Blackburn T (1889) Further notes on Australian Coleoptera, with descriptions of new genera and species. *Proceedings of the Linnean Society of New South Wales* **3**, 1387–1506.

Blackburn T (1890a) Further notes on Australian Coleoptera, with descriptions of new genera and species. Part VII. *Transactions of the Royal Society of South Australia* **13**, 82–93.

Blackburn T (1890b) Notes on Australian Coleoptera, with descriptions of new species. Part V. *Proceedings of the Linnean Society of New South Wales* **4**, 1247–1276. doi:10.5962/bhl.part.15084

Blackburn T (1890c) Notes on Australian Coleoptera, with descriptions of new species. Part VI. *Proceedings of the Linnean Society of New South Wales* **5**, 147–156. doi:10.5962/bhl.part.18628

Blackburn T (1890d) Notes on Australian Coleoptera, with descriptions of new species. Part VII. *Proceedings of the Linnean Society of New South Wales* **5**, 303–366. doi:10.5962/bhl.part.18641

Blackburn T (1890e) Notes on Australian Coleoptera, with descriptions of new species. Part VIII. *Proceedings of the Linnean Society of New South Wales* **5**, 553–592. doi:10.5962/bhl.part.18655

Blackburn T (1892a) Further notes on Australian Coleoptera, with descriptions of new genera and species. Part XII. *Transactions of the Royal Society of South Australia* **15**, 207–261.

Blackburn T (1892b) Notes on Australian Coleoptera, with descriptions of new species. *Proceedings of the Linnean Society of New South Wales* **7**, 283–300. doi:10.5962/bhl.part.26052

Blackburn T (1894) Notes on Australian Coleoptera with descriptions of new species. Part XIII. *Proceedings of the Linnean Society of New South Wales* **8**, 185–208.

Blackburn T (1895) Further notes on Australian Coleoptera with descriptions of new genera and species. Part XVII. *Transactions of the Royal Society of South Australia* **19**, 27–60.

Blackburn T (1896) Further notes on Australian Coleoptera, with descriptions of new genera and species. Part XX. *Transactions of the Royal Society of South Australia* **20**, 233–259.

Blackburn T (1897) Further notes on Australian Coleoptera, with descriptions of new genera and species. Part XXI. *Transactions of the Royal Society of South Australia* **21**, 28–39.

Boisduval JA (1835) Faune Entomologique de l'Océan Pacifique, avec l'illustration des insects nouveaux recueillis pendent de voyage. 2me Partie. Coléoptères et autres Ordres. In, *Voyage de Découvertes de l'Astrolabe exécuté par ordre du Roi, pendant les annés 1826–1829, sous le commandement de M.J. Dumont d'Urville*. Vol. 7. (Eds D D'Urville and JS César) J. Tastu, Paris.

Bouchard P, Bousquet Y, Davies AE, Alonso-Zarazaga MA, Lawrence JF, Lyal CHC, Newton AF, Reid CAM, Schmitt M, Slipinski SA, Smith ABT (2011) Family names in Coleoptera (Insecta). *ZooKeys* **88**, 1–972. doi:10.3897/zookeys.88.807

Browne DJ, Scholtz CH (1998) Evolution of the scarab hind wing articulation and wing base: a contribution towards the phylogeny of the Scarabaeoidea (Coleoptera). *Systematic Entomology* **23**, 307–326. doi:10.1046/j.1365-3113.1998.00059.x

Burmeister HCC (1847) *Handbuch der Entomologie. Coleoptera Lamellicornia Xylophila et Pectinicornia.* Vol. 5. TCF Enslin, Berlin.

Carne PB (1957a) *Cyclocephala signaticollis* Burmeister, an introduced pasture scarab (Coleoptera). *Proceedings of the Linnean Society of New South Wales* **81**, 217–221.

Carne PB (1957b) *Systematic Revision of the Australian Dynastinae*. Division of Entomology, CSIRO, Melbourne.

Carne PB (1976) *Cheiroplatys volsellus* sp. n. and notes on related species (Coleoptera: Scarabaeidae: Dynastinae). *Journal of the Australian Entomological Society* **15**, 85–88. doi:10.1111/j.1440-6055.1976.tb02078.x

Carne PB (1978) *Dasygnathus blattocomes* sp. n. (Coleoptera: Scarabaeidae: Dynastinae). *Journal of the Australian Entomological Society* **17**, 91–93. doi:10.1111/j.1440-6055.1978.tb01749.x

Carne PB (1980) *Pseudoryctes storeyi* sp. n. and new records of other species of *Pseudoryctes* Sharp. (Coleoptera: Scarabaeidae: Dynastinae). *Journal of the Australian Entomological Society* **19**, 255–258. doi:10.1111/j.1440-6055.1980.tb00981.x

Carne PB (1981) *Cryptoryctes minchami* sp. n. and notes on other *Cryptoryctes* spp. (Coleoptera: Scarabaeidae: Dynastinae). *Journal of the Australian Entomological Society* **20**, 249–252. doi:10.1111/j.1440-6055.1981.tb01042.x

Carne PB (1985a) *Neonastes uptoni* sp. n. from the Northern Territory, with notes on the holotype and distribution of *N. glabricollis* (Macleay) (Coleoptera: Scarabaeidae: Dynastinae). *Journal of the Australian Entomological Society* **24**, 69–72. doi:10.1111/j.1440-6055.1985.tb00187.x

Carne PB (1985b) A new genus and species of Dynastinae from North Queensland (Coleoptera: Scarabaeidae). *Journal of the Australian Entomological Society* **24**, 75–76. doi:10.1111/j.1440-6055.1985.tb00189.x

Carne PB, Allsopp PG (1987) *Novapus macfarlandi* sp. n. and notes on other species of *Novapus* Sharp (Coleoptera: Scarabaeidae: Dynastinae). *Journal of the Australian Entomological Society* **26**, 309–312. doi:10.1111/j.1440-6055.1987.tb01973.x

Castelnau FL de Laporte (1840) *Histoire Naturelle des Insectes Coléoptères, ave une introduction renfemant l'anatomie et la physiologie des animaux articulés, par M Brullé.* Vol. 2. P Duménil, Paris.

Cumpston D (1940) On the external morphology and biology of *Heteronychus sanctae-helenae* Blanch. and *Metanastes vuglivagus* Olliff (Col. Scarabaeidae, Dynastinae). *Proceedings of the Linnean Society of New South Wales* **65**, 289–300.

Daniels G (1999) An overlooked generic name in the Dynastinae (Coleoptera: Scarabaeidae). *Australian Entomologist* **26**(1), 28.

Dechambre R-P (1999a) *Enracius longipes* n. gen., n. sp., nouveau Dynastide d'Australie (Col. Scarabaeoidea Dynastidae). *Bulletin de la Société Entomologique de France* **104**(1), 37–38.

Dechambre R-P (1999b) Note synonymique sur le genre *Phyllognathus* Eschscholtz, 1830 (Col. Scarabaeoidea Dynastidae). *Bulletin de la Société Entomologique de France* **104**(2), 174.

Dechambre R-P (2003a) Contribution á la connaissance des Pentodontini australiens (Coleoptera, Dynastidae). *Coléoptères* **9**(27), 377–382.

Dechambre R-P (2003b) Le genre *Onychionyx* Arrow, 1914 (Coleoptera, Dynastidae). *Coléoptères* **9**(16), 245–250.

Dechambre R-P (2004) Note synonymique sur le genre *Corynophyllus* Hope, 1845 (Col. Dynastidae). *Bulletin de la Société Entomologique de France* **109**(2), 168.

Dechambre R-P (2005a) *Les Coléoptères du Monde/The Beetles of the World. 30: Dynastidae australiens et océaniens.* Hillside Books, Canterbury, UK.

Dechambre R-P (2005b) Les *Papuana* du groupe *fortepunctata* Heller, 1913 (Coleoptera, Dynastidae). *Coléoptères* **11**(9), 81–88.

Dechambre R-P, Drumont A (2004) Le genre *Haploscapanes* Arrow, 1908 (Coleoptera, Dynastidae). *Coléoptères* **10**(16), 197–203.

Endrödi S (1971a) Monographie der Dynastinae 2. Tribus: Oryctoderini (Coleoptera: Lamellicornia: Melolonthidae). *Pacific Insects* **13**, 207–241.

Endrödi S (1971b) Eine neue australische Pentodotine, *Teinogenys demarzi* n. sp. (Col. Melolonthidae). *Entomologische Arbeiten aus dem Museum G. Frey Tutzing bei München* **22**, 105–108.

Endrödi S (1974) Monographie der Dynastinae 4. Tribus: Pentodontini von Australien und Neu-Zeeland (Coleoptera). *Folia Entomologica Hungarica* **27**, 17–70.

Endrödi S (1977) Monographie der Dynastinae VIII. Tribus: Phileuriini (Coleoptera: Lamellicornia). *Entomologische Abhandlungen* **41**, 93–134.

Endrödi S (1978) *Oryctoderinus walfordorum* gen. et sp. n. (Coleoptera: Scarabaeidae: Dynastinae). *Journal of the Australian Entomological Society* **17**, 159–161. doi:10.1111/j.1440-6055.1978.tb02225.x

Endrödi S (1985) *The Dynastinae of the World*. W. Junk, Dordrecht.

Erichson WF (1842) Beitrag zur Insecten-fauna von Vandiemensland, mit besonderer Berucksichtigung der geographischen Verbreitung der Insecten. *Archiv für Naturgeschichte* **8**, 83–287, pls 4, 5.

Fabricius JC (1775) *Systema Entomologiae, sistens Insectorum Classes, Ordines, Genera, Species, adiectis Synonymis, Locis, Descriptionibus, Observationibus*. Kortius, Flensburgi et Lipsiae.

Fairmaire L (1877) Trois nouvelles espèces de Scarabéides du genre *Oryctodus* oisduval. *Annales de la Société Entomologique de France* **7**(5), 158.

Germar EF (1848) Beiträge zur Insektenfauna von Adelaide. *Linnaea Entomologica* **3**, 153–247.

Guérin-Méneville FÉ (1830) Insectes. In *Voyage autour du monde, exécuté par ordre du Roi, sur la corvette de S. M. La Coquille, pendant les années 1822–25*. (Ed. LI Duperrey) Arthus Bertrand, Paris.

Guérin-Méneville FÉ (1831) Insectes. In *Voyage autour du monde, exécuté par ordre du Roi, sur la corvette de sa Majest., La Coquille, pendant les années 1822, 1823, 1824 et 1825 sous le ministère de S.E.M. Le Marquis de Clermont-Tonnerre, et publié sous les auspices de son Excellence M. Le Cte De Chabrol, Ministre de la Marine et des Colonies. Histoire naturelle, zoologie*. (Ed. LI Duperrey) Pls 1–21. A. Bertrand, Paris.

Gunter NL, Weir TA, Slipinski A, Bocak L, Cameron SL (2016) If dung beetles (Scarabaeidae: Scarabaeinae) arose in association with dinosaurs, did they also suffer a mass co-extinction at the K-PG boundary? *PLOS One* **11**, e0153570.

Hope FW (1837) *The Coleopterist's Manual, containing the Lamellicorn Insects of Linnaeus and Fabricius*. Vol. 1L. H.G. Bohn, London.

Hope FW (1845) Descriptions of some new species of Coleoptera from Adelaide in New Holland. *The Transactions of the Entomological Society of London* **4**, 100–113[pls VI–VII.]. doi:10.1111/j.1365-2311.1845.tb01345.x

Howden HF (1982) Larval and adult characters of *Frickius* Germain, its relationship to the Geotrupini, and a phylogeny of some major taxa in the Scarbaeioidea (Insecta: Coleoptera). *Canadian Journal of Zoology* **60**, 2713–2724. doi:10.1139/z82-347

Howden HF, Vladislav M (2005) *Cryptoryctes tricornutus* sp. nov. from the Northern Territory, Australia (Scarabaeidae, Dynastinae, Oryctini). *Canadian Entomologist* **137**, 274–277. doi:10.4039/n04-102

Hunt T, Bergsten J, Levkanicova Z, Papadopoulou A, John OS, Wild R, Hammonf PM, Ahrens D, Balke M, Caterino MS, Gómez-Zurita J, Ribera I, Barraclough TG, Bocakova M, Bocak L, Vogler AP (2007) A comprehensive phylogeny of beetles reveals the evolutionary origins of a superradiation. *Science* **318**, 1913. doi:10.1126/science.1146954

Jameson ML (1998) Phylogenetic analysis of the subtribe Rutelina and revision of *Rutela* generic groups (Coleoptera: Scarabaeoidea: Rutelinae: Rutelini). *Bulletin of the University of Nebraska State Museum* **14**, 1–184.

Jameson ML, Oishi DE, Ratcliffe BC, McQuate GT (2009) Two additional invasive scarabaeoid beetles (Coleoptera: Scarabaeidae: Dynastinae) in Hawaii. *Proceedings of the Hawaiian Entomological Society* **41**, 25–30.

Jin H, Yonezawa T, Zhong Y, Kishino H, Hasegawa M (2016) Cretaceous origin of giant rhinoceros beetles (Dynastini; Coleoptera). *Genes & Genetic Systems* **91**, 209–215. doi:10.1266/ggs.16-00003

Kirby W (1818) XXVIII. A description of several new species of insects collected in New Holland by Robert Brown, Esq FRS Lin. Linn. Soc. *Transactions of the Linnean Society of London* **12**, 454–482 [pls XXI–XXIII.]. doi:10.1111/j.1095-8339.1817.tb00240.x

Krell FT (2007) Catalogue of fossil Scarabaeoidea (Coleoptera: Polyphaga) of the Mesozoic and Tertiary. Available at http://museum.unl.edu/research/entomology/fossilrecord-web.pdf

Krell FT, Hangay G (1998) The African rhinoceros beetle *Temnorhynchus retusus* (Fabricius) established in eastern Australia (Coleoptera: Scarabaeidae: Dynastinae). *Australian Journal of Entomology* **37**, 312–314. doi:10.1111/j.1440-6055.1998.tb01588.x

Lea AM (1917) Descriptions of new species of Australian Coleoptera. Part XIII. *Proceedings of the Linnean Society of New South Wales* **42**, 545–582. doi:10.5962/bhl.part.4862

Lea AM (1919a) Descriptions of new species of Australian Coleoptera. Part XIV. *Proceedings of the Linnean Society of New South Wales* **43**, 715–746.

Lea AM (1919b) Notes on some miscellaneous Coleoptera, with descriptions of new species. Part V. *Transactions of the Royal Society of South Australia* **43**, 166–261.

Lea AM (1920) Descriptions of new species of Australian Coleoptera. Part XIV. *Proceedings of the Linnean Society of New South Wales* **44**, 735–760.

Lea AM (1926) Notes on some miscellaneous Coleoptera, with descriptions of new species. Part VI. *Transactions of the Royal Society of South Australia* **50**, 45–84.

Macleay WS (1819) *Horae Entomologicae: or essays on the Annulose animals.* Vol. I, Pt. 1. S Bagster, London.

Macleay WJ (1871) Notes on a collection of insects from Gayndah. *Proceedings of the Linnean Society of New South Wales* **2**, 79–205.

Macleay WJ (1887) The insects of the Cairns district, northern Queensland. *Proceedings of the Linnean Society of New South Wales* **2**, 213–238. doi:10.5962/bhl.part.29175

Macleay WJ (1888) The insects of King's Sound and its vicinity. Part II. The Lamellicornes. *Proceedings of the Linnean Society of New South Wales* **3**, 897–924.

McKenna DD, Wild AL, Kanda K, Bellamy CL, Beutel RG, Caterino MS, Farnum CW, Hawks DC, Ivie MA, Jameson ML, Leschen RAB, Marvaldi AE, Mchugh JV, Newton AF, Robertson JA, Thayer MK, Whiting MF, Lawrence JF, Maddison A, Slipinski DR, Farrell BD (2015) The beetle tree of life reveals that Coleoptera survived end-Permian mass extinction to diversify during the Cretaceous terrestrial revolution. *Systematic Entomology* **40**(4), 835–880. doi:10.1111/syen.12132

Mcquillan PB (1985) The identification of root-feeding cockchafer larvae (Coleoptera: Scarabaeidae) found in pastures in Tasmania. *Australian Journal of Zoology* **33**, 509–546. doi:10.1071/ZO9850509

Montrouzier P (1860) Essau sur la faune entomologiqu de la Nouvelle-Calédonie (Balade) et des îles des Pins, Art, Lifui, etc. *Annales de la Société Entomologique de France* **8**, 229–308.

Olliff AS (1889) The insect fauna of Lord Howe Island. *Australian Museum Memoirs* **2**, 77–98. doi:10.3853/j.0067-1967.2.1889.482

Özdikmen H (2009) Nomenclatural changes for five preoccupied scarab beetle genus group names (Coleoptera: Scarabaeidae). *Munis Entomology and Zoology Journal* **4**(1), 139–147.

Reiche LJ (1860) Remarqués entomologiques description d'une novella espéce de Coléoptère. *Annales de la Société Entomologique de France* **8**, 331–333.

Ritcher PO (1966) *White Grubs and Their Allies. A Study of North American Scarabaeoid Larvae*. Oregon State University Press (Studies in Entomology No. 4), Corvallis, Oregon.

Rowland JM (2011) Notes on nomenclature in *Xylotrupes* Hope (Scarabaeidae: Dynastinae: Dynastini). *Insecta Mundi* **0176**, 1–10.

Rowland JM, Miller KB (2012) Phylogeny and systematic of the giant rhinoceros beetles (Scarabaeidae: Dynastini). *Insecta Mundi* **0263**, 1–15.

Sharp D (1873) Description de nouveaux genres et de nouvelles espéce de Coléoptères Lamellicornes. *Revue et Magasin de Zoologie (Paris)* **1**, 262–273.

Sharp D (1875) Description de Coléoptères (Lamellicornes) nouveaux de l'Australie. *Revue et Magasin de Zoologie (Paris)* **3**, 37–50.

Silvestre G (2003) Les *Xylotrupes* de Nouvelle-Guinée, d'Australie et d'Océanie (Coleoptera, Dynastidae). *Coléoptères* **9**(12), 151–162.

Silvestre G (2006) Quatre nouveaux *Dipelicus* Hope, 1845 (Coleoptera, Dynastidae). *Coléoptères* **12**(33), 435–443.

Silvestre G (2012) Révision des *Dipelicus* du groupe *triangularis* (Montrouzier, 1855) (Coleoptera, Dynastidae). *Coléoptères* **18**(6), 31–55.

Smith ABT (2006) A review of the family-group names for the superfamily Scarabaeoidea (Coleoptera) with corrections to nomenclature and a current classification. *Coleopterists Society Monographs* **5**, 144–204.

Smith ABT, Hawks DC, Heraty JM (2006) An overview of the classification and evolution of the major scarab beetle clades (Coleoptera: Scarabaeiodea) based on preliminary molecular analyses. *Coleopterists Society Monographs* **5**, 35–46.

Sternberg C (1910) Neue Dynastiden-Arten. *Annales de la Société de Belgique (Comptes-rendes)* **54**, 91–102.

Szent-Ivany JJH (1969) The distribution, systematics and feeding habits of the elephant beetle, *Xylotrupes gideon* (L.) in New Guinea and in the Solomon Islands. *Transactions of the Papua New Guinea Scientific Society* **10**, 14–18.

Westwood JO (1841) On the Australian genus *Cryptodus,* and upon *Parastasia,* the Asiatic representatives of the Rutellidae. *Proceedings of the Entomological Society of London* **1841**, 34–35.

Westwood JO (1856) Descriptions of the species of the Australian lamellicorn genus *Cryptodus. Proceedings of the Entomological Society of London* **4**, 1–7.

White A (1841) Notes on some insects from King George's Sound, collected and presented to the British Museum by Captain George Grey. In *Journals of two expeditions of discovery in North-west and Western Australia during the years 1837, 38, and 39, under the authority of Her Majesty's government. Describing many newly discovered, important, and fertile districts, with observations on the moral and physical condition of the aboriginal inhabitants, & c. & c.* Vol. 2. (Ed. G Grey) pp. 450–482. T. & W. Boone, London.

34. SCARABAEIDAE: CETONIINAE LEACH, 1815

Christian H. Moeseneder, Tom A. Weir, Cate Lemann and Paul M. Hutchinson

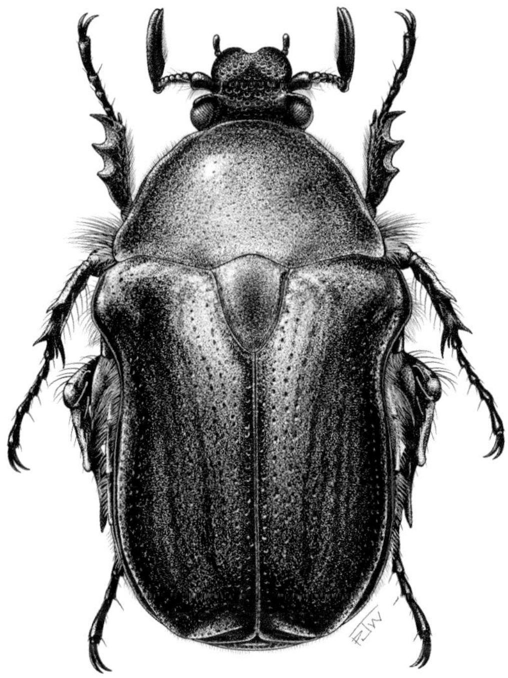

Fig. 34.1. *Bisallardiana philippei* (Allard).

Common names. Flower beetles, flower chafers.

Introduction. In Australia the cosmopolitan subfamily Cetoniinae currently contains 141 species in 3 tribes and 37 genera. 68% of genera and 90% of species are endemic to this continent. Worldwide there are 4273 described cetoniine species in 485 genera (Krajčík 2012).

The Cetoniinae are a conspicuous group of insects and several species are known to the public in flight and as flower-visitors in suburban backyards, for example the 'Cowboy Beetle' (*Chondropyga dorsalis* (Donovan)) and the 'Fiddler Beetle' (*Eupoecila australasiae* (Donovan)). Adult beetles range in size from less than 5 mm (*Microvalgus* Kraatz) to 45 mm (*Dilochrosis* Thomson), but the majority are medium-sized beetles. While a few species are well known and reared for education and by hobbyists, little or nothing is understood of the biology of most Australian cetoniines. Some species are so rarely collected that only several specimens have ever been found and the females of six species are unknown. In several large genera, such as *Microvalgus*, few biological or ecological observations of Australian species have been published.

Cetoniines occur in all Australian states, with the tropical north Queensland region especially rich in species and Tasmania particularly poor and represented only by the genus *Microvalgus*. Krikken (1984) suggested the following biogeography for the Australian cetoniines: the Cetoniini and *Charitovalgus* Kolbe (14 species) invaded from the Oriental region, *Microvalgus* (16 species) represent an ancient Gondwanan element (with extant members in Africa) while the Schizorrhinini (111 species) evolved *in situ* and dispersed from there into the Oriental regions, where they are now highly speciose and occur as far as Sri Lanka. Some Australian cetoniines, namely *Dilochrosis atripennis* (Macleay) and *Dilochrosis brownii* (Kirby) have spread over a large part of the continent and are found along the East Coast but also in WA. From the arid centre of Australia few species are known (Moeseneder & Hutchinson 2016) but current collecting effort is yielding new species in such poorly collected regions. No cetoniine species are known to have been introduced by humans to Australia. Within the continent, however, *Protaetia fusca* (Herbst) and *Glycyphana stolata* (Fabricius) were unintentionally transported from their original eastern states range into WA where they have become established.

As an attractive group, cetoniines have received much attention but also turmoil from the beginning of taxonomy. The placement and grouping of species within the Scarabaeoidea and the names of taxa were subject to many changes. Linneaus (1758) grouped the species which are now in the subfamily Cetoniinae with all other Scarabaeoidea in *Scarabaeus*. The subfamily was named by Fabricius (1775) who created *Cetonia* for all current Cetoniinae. Donovan (1805) described the first species in endemic genera. Leach (1815) placed his family Cetonida under tribe Scarabaeides. The genus *Schizorhina* which held most subsequently described Australian cetoniines was described by Kirby (1825). W. S. Macleay (1838) created an elaborate system with four levels under Cetonidae, which included, for example, Insulares, a taxon containing Madagascan and Australasian species; *Valgus* Scriba within the genus *Trichinus* Kirby; and the Australian cetoniine species with a pronotal lobe in *Gymnetinus* Macleay. Burmeister (1842) presented a key to the known species of Schizorrhinidae (it included species from Africa, Madagascar and Australia) and Cetoniades genuini (all current Cetoniini). W. J. Macleay (1863), created subsection Lenosoma within *Cetonia* F. for *C. fulgens* Macleay and other small species. Van de Poll (1886) recognised the consistency of male genitalia within genera. Gemminger & von Harold (1869) listed all Australian species under genus *Schizorhina* Kirby. Thomson (1878, 1880) described genera in Schizorhina. Kraatz (1880) dealt largely with classification, often based on the writing of other authors rather than his own inspection, and created more genera for the 40 diverse species remaining in genus *Schizorhina* Kirby [as *Schizorrhina* (sic)]; ten of these are still valid. Schoch's important papers (Schoch 1894, 1895, 1896) presented keys to genera and a catalogue of cetoniines based mainly on distribution and shape and extent of mesoventral process. In the Schizorrhinini he separated those with a lancet-shaped process (Hemipharidae) and those without (Diaphonidae). In the Cetonini he created the subtribe Glycyphanidae which included *Protaetia* Burmeister and *Glycyphana* Burmeister. Lea (1914) inspected a large amount of material, providing distribution information and in depth morphological work, often based on colour and unusual specimens, but he ignored male genital construction entirely.

This is the only work that dealt with Australian *Microvalgus*, naming 11 of the 17 species. Schenkling presented a catalogue (Schenkling 1921) based on Schoch's divisions and listed Valginae genera without tribes. Bacchus (1974) revised the Australian *Glycyphana*. Krikken (1978) provided morphological diagnosis for the Valgini (as Valginae) and supported Arrow (1910) in his division of the valgines into two groups. Early workers who described the most species (numbers of those still valid in parentheses) were: Lea (28), Janson (15), W. J. Macleay (15), Blackburn (6) and Thomson (6). Taxonomic work after Schenkling is discussed in the classification section.

In recent times, Bacchus (1974) published a review of the Australian species of *Glycyphana*. Since then several authors provided publications and online resources about the Australian cetoniines, most of general nature, elaborately figured and excluding Valgini: the *Beetles of the World* series by Rigout & Allard (1997) and Allard (1995a, 1995b) on Schizorhinini with monographs on *Lomaptera* Gory and Percheron, *Ischiopsopha* Gestro and *Mycterophallus* van de Poll, (the three books in this series provide a basic overview but are often erroneous, type descriptions are brief and ambiguous and their images have been manipulated); Sakai & Nagai's (1998) cetoniine volume in the Mushi-Sha series; a history of the research of Australian cetoniines by Hangay (2001); Reid & Bulbert's (2002) Lucid Key to the flower chafers of NSW, the ABRS AFD section Cassis *et al.* (2002) and a field guide by Golding (2009) the images of which may be useful. Moeseneder & Hutchinson are current workers of the group. In the last 20 years a dozen new Australian species have been described and at least as many await description.

The most recent work on Australian Cetoniinae (Krikken 2018) was published too late for its complete incorporation into the present chapter. Two new genera are described in that paper: *Axillonia* Krikken with the type *Clithria bacchusi* Allard, 1995, and *Territonia* Krikken, with the type *T. campbelli* Krikken, 2018. In the key below, *Axillonia bacchusi* keys out at couplet 17 and *T. campbelli* may key out with *Clithria* or related genera.

Biology. The majority of Australian Cetoniinae are anthophagous, a trait which evolved ~62 Ma in the Scarabaeoidea and gave rise to the diversity in the Cetoniinae (Ahrens *et al.* 2014). In these flower-visiting species, both sexes are volant, sexual dimorphism is not pronounced and mating usually occurs on flowers. Adults are important pollinators of many flowering trees and bushes, especially those in the white-flowering Myrtaceae (*Angophora* Cav., *Leptospermum* J.R.Forst. & G.Forst., *Eucalyptus* L'Her. and Melaleuca L.). The cetoniines consume pollen and nectar when feeding, but some, such as *Dilochrosis balteata* Vollenhoven and *Protaetia fusca* may consume or damage the flower structures. In the wild, a few cetoniines feed on ripe fruit (not in numbers to be regarded as pests) and captured specimens consume offered honey. Some species have been recorded feeding at sap flows (*Dilochrosis brownii* in WA). Adult peak activity times are from October to May and coincide with the flowering of trees which is determined by rainfall. Mating is generally opportunistic and without courtship in the species that are flower-visiting but there are exceptions.

In the ecology section of each genus, records of the host plants on which adult cetoniines have been recorded were compiled from Alderson (1976), Armstrong (1979), Bacchus (1974), Froggatt (1894, 1907, 1914), Golding (2009), Hasenpusch & Moeseneder (2010), Hawkeswood (1981, 1982, 2002), Hawkeswood & Turner (2003, 2007), Hiller (1990), Matthews (1984), Moeseneder *et al.* (2014), Moeseneder & Cook (2014), Moeseneder & Hutchinson (2012), Moore (1987), Reid & Bulbert (2002), Simpson (1990), Tepper (1887), Webb (1987), Williams & Adam (1998) and Zietek (2008).

In some genera (*Tapinoschema* Thomson, most *Chondropyga* Kraatz, *Diaphonia* Newman, *Grandaustralis* Hutchinson & Moeseneder, *Navigator* Moeseneder & Hutchinson and some *Pseudoclithria* van de Poll) adults are infrequently or never encountered on flowers. The males have enlarged antennal clubs and are observed flying most likely in search of sedentary females. The females of six species are not known (all are known in the flower-visiting group). Sexual dimorphism is pronounced with females having heavier build and darker coloration. Usually males are found more commonly.

Most Australian cetoniines can be caught by netting specimens on flowering trees with a net on a long pole. However in the last few decades, flight intercept traps have yielded rare and also undescribed species, especially in the group of the non-flower visiting taxa. Fruit-bait traps have also been successfully used in catching some species. The species in genera *Lomaptera* and *Trichaulax* Kraatz are very strong and fast flyers and can be exceedingly difficult to catch since they often frequent the uppermost flowers on mature trees.

The life history of few Australian cetoniine species is known. Froggatt (1894), Alderson (1976), Moore (1987), Matthews (1984), Moeseneder & Cook (2014), Moeseneder & Hutchinson (2016) and Tepper (1887) provided observations about several cetoniine species. A few attractive species such as those in the genera *Dilochrosis*, *Trichaulax* and *Ischiopsopha* have been reared for commercial sale, by zoos and for education.After copulation, females lay between 15 and 40 eggs in suitable substrate. Publications and observations show that in the great majority of species larvae develop in decaying, never living, wood where they are generally beneficial as organic recyclers by breaking down decaying plant material. Often larvae inhabit dead branches and tree trunks on the ground. When larvae have been found in standing living trees, they are encountered within the decaying wood areas in rotten branches, in the core of the tree or openings and splits at the base of the tree, frequently near the boundary layer to living wood (*Trichaulax*, *Lyraphora, Chlorobapta*). In some (*Pseudoclithria*, *Navigator*, *Tapinoschema* and one *Dilochrosis* sp.), however, larvae are found freely in soil under leaf litter, unassociated in sand or in decomposing plant litter. The Australian cetoniines show some particularities which may be associated with the variable and often sparse rainfall in Australia. At least some species are highly tolerant to desiccation, which was shown to cause severe shrinkage and resultant torpor in *Dilochrosis atripennis* and *Tapinoschema digglesii*. Normal growth and activity were rapidly resumed when exposed to a more humid environment and normally developed adults hatched (CHM). Larval development requires 1–3 years in 3 instars. A link to ants or termites during larval stage for several, particularly *Valgini*, has been speculated by some authors but not proven. Cocoons are constructed from humus, soil or decaying wood substrate and incorporate a lesser or greater amount of the larva's faecal pellets. In many species they remain loose in the substrate, in others they are cemented to the tree's decaying or living wood. Australian cetoniines pupate early and often spend many months in this stage, most likely as a further adaptation to arid conditions.

Predation and parasitism of Australian cetoniine larvae, pupae and adults by insects (usually flies), spiders, birds, mammals and fish has been documented by Alderson (1976), Froggatt (1894), Hawkeswood (2007), Hiller (1990), Moeseneder *et al.* (2014), CHM and PMH pers. obs., Reid & Bulbert (2002) and Simpson (1990).

Australian cetoniines are not of great economic importance to humans. In recent years, however, *Bisallardiana philippei* (Fig. 34.1) and an undescribed, similar species have been observed in Queensland entering hives of the non-native Western honey bee (*Apis mellifera* L.) in large numbers and causing damage to cells, consuming honey and stressing the colony.

Classification. Recent classification work which is relevant to Australian cetoniines began with Krikken (1984) who provided a suprageneric key to the Cetoniinae based on adult morphology. He raised the Cetoniinae to family rank, as Cetoniidae, with subfamilies Cetoniinae, Trichiinae and Valginae and transferred the genus *Pseudoclithria* into the Xiphoscelidini – both taxonomic arrangements were generally not accepted by subsequent authors. Krikken also transferred the Australian genera with a developed medial pronotal lobe from the Gymnetini to the Schizorhinini, a decision which found acceptance by all subsequent authors.

Cladistic analysis of the Scarabaeoidea (Browne & Scholtz 1995, 1998, 1999) concluded that Cetoniinae and Valginae are well defined subfamilies and remain in the Scarabaeidae. In a molecular phylogenetic review of the major Scarabaeoidea families and subfamilies of the Scarabaeidae, Smith *et al.* (2006) decided to retain their previous placement and align them with the other phytophagous groups (Melolonthinae, Dynastinae, Rutelinae) despite evidence that Cetoniinae branch off at the base of the Scarabaeidae. The preliminary findings followed those of Browne & Scholtz (1998) but Smith *et al.* (2006) listed Trichiini and Valgini at tribal level.

To date, the most comprehensive molecular phylogenetic study of the subfamily sampled 125 species representing ten of the twelve currently recognised tribes and recovered Cetoniinae as a monophyletic linieage sister to Rutelinae + Dynsatinae (Šípek *et al.* 2016). The subfamily was further divided into two clades comprised of (i) Valgini, *Microvalgus* (Microvalgini) and Trichiini in part and (ii) Cetoniini, Crematoscheilini, Diplognathini, Goliathini, Gymnetini, Osmodermatini, Schizorhinini, Stenotarsini, Taenioderini, and Trichiini in part (Šípek *et al.* 2016). Their results highlight the need for taxonomic changes of the tribal classification as only Cremastocheilini was recovered as a monophylum and all other tribes recovered as poly- or paraphyletic (Šípek *et al.* 2016).

Of the 12 tribes currently included within the Cetoniinae (Smith 2006, based on Browne & Scholtz 1995, Ratcliffe & Jameson 2002), only Valgini, Cetoniini and Schizorhinini are represented in the Australian fauna. Sixteen Australian genera representing all three of these tribes were included in the phylogeny of Gunter *et al.* (2016), however within this study the clade representing Cetoniinae was primarily comprised of Australian taxa (17 of 22 species) so limited extrapolations can be made regarding the overall tribal relationships. The topology was consistent with classification in three distinct lineages, confirming *Glycyphana* and *Protaetia* (Cetoniini) in close relationship to Cetoniini from other continents, and distinct from the monophyletic clades containing thirteen Australian Schizorhinini genera or *Microvalgus* (Valgini) (Gunter *et al.* 2016).

Catalogues at species level are Krajčík (2012) and Cassis *et al.* (2002). The Australian cetoniine fauna is very diverse and, with 3.8 described species per genus (Passalidae 3.9, Rutelinae 5.0, Dynastinae 5.3, and Lucanidae 5.5), Krikken's (1984) advice not to lump genera in the Cetoniinae due to the danger of obscuring complexity seems well placed.

No phylogenetic study or key to the entirety of Australian cetoniines as they are known in recent times has been published (but see Introduction above). The classification of genera presented herein is based on Smith (2006) and Bouchard *et al.* (2011). At this time *Clithria* Burmeister, *Pseudoclithria* and *Diaphonia* are polyphyletic and in need of revision, hence these genera resolve to several places in the key.

Schochidia Berg has been listed from Australia in the past. No specimens have been found in Australian museums by the authors and all specimens recorded in literature were captured in New Guinea. Krikken (1983) also noted Australia to be an incorrect location for this species and *Schochidia* is hence not included.

Characteristics. Mandibles and labrum concealed beneath clypeus, not visible from above. Antennal insertion exposed beside clypeus in dorsal view. Antenna 10-segmented; club 3-segmented. Frontoclypeal suture absent. Mesepimeron protrudent at base of elytra (except in Valgini). Elytral subhumeral emargination present (except in Valgini). Propygidium fused with penultimate abdominal ventrite and usually concealed by elytra (except in Valgini). Pygidium exposed. Procoxa conical and ventrally projecting. Mesocoxa usually separated (except in Valgini) and with or without development of an interposed mesoventral process. Mesocoxae aligned transversely. Metacoxa approximate (except in Valgini). Tarsal claws usually simple and subequal length but not independently articulate.

Secondary sexual characteristics. In ~90% of Australian Schizorhinini species, males have longitudinally, transversely or broadly impressed tergites, modified legs and/or enlarged antennal clubs. Male genitalia may bear unique ventral structures or pilose patches in some species. Females possess fewer diagnostic characters but are generally wider and stockier. They may be similar to males in general appearance or so morphologically dissimilar they have at times been described as separate species (*Diaphonia dispar* Newman). Females can resemble congeneric species to a degree that they are difficult to determine or indeterminable (*Metallesthes* Kraatz and *Hemichnoodes* Kraatz) without extensive knowledge or when not collected in conjunction with males.

Melanic forms occur in a large number of Australian Schizorrhinini. Where possible the generic key provided here does not rely on colour as a primary diagnostic character.

Key to the tribes, subtribes and genera of Australian Cetoniinae

Notes on key characters. Specimen lengths were measured from the clypeal apex to pygidial apex. Characters in the key and generic descriptions for taxa which also occur outside the Australian region (*Protaetia, Glycyphana, Dilochrosis, Eupoecila* Burmeister, *Ischiopsopha, Mycterophallus, Lomaptera, Chalcopharis* Heller, *Poecilopharis* Kraatz, *Microvalgus, Charitovalgus*) consider only species which occur in Australia and not extralimital species.

Some genera key out more than once, allowing for variation between species and difficulty in interpreting certain characters in the key as well as indicating the need for revision (eg *Pseudoclithria, Diaphonia, Aphanesthes, Bisallardiana, Clithria*). Males usually have longitudinally, transversely or circularly impressed abdominal ventrites or abdomen concave in lateral view, modified legs and/or enlarged antennal clubs. Females are generally wider and stockier with abdomen convex or straight in lateral view, but have fewer diagnostic characters. Costae (Figs 34.16sc–sutural costa; 34.16c1–first costa; 34.16c2–second costa) here are defined as longitudinal raised areas on the elytra which may be broad or narrow, distinct or indistinct or absent. Posthumeral emargination can be arcuate (Fig. 34.21pe) or sinuate (Fig. 34.20pe); if arcuate, the elytra are distinctly constricted in dorsal view at level of posthumeral emargination. Sculpture, setae, pilosity, colour and other non-diagnostic characters are omitted from all drawings.

1. Elytra with posthumeral emarginations absent (Fig. 34.13pe); elytra together often square, disc almost flat (Fig. 34.13); mesepimera barely or not visible between pronotum and elytra when viewed dorsally; propygidium fully exposed or nearly so (Fig. 34.13pp); metacoxae widely separated; body usually covered with squamose setae or scales (except in *Microvalgus glaber*); 6 mm or less in length (not including female ovipositor, when present)….**Valgini** **2**

– Elytra with post humeral emarginations usually distinct (Figs 34.7pe, 34.8pe, 34.10pe, 34.11pe, 34.12pe); elytra together longer than wide, disc somewhat convex; mesepimera clearly visible when viewed dorsally, protruding between pronotum and elytra to a greater or lesser degree, sometimes concealed under dense setae (Figs 34.5me, 34.6me, 34.7me, 34.8me, 34.16me); propygidium concealed or at most only partially visible in intact specimens (Fig. 34.10pp); metacoxae close together (Fig. 34.33mx); body not covered with squamose setae or scales; greater than 6 mm in length **3**

2(1). Anterolateral angles of pronotum not obvious; pronotal hypomera somewhat rectangular (Fig. 34.2ph), anterior edge vertical and embracing at least base of eye; pronotum with paramedian ridges and lateral impressions anteriorly and raised above plane of elytra; protibiae with 5 denticles or teeth laterally; females with distinct ovipositor, about as long as pronotum; 5–6 mm in length (not including ovipositor); north QLD .... ***Charitovalgus* Kolbe**

– Anterolateral angles of pronotum obvious, acute (Fig. 34.3aa); pronotal hypomera not rectangular (Fig. 34.3ph), anterior edge not vertical, exposing most of eye; pronotum without ridges and in same plane as elytra; protibiae with 5 or fewer denticles or teeth laterally, if with 5, then second and fourth are deflexed; 2–4 mm in length; QLD, NSW, VIC, SA, TAS, WA........ ***Microvalgus* Kraatz**

3(1). Pronotum with basolateral angles broadly rounded (Figs 34.18ba, 34.19ba), usually more than 90 degrees, more or less shifted forward relative to base of elytra; mesoscutellum with rounded or broad apex (Figs 34.18sl, 34.19sl); mesoventral process transverse, markedly constricted between mesocoxae (Fig. 34.33m/ba), expanded anteriorly with acute lateral angles (Fig. 34.33m/la), anterior margin more or less straight or broadly rounded, barely extending anterior of mesocoxae (Fig. 34.33m), sometimes with row of punctures or setae adjacent to apex (Fig. 34.33i); body dorsally velutinous (except in *Protaetia andrewsi* from Christmas Island, and some *Protaetia acuminata* from Cocos-Keeling Islands), and with at least a few white or coloured embedded maculae formed by microtrichiae….**Cetoniini: Cetoniina** **4**

– Pronotum with basolateral angles more or less distinct, usually less than 90 degrees (except *Lyraphora, Octocollis, Neoclithria, Grandaustralis, Phyllopodium,* some *Chlorobapta,* some *Pseudoclithria* all of which have the mesoventral process other than transverse), more closely applied to base of elytra (Figs 34.4ba, 34.6ba, 34.14ba, 34.15ba, 34.16ba, 34.17ba); mesoscutellum, if visible, with apex usually sharply acute or indistinctly rounded (Figs 34.7sl, 34.14sl, 34.16sl), sometimes rounded (*Bisallardiana* and female *Poecilopharis leai* which have mesoventral process elongate); mesoventral process of various shapes (Figs 34.33a, b, d–k, m), sometimes absent (Figs 34.33c, l), sometimes transverse (some *Chondropyga* and some *Diaphonia*) (Fig. 34.33f) but not as above; body rarely velutinous (except some *Lyraphora*) or rarely with maculae formed by microtrichiae (except some *Bisallardiana*) ….**Schizorhinini** **5**

4(3). Clypeus distinctly emarginate, not distinctly upturned apically (Fig. 34.18ca); basal emargination of pronotum shallow, smoothly rounded, without distinct lateral angles (Fig. 34.18se); pronotal lateral margins broadly rounded, often parallel basally (Fig. 34.18mg); elytra with posthumeral emargination deep, arcuate; dorsal maculation generally as in Fig. 34.18ma; 10–16 mm in length; widespread – all Australian mainland states .... ***Glycyphana* Burmeister**

– Clypeus truncate, distinctly upturned apically (Fig. 34.19ca); basal emargination of pronotum deeper, lateral angles usually distinct (Fig. 34.19se); pronotal lateral margins more linear, not parallel basally (Fig. 34.19mg); elytra with posthumeral emargination shallower, sinuate; dorsal maculation generally as in Fig. 34.19ma; 13–18 mm in length; widespread – all Australian mainland states, Christmas Island, Cocos-Keeling Islands........ ***Protaetia* Burmeister**

5(3). Pronotum with basal lobe covering mesoscutellum, entirely or nearly so (Figs 34.15bl, 34.34a, b)….**Lomapterina** **6**

– Pronotum without basal lobe, mesoscutellum fully exposed (Figs 34.7sl, 34.14se/sl, 34.16se/sl)….**Schizorhinina** **8**

6(5). Pronotal basal lobe with notched apex; mesoscutellum apex visible (Figs 34.15, 34.34a); abdominal ventrites 2 and 3 with lateral, distinct patches bearing fine diagonal striations (Fig. 34.41); 19–30 mm in length; northern Australia – QLD, NT, WA .... ***Ischiopsopha* Gestro**

– Pronotal basal lobe without notched apex (Fig. 34.34b), apex of mesoscutellum not or only barely visible; abdominal ventrites 2 and 3 laterally without such distinct striated patches **7**

7(6). Elytra with preapical umbone obvious (Fig. 34.38au); metaventrite, metacoxae and abdominal ventrites without or with only a few setae; elytral apex somewhat truncate, not sinuate before suture (Fig. 34.38); 16–25 mm in length; QLD .... ***Lomaptera* Gory and Percheron**

– Elytra with preapical umbone not obvious (Fig. 34.37au); metaventrite, metacoxae and abdominal ventrites laterally with abundant pale setae; elytral apex somewhat sinuate before suture (Fig. 34.37); 23–30 mm in length; QLD .... ***Mycterophallus* van de Poll**

8(5). Mesoventral process absent or greatly reduced, not surpassing anterior edge of mesocoxae in lateral view (Figs 34.33c, l) **9**

– Mesoventral process present, variously shaped, surpassing anterior edge of mesocoxae in lateral view (Figs 34.33a–b, d–h, j–k) **14**

9(8). Pronotum widest at mid length, appearing somewhat circular AND tarsi of all legs with segments 1 to 5 combined longer than tibiae AND male antennal club greatly enlarged AND body non-metallic dorsally; 16–19 mm in length; north QLD; known from males only .... ***Octocollis* Moeseneder & Hutchinson**

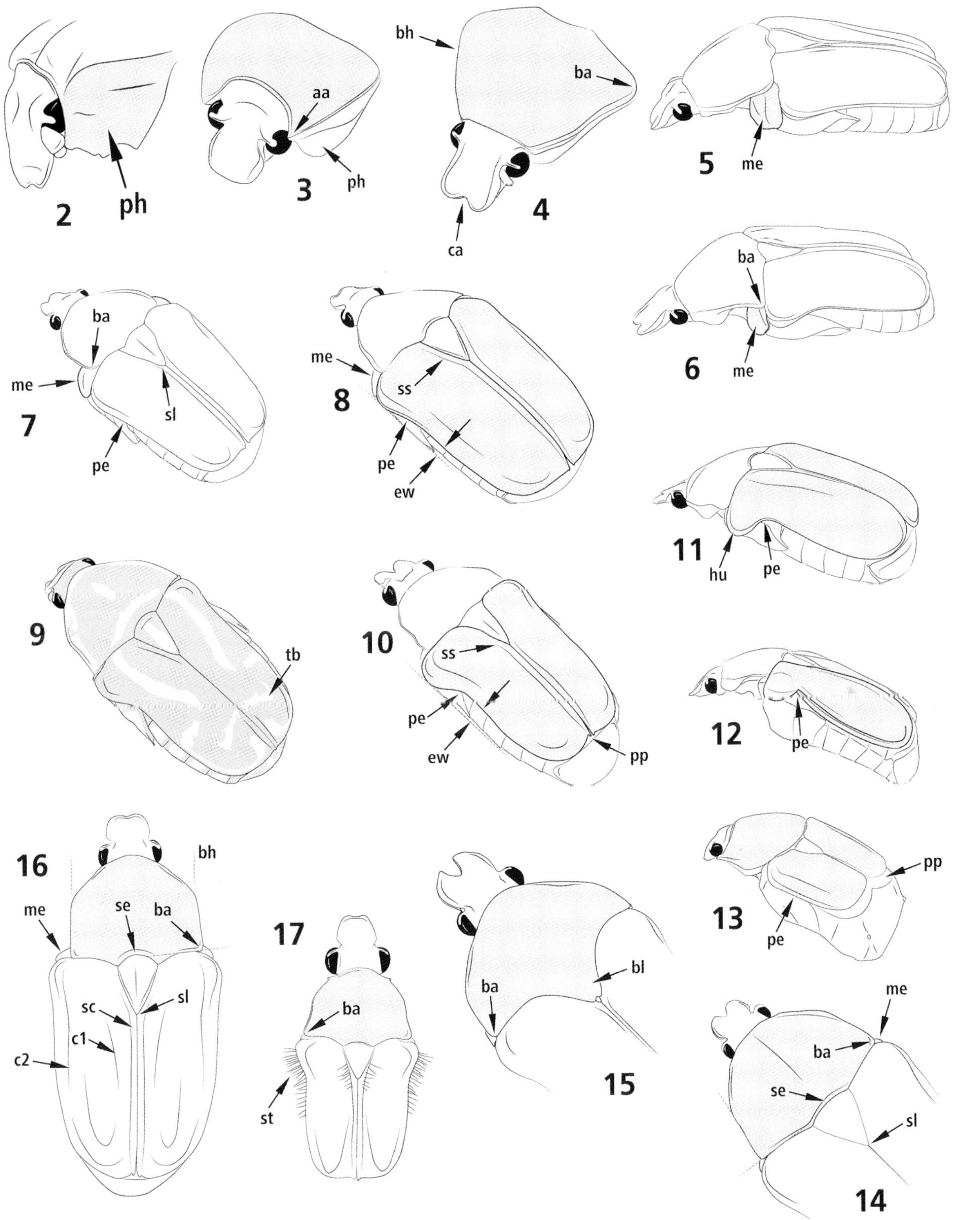

Figs 34.2–34.17 Characters of Australian Cetoniinae genera. Taxa: 2, *Charitovalgus quinquedentatus* (Lea) female; 3, *Microvalgus* sp. male; 4, *Aphanesthes pullata* (Janson) male; 5, *Lyraphora obliquata* (Westwood) male; 6, *Aphanesthes trapezifera* (Thomson) female; 7, *Lyraphora obliquata* male; 8, *Chondropyga dorsalis* (Donovan) male; 9, *Eupoecila australasiae* (Donovan) male; 10, *Clithria eucnemis* (Burmeister) male; 11, *Bisallardiana* sp.; 12, *Clithria bacchusi* Allard male; 13, *Microvalgus* sp.; 14, *Hemichnoodes mniszechi* (Janson) female; 15, *Ischiopsopha wallacei* (Thomson) male; 16, *Chlorobapta frontalis* (Donovan) male; 17, *Lenosoma fulgens* (Macleay) male.

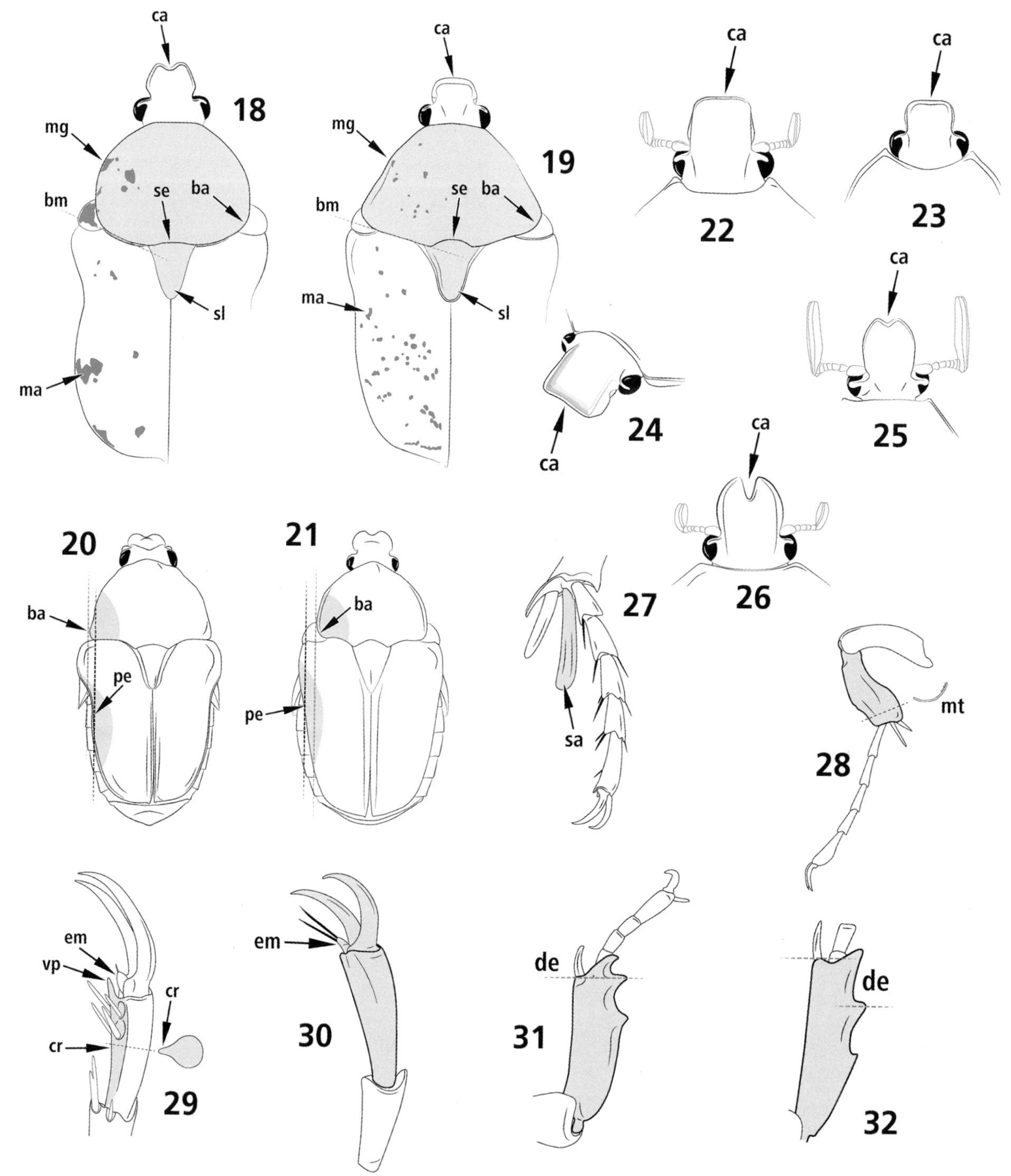

Figs 34.18–34.32. Characters of Australian Cetoniinae genera. 18, *Glycyphana (Glycyphaniola) stolata* (Fabricius) female; 19, *Protaetia* (*Pseudourbania) andrewsi* Gahan male; 20, *Clithria albersi* Kraatz female; 21, *Lyraphora velutina* (Macleay) female; 22, *Stenopisthes frenchi* (Blackburn) male; 23, *Poecilopharis leai* Schürhoff male; 24, *Chalcopharis lansbergei* (Gestro) male; 25, *Diaphonia antoinei* Allard male; 26, *Dilochrosis atripennis* (Macleay) female; 27, *Navigator* sp.; 28, *Clithria eucnemis* male; 29, *Aphanesthes pullata*; 30, *Chondropyga* sp.; 31, *Poecilopharis leai* Schurhoff male; 32, *Macrotina satanas* (Schürhoff) male.

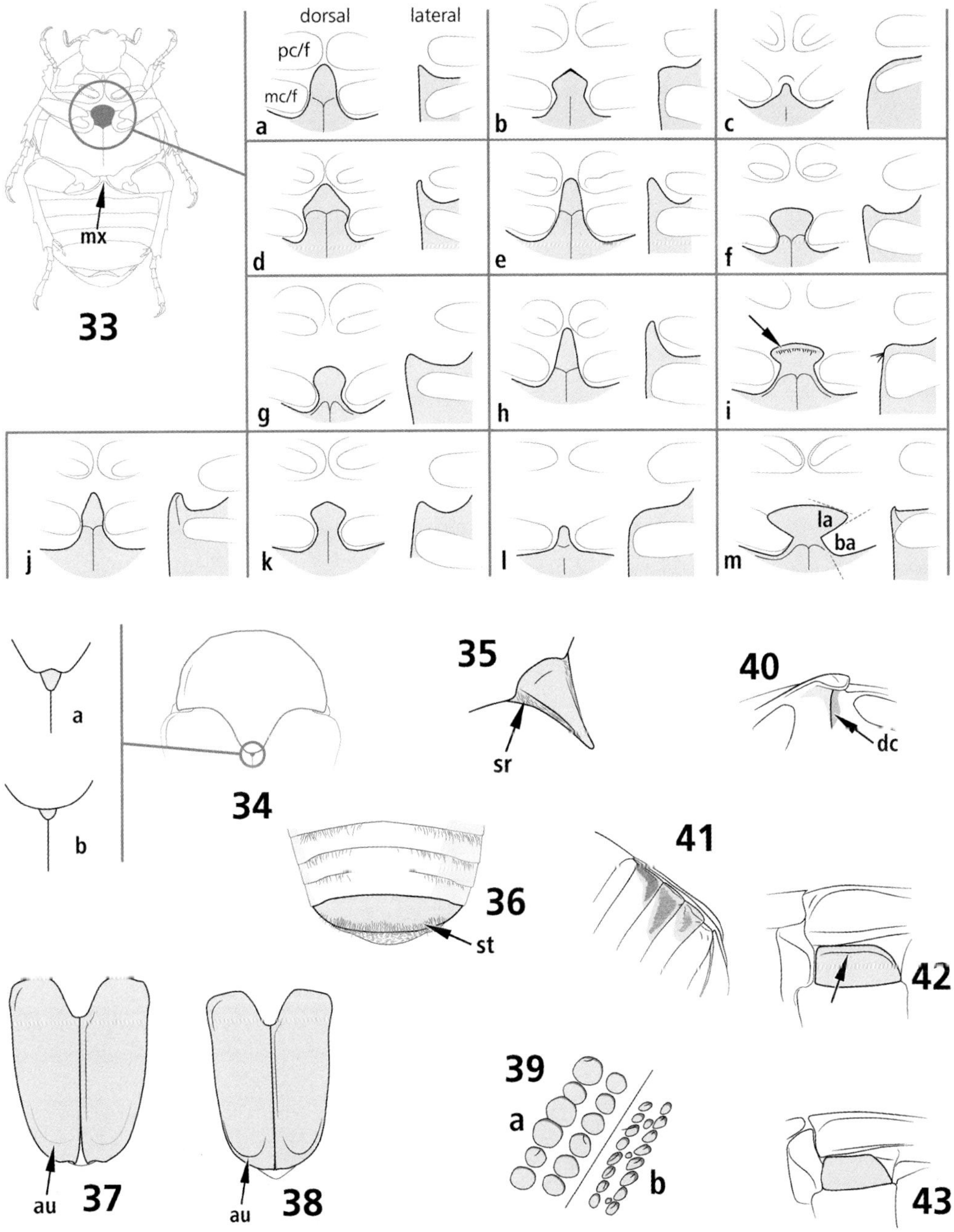

Figs 34.33–34.43. Characters of Australian Cetoniinae genera. 33, *Aphanesthes succinea* (Hope) female; 33a, *Evanides bakewellii* (White) female; 33b, *Micropoecila cincta* (Gory & Percheron); 33c, *Grandaustralis boomerang* Hutchinson & Moeseneder male; 33d, *Schizorhina atropunctata* (Kirby) male; 33e, *Dilochrosis brownii* (Kirby) male; 33f, *Diaphonia antoinei* Allard male, 33 g, *Aphanesthes trapezifera* male; 33h, *Dilochrosis atripennis* (Macleay) female; 33i, *Glycyphana (Glycyphaniola) stolata* (Fabricius) male; 33j, *Bisallardiana* sp.; 33k, *Chlorobapta bestii* (Westwood) female; 33l, *Pseudoclithria* undescribed sp. male; 33m, *Protaetia fusca* (Herbst) male; 34a, *Ischiopsopha wallacei* (Thomson) female; 34b, *Lomaptera australis* Wallace female; 35, *Clithria tibiale* (Macleay) male; 36, *Chondropyga dorsalis* female; 37, *Mycterophallus duboulayi* (Thomson) male; 38, *Lomaptera australis* female; 39a, *Tapinoschema digglesii* (Janson); 39b, *Metallesthes metallescens* (White); 40, *Eupoecila intricata* Lea; 41, *Ischiopsopha wallacei* male; 42, *Dilochrosis brownii*; 43, *Hemipharis insularis* (Gory & Percheron). Character abbreviations: aa–pronotal anterolateral angle, au–apical umbone, ba–basolateral angle or basal angle, bh–pronotum basal half, bl–pronotal basal lobe, bm–pronotal basolateral margin, c1–first costa, c2–second costa, ca–clypeus apex, dc–dorsal carina, cr–carina, de–dentition, el–elytral lateral margin, em–empodium, hu–humeral umbone, la–lateral angle, ma–macula, mc/f–mesocoxa and mesofemur, me–mesepimeron, mg–margin, mt–metatibia, mx–metacoxae, pc/f–procoxa and profemur, pe–posthumeral emargination, ph–pronotal hypomera, pp–propygidium, sa–spatulate spur, sc–sutural costa, se–scutellar emargination, sl–scutellum, sp–spine, sr–striae, ss–sutural striae, st–setae, tb–transverse band, vp–ventrodistal projection.

– Pronotum widest across posterolateral angles, more tapered anteriorly; tarsi of all legs with segments 1 to 5 combined usually at most as long as tibiae (*Navigator fossor* and *N. pixii* have all legs with segments 1 to 5 combined longer than tibiae); male antennal club variable; body metallic or matt dorsally **10**

10(9). Abdominal ventrites with distinctive colour pattern: 1 to 5 yellow and 6 entirely black; 15 mm in length; WA; known from males only (*Diaphonia kerleyi*) .... ***Diaphonia* Newman (part)**

– Abdominal ventrites with colour pattern otherwise **11**

11(10). Elytra distinctly constricted in dorsal view at level of posthumeral emargination; humeral umbone distinct, protruding; posthumeral emargination deep, arcuate (Fig. 34.11pe); 12–21 mm in length; QLD, NSW, WA, NT .... ***Bisallardiana* Antoine (part)**

– Elytra not so constricted in dorsal view at level of posthumeral emargination; humeral umbone distinct or indistinct but not protruding; posthumeral emargination shallow, sinuate (Fig. 34.8pe) **12**

12(11). Elytra without distinct raised costae; clypeus with sharp, raised, vertical lateral and anterior margins (less so in female), clypeal apicolateral angles square; mesotibiae with 2 distinct teeth laterally; male antennal club greatly enlarged; sexually dimorphic with regard to colour: male pronotum, elytra and pygidium burnt orange; female all black; non-metallic; 18.5–25.5 mm in length; south-west WA .... ***Grandaustralis* Hutchinson & Moeseneder**

– Elytra with 1 or 2 distinct raised costae; clypeus without sharp, raised, vertical lateral and anterior margins, OR lacking raised margins, clypeal apicolateral angles more rounded; mesotibiae usually with 1 tooth externally (2 in the metallic *Navigator fossor*); male antennal club enlarged or not; not dimorphic with regard to colour; metallic or matt dorsally; generally smaller species 10–19 mm in length **13**

13(12). Elytra with 2 distinct, narrow costae; colour black or brown, with metallic copper, blue or green reflections (less so in old specimens); pronotum coarsely punctate; male antennal club greatly enlarged; female metatibial inner spur widened apically (Fig. 34.27sa); 10–19 mm in length; QLD, SA, VIC, WA .... ***Navigator* Moeseneder & Hutchinson**

– Elytra with 1 or 2 distinct costae, first costa usually wide with somewhat defined margins; colour black, brown or yellow, without metallic reflections; pronotum finely punctate (coarsely in *P. mastersii*); male antennal clubs not greatly enlarged (except *P. adusta*); female metatibial inner spur parallel sided or tapering apically; 10–19 mm in length; QLD, SA, WA .... ***Pseudoclithria* van de Poll (part)**

14(8). Elytra with prominent specialised setae (as well as various other setae): EITHER rows of few erect setae above posthumeral emargination (Fig. 34.17st) and on sutural interval either side of mesoscutellum OR prominent tufts of setae on apical declivity between suture and costa; smaller species 11–18 mm in length **15**

– Elytra without such specialised setae as above (except *Trichaulax* which has elytra with extensive rows of dense setae including on apical declivity) **16**

15(14). Elytra with rows of erect setae above posthumeral emargination and on sutural interval either side of mesoscutellum (Fig. 34.17st); body metallic green, red, coppery or blue; 11–16 mm in length; QLD, NSW .... ***Lenosoma* Kraatz**

– Elytra with tufts of setae on apical declivity between suture and costa; body not metallic, black or brown, with or without maculae; 11–18 mm in length; QLD .... ***Storeyus* Hasenpusch & Moeseneder**

16(14). Pygidium with a setose longitudinal groove, broadening apically to a flattened setose depression in female; pronotum with parallel sides in basal half; elytra black with yellow maculae; males with broad, parallel-sided metatibiae and elongate metatarsi; narrow species; 11–14 mm in length; QLD .... ***Neoclithria* van de Poll**

– Pygidium without a setose longitudinal groove, at most with a flattened setose impression; sides of pronotum variable; elytra colour variable; male metatarsi not elongate; body not noticeably narrow **17**

17(16). Elytra with posthumeral emargination deep and sharp, angled at 90 degrees (Fig. 34.12pe); males with a tooth on inner claw of protarsi; elytra colour yellow with black maculae; 10–16 mm in length; north QLD....(*Clithria bacchusi*) .... ***Clithria* Burmeister (part)**

– Elytra with posthumeral emargination more evenly rounded, arcuate or sinuate, much greater than 90 degrees (Figs 34.8pe, 34.10pe, 34.11pe); males without a tooth on inner claw of protarsi; elytral colour variable **18**

18(17). Mesoventral process large, wide, flat, spade-shaped with angled apex, reaching at least posterior margin of procoxae (Fig. 34.33d, but not as in 34.33b or 34.33k) **19**

– Mesoventral process otherwise shaped (Figs 34.33a–b, e–h, j–k) **20**

19(18). Elytra impunctate with 2 wide costae developed, interstices and/or lateral margins grooved and densely setose; elytral margins smooth; dorsally entirely black, dark maroon or green or black elytra with reddish-brown pronotum; 16–39 mm in length; QLD, NSW, VIC, SA, WA .... ***Trichaulax* Kraatz**

– Elytra with shallow punctures, coalesced laterally, without costae or dense setae; elytral margins somewhat serrate except near base; yellow to yellow-brown in colour, sometimes with a green hue; 22–28 mm in length; QLD, NSW .... ***Schizorhina* Kirby**

20(18). Embedded maculae formed by microtrichiae present at least on pygidium (rarely absent) and lateral abdominal ventrites; colour black; elytra distinctly constricted in dorsal view at level of posthumeral emargination (Fig. 34.11); 12–18 mm in length; QLD, NT (*Bisallardiana decorticata* and *B. obscura*) .... ***Bisallardiana* Antoine (part)**

– Elytra without embedded maculae formed by microtrichiae, although other coloured maculae not formed by microtrichiae may be present; colour variable; elytral constriction variable **21**

21(20). Elytra with dense, deep, large, simple, round punctures arranged in longitudinal rows (Fig. 34.39a); sutural striae deep, and punctate, from apex of mesoscutellum to near apex of elytra; elytra without costae; colour EITHER pronotum burnt orange and elytra burnt orange with black elytral sutural maculae tapering posteriorly OR pronotum burnt orange and metallic blue elytra; 19–29 mm in length; QLD, SA, WA .... ***Tapinoschema* Thomson (part)**

– Elytra with punctures of different types but never entirely with dense, deep, large, round, simple punctures (*Metallesthes* has dense punctures on elytra, but these are smaller and elongate (Fig. 34.39b) and the elytra have 2 narrow costae); with or without costae; colour variable **22**

22(21). Mesoventral process EITHER tapering from base to apex (Figs 34.33a, e, h) OR parallel sided basally and then tapering to apex (Fig. 34.33j) OR widened medially and then tapering to apex; apex sharply or bluntly pointed; process usually with a distinct dorsal carina or keel (Fig. 34.40dc) (absent in *Neorrhina, Evanides, Macrotina* and *Stenopisthes*) **23**

– Mesoventral process not as above, either expanded, truncate, rounded or angled, sometimes parallel sided with rounded apex (Figs 34.33f, g, k); process without a distinct dorsal carina or keel (present in *Micropoecila* (Fig. 34.33b) **33**

23(22). Elytra distinctly constricted in dorsal view at level of posthumeral emargination; humeral umbone distinct, protruding (Fig. 34.11hu); posthumeral emargination, deep, arcuate (Fig. 34.11pe) **24**

– Elytra not so constricted in dorsal view at level of posthumeral emargination; humeral umbone distinct or not distinct, but not protruding; posthumeral emargination shallow, sinuate (Fig. 34.8pe) **27**

24(23). Elytra yellow with black maculae generally in the form of spots; legs all yellow; sutural striae present from elytral apices past scutellum usually to pronotal base as a fine but distinct line; 13–18 mm in length; QLD, NSW, VIC, WA, NT .... ***Neorrhina* Thomson**

– Elytra otherwise coloured; legs brown, black or bicoloured; sutural striae variable, not as above, often shorter, if they extend past scutellum apex then not to pronotal base and only as punctures or dashes **25**

25(24). Pronotum, elytra and pygidium with extensive, complex, hieroglyph-like markings composed of bands, lines and spots (as Fig. 34.9); elytra with post-medial sinuous transverse band (Fig. 34.9tb); 14–24 mm in length; QLD, NSW, VIC, SA, WA .... ***Eupoecila* Burmeister**

– Pronotum, elytra and pygidium without extensive hieroglyph-like markings **26**

26(25). Mesoscutellum somewhat rounded apically; mesoventral process with a dorsal carina or keel (sometimes concealed by setae); 12–21 mm in length; QLD, NSW, WA, NT .... ***Bisallardiana* Antoine (part)**

– Mesoscutellum sharply pointed apically; mesoventral process without a dorsal carina or keel; 19–27 mm in length; QLD, WA .... ***Aphanesthes* Kraatz (part)**

27(23). Apex of clypeus truncate or slightly emarginate (Figs 34.22ca, 34.23ca, 34.24ca) **28**

– Apex of clypeus with a deep notch (Fig. 34.26ca) or distinctly emarginate (Figs 34.4ca, 34.25ca) **31**

28(27). Clypeus elongate (Fig. 34.22); base of second protibial tooth posterior to base of apical spur (Fig. 34.32de); (often only one tooth present in male *Stenopisthes*); mesoventral process without a dorsal carina or keel **29**

– Clypeus quadrate (Figs 34.23, 34.24) or wider than long; second protibial tooth in line with base of apical spur (Fig. 34.31de); mesoventral process with a dorsal carina or keel (Fig. 34.40dc) **30**

29(28). Dorsally entirely metallic green with coppery reflections; clypeus with apex truncate (Fig. 34.22ca); metafemora with a few sparse setae on posterior margin; elytra narrowed apically, sides evenly curved from base to apical umbone; male antennal club not greatly enlarged; 26–33 mm in length; north QLD .... ***Stenopisthes* Moser**

– Dorsally black or black and pale brown; clypeus with apex slightly emarginate; metafemora with dense fringes of setae on posterior margin; elytra not much narrowed apically, nearly straight from base to apical umbone; male antennal club greatly enlarged; 20–26 mm in length; north QLD .... ***Macrotina* Strand**

30(28). Clypeus with sharply raised apical and lateral margins; profemora and protibiae with long dense inner fringes of setae; dorsally entirely glossy brass coloured; elytra with sutural striae impressed only apically; 30–37 mm in length; northern WA; extralimital to New Guinea .... ***Chalcopharis* Heller**

– Clypeus with margins not sharply raised; profemora and protibiae with shorter, sparser setae; dorsally glossy red-brown or coppery with multiple yellow or orange maculae; elytra with sutural striae impressed from base to apex in the one Australian species; 17–20 mm in length; north QLD; extralimital to New Guinea, Fiji .... ***Poecilopharis* Kraatz**

31(27). Dorsally entirely glossy green or brassy; sides of elytra striate at least in apical half; metanepisternum without a dorsal groove (Fig. 34.43); 19–23 mm in length; QLD, NT, WA .... ***Hemipharis* Burmeister**

– Dorsally variously coloured but never entirely green; sides of elytra not striate; metanepisternum with a dorsal groove (Fig. 34.42) or short row of chained punctures (male *Evanides rufolatera*) **32**

32(31). Clypeus with raised lateral margins, apex with a deep V-shaped notch and pointed apices (Fig. 34.26ca); mesoventral process usually with a dorsal carina or keel; male without long curved setae on apical third of metatibiae; 22–45 mm in length; QLD, NSW, SA, WA .... ***Dilochrosis* Thomson**

– Clypeus without raised lateral margins, apex with a nearly right-angle shaped notch and rounded apices; mesoventral process without a dorsal carina or keel; male with long curved setae on apical third of metatibiae; 30–40 mm in length; QLD, NSW, VIC .... ***Evanides* Thomson**

33(22). Pronotum punctate AND all legs with tarsomere 5 having a narrow ventroapical projection (as Fig. 34.29vp) (not to be confused with the empodium which arises between the claws (Fig. 34.29em)) AND elytra setose with irregular, often coalesced punctures (Fig. 34.39b) AND 2 narrow indistinct costae AND post humeral emargination sinuate; dorsally black with or without metallic reflections; 15–25 mm in length; QLD, NSW, VIC, SA, WA .... ***Metallesthes* Kraatz**

– Without ALL of the above characters in combination. Some *Bisallardiana* have a punctate pronotum but have the elytra distinctly constricted in dorsal view at level of posthumeral emargination and the emargination arcuate. *Hemichnoodes*, *Micropoecila*, some *Aphanesthes* and some *Diaphonia* have the narrow ventroapical projection on tarsomere 5, but without setose elytra. Some *Chondropyga* have a punctate pronotum but lack the narrow ventroapical projection on tarsomere 5 and setose elytra **34**

34(33). Width of base of pronotum in relation to width of elytra across posthumeral emarginations wide – a line from basolateral angles of pronotum (Fig. 34.20ba) and parallel to elytral suture passes through or outside of posthumeral emargination (Fig. 34.20pe); mesepimeron usually only slightly visible when viewed dorsally (Fig. 34.14me) **35**

– Width of base of pronotum in relation to width of elytra across posthumeral emarginations narrower – a line from basolateral angles of pronotum (Fig. 34.21ba) and parallel to elytral suture passes inward of posthumeral emargination (Fig. 34.21pe); mesepimeron may or may not be clearly visible when viewed dorsally (Figs 34.8me, 34.16me), sometimes obscured by dense setae **40**

35(34). Mesoscutellum with distinct lateral grooves that are striate and extending to base (Fig. 34.35sr); head, pronotum and elytra with long setae (longer and denser in males); elytra dorsally distinctly constricted at level of posthumeral emargination; posthumeral emargination deep, arcuate; males with metatibiae expanded, more or less twisted (Fig. 34.28) and with specialised setae on inner side; colour various combinations of black, yellow and orange; smaller species 12–19 mm in length; QLD, NSW, VIC .... ***Clithria* Burmeister (part)**

– Mesoscutellum without distinct striate lateral grooves; other characters not all present in combination as above. *Bisallardiana* that come out in couplet 38 have elytra dorsally distinctly constricted at the level of post humeral emargination and posthumeral emargination deep, arcuate, but lack the lateral grooves on mesoscutellum **36**

36(35). Mesoventral process somewhat expanded laterally and apex truncated or slightly arcuate (Fig. 34.33f) **37**

– Mesoventral process usually simple, round, if somewhat expanded then apex rounded (Fig. 34.33g) **38**

37(36). Legs bicoloured with femora yellow to orange and tibiae brown-black; dorsally usually uniform yellow-orange on pronotum and elytra; antennal club in both sexes shorter than antennomeres 1 to 7 combined; 18–27 mm in length; WA (*Aphanesthes succinea*) .... ***Aphanesthes* Kraatz (part)**

– Legs not bicoloured (except *Diaphonia lateralis* which is dorsally black with yellow margins to pronotum and elytra); dorsally maybe otherwise coloured on pronotum and elytra; male antennal club as long as or longer than antennomeres 1–7 combined; 17–24 mm in length; QLD, NSW, SA, WA .... ***Diaphonia* Newman (part)**

38(36). Elytra dorsally distinctly constricted at level of post humeral emargination; posthumeral emargination deep, arcuate (Fig. 34.11pe); 12–21 mm in length; QLD, NSW, NT, WA .... ***Bisallardiana* Antoine (part)**

– Elytra not so constricted in dorsal view at level of post humeral emargination; posthumeral emargination shallow, sinuate (Figs 34.8pe, 34.10pe) **39**

39(38). Elytra at most with fine indistinct punctures in rows, inner costae, if present, not reaching base; pronotum with fine punctures, sparse medially; dorsally pronotum and elytra yellow-orange with black central markings on pronotum to all black; 21–24 mm in length; QLD; common (some *Aphanesthes trapezifera* and *A. pullata*) .... ***Aphanesthes* Kraatz (part)**

– Elytra with noticeable large dense punctures often in rows, inner costae reaching base; pronotum with coarser, evenly spaced punctures, less so on midline; in male dorsally pronotum orange brown with black central maculae and elytra orange brown with black sutural stripe widening apically to form an anchor shaped mark, in female dorsally black with some red wash at central base of each elytron; 22–27 mm in length: SA; very rare (*Diaphonia dispar*) .... ***Diaphonia* Newman (part)**

40(34). Pygidium uniformly covered with with dense, long pilosity; pronotum and elytra glabrous or with very fine short setae; apex of clypeus with a deep V-shaped notch and somewhat pointed apices (Fig. 34.26); elytra mostly uniform coloured; 23–39 mm in length; NSW, VIC, SA, WA .... ***Hemichnoodes* Kraatz (part)**

– Pygidium not uniformly covered with dense, long pilosity (male of *Phyllopodium palmatum* has pygidium with long setae on either side of a glabrous midline, but also has distinct broad, flat protarsi), but if setae present then pronotum and elytra also with setae different to above; apex of clypeus not as above, either truncate, sinuate or with a shallower notch and rounded apices (Figs 34.21, 34.25); elytral colour variable **41**

41(40). Mesoventral process medially with a right angle, a vertical dorsal keel and glabrous ventrally (Fig. 34.33b); elytra without costae; dorsally shiny brown with large central black maculae on pronotum and elytra; 18–23 mm in length; NSW, SA, WA .... ***Micropoecila* Kraatz**

– Mesoventral process medially without a right angle and vertical dorsal keel, with or without setae; elytra with costae or longitudinal punctures; dorsally colour not as above; QLD, SA, WA, NSW **42**

42(41). Clypeus subquadrate, apex truncate and raised, anterolateral angles square; pronotum and elytra with setae; male with pronotum black and elytra orange; female all black; 15–20 mm in length; south-west WA (*Diaphonia bacchusi*) .... ***Diaphonia* Newman (part)**

– Clypeus more elongate, apex sinuate or emarginate, if somewhat subquadrate then anterolateral angles rounded; pronotum and elytra with or without setae; colour variable **43**

43(42). Mesepimera enlarged and swollen, produced anteriorly (Figs 34.5me, 34.7me); pronotum with basolateral angles broadly rounded (Fig. 34.7ba), sometimes greater than 90 degrees, more or less shifted forwards and overlaying mesepimera; colour dorsally yellow to orange with black maculae, shiny or velutinous; 15–24 mm in length; QLD, NSW, VIC .... ***Lyraphora* Kraatz**

– Mesepimera visible but not enlarged and swollen as above (Figs 34.8me, 34.16me), if larger, then pronotal basolateral angles not broadly rounded; pronotum with basolateral angles variously rounded, usually less than 90 degrees and usually more closely applied to base of elytra (Fig. 34.16ba) (except *Chlorobapta* and some *Chondropyga* e.g. Fig. 34.16ba); colour variable but never velutinous **44**

44(43). Mesoscutellum with distinct lateral grooves, bearing either striations (Fig. 34.35sr) or stout setae; males with either flattened expanded protarsi OR expanded and more or less twisted metatibiae **45**

– Mesoscutellum usually without distinct lateral grooves, if present then neither striate nor setose (a few weak striae in some *Aphanasthes trapezifera*); males with protarsi and metatibiae not as above **46**

45(44). Lateral grooves of mesoscutellum striate and extending to base (Fig. 34.35sr), sometimes with fine setae as well; head, pronotum and elytra with long setae; elytra dorsally distinctly constricted at level of posthumeral emargination; posthumeral emargination deep, arcuate; elytra without distinct costae; males with metatibiae expanded, more or less twisted (Fig. 34.28) and with a tuft of specialised setae on inner side; colour various combinations of black, yellow and orange; smaller species, 12–19 mm in length; QLD, NSW, VIC .... ***Clithria* Burmeister (part)**

– Lateral grooves of mesoscutellum with stout setae, not extending to base; head, pronotum and elytra with only short setae; elytra not so constricted at level of posthumeral emargination; posthumeral emargination shallow, sinuate; elytra with at least one distinct costa; males with expanded, flattened protarsi and curved metatibiae; head, pronotum and basal third of elytra black – rest of elytra orange; larger species, 22–26 mm in length; QLD, NSW, VIC, SA .... ***Phyllopodium* Schoch**

46(44). Head both deeply and densely punctate and in male, setose, leaving a raised, glabrous, V-shaped area on vertex of male, a node on female; elytra with 1 distinct raised costa; colour brown and black with yellow maculae on abdominal ventrites or entirely dark brown; 15–18 mm in length; QLD (*Pseudoclithria hirticeps*) .... ***Pseudoclithria* van de Poll (part)**

– Head usually not both deeply and densely punctate and setose, if long setae present (some male *Chondropyga* and *Tapinoschema lacunosa*), then without a raised, glabrous, V-shaped area or node on vertex; elytra with or without costae; colour variable; abdominal ventrites with or without yellow maculae **47**

47(46). Elytra with simple, shallow, variously shaped, often coalesced punctures, each with a seta and loosely arranged in longitudinal rows; head, pronotum, scutellum, sutural area of elytra and pygidium with long setae; dense long setae present on ventrites and legs; lateral margins of clypeus sharply raised and expanded; dark brown–black; 18–20 mm in length; WA, SA (*Tapinoschema lacunosa*) .... ***Tapinoschema* Thomson (part)**

– Elytra without punctures and setae as above; scutellum, elytra, and pygidium without long setae, although short setae may be present in some areas; ventrites usually with sparser and shorter setae (most species of *Chondropyga* with longer setae concentrated apically on ventrite 5 (Fig. 34.36st)); lateral margins of clypeus not sharply raised **48**

48(47). Elytra light brown with 2 longitudinal dark brown lines marking the position of the 2 costae, inner one more raised than outer; pronotum with a distinct narrow impunctate raised longitudinal midline in anterior half; head black, pronotum entirely black in males and black with 4 orange maculae in females; 23–25 mm in length; central QLD (*Chondropyga insignicosta*)... ***Chondropyga* Kraatz (part)**

– Elytra otherwise coloured, if light brown then without 2 longitudinal dark brown lines; with or without costae; pronotum without a distinct impunctate raised midline anteriorly; rarely with head and pronotum black (except *Chondropyga frenchi* and WA specimens of *Chlorobapta frontalis*).. **49**

49(48). Pronotum with basal half parallel or sub parallel (Fig. 34.16bh), especially in female, basolateral angles often greater than 90 degrees; elytral disc somewhat flattened; abdominal ventrites with extensive yellow maculae centrally, interrupted in the middle; colour dorsally black with distinctively shaped complex green, yellow or blue maculae; males with somewhat shortened mesotibiae and asymmetrical mesotarsal claws; 18–25 mm in length; all Australian states except Tasmania... ***Chlorobapta* Kraatz**

– Pronotum more evenly curved from basolateral angles to apex (Fig. 34.4bh), basolateral angles less than 90 degrees; elytral disc more convex; abdominal ventrites not with medially interrupted extensive yellow central maculae; colour dorsally not as above; males with mesotibiae not shortened and mesotarsal claws symmetrical... **50**

50(49). Elytra with sutural striae distinct from base to apex, curving around mesoscutellum (Fig. 34.8ss); abdominal ventrite 5 with rows of long setae concentrated apically (Fig. 34.36st); mesotarsomere 5 and metatarsomere 5 ventroapically truncate or slightly concave (Fig. 34.30); empodium of at least metatarsi distinct, with 2 or 3 conspicuous long setae (Fig. 34.30em); metatarsal claws more curved (Fig. 34.30); 17–32 mm in length; QLD, NSW, VIC... ***Chondropyga* Kraatz (part)**

– Elytra with sutural striae not reaching base, present only from apex of mesoscutellum to apex of elytra (Fig. 34.10ss); abdominal ventrite 5 with setae more generally distributed; mesotarsomere 5 and metatarsomere 5 with a pointed ventroapical projection usually arising from a keel-like carina (Fig. 34.29vp, cr); empodium indistinct, with 2 insignificant short setae (Fig. 34.29em); metatarsal claws less curved (Fig. 34.29)... **51**

51(50). Mesosternal process somewhat expanded and truncated or with a slightly arcuate apex (Fig. 34.33f); males with antennal club elongate, equal to or longer than antennomeres 1–7 combined; 17–24 mm in length; QLD, WA; rare (*Diaphonia antoinei* and *D. vicina*)... ***Diaphonia* Newman (part)**

– Mesosternal process with simple rounded apex (Fig. 34.33g); males with antennal club shorter, less than antennomeres 1–7 combined; 17–21 mm in length; QLD; common (*Aphanesthes pullata*, some *A. trapezifera*)... ***Aphanesthes* Kraatz (part)**

Fig. 34.44. *Navigator fossor* (Lea), male.

Classification of the Australian genera

Tribe Cetoniini Leach, 1815

Characteristics. Head usually unarmed. Anterolateral angles of clypeus usually rounded. Pronotum with basolateral angles broadly rounded and basolateral margin angled anteriorly. Elytral posthumeral emargination usually distinct. Mesocoxae separated and mesoventral process transverse.

Subtribe Cetoniina Leach, 1815

Characteristics. Pronotum with distinct posterolateral angles, scutellar emargination usually present. Scutellum with rounded or broad apex.

Glycyphana Burmeister, 1842 (Pl. 6C, 73I)

Type species. *Cetonia horsfieldii* Hope, 1831.

Characteristics. Small, 10–15 mm. Clypeus with emargination sinuate. Pronotal basolateral angles broadly rounded; basolateral margin angled anteriorly. Scutellum with rounded apex. Elytra, posthumeral emargination arcuate; humeral umbone protrudent. Mesoventral process transverse. Metacoxae usually separated. Body dorsally dull satin textured and light tomented with many small maculae in most. Not sexually dimorphic.

Australian species. Nine in two subgenera, 2 subspecies. Genus total ~104.

Distribution. All Australian mainland states. *Glycyphana (Glycyphaniola) stolata* (Fabricius) has been introduced in WA and has become established there. Rest of genus: tropical and subtropical Asia and Australasia.

Biology. All *Glycyphana* are known to be flower visiting and recorded from many taxa of flowering tree where some species occur in great numbers: *Angophora*, *Leptospermum*, *Terminalia* L., *Melaleuca*, *Eucalyptus*, *Euroschinus* Hook.f., *Kunzea* Rchb., *Tristaniopsis* Brongn. & Gris, *Helichrysum* Mill., *Pittosporum* Banks ex Gaertn., *Viburnum* L., *Alphitonia* Endl., *Aleurites* J.R.Forst. & G.Forst., *Syzygium* Gaertn., *Buckinghamia* F.Muell., *Sannantha* Peter G.Wilson, *Ligustrum* L. A life cycle has not been published for any Australian species but observations show that *Glycyphana* larvae live in rotting, dead or living trees, in sand or in the soil under leaf litter. *G. stolata* has been reared from decaying mulch of *Eucalyptus gomphocephala* DC. and in potting medium of potted palms (PMH).

Notes. *Glycyphana* contains two subgenera. Most species are common.

Key to species. Mikšić (1970, 1971); Bacchus (1974).

Subgenus *Caloglycyphana* Mikšić, 1968

Characteristics. Dorsally dull brown or black with multiple white or yellow tomented maculae and red maculation. Male genitalia, parameres gradually tapering to arcuate apex, preapically with arcuate suture giving appearance of secondary apex, dorsal cleft narrow.

Australian species. Two.

Subgenus *Glycyphaniola* Mikšić, 1968

Characteristics. Dorsally dull green to brown with multiple small white tomented maculae. Male genitalia, parameres very broad and short, either broadly arcuate to apex or truncate apex, basally fused, dorsal cleft absent or with apical notch.

Australian species. Seven.

Protaetia Burmeister, 1842 (Pl. 74D, 75J)

Type species. *Protaetia spectabilis* Schaum, 1841.

Characteristics. Small, 13–18 mm. Dorsally brown with greenish reflections and multiple small tomented maculae. Clypeus with apex linear. Pronotal basolateral angles broadly rounded; basolateral margin angled anteriorly. Scutellum with broad and rounded apex. Elytra, posthumeral emargination arcuate, humeral umbone indistinctly protrudent. Mesoventral process transverse. Metacoxae usually touching. Body dorsally dull, satin textured and light tomented with many small maculae in most. Not sexually dimorphic. At this time the 4 subgenera of *Protaetia* which occur in Australia can only be differentiated by male genitalia.

Australian species. Four. Genus total ~345.

Distribution. All Australian mainland states, Christmas Island and Cocos (Keeling) Islands. *Protaetia (Protaetia) fusca* (Herbst) was found only in the eastern states of Australia but has now been introduced to all mainland states. *Protaetia (Pseudourbania) andrewsi* Gahan is found on Christmas Island, *Protaetia (Netociomima) advena* Janson is recorded from Cape York and *Protaetia (Miksicus) acuminata* (Fabricius) from the Cocos-Keeling Islands, extralimital in Indonesia and South-east Asia. Rest of genus: south-east Florida, Hawaii, Europe, Asia, Indonesia and Solomon Islands.

Biology. In Australia adult *Protaetia* can be found on the flowers of most trees, especially *Angophora* and *Eucalyptus*, but also on flowers of citrus, avocados, figs, peaches, roses and *Terminalia* (Simpson 1990) and on tree sap and ripe fruit. Simpson (1990) provided extensive observations on the life cycle of *Protaetia fusca* in captivity. Arrow (1910) wrote that *P. fusca* was observed entering nets of *Trigona* Jurine, 1807 bees in search of honey and the species is known as a pest in Hawaii (Woodruff 2006).

Notes. *Protaetia* in Australia contains four species, each in a separate subgenus. The genus *Miksicus* Özdikmen and Turgut, 2009 has generally not been accepted and is listed in Krajčík (2012) as a subgenus of *Protaetia*.

Key to species. Mikšić (1987) and Arrow (1910) on Palearctic and Oriental *Protaetia*, which includes *P. (Protaetia) fusca*, *P. (Pseudourbania) andrewsi and P. (Miksicus) acuminata*.

Tribe Schizorhinini Burmeister, 1842

Characteristics. Head without horns. Clypeus apical margin either linear to usual bilobate form. Pronotal base triconcave, often with basomedian lobe extended. Pronotal posterolateral angles usually 90 degrees or less, very distinct. Mesepimera usually concealed under posterolateral corners. Elytral posthumeral emargination usually sinuate, sometimes arcuate or angulate. Pygidium usually very broad, usually deflexed creating dorsal and ventral zones (ventral zone often invisible in dorsal view); derm usually with conspicuous, subconcentric striolation. Mesocoxae separated, mesoventral process present, in some absent or reduced.

Subtribe Lomapterina Burmeister, 1842

Characteristics. Pronotum with basomedian lobe extended, covering scutellum.

Ischiopsopha Gestro, 1874 (Pl. 74I)

Type species. *Cetonia bifasciata* Quoy & Gaimard, 1824.

Characteristics. Medium to large, 20–27 mm. Dorsally glossy, green. Pronotum with basal lobe extended; apex notched, rendering apex of scutellum visible. Pygidium in dorsal view; arcuate in males and females. Not sexually dimorphic. Male genitalia, parameres not apically fused, forming incomplete sphere, tongue absent.

Australian species. Five. Genus total 81 (Krajčík 2012).

Distribution. North-west WA, NT, northern QLD. Outside Australia: New Guinea, Java, Solomon Islands.

Biology. Adults on flowers of *Melaleuca*, *Ligustrum* and *Schefflera* J.R.Forst. & G.Forst.

References. Allard (1995b).

Key to species. Mikšić (1978).

Notes. Species-rich genus, often difficult to differentiate. *Ischiopsopha* is divided into 2 subgenera; all Australian species are in *Ischiopsopha* s. str.

Lomaptera Gory and Percheron, 1833 (Pl. 74B)

Type species. *Cetonia papua* Guérin-Méneville, 1830.

Characteristics. Medium to large, 16–24 mm. Dorsally glossy, green or brown. Pronotum with basal lobe extended; apex not notched, covering scutellum. Pygidium in dorsal view; male arcuate, female conical. Sexually dimorphic. Male genitalia, parameres fused apically forming oval shape and with bifurcate tongue.

Australian species. Seven in two subgenera, two subspecies. Genus total 112.

Distribution. QLD, New Guinea and neighbouring islands.

Biology. Adults feed on the flowers of palms. de Baar (1988) reported finding *Lomaptera* larvae in the mulch in trunks of rotten coconut palms throughout the Torres Strait. Adults are very apt and fast flyers.

References. Rigout & Allard (1997); Valck Lucassen (1961).

Key to species. Valck Lucassen (1961).

Mycterophallus van de Poll, 1886 (Pl. 73H)

Type species. *Lompatera validipes* Thomson, 1857.
Lomapteroides Schoch, 1898
Neophonia Kraatz, 1885
Mucterophallus van de Poll, 1886

Characteristics. Medium to large, 25–27 mm. Dorsally glossy metallic green with pale margin. Pronotum with basal lobe extended; apex not notched. Pygidium in dorsal view; arcuate in males and females. Not sexually dimorphic. Male genitalia, parameres not fused apically, forming incomplete sphere, tongue present.

Australian species. One, *Mycterophallus duboulayi* (Thomson). Genus total five (Krajčík 2012)

Distribution. QLD, New Guinea.

Biology. Adults on flowers of *Schefflera*, *Ligustrum*, *Sannantha* and *Murraya* J.Koenig ex L.

References. Allard (1995b).

Subtribe Schizorhinina Burmeister, 1842

Characteristics. Pronotum with basomedian lobe not extended over scutellum.

Aphanesthes Kraatz, 1880 (Pl. 73F)

Type species. *Eupoecila pullata* Janson, 1873.
Ablacopous Thomson, 1880
Ablacopus Schoch, 1895
Anthracopharis Schoch, 1895
Poecilocephala Kraatz, 1880

Characteristics. Medium to large, 19–27 mm. Dorsally brown and black, some melanic. Variable genus with several undescribed species. Clypeus elongate, apical emargination arcuate. Pronotal base wide, basolateral angles distinct. Elytra, posthumeral emargination arcuate, costae indistinct. Mesoventral process either transverse, elongate or bud-shaped. Not sexually dimorphic. Male genitalia, parameres attenuate, converging at apices, dorsal cleft wide. *A. pullata* and *A. trapezifera* Thomson occur in Queensland and are highly variable in size, coloration and shape. At least three forms are known in each species. Some differ to such a degree that they are likely the most variable of Australian cetoniines and easily confused with each other or with other cetoniines such as *Bisallardiana* Antoine, 2003.

Australian species. Three.

Distribution. *A. succinea* Hope in WA, *A. pullata* and *A. trapezifera* in QLD.

Biology. Adults have been collected on flowers of *Eucalyptus* and *Melaleuca*. The larval biology is unknown.

Notes. The genus is in need of revision and several undescribed species are known.

Bisallardiana Antoine, 2003 (Pl. 73C, 75A)

Type species. *Cetonia gymnopleura* Fischer von Waldheim, 1823.
Cacochroa Kraatz, 1880
Camilla Thomson, 1880
Allardiana Rigout, 1994

Characteristics. Small, 12–21 mm. Dorsally brown, yellow or black, several with white, yellow or black maculation. Clypeus elongate, apical emargination arcuate. Elytra, posthumeral emargination arcuate, humeral umbone protruding, costae indistinct. Mesoventral process elongate attenuate (length variable). *B. decorticata* (Macleay) and *B. obscura* (Blackburn) with tomentose fovea. Not sexually dimorphic. Males with slightly enlarged antennal clubs. Male genitalia, parameres narrow, converging to often broader apices, dorsal cleft very wide, phallobase ventrally often with setal tufts or sclerotised structures.

Australian species. Nine and one subspecies.

Distribution. WA, NT, QLD, NSW.

Biology. Most *Bisallardiana* are common and found on the flowers of trees and bushes, namely *Angophora*, *Leptospermum*, *Eucalyptus*, *Terminalia*, *Waterhousea* B.Hyland, *Lomatia* R.Br., *Bursaria* Cav. and *Actinotus* Labill. Larvae of the described species have not been mentioned in literature but larvae of one currently undescribed species in WA were found in sand under *Eucalyptus* leaf litter (PMH).

Notes. Rigout (1994) replaced the name *Camilla*, which he wrote to be preoccupied in three instances, with *Allardiana*. *Allardiana*, however, was also a preoccupied name and Antoine (2003) provided its current name. The genus is in need of revision (Moeseneder & Hutchinson pers. comm.) and there are at least six undescribed species. The existence and shape of structures of the ventral phallobase and its pilosity is useful in identification.

Chalcopharis Heller, 1903 (Pl. 73L)

Type species. *Chalcopharis nigroaenea* Heller, 1903.

Characteristics. Large, 30–37 mm. Dorsally glossy brass. Clypeus transverse rectangular, apical emargination sinuate, margins raised. Pronotum with basal lobe slightly extended posteriorly. Elytra, humeral and apical umbones indistinct, posthumeral emargination sinuate, non-costate. Mesoventral process elongate. Protibial tridenticles grouped in apical half, second in line with base of apical spur. Metacoxae spinose. Profemur and protibia with dense, long setal brush. Not sexually dimorphic. Male genitalia, parameres on vertical plane, broad, equal width to apex, dorsal cleft narrow.

Australian species. One, *Chalcopharis lansbergei* Gestro; this and three other species occur in New Guinea.

Distribution. North coastal WA. Rest of genus: New Guinea.

Biology. Unknown.

Notes. No Australian specimens of *C. lansbergei* were seen by the authors.

Chlorobapta Kraatz, 1880 (Pl. 7G, 73J)

Type species. *Schizorhina bestii* Westwood, 1842.

Characteristics. Medium-sized, 18–25 mm. Dorsally black with green, yellow or blue maculae. Clypeus subquadrate or elongate, apical emargination sinuate. Pronotum narrow and parallel in basal half. Elytra, posthumeral emargination sinuate, costae distinct. Mesoventral process short, strongly divergent fungiform or stout spade shape. Not sexually dimorphic. Males with enlarged antennal clubs and leg modifications – often with stout mesotarsomeres with uneven claws and/or metatibia unusually shaped with or without long setal brushes. Male genitalia, parameres elongate, narrow broadening to apex, dorsal cleft narrow.

Australian species. Five.

Distribution. All Australian mainland states. *C. goerlingi* Schürhoff occurs in WA, *C. frontalis* is found in Eastern states and NT; other forms, such as in WA, require inspection. All remaining species, *C. bestii*, *C. tibialis* Lea and the rare *C. hirtipes* Lea, in the eastern states.

Biology. Adult *Chlorobapta* are found less often on flowers of trees than many other cetoniines. Records show captures on *Melaleuca* and *Eucalyptus*. Larvae of *C. frontalis* were found in decaying hollow areas, below or above ground of standing live *Angophora* or *Eucalytpus* trees, often in the matrix of termite nests. Larval stage lasts 1–3 years (CHM and PMH pers. obs. and Tepper 1887). *Chlorobapta hirtipes* larvae were encountered in large, hollow standing *Eucalytpus* together with *Evanides bakewellii*. *Chlorobapta goerlingi* larvae found in decaying *Eucalytpus* tuber (CHM and PMH).

Chondropyga Kraatz, 1880 (Pl. 73D)

Type species. *Diaphonia gulosa* Janson, 1873.

Characteristics. Medium to large, 17–29 mm. Variable genus.

Group I: Medium, 20–25 mm. Dorsally black and brown with maculated costae. Clypeus elongate, apical emargination sinuate. Pronotum with distinct midline. Elytra, humeral umbone protrudent, posthumeral emargination arcuate, bicostae maculated. Mesoventral process slightly divergent to arcuate apex. Not sexually dimorphic. Male genitalia, parameres parallel, gradually widening to two-thirds length, then abruptly narrowing to twisting apical third, at two-thirds length in lateral view bends through 90 degrees, dorsal cleft narrow.

Group II: Medium to large, 17–29 mm. Dorsally yellow or brown with pronotal and elytral macula. Clypeus elongate, apical emargination sinuate. Pronotum wide across base however mesepimeron distinct and slightly anteriorly protrudent. Elytra subhumeral emargination sinuate, costae indistinct. Mesoventral process divergent with truncate apex or fungiform. Abdominal ventrite 6 apically with long setae. Pygidium very transverse. Not sexually dimorphic. Males of 2 of 4 species have enlarged antennal clubs. Male genitalia, parameres parallel elongate attenuate to acute apex, dorsal cleft narrow.

Australian species. Six.

Distribution. QLD, NSW, VIC.

Biology. Some *Chondropyga* species are often seen in flight. Almost all captures of *C. allardi* Rigout and Allard and *C. insignicosta* Hutchinson & Moeseneder have been with flight intercept traps. These two species inhabit a small area of rainforest in east-central Queensland. Adult *Chondropyga* feed on flowers of *Angophora*, *Eucalyptus*, *Euroschinus*, *Waterhousea* and *Ligustrum* but also on the fruits of apricot and peach. Alderson (1976) published the biology of *C. dorsalis* (Donovan) including descriptions of larva and pupa and mentions larvae in soil under logs. Moore (1987) reported that he found *C. dorsalis* larvae in garden compost heaps.

Notes. *Chondropyga insignicosta* is known only from approximately ten males and one female (Hutchinson &

Moeseneder 2013a). *Chondropyga suturata* Nonfried is excluded since it was incorrectly placed in this genus by Allard (1995a).

Clithria Burmeister, 1842 (Pl. 73A)

Type species. *Diaphonia (Clithria) eucnemis* Burmeister, 1842.

Characteristics. Small, 12–19 mm. Dorsally light brown or black with black, light or orange macula. Clypeus elongate, apical emargination sinuate or arcuate. Elytra, humeral umbone protrudent, posthumeral emargination arcuate or angulate, costae indistinct. Mesoventral process short, divergent, with arcuate apex. Sexually dimorphic. Males with either flabellate metatibia or basal denticle on claw of proleg. Male genitalia, parameres variable usually uniform width along length, may be divergent, parallel or convergent to apices. Apices with ventral lobes or phallobase with ventral sclerotisation or structure. *Clithria bacchusi* Allard is untypical for the genus.

Australian species. Four.

Distribution. QLD, NSW, VIC.

Biology. As adults, these small yellow-marked beetles feed on the flowers of trees. No other information has been published on the biology of any *Clithria* species.

Notes. The genus in need of revision (Hutchinson and Moeseneder pers. comm.).

Diaphonia Newman, 1840 (Pl. 73G)

Type species. *Diaphonia dispar* Newman, 1840.

Dysectoda Kraatz, 1880

Dysdiatheta Kraatz, 1880

Melobastes Thomson, 1880

Characteristics. Group I (near *D. kerleyi* Rigout and Allard, 1997): Small, 17 mm. Dorsally brown, abdominal ventrite 6 black, remainder brown. Clypeus elongate, apical emargination sinuate. Elytra, posthumeral emargination sinuate, exposing sternites, costa distinct. Mesoventral process reduced. Pygidium transverse. Protibia with 2 elongate acute denticles, metacoxae non-spinose. Sexual dimorphism unknown as females unknown. Male genitalia, parameres narrow, converging apically, dorsal cleft wide.

Group II: Medium-sized, 17–24 mm. Dorsally brown often with dark maculation, females often black or black with pale margin. Clypeus elongate, apical emargination sinuate or arcuate. Pronotum wide at base. Elytra, humeral umbone protrudent or indistinctly so, posthumeral emargination sinuate or arcuate, costae indistinct. Mesoventral process fungiform or divergent with arcuate apex. Sexually dimorphic. Male antennal clubs enlarged, females stockier and often melanic. Male genitalia, parameres narrowing then converging to apex, dorsal cleft wide.

Group III (near *D. bacchusi* Rigout and Allard): Small, 15–19 mm. Dorsally brown and black, females black. Clypeus subquadrate, apex linear and raised. Elytra, humeral umbone not protrudent, posthumeral emargination sinuate, costae indistinct. Mesoventral process short with apex arcuate. Sexually dimorphic. Male antennal club not enlarged, protibia unidentate; females black, stockier and foretibia tridentate. Male genitalia, parameres narrowing then widening to converging apices, dorsal cleft wide.

Australian species. 11.

Distribution. WA, SA, QLD, NSW.

Biology. Adult males generally fly in search of sedentary females and infrequently visit flowers, namely *Eucalyptus*. Larval biology is not known for any *Diaphonia*. Merely Blackburn (1893) writes that Tepper found *D. dispar* Newman pupae at the base of a tree. Detailed observations are needed to understand this unusual group of cetoniines.

Notes. *Diaphonia* contains some very untypical species and is in need of extensive revision (CHM and PMH pers. comm.). The number of synonyms at generic level are an indication of the difficulties which authors have had with, and still exist, in the group. Included in the species count is *Chondropyga suturata* (Nonfried) since it was originally described in *Diaphonia*. The genus currently contains two species which are better placed elsewhere: *D. bacchusi* and *D. kerleyi*. The females of some described species are unknown and at least two further undescribed species are awaiting description. At the moment *Diaphonia* are difficult to separate from *Aphanesthes* without utilising primary sexual and ecological characters.

Dilochrosis Thomson, 1878 (Pl. 73K)

Type species. *Cetonia flammula* Blanchard, 1853.

Phaeopharis Kraatz, 1880

Lethosesthes Thomson, 1880

Characteristics. Large, 22–45 mm. Dorsally yellow to brown with black maculae, or black with or without red maculae. Clypeus elongate, apical emargination deep v-shaped notch with acute apices and lateral margins raised. Elytra, posthumeral emargination sinuate, non-costate. Mesoventral process elongate. Not sexually dimorphic. Male genitalia, parameres even width to converging apices, dorsal cleft wide. *Dilochrosis brownii* (Kirby) of the Torres Strait are consistently different morphologically from the nominal form. In WA, *D. brownii* has another localised form which bears a broad, black, basal elytral band.

Australian species. Four. Six further species outside Australia.

Distribution. WA, SA, NT, QLD, NSW, New Guinea, Moluccas, Solomon Islands. *D. balteata* (Vollenhoven) is found in Australia and in New Guinea. *D. walteri* Lea is rare and found in WA, *D. atripennis* (Macleay), and *D. brownii* are found in eastern states and WA.

Biology. Adult beetles have been collected on flowers of *Eucalyptus* and *Terminalia* but some also feed on fruit or sap, specifically sap of *Codonocarpus* Endl. (PMH). The larvae of *D. walteri* were encountered in hollow, standing *Eucalyptus* stems and tubers and in soil beneath leaf litter (CHM and PMH pers. obs.), and *D. brownii* were reared to adult on leaf litter (PMH). de Barr (1988) found *D. brownii* larvae in rotten branches on the ground on several Torres Strait islands. Australian *Dilochrosis* species are available commercially and reared by individuals, schools and zoos.

Notes. The species *D. bakewellii* White, and *D. rufolatera* Lea, which have been included in *Dilochrosis* historically, are here listed under genus *Evanides* since they were correctly separated from *Dilochrosis* by Thomson (1880).

Eupoecila Burmeister, 1842 (Pl. 73E)

Type species. *Cetonia australasiae* Donovan, 1805.
Neophonia Thomson, 1878

Characteristics. Medium-sized, 16–24 mm. Dorsally black or black and brown with multiple yellow maculae or green with black maculae. Clypeus elongate, widest at midlength, apical emargination deeply arcuate or with v-shaped notch. Pronotal base wide. Elytra, posthumeral emargination arcuate, non-costate. Mesoventral process elongate. Abdominal ventrites with centres maculated. Not sexually dimorphic. Male genitalia, parameres narrow converging preapically, then parallel and widening at apex; dorsal cleft very wide, phallobase ventrally with or without setae.

Australian species. Four. One species in Indonesia.

Distribution. WA, SA, QLD, NSW, VIC, *Eupoecila miskini* Janson has recently been collected from Aru Islands, Indonesia; other locations are unconfirmed. *Eupoecila australasiae* is well known since it is very common and occurs in backyards in the eastern states.

Biology. All *Eupoecila* are flower-visiting. Alderson (1976) published the life cycle of *E. australasiae* and descriptions of its larvae and pupae. The larvae develop in a wide variety of rotten wood, such as felled trees, rotting *Xanthorrhoea* Sol. ex Sm. (Froggatt 1907), but also in other dead wood such as fence posts (Illidge 1917). Adult *Eupoecila* are found on flowers of *Angophora*, *Eucalyptus, Alphitonia*, *Melaleuca*, *Waterhousea*, *Xanthorrhoea*, *Bursaria*, *Leptospermum*, *Juniperus* L., *Acmenea* DC., *Tristaniopsis* Brongn. & Gris, *Terminalia*.

Evanides Thomson, 1880 (Pl. 76E)

Type species. *Schizorhina bakewellii* White, 1859.

Characteristics. Large, 30.5–37 mm. Dorsally red-brown with black maculae or black with orange maculae. Clypeus elongate, apical emargination u-shaped with rounded apices, lateral margins not raised. Elytra, posthumeral emargination sinuate, non-costate. Mesoventral process elongate. Not sexually dimorphic. Males with enlarged antennal clubs, metatibia arcuate and with tuft of long setae near apex. Male genitalia, parameres broad, even width and parallel to apex, dorsal cleft narrow.

Australian species. Two.

Distribution. QLD, NSW, VIC.

Biology. *Evanides bakewellii* (White) larvae develop in large, hollow, standing *Eucalyptus* in the rainforests of QLD, NSW and VIC. The adults are rarely found on the flowers of trees. The development of *E. rufolatera* (Lea) is unknown and it occurs only in a small area of northern QLD. Most adults have been found dead in water-carrying structures.

Notes. *Evanides*, an endemic genus, was described by Thomson (1880) for *Dilochrosis bakewellii*. The genus was not accepted by subsequent authors, except for Bergé (1884). Inspection of larvae and adults confirmed that they need to be in a separate genus. Both species are large and rare. Few collections have specimens of *E. bakewellii*. Only very few specimens of *E. rufolatera* are known.

Grandaustralis Hutchinson and Moeseneder, 2013 (Pl. 74A)

Type species. *Grandaustralis boomerang* Hutchinson & Moeseneder, 2013.

Characteristics. Medium-sized, 18.5–25.5 mm. Dorsally, males brown and black, females black. Clypeus elongate trapezoidal, widest apically, apical margin linear. Pronotal basolateral angles greater than 90 degrees, distinct angles, parallel in basal half. Elytra, posthumeral emargination sinuate, elytra distinctly punctate. Mesoventral process absent. Tarsomeres equal to tibial length. Sexually dimorphic. Male with greatly enlarged antennal club, female stockier and black. Male genitalia, parameres equal width along length, dorsal cleft and parameres divergent at midlength.

Australian species. One, *Grandaustralis boomerang* Hutchinson & Moeseneder.

Distribution. WA, localised.

Biology. Adults of both sexes are generally not found on flowers but there is a solitary record of a specimen collected on flowering *Eucalyptus*. Females are sedentary and males are often seen in flight. Hutchinson & Moeseneder (2013b) provided some observations on the adults.

Notes. Only two female specimens are known.

Hemichnoodes Kraatz, 1880 (Pl. 8C, 74F)

Type species. *Diaphonia mniszechii* Janson, 1873.

Characteristics. Large, 23–39 mm. Dorsally straw yellow with or without pronotal maculae. Clypeus elongate, apical emargination deep v-shaped notch, apices acute.

Elytra, humeral umbone protrudent, posthumeral emargination arcuate, costae indistinct. Mesoventral process fungiform or compact spade shape with arcuate apex. Pygidium pilose. Mesotarsomeres and metatarsomeres with long bristles. Sexually dimorphic. Male tibia and tarsomeres elongate, mesotibia in cross-section c-shaped, metatibia compressed and apex sublinear. Female stockier. Male genitalia, parameres very broad and sculptured, abruptly constricted to narrow setose apex, or wide along length to converging widening apices. In both forms apices setose, dorsal cleft narrow and phallobase dorsally with wide deep groove.

Australian species. Three.

Distribution. WA, SA, NSW, VIC.

Biology. Development of *H. mniszechii* has been documented in *Eucalyptus*. Adult *Hemichnoodes* were collected feeding on flowers of *Eucalyptus* (Hawkeswood 1982; PMH). *H. nigriceps* (Blanchard 1850) males frequently flying between *Eucalyptus* close to the ground (PMH).

Hemipharis Burmeister, 1842 (Pl. 74L)

Type species. *Schyzorhina insularis* Gory & Percheron, 1833.

Characteristics. Medium-sized, 19–23 mm. Dorsally glossy green with brassy forms. Clypeus subquadrate, apex with wide arcuate emargination with lobes upturned. Pronotum with basal lobe slightly extended posteriorly. Elytra, humeral umbone indistinct, posthumeral emargination sinuate, non-costate. Mesoventral process elongate. Not sexually dimorphic. Male genitalia, parameres on vertical plane, converging midlength then diverging with fold (collar) to apex forming figure '8' dorsal cleft.

Australian species. One, *Hemipharis insularis* (Gory & Percheron).

Distribution. WA, Melville Island, NT, QLD.

Biology. Adults on flowers of *Terminalia*. Also observed in large agglomerations on fruiting branches of palms.

Lenosoma Kraatz, 1880 (Pl. 74E)

Type species. *Schizorhina* (*Cetonia*) *fulgens* Macleay, 1863.

Characteristics. Small, 11–14 mm. Dorsally metallic green or copper. Clypeus elongate, apical emargination sinuate. Pronotum with shallow paramedian depressions. Elytra with humeral umbone protrudent, posthumeral emargination arcuate, costae distinct, specialised sparse long setae present aside scutellum and subhumeral emargination. Mesoventral process wide and apically arcuate. Metacoxa distinctly spinose. Not sexually dimorphic. Male genitalia, parameres convergent to broad apices, dorsal cleft narrow.

Australian species. One, *Lenosoma fulgens* (Macleay).

Distribution. QLD, NSW.

Biology. Adults on flowers of *Melicope* J.R.Forst. & G.Forst. and *Barklya* F.Muell., *Bursaria* (Hasenpusch & Moeseneder 2010).

Lyraphora Kraatz, 1880 (Pl. 74C)

Type species. *Schizorhina obliquata* Westwood, 1842.

Platedelosis Kraatz, 1880

Characteristics. Medium-sized, 15–24 mm. Dorsally shiny or matt yellow with black maculae. Clypeus elongate, apical emargination sinuate or acuate. Pronotum narrow at base, clearly exposing anteriorly protrudent mesepimera. Elytra, posthumeral emargination sinuate, costae indistinct. Mesoventral process parallel or divergent with apex arcuate. Metatibia apex with 3 acute spines. Not sexually dimorphic in most. Males with enlarged antennal clubs. Male genitalia, parameres narrow, equal width, parallel to apex, dorsal cleft narrow.

Australian species. Four.

Distribution. QLD, NSW, VIC.

Biology. Larvae of *L. obliquata* were found in hollow of standing *Callitris columellaris* several metres above ground (CHM). Adults on flowers of *Backhousia* Hook. & Harv., *Buckinghamia* F.Muell., *Commersonia* J.R.Forst. & G.Forst., *Barklya* and *Sannantha*.

Macrotina Strand, 1934 (Pl. 74H)

Type species. *Macrotis satanas* Schürhoff, 1933.

Macrotis Schürhoff, 1933

Characteristics. Medium-sized, 22–26 mm. Dorsally black or black and pale brown. Clypeus elongate. Elytra, posthumeral emargination sinuate, costa indistinct. Mesoventral process elongate, spearhead-shaped. Metacoxa spinose. Sexual dimorphism indistinct. Male antennal clubs enlarged. Male genitalia not examined.

Australian species. One, *Macrotina satanas* (Schürhoff).

Distribution. QLD.

Biology. Larvae found in decaying *Xanthorrhea*.

Notes. A rarely collected species, especially females.

Metallesthes Kraatz, 1880 (Pl. 7H, 74K)

Type species. *Diaphonia metallescens* White, 1859.

Characteristics. Medium-sized, 15–22 mm. Dorsally black with or without metallic reflections. Clypeus subquadrate to elongate, apical emargination sinuate or arcuate. Pronotum punctate with narrow impunctate midline. Elytra, posthumeral emargination sinuate, bicostate distinct or as narrow ridge, deep and large punctation. Mesoventral process either spade-shaped or parallel elongate. Protarsomere 5 with ventrodistal process. Not sexually dimorphic. Males with metatibia compressed with apical margin linear. Male genitalia, parameres narrow and even width to apex, dorsal cleft narrow, phallobase dorsally with wide, deep groove. *Metallesthes metallescens* and *M. unicolor* (Macleay) are very similar. Females of these species are difficult to differentiate.

Australian species. Three.

Distribution. Southern WA, SA, south-eastern QLD, NSW, VIC.

Biology. Adults collected on flowers of *Eucalyptus*, *Melaleuca*, *Grevillea* R.Br. ex Knight, *Xylomelum* Sm., *Nuytsia* R.Br. ex G.Don, *Leptospermum*, *Triodia* R. Br., *Xanthorrhoea*, *Styphelia* Sm. and *Angophora*. Larvae in roots and base of *Eucalyptus* (Tepper 1887).

Notes. Recent revision and some ecology by Moeseneder *et al.* (2014).

Key to species. Moeseneder *et al.* (2014).

Micropoecila Kraatz, 1880 (Pl. 6B, 74J)

Type species. *Schyzorhina cincta* Gory & Percheron, 1833.

Characteristics. Medium-sized, 18–23 mm. Dorsally black with brown margin. Clypeus subquadrate, apical emargination sinuate. Pronotum wide at base, semicircular, quite linear across base. Elytra, humeral and apical umbones indistinct, posthumeral emargination sinuate, non-costate. Mesoventral process short, diamond-shaped. Not sexually dimorphic. Male genitalia, parameres narrowing, converging apically, dorsal cleft wide.

Australian species. One, *Micropoecila cincta* (Gory & Percheron).

Distribution. WA, SA, NSW.

Biology. Adults infrequently on flowers of *Melaleuca*, *Eucalyptus* and *Angophora*. Froggatt (1896) described the larvae and provided ecological notes such as the larval occurrence in decaying trunks of *Xanthorrhoea*.

Navigator Moeseneder & Hutchinson, 2016 (Fig. 34.44)

Type species. *Pseudoclithria fossor* Lea, 1914.

Characteristics. Small to medium, 13–18 mm. Dorsally metallic blue, green, brown or gold. Clypeus subquadrate or elongate, apical emargination linear or sinuate; vertex raised medially. Elytra, humeral umbone non-protrudent, posthumeral emargination sinuate; costae distinctly reduced to narrow ridge. Mesosternal process reduced. Protibia with 2 elongate acute denticles (3 in females); metacoxa non-spinose. Sexual dimorphism distinct, male with enlarged antennal club, females with extended pygidium and very distended abdomen. Male genitalia, parameres narrow, converging to wider apices; dorsal cleft wide.

Australian species. Four.

Distrtibution. WA, SA, VIC, QLD.

Biology. Adults rarely on flowers of *Eucalyptus* and *Melaleuca*. Males often in flight searching for females which are mostly sedentary. Larvae found beneath *Acacia* Mill. leaf litter.

Notes. The genus separates *Pseudoclithria fossor* Lea and *P. ruficornis* (Westwood) from *Pseudoclithria* and contains two newly described species, *N. pixii* (Moeseneder & Hutchinson) and *N. interior* (Moeseneder & Hutchinson).

Key to species. Moeseneder & Hutchinson (2016).

Neoclithria van de Poll, 1886 (Pl. 75B)

Type species. *Cetonia incana* Macleay, 1863.

Characteristics. Small, 13–14 mm. Dorsally black with yellow maculae. Clypeus elongate, apical emargination sinuate. Pronotum parallel in basal half. Elytra, humeral umbone protrudent, posthumeral emargination arcuate, costae indistinct. Mesoventral process short with arcuate apex. Metacoxa non-spinose. Pygidium with shallow longitudinal setose impression. Not sexually dimorphic. Male genitalia, parameres narrowing and converging abruptly to preapex, then linear and parallel, dorsal cleft wide basally, phallobase ventrally with pair of setose structures.

Australian species. One, *Neoclithria eburneoguttata* (Blanchard).

Distribution. QLD.

Biology. Adults on flowers.

Neorrhina Thomson, 1878 (Pl. 3D, 75C)

Type species. *Eupoecila ochracea* Westwood, 1854.

Polystigma Kraatz, 1880

Characteristics. Small, 13–18 mm. Dorsally yellow with black maculae. Clypeus elongate, apical emargination sinuate or arcuate. Pronotal basolateral angles produced. Elytra, humeral umbone protrudent, posthumeral emargination arcuate, non-costate. Mesoventral process elongate. Metatibia trispinose. Not sexually dimorphic. Male genitalia, parameres elongate attenuate then abruptly broadening to setose apex. Phallobase ventrally with elongate medial sclerotisation.

Australian species. Three.

Distribution. WA, NT, QLD, NSW, VIC.

Biology. Adults have been collected on flowers of these trees and bushes: *Angophora*, *Acmenea*, *Alphitonia*, *Bursaria*, *Eucalyptus*, *Leptospermum*, *Tristaniopsis*, *Waterhousea*, *Ligustrum*, *Sannantha*, *Murraya*, *Buckinghamia*.

Notes. *Neorrhina calopyga* (Lea) is known only from few specimens.

Octocollis Moeseneder & Hutchinson, 2012 (Pl. 75G)

Type species. *Octocollis setosus* Moeseneder & Hutchinson, 2012.

Characteristics. Medium-sized, 16–19 mm. Dorsally brown and black with entirely black forms. Clypeus subquadrate, apical emargination sinuate. Pronotal basolateral angles greater than 90 degrees, rounded, widest medially. Elytra, posthumeral emargination sinuate, elytra distinctly punctate. Mesoventral process absent. Legs slender, tarsomeres longer than tibia. Sexual dimorphism unknown as females unknown. Male antennal clubs very enlarged. Male genitalia, parameres parallel along length with additional apical process, dorsal cleft narrow.

Australian species. One, *Octocollis setosus* Moeseneder & Hutchinson.

Distribution. QLD, a small area west of the Great Dividing Range and west of Paluma.

Biology. Adult males have been found on flowers of *Eucalyptus similis* Maiden growing on rocky ridges. The species occurs very localised and is rare. More frequent in the past (Moeseneder & Hutchinson 2012).

Notes. Females are unknown.

Phyllopodium Schoch, 1895 (Pl. 75H)

Type species. *Schizorrhina (Diaphonia) palmata* Schaum, 1848.

Characteristics. Medium-sized, 22–26 mm. Dorsally apically black, posteriorly yellow with black maculae. Clypeus elongate, apical emargination sinuate. Scutellum with lateral setose grooves. Elytra, posthumeral emargination sinuate, bicostate. Mesoventral process short, spade-shaped. Sexually dimorphic. Male legs highly modified, protarsomere depressed, mesoclaws unequal, metatibia arcuate. Male genitalia, parameres narrow at base, equal width and parallel to apex, dorsal cleft narrow. Phallobase with unique pair of articulated appendages with long setal brush along length.

Australian species. One, *Phyllopodium palmatum* (Schaum).

Distribution. SA, QLD, NSW, VIC.

Biology. Adults on flowers of *Angophora.* Adults have extensive courtship behaviour. Males bear unique, articulated genital brushes and have unique broad, flattened protarsi. Both sexes of *P. palmatum* stridulate, and are the only Australian cetoniines known to do so. Biology and behaviour by Moeseneder & Cook (2014). *Phyllopodium palmatum* is uncommon but occurs over a wide range.

Poecilopharis Kraatz, 1880 (Pl. 75I)

Type species. *Schizorhina bouruensis* Wallace, 1867.

Characteristics. Medium-sized, 17–20 mm. Dorsally glossy red-brown with multiple yellow maculae. Clypeus transverse rectangular, apical margin linear, margins not raised. Pronotum with basal lobe slightly extended posteriorly. Elytra, humeral and apical umbones indistinct, posthumeral emargination sinuate, non-costate. Mesoventral process elongate. Protibia tridenticles grouped in apical half, second in line with base of apical spur. Metacoxae spinose. Not sexually dimorphic. Male genitalia, parameres on vertical plane, broad, equal width to folded apex, dorsal cleft narrow.

Australian species. One, *Poecilopharis leai* Schürhoff. Genus total 36.

Distribution. QLD, New Guinea region, Solomon Islands, Fiji.

Biology. Adults are flower feeding.

Pseudoclithria van de Poll, 1886 (Pl. 75D)

Type species. *Schizorhina hirticeps* Macleay, 1871 [= *Clithria bicostata* Kraatz, 1885].

Characteristics. Small, 11–18 mm. The type species differs morphologically from the other species in the genus: 15–18 mm. Dorsally black and brown or melanic. Clypeus subquadrate, apical emargination sinuate; vertex raised medially. Elytra, humeral umbone protrudent, posthumeral emargination sinuate; distinct unicostate. Mesoventral process short, bud-shaped. Abdominal sternite 6 with row of setae. Protibia with 2 acute elongate denticles. Metacoxa distinctly spinose. Sexually dimorphic. Male with enlarged antennal clubs, female with very transverse pygidium. Male genitalia, parameres long, narrow, attenuate; dorsal cleft indistinct.

The remainder of the species are as follows: 10–17 mm. Dorsally black and brown or black. Clypeus subquadrate or elongate, apical emargination sinuate. Elytra, humeral umbone non-protrudent, posthumeral emargination sinuate, costae distinct but usually reduced to narrow ridge. Mesoventral process reduced. Protibia with 2 or 3 acute elongate denticles. Metacoxa non-spinose. Not sexually dimorphic. Male genitalia, parameres narrow, converging to wider apices; dorsal cleft wide.

Australian species. Nine.

Distribution. WA, SA.

Biology. Larvae of one undescribed species has been found free-living in sand; adults found in coastal sand dune ridges. Some species have recently been captured with flight intercept traps (CHM and PMH pers. obs.)

Notes. The genus is poorly represented in collections and in urgent need of revision (CHM and PMH pers. comm.). Polyphyletic group. Keys to several places in the generic key. *Pseudoclithria* is a catch-all genus which contains several distinct genera and there are a further seven undescribed species.

Schizorhina Kirby, 1825 (Pl. 75E)

Type species. *Cetonia atropunctata* Kirby, 1818.
Schyzorhina Gory and Percheron, 1833
Schizorrhina Burmeister, 1842

Characteristics. Large, 25–27 mm. Dorsally light yellow to yellow-brown, sometimes with green hue. Clypeus elongate, apical emargination with deep v-shaped notch. Pronotum basolateral angles produced. Elytra, humeral and apical umbones indistinct, subhumerally explanate covering metacoxae, non-costate. Mesoventral process large, spade-shaped. Tarsomeres elongate. Not sexually dimorphic. Male genitalia, parameres basally abruptly twisting into vertical plane and parallel to expanded setose apex. Dorsal cleft very narrow. Both species are very similar but parameres are diagnostic.

Australian species. Two.

Distribution. QLD, NSW.

Biology. Adults on flowers of *Angophora*. Larvae in forest humus (Moore 1987). Both species frequent woodlands.

Stenopisthes Moser, 1913 (Pl. 75L)

Type species. *Dilochrosis frenchi* Blackburn, 1894.

Characteristics. Large, 26–31 mm. Dorsally metallic green with coppery reflections. Clypeus elongate, apical margin linear. Pronotum subparallel in basal half. Elytra, humeral umbone indistinct, posthumeral emargination absent, metacoxae covered in dorsal view, non-costate. Mesoventral process elongate. Not sexually dimorphic. Male genitalia, parameres equal width along length, converging at apex, dorsal cleft wide.

Australian species. One, *Stenopisthes frenchi* (Blackburn).

Distribution. QLD.

Biology. Flower visiting but also on fruit.

Storeyus Hasenpusch & Moeseneder, 2010 (Pl. 75F)

Type species. *Schizorhina (Cetonia) fasciculata* Macleay, 1863.

Characteristics. Small, 12–18 mm. Dorsally brown or black. Clypeus elongate, apical emargination arcuate. Pronotum distinctly setose either side of midline. Elytra, humeral umbone protrudent, posthumeral emargination arcuate. Costae distinct, specialised tufts of setae on apical declivity. Mesoventral process broad, short and apex arcuate. Metacoxa distinctly spinose. Not sexually dimorphic. Male antennal club enlarged, metatibia arcuate with long thick setal brush along length. Male genitalia, parameres very long (subequal to length of phallobase), narrow and parallel, dorsal cleft reduced to base.

Australian species. Two.

Distribution. QLD, NSW.

Biology. Adults found on flowers of *Syzygium*, *Cardwellia* F.Muell., *Symplocos* Jacq., *Angophora*, *Barklya*, *Backhousia* and *Waterhousea* (Hasenpusch & Moeseneder 2010).

Notes. *Storeyus* are very poorly represented in collections.

Tapinoschema Thomson, 1880 (Pl. 75K)

Type species. *Schizorhina impar* Macleay, 1863.

Characteristics. Medium to large, 19–28 mm. Dorsally glossy black or orange-red and dark metallic blue. Clypeus subquadrate to elongate, apical emargination sinuate or arcuate. Pronotum punctate, midline indistinct. Elytra, posthumeral emargination sinuate, costa indistinct, deep and large punctation. Mesoventral process either parallel or attenuate elongate. Not sexually dimorphic. Male with elongated antennal clubs. Male genitalia, parameres narrow at base and gradually widening to apex.

Australian species. Three.

Distribution. WA, SA, QLD.

Biology. Larvae of *T. digglesii* (Janson) have been found under composting plant litter and *T. lacunosa* (Janson) larvae in sand under leaf litter. Adults are infrequently seen on flowers of *Melaleuca*. *Tapinoschema digglesii* occur localised, females are sedentary and males have been observed, at times in large numbers, in flight searching for females.

Trichaulax Kraatz, 1880 (Pl. 76G)

Type species. *Cetonia philipsii* Schreibers, 1802.

Characteristics. Large, 21–35 mm. Dorsally black, brown and black or green. Clypeus elongate, apical emargination arcuate or with v-shaped notch. Elytra, humeral and apical umbones distinct, posthumeral emargination sinuate, bicostate, costal interstices and/or elytral margins grooved and densely setose. Mesoventral process large, spade-shaped. Tarsomeres elongate. Not sexually dimorphic. Male genitalia, parameres broad, converging to slightly narrowing setose apex. Dorsal cleft narrow or wide.

Australian species. Six. Genus total eight.

Distribution. WA, SA, QLD, NSW, VIC, extralimital: New Guinea.

Biology. Adults on flowers of *Bursaria*, *Eucalyptus*, *Angophora*, *Euodia* J.R.Forst. & G.Forst., *Leptospermum*, *Bursaria* and *Corymbia* K.D.Hill & L.A.S.Johnson. *Trichaulax* are very strong flyers which often prefer the highest flowers on trees. Biology has been published by Hiller (1990), Cantrell (1979), Illidge (1917) and Zietek (2008). Larvae collected in hollows of live or dead *Eucalyptus*, *Angophora* and *Flacourtia* Comm. ex L'Her.

Key to species. Hiller (1990).

Notes. The status of *T. kirbyi* (Thomson) is unresolved.

Tribe Valgini Mulsant, 1842

Characteristics. Pronotal base never with strongly developed basomedian lobe, disc without any strong projections. Mesepimeron only slightly protruding anteriorly. Elytral posthumeral emargination absent. Elytral disc flat, non-costate. Propygidium broadly exposed. Foretibia with 3–5 denticles. Mesocoxae rarely separated by distinct mesoventral process. Metacoxae widely separated.

Key to genera. Krikken (1978).

Subtribe Valgina Mulsant, 1842

Characteristics. Pronotum with longitudinal ridges and/or tubercles, occasionally with callosities and impressions. Body usually abundantly squamose and/or setose.

Charitovalgus Kolbe, 1904 (Pl. 73B)

Type species. *Valgus pulcher* Kraatz, 1883.

Characteristics. Very small, less than 5 mm. Dorsally brown to black with brown scales. Pronotum with anterolateral angles extended beside head; derm with paramedial longitudinal ridges and lateral impressions. Body with squamose setae present. Protibia with 5 denticles.

Australian species. One, *Charitovalgus quinquedentatus* (Lea). Genus total 10.

Distribution. *C. quinquedentatus* in QLD, Rest of genus: Andaman Islands, S.E. Asia, Taiwan, Japan, Indonesia and New Guinea.

Biology. Biology of the single Australian species is unknown.

Subtribe Microvalgina Kolbe, 1904

Characteristics. Pronotum evenly convex. Mesoventrite posteriorly interposed between mesocoxae. Body usually sparsely squamose or setose.

Microvalgus Kraatz, 1883 (Pl. 3C, 74G)

Type species. *Valgus lapeyrouse* Gory and Percheron, 1833.

Characteristics. Very small, less than 5 mm. Dorsally dark brown to black with either yellow, red-brown or black elytra and scales white, yellow or grey. Pronotum with anterolateral angles extended beside head; derm without ridges or impressions. Body usually with squamose setae present, rarely absent. Protibia with less than 5 denticles.

Australian species. 16 and two subspecies. Genus total ~51.

Distribution. WA, QLD, NSW, VIC, TAS. Rest of genus: Africa.

Biology. Adults recorded on *Angophora* Cav. and *Leptospermum* J.R.Forst. & G.Forst. Some specimens were collected from bark. Larvae are possibly inhabitants of termite nests (Cassis & Weir 1992; Matthews 1984) but biology has not been described.

Notes. Ten *Microvalgus* species were described by Lea (1914). *M. vagans* Lea has 2 subspecies.

Key to species. Lea (1914).

Acknowledgments

J. Hasenpusch of the Australian Insect Farm, Innisfail, Queensland for the donation of a considerable number of specimens which provided a larger picture of the variability and distribution of species and for ecological observations; Andrew Mitchell of the Australian Museum for genetics work; Christine Lambkin, Susan Wright and Geoff Monteith of the Queensland Museum for loans and Justin Bartlett of Biosecurity Queensland, Department of Agriculture and Fisheries, for general advice.

References

Ahrens D, Schwarzer J, Vogler AP (2014) The evolution of scarab beetles tracks the sequential rise of angiosperms and mammals. *Proceedings. Biological Sciences* **281**, 1–10. doi:10.1098/rspb.2014.1470

Alderson J (1976) Behaviour and larvae of two rose chafer beetles *Eupoecila australasiae* (Don), *Diaphonia dorsalis* (Don) (Coleoptera: Scarabaeidae, Cetoniinae). *Victorian Naturalist* **93**, 86–91.

Allard V (1995a) *Les Coléoptères du Monde. Volume 23: Schizorhinini 1.* Sciences Nat, Venette, France.

Allard V (1995b) *Les Coléoptères du Monde. Volume 24: Schizorhinini 2.* Sciences Nat, Venette, France.

Antoine P (2003) *Bisallardiana*, nom nouveau de remplacement pour *Allardiana* Rigout, 1994 (Coleoptera, Cetoniidae). *Revue française d'Entomologie* (N.S.) **25**, 112.

Armstrong J (1979) Biotic pollination mechanisms in the Australian flora – a review. *New Zealand Journal of Botany* **17**, 467–508. doi:10.1080/0028825X.1979.10432565

Arrow GJ (1910) *The fauna of British India, including Ceylon and Burma. Coleoptera, Lamellicornia (Cetoniina and Dynastinae).* Taylor and Francis, London.

Bacchus ME (1974) A revision of the Australian species of the genus *Glycyphana* Burmeister (Coleoptera: Scarabaeidae). *Journal of the Australian Entomological Society* **13**, 111–128. doi:10.1111/j.1440-6055.1974.tb02162.x

Bergé A (1884) Énumeration des Cetonides derits depuis la publication du catalogue de MM. Gemminger & de Harold. *Annales de la Société Entomologique de Belgique* **28**, 113–163.

Blackburn T (1893) Notes on Australian Coleoptera, with description of new species. *Proceedings of the Linnean Society of New South Wales* **8**, 245–286.

Blanchard CE (1850) *Muséum d'Histoire Naturelle de Paris. Catalogue de la Collection Entomologique. Classe des Insectes. Ordre des Coléoptères* (Vol. 1). Gide et Baudry, Paris.

Bouchard P, Bousquet Y, Davies AE, Alonso-Zarazaga MA, Lawrence JF, Lyal CHC, *et al.* (2011) Family-group names in Coleoptera (Insecta). *ZooKeys* **88**, 1–972. doi:10.3897/zookeys.88.807

Browne DJ, Scholtz CH (1995) Phylogeny of the families of Scarabaeoidea (Coleoptera) based on characters of the hindwing articulation, hindwing base and wing venation. *Systematic Entomology* **20**, 145–173.

Browne DJ, Scholtz CH (1998) Evolution of the scarab hindwing articulation and wing base: a contribution toward the phylogeny of the Scarabaeidae (Scarabaeoidea: Coleoptera). *Systematic Entomology* **23**, 307–326.

Browne DJ, Scholtz CH (1999) A phylogeny of the families of Scarabaeoidea (Coleoptera). *Systematic Entomology* **24**, 51–84.

Burmeister H (1842) *Handbuch der Entomologie. Dritter Band. Coleoptera Lamellicornia Melitophila.* 827 pp. Enslin, Berlin.

Cantrell B (1979) Breeding site of *Trichaulax philipsii* (Coleoptera: Scarabaeidae). *News Bulletin of the Entomological Society of Queensland* **6**, 127–128.

Cassis G, Weir TA (1992) Cetoniinae. In *Zoological Catalogue of Australia. Coleoptera: Scarabaeoidea.* (Ed. WWK Houston) pp. 30–40. AGPS, Canberra.

Cassis G, Weir TA, Calder AA (2002) Cetoniinae. Australian Faunal Directory. In. Australian Biological Resources Study, Canberra. Available from: http://www.environment.gov.au/biodiversity/abrs/online-resources/fauna/afd/taxa/Cetoniinae

de Baar M (1988) Insects collected during a trip to Torres Strait 27 March to 10 April, 1987. *The Entomological Society of Queensland News Bulletin* **15**(9), 107–117.

Donovan E (1805) *An epitome of the natural history of the insects of New Holland, New Zealand, New Guinea, Otaheite, and other islands in the Indian, Southern, and Pacific Oceans: including the figures and descriptions of one hundered and fifty-three species*. Bensley, London.

Fabricius JC (1775) *Systema Entomologiae, sistens insectorum classes, ordines genera, species, adiectus synonymus, locis, descriptionibus, observations*. Officina Libraria Kortii, Flensburg and Leipzig, Germany.

Fabricius JC (1781) *Species insectorum, exhibentes eorum differentias specificas, synonyma auctorum, loca natalia, metamorphosin, adiectis observationibus, descriptionibus. Tom.I.* Carol Ernest Bohhn, Hamburg and Kiel, Germany.

Fischer von Waldheim G (1823) Coleoptera quaedam exotica descripta. *Mémoires de la Société Naturalistes de Moscou* **6**, 254–267.

Froggatt WW (1894) On the life-histories of Australian Coleoptera. *Proceedings of the Linnean Society of New South Wales* **9**, 113–125. doi:10.5962/bhl.part.18097

Froggatt WW (1896) The entomology of the grass-trees (Xanthorrhoea). *Proceedings of the Linnean Society of New South Wales* **21**, 74–87. doi:10.5962/bhl.part.8462

Froggatt WW (1907) *Australian Insects*. W. Brooks & Co., Sydney.

Froggatt WW (1914) Insects associated with *Angophora cordifolia*. *Australian Naturalist* **3**, 18–20.

Gemminger M, von Harold B (1869) *Catalogus Coleopterorum hucusque descriptorum synonymicus et systematicus Tom. IV. Scarabaeidae*. pp. 979–1346. Monachii. Sumptu E. H. Gummi, Paris.

Gestro R (1876) Appendice all'enumerazione dei Cetonidi raccolti nell' Archipelago Malese e nella Papuasia dai signori G. Doria, O. Beccari e L.M. D'Albertis. *Annali del Museo Civico di Storia Naturale di Genova* **9**, 83–110.

Golding MR (2009) *A pictorial field guide to the beetles of Australia. Part 5: Cetoniidae*. Total Digital Solutions, Burswood, Western Australia.

Gunter NL, Weir TA, Slipinski A, Bocak L, Cameron SL (2016) If dung beetles (Scarabaeindae: Scarabaeinae) arose in association with dinosaurs, did they also suffer a mass co-extinction at the K-PG boundary? *Public Library of Science ONE* **11**, e0153570. doi:10.137/journal.pone.0153570

Hangay G (2001) Notes on Australian Cetoniinae research. *Cetoniimania* **1**, 122–125.

Hasenpusch J, Moeseneder C (2010) *Storeyus*, a new genus and *Storeyus pseudodipterus*, a new species of Cetoniinae (Coleoptera: Scarabaeidae) from Australia, with a redescription of *Lenosoma* Kraatz. *Australian Entomologist* **36**, 221–234.

Hawkeswood TJ (1981) Insect pollination of *Angophora woodsiana* F. M. Bail. (Myrtaceae) at Burbank, south-east Queensland. *Victorian Naturalist* **98**, 120–129.

Hawkeswood TJ (1982) Notes on insect pollination of two species of *Eucalyptus* (Myrtaceae) from South-west Western Australia. *Victorian Naturalist* **99**, 28–37.

Hawkeswood TJ (2002) New adult host plant records for *Glycyphana stolata* (Fabricius, 1781) (Coleoptera: Scarabaeidae) from Australia. *Entomologische Zeitschrift* **112**, 179–180.

Hawkeswood TJ, Turner JR (2003) A late occurrence of the Australian cetonid beetle, *Polystigma punctata* (Donovan)(Coleoptera: Scarabaeidae: Cetoniinae) and comments on the late-flowering of its food plant, *Bursaria spinosa* Cav. (Pittosporaceae) in the Sydney area, New South Wales, Australia. *Calodema* **1**, 13–14.

Hawkeswood TJ, Turner JR (2007) Record of pollination of *Lomatia silaifolia* (Sm.) R. Br. (Proteaceae) by the cetonid beetle *Bisallardiana gymnopleura* (Fischer von Waldheim, 1823) (Coleoptera: Scarabaeidae) with comments on some entomological observations of Denham & Whelan (2000) in the Australian Journal of Botany. *Calodema* **52**, 1–5.

Hawkeswood TJ (2007) A record of Eriophora transmarinus (Keyserling, 1865) (Arachnida: Araneidae) feeding on the large cetonid beetle, Diaphonia dorsalis (Donovan, 1805) (Coleoptera: Scarabaeidae). *Calodema Supplementary Paper* **17**, 1–2.

Hiller A (1990) Notes on the systematics and biology of *Trichaulax* Kraatz (Coleoptera: Scarabaeidae: Cetoniinae). *Australian Entomological Magazine* **17**, 117–130.

Hope FW (1844) On some nondescript lamellicorn beetles. *The Transactions of the Entomological Society of London* **3**, 279–283.

Hutchinson P, Moeseneder C (2013a) *Chondropyga insignicosta*, a new species of Cetoniinae (Coleoptera: Scarabaeidae) from Queensland, Australia. *Zootaxa* **3710**, 281–288. doi:10.11646/zootaxa.3710.3.6

Hutchinson P, Moeseneder C (2013b) *Grandaustralis*, a new genus and *Grandaustralis boomerang*, a new species of Cetoniinae (Coleoptera: Scarabaeidae) from Western Australia. *Zootaxa* **3669**, 17–26. doi:10.11646/zootaxa.3669.1.2

Illidge R (1917) Life history of *Trichaulax marginipennis* and notes on other Cetonidae, etc. *Queensland Naturalist* **2**, 50–53.

Kirby W (1825) A description of such genera and species of insects, alluded to in the "Introduction to Entomology" of Messrs. Kirby and Spence, as appear not to have been before sufficiently noticed or described. *Transactions of the Linnean Society of London* **14**, 563–572. doi:10.1111/j.1095-8339.1823.tb00103.x

Kraatz G (1880) Genera Cetonidarum Australiae. *Deutsche Entomologische Zeitschrift* **24**, 177–214.

Krajčík M (2012) Checklist of the world Scarabaeoidea. *Animma.X* **5**, 1–278.

Krikken J (1978) Valgine beetles: A preliminary review of the genera, with descriptions of two novelties. *Zoölogische Mededeelingen* **53**, 153–164.

Krikken J (1983) A new genus and species of cetoniine beetle (Col., Scarabaeoidea) from New Guinea. *Entomologist's Monthly Magazine* **119**, 137–139.

Krikken J (1984) A new key to the suprageneric taxa in the beetle family Cetoniidae, with annotated lists of the known genera. *Zoölogische Verhandelingen* **210**, 1–75.

Krikken J (2018) Two new genera of Australian flower chafers, with an annotated overview of their relatives (Coleoptera: Scarabaeidae: Cetoniinae). *Haroldius* **5**, 1–36.

Lea AM (1914) Notes on Australian Cetonides; with a list of species and descriptions of some new ones. *Transactions and Proceedings of The Royal Society of South Australia* **38**, 132–218.

Leach WE (1815) *Entomology*. Vol. 9. W. Blackwood, J. Waugh, Edinburgh.

Linnaeus C (1758) *Systema Naturae per regna tria naturae, secundum classes, ordines, genera, species, cum characteribus, differentiis, synonymis, locis*. Editio decima, reformata. Laurentius Salvius, Stockholm.

Macleay WS (1838) *Illustrations of the zoology of South Africa, consisting chiefly of figures and descriptions of the objects of natural history collected during an expedition into the interior of South Africa, in the years 1834, 1835, and 1836; fitted out by "The Cape of Good Hope association for exploring Central Africa": together with a summary of African zoology, and an inquiry into the geographical ranges of species in that quarter of the globe*. Smith, Elder and Company, London.

Macleay W (1863) Description of twenty new species of Australian Coleoptera, belonging to the families Cicindelidae and Cetoniidae. *Transactions of the Entomological Society of New South Wales* **1**, 9–21.

Matthews EG (1984) *A guide to the genera of beetles of South Australia. Part 3. Polyphaga: Eucinetoidea, Dascilloidea and Scarabaeoidea*. Adelaide: South Australian Museum.

Mikšić R (1970) Revision der Gattung *Glycyphana* Burmeister (Coleoptera, Cetoniinae) I. Teil. *Zoölogische Verhandelingen* **107**, 1–112.

Mikšić R (1971) Revision der Gattung *Glycyphana* Burmeister (Coleoptera, Cetoniinae) II. Teil. *Zoölogische Verhandelingen* **118**, 1–153.

Mikšić R (1978) Revision der Gattung *Ischiopsopha*. *Entomologische Abhandlungen und Berichte aus dem Staatlichen Museum für Tierkunde in Dresden* **41**, 235–286.

Mikšić R (1987) *Monographie der Cetoniinae der Paläarktischen und Orientalischen Region. Coleoptera: Lamellicornia. Band 4*. Grafički zavod Hrvatske, Zagreb, Yugoslavia.

Moeseneder C, Cook L (2014) Captive observations on mating, stridulation and male genital brushes of the Australian flower chafer *Phyllopodium palmatum* (Schaum, 1848) (Coleoptera: Scarabaeidae: Cetoniinae). *Australian Entomologist* **41**, 77–90.

Moeseneder C, Hutchinson P (2012) *Octocollis*, a new genus and *Octocollis setosus*, a new species of Cetoniinae (Coleoptera: Scarabaeidae) from Queensland, Australia. *Zootaxa* **3557**, 40–48.

Moeseneder C, Hutchinson PM (2016) *Navigator,* a new endemic genus of Cetoniinae (Coleoptera: Scarabaeidae) from Australia with descriptions of two new species and behavioural studies. *Zootaxa* **4173**, 530–556. doi:10.11646/zootaxa.4173.6.2

Moeseneder C, Hutchinson P, Lambkin C (2014) Revision of the genus *Metallesthes* Kraatz and description of *Metallesthes anneliesae*, a new species of Cetoniinae (Coleoptera: Scarabaeidae) from Queensland and New South Wales, Australia. *Zootaxa* **3881**, 301–327. doi:10.11646/zootaxa.3881.4.1

Moore B (1987) *A Guide to the Beetles of South-Eastern Australia*. Australian Entomological Press, Greenwich, NSW.

Quoy JRC, Gaimard JP (1824) *Voyage autour du monde: entrepris par ordre du Roi,* éxecuté *sur les Corvettes de SM L'Uranie et de la Physicienne 1817–1820. Zoologie*. Chez Pillet Ainé, Paris.

Ratcliffe BC, Jameson ML (2002) Superfamily Scarabaeoidea Latreille 1802. In *American Beetles. Polyphaga: Scarabaeoidea through Curculionoidea. Volume 2*. (Eds RH Arnett, MC Thomas, PE Skelley and JH Frank) pp. 1–5. CRC Press, Boca Raton, Florida.

Reid C, Bulbert M (2002) *Flower Chafers of NSW: How to Identify the Cetoniinae of New South Wales*. Australian Museum, Sydney.

Rigout J (1994) Un nouveau nom pour un genre de cetoines Australiennes. *Bulletin de la Société Sciences Nat* **82**, 39.

Rigout J, Allard V (1997) *Schizorhinini 3. The Beetles of the World.* Vol. 25. Hillside Books, Canterbury, Great Britain.

Sakai K, Nagai S (1998) *The Cetoniine Beetles of the World.* Vol. 3. Mushi-Sha, Japan.

Schenkling W (1921) *Coleopterorum Catalogus*. Pars 72. Scarabaeidae: Cetoninae. Dr. W. Junk Verlag für Naturwissenschaften, Berlin.

Schoch G (1894) Ueber die Systematik der Cetoniden. *Mitteilungen der Schweizerischen Entomologischen Gesellschaft* **9**, 164–225.

Schoch G (1895) *Die Genera und Species meiner Cetonidensammlung. I. Teil: Trib. Goliathidae, Madagassae, Schizorhinidae*. E. Zwingli, Zürich.

Shoch G (1896) Lamellicornia melitophila: *Catalogus systematicus Cetonidarum at Trichiidarum ad huc cognitarum*. W. Junk, Zürich.

Šipek P, Fabrizi S, Eberle J, Ahrens D (2016) A molecular phylogeny of rose chafers (Coleoptera: Scarabaeidae: Cetoniinae) reveals a complex and concerted morphological evolution related to their flight mode. *Molecular Phylogenetics and Evolution* **101**, 163–175.

Simpson GB (1990) Immature stages of *Protaetia fusca* (Herbst) (Coleoptera: Scarabaeidae: Cetoniinae) with notes on biology. *Journal of the Australian Entomological Society* **29**, 67–73. doi:10.1111/j.1440-6055.1990.tb00317.x

Smith ABT (2006) A review of the family-group names for the superfamily Scarabaeoidea (Coleoptera) with corrections to nomenclature and current classification. *Coleopterists Society Monographs* **5**, 144–204.

Smith ABT, Hawks DC, Heraty JM (2006) An overview of the classification and evolution of the major scarab beetle clades (Coleoptera: Scarabaeoidea) based on preliminary molecular analyses. *The Coleopterists Bulletin* **60**, 35–46.

Tepper JG (1887) *Common native insects of South Australia. Part 1. Coleoptera or beetles*. E.S. Wigg and Son, Adelaide, South Australia, Australia.

Thomson J (1857) *Wallace. Voyage dans l'Asie orientale. Fragments entomologiques renfermant la description de Coléoptères nouveaux ou rares. Tome premier.* pp. 425–429. Bureau du Trésorier de la Société entomologique de France, Paris.

Thomson J (1878) *Typi Cetonidarum suivis de typi monommidarum et de typi nilionidarum Musaei Thomsoniani*. E. Deyrolle, Paris.

Thomson J (1880) Diagnoses de genres nouveaux de la famille des Cétonides. *Le Naturaliste* **2**, 293–294.

Valck Lucassen FT (1961) *Monographie du genre Lomaptera Gory & Percheron (Coleoptera, Cetonidae)*. 299 pp. Nederlandsche Entomologische Vereeniging, Amsterdam.

van de Poll JRH (1886) Ueber den systematischen Werth der Forceps-Bildung von *Mycterophallus* v. d. Poll. *Deusche Entomologische Zeitschrift* **30**, 301–304.

Webb GA (1987) Beetle visitors to flowers of *Angophora hispida* (Sm.) D. Blaxell (Myrtaceae) and some other *Angophora* in the Sydney region. *Australian Zoologist* **23**, 59–62. doi:10.7882/AZ.1987.004

Westwood JO (1842) *Arcana entomologica; or illustrations of new, rare, and interesting insects*. Vol. 1. pp. 103–104. William Smith, London.

Williams GA, Adam P (1998) Pollen loads collected from large insects in Australian subtropical rainforests. *Proceedings of the Linnean Society of New South Wales* **120**, 49–67.

Woodruff RE (2006) The Asian mango flower beetle, *Protaetia fusca* (Herbst), and *Euphoria sepulcralis* (Fabricius) in Florida and the West Indies (Coleoptera: Scarabaeidae: Cetoniinae). *Insecta Mundi* **20**, 227–231.

Zietek R (2008) Flower Chafer – *Trichaulax marginipennis* (Macleay). *In: Metamorphosis Australia*. Butterfly and Other Invertebrates Club Inc., Queensland, pp. 4–5.

35. BUPRESTIDAE LEACH, 1815

John F. Lawrence and Cate Lemann

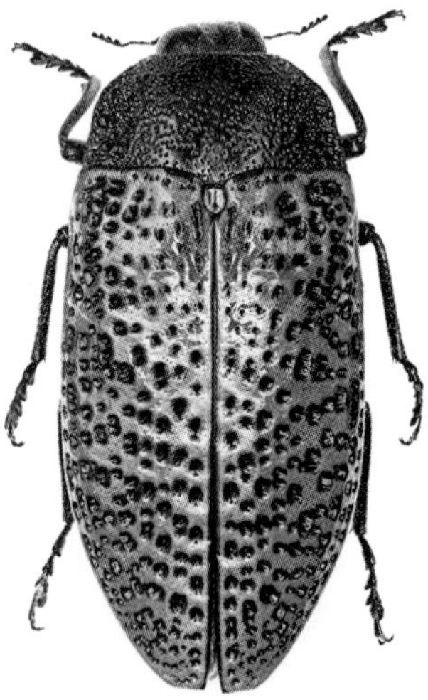

Fig. 35.1. *Stigmodera maculata* (Donovan), habitus.

Common names. Metallic wood-boring beetles, jewel beetles, flat-headed wood-borers.

Introduction. The Buprestidae is a cosmopolitan family with ~520 genera and 15 000 species currently placed in six subfamilies and 87 tribes and subtribes (Bellamy 2008a–d, 2009). The present treatment includes four subfamilies, 30 tribes and subtribes and 74 genera, with two groups of genera not placed in subtribes. There is little doubt that the family Buprestidae is sister to the Schizopodidae, as first noted by Forbes (1926, 1942) and argued by Nelson & Bellamy (1991, 2002), Bellamy & Nelson (2002) and Bellamy (2003). The latter has been considered a subfamily of Buprestidae by Lawrence & Newton (1995), Lawrence *et al.* (2011) and Lawrence & Ślipiński (2013), but the two positions do not really differ phylogenetically and Schizopodidae is used in the current world catalogue (Bellamy 2008a). Both Schizopodidae and Buprestidae are known from middle Jurassic fossils of Kazakhstan and China and the latter group is relatively common throughout the late Mesozoic and Cenozoic (Ponomarenko 1971; Alexeev 1993, 1995, 2009; Hörnschemeyer & Wedmann 1994; Wedmann & Hörnschemeyer 1994; Yu *et al.* 2013; Cai *et al.* 2015). Records of Triassic Buprestidae based on single elytra from Australia (Dunstan 1923) and larval tunnels from Arizona (Walker 1938) need confirmation.

Crowson (1982) considered the buprestoids to be most closely related to the Dryopidae and related families, and in the morphological analysis of Lawrence *et al.* (2011), they form a clade with Dryopidae, Limnichidae, Lutrochidae Elmidae and Heteroceridae and sister to Byrrhidae, but excluding the psephenoid families, as defined by Lawrence (1988). An analysis based on DNA sequences from eight nuclear genes and utilising the same taxa as in the above study (McKenna *et al.* 2015), Buprestoidea was sister to Byrrhoidea in the broad sense (including the psephenoid families), a result diffrering only slightly from those of Hunt *et al.* (2007) and Bocak *et al.* (2014).

The subfamily, tribal and subtribal classification of the family Buprestidae is still in a state of flux and various classifications have been proposed by Kerremans (1893, 1906–1913), Obenberger (1923, 1928), Théry in Carter (1929), Cobos (1957, 1978, 1979, 1980, 1986), Nelson (1981), Toyama (1987), Bellamy (1988, 2003), Hołyński (1988, 1993), Nelson & Bellamy (1991, 2002), Bílý (2000), Kolibáč (2000), Kubáň *et al.* (2001), Volkovitsh (2001), Volkovitsh & Bílý (2015) and Bellamy & Volkovitsh (2016). In the recent molecular study by Evans *et al.* (2015), Polycestinae, as recognised here, was found to be monophyletic, with the exception of the tribe Haplostethini, which formed a clade with the exotic Julodinae, and Agrilinae formed a strongly supported clade; both Buprestinae and Chrysochroinae, however, were polyphyletic, with chrysochroine genera falling into different clades with buprestine genera. Within the Agrilinae, neither Agrilini nor Trachyini were monophyletic and the several genera of leaf-mining agrilines did not form a monophyletic group.

The first key to Australian subfamilies, tribes and genera was that of Théry in Carter (1929), while Matthews (1985) provided an illustrated key to genera occurring in South Australia and Bellamy (1986) published a revised subfamily and tribal key and a list of included Australian genera. Numerous modifications have been made in the last 30 years based on adults, larvae and more recently DNA sequence data. The Australian fauna was catalogued by Bellamy (2002) and the world fauna by Bellamy (2008a–d, 2009).

Characteristics. Adults. 1.5–65 mm long and usually elongate with sides subparallel or converging posteriorly, slightly flattened to strongly convex (short, broad and flattened in leaf-miners); colour reddish-orange to black, but often bicolored, forming light and dark pattern, or with metallic coloration. Upper surfaces usually subglabrous but sometimes clothed with short, decumbent to erect, fine hairs, rarely with thickened or scale-like setae forming patches on elytra. Head somewhat flattened, deeply inserted into prothorax, strongly declined, so that mouthparts are ventrally oriented, almost always with distinct median endocarina. Eyes large, entire, vertically oval, not protuberant, finely facetted, without interfacetal setae; ommatidium of the eucone type. Antennal insertions exposed, usually widely separated, but sometimes enclosed within antennal fossae, which may be narrowly separated in Agrilinae. Frontoclypeal suture absent; clypeus short and usually anteriorly emarginate, sometimes with membranous anteclypeus visible within emargination. Antennae 11-segmented and usually serrate from antennomere 4 (very rarely pectinate). Labrum visible, free, small, transverse. Mandible small, short and broad; unidentate or bidentate, without mola or prostheca. Maxilla almost always with distinct, setose galea and lacinia, rarely with single lobe; lacinia without uncus; apical palpomere cylindrical or fusiform. Gular sutures usually widely separated, sometimes narrowly separated or absent. Corpotentorium slender and arched; anterior tentorial arms not expanded

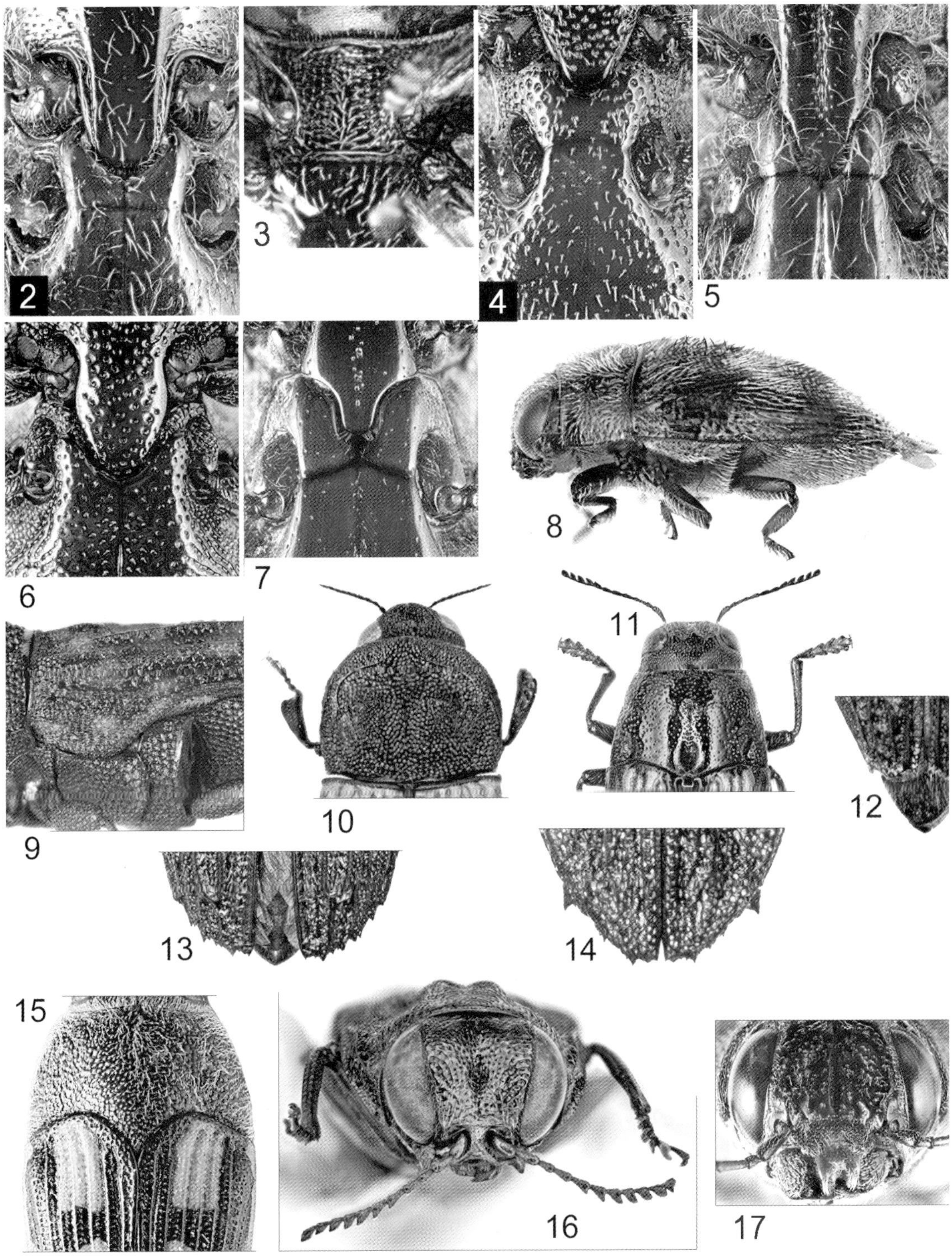

Figs 35.2–35.17. 2, *Prospheres aurantiopictus* (Laporte & Gory), prosternal process, meso- and metaventrite, ventral; 3, *Paratrachys australis* Bellamy & Williams, prosternal process, meso- and metaventrite, ventral; 4, *Polycesta mastersi* Macleay, prosternal process, meso- and metaventrite, ventral; 5, *Buprestis aurulenta* Linneaus, prosternal process, meso- and metaventrite, ventral; 6, *Diphucrania leucosticta* (Kirby), prosternal process, meso- and metaventrite, ventral; 7, *Iridotaenia bellicosa* (Blackburn), prosternal process, meso- and metaventrite, ventral; 8, *Paratrachys australis* Bellamy & Williams, habitus, lateral; 9, *Xyroscelis bumanna* Williams & Watkins, elytral base, metathorax, metacoxa, lateral; 10, *Xyroscelis crocata* (Gory & Laporte), head and prothorax, dorsal; 11, *Prospheres aurantiopictus* (Laporte & Gory), head and prothorax, dorsal; 12, *Prospheres aurantiopictus* (Laporte & Gory), elytral apices and pygidium, dorsal; 13, *Strigoptera bimaculata* (Linnaeus), elytral apices, dorsal; 14, *Polycesta mastersi* Macleay, elytral apices, dorsal; 15, *Astraeus flavopictus* Laporte & Gory, pronotum and elytral bases, dorsal; 16, *Aaaba nodosus* (Deyrolle); head, frontal; 17, *Austrochalcophora subfasciata* (Carter), head, frontal.

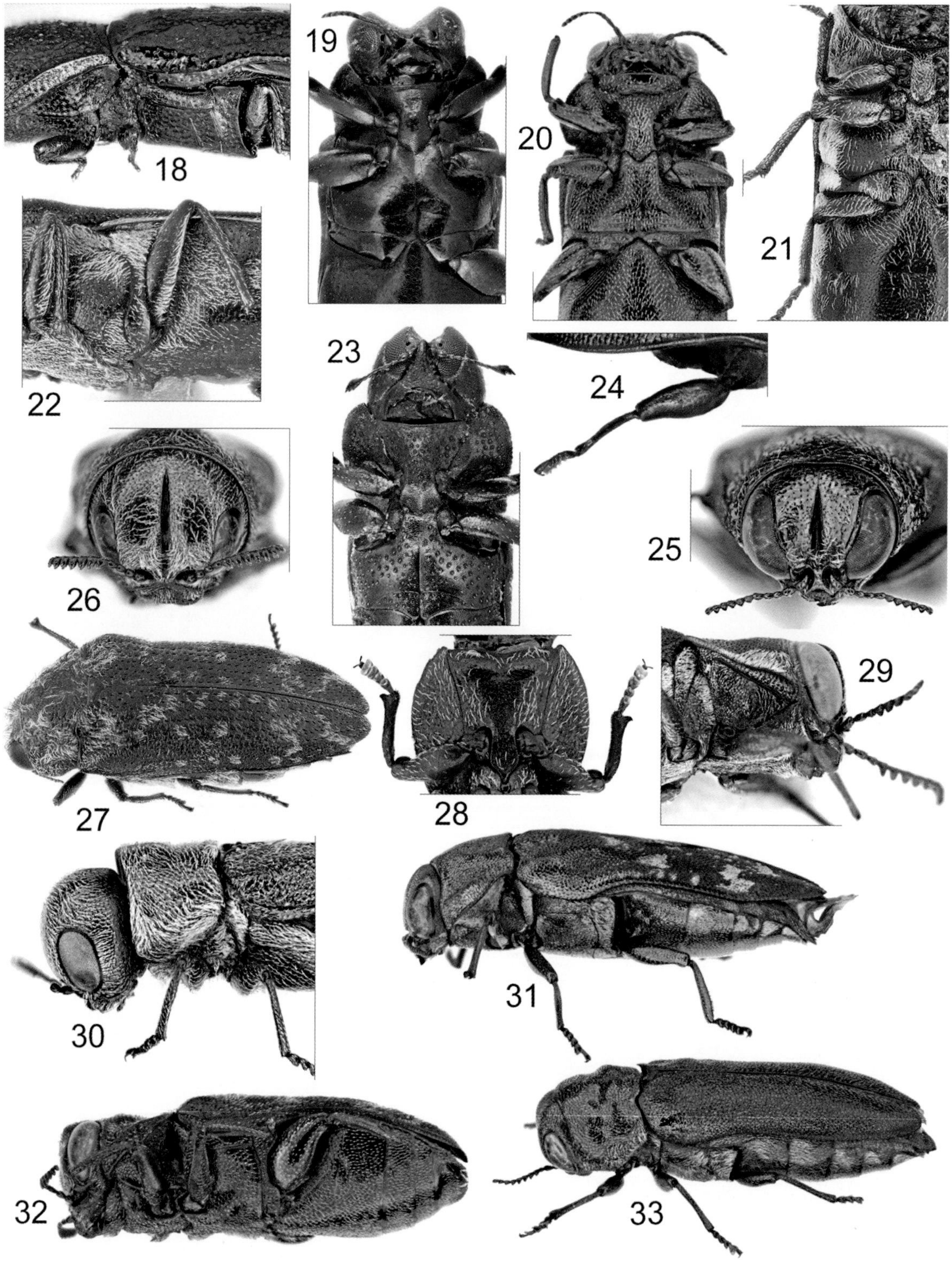

Figs 35.18–35.33. 18, *Helferella frenchi* (Théry), pronotum and pterotorax, lateral; 19, *Endelus* sp., head, thorax, base of abdomen, ventral; 20, *Sambus australis* Bellamy & Peterson, head, thorax, base of abdomen, ventral; 21, *Meliboeithon intermedium* (Kerremans), thorax, base of abdomen, ventral; 22, *Agrilus hypoleucus* Gory & Laporte; mesothorax, metathorax, base of abdomen, ventrolateral; 23, *Aphanisticus endeloides* Carter, head and thorax, ventral; 24, *Anthaxomorphus bougainvillensis* Williams & Weir, hind leg, lateral; 25, *Ethonion fissiceps* (Kirby); head and pronotum, anterior; 26, *Meliboeithon intermedium* (Kerremans), head and pronotum, anterior; 27, *Ethonion* sp., habitus, dorsolateral; 28, *Synechocera tasmanica* Théry, prothorax, ventral; 29, *Agrilus hypoleucus* Gory & Laporte, head and prothorax, lateral; 30, *Paracephala pistacina* (Hope), head, thorax, elytral base, lateral; 31, *Neospades rugiceps* (Thomson), habitus, lateral; 32, *Sambus australis* Bellamy & Peterson, habitus, ventrolateral; 33, *Dinocephalia cyanipennis* (Blackburn), habitus, lateral.

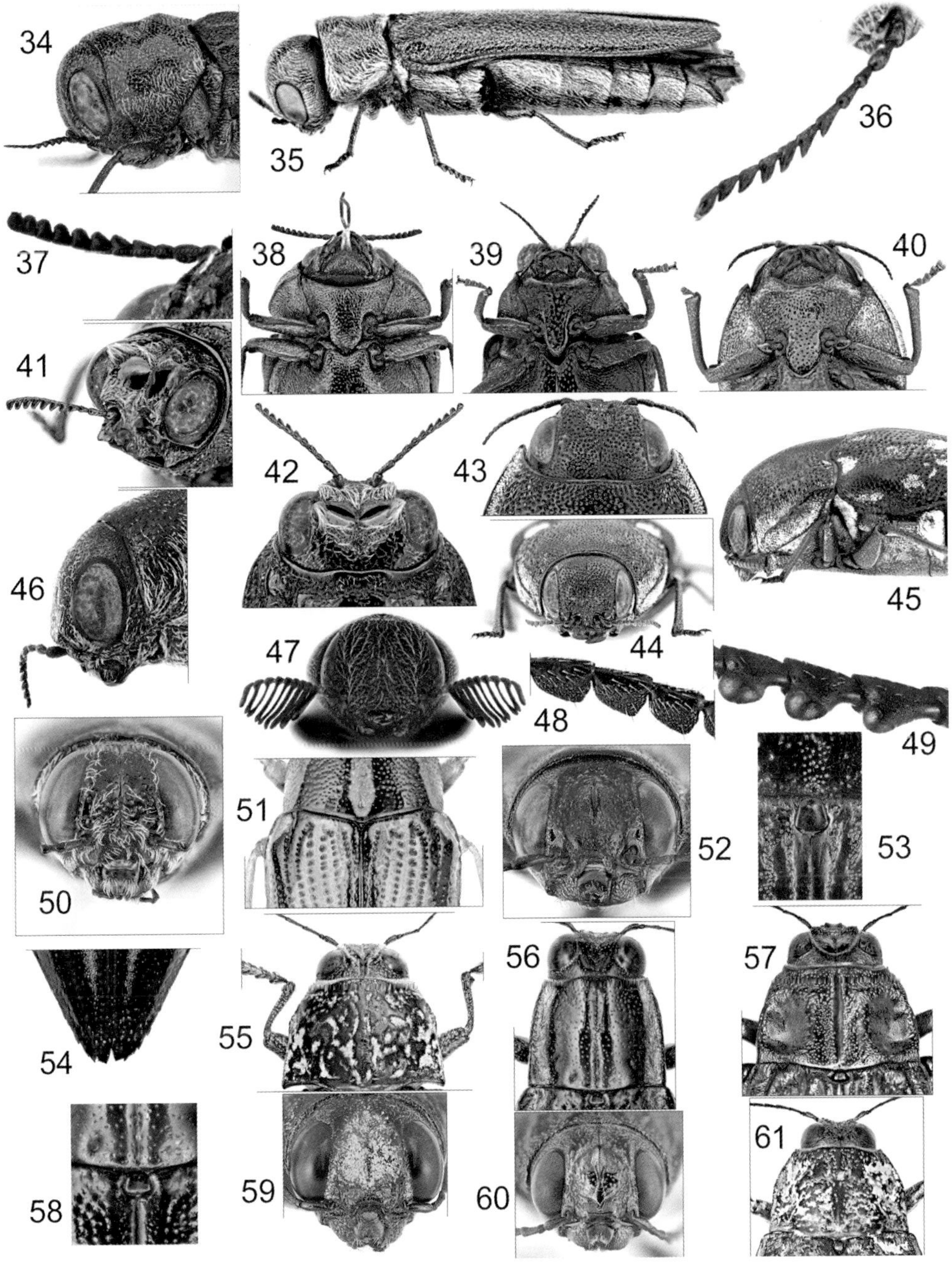

Figs 35.34–35.61. 34, *Dinocephalia cyanipennis* (Blackburn), head and prothorax, anterolateral; 35, *Paracephalia pistacina* (Hope), habitus, lateral; 36, *Aaaba nodosus* (Deyrolle); antenna; 37, *Pachycisseis bicolor* (Gory & Laporte), antenna; 38, *Pachycisseis bicolor* (Gory & Laporte), head, prothorax, mesothorax, ventral; 39, *Hypocisseis latipennis* (Macleay), head, prothorax, mesothorax, ventral; 40, *Diphucrania* sp., head, prothorax, mesothorax, ventral; 41, *Hypocisseis latipennis* (Macleay), head, anterodorsolateral; 42, *Hypocisseis latipennis* (Macleay), head and pronotum, dorsal; 43, *Diphucrania leucosticta* (Kirby), head and pronotum, dorsal; 44, *Diphucrania leucosticta* (Kirby), head and pronotum, anterior; 45, *Diphucrania leucosticta* (Kirby), head, thorax and elytral base, lateral; 46, *Stanwatkinsius perplexus* (Blackburn), head, lateral; 47, *Australorhipis aphanochila* Bellamy, head, anterior; 48, *Chysodema aurofoveata* (Guérin-Méneville), three antennomeres; 49, *Temognatha variabilis* (Donovan), three antennomeres; 50, *Cyroides imperialis* (Fabricius), head, anterior; 51, *Diadoxus erythrurus* (White), bases of pronotum and elytra, dorsal; 52, *Iridotaenia albivittis* (Hope), head, anterior; 53, *Chalcophorotaenia martinii* (Saunders); scutellar shield, dorsal; 54, *Cyphogastra pistor* (Laporte & Gory), elytral apices, dorsal; 55, *Paeudotaenia salamandra* (Thomson); head and pronotum, dorsal; 56, *Iridotaenia belicosa* (Blackburn), head and pronotum, dorsal; 57, *Chrysodema aurofoveata* (Guérin-Méneville), head and pronotum, dorsal; 58, *Iridotaenia belicosa* (Blackburn), scutellar shield, dorsal; 59, *Metataenia aurofoveata* (Saunders), head, anterior; 60, *Chalcophorotaenia martinii* (Saunders), head, anterior; 61, *Austrophorella quadrisignata* (Saunders), head and pronotum, dorsal.

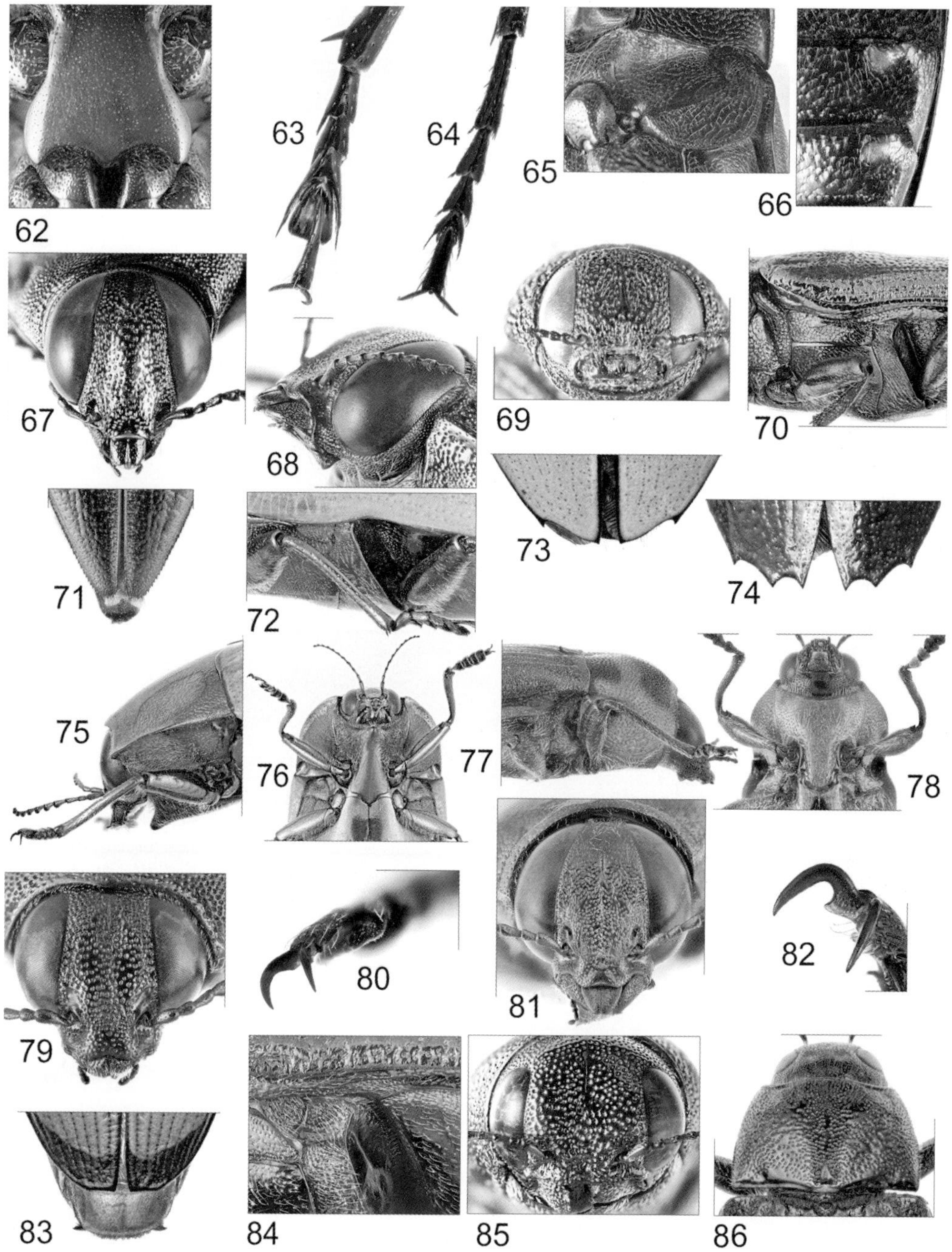

Figs 35.62–35.86. 62, *Belinota prasina* (Thunberg), apex of prosternal process, ventral; 63, *Belinota prasina* (Thunberg), tarsus; 64, *Merimna atrata* (Gory & Laporte), tarsus; 65, *Merimna atrata* (Gory & Laporte), mesocoxa, mesanepisternum, mesepimeron; 66, *Merimna atrata* (Gory & Laporte), left edges of abdominal ventrites 2 and 3; 67, *Selagis caloptera* (Boisduval), head, anterior; 68, *Temognatha variabilis* (Donovan), head, lateral; 69, *Microcastalia globithorax* (Thomson), head, anterior; 70, *Selagis* sp., mesothorax, metathorax, elytral base, lateral; 71, *Selagis aurifera* (Gory & Laporte), elytral apices, dorsal; 72, *Calodema regale* (Gory & Laporte), mesothorax and metathorax, lateral; 73, *Calodema regale* (Gory & Laporte), elytral apices, dorsal; 74, *Metaximorpha gloriosa* Blackburn, elytral apices, dorsal; 75, *Calodema regale* (Gory & Laporte), head and prothorax, lateral; 76, *Calodema regale* (Gory & Laporte), head and thorax, ventral; 77, *Castiarina pertii* (Gory & Laporte), head, thorax and elytral base, lateral; 78, *Castiarina pertii* (Gory & Laporte), head and thorax, ventral; 79, *Castiarina pictipennis* (Saunders), head, anterior; 80, *Castiarina pertii* (Gory & Laporte), pretarsal claws; 81, *Temognatha variabilis* (Donovan), head, anterior; 82, *Calotemognatha yarelli* (Gory & Laporte), pretarsal claws; 83, *Calotemognatha yarelli* (Gory & Laporte), Elytral apices and abdominal apex, dorsal; 84, *Neobuprestis frenchi* (Blackburn), metathorax and abdominal base, ventrolateral; 85, *Bubastes sphaenoida* Laporte & Gory, head, anterior; 86, *Neobuprestis frenchi* (Blackburn), head and pronotum, dorsal.

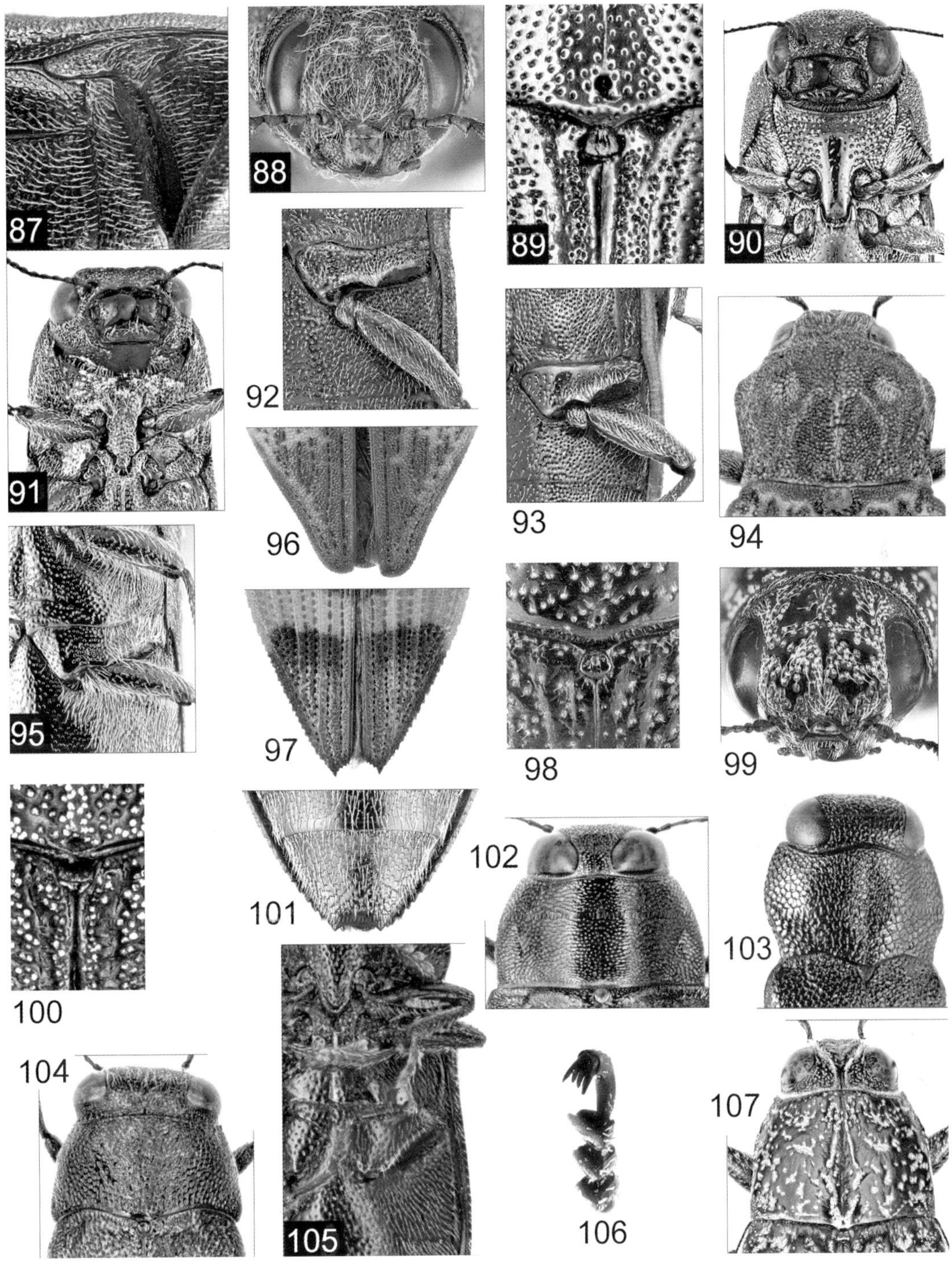

Figs 35.87–35.107. 87, *Torresita cuprifera* (Kirby), metathorax and abdominal base, ventrolateral; 88, *Buprestis aurulenta* Linnaeus, head, anterior; 89, *Bubastes sphaenoida* Laporte & Gory, pronotum, scutellum, elytral bases, dorsal; 90, *Bubastes sphaenoida* Laporte & Gory, head, prothorax, mesothorax, ventral; 91, *Euryspilus chalcodes* (Gory & Laporte), head, prothorax, mesothorax, ventral; 92, *Nascio* sp., metacoxa and abdominal base, ventral; 93, *Nascioides* sp., metacoxa and abdominal base, ventral; 94, *Nascio vetusta* (Boisduval), head and prothorax, dorsal; 95, *Neobubastes flavovittata* Carter, metaventrite, metacoxa and abdominal base, ventral; 96, *Nascio* sp., elytral apices, dorsal; 97, *Nascioides* sp., elytral apices, dorsal; 98, *Notobubastes orientalis* Carter, bases of pronotum and elytra, scutellar shield, dorsal; 99, *Notobubastes aurosulcata* Carter, head, anterior; 100, *Microcastalia* sp., bases of pronotum and elytra, scutellar shield, dorsal; 101, *Melobasis* sp., abdominal apex, ventral; 102, *Torresita cuprifera* (Kirby), head and pronotum, dorsal; 103, *Notographus* sp., head and pronotum, dorsal; 104, *Pseudanilara purpureicollis* (Macleay), head and pronotum, dorsal; 105, *Maoraxia auroimpressa* (Carter), thorax and abdominal base, ventral; 106, *Ethonion* sp., apical tarsomeres and pretarsal claws; 107, *Pseudotaenia salamandra* (Thomson), head and pronotum, dorsal.

mesally. Cervical sclerites well developed. Pronotum slightly to strongly transverse, not distinctly narrower than combined elytral bases; lateral carinae usually complete or incomplete anteriorly, occasionally absent, simple; disc usually simple, sometimes with median groove or curved, oblique costae; sublateral carinae present in Haplostethini and some Agrilinae. Prosternum usually moderately long, sometimes with short, broad chin piece. Prosternal process complete, usually parallel-sided or gradually narrowed apically, but sometimes abruptly narrowed at apex. Notosternal suture complete, with antennal groove in *Xyroscelis*; hypomeron with short antennal groove in *Aphanisticus*. Procoxae subglobular or slightly transverse, not projecting, without concealed lateral extensions. Procoxal cavities moderately widely separated, externally broadly open, internally open. Trochantin broadly exposed, and more or less abutting prosternum and hypomeron. Elytra narrowly elongate to short and broad; sides usually subparallel anteriorly and converging posteriorly (more or less converging from base to apex in *Paratrachys* and Trachyini); apices conjointly rounded, independently rounded, acute or biacute; lateral edges often serrate posteriorly; disc with irregular punctation or with 10 puncture rows; epipleuron gradually or abruptly narrowed and usually incomplete. Anterior edge of mesoventrite with paired, slightly to strongly declined procoxal rests; discrimen more or less developed or obliterated by large, deep mesoventral cavity. Mesocoxae not projecting, with exposed trochantins. Mesocoxal cavities widely separated, subcircular or longitudinally oval, partly closed laterally by mesanepisterna and mesepimera. Mesoventral process not divided, more or less truncate or emarginate and either fused to or abutting metaventral process. Metaventrite with discrimen usually complete to apex of metaventral process and with moderately long, distinctly curved, transverse (katepisternal) suture, which crosses midline well in front of metacoxae. Exposed portion of metanepisternum usually relatively short (elongate and slender in Haplostethini and Aphanisticini; metanepisternum concealed in *Paratrachys*); metepimeron exposed or concealed. Metacoxae contiguous, extending laterally to meet elytra and usually as broad laterally as mesally; metacoxal plates well developed. Metendosternite with moderately long stalk, short lateral arms and short to moderately long anterior process with approximate anterior tendons; laminae and ventrolateral processes absent. Hind wing moderately elongate; radial bar extending well beyond end of radial cell with long, but weak bending zone; apical field relatively short, radial cell usually narrowly elongate with inner edge slightly oblique, sometimes absent; cross-vein r3 either absent or fused basally with r4; R-M loop forming acute angle; basal portion of RP very short or absent, apical extension usually present; median spur usually straight and extending to wing margin; medial field with 6 free veins and no wedge cell, 5 free veins and an apically acute wedge cell, or witgh venation somewhat reduced; MP_{3+4} sometimes with cross-vein and basal spur, but often detached; anal lobe usually well developed with weak embayment weak or absent. Legs usually slender; femora flattened and expanded to receive tibiae in *Aphanisticus*; tibiae flattened and expanded to receive tarsi in *Xyroscelis*; outer edge simple; tibial spurs usually paired on all legs. Tarsi 5–5–5; membranous lobes present on at least some preapical tarsomeres. Claws simple, empodium not apparent. Abdomen usually with five ventrites, the basal two of which are connate; intercoxal process usually acute (broadly rounded or angulate in Trachyini); suture between ventrites 1 and 2 sometimes obliterated. Paired small, oval impressions (cuticularia) frequently present at or near joints between ventrites. All tergites usually heavily sclerotised, but occasionally very lightly so, with as many as 5 pairs of cuticularia similar to those on ventrites. Functional spiracles present on segments I–VIII, those on II–VIII located on tergites or laterotergites, that on VIII sometimes reduced in size. Anterior edge of sternite VIII in male without median strut. Anterior edge of segment IX in male with emarginate subgenital plate; tergite IX either deeply emarginate and X well developed and free, or IX partly or completely fused to X. Aedeagus of trilobate type, with phallobase highly reduced and more or less fused to parameres, which form a flattened tube more or less enveloping penis. Sternite VIII in female without spiculum ventrale. Ovipositor usually elongate, more or less flattened and lightly sclerotised, except for bacula; proctiger usually more or less evenly sclerotised, usually with 2 short, basolateral struts articulating with apex of paraproctal bacula; the latter, in turn, abruptly curved or angulate ventrally near anterior end and articulated with coxital bacula; coxites located mesoventrally, undivided and usually slightly expanded at apex, with 2 pairs of longitudinal bacula and with short, palpiform, terminal or subterminal styli. Ovipositor sometimes short, broad and somewhat reduced (Good 1925; Carnaby 1986; Bellamy 1986, 1988, 1997, 2002, 2003; Williams & Weir 1987, 1988; Gardner 1989; Bílý 2000; Kolibáč 2000; Kubáň *et al.* 2001; Jendek 2001; Curletti 2001; Volkovitsh 2001; Bellamy & Nelson 2002; Barker 2006; Bellamy & Weir 2008; Bellamy & Volkovitsh 2016).

Larvae. Elongate, usually more or less parallel-sided, except for thorax, which is much wider than abdomen (less elongate, widest anteriorly and gradually narrowing posteriorly in Trachyini and in other leaf-mining forms), subcylindrical to slightly flattened, except for thorax, which is usually distinctly flattened. Very lightly pigmented, except for mouth frame and mandibles, but often with yellow dorsal and ventral prothoracic plates and sometimes patches of small sclerites; with paired, sclerotised, acute processes on segment X in *Agrilus*; vestiture of short to moderately long suberect, fine hairs. Head prognathous and deeply retracted into prothorax, with only short anterior portion visible; slightly rounded laterally and strongly flattened; posterior edge deeply emarginate dorsally and ventrally. Epicranial stem absent; frontal arms V-shaped, contiguous at base; median endocarina extending to frontoclypeal suture. Stemmata usually absent (2 on each side in some Trachyini). Antennae highly reduced, either 3-segmented, with conical sensorium on antennomere 2 much longer than antennomere 3, or 2-segmented, with sensorium at apex of antennomere 2. Frontoclypeal suture present. Labrum free, lightly sclerotised. Mandibles robust, usually bidentate or tridentate; mola and prostheca absent. Ventral mouthparts protracted; cardines more or less distinct, sclerotised or not, separated from each other by labium; stipes not or only slightly longer than wide; maxillary articulating area absent; maxilla with articulated mala; palp 2-segmented. Labium consisting of prementum and postmentum or a single labial plate; ligula broadly rounded, undivided;

labial palps 1-segmented and widely separated or absent. Hypopharyngeal bracon and sclerome absent; hypostomal rods long and diverging; gular region short; sutures narrowly separated. Prothorax usually longer than meso- and metathorax combined (not so in leaf-mining groups) and moderately to strongly flattened; often with moderately sclerotised dorsal and ventral plates, usually with V- or Y-shaped dorsal endocarina and single median ventral endocarina, occasionally with single median endocarina both dorsally and ventrally and sometimes with additional small sclerites; mesothorax strongly transverse and usually shorter than metathorax, which may be transverse or subquadrate, sometimes with ventrolateral ampullae. Thoracic legs almost always absent (vestiges occasionally present). Abdominal segments I to IX more or less similar in length, often subdivided by transverse folds (strongly transverse, undivided and somewhat expanded laterally in leaf-mining forms), without patches or rows of asperities or lateral processes; tergum IX shorter than VIII, subterminal, without urogomphi; sternum IX well developed and simple; segment X usually well developed and posteriorly oriented with paired, oval lobes lying on each side of vertical anal opening (segment X somewhat reduced in leaf-mining forms); with posteriorly projecting, acute processes in *Agrilus*. Spiracles cribriform, usually reniform with spiracular scar anterior to sieve plate, but sometimes reduced and so resembling multiforous or uniforous type. Spiracular closing apparatus present (Böving & Craighead 1931; Rees 1941; Hawkeswood 1985; Bílý 1972, 1975, 1986, 1989, 1992, 1999; Volkovitsh and Hawkeswood 1987, 1990, 1993, 1994, 1999; Lawrence 1991; Bellamy & Nelson 2002; Bílý & Volkovitsh 2001, 2003, 2005; Volkovitsh *et al.* 2003; Bellamy & Volkovitsh 2016; Bílý *et al.* 2008, 2013; Grebennikov 2013; Volkovitsh & Bílý 2015).

Biology. Adults are often active in hot weather, will readily fly in sunlight and feed on foliage or pollen and nectar; some, such as the stigmoderines, have mouthparts adapted for nectar-feeding. Species of *Xyroscelis*, however, feed on the sap of the host plant. Larvae usually feed in wood (often in the phloem just beneath the bark) or in root systems of trees or shrubs, but some feed in herbaceous stems or galls or mine leaves. Eggs are usually laid in cracks in bark, on or in root crowns or directly on plant tissue, and larvae chew into the substrate, but some species deposit eggs in soil adjacent to roots. Species of *Castiarina* are often aposematically coloured and form part of a mimicry ring with species of Lycidae, Pyrochroidae, Oedemeridae, Belidae and some other families; these and other buprestids contain bitter chemicals called buprestins (Brown *et al.* 1985; Moore & Brown 1985, 1989). Further details on buprestid biology may be found in general works, such as the introduction to the Australian buprestid catalogue (Bellamy 2002), the first volume of this series (Lawrence & Ślipiński 2013) and DeGruyter's Handbook of Zoology (Bellamy & Volkovitsh 2016), in the numerous references cited by Bellamy *et al.* (2013) and Evans *et al.* (2015), or included in the generic sections below.

Key to the Australian genera

1. Mesoventral cavity formed by mesoventrite only, not extending into metaventrite, so that apex of prosternal process in intact specimen always separated from metaventrite by at least a slender strip of mesoventral cuticle (Figs 35.2–35.4) **2**

– Mesoventral cavity formed in part by metaventrite, so that apex of prosternal process in intact specimen is in direct contact with metaventrite (Figs 35.5–35.7) **6**

2(1). Body short, broad and flattened (Pl. 81C; Fig. 35.8); total length less than 3.5 mm and less than 1.75 times greatest width; exposed portion of mesoventrite a very slender, transverse strip (Fig. 35.3); upper surfaces finely punctate, smooth and shiny; metanepisternum completely concealed by elytral epipleura (Fig. 35.8) .... ***Paratrachys* Saunders**

– Body longer and narrower, not flattened; total length more than 4 mm and more than 2.5 times greatest width; exposed portion of mesoventrite greater; IF metanepisternum concealed by elytral epipleuron (Fig. 35.9), THEN upper surfaces coarsely punctate with pronotal carinae and elytral costae (Pl. 83A) **3**

3(2). Metanepisternum concealed by elytral epipleuron (Fig. 35.9); body subcylindrical, 10 mm or less in length; pronotum (Pl. 83A; Fig. 35.10) strongly convex, with two pairs of curved or sinuate dorsal carinae; all tibiae with apical half expanded and flattened; elytra with regular rows of large window punctures and with alternate intervals costate .... ***Xyroscelis* Saunders**

– Metanepisternum not concealed by elytral epipleuron; body somewhat flattened, almost always more than 10 mm in length; pronotum without dorsal carinae **4**

4(3). Pronotum (Pl. 82N; Fig. 35.11) widest at base, where there are 3 distinct pits, one at midline and two sublateral; all elytral intervals equally convex, not carinate (except in *P. norfolkensis*); elytra strongly narrowed posteriorly (Pl. 82N), the apex of each bearing two small spines, one at outer edge and another at inner edge (Fig. 35.12) .... ***Prospheres* Saunders**

– Pronotum widest at middle (Pl. 82C, E), without distinct basal pits; alternate elytral intervals strongly carinae (Pl. 82B, C, E); elytra less strongly narrowed posteriorly, the apex of each with numerous small spines (Fig. 35.13) **5**

5(4). Posterior edge of pronotum strongly biemarginate forming median lobe which extends between anterior edges of elytra (Pl. 82C, E); scutellar shield narrow and elongate, somewhat depressed between elytral bases; elytral apices more or less rounded; colour black or blue, usually with large red macula on each elytron (Pl. 82E) .... *Strigoptera* Dejean

– Posterior edge of pronotum not or only very weakly biemarginate, without median lobe (Pl. 82B); scutellar shield short, broad and posteriorly widened, not depressed; elytral apices obliquely truncate with acute sutural angle (Fig. 35.14); colour not as above .... ***Polycesta* Serville**

Mesoventral cavity partly formed by metaventrite (6–end)

6(2). Posterior edge of pronotum deeply biemarginate forming acute medial lobe which completely conceals scutellar shield (Pl. 77B; Fig. 35.15) .... ***Astraeus* Laporte & Gory**

– Posterior edge of pronotum not or only slightly biemarginate; scutellar shield almost always exposed **7**

7(6). Distance between antennal fossae less than half the shortest distance between eyes (Fig. 35.16); length never greater than 25 mm; lateral pronotal carinae complete and usually exposed from above or rarely absent; prescutellar pit absent **8**

– Distance between antennal fossae almost always more than half the shortest distance between eyes (Fig. 35.17) (IF distance less than this THEN length more than 30 mm (*Julodimorpha*) OR pronotum strongly convex with prescutellar pit and incomplete lateral carinae concealed from above (Pl. 78J)) **30**

Antennal fossae close together (8–29)

8(7). Pronotum with two complete lateral carinae (Fig. 35.18); total length less than 3.5 mm; mesanepisternum extending laterally as far as lateral edge of mesepimeron; anterior edge of each elytron with submarginal transverse groove; body somewhat flattened and body parallel-sided (Pl. 79D) .... ***Helferella* Cobos**

– IF pronotum with two lateral carinae THEN total length greater than 3.5 mm; mesanepisternum not extending laterally as far as lateral edge of mesepimeron; anterior edge of elytron without transverse groove. IF body somewhat flattened, THEN elytra not parallel-sided.... **Agrilinae 9**

9(8). Mesocoxae almost always separated by a distance distinctly greater than that separating procoxae (Fig. 35.19); IF pro- and mesocoxae separated by similar distance THEN body less than 4.5 mm long and more than 3.5 times as long as wide, with terminal 4 or 7 antennomeres enlarged forming a club; anterior edges of metacoxae not or scarcely concave (Fig. 35.19); tarsi very short **10**

– Mesocoxae separated by approximately the same width as the procoxae (Figs 35.20–35.21); body length usually greater than 4.5 mm and antennae not clubbed; anterior edges of metacoxae usually strongly concave, their anterolateral corners extending anteriorly between metaventrite and lateral prolongation of abdomen (Figs 35.20, 35.22); tarsi more or less elongate **17**

10(9). Length more than 2.3 times width **11**

– Length less than 2 times width **13**

11(10). Form cylindrical (Pl. 79A); frontovertex weakly convex; sides of pronotum not explanate, its posterior edge more less truncate; length less than 3 mm; colour black, not metallic; widely distributed .... ***Germarica* Blackburn**

– Form depressed (Pl. 77K, 79C); frontovertex deeply excavate; sides of pronotum more or less explanate, its posterior edge distinctly biemarginate; length greater than 3 mm; colour metallic **12**

12(11). Antennal serrations beginning on antennomere 8, so that 8–11 form a club, which fits into a cavity at the anterior edge of the pronotal hypomeron; pronotum more than 0.6 times as long as wide (Pl. 77K; Fig. 35.23) .... ***Aphanisticus* Latreille**

– Antennal serrations beginning before antennomere 8, terminal antennomeres not forming club; pronotum less than 0.6 times as long as wide (Pl. 79C) .... ***Endelus* Deyrolle**

13(10). Width of scutellar shield more than 0.25 times as great as thoracic width (Pl. 79F, 81B); tibiae flattened, expanded and excavate to receive tarsi at rest; elytra with pattern of transverse zigzag patterns composed or silver or silver and red thickened setae; antennae fitting into hypomeral grooves **14**

– Width of scutellar shield less than 0.15 times as great as thoracic width (Pl. 77C, 79E, 83K); tibiae neither flattened nor excavate; elytra without zigzag patterns; hypomeral grooves absent **15**

14(13). Antennae strongly clavate, antennomeres 6–10 strongly expanded and flattened; posterior pronotal angles acute; outer edges of meso- and metatibiae sharply angulate (Pl. 81B); length 4 mm; north-western WA (possible introduction) .... ***Pachyschelus* Solier**

– Antennae weakly clavate, antennomeres 6–10 weakly, asymmetrically expanded; posterior pronotal angles more or less obtuse; outer edges of meso- and metatibiae more weakly expanded, somewhat rounded (Pl. 79F); QLD (introduced for biological control) .... ***Hedwigiella* Obenberger**

15(13). Sides of elytra slightly sinuate; pronotum widest at middle; metafemora distinctly inflated (Pl. 77C; Fig. 35.24); metacoxal plates incomplete laterally; length ~3.4 mm; northern QLD .... ***Anthaxomorphus* Deyrolle**

– Sides of elytra straight or slightly rounded; pronotum widest posteriorly (Pl. 79E, 83K); metafemora not inflated; metacoxal plates complete **16**

16(15). Form more or less triangular, body widest at junction of pronotum and elytra and sides of elytra gradually converging apically; anterior edge of pronotum distinctly concave, with sharp anterior angles; each elytron with longitudinal carina extending posteriorly from humeral callus (Pl. 79E); metacoxal plates distinctly wider mesally than laterally .... ***Habroloma* Thomson**

– Form more or less oval, body widest behind junction of pronotum and elytra and sides of elytra subparallel anteriorly, curved and gradually converging posteriorly; anterior edge of pronotum more or lessd truncate, without distinct anterior angles; elytron without longitudinal carina (Pl. 83K); metacoxal plates about as wide mesally as laterally .... ***Trachys* Fabricius**

17(9). Frontovertex with narrow longitudinal groove (Figs 35.25–35.26) **18**

– Frontovertex without such a groove, but may be broadly, longitudinally depressed between eyes (Pl. 83E) **21**

18(17). Pretarsal claws bifid (Fig. 35.106); dorsal setae in patches or fasciae (Pl. 79G) .... ***Ethonion* Kubán**

– Pretarsal claws appendiculate or only swollen at base; dorsal setae regular and sparsely distributed (Pl. 80D) **19**

19(18). Body subcylindrical; pronotum widest at base with sides narrowing to apex (Pl. 80D) .... ***Meliboeithon* Obenberger**

– Body strongly flattened; sides of pronotum arcuate, widest at middle (Pl. 83E, 83I) **20**

20(19). Tibiae straight; outer apical angle of protibia produced to form tooth (Fig. 35.28); pronotal disc more or less even; lateral pronotal carinae simple or with weak foveate impressions; elytral vestiture not forming pattern (Pl. 83I) .... ***Synechocera* Deyrolle**

– Tibiae outwardly curved; outer apical angle of protibia simple; pronotal disc with distinct prominences and depressions; elytra with thicker white setae forming a pattern including zigzag transverse fasciae (Pl. 83E) .... ***Toxoscelus* Deyrolle**

21(17). Lateral edge of pronotum with complete lateral carina plus a ventral hypomeral carina (Fig. 35.29) and often also a supralateral discal carina which may be short or almost reaching anterior edge of pronotum; metatarsomere 1 as long as 2–4 combined; scutellar shield usually with transverse carina (Pl. 77J) .... ***Agrilus* Curtis**

– Lateral edge of pronotum without carina (Fig. 35.30) or with complete lateral carina as well as a supralateral discal carina (Fig. 35.31) which may be continuous or broken, long or short, metatarsomere 1 not longer than 2 and 3 combined; scutellum without transverse carina **22**

22(21). Metafemora strongly swollen (Fig. 35.32); vertex with pair of projecting lobes; elytra with zigzag setal pattern (Pl. 82I) .... ***Sambus* Deyrolle**

– Metafemora not strongly swollen; vertex without paired projecting lobes; elytra without zigzag setal pattern **23**

23(22). Lateral pronotal carinae absent; body more or less cylindrical; posterior edges of eyes located some distance from anterior edge of pronotum **24**
– Lateral pronotal carinae present; body more or less flattened; posterior edges of eyes contiguous or nearly so with anterior edge of pronotum **25**
24(23). Elytron epipleuron not separated from disc by distinct carina (Fig. 35.33); antennae serrate from antennomere 4 (Fig. 35.34) (Pl. Dinocephalia) .... ***Dinocephalia* Obenberger**
– Elytral epipleuron separated from disc by distinct carina (Fig. 35.35); antennae serrate from antennomere 5 (Pl. 81H)........***Paracephala* Saunders**
25(23). Pretarsal claws bifid; metepimeron concealed beneath lateral projection of elytron (Fig. 35.31; Pl. 81A).......... ***Neospades* Blackburn**
– Pretarsal claws appendiculate or at most swollen at base **26**
26(25). Antennae serrate from antennomere 5 (Fig. 35.36); body slightly flattened or subcylindrical (Pl. 77I) .... ***Aaaba* Bellamy**
– Antennae serrate from antennomere 4 (Fig. 35.37); body dorsoventrally flattened **27**
27(26). Anterior edge of prosternum very broadly, more or less evenly emarginate (Fig. 35.38); maxillary palps elongate in male (Pl. 81M).......... ***Pachycisseis* Théry**
– Anterior edge of prosternum weakly triemarginate, lateral emarginations narrow and median one broad with curved base (Figs 35.39–35.40); maxillary palps short, equal in both sexes **28**
28(27). Frontovertex broadly depressed between eyes, with surface punctation reduced or absent (Figs 35.41–35.42); pronotum uneven, with elevated tubercles or costae; sublateral carinae bisinuate, short or absent (Pl. 80G) .... ***Hypocisseis* Thomson**
– Frontovertex flat or only slightly depressed, with surface evenly punctate (Figs 35.43–35.44); pronotum more or less convex with sublateral carinae straight or only slightly curved and complete or almost complete to anterior edge (Fig. 35.45).......... **29**
29(28). Clypeal region not or only weakly declined (Figs 35.43–35.44); pretarsal claws symmetrical; elytral surfaces usually with thickened white setae forming spots or fasciae (Pl. 79B); ovipositor without groups of curved ventral spines; length often greater than 6.5 mm .... ***Diphucrania* Dejean**
– Clypeal region strongly declined, forming a distinct angle with frontal ridge (Fig. 35.46); pretarsal claws asymmetrical; elytral surfaces never with white spots, often transversely rugulose (Pl. 82D); ovipositor ventrally with two groups of curved spines; length rarely greater than 6.5 mm .... ***Stanwatkinsius* Barker & Bellamy**

Antennal fossae almost always more widely separated (30–end)

30(7). Prosternum in front of coxae shorter than diameter of procoxal cavity; mesoventral cavity shallow, without steep sides posteriorly; body strongly convex, subcylindrical and more than 30 mm in length (Pl. 80A); elytral punctation partly seriate, but irregular in places, with moderately convex intervals interrupted along their lengths by impressions and punctures .... ***Julodimorpha* Gemminger & Harold**
– Prosternum in front of coxae distinctly longer than diameter of procoxal cavity; mesoventral cavity deeper with steeply sloping sides; body less convex, at least slightly flattened; IF body length greater than 30 mm THEN elytral punctation more regularly seriate or striate (Pl. 80B) **31**
31(30). Antennomeres 3–11 unipectinate in both sexes, with rami much longer in male and with sensilla generally distributed over surface of each ramus (Fig. 35.47); labrum concealed beneath clypeus; pronotum with small posteromesal fossa just in front of scutellum (Pl. 77E); length 5–6.1 mm .... ***Australorhipis* Bellamy**
– Antennae never unipectinate; IF rarely bipectinate in male THEN antennae 12-segmented and length greater than 8.5 mm; without other features combined **32**
32(31). Antennal sensilla generally distributed, at least over dorsal surfaces of serrate or lobed antennomeres (Fig. 35.48), in some cases also concentrated in depressions or foveae on ventral or apical surfaces **33**
– Antennal sensilla always concentrated and often in well defined, apical or ventral depressions (foveae) of serrate or lobed antennomeres (Fig. 35.49) **44**

Antennal sensilla generally distributed, at least over dorsal surfaces (33–43)

33(32). Anterior edge of frontoclypeus with broad emargination extending laterally as far as antennal fossae and exposing a wider portion of the membranous anteclypeus (Fig. 35.50); scutellar shield absent or highly reduced and apically acute (Fig. 35.51); elytral apices rounded, unidentate or bidentate; colour never metallic, either black or bicolored (black and yellow or red); without patches of glandular setae or yellow flocculence **34**
– Anterior edge of frontoclypeus with narrower (but often deeper) emargination ending laterally well before antennal fossae and concealing or exposing narrower portion of membranous anteclypeus (Fig. 35.52); scutellar shield well developed (Fig. 35.53); elytral apices multidentate (Fig. 35.54); colour always more or less metallic, often green; surfaces almost always with some patches of glandular setae producing yellow flocculence (which may be rubbed off or washed off in some specimens) (Pl. 77L, 82K–M) **36**
34(33). Scutellar shield present but minute, not sunken below level of elytra; elytral apices unidentate (Fig. 35.51); colour yellow with dark markings (Pl. 79L); length 5–22 mm .... ***Diadoxus* Saunders**
– Scutellar shield absent or sunken beneath level of elytral bases; elytral apices bidentate or rounded.......... **35**
35(34). Elytra with deeply impressed striae and convex, costa-like intervals; elytral apex bidentate; colour black or dark brown; upper surface matt (Pl. 77G); length 17 mm or less .... ***Araucariana* Levey**
– Elytra with rows of very fine punctures and flat intervals; elytral apices simple and rounded; colour black and red; upper surface glossy (Pl. 78G); length at least 18 mm.......... ***Cyrioides* Carter**
36(33). Intercoxal process of abdominal ventrite 1 with prominent, posteriorly rounded plate extending slightly beyond posterior edge of ventrite; sides of prothorax parallel posteriorly but abruptly narrowed at anterior fourth (Pl. 78H–I); pronotal disc with two large, posterolateral, C-shaped impressions and two anterior rounded ones; elytra without costae .... ***Cyphogastra* Deyrolle**
– Intercoxal process of abdominal ventrite 1 without prominent plate; IF sides of prothorax abruptly narrowed anteriorly THEN pronotal disc with numerous, irregular impressions and reliefs (Fig. 35.55).......... **37**
37(36). Pronotal disc with distinct median groove (Fig. 35.56); pronotal punctation relatively fine and regular OR IF somewhat irregular THEN upper surfaces without setose impressions or yellow flocculence (Fig. 35.56) **38**
– Pronotal disc without median groove (sometimes with paired median grooves defining a longitudinal smooth relief (Fig. 35.57); pronotal punctation varying considerably in size and distribution; upper surfaces with setose impressions filled with yellow flocculence (Pl. 82K).......... **40**

38(37). Upper surfaces without setose impressions or yellow flocculence; sides of pronotum subparallel, very slightly converging anteriorly; each elytron with 5 costae (Pl. 81E); scutellar shield more or less oval .... ***Paracupta* Deyrolle**

– Upper surfaces with setose impressions and yellow flocculence; sides of pronotum distinctly converging anteriorly; elytra not costate; scutellar shield distinctly widened posteriorly with truncate apex (Fig. 35.58) **39**

39(38). Frontovertex with broad, shallow setose impression filled with yellow flocculence (Fig. 35.59); pronotal disc with 2 oval setose impressions and elytra each with two setose impressions anteriorly and at middle and a linear oblique setose impression extending from middle almost to apex (Pl. 80F); tarsi yellow .... ***Metatenia* Théry**

– Frontovertex with narrow, deep, longitudinal groove; pronotal disc with pair of slender, sublateral setose impressions and each elytron with a similar impression, the two sometimes meeting (Fig. 35.52); tarsi dark and metallic (Pl. 80L) .... ***Iridotaenia* Deyrolle**

40(37). Frontovertex with very broad impression almost reaching eyes on either side and a short, deep median groove in its centre (Fig. 35.60); pronotal disc with more regular punctures and two or four larger impressions; elytra with at least weak longitudinal costae **41**

– Frontovertex with narrow median impression or groove, well separated from eyes; pronotal disc with numerous, highly irregular, setose impressions; elytra without distinct costae **43**

41(40). Pronotal disc with median punctate furrow and sublateral carinae extending anteriorly to just beyond middle; elytral sublateral setose impression extending from beneath humerus almost to apex (Pl. 78C) .... ***Chalcophorotaenia* Obenberger**

– Pronotal disc with raised median impunctate ridge (Fig. 35.57) **42**

42(41). Pronotal disc with one pair of large, oval, setose basolateral impressions, each defined laterally by a sharp carina (Fig. 35.57); elytra with 5 or more distinct costae (Pl. 79K) .... ***Chrysodema* Laporte & Gory**

– Pronotal disc with 1 or 2 lateral impressions or areas of densely packed setose; punctures, not defined by lateral carina; elytral costae weakly defined (at least mesally) and each elytral disc with sublateral linear setose impression and at least 2 ovate setose impressions (Pl. 77M).......... ***Austrochalcophora* Bellamy**

43(40). Setose depressions on head more or less scattered over surface; scutellar shield distinctly longer than wide (Fig. 35.61; Pl. 77L).......... ***Austrophorella* Kerremans**

– Setose depressions on head concentrated in two oblique linear bands at inner edges of eyes and along median, longitudinal groove (Fig. 35.107); scutellar shield not longer than wide (Pl. 82K–M).......... ***Pseudotaenia* Kerremans**

Antennal sensilla always concentrated and usually in foveae (44–end)

44(32). Apex of prosternal process biemarginate forming three more or less equal lobes, two lateral and one apical (Fig. 35.62); antennomere 3 more than 1.5 times as long as 4; posterior edge of pronotum deeply biemarginate to receive pair of sharp elytral lobes on either side of scutellar shield, which is longer than wide and sharply acute posteriorly (Pl. 78F) **45**

– Apex of prosternal process without or with shorter lateral lobes; antennomere 3 not or only slightly longer than 4; posterior edge of pronotum not or only shallowly biemarginate and anterior elytral lobes not as sharply defined **46**

45(44). Tarsomeres 1–3 each deeply emarginate with sides produced to form a pair of slender processes bearing a tight cluster of spine like setae, those on tarsomere 3 usually extending beyond the apex of tarsomere 4 (Fig. 35.63); eyes strongly converging dorsally; pronotal disc with 2 large, obliquely transverse impressions; scutellar shield at least 1.5 times as long as wide and about a fifth as long as elytra, which lack golden impressions (Pl. 78F); profemur not expanded.......... ***Belionota* Eschscholtz**

– Tarsomeres 1–3 without paired, apicolateral acute processes; eyes slightly converging; pronotal disc without large, transverse impressions; scutellar shield only slightly longer than wide and less than a tenth as long as elytra, each of which bears 3 golden impressions; profemur expanded internally to form angulate process (Pl. 78F) .... ***Chrysobothris* Eschscholtz**

46(44). Mesepimeron expanded laterally and partly overlapping elytral epipleuron (Fig. 35.65); abdominal ventrites 2 and 3 each with pair of lateral yellow impressions (Fig. 35.66); tarsomeres 1–3 each deeply emarginate with sides produced to form paired processes similar to those in *Belionota* (Fig. 35.64); dorsal surfaces entirely black (Pl. 80H); widely distributed .... ***Merimna* Saunders**

– Tarsomeres 1–3 never with paired, slender processes; mesepimeron not laterally expanded, not overlapping epipleuron; abdominal ventrites without paired yellow impressions **47**

47(46). Labrum well sclerotised, overlapping mandibles, sometimes anteriorly narrowed, longitudinally grooved, and extending to mandibular apex; frontoclypeus produced forward to form muzzle (Figs 35.67–35.68); mentum not strongly transverse and usually well sclerotised **48**

– Labrum transverse and usually weakly sclerotised; with or without longitudinal groove, frontoclypeus not produced forward to form muzzle (Fig. 35.69); mentum strongly transverse and at least apically membranous **54**

Labrum well sclerotised, moderately long; frontoclypeus forming muzzle (48–53)

48(47). Anterolateral angles of first abdominal ventrite projecting forward to partly conceal metepimeron (Fig. 35.70); elytral apices rounded and posterolateral edges of elytra finely but sharply denticulate (Fig. 35.71); pronotum with weak median groove; frontovertex broadly impressed; labrum longitudinally grooved; length usually less than 15 mm; pygidium heavily sclerotised and usually exposed (Fig. 35.71; Pl. 82J) .... ***Selagis* Dejean**

– Anterolateral angles of first abdominal ventrite not projecting forward or, if so, not partly concealing metepimeron (Fig. 35.72); elytral apices rounded, truncate, unidentate or bidentate, but posterolateral edges of elytra without or with very fine and weak denticulations (Figs 35.73–35.74); pronotum more or less even, not grooved; frontovertex not impressed; pygidium weakly sclerotised and usually concealed; length often more than 15 mm (Pl. 78M, 80B) **49**

49(48). Anterior portion of prosternum strongly tumid forming subconical process, which projects beyond anterior edge (Figs 35.75–35.76) **50**

– Prosternum flat or slightly convex without anteriorly projecting process (Figs 35.77–35.78) **51**

50(49). Posterior edge of pronotum strongly biemarginate with very prominent medial lobe; scutellar shield small, elongate and apically acute; elytral punctation seriate; elytral apices bidentate and non-overlapping (Pl. 78M) .... ***Calodema* Gory & Laporte**

– Posterior edge of pronotum weakly bisinuate, without prominent medial lobe; scutellar shield larger, about as long as wide and apically rounded; elytral punctation striate; elytral apices tridentate and overlapping (Pl. 80B) .... ***Metaxymorpha* Parry**

51(49). Elytra with large, more or less irregularly spaced foveoles (Pl. 82F, H).......... ***Stigmodera* Eschscholtz**

– Elytra striate or striate-punctate, sometimes with costae (Pl. 78D–E, K–L, 83G, J, M– N) **52**

52(51). Frontovertex almost always without hairs (Fig. 35.79) (present in the Tasmanian *C. setifera* Barker, *C. jubata* (Blackburn), *C. macquillani* (Barker) and *C. rudis* (Carter)); length always less than 29 mm and usually less than 20 mm; pretarsal claws always simple, without basal tooth (Fig. 35.80); mesoventral cavity strongly transverse (Pl. 78K–L, 83G, J, M–N) .... ***Castiarina* Gory & Laporte**

– Frontovertex with hairs (Fig. 35.81); length at least 14 mm and usually greater than 20 mm; pretarsal claws with basal tooth in about three-fourths of the species (Fig. 35.82); mesoventral cavity as long as wide **53**

53(52). Ovipositor with pair of heavily sclerotised projections (often externally visible in intact specimens) (Fig. 35.83; Pl. 78D–E)................................... .... ***Calotemognatha* Peterson**

– Ovipositor without sclerotised projections; without other features combined (Pl. 83C) .... ***Temognatha* Solier**

Labrum strongly transverse, weakly sclerotised; frontoclypeus not forming muzzle (54–end)

54(47). Anterolateral angle of first abdominal ventrite not extending anteriorly beyond metacoxa and not abutting or overlapping metepimeron (Fig. 35.84); clypeus expanded laterally, closing antennal cavities anteriorly (Fig. 35.85); pronotal disc without distinct longitudinal groove or impression (Fig. 35.86) **55**

– Anterolateral angle of first abdominal ventrite extending anteriorly beyond metacoxa and abutting or overlapping metepimeron (Fig. 35.87); clypeus usually not expanded, so that antennal cavities are anteriorly open (Fig. 35.88); IF cavities anteriorly closed THEN and pronotal disc with median longitudinal groove or impression (Pl. 78A–B).. **60**

Ventrite 1 not extending anteriorly (55–59)

55(54). Body subcylindrical; pronotal disc evenly convex, with lateral carinae concealed from above; elytra weakly costate with fine, confused punctures between the costae (Pl. 78J, 79M); elytral apices truncate or slightly concave, often forming paired teeth or spines....................................... **56**

– Body somewhat flattened; pronotal disc usually somewhat uneven, with weak impressions, but sometimes highly irregular (Pl. 80M, 81K); IF lateral pronotal carinae concealed from above, THEN elytral punctures forming distinct striae; IF elytra with fine, confused punctures, THEN costae more strongly developed; elytral apices more or less rounded **57**

56(55). Protnoum almost always with prescutellar pit (Fig. 35.89); body somewhat stouter (Pl. 78J); apex of prosternal process truncate (Fig. 35.90), without or with very weak lateral processes... ***Bubastes* Laporte & Gory**

– Pronotum without prescutellar pit (Pl. 79M); body more slender; prosternal process apically subacute with distinct lateral processes (Fig. 35.91)... .... ***Euryspilus* Lacordaire**

57(55). Posterior edge of metacoxal plates dentate or at least markedly angulate at middle (Figs 35.92–35.93); pronotal disc with either several large and small impression (Fig. 35.94) OR a small prescutellar pit and two weak, elongate, oblique posterolateral impressions **58**

– Posterior edge of metacoxal plates straight or feebly convex at middle (Fig. 35.95); pronotal disc with neither combination of features.............. **59**

58(57). Posterolateral edges of elytra entire (Fig. 35.96); eyes neither prominent nor projecting; body robust, not metallic; pronotum with several large to small, rounded impressions (usually three anterior and three posterior) (Pl. 80M)... ***Nascio* Laporte & Gory**

– Posterolateral edges of elytra serrate or spinose (Fig. 35.97); eyes prominent and projecting; body more elongate, metallic; pronotum with prescutellar pit and a pair of weak, elongate, oblique posterolateral impressions (Pl. 81F, 83H) .... ***Nascioides* Kerremans**

59(57). Elytral punctures forming distinct striae, at least mesally; pronotal disc strongly, evenly convex, the lateral carinae not visible from above; posterolateral edges of elytra denticulate (Pl. 81G) .... ***Neobubastes* Blackburn**

– Elytra with confused punctures and distinct longitudinal costae; pronotal disc less convex and at least slightly irregular, usually with a weak median groove and/or a pair of weak posterolateral impressions adjacent to posterior edge (rarely with a pair of longitudinal paramedian ridges); lateral pronotal carinae usually partly visible from above (Pl. 81I, 81K)... ***Neobuprestis* Kerremans**

Ventrite 1 extending anteriorly (60–end)

60(54). Clypeus expanded laterally, closing antennal cavities anteriorly (Fig. 35.88); elytra with distinct costae separated by irregularly distributed, nonseriate punctures; pronotal disc with more or less complete longitudinal groove or broad impression (Pl. 78A–B, 79I); frontovertex with median carina; length more than 15 mm **61**

– Clypeus not expanded laterally, antennal cavities anteriorly open (Fig. 35.69); IF elytra costate THEN frontovertex without median carina; pronotal median groove, if present, weak, narrow and incomplete (Pl. 81D, 83F).. **62**

Antennal cavities anteriorly closed by laterally expanded clypeus (61)

61(60). Pronotum with large, broad, longitudinal depression (Pl. 78A–B); tarsomeres 1 and 2 with reduced adhesive pads (much less than half tarsomere length); colour non-metallic; widely distributed .... ***Burnsiellus* Levey & Bellamy**

– Pronotum wit weak, narrow longitudinal depression; adhesive pads on tarsomeres 1 and 2 more than half tarsomere length; colour brightly metallic, usually green and coppery-red (Pl. 79I); introduced in timber from U.S.. .***Buprestis* Linnaeus**

Antennal cavities anteriorly open (62–end)

62(60). Elytral punctation either striate or costate, with punctures in intervals more or less irregularly distributed and almost always containing a white flocculence (Pl. 80J–K, 81D, 83F); posterolateral edges of elytra not denticulate; pronotal base narrower than distance between elytral humeri; antennae very short, not extending beyond anterior third of prothorax **63**

– Elytral punctation usually neither striate nor costate and without white flocculence; IF elytra striate or costate THEN posterolateral edges of elytra strongly denticulate AND pronotal base not narrower than distance between elytral humeri (Pl. 80I); antennae usually extending at least to middle of prothorax **64**

63(62). Elytra striate (Pl. 81D); scutellar shield more or less circular or pentagonal, sometimes with longitudinal groove (Fig. 35.98); anterior edge of clypeus with sharply angulate emargination (Fig. 35.99; Pl. 81D, 83F); Central QLD, WA.. .... ***Notobubastes* Carter & *Buprestina* Obenberger**

– Elytra costate (Pl. 80J); scutellar shield transverse and more or less truncate or with transverse groove (Fig. 35.100); anterior edge of clypeus with deep, rounded emargination (Fig. 35.69); VIC, SA, WA (Pl. 80K).. ***Microcastalia* Heller**

64(62). Last abdominal ventrite almost always excised apically between two strong spines (Fig. 35.101); posterolateral edges and apices of elytra usually sharply denticulate; pronotal disc neither transversely rugulose nor covered with polygonal cells; length usually greater than 5 mm and body usually brightly coloured (Pl. 80C, 80I); .... ***Melobasis* Gory & Laporte**

– Last abdominal ventrite rounded or only feebly excised apically; posterolateral edges and apices of elytra not or only finely denticulate; IF length greater than 5 mm THEN pronotal disc transversely rugulose or covered with polygonal cells **65**

65(64). Pronotal disc transversely rugulose; base of pronotum truncate (Fig. 35.102); elytra widest at posterior third, with apicolateral edges somewhat explanate (Pl. 83B, 83L); eyes dorsally converging; length greater than 7 mm .... ***Torresita* Gemminger & Harold**

– Pronotal disc not transversely rugulose; elytra not enlarged posteriorly; eyes not dorsally converging; IF base of pronotum truncate THEN length less than 7 mm AND sides of elytra subparallel and not explanate **66**

66(65). Pronotal disc with polygonal cells, sometimes partly transformed into wrinkles, but without distinct punctures (Fig. 35.103) **67**

– Pronotal disc punctate without polygonal cells (Pl. 81J) **72**

Pronotum with polygonal cells (67–71)

67(66). Pygidium completely or almost completely exposed; pronotum strongly transverse with posterior angles obtuse; elytra more than 3.6 times as long as pronotum with broadly rounded, well separated apices (Pl. 77A); southern QLD .... ***Barakula* Peterson & Bellamy**

– Pygidium completely concealed or barely exposed; pronotum usually less strongly transverse; posterior pronotal angles right or acute; elytra less than 3.5 times as long as pronotum **68**

68(67). Base of pronotum truncate or barely biemarginate (Pl. 77H, 83D) **69**

– Base of pronotum distinctly biemarginate (Figs 35.103–35.104). **70**

69(68). Head across eyes wider than prothorax; pronotum more than 0.75 times as long as wide, widest posteriorly with sides straight and slightly narrowed anteriorly; colour metallic green (Pl. 83D) .... ***Theryaxia* Carter**

– Head across eyes not wider than prothorax; pronotum less than 0.7 times as long as wide, widest at or before middle with sides rounded at least anteriorly (Pl. 77H) .... ***Anilara* Saunders**

70(68). Pronotum widest at or near base; posterior pronotal angles more or less acute; anterior edge of pronotum straight or slightly convex; distance between posterior pronotal angles about the same as that between elytral humeri) (Fig. 35.104; Pl. 82G) .... ***Pseudanilara* Théry**

– Pronotum widest at or before middle and at least slightly constricted posteriorly; posterior pronotal angles more or less right; anterior edge of pronotum concave or biconcave; distance between posterior pronotal angles less than 0.9 times that between elytral humeri (Pl. 77D, 81L) **71**

71(70). Pronotum widest before middle and strongly constricted posteriorly (Fig. 35.103); pronotal disc with weak longitudinal impression or flat area anteriorly and narrow longitudinal groove posteriorly; eyes large and posterodorsally convergent (Pl. 81L) .... ***Notographus* Thomson**

– Pronotum widest at about middle and only weakly constricted posteriorly; pronotal disc more or less evenly convex; eyes only slightly convergent posteriorly (Pl. 77D) .... ***Anthaxoschema* Obenberger**

Pronotum without polygonal cells (72)

72(66). Metacoxal plate laterally narrowed to one-fifth or less of its width near midline (Fig. 35.105); pretarsal claws toothed; prosternal process not expanded near apex (Fig. 35.105; Pl. 80E) .... ***Maoraxia* Obenberger**

– Metacoxal plate nearly parallel-sided; pretarsal claws simple; prosternal process laterally expanded near apex (Pl. 81J) .... ***Neocuris* Saunders**

Classification of Australian genera

As mentioned above, buprestid classification is still undergoing revision, and although considerable time and effort were put into the preparation of the key, the authors, who are not a specialists on this family, do not feel competent to include subfamilal, tribal, subtribal or generic diagnoses. Unless otherwise indicated adult and larval host records in the following sections have been extracted from Bellamy *et al.* (2013) and much of the distributional data is based on Bellamy (2008a–d).

Subfamily Polycestinae Lacordaire, 1857

Tribe Haplostethini LeConte, 1861

Helferella Cobos, 1957 (Pl. 79D)

Type species. *H. dianae* Cobos, 1957.

Australian species. Nine.

Distribution. *Helferella* occurs in the Philippines, Fiji, Papua New Guinea and Australia. Eight of the Australian species occur mainly along the eastern coast of Australia from southern VIC to northern QLD, with a single species recorded from WA.

Biology. Little is known about the biology or immature stages of Haplostethini; however, a North American *Mastogenius* has been reported to breed in 'small branches of oak' (Baker 1972) and a larva of *M. texanus* Bellamy was found in a dead twig of *Quercus pungens* (Fagaceae) (Volkovitsh & Bílý 2015). Adults of *Helferella manningensis* Williams & Weir were observed on foliage of *Guioa semiglauca* (Sapindaceae), but the larval habitat is unknown.

References. Cobos (1957); Williams & Weir (1987, 1988).

Tribe Paratrachyini Cobos, 1980

Paratrachys Saunders, 1873 (Pl. 81C)

Type species. *P. hederae* Saunders, 1873.

Friendiella Holynski, 1992 (subgenus).

Australian species. Two.

Distribution. *Paratrachys* is distributed throughout the eastern Palaearctic and Oriental regions and extends to Indonesia, Papua New Guinea and Australia. The two Australian species, *P. queenslandica* Bellamy & Williams and *P. australia* Bellamy & Williams, occur along the eastern coast of Australia, from northern QLD and central NSW, respectively.

Biology. Hołyński (1992) sugested that *Paratrachys* species may have coevolved with several species of *Ficus* (Moraceae). Larvae of the two Australian species are unknown, but those of the Japanese *P. hederae* Saunders are known to be miners in the leaves of *Ficus pumila* (Fagaceae) (Volkovitsh & Bílý 2015).

References. Bílý (1989); Hołyński (1992); Bellamy & Williams (1995).

Tribe Xyroscelidini Cobos, 1955

Xyroscelis Saunders, 1868 (Pl. 83A)

Type species. *Amorphosoma crocatum* Gory & Laporte, 1839.

Australian species. Two.

Distribution. *Xyroscelis bumana* Williams & Watkins is known from south-east coastal NSW and *X. crocata* (Gory & Laporte, 1839) from south-western WA.

Biology. Adults of *Xyroscelis* crocata have been observed feeding on the sap of *Macrozamia riedlei* (Cycadales: Zamiaceae) and *Xyroscelis* larvae feed within *Macrozamia* fronds (Bellamy 1997).

References. Williams & Watkins (1986); Volkovitsh & Bílý (2015).

Tribe Polycestini Lacordaire, 1857

Polycesta Serville in Dejean, 1833 (Pl. 82B)

Type species. *Buprestis porcata* Fabricius, 1775.

Australian species. One.

Distribution. *Polycesta* is a cosmopolitan genus, and the single Australian species, *P. mastersi* Macleay, is known from the Murray–Darling basin, QLD

Biology. Biology of the Australian species is unknown, but larvae of exotic *Polycesta* species are known to feed in the wood of *Quercus* (Fagaceae) and *Bursaria* (Bursariaceae) (Volkovitsh & Bílý 2015).

References. Bílý (1989).

Strigoptera Dejean, 1833 (Pl. 82C, E)

Type species. *Buprestis bimaculata* Linnaeus, 1758.

Castalia Gory & Laporte, 1839, not Lamarck, 1819.

Australian species. One.

Distribution. The genus *Strigoptera* occurs in the Oriental, Afrotropical and Australasian regions; the Australian species, *S. bimaculata* (Linnaeus), occurs in New Guinea, as well as NT and nothern QLD.

Biology. Adults have been recorded on leaves of a mangrove, *Camptostemon schulzii* (Bombacaceae) and larvae of the same species were collected near Darwin, NT from the mangrove *Ceriops tagal* (Rhizophoraceae).

References. Hawkeswood (1988).

Tribe Prospherini Cobos, 1980

Prospheres Saunders, 1868 (Pl. 82N)

Type species. *Buprestis aurantiopicta* Laporte & Gory, 1837.

Australian species. Two.

Distribution. *Prospheres* species are known from New Caledonia, Papua New Guinea and Australia. The Australian *P. aurantiopictus* (Laporte & Gory) occurs in from VIC to northern QLD, as well as on Norfolk Island, where a second species, *P. norfolkensis* Levey also occurs.

Biology. Larvae of *P. aurantiopicta* are usually found in the wood of *Araucaria cunninghamii* (Araucariaceae).

References. Levey (1978b); Volkovitsh & Hawkeswood (1999).

Tribe Astraeini Cobos, 1980

Astraeus Laporte & Gory, 1837 (Pl. 77B)

Type species. *A. flavopictus* Laporte & Gory, 1837.

Depollus Barker 1975 (subgenus).

Australian species. 50.

Distribution. The genus *Astraeus* is widely distributed in Australia, and also occurs in New Guinea and New Caledonia.

Biology. Adults are most often found on foliage of *Casuarina* and *Allocasuarina* (Casuarinaceae), but there are adult records from several genera of Fabaceae, Myrtaceae and Proteaceae. Larvae of *A. crassus* Van de Poll were found in branches of *Bursaria spin*osa (Pittosporaceae). *Astraeus* adults have a spring mechanism involving the release of the elytra.

References. Barker (1975); Turner & Hawkeswood (1996).

Subfamily Chrysochroinae Laporte, 1835

Austrochalcophora Bellamy, 2006 (Pl. 77M)

Type species. *Chalcophora subfasciata* Carter, 1916.

Australian species. One.

Distribution. The single species, *A. subfasciata* (Carter), is known from NT and northern WA.

Biology. Larvae of *A. subfasciata* (Carter) found in the wood of *Acacia melanoxylon* (Fabaceae) and adults occur on its foliage.

References. Bellamy (2006).

Austrophorella Kerremans, 1903 (Pl. 77L)

Type species. *Chalcotaenia quadrisignata* Saunders, 1872.

Australian species. One.

Distribution. The only species, *Austrophorella quadrisignata* (Saunders), is known from NSW and QLD.

Biology. Adults have been recorded from the foliage of *Alphitonia excelsa* and *A. petriei* (Rhamnaceae) and larvae were found in stems of the former.

References. Bellamy (2006).

Chalcophorotaenia Obenberger, 1928 (Pl. 78C)

Type species. *Chalcotaenia quadriimpressa* Waterhouse, 1875.

Australian species. 13.

Distribution. *Chalcophorotaenia* species occur mainly in the northern and western part of the continent.

Biology. Adults have been recorded on foliage of *Pleiogynium cerasiferum* (Anacardiaceae), as well as *Eucalyptus* and *Melaleuca* (Myrtaceae). Larvae are known from *Gastrolobium grandiflorum* (Fabaceae).
References. Obenberger (1928); Belamy (2006).

Chrysodema Laporte & Gory, 1835 (Pl. 79H, K)

Type species. *C. sonnerati* Laporte & Gory, 1835.
Pseudochrysodema Saunders, 1874 (subgenus).
Australian species. Three.
Distribution. *Chrysodema* species occur in the Palaearctic, Oriental and Australasian regions. *Chrysodema aurofoveata* (Guérin-Méneville) and *C. furcata* Kerremans are known from northern QLD and NT and *C. simplex* Waterhouse from Christmas Island.
References. Hołyński (1993, 2014d).

Cyphogastra Deyrolle, 1864 (Pl. 78H–I)

Type species. *Chrysodema semipurpurea* Laporte & Gory, 1835.
Australian species. Three.
Distribution. The genus *Cyphogastra* occurs in South-east Asia, the Philippines, Indonesia, Papua New Guinea, New Caledonia, Australia and several Pacific islands. The Australian species, *C. farinosa* (Fabricius), *C. pistor* (Laporte & Gory) and *C. quadrivittata* Carter, occurr across northern Australia from QLD to northern WA
Biology. Larvae of *C. farinosa* (Fabricius) are known from *Elaeocarpus angustifolius* (Elaeocarpaceae) and *Barringtonia calyptrata* (Lecythidaceae), and those of *C. pistor* (Laporte & Gory) from *Terminalia* spp. (Combretaceae), *Aleurites moluccana* (Euphorbiaceae) and *Alphitonia petriei* (Rhamnaceae). The exotic species *C. bruijni* Lansberge was recorded from introduced rubber trees (*Hevea brasiliensis*) (Euphorbiaceae) in Papua New Guinea.
References. Hawkeswood (2003).

Pseudotaenia Kerremans, 1903 (Pl. 82K–M)

Type species. *Chalcotaenia salamandra* Thomson 1879.
Austrophorella Toyama, 1986, not Kerremans, 1903.
Australian species. Seven.
Distribution. Restricted to Australia, with species from NSW, QLD, SA and WA.
Biology. Adults have usually been found on foliage of various species of *Acacia* (Fabaceae) and larvae usually in *Acacia* wood.
References. Hawkeswood (1983); Toyama (1986); Pullen (1987); Bellamy (2006).

Iridotaenia Deyrolle, 1864 (Pl. 80L)

Type species. *Chrysodema sumptuosa* Laporte & Gory, 1835.
Platymroczkowskia Hołyński 1997
Australian species. Three.
Distribution. The genus *Iridotaenia* occurs in the Oriental, Afrotropical and Australasian regions. Australian species are known from eastern Australia (VIC to northern QLD).
Biology. Adults have been found on foliage of *Eucalyptus* and *Xanthostemon* (Myrtaceae); larvae of *S. bellicosa* (Blackburn) have been collected in wood of *Mallotus paniculatus* (Euphorbiaceae).
References. Hołyński (1997, 2014b).

Paracupta Deyrolle, 1864 (Pl. 81E)

Type species. *Buprestis xanthochroa* Boisduval, 1835.
Chalcotaenia Deyrolle, 1864.
Australian species. One.
Distribution. Species of *Paracupta* are known from South-east Asia, Indonesia, Papua New Guinea, New Caledonia, the Solomon Islands, Vanuatu, Fiji and other Pacific islands, as well as Australia. *Paracupta* (*Chalcotaenia*) *lambertii* (Laporte & Gory) is known from NSW.
References. Hołyński (2014a).

Metataenia Théry, 1923 (Pl. 80F)

Type species. *M. quadrimaculata* Théry, 1923.
Australian species. Two.
Distribution. *Metataenia* occurs in the Oriental and Australasian regions.
Biology. *Metataenia aurofoveata* (Saunders), placed in *Paracupta* in the most recent catalogue (Bellamy 2008a), occurs in QLD, WA and Papua New Guinea, while *M. meecki* (Kerremans) occurs in QLD, Papua New Guinea and New Caledonia.
References. Hołyński (2014c).

Buprestinae Leach, 1815

Tribe Maoraxiini Hołyński, 1984

Maoraxia Obenberger, 1937 (Pl. 80E)

Type species. *Maoriella novaezeelandiae* Obengerger, 1924.
Maoriella Obenberger, 1924, not Attems, 1903.
Australian species. Three.
Distribution. Species of *Maoraxia* are known from the Philippines, New Caledonia, the Solomon Islands, Fiji, Tonga, New Zealand and Australia. *Maoraxia auroimpressa* (Carter) occurs in NSW and QLD, *M. roseocuprea* Bellamy & Peterson on Lord Howe I. and *M. storeyi* Williams & Bellamy in northern QLD.
Biology. Adults *of M. auroimpressa* (Carter) have been recorded from foliage of *Elaeodendron* (Celastraceae), *Podocarpus* (Podocarpaceae), *Acronychia* (Rutaceae), *Schizomeria* (Cunoniaceae), *Syzygium* (Myrtaceae) and *Heritiera* (Sterculiaceae), and were found emerging from *Alectryon* and *Guioa* (Sapindaceae). Larvae have been found in *Elaeodendron australe*.

References. Bellamy & Williams (1985); Bílý & Volkovitsh (2005).

Tribe Julodimorphini Kerremans, 1902

Julodimorpha Gemminger & Harold, 1869 (Pl. 80A)

Type species. *Stigmodera bakewelli* White, 1859.
Australian species. Two.
Distribution. *Julodimorpha bakewelli* (White) is known from NSW, VIC and SA, while *J. saundersi* Thomson occurs in southern WA.
Biology. Adults of *J. saundersi* have been observed ovipositing in *Calothamnus* (Myrtaceae) (Hawkeswood and Peterson 1982). Larvae are apparently free-living in soil and the larva described by Bílý *et al.* (2013) was found in the soil within a plant assemblage including various Proteaceae and Myrtaceae, including a species of *Calothamnus*. The unusual behaviour of male *Julodimorpha*, who attempt copulation with 'stubby' beer bottles, was described by Gwynne & Rentz (1983) and commented on by Hawkeswood (2005).
References. Bellamy & Weir (2008).

Tribe Nascionini Hołyński, 1988

Nascio Laporte & Gory, 1837 (Pl. 80M)

Type species. *Buprestis vetusta* Boisduval, 1835.
Australian species. Four.
Distribution. Three *Nascio* species occur along the east coast and one is from southern WA.
Biology. *Nascio* species have been associated with a variety of species of *Eucalyptus* and some related Myrtaceae; adults feed on the foliage and larvae have been recorded from seven *Eucalyptus* species, one *Corymbia* and one *Metrosideros*.
References. Turner & Hawkeswood (1995b).

Nascioides Kerremans, 1903 (Pl. 81F, 83H)

Type species. *Stigmodera parryi* Hope, 1843.
Australian species. 18.
Distribution. Australian *Nascioides* occur throughout the eastern part of the continent, from TAS to northern QLD, and in New Caledonia and New Zealand.
Biology. The common *Nascioides parryi* (Hope) is associated primarily with *Eucalyptus* species but is also known to breed in *Leptospermum*. Larvae of the Tasmanian *N. quadrinotatus* (Van de Poll) occur in *Nothofagus cunninghamii*, while the New England *N. tillyardi* (Carter) breeds in *Nothogagus moorei*. Larval host plants of *N. falsomultisemus* Williams include species of *Doryphora* (Monimiaceae), *Ceratopetalum* and *Schizomeria* (Cunionaceae) and *Sloanea* (Elaeocarpaceae).
References. Williams (1985); Williams & Watkins (1985); Hawkeswood (2007b).

Tribe Epistomentini Levey, 1978a

Araucariana Levey, 1978a (Pl. 77G)

Type species. *A. queenslandica* Levey, 1978a.
Australian species. One.
Distribution. *Araucariana queenslandica* Levey occurs in QLD
Biology. *A. queenslandica* breeds in *Lophozonia cunninghamii* (Araucariaceae).
References. Levey (1978a).

Cyrioides Carter, 1920 (Pl. 78G)

Type species. *C. sexspilota* Carter, 1920.
Cyria Serville in Dejean, 1833, not Leach, 1818.
Cyriola Strand, 1928.
Xenocyria Obenberger, 1947.
Australian species. Six.
Distribution. Four species of *Cyrioides* occur in eastern Australia and two in southern WA.
Biology. Most species occur in *Banksia* species (Proteaceae), but larvae of the northern QLD species *C. cincta* (Carter) have been found in *Davidsonia* (Cunionaceae).
References. Levey (1978a).

Diadoxus Saunders, 1868 (Pl. 79L)

Type species. *Buprestis scalaris* Gory & Laporte, 1837.
Australian species. Three.
Distribution. All three species of *Diadoxus* occur in the Murray–Darling Basin (NSW, VIC, SA), but *D. erythrurus* (White) also occurs in coastal VIC, NSW and QLD.
Biology. These beetles are associated primarily with Cupressaceae (*Callitris* and introduced *Cupressus*) and Casuarinaceae, although a few other trees have been recorded as larval host plants.
References. Hadlington & Gardner (1959); Levey (1978a).

Tribe Buprestini Leach, 1815

Buprestis Linnaeus, 1758 (Pl. 79I)

Type species. *Buprestis octoguttata* Linnaeus, 1758.
Cypriacis Casey, 1909.
Australian species. One (introduced).
Distribution. The genus *Buprestis* occurs primarily in the Holarctic, Neotropical and Oriental regions. *Buprestis aurulenta* Linnaeus, is known from NSW and VIC, where it was introduced from North America.
Biology. Larvae always occur in conifers of various kinds, including species of *Pinus*, *Abies*, *Pseudotsuga*, *Picea* and *Larix*.

Burnsiellus Levey & Bellamy, 2013 (Pl. 78A–B)

Type species. *Neobuprestis trisulcata* Carter, 1932.
Australian species. Four.

Distribution. *Burnsiellus* includes one species from northern QLD and three from NSW, VIC, SA and WA.

Biology. An adult of *B. triculcata* (Carter) was observed emerging from a large *Acacia* species, and specimens of *B. marmoratus* (Blackburn) emerged from a billet of *Melaleuca lanceolate*.

References. Levey & Bellamy (2013).

Neobuprestis Kerremans, 1903 (Pl. 81I, K)

Type species. *Strigoptera australis* Blackburn, 1892.
Balthasarella Obenberger,1958 (subgenus).

Australian species. Three.

Distribution. Species of *Neobuprestis* are known from VIC, NSW, SA and WA.

References. Levey & Bellamy (2013).

Tribe Bubastini Obenberger, 1920

Bubastes Laporte & Gory, 1836 (Pl. 78J)

Type species. *Bubastes sphaenoida* Laporte & Gory, 1836.
Neraldus Théry, 1911.

Australian species. 40.

Distribution. *Bubastes* species are widely distributed in continental Australia.

Biology. Adults of this genus have been found on foliage of *Melaleuca* (Myrtaceae).

Euryspilus Lacordaire, 1857 (Pl. 79M)

Type species. *Eurybia chalcodes* Gory & Laporte, 1838.
Eurybia Gory & Laporte, 1838, not Latreille, 1809.
Neurybia Théry, 1911.

Australian species. Four.

Distribution. Three species have been recorded only from WA, and a fourh occurs in SA and southern WA.

Biology. Adults of *E. viridis* Carter feed on leaves of various sedges (Cyperaceae).

Tribe Stigmoderini Lacordaire, 1857

Calodema Gory & Laporte, 1838 (Pl. 78M)

Type species. *Stigmodera* (*Calodema*) *regale* Gory & Laporte, 1838.

Australian species. Three.

Distribution. Eastern Indonesia (Papua), Papua New Guinea, north-eastern and eastern Australia. Two Australian species are restricted to Cape York Peninsula, while *C. regalis* (Gory & Laporte) occurs from Cape York to northern NSW.

Biology. Adults occur on the flowers of several tree species, primarily in the families Myrtaceae, Proteaceae and Rutaceae. Larvae apparently belonging to the New Guinean *Calodema ribbei* Neervoort van de Poll were found in stems and branches of a rainforest tree called Tafaia.

References. Gardner (1989); Nylander (2008).

Calotemognatha Peterson, 1991 (Pl. 78D–E)

Type species. *Stigmodera laevicollis* Saunders, 1868.

Australian species. Three.

Distribution. Known from VIC to southern WA.

Biology. Adults have been found primarily on flowers of *Eucalyptus* and *Melaleuca* (Myrtaceae).

Note. Because the only character separating this genus from *Temognatha* involves a feature of the ovipositor not always visible without dissection, it is possible that the status of this genus or at least its limits may change, as more female genitalia are examined.

References. Peterson (1991); Peterson & Bellamy (2000).

Castiarina Gory & Laporte, 1838 (Pl. 78K–L, N, 83G, J, M–N)

Type species. *Stigmodera pertii* Gory & Laporte, 1838.
Hypostigmodera Blackburn, 1892.
Archaeozodes Obenberger, 1920.
Petersonia Hawkeswood, 2007.

Australian species. 475.

Distribution. *Castiarina* species are widely distributed throughout Australia.

Biology. Adults visit the flowers of a large number of plants, and larval hosts include species of *Acacia* (Fabaceae), *Cleroderndrum* (Verbenaceae), *Dicrastylis* (Lamiaceae), *Dodonaea* (Sapindaceae), *Eucalyptus* (Myrtaceae) and *Muellerina* (Loranthaceae). Species of *Castiarina* are often aposematically coloured and form part of a mimicry ring with species of Lycidae and several other families; these and other members of the family contain bitter chemicals called buprestins (Brown *et al.* 1985; Moore & Brown 1985, 1989).

Note. Males of two unusual species, *C. variegata* (Blackburn) from NSW and QLD and *C. shelleybarkeri* Nylander (Nylander 2006) from Papua New Guinea (Pl. 78L, N), have bipectinate antennae The former was originally placed in a new genus *Hypostigmodera* (Blackburn 1892) but returned to *Castiarina* by Gardner (1989). The latter was made the type of a new genus *Petersonia* by Hawkeswood (2007a) but returned to *Castiarina* by Bellamy & Nylander (2007).

References. Carnaby (1986); Gardner (1989); Barker (1979, 2006).

Metaxymorpha Parry, 1848 (Pl. 80B)

Type species. *Stigmodera* (*Metaxymorpha*) *grayii* Parry 1848.
Semnopharus van de Poll, 1886.

Australian species. Five.

Distribution. Eastern Indonesia (Papua), Papua New Guinea, northern and eastern Australia. Australian species occur primarily in north QLD, with one in NT and another extending into northern NSW.

Biology. Adults are known from the flowers of several different Myrtaceae, Proteaceae, Elaeocarpaceae, Sapindaceae, Rutaceae and Rubiaceae. Larvae of *M. gloriosa* Blackburn occur in *Casuarina cunninghamiana* (Casuarinaceae) and species of *Cyclophyllum* (Rubiaceae), *Guioa* and *Sarcopteryx* (Sapindaceae).
References. Gardner (1989); Nylander (2008).

Stigmodera Eschscholtz, 1829 (Pl. 82F, H)

Type species. *Buprestis macularia* Donovan, 1805.
Australian species. Seven.
Distribution. Distributed primarily in the southern part of the continent, from VIC, NSW, SA, and southern WA, with three species extending into QLD.
Biology. Larvae occur in species of *Agonis*, *Melaleuca* and *Leptospermum* (Myrtaceae) and adults visit the flowers of these and other myrtaceous genera.
References. Gardner (1989).

Temognatha Solier, 1933 (Pl. 83C)

Type species. *Buprestis variabilis* Donovan, 1805.
Stigmoderoides Théry, 1937.
Australian species. 85+.
Distribution. Widely distributed in Australia.
Biology. Adults have been taken on the flowers of several different Myrtaceae, Casuarinaceae, Pittosporaceae (*Bursaria*) and Loranthaceae (*Nuytsia*), and larval host plants include members of the Casuarinaceae and Myrtaceae. Oviposition and preoviposition behaviour was discussed by Peterson (1996).
References. Carnaby (1986); Gardner (1989).

Tribe Pterobothrini Volkovitsh, 2001

Neobubastes Blackburn, 1892 (Pl. 81G)

Type species. *N. aureocincta* Blackburn, 1892.
Castelnaudia Obenberger 1923, not Tschitscherine, 1891.
Castelnaudina Obenberger, 1924.
Eububastes Obenberger, 1928.
Australian species. Four.
Distribution. Three species are known only from WA, while *N. aureocincta* Blackburn occurs also in SA, VIC, NT and QLD.
Biology. Adults of *Neobubastes nickerlli* (Obenberger) have been recorded feeding on foliage of *Acacia* (Fabaceae), and specimens of *N. obscura* Bílý & Powell have been reared or cut from trunks of two species of *Melaleuca*.
References. Bílý & Powell (2017).

Tribe Uncertain

Buprestina Obenberger, 1923

Type species. *B. prosternalis* Obenberger, 1923.
Australian species. One.
Distribution. The only included species was described from QLD (no exact locality).
Note. *Buprestina prosternalis* Obenberger appears to be closely related to *Notobubastes orientalis* Carter and these two genera may not be distinct (in which case the Obenberger name would have precedence). The two genera are keyed out in the same couplet in the generic key.
References. Obenberger (1923).

Microcastalia Heller, 1891 (Pl. 80J–K)

Type species. *Castalia globithorax* Thomson, 1878.
Bubastodes Blackburn, 1892.
Australian species. Two.
Distribution. *Microcastalia globithorax* (Thomson) is known from VIC and SA, and *M. scintillans* (Carter) occurs in southern WA.

Notobubastes Carter, 1924 (Pl. 81D, 83F)

Type species. *N. occidentalis* Carter, 1924.
Australian species. Four.
Distribution. Three species occur in WA and a fourth in northern QLD.

Tribe Curidini Hołyński, 1988

Subtribe Curidina Hołyński, 1988

Selagis Dejean, 1836 (Pl. 82J)

Type species. *Buprestis caloptera* Boisduval, 1835.
Curis Gory & Laporte, 1838.
Neocuropsis Obenberger, 1923
Australian species. 24.
Distribution. *Selagis* species are widely distributed in Australia.
Biology. Adults occur on the flowers of Myrtaceae (*Angophora*, *Corymbia*, *Eucalyptus*, *Kunzea*, *Leptospermum*, *Melaleuca*, *Syzygium*), Casuarinaceae (*Allocasuarina*) and Pittoporaceae (*Bursaria*), while larvae have been recorded from *Allocasuarina* and *Eucalyptus*.

Subtribe Neocuridina Hołyński, 1988

Neocuris Saunders, 1868 (Pl. 81J)

Type species. *Anthaxia fortnumi* Hope, 1846.
Australian species. 30.
Distribution. *Neocuris* species are known from VIC, NSW, QLD, SA, VIC, WA and NT.
Biology. Adults may be found on the flowers of various Myrtaceae (*Angophora*, *Backhousia*, *Baeckea*, *Eucalyptus*, *Kunzea*, *Leptospermum*, as well as Pittosporaceae (*Bursaria*), Loranthaceae (*Nuytsia*), Asteraceae (*Helichrysum*), Apiaceae (*Actinotus*) and Epacridaceae (*Epacris*); larvae are known from a species of *Pultenea* (Fabaceae).
References. Volkovitsh & Hawkeswood (1987).

Pseudanilara Théry, 1911 (Pl. 82G)

Type species. *Pseudanilara roberti* Théry, 1911.
Neotorresita Obenberger, 1923
Australian species. 11.
Distribution. Species of *Pseudanilara* are known from TAS, NSW, QLD and WA.
Biology. Adults of *P. cupripes* (Macleay) have been found on flowers of *Corymbia* and *Syzygium* (Myrtaceae), while those of two other species were taken on leaves of dying eucalypts. Larvae of the same species have been recorded from *Ceratopetalum* (Cunoniaceae) and from the myrtaceous genera *Eucalyptus*, *Kunzea*, *Rhodomyrtus* and *Syzygium*.
References. Obenberger (1923); Williams (1985).

Theryaxia Carter, 1928 (Pl. 83D)

Type species. *Theryaxia suttoni* Carter, 1928.
Australian species. One.
Distribution. *Theryaxia suttoni* Carter, occurs in NSW and QLD.
Biology. Larvae have been recorded from *Callitris* (Cupressaceae).

Torresita Gemminger & Harold, 1869 (Pl. 83B, L)

Type species. *Buprestis chrysochloris* Laporte & Gory, 1837.
Plagiope Saunders, 1868, not Erichson 1845.
Australian species. Two.
Distribution. *Torresita cuprifera* (Kirby) and *T. parallela* Kerremans both occurr in the south-eastern part of the continent.
Biology. Adults of *T. cuprifera* have been recorded from the flowers of various Myrtaceae (*Angophora*, *Eucalyptus*, *Kunzea*, *Leptospermum*, *Syzygium*), Cunoniaceae (*Ceratopetalum*) and Rhamnaceae (*Alphitonia*). The only known larval host plant is *Rhodomyrtus psidioides* (Myrtaceae).
References. Williams (1985).

Subtribe Anilarina Bílý, 2000

Anilara Saunders, 1868 (Pl. 77H)

Type species. *Anthaxia adelaidae* Hope, 1846.
Australian species. 30.
Distribution. Species of *Anilara* are widely distributed in Australia.
Biology. Larval host plants of *Anilara* species include species of *Eucalyptus* and *Melaleuca* (Mytraeae) and *Flindersia* (Flindersiaceae).
References. Obenberger (1923); Bílý (2000).

Anthaxoschema Obenberger, 1923 (Pl. 77D)

Type species. *A. terraereginae* Obenberger, 1923.
Australian species. Two.
Distribution. *Anthaxoschema carteri* Théry is known from WA and *A. terraereginae* Obenberger from QLD.
References. Obenberger (1923); Théry (1945).

Australorhipis Bellamy, 1986 (Pl. 77E)

Type species. *A. aphanochila* Bellamy, 1986.
Australian species. One.
Distribution. *Austrlorhipis aphanochila* Bellamy is known only from the vicinity of Fowler's Gap in western NSW.
References. Bellamy (1986).

Barakula Peterson, 2000 (Pl. 77A)

Type species. *B. petersonorum* Peterson.
Australian species. One.
Distribution. *B. petersonorum* Peterson occurs in central QLD.
References. Peterson in Peterson & Bellamy (2000).

Notographus Thomson, 1879 (Pl. 81L)

Type species. *Notographus sulcipennis* Thomson, 1879.
Notograptus W. J. Macleay, 1872, not Günther, 1867.
Australian species. Four.
Distribution. Species of *Notographus* are known from QLD, VIC and WA.
References. Obenberger (1923).

Tribe Melobasini Bílý, 2000

Melobasis Gory & Laporte, 1837 (Pl. 80C, I)

Type species. *Buprestis cupriceps* Kirby, 1818.
Briseis Saunders, 1871 (subgenus).
Diceropygus Deyrolle, 1865 (subgenus).
Australian species. 133.
Distribution. *Melobasis* occurs in the Oriental and Australasian regions and is widely distributed in Australia.
Biology. Many species are associated as both larvae and adults with genera of Fabaceae, including *Acacia*, *Bossiacea*, *Jacksonia*, *Oxylobium*, *Pultenea* and *Viminaria*, but several other larval hosts have been recorded, including species of *Eucalyptus* and *Corymbia* (Myrtaceae), *Alphitonia* (Rhamnaceae), *Doryphora* (Atherospermataceae), *Nothofagus* (Fagaceae), and *Santalum* (Santalaceae).
References. Williams (1985); Volkovitsh & Hawkeswood (1994); Bílý (2000); Levey (2012, 2016).

Tribe Melanophilini Bedel, 1921

Merimna Saunders, 1868 (Pl. 80H)

Type species. *Chrysobothris atrata* Gory & Laporte, 1838.
Australian species. One.
Distribution. *Merimna atrata* (Gory & Laporte) ocurs in VIC, SA, QLD, NT, WA and Papua New Guinea.

Biology. *Merimna atrata* is commonly called the Australian fire beetle and is attracted in large numbers to so-called hot spots, areas of burning wood, on which they will eventually lay their eggs. The species has been associated with *Angophora hispida* (Myrtaceae) and a female was observed ovipositing into the smouldering bark of *Corymbia calophylla*. Larval host plants include the following Myrtaceae: *Baeckia frutescens*, *Corymbia calophylla*, *Eucalyptus gomphocephala*, *E. marginata* and *Melaleuca rhaphiophylla*. The ability to sense hot spots is due to the presence on the beetle abdomen of pairs of infrared receptors described and discussed in detail by H. and A. Schmitz and colleagues listed below.

References. Hawkeswood & Peterson (1982); Williams & Williams (1983); Schmitz *et al.* (2000, 2001); Schmitz and Trenner (2003); Mainz *et al.* (2004); Hawkeswood (2007e); Kitchin (2009); Schmitz *et al.* (2015); Schneider & Schmitz (2013, 2014).

Tribe Actenodini Kerremans, 1890

Belionota Eschscholtz, 1829 (Pl. 78F)

Type species. *B. saggitaria* Eschscholtz, 1829.

Australian species. One.

Distribution. The genus *Belionota* occurs in the Afrotropical, Oriental and Australasian regions. The only Australian species, *B. prasina* (Thunberg), occurs in China, India, and Sri Lanka, the Mascarines, Seychelles, the Philippines, South-east Asia, Indonesia and Papua New Guinea, as well as in northern QLD.

Biology. Larvae of *B. prasina* have been recorded from the genera *Mangifera* (Anacardiaceae), *Ceiba* (Bombacaceae) and *Delonix* (Fabaceae).

References. Théry (1929); Bellamy (1991).

Tribe Chrysobothrini Gory & Laporte, 1838

Chrysobothris Eschscholtz, 1829 (Pl. 77F)

Type species. *Buprestis chrysostigma* Linnaeus 1758.

Australian species. 17.

Distribution. *Chrysobothris* is a cosmopolitan genus, which is widely distributed in Australia.

Biology. Larval host plants include species of *Acacia* (Fabaceae), *Allocasuarina* (Casuarinaceae), *Rapanea* (Myrsinaceae) and *Backhousia* (Myrtaceae).

References. Benoit (1964); Bílý (1975); Williams (1985); Hawkeswood & Turner (1997).

Subfamily Agrilinae Laporte, 1835

Tribe Coraebini Bedel, 1921

Subtribe Ethoniina Majer, 2001

Ethonion Kubáň, 2001 (Pl. 79G)

Type species. *Buprestis fissiceps* Kirby, 1818.
Ethon Lacordaire, 1857, not Gory & Laporte, 1839.

Australian species. Eight.

Distribution. Species of *Ethonion* are distributed throughout the southern part of Australia.

Biology. *Ethonion* species are almost always associated with Fabaceae. Adults have been found on flowers of *Davesia*, *Dillwynia*, *Jacksonia*, *Pultenea*, as well as species of *Leptospermum* (Myrtaceae). Larvae form galls in *Dillwynia* and *Pultenea* species.

References. Majer in Kubáň *et al.* (2001); Kubáň in Kubáň *et al.* (2001).

Subtribe Cisseisina Majer, 2001

Diphucrania Dejean, 1833 (Pl. 79B)

Type species. *Buprestis leucosticta* Kirby, 1818.
Cisseis Gory & Laporte, 1839.
Ethon Gory & Laporte, 1939.

Australian species. 112.

Distribution. *Diphucrania* is widely distributed throughout Australia and also occurs in Papua New Guinea, Indonesia, New Caledonia, the Solomon Islands and the Philippines; a single species has been introduced into Spain.

Biology. Adults occur on the foliage of several different Fabaceae and larval host plants include members of the genera *Bossiaea*, *Dillwynia*, *Pultenea* and several *Acacia* species (Fabaceae), as well as *Xanthorrhoea* (Xanthorrhoeaceae) and *Elattostachys* (Sapindaceae).

References. Turner & Hawkeswood (1995a); Majer in Kubáň *et al.* (2001); Hooper *et al.* (2010).

Hypocisseis Thomson, 1879 (Pl. 80G)

Type species. *Hypocisseis laticornis* Thomson, 1879.
Maschalix Waterhouse, 1887.
Cisseoides Kerremans, 1893.

Australian species. 10.

Distribution. *Hypocisseis* occurs in the Philippines, Indonesia and Papua New Guinea and is relatively widely distributed in Australia.

Biology. Adults are known from the foliage of *Alphitonia* (Rhamnaceae), *Amyema* (Loranthaceae), *Maytenus* (Celastraceae), *Casuarina* (Casuarinaceae), *Leptospermum* (Myrtaceae) and *Hakea* (Proteaceae). The larval host plant of *H. suturalis* (Saunders) is *Exocarpos cupressiformis* (Santalaceae).

Neospades Blackburn, 1887 (Pl. 81A)

Type species. *Coraebus chrysopygius* Germar, 1848.

Australian species. 11.

Distribution. *Neospades* occurs in VIC, NSW QLD, WA and NT.

Biology. Adults have been found on flowers or foliage of various plants, including species of *Acacia*, *Atylosia*, *Cantharospermum*, *Jacksonia* and *Tephrosia* (Fabaceae),

Hibiscus and *Sida* (Malvaceae), *Solanum* (Solanaceae), and *Corymbia* and *Eucalyptus* (Myrtaceae). There are no larval host records.

References. Hawkeswood (2007d).

Pachycisseis Théry in Carter, 1929 (Pl. 81M)

Type species. *Ethon bicolor* Gory & Laporte, 1839.

Australian species. One.

Distribution. *Pachycisseis bicolor* (Gory & Laporte), the single included species, is known from NSW, VIC SA and NT.

Biology. Adults of *P. bicolor* have been recorded from the flowers and leaves of *Acacia* (Fabaceae) and *Leptospermum* (Myrtaceae).

References. Carter (1929).

Stanwatkinsius Barker & Bellamy, 2001 (Pl. 82D)

Type species. *Cisseis perplexa* Blackburn, 1891.

Australian species. 16.

Distribution. *Stanwatkinsius* is widely distributed in Australia.

Biology. Adults of this genus occur on the foliage of *Allocasuarina* (Casuarinaceae) and Proteaceae in the genera *Grevillea* and *Hakea*.

References. Barker & Bellamy (2001).

Aaaba Bellamy, 2002 (Pl. 77I)

Type species. *Alcinous nodosus* Deyrolle, 1865.

Alcinous Deyrolle, 1865, not Costa, 1812.

Australian species. Two.

Distribution. The genus *Aaaba* is restricted to Australia, *A. fossicollis* (Kerremans) and *A. nodosus* (Deyrolle), both occurring in VIC, NSW and QLD.

Biology. Adults of the two Australian species occur on the leaves and canes of *Rubus* spp. (Rosaceae) and larvae have also been recorded from *Rubus*.

References. Bellamy (2002, 2003).

Subtribe Toxoscelina Majer, 2001

Toxoscelus Deyrolle, 1865 (Pl. 83E)

Type species. *T. undatus* Deyrolle, 1865.

Australian species. Two.

Distribution. *Toxoscelus* species occur throughout the eastern Palaearctic, Oriental and Australasian regions *Toxoscelus eylandtus* Bellamy & Peterson and *T. queenslandicus* Bellamy & Peterson are known from northern NT and QLD, respectively.

Biology. Little is known about the biology of this group. The Japanese species *T. nakajumai* Ohmomo is known to be associated with *Distylium* (Hamamelidaceae).

References. Bellamy & Peterson (2000b); Majer in Kubáň *et al.* (2001); Ohmomo (2002).

Subtribe Synechocerina Majer, 2001

Synechocera Deyrolle, 1865 (Pl. 83I)

Type species. *Buprestis deplana* Fabricius, 1801.

Australian species. 10.

Distribution. Species of *Synechocera* are known from TAS, VIC, NSW, SA, WA, QLD and Indonesia (Ambon).

Biology. Adults occur on foliage of *Xanthorrhoea* (Xanthorhoeaceae), *Leptospermum* (Myrtaceae) and *Gahnia* (Cyperaceae), while *Gahnia* sp. is also recorded as a larval host plant.

References. Bellamy (1987a), Majer in Kubáň *et al.* (2001).

Subtribe Uncertain

Dinocephalia Obenberger, 1923 (Pl. 79J)

Type species. *D. gigantea* Obenberger, 1923.

Pseudosynechocera Obenberger, 1943.

Australian species. Seven.

Distribution. The genus *Dinocephalia* is widely distributed in Australia.

Biology. Adults have been recorded from the foliage of *Casuarina* and *Allocasuarina* (Casuarinaceae), *Dillwynia* and some related Fabaceae, and *Leptospermum* (Myrtaceae). Larvae of *D. cyanipennis* (Blackburn) live in galls of *Allocasuarina distyla*.

References. Obenberger (1923); Bellamy (1988).

Meliboeithon Obenberger, 1920 (Pl. 80D)

Type species. *M. fissus* Obenberger, 1920.

Australian species. Six.

Distribution. The genus *Meliboeithon* is known from VIC, NSW, QLD and WA.

Biology. Adults have been recorded feeding on flowers of *Juncus* (Juncaceae) and one species was collected from swellings on the lower parts of stems of *Oxylobium* (Fabaceae).

References. Bellamy (1988); Hawkeswood (2006).

Paracephala Saunders, 1868 (Pl. 81H)

Type species. *Agrilus pistacinus* Hope, 1846.

Australian species. Seven.

Distribution. Species of *Paracephala* are known from VIC, NSW, SA, QLD and WA.

Biology. Adults are usually found on flowers of grasses, such as *Chrysopogon* and *Stipa* (Poaceae).

References. Bellamy (1988).

Tribe Agrilini Laporte, 1835

Agrilus Curtis, 1825 (Pl. 77J)

Type species. *Buprestis viridis* Linnaeus, 1758.

Agrartus Curletti, 2001 (subgenus).

Pinarinus Curletti, 2001 (subgenus).

Australian species. 54.

Distribution. The genus *Agrilus* is cosmopolitan in distribution.

Biology. Adults of many species occur on foliage of *Acacia* (Fabaceae) and Casuarinaceae. Larvae have been found in various *Acacia* species, but also in *Drypetes* (Euphorbiaceae), *Eucalyptus* and *Leptospermum* (Myrtaceae), *Allocasuarina* (Casuarinaceae), *Hibiscus* (Malvaceae) and *Citrus* (Rutaceae). *Agrilus hyperici* (Creutzer) was introduced into Australia in 1940 for the control of St. John's wort (*Hypericum perforatum*, Clusiaceae).

References. Williams (1985); Volkovitsh & Hawkeswood (1990); Briese (1991); Curletti (2001).

Sambus Deyrolle, 1865 (Pl. 82I)

Type species. *S. lafertei* Deyrolle, 1865.

Australian species. One.

Distribution. The genus *Sambus* occurs in the Oriental, Afrotropical and Australasian regions. The single Australian species, *S. australis* Bellamy & Peterson occurs in northern QLD.

Biology. Adults of a Philippine species, *S. isis* Jackman, were associated with a species of *Ficus*.

References. Jackman (1987); Bellamy & Peterson (2000a).

Tribe Aphanisticini Jacquelin DuVal, 1863

Subtribe Aphanisticina Jacquelin DuVal, 1863

Aphanisticus Latreille, 1829 (Pl. 77K)

Type species. *Buprestis emarginatus* Olivier, 1790.

Australian species. Three.

Distribution. *Aphanisticus* species occur in the Palaearctic, Afrotropical, Oriental and Australasian regions. The three Australian species occur mainly in northern QLD.

Biology. In Europe, species in this genus are known to develop in stlks of grasses and rushes. *Aphanisticus cochinchinae seminulum* Obenberger is a well known pest of sugarcane in Hawaii, and the species has been introduced into Fiji, Florida, Texas, Honduras and Panama, but has not been reported from Australia.

References. Chang & Oto (1984); Bílý (1986, 1999); Heu (1987); Wellso & Jackman (1995); Peck & Thomas (1998); MacRae & Nelson (2003); Bellamy (2007).

Endelus Deyrolle, 1865 (Pl. 79C)

Type species. *E. cupido* Deyrolle, 1865.

Australian species. Two.

Distribution. Endelus species are known from the Oriental and Australasian regions. The Australian species of *Endelus* occur in QLD.

Biology. *Endelus* species are known to breed in ferns, but no hosts have been recorded in Australia.

References. Bellamy (2007); Kalashian (2013).

Subtribe Anthaxomorphina Hołyński, 1993

Anthaxomorphus Deyrolle, 1865 (Pl. 77C)

Type species. *A. papuanus* Deyrolle, 1865.

Australian species. One.

Distribution. The genus *Anthaxomorphus* occurs in the Afrotropical, Oriental and Australasian regions. The single Australian species, *A. queenslandicus* Williams & Weir, occurs in northern QLD.

Biology. Nothing is known on the biology of the Australian *Anthaxomorphus*, but two South African species have been reared from the leaves of *Ficus* sp. (Bellamy 1987b).

References. Williams & Weir (1992); Hołyński (1993).

Subtribe Germaricina Cobos, 1979

Germarica Blackburn, 1887 (Pl. 79A)

Type species. *G. casuarinae* Blackburn, 1887.

Australian species. Three.

Distribution. The three described species of *Germarica* occur in eastern Australia.

Biology. Adults occur on the foliage of various Casuarinaceae. Although larvae have not yet been identified, it is possible that they feed in the small, needle-like branchlets of this tree, given the small size of the adults.

References. Obenberger (1923); Cobos (1979).

Tribe Trachyini Laporte, 1835

Subtribe Trachyina Laporte, 1835

Habroloma Thomson, 1864 (Pl. 79E)

Type species. *Buprestis nanus* Paykull, 1799.

Australian species. Seven.

Distribution. The genus *Habroloma* is widely distributed throughout the Palaearctic, Afrotropical, Oriental and Australasian regions.

Biology. Adults of *H. australis* (Macleay) have been observed on several occasions on the leaves of *Atylosia* (Fabaceae). Larvae of *H. socialis* (Lea) mine the leaves of *Kennedia* (Fabaceae). A species of *Habroloma* from India is known to have a close association with *Oecophylla* ants; larvae mine the leaves comprising the ant nest and adults are not killed by the ants but merely thrown out of the nest along with dead ants.

References. Hawkeswood (2007c); Bílý *et al.* (2008).

Trachys Fabricius, 1801 (Pl. 83K)

Type species. *Buprestis pygmaeus* Fabricius, 1801.

Australian species. One.

Distribution. The genus *Trachys* is widely distributed in the Palaearctic, Oriental, Afrotropical and Australasian regions with one species introduced into North America. *Trachys blackburni* Kerremans was described from NSW and also occurs in Papua New Guinea.

Biology. In Europe, species of *Trachys* mine leaves of several dicotyledonous plants, including members of the Betulaceae (*Carpinus, Corylus*), Caprifoliaceae (*Succisa, Scabiosa*), Lamiaceae (Stachys), Malvaceae (*Althaea*), Rosaceae (*Fragaria, Potentilla*), Salicaceae (*Salix*) and Ulmaceae (Ulmus). Several species of *Trachys* have been described from Australia, but all except *Trachys blackburni* have since been transferred to *Habroloma*.

References. Kerremans (1896); Bílý (1999).

Subtribe Pachyschelina Böving & Craighead, 1931

Hedwigiella Obenberger, 1941 (Pl. 79F)

Type species. *H. jureceki* Obenberger, 1941.

Australian species. One (introduced).

Distribution. The genus *Hedwigiella* contains three species: *H. laticeps* (Waterhouse) and *H. cbellamyi* Hespenheide from Central America and *H. jureceki* from Brazil, Argentina and Paraguay. The latter species occurs in QLD.

Biology. *Hedwigiella jureceki* Obenberger occurs naturally in Argentina, Brazil and Paraguay and has been introduced into Australia for the control of Cat's Claw Creeper, *Macfadyena unguiscati* (L.) A. H. Gentry (Bignoniaceae).

References. Dhileepan *et al.* (2012); Snow & Dhileepan (2014); Hespenheide (2014).

Pachyschelus Solier, 1833 (Pl. 81B)

Type species. *P. scutellatus* Solier, 1833.

Australian species. One?

Distribution. About 260 species of *Pachyschelus* occur in the New World from the eastern and southern United States to Argentina and 15 are known from India and south-eastern Asia to Indonesia. An undescribed species or described Oriental species of *Pachyschelus* has been seen from Broome, WA, collected in Nov. 1946 by 'WdB' and originally deposited in the J. G. Brooks collection (Australian National Insect Collection).

Biology. *Pachyschelus* larvae are known to mine the leaves of *Croton* spp. (Euphorbiaceae) and some species of Fabaceae.

References. Hespenheide & Kim (1992); Queiroz (2002); Hespenheide & Bellamy (2004); Grebennikov (2013).

Excluded genera

The following genera have been excluded from the above list and from the generic key, since they appear to be based on mislabelled type specimens or represent one-time introductions.

***Nesotrinchus* Obenberger, 1924** (Chrysochroinae: Poecilonotini). *Nesotrinchus australicus* (Kerremans, 1903). Is synonymous with *N. simondsi* Obenberger, 1924, from Fiji and the Australian type locality is probably in error (Bílý *et al.* 2009).

***Lampetis* Dejean, 1833** (Chrysochroinae: Dicerini: Dicerina: *Lampetis* generic group). *Lampetis* is an Oriental genus and it is likely that both *L. corruscans* (Carter) and *L. fastuosa* (Fabricius) are either based on mislabelled types or represent one-time introductions (Bellamy 2002).

***Anthaxia* Eschscholtz, 1829** (Buprestinae: Anthaxiini). *Anthaxia novaehollandiae* Obenberger, 1928 was probably based on a mislabelled type. The species is synonymous with *Anthaxia splendida* Chevrolat from South Africa and Zimbabwe (Bellamy 2008c).

***Amorphosoma* Laporte, 1835** (Agrilinae: Coraebini). *Amorphosoma tasmanicum* Germar is probably based on a mislabelled type. The species is synonymous with *A. penicillatum* (Klug) from Argentina and Brazil (Bellamy 2008d).

***Coraebus* Gory & Laporte, 1839** (Agrilinae: Coraebini). *Coraebus longipennis* Deyrolle, 1864, is based on a mislabelled type or one-time introduction (Bellamy 2002). The species occurs in Indonesia and is only doubtfully recorded from Australia.

***Polyonychus* Chevrolat, 1838** (Agrilinae: Coraebini). *Polyonychus mucidus* Chevrolat is an Indian species and the Australian record (as *Discoderes torridus* Blackburn, 1891) is based on a mislabelled type of a one-time introduction (Bellamy 2002).

Acknowledgments

We are grateful to Svatopluk Bílý, Mark Volkovitsh and Allen Sundholm for providing information on buprestid relationships and reviewing an early version of the key and to Svatopluk Bílý for providing an image of the holotype of *Anthaxomorpha terraereginae* Obenberger (Pl. 77D). Geoff Monteith shared information relating to the use of the ovipositor in buprestid classification, but we were unable to follow up on this. Justin Bartlet and Kunjothapam Dhileepan of Biosecurity Queensland, DAF provided specimens of *Hedwigiella jureceki* Obenberger.

References

Alexeev AV (1993) Jurassic and Lower Cretaceous Buprestidae (Coleoptera) from Eurasia. *Paleontological Journal* **27**(1A), 9–34.

Alexeev AV (1995) New taxa of metallic wood-boring beetles (Coleoptera, Buprestidae) from the Mesozoic of Russia, Kazakhstan, and Mongolia. *Paleontologicheskii Zhurnal* 1995 (4), 310–317 (in Russian; translation in *Paleontological Journal* **30** (3), 310–317, 1996).

Alexeev AV (2009) New jewel beetles (Coleoptera: Buprestidae) from the Cretaceous of Russia, Kazakhstan, and Mongolia. *Paleontologicheskii Zhurnal* **2009** (3), 42–45 (in Russian; translation in *Paleontological Journal* **43** (3), 277–281, 1996).

Baker WL (1972) Eastern Forest Insects. U. S. Department of Agriculture (Miscellaneous Publications No. 1175), Washington, D. C., 642 pp.

Barker S (1975) Revision of the genus *Astraeus* Laporte & Gory (Coleoptera: Buprestidae). *Transactions of the Royal Society of South Australia* **99**, 105–142.

Barker S (1979) New species and a catalopgue of *Stigmodera* (*Castiarina*) (Coleoptera: Buprestidae). *Transactions of the Royal Society of South Australia* **103**, 1–23.

Barker S (2006) *Castiarina: Australia's Richest Jewel Beetle Genus.* Australian Biological Resources Study, Canberra.

Barker S, Bellamy CL (2001) *Stanwatkinsius*, a new genus of Australian jewel beetles (Coleoptera: Buprestidae: Agrilinae) with a key to known species. *Transactions of the Royal Society of South Australia* **125**(1), 1–14.

Bellamy CL (1986) The higher classification of Australian Buprestidae, with the description of a new genus and species (Coleoptera). *Australian Journal of Zoology* **34**, 583–600. doi:10.1071/ZO9860583

Bellamy CL (1987a) A revision of the genus *Synechocera* Deyrolle (Coleoptera: Buprestidae: Agrilinae). *Invertebrate Taxonomy* **1**, 17–34. doi:10.1071/IT9870017

Bellamy CL (1987b) Two new species of *Anthaxomorphus* Deyrolle from southern Africa with a new synonym and a world checklist (Coleoptera: Buprestidae: Trachyinae). *Coleopterists Bulletin* **41**(3), 287–295.

Bellamy CL (1988) The classification and phylogeny of Australian Coraebini with a revision of the genera *Paracephala, Meliboeithon* and *Dinocephalia. Invertebrate Taxonomy* **2**, 413–453. doi:10.1071/IT9880413

Bellamy CL (1991) New species of *Phrixia* Deyrolle and *Belionota* Eschscholtz from Luzon, Philippines with checklists of both genera (Coleoptera: Buprestidae). *Coleopterists Bulletin* **45**(4), 341–349.

Bellamy CL (1997) Phylogenetic relationships of *Xyroscelis* (Coleoptera: Buprestidae. *Invertebrate Taxonomy* **11**, 569–574. doi:10.1071/IT94026

Bellamy CL (2002) *Zoological Catalogue of Australia. Volume 29.5. Coleoptera: Buprestoidea.* CSIRO Publishing and Australian Biological Resources Study, Canberra, ACT.

Bellamy CL (2003) An illustrated summary of the higher classification of the superfamily Buprestoidea (Coleoptera). *Supplementum* **10**, 1–197.

Bellamy CL (2006) Studies on the Australian Chalcophorini: a new genus for *Chalcophora subfasciata* Carter, 1916 and a review of the *Pseudotaenia* Kerremans, 1893 generic-group (Coleoptera: Buprestidae). *Zootaxa* **1206**, 23–46.

Bellamy CL (2007) The genera *Aphanisticus* Latreille and *Endelus* Deyrolle in Fiji (Coleoptera: Buprestidae: Aphanisticini). In *Fiji Arthropods VII* (Eds. NL Evenhuis & DJ Bickel) pp. 13–25.

Bellamy CL (2008a) *A World Catalogue and Bibliography of the Jewel Beetles (Coleoptera: Buprestoidea), Volume 1: Introduction; Fossil Taxa; Schizopodidae; Buprestidae: Julodinae – Chrysochroinae: Poecilonotini.* pp. 1–625. Pensoft (Series Faunistica No. 76), Sofia.

Bellamy CL (2008b) *A World Catalogue and Bibliography of the Jewel Beetles (Coleoptera: Buprestoidea), Volume 2: Chrysochroinae: Sphenopterini through Buprestinae: Stigmoderini.* pp. 626–1259. Pensoft (Series Faunistica No. 77), Sofia.

Bellamy CL (2008c) *A World Catalogue and Bibliography of the Jewel Beetles (Coleoptera: Buprestoidea), Volume 3: Buprestinae: Pterobothrini through Agrilinae: Rhaeboscelina.* pp. 1260–1931. Pensoft (Series Faunistica No. 78), Sofia.

Bellamy CL (2008d) *A World Catalogue and Bibliography of the Jewel Beetles (Coleoptera: Buprestoidea), Volume 4: Agrilinae: Agrilina through Trachyini.* pp. 1932–2684. Pensoft (Series Faunistica No. 79), Sofia.

Bellamy CL (2009) *A World Catalogue and Bibliography of the Jewel Beetles (Coleoptera: Buprestoidea), Volume 5: Appendices, Bibliography, Indices. pp. 2685–3264.* Pensoft (Series Faunistica No. 80), Sofia.

Bellamy CL, Nelson GH (2002) 41. Buprestidae Leach 1815. In *American Beetles. Volume 2. Polyphaga: Scarabaeoidea through Curculionoidea.* (Eds RH Arnett Jr, MC Thomas, PE Skelley, JH Frank) pp. 98–112. CRC Press, Boca Raton, Florida.

Bellamy CL, Nylander U (2007) New genus-group synonymy in Stigmoderini (Coleoptera: Buprestoidea). *Coleopterists Bulletin* **61**(3), 423–427. doi:10.1649/0010-065X(2007)61[423:NGSISC]2.0.CO;2

Bellamy CL, Peterson M (2000a) Contribution to the taxonomy of Australian Buprestidae (Coleoptera) Part I. New synonymy, combinations, and names. *Folia Heyrovskyana* **8**(2), 73–100.

Bellamy CL, Peterson M (2000b) Contribution to the taxonomy of Australian Buprestidae (Coleoptera) Part II. Three new species and a new coraebine generic member of the Australian fauna. *Folia Heyrovskyana* **8**(2), 101–108.

Bellamy CL, Volkovitsh MG (2016) 18. Buprestoidea, 18.1. Buprestidae Leach, 1815. In *Handbuch der Zoologie/Handbook of Zoology. BandVolume IV Arthropoda: Insecta. Teilband/Part 38. Coleoptera, Beetles. Volume 1: Morphology and Systematics (Archostemata, Adephaga, Myxophaga, Polyphaga partim).* 2nd Edition. (Eds RG Beutel and RAB Leschen) pp. 543–552. Walter de Gruyter, Berlin.

Bellamy CL, Weir T (2008) The reinstatement of *Julodimorpha saundersii* Thomson 1879 (Coleoptera: Buprestidae) as a valid species. *Zootaxa* **1751**, 46–54.

Bellamy CL, Williams GA (1985) A revision of the genus *Maoraxia* with a new synonym in *Acmaeodera* (Coleoptera: Buprestidae). *International Journal of Entomology* **27**, 147–161.

Bellamy CL, Williams GA (1995) The first Australian *Paratrachys* (Coleoptera: Buprestidae), with comments on the higher classification of the genus. *Invertebrate Taxonomy* **9**, 1265–1276. doi:10.1071/IT9951265

Bellamy CL, Williams GA, Hasenpusch J, Sundholm A (2013) A summary of the published data on host plants and morphology of immature stages of Australian jewel beetles (Coleoptera: Buprestidae), with additional new records. *Insecta Mundi* **0293**, 1–172.

Benoit P (1964) Comparative morphology of some *Chrysobothris* larvae (Coleoptera: Buprestidae) of eastern Canada. *Canadian Entomologist* **96**(8), 1107–1117. doi:10.4039/Ent961107-8

Bílý S (1972) Larvae of *Ptosima flavoguttata* (Illiger) (Coleoptera, Buprestidae). *Acta Entomologica Bohemoslovaca* **69**, 18–22.

Bílý S (1975) Larvae of European species of the genus *Chrysobothris* Eschsch. (Coleoptera, Buprestidae). *Acta Entomologica Bohemoslovaca* **72**, 418–424.

Bílý S (1986) Descriptions of adult larvae of *Thrincopyge alacris* LeConte and *Aphanisticus cochinchinae seminulum* Obenberger (Coleoptera, Buprestidae). In *Entomological Papers presented to Y. Kurosawa on the Occasion of his Retirement, March 20, 1986.* (Ed. I Ueno) pp. 198–201. Coleopterist's Association of Japan, Tokyo.

Bílý S (1989) Descriptions of last instar larvae of *Polycesta porcata* and *Paratrachys hederae hederae* (Coleoptera: Buprestidae). *Acta Entomologica Bohemoslovaca* **86**, 61–66.

Bílý S (1992) The larva of *Trachys troglodytes* (Coleoptera: Buprestidae) and a key to the trachyine larvae of Scandinavia. *Entomologica Scandinavica* **23**(4), 415–418. doi:10.1163/187631292X00209

Bílý S (1999) Larvae of buprestid beetles (Coleoptera: Buprestidae) of Central Europe. *Acta Entomologica Musei Nationalis Pragae. Supplementum* 9, 45 pp., 33 pls.

Bílý S (2000) A new concept of Anthaxiini (Coleoptera: Buprestidae). *Folia Heyrovskyana* **8**(2), 109–114.

Bílý S, Powell M (2017) A new species of the genus *Neobubastes* Blackburn, 1892 from Western Australia (Coleoptera: Buprestidae: Buprestinae: Pterobothrini). *Zootaxa* **4329**(1), 86–90. doi:10.11646/zootaxa.4329.1.5

Bílý S, Volkovitsh MG (2001) Larvae of some tropical genera of Buprestidae (Coleoptera: Buprestidae). *Elytron* **15**, 49–73.

Bílý S, Volkovitsh MG (2003) Larvae of Australian Buprestidae (Coleoptera). Part 1. Genera *Austrophorella* and *Pseudotaenia. Acta Societatis Zoologicae Bohemoslovenicae* **67**, 99–114.

Bílý S, Volkovitsh MG (2005) Larvae of Australian Buprestidae (Coleoptera). Part 3. Genera *Maoraxia* and *Anthaxoschema* with a review of larval characters of known anthaxiine taxa. *Folia Heyrovskyana. Serie A* **13**(1–2), 7–26.

Bílý S, Fikáček M, Šipek P (2008) First record of myrmecophily in buprestid beetles: immature stages of *Habroloma myrmecophila* sp. nov. (Coleoptera: Buprestidae) associated with *Oecophylla* ants (Hymenoptera: Formicidae). *Insect Systematics & Evolution* **39**, 121–131. doi:10.1163/187631208788784084

Bílý S, Kubáň V, Volkovitsh MG (2009) A study on the tribe Poecilonotini, with a revision of the subtribe Nesotrinchina subtrib. nov. and the description of a new genus and species from Papua New Guinea (Coleoptera: Buprestidae: Chrysochroinae). *Acta Entomologica Musei Nationalis Pragae* **49**(2), 729–767.

Bílý S, Volkovitsh M, Peterson M (2013) Larvae of Australian Buprestidae (Coleoptera). Part 4. Genus *Julodimorpha. Zootaxa* **3637**(3), 341–354. doi:10.11646/zootaxa.3637.3.6

Blackburn T (1892) Further notes on Australian Coleoptera with descriptions of new genera and species. *Transactions of the Royal Society of South Australia* **15**, 207–261.

Bocak L, Barton C, Crampton-Platt A, Chesters D, Ahrens D, Vogler AP (2014) Building the Coleoptera tree-of-like for >8000 species: composition of public DNA data and fit with Linnaean classification. *Systematic Entomology* **39**, 97–110. doi:10.1111/syen.12037

Böving AG, Craighead FC (1931) An illustrated synopsis of the principal larval forms of the order Coleoptera. *Entomologica Americana* (N.S.) **11** (1930), 1–351.

Briese DT (1991) Current status of *Agrilus hyperici* (Coleoptera: Buprestidae) released in Australia in 1940 for the control of St. John's wort: Lessons for insect introductions. *Biocontrol Science and Technology* **1**(3), 207–215. doi:10.1080/09583159109355200

Brown VV, Jones AJ, Lacey MJ, Moore BP (1985) The chemistry of buprestins a and b. Bitter principles of jewel beetles (Coleoptera: Buprestidae). *Australian Journal of Chemistry* **38**, 197–206. doi:10.1071/CH9850197

Cai C, Ślipiński A, Huang D (2015) First false jewel beetle (Coleoptera: Schizopodidae) from the Lower Cretaceous of China. *Cretaceous Research* **52**(B), 490–494.

Carnaby K (1986) *Jewel Beetles of Western Australia.* Privately published, Wilga, WA, 69 pp.

Carter HJA (1929) A checklist of the Australian Buprestidae. With tables and keys to sub-families, tribes, and genera by André Théry. *Australian Zoologist* **5**, 265–304.

Chang V, Oto AK (1984) A new insect pest of sugarcane found in Hawaii. *Annual Report of the Hawaiian Sugar Planters' Association* **1984**, 36–37.

Cobos A (1957) Un genero y does nuevas especies de Buprestidae de Nueva Guinea. *Coleopterists Bulletin* **10**(6), 91–96.

Cobos A (1978) Ensayo sobre los generous de la subfamilia Polycestinae (Coleoptera, Buprestidae). Parte I. *Eos (Washington, D.C.)* **54**, 15–94.

Cobos A (1979) Revisión de la subfamilia Trachyinae a niveles supraespecificos (Coleoptera, Buprestidae). *Acta Entomologica Bohemoslovaca* **76**, 414–430.

Cobos A (1980) Ensayo sobre los géneros de la subfamilia Polycestinae (Coleoptera, Buprestidae). Parte II. *Eos (Washington, D.C.)* **55–56**, 23–94.

Cobos A (1986) *Fauna Iberica de Coleopteros Buprestidae.* Consejo Superior de Investigaciones Cientificas, Madrid, 364 pp.

Crowson RA (1982) On the dryopoid affinities of Buprestidae. *Coleopterists Bulletin* **36**, 22–25.

Curletti G (2001) The genus *Agrilus* in Australia. *Jewel Beetles* **9**, 76 pp.

Dhileepan K, Taylor DBJ, Lockett C, Treviño M (2012) Cat's claw creeper leaf-mining jewel beetle *Hylaeogena jureceki* Obenberger (Coleoptera: Buprestidae), a host-specific biological control agent for *Dolichandra unguis-cati* (Bignoniaceae) in Australia. *Australian Journal of Entomology* **52**(2), 175–181. doi:10.1111/aen.12014

Dunstan B (1923) *Mesozoic Insects of Queensland. Part I. – Introduction and Coleoptera. Queensland Geological Survey Publication No. 273.* Queensland Department of Mines, Brisbane.

Evans AM, McKenna DD, Bellamy CL, Farrell BD (2015) Large-scale molecular phylogeny of metallic wood-boring beetles (Coleoptera: Buprestoidea) provides new insights into relationships and reveals multiple evolutionary origins of the larval leaf-mining habit. *Systematic Entomology* **40**(2), 385–400. doi:10.1111/syen.12108

Forbes WTM (1926) The wing folding patterns of the Coleoptera. *Journal of the New York Entomological Society* **34**, 42–115, pls. 7–18.

Forbes WTM (1942) The wing of the Schizopini (Coleoptera: Dascillidae). *Entomological News* **53**, 101–102.

Gardner JA (1989) Revision of the genera of the tribe Stigmoderini (Coleoptera: Buprestidae) with a discussion of phylogenetic relationships. *Invertebrate Taxonomy* **3**, 291–361. doi:10.1071/IT9890291

Good HG (1925) Wing venation of the Buprestidae. *Annals of the Entomological Society of America* **18**, 251–272, pls. 8–11.

Grebennikov VV (2013) Life in two dimensions or keeping your head down: Lateral exuvial splits in leaf-mining larvae of *Pachyschelus* (Coleoptera: Buprestidae) and *Cameraria* (Lepidoptera: Gracillariidae). *European Journal of Entomology* **110**(1), 165–172. doi:10.14411/eje.2013.024

Gwynne DT, Rentz DCF (1983) Beetles on the bottle: male buprestids mistake stubbies for females (Coleoptera). *Journal of the Australian Entomological Society* **22**, 79–80. doi:10.1111/j.1440-6055.1983.tb01846.x

Hadlington P, Gardner MJ (1959) *Diadoxus erythrurus* (White) (Coleoptera-Buprestidae) attack on fire-damaged *Callitris* trees. *Proceedings of the Linnean Society of New South Wales* **84**, 325–331.

Hawkeswood TJ (1983) Notes on the biology and distribution of a rare jewel beetle *Pseudotaenia waterhousei* (Van de Poll) (Coleoptera: Buprestidae). *Victorian Naturalist* **100**, 238–241.

Hawkeswood TJ (1985) The larva of *Diadoxus erythrurus* (White) (Coleoptera: Buprestidae). *Australian Entomological Magazine* **12**, 23–28.

Hawkeswood TJ (1988) A review of larval host records for twelve Australian Buprestidae (Coleoptera). *Giornale Italiano di Entomologia* **4**, 81–88.

Hawkeswood TJ (2003) Some notes on the biology and a new host plant of *Cyphogastra bruijni* Lansberge (Coleoptera: Buprestidae) in Papua New Guinea. *Spilopyra* **7**, 1–4.

Hawkeswood TJ (2005) Review of the biology and host-plants of the Australian jewel beetle *Julodimorpha bakewelli* (White, 1859) (Coleoptera: Buprestidae). *Calodema* **3**, 3–5.

Hawkeswood TJ (2006) Review of the biology of the Australian genus *Meliboeithon* Obenberger, 1920 (Coleoptera: Buprestidae). *Calodema* **8**, 3–4.

Hawkeswood TJ (2007a) *Petersonia* gen. n. (Coleoptera: Buprestidae) from Papua New Guinea and a redescription of *Petersonia shelleybarkeri* (Nylander, 2006). *Calodema Supplementary Paper* **9**, 1–5.

Hawkeswood TJ (2007b) Review of the biology of the Australian species of the genus *Nascioides* Kerremans, 1903 (Coleoptera: Buprestidae). *Calodema Supplementary Paper* **22**, 1–7.

Hawkeswood TJ (2007c) Review of the biology of two Australian species of the genus *Habroloma* Thomson, 1864 (Coleoptera: Buprestidae). *Calodema Supplementary Paper* **4**, 1–2.

Hawkeswood TJ (2007d) Review of the biology of the genus *Neospades* Blackburn, 1887 (Coleoptera: Buprestidae). *Calodema Supplementary Paper* **19**, 1–3.

Hawkeswood TJ (2007e) Review of the biology of the genus *Merimna* Saunders, 1868 (Coleoptera: Buprestidae). *Calodema* **9**, 12–13.

Hawkeswood TJ, Peterson M (1982) A review of larval host records for Australian jewel beetles (Coleoptera: Buprestidae). *Victorian Naturalist* **99**, 240–251.

Hawkeswood TJ, Turner JR (1997) The taxonomy, biology and distribution of the Australian jewel beetle *Chrysobothris saundersi* Macleay. *Spixiana* **20**(2), 141–149.

Hespenheide HA (1992) Clutch size, survivorship, and biology of larval *Pachyschelus psychotriae* Fisher (Coleoptera: Buprestidae). *Annals of the Entomological Society of America* **85**(1), 48–52. doi:10.1093/aesa/85.1.48

Hespenheide HA (2014) A reconsideration of *Hylaeogena* Obenberger, 1923 (Coleoptera: Buprestidae), with description of a new genus and new species from Mexico and Central America. *Coleopterists Bulletin* **68**(1), 21–30. doi:10.1649/0010-065X-68.1.21

Hespenheide HA, Bellamy CL (2004) The first Antillean *Pachyschelus*, and a new *Leiopleura*, from Hispanola (Coleoptera: Buprestidae). *Folia Heyrovskyana* **12**(2–3), 105–112.

Hespenheide HA, Kim CO (1992) Clutch size, survivorship, and biology of larval *Pachyschelus psychotriae* Fisher (Coleoptera: Buprestidae). *Annals of the Entomological Society of America* **85**(1), 48–52. doi:10.1093/aesa/85.1.48

Heu R (1987) *Aphanisticus cochinchinae seminulum* Obenberger. *Proceedings of the Hawaiian Entomological Society* **27** (1986), 7.

Hołyński RB (1988) Remarks on the general classification of Buprestidae Leach as applied to Maoraxiina. *Folia Entomologica Hungarica* **46**, 49–54.

Hołyński RB (1992) A review of the genus *Paratrachys* Snd. (Coleoptera: Buprestidae). *Annals of the Upper Silesian Museum, Entomology* **3** (1991), 115–136.

Hołyński RB (1993) A reassessment of the internal classification of the Buprestidae Leach (Coleoptera). *Crystal. Publications of the Natural Science Foundation at Göd. Series Zoologica* **1**, 1–42.

Hołyński RB (1997) *Mroczkowskia*-knot and the evolution of the subtribe Chrysochroina (Coleoptera: Buprestidae). *Annales Zoologici, Warszawa* **47**(1/2), 179–188.

Hołyński RB (2014a) Description of three new subgenera and two new species of *Paracupta* Deyr. (Coleoptera: Buprestidae). *Genus* **25**(3), 403–414.

Hołyński RB (2014b) A new species of the genus *Iridotaenia* Deyr. (Coleoptera: Buprestidae). *Genus* **25**(3), 387–391.

Hołyński RB (2014c) New subgenus and three new species of the genus *Metataenia* Théry (Coleoptera: Buprestidae). *Genus* **25**(3), 391–402.

Hołyński RB (2014d) Supplementary remarks on the taxonomic structure of the genus *Chrysodema* C.G. (Coleoptera: Buprestidae). *Genus* **25**(3), 373–375.

Hooper RJ, Wills A, Shearer BL, Sivasithamparam K (2010) A redescription and notes on biology of *Cisseis fascigera* Obenberger (Coleoptera: Buprestidae) on declining *Eucalyptus wandoo* in south-western Australia. *Australian Journal of Entomology* **49**, 234–244. doi:10.1111/j.1440-6055.2010.00755.x

Hörnschemeyer T, Wedmann S (1994) Fossile Prachkäfer (Coleoptera: Buprestidae: Buprestinae und Agrilinae) aus dem Mitteleozän der Grube Messel bei Darmstadt. Teil 1. *Courier Forschungsinstitut Senckenberg* **170**, 85–136.

Hunt T, Bergsten J, Levkanicova Z, Papadopoulou A, St. John O, Wild R, Hammond PM, Ahrens D, Balke M, Caterino MS, Gómez-Zurita J, Ribera I, Barraclough TG, Bocakova M, Bocak L, Vogler AP (2007) A comprehensive phylogeny of beetles reveals the evolutionary origins of a superradiation. *Science* **318**, 1913–1916. doi:10.1126/science.1146954

Jackman JA (1987) A new species of *Sambus* from Luzon Island, Philippines, with notes on species in other genera (Buprestidae: Coleoptera). *Coleopterists Bulletin* **41**(1), 27–33.

Jendek E (2001) A comparative study of the abdomen in the family Buprestidae (Coleoptera). *Acta Musei Moraviae. Scientiae Biologicae, Brno* **86**, 1–41.

Kalashian MYu (2013) Two new species of the genus *Endelus* Deyrolle, 1864 (Coleoptera: Buprestidae) from Malaysia and the Philippines with notes on the synonymy of some species of the subgenus *Endelus* s. str. *Caucasian Entomological Bulletin* **9**(1), 83–88.

Kerremans C (1893) Essai de groupement des Buprestidae. *Annales de la Société Entomologique de Belgique* **37**, 94–122.

Kerremans C (1896) Trachydes nouveaux. *Annales de la Société Entomologique de Belgique* **40**, 306–333.

Kerremans C (1906–1913) *Monographie des Buprestides - Vol. I–VII* - V. Dulan et Co., London.

Kitchin DR (2009) Notes on the biology of *Merimna atrata* (Gory & Laporte) (Coleoptera: Buprestidae). *Australian Entomologist* **36**(1), 1–2.

Kolibáč J (2000) Classification and phylogeny of the Buprestoidea (Insecta: Coleoptera). *Acta Musei Moraviae. Scientiae Biologicae, Brno* **85**, 113–184.

Kubáň V, Majer K, Kolibáč J (2001) Classification of the tribe Coraebini Bedel, 1921 (Coleoptera, Buprestidae, Agrilinae). *Acta Musei Moraviae. Scientiae Biologicae, Brno* **85**, 185–287.

Lawrence JF (1988) Rhinorhipidae, a new beetle family from Australia, with comments on the phylogeny of the Elateriformia. *Invertebrate Taxonomy* **2** (1987), 1–53.

Lawrence JF (1991) Buprestidae (Buprestoidea) (including Schizopodidae). In *Immature Insects. Volume 2.* (Ed. FW Stehr) pp. 386–388. Kendall/Hunt, Dubuque, Iowa.

Lawrence JF, Newton AF Jr (1995) Families and subfamilies of Coleoptera (with selected genera, notes, references and data on family-group names). In *Biology, Phylogeny, and Classification of Coleoptera. Papers Celebrating the 80th Birthday of Roy A. Crowson.* (Eds J Pakaluk and SA Ślipiński) pp. 779–1006. Muzeum I Instytut Zoologii Polska Akademia Nauk, Warszawa.

Lawrence JF, Ślipiński A (2013) *Australian Beetles: Morphology, Classification and Keys. Volume 1.* CSIRO Publishing, Collingwood, Victoria.

Lawrence JF, Ślipiński A, Seago AE, Thayer MK, Newton AF, Marvaldi AE (2011) Phylogeny of the Coleoptera based on morphological characters of adults and larvae. *Annales Zoologici* **61**(1), 1–217. doi:10.3161/000345411X576725

Levey B (1978a) A new tribe, Epistomentini, of Buprestidae (Coleoptera) with a redefinition of the tribe Chrysochroini. *Systematic Entomology* **3**, 153–158. doi:10.1111/j.1365-3113.1978.tb00111.x

Levey B (1978b) A taxonomic revision of the genus *Prospheres* (Coleoptera: Buprestidae). *Australian Journal of Zoology* **26**, 713–726. doi:10.1071/ZO9780713

Levey B (2012) A revision of the Australian species of the genus *Melobasis* Laporte & Gory 1837 (Coleoptera: Buprestidae). Part I (Introductory material, key to species groups and keys to species of the *thoracica, pusilla, formosa, propinqua* & *gloriosa* species-groups). *Zootaxa* **3464**, 1–107.

Levey B (2016) *Melobasis bilyi*, an interesting new species of the *M. pusilla* species-group from Queensland, Australia (Coleoptera: Buprestidae). *Acta Entomologica Musei Nationalis Pragae* **56**(1), 23–27.

Levey B, Bellamy CL (2013) A taxonomic revision of *Neobuprestis* Kerremans (Coleoptera: Buprestidae) with the description of a new genus and two new species. *Zootaxa* **3681**(3), 225–240. doi:10.11646/zootaxa.3681.3.2

MacRae TC, Nelson GH (2003) Distributional and biological notes on Buprestidae (Coleoptera) in North and Central America and the West Indies, with vaidation of one species. *Coleopterists Bulletin* **57**(1), 57–70. doi:10.1649/0010-065X(2003)057[0057:DABNOB]2.0.CO;2

Mainz T, Schmitz A, Schmitz H (2004) Variation in number and differentiation of the abdominal infrared receptors in the Australian "fire-beetle" *Merimna atrata* (Coleoptera, Buprestidae). *Arthropod Structure & Development* **33**, 419–430. doi:10.1016/j.asd.2004.06.004

Matthews EG (1985) *A Guide to the Genera of Beetles of South Australia. Part 4. Polyphaga: Byrrhoidea, Buprestoidea, Dryopoidea, Elateroidea, Cantharoidea, Derodontoidea, Bostrichoidea.* South Australian Museum (Special Education Bulletin Series, No. 7), Adelaide.

McKenna DD, Wild AL, Kanda K, Bellamy CL, Beutel RG, Caterino MS, Farnum CW, Hawks DC, Ivie MA, Jameson ML, Leschen RAB, Marvaldi AE, McHugh JV, Newton AF, Robertson JA, Thayer MK, Whiting MF, Lawrence JF, Ślipiński A, Maddison DR, Farrell BD (2015) The beetle tree of life reveals Coleoptera survived end Permian mass extinction to diversify during the Cretaceous terrestrial revolution. *Systematic Entomology* **40**, 35–60. doi:10.1111/syen.12093

Moore BP, Brown WV (1985) The buprestins: bitter principles of jewel beetles (Coleoptera: Buprestidae). *Journal of the Australian Entomological Society* **24**, 81–85. doi:10.1111/j.1440-6055.1985.tb00191.x

Moore BP, Brown WV (1989) Graded levels of chemical defence in mimics of lycid beetles of the genus *Metriorrhynchus* (Coleoptera). *Journal of the Australian Entomological Society* **28**, 229–233. doi:10.1111/j.1440-6055.1989.tb00889.x

Nelson GH (1981) A new tribe, genus, and species of North American Buprestidae with consideration of subfamilial and tribal categories. *Coleopterists Bulletin* **35**(4), 431–450.

Nelson GH, Bellamy CL (1991) A revision and phylogenetic re-evaluation of the family Schizopodidae (Coleoptera, Buprestoidea). *Journal of Natural History* **25**, 985–1026. doi:10.1080/00222939100770651

Nelson GH, Bellamy CL (2002) 40. Schizopodidae LeConte 1861. In *American Beetles. Volume 2. Polyphaga: Scarabaeoidea through Curculionoidea.* (Eds RH Arnett Jr, MC Thomas, PE Skelley and JH Frank) pp. 95–97. CRC Press, Boca Raton, Florida.

Nylander U (2006) The New Guinean species of the genus *Castiarina* Gory & Laporte (Coloeptera: Buprestidae, Stigmoderini). *Lambillionea. Revue Internationale d. Entomologie* **106**(Supplement 2), 3–13.

Nylander U (2008) Review of the genera of *Calodema* and *Metaxymorpha* (Coleoptera: Buprestidae: Stigmoderini). *Folia Heyrovskyana Supplementum* **13**, 84 pp.

Obenberger J (1923) Une série de nouveaux genres de Buprestides (Coleoptera). *Acta Entomologica Musei Nationalis Pragae* **1**(2), 13–44.

Obenberger J (1928) Opuscula Buprestologica I. Beiträge zur Kenntnis der Buprestiden (Col.). *Archiv für Naturgeschichte* **92** (A) 9–11 (1926), 1–350.

Ohmomo S (2002) Buprestid beetles (Coleoptera: Buprestidae) new to Japanese fauna (Part 1). *Entomological Review of Japan* **57**(2), 155–164.

Peck SB, Thomas MC (1998) A distributional checklist of the beetles (Coleoptera) of Florida. *Arthropods of Florida and Neighboring Land Areas* **16**, 1–180.

Peterson M (1991) A new southern Australian subgenus of *Temognatha* Solier (Coleoptera: Buprestidae: Buprestinae), with nomenclatural notes on Solier's generic name. *Records of the Western Australian Museum* **15**(1), 117–127.

Peterson M (1996) Aspects of female reproductive biology of two southwestern Australian *Temognatha* species (Coleoptera: Buprestidae). *Records of the Western Australian Museum* **18**, 203–208.

Peterson M, Bellamy CL (2000) Contribution to the taxonomy of Australian Buprestidae (Coleoptera) Part III. A new genus and species of buprestid from Queensland and comments about its placement. *Folia Heyrovskyana* **8**(2), 127–131.

Ponomarenko AG (1971) On the systematic position of some beetles from the Solenhofen shales of Bavaria. *Paleontologicheskii Zhurnal* **1971** (1), 67–81 (in Russian; translation in *Paleontological Journal* **1971** (1), 62–75.

Pullen KR (1987) *Pseudotaenia waterhousei* (V. d. Poll) (Coleoptera: Buprestidae) in New South Wales. *Australian Entomological Magazine* **14** (1, 2), 23–28.

Queiroz JM (2002) Distribution, survivorship and mortaliyy sources in immature stages of the Neotropical leaf miner *Pachyschelus coeruleipennis* Kerremans (Coleoptera: Buprestidae). *Brazilian Journal of Biology* **62**(1), 69–76. doi:10.1590/S1519-69842002000100009

Rees BE (1941) First instar larvae of *Buprestis rusticorum* (Kby.) and *Schizopus sallei* Horn, with notes on the classification of *Schizopus. Proceedings of the Entomological Society of Washington* **43**, 210–222, pl. 22–23.

Schmitz A, Schneider ES, Schmitz H (2015) Behaviour of the Australian 'fire-beetle' *Merimna atrata* (Coleoptera: Buprestidae) on burnt areas after bushfires. *Records of the Western Australian Museum* **30**, 001–011.

Schmitz H, Trenner S (2003) Electrophysiological characterization of the multipolar thermoreceptors in the "fire-beetle" *Merimnus atrata* and comparison with the infrared sensilla of *Melanophila acuminata* (both Coleoptera, Buprestidae). *Journal of Comparative Physiology. A, Neuroethology, Sensory, Neural, and Behavioral Physiology* **189**, 715–722. doi:10.1007/s00359-003-0447-6

Schmitz H, Schmitz A, Bleckmann H (2000) A new type of infrared organ in the Australian "fire-beetle" *Merimna atrata* (Coleoptera: Buprestidae). *Naturwissenschaften* **87**, 542–545. doi:10.1007/s001140050775

Schmitz H, Schmitz A, Bleckmann H (2001) Morphology of a thermosensitive multipolar neuron in the intrared organ of *Merimna atrata* (Coloeoptera, Buprestidae). *Arthropod Structure & Development* **30**, 99–111. doi:10.1016/S1467-8039(01)00028-7

Schneider ES, Schmitz H (2013) Bimodal innervation of the infrared organ of *Merimna atrata* (Coleoptera, Buprestidae) by thermo- and mechanosensory units. *Arthropod Structure & Development* **42**, 135–142. doi:10.1016/j.asd.2012.11.001

Schneider ES, Schmitz H (2014) Thermomechanical properties of the stimulus transducing cuticle in the infrared organ of *Merimna atrata* (Coleoptera, Buprestidae). *Journal of Morphology* **275**(9), 991–1003. doi:10.1002/jmor.20276

Snow EL, Dhileepan K (2014) The jewel beetle *(Hylaeogena jureceki*): a new biological control for cat's claw creeper (*Dolichandra unguis-cati*) in Queensland. In *Nineteenth Australasian Weeds Conference.* (Ed. M. Baker) pp. 50–54. Tasmanian Weed Society, Hobart.

Théry A (1929) Révision monographie des espéces du genre *Belionota. Annales de la Société Linnéenne de Lyon* **74**, 132–147.

Théry A (1945) Descriptions de deux Buprestides des faunes australienne et mélanésienne (Col.). *Bulletin de la Société Entomologique de France* **50**(4), 45–47.

Toyama M (1986) The buprestid genus *Chalcophorella* Kerremans and its related genera. *Entomological Papers Presented to Yoshihiko Kurasawa on the Occasion of his Retirement, March 20, 1986*, pp. 188–193.

Toyama M (1987) The systematic position of some buprestid genera (Coleoptera, Buprestidae). *Elytra* **15**, 1–11.

Turner JR, Hawkeswood TJ (1995a) A new larval host plant and other biological data for the Australian jewel beetle *Cisseis acuducta* (Kirby) (Coleoptera: Buprestidae). *Giornale Italiano di Entomologia* **7**, 273–277.

Turner JR, Hawkeswood TJ (1995b) A note on *Nascio vetutsta* (Boisduval) (Coleoptera: Buprestidae) from Australia. *Australian Zoologist* **30**(1), 65–67. doi:10.7882/AZ.1995.009

Turner JR, Hawkeswood TJ (1996) A note on the larval host plant and biology of the Australian jewel beetle *Astraeus crassus* Van de Poll (Coleoptera: Buprestidae). *Mauritiana* **16**(1), 75–79[Altenburg].

Volkovitsh MG (2001) The comparative morphology of antennal structures in Buprestidae (Coleoptera): evolutionary trends, taxonomic and phylogenetic implications. Part 1. *Acta Musei Moraviae. Scientiae Biologicae, Brno* **86**, 43–169.

Volkovitsh MG, Bílý S (2015) Larvae of Australian Buprestidae (Coleoptera). Part 5. Genera *Astraeus* and *Xyroscelis* with notes on larval characters of Australian polycestine taxa. *Acta Entomologica Musei Nationalis Pragae* **55**(1), 173–202.

Volkovitsh MG, Hawkeswood TJ (1987) The larva of *Neocuris gracilis* Macleay (Coleoptera: Buprestidae). *Zoologischer Anzeiger* **219**(5/6), 274–282.

Volkovitsh MG, Hawkeswood TJ (1990) The larvae of *Agrilus austrlasiae* Laporte & Gory and *Ethon affine* Laporte & Gory (Insecta: Coleoptera: Buprestidae). *Spixiana* **13**(1), 43–59.

Volkovitsh MG, Hawkeswood TJ (1993) The larvae of *Anilara antiqua* Thèry and *Anilara nigrita* Kerremans (Insecta, Coleoptera, Buprestidae). *Spixiana* **16**(2), 157–167.

Volkovitsh MG, Hawkeswood TJ (1994) The larvae of *Melobasis* (*Melobasis*) *vertebralis* Carter (Coleoptera: Buprestidae). *Giornale Italiano di Entomologia* **7**, 11–27.

Volkovitsh MG, Hawkeswood TJ (1999) The larva of *Prospheres aurantiopicta* (Laporte & Gory) with comments on the larval characteristics of polycestoid taxa (Insecta, Coleoptera, Buprestidae). *Mauritiana. Altenburg* **17**, 295–314.

Volkovitsh MG, Bílý S, Hasenpush J (2003) Larvae of Australian Buprestidae (Coleoptera). Part 2. Genus *Metaxymorpha. Folia Heyrovskyana* **11**(3–4), 203–216.

Walker MV (1938) Evidence of Triassic insects in the Petrified Forest National Monument, Arizona. *Proceedings of the United States*

National Museum **85**(3033), 137–141. doi:10.5479/si.00963801.85-3033.137

Wedmann S, Hörnschemeyer T (1994) Fossile Prachkäfer (Coleoptera: Buprestidae: Buprestinae und Agrilinae) aus dem Mitteleozän der Grube Messel bei Darmstadt.Teil 2. *Courier Forschungsinstitut Senckenberg* **170**, 137–187.

Wellso SG, Jackman JA (1995) The first record of *Aphanisticus cochinchinae seminulum* Obenberger (Coleoptera: Buprestidae), a potential sugarcane pest in the Western Hemisphere. *Coleopterists Bulletin* **49**(3), 287–288.

Williams GA (1985) New larval foods plants for some Australian Buprestidae and Cerambycidae (Coleoptera). *Australian Entomological Magazine* **12**, 41–46.

Williams G, Watkins S (1985) A new species of *Nascioides* Kerremans (Coleoptera: Buprestidae). *Journal of the Australian Entomological Society* **24**, 255–259. doi:10.1111/j.1440-6055.1985.tb00239.x

Williams G, Watkins S (1986) A revision of the genus *Xyroscelis* Saunders (Coleoptera: Buprestidae). *Journal of the Australian Entomological Society* **25**, 295–299. doi:10.1111/j.1440-6055.1986.tb01118.x

Williams GA, Weir TA (1987) Four new species and new records of Australian Mastogeniinae (Coleoptera: Buprestidae). *Journal of the Australian Entomological Society* **26**, 153–159. doi:10.1111/j.1440-6055.1987.tb00277.x

Williams GA, Weir TA (1988) Further new species of Australian Mastogeniinae (Coleoptera: Buprestidae). *Journal of the Australian Entomological Society* **27**, 179–181. doi:10.1111/j.1440-6055.1988.tb01518.x

Williams GA, Weir TA (1992) Two new species of *Anthaxomorphus* Deyrolle (Coleoptera: Buprestidae): Trachyinae) from the Australasian region. *Journal of the Australian Entomological Society* **31**, 227–230. doi:10.1111/j.1440-6055.1992.tb00494.x

Williams GA, Williams T (1983) A list of the Buprestidae (Coleoptera) of the Sydney Basin, New South Wales, with adult food plant records and biological notes on food plant associations. *Australian Entomological Magazine* **9**(6), 81–93.

Yu YL, Ślipiński A, Shih CG, Pang H, Ren D (2013) A new fossil jewel beetle (Coleoptera: Buprestidae) from the Early Cretaceous of Inner Mongolia, China. *Zootaxa* **3637**, 355–360. doi:10.11646/zootaxa.3637.3.7

36. TENEBRIONIDAE LATREILLE, 1802

Eric G. Matthews and John F. Lawrence

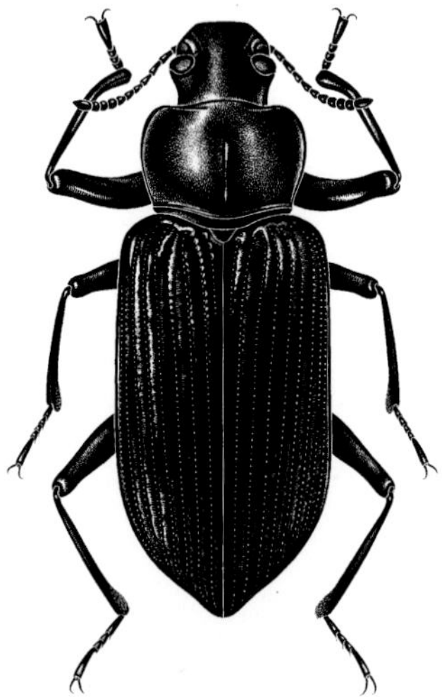

Fig. 36.1 *Promethis nigra* (Blessig)

Common names. Darkling beetles, Schwarzkäfer, жуки чернотелки.

Family synonyms. Alleculidae Laporte, 1840; Lagriidae Latreille, 1825; Nilionidae Oken, 1844, Pimeliidae Latreille, 1802; Tentyriidae Eschscholtz, 1831.

Introduction. Tenebrionidae is the seventh largest family of Coleoptera, with ~2300 genera and 20 000 species. The Australian fauna presently includes 213 genera and 1641 named species. The current higher classification of the family is based mainly on the morphological studies of Watt (1975), Doyen & Lawrence (1979), Tschinkel & Doyen (1980), Doyen & Tschinkel (1982), Doyen *et al.* (1990) and Doyen (1994), resulting in the formal classification of Matthews & Bouchard (2008) for Australia, as well as Matthews *et al.* (2010) and Bouchard *et al.* (2005, 2011) for the world. Nine subfamilies and just under 100 tribes are recognised, of which eight subfamilies and 41 named tribes occur in Australia, as well as three apparent but unnamed tribes.

The principal morphological structures on which higher classification is based are the internal female reproductive system and the ovipositor, the labrum-epipharynx, the aedeagus, the defensive glands, and the antennal sensoria (Medvedev 1977). All except the last are internal and require dissection to be seen. The key and brief characterisations below are based only on external characters. For descriptions of internal characters see Matthews & Bouchard (2008).

More recently, there have been phylogenetic reconstructions based on DNA sequencing: Kergoat *et al.* (2014a, b), Gunter *et al.* (2014) and McKenna *et al.* (2015) for Tenebrionoidea as a whole, as well as Kanda *et al.* (2015) mainly for Lagriinae. Tenebrionidae always recovers as a monophyletic group. At least one of the above studies supports the monophyly of Zolodininae, Lagriinae, Phrenapatinae, Alleculinae and Stenochiinae, but Pimeliinae recovers as two paraphyletic lineages, while Tenebrioninae and Diaperinae are broadly polyphyletic. Evidently, there will have to be a partial reorganisation of the higher classification of the family which correlates morphological and molecular information, undoubtedly leading to an increase in the number of tribes and subfamilies.

Indications are that there is a fundamental dichotomy at the base of Tenebrionidae: Pimeliinae versus non-Pimeliinae (Kergoat *et al.* 2014a, b; Kanda *et al.* 2015). The two branches are about equal in diversity. This dichotomy had already been noted by Doyen (1972), who proposed that there should be two families: Tentyriidae (= Pimeliidae) and Tenebrionidae, based mainly on the primitive absence of defensive glands, inverted aedeagus and arid habitats of the former, versus the opposites for the latter. The idea of separate families has not been supported subsequently, but this evident basal split raises the possibility that the family arose in an arid environment, a hypothesis supported by the molecular dating analysis of Kergoat *et al.* (2014a) which proposes that the family originated in the mid Jurassic as a group adapted to the arid conditions which predominated at that time. Also, a xerophilic adaptation which seems to be basal is the ability to interlock lateral and apical grooves on the abdominal ventrites with the elytral edges, thus forming a sealed subelytral cavity which minimises respiratory water loss. These grooves are present in Pimeliinae, Zolodininae, Lagriinae and Palorini (Tenebrioninae), as well as some others discussed by Bouchard & Gorb (2000). Palorini have an inverted aedeagus and may be in a key position in tenebrionid phylogeny, but so far they have been omitted from molecular analyses.

Matthews & Bouchard (2008) suggested on morphological grounds that Zolodininae is the most basal tenebrionid group mainly because it combines primitive glandlessness and an inverted aedeagus with a generalised saprophagous larva (Pimeliinae have modified soil-dwelling larvae). However, in the molecular phylogeny of Kanda *et al.* (2015) Zolodininae forms the second branch (after Pimeliinae), not the first. If this is correct, Zolodininae (which occurs in forests) represents the first venture of the family into mesic biomes and is the sister group of the remaining tenebrionids (other than Pimeliinae). These non-pimeliines, in turn, clearly divide into two main branches after Zolodininae: the lagrioids and the tenebrionoids. The lagrioid branch contains only Lagriinae and the Neotropical Nilioninae, not also Phrenapatinae as proposed by Matthews & Bouchard (2008). Phrenapatinae falls among the tenebrionoids in molecular studies, and the New Zealand phrenapatine *Aphtora* Bates, 1872 has a tenebrionine type of female tract (Matthews 2003a). For a discussion of the three main branches of the family – pimelioid, lagrioid and tenebrionoid – see Matthews & Bouchard (2008) and Matthews *et al.* (2010).

When defensive glands did evolve in the family, they did so independently in the lagrioids on the one hand and the tenebrionoids on the other (Tschinkel & Doyen 1980). The most basal lagrioids, Belopini and Cossyphini, are primitively glandless (and also somewhat xerophilic). According to the molecular phylogeny of Lagriinae in Aalbu *et al.* (2017), defensive glands first appeared in the Eschatoporiini-Adeliini clade in positions which do not correspond to that seen in the remaining glanded Lagriinae (where they open between abdominal sternites 7/8). In Eschatoporiini they open between tergites 7/8 (Kanda 2016) and in Adeliini always between sternites 8/9, although a few Adeliini also have small glands in the sternite 7/8 position (Matthews 1998). In the remaining Lagriinae, the next branch (Laenini) has no glands at all (persistence of an ancestral state?), while the rest have defensive glands fixed in the sternite 7/8 position, which is also where they occur in all tenebrionoids (if not lost). Primitively glandless tenebrionoids are not known.

A secondary return to xerophily occurred twice in the lagrioids (Chaerodini, some Adeliini) and nine times in the tenebrionoids, represented by the seven tenebrionoid clades shown returning to arid habitats by Kergoat *et al.* (2014a, Fig. 2) (Melanimonini, Heterotarsina, Opatrini, phaleriine Diaperinae, Crypticini, Pedinini and Blaptini) plus two others in Australia: many Heleini and some Cnodalonini, which became secondarily xerophilic as a result of the desertification of the continent in the Cenozoic.

Biogeography. The Australian tenebrionid fauna has two main historical components: one which has been present on the continent by vicariance since it was a part of Gondwana in the Mesozoic, and the other which dispersed from South-east Asia during the Cenozoic. The Gondwanan component, in turn, has two taxonomic elements which differ in their adaptation to arid conditions, their distribution on the continent, and the locations of their overseas sister groups.

The first of the Gondwanan (Mesozoic) elements consists of taxa which do not occur in humid forests: Cnemeplatiini and Vacronini (Pimeliinae), Belopini and Chaerodini (Lagriinae), Hyociini, Ectychini and Trachyscelini (Diaperinae), plus the genus *Hangaya* Matthews & Merkl, 2015 (Tenebrioninae), comprising 21 genera altogether. All these are found primarily or entirely west of the Great Divide in more or less arid environments, as well as along sea shores. Their relatives do not occur in southern hemisphere forests but in northern hemisphere arid and saline areas (Cnemeplatiini, Vacronini, Belopini and Phaleriini, which is the sister group of the three diaperine tribes mentioned above) and New Zealand coastal sands (Cnemeplatiini and Chaerodini). Matthews (2000) called this biogeographic element 'Tethyan' on the assumption that before the break-up of Gondwana its ancestors lived mainly along the coastal dunes of the Tethys Sea, the southern coast of which stretched from what is now New Zealand across northern Australia, India and the Mediterranean to the Caribbean. In Australia, which was covered by dense forest in the Mesozoic, the coastal dunes would have been about the only xeric habitats present; they are edaphic deserts which pre-adapt their inhabitants to move into deserts proper when the latter appear (Byrne *et al.* 2008). The Tethyan element contains the most geographically and taxonomically isolated Australian groups: Hyociini and Ectychini, which are endemic tribes, *Hangaya* whose relationship is unknown, Chaerodini which occurs only on Australian and New Zealand beaches, and Vacronini which is known only from central Australia and western North America.

The second Gondwanan element, which Matthews (2000) called Austral Gondwanan although the term Notogean is more commonly used, occurs mainly in forests of various types in the east and south of the continent, and comprises Zolodininae, Adeliini (Lagriinae), Phrenapatinae, Palorini, Heleini, Titaenini and Trachelostenini (Tenebrioninae), as well as Alleculinae, 103 genera altogether. Other forest-inhabiting groups in Australia have close Indonesian relatives and are considered to be Indo-Malayan (see below), not Notogean. The Notogean taxa are found mainly east of the Great Divide and along the southern coast, or at least (in the cases of Adeliini, Palorini and Heleini) that is where their apparently basal taxa live. The more recently evolved taxa of these tribes adapted to arid conditions in the central and western parts of Australia (probably *in situ* from mesophilic ancestors already present there), following isolation of the continent and desertification in the Cenozoic. Heleina in particular became very diverse in the centre and west with a concentration in the Murchison bioregion (Matthews 1993, map fig. 114), where they occupy xeric niches which elsewhere in the world would be filled by Pimeliinae. Palorini adapted to the west by becoming termitophilous; Alleculinae are mainly peripheral but have also evolved some xerophilic taxa characterised by 'stomatiform' punctures on the elytra (see under *Metistete* Pascoe, 1866 in the classification section).

The recognised overseas sister groups of the Notogean element are, in contrast to those of the Tethyan, largely in southern hemisphere forest biomes: New Zealand (Zolodininae, Adeliini, Palorini, Heleini and Titaenini), New Caledonia (Adeliini and Titaenini) and southern Chile (Adeliini and Trachelostenini), while Phrenapatinae and Alleculinae occur in all three of those places as well as elsewhere in the world. Palorini also reached Asia, perhaps as a result of the northward drift of India. In the most recent area to have separated, Chile, we see groups which are the most closely related to those of Australia including two instances of shared genera (*Licinoma* Pascoe, 1869 of Adeliini and *Archeophthora* Kaszab, 1978 of Phrenapatinae).

New Caledonia, long believed to have a relict continental biota like New Zealand, is in fact an oceanic island populated by overseas colonisation events beginning in the Oligocene (Grandcolas *et al.* 2008), followed by local radiations. Adeliini in particular diversified extensively in New Caledonia but there were apparently just two founders: one from the *Adelium* group of Matthews (1998) resulting in one endemic genus, *Neoadelium* Carter, 1908 with five species, and the other from the *Pseudobyrsax* group (containing the Australian *Monteithium* Matthews, 1998) which proliferated into five genera and 16 species (Matthews 1998, 2003b). Also, at least one species of *Titaena* Erichson, 1842 (Titaenini) found its way to New Caledonia and subsequently radiated into no less than 55 local species (currently placed in *Callismilax* Bates, 1874) (Kaszab 1982d). The other tenebrionid genera in New Caledonia appear to be of Indo-Malayan origin. New Guinea, although geologically a part of the Australian plate, has very few Gondwanan Tenebrionidae, none of which belongs to endemic genera.

The Gondwanan genera in Australia are mostly concentrated in just ten relatively small areas of endemism distributed throughout the continent (mapped in Matthews & Bouchard 2008, fig. 4).

The second major component mentioned above (the Cenozoic one) consists of numerous groups which dispersed from south-east Asia, and which are listed in Matthews & Bouchard

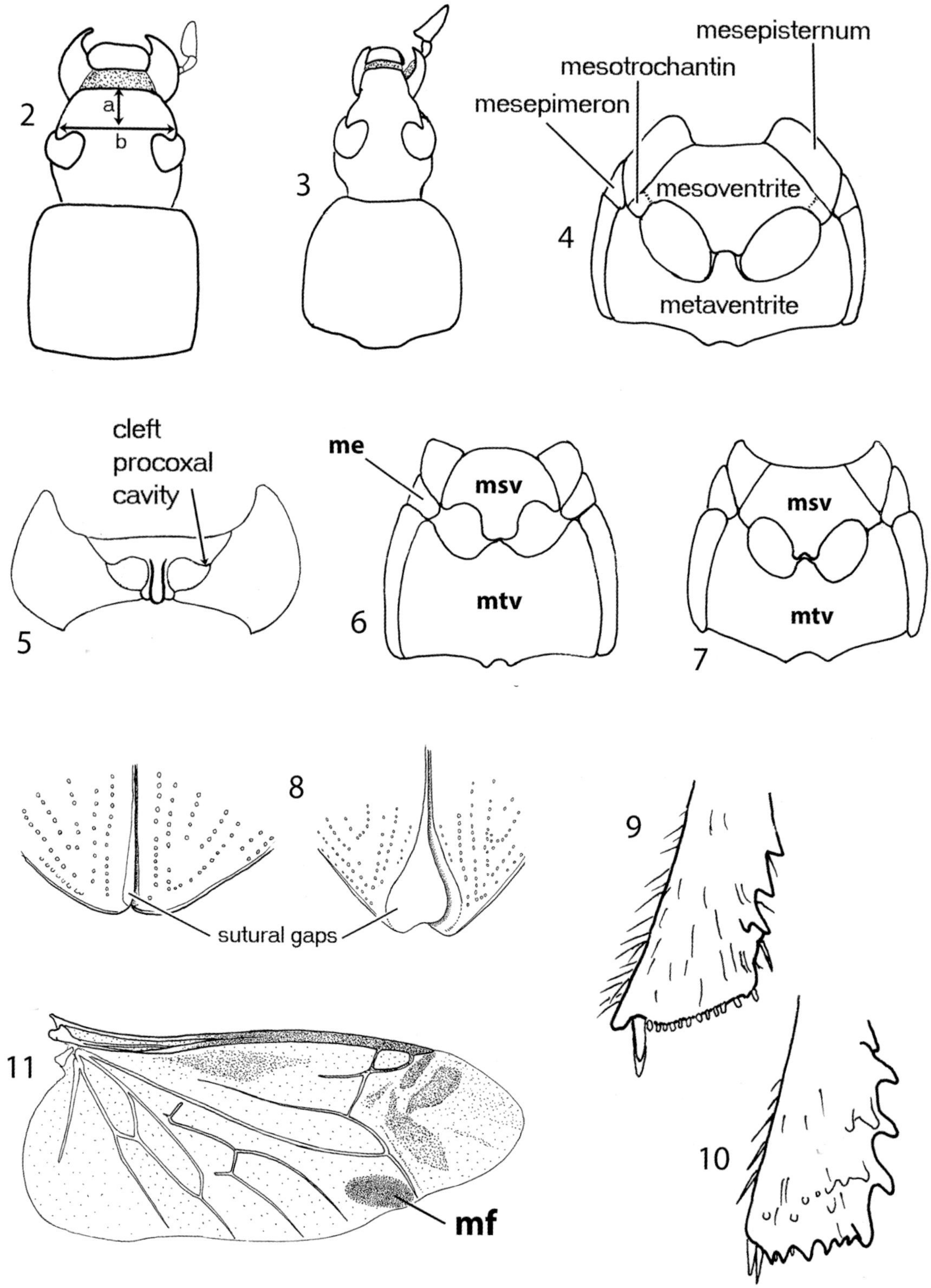

Figs 36.2– 36.11. 2, Head and pronotum of *Lepturidea* indicating frontal index (a/b) of ~0.3; 3, Head and pronotum of *Atoichus*, frontal index ~0.7; 4, Ventral sclerites of meso- and metathorax of *Dioedus*; 5, Ventral sclerites of prothorax of *Triplehornia;* 6, Ventral sclerites of meso- and metathorax of *Scotoderus* showing open mesocoxal cavities. **me** – mesepimeron, **msv** – mesoventrite; **mtv** – metaventrite; 7, Ditto of *Diaclina*, showing closed mesocoxal cavities (meso- and metaventrites meet); 8, Elytral apices of two species of *Dimorphochilus* showing sutures separated by a membranous gap; 9, End of metatibia of *Uloma,* apical edge more or less straight and lined with peg-like setae; 10, End of metatibia of *Achthosus,* apical edge irregularly serrate; 11, Hind wing of *Pachycoelia*. **mf** – medial fleck.

(2008, table 3) as 'Indo-Malayan'. In that list, Phrenapatinae and Alleculinae should be deleted as they were undoubtedly present in Gondwanan Australia long before the break-up. The distribution of the Indo-Malayan groups displays a typical dispersal pattern of maximum diversity in the north gradually diminishing southward. The taxa involved are all closely related to those of South-east Asia and New Guinea: three tribes of Lagriinae, eight tribes of Tenebrioninae, eight tribes of Diaperinae, and all Stenochiinae, 79 genera altogether. Most of these are forest inhabiting because they came from Asian tropical forests, but two of the tenebrionine tribes, Opatrini and Pedinini, are xerophilic and are found along sea shores and inland west of the Divide, while three genera of Stenochiinae-Cnodalonini were able to adapt to arid conditions after arrival in Australia and also inhabit the centre, namely *Hypaulax* Bates,1868, *Omolipus* Pascoe, 1860 and *Oectosis* Pascoe, 1869.

In addition to the groups mentioned above there are 19 tenebrionid species introduced through human commerce, belonging to 11 different tribes and 14 genera, three of which also include native species (see list in Matthews & Bouchard 2008, Table 4).

Characteristics. Adults. Length 1.5–36 (usually 2.5–25) mm. Body of varied form but often elongate and moderately convex; colour often dark brown to black, especially in more arid regions, but sometimes bicolored or metallic, especially in more humid environments. Eyes usually vertically oriented and emarginate anteriorly or anteriorly and posteriorly. Antennal insertions almost always concealed from above (exposed in Leiochrinini and some Lagriini). Labrum free, usually transverse, sometimes subquadrate to elongate, with or without exposed basal membrane. Antennae usually 11-segmented, but with 10 or fewer in several lineages, usually filiform to clavate, with apical antennomeres often bearing compound (stellate) sensoria, or placoid sensoria in Ulomini. Mandibles almost always with well developed mola and distinct prostheca, and apical maxillary and labial palpomeres cylindrical to subtriangular or securiform. Cervical sclerites absent. Pronotum usually as wide at base as combined elytral bases, with lateral pronotal carinae almost always complete and usually simple. Prosternal process complete and almost always dorsally curved behind coxae. Procoxae rarely projecting below sternum; trochantins concealed. Procoxal cavities almost always externally closed (open in *Tanylypa* Pascoe, 1869) and internally closed (open in *Tanylypa*, some Tenebrioninae and a few Diaperinae). Elytra usually irregularly punctate or with nine distinct puncture rows or striae (ten rows in some Lagriinae, *Tanylypa* and *Toxicum* Latreille, 1802 and multiple rows in Australian Vacronini and most Heleini). Mesoventrite often on different plane than metaventrite, its anterior edge without procoxal rests. Mesocoxae with exposed or concealed trochantins. Mesocoxal cavities narrowly to widely separated, laterally open or closed, with trochantins exposed or concealed. Metacoxae contiguous to widely separated. Metendosternite usually without laminae or anterior process, tendons on lateral arms. Legs with trochanterofemoral joint almost always strongly oblique, with base of femur in contact with coxa. Tarsi usually 5–5–4 but reduced in several groups; tarsomeres usually simple but sometimes with penultimate tarsomere lobed beneath; pretarsal claws usually simple (pectinate or serrate in almost all Alleculinae). Abdomen with five ventrites, the first three of which are connate; membrane between sternites V and VI and VI and VII usually visible in Australian species. Paired abdominal defensive glands usually present. Aedeagus usually of uninverted tenebrionoid type, with dorsal tegmen (inverted in *Tanylypa*, Pimeliinae and Palorini); parameres partly or entirely fused together forming apicale, which lacks acccessory lobes and usually has a pair of internal sclerotised rods or lobes (alae).

Larvae. Body almost always elongate, more or less parallel-sided and subcylindrical to slightly flattened (rarely broadly ovate or strongly flattened); dorsal surfaces usually more or less evenly sclerotised and yellow to dark brown in colour (sometimes membranous or with light and dark pigment pattern); generally smooth (occasionally with large, shallow punctures); vestiture almost always of fine hairs or setae (rarely densely hirsute). Head almost always prognathous and protracted. Epicranial stem usually moderately long (rarely absent); frontal arms usually V-shaped or U-shaped and contiguous at base. Frontoclypeal suture always distinct. Labrum free. Antennae usually well developed, 3-segmented, with dome-like or C-shaped sensorium, 2-segmented in Lagriinae. Mandibles usually with well developed mola, which is simple, irregularly ridged or with fine transverse ridges. Ventral mouthparts retracted, with well developed maxillary articulating area; mala rounded or truncate. Hypopharyngeal sclerome almost always consisting of molar-like sclerite (absent in *Spiloscapha* Bates, 1873). Hypostomal rods absent. Thoracic and abdominal terga without patches or rows of asperities, transverse plicae or lateral processes. Prothoracic venter usually without distinct plates; postcoxalia greatly enlarged and meeting at midline in some Pimeliinae. Legs well developed; fore legs enlarged in some groups. Tergum IX excluding appendages shorter than to slightly longer than VIII, usually extending onto ventral surface, not forming articulated plate, usually either without urogomphi (tergum IX posteriorly rounded or subconical) or with pair of simple, acute urogomphi with no pit between them; sternum IX exposed, simple, segment X often reduced and concealed or with distinct pair of pygopods. Spiracles annular or annular-multiforous, with peritreme sometimes crenulate. More complete adult and larval descriptions may be found in Matthews *et al.* (2010), Lawrence & Ślipiński (2013) and references contained therein.

Biology. Tenebrionidae are primarily scavengers, feeding on dead material of plant or fungal origin, and they have been able to survive better than most beetles in arid habitats throughout the world. Many species occur under rocks or in leaf litter, with the larvae living in the soil or scavenging on the ground at night, while others feed in rotten wood (usually as larvae or occasionally as adults). Other groups, particularly in forested areas occur on surfaces at night feeding on lichens or algae, or inhabit the fruiting bodies of wood-rotting fungi (particularly Polyporales and Hymenochaetales). Some subterranean larvae (false wireworms) feed on the roots of plants including commercial crops, and several species (mealworms and flour beetles) have become pests of stored products. The chemical composition of the adult defensive secretions was analysed by Brown *et al.* (1992). For further information, see Matthews & Bouchard (2008), Matthews *et al.* (2010) and Lawrence & Ślipiński (2013).

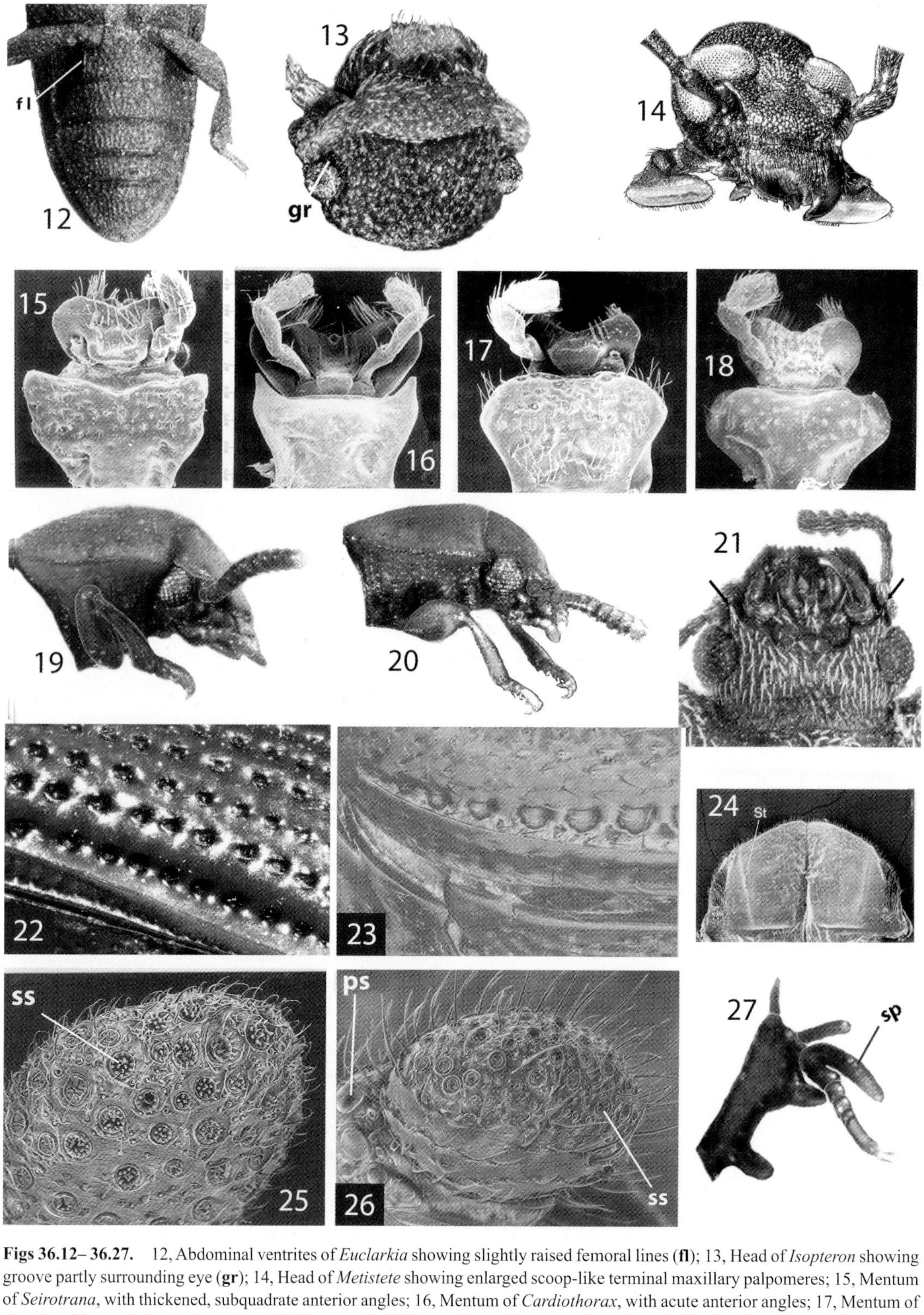

Figs 36.12– 36.27. 12, Abdominal ventrites of *Euclarkia* showing slightly raised femoral lines (**fl**); 13, Head of *Isopteron* showing groove partly surrounding eye (**gr**); 14, Head of *Metistete* showing enlarged scoop-like terminal maxillary palpomeres; 15, Mentum of *Seirotrana*, with thickened, subquadrate anterior angles; 16, Mentum of *Cardiothorax*, with acute anterior angles; 17, Mentum of *Apasis*, with rounded anterior angles; 18, Mentum of *Adelium*, with swollen central portion; 19, Head and prothorax of *Pseudeba*, dorsal edge of eye below edge of gena; 20, Head and prothorax of *Palorus*, eye encroaching on gena; 21, Head of *Csiro* in ventral view, arrows point to subgenal spines; 22, *Metistete*, part of lateral edge of elytron showing deep punctures with thickened lateral lips (stomatiform punctures); 23, *Metistete*, humeral edge of elytron showing enlarged stomatiform punctures confined to lateral row. © GC; 24, *Adelium auratum*, abdominal tergite VII showing stridulatory ridges (**St**); 25, Terminal antennomere of *Strongylium* showing compound stellate sensoria (**ss**). © GC; 26, Terminal and subterminal antennomeres of *Uloma* showing plate-like placoid sensoria (**ps**) and stellate sensoria (**ss**). © GC; 27, Protibia and tarsus of *Thorictosoma*. **sp** – spur-like process of basal tarsomere.

Key to the genera of Australian Tenebrionidae

1. Abdomen without visible membranes (there may be grooves showing smooth cuticle) **2**
– Abdomen with membranes visible between ventrites 3 to 5, sometimes sunken and difficult to see **17**

No visible abdominal membranes 2–16

2(1). Basal ventrite of abdomen with a pair of raised femoral lines which are either longitudinal and parallel (Fig. 36.12), or diagonal and bordering intercoxal process … **BELOPINI** **3**
– Without femoral lines **6**
3(2). Tarsal segmentation 5–5–4. Antennae with 11 segments. Head and pronotum with numerous small tubercles. Eyes absent. L ~3 mm. S WA .... ***Eulea* Carter** (Fig. 36.28, Pl. 90I)
– Tarsal segmentation 4–4–4. Surfaces not tuberculate. Eyes present **4**
4(3). Dorsal surfaces strongly carinate and glabrous. Posterior canthus of head encroaching on outline of eye. Antennae clubbed **5**
– Pronotum smooth, elytra with fine carinae and short erect setae arranged in rows. Eyes simply rounded. Antennae gradually enlarged apically except for terminal segment, which is minute. L ~1.5 mm. S WA .... ***Doyenia* Matthews & Lawrence** (Fig. 36.29, Pl. 89I)
5(4). Antennae with three segments, terminal one large and cylindrical. L 2.5–3.5 mm. S WA .... ***Euclarkia* Lea** (Fig. 36.30, Pl. 90L)
– Antennae with eight segments, terminal one forming a club. L 3–4 mm. Southern arid Australia .... ***Kershawia* Lea** (Fig. 36.31, Pl. 92I)
6(2). Postcoxal bridges meeting along midline behind prosternal process. Wide pronotal and elytral flanges entirely covering head and legs. L 4–6 mm. Northern Australia, Eurasia, Africa .... **COSSYPHINI: *Cossyphus* Olivier** (Fig. 36.33, Pl. 88A)
– Postcoxal bridges not meeting at midline behind procoxae. No flanges **7**
7(6). Procoxal cavities externally open behind. L 7–12 mm. TAS .... **ZOLODININAE: *Tanylypa* Pascoe** (Fig. 36.32, Pl. 99H)
– Procoxal cavities externally closed behind **8**
8(7). Mentum large, entirely concealing maxillae except for palpi. Mandibles with large dorsal lobe which flanks clypeus at rest. Clypeus trilobed. L 7–8 mm. S SA, South America .... **TRILOBOCARINI: *Salax* Guérin-Méneville** (Fig. 36.34, Pl. 97C)
– Mentum smaller, exposing stipes and cardo. Mandibles without lobe flanking clypeus. Clypeus not trilobed **9**
9(8). Eyes vestigial or absent. Antennae short with clubs received in cavities under anterior angles of prothorax. Scutellum not visible … **CNEMEPLATIINI** **10**
– Eyes fully developed. Antennae longer, not received in prothoracic cavities. Scutellum visible **11**
10(9). Middle of outer edge of protibiae with a prominent tooth. Basal protarsomere expanded into a long spur-like structure (Fig. 36.27). Elytral bases with beading paralleling pronotal base. Prosternum medially strongly convex. L 2–3 mm. Southern arid Australia .... ***Thorictosoma* Lea** (Fig. 36.35, Pl. 99D)
– Protibiae unarmed. Protarsi unmodified. Elytral bases without beading. Prosternum nearly flat. L 2–3 mm. Northern arid Australia .... ***Wattiana* Matthews & Lawrence** (Fig. 36.36, Pl. 101B)
11(9). Antennae with last two to four segments enlarged as distinct club. Elytral punctures, if seriate, forming 10 or fewer complete rows. Middle of clypeal margin straight or arcuately produced, sometimes horned … **TOXICINI** **12**
– Antennae without club, with apical segment smaller than subapical. Elytra with at least 16 rows of quadrate or round window punctures. Middle of clypeal margin straight or slightly excised … **VACRONINI** **16**
12(11). Dorsal surfaces smooth, nearly glabrous. Antennae clearly 11-segmented … **TOXICINA** **13**
– Dorsal surfaces very uneven, densely covered with short squamiform setae. Last two segments of antennal club more or less fused, antennae therefore sometimes appear to be 10-segmented … **DYSANTINA** **14**
13(12). Eyes completely divided by genal canthus. Supraorbital horns of males glabrous, head without clypeal horns. L 8–11 mm. E NSW, SE QLD, Eurasia, Africa .... ***Cryphaeus* Klug** (Fig. 36.38, Pl. 88J)
– Eyes not completely divided. Supraorbital horns of male fringed with yellow hairs, with a pair of horns on clypeus. L 8–17 mm. S, E and N Australia, Eurasia, Africa .... ***Toxicum* Latreille** (Fig. 36.39, Pl. 100G, H)
14(12). Metaventrite shorter along midline than first abdominal ventrite. No projection above eye. Hind body oval in outline. Dorsal projections not surmounted by shiny tubercles. L 7–11 mm. NE NSW, E QLD .... ***Mychestes* Pascoe** (Fig. 36.37, Pl. 94E)
– Metaventrite longer along midline than first abdominal ventrite. Prominent ridge or cone above eye. Hind body parallel in outline. At least some dorsal projections with apical shiny tubercles **15**
15(14). Hind edge of pronotum more or less emarginate opposite scutellum. L 6–8 mm. E VIC, E NSW, N QLD .... ***Orcopagia* Pascoe** (Fig. 36.40, Pl. 95D)
– Hind edge of pronotum medially straight. L ~5 mm. NE NSW, SE QLD .... ***Ilyxerus* Pascoe** (Fig. 36.41, Pl. 91L)
16(11). Mandibles unidentate apically, projecting so as to leave a gap between them before labrum when closed. Apical maxillary palpal segments narrowly triangular. Prothorax widest at base, narrowing anteriorly, surface of pronotum finely rugose. Antennae dimorphic, those of male widened and serrate except for small apical segment, reaching middle of elytra, those of female simple and reaching only to basal 1/4 of elytra. L 10–11 mm. Central Australia .... ***Lixionica* Blackburn** (Fig. 36.42, Pl. 93I)
– Mandibles narrowly truncate apically, not projecting, not leaving visible gap when closed. Apical maxillary palpal segments broadly securiform. Prothorax widest in middle, surface of pronotum granulate or vermiculate. Antennae not dimorphic, segments subcylindrical to obconic, except for small apical two or three segments, reaching basal 1/4 of elytra or less. L 7.5–12 mm. SA, S NT, NW VIC, W NSW .... ***Exangeltus* Blackburn** (Fig. 36.43, Pl. 90D)

Abdominal membranes visible 17–end

17(1). Procoxal cavities with lateral margins more or less cleft, exposing some of lateral, narrowed part of coxa (Fig. 36.5). Epipleura ending before elytral apices. Form of beetle convex oval **18**
– Procoxal cavities with lateral margins broadly rounded or nearly straight, at most very shallowly cleft, lateral part of coxa always concealed. Epipleura and form variable **22**

Procoxal cavities cleft, epipleura incomplete 18–22

18(17). Antennae with 10 segments, of which the apical four form a club. Elytra with more than 15 puncture rows. Beetle uniformly dark with greenish reflections. L 2–3 mm. E QLD, SE Asia .... ***Triplehornia* Matthews & Lawrence** (Fig. 36.47, Pl. 100K)

– Antennae with 11 segments, gradually expanding or with 7-segmented club. Elytra with fewer or no puncture rows. Surfaces patterned or nitid, without reflections **19**

19(18). Mesocoxal cavities closed in part by mesepimera (as in Fig. 36.6) (considered 'open'). Body less convex, oval. Elytra striate. E NSW, New Guinea, E Asia ... **SCAPHIDEMINI** .... ***Spiloscapha* Bates** (Fig. 36.45, Pl. 98E)

– Mesocoxal cavities closed by ventrites only (as in Fig. 36.7) (considered 'closed'). Body more or less hemispherical. Elytra without striae ... **LEIOCHRININI** **20**

20(19). Tarsomeres simple, not lobed. Antennae short, moderately flattened, segment 4 as wide as 5, segment 3 as wide distally as base of 4. Legs relatively short, tibiae without keel on outer edge. L 2–3 mm. NE NSW, SE QLD, New Guinea, Asia .... ***Derispia* Lewis** (Fig. 36.44, Pl. 88L)

– Tarsomeres (except apical) strongly lobed below. Antennae long, segments subcylindrical, segments 3 and 4 narrower than 5. Legs longer, tibiae with sharp median keel on outer edge **21**

21(20). Pronotum covering head at rest, its anterior edge broadly rounded. L 4.5–5.5 mm. NE QLD, New Guinea, Asia .... ***Leiochrinus* Westwood** (Fig. 36.46, Pl. 92L)

– Head free beyond hind edge of eyes. Anterior edge of pronotum deeply excised. L 2–3.5 mm. E QLD, New Guinea, Asia .... ***Leiochrodes* Westwood** (Fig. 36.48, Pl. 92E)

Procoxal cavities rounded 22–end

22(17). Tarsal claws pectinate (comb-like), denticulate, serrate or crenulated. In some cases denticles are minute and clustered at bases of claws, which appear to be simple except at high magnification ... **ALLECULINAE** **23**

– Tarsal claws simple **45**

Tarsal claws pectinate or denticulate 23–44

23(22). Tarsomeres subparallel, not lobed. Claws dentate for basal half only. L 8 mm. WA .... ***Litopous* Matthews** (Fig. 36.49, Pl. 101J)

– Penultimate tarsomeres more or less lobed (minimally in *Taxes*). Dentition of claws variable **24**

24(23). Outer edge of protibia with a serrated row of minute spinules, not always present in *Nocar*. Form of beetle compact and oval or oblong, lateral margins of prothorax prominent, its base almost or quite as wide as elytral bases, head short and legs relatively short, less than 1/4 of femoral length visible beyond body outline from above. Mandibles nearly always apically bidentate, otherwise truncate or broadly rounded. L of beetle not exceeding 9 mm **25**

– Outer edge of protibia without spinules or with only a few irregularly spaced ones, occasionally with a row of spinules, in which case head is elongate before eyes and mandibles are long and apically unidentate. Form of beetle variable, may be as above but usually more elongate, lateral margins of prothorax may not project at all, its base often narrower than elytra, head short or long, legs usually longer than above. Mandibles as above or apically acuminate. L of beetle may be over 9 mm **29**

25(24). Upper surfaces glabrous or nearly so. Pronotal punctures separated by more than one puncture width, interpunctural surface smooth. Prothorax basally not quite as wide as elytral bases. Forespur sometimes enlarged. Apical maxillary palpomere cultriform. L 4.7–6.6 mm. Throughout Australia .... ***Scaletomerus* Blackburn** (Fig. 36.50, Pl. 97G)

– Upper surfaces densely setose with short to long setae. Pronotal punctures separated by less than one puncture width, interpunctural surface often shagreened. Prothorax often as wide basally as elytral bases. Other characters variable **26**

26(25). Lobes of penultimate tarsomeres much reduced, not wider than segments, with long bristles on plantar surfaces. Teeth confined to basal half of claws. Apical maxillary palpomeres securiform. Mandibles sometimes apically truncate or bluntly rounded. Elytral stria 9 frequently deeper than the others and with larger punctures. Form of beetle usually oblong with parallel sides. L 3.5–6.8 mm. Northern and central Australia .... ***Taxes* Champion** (Fig. 36.51, Pl.99B)

– Lobes of penultimate tarsomeres readily apparent and wider than segments. Teeth present along most of claw length. Palpomeres variable. Mandibles distinctly bidentate. Elytral stria 9 similar to the others or only a little deeper. Form of beetle oval or oblong **27**

27(26). Protibiae apically expanded. Antennomeres elongate, ~3 times as long as wide. Apical maxillary palpomeres securiform. Elytral intervals minutely tuberculate. Form of beetle oblong, prothorax widest in middle. Colour rufocastaneous with paler legs. L 7.0–9.0 mm. QLD .... ***Barycistela* Blackburn** (Fig. 36.61, Pl. 86E)

– Protibiae subparallel or only feebly widening apically. Antennomeres shorter, ~2 times as long as wide or less. Apical maxillary palpomeres cultriform. Elytral intervals minutely rugose or smooth. Form of beetle usually oval with semicircular prothorax widest at base, occasionally oblong and parallel-sided. Colour variable **28**

28(27). Elytral striae subobsolete on disc, not discernible from general surface punctation although some punctures arranged in rows. Apical labial palpomeres not bilobed. Male metatibiae dilated with a deep smooth concavity on inner faces. L ~4 mm. Northern Australia .... ***Otys* Champion** (Fig. 36.59)

– All elytral striae impressed. Three species with apical labial palpomere bilobed. Without sexual dimorphism. L 4.6–9.0 mm. Throughout Australia .... ***Nocar* Blackburn** (Fig. 36.52, Pl. 95B)

29(24). Apices of mandibles acuminate and unidentate (a second small tooth may be present on lower edge but remote from apex). If with two minute apical teeth (*Euomma*), head moderately prolonged, as are labrum and maxillary palpi. Apical maxillary palpomere cultriform or strongly asymmetrically triangular **30**

– Mandibles clearly bidentate with both teeth apical or nearly so, or truncate, usually with external apical groove. Head shorter. Apical maxillary palpomere broadly securiform or triangular, usually more or less equilateral, or, if asymmetrical, with lower edge more prominent than upper (scoop-like) **37**

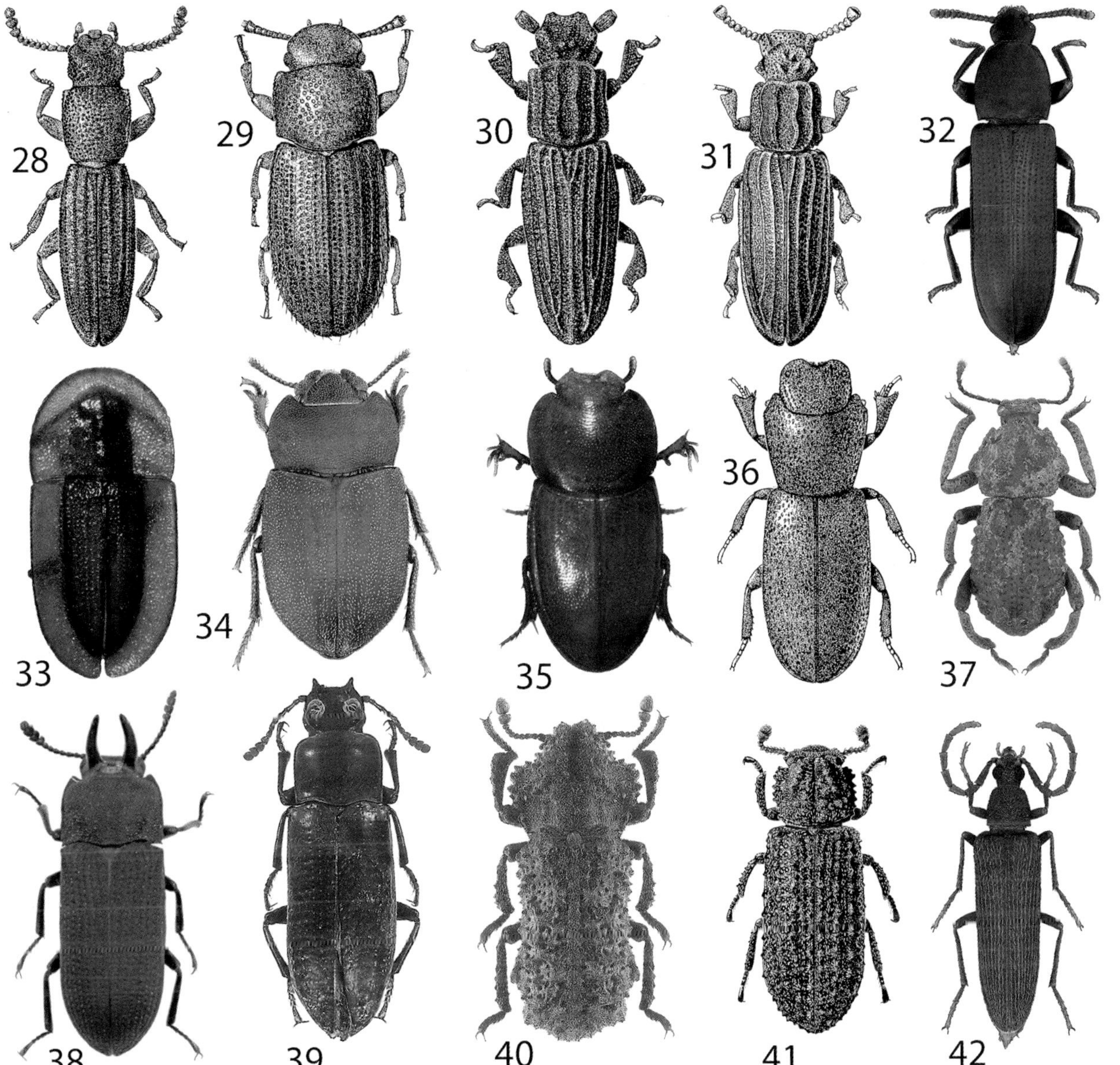

Figs 36.28– 36.46. 28, *Eulea*; 29; *Doyenia* © AZ; 30, *Euclarkia*; 31, *Kershawia*; 32, *Tanylypa* © GC; 33, *Cossyphus*; 34, *Salax* © GC; 35, *Thorictosoma*; 36, *Wattiana* © AZ; 37, *Mychestes* © GC; 38, *Cryphaeus*; 39, *Toxicum*; 40, *Orcopagia*; 41, *Ilyxerus*; 42, *Lixionica* © GC; 43, *Exangeltus* © GC; 44, *Derispia*; 45, *Spiloscapha*; 46, *Leiochrinus*.

30(29). Frons and clypeus not produced, frontal index (a/b in Fig. 36.2) not more than 0.4, clypeolabral membrane as long as half length of labrum or longer. Prothorax subparallel for basal half at least. Eyes usually moderate, separated by distance about equal to one eye width, but may be larger. Integument sometimes brilliantly metallic. One unnamed species with simple claws. L 5.0–16.0 mm. Throughout Australia .... ***Lepturidea*** **Fauvel** (Fig. 36.57, Pl. 84H)

– Frons and especially clypeus at least a little more produced (e.g. Fig. 36.3), clypeolabral membrane often shorter than half length of labrum. Other characters variable **31**

31(30). Procoxae contiguous **32**

– Procoxae separated by narrow prosternal process **33**

32(31). Head abruptly constricted basally, forming distinct neck. Antennomeres subtriangular. Colour black combined with rufous, flavous or testaceous areas. L 4.0–6.0 mm. SE QLD, NSW, VIC, SA .... ***Atoichus*** **Carter** (Fig. 36.53, Pl. 85E)

– Head gradually and only slightly narrowing basally. Antennomeres more elongate. Colour as above, or entirely testaceous. L 4.0–6.0 mm. NSW, VIC, TAS .... ***Pemanoa*** **Buck** (Fig. 36.54, Pl. 96H)

33(31). L of beetle > 11 mm. Glabrous and nitid. Head moderately to strongly elongate, frons with subparallel or anteriorly diverging grooves. Colour uniformly castaneous or in *T. pulcher* (Carter) with metallic blue-green reflections. Male protibiae sometimes sinuate on inner edges. L 11.7–19.0 mm. Throughout Australia .... ***Tanychilus*** **Newman** (Fig. 36.58, Pl. 99G)

– L of beetle 11 mm or less. With short recumbent setae, rarely glabrous. Other characters variable **34**

34(33). Middle of pronotum with a pair of discal impressions. Head short before eyes. Integument sometimes with metallic reflections, violet on elytra and green on head and prothorax. L 7.5–9.5 mm. TAS .... ***Nypsius*** **Champion** (Fig. 36.55, Pl. 95F)

– Disc of pronotum without impressions, or if with a pair of basal impressions, integument not metallic or head longer, or both **35**

35(34). Only first two to four elytral striae evident. Head strongly prolonged. Colour entirely black or black with yellow elytra, legs rufous or testaceous. L 4.9–5.7 mm. NSW, VIC .... ***Neocistela*** **Borchmann** (Fig. 36.63, Pl. 94C)

– All elytral striae distinct. Head moderately prolonged. Colour fuscous with pale elytral disc patches, or entirely fuscous or testaceous **36**

36(35). Claws pectinate, although teeth may be small. Eyes variable in size, often large, separated dorsally by a distance equal to much less than one eye width. Head moderately long, apical maxillary palpomere elongate-cultriform, more strongly so in males. Prothorax about as long as wide, sides broadly rounded anteriorly, rarely transverse with rounded sides. Male sexual dimorphism variously involving antennae, maxillary palpi, legs or apical ventrite. L 4.0–10.0 mm. Throughout Australia .... ***Euomma*** **Boheman** (Fig. 36.56, Pl. 90F, 91G)

– Claws only feebly crenulate on inner edges, or completely simple. Eyes small, separated dorsally by distance equal to approximately one eye width. Head shorter, apical maxillary palpomere moderately cultriform. Prothorax a little transverse with rounded sides. Male with shallow apical emargination on inner edge of mesotibia. L 4.0–5.6 mm. VIC, S WA .... ***Hemicistela*** **Blackburn** (Pl. 91I)

37(29). Basal metatarsomeres more or less widened relative to other segments, more strongly so in male. Prothorax without or with incomplete lateral marginal beading, its base sometimes narrower than apex. Meso- and metatibiae strongly evenly curved in male. Colour uniformly rufous. L 3.5–4.5 mm. N WA, NT .... ***Jophon*** **Champion** (Fig. 36.62, Pl. 92F)

– Basal metatarsomeres not widened. Prothorax with complete lateral marginal beading or carina, its base as wide as apex or wider. Tibiae not evenly curved in male, may be expanded or sinuate, or simply straight. Colour variable **38**

38(37). With one or both of the following character states: (1) At least some elytral punctures stomatiform (deep with thickened lateral edges resembling lips or sometimes tubercles, Fig. 36.22), and containing a single very fine recumbent seta. Often stomatiform punctures confined to an enlarged and impressed segment of stria 9 just behind humerus (Fig. 36.23), or all elytral punctures may be stomatiform (Fig. 36.22). Sometimes punctures are almost closed over, leaving only slits. (2) Apical maxillary palpomere more or less enlarged (more so in males) and scoop-like (Fig. 36.14), the inner end sometimes curled over **39**

– Elytral punctures simple, rounded. Punctures of stria 9 behind humerus may be more impressed than others, but not enlarged or modified. Apical maxillary palpomere securiform but not greatly enlarged **41**

39(38). Occiput of head deeply transversely excavate and receiving prolongation of anterior part of pronotal disc. Males often with distally enlarged antennomeres. L 5–6 mm. NW WA .... ***Scaphinion*** **Matthews** (Fig. 36.60, Pl. 101E)

– Occiput not excavate. Front of pronotum not prolonged. Antennae not enlarged **40**

40(39). Beetle generally fully winged, but females of some species flightless. Eyes moderate to large, separated by two eye widths or less. Elytral surfaces rarely tuberculate, sometimes carinate. L 8.4–24.0 mm. Throughout Australia .... ***Metistete*** **Pascoe** (Fig. 36.64, Pl. 94G)

– Beetle flightless. Eyes small, separated by two to three eye widths. Elytral surfaces densely minutely tuberculate. L 8–12 mm. NT, WA .... ***Simarus*** **Borchmann** (Fig. 36.65, Pl. 98H)

41(38). Beetle flightless, metaventrite shorter than first abdominal ventrite and intercoxal process of latter broadly rounded or truncate. Males often with notch or tooth on inner edge of pro- and mesotibiae and metatibia usually flattened or distorted **42**

– Beetle winged, or rarely flightless in female only, metaventrite longer and intercoxal process of first abdominal ventrite acute. Dimorphism rarely involving meso- or metatibiae **43**

42(41). Elytra with large pustules, at least posteriorly, each nearly always bearing one long erect seta. Pronotum densely vermiculate-punctate, laterally feebly convex and meeting hypomeron at a distinct angle. Colour usually uniformly castaneous. L 6.0–10.0 mm. WA, central Australia .... ***Notocistela*** **Carter** (Fig. 36.66, Pl. 95G)

– Elytra without pustules. Pronotum usually with simple punctures, laterally very convex and meeting hypomeron at a very obtuse angle or none at all. Surfaces nitid or with long erect setae. Body castaneous, legs usually pale testaceous or flavous, often with apices of femora dark. L 5.6–12.7 mm. Throughout Australia .... ***Melaps*** **Carter** (Fig. 36.67, Pl. 93G)

43(41). Thin inner lamina of elytral sutures at least slightly expanded apically where there is often a sutural gap (Fig. 36.8). Head moderately short before eyes. Eyes small to moderate. Prothorax narrower than elytral bases. Beetle usually glabrous, rarely with very short setae, castaneous, usually with paler legs. Males with sternite VIII divided and forcipate, and inner edge of protibiae more or less sinuate. L 8.7–15.0 mm. Throughout Australia .... ***Dimorphochilus*** **Borchmann** (Fig. 36.68, Pl. 89F)

– Inner lamina of elytral sutures narrow throughout, not expanded apically, no sutural gap. Head short to very short before eyes. Eyes generally larger. Other characters variable **44**

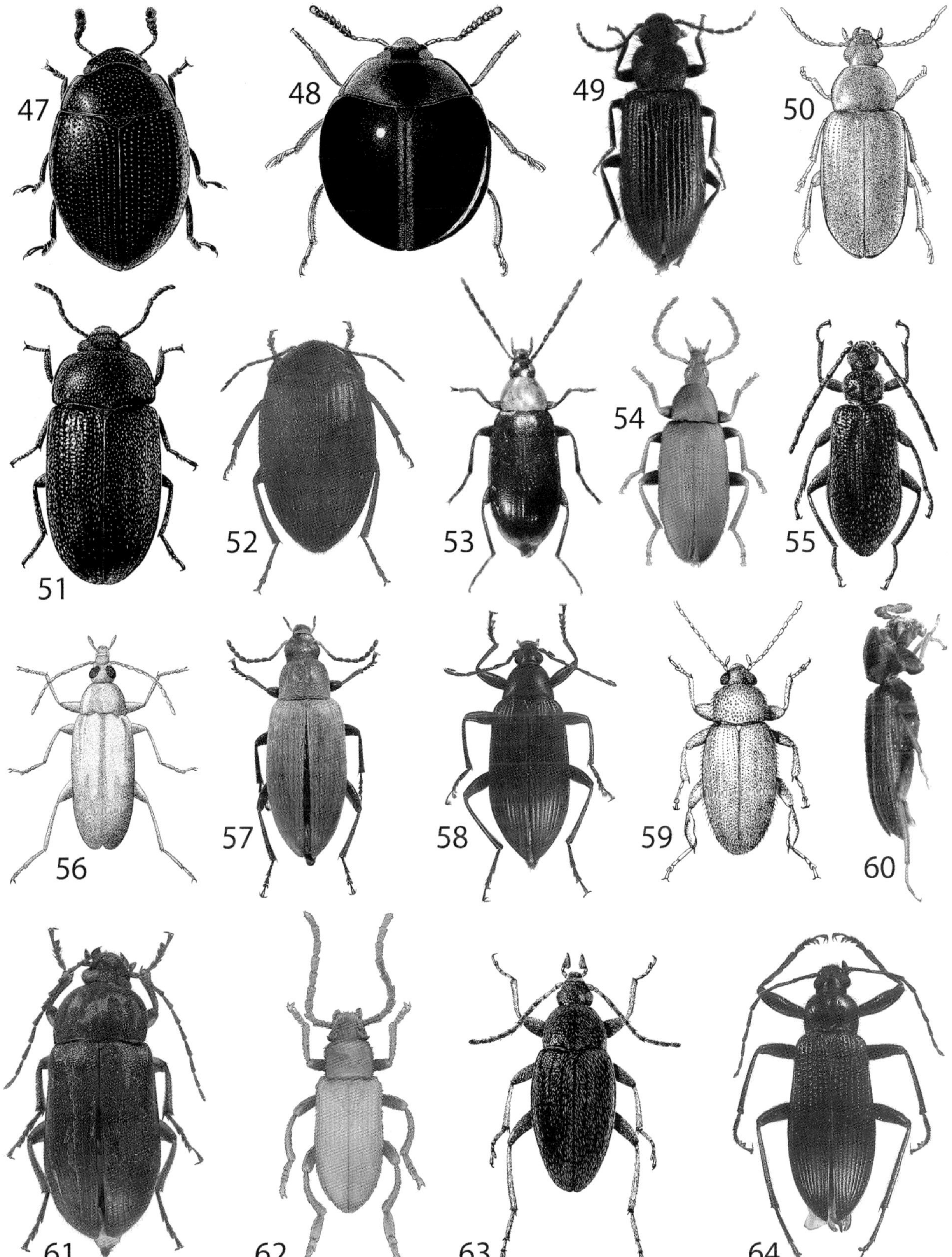

Figs 36.47– 36.64. 47, *Triplehornia* © AZ; 48, *Leiochrodes*; 49, *Litopous* © EGM; 50, *Scaletomerus* © SAM; 51, *Taxes*; 52, *Nocar* © GC; 53, *Atoichus* © SAM; 54, *Pemanoa* © GC; 55, *Nypsius*; 56, *Euomma* © SAM; 57, *Lepturidea*; 58, *Tanychilus*; 59, *Otys;* 60, *Scaphinion* © EGM; 61, *Barycistela* © GC; 62, *Jophon* © GC; 63, *Neocistela*; 64, *Metistete*.

44(43). Antennae dimorphic, apical segment longer than subapical in males, subequal in females, middle segments in male may be enlarged and glabrous below, antennomeres subtriangular and black except in *O. mastersi* Macleay, where they are obconic and fuscous or flavous. Body and legs with very long setae some of which are erect, at right angle to body surface. Integument strongly nitid, often bicoloured. Tibiae of males not dimorphic. L 4.8–8.0 mm. QLD, NSW .... ***Ommatophorus* Macleay** (Fig. 36.69, Pl. 95J)

– Antennae filiform or serrate in both sexes, if latter only a little wider in males. Surfaces usually more or less densely setose with long erect or short recumbent setae, latter in larger species often pale or white and forming patches or V-shaped patterns on elytra, smaller species with uniformly short dense recumbent setae. Males often with protibiae dentate on inner edges, rarely with flattened, twisted and widened metatibiae, one unnamed species with modified mesotibia and long spur on profemur. Colour of larger species piceous, sometimes with rufous elytra, smaller species fuscous or rufous often with legs flavous with dark femoral apices. L 5.0–17.0 mm. Throughout Australia .... ***Homotrysis* Pascoe** (Fig. 36–70, Pl. 91F, 101G)

Claws simple 45–end

45(22). Labrum subquadrate or elongate, strongly projecting beyond edge of clypeus, and penultimate tarsomeres cupuliform or lobed (except in some Adeliini). Antennae with simple hairlike sensilla ... **LAGRIINAE** **46**

– Without above combination. Labrum transverse, or if subquadrate (Phrenapatinae, Palorini, Bolitophagini, Opatrini, Pedinini), tarsomeres are linear except sometimes on protarsi only. Antennal sensilla variable **85**

Tarsomeres cupuliform or lobed. Labrum subquadrate and projecting 46–84

46(45). Meso- and metafemora massive, ovoid. Meso- and metatibiae apically expanded, densely set with stout spines. Antennae short, clavate. Elytra without striae. L 5–7 mm. On beaches ... **CHAERODINI** .... ***Sphargeris* Pascoe** (Fig. 36.71, Pl. 98C)

– Meso- and metafemora and tibiae subcylindrical, with few spines or stout setae. Other characters variable **47**

47(46). Procoxae contiguous and postcoxal bridges meeting behind them ...**LAGRIINI-LAGRIINA** **48**

– Procoxae and postcoxal bridges separated by prosternal process **53**

48(47). Pronotum smooth, impunctate, surface uneven. Femora and tibiae flattened dorsoventrally. Elytra fulvous, rest of body bluish black. Antennae of males thick, segments 7 and 9 strongly expanded at inner side, segments 8 and 10 simple, segment 11 bent, longer than preceding six segments combined. L 9.5–13 mm. N QLD, New Guinea, Philippines, Moluccas .... ***Acerogria* Borchmann** (Fig. 36.72, Pl. 84I)

– Pronotum distinctly punctate, surface more or less evenly convex. Femora and tibiae less flattened. Colour pattern not as above. Antennae of males with segments simple, or deformed in different ways from above, segment 11 nearly straight, never longer than five preceding segments combined, usually much shorter **49**

49(48). Antennae of males with segments simple, like those of females **50**

– Antennae of males with segment 9, or 9 and 10, more or less expanded at outer edges **52**

50(49). Antennomere 1 of males considerably, of females slightly, longer than half of interantennal distance. Antennomeres (at least basal four) with some hairs as long as or longer than length of segments. Size larger, L 9.5–14 mm. SE Australia .... ***Metriolagria* Merkl** (Fig. 36.73, Pl. 94F)

– Antennomere 1 of both sexes much shorter than half of interantennal distance. Antennomeres with hairs much shorter than length of segments. Size smaller, not over 12 mm in length **51**

51(50). Antennae thick and short, segment 3 not longer than segment 1, segment 2 not longer than wide. Elytra subparallel-sided, barely dilated posteriorly in both sexes, humeral angle pronounced. L 7.5–9 mm. NT, N QLD .... ***Xenolagria* Merkl** (Fig. 36.74, Pl. 101K)

– Antennae slenderer and longer, segment 3 usually longer than 1, segment 2 longer than wide. Elytra distinctly dilated posteriorly at least in females, humeral angle not conspicuous. L 6–12 mm. NT, QLD, Eurasia, Africa .... ***Lagria* Fabricius** (Pl. 92J)

52(49). Antennae of males with segment 1 very elongate, longer than interantennal distance, those of females longer than half of interantennal distance. Male pronotum slightly transverse, sometimes as long as wide, but broader than head. Size larger, body robust, legs strong. L 7–16 mm. Throughout Australia, peripheral .... ***Ecnolagria* Borchmann** (Fig. 36.75, Pl. 89G)

– Antennae of males with segment 1 less elongate, only longer than half of interantennal distance, those of females shorter. Male pronotum at least as long as wide, usually longer, narrower than head. Size smaller, body slender, legs gracile. L 6–10 mm. N QLD .... ***Stenolagria* Merkl** (Fig. 36.76, Pl. 98F)

53(47). Apical antennomere two to four times longer than wide. Lateral pronotal margins absent. Head abruptly narrowed just behind eyes, forming distinct neck ... **LAGRIINI-STATIRINA** **54**

– Apical antennomere not more than twice as long as wide. Lateral pronotal margins nearly always distinct. Head behind eyes variable **55**

54(53). Last antennomere not longer than preceding two segments combined. Head about as broad as prothorax. Pronotum smooth, not punctate. L 7–10.5 mm. N QLD, New Guinea, S Asia, Africa .... ***Casnonidea* Fairmaire** (Fig. 36.77, Pl.87F)

– Last antennomere longer than preceding two segments combined. Head distinctly narrower than prothorax. Pronotum punctate. L 8–10 mm. N QLD, New Guinea, SE Asia, Africa .... ***Sora* Walker** (Fig. 36.78, Pl. 98G)

55(53). Elytra without striae or puncture rows and with smooth surface. No stridulatory files on abdomen **56**

– Elytra with 10 striae including one along epipleuron or, if estriate, with tuberculate, vermiculate, or verrucose surface. Tergite VII of abdomen often with a pair of stridulatory files (Fig. 36.24 St) ... **ADELIINI** **60**

56(55). Mesocoxal cavities partly closed by mesepimera. Surfaces densely covered with long semirecumbent setae. Prothorax with sides explanate, edges and hind angles strongly dentate. L 5–8 mm. NSW, N Australia, New Guinea ... **GONIADERINI** .... ***Pseudolyprops* Fairmaire** (Fig. 36.79, Pl. 97J)

– Mesocoxal cavities closed by ventrites only. Surfaces glabrous or with short recumbent setae. Prothorax with sides entire or crenulate and hind angles subquadrate or obtuse ... **LUPROPINI** **57**

57(56). Front edge of clypeus produced and anteriorly very shallowly emarginate. Tarsi short and broad, all preapical segments transverse. Antennae short, not reaching middle of pronotum. L 3–4 mm. S WA, on beaches .... ***Mesotretis* Bates** (Fig. 36.80, Pl. 93K)

– Front edge of clypeus produced or not, but never emarginate. Tarsi narrower, at least some preapical segments longer than wide. Antennae reaching beyond middle of pronotum **58**

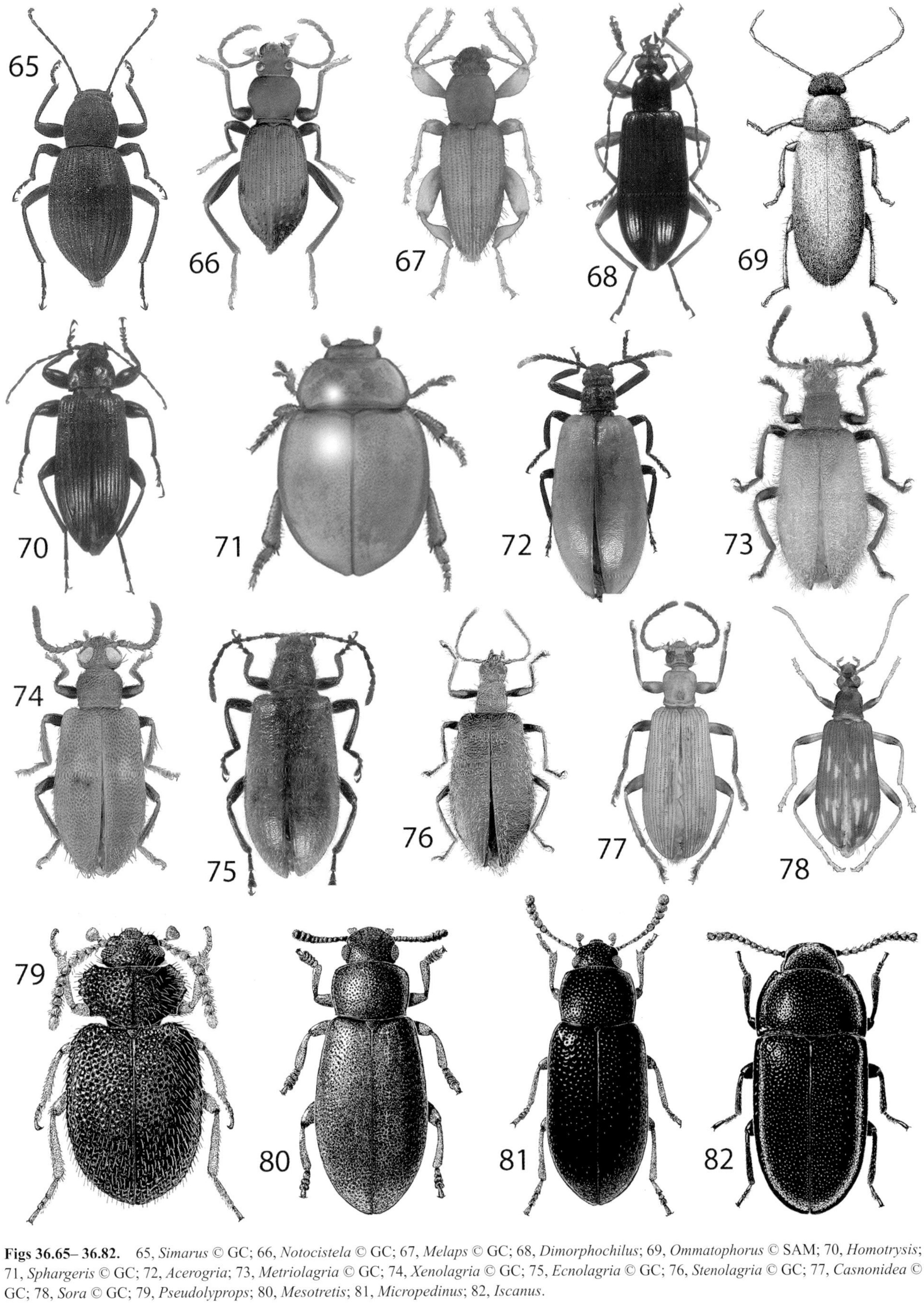

Figs 36.65– 36.82. 65, *Simarus* © GC; 66, *Notocistela* © GC; 67, *Melaps* © GC; 68, *Dimorphochilus*; 69, *Ommatophorus* © SAM; 70, *Homotrysis*; 71, *Sphargeris* © GC; 72, *Acerogria*; 73, *Metriolagria* © GC; 74, *Xenolagria* © GC; 75, *Ecnolagria* © GC; 76, *Stenolagria* © GC; 77, *Casnonidea* © GC; 78, *Sora* © GC; 79, *Pseudolyprops*; 80, *Mesotretis*; 81, *Micropedinus*; 82, *Iscanus*.

58(57). Labial palpi with terminal segment larger than subterminal. Antennae from 7th segment to end gradually widened. L 2.5–4 mm. WA, N QLD, W Pacific, on beaches .... ***Micropedinus* Lewis** (Fig. 36.81, Pl. 94B)

– Labial palpi with subterminal segment enlarged, terminal one small and linear. Antennae a little more abruptly widened from segments 8 or 9, last 3 or 4 segments of about equal width **59**

59(58). Pronotal base not or barely narrower than elytral bases. Humeri angulate, not rounded. Basal membrane of labrum covered by produced edge of clypeus. L 3–4 mm. Norfolk Island, other Pacific islands .... ***Iscanus* Fauvel** (Fig. 36.82, Pl. 92C)

– Pronotal base slightly narrower than elytral bases. Humeri rounded. Basal membrane of labrum exposed. L 3–4.5 mm. N QLD, Lord Howe and Norfolk islands, New Zealand, pantropical .... ***Lorelus* Sharp** (Fig. 36.83, Pl. 93C)

60(55). Dorsal surfaces capable of holding a layer of soil (squalid), often scabrous or tuberculate. Setae, if present, dense, short or minute and usually recumbent and squamiform or thickened. Prosternal process often bilobate. Base of prothorax not overlapping elytral bases, epipleuron not dentate at humerus and mentum not carinate or vermiculate **61**

– Dorsal surfaces usually smoother. Setae, if present, fine and often erect, at least in part, rarely entirely short and thickened. Prosternal process bilobate only in a few *Adelium*. If surfaces squalid, base of prothorax overlapping elytral bases (*Yarranum*, *Seirotrana*), epipleuron dentate at humerus (*Daedrosis*), or mentum vermiculate or longitudinally carinate medially (*Nototrintus*, *Yarranum*) **65**

61(60). Eyes at least partly surrounded by a groove which connects with a pit on gena (Fig. 36.13 gr), sometimes indistinct. Eyes often divided, or at least constricted. Anterior edge of clypeus arcuately emarginate. Pronotum evenly convex or with shallow depressions on disc. Front margin of prosternum not raised into a collar. Metaventrite not swollen. Some species winged (metaventrite longer than first abdominal ventrite). L 5–13 mm. Throughout Australia .... ***Isopteron* Hope** (Fig. 36.84, Pl. 92H)

– Eyes not bordered by a groove or if shallow one present, pit on gena, when present, not connected to eye groove. Eyes only a little constricted at most. Anterior edge of clypeus usually straight in middle, but may be deeply emarginate in some *Adelodemus*. Pronotum with shallow median transverse depression or other surface irregularities. Front margin of prosternum sometimes raised into a collar. Metaventrite often swollen in middle. Wings absent **62**

62(61). Gula with very deep groove varying in shape between species, rarely just a median pit. Frontoclypeal suture deeply impressed or grooved. Lateral edges of prothorax, at least anteriorly, not dentate, may be lobed. Prosternal process bifurcate or at least weakly bilobate. L 9–14 mm. E QLD .... ***Adelodemus* Haag-Rutenberg** (Fig. 36.88, Pl. 84A)

– Gula without groove (except in *Monteithium storeyi* Matthews & Monteith), although small anteromedian pit may be present. Frontoclypeal suture only moderately or shallowly impressed. Lateral edges of prothorax often more or less dentate or ornate. Prosternal process bifurcate only in *Monteithium* **63**

63(62). Prosternal process unmodified, not forming prominent projection. Elytra with continuous lateral carina outside 8th stria, forming pseudepipleuron. Epipleura proper narrow throughout. Tarsi short, segments not lobed. L ~3 mm. QLD wet tropics .... ***Bolusculus* Matthews** (Pl. 86L)

– Prosternal process strongly projecting subapically. Elytra without pseudepipleura. Epipleura variable. Tarsi with penultimate segment lobed **64**

64(63). Sides of prothorax with highly ornate outgrowths. Eyes oval, ~1.5 × as long as wide. Last four antennomeres forming weak tomentose club after cupuliform segment 7. Prosternal process very broad and deeply bifurcate. Elytral crests capped by tubercles. Penultimate tarsomeres with rounded lobe. L 5–6 mm. QLD wet tropics .... ***Monteithium* Matthews** (Fig. 36.86, Pl. 94K)

– Sides of prothorax dentate or crenulate only. Eyes transverse, ~2 × as long as wide. Apical antennomeres similar to remainder. Prosternal process of normal width, nodiform, not bifurcate. Elytral crests crenulate or simple. Penultimate tarsomeres acuminate or asymmetrical beneath. L 8–9 mm. QLD wet tropics .... ***Bellendenum* Matthews** (Fig. 36.85)

65(60). With rounded pseudepipleural carina outside 7th interval, and with epipleuron proper entirely narrow, effaced or indistinctly outlined anteriorly. Elytral striae either effaced, irregularly confluent, or geminate, often with large ocellate impressions on intervals. L 9–19 mm. E Australia, TAS .... ***Coripera* Pascoe** (Fig. 36.87, Pl. 88H)

– Without pseudepipleuron (similar carina may be present, but corresponds to 9th interval) or, if present (some *Adelium*), epipleuron complete and widened anteriorly. Elytral striae rarely as above **66**

66(65). Mentum with anterior angles strongly produced and acute (Fig. 36.16). Frontal grooves often curved inward posteriorly, usually meeting on occiput. In most species prothorax with hind edge clearly narrower than front edge, usually with explanate sides delimited from disc by shallow groove; some species with subparallel prothorax. Elytral striae impunctate, consisting of just impressed lines or rarely irregular foveae. Males often with modified legs, without expanded protarsi. L 9–23 mm. E Australia except TAS .... ***Cardiothorax* Motschulsky** (Fig. 36.95, Pl. 87C)

– If anterior angles of mentum produced, apices less acute. Males very rarely with modified legs, usually with expanded protarsi. Other characters variable **67**

67(66). Epipleura relatively narrow for most or all of their length, their edges passing under humeri such that they are not visible from directly above (except at elytral bases in some species), and beetle over 14 mm long. Apical antennomere nearly always subparallel. Prothorax much narrower than hind body, as long as wide. Elytral striae (if present) impunctate **68**

– Epipleura gradually or abruptly widened anteriorly, visible from above at humeri, or if entirely narrow, beetle smaller (not over 11 mm long) and prothorax not much narrower than hind body. Apical antennomeres oval (linear only in some *Adelium*). Elytral striae variable **69**

68(67). Third antennomere longer than next two combined. Elytra with contiguous bullae, without striae except for sutural one. L 15–17 mm. QLD wet tropics .... ***Bluops* Carter** (Fig. 36.93, Pl. 86F)

– Third antennomere shorter. Elytra striate, without bullae. L 15–27 mm. E Australia except TAS .... ***Blepegenes* Pascoe** (Fig. 36.94, Pl. 86B, 86G)

69(67). Eyes reduced, oval, oblong or nearly round in outline, their long (transverse) axis not more than 2 × short (longitudinal) axis at widest point...... **70**

– Eyes transverse, their long axis more than 2 × short one **75**

70(69). Penultimate tarsomeres forming large lobes which subtend ~1/2 or more of claw segments. Elytra inflated and with projecting taper posteriorly. Epipleural carinae dentate at humeri in QLD species. L 3–4 mm. QLD and NSW rainforests. .... ***Diaspirus* Matthews** (Fig. 36.92, Pl. 89L)

– Penultimate tarsomeres without or with shorter lobes. Elytral edges almost evenly curved to apices. Epipleural carinae not dentate except in *Epomidus* **71**

71(70). Lateral pronotal and epipleural carinae absent, prothorax without angles, globiform. Surfaces covered with dense pile of erect setae and elytral intervals minutely tuberculate. L 3–5 mm. QLD wet tropics .... ***Apocryphodes* Matthews** (Fig. 36.96, Pl. 85F)

– Lateral pronotal and epipleural carinae, and at least posterior pronotal angles, present. Other characters variable **72**

72(71). Epipleural carinae at humeri forming strong projecting lobes visible from above. Dorsal surfaces with numerous erect setae. L ~4 mm. QLD wet tropics .... ***Epomidus* Matthews** (Fig. 36.91, Pl. 90G)

– Epipleural carinae even around edges of humeri. Setae sparser **73**

73(72). Bases of elytra depressed between humeral angles and receiving prothoracic base, causing angles to project forward. Intercoxal process of first ventrite truncate. L 5–9 mm. QLD wet tropics .... ***Dicyrtodes* Matthews** (Fig. 36.89, Pl. 89K)

– Bases of elytra convex, humeral angles not projecting forward. Intercoxal process of first ventrite rounded in outline **74**

74(73). Underside of pronotum (hypomeron) and prosternum impunctate. Elytra often with minute tubercles laterally and posteriorly. Eyes oblong in outline. L 7–11 mm. TAS .... ***Diemenoma* Matthews (part)** (Fig. 36.90, Pl. 89E)

– Hypomeron and prosternum punctate. Elytra without tubercles. Eyes round in outline, appearing more or less protuberant in dorsal view because of constriction of genal edges before them. L 3–8 mm. E and S Australia, TAS .... ***Brycopia* Pascoe** (Pl. 87B)

75(69). Epipleural carinae at humeri forming prominent angles or lobes visible from above in all but *D. puncticeps* (Carter), where epipleural carinae are merely sinuate at humeri. Frontal grooves absent. Anterior margination of pronotum effaced in middle and hypomeron coarsely punctate. Beetle largely glabrous. L 7–12 mm. SE Australia .... ***Daedrosis* Bates** (Fig. 36.98, Pl. 88I)

– Epipleural carinae evenly curved, not forming angles, or if angles present (some *Leptogastrus*), beetle less than 7 mm long. If epipleural carinae sinuate at humeri and pronotal margination and hypomeron as above (*Licinoma*), frontal grooves present **76**

76(75). Base of prothorax narrowed and bases of elytra rounded off, the two well separated by an exposed condyle formed by base of mesothorax, giving body pedunculate appearance. Hind body subparallel. Rounded base and sides of prothorax sometimes with prominent acute teeth. Dorsal surfaces densely and very coarsely punctate with numerous erect bristles. Frontal grooves absent. Apical antennomeres usually more than twice as massive as subapical. Elytral epipleural carina narrow basally, passing under humerus, angulate anteriorly in one species only. L 4–11 mm. E and SW Australia .... ***Leptogastrus* Macleay** (Fig. 36.106)

– Body not pedunculate, or if so broader and hind body not subparallel. Dorsal surfaces and frontal grooves variable. Apical antennomeres about twice as massive as subapical, or less. Elytral epipleura narrow basally only in *Dorrigonum*, not dentate **77**

77(76). Base of prothorax capable of at least slightly overlapping elytral bases. Elytra with at least a trace of elongated pustules or interrupted carinae... **78**

– Base of prothorax not overlapping elytral bases, although sometimes coming close. Elytra smooth **79**

78(77). Mentum flat or medially depressed, coarsely punctate with thickened edges and produced, prominent anterolateral processes which overhang and obscure true front angles (Fig. 36.15). Pronotum often with punctures of two sizes. Intercoxal process of first ventrite broadly truncate. Depressed areas of male ventrites almost always longitudinally corrugated. Legs without yellowish bands but apices of femora sometimes yellow. L 9–19 mm. E Australia, TAS .... ***Seirotrana* Pascoe** (Fig. 36.99, Pl. 98I)

– Mentum with irregular, tuberculate or vermiculate surface, without anterolateral processes. Intercoxal process of first ventrite rounded apically. Depressed areas of male ventrites smooth. Legs with yellowish bands. L 7–13 mm. VIC, S NSW .... ***Yarranum* Matthews** (Fig. 36.100, Pl. 101L)

79(77). Mentum with broadly rounded anterior angles and thickened anterior edge, often appearing swollen across its width but not more convex medially (Fig. 36.17). Male (with expanded protarsi) has first two medially concave ventrites often with pair of short inconspicuous longitudinal ridges posteriorly or shallow median channel in some individuals. Larger beetles (L 12–23 mm). SE Australia .... ***Apasis* Pascoe** (Fig. 36.101, Pl. 85H)

– Mentum with anterior angles variable but not thickened at anterior edge, flat, or if convex, in middle only. Male ventrites either unmodified or, if concave, without ridges or channel. Size and elytral striae variable **80**

80(79). Elytral epipleura relatively narrow basally, their carinate edges passing under crown of humeri. Apical antennomeres contrastingly flavous. Basal tarsomeres of metatarsi about twice as long as claw segments. L 7–10 mm. NE NSW, S QLD .... ***Dorrigonum* Matthews** (Fig. 36.102, Pl. 89J)

– Epipleura strongly widened at base, their carinate edge level with crown of humerus. Apical antennomeres concolorous or at most a little paler than rest. Basal metatarsomeres less than twice as long as claw segments **81**

81(80). Mentum more or less convex anteriorly, especially in middle (Fig. 36.18). Prothorax nearly always transverse, its ratio of width to length in middle ~3:2, its sides usually arcuately or sinuately produced, occasionally subquadrate, rarely narrowed at base. Frontal grooves present except in a few species. Occipital groove absent. Hind body often notably wider than prothorax. Penultimate tarsomeres distinctly lobed except in a few species. Basal metatarsomeres usually longer than claw segments. L 5–25 mm. Throughout Australia .... ***Adelium* Kirby** (Fig. 36.103, Pl. 84L)

– Mentum slightly or not convex anteriorly, may be longitudinally carinate. Prothorax less or not transverse (except in some *Nolicima*), sides less produced. Frontal grooves present or absent. Occipital groove usually present. Hind body only slightly wider than prothorax. Penultimate tarsomeres usually not lobed, subcupuliform. Basal metatarsomeres variable **82**

82(81). Prothorax distinctly wider in front than behind, with sharply angulate posterior angles, sides more or less sinuate just before them. Disc with median longitudinal sulcus. Elytral striae impunctate except in *N. asperatus* (Carter) (which has alternately carinate/granulate intervals). Epipleura with upper carinate edges reflexed and somewhat elevated above discal surface at humeri. Frontal grooves present (obscured by rugosity in *N. asperatus*). L 11–18 mm. NE NSW, S QLD .... ***Nototrintus* Carter** (Fig. 36.104, Pl. 95E)

– Prothorax subquadrate or rounded, at most a little widened anteriorly, hind angles more rounded, sides not sinuate. Disc with or without sulcus. Elytral striae rarely impunctate. Edges of epipleura at humeri not elevated. Frontal grooves present or absent. Length not more than 13 mm **83**

83(82). Anterior edge of pronotum with margination effaced, at least in middle. Hypomera with deep coarse punctures. Edges of epipleura, when viewed from side, sinuate, rising somewhat at humeri and descending to bases. Frontal grooves present and prolonged backward beyond level of front of eyes. L 6–12 mm. E Australia except TAS, Chile .... ***Licinoma* Pascoe** (Fig. 36.105, Pl. 93J)

– Pronotal margination complete. Hypomera impunctate or nearly so. Edges of epipleura, when seen from side, nearly always straight, running level across humeri to elytral bases. Frontal grooves often absent or incomplete, rarely reaching eyes **84**

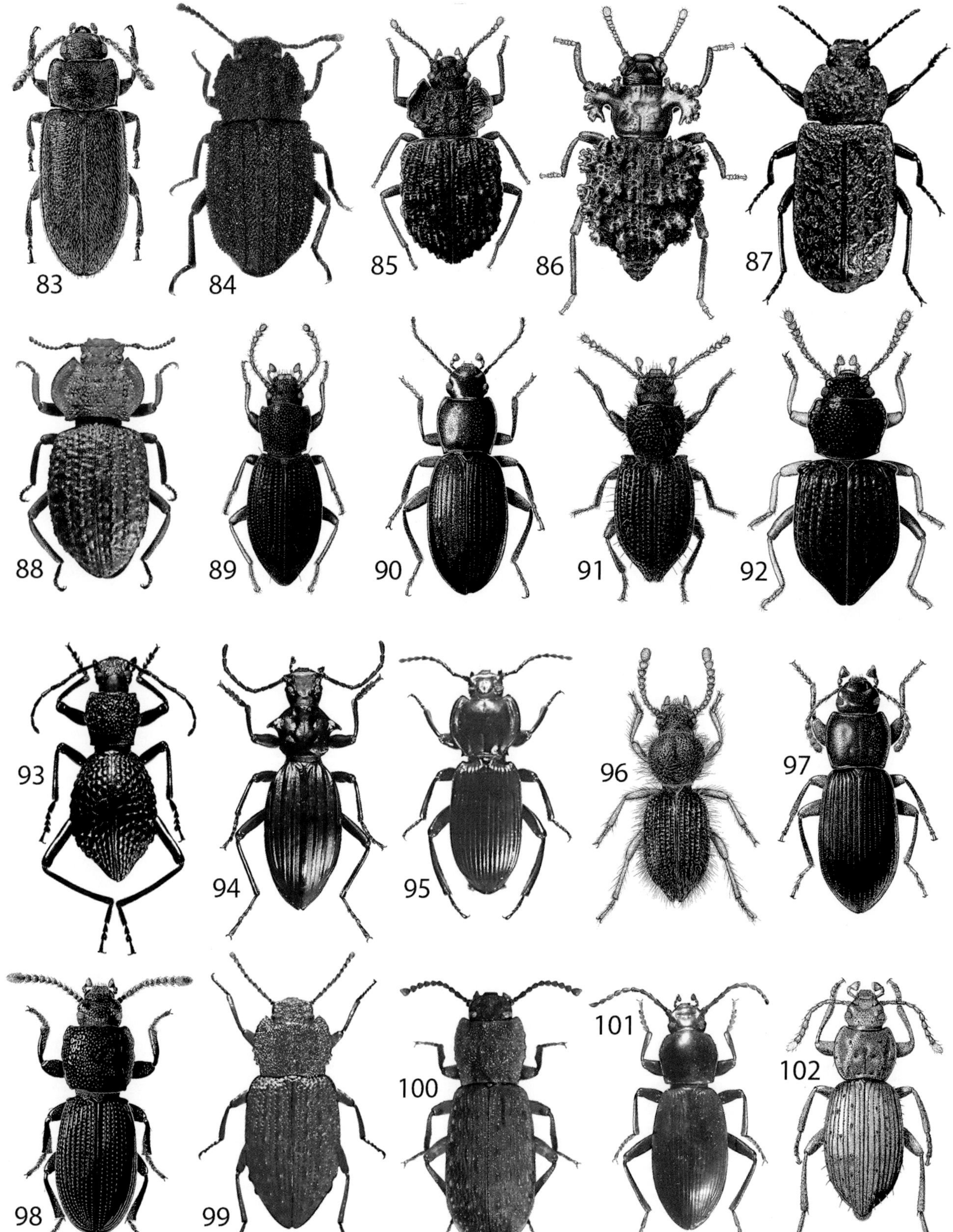

Figs 36.83–36.102. 83, *Lorelus*; 84, *Isopteron*; 85, *Bellendenum* © QM; 86, *Monteithium* © QM; 87, *Coripera*; 88, *Adelodemus*; 89, *Dicyrtodes* © GT; 90, *Diemenoma*; 91, *Epomidus* © GT; 92, *Diaspirus* © GT; 93, *Bluops*; 94, *Blepegenes*; 95, *Cardiothorax*; 96, *Apocryphodes* © GT; 97, *Nolicima*; 98, *Daedrosis*; 99, *Seirotrana*; 100, *Yarranum*; 101, *Apasis*; 102, *Dorrigonum*.

84(83). Eyes when seen from side with long (transverse) axis ~2 × longer than short one. Elytra often with minute tubercles laterally and posteriorly. L 7–11 mm. TAS .... ***Diemenoma* Matthews (part)** (Fig. 36.90, Pl. 89E)

– Eyes with long axis 3–4 longer than short one. Elytra not tuberculate. L 5–13 mm. E and SW Australia, TAS .... ***Nolicima* Matthews** (Fig. 36.97)

Tarsomeres not lobed, or if so, labrum transverse 85–end

85(45). Protibiae with outer apical angle strongly produced forming a rounded process; tibial spurs as long as first three tarsomeres combined. Meso- and metatibiae strongly clavate and spinose. Antennae with well developed 5-segmented club, the first four segments of which are strongly transverse. L 2–3 mm. Australia, Lord Howe Island, S Asia, Mediterranean, Gulf of Mexico, on beaches … **TRACHYSCELINI** .... ***Trachyscelis* Latreille** (Fig. 36.107, Pl. 100A)

– Without above characters in combination **86**

86(85). Eyes round or oval, or sometimes transverse but never emarginate **87**

– Eyes at least slightly emarginate, reniform, or divided by frontal canthus, rarely absent **110**

Eyes round or oval, not emarginate 87–109

87(86). Antennae with less than 11 segments **88**

– Antennae 11-segmented **91**

88(87). Antennae not clubbed, 10-segmented, tapering apically with terminal segment smallest. Pronotum dorsally with very large and deep pit posteriorly on either side of a pair of carinae. Tarsi 4–4–4. L ~2.5 mm. NW WA … **PALORINI (part)** .... ***Eutermicola* Lea** (Fig. 36.108, Pl. 101I)

– Antennae strongly clubbed. Other characters variable **89**

89(88). Antennae 10-segmented with 2-segmented club received in pockets beneath eyes and on underside of pronotum. A deep irregular groove across base of pronotum. Tarsi 4–4–4. L ~2 mm. ACT, VIC, Pacific region, Caribbean … **ARCHAEOGLENINI** .... ***Archaeoglenes* Broun** (Fig. 36.109, Pl. 85L)

– Without above combination … **GNATHIDIINI** **90**

90(89). Antennae 10-segmented with club of three or four segments. Surfaces largely glabrous. Lateral margins of pronotum simple. Tarsi 5–5–4. L 2.5–4 mm. S QLD, N NSW, VIC, Lord Howe Island, Asia-Pacific region … **GNATHIDIINA** .... ***Menimus* Sharp** (Fig. 36.110, Pl. 93A)

– Antennae 6-segmented with large 1-segmented club. Dorsal surfaces with dense erect setae. Lateral margins of pronotum widened and serrate. Tarsi 4–4–4. L 1.5–2 mm. S QLD, NSW, pantropical … **ANOPIDIINA** .... ***Tyrtaeus* Champion** (Fig. 36.111, Pl. 100B)

91(87). Tarsal segmentation actually or apparently 4–4–4 .... **92**

– Tarsal segmentation clearly 5–5–4 .... **100**

92(91). Lateral pronotal margins absent although pronotal surface is abruptly folded laterally. Antennae with a 4- or 5-segmented club. L ~2 mm. Cosmopolitan … **MYRMECHIXENINI** .... ***Myrmechixenus* Chevrolat** (Fig. 36.112, Pl. 94D)

– Lateral pronotal margins present. Antennae clubbed or not **93**

93(92). Antennae with a 3-segmented club, with stellate sensoria (as in Fig. 36.25 ss). Labrum transverse … **HYOCIINI (part)** **94**

– Antennae gradually enlarging distally or with weak 5-segmented club, with simple hairlike sensilla only. Labrum subquadrate … **PALORINI (part)** **96**

94(93). Anterior margin of eyes separated from gena by deep groove continued along inner eye margin. With acute subgenal processes below antennal bases (Fig. 36.21). Sides of pronotum explanate, thin and shiny. L 2–2.3 mm. Pilbara, WA. …**HYOCIINA (part)** .... ***Csiro* (*Millstreamia*) Medvedev & Lawrence** (Fig. 36.113)

– Eyes not separated from gena by groove. Subgenal processes absent. Sides of pronotum not shiny, although more or less flattened … **BRITTONINA** **95**

95(94). Outer margin of protibiae with two teeth separated by broad notch. Antennae short, club compact, the segments strongly transverse, apical setae present on segments 9–11, rest of club surface bare. Pronotum very little flattened along lateral margins, covered with scattered and barely distinguishable setae. Elytra without dark spots. L 1.6–1.9 mm. NT .... ***Brittona* Medvedev & Lawrence** (Fig. 36.114, Pl. 86J)

– Protibiae uniformly widened towards apex, outer margin without teeth but with sharp spinules. Antennomeres 3–8 weakly transverse or oblong, club looser, its segments uniformly covered with recumbent setae. Pronotum broadly flattened laterally, mesally either convex, a little flattened, or slightly depressed. Setal vestiture of pronotum well developed. Elytra with black spots, rarely absent. L 1.9–2.1 mm. N Australia .... ***Magela* Medvedev & Lawrence** (Fig. 36.115, Pl. 93D)

96(93). Antennae long, reaching base of pronotum, all segments longer than wide. Anterior (usually) and posterior angles of pronotum dentate. Form strongly flattened. Colour piceous with paler or rufous elytral areas. L 3–3.5 mm. E NSW, E QLD, N NT, Eurasia, Africa .... ***Platycotylus* Olliff** (Fig. 36.116, Pl. 96I)

– Antennae much shorter, with at least some segments wider than long. Angles of pronotum not dentate. Form more convex. Colour uniformly fuscous **97**

97(96). Elytra with interval 7 (humeral interval) carinate. L 3.6–4.3 mm. N QLD, cosmopolitan .... ***Ulomina* Baudi di Selve** (Fig. 36.117)

– Elytra either without carinae or with more than one carina **98**

98(97). Apical maxillary palpomere a little widened. Antennae loosely articulated. L 3–3.5 mm. Edges of genae angulate before eyes. N WA, N NT .... ***Austropalorus* Halstead** (Fig. 36.118, Pl. 85I)

– Apical maxillary palpomere oval or acuminate. Antennae usually compact, rarely loose with 5-segmented club. Edges of genae rounded or nearly straight before eyes **99**

99(98). Dorsal margin of eye level with or above side margin of gena (Fig. 36.20). Elytra not carinate, with interstrial punctures usually much finer. Tarsi 5–5–4, but basal pro- and mesotarsomeres very short, difficult to see. L 2.0–3.2 mm. Cosmopolitan .... ***Palorus* Mulsant** (Fig. 36.119, Pl. 96J)

– Dorsal margin of eye lower than side margin of gena (Fig. 36.19). Elytra carinate, or if punctate, with interstrial punctures comparatively large, equal to 1/4–2/3 diameter of strial punctures. Tarsi 5–5–4 or 4–4–4. L 2.3–2.8 mm. N WA, NE QLD .... ***Pseudeba* Blackburn** (Fig. 36.120, Pl. 97L)

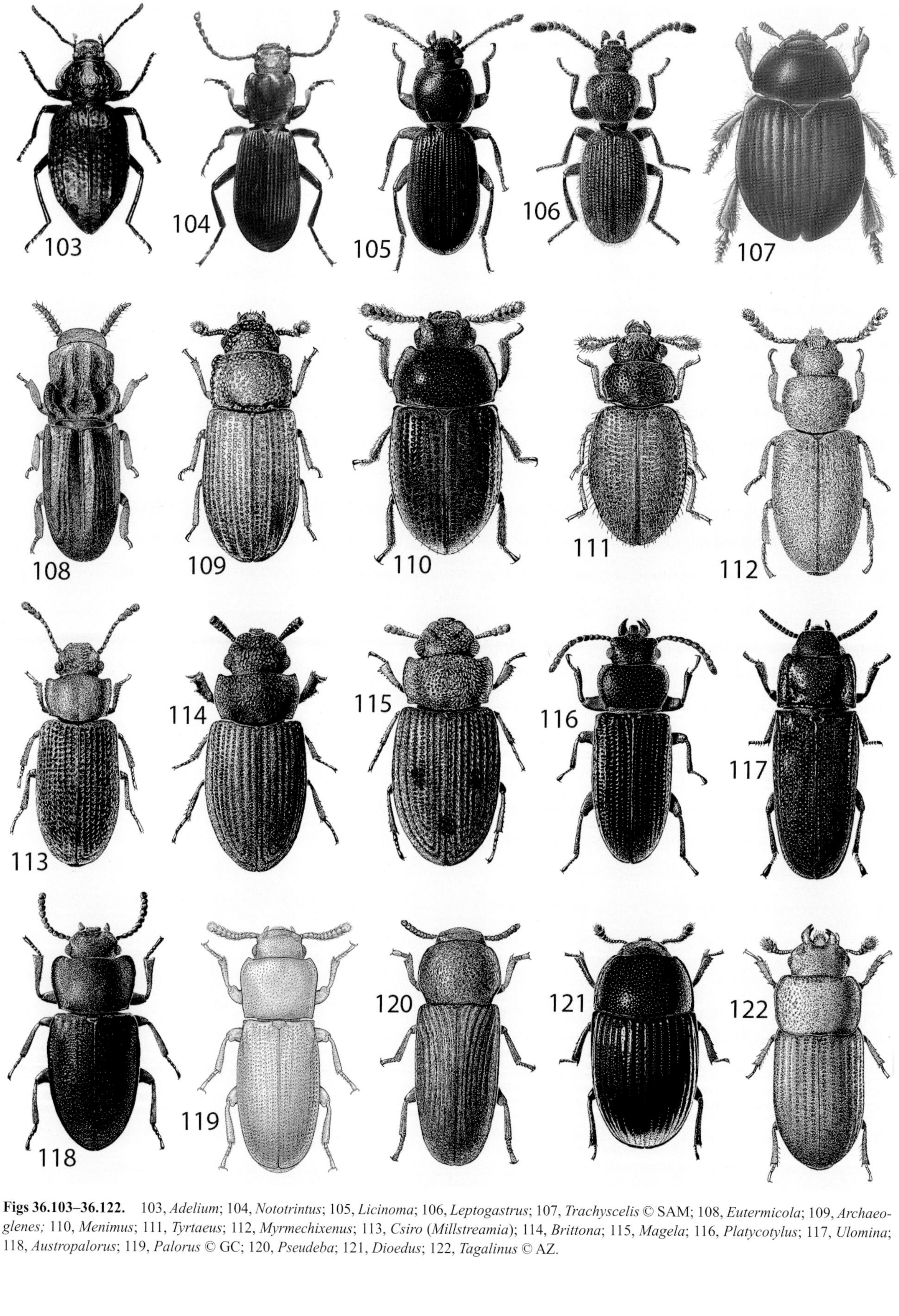

Figs 36.103–36.122. 103, *Adelium*; 104, *Nototrintus*; 105, *Licinoma*; 106, *Leptogastrus*; 107, *Trachyscelis* © SAM; 108, *Eutermicola*; 109, *Archaeoglenes;* 110, *Menimus*; 111, *Tyrtaeus*; 112, *Myrmechixenus*; 113, *Csiro* (*Millstreamia*); 114, *Brittona*; 115, *Magela*; 116, *Platycotylus*; 117, *Ulomina*; 118, *Austropalorus*; 119, *Palorus* © GC; 120, *Pseudeba*; 121, *Dioedus*; 122, *Tagalinus* © AZ.

100(91). Mesotrochantin well developed and exposed, forming lateral wall of coxal cavity (Fig. 36.4), which is not closed laterally by meeting of meso- and metaventrites. Antennal clubs strongly developed, bearing only simple, hairlike sensilla. Underside of head with pockets for reception of antennae. Protibiae with two apical spurs ... **PENETINI** **101**

– Mesotrochantin not visible, mesocoxal cavities closed laterally by meeting of meso- and metaventrites (as in Fig. 36.7). Antennal clubs weakly developed, bearing compound sensoria. Underside of head without pockets. Protibiae without or with only one apical spur ... **HYOCIINI (part)** **105**

101(100). Antennal club 2-segmented, antennal pockets sharply defined or not **102**

– Antennal club 3- or 4-segmented, antennal pockets sharply defined **103**

102(101). Stria 8 of elytra scarcely shorter than 7. Impressions on underside of head for reception of antennae not sharply defined. Edges of genae simple. L 2.2–4.8 mm. N QLD, Lord Howe and Norfolk islands, Pacific region, N and S America .... ***Dioedus* LeConte** (Fig. 36.121, Pl. 89A)

– Stria 8 very short, beginning well away from base and barely going beyond middle of elytra. Underside of head with sharply defined grooves to receive antennae. Anterior edges of genae with a small angular projection. L 3.2–4.5 mm. N QLD, New Caledonia .... ***Tagalinus* Kaszab** (Fig. 36.122, Pl. 99J)

103(101). Genae distinctly margined between eyes and clypeal suture. Margin between clypeus and genae, where curved frontal groove runs into it, emarginate. Antennal club 3- or 4-segmented. Left mandible with three sharp teeth. L 3.5–4.5 mm. E Australia, TAS, Sumatra, Philippines .... ***Scolytocaulus* Fairmaire** (Fig. 36.123, Pl. 98K)

– Genae without trace of margin. Clypeal and genal edges more or less confluent. Antennal club 3-segmented. Left mandible with two or three teeth **104**

104(103). Mandibles with three teeth. Pronotum and elytra moderately convex. Striae 7 and 8 subequal in length. Winged. L ~4 mm. SE Australia .... ***Pseudophthora* Kaszab** (Fig. 36.124, Pl. 97K)

– Mandibles with two teeth. Pronotum and elytra very convex, subcylindrical. Stria 8 anteriorly much shorter than 7. Wingless. L 2.3–3 mm. TAS, NSW, Chile .... ***Archeophthora* Kaszab** (Fig. 36.125, Pl. 85J)

105(100). Protibiae very broad (length not more than twice width), outer edges with two broad teeth (apical and middle) divided by deep arcuate emargination. Sides of pronotum and elytra covered with long hairs. Intercoxal process of ventrite 1 angulate. Antennae short, barely reaching beyond anterior edge of pronotum, with setae only on apical edges of club segments. L 2.1–3.1 mm. NT and N SA ... **UPTONINA** .... ***Uptona* Medvedev & Lawrence** (Fig. 36.126, Pl. 101A)

– Protibiae broad or not, but more than twice as long as wide, the outer edge without large teeth, but may have spines. Sides of pronotum and elytra covered with stout recumbent setae. Intercoxal process of first ventrite broadly rounded apically. Antennae long or short, segments more or less densely setose ... **HYOCIINA (part)** **106**

106(105). Strong subgenal spines present before eyes (Fig. 36.21). Margins of pronotum explanate. Anterior margin of eye often separated from gena by deep groove. Body usually black or dark grey, sometimes brownish. L 1.6–3.4 mm. Throughout Australia .... ***Csiro s. str.* Medvedev & Lawrence** (Fig. 36.127)

– No subgenal spines. Other characters variable **107**

107(106). Antennae and body black. Dorsal vestiture indistinct, consisting of very short setae on elytral intervals. Occipital ridge running along hind edge of eye bent forward below. Sides of elytra distinctly arcuately convex. Pronotal base opposite scutellum and near hind angles deeply impressed. Posterior angles smooth. L 2–2.3 mm. N QLD, Pacific islands .... ***Parahyocis* Kaszab** (Fig. 36.128, Pl. 96K)

– Antennae entirely or at least basally pale, pronotum and elytra, or elytra alone, pale, if body entirely black dorsally, then antennae entirely pale. Vestiture usually well developed on body, always distinct on elytral intervals. Occipital surface without distinct ridge, or it latter present (*Hyocis*), not continued forward along lower eye margin **108**

108(107). Disc of pronotum with median longitudinal depression or groove, sometimes indistinct or confined to hind edge, and apical spur of protibiae not downwardly and inwardly uncinate. Outer edges of protibiae spinulose. L 2.4–3.2 mm. Australia wide .... ***Hyocis s. str.* Pascoe** (Fig. 36.129, Pl. 91H)

– Disc of pronotum without median longitudinal groove, evenly convex, but with depression on base opposite scutellum. Outer edges of protibiae entire, without teeth or spines **109**

109(108). 9th (outer) stria not continued forward to elytral base, leaving gap equivalent to about four punctations. Eyes weakly transverse (ratio of height to length ~9:8), scarcely encroaching on lower surface of head. Apical spur of protibia short, weakly bent downward. Pronotal base with deep depression opposite scutellum. L 1.7 mm. S WA .... ***Hyocis* (*Nannohyocis*) Medvedev & Lawrence** (Fig. 36.130)

– 9th stria running across humerus to elytral base. Eyes large, a little more transverse, covering lateral and lower surfaces of head. Apical spur of protibia usually downwardly and inwardly uncinate. Pronotal base with indistinct depression opposite scutellum. L 2.1–4.1 mm. NT, N QLD, NSW, N SA .... ***Hyocis* (*Neohyocis*) Medvedev & Lawrence**. (Fig. 36.131)

Eyes emarginate, divided, or absent 110–end

110(86). Clypeus with anterior margin deeply excised **111**

– Clypeus with anterior margin produced, straight, or very shallowly excised **126**

Clypeus deeply excised 111–125

111(110). Clypeolabral membrane broadly exposed at base of labrum. Protibiae slender **112**

– Clypeolabral membrane concealed under clypeus. Protibiae more or less apically expanded, often dentate on outer edges **114**

112(111). Apical antennomeres bearing numerous compound sensoria which appear as pale round spots at lower magnifications. Elytra with nine striae and a scutellary striole, without costae. L 25–34 mm. W Hemisphere, introduced into Australia ... **TRIBE UNCERTAIN** .... ***Zophobas* Dejean** (Fig. 36.132, Pl. 101C)

– Antennomeres with simple hairlike sensilla only. Elytra without distinct striae, more or less costate **113**

113(112). Tibial spurs longer than basal tarsomeres. Eyes small, dorsally separated by 2.5–5 eye widths. Males with meso- and metafemora unmodified or with their hind margins angularly expanded near bases. L 10–26 mm. Arid zone, mainly in caves .... **HELEINI-HELEINA (Part)** .... ***Brises*** **Pascoe** (Fig. 36.133, Pl. 86A)
– Tibial spurs much shorter. Eyes large, dorsally separated by about one eye width. Male with meso- and metafemora having slightly concave posterior faces bearing rows of tubercles at upper and lower edges. L 21–27 mm. SE SA, VIC, E NSW, E QLD ... **HELEINI-CYPHALEINA (Part)** .... ***Prophanes*** **Westwood** (Fig. 36.134, Pl. 97F)
114(111). Dorsal surfaces, including eyes, completely covered with variegated scales. L 3.5–5 mm. S WA, E Australia, world-wide, coastal ... **PEDININI-LEICHENINA** .... ***Leichenum*** **Blanchard** (Fig. 36.135, Pl. 92B)
– Scales, if present, not wholly covering surfaces and of uniform colour, not present on eyes **115**
115(114). No trace of elytral striae **116**
– Elytral striae present, at least in part **117**
116(115). Dorsal surfaces with long fine adpressed silken setae. Tibiae without any preapical teeth. L 7–9 mm. TAS, coastal ... **HELEINI (HELEINA) (Part)** .... ***Edylius*** **Champion** (Fig. 36.136, Pl. 90B)
– Dorsal surfaces mostly glabrous with lateral fringe of long bristles. All tibiae with preapical teeth or prominent tubercles on outer edges. L 4–5 mm. S WA, S SA, Eurasia, Africa, N. America ... **MELANIMONINI** .... ***Cheirodes*** **Gené** (Fig. 36.137, Pl. 88B)
117(115). Elytral epipleura complete to apices. Males with strongly distorted protibiae and expanded protarsi. L 11–20 mm. Southern India, Sri Lanka, Cocos-Keeling islands ... **PEDININI-PLATYNOTINA** .... ***Menearchus*** **Carter** (Fig. 36.142, Pl. 93H)
– Elytral epipleura terminating before apices. Protibiae not distorted, protarsi expanded or not **118**
118(117). Pro- and usually mesotarsomeres 2 and 3 wider than rest, more strongly so in males. Dorsal surfaces glabrous ... **OPATRINI-HETEROTARSINA** **119**
– Tarsomeres not expanded. Dorsal surfaces with adpressed or semi-erect setae, usually squamiform **121**
119(118). Metaventrite subequal in length to first abdominal ventrite. Outer edges of protibiae apically produced into broadly rounded lobes. L 6–8 mm. S and E coasts .... ***Scymena*** **Pascoe** (Fig. 36.141, Pl. 98J)
– Metaventrite longer than first abdominal ventrite. Outer edges of protibiae not appreciably produced **120**
120(119). Elytral striae evenly impressed throughout. L 5–8 mm. All coasts of Australia, Asia and Africa .... ***Heterocheira*** **Dejean** (Fig. 36.140, Pl. 91A)
– Striae at bases of elytra more shallowly impressed or absent. L 4–5 mm. N WA, N NT, Asia, Africa, coastal .... ***Diphyrrhynchus*** **Fairmaire** (Fig. 36.139, Pl. 89C)
121(118). Terminal maxillary palpomere securiform. Antennae long, reaching to about middle of prothorax. Protibiae not strongly expanded, without preapical teeth, their undersurface not tuberculate. Scutellum fully exposed behind pronotal base ... **OPATRINI-OPATRINA** **122**
– Terminal maxillary palpomere triangular or oval. Antennae short, reaching to only just beyond front edge of prothorax. Protibiae strongly expanded apically, with one or more preapical teeth or lobes, their undersurface with at least some tubercles. Scutellum largely or wholly overlapped by pronotal base ... **OPATRINI-AMMOBIINA** **124**
122(121). Eyes completely, or almost completely, divided by genal canthus. Head surface even, with no trace of frontoclypeal suture. Mentum with prominent median keel. L 6–8 mm. Arid Australia, Asia, Africa .... ***Mesomorphus*** **Miedel** (Fig. 36.138, Pl. 93F)
– Eyes constricted but not divided. Frontoclypeal suture variable. Mentum without median keel or with a low one **123**
123(122). Each elytral interval with a single row of tubercles, each of which bears a long, semi-erect squamiform seta, strial punctures large and deep. L 5–7 mm. NE SA, Eurasia, Africa .... ***Scleropatroides*** **Löbl & Merkl** (Fig. 36.147, Pl. 98A)
– Each elytral interval with multiple rows of shorter squamiform or ciliate setae, or setae absent. Strial puncture small or absent. L 6–11 mm. Throughout Australia, Eurasia, Africa .... ***Gonocephalum*** **Solier** (Fig. 36.143, Pl. 91K)
124(121). Abdominal ventrites with minute tubercle at base of each squamiform seta. Elytra with at least a few minute tubercles. L 5–7 mm. Arid and coastal Australia, Eurasia, Africa .... ***Caedius*** **Blanchard** (Fig. 36.144, Pl. 87K)
– Ventrites with only punctures at bases of setae. Elytra without tubercles **125**
125(124). Width of protibia behind lateral apical tooth less than length of tooth. Scutellum not visible. L 3–4 mm. N WA, N QLD, N SA .... ***Sobas*** **Pascoe** (Fig. 36.145, Pl. 98B)
– Width of protibia behind apical tooth much more than length of tooth, tibia strongly dilated. A minute scutellum visible. L 4–5 mm. All Australian coasts .... ***Caediomorpha*** **Blackburn** (Fig. 36.146, Pl. 87J)

Clypeus not deeply excised 126–end

126(110). Head in repose vertical or nearly so (hypognathous). Prosternum in front of coxae less than twice as long as postcoxal bridge **127**
– Head in repose porrect. Prosternum in front of coxae more than twice as long as postcoxal bridge **135**

Hypognathous 127–134

127(126). Clypeolabral membrane hidden. Epipleura ending at about anterior edge of 5th ventrite. Antennae with simple, hair-like sensilla only. L 8–13 mm. E Australia, TAS ... **TITAENINI** .... ***Titaena*** **Erichson** (Fig. 36.148, Pl. 99L)
– Clypeolabral membrane fully exposed. Epipleura complete or not. Apical segments of antennae bearing compound sensoria consisting of clusters of very short sensilla (Fig. 36.25 ss) **128**
128(127). Epipleura terminating before elytral apices. Eyes small, separated by more than two eye diameters in front view. Body subglobose, colour light brown. L 4–5 mm. NE QLD ... **CNODALONINI (part)** .... ***Cuemus*** **Bouchard** (Fig. 36.149, Pl. 88E)
– Epipleura complete to apices. Eyes larger. Body less convex, colour metallic, patterned, or black **129**
129(128). Head less adpressed to prosternum (portion of prosternum in front of coxae at least as long as postcoxal bridge) ... **STENOCHIINI** **130**
– Head closely adpressed to prosternum (portion of prosternum in front of coxae shorter than postcoxal bridge) ... **AMARYGMINI** **131**
130(129). Wingless, metaventrite along midline shorter than first abdominal ventrite. L 7–8 mm. N WA, N NT .... ***Eutherama*** **Carter** (Fig. 36.150, Pl. 90C)
– Winged, metaventrite along midline longer than first abdominal ventrite. L 8–15 mm. N QLD, N NT, Pantropical .... ***Strongylium*** **Kirby** (Fig. 36.151, Pl. 99F)

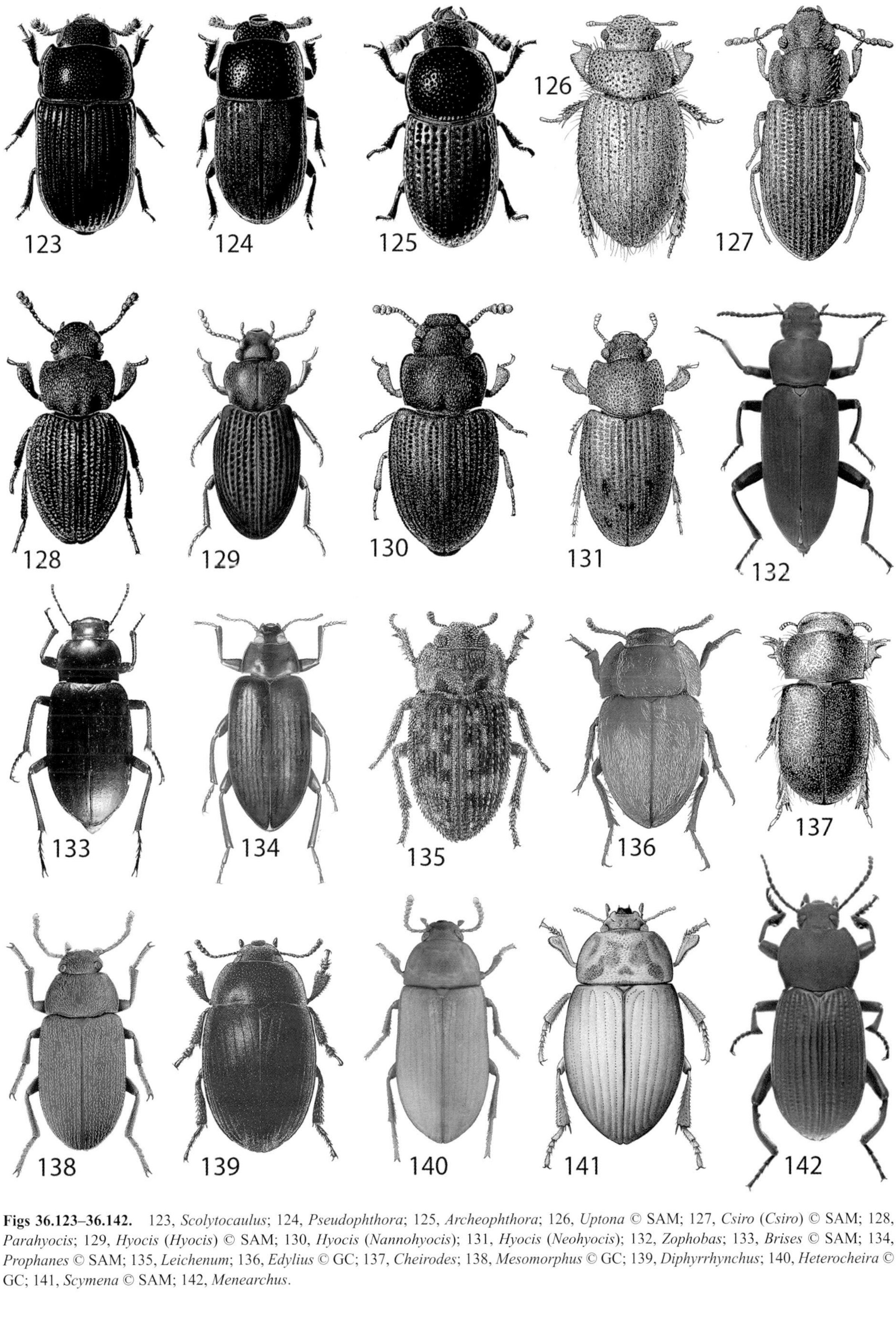

Figs 36.123–36.142. 123, *Scolytocaulus*; 124, *Pseudophthora*; 125, *Archeophthora*; 126, *Uptona* © SAM; 127, *Csiro* (*Csiro*) © SAM; 128, *Parahyocis*; 129, *Hyocis* (*Hyocis*) © SAM; 130, *Hyocis* (*Nannohyocis*); 131, *Hyocis* (*Neohyocis*); 132, *Zophobas*; 133, *Brises* © SAM; 134, *Prophanes* © SAM; 135, *Leichenum*; 136, *Edylius* © GC; 137, *Cheirodes*; 138, *Mesomorphus* © GC; 139, *Diphyrrhynchus*; 140, *Heterocheira* © GC; 141, *Scymena* © SAM; 142, *Menearchus*.

131(129). Metaventrite subequal in length along midline to first abdominal ventrite (beetle flightless). Lateral margin of pronotum largely obliterated. Elytral striae consisting of large, deep punctures joined by longitudinal impressed lines. Apex of right mandible truncate, that of left bifid. L 17–20 mm. Central QLD .... ***Axynaon* Blackburn** (Fig. 36.152, Pl. 85B)
– Metaventrite longer (beetle winged). Lateral margin of pronotum complete. Elytral striae with smaller or shallower punctures. Both mandibles either truncate or bifid **132**
132(131). Beetle with dense, long, erect pilosity. Mandibles apically truncate. L 10–11 mm. Central WA .... ***Trichamarygmus* Carter** (Fig. 36.153, Pl. 100F)
– Beetle glabrous. Mandibles variable **133**
133(132). Mandibles apically bifid, outwardly with distal sulcus arising between teeth. Body form usually oval, rarely subparallel. Elytra usually distinctly striate, but sometimes only punctate. L 6–15 mm. Throughout Australia, mostly N, Eurasia, Africa .... ***Amarygmus* Dahm** (Fig. 36.155, Pl. 84E)
– Mandibles apically simple, without sulcus. Body subparallel. Elytra variable **134**
134(133). Mandibles apically narrowly rounded. Elytra deeply striate with strongly convex intervals. L 5 mm or less. NT and N QLD .... ***Isopteroplonyx* Bremer** (Fig. 36.156, Pl. 92G)
– Mandibles apically truncate or broadly rounded. Elytra sometimes with striae impressed, but usually with only superficial and minute punctures, intervals less convex. L 6–23 mm. Throughout Australia, mostly S .... ***Chalcopteroides* Strand** (Fig. 36.154, Pl. 88F)

Head porrect 135–end

135(126). Antennae with compound stellate sensoria consisting of clusters of minute sensilla (Fig. 36.25), at lower magnifications appearing as whitish spots **136**
– Antennae with simple hairlike sensilla only **174**

Antennae with stellate sensoria 136–173

136(135). Elytra exposing hind edge of 7th abdominal tergite, their apices arcuately cut away with epipleura incomplete **137**
– Elytra completely covering abdomen, epipleura complete or not **142**
137(136). Form elongate-cylindrical. Protibiae with outer edges entire. Antennae without placoid sensoria on distal segments. L 1.5–6 mm. N Australia, world-wide … **HYPOPHLAEINI** .... ***Corticeus* Piller & Mitterpacher** (Fig. 36.157, Pl. 88D)
– Form oblong. Protibiae with outer edges dentate or spinulose. Antennae with platelike placoid sensoria as well as stellate ones (Fig. 36.26 ps) … **ULOMINI** **138**
138(137). Outer edge of protibiae bearing a row of numerous small spines. Clypeo-labral membrane broadly exposed, labrum in same plane as clypeus. Antennomeres 6–10 only a little wider than long. L 6.0–6.5 mm. NT, SE Asia .... ***Cenoscelis* Wollaston** (Fig. 36.158, Pl. 87A)
– Outer edge of protibiae strongly toothed. Clypeo-labral membrane covered, or exposed in middle only, labrum in lower plane than clypeus. Antennomeres 6–10 much wider than long **139**
139(138). Eyes absent. L 4–4.5 mm. NE QLD .... ***Typhluloma* Lea** (Fig. 36.159, Pl.100D)
– Eyes developed. L > 5 mm **140**
140(139). Apical edge (truncation) of meso- and metatibiae toothed ventrally with only a few spine-like bristles between teeth (Fig. 36.10). Both sexes similar, with produced clypeal lamella. L 14–23 mm. E Australia, TAS, N NT .... ***Achthosus* Pascoe** (Fig. 36.160, Pl. 84C)
– Truncation of meso- and metatibiae with ventral row of short spine- or peg-like bristles, not toothed (Fig. 36.9). Males and females often with different cephalic and pronotal sculpturing. (*Uloma*) **141**
141(140). Metaventrite at least as long as half length of mesocoxae (beetle winged). Elytra parallel or a little oval. L 6.5–15 mm. E and N Australia, TAS, world-wide .... ***Uloma s. str.* Dejean** (Fig. 36.161, Pl. 100C)
– Metaventrite much shorter than half length of mesocoxae (beetle flightless). Elytra short, broadly oval. L 5.5–5.8 mm. N QLD .... ***Uloma* (*Apterulomoides* Kaszab)** (Fig. 36.162)
142(136). Mesocoxal cavities closed by ventrites only (as in Fig. 36.7) **143**
– Mesocoxal cavities closed in part by mesepimera (as in Fig. 36.6) **155**

Mesocoxal cavities closed by ventrites only 143–154

143(142). Clypeolabral membrane concealed under clypeus **144**
– Clypeolabral membrane exposed before clypeus **147**
144(143). Epipleura terminating abruptly before elytral apices. Prosternal process not acutely prolonged. Male with distorted head and hornlike outgrowths on mandibles. L 3–4 mm. Cosmopolitan … **DIAPERINI (part)** .... ***Gnatocerus* Thunberg** (Fig. 36.163, Pl. 90A)
– Epipleura continuing to elytral apices. Prosternal process acutely prolonged and received in deep V-shaped concavity of mesoventrite **145**
145 (144). Elytra pale with dark markings. L 2–3 mm. N Australia, world-wide. … **CRYPTICINI** .... ***Ellipsodes* Wollaston** (Fig. 36.164, Pl. 90H)
– Elytra uniformly dark. L > 4 mm. … **ALPHITOBIINI** **146**
146(145). Middle of clypeal margin shallowly excised. Apical six antennomeres abruptly enlarged. Mentum with broad median flat area bounded laterally by vertical edges. Prosternal process arcuate in profile, apically rounded. Surfaces coarsely punctate. L 4.5–6 mm. Cosmopolitan .... ***Alphitobius* Stephens** (Fig. 36.165, Pl. 84B)
– Middle of clypeal margin straight. Antennae gradually enlarging distally. Mentum with median rounded convexity. Prosternal process subhorizontal in profile, apically acute. Surfaces very finely punctate. L 5–6 mm. NE QLD, Eurasia.... ***Diaclina* Jacquelin du Val** (Fig. 36.166, Pl. 89H)
147(143). Protibiae enlarged and flattened apically. Meso- and metatibiae bearing dense, very coarse erect spines. L ~8 mm. Port Phillip Bay, VIC, world-wide on beaches … **PHALERIINI** .... ***Phaleria* Latreille** (Fig. 36.167, Pl. 96E)
– Protibiae subcylindrical, slightly or not enlarged at apices. Meso- and metatibiae bearing fine setae or prostrate bristles … **DIAPERINI (most)**. **148**

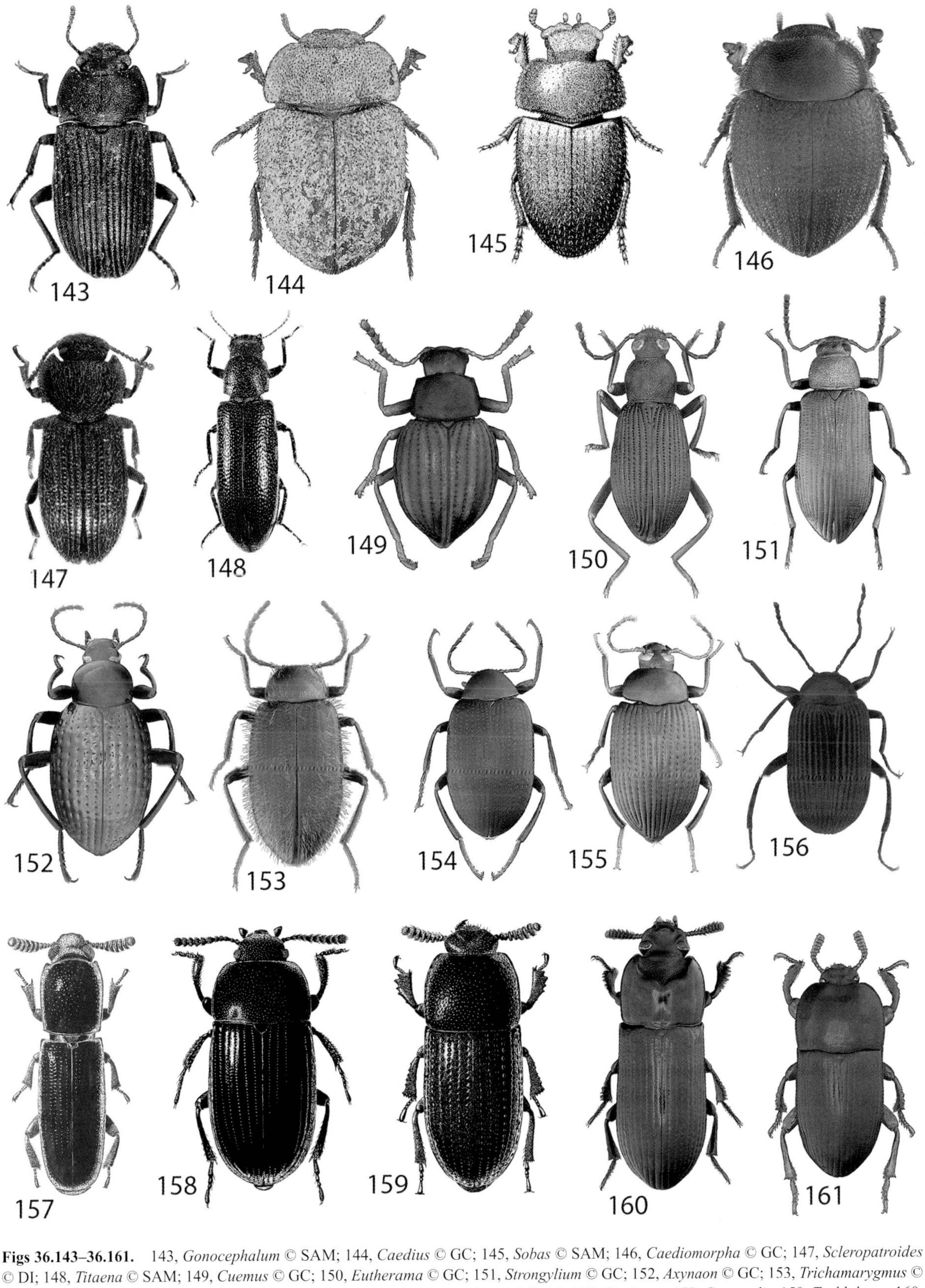

Figs 36.143–36.161. 143, *Gonocephalum* © SAM; 144, *Caedius* © GC; 145, *Sobas* © SAM; 146, *Caediomorpha* © GC; 147, *Scleropatroides* © DI; 148, *Titaena* © SAM; 149, *Cuemus* © GC; 150, *Eutherama* © GC; 151, *Strongylium* © GC; 152, *Axynaon* © GC; 153, *Trichamarygmus* © GC; 154, *Chalcopteroides* © GC; 155, *Amarygmus* © GC; 156, *Isopteroplonyx* © SAM; 157, *Corticeus*; 158, *Cenoscelis*; 159, *Typhluloma*; 160, *Achthosus* © GC; 161, *Uloma* (*Uloma*) © GC.

148(147). Antennae with capitate club of five subequal segments. Apical maxillary palpomeres fusiform. Elytra not striate, without large punctures. L 2–2.5 mm. NE NSW, E QLD, N NT, world-wide .... ***Pentaphyllus* Dejean** (Fig. 36.168, Pl. 96L)
– Antennae gradually enlarged apically or with only feeble club. Apical maxillary palpomeres asymmetrical. Elytra striate or with lines of large punctures **149**
149(148). Elytral epipleura terminating before apices, opposite base or middle of 5th ventrite **150**
– Elytral epipleura complete to apices or at least to point beyond middle of 5th ventrite **152**
150(149). Form extremely flattened, with elytral surface becoming abruptly vertical outside 7th stria. L 4.5–5 mm. NWA, Asia, W Hemisphere .... ***Adelina* Dejean** (Fig. 36.169, Pl. 84K)
– Form more convex, elytral surface evenly rounded **151**
151(150). Tarsomeres subparallel. Antennomeres 4–10 gradually widening. Form of beetle elongate-oblong. L ~3 mm. NT, QLD, NSW, VIC, SE SA........ ***Palembomimus* Matthews & Lawrence** (Fig. 36.170, Pl. 96A)
– Penultimate tarsomeres cupuliform. Last four antennomeres strongly widening towards apex. Form oval. L ~4 mm. N QLD, Pacific .... ***Sciophagus* Sharp** (Fig. 36.171, Pl. 97D)
152(149). Outline of head indented at frontoclypeal suture. Dorsum covered with minute fine recumbent setae. Elytra bicoloured. Male with swollen clypeus. L ~2 mm. Cosmopolitan........ ***Alphitophagus* Stephens** (Fig. 36.172, Pl. 84G)
– Outline of head describing a more or less even curve. Surfaces glabrous. Elytral colour variable. Without swollen clypeus. Size larger........ **153**
153(152). Antennae subserrate. L 8–10 mm. E QLD, Asia, Africa .... ***Ceropria* Laporte & Brullé** (Fig. 36.173, Pl. 87I)
– Antennomeres symmetrical. Beetle smaller........ **154**
154(153). Base of pronotum completely margined. Intercoxal process of prosternum declivous behind coxae with small elevated apex, or entirely feebly elevated. Form often flattened and subparallel. L 2.5–5 mm. Throughout Australia, Asia, Africa, N America .... ***Ulomoides* Blackburn** (Fig. 36.174, Pl. 100E)
– Base of pronotum usually without margin, or this confined to middle. Intercoxal process of prosternum more elevated, its posterior face high and vertical when viewed from side. Form usually more convex and oval. L 3–7 mm. Throughout Australia, world-wide........ ***Platydema* Laporte & Brullé** (Fig. 36.175, Pl. 96C)

Mesocoxal cavities closed in part by mesepimera (open) 155–end

155(142). Pronotum and elytra clothed with long erect bristles. Tarsi with spines on ventral surfaces. Posterior pronotal angles excised and acutely dentate behind excision. Elytra often with metallic blue reflections … **ECTYCHINI** **156**
– Pronotum and elytra glabrous or nearly so. Tarsi and apices of tibiae densely set with fine setae on ventral surfaces, usually appearing pilose or tomentose. Posterior pronotal angles not excised. Colour variable … **CNODALONINI (except *Cuemus*)** **157**
156(147). Antennae perfoliate (consisting of loosely-jointed disc-like segments), especially towards apices, usually long enough to reach or surpass base of prothorax. L 4.5–8.0 mm. WA and NT .... ***Ectyche* Pascoe** (Fig. 36.176, Pl. 90E)
– Antennae more compact and shorter. L 3.5–4.5 mm. WA and SA .... ***Micrectyche* Bates** (Fig. 36.177, Pl. 94L)
157(155). Clypeolabral membrane narrowly exposed, clypeus very short and crescentic in shape, its anterior edge broadly excised. L 7–9 mm. NE QLD, New Guinea .... ***Neotheca* Carter** (Fig. 36.178, Pl. 94I)
– Clypeolabral membrane concealed or almost so (if membrane is widely exposed across width of clypeus, you have a stenochiine; go back to couplet 129), clypeus longer and with anterior edge a little produced, straight or slightly concave **158**
158(157). Dorsal surfaces strongly tuberculate, vermiculate and carinate, lateral edges of pronotum serrate or undulate, dorsal edges of eyes usually raised into low crests. L 6–9 mm. NE QLD, New Guinea, Asia, Pacific islands, Africa .... ***Bradymerus* Perrroud** (Fig. 36.187, Pl. 86I)
– Dorsal surfaces smooth, or if partly carinate (some *Apterotheca*) lateral edges of pronotum not serrate, may be feebly undulate, no crests above eyes **159**
159(158). Hind edge of pronotum not, or not completely, margined (sometimes with very fine margin in *Campolene*)........ **160**
– Hind edge of pronotum fully margined **164**
160(159). Prosternal process strongly produced posteriorly and received in deep V-shaped concavity of mesoventrite **161**
– Prosternal process less or not produced, mesoventrite without concavity or with shallow one **163**
161(160). Prothorax transverse, about twice as wide as long, the anterior angles strongly produced. Preapical tarsomeres lobed. L 4–5 mm. N NSW, E QLD, New Guinea, Asia........ ***Androsus* Gebien** (Fig. 36.179, Pl. 85C)
– Prothorax less transverse, anterior angles subquadrate or rounded. Tarsomeres simple. L > 5 mm **162**
162(161). Surface of elytral apices evenly convex. L 7–9 mm. N QLD, New Guinea .... ***Chariotheca* Pascoe** (Fig. 36.180, Pl. 88C)
– Surface of elytral apices with interstria 7 distinctly more convex than neighbouring interstriae, forming obtuse angle between dorsal surface and sides of elytra. L 10–12 mm. Lord Howe and Norfolk islands .... ***Metisopus* Bates** (Fig. 36.181, Pl. 94A)
163(160). Legs rufous or fuscous, contrasting with bluish-black body. Tibiae distinctly bowed. Pronotal disc more convex anteriorly, with deep groove along lateral edges. Surfaces smooth. Reduced wings present. L 8–10 mm. N NSW, SE QLD .... ***Campolene* Pascoe** (Fig. 36.182, Pl. 87H)
– Legs concolorous with body. Tibiae less bowed, pronotal disc flatter. Surfaces either smooth, tuberculate, or carinate (on elytra). Wings absent. 7–16 mm. NE QLD .... ***Apterotheca* Gebien** (Fig. 36.183, Pl. 85K)
164(159). Fold between pronotal disc and hypomeron in cross section forming obtuse angle or even curve, lateral margin fine and hind angles obsolescent...... **165**
– Fold between pronotal disc and hypomeron in cross section forming right or acute angle, lateral margin and hind angles more prominent...... **167**
165(164). Femora clavate. Elytral epipleura complete to apices. Tarsomeres simple. L 10–16 mm. E Australia, NT, Asia, Pacific islands, Africa .... ***Derosphaerus* Thomson** (Fig. 36.184, Pl. 89D)
– Femora linear or fusiform. Epipleura incomplete. Preapical tarsomeres lobed or cupuliform **166**
166(165). Pronotum longer than wide. Beetle winged (metaventrite longer than first abdominal ventrite). Elytral striae fully impressed. Anterior edges of abdominal ventrites completely crenulated. L 7–10 mm. S WA, SA, VIC .... ***Oectosis* Pascoe** (Fig. 36.185, Pl. 95L)
– Pronotum about as long as wide. Beetle wingless (metaventrite shorter than first abdominal ventrite). Elytra nearly always without, or with superficial, striae joining deep punctures. Anterior edges of abdominal ventrites crenulate at sides only. L 6–16 mm. NSW, QLD, NT, N WA........ ***Omolipus* Pascoe** (Fig. 36.186, Pl. 95I)

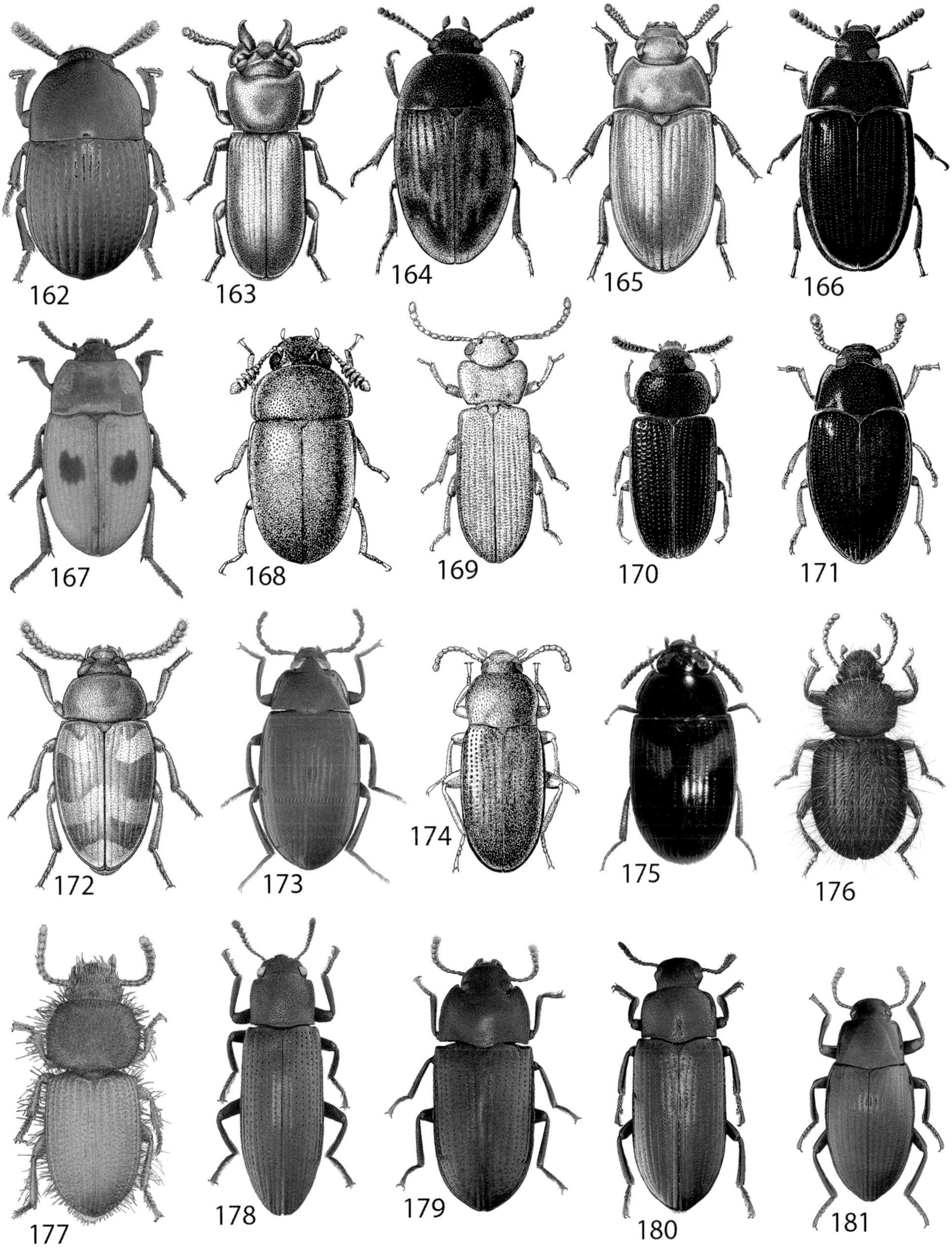

Figs 36.162– 36.181. 162, *Uloma* (*Apterulomoides*) © GC; 163, *Gnatocerus* male © GC; 164, *Ellipsodes*; 165, *Alphitobius* © GC; 166, *Diaclina*; 167, *Phaleria* © GC; 168, *Pentaphyllus*; 169, *Adelina*; 170, *Palembomimus* © AZ; 171, *Sciophagus*; 172, *Alphitophagus* © GC; 173, *Ceropria* © GC; 174, *Ulomoides*; 175, *Platydema*; 176, *Ectyche* © SAM; 177, *Micrectyche* © GC; 178, *Neotheca* © GC; 179, *Androsus* © GC; 180, *Chariotheca* © GC; 181, *Metisopus* © GC.

167(164). Prosternal process approximately as wide as a procoxa, nearly straight or feebly arched in profile **168**

– Prosternal process narrower, more strongly arched or sinuate in profile **169**

168(167). Wingless (metaventrite shorter than first abdominal ventrite). Gula nearly always with very deep transverse groove. Mentum usually glabrous, rarely pilose. L 12–26 mm. Arid Australia .... ***Hypaulax* Bates** (Fig. 36.188, Pl. 91D)

– Winged (metaventrite longer than first abdominal ventrite). Gula without a deep transverse groove, may have very shallow one. Mentum often with very long and dense pilosity. L 15–28 mm. E Australia including TAS, New Guinea, Pacific islands, Asia .... ***Promethis* Pascoe** (Figs 36.1, 36.189, Pl. 97E)

169(167). Wingless. L 11–16 mm. E QLD, Lord Howe Island .... ***Hydissus* Pascoe** (Fig. 36.190, Pl. 91E)

– Winged **170**

170(169). Pronotum with deep basal transverse marginal sulcus, the margin usually interrupted medially. L 7–9 mm. NE NSW, SE QLD, New Guinea, Pacific islands, SE Asia .... ***Scotoderus* Perroud** (Fig. 36.191, Pl. 98D)

– Pronotum without basal sulcus **171**

171(170). Frons with a pair of short, shallow longitudinal grooves or rugae between eyes (except in *Z. socius* (Carter)). Posterior pronotal angles acute or produced, sometimes overlapping humeri. Hypomeron more or less rugose. Tarsomeres simple. L 9–19 mm. E Australia including TAS, NT, New Guinea, Borneo .... ***Zophophilus* Fairmaire** (Fig. 36.192, Pl. 101D)

– Without frontal grooves. L < 9 mm. Other characters variable **172**

172(171). Elytra with brilliant transverse violet and gold colour pattern. Prothorax more than twice as wide as long. L 5–6 mm. NE QLD, New Guinea, SE Asia .... ***Espites* Pascoe** (Fig. 36.193, Pl. 100I)

– Elytra uniform in colour or with greenish longitudinal stripes. Prothorax twice as wide as long or less **173**

173(172). Lateral edges of pronotum a little uneven, slightly undulate or crenulate, the disc more convex anteriorly and coarsely punctate, hypomera coarsely punctate. Some elytral intervals laterally stepped. Tarsomeres simple. L 6–8 mm. E Australia including TAS, SE Asia, Pacific islands .... ***Tetragonomenes* Chevrolat** (Fig. 36.194, Pl. 99A)

– Lateral edges of pronotum simple, the disc evenly feebly convex and finely punctate, hypomera indistinctly or not punctate. Elytral intervals evenly convex. Preapical tarsomeres lobed. L 6–8 mm. E Australia including TAS, NT .... ***Kaszaba* Matthews & Doyen** (Fig. 36.195, Pl. 92K)

Antennae with simple hairlike sensilla only 174–end

174(135). Eyes divided by genal canthus. L 4.5–6 mm. E NSW, S WA ... **TENEBRIONINI (part)** .... ***Paratoxicum* Champion** (Fig. 36.196, Pl. 96D)

– Eyes emarginate but entire, or if divided (*Dysarchus*), beetle more than 9 mm long **175**

175(174). Elytral epipleura abbreviated, not reaching apices **176**

– Epipleura complete to elytral apices **177**

176(175). Prosternum very narrow and depressed between procoxae, so that latter are strongly projecting and nearly contiguous. Dorsal surfaces smooth with patches of adpressed white setae. L 6.5–8.5 mm. TAS, NE NSW, NE QLD ... **TRACHELOSTENINI** .... ***Leaus* Matthews & Lawrence** (Fig. 36.197, Pl. 92A)

– Prosternum a little wider and not depressed, procoxae only moderately projecting and clearly separated. Dorsal surfaces scabrous with evenly distributed adpressed stout setae. L ~4 mm. Central Australia ... **TRIBE UNKNOWN** .... ***Hangaya* Matthews & Merkl** (Fig. 36.198, Pl. 91B)

177(175). Surfaces either cribrate (with contiguous large punctures), or tuberculate. Antennae short with terminal six or seven segments strongly widened. Tibiae longitudinally carinate ... **BOLITOPHAGINI** **178**

– Without above combination **179**

178(177). Elytra costate, surfaces not tuberculate or horned. Tarsi 4–4–4. L 2.5–4 mm. NE NSW, NE QLD, NT, Eurasia, N America .... ***Rhipidandrus* LeConte** (Fig. 36.199, Pl. 97I)

– Dorsal surfaces strongly tuberculate. Male with cephalic horns. Tarsi 5–5–4, or 4–4–4 in smaller species. L 4–13 mm. E and N Australia, Asia .... ***Byrsax* Pascoe** (Fig. 36.200, Pl. 87L)

179(177). L 2–6 mm ... **TRIBOLIINI** **180**

– L 6 mm or more **182**

180(179). Antennae with 4-segmented club. Elytral striae more or less indistinct, epipleura vertical. Form subcylindrical. L 2.5–6 mm. Cosmopolitan .... ***Lyphia* Mulsant & Rey** (Fig. 36.201, Pl. 93L)

– Antennal club, if present, of 3, 5 or 7 segments. Elytra distinctly striate, epipleura more horizontal. Form flatter **181**

181(180). Elytral intervals usually carinate, at least laterally, rarely smooth. Clypeus much smaller in area than frons, subtrapezoidal. L 2.5–4.5 mm. Cosmopolitan .... ***Tribolium* Macleay** (Fig. 36.202, Pl. 100L)

– Elytral intervals smooth. Clypeus very large, occupying almost as much area as frons, subrectangular. L 2.5–3 mm. Cosmopolitan .... ***Latheticus* Waterhouse** (Fig. 36.203, Pl. 92D)

182(179). Anterior margin of labrum emarginate and its fringe medially interrupted. Elytral apices prolonged into a long process (mucro). L ~40 mm. S SA, Eurasia, Africa ... **BLAPTINI** .... ***Blaps* Fabricius** (Fig. 36.204, Pl. 86H)

– Anterior margin of labrum entire or, if emarginate, its fringe not interrupted. Elytral mucro rarely present. L not over 35 mm **183**

183(182). Outer faces of all tibiae sulcate and surfaces black ... **HELEINI-ASPHALINA (part)** **184**

– Outer faces of tibiae not sulcate, or if so (some Cyphaleina), surfaces metallic green **185**

184(183). Elytra costate, finely tuberculate between costae. Frontoclypeal suture deeply impressed. Wingless. L 18–19 mm. NE NSW .... ***Sloanea* Carter** (Fig. 36.205, Pl. 98L)

– Elytra striate, smooth. Frontoclypeal suture shallowly impressed. Winged. L 15–25 mm. SE SA, VIC, TAS, E NSW, SE QLD .... ***Meneristes* Pascoe** (Fig. 36.206, Pl. 93B)

185(183). Elytra with nine (primary) striae and a scutellary striole. Surfaces black **186**

– Elytra without striae or with more than nine (supernumerary), rarely with nine (some *Platyphanes*), in which case integument is metallic ... **HELEINI-CYPHALEINA (Part)** **188**

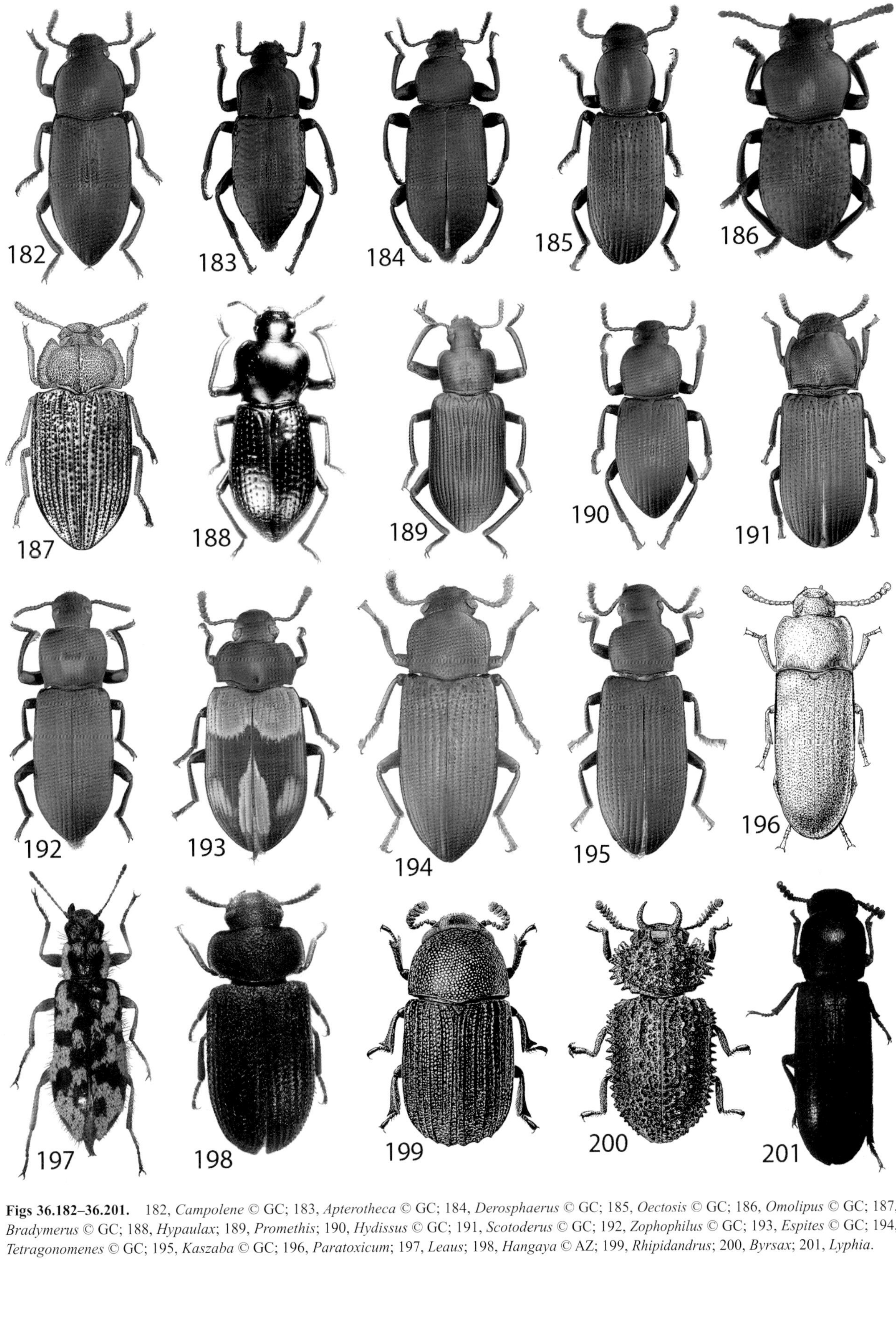

Figs 36.182–36.201. 182, *Campolene* © GC; 183, *Apterotheca* © GC; 184, *Derosphaerus* © GC; 185, *Oectosis* © GC; 186, *Omolipus* © GC; 187, *Bradymerus* © GC; 188, *Hypaulax*; 189, *Promethis*; 190, *Hydissus* © GC; 191, *Scotoderus* © GC; 192, *Zophophilus* © GC; 193, *Espites* © GC; 194, *Tetragonomenes* © GC; 195, *Kaszaba* © GC; 196, *Paratoxicum*; 197, *Leaus*; 198, *Hangaya* © AZ; 199, *Rhipidandrus*; 200, *Byrsax*; 201, *Lyphia*.

186(185). Wings without medial fleck. L 12–18 mm. Cosmopolitan ... **TENEBRIONINI (part)**........................ ***Tenebrio* Linnaeus** (Fig. 36.207, Pl. 99K)
– Wings with medial fleck (Fig. 36.11 mf) ... **HELEINI-ASPHALINA (part)**........................ **187**
187(186). Posterior edge of pronotum not margined, base without short oblique sulcus or depression on either side of middle, with posterior angles acute. Wingless. L 17–21 mm. NE NSW, SE QLD........................ ***Asphalus* Pascoe** (Fig. 36.208, Pl. 85D)
– Posterior edge of pronotum margined, base on either side of middle usually with short, more or less developed anterior extension of marginal groove, with posterior angles subquadrate. Winged. L 6–13 mm. SE SA, VIC, TAS, E NSW, E QLD........................ ***Bassianus* Matthews & Doyen** (Fig. 36.209, Pl. 86C)
188(185). Surfaces with at least a few white adpressed setae forming patches in shallow depressions on elytra, sometimes largely rubbed off. Pronotum with deep median longitudinal groove. L 13–22 mm. E Australia, TAS........................ ***Pachycoelia* Boisduval** (Fig. 36.210, Pl. 93E)
– No white adpressed setae. Pronotum without median groove........................ **189**
189(188). Clypeolabral membrane at least slightly exposed before clypeus when viewed from directly above, sometimes only in middle, or if concealed (many *Nyctozoilus*), scutellum strongly transverse........................ **190**
– Clypeolabral membrane concealed when viewed from directly above (partly exposed in some *Platyphanes*). Scutellum subequilateral........................ **200**

Clypeolabral membrane exposed 190–199

190(189). Wings reduced or absent. Metaventrite shorter than first abdominal ventrite (except in *Amphianax*). Lateral margins of pronotum thickened. Scutellum usually transverse. Eyes small, separated across top of head by 2.5–5 eye widths (*Nyctozoilus* group)........................ **191**
– Wings fully developed. Metaventrite about as long as first abdominal ventrite. Margins of pronotum rarely markedly thickened. Scutellum subequilateral. Eyes usually larger........................ **195**
191(190). Wings present but reduced, metaventrite about as long as first abdominal ventrite. Scutellum forming a subequilateral triangle. L 12–19 mm. Southern Australia........................ ***Amphianax* Bates** (Fig. 36.211, Pl. 85G)
– Wings entirely absent, metaventrite shorter. Scutellum wider than long........................ **192**
192(191). Scutellum not more than twice as wide as long. Sutures between prosternum and hypomera deeply impressed in most species, forming a groove. Form elongate-oblong with prothorax as wide as elytral bases. Without prominent teeth or tusks on submentum. Elytral surface with prominent costae, rugose between costae. L 17–24 mm. SE and SW Australia........................ ***Byallius* Pascoe** (Fig. 36.212, Pl. 87G)
– Scutellum 2–3 times as wide as long and often swollen. Sutures between prosternum and hypomera rarely impressed, in which case body form is not elongate. Other characters variable........................ **193**
193(192). Side of pronotum irregularly crenulate or lobate, discal surface uneven, usually vermiculate. Middle of frons and base of pronotum with transverse and/or longitudinal grooves. Gular grooves or pits, if present, confined to base of head. Submentum without prominent tusks, may have small angulation of margin. Elytral surface strongly sculptured. Form more or less elongate. L 12–20 mm. NE NSW, SE QLD........................ ***Styrus* Bates** (Fig. 36.216, Pl. 99E)
– If side of pronotum crenulate form of beetle is not elongate. Pronotum not grooved. Gular grooves, if present, often extending to submentum, which often bears a pair of prominent obtuse projections. Elytral surface variable........................ **194**
194(193). Prothorax abruptly constricted just before hind angles, which are therefore dentiform. Head above with deep grooves (sometimes only depressions) extending forward from eyes and marking inner edges of genae, which are small, laterally compressed and pad-like. Apical maxillary palpomere triangular. Body form usually parallel and elongate (broadened in *A. frenchi* Carter). L 17–25 mm. S WA, N SA, S NT, W VIC........................ ***Agasthenes* Bates** (Fig. 36.213, Pl. 84J)
– Lateral prothoracic margin posteriorly sinuate at most. No deep groove before eyes although sometimes a depression there, and genae not compressed. Apical maxillary palpomere may be securiform or cultriform. Body form usually broadly oval. L 11–26 mm. Throughout Australia........................ ***Nyctozoilus* Guérin-Méneville** (Fig. 36.214, Pl. 95C, 95H)
195(190). Male without setose line on inner margin of metatibiae and with pro- and mesotarsi widened. Wings without medial fleck (except in *Amarygmimus*). Penultimate two abdominal ventrites usually with coriaceous expansions of intersegmental membranes........................ **196**
– Male with tomentose or setose lines on inside edge of metatibiae at least, sometimes also with widened protarsi, rarely without sexual dimorphism. Wings usually with medial fleck. Ventrites without coriaceous edges........................ **199**
196(195). Pro- and mesotibiae, at least, often sulcate on outer faces. Body form broadly oval or subcircular, with legs and antennae relatively short........................ **197**
– Tibiae not sulcate. Body form narrower and appendages usually longer........................ **198**
197(196). All tibiae deeply sulcate on outer faces. Body form very broad, with legs largely hidden underneath. Elytral epipleura very wide and excavated under humeri. Males with enlarged apical maxillary palpomeres. L 12–16 mm. VIC, E NSW, E QLD, New Guinea........................ ***Hemicyclus* Westwood** (Fig. 36.215, Pl. 91J)
– Only pro- and mesotibiae shallowly sulcate in some species. Body form less broad, not largely covering legs. Epipleura narrower and not excavated. Male with apically dilated and flattened metatibiae. L 7–12 mm. E Australia........................ ***Bolbophanes* Carter** (Fig. 36.217, Pl. 86D)
198(196). Pronotum evenly convex, in side profile continuing outline of elytra. Third antennomere only 1.5 × as long as 4th and antennae barely reaching base of prothorax. Legs short, apices of femora only just extending beyond body outline. L 11–14 mm. S WA........................ ***Amarygmimus* Bates** (Fig. 36.218, Pl. 84F)
– Pronotum more flattened than bases of elytra. Third antennomere twice as long as 4th and antennae able to reach well beyond base of prothorax. Femora projecting well beyond body outline. L 11–35 mm. Throughout Australia, peripheral........................ ***Cyphaleus* Hope** (Fig. 36.220, Pl. 88G, 101F, 101H)
199(195). Male with metatibial tomentum only. Pronotal surface matt. Elytral surface relatively smooth apart from coarse punctures and lateral row of small tubercles near bases, sericeous, uniformly greenish or fuscous. Epipleura not grooved. L 15–18 mm. E VIC, E NSW, SE QLD........................ ***Phanechloros* Matthews & Bouchard** (Fig. 36.219, Pl. 96G)
– Male with row of tubercles along inner face of protibiae, as well as metatibial tomentum. Elytral surfaces usually uneven, vermiculate or tuberculate (but without small lateral tubercles), often with cupreous or golden reflections on raised areas. Epipleura usually with fine groove along inner edge posteriorly. L 6–16 mm. E Australia........................ ***Atoreuma* Gebien** (Fig. 36.221, Pl. 85A)

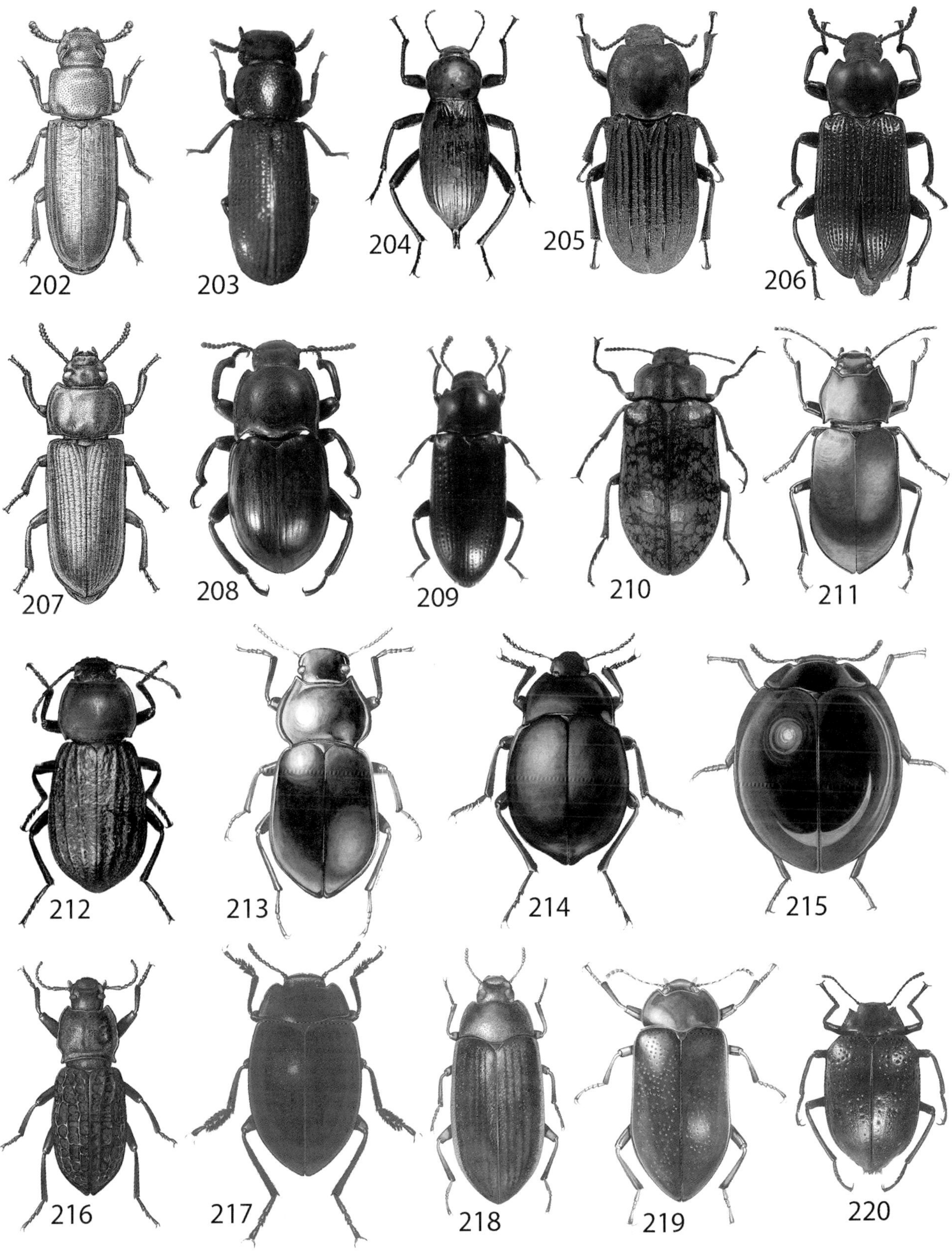

Figs 36.202–36.220. 202, *Tribolium* © GC; 203, *Latheticus*; 204, *Blaps* © SAM; 205, *Sloanea* © GC; 206, *Meneristes* © GC; 207, *Tenebrio* © GC; 208, *Asphalus* © GC; 209, *Bassianus* © GC; 210, *Pachycoelia* © SAM; 211, *Amphianax* © SAM; 212, *Byallius*; 213, *Agasthenes* © SAM; 214, *Nyctozoilus* © SAM; 215, *Hemicyclus* © SAM; 216, *Styrus* © SAM; 217, *Bolbophanes* © SAM; 218, *Amarygmimus* © SAM; 219, *Phanechloros* © SAM; 220, *Cyphaleus* © QM.

Clypeolabral membrane concealed 200–end

200(189). Metaventrite as long as or longer than first abdominal ventrite. Dorsal part of eyes rounded or subquadrate, each usually occupying 1/4 or more of distance along a transverse line across head (sometimes less). Pronotum and elytra with or without lateral flanges. Beetle fully winged, rarely with reduced or absent wings in some *Pterohelaeus* **201**

– Metaventrite shorter. Dorsal part of eyes either slit-like or, if rounded, occupying less than 1/4 of distance across head. Beetle wingless **209**

Winged, metaventrite as long as or longer than first ventrite 201–208

201(200). Pronotum and elytra without foliate lateral extensions (flanges) ... **HELEINA-CYPHALEINA (Part)** **202**

– Pronotum at least, and often also elytra, with flanges partially or entirely covering legs ... **HELEINI-HELEINA** (Part) **206**

202(201). Epipleura with groove on inside edge of posterior 2/5. Pronotum subquadrate, narrower than elytral bases with anterior angles strongly produced. Dorsal surfaces densely punctate, or asperate on elytra. Pro- and mesotibiae of males subapically dentate or expanded. L 9–11 mm. S WA, SA to SE QLD .... ***Mithippia* Pascoe** (Fig. 36.239, Pl. 94H)

– Epipleura without groove, or with a partial one in *Platyphanes*. Dorsal surfaces usually more sparsely punctate. Male tibiae not dentate or expanded **203**

203(202). Elytral epipleura with outer edges partly effaced and undulate anteriorly. Disc just above epipleura without tubercles. Femora clavate, usually with white setose fringes. Males with more or less elongate apical antennomeres. L 10–14 mm. E QLD .... ***Paraphanes* Macleay** (Fig. 36.222, Pl. 96F)

– Epipleura with outer edges entire. Grooves just above epipleura often containing a row of minute tubercles. Femora not clavate and without fringes. Apical antennomere of male not elongate **204**

204(203). Elytral punctures arranged in rows or striae (scarcely discernible in *P. browni* Carter). Antennae short, with 3rd segment less than 2 × as long as 4th. L 14–27 mm. S WA, E Australia, TAS .... ***Platyphanes* Westwood** (Fig. 36.223, Pl. 97B)

– Elytral punctures confused. Antennae longer, with 3rd segment about twice as long as 4th. Beetles rarely more than 11 mm long **205**

205(204). Mandibles with wide, deep channel along outer edge dorsally. Pronotum with only a narrow gutter between lateral edge and disc, which is evenly convex. Intercoxal process of abdomen acutely rounded. Wings without medial fleck. L 9–12 mm. S WA, NE NSW, SE QLD .... ***Mitrothorax* Carter** (Fig. 36.224, Pl. 94J)

– Mandibles with shallow dorsal groove at most. Sides of pronotum broadly concave. Intercoxal process broadly rounded. Wings with medial fleck. L 9–14 mm. TAS, E Australia .... ***Olisthaena* Erichson** (Fig. 36.226, Pl. 95A)

206(201). Lateral edges of prothorax with a dense fringe of setae. Elytral flanges present but not evident from above, turned down. Colour a uniform nitid brown. L 8–10 mm. Southern Australia .... ***Camponotiphilus* Lea** (Fig. 36.227, Pl. 87D)

– Lateral edges without setae. Elytral flanges, if present, rarely turned down (some *Emcephalus*). Colour black or variegated brown **207**

207(206). Upper spur of protibia (forespur) longer than basal protarsomere, except in one species. Dorsal surfaces often a little uneven and colour generally brown, sometimes with darker maculae or pale patterning. Elytral flanges narrow to moderately wide. L 9–16 mm. E and N Australia .... ***Ospidus* Pascoe** (Fig. 36.228, Pl. 96B)

– Protibial spurs not longer than basal protarsomere. Dorsal surfaces usually black; if as above, elytral flanges very wide **208**

208(207). Protibial spurs less than 1/4 length of basal protarsomere. Wings with medial fleck. Apical maxillary palpomeres usually enlarged, cultriform. Colour often brownish with darker reticulate raised areas, sometimes evenly dark brown. Most species floccose in life, entirely covered with white waxy filaments. L 11–22 mm. E Australia, New Guinea .... ***Emcephalus* Kirby** (Fig. 36.231, Pl. 90J. 90K)

– Protibial spurs ~1/2 length of basal protarsomere. Wings without medial fleck. Apical maxillary palpomeres variable. Colour piceous, rarely brown, without reticulate ridges. Not floccose, but sometimes pruinose. L 6–25 mm. Throughout Australia, New Guinea .... ***Pterohelaeus* Brême** (Fig. 36.230, Pl. 97H)

Wingless, metaventrite shorter than first ventrite 209–end

209(200). Apices of mandibles apically unidentate, with prominent apical tooth and smaller tooth distant from apex **210**

– Apices of mandibles apically bidentate **212**

210(209). Protibiae unmodified. Surfaces glabrous. L 8–12 mm. Central WA .... ***Aglypta* Gebien** (Fig. 36.235, Pl. 84D)

– Protibiae flattened and dentate or crenulate on outer edges. Surfaces with long pilosity or short, somewhat squamiform setae **211**

211(210). Protibiae with long outwardly directed apical spur. Intercoxal process of first abdominal ventrite narrowly triangular. L 6–16 mm. Eastern arid zone .... ***Trichosaragus* Blackburn** (Fig. 36.232, Pl. 100J)

– Protibiae with very short apical spur. Intercoxal process broadly rounded. L 9–16 mm. Western arid zone .... ***Onotrichus* Carter** (Fig. 36.229, Pl. 95K)

212(209). Eyes divided. Protibiae, when not worn, with single strong preapical tooth on outer edge and large apical spur fused to tibia. Mentum enlarged. Antennae with sensilla confined to small patches and terminal segment reduced. Males with much enlarged maxillary and labial palpi. L 10–23 mm. S WA, S SA, SE QLD .... ***Dysarchus* Pascoe** (Fig. 36.236, Pl. 89B)

– Eyes entire. Protibia without single lateral tooth, may have a row of denticles. Forespur not fused to tibia. Mentum not enlarged. Antennal sensilla spread over most of distal segment surfaces, terminal segment not reduced. No sexual dimorphism in palpi **213**

213(212). Pronotal flanges extended forward and inward to approach one another or meet above or in front of head, rarely with anterior angles of flanges merely acute, widely separated. Elytral flanges generally wide. Pronotal disc usually with a median costa or spine. L 10–37 mm. Arid and subarid Australia .... ***Helea* Latreille** (Fig. 36.237, Pl. 91C)

– Pronotal flanges not extended as much. Other characters variable **214**

214(213). Disc of pronotum usually sculptured, either with median ridge (at least a trace), pair of gibbosities, or single strong convexity. Elytral flanges generally well developed, making up 1/4 to more than 1/2 total width of beetle **215**

– Disc of pronotum more or less evenly convex. Elytral flanges narrower, rarely absent **216**

215(214). Disc of pronotum gibbous or binodose, except in *C. ovalis* (Macleay). Prosternal process prominent, received in a deep concavity or against a high step of mesoventrite. Forespur not elongate. Elytral suture little or not raised. L 18–25 mm. E mainland Australia ..***Cillibus*** **Matthews** (Fig. 36.225, Pl. 88K)

– Disc of pronotum usually with median carina. Prosternal process flattened, not received in mesoventrite. Forespur usually long. Elytral suture nearly always strongly raised, forming median carina. L 8–27 mm. WA ....***Sympetes*** **Pascoe** (Fig. 36.234, Pl. 99C)

216(214). Forespur enlarged, longer than first tarsomere (may be worn down to short broad stub), dorsal surfaces tuberculate and costate, or densely squamose. L 10–20 mm. Southern Australia .... ***Saragus*** **Erichson** (Fig. 36.238, Pl. 97A)

– Forespur not enlarged. Dorsal surfaces generally not tuberculate with some exceptions, not squamose **217**

217(216). Intercoxal process of first ventrite broadly subquadrate. Elytra each with two or three costae, sometimes zig-zagged, inner costa more prominent and forming with opposite one a raised flattened longitudinal median surface. Without sexual dimorphism. L 12–20 mm. N Australia ....***Boreosaragus*** **Matthews** (Fig. 36.233, Pl. 86K)

– Intercoxal process of first ventrite U-shaped. Elytra smooth or costate, but not forming raised median area. Males often with dilated protarsi. L 8–22 mm. S and E Australia, Lord Howe Island .... ***Celibe*** **Boisduval** (Fig. 36.240, Pl. 87E).

Classification of the Australian Genera

Subfamily Pimeliinae Latreille, 1802

Characteristics. Antennae 11-segmented with simple sensilla only. Procoxal cavities closed. Hind wings, if present, without medial fleck. Tarsi 5–5–4, not lobed. Abdomen without visible intersegmental membranes. Defensive glands absent. Edge of last abdominal ventrite grooved to receive elytral edges. Aedeagus inverted (with tegmen ventral).

Distribution. World-wide in arid regions.

Australian taxa. Two native tribes (Cnemeplatiini and Vacronini) and one introduced one (Trilobocarini). Two additional tribes which do not now occur in Australia and which are not further discussed here (Tentyriini and Pimeliini) are represented respectively by a species of *Stenosida* Solier, 1835 described by Medvedev (1995) which is based on mislabelled specimens, and a species of *Trachyderma* Latreille, 1828 reported as *Ocnera* Fischer, 1822 by Lea (1931) which has not survived following its original appearance in SA (see Matthews & Bouchard 2008 for details).

Note. Australia is exceptional among continents of the world in having very few indigenous Pimeliinae. Elsewhere they are the dominant element of the tenebrionid fauna of arid regions. Those which do occur here represent early branches of the subfamily in the cladograms of Doyen (1994). Matthews *et al.* (2010: 602) postulated that the strictly xerophilic Pimeliinae could not live in densely-forested Australia while it was a part of Gondwana, except on coastal dunes (the New Zealand representative of Cnemeplatiini inhabits dunes), and by the time deserts formed the continent was isolated, allowing other groups, particularly Heleini (Tenebrioninae), to move into the xeric niches which appeared.

Tribe Cnemeplatiini Jacquelin du Val, 1861

Subtribe Thorictosomatina Watt, 1992

Characteristics. Oblong, moderately convex, glabrous or with minute setae, rufous or fuscous. Eyes reduced to narrow vertical slits concealed by front of prothorax. Labrum moderately transverse and bilobed. Edge of clypeus deeply arcuately emarginate. Antennae very short, held in cavities of head and prothorax, segments transverse, cupuliform, very slightly widening distally, last three forming club. Mentum enlarged, oval, partially covering maxillae. Mesocoxal cavities fully closed by broad meeting of ventrites. Elytra estriate but with large subsurface punctures arranged in rows. Scutellum not visible. Wings absent. Legs fossorial, all tibiae broadly expanded apically. [Watt 1992].

Distribution. The tribe has a world-wide distribution including New Zealand. The two Australian genera occur in the arid zone mainly in western and central Australia.

Australian taxa. Two genera.

Thorictosoma Lea, 1919 (Pl. 99D; Figs 36.27, 36.35)

Type species. *Thorictosoma ectatommae* Lea, 1919.

Characteristics. Length 2–3 mm. Antennal club weak. Prothorax subquadrate with arcuate sides. Protibiae (Fig. 36.27) with strong outer basal tooth and two enlarged apical spurs, basal protarsomere with long spur-like process.

Australian species. Two: *T. ectatommae* and *T. tibiale* Lea, 1919.

Distribution. The southern part of arid Australia.

Biology. The type species was taken from a nest of *Rhytidoponera* (as *Ectatomma*) and Lea (1919) believed that *T. tibiale* was 'probably from a nest of ants', but numerous specimens of the genus have been collected subsequently only in pitfall traps.

Wattiana Matthews & Lawrence, 2005 (Pl. 101B; Fig. 36.36)

Type species. *Wattiana greensladei* Matthews & Lawrence, 2005.

Characteristics. Length 2.0–2.4 mm. Antennal club strong. Prothorax about as long as wide, widest at front angles, sides straight, narrowing to base. Protibiae gradually, strongly expanding to apices which are outwardly dentate, without preapical teeth. Tarsomeres simple.

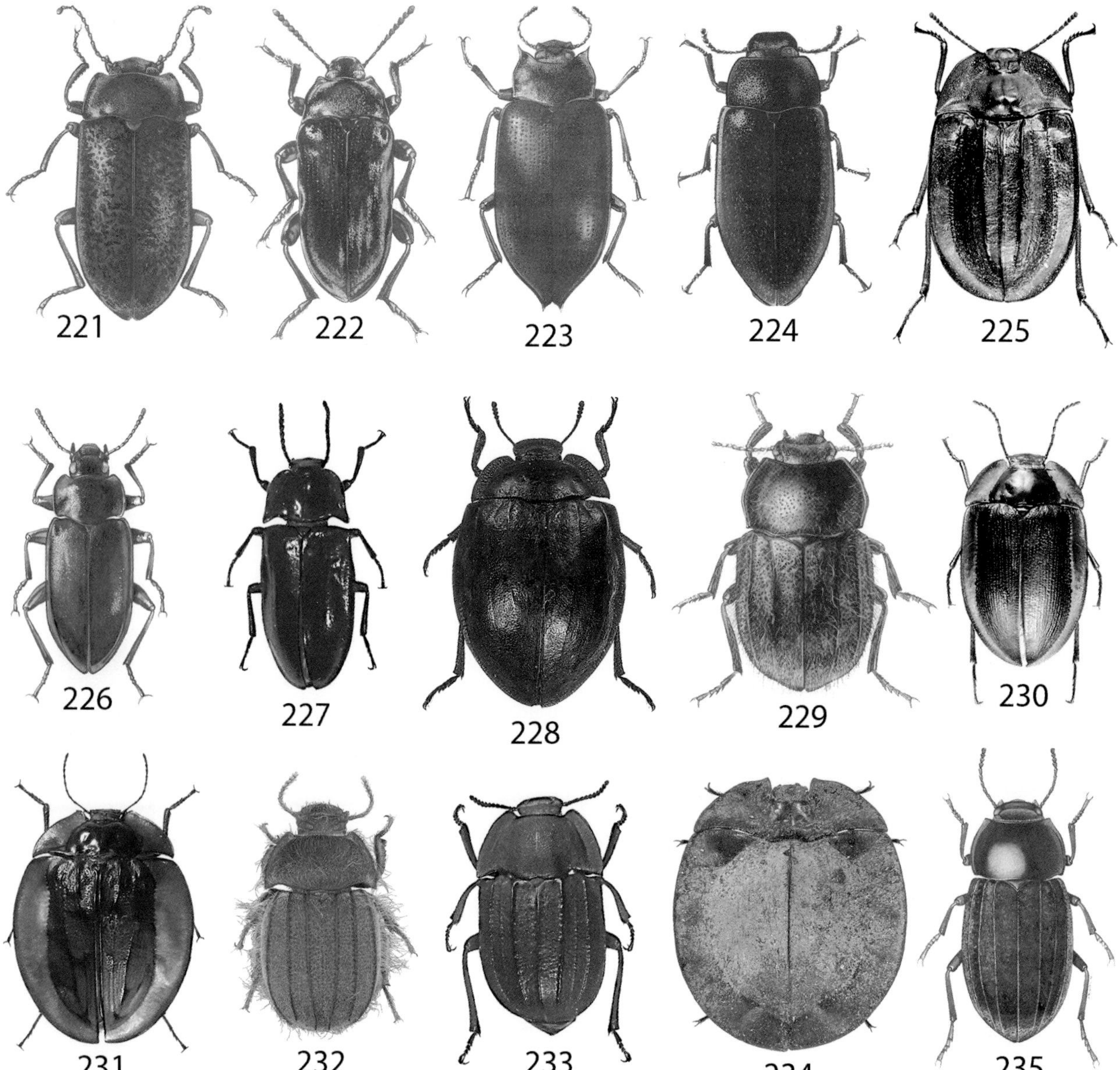

Figs 36.221–36.240. 221, *Atoreuma* © SAM; 222, *Paraphanes* © SAM; 223, *Platyphanes* © SAM; 224, *Mitrothorax* © SAM; 225, *Cillibus* © SAM; 226, *Olisthaena* © SAM; 227, *Camponotiphilus* © SAM; 228, *Ospidus* © SAM; 229, *Onotrichus* © SAM; 230, *Pterohelaeus* © SAM; 231, *Emcephalus* © SAM; 232, *Trichosaragus* © GC; 233, *Boreosaragus* © SAM; 234, *Sympetes* © GC; 235, *Aglypta* © SAM; 236, *Dysarchus* © SAM; 237, *Helea*; 238, *Saragus*; 239, *Mithippia* © SAM; 240, *Celibe* © SAM.

Australian species. One: *W. greensladei*.

Distribution. Only two specimens are known: the holotype of *W. greensladei* from extreme south-western QLD and a paratype from the north of the NT, without biological data.

Tribe Vacronini Gebien, 1910

Characteristics. Elongate with long slender legs, with numerous flattened recumbent setae, fuscous. Eyes moderate-sized to large and convex, slightly reniform. Labrum strongly transverse. Edge of clypeus nearly straight. Antennae long, extending beyond base of prothorax, first two segments very short, rest linear to elongate-obconic, or flattened and serrate, except for last one or two segments which are very short and subtriangular. Mentum transversely quadrate, not covering maxillae. Mesocoxal cavities open. Elytra without striae, with numerous rows of punctures and short curved recumbent setae. Wings fully developed. Legs not fossorial.

Distribution. This plesiomorphic tribe occurs in central Australia and in the western United States of America (Doyen 1994). Matthews (2000) suggested that this anomalous distribution is the result of the fragmentary survival of a group once widespread in an arid northern part of Gondwana in the Mesozoic.

Australian taxa. Two genera.

Biology. Nothing is known about the biology of Australian Vacronini and larvae have not been described.

Exangeltus Blackburn, 1897 (Pl. 90D; Fig. 36.43)

Type species. *E. angustus* Blackburn, 1897.

Exangelutus Kulzer, 1964

Characteristics. Length 7.5–12 mm. Mandibles truncate and not projecting. Apical maxillary palpomere securiform. Antennae simple and not reaching beyond basal 1/4 of elytra in both sexes. Pronotum subquadrate with rounded or feebly angular sides. Prosternum about as long before coxae as behind them. Elytra with multiple rows of large window punctures.

Australian species. Four.

Distribution. NT, WA, SA, QLD in arid areas.

Lixionica Blackburn, 1896 (Pl. 93I; Fig. 36.42)

Type species. *L. costatipennis* Blackburn, 1896.

Characteristics. Length 10–11 mm. Apical maxillary palpomere narrowly triangular or fusiform. Antennae of males reaching middle of elytra, the segments widened and serrate. Antennae of females shorter, segments simple. Mandibles unidentate and strongly porrect. Pronotum elongate with anteriorly converging sides. Prosternum much longer before coxae than behind them. Elytra with small round punctures.

Australian species. One named: *L. costatipennis*.

Distribution. Originally found in the extreme north of SA at Stevenson Creek during the 1894 Horn Expedition, *Lixionica* has since been collected further west near the Mann Ranges on the SA–NT border, as well as much further west along the Canning Stock Route in central WA. The WA specimens look a little different from the SA ones and may represent a different species.

Tribe Trilobocarini Lacordaire, 1859

Salax Guérin-Méneville, 1834 (Pl. 97C; Fig. 36.34)

Type species. *S. lacordairei* Guérin-Méneville, 1834 (Argentina).

Characteristics. Broadly oblong with numerous very short squamiform setae. Length 7–8 mm. Labrum prominent and rounded, not transverse. Clypeus trilobed. Antennae short, reaching only to basal third of pronotum, slightly widening distally with terminal segment smaller than subterminal. Mandibles with large dorsal lobes which flank clypeus. Mentum large, entirely concealing maxillae except for palpi.

Australian species. One unnamed. Following its discovery in SA specimens were sent to a specialist in Argentina who noted that the species was different from the only described one (the generitype) and it could not be matched with any available collected material.

Distribution. The genus is known from Argentina and Chile in saline humid areas (Vidal & Guerrero 2007). It first appeared in SA in 1972 among the stomach contents of a dotterel taken in a partially flooded mid-north pasture. Since then it has turned up in eight coastal localities around Adelaide and on Yorke and Eyre peninsulas.

Subfamily Zolodininae Watt, 1975

Characteristics. Elongate-oblong, somewhat flattened, glabrous, piceous. Length 12–15 mm. Eyes weakly reniform. Edge of clypeus nearly straight. Clypeo-labral membrane not exposed. Labrum subquadrate. Antennae reaching only to about middle of prothorax, segments gradually and weakly expanding apically, cupuliform, apical segment not enlarged. Mandibles bidentate. Maxillary palpi with apical segment subparallel. Mentum transverse-trapezoidal, narrower at base, not carinate. Pronotum elongate, tapering anteriorly, feebly evenly convex. Prosternum very long before coxae. Procoxal cavities open both internally and externally. Mesocoxal cavities open. Elytra elongate, smooth, with scutellary striole and 10 striae. Wings fully developed, with medial fleck. Legs slender, not fossorial, preapical tarsomeres a little cupuliform. Abdomen without visible intersegmental membranes. Edge of last abdominal ventrite grooved to receive elytral edges. Defensive glands absent. Aedeagus inverted.

Distribution. There are two monotypic genera: *Tanylypa* Pascoe, 1869 in Tasmania, and *Zolodinus* Blanchard, 1853 in New Zealand.
Australian taxa. One genus and species.

Tanylypa Pascoe, 1869 (Pl. 99H; Fig. 36.32)

Type species. *T. morio* Pascoe, 1869.
Characteristics. As for the subfamily.
Distribution. TAS, mainly in *Nothofagus* forest.
Australian species. One: *T. morio*.
Biology. Larvae and adults have been collected from rotten wood.

Subfamily Lagriinae Latreille, 1825

Characteristics. Labrum subquadrate to elongate. Antennae with simple sensilla only. Procoxal cavities closed externally and internally. Wings without medial fleck. Elytral striae, when present, 10 plus scutellary striole. Abdomen with exposed intersegmental membranes between last three ventrites except in Belopini and Cossyphini. Tarsomeres frequently lobed or cupuliform. Edge of last abdominal ventrite grooved to receive elytral edges. Aedeagus not inverted.
Distribution. The subfamily is found throughout the world.
Note. Representative genera of Lagriinae have been subjected to DNA sequencing in three recent studies (Kergoat *et al.* 2014b; Kanda *et al.* 2015; Aalbu *et al.* 2017), sampling 12, 28 and 29 genera respectively. In all three studies the subfamily as currently understood appears as a monophyletic group and, except for Lupropini, all tribes also recover as monophyletic. Cossyphini and Belopini form basal branches of the subfamily, a result which agrees with conclusions drawn from morphological data (Matthews & Bouchard 2008).
Australian taxa. Seven tribes.

Tribe Cossyphini Latreille, 1802

Characteristics. Oblong-oval, strongly flattened, with very wide pronotal and elytral flanges covering head and legs, glabrous, fuscous. Length 4–6 mm. Eyes round, not emarginate. Edge of clypeus arcuately excised. Clypeolabral membrane concealed. Antennae reaching base of prothorax, segments gradually and strongly widened distally, apical segment transversely oval. Extensions of hypomeron meet behind procoxae and are overlapped by prosternal process. Mesocoxal cavities closed. Wings delicate, reduced. Legs not fossorial, tarsomeres parallel, not lobed or cupuliform. Intersegmental membranes of abdominal ventrites not visible. Defensive glands absent. Outer edge of last visible ventrite not grooved as such, but with fine ridge probably representing inner edge of former groove [Scupola 2000].
Distribution. Africa and Eurasia eastward to South-east Asia, northern Australia.
Australian taxa. One genus and species.

Cossyphus Olivier, 1791 (Pl. 88A; Fig. 36.33)

Type species. *Lampyris depressa* Fabricius, 1781 (India).
Acontodactylus Desbrocher des Loges, 1894 (subgenus).
Characteristics. As for the tribe.
Australian species. One: *C. odewahni* Pascoe, 1866.
Distribution. As for the tribe. In Australia found mainly in the north, extending southward to central NSW, habitat and biology unknown.

Tribe Belopini Reitter, 1917

Characteristics. Elongate-oblong, moderately convex, glabrous or with short setae. Clypeo-labral membrane concealed. Maxillary palpi with apical segment ovoid, fusiform or obliquely truncate. Mentum more or less widened, covering bases of maxillae. Pronotum elongate-quadrangular or subquadrate. Prosternum long before coxae. Mesocoxal cavities open, trochantins visible. Wings fully developed. Abdomen with femoral lines on first ventrite which are either parallel (Fig. 36.12) or divergent posteriorly (Matthews & Bouchard 2008 Fig. 9D), without visible intersegmental membranes. Tarsomeres not lobed, subcylindrical. Defensive glands absent.
Distribution. The tribe occurs in semi-arid and saline parts of Asia, the Mediterranean, and the Americas. In Australia it occurs mainly in the south-west of the continent where three of the four genera are confined, but *Kershawia* Lea, 1904, while also found there, extends its range eastward to the drier parts of the NSW coast.
Australian taxa. Four genera.
Biology. Three of the four Australian genera have been found exclusively in ants' nests. No larvae of Australian Belopini have been described.
Note. The absence of defensive glands and visible abdominal intersegmental membranes, with consequent central hinging of the ventrites, and the enlarged mentum, are pimeliine features. Doyen (1994) selected Belopini as an outgroup for some of his pimeliine cladograms. It is to be noted also that while Australian Belopini basically conform to the characteristics of the tribe, as first noted by Doyen *et al.* (1990), the presence of femoral lines on the first abdominal ventrite, as well as myrmecophily (both unknown in Belopini elsewhere), suggest long isolation.

Doyenia Matthews & Lawrence, 2005 (Pl. 89I; Fig. 36.29)

Type species. *D. crematogastri* Matthews and Lawrence, 2005.
Characteristics. Length ~1.5 mm. Eyes entire. Edge of clypeus not excised. Antennae with nine segments, segment 8 enlarged, 9 small. Pronotum smooth, elytra with fine serrate carina outside stria 8 and bristles aligned in rows. Tibiae slightly expanded distally. Tarsi 5–5–4.

Distribution. Southern WA, in nests of *Crematogaster*.
Australian species. One: *D. crematogastri*.

Euclarkia Lea, 1919 (Pl. 90L; Figs 36.12, 36.30)

Type species. *E. costata* Lea, 1919.
Characteristics. Length 2.5–3.5 mm. Eyes partly divided by posterior canthus. Edge of clypeus arcuately excised. Antennae with three segments, terminal one large and cylindrical. Pronotum and elytra with prominent carinae. All tibiae broadly expanded distally. Tarsi 4–4–4.
Australian species. One: *E. costata*.
Distribution. Southern WA. In nests of *Iridomyrmex*.

Eulea Carter, 1937 (Pl.90I; Fig. 36.28)

Type species. *E. caeca* Carter, 1937.
Characteristics. Length ~3 mm. Eyes absent. Edge of clypeus arcuately excised. Antennae with 11 segments, last three forming feeble club. Head and pronotum densely tuberculate. Elytra with prominent carinae. Tibiae slightly expanded distally. Tarsi 5–5–4.
Australian species. One: *E. caeca*.
Distribution. Southern WA. Habitat unknown.

Kershawia Lea, 1904 (Pl. 92I; Fig. 36.31)

Type species. *K. rugiceps* Lea, 1904.
Characteristics. Length 3–4 mm. Eyes partly divided by posterior canthus. Edge of clypeus feebly excised. Antennae with eight segments, terminal one enlarged. Pronotum and elytra with prominent carinae. All tibiae broadly expanded distally. Tarsi 4–4–4.
Australian species. One: *K. rugiceps*.
Distribution. Throughout southern Australia, in nests of *Iridomyrmex*.

Tribe Adeliini Kirby, 1828

Characteristics. Oblong or oval, from smooth and clean (carabiform) to strongly rugose or tuberculate and squalid (opatriform), nearly glabrous to very densely setose. Colour generally fuscous to piceous, sometimes with more or less brilliant metallic reflections, without pigment patterns on elytra, but sometimes on legs. Eye shape varying from round and entire through reniform to fully divided. Middle of clypeal margin from slightly produced to deeply arcuately excised. Clypeo-labral membrane concealed. Maxillary palpi with apical segment strongly triangular. Antennae reaching to base of prothorax, rarely shorter, segments gradually enlarging to apex, cylindrical, moniliform or subtriangular, apical segment more massive than subapical but not greatly enlarged. Lateral edges of pronotum moderately carinate to strongly foliate, often variously irregular, absent in *Apocryphodes* Matthews 1998. Posterior edge slightly overlapping elytral bases in 15 genera. Prosternum long before procoxae. Hind wings absent except in some *Isopteron* Hope, 1840. Mesocoxal cavities open, trochantin visible. Metaventrite very short except in winged species. Legs slender, not dentate or fossorial. Tarsi slender except protarsi of many males, penultimate segment lobed or not but even if not, insertion of claw segment is above centre of face of penultimate segment, which is at least slightly cupuliform. Tergite VII with pair of stridulatory files (Fig. 36.24) in 21 genera. Pair of very long eversible sacs of defensive glands always present, opening in membrane between segments VIII and IX [Matthews 1998].
Distribution. Adeliini occur in Australia, New Caledonia, New Zealand and southern Chile, mainly in temperate and tropical rainforests, but some genera in Australia have adapted to sclerophyll forest, and one genus, *Isopteron*, occurs in the arid zone.
Australian taxa. Twenty-five genera.
Biology. Most Adeliini occur under logs or in leaf litter and other accumulations of dead plant matter on which both adults and larvae feed; some species, however, may be found in rotten wood, and at least one species of *Isopteron* has been recorded feeding on germinating wheat (Watt 1975; Allsopp 1986a). Adult adeliines display a defensive behaviour which consists of a disturbance sound produced by an elytro-tergital stridulatory mechanism, followed by the eversion of a pair of large abdominal glands containing defensive chemicals (Eisner *et al.* 1974). These glands are not homologous with the defensive glands of other Tenebrionidae but appear to be homologous with the colleterial glands seen in females of many Lagriinae (Tschinkel & Doyen 1980: 327) such as Goniaderini (Matthews & Bouchard 2008, Fig. 22D).
Note. On molecular evidence the sister group of Adeliini is not Laenini as believed by Matthews (1998) but the cavernicolous Eschatoporiini of California (Aalbu *et al.* 2017). The two together form the most basal branch of the glanded Lagriinae, Laenini then forming the next branch.

Adelium Kirby, 1819 (Pl. 11A, B, 84L; Figs 36.18, 36.24, 36.103)

Type species. *Carabus porcatus* Fabricius, 1775.
Tropidopterus Blanchard, 1845
Dystalica Pascoe, 1869
Rues Casey, 1891
Characteristics. Oblong, dorsal surfaces smooth, glabrous or setose, some species with greenish reflections. Length 5–25 mm. Eyes transverse. Mentum more or less convex anteriorly, especially in middle (Fig. 36.18). Epipleura gradually or abruptly widened anteriorly, visible from above at humeri. Base of prothorax not overlapping

elytral bases. Prothorax nearly always transverse, its ratio of width to length in middle ~3:2, its sides usually arcuately or sinuately produced, occasionally subquadrate, rarely strongly dentate. Hind body often notably wider than prothorax. Penultimate tarsomeres distinctly lobed except in a few species. Basal metatarsomeres usually longer than claw segments.

Distribution. Throughout Australia in sclerophyll and rain forest leaf litter.

Australian species. 92 named and numerous unnamed.

Biology. Larvae and adults occur under rocks and in leaf litter. The flattened and often heavily sclerotised larvae are capable of moving rapidly among leaf litter and other ground debris. In SA and WA *Adelium brevicorne* Blessig, 1861 has become a synanthropic species common in gardens and other disturbed areas and may cause damage to canola and strawberry crops.

Note. Matthews (1998) writes of *Adelium*: 'This is a very difficult genus to recognise because it has no generic autapomorphies and because many shared character states to be found in the Adeliini reappear somewhere within *Adelium* ...'. Some species have the habitus of *Seirotrana*, *Adelodemus* and *Brycopia*. Many pilose species were described in *Brycopia* Pascoe, 1869 and resemble the latter, but can be separated mainly by their transverse eyes. *Brycopia* has round or oval eyes.

Key to species. Carter (1908a).

Adelodemus Haag-Rutenberg, 1878 (Pl. 84A; Fig. 36.88)

Type species. *A. asperulus* Haag-Rutenberg, 1878.

Apostethus Pascoe, 1882

Characteristics. Elongate-oblong, dorsal surfaces capable of holding a layer of soil (squalid), scabrous or tuberculate. Length 9–14 mm. Eyes transverse. Anterior edges of clypeus and labrum deeply emarginate. Frontoclypeal suture deeply impressed or grooved. Gula with deep groove varying in shape between species, rarely just a median pit. Pronotum with shallow median transverse depression and broadly deplanate lateral edges which are posteriorly sinuate to deeply excised. Front edge of prosternum raised into a collar. Prosternal process apically bifurcate or at least weakly bilobate.

Australian species. Three named plus several unnamed.

Distribution. Coastal mountains of QLD in dense forest.

Key to species. Kulzer (1964).

Apasis Pascoe, 1869 (Pl. 85H; Figs 36.17, 36.101)

Type species. *A. howitti* Pascoe, 1869.

Characteristics. Elongate-oblong, dorsal surfaces smooth. Length 12–23 mm. Eyes transverse. Mentum with broadly rounded anterior angles and thickened anterior edge, often appearing swollen across its width but not more convex medially (Fig. 36.17). Base of prothorax not overlapping elytral bases. Elytra with continuous impunctate striae except in *A. ruptus* (Pascoe, 1869) where they are interrupted, and *A. apasioides* (Carter, 1920) where they are punctate, the epipleura gradually widened anteriorly, their carina visible from above at humeri. Males (with expanded protarsi) have the first two abdominal ventrites medially concave, often with a pair of short longitudinal ridges posteriorly or a shallow median channel.

Australian species. Four.

Distribution. South-east SA, VIC, south-east NSW including higher elevations, in eucalypt forest.

Biology. The subcylindrical and lightly sclerotised larvae feed in white rotten wood.

Apocryphodes Matthews 1998 (Pl. 85F; Fig. 36.96)

Type species. *A. thompsoni* Matthews, 1998.

Characteristics. Form pedunculate (with globose prothorax separated by a narrow waist from oval hind body), densely pilose. Length 3–5 mm. Eyes reduced, oval. Apical antennomeres enlarged. Lateral pronotal and epipleural carinae absent. Epipleura entirely narrow. Elytral intervals minutely tuberculate. One unnamed species with stout spine on elytral humerus.

Australian species. *A. thompsoni* plus four unnamed species.

Distribution. QLD wet tropics, in rainforest.

Bellendenum Matthews, 1998 (Fig. 36.85)

Type species. *B. gonyxuthum* Matthews, 1998.

Characteristics. Oblong, dorsal surfaces capable of holding a layer of soil (squalid), rugose and tuberculate. Length 8–9 mm. Eyes transverse. Sides of pronotum expanded into irregularly shaped lobes. Prosternal process nodiform, not bifurcate. Penultimate tarsomeres ventrally prolonged into acute or asymmetrical lobes.

Australian species. *B. gonyxuthum* plus three unnamed species.

Distribution. QLD wet tropics, in rainforest.

Blepegenes Pascoe 1868 (Pl. 86B, G; Fig. 36.94)

Type species. *B. aruspex* Pascoe, 1868.

Ceradelium Preudhomme de Borre, 1868.

Characteristics. Pedunculate, dorsal surfaces smooth, glabrous. Length 15–27 mm. Epipleura narrow, passing under humeri and inflected elytral sides. Prothorax as long as wide, narrower than oval hind body, with rounded anterior and posterior angles and lateral edges bearing a strong spine in all but one species. Elytral striae impunctate. Legs long and slender.

Australian species. Seven named plus one unnamed.

Distribution. Eastern VIC to south-eastern QLD, then in widely separated areas in central and northern QLD, in montane rainforest.

Biology. The subcylindrical and lightly sclerotised larvae feed in white rotten wood.

Key to species. Neboiss (1962).

Bluops Carter, 1914 (Pl. 86F; Fig. 36.93)

Type species. *B. verrucosus* Carter, 1914.

Characteristics. Pedunculate, dorsal surfaces glabrous, black with bluish legs and antennae. Length 15–27 mm. Third antennomere longer than next two combined. Epipleura narrow, passing under humeri and inflected elytral sides. Prothorax subquadrate, much narrower than broadly oval hind body, with right anterior and posterior angles. Elytra covered with large contiguous bullae, without striae except for sutural one. Legs long and slender.

Australian species. One: *B. verrucosus*.

Distribution. QLD wet tropics, in rainforest.

Bolusculus Matthews, 1998 (Pl. 86L)

Type species. *B. arcanus* Matthews, 1998.

Characteristics. Subquadrate, dorsal surfaces capable of holding a layer of soil (squalid), tuberculate. Length ~3 mm. Lateral edges of prothorax explanate and irregularly lobed. Prosternal process simple, not produced or bilobate. Elytra with prominent tubercles in rows, with continuous lateral carina outside 8th stria, forming pseudepipleuron. Epipleura proper narrow throughout. Tarsi short, segments not lobed.

Australian species. One: *B. arcanus*.

Distribution. QLD wet tropics, in rainforest.

Brycopia Pascoe, 1869 (Pl. 87B)

Type species. *B. pilosella* Pascoe, 1869.

Dinoria Pascoe, 1869

Characteristics. Oblong, dorsal surfaces smooth, from nearly glabrous to densely pilose, some species with pale elytral sides and/or pale legs. Length 3–8 mm. Eyes reduced, oval, oblong or nearly round in outline, appearing more or less protuberant in dorsal view because of constriction of genal edges before them. Hypomeron and prosternum punctate. Epipleura widened anteriorly, visible from above at humeri.

Australian species. 12.

Distribution. TAS, where half the known species occur, VIC, south-eastern NSW mostly at higher elevations, southern SA and WA. In southern beech and eucalypt forest litter and on moss and lichens.

Cardiothorax Motschulsky, 1860 (Pl. 87C; Figs 36.16, 36.95)

Type species. *Thoracophorus walckenaeri* Hope, 1840.

Thoracophorus Hope, 1840

Prosodes Germar, 1848

Atryphodes Pascoe, 1866

Otrintus Pascoe, 1866

Characteristics. Elongate-oblong, dorsal surfaces glabrous. Length 9–23 mm. Mentum with anterior angles strongly produced and acute (Fig. 36.16). Frontal grooves often curved inward posteriorly, usually meeting on occiput. In most species prothorax with hind edge clearly narrower than front edge, usually with explanate sides delimited from disc by shallow grooves; some species (formerly in *Otrintus*) with subparallel prothorax. Elytral striae impunctate, often effaced laterally, consisting of just impressed lines or rarely irregular foveae, intervals sometimes carinate. Males often with modified legs, without expanded protarsi.

Australian species. 66 named and several unnamed.

Distribution. Eastern Australia from south-eastern SA to northern QLD, not known from TAS, primarily in eucalypt forest.

Biology. The long-legged larvae have long, slender antennae and urogomphi and occur in leaf litter and under logs. According to label data, a larva was seen using the long urogomphi as a means of defence, but no further data are available on this possible behaviour.

Key to species. Carter (1914a).

Coripera Pascoe, 1866 (Pl. 88H; Fig. 36.87)

Type species. *Adelium deplanatum* Boiduval, 1835.

Characteristics. Oblong, glabrous, surfaces uneven, sometimes tuberculate. Length 9–19 mm. Pronotal base closely applied to elytral bases. With rounded pseudepipleural carina outside 7th interval, and epipleuron proper entirely narrow, effaced or indistinctly outlined anteriorly. Elytral striae either effaced, irregularly confluent, or geminate, intervals often with large ocellate impressions.

Australian species. Nine named, with at least four unnamed in the wet tropics of QLD.

Distribution. Eastern Australia from TAS to north QLD, in wet sclerophyll and rainforest.

Key to species. Carter (1935).

Daedrosis Bates, 1868 (Pl. 88I; Fig. 36.98)

Type species. *D. crenatostriata* Bates, 1868.

Macroperas Carter, 1914

Characteristics. Elongate-oblong, glabrous, surfaces smooth or a little rugose. Length 7–12 mm. Terminal antennomere enlarged, oval. Sides of pronotum irregularly undulate. Anterior margination of pronotum effaced in middle and hypomeron coarsely punctate. Epipleural carina dentate or lobate at humeri except in one species.

Australian species. Four named and one unnamed.

Distribution. Southern SA, southern VIC, south-eastern NSW in montane eucalypt forest.

Diaspirus Matthews 1998 (Pl. 89L; Fig. 36.92)

Type species. *D. bellendenus* Matthews, 1998.

Characteristics. Oval, hind body much broader than prothorax. Length 3–6 mm. Eyes reduced, oval. Pronotum with base narrower than apex, sides irregularly sinuate. Elytra inflated and with projecting taper posteriorly. Epipleural carinae dentate at humeri in QLD species. Penultimate tarsomeres forming large lobes which subtend ~1/2 or more of claw segments.

Australian species. *D. bellendenus* and *D. crenaticollis* (Carter, 1908) plus one unnamed.

Distribution. North-eastern NSW, northern QLD, in rainforest.

Dicyrtodes Matthews, 1998 (Pl. 89K; Fig. 36.89)

Type species. *D. arneius* Matthews, 1998.

Characteristics. Elongate oval, glabrous except for a few long erect setae. Length 5–9 mm. Eyes reduced, oval. Epipleural carinae not dentate at humeri. Bases of elytra depressed between humeral angles and receiving prothoracic base there.

Australian species. *D. arneius* and two unnamed.

Distribution. Northern QLD in rainforest.

Diemenoma Matthews, 1998 (Pl. 89E; Fig. 36.90)

Type species. *Adelium commodum* Pascoe, 1869.

Characteristics. Elongate oblong-oval, glabrous with very few fine semi-recumbent setae. Length 7–11 mm. Eyes oblong in outline. Hypomeron and prosternum impunctate. Epipleural carinae not dentate at humeri. Elytra often with minute tubercles laterally and posteriorly.

Australian species. Five.

Distribution. TAS in forest litter and rotting logs.

Dorrigonum Matthews, 1998 (Pl. 89J; Fig. 36.102)

Type species. *Licinoma umbilicata* Carter, 1924.

Characteristics. Elongate oval, surfaces smooth, sometimes sericeous, with scattered long erect setae, otherwise glabrous or with minute recumbent setae. Length 7–10 mm, Terminal antennomere whitish, paler than rest of antenna. Hypomeron impunctate or with shallow punctures. Elytral epipleura narrow, not visible from above. Basal tarsomeres of metatarsi about twice as long as claw segments.

Australian species. *D. umbilicatum* and *D. longipes* (Macleay, 1872) plus one unnamed.

Distribution. North-eastern NSW, south-eastern QLD, in rainforest litter.

Epomidus Matthews, 1998 (Pl. 90G; Fig. 36.91)

Type species. *E. prionodes* Matthews, 1998.

Characteristics. Oblong, dorsal surfaces with numerous erect setae. Length 3–4 mm. Eyes reduced, oval. Frontoclypeal suture deeply grooved, genae tumid. Epipleural carinae at humeri forming strong projecting lobes visible from above.

Australian species. *E. prionodes* and one unnamed.

Distribution. Northern QLD in rainforest.

Isopteron Hope, 1840 (Pl. 92H; Figs 36.13, 36.84)

Type species. *I. australe* Hope, 1840.

Cestrinus Erichson, 1842
Asida Hope, 1843
Isopterum Agassiz, 1846
Apatelus Mulsant & Rey, 1859
Prionotus Mulsant & Rey, 1859
Achora Pascoe, 1869
Priothorax Gebien, 1910

Characteristics. Elongate-oblong, squalid, scabrous or tuberculate, with dense short recumbent squamiform or ciliate setae. Length 5–13 mm. Eyes at least partly surrounded by a groove which connects with a pit on gena (Fig. 36.13 gr), sometimes indistinct. Eyes often divided, or at least constricted. Clypeus arcuately emarginate. Prosternal process often apically bilobate. Pronotum evenly convex or with shallow depressions on disc. Some species winged (metaventrite longer than first abdominal ventrite).

Distribution. Widely distributed throughout Australia in arid and semi-arid habitats.

Australian species. 15.

Biology. This genus is unique among Adeliini in having adapted to desert conditions, with soil-inhabiting larvae (Allsopp 1986a), and is the only adeliine genus containing species with fully developed hind wings.

Key to species. Kulzer (1964).

Leptogastrus Macleay, 1872 (Fig. 36.106)

Type species. *L. mastersi* Macleay, 1872.

Characteristics. Base of prothorax narrowed and bases of elytra rounded off, the two well separated by an exposed condyle formed by base of mesothorax, giving body pedunculate appearance. Length 4–11 mm. Sometimes with metallic bronze or blue reflections. Dorsal surfaces densely and very coarsely punctate with numerous erect bristles. Eyes strongly transverse. Apical antennomeres usually more than twice as massive as subapical. Rounded base and sides of prothorax sometimes with large acute teeth in some unnamed species. Elytral epipleural carina passing under humerus, angulate anteriorly in one species only.

Australian species. Six named plus five unnamed.

Distribution. Eastern coastal Australia from VIC to northern QLD including Torres Strait islands, one species in southern WA.

Key to species. Carter (1920).

Licinoma Pascoe, 1869 (Pl. 93J; Fig. 36.105)

Type species. *L. nitida* Pascoe, 1869.

Characteristics. Elongate-oblong, smooth, glabrous with a few semi-erect setae, rarely densely setose. Lengh 6–12 mm. Eyes transverse. Pronotum subquadrate, disc finely to coarsely punctate, anterior margination effaced in middle, hypomeron coarsely punctate. Epipleura strongly widened anteriorly, visible from above at humeri. Hind body only slightly wider than prothorax. Penultimate tarsomeres not lobed, subcupuliform.

Australian species. 12.

Distribution. Southern SA, eastern VIC and NSW, south-eastern and north coastal QLD. One species in southern Chile.

Key to species. Carter (1927).

Monteithium Matthews, 1998 (Pl. 94K; Fig. 36.86)

Type species. *M. ascetum* Matthews, 1998.

Characteristics. Dorsal surfaces capable of holding a layer of soil (squalid), tuberculate, largely glabrous. Length 5–6 mm. Eyes oval, ~1.5 × as long as wide. Last four antennomeres forming weak tomentose club after cupuliform segment 7. Pronotum with highly ornate lateral outgrowths and rounded median depression. Prosternal process very broad and deeply bifurcate. Elytral crests capped by tubercles. Penultimate tarsomeres with rounded lobe.

Australian species. Two: *M. ascetum* and *M. storeyi* Matthews & Monteith, 2010.

Distribution. QLD wet tropics, in rainforest at altitudes above 800 m.

Nolicima Matthews, 1998 (Fig. 36.97)

Type species. *Cardiothorax angusticollis* Carter, 1906.

Characteristics. Elongate-oblong, smooth, nearly glabrous. Length 5–13 mm. Eyes strongly transverse. Pronotum subquadrate or transverse, disc finely punctate often with shallow median longitudinal impressed line and 2 or 4 shallow depressions, margination complete, hypomera impunctate. Hind body only slightly wider than prothorax. Elytral striae finely punctate, epipleura widened anteriorly, their edges, when seen from side, nearly always straight, running level across humeri to elytral bases. Penultimate tarsomeres subcupuliform to lobed.

Australian species. 12 named and three unnamed.

Distribution. TAS to southern QLD, northern QLD, southern WA, in forest litter.

Nototrintus Carter, 1924 (Pl. 95E; Fig. 36.104)

Type species. *Otrintus jacksoni* Carter, 1905.

Characteristics. Elongate-oblong, smooth or rugose, glabrous with a few erect setae. Length 11–18 mm. Eyes transverse. Prothorax distinctly wider in front than behind, with sharply angulate posterior angles, sides more or less sinuate just before them. Disc with median longitudinal sulcus. Elytral striae impunctate except in *N. asperatus* (Carter, 1916) (which has alternately carinate/granulate intervals). Hind body only slightly wider than prothorax. Epipleura with upper carinate edges reflexed and somewhat elevated above discal surface at humeri.

Australian species. Five.

Distribution. From Barrington Tops, NSW to south-eastern QLD.

Seirotrana Pascoe, 1866 (Pl. 98I; Figs 36.15, 36.99)

Type species. *Adelium catenulatum* Boisduval, 1835.

Characteristics. Elongate-oval, surfaces more or less uneven, with bullae or short carinae on elytra, sometimes squalid, sparsely setose, sometimes with bronze reflections and apices of femora yellow. Length 9–19 mm. Eyes transverse. Mentum (Fig. 36.15) flat or medially depressed, coarsely punctate with thickened edges and produced, prominent anterolateral processes which overhang and obscure true front angles. Base of prothorax capable of at least slightly overlapping elytral bases. Visible part of scutellum distinctly transverse. Intercoxal process of first ventrite broadly truncate. Depressed areas of male ventrites almost always longitudinally corrugated [Carter 1908b].

Australian species. 30.

Distribution. South-eastern SA, northern TAS, VIC, eastern NSW, south-eastern QLD north to Mossman.

Biology. Larvae occur in leaf litter.

Key to species. Carter (1933).

Yarranum Matthews, 1998 (Pl. 101L; Fig. 36.100)

Type species. *Seirotrana crenicollis* Pascoe, 1869.

Characteristics. Oblong, squalid, surfaces matt with shiny elevations, with scattered short, fine erect setae, femora with yellowish bands. Length 7–13 mm. Eyes transverse and medially constricted. Mentum with irregular, tuberculate or vermiculate surface, without anterolateral processes. Pronotum subquadrate, discal surface densely tuberculate, lateral edges crenate, base slightly overlapping elytral bases. Intercoxal process of first ventrite rounded apically. Depressed areas of male ventrites smooth.

Australian species. One named, *Y. crenicolle*, possibly a species complex, plus one unnamed. Closely related to the Chilean genus *Gondvanadelium* Kaszab, 1981 (Matthews 1998).

Distribution. South-central and montane VIC, ACT.

Tribe Goniaderini Lacordaire, 1859

Characteristics. Oblong, somewhat flattened, with fine erect pilosity, fuscous. Length 5–7 mm. Eyes large, reniform. Edge of clypeus more or less straight. Clypeo-labral membrane concealed. Maxillary palpi with apical segment triangular. Mentum trapezoidal, more or less flat, coarsely punctate. Antennae reaching base of prothorax, segments short, slightly obconic, very gradually enlarging to apex. Pronotum with sides explanate, edges strongly irregularly dentate. Mesocoxal cavities open. Elytra estriate, densely confusedly punctate. Wings fully developed. Legs not fossorial, preapical tarsomeres cupuliform. Defensive glands absent but colleterial glands of segments VIII/IX possibly adapted for defensive purposes as in Adeliini (Matthews 1998).

Distribution. Goniaderini have a pantropical distribution throughout the world.

Australian taxa. One genus.

Biology. Found in rainforest floor litter.

Pseudolyprops Fairmaire, 1882 (Pl. 97J; Fig. 36.79)

Type species. *P. dilaticollis* Fairmaire, 1882 (Sumatra).

Trichulodes Carter, 1914

Characteristics. As for tribe.

Australian species. Two named: *P. punctatus* (Carter, 1914) and *P. australiae* Carter, 1930, plus one unnamed.

Distribution. Northern NT, northern QLD, recently found in northern NSW, Sumatra, India.

Tribe Lupropini Lesne, 1926

Characteristics. Oblong, somewhat flattened to moderately convex, smooth, glabrous or setose, fuscous to piceous. Eyes ovoid, not emarginate. Edge of clypeus slightly to moderately produced. Mentum trapezoidal, narrow basally, with weak median costa. Pronotum quadrangular or narrowing anteriorly, with simple lateral margins. Mesocoxal cavities closed by ventrites only except in *Micropedinus* Lewis, 1894, in which they are barely open. Elytra estriate, densely confusedly punctate. Wings fully developed except in one unnamed species of *Lorelus* Sharp, 1876. Legs slender, not fossorial, preapical tarsomeres widened, lobed or cupuliform. Defensive glands absent in Australian taxa.

Distribution. The tribe has a pantropical distribution but is most diverse in the Pacific region.

Australian taxa. Four genera (Kaszab 1982e).

Iscanus Fauvel, 1904 (Pl. 92C; Fig. 36.82)

Type species. *I. kuniensis* Fauvel, 1904 (New Caledonia).

Araucaricola Lea, 1929

Characteristics. Oblong-oval, glabrous. Length 3–4 mm. Clypeo-labral membrane concealed. Front edge of clypeus more or less straight. Labial palpi with subterminal segment enlarged, terminal one smaller and linear. Antennae reaching beyond middle of pronotum, last three segments slightly enlarged. Pronotal base not or barely narrower than elytral bases. Humeri angulate. Tarsi narrow, at least some preapical segments longer than wide.

Australian species. Two: *I. ebeninus* (Lea, 1929) and *I. glaber* (Lea, 1929).

Distribution. Norfolk Island, some other Pacific islands.

Biology. Found in rotten wood.

Key to species. Kaszab (1982c).

Lorelus Sharp, 1876 (Pl. 93C; Fig. 36.83)

Type species. *L. priscus* Sharp, 1876 (New Zealand).

Characteristics. Elongate oblong, densely setose with short, fine recumbent setae. Length 3–4.5 mm. Clypeo-labral membrane exposed. Front edge of clypeus more or less straight. Labial palpi with subterminal segment enlarged, terminal one smaller and linear. Antennae reaching beyond middle of pronotum, a little more abruptly widened from segments 8 or 9, last three or four segments of about equal width. Pronotal base slightly narrower than elytral bases. Humeri rounded. Tarsi narrow, at least some preapical segments longer than wide.

Australian species. Four.

Distribution. Lord Howe and Norfolk islands, southern QLD, Pacific islands, pantropical.

Biology. Found in litter and on vegetation.

Key to species. Kaszab (1982e).

Mesotretis Bates, 1872 (Pl. 93K; Fig. 36.80)

Type species. *M. ferruginea* Bates, 1872.

Characteristics. Elongate oval, with short, very fine, dense recumbent setae. Length 3–4 mm. Clypeo-labral membrane concealed. Front edge of clypeus produced and anteriorly very shallowly emarginate. Labial palpi with subterminal segment enlarged, terminal one smaller and linear. Antennae short, not reaching middle of pronotum, last three segments forming weak club. Pronotal base slightly narrower than elytral bases. Humeri rounded. Tarsi short and broad, all preapical segments transverse.

Australian species. *M. ferruginea*, the only species in the genus.
Distribution. South-western WA.
Biology. Found on beaches under stranded seaweed.

Micropedinus Lewis, 1894 (Pl. 94B; Fig. 36.81)

Type species. *M. algae* Lewis, 1894 (Japan).
Notoprataeus Carter, 1924
Characteristics. Oblong, glabrous. Length 2.5–4 mm. Clypeo-labral membrane concealed. Front edge of clypeus more or less straight. Antennae reaching base of pronotum, from 7th segment to end gradually widened. Labial palpi with terminal segment larger than subterminal. Pronotal base slightly narrower than elytral bases. Humeri rounded. Tarsi narrow, at least some preapical segments longer than wide.
Australian species. Two: *M. glaber* (Carter, 1933) and *M. inconstans* (Lea, 1917).
Distribution. WA, northern QLD, western Pacific.
Biology. Found on beaches under stranded seaweed.

Tribe Lagriini Latreille, 1825

Subtribe Lagriina Latreille, 1825

Characteristics. Ovoid, moderately convex, with forebody much smaller than hind body, densely setose, bronze or metallic blue-green. Eyes reniform, deeply invaded by genal canthus. Edge of clypeus arcuately excised, revealing clypeolabral membrane. Antennae reaching beyond base of prothorax, segments gradually enlarging distally, cupuliform, apical segment more or less elongate. Maxillary palpi with apical segment cultriform. Mentum small, subquadrate, smooth. Pronotum without, or with weak, lateral margins. Lobes of hypomeron meet behind procoxae. Mesocoxal cavities open. Elytra estriate and confusedly punctate. Wings fully developed. Legs slender, not fossorial, preapical tarsomeres obconic except penultimate which is very broadly bilobed. Defensive glands present. [Merkl 1987].
Distribution. Eurasia, Africa, New Guinea, Australia and the Americas, most diverse in the Afrotropical and Indo-Malayan regions. In Australia there are four centres of distribution (Merkl 1987): northern QLD, north of the NT, south-eastern Australia, and south-western WA.
Australian taxa. Six genera.

Acerogria Borchmann, 1936 (Pl. 84I; Fig. 36.72)

Type species. *Cerogria dohrni* Borchmann, 1911 (Philippines).
Birolagria Pic, 1956
Characteristics. Elytra yellowish, rest of body bluish black. Length 9.5–13 mm. Antennae of males thick, segments 7 and 9 strongly expanded at inner side, segments 8 and 10 simple, segment 11 bent, longer than preceding six segments combined. Pronotum smooth, impunctate, surface uneven. Femora and tibiae flattened dorsoventrally.
Australian species. One: *A. oriunda* (Borchmann, 1924).
Distribution. Northern QLD, New Guinea, the Philippines and Moluccas.

Ecnolagria Borchmann, 1915 (Pl. 9A, B, 89G; Fig. 36.75)

Type species. *Lagria grandis* Gyllenhal,1817.
Characteristics. Bronze or reddish with green or violet reflections. Length 7–16 mm. Antennae of males with segment 1 very elongate, longer than interantennal distance and segment 9, or 9 and 10, more or less expanded at outer edges. Antennae of females longer than half of interantennal distance. Pronotum distinctly punctate, in males slightly transverse, sometimes as long as wide, but broader than head. Surface more or less evenly convex.
Australian species. Eight.
Distribution. TAS to northern QLD, SA, southern WA.
Biology. *E. grandis* adults are seen in large numbers on the leaves of understorey plants in woodlands. Larvae feed on fallen leaves.
Key to species. Merkl (1987).

Lagria Fabricius, 1775 (Pl. 92J)

Type species. *Chrysomela hirta* Linnaeus, 1758 (Europe).
Lachna Billberg, 1820
Characteristics. Blue, green or black with bronze reflections, antennae usually partly yellowish. Length 6–12 mm. Antennae of males with segments simple, like those of females, with hairs much shorter than length of segments. Antennomere 1 of both sexes much shorter than half of interantennal distance, segment 3 usually longer than 1, segment 2 longer than wide. Pronotum distinctly punctate, surface more or less evenly convex. Elytra distinctly dilated posteriorly at least in females, humeral angle not conspicuous.
Australian species. Eight.
Distribution. NT, QLD, Eurasia, Africa.
Key to species. Merkl (1987).

Metriolagria Merkl, 1987 (Pl. 94F; Fig. 36.73)

Type species. *Lagria affinis* Boisduval, 1835.
Characteristics. Yellowish to blackish brown. Length 9.5–14 mm. Antennomeres (at least basal four) with some hairs as long as or longer than length of segments. Antennomere 1 of males considerably, of females slightly, longer than half of interantennal distance. Pronotum distinctly punctate, surface more or less evenly convex.
Australian species. Two: *M. affinis* and *M. formicicola* (Lea, 1895).
Distribution. SA, VIC, NSW to southern QLD.

Stenolagria Merkl, 1987 (Pl. 98F; Fig. 36.76)

Type species. *S. matthewsi* Merkl, 1987.
Characteristics. Brownish to black with weak metallic sheen. Body slender, legs gracile. Length 6–10 mm. Antennae of males with segment 1 less elongate, only longer than half of interantennal distance, those of females shorter. Pronotum distinctly punctate. Male pronotum at least as long as wide, usually longer, narrower than head.
Australian species. Two: *S. leai* (Borchmann, 1936) and *S. matthewsi*.
Distribution. Northern QLD in rainforest.

Xenolagria Merkl, 1987 (Pl. 101K; Fig. 36.74)

Type species. *Lagria tincta* Blackburn, 1889.
Characteristics. Reddish to blackish brown with vague light markings on elytra. Length 7.5–9 mm. Antennae thick and short, segment 3 not longer than segment 1, segment 2 not longer than wide, without conspicuously long vestiture. Antennomere 1 of both sexes much shorter than half of interantennal distance. Pronotum distinctly punctate, surface more or less evenly convex. Elytra subparallel-sided, barely dilated posteriorly in both sexes, humeral angle pronounced.
Australian species. One: *X. tincta*.
Distribution. Northern QLD, north of NT.

Tribe Lagriini Latreille, 1825

Subtribe Statirina Blanchard, 1845

Characteristics. As for the subtribe Lagriina but with the clypeolabral membrane concealed, head abruptly narrowed just behind eyes, forming distinct neck, the pronotum more or less constricted before the hind angles, the lobes of the hypomeron not meeting (separated by the prosternal process), and the elytra with 10 distinct striae. [Merkl 1986].
Distribution. North and South America, Africa, southern Asia, Japan, Taiwan, Indonesia, New Guinea, Bismarck Archipelago, Solomon Islands and Australia, most diverse in the rainforests of South America and South-east Asia. In Australia they are poorly represented and restricted to the north of the continent.
Australian taxa. Two genera.

Casnonidea Fairmaire 1882 (Pl.87F; Fig. 36.77)

Type species. *C. holomelaena* Fairmaire, 1882 (Sumatra).
Synatractus Macleay, 1887
Characteristics. Head about as broad as prothorax. Length 7–10.5 mm. Apical antennomere two to four times longer than wide but not longer than preceding two segments combined. Pronotum smooth, not punctate.
Australian species. Six.
Distribution. Northern QLD, New Guinea, southern Asia, Africa, Madagascar.
Key to species. Merkl (1986).

Sora Walker, 1859 (Pl. 98G; Fig. 36.78)

Type species. *S. marginata* Walker, 1859 (Sri Lanka).
Nemostira Fairmaire, 1868
Characteristics. Head distinctly narrower than prothorax. Length 8–10 mm. Apical antennomere longer than preceding two segments combined. Pronotum punctate.
Australian species. One: *S. lawrencei* Merkl, 1986.
Distribution. Northern QLD, New Guinea, South-east Asia, Africa, Madagascar, in rainforest.

Tribe Chaerodini Doyen, Matthews & Lawrence, 1990

Characteristics. Hemispherical, glabrous, flavous or rufous. Length 5–7 mm. Eyes ovoid, small. Edge of clypeus straight. Labrum subquadrate. Clypeolabral membrane concealed. Maxillary palpi with apical segment narrowly triangular. Mentum small, narrower at base. Antennae very short, not reaching beyond middle of prothorax, segments very transverse, distal five forming an asymmetrical club, apical segment very small. Pronotum transverse, evenly and strongly convex. Prosternum short before procoxae. Mesocoxal cavities open. Elytra evenly very convex, estriate, confusedly punctate. Wings absent. Legs strongly fossorial, protibiae with very large rounded outer apical lobe, preapical tarsomeres cupuliform. Membrane visible between last three abdominal ventrites. Defensive glands absent.
Distribution. Australian and New Zealand sea shores. In New Zealand the tribe is represented by a closely related genus, *Chaerodes* White, 1846, with two species.
Australian taxa. One monotypic genus.
Biology. Chaerodini inhabit beach sands and feed on stranded seaweed as both adults and larvae.

Sphargeris Pascoe, 1860 (Pl. 98C; Fig. 36.71)

Type species. *S. physodes* Pascoe, 1860.
Characteristics. As for the tribe.
Australian species. One: *S. physodes*.
Distribution. All coastal beaches.

Subfamily Phrenapatinae Solier, 1834

Characteristics. Eyes small, entire. Mandibles usually apically tridentate. Antennae clubbed. Elytral striae nine or more, without scutellary striole. Wings present. Mesepisternum not reaching mesocoxal cavities. Tarsomeres not expanded or lobed. Abdomen with membrane visible after ventrites 3 and 4. Defensive glands absent. Outer edge of last ventrite with broad groove to receive elytral edges. [Doyen & Lawrence 1979].

Distribution. Pantropical, mostly on islands. The New Zealand genus *Aphtora* Bates, 1872 is a phrenapatine on both morphological and molecular evidence (Matthews 2003a; Kergoat *et al.* 2014b) but its tribal position is uncertain.

Australian taxa. Two tribes.

Biology. Phrenapatinae are found in forests in decaying vegetation and sometimes fruiting bodies of fungi (Doyen & Lawrence 1979). Leschen & Cutler (1994) report that calcium carbonate crystals are deposited inside the cuticle of Penetini and Archaeoglenini, something otherwise unknown in Coleoptera, and Doyen & Lawrence (1979) mention a surface encrusting system in the same two tribes, whereby a hardened material secreted by epidermal glands collects in cuticular surface depressions.

Tribe Archaeoglenini Watt, 1975

Characteristics. Oblong, moderately convex, glabrous, flavous. Total length ca 2 mm. Edge of clypeus very shallowly emarginate. Labrum prominent. Mandibles tridentate. Antennae short, not reaching base of head, received in grooves beneath eyes, composed of 10 segments (sometimes nine elsewhere) of which last two are strongly enlarged into a club. Pronotum subrectangular with lateral edges explanate and with thickened crenulate margins, disc with a deep, irregularly defined basal depression narrowing medially. Hypomera excavated anteriorly to receive antennal clubs. Mesocoxal cavities closed. Elytra with 14 rows of punctures (usually 9–11 elsewhere), epipleura wide throughout. Wings fully developed (in Australia). Legs with tibiae linear, tarsal formula 4–4–4 (sometimes 3–3–3 or 5–5–4 elsewhere). [Doyen & Lawrence 1979; Kaszab 1982a; Iwan *et al.* 2015].

Distribution. VIC, ACT, NSW, New Zealand, Indian Ocean islands, Japan, most Pacific islands, Jamaica, Mexico, Central America and northern South America. Thirteen of the 17 known species of *Archaeoglenes* are found exclusively on islands (Iwan *et al.* 2015).

Australian taxa. There is only one genus in the tribe: *Archaeoglenes* Broun, 1893.

Archaeoglenes Broun, 1893 (Pl. 85L; Fig. 36.109)

Type species. *A. costipennis* Broun, 1893 (New Zealand).

Characteristics. As for the tribe.

Australian species. One: *A. australis* Lawrence, 1979.

Distribution. ACT; Wilson's Promontory, VIC; Kiama, NSW.

Tribe Penetini Lacordaire, 1859

Characteristics. Broadly oblong, convex, glabrous, fuscous to piceous. Edge of clypeus shallowly arcuately emarginate or straight. Labrum with anterior edge arcuate, its base sharply constricted. Mandibles apically with two or three teeth. Antennae short, extending only a little beyond base of head, partly held in sharply defined grooves beneath eyes except in *Dioedus*, 11-segmented and expanded into a 2- to 4-segmented capitate club. Pronotum evenly convex, usually with widened lateral gutters. Prosternum short before coxae. Mesocoxal cavities closed in part by mesotrochantin which appears to be attached to mesoventrite (Fig. 36.4). Elytra strongly convex, smooth. Legs with tibiae somewhat expanded apically, anterior ones outwardly toothed. Tarsal formula 5–5–4. [Kaszab 1977a].

Distribution. There are more than 25 genera of Penetini world-wide, mainly tropical and frequently occurring on islands (Doyen & Lawrence 1979).

Australian taxa. Five genera.

Biology. Larvae feed in rotten wood or other dead plant matter.

Archeophthora Kaszab, 1978 (Pl. 85J; Fig. 36.125)

Type species. *Brachycilibe tasmanica* Carter, 1919.

Characteristics. Form very convex, subcylindrical. Length 2.3–3 mm. Clypeal and genal edges more or less confluent. Left mandible with two teeth. Genae without margin. Antennal club 3-segmented, antennae held in sharply defined grooves beneath eyes. Strial punctures large, stria 8 anteriorly much shorter than 7. Wingless.

Australian species. One: *A. tasmanica*.

Distribution. TAS, NSW, southern Chile.

Dioedus LeConte, 1862 (Pl. 89A; Figs 36.4, 36.121)

Type species. *D. punctatus* LeConte, 1862 (North America).

Arrhabaeus Champion, 1886

Brachycilibe Carter, 1911

Tagalus Gebien, 1914

Characteristics. Oblong-oval, strongly convex. Length 2.2–4.8 mm. Edges of genae simple. Antennal club 2-segmented. Impressions on underside of head for reception of antennae not sharply defined. Stria 8 of elytra scarcely shorter than 7. Winged.

Australian species. Two: *D. antennatus* (Carter, 1911) and *D. araucariae* (Lea, 1929).

Distribution. Northern QLD, Lord Howe and Norfolk islands, other Pacific islands, North and Central America.

Key to species. Kaszab (1978).

Pseudophthora Kaszab, 1970 (Pl. 97K; Fig. 36.124)

Type species. *P. laena* Kaszab, 1970 (New Guinea).

Characteristics. Oblong, moderately convex. Length ~4 mm. Left mandible with three teeth. Genae without trace of margin. Clypeal and genal edges more or less confluent. Antennal club 3-segmented, antennae held in sharply defined grooves beneath eyes. Striae 7 and 8 subequal in length. Winged.

Australian species. One: *P. wilsoni* Kaszab, 1978.

Distribution. VIC, other species in New Guinea and Micronesia.

Scolytocaulus Fairmaire, 1869 (Pl. 98K; Fig. 36.123)

Type species: *S. bouchardi* Fairmaire, 1896 (Sumatra).
Platycilibe Carter, 1911
Picnotagalus Kaszab, 1939

Characteristics. Oblong, moderately convex. Length 3.5–4.5 mm. Left mandible with three sharp teeth. Genae distinctly margined between eyes and clypeal suture. Margin between clypeus and genae, where curved frontal line runs into it, emarginated. Antennal club 3- or 4-segmented, antennae held in sharply defined grooves beneath eyes. Winged.

Australian species. Three.

Distribution. Eastern Australia, TAS, Sumatra, Philippines.

Key to species. Kaszab (1978).

Tagalinus Kaszab, 1977a (Pl. 99J; Fig. 36.122)

Type species. *Uloma lifuana* Montrouzier, 1860 (New Caledonia).

Characteristics. Oblong, somewhat flattened. Length 3.2–4.5 mm. Anterior edges of genae with a small angular projection. Antennal club 2-segmented. Underside of head with sharply defined grooves to receive antennae. Stria 8 very short, beginning well away from base and barely going beyond middle of elytra. Winged.

Australian species. One: *T. australis* Matthews & Lawrence, 2005.

Distribution. Northern QLD, New Caledonia.

Subfamily Tenebrioninae Latreille, 1802

Characteristics. Labrum usually transverse (but subquadrate in Palorini, Bolitophagini, Ulomini, Opatrini, Pedinini and some Toxicini). Antennae either with simple sensilla only or with complex sensoria in Alphitobiini, Ulomini, Amarygmini and the genus *Zophobas* Dejean, 1834. Elytral striae 9 (10 in some Toxicini and Bolitophagini) plus scutellary striole. Hind wings frequently with medial fleck. Claws simple. Membranes visible between last three abdominal ventrites except in Toxicini. Dorsal edge of last abdominal ventrite evenly rounded except in Palorini, where it is grooved. Defensive glands present except in *Eutermicola* Lea, 1916 (Palorini). Aedeagus not inverted except in Palorini.

Distribution. World-wide.

Australian taxa. Fifteen named and two unnamed tribes.

Note. This is one of the subfamilies shown to be polyphyletic by Kergoat *et al*. (2014b), who recovered five widely separated lineages.

Tribe Palorini Matthews, 2003

Characteristics. Oblong, parallel-sided, either subcylindrical, moderately convex, or somewhat flattened, subglabrous, fuscous to piceous, sometimes with partly rufous elytra. Eyes entire, oval or subcircular. Edge of clypeus more or less straight in middle to shallowly emarginate, sexually dimorphic in *Platycotylus* Oliff, 1883. Labrum subcircular or subquadrate, prominent. Antennae usually short, compressed and cylindrical, not reaching beyond middle of prothorax, with very short first segment, but antennae much longer in *Platycotylus*, segments either becoming smaller distally or gradually enlarging distally, not clubbed, with 11 segments except in *Eutermicola*, where there are 10. Pronotum subquadrate or narrowing basally, in *Eutermicola* strongly sculptured. Mesocoxal cavities narrowly closed by ventrites or open (closed by mesepimera). Scutellum transverse. Elytra with nine striae, often obscured laterally, with or without scutellary striole, epipleura complete to apices but very narrow there, sometimes entirely narrow. Wings present, with reduced venation except in *Platycotylus*. Legs not fossorial, tarsomeres subparallel and cylindrical, basal segment of pro- and mesotarsi very small, difficult to see, or absent altogether, therefore tarsal formula actually or apparently 4–4–4. Abdomen with membranes visible after ventrites 3 and 4. Groove along edge of apical ventrites present, but only laterally in *Platycotylus* and absent in *Pseudeba* Blackburn, 1903 and *Eutermicola*. Defensive glands present except in *Eutermicola*. Bands of dense subcuticular pits seen on some ventrites internally in males of most taxa. Aedeagus with very short basale, inverted. [Halstead 1967; Matthews 2003a; Masumoto & Grimm 2004].

Distribution. New Zealand, Australia, Africa, Madagascar, the Oriental biogeographic region, Europe, Japan, and the Pacific. Absent from the New World except for introduced species. There is one fossil genus, as well as a fossil species of *Palorus* Mulsant, 1854, known from Eocene Baltic amber (Alekseev & Nabozhenko 2015, 2017). An unnamed genus with compact body and expanded tibiae occurs in western India.

Australian taxa. Six genera.

Biology. The probable original habitat is in dead trees under bark, where many palorines are found in association with wood-boring beetles (bostrichids, scolytines and platypodines); others occur in the food-storage chambers of social insects and of man. In Europe *Palorus depressus* (Fabricius, 1790) has been able to extend its range as far north as Scandinavia only by moving into nests of *Formica* (Halstead 1967). In northern Australia *Pseudeba* and *Eutermicola* live in termite colonies.

Note. The New Zealand *Ulomotypus* Broun, 1886 is the most plesiomorphic palorine, not showing the much-reduced spermathecal structure, wing venation and basale of the aedeagus seen in most other genera (Matthews 2003a). It indicates that the tribe originated in Gondwana but then dispersed into Eurasia and Africa, perhaps by way of India. Aedeagal inversion is basal in Tenebrionidae; this suggests that the tribe represents a very early offshoot of the tenebrionoid branch.

Austropalorus Halstead, 1967 (Pl. 85I; Fig. 36.118)

Type species. *A. planatus* Halstead, 1967.

Characteristics. Moderately depressed. Length 3–3.5 mm. Apical maxillary palpomere a little widened apically. Antennae loosely articulated, slender. Pronotum more or less narrowing basally. Pronotum and elytra without carinae. Tarsi 5–5–4.

Australian species. Two: *A. demarzi* Halstead, 1967 and *A. planatus*.

Distribution. Northern WA, northern NT.

Eutermicola Lea, 1916 (Pl. 101I; Fig. 36.108)

Type species. *E. sculpticollis* Lea, 1916.

Characteristics. Convex. Length ~2.5 mm. Antennae compact, 10-segmented, tapering apically with terminal segment smallest. Pronotum dorsally with very large and deep pit posteriorly on either side of a pair of strong sinuate carinae. Elytra with a pair of prominent carinae flanking suture and a sharp carina emerging from humeral angles. Tarsi 4–4–4.

Australian species. One: *E. sculpticollis*.

Distribution. Derby, northern WA. In the original description Lea (1916) mentions seven specimens collected by W.D. Dodd, one of which went to the WA Museum. The holotype and one paratype are in the SA Museum and there is one paratype in ANIC, but the present location of the other three is unknown. The taxon has not been collected again.

Biology. Termitophilous. Some specimens are mounted with termites identified as *Eutermes palmae* (Froggatt MS).

Palorus Mulsant, 1854 (Pl. 96J; Figs 36.20, 36.119)

Type species. *Hypophloeus depressus* Fabricius, 1790 (Europe).

Characteristics. Convex. Length 2.0–3.2 mm. Apical maxillary palpomere oval or acuminate. Antennae rather compact and tightly articulated. Dorsal margin of eyes level with or above side margin of gena (Fig. 36.20). Pronotum and elytra not carinate, with interstrial punctures fine. Tarsi 5–5–4, but basal pro- and mesotarsomeres very short, difficult to see.

Australian species. 11, of which four are endemic, three are shared with South-east Asia, New Guinea, New Caledonia and other Pacific islands, and four are cosmopolitan pests of stored products.

Distribution. There are ~40 species found naturally in Africa, Europe, southern and eastern Asia and some Pacific islands. Absent from New Zealand. In Australia the native species are found in forests from SA to northern QLD, with one, *P. austrinus* Champion, 1896, occurring in northern WA and the NT.

Key to species. Halstead (1967).

Platycotylus Olliff, 1883 (Pl. 96I; Fig. 36.116)

Type species. *P. inusitatus* Olliff, 1883 (Borneo).

Thurea Ferrer, 1998

Characteristics. Strongly flattened. Colour piceous with paler or rufous elytral areas. Length 3–3.5 mm. Edge of clypeus bidentate in males. Antennae long, reaching beyond middle of pronotum, all segments longer than wide. Posterior, and usually anterior, angles of pronotum dentiform. Pronotum and elytra not carinate. Tarsi 5–5–4. [Merkl 1992].

Australian species. One named, *P. nitidulus* (Macleay, 1872) and at least two unnamed in NSW and SA.

Distribution. Northern QLD, NT, eastern NSW, southern SA, Eurasia, Africa.

Pseudeba Blackburn, 1903 (Pl. 97L; Figs 36.19, 36.120)

Type species. *P. novica* Blackburn, 1903.

Characteristics. Moderately depressed. Length 2.3–2.8 mm. Dorsal margin of eyes lower than side margin of gena (Fig. 36.19). Apical maxillary palpomere oval or acuminate. Antennae compact. Pronotum smooth, elytral striae with finely carinate edges in *P. novica*, interstriae with large shallow punctures in *P. watsoni* Britton, 1973 and *P. fossor* Halstead, 1967. Tarsi 5–5–4 or 4–4–4.

Australian species. Four.

Distribution. Northern WA, northern QLD, north-central NSW.

Biology. *P. novica* was taken in mounds of *Nasutitermes magnus* (Froggatt) and *P. watsoni* was found in the food-storage chambers of *Drepanotermes perniger* (Froggatt) (Britton 1973).

Key to species. Halstead (1967).

Ulomina Baudi di Selve, 1876 (Fig. 36.117)

Type species. *U. carinata* Baudi di Selve, 1876 (Italy).

Coelopalorus Blair, 1930.

Characteristics. Moderately depressed. Length 3.6–4.3 mm. Antennae compact. Apical maxillary palpomere oval or acuminate. Pronotum with deep lateral longitudinal depressions. Elytral interval 7 with strong carina extending from humerus almost to apex. Protibiae with row of scaliform spinules beneath external and apical margins. Tarsi 5–5–4.

Australian species. One: *U. carinata* (= *Coelopalorus foveicollis* Blair, 1930).

Distribution. There are two species in the genus: *U. carinata* found in southern Asia including the Cocos Keeling islands, Taiwan, the Philippines, Hawaii, and northern QLD, introduced with imported products into Europe, Kenya, Trinidad, and the USA, and *U. martinae* Scupola, 2002 (= *Coelopalorus carinatus* Blair, 1930) found in

India, Sri Lanka, South-east Asia and Hawaii (Halstead 1967; Scupola 2002).

Biology. Found in tunnels of bostrichine beetles and in stored or imported vegetable products.

Tribe Toxicini Oken, 1843

Characteristics. Edge of clypeus slightly arcuately produced to shallowly emarginate. Eyes strongly transverse, reniform. Labrum only a little transverse. Clypeolabral membrane not visible. Antennae of moderate length, not or barely reaching base of prothorax, segments subequal in width except last two to four, which are expanded into a club, with simple setiform sensilla only. Mandibles short, bidentate, molar surface without fine striations. Maxillary palpi with ovoid or truncate apical segment, slightly expanded. Mentum trapezoidal or apically expanded and bilobed. Mesocoxal cavities open. Elytral epipleura complete to apex. Legs not fossorial, tarsomeres not lobed. Abdomen without visible external membranes but defensive glands present.

Distribution. Pantropical, eastern Australia.

Australian taxa. Two subtribes.

Biology. Larvae occur in the fruiting bodies of higher fungi and in decaying wood (Kompantseva 1999).

Subtribe Dysantina Gebien, 1922

Characteristics. Elongate oblong or oval with strongly sculptured surfaces and short squamiform setae able to hold a layer of soil. Last two segments of antennal club more or less fused, antennae therefore appear to be 10-segmented. Pronotum more or less sculptured with irregularly crenate sides. Elytral striae not visible. Wings present or absent.

Distribution. Eastern VIC north to northern QLD, New Caledonia, West Indies, Brazil, in humid forests.

Australian taxa. Three genera.

Ilyxerus Pascoe, 1866 (Pl. 91L; Fig. 36.41)

Type species. *I. asper* Pascoe, 1866.

Characteristics. Elongate-oblong, hind body parallel-sided. Length ~5 mm. Prominent crest above eyes. Hind edge of pronotum medially straight. Metaventrite longer along midline than first abdominal ventrite. Some dorsal projections with apical shiny tubercle.

Australian species. One: *I. asper*.

Distribution. North-eastern NSW, south-eastern QLD.

Mychestes Pascoe, 1870 (Pl. 94E; Fig. 36.37)

Type species. *M. lignarius* Pascoe, 1870.

Characteristics. Oblong-oval, hind body oval in outline. Length 7–11 mm. No projection above eye. With very prominent protuberances on pronotal and elytral surfaces. Metaventrite shorter along midline than first abdominal ventrite.

Australian species. Six.

Distribution. Eastern NSW, eastern QLD.

Key to species. Carter (1937).

Orcopagia Pascoe, 1869 (Pl. 95D; Fig. 36.40)

Type species. *O. monstrosa* Pascoe, 1869.

Characteristics. Elongate-oblong, hind body parallel-sided. Length 6–8 mm. Prominent crest above eyes. Hind edge of pronotum more or less emarginate opposite scutellum. Metaventrite longer along midline than first abdominal ventrite. Each dorsal projection with apical shiny tubercle.

Australian species. Two: *O. monstrosa* and *O. regularis* Carter, 1914.

Distribution. Eastern VIC, eastern NSW, northern QLD.

Biology. Larvae feed in rotten wood or other dead plant matter.

Subtribe Toxicina Oken, 1843

Characteristics. Elongate and parallel-sided, smooth, glabrous, piceous, matt. Length 6–18 mm. Males of many species with curved frontal horns. Pronotum subquadrate with smooth disc. Elytra usually with nine striae and scutellary striole, sometimes with part of additional stria laterally. Wings present, with medial fleck.

Distribution. Southern, eastern and northern Australia, southern Eurasia, Africa, in forests. Absent from the New World.

Australian taxa. Two genera.

Cryphaeus Klug, 1833 (Pl. 88J; Fig. 36.38)

Type species. *C. aries* Klug, 1833 (Africa).

Anthracias Redtenbacher, 1858

Characteristics. Length 8–11 mm. Eyes completely divided by genal canthus. Antennal club 3-segmented. Supraorbital horns of males glabrous, no clypeal horns. [Grimm 2013].

Australian species. One, *C. wachteli* Grimm, 2013.

Distribution. Eastern NSW, south-eastern QLD, Eurasia, Africa.

Biology. *Cryphaeus wachteli* has been collected in fruiting bodies of *Coriolopsis strumosus*.

Toxicum Latreille, 1802 (Pl. 100G, H; Fig. 36.39)

Type species. *T. richesianum* Latreille, 1804 (India).

Trestonia Rafinesque, 1815

Characteristics. Length 8–17 mm. Eyes not completely divided. Antennal club 4-segmented. Supraorbital horns of male fringed with long yellow hairs, with a pair of horns on clypeus.

Australian species. Six.
Distribution. S, E and N Australia, Eurasia, Africa.
Biology. *Toxicum gracile* Pascoe, 1883 was associated with *Phaeotrametes decipiens* in WA, a species of *Toxicum* from NT was found in fruiting bodies of *Coriolopsis asper*, and *T. antilope* Arrow from Christmas Island was breeding in a species of *Coriolopsis*.
Key to species. Carter (1914b).

Tribe Bolitophagini Kirby, 1837

Characteristics. Oblong-oval, strongly convex, with squamiform setae, often strongly sculptured and tuberculate, fuscous, matt. Eyes large, oval, minimally excised by frontal canthus. Edge of clypeus feebly produced. Labrum subquadrate. Antennae short to moderate, not or barely reaching base of prothorax, segments gradually and strongly expanding apically to prominent asymmetrical club involving about seven segments, with simple sensilla only. Maxillary palpi with apical segment fusiform. Mentum trapezoidal. Pronotum strongly tuberculate with sides serrate. Prosternum very short before coxae. Mesocoxal cavities open. Elytral epipleura complete to apices. Wings present, without medial fleck. Legs not fossorial, tibiae with longitudinal carinae, subapical tarsomeres very short, obliquely truncate. Tarsal formula 5–5–4, or 4–4–4 in species of small size.
Distribution. Pantropical in forests. In Australia they occur mainly in the north-east and east.
Australian taxa. Two genera.
Biology. Bolitophagini are specialised feeders in the fruiting bodies of higher fungi in which both larvae and adults are able to tunnel.

Byrsax Pascoe, 1860 (Pl. 87L; Fig. 36.200)

Type species. *B. coenosus* Pascoe, 1860 (Singapore).
Characteristics. Oblong. Length 4–13 mm. Dorsal surfaces strongly tuberculate. Sides of pronotum and elytra strongly dentate. Males often with a pair of curved cephalic horns. Tarsi 5–5–4, or 4–4–4 in smaller species.
Australian species. Three.
Distribution. Eastern and northern Australia, Asia.
Biology. The three Australian species each occur in fruiting bodies of different groups of polypore fungi: *B. macleayi* Pascoe, 1866 in species of *Ganoderma*, *B. egenus* Pascoe, 1866 in species of *Phellinus* and the related *Inonotus*, and *B. pinnaticollis* Carter, 1914 in *Antrodiella*, *Nigroporus* and *Rigidoporus*. *B. macleayi* appears to produce a distinctive defensive chemical, since adults placed in alcohol turn the liquid bright purple.

Rhipidandrus LeConte, 1862 (Pl. 97I; Fig. 36.199)

Type species. *Xyletinus flabellicornis* Sturm, 1826 (North America).
Characteristics. Oblong. Length 2.5–4 mm. Elytra costate, surfaces not tuberculate or horned. Sides of pronotum and elytra simple. Tarsi 4–4–4.
Australian species. Five.
Distribution. North-eastern NSW, north-eastern QLD, NT, Eurasia, North America.
Key to species. Merkl & Kompantseva (1996).
Biology. *Rhipidandrus* species normally occur in species of *Ganoderma*, but at least *R. crowsoni* Merkl and Kompantseva, 1996 has also been associated with *Nigroporus vinosus* and *Fomitopsis dochmia*.

Tribe Tenebrionini Latreille, 1802

Characteristics. Elongate-oblong, parallel-sided, glabrous, piceous, Length 6–18 mm. Eyes transverse, reniform or divided by genal canthus. Edge of clypeus slightly convex in middle. Labrum transverse, clypeolabral membrane not exposed. Antennae not reaching base of prothorax. Antennomeres moniliform or subtriangular, gradually becoming more transverse distally, with simple sensilla only. Maxillary palpi with apical segment suboval or subtriangular. Pronotum subquadrate, disc evenly and feebly convex. Prosternum long before coxae. Mesocoxal cavities open. Elytra smooth with nine striae and scutellary striole, epipleura complete to apices. Wings present, without medial fleck. Legs not fossorial, tarsomeres subparallel.
Distribution. World-wide.
Australian taxa. Two genera.

Paratoxicum Champion, 1894 (Pl. 96D; Fig. 36.196)

Type species. *P. iridescens* Champion, 1894
Schizophthalmotribolium Kaszab, 1940
Characteristics. Surfaces piceous with a velvety, slightly iridescent texture. Eyes divided by genal canthus. Length ~6 mm.
Australian species. Three.
Distribution. Eastern NSW, southern WA.

Tenebrio Linnaeus, 1758 (Pl. 99K; Fig. 36.207)

Type species: *T. molitor* Linnaeus, 1758.
Characteristics. Surfaces matt, piceous. Eyes not divided. Length 12–18 mm.
Australian species. Two: *T. molitor* and *T. obscurus* Fabricius, 1793.
Distribution. The genus occurs naturally in Africa and southern Asia, with the two species named above introduced by human commerce throughout the world.
Biology. The larvae known as mealworms are widely cultured in milled grain as pet food, but *Tenebrio* is rarely found naturally in Australia.

Tribe Uncertain – Tenebrioninae

Zophobas Dejean, 1834 (Pl. 101C; Fig. 36.132)

Type species: *Helops morio* Fabricius, 1776.

Characteristics. Surfaces matt, piceous, glabrous. Length 23–34 mm. Clypeus elongate and emarginate, deeply so in male, shallowly in female, broadly exposing middle of clypeolabral membrane. Antennae reaching base of pronotum, with numerous compound sensoria on last five segments. Pronotum a little transverse, sides evenly convex, anterior angles rounded, posterior angles right. Mesocoxal cavities open. Elytra tapering apically, smooth with nine striae and scutellary striole, epipleura complete to apices. Wings present, without medial fleck. Legs not fossorial, tarsomeres subparallel, ventrally with dense, long orange setae. Last abdominal ventrite with deep marginal groove.

Australian species. One, *Z. morio*. Ferrer (2006) separated the two closely related species *Z. morio* and *Z. atratus* (Fabricius, 1775) which had previously been synonymised. Specimens sold in Australia agree with Ferrer's description of *morio*.

Distribution. *Zophobas*, with 21 named species, occurs mainly in the Neotropical region. Introduced into Africa and Australia (Ferrer 2011).

Biology. Larvae of *Z. morio*, known as superworms or giant mealworms, are cultured in oat bran and widely sold in Australia as pet food. The natural food of the genus is very diverse, consisting of rotten wood, bat guano, decaying vegetation, and carrion (Ferrer 2011).

Note. The tribal position of this genus must be considered uncertain. In all current works it is placed in Tenebrionini but differs profoundly from *Tenebrio* and related genera in both the adult and larval stages. Adults have compound sensoria on the antennae and exhibit differences in the mouthparts, defensive gland reservoirs, female tract and aedeagus. Larvae differ from those of *Tenebrio* in several ways, among which are the following: (1) tergum IX is subconical, slightly flattened above and bears four short, articulated spines – one on each side of midline and one on each side slightly posterior to these (in *Tenebrio* without spines and with a pair of fixed urogomphi); (2) antennomere 3 very small and sensorium at the apex of antennomere 2 is large and consists of numerous circular discs (in *Tenebrio* antennomere 3 is moderately long and the sensorium is C-shaped); and (3) fore legs are enlarged (not so in *Tenebrio*). The combination of enlarged fore legs and a subconical ninth tergum with a variable number of spines (often four) occurs in members of several tenebrionine tribes, including Amphidorini, Blaptini, Opatrini, Pedinini and Cerenopini, but the sensorium may be unique.

Tribe Unnamed – Tenebrioninae

Hangaya Matthews & Merkl, 2015 (Pl. 91B; Fig. 36.198)

Type species. *H. enigmatica* Matthews & Merkl, 2015.

Characteristics. Oblong, surfaces scabrous with numerous sparse, stout semi-recumbent setae. Length ~4.3 mm. Eyes large, reniform. Edge of clypeus evenly convex, labrum strongly transverse, not exposed. Antennae gradually expanding, reaching only to about middle of pronotum, with simple sensilla only. Pronotum transverse with broadly rounded anterior angles and sinuate posterior angles. Elytra with nine shallow punctate striae, without scutellary striole, intervals feebly convex with one or two rows of inwardly inclined, semirecumbent setae, epipleura incomplete, not reaching apices. Mesocoxal cavities barely open. Legs cursorial, tarsomeres simple.

Australian species. One: *H. enigmatica*.

Distribution. Known only from one locality in central NT in the Central Ranges xeric scrub habitat.

Note. Characteristics of the single known species are mostly the basal ones of Tenebrioninae with very few apomorphies, the two most salient being abbreviated elytral epipleura, which in the subfamily are otherwise seen in Opatrini, Titaenini, and Trachelostenini (all of which differ from *Hangaya*), and annulated defensive gland reservoirs (Matthews & Merkl 2015, fig. 12), only known in the Scaurini among the Tenebrioninae (Tschinkel & Doyen 1980) and in the Cnodalonini (Stenochiinae), but *Hangaya* bears no resemblance to any member of the Scaurini or Cnodalonini. The closest relative of this genus remains unknown at present.

Tribe Heleini Fleming, 1821

Characteristics. Oblong, ovoid or subcircular in outline. Mandibles apically bidentate, unidentate or truncate. Maxillary palpi with apical segment more or less triangular. Antennae variable in length, from not reaching base of prothorax to reaching well beyond it, segments basally more or less elongate becoming shorter and subtriangular distally, only a little widened there, with simple sensilla only. Prosternum moderately long before coxae, its posterior process usually received in concavity of mesoventrite. Mesocoxal cavities open. Elytral epipleura complete to apices. Wings present or absent.

Distribution. Australia, New Guinea, Ceram, Lord Howe Island, New Zealand. The New Zealand *Cerodolus* Sharp, 1886, currently placed in Titaenini, belongs in Heleini according to both molecular and adult morphological evidence (Kergoat *et al.* 2014b; Matthews & Lawrence 2015). *Cerodolus* and the closely related *Pseudhelops* Guérin-Méneville, 1841 are not related to any Australian heleines and seem to represent a basal branch of the tribe present in New Zealand by vicariance, whereas the other heleine genus there, *Mimopeus* Pascoe, 1866 is indistinguishable from the Australian *Celibe* Boisduval, 1835 and clearly arrived later by overseas dispersal (Matthews 1993).

Australian taxa. Three subtribes.

Biology. Heleini occur mainly in forests but have adapted to Australian arid habitats four times independently, once in Cyphaleina and three times in Heleina (Matthews 2000). They underwent an extraordinary adaptive radiation on the Australian continent during the time that it was isolated, managing to occupy virtually every ecological niche that is open to the Tenebrionidae and thus evolving into forms that seem equivalent to Pimeliinae, Stenochiinae and Opatrini elsewhere.

Subtribe Asphalina Matthews & Lawrence, 2005

Characteristics. Oblong, glabrous, black. Length 6–21 mm. Clypeolabral membrane not visible from above. Pronotum subquadrate to a little transverse with acute hind angles, the sides not explanate. Elytra smooth with nine striae and a scutellary striole except in *Sloanea* Carter, 1916 where striae are partly replaced by costae. Wings present, with medial fleck, or absent. Legs not fossorial.

Distribution. South-east SA, VIC, TAS, eastern NSW, south-eastern and north-eastern QLD, in forests.

Australian taxa. Four genera.

Biology. Larvae feed in rotten wood.

Asphalus Pascoe, 1868 (Pl. 85D; Fig. 36.208)

Type species. *A. ebeninus* Pascoe, 1868.

Characteristics. Posterior edge of pronotum strongly sinuate with prominent acute posterior angles, anterior angles acute. Length 17–21 mm. Elytra smooth, striate. Wingless. Tibiae not sulcate, strongly curved.

Australian species. Two: *A. ebeninus* and *A. striatus* Carter, 1911.

Distribution. North-eastern NSW, south-eastern QLD.

Key to species. Carter (1914b).

Bassianus Matthews & Doyen, 1989 (Pl. 86C; Fig. 36.209)

Type species. *Tenebrio colydioides* Erichson, 1842.

Characteristics. Posterior edge of pronotum straight or a little sinuate, posterior angles subquadrate or subacute, anterior angles strongly projecting. Length 6–13 mm. Elytra smooth, striate. Winged. Tibiae not sulcate, not strongly curved.

Australian species. Four.

Distribution. TAS, southern and eastern Australia from SA to northern QLD.

Key to species. Matthews & Doyen (1989).

Meneristes Pascoe, 1869 (Pl. 93B; Fig. 36.206)

Type species: *M. laticollis* Pascoe, 1869.

Asiris Motschulsky, 1872

Characteristics. Anterior and posterior angles of pronotum acute. Length 15–25 mm. Elytra smooth, striate. Winged. Tibiae strongly curved, the outer faces sulcate.

Australian species. Five.

Distribution. TAS, southern and eastern Australia from SA to south-eastern QLD.

Key to species. Carter (1914b).

Sloanea Carter, 1916 (Pl. 98L; Fig. 36.205)

Type species. *S. costata* Carter, 1916.

Characteristics. Anterior angles of pronotum rounded, posterior angles subacute. Length 18–19 mm. Elytra not striate, with seven nitid costae separated by matt rugose intervals. Wingless. Tibiae straight, the outer faces sulcate.

Australian species. One: *S. costata*.

Distribution. North-eastern NSW.

Subtribe Cyphaleina Lacordaire, 1859

Characteristics. From oblong to subcircular, fuscous to piceous or more or less metallic, glabrous or setose. Clypeolabral membrane exposed in 13 of the 20 genera. Sides of pronotum explanate or not, but if so not serrate on posterior edges. Prosternum apically prolonged and received in a deep concavity of mesoventrite except in three genera where the concavity is shallow. Elytra never with lateral flanges. Males often with patch or line of tomentum or longer setae on inner edges of meso and metatibiae, and often on femora. Legs not fossorial. [Matthews 1992].

Distribution. Australia, throughout the continent.

Australian taxa. Twenty genera. Since the revision of Matthews (1992), treating 19 genera, *Pachycoelia* Boisduval, 1835 was returned to the subtribe (as *Lepispilus* Westwood, 1841) by Matthews & Bouchard (2008).

Biology. Larvae of most Cyphaleina are found in rotten wood (Watt 1975), but some of those of the arid-zone genera may be soil-inhabiting although this has not been confirmed. The subtribe is divided ecologically into two groups: a winged mesophilic one of 13 genera occurring in forest habitats, and a flightless xerophilic one (informally called the *Nyctozoilus* group) of seven related genera occurring in arid shrubland and grassland areas: *Agasthenes* Bates, 1873, *Aglypta* Gebien, 1908, *Amphianax* Bates, 1873, *Byallius* Pascoe, 1869, *Nyctozoilus* Guérin-Méneville, 1830, *Onotrichus* Carter, 1911 and *Styrus* Bates, 1873.

Agasthenes Bates, 1873 (Pl. 84J; Fig. 36.213)

Type species. *A. westwoodi* Bates, 1873.

Characteristics. Body form usually parallel and elongate (broadened in *A. frenchi* Carter, 1909), smooth, matt, glabrous, piceous. Length 17–25 mm. Clypeolabral membrane exposed. Head above with deep grooves (sometimes only depressions) extending forward from eyes and marking inner edges of genae, which are small, laterally compressed and pad-like. Submentum with prominent tusk- or knob-like processes on anterior edge laterally, with or without gular groove. Prothorax abruptly constricted just before hind angles, which are therefore dentiform. Lateral margins of pronotum thickened. Elytra smooth, striate. Scutellum 2–3 times as wide as long and often swollen. Wingless.

Australian species. Three named and two unnamed.

Distribution. Western and southern WA, SA, southern NT, western VIC, arid zone.

Aglypta Gebien, 1908 (Pl. 84D; Fig. 36.235)

Type species. *A. octocostata* Gebien, 1908.

Onoglypta Carter, 1926

Characteristics. Oblong, strongly convex, matt, piceous, glabrous. Length 8–12 mm. Clypeolabral membrane concealed. Apices of mandibles apically unidentate, with prominent apical tooth and smaller tooth distant from apex. Elytra without striae, sometimes with costae. Wingless.

Australian species. Four.

Distribution. A restricted area of central WA inland from Geraldton around Mullewa, Cue and Kookynie.

Amarygmimus Bates, 1873 (Pl. 84F; Fig. 36.218)

Type species: *A. duboulayi* Bates, 1873.

Amarygmomimus Rye, 1875

Characteristics. Oblong, nitid, with short erect setae, fuscous with violet and green reflections. Length 11–14 mm. Eyes large. Antennae short, barely reaching base of prothorax. Pronotum transverse, evenly convex, angles not projecting. Legs short, apices of femora only just extending beyond body outline. Wings present, with medial fleck.

Australian species. One: *A. duboulayi*, known from only three specimens, the last collected in 1906.

Distribution. South-western WA, without habitat data.

Amphianax Bates, 1873 (Pl. 85G; Fig. 36.211)

Type species: *A. subcoriaceus* Bates, 1873.

Characteristics. Oblong, matt, piceous. Length 12–19 mm. Clypeolabral membrane exposed. Eyes small, separated across top of head by 3–5 eye widths. Pronotum with acute angles, impunctate, the lateral margins thickened. Scutellum forming a subequilateral triangle. Elytra smooth, without striae, with only a few scattered punctures. Wings vestigial.

Australian species. One: *A. subcoriaceus*.

Distribution. Western VIC, southern SA and WA, in mallee.

Atoreuma Gebien, 1941 (Pl. 85A; Fig. 36.221)

Type species. *Toreuma cupreum* Carter, 1913.

Toreuma Carter, 1913

Eutoreuma Carter, 1914

Characteristics. Oblong, matt or nitid, usually with metallic reflections. Glabrous or densely setose. Length 6–15 mm. Eyes large. Clypeolabral membrane exposed. Pronotum transverse, the angles obtuse. Scutellum subequilateral. Elytral surfaces usually uneven, vermiculate or tuberculate, often with cupreous or golden reflections on raised areas. Epipleura usually with fine groove along inner edge posteriorly. Wings present, usually with medial fleck. Male with row of tubercles along inner face of protibiae, as well as metatibial tomentum, sometimes also with widened protarsi, rarely without sexual dimorphism.

Australian species. Four named, five unnamed.

Distribution. Coastal NSW and QLD, from near Sydney to Cape York, in tall open forest and rainforest.

Bolbophanes Carter, 1913 (Pl. 86D; Fig. 36.217)

Type species. *Paraphanes dumbrelli* Lea, 1895.

Pseudobolbophanes Kulzer, 1954

Characteristics. Broadly oval, strongly convex, nitid, metallic green or purple, glabrous or with recumbent setae. Length 7–12 mm. Eyes large. Clypeolabral membrane exposed. Pronotum transverse, finely punctate, may have dense recumbent setae, the angles obtuse. Scutellum sunken, equilaterally triangular. Elytra smooth, densely punctate, glabrous. Wings present, with medial fleck. Penultimate two abdominal ventrites with coriaceous expansions of intersegmental membranes. Pro- and mesotibiae often sulcate on outer faces. Male with pro- and mesotarsi widened, and metatibiae apically dilated and flattened.

Australian species. Six.

Distribution. Eastern Australia from VIC to northern QLD, in sclerophyll forests.

Byallius Pascoe, 1869 (Pl. 87G; Fig. 36.212)

Type species. *B. reticulatus* Pascoe, 1869.

Characteristics. Elongate-oblong, strongly convex, matt, piceous, glabrous. Length 17–24 mm. Eyes small, separated across top of head by 2.5–5 eye widths. Clypeolabral membrane exposed. Submentum without prominent teeth or tusks. Pronotum as wide as elytral bases, its lateral margins thickened. Sutures between prosternum and hypomera deeply impressed, forming a groove, except in WA species. Elytra with three prominent costae, surface rugose or vermiculate between costae. Scutellum not more than twice as wide as long. Wingless.

Australian species. Eight named and one unnamed.

Distribution. VIC, south-eastern NSW primarily in the Australian Alps, south-eastern SA, in sclerophyll forest. An unnamed species is known from the south-east of WA on Mount Ragged and around Lake Daringdella, near Israelite Bay.

Key to species. Carter (1933).

Cyphaleus Hope, 1840 (Pl. 88G, 101F, H; Fig. 36.220)

Type species. *Helops rugosus* Gray, 1832.

Chartopteryx Westwood, 1841.

Oremasis Pascoe, 1866

Altes Pascoe, 1869

Anausis Bates, 1873

Apomestris Bates, 1873

Trisilus Haag-Rutenberg, 1878

Characteristics. Oblong or oval, sometimes with spiniform anterior pronotal angles and elytral apices, often strongly convex across elytral bases, usually nitid bronze or bluish, rarely matt brown, glabrous or setose. Length 11–35 mm. Clypeolabral membrane exposed. Third antennomere twice as long as 4th and antennae able to reach well beyond base of prothorax. Pronotum more flattened than bases of elytra. Winged, without medial fleck. Femora projecting well beyond body outline. Tibiae not sulcate. Male without setose line on inner margin of metatibia and with pro- and mesotarsi widened. Penultimate two abdominal ventrites with coriaceous expansions of intersegmental membranes.

Australian species. 29 named. Habitus images of 18 species of *Cyphaleus* are presented by Matthews & Bouchard (2008).

Distribution. Twenty-six species occur in VIC, eastern NSW, eastern QLD, and northern NT in forests. Three other, closely-related species, *C. metallescens* (Westwood, 1849), *blackburni* (Carter, 1913) and *macleayi* (Bates, 1873) occur outside these areas in the arid zone from western NSW to WA (Matthews 1992, map fig. 105).

Hemicyclus Westwood, 1841 (Pl. 91J; Fig. 36.215)

Type species. *H. grandis* Westwood, 1841.

Cyclophanes Carter, 1913

Characteristics. Hemispherical, strongly convex, nitid or sericeous, metallic green or purple, glabrous. Length 12–16 mm. Clypeolabral membrane exposed. Males with enlarged apical maxillary palpomeres. Antennae short. Elytral epipleura very wide and excavated under humeri. Winged, without medial fleck. Legs largely hidden under body, all tibiae deeply sulcate on outer faces. Penultimate two abdominal ventrites with coriaceous expansions of intersegmental membranes.

Australian species. Six.

Distribution. VIC, eastern NSW, south-eastern QLD, northern QLD, in tall open forest and rainforest.

Key to species. Carter (1913a).

Mithippia Pascoe, 1869 (Pl. 94H; Fig. 36.239)

Type species. *M. aurita* Pascoe, 1869 (= *Tarpela catenulata* Allard, 1877).

Characteristics. Elongate-oblong, alutaceous, fuscous, glabrous. Length 9–11 mm. Dorsal surfaces densely, coarsely punctate, or asperate on elytra. Clypeolabrtal membrane concealed. Pronotum subquadrate, narrower than elytral bases with anterior angles strongly produced. Elytra without distinct striae, densely, coarsely punctate with punctures arranged in rows or reticulate pattern. Epipleura with groove on inside edge of posterior 2/5. Winged, without medial fleck. Pro- and mesotibiae of males subapically dentate or expanded, two species with spine on fore trochanter.

Australian species. Two named: *M. aurita* and *M. jansoni* Bates, 1873, and one unnamed.

Distribution. South-western WA (*M. jansoni*) and south-eastern SA, eastern VIC and NSW north to QLD border (the other two species), in sclerophyll forest.

Key to species. Carter (1913a).

Mitrothorax Carter, 1914 (Pl. 94J; Fig. 36.224)

Type species. *Mitrephorus convexicollis* Carter, 1913.

Ctimene Bates, 1873

Mitrephorus Carter, 1913

Timeneca Carter, 1914

Characteristics. Elongate-oval, fuscous, alutaceous, with sparse minute semi-erect ciliae. Length 9–12 mm. Mandibles with wide, deep channel along outer edge dorsally. Pronotum transverse, the angles obtuse. Elytral punctures confused, deep grooves just above epipleura containing a row of minute tubercles. Winged, without medial fleck.

Australian species. Two: *M. convexicollis* and *M. breweri* (Bates, 1873).

Distribution. North-eastern NSW to south-eastern QLD (*M. convexicollis*) and south-western WA (*M. breweri*), in forests.

Nyctozoilus Guérin-Méneville, 1830 (Pl. 10C, H, 95C, H; Fig. 36.214)

Type species. *N. obesus* Guérin-Méneville, 1830.

Sphenogenius Solier, 1848

Onosterrhus Pascoe, 1866

Hypocilibe Bates, 1872
Aethalides Bates, 1873
Ononyctus Carter, 1914

Characteristics. Oval, rarely elongate, matt, piceous, glabrous, ciliate or squamose. Length 11–26 mm. Eyes small, separated across top of head by 3–5 eye widths. Clypeolabral membrane exposed or concealed. Submentum simple or sometimes with groove, often with a pair of tusk-like projections. Pronotum transverse, evenly convex, the lateral margins thickened, the angles subquadrate or acute. Scutellum 2–3 times as wide as long and often swollen. Elytral surface smooth, rugose or costate. Wingless. [Carter 1911b].

Australian species. 71 named.

Distribution. All states except TAS, primarily in inland arid areas.

Biology. Adults have been seen on the ground surface in a mallee area at night nibbling on moulds growing on fallen branches.

Key to species. Carter (1925) (part).

Olisthaena Erichson, 1842 (Pl. 95A; Fig. 36.226)

Type species. *O. nitida* Erichson, 1842.
Olisthoena Fairmaire, 1849
Hectus Pascoe, 1869
Decialma Pascoe, 1869
Aphectus Carter, 1926

Characteristics. Oblong-oval, nitid, glabrous. Length 9–14 mm. Clypeolabral membrane concealed. Sides of pronotum broadly convex. Elytral punctures confused, grooves just above epipleura often containing a row of minute tubercles. Winged, with medial fleck.

Australian species. Four.

Distribution. TAS, south-eastern SA, VIC, coastal NSW to south-eastern QLD, northern QLD, in sclerophyll forest.

Onotrichus Carter, 1911 (Pl. 95K; Fig. 36.229)

Type species. *O. lateralis* Carter, 1911.

Characteristics. Oblong, matt, piceous, with long pilosity or short setae. Length 9–16 mm. Clypeolabral membrane concealed. Mandibles unidentate. Pronotum evenly convex, the lateral edges thickened. Scutellum broadly triangular. Elytra irregularly punctate and pilose, or indistinctly rugulose with short erect setae, with three or four costae. Wingless. Protibiae with outer edges crenulate. Intercoxal process of first abdominal ventrite broadly rounded.

Australian species. Two named: *O. lateralis* and *O. minor* Carter, 1914, and two unnamed.

Distribution. Central WA inland from Geraldton and Shark Bay.

Pachycoelia Boisduval, 1835 (Pl. 10G, 93E; Figs 36.11, 36.210)

Type species. *P. sulcicollis* Boisduval, 1835.
Lepispilus Westwood, 1841
Lepidospilus Agassiz, 1846
Tyndarisus Pascoe, 1869

Characteristics. Oblong, smooth, fuscous usually with adpressed white setae forming patches. Length 13–22 mm. Clypeolabral membrane concealed. Antennae short. Pronotum transverse with deep median longitudinal sulcus and often setose white borders. Elytra without striae, with shallow depressions containing white setae, sometimes uniformly fuscous. Winged, with medial fleck (Fig. 36.11 mf). Often with strong sexual dimorphism, males with narrower bodies and larger eyes than females.

Australian species. Four.

Distribution. TAS, VIC, eastern NSW, south-eastern QLD, mostly montane. The type locality for *P. rotundicollis* (Blackburn, 1889), 'C.A.' (Central Australia?) is misleading, as the species is only known from south-eastern QLD and the Blue Mountains (Carter 1932).

Biology. Adults of *Pachycoelia*, most of which have a mottled appearance, are unique in the tribe in being flower-visiting. Their larvae feed on wood mould (Watt 1975).

Note. *Pachycoelia* is listed as a preoccupied name in Matthews & Bouchard (2008) but that is incorrect.

Key to species. Carter (1913a).

Paraphanes Macleay, 1887 (Pl. 96F; Fig. 36.222)

Type species. *P. nitidus* Macleay, 1887.

Characteristics. Oblong, nitid, fuscous or rufous with metallic green reflections, mostly glabrous with a few fine recumbent setae. Length 10–14 mm. Clypeolabral membrane concealed. Antennae relatively long, especially in males. Pronotum transverse, anterior angles subquadrate, posterior angles acute, disc with uneven surface. Elytral epipleura with outer edges partly effaced and undulate anteriorly. Winged, with medial fleck. Femora clavate, usually with white setose fringes.

Australian species. One named, *P. nitidus* and one unnamed differing in sexual characters.

Distribution. North-eastern QLD (*P. nitidus*), central and south-eastern QLD (an unnamed species), in rainforest.

Phanechloros Matthews & Bouchard, 2008 (Pl. 96G; Fig. 36.219)

Type species. *Platyphanes punctipennis* Carter, 1911.
Chlorophanes Matthews, 1992

Characteristics. Oblong, sericeous with matt pronotum, greenish or black, glabrous. Length 15–18 mm. Clypeolabral membrane exposed. Pronotum transverse with angles subquadrate, disc evenly convex. Elytral surface smooth alutaceous with scattered coarse punctures and lateral row of small tubercles near bases. Male with tomentose line on inside edge of metatibiae. Winged, with faint medial fleck.

Australian species. One: *P. punctipennis*.

Distribution. VIC and eastern NSW, in sclerophyll forest.

Platyphanes Westwood, 1849 (Pl. 97B; Fig. 36.223)

Type species. *P. gibbosus* Westwood, 1849.

Opigenia Pascoe, 1869

Laonicus Haag-Rutenberg, 1878

Characteristics. Oblong or oval, piceous usually with metallic blue or green reflections, glabrous or pilose. Length 14–27 mm. Clypeolabral membrane usually concealed but sometimes slightly exposed. Pronotum transverse or subquadrate, some species with acuminate anterior angles, discal surface even or uneven. Elytral punctures arranged in rows or striae (scarcely discernible in *P. browni* Carter, 1914), grooves just above epipleura containing a row of minute tubercles, apices sometimes spiniform. Winged, with medial fleck.

Australian species. 25 named and five unnamed.

Distribution. East of the Great Divide from south-eastern SA to northern QLD in sclerophyll forest, with two rarely collected species in south-central WA in semi-arid woodland.

Key to species. Carter (1913a).

Prophanes Westwood, 1849 (Pl. 97F; Fig. 36.134)

Type species. *P. aculeatus* Westwood, 1849.

Lygestira Pascoe, 1866

Maerodes Waterhouse, 1877

Moerodes Rye, 1879

Characteristics. Oblong, fuscous without metallic reflections, glabrous or with dense, very short fine setae. Length 21–27 mm. Clypeolabral membrane exposed in middle, laterally covered by extensions of clypeal margin. Pronotum transverse, anterior angles rounded, or acuminate in the type species. Elytra densely, evenly punctate with shallow confused punctures and traces of costae, apices acuminate in type species. Winged, without medial fleck. Male with meso- and metafemora having slightly concave posterior faces bearing rows of tubercles at upper and lower edges.

Australian species. Three.

Distribution. South-eastern SA, VIC, eastern NSW, south-eastern QLD, northern QLD in sclerophyll forest and rainforest.

Styrus Bates, 1873 (Pl. 99E; Fig. 36.216)

Type species. *S. elongatulus* Bates, 1873.

Characteristics. Elongate-oval, strongly convex with uneven surfaces, matt, piceous. Length 12–20 mm. Clypeolabral membrane largely exposed by broadly emarginate edge of clypeus. Middle of frons with transverse groove. Pronotum with sides irregularly crenulate or lobate, angles produced, discal surface uneven, usually vermiculate, base with transverse groove. Scutellum 2–3 times as wide as long and often swollen. Elytral surfaces strongly sculptured with prominent irregular reticulation. Wingless.

Australian species. Five.

Distribution. North-eastern NSW and south-eastern QLD on the western slopes of the Great Divide in dry sclerophyll forest.

Subtribe Heleina Fleming, 1821

Characteristics. Oblong, ovoid or subcircular in outline, glabrous or pilose, piceous or fuscous. Edge of clypeus either more or less straight in middle or excised in *Brises* Pascoe, 1869 and some *Pterohelaeus* Brême, 1842. Eyes transverse, reniform or slit-like. Clypeolabral membrane concealed except in *Brises*. Antennal sensilla can be clustered in New Guinean *Emcephalus* Kirby, 1828 (Matthews 1993, figs 74, 75), but are otherwise simple. Pronotum usually explanate laterally. Elytra estriate or with supernumerary striae, epipleura complete to apices. Wings present or absent, with or without medial fleck. Legs either unmodified and cursorial, or protibiae variously modified for digging, if the latter, then often with enlarged spur of protibia (forespur). [Matthews 1993].

Distribution. Australia, New Guinea, Ceram, Lord Howe Island, New Zealand.

Australian taxa. Fourteen genera.

Biology. Known larvae of Heleina are soil-inhabiting and the group as a whole is more xerophilic than the other two subtribes, occurring in open forest, shrubland and grassland and being absent from rainforest. Matthews (1993) recognised three clades (genus groups) of which the most basal contains the very large genus *Pterohelaeus* and the closely-related *Celibe* Boisduval, 1835, as well as two other genera, while the other two cluster around *Saragus* Erichson, 1842 and *Helea* Latreille, 1804 respectively. Matthews (2000) proposed that some members of each of the three clades adapted to desert conditions independently, each differing mainly in the choice of adult daytime shelters (Heleina are rarely day-active). The single desert-inhabiting member of the *Pterohelaeus* clade is *Brises*, which lives almost entirely in desert caves. Members of the *Saragus* clade, containing the geophilous genera *Dysarchus* Pascoe, 1866, *Boreosaragus* Matthews, 1993, *Saragus* and *Trichosaragus* Blackburn, 1890 are able to dig their own burrows for shelter using the protibiae usually equipped with an enlarged forespur that often shows signs of wear. Members of the *Saragus* clade are not seen foraging on the ground surface. *Ospidus* Pascoe, 1866 and *Edylius* Champion, 1894 seem to belong to the *Saragus* clade but the former is arboreal in the adult stage and the latter lives on coastal sand dunes. The *Helea* clade, characterised by greatly developed pronotal and elytral flanges which seem to have evolved from the narrower flanges of *Pterohelaeus*, has *Emcephalus*, an arboreal mesophilic genus, plus three geophilous xerophilic genera: *Helea*, *Sympetes* Pascoe, 1866 and *Cillibus* Matthews, 1993. These are unable to dig and utilise mainly mammal burrows for daytime shelter, emerging at night to forage on the surface.

Boreosaragus Matthews, 1993 (Pl. 86K; Fig. 36.233)

Type species. *Saragus lugubris* Lea, 1897.

Characteristics. Oblong, convex, matt, either glabrous, minutely bristled or densely squamose. Length 12–20 mm. Pronotum transverse, as wide as elytral bases, with moderately wide lateral flanges, disc evenly convex, angles subquadrate. Elytra each with two or three costae, sometimes zigzagged, inner costa more prominent and forming with opposite one a characteristic raised flattened longitudinal median surface. Wingless. Protibiae linear, forespur not enlarged. Intercoxal process of first abdominal ventrite broadly subquadrate.

Australian species. 17 named, at least two unnamed.

Distribution. Australia north of about the 30th parallel, in arid and monsoonal areas.

Brises Pascoe, 1869 (Pl. 86A; Fig. 36.133)

Type species. *B. trachynotoides* Pascoe, 1869.

Ephidonius Pascoe, 1869

Characteristics. Oblong with prothorax narrower than elytral bases, entirely piceous, glabrous or with minute setae. Length 10–26 mm. Anterior edge of clypeus deeply emarginate in middle, exposing middle part of clypeolabral membrane. Pronotum subquadrate or a little transverse. Elytral surfaces with striae ill-defined or absent, usually costate, or sometimes smooth or with small tubercles. Wings fully developed except in two species, where they are reduced, without medial fleck. Legs long, tibial spurs longer than basal tarsomeres.

Australian species. Eight named, with several more unnamed in northern caves.

Distribution. Arid parts of WA, NT, SA, NSW, and QLD.

Biology. Specimens of *Brises* have been found only in desert caves except for *B. blairi* Carter, 1914 and *B. parvicollis* (Blackburn, 1895) which are found in mammal burrows and wandering on sand dunes in the arid zone at night or on cloudy days. In caves they feed on the faeces of bats, other mammals, swallows and raphidophorid crickets. In northern Australia each species of the genus is endemic to a single cave. *Brises* does not occur in caves outside the arid zone (Matthews 1986).

Key to species: Matthews (1986).

Camponotiphilus Lea, 1914 (Pl. 87D; Fig. 36.227)

Type species. *C. fimbricollis* Lea, 1914.

Campanotiphilus Carter, 1926

Characteristics. Elongate-oblong, flattened, nitid, reddish brown, with very fine erect setae on elytra. Length 8–10 mm. Pronotum transverse, anterior angles rounded, posterior acute, lateral flanges with a dense fringe of setae. Elytra smooth, impunctate, with flanges present but not evident from above, turned down. Winged, without medial fleck.

Australian species. One: *C. fimbricollis*.

Distribution. South-western WA, north-western VIC, inland southern QLD.

Biology. The type and several other specimens are recorded as occurring in nests of two species of *Camponotus* ants (Lea 1914), while other specimens have no host data.

Celibe Boisduval, 1835 (Pl. 87E; Fig. 36.240)

Type species. *C. australis* Boisduval, 1835.

Cilibe Dejean, 1836

Characteristics. Oval, piceous, glabrous, densely punctate, smooth, rugulose, granulate or rarely costate. Length 8–22 mm. Mandibles bidentate. Disc of pronotum evenly convex, the flanges moderately wide, base as wide as elytral bases. Elytra smooth or variously sculptured, but not forming raised median area, the flanges narrow or rarely absent. Wingless. Forespur of protibia not elongate, male often with dilated protarsi.

Australian species. 33 named, but many are synonyms.

Distribution. Southern WA, SA, VIC, TAS, eastern NSW, south-eastern QLD, one species in northern QLD, Lord Howe Island, mainly in coastal habitats. The 23 species of the New Zealand genus *Mimopeus* Pascoe, 1866 differ from *Celibe* only in the structure of the aedeagus (Matthews 1993).

Cillibus Matthews, 1993 (Pl. 88K; Fig. 36.225)

Type species. *Saragus blackburni* Macleay, 1888.

Characteristics. Oblong-oval, piceous or rufous, convex or flattened dorsally, sericeous or matt, glabrous or with short curved squamae which may be clustered. Length 18–25 mm. Mandibles truncate, or bidentate in *C. blackburni*. Disc of pronotum gibbous or binodose, except in *C. ovalis* (Macleay, 1872) where it is only feebly convex. Elytra each with one strong carina inwardly and three rows of tubercles laterally, except in *C. ovalis* which has smooth elytra, flanges narrow to broad, usually reflexed, their outer edges sometimes with vertical or thickened border. Wingless. Protibial spurs very small.

Australian species. Four named, one unnamed.

Distribution. Western VIC and eastern SA, eastern NSW to central coastal QLD in sandy areas, foraging on the ground surface.

Dysarchus Pascoe, 1866 (Pl. 89B; Fig. 36.236)

Type species. *D. odewahni* Pascoe, 1866.
Saragodinus Bates, 1872
Saragella Carter, 1937
Characteristics. Oblong-oval, fuscous, more or less convex, rugose, tuberculate and costate, glabrous or with short erect setae. Length 10–23 mm. Eyes divided. Males with much enlarged maxillary and labial palpi. Mentum enlarged, covering most of maxillae. Pronotum with sides explanate, base narrower than elytral bases. Elytra without flanges, with costae or rows of tubercles. Wingless. Protibiae, when not worn, with single strong preapical tooth on outer edge and large apical spur which is fused to the tibia.
Australian species. Seven named and at least seven unnamed.
Distribution. Southern WA where most species occur, south-eastern SA, south-eastern QLD.
Biology. The species of *Dysarchus* appear to be subterranean, as is shown by their large forespur fused to the protibia, which is often extremely worn, enlarged mentum protecting the mouthparts, small divided eyes, and modifications of the aedeagus and ovipositor discussed by Matthews (1993). The enlarged palpi of the males suggest that they seek females that remain in their burrows. The species in collections are largely undescribed and each is often known only from a single locality.

Edylius Champion, 1894 (Pl. 90B; Fig. 36.136)

Type species. *E. canescens* Champion, 1894.
Characteristics. Oval, smooth, finely punctate with long, fine adpressed silken setae. Length 7–9 mm. Clypeus with anterior margin deeply excised. Pronotum with wide lateral flanges, disc evenly convex, base as wide as elytral bases. Elytra smooth, estriate, without flanges. Wingless. Protibiae apically dentate, without any preapical teeth.
Australian species. One: *E. canescens*.
Distribution. TAS, on coastal sand dunes.
Note. *Edylius* was omitted from the Heleini and placed in Pedinini or Melanimonini (because of its deeply emarginate clypeus) in all works before Matthews & Bouchard (2008), who pointed to its unmistakeably heleine characteristics.

Emcephalus Kirby, 1828 (Pl. 9D, 90J, K; Fig. 36.231)

Type species. *E. gibbosus* Kirby, 1828.
Encephalus Brême, 1842
Ellaemus Pascoe, 1866
Encara Gemminger & Harold, 1870
Euhelaeus Gebien, 1920
Characteristics. Broadly oval to subcircular, glabrous or with fine recumbent setae or squamae, sometimes granulose. Colour often brownish with darker reticulate raised areas, sometimes evenly dark brown. Length 11–22 mm. Most species floccose in life, entirely covered with white waxy filaments (Pl. 90K). Apical maxillary palpomeres usually enlarged, cultriform. Pronotum and elytra with wide flanges covering legs, hind edges of pronotal flanges sometimes serrate, as wide as elytral bases. Protibial spurs very short, not longer than basal protarsomeres. Winged, with medial fleck.
Australian species. Six named, three unnamed.
Distribution. Eastern VIC to south-eastern QLD, northern QLD, NT, New Guinea, New Britain, Ceram.
Biology. Adults are found in the daytime on the trunks of eucalypts where their covering of white filaments provides camouflage.
Key to species. Carter (1911a).

Helea Latreille, 1804 (Pl. 10B, 91C; Fig. 36.237)

Type species. *Heleus perforatus* Latreille, 1817.
Heleus Latreille, 1817
Helaeus Kirby, 1819
Elaeus Gemminger & Harold, 1870
Characteristics. Oval to subcircular, sericeous or nitid, glabrous, squamose or setose. Length 10–37 mm. Dorsal eye lobes obliquely transverse. Pronotal flanges extended forward and inward to approach one another or meet above or in front of head, their hind edges serrate, as wide as elytra bases, disc usually with a median costa or spine. In some individuals of *H. haagi* (Dohrn, 1881) anterior angles of pronotum acute but not extended. Elytral flanges generally wide, covering legs, their outer edges simple or forming raised borders. Wingless.
Australian species. 50 named, at least five unnamed.
Distribution. Arid and subarid Australia.
Biology. Adults of *Helea* find shelter in the daytime, often in mammal burrows, and emerge at night to forage widely over the ground, nibbling at plant matter but also ingesting any dead arthropods encountered, returning to the home burrow at dawn. The wide flanges act as shields to protect the appendages against attack (Matthews 1993 and references therein).
Key to species. Carter (1910).

Ospidus Pascoe, 1866 (Pl. 96B; Fig. 36.228)

Type species. *O. chrysomeloides* Pascoe, 1866.
Characteristics. Oblong to broadly oval, moderately to strongly convex, glabrous or with dense, short recumbent setae forming patterns, cupreous, bronzed or fuscous, sometimes with white patches formed by a waxy pruinosity. Length 9–16 mm. Dorsal surfaces often a little uneven. Pronotum with flanges partially or entirely covering legs, the hind edges serrate, as wide as elytra bases. Elytra

laterally extended to mostly cover legs, flanges narrow to moderately wide. Upper spur of protibia (forespur) longer than basal protarsomere, except in one species. Winged, with medial fleck.

Australian species. Four named, two unnamed.

Distribution. Extreme north-eastern NSW, eastern QLD to Cape York, northern NT.

Biology. Collecting data indicate that *Ospidus* adults occur on the foliage of eucalypts together with paropsine chrysomelids, which they resemble.

Key to species. Carter (1920).

Pterohelaeus Brême, 1842 (Pl. 97H; Fig. 36.230)

Type species. *Helaeus (Pterohelaeus) walkeri* Brême, 1842.

Barytipha Pascoe, 1869

Pterelaeus Gemminger & Harold, 1870

Pezohelaeus Gebien, 1920

Characteristics. Oblong to broadly oval, rarely subcircular, flattened to strongly convex, matt or nitid, glabrous or setose, sometimes pruinose, piceous. Length 6–25 mm. Front edge of clypeus from feebly convex to distinctly concave. Pronotal disc usually evenly convex, flanges narrow to wide, anterior angles sometimes acuminate, hind edges usually serrate or denticulate, as wide as elytral bases. Elytra with disc moderately convex, even or with indistinct costae, with numerous puncture rows, flanges narrow to wide, rarely absent. Protibial spurs ~1/2 length of protarsomeres. Most species winged, without medial fleck.

Australian species. 102 named.

Distribution. Throughout Australia, New Guinea.

Biology. Adults are frequently found under loose bark of living and dead trees mostly in sclerophyll forest. At night, individuals move about on the outside of trunks. Larvae of at least some species, known as false wireworms, occur in soil where they may damage the roots of crops (Allsopp 1986b and previous works).

Keys to species. Carter (1910, 1911a).

Saragus Erichson, 1842 (Pl. 97A; Fig. 36.238)

Type species. *Cilibe costatus* Solier, 1848 (misidentified by Erichson as *Silpha laevicollis* Fabricius, 1775, now in *Boreosaragus*).

Cyclosattus Casey, 1892

Characteristics. Broadly oval, moderately to very convex, glabrous, smooth or sculptured with granules and interrupted costae, piceous. Length 10–20 mm. Dorsal eye lobes narrowed, obliquely transverse. Disc of pronotum more or less evenly convex, flanges narrow to broad, anterior angles rounded, posterior angles prolonged and acute, base as wide as elytral bases. Elytra smooth or granulate and costate, flanges narrow to broad. Forespur enlarged, longer than first tarsomere. Wingless.

Australian species. 14 named and five unnamed.

Distribution. Widely distributed in TAS and southern NSW, VIC, SA, and WA, extending north along the coast of WA into southern NT, in open forest and shrubland. Allopatric with *Boreosaragus*, which occupies areas to the north of this.

Key to species. Carter (1911a).

Sympetes Pascoe, 1866 (Pl. 11G, 99C; Fig. 36.234)

Type species. *S. macleayi* Pascoe, 1866 (= *Encephalus tricostellus* White, 1891).

Characteristics. Subcircular, flattened, matt or nitid, glabrous or with very short setae, often covered with a layer of soil, fuscous sometimes with flanges flavous. Length 8–27 mm. Disc of pronotum usually with median carina and very broad flanges with anterior angles rounded, subquadrate or acuminate, hind angles acute, hind edge as wide as elytral bases, not serrate. Prosternal process flattened, not received in mesoventrite. Elytral suture nearly always strongly raised, forming median carina, flanges very wide. Forespur usually long.

Australian species. 15 named.

Distribution. WA, in the south and west and along the north coast, in open sandy areas.

Biology. This is a nocturnal forager like *Helea* and has been observed coming in numbers to human faeces.

Key to species. Carter (1910).

Trichosaragus Blackburn, 1890 (Pl. 100J; Fig. 36.232)

Type species. *T. pilosellus* Blackburn, 1890.

Characteristics. Broadly oblong, strongly convex, with long pilosiity or shorter setae or squamae, rufous or piceous. Length 6–16 mm. Apices of mandibles usually unidentate, with prominent apical tooth and smaller tooth distant from apex, but bidentate in *T. granulatus* Carter, 1926. Pronotum with disc strongly, evenly convex, with narrow flanges, lateral edges crenulate. Elytra without flanges, with three costae or rows of tubercles and usually granulate surface between costae. Protibiae flattened and dentate or crenulate on outer edges, with long outwardly directed apical spur. Wingless.

Australian species. Three named, one unnamed.

Distribution. Eastern SA, western NSW, south-western QLD in the eastern part of the arid zone.

Tribe Titaenini Fauvel, 1905

Characteristics. Elongate, parallel-sided, subcylindrical, with fine erect setae, head held vertically, fuscous with greenish or bluish reflections. Length 5–12 mm. Eyes rounded with only minor encroachment by frontal canthus. Clypeolabral membrane concealed under clypeus. Labrum transverse. Edge of clypeus almost straight medially. Antennae short, not reaching base of prothorax, segments subtriangular, only slightly widening distally, with simple sensilla. Mandibles apically bidentate. Terminal maxillary palpomere securiform. Pronotum subcylindrical with fine lateral margins. Prosternum very short before coxae. Mesocoxal cavities open. Elytra either with nine striae and scutellary striole, with only partial striation, or estriate with confused punctation, epipleura abbreviated, not reaching apices of elytra. Legs not fossorial, tarsomeres subparallel. Winged, with medial fleck. The above description applies to Australia and New Caledonia only; New Zealand members of the tribe are very different.

Distribution. Eastern Australia, New Caledonia, New Zealand, Subantarctic islands.

Australian taxa. One genus.

Biology. Watt (1975) reported that *Artystona* Bates, 1873 and some other genera of Titaenini in New Zealand feed on lichens at night, and larvae of *Titaena* Erichson, 1842 in Australia have also been observed feeding externally on bark at night, suggesting that the tribe as a whole feeds on lichens.

Note. The New Caledonian *Callismilax* Bates, 1874, not distinguishable from *Titaena*, was tentatively recorded as occurring in Australia by Doyen *et al.* (1990) on the basis of specimens originally misidentified by Carter as a species of *Thesilea* Haag-Rutenberg, 1878, a genus which does not occur in Australia.

Titaena Erichson, 1842 (Pl. 99L; Fig. 36.148)

Type species. *T. alcyonea* Erichson, 1842.
Characteristics. As for the tribe.
Australian species. Nine.
Distribution. From TAS to central QLD.

Tribe Trachelostenini Lacordaire, 1859

Characteristics. Elongate, parallel-sided, a little flattened, with a few long, fine erect setae and also adpressed and flattened white ones forming patterns. Total length 4–9 mm. Eyes rounded with only minor encroachment by frontal canthus. Edge of clypeus almost straight medially. Labrum transverse, clypeolabral membrane narrowly exposed. Antennae short, not reaching base of prothorax, segments subtriangular, only slightly widening distally, with simple sensilla. Terminal maxillary palpomere securiform. Pronotum with lateral margin uneven or denticulate. Prosternum moderately long before coxae. Procoxae very prominent, approximated and covering prosternum. Mesocoxal cavities open. Elytra smooth with incomplete puncture rows, epipleura abbreviated, not reaching apices. Legs not fossorial, tarsomeres subparallel. Winged, without medial fleck. [Matthews & Lawrence 2015].

Distribution. The tribe has one genus in eastern Australia and two genera in southern Chile.

Australian taxa. One genus.

Note. In their original description of *Leaus*, Matthews & Lawrence (1992) noted that its most anomalous feature is the strongly projecting and approximated procoxae, otherwise seen in the Chilean genus *Trachelostenus* Solier, 1851, then in a separate family Trachelostenidae. Later, Lawrence & Slipiński (2013) noted that the larvae of these two genera are very similar. Then in the molecular phylogeny of McKenna *et al.* (2015) the two recover as sister groups within Tenebrionidae. Finally, Matthews & Lawrence (2015) formally brought the two genera together in a tenebrionid tribe Trachelostenini based on compelling larval and adult synapomorphies, adding a third genus, the Chilean *Myrmecodema* Gebien, 1943.

Leaus Matthews & Lawrence, 1992 (Pl. 10A, 92A; Fig. 36.197)

Type species. *L. tasmanicus* Matthews & Lawrence, 1992.
Characteristics. As for the tribe.
Australian species. Four named and one unnamed.
Distribution. TAS, northern NSW to northern QLD.
Biology. Larvae have been collected from tree trunks by pyrethrum knockdown. They are probably external feeders, perhaps on lichens, considering the patterned appearance of their dorsal surfaces.
Key to species. Matthews & Lawrence (2015).

Tribe Alphitobiini Reitter, 1917

Characteristics. Ovoid, somewhat flattened, glabrous, piceous. Eyes transverse, reniform, more or less deeply divided by frontal canthus. Edge of clypeus medially straight or shallowly emarginate. Labrum transverse, largely covered by clypeus, clypeolabral membrane hidden. Antennae not reaching base of prothorax, antennomeres gradually asymmetrically widening distally, with compound sensoria. Maxillary palpi with apical segment fusiform or suboval. Pronotum transverse, widest basally, evenly convex; prosternum moderate before coxae. Prosternal process acutely prolonged and received in deep V-shaped concavity of mesoventrite. Mesocoxal cavities closed or almost closed by ventrites only. Elytra with nine striae and scutellary striole, epipleura complete to apices. Winged, without medial fleck. Legs not fossorial, tarsomeres simple.

Distribution. World-wide.

Australian taxa. Two genera.

Alphitobius Stephens, 1829 (Pl. 84B; Fig. 36.165)

Type species. *Helops picipes* Panzer, 1794 (= *Opatrum laevigatum* Fabricius, 1781) (Germany).
Microphyes Macleay, 1872

Characteristics. Surfaces coarsely punctate. Length 4.5–6 mm. Middle of clypeal margin shallowly excised. Apical six antennomeres abruptly enlarged. Mentum with broad median flat area bounded laterally by vertical edges. Prosternal process arcuate in profile, apically rounded.

Australian species. Two: *A. laevigatus* and *A. diaperinus* (Panzer, 1797).

Distribution. Cosmopolitan, introduced into Australia with commerce, the larvae known as lesser mealworms.

Biology. Found primarily in aviaries, where the larvae and adults feed on spilled grain and carrion, and are occasionally predatory. The biology of *Alphitobius* is discussed at length by Matthews *et al.* (2010: 592).

Diaclina Jacquelin du Val, 1861 (Pl. 89H; Figs 36.7, 36.166)

Type species. *Tenebrio chrysomelinus* Rossi, 1790 (Italy).

Characteristics. Surfaces very finely punctate. L 5–6 mm. Middle of clypeal margin straight. Antennae gradually enlarging distally. Mentum with median rounded convexity. Prosternal process subhorizontal in profile.

Australian species. One: *D. nitida* (Carter, 1908).

Distribution. North-eastern QLD, New Guinea, Eurasia, under bark and in fungi.

Tribe Triboliini Gistel, 1848

Characteristics. Oblong, more or less elongate, parallel-sided, moderately convex to subcylindrical, glabrous, fuscous. Total length 2–6 mm. Eyes reniform. Edge of clypeus nearly straight in middle or shallowly emarginate. Labrum transverse, clypeolabral membrane concealed. Antennae short, not reaching base of prothorax, segments gradually or (usually) abruptly enlarging distally to form club, with simple setiform sensilla. Maxillary palpi with apical segment fusiform. Prosternum long before coxae. Mesocoxal cavities open. Elytra with nine striae and scutellary striole, epipleura complete to apices. Winged, without medial fleck, apical membrane ~2⁄5 of wing length. Legs not fossorial, subapical tarsomeres somewhat cupuliform.

Distribution. World-wide.

Australian taxa. Three genera.

Latheticus Waterhouse, 1880 (Pl. 92D; Fig. 36.203)

Type species. *L. oryzae* Waterhouse, 1880 (Europe).

Characteristics. Elongate oblong, moderately convex, castaneous. Length 2.5–3 mm. Clypeus very large, occupying almost as much area as frons, subrectangular. Antennal club of five segments. Elytra striate, the intervals smooth.

Australian species. One: *L. oryzae*.

Distribution. Cosmopolitan, in stored grain.

Lyphia Mulsant & Rey, 1859 (Pl. 93L; Fig. 36.201)

Type species. *L. ficicola* Mulsant & Rey, 1859 (Italy).

Lindia Blackburn, 1888

Characteristics. Elongate subcylindrical, castaneous. Length 2.5–6 mm. Clypeus much smaller in area than frons, subtrapezoidal. Antennae with four-segmented club. Elytral striae indistinct, epipleura vertical.

Australian species. Three.

Distribution. Cosmopolitan, under bark.

Tribolium Macleay, 1825 (Pl. 100L; Fig. 36.202)

Type species. *Colydium castaneum* Herbst, 1797.

Leanum Uyttenboogaart, 1934 (subgenus)

Characteristics. Oblong, moderately convex, castaneous. Length 2.5–4.5 mm. Clypeus much smaller in area than frons, subtrapezoidal. Antennal club of three, five or seven segments. Elytra striate, intervals usually carinate, at least laterally, rarely smooth. [Hinton 1948].

Australian species. Six.

Distribution. Eastern Australia from VIC to north QLD, cosmopolitan.

Biology. The two cosmopolitan flour beetles, *T. castaneum* and *T. confusum* Jacquelin du Val, 1862 occur in stored grain products and have been widely cultured for experimental purposes. The native *T. apiculum* Neboiss, 1962 was found in a nest of *Trigona* bees, while *T. myrmecophilum* Lea, 1904 was found with *Iridomyrmex nitidus* Mayr.

Key to species. Neboiss (1962).

Tribe Amarygmini Gistel, 1848

Characteristics. Ovoid, convex, smooth, usually glabrous, usually with metallic reflections. Eyes reniform. Edge of clypeus feebly produced or nearly straight in middle. Labrum transverse, clypeolabral membrane exposed for its full width. Antennae reaching beyond base of prothorax, filiform, only slightly widening distally, with numerous compound sensoria on distal segments. Pronotum widest basally, evenly convex. Portion of prosternum in front of coxae shorter than postcoxal bridge, allowing head to be held vertically against thorax (hypognathous). Mesocoxal cavities open. Elytra with nine striae and scutellary striole, epipleura complete to apices. Wings present (except in *Axynaon* Blackburn, 1897), without medial fleck. Legs slender, not fossorial, tarsomeres not lobed. [Bremer & Lillig 2014].

Distribution. Mostly Old World tropics, throughout Australia.

Australian taxa. Five genera.

Biology. Larvae of Amarygmini are mostly found in decaying wood and adults frequently occur under bark in the daytime, emerging at night when they can be beaten from tree branches or fall into pitfall traps. Larvae of at least some *Chalcopteroides* Strand, 1935 are soil inhabiting. Larvae of some *Amarygmus* Dalman, 1923 feed on lichens at night. *Axynaon* breeds in the compacted dead grass that accumulates in deep soil cracks. *Isopteroplonyx* Bremer, 2006 is termitophilous.

Amarygmus Dalman, 1823 (Pl. 11D, 84E; Fig. 36.155)

Type species. *Chrysomela micans* Fabricius, 1794 (Sunda Is).
Dietysus Pascoe, 1866
Elixota Pascoe, 1866
Eurypera Pascoe, 1870
Aphyllocerus Fairmaire, 1881
Anacycus Fairmaire, 1896
Platolenes Gebien, 1913
Pseudamarygmus Pic, 1915
Podamarygmus Carter, 1928
Characteristics. Usually oval, rarely subparallel, convex, glabrous, usually fuscous or black, but may be metallic green or purple. Length 6–15 mm. Mandibles apically bifid, outwardly with distal sulcus arising between teeth. Lateral margins of pronotum complete. Elytra usually distinctly striate, but sometimes only punctate. Metaventrite long (beetle winged). Tarsal vestiture fulvous. [Bremer 2001a, b, 2005a, b, 2006a, b].
Australian species. 54.
Distribution. Australia, mostly in the north, New Guinea, Eurasia, Africa.
Key to species. Carter (1913b).

Axynaon Blackburn, 1897 (Pl. 85B; Fig. 36.152)

Type species. *A. championi* Blackburn, 1897.
Catopherus Carter, 1918
Characteristics. Oblong-oval, glabrous, metallic green. Length 17–20 mm. Apex of right mandible truncate, that of left bifid. Lateral margins of pronotum largely obliterated. Elytral striae consisting of large, deep punctures joined by longitudinal impressed lines. Metaventrite subequal in length along midline to first abdominal ventrite (beetle flightless).
Australian species. One: *A. championi*.
Distribution. Central QLD.

Chalcopteroides Strand, 1935 (Pl. 88F; Fig. 36.154)

Type species. *Chalcopterus iridicolor* Blessig, 1861.
Chalcopterus Blessig, 1861
Characteristics. Oblong, glabrous, usually metallic blue or green. Length 6–23 mm. Mandibles apically truncate or broadly rounded, without sulcus. Lateral margins of pronotum complete. Elytra sometimes with striae impressed, but usually with only superficial and minute punctures. Metaventrite long (beetle winged). Tarsal vestiture mostly black.
Australian species. 114 named.
Distribution. Throughout Australia, more numerous in the south and west than *Amarygmus*.
Key to species groups. Watt (1989).
Key to species. Carter (1913b).

Isopteroplonyx Bremer, 2006 (Pl. 92G; Fig. 36.156)

Type species. *Amarygmus termitophilus* Lea, 1910.
Characteristics. Oblong oval, glabrous, bluish black. Length 5 mm or less. Mandibles apically narrowly rounded. Lateral margins of pronotum complete. Elytra deeply striate with strongly convex intervals. Metaventrite long (beetle winged).
Australian species. One: *I. termitophilus*.
Distribution. Northern NT and northern QLD, found in nests of *Eutermes* (Bremer 2006b).

Trichamarygmus Carter, 1913 (Pl. 100F; Fig. 36.153)

Type species. *T. pilosus* Carter, 1913.
Characteristics. Oblong, bluish, with dense, long, erect pilosity. Length 10–11 mm. Mandibles apically truncate. Not distinguishable from *Chalcopteroides* except for pilosity.
Australian species. One: *T. pilosus*.
Distribution. Central coastal WA.

Tribe Ulomini Blanchard, 1845

Characteristics. Oblong, parallel-sided, convex, glabrous, piceous. Eyes reniform (absent in *Typhluloma* Lea, 1912). Edge of clypeus more or less straight in middle. Labrum trapezoidal, narrower basally. Antennae short, reaching halfway or less to base of prothorax, segments gradually widening distally, transverse, distal edges of preapical segments bearing a circle of placoid (plate-like) sensoria (Fig. 36.26 ps), apical segment with stellate compound sensoria. Maxillary palpi with apical segment obliquely securiform. Pronotum subquadrate, base as wide as elytral bases, disc evenly convex or often anteriorly excavate in males. Prosternum moderately long before coxae. Mesocoxal cavities closed. Elytra with nine striae and scutellary striole, epipleura not reaching apices, latter excised uncovering part of last tergite. Winged (except *U.* (*Apterulomoides* Kaszab, 1982)), without medial fleck. Legs with pro- and mesotibiae fossorial, flattened and with dentate or spinulose outer edges, tarsomeres not lobed.
Distribution. World-wide in temperate and tropical forests.
Australian taxa. Four genera.
Biology. Ulomini feed on well rotted wood.

Achthosus Pascoe, 1863 (Pl. 10D, 84C, Figs 36.10, 36.160)

Type species. *A. westwoodi* Pascoe, 1863.
Acthosus Carter, 1926

Characteristics. Length 14–23 mm. Clypeolabral membrane covered, labrum in lower plane than clypeus. Antennomeres 6–10 much wider than long. Outer edges of protibiae strongly toothed. Apical end (truncation) of meso- and metatibiae toothed ventrally with only a few peg-like bristles between teeth (Fig. 36.10). Both sexes similar, with produced clypeal lamella and sculptured pronotum. [Kaszab 1982b]

Australian species. One: *A. westwoodi*.

Distribution. TAS, eastern Australia, north of NT.

Cenoscelis Wollaston, 1867 (Pl. 87A; Fig. 36.158)

Type species. *C. tibialis* Wollaston, 1867 (Cape Verde).

Characteristics. Length 6.0–6.5 mm. Clypeolabral membrane broadly exposed, labrum in same plane as clypeus. Antennomeres 6–10 only a little wider than long. Outer edges of protibiae not toothed, bearing a row of numerous small spines.

Australian species. One undetermined (Matthews & Scupola 2003).

Distribution. North of the NT, Asia, Africa.

Typhluloma Lea, 1912 (Pl.100D; Fig. 36.159)

Type species. *T. inops* Lea, 1912.

Characteristics. Length 4–4.5 mm. Eyes absent. Clypeolabral membrane covered, labrum in lower plane than clypeus. Antennomeres 6–10 much wider than long. Outer edges of protibiae strongly toothed.

Australian species. One: *T. inops*.

Distribution. Northern QLD.

Biology. No definite information is available on the habitat of *Typhluloma*, but according to Lea (1912) it might be a myrmecophile.

Uloma Dejean, 1821 (Pl. 100C; Figs 36.9, 36.26, 36.161)

Type species. *Tenebrio culinaris* Linnaeus, 1758 (Europe).

Prioscelida White, 1846

Apterulomoides Kaszab, 1982 (subgenus)

Characteristics. Length 6.5–15 mm. Clypeolabral membrane covered, or exposed in middle only, labrum in lower plane than clypeus. Antennomeres 6–10 much wider than long. Outer edges of protibiae strongly toothed. Apical end (truncation) of meso- and metatibiae with ventral row of short spine- or peg-like bristles, not toothed (Fig. 36.9). Males and females often with different cephalic and pronotal sculpturing.

Australian species. 27.

Distribution. TAS, eastern and northern Australia, worldwide.

Key to species. Kaszab (1982b).

Tribe Blaptini Leach, 1815

Characteristics. Elongate-oval, black. Length more than 40 mm. Eyes emarginate. Anterior margin of labrum emarginate, its fringe medially interrupted. Antennae filiform, segment 3 very long, last four segments moniliform, terminal one conical. Pronotum sub-globose. Elytra indistinctly carinate, their apices strongly prolonged into a tapering process (mucro). Legs long, tarsomeres not expanded.

Distribution. Widespread in Eurasia and Africa, introduced into the United States and Australia.

Australian taxa. One introduced genus and species.

Blaps Fabricius, 1775 (Pl. 86H; Fig. 36.204)

Type species. *Tenebrio mortisagus* Linnaeus, 1758 (Europe).

Characteristics. As for the tribe.

Australian species. One: *B. polychresta* (Forskål, 1775) introduced from North Africa.

Distribution. As for the tribe.

Note. *Blaps polychresta* first appeared near Wallaroo, SA in 1930 and was reported as *Blaps gigas* (Linnaeus, 1767) by Lea (1930), who stated that it had arrived on a grain ship returning from Egypt. Subsequently the species has become widespread in urban and agricultural areas throughout southern SA. It feeds on spilled grain, lucerne cubes and rabbit faecal pellets (Matthews 1975).

Tribe Melanimonini Seidlitz, 1894

Characteristics. Oblong, strongly convex, fuscous, mostly glabrous with lateral fringe of long bristles. Length 4–5 mm. Clypeus with anterior margin deeply excised. Labrum subquadrate. Antennae very short, last four segments slightly englarged. Mesocoxal cavities barely open. Elytra with no trace of striae, epipleura complete to apices. Protibiae fossorial, outer edge with strong tooth at apex and in middle, a small tooth near base.

Distribution. Widespread in Eurasia, Africa and North America.

Australian taxa. One introduced genus and species.

Cheirodes Gené, 1839 (Pl. 88B; Fig. 36.137)

Type species. *C. sardous* Gené, 1839 (Sardinia).

Anaemia Laporte, 1840

Characteristics. As for the tribe.

Australian species. One: *C. sardous*.

Distribution. Eurasia, Africa, introduced into WA and SA.

Note. The Mediterranean *C. sardous* first appeared in WA *c.* 1920 and was named *Anaemia caulobioides* Carter, 1920. It has now spread throughout southern WA, as well as southern SA where it appeared in 1960. It presently occurs in large numbers feeding on dried cow dung.

Tribe Pedinini Eschscholtz, 1829

Subtribe Leichenina Mulsant, 1854

Characteristics. Oblong, not very convex, covered with variegated scales. Total length ~5 mm. Eyes only feebly emarginate, with erect scales between facets. Edge of clypeus deeply excised in middle. Antennae with weak 4-segmented club, with simple sensilla only. Mesocoxal cavities closed. Elytra striate, with complete epipleura. Winged, without medial fleck. [Medvedev 1973].

Distribution. Africa, Eurasia, the Pacific, south-eastern United States (introduced), on sea shores.

Australian taxa. One genus and species.

Leichenum Dejean, 1834 (Pl. 92B; Fig. 36.135)

Type species. *Opatrum pictum* Fabricius, 1801 (Italy).

Endothina Carter, 1924

Endothyna Gebien, 1939

Characteristics. As for the subtribe.

Australian species. One: *L. canaliculatum* (Fabricius, 1798).

Distribution. All Australian coasts, southern Eurasia, Taiwan, Japan, Africa.

Subtribe Platynotina Mulsant & Rey, 1853

Characteristics. Oblong, not very convex, glabrous, entirely black. Total length 16–19 mm. Eyes strongly transverse and deeply excised. Antennae long, reaching almost to base of pronotum, segments 7–11 slightly widened, with simple sensilla only. Clypeus moderately excised. Mentum with raised flattened median prominence and triangular lateral lobes. Mesocoxal cavities open. Elytra deeply striate with strongly convex intervals, with complete epipleura. Wings reduced. Male with protarsi widened, inner edge of protibiae deeply excised, distally with tuft of setae. [Iwan 2002; Ferrer 2006].

Distribution. Africa, southern Asia, Western Hemisphere.

Australian taxa. One genus and species.

Menearchus Carter, 1920 (Pl. 93H; Fig. 36.142)

Type species. *Helops nigrita* Fabricius, 1776 (= *Tenebrio dispar* Herbst, 1797) (India).

Characteristics. As for the subtribe.

Australian species. One: *M. nigritus* (= *impressosulcatus* Carter, 1920).

Distribution. Southern India, Sri Lanka, Cocos-Keeling islands.

Tribe Opatrini Brullé, 1832

Characteristics. Oblong-oval, moderately to strongly convex. Total length 5–11 mm. Antennae slightly and gradually expanded distally, with simple setiform sensilla only. Edge of clypeus deeply excised in middle. Labrum subquadrate. Mesocoxal cavities open. Elytra striate, with incomplete epipleura. Protibiae fossorial, apically more or less expanded. [Iwan & Kamiński 2016].

Distribution. World-wide, in more or less arid habitats.

Australian taxa. Three subtribes.

Subtribe Heterotarsina Blanchard, 1845

Characteristics. Dorsal surfaces glabrous. Protibiae without lateral teeth. Basal three protarsomeres strongly expanded in both sexes and densely setose beneath. Wings with medial fleck.

Distribution. Heterotarsina are found along sea shores in the Ethiopian, Indo-Malayan and Australian regions, and are absent from the Palaearctic and Western Hemisphere.

Note. In the molecular phylogeny of Kergoat *et al.* (2014b) this group (represented by *Diphyrrhynchus* Fairmaire, 1849) recovers separately from other Opatrini as the sister group of Bolitophagini. Also, Matthews & Bouchard (2008) noted that this subtribe (as Heterocheirina) differs significantly from the other Opatrini examined in the structure of the female tract, wings, and aedeagus. These combined results indicate that Heterotarsina should be elevated to tribal status.

Australian taxa. Three genera.

Diphyrrhynchus Fairmaire, 1849 (Pl. 89C; Fig. 36.139)

Type species. *D. chalceus* Fairmaire, 1849 (Tonga).

Neoabantis Gebien, 1910 (subgenus).

Characteristics. Fuscous. Length 4–5 mm. Protibiae evenly and strongly widened, without apical lobes. Striae at bases of elytra shallowly impressed or absent. Metaventrite longer than first abdominal ventrite. Wings reduced.

Australian species. Two: *D. ellipticus* Champion, 1894 and *D. nicobaricus* Redtenbacher, 1868.

Distribution. Northern and central WA, northern NT, Asia and Africa, coastal.

Heterocheira Dejean, 1836 (Pl. 91A; Fig. 36.140)

Type species. *Uloma australis* Boisduval, 1835.

Heterochira Agassiz, 1846

Characteristics. Fuscous. Length 5–8 mm. Protibiae weakly widened apically, without apical lobes. Elytral striae evenly impressed throughout. Metaventrite longer than first abdominal ventrite. Wings developed.

Australian species. Two: *H. australis* and *H. tropica* Carter, 1924.

Distribution. All coasts of Australia, southern Asia and Africa.

Scymena Pascoe, 1866 (Pl. 98J; Fig. 36.141)

Type species. *S. variabilis* Pascoe, 1866.
Characteristics. Pale castaneous with darker markings on pronotum. Length 6–8 mm. Protibiae apically expanded into a prominent rounded lobe. Elytral striae evenly impressed throughout. Metaventrite subequal in length to first abdominal ventrite. Wings vestigial.
Australian species. Two: *S. amphibia* Pascoe, 1870 and *S. variabilis*.
Distribution. Southern WA, SA, under stranded seaweed.

Subtribe Ammobiina Desbrochers des Loges, 1902

Characteristics. Dorsal surfaces with small adpressed or semi-erect setae, usually squamiform. Eyes reniform. Apical maxillary palpomere triangular or oval. Pronotum without basal depressions. Protibiae strongly expanded apically with one or more preapical teeth or lobes, tarsomeres not expanded. Scutellum largely or wholly overlapped by pronotal base. Wings without medial fleck.
Distribution. World-wide.
Australian taxa. Three genera.

Caediomorpha Blackburn, 1888 (Pl. 87J; Fig. 36.146)

Type species. *Morychus heteromerus* King, 1869 (= *Caediomorpha australis* Blackburn, 1888).
Characteristics. Broadly oval, strongly convex. Length 4–5 mm. Antennae very short, compact. Terminal maxillary palpomere triangular or oval. Protibiae strongly expanded apically, with one or more preapical rounded lobes, their undersurface with at least some basal tubercles. Scutellum largely overlapped by pronotal base.
Australian species. One: *C. heteromera*.
Distribution. All Australian coasts on beaches and sand dunes.

Caedius Blanchard, 1845 (Pl. 87K; Fig. 36.144)

Type species. *Opatrum sphaeroides* Hope, 1842.
Characteristics. Broadly oval, moderately convex. Length 5–7 mm. Antennae very short, reaching little beyond front edge of pronotum. Terminal maxillary palpomere triangular or oval. Base of prothorax feebly subangulate or bisinuate. Protibiae with prominent teeth or lobes on outer edges, apical one projecting more than distance equal to width of tibia behind it. Scutellum overlapped by pronotal base. Elytra with at least a few minute tubercles. Abdominal ventrites with minute tubercle at base of each squamiform seta.
Australian species. Five.
Distribution. Arid and coastal Australia, Asia, Africa.
Key to species. Kulzer (1964).

Sobas Pascoe, 1863 (Pl. 98B; Fig. 36.145)

Type species. *Trigonotarsus australis* Hope, 1843.
Trigonotarsus Hope, 1843
Pseudocaedius Blackburn, 1890
Characteristics. Elongate-oval, moderately convex. Length 3–4 mm. Terminal maxillary palpomere triangular or oval. Antennae very short, reaching little beyond front edge of pronotum. Base of prothorax evenly arcuate. Protibiae with a prominent apical tooth on outer edge, a smaller tooth close to base. Scutellum wholly sunken. Elytra without tubercles.
Australian species. Two: *S. australis* and *S. minor* Carter, 1937.
Distribution. Western WA, northern SA, NT.

Subtribe Opatrina Brullé, 1832

Characteristics. Dorsal surfaces with small adpressed or semi-erect setae, usually squamiform. Eyes reniform or divided. Apical maxillary palpomere securiform. Pronotum with a pair of basal depressions. Protibiae not strongly expanded, without preapical teeth, tarsomeres not expanded. Scutellum well developed behind pronotal base. Wings without medial fleck.
Distribution. African, Eurasian and Australian regions. Absent from the Western Hemisphere.
Australian taxa. Three genera.

Gonocephalum Solier, 1834 (Pl. 11F, 91K; Fig. 36.143)

Type species. *Opatrum pygmaeum* Steven, 1829 (Europe).
Hopatrum auctorum (*partim*)
Characteristics. Oblong, moderately convex. Length 6–11 mm. Eyes constricted but not divided. Antennae long, reaching beyond middle of pronotum. Terminal maxillary palpomere securiform. Protibiae not strongly expanded, without preapical teeth. Scutellum fully exposed behind pronotal base. Each elytral interval with multiple rows of short squamiform or ciliate setae. Strial punctures small or absent. [Iwan *et al.* 2010, 2011].
Australian species. 18.
Distribution. Arid and semi-arid Australia, Eurasia and Africa.
Biology. One or more species of *Gonocephalum* are major pests of canola crops where zero-tillage practices are adopted, allowing stubble retention. Larvae are surface feeders and may reach densities of 2000/m^2, attacking seedlings (B.H. Rohitha pers. comm. 1980).
Key to species. Blackburn (1907) (as *Hopatrum*).

Mesomorphus Miedel, 1880 (Pl. 93F; Fig. 36.138)

Type species. *Opatrum murinus* Baudi di Selve, 1876 (Egypt).
Hopatromorpha Blackburn, 1907

Characteristics. Oblong, moderately convex. Length 6–8 mm. Eyes completely, or almost completely, divided by genal canthus. Head surface even, with no trace of frontoclypeal suture. Terminal maxillary palpomere securiform. Mentum with prominent median keel. Antennae long, reaching beyond middle of pronotum. Protibiae not strongly expanded, without preapical teeth. Scutellum fully exposed behind pronotal base. Each elytral interval with multiple rows of short squamiform or ciliate setae. Strial punctures small or absent.

Australian species. Seven.

Distribution. Arid and semi-arid Australia, Eurasia and Africa.

Key to species. Ferrer (2008).

Scleropatroides Löbl & Merkl, 2003 (Pl. 98A; Fig. 36.147)

Type species. *Opatrum hirtulum* Baudi di Selve, 1876 (Central Asia).

Characteristics. Oblong, moderately convex, dorsal surfaces scabrous. Length 5–7 mm. Eyes reniform. Terminal maxillary palpomere securiform. Antennae long, reaching beyond middle of pronotum. Protibiae not strongly expanded, without preapical teeth. Scutellum fully exposed behind pronotal base. Elytral intervals with a single row of tubercles, each bearing a long, semi-erect squamiform seta, strial punctures large and deep. [Iwan & Matthews 2015].

Australian species. One: *S. triplehorni* Iwan & Matthews, 2015.

Distribution. Africa, central and southern Asia, central Australia.

Subfamily Alleculinae Laporte, 1840

Tribe Alleculini Laporte, 1840

Subtribe Alleculina Laporte, 1840

Characteristics. Either elongate-oblong with narrow prothorax, or ovoid. Eyes reniform. Labrum strongly transverse. Maxillary palpi with apical segment either securiform or cultriform. Antennae reaching beyond base of prothorax, with numerous stellate compound sensoria, usually filiform but rarely serrate or otherwise modified, at least in males. Prosternum before coxae about as long as behind them. Mesocoxal cavities open. Elytral striae nine plus scutellary striole. Winged or flightless, without medial fleck. Legs slender, not fossorial, preapical tarsomeres strongly lobed except in *Taxes* Champion, 1895 and *Litopous* Matthews, 2012, claws pectinate or serrate with rare exceptions discussed below under notes. [Carter 1915b, 1930; Matthews & Bouchard 2008; Matthews 2012].

Distribution. World-wide.

Australian taxa. Twenty-three genera.

Biology. Larvae of Alleculini are found mainly in rotten wood but also in soil (Watt 1975; Lawrence & Britton 1994). Adults are associated with plants and the pectinate claws assist walking on foliage (Watt 1975). Species of *Lepturidea* Fauvel, 1862 and *Tanychilus* Newman, 1838 have been observed coming to flowers, and a brush-like development of the maxillary galea, seen in all genera examined, seems to be an adaptation for collecting pollen. Canopy fogging of rainforest trees in Queensland has yielded very large numbers of small, setose species of *Homotrysis* Pascoe, 1866 that could not be identified to species.

Notes. Three taxa from WA appear to have simple claws: an unnamed species of *Lepturidea* Fauvel, 1862 mentioned by Matthews (2012), the WA specimens of *Hemicistela* Blackburn, 1891, and an unnamed genus represented by a single specimen from Coolgardie in the SA Museum labelled by H. J. Carter as 'near *Strongylium cylindricus* Carter', an unpublished name. However, a close examination of all these taxa at high magnification reveals minute remnants of teeth or crenulations at least at the bases of the claws.

Bousquet *et al.* (2015) published a catalogue of genus-group names in Alleculinae which requires some changes to be made to the catalogue in Matthews & Bouchard (2008). In particular, *Lepturidea* Fauvel, 1862 has to replace *Aethyssius* Pascoe, 1863, and the type species designations in Matthews & Bouchard (2008) for several genera are invalid because type species had already been designated for these genera by Lucas (1920). In many cases the species involved are the same in both works, but for *Nocar* Blackburn, 1891, *Nypsius* Champion, 1895 and *Otys* Champion, 1895 the generitypes listed below are now different from those in Matthews & Bouchard (2008). These changes have been incorporated into the Australian Faunal Directory.

Atoichus Carter, 1915 (Pl. 85E; Figs 36.3, 36.53)

Type species. *Licymnius bicolor* Blackburn, 1893.

Characteristics. Elongate-oval, mostly glabrous with a few minute setae. Colour black combined with rufous, flavous or testaceous areas. Length 4.0–6.0 mm. Head very long before eyes (frontal index 0.7–1.1), abruptly constricted basally forming distinct neck. Apices of mandibles unidentate. Apical maxillary palpomere cultriform. Procoxae contiguous. Winged.

Australian species. Four

Distribution. South-eastern QLD, NSW, VIC, SA.

Barycistela Blackburn, 1891 (Pl. 86E; Fig. 36.61)

Type species. *B. robusta* Blackburn, 1891.

Characteristics. Oblong, prothorax widest in middle, base almost as wide as elytral bases, with dense minute setae, rufocastaneous with paler legs. Length 7.0–9.0 mm. Head

short before eyes. Mandibles bidentate. Apical maxillary palpomere securiform. Antennomeres elongate, ~3 times as long as wide. Elytral intervals minutely tuberculate. Protibiae apically expanded. Winged.
Australian species. One: *B. robusta*.
Distribution. QLD.

Dimorphochilus Borchmann, 1908 (Pl. 89F; Figs 36.8, 36.68)

Type species. *D. apicalis* Borchmann, 1908.
Characteristics. Elongate oblong with prothorax narrower than elytral bases, glabrous, castaneous, with paler legs. Length 8.7–15.0 mm. Head moderately short before eyes (frontal index 0.5). Mandibles apically bidentate. Apical maxillary palpomere securiform. Thin inner lamina of elytral sutures at least slightly expanded apically (Fig. 36.8). Winged. Males with last abdominal ventrite divided and forcipate, and inner edge of protibiae more or less sinuate.
Australian species. Five named, two unnamed. *D. diversicollis* Borchmann, 1908 was listed as a synonym in Matthews & Bouchard (2008) but resurrected by Matthews (2012), who discovered in the WA Museum three specimens of the original type series, one of which was made a lectotype. It had previously been thought that the entire type series was in Hamburg and was destroyed during WW II.
Distribution. WA, SA, NSW, QLD.

Euomma Boheman, 1858 (Pl. 90F, 91G; Fig. 36.56)

Type species. *E. lateralis* Boheman, 1858.
Apellatus Pascoe, 1863.
Characteristics. Elongate-oval, glabrous, fuscous with pale elytral disc areas. Length 4.0–10.0 mm. Head long before eyes (frontal index 0.6–0.7). Eyes large, separated dorsally by a distance equal to much less than one eye width. Mandibles acute with two minute apical teeth. Apical maxillary palpomere elongate-cultriform, more strongly so in males. Prothorax about as long as wide, sides broadly rounded anteriorly. Winged.
Australian species. 12.
Distribution. Throughout the continent.

Hemicistela Blackburn, 1891 (Pl. 91I)

Type species. *H. discoidalis* Blackburn, 1891.
Characteristics. Elongate-oblong, glabrous, fuscous with anterior 2/3 of elytral disc pale. Length 4.0–5.6 mm. Head short before eyes (frontal index 0.5), apical maxillary palpomere moderately cultriform. Prothorax a little transverse with rounded sides. Claws only feebly crenulate on inner edges, or completely simple. Winged. Male with shallow apical emargination on inner edge of mesotibia.
Australian species. One: *H. discoidalis*.
Distribution. VIC, Southern WA.

Homotrysis Pascoe, 1866 (Pl. 91F, 101G; Fig. 36.70)

Type species. *Allecula tristis* Germar, 1848.
Hybrenia Pascoe, 1866
Characteristics. Elongate-oblong. Length 5.0–17.0 mm. Surfaces usually more or less densely setose with long erect or short recumbent setae, latter in larger species often pale or white and forming patches or V-shaped patterns on elytra, smaller species with uniformly short dense recumbent setae. Colour of larger species piceous, sometimes with rufous elytra, smaller species fuscous or rufous often with legs flavous with dark femoral apices. Head short to very short before eyes (frontal index 0.3–0.5). Mandibles bidentate. Apical maxillary palpomere securiform. Antennae filiform or serrate, if latter a little wider in males. Elytral punctures simple, rounded. Winged.
Australian species. 46 named, many unnamed.
Distribution. Throughout Australia.

Jophon Champion, 1895 (Pl. 92F; Fig. 36.62)

Type species. *Iophon myrmecophilus* Champion, 1895.
Iophon Champion, 1895
Jophon Kolbe, 1896 (see note below)
Characteristics. Oblong, densely setose with short setae, entirely rufous. Length 3.5–4.5 mm. Head short before eyes (frontal index 0.3–0.5). Mandibles unidentate. Apical maxillary palpomere cultriform. Prothorax without or with incomplete lateral marginal beading, its base sometimes narrower than apex. Basal metatarsomeres more or less widened relative to other segments, more strongly so in males. Winged. Meso- and metatibiae strongly evenly curved in males.
Australian species. One: *J. myrmecophilus*.
Distribution. NT.
Note. Champion (1895) named this genus *Iophon*, a preoccupied name. Subsequently Kolbe (1896) incorrectly altered it to *Jophon* and it has been known by the latter name ever since, becoming a modified spelling of Champion's name in prevailing usage and therefore available under Article 33.3.1 of the ICZN. It is to be attributed to Champion not Kolbe.

Lepturidea Fauvel, 1862 (Pl. 9C, 84H; Figs 36.2, 36.57)

Type species. *Lepturidea deplanchei* Fauvel, 1862.
Atractus Boisduval, 1835
Aethyssius Pascoe, 1863
Chromomaea Pascoe, 1866
Alcmeonis Bates, 1868

Licymnius Bates, 1868
Anaxo Bates, 1868
Alcmaeonis Blackburn, 1893
Neoatractus Borchmann, 1909
Chromomoea Carter, 1915

Characteristics. Elongate-oblong with prothorax narrower than elytral bases, densely setose with short recumbent setae, colour very variable, often metallic green. Length 5.0–16.0 mm. Head very short before eyes (frontal index not more than 0.4), clypeolabral membrane as long as half length of labrum or longer. Apices of mandibles acuminate and unidentate. Apical maxillary palpomere cultriform or strongly asymmetrically triangular. A few species with simple claws. Prothorax subparallel for basal half at least. Winged.

Australian species. 56 named.

Distribution. Throughout the continent.

Litopous Matthews, 2012 (Pl. 101J; Fig. 36.49)

Type species. *L. baehri* Matthews, 2012.

Characteristics. Oblong, densely setose with long erect bristles on all dorsal surfaces and legs, fuscous. Length ~8 mm. Head short before eyes. Mandibles bidentate. Apical maxillary palpomere securiform. Antennae with only a few compound sensoria on terminal three antennomeres. Tarsomeres bearing long bristles only, without any lobes. Claws not pectinate, with only small teeth on basal half. Winged.

Australian species. One: *L. baehri*.

Distribution. Inland from Geraldton in south-central WA.

Note. This genus is unique among Australian Alleculinae in that it does not have any lobed or cupuliform tarsomeres, hence it does not technically belong in the subtribe Alleculina to which all other Australian genera have been assigned; it was described as *incertae sedis* in Alleculini. One other Australian genus, *Taxes* Champion, 1895 (q.v.) has greatly reduced (but still present) tarsal lobes and shares with *Litopous* the semi-dentate claws, but the two genera are otherwise very different. Our knowledge of the Alleculinae is not sufficient to allow us to identify the closest relative of *Litopous*.

Melaps Carter, 1908 (Pl. 93G; Fig. 36.67)

Type species. *M. cisteloides* Carter, 1908.
Oocistela Borchmann, 1908

Characteristics. Elongate-oval, nitid or with long erect setae, brown with paler legs. Length 5.5–13 mm. Head short before eyes. Mandibles bidentate. Apical maxillary palpomere broadly securiform. Pronotum laterally very convex and meeting hypomeron at a very obtuse angle or none at all. Elytral punctures simple, rounded. Metaventrite shorter than first abdominal ventrite and intercoxal process of latter broadly rounded or truncate. Wings absent. Males often with notch or tooth on inner edge of pro- and mesotibiae and metatibiae usually flattened or distorted.

Australian species. Eight.

Distribution. Throughout Australia.

Metistete Pascoe, 1866 (Pl. 94G; Figs 36.14, 36.22, 36.23, 36.64)

Type species. *Tanychilus gibbicollis* Newman, 1838.
Lisa Haag-Rutenberg, 1879

Characteristics. Elongate oblong or oval, glabrous, uniformly fuscous. Length 8.5–24.0 mm. Head short before eyes. Mandibles bidentate or truncate. Apical maxillary palpomere very broadly securiform (Fig. 36.14), enlarged in males. Elytra usually (but not always) with punctures stomatiform (deep with thickened lateral edges resembling lips or sometimes tubercles) containing a single very fine recumbent seta. Often stomatiform punctures are confined to an enlarged and impressed segment of stria 9 just behind humerus (Fig. 36.23), or all elytral punctures may be stomatiform (Fig. 36.22). Sometimes punctures are almost closed over, leaving only slits. Beetle generally fully winged, but females of some species flightless.

Australian species. 46. Matthews (2012) added three existing names to *Metistete* and removed one to *Homotrysis*.

Distribution. Throughout Australia.

Neocistela Borchmann, 1909 (Pl. 94C; Fig. 36.63)

Type species. *Pseudocistela ovalis* Blackburn, 1891.
Pseudocistela Blackburn, 1891

Characteristics. Elongate-oval, with dense short recumbent setae, entirely black or black with yellow elytra, legs rufous or testaceous. Length 4.9–5.7 mm. Head strongly prolonged (frontal index 0.7–1.1). Apices of mandibles unidentate. Apical maxillary palpomere cultriform. Only first two to four elytral striae evident. Winged.

Australian species. One: *N. ovalis*.

Distribution. Alpine VIC and NSW.

Nocar Blackburn, 1891 (Pl. 95B; Fig. 36.52)

Type species. *Cistela depressiuscula* Macleay, 1872.

Characteristics. Oval with base of semicircular pronotum as wide as elytral bases, densely setose with short recumbent setae, uniformly fuscous. Length 4.6–9.0 mm. Head short. Mandibles bidentate. Apical maxillary palpomeres cultriform. Three species with apical labial palpomeres bilobed. Legs relatively short, less than 1/4 of femoral length visible beyond body outline from above. Winged.

Australian species. 10. Matthews & Bouchard (2008) list nine species, but *Nocar subfasciatus* Carter, 1920 was returned to this genus by Matthews (2012).

Distribution. Throughout Australia.

Notocistela Carter, 1915 (Pl. 95G; Fig. 36.66)

Type species. *N. tibialis* Carter, 1915.

Characteristics. Elongate-oval, with short dense recumbent setae and some long erect setae on elytra. Length 6.0–10.0 mm. Head short before eyes. Mandibles bidentate. Apical maxillary palpomere broadly securiform. Pronotum densely vermiculate-punctate, laterally rounded. Elytra with large pustules, at least posteriorly, each bearing one long erect seta. Metaventrite shorter than first abdominal ventrite. Flightless.

Australian species. Three.

Distribution. WA, central Australia.

Nypsius Champion, 1895 (Pl. 95F; Fig. 36.55)

Type species. *N. aeneopiceus* Champion, 1895.

Characteristics. Elongate-oblong, prothorax much narrower than elytral bases, with short dense recumbent setae, sometimes with metallic reflections, violet on elytra and green on head and prothorax. Length 7.5–9.5 mm. Head short before eyes. Mandibles unidentate. Apical maxillary palpomere triangular. Middle of pronotum with a pair of discal impressions. Winged.

Australian species. Two: *N. aeneopiceus* and *N. foveatus* Champion, 1895.

Distribution. TAS.

Ommatophorus Macleay, 1872 (Pl. 95J; Fig. 36.69)

Type species. *O. mastersi* Macleay, 1872.

Characteristics. Elongate-oval, strongly nitid, often bicoloured, body and legs with very long setae some of which are erect and at right angles to body surface. Length 4.8–8.0 mm. Head very short before eyes (frontal index 0.3). Eyes very large. Mandibles bidentate. Apical maxillary palpomere broadly securiform. Antennae dimorphic, apical segment longer than subapical in males, subequal in females; in males middle segments may be enlarged and devoid of sensoria below. Winged.

Australian species. Four.

Distribution. NSW, QLD.

Otys Champion, 1895 (Fig. 36.59)

Type species. *O. harpalinus* Champion, 1895 (= *Scaletomerus proximus* Blackburn, 1891).

Characteristics. Elongate-oval, densely setose with short semi-erect setae, uniformly castaneous. Length ~4 mm. Head short before eyes. Mandibles bidentate. Apical maxillary palpomeres cultriform. Elytral striae obsolete on disc, not discernible from general surface punctation although some punctures arranged in rows. Outer edge of protibia with a serrated row of minute spinules. Male metatibiae dilated and twisted with a deep smooth concavity on inner faces. Winged.

Australian species. One: '*Otys' armatus* Champion, 1895.

Distribution. Northern WA.

Note. Matthews & Bouchard (2008) designated *O. armatus* as the type species of this genus, but Lucas (1920) had already designated *O. harpalinus*, which was subsequently shown to be a synonym of *Scaletomerus proximus*, therefore the name *Otys* is unfortunately a synonym of *Scaletomerus*. A new name will be required for the distinct genus here called *Otys*.

Pemanoa Buck, 1955 (Pl. 96H; Fig. 36.54)

Type species. *P. mixta* Buck, 1955.

Characteristics. Elongate-oval, with dense short semi-recumbent setae, either uniformly testaceous or bicoloured, black with rufous or flavous areas. Length 4.0–6.0 mm. Head very long before eyes (frontal index 1.0), gradually and only slightly narrowing basally. Mandibles unidentate. Apical maxillary palpomere cultriform. Winged.

Australian species. Four.

Distribution. TAS, VIC, NSW.

Scaletomerus Blackburn, 1891 (Pl. 87G; Fig. 36.50)

Type species. *S. harpaloides* Blackburn, 1891.

Otys Champion, 1895

Characteristics. Oblong, glabrous, uniformly castaneous. Length 4.7–6.6 mm. Head short before eyes. Mandibles bidentate. Apical maxillary palpomere cultriform. Prothorax basally not quite as wide as elytral bases. Outer edge of protibia with a serrated row of minute spinules, the spur sometimes enlarged. Winged.

Australian species. Four

Distribution. Throughout Australia.

Scaphinion Matthews, 2012 (Pl. 101E; Fig. 36.60)

Type species. *S. clavatus* Matthews, 2012.

Characteristics. Elongate-oblong, bearing dense long recumbent setae, fuscous. Length 5–6 mm. Occiput of head deeply transversely excavate and receiving prolongation of anterior part of pronotal disc. Mandibles bidentate. Apical maxillary palpomere broadly securiform. Males of type species with distal antennomeres very strongly enlarged, less or not enlarged in other two species (Matthews 2012, fig. 4). Elytral punctures uniformly large, deep, with side edges thickened (stomatiform). Winged.

Australian species. *S. clavatus* and two unnamed.

Distribution. The Kimberley District of northern WA.

Simarus Borchmann, 1909 (Pl. 98H; Fig. 36.65)

Type species. *Ismarus godeffroyi* Haag-Rutenberg, 1879.

Ismarus Haag-Rutenberg, 1878

Characteristics. Elongate-oval, with dense very short setae, fuscous. Length 8–12 mm. Head short before eyes.

Mandibles bidentate. Apical maxillary palpomere very broadly securiform, enlarged in males. Eyes small, separated by two to three eye widths. Elytral surfaces densely minutely tuberculate, the punctures stomatiform (see *Metistete* above). Flightless

Australian species. Four.

Distibution. WA, NT, arid areas.

Tanychilus Newman, 1838 (Pl. 99G; Fig. 36.58)

Type species. *T. striatus* Newman, 1838.

Characteristics. Elongate-oblong, glabrous and nitid, uniformly castaneous or in *T. pulcher* (Carter) with metallic blue-green reflections. Length 11.7–19.0 mm. Head moderately to strongly elongate before eyes (frontal index 0.5–0.9), frons with subparallel or anteriorly diverging grooves. Mandibles unidentate. Apical maxillary palpomere cultriform. Male protibiae sometimes sinuate on inner edges. Winged.

Australian species. Six.

Distribution. Throughout Australia, New Zealand.

Taxes Champion, 1895 (Pl. 99B; Fig. 36.51)

Type species. *T. depressus* Champion, 1895.

Characteristics. Oblong, with dense minute setae or dense long setae, fuscous. Length 3.5–6.8 mm. Head short. Mandibles sometimes apically truncate or bluntly rounded. Apical maxillary palpomere securiform. Elytral stria 9 frequently deeper than the others and with larger punctures. Penultimate tarsomeres with long bristles, with lobes much reduced. Teeth confined to basal half of claws. Winged.

Australian species. Two named, at least one unnamed. *Nocar subfasciatus* Carter, 1920, listed erroneously as a species of *Taxes* in Matthews & Bouchard (2008), is a *Nocar* (Matthews 2012)

Distribution. Northern and central Australia.

Subfamily Diaperinae Latreille, 1802

Characteristics. Form usually compact, ovoid or subhemispherical but sometimes more elongate or flattened, with relatively short antennae and legs. Labrum transverse. Antennae with complex stellate sensoria. Elytral striae nine, with or without scutellary striole, striae sometimes effaced. Hind wings always present, without medial fleck. Claws simple.

Distribution. World-wide.

Australian taxa. 12 tribes.

Notes. The Diaperinae were divided by Matthews & Bouchard (2008) into two informal groups: the 'phaleriine Diaperinae' and the 'diaperine Diaperinae'. The former live in edaphic or climatic deserts and include Crypticini, Trachyscelini, Phaleriini, Hyociini and Ectychini, and the latter live mainly in forests and include all the other Diaperinae. Morphologically these two groups differ profoundly in labral, ovipositor, and especially internal female tract structure.

In the phylogeny of Kergoat *et al.* (2014b), based on DNA sequencing, Diaperinae as now understood do not recover as monophyletic but appear as several separate branches. The above-mentioned phaleriine-diaperine division is largely supported in that Trachyscelini, Hyociini and Phaleriini (but not Crypticini) cluster together, separately from other Diaperinae (Ectychini were not sampled), as the sister group of Alleculinae. Crypticini form part of a cluster also including *Pentaphyllus* Dejean, 1821 and Scaphidemini.

Tribe Trachyscelini Blanchard, 1845

Characteristics. Ovoid, strongly convex, glabrous, fuscous or rufous. Length 2–3 mm. Eyes transverse, oval, not emarginate. Antennae extremely short, not reaching to base of head, last five segments expanded into a broad club. Clypeolabral membrane not exposed. Pronotum widest at base, disc evenly convex. Mesocoxal cavities open. Elytral epipleura not reaching apices. Legs strongly fossorial, stout, with distally expanded tibiae and peg-like setae, tarsomeres subtriangular.

Distribution. Mediterranean, southern Asia, Australia, introduced around Gulf of Mexico, on coastal sand dunes.

Australian taxa. One genus.

Trachyscelis Latreille, 1809 (Pl. 100A; Fig. 36.107)

Type species. *T. aphodioides* Latreille, 1809 (Europe).

Characteristics. As for the tribe.

Australian species. Four.

Distribution. All mainland Australian shores, Lord Howe Island.

Tribe Phaleriini Blanchard, 1845

Characteristics. Oval, strongly convex, glabrous, rufous with a darker macula in middle of each elytron. Length ~8 mm. Eyes transverse, oval. Antennae reaching to about middle of pronotum, gradually widened apically. Clypeolabral membrane exposed before clypeus. Pronotum widest at base, disc evenly convex. Mesocoxal cavities closed. Protibiae enlarged and flattened apically. Meso- and metatibiae bearing dense erect spines.

Distribution. World-wide, on marine beaches.

Australian taxa. One introduced genus and species.

Phaleria Latreille, 1802 (Pl. 96E; Fig. 36.167)

Type species. *Tenebrio cadaverinus* Fabricius, 1793 (North Sea).

Characteristics. As for the tribe.

Australian species. One: *P. bimaculata* (Linnaeus, 1767).

Distribution. The species named above, common on beaches around the Mediterranean, has become established in Port Phillip Bay, VIC where it was first found in 1940 (Matthews & Queale 1997).

Biology. *Phaleria* species feed on dead animal matter and faeces on marine beaches.

Tribe Hyociini Medvedev & Lawrence, 1982

Characteristics. Elongate-oval, more or less convex, sometimes sculptured or tuberculate, with numerous adpressed pale setae, fuscous or piceous, elytra sometimes mottled. Eyes transversely oval or rounded, not emarginate. Labrum with clypeolabral membrane hidden. Antennomeres of about equal width except last three which are expanded into a weak, loose or compact club. Pronotum transverse, often explanate laterally, sides rounded in outline, often narrower basally. Mesocoxal cavities closed. Elytra without scutellary striole, epipleura complete to apices. Wings sometimes greatly reduced. Legs more or less fossorial, with protibia widened and outwardly spinulose or dentate, bearing at most one apical spur, tarsomeres not lobed, tarsi sometimes 4–4–4.

Distribution. Hyociini occur throughout Australia, the areas of greatest diversity and endemism being the northern and western parts of the continent (Medvedev & Lawrence 1987). The tribe is an exclusively Australian group apart from one species of *Parahyocis* Kaszab, 1955 which occurs on some Pacific islands, probably as a result of recent dispersal.

Australian taxa. Six genera and five subgenera arranged in three subtribes. The subtribes are not separately characterised here but are indicated in the key.

Biology. Hyociini live on coastal dunes and beaches and in the arid interior of the continent. The group as a whole is adapted to life in the soil and displays various degrees of fossorial modification of the protibiae, from minimal in *Csiro* Medvedev & Lawrence, 1986 to highly developed in *Uptona* Medvedev & Lawrence, 1986. The soil-inhabiting larva of a species of *Hyocis* Pascoe, 1866 was described by Lawrence & Medvedev (1982).

Brittona Medvedev & Lawrence, 1986 (Pl. 86J; Fig. 36.114)

Type species. *B. minuta* Medvedev & Lawrence, 1986.

Characteristics. Uniformly fuscous. Length 1.6–1.9 mm. Eyes subcircular, not separated from gena by slit-like groove. Antennae short, with compact club, the segments strongly transverse, apical setae present on segments 9–11, rest of club surface bare. Pronotum little flattened along lateral margins, covered with scattered and barely distinguishable setae. Sides of pronotum not shiny, although more or less flattened. Outer margin of protibiae with two teeth separated by broad notch. Tarsi 4–4–4.

Australian species. One: *B. minuta*.

Distribution. NT.

Csiro Medvedev & Lawrence, 1984 (Pl. 99I; Figs 36.21, 36.113, 36.127)

Type species. *Hyocis subparallela* Champion, 1894.
Millstreamia Medvedev & Lawrence, 1984 (subgenus)

Characteristics. Uniformly fuscous. Length 1.6–3.4 mm. Eyes subcircular, their anterior margin separated from gena by deep groove continued along inner eye margin. With acute subgenal process below antennal bases (Fig. 36.21). Sides of pronotum explanate, thin and shiny, Protibiae more than twice as long as wide, apically almost parallel-sided, outer edge without large teeth. Tarsi 5–5–4 (*Csiro* s. str.) or 4–4–4 (*Millstreamia*).

Australian species. 10.

Distribution. Throughout Australia (*Csiro* s. str.) and the Pilbara Region of WA (*Millstreamia*).

Key to species. Medvedev & Lawrence (1984).

Hyocis Pascoe, 1866 (Pl. 91H; Figs 36.129, 36.130, 36.131)

Type species. *H. bakewelli* Pascoe, 1866.
Nannohyocis Medvedev & Lawrence, 1983 (subgenus)
Neohyocis Medvedev & Lawrence, 1983 (subgenus)

Characteristics. Fuscous, sometimes with dark spots on elytra, antennae entirely or at least basally pale, pronotum and elytra, or elytra alone, pale, if body entirely black dorsally, then antennae entirely pale. Length 1.7–4.1 mm. Eyes weakly to strongly transverse. Disc of pronotum either with median longitudinal groove, sometimes indistinct, or with depression on base opposite scutellum. Tarsi 5–5–4.

Australian species. 15.

Distribution. Throughout Australia.

Biology. The monotypic subgenus *Nannohyocis* of southern WA is myrmecophilous, the ant host being recorded by Carter (1921) as *Iridomyrmex conifera* Clark.

Keys to species. Lawrence & Medvedev (1982); Medvedev & Lawrence (1983).

Magela Medvedev & Lawrence, 1986 (Pl. 93D; Fig. 36.115)

Type species. *M. uptoni* Medvedev & Lawrence, 1986.

Characteristics. Elytra with black spots, rarely absent. Length 1.9–2.1 mm. Eyes subcircular. Pronotum broadly flattened laterally. Protibiae uniformly widened towards apex, outer margin without teeth but with sharp spinules. Tarsi 4–4–4.

Australian species. Two: *M. uptoni* and *M. cordicollis* Medvedev & Lawrence, 1986.

Distribution. Western QLD, NT.

Parahyocis Kaszab, 1955 (Pl. 96K; Fig. 36.128)

Type species. *Hyocis championi* Fauvel, 1904 (New Caledonia).

Characteristics. Antennae and body black. Length 2–2.3 mm. Dorsal vestiture indistinct, consisting of very short setae on elytral intervals. Eyes subcircular. Occipital ridge running along hind edge of eye bent forward below. Sides of elytra convex. Pronotal base opposite scutellum and near hind angles deeply impressed. Protibiae broadly expanded. Tarsi 5–5–4.

Australian species. One: *P. kaszabi* Medvedev & Lawrence, 1984.

Distribution. Northern QLD, New Caledonia, Vanuatu, Samoa.

Uptona Medvedev & Lawrence, 1986 (Pl. 101A; Fig. 36.126)

Type species. *Uptona pallida* Medvedev & Lawrence, 1986.

Characteristics. Rufous to black, sides of pronotum and elytra covered with long hairs. Length 2.1–3.1 mm. Antennae short, barely reaching beyond anterior edge of pronotum, with setae only on apical edges of club segments. Protibiae extremely broad (length not more than twice width), outer edges with two broad teeth (apical and middle) divided by deep arcuate emargination. Tarsi 5–5–4.

Australian species. Four.

Distribution. NT and northern SA.

Key to species. Medvedev & Lawrence (1986).

Tribe Ectychini Doyen, Matthews & Lawrence, 1990

Characteristics. Bipartite-oval or oblong, convex, pilose with long erect setae, fuscous or piceous, sometimes with slight bluish reflections. Length 3–8 mm. Eyes transverse or slightly reniform. Labrum with clypeolabral membrane not exposed. Antennae reaching base of prothorax, segments gradually enlarging distally, usually loose-jointed, with complex sensoria on distal edges of last four segments. Pronotum transverse with rounded sides, with narrow base nearly always excised and acutely dentate at posterior angles, disc weakly, evenly convex. Prosternum much longer before coxae than postcoxal bridge. Mesocoxal cavities open. Elytra with nine striae, with or without scutellary striole, epipleura complete to apices. Wings absent, or vestigial without venation. Legs not fossorial, tibiae subparallel or only slightly expanded distally, tarsomeres not lobed.

Distribution. Australia only, in arid areas and offshore islands.

Australian taxa. Two genera. A species of the tribe in the WA Museum which might belong to an unnamed genus (because it does not have excised posterior pronotal angles) was found in a cave on the North West Cape.

Ectyche Pascoe, 1869 (Pl. 90E; Fig. 36.176)

Type species. *E. erebea* Pascoe, 1869.

Characteristics. Length 4.5–8.0 mm. Antennae perfoliate (consisting of loosely-jointed disc-like segments), especially towards apices, usually long enough to reach or surpass base of prothorax. Otherwise as for the genus.

Australian species. Eight.

Distribution. WA, SA, NT, more common on offshore islands.

Micrectyche Bates, 1873 (Pl. 94L; Fig. 36.177)

Type species. *M. intermedia* Bates, 1873.

Micretyche Carter, 1924.

Characteristics. Length 3.5–4.5 mm. Antennae more compact and shorter. Otherwise as for the genus.

Australian species. Three.

Distribution. Southern WA, coastal.

Tribe Crypticini Brullé, 1832

Characteristics. Ovoid, moderately convex, dorsal surfaces covered with short fine setae, rufous or partly flavous with darker elytral markings. Length 2–3 mm. Eyes very transverse, only slightly emarginate anteriorly. Labrum with clypeolabral membrane not exposed. Antennae with apical five segments forming weak club, with a few stellate sensoria on distal edges. Pronotum widest at base, evenly convex. Prosternal process acutely prolonged and received in cavity of mesoventrite. Mesocoxal cavities closed. Epipleura complete to apices. Legs not distinctly fossorial, but pro- and mesotibiae evenly widening distally and their outer edges spinulose, tarsomeres subparallel.

Distribution. World-wide.

Australian taxa. One genus, revised by Kaszab (1975) (as *Microcrypticus*).

Ellipsodes Wollaston, 1854 (Pl. 90H; Fig. 36.164)

Type species. *Sphaeridium glabratum* Fabricius, 1793 (Madeira).

Microcrypticus Gebien, 1920 (*partim*)

Anthrenopsis Koch, 1950 (subgenus)

Characteristics. As for the tribe.

Australian species. Two: *E. scriptus* (Lewis, 1894) and *E. ziczac* (Motschulsky, 1873).

Distribution. NT, northern QLD, Africa, southern Asia, Japan, USA (introduced), in dry habitats.

Biology. Larvae have been collected in large numbers in leaf litter.

Tribe Hypophlaeini Billberg, 1820

Characteristics. Elongate, parallel-sided, subcylindrical, glabrous, piceous or fuscous. Total length 1.5–6 mm. Eyes reniform. Labrum with clypeolabral membrane exposed for its entire width. Antennae not reaching beyond middle of prothorax, last seven segments widened into a club, with complex (multipronged) sensoria. Pronotum elongate and subparallel. Prosternum very long before coxae. Mesocoxal cavities closed. Elytra with nine striae and scutellary striole or estriate, epipleura not complete to apices, which are roundly truncate to expose last visible tergite. Wings present. Legs not fossorial, tarsomeres not lobed. Last visible abdominal tergite partially exposed and heavily sclerotised, forming pygidium.

Distribution. This is a pantropical group that is poorly represented in Australia, where it occurs only in northern QLD and the NT, and is represented there by three species of the world-wide genus *Corticeus* Piller & Mitterpacher, 1783 which are conspecific with, or closely related to, species that occur in New Guinea and the Pacific region [Bremer 1990, 1993, 2013, 2015; Bremer & Lillig 2017].

Australian taxa. One genus.

Biology. Hypophlaeini are found under the bark of trees in association with scolytine larvae either feeding on the latter (in a rare instance of predation in Tenebrionidae), or feeding also, or exclusively, on frass and fungi (Doyen & Lawrence 1979).

Corticeus Piller & Mitterpacher, 1783 (Pl. 88D; Fig. 36.157)

Type species. *C. unicolor* Piller & Mitterpacher, 1783 (Europe).

Hypophlaeus Fabricius, 1790

Stenophloeus Blair, 1921 (subgenus)

Cnemophloeus Bremer, 1998 (subgenus)

Characteristics. As for the tribe.

Australian species. Three. The name *C. hackeri* (Carter, 1928) listed as a valid species in Matthews & Bouchard (2008) was shown by Bremer (2013) to be a synonym of *C. cylindricus* (Reitter, 1877) known from Fiji and New Guinea.

Distribution. Northern QLD, northern NT, world-wide.

Tribe Myrmechixenini Jacquelin du Val, 1858

Characteristics. Oblong, fuscous, smooth with dense, short recumbent setae. Length ~2 mm. Eyes round. Antennae 11-segmented with a 4- or 5-segmented club. Lateral pronotal margins absent. Elytra without striae. Tarsi 4–4–4. [Doyen & Lawrence 1979].

Distribution. World-wide.

Australian taxa. One introduced genus and species.

Myrmechixenus Chevrolat, 1835 (Pl. 94D; Fig. 36.112)

Type species. *M. subterraneus* Chevrolat, 1835 (France).

Characteristics. As for the tribe.

Australian species. One: *M. vaporariorum* Guérin-Méneville, 1843.

Distribution. Introduced into Australia. Doyen *et al.* (1990) were the first to record this genus and species from Australia on the basis of specimens collected throughout the country, represented by a widespread European species also found in Africa, occurring in stored food products.

Tribe Gnathidiini Gebien, 1921

Subtribe Anopidiina Jeannel & Paulian, 1945

Characteristics. Oval, moderately convex, with numerous long erect setae, fuscous or piceous. Total length 1.5–2 mm. Eyes entire, small. Labrum with clypeolabral membrane exposed. Antennae reaching to only about middle of prothorax, with six segments including a large apical club showing traces of three or four fused segments. Pronotum transversely subrectangular, widest in middle, disc evenly convex, lateral margins widened and serrate. Mesocoxal cavities closed. Elytra estriate, epipleura not reaching apices. A sulcus next to epipleural carina which curves around humerus. Wings present, with reduced venation. Legs not fossorial, tarsomeres not lobed, segmentation 4–4–4. [Doyen *et al.* 1990].

Distribution. Oriental and Neotropical regions, eastern Australia.

Australian taxa. One genus and species.

Tyrtaeus Champion, 1913 (Pl. 100B; Fig. 36.111)

Type species. *T. rufus* Champion, 1913 (Guatemala).

Characteristics. As for the subtribe.

Australian species. One: *T. bicoloratus* Matthews & Lawrence, 2005.

Distribution. Pantropical, Southern QLD, NSW.

Note. At least one undescribed species is known from Christmas Island.

Subtribe Gnathidiina Gebien, 1921

Characteristics. Oblong, moderately convex, nearly glabrous with a few lateral elytral setae. Total length 2.5–4 mm. Eyes entire, small. Labrum with clypeolabral membrane not exposed. Antennae reaching to only about middle of prothorax, with 10 segments of which apical three or four are expanded into a club. Pronotum transversely trapezoidal, widest at base, disc evenly convex, lateral margins simple. Mesocoxal cavities closed. Elytra with nine sometimes indistinct striae and trace of scutellary striole, epipleura not reaching apices. A sulcus next to epipleural carina which curves around humerus, carina usually angulate at humerus. Wings developed or vestigial. Legs not fossorial, tarsomeres not lobed, segmentation 5–5–4. [Kaszab 1978; Doyen & Lawrence 1979].

Distribution. Widespread in the Asia-Pacific.

Australian taxa. One genus.

Biology. Adults and larvae are found in vine scrub and rainforest in forest-floor litter, on tree trunks, on moss, in rotten wood, and in polypore fungi (Watt 1975).

Menimus Sharp, 1876 (Pl. 93A; Fig. 36.110)

Type species. *M. batesi* Sharp, 1876 (New Zealand).
Ceramba Fauvel, 1904
Paita Fauvel, 1904
Microcilibe Carter, 1919
Tjikoraia Pic, 1921
Neomenimus Kaszab, 1939
Characteristics. As for the subtribe.
Australian species. Four.
Distribution. Northern NSW to northern QLD, Lord Howe Island, tropical Asia, Pacific islands.
Key to species. Kaszab (1978).

Tribe Unnamed – Diaperinae

Characteristics. Broadly oval, strongly convex, glabrous, nitid with greenish reflections. Total length 2.3–2.9 mm. Eyes well developed, reniform. Labrum transverse, clypeolabral membrane not exposed. Antennae 10-segmented, distal four segments enlarged into a club. Pronotum transverse, strongly and evenly convex with ventrally concave sides covering legs. Procoxal cavities expanded laterally where they form an acutely angled cleft (Fig. 36.5). Mesoventrite not excavated, mesocoxal cavities closed. Elytra evenly, strongly convex, epipleura terminating before apices, at humeral angles greatly widened to form deflexed flange covering legs. Scutellum strongly transversely triangular. Wings present. Legs short, largely covered by pronotal and elytral edges. Tarsi with preapical segments short, ventrally lobed, claw segments elongate, in meso- and metatarsi longer than remaining segments combined, claws basally subdentate. [Matthews & Lawrence 2005; Grimm 2011].
Distribution. Malaysia, Philippines, northern QLD.
Australian taxa. One genus and species.

Triplehornia Matthews & Lawrence, 2005 (Pl. 100K; Figs 36.5, 36.47)

Type species. *T. metallica* Matthews & Lawrence, 2005.
Characteristics. As for the tribe.
Australian species. One: *T. metallica*.
Distribution. Northern QLD, in rain forest.

Tribe Leiochrinini Lewis, 1894

Characteristics. Hemispherical, glabrous, variously coloured and patterned, nitid. Eyes transverse or reniform. Labrum with clypeolabral membrane exposed. Antennae reaching well beyond base of prothorax, last seven or eight segments expanded into a weak parallel club with relatively few large complex sensoria. Pronotum transverse, semicircular, broad at base, sometimes covering head, notal cowling excised at angle leaving much of procoxa exposed. Prosternum very short before coxae, its process expanded behind them. Mesocoxal cavities narrowly closed by ventrites, mesepimeron almost reaching them. Elytra without distinct striae (in Australia), very smooth, epipleura not reaching apices, widened anteriorly. Wings present, with greatly reduced venation. Legs not fossorial, preapical tarsomeres either parallel or strongly lobed. Ventral intercoxal process of abdomen very wide and truncate. [Kaszab 1946, 1961a, b, c].
Distribution. Leiochrinini are inhabitants of Old World, mainly tropical, forests and are most diverse in the Oriental Region. In Australia they are confined to the wetter parts of QLD and northern NSW.
Australian taxa. Three genera.
Biology. Larvae and adults of Leiochrinini have been observed feeding exposed on lichens, algae and mosses. The coccinelloid form and colours of the adults are unmistakeable indicators of exposed feeding.

Derispia Lewis, 1894 (Pl. 9E, 88L; Fig. 36.44)

Type species. *Diaperis maculipennis* Marseul, 1876 (Japan).
Characteristics. Head not covered by pronotum. Length 2–3 mm. Antennae short, moderately flattened, segment 4 as wide as 5, segment 3 as wide distally as base of 4. Legs relatively short, tibiae without keel on outer edge, tarsomeres simple, not lobed.
Australian species. Three.
Distribution. Northern NSW to northern QLD.
Biology. Larvae and adults of *Derispia variabilis* Carter, 1930 have been observed feeding on algae covering a stone wall in southern QLD
Key to species. Kaszab (1961b).

Leiochrinus Westwood, 1883 (Pl. 92L; Fig. 36.46)

Type species. *L. fulvicollis* Westwood, 1883.
Liochrinus Alluaud, 1900.
Characteristics. Pronotum covering head at rest, its anterior edge broadly rounded. Length 4.5–5.5 mm. Antennae long, segments subcylindrical, segments 3 and 4 narrower than 5. Legs longer, tibiae with sharp median keel on outer edge, tarsomeres (except apical) strongly lobed below.
Australian species. One: *L. fulvicollis*.
Distribution. North-eastern QLD, New Guinea, Asia.

Leiochrodes Westwood, 1883 (Pl. 92E; Fig. 36.48)

Type species. *L. discoidalis* Westwood, 1883 (Malacca).
Characteristics. Head not covered by pronotum. Length 2–3.5 mm. Anterior edge of pronotum deeply excised. Antennae long, segments subcylindrical, segments 3 and 4 narrower than 5. Legs longer, tibiae with sharp median keel on outer edge, tarsomeres (except apical) strongly lobed below.

Australian species. One: *L. suturalis* Westwood, 1883.
Distribution. Eastern QLD, New Guinea, Asia.

Tribe Scaphidemini Reitter, 1922

Characteristics. Oval, convex, glabrous, with black and red colour pattern. Total length 3–5 mm. Eyes reniform. Labrum with clypeolabral membrane exposed. Antennae reaching beyond base of prothorax, segments from 4 onward gradually expanded distally, their distal edges with a row of complex sensoria. Pronotum widest at base, disc evenly convex. Prosternum moderate before coxae, about twice as long as postcoxal bridge. Pronotal cowling excised at angle, leaving much of narrow part of procoxa exposed. Mesocoxal cavities open. Elytra with striae indistinct laterally, epipleura terminating before apices. Wings present. Legs not fossorial, tarsomeres not lobed. Abdominal intercoxal process widely truncate. [Doyen *et al.* 1990; Schawaller 1997, 2004].
Distribution. There are five genera in the tribe: three found in north-eastern Asia, one, *Scaphidema* Redtenbacher, 1849 which has a Holarctic distribution extending (unusually for a tenebrionid) into boreal and Arctic regions, and *Spiloscapha* Bates, 1873.
Australian taxa. One genus and species.

Spiloscapha Bates, 1873 (Pl. 98E; Fig. 36.45)

Type species. *Platydema thalloides* Pascoe, 1869 (= *S. crassicornis* Bates, 1873).
Characteristics. As for the tribe.
Australian species. One: *S. thalloides*.
Distribution. *S. thalloides* occurs in eastern NSW. There are also some 25 other species of the genus in New Guinea and southern Asia as far north as Taiwan (Schawaller 1997, 2004).
Biology. Larvae of *S. thalloides* have been seen feeding on the surface of a fungus growing on a log. The larva is bicoloured, suggesting external feeding with chemical defence, and cuticular glands are present on the head and most body segments.

Tribe Diaperini Latreille, 1802

Characteristics. Oval or subparallel, convex or rarely strongly flattened, glabrous or with minute recumbent setae, colour uniform or with variously patterned elytra. Eyes reniform. Labrum with clypeolabral membrane exposed except in *Gnatocerus* Thunberg, 1814. Antennae reaching beyond base of prothorax, segments gradually or abruptly widening distally, with complex stellate sensoria on apical 5–7 segments. Prosternum short or moderate before coxae, 2–3 times as long as postcoxal bridge. Mesocoxal cavities closed by ventrites only or nearly so. Elytra with nine striae and scutellary striole but estriate in *Pentaphyllus* Dejean, 1821. Wings fully developed. Legs not fossorial, tarsomeres not lobed except in *Sciophagus* Sharp, 1885. [Doyen 1984].
Distribution. World-wide.
Australian taxa. Two subtribes.
Biology. Diaperini generally feed in fungi growing on decaying wood. Introduced *Gnatocerus* Thunberg, 1814 and *Alphitophagus* Stephens, 1832 may occasionally be found in mouldy stored grains.
Note. The current concept of the tribe Diaperini and its two subtribes is based entirely on internal characters (Doyen 1984), which are not described in the present work. They involve the structure of the female tract, the presence or absence of aedeagal clavae, and the symmetry of the epipharyngeal setal pattern, see Matthews & Bouchard (2008).

Subtribe Adelinina LeConte, 1862

Characteristics. As defined by Doyen (1984) this subtribe is based entirely on the presence of clavae (also called laciniae) on the aedeagus. These take the form of hooks or elaborate structures attached to the penis (Matthews & Lawrence 2005, Figs 13, 15; Matthews & Bouchard 2008, Fig. 98). Externally, Adelinina (except for *Alphitophagus* Stephens, 1832) have incomplete elytral epipleura.
Distribution. Adelinina are primarily a Western Hemisphere group which has dispersed around the world, but *Palembomimus* Matthews & Lawrence, 2005 is confined to Australia.
Australian taxa. Four genera.

Adelina Dejean, 1835 (Pl. 84K; Fig. 36.169)

Type species. *Cucujus planus* Fabricius, 1801 (North America).
Doliema Pascoe, 1860
Characteristics. Form extremely flattened, subparallel, glabrous, pale brown. Length 4.5–5 mm. Antennae gradually enlarged apically or with only feeble club. Prothorax narrowest at base. Elytral surface becoming abruptly vertical outside 7th stria [Merkl 1992].
Australian species. One: *A. platisoides* (Pascoe, 1860).
Distribution. Northern WA, Asia, Western Hemisphere.
Biology. *Adelina* lives under the bark of dead trees.

Alphitophagus Stephens, 1832 (Pl. 84G; Fig. 36.172)

Type species. *A. quadripustulatus* Stephens, 1832 (Britain).
Characteristics. Oval, moderately convex, covered with minute fine recumbent setae, elytra bicoloured. Length ~2 mm. Outline of head indented at frontoclypeal suture. Male with swollen clypeus. Antennae gradually enlarged apically or with only feeble club.
Australian species. One: *A. bifasciatus* (Say, 1824).
Distribution. South-eastern SA and VIC (introduced), cosmopolitan in mouldy stored grains.

Gnatocerus Thunberg, 1814 (Pl. 90A; Fig. 36.163)

Type species. *G. ruber* Thunberg, 1814 (= *Trogosita cornuta* Fabricius, 1798).

Gnathocerus Agassiz, 1846

Characteristics. Oblong, moderately convex, glabrous, uniformly fuscous. Length 3–4 mm. Clypeolabral membrane concealed under clypeus. Posterior angles of pronotum acute. Male with distorted head and hornlike outgrowths on mandibles.

Australian species. One: *G. cornutus*.

Distribution. Cosmopolitan, introduced into Australia, in mouldy stored grains.

Palembomimus Matthews & Lawrence, 2005 (Pl. 96A; Fig. 36.170)

Type species. *Platydema deplanatum* Champion, 1894.

Characteristics. Oblong, parallel-sided, flattened, densely and coarsely punctate, minutely setulose, uniformly fuscous or with rufous fasciae on elytra. Length 2.7–3.6 mm. Antennomeres 4–10 gradually widening.

Australian species. Two: *P. bicinctus* (Champion, 1894) and *P. deplanatus*.

Distribution. South-east SA, VIC, NSW, QLD, NT.

Subtribe Diaperina Latreille, 1802

Characteristics. Epipleura complete to elytral apices except in *Sciophagus* Sharp, 1885. Aedeagal clavae absent. Otherwise as for tribe.

Distribution. World-wide.

Australian taxa. Five genera.

Ceropria Laporte & Brullé, 1831 (Pl. 87I; Fig. 36.173)

Type species. *Helops indutus* Wiedemann, 1819 (India).

Characteristics. Oval, convex, glabrous, uniformly fuscous. Length 8–10 mm. Antennae gradually enlarged apically, subserrate. Elytra striate.

Australian species. Two: *C. maculata* Gebien, 1911 and *C. peregrina* Pascoe, 1866.

Distribution. Eastern QLD, Asia, Africa.

Biology. Species of *Ceropria* usually occur on softer basidiomycete fruiting bodies, such as those of Auriculariaceae and Thelephoraceae.

Pentaphyllus Dejean, 1821 (Pl. 96L; Fig. 36.168)

Type species. *Mycetophagus testaceus* Hellwig, 1792 (Europe).

Characteristics. Oval, strongly convex, fuscous. Length 2–2.5 mm. Antennae with large club of five subequal segments. Elytra not striate, without large punctures.

Australian species. Three.

Distribution. North-eastern NSW, eastern QLD, north of NT, world-wide.

Biology. *Pentaphyllus exilis* (Champion, 1894) is known to breed in fruiting bodies of *Ganoderma*.

Platydema Laporte & Brullé, 1831 (Pl. 96C; Fig. 36.175)

Type species. *P. dejeani* Laporte & Brullé, 1831 (Europe).

Characteristics. Oval, convex, glabrous, uniformly fuscous or frequently with rufous fascies on elytra. Length 3–7 mm. Antennomeres symmetrical, gradually enlarged apically. Base of pronotum usually without margin, or this confined to middle. Intercoxal process of prosternum more elevated, its posterior face high and vertical when viewed from side. Males frequently with a pair of frontal horns.

Australian species. 16.

Distribution. Throughout Australia, world-wide.

Biology. *Platydema* species are associated with a wide variety of basidiomycete fungi.

Key to species. Carter (1918) (including *Ulomoides*).

Sciophagus Sharp, 1885 (Pl. 97D; Fig. 36.171)

Type species. *Uloma pandanicola* Boisduval, 1835 (Pacific islands).

Pachycerus Montrouzier, 1860

Characteristics. Oval, moderately convex, glabrous, black. Length ~4 mm. Last four antennomeres strongly widening apically. Elytral epipleura terminating before apices. Penultimate tarsomeres cupuliform.

Australian species. One: *S. flavipes* Gebien, 1920.

Distribution. Northern QLD, New Guinea, Pacific region.

Ulomoides Blackburn, 1888 (Pl. 100E; Fig. 36.174)

Type species. *U. humeralis* Blackburn, 1888.

Characteristics. Elongate-oval or oblong, a little flattened, glabrous, uniformly fuscous or with rufous areas on elytra. Length 2.5–5 mm. Antennomeres symmetrical, gradually enlarged apically. Base of pronotum completely but very finely margined. Intercoxal process of prosternum declivous behind coxae with small elevated apex, or uniformly feebly elevated.

Australian species. 12.

Distribution. Throughout Australia, Asia, Africa, North America.

Biology. At least two species, *U. pascoei* (Macleay, 1872) and *U. heroni* (Carter, 1929), are frequently found in nests of five species of finches, both as adults and larvae (Matthews & Bouchard 2008 and references therein). Two species, *U. australis* (Carter, 1932) and *U. xamiaphila* (Carter, 1920) were found in male cones of *Cycas* (Ornduff 1992).

Key to species. Carter (1918) (including *Platydema*).

Subfamily Stenochiinae Kirby, 1837

Characteristics. Eyes reniform. Labrum transverse. Antennae with scattered complex stellate sensoria on last 5–7 segments (Fig. 36.25 ss). Mesocoxal cavities open. Elytral striae nine plus scutellary striole, sometimes effaced. Wings usually present, rarely absent. Legs not fossorial. Claws simple. Dorsal edge of last abdominal ventrite with angular inner edge, or simply rounded.

Distribution. Stenochiinae are a pantropical lineage with groups of closely related genera that have evolved in the Neotropical, Asian and African regions. In Australia they clearly belong to the northern invasive component.

Australian taxa. Two tribes.

Tribe Cnodalonini Oken, 1843

Characteristics. Oblong or ovoid, convex or moderately flattened, glabrous, fuscous, piceous or often with brilliant metallic reflections, rarely tuberculate. Distance between eyes in dorsal view greater than width of one eye. Clypeolabral membrane concealed except in *Cuemus* Bouchard, 2000 and *Neotheca* Carter, 1930. Antennae not or barely reaching base of prothorax (reaching middle of elytra in *Cuemus*), segments gradually enlarging distally or rarely with weak 6- or 7-segmented club. Prosternum one to three times as long before coxae as postcoxal bridge. Elytral epipleura uusually incomplete, terminating opposite base of last ventrite, sometimes complete to apices, often excavated posteriorly. Tarsomeres parallel, cupuliform, or strongly lobed, their plantar surfaces with dense tomentum or with parallel rows of usually yellow setae. Apices of tibiae with similar setae. Apical tarsomere of hind legs subequal to or longer than first tarsal segment. [Bouchard & Yeates 2001].

Distribution. Pantropical. In Australia mostly in humid northern Queensland, although four genera have penetrated as far as TAS and the flightless *Hypaulax* Bates, 1868, *Oectosis* Pascoe, 1869 and *Omolipus* Pascoe, 1860 have adapted to more arid areas.

Australian taxa. 19 genera.

Biology. The immature stages of Cnodalonini develop in decaying wood (Bouchard & Yeates 2001). Pupae have defensive structures and behaviours described by Bouchard & Steiner (2004). Larvae of species that occur in arid habitats (*Hypaulax* and *Omolipus*) are also associated with dead or decaying woody tissue.

Androsus Gebien, 1920 (Pl. 85C; Fig. 36.179)

Type species. *Chariotheca violacea* Pascoe, 1887.

Characteristics. Parallel sided, smooth, with metallic green reflections. Length 4–5 mm. Prothorax transverse, about twice as wide as long, the anterior angles strongly produced. Hind edge of pronotum not margined. Prosternal process almost straight in profile, produced posteriorly and received in deep V-shaped concavity of mesoventrite. Elytral epipleura complete. Preapical tarsomeres lobed. Winged.

Australian species. Four.

Distribution. Northern NSW, eastern QLD, New Guinea, Asia.

Key to species. Kulzer (1951).

Apterotheca Gebien, 1920 (Pl. 85K; Fig. 36.183)

Type species. *Chariotheca besti* Blackburn, 1894.

Austropeus Carter, 1924

Caxtonana Buck, 1960

Characteristics. Smooth or strongly tuberculate or carinate on elytra, metallic green or bronze, or uniformly fuscous. Length 7–16 mm. Prosternal process arcuate or sinuate in profile and not produced. Hind edge of pronotum not margined. Epipleura incomplete. Tarsomeres simple. Male hind tibiae with apical teeth. Wingless.

Australian species. 44.

Distribution. Northern QLD.

Key to species. Bouchard (2002).

Bradymerus Perroud, 1864 (Pl. 86I; Fig. 36.187)

Type species. *B. tuberculatus* Perroud, 1864 (New Caledonia).

Isostira Pascoe, 1870.

Characteristics. Dorsal surfaces strongly tuberculate and carinate. Length 6–9 mm. Dorsal edges of eyes usually raised into low crests. Pronotum with sides explanate, lateral edges serrate or undulate, disc densely punctate with median longitudinal sulcus, the hind edge fully margined. Epipleura complete. Metaventral process bordered. Tarsomeres simple. Winged.

Australian species. Two: *B. crenatus* (Pascoe, 1870) and *B. raucipennis* (Blackburn, 1892).

Distribution. Northern QLD, New Guinea, Pacific islands, Asia, Africa.

Campolene Pascoe, 1863 (Pl. 87H; Fig. 36.182)

Type species. *C. nitida* Pascoe, 1863.

Characteristics. Smooth, legs rufous or fuscous contrasting with bluish-black body. Length 8–10 mm. Pronotal disc more convex anteriorly, the hind edge either not margined or with very fine margin, lateral edges with deep grooves. Prosternal process not produced, convex in lateral view. Tibiae distinctly bowed. Tarsomeres simple. Male hind tibiae with apical teeth. Wings reduced.

Australian species. Three.

Distribution. Northern NSW, South-eastern QLD.

Chariotheca Pascoe, 1860 (Pl. 88C; Fig. 36.180)

Type species. *C. coruscans* Pascoe, 1860 (Moluccas).
Characteristics. Oblong, smooth, metallic purple or green. Length 7–9 mm. Pronotum nearly as wide as elytral bases, in lateral view continuing outline of elytra, hind edge not margined. Prosternal process straight in profile, received in deep concavity of mesoventrite. Tarsomeres simple. Winged.
Australian species. Three.
Distribution. Northern QLD, New Guinea, Maluku.

Cuemus Bouchard, 2000 (Pl. 88E; Fig. 36.149)

Type species. *C. monteithi* Bouchard, 2000.
Characteristics. Subglobose, light to reddish brown. Length 4–5 mm. Head held vertical, hypognathous. Eyes small, separated by about three eye diameters. Clypeolabral membrane exposed, clypeus produced beyond front edges of genae. Length of prosternum before coxae equal to less than diameter of a coxa. Epipleura terminating before elytral apices. Tarsomeres simple. Wingless.
Australian species. Two: *C. cooki* Bouchard, 2000 and *C. monteithi*, closely related to two Fijian genera.
Distribution. Northern QLD, in rainforest at higher altitudes.

Derosphaerus Thomson, 1858 (Pl. 89D; Fig. 36.184)

Type species. *D. globicollis* Thomson, 1858 (Africa).
Encyalesthus Motschulsky, 1860
Pachyurgus Leconte, 1862
Cholipus Pascoe, 1866
Notiolesthus Motschulsky, 1872
Falsocyalesthus Pic, 1923
Characteristics. Smooth, prothorax subglobose, narrower than elytral bases, uniformly fuscous or piceous. Length 10–16 mm. Fold of pronotal disc and hypomeron in cross section forming obtuse angle or even curve, lateral margin fine and hind angles obsolescent, hind edge fully margined. Epipleura complete. Femora clavate. Tarsomeres simple. Winged.
Australian species. Seven.
Distribution. VIC, NSW, QLD, NT, New Guinea, Pacific islands, Asia, Africa.
Key to species. Kaszab (1987).

Espites Pascoe, 1882 (Pl. 100I; Fig. 36.193)

Type species. *E. basalis* Pascoe, 1882.
Characteristics. Smooth, pronotum blue, elytra with brilliant transverse violet and gold colour pattern. Length 5–6 mm. Prothorax more than twice as wide as long. Hind edge of pronotum fully margined. Prosternal process not produced. Tarsomeres simple. Winged.
Australian species. One: *E. basalis*.
Distribution. Northern QLD, New Guinea.

Hydissus Pascoe, 1869 (Pl. 91E; Fig. 36.190)

Type species. *Nyctobates feronioides* Pascoe, 1866.
Hydisus Gemminger & Harold, 1870
Characteristics. Smooth, fuscous. Length 11–16 mm. Hind edge of pronotum fully margined. Prosternal process not produced. Tarsomeres simple. Wingless.
Australian species. Two: *H. feronioides* and *H. vulgaris* (Olliff, 1889).
Distribution. Eastern QLD, Lord Howe Island.
Key to species. Carter (1914b).

Hypaulax Bates, 1868 (Pl. 10E, 91D; Fig. 36.188)

Type species. *H. marginata* Bates, 1868
Chileone Bates, 1868
Characteristics. Smooth, piceous. Length 12–26 mm. Mentum sometimes densely pilose. Gula nearly always with very deep transverse groove. Prosternal process nearly straight in profile, as wide as a procoxa. Hind edge of pronotum fully margined. Epipleura complete. Tarsomeres simple. Males with curved protibiae. Wingless.
Australian species. 16.
Distribution. Australian arid zone.
Note. *Hypaulax* is unusual among the Stenochiinae in that it has extensively adapted to arid conditions. It appears to have evolved polyphyletically from different species of the closely related mesophilic *Promethis* Pascoe, 1869 (Matthews & Bouchard 2008).
Key to species. Carter (1914b).

Kaszaba Matthews & Doyen, 1989 (Pl. 92K; Fig. 36.195)

Type species. *Tenebrio corvinus* Erichson, 1842.
Characteristics. Smooth, fuscous or with bluish or greenish areas. Length 6–8 mm. Prosternal process not produced. Lateral edges of pronotum simple, the disc evenly convex and finely punctate, hypomeron indistinctly or not punctate, hind edge fully margined. Epipleura incomplete. Preapical tarsomeres lobed. Winged. Anterior face of male profemur with small tomentose patch.
Australian species. Four.
Distribution. TAS, eastern Australia, NT.
Key to species. Matthews & Doyen (1989).

Metisopus Bates, 1873 (Pl. 94A; Fig. 36.181)

Type species. *M. purpureipennis* Bates, 1873.
Characteristics. Smooth, fuscous with bronze and violet reflections. Length 10–12 mm. Prosternal process almost straight in profile, produced posteriorly and received in deep V-shaped concavity of mesoventrite. Hind edge of

pronotum not margined. Surface of elytral apices with interstria 7 distinctly more convex than neighbouring interstriae, forming obtuse angle between dorsal surface and sides of elytra. Tarsomeres simple. Winged.

Australian species. Two: *M. curtulus* (Olliff, 1889) and *M. purpureipennis*.

Distribution. Lord Howe Island (*M. curtulus*) and Norfolk Island (*M. purpureipennis*).

Neotheca Carter, 1930 (Pl. 94I; Fig. 36.178)

Type species. *N. fusca* Carter, 1930 (New Guinea).

Characteristics. Smooth, fuscous. Length 7–9 mm. Clypeolabral membrane narrowly exposed, clypeus very short and crescentic in shape, its anterior edge broadly excised. Prosternal process not prolonged. Hind edge of pronotum fully margined. Elytral epipleura complete. Tarsomeres simple. Winged.

Australian species. One: *N. fusca*.

Distribution. North-east QLD, New Guinea.

Oectosis Pascoe, 1869 (Pl. 95L; Fig. 36.185)

Type species. *Upis cylindrica* Germar, 1848.

Characteristics. Smooth, pronotum longer than wide, fuscous. Length 7–10 mm. Prosternal process not prolonged. Fold of pronotal disc and hypomeron in cross section forming obtuse angle or even curve, lateral margins fine and hind angles obsolescent, hind edge fully margined. Elytral striae impressed, epipleura incomplete. Anterior edges of abdominal ventrites strongly and completely crenulated. Tarsomeres lobed. Winged.

Australian species. One: *O. cylindrica*.

Distribution. Southern WA, SA, VIC.

Omolipus Pascoe, 1860 (Pl. 95I; Fig. 36.186)

Type species. *O. corvus* Pascoe, 1860.

Characteristics. Smooth, pronotum about as long as wide, piceous or metallic blue. Length 6–16 mm. Prosternal process not prolonged. Fold of pronotal disc and hypomeron in cross section forming obtuse angle or even curve, lateral margins fine and hind angles obsolescent, hind edge fully margined. Elytra without, or with superficial, striae joining deep punctures, but those of *O. punctatostriatus* Carter, 1920 deeply striate, and those of *O. laevis* Pascoe, 1869 almost impunctate. Epipleura incomplete. Anterior edges of abdominal ventrites moderately crenulate at sides only. Tarsomeres lobed. Wingless.

Australian species. 16.

Distribution. NSW, QLD, NT, WA in semi-arid areas.

Promethis Pascoe, 1869 (Pl. 97E; Figs 36.1, 36.189)

Type species. *Upis angulata* Erichson, 1842.
Mederis Motschulsky, 1872
Setenis Motschulsky, 1872
Pediris Motschulsky, 1872
Pseudobates Fairmaire, 1882

Characteristics. Smooth, prothorax narrower than elytral bases, legs long, piceous. Length 15–28 mm. Mentum often with very long and dense pilosity. Prosternal process nearly straight in profile, as wide as a procoxa. Hind edge of pronotum fully margined. Epipleura complete. Tarsomeres simple. Male with curved protibiae, hind tibiae with apical teeth. Winged. [Kaszab 1988a, b].

Australian species. 14.

Distribution. Eastern Australia including TAS, New Guinea, Lord Howe Island, other Pacific islands, Asia.

Keys to species. Carter (1914b); Kaszab (1988a).

Scotoderus Perroud, 1864 (Pl. 98D; Figs 36.6, 36.191)

Type species. *Tenebrio (Iphthinus) cancellatus* Montrouzier, 1860 (New Caledonia).
Dechius Pascoe, 1866
Gonespites Gebien, 1920
Pelecypalpus Hinton, 1947

Characteristics. Smooth, fuscous. Length 7–9 mm. Prosternal process arched or sinuate in profile, not prolonged. Pronotum with deep basal transverse marginal sulcus, the margin usually interrupted medially, with deep sulci along lateral margins, anterior angles prominent. Metaventral process bordered. Tarsomeres simple. Winged.

Australian species. Three.

Distribution. North-eastern NSW to northern QLD, New Guinea, Pacific islands, South-east Asia.

Tetragonomenes Chevrolat, 1878 (Pl. 99A; Fig. 36.194)

Type species. *T. semiviridis* Chevrolat, 1878 (Moluccas).
Tetragonomecus Rye, 1880
Obriomaia Gebien, 1927

Characteristics. Smooth, fuscous with greenish reflections on elytra. Length 6–8 mm. Prosternal process arched or sinuate in profile, not produced. Lateral edges of pronotum uneven, slightly undulate or crenulate, the disc more convex anteriorly and coarsely punctate, hypomeron coarsely punctate, hind edge fully margined. Elytra with some intervals laterally carinate, epipleura incomplete. Metaventral process bordered. Tarsomeres simple. Winged.

Australian species. Five.

Distribution. TAS, VIC, NSW, QLD, South-east Asia, Pacific islands.

Key to species. Kulzer (1951).

Zophophilus Fairmaire, 1881 (Pl. 101D; Fig. 36.192)

Type species. *Z. curticornis* Fairmaire, 1881 (Neupommern, Bismarck Archipelago).
Sphenothorax Gebien, 1906
Teremenes Carter, 1914
Characteristics. Smooth, fuscous, Length 9–19 mm. Frons with a pair of short, shallow longitudinal grooves or rugae between eyes (except in *Z. socius* (Carter, 1914)). Prosternal process not produced. Posterior pronotal angles acute or produced, sometimes overlapping humeri. Hind edge of pronotum fully margined. Hypomeron more or less rugose. Elytral epipleura complete. Tarsomeres simple. Male with curved protibiae, hind tibiae with apical teeth. Winged.
Australian species. Six.
Distribution. Eastern Australia, TAS, NT, New Guinea, Borneo.
Key to species. Carter (1914b).

Tribe Stenochiini Kirby, 1837

Characteristics. Elongate-oblong, usually parallel-sided with forebody narrower than hind, convex, glabrous or rarely with adpressed thickened pale setae, colour and sculpturing variable. Clypeolabral membrane widely exposed, with lateral edges attached to corners of clypeus. Antennal segments elongate and subparallel, segments 7–11 a little wider and bearing numerous scattered stellate sensoria. Pronotum subquadrate with lateral margins not prominent, sometimes subdentate. Prosternum less than twice as long before coxae as postcoxal bridge. Elytral striation complete, epipleura complete to apices. Wings developed except in *Eutherama* Carter, 1914. Legs slender, tarsomeres elongate, cylindrical, no tomentum on apices of tibiae. Apical tarsomere of hind legs shorter than length of first tarsal segment. [Kaszab 1977b].
Distribution. Stenochiini, like Cnodalonini, are pantropical with only small numbers of species at higher latitudes. In Australia they are clearly recent northern invaders with species concentrated in the far north of the continent. Absent from New Caledonia, New Zealand and Vanuatu.
Australian taxa. Two genera.
Biology. Immature stages of Stenochiini develop in decaying wood.

Eutherama Carter, 1914 (Pl. 90C; Fig. 36.150)

Type species. *E. cyaneum* Carter, 1914.
Characteristics. Elongate oval, elytra blue or violet. Length 7–8 mm. Wingless, metaventrite along midline shorter than first abdominal ventrite.
Australian species. Two: *E. cyaneum* and *E. coeruleum* Carter, 1939.
Distribution. Northern WA, northern NT.

Strongylium Kirby, 1819 (Pl. 99F; Figs 36.25, 36.151)

Type species. *S. chalconotum* Kirby, 1819 (Brazil).
Anomoearthrum Mäklin, 1864
Ebenolus Fairmaire, 1897
Notostrongylium Carter, 1915
Microstrongylium Pic, 1917
Falsolophocnemis Pic, 1917
Bionesus sensu Kulzer, 1966
Characteristics. Subparallel, fuscous to piceous, nearly always with metallic reflections, sometimes with rufous areas on pronotum and/or elytra, rarely with rugose and sculptured integument. Length 8–15 mm. Eyes moderate in size to very large. Winged, metaventrite along midline longer than first abdominal ventrite.
Australian species. 27.
Distribution. VIC, eastern NSW, QLD, NT, northern WA, Torres Strait islands, some Pacific islands, pantropical.
Keys to species. Carter (1915a); Kulzer (1966); Kaszab (1977b).

Acknowledgments

The authors wish to thank Vratislav Bejšák for checking all references in the catalogue of tenebrionid names published in Matthews & Bouchard (2008) and uncovering several errors, also Patrice Bouchard (Agriculture and Agri-food Canada) for an opinion on the priority of *Pachycoelia* Boisduval. For information on *Zophobas* Dejean they thank Maxim Nabozhenko (Russian Academy of Sciences), Aaron Smith (Northern Arizona Universirty) and Warren Steiner (Smithsonian Institution). They are also most grateful to Cate Lemann and Anne Hastings (Australian National Insect Collection) for producing nearly all the excellent coloured habitus images.

Many illustrations have been published previously and the following copyright holders, as indicated in the figure captions, are thanked for giving permission to reproduce them here. The relevant artists and photographers involved (if not the senior author) are listed below for each institution: AZ – Annales Zoologici; DI – Dariusz Iwan; GC – Government of Canada (A. Davies, R. Idema, G. Sato); GT – G. Thompson; QM – Queensland Museum (G. Thompson); AM – The South Australian Museum (K. Ashby, K. Bowshall-Hill, J. Forrest, J. Thurmer, A. Tindall).

References

Aalbu RL, Kanda K, Smith AD (2017) Reinstatement of Eschatoporiini Blaisdell, 1906, a unique tribe of blind cavernicolous Tenebrionidae from California, with a new species from Napa County (Coleoptera, Tenebrionidae, Lagriinae). *ZooKeys* **688**, 135–149. doi:10.3897/zookeys.688.13575

Alekseev VI, Nabozhenko M (2015) A new fossil tenebrionid beetle of the tribe Palorini from Eocene Baltic amber. *The Coleopterists Society Monographs* **14**(4), 127–130.

Alekseev VI, Nabozhenko M (2017) *Palorus platycotyloides* sp. n., the second fossil representative of the tribe Palorini (Coleoptera: Tenebrionidae) from Baltic Amber. *Acta Zoologica Bulgarica* **69**(2), 167–170.

Allsopp PG (1986a) Identification of an additional false wireworm *Cestrinus trivialis* Erichson (Coleoptera: Tenebrionidae) from south Queensland. *Journal of the Australian Entomological Society* **25**, 137–140. doi:10.1111/j.1440-6055.1986.tb01093.x

Allsopp PG (1986b) Development, longevity and fecundity of the false wireworms *Pterohelaeus darlingensis* and *P. alternatus* (Coleoptera: Tenebrionidae) 4. Models for larval and pupal development under fluctuating temperatures. *Australian Journal of Zoology* **34**, 815–825. doi:10.1071/ZO9860815

Blackburn T (1907) Further notes on Australian Coleoptera with descriptions of new genera and species. Part XXXVII. *Transactions of the Royal Society of South Australia* **31**, 231–299.

Bouchard P (2002) Phylogenetic revision of the flightless Australian genus *Apterotheca* Gebien (Coleoptera: Tenebrionidae: Coelometopinae). *Invertebrate Systematics* **16**, 449–554. doi:10.1071/IT01029

Bouchard P, Gorb SN (2000) The elytra-to-body binding mechanism of the flightless rainforest species *Tabarus montanus* Kaszab (Coleoptera: Tenebrionidae). *Arthropod Structure & Development* **29**, 323–331. doi:10.1016/S1467-8039(01)00013-5

Bouchard P, Steiner WE, Jr (2004) First descriptions of Coelometopini pupae (Coleoptera: Tenebrionidae) from Australia, Southeast Asia and the Pacific region, with comments on phylogenetic relationships and antipredator adaptations. *Systematic Entomology* **29**, 101–114. doi:10.1111/j.1365-3113.2004.00230.x

Bouchard P, Yeates DK (2001) Phylogenetic relationships of the Australasian Coelometopini (Coleoptera: Tenebrionidae); a quantitative cladistic analysis with a review of biology. *Organisms, Diversity & Evolution* **1**, 17–43. doi:10.1078/1439-6092-00003

Bouchard P, Lawrence JF, Davies AE, Newton AF (2005) Synoptic classification of the world Tenebrionidae (Insecta: Coleoptera) with a review of family-group names. *Annales Zoologici* **55**(4), 499–530.

Bouchard P, Bousquet Y, Davies AE, Alonso-Zarazaga MA, Lawrence JF, Lyal CHC, Newton AF, Reid CAM, Schmitt M, Ślipiński A, Smith BT (2011) Family-group names in Coleoptera (Insecta). *ZooKeys* **88**, 1–972. doi:10.3897/zookeys.88.807

Bousquet Y, Bouchard P, Campbell JM (2015) Catalogue of genus-group names in Alleculinae (Coleoptera: Tenebrionidae). *The Coleopterists Society Monograph Number* **14**, 131–151.

Bremer HJ (1990) Eine neue *Corticeus*-Art aus Brasilien, sowie Anmerkungen zur Synonymie einiger orientalischer und papuanisch-australischer *Corticeus*-Art (Coleoptera, Tenebrionidae, Hypophloeini). *Entomofauna* **11**, 313–320.

Bremer HJ (1993) Neue Arten des Genus *Corticeus* Piller & Mitterpacher, 1783, aus der papuanisch-melanesischen Region. II. Mitteilung. *Entomofauna* **14**(32), 509–528.

Bremer HJ (2001a) Revision der Gattung *Amarygmus* Dalman, 1823 und verwandter Gattungen. I. Allgemeine Bemerkungen, Status einiger Gattungen *affine Amarygmus* Dalman, 1823; neue Kombinationen von Arten der Gattung *Amarygmus* Dalman (Coleoptera: Tenebrionidae: Alleculinae: Amarygmini). *Coleoptera* **5**, 57–80.

Bremer HJ (2001b) Revision of the Genus *Amarygmus* Dalman, 1823 and related genera. VI. Catalogue of already described species of *Amarygmus* Dalman (Coleoptera: Tenebrionidae: Amarygmini). *Coleoptera* **5**, 173–338.

Bremer HJ (2005a) Revision der Gattung *Amarygmus* Dalman, 1823 sowie verwandter Gattungen, XXVIII. Angaben zu *Amarygmus*-Arten, die von Fabricius, Weber, Wiedemann, Hope und Pascoe beschriebenen wurden. *Spixiana* **28**(1), 41–89.

Bremer HJ (2005b) Revision der Gattung *Amarygmus* Dalman, 1823 sowie verwandter Gattungen, XXXIV. Anmerkungen zu den Genera *Amarygmus* Dalman, *Becvaramarygmus* Masumoto, *Eumolpamarygmus* Pic, *Lobatopezus* Pic, *Oogeton* Kaszab und *Pyanirygmus* Pic (Insecta, Coleoptera, Tenebrionidae, Amarygmini, Chrysomelidae, Eumolpinae). *Spixiana* **28**(3), 199–221.

Bremer HJ (2006a) Revision der Gattung *Amarygmus* Dalman, 1823 sowie verwandter Gattungen. XXXVII. Nachbeschreibungen und Abbildungen australischer *Amarygmus*-Arten, die von Blackburn beschriebenen wurden (Insecta, Coleoptera, Tenebrionidae, Amarygmini). *Spixiana* **29**(1), 31–50.

Bremer HJ (2006b) Revision der Gattung *Amarygmus* Dalman, 1823 und verwandter Gattungen XXXIX. Eine neue Gattung der Amarygmini; Anmerkungen zu den Gattungen *Amarygmus* Dalman, *Hoplobrachium* Fairmaire, *Eupezoplonyx* Pic und *Eumolparagmarygmus* Bremer (Col; Tenebrionidae, Amarygmini). *Acta Coleopterologica* **XXII**(1), 5–13.

Bremer HJ (2013) The species of *Corticeus* Piller et Mitterpacher, 1783 of the Papuan faunal area – Part III (Coleoptera: Tenebrionidae, Hypophloeini). *Entomologische Zeitschrift* **123**(4), 251–255.

Bremer HJ (2015) The species of *Corticeus* Piller & Mitterpacher, 1783 of the Papuan faunal area. – Part IV (Coleoptera; Tenebrionidae; Hypophlaeini). *Mitteilungen der Münchner Entomologischen Gesellschaft* **105**, 5–39.

Bremer HJ, Lillig M (2014) World catalogue of Amarygmini, Rhysopaussini and Falsocossyphini (Coleoptera: Tenebrionidae). *Münchner entomologische Gesellschaft* **104**(Suppl.), 3–176.

Bremer HJ, Lillig M (2017) Remarks on the Genera of Hypophlaeini and the Subgenera of *Corticeus* Piller & Mitterpacher, 1783, with descriptions of new species of *Corticeus* of the Oriental Region (Coleoptera: Tenebrionidae, Hypophlaeini. *Entomologische Zeitschrift* **127**(2), 67–75.

Britton EB (1973) *Pseudeba watsoni* sp. n. (Coleoptera: Tenebrionidae: Ulominae) from the food stores of harvester termites. *Journal of the Australian Entomological Society* **12**, 230–232. doi:10.1111/j.1440-6055.1973.tb01664.x

Brown WV, Doyen JT, Moore BP, Lawrence JF (1992) Chemical composition and taxonomic significance of defensive secretions of some Australian Tenebrionidae (Coleoptera). *Journal of the Australian Entomological Society* **31**, 79–89. doi:10.1111/j.1440-6055.1992.tb00461.x

Byrne M, Yeates DK, Joseph L, Kearney M, Bowler J, Williams MAJ, Cooper S, Donellan SC, Keogh JS, Leys R, Melville J, Murphy DJ, Porch N, Wyrwoll K-H (2008) Birth of a biome: insights into the assembly and maintenance of the Australian arid zone biota. *Molecular Ecology* **17**, 4398–4417. doi:10.1111/j.1365-294X.2008.03899.x

Carter HJ (1908a) Revision of the Australian species of *Adelium*. *Proceedings of the Linnean Society of New South Wales* **33**, 257–285.

Carter HJ (1908b) Revision of the genus *Seirotrana*, together with descriptions of new species of other Australian Coleoptera. *Proceedings of the Linnean Society of New South Wales* **33**, 392–422.

Carter HJ (1910) Revision of *Sympetes* and *Helaeus*: with descriptions of new species of Tenebrionidae (Coleoptera). *Proceedings of the Linnean Society of New South Wales* **35**, 77–134. doi:10.5962/bhl.part.25544

Carter HJ (1911a) Revision of *Pterohelaeus* (continued) and of *Saragus*; with descriptions of new species of Australian Tenebrionidae. *Proceedings of the Linnean Society of New South Wales* **36**, 179–223.

Carter HJ (1911b) Revision of the Nyctozoilides – genera and species (Fam. Tenebrionidae). *Annals of the Queensland Museum* No. 10, 136–166.

Carter HJ (1913a) Revision of Australian species of the subfamilies Cyphaleinae and Cnodaloninae. *Proceedings of the Linnean Society of New South Wales* **38**, 61–105. doi:10.5962/bhl.part.13554

Carter HJ (1913b) Notes and tabulation of the Australian Amarygminae (Family Tenebrionidae), with descriptions of new species. *Transactions of the Royal Society of South Australia* **37**, 6–47.

Carter HJ (1914a) Notes on Australian Tenebrionidae, with descriptions of new species. *Transactions of the Royal Society of South Australia* **38**, 369–406.

Carter HJ (1914b) Revision of the subfamily Tenebrioninae, Family Tenebrionidae. *Proceedings of the Linnean Society of New South Wales* **39**, 44–86. doi:10.5962/bhl.part.2275

Carter HJ (1915a) The Australian Strongyliinae and other Tenebrionidae, with descriptions of new genera and species. *Proceedings of the Linnean Society of New South Wales* **40**, 522–539. doi:10.5962/bhl.part.18883

Carter HJ (1915b) Revision of the Australian Cistelidae. *Proceedings of the Royal Society of Victoria* **28**, 52–104.

Carter HJ (1918) Some new Heteromera, and one *Stigmodera* from Tropical Australia. *Proceedings of the Linnean Society of New South Wales* **42** [1917], 701–719.

Carter HJ (1920) Notes on some Australian Tenebrionidae, with descriptions of new species; – also of a new genus and species of Buprestidae. *Proceedings of the Linnean Society of New South Wales* **45**, 222–249. doi:10.5962/bhl.part.19543

Carter HJ (1921) Australian Coleoptera: notes and new species. *Proceedings of the Linnean Society of New South Wales* **46**, 301–323. doi:10.5962/bhl.part.14020

Carter HJ (1925) Revision of the Australian species of *Chrysobothris* (Fam. Buprestidae), together with notes and descriptions of new species of Coleoptera. *Proceedings of the Linnean Society of New South Wales* **50**, 225–244.

Carter HJ (1927) Australian Coleoptera: notes and new species. No. v. *Proceedings of the Linnean Society of New South Wales* **52**, 222–234.

Carter HJ (1930) Check list of the Australian Cistelidae. Order Coleoptera. *Australian Zoologist* **6**, 269–276.

Carter HJ (1932) New Guinea and Australian Coleoptera. Notes and new species. No. ii. *Proceedings of the Linnean Society of New South Wales* **57**, 101–115.

Carter HJ (1933) Australian Coleoptera. Notes and new species. viii. *Proceedings of the Linnean Society of New South Wales* **58**, 159–180.

Carter HJ (1935) Australian Coleoptera. Notes and new species. ix. *Proceedings of the Linnean Society of New South Wales* **60**, 179–193.

Carter HJ (1937) Some new Tenebrionidae in the South Australian Museum, together with notes and descriptions of other Australian Coleoptera. *Transactions of the Royal Society of South Australia* **61**, 121–144.

Champion GC (1895) On the Heteromerous Coleoptera collected in Australia and Tasmania by Mr. James J. Walker, R.N., F.L.S., during the voyage of H.M.S. "Penguin", with descriptions of new genera and species. Part ii. *The Transactions of the Entomological Society of London* **1895**, 213–275.

Doyen JT (1972) Familial and subfamilial classification of the Tenebrionidae (Coleoptera) and revised generic classification of the Coniontini. *Quaestiones Entomologicae* **8**, 357–376.

Doyen JT (1984) Reconstitution of the Diaperini of North America, with new species of *Adelina* and *Sitophagus* (Coleoptera: Tenebrionidae). *Proceedings of the Entomological Society of Washington* **86**, 777–789.

Doyen JT (1994) Cladistic relationships among Pimeliine Tenebrionidae (Coleoptera). *Journal of the New York Entomological Society* [1993] **101**, 443–514.

Doyen JT, Lawrence JF (1979) Relationships and higher classification of some Tenebrionidae and Zopheridae (Coleoptera). *Systematic Entomology* **4**, 333–377. doi:10.1111/j.1365-3113.1979.tb00619.x

Doyen JT, Tschinkel WR (1982) Phenetic and cladistic relationships among tenebrionid beetles (Coleoptera). *Systematic Entomology* **7**, 127–183. doi:10.1111/j.1365-3113.1982.tb00129.x

Doyen JT, Matthews EG, Lawrence JF (1989) Classification and annotated checklist of the Australian genera of Tenebrionidae (Coleoptera). *Invertebrate Taxonomy* **3**, 229–260. doi:10.1071/IT9890229

Eisner T, Aneshansley D, Eisner M, Rutowski R, Chong B, Meinwald J (1974) Chemical defense and sound production in Australian tenebrionid beetles (*Adelium* spp.). *Psyche* **81**, 1189–1208.

Ferrer J (2006) Description d'un genre nouveau et notes synonymiques et systématiques sur les genres *Alobates* MOTSCHOULSKY, 1872 et *Acanthobas* GEBIEN, 1928 (Coleoptera, Tenebrionidae, Tenebrionini). *Zeitschrift für Entomologie* **27**(19), 229–240.

Ferrer J (2008) Les espèces australiennes du genre *Mesomorphus* Miedel, 1880, et description d'une nouvelle espèce d'Afrique du Sud (Coleoptera, Tenebrionidae, Opatrini). *Bulletin de la Société Entomologique de France* **113**(3), 287–300.

Ferrer J (2011) Revisión del género *Zophobas* Dejean, 1834 (Coleoptera, Tenebrioninae, Tenebrionini). *Boletín de la Sociedad Entomológica Aragonesa (S.E.A.)* **48**, 287–319.

Grandcolas P, Murienne J, Robiliard T, Desutter-Grandcolas L, Jourdan H, Guilbert E, Deharveng L (2008) New Caledonia: a very old Darwinian island? *Philosophical Transactions of the Royal Society of London (B). Biological Sciences* **363**, 3309–3317. doi:10.1098/rstb.2008.0122

Grimm R (2011) Two new species of the genus *Triplehornia* Matthews & Lawrence from Borneo and the Philippines (Coleoptera: Tenebrionidae). *Stuttgarter Beiträge zur Naturkunde A. Neue Serie* **4**, 259–261.

Grimm R (2013) The genus *Cryphaeus* Klug, 1833 in Australia, with description of a new species (Coleoptera: Tenebrionidae: Toxicini). *Australian Entomologist* **40**(4), 243–247.

Gunter NL, Levkaničová Z, Weir TH, Ślipiński A, Stephen L, Cameron SL, Bocak L (2014) Towards a phylogeny of the Tenebrionoidea (Coleoptera). *Molecular Phylogenetics and Evolution* **79**, 305–312. doi:10.1016/j.ympev.2014.05.028

Halstead DGH (1967) A revision of the genus *Palorus* (*sens. lat.*) (Coleoptera: Tenebrionidae). *Bulletin of the British Museum (Natural History). Historical Series* **19**, 59–148. doi:10.5962/bhl.part.28299

Hinton HE (1948) A synopsis of the genus *Tribolium* Macleay, with some remarks on the evolution of its species-groups (Coleoptera, Tenebrionidae). *Bulletin of Entomological Research London* **39**, 13–55. doi:10.1017/S0007485300024287

Iwan D (2002) Catalogue of the world Platynotini (Coleoptera: Tenebrionidae). *Genus* **13**, 219–323.

Iwan D, Kamiński MJ (2016) Toward a natural classification of opatrine darkling beetles: comparative study of female terminalia. *Zoomorphology* doi:10.1007/s00435-016-0328-5.

Iwan D, Matthews EG (2015) *Scleropatroides* Löbl and Merkl (Coleoptera: Tenebrionidae: Opatrini) discovered in Australia, with description of a new species and discussion of related genera. *The Coleopterists Society Monograph Number* **14**, 115–121.

Iwan D, Ferrer J, Raś M (2010) Catalogue of the World *Gonocephalum* Solier, 1834 (Coleoptera, Tenebrionidae, Opatrini). Part 1. List of the species and subspecies. *Annales Zoologici* **60**(2), 245–304. doi:10.3161/000345410X516920

Iwan D, Merkl O, Ferrer J (2011) Catalogue of the world *Gonocephalum* Solier, 1834 (Coleoptera: Tenebrionidae: Opatrini). Part 2. Comments, additions and references. *Annales Zoologici* **61**(2), 259–276. doi:10.3161/000345411X584762

Iwan D, Kamiński MJ, Raś M (2015) Amphi-Indian Ocean disjunction in the trans-Pacific genus *Archaeoglenes* Broun (Coleoptera: Tenebrionidae: Phrenapatinae): New taxonomic and distributional data. *The Coleopterists Society Monograph Number* **14**, 152–166.

Kanda K (2016) *Grabulax* gen. nov *darlingtoni* sp. nov. (Tenebrionidae: Lagriinae: Laenini) from Colombia, with reevaluation of New World Laenini. *Annales Zoologici* **66**(4), 557–562. doi:10.3161/00034541ANZ2016.66.4.007

Kanda K, Pflug JM, Sproul JS, Dasenko MA, Maddison DR (2015) Successful recovery of nuclear protein-coding genes from small insects in museums using illumina sequencing. *PLoS One* **10**(12), e0143929 doi:10.1371/journal.pone.0143929.

Kaszab Z (1946) Monographie der Leiochrinen. *Naturwissenschaftliche Monographien* **3**. Ungarisches naturwissenschaftliches Museum, Budapest.

Kaszab Z (1961a) Neue Arten der Gattung *Leiochrodes* Westwood (Coleoptera: Tenebrionidae). *Acta Zoologica Academiae Scientiarum Hungaricae* **7**, 433–466.

Kaszab Z (1961b) Revision der tenebrioniden-Gattung *Derispia* Lewis (Coleoptera). *Acta Zoologica Academiae Scientiarum Hungaricae* **7**, 139–184.

Kaszab Z (1961c) Beiträge zur Kenntnis der Tenebrioniden-Tribus Leiochrini (Coleoptera). *Annales Historico-Naturales Musei Nationalis Hungarici* **53**, 357–380.

Kaszab Z (1975) Die Arten der Tenebrionidae-Gattung *Microcrypticus* Gebien 1920 (Coleoptera). *Folia Entomologica Hungarica* **XXVIII**(1), 99–104.

Kaszab Z (1977a) Die Phrenapatinen des papuanisch-pazifischen Gebietes (Coleoptera: Tenebrionidae). *Acta Zoologica Academiae Scientiarum Hungaricae* **23**, 299–339.

Kaszab Z (1977b) Die Tenebrioniden des papuanischen Gebietes. I. Strongyliini (Coleoptera: Tenebrionidae). *Pacific Insects Monographs* **33**, 1–219.

Kaszab Z (1978) Australische und südpazifische Tenebrioniden der Tribus Phrenapatini und Gnathidiini (Coleoptera) sowie synonymische Bemerkungen. *Annales Historico-Naturales Musei Nationalis Hungarici* **70**, 163–177.

Kaszab Z (1982a) Die papuanisch-pazifischen Arten der Gattung *Archeoglenes* Broun, 1893 (Coleoptera: Tenebrionidae). *Acta Zoologica Academiae Scientiarum Hungaricae* **28**, 45–55.

Kaszab Z (1982b) Revision der australischen *Uloma*-Arten (Coleoptera: Tenebrionidae). *Acta Zoologica Academiae Scientiarum Hungaricae* **28**, 233–291.

Kaszab Z (1982c) Die Gattung *Iscanus* Fauvel (Coleoptera, Tenebrionidae) aus dem Pazifik. *Folia Entomologica Hungarica* **XLIII**(1), 59–62.

Kaszab Z (1982d) Die Tenebrioniden Neucaledoniens und der Loyauté-Inseln (Coleoptera). *Folia Entomologica Hungarica* **XLIII**(2), 1–294.

Kaszab Z (1982e) Die papuanisch-pazifischen Arten der Gattung *Lorelus* Sharp, 1876 (Coleoptera, Tenebrionidae). *Annales Historico-Naturales Musei Nationalis Hungarici* **74**, 151–191.

Kaszab Z (1987) Die papuanisch-australischen Arten der Gattung *Derosphaerus* Thomson, 1858 (Coleoptera: Tenebrionidae). *Acta Zoologica Hungarica* **33**, 41–85.

Kaszab Z (1988a) Katalog und Bestimmungstabelle der Gattung *Promethis* Pascoe, 1869 (Coleoptera, Tenebrionidae). *Acta Zoologica Academiae Scientiarum Hungaricae* **34**, 67–170.

Kaszab Z (1988b) Faunistische Angaben der Gattung *Promethis* Pascoe, 1869 (Coleoptera, Tenebrionidae). *Folia Entomologica Hungarica* **XLIX**, 55–116.

Kergoat GJ, Bouchard P, Clamens A-L, Abbate JL, Jourdan H, Jabdour-Zahab R, Genson G, Soldati L, Condamine L (2014a) Cretaceous environmental changes led to high extinction rates in a hyperdiverse beetle family. *BMC Evolutionary Biology* **14**, 220 doi:10.1186/s12862-014-0220-1.

Kergoat GJ, Soldati G, Clamens A-L, Jourdan H, Jabdour-Zahab R, Genson G, Bouchard P, Condamine FL (2014b) Higher level molecular phylogeny of darkling beetles (Coleoptera: Tenebrionidae). *Systematic Entomology* **39**, 486–499 doi:10.1111/syen.12065.

Kolbe HJ (1896) Bericht über die wissenschaftlichen Leistungen im Gebiete der Entomologie während des Jahres 1895. Coleoptera. *Archiv für Naturgeschichte* **62**, 427–654.

Kompantseva TV (1999) Larvae of tenebrionid beetles of the tribe Toxicini and notes on the systematic position of this group in the family Tenebrionidae (Coleoptera). *Entomologicheskoe Obozrenie* **78**, 636–647.

Kulzer H (1951) Fünfter Beitrag zur Kenntnis der Tenebrioniden. *Entomologische Arbeiten aus dem Museum Georg Frey* **2**, 461–573.

Kulzer H (1964) Über neue Tenebrionidenarten (Col.). 27. Beitrag zur Kenntnis der Tenebrioniden. *Entomologische Arbeiten aus dem Museum Georg Frey* **15**, 221–276.

Kulzer H (1966) Australische und papuanische Strongyliini. 29. Beitrag zur Kenntnis der Tenebrioniden. *Entomologische Arbeiten aus dem Museum Georg Frey* **17**, 338–396.

Lawrence JF, Britton EB (1994) *Australian Beetles*. Melbourne University Press, Carlton.

Lawrence JF, Medvedev GS (1982) A new tribe of darkling beetles (Coleoptera: Tenebrionidae) from Australia and its systematic position. [Tribal name to be attributed to Medvedev and Lawrence]. *Entomologicheskoye Obozreniye* **61**(3), 548–571 (in Russian; English translation in *Entomological Review* **61**(3), 85–107).

Lawrence JF, Ślipiński A (2013) *Australian beetles. Volume 1: morphology, classification and keys*. CSIRO Publishing, Melbourne.

Lea AM (1912) Descriptions of new species of Australian Coleoptera. Part ix. *Proceedings of the Linnean Society of New South Wales* **34** [1911], 426–478.

Lea AM (1914) Notes on some miscellaneous Coleoptera, with descriptions of new species. *Transactions of the Royal Society of South Australia* **38**, 249–344.

Lea AM (1916) Notes on some miscellaneous Coleoptera, with descriptions of new species. Part ii. *Transactions of the Royal Society of South Australia* **40**, 272–436.

Lea AM (1919) Notes on some miscellaneous Coleoptera, with descriptions of new species. Part v. *Transactions of the Royal Society of South Australia* **43**, 166–261.

Lea AM (1930) On Australian Coleoptera. Part vi. *Records of the South Australian Museum* **4**, 243–273.

Lea AM (1931) Australian Coleoptera. *Records of the South Australian Museum* **4**, 365–408.

Leschen RAB, Cutler B (1994) Cuticular calcium in beetles (Coleoptera: Tenebrionidae: Phrenapetinae). *Annals of the Entomological Society of America* **87**, 918–921. doi:10.1093/aesa/87.6.918

Lucas R (1920) *Catalogus alphabeticus generum et subgenerum Coleopterorum orbis terrarum totius*. R. Stricker, Berlin.

Masumoto K, Grimm R (2004) A new genus and species of the Palorinae (Coleoptera: Tenebrionidae) from Japan. *Entomological Review of Japan* **59**, 127–130.

Matthews EG (1975) The Mediterranean beetle *Blaps polychresta* Forskål in South Australia (Tenebrionidae). *South Australian Naturalist* **49**, 35–39.

Matthews EG (1986) Revision of the troglophilic genus *Brises* Pascoe, with a discussion of the Cyphaleini (Coleoptera, Tenebrionidae). *Records of the South Australian Museum* **19**, 77–90.

Matthews EG (1992) Classification, relationships and distribution of the genera of Cyphaleini (Coleoptera: Tenebrionidae). *Invertebrate Taxonomy* **6**, 437–522. doi:10.1071/IT9920437

Matthews EG (1993) Classification, relationships and distribution of the genera of Heleini (Coleoptera: Tenebrionidae). *Invertebrate Taxonomy* **7**, 1025–1095. doi:10.1071/IT9931025

Matthews EG (1998) Classification, phylogeny and biogeography of the genera of Adeliini (Coleoptera: Tenebrionidae). *Invertebrate Taxonomy* **12**, 685–824. doi:10.1071/IT97008

Matthews EG (2000) Origins of Australian arid-zone tenebrionid beetles. *Invertebrate Taxonomy* **14**, 941–951. doi:10.1071/IT00021

Matthews EG (2003a) *Ulomotypus* Broun, a member of the new subfamily Palorinae, with remarks on *Aphtora* Bates and *Demtrius* Broun (Coleoptera, Tenebrionidae). *New Zealand Entomologist* **26**, 7–14. doi:10.1080/00779962.2003.9722104

Matthews EG (2003b) *Aoupinia*, a remarkable new genus of Adeliini from New Caledonia (Coleoptera: Tenebrionidae). *Memoirs of the Queensland Museum* **49**(1), 441–445.

Matthews EG (2012) Australian Alleculinae: new genera, new combinations, and a lectotype designation (Coleoptera: Tenebrionidae). *Psyche* **2012**, 814865.

Matthews EG, Bouchard P (2008) *Tenebrionid Beetles of Australia. Descriptions of Tribes, Keys to Genera, Catalogue of Species*. Australian Biological Resources Study, Canberra.

Matthews EG, Doyen JT (1989) A reassessment of the Australian species of *Menephilus* Mulsant (Coleoptera: Tenebrionidae) with descriptions of two new genera and a larva and pupa. *Records of the South Australian Museum* **23**, 39–50.

Matthews EG, Lawrence JF (1992) A new genus and species of Heleini from Tasmania (Coleoptera: Tenebrionidae). *Journal of the Australian Entomological Society* **31**, 311–316. doi:10.1111/j.1440-6055.1992.tb00513.x

Matthews EG, Lawrence JF (2005) New taxa, new synonymy and new generic records for Australian Tenebrionidae (Coleoptera). *Annales Zoologici* **55**(4), 531–547.

Matthews EG, Lawrence JF (2015) Trachelostenini *sensu novo*: redescriptions of *Trachelostenus* Solier, *Myrmecodema* Gebien and *Leaus* Matthews & Lawrence, based on adults and larvae, and descriptions of three new species of *Leaus* (Coleoptera: Tenebrionidae). *Zootaxa* **4020**(2), 289–312. doi:10.11646/zootaxa.4020.2.4

Matthews EG, Merkl O (2015) *Hangaya enigmatica*, a new genus and species of Tenebrionidae from Central Australia (Coleoptera). *Annales Zoologici* **65**(3), 479–482. doi:10.3161/00034541ANZ2015.65.3.004

Matthews EG, Queale LF (1997) Littoral insects (Phylum Arthropoda: Class Insecta). In *Marine invertebrates of southern Australia. Part III.* (Eds SA Shepherd and M Davies) pp. 1073–1091. South Australian Research and Development Institute, Adelaide.

Matthews EG, Scupola A (2003) Entomological investigations in Australia by the Natural History Museum of Turin: Coleoptera, Tenebrionidae. *Museo Regionale di Scienze Naturali Monografie* **35**, 281–302.

Matthews EG, Lawrence JF, Bouchard P, Steiner WE, Jr, Ślipiński A (2010) 11.14. Tenebrionidae Latreille, 1802. In *Handbuch der Zoologie/Handbook of Zoology. Band/Volume IV Arthropoda: Insecta Teilband/Part 38. Coleoptera, Beetles. Volume 2. Morphology and Systematics (Polyphaga partim).* (Eds RAB Leschen, RG Beutel and JF Lawrence) pp. 574–659. W de Gruyter, Berlin.

McKenna DD, Wild AL, Kanda K, Bellamy CL, Beutel RG, Caterino MS, Farnum CW, Hawks DC, Ivie MA, Jameson ML, Leschen RAB, Marvaldi AE, McHugh JV, Newton AF, Robertson JA, Thayer MK, Whiting MF, Lawrence JF, Ślipiński A, Maddison DR, Farrell BD (2015) The beetle tree of life reveals that Coleoptera survived end-Permian mass extinction to diversify during the Cretaceous terrestrial revolution. *Systematic Entomology* **40**, 835–880 doi:10.1111/syen.12132.

Medvedev GS (1973) Position of the genera *Leichenum* Dej. and *Idisia* Pasc. (Coleoptera, Tenebrionidae) in the system and a description of a new genus from northern Karakums. *Entomologicheskoe Obozrenie* **52**, 644–650 (in Russian; English translation in *Entomological Review* **52**, 428–433).

Medvedev GS (1977) Taxonomic significance of the antennal sensilla of darkling beetles (Coleoptera, Tenebrionidae). *Trudy Vsesoyuznogo Entomologicheskogo Obshchestva* **58**, 61–86 [in Russian].

Medvedev GS (1995) New species of darkling beetles of the genus *Stenosida* Sol. (Coleoptera, Tenebrionidae) from Australia and Indonesia. *Trudy Zoologicheskogo Instituta RAN* **258**, 51–58 [in Russian].

Medvedev GS, Lawrence JF (1983) Beetles of the tribe Hyocini (Coleoptera, Tenebrionide) of Australia. I. Subgenera *Nannohyocis* subgen. n. and *Neohyocis* subgen. n. (genus *Hyocis* Pasc.). *Entomologicheskoe Obozrenie* **62**(3), 569–582 (in Russian; English translation in *Entomological Review* **62**(3), 117–130).

Medvedev GS, Lawrence JF (1984) Beetles of the tribe Hyocini (Coleoptera, Tenebrionidae) of Australia. II. Genera *Parahyocis* Kasz. and *Csiro* gen. n. *Entomologicheskoe Obozrenie* **63**(3), 561–581 (in Russian; English translation in *Entomological Review* **63**(3), 129–148).

Medvedev GS, Lawrence JF (1986) Tenebrionidae of the tribe Hyocini (Coleoptera, Tenebrionidae) of Australia. III. Brittonina and Uptonina, new subtribes. *Entomologicheskoe Obozrenie* **65**(3), 574–591 (in Russian; English translation in *Entomological Review* **65**(3), 161–179).

Medvedev GS, Lawrence JF (1987) Tenebrionid beetles of the tribe Hyocini (Coleoptera, Tenebrionidae) of Australia. IV. A key to genera and subgenera. *Entomologicheskoe Obozrenie* **66**(4), 803–816 (in Russian; English translation in *Entomological Review* **67**(3) [1988], 159–171).

Merkl O (1986) A review of the Australian species of the subtribe Statirina (Coleoptera, Tenebrionidae: Lagriini). *Annales Historico-Naturales Musei Nationalis Hungarici* **78**, 187–199.

Merkl O (1987) A review of the Australian species of the subtribe Lagriina (Coleoptera, Tenebrionidae: Lagriini). *Annales Historico-Naturales Musei Nationalis Hungarici* **79**, 121–166.

Merkl O (1992) Tenebrionidae (Coleoptera) from Laos and Vietnam, with reclassification of Old World "*Doliema*". *Acta Zoologica Hungarica* **38**, 261–280.

Merkl O, Kompantseva TV (1996) Old World *Rhipidandrus* LeConte: synonymies, faunistics, identification key and description of two new species from Australia (Coleoptera: Tenebrionidae). *Acta Zoologica Academiae Scientiarum Hungaricae* **42**, 89–109.

Neboiss A (1962) Notes on distribution and descriptions of new species (Orders: Odonata, Plecoptera, Orthoptera, Trichoptera and Coleoptera). *Memoirs of the National Museum of Victoria* **25**, 243–257. doi:10.24199/j.mmv.1962.25.13

Ornduff R (1992) Features of coning and foliar phenology, size classes, and insect associates of *Cycas armstrongi* (Cycadaceae) in the Northern Territory, Australia. *Bulletin of the Torrey Botanical Club* **119**, 39–43. doi:10.2307/2996918

Schawaller W (1997) The genus *Spiloscapha* Bates (Coleoptera: Tenebrionidae), with descriptions of new species from the Oriental and Papuan Regions. *Stuttgarter Beiträge zur Naturkunde* **566**, 1–15.

Schawaller W (2004) New species and records of the genus *Spiloscapha* Bates (Coleoptera: Tenebrionidae) from the Oriental and Papuan Regions. *Stuttgarter Beiträge zur Naturkunde* **661**, 1–10.

Scupola A (2000) Revisione della tribu Cossyphini Latreille, 1802. Parte I. Introduzione e genere *Cossyphus* Olivier, 1791 (Coleoptera, Tenebrionidae). *Atti del Museo Civico di Storia Naturale di Trieste* **48**, 185–249.

Scupola A (2002) A proposito di *Ulomina carinata* Baudi di Selve, 1876 (Coleoptera, Tenebrionidae). *Bollettino del Museo Regionale di Scienze Naturali di Torino* **19**, 185–189.

Tschinkel WR, Doyen JT (1980) Comparative anatomy of the defensive glands, ovipositors and female genital tubes of tenebrionid beetles (Coleoptera). *International Journal of Insect Morphology & Embryology* **9**, 321–368. doi:10.1016/0020-7322(80)90009-4

Vidal G-HP, Guerrero GM (2007) *Los Tenebriónidos de Chile*. Ediciones Universidad Católica de Chile, Santiago.

Watt JC (1975) A revised subfamily classification of Tenebrionidae (Coleoptera). *New Zealand Journal of Zoology* **1** [1974], 381–452.

Watt JC (1989) The identity of two Fabrician species of Amarygmini (Coleoptera: Tenebrionidae) from Australia with a key to species groups and some species of *Chalcopteroides* Strand. *Journal of the Australian Entomological Society* **28**, 115–123. doi:10.1111/j.1440-6055.1989.tb01207.x

Watt JC (1992) Relationships of *Actizeta* and Cnemeplatiini (Coleoptera: Tenebrionidae). *Systematic Entomology* **17**, 287–299. doi:10.1111/j.1365-3113.1992.tb00339.x

Addendum

While this chapter was in preparation the homonym *Agasthenes* Bates, 1873 was replaced by *Batessia* Ponting, 2018.

Ponting J (2018) A replacement name for *Agasthenes* Bates, 1873 (Coleoptera: Tenebrionidae), junior homonym of *Agasthenes* Förster, 1869 (Hymenoptera: Ichneumonidae). *Records of the Western Australian Museum* **33**, 131–132.

Plate 1. Beetles in their natural habitats. A, *Omma stanleyi* Newman (Ommatidae) © Owen Kelly; B, *Distocupes varians* (Lea) (Cupedidae) © Kristi Ellingsen; C, *Cybister tripunctatus* Olivier (Dytiscidae) © Kristi Ellingsen; D, *Rhantus suturalis* (McLeay) (Dytiscidae) © Kristi Ellingsen; E, *Leoglymmius lignarius* (Oliff) (Carabidae: Rhysodini) © Simon Grove; F, *Calosoma (Australodrepa) schayeri* Erichson (Carabidae: Carabini) © Jiří Lochman; G, *Mecynognathus damelii* Macleay (Carabidae) © Owen Kelly; H, Carabidae: *Pamborus alternans* Latreille (Carabidae) © Zhenhua Liu.

Plate 2. Beetles in their natural habitats. Carabidae (A–G), Dytiscidae (H). A, *Scaraphites lenaeus* Westwood (Scaritini); B, *Philoscaphus tuberculatus* (Macleay) (Scaritini); C, *Craspedophorus insignis* (Schaum) (Panagaeini); D, *Megacephala castelnaui* Sloane (Cicindelini); E, *Carenum* sp. (Scaritini); F, *Pheropsophus verticalis* (Dejean) (Brachinini); G, Carabidae: *Eurylychnus blagravei* (Laporte) (Broscini); H, *Eretes australis* (Erichson). A–G © Jiří Lochman; H, © Nick Monaghan.

Plate 3. Beetles in their natural habitats. Scarabaeidae (A–D, H), Lucanidae (E-G). A, *Anoplognathus montanus* Macleay (Rutelinae); B, *Anoplognathus pallidicollis* Blanchard, male (Rutelinae); C, *Microvalgus* sp. (Cetoniinae); D, *Neorrhina punctatum* (Donovan) (Cetoniinae); E, *Ceratognathus* sp.; F, *Lissotes furcicornis* Westwood; G, *Syndesus cornutus* (Fabricius); H, *Anoplognathus* sp. (Rutelinae). A–D, F © Nick Monaghan; E, G, H © Zhenhua Liu.

Plate 4. Beetles in their natural habitats. Scarabaeidae (A), Geotrupidae (B), Lucanidae (C–H). A, *Liparetrus* sp. (Melolonthinae); B, *Elephastomus proboscideus* (Schreibers) (Bolboceratinae); C, *Ceratognathus westwoodi* Thomson; D, *Lissapterus obesus* Bowmans; E, *Ryssonotus nebulosus* (Kirby); F, *Ryssonotus nebulosus* (Kirby), pupa; G, *Ryssonotus nebulosus* (Kirby), larva; H, *Prosopocoilus torresensis* (Deyrolle), minor male. A, © Nick Monaghan; B, D–H © Owen Kelly; C, © Simon Grove.

Plate 5. Beetles in their natural habitats. A, *Macrohelodes* sp. (Scirtidae); B, *Repsimus manicatus* (Swartz) (Scarabaeidae: Rutelinae); C, *Hydrophilus* sp. (Hydrophilidae); D, *Aulacocyclus* sp. (Passalidae); E, *Chlamydopsis* sp. (Histeridae); F, Passalidae, larva (Passalidae); G, *Omorgus* sp. (Trogidae); H, *Diphucephala colaspidoides* (Gyllenhal) (Scarabaeidae: Melolonthinae). A, © Kristi Ellingsen; B–D, F © Owen Kelly; E, H, © Nick Monaghan.

Plate 6. Beetles in their natural habitats. Scarabaeidae. A, *Calloodes grayianus* (White) (Rutelinae); B, *Micropoecila cincta* (Gory et Percheron) (Cetoniinae); C, *Glycyphana (Glycyphaniola) stolata* (Fabricius) (Cetoniinae); D, *Copris hispanus* Linnaeus (Scarabaeinae); E, *Cryptodis* sp. (Dynastinae); F, *Diphucephala* sp. (Melolonthinae); G, *Liparetrus luridipennis* Macleay (Melolonthinae); H, *Lepidiota grisea* Britton (Melolonthinae). A–H © Jiří Lochman.

Plate 7. Beetles in their natural habitats. A, *Blackburnium sloanei* (Blackburn) (Geotrupidae: Bolboceratinae); B, *Onthophagus ferox* Harold (Scarabaeidae: Scarabaeinae); C, *Lamprima aurata* Latreille (Lucanidae); D, *Bolborhachium* sp. (Geotrupidae: Bolboceratinae); E, *Pharochilus dilatatus* (Dalman) (Passalidae); F, *Anoplognathus sp.* (Scarabaeidae: Rutelinae); G, *Chlorobapta frontalis* (Donovan) (Scarabaeidae: Cetoniinae); H, *Metallesthes metallescens* (White) (Scarabaeidae: Cetoniinae). A–H © Jiří Lochman.

Plate 8. Beetles in their natural habitats. Scarabaeidae (A–G) and Trogidae (H). A, *Phyllotocus* sp. (Melolonthinae); B, *Mimadoretus niveosquamosus* Lea (Rutelinae); C, *Hemichnoodes mniszechi* (Janson) (Cetoniinae); D, *Colpochila* sp. (Melolonthinae); E, *Xylonichus eucalypti* Boisduval (Melolonthinae); F, *Xylotrupes australicus* Thomson (Dynastinae); G, *Onitis aygulus* (Fabricius) (Scarabaeinae); H, *Omorgus (Omorgus) tatei* (Blackburn). A–H © Jiří Lochman.

Plate 9. Beetles in their natural habitats. Tenebrionidae (A–E) and Scarabaeidae (F–H). A, *Ecnolagria* sp.; B, *Ecnolagria grandis* (Gyllenhal); C, *Lepturidea* sp.; D, *Emcephalus* sp.; E, *Derispia variabilis* (Carter); F, *Xylotrupes australicus* Thomson; G, *Digitonthophagus gazella* (Fabricius); H, *Lepanus* sp. on bird dropping. © Nick Monaghan.

Plate 10. Tenebrionidae. A, *Leaus tasmanicus* Matthews & Lawrence; B, *Helea squamosa* (Pascoe); C, *Nyctozoilus* sp.; D, *Achthosus westwoodi* Pascoe; E, *Hypaulax* sp.; F, *Boreosargus* sp.; G, *Pachyocelia rotundicollis* (Blackburn) NM; H, *Nyctozoilus major* (Blackburn). A © Simon Grove; B-F, H © Owen Kelly; H © Nick Monaghan.

Plate 11. Tenebrionidae. A, *Adelium augurale* Pascoe; B, *Adelium sp.* Larva; C, *Cyphaleus ducalis* Carter; D, *Amarygmus* sp.; E, *Cyphaleus imperialis* (Carter); F, *Gonocephalum* sp.; G, *Sympetes* sp.; H, *Nyctozoilus bos* (Carter). All © Jiří Lochman.

Plate 12. Buprestidae. A, *Melobasis cuprifera* (Laporte & Gory); B, *Merimna atrata* (Gory & Laporte); C, *Ethonion* sp. (Moore River NP); D, *Julodimorpha saundersii* Thomson; E, *Castiarina forresti* (Barker); F, *Diphucrania stigmata* (Gory & Laporte); G, *Cyrioides elateroides* (Saunders); H, *Diadoxus erythrurus* (White). A, B, D–H © Jiří Lochman; C © Marie Lochman.

Plate 13. Buprestidae. A, *Cyphogastra* sp.; B, *Pseudotaenia superba* (Saunders); C, *Stigmodera cancellata* (Donovan); D, *Stigmodera roei* Saunders; E, *Temognatha duponti* (Boisduval); F, *Temognatha bruckii* (Thomson); G, *Castiarina goerlingi* (Carter); H, *Agrilus (Agrilus) australasiae* Gory & Laporte. A–G © Jiří Lochman; H © Simon Grove.

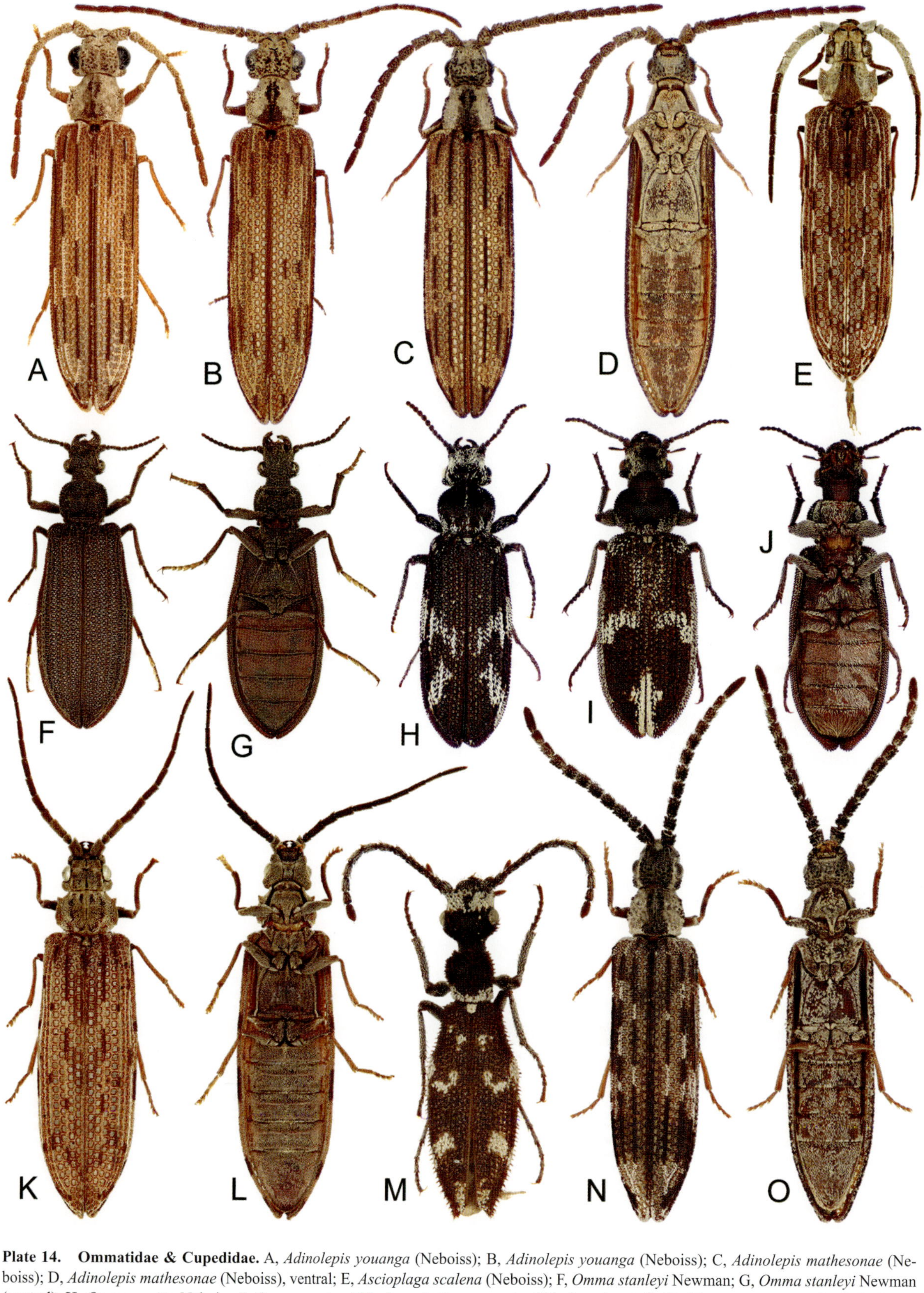

Plate 14. Ommatidae & Cupedidae. A, *Adinolepis youanga* (Neboiss); B, *Adinolepis youanga* (Neboiss); C, *Adinolepis mathesonae* (Neboiss); D, *Adinolepis mathesonae* (Neboiss), ventral; E, *Ascioplaga scalena* (Neboiss); F, *Omma stanleyi* Newman; G, *Omma stanleyi* Newman (ventral); H, *Omma sagitta* Neboiss; I, *Omma mastersi* Macleay; J, *Omma mastersi* Macleay (ventral); K, *Distocupes varians* (Lea); L, *Distocupes varians* (Lea); M. *Omma rutherfordi* Lawrence; N. *Adinolepis eumana* (Neboiss); O. *Adinolepis eumnana* (Neboiss), ventral.

Plate 15. Dytiscidae. A, *Australphilus saltus* Watts; B, *Clypeodytes feryi* Hendrich & Wang; C, *Batrachomatus daemeli* (Sharp); D, *Copelatus nigrolineatus* Sharp; E, *Austrodytes insularis* (Hope); F, *Carabhydrus niger* Watts; G, *Hyderodes schuckardi* Hope, female; H, *Hyderodes schuckardi* Hope, male; I, *Chostonectes gigas* (Boheman); J, *Barretthydrus geminatus* Lea; K, *Eretes australis* (Erichson); L, *Exocelina australiae* (Clark).

Plate 16. Dytiscidae. A, *Leiodytes migrator* (Sharp); B, *Megaporus howitti* (Clark); C, *Paroster ursulae* Hendrich & Balke; D, *Petrodessus conatus* Miller; E, *Paroster plutonicensis* (Watts & Humphreys); F, *Platynectes* sp.; G, *Limbodessus raesideensis* (Watts & Humphreys); H, *Necterosoma penicillatum* (Clark); I, *Onychohydrus scutellaris* (Germar); J, *Rhantus suturalis* (W. S. Macleay); K, *Rhantaticus congestus* (Klug); L, *Lancetes lanceolatus* (Clark).

Plate 17. Dytiscidae (A–F), Noteridae (G–I, L), Hygrobiidae (J), Haliplidae (K). A, *Paroster caecus* (Watts); B, *Sekaliporus kreigi* Watts; C, *Sternopriscus tasmanicus* Sharp; D, *Tiporus undecimaculatus* (Clark); E, *Sternhydrus* sp.; F, *Spencerhydrus pulchellus* Sharp; G, *Notomicrus tenellus* (Clark); H, *Neohydrocoptus subfasciatus* (Sharp); I, *Canthydrus bovillae* Blackburn; J, *Hygrobia maculata* Britton; K, *Haliplus (Neohaliplus) bistriatus* Wehncke; L, *Sternocanthus australasiae* (Wehncke).

Plate 18. Carabidae. A, *Adelotopus dytiscides* Newman (Pseudomorphini); B, *Abacetus angustior* Macleay (Abacetini); C, *Aenigma iris* Newman (Helluonini); D, *Adelotopus rufozonatus* Baehr (Pseudomorphini); E, *Adotela apicalis* (Sloane) (Broscini); F, *Amblystomus laetus* (Blackburn) (Harpalini); G, *Anthracus exactus* (Darlington) (Harpalini); H, *Aephnidius adeliodes* W. S. Macleay (Cyclosomini); I, *Acrogenys hirsuta* Macleay (Zuphiini); J, *Agonica victoriensis* Moore (Agonicini); K, *Acupalpus tachoides* (Sloane) (Harpalini); L, *Agonocheila curtula* (Erichson) (Lebiini).

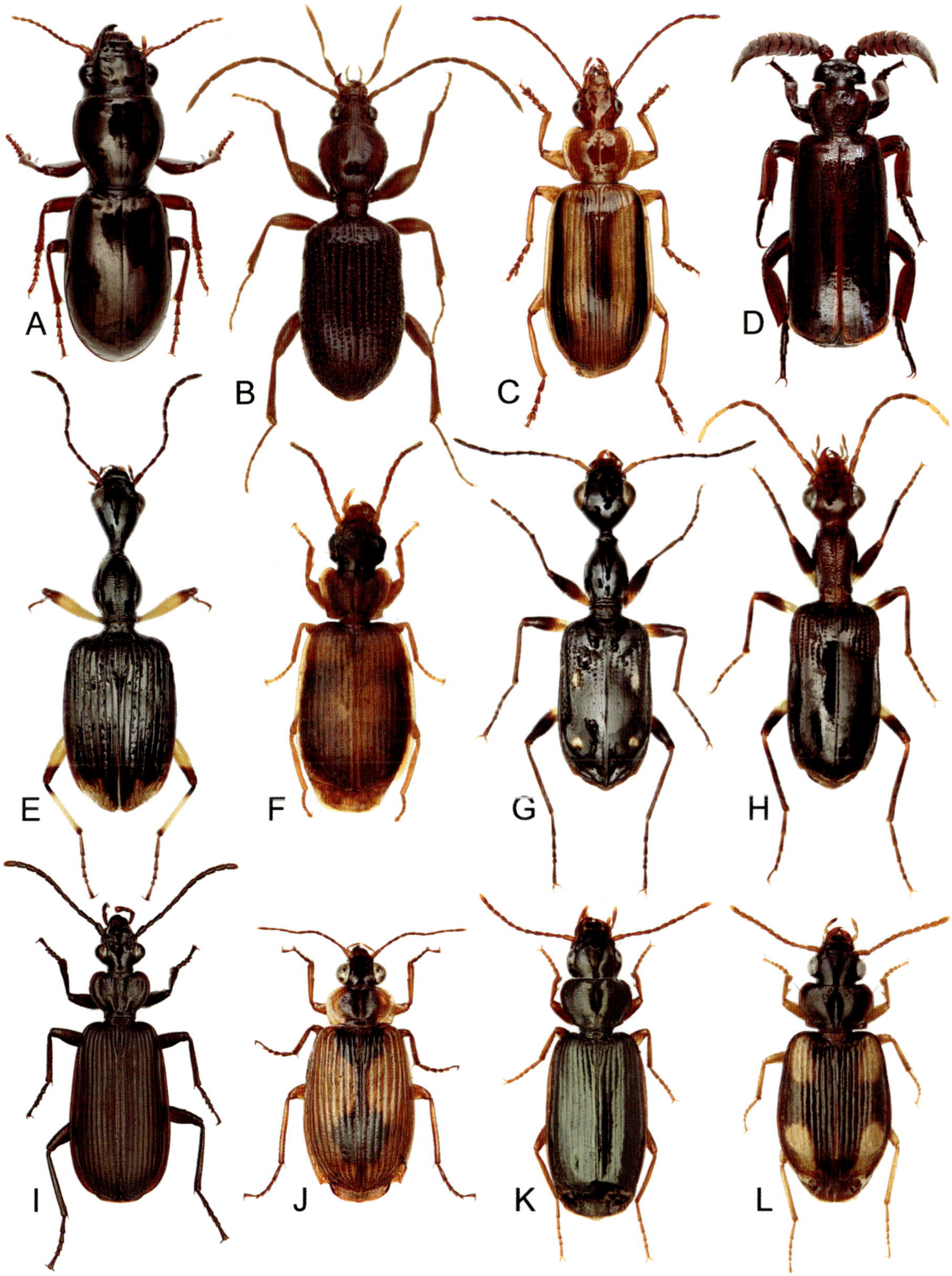

Plate 19. Carabidae. A, *Anheterus gracilis* (Germain) (Broscini); B, *Apotomus australis* Laporte (Aptotomiini); C, *Amblytelus curtus curtus* (Fabricius) (Psydrini); D, *Arthropterus wilsoni* (Westwood) (Paussini); E, *Aulacolius triordinatus* Sloane (Odacanthini); F, *Anomotariella hippocrepis* Baehr (Lebiini); G, *Archicolliuris splendissimus* Baehr (Odacanthini); H, *Basistichus micans* (Macleay) (Odacanthini); I, *Ametroglossus ater* (Macleay) (Helluonini); J, *Aristolebia mucronata* (Sloane) (Lebiini); K, *Anomotarus illawarrae* (Macleay) (Lebiini); L, *Barrymooreana quadrimaculata* Baehr (Lebiini).

Plate 20. Carabidae. A, *Calosoma schayeri* Erichson (Carabini); B, *Batoscelis oblongus* (Dejean) (Harpalini); C, *Calyptogonia atra* Sloane (Migadopini); D, *Brachidius crassicornis* Chaudoir (Cratocerini); E, *Brithysternum macleayi* Sloane (Broscini); F, *Castelnaudia marginifera* (Chaudoir) (Pterostichini); G, *Caelostomus* sp. (Drimostomatini); H, *Catadromus australis* Laporte (Pterostichini); I, *Cainogenion ephippiatum* (Newman) (Pseudomorphini); J, *Bembidion (Sloanephila) jacksoniense* Guérin-Méneville (Bembidiini); K, *Carenum elegans* Macleay (Scaritini); L, *Caphora humilis* Schmidt-Goebel (Cyclosomini).

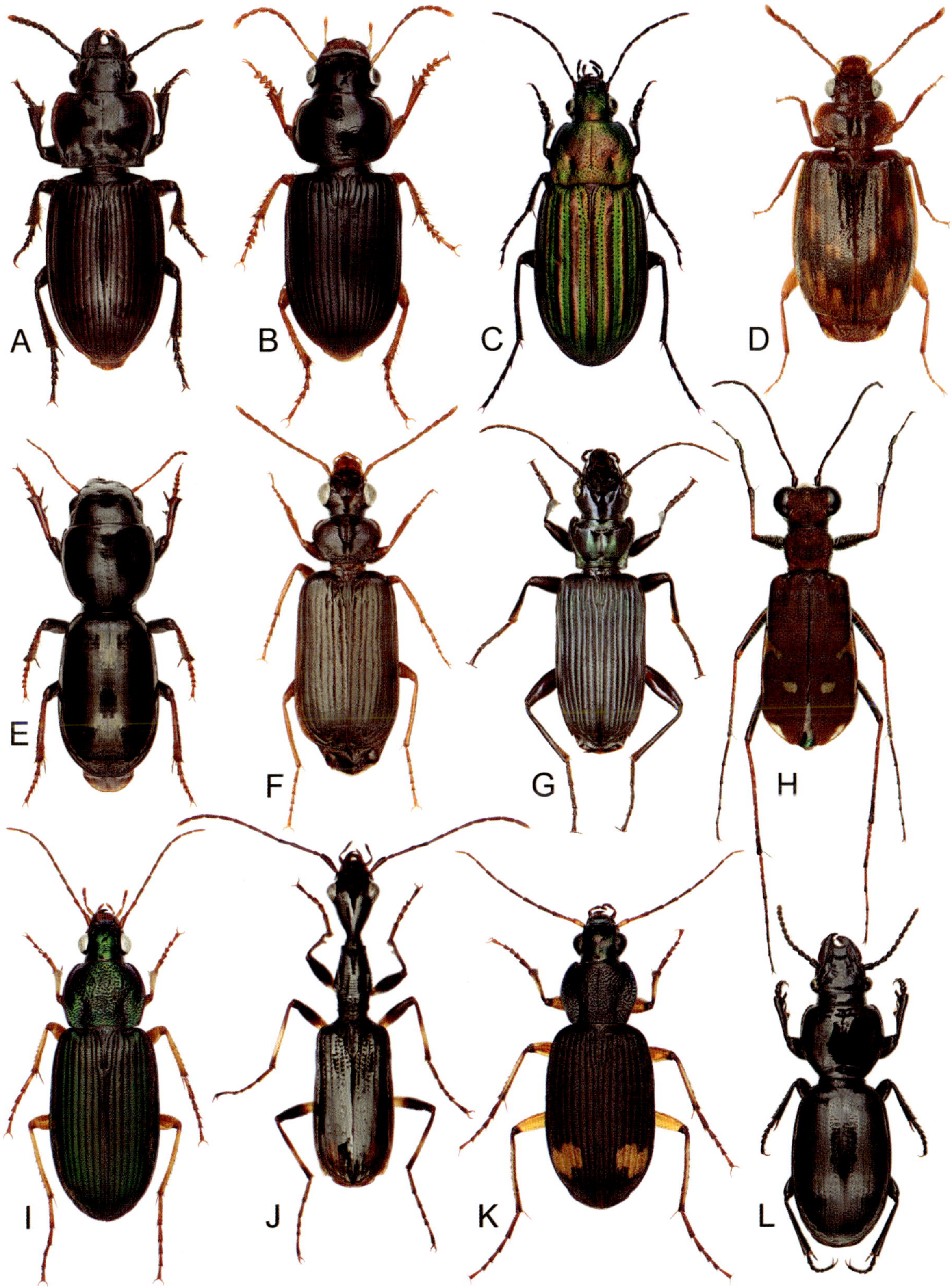

Plate 21. Carabidae. A, *Celanida montana* Laporte (Psydrini); B, *Cenogmus castelnaui* Csiki (Harpalini); C, *Chlaenius greyanus* White (Chlaeniini); D, *Cheilagona strictica* (Blackburn) (Lebiini); E, *Cerotalis* sp. (Broscini); F, *Celaenephes linearis* (Walker) (Lebiini); G, *Catascopus chaudoiri* Laporte (Lebiini); H, *Cicindela mastersi catoptriola* Horn (Cicindelini); I, *Chlaenius ophonoides* Fairmaire (Chlaeniini); J, *Clarencia aliena* Pascoe (Odacanthini); K, *Chlaenius flaviguttatus* W. S. Macleay (Chlaeniini); L, *Chylnus ater* (Putzeys) (Broscini).

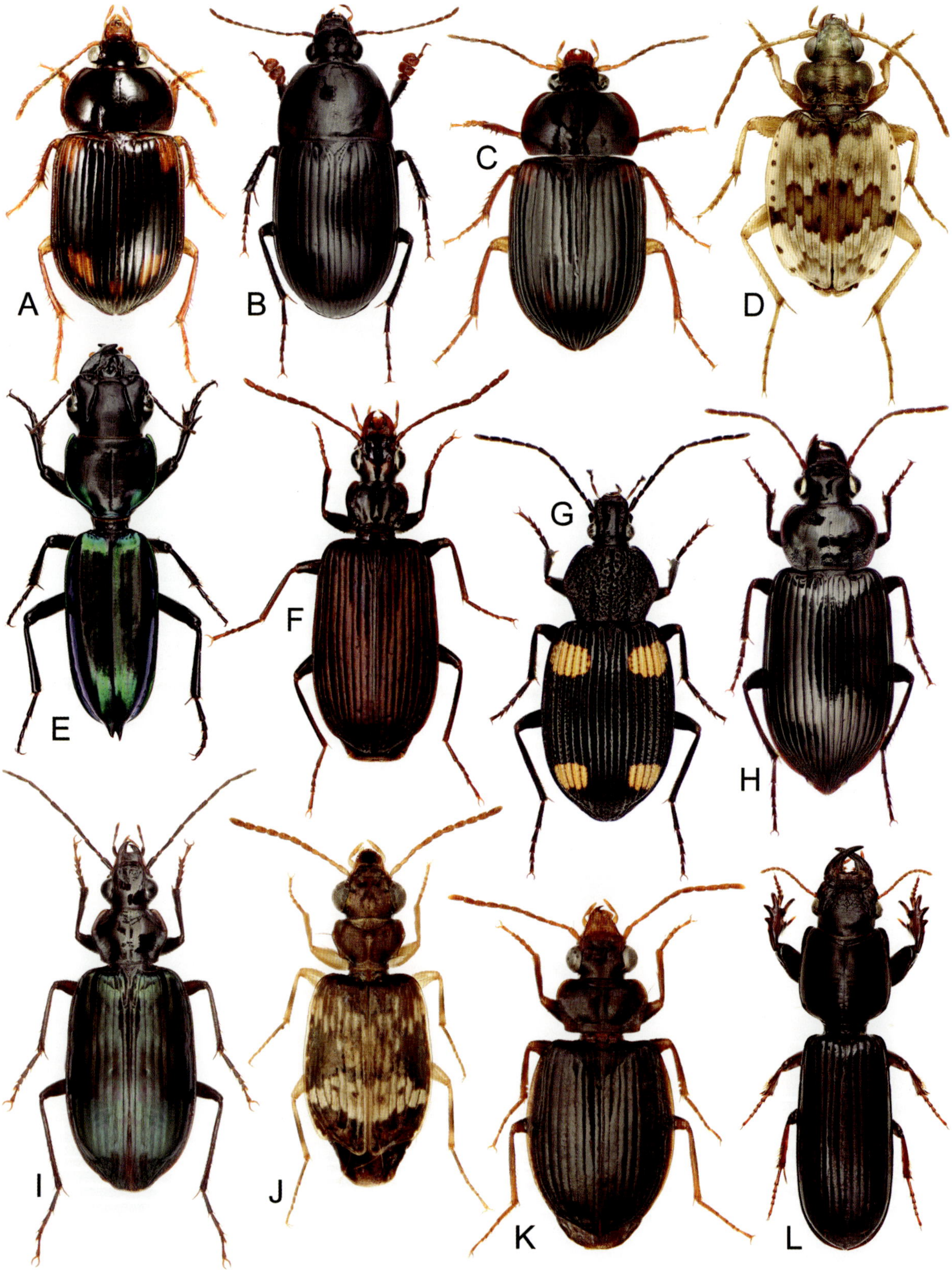

Plate 22. Carabidae. A, *Cosmodiscus rubripictus* Sloane (Abacetini); B, *Coptocarpus australis* (Dejean) (Oodini); C, *Cosmodiscus rubripictus* Sloane (Abacetini); D, *Tetragonoderus undatus* (Dejean) (Cyclosomini); E, *Conopterum leai* (Sloane) (Scaritini); F, *Coptoglossus porphyriacus* (Sloane) (Lebiini); G, *Craspedophorus rockhamptonensis* (Laporte) (Panagaeini); H, *Coleolissus papua* Darlington (Harpalini); I, *Metacolpodes truncatellus* (Fairmaire, 1881) (Platynini); J, *Coptodera fasciolata* (Macleay) (Lebiini); K, *Coptodera oxyptera* Chaudoir (Lebiini); L, *Clivina planiceps* (Putzeys) (Scaritini).

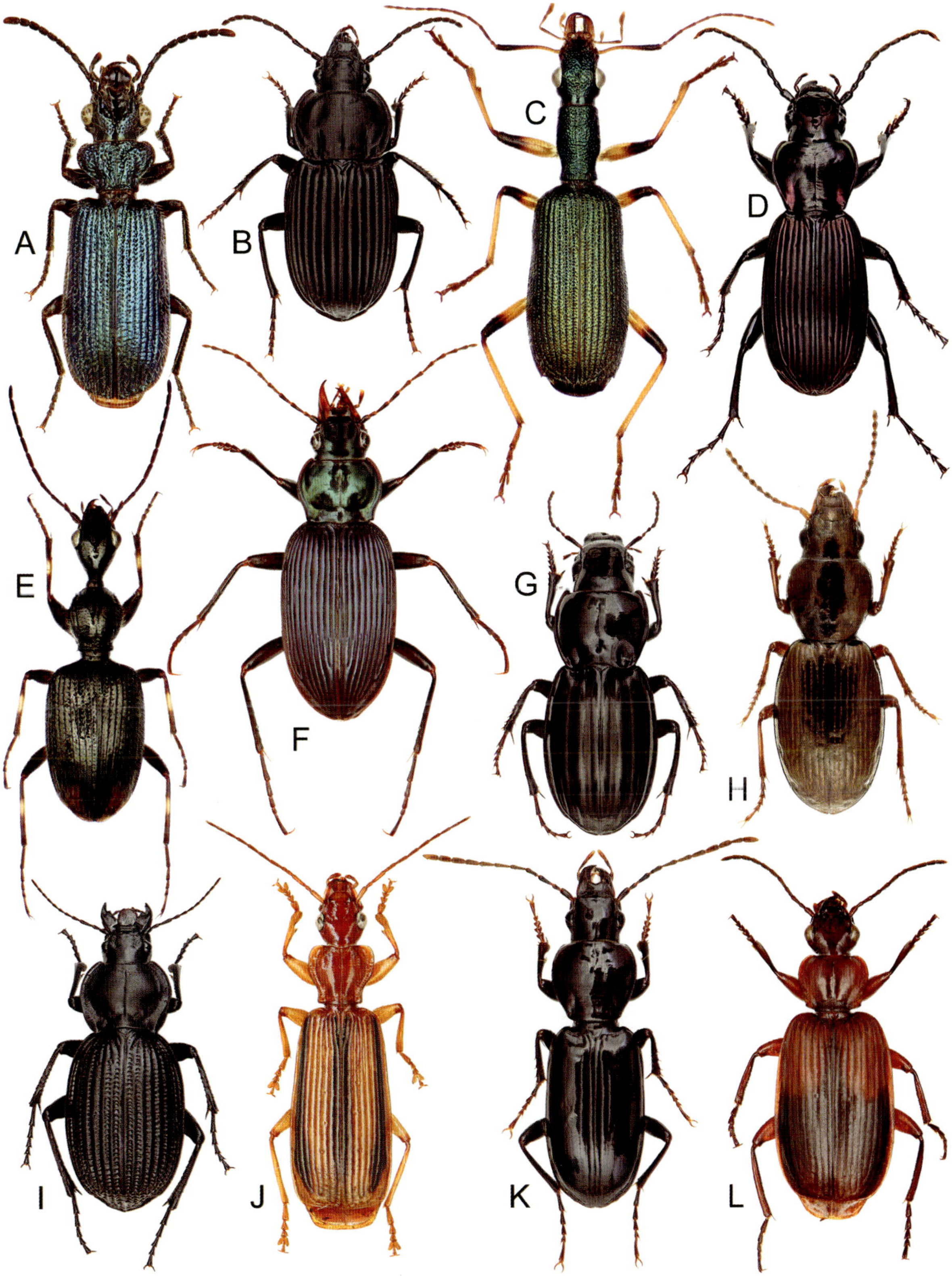

Plate 23. Carabidae. A, *Dicranoglossus resplendens* (Laporte) (Helluonini); B, *Cratogaster occidentalis* Macleay (Pterostichini); C, *Dendrocellus smaragdinus* Chaudoir (Dryptini); D, *Cratoferonia phylarchus* (Sloane) (Pterostichini); E, *Deipyrodes palustris* (Sloane) (Odacanthini); F, *Decogmus chalybaeus* Sloane (Migadopini); G, *Delinius essingtoni* Westwood (Pterostichini); H, *Acallistus cuprescens* (Sloane) (Broscini); I, *Cuneipectus foveatus* Sloane (Pterostichini); J, *Demetrida grandis* (Chaudoir) (Lebiini); K, *Darodilia mandibularis* Laporte (Pterostichini); L, *Diabaticus australis* (Erichson) (Lebiini).

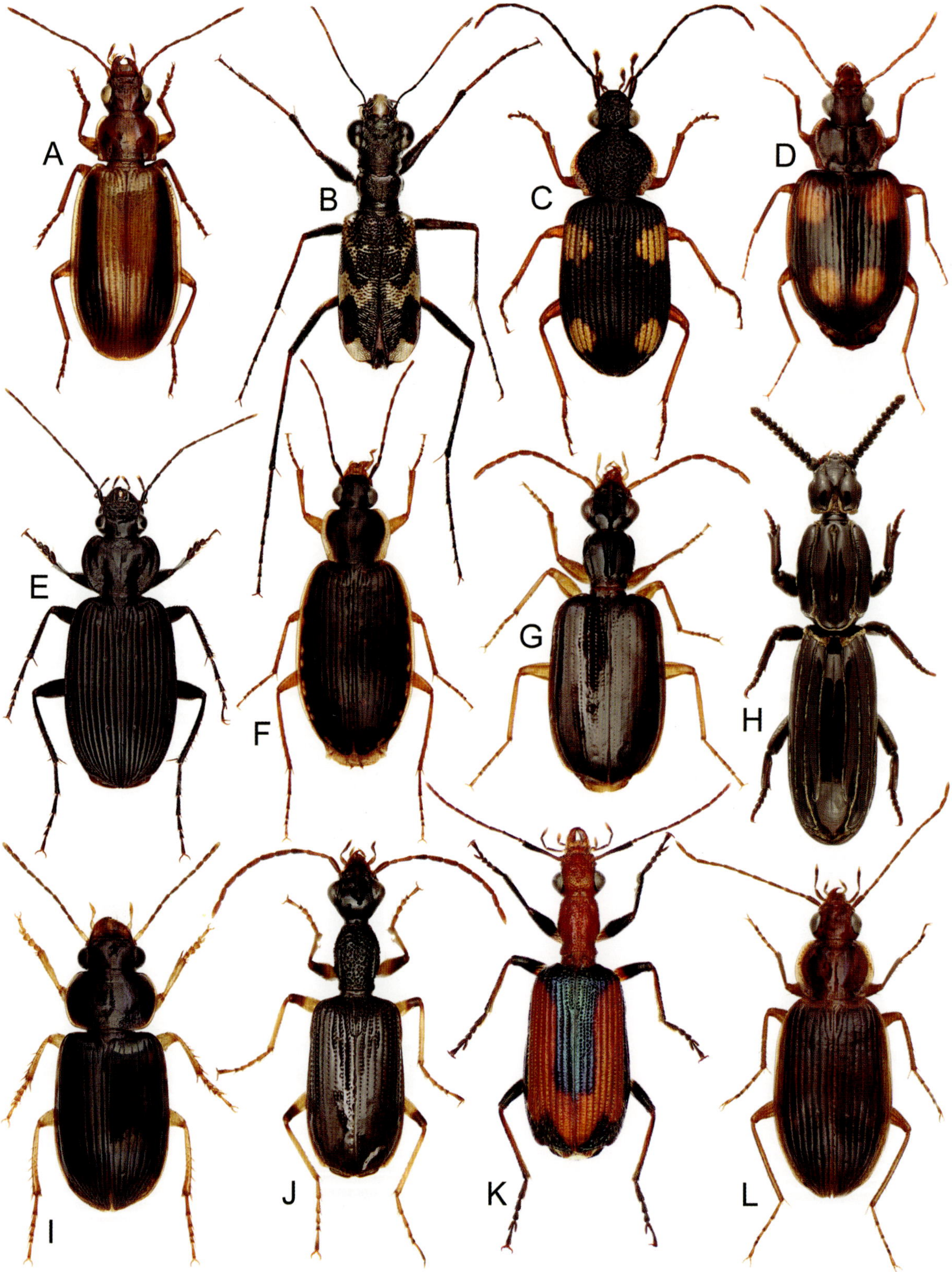

Plate 24. Carabidae. A, *Dystrichothorax bicolor* Blackburn (Psydrini); B, *Distipsidera undulata* Westwood (Cicindelini); C, *Adischissus notulatus* (Fabricius) (Panagarini); D, *Dolichoctis striatus* Schmidt-Goebel (Lebiini); E, *Dicrochile goryi* (Boisduval) (Licinini); F, *Dilonchus pictus* Darlington (Licinini); G, *Dicraspeda nitida* (Sloane) (Odacanthini); H, *Rhyzodiastes* (*Apotomus*) *mirabilis* (Lea) (Rhysodini);. I, *Egadroma piceus* (Guérin-Méneville) (Harpalini); J, *Dicraspeda obscura* (Laporte) (Odacanthini); K, *Drypta mastersi* Malceay (Dryptini); L, *Dicranoncus queenslandicus* (Sloane) (Platynini).

Plate 25. Carabidae. A, *Eurylychnus blagravei* (Laporte) (Broscini); B, *Eudalia macleayi* Bates (Odacanhthini); C, *Eurystomis castelnaui* Chaudoir (Pterostichini); D, *Epilectus mastersi* (Macleay) (Scaritini); E, *Euplynes apicalis* Darlington (New Guinea) (Platynini); F, *Cerabilia* (*Feronista*) *amaroides* Moore (Abacetini); G, *Tachyura bipustulatus* (Macleay) (Bembidiini); H, *Endynomena pradieri* (Fairmaire) (Cocos-Keeling Is.) (Lebiini); I, *Epimicodema mastersii* (Macleay) (Helluonini); J, *Epelyx walkeri* Baehr (Psydrini); K, *Euryscaphus dilatatus* Macleay (Scaritini); L, *Euthenarus comes* Sloane (Harpalini).

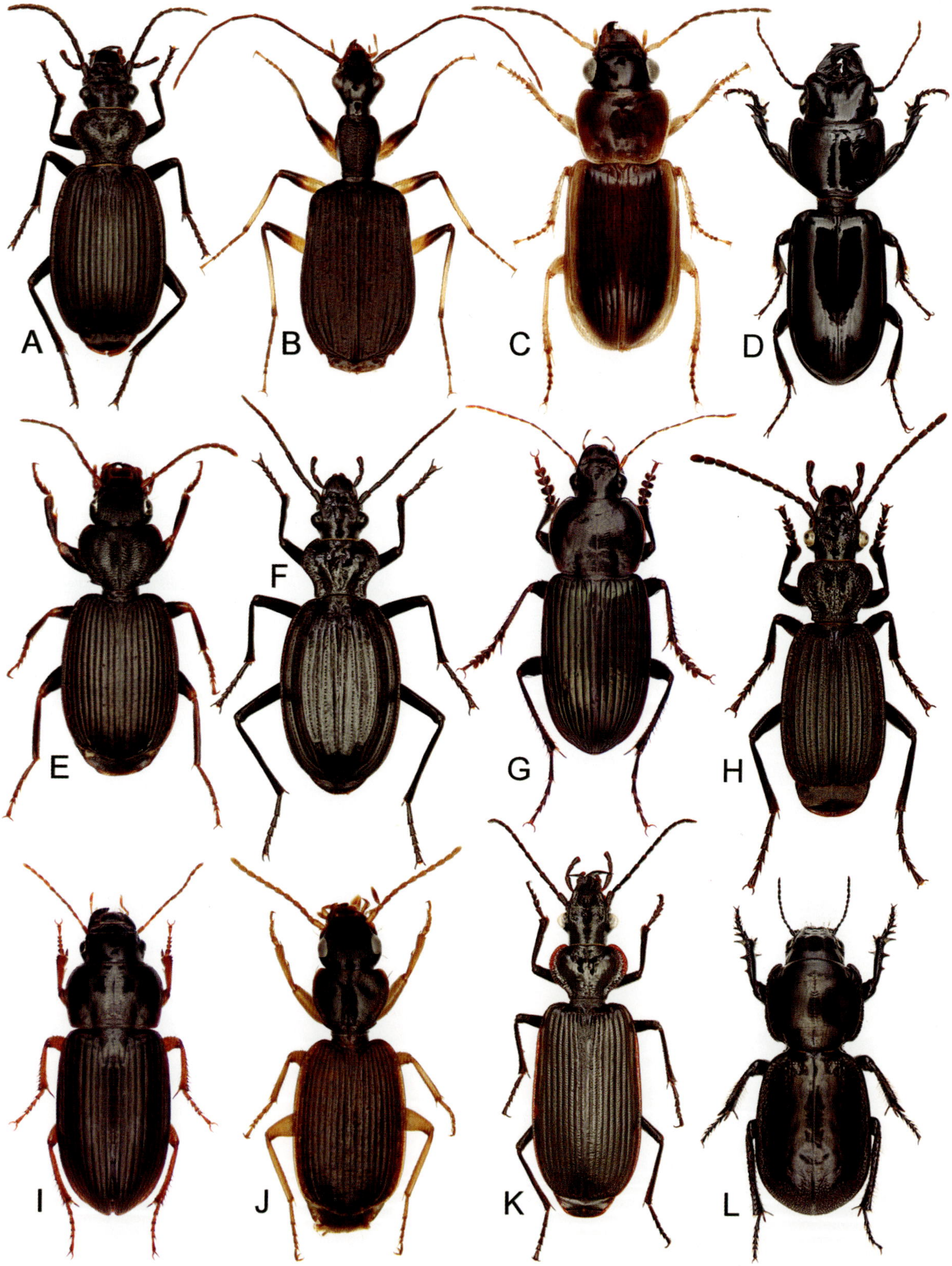

Plate 26. Carabidae. A, *Helluapterus niger* Sloane (Helluonini); B, *Gestroania storeyi* Baehr (Odacanthini); C, *Haplaner* sp. (Harpalini); D, *Geoscaptus* sp. (Scaritini); E, *Habutarus crassiceps* (Macleay) (Lebiini); F, *Helluarchus robustus* Sloanc(Helluonini); G, *Gnathaphanus latus* Sloane(Harpalini); H, *Helluo costatus* Bonelli (Helluonini); I, *Harpalus fulvicornis* (Thunberg) (Harpalini); J, *Habutarus calderi* Baehr (Lebiini); K, *Gigadema mandibulare* Blackburn (Helluonini); L, *Gnathoxys granularis* Westwood (Broscini).

Plate 27. Carabidae. A, *Laccocenus ambiguus* Sloane (Psydrini); B, *Helluosoma atrum* Laporte (Helluonini); C, *Idacarabus troglodytes* Lea (Zolini); D, *Holcoderus caeruleipennis* Sloane (Lebiini); E, *Homethes guttifer* Germar (Pentagonicini); F, *Hormacrus latus* Sloane (Licinini); G, *Hyperion schroetteri* (Schreibers) (Morionini); H, *Helluodema unicolor* (Hope) (Helluonini); I, *Hypharpax krefti* (Laporte) (Harpalini); J, *Laccopterum darwiniense* (Malceay) (Scaritini); K, *Helluonidius aterrimus* (Macleay) (Helluonini); L, *Pericompsus australis* (Schaum) (Bembidiini).

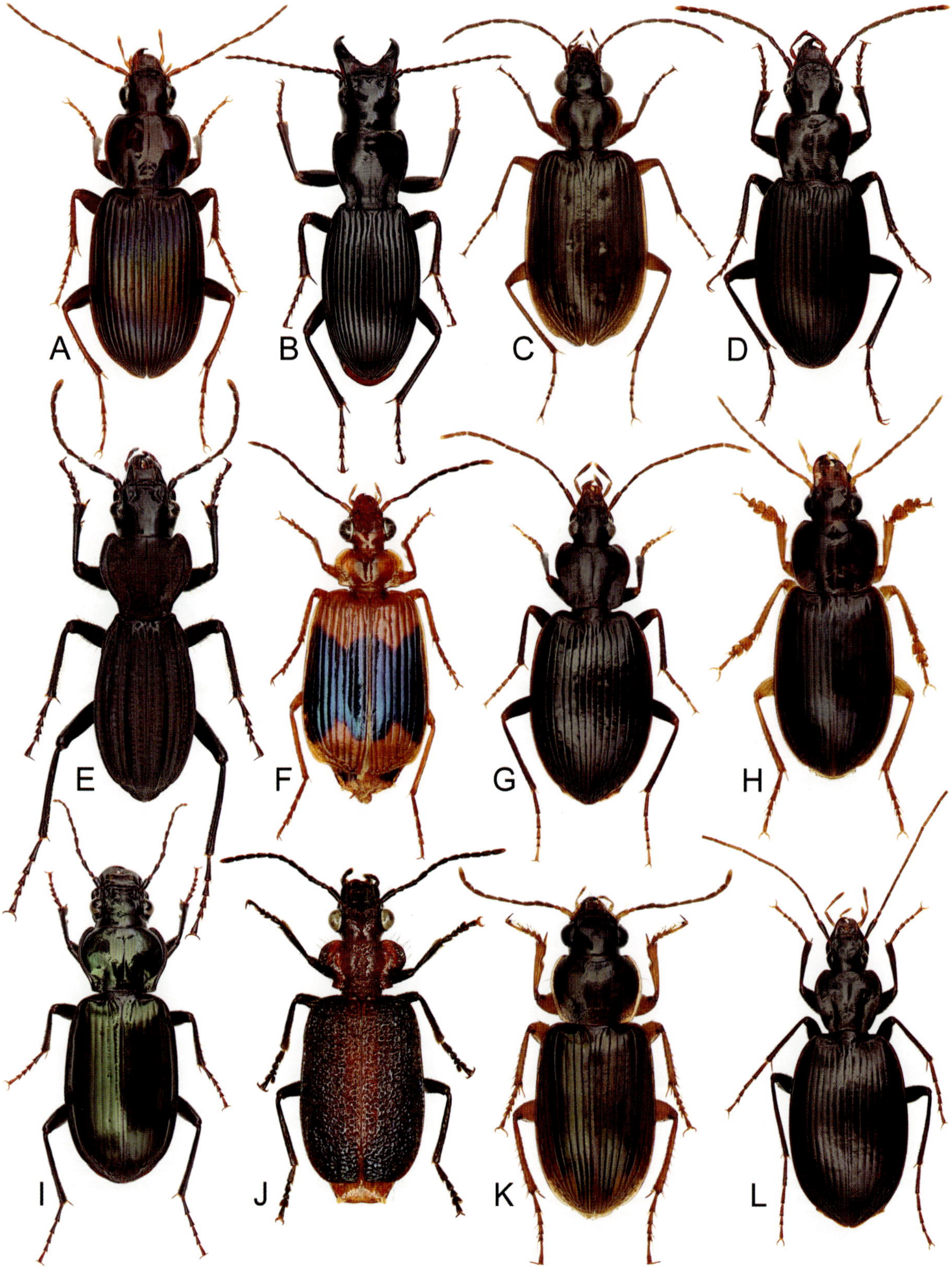

Plate 28. Carabidae. A, *Loxandrus subgagatinus* (Laporte) (Abacetini); B, *Liopasa crepera* Tschitschérine (Pterostichini); C, *Lorostema bothriophora* (Redtenbacher) (Platynini); D, *Laemostenus complanatus* (Dejean) (Sphodrini); E, *Loxogenius opacipennis* (Malceay) (Pterostichini); F, *Lebia bicolor* (Sloane) (Lebiini); G, *Lacordairia proxima* Laporte (Licinini); H, *Lecanomerus verticalis* (Erichson) (Harpalini); I, *Lesticus chloronotus* Chaudoir (Pterostichini); J, *Lachnoderma cinctum* Macleay (Lebiini); K, *Loxoncus marginatus* (Macleay) (Harpalini); L, *Lestignathus cursor* Erichson (Licinini).

Plate 29. Carabidae. A, *Melisodera picipennis* Westwood (Psydrini); B, *Migadopiella octoguttata* Baehr (Migadopini); C, *Microferonia cinctipennis* Sloane (Licinini); D, *Megacephala australis* (Chaudoir) (Cicindelini); E, *Megadromus eborensis* Moore (Pterostichini); F, *Meonis semistriatus* Sloane (Psydrini); G, *Mecyclothorax cordicollis* (Sloane) (Psydrini); H, *Mecynognathus dameli* Macleay (Pterostichini); I, *Megalopaussus amplipennis* Lea, female (Paussini); J, *Microlestodes atrifasciatus* (Sloane) (Lebiini); K, *Megalopaussus amplipennis* Lea, male (Paussini); L, *Adischissus quadrimaculatus* (Cziki) (Panagaeini).

Plate 30. Carabidae. A, *Moriomorpha adelaidae* Laporte (Psydrini); B, *Myrmecodemus formicoides* (Sloane) (Odacanthini); C, *Moriodema mcoyei* Laporte (Psydrini); D, *Mochtherus tetraspilotus* (W.S. Macleay) (Lebiini); E, *Monocentrum megacephalum* (Hope) (Scaritini); F, *Mochtheroides niger* Jedlicka (Lebiini); G, *Mystropomus subcostatus* Chaudoir (Ozaenini); H, *Morion longicollis* Macleay (Morionini); I, *Neocarenum elongatum* (Macleay) (Scaritini); J, *Miscelus unicolor* Putzeys (Lebiini); K, *Nanodiodes* sp. (Oodini); L, *Minuthodes queenslandica* (Sloane) (Lebiini).

Plate 31. Carabidae. A, *Notonomus (Conchitella) clivinoides* (Moore) (Pterostichini); B, *Nickerlea sloanei* (Lea) (Cicindelini); C, *Notabax monteithi* Moore (Pterostichini); D, *Notiobia denisonensis* (Laporte) (Harpalini); E, *Neoeudalia nigra* (Sloane) (Odacanthini); F, *Neoscaphus simplex* Sloane (Scaritini); G, *Neonomius laevicollis* (Sloane) (Psydrini); H, *Notolestus sulcipennis* (Macleay) (Pterostichini); I, *Kaveinga (Angekiva) frontalis* (Grouvelle) (Rhysodini); J, *Neohelluo angulicollis* Sloane (Helluonini); K, *Notagonum tasmanicum* Baehr (Platynini); L, *Notonomus (Acanthoferonia) ferox* Moore (Pterostichini).

Plate 32. Carabidae. A, *Notonomus (Leiradira) aurifer* Darlington (Pterostichini); B, *Notonomus (Loxodactylus) carinulatus* (Chaudoir) (Pterostichini); C, *Oodes modestus* Laporte (Oodini); D, *Ophionea thouzeti* Laporte (Odacanthini); E, *Nurus imperialis* (Sloane) (Pterostichini); F, *Notonomus aeques* (Laporte) (Pterostichini); G, *Notonomus triplogenioides* (Chaudoir) (Pterostichini); H, *Notospeophonus castaneus* Moore (Harpalini); I, *Notonomus* sp. (Pterostichini); J, *Notoplatynus hilaris* (Oliff) (Platynini); K, *Omestes torta* Andrewes (Licinini); L, *Notonomus (Leiradira) auricollis* (Laporte) (Pterostichini).

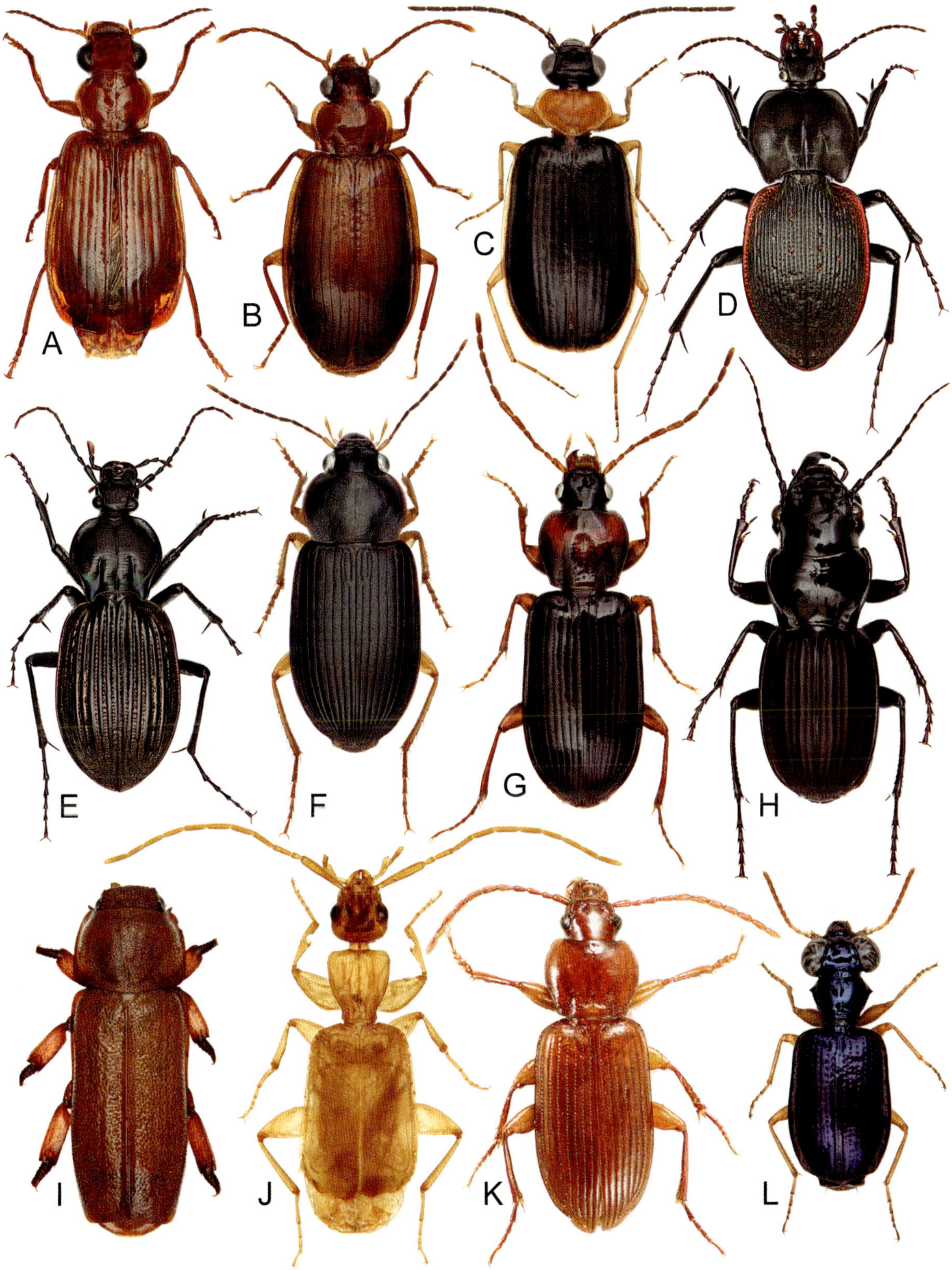

Plate 33. Carabidae. A, *Parena picea* (Macleay) (Lebiini); B, *Paratrichothorax brevistylus* Baehr (Psydrini); C, *Pentagonica dichroa* Sloane (Pentagonicini); D, *Pamborus guerini* (Gory) (Pamborini); E, *Pamborus alternans* Latreille (Pamborini); F, *Parophonus opacus* (Macleay) (Harpalini); G, *Pediomorphus ruficollis* Sloane (Abacetini); H, *Paranurus dilaticeps* (Chaudoir) (Pterostichini); I, *Paussotropus cylindricus* (Chaudoir) (Pseudomorphini); J, *Parazuphium mastersii* (Laporte) (Zuphiini); K, *Pediomorphus obtusus* Will (Abacetini); L, *Parascopodes cyaneus yorkianus* Baehr (Pentagonicini).

Plate 34. Carabidae. A, *Philophloeus australis* (Dejean) (Lebiini); B, *Physolaesthus caviceps* (Andrewes) (Licinini); C, *Percodermus niger* Sloane (Zolini); D, *Phorticosomus franzi* Baehr (Harpalini); E, *Pheropsophus verticalis* (Dejean) (Brachinini); F, *Phloeocarabus semivittatus* (Macleay) (Lebiini); G, *Physolaesthus grandipalpis* Macleay (Licinini); H, *Philoscaphus* sp. (Scaritini); I, *Percolestus blackburni* Sloane (Broscini); J, *Phaenaulax nanus* (Sloane) (Pterostichini); K, *Platysphyrus tibialis* Sloane (Scaritini); L, *Perigona litura* (Perroud & Montrouzier) (Perigonini).

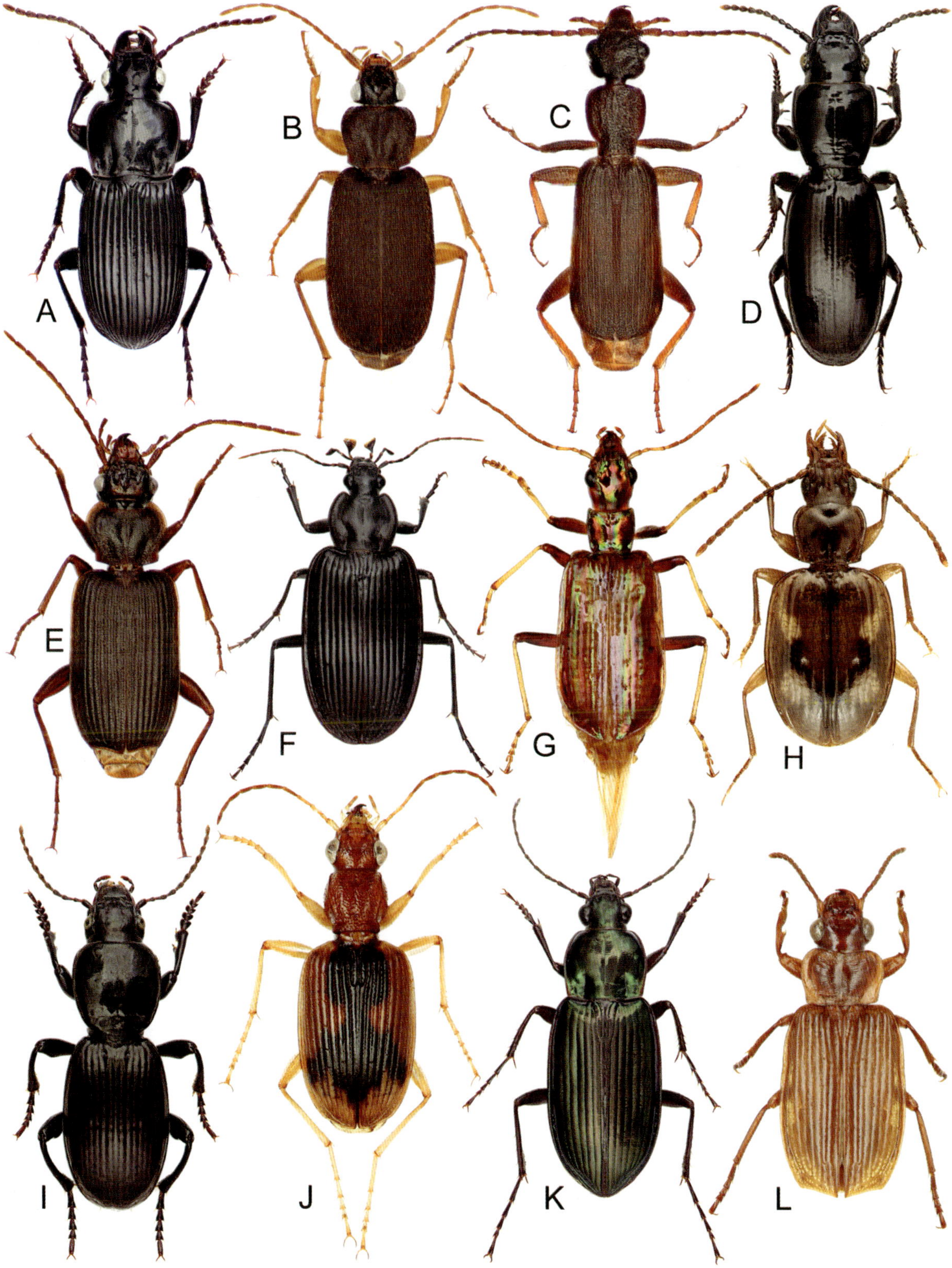

Plate 35. Carabidae. A, *Prosopogmus chalybeipennis* (Chaudoir) (Pterostichini); B, *Planetes magelae* Baehr (Zuphiini); C, *Platyhelluo weiri* Baehr (Helluonini); D, *Percosoma montanum* (Laporte) (Broscini); E, *Pogonoglossus porosus* (Sloane) (Physocrotaphini); F, *Platylytron amplipenne* Macleay (Licinini); G, *Plagiotelum opalescens* Oliff (Ctenodactylini); H, *Bothynotrechus lynx* Moore (Trechini); I, *Promecoderus bassii* Laporte (Broscini); J, *Porocara punctata* Sloane (Odacanthini); K, *Platycoelus melliei* (Montrouzier) (Pterostichini); L, *Plochionus pallens* (Fabricius) (Lebiini).

Plate 36. Carabidae. A, *Rhaebolestes walkeri* Sloane (Psydrini); B, *Pseudagonica obscuripes* Baehr (Agonicini); C, *Raphetis gracilis* Moore (Psydrini); D, *Rhysocara crassa* Sloane (Scaritini); E, *Pterogmus rufipes* Sloane (Psydrini); F, *Pseudaptinus fulvus* (Laporte) (Zuphiini); G, *Cicindela (Grandopronotalia) browni* Sloane (Cicindelini); H, *Rhabdotus reflexus* (Chaudoir) (Pterostichini); I, *Renneria kamouni* Baehr (Odacanthini); J, *Pseudoceneus iridescens* (Laporte) (Pterostichini); K, *Austropseudomorpha insignis* (Sloane) (Pseudomorphini); L, *Pterocyrtus striatulus* Sloane (Zolini).

Plate 37. **Carabidae.** A, *Setodyschirius wilsoni* (Sloane) (Scaritini); B, *Sarothrocrepis luctosa* (Newman) (Cyclosomini); C, *Sarothrocrepis elegans* (Blackburn) (Cyclosomini); D, *Secatophus australis* (Hope) (Pterostichini); E, *Rhytisternus cyathoderus* (Chaudoir) (Pterostichini); F, *Sarticus dixoni* Sloane (Pterostichini); G, *Rhysopleura orbicollis* (Sloane) (Cicindelini); H, *Scopodes boops* Erichson (Pentagonicini); I, *Setalimorphus punctiventris* Sloane (Pterostichini); J, *Rugitarus puellarum puellarum* Baehr (Lebiini); K, *Scaraphites rotundipennis* (Dejean) (Scaritini); L, *Setalis niger* Laporte (Pterostichini).

Plate 38. Carabidae. A, *Steganomma porcatum* Macleay (Scaritini); B, *Simodontus grandiceps* Sloane (Pterostichini); C, *Siagonyx blackburni* Sloane (Licinini); D, *Sitaphe rotundata* Moore (Psydrini); E, *Stricklandiana nigra* Sloane (Lebiini); F, *Speotarus princeps* Moore (Lebiini); G, *Sloaneana tasmaniae* (Sloane) (Zolini); H, *Syleter papua* Darlington (Scaritini); I, *Sphallomorpha albopicta* Newman (Pseudomorphini); J, *Stichonotus piceus* Sloane (Migadopini); K, *Syntomus quadripunctatus* (Schmidt-Goebel) (Lebiini); L, *Tasmanorites brevinotatus* (Sloane) (Trechini).

Plate 39. Carabidae. A, *Theprisa convexa* (Sloane) (Psydrini); B, *Trichisia azurea* (Chaudoir) (Panagaeini); C, *Trichotichnus straneoi* (Louwerens) (Harpalini); D, *Teropha besti* (Sloane) (Pterostichini); E, *Trichocarenum* sp. (Scaritini); F, *Zuphium australe* Chaudoir (Zuphiini); G, *Violagonum violaceum* (Chaudoir) (Platynini); H, *Tropidotrechus victoriae* (Blackburn) (Trechini); I, *Zeodera atra* Laporte (Abacetini); J, *Trigonothops flavofasciata* Chaudoir (Lebiini); K, *Trichosternus nudipes* Darlington (Pterostichini); L, *Teraphis helmsi* (Sloane) (Psydrini).

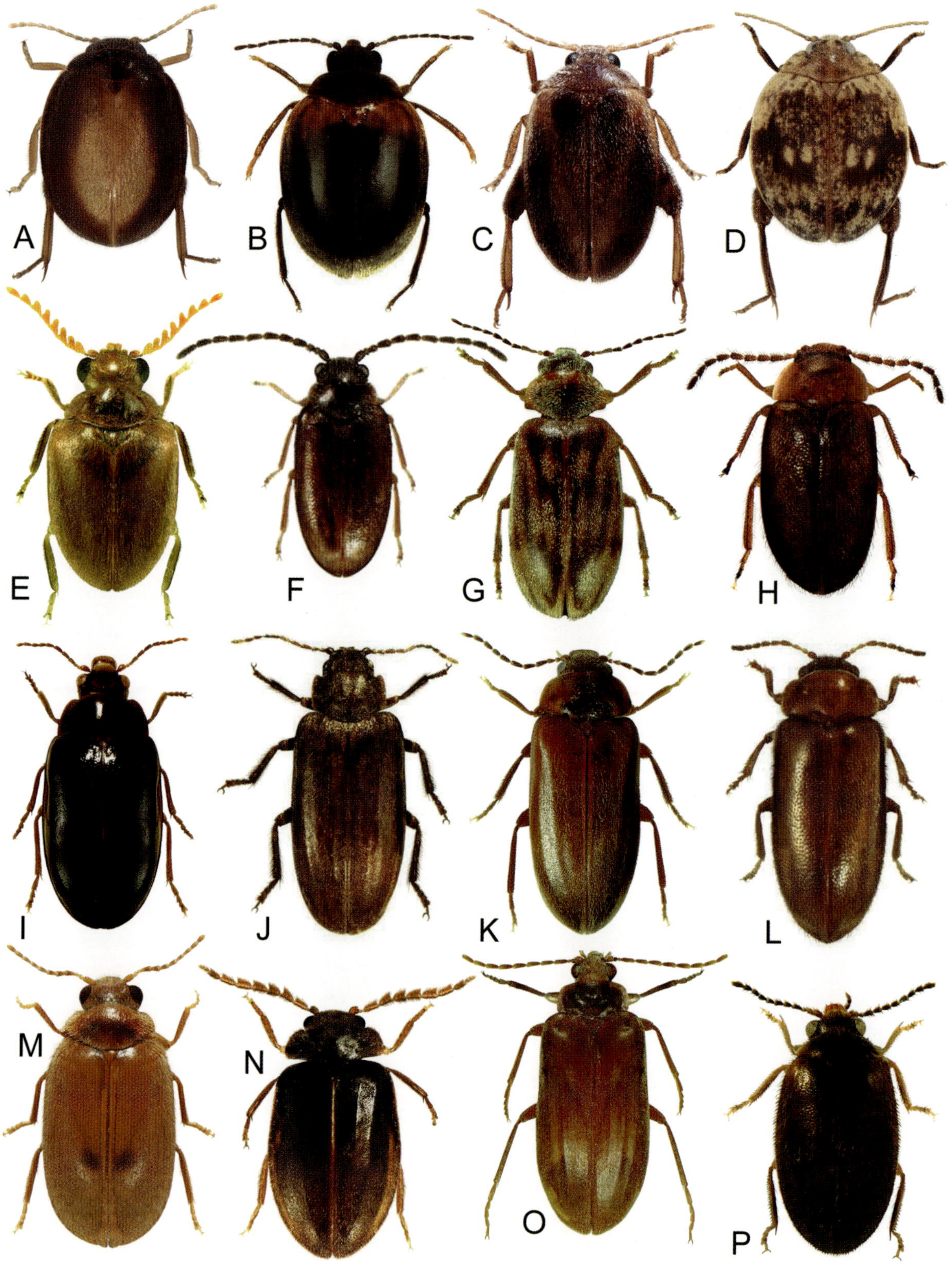

Plate 40. Scirtidae. A, *Exomoscirtes ruforotundus* (Watts); B, *Eurycyphon fulvus* Watts; C, *Scirtes emmaae* Watts; D, *Ora floccosa* Watts; E, *Prionocyphon tolgaensis* Watts, male; F, *Petrocyphon tasmanicus* Watts, male; G, *Accolabass maculatus* (Waterhouse); H, *Leptocyphon furcalonga* Zwick; I, *Macrohelodes princeps* (Blackburn); J, *Daploeuros tasmanicus* Watts; K, *Heterocyphon thunguttiensis* Watts; L, *Chameloscyphon huonensis* Watts; M, *Nothocyphon lindensis* (Blackburn); N, *Macrodascillus scalaris* (Lea); O, *Peneveronatus macedonensis* (Armstrong); P, *Austrocyphon adelaidae* (Blackburn).

Plate 41. Scirtidae larvae (A–D), Histeridae (E–Q). A, *Macrodascillus scalaris* (Lea); B, *Austrocyphon adelaidae* (Blackburn); C, *Nothocyphon lindensis* (Blackburn); D, *Peneveronatus macedonensis* Armstrong; E, *Arbolister termitophilus* Mazur; F, *Ectatommiphila opaca* (Lea), frontal view; G, *Ectatommiphila opaca* (Lea); H, *Epiechinus costatus* (Macleay); I, *Chlamydopsis* sp.; J, *Dendrophilus xavieri* Marseul; K, *Chlamydopsis* sp.; L, *Epierus* sp.; M. *Asterix* sp.; N, *Carcinops (Carcinops) pumilio* (Erichson); O, *Eurylister bakewelli* (Marseul); P, *Eblisia* sp.; Q, *Eblisia* sp.

Plate 42. Histeridae. A, *Hister nomas* Erichson; B, *Hister ambulator* Thayer; C, *Orectoscelis blackburni* Dégallier & Caterino; D, *Platysoma (Kanaarister) latisternus* (Marseul); E, *Pachylomalus mus* (Marseul); F, *Onthophilus australis* Helava & Howden; G, *Hololepta* sp.; H, *Platylomalus umbilicatus* (Marseul); I, *Pacifister* sp.; J, *Parepierus* sp.; K, *Pheidoliphila arriagadai* Dégallier & Caterino; L, *Platysoma (Platylister) comes* (Marseul); M. *Platysoma (Popinus) querulus* (Marseul); N. *Eblisia* sp.; O. *Platysoma (Cylister) bipunctatum* Lewis; P. *Platysoma (Platysoma) paugami* (LeGuillou).

Plate 43. Georissidae (A–B), Hydrochidae (C–D), Spercheidae (E), Hydrophilidae. Hydrophilinae: Amphiopini (F), Berosini (G–K) and Laccobiini (L–N), dorsal and lateral views. A, *Georissus* cf. *kingii* MacLeay; B, *Georissus* sp.; C, *Hydrochus horni* Blackburn; D, *Hydrochus burdekinensis* Watts; E, *Spercheus platycephalus* MacLeay; F, *Amphiops australicus* Blackburn; G, *Allocotocerus punctatus* (Blackburn); H, *Regimbartia attenuata* (Fabricius); I, *Berosus veronicae* Watts; J, *Berosus pulchellus* MacLeay; K, *Berosus involutus* (MacLeay); L, *Laccobius* (*Notoberosus*) *zietzi* (Blackburn); M, *Laccobius* (*Hydroxenus*) *clarus* Gentili; N, *Paracymus pygmaeus* (MacLeay).

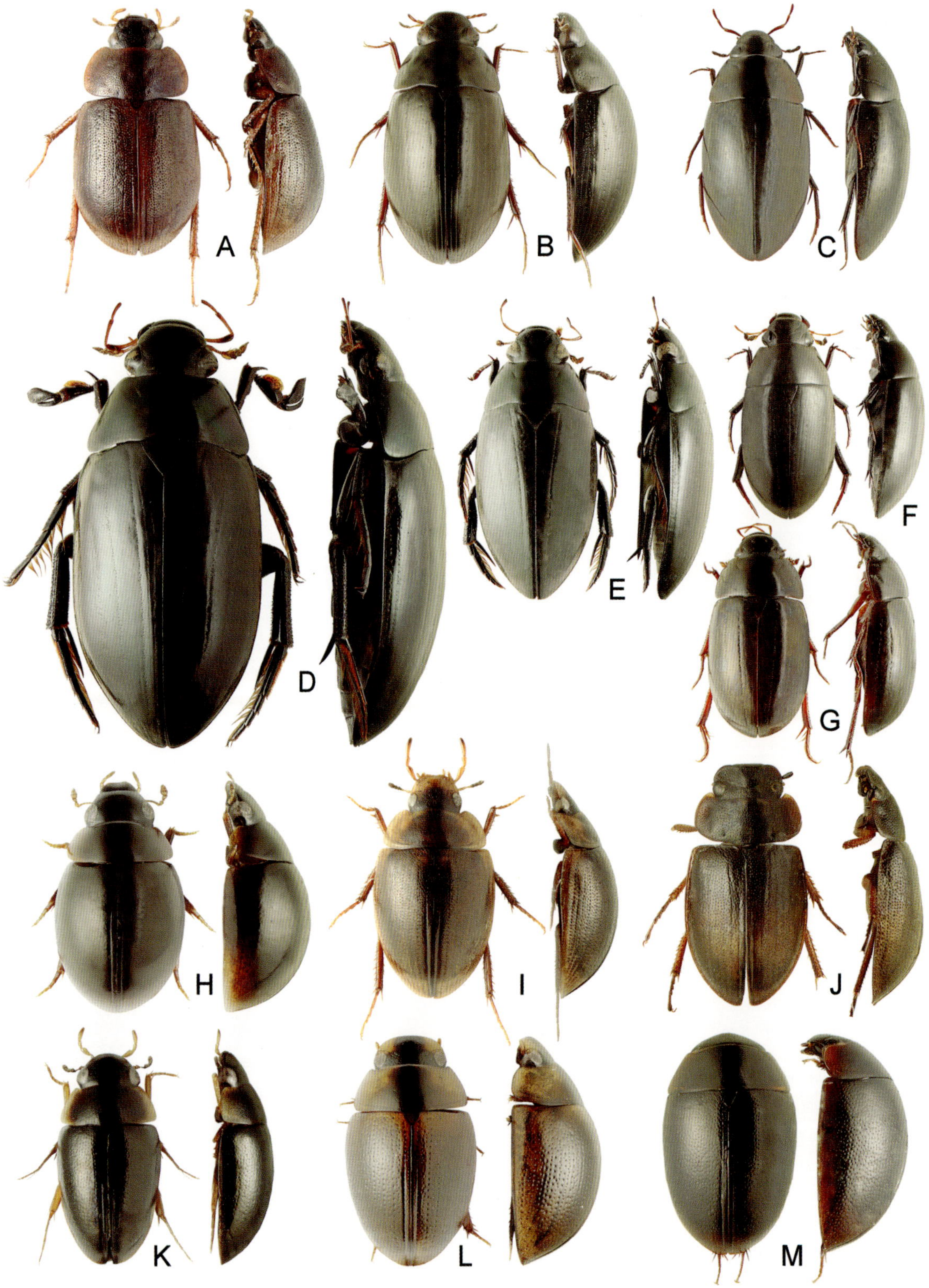

Plate 44. Hydrophilidae. Hydrobiusini (A–B), Hydrophilini (C–G) and Chaetarthriinae (H–M), dorsal and lateral views. A, *Hybogralius harmeyeri* (Régimbart); B, *Limnoxenus zealandicus* (Broun); C, *Sternolophus marginicollis* (Hope); D, *Hydrophilus macronyx* (Régimbart); E, *Hydrophilus bilineatus* (MacLeay); F, *Hydrobiomorpha bovilli* Blackburn; G, *Hydrobiomorpha helenae* Blackburn; H, *Chaetarthria nigerrima* (Blackburn); I, *Notohydrus australis* (Blackburn); J, *Phelea breviceps* Hansen; K, *Crenitulus nitens* (Gentili); L, *Anacaena lindi* (Blackburn); M, *Anacaena littoralis* (Orchymont).

Plate 45. Hydrophilidae: Enochrinae (A–B), Acidocerinae (C–F) and Rygmodinae (G–O), dorsal and lateral views. A, *Enochrus* (*Methydrus*) *eyrensis* (Blackburn); B, *Enochrus* (*Hydatotrephis*) *mastersi* (MacLeay); C, *Chasmogenus* sp.; D, *Agraphydrus coomani* (Orchymont); E, *Helochates* (*Hydrobaticus*) *percyi* Watts; F, *Helochares* (s.str.) *foveicollis* (Montrouzier); G, *Austrotypus nothofagi* Fikáček, Minoshima & Newton; H, *Borborophorus tuberculus* Hansen; I, *Coelostomopsis picea* Hansen; J, *Eurygmus helocharoides* Hansen; K, *Petasopsis brevitarsis* Hansen; L, *Pseudohydrobius flavus* Lea; M, *Pseudohydrobius floricola* Blackburn; N, *Rygmostralia brunnea* Orchymont; O, *Rygmostralia* sp.

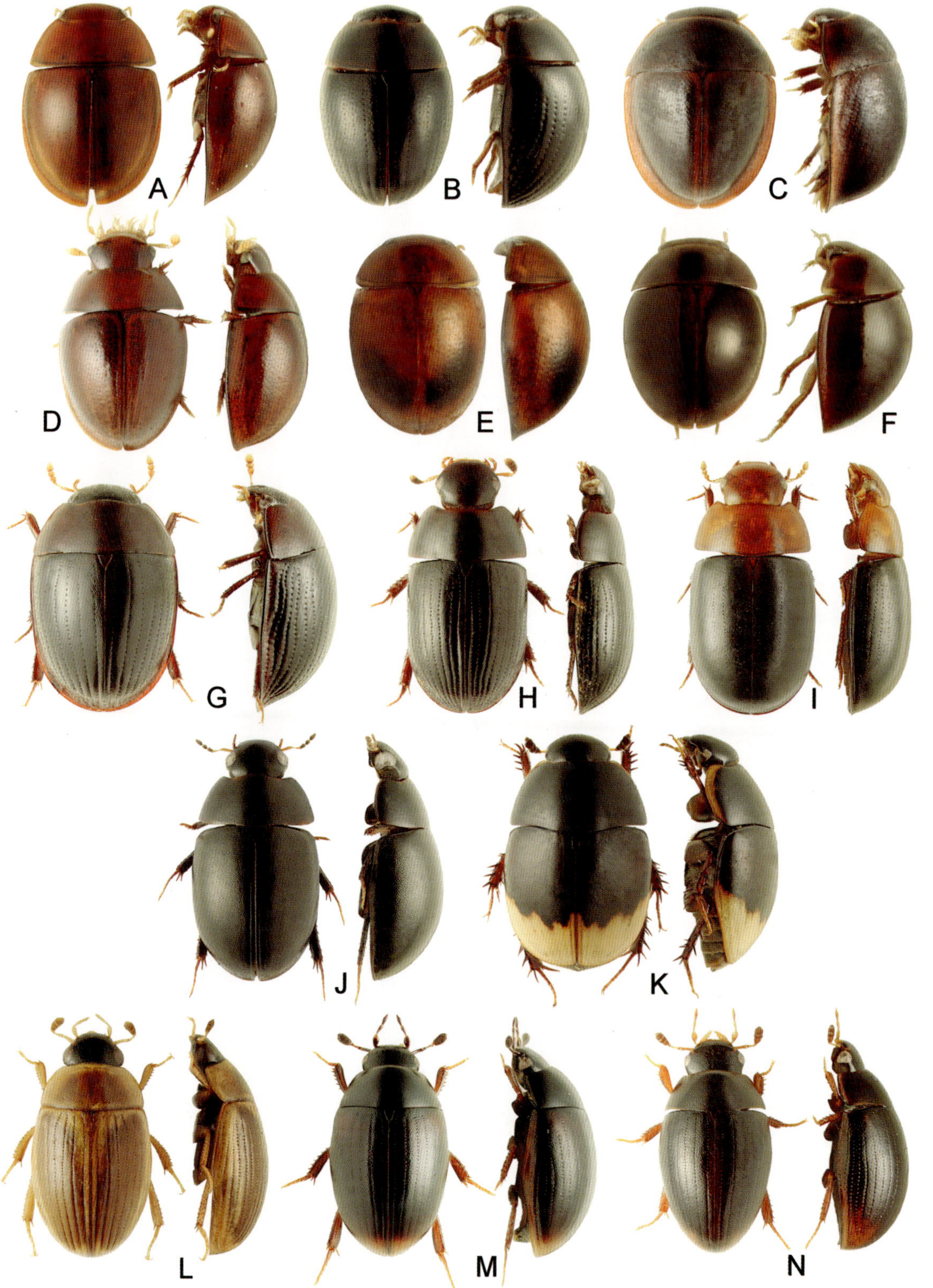

Plate 46. Hydrophilidae. Sphaeridiinae, dorsal and lateral views: A–F, Omicrini. A, *Mircogioton* sp.; B, *Noteropagus* sp.; C–D, *Paromicrus* sp.; E–F, *Psalitrus* sp. G–J, Coelostomatini: G, *Dactylosternum marginale* (Sharp); H, *Dactylosternum abdominale* (Fabricius); I, *Dactylosternum dytiscoides* Fabricius); J, *Coelostoma fabricii* (Montrouzier). K, Sphaeridiini: *Sphaeridium discolor* Orchymont. L–N, introduced species of Megasternini: L, *Cercyon* (*Paracycreon*) *laminatus* Sharp; M, *Cercyon* (s.str.) *haemorrhoidalis* (Fabricius); N, *Cercyon* (*Paracercyon*) *analis* (Paykull).

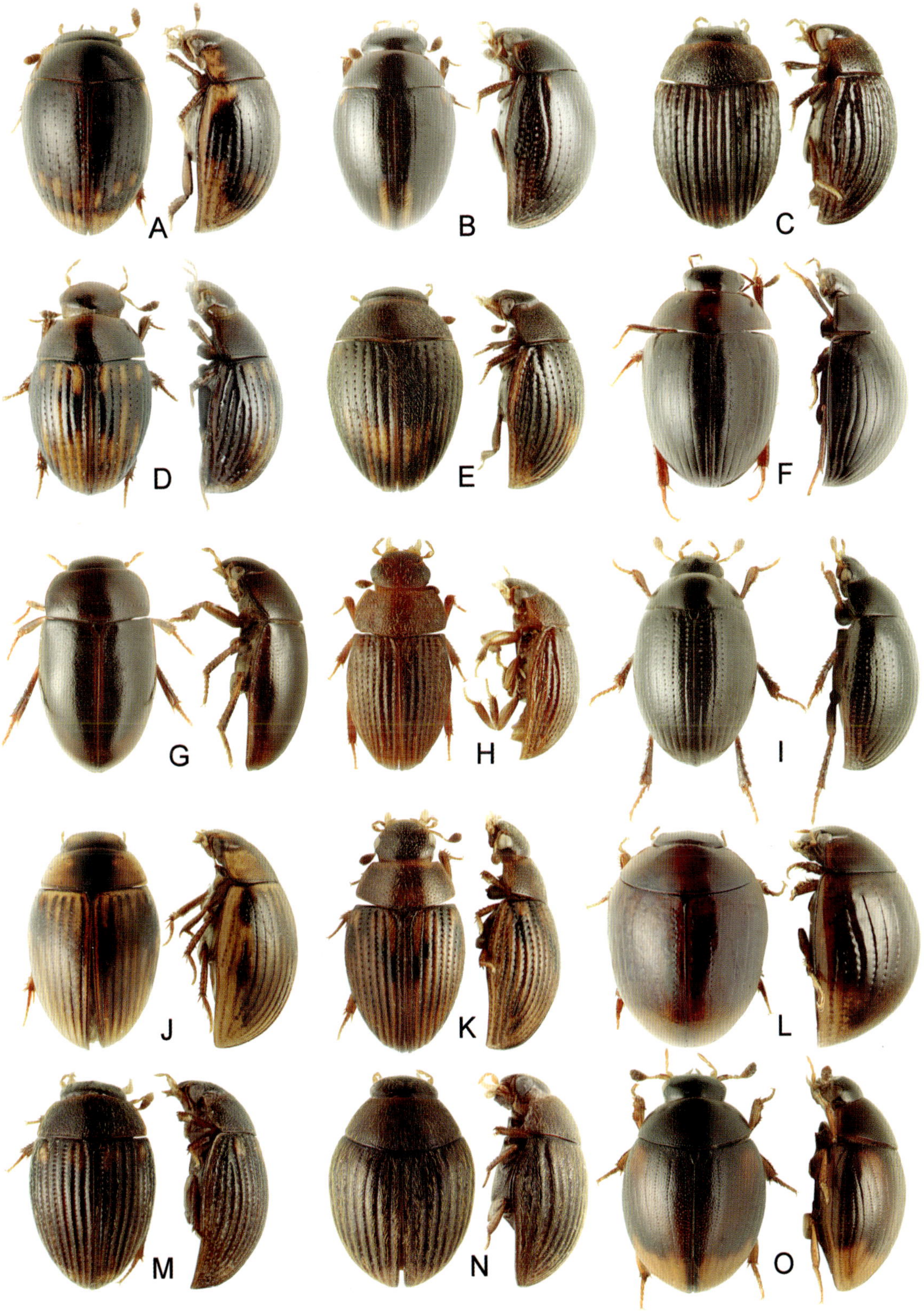

Plate 47. Hydrophilidae. Megasternini, dorsal and lateral views. A, *Australocyon variegatus* Hansen; B, *Australocyon flavolineatus* Hansen; C, *Cenebriophilus costatus* Hansen; D, *Chledocyon marmoratus* Hansen; E, *Chledocyon semiopacus* Hansen; F, *Cetiocyon* sp.; G, *Cercyodes kingensis* (Blackburn); H, *Ceronocyton obscurum* Hansen; I, *Ercycodes fossus* (Blackburn); J, *Merosoma* sp.; K, *Notocercyon ornatus* Blackburn; L, *Pilocnema maculiapex* Hansen; M, *Pseudoosternum maculatum* Hansen; N, *Cryptopleurum* sp.; O, *Pachysternum capense* (Mulsant).

Plate 48. Geotrupidae. A, *Bolbobaenius planiceps* (Macleay), male; B, *Blackburnium angulicorne* (Macleay), male; C, *Blackburnium angulicorne* (Macleay), male (lateral); D, *Blackburnium cavicolle* (Macleay), female; E, *Eucanthus felshei* Boucomont; F, *Geotrupes spiniger* Marsham; G, *Bolborhachium tricavicolle* (Lea), male; H, *Bolborhachium tricavicolle* (Lea), male (lateral); I, *Bolboleaus quadriarmigerus* (Howden), male; J, *Bolboleaus quadriarmigerus* (Howden), male (lateral); K, *Blackbolbus taurus* (Westwood), female; L, *Blackbolbus taurus* (Westwood), male; M, *Australobolbus loweri* (Blackburn), female; N, *Bolborhachium recticorne* (Guérin-Méneville), female; O, *Bolborhachium recticorne* (Guérin-Méneville), male; P, *Bolborhachium recticorne* (Guérin-Méneville), male (lateral).

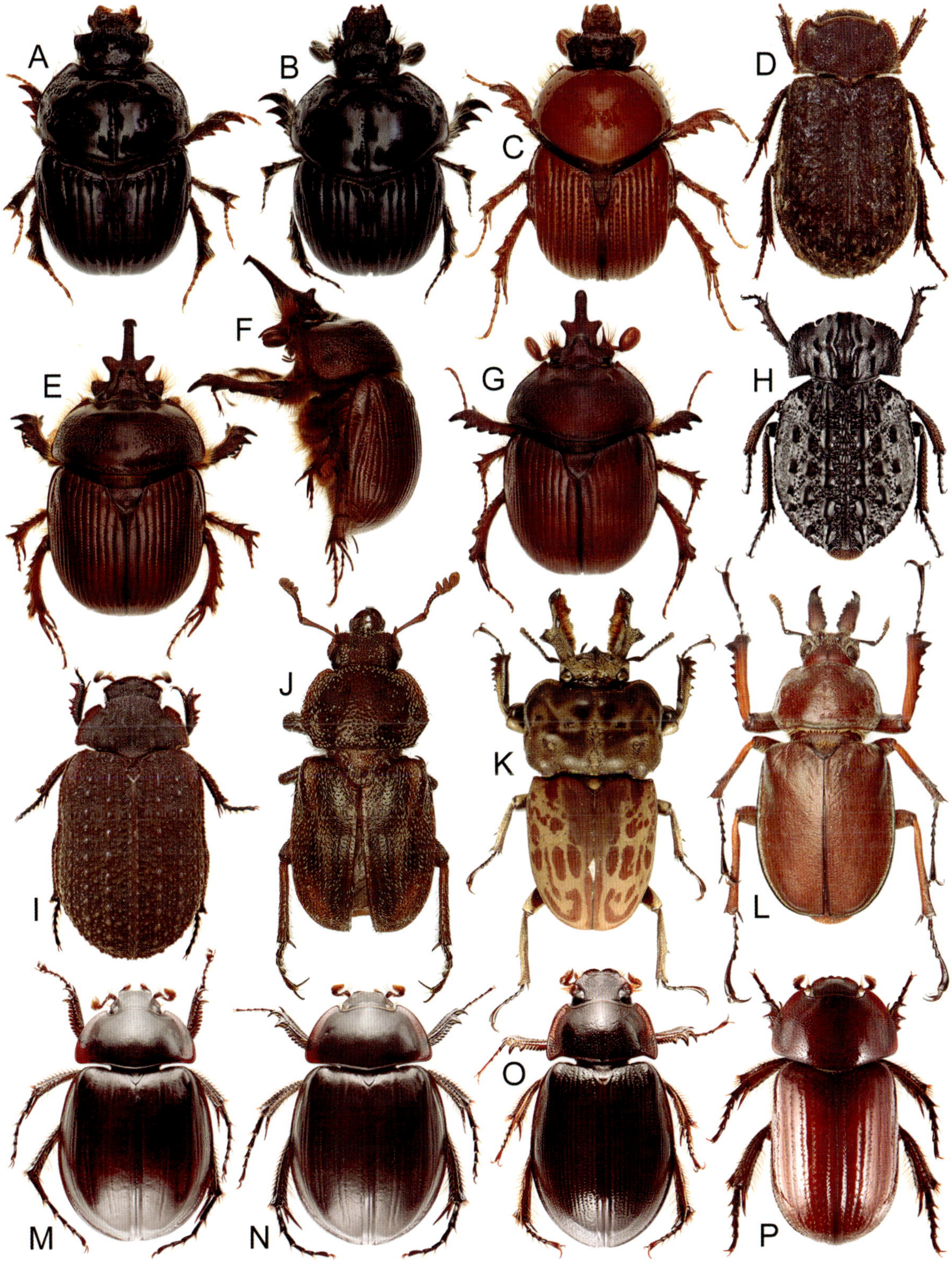

Plate 49. Geotrupidae (A–C, E–G), Trogidae (D, H–I), Lucanidae (J–L), Hybosoridae (M–P). A, *Stenaspidius ruficornis* (Boucomont), female; B, *Stenaspidius nigricornis* (Boucomont), male; C, *Gilletinus bipagus* Howden, female; D, *Trox scaber* (Linnaeus); E, *Elephastomus proboscideus* (Schreibers), male; F, *Elephastomus proboscideus* (Schreibers), male (lateral); G, *Elephastomus gellarus* Carne, male; H, *Omorgus tatei* (Blackburn); I, *Omorgus amictus* (Haaf); J, *Cacostomus subvittatus* (Moore); K, *Ryssonotus nebulosus* (Kirby); L, *Australognathus queenslandicus* (Moore); M, *Liparochrus hackeri* Blackburn, male; N, *Liparochrus hackeri* Blackburn, female; O, *Phaeochrous emarginatus* Laporte; P, *Antiochrus* sp.

Plate 50. Lucanidae. A, *Ceratognathus niger* Westwood; B, *Hololamprima crenulata* Macleay; C, *Cacostomus squamosus* Newman; D, *Lissapterus grammicus* (Lea); E, *Phalacrognathus muelleri* (Macleay); F, *Hoplogonus bornemisszai* Bartolozzi; G, *Lamprima aurata* Latreille male; H, *Lissotes rudis* Lea; I, *Figulus nitens* Waterhouse; J, *Aegus jansoni* Boileau; K, *Lamprima aurata* Latreille, female.

Plate 51. Passalidae (A), Scarabaeidae: Melolonthinae (B), Aphodiinae (C–L). A, *Analaches australiensis* (Stoliczka); B, *Phaenognatha aequistriata* Arrow; C, *Leiopsammodius rufus* (Rakovič); D, *Australammoecius goyderensis* (Blackburn); E, *Australaphodius frenchi* (Blackburn); F, *Australoxenella wurrook* Storey & Howden; G, *Saprus griffithi* Blackburn; H, *Neotrichiorhyssemus hirsutus* (Clouët); I, *Odontolochus weiri* Stebnicka & Howden; J, *Rhyssemus inscitus* (Walker); K, *Ozodius neglectus* (Schmidt); L, *Psammorpha lawrencei* Stebnicka.

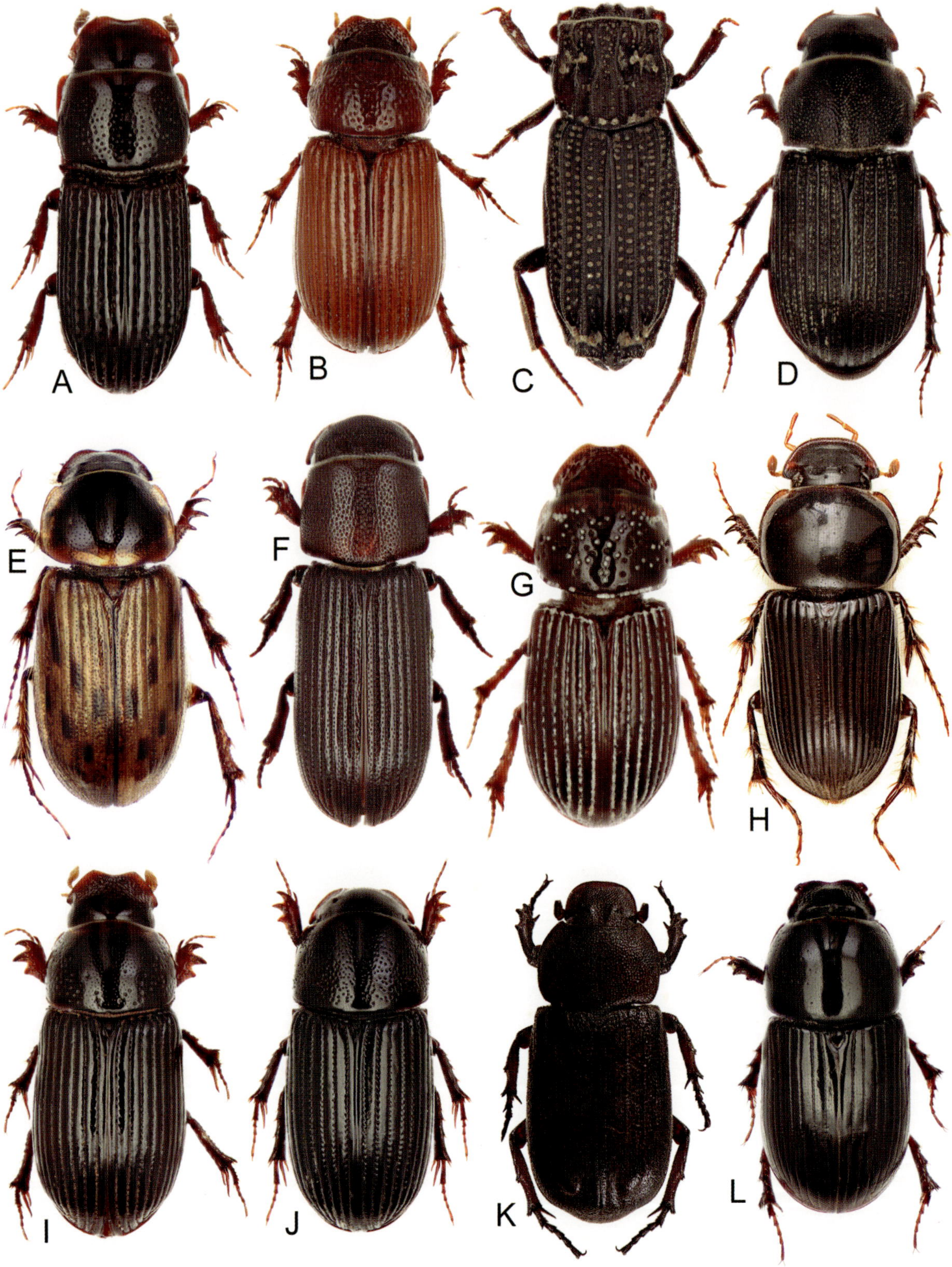

Plate 52. Scarabaeidae: Aphodiinae (A–I, L), Dynastinae (K). A, *Saprosites nitidicollis* (Macleay); B, *Aphodopsammobius zietzi* (Blackburn); C, *Rhyparus helophoroides* Fairmaire; D, *Harmogaster geminata* Schmidt; E, *Candezeolus pseudocandezei* Stebnicka & Howden; F, *Cnematoplatys tozerensis* Stebnicka & Howden; G, *Platyomus tibialis* (Fabricius); H, *Acrossidius* sp.; I, *Parataenius simulator* (Harold); J, *Ataenius australasiae* (Boheman); K, *Cryptodus paradoxus* Macleay; L, *Aphodius granarius* (Linnaeus).

Plate 53. Scarabaeidae: Aphodiinae (A–D, L), Dynastinae (E–K). A, *Euparotrix squamosa* (Lea); B, *Acrossidius tasmaniae* (Hope); C, *Podotenus suberosus* (Blackburn); D, *Gongrolophus storeyi* Stebnicka & Howden; E, *Haploscapanes australicus* (Arrow) (lateral); F, *Pseudoryctes mullerianus* (White); G, *Haploscapanes australicus* (Arrow); H, *Xylotrupes australicus australicus* Thomson; I, *Papuana woodlarkiana* (Montrouzier); J, *Xylotrupes australicus australicus* Thomson (lateral); K, *Heteronychus arator* (Fabricius); L, *Proctophanes sculptus* (Hope).

Plate 54. Scarabaeidae: Dynastinae. A, *Oryctoderinus walfordorum* Endrödi; B, *Adoryphorus coulonii* (Burmeister); C, *Neonastes glabricollis* (Malceay); D, *Trissodon australasiae* (Hope); E, *Carneodon pecuarius* (Reiche); F, *Teinogenys nitidus* Sharp; G, *Neocorynophyllus aputaringus* Carne; H, *Temnorrhynchus retusus* (Fabricius); I, *Enracius longipes* Dechambre; J, *Anomalomorpha anthracina* Arrow; K, *Metanastes vulgivagus* (Olliff); L, *Oryctoderus latitarsis* Boisduval; M, *Pentodon algerinum* (Herbst); N, *Phylliocephala nigrohirta* Blackburn; O, *Epironastes limbatus* Carne; P, *Neocnecus punctata* (Sharp).

Plate 55. Scarabaeida: Dynastinae (A–O), Aphodiinae (P). A, *Phylliocephala interioris* (Blackburn); B, *Dasygnathus dejeani* W.S. Macleay; C, *Cavonus armatus* Sharp; D, *Dipelicus optatus* (Sharp); E, *Novapus crassus* Sharp; F, *Hyphoryctes maculatus* Blackburn; G, *Corynophyllus fortnumi* Hope; H, *Pseudocavonus antennalis* Blackburn; I, *Semanopterus subcostatus* (Laporte); J, *Carneoryctes brittoni* (Carne); K, *Cyclocephala signaticollis* Burmeister; L, *Haploscapanes barbarossa* (Fabricius); M, *Neodasygnathus juba* (Kirby); N, *Novapus adelaidae* Blackburn; O, *Cheiroplatys latipes* (Guérin-Méneville); P, *Tesarius sulcipennis* (Lea).

Plate 56. Scarabaeidae: Dynastinae (A, C, G), Melolonthinae (B, D, F, H–K); Hybosoridae (E). A, *Oryctoderus latitarsis* Boisduval; B, *Anacanthodes inermis* (Blackburn); C, *Eophileurus dentatus* (Blackburn); D, *Hadrops flavus* (Lea); E, *Pterorthochaetes danielsi* Ballerio; F, *Evbrittonia longitarsis* Szito; G, *Xylotrupes* sp.; H, *Leonotus pilosicollis* (Lea); I, *Acheilo clypeatus* Britton; J, *Allara insularis* Britton; K, *Adossa sylvicola* Britton.

Plate 57. Scarabaeidae: Rutelinae (A–G, I, L–P); Melolonthinae (H, J, K). A, *Parastasia terrareginae* Kuijten; B, *Bilobatus luridipennis* (Waterhouse); C, *Wambo puticasus* Allsopp; D, *Dungoorus frater* Smith; E, *Mimadoretus niveosquamosus* Lea; F, *Amblochilus bicolor* Blanchard; G, *Amblyterus simplicitarsus* Carne; H, *Homolotropus metallicus* Britton; I, *Exochogenys nigripennis* (Blanchard); J, *Liparetrus karallus* Britton; K, *Macleayella singularis* (Blackburn); L, *Pseudoschizognathus lajoyi* Ohaus; M, *Saulostomus villosus* Waterhouse; N, *Mimadoretus flavomaculatus* (Macleay); O, *Trioplognathus griseopilosus* (Ohaus); P, *Schizognathus* sp.

Plate 58. Scarabaeidae: Melolonthinae. A, *Aphanesia greyi* Britton; B, *Aneucomides coloratus* Blackburn; C, *Callabonnica propria* Blackburn; D, *Ancylonyx* sp.; E, *Automolius depressus* (Blanchard); F, *Aplopsis lineoligera* Blanchard; G, *Anthotocus luridus* (Macleay); H, *Biphyllocera kirbyana* White; I, *Antitrogus tasmanicus* Burmeister; J, *Byrrhomorpha verres* Blackburn; K, *Brittonius castaneus* (Britton); L, *Astibicola bicolor* Britton.

Plate 59. Scarabaeidae: Melolonthinae (A–B, E–H, K–L), Aphodiinae (D), Dynastinae (C, I–J). A, *Apogonia* sp.; B, *Idanastes abditus* Britton; C, *Onychionyx australicus* Dechambre; D, *Airapus obscurus* (Macleay); E, *Hypolepida wilsoni* Britton; F, *Maechidinus sculptilis* Britton; G, *Glossocheilifer labialis* Blackburn; H, *Pseudoheteronyx creber* Blackburn; I, *Metanastes bicornis* Lea; J, *Pimelopus porcellus* Erichson; K, *Neoheteronyx cribrifrons* (Lea); L, *Maechidius variolosus* Macleay.

Plate 60. Scarabaeidae: Melolonthinae. A, *Chariochilus costipennis* Moser; B, *Allothnonius brooksi* Britton; C, *Cheirora interstitialis* (Blackburn); D, *Cheirodontus bilobus* (Lea); E, *Colobostoma* sp.; F, *Cheiragra ruficollis* Macleay; G, *Chilodiplus weiri* Allsopp; H, *Colpochila andersoni* Britton; I, *Cheilo liparetroides* Britton; J, *Colpochelyne dux* Britton; K, *Colpochilodes peregrina* Britton; L, *Colymbomorpha vittata* Britton.

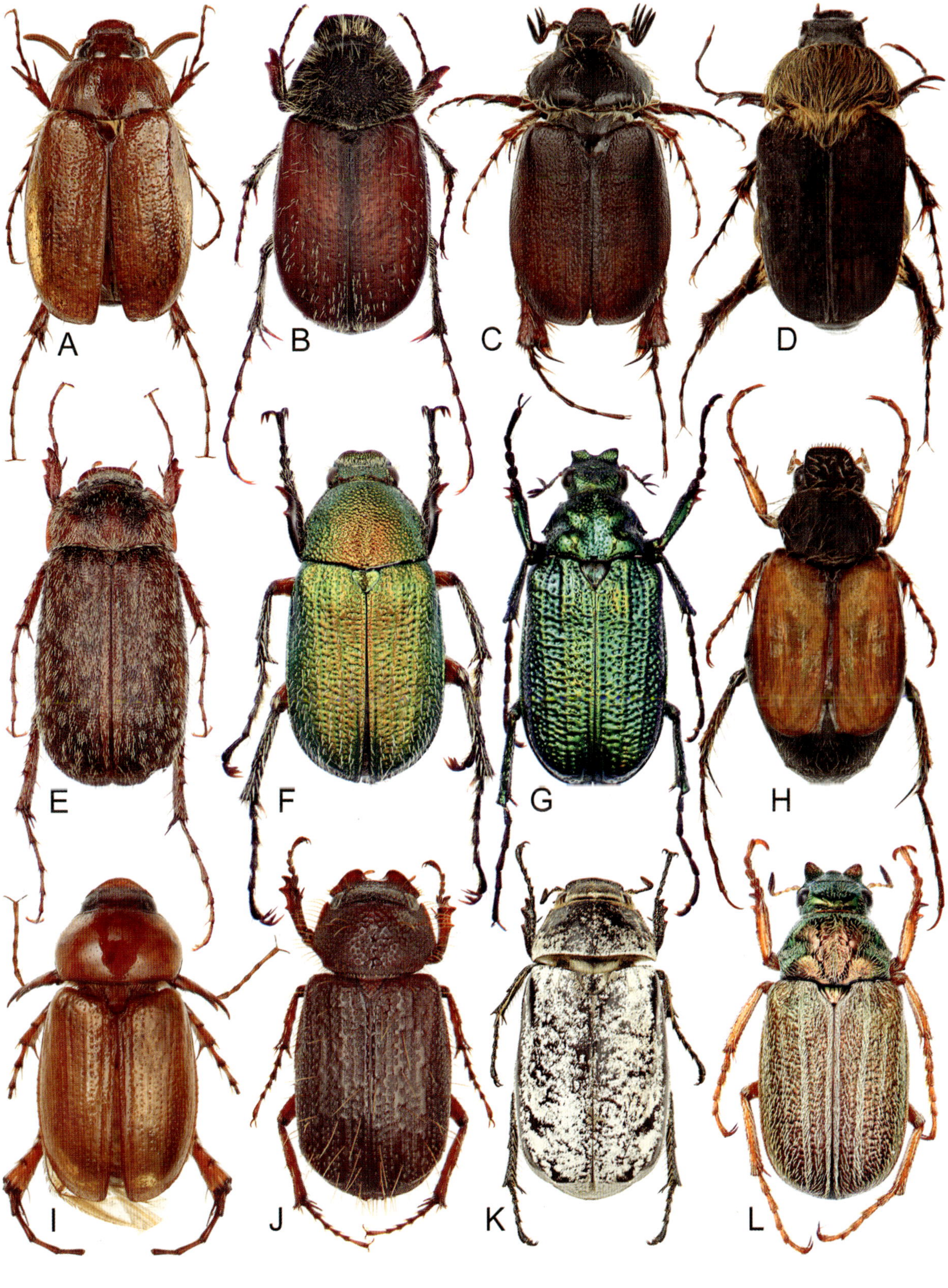

Plate 61. Scarabaeidae: Melolonthinae. A, *Dysphanochila pilosipennis* Blackburn; B, *Deuterocaulobius villosus* (Guillou); C, *Dikellites abditus* Britton; D, *Enamillus striatus* Sharp; E, *Eurychelus marmoratus* Blanchard; F, *Cunderdinia orientalis* Britton; G, *Diphucephala elegans* Blackburn; H, *Comophorina testaceipennis* (Blanchard); I, *Engyopsina spectans* (Blackburn); J, *Epholcis bilobiceps* (Fairmaire); K, *Dermolepida albohirtum* (Waterhouse); L, *Diphucephala lineata* Boisduval.

Plate 62. Scarabaeidae: Melolonthinae. A, *Holorhopaea sagata* Britton; B, *Harpechys chilo* Britton; C, *Exostethus brunneus* Britton; D, *Ictigaster ruficollis* (Lea); E, *Megacoryne carnabyorum* Britton; F, *Heteronyx australis* Guériin-Méneville; G, *Luftius parva* (Szito); H, *Hadropechys victoriae* (Blackburn); I, *Lepidiota caudata* Blackburn; J, *Gnaphalopoda proxima* Britton; K, *Frenchella hirticollis* Blackburn; L, *Megarhopaea gigas* (Lea).

Plate 63. Scarabaeidae: Melolonthinae. A, *Nitorellus splendidus* (Lea); B, *Pachygastra tasmanica* Germar; C, *Microthopus castanopterus* Burmeister; D, *Neophyllotocus dispar* (Blackburn); E, *Microcoenus nanus* Britton; F, *Opsitocus obscurus* (Lea); G, *Nosphisthis perkinsi* Blackburn; H, *Ophropyx hispida* (Blackburn); I, *Neso flavipennis* (Macleay); J, *Odontotonyx ruficeps* Lea; K, *Metatrogus septuosus* Britton; L, *Othnonius batesi* Olliff.

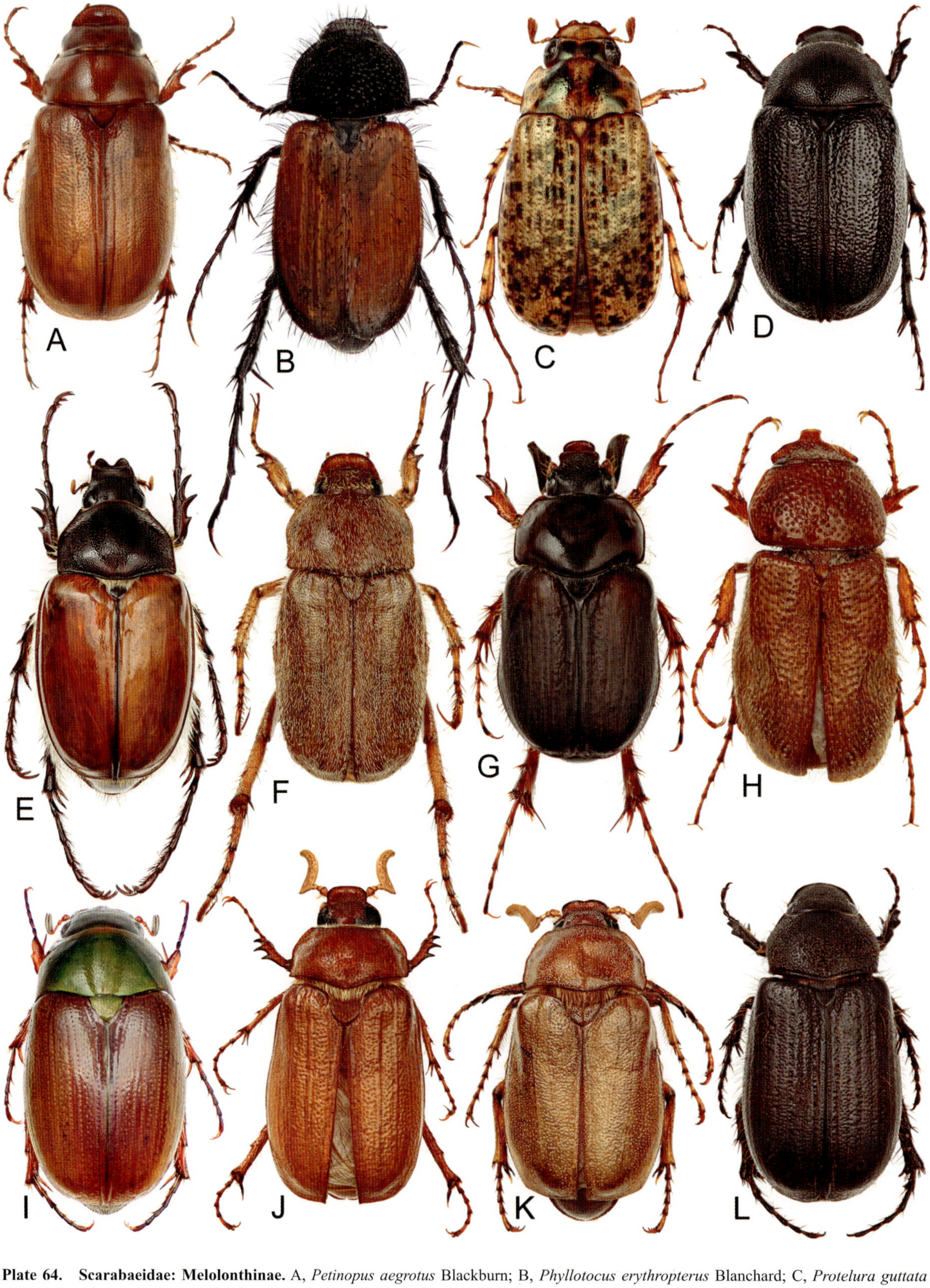

Plate 64. Scarabaeidae: Melolonthinae. A, *Petinopus aegrotus* Blackburn; B, *Phyllotocus erythropterus* Blanchard; C, *Protelura guttata* Britton; D, *Phorine anomala* (Blackburn); E, *Pachytricha castanea* Hope; F, *Plectris aliena* Chapin; G, *Prochelyna heterodoxa* Burmeister; H, *Proborhinus cornutus* (Blackburn); I, *Phyllococerus purpurascens* Waterhouse; J, *Pararhopaea callabonensis* (Blackburn); K, *Pseudoholophylla cavifrons* (Lea); L, *Paronyx setifera* Britton.

Plate 65. Scarabaeidae: Melolonthinae. A, *Scitala aureorufa* (Blanchard); B, *Sericesthis brooksi* Britton; C, *Phyllotocidium bimaculiflavum* Lea; D, *Sphaeroscelis pectoralis* Burmeister; E, *Sarothromerus aranda* Allsopp; F, *Colobostoma hirsuta* (Frey); G, *Synchilus gisleni* Britton; H, *Sphyrocallus desertus* Allsopp; I, *Stenochelyne noctis* Britton; J, *Sciton* sp.; K, *Scitaloides malanda* Britton; L, *Rhopea magnicornis* Blackburn.

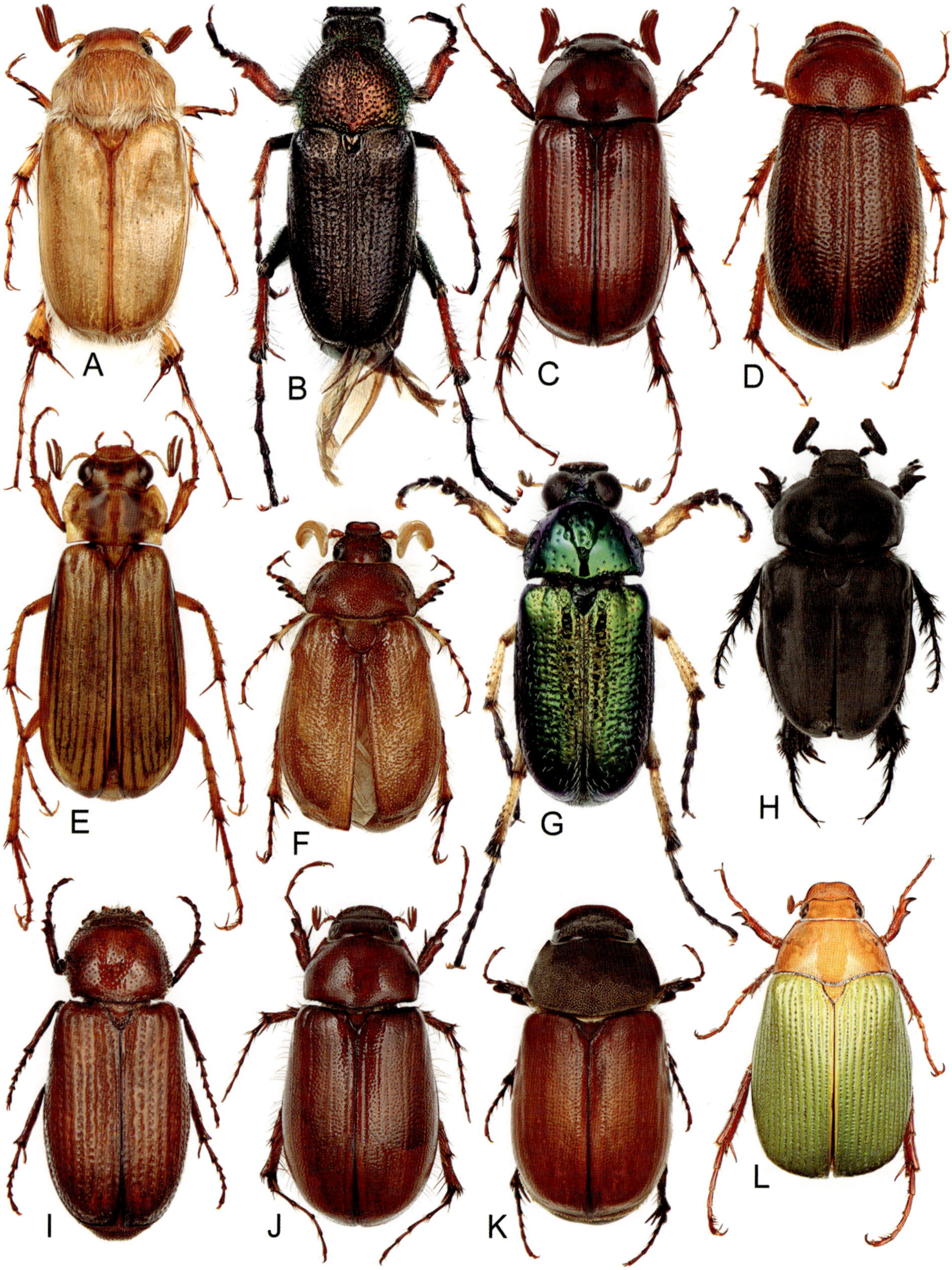

Plate 66. Scarabaeidae: Melolonthinae. A, *Xyridea hirticollis* (Blackburn); B, *Watkinsia bella* Britton; C, *Xyrine inusitatus* Britton; D, *Webbella labralis* (Blackburn); E, *Telura alta* Britton; F, *Zietzia geologa* Blackburn; G, *Watkinsia megalops* Britton; H, *Systellopus obtusus* Sharp; I, *Termitophilus spadix* Britton; J, *Xyroa nitida* Britton; K, *Xyrodes calorata* (Blackburn); L, *Xylonichus eucalypti* Boisduval.

Plate 67. Scarabaeidae: Rutelinae. A, *Calloodes atkinsoni* Waterhouse; B, *Anoplognathus aureus* Waterhouse; C, *Calloodes rayneri* Macleay; D, *Calloodes grayianus* (White); E, *Anoplognathus abnormis* Macleay; F, *Anoplostethus laetus* Rothschild & Jordon; G, *Epichrysus lamprimoides* (White); H, *Anoplognathus aeneus* Waterhouse; I, *Calloodes frenchi* Ohaus; J, *Repsimus manicatus* (Swartz); K, *Anoplognathus punctulatus* Olliff; L, *Anomala antiqua* (Gyllenhal).

Plate 68. Scarabaeidae: Rutelinae (A–B, F–H), Scarabaeinae (C–E, I–L). A, *Mesystoechus ciliatus* Waterhouse; B, *Eosaulostomus halei* Carne; C, *Demarziella metallica* (Carter); D, *Demarziella geminata* (Macleay); E, *Aptenocanthon hopsoni* (Carter); F, *Repsimus manicatus* (Swartz) male; G, *Repsimus aeneus* (Fabricius); H, *Anoplognathus* sp.; I, *Amphistomus complanatus* Matthews; J, *Aulacopris reichei* White; K, *Boletoscapter cornutus* (Macleay) female; L, *Boletoscapter cornutus* (Macleay) male.

Plate 69. Scarabaeidae: Scarabaeinae. A, *Bubas bison* (Linnaeus), female (lateral); B, *Digitonthophagus gazella* (Fabricius), male (lateral); C, *Copris hispanus* Linnaeus, female (lateral); D, *Copris hispanus* Linnaeus, male (lateral); E, *Digitonthophagus gazella* (Fabricius), female (lateral); F, *Bubas bison* (Linnaeus), male (lateral); G, *Cephalodesmius armiger* Westwood; H, *Bubas bison* (Linnaeus), male; I, *Copris hispanus* Linnaeus, male; J, *Diorygopyx niger,* Matthews, male; K, *Canthonosoma macleayi* (Harold); L, *Bubas bison* (Linnaeus), female; M, *Digitonthophagus gazella* (Fabricius), female; N, *Coproecus hemisphaericus* (Guérin-Méneville); O, *Copris hispanus* Linnaeus, female; P, *Coptodactyla glabricollis* (Hope), male; Q, *Coptodactyla glabricollis* (Hope), female; R, *Digitonthophagus gazella* (Fabricius), male.

Plate 70. Scarabaeidae: Scarabaeinae. A, *Onthophagus propinquus* Macleay, male (lateral); B, *Onitis alexis* Klug, male (lateral); C, *Onitis alexis* Klug, male (lateral); D, *Onitis alexis* Klug, female (lateral); E, *Onthophagus propinquus* Macleay, female (lateral); F, *Matthewsius rossi* (Matthews); G, *Euoniticellus intermedius* (Reiche), male; H, *Euoniticellus intermedius* (Reiche), female; I, *Liatongus militaris* (Laporte); J, *Onitis alexis* Klug, male; K, *Onitis alexis* Klug, female; L, *Onthophagus propinquus* Macleay, male; M. *Labroma umbratilis* Matthews; N, *Monoplistes curvipes* Lea; O. *Onthophagus propinquus* Macleay, female; P. *Lepanus storeyi* Weir & Monteith; Q. *Mentophilus hollandiae* (Laporte).

Plate 71. Scarabaeidae: Scarabaeinae (A–C, E–G), Histeridae (D, H–L). A, *Pseudignambia squamata* (Matthews); B, *Thyregis kershawi* Blackburn; C, *Temnoplectron boucomonti* Paulian; D, *Aulacosternus* sp.; E, *Tesserodon tenebroides* Matthews; F, *Neosisyphus spinipes* (Thunberg); G, *Sauvagesinella becki* (Paulian); H, *Stictostix parra* (Marseul); I, *Tribalus australis* (Macleay); J, *Teretrius melburnius* Marseul; K, *Sunilis robustus* (Schmidt); L, *Saprinus splendens* (Paykull).

Plate 72. Passalidae (A–D, I–L), Lucanidae (E–G, H). A, *Pharochilus dilatatus* (Dalman); B, *Aulacocyclus deyrolli* Kaup; C, *Labienus ptox* (Kaup); D, *Gonatas* sp.; E, *Syndesus cornutus* (Fabricius); F, *Prosopocoilus torresensis* (Deyrolle); G, *Dorcus wickhami* (Waterhouse); H, *Lamprima aurata* Latreille; I, *Leptaulax* sp.; J, *Austropassalus hultgreni* Mjöberg; K, *Mastachilus australasicus* (Percheron); L, *Protomocoelus australis* (Boisduval).

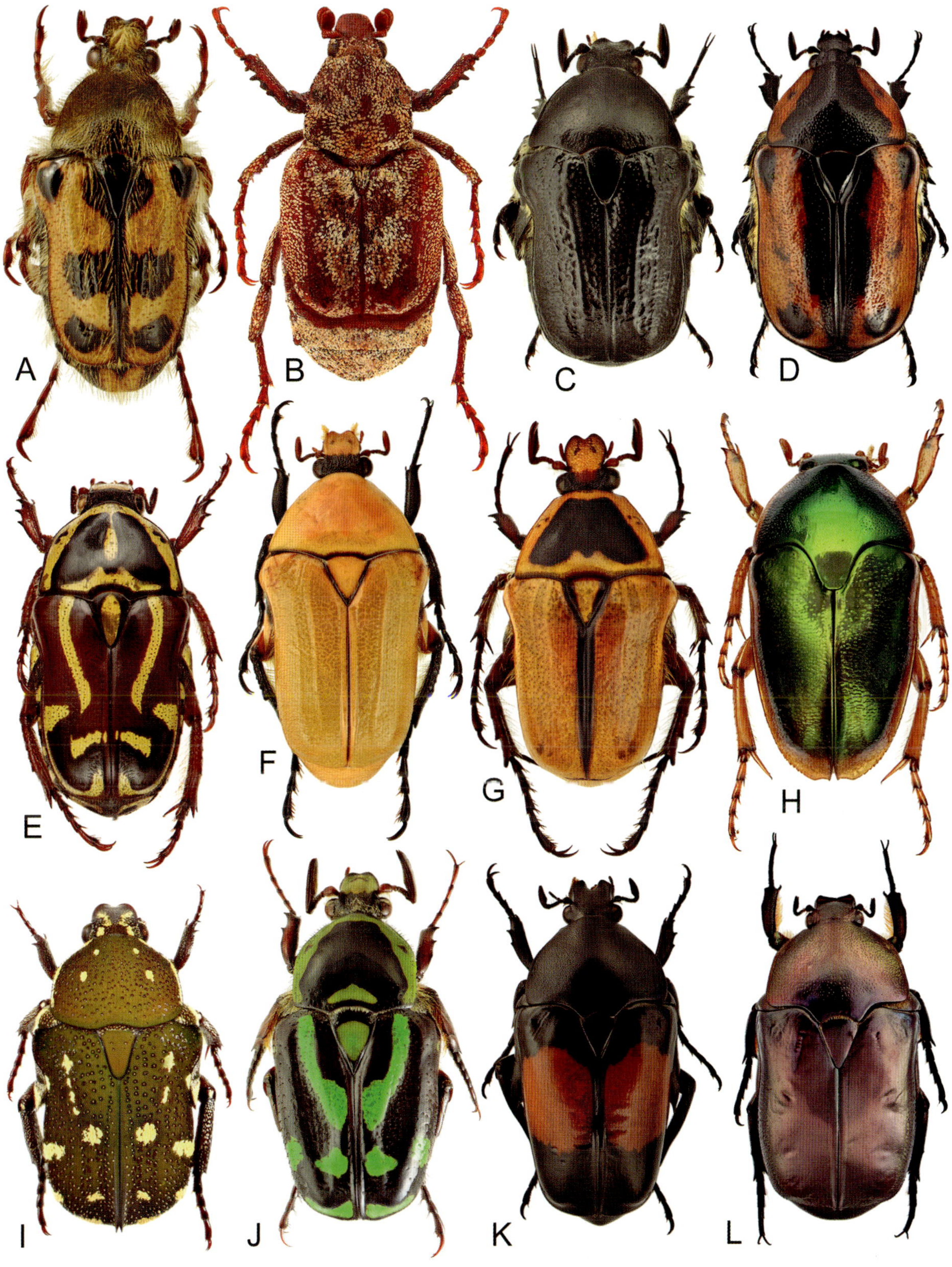

Plate 73. Scarabaeidae: Cetoniinae. A, *Clithria eucnemis* (Burmeister); B, *Charitovalgus quinquedentatus* (Lea); C, *Bisallardiana philippei* (Allard); D, *Chondropyga dorsalis* (Donovan); E, *Eupoecila australasiae* (Donovan); F, *Aphanesthes succinea* (Hope); G, *Diaphonia antoinei* Allard; H, *Mycterophallus duboulayi* (Thomson); I, *Glycyphana stolata* (Fabricius); J, *Chlorobapta frontalis* (Donovan); K, *Dilochrosis balteata* (Vollenhoven); L, *Chalcopharis lansbergei* (Gestro).

Plate 74. Scarabaeidae: Cetoniinae. A, *Grandaustralis boomerang* Hutchinson & Moeseneder; B, *Lomaptera cinnamomea* Thomson; C, *Lyraphora obliquata* (Westwood); D, *Protaetia acuminata* (Fabricius); E, *Lenosoma fulgens* (Macleay); F, *Hemichnoodes mniszechi* (Janson); G, *Microvalgus* sp.; H, *Macrotina satanas* Schürhoff; I, *Ischiopsopha wallacei* (Thomson); J, *Micropoecila cincta* (Gory & Percheron); K, *Metallesthes metallescens* (White); L, *Hemipharis insularis* (Gory & Percheron).

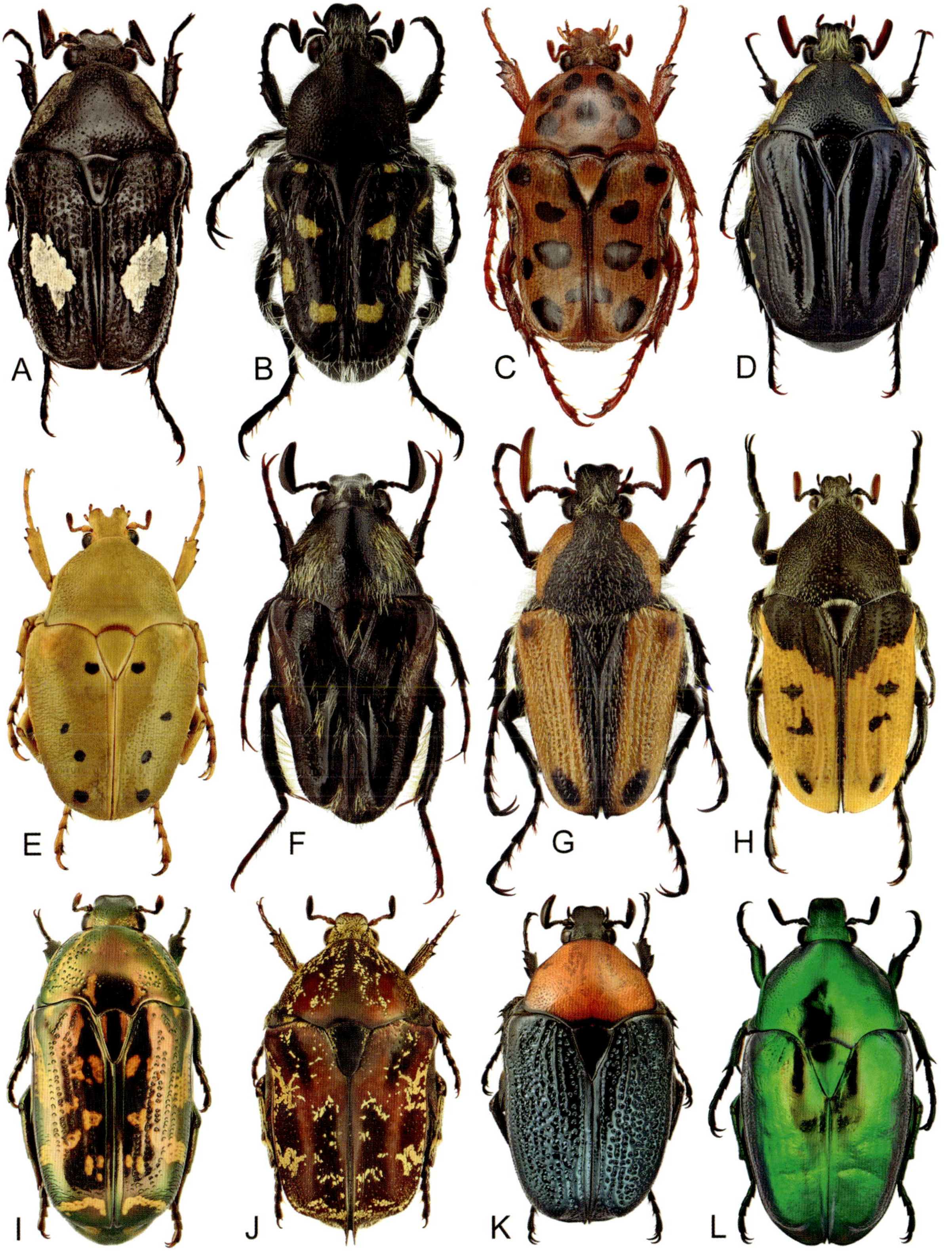

Plate 75. Scarabaeidae: Cetoniinae. A, *Bisallardiana decorticata* (Macleay); B, *Neoclithria eburneoguttata* (Blanchard); C, *Neorrhina punctatum* (Donovan); D, *Pseudoclithria hirticeps* (Macleay); E, *Schizorhina atropunctata* (Kirby); F, *Storeyus pseudodipterus* Hasenpusch & Moeseneder; G, *Octocollis setosus* Moeseneder & Hutchinson; H, *Phyllopodium palmatum* (Schaum); I, *Poecilopharis leai* Schürhoff; J, *Protaetia fusca* (Herbst); K, *Tapinoschema digglesii* (Janson); L, *Stenopisthes frenchi* (Blackburn).

Plate 76. Lucanidae (A–B, I, L), Geotrupidae (C), Rutelinae (D, F, H, J–K), Cetoniinae (E, G). A, *Safrina polita* (Carter); B, *Pseudodorcus nitidus* Didier; C, *Geotrupes spiniger* Marsham, ventral; D, *Anoplognathus smaragdinus* Ohaus; E, *Evanides bakewellii* (White); F, *Anoplognathus parvulus* Waterhouse; G, *Trichaulax philipsii* (Schreibers); H, *Anoplognathus velutinus* Boisduval; I, *Dorculus* sp.; J, *Anoplognathus viridiaeneus* (Donovan); K, *Anoplognathus rubiginosus* Macleay; L, *Mitophyllus ocularis* (Carter).

Plate 77. Buprestidae. A, *Barakula petersonorum* Peterson; B, *Astraeus flavopictus* Laporte & Gory; C, *Anthaxomorphus bogainvillensis* Williams & Weir (Papua New Guinea); D, *Anthaxoschema terraereginae* Obenberger; E, *Australorhipis aphanochila* Bellamy; F, *Chrysobothris australasiae* Hope; G, *Araucariana queenslandica* Levey; H, *Anilara adelaidae* (Hope); I, *Aaaba nodosus* (Deyrolle); J, *Agrilus hypoleucus* Gory & Laporte; K, *Aphanisticus endeloides* Carter; L, *Austrophorella quadrisignata* (with yellow flocculence); M, *Austrochalcophora subfasciata* (Carter).

Plate 78. Buprestidae, dorsal (except for B and N). A, *Burnsiellus marmorata* (Blackburn); B, *Burnsiellus marmorata* (Blackburn), lateral; C, *Chalcophorotaenia quadriimpressa* (Waterhouse); D, *Calotemognatha yarrelli* (Gory & Laporte); E, *Calotemognatha varicollis* (Carter); F, *Belionota prasina* (Thunberg); G, *Cyrioides imperialis* (Fabricius); H, *Cyphogastra farinosa* (Fabricius); I, *Cyphogastra pistor* (Laporte & Gory); J, *Bubastes sphaenoida* Laporte & Gory; K, *Castiarina pertyi* (Gory & Laporte); L, *Castiarina shellybarkeri* Nylander; M, *Calodema regale* (Gory & Laporte); N, *Castiarina shellybarkeri* Nylander, ventral.

Plate 79. Buprestidae. A, *Germarica liliputana* (Thomson); B, *Diphucrania leucosticta* (Kirby); C, *Endelus* sp. (New Britain); D, *Helferella frenchi* (Théry); E, *Habroloma australis* (Macleay); F, *Hedwidgiella jurecki* Obenberger; G, *Ethonion fissiceps* (Kirby); H, *Chrysodema aurofoveata* (Guérin-Méneville); I, *Buprestis aurulenta* (Linnaeus); J, *Dinocephalia cyanipennis* (Blackburn); K, *Chrysodema subfasciata* (Carter); L, *Diadoxus erythrurus* (White); M, *Euryspilus chalcodes* (Gory & Laporte).

Plate 80. Buprestidae. A, *Julodimorpha bakewelli* (White); B, *Metaxymorpha gloriosa* Blackburn; C, *Melobasis cuprifera* (Gory & Laporte); D, *Meliboeithon intermedium* (Kerremans); E, *Maoraxia auroimpressa* Carter); F, *Metataenia aurofoveata* (Saunders); G, *Hypocisseis latipennis* (Macleay); H, *Merimna atrata* (Gory & Laporte); I, *Melobasis cupriceps* (Kirby); J, *Microcastalia globithorax* (Thomson); K, *Microcastalia* sp. (with white flocculence); L, *Iridotaenia bellicosa* (Blackburn); M, *Nascio vetusta* (Boisduval).

Plate 81. Buprestidae. A, *Neospades rugiceps* (Thomson); B, *Pachyschelus* sp. (Broome, WA); C, *Paratrachys australia* Bellamy & Williams; D, *Notobubastes orientalis* Carter; E, *Paracupta lambertii* (Laporte & Gory); F, *Nascioides parryi* (Hope); G, *Neobubastes aureocincta* Blackburn; H, *Paracephala pistacina* (Hope); I, *Neobuprestis frenchi* (Blackburn); J, *Neocuris carnabyae* Barker; K, *Neobuprestis perroni* (Gory & Laporte); L, *Notographus sulcipennis* (Macleay); M, *Pachycisseis bicolor* (Gory & Laporte).

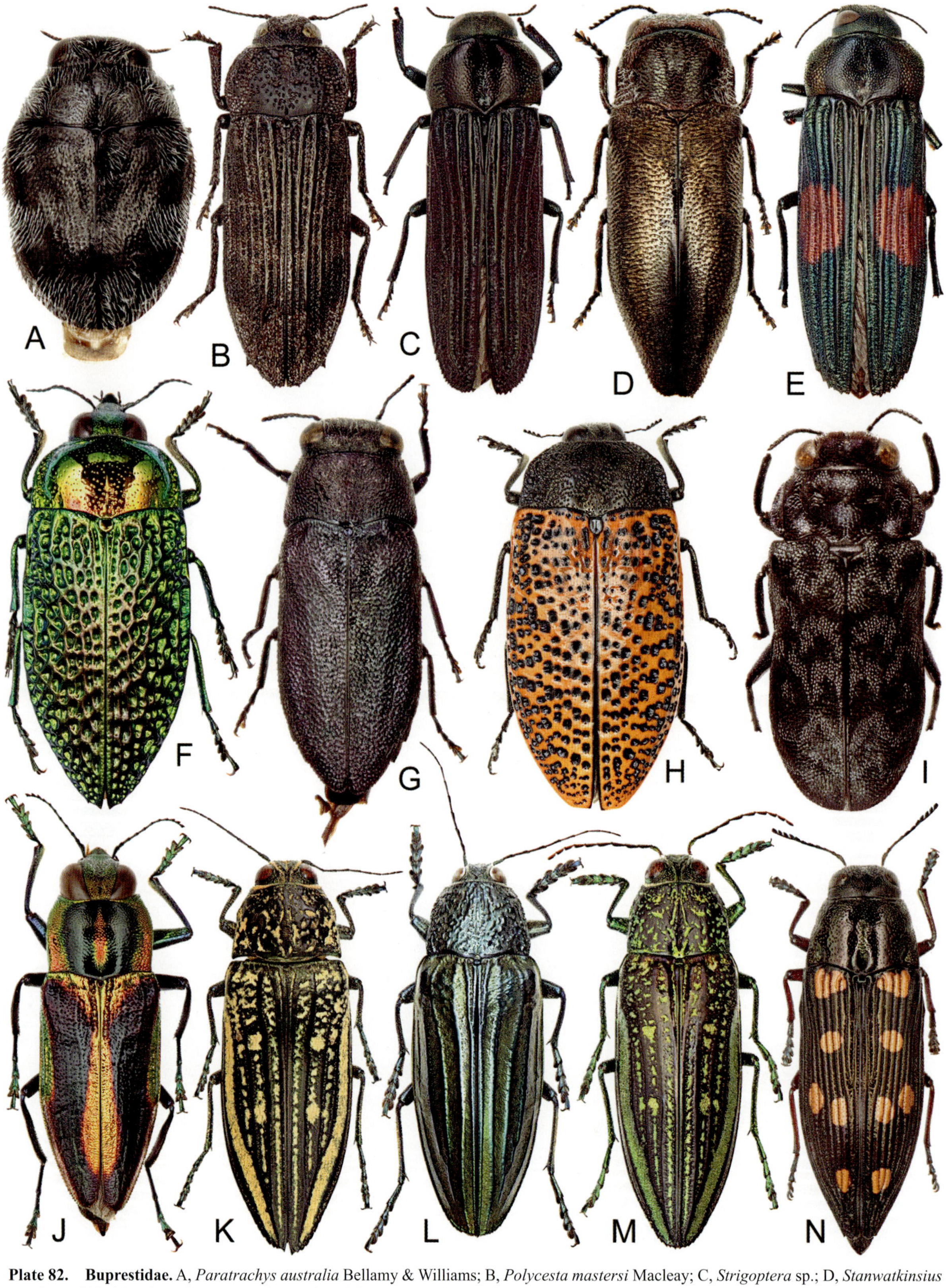

Plate 82. Buprestidae. A, *Paratrachys australia* Bellamy & Williams; B, *Polycesta mastersi* Macleay; C, *Strigoptera* sp.; D, *Stanwatkinsius perplexus* (Blackburn); E, *Strigoptera bimaculata* (Linnaeus); F, *Stigmodera gratiosa* Chevrolat; G, *Pseudanilara purpureicollis* (Macleay); H, *Stigmodera macularia* (Donovan); I, *Sambus australis* Bellamy & Peterson; J, *Selagis caloptera* (Boisduval); K, *Pseudotaenia salamandra* (Thomson) (with yellow flocculence); L, *Pseudotaenia waterhousei* (van de Poll); M, *Pseudotaenia salamandra* (Thomson); N, *Prospheres aurantiopictus* (Laporte & Gory).

Plate 83. Buprestidae. A, *Xyroscelis crocata* (Gory & Laporte); B, *Torresita cuprifera chrysochloris* (Laporte & Gory); C, *Temognatha variabilis* (Donovan); D, *Theryaxia suttoni* Carter; E, *Toxoscelis queenslandicus* Bellamy & Peterson; F, *Notobubastes aurosulcata* (with white flocculence); G, *Castiarina dilatata* (Carter); H, *Nascioides quadrinotatus* (van de Poll); I, *Synechocera deplana* (Fabricius); J, *Castiarina dimidiata* (Carter); K, *Trachys azurea* Deyrolle (Indonesia: Aru, Misool); L, *Torresita cuprifera* (Kirby); M, *Castiarina deyrollei* (Thomson); N, *Castiarina distincta* (Saunders).

Plate 84. Tenebrionidae. A, *Adelodemus excisicollis* Carter 13 mm; B, *Alphitobius diaperinus* (Panzer) 7 mm; C, *Achthosus westwoodi* Pascoe 16 mm; D, *Aglypta octocostata* Gebien 12 mm; E, *Amarygmus aeger* Blackburn 12.5 mm; F, *Amarygmimus duboulayi* Bates 11.5 mm; G, *Alphitophagus bifasciatus* (Say) 2.5 mm; H, *Lepturidea viridis* (Boisduval) 10 mm; I, *Acerogria oriunda* (Borchmann) 13 mm; J, *Agasthenes* sp. 21 mm; K, *Adelina platisoides* (Pascoe) 4.5 mm; L, *Adelium augurale* Pascoe 20 mm.

Plate 85. Tenebrionidae. A, *Atoreuma* sp. 7 mm; B, *Axynaon championi* Blackburn 20 mm; C, *Androsus* sp. 6.5 mm; D, *Asphalus ebeninus* Pascoe 18 mm; E, *Atoichus bicolor* (Blackburn) 6 mm; F, *Apocryphodes thompsoni* Matthews PT ANIC 5.5 mm; G, *Amphianax subcoriaceus* Bates 19 mm; H, *Apasis howitti* Pascoe 19 mm; I, *Austropalorus planatus* Halstead HT ANIC 2.5 mm; J, *Archeophthora tasmanica* (Carter) 2.5 mm; K, *Apterotheca costata* (Buck) 16 mm; L, *Archaeoglenes australis* Lawrence 2.5 mm.

Plate 86. **Tenebrionidae.** A, *Brises acuticornis* (Pascoe) 15 mm; B, *Blepegenes aruspex* Pascoe 19 mm; C, *Bassianus sydneyanus* (Blackburn) 11 mm; D, *Bolbophanes dumbrelli* (Lea) 11 mm; E, *Barycistela robusta* Blackburn 9.5 mm; F, *Bluops verrucosus* Carter 16 mm; G, *Blepegenes blepegenioides* (Carter) 18 mm; H, *Blaps polychresta* (Forskål) 41 mm; I, *Bradymerus raucipennis* (Blackburn) 8 mm; J, *Brittona minuta* Medvedev and Lawrence 2 mm; K, *Boreosaragus tricarinatus* (Blackburn) 15 mm; L, *Bolusculus arcanus* Matthews 3 mm.

Plate 87. Tenebrionidae. A, *Cenoscelis* sp. 6 mm; B, *Brycopia pilosella* Pascoe 7 mm; C, *Cardiothorax australis* Carter 14 mm; D, *Camponotiphilus fimbricollis* Lea 11mm; E, *Celibe limbata* (Pascoe) 12.5 mm; F, *Casnonidea variabilis* (Macleay) 12 mm; G, *Byallius revolutus* (Carter) 21 mm; H, *Campolene nitida* Pascoe 9.5 mm; I, *Ceropria peregrina* Pascoe 11 mm; J, *Caediomorpha heteromera* (King) 5 mm; K, *Caedius sphaeroides* (Hope) 6.5 mm; L, *Byrsax macleayi* Pascoe 11.5 mm.

Plate 88. Tenebrionidae. A, *Cossyphus odewahni* Pascoe 7 mm; B, *Cheirodes sardous* Gené 5 mm; C, *Chariotheca* sp. 9.5 mm; D, *Corticeus* sp. 4 mm; E, *Cuemus monteithi* Bouchard 5.5 mm; F, *Chalcopteroides cyanopterus* (Hope) 13 mm; G, *Cyphaleus cupricollis* Macleay 15.5 mm; H, *Coripera geminata* Lea 14.5 mm; I, *Daedrosis crenatostriata* Bates 10.5 mm; J, *Cryphaeus wachteli* Grimm 9.5 mm; K, *Cillibus incisus* (Pascoe) 24.5 mm; L, *Derispia variabilis* (Carter) 4 mm.

Plate 89. Tenebrionidae. A, *Dioedus antennatus* (Carter) 5.5 mm; B, *Dysarchus* sp. 14 mm; C, *Diphyrrhynchus ellipticus* Champion 3 mm; D, *Derosphaerus excisipes* (Carter) 12.5 mm; E, *Diemenoma commoda* (Pascoe) 10.5 mm; F, *Dimorphochilus pascoei* (Macleay) 11.5 mm; G, *Ecnolagria rufescens* (Boisduval) 16 mm; H, *Diaclina nitida* (Carter) 6 mm; I, *Doyenia cremastogastri* Matthews & Lawrence PT ANIC 2 mm; J, *Dorrigonum umbilicatum* (Carter) 8.5 mm; K, *Dicyrtodes arneius* Matthews PT ANIC 5.5 mm; L, *Diaspirus bellendenus* Matthews PT ANIC 4.5 mm.

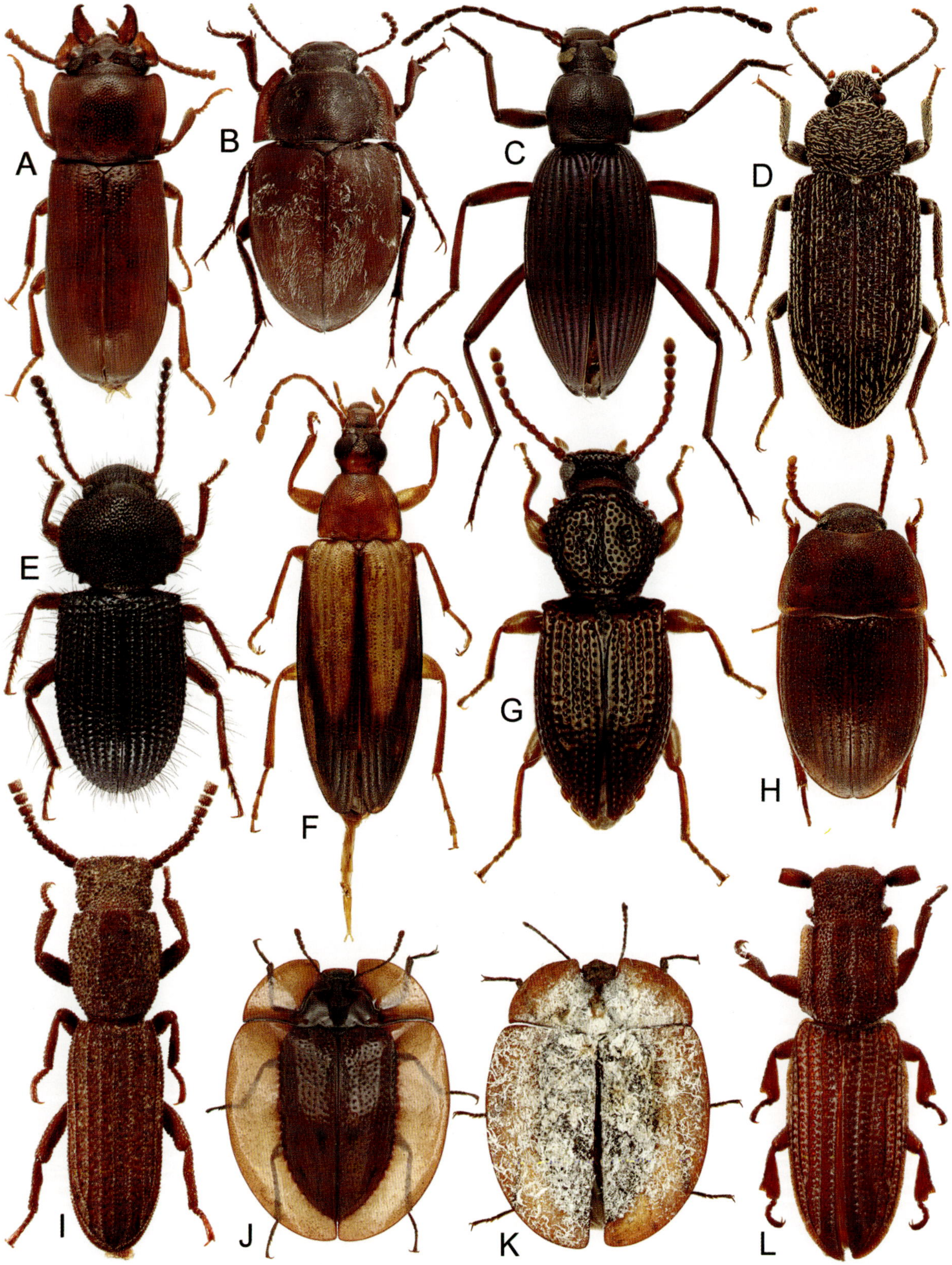

Plate 90. Tenebrionidae. A, *Gnatocerus cornutus* (Fabricius) 4 mm; B, *Edylius canescens* Champion 9 mm; C, *Eutherama cyaneum* Carter 7 mm; D, *Exangeltus rufipennis* Carter 10.5 mm; E, *Ectyche bicolor* Carter 5.5 mm; F, *Euomma lateralis* Bohemann 9 mm; G, *Epomidus prionodes* Matthews 4.5 mm; H, *Ellipsodes scriptus* (Lewis) 3 mm; I, *Eulea caeca* Carter 4 mm; J, *Emcephalus floccosus* (Pascoe) with floccosity removed 15 mm; K, *Emcephalus floccosus* (Pascoe) as it appears in life 15 mm; L, *Euclarkia costata* Lea 3.5 mm.

Plate 91. Tenebrionidae. A, *Heterocheira* sp. 7 mm; B, *Hangaya enigmatica* Matthews & Merkl HT SAMA 4.5 mm © AZ; C, *Helea consularis* (Pascoe) 22 mm; D, *Hypaulax deyrollei* (Bates) 19.5 mm; E, *Hydissus feronioides* (Pascoe) 13 mm; F, *Homotrysis cisteloides* (Newman) 15.5 mm; G, *Euomma testacea* (Carter) 6 mm; H, *Hyocis* (*H.*) *bakewelli* Pascoe 2.5 mm; I, *Hemicistela discoidalis* Blackburn 5 mm; J, *Hemicyclus reaumuri* (Laporte) 14 mm; K, *Gonocephalum arenarium* (Fabricius) 8.5 mm; L, *Ilyxerus asper* Pascoe 7.5 mm.

Plate 92. Tenebrionidae. A, *Leaus tasmanicus* Matthews & Lawrence 8.5 mm; B, *Leichenum canaliculatum* (Fabricius) 5 mm; C, *Iscanus glaber* (Lea) 5.5 mm; D, *Latheticus oryzae* Waterhouse 3 mm; E, *Leiochrodes suturalis* Westwood 3.5 mm; F, *Jophon myrmecophilus* Champion 4.5 mm; G, *Isopteroplonyx termitophilus* (Lea) HT SAMA 5 mm © SAMA, H, *Isopteron aversum* (Pascoe) 8.5 mm; I, *Kershawia rugiceps* Lea 5 mm; J, *Lagria cyanea* Macleay 10.5 mm; K, *Kaszaba coerulescens* (Haag-Rutenberg) 11 mm; L, *Leiochrinus fulvicollis* Westwood 4.5 mm.

Plate 93. Tenebrionidae. A, *Menimus castaneus* (Carter) 3 mm; B, *Meneristes australis* (Boisduval) 18 mm; C, *Lorelus fumatus* (Lea) 5.5 mm; D, *Magela cordicollis* Medvedev & Lawrence PT ANIC 2 mm; E, *Pachycoelia ocularis* (Carter) 20 mm; F, *Mesomorphus darlingensis* (Blackburn) 9.5 mm; G, *Melaps victoriae* Carter 8.5 mm; H, *Menearchus nigritus* (Fabricius) 19 mm; I, *Lixionica costatipennis* Blackburn 11 mm © GC; J, *Licinoma coerulea* Carter 13.5 mm; K, *Mesotretis ferruginea* Bates 5.5 mm; L, *Lyphia australis* Gebien 3.5 mm.

Plate 94. Tenebrionidae. A, *Metisopus purpureipennis* Bates 11 mm; B, *Micropedinus* sp. (Japan) 4 mm; C, *Neocistela ovalis* (Blackburn) 5.5 mm; D, *Myrmechixenus vaporariorum* Guérin-Méneville 2 mm; E, *Mychestes lignarius* Pascoe 8 mm; F, *Metriolagria affinis* (Boisduval) 11.5 mm; G, *Metistete grandis* (Borchmann) 25 mm; H, *Mithippia aurita* Pascoe 9.5 mm; I, *Neotheca fusca* Carter 8.5 mm; J, *Mitrothorax convexicollis* (Carter) 9.5 mm; K, *Monteithium storeyi* Matthews 6 mm; L, *Micrectyche ferruginea* Bates 4 mm.

Plate 95. **Tenebrionidae**. A, *Olisthaena anthracina* (Pascoe) 11 mm; B, *Nocar depressiusculus* (Macleay) 6.5 mm; C, *Nyctozoilus denticollis* Carter 18 mm; D, *Orcopagia monstrosa* Pascoe 8.5 mm; E, *Nototrintus asperatus* (Carter) 16 mm; F, *Nypsius foveatus* Champion 10 mm; G, *Notocistela* sp. 10.5 mm; H, *Nyctozoilus bos* (Carter) 16.5 mm; I, *Omolipus* sp. 14.5 mm; J, *Ommatophorus mastersi* Macleay 6.5 mm; K, *Onotrichus minor* Carter 11 mm; L, *Oectosis* sp. 9 mm.

Plate 96. Tenebrionidae. A, *Palembomimus bicinctus* (Champion) 3.5 mm; B, *Ospidus chrysomeloides* Pascoe 12.5 mm; C, *Platydema fuligineum* (Pascoe) 6 mm; D, *Paratoxicum iridescens* Champion 7 mm; E, *Phaleria bimaculata* (Linnaeus) 8 mm; F, *Paraphanes nitidus* Macleay 11.5 mm; G, *Phanechloros punctipennis* (Carter) 15.5 mm; H, *Pemanoa* sp. 6.5 mm; I, *Platycotylus nitidulus* (Macleay) 3.5 mm; J, *Palorus intermedius* Halstead 2.5 mm; K, *Parahyocis championi* (Fauvel) 2.5 mm (New Caledonia); L, *Pentaphyllus exilis* (Champion) 2.5 mm.

Plate 97. Tenebrionidae. A, *Saragus costatus* (Solier) 15.5 mm; B, *Platyphanes frenchi* Carter 19.5 mm; C, *Salax* sp. 7.5 mm; D, *Sciophagus flavipes* Gebien 4.5 mm; E, *Promethis quadricollis* Pascoe 19.5 mm; F, *Prophanes aculeatus* Westwood 23 mm; G, *Scaletomerus politus* (Macleay) 5 mm; H, *Pterohelaeus arcanus* Pascoe 21 mm; I, *Rhipidandrus simpsoni* (Waterhouse) 4.5 mm; J, *Pseudolyprops punctatus* (Carter) 6 mm; K, *Pseudophthora wilsoni* Kaszab 4.5 mm; L, *Pseudeba watsoni* Britton 3.5 mm.

Plate 98. Tenebrionidae. A, *Scleropatroides triplehorni* Iwan & Matthews 5 mm PT SAMA © DI; B, *Sobas australis* (Hope) 5 mm; C, *Sphargeris physodes* Pascoe 8.5 mm; D, *Scotoderus cancellatus* (Montrouzier) 8.5 mm; E, *Spiloscapha thalloides* (Pascoe) 5 mm; F, *Stenolagria leai* (Borchmann) 11.5 mm; G, *Sora lawrencei* Merkl 12.5 mm; H, *Simarus godeffroyi* (Haag-Rutenberg) 7 mm; I, *Seirotrana catenulata* (Boisduval) 16.5 mm; J, *Scymena variabilis* Pascoe 6 mm; K, *Scolytocaulus* sp. 5.5 mm; L, *Sloanea costata* Carter 20 mm.

Plate 99. Tenebrionidae. A, *Tetragonomenes ocularis* (Kulzer) 7.5 mm; B, *Taxes alphitobioides* Champion 4.5 mm; C, *Sympetes tricostellus* (White) 24.5 mm; D, *Thorictosoma* sp. 3 mm; E, *Styrus batesi* Haag-Rutenberg 16 mm; F, *Strongylium neboissianum* Kaszab 10.5 mm; G, *Tanychilus dubius* Newman 17.5 mm; H, *Tanylypa morio* Pascoe 12.5 mm; I, *Csiro matthewsi* Medvedev & Lawrence, 2 mm; J, *Tagalinus* sp. 4 mm; K, *Tenebrio obscurus* Fabricius 17.5 mm; L, *Titaena columbina* Erichson 9.5 mm.

Plate 100. Tenebrionidae. A, *Trachyscelis ciliaris* Champion 3.5 mm; B, *Tyrtaeus* sp. 2 mm; C, *Uloma* (*U.*) *queenslandica* Kaszab 11 mm; D, *Typhluloma inops* Lea 4.5 mm; E, *Ulomoides heroni* (Carter) 5 mm; F, *Trichamarygmus pilosus* Carter 11 mm; G, *Toxicum punctipenne* Pascoe female 12.5 mm; H, *Toxicum punctipenne* Pascoe male 12 mm; I, *Espites basalis* Pascoe 7.5 mm; J, *Trichosaragus pilosellus* Blackburn 9 mm; K, *Triplehornia metallica* Matthews & Lawrence PT ANIC 2.5 mm; L, *Tribolium confusum* Jacquelin du Val 4 mm.

Plate 101. Tenebrionidae. A, *Uptona pallida* Medvedev & Lawrence PT ANIC 3 mm; B, *Wattiana greensladei* Matthews & Lawrence HT ANIC 3 mm; C, *Zophobas morio* (Fabricius) 27 mm; D, *Zophophilus* sp. 16 mm; E, *Scaphinion clavatus* Matthews HT WAMA © Psyche 6 mm, lateral. F, *Cyphaleus aereus* Waterhouse 20.5 mm; G, *Homotrysis canescens* (Hope) 17.5 mm; H, *Cyphaleus mastersi* Pascoe 21 mm; I, *Eutermicola sculpticollis* Lea 3.5 mm; J, *Litopous baehri* Matthews HT ANIC © Psyche 8 mm; K, *Xenolagria tincta* (Blackburn) 8.5 mm; L, *Yarranum crenicolle* (Pascoe) 13 mm.

Plate 102. Dytiscidae. A, *Allodessus bistrigatus* (Clark); B, *Antiporus gilbertii* (Clark); C, *Antiporus blakeii* (Clark); D, *Brancuporus gottwaldi* (Hendrich); E, *Gibbidessus pictipes* (Lea); F, *Kakadudessus tomweiri* Hendrich & Balke; G, *Hydroglyphus trifasciatus* (Watts); H, *Hydroglyphus daemeli* (Sharp); I, *Hydaticus quadrivittatus* Blanchard; J, *Hydaticus consanguineus* Aubé; K, *Hydrovatus opacus* Sharp.

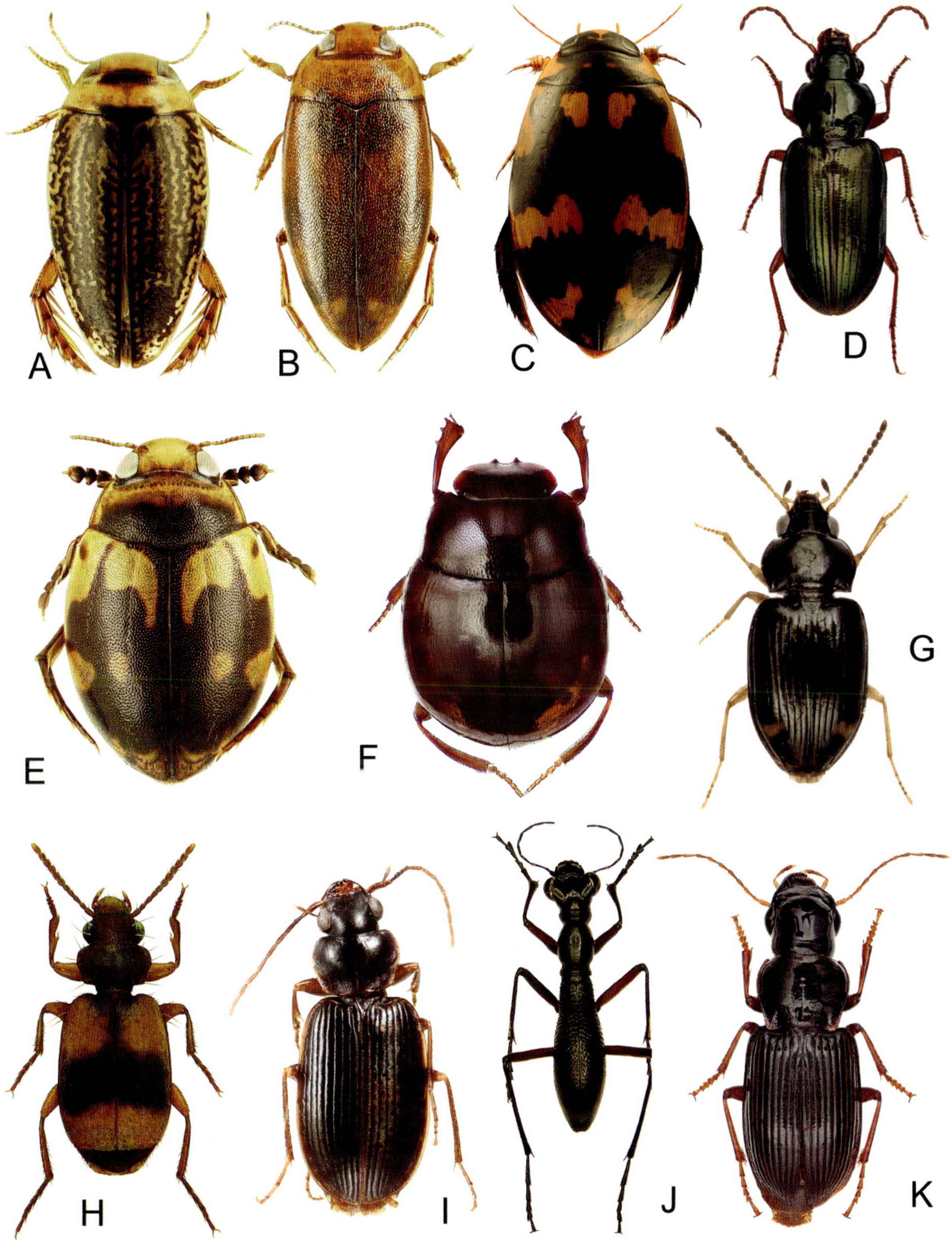

Plate 103. Dytiscidae (A–C, E), Carabidae (D, G–K), Scarabaeidae (F). A, *Laccophilus clarki* Sharp; B, *Neobidessodes grossus* (Zimmermann); C, *Sandracottus bakewelli* (Clark); D, *Pogonus cardiotrachelus* Chaudoir (Pogonini); E, *Hyphydrus elegans* (Montrouzier); F, *Monteithocanthon peckorum* Gunter & Weir; G, *Elaphropus bipistulatus* (Macleay) (Bembidiini); H, *Somotrichus unifasciatus* (Dejean) (Lebiini); I, *Badister* (*Baudia*) *sundaicus* Andrewes (Licinini); J, *Tricondyla aptera* (Olivier) (Cicindelini); K, *Nornalupia impuncta* Kataev (Harpalini).

ERRATA FOR VOLUME 1

FIGURE CAPTIONS

Fig. 4C. Change '*Selagis* sp.' to '*Diphucrania* sp.'

Fig. 54C. Change '*Distocupes varians* (Lea)' to '*Adinolepis matthesonae* (Neboiss)'. This misidentification was carried over from the first edition of *Australian Beetles* (Lawrence & Britton 1994). The original image in the revised edition of *Insects of Australia* (Lawrence & Britton (1991) was properly identified, although misspelled as '*Cupes mathesoni*'.

Fig. 73I. Change '*Macleayana vesca* (Olliff)' to '*Macleayania vesca* (Olliff)'.

1. INTRODUCTION. ADULT MORPHOLOGY. *Female Genitalia*

P. 61, col. 2, para 4, line 4: change 'baculi' to 'bacula'. Although baculus, baculi is sometimes used for the various struts in the ovipositor, the correct term is baculum, a Latin term with the plural bacula.

2. CLASSIFICATION AND KEYS

ADULTS

11(10). Antennae 11-segmented .... **Geotrupidae**

– Antennae with fewer than 11 antennomeres........ **11a**

11a(11). Body strongly contractile, capable of being rolled into the shape of a ball (Fig. 60M); eye completely divided into two parts by ocular canthus; metatibiae and to a lesser extent mesotibiae and protibiae expanded and flattened, with carinate outer edge; length less than 5 mm........ **Hybosoridae: Ceratocanthinae**

– Body not contractile, or IF moderately so THEN eye not completely divided; metatibiae not flattened and carinate and length greater than 5 mm........ **12**

14(12). Change 'not capable' to 'usually not capable'.

15(14)........ **Hybosoridae (major part)**

39(38). Insert after 'Figs': '13C,'

41(40). Tarsi 5-5-5

42(41). Apical 3 (or apparently 4) antennomeres much longer than wide, each one longer than antennomeres 3–6 combined; head with median ocellus; elongate, slender, soft-bodied beetles with 7 abdominal ventrites and all coxae conical and projecting........ **Dermestidae: Trinodinae (*Thylodrias* males) (part)**

– Apical antennomeres not much longer than wide; without other features combined........ **42a**

42a(42). Antennae 10-segmented; scape in male often enlarged and modified; soft-bodied and brightly coloured (red or yellow with blue or black)........ **Melyridae: Malachiinae (part)**

– Antennae 11-segmented (or rarely with more than 10 antennomeres)........ **43**

47(45). End of line, replace '**8**' with '**48**'.

74 (40). Club segments not or barely wider than preceding 4 antennomeres but each one longer than 3–6 combined; head with median ocellus; elongate, slender, soft-bodied beetles with 7 abdominal ventrites and all coxae conical and projecting........ **Dermestidae: Trinodinae (*Thylodrias* males) (part)**

– Last 3 to 6 antennomeres forming club not as above; head without median ocellus; without other features combined........ **74a**

74a(74). Antennae 7- to 9-segmented........ **75**

– Antennae 10-segmented........ **77**

– Antennae 11-segmented........ **80**

Change 77(74) to 77(74a).

78(77). Mandible with dorsal tubercle and setose cavity etc........ **Sphindidae: *Notosphindus***

– Mandible without dorsal tubercle or setose cavity; frontoclypeal suture absent or indistinct; metathoracic discrimen absent........ **78a**

78a(78). Head prognathous or slightly declined etc........ **Trogossitidae: Lophocaterinae (part)**

– Head hypognathous; transverse occipital ridge present (Fig. 44G); prosternal process incomplete, terminating between coxae; outer edge of mesotibia simple; body strongly convex, more or less globose........ **79**

79(78a). Head with median ocellus; apical antennomere only slightly longer than preapical one; protibiae not strongly expanded; upper surfaces clothed with erect, dehiscent spines; abdominal ventrite 1 with paired diverging postcoxal lines; mesocoxae separated by distance much greater than shortest siameter of mesocoxal cavity........ **Dermestidae: Trinodinae (part)**

– Head without median ocellus; apical antennomere as long as preceding 3 combined; protibiae strongly expanded subapically; upper surfaces clothed with moderately long, decumbent fine hairs; abdominal ventrite 1 without paired postcoxal lines; mesocoxae separated by distance less than shortest diameter of mesocoxal cavity........ **Trogossitidae: Rentoniinae (*Globorentonium*)**

80(74a). Pronotal hypomera and sides of pronotal disc strongly tuberculate, main portion of pronotal disc finely punctate and separated from tuberculate regions by pair of longitudinal furrows; dorsal vestiture consisting of erect, dehiscent spines (inclined at various angles and often partly missing); antennal club 4- to 6-segmented, but not clearly defined and with most antennomeres variously thickened in male........ **Dermestidae: Trinodinae (*Trichelodes*)**

– Pronotal hypomeron and sides of pronotal disc not strongly tuberculate; IF pronotal disc with paired longitudinal furrows and dorsal surfaces clothed with dehiscent spines THEN antennal club 1- or 2-segmented AND head with median ocellus........ **80a**

80a(80). Abdomen with 6 or 7 ventrites **81**
– Abdomen with fewer than 6 ventrites **84**
Change 81(80) to 81(80a) and in second line change 'betles' to 'beetles'
Change 84(80) to 84(80a).
109(105). Change 'entrites' to 'ventrites'.
138(137). Antennal insertions concealed from above (Fig. 44A); prosternal process complete, extending behind coxae (Fig. 25C); ventrites 1 to 3 connate (Fig. 25K)... **Tenebrionidae (part)**
– Antennal insertions exposed from above **138a**
138a(138). Prosternal process complete, extending behind coxae; ventrites 1–3 connate (Fig. 25K) .... **Tenebrionidae: Lagriinae: Statirini**
– Prosternal process incomplete, terminating between coxae (Fig. 18D); postcoxal lobes of hypomera meeting behind coxae **138b**
138b(138a). All ventrites free; anterior edge of prothorax with distinct collar .... **Anthicidae: Eurygeniinae (part)**
– Ventrites 1–3 connate; anterior edge of prothorax without collar .... **Tenebrionidae: Lagriinae: Lagriini**
191(190). Change 'bewteen' to 'between'.

3. FAMILY TREATMENTS

28. BUPRESTIDAE

Pp. 220–223, throughout text: change '*Cisseis*' to '*Diphucrania*'.

P. 223, col 2, lines 10–11: change '*Cisseis* Gory & Laporte' to '*Diphucrania* Dejean'.

29. BYRRHIDAE

P. 225, col. 1, last line of para 2: move '2002a, 2005' to a position after '1987,' (these are references of Johnson, not Szpojda).

50. PTINIDAE

P. 283, para 2, line 1: change 'Anobiids' to 'Ptinids'.

79. NITIDULIDAE

P. 327, para 2, line 2, change '44500' to '4500'.

INDEX OF GENERIC AND FAMILY GROUP NAMES